WATER QUALITY & TREATMENT

ABOUT THE AMERICAN WATER WORKS ASSOCIATION

American Water Works Association is the authoritative resource for knowledge, information, and advocacy to improve the quality and supply of water in North America and beyond. AWWA is the largest organization of water professionals in the world. AWWA advances public health, safety, and welfare by uniting the efforts of the full spectrum of the entire water community. Through our collective strength we become better stewards of water for the greatest good of the people and the environment.

American Water Works Association
6666 W. Quincy Ave.
Denver, Colorado 80235
303.794.7711
www.awwa.org

WATER QUALITY & TREATMENT

A Handbook on Drinking Water

American Water Works Association

James K. Edzwald, Editor

Sixth Edition

McGraw Hill

New York Chicago San Francisco Lisbon London Madrid
Mexico City Milan New Delhi San Juan Seoul
Singapore Sydney Toronto

The McGraw-Hill Companies

Cataloging-in-Publication Data is on file with the Library of Congress

Water Quality & Treatment: A Handbook of Drinking Water, Sixth Edition

5 6 7 8 9 10 QVS/QVS 21 20 19 18 17

ISBN 978-0-07-163011-5
MHID 0-07-163011-2

The pages within this book were printed on acid-free paper.

Cover photo: Modesto Regional Water Treatment Plant (Modesto, CA). Photo courtesy of Black & Veatch Corporation.

AWWA Team
Executive Director
David LaFrance

Deputy Executive Directors
Paula MacIlwaine, Tom Curtis

Publications Manager
Gay Porter De Nileon

McGraw-Hill Team
Sponsoring Editor
Joy Bramble

Acquisitions Coordinator
Alexis Richard

Editorial Supervisor
David E. Fogarty

Project Manager
Deepti Narwat, Glyph International

Copy Editors
James Madru, Carol Loomis

Proofreader
Surendra Nath Shivam, Glyph International

Indexer
Robert Swanson

Production Supervisor
Pamela A. Pelton

Composition
Glyph International

Art Director, Cover
Jeff Weeks

ABOUT THE EDITOR

James K. Edzwald is Professor Emeritus of the Department of Civil and Environmental Engineering at the University of Massachusetts, Amherst. He earned his B.S. and M.S. degrees in Civil Engineering and Environmental Health Engineering from the University of Maryland, and a Ph.D. in Water Resources Engineering from the University of North Carolina, Chapel Hill. He also held faculty positions at the University of Missouri, Clarkson University, and Rensselaer Polytechnic Institute. His research interests include water supply, drinking water treatment, and aquatic chemistry. Professor Edzwald has authored or coauthored over 150 publications on water quality and treatment. He is a recipient of the 2004 A.P. Black Award from AWWA for his contributions in water supply research and a recipient of the 2009 Founders' Award from the Association of Environmental Engineering and Science Professors for his contributions to environmental engineering education and practice. He is a registered professional engineer in New York.

CONTENTS

Preface xv
Acknowledgments xvii

Chapter 1. Drinking Water Standards, Regulations, and Goals
J. Alan Roberson, P.E., and Eric G. Burneson, P.E. **1.1**

Regulatory History Prior to the 1974 SDWA / *1.2*
Evolution of the SDWA / *1.3*
The Risk Management and Standard-Setting Processes / *1.8*
Current Drinking Water Regulations / *1.19*
Role of State Agencies / *1.26*
Peer Review, Outside Consultation, and Public Involvement / *1.30*
Other Countries and International Standards / *1.32*
Outlook for the Future / *1.33*
The Internet as a Resource / *1.34*
Disclaimer / *1.35*
Abbreviations / *1.35*
References / *1.37*

Chapter 2. Health and Aesthetic Aspects of Drinking Water
Gloria B. Post, Ph.D., D.A.B.T., Thomas B. Atherholt, Ph.D., and Perry D. Cohn, Ph.D., M.P.H. **2.1**

Waterborne Disease / *2.3*
Pathogenic Organisms / *2.6*
Indicators of Water Quality / *2.19*
Toxicological Evaluation of Drinking Water Contaminants / *2.23*
Risk Assessment of Drinking Water Contaminants / *2.27*
Inorganic Constituents / *2.33*
Organic Constituents / *2.44*
Disinfectants and Disinfection By-Products / *2.59*
Radionuclides / *2.69*
Aesthetic Quality / *2.72*
Preparedness and Health / *2.75*
Final Comment / *2.76*
Internet Resources / *2.76*
Abbreviations / *2.77*
Acknowledgments / *2.78*
References / *2.79*

Chapter 3. Chemical Principles, Source Water Composition, and Watershed Protection *James K. Edzwald, Ph.D., B.C.E.E., and John E. Tobiason, Ph.D., B.C.E.E.* **3.1**

Introduction / *3.2*
Chemical Principles and Concepts / *3.2*
Source Water Composition / *3.24*
Particles / *3.42*
Natural Organic Matter / *3.58*
Source Water Selection and Protection / *3.67*
Abbreviations / *3.71*
Notation for Equations / *3.72*
References / *3.72*

Chapter 4. Hydraulic Characteristics of Water Treatment Reactors and Their Effects on Treatment Efficiency *Desmond F. Lawler, Ph.D., P.E.* **4.1**

Introduction / *4.2*
Continuous Flow Reactors: Ideal and Non-Ideal Flow / *4.2*
Tracer Studies / *4.3*
Choosing a Step or Pulse Input Tracer Test / *4.11*
Mathematical Models for Non-Ideal Flow / *4.16*
Computational Fluid Dynamics / *4.24*
Reaction Rate Expressions / *4.26*
Reactions in Continuous Flow Systems at Steady State: Combining Hydraulics and Reaction Kinetics / *4.33*
Reactors in Water Treatment and Their Hydraulics Characteristics / *4.43*
Summary / *4.47*
Abbreviations / *4.47*
Notation for Equations / *4.47*
References / *4.49*

Chapter 5. Overview of Water Treatment Processes *Doug Elder, P.E., and George C. Budd, Ph.D., P.E.* **5.1**

Introduction / *5.2*
Source Water Quality Considerations (Chap. 3) / *5.2*
Characteristics and General Capabilities of Unit Processes / *5.4*
Distribution System Considerations (Chaps. 19–21) / *5.25*
Treatment Process Residuals Management (Chap. 22) / *5.27*
Other Considerations / *5.28*
Treatment Process Configurations / *5.29*
Abbreviations / *5.36*
References / *5.38*

Chapter 6. Gas–Liquid Processes: Principles and Applications *David W. Hand, Ph.D., David R. Hokanson, Ph.D., P.E., B.C.E.E., and John C. Crittenden, Ph.D., P.E., D.E.E., N.A.E.* **6.1**

Introduction / *6.1*
Theory of Gas Transfer / *6.2*
Packed Towers / *6.14*
Diffused or Bubble Aeration / *6.41*

Surface Aeration / 6.52
Spray Aerators / 6.56
Notation for Equations / 6.61
References / 6.63

Chapter 7. Chemical Oxidation ***Philip C. Singer, Ph.D., P.E., B.C.E.E., and David A. Reckhow, Ph.D.*** **7.1**

Introduction / 7.1
Principles of Oxidation / 7.2
Oxidants Used in Water Treatment / 7.15
Applications of Oxidation Processes to Water Treatment Processes / 7.31
Abbreviations / 7.47
Notations for Equations / 7.48
References / 7.48

Chapter 8. Coagulation and Flocculation ***Raymond D. Letterman, Ph.D., P.E., and Sotira Yiacoumi, Ph.D.*** **8.1**

Introduction / 8.2
Contaminants / 8.3
Stability of Particle Suspensions / 8.6
Destabilization Mechanisms / 8.14
Coagulants / 8.15
The Rapid Mixing and Flocculation Processes / 8.50
Abbreviations / 8.71
Equation Notation / 8.71
References / 8.72

Chapter 9. Sedimentation and Flotation ***Ross Gregory, MPhil, C.Eng., C.WEM, and James K. Edzwald, Ph.D., B.C.E.E.*** **9.1**

Modern History of Sedimentation / 9.2
Sedimentation Theory / 9.3
Operational and Design Considerations for Sedimentation / 9.26
Introduction to Dissolved Air Flotation / 9.45
Fundamentals of Dissolved Air Flotation / 9.48
Operational and Design Considerations for Flotation / 9.65
Applications / 9.84
Abbreviations / 9.89
Notation for Equations / 9.89
References / 9.92

Chapter 10. Granular Media Filtration ***John E. Tobiason, Ph.D., P.E., B.C.E.E., John L. Cleasby, Ph.D., P.E., B.C.E.E., Gary S. Logsdon, D.Sc., P.E., B.C.E.E., and Charles R. O'Melia, Ph.D., P.E.*** **10.1**

Overview of Particle Filtration Processes / 10.2
Granular Media Filtration Process Description / 10.9
Media Filtration Theory and Modeling / 10.26

Rapid-Rate Filter Performance / *10.39*
Flow Control in Filtration / *10.49*
Backwashing and Maintenance of Filter Media / *10.53*
Direct Filtration / *10.70*
Pressure Granular Bed Filters / *10.74*
Slow Sand Filtration / *10.76*
Precoat Filtration / *10.87*
Dedication / *10.95*
Abbreviations / *10.95*
Equation Notation / *10.96*
References / *10.97*

Chapter 11. Membranes ***Steven J. Duranceau, Ph.D., P.E., and James S. Taylor, Ph.D., P.E.*** **11.1**

Size Ranges for Membrane Processes / *11.4*
Classifications and Configurations of Membrane Processes / *11.9*
RO-NF Configuration / *11.18*
Membrane Properties and Rejection Characteristics / *11.26*
Bubble-Point Direct Testing / *11.35*
Mass Transport and Separation / *11.37*
Integrated MF and UF Process Applications and Process Design / *11.57*
NF and RO Process Concepts and Design Criteria / *11.65*
Residuals Disposal and Concentrate Management / *11.91*
Pilot Plant Testing / *11.94*
Abbreviations / *11.96*
Notation for Equations / *11.98*
References / *11.99*

Chapter 12. Ion Exchange and Adsorption of Inorganic Contaminants ***Dennis Clifford, Ph.D., P.E., B.C.E.E., Thomas J. Sorg, P.E., B.C.E.E., and Ganesh L. Ghurye, Ph.D., P.E., B.C.E.E.*** **12.1**

Overview / *12.2*
Introduction and Theory of Ion Exchange / *12.2*
Applications of IX and Adsorption / *12.27*
IX Modeling Using EMCT / *12.74*
Waste Disposal / *12.82*
Summary / *12.82*
Abbreviations / *12.87*
Notation for Equations / *12.89*
References / *12.90*

Chapter 13. Precipitation, Coprecipitation, and Precipitative Softening ***Stephen J. Randtke, Ph.D., P.E.*** **13.1**

Introduction / *13.1*
Principles / *13.2*
Precipitative Softening / *13.12*
Other Applications / *13.70*
Acknowledgments / *13.74*
Abbreviations / *13.74*
Notation for Equations / *13.75*
References / *13.76*

Chapter 14. Adsorption of Organic Compounds by Activated Carbon *R. Scott Summers, Ph.D., Detlef R. U. Knappe, Ph.D., and Vernon L. Snoeyink, Ph.D.* **14.1**

Adsorption Overview / *14.2*
Adsorbent Characteristics / *14.4*
Adsorption Theory / *14.8*
GAC Adsorption Systems / *14.37*
Performance of GAC Systems / *14.39*
GAC Performance Estimation / *14.59*
Powdered Activated Carbon (PAC) Adsorption / *14.77*
Thermal Reactivation of GAC / *14.85*
Adsorption of Organic Matter by Other Adsorbents / *14.87*
Abbreviations / *14.89*
Notation for Equations / *14.91*
References / *14.91*

Chapter 15. Natural Treatment Systems *Saroj K. Sharma, Ph.D. and Gary Amy, Ph.D.* **15.1**

Introduction / *15.2*
River (RBF) and Lake (LBF) Bank Filtration / *15.3*
Artificial Recharge and Recovery (ARR) / *15.7*
Subsurface Groundwater Treatment / *15.10*
Soil Aquifer Treatment (SAT) for Indirect Potable Reuse / *15.12*
Water Quality Improvements in Natural Treatment Systems / *15.15*
Design and Operation of Natural Water Treatment Systems / *15.20*
Selected Case Studies of Natural Treatment Systems / *15.23*
Abbreviations / *15.28*
References / *15.29*

Chapter 16. Water Reuse for Drinking Water Augmentation *Jörg E. Drewes, Ph.D., and Stuart J. Khan, Ph.D.* **16.1**

Introduction to Potable Reuse / *16.2*
Source Water Characteristics / *16.5*
System Reliability and Health Risk Considerations / *16.17*
Design of Potable Reuse Schemes / *16.23*
Monitoring Strategies for Process Performance and Compliance / *16.33*
Regulations and Guidelines for Drinking Water Augmentation / *16.36*
Public Perception to Indirect Potable Reuse / *16.39*
Abbreviations / *16.42*
References / *16.43*

Chapter 17. Chemical Disinfection *Charles N. Haas, Ph.D.* **17.1**

Introduction / *17.1*
History of Disinfection / *17.3*
Regulatory Issues for Disinfection / *17.4*
Disinfectants and Theory of Disinfection / *17.5*
Assessment of Microbial Quality (Indicators) / *17.19*
Disinfection Kinetics / *17.21*
Mode of Action of Disinfectants / *17.27*

Disinfectant Residuals for Posttreatment Protection / *17.30*
Design and Application of Technologies / *17.31*
Relative Comparisons / *17.41*
Abbreviations / *17.42*
Notation for Equations / *17.43*
References / *17.43*

Chapter 18. Ultraviolet Light Processes ***Karl G. Linden, Ph.D., and Erik J. Rosenfeldt, Ph.D., P.E.*** **18.1**

Introduction to Ultraviolet Light Processes / *18.1*
Fundamentals of UV Light / *18.4*
UV Disinfection / *18.10*
UV Photolysis / *18.17*
UV Advanced Oxidation Processes (AOPS) / *18.24*
Abbreviations / *18.34*
Notation for Equations / *18.36*
References / *18.36*

Chapter 19. Formation and Control of Disinfection By-Products ***David A. Reckhow, Ph.D., and Philip C. Singer, Ph.D., P.E., B.C.E.E.*** **19.1**

Introduction / *19.1*
Formation of Disinfection (and Oxidation) By-Products / *19.2*
Control of Oxidation/Disinfection By-Products / *19.27*
Disinfection By-Products in the Distribution System / *19.43*
Abbreviations / *19.46*
References / *19.48*

Chapter 20. Internal Corrosion and Deposition Control ***Michael R. Schock and Darren A. Lytle*** **20.1**

Introduction / *20.2*
Corrosion, Passivation, and Immunity / *20.3*
Physical Factors Affecting Corrosion and Metals Release / *20.20*
Chemical Factors Affecting Corrosion / *20.25*
Corrosion of Specific Materials / *20.40*
Direct Methods for the Assessment of Corrosion / *20.66*
Corrosion Control Alternatives / *20.73*
Water Sampling for Corrosion Control / *20.77*
Acknowledgments / *20.81*
Disclaimer / *20.82*
Abbreviations / *20.82*
Notation for Equations / *20.82*
References / *20.83*

Chapter 21. Microbiological Quality Control in Distribution Systems ***Mark W. LeChevallier, Ph.D., Marie-Claude Besner, Ph.D., Melinda Friedman, P.E., and Vanessa L. Speight, Ph.D., P.E.*** **21.1**

Microbial Risks from Distribution System Contamination / *21.2*
Microbes in Distribution Systems / *21.7*

Factors Contributing to Microbial Occurrences in Distribution Systems / *21.21*
Monitoring Distribution Systems / *21.34*
Engineering and Design of Distribution Systems / *21.43*
Controlling Microbial Occurrences in Distribution Systems / *21.47*
Final Comments / *21.65*
Abbreviations / *21.66*
References / *21.67*

Chapter 22. Water Treatment Plant Residuals Management
David A. Cornwell, Ph.D., P.E., B.C.E.E., and Damon K. Roth, P.E. **22.1**

Introduction / 22.2
Coagulation and Lime-Softening Residuals / 22.4
Thickening / 22.16
Non-Mechanical Dewatering / 22.21
Mechanical Dewatering / 22.29
Spent Filter Backwash Treatment / 22.37
Recycle / 22.40
Membrane Residuals / 22.43
Ion Exchange and Inorganic Adsorption Process Residuals / 22.54
Residuals Containing Arsenic / 22.62
Residuals Containing Radioactivity / 22.66
Ultimate Disposal and Utilization of Solids / 22.70
Abbreviations / 22.75
Notation for Equations / 22.76
References / 22.77

Appendix A. Chemical Elements **A.1**

Appendix B. Useful Constants **B.1**

Appendix C. Conversion Factors **C.1**

Appendix D. Properties of Water and Gases **D.1**

Index I.1

PREFACE

This sixth edition of *Water Quality & Treatment: A Handbook on Drinking Water* serves as a handbook for scientists, engineers, and other professionals who study and work in drinking water—particularly the quality of water supplies, the quality of treated drinking water, and water treatment processes. It is meant as a resource for those in academics (professors and students), consulting engineering practice, water utilities, federal and state regulatory agencies, and the water process and chemical industries. The book emphasizes principles (theory) and applications (practice). It serves as a companion to the book on design, AWWA-ASCE, *Water Treatment Plant Design* (the fifth edition is in preparation, with expected publication in late 2011).

This book is an activity of the American Water Works Association's (AWWA's) Water Quality and Technology Division (WQTD). James K. Edzwald served as the technical editor and worked with the authors of the chapters in preparing the book. An ad hoc committee of the WQTD consisting of James P. Malley, Jr., Marilyn M. Marshall, and Dixie Fanning provided advice to the technical editor throughout the preparation of this book.

Water Quality & Treatment, sixth edition, differs greatly from the fifth edition—published in 1999; it contains significant revisions, updating of material, and new chapters. Five new chapters expand the scope of this book: Chapter 4, "Hydraulic Characteristics of Water Treatment Reactors and Their Effects on Treatment Efficiency," Chapter 15, "Natural Treatment Systems," Chapter 16, "Water Reuse for Drinking Water Augmentation," Chapter 18, "Ultraviolet Light Processes," and Chapter 19, "Formation and Control of Disinfection By-Products." A sixth chapter, Chapter 3, "Chemical Principles, Source Water Composition, and Watershed Protection," replaces one from the fifth edition on source water quality management, and it is essentially another new chapter in that it contains new material on chemical principles and additional material on source water quality.

Since publication of the fifth edition, the drinking water field has faced new regulations and concerns about the health effects of some new and previously known contaminants. Furthermore, in the last 10 years we have seen the development of new technologies and refinements of older technologies that are now covered in this edition. The sixth edition covers the health effects and treatment technologies to remove some contaminants not covered previously, such as nanoparticles, endocrine disrupting compounds, and pathogens; it contains updated material on many other contaminants, such as disinfection by-products, arsenic, and pathogens, including viruses and protozoan cysts such as *Cryptosporidium*; and it addresses subjects not adequately covered in the prior edition, such as water reuse, ultraviolet light processes, and natural treatment systems.

Several other new features are notable in this sixth edition. The International System of Units (SI) is used with U.S. units in parenthesis where appropriate. This makes the book useful to professionals outside the United States and to those within the United States working on water projects around the world. Each chapter has its own table of contents to aid readers in finding subject matter within chapters. Four new appendices provide quick references for atomic numbers and masses, physical and chemical constants, unit conversion factors, and the physical properties of water and gases.

The book is organized beginning with five foundation chapters that contain material on drinking water standards and regulations (Chap. 1); health effects (Chap. 2); chemical principles, source water composition, and watershed protection (Chap. 3);

hydraulics of treatment processes (Chap. 4); and an overview of water treatment processes (Chap. 5). This is followed by coverage of various water treatment processes in Chapters 6 through 14 that present principles and applications of the removal of various contaminants from water supplies. Chapter 15 covers natural treatment systems such as river bank filtration, and Chapter 16 deals with water reuse. Chapters 17 and 18 follow with disinfection and ultraviolet light processes, including disinfection and advanced oxidation processes. Chapters 19, 20, and 21 cover disinfection by-products, corrosion, and microbiological quality in distribution systems, respectively. Chapter 22 ends the book with the properties, treatment, and management of water treatment residuals.

James K. Edzwald
Editor
Professor Emeritus, University of Massachusetts

James P. Malley, Jr.
Chairman of the Board of Trustees, AWWA Water Quality and Technology Division
Professor, University of New Hampshire

ACKNOWLEDGMENTS

The sixth edition of *Water Quality & Treatment: A Handbook on Drinking Water* is a valuable resource for the drinking water field that is made possible through the efforts of many people. First and foremost, the quality of the book is due to the efforts of the 45 authors who prepared 22 chapters in the book.

Revision of the book began with an assessment of the fifth edition. Several professionals from water utilities, consulting engineering firms, and academics were asked to review the fifth edition and to make recommendations for new material for inclusion in the sixth edition. I wish to thank the following: William C. Becker (Hazen and Sawyer), William D. Bellamy (CH2M Hill), Steve Bishop (Metcalf and Eddy), Howard Dunn (Vice President of Operations and Technology, Aquarion Water Company of Connecticut), Harold T. Glaser (Kennedy Jenks), Raymond D. Letterman, (Syracuse University and Technical Editor of the fifth edition), Michael J. MacPhee (Malcolm Pirnie), Charles R. O'Melia (Johns Hopkins University), Vernon L. Snoeyink (University of Illinois), and John P. Walsh (formerly, Director of Operations and Distribution, Aquarion Water Company of Connecticut, now with Environmental Partners Group).

I am grateful to the reviewers who commented on draft chapters and provided comments for the authors for improving their chapters. They are Robert Andrews, Brian Arbuckle, Takashi Asano, Khalil Z. Atasi, Benoit Barbeau, William Ball, Tim Bartrand, William Becker, Ernest Blatchley III, James Bolton, Anne Camper, Sarah Clark, Joseph Cotruvo, James Crook, Brian Dempsey, Francis DiGiano, Bruce Dorvak, Jörg E. Drewes, Nicholas Dugan, Marc Edwards, Doug Elder, Tom Gillogly, Thomas Grischek, Johannes Haarhoff, Robert Howd, Kerry Howe, Michael Kavanaugh, William Knocke, Yann Le Gouellec, France Lemieux, Gary Logsdon, Michael J McGuire, James P. Malley, Jr., Margaret H. Nellor, Eva C. Nieminski, John Novak, David Pernitsky, David Reckhow, Michael Semmens, Sukalyan Sengupta, Robert Sharp, Jim Taft, Ian Watson, Paul Westerhoff, and Yuefeng Xie.

This book project was initiated by James P. Malley, Jr., Marilyn, M. Marshall, and Dixie Fanning, members of the ad hoc committee representing the Water Quality and Technology Division of AWWA. Their advice was invaluable, and I thank them. I am particularly indebted to Jim Malley for his leadership. He was also always there to give advice and help me over the hurdles. Finally, I thank the staff with AWWA Publications and with McGraw-Hill for their work in producing the book. A special thanks goes to Gay Porter De Nileon, AWWA Publications Manager, who provided essential support from AWWA; without her assistance the book could not have been completed.

James K. Edzwald
Editor
Professor Emeritus, University of Massachusetts

CHAPTER 1
DRINKING WATER STANDARDS, REGULATIONS, AND GOALS

J. Alan Roberson, P.E.
Director of Security and Regulatory Affairs
American Water Works Association
Washington, D.C., United States

Eric G. Burneson, P.E.
Targeting and Analysis Branch Chief, Office of Ground Water and Drinking Water
U.S. Environmental Protection Agency[1]
Washington, D.C., United States

REGULATORY HISTORY PRIOR TO THE 1974 SDWA **1.2**
Early History 1.2
The U.S. Public Health Standards 1.3
EVOLUTION OF THE SDWA **1.3**
Setting the Stage for the SDWA 1.3
The First 1974 SDWA 1.4
The 1986 SDWA Amendments 1.6
The 1996 SDWA Amendments 1.7
The Bioterrorism Act of 2002 1.7
THE RISK MANAGEMENT AND STANDARD-SETTING PROCESSES **1.8**
Contaminant Candidate List 1.8
Regulatory Determinations 1.9
Unregulated Contaminant Monitoring Regulations 1.14
National Primary Drinking Water Regulations 1.15
National Primary Drinking Water Regulation Review 1.16
National Secondary Drinking Water Regulations 1.16
Health Advisories and Other Actions 1.16
CURRENT DRINKING WATER REGULATIONS **1.19**
Individual Rules 1.19
The Traditional and Negotiated Rulemaking Processes 1.25
Going beyond the Regulations 1.26
ROLE OF STATE AGENCIES **1.26**
State Agencies and USEPA as Co-regulators 1.26
Primacy 1.27
New State Programs from the 1996 SDWA Amendments 1.28
State Standards 1.29
PEER REVIEW, OUTSIDE CONSULTATION, AND PUBLIC INVOLVEMENT **1.30**
The National Academy of Sciences 1.30
The Science Advisory Board 1.30
The National Drinking Water Advisory Council 1.30
The Office of Management and Budget 1.31
Public Involvement 1.31

[1]See Disclaimer section.

The Consumer Confidence Report and the Public Notification Rule..... 1.31
OTHER COUNTRIES AND INTERNATIONAL STANDARDS **1.32**
Canada 1.32
Australia 1.32
European Union 1.33
World Health Organization 1.33
OUTLOOK FOR THE FUTURE **1.33**
THE INTERNET AS A RESOURCE **1.34**
DISCLAIMER **1.35**
ABBREVIATIONS **1.35**
REFERENCES **1.37**

The initial Safe Drinking Water Act (SDWA) was signed into law on Dec. 16, 1974 (PL 93-523). The 1974 SDWA established the national regulatory structure by which the U.S. Environmental Protection Agency (USEPA), state and local regulatory agencies, and water utilities work together to ensure safe drinking water. This chapter presents the history of drinking water regulations leading to the 1974 SDWA, subsequent SDWA amendments in 1986 and 1996, and the history of the drinking water regulations that resulted from the 1974 SDWA and the 1986 and 1996 amendments. The current risk management and standard-setting processes are discussed, along with the roles of the states and the public in the standard-setting process. Standards developed at the state level, as well as international standards, are also discussed.

REGULATORY HISTORY PRIOR TO THE 1974 SDWA

Early History

Protection of drinking water quality goes back several hundred years. Scientific and medical advances in the 1800s, along with the need to provide basic sanitation in the rapidly urbanizing cities, laid the foundation for today's drinking water field. Philadelphia was one of the first cities in the United States to provide piped drinking water; drinking water first flowed through mains of the Philadelphia Water Department in 1801 (Philly H_2O, 2008).

Connecting disease epidemics with centralized water systems was a major step in public health protection (McGuire, 2006). As part of the major cholera outbreak in London and the investigation into the area surrounding the Broad Street Pump in 1854, Dr. John Snow concluded that cholera was a waterborne disease. He removed the pump handle and no further epidemics occurred in the area surrounding that well. At that point, safe drinking water and basic sanitation started to become part of basic public health protection (Johnson, 2006). The Centers for Disease Control and Prevention (CDC) has recognized conventional drinking water treatment, i.e., the traditional multibarrier approach of using the best available source, treating the water appropriately by using filtration and disinfection, and maintaining distribution system integrity, as one of the 10 great public health improvements of the twentieth century (under the umbrella of infectious disease control) (CDC, 1999).

The first federal action taken regarding drinking water quality was passage of the Interstate Quarantine Act in 1893 (U.S. Statutes, 1893). This legislation gave the Surgeon

General of the U.S. Public Health Service (USPHS) the authority to develop and enforce regulations to prevent the introduction and transmission of communicable diseases. Interstate quarantine regulations followed the next year.

The first national drinking water regulation was adopted in 1912, a result of the nation's growing railroad network, and prohibited the use of the "common cup" on interstate train carriers (Roberson, 2006). Bringing the interstate transport challenges into current times for airlines, USEPA conducted a stakeholder effort in 2006–2007 to develop and aircraft drinking water rule. As a result of this effort, USEPA finalized a regulation for drinking water on aircraft (USEPA, 2009a).

The U.S. Public Health Standards

The "common cup" regulatory framework was soon found to be deficient, as the "common cup" regulation could only protect public health if the water placed in the cup was safe. The task of developing these standards fell to the USPHS, which at that time was an agency within the U.S. Treasury Department. On Oct. 14, 1914, the Secretary of the Treasury promulgated Standards for Purity of Drinking Water Supplied to the Public by Common Carriers in Interstate Commerce, the first national drinking water standards (AWWA, 1990). These standards, known as the "Treasury Standards," were limited to the bacteriological quality of the water.

Even though the Treasury Standards were legally binding on interstate carriers, many state and local governments adopted these standards as guidelines for their water systems. States used these standards to develop their own regulations and provided regulatory oversight for systems in their states. These standards were the start of federal, state, and local cooperation in protecting drinking water quality at the community level that continues to this day.

The Treasury Standards were revised in 1925 by the USPHS to strengthen the bacteriological quality requirements and to add basic physical and chemical standards (USPHS, 1925). These standards were revised again in 1942, 1946, and 1962 (USPHS, 1943, 1946, 1962). The 1962 standards were the most comprehensive, covering 28 constituents, and were used by all 50 states either as standards or guidelines. However, depending on the state regulations, these standards were not legally enforceable for many systems and were only legally binding for those systems that supplied water to the interstate carriers.

EVOLUTION OF THE SDWA

Setting the Stage for the SDWA

Public concern and media attention about the presence of contaminants in the environment continued to grow in the late 1960s and early 1970s. The modern environmental movement began at this time, and the public concern and media attention translated to pressure on the federal government to act. From a federal perspective, the government wanted to keep up to date with the newest scientific developments in drinking water research and incorporate the latest results into the USPHS standards.

In 1969, the USPHS Bureau of Water Hygiene started the Community Water System Survey (CWSS) in an effort to revisit the 1962 standards and conducted a review of water systems to determine how many met these standards. The USPHS surveyed approximately 1000 public water systems (PWSs) that, at the time, served approximately 12 percent of the population. Released in 1970, the survey results showed that 41 percent of the systems did not meet the 1962 guidelines (USPHS, 1970). Many systems were deficient in one or more components of the multibarrier approach (source water protection, filtration, disinfection, and protecting the integrity of the distribution system).

Soon thereafter, drinking water researchers in both the United States and Europe were conducting their own surveys that began to raise public awareness. Analytical methods that allowed for better separation and quantification of organic chemicals had improved. A 1972 study of the Mississippi River, which supplies New Orleans, found 36 synthetic organic chemicals (SOCs) (USEPA, 1972). In addition, researchers in the United States and the Netherlands published their seminal work on disinfection by-products (DBPs) with the discovery of trihalomethanes (THMs) (Bellar et al., 1974; Rook, 1974).

Building on these scientific reports, several national media stories raised consumers' concern about drinking water safety and put pressure on Congress for legislative action. The initial congressional hearings on drinking water were held in 1971 and 1972. Like most major legislation, there was substantial debate on the best legislative approach, and more than one session of Congress was needed to pass the initial SDWA. After four years of work, Congress passed the first SDWA in November 1974, which was signed into law on Dec. 16, 1974 (PL 93-523).

The First 1974 SDWA

The 1974 SDWA established a partnership between the states and the federal government for the implementation of the drinking water program that continues to the present. This legislation dramatically changed the federal-state regulatory relationship. Under the SDWA, USEPA conducts the necessary research and analyses and establishes National Primary Drinking Water Regulations (NPDWRs). NPDWRs are legally enforceable standards that apply to PWSs, which are defined by the SDWA as having at least 15 service connections or regularly serving 25 residents. It should be noted that systems with fewer than 15 service connections and private wells are not covered by the SDWA and the resultant NPDWRs. These regulations protect public health by limiting the levels of contaminants in drinking water using maximum contaminant levels (MCL) or treatment techniques (TT) if analytical techniques are not economically or technologically feasible for the specific contaminant.

If individual states or American Indian tribal nations pass their own regulations that are at least as stringent as the federal regulations and have programs and enforcement authorities to ensure that PWSs within the state are in compliance with the regulations, USEPA will delegate primacy to the state or tribe. Currently, 49 states and 1 tribe have primacy and oversee PWSs (with some federal assistance and oversight).

PWSs have the ultimate responsibility for compliance with these regulations, including specific requirements for monitoring and reporting. Failure to meet any of these requirements can result in enforcement actions and, in some cases, penalties. Before the 1974 SDWA was passed, national drinking water standards were not enforceable, except for the coliform standard for interstate carriers, i.e., trains, airplanes, buses, and ships.

Soon after passage of the 1974 SDWA, USEPA published the first two national drinking water regulations (Table 1-1): the National Interim Primary Drinking Water Regulations (NIPDWRs), using the USPHS standards as the starting point; and the Total Trihalomethanes (TTHM) Rule. These two rules increased the number of regulated contaminants to 23 (Fig. 1-1).

The TTHM Rule was the first national primary drinking water regulation for which USEPA prepared detailed assessments of toxicology and health risk, occurrence and exposure, analytical methods, treatment technologies, and economic impacts. Many of the policies and procedures used to develop the economic analyses, occurrence estimates, and technologies and costs for the TTHM Rule formed the foundation of the current regulatory development process.

TABLE 1-1 National Primary Drinking Water Regulations

Promulgation date	Regulation	Reference
Dec. 24, 1975	National Interim Primary Drinking Water Regulations	FR* 40:248:59566
Nov. 29, 1979	Total Trihalomethanes	FR 44:231:68624
April 2, 1986	Fluoride	FR 51:63:11396
July 8, 1987	Phase I Volatile Organic Chemicals	FR 52:130:25690
June 29, 1989	Surface Water Treatment Rule	FR 54:124:27486
June 29, 1989	Total Coliform Rule	FR 54:124:27544
Jan. 20, 1991	Phase II Synthetic Organic Chemicals (SOCs) and Inorganic Chemicals (IOCs)	FR 56:20:3526
June 7, 1991	Lead and Copper Rule	FR 56:110:26460
July 17, 1992	Phase V SOCs and IOCs	FR 57:138:31776
Dec. 16, 1998	Stage 1 Disinfection By-Products Rule	FR 63:241:69389
Dec. 16, 1998	Interim Enhanced Surface Water Treatment Rule	FR 63:241:69477
Dec. 7, 2000	Radionuclides	FR 65:236:76707
Jan. 22, 2001	Arsenic	FR 66:14:6975
June 8, 2001	Filter Backwash Recycling Rule	FR 66:111:31085
Jan. 14, 2002	Long Term 1 Enhanced Surface Water Treatment Rule	FR 67:91:1844
Jan. 4, 2006	Stage 2 Disinfection By-Products Rule	FR 71:2:387
Jan. 5, 2006	Long Term 2 Enhanced Surface Water Treatment Rule	FR 71:3:653
Nov. 8, 2006	Ground Water Rule	FR 71:216:65573

*FR – *Federal Register*

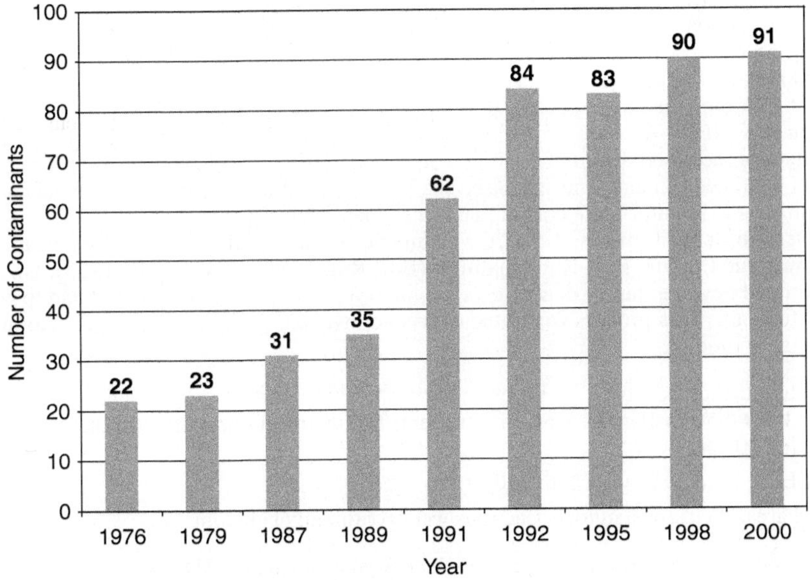

FIGURE 1-1 Number of regulated contaminants from 1976 through 2000. (*Source: www.epa.gov/safewater/contaminants/pdfs/contam_timeline.pdf.*)

However, these analyses require a significant amount of data, and many complex technical and policy issues must be debated and resolved in order to complete these analyses. To effectively utilize taxpayer dollars and adhere to the SDWA goals, USEPA should target its drinking water research and regulatory development efforts on the contaminants that present the greatest health risk. Consequently, in 1983, USEPA, in collaboration with the Awwa Research Foundation (AwwaRF, now known as the Water Research Foundation), conducted a series of workshops with a variety of national drinking water experts to discuss the following questions (AwwaRF, 1983):

- What are the most important contaminants to regulate?
- Are there robust and reliable analytical methods for analyzing these contaminants?
- What are the health effects of these contaminants and what is the exposure?
- What treatment technologies work for these contaminants and what do these technologies cost?
- How would water utilities monitor and report compliance to the states?

Answering these questions with limited data for a large number of potential contaminants is not easy (these questions are still relevant today for drinking water risk management and regulatory development). After these workshops, USEPA continued to collect data on health effects, analytical methods, occurrence, and treatment technologies but did not issue any new national drinking water regulations until 1986.

The 1986 SDWA Amendments

Frustrated by USEPA's lack of regulatory progress (progress being defined as an increasing number of regulations), Congress amended the SDWA in 1986 (PL 99-339). The amendments placed USEPA on a "regulatory treadmill" with requirements to regulate a specific list of 83 contaminants in the first five years and then 25 new contaminants every three years thereafter. On the basis of these statutory requirements, the number of regulated contaminants would have exceeded 250 in 2007.

USEPA increased its regulatory development process in the late 1980s and early 1990s. Seven new NPDWRs were promulgated between 1986 and 1992 (see Table 1-1). These regulations increased the number of regulated contaminants to 84 (see Fig. 1-1). The number of regulated contaminants increased sharply in 1991 and 1992, and the financial burden for utilities to monitor these contaminants also increased substantially.

Despite its best efforts, USEPA was unable to meet multiple regulatory deadlines and was sued by the Bull Run Coalition (Bull Run Coalition v. Reilly, 1993). USEPA negotiated new regulatory deadlines, then missed those new deadlines, and had to renegotiate again. This process frustrated everyone involved in the regulatory development process, including:

- Water utilities, because they never knew when new regulations were coming out and did not know how to plan for capital investments for treatment improvements that would last 50 to 100 years
- USEPA, because it was continually being sued
- Congress, because statutory deadlines were continually being missed

Throughout the early 1990s, pressure increased to amend the SDWA and allow USEPA to jump off the regulatory treadmill and more appropriately focus its limited resources. Congress began holding hearings and debating potential SDWA amendments in the

103rd Congress in 1993 and 1994 and ultimately passed the 1996 SDWA Amendments (PL 104-208) in the 104th Congress.

The 1996 SDWA Amendments

The 1996 SDWA Amendments can be divided into the following areas:

- A new standard-setting process with specific statutory language on how to select contaminants for potential regulation and then how to set the regulation.
- Priority regulations with specific deadlines for contaminants such as arsenic, sulfate, and radon and the Microbial/Disinfection By-Product (M/DBP) cluster.
- New state programs for source water assessments, capacity development, operator certification, and a drinking water state revolving loan fund.
- New public information programs, such as the Consumer Confidence Report (CCR) for utilities, and revision of the Public Notification Regulation (PNR) by USEPA.

USEPA promulgated nine new or revised NPDWRs between 1998 and 2006 (see Table 1-1). These regulations increased the number of regulated contaminants to 91 (see Fig. 1-1). The nine NPDWRs promulgated by USEPA since the 1996 SDWA Amendments are primarily new or expanded treatment technique requirements. Therefore, although the number of contaminants with MCLs has not increased significantly, the complexity of the treatment techniques, i.e., the more complex turbidity requirements in the Interim and Long Term 1 Enhanced Surface Water Treatment Rules, and more advanced compliance treatment technologies, i.e., ion exchange for arsenic removal, have significantly increased costs for many PWSs.

The Bioterrorism Act of 2002

Prior to September 11, 2001, water security had not been a significant problem for water utilities and there were no legislative or regulatory requirements. After 9/11, Congress reacted to address security concerns for critical infrastructure (CI), with the water sector (both drinking water and wastewater) being one of the CI sectors. To address water security concerns, the SDWA was amended through the Public Health Security and Prevention Preparedness Act of 2002 (the Bioterrorism Act, PL 107-188). The legislation required water utilities serving more than 10,000 people to meet five new statutory requirements: (1) conduct a vulnerability assessment (VA); (2) submit the VA to USEPA (USEPA had statutory requirements to develop policies and procedures for protection of the VAs that were submitted); (3) certify to USEPA that the VA was properly conducted and met the requirements of the Bioterrorism Act; (4) conduct or revise the utility emergency response plan (ERP) based on the knowledge derived from the VA; and (5) certify to USEPA that the new or revised ERP has been completed.

Although not part of the SDWA, the Homeland Security Act (PL 107-296) was also passed in 2002 and created the Department of Homeland Security (DHS) by merging parts of 22 different federal agencies into one. DHS has the overall responsibility for homeland security, and USEPA has been designated as the lead agency for the water sector. DHS created an overall risk management framework for critical infrastructure through the National Infrastructure Protection Plan (NIPP) (DHS, 2006). Under the NIPP framework, the water sector developed its own Water Sector-Specific Plan (SSP). The Water SSP, along with the other SSPs, was released by DHS in 2007 (DHS, 2007). See other publications for more detail on water security issues (States, 2010; Roberson and Morley, 2006; Mays, 2004).

THE RISK MANAGEMENT AND STANDARD-SETTING PROCESSES

The 1996 SDWA Amendments established a scientific, risk-based approach to targeting, assessing, and managing health risks from contaminants in PWSs. This approach targets research, assessment, and regulatory activities on the contaminants that have the greatest likelihood of presenting health risks from drinking water. The amendments also recognized that over time, better information becomes available and requires USEPA to regularly reassess and reprioritize its risk management efforts.

The mechanisms required by the SDWA for USEPA to gather and assess data to prioritize contaminants for risk management actions include: (1) the Contaminant Candidate List (CCL), (2) the Unregulated Contaminant Monitoring Rules (UCMRs), (3) regulatory determinations, and (4) the review of NPDWRs (six-year review). The risk management actions that SDWA authorizes include: (1) NPDWRs, (2) National Secondary Drinking Water Regulations, and (3) Health Advisories and Other Actions.

This section discusses each of these targeting and risk management processes. Figure 1-2 provides an overview of how these different processes fit together in the development of regulations.

Contaminant Candidate List

The Contaminant Candidate List (CCL) is developed by USEPA as a listing of priority contaminants for regulatory decision making and information collection. The SDWA requires that every five years, USEPA publish a list of unregulated contaminants that are known or anticipated to occur in PWSs and that may require regulation. In developing a CCL, USEPA must consider the contaminants identified in section 101(14) of the Comprehensive Environmental Response, Compensation, and Liability Act of 1980 (CERCLA, or Superfund) and substances registered as pesticides under the Federal Insecticide, Fungicide, and Rodenticide Act (FIFRA). USEPA must also consult with the scientific community and request and consider public comment on a draft list.

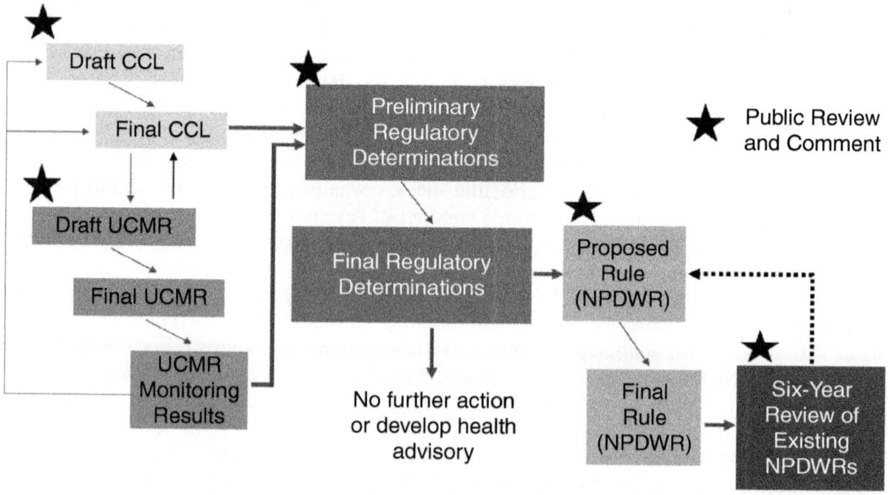

FIGURE 1-2 Overview of SDWA regulatory processes.

USEPA published the first CCL (CCL1) in 1998 (USEPA, 1998a). CCL1 contained 50 chemicals and 10 microbial contaminants. USEPA consulted with the scientific community to develop a process to identify CCL1 contaminants. The process used a combination of expert judgment for microbial contaminants and screening and evaluation criteria to identify chemical contaminants.

In response to comments that a more comprehensive and reproducible approach was needed for selecting contaminants for future CCLs, USEPA sought advice from the National Academies of Science–National Research Council (NRC). The NRC recommended that USEPA continue to use expert judgment and public involvement to identify future contaminants for the CCLs (NRC, 2001a). The NRC also recommended that USEPA first screen a broad universe of contaminants of potential concern to identify a preliminary CCL (PCCL) based on available health risk data and likelihood of occurrence in drinking water. Then USEPA would assess the PCCL contaminant data in a more detailed manner, using classification tools and expert judgment to evaluate the likelihood that specific contaminants could occur in drinking water at levels and at frequencies that pose a public health risk.

To ensure broad stakeholder input, USEPA also consulted with the National Drinking Water Advisory Council (NDWAC) on its implementation of the NRC-recommended CCL process. The NDWAC endorsed the NRC recommendations, which it described as a three-step process, as depicted in Fig. 1-3 (NDWAC, 2004). The NDWAC provided specific recommendations for implementing each step. Because of differences in the information available for microbes and chemicals, the NDWAC recommended these contaminants be evaluated in parallel procedures. The NDWAC also recommended that USEPA move forward using a step-wise adaptive management approach to build upon advances in technology and the experience it has gained in developing previous CCLs.

USEPA did not implement the NRC and NDWAC recommendations for the second CCL (CCL2) published in 2005 because the recommended processes would not have been completed in time (USEPA, 2005a). However, the agency described the improved process it would implement for future CCLs. CCL2 consisted of the 51 (42 chemical and 9 microbial) contaminants for which USEPA had not yet made regulatory determinations.

USEPA published the draft of the third CCL (CCL3) for public comment in February 2008 (USEPA, 2008a). This draft included 104 contaminants—93 chemicals and 11 microbiological contaminants. The draft CCL3 was developed using the NRC/NDWAC-recommended process to evaluate approximately 7500 chemical and microbial contaminants. USEPA also considered the contaminant nominations and information received from the public in preparing the draft CCL3. The final CCL3 was published in October 2009 and included 116 contaminants (104 chemicals and 12 microbiological contaminants), as listed in Table 1-2 (USEPA, 2009b).

Research plays a significant role in filling the data gaps identified in the CCL process. The final CCL3 contained a table on regulatory determination data/information needs for each contaminant, broken down into health effects, occurrence, and need for analytical methods (USEPA, 2009b). This table shows the depth and breadth of the potential research agenda for USEPA's drinking water program, as there are numerous data/information needs for the CCL3 contaminants.

Regulatory Determinations

A regulatory determination is a decision made by USEPA on whether to initiate a national primary drinking water rulemaking for a contaminant. The SDWA requires that every five years USEPA make regulatory determinations for at least five contaminants on the CCL. Section 1412(b)(1) of the SDWA specifies three criteria that must be met for USEPA to make a determination to develop a national regulation for a contaminant: "(1) the contaminant may

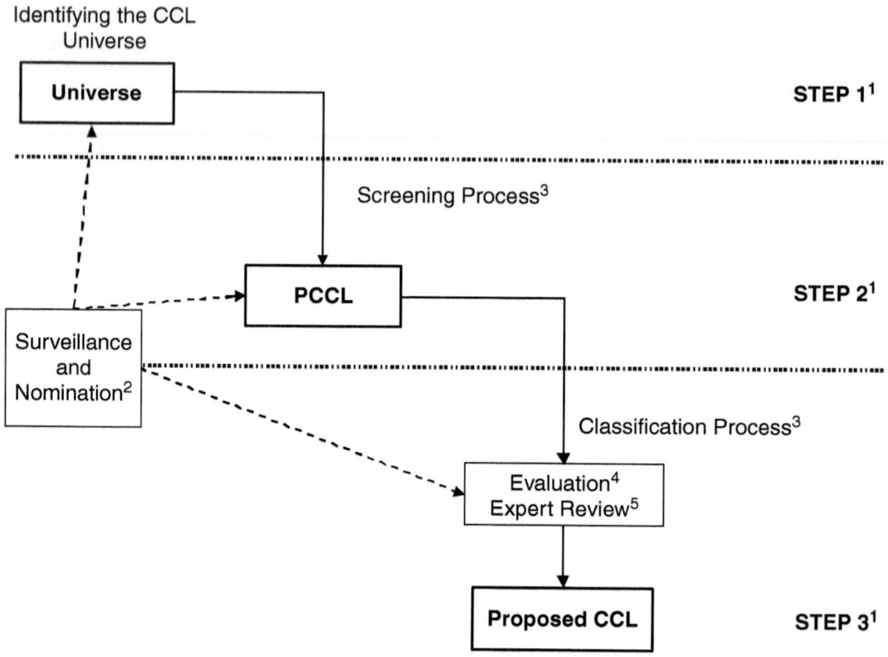

Notes:

1. Steps are sequential, as are components of each step, with the exception of surveillance and nomination. This generalized process is applicable to both chemical and microbial contaminants, though the specific execution of particular steps may differ in practice.

2. Surveillance and nomination provide an alternative pathway for entry into the CCL process for new and emerging agents, in particular. Most agents would be nominated to the CCL Universe. Depending on the timing of the nomination and the information available, a contaminant could move onto the PCCL or CCL, if justified.

3. Expert judgment, possibly including external expert consultation, will be important throughout the process, but particularly at key points, such as reviewing the screening criteria and process from the Universe to the PCCL; assessing the training data set and classification algorithm performance during development of the PCCL to CCL classification step.

4. After implementing the classification process, the prioritized list of contaminants would be evaluated by experts, including a review of the quality of information.

5. The CCL classification process and draft CCL list would undergo a critical Expert Review by us EPA and by outside experts before the CCL is proposed.

FIGURE 1-3 Overview of CCL process recommended by NDWAC Work Group and incorporated by USEPA into CCL3. (*Source: National Drinking Water Advisory Council, 2004; www.epa.gov/safewater/ndwac/pdfs/report_ccl_ndwac_07-06-04.pdf.*)

have an adverse effect on the health of persons; (2) the contaminant is known to occur or there is a substantial likelihood the contaminant will occur in public water systems with a frequency and at levels of public health concern; and (3) in the sole judgment of the Administrator, regulation of the contaminant presents a meaningful opportunity for health risk reductions for persons served by public water systems."

TABLE 1-2 Third Contaminant Candidate List

Microbial Contaminants (12)	
Adenovirus	
Caliciviruses	
Campylobacter jejuni	
Enterovirus	
Escherchia coli (O157)	
Helicobacter pylori	
Hepatitis A virus	
Legionella pneumophila	
Mycobacterium avium	
Naegleria fowleri	
Salmonella enterica	
Shigella sonnei	
Chemical Contaminants (104)	
Common name—registry name	CASRN*
alpha-Hexachlorocyclohexane	319-84-6
1,1,1,2-Tetrachloroethane	630-20-6
1,1-Dichloroethane	75-34-3
1,2,3-Trichloropropane	96-18-4
1,3-Butadiene	106-99-0
1,3-Dinitrobenzene	99-65-0
1,4-Dioxane	123-91-1
17alpha-estradiol	57-91-0
1-Butanol	71-36-3
2-Methoxyethanol	109-86-4
2-Propen-1-ol	107-18-6
3-Hydroxycarbofuran	16655-82-6
4,4'-Methylenedianiline	101-77-9
Acephate	30560-19-1
Acetaldehyde	75-07-0
Acetamide	60-35-0
Acetochlor	34256-82-1
Acetochlor ethanesulfonic acid (ESA)	187022-11-3
Acetochlor oxanilic acid (OA)	184992-44-4
Acrolein	107-02-8
Alachlor ethanesulfonic acid (ESA)	142363-53-9
Alachlor oxanilic acid (OA)	171262-17-2
Aniline	62-53-3
Bensulfide	741-58-2
Benzyl chloride	100-44-7
Butylated hydroxyanisole	25013-16-5
Captan	133-06-2
Chloromethane (Methyl chloride)	74-87-2
Clethodim	110429-62-4
Cobalt	7440-48-4
Cumene hydroperoxide	80-15-9
Cyanotoxins (3)	
Dicrotophos	141-66-2
Dimethipin	55290-64-7
Dimethoate	60-51-5
Disulfoton	298-04-4

(*Continued*)

TABLE 1-2 Third Contaminant Candidate List (*Continued*)

Diuron	330-54-1
Equilenin	517-09-9
Equilin	474-86-2
Erythromycin	114-07-8
Estradiol (17-beta estradiol)	50-28-2
Estriol	50-27-1
Estrone	53-16-7
Ethinyl Estradiol (17-alpha ethynyl estradiol)	57-63-6
Ethoprop	13194-48-4
Ethylene glycol	107-21-1
Ethylene oxide	75-21-8
Ethylene thiourea	96-45-7
Fenamiphos	22224-92-6
Formaldehyde	50-00-0
Germanium	7440-56-4
HCFC-22	75-45-6
Hexane	110-54-3
Hydrazine	302-01-2
Mestranol	72-33-3
Methamidophos	10265-92-6
Methanol	67-56-1
Methyl bromide (Bromomethane)	74-83-9
Methyl-*tert*-butyl-ether (MTBE)	1634-04-4
Metolachlor	51218-45-2
Metolachlor ethanesulfonic acid (ESA)	171118-09-5
Metolachlor oxanilic acid (OA)	152019-73-3
Molinate	2212-67-1
Molybdenum	7439-98-7
Nitrobenzene	98-95-3
Nitroglycerin	55-63-0
N-methyl-2-pyrrolidone	872-50-4
N-nitrosodiethylamine (NDEA)	55-18-5
N-nitrosodimethylamine (NDMA)	62-75-9
N-nitroso-di-n-propylamine (NDPA)	621-64-7
N-nitrosodiphenylamine	86-30-6
N-nitrosopyrrolidine (NPYR)	930-55-2
Norethindrone (19-Norethisterone)	68-22-4
n-Propylbenzene	103-65-1
o-Toluidine	95-53-4
Oxirane, methyl-	75-56-9
Oxydemeton-methyl	301-12-2
Oxyfluorfen	42874-03-3
Perchlorate	14797-73-0
Perfluorooctane sulfonic acid (PFOS)	1763-23-1
Perfluorooactanoic acid (PFOA)	335-67-1
Premethrin	52645-53-1
Profenofos	41198-08-7
Quinoline	91-22-5
RDX (Hexahydro-1,3,5-trinitro-1,3,5-triazine)	121-82-4
sec-Butylbenzene	135-98-8
Strontium	7440-24-6

(*Continued*)

TABLE 1-2 Third Contaminant Candidate List (*Continued*)

Tebuconazole	107534-96-3
Tebufenozide	112410-23-8
Tellurium	13494-80-9
Terbufos	13071-79-9
Terbufos sulfone	56070-16-7
Thiodicarb	59669-26-0
Thiophanate-methyl	23564-05-8
Toluene diisocyanate	26741-62-5
Tribufos	78-48-8
Triethylamine	121-44-8
Triphenyltin hydroxide (TPTH)	76-87-9
Urethane	51-79-6
Vanadium	7440-62-2
Vinclozolin	50471-44-8
Ziram	137-30-4

*Chemical Abstracts Service Registry Number

USEPA developed a comprehensive approach for evaluating these criteria with significant input from the NRC (NRC, 1999a, 1999b) and the NDWAC (USEPA, 2003a). To evaluate the first criterion, USEPA evaluates best available, peer-reviewed assessments[2] to characterize the health effects that may result from consuming the contaminant in drinking water (USEPA, 2008b). From this information, USEPA estimates a health reference level (HRL) that takes into account the potential for other routes of exposure (e.g., food). To evaluate the second criterion, USEPA analyzes data from nationally representative occurrence studies[3] and compares these data to the HRL to determine the frequency at which PWSs exceed this level of concern. To evaluate the third statutory criterion, USEPA evaluates the potential health risks in the populations above the health reference level. USEPA also evaluates the nondrinking water route of exposure to determine if removing the contaminant from drinking water will significantly reduce the population's exposure to the contaminant.

USEPA has made regulatory determinations for 20 contaminants, 9 from CCL1 and 11 from CCL2, as listed in Table 1-3 (USEPA, 2000, 2008b). For all of these contaminants, USEPA has made a determination not to regulate them because they did not occur frequently in PWSs at levels of health concern and/or there was not a meaningful opportunity for health risk reduction through a national primary drinking water rule.

USEPA requested comment on a preliminary regulatory determination to not regulate perchlorate in October 2008 (USEPA, 2008c). In August 2009, USEPA published a supplemental request for comments on alternative analysis of the perchlorate regulatory determination (USEPA, 2009c). In this notice, USEPA presented a broader range of alternatives for interpreting the available data on: (1) the level of concern, (2) the frequency of occurrence of perchlorate in drinking water, and (3) the opportunity for health-risk reduction through a national perchlorate standard.

[2]USEPA has relied upon peer-reviewed risk assessments from the agency's Integrated Risk Information System (IRIS) or the pesticide reregistration eligibility decisions (RED), as well as from the National Academy of Sciences (NAS) or the Agency for Toxic Substances and Disease Registry (ATSDR).

[3]USEPA has relied on data from the Unregulated Contaminant Monitoring (UCM) Program, the National Inorganic and Radionuclide Survey (NIRS), and the first Unregulated Contaminant Monitoring Rule (UCMR1) in making its regulatory decisions for CCL1 and CCL2 contaminants.

TABLE 1-3 Contaminants Not Regulated by First and Second Regulatory Determinations

First regulatory determinations (9)	Second regulatory determinations (11)
Managanese	Boron
Sodium	Dachtal mono-acid degradate
Sulfate	Dachtal di-acid degradate
Aldrin	1,1-dichloro-2,2-bis(p-chlorophenyl)ethylene
Dieldrin	1,3-dichloropropene
Metribuzin	2,4-dinitrotoluene
Hexachlorobutadiene	2,6-dinitrotoluene
Napthalene	*s*-ethyl dipropulthiocarbamate
Acanthamoeba	Fonofos
	Terbacil
	1,1,2,2-tetrachloroethane

Unregulated Contaminant Monitoring Regulations

Unregulated contaminant monitoring regulations (UCMRs) require the collection of drinking water contaminant occurrence data that can be used by USEPA to identify contaminants for the CCL, to support regulatory determinations, and to develop national primary drinking water regulations. The 1986 SDWA Amendments provided authority for USEPA to gather information on unregulated contaminants. USEPA included unregulated contaminant monitoring (UCM) requirements in the Phase I and Phase II regulations. The UCM monitoring continued until the 1996 SDWA Amendments required substantial revisions to the program. Under the 1996 SDWA Amendments, USEPA is required to: (1) publish a list of not more than 30 unregulated contaminants every five years, i.e., the UCMR; (2) identify a representative sample of PWSs serving 10,000 or fewer people to monitor, with USEPA paying the cost of analyzing samples from those systems; (3) place the monitoring data in the National Contaminant Occurrence Database; and (4) notify consumers that the monitoring results are available.

USEPA selects contaminants for the UCMR by evaluating contaminants that have been targeted through prioritization processes (i.e., the CCL). The agency identifies additional contaminants through an evaluation of current research on occurrence and health-effects risk factors. USEPA does not list contaminants that do not have an analytical reference standard or contaminants whose analytical methods are not ready for widespread use under UCMR.

The first UCMR (UCMR1), which was promulgated in 1999 (USEPA, 1999a), listed 12 contaminants for assessment monitoring (List 1) at all large water systems (serving more than 10,000 people) and a representative sample of 800 small water systems (serving fewer than 10,000 people). UCMR1 also listed 14 contaminants for screening monitoring (List 2) at 300 randomly selected large and small water systems. Surface water systems were required to collect four quarterly samples, and groundwater systems were required to collect two semiannual samples. Monitoring data for UCMR1 were reported to USEPA from 2001 to 2005 and are available on the Internet at www.epa.gov/safewater/ucmr/data.html#2.

USEPA promulgated the second UCMR (UCMR2) in 2007 (USEPA, 2007a). The UCMR2 lists 10 contaminants for assessment monitoring (List 1) at all large systems and at 800 selected small systems. UCMR2 also requires screening monitoring (List 2) for 15 contaminants from all very large systems (serving more than 100,000 people) and from 600 selected medium and small systems. Monitoring for the UCMR2 is to be performed during a 12-month period from January 2008 to December 2010.

National Primary Drinking Water Regulations

National Primary Drinking Water Regulations (NPDWRs), which are legally enforceable standards that apply to PWSs, protect public health by limiting the levels of contaminants in drinking water. NPDWRs take the form of MCLs or TTs. An MCL is the maximum permissible level of a contaminant in water that is delivered to any user of a PWS. A treatment technique is an enforceable procedure or level of technological performance that PWSs must follow to ensure control of a contaminant. Examples of TT rules are the Surface Water Treatment Rule (disinfection and filtration for inactivation/removal of target pathogens) and the Lead and Copper Rule (optimized corrosion control). More details on the individual rules are found in the next section.

To propose a new or revised NPDWR, the 1996 SDWA Amendments require USEPA to undertake a number of steps, including:

- Establish a maximum contaminant level goal (MCLG). The MCLG is the maximum level of a contaminant in drinking water at which no known or anticipated adverse effect on the health of persons would occur, allowing for an adequate margin of safety. MCLGs are nonenforceable public health goals.
- Set the MCL as close as feasible to the MCLG. The feasible level is the level that may be achieved with the use of the best available technology, TTs, and other means that USEPA finds (after examination for efficiency under field conditions and not solely under laboratory conditions) are available, taking cost into consideration. When there is no reliable method that is economically and technically feasible to measure a contaminant, USEPA establishes a TT for control of that contaminant.
- Prepare a health-risk reduction cost analysis (HRRCA) that includes estimates of the quantifiable and nonquantifiable costs and benefits of the regulatory alternatives, including the feasible level that is closest to the MCLG.
- Determine if the costs justify the benefits at the feasible level. If not, USEPA may set the MCL at a level that maximizes health risk reduction benefits at a cost that is justified by the benefits.
- List the technologies that achieve compliance with the MCL or TT. USEPA can update the list at any time after promulgating a standard to list new or innovative technologies that achieve compliance with a standard.
- Identify affordable small-system compliance technologies. If none are available, USEPA must identify small-system variance technologies that remove the contaminant to the maximum extent affordable and are protective of public health.

The SDWA specifies that USEPA use the "the best available, peer-reviewed science" in the decision-making processes (i.e., CCL, regulatory determinations, and developing NPDWRs). The SDWA also specifies that USEPA must propose a NPDWR within 24 months of making a determination to regulate a contaminant and promulgate a final regulation within 18 months of proposal. At the Administrator's discretion and public notification, USEPA can extend this deadline for the final rule by up to nine months.

Although USEPA has not yet made a determination to regulate a CCL contaminant because none of the CCL contaminants have met the three SDWA criteria previously discussed, the agency has implemented the standard-setting processes required by the 1996 SDWA Amendments in developing the priority regulations (e.g., Arsenic, Radionuclides, The Microbial and Disinfection By-Products Rules).

National Primary Drinking Water Regulation Review

The National Primary Drinking Water Regulation Review, or six-year review, is an evaluation by USEPA of the available information on health effects, analytical methods, treatment technologies, and any other factors for existing NPDWRs to determine if revisions are appropriate. The SDWA requires that USEPA review and revise as appropriate each NPDWR every six years. The SDWA also requires that each revision shall maintain or provide for greater protection of public health.

USEPA has developed a protocol based upon input from the NDWAC to systematically evaluate NPDWRs to determine if a revision presents a meaningful opportunity to improve the level of public health protection or to achieve cost savings while maintaining or improving the level of health protection, as shown in Fig. 1-4 (USEPA, 2003b). In carrying out the six-year review, USEPA compiles the available, peer-reviewed information on health effects, analytical feasibility, and treatment for regulated contaminants to determine if the data indicate a need to reevaluate a contaminant's NPDWR. If no new data are available, USEPA assumes the existing NPDWRS remain appropriate. However, if new data are available, USEPA determines whether changes in the NPDWR for that contaminant are warranted.

For example, if the current MCL for a contaminant was set at the level of analytical feasibility and a new or improved analytical method is now available for a contaminant, USEPA then determines if the lower analytical quantitation level is feasible. USEPA also determines if there is a meaningful opportunity to improve public health by changing the standard. For example, based on contaminant occurrence data collected from states, USEPA will estimate the population served by systems where the concentration of the contaminant exceeds the potentially lower new standard.

USEPA completed the first six-year review in 2003 (USEPA, 2003c); the agency reviewed NPDWRs for 69 contaminants and concluded that it was appropriate to revise one NPDWR, the Total Coliform Rule (TCR) at that time. The TCR revisions are discussed later in this chapter.

National Secondary Drinking Water Regulations

National Secondary Drinking Water Regulations, referred to as secondary maximum contaminant levels (SMCLs), are nonenforceable guidelines for contaminants that may cause cosmetic effects in consumers (e.g., skin discoloration) or aesthetic effects in drinking water (e.g., taste, odor, or color) (USEPA, 1992). PWSs are not required to comply with SMCLs unless their states have chosen to adopt them as an enforceable standard. The SDWA defines an SMCL as a regulation that USEPA determines is "requisite to protect the public welfare." USEPA has established SMCLs for 15 contaminants, as shown in Table 1-4 (USEPA, 1979, 1986, 1991).

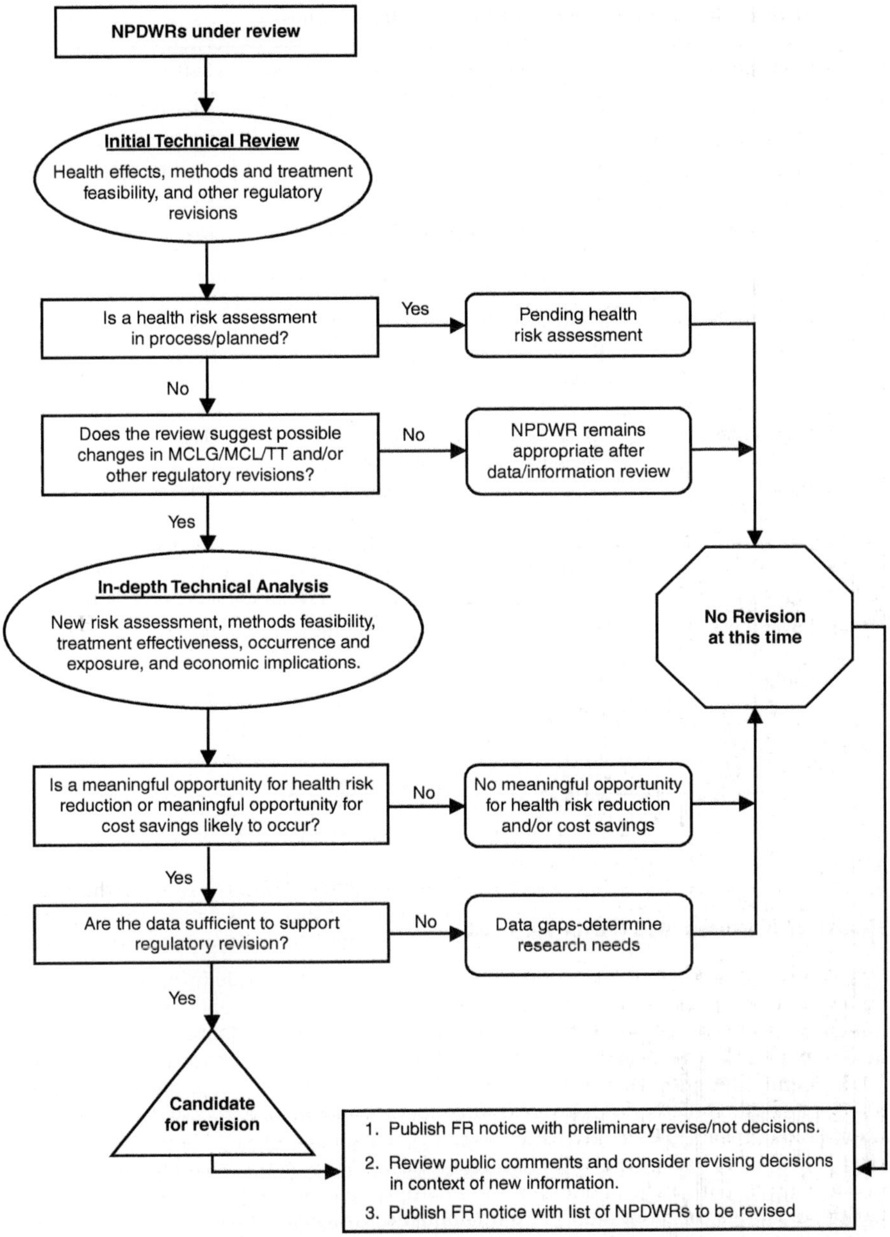

FIGURE 1-4 Overview of the six-year review protocol and making the revise/not revise decision. (*Source: EPA Protocol for the Review of Existing National Primary Drinking Water Regulations, EPA-815-R-03-002, 2003; www.epa.gov/safewater/standard/review/pdfs/support_6yr_protocal_final.pdf.*)

TABLE 1-4 National Secondary Drinking Water Standards

Contaminant	Effect(s)	SMCL, mg/L
Aluminum	Colored water	0.05–0.2
Chloride	Salty taste	250
Color	Visible tint	15 color units
Copper	Metallic taste; blue-green stain	1.0
Corrosivity	Metallic taste; corrosion; fixture staining	Noncorrosive
Fluoride	Tooth discoloration	2
Foaming agents	Frothy, cloudy; bitter taste; odor	0.5
Iron	Rusty color; sediment; metallic taste; reddish or orange staining	0.3
Manganese	Black to brown color; black staining; bittermetallic taste	0.05
Odor	"Rotten egg," musty, or chemical smell	3 TON
pH	Low pH: bitter metallic taste, corrsosion high pH: slippery feel, soda taste, deposits	6.5–8.5
Silver	Skin discoloration; greying of the white part of the eye	0.10
Sulfate	Salty taste	250
Total dissolved solids (TDS)	Hardness; deposits; colored water; staining; salty taste	500
Zinc	Metallic taste	5

Source: Letterman, 1999

Health Advisories and Other Actions

Health advisories are documents prepared by USEPA that provide information on contaminants that can cause human health effects and are known or anticipated to occur in drinking water. Health advisories provide nonenforceable guidance values (HA values) based on noncancer health effects for different durations of exposure (e.g., 1-day, 10-day, and lifetime). Health advisories also provide technical guidance on health effects, analytical methodologies, and treatment technologies associated with drinking water contaminants. These advisories were first developed in 1987 before USEPA had established many of the NPDWRs in effect today. USEPA has issued health advisories in association with its regulatory determinations, and has recently issued or revised health advisories for more than 170 contaminants (USEPA, 2006a).

In addition to health advisories, USEPA takes other actions to address concerns associated with drinking water contaminants. These actions include issuing drinking water advisories (DWAs), which are similar to a health advisories in that they provide a nonenforceable guidance value. However, the DWA value is based on aesthetic values (taste, odor, and color). USEPA has published DWAs for sulfate, methyl-*tert*-butyl-ether (MTBE), and sodium (USEPA, 2006a).

CURRENT DRINKING WATER REGULATIONS

Individual Rules

USEPA finalized 18 NPDWRs between 1975 and 2006 (see Table 1-1). Typically, minor technical corrections are needed after final rule promulgation, and a separate *Federal Register* notice is issued for these corrections. The number of regulated contaminants has varied from 90 to 91 since the 1996 SDWA Amendments (see Fig. 1-1). The secondary standards are listed in Table 1-4, and the MCLGs and MCLs for the primary drinking water standards are listed in Table 1-5.

The 1996 SDWA Amendments mandated specific deadlines for a handful of regulations that were known as the "priority" regulations. These regulations address the following:

- *Disinfectants and disinfection by-products.* This set of regulations was to be finalized in accordance with the regulatory schedule listed in the 1994 proposed Information Collection Rule (ICR). The Stage 1 Disinfection By-Product Rule (DBPR) and Interim Enhanced Surface Water Treatment Rule (IESWTR) were promulgated in 1998, the Long Term 1 Enhanced Surface Water Treatment Rule (LT1ESWTR) was promulgated in 2002, and the Stage 2 DBPR and Long Term 2 Enhanced Surface Water Treatment Rule (LT2ESWTR) were promulgated in 2006.
- *Arsenic.* The arsenic regulation was to be proposed by Jan. 1, 2000, and finalized by Jan. 1, 2001. The arsenic regulation was promulgated in 2001.
- *Sulfate.* Regulatory determination was to be made for sulfate by August 2001 as part of the first regulatory determinations. USEPA made a final determination not to regulate sulfate in July 2003 (USEPA, 2003a).
- *Filter backwash.* A regulation to address filter backwash was to be finalized by August 2000. This rule was promulgated in 2001.
- *Radon.* The radon regulation was to be proposed by August 1999 and finalized by August 2000. USEPA proposed a radon in drinking water rule in November 1999 but has not promulgated a final rule (USEPA, 1999b).
- *Groundwater.* A rule to address potential groundwater contamination was to be finalized sometime between August 1999 and publication of the final Stage 2 DBPR. The Ground Water Rule (GWR) was promulgated in 2006.

USEPA has also promulgated other national drinking water regulations that are not NPDWRs. Many of these regulations mandate specific monitoring, such as the ICR, and the first and second Unregulated Contaminant Monitoring Rule (UCMR1 and UCMR2). This category of regulations also includes the Consumer Confidence Report (CCR), a report on water quality that must be sent annually to customers. The CCR is an important part of the improved public education component of the 1996 SDWA Amendments. Table 1-6 lists other significant national drinking water regulations.

As previously discussed, USEPA has taken other regulatory actions, such as CCLs and regulatory determinations, that are not regulations in the sense that compliance by PWSs is not required. However, these actions form the foundation of the regulatory development process by identifying contaminants that may require regulation.

For most regulations, USEPA develops a variety of publications in order to provide compliance assistance for water utilities and state agencies. These publications typically include fact sheets on the regulations and a variety of guidance manuals that provide detailed information on technical issues that cannot be found in USEPA's regulatory language and preamble. It should be noted that these guidance manuals are suggestions and

TABLE 1-5 National Primary Drinking Water Standards

Contaminant	MCLG, mg/L	MCL, mg/L	Potential health effects	Sources of drinking water contamination
Fluoride Rule				
Fluoride	4.0	4.0	Skeletal and dental fluorosis	Natural deposits; fertilizer, aluminum industries; drinking water additive
Phase I Volatile Organics				
Benzene	zero	0.005	Cancer	Some foods; gas, drugs, pesticide, paint, plastic industries
Carbon tetrachloride	zero	0.005	Cancer	Solvents and their degradation products
p-Dichlorobenzene	0.075	0.075	Cancer	Room and water deodorants and mothballs
1,2-Dichloroethane	zero	0.005	Cancer	Leaded gas, fumigants, paints
1,1-Dichloro-ethylene	0.007	0.007	Cancer, liver, kidney effects	Plastics, dyes, perfumes, paints
Trichloro-ethylene	zero	0.005	Cancer	Textiles, adhesives, metal degreasers
1,1,1-Tri-chloroethane	0.2	0.2	Liver, nervous system effects	Adhesives, aerosols, textiles, paints, inks, metal degreasers
Vinyl chloride	zero	0.002	Cancer	May leach from PVC pipe; formed by solvent breakdown
Surface Water Treatment Rule and Total Coliform Rule				
Giardia lamblia	zero	TT	Gastroenteric disease	Human and animal fecal wastes
Legionella	zero	TT	Legionnaire's disease	Natural waters; can grow in water heating systems
Heterotrophic plate count	N/A	TT	Indicates water quality, effectiveness of treatment	
Total coliform	zero	< 5%+	Indicates gastroenteric pathogens	Human and animal fecal wastes
Turbidity	N/A	TT	Interferes with disinfection	Soil runoff
Viruses	zero	TT	Gastroenteric disease	Human and animal fecal wastes
Phase II Rule Inorganics				
Asbestos (>10 μm)	7 MFL	7 MFL	Cancer	Natural deposits; asbestos cement in water systems
Barium	2	2	Circulatory system effects	Natural deposits; pigments, epoxy sealants, spent coal
Cadmium	0.005	0.005	Kidney effects	Galvanized pipe corrosion; natural deposits; batteries, paints
Chromium 0.1 (total)	0.1	0.1	Liver, kidney, circulatory disorders	Natural deposits; mining, electroplating, pigments
Mercury (inorganic)	0.002	0.002	Kidney, nervous system disorders	Crop runoff; natural deposits; batteries, electrical switches
Nitrate	10	10	Methemoglobulinemia	Animal waste, fertilizer, natural deposits, septic tanks, sewage

Nitrite	1	1	Methemoglobulinemia	Same as nitrate; rapidly converted to nitrate
Nitrate + nitrite	10	10		
Selenium	0.05	0.05	Liver damage	Natural deposits; mining, smelting, coal/oil combustion
Phase II Rule Organics				
Acrylamide	zero	TT	Cancer, nervous system effects	Polymers used in sewage/waste-water treatment
Alachlor	zero	0.002	Cancer	Runoff from herbicide on corn, soybeans, other crops
Aldicarb	delayed	delayed	Nervous system effects	Insecticide on cotton, potatoes, other crops; widely restricted
Aldicarb sulfone	delayed	delayed	Nervous system effects	Biodegradation of aldicarb
Aldicarb sulfoxide	delayed	delayed	Nervous system effects	Biodegradation of aldicarb
Atrazine	0.003	0.003	Mammary gland tumors	Runoff from use as herbicide on corn and noncropland
Carbofuran	0.04	0.04	Nervous, reproductive system effects	Soil fumigant on corn and cotton; restricted in some areas
Chlordane	zero	0.002	Cancer	Leaching from soil treatment for termites
Chlorobenzene	0.1	0.1	Nervous system, liver effects	Waste solvent from metal degreasing processes
2,4-D	0.07	0.07	Liver and kidney damage	Runoff from herbicide on wheat, corn, rangelands, lawns
o-Dichloroethylene	0.1	0.1	Liver, kidney, blood cell damage	Paints, engine cleaning compounds, dyes, chemical
cis-1,2-Dichloroethylene	0.07	0.07	Liver, kidney, nervous, circulatory system effects	Waste industrial extraction solvents
trans-1,2-Dichloroethylene	0.1	0.1	Liver, kidney, nervous, circulatory system effects	Waste industrial extraction solvents
1,2-Dibromo-3-chloropropane	zero	0.0002	Cancer	Soil fumigant on soybeans, cotton, pineapple, orchards
1,2-Dichloropropane	zero	0.005	Liver, kidney effects; cancer	Soil fumigant; waste industrial solvents
Epichlorohydrin	zero	TT	Cancer	Water treatment chemicals; waste epoxy resins, coatings
Ethylbenzene	0.7	0.7	Liver, kidney, nervous system effects	Gasoline; insecticides; chemical manufacturing wastes
Ethylene dibromide	zero	0.00005	Cancer	Leaded gas additives; leaching of soil fumigant
Heptachlor	zero	0.0004	Cancer	Leaching of insecticide for termites, very few crops
Heptachlor epoxide	zero	0.0002	Cancer	Biodegradation of heptachlor
Lindane	0.0002	0.0002	Liver, kidney, nervous system, immune system, circulatory system effects	Insecticide on cattle, lumber, gardens; restricted in 1983
Methoxychlor	0.04	0.04	Growth; liver, kidney, nervous, system effects	Insecticide for fruits, vegetables, alfalfa, livestock, pets

(Continued)

TABLE 1-5 National Primary Drinking Water Standards (*Continued*)

Contaminant	MCLG, mg/L	MCL, mg/L	Potential health effects	Sources of drinking water contamination
Pentachlorophenol	zero	0.001	Cancer; liver, kidney effects	Wood preservatives, herbicide, cooling tower wastes
PCBs	zero	0.0005	Cancer	Coolant oils from electrical transformers; plasticizers
Styrene	1	1	Liver, nervous system effects	Plastics, rubber, resin, drug damage industries; leachate from city landfills
Tetrachloroethylene	zero	0.005	Cancer	Improper disposal of dry cleaning and other solvents
Toluene	1	1	Liver, kidney, nervous system, circulatory system effects	Gasoline additive; manufacturing and solvent operations
Toxaphene	zero	0.003	Cancer	Insecticide on cattle, cotton, soybeans; cancelled in 1982
2,4,5-TP (Silvex)	0.05	0.05	Liver, kidney damage	Herbicide on crops, rights-of-way, golf courses; cancelled in 1983
Xylenes (total)	10	10	Liver, kidney, nervous system effects	By-product of gasoline refining; paints, inks, detergents
Lead and Copper Rule				
Lead	zero	TT[#]	Kidney, nervous system effects	Natural/industrial deposits; plumbing solder, brass alloy faucets
Copper	1.3	TT[##]	Gastrointestinal irritation	Natural/industrial deposits; wood preservatives, plumbing
Phase V Inorganics				
Antimony	0.006	0.006	Cancer	Fire retardents, ceramics, electronics, fireworks
Beryllium	0.004	0.004	Bone, lung damage	Electrical, aerospace, defense industries
Cyanide	0.2	0.2	Thyroid, nervous system damage	Electroplating, steel, plastics, mining, fertilizer
Nickel	0.1	0.1	Heart, liver damage	Metal alloys, electroplating, batteries, chemical production
Thallium	0.0005	0.002	Kidney, liver, brain, intestinal effects	Electronics, drugs, alloys, glass
Phase V Organics				
Adipate, (di(2-ethylhexyl))	0.4	0.4	Decreased body weight	Synthetic rubber, food packaging, cosmetics
Benzo(a)pyrene (PAHs)	zero	0.0002	Cancer	Coal tar coatings; burning organic matter; volcanoes, fossil fuels
Dalapon	0.2	0.2	Liver, kidney effects	Herbicide on orchards, beans, coffee, lawns, roads, railways
Di(2-ethylhexyl) phathalate	zero	0.006	Cancer	PVC and other plastics
Dichloromethane	zero	0.005	Cancer	Paint stripper, metal degreaser, propellant, extractant

Di(2-ethylhexyl) phathalate	zero	0.006	Cancer	PVC and other plastics
Dinoseb	0.007	0.007	Thyroid, reproductive organ damage	Runoff of herbicide from crop and noncrop applications
Diquat	0.02	0.02	Liver, kidney, eye effects	Runoff of herbicide on land and aquatic weeds
Dioxin	zero	3×10^{-8}	Cancer	Chemical production by-product; impurity in herbicides
Endothall	0.1	0.1	Liver, kidney, gastrointestinal effects	Herbicide on crops, land/aquatic weeds; rapidly degraded
Glyphosate	0.7	0.7	Liver, kidney damage	Herbicide on grasses, weeds, brush
Hexachlorobenzene	zero	0.001	Cancer	Pesticide production waste by-product
Hexachlorocyclopentadiene	0.05	0.05	Kidney, stomach damage	Pesticide production intermediate
Oxamyl (Vydate)	0.2	0.2	Kidney damage	Insecticide on apples, potatoes, tomatoes
Picloram	0.5	0.5	Kidney, liver damage	Herbicide on broadleaf and woody plants
Simazine	0.004	0.004	Cancer	Herbicide on grass sod, some crops, aquatic algae
1,2,4-Trichlorobenzene	0.07	0.07	Kidney, liver damage	Herbicide production; dye carrier
1,1,2-Trichloroethane	0.003	0.003	Kidney, liver, nervous system damage	Solvent in rubber, other organic products; chemical production wastes
Disinfectants and Disinfection By-Products				
Bromate	zero	0.010	Cancer	Ozonation by-product
Bromodichloromethane	zero	see TTHMs	Cancer, kidney, liver, reproductive effects	Chlorination by-product
Bromoform	zero	see TTHMs	Cancer, kidney, liver, nervous system effects	Chlorination by-product
Chlorine	4 (as Cl_2) (MRDLG)	4.0 (as Cl_2) (MRDL)		
Chloramies	4 (as Cl_2) (MRDLG)	4.0 (as Cl_2) (MRDL)		
Chlorine dioxide	0.8 (as ClO_2) (MRDLG)	0.8 (as ClO_2) (MRDL)		
Chloral hydrate	0.04	TT	Liver effects	Chlorination by-product
Chlorite	0.08	1.0	Developmental neurotoxicity	Chlorine dioxide by-product
Chloroform	zero	see TTHMs	Cancer, kidney, liver reproductive effects	Chlorination by-product
Dibromochloromethane	0.06	see TTHMs	Nervous system, kidney, liver, reproductive effects	Chlorination by-product
Dichloroacetic acid	zero	see HAA5	Cancer, reproductive, developmental effects	Chlorination by-product

(Continued)

TABLE 1-5 National Primary Drinking Water Standards (*Continued*)

Contaminant	MCLG, mg/L	MCL, mg/L	Potential health effects	Sources of drinking water contamination
Disinfectants and Disinfection By-Products				
Haloacetic acids (HAA5)**	zero	0.060	Cancer and other effects	Chlorination by-product
Trichloroacetic acid	0.3	see HAA5	Liver, kidney, spleen, mental effects	Chlorination by-product
Total trihalomethanes (TTHMs)	zero	0.010	Cancer and other effects	Chlorination by-product
Radionuclides				
Beta/photon emitters	zero	4 mrem/yr	Cancer	Natural and man-made deposits
Alpha particles	zero	15 pCi/L	Cancer	Natural deposits
Radium 226 and 228	zero	5 pCi/L	Bone cancer	Natural deposits
Uranium	zero	0.03	Cancer	Natural deposits
Arsenic				
Arsenic	zero	0.010	Skin, nervous system toxicity; cancer	Natural deposits; smelters, glass, electronics wastes; orchards

Notes:
Action level = 0.015 mg/L
Action level = 1.3 mg/L
TT = treatment technique requirement
MFL = million fibers per liter
*For water systems analyzing at least 40 samples per month, no more than 5.0 percent of the monthly samples may be positive for total coliforms. For systems analyzing fewer than 40 samples per month, no more than one sample per month may be positive for total coliforms.
**Sum of the concentrations of mono-, di-, and trichloroacetic acids and mono- and dibromoacetic acids.
Alternatives allowing public water systems the flexibility to select compliance options appropriate to protect the population served were proposed.

TABLE 1-6 Other Significant National Drinking Water Regulations

Promulgation date	Regulation	Reference
July 19, 1979	Secondary Standards	FR* 44:140:42195
May 14, 1996	Information Collection Rule	FR 61:94:24354
Aug. 14, 1998	Variance and Exemption Rule	FR 63:157:43833
Aug. 16, 1998	Consumer Confidence Report Rule	FR 63:160:44511
Sept. 17, 1999	First Unregulated Contaminant Monitoring Rule	FR 64:180:50555
May 4, 2000	Public Notification Rule	FR 65:87:25891
Jan. 4, 2007	Second Unregulated Contaminant Monitoring Rule	FR 72:2:367
Oct. 10, 2007	Short-Term Revisions to the Lead and Copper Rule	FR 72:195:57781

* FR – *Federal Register*

guidance and do not have the same enforcement implications as the regulatory language. More information on USEPA guidance manuals can be found on the USEPA Web site listed in the section "The Internet as a Resource."

The Traditional and Negotiated Rulemaking Processes

In the traditional rulemaking process, USEPA publishes the proposed regulation in the *Federal Register* and accepts comments from the public on various aspects of the regulation. Then, USEPA evaluates the comments, develops a comment response document, and finalizes the regulation. The 1996 SDWA Amendments allow PWSs three years after the regulation has been finalized for compliance, and the compliance deadline can be extended, in conjunction with the state primacy agency, by an additional two years if additional time is needed for capital improvements.

For some of its more recent and more complex rulemakings, USEPA has used a negotiated rulemaking process to develop the substance of the regulation. In the late 1980s and early 1990s, USEPA considered revising the 1979 TTHM Rule but realized the enormous regulatory challenge in simultaneously balancing the risks from DBPs against the microbial risks. This risk–risk balancing did not lend itself to the traditional rulemaking process in which the national standard is set at the appropriate risk level.

From 1992 to 1993, USEPA used a negotiated rulemaking process to develop the proposed ICR, the proposed Stage 1 DBPR, and the proposed IESWTR. Later, for more regulatory negotiations, a Federal Advisory Committee was formed under the auspices of the Federal Advisory Committee Act (FACA) to provide regulatory recommendations through additional Agreement in Principles (AIPs) to USEPA. The members of these committees represented a broad range of stakeholders in the drinking water community including utilities, USEPA, state regulators, tribes, local government, public health authorities, consumer advocates, and environmental advocates. To marshal the widely divergent views, a facilitator was hired by USEPA to run the face-to-face meetings.

For the Microbial/Disinfection By-Product (M/DBP) cluster of regulations, USEPA used negotiated rulemaking/FACA processes in:

- 1992 to 1993 for the proposed ICR, the proposed Stage 1 DBPR, and the IESWTR
- 1997 for finalizing the Stage 1 DBPR and the IESWTR because the ICR was not completed in time for those regulations as originally anticipated in 1992 to 1993
- 1999 to 2000 for the proposed Stage 2 DBPR and the LT2ESWTR

As previously discussed, USEPA decided in 2003 to revise the TCR in its first six-year review of existing drinking water regulations. Due to the inherent complexities with the myriad of distribution systems, USEPA, in 2007, elected to form the Total Coliform Rule Distribution System Advisory Committee (TCRDSAC). The TCRDSAC met from July 2007 to September 2008 to develop proposed revisions to the TCR and to develop recommendations for distribution system research. In September 2008, the members of the TCRDSAC signed the AIP that will be used as the foundation for proposed TCR revisions (TCRDSAC AIP, 2008).

These Federal Advisory Committees are both time and resource intensive for all committee members. The face-to-face meetings typically take two days and are sometimes preceded by a Technical Work Group (TWG) meeting. The TWG typically provides substantial data analysis and technical support for the committee. For example, the TWG conducted many variations of analyses of the ICR data that were used to support the development of the Stage 2 DBPR and the LT2ESWTR. The output from some of these analyses was eventually compiled into an ICR data analysis book that was edited by three TWG members (McGuire et al., 2002). Even in cases where USEPA serves as a Federal Advisory Committee (FAC) and signs the AIP, USEPA still has the responsibility for developing the proposed preamble and regulatory language, conducting the analyses previously discussed, and considering public comments before issuing a final regulation.

Going Beyond the Regulations

Systems sometimes develop their own water quality goals that are stricter than the regulations. These goals might be for particular contaminants that occur in their source water. One reason that systems adopt this approach is because they want to be leaders in the drinking water community and serve their customers the highest quality water possible. Another reason is that systems do not want to have any violations for any reason and want to have a considerable margin of safety relative to the standards.

Programs have been developed to assist utilities in going beyond the regulations to work toward optimizing treatment. The Partnership for Safe Water is a voluntary cooperative effort between USEPA, AWWA, and other drinking water organizations (AWWA, 2009). More than 200 surface water utilities serving more than 85 million people involved in the Partnership have improved the quality of their drinking water by optimizing system operations and improving filtration performance. Another similar collaborative program between USEPA and several state primacy agencies is the Area-Wide Optimization Program (AWOP) (Barr, 2007). AWOP started in 1989 and now operates in 21 states and focuses on treatment optimization.

ROLE OF STATE AGENCIES

State Agencies and USEPA as Co-regulators

USEPA and states are co-regulators under the SDWA. USEPA assesses contaminants and establishes NPDWRs. State agencies that obtain primacy ensure water systems comply with the drinking water standards. USEPA provides technical assistance and oversight to states in their efforts to ensure compliance with the regulations. States' experience in implementing NPDWRs provide a valuable perspective when developing new regulations. State agencies work closely with USEPA in a state-USEPA workgroup in the initial drafting of a proposed regulation, but USEPA is ultimately responsible for issuing both the proposed and final regulations.

Primacy

If state and tribal regulations are passed that are at least as strict as the federal standards and oversight programs and enforcement mechanisms are in place, primacy for PWS supervision is retained by the state or tribe. Some state agencies further delegate these responsibilities to county or local health departments. All states except Wyoming have primacy. For each new regulation, states and tribes must submit a primacy package on those regulations to USEPA for review and approval.

State agency oversight of PWSs goes beyond simply ensuring compliance with the numerous standards. Sanitary surveys are a critical oversight activity that entails looking at all aspects of a utility's operations and management proactively in order to get ahead of any potential compliance problems. The identification and correction of significant deficiencies are now required under the IESWTR, LT1ESWTR, and GWR and provide a significant new tool for state regulators to use in bringing systems back into compliance.

Other state oversight activities include: (1) review and approval of construction plans and specifications, typically for treatment plants and larger transmission and distribution piping; (2) technical assistance, ranging from assisting with treatment evaluation and selection to assisting with evaluation of water rates to filling out the required paperwork and everything in between; and (3) sample collection and analysis (in some states). These activities help move the PWSs towards compliance.

The primacy agencies make the initial compliance determination. Systems may incur violations not only for MCLs and TTs but also for monitoring and reporting (M&R). The primacy agency works with systems to bring them back into compliance However, if the systems remain out of compliance, USEPA can step in to enforce the regulations.

The violations listed in the state-level data management system are transferred to the USEPA data system, or Safe Drinking Water Information System–Federal (SDWIS-FED). From 1997 to 2007, national compliance with the NPDWRs, based on population served, has increased slightly from 86.5 percent in 1997 to 91.5 percent in 2007 (USEPA, 2008d). The highest rate was in 2003 with 93.6 percent compliance. The national compliance rate is important to USEPA, because USEPA uses this metric to meet its obligations under the Government Performance and Results Act (GPRA) to measure how well they are performing. USEPA set an objective in its strategic plan that by 2008, 95 percent of the population served by community water systems would meet all applicable health-based standards.

USEPA's regional offices also play a role in the national drinking water program. USEPA divides the states and territories into 10 regions, as shown on Fig. 1-5. These regional offices assist state and tribal agencies that hold primacy with their regulatory oversight. In USEPA Region 3 (the District of Columbia) and USEPA Region 8 (Wyoming), the regional office directly implements the drinking water program.

USEPA provides funding to the primacy agencies for their drinking water programs through Public Water System Supervision (PWSS) grants. Since 2004, the total amount of PWSS grants has effectively been held constant (USEPA, 2008e), despite the increasing complexity of regulations such as the Filter Backwash Recycling Rule (FBRR), the IESWTR, and the LT1ESWTR. While some state agencies charge a fee for service to the utilities and/or their consultants for some of the previously mentioned activities, other states are prohibited from charging any fees. States agencies' resources continue to be challenged. In 2003, the Association of State Drinking Water Administrators (ASDWA) released a report on the funding gap for the states' drinking water programs (ASDWA, 2003). This report found a gap in 2002 of approximately $230 million between the funds expended at the state level for drinking water programs and the estimated $535 million needed. The funding shortfall was estimated to increase to approximately $369 million by 2006, and this shortfall will continue to grow as the number of regulations increases, the population increases, and water systems expand.

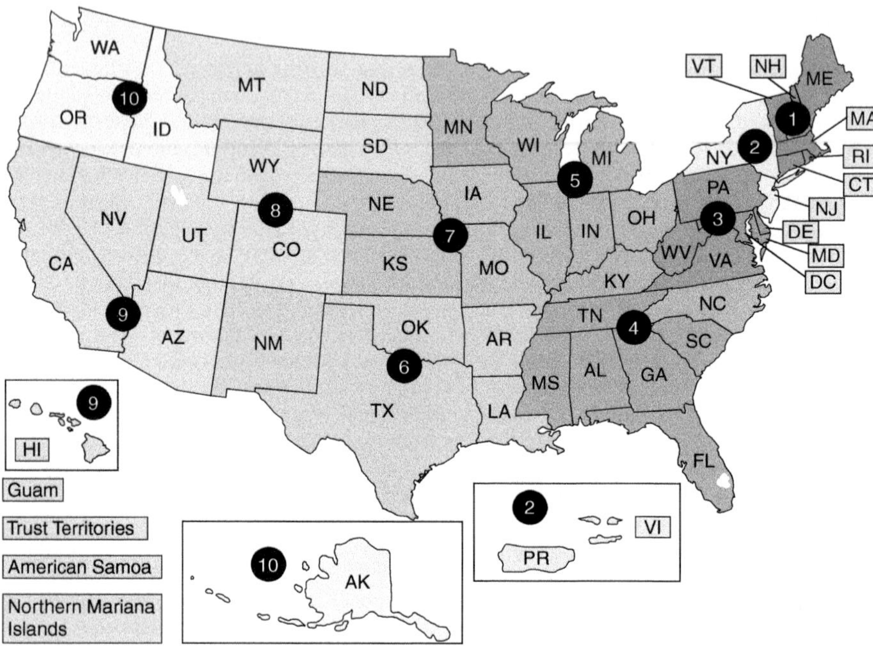

FIGURE 1-5 Map of USEPA regions. (*Source: www.epa.gov/epafiles/samples/map.html#eparegions.*)

New State Programs from the 1996 SDWA Amendments

State agencies' resources have been squeezed further by four new state programs that resulted from the 1996 SDWA Amendments: (1) source water assessments, (2) capacity development, (3) Drinking Water State Revolving Funds, and (4) operator certification.

As part of the 1996 SDWA Amendments, USEPA was required to develop source water protection guidance. Within 18 months of publication of this guidance, state agencies were required to develop source water assessment programs. The programs were to delineate the source water areas for PWSs, identify potential sources of contamination, and assess the susceptibility of the system to contamination. State agencies were required to complete all source water assessments by 1999. State agencies and utilities were further encouraged to develop voluntary source water protection programs.

USEPA was also required to review existing state assessment programs of the technical, financial, and managerial capacity of PWSs, and to publish guidance on how to strengthen these programs. State agencies must have the legal authority to require systems to demonstrate adequate capacity and a strategy to identify and assist systems that have capacity deficiencies. The objective of these capacity development programs is to ensure system viability for the long term.

The creation of the Drinking Water State Revolving Fund (DWSRF) provided a new funding mechanism for PWSs to construct necessary improvements. Most state agencies established drinking water state SRFs that are parallel to their existing Wastewater SRFs. Congress, which annually appropriates funding for the DWSRFs, started the fund with an appropriation of $1.2 billion. Between 1997 and 2007, the DWSRF provided $12.6 billion in funding for 5,555 projects; 39 percent of this funding has gone to systems serving fewer than 10,000 people (USEPA, 2007b). The American Recovery and Reinvestment Act

(ARRA) of 2009 provided additional funding of $2 billion for the DWSRF program on a one-time basis (USEPA, 2009d).

USEPA is required by the 1996 SDWA Amendments to conduct a national needs survey every four years that is used to establish the percent allocated to each state for its SRF. The 2007 Drinking Water Needs Survey identified a 20-year (2007–2026) capital investment need of $334.8 billion for PWSs eligible for SRF funding, noting that private systems and projects resulting from population and system growth are not eligible for SRF funding (USEPA, 2005b). The majority, approximately 60 percent ($200.8 billion), of the total national need is for transmission and distribution projects. Approximately $52 billion (16 percent) can be attributed to the NPDWRs. Each state agency develops its own process for identifying and prioritizing projects for potential SRF funding by developing a state-level intended use plan (IUP).

One notable feature of the drinking water SRF is the ability of state agencies to use a portion of their SRF funding for these new programs. State agencies can use up to 10 percent for source water protection, capacity development, and operator certification, as well as for the overall drinking water program. Up to another 15 percent can be used by systems for preparation of source water assessments, source water protection loans, capacity development, and wellhead protection. It should be noted that these "set-asides" take funding away from needed construction improvements. Between 1997 and 2007, state agencies "set aside" $1.4 billion from the DWSRF.

Most state agencies have had operator certification programs for many years, but these programs varied significantly from state to state. In 1999, USEPA developed guidelines for these programs, with eight elements that operator certification programs must contain. The objective was to not have any state agencies lower their standards for operator certification but rather to establish a national base level for these programs. These guidelines also mandated the development of programs for certifying distribution system operators, which was an expansion of traditional certification programs for treatment plant operators. The 1996 SDWA Amendments also provided for grants to state agencies to reimburse operators of systems serving fewer than 3300 people for training and certification costs.

The 1996 SDWA Amendments provided for some type of earmarked allocation to states for these programs. These allocations helped the states meet the requirements of these new programs. However, combined with the new regulations and their increasing complexity, state resources continue to be challenged.

State Standards

States and tribes must pass their own regulations that are at least as strict as the federal standard in order to retain primacy, but some states pass stricter standards or develop their own standards that are not regulated by USEPA. California, New York, New Jersey, and Massachusetts are some of the states that have more or stricter standards than those of USEPA. Tracking the development of state standards across all states can be a substantial effort. ASDWA has links to each state's drinking water program on its Web site (see "The Internet as a Resource") (ASDWA, 2008). For information about a drinking water standard for a particular state or tribe, the appropriate state drinking water program agency should be contacted directly.

Even when the regulations adopted by primacy agencies are no more stringent than the federal regulations, differences in implementation at the state level can make them operationally different from the system's perspective. For example, under the TCR, the state of Ohio used to report missing repeat samples as an MCL violation as opposed to an M&R violation, which caused them to have a disproportionately high number of MCL violations. Other states have differing definitions of significant deficiencies under sanitary surveys previously discussed; consequently, in many cases, even when the numerical standards are the same, variation in implementation can make regulations look different from state to state.

PEER REVIEW, OUTSIDE CONSULTATION, AND PUBLIC INVOLVEMENT

USEPA regularly conducts peer review of its reports and assessments and also consults with outside parties to provide recommendations on specific technical and science policy issues to incorporate into its regulatory development process. The parties that USEPA uses for each consultation and the specific technical and science policy issues USEPA asks for input on are specific to each regulation. The intent of these outside consultations is to provide recommendations and advice to USEPA in meeting the SDWA requirement and to use the "...best available, peer-reviewed science..." in its regulations, as required by Section 1412(b)(3)(A) of the SDWA. USEPA also has to meet its own internal guidelines for peer review and information quality (USEPA, 2002, 2006b).

The National Academy of Sciences

USEPA has repeatedly gone to the National Academy of Sciences (NAS) for advice and recommendations on scientific and technical issues to apply into its regulatory development process. Some NAS studies have been mandated by Congress, while others were solely at USEPA's request. As a result of the 1974 SDWA, the NRC of the NAS conducted a study to assess human exposure via drinking water and the toxicology of contaminants in drinking water. USEPA published the NAS study recommendations in 1977 and used this report as the basis for revising its drinking water regulations (USEPA, 1977).

More recently, the NRC provided recommendations to USEPA on the CCL and regulatory determination processes (NRC, 1999a, 1999b, 2001a). The NRC has also assessed the health effects of fluoride, radon, arsenic, and copper and reviewed and provided recommendations for USEPA's fluoride standards (NRC, 1993, 1999c, 1999d, 2000, 2001b, 2005, 2006).

The Science Advisory Board

Congress established USEPA's Science Advisory Board (SAB) in 1978 to provide guidance regarding scientific and technical issues to support the agency throughout its regulatory development process. The SAB develops reports and advisories that review the approaches to regulatory science, reviews research programs, critiques analytical methods such as mathematical modeling, and provides other science policy and technical advice as necessary.

The USEPA Administrator appoints members of the SAB and nominations from the public are solicited for the SAB and the SAB standing committees. The SAB currently has six standing committees, including a Drinking Water Committee. Ad hoc committees and panels are formed as needed to address specific topics. Section 1412 of the SDWA mandates USEPA to consult with the scientific community, including the SAB, as part of its development of draft CCLs. USEPA regularly consults with the SAB Drinking Water Committee on a wide variety of drinking water regulatory issues.

The National Drinking Water Advisory Council

Section 1446 of the SDWA mandates the formation of the NDWAC. This section also details representation of the general public, state and local agencies, and private groups for the 15 NDWAC members; the members' terms; and the rotation of the members' terms.

NDWAC members are appointed by the USEPA Administrator. The NDWAC provides policy guidance to USEPA on individual regulations and other regulatory issues.

The NDWAC typically sends letters to the USEPA Administrator after its review of a specific issue. The council often forms a working group to review more complex issues and make more detailed policy recommendations to USEPA. These working groups are usually comprised of two to four NDWAC members plus additional subject matter experts representing a broad range of stakeholders' perspectives. Past NDWAC working groups included Benefits, CCR, Drinking Water SRF, Health Care Providers, Occurrence and Contaminant Selection, Operator Certification, Right-to-Know, Small Systems, Small System Affordability, Source Water, Arsenic Cost, CCL and Six-Year Review, CCL Classification Process, Research, Water Security, and Public Education Requirements of the LCR.

The Office of Management and Budget

Under Executive Order (EO) 12866, the Office of Information and Regulatory Affairs (OIRA) in the Office of Management and Budget (OMB) conducts a regulatory review of each proposed and final regulation by any federal agency, including USEPA. These reviews are typically conducted with a 90-day deadline, and completion of these reviews typically signifies that the proposed or final regulations will be published soon thereafter in the *Federal Register*. These reviews are conducted to ensure that the agencies are following other Executive Orders (EOs) that provide guidance on the use of benefit–cost analysis (BCA) in the regulatory development process.

Public Involvement

The public plays an important role in the provision of safe drinking water. Because the public is the ultimate consumer of the treated drinking water, it has a vested interest in the USEPA regulatory development process, as well as state and USEPA oversight of the utilities. At the national level, the public can play a role and have some influence on Congress in future SDWA reauthorizations and other legislation that might impact drinking water quality. Members of Congress are always willing to listen to the public regarding issues (such as drinking water) that are important to their constituents that ultimately vote to elect or re-elect them to office. The public can play a role in USEPA's regulatory development process by participating in USEPA's stakeholder efforts as previously described. The public can also submit comments on USEPA's proposed regulatory actions; these comments are published in the *Federal Register* (see the stars in Fig. 1-2). The public can also play a role in the development of state-level drinking water regulations by providing input at public hearings and providing comments on the states' proposed regulations. At the local level, the public can attend board meetings of their local utility and provide input during any public comment period at rate hearings.

The Consumer Confidence Report and the Public Notification Rule

Increased consumer education was one improvement that resulted from the 1996 SDWA Amendments. One important component of that increased consumer education was that USEPA must develop a new regulation that requires utilities to prepare annual water quality reports and for these reports to be delivered to their customers. These reports are known as Consumer Confidence Reports (CCRs), and the initial CCR regulation was based on a report from an NDWAC workgroup (USEPA, 1998b).

The CCR regulation requires utilities to develop a water quality report that includes the following:

- Identification of water source and treatment characterization
- Susceptibility of the source water to contamination based on the source water assessment and how to get a copy of the assessment
- Contaminants that are found in the water and their likely sources
- The levels of the contaminants, a comparison to the standard, and the health effects
- An educational statement for vulnerable populations about *Cryptosporidium* and additional information statements on nitrate, arsenic, and lead
- Sources of additional information, such as the utility's main phone number and USEPA's SDWA Hotline

The CCR regulation also requires utilities to send the CCR to customers annually, by July 1 of each year, and to provide methods to get the report in languages other than English if English is not the primary language for a significant portion of customers.

Revisions of the PNR provide another avenue for consumer education. Public notification has always been a part of the NPDWRs; however, over the years, this component has evolved with each new NPDWR. USEPA conducted a stakeholder effort from 1998 to 1999 to revise to the PNR and published the revised PNR in 2000 (USEPA, 2000). The objectives of the revised PNR are to provide faster notice in the event of an emergency, provide fewer notices overall in order to minimize confusion, and ensure that public notices are easy to understand by using clearer language and less scientific and technical jargon. USEPA and the stakeholders developed new guidance for the revised PNR that includes several templates that utilities can use for compliance.

OTHER COUNTRIES AND INTERNATIONAL STANDARDS

Canada

In Canada, drinking water guidelines are developed by Health Canada; if they choose to, the provinces and territories use these guidelines as a basis for developing enforceable standards. This regulatory approach is different from the U.S. regulatory approach in that the provinces and territories are not required to adopt the Health Canada guidelines. Health Canada's Water Quality and Health Bureau takes the lead by developing *Guidelines for Canadian Drinking Water Quality* through a collaborative effort with the provinces, territories, utilities, and the public. Development of the Guidelines generally parallels USEPA's regulatory development process, with the initial identification of substances that have been found in source waters and are known or suspected to be harmful. A maximum acceptable concentration (MAC), which is comparable to USEPA's MCL, is then established. More information on the Canadian guidelines can be found at the Web site listed in the section "The Internet as a Resource."

Australia

Australia's regulatory approach warrants discussion because it is based on a preventive risk management approach for water supplies. This approach was jointly developed by the Cooperative Research Centres and the National Health and Medical Research Council

(NHMRC). The 2004 *Australian Drinking Water Guidelines* (ADWG) were developed by the NHMRC, in collaboration with the Natural Resource Management Ministerial Council (NRMMC). The guidelines incorporate a "Framework for the Management of Drinking Water Quality" and provide guidance on what constitutes safe drinking water. This risk-based approach takes a holistic view of the watershed and/or aquifer, treatment plant, and distribution system; analyzes for any potential weaknesses or vulnerabilities; and then develops a plan to address these potential weaknesses or vulnerabilities. This approach is comparable to the Hazard Analysis and Critical Control Point (HACCP) framework that is used in the United States for food production and supply. A report by the Awwa Research Foundation (now the Water Research Foundation) evaluated application of the HACCP framework for distribution system protection at two water utilities. However, this regulatory approach does not easily fit within the current SDWA framework of contaminant-by-contaminant regulation (AwwaRF, 2006). More information on the Australian guidelines can be found at their Web site (see "The Internet as a Resource").

European Union

The European Union (EU) is a voluntary economic alliance consisting of several European countries. The Drinking Water Directives (DWDs) for drinking water quality developed by the EU must be adopted by the individual countries in order to remain an EU member state. Comparable to regulations established by USEPA, the enforceable drinking water standards developed by the EU member states must be at least as strict as the limits established in the DWD. The member states can regulate additional parameters beyond the parameters in the DWD, but cannot adopt stricter standards. The DWD, Council Directive 98/83/EC, addresses the quality of water intended for human consumption and lists 48 parameters that must be monitored and tested regularly. The European Commission started a review of Directive 98/83/EC in 2003 with a large public consultation. Member states report back to the European Commission every three years with their monitoring results. Generally, the WHO guidelines for drinking water are used as the basis for the DWD standards. More information the EU standards can be found at their Web site (see "The Internet as a Resource").

World Health Organization

The World Health Organization (WHO) is a specialized agency within the United Nations with the primary responsibility of protecting international public health. The WHO produces international guidelines for drinking water quality that are used as the basis for regulation and standard setting by both developing and developed countries. Every country considers a broad range of environmental, social, cultural, and economic factors when determining which guidelines to adopt as their own standards. The 2004 *Guidelines for Drinking Water Quality* introduced the concept of Water Safety Plans, which shifts the regulatory focus from control of the drinking water at the end of the treatment plant or at the tap toward managing drinking water quality from the watershed, through the treatment plant, and through the distribution system. More information on the WHO guidelines can be found at their Web site (see "The Internet as a Resource").

OUTLOOK FOR THE FUTURE

The SDWA is the starting point for U.S. drinking water regulations. With competing legislative priorities in Congress, it is not likely that the SDWA will be significantly amended anytime soon. The risk management and regulatory development process used by USEPA will continue to evolve as more research is conducted and more contaminants are evaluated as potential candidates for regulation. As analytical methodologies continue to improve the ability to measure extremely low levels of contaminants, previously undetected contaminants will likely be found in some drinking water supplies. These emerging contaminants may generate public concern because of the uncertainty of their impact on public health at these low levels. Emerging contaminants such as endocrine disruptors, pharmaceuticals, and personal care products will need to be addressed through USEPA's regulatory development processes, using both the SDWA and other environmental statutes, such as the Clean Water Act (CWA) and Food Quality Protection Act (FQPA). States and PWSs will continue to be challenged by more regulations with more inherent complexities that will continue to stress limited resources. The public will continue to be interested in the safety of their drinking water, based on their own growing interest and the stories that will appear in the media.

The U.S. population will continue to grow. Currently, most population growth is projected to be in the southern half of the United States where water quantity and water quality problems will continue to increase. Utilities, regulators, and the public must remain vigilant in protecting existing source waters; finding new sources of water; and raising enough capital to build the appropriate transmission, treatment, and distribution facilities. The future will pose many challenges to keeping water safe and secure.

THE INTERNET AS A RESOURCE

Since the mid-1990s, the Internet has become an increasingly valuable resource for information on drinking water regulations. But one must be careful, because as with many issues, information overload can occur. In 2008, more than 8 million hits resulted when "drinking water regulations" was searched on Google™.

Many government Web sites are now the most direct source for copies of *Federal Register* notices, guidance documents, background documentation, NDWAC reports, and other information. The starting point for any research on U.S. drinking water regulations is the USEPA's Office of Ground Water and Drinking Water Web site (water.epa.gov/drink/index.htm). This Web site is a virtual treasure trove of drinking water regulatory information and has links to electronic copies of *Federal Register* notices dating back to May 1996.

Copies of *Federal Register* notices can also be found on the Government Printing Office's Web site (www.gpoaccess.gov/fr/). Dating back to 2000, the table of contents can be searched, but he publication date of the specific *Federal Register* notice is needed in order to search the table of contents.

Background documentation, such as the economic analysis (EA) and the technology and cost (T&C) documents, can be found through a federal government clearinghouse Web site (www.regulations.gov). For example, searching by the docket number greatly simplifies the search for background documentation on a proposed drinking water regulation. Without the docket number, searching through this Web site can be challenging.

Other useful government Web sites include:

- Office of Management and Budget (OMB); www.whitehouse.gov/omb/inforeg_default
- U.S. Geological Survey (USGS); http://water.usgs.gov/
- U.S. Department of Health and Human Services, Centers for Disease Control and Prevention (CDC); www.cdc.gov/healthywater/drinking/index.html
- U.S. Department of Homeland Security (DHS), Office of Infrastructure Protection; www.dhs.gov/xabout/structure/gc_1185203138955.shtm

- U.S. Department of Agriculture (USDA) Rural Development Utilities Programs, Water and Environmental Programs; www.usda.gov/rus/water
- The following drinking water associations can also be a useful source of information:
  - American Water Works Association (AWWA); www.awwa.org
  - Association of State Drinking Water Administrators (ASDWA); www.asdwa.org. Web sites for state drinking water programs can generally be found by searching on the state name followed by "drinking water".
  - Association of Metropolitan Water Agencies (AWMA); www.amwa.net
  - National Rural Water Association (NRWA); www.nrwa.org
  - National Association of Water Companies (NAWC); www.nawc.com
  - Water Research Foundation; www.waterresearchfoundation.org

Government Web sites (outside of the United States) that are useful sources of information on international drinking water standards include:

- Health Canada; www.hc-sc.gc.ca/ewh-semt/water-eau/drink-potab/index-eng.php
- European Union; http://ec.europa.eu/environment/water/water-drink/index_en.html
- World Health Organization; www.who.int/water_sanitation_health/dwq/en
- Australia; www.nhmrc.gov.au/publications/synopses/eh19syn.htm

DISCLAIMER

This chapter does not necessarily reflect the views and policies of USEPA nor does the mention trade names or products constitute endorsement or recommendation for their use.

ABBREVIATIONS

ADWG	Australian Drinking Water Guidelines
AIP	agreement in principle
AMWA	Association of Metropolitan Water Agencies
ARRA	American Recovery and Reinvestment Act
ASDWA	Association of State Drinking Water Administrators
ATSDR	Agency for Toxic Substances and Disease Registry
AWOP	area-wide optimization program
AWWA	American Water Works Association
AwwaRF	Awwa Research Foundation, now known as the Water Research Foundation
BCA	benefit–cost analysis
CCL	Contaminant Candidate List
CCR	Consumer Confidence Report
CDC	Centers for Disease Control and Prevention
CERCLA	Comprehensive Environmental Response, Compensation, and Liability Act
CI	critical infrastructure

CWA	Clean Water Act
CWSS	Community Water System Survey
DBP	disinfection by-product
DBPR	Disinfection By-Product Rule
DHS	Department of Homeland Security
DWA	Drinking Water Advisory
DWD	Drinking Water Directive
DWSRF	Drinking Water State Revolving Fund
EA	economic analysis
EO	Executive Order
ERP	emergency response plan
EU	European Union
FACA	Federal Advisory Committee Act
FBRR	Filter Backwash Recycling Rule
FIFRA	Federal Insecticide, Fungicide, and Rodenticide Act
FQPA	Food Quality Protection Act
HA	Health Advisory
HACCP	Hazard Analysis and Critical Control Points
HRRCA	health risk reduction cost analysis
ICR	Information Collection Rule
IESWTR	Interim Enhanced Surface Water Treatment Rule
IOC	inorganic chemical
IRIS	Integrated Risk Information System
IUP	intended use plan
GPRA	Government Performance and Results Act
GWR	Ground Water Rule
LCR	Lead and Copper Rule
LT1ESWTR	Long Term 1 Enhanced Surface Water Treatment Rule
LT2ESWTR	Long Term 2 Enhanced Surface Water Treatment Rule
M&R	monitoring and report
MAC	maximum acceptable concentration
MCL	maximum contaminant level
MCLG	maximum contaminant level goal
M/DBP	microbial/disinfection by-product
MRDL	maximum residual disinfectant level
MRDLG	maximum residual disinfectant level goal
MTBE	methyl-*tert*-butyl-ether
NAS	National Academy of Sciences
NAWC	National Association of Water Companies
NDWAC	National Drinking Water Advisory Council
NHMRC	National Health and Medical Research Council

NIPP	National Infrastructure Protection Plan
NIRS	National Inorganic and Radionuclide Survey
NPDWRs	National Primary Drinking Water Regulations
NRC	National Research Council
NRMMC	Natural Resource Management Ministerial Council
NRWA	National Rural Water Association
OIRA	Office of Information and Regulatory Affairs
OMB	Office of Management and Budget
PNR	Public Notification Rule
PWSS	public water system supervision
RED	Registration Eligibility Decision
SAB	Science Advisory Board
SDWA	Safe Drinking Water Act
SDWIS–FED	Safe Drinking Water Information Systems–Federal
SMCL	secondary maximum contaminant level
SOC	synthetic organic chemical
SSP	sector-specific plan
SWTR	Surface Water Treatment Rule
T&C	technology and cost
TCR	Total Coliform Rule
TCRDSAC	Total Coliform Rule/Distribution System Advisory Committee
TON	threshold odor number
TT	treatment technique
TTHMs	total trihalomethanes
TWG	Technical Work Group
UCM	unregulated contaminant monitoring
UCMR	Unregulated Contaminant Monitoring Rule
USDA	U.S. Department of Agriculture
USEPA	U.S. Environmental Protection Agency
USGS	U.S. Geological Survey
USPHS	U.S. Public Health Service
VA	vulnerability assessment
VOC	volatile organic chemical
WHO	World Health Organization

REFERENCES

ASDWA (Association of State Drinking Water Administrators) (2003). Public Health Protection Threatened by Inadequate Resources for State Programs: An Analysis of State Drinking Water Programs Resources, Needs, and Barriers, Washington, D.C.: Association of State Drinking Water Administrators.

ASDWA (2008). www.asdwa.org/index.cfm?fuseaction=Page.viewPage&pageId=487 (accessed Jan. 5, 2009).

AWWA (American Water Works Association) (1990). *Water Quality & Treatment, Fourth Edition.*

AWWA (American Water Works Association) (2009). Partnership for Safe Water. www.awwa.org/Resources/PartnershipforSafeWater.cfm?ItemNumber=3790&navItemNumber=33971 (accessed October 20, 2009).

AwwaRF (Awwa Research Foundation) (1983). Final Report—Workshops on the Revised Drinking Water Regulations. Denver, Colo.: Water Research Foundation.

AwwaRF (2006). *Application of HACCP for Distribution System Protection.* Denver, Colo.: Water Research Foundation.

Barr, P. (2007). Treatment Optimization : Providing More Effective Multiple-Barrier Protection. *Jour. AWWA*, 99 (12), 40–43.

Bellar, T.A., Lichtenberg, J.J., and Kroner, R.C. (1974). The Occurrence of Organohalides in Chlorinated Drinking Water. *Jour. AWWA*, 66 (12), 703–706.

Bull Run Coalition v. Reilly (1993). 1 F3d 1246. 9th Circuit, 1993.

CDC (Centers for Disease Control and Prevention) (1999). Ten Great Public Health Achievements—United States, 1900–1999. *Morbidity & Mortality Weekly Rept.*, 48(12), 241.

DHS (Department of Homeland Security) (2006). National Infrastructure Protection Plan. www.dhs.gov/nipp (accessed Jan. 5, 2009).

DHS (2007). Sector-Specific Plans (SSPs). www.dhs.gov/xprevprot/programs/gc_1179866197607.shtm (accessed Jan. 5, 2009).

Johnson, S. (2006). *The Lost Map*. New York: Riverhead Books.

Mays, L.M. (2004). *Water Supply Systems Security.* New York: McGraw-Hill.

McGuire, M.J. (2006). Eight Revolutions in the History of US Drinking Water Disinfection. *Jour. AWWA*, 98 (3), 123–149.

McGuire, M.J., McLain, J.L., Obolensky, A., eds. (2002) *Information Collection Rule Data Analysis*. Denver, Colo.: Water Research Foundation.

National Drinking Water Advisory Council (2004). Report on the CCL Classification Process. www.epa.gov/safewater/ccl/basicinformation.html#eleven (accessed Jan. 5, 2009).

NRC (National Research Council) (1993). *Health Effects of Ingested Fluoride.* Washington, D.C.: National Academy Press.

NRC (1999a). *Setting Priorities for Drinking Water Contaminants.* Washington, D.C.: National Academy Press.

NRC (1999b). *Identifying Future Drinking Water Contaminants.* Washington, D.C.: National Academy Press.

NRC (1999c). *Risk Assessment of Exposure to Radon in Drinking Water.* Washington, D.C.: National Academy Press.

NRC (1999d). *Arsenic in Drinking Water.* Washington, D.C.: National Academy Press.

NRC (2000). *Copper in Drinking Water.* Washington, D.C.: National Academy Press.

NRC (2001a). *Classifying Drinking Water Contaminants for Regulatory Consideration.* Washington, D.C.: National Academy Press.

NRC (2001b). *Arsenic in Drinking Water: 2001 Update.* Washington, D.C.: National Academy Press.

NRC (2005). *Health Implications of Perchlorate Ingestion.* Washington, D.C.: National Academy Press.

NRC (2006). *Fluoride in Drinking Water: A Scientific Review of EPA's Standards.* Washington, D.C.: National Academy Press.

Philly H_2O (2008). The History of Philadelphia's Watersheds and Sewers. www.phillyh2o.org/backpages/MSB_Water.htm#01 (accessed Oct. 24, 2008).

Roberson, J.A. (2006). From Common Cup to *Cryptosporidium*: A Regulatory Evolution. *Jour. AWWA*, 98 (3), 198–207.

Roberson, J.A., and Morley, K.M. (2006). Water Security: Shifting to an All-Hazards Resiliency Approach. *Jour. AWWA*, 98 (5), 46–47.

Rook, J.J. (1974). Formation of Haloforms During Chlorination of Natural Water. *Water Treatment & Examination*, 23 (2), 234–243.

States, S. (2010). *Security and Emergency Planning for Water and Wastewater Utilities*. Denver, Colo.: American Water Works Association.

TCRDSAC AIP (2008). www.awwa.org/files/AIP.pdf (accessed Jan. 5, 2009).

USEPA (U.S. Environmental Protection Agency) (1972). Industrial Pollution of Lower Mississippi River in Louisiana. Surveillance and Analysis Division. Dallas, Texas: USEPA, Region VI.

USEPA (1977). Recommendations of the National Academy of Sciences. *Federal Register*, 42:110:35764.

USEPA (1979). National Secondary Drinking Water Regulations. Final Rule. *Federal Register*, 44:140:42195.

USEPA (1986). National Primary and Secondary Drinking Water Regulations. Final Rule. *Federal Register*, 51:63:11396.

USEPA (1991). National Primary Drinking Water Regulations. Synthetic Organic Chemicals and Inorganic Chemicals; Monitoring for Unregulated Contaminants; National Primary Drinking Water Regulation Implementation; National Secondary Drinking Water Regulations. Final Rule. *Federal Register*, 56:20:3526.

USEPA (1992). Secondary Drinking Water Regulations: Guidance for Nuisance Chemicals. EPA 810/K-92-001. www.epa.gov/safewater/consumer/2ndstandards.html (accessed Jan. 5, 2009).

USEPA (1998a). Announcement of the Contaminant Candidate List. *Federal Register*, 63:40:10273.

USEPA (1998b). Consumer Confidence Reports; Final Rule. *Federal Register*, 63:160:44511.

USEPA (1999a). Revisions to the Unregulated Contaminant Monitoring Regulation for Public Water Systems; Final Rule. *Federal Register*, 64:180:50555.

USEPA (1999b). National Primary Drinking Water Regulations; Radon-222; Proposed Rule. *Federal Register*, 64:211:59245.

USEPA (2000). Public Notification Rule. *Federal Register*, 65:87:25891.

USEPA (2002). Guidelines for Ensuring and Maintaining the Quality, Objectivity, Utility, and Integrity of Information Disseminated by the Environmental Protection Agency. EPA/260R-02-08. www.epa.gov/QUALITY/informationguidelines/ (accessed Aug. 6, 2009).

USEPA (2003a). Announcement of Regulatory Determinations for Priority Contaminants on the Drinking Water Contaminant Candidate List. *Federal Register*, 68:138:42897.

USEPA (2003b). EPA Protocol for the Review of Existing National Primary Drinking Water Regulations. EPA 815-R-03-002. www.epa.gov/safewater/review.html (accessed Jan. 5, 2009).

USEPA (2003c). National Primary Drinking Water Regulations; Announcement of Completion of EPA's Review of Existing Drinking Water Standards. *Federal Register*, 68:138:42907.

USEPA (2005a). Drinking Water Contaminant Candidate List 2; Final Notice. *Federal Register*, 70:36:9071.

USEPA (2005b). Drinking Water Infrastructure Needs Survey and Assessment. www.epa.gov/safewater/needssurvey/index.html (accessed Jan. 5, 2009).

USEPA (2006a). Drinking Water Health Advisories; 2006 Edition. EPA 822-R-06-013. www.epa.gov/waterscience/criteria/drinking (accessed Jan. 5, 2009).

USEPA (2006b). U.S. Environmental Protection Agency Peer Review Handbook. EPA/100/B-06/002. www.epa.gov/OSA/spc/2peerrev.htm (accessed Aug. 6, 2009).

USEPA (2007a). Unregulated Contaminant Monitoring Regulation (UCMR) for Public Water Systems Revisions. *Federal Register*, 72:2:367.

USEPA (2007b). Drinking Water State Revolving Fund, 2007 Annual Report. www.epa.gov/safewater/dwsrf/pdfs/report_dwsrf_annual_2007.pdf (accessed Jan. 5, 2009).

USEPA (2008a). Drinking Water Contaminant Candidate List 3; Draft Notice. *Federal Register,* 73:35:9627.

USEPA (2008b). Drinking Water: Regulatory Determinations Regarding Contaminants on the Second Drinking Water Contaminant Candidate List. *Federal Register,* 73:147:44251.

USEPA (2008c). Preliminary Regulatory Determination on Perchlorate. *Federal Register,* 73:198:60262.

USEPA (2008d). FACTOIDS: Drinking Water and Ground Water Statistics for 2008. www.epa.gov/safewater/databases/pdfs/data_factoids_2008.pdf (accessed Jan. 5, 2009).

USEPA (2008e). PWSS Formula Grant Allotments for the Past 5 Years. www.epa.gov/safewater/pws/grants/pdfs/summary_pwss_allotments_historical.pdf (accessed Jan. 5, 2009).

USEPA (2009a). National Primary Drinking Water Regulations; Drinking Water Regulations for Aircraft Public Water Systems; Final Rule. *Federal Register,* 74:200:53589.

USEPA (2009b). Drinking Water Contaminant Candidate List 3—Final. *Federal Register,* 74:194:51850.

USEPA, (2009c). Perchlorate Supplemental Request for Comments, Notice. *Federal Register,* 74:159:41883.

USEPA (2009d). EPA Information Related to the American Recovery and Reinvestment Act of 2009 (Recovery Act). www.epa.gov/recovery/basic.html#overview (accessed Aug. 6, 2009).

USPHS (U.S. Public Health Service) (1925). Report of the Advisory Committee on Official Water Standards. *Public Health Rept.*, 40:693.

USPHS (1943). Public Health Service Drinking Water Standards and Manual of Recommended Water Sanitation Practice. *Public Health Rept.*, 58:69.

USPHS (1946). Public Health Service Drinking Water Standards. *Public Health Rept.*, 61:371.

USPHS (1962). Drinking Water Standards. *Federal Register,* 27:42:2152.

USPHS (1970). Community Water Supply Study: Significance of National Findings. NTIS PB215198/BE. Springfield, Va.

U.S. Statutes (1893). Interstate Quarantine Act of 1893 Chapter 114, *U.S. Statutes at Large*, 27, February 15, 1893:449.

CHAPTER 2

HEALTH AND AESTHETIC ASPECTS OF DRINKING WATER*

Gloria B. Post, Ph.D., D.A.B.T.
Research Scientist
Office of Science
New Jersey Department of Environmental Protection
Trenton, New Jersey, United States

Thomas B. Atherholt, Ph.D.
Research Scientist
Office of Science
New Jersey Department of Environmental Protection
Trenton, New Jersey, United States

Perry D. Cohn, Ph.D., M.P.H.
Research Scientist
Consumer and Environmental Health Service
New Jersey Department of Health and Senior Services
Trenton, New Jersey, United States

WATERBORNE DISEASE **2.3**
PATHOGENIC ORGANISMS **2.6**
Bacteria 2.7
Viruses 2.11
Protozoa 2.15
Cyanobacteria 2.18
INDICATORS OF WATER QUALITY ... **2.19**
Coliform Bacteria 2.19
Enterococcus 2.21
Coliphage 2.21
Other Fecal Indicator Organisms 2.22
Heterotrophic Bacteria 2.22
Microscopic Particulate Analysis 2.22
Turbidity 2.22
TOXICOLOGICAL EVALUATION OF DRINKING WATER CONTAMINANTS **2.23**
Dose-Response in Toxicology 2.23
Types of Toxicological Effects 2.25
Toxicology Studies for Evaluation of Drinking Water Contaminants 2.25
RISK ASSESSMENT OF DRINKING WATER CONTAMINANTS **2.27**
Noncarcinogen Risk Assessment 2.28
Carcinogen Risk Assessment 2.29
Risk Assessment of Suggestive or Possible Human Carcinogens 2.30
Additional Considerations in MCL Development 2.31
Sources of Information for Health-based Drinking Water Values 2.31
Microbial Risk Assessment 2.31
INORGANIC CONSTITUENTS **2.33**
Aluminum 2.33

*The views expressed herein are those of the authors and do not necessarily represent the policies of the New Jersey Department of Environmental Protection or the New Jersey Department of Health and Senior Services.

Arsenic 2.34
Asbestos 2.35
Boron 2.35
Cadmium 2.36
Chromium 2.36
Copper 2.37
Fluoride 2.38
Hardness 2.38
Iron 2.39
Lead 2.39
Manganese 2.40
Mercury 2.40
Nitrate and Nitrite 2.41
Perchlorate 2.42
Selenium 2.43
Sodium 2.43
Sulfate 2.44
ORGANIC CONSTITUENTS 2.44
Volatile Organic Chemicals 2.45
Pesticides 2.50
Herbicides 2.52
Insecticides 2.54
Fungicides 2.56
Additional and Emerging Drinking Water Contaminants 2.56
Chemicals in Treatment Additives, Linings, and Coatings 2.58
DISINFECTANTS AND DISINFECTION BY-PRODUCTS 2.59
General Background 2.59
Health Basis for DBP Regulation 2.60
Approaches for Evaluation of Health Effects of DBPs 2.60
Disinfectants and Inorganic DBPs 2.64
Organic Disinfection By-products 2.66
RADIONUCLIDES 2.69
Introduction 2.69
Health Effects of Radionuclides 2.70
General Considerations for Drinking Water Standards for Radionuclides 2.71
Specific Radionuclides as Drinking Water Contaminants 2.71
AESTHETIC QUALITY 2.72
Taste and Odor 2.73
Turbidity and Color 2.74
Mineralization 2.75
Hardness 2.75
Staining 2.75
PREPAREDNESS AND HEALTH 2.75
FINAL COMMENT 2.76
INTERNET RESOURCES 2.76
ABBREVIATIONS 2.77
ACKNOWLEDGMENTS 2.78
REFERENCES 2.79

Microbiological and chemical contaminants in drinking water may cause acute or chronic health effects or undesirable aesthetic properties when present at excessive concentrations. Hence, health and aesthetic concerns are the principal motivations for water treatment. This chapter summarizes the approaches used to develop the health and aesthetic basis for drinking water standards and guidance, and the health and aesthetic effects of the major groups of contaminants found in drinking water. Following an introductory discussion of waterborne disease outbreaks, specific pathogenic organisms and indicator organisms are discussed. For chemical contaminants, basic concepts of toxicology and risk assessment for chemical and microbial contaminants are presented, followed by separate sections on inorganic constituents, organic compounds, disinfectants and disinfection by-products (DBPs), and radionuclides. Emphasis is placed on contaminants that occur more frequently and at levels that are of concern for human health as well as emerging contaminants of recent interest. Taste and odor, turbidity, color, mineralization, and hardness are discussed in the final section, which is devoted to aesthetic quality.

Because of space considerations, citations have been limited to those most generally applicable, especially those from U.S. governmental organizations such as the U.S. Environmental Protections Agency (USEPA) and the Centers for Disease Control and Prevention (CDC). A general reference for toxicology is Klaasen (2007). For risk assessment, general references are the USEPA risk assessment documents listed in the guidance section of the USEPA Integrated Risk Information System (IRIS) Website. Useful sources of information on health effects, occurrence, and environmental fate and transport of specific contaminants include the National Academy of Sciences series on drinking water, the Toxicological Profiles produced by the Agency for Toxic Substances and Disease Registry (ATSDR) of the CDC, the USEPA IRIS database, the World Health Organization (WHO) Guidelines for Water Quality, which will be updated in 2010, and the National Library of Medicine's Toxnet/Hazardous Substances Data Bank (HSDB) and PubMed databases. The Internet websites for these resources are provided at the end of this chapter.

As mentioned in Chap. 1, some states have established standards for some contaminants that are more stringent than the federal standards. More stringent standards may be based on different interpretation of the health effects information, improved capabilities of available analytical methods, or a more favorable cost-benefit analysis, reflecting the local conditions. States may also establish standards or guidance for additional contaminants not regulated on a national basis.

For chemicals, emphasis has been given to potential health effects from long-term exposure to concentrations that may occur in drinking water rather than to the effects of acute poisoning by much higher concentrations. Where the effects are markedly different between routes of exposure, effects from oral exposure rather than inhalation are emphasized. This chapter is intended as an introductory overview of information on health effects and occurrence of drinking water contaminants. The cited references and public health officials should be consulted when evaluating specific situations of contamination. This chapter does not address acute poisoning situations. Effects of acute poisoning are addressed in other handbooks and Internet Websites. If such poisoning has occurred, local poison control authorities should be consulted.

WATERBORNE DISEASE

From 1993 through 2006, the CDC reported 203 waterborne disease outbreaks in the United States and its territories, with a total of 418,894 actual or estimated cases of illness (Table 2-1). The responsible agent was not identified in 50 (25 percent) of these outbreaks, while microbes caused 119 (59 percent) and chemicals caused 34 (17 percent) of the outbreaks. The data include an estimated 403,000 illnesses from the 1993 *Cryptosporidium* disease outbreak in Milwaukee, Wis., but this number has been questioned due to recall bias (Hunter and Syed, 2002). The data also include 26 disease outbreaks (156 cases and 12 deaths) caused by *Legionella* following exposure to water intended for drinking for the period 2001 through 2006. *Legionella* data were not available prior to 2001; thus the outbreak data are underrepresented in this regard. Much of the disease burden involves gastrointestinal (GI) illness, but a variety of other, more serious illnesses also occur (e.g., pneumonia, hepatitis). This reporting period was selected because 1993 was the year when compliance with most provisions of the USEPA Surface Water Treatment Rule was required (USEPA, 1989a). All provisions of the USEPA Total Coliform Rule (TCR) were in effect at the start of 1991 (USEPA, 1989b). The last year for which data are available is 2006.

If outbreaks caused by a source other than a community or a noncommunity water system (e.g., illnesses caused by contaminated bottled water or container, consuming non-potable water, water from private systems, or water from systems serving fewer than 15 residences) are excluded from the data, there were 144 outbreaks affecting 417,468 persons, including 23 outbreaks (16 percent) caused by chemicals affecting 281 persons. Excluding the caseload from the large 1993 Milwaukee *Cryptosporidium* outbreak, the outbreaks due to *Legionella*, as well as the outbreaks in U.S. territories, between 1993 and 2006 the number of disease outbreaks and cases of illness caused by exposure to contaminated water linked to a community or non-community water system in the United States was 122 and 14,324, respectively. This equates to a 14-year

TABLE 2-1 Agents Causing Waterborne Disease Outbreaks in the United States and Territories, 1993–2006[a,b]

Agent	Number of outbreaks	Footnote	Percent	Number of cases of illness
Unknown	**50**		**25**	**2761**
Bacteria	**63**		**31**	**3423**
Legionella spp.[c]	26	d,e-1		156
E. coli O157:H7	11			267
Campylobacter spp.	9			447
Shigella spp.	5	d		439
Salmonella spp.	4	d		928
Vibrio cholerae	2	e-2		114
Mixed bacteria	6	f		1072
Viruses	**17**		**8**	**3636**
Norovirus	15			3612
Hepatitis A	2			24
Parasites	**35**		**17**	**407001**
Giardia intestinalis (*G. lamblia*)	22			1908
Cryptosporidium parvum	10			404728
Entamoeba histolytica	1	e-3		59
Naegleria fowleri	1			2
Mixed parasite	1	g		304
Mixed Microbe Groups	**4**	h	**2**	**1698**
Chemicals	**34**		**17**	**300**
Copper	11			167
Nitrate/nitrite	6			21
Sodium hydroxide	4			38
Lead	3			3
Fluoride	2			43
Soap/cleaning product	2			15
Other chemical	6	i		13
Total	**203**		**100**	**418819**

[a]An outbreak is defined as: (1) two or more persons experience a similar illness after consumption or use of water intended for drinking, and (2) epidemiologic evidence implicates the water as a source of illness. A single case of chemical poisoning constitutes an outbreak if a laboratory study indicates that the water has been contaminated by the chemical.

[b]Data are from Kramer et al. (1996), Levy et al. (1998), Barwick et al. (2000), Lee et al. (2002), Blackburn et al. (2004), Liang et al. (2006), and Yoder et al. (2008).

[c]Legionella data are from 2001–2006. Only includes illness data for water intended for drinking.

[d]*Legionella pneumophila* (20), *L. micdadei* (1), *Legionella* spp. (5); *Shigella sonnei* (4) and *Shigella flexneri* (1); *Salmonella* Typhimurium (3), *S.* Bareilly (1).

[e]U.S. territories: 1 = one outbreak from the Virgin Islands; 2 = Mariana Islands (Saipan) and Marshall Islands; 3 = Palau.

[f]1996: *Plesiomonas shigelloides* and *Salmonella* Hartford. 1999: *E. coli* O157:H7 and *Campylobacter jejuni.* 2000: *C. jejuni, E. coli* O157:H7, and *E. coli* O111. 2001: *C. jejuni* and *Yersinia enterocolitica.* 2003: *C. jejuni* and *Shigella* spp. 2006: *E. coli* O157:H7, *E. coli* O145, and *C. jejuni.*

[g]1994: *Giardia intestinalis* and *Entamoeba histolytica.*

[h]2002: *C. jejuni, Entamoeba* spp., and *Giardia* spp. 2004a: *C. jejuni, Campylobacter lari, Cryptosporidium* spp., and *Helicobacter canadensis.* 2004b: *C. jejuni,* norovirus, and *Giardia intestinalis.* 2006: *Norovirus* (G1, G2) and *C. jejuni.*

[i]One outbreak each due to: ethylene glycol; bromate and disinfection by-products; ethyl benzene, toluene, and xylene; gasoline by-products; chlorine; unknown chemical.

average of 9 outbreaks and 1023 illnesses per year. Of the 63 (52 percent) outbreaks (6193 cases) that had a community water system as the source, a review of the outbreak details revealed that at least 26 (41 percent), likely more, were caused by events that occurred after delivery of water to the customer. Although 63 percent of the 144 outbreaks involved public water systems (PWSs) using groundwater sources, 91 percent of the PWSs in the United States use groundwater sources (www.epa.gov/safewater/databases/pdfs/data_factoids_2008.pdf). Furthermore, although the USEPA Ground Water Rule (GWR) treatment requirements did not become effective until December 2009 (USEPA, 2006b), many states have had treatment requirements in place for many years for PWS using groundwater sources.

Disease outbreak reporting is voluntary, and cases must be diagnosed and characterized as waterborne. As such, the number or outbreaks and cases are underrepresented in some states, given their respective populations. The summer peak in outbreaks is explained by the many outbreaks that occur at seasonal venues such as summer camps, fairs, and resorts (Table 2-2). Harder to explain is the smaller number of outbreaks that occurred in winter as opposed to the other seasons at year-round venues such as homes, communities, and various facilities.

TABLE 2-2 U.S. Disease Outbreaks by Source Venue and Season, 1993–2006*

		Season‡			
Source venue†	Total	Winter	Spring	Summer	Fall
Year-round					
Homes	78	11	21	27	19
Facilities	64	10	11	22	21
Total	142	21	32	49	40
Seasonal					
Recreation	43	0	3 (May)	36	4 (3 Sep.)
Winter recreation	4	3	1 (Mar.)	0	0

*Data are from Kramer et al. (1996), Levy et al. (1998), Barwick et al. (2000), Lee et al. (2002), Blackburn et al. (2004), Liang et al. (2006), and Yoder et al. (2008). Outbreaks due to contaminated containers (11), unknown season (2), or multiple seasons (1) not included.

†Homes = homes, apartments, condominiums, mobile home parks, communities. Facilities = restaurants (including golf course restaurants), hotels, inns, schools, churches, offices, factories, stores, medical care centers, other. Recreation = camps, fairs/fairgrounds, parks, sport camps/events/venues, resorts/resort areas (except winter resorts).

‡Winter = Dec. to Feb.; spring = Mar. to May; summer = Jun. to Aug. fall = Sep. to Nov.

Disease outbreak data do not include sporadic or endemic cases of illness. For several reasons, the number of reported outbreaks and illnesses is a small proportion of the total waterborne disease burden in the United States. The amount of GI illness attributable to public drinking water systems in the United States each year has been estimated to be between 4 and 33 million cases (Colford et al., 2006; Messner et al., 2006). If disease endpoints other than GI illness were included, even higher numbers of cases would result. It should also be noted that a portion of the total population exposed to any disease-causing agent does not become ill. People whose infections are asymptomatic may become "carriers," capable of spreading disease through the shedding of the infectious agent in their feces and other excretions. This topic is discussed further under Microbial Risk Assessment in the section on Risk Assessment of Drinking Water Contaminants.

PATHOGENIC ORGANISMS

Pathogenic organisms are microorganisms capable of causing disease after getting past host barriers and defenses and multiplying within the host. A number of characteristics and circumstances determine whether or not a microorganism is a pathogen or a harmless or even beneficial member of a host's microbial community (Falkow, 1997). For example, uptake of a bacterial virus or "plasmid" carrying "virulence genes" can convert some types of harmless bacteria into pathogens. Readers who desire information on waterborne pathogens beyond what can be included in this chapter are encouraged to consult the most recent edition of *Waterborne Pathogens* (AWWA, 2006), as well as other recent reviews (e.g., Szewzyk et al., 2000; Leclerc, Schwartzbod, and Dei-Cas, 2002; Cotruvo et al., 2004; Craun et al., 2006). Recent pathogen-specific reviews are cited throughout this section.

The three categories of pathogenic microorganisms that can be transmitted by water are bacteria, viruses, and protozoan parasites (Table 2-1). Disease agents in these categories were responsible for 63 (31 percent), 17 (8 percent), and 35 (17 percent) of the disease outbreaks, respectively (multiple groups were involved in four outbreaks). Due to diagnostic limitations, many of the outbreaks caused by an unknown agent are likely caused by a virus. Although a variety of fungi have been identified in water (Hageskal et al., 2006), there are no documented waterborne disease outbreaks due to fungi in the United States, and there is little if any evidence of potential adverse health effects due to waterborne fungi. Many fungi are pathogenic to humans, but exposure is mostly by airborne inhalation or direct dermal contact. A fourth category, cyanobacteria (blue-green algae), is covered separately. Cyanobacteria are not pathogens. However, excessive growth (blooms) in rivers, lakes, and reservoirs can result in the release of taste- and odor-causing chemicals that can be difficult and expensive to remove during treatment. Some types of cyanobacteria can also produce toxic chemicals, which have harmed wild and domestic animals and, on rare occasions, humans (Carmichael, 2001).

A term increasingly used is *emerging pathogen* or *emerging infectious disease*. The CDC defines emerging diseases as those in which "the incidence in humans has increased in the past two decades or threatens to increase in the near future" (www.cdc.gov/ncidod/EID/about/background.htm). Emerging diseases include those resulting from: (1) a new pathogen created as a result of a change or evolution of an existing organism; (2) a known pathogen spreading to a new geographic area or population; (3) a previously unrecognized pathogen, including those appearing in areas undergoing ecologic transformation; and (4) previously controlled infections reemerging as a result of antimicrobial resistance or breakdowns in public health measures. Some of the pathogens discussed below are considered to be emerging pathogens.

Every biological species except viruses bears a two-word italicized name in Latin. The first word, always capitalized, is the genus (e.g., *Escherichia*). The second name is the species name, not capitalized, and there is usually more than one species for any given genus (e.g., *Escherichia coli* [*E. coli*], *Escherichia hermannii*, etc.). Species are further differentiated as to the types of antigens (protein and carbohydrate molecules) on their surface to which host immune systems respond (serotypes, e.g., *E. coli* O157:H7). For the genus *Salmonella*, what appear to be species names are actually serotypes with the serotype capitalized (e.g., *Salmonella enterica* serotype Typhimurium or simply *Salmonella* Typhimurium). Species can also be differentiated by genetic differences ("genotypes"), their mode of action (e.g., enteroinvasive *E. coli* (EIEC), enterohemorrhagic *E. coli* (EHEC), etc.), or other characteristics.

In nature in general and in source waters and drinking water distribution systems in particular, microbes exist primarily in "biofilms": diverse communities of protozoa, bacteria (and their secretions), and viruses. Microbial interactions in this environment are constant. For example, viruses (phage) infect bacteria, protozoans feed on bacteria and each other, and bacteria compete with one another. Some bacteria have evolved mechanisms to survive and multiply within protozoa, most notably, *Legionella*, but also other bacterial pathogens (Harb, Gao, and Abu, 2000; Snelling et al., 2006a). Microbial characteristics in this environment, such as survival and resistance to disinfection, can be quite different than when grown in isolation (King et al., 1988; Eboigbodin, Seth, and Biggs, 2008). For example, in the natural environment, the function of antibiotic compounds is not to kill or inhibit the growth of other organisms (with some exceptions) but rather to serve as signaling molecules to communicate and interact with other microbes (Mlot, 2009).

Bacteria

Bacteria are single-celled prokaryotic microorganisms, meaning that their cellular architecture is different from that of the eukaryotic cells of protozoa and animals or of the archaea, which are bacteria-like microorganisms with their own unique architecture. A few types such as *Clostridium* and *Bacillus* can develop into resting endospores that can withstand extreme environmental stresses. Bacteria exist in a variety of shapes, but two common shapes are spheres (cocci) and round-ended rods (bacilli). Typical sizes range from 0.5 to 8 μm, but smaller and larger sizes exist. Some bacteria exist in colonies as clumps, sheets, or chains.

Bacteria reproduce by binary fission. Of the three main groups of pathogens, bacteria are the only group with members capable of growth (reproduction) outside of their host. Many bacteria, including many clinically important pathogens, can be grown in the laboratory. Bacteria occur in feces in high concentrations (e.g., 10^7 to 10^{11} per gram). Some pathogens are aerobes requiring oxygen for growth, while others are anaerobes that are killed in the presence of oxygen (but anaerobic *Clostridia* can form endospores that remain viable in the presence of oxygen). Another group, facultative anaerobes, can thrive in the presence or absence of oxygen. Of the many bacteria in nature, only a few cause disease. Many bacterial infections can be treated with antibiotics. Because of the extensive use of antibiotics in hospitals and farms over many years, a number of bacterial pathogens have developed resistance to many antibiotics. Bacterial pathogens of current interest are described below.

Enteric Bacteria. Until the early 1900s, enteric bacteria were responsible for many waterborne disease epidemics worldwide. Epidemics of typhoid fever (caused by *Salmonella* Typhi), dysentery (*Shigella dysenteriae*), and cholera (*Vibrio cholerae*) were not uncommon and resulted in considerable mortality. Following the advent of clean water technologies (Cutler and Miller, 2005) and improvements in hygiene in the early 1900s, disease outbreaks due to these bacteria no longer occur in the United States, but they continue to occur in countries with inadequate sanitation. Although self-limiting gastrointestinal illness is the primary symptom in most cases of illness caused by enteric bacteria, other types of illnesses occur, and some outbreaks have resulted in hospitalizations and mortality.

Enteric bacteria are members of the family Enterobacteriaceae. Members of this family are foodborne as well as waterborne pathogens that are able to infect exposed persons by the fecal-oral route. For waterborne outbreaks, this usually means drinking water that is contaminated with human or animal feces from which these pathogens are derived. Between 1993 and 2006, species of enteric bacteria in the genera *Escherichia*, *Shigella*,

Salmonella, *Plesiomonas*, and *Yersinia* were identified as causative agents in 26 waterborne disease outbreaks in the United States (Table 2-1). Although all of these pathogens can cause gastroenteritis, infection with enterohemorrhagic *E. coli* O157:H7 can result in bloody diarrhea. In susceptible persons, particularly children and the elderly, this infection can sometimes progress to hemolytic-uremic syndrome whereby kidney failure can result in serious illness or death (Yoon and Hovde, 2008). Cattle and other ruminants are significant sources of *E. coli* O157:H7. A few other *E. coli* strains can also cause hemorrhagic disease, and strains of enterotoxigenic *E. coli* can cause a mild to severe watery diarrheal illness in young children and travelers (Hunter, 2003).

Enteric-like Bacteria. Enteric-like bacteria are found in feces and transmitted by the fecal-oral route of exposure, but they are also found in various aquatic niches. Drinking water–related genera in this group include *Vibrio* and *Aeromonas*. *Vibrio cholerae* can cause gastroenteritis. Two toxin-producing serotypes, *V. cholerae* O1 and O139, can cause cholera, a disease characterized by profuse watery diarrhea leading to dehydration and loss of electrolytes. In the developing world, lack of access to rehydration therapy and medical care are causes of considerable illness and death from cholera. *V. cholerae* caused disease outbreaks in two tropical U.S. territories (Kramer et al., 1996; Blackburn et al., 2004).

Aeromonas hydrophyla and other aeromonads are commonly found in the environment, including distribution system biofilms. Only a few strains of *A. hydrophyla* are human opportunistic pathogens. They can cause gastroenteritis as well as a number of other more serious illnesses that can develop following wound infections. Waterborne transmission is possible but not proven (Borchardt, Stemper, and Standridge, 2003).

Treated water from a properly operated water treatment system (i.e., in compliance with all pathogen-related regulations) should be free of enteric and enteric-like bacteria. These organisms are inactivated by all disinfectants commonly used in water treatment. Disease outbreaks caused by these bacteria have only occurred after consumption of water that was untreated, inadequately treated, or contaminated following treatment, such as by cross-connections or back-siphonage within the distribution system or by use of a contaminated container.

Campylobacter. *Campylobacter* spp. are genetically related to *Helicobacter* spp. (see below) and, like *Helicobacter*, have a helical shape but are otherwise like enteric bacteria in terms of being a leading cause of foodborne and waterborne gastroenteritis (Moore et al., 2005). Between 1993 and 2006, species of *Campylobacter* were involved in 18 outbreaks, most involving the thermophilic species *C. jejuni* (Table 2-1). Other less frequent human infecting species include *C. coli* and *C. upsaliensis*. *Campylobacter* was also the agent responsible for most bacterial outbreaks in Canada (Schuster et al., 2005). Although infection with *Campylobacter* leads mostly to GI illness, in some cases, it can lead to Reiter's syndrome, which is an arthritic disease affecting the joints (~1 percent of cases); Guillain-Barré and Miller Fisher syndromes, which cause acute neuromuscular paralyses (~0.1 percent of cases); and other diseases (Peterson, 1994; Dorrell and Wren, 2007). Men are more susceptible to campylobacteriosis than women (Strachan et al., 2008). Domestic and wild birds are a significant source of *Campylobacter*, but a number of other warm-blooded animals including pets can also be sources. *Campylobacter* can remain viable in environmental waters for extended periods of time in a nonculturable state, and they can also form biofilms and infect free-living protozoa. As with enteric bacteria, *Campylobacter* is amenable to conventional water treatment and disinfection. A related genus, *Arcobacter*, contains several emerging foodborne and waterborne pathogenic species (especially *A. butzleri*) (Ho, Lipman, and Gaastra, 2006; Snelling et al., 2006b; Miller et al., 2007). *Arcobacter* have been identified in water samples collected during two U.S. waterborne disease outbreaks involving *Campylobacter* and other agents (Figueras, 2008).

Helicobacter. This genus currently contains many named and unnamed gastric, intestinal, and hepatic species. The human-infecting species, *Helicobacter pylori* (*H. pylori*), is found in the stomach where it can cause gastritis. Most infected persons remain asymptomatic, but duodenal or gastric ulcer may develop in about 5 to 10 percent of cases. *H. pylori* also causes mucosa-associated lymphoid tissue (MALT) lymphoma and is a major risk factor for gastric adenocarcinoma (Kusters, van Vliet, and Kuipers, 2006).

In developed countries the mode of transmission of *H. pylori* appears to be mostly by person-to-person from *H. pylori*-containing vomitus, saliva, stools, or medical equipment. Some species, *H. pylori* included, can infect both humans and wild animals. Hence, animals may represent a source of spread among humans (Solnick and Schauer, 2001; On, Hynes, and Wadstron, 2002; Bellack et al., 2006).

The possibility of waterborne transmission of *H. pylori* is an area of active research. Once excreted into the environment, *H. pylori* quickly transits to a nonculturable *coccoid cell* state, which may (Cellini et al., 1994; She et al., 2003) or may not (Eaton et al., 1995; Kusters et al., 1997) be infectious. Some *Helicobacter* species can persist in water in a culturable state for only a few hours, but *H. pylori* can survive for days to weeks (Shahamat et al., 1993; Konishi et al., 2007; Azevedo et al., 2008), a period sufficient for exposure and infection in persons drinking untreated, fecal-contaminated water (Baker and Hegarty, 2001; Rolle-Kampczyk et al., 2004). *H. pylori* has been detected in surface and groundwaters by several investigators but has never been cultured from these sources, perhaps because of limitations of the assays used (She et al., 2003; Watson et al., 2004). *H. pylori* may gain entrance to drinking water distribution systems through either infiltration or treatment deficiencies and has been detected in distribution system biofilms in a nonculturable state (Park, MacKay, and Reid, 2001; Watson et al., 2004; Giao et al., 2008). *H. pylori* can also infect the protozoan *Acanthamoeba* (see below), remaining culturable for up to eight weeks (Winieka-Krusnell et al., 2002).

H. pylori is readily killed by concentrations of disinfectants typically used in drinking water treatment. As a possible corollary to this, infection rates of *H. pylori* in the United States (20 percent of people under age 40 and 50 percent over age 60) are decreasing to the point where some models show the organism gradually disappearing from the population over the next century (Rupnow et al., 2000).

Legionella. *Legionella* bacteria can cause legionellosis, which can be either a self-limiting febrile illness called Pontiac fever or a serious type of pneumonia called Legionnaires' disease with symptoms that are indistinguishable from pneumococcal pneumonia (Fields, Benson, and Besser, 2002). The primary human exposure route is thought to be the inhalation of aerosols of water containing high concentrations of *Legionella* such as during showering. Aerosol inhalation of *Legionella* containing waters from cooling towers and whirlpool spas are other common, nondrinking water–related exposure routes, but disease outbreaks derived from these sources are not included in the data in Table 2-1. Waterborne ingestion of low levels of *Legionella* is common, perhaps universal, but is not associated with disease.

The CDC began reporting disease outbreaks caused by *Legionella* bacteria in 2001. In the six-year reporting period 2001 through 2006, there were 26 outbreaks involving 156 cases and 12 deaths due to *Legionella* from "water intended for drinking" (Table 2-1) (Blackburn et al., 2004; Liang et al., 2006; Yoder et al., 2008). Thus, in the United States, *Legionella* causes more drinking water–related outbreaks (but not more cases), and probably more serious illness and mortality, than any other microorganism. An estimated 8000 to 18,000 persons are hospitalized with legionellosis each year in the United States (Marston et al., 1997).

Mortality rates from Legionnaires' disease used to be significant (15–30 percent) but have been falling since 1980 because of better and faster diagnosis and more widespread use of prophylactic antibiotic therapy (Diederen, 2008). Although a number of different *Legionella* species are frequently found in water distribution systems, about 91 percent of confirmed cases of community-acquired Legionnaire's disease are caused by *L. pneumophila,* and, of this species, 84 percent were *L. pneumophila* serogroup 1 (Yu et al., 2002). Of exposed individuals who become ill, most are elderly, smokers, or have weakened immune systems or underlying illnesses. About 20 other species have been implicated in human disease (Riffard et al., 2004).

Legionella are naturally occurring bacteria (not the result of fecal pollution) widely distributed in fresh water, including groundwater and wet soil (Riffard et al., 2004), mostly as a parasite of protozoa. Growth may occur in the environment in the presence of algae and cyanobacteria (Fliermans, 1996; WHO, 2002). Small numbers of these organisms can survive chlorination and have been found in distribution and plumbing system biofilms (Rogers and Keevil, 1992).

The organism can multiply within many different types of protozoan hosts (e.g., *Acanthamoeba* spp. and *Hartmanella* spp.), but only at temperatures above 20°C (Fliermans, 1996; Snelling et al., 2006a; Ohno et al., 2008). Environments such as hot water systems, showerheads, taps, and respiratory ventilators favor *Legionella* growth (Diederen, 2008). It is believed that the ability to multiply within protozoan hosts enables *L. pneumophila* to evade disinfection and to infect lung macrophages (phagocytic white blood cells) following inhalation.

It has been suggested that control actions should be taken whenever concentrations of *Legionella* exceed 0.1 colony-forming unit (CFU) per mL or when heterotrophic bacteria (see section on Indicators of Water Quality) exceed 10,000 CFU per mL (Health and Safety Commission, 2000; WHO, 2007). Some protozoan hosts can withstand high concentrations of chlorine and high temperatures for brief periods, and *Legionella* embedded within biofilms are also quite resistant to disinfection (Thomas et al., 2004; Loret et al., 2005). Thus, it is difficult to eliminate *Legionell*a from plumbing systems. *Legionella* can be controlled in hot water storage tanks by keeping the temperature at 60°C, but high temperatures present a risk for scalding. The installation of thermostatically controlled mixing valves can eliminate the scalding risk (WHO, 2006). High concentrations of chlorine can eliminate *Legionella*, but such doses can corrode pipes. Therefore, most remediation contractors employ superheating to eliminate *Legionella* from building plumbing systems. However, some species can withstand 70°C for up to an hour and remain infective, so recolonization of plumbing systems following heat-shock treatment may occur (Allegra et al., 2008).

Mycobacteria. As with *Legionella*, *Mycobacteria* are free-living bacteria in water and soil (including plant potting soil) that often colonize water distribution systems. As with *Legionella*, *Mycobacteria* have been shown to be able to survive within a number of protozoa (Vaerewijck et al., 2005; Mura et al., 2006; Pagnier et al., 2009). Environmental mycobacteria (EM), including the so-called *Mycobacterium avium* complex (*M. avium* and *M. intracellulare*; MAC), can be waterborne pathogens under certain conditions (Pedley et al., 2004; Vaerewijck et al., 2005). MAC can infect the lungs or GI tract, particularly those of AIDS patients. Symptoms can include fatigue, cough, weight loss, and fever. In AIDS patients, the bacterium can cause debilitating disease and death. Some species of EM, MAC organisms in particular, have been isolated from water taps in medical facilities with the same species isolated from AIDS patients in those facilities; however, the degree of risk for either AIDS patients or healthy persons is unknown. The introduction of antiretroviral therapy for AIDS patients has reduced the amount of disease due to MAC and other mycobacteria (Jones et al., 1999). Many EM other than MAC organisms (e.g., *M. gordonae, M. kansasii, M. abscessus, M. fortuitum, M. xenopi*) have been found in water distribution systems, often more frequently than MAC organisms, and some of these organisms have also been categorized as *opportunistic pathogens* associated with human disease.

Mycobacteria are resistant to various environmental stresses and water treatment disinfectants compared to most other waterborne bacteria and viruses (LeChevallier, 2006; Shin et al., 2008). The ability of *Mycobacteria* to persist in water distribution systems is associated with a complex cell wall, survival or growth at warm water temperatures, association with biofilms, and survival within protozoa. The role of protozoa in the persistence and virulence of EM in distribution systems and the resistance of EM to disinfection requires additional study.

Opportunistic Pathogens. For information on *Acinetobacter, Enterobacter, Klebsiella, Pseudomonas*, and other bacteria groups with species that are potentially pathogenic in persons with weakened immune systems or in other susceptible subpopulations such as burn patients, the reader is referred to AWWA (2006), Leclerc, Schwaartzbod, and Dei-Cas (2002), and Szewzyk et al. (2000). Increasing levels of antibiotic resistance are a growing problem in members of these and other genera (Taubes, 2008).

Viruses

Along with prions (misfolded proteins that can cause disease in some animals including humans) and viroids (small, circular strands of RNA that can infect some plants), viruses are among the smallest, most abundant infectious agents on earth. They range in size from 0.02 to 0.3 μm. They consist of a core of nucleic acid (either DNA or RNA) surrounded by a protein capsid or shell. Some viruses also have an outer host-derived lipid membrane. They depend on host cells for reproduction and have no metabolism of their own. Enteric viruses comprise a number of classes of viruses that infect the intestinal tract of humans and animals and are excreted in their feces (Carter, 2005; Nwachuku and Gerba, 2006). Most pathogenic waterborne viruses can cause acute GI disease, although some cause more severe illnesses. Enteric viruses that infect humans generally do not infect other animals and vice versa, although there are a few exceptions to this rule. Viruses that have caused or have the potential to cause waterborne disease are discussed next.

Norovirus and Other Calicivirus. Of the four groups that comprise the Calicivirus family (Caliciviridae), two can infect humans: noroviruses (NVs) and sapoviruses (SVs). Noroviruses were previously called small round-structured viruses (SRSV), human caliciviruses (HCV), or Norwalk-like viruses. Sapoviruses were previously called Sapporo-like viruses. With the exception of *Legionella pneumophila* and *Giardia lamblia* (see below), NVs have caused more identified waterborne disease outbreaks than any other biological agent. Because caliciviruses cannot be grown in the laboratory and because these viruses are highly infectious, many outbreaks of unknown etiology are likely caused by NVs (Schaub and Oshiro, 2000). In addition, because illness symptoms, acute gastroenteritis, are generally mild compared to those of some other agents, it is likely that many other NV outbreaks are never identified and thousands of additional cases are never reported. Indeed, NVs are estimated to cause 93 percent of the nonbacterial gastroenteritis in the United States (Fankhauser et al., 2002). NVs likely cause more cases of foodborne illness than any other pathogen.

Between 1993 and 2006, there were 15 identified waterborne disease outbreaks (3612 cases) in the United States due to NVs. Of the 12 outbreaks in which a cause was identified, 9 involved the consumption of untreated contaminated water (including two in which a chlorinator was not functioning), two followed a back-siphonage or cross-contamination event (with other possible causes in one), and one involved an improperly cleaned ice-making machine. Transmission is mainly by the fecal-oral route, with person-to-person spread also playing an important role. Airborne transmission over short distances and spread by contaminated objects (fomites) may also facilitate spread during outbreaks (Parashar et al., 2001). Outbreaks due to the consumption of contaminated foods, particularly molluscan shellfish, are common. Historically, NV disease was often referred to as "winter vomiting disease," reflecting the seasonal outbreak peak and one of the most common illness symptoms in addition to diarrhea.

Rotavirus. Although no waterborne disease outbreaks due to rotaviruses (RVs) have occurred in the United States in the past 14 years, outbreaks have occurred in earlier years and in other countries as well. RVs are responsible for a significant amount of disease, primarily among infants and young children (Gray et al., 2008). Symptoms in children include vomiting followed by watery diarrhea lasting four to seven days and fever. Most children are infected by age 2 to 3. Adults are susceptible, but in many cases, adults are asymptomatic, though still infectious. The annual pattern of infection in the United States, a winter peak with disease spreading from the west coast in late fall to the east coast in the spring, appears to be due in large part to interstate differences in birth rates (Pitzer et al., 2009). Dehydration can be rapid but can be combated with rehydration therapy. In the developing world, lack of access to rehydration therapy and medical care results in 600,000 to 900,000 deaths each year, which is 6 percent of all mortality of children under age 5 (Bass, Pappano, and Humiston, 2007). Death due to RV in the United States is rare (20–60 per year), but an estimated 477,000 medical care visits and 55,000 to 70,000 hospitalizations each year are a significant burden on health care resources and the economy (Parashar, Alexander, and Glass, 2006).

RVs are primarily transmitted by the fecal-oral route, but there is growing evidence of respiratory transmission as well. RVs have a low infectious dose (Ward et al., 1986) and can persist on surfaces for long periods of time and on hands for an hour or more, likely contributing to RVs apparent hand-to-mouth mode of transmission (Cook et al., 2004). RV is capable of surviving for weeks to months in surface and groundwaters in an infectious state (Caballero et al., 2004; Espinosa et al., 2008). Infection from livestock and pets has been reported but is uncommon (Cook et al., 2004).

Because natural infection confers significant immunity and because two U.S. Food and Drug Administration-licensed RV vaccines are available, the CDC now recommends routine vaccination for infants (Parashar, Alexander, and Glass, 2006; Cortese and Parashar, 2009). The vaccines protect against most, but not all, RV serotypes (Gentsch et al., 2005). Vaccine use has resulted in delayed onset and reduced severity of RV illness among children in the United States (Staat et al., 2008).

Hepatitis A Virus. Hepatitis A virus (HAV) is the sole member of the genus *Hepatovirus* within the small RNA virus family. It is the cause of most cases of infectious hepatitis, an inflammatory disease of the liver (Cuthbert, 2001). In countries with poor sanitation, most people are infected with HAV as children. Disease symptoms in children are either mild or absent and subsequent immunity is lifelong. In the United States and other industrialized countries, most people reach adulthood having never been exposed to HAV. Disease symptoms are generally more severe when the first HAV infection occurs in adulthood rather than childhood (www.who.int/mediacentre/factsheets/fs328/en/index.htmL). Hepatitis can involve liver inflammation, with symptoms of dark urine and subsequent jaundice (yellowing of skin and eyes due to the inhibition of hemoglobin breakdown) and death in rare cases. Acute disease lasts 2 to 4 weeks and full recovery can take 8 to 10 weeks, with excessive fatigue a common postinfection problem. Relapse occurs in some people. Between 1993 and 2006, two drinking water–related outbreaks occurred in the United States involving the consumption of untreated water from contaminated residential sources.

HAV is spread by the fecal-oral route and is quite stable under a variety of environmental stresses (Nasser, 1994). Travel remains a significant risk factor, as is close contact with an HAV-infected person. Food, especially contaminated shellfish, and waterborne etiologies are involved in about 7.5 percent of all cases. Considerable research is being directed at improved detection methods for this difficult-to-grow virus (Konduru and Kaplan, 2006).

Hepatitis A incidence has declined by 90 percent in the United States (to 1.2 cases per 100,000 persons) between 1995, the first year HAV vaccines were available, and 2006, especially in states where routine vaccination was recommended beginning in 1999 (Wesley, Grytdal, and Gallagher, 2008). However, with underreporting and asymptomatic cases taken into account, there are still an estimated 32,000 new infections each year. In 2006, it was recommended that children ages 12 to 23 months in all states receive an HAV vaccine. This practice should result in the eventual elimination of indigenous transmission in the United States (Fiore, Wasley, and Bell, 2006).

Hepatitis E Virus. A member of the small RNA virus family, Hepatitis E virus (HEV) is the sole member of the genus *Hepevirus* (Panda, Thakral, and Rehman, 2007). Like HAV, HEV is a fecal-oral pathogen. HEV is transmitted mostly by contaminated water in developing countries (Worm, van der Poel, and Brandstatter, 2002). Infection without illness is common. The clinical disease, hepatitis, has symptoms similar to those caused by HAV and other hepatitis viruses, except in pregnant women and their unborn children. In pregnant women, fulminant hepatitis is more common, and death rates can approach 25 percent in the third trimester (Kumar et al., 2004). The most susceptible age groups are young adults. The few cases that occur in the United States and other developed countries are mostly contracted during travel to countries in the developing world where HEV infection is endemic. Sporadic, nontravel-related cases occur mostly in men over age 50 who often have underlying medical conditions (Borgen et al., 2008). Sources of sporadic infections are unknown, but exposures to HEV-containing pigs, pig products, or other domestic or wild animals are possibilities.

Enterovirus. Enteroviruses are a group of more than 60 different types of small RNA viruses that commonly infect humans. They were formerly classified numerically within the groups Polioviruses, Coxsackieviruses, Echoviruses, and Enteroviruses but are now classified as either Polioviruses or Enterovirus groups A, B, C, and D (HEV-A, HEV-B, HEV-C, and HEV-D) but with traditional names kept for previously established serotypes. Echoviruses 22 and 23 have been reclassified as a new genus, *Parechovirus*, as parechoviruses 1 and 2, respectively. Enteroviruses are fecal-oral waterborne pathogens that are associated with a wide range of serious illnesses, including paralysis, aseptic meningitis, myocarditis, encephalitis, and neonatal sepsis. Less serious illnesses include conjunctivitis, mouth and throat sores, sharp abdominal pain, rashes, fever, and respiratory and GI illnesses. Most enterovirus detections occur from June through October. Infections can be acquired through contact with secretions from infected persons, contaminated surfaces, or contaminated water. Transmission to newborns from infected mothers also occurs. Unlike many other viral pathogens, many enteroviruses can be grown in laboratory cultures.

Polioviruses (serotypes 1, 2, and 3) were major worldwide pathogens until the advent of vaccines in 1955 and 1961 and the WHO's effort beginning in 1988 to eradicate these viruses. Acute flaccid paralysis (AFP), which develops in about 1 percent of all infections, mostly in children, is now due to serotypes 1 and 3. Serotype 2 was eradicated, but type 2 vaccine derived strains are now circulating in at least two countries (see below; Shulman et al., 2009; Roberts, 2009a). AFP still occurs in 18 countries, with India, Nigeria, Pakistan, and Afghanistan being the ultimate source of the viruses in the other 14 countries (CDC, 2007; Roberts, 2009b). Outbreaks of AFP due to mutated forms of the live but nonvirulent oral vaccine derived polio virus (VDPV) also occur in these and other countries. Oral polio vaccines are no longer used in the United States and some other countries. The last cases of naturally acquired polio in the United States occurred in 1979, but four nonparalytic cases of VDPV in unvaccinated children were documented in the United States in 2005 (Bahta et al., 2005). A few enteroviruses other than polioviruses can also cause AFP (Dufresne and Gromeier, 2006; Khetsuriani et al., 2006).

About 15 types of enteroviruses (mostly HEV-B but also HEV-A serotypes) predominate in the United States, with echoviruses 9, 11, 33, and 6 and coxsackievirus B5 accounting for almost half of the 52,812 enteroviruses detected between 1970 and 2005 (Khetsuriani et al., 2006). Illness occurs most often in infants and young children in both epidemic and endemic patterns of infection. About 3.3 percent of the cases, mostly involving infants, are fatal, with fatality rates of 10 to 11 percent for coxsackievirus B4 and parechovirus 1. No disease outbreaks due to drinking water have been recorded in the United States since 1993, but recreational waterborne outbreaks have occurred.

Adenovirus. Adenoviruses (Ads) are among the few types of enteric viruses with a DNA core (Fong and Lipp, 2005). There are 51 numbered serotypes (e.g., Adenovirus 1, or Ad1) that can infect humans. As with many other pathogens, infants are most susceptible among persons with normal immune system function. Exposure is common, perhaps universal (Langley, 2005), and Ads are a common cause of childhood gastroenteritis. Persons with weakened immune systems are most susceptible to infection, with high rates of severe illness and mortality.

Some serotypes are shed in feces but cause respiratory (e.g., Ads 1 through 7) or eye disease (e.g., Ads 3, 7, and 14). These types can also spread via aerosol, direct contact, or sexual routes (Jiang et al., 2006; Langley, 2005). Ads 12, 18, and 31 may cause diarrhea on occasion, but infection usually results in inapparent illness in infants. Other serotypes are associated with GI illness but have not been proven to cause illness. However, there is probably sufficient evidence that Ad 40 and Ad 41 are waterborne pathogens that can cause diarrhea in infants.

Ads can be shed in the feces for months to years following infection. Ads are among the most frequently observed viruses in sewage and environmental waters (Katayama et al., 2008; Rodriguez, Gundy, and Gerba, 2008). Many Ads are difficult if not impossible to grow in laboratory cultures. Hence, much recent research has focused on the development of detection and quantitation assays.

Astrovirus. Astroviruses are members of the small RNA virus family that have a star-like appearance when viewed by electron microscopy (Mendez and Arias, 2007). Astroviruses are divided into two groups, Avastroviruses, which infect birds, and Mamastroviruses, which infect mammals. The human-infecting astroviruses (HAstV) are part of the latter group. HAstVs are a significant cause of mild, self-limiting GI illness throughout the world, especially among infants and young children (Walker and Mitchell, 2003). The elderly and persons with weakened immune systems are also at increased risk of illness. About 2 to 6 percent of diarrheal illness is caused by HAstVs in developed countries. Seasonal peaks occur in the United States, with a winter peak in western states and a late spring or early summer peak in the east. Water and foodborne transmission is likely but not proven. HAstVs are relatively stable in the environment (Abad et al., 2001; Espinosa et al., 2008). Secondary wastewater treatment appears effective in eliminating infectious HAstVs from sewage even without disinfection (El-Senousy et al., 2007).

Other Viruses. New detection methods have led to the discovery of other enteric viruses (Check, 2009). The reader should consult Gerba (2006) and Wilhemi, Roman, and Sanchez-Fauquier (2003) for additional information.

Protozoa

Protozoa, or more precisely protista, are single-celled eukaryotic organisms that vary greatly in size, but most are larger than bacteria. Unlike algae, they cannot perform photosynthesis. They are common in natural waters, moist soils, and as symbionts or parasites of animals and plants (Fayer, 1994; Marshall et al., 1997). Protozoans feed on bacteria, but some pathogenic bacteria such as *Legionella* (see above) have evolved to survive and multiply within these hosts (Snelling et al., 2006a). Protozoa often occur in the environment as resistant spores, cysts, or oocysts. These are dormant stages in the life cycle of the parasite that are resistant to adverse environmental conditions and allow the organism to survive until conditions improve or to facilitate its spread from one host to another. Such cysts and spores are often more resistant to chlorine and other oxidizing disinfectants than viruses and most bacteria with the exception of *Bacillus* and *Clostridium* endospores. Protozoa are found in surface waters and groundwaters directly influenced by surface waters. Worldwide, they have caused hundreds of disease outbreaks (Karanis, Kourenti, and Smith, 2007). A few of the pathogenic protozoa are described next.

Parasitic Protozoa. *Giardia lamblia* (syn. *Giardia intestinalis*, *Giardia duodenalis*) is the most common cause of diarrheal disease worldwide (Adam, 2001) and the second leading cause of drinking water disease outbreaks in the United States after *Legionella*. Between 1993 and 2006, *Giardia* caused or contributed to 25 outbreaks (Table 2-1). *Giardia* has also caused foodborne outbreaks on rare occasions. *Giardia lamblia* can infect many other animals in addition to humans; thus, animal wastes can be a source of human-infecting *Giardia*. The organism forms thick-walled oval cysts (6 to 10 μm wide and 10 to 15 μm long) in the host intestine prior to excretion in feces.

Exposure is by the ingestion of fecal-contaminated water, through direct fecal-oral contact (including touching contaminated objects) or, less commonly, from contaminated food. Summer peaks in disease incidence are due to the fact that *Giardia*, like *Cryptosporidium* (see below), is also a frequent cause of recreational disease outbreaks (Yoder and Beach, 2007). Groups at increased risk for contracting giardiasis include travelers to developing countries, young children, and homosexual males (Wolfe, 1992). Giardiasis is usually a self-limiting disease, but subacute or chronic infection can occur in some people. Most infections are without symptoms. Disease outbreaks occur only when untreated or insufficiently treated water is ingested.

Cryptosporidium includes a diverse group of organisms that, like some other protozoa, forms oocysts (thick-walled zygotes) in the intestines of infected host animals prior to excretion in feces. Oocysts are small in size, 4 to 5 μm in diameter. Oocysts from 13 named species and many more unnamed species are commonly found in the environment and they can infect a wide range of vertebrate hosts (Xiao et al., 2004). Each species has a limited host range but few, if any, are single host-specific. *C. parvum* (previously, bovine genotype, or genotype 2) and *C. hominis* (previously, *C. parvum* genotype 1; Morgan-Ryan et al., 2002) cause most human cases of cryptosporidiosis, and these organisms are second to *Giardia lamblia* in the number of disease outbreaks caused by protozoa. Between 1993 and 2006, 10 outbreaks were caused by these organisms, including the 1993 outbreak in Milwaukee, Wis., that sickened up to 25 percent of the city's population of 1.61 million. This outbreak together with the finding that most existing water treatment processes did not completely remove *Cryptosporidium* oocysts from the treated water (Aboytes *et al.*, 2004) resulted in promulgation of the Enhanced Surface Water Treatment Rules in the United States (ESWTRs, see Chap. 1). The ESWTRs strengthen protection against infection from *Cryptosporidium* and other pathogens while minimizing the health risks from disinfection by-products (DBPs, see below). Humans and farm animals, especially young cattle, are the major sources of these two species. Ingestion of water contaminated with oocysts is the primary mode of contracting cryptosporidiosis, but person-to-person transmission and contact with fecal-contaminated surfaces are also important causes, especially in child and elderly care centers. A few foodborne outbreaks have also been recorded.

Cryptosporidiosis is a self-limiting diarrheal disease in most people. The disease can be more severe in persons with advanced AIDS and other diseases involving impaired T-cell function (Hunter and Nichols, 2002). The introduction of antiretroviral therapy for AIDS patients has reduced the incidence of severe cryptosporidiosis (Jones et al., 1999).

Method 1623, which is used to enumerate *Cryptosporidium* in water (USEPA, 2001a), cannot differentiate species that can infect humans from those that infect other animals but not humans. Studies of surface waters using methods capable of differentiating species have shown that many oocysts detected in surface waters using Method 1623 are not likely to infect humans (e.g., Xiao et al., 2000; Yang et al., 2008).

The oocysts are not affected by chlorine concentrations used in water treatment, and their high chlorine resistance is the reason why *Cryptosporidium* is the most frequently recognized cause of outbreaks of gastroenteritis in treated recreational waters, such as swimming pools, water parks, and splash fountains (Yoder and Beach, 2007).

Isospora are a large group of coccidian parasites in the phylum Apicomplexa, which also includes *Cryptosporidium* (Lindsay, Dubey, and Blagburn, 1997). The species that can cause diarrheal disease in humans, *Isospora belli*, is a common parasite, especially in warm climates. This species produces large oval oocysts 10 to 19 μm wide and 20 to 33 μm long that resemble oocysts of *Toxoplasma* (see below) and, like *Toxoplasma* but unlike *Cryptosporidium*, require 24 hours or more to sporulate and become infectious for humans and other animals following excretion to the environment. Human exposure is by contaminated food or water. The disease is usually self-limiting, but the disease isosporiasis can be chronic and more severe in infants, young children, and especially persons with weakened immune systems such as AIDS patients.

Toxoplasma gondii is a common coccidian parasite of warm-blooded animals (most being *intermediate hosts*) that infects about 16 percent of the U.S. population ages 12 to 49 (Jones, Kruszon-Moran, and Wilson, 2003). Infection occurs by ingesting raw or undercooked meat or contact with soil, sand boxes, cat litter boxes, or water that has been contaminated by feces of domestic or wild cats (Tenter, Heckeroth, and Weiss, 2000; Dubey, 2004). Only cats (the *definitive host*) can excrete *T. gondii* oocysts, which are 10 to 12 μm in diameter and require 1 to 21 days in the environment to sporulate and become infectious. Few symptoms occur in most healthy persons, but cysts can remain permanently in nerve, muscle, and other tissues. Fetal exposure from an infected mother can result in abortion or blindness, mental retardation, or epilepsy in the infant or it can remain a subclinical infection. Toxoplasmosis in AIDS patients and persons with weakened immune systems, following reactivation of tissue cysts, is the most frequent cause of severe or fatal encephalitis. The introduction of antiretroviral therapy for AIDS patients has reduced the occurrence of toxoplasmosis of the brain (Jones et al., 1999). Waterborne outbreaks are rare, but three outbreaks outside of the United States have been documented (Villena et al., 2004). Resistant oocysts can survive for months in the environment. Conventional filtration is usually effective in removing oocysts, but oocysts may be difficult to inactivate using disinfection during water treatment (Tenter, Heckeroth, and Weiss, 2000; Wainwright et al., 2007).

Entamoeba histolytica is a human-specific pathogen that causes amoebic dysentery following ingestion of cysts from contaminated water or food (Haque et al., 2003). The disease prevalence worldwide is fourth behind AIDS, malaria, and shistosomiasis (Keene, 2006). In the United States, at-risk groups include male homosexuals, travelers and recent immigrants, and institutionalized populations. Infection can result in no symptoms in most cases to mild gastroenteritis and, less frequently, acute bloody diarrhea, fever, sometimes involving liver abcesses (predominantly in males), or a fulminant disease with significant mortality rates (Stanley, 2003; Stauffer and Ravdin, 2003). In 2002, there was an outbreak in the

U.S. territory of Palau involving 59 persons who drank untreated surface water. Waterborne disease outbreaks due to this organism are infrequent in the United States. *E. histolytica* contributed to a 1994 outbreak (424 cases) caused mainly by *Giardia*, which was due to a cross-contamination event at a Tennessee prison. A 2002 disease outbreak occurred at a New York apartment building in which a nonspeciated *Entamoaeba* was identified along with *Campylobacter* and *Giardia* (morphologically indistinguishable species *E. dispar* and *E. moshkovskii* are nonpathogenic). This outbreak involved 27 persons who became ill after sewage contaminated a basement well.

Cyclospora cayetanensis has caused 11 or more known foodborne diarrheal disease outbreaks in the United States and Canada and is linked to imported foods such as raspberries and leaf vegetables including basil (Herwaldt, 2007). This organism is also suspected, but not proven, to be a cause of a few waterborne outbreaks, including a U.S. outbreak in 1990 (Huang et al., 1995). The organism is shed in the feces of infected persons as oocysts, which do not sporulate (become infectious) until days or weeks following excretion. The oocysts are 8 to 10 μm in size. The oocysts are usually removed by conventional treatment.

Microsporidia are a large group (~1200 species) of bacteria-sized parasitic organisms related to fungi (Hirt et al., 1999) and with spores possessing a characteristic polar filament or tube. They can infect a large number of animals and are shed in urine, feces, and other secretions (Didier and Weiss, 2006). Fourteen species have been shown to cause disease, sometimes chronic, in humans, with the most significant being *Enterocytozoon bieneusi* (causing diarrhea) and three species of *Encephalitozoon* (*E. intestinalis, E. hellem*, and *E. cuniculi*; causing disseminated infections). Prior to the advent of antiretroviral therapies, human-infecting microsporidia most often caused illness in persons with AIDS (Weber et al., 1994). Other at-risk groups include travelers, children, the elderly, and organ transplant patients. Microsporidiosis is likely a zoonotic disease, and waterborne transmission is possible but not proven.

Free-living Protozoa. Free-living protozoa that can also infect suitable hosts and cause illness are called amphizoic pathogens. *Naegleria fowleri* is a thermophilic organism commonly found in moist soils, water, and decaying vegetation that forms a dormant cyst at colder temperatures. Though commonly found in surface waters, *N. fowleri* rarely causes disease, with a 14-year total of 29 cases (2 per year) in the United States. Unfortunately, the disease, primary amoebic meningoencephalitis (PAM), is almost always fatal, with death occurring within 72 hours after symptoms (similar to viral and bacterial meningitis) first appear. Also, unlike encephalitis due to *Acanthamoeba* (see below), many cases of PAM involve young people. Most cases occur following swimming in warm fresh waters, with the transmission route being the nasal passage leading to the brain. In 2002, two young, unrelated individuals contracted fatal PAM in Arizona after being exposed to, but not consuming, untreated well water from the same water system (Marciano-Cabral et al., 2003). In this outbreak, *N. fowleri* was detected in the water storage tank and in refrigerator filter water. The DNA of *N. fowleri* has been detected in 11 (7.7 percent) of 143 wells in Arizona. The average water temperature of the positive wells was 29°C (Blair et al., 2008).

Acanthamoeba spp. is a group of amoeba commonly found in air, soil, and water including swimming pools, spas, and drinking water (Nwachuku and Gerba, 2004). Exposure to these organisms by way of the eyes, nose, lungs, or breaks in the skin is common, but illness is rare. A few species can cause inflammation of the cornea in some people, most often in contact lens wearers exposed to water containing these organisms. These and other species, as well as the related organism *Balamuthia mandrillaris*, can also cause granulomatous amoebic encephalitis (GAE), a fatal disease in persons with weakened immune systems (Martinez and Visvesvara, 1997; Marciano-Cabral and Cabral, 2003). In addition to being pathogenic, *Acanthamoeba* can be hosts for a number of pathogenic bacteria, including important waterborne pathogens such as *Legionella, Mycobacteria, Helicobacter pylori*, and enteric and similar bacteria (see above and Snelling et al., 2006a). These bacterial–protozoan interactions enhance the survival capacity (Laskowski-Arce and Orth, 2008) and the antibiotic and disinfectant resistance of the bacteria, as well as the virulence of both entities. Control of these bacteria in drinking water distribution systems requires control of *Acanthamoeba* and other protozoan hosts (e.g., *Hartmannella*). The sizes of the amoeboid and cyst stages (13 to 40 μm) are fairly large, and effective removal by conventional treatment is likely (Marciano-Cabral and Cabral, 2003).

Other Protozoa. Other protozoa that have been identified as the cause of one or two waterborne diarrheal disease outbreaks worldwide include *Balantidium coli* and *Blastocystis hominis* (Karanis, Kourenti, and Smith, 2007). Feces from pigs, humans, and primates are sources of *B. coli* cysts, which are large (50 to 70 μm in size). Removal by conventional treatment is expected. Most human infections occur in warm climates. *B. hominis* have been observed in many types of animals, but its role as a disease agent is still under study. *B. hominis* cysts (*central body forms*) are 6 to 40 μm in size. More information on these parasites, as well as few multicellular parasites of soil and water, can be found in the AWWA Manual 48 (AWWA, 2006).

Cyanobacteria

Cyanobacteria are often referred to as blue-green algae, but they are photosynthetic bacteria and not algae. They can be unicellular or filamentous (colonial) organisms. They are not pathogens, but excessive growth of cyanobacteria in source waters can lead occasionally to the production of chemicals with undesirable tastes or odors (see Aesthetic Quality section) and toxic chemicals (Chorus and Bartram, 1999).

A number of free-floating, toxin-producing strains of cyanobacteria (e.g., *Microcystis, Oscillatoria, Anabaena, Nostoc*) can produce several classes of toxic chemicals, primarily liver toxins and nerve toxins (Haider et al., 2003). The liver toxins (hepatotoxins) include cylindrospermopsin (CYL) and families of microcystin (60 types) and nodularin (6 types). In addition to the liver, CYL can also affect the kidneys, thymus, and heart. The neurotoxins include a family of anatoxin (3) and saxitoxin (20) compounds (Briand et al., 2003) and β-*N*-methylamino-L-alanine (BMAA) (Cox et al., 2005). The function of these toxins in nature is uncertain. They may protect their producers from protozoan predators and/or give a competitive growth advantage over nontoxin-producing cyanobacteria and algae (Wiegand and Pflugmacher, 2005; Welker and von Döhren, 2006), but other functions are possible (Mlot, 2009).

Cyanotoxins have been linked to multiple episodes of illness and death in wild and domestic animals and, on rare occasions, have caused illness and death in humans (Carmichael, 2001). In 1996, intravenous exposure to hepatotoxins at a dialysis center in Brazil caused the death of 76 of 131 people (Carmichael et al., 2001). These and other toxins can also cause respiratory and allergic reactions and gastroenteritis in humans. BMAA is hypothesized to be associated with motor neuron diseases such as Alzheimer's and Parkinson's disease. Some of these toxins also have cancer-promoting effects. Thus, even low concentrations in drinking water are of concern. The WHO has set a guideline of 1 μg/L for microcystin-LR and is proposing the same concentration for CYL in drinking water (Rodriguez et al., 2007). Currently, there is insufficient information to set acceptable levels for any of the other microcystin toxins (e.g., microcystin-LA, -YR, -YM, etc.) or for any of the other hepatotoxins or neurotoxins.

These toxins are produced following algal bloom events in eutrophic lakes and rivers. The toxins are cell-bound and are released upon death or aging of the organisms. As such, algicide treatment of cyanobacterial blooms in lakes and reservoirs or chemical disinfection can increase waterborne toxin levels. In addition to potential drinking water exposure, human exposure can occur during water recreation or consumption of cyanobacterial health food products. Conventional water treatment has been shown to reduce microcystin concentrations in U.S. waters by 90 to 99.9 percent (Karner et al., 2001; Haddix, Hughley, and LeChevallier, 2007). Chlorine disinfection alone may not be sufficient to reduce high concentrations of anatoxin-a to safe levels, and chloramine disinfection may not be effective against any of the cyanotoxins (Rodriguez et al., 2007).

INDICATORS OF WATER QUALITY

Treated as well as untreated sewage likely contains some level of human pathogens most if not all of the time. However, pathogen occurrence in fecal wastes from individuals, septic tank effluents, or nonhuman animal wastes can be sporadic. In addition, different pathogens can occur at different times and, as already mentioned, many pathogens are difficult or impossible to grow in laboratory cultures. Almost all pathogens can now be detected by molecular tests, but many of these tests are such that the viability or infectivity of the detected pathogen is unknown. In addition, many of these methods are not able to detect low but infectious concentrations. Finally, many pathogen tests are difficult, costly, and time-consuming. For all of these reasons, the testing of water samples for the presence of pathogenic organisms is problematic (Allen, Clancy, and Rice, 2000).

Instead, the sanitary quality of water is determined by testing for one or more *indicator* organisms or groups of organisms. The presence of these organisms indicates that the tested water contains or may contain fecal pollution and hence the possible presence of pathogenic organisms. The detection methods for indicator organisms should be fast, easy to perform, and inexpensive compared to pathogen tests, and this is the case. Indicator organisms should always occur in untreated fecal wastes in concentrations that are higher than those of pathogens, and this is also the case. Ideally, they should mimic pathogens in terms of transport and survival in environmental waters, but this is not always the case. Ideally, they should only be found in fecal wastes, but all indicator organism groups in use today also have nonfecal sources to varying extents. Current indicator organisms and organism groups used as fecal pollution indicators in U.S. drinking water regulations are shown in Table 2-3. The methods used to detect these organisms are culture based and standardized (APHA, AWWA, and WEF, 1998; USEPA, 2001b, 2001c). Although many molecular tests in use cannot distinguish living from dead organisms, other types of molecular tests are under development that may be able to discriminate between live and dead indicator organisms (e.g., Gedalanga and Olson, 2009).

Coliform Bacteria

Total Coliform. Total coliform (TC) bacteria comprise many members of the family Enterobacteraceae (Leclerc et al., 2001). TC bacteria are those that can grow in selective media at 35°C and ferment lactose or possess a β-galactosidase enzyme. As an indicator of fecal contamination, the TC bacteria test has an acknowledged weakness and perhaps an underappreciated strength. On the one hand, the TC group of bacteria are unreliable indicators of fecal contamination because many members are capable of growth and long-term persistence (having a nonfecal origin) in many environments, including water distribution systems. On the other hand, there are more TC bacteria in untreated fecal waste than any of the other fecal indicators or indicator groups, making the TC test the most sensitive (albeit least specific) of all indicator tests (Atherholt et al., 2003). Because of this sensitivity, the

TABLE 2-3 Indicator Organisms Specified in U.S. Regulations

Indicator	Group composition*	U.S. regulatory use†
Coliform Bacteria		
Total coliforms	***Escherichia coli, Enterobacter, Klebsiella, Citrobacter***, *Serratia, Leclercia, Yersinia*, others‡	TCR
Fecal coliforms	***Escherichia coli***, *Enterobacter*, and *Klebsiella*	TCR
Escherichia coli		TCR; LT2ESWTR; GWR
Enterococcus Bacteria	***Enterococcus faecalis, E. faecium, E. casseliflavus, E. hirae, E. mundtii***, *E. durans, E. gallinarum*, others	GWR
Coliphage (Viruses That Infect Coliform Bacteria)		
Somatic coliphage	Myoviridae (e.g., T2,T4,T6); Podoviridae (e.g., T3,T7); Siphoviridae (e.g., λ, Φ80, T5); Microviridae (e.g., ΦX174, S13)	GWR
F+ RNA coliphage	Leviviridae: serogroups I (e.g., MS-2, f2); II (e.g., GA); III (e.g., Qβ, VK); IV (e.g., SP, FI)	GWR

*Major assay constituents in bold type.

†TCR = Total Coliform Rule (USEPA, 1989b); LT2ESWTR = Long Term 2 Enhanced Surface Water Treatment Rule (USEPA, 2006a); GWR = Ground Water Rule (USEPA, 2006b).

‡Multiple species of each genera listed plus *Hafnia, Buttiauxella, Kluyvera, Pantoea, Rahnella.*

TCR (USEPA, 1989b) relies on the TC bacteria test as the initial test to detect the possible presence of fecal contamination in delivered water, as well as to assess water treatment effectiveness and the integrity of the distribution system. Distribution system waters that are free of TC bacteria should have no or minimal levels of pathogens.

Fecal Coliform. Under the TCR, if the TC test result is positive, that sample is then further tested for the presence of either fecal coliform (FC) bacteria or *E. coli*. The FC group of organisms is a subset of the TC group that can grow in selective media at 44.5°C and ferment lactose. The FC group has far fewer nonfecal sources than the TC group but more so than *E. coli*. In most environmental waters, the majority of FC bacteria are *E. coli*. However, other coliform bacteria can, on occasion, comprise a significant proportion of the FC group, with wastewaters from pulp and paper mills the most well-known example. Nonfecal *Klebsiella* can dominate the composition of the FC group in these wastewaters. Because of this lack of specificity and because there are now fast and easy assays available to detect *E. coli*, recommendations have been made to eliminate the FC test from the forthcoming revised TCR (Total Coliform Rule/Distribution System Federal Advisory Committee, 2008).

E. coli. Until recently, *E. coli* was considered unique to fecal wastes with no environmental sources. Nonfecal strains of *E. coli* were first observed in warm water environments and later in temperate waters and sediments as well (Standridge, 2008). Despite this fact, the presence of *E. coli* is still considered to be the most fecal specific of all currently used indicator tests. Most *E. coli* possess a β-glucuronidase as well as a β-galactosidase enzyme and can grow in selective media and ferment lactose at 44.5°C. *E. coli* lack a urease enzyme, allowing this species to be distinguished from closely related enteric organisms in some tests. The *E. coli* test is the indicator used by small systems in addition to or in place of testing for *Cryptosporidium* in the Long Term 2 Enhanced Surface Water Treatment Rule (USEPA, 2006a). In addition, the GWR allows the use of an *E. coli* test (or an enterococcus or coliphage test depending on state specifications) to examine untreated groundwater following a positive TC test result in the distribution system, per TCR test requirements (USEPA, 2006b).

Enterococcus

Enterococci grow on selective medium at 41°C and possess a β-glucosidase enzyme (some closely related organisms possess this enzyme as well). Fecal wastes contain concentrations of enterococcus bacteria comparable to that of *E. coli*. However, there can be significant nonfecal sources of enterococcus in many types of environmental waters (Leclerc, Devriese, and Mossel, 1996; Muller et al., 2001; Moore, Guzman, and McGee, 2008). The GWR allows the use (depending on state specifications) of an enterococcus test to examine untreated groundwater following a positive TC test result in the distribution system under the TCR (USEPA, 2006b).

Coliphage

Coliphages are viruses that infect *E. coli* and other coliform bacteria (Leclerc et al., 2000). There are two categories of indicator coliphage, or simply phage: (1) somatic phage that infect *E. coli* through their cell wall and (2) male-specific (F+ or F-specific) phages that gain entry through hair-like structures called pili that only male, or F-containing, *E. coli* possess. The test for the two coliphage groups requires two different host bacteria for maximum recoveries (there is a universal host that will detect both coliphage groups, but at lower recovery efficiencies), making the USEPA-approved tests for coliphage slightly more complex and costly compared to the bacterial indicator tests. Coliphage are almost always found in treated as well as untreated sewage, but they are present in the feces of only a few percent of humans at any given time. Thus, coliphage may not be the best indicators for some types of pollution sources. In addition, there are other issues related to the use of coliphage as fecal pollution indicators that warrant caution (Leclerc et al., 2000).

Because coliphage are viruses and more closely mimic HEV pathogens in terms of environmental transport and survival, they are being investigated as a potentially more reliable indicator of fecal pollution in general and of HEV pollution in particular, especially in some groundwaters. The GWR allows the use of coliphage in place of *E. coli* or enterococci (depending on state specifications) to indicate fecal pollution in groundwater (USEPA, 2006b), and the GWR Source Water Monitoring Methods Guidance Manual states that coliphage should be used when the wells that are targeted for assessment monitoring are located in either sand aquifers or tropical settings (USEPA, 2008a). However, a study of hundreds of wells across the United States showed that at sites where more than one sample was collected, if the sample tested positive at some point in time for HEV, it also tested positive for bacterial indicators (Abbaszadegan, LeChevallier, and Gerba, 2003). A study of Canadian groundwater found sampling for bacterial indicators to be the best approach for detecting HEV contamination (Locas et al., 2008). A few studies, however, have found HEV in some groundwater in the absence of fecal indicators (e.g., Borchardt et al., 2003). However, it should be noted that in studies determining the presence of HEV and fecal pollution indicators, sample (or equivalent sample) test volumes for HEV are typically 1 L or more, often 10 to 100 L, while volumes for bacterial indicator organisms are standardized at 100 mL (0.1 L).

Other Fecal Indicator Organisms

No other fecal pollution indicators are currently specified in U.S. drinking water regulations. The use of *Clostridium perfringens* endospores is under investigation as a more suitable indicator than bacteria for predicting the presence of viral and protozoan pathogens and for assessing water treatment efficiency. Due to their extreme longevity in the environment, *C. perfringens* endospores have often been monitored for evidence of human wastes or sewage sludge in ocean and other sediments. Similarly, *Bacillus* spp. endospores have been studied as a possible indicator of water treatment efficiency, with emphasis on their ability to predict the removal of protozoan pathogens. The anaerobic genera *Bacteroides* and *Bifidobacteria*, and the phage that infect these bacteria, are also under investigation as fecal pollution indicators. These bacteria are among the dominant members of the fecal waste microbial community. However, because all but very low oxygen concentrations are toxic to these bacteria and because they are not capable of producing resistant endospores, they survive in the environment for only short periods of time. *Bacteroides* and *Bifidobacteria* are under consideration, along with other bacteria, coliphage, enteric viruses, and some chemicals, as tools for microbial source tracking studies (Scott et al., 2002; USEPA, 2005a).

Heterotrophic Bacteria

Heterotrophic bacteria are members of a large group of bacteria that use organic carbon for energy and growth (Bartram et al., 2003). In the United States, laboratories often measure heterotrophic bacteria using the heterotrophic plate count test (HPC) (APHA, AWWA, and WEF, 1998). Because of its lack of specificity, HPC has not been used to assess the likelihood of waterborne disease; a specific HPC level might contain many, few, or no pathogens. The significance of HPC lies in its indication of poor general biological quality of drinking water. A sudden significant increase in the HPC can suggest a problem with treatment, including poor disinfection practice. Under USEPA regulations, systems that have no detectable disinfectant in the distribution system may claim, for the purposes of the regulation, that disinfectant is present if the HPC does not exceed 500 colonies/mL (see Geldreich et al., 1972).

Microscopic Particulate Analysis

Microscopic particulate analysis (MPA) is a tool for examining groundwater samples to determine if they are under the direct influence of surface water and for evaluating the efficiency of filter treatment processes for systems using surface waters. The method consists of a microscopic examination of the water for the presence of plant debris, pollen, rotifers, crustaceans, amoebas, nematodes, insects/larvae, algae (including diatoms), coccidia (e.g., *Cryptosporidium*), and *Giardia* cysts. Microscopic particulate analysis guidance for groundwater defines five risk categories, based on the concentration of each of these bioindicators. The concentration of this material should be low in true groundwaters. The MPA is described in two USEPA documents: the first for use in groundwaters (USEPA, 1992a) and the second for use by surface water systems to measure the effectiveness of steps taken to optimize filter performance (USEPA, 1996).

Turbidity

Turbidity is a nonspecific measure of the amount of particulate matter in water (e.g., clay, silt, finely divided organic and inorganic matter, microorganisms). Details about the use, measurement, and limitations of turbidity are found in Chap. 3. Adequate turbidity removal involves partial removal of pathogens in the source water, since most pathogens as well as nonpathogenic bacteria such as iron bacteria, nitrifying bacteria, and sulfur bacteria tend to aggregate with particles, each other, and the chemical coagulants used in the treatment process. High turbidity levels can reduce the efficiency of disinfection by creating a disinfectant demand. The particles may also provide absorption sites for toxic substances in the water, protect pathogens and indicator organisms from disinfection by adsorbing or encasing them, and interfere with the total coliform analysis (USEPA, 1995; LeChevallier, Evans, and Seidler, 1981). Simple analytical methods for turbidity are available (APHA, AWWA, and WEF, 1998). Aesthetic issues related to turbidity are discussed in the section on Aesthetic Quality.

Particle counts in water can also be used as a general indicator of water quality, and this is discussed in Chap. 3.

TOXICOLOGICAL EVALUATION OF DRINKING WATER CONTAMINANTS

Toxicology, the study of the adverse effects of chemicals on living organisms, provides a means of evaluating and understanding the biological effects of drinking water contaminants. Human epidemiology studies usually cannot define cause-and-effect relationships, and sufficient data on duration and magnitude of human exposure do not exist for most contaminants. For these reasons, data from laboratory animal studies are typically used to assess a contaminant's potential risk to humans. In such studies, the doses and the length of exposure to the chemical of concern are known, providing data on the effects and dose-response of the chemical under controlled conditions. The underlying assumption is that effects observed in animals may occur in humans, unless information to the contrary is available.

However, epidemiology, the study of the distribution and determinants of diseases and injuries in human populations, can also provide information about the adverse effects of chemicals on humans. Exposure to substances including arsenic, benzene, vinyl chloride, asbestos, cigarette smoke, and radiation was first linked to the induction of cancer by epidemiologic studies. Although a complete review of toxicological and epidemiological principles is beyond the scope of this chapter, basic concepts needed to understand the information follow.

Chemicals can cause clearly deleterious effects as well as biological changes (e.g., changes in certain enzyme levels) that may or may not be considered adverse. Some adverse health effects in organisms occur immediately (within 24–48 hours after exposure), but others are delayed (e.g., 5–40 years or more for cancer latency in humans). Adverse effects may be reversible or irreversible, depending upon their nature, severity, and the organ affected. Often more is known about the immediate effects of single or short-term high doses than the delayed effects of long-term, low-dose exposure. Unless otherwise noted, the emphasis in this chapter is on the latter, since long-term, low-dose effects are more relevant to most drinking water contamination issues.

Dose-Response in Toxicology

The underlying principle of toxicology is that any chemical may have harmful effects and that the response of a living organism to a chemical depends upon the dose. Dose is typically expressed as weight of chemical administered per body weight per day, such as milligrams per kilogram per day (mg/kg/day). As dose is increased, effects are usually expected to occur in a greater percentage of the exposed group of people or experimental animals. This is termed the *dose-response relationship.*

The concept of dose-response is important because simply identifying the toxicological effects of a chemical (known as *hazard identification*) does not provide the complete information needed to assess human health risk. Before an evaluation (*risk characterization*) can be made regarding the public health significance of exposure to a substance, the dose-response relationship and extent of human exposure must also be known.

Usually, a clear dose-response relationship exists, with effects increasing over a certain dose range below which there is a threshold dose at which no effects are seen (Fig. 2-1). For some effects, particularly cancers which occur through a genotoxic mode of action, it is assumed that exposure to any dose results in some level of risk; thus, there is no threshold below which no risk exists. Nonmonotonic (U-shaped or inverted U-shaped) dose-response relationships can also occur. For these, lower doses can produce greater effects than higher doses over a portion of the dose range, particularly for some endocrine and neurotoxic effects. Vitamins can also exhibit this type of dose-response relationship, as toxicity can result both from deficiency at low doses and from excessive exposure at high doses. Development of approaches to address these types of dose-response relationships is a current issue for risk assessment of environmental contaminants (Myers, Zoeller, and vom Saal, 2009).

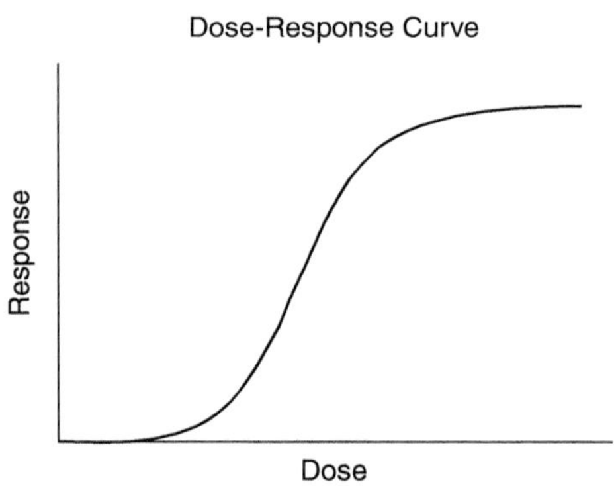

FIGURE 2-1 Typical dose-response curve for a chemical with a threshold for response.

The response of a person or an animal to a chemical depends on both toxicodynamic and toxicokinetic factors. Toxicodynamics is concerned with the biochemical and physiological effects of chemicals and the mechanisms of action by which these effects occur. Toxicokinetics relates to the fate of a substance in the body, including absorption, distribution, metabolism, and excretion. In other words, toxicodynamics refers to what a chemical does to the body, while toxicokinetics refers to what the body does to the chemical. Both toxicokinetic and toxicodynamic parameters may vary between species as well as among members of the same species, including experimental animal groups and human populations. Thus, humans may be more or less sensitive to a chemical than experimental animals, and some exposed people may be more sensitive than others.

Specific subgroups within the human population may be more sensitive to the toxicity of chemicals due to differences in toxicodynamics and/or toxicokinetics. These subpopulations include fetuses, children, the elderly, and individuals with certain genetic variants, diseases, or medical conditions. Effects occurring during fetal development or infancy (e.g., inhibition of thyroid function) may cause permanent damage, while the same effect in older individuals may have less serious consequences that are reversible when exposure ceases. Additionally, fetuses and children may be more sensitive because of physiological or biochemical differences from adults. Toxicokinetic differences may also increase the sensitivity of these subgroups. For example, GI absorption of lead and mercury salts is as much as 10 to 20 times greater in infants, and up to several-fold greater in pregnant women, than in other adults (CDC, 1991; ATSDR, 1999). The ability to metabolize or excrete certain toxic chemicals can be diminished in the elderly, increasing their sensitivity to these chemicals.

Types of Toxicological Effects

A variety of adverse health effects are possible. The general types of effects, listed below, are those that are most relevant to exposure to contaminants found in drinking water.

1. *Systemic toxicity.* Causing a deleterious response in a biological system, disrupting function, or producing death. These effects may result from acute (short-term), subchronic (intermediate-term), or chronic (long-term) exposure. The term *systemic toxicity* is often used to refer to noncancer effects in the context of toxicity and risk assessment of environmental contaminants. Examples of systemic toxicity include weight loss, anemia, damage to the liver or kidney, and decreased thyroid hormone production. It is assumed that there is a threshold dose for most systemic effects below which no effects are expected to occur.
2. *Carcinogenicity (oncogenicity).* Causing or inducing uncontrolled growth of aberrant cells into neoplastic tumors. Chemicals may cause tumor formation by initiating the series of changes in DNA necessary for carcinogenesis, and/or by promoting the growth of initiated cells. These chemicals are referred to as *initiators* and *promotors*, respectively. Chemicals that cause benign tumors, in addition to those that cause malignant tumors, are usually assumed to pose a cancer risk to humans. In assessing the risks of exposure to chemicals that cause cancer, it is usually assumed that no threshold exists. This means that there is some risk of cancer from exposure to any amount of the chemical.
3. *Mutagenicity.* Causing heritable alteration of the genetic material (DNA) within living cells. A mutation is typically defined as a change in the genetic code of a specific gene that results in a change in a cellular or biochemical characteristic. Many carcinogens are also mutagens, and vice versa. Mutagenicity is a subset of the effects included in genotoxicity.
4. *Genotoxicity (clastogenicity).* Causing alterations or damage to the genetic material (DNA) in living cells, such as deletions of portions of or entire chromosomes, but also including other phenomena, such as creation of micronuclei, sister chromatid exchanges, and unscheduled DNA synthesis (due to repair processes). Some changes may be lethal to the cell. Genotoxicity encompasses mutagenicity as well as other types of damage to DNA. Many carcinogens are also genotoxic.
5. *Teratogenicity.* Causing nonhereditary congenital malformations (birth defects) in offspring as a result of exposure during gestation.
6. *Developmental toxicity.* Adverse effects resulting from exposure to a chemical during development, which includes exposure to either parent prior to conception, exposure during gestation, or exposure after birth until puberty. Developmental effects may become evident at any time of life. For example, some effects resulting from prenatal exposures are not seen until puberty or middle age. Teratogenic effects are a subset of the effects in this category.
7. *Reproductive toxicity.* Adverse effects on the male or female reproductive system resulting from exposure to a chemical. Examples include effects on sexual development, mating behavior, fertility, pregnancy, and lactation.

Toxicology Studies for Evaluation of Drinking Water Contaminants

As stated above, the focus of this section is on effects of chronic (lifetime, which is about two years in rodents) or subchronic (up to 10 percent of the lifespan, 2 to 13 weeks in rodents) exposures, rather than acute effects resulting from brief exposures to high doses.

For assessment of developmental effects, studies involving exposure during gestation are relevant. Chemicals may be administered to animals in drinking water or the diet, by gavage (administered through a tube into the stomach as a bolus dose dissolved in oil or water), dermally, by inhalation, or by intraperitoneal (abdominal), intravenous, or intramuscular injection. For assessment of drinking water contaminants, oral studies, particularly those in which exposure is through drinking water or the diet, are preferred, because these dosing methods best simulate human exposures through drinking water. Studies in which chemicals are administered by injection are not used as the basis for quantitative risk assessment of drinking water contaminants, as this route is not relevant to environmental exposure.

Noncancer Effects. For noncarcinogenic endpoints, studies examine mortality, body weight, and pathological changes in organs and blood, including absolute organ weights and the ratio of organ weights to body weight, microscopic appearance, and enzyme activities. Among the organs frequently affected are the liver and kidney because of their roles in metabolism and excretion of toxic chemicals. Neurologic function, effects on the immune system, behavioral changes, changes in hormone levels, and other specific endpoints may also be assessed if relevant. In recent years, increased emphasis has been given to studies of teratogenicity and developmental and reproductive toxicity (such as changes in fertility, birth weight, and survival), as the period of development is now known to be the most sensitive life stage for many chemicals.

Carcinogenic Effects. Assessment of a chemical's carcinogenic potential is primarily based on the results of two types of studies: chronic (lifetime) animal studies and tests of mutagenicity and genotoxicity. Many chronic studies have been conducted by the National Cancer Institute (NCI) or National Toxicology Program (NTP), and reports of these studies are available on the NTP Website. In a chronic rodent study, rats, mice, or hamsters are exposed to the test chemical for two years. Doses are determined by subchronic range-finding studies conducted prior to the chronic study. Doses ideally include the maximum tolerated dose (MTD), often defined as the dose reducing body weight by about 10 percent, and one half and one quarter of the MTD. A control (unexposed) group is also included. As mentioned above, oral studies are preferred for assessment of drinking water contaminants. However, if no appropriate oral data are available, inhalation data may be used.

After the end of the exposure period, animals are killed and necropsied. A wide variety of organs or tissues are examined for tumors both visually and microscopically. The data on tumor incidence in each of the tissues evaluated are analyzed to determine whether there is a statistically significant increase in the incidence of malignant tumors or total tumors (malignant and benign) in the experimental groups in any of the tissues compared to background incidence in the control group. There has been some disagreement about the human relevance of the results of cancer studies in rodents, particularly liver cancers in certain mouse strains following very high doses of the test chemical. However, default assumptions used in cancer risk assessment are that, in the absence of specific data to the contrary, carcinogenicity observed in animals indicates the potential for carcinogenic effects in humans, and that a carcinogenic chemical may cause tumors in different organs in animals and humans.

The doses used in animal studies are much higher than the doses to which humans are exposed in the environment for several reasons. High doses are used in order to increase the number of transformed cells within the target tissues. In addition, with the numbers of animals used in these studies (typically 50 per sex per dose group), an increased tumor incidence of at least several percent must occur in order to detect a statistically significant difference between treated and control groups. High doses must be given to produce increased tumor frequencies of this magnitude. In contrast, the cancer risk levels of interest in humans exposed to low doses of environmental contaminants are much lower, down to the 1 in one million range.

Genotoxicity and Mutagenicity. The Ames test is one of many *in vitro* screening systems for mutagenicity and other types of genotoxic effects. Such effects are considered to be an indicator of carcinogenic potential. Strains of the bacterium *Salmonella* Typhimurium that can detect different types of mutations, such as frame shift mutations and base pair substitution mutations, have been developed. These strains require the amino acid histidine to grow. Through mutation of the relevant gene, they can become histidine independent and thus able to grow in the absence of histidine. Mutagenic potency is determined by comparing the number of bacteria that become histidine independent when exposed to the test chemical to the number that become histidine independent spontaneously under control (unexposed) conditions. Usually, a chemical is tested with and without the addition of rodent liver metabolic enzymes and necessary cofactors in order to detect chemicals that are not themselves mutagenic but are metabolized, or activated, in the body to mutagenic metabolites.

To characterize mutagenicity, a battery of *in vivo* (whole animal) and *in vitro* (in cell cultures, bacteria, etc.) assays using other microorganisms (e.g., yeast and fungi), fruit flies (*Drosophila*), and mammalian cells are often used. Because many known carcinogens test positive in these assays, they are useful as screening tools. Endpoints evaluated include mutagenicity, DNA strand breaks, chromosomal aberrations, and ability of a chemical to react with the DNA (adduct formation), as well as assays of other phenomena whose meaning are not fully understood but appear to be correlated with carcinogenic potential (e.g., sister chromatid exchange and micronuclei formation). Some assays appear to be more sensitive to certain types of carcinogens.

The animal tests currently used for the assessment of noncancer and cancer effects of chemicals of environmental concern are expensive and time consuming, so that the number of chemicals that can be tested each year is limited by practical considerations. Additionally, reducing the use of animals in toxicity studies addresses concerns related to animal rights and animal welfare. A recent report by the National Research Council (NRC) of the National Academy of Sciences (NRC, 2007), commissioned by USEPA, envisaged a shift from the current whole animal-based testing systems to testing founded primarily on *in vitro* methods that evaluate changes in biological processes using cells, cell lines, or cellular components. However, many questions and issues must be addressed for this vision to become a reality, and it is unclear whether the *in vitro* methods advocated by this report will ever be capable of detecting all or most toxic effects of concern.

RISK ASSESSMENT OF DRINKING WATER CONTAMINANTS

The human health basis (maximum contaminant level goals, MCLGs) for the drinking water standards (maximum contaminant levels [MCLs]), as well as the drinking water guidance (Health Advisories), developed by USEPA and other agencies are derived through risk assessment approaches. Recent general references for risk assessment of environmental contaminants are Nielsen, Ostergaard, and Larsen (2008) and, specifically for risk assessment of drinking water contaminants, Howd and Fan (2008). As discussed in Chap. 1, MCLGs are nonenforceable, health-based goals that are to "be set at a level at which no known or anticipated adverse effect on human health occurs and that allows for an adequate margin of safety." MCLGs are considered along with other factors to develop the MCL, which is an enforceable regulatory standard.

Guidance documents providing details on risk assessment approaches are linked from the USEPA IRIS Website. The overall framework for the risk assessment approach used by USEPA and other agencies was developed by the NRC (1983). The four steps of this approach are:

1. *Hazard identification*—qualitative determination of the adverse effects that a constituent may cause
2. *Dose-response assessment*—quantitative determination of the effects that a constituent may cause at different doses
3. *Exposure assessment*—determination of the route and amount of human exposure
4. *Risk characterization*—description of risk assessment results, underlying assumptions, and uncertainties

Many uncertainties are involved in the extrapolation of data from experimental animals to potential effects in humans and from the high doses typically used in the animal studies to the much lower exposures usually occurring from drinking water contaminants. Because of these uncertainties, it is necessary to use assumptions in the risk assessment process. As the risk assessment process is intended to be protective of public health, the assumptions used in the presence of uncertainty are meant to be reasonable but conservative. If data become available to address a source of uncertainty, the risk assessment can be revised to replace the assumption with information based on the data.

A chemical's classification as a carcinogen or a noncarcinogen is important, since a different general risk assessment approach is used for noncarcinogens than for most carcinogens. Contaminants are classified as to their weight of evidence for carcinogenicity according to categorization schemes. USEPA adopted new Guidelines for Cancer Risk Assessment in 2005 to replace the previous 1986 guidelines. However, assessments under the 1986 guidelines are still in place for many chemicals that have not been reassessed under the new USEPA (2005b) guidelines.

Chemicals treated as carcinogens are those classified as "carcinogenic to humans" or "likely to be carcinogenic to humans" under the 2005 guidelines, as "known human carcinogens" (Group A) or "probable human carcinogens" (Group B1 and Group B2) under the 1986 guidelines, and as Group 1 or Group 2A by the International Agency for Research on Cancer (IARC). Chemicals treated as noncarcinogens are those classified as having "inadequate evidence" or as "not likely to be carcinogenic" (USEPA, 2005b), as "not classifiable" (Group D) or "not carcinogenic to humans" (Group E) (USEPA, 1986a), or as Group 3 or Group 4 by IARC. Specific approaches are used to address chemicals classified as having "suggestive evidence" (USEPA, 2005b), "possible human carcinogens" (Group C) (1986a), or Group 2B by IARC.

Noncarcinogen Risk Assessment

Point of Departure. For most noncarcinogenic effects, it is assumed that there is a threshold dose below which no adverse effects occur. A Reference Dose (RfD) at which no adverse effects are expected from lifetime exposure is developed from appropriate animal or, less commonly, human data. The RfD is derived by applying appropriate uncertainty factors to a dose (in mg/kg/day) chosen from the dose range used in the study or modeled from the study data. This dose, called the point of departure, can be the highest dose at which no effects have been observed (no observed adverse effect level, or NOAEL), the lowest dose at which adverse effects have been observed (lowest observed adverse effect level, or LOAEL), or a benchmark dose derived by modeling the dose-response data to predict a dose at which minimal effects (e.g., 5 percent or 10 percent change in the parameter of interest) will occur. When data from several appropriate studies are available, the highest NOAEL that is below the lowest LOAEL from the various studies is chosen as the point of departure for RfD development.

Uncertainty Factors. The uncertainty factors (formerly called "safety factors") used in RfD development account for factors such as: differences in sensitivity between humans and experimental animals, variability in sensitivity within the human population, less than lifetime duration of a study, need to extrapolate from an LOAEL to an NOAEL, or gaps in the toxicology database for the chemical. Uncertainty factors of 10 each are often used, but factors of 3 or other values can be chosen if warranted by the data. USEPA recommends that the total uncertainty factor (the product of the individual uncertainty factors) not exceed 3000.

$$\text{RfD(mg/kg/day)} = \frac{\text{NOAEL, LOAEL, or benchmark dose (mg/kg/day)}}{\text{product of appropriate uncertainty factors}} \quad (2\text{-}1)$$

Exposure Assumptions. From the RfD, a drinking water equivalent level (DWEL) is calculated. The DWEL represents a concentration at which adverse health effects are not anticipated to occur from a lifetime of exposure, assuming 100 percent of exposure to the contaminant comes from drinking water.

$$\text{DWEL (mg/L)} = \frac{\text{RfD (mg/kg/day)} \times \text{body weight (kg)}}{\text{ingestion rate (L/day)}} \quad (2\text{-}2)$$

Exposure assumptions for adults (body weight of 70 kg, and a drinking water ingestion rate of 2 L/day) are often used to develop health-based drinking water concentrations protective for lifetime exposure, such as MCLGs and Lifetime Health Advisories. If the assessment is based on effects relevant to infants (e.g., methemoglobinemia from exposure to nitrate), then assumptions appropriate for infants, such as a body weight of 10 kg and a consumption rate of 1 L/day are assumed. Assumptions for infants or children are also used by USEPA to develop Health Advisories for short-term (1 day, 10 days) or longer-term (7 years) exposure durations.

Relative Source Contribution. In developing MCLGs or Health Advisories for noncarcinogenic contaminants, contributions from nondrinking water sources of exposure, including air and food, are also taken into account so that total exposure from all sources does not exceed the RfD. In order to do this, a relative source contribution (RSC) factor of between 20 and 80 percent is applied to the DWEL (USEPA, 2000a). When the necessary data on air and food exposures are not available to develop a contaminant-specific RSC, as is the case for most contaminants, a conservative default value for the RSC of 20 percent is used. If sufficient data are available on non-water sources of exposure, a contaminant-specific RSC is derived. If drinking water contributes between 20 and 80 percent of total exposure at the DWEL, the exposure contributions from food and air are subtracted from the exposure at the DWEL to calculate the MCLG or Health Advisory. If drinking water contributes 80 percent or more to total exposure, a "ceiling" RSC value of 80 percent is used to protect individuals whose non-water exposures may be higher than that indicated by available data. If drinking water contributes less than 20 percent of total exposure, a "floor" RSC value of 20 percent is generally used. In these cases where drinking water contributes a relatively small portion of total exposure, efforts are best directed toward reducing the exposures from other sources, as further reductions of the drinking water standard or guidance based on an RSC of less than 20 percent will not result in a significant decrease in total exposure to the contaminant. Once the RSC has been determined, the MCLG or Health Advisory is calculated as follows:

$$\text{MCLG or Health Advisory (mg/L)} = \text{DWEL (mg/L)} \times \text{RSC (\%)} \quad (2\text{-}3)$$

Carcinogen Risk Assessment

Nonthreshold versus Threshold Assumption. In assessing the risks of exposure to chemicals that cause cancer, it is usually assumed that some risk of cancer results from exposure to any amount of the chemical. The default assumption recommended in the USEPA

Guidelines for Carcinogen Risk Assessment (USEPA, 2005b) is that there is no threshold for carcinogenicity unless a threshold mode of action has been clearly demonstrated.

For a few chemicals, it has been shown that cancer occurs secondary to systemic effects. In these cases, it is assumed that a threshold for carcinogenicity exists and an RfD approach similar to that described above for noncarcinogens is appropriate. Examples of this approach are chloroform, for which tumors in rodents appear to occur only after cell damage and regenerative growth, and antithyroid agents that cause sustained elevations in levels of thyroid-stimulating hormone (TSH), resulting in continued stimulation of the thyroid and the development of thyroid tumors.

Because of the lack of evidence for a threshold for most carcinogens, USEPA sets MCLGs for most carcinogenic contaminants at "zero" as an aspirational goal, as a matter of regulatory policy (USEPA, 1987a, 1991a). This expresses the ideal that drinking water should not contain carcinogens. In this context, zero is not a number (as in 0 mg/L) but is a concept (as in "none").

Estimation of Cancer Risk. Under the current USEPA (2005b) risk assessment guidelines, cancer risk is usually estimated quantitatively by linear extrapolation from a benchmark dose for tumor data from animal studies. In some cases, such as benzene, risk estimates come from human epidemiological data. The carcinogenic potential of a chemical is expressed quantitatively as a slope factor or potency factor, in which cancer risk per lifetime daily dose is given in units of per milligrams per kilogram per day [mg/kg/day]–1. The slope factor used for risk assessment is the 95th percentile upper confidence limit on the modeled slope factor. Because of the conservative assumptions and approaches used in cancer risk assessment, the risk estimates are considered to be upper bounds on the actual risk, and the actual risk is likely to be lower and could be as low as zero. The slope factor or potency factor is related to the dose (in mg/kg/day) and the lifetime cancer risk, which is unitless, as follows:

$$\text{Risk (unitless)} = \text{dose (mg/kg/day)} \times \text{slope factor (mg/kg/day)}^{-1} \qquad (2\text{-}4)$$

Health-based Drinking Water Concentration. The drinking water concentration at a given cancer risk level (10^{-4}, 10^{-5}, 10^{-6}) can be calculated from the following equation, using assumptions of 70 kg for adult body weight and 2 L/day for ingestion rate:

$$\text{Drinking water conc. (mg/L)} = \frac{\text{risk level } (10^{-x}) \times \text{body weight (kg)}}{\text{slope factor (mg/kg/day)}^{-1} \times \text{ingestion rate (L/Day)}} \qquad (2\text{-}5)$$

The actual regulatory standard, or MCL, is based on practical considerations, such as analytical limitations and treatment costs, with the goal that the lifetime cancer risk should remain below 1 in 10,000.

Risk Assessment of Suggestive or Possible Human Carcinogens

For certain contaminants, some evidence for carcinogenicity exists, but the weight of evidence is not sufficient for classification as "likely to be carcinogenic to humans" (USEPA, 2005b) or "probable human carcinogen" (Group B2) (USEPA, 1986a). These contaminants are classified as "suggestive carcinogens" (USEPA, 2005b) or "possible human carcinogens" (Group C) (USEPA, 1986a). The risk assessment approach used for such chemicals is a science policy decision. The USEPA drinking water MCLGs and Health Advisories for contaminants in these categories are based on the RfD for systemic effects, with the incorporation of an additional uncertainty factor to account for the evidence of possible carcinogenicity. The uncertainty factor is usually 10, but other values from 1 to 10 may be used as appropriate. The same equation used to develop MCLGs for noncarcinogens (above) is used. Less commonly, if there are insufficient data to develop an RfD, the MCLG is based on the cancer slope factor and a lifetime risk level of 10^{-5} (1 in 100,000). Within USEPA, other programs may use other approaches to address chemicals in this category. For example, in the Superfund program for cleanup of contaminated sites, health-based criteria for these chemicals are based on a lifetime cancer risk level of 10^{-6} (1 in 1,000,000).

Additional Considerations in MCL Development

As discussed in more detail in Chap. 1, regulatory standards for drinking water contaminants (MCLs) consider other factors in addition to health effects alone. Factors considered may include analytical limitations, availability of treatment technology, costs, and others. For example, the public health benefit of reducing infectious disease from the use of chemical disinfectants is balanced against the cancer risks of the disinfection by-products formed during the disinfection process. For carcinogens, the MCLG is set at zero as a matter of policy, and the MCL is determined by considering the factors mentioned above. For some noncarcinogens, the MCL is also set above the MCLG due to practical considerations.

Sources of Information for Health-based Drinking Water Values

As discussed in Chap. 1, a useful resource for information on health-based assessments of drinking water contaminants is the USEPA Drinking Water Standards and Health Advisories tables (USEPA, 2006c; USEPA, 2009a), which are periodically updated. In addition to the regulatory MCLs, MCLGs, and the various short-term and long-term Health Advisories for children and adults, the tables summarize the RfDs and DWELs for contaminants with risk assessments based on noncarcinogenic end points. As the RfD is intended to provide a wide margin of safety, small exceedances of an MCL usually do not represent immediate concern (but see sections on nitrates and lead for examples of exceptions to this generalization). The tables also provide the carcinogenicity classification and the concentrations of compounds corresponding to a 10^{-4} (1 in 10,000) incremental lifetime cancer risk as estimated by USEPA. These tables also provide secondary MCLs (SMCLs) based on aesthetic effects, as well as useful definitions of terms used in drinking water risk assessment and regulation.

Information on health effects and occurrence of unregulated drinking water contaminants that have been considered for regulation by USEPA can be found on the USEPA Office of Water Contaminant Candidate List (CCL) website. Additional information on health effects may be found in the general references mentioned at the beginning of this chapter and in the sections on individual contaminants. Within this chapter, for consistency, MCLGs and Health Advisories developed by the USEPA Office of Water are generally provided. However, it should be noted that the risk assessment basis for some older MCLs may not agree with the more current information provided in IRIS.

Additionally, as noted above, some states have established standards that are more stringent than the federal standards for some contaminants. States have also established standards or guidance for contaminants that have not been addressed by USEPA.

Microbial Risk Assessment

Epidemiological studies attempt to directly measure the amount of infection or illness due to the consumption of drinking water in sample populations (Colford et al., 2006; Messner et al., 2006). Determining the level of such illness through these studies is difficult, expensive, and time consuming. In addition, these studies have low sensitivity (i.e., the ability to measure small increases in health risk). Microbial risk assessment (MRA) is a useful alternative or complement to epidemiology studies when attempting to assess the extent of infection (e.g., Eisenberg et al., 2006). In contrast to epidemiological studies, MRA studies are inexpensive and can be conducted comparatively quickly. MRA may be the sole means to obtain health-risk information when epidemiologically derived data are limited or absent.

Drinking water MRA has been an area of research for over 25 years. It is a formal process that translates the occurrence of pathogens in drinking water and the volume of water consumed into a probability of infection or illness (Haas, Rose, and Gerba, 1999). There are typically four steps involved in MRA: (1) an important subset of pathogens is identified; (2) human exposure, from ingestion or inhalation, is characterized; (3) pathogen dose-response information is obtained from existing data from clinical trials or disease surveillance studies; and (4) a range of possible risk estimates is mathematically determined. Depending on data availability and the type of information needed, different statistical approaches employing different types of MRA models are used.

The many uses of MRAs include providing information for setting regulatory standards, analyzing and improving water treatment, characterizing the mode of disease transmission during an outbreak, and identifying additional information needs. As one example, MRA is the basis of some U.S. drinking water treatment regulations whereby treatment requirements (e.g., specified levels of disinfection and filtration) are specified. These treatments achieve specified levels of removal or inactivation (e.g., 99.9 percent, 99.99 percent) of target organisms or organism groups in source waters such that the resulting level of "acceptable risk" from consuming the treated water is not more than 1 infection in 10,000 exposed persons per year (e.g., USEPA, 1989a, 2006a). The term *acceptable risk* is controversial, and alternative risk endpoints, such as *disability adjusted life years* (the number of days lost due to sickness or death), have been used by other organizations (LeChevallier and Buckley, 2007).

Microorganisms have attributes that make MRA more complicated than chemical risk assessment (ILSI, 1996). For example, although pathogen infectivity information can sometimes be obtained from disease outbreak data, the most reliable dose-response information is obtained from feeding studies whereby small numbers of healthy adults each ingest a different amount of a single pathogen. The number of infections or illnesses that result from each dose determines the infective potential for that pathogen (e.g., the dose that is infectious for 50 percent of the exposed population, namely the ID_{50}). However, for ethical and economic reasons, information from human feeding studies is available for only a few strains of pathogens. Furthermore, the general population includes children, the elderly, pregnant women, and persons with weakened immune systems, and these "sensitive subpopulations" likely have susceptibilities to pathogens that are different from adults with normal immune system function.

There are thousands of species and strains of pathogens. Different species and different strains of the same species (see section on Pathogenic Organisms) can vary greatly in both dose-response (see above) as well as the severity of symptoms following infection. For example, for three different strains of *Cryptosporidium parvum*, the ID_{50} is 2066, 132, and 12 oocysts (Messner, Chappell, and Okhuysen, 2001). Because of this, MRA is usually based on analysis of only a few pathogens (or a combination of characteristics of related pathogens, such as enteric viruses). Such pathogens have a low infectious dose, the potential to cause serious disease symptoms, occur most frequently in source water, and/or have properties that make them difficult to treat.

Additional issues include the fact that the types of pathogens present and the concentration and distribution of each at a source water location can vary greatly over seasons as well as much shorter periods of time and within short distances within the water column (e.g., Wilkes et al., 2009). There are also a host of methodological problems involved in detecting pathogens, determining their infective potential, and determining their respective concentrations in source waters and in treated waters where they are either absent or present at low levels (see Indicators of Water Quality section; Regli et al., 1991; Emelko, Schmidt, and Roberson, 2008). MRA is a field of study in the early stages of development. Given all of the pathogen-specific issues that underlie MRA-derived information, the best use of MRA data at the present time is perhaps to focus on the uncertainty or limits of the information obtained.

INORGANIC CONSTITUENTS

Inorganic constituents can be present in natural waters, in contaminated source waters, or in some cases may result from contact of water with piping or plumbing materials. Lead, copper, zinc, and asbestos are constituents that can originate from distribution and plumbing systems.

Inorganic chemicals found in drinking water represent a variety of health concerns. Some are known or suspected carcinogens, such as arsenic and lead. Numerous inorganic chemicals are essential to human nutrition at low doses, yet demonstrate adverse health effects at higher doses. These include chromium, copper, manganese, molybdenum, nickel, selenium, zinc, and sodium and are reviewed by the Institute of Medicine Food and Nutrition Board (2006). Several inorganic chemicals, including sodium, barium, and lead, have been associated with increased blood pressure. Numerous reports have also shown an inverse relationship between water hardness and hypertensive heart disease, and this topic is under continuing investigation.

Health aspects of inorganic constituents of interest in drinking water are summarized in this section. Inorganic disinfectants and DBPs are discussed in a later section.

Aluminum

Aluminum occurs naturally in nearly all foods, and the average dietary intake is about 20 mg/day. Aluminum salts are widely used in antiperspirants, soaps, cosmetics, and food additives. Additionally, cans for carbonated beverages are made of aluminum. Aluminum occurs in both raw and treated drinking waters, but it is estimated that drinking water typically represents only a small fraction of total aluminum intake (Reiber and Kukull, 1996).

Aluminum shows low acute toxicity but is a neurotoxicant when administered under certain conditions, as reviewed by California EPA (2001) and ATSDR (2008a). Chronic high-level exposure data are limited but indicate that aluminum affects phosphorus absorption, resulting in weakness, bone pain, and anorexia. Carcinogenicity, mutagenicity, and teratogenicity tests have all been negative. Associations between aluminum and two neurological disorders, Alzheimer's disease and dementia associated with kidney dialysis, have been studied. The available data suggest that aluminum is not likely the causative agent in the development of Alzheimer's disease. However, aluminum may play a role by acting as a cofactor in the chain of pathological events resulting in Alzheimer's disease (ATSDR, 2008a). Dialysis dementia has been reasonably documented to be caused by aluminum (Ganrot, 1986; Shovlin et al., 1993). Kidney dialysis machines now use water with a very low concentration of aluminum to prevent this exposure in the patients.

Aluminum was included on the original list of 83 contaminants to be regulated under the 1986 SDWA amendments. USEPA removed aluminum from the list because it was concluded that there was no evidence that aluminum ingested in drinking water poses a health risk (USEPA, 1988a). The USEPA secondary maximum contaminant level (SMCL) for aluminum of 50 to 200 μg/L is based on prevention of discoloration of water and is intended to encourage removal of coagulated material before treated water enters the distribution system.

Arsenic

Arsenic concentrations in U.S. drinking waters are typically low. However, an estimated 5000 community systems (out of 70,000) using groundwater and 370 systems (out of 6000) using surface water had concentrations above 5 μg/L (Reid, 1994). Elevated arsenic concentrations occur primarily in the Western states, but are also found in the Midwest and the geological Piedmont Region of the East Coast. Weathering of rocks containing arsenic and the smelting of nonferrous metal ores, especially copper, account for most of the arsenic in water supplies. Until the 1950s, arsenical compounds were also major agricultural insecticides. Arsenic exists in drinking water primarily in the inorganic state, and arsenite (III) and arsenate (V) are the most prevalent toxic forms found in drinking water (NRC, 1999a; USEPA, 2000b).

Arsenic may be a trace dietary requirement in certain farm animals, but not in humans, and is present in many foods such as meat, fish, poultry, grain, and cereals (NRC, 1999a). Market basket surveys suggest that the daily adult intake of all forms of arsenic is about 50 μg, of which 1 to 20 μg is inorganic (ATSDR, 2007). In fish and shellfish, it is present in organic arsenical forms, which are thought to be much less toxic than inorganic arsenic. In grains such as rice, arsenic is present primarily in the inorganic form.

The toxicity of arsenic has been thoroughly reviewed (NRC, 1999a, 2001; ATSDR, 2007; USEPA, 2000b). The toxic endpoints of concern for arsenic in drinking water relate to chronic effects of arsenic; levels found in drinking water are below those resulting in acute effects. Acute effects resulting from accidental or intentional exposure to large doses of arsenic involve the skin, nervous system, GI system, blood, and kidneys. The lethal dose for poisoning incidents has been reported to be 1 to 4 mg/kg (USEPA, 2000b).

Effects of chronic exposure of humans to lower levels of arsenic include both systemic (noncancer) toxicity and cancer. Much of the data on these effects comes from studies of populations with elevated levels of arsenic in their drinking water in Taiwan, Bangladesh, Mongolia, South America, and the western United States. Systemic (noncancer) effects of arsenic include dermatological toxicity, peripheral vascular disease, and increased risk of cardiovascular disease, diabetes, and gastroenterological disease. The most prominent systemic effects of long-term arsenic exposure involve the skin and include hyperpigmentation and development of keratoses. Blackfoot disease, a severe peripheral vascular insufficiency that can cause gangrene of the extremities, has been reported in Taiwanese individuals exposed to arsenic.

Arsenic is one of a relatively small number of chemicals that has been classified by USEPA as a Known Human Carcinogen (Group A), based on human epidemiological data. Unlike most other carcinogens of environmental concern, arsenic does not induce cancer in the traditional animal testing models, perhaps due to differences in metabolism between the test animals and humans. However, gestational exposure of rodents to arsenic was shown to increase tumor incidence in adulthood (Liu and Waalkes, 2008). Exposure to arsenic through drinking water has long been linked to an increased risk of skin cancer in humans. More recently, an elevated risk of internal cancers has been demonstrated in populations from Taiwan, Japan, Argentina, and Chile. The organs in which cancer rates were reported to be elevated include bladder, lung, liver, and kidney (reviewed in NRC, 1999a, 2001; ATSDR, 2007). Ingestion of arsenical medicines and other arsenic exposures have also been associated with several internal cancers, but several small studies of communities in the United States with relatively high arsenic levels in drinking water have failed to demonstrate any health effects. The interactive effect of smoking and the relative absence of tobacco use in those communities may have played a role in the apparent lack of effect. The NRC (2001) concluded that it would be difficult to detect increases in cancer risk from arsenic in the U.S. population, even at the highest of the possible estimated risk levels, since the arsenic concentrations present in U.S. drinking water are not expected to cause a cancer incidence that is statistically significantly elevated over the relatively high background rates of the cancers of concern.

The mechanism of action by which arsenic causes cancer has not been definitively elucidated. Arsenic appears to impact a number of cellular processes involved with DNA repair and chromosome replication, leading to the hypothesis that it causes cancer by indirect mechanisms, rather than by directly interacting with DNA. However, it has been shown (Mass et al., 2001) that methylated trivalent forms of arsenic, which can be formed through metabolic processes from inorganic arsenic, can act directly to damage DNA *in vitro*. Several studies have suggested that the primary carcinogenic mechanism involves the metabolites, mono- and dimethylarsenous acid.

In 2001, USEPA (2001d) decreased the MCL for arsenic to 10 μg/L, based on cancer risks, replacing the existing MCL of 50 μg/L based on a 1942 U.S. Public Health Service standard. As part of the MCL development process, cancer risk estimates for arsenic in drinking water were developed by the NRC (1999a, 2001) from bladder and lung cancer data from Taiwanese epidemiology studies. The lifetime cancer risk at the MCL of 10 μg/L is estimated to be 3 to 4 per 1000, which is high relative to the risks of most other regulated drinking water contaminants.

Asbestos

Asbestos is the name for a group of naturally occurring, hydrated silicate minerals with fibrous appearance. Included in this group are the minerals chrysotile, crocidolite, anthophyllite, and some of the tremolite-actinolite and cummingtonite-grunerite series. All except chrysotile fibers are known as amphibole. Most commercially mined asbestos is chrysotile. Asbestos occurs in water exposed to natural deposits of these minerals, asbestos mining discharges, and asbestos-cement pipe (USEPA, 1985a).

The physical dimensions of asbestos fibers, rather than the type of fiber, are more important in determining the health effects, with the shorter, thinner fibers more highly associated with cancers by inhalation. Human occupational and laboratory animal inhalation exposures are associated with a specific type of cancer, mesothelioma, in lung, pleura, and peritoneum (USEPA, 1985a). An NTP study also observed benign GI neoplasms in rats exposed in the diet to intermediate-range fibers (65 percent of the fibers larger than 10 μm, 14 percent larger than 100 μm) for their lifetime (USEPA, 1985a). Human epidemiologic studies of asbestos in drinking water have had inconsistent results, but there are suggestions of elevated risk for gastric, kidney, and pancreatic cancers (Cantor, 1997).

The USEPA MCL and MCLG for asbestos fibers >10 μm is 7×10^6 fibers/L. This is based on the estimate that drinking water exposure to 7×10^6 fibers (of size >10 μm/L) will not result in an increased lifetime cancer risk of more than 1×10^{-6} (1 in 1,000,000). This estimate is based on the results of the NTP oral rat study (USEPA, 1991a).

Boron

Boron may be present in water from natural or anthropogenic sources. It is found in rock formations, especially in the western United States, and in seawater at about 4.5 mg/L. It is used industrially, in pesticides, and in detergents and soaps and is released from coal-burning power plants (USEPA, 2008b).

Boron is detected at some concentration in almost all surface and groundwaters tested, with higher concentrations generally found in groundwater. In a study of 228 groundwater and 113 surface water samples, it was detected in groundwater above 0.7 mg/L in 8.8 percent and above 1.4 mg/L in 3.1 percent of samples but was not detected above 0.7 mg/L in any surface water sample. Boron is often present at milligram per liter concentrations in the finished drinking water produced by desalination of seawater (NRC, 2008).

Acute exposure to high levels of boron causes GI effects and skin redness. It is considered a possible trace mineral nutrient for humans (IOM, 2006). The primary effects of concern from long-term exposure to low levels of boron are testicular toxicity and developmental effects. Testicular effects including atrophy, inhibition of sperm formation, decreased fertility, and sterility have been seen in several animal species. In developmental studies with rats, mice, and rabbits, oral exposure to boric acid resulted in decreased pregnancy rates, increased prenatal mortality, decreased fetal weights, and increased malformations in fetuses and pups. Boron is not considered mutagenic and is not considered to be carcinogenic by USEPA (USEPA, 2008b).

USEPA (2008b) recently decided not to develop an MCL for boron as part of its CCL process. The USEPA RfD for boron is 0.2 mg/kg/day based on developmental effects in rats. USEPA Drinking Water Health Advisories for boron have recently been updated (USEPA, 2008c). The Ten-day Health Advisory of 3 mg/kg/day for children is based on testicular toxicity in a subchronic dietary study in rats. The Longer-term Health Advisory of 2 mg/L for children is based on testicular toxicity in a chronic rat study. The Longer-term Health Advisory for adults and the Lifetime Health Advisory of 5 mg/L are based on developmental effects in the fetus and exposure assumptions for pregnant women (USEPA, 2008c).

Cadmium

Cadmium enters the environment from a variety of industrial applications and may enter consumers' tap water by corrosion of galvanized pipe and solder used in hot water heaters. Various foods and tobacco smoke also contribute to cadmium exposure.

Cadmium can cause kidney dysfunction, hypertension, anemia, and altered liver function (USEPA, 1985a). Chronic occupational exposure has resulted in renal dysfunction and neuropsychological impairments. The concentration of cadmium in the kidney increases over time, and renal toxicity is related to the concentration in the kidney. Cadmium has been shown to induce testicular and prostate tumors in laboratory animals injected subcutaneously (ATSDR, 1993). It is considered to be a probable human carcinogen (Group B1) by the inhalation route based on limited evidence in workers but is not classifiable as to carcinogenicity by the oral route. It is regulated based on its renal toxicity in humans, based on the highest cadmium concentration in the kidney, 200 mg/kg, which is not associated with significant proteinuria (USEPA, 1985a). A toxicokinetic model was used to determine the NOAEL of 0.005 mg/kg/day, which is the chronic dose from drinking water that would result in a concentration of 200 mg/kg in the kidney. An uncertainty factor of 10, appropriate for a NOAEL from human data, and an RSC factor of 25 percent were applied to develop the MCLG and an MCL of 5 μg/L (USEPA, 1985a, 1991a).

Chromium

Chromium may occur in drinking water in its hexavalent or trivalent states. The process of water disinfection can affect its oxidation state, increasing the proportion in the hexavalent state (Lai and McNeill, 2006). Chromium can occur naturally in water or result from sources such as wastes from chromate production or electroplating operations, mining, and garbage and fossil fuel combustion (USEPA, 1985a).

The toxic potency of chromium is dependent on its oxidation state, with Cr(VI) being much more potent than Cr(III). The stomach has a substantial capacity to reduce Cr(VI) to Cr(III), but kinetic factors which compete with the reduction capacity make it likely that absorption of unreduced Cr(VI) occurs after ingestion of even relatively small doses of Cr(VI) (O'Flaherty et al., 2001; NJDEP, 2009). Cr(VI) compounds rapidly enter cells, while Cr(III) compounds enter much more slowly. Once it is taken into cells, Cr(VI) is rapidly reduced to Cr(III), with Cr(V) and Cr(IV) as intermediates. Within the cell, Cr(III) may react with peptides, proteins, and DNA, resulting in DNA-protein crosslinks, DNA strand breaks, and alterations in cellular signaling pathways, which may contribute to toxicity and carcinogenicity of chromium compounds (ATSDR, 2008b).

Cr(III) is an essential nutrient for normal glucose, protein, and fat metabolism; has low toxicity; and is poorly absorbed. Currently, the biological target for the essential effects of Cr(III) is unknown. The Institute of Medicine Food and Nutrition Board (2001) of the NAS determined an adequate intake (e.g., a level typically consumed by healthy individuals) of 20 to 45 μg Cr(III)/day for adolescents and adults.

Cr(VI) is a known human carcinogen (Group A) by the inhalation route of exposure, as epidemiology studies have shown it to cause lung tumors in exposed workers. Until recently, inadequate data were available to assess its oral carcinogenicity. The MCL and MCLG of 100 μg/L for total chromium are based on an NOAEL of 2.4 mg/kg/day in a one-year rat drinking water study of hexavalent chromium, an uncertainty factor of 500, and the assumption that humans are exposed to 0.1 mg/day in the diet. However, a recent NTP (2008a) drinking water study has shown clear evidence of oral carcinogenicity based on gastrointestinal tumors in mice and oral tumors in rats. If the MCLG and MCL are reevaluated based on a slope factor developed from these data (California EPA, 2009; NJDEP, 2009), a much lower drinking water standard will likely result.

Copper

Copper is found commonly in drinking water (USEPA, 1985a). The principal sources in water supplies are from the addition of copper salts to control algae in lakes and reservoirs. Additionally, low levels (generally, below 20 μg/L) can derive from rock weathering, and some can come from industrial sources. In treated drinking waters, sources of copper are from water corrosion of brass and copper pipes and fixtures.

Copper is a nutritional requirement, and deficiency leads to anemia, skeletal defects, nervous system degeneration, and reproductive abnormalities. The daily requirements are estimated as 0.9 mg/day for adults, 0.34 mg/day for 1- to 3-year-olds, and 0.2 mg/day for infants (Institute of Medicine Food and Nutrition Board, 2001).

Copper in excess of nutritional requirements is excreted. However, at high doses, copper can cause acute effects such as GI disturbances, damage to the liver and kidneys, and anemia. The lowest dose at which adult GI tract irritation was seen was 5.3 mg/day (USEPA, 1991b). A study of 4- to 6-month-old infants found that growth and the incidence of diarrhea were not affected by ingesting 2 mg/L in milk formula (Olivares et al., 1998). The Institute of Medicine tolerable intake level for 1- to 3-year-olds was set at 1 mg/day. Those with Wilson's disease, an inherited condition in about 1 in 40,000 individuals that results in copper accumulation in the body, are more sensitive to the effects of copper in water and food and may require medical intervention. Exposure of mice via subcutaneous injection yielded tumors; however, oral exposure in several studies did not. Mutagenicity tests have been negative.

Copper is regulated under special provisions of the Lead and Copper Rule (USEPA, 1991b). Under the rule, if more than 10 percent of the samples from residential taps most at risk for copper contamination exceed the action level of 1.3 mg/L, then water purveyors must take action to minimize corrosion. An SMCL of 1.0 mg/L, based on taste, applies to water entering the distribution system from the water treatment plant. Residents in homes that use copper pipe are often advised to drink water only from the cold tap and to flush standing water from the pipes each morning before drinking it or using it to make beverages; these measures usually also decrease the concentration of lead in the drinking water.

Fluoride

Fluoride occurs naturally in some soils and rocks and in many water supplies. Drinking water is the source of greatest exposure in most individuals; other exposure sources include food, dental products, and pesticides. To protect against the development of dental caries, fluoride is widely added at 0.7 to 1.2 mg/L to public water supplies in the United States that do not contain it naturally at the desired level. In 2000, it was estimated that approximately 162 million people in the United States used artificially fluoridated water (NRC, 2006a). Acute effects of overdosing in the range of 20 to 200 mg/L, which can occur due to equipment failure in fluoridating systems, can cause nausea, diarrhea, abdominal pain, headache, and dizziness (CDC, Division of Oral Health, 1992).

The USEPA SMCL of 2 mg/L is intended as a guideline for areas that have high levels of naturally occurring fluoride. The SMCL is based on protection from dental fluorosis (tooth discoloration and/or pitting) caused by excess fluoride exposures during the formative period prior to eruption of teeth in children. This effect is considered by USEPA to be a cosmetic effect rather than an adverse health effect. Chronic high-level exposure to fluoride can increase the risk of fractures and lead to skeletal fluorosis. In this condition, fluoride accumulates in the bone over many years, causing stiffness and pain in the joints and eventually changes in bone structure, calcification of ligaments, and pain. Evidence of a relationship between fluoride exposure and an increased incidence of bone cancer (osteosarcoma) in young males is mixed but cannot yet be ruled out.

The MCL and MCLG for fluoride are 4 mg/L, based on prevention of bone disease. In 1992, it was estimated that 1.4 million people in the United States used drinking water with natural fluoride concentrations of 2.0 to 3.9 mg/L, and just over 200,000 people used drinking water with concentrations equal to or exceeding 4 mg/L (NRC, 2006a). A recent NRC (2006a) report concluded that the MCL and MCLG of 4 mg/L should be decreased to protect children from developing severe enamel fluorosis and to reduce the risk of bone fracture and possibly skeletal fluorosis, particularly in subpopulations that accumulate fluoride in their bones.

Hardness

Hardness is due to calcium and magnesium present in water and is expressed as an equivalent concentration as calcium carbonate ($CaCO_3$). Other divalent and trivalent ions can contribute to hardness but are almost always less than 1 mg/L and do not make a significant contribution. Although no distinctly defined levels exist, waters with less than 75 mg/L $CaCO_3$ are generally considered soft, those with between 75 and 150 mg/L $CaCO_3$ are said to be moderately hard, those with 150 to 300 mg/L $CaCO_3$ are hard, and waters with more than 300 mg/L $CaCO_3$ are classified as very hard.

An inverse relationship has been observed between the incidence of cardiovascular disease and the amount of hardness in the water (Catling et al., 2008). Hypotheses outlined below suggest a protective effect from either the major or minor constituents of hard water or, conversely, a harmful effect from corrosion by-products found at higher concentrations in soft water.

Many investigators attribute a protective effect for cardiovascular disease to the presence of calcium and magnesium (reviewed by Marx and Neutra, 1997; McCarron, 1998; Monarca et al., 2006). A moderate increase in calcium in the diet has been observed to decrease levels of circulating cholesterol; this is speculated to be a factor in relating water hardness and cardiovascular disease. Magnesium is theorized to protect against lipid deposits in arteries, to reduce cardiac irritability and damage, and may also have anticoagulant properties that could protect against cardiovascular diseases by inhibiting blood clot formation.

A few other studies suggest that minor constituents often associated with hard water may exert a beneficial effect on the cardiovascular system. Candidate trace elements include vanadium, lithium, manganese, and chromium. On the other hand, other investigators suggest that certain trace metals found in higher concentrations in soft water due to corrosion, such as cadmium, lead, copper, and zinc, may be involved in the causation of cardiovascular disease.

For each of these hypotheses, it is uncertain whether drinking water provides enough of these elements to have any significant impact on the pathogenesis of cardiovascular disease. For example, hard water generally supplies less than 10 percent of the total dietary intake of calcium and magnesium, and water provides even smaller proportions of the total intake of the various suspect trace metals. Given the level of uncertainty on this subject, USEPA currently has no policy with respect to the hardness or softness of public water supplies. However, USEPA strongly supports corrosion control measures, some of which increase hardness, to reduce exposure to lead.

Iron

Iron is commonly found in the soluble reduced form (FeII) in groundwater and at the bottom of reservoirs where reducing conditions exists. Iron can also enter drinking water from corrosion of iron or steel pipes and storage tanks.

Because there is a nutritional requirement of 10 to 12 mg per day of iron for healthy adult men and 10 to 15 mg per day for women (NRC, 1989), it is unlikely that health problems will arise even from the maximum concentrations found in drinking water. However, in individuals genetically susceptible to hemochromatosis, excess iron can accumulate in the body, resulting in liver, pancreatic, and heart dysfunction and failure after long-term high exposures (Motulsky, 1988). About 1 person in 200 is potentially at risk for this condition, though the actual incidence is much lower (Walker et al., 1998).

The SMCL for iron of 0.3 mg/L is based on discoloration of laundry and a metallic taste that becomes noticeable in the range of 0.1 to 1.0 mg/L (see Aesthetic Quality later in this chapter).

Lead

Lead occurs in drinking water primarily from corrosion of lead pipe and of solder and faucets constructed with leaded brass, especially in areas with soft or acidic water. However, the primary source of exposure for children, who are more sensitive than adults to lead toxicity, is lead-based paint in older homes. Because of concerns about the health effects of lead, it is no longer used in gasoline, residential paint, or solder for food cans and water pipes.

Lead accumulates in bone over the lifetime; most lead in the body is stored in bone. The main target for lead toxicity is the nervous system, both in adults and children. Infants and young children absorb ingested lead more readily than older children and young adults (National Academy of Sciences Safe Drinking Water Committee, 1982). Of great public health concern, low levels of exposure during prenatal development, in infancy, or in early childhood causes neurodevelopmental effects, such as delays in mental and physical development and decreases in intelligence, which may be permanent. Higher blood lead levels are associated with many other effects, including interference with heme synthesis necessary for formation of red blood cells, anemia, kidney damage, miscarriage, impaired sperm production, interference with vitamin D metabolism, and elevations in blood pressure in adults (USEPA, 1988a). In its IRIS database of risk assessment information, USEPA has classified lead as a probable human carcinogen because some lead compounds cause renal tumors in rats.

Exposure to lead is measured by the blood lead level. The CDC action level for blood lead in children is 10 μg/dL, but no threshold blood lead level for neurodevelopmental effects in children, such as slight IQ decrements, has been determined. At this low blood level, the lead from plumbing becomes more significant as a potential source, particularly when other sources are reduced or eliminated.

USEPA has established an action level instead of an MCL for lead (USEPA, 1991b). It requires that water purveyors sample residential taps that have been determined to be the most likely to have a lead corrosion problem. If more than 10 percent of the residential taps sampled are above the action level of 15 μg/L, treatment measures are required to reduce corrosive properties of the water and, thus, the lead levels. The MCLG for lead is set at zero because no threshold is known for neurodevelopmental effects, to affirm the overall USEPA goal of reducing total lead exposures, and because lead is classified as a probable human carcinogen.

Manganese

Manganese is found in the soluble reduced form (MnII) in groundwater and at the bottom of reservoirs where reducing conditions exists. Man-made sources include discarded batteries, steel alloy production, and agricultural products. Manganese is removed from water primarily due to aesthetic reasons (taste and staining), with a SMCL of 0.05 mg/L.

Manganese is an essential nutrient, fulfilling a catalytic role in various cellular enzymes, but deficiencies are rare because it is found in many foods. The greatest source of exposure is the diet (0.7 to 10.9 mg/day) (USEPA, 2004), while the median drinking water concentration is 10 μg/L. The adequate daily intake for infants up to 6 months old is thought to be 3 μg/day, based on levels in breast milk, with higher requirements for older individuals, up to 2.3 mg/day in adult males (Institute of Medicine, 2001).

Occupational inhalation exposure can lead to manganism, an irreversible, slow-onset neurotoxic disease with symptoms similar to Parkinson's disease. Manganese is generally considered to be nontoxic orally, and human epidemiology studies of its neurotoxic potential through oral exposure are inconclusive. Rodents are not considered to be a good model for manganese toxicity in humans due to differences in accumulation in the brain and higher dietary requirements. Manganese is not considered to be carcinogenic. The USEPA Lifetime Health Advisory (2004) of 0.3 mg/L is based on the estimation of a safe level from the reported upper limit of dietary exposures in adults. It includes a modifying factor of 3 to account for possible higher bioavailability from water than from food. A recent publication (Ljung and Vahter, 2007) has raised questions as to whether the WHO guideline for manganese in drinking water of 0.4 mg/L is protective in children, based on data on greater absorption, slower excretion, and greater susceptibility to neurotoxicity in infants and children than in adults.

Mercury

Mercury occurs primarily as an inorganic salt in water and in the organic form, methylmercury, in sediments and fish (USEPA, 1985a). Mercury enters the environment from the burning of fossil fuels, incineration of products containing mercury, past uses in pesticides, and leaching of organic mercury from antifungal outdoor paints, as well as natural sources.

Inorganic mercury is poorly absorbed in the adult GI tract, does not readily penetrate cells, and is not as toxic as methylmercury. Calomel, mercurous chloride, was used in the past as a laxative. However, the absorption of inorganic mercury can be much higher in infants and young children than in older individuals. The most sensitive toxic endpoint for inorganic mercury in adult laboratory animals is the induction of an autoimmune disease of the kidney (ATSDR, 1999).

Inorganic mercury released into the environment enters water bodies where it is transformed by bacteria in sediments to methylmercury, which bioaccumulates through the food chain. Organic forms such as methylmercury are readily absorbed in the GI tract and can easily enter the central nervous system (CNS). Methylmercury is neurotoxic, and exposure can result in mental and motor dysfunctions or death. It can cross the placental barrier and reach the fetus, which is sensitive to its neurodevelopmental toxicity. Thus, pregnant women are the subpopulation of greatest concern for exposure to methylmercury. Fish contaminated with methylmercury caused the well-known mercury poisonings in Japan's Minamata Bay, characterized by deaths, CNS disorders, and mental retardation. Lower levels of exposure, such as those resulting from consuming large amounts of fish containing mercury, result in neurodevelopmental effects in children and neurological effects in adults. However, drinking water is not a significant source of methylmercury exposure. Metallic (elemental) mercury is also neurotoxic when inhaled but is not found at significant levels in water.

USEPA has classified inorganic mercury as a possible human carcinogen (Group C) based on a chronic oral study in rats (NTP, 1993) that showed some evidence of carcinogenicity in male rats and equivocal evidence in female rats and male mice. *In vitro* mutagenicity and genotoxicity studies of inorganic and organic mercury gave mixed results (ATSDR, 1999). The MCLG and MCL of 2 μg/L are based on kidney effects of the inorganic mercury, which is the form of mercury found in water (USEPA, 1991a).

Nitrate and Nitrite

Nitrate is one of the major anions in natural waters, and concentrations can be greatly elevated due to leaching of nitrogen from farm fertilizer, feedlots, and septic tanks. The mean concentration of nitrate nitrogen (NO_3^- as N) in a typical surface water supply is around 0.2 to 2 mg/L, and individual wells can have significantly higher concentrations. Adult dietary nitrate intake is approximately 20 mg/day, mostly from vegetables such as lettuce, celery, beets, and spinach (National Academy of Sciences Committee on Nitrite and Alternative Curing Agents in Food, 1981).

Nitrite does not occur typically in natural waters at significant levels, except under reducing conditions or by oxidation of ammonia. Sodium nitrite was widely used for cured meats, pickling, and beer, but such usage has been sharply curtailed. Rarely, water supplies in buildings have been contaminated with nitrite by faulty cross-connections or procedures during boiler cleaning with nitrous acid.

Nitrite, or nitrate converted to nitrite in the body, can undergo two chemical reactions that can cause adverse health effects: induction of methemoglobinemia, especially in infants under six months of age, and the potential formation of carcinogenic nitrosamides and nitrosamines (National Academy of Sciences Safe Drinking Water Committee, 1977; USEPA, 1989c).

Methemoglobin, normally present at 1 to 3 percent in the blood, is the oxidized form of hemoglobin and cannot act as an oxygen carrier. Certain substances, such as the nitrite ion, oxidize hemoglobin to methemoglobin (National Academy of Sciences Safe Drinking Water Committee, 1977). In adults, only a small percentage of ingested nitrate is reduced to nitrite by bacteria in the mouth. However, in infants, the relatively alkaline conditions in the stomach allow bacteria there to reduce up to 100 percent of the nitrate to nitrite, particularly when a concurrent GI infection is present. Furthermore, infants do not have the same capability as adults to convert methemoglobin back to hemoglobin. When the concentration of methemoglobin in the blood reaches 5 to 10 percent, symptoms can include lethargy, shortness of breath, and a bluish skin color (blue-baby syndrome). Anoxia and death can occur at high exposures to nitrites or nitrates.

Carcinogenic nitrosamines and nitrosamides are formed when nitrite reacts in the stomach with secondary amines, such as the amino acids from dietary protein (IARC, 1978). This process is inhibited by antioxidants in the diet, such as vitamin C. A recent series of epidemiological case-control studies conducted in Iowa by the National Cancer Institute and others (reviewed by Ward et al., 2005, 2007, 2008) suggest a relationship between nitrate and cancer in various tissues in the subset of people consuming more red meat and less vitamin C than the average.

The data on the role of nitrates in developmental effects, such as birth defects, are regarded as being inconclusive (Ward et al., 2005).

The MCLGs and MCLs are 10 mg/L for nitrate measured as nitrogen (equivalent to 45 mg/L nitrate) and 1 mg/L for nitrite measured as nitrogen (USEPA, 1991a). In addition, the MCL for total nitrate-N and nitrite-N is 10 mg/L. While most MCLGs are based on exposure assumptions for adults (70 kg body weight and 2 L/day water consumption), the MCLGs for nitrate and nitrite are based on exposure assumptions for an infant (10 kg body weight and 1 L/day water consumption), as they are most sensitive to the effects of these compounds.

Perchlorate

Perchlorate is a chemical with diverse uses in flares, fireworks, rocket fuel, and explosives and it occurs naturally in fertilizers from certain regions, such as Chile. It also has been used clinically to treat hyperthyroidism because it blocks the uptake of iodine into the thyroid gland, resulting in reduced production of the iodine-containing hormones, thyroxine (T4) and triiodothyronine (T3). In a normal, healthy individual, reduced levels of thyroid hormones are prevented to a large extent by a negative feedback loop involving stimulation of the thyroid by the pituitary hormone, thyroid stimulating hormone (TSH). Thyroid hormones are critical in fetal and infant brain development, and deficiency during these periods can cause irreversible developmental effects. Thus, the primary concern about environmental exposures to perchlorate is its potential to interfere with thyroid function in fetuses and infants.

The NRC (2005) recommended an RfD based on the small but detailed study by Greer et al. (2002), which measured the inhibition of iodine uptake by the thyroid and the levels of thyroid hormones in adult volunteers dosed with perchlorate. In this study, a dosage of 7 μg/kg/day did not affect thyroidal iodine uptake. Reduced iodine uptake is not considered to be an adverse effect and is a full biological step away from a reduction in thyroxine or a homeostatic increase in TSH. Neither of these effects occurred even at the highest dose of perchlorate in the study, 500 μg/kg/day.

Based on this study, the NRC established an RfD of 0.0007 mg/kg/day, which was adopted by the USEPA IRIS database in 2006. This RfD incorporates an uncertainty factor of 10 to protect sensitive subpopulations. Additional uncertainty factors were not considered necessary since the observed effect was not in and of itself considered adverse. The protectiveness of this RfD is supported by other data indicating the absence of a thyroid hormone effect on newborns and pregnant women in locations with perchlorate in the drinking water.

The DWEL for a 70-kg adult based on this RfD is 24.5 μg/L of perchlorate. The recent publication of U.S. Food and Drug Administration data on exposure to perchlorate in foods

(Murray et al., 2008) indicates that an RSC of 0.6 is appropriate, instead of the default RSC of 0.2, suggesting a health-based drinking water level protective for adults of 15 μg/L.

Recent physiologically based pharmacokinetic modeling (USEPA, 2008d) indicates that infants who are bottle-fed (mostly with formula reconstituted from powder) receive a dose (mg/kg/day) approximately six times higher than the adult dose at the same drinking water concentration. These findings suggest that the drinking water concentration of 15 μg/L based on adult exposure is not sufficiently protective of infants. USEPA (2008e) has recently developed an Interim Health Advisory of 15 μg/L for perchlorate. Perchlorate is one of four contaminants listed on the USEPA CCL3 which the USEPA Science Advisory Board recommended as a priority for consideration for regulation based on available data on toxicity, occurrence, and treatability (USEPA, 2009b). At the time of this writing, USEPA is reevaluating an earlier proposal not to establish an MCL for perchlorate and has requested comments on this issue (USEPA, 2009c).

The USEPA Office of Water perchlorate Website should be consulted for further updates. http://www.epa.gov/safewater/contaminants/unregulated/perchlorate.htmL. California, Massachusetts, and New Jersey have proposed or adopted a perchlorate drinking water standard (Zewdie et. al., 2009).

Selenium

Selenium is an essential dietary element with most human intake coming from food. The NRC (1989) has recommended that the daily diet include between 55 and 75 μg selenium, with the higher end of this range recommended for adult males and pregnant or nursing females. Selenium is a key component of glutathione peroxidase, a vital thyroid enzyme (Corvilain, Contempre, and Longombe, 1993). In humans exposed to high dietary levels, dermatitis, hair loss, abnormal nail formation and loss, diarrhea, liver degeneration, fatigue, peripheral nervous system abnormalities, and a garlic odor have been observed (reviewed by Poirier, 1994 and by Patterson and Levander, 1997). The USEPA IRIS RfD of 0.005 mg/kg/day is based on the NOAEL for selenium toxicity in an area of China with high levels in the environment, and includes an uncertainty factor of 3 to account for sensitive individuals.

Selenium reacts *in vivo* with other elements to protect against heavy metal toxicity. Naturally occurring selenium compounds have not been shown to be carcinogenic in animals, and selenium may inhibit tumor formation (USEPA, 1985a; Patterson and Levander, 1997), although the data on this effect are controversial (Clark and Alberts, 1995; Navarro et al., 2007). Selenium has not been found to be teratogenic in mammals (Poirer, 1994).

An MCLG and an MCL of 50 μg/L (USEPA, 1991a) have been established based on the RfD described above.

Sodium

Sodium is a naturally occurring constituent of drinking water. A survey of 2100 finished waters conducted between 1963 and 1966 by the U.S. Public Health Service found concentrations ranging from 0.4 to 1900 mg/L, with 42 percent of samples above 20 mg/L and 4 percent above 250 mg/L (White et al., 1967). Sodium levels at the tap can be increased by water softeners, which can add approximately 1 mg sodium for every 2 mg hardness removed.

There is strong evidence for an at-risk population that is predisposed to high blood pressure (hypertension) from dietary sodium (Institute of Medicine Panel on Dietary Reference Intakes for Electrolytes and Water, 2004). Older persons, African-Americans, and those with hypertension, diabetes, or kidney disease are more likely than others to be members of this at-risk population. Hypertension is of concern because it increases the risk of coronary heart disease, stroke, and certain other diseases (Institute of Medicine Panel on Dietary Reference Intakes for Electrolytes and Water, 2004).

Renin, angiotensin, and aldosterone are hormones that form a system that regulates cardiac output and blood pressure in response to changes in physiological factors, including sodium levels. This system can become less responsive with advancing age, resulting in increased sensitivity of blood pressure to sodium intake (Institute of Medicine Panel on Dietary Reference Intakes for Electrolytes and Water, 2004). The Institute of Medicine report concluded that "overall, the available evidence on the effects of sodium reduction on blood pressure in children is limited and inconsistent."

Food is the major source of sodium. The suggested adult maximum daily intake is 2300 mg (American Heart Association, 2008), and the Institute of Medicine (2004) suggested that 1200 mg/day is an adult minimum requirement. Typical adult intake is 2000 to 10,000 mg/day (Institute of Medicine, 2004), with an average intake of 3000 to 5000 mg/day (USDA, 2009); physically active persons may require higher levels. Drinking water at a typical concentration of 20 mg/L contributes about 1 percent of the minimum requirement, assuming consumption of 2 L/day. The USEPA guidance level for sodium of 30 to 60 mg/L is based on taste (see Taste and Odor in section on Aesthetic Quality), and consuming 2 L per day of water with sodium at the high end of this range would contribute approximately 5-10 percent of the minimum requirement.

Sulfate

Sulfate is a naturally occurring anion. High concentrations of sulfate in drinking water may cause transitory diarrhea (USEPA, 1990a). Most adults experience a laxative effect when sulfate concentrations are above 1000 mg/L, whereas medical case reports indicate that bottle-fed infants develop diarrhea at levels above 600 mg/L. Acute diarrhea can cause dehydration, particularly in infants and young children who may already have a microbial diarrheal condition. Adults living in areas having high sulfate concentrations in their drinking water easily adjust, with no ill effects. The SMCL for sulfate is 250 mg/L based on taste (see section on Aesthetic Quality). The U.S. Army standard for soldiers drinking 5 L/day of water is 300 mg/L.

ORGANIC CONSTITUENTS

Organic compounds in water derive from three major sources: (1) the breakdown of naturally occurring organic materials, (2) man-made chemicals from domestic and commercial activities, and (3) treatment additives and chemicals formed during reactions that occur during water treatment and transmission. Naturally occurring materials are the greatest source of organic compounds in water (see Chap. 3). They are composed of humic materials from plants and algae, microorganisms and their metabolites, and high-molecular-weight aliphatic and aromatic hydrocarbons. These natural organics are typically not of health concern, although some are nuisance constituents such as odiferous metabolites (see section on Aesthetic Quality). A few of the high-molecular-weight natural aliphatic and aromatic hydrocarbons may have adverse health effects. In addition, humics serve as precursors (reactants) in the formation of trihalomethanes (THMs), haloacetic acids (HAAs), and other organohalogen DBPs during disinfection (see Chap. 19).

Man-made organic chemicals from domestic and commercial activities enter water supplies through wastewater discharges, agricultural and urban runoff, leachate from contaminated soils, and air deposition to surface waters or to soils followed by migration to groundwater. Many of the organic drinking water contaminants with potential for adverse health effects are part of this group. They include pesticides (such as atrazine and aldicarb), solvents and metal degreasers (such as trichlorobenzene, tetrachloroethylene, trichloroethylene, and trichloroethane), and emerging contaminants such as pharmaceuticals and perfluorinated chemicals.

Organic contaminants formed during water disinfection include by-products such as THMs (e.g., chloroform) and HAAs (e.g., di- and trichloroacetic acids). Other compounds such as acrylamide are components of high molecular weight polymers used as floc aids and filter aids used in water treatment and epichlorohydrin, a component of one group of cationic polymers used as a coagulant. During finished water transmission, undesirable components of pipes, coating, linings, and joint adhesives, such as PAHs, epichlorohydrin, and solvents, can leach into water. Health effects of some organic chemicals that are of concern as drinking water contaminants are summarized below.

Volatile Organic Chemicals

The term *volatile organic chemical* (VOC) refers to the characteristic evaporative properties (or the vapor pressure) of these compounds. In general, VOCs are at least slightly water soluble, and three broad groups of VOCs have been found in drinking water. One group includes compounds found in petroleum products, including aromatics such as benzene, toluene, ethylbenzene, and xylenes, as well as gasoline additives such as methyl-*tert*-butyl-ether. Typical sources are leaking fuel oil and gasoline tanks. Another group is the halogenated VOCs, used as solvents and degreasers in industrial and commercial facilities including dry cleaners and automobile repair garages. Their former use as septic tank cleaners also accounts for many instances of private well contamination. Some halogenated VOCs (e.g., ethylene dibromide, dichloropropanes, and 1,2,3-trichloropropane) have also been used as pesticide fumigants. The third group includes some of the chlorinated organic DBPs, particularly the THMs. Some THMs have also entered groundwater through industrial contamination.

As VOCs easily volatilize from hot water used in the home, inhalation is an important route of exposure to these chemicals from drinking water, in addition to ingestion. The air in bathrooms can represent a major source of exposure because of volatilization from shower and bath water (McKone, 1987; Jo, Weisel, and Lioy, 1990; Maxwell, Burmaster, and Ozonoff, 1991). In addition, dermal absorption (through the skin) can occur from bath water because the VOCs are lipophilic. The exposures to VOCs in drinking water from the combined dermal and inhalation routes can equal the exposure received through ingestion (McKone, 1987).

Their lipophilic characteristic also enable VOCs to enter the brain from the blood. At high exposure concentrations, VOCs can cause reversible effects, including dizziness, nausea, and cardiac depression. Chronic inhalation of high levels of many VOCs causes CNS toxicity and may also produce irritation of the eyes, nose, and throat, but these effects are not relevant to drinking water contamination. Many VOCs cause toxicity to the liver and/or the kidney after chronic exposure, and some VOCs have been found to cause cancer in humans and/or laboratory animals.

Benzene and Other Aromatic Gasoline Components (Ethylbenzene, Toluene, and Xylene). The group of aromatic hydrocarbons that are found in petroleum products—benzene, toluene, ethylbenzene, and xylene, are known as the BTEX compounds. Drinking water can be contaminated with these chemicals from leaking underground storage tanks and other sources of petroleum contamination. Benzene, which is toxic to the bone marrow and causes leukemia, is by far the most toxic of this group of structurally related compounds.

Benzene is produced by petroleum refining, coal tar distillation, coal processing, and coal coking. It is also widely used as a chemical intermediate. Gasoline in the United States has contained anywhere from 0.5 to 5 percent benzene by volume. By 2011, USEPA will require that gasoline made or imported to the United States have an average annual benzene content of no higher than 0.62 percent (USEPA, 2008f).

Chronic exposure to sufficient levels of benzene, such as have occurred in occupational settings, causes toxicity to the hematopoietic (blood-forming) system. The target organ is the bone marrow, and effects include decreased numbers of red blood cells, platelets, or white blood cells (anemia, thrombocytopenia, or leukopenia) and leukemia. It is considered to be a human carcinogen. The mechanism of toxicity has not been totally elucidated, but toxicity is known to result from metabolites of benzene, rather than from the parent compound. By extrapolation from occupational studies, a drinking water concentration of 0.6 to 2.2 μg/L benzene has been calculated to result in an excess lifetime cancer risk of 1 in 1,000,000 (USEPA 1998a, 1999). The federal MCL of 5 μg/L for benzene is based on carcinogenic potential. Many states have promulgated stricter standards.

Ethylbenzene, toluene, and xylene are components of gasoline and fuel oil that are less toxic than benzene. Ethylbenzene is also used as a solvent and in the production of styrene. In a two-year inhalation study (NTP, 1999), kidney tumors in male and female rats, testicular tumors in male rats, lung tumors in male mice, and liver tumors in female mice occurred at an increased incidence. The relevance of these findings to human carcinogenic potential has not yet been evaluated by USEPA, and it is treated as a noncarcinogen and classified as Group D by the USEPA IRIS risk assessment database and in the development of the MCLG (USEPA, 1987a). The MCL and MCLG of 0.7 mg/L and the IRIS RfD of 0.1 mg/kg/day are based on the NOAEL for liver and kidney effects in a one-year gavage study in rats.

Toluene is part of the aromatic fraction in gasoline and is used as a starting material in the production of benzene and other chemicals and as a solvent for paints, coatings, gums, glues, and resins. CNS toxicity is the primary effect of concern from acute or chronic inhalation exposure (ATSDR, 2000a). It is not considered to be a carcinogen. The RfD in the USEPA IRIS database of 0.08 mg/kg/day and the MCL and MCLG of 1 mg/L are based on increased kidney weight in a subchronic rat gavage study.

Xylenes occur as three isomers, ortho-, meta-, and para-, that are treated as one in the USEPA health effects evaluation. Xylenes are part of the aromatic fraction that has been used in gasoline and are used as solvents and in the synthesis of many organic chemicals, pharmaceuticals, and vitamins. Like other chemicals in this group, high levels of exposure cause CNS toxicity. In a two-year gavage study (NTP, 1986), neither cancer nor other specific organ toxicity was observed in rats or mice. The MCL and MCLG of 10 mg/L for xylene, as well as the IRIS RfD, are based on the NOAEL for body weight loss in this study.

Chlorinated Volatile Organic Chemicals. Chlorinated solvents are among the most commonly encountered organic drinking water contaminants. Information on important members of this group of chemicals is summarized below.

Carbon Tetrachloride. Carbon tetrachloride has been used as a dry cleaning agent, solvent, reagent in chemical synthesis, fire extinguisher fluid, and grain fumigant, but its primary use was in chlorofluorocarbon (CFC) production. Annual production has declined since its use in consumer products was banned in the 1970s. Its production and import are currently being phased out as an ozone-depleting substance under the Clean Air Act and the Montreal Protocol.

The liver and kidney are the organs that are most sensitive to the effects of carbon tetrachloride from both acute and long-term exposure. Hepatic effects include fatty infiltration, release of liver enzymes, inhibition of enzyme activity, and, ultimately, necrosis.

Mutagenicity and genotoxicity tests using bacteria, mammalian cells, and short-term *in vivo* rat assays were mostly negative. Chronic exposures produced liver cancers in mice, rats, and hamsters, and it is classified as a probable human carcinogen (Group B2) by USEPA. No teratogenic effects have been demonstrated (ATSDR, 1994). The federal MCL for carbon tetrachloride of 5 μg/L is based on carcinogenic potential.

Dichlorobenzenes (ortho-, meta-, para-, or o-, m-, p-). o-Dichlorobenzene is used primarily in the production of organic chemicals, pharmaceuticals, pesticides, and dyes and as a deodorant in industrial wastewater treatment, and m-dichlorobenzene is used to make herbicides. Large amounts of p-dichlorobenzene are used as an insecticidal fumigant, lavatory deodorant, and for mildew control. Dichlorobenzenes have been found in the milk of nursing mothers in several states (USEPA, 1985a).

The primary toxic effects of o- and p-dichlorobenzenes are on the blood, lung, kidney, and liver. Since similar data are not available for the meta-isomer, assessments for this isomer are based on data for o-dichlorobenzene. There is no evidence of carcinogenicity for o-dichlorobenzene, and USEPA treats it as a noncarcinogen (Group D). The RfD is based on the NOAEL for liver necrosis in rats in a subchronic gavage study (NTP, 1985). The MCL and MCLG for o-dichlorobenzene are 0.6 mg/L, based on an RfD of 0.9 mg/kg/day. The Lifetime Health Advisory for m-dichlorobenzene is also 0.6 mg/L based on the same RfD.

In a chronic gavage study of p-dichlorobenzene (NTP, 1987a), treatment-related increases were observed in the incidence of renal tubular cell adenocarcinomas in male rats and hepatocellular adenomas in male and female mice. Rare hepatoblastomas also occurred in male mice. Based on these data, p-dichlorobenzene is classified as a possible human carcinogen (Group C). The MCL and MCLG are 0.075 mg/L, based on the NOAEL for kidney lesions in rats (USEPA, 1987a, 1987b).

1,2-Dichloroethane (Ethylene Dichloride). 1,2-Dichloroethane is or has been used in the manufacture of vinyl chloride, as a lead scavenger in gasoline, as an insecticidal fumigant, as a constituent of paint varnish and finish removers, as a metal degreaser, in soap and scouring compounds, in wetting agents, and in ore flotation. It is environmentally persistent and is slightly water-soluble.

In humans, high doses of 1,2-dichloroethane cause toxicity to the CNS, circulatory system, kidney, and possibly the liver. Chronic doses in laboratory animals lead to hepatic and renal changes. 1,2-Dichloroethane is metabolized to reactive intermediates that are mutagenic and genotoxic in many microbial and mammalian test systems (ATSDR, 2001). In a chronic gavage study, it increased the incidence of tumors of the forestomach, circulatory system, and mammary glands in rats and of the uterus, mammary glands, and lung in mice (NTP, 1978). It is classified as a probable human carcinogen (Group B2). The MCLG is zero and the MCL is 5 μg/L.

1,1-Dichloroethylene (Vinylidene Chloride). 1,1-Dichloroethylene is used primarily as an intermediate in the synthesis of copolymers for food packaging films, adhesives, and coatings. It is formed by the hydrolysis of 1,1,1-trichloroethane.

Acute oral or inhalation exposures to 1,1-dichloroethylene cause toxicity to the liver, kidneys, and lung (USEPA, 2002a). Liver is the primary site of toxicity after chronic exposures, and kidney toxicity occurs in male mice. The organ damage is believed to result from metabolism of the parent compound to cytotoxic metabolites. Most genotoxicity assays in mammalian systems were negative. In the presence of mammalian enzymes that metabolize foreign chemicals, 1,1-dichloroethylene was mutagenic in bacterial and mammalian cell assays (USEPA, 2002a). Carcinogenicity test results have been equivocal. One inhalation study with mice and rats appeared to produce malignant kidney tumors in rats and malignant mammary tumors and leukemia in mice. The results of many other inhalation and feeding studies were negative, with the possible exception of mammary tumors. However, the design of many of these studies was not optimal for detecting carcinogenic effects (USEPA, 2002a). 1,1-Dichloroethylene is classified as a suggestive carcinogen by USEPA

(2002a). The MCLG and MCL are 7 μg/L based on liver effects in rats in a chronic drinking water study.

1,2-Dichloroethylenes. 1,2-Dichloroethylenes exist as cis- and trans-isomers. Both isomers are used, alone or in combination, as solvents and chemical intermediates. They are moderately water soluble.

The 1,2-dichloroethylenes are not highly toxic. Little toxicity was seen in a subchronic dietary study of the trans-isomer at doses up to 8000 mg/kg/day in mice and 3000 mg/kg/day in rats (NTP, 2002). No genotoxic or mutagenic effects were observed for trans-1,2-dichloroethylene, though cis-1,2-dichloroethylene was positive in some assays. They are not considered to be carcinogenic and are classified as Group D by USEPA. The MCLGs for both isomers are based on liver effects in animal studies. The MCLG and MCL for cis-1,2-dichloroethylene are 70 μg/L and for trans-1,2-dichloroethylene are 100 μg/L.

Methylene Chloride (Dichloromethane). Methylene chloride has been widely used in the manufacture of paint removers, urethane foam, pesticides and other chemical products, degreasers, cleaners, pressurized spray products, and fire extinguishers.

Increased levels of carbon monoxide bound to hemoglobin (carboxyhemoglobin) have been observed in people acutely poisoned by methylene chloride through inhalation or oral exposures (ATSDR, 2000b). Rats maintained on drinking water containing 130 mg/L methylene chloride for 91 days showed no adverse effects. Methylene chloride was mutagenic in bacterial assays but was negative in most mammalian cell assays. Chronic exposure through inhalation caused increased mammary gland tumors in rats and increased lung and liver tumors in mice. A chronic drinking water study showed increased liver tumors in female mice (ATSDR, 2000b). Methylene chloride is classified as a probable human carcinogen (Group B2). The MCL of 5 μg/L is based on carcinogenicity.

Tetrachloroethylene (Perchloroethylene). Tetrachloroethylene is used as a solvent in textile processing, metal degreasing, and dry cleaning and as a precursor in the manufacture of fluorocarbons. It was also used as a septic tank cleaner. USEPA (2010) recently estimated that over 600,000 people (>0.3 percent) served by public water supplies receive water with tetrachloroethylene at or above the MCL of 5 ug/L, and over 12,000,000 people (>5 percent) served by public water supplies receive water with concentrations at or above the Estimated Quantitation Level of 0.5 ug/L. Contamination in private wells has been as high as several milligrams per liter.

A recent draft review of the toxicity of tetrachloroethylene (USEPA, 2008g) concluded that inhalation causes CNS effects in both humans and animals, with effects on both vision and cognition. The doses at which these effects were seen are likely above those relevant to exposures from drinking water. The liver and kidney are affected by both oral and inhalation exposures. USEPA (2008g) concluded that tetrachloroethylene is "likely to be carcinogenic to humans" by both inhalation and oral exposure. This conclusion was based on findings of carcinogenicity in 10 of 10 chronic rodent studies and consistent association with excess cancer risks in human epidemiologic studies, although a causal mechanism has yet to be definitively established. The MCLG for tetrachloroethylene is zero, based on carcinogenicity, and the MCL is 5 μg/L.

1,1,1-Trichloroethane (Methyl Chloroform). Because 1,1,1-trichloroethane is less toxic than many other chlorinated solvents, it was previously commonly used in many household products and industrial applications. However, 1,1,1-trichloroethane is an ozone-depleting substance, and its production is currently being phased out under the Clean Air Act and the Montreal Protocol (ATSDR, 2006).

As with most VOCs, exposure to high levels of 1,1,1-trichloroethane orally or by inhalation causes CNS toxicity such as behavioral effects, unconsciousness, and death due to respiratory depression. Some recent animal studies suggest that long-term inhalation can cause permanent brain damage (ATSDR, 2006; USEPA, 2007a).

The USEPA MCLG and MCL of 200 μg/L are based on minimal liver effects in mice exposed by inhalation for 14 weeks (USEPA, 1984), as no appropriate oral study was available when these values were developed. It is considered a noncarcinogen (Group D). More recently, the NTP conducted a subchronic study in which 1,1,1-trichloroethane was administered to mice and rats in microcapsules in the feed (NTP, 2000). This dosing regimen was designed to simulate drinking water exposure. The most sensitive endpoint was decreased body weight in mice compared to vehicle controls. Evaluation of the results of this study will likely lead to a recommendation of a health-based drinking water concentration much higher than the current MCLG of 200 μg/L.

Trichloroethylene (Trichloroethene). Trichloroethylene (TCE) is used primarily in metal degreasing. Its past use in dry cleaning was largely discontinued in the 1950s when it was replaced by tetrachloroethylene. Other uses include adhesives, paint strippers, paints, and varnishes. Other uses in the production of foods, pesticides, drugs, and cosmetics have been discontinued.

TCE is the most frequently reported organic contaminant in groundwater and is a common contaminant at hazardous waste sites (ATSDR, 1997). USEPA (2010) recently estimated that over 400,000 (~0.2 percent) of people served by public water supplies receive water with tetrachloroethylene at or above the MCL of 5 ug/L, and about 12,000,000 (>5 percent) of people served by public water supplies receive water with concentrations at or above the Estimated Quantitation Level of 0.5 ug/L. Trichloroethylene has been found at concentrations up to several milligrams per liter in private wells.

TCE causes toxicity to the central nervous system, the kidney, the liver, the immune system, the male reproductive system, and the developing fetus (USEPA, 2009). It is metabolized through two separate pathways: oxidation by cytochrome P-450 to metabolites thought responsible for liver toxicity, and conjugation with glutathione by glutathione S-transferase leading to formation of reactive metabolites thought responsible for kidney toxicity.

USEPA (2009d) conducted a meta-analysis of 14 high-quality human epidemiology studies, and classified TCE as "Carcinogenic to Humans" by all routes of exposure based on convincing evidence of a causal association between TCE exposure and kidney cancer. USEPA (2009d) also concluded that there is compelling evidence that TCE causes lymphoma in humans, and more limited evidence for lymphoma and biliary tract cancer. In chronic animal studies, TCE caused kidney tumors in rats, but not mice or hamsters, and liver tumors in mice. The data for several other types of tumors were weaker. The reactive metabolites formed through the glutathione conjugation pathway are clearly mutagenic, leading USEPA (2009d) to conclude that the kidney tumors, thought to result from these metabolites, occur through a mutagenic mode of action. IARC classified TCE as a probable human carcinogen based on occupational and environmental epidemiology (IARC, 1996). The MCLG is zero, based on carcinogenicity, and the MCL is 5 μg/L.

Vinyl Chloride (Monochloroethene). Vinyl chloride is used in the production of polyvinyl chloride (PVC) resins, which are widely used in construction, including drinking water pipes, as well as in many consumer and industrial products and in the building and construction industries. Its use in propellants and aerosols was banned in 1974.

The association between occupational exposure to vinyl chloride and the development of liver angiosarcomas is one of the best-characterized cases of chemical-induced carcinogenicity in humans. Liver angiosarcomas are an extremely rare tumor, with only 20 to 30 cases per year reported in the United States, and many of the reported cases are associated with vinyl chloride exposure. Exposure has also been associated with increased death due to primary liver cancer and may be associated with other types of cancer (USEPA, 2000c). Angiosarcoma and liver cancer are also found in rats, mice, and hamsters administered vinyl chloride via the oral and inhalation routes. The genotoxic mode of action of vinyl chloride is well characterized, involving a reactive metabolite that forms DNA adducts that lead to tumor formation. Vinyl chloride is classified as a known human carcinogen by the oral and inhalation route. The MCLG is zero, based on carcinogenicity, and the MCL is 2 μg/L.

Methyl-*Tert*-Butyl-Ether and Other Gasoline Oxygenates. Methyl-*tert*-butyl-ether (MTBE) is an oxygenate that has been widely used both as a gasoline octane enhancer and for meeting USEPA Clean Air Act oxyfuel requirements. Premium gasoline has contained 5 to 8 percent MTBE to boost octane. Under the Clean Air Act, gasoline containing 15 percent MTBE was used in the Wintertime Oxygenated Fuels Program to reduce carbon monoxide formation in nonattainment areas, and gasoline containing 11 percent MTBE was used in the Federal Reformulated Gasoline Program to reduce formation of ozone precursors in nonattainment areas. In 1995, MTBE was the third most-produced organic chemical in the United States, behind ethylene and propylene, with 17.6 billion pounds produced (NJDEP, 2001).

MTBE is one of the most commonly detected groundwater contaminants and its use in gasoline is now being phased out in many states because of its potential to contaminate water supplies. For example, it has been detected in about 20 percent of public water supplies and private wells tested in New Jersey (NJDEP, 2001). MTBE is more resistant to degradation and moves through the soil to groundwater much faster than the other common groundwater contaminants from gasoline, such as benzene, toluene, ethylbenzene, and xylene (BTEX compounds). Sources of MTBE in groundwater and surface water include leaking underground storage tanks and pipelines, accidental spills, stormwater runoff, the use of gasoline engines in boats, and air deposition (NJDEP, 2001).

Individuals inhaling vapors from gasoline containing MTBE at the gasoline pump and in automobiles have reported symptoms such as headaches, nausea, and respiratory and eye irritation. In water, MTBE has a noticeable odor at 20 to 40 μg/L (USEPA, 1997a). It was not highly toxic to rats exposed by gavage for several months (Robinson, Bruner, and Olson, 1990). Although MTBE is not mutagenic or genotoxic, exposure to high levels by inhalation (8000 ppm) or by gavage (1000 mg/kg) was associated with the development of lymphoma and leukemia, as well as liver, renal, and testicular cancers in rodents (Burleigh-Flayer, Chun, and Kintigh, 1992; Belpoggi, Soffritti, and Maltoni, 1995). The relevance of these cancers to human health is not clear. A chronic study intended to evaluate carcinogenicity through the drinking water route is currently underway (Bermudez, Parkinson, and Dodd, 2009).

USEPA has not developed an MCL for MTBE, although it has been considered as a candidate for regulation in the CCLs developed in 1998, 2005, and 2008. MTBE is one of four contaminants listed on the USEPA CCL3, which the USEPA Science Advisory Board recommended as a priority for consideration for regulation based on available data on toxicity, occurrence, and treatability (USEPA, 2009b). Because of its widespread occurrence in drinking water and groundwater, at least 42 states have developed a standard or guidance for drinking water and/or groundwater (Delta Environmental Consultants, 2005a). USEPA has developed an advisory of 20 to 40 μg/L based on taste and odor, which is intended to provide a health-protective margin of exposure for both cancer and noncancer health effects (USEPA, 1997a).

In addition to MTBE, a number of other ethers and alcohols have been used as gasoline oxygenates. These include di-isopropyl ether (DIPE), ethyl *tert*-butyl ether (ETBE), *tert*-amyl methyl ether (TAME), *tert*-butyl alcohol (TBA), and ethanol. USEPA has not developed MCLs for these oxygenates, and discussion of their health effects is beyond the scope of this chapter. Many states have developed standards or guidance values for drinking water and/or groundwater for one or more of these contaminants (Delta Environmental Consultants, 2005b).

Pesticides

Pesticides include a wide range of compounds such as insecticides, herbicides, nematicides, rodenticides, and fungicides. These categories are divided into chemical families.

Chemicals within the same family typically result in related types of health effects, as discussed below.

Pesticides display a range of solubility in water, and many bind tightly to organic material in soil particles, slowing entry into surface or groundwater. Many of these compounds are chlorinated, increasing their persistence for agricultural uses as well as in the general environment. Fumigant pesticides are relatively water soluble, resulting in contamination of drinking water sources. In contrast, the lipophilic organochlorine pesticides, such as lindane, chlordane, and DDT (and its degradates/metabolites, DDD and DDE), are rarely seen above trace levels in public water systems, especially since their uses were cancelled. However, they are often found in river sediments, and they bioaccumulate in lipids (fat) through the food chain.

For the great majority of people, food represents a greater source of exposure to pesticides than drinking water (MacIntosh et al., 1996). Pesticides can reach drinking water concentrations of major health concern after accidental spills or in wells located close to application or mixing sites. Occurrence of pesticides in water sources varies with land use (e.g., low in areas with undeveloped land), season, and periods of high runoff. The U.S. Geological Survey (USGS) has conducted several surveys of water quality that have included pesticides. The National Ambient Water Quality Assessment (NAWQA) survey, conducted in 1992 through 2001, detected pesticides in 55 to 60 percent of shallow wells (mostly non-potable) and almost 100 percent of streams in agricultural and urban areas (Gilliom et al., 2006). Most detections were in the low part per trillion (0.001 μg/L) range. Two triazine and three acetanilide herbicides (see below) were associated with agricultural land use, while another triazine, two substituted-urea herbicides, 2,4-D (high-use lawn herbicide), and organophosphate and carbamate insecticides were found in urban settings. Health-based levels were exceeded for one or more pesticide in 1 percent of shallow wells in agricultural areas and 5 percent in urban areas and in 7 to 10 percent of streams. The NAWQA Source Water Quality Assessment, which sampled during 2002–2005, detected a variety of pesticides in streams and finished water. Maximum concentrations of two triazines, one acetanilide, and one substituted-urea herbicide were close to or exceeded levels of health concern in some samples (Kingsbury, Delzer, and Hopple, 2008). Data from the USEPA Unregulated Contaminant Monitoring Rule indicate that pesticides were detected at a point of entry in less than 1 percent of public drinking water systems and were detected above an MCL or Health Advisory at a point of entry in less than 0.1 percent of public water systems. Most detections above a Health Advisory were of the herbicide dacthal or its metabolites (USEPA, 2001e, 2008g).

Cancellation or severe restriction of use during the 1970s and 1980s has reduced the levels of some pesticides in the environment (USGS, 2006). These pesticides include the herbicide 2,4,5-TP (Silvex); certain fumigant nematicides (e.g., dibromochloropropane (DBCP), 1,2-dichloropropane, 1,2,3-trichloropropane, and ethylene dibromide [EDB]); and most organochlorine insecticides (chlordane, DDT, dieldrin, endrin, heptachlor, lindane, and toxaphene). Nevertheless, in monitoring done under the Unregulated Contaminant Monitoring Rule, some were found at or above levels of potential health concern.

The health effects associated with pesticides vary among the specific chemicals. Effects can be crudely differentiated into acute effects that occur from short-term exposure to high levels and chronic effects from long-term, low-level exposure. The former can include liver and kidney damage; major interference with nervous, immune, and reproductive system functions; and birth defects. Less severe effects on the nervous system are frequently expressed as "nonspecific" symptoms such as dizziness, nausea, and fatigue. Effects that may occur from long-term, low-level exposures include increased risk of birth defects and cancer risk. Uncertainties exist regarding the effects of exposure to mixtures of pesticides. Most pesticide degradates and metabolites are less toxic than their parent pesticide, but some have similar or greater toxicities.

For most pesticides, estimates of the risks of cancer and birth defects are based on studies in laboratory animals. However, epidemiologic studies have found increased incidence of soft-tissue sarcoma and lymphoid cancers among farmers (reviewed by Dich et al., 1997). These effects may be related to the past presence of dioxin and other organochlorine chemicals in some pesticides as by-products of manufacturing (Kogevinas et al., 1997). Some studies have linked DDT and its metabolites to breast cancer, but the data are mostly negative (reviewed by Calle et al., 2002). Childhood cancers have also been linked to general pesticide exposure, particularly parental exposure during pregnancy and nursing (Infante-Rivard and Weichenthal, 2007), but these studies are not considered conclusive. The National Cancer Institute's Agricultural Health Study (2008) (recently updated by Blair and Beane Freeman (2009)) of agricultural pesticide applicators provides more recent evidence of a link between cancer and high-intensity exposure to certain pesticides. Prostate cancer was associated specifically with organochlorine pesticides, and an increased incidence among those with family histories of prostate cancer was also associated with exposure to a wider variety of pesticides, particularly certain organophosphate insecticides (Alavanja et al., 2003).

Epidemiological studies of reproductive and developmental effects of pesticides in current use, including organophosphates, carbamates, pyrethroids, ethylenebisdithiocarbamates, and chlorophenoxy herbicides, were reviewed by Bjørling-Poulsen et al. (2008). Many of these studies suggest that maternal exposure to pesticides is linked to adverse reproductive outcome, but exposures to mixtures of pesticides and the potential for recall bias preclude making definitive linkages between specific chemicals and effects.

Some pesticides affect endocrine systems, such as the effect of the fungicide vinclozolin on the sex steroid system and the effect of the triazine and acetanilide herbicides on the thyroid (Kavlock, 2001). Considerable research is being conducted on the relationship between these endocrine effects and cancer and reproductive toxicity. There is strong evidence that environmental exposures to pesticides can cause endocrine effects in nontarget wildlife species.

Families of pesticides typically share common toxicological mechanisms. USEPA documents, which address cumulative risk assessment from pesticides in general and specifically for organophosphates, N-methyl carbamates, triazines, and chloroacetanilides, are found at http://www.epa.gov/oppsrrd1/cumulative/.

Herbicides

Herbicides are not highly acutely toxic to mammals including humans, but some may be weakly carcinogenic. Determination of the carcinogenic potential of chlorinated aromatic herbicides was complicated by contamination due to the coproduction of carcinogenic polychlorinated aromatics during earlier manufacturing. Most notably, 2,4,5-trichlorophenoxyacetic acid (a component of Agent Orange, used as a Vietnam War defoliant) was contaminated with dioxin (2,3,7,8-TCDD). 2,4,5-T was the first of the herbicides to be discontinued. Dacthal was also previously similarly contaminated but is still in use since these contaminants are not present during current production (see below). These contaminant by-products are not mobile in soils and are generally not found in drinking water.

Triazine Herbicides (Atrazine, Metribuzin, Prometon, Simazine). USEPA (2006d) concluded that "certain triazine pesticides have been identified by a common mechanism group, including atrazine, simazine, propazine, and their chlorinated degradates desethyl-s-atrazine (DEA), desisopropyl-s-atrazine (DIA), and diaminochlorotriazine (DACT). These triazine pesticides share a common mechanism of toxicity. In laboratory studies in rats, at experimental dose levels higher than those encountered in the environment, the triazines have the ability to potentially cause neuroendocrine developmental and reproductive effects that may be relevant to humans. Specifically, these pesticides may disrupt part of the central nervous system (the hypothalamic-pituitary-gonadal (HPG) axis). This endocrine disruptive effect causes increased mammary tumors in Sprague-Dawley rats but may not be relevant to the potential risk of human breast cancer. Tumors were not found in mice treated with these chemicals.

Other effects have also been observed for specific triazine herbicides, as noted below.

1. *Atrazine.* Specific organ toxicity or teratogenicity was not observed in a variety of rodent studies, but cardiac pathologies were seen in dogs (ATSDR, 2003). Atrazine appears to be weakly to moderately mutagenic and genotoxic (ATSDR, 2003), and there is some evidence linking it with lymphomas among those with occupational exposures (Rusiecki et al., 2004), but there is little data on carcinogenicity at environmentally relevant doses. There is also some epidemiological evidence linking atrazine at environmentally relevant levels to reduced fetal growth and certain birth defects (Villaneueva et al., 2005; Mattix et al., 2007; Ochoa-Acuna et al., 2009; Waller et al., 2010). The NOAEL from a rat two-generation reproduction study is the basis for the USEPA DWEL of 200 μg/L.
2. *Metribuzin.* Effects seen after long-term exposure include thyroid enlargement in rats, kidney and liver enlargement in mice, but no apparent occurred toxicity in dogs (USEPA, 2008i). No teratogenicity was observed. There is little evidence for mutagenicity or carcinogenicity. The Lifetime Health Advisory is 70 μg/L.
3. *Prometon.* This chemical appears to have relatively low toxicity, but due to "low confidence" in the published studies, USEPA is conducting further investigations (USEPA, 2008i). There was little evidence of organ toxicity or teratogenicity and no data on mutagenicity and carcinogenicity. The Lifetime Health Advisory level is 100 μg/L.
4. *Simazine.* In a two-year rat feeding study, decreases in weight gain and hematologic parameters were observed (USEPA, 2008i). A recent one-year dog feeding study also identified a hematological effect. No adverse effects on female reproductive capacity were observed after a two-generation rat feeding study, and no developmental effects, aside from delayed bone formation, have been seen. Mutagenicity and genotoxicity tests yielded mixed results. The MCL is 4 μg/L.

Acetanilides (Alachlor and Metolachlor)

1. *Alachlor.* Chronic effects noted in laboratory feeding studies include hepatotoxicity and hemolytic anemia in dogs and hepatotoxicity, nephritis, retinal degeneration, and degenerative changes in a variety of other organs in rats (USEPA, 2008i). Few developmental or teratogenic effects have been observed. Chronic feeding studies using mice have demonstrated that alachlor causes lung tumors, and studies in rats have demonstrated stomach, thyroid, and nasal tumors (Heydens, Lamb, and Wilson, 2001). Stomach tumors appear to arise from nonthreshold effects on the gastric mucosa. Thyroid tumors were probably mediated by a thyroidal-pituitary homeostatic feedback threshold mechanism. Nasal tumors appear to be related to a metabolite produced at high levels in the rat but not in mice or primates. The Agricultural Health Study reported an increased incidence of lymphohematopoietic cancers, such as leukemias and multiple myelomas, among agricultural pesticide users (Lee et al., 2004). Alachlor and its metabolites are moderately genotoxic and weakly mutagenic. The USEPA (2006c) determined the drinking water concentration resulting in a 1 in 10,000 lifetime cancer risk as 40 μg/L and established the MCL at 2 μg/L.
2. *Metolachlor.* Long-term exposure to high doses in rats caused liver and kidney enlargement and atrophy of seminal vesicle and testes in males. Reproductive and teratogenic effects were not seen. Evidence of increased liver tumors was seen in female rats, but male rats and mice showed no increase in tumors and mutagenicity tests were negative (USEPA, 2008h). An MCL has not been set, and the Lifetime Health Advisory is 700 μg/L.

Chlorophenoxyacetic Acids 2,4-D (2,4-Dichlorophenoxyacetic Acid). In a 90-day feeding study, 2,4-D caused hematologic (reduced hemoglobin and red blood cells), hepatic, and renal toxicity in rats, although mice were less sensitive (USEPA, 2008h). There was no reproductive or teratogenic toxicity. Tests in most microbial systems showed no mutagenic activity, and *in vivo* animal assays have also been largely negative. Teratogenicity data are largely negative, and no dose-dependent tumor formation has been reported. There has been considerable epidemiologic analysis, and the results have been mixed though mostly negative (Garabrant and Philbert, 2002). Despite its very widespread use of 2,4-D, co-exposures to other agricultural chemicals prevent making definitive associations with any effects. The MCL is 70 μg/L. Metabolites of 2,4-D in drinking water may also be of concern, but this has not yet been addressed in regulation.

Other Herbicides

1. *Dacthal (DCPA).* Chronic feeding studies in rats have resulted in lung, liver, kidney, and thyroid pathology as well as alterations in thyroid hormone levels (USEPA, 2008j). There were no adverse reproductive or teratological effects. The thyroid tumors may be caused by a mechanism for which a threshold exists involving stimulation of the thyroid through the thyroid-pituitary feedback system. The liver cancer data are difficult to interpret due in part to contamination of the test compound by dioxin and hexachlorobenzene, but the lung and kidney tumors are not explained by the contaminants. Little information is available on toxicity of dacthal's metabolites, but they appear to be much less toxic than the parent compound. Dacthal was classified as a possible human carcinogen by USEPA, and the Lifetime Health Advisory is 70 μg/L.
2. *Diuron.* Diuron has low acute toxicity but causes red blood cell damage (hemolytic anemia) and compensatory regeneration and developmental effects on bone growth at high doses. It has low mutagenic and genotoxic potential but caused bladder tumors in rats and mammary carcinomas in female mice given high doses in chronic studies (USEPA, 2003a). It is classified as likely to be carcinogenic to humans by USEPA, and the drinking water concentration at the 1 in 10,000 lifetime is given as 200 μg/L. Its major aquatic metabolite resembles a related herbicide that is also considered a likely carcinogen.
3. *Glyphosate.* Glyphosate is widely used to control weeds, grasses, and other plants. It appears to be a relatively low-toxicity herbicide in terms of human health but can affect liver, kidney, and reproductive health with long-term exposure at high levels. The MCLG and MCL have been set by USEPA at 700 μg/L.

Insecticides

Insecticide chemical families of concern, based on toxicity and presence in source water, include organophosphates (e.g., chlorpyrifos and diazinon), which are related to chemical warfare nerve agents, and carbamates (e.g., carbaryl, carbofuran, and oxamyl). These chemicals inhibit the neuromuscular junction enzyme, acetylcholinesterase, in insects. While use of these insecticides has been increasingly restricted, many are still used. Because of biochemical similarities between insect and vertebrate nervous systems, these classes of insecticides have a high potential to cause acute and medium-term health effects in sensitive individuals. In humans and other animals, acetylcholinesterase inhibition causes a buildup of acetylcholine, affecting both the skeletal muscles and the nerves and smooth muscles of the autonomic nervous system. Effects include weakness, sweating, increased intestinal motility (diarrhea-like symptom), salivation, tears, and constriction of the pupils. The effects end within hours after termination of exposure. Exposure to these insecticides can be monitored by measuring inhibition of a related cholinesterase in blood. In general, the levels found in public drinking water probably do not constitute an acute threat to health. However, recent human and laboratory animal studies indicate that carbamate and organophosphate insecticides, especially chlorpyrifos, may also cause neurodevelopmental effects (Eskenazi et al., 2007).

Carbamates (Aldicarb, Aldicarb Sulfoxide, and Aldicarb Sulfone). Aldicarb is the most acutely neurotoxic of any of the widely used pesticides, and its use has been cancelled. Aldicarb has not been shown to be a mutagen or a carcinogen in animal studies, but its use was associated with increased colon cancer in the Agricultural Health Study (Lee et al., 2007). The MCLs for aldicarb and its metabolites aldicarb sulfoxide and aldicarb sulfone are 3, 4, and 2 μg/L, respectively. Because these related compounds share a common mechanism of action, their toxicity in a mixture is considered to be additive.

In addition to neurotoxic effects, laboratory animals exposed to carbofuran exhibited testicular degeneration. Based on limited evidence, it does not appear to be carcinogenic. The MCL is 40 μg/L.

Oxamyl is mildly neurotoxic. No histopathological changes were noted in organs of exposed animals, and no reproductive or developmental effects have been seen (USEPA, 1992b). It is classified as a noncarcinogen. The MCL is 200 μg/L.

Organophosphates (Diazinon and Chlorpyrifos). The metabolite diazoxon accounts for the major effects of diazinon, including neurotoxicity. Diazinon is moderately mutagenic but is not carcinogenic. No MCL has been set by USEPA, and the Lifetime Health Advisory is 1 μg/L.

In addition to inhibition of acetylcholinesterase, chlorpyrifos causes neurotoxicity through other mechanisms. It was associated with prostate, lung, and rectal cancer in the Agricultural Health Study (Alavanja et al., 2003; Lee et al., 2007). Limited data from studies of carcinogenicity in laboratory animals are negative, though mutagenicity and genotoxicity tests show mixed results. No MCL has been set by USEPA, and the Lifetime Health Advisory is 2 μg/L.

Organochlorines. A variety of organochlorines were previously used as pesticides because of their neurotoxic effects. Their mode of action appears to involve effects on the neurotransmitter gamma-amino butyric acid and effects on the sodium channel in nerves of the insect nervous system. However, their use has been restricted and/or cancelled due to their environmental persistence. Other concerns with these compounds included development of resistance by insects (e.g., DDT and malarial mosquitoes), toxicity to wildlife (e.g., DDT and malformed bird eggs), and human carcinogenic potential. Several are potent carcinogens. For example, the drinking water concentration resulting in a 1 in 10,0000 lifetime cancer risk for aldrin and dieldrin is estimated as only 0.2 μg/L (USEPA, 2008i).

Soil Fumigants/Nematicides. A variety of chlorinated short alkanes, 1,2-dibromo-3-chloropropane (DBCP), 1,2-dichloropropane, 1,3-dichloropropene, 1,2,3-trichloropropane, and ethylene dibromide (EDB), have been used as fumigant nematicides. Use of several of these has been phased out, as noted above. These chemicals have also been used for other purposes; for example, 1,2,3-trichloropropane was used as an industrial solvent and degreasing agent and is still used as an intermediate in the production of other chemicals. DBCP caused reproductive toxicity in occupationally exposed men, including decreased or absent sperm production, testicular atrophy, hormonal changes, and decreased proportion of male offspring (ATSDR, 1992). Both DBCP and EDB caused male reproductive toxicity in animal studies (USEPA, 2008i). In general, they are classified as likely carcinogens, causing tumors in multiple organs in animal studies through a mutagenic mode of action (NTP, 2005a; USEPA, 2006c). In particular, 1,2,3-trichloropropane has been shown to be especially potent in laboratory animals, with the drinking water concentration resulting in a 1 in 10,000 lifetime cancer risk of less than 1 μg/L. The MCLs for these chemicals are 5 μg/L for 1,2-dichloropropane, 0.2 μg/L for DBCP, and 0.05 μg/L for EDB (USEPA, 2006c). There is no MCL for 1,3-dichloropropene or 1,2,3-trichloropropane. 1,2,3-Trichloropropane is considered to be an emerging contaminant based on occurrence and potential health effects. Several states are in the process of developing a drinking water standard for it, and it is currently being considered by USEPA for MCL development through the CCL process.

Fungicides

Although no fungicides regarded as significant health threats are found at notable levels in water, ethylene thiourea (ETU), which is a water-soluble breakdown product of the ethylene bisdithiocarbamate (EBDC) chemical family (e.g., maneb, mancozeb, and zineb) is important. Uses of many of the EBDC fungicides were restricted in the late 1980s due to concerns about residues on food crops and risks from occupational exposure, but their use is still widespread.

ETU caused thyroid toxicity in adult rodents, resulting in reduced levels of thyroid hormone. Developmental toxicity including retarded CNS development may be related to thyroid toxicity (USEPA, 2008i). Liver and thyroid tumors were increased in mice and rats exposed to ETU, but the results of mutagenicity and genotoxicity testing were mixed (reviewed by Dearfield, 1994). Thyroid tumors occurred in the context of thyroid hyperplasia resulting from the increased level of thyroid stimulating hormone production by the pituitary in response to the low level of thyroid hormone. There is no MCL and MCLG for ETU. It is listed as a probable human carcinogen by the USEPA Office of Water, and the drinking water concentration at the 1 in 10,000 cancer risk level is given as 20 μg/L.

Additional and Emerging Drinking Water Contaminants

Endocrine Disruptors. A recent area of concern is the occurrence in water of low levels of a wide variety of environmental contaminants that can affect the endocrine system (Snyder et al., 2008; Diamante-Kandarakis et al., 2009). In particular, environmental contaminants that affect the function of thyroid and steroid hormones are known to interfere with development and reproduction in wildlife, and such effects may also be occurring in humans (Colburn, vom Saal, and Soto, 1993; Porterfield, 1994; Toppari et al., 1996). Excreted steroids of both endogenous and pharmaceutical origin, various pesticides, organochlorines (e.g., dioxin, certain PCB congeners, DDT and metabolites), chemicals found in plastics such as phthalates and bisphenol A, and detergent compounds are among the chemicals of concern, in large part because of their documented effects in laboratory animals and on wildlife. These chemicals can enter waterways through known point sources such as wastewater treatment plants and agricultural runoff, as well as nonpoint sources such as atmospheric deposition and stormwater discharge. An AwwaRF (2003) project demonstrated that most surface waters and many groundwater sources exhibited measurable estrogenic activity, although most finished drinking waters did not exhibit activity.

Pharmaceuticals and Personal Care Products. A related area of concern is the plethora of chemicals found at trace concentrations in surface waters, stemming from pharmaceutical, food, and household products (USGS, 2002; Focazio et al., 2008). These chemicals enter waterways primarily through incomplete removal in wastewater treatment plants. Medications enter wastewater both by excretion from the body and by disposal of unused medications by flushing down the toilet. Available occurrence studies and risk assessments indicate that medicines occur in finished drinking water at levels well below therapeutic doses (Focazio et al., 2008; Schulman et al., 2002; Schwab et al., 2005; Webb et al., 2003).

Mixtures of Organic Chemicals. Groundwater studies have also found trace levels of a variety of organic chemicals at low levels, some of which may be only tentatively identified (NJDEP, 2003). The potential for human health effects from exposure to low levels of diverse chemical mixtures has yet to be determined (Trubo, 2005). A potential approach to address such mixtures of contaminants that lack specific health effects information is to institute treatment technologies, such as activated carbon, which will remove organic chemicals in general.

Perfluorinated Chemicals. The organic chemicals classified as persistent, bioaccumulative, and toxic (PBT) by USEPA are well known for their toxicity to humans and wildlife, persistence in the environment, and bioaccumulation through the food chain. These chemicals, which include dioxins and furans, polychlorinated biphenyls, and pesticides such as chlordane, DDT and related compounds, and aldrin/dieldrin, as well as some of more recent concern such as the flame retardant polybrominated diphenyl ethers (PBDEs), are not primarily significant as drinking water contaminants. This is because these chemicals tend to have high octanol/water partition coefficients, and thus a high affinity for sediments and a low solubility in water. The primary source of human exposure to these compounds is through the diet, from lipids in fish, meat, and dairy products, and they tend to concentrate in fat in the body.

Recently, another group of persistent and toxic chemicals that are of concern in drinking water has emerged. These are perfluorochemicals (PFCs) such as perfluorooctanoic acid (PFOA, C8), perfluorooctane sulfonate (PFOS), perfluorobutanoic acid (PFBA), perfluoropentanoic acid (PFPeA), perfluorohexane sulfonate (PFHxS), perfluorohexanoic acid (PFHxA), and other related chemicals. In contrast to the PBT chemicals mentioned above, PFCs are highly water soluble. They are used in the production of water- and stain-resistant products including cookware and clothing, as well as in fire-fighting foams. They also arise from their breakdown in the environment and in the body of related chemicals, fluorotelomer alcohols, which are used in consumer products such as grease-proof food wrappers and stain-resistant carpet treatments. They do not degrade in the environment and have been found at low concentrations in drinking water not known to be impacted by discharge from a facilities that manufacture or use them (Mak et al., 2009; Post et al., 2009). The use of PFOS has been phased out, and the use of PFOA is currently being phased out by the manufacturers, and efforts are ongoing to replace them with less toxic shorter chain compounds in the future.

The environmental occurrence and toxicology of several PFCs was reviewed by Lau et al. (2007). The discussion below focuses on PFOA, the PFC that has received the greatest attention as a drinking water contaminant. PFOA is found in the serum of almost all individuals, at a mean concentration in the United States of about 4 μg/L, with a human serum half-life of several years (Lau et al., 2007). In communities with water supplies contaminated with PFOA, the median serum concentration contributed from drinking water is about 100 times the concentration in the drinking water source (Emmett et al., 2006; Post et al., 2009). Thus, low levels in drinking water may add substantially to the general population's exposure from the diet, consumer products, and other sources. For example, a drinking water concentration of 0.02 μg/L would be expected to increase serum concentrations by about 2 μg/L, an increase of 50 percent over the background serum concentration of about 4 μg/L.

In animal studies, PFOA causes developmental effects and adverse effects on the liver, immune system, and lipid metabolism (Lau et al., 2007). Increased tumor incidence was seen in chronic rat studies, and it has been classified as a likely human carcinogen by the USEPA Science Advisory Board (USEPA, 2006e). Recent studies in communities whose drinking water was contaminated by industrial discharge have found associations between serum levels of PFOA and elevated cholesterol and uric acid, changes in immune function, and possibly other biological parameters (Fletcher, Steenland, and Savitz, 2009; Steenland, Fletcher, and Savitz, 2008, 2009; C8 Health Project, 2009).

PFOA is one of four contaminants listed on the USEPA CCL3, which the USEPA Science Advisory Board recommended as a priority for consideration for regulation based on available data on toxicity, occurrence, and treatability (USEPA, 2009b).

Nanoparticles. A final category of emerging contaminants is nanoparticles or nanomaterials, which are defined as particles or materials with at least one dimension smaller than 100 nm. The toxicological effects of nanoparticles are believed to depend on their physicochemical parameters such as particle size and size distribution, agglomeration state, shape, crystal structure, chemical composition, surface area, surface chemistry, surface charge, and porosity, rather than solely on their chemical composition (Oberdörster et al., 2005). Nanoparticles are currently used in many consumer products, including clothing, laundry detergents, and washing machines and in drug delivery systems for pharmaceuticals. Their possible use in groundwater remediation and drinking water treatment is being investigated. These and other applications could potentially lead to their presence in drinking water. Future research will provide information on the occurrence and potential for human health risks from nanoparticles in drinking water; currently, little is known about these questions (Boxall, Tiede, and Chaudry, 2007).

Chemicals in Treatment Additives, Linings, and Coatings

In the United States, NSF International (www.nsf.org) develops standards for all types of materials and equipment used in drinking water treatment, distribution systems, and storage facilities. These standards are based on prevention of leaching at unsafe levels of these chemicals into drinking water. Chemicals evaluated by NSF International include solvents, monomers, and heavy metals. A few examples of additives, linings, and coatings are given here.

Acrylamide. The most important source of drinking water contamination by acrylamide is the use of polyacrylamide flocculants containing residual levels of acrylamide monomer. They are also used as grouting agents in the construction of drinking water reservoirs and wells (WHO, 2003). It has recently been discovered that high levels of acrylamide are formed during the frying or baking of a variety of foods, particularly those containing carbohydrates, such as cereals, french fried potatoes, potato chips, breads, pastries, and coffee (WHO, 2005).

Acrylamide's principal toxic effect is peripheral neuropathy (damage to the nerves in the arms or legs). Neurotoxicity is seen in animal studies and in humans exposed via the dermal, oral, or inhalation routes. Acrylamide is metabolized to a reactive compound that binds to DNA and hemoglobin. DNA adducts in many organs, and hemoglobin adducts are detected in animals given acrylamide or its metabolite. Acrylamide in drinking water was carcinogenic to rats in two studies, causing tumors of the thyroid, scrotum, adrenal and mammary glands, central nervous system, oral cavity, uterus, clitoris, and pituitary (WHO, 2005). In reproduction studies, male rodents showed reduced fertility, dominant lethal effects, and adverse effects on sperm count and morphology.

Acrylamide is regulated by a treatment technique in lieu of an MCL. When polyacrylamide is used in drinking water systems, the combination of polymer dose and monomer level may not exceed 0.05 percent monomer dosed at 1 mg/L polymer (USEPA, 1991a). In its recent review of its drinking water standards, USEPA (2010) concluded that improvements in manufacturing capabilities have reduced the residual monomer content in acrylamide-based polymeric coagulants aids, and these changes would support a decrease in the treatment technique for acrylamide.

Epichlorohydrin. Epichlorohydrin (ECH) is a halogenated alkyl epoxide. It is used as a raw material in the manufacture of a variety of plastics, including cationic polymers used as coagulants in potable water treatment and as a solvent for lacquers sometimes used to coat the interiors of water tanks and pipes.

Epichlorohydrin is mutagenic in bacterial and mammalian cells and is carcinogenic, causing forestomach tumors when given orally and nasal tumors by inhalation (IARC, 1999). Epichlorohydrin is regulated by a treatment technique in lieu of an MCL. When epichlorohydrin is used in drinking water systems, the combination of dose and monomer level may not exceed 0.01 percent monomer dosed at 20 mg/L polymer (USEPA, 1991a). In its recent review of its drinking water standards, USEPA (2010) concluded that improvements in manufacturing capabilities have reduced the residual monomer content in epichlorohydrin-based polymeric coagulants aids, and these changes would support a decrease in the treatment technique for epichlorohydrin.

Polynuclear Aromatic Hydrocarbons. Polynuclear (or polycyclic) aromatic hydrocarbons (PAHs) are a diverse class of compounds consisting of substituted and unsubstituted polycyclic and heterocyclic aromatic rings. They are formed in the pyrolysis of naturally occurring hydrocarbons and found as constituents of coal, coal tar, petroleum, fuel combustion products, and cigarette smoke (National Academy of Sciences Safe Drinking Water Committee, 1982). Polynuclear aromatic hydrocarbons can enter drinking water by leaching from tar or asphalt linings of distribution pipelines, but their solubility is limited.

PAHs vary in their toxicity, and some are carcinogenic while others are not. Benzo(a)pyrene is the most thoroughly studied PAH. It causes cancer in animals exposed orally or by inhalation. Benzo(a)pyrene is genotoxic and is metabolized to a reactive compound that forms adducts with DNA (California EPA, 1997; National Academy of Sciences Safe Drinking Water Committee, 1982). The MCL for benzo(a)pyrene is 0.2 μg/L.

DISINFECTANTS AND DISINFECTION BY-PRODUCTS

General Background

The disinfection of drinking water, although reducing the risk of waterborne disease, creates other potential risks through the formation of compounds known as disinfection by-products (DBPs) during the disinfection process. The occurrence, genotoxicity, and carcinogenicity of these compounds were recently comprehensively reviewed by Richardson et al. (2007). Organic DBPs are formed through the reaction of disinfectants with organic matter in source waters; their presence was first recognized in the 1970s when trihalomethanes (THMs) were detected in chlorinated drinking water. At this time, more than 600 DBPs have been identified, including haloacetic acids (HAAs), haloacetonitriles, haloketones, haloaldehydes, chloropicrin, cyanogen chloride, MX and related compounds, chlorophenols, nitrosamines, and others. These compounds are usually present in drinking water at relatively low concentrations, ranging from nanograms per liter to approximately 100 μg/L. For many of these compounds, occurrence data and health effects information are limited or unavailable. In addition to the identified DBPs, a large portion of the total organic halide formed during chlorination and the assimilable organic carbon formed during ozonation remains uncharacterized. The formation and control of DBPs is covered in Chap. 19.

It is important to recognize that, as with all substances, the human health risk from DBPs in drinking water is a function of both their concentration and their toxic potency. Thus, highly toxic DBPs, such as MX, iodo-trihalomethanes, and others, which are present at very low concentrations, may pose an equal or greater risk than less toxic DBPs present at much higher concentrations. Additionally, the identified and uncharacterized DBPs in the complex mixtures present in drinking water may interact toxicologically in ways that have not yet been characterized. A final consideration relevant to assessing the risks of DBPs in drinking water relates to exposure pathways. Volatile DBPs enter household air, especially with hot water use in showers, and inhalation and dermal exposures can be more significant than ingestion.

Health Basis for DBP Regulation

Under the Stage 1 Disinfectants/Disinfection By-Products Rule finalized by USEPA in 1998 (USEPA, 1998b), public water systems are required to meet limits for 11 DBPs: total trihalomethanes (TTHMs; chloroform, bromodichloromethane, dibromochloromethane, and bromoform), the sum concentration for five HAAs (mono-, di-, and trichloroacetic acids and mono- and dibromoacetic acids), bromate, chlorite and three disinfectants: chlorine, chlorine dioxide, and chloramines. The rule establishes maximum residual disinfectant level goals (MRDLGs) and maximum residual disinfectant levels (MRDLs) for disinfectants. The MRDLGs are nonenforceable goals, analogous to MCLGs and based on potential health effects, without consideration of disinfectant benefits of the chemical. The MRDLs are enforceable standards, analogous to MCLs, which are set as close to the MRDLGs as feasible and which recognize the benefits of adding a disinfectant to water. MCLGs and MCLs are established for DBPs. A RSC factor of 0.8 (80 percent), instead of the default value of 20 percent, was used in setting MRDLGs and MCLGs for noncarcinogenic disinfectants and DBPs, as sufficient data were available to show that food and ambient air are not significant sources of exposure (USEPA, 1994a). As a treatment control technology, USEPA has a coagulation requirement to control DBPs by removing total organic carbon (TOC) as a collective surrogate for DBP precursors—see Chap. 19. In 2006, the Stage 2 D/DBP Rule, which contains additional implementation requirements to ensure that the regulatory limits are not exceeded at specific locations within a drinking water distribution system, was finalized (USEPA, 2006f).

Approaches for Evaluation of Health Effects of DBPs

The primary toxic endpoints of concern from chronic exposure to low levels of DBPs in drinking water are cancer and reproductive/developmental effects. However, chronic carcinogenicity bioassays and/or developmental and reproductive studies in experimental animals have been conducted for only a relatively small number of DBPs. Discussion of the health effects of some of these disinfectants and DBPs follows. Many emerging DBPs have been tested only *in vitro* for genotoxicity and/or cytotoxicity, while for others, no toxicity data exist. These data are summarized in Table 2-4 (from Richardson et al., 2007). Some overall generalizations can be made about the biological activity of DBPs. For analogous DBPs, genotoxicity and cytotoxic potency are generally greatest in iodinated compounds, followed by brominated compounds, with chlorinated compounds the least potent. Similarly, DBPs containing nitrogen are generally more active than their analogues that do not contain nitrogen (Richardson et al., 2007).

As long-term animal testing is not feasible for the vast number of DBPs and mixtures of DBPs that may be present in drinking water, several different approaches have been proposed or applied for evaluation of the health risks of DBPs in drinking water. For example, DBPs were prioritized for further toxicology studies based on mechanism-based structure-activity relationships and genotoxicity data (Woo et al., 2002). An occurrence survey was conducted for the 50 DBPs that received the highest priority ranking in this study, including iodinated THMs, other mono- and di- halomethanes, a haloacid, haloacetonitriles, haloacetates, haloketones, halonitromethanes, haloamides, MX and other halogenated furanones, and carbonyls (Krasner et al., 2006). An approach using exposure assessment modeling, physiologically

TABLE 2-4 Summary of Occurrence, Genotoxicity, and Carcinogenicity of Regulated and Unregulated DBPs

DBP	Occurrence[a]	Genotoxicity[b]	Carcinogenicity
Regulated DBPs			
THMs			
Chloroform	*****	–	+
Bromodichloromethane	****	+	+
Chlorodibromomethane	****	+	+
Bromoform	****	+	+
HAAs			
Chloroacetic acid	***	+	–
Bromoacetic acid	***	+	
Dichloroacetic acid	*****	+	+
Dibromoacetic acid	*****	+	+
Trichloroacetic acid	*****	–	+
Oxyhalides			
Bromate	***	+	+
Chlorite	******	–	–[c]
Unregulated DBPs			
Halonitromethanes			
Chloronitromethane	**	+	
Bromonitromethane	**		+
Dichloronitromethane	**	+	
Dibromonitromethane	***	+	
Bromochloronitromethane	**	+	
Trichloronitromethane (chloropicrin)	****	+	
Bromodichloronitromethane	***	+	
Dibromochloronitromethane	***	+	
Tribromonitromethane	***	+	
Iodo-acids			
Iodoacetic acid	***	+	
Bromoiodoacetic acid	***	+	
(Z)-3-Bromo-3-iodopropenoic acid	**		
(E)-3-Bromo-3-iodopropenoic acid	**		
2-Iodo-3-methylbutenedioic acid	***	+	
Other halo-acids			
Bromochloroacetic acid	****		+
Bromodichloroacetic acid	****		+
Dibromochloroacetic acid	****		+
Tribromoacetic acid	****	+	
Iodo-THMs and other unregulated THMs			
Dichloroiodomethane	***		
Bromochloroioomethane	***		
Dibromoiodomethane	***		
Chlorodiiodomethane	***		

(*Continued*)

TABLE 2-4 Summary of Occurrence, Genotoxicity, and Carcinogenicity of Regulated and Unregulated DBPs (*Continued*)

DBP	Occurrence[a]	Genotoxicity[b]	Carcinogenicity
Iodo-THMs and other unregulated THMs			
Bromodiiodomethane	***		
Iodoform	***	+	
Dichloromethane	***	+	
Bromochloromethane	ND	+	
Dibromomethane	ND/**	+	
MX compounds			
MX	**	+	+
Red-MX	*	+	
Ox-MX	*	+	
EMX	*	+	
ZMX	*	+	
Mucochloric acid	**	+	
BMX-1	**	+	
BMX-2	*	+	
BMX-3	*	+	
BEMX-1	**	+	
BEMX-2	**	+	
BEMX-3	**	+	
Haloamides			
Chloroacetamide	***	+	
Bromoacetamide	***	+	
Iodoacetamide		+	
Dichloroacetamide	***	+	
Bromochloroacetamide	***	+	
Dibromoacetamide	***	+	
Bromoiodoacetamide	***	+	
Trichloroacetamide	***	+	
Bromodichloracetamide	***	+	
Dibromochloracetamide	***	+	
Tribromoacetamide	***	+	
Diiodoacetamide		+	
Chloroiodoacetamid		+	
Haloacetonitriles			
Chloroacetonitrile	***	+	
Bromoacetonitrile	***	+	
Iodoacetonitrile		+	
Dichloroacetonitrile	***	+	
Bromochloroacetonitrile	***	+	
Dibromoacetonitrile	***	+	On test
Trichloroacetonitrile	***	+	
Bromodichloroacetonitrile	***		
Dibromochloroacetonitrile	***		
Tribromoacetonitrile	***		
Halopyrroles			
2,3,5-Tribromopyrrole	**	+	

(*Continued*)

TABLE 2-4 Summary of Occurrence, Genotoxicity, and Carcinogenicity of Regulated and Unregulated DBPs (*Continued*)

DBP	Occurrence[a]	Genotoxicity[b]	Carcinogenicity
Nitrosamines			
NDMA	**	+	+
N-Nitrosopyrrolidine	*	+	+
N-Nitrosopiperidine	*	+	+
N-Nitrosopiperidine	*	+	+
N-Nitrosodiphenylamine	*	+	+
Aldehydes			
Formaldehyde	***	+	+
Acetaldehyde	***	+	+
Chloroacetaldehyde	***		+
Dichloroacetaldehyde	***		
Bromochloroacetaldehyde	***		
Trichloroacetaldehyde (chloral hydrate)	****	+	+
Tribromoacetaldehyde	***		
Other DBPs			
Chlorate	******	+	+

[a] Key to occurrence symbols: *low-ng/L levels; **ng/L to sub-ug/L levels; ***sub- to low-ug/L levels; ****low-ug/L levels; *****low- to mid- ug/L levels; ******high ug/L levels; entries left blank have no occurrence data available; bromine-containing DBPs formed only when source waters contain natural bromide (occurrence lower than shown if low bromide levels in source waters).

[b] Symbols represent weight of evidence for the genotoxicity data. In general, where a compound was genotoxic in several studies in the same assay or was genotoxic in several different assays, it was declared "+" in the table even if the compound was negative in other assays.

[c] Based on 85-week studies.

Source: from Richardson, S.D., Plewa, M.J., Wagner, E.D., Schoeny, R., and Demarini, D.M., "Occurrence, genotoxicity, and carcinogenicity of regulated and emerging disinfection by-products in drinking water: a review and roadmap for research," Mutation Research, vol. 636, pp. 178–242, Copyright (2007), with permission from Elsevier.

based pharmacokinetic modeling, and cumulative relative potency factors has been proposed to assess the risks of multiroute exposures to mixtures of DBPs (Teuschler et al., 2004). Another approach involves studying the toxicity of the four regulated THMs individually, in binary combinations, and as the mixture of all four compounds (Simmons et al., 2004).

Additionally, the toxicity of undefined complex mixtures containing both known and uncharacterized DBPs may be assessed through studies of concentrates of finished drinking water. For example, the Four-Lab Study was a multicomponent study of the mixture of DBPs present in surface water treated either by chlorination or ozonation followed by posttreatment chlorination (Simmons et al., 2008). The finished water was concentrated approximately 130-fold by reverse osmosis. This concentrate was tested for mutagenicity in *Salmonella* (Claxton et al., 2008), cytotoxicity and effects on gene expression in cultured rat hepatocytes (Crosby et al., 2008), and developmental effects in rats given the concentrates as their drinking water on days 6 through 16 of gestation (Narotsky et al., 2008). In these studies, mutagenicity was similar to that seen in previous studies of drinking water concentrates; differences in gene expression were seen in concentrates of chlorinated and ozonated waters and no adverse developmental effects were seen in the offspring of rats dosed during pregnancy.

In addition to toxicology studies on the effects of disinfectants and DBPs in experimental animals, epidemiology studies have evaluated the association of exposure to chlorinated surface water with several adverse outcomes: cancer, cardiovascular disease, and adverse reproductive outcomes, including birth defects. A number of studies have found small increases in bladder, colon, and rectal cancers, linked with duration of exposure and volume of water consumed or with noningestion exposures to water such as showering and swimming. However, other studies have observed no association (Richardson et al., 2007). A pooled analysis of several recent well-designed case-control studies has also observed a small, but statistically significant, dose-related, elevated risk of bladder cancer after adjustment for confounding factors (Villanueva et al., 2004). Increased incidence of spontaneous abortions, decreased birth weight, prematurity, intrauterine growth retardation, and oral cleft and neural tube defects have also been reported with exposure to DBPs, while other studies were negative for these effects (Rice et al., 2008). Thus, there is still considerable debate in the scientific community on the significance of these findings.

Discussion of the health effects of some important disinfectants and DBPs follows. Because of space limitations, it is not possible to include discussions of all categories of DBPs that have been detected in finished drinking water.

Disinfectants and Inorganic DBPs

Chlorine. The forms of chlorine used in water treatment and its chemistry are covered in Chap. 17.

No evidence of reproductive or developmental effects have been reported for chlorine in drinking water, and no systemic effects were observed in rodents following oral exposure to chlorine as hypochlorite over a two-year period (USEPA, 1994b, 1997b). Chlorine in water was not carcinogenic in chronic rodent studies. Chlorinated water has been shown to be mutagenic to bacterial and mammalian cells, but assessment of the mutagenic potential of chlorine is confounded by the reactive nature of chlorine and by the formation of mutagenic reaction products during *in vitro* testing. The MRDLG and an enforceable MRDL of 4.0 mg/L for residual free chlorine in water are based on the NOAEL in a chronic rat drinking water study.

Chloramines. The chemistry of chloramination and discussion of its use are covered in Chap. 17. Inorganic chloramines can occur as mono-, di-, and trichloramine. The form that dominates depends on dosing of chlorine and ammonia and on pH—see Chap. 17—but it is common to find monochloramine as the dominant species for water plants practicing chloramination. It is desirable to minimize the formation of di- and trichloramines, since they produce an undesirable odor.

Human health effects of monochloramine in drinking water have been observed only in hemodialysis patients. Chloramines in dialysis water cause oxidation of hemoglobin to methemoglobin and denaturation of hemoglobin. For this reason, it is a good practice for water treatment utilities to notify dialysis centers when using chloramination. Monochloramine in drinking water at concentrations up to 24 mg/L (short term) and 5 mg/L (for 12 weeks) caused no effects in healthy volunteers (USEPA, 1994c, 1997b). Two lifetime rodent studies showed minimal toxicity, such as decreased body and organ weights, along with some liver effects (weight changes, hypertrophy, and changes in chromatid patterns). These effects appear to be related to decreased water consumption, due to the unpalatability of higher concentrations of chloramines to the test animals, and thus may not be indicative of toxicity of the chloramines being tested (USEPA, 1994c, 1997b). The MRDLG and MRDL are 4.0 mg/L, measured as chlorine residual in water based on the NOAEL in a chronic drinking water study in rats (USEPA, 1997b, 1998b).

Chlorine Dioxide, Chlorite, and Chlorate. Chlorine dioxide (ClO_2) is fairly unstable and dissociates rapidly into chlorite and chloride in water (USEPA, 1994d). The primary concerns with using chlorine dioxide as a disinfectant are the toxic effects attributed to residual chlorite and chlorate. Because chlorine dioxide converts to chlorite *in vivo*, health effects attributed to chlorine dioxide are assumed to be the same as for chlorite. The health effects of chlorine dioxide, chlorite, and chlorate have each been evaluated separately in humans and in laboratory animals.

Chlorine dioxide. The major toxic effects seen in animals exposed to chlorine dioxide and its by-products, chlorite and chlorate, in drinking water are oxidative damage to red blood cells, decreased thyroxine levels, and delayed neurodevelopment from exposures both after birth and in a two-generation study (USEPA, 1994d). Studies evaluating developmental or reproductive effects have found decreases in the number of implants and live fetuses per dam in female rats given chlorine dioxide in drinking water before mating and during pregnancy. Mutagenicity tests in mice and bacterial systems were negative (USEPA, 1994d, 1997b). No studies on the carcinogenic properties of orally ingested ClO_2 are available, but exposure to concentrates of water treated with ClO_2 did not increase the incidence of lung tumors in mice, nor was any initiating activity observed in mouse skin or rat liver bioassays (USEPA, 1994d). Chlorine dioxide is considered a noncarcinogen (Group D). The MRDLG and an MRDL of 0.8 mg/L are based on decreased thyroxine and behavioral effects in rat pups exposed during gestation and lactation (USEPA, 1994d, 1998b).

Chlorite. Chlorite ion is relatively stable and degrades slowly to chloride (USEPA, 1994d). Rats exposed in drinking water had decreased red blood cells and hemoglobin, and these effects are thought to be due to oxidative stress from glutathione depletion. In several studies, chlorite was associated with a decrease in the growth rate of rat pups between birth and weaning. Behavioral effects indicating delayed neurodevelopment are seen in rat pups exposed perinatally to chlorite (USEPA, 1994d, 1998c, 1998d). Mutagenicity testing of chlorite in mouse assays was negative. As with chlorine dioxide, no clear tumorigenic activity has been observed in animals given oral doses of chlorite, and it is considered to be a noncarcinogen (Group D) (USEPA, 1994d). The MCLG of 0.8 mg/L is based on behavioral effects in rat pups exposed prenatally and after birth. The MCL is 1.0 mg/L (USEPA, 1998d).

Chlorate. Chlorate occurs as a stable by-product in drinking water disinfected with chlorine dioxide. Incidents of chlorate poisoning were reported during its previous use as an herbicide. Effects in people, who consumed large amounts included methemoglobinemia, hemolysis, and renal failure, with the lowest fatal dose reported as 15 g (USEPA, 1994d). Data were considered inadequate to develop an MCLG for chlorate in the USEPA Stage 1 D/DBP Rule (USEPA, 1998b). More recently, a chronic drinking water study and mutagenicity testing were conducted by the NTP (2005b). Effects seen in this study included thyroid follicular cell hypertrophy in both sexes of rats and female mice, bone marrow hyperplasia in male rats and female mice, and hematopoietic cell proliferation in male rats. Thyroid tumors were increased in both sexes of rats, and pancreatic tumors were increased in female mice. Chlorate was not mutagenic in several bacterial strains and did not increase micronuclei in red blood cells of mice.

Bromine. Bromine has been used in swimming pools for disinfection and in cooling towers, but it is not used in drinking water. Bromine can, however, be found in drinking water produced through desalination of seawater. It is formed by the reaction between bromide in seawater and free chlorine, which is often used as a biocide in the intake and pretreatment systems of seawater desalination plants. In water at pH 8.6, bromine exists primarily as hypobromous acid (HOBr). In this uncharged form, it may pass through reverse osmosis membranes and be present in the drinking water that is produced (NRC, 2008). Below pH 6, Br_2, Br_3, bromine chloride, and other halide complexes occur, and in the presence of ammonia or organic amines, bromoamines are formed.

Most health effects information is on bromide salts because of their pharmaceutical use (National Academy of Sciences Safe Drinking Water Committee, 1980). In humans and/or experimental animals, sedation, GI effects, and thyroid hyperplasia may occur from high doses of bromide salts.

Iodine. Iodine has been used to disinfect drinking water in field situations, such as camping and the military, and in swimming pools. It can occur in water as iodine (I_2), hypoiodous acid (HOI), iodate (IO_3), or iodide (I^-). Iodoamines do not form to any appreciable extent. Iodine is an essential trace element that is required for synthesis of the thyroid hormone, with an estimated adult requirement of 80 to 150 μg/day. Most intake of iodine is from food, especially seafood, and, in the United States, table salt that has been supplemented with potassium iodide (National Academy of Sciences Safe Drinking Water Committee, 1987) to prevent goiters (thyroid enlargement) or mental retardation that can occur in populations lacking sufficient iodine in their diet during gestation or in early childhood. High doses of iodine cause irritation of the GI tract. A dose of 2 to 3 g may be fatal, although acute iodine poisoning is rare. ATSDR (2004) has developed a chronic oral minimum risk level of 0.01 mg/kg/day for iodide, based on the rate of occurrence of subclinical hypothyroidism in children in two areas of China with high or low iodide levels in drinking water.

Ozone. Ozone is used as an oxidant and disinfectant in drinking water treatment. Its chemistry and use are covered in Chaps. 7 and 17. Ozone is unstable and is dissipated largely in the ozonation unit, so that no residual ozone is present in the finished water leaving the treatment plant. Thus, concerns about its safety are related to its reactions in the formation of by-products such as bromate and formaldehyde and not to the ozone itself (Bull and Kopfler, 1991).

Bromate. Bromate is formed in water following the use of ozone in water containing bromide ion (USEPA, 1994f; Bonacquisti, 2006). Coverage of the chemistry of formation and control is found in Chap. 19.

Noncancer effects in rats exposed to bromate in drinking water for 100 weeks included decreased survival and body weight; increased liver, kidney, thyroid, and spleen weights; and kidney lesions (USEPA, 2008i). Mice were less sensitive than rats in this study. Renal effects have also been seen in several other chronic oral studies in rats. Bromate was shown to be mutagenic in several test systems (USEPA, 1993, 1998e), and effects on sperm density in male rats were seen after 35 days of exposure (Wolfe and Kaiser, 1996). Bromate in drinking water was carcinogenic to male and female rats in chronic studies, with increased incidences of kidney and thyroid tumors and peritoneal mesothelioma (USEPA, 2008i). It is classified as a probable human carcinogen (Group B2) under the 1986 USEPA cancer risk assessment guidelines and as a likely carcinogen under the more recent 2005 guidelines (USEPA, 2008i). The MCLG is zero, based on carcinogenicity, and the MCL is 0.01 mg/L (USEPA, 1994a).

Organic Disinfection By-products

Factors influencing the formation of DBPs are discussed in detail in Chap. 19. Discussion of the health effects of some important organic DBPs that have been identified in treated drinking water follows.

Trihalomethanes. The primary THMs include chloroform, dibromochloromethane, dichlorobromomethane, and bromoform, but iodinated forms have also been detected. Because all THMs are volatile, inhalation exposure occurs during showering and other household uses of water. For most drinking waters, chloroform is by far the most prevalent, but higher bromide concentrations in the water supply can significantly shift the relative proportions of the THMs. The MCL for TTHMs is 80 μg/L, based on a running annual average of quarterly samples. Health effects information for individual THMs follows.

Chloroform. Chloroform was formerly used as an anesthetic, and inhalation of sufficient concentrations can cause CNS depression. The primary target organs in animal studies are the liver, kidney, and nasal epithelium (USEPA, 2001f). Developmental effects are likely secondary to toxicity to the mother (USEPA, 2001f). The effects on the liver and the kidney are believed to result from cellular toxicity of a reactive metabolite, phosgene. Chloroform induced liver tumors in mice when administered by gavage in corn oil, but not via drinking water, and induced kidney tumors in male rats, regardless of the carrier vehicle (oil or drinking water). It is not a strong mutagen, and neither it nor its metabolites bind readily to DNA. USEPA (2001f) has concluded that it does not cause cancer primarily through a mutagenic mode of action and that its carcinogenicity is secondary to cytotoxicity and regenerative hyperplasia. Doses that are not cytotoxic are considered unlikely to result in increased risk of cancer. Thus, chloroform is one of the few chemicals for which USEPA has concluded that there is a threshold for carcinogenicity. Prior to the USEPA (2001f) assessment, an MCLG of zero was adopted by USEPA in 1998, based on the default assumption that exposure to any level of a carcinogen results in some increase in cancer risk. The current MCLG is 0.07 mg/L and is based on an RfD of 0.01 mg/kg/day that is protective for the effects considered to be precursors to tumor formation.

Bromodichloromethane. In mice and rats, the target organs for bromodichloromethane (BDCM) toxicity are the kidney, liver, and thyroid (NTP, 2006). BDCM also caused pregnancy loss in rats, possibly through disruption of the secretion of luteinizing hormone (Bielmeier et al., 2004). The results of mutagenicity studies are mixed, and the overall results suggest BDCM is genotoxic (NTP, 2006). In a chronic study in which BDCM was given by corn oil gavage, there was an increased incidence of kidney tumors in male mice, liver tumors in female mice, and kidney and large intestine tumors in male and female rats (NTP, 1987b). Based on these results, BDCM was classified as a probable human carcinogen (Group B2) with an MCLG of zero (USEPA, 1998b). BDCM was shown to induce aberrant crypt foci, thought to be precursors of cancers of the large intestine, equally when administered by drinking water or corn oil gavage (Geter et al., 2004). However, in a more recent chronic drinking water study in male rats and female mice, BDCM did not cause benign or malignant tumors at doses up to 700 mg/L for two years (NTP, 2006).

Dibromochloromethane. Aside from its occurrence as a DBP, dibromochloromethane (DBCM) is used as a chemical intermediate in the manufacture of fire-extinguishing agents, aerosol propellants, refrigerants, and pesticides (USEPA, 1994a). In mice and rats, DBCM causes changes in kidney, liver and serum enzyme levels and decreased body weight (USEPA, 1994a). A multigeneration reproductive study of mice treated with DBCM in drinking water showed maternal toxicity (weight loss, liver pathology) and fetal toxicity (USEPA, 1994e). In a chronic study in which DBCM was given by gavage in corn oil, benign and malignant liver tumors were significantly increased in female and male mice, while rats did not develop cancer (USEPA, 1994e). Results of mutagenicity studies are inconclusive (USEPA, 1994e). USEPA has classified DBCM as a possible human carcinogen (Group C). The MCLG of 0.6 mg/L is based on the NOAEL for liver effects in a subchronic gavage study, with an additional uncertainty factor of 10 for possible carcinogenicity (USEPA, 1998b).

Bromoform. Bromoform occurs less frequently as a DBP in drinking water than the other THMs. Liver is the primary target for bromoform toxicity, with more severe toxicity seen with a bolus dose given by gavage than with drinking water exposure (ATSDR, 2005). Results of *in vitro* and *in vivo* genotoxicity studies are mixed (USEPA, 1994e). Bromoform administered by corn oil gavage to rats produced an increased incidence of benign and malignant tumors of the large intestine and rectum (USEPA, 1994e). It is classified as a probable human carcinogen (Group B2) with an MCLG of zero.

Haloacetic Acids. Haloacetic acids (HAAs) consist of nine highly water-soluble compounds and, as a group, are the next most prevalent DBPs after THMs. The five regulated HAAs are mono-, di-, and trichloroacetic acids and mono- and dibromoacetic acid. Dichloroacetic and trichloroacetic acids are the most prevalent HAAs. In high-bromide waters, the brominated species such as bromodichloroacetic acid and bromochloroacetic acid are more prevalent, and toxicity of these has been studied recently (WHO, 2004). The MCL for the total of five HAAs is 60 μg/L, based on a running annual average of quarterly samples.

Dichloroacetic Acid. In addition to its formation as a DBP, dichloroacetic acid (DCA) is used as a chemical intermediate and as an ingredient in pharmaceuticals and medicine. It was used experimentally in the past to treat diabetes and hypercholesterolemia in human patients (USEPA, 1994a) and it is also a metabolite of trichloroethylene.

The health effects of DCA have been reviewed extensively by USEPA (2003b). DCA inhibits enzymes involved in intermediary metabolism and cholesterol synthesis and thus decreases levels of glucose, pyruvate, and lactate, as well as cholesterol. It has been used experimentally to treat lactic acidosis, diabetes, and familial hypercholesteremia. DCA causes liver enlargement associated with hepatic glycogen deposition. It causes increased serum liver enzymes in mice, rats, dogs, and humans, but liver necrosis has been observed only in mice. DCA causes male reproductive effects, including decreased testicular weight and viable sperm production, and dogs are the most sensitive species for these effects. DCA exposure in pregnant female rats causes impaired fetal maturation and soft tissue anomalies (primarily of cardiac origin) in the offspring. Neurological effects including sedation and peripheral neuropathy have occurred in humans treated with DCA, and neurological toxicity has been reported in dogs and rats, but not mice. DCA caused increased liver tumors in male and female mice and male rats in several drinking water studies. Although DCA has been found to be mutagenic, responses generally occur at relatively high doses. USEPA (1994a, 1998b) has classified DCA as a probable human carcinogen (Group B2) with an MCLG of zero. More recently (USEPA, 2003b), it has been classified as a likely human carcinogen under the newer USEPA classification scheme.

Trichloroacetic Acid. Trichloroacetic acid (TCA) given in drinking water to male rats caused decreased spleen weight and increased liver and kidney weights (USEPA, 1994f, 1997b). Several studies have shown that TCA, like DCA, can produce developmental malformations in rats, particularly in the cardiovascular system (USEPA, 1994f, 1997b). Trichloroacetic acid was negative in bacterial mutagenicity tests but it produced bone marrow chromosomal aberrations and sperm abnormalities in mice and induced single-strand DNA breaks in rats and mice exposed by gavage. Overall, the available studies suggest that TCA does not cause cancer through a mutagenic mechanism (USEPA, 1994f, 1997b). TCA has induced liver tumors in mouse studies but it was not carcinogenic in male rats (USEPA, 1994f, 1997b). USEPA (1998b) has classified TCA as a possible human carcinogen (Group C) with an MCLG of 0.3 mg/L.

Brominated Acetic Acids. While bromoacetic acid, dibromochloroacetic acid, and tribromoacetic acid were negative in studies of reproductive toxicity in male rats, dibromoacetic acid is a male reproductive toxicant in both rats and rabbits (Melnick et al., 2007; Veeramachaneni, Palmer, and Klinefelter, 2007), causing such effects as abnormal sperm, decreased sperm count and motility, and decreased testis and epididymis weight. Bromodichloroacetic acid was found to cause reproductive toxicity in female rats (NTP, 1998).

Dibromoacetic acid in drinking water caused liver toxicity in subchronic studies in rats and mice. In chronic studies, it caused neoplasms at multiple sites in rats and mice, including mononuclear cell leukemia and mesotheliomas in rats, and liver and lung tumors in mice. Liver tumors were significantly increased in male mice at the lowest dose, approximately 5 mg/kg/day (Melnick et al., 2007). It also caused neurotoxicity in rats when administered by drinking water (Moser et al., 2004). The lowest dose in the study, 200 mg/L, caused neurobehavioral changes, while higher doses also caused neuropathological changes.

Bromochloroacetic acid was also carcinogenic to rats and mice (NTP, 2008b). It caused mesotheliomas in male rats, large intestinal adenomas in male and female rats, mammary gland fibroadenomas in female rats, and liver tumors in male and female mice.

3-Chloro-4-(Dichloromethyl)-5-Hydroxy-2(5H)-Furanone (MX). The organic DBP MX appears to be produced in water treated with chlorine and has been found in drinking water in several countries (Bull and Kopfler, 1991). It is one of the most potent mutagens in the Ames test (Bull and Kopfler, 1991). In a Finnish study (Smeds et al., 1997), it accounted for a large portion (11–70 percent) of mutagenicity in chlorinated drinking waters. Male and female rats given MX in drinking water developed cancer at multiple sites at all doses (Komulainen et al., 1997). McDonald and Komulainen (2005) provide a comprehensive review of occurrence, mutagenicity, and carcinogenicity of MX.

Nitrosamines. The reaction of amines with chlorine or chloramines results in the production of nitrosamines. Nitrosodimethylamine (NDMA) has received the most attention as an emerging drinking water contaminant. In addition to its formation as a DBP, it can form through the degradation of dimethylhydrazine, a component of rocket fuel, and has been found in water supplies near aerospace facilities (Howd and Fan, 2008; WHO, 2008). However, most exposure to nitrosamines comes from dietary sources. Exposure to exogenous NDMA is greater from food than from drinking water, and NDMA is also formed by nitrosation of proteins in food in the stomach (Fristachi and Rice, 2007), as discussed under nitrate.

The nitrosamines are considered probable human carcinogens. The drinking water concentration of NDMA estimated to result in a 1 in 10,000 lifetime cancer risk level is 0.07 μg/L (USEPA, 2008i), but no federal MCL has been promulgated. California (Howd and Fan, 2008) and Massachusetts have set 0.01 μg/L as a notification or guidance level. NDMA is one of four contaminants listed on the USEPA CCL3, which the USEPA Science Advisory Board recommended as a priority for consideration for regulation based on available data on toxicity, occurrence, and treatability (USEPA, 2009b).

RADIONUCLIDES

Introduction

Radionuclides are unstable atoms that spontaneously decay, releasing a portion of their nucleus as radioactivity in the form of one of two different types of particles with mass:

- *Alpha radiation:* Positively charged particles made up of two protons and two neutrons (the same as a helium nucleus)
- *Beta radiation:* Electrons or positrons (the positively charged antiparticle of electrons)

Any additional released energy not contained by these particles is released as gamma radiation, which is high-energy, electromagnetic, wave-type energy (equivalent to photons with energies above 100,000 electron volts). All of these constitute ionizing radiation that can pass through tissues to produce reactive oxygen species that damage all cellular constituents, including DNA.

Elements heavier than lead, such as radon, radium, thorium, and uranium isotopes, decay by the release of alpha, alpha and gamma, or beta and gamma emissions. Radionuclides that are lighter than lead, such as radioactive isotopes of iodine, strontium, and potassium, decay by beta and/or gamma emissions. Potassium-40 is the naturally occurring long-lived isotope that is the most significant ingested source of beta radiation. However, most of the lighter radionuclides are man-made.

Radioactive decay results in the formation of a lighter isotope or element. The isotope that decays is called the parent, and the new element is called the progeny or daughter. Each naturally occurring isotope heavier than lead belongs to one of three series of stepwise decay pathways stemming from a long-lived isotope. For example, radium-226 (Ra-226) and radon-222 (Rn-222) are 2 of the 14 progenies of uranium-238 (U-238). Eventually, the progenies in this series decay to the stable isotope lead-206. Radium-228 is 1 of 10 progenies of the thorium-232 (Th-232) series that decays eventually to the stable isotope lead-208. The parent of the third series is U-235.

Different isotopes decay at different rates. The half-life of an isotope is the time required for one half of the atoms present to decay. The half-lives of various isotopes range from fractions of a second to billions of years (U-238 and Th-232). Isotopes with very short half-lives are not significant health risks because they do not survive transport through drinking water distribution systems; however, their progenies may cause significant biological effects.

Radioactivity is generally reported in units of curies (Ci). The dose as a measure of absorbed energy is reported as rads, while dose as a measure of the resulting biological damage is reported as rems. One curie equals 3.7×10^{10} nuclear disintegrations per second (the radioactivity in 1 g of radium, by Marie Curie's definition). A commonly used unit is the picocurie (pCi), which equals 10^{-12} Ci. By comparison to radium, 1 g of the relatively stable uranium-238 has 0.36×10^{-6} Ci of radioactivity. In the international system (SI), a Becquerel is equal to one disintegration per second, or 27 pCi. (Rads and rems are discussed below.)

Based on occurrence and health effects, the radionuclides of most concern in drinking water are Ra-226, Ra-228, U-238, and Rn-222 (USEPA, 1988b). Ra-224 and polonium-210 are also frequently found in drinking water. These are all naturally occurring alpha emitters, except for radium-228, which is a beta emitter.

Health Effects of Radionuclides

Each type of radiation interacts differently with cells in the human body. Alpha particles have little ability to penetrate the skin, traveling at most 50 to 70 microns. However, ingested alpha particles can be very damaging to nearby cells, due to both the electric charge they carry and to the energy imparted by their relatively high mass. Damage to DNA (particularly double-strand breaks) is a typical feature of their effect. The smaller mass of beta particles allows greater penetration but causes less damage to organs and tissue. Gamma radiation has tremendous penetrating power but has limited effect on the body at low levels.

A rad quantifies the energy absorbed by tissue or matter, without regard to the amount of damage caused. Thus, a rad of alpha particles creates more damage than a rad of beta particles, while a rad of gamma radiation causes the least damage. In contrast, a rem quantifies radiation in terms of its dose effect, such that equal doses expressed in rems produce the same biological effect regardless of the type of radiation involved. Numerically, a rem is equal to the absorbed dose (in rads) times a quality factor (Q) that is specific to the type of radiation and reflects its biological effects. (In the newer SI units, a Gray is equal to 100 rad and a Sievert is equal to 100 rem.)

The chemical nature of the radioactivity is also important. For example, radioactive iodine can cause thyroid cancer, since iodine accumulates in the thyroid gland and is incorporated there into the hormone, thyroxine, which is a tetra-iodinated amino acid. The half-lives of different isotopes can also play a role in the disposition of the radionuclide in the body and thus its effect, such as the differing distribution patterns of different radium isotopes in bone. In addition, after a decay step, the daughter element formed may have chemical characteristics that are different than the parent isotope (see Radium, below).

Humans receive an average annual dose of radiation of about 200 to 300 mrem from all sources, and USEPA estimates that drinking water contributes about 0.1 to 3 percent of a person's annual radiation dose. Local conditions can alter this considerably, however. For example, the level of 10,000 pCi/L of radon found in some water systems corresponds to an annual effective dose equivalent of 100 mrem/yr (USEPA, 1991c).

Radioactivity can cause developmental and teratogenic effects, heritable genetic effects, and somatic effects including carcinogenesis. These outcomes result from the interactions of the various cellular effects of radiation, including cell death, induction of cell division, and repair of damage to DNA. High doses lead primarily to cell death, while low doses can stimulate repair processes that can save a cell or return a cell to normal homeostasis and cell-cycle control. Events affecting neighboring cells can also lead to changes in cells not directly impacted by radiation ("bystander effect"). The carcinogenic effects of nuclear radiation (alpha, beta, and gamma) are thought to be due to the ionization of cellular constituents leading to oxidative changes in DNA. All radionuclides are considered to be human carcinogens, although their primary target organs and relative efficiencies as carcinogens differ.

General Considerations for Drinking Water Standards for Radionuclides

The linear nonthreshold approach to extrapolating cancer risk is currently the most widely accepted risk assessment model (NRC, 2006b) for radionuclides. With this approach, lifetime risk is proportional to total exposure, but age during exposure is a modifier. Exposure during early childhood can increase lifetime risk more than predicted by simple proportional increase in years at risk.

Under the original National Interim Primary Drinking Water Regulations, MCLs were set for gross alpha emitters (not including radon) at 15 picoCuries per liter (pCi/L), for Ra-226 and Ra-228 combined at 5 pCi/L, and for gross beta emitters at 4 mrem/yr. These standards were retained under the revised regulations (USEPA, 2000d). If testing is done within 48 to 72 hours after the sample is taken, alpha radiation from the short-lived Ra-224 is also detected (see below). The gross alpha standard is meant as a trigger for further testing and is the only standard for detecting alpha radiation from unregulated emitters, such as polonium-210. USEPA has not promulgated a drinking water standard for radon.

Information on specific radionuclides is presented below. The reader can also refer to the Radionuclides Notice of Data Availability Technical Support Document (USEPA, 2000d) for more background on radionuclides and their risks.

Specific Radionuclides as Drinking Water Contaminants

Radium. Radium accumulates in bone since it distributes in the body similarly to another divalent cation, calcium. It causes bone sarcomas, particularly osteosarcoma. Ra-226 also induces head carcinomas because it decays into radon, which as a gas can enter the nasal sinuses. The half-lives of Ra-224, Ra-226, and Ra-228 are 3.6 days, 1600 years, and 5.8 years, respectively.

Health effects data for radium come from studies of radium watch-dial painters, as well as from persons medically treated with radium injections. In addition, several studies of radium in drinking water indicated that environmentally relevant concentrations are associated with measurable increases in osteosarcoma incidence in humans. The USEPA estimated that at the MCL for combined Ra-226 and Ra-228 of 5 pCi/L, the lifetime cancer risk is about 1 to 2 per 10,000 (USEPA, 2000e).

Radon. Radon is a naturally occurring noble (or inert) gas formed from the radioactive decay of radium-226, which is primarily a decay product of uranium-238 in the earth's crust. Some of the radon released through radioactive decay moves through air- or water-filled pores in the soil to the soil surface and enters the air in buildings or outdoors, while some remains below the surface and dissolves in groundwater. It quickly volatilizes to air from water, and the radon from drinking water can significantly contribute to the radon level of indoor air within a home. Because the half-life of the primary radon isotope, Rn-222, is only 3.8 days, there is significant decay to nongaseous daughters that attach to dust and other particles in air. Small, breathable particles carrying the attached radionuclides can lodge in lung tissue. For each radon atom that reaches the lungs, up to eight progeny will contribute to the radiation load, mostly as alpha radiation.

Direct association has been shown between inhalation of Rn-222 by uranium miners and lung cancer, as well as a small increased risk of other cancers (NRC, 1999b). The effect of radon was increased among smokers. There is also strong evidence for a significant lung cancer risk from radon in residential indoor air (Krewski et al., 2006). It has been estimated that volatilization during showering releases most of the radon dissolved in water and that with typical water use (especially showering), 10,000 pCi/L of radon in water results in an increase of 1 pCi/L in the air in a typical house (Hopke et al., 1996). To put this into perspective, the USEPA residential indoor air guidance level is 4 pCi/L (of air), above which remedial measures are recommended; the approximate level in outdoor air is 0.4 pCi/L. Inhalation of volatilized radon from indoor air is the primary route of exposure to radon in drinking water, with ingestion of radon dissolved in drinking water contributing a minor portion of exposure. The average lifetime cancer risk due to inhalation and ingestion exposures from 100 pCi/L in drinking water was estimated to be approximately 7 per 100,000 (NRC, 1999c).

The health risks from radon in drinking water can exceed those of most regulated drinking water contaminants. For example, radon was found in public drinking water supplies in New Jersey at concentrations ranging from nondetectable to 41,000 pCi/L, with an average concentration of 921 pCi/L. In New Jersey private wells, it was found at 50 to 170,000 pCi/L with an average concentration of 5040 pCi/L (New Jersey Drinking Water Quality Institute, 2009). The estimated lifetime cancer risks at these average concentrations exceed the upper end of USEPA's target risk range for carcinogens in drinking water of 1 in 10,000 (10^{-4}). A drinking water regulation for radon was proposed by USEPA in 1999, but it has not been adopted.

Uranium. Uranium accumulates in the bones similarly to radium. The USEPA classifies it as a human carcinogen, and it also causes kidney toxicity in humans, leading to kidney inflammation and cell death in the renal tubules. The MCL was set at 30 μg/L (approximately corresponding to 27 pCi/L), based on these noncancer kidney effects, since they were deemed to be more critical than carcinogenicity (USEPA, 2000e). The estimated lifetime cancer risk at the MCL is approximately 1 in 10,000. It is notable that uranium represents the first instance of an MCL being set above the MCLG (20 μg/L) indicated by the RfD because of economic considerations. Health Canada has an interim maximum acceptable concentration of 20 μg/L (http://www.hc-sc.gc.ca/ewh-semt/pubs/water-eau/uranium/guideline-recommandation-eng.php).

AESTHETIC QUALITY

Aesthetic complaints are regarded as important signals of changes in water quality and can potentially serve as an early warning of the presence of contaminants to alert water utility staff. For example, in the days leading up to the 1993 *Cryptosporidium* outbreak in

Milwaukee, Wis., there was an increase in the number of water quality complaints (Proctor, Blair, and Davis, 1998). USEPA advises water systems to include programs for monitoring these complaints in preparedness planning. Taste, odor, color, turbidity, and hardness are discussed below.

Taste and Odor

Certain generalizations can be made about the taste and odor of water. Foremost is that the two sensations can influence one another. Taste tends to reflect inorganic constituents, while odor mainly reflects organic constituents, often at much lower water concentrations. Taste is often influenced by temperature and pH. Both taste and odor can be difficult to measure properly, and efforts must be made to separate the two sensory paths. There can be a wide variation of taste and odor thresholds among individuals, which may exceed the range reported by taste panels. Many individuals adapt to certain tastes, so that the tastes are no longer noticed. Methods for identification and treatment of taste and odor problems are discussed extensively by Mallevialle and Suffet (1987) and in Manual 7 of the American Water Works Association (2004).

Taste and Odor of Inorganic and Organic Contaminants. Taste problems in water derive in part from salts, total dissolved solids (TDS), and the presence of specific metals such as iron, copper, manganese, and zinc. The prevalent anions can often have a larger effect on taste. In general, waters with TDS less than 1200 mg/L are minimally acceptable to consumers, although levels below about 600 mg/L are preferable (Mallevialle and Suffet, 1987). The USEPA SMCL for TDS is 500 mg/L. Specific salts may be more significant in terms of taste, notably magnesium chloride and magnesium bicarbonate. The sulfate salts, magnesium sulfate and calcium sulfate, on the other hand, have been found to be relatively inoffensive (Mallevialle and Suffet, 1987). Sulfate may also impart taste at levels above 300 to 400 mg/L, and concentrations of chloride above 250 mg/L may give water a salty taste. Fluoride can also cause a distinct taste above about 2.4 mg/L. Testing of metals in drinking water (James M. Montgomery Consulting Engineers, Inc., 2005) showed the following taste thresholds (in mg/L): Zn at 4 to 9; Cu at 2 to 5; Fe at 0.04 to 0.1; and Mn at 4 to 30. However, individual taste thresholds may vary widely; for example, the taste threshold for Fe spans a concentration range of over 6000-fold (Cohen et al., 1960). The USEPA Drinking Water Advisory (USEPA, 2003c) for sodium of 30 to 60 mg/L is based on taste.

Objectionable tastes and odors may also occur in water contaminated with synthetic organics or as a result of water treatment or from coatings used inside tanks and pipes. Odor thresholds of some common solvents identified in groundwater are shown in Table 2-5. Many consumers object to the taste of chlorine, which has an average taste threshold of about 0.2 (0.02–0.3) mg/L at neutral pH, whereas the disinfectant monochloramine has been found to

TABLE 2-5 Odor Thresholds of Chlorinated Solvents

Solvent	Detection odor threshold, mg/L
1,4-Dichlorobenzene	0.0003
Trichloroethylene	0.5
Tetrachloroethylene	0.3
Carbon tetrachloride	0.2

Source: Van Gemert, L.J. and Nettenbreijer, A.H. eds. 1977. *Compilation of Odour Threshold Values in Air and Water.* Natl. Inst. for Wtr. Supply, Voorburg, Netherlands, and Centr. Inst. for Nutr. & Food Res. TNO, Zeist, Netherlands.

have a higher taste threshold of 0.48 mg/L (Mallevialle and Suffet, 1987). Di- and trichloramine have stronger odors than monochloramine. Although chlorine can reduce many taste and odor problems by oxidizing the offending compounds, the compounds formed when it reacts with organics may have objectionable taste or odor. Most notorious is the phenolic odor resulting from reactions between chlorine and phenols. The DBP chloroform has been detected by smell at 0.1 mg/L (Mallevialle and Suffet, 1987). The formation of aldehydes from chlorinated amino acids has also resulted in problem odors (Gittleman and Luitweiler, 1996). Other oxidants can also be used to remove chemicals with objectionable taste or odor.

Taste and Odor from Microorganisms. Excessive growth of certain types of filamentous cyanobacteria (e.g., *Anabaena, Oscillatoria, Phormidium*), filamentous Actinomycete bacteria (e.g., *Streptomyces* and *Nocardia)*, fungi, amoeba, and other organisms in source waters can lead to the production of taste and odor chemicals (Juttner and Watson, 2007). The most common taste and odor chemicals produced by these organisms include geosmin and 2-methylisoborneol (MIB). Geosmin and MIB are "earthy" smelling, volatile compounds. In addition to production in the water column, these compounds can also be produced by terrestrial organisms and then washed into a water body by rainfall. Organisms produce these compounds in both dissolved and protein-bound forms. Cell degradation or predator grazing can cause the release of the bound fraction to the dissolved form. Both geosmin and MIB are relatively stable in water. There are different methods to remove these compounds such as oxidation (Chap. 7) and adsorption on activated carbon (Chap. 14). Chlorination or the use of algicides in lakes and reservoirs can increase levels of these compounds due to cell disruption. Organism removal by filtration, sedimentation, or dissolved air flotation should precede the use of oxidants when treating geosmin- and MIB-containing waters.

In groundwaters and in some distribution systems, a highly unpleasant odor of hydrogen sulfide may occur as the result of anaerobic bacterial action on sulfates (Mallevialle and Suffet, 1987). This rotten egg odor can be detected at ≤ 0.1 μg/L. The bacterium most often responsible for hydrogen sulfide production is *Desulfovibrio desulfuricans.* Other bacterially produced sulfur compounds creating swampy and fishy tastes and odors in distribution systems include dimethylpolysulfides and methylmercaptan. Various types of *Pseudomonas* bacteria can also produce obnoxious sulfur compounds.

Maintenance of a chlorine residual in the distribution system is an important measure in controlling growth of aesthetically significant microorganisms, many of which grow in biofilm on pipes (see Opportunistic Pathogens in the section Pathogenic Organisms). However, if nonpathogenic constituents, such as iron bacteria, nitrifying bacteria, or sulfur bacteria, have built up excessively in the distribution system, the chlorine demand can be significantly increased (American Water Works Association, 2004). The subsequent decreased chlorine residual can allow increased growth of microorganisms in the distribution system, especially in low-flow areas.

Turbidity and Color

Two important aesthetic aspects of water quality are color and turbidity. High-quality drinking water should have no noticeable color and a low level of turbidity. Typical finished waters have color unit (CU) values of less than 5 and turbidities under 1 nephelometric turbidity unit (ntu). Under the Long Term 1 Enhanced Surface Water Treatment Rule (USEPA, 2002b), drinking water cannot exceed 1 ntu and must be under 0.3 ntu in 95 percent of each month's tests. The sources of color and turbidity are covered in Chap. 3.

Suspended particles can be a significant health concern if heavy metals and hydrophobic chemicals (e.g., many pesticides) are adsorbed to them. The association of turbidity with infectious disease is described in the Indicators of Water Quality section of this chapter.

Color and turbidity removals are typically achieved by the processes of coagulation, flocculation, clarification (sedimentation or dissolved air flotation), and filtration.

Mineralization

Waters with high levels of salts, measured as TDS, may be less palatable to consumers. For example, as discussed earlier in this chapter, sodium and sulfate can impart a taste to the water. The presence of certain salts, such as high levels of sulfate, may have a transitory laxative effect on the consumer.

High salts can also adversely affect industrial cooling operations, boiler feed, and specific processes requiring softened or demineralized waters, such as food and beverage industries and electronics firms. Typically, high salts will force these industries to pretreat their water. High levels of chloride and sulfate can accelerate corrosion of metals in both industrial and consumer systems.

Removal of salts requires expensive treatment, such as demineralization by ion exchange, nanofiltration, or reverse osmosis, and is not usually done for drinking water. Water with excessive levels of salt may be blended with source water with a lower salt level to reduce the salt concentration to acceptable levels.

Hardness

Possible health effects and association were discussed in the section on Inorganic Constituents. Hardness affects the capacity of the water for precipitating soap—the harder the water the more difficult it is to form lather. It is this aspect of hard water that is the most perceptible to consumers.

Staining

Staining of laundry and household fixtures can occur from waters with high iron, manganese, or copper in solution. In oxygenated surface waters, typical concentrations of total iron (mostly the ferric form) are low, and treated waters are well below the SMCL of 0.3 mg/L. From reservoir bottoms deficient in oxygen and from groundwater, the occurrence of ferrous iron at concentrations of 1.0 to 10 mg/L is common. If reduced Fe is carried into the distribution system, the presence of oxygen or chlorine will oxidize Fe producing a red-brown ferric hydroxide precipitate that stains fixtures and laundry. Manganese is often present in waters from the bottom of reservoirs and groundwaters and can cause staining problems—dark brown to black precipitate. As discussed in Chap. 7, drinking water treatment plants attempt to reduce Mn to 15 μg/L or less to prevent chronic staining problems. This is less than the SMCL of 0.05 mg/L.

Excess copper in water may create blue stains.

PREPAREDNESS AND HEALTH

Preparedness is a cross-cutting issue. Threats from the potential release of biological, chemical and radiological agents considered by agencies involved in homeland security can pose acute to long-term health risks. Publicly available lists of agents of concern have been in existence for many years because of past public discussions about preparedness and prohibition of the use of these agents as weapons by militaries (NRC, 1995; Christen, 2001), but potential threats range from the "hardware store" variety to esoteric agents. A subset of these agents may be relevant to drinking water because of resistance to treatment and disinfection or because of the potential for intentional intrusion into a system. It is suggested that the reader make use of the various reports and Websites provided at the end of the section for more detail and updates.

The biological agents of concern for drinking water include anthrax because of its resistance to disinfection (Rice et al., 2005). Other agents, including those that have caused disease outbreaks in the past, could pose a hazard if the disinfectant residual is low or compromised. Bacterial or fungal toxins also have received attention as potential threats.

Chemical warfare agents of concern include the nerve agents that inhibit human acetylcholinesterase, which is a critical enzyme located at the neuromuscular junction, in the autonomic nervous system and the brain. Inhibition causes a dangerous buildup of the neurotransmitter, acetylcholine (see discussion in the introduction to the section on Insecticides). Insecticide inhibitors of acetylcholinesterase can also pose a risk for some individuals and exposure scenarios, but many insecticides have a toxicity profile that affects humans much less than insects. Other chemicals of concern include well-known poisons, such as cyanide, which inhibits the ability of animal cells to respire, arsenic, and thallium, as well as protein based plant- and animal-derived toxins, like ricin.

Among the various sources of information are the Water Contamination Information Tool (USEPA, 2007b); USEPA National Homeland Security Research Center, including its initiative to establish Provisional Advisory Levels for hazardous agents (USEPA, 2008k); the CDC (2008) Website for Bioterrorism Agents and Diseases; and the Water Information Sharing and Analysis Center (http://www.waterisac.org). In addition, the USEPA has developed a number of emergency response tools, including the Response Protocol Toolbox Modules, found at: http://cfpub.epa.gov/safewater/watersecurity/home.cfm?program_id=8.

FINAL COMMENT

Health and aesthetic aspects of water quality are the driving forces behind water quality regulations and water treatment practice. Because of the complexity of the studies summarized in this chapter, the reader is urged to review the cited literature, Health Advisories and IRIS documents from USEPA, the Toxicological Profile series from the ATSDR, and other information sources provided below for more details on any contaminant of particular interest. Furthermore, because new information on organisms causing waterborne disease and chemical contaminants is constantly becoming available, review of the literature since the publication of this chapter is recommended to supplement the information herein during the process of decision making.

INTERNET RESOURCES

The following Internet Websites are useful sources of information on toxicity and risk assessment of drinking water contaminants:

USEPA Drinking Water Health Advisories, Standards, and Health Effects Support Documents for Regulatory Determinations, www.epa.gov/waterscience/criteria/drinking/

USEPA MCLs and MCLGs, www.epa.gov/safewater/contaminants/index.html

USEPA Drinking Water Contaminant Candidate List, www.epa.gov/OGWDW/ccl/index.htmL

USEPA IRIS Database, www.epa.gov/iris/

ATSDR Toxicological Profiles, www.atsdr.cdc.gov/toxpro2.html

WHO Guidelines for Drinking Water Quality, www.who.int/water_sanitation_health/dwq/guidelines/en

California EPA Drinking Water Public Health Goals, http://oehha.ca.gov/water/phg/index.htmL

New Jersey Drinking Water Quality Institute MCL Recommendation Support Documents, www.state.nj.us/dep/watersupply/njdwqinstitute.htm

IARC Monographs on the Evaluation of Carcinogenic Risks to Humans, http://monographs.iarc.fr

National Academies Press (many books on health effects of drinking water contaminants available online), www.nap.edu

National Toxicology Program Study Reports, http://ntp.niehs.nih.gov/?objectid=7DA86165-BDB5-82F8-F7E4FB36737253D5

National Library of Medicine TOXNET Toxicology Data Network, http://toxnet.nlm.nih.gov

National Library of Medicine PubMed Database, www.ncbi.nlm.nih.gov/pubmed

ABBREVIATIONS

AFP acute flaccid paralysis
ATSDR Agency for Toxic Substances and Disease Registry
BMAA β-*N*-methylamino-L-alanine
CCL Contaminant Candidate List
CDC Centers for Disease Control and Prevention
CFU colony forming unit
CNS central nervous system
CU color unit
CYL cylindrospermopsin
DBP disinfection by-product
DWEL drinking water equivalent level
FC fecal coliform
GI gastrointestinal
GWR Ground Water Rule
HAA haloacetic acid
HSDB Hazardous Substances Data Bank of the National Library of Medicine's Toxicology Data Network
HPC heterotrophic plate count
IARC International Agency for Research on Cancer
ID_{50} Dose that is infectious to 50 percent of the exposed population

IRIS	Integrated Risk Information System
LD_{50}	Dose that is lethal to 50 percent of the exposed population
LOAEL	lowest observed adverse effect level
MCL	maximum contaminant level
MCLG	maximum contaminant level goal
MIB	2-methylisoborneol
MRA	microbial risk assessment
MRDL	maximum residual disinfectant level
MRDLG	maximum residual disinfectant level goal
MTD	maximum tolerated dose
NOAEL	no observed adverse effect level
NCI	National Cancer Institute
NRC	National Research Council
NTP	National Toxicology Program
ntu	nephelometric turbidity unit
PAH	polynuclear aromatic hydrocarbon
PALs	Provisional Advisory Levels
PAM	primary amoebic meningoencephalitis
PWS	public water system
RfD	reference dose
RSC	relative source contribution
SDWA	Safe Drinking Water Act
SMCL	secondary maximum contaminant level
TC	total coliform
TCR	Total Coliform Rule
TDS	total dissolved solids
THM	trihalomethane
TSH	thyroid-stimulating hormone
USEPA	U.S. Environmental Protection Agency
VOC	volatile organic chemical
WHO	World Health Organization

ACKNOWLEDGMENTS

The authors thank the editor of this book and the reviewers of this chapter for their thorough review and comments. We also thank Dr. Keith Cooper, Rutgers University; Ms. Sandra Krietzman, New Jersey Department of Environmental Protection; Dr. Mark LeChevallier, American Water Works Company, Inc.; and Dr. Francis Mègraud, Institut National de la Santé et de la Recherche Mèdicale (INSERM), Bordeaux, France, for their helpful input on the sections relating to their areas of expertise.

REFERENCES

Abad, F.X., C. Villena, S. Guix, S. Caballero, R.M. Pinto, and A. Bosch (2001). Potential role of fomites in the vehicular transmission of human astroviruses, *Appl. Environ. Microbiol.*, 67: 3904–3907.

Abbaszadegan, M., M. LeChevallier, and C. Gerba (2003). Occurrence of viruses in US goundwaters, *J. AWWA*, 95(9): 107–120.

Aboytes, R., G.D. DiGiovanni, F.A. Abrams, C. Rheinecker, W. McElroy, N. Shaw, and M.W. LeChevallier (2004). Detection of infectious *Cryptosporidium* in filtered drinking water, *J. AWWA*, 96(9): 88–98.

Adam, R.D. (2001). Biology of *Giardia lamblia, Clin. Microbiol.* Rev., 14: 447–475.

Agency for Toxic Substances and Disease Registry (ATSDR) (1992). Toxicological Profile for 1,2-Dibromo-3-Chloropropane. Atlanta, Ga.: Agency for Toxic Substances and Disease Registry.

ATSDR (1993). Toxicological Profile for Cadmium. TP/92–06. Atlanta, Ga.: Agency for Toxic Substances and Disease Registry.

ATSDR (1994). Toxicological Profile for Carbon Tetrachloride. TP/93-02. Atlanta, Ga.: Agency for Toxic Substances and Disease Registry.

ATSDR (1997). Toxicological Profile for Carbon Trichloroethylene. Atlanta, Ga.: Agency for Toxic Substances and Disease Registry.

ATSDR (1999). Toxicological Profile for Mercury. PB/99/142416. Atlanta, Ga.: Agency for Toxic Substances and Disease Registry.

ATSDR (2000a). Toxicological Profile for Toluene. Atlanta, Ga.: Agency for Toxic Substances and Disease Registry.

ATSDR (2000b). Toxicological Profile for Methylene Chloride. Atlanta, Ga.: Agency for Toxic Substances and Disease Registry.

ATSDR (2001). Toxicological Profile for 1,2-Dichloroethane. Atlanta, Ga.: Agency for Toxic Substances and Disease Registry.

ATSDR (2003). Toxicological Profile for Atrazine. Atlanta, Ga.: Agency for Toxic Substances and Disease Registry.

ATSDR (2004). Toxicological Profile for Iodine. Atlanta, Ga.: Agency for Toxic Substances and Disease Registry.

ATSDR (2005). Toxicological Profile for Bromoform and Dibromochloromethane. Atlanta, Ga.: Agency for Toxic Substances and Disease Registry.

ATSDR (2006). Toxicological Profile for 1,1,1-Trichloroethane. Atlanta, Ga.: Agency for Toxic Substances and Disease Registry.

ATSDR (2007). Toxicological Profile for Arsenic. TP/92-02. Atlanta, Ga.: Agency for Toxic Substances and Disease Registry.

ATSDR (2008a). Toxicological Profile for Aluminum. Update. Atlanta, Ga.: Agency for Toxic Substances and Disease Registry.

ATSDR (2008b). Draft Toxicological Profile for Chromium. Update. Atlanta, Ga.: Agency for Toxic Substances and Disease Registry.

Alavanja, M.C., C. Samanic, M. Dosemeci, J. Lubin, R. Tarone, C.F. Lynch, C. Knott, K. Thomas, J.A. Hoppin, J. Barker, J. Coble, D.P. Sandler, and A. Blair (2003). Use of agricultural pesticides and prostate cancer risk in the Agricultural Health Study cohort, *Amer. J. Epidem.*, 157(9): 800–814.

Allegra, S., F. Berger, P. Berthelot, F. Grattard, B. Pozzetto, and S. Riffard (2008). Use of flow cytometry to monitor *Legionella* viability, *Appl. Environ. Microbiol.*, 74: 7813–7816.

Allen, M.J., J.L. Clancy, and E.W. Rice (2000). The plain, hard truth about pathogen monitoring, *J. AWWA*, 92(9): 64–76.

American Heart Association (2008). Sodium. AHA Recommendation.http://www.americanheart.org/presenter.jhtml?identifier=4708. Last accessed May 2008.

American Public Health Association, American Water Works Association, Water Environment Federation (APHA, AWWA, and WEF) (1998). *Standard Methods for the Examination of Water and Wastewater*, 20th ed. Washington, D.C.: American Public Health Association.

American Water Works Association (2004). M7, *Problem Organisms in Water: Identification and Treatment*, 3rd ed. AWWA Manual of Water Supply Practices. Denver, Colo.: American Water Works Association.

American Water Works Association (2006). M48, *Waterborne Pathogens,* 2nd ed. AWWA Manual of Water Supply Practices. Denver, Colo.: American Water Works Association.

AWWA Research Foundation (2003). *Assessment of Waters for Estrogenic Activity.* Denver, Colo.: Water Research Foundation.

Atherholt, T., E. Feerst, B. Hovendon, J. Kwak, and J.D. Rosen (2003). Evaluation of indicators of fecal contamination in groundwater, *J. AWWA,* 95(10): 119–131.

Azevedo, N.F., C. Almeida, I. Fernandes, L. Cerqueira, S. Dias, C.W. Keevil, and M.J. Vieira (2008). Survival of gastric and enterohepatic *Helicobacter* spp. in water: implications for transmission, *Appl. Environ. Microbiol.*, 74: 1805–1811.

Bahta, L., J. Bartkus, J. Besser, N. Crouch, E. Cebelinski, K. Ehresmann, S. Fuller, K. Harriman, J. Harper, H. Hull, R. Lynfield, C. Miller, J. Rainbow, M. Sullivan, G. Wax, P. Ackerman, and A. Parker (2005). Poliovirus infections in four unvaccinated children—Minnesota, August–October 2005, *Morbidity and Mortality Weekly Rept.,* 54(41): 1053–1055.

Baker, K.H., and J.P. Hegarty (2001). Presence of *Helicobacter pylori* in drinking water is associated with clinical infection, *Scand. J. Infect. Dis.*, 33: 744–746.

Bartram, J., J. Cotruvo, M. Exner, C. Fricker, and A. Glasmacher, eds. (2003). *Heterotrophic Plate Counts and Drinking-water Safety,* London: World Health Organization, IWA Publishing.

Barwick, R.S., D.A. Levy, G.F. Craun, M.J. Beach, and R.L. Calderon (2000). Surveillance for water-borne-disease outbreaks—United States, 1997–1998, *Morbidity and Mortality Weekly Rept.*, 49(SS-4): 1–35.

Bass, E.S., D.A. Pappano, and S.G. Humiston (2007). Rotavirus, *Pediatr. Rev.,* 28: 183–191.

Bellack, N.R., M.W. Koehoorn, Y.C. MacNab, and M.G. Morshed (2006). A conceptual model of water's role as a reservoir in *Helicobacter pylori* transmission: a review of the evidence, *Epidemiol. Infect.,* 134: 439–449.

Belpoggi, F., M. Soffritti, and C. Maltoni (1995). Methyl-tertiary-butyl ether (MTBE)—A gasoline additive—causes testicular and lymphohaematopoietic cancers in rats, *Toxicol. Industr. Health,* 11: 119–149.

Bermudez, E., H. Parkinson, and D.E. Dodd (2009). Methyl tertiary butyl ether (MTBE): One-year toxicity drinking water study in Wistar rats. Presented at 2009 Annual Meeting of the Society of Toxicology, Abstract 1481.

Bielmeier, S.R., D.S. Best, and G. Narotsky (2004). Serum hormone characterization and exogeneous hormone rescue of bromodichloromethane-induced pregnancy loss in the F344 rat. *Toxicol. Sci.*, 77: 101–108.

Bjørling-Poulsen M., H.R. Andersen, and P. Grandjean (2008). Potential developmental neurotoxicity of pesticides used in Europe. *Environ. Health,* 7:50.

Blackburn, B.G., G.F. Craun, J.S. Yoder, V. Hill, R.L. Calderon, N. Chen, S.H. Lee, D.A. Levy, and M.J. Beach (2004). Surveillance for waterborne-disease outbreaks associated with drinking water–United States, 2001–2002, *Morbidity and Mortality Weekly Rept.*, 53(SS-8): 23–45.

Blair A and Beane Freeman L (2009). Epidemiologic Studies of Cancer in Agricultural Populations: Observations and Future Directions. *J Agromedicine*. 2009; 14(2): 125–131.

Blair, B., P. Sarkar, K.R. Bright, F. Marciano-Cabral, and C.P. Gerba (2008). *Naegleria fowleri* in well water, *Emerg. Infect. Dis.*, 14: 1499–1501.

Bonacquisti, T.P. (2006). A drinking water utility's perspective on bromide, bromate, and ozonation, *Toxicology*, 221: 145–148.

Borchardt, M.A., P.D. Bertz, S.K. Spencer, and D.A. Battigelli (2003). Incidence of enteric viruses in groundwater from household wells in Wisconsin, *Appl. Environ. Microbiol.*, 69: 1172–1180.

Borchardt, M.A., M.E. Stemper, and J.H. Standridge (2003). *Aeromonas* isolates from human diarrheic stool and groundwater compared by pulsed-field gel electrophoresis, *Emerg. Infect. Dis.*, 9: 224–228.

Borgen, K., T. Herremans, E. Duizer, H. Vennema, S. Rutjes, A. Bosman, A.M. de Roda Husman, and M. Koopmans (2008). Non-travel related Hepatitis E virus genotype 3 infections in the Netherlands: a case series 2004–2006, *BMC Infect. Dis.*, vol. 8, article 61, doi:10.1186/1471-2334-8-61.

Boxall, A.B.A., K. Tiede, and Q. Chaudry (2007). Engineered nanomaterials in soils and water: how do they behave and could they pose a risk to human health? *Nanomedicine,* 2:919–927.

Briand, J.-F., S. Jacquet, C. Bernard, and J.-F. Humbert (2003). Health hazards for terrestrial vertebrates from toxic cyanobacteria in surface water ecosystems, *Vet. Res.*, 34: 361–377.

Bull, R.J., and F.C. Kopfler (1991). *Health Effects of Disinfectants and Disinfectant By-Products.* Denver, Colo.: Water Research Foundation.

Burleigh-Flayer, H.D., J.S. Chun, and W.J. Kintigh (1992). MTBE: Vapor Inhalation Oncogenicity Study in CD-1 Mice. BRRC Report No. 91N0013A. Export, Pa.: Bushy Run Research Center.

C8 Health Project (2009). WVU Data Hosting Website. West Virginia University School of Medicine, Department of Community Medicine. www.hsc.wvu.edu/som/cmed/c8. Last accessed Aug. 1, 2009.

Caballero, S., F.X. Abad, F. Loisy, F.S. Le Guyader, J. Cohen, R.M. Pinto, and A. Bosch (2004). Rotavirus virus-like particles as surrogates in environmental persistence and inactivation studies, *Appl. Environ. Microbiol.,* 70: 3904–3909.

California Environmental Protection Agency (1997). Public Health Goal for Benzo(a)pyrene in Drinking Water. Office of Environmental Health Hazard Assessment. December 1997.

California Environmental Protection Agency (2001). Public Health Goal for Aluminum in Drinking Water. Office of Environmental Health Hazard Assessment. April 2001.

California Environmental Protection Agency (2009). Draft Public Health Goal for Hexavalent Chromium in Drinking Water. Office of Environmental Health Hazard Assessment. August 2009.

Calle, E.E., H. Frumkin, S.J. Henley, D.A. Savitz, and M.J. Thun (2002). Organochlorines and breast cancer risk, *CA: A Cancer Journal for Clinicians*, 52(5): 301–309.

Cantor, K.P. (1997). Drinking water and cancer. *Cancer Causes Control,* 8: 292–308.

Carmichael, W.W. (2001). Health effects of toxin-producing cyanobacteria: 'the CyanoHABs', *Human Ecol. Risk Assess.*, 7: 1393–1407.

Carmichael, W.W., S.M.F.O. Azevedo, J.S. An, R.J.R. Molica, E.M. Jochimsen, S. Lau, K.L. Rinehart, G.R. Shaw, and G.K. Eaglesham (2001). Human fatalities from cyanobacteria: chemical and biological evidence for cyanotoxins, *Environ. Health Persp.,* 109: 663–668.

Carter, M.J. (2005). Enterically infecting viruses: pathogenicity, transmission and significance for food and waterborne infection, *J. Appl. Microbiol.,* 98: 1354–1380.

Catling, L.A., I. Abubakar, I.R. Lake, L. Swift, and P.R. Hunter (2008). A systematic review of analytical observational studies investigating the association between cardiovascular disease and drinking water hardness. *J. Water Health.,* 6(4): 433–442.

Cellini, L., N. Allocati, D. Angelucci, T. Iezzi, E. Di Campli, L. Marzio, and B. Dainelli (1994) Coccoid *Helicobacter pylori* not culturable *in vitro* reverts in mice, *Microbiol. Immunol.*, 38: 843–850.

Centers for Disease Control and Prevention (CDC) (2007). Laboratory surveillance for wild and vaccine-derived polioviruses–worldwide, January 2006–June 2007, *Morbidity and Mortality Weekly Rept.,* 56(37): 965–969.

CDC (1991). Preventing Lead Poisoning in Young Children. Atlanta, Ga.: U.S. Department of Health and Human Services.

CDC, Division of Oral Health. (1992). Fact Sheet. Unintentional High Fluoride Concentration: Acute Adverse Health Events. June 11. Atlanta, Ga.: Centers for Disease Control National Center for Preventive Services.

CDC (2008). Bioterrorism Agents and Diseases. U.S. Centers for Disease Control and Prevention Emergency Preparedness and Response. www.bt.cdc.gov/Agent/Agentlist.asp.

Check, W. (2009). DNA sequencing grows virtuosic—and deep, *Microbe*, 4(1): 18–22.

Chorus, I., and J. Bartram, eds. (1999). *Toxic Cyanobacteria in Water: A Guide to Their Public Health Consequences, Monitoring and Management.* London: World Health Organization, E & FN Spon.

Christen, C. (2001). Bioterrorism and waterborne pathogens: how big is the threat? *Environmental Science and Technology,* 35(19): 396A–397A.

Clark, L.C., and D.S. Alberts (1995). Selenium and cancer: risk or protection?, *J. Nat. Cancer Inst.*, 87: 473–475.

Claxton, L.D., R. Pegram, K.M. Schenck, J.E. Simmons, and S.H. Warren (2008). Integrated disinfection by-products research: salmonella mutagenicity of water concentrates disinfected by chlorination and ozonation/postchlorination. *J. Toxicol. Environ. Health A*, 71: 1187–1194.

Cohen, J.M., L.J. Kamphake, E.K. Harris, and R.L. Woodward (1960). Taste threshold concentrations of metals in drinking water, *J. AWWA,* 52: 660–670.

Colburn, T., F.S. vom Saal, and A.M. Soto (1993). Developmental effects of endocrine-disrupting chemicals in wildlife and humans. *Environ. Health Persp.*, 101: 378–384.

Colford, J.M., Jr., S. Roy, M.J. Beach, A. Hightower, S.E. Shaw, and T.W. Wade (2006). A review of household drinking water intervention trials and an approach to the estimation of endemic waterborne gastroenteritis in the United States, *J. Wat. Health*, 4(suppl. 2): 71–88.

Cook, N., J. Bridger, K. Kendall, M.I. Gomara, L. El-Attar, and J, Gray (2004). The zoonotic potential of Rotavirus, *J. Infect.,* 48: 289–302.

Cortese, M.M. and U.D. Parashar (2009). "Prevention of Rotavirus gastroenteritis among infants and children: recommendations of the Advisory Committee on Immunization Practices (ACIP), *Morbidity and Mortality Weekly Rept.,* 58(RR-2): 1–25.

Corvilain, B., B. Contempre, A.O. Longombe, et al. (1993). Selenium and the thyroid: how the relationship was established, *Am. J. Clin. Nutr.*, 57(suppl. 2): 244S–248S.

Cotruvo, J.A., A. Dufour, G. Rees, J. Bartram, R. Carr, D.O. Cliver, G.F. Craun, R. Fayer, and V.P.J. Gannon, eds. (2004). *Waterborne Zoonoses: Identification, Causes and Control*, London : World Health Organization, IWA Publishing.

Cox, P.A., S.A. Banack, S.J. Murch, U. Rasmussen, G. Tien, R.R. Bidigare, J.S. Metcalf, L.F. Morrison, G.A. Codd, and B. Bergman (2005). Diverse taxa of cyanobacteria produce β-*N*-methylamino-L-alanine, a neurotoxic amino acid, *Proc. Acad. Natl. Sci.,* 102: 5074–5078.

Craun, M.F., G.F. Craun, R.L. Calderon, and M.J. Beach (2006). Waterborne outbreaks reported in the United States, *J. Wat. Health,* 4(suppl. 2): 19–30.

Crosby, L.M., J.E. Simmons, W.O. Ward, T.M. Moore, K.T. Morgan, and A.B. Deangelo (2008). Integrated disinfection by-products (DBP) mixtures research: gene expression alterations in primary rat hepatocyte cultures exposed to DBP mixtures formed by chlorination and ozonation/postchlorination, *J. Toxicol. Environ. Health A,* 71: 1195–1215.

Cuthbert, J.A. (2001). Hepatitis A: old and new, *Clin. Microbiol. Rev.,* 14: 38–58.

Cutler, D. and G. Miller (2005). The role of public health improvements in health advances: the twentieth-century United States, *Demography*, 42(1): 1–22.

Dearfield, K.L. (1994). Ethylene thiourea (ETU). A review of the genetic toxicity studies. *Mutat. Res.*, 317: 111–132.

Delta Environmental Consultants (2005a). MTBE Groundwater Action/Cleanup Levels for LUST Sites: Current and Proposed. St. Paul, Minn. *www.epa.gov/oust/mtbe/mtbemap.pdf*. Accessed 6/21/10.

Delta Environmental Consultants (2005b). Groundwater Oxygenate Cleanup Levels for LUST Sites. St. Paul, Minn. *http://www.epa.gov/swerust1/mtbe/oxytable.pdf*. Accessed 6/21/20.

Diamanti-Kandarakis, E., J-P. Bourguignon, L.C. Giudice, R. Hauser, G.S. Prins, A.M. Soto, R.T. Zoeller, and A.C. Gore (2009). Endocrine-disrupting chemicals: an Endocrine Society scientific statement, *Endocrine Reviews*, 30(4): 293–342.

Dich, J., S. Hoar Zahm, A. Hanberg, and H.-O. Adami (1997). Pesticides and cancer, *Cancer Causes and Control*, 8: 420–423.

Didier, E.S., and L.M. Weiss (2006). Microsporidiosis: current status, *Curr. Opin. Infect. Dis.,* 19: 485–492.

Diederen, B.M.W. (2008) *Legionella* spp. and Legionnaires' disease, *J. Infect.,* 56: 1–12.

Dorrell, N., and B.W. Wren (2007). The second century of *Campylobacter* research: recent advances, new opportunities and old problems, *Curr. Opin. Infect. Dis.*, 20: 514–518.

Dubey, J.P. (2004). Toxoplasmosis–A waterborne zoonosis, *Vet. Parasitol.*, 126: 57–72.

Dufresne, A.T., and M. Gromeier (2006). Understanding polio: new insights from a cold virus, *Microbe*, 1: 13–18.

Eaton, K.A., C.E. Catrenich, K.M. Makin, and S. Krakowka (1995). Virulence of coccoid and bacillary forms of *Helicobacter pylori* in gnotobiotic piglets, *J. Infect. Dis.,* 171: 459–462.

Eboigbodin, K.E., A. Seth, and C.A. Biggs (2008). A review of biofilms in domestic plumbing, *J. AWWA,* 100(10): 131–137.

Eisenberg, J.N.S., A. Hubbard, T.J. Wade, M.D. Sylvester, M.W. LeChevallier, D.A. Levy, and J.M. Colford Jr. (2006). Inferences drawn from a risk assessment compared directly with a randomized trial of a home drinking water intervention, *Environ. Health Persp.*, 114(8): 1199–1204.

El-Senousy, W.M., S. Guix, I. Abid, R.M. Pinto, and A. Bosch (2007). Removal of Astrovirus from water and sewage treatment plants, evaluated by a competitive reverse transcription-PCR, *Appl. Environ. Microbiol.,* 73: 164–167.

Emelko, M.B., P.J. Schmidt, and J.A. Roberson (2008). Quantification of uncertainty in microbial data–Reporting and regulatory implications, *J. AWWA*, 100(3): 94–104.

Emmett, E.A.; F.S. Shofer, H. Zhang, D. Freeman, C. Desai, and L.M. Shaw, L.M. (2006). Community exposure to perfluorooctanoate: relationships between serum concentrations and exposure sources. *J. Occup. Environ. Med.*, 48: 759–770.

Eskenazi, B., A.R. Marks, A. Bradman, K. Harley, D.B. Barr, C. Johnson, N. Morga, and N.P. Jewell (2007). Organophosphate pesticide exposure and neurodevelopment in young Mexican-American children. *Environ. Health Persp.*, 115(5): 792–798.

Espinosa, A.C., M. Mazari-Hiriart, R. Espinosa, L. Maruri-Avidal, E. Mendez, and C.F. Arias (2008). Infectivity and genome persistence of Rotavirus and Astrovirus in groundwater and surface water, *Water Research*, 42: 2618–2628.

Falkow, S. (1997). What is a pathogen?, *ASM News*, 63(7): 359–365.

Fankhauser, R.L., S.S. Monroe, J.S. Noel, C.D. Humphrey, J.S. Bresee, U.D. Parashar, T. Ando, and R.I. Glass (2002). Epidemiologic and molecular trends of 'Norwalk-like viruses' associated with outbreaks of gastroenteritis in the United States, *J. Infect. Dis.*, 186: 1–7.

Fayer, R. (1994). Foodborne and waterborne zoonotic protozoa, pp. 331–362, in Y.H. Hui, J.R. Gorham, K.D. Murrell, and D.O. Cliver (eds.), *Foodborne Disease Handbook,* vol. 2, New York: Marcel Dekker, Inc.

Fields, B.S., R.F. Benson, and R.E. Besser (2002). *Legionella* and Legionnaire's disease: 25 years of investigation, *Clin. Microbiol. Rev.*, 15: 506–526.

Figueras, M.J. (2008). *Arcobacter* spp., a poorly known group of bacteria already associated with two well-water outbreaks in the USA. EPA Symposium of Groundwater-Borne Infectious Disease, Etiologic Agents and Indicators, Washington, D.C., Dec. 2–4, 2008.

Fiore, A.E., A. Wasley, and B.P. Bell (2006). Prevention of hepatitis A through active or passive immunization: recommendations of the Advisory Committee on Immunization Practices (ACIP), *Morbidity and Mortality Weekly Rept.,* 55(RR-7): 1–23.

Fletcher, T., K. Steenland, and D. Savitz (2009). Status Report: PFOA and immune biomarkers in adults exposed to PFOA in drinking water in the mid Ohio valley. C8 Science Panel. Mar. 16, 2009.

http://c8sciencepanel.org/pdfs/Status_Report_C8_and_Immune_markers_March2009.pdf. Accessed 6/21/10.

Fliermans, C.B. (1996). Ecology of *Legionella*: from data to knowledge with a little wisdom, *Microb. Ecol.*, 32: 203–228.

Focazio, M.J., D.W. Kolpin, K.K. Barnes, E.T. Furlong, M.T. Meyer, S.D. Zaugg, L.B. Barber, and M.E. Thurman (2008). A national reconnaissance for pharmaceuticals and other organic wastewater contaminants in the United States—(II) Untreated drinking water sources. *Sci. Total. Environ.*, 402: 201–216.

Fong, T.-T., and E.K. Lipp (2005). Enteric viruses of humans and animals in aquatic environments: health risks, detection, and potential water quality assessment tools, *Micro. Mol. Biol. Rev.*, 69: 357–371.

Fristachi, A., and G. Rice (2007). Estimation of the total daily oral intake of NDMA attributable to drinking water, *J. Water and Health*, 5(3): 341–355.

Ganrot, P.O. (1986).Metabolism and possible health effects of aluminum, *Environ. Health Persp.*, 65: 363–441.

Garabrant, D.H. and M.A. Philbert. (2002). Review of 2,4-dichlorophenoxyacetic acid (2,4-D) epidemiology and toxicology, *Critical Reviews in Toxicology*, 32(4): 233–257.

Gedalanga, P.R., and B.H. Olson. (2009). Development of a quantitative PCR method to differentiate between viable and nonviable bacteria in environmental water samples, *Appl. Microbiol. Biotechnol.*, 82: 587–596.

Geldreich, E.E., H.D. Nash, D.J. Reasoner, and R.H. Taylor (1972). The necessity of controlling bacterial populations in potable waters: community water supply, *J. AWWA*, 64(9): 596–602.

Gentsch, J.R., A.R. Laird, B. Bielfelt, D.D. Griffin, K. Banyai, M. Ramachandran, V. Jain, N.A. Cunliffe, O. Nakagomi, C.D. Kirkwood, T.K. Fischer, U.D. Parashar, J.S. Bresee, B. Jiang, and R.I. Glass (2005). Serotype diversity and reassortment between human and animal rotavirus strains: implications for Rotavirus vaccine programs, *J. Infect. Dis.*, 192(suppl. 1): S146–S158.

Gerba, C.P. (2006). Emerging Viruses," pp. 263–265, in M48, *Waterborne Pathogens.* Denver, Colo.: American Water Works Association.

Geter, D.R., M.H. George, T.M. Moore, S. Kilburn, G. Huggins-Clark, and A.B. DeAngelo (2004). Vehicle and mode of administration effects on the induction of aberrant crypt foci in the colons of male F344/N rats exposed to bromodichloromethane. *J. Toxicol. Environ. Health A,* 67: 23–29.

Giao, M.S., N.F. Azevedo, S.A. Wilks, M.J. Vieira, and C.W. Keevil (2008). Persistence of *Helicobacter pylori* in heterotrophic drinking water biofilms, *Appl. Environ. Microbiol.*, 74: 5898–5904.

Gilliom, R.J., J.E. Barbash, C.G. Crawford, P.A. Hamilton, J.D. Martin, N. Nakagaki, L.H. Nowell, J.C. Scott, P.E. Stackelberg, G.P. Thelin, and D.M. Wolock (2006). The quality of our nation's waters—Pesticides in the nation's streams and groundwater, 1992–2001. U.S. Geological Survey Circular 1291.

Gittleman, T.S., and P. Luitweiler (1996). Tastes and Odors in Treated Water from Algae, Urea, and Industrial Chemicals in Source Waters. In Proc. AWWA Water Quality and Treatment Conference. Denver, Colo.: American Water Works Association.

Gray, J., T. Vesikari, P. Van Damme, C. Giaquinto, J. Mrukowicz, A. Guarino, R. Dagan, H. Szajewska, and V. Usonis (2008). Rotavirus, *J. Pediatr. Gastroenterol. Nutr.*, 46(suppl. 2): S24–S31.

Greer, M.A., G. Goodman, R.C. Pleus, and S.E. Greer (2002). Health effects assessment for environmental perchlorate contamination: the dose response for inhibition of thyroidalradioiodine uptake in humans, *Envir. Health Perspectives*, 110: 927–937.

Haas, C.N., J.B. Rose, and C.P. Gerba (1999). *Quantitative Microbial Risk Assessment.* New York: John Wiley & Sons.

Haddix, P.L., C.J. Hughley, and M.W. LeChevallier (2007). Occurrence of microcystins in 33 US water supplies, *J. AWWA*, 99(9): 118–125.

Hageskal, G., A.K. Knutsen, P. Gaustad, G. Sybren de Hoog, and I. Skaar (2006) Diversity and significance of mold species in Norwegian drinking water, *Appl. Environ. Microbiol.*, 72: 7586–7593.

Haider, S., V. Naithani, P.N. Viswanathan, and P. Kakkar (2003). Cyanobacterial toxins: a growing environmental concern, *Chemosphere*, 52: 1–21.

Haque, R., C.D. Huston, M. Hughes, E. Houpt, and W.A. Petri, Jr. (2003). Amebiasis, *New Engl. J. Med.*, 348: 1565–1573.

Harb, O.S., L.-Y. Gao, and Y. Abu (2000). From protozoa to mammalian cells: a new paradigm in the life cycle of intracellular bacterial pathogens, *Environ. Microbiol.*, 2: 251–265.

Health and Safety Commission (2000). *Legionnaires' Disease. The Control of Legionella Bacteria in Water Systems. Approved Code of Practice & Guidance–L8*, 3rd ed. Norwich, UK: HSE Books.

Herwaldt, B.L. (2007). *Cyclospora cayetanensis*: a review, focusing on the outbreaks of cyclosporiasis in the 1990s, *Clin. Infect. Dis.*, 31: 1040–1057.

Heydens, W.F., I.C. Lamb, and A.G.E. Wilson (2001). Chloroacetanilides, in R.I. Krieger, ed., *Handbook of Pesticide Toxicology*, vol. 1, 2nd ed. New York: Academic Press.

Hirt, R.P., J.M. Logsdon, Jr., B. Healy, M.W. Dorey, W.F. Doolittle, and T.M. Embley (1999). Microsporidia are related to fungi: evidence from the largest subunit of RNA polymerase II and other proteins, *Proc. Natl. Acad. Sci.*, 96: 580–585.

Ho, H.T.K., L.J.A. Lipman, and W. Gaastra (2006) *Arcobacter*, what is known and unknown about a potential foodborne zoonotic agent!, *Vet. Microbiol.*, 115(1–3): 1–13.

Hopke, P.K., D. Vandana, B. Fitzgerald, T. Raunemaa, and K. Kuuspalo (1996). *Critical Assessment of Radon Progeny Exposure While Showering in Radon-Bearing Water*. Denver, Colo.: Water Research Foundation.

Howd, R.A. and A.M. Fan (2008). *Risk Assessment for Chemicals in Drinking Water*. Hoboken, NJ: Wiley.

Huang, P., J.T. Weber, D.M. Sosen, P.M. Griffin, E.G. Long, J.J. Murphy, F. Kocka, C. Peters, and C. Kallick (1995). The first reported outbreak of diarrheal illness associated with *Cyclospora* in the United States, *Ann. Internal Med.*, 123: 409–414.

Hunter, P.R. (2003). Drinking water and diarrheal disease due to *Escherichia coli*, *J. Water and Health*, 1(2) 65–72.

Hunter, P.R. and G. Nichols (2002). Epidemiology and clinical features of *Cryptosporidium* infection in immunocompromised patients, *Clin. Microbiol. Rev.*, 15: 145–154.

Hunter, P.R., and Q. Syed. (2002). A community survey of self-reported gastroenteritis undertaken during an outbreak of cryptosporidiosis strongly associated with drinking water after much press interest, *Epidemiol. Infect.*, 128: 433–438.

IARC (International Agency for Research on Cancer) (1978). *IARC Monographs on the Evaluation of Carcinogenic Risks to Humans. Some N-Nitroso Compounds*. Lyon, France: World Health Organization.

IARC (1996). *IARC Monographs on the Evaluation of Carcinogenic Risks to Humans. Dry Cleaning. Some Chlorinated Solvents and Other Industrial Chemicals,* vol. 63. Lyon, France: World Health Organization.

IARC (1999). *IARC Monographs on the Evaluation of Carcinogenic Risks to Humans. Re-evaluation of Some Organic Chemicals, Hydrazine and Hydrogen Peroxide,* vol. 71. Lyon, France: World Health Organization.

ILSI Risk Science Institute Pathogen Risk Assessment Working Group (1996). A conceptual framework to assess the risks of human disease following exposure to pathogens, *Risk Anal.*, 16(6): 841–848.

Infante-Rivard, C., and S. Weichenthal (2007). Pesticides and childhood cancer: an update of Zahm and Ward's 1998 review. *J. Toxicol. Environ. Health* B, 10: 81–99.

Institute of Medicine, Food and Nutrition Board (2001). *Dietary Reference Intakes for Vitamin A, Vitamin K, Arsenic, Boron, Chromium, Copper, Iodine, Iron, Manganese, Molybdenum, Nickel, Silicon, Vanadium and Zinc*. Food and Nutrition Board, Institute of Medicine. Washington, D.C.: National Academies Press.

Institute of Medicine, Food and Nutrition Board (2006). *Dietary Reference Intakes Research Synthesis Workshop Summary*. Washington, D.C.: National Academies Press.

Institute of Medicine, Panel on Dietary Reference Intakes for Electrolytes and Water (2004). *Dietary Reference Intakes for Water, Potassium, Sodium, Chloride, and Sulfate*. Washington, D.C.: National Academies Press.

International Life Sciences Institute (1997). *An Evaluation of EPA's Proposed Guidelines for Carcinogen Risk Assessment Using Chloroform and Dichloroacetate as Case Studies: Report of an Expert Panel*. Washington, D.C.: International Life Sciences Institute.

James M. Montgomery Consulting Engineers, Inc. (2005). *Water Treatment Principles and Design*, 2nd ed. Hoboken: John Wiley & Sons.

Jiang, S.C. (2006). Human adenoviruses in water: occurrence and health implications: a critical review, *Environ. Sci. Technol.*, 40: 7132–7140.

Jo, W.K., C.P. Weisel, and P.J. Lioy (1990). Routes of chloroform exposure and body burden from showering with chlorinated tap water, *Risk Anal.*, 10: 575–580.

Jones, J.L., D.L. Hanson, M.S. Dworkin, D.L. Alderton, P.L. Fleming, J.E. Kaplan, and J. Ward (1999). Surveillance for AIDS-defining opportunistic illnesses, 1992–1997, *Morbidity and Mortality Weekly Rept.*, 48(SS-2): 1–22.

Jones, J.L., D. Kruszon-Moran, and M. Wilson (2003). *Toxoplasma gondii* infection in the United States, 1999–2000, *Emerg. Infect. Dis.*, 9: 1371–1374.

Juttner, F., and S.B. Watson (2007). Biochemical and ecological control of geosmin and 2-methylisoborneol in source waters, *Appl. Environ. Microbiol.*, 73: 4395–4406.

Karanis, P., C. Kourenti, and H. Smith (2007). Waterborne transmission of protozoan parasites: a worldwide review of outbreaks and lessons learnt, *J. Water and Health*, 5: 1–38.

Karner, D.A., J.H. Standridge, G.W. Harrington, and R.P. Barnum (2001). Microcystin algal toxins in source and finished drinking water, *J. AWWA*, 93(8): 72–81.

Katayama, H., E. Haramoto, K. Oguma, H. Yamashita, A. Tajima, H. Nakajima, and S. Ohgaki (2008). One-year monthly quantitative survey of Noroviruses, Enteroviruses, and Adenoviruses in wastewater collected from six plants in Japan, *Water Research*, 42: 1441–1448.

Kavlock, R.J. (2001). Pesticides as endocrine-disrupting chemicals, in R.I. Krieger, ed., *Handbook of Pesticide Toxicology*, vol. 1, 2nd ed. New York: Academic Press.

Keene, W.E. (2006). *Entamoeba histolytica,* pp. 203–208, in M48, *Waterborne Pathogens*. Denver, Colo.: American Water Works Association.

Khetsuriani, N., A. LaMonte-Fowlkes, M.S. Oberste, and M.A. Pallansch (2006). Enterovirus surveillance–United States, 1970–2005, *Morbidity and Mortality Weekly Rept.*, 55(SS-8): 1–20.

King, C.H., E.B. Shotts, Jr., R.E. Wooley, and K.G. Porter (1988). Survival of coliforms and bacterial pathogens within protozoa during chlorination, *Appl. Environ. Microbiol.*, 54: 3023–3033.

Kingsbury, J.A., G.C. Delzer, and J.A. Hopple (2008). Anthropogenic Organic Compounds in Source Water of Nine Community Waters Systems that Withdraw from Streams, 2002–2005. Scientific Investigations Report 2008-5208. U.S. Geological Survey. National Water-Quality-Assessment Program.

Klaasen, C.D., ed. (2007). *Casarett and Doull's Toxicology: The Basic Science of Poisons, 7th Edition*. New York: McGraw Hill.

Kogevinas, M., H. Becher, T. Benn, P.A. Bertazzi, P. Boffetta, H. Bas Bueno-de-Mewquita, D. Cogon, D. Colin, D. Flesch-Janys, M. Fingerhut, L. Green, T. Kauppinen, M. Littorin, E. Lynge, J.D. Matthews, M. Neuberger, N. Pearce, and R. Saracci (1997). Cancer mortality in workers exposed to phenoxy herbicides, chlorophenols, and dioxins, *Amer. J. Epidem.*, 145: 1061–1075.

Komulainen, H., V. Kosma, S. Vaittinen, T. Vartiainen, E. Korhonen, S. Lotojonen, R. Tuominen, and J. Tuomisto (1997). Carcinogenicity of the drinking water mutagen 3chloro-4(dichloromethyl)-5-hydroxy-2(5H)-furanone in the rat, *J.National Cancer Institute*, 89(12): 848–856.

Konduru, K., and G.G. Kaplan (2006) Stable growth of wild-type hepatitis A virus in cell culture, *J. Virol.*, 80: 1352–1360.

Konishi, K., N. Saito, E. Shoji, H. Takeda, M. Kato, M. Asaka, and H.-K. Ooi (2007). *Helicobacter pylori*: longer survival in deep ground water and sea water than in a nutrient-rich environment, *APMIS*, 115: 1285–1291.

Kramer, M.H., B.L. Herwaldt, G.F. Craun, R.L. Calderon, and D.D. Juranek (1996). Surveillance for waterborne-disease outbreaks–United States, 1993–1994, *Morbidity and Mortality Weekly Rept.*, 45(SS-1): 1–33.

Krasner, S.W., H.S Weinberg, S.D. Richardson, S.J. Pastor, R. Chinn, M.J. Sclimenti, G.D. Onstad, and A.D. Thruston Jr. (2006). Occurrence of a new generation of disinfection by-products, *Environ. Sci. Technol.*, 40: 7112–3.

Krewski, D., J.H. Lubin, J.M. Zielinski, M. Alavanja, V.S. Catalan, R.W. Field, J.B. Klotz, E.G. Létourneau, C.F. Lynch, J.L. Lyon, D.P. Sandler, J.B. Schoenberg, D.J. Steck, J.A. Stolwijk, C. Weinberg, and H.B. Wilcox (2006). A combined analysis of North American case-control studies of residential radon and lung cancer, *J. Toxicol. Environ. Health A*, 69(7): 533–597.

Kumar, A., M. Benimal, P. Kar, J.B. Sharma, and N.S. Murthy (2004). Hepatitis E in pregnancy, *Int. J. Gynaecol. Obstet.*, 85: 240–244.

Kusters, J.G., M.M. Gerrits, J.A.G. Van Strijp, and C.M.J.E. Vandenbroucke-Grauls (1997). Coccoid forms of *Helicobacter pylori* are the morphologic manifestation of cell death, *Infect. Immun.*, 65: 3672–3679.

Kusters, J.G., A.H.M. van Vliet, and E.J. Kuipers (2006). Pathogenesis of *Helicobacter pylori* infection, *Clin. Microbiol. Rev.*, 19(3): 449–490.

Lai, H., and L.S. McNeill (2006). Chromium redox chemistry in crinking water systems. *J. Environmental Engineering*, 132: 842–851.

Langley, J.M. (2005). Adenoviruses, *Pediatr. Rev.*, 26: 244–249.

Laskowski-Arce, M.A., and K. Orth (2008). *Acanthamoeba castellanii* promotes the survival of *Vibrio parahaemolyticus*, *Appl. Environ. Microbiol.*, 74: 7183–7188.

Lau, C., K. Anitole, C. Hodes, D. Lai, A. Pfahles-Hutchens, and J. Seed, (2007). Perfluoroalkyl acids: A review of monitoring and toxicological findings, *Toxicol. Sci.*, 99: 366–394.

LeChevallier, M.W. (2006). *Mycobacterium avium* complex, pp. 125–130, in M48, *Waterborne Pathogens*. Denver, Colo.: American Water Works Association.

LeChevallier, M.W., and M. Buckley (2007). *Clean Water: What Is Acceptable Microbial Risk?* Washington, D.C.: American Academy of Microbiology.

LeChevallier, M.W., T. M. Evans, and R.J. Seidler (1981). Effect of turbidity on chlorination efficiency and bacterial persistence in drinking water, *Appl. Environ. Microbiol.*, 42: 159–167.

Leclerc, H., L.A. Devriese, and D.A.A. Mossel (1996). Taxonomical changes in intestinal (faecal) enterococci and streptococci: consequences on their use as indicators of faecal contamination in drinking water, *J. Appl. Microbiol.*, 81: 459–466.

Leclerc, H., S. Edberg, V. Pierzo, and J.M. Delattre (2000). Bacteriophages as indicators of enteric viruses and public health risk in groundwaters, *J. Appl. Microbiol.*, 88: 5–21.

Leclerc, H., D.A.A. Mossel, S.C. Edberg, and C.B. Struijk (2001). Advances in the bacteriology of the coliform group: their suitability as markers of microbial water safety, *Ann. Rev. Microbiol.*, 55: 201–234.

Leclerc, H., L. Schwartzbod, and E. Dei-Cas (2002). Microbial agents associated with waterborne diseases, *Crit. Rev. Microbiol.*, 28(4): 371–409.

Lee, S.H., D.A. Levy, G.F. Craun, M.J. Beach, and R.L. Calderon (2002). Surveillance for waterborne-disease outbreaks–United States, 1999–2000, *Morbidity and Mortality Weekly Rept.*, 51(SS-8): 1–47.

Lee W.J., J.A. Hoppin, A. Blair, J.H. Lubin, M. Dosemeci, D.P. Sandler, and M.C. Alavanja (2004). Cancer incidence among pesticide applicators exposed to alachlor in the Agricultural Health Study, *Amer. J. Epidem.*, 159(4): 373–380.

Lee, W.J., D.P. Sandler, A. Blair, C. Samanic, A.J. Cross, and M.C. Alavanja (2007). Pesticide use and colorectal cancer risk in the Agricultural Health Study. *International Journal of Cancer,* 121(2): 339–346.

Levy, D.A., M.S. Bens, G.F. Craun, R.L. Calderon, and B.L. Herwaldt (1998). Surveillance for waterborne-disease outbreaks–United States, 1995–1996, *Morbidity and Mortality Weekly Rept.*, 47(SS-5): 1–34.

Liang, J.L., E.J. Dziuban, G.F. Craun, V. Hill, M.R. Moore, R.J. Gelting, R.L. Calderon, M.J. Beach, and S.L. Roy (2006). Surveillance for waterborne disease outbreaks associated with drinking water and water not intended for drinking—United States, 2003–2004, *Morbidity and Mortality Weekly Rept.*, 55(SS-12): 31–65.

Lindsay, D.S., J.P. Dubey, and B.L. Blagburn (1997). Biology of *Isospora* spp. from humans, nonhuman primates, and domestic animals, *Clin. Microbiol. Rev.*, 10: 19–34.

Liu, J., and Waalkes, M.P. (2008). Liver is a target of arsenic carcinogenesis. *Tox. Sci.* 105: 24–32.

Ljung, K., and M. Vahter (2007). Time to re-evaluate the guideline value for manganese in drinking water*? Environ. Health Persp.*, 115: 1533–1538.

Locas, A., C. Barthe, A.B. Margolin, and P. Payment (2008). Groundwater microbiological quality in Canadian drinking water municipal wells, *Can. J. Microbiol.*, 54: 472–478.

Loret, J.F., S. Robert, V. Thomas, A.J. Cooper, W.F. McCoy, and Y. Levi (2005). Comparison of disinfectants for biofilm, protozoa and *Legionella* control *J. Water and Health*, 3: 423–433.

MacIntosh, D.L., J.D. Spengler, H. Ozkaynak, L.-h. Tsai, and P.B. Ryan (1996). Dietary exposures to selected metals and pesticides, *Environ. Health Persp.*, 104: 202–209.

Mallevialle, J., and F.H. Suffet, eds. (1987). *Identification and Treatment of Tastes and Odors in Drinking Water*. Awwa Research Foundation and Lyonnaise des Eaux. Denver, Colo.: Water Research Foundation.

Marciano-Cabral, F., and G. Cabral (2003). *Acanthamoeba* spp. as agents of disease in humans, *Clin. Microbiol. Rev.*, 16: 273–307.

Marciano-Cabral, F., R. MacLean, A. Mensah, and L. LaPat-Polasko (2003). Identification of *Naegleria fowleri* in domestic water sources by nested PCR, *Appl. Environ. Microbiol.*, 69: 5864–5869.

Marshall, M.M., D. Naumovitz, Y. Ortega, and C.R. Sterling (1997). Waterborne protozoan pathogens, *Clin. Microbiol. Rev.*, 10: 67–85.

Marston, B.J., J.F. Plouffe, T.M. File, Jr., B.A. Hackman, S.-J. Salstrom, H.B. Lipman, M.S. Kolczak, and R.F. Breiman (1997). Incidence of community-acquired pneumonia requiring hospitalization. Results of a population-based active surveillance study in Ohio. The Community-Based Pneumonia Incidence Study Group, *Arch. Intern. Med.*, 157(15): 1709–1718.

Martinez, A.J., and G.S. Visvesvara (1997). Free-living, amphizoic and opportunistic amebas, *Brain Pathol.*, 7: 583–598.

Marx, A., and R.R. Neutra (1997). Magnesium in drinking water and ischemic heart disease, *Epidem. Rev.*, 19: 258–272.

Mass, M.J., A. Tennant, B.C. Roop, A. Cullen, M. Styblo, D.J. Thomas, and A.D. Kligerman (2001). Methylated trivalent arsenic species are genotoxic, *Chem. Res. Toxicol.* 14: 355–361.

Mattix K.D., P.D. Winchester, L.R. Scherer (2007). Incidence of abdominal wall defects is related to surface water atrazine and nitrate levels. *J Pediatr Surg*. 42(6):947–9.

Maxwell, N.I., D.E. Burmaster, and D. Ozonoff (1991). Trihalomethanes and maximum contaminant levels: the significance of inhalation and dermal exposures to chloroform in household water, *Regul. Toxicol. Pharmaco.* 14: 297–312.

McCarron, D.A (1998). Importance of dietary calcium in hypertension, *J. Am. Coll. Nutr.*, 17: 97–99.

McDonald, T.A., and H. Komulainen (2005). Carcinogenicity of the chlorination disinfection by-product MX. *J. Environ. Sci. Health C. Environ. Carcinog. Ecotoxicol. Rev.*, 23: 163–214.

McKone, T.E. (1987). Human exposure to volatile organic compounds in household tap water: the indoor inhalation pathway, *Environ. Sci. Technol.*, 21: 1194–2101.

Melnick, R.L., A. Nyska, P.M. Foster, J.H. Roycroft, and G.E. Kissling (2007). Toxicity and carcinogenicity of the water disinfection by-product, dibromoacetic acid, in rats and mice, *Toxicology,* 230: 126–136.

Mendez, E., and C.F. Arias (2007). Astroviruses, pp. 981–1000, in *Field's Virology*, 5th ed. Philadelphia: Lippincott, Williams, and Wilkins.

Messner, M.J., C.L. Chappell, and P.C. Okhuysen (2001). Risk assessment for *Cryptosporidium*: a hierarchical Bayesian analysis of human dose response data, *Wat. Res.*, 35(16): 3934–3940.

Messner, M., S. Shaw, S. Regli, K. Rotert, V. Blank, and J. Soller (2006). An approach for developing a national estimate of waterborne disease due to drinking water and a national estimate model application, *J. Water and Health,* 4(suppl. 2): 201–240.

Miller, W.G., C.T. Parker, M. Rubenfield, G.L. Mendz, M.M.S.M. Wosten, D.W. Ussery, J.F. Stolz, T.T. Binnewies, P.F. Hallin, G. Wang, J.A. Malek, A. Rogosin, L.H. Stanker, and R.E. Mandrell (2007). The complete genome sequence and analysis of the epsilonproteobacterium *Arcobacter butzleri*, *PLoS ONE,* 2(12): e1358. doi:10.1371/journal.pone.0001358.

Mlot, C. (2009). Antibiotics in nature: beyond biological warfare, *Science*, 324: 1637–1639.

Monarca, S., F. Donato, I. Zerbini, R.L. Calderon, and G.F. Craun (2006). Review of epidemiological studies on drinking water hardness and cardiovascular diseases. *Eur. J. Cardiovasc. Prev. Rehabil.*, 13(4): 495–506.

Moore, D.F., J.A. Guzman, and C. McGee (2008). Species distribution and antimicrobial resistance of *Enterococci* isolated from surface and ocean water, *J. Appl. Microbiol.*, 105(4): 1017–1025.

Moore, J.E., D. Corcoran, J.S.G. Dooley, S. Fanning, B. Lucey, M. Matsuda, D.A. McDowell, F. Megraud, B.C. Millar, R. O'Mahony, L. O'Riordan, M. O'Rourke, J.R. Rao, P.J. Rooney, A. Sails, and P. Whyte (2005). *Campylobacter*, *Vet. Res.*, 36: 351–382.

Morgan-Ryan, U.M., A. Fall, L.A. Ward, N. Hijjawi, I. Sulaiman, R. Fayer, R.C. Thompson, M. Olson, A. Lal, and L. Xiao (2002). *Cryptosporidium hominis* n. sp. (Apicomplexa: Cryptosporidiidae) from *Homo sapiens*, *J. Eukaryot. Microbiol.*, 49: 433–440.

Moser, V.C., P.M. Phillips, A.B. Levine, K.L. McDaniel, R.C. Sills, B.S. Jortner, and M.T. Butt (2004). Neurotoxicity produced by dibromoacetic acid in drinking water of rats, *Toxicol. Sci.,* 79: 112–22.

Motulsky, A.G. (1988). Hemochromatosis (iron storage disease), in *Cecil Textbook of Medicine*, J.B. Wyngaarden and L.H. Smith, eds. Philadelphia: W.B. Saunders Co.

Muller, T., A. Ulrich, E.-M. Ott, and M. Muller (2001). Identification of plant-associated enterococci, *J. Appl. Microbiol.*, 91: 268–278.

Mura, M., T.J. Bull, H. Evans, K. Sidi-Boumedine, L. McMinn, G. Rhodes, R. Pickup, and J. Hermon-Taylor (2006). Replication and long-term persistence of bovine and human strains of *Mycobacterium avium* subsp. *paratuberculosis* within *Acanthamoeba polyphaga*, *Appl. Environ. Microbiol.*, 72: 854–859.

Murray, C.W. III, S.K. Egan, H. Kim, N. Beru, and P.M. Bolger. (2008). US Food and Drug Administration's Total Diet Study: dietary intake of perchlorate and iodine. *Journal of Exposure Science and Environmental Epidemiology*, 18: 571–580.

Myers, J.P., R.T. Zoeller, and F.S. vom Saal (2009). A clash of old and new scientific concepts in toxicology, with important implications for public health. *Environ. Health Persp.* Doi: 10.1289/ehp.0900887. Online July 30, 2009.

Narotsky, M.G., D.S. Best, E.H. Rogers, A. McDonald, Y.M. Sey, and J.E. Simmons (2008). Integrated disinfection by-products mixtures research: assessment of developmental toxicity in Sprague-Dawley rats exposed to concentrates of water disinfected by chlorination and ozonation/postchlorination. *J. Toxicol. Environ. Health A*, 71: 1216–1221.

Nasser, A.M. (1994). Prevalence and fate of hepatitis A virus in water, *Crit. Rev. Environ. Sci. Technol.*, 24: 281–323.

National Academy of Sciences Committee on Nitrite and Alternative Curing Agents in Food. (1981). *The health effects of nitrate, nitrite, and N-nitroso compounds*. Washington, D.C.: National Academies Press.

National Academy of Sciences Safe Drinking Water Committee. (1977). Drinking water and health. Washington, D.C.

National Academy of Sciences Safe Drinking Water Committee. (1980). *Drinking Water and Health*, vol. 2. Washington, D.C.: National Academies Press.

National Academy of Sciences Safe Drinking Water Committee. (1982). *Drinking Water and Health*, vol. 4. Washington, D.C.: National Academies Press.

National Academy of Sciences Safe Drinking Water Committee. (1987). *Drinking Water and Health*, vol. 7. Washington, D.C.: National Academies Press.

National Cancer Institute. (2008). Agricultural Health Study. http://aghealth.nci.nih.gov/.

Navarro Silvera SA, Rohan TE. (2007). Trace elements and cancer risk: a review of the epidemiologic evidence, Cancer Causes Control., 18(1):7-27.

National Research Council (NRC) (1983). Committee on the Institutional Means for Assessment of Risks to the Public. *Risk Assessment in the Federal Government: Managing the Process.* Washington, D.C.: National Academies Press.

NRC (1989). *Recommended Dietary Allowances*, 10th ed. Washington, D.C.: National Academies Press.

NRC, Board on Environmental Studies and Toxicology. Board on Environmental Studies and Toxicology (1995). *Guidelines for Chemical Warfare Agents in Military Field Drinking Water.* Washington, D.C.: National Academies Press.

NRC, Subcommittee on Arsenic in Drinking Water (1999a). *Arsenic in Drinking Water.* Washington, D.C.: National Academies Press.

NRC, Committee to Assess Health Risks from Exposure to Low Levels of Ionizing Radiation (1999b). *Health Effects of Exposure to Radon: BEIR VI.* Washington, D.C.: National Academies Press.

NRC, Committee on Risk Assessment of Exposure to Radon in Drinking Water (1999c). *Risk Assessment of Radon in Drinking Water.* Washington, D.C.: National Academies Press.

NRC, Subcommittee to Update the 1999 Arsenic in Drinking Water Report (2001). *Arsenic in Drinking Water.* Washington, D.C.: National Academies Press.

NRC, Board on Environmental Studies and Toxicology (2005). *Health Implications of Perchlorate Ingestion.* Washington, D.C.: National Academies Press.

NRC, Committee on Fluoride in Drinking Water (2006a). *Fluoride in Drinking Water—A Scientific Review of EPA's Standards.* Washington, D.C.: National Academies Press.

NRC, Committee to Assess Health Risks from Exposure to Low Levels of Ionizing Radiation (2006b). *Health Risks from Exposure to Low Levels of Ionizing Radiation: BEIR VII—Phase 2.* Washington, D.C.: National Academies Press.

NRC, Committee on Toxicity Testing and Assessment of Environmental Agents (2007). *Toxicity Testing in the 21st Century: A Vision and a Strategy.* Washington, D.C.: National Academies Press.

NRC (2008). *Desalination, A National Perspective.* Washington, D.C.: National Academies Press.

National Toxicology Program (NTP) (1978). Bioassay of 1,2-Dichloroethane for Possible Carcinogenicity (CAS No. 107-06-2). TR-55.

NTP (1985). Toxicology and Carcinogenesis Studies of 1,2-Dichlorobenzene (o-Dichlorobenzene) (CAS No. 95-50-1) in F344/N Rats and B6C3F1 Mice (Gavage Studies). TR-255.

NTP (1986). Toxicology and Carcinogenesis Studies of Xylenes (Mixed) (60% m-Xylene, 14% p-Xylene, 9% o-Xylene, and 17% Ethylbenzene) (CAS No. 1330-20-7) in F344/N Rats and B6C3F1 Mice (Gavage Studies).

NTP (1987a). Toxicology and Carcinogenesis Studies of 1,4-Dichlorobenzene (CAS No. 106-46-7) in F344/N Rats and B6C3F1 Mice (Gavage Studies). TR-319.

NTP (1987b). Toxicology and Carcinogenesis Studies of Bromodichloromethane (CAS No. 75-27-4) in F344/N Rats and B6C3F1 Mice (Gavage Studies). TR-321.

NTP (1993). Toxicology and Carcinogenesis Studies of Mercuric Chloride (CAS No. 7487-94-7) in F344 Rats and B6C3F1 Mice (Gavage Studies). TR 408.

NTP (1998). Short Term Reproductive and Developmental Toxicity of Bromochloroacetic Acid (CAS No. 5589-96-8) Administered in the Drinking Water to Sprague-Dawley Rats. NTP Study Number: RDGT96001.

NTP (1999). Toxicology and Carcinogenesis Studies of Ethylbenzene (CAS No. 100-41-4) in F344/N Rats and B6C3F1 Mice (Inhalation Studies). TR-466.

NTP (2000). NTP Technical Report on the Toxicity Studies of 1,1,1-Trichloroethane (CAS No. 71-55-6) Administered in Microcapsules in Feed to F344/N Rats and B6C3F1 Mice. TOX-41.

NTP (2002). NTP Technical Report on the Toxicity Studies of trans-1,2-Dichloroethylene (CAS No. 156-60-5) Administered in Microcapsules in Feed to F344/N Rats and B6C3F1 Mice. TOX-55.

NTP (2005a). 1,2,3-Trichloropropane, Report on Carcinogens, 11th ed.; U.S. Department of Health and Human Services, Public Health Service, National Toxicology Program.

NTP (2005b). Toxicology and Carcinogenesis Studies of Sodium Chlorate (CAS No. 7775-09-9) in F344/N Rats and B6C3F1 Mice (Drinking Water Studies). TR-517.

NTP (2006). Toxicology and Carcinogenesis Studies of Bromodichloromethane (CAS No. 75-27-4) in Male F344/N Rats and Female B6C3F1 Mice (Drinking Water Studies). TR-532.

NTP (2008a). Toxicology and Carcinogenesis Studies of Sodium Dichromate Dihyrate (CAS No. 7789-12-0) in F344/N Rats and B6C3F1 Mice (Drinking Water Studies). NTP TR-546. NIH Publication No. 08-5887.

NTP (2008b). Toxicology and Carcinogenesis Studies of Bromochloroacetic Acid (CAS No. 5589-96-8) in F344/N Rats and B6C3F1 Mice (Drinking Water Studies). Draft. TR-549.

New Jersey Department of Environmental Protection (NJDEP) (2001). Division of Science, Research and Technology. MTBE in New Jersey's Environment. http://www.state.nj.us/dep/dsr/mtbe/MTBE-NJ.PDF. Trenton, N.J. Accessed 6/21/10.

NJDEP (2003). Division of Science, Research and Technology. The Characterization of Tentatively Identified Compounds (TICs) in Samples from Public Water Systems in New Jersey. http://www.state.nj.us/dep/dsr/TIC-report.pdf. Trenton, N.J. Accessed 6/21/10.

NJDEP (2009). Division of Science, Research and Technology. Derivation of Ingestion-Based Soil Remediation Criteria for Cr+6 Based on the NTP Chronic Bioassay Data for Sodium Dichromate Dihydrate. http://www.state.nj.us/dep/dsr/chromium/soil-cleanup-derivation.pdf. Trenton, N.J. Accessed 6/21/10.

New Jersey Drinking Water Quality Institute (2009). Maximum Contaminant Level Recommendation Document on Radon-222. Trenton, N.J. http://www.state.nj.us/dep/watersupply/radon_report_dwqi_2_17_09.pdf. Accessed 6/21/10.

Nielsen, E., G. Ostergaard, and J.C. Larsen (2008). *Toxicological Risk Assessment of Chemicals*. New York: Informa Healthcare.

Nwachuku, N., and C.P. Gerba (2004). Health effects of *Acanthamoeba* spp. and its potential for waterborne transmission, *Rev. Environ. Contam. Toxicol.*, 180: 93–131.

Nwachuku, N. and C.P. Gerba (2006). Health risks of enteric viral infections in children, *Rev. Environ. Contam. Toxicol.*, 186: 1–56.

Oberdörster, G., A. Maynard, K. Donaldson, V. Castranova, J. Fitzpatrick, K. Ausman, J. Carter, B. Karn, W. Kreyling, D. Lai, S. Olin, N. Monteiro-Riviere, D. Warheit, and H. Yang (2005). Principles for characterizing the potential human health effects from exposure to nanomaterials: elements of a screening strategy. *Part Fibre Toxicol.*, 2: 8.

Ochoa-Acuña H., J. Frankenberger, L. Hahn, C. Carbajo (2009). Drinking-water herbicide exposure in Indiana and prevalence of small-for-gestational-age and preterm delivery. *Environ Health Perspect.* 117(10):1619–24.

O'Flaherty, E.J., B.D. Kerger, S.M. Hays, and D.J. Paustenbach (2001). A physiologically based model for the ingestion of chromium(III) and chromium(VI) by humans, *Toxicol. Sci.*, 60: 196–213.

Ohno, A., N. Kato, R. Sakamoto, S. Kimura, and K. Yamaguchi (2008). Temperature-dependent parasitic relationship between *Legionella pneumophila* and a free-living amoeba (*Acanthamoeba castellanii*), *Appl. Environ. Microbiol.*, 74: 4585–4588.

Olivares, M., F. Pizarro, H. Speisky, B. Lonnerdal, and R. Uauy (1998). Copper in infant nutrition: safety of the World Health Organization provisional guideline value for copper content in drinking water, *J. Pediatr. Gastroenterol. Nutr.*, 26(25): 1–257.

On, S.L.W., S. Hynes, and T. Wadstrom (2002). Extragastric *Helicobacter* species, *Helicobacter*, 7(suppl. 1): 63–67.

Pagnier, I., M. Merchat, D. Raoult, and B. La Scola (2009). Emerging *Mycobacteria* spp. in cooling towers, *Emerg. Infect. Dis.*, 15: 121–122.

Panda, S.K., D. Thakral, and S. Rehman. (2007). Hepatitis E virus, *Rev. Med. Virol.*, 17: 151–180.

Parashar, U.D., J.P. Alexander, and R.I. Glass (2006). Prevention of Rotavirus gastroenteritis among infants and children: recommendations of the Advisory Committee on Immunization Practices, *Morbidity and Mortality Weekly Rept.*, 55(RR-12): 1–13.

Parashar, U.D., E.S. Quiroz, A.W. Mounts, S.S. Monroe, R.L. Fankhauser, T. Ando, J.S. Noel, S.N. Bulens, R.S. Beard, J.-F. Li, J.S. Bresee, and R.I. Glass (2001). 'Norwalk-like viruses' public health consequences and outbreak management, *Morbidity and Mortality Weekly Rept.*, 50(RR-9): 1–17.

Park, S.R., W.G. MacKay, and D.C. Reid (2001). *Helicobacter* sp. recovered from drinking water biofilm sampled from a water distribution system, *Water Res.*, 35: 1624–1626.

Patterson, B.H., and O.A. Levander (1997). Naturally occurring selenium compounds in cancer chemoprevention trials: a workshop summary, *Cancer Epidem. Biomarkers Prev.*, 6: 63–69.

Pedley, S., J. Bartrum, G. Rees, A. Dufour, and J.A. Cotruvo, eds. (2004). *Pathogenic Mycobacteria in Water: A Guide to Public Health Consequences, Monitoring and Management.* London: World Health Organization, IWA Publishing.

Peterson, M.C. (1994). Clinical aspects of *Campylobacter jejuni* infections in adults, *West J. Med.*, 161: 148–152.

Pitzer, V.E., C. Viboud, L. Simonsen, C. Steiner, C.A. Panozzo, W.J. Alonso, M.A. Miller, R.I. Glass, J.W. Glasser, U.D. Parashar, and B.T. Grenfell (2009). Demographic variability, vaccination, and the spatiotemporal dynamics of Rotavirus epidemics, *Science*, 325: 290–294.

Poirier, K. (1994). Summary of the derivation of the reference dose for selenium. in *Risk Assessment of Essential Elements*, W. Mertz, C.O. Abernathy, and S.S. Olin, eds. Washington, D.C.: International Life Sciences Institute Press.

Porterfield, S.P. (1994). Vulnerability of the developing brain to thyroid abnormalities: environmental insults to the thyroid system, *Environ. Health Persp.*, 102(Suppl 2): 125–130.

Post, G.B., J.B. Louis, K.R. Cooper, B.J. Boros-Russo, and R.L. Lippincott (2009). Occurrence and potential significance of perfluorooctanoic acid (PFOA) detected in New Jersey public drinking water systems, *Environ. Sci. Technol.*, 43: 4547–4554.

Proctor, M.E., K.A. Blair, and J.P. Davis (1998). Surveillance data for waterborne illness detection: an assessment following a massive waterborne outbreak of *Cryptosporidium* infection, *Epidemiol. Infect.*, 120: 43–54.

Regli, S., J.B. Rose, C.N. Haas, and C.P. Gerba (1991). Modeling the risk from *Giardia* and viruses in drinking water, *J. AWWA*, 83(11): 76–84.

Reiber, S.H., and W.A. Kukull. (1996). *Aluminum, Drinking Water, and Alzheimer's Disease*. Denver, Colo.: Awwa Research Foundation.

Reid, J. (1994). Arsenic occurrence: USEPA seeks a clearer picture, *J. AWWA*, 86: 44–51.

Rice, E.W., N.J. Adcock, M. Sivaganesan, and L.J. Rose (2005). Inactivation of spores of *Bacillus anthracis* Sterne, *Bacillus cereus*, and *Bacillus thuringiensis* subsp. israelensis by chlorination, *Appl. Environ.Microbiol.*, 71: 5587–5589.

Rice, G., L.K. Teuschler, T.F. Speth, S.D. Richardson, R.J. Miltner, K.M. Schenck, C. Gennings, E.S. Hunter, III, M.G. Narotsky, and J.E. Simmons (2008). Integrated disinfection by-products research: assessing reproductive and developmental risks posed by complex disinfection by-product mixtures, *J. Toxicol. Environ. Health A*, 71: 222–234.

Richardson, S.D., M.J. Plewa, E.D. Wagner, R. Schoeny, and D.M. Demarini (2007). Occurrence, genotoxicity, and carcinogenicity of regulated and emerging disinfection by-products in drinking water: a review and roadmap for research, *Mutat. Res.*, 636: 178–242.

Riffard, S., S. Springthorpe, L. Filion, S.A. Sattar, T. Brooks, R. Osiki, M. Lee, and D. Abrial (2004). *Occurrence of Legionella in Groundwater.* Denver, Colo.: Water Research Foundation.

Roberts, L. (2009a). Type 2 Poliovirus back from the dead in Nigeria, *Science*, 325: 660–661.

Roberts, L. (2009b). Polio: looking for a little luck, *Science*, 323: 702–705.

Robinson, M., R.H. Bruner, and G.R. Olson (1990). Fourteen and ninety day toxicity studies of methyl tertiary-butyl ether in Sprague-Dawley rats. *Journal of the American College of Toxicology*, 9: 525–540.

Rodriguez, E., A. Sordo, J.S. Metcalf, and J.L. Acero (2007). Kinetics of the oxidation of cylindrospermopsin and anatoxin-a with chlorine, monochloramine and permanganate, *Water Res.*, 41: 2048–2056.

Rodriguez, R.A., P.M. Gundy, and C.P. Gerba (2008). Comparison of BGM and PLC/PRC/5 cell lines for total culturable viral assay of treated sewage, *Appl. Environ. Microbiol.*,74: 2583–2587.

Rogers, J., and C.W. Keevil (1992). Immunogold and fluorescein immunolabelling of *Legionella pneumophila* within an aquatic biofilm visualized by using episcopic differential interference contrast microscopy, *Appl. Environ. Microbiol.*, 58: 2326–2330.

Rolle-Kampczyk, U.E., G.J. Fritz, U. Diez, I. Lehmann, M. Richter, and O. Herbarth (2004). Well water–One source of *Helicobacter pylori* colonization, *Int. J. Hyg. Environ. Health*, 207: 363–368.

Rupnow, M.F.T., R.D. Shachter, D.K. Owens, and J. Parsonnet (2000). A dynamic transmission model for predicting trends in *Helicobacter pylori* and associated diseases in the United States, *Emerg. Infect. Dis.*, 6: 228–236.

Rusiecki J.A., A. De Roos, W.J. Lee, M. Dosemeci, J.H. Lubin, J.A. Hoppin, A. Blair, M.C. Alavanja (2004). Cancer incidence among pesticide applicators exposed to atrazine in the Agricultural Health Study. *J Natl Cancer Inst.*, 96(18):1375–82.

Schaub, S.A., and R.K. Oshiro (2000). Public health concerns about Caliciviruses as waterborne contaminants, *J. Infect. Dis.*, 181:suppl. 2: S374–S380.

Schulman, L.J., E.V. Sargent, B.D. Naumann, E.C. Faria, D.G. Dolan, and J.P. Wargo (2002). A human health risk assessment of pharmaceuticals in the aquatic environment, *Human Ecol. Risk Assess.*, 8(4): 657–680.

Schuster, C.J., A.G. Ellis, W.J. Robertson, D.F. Charron, J.J. Aramini, B.J. Marshall, and D.T. Medeiros (2005). Infectious disease outbreaks related to drinking water in Canada, 1974–2001, *Can. J. Pub. Health*, 96: 254–258.

Schwab, B.W., E.P. Hayes, J.M. Fiori, F.J. Mastrocco, N.M. Roden, D. Cragin, R.D. Meyerhoff, V.J. D'Aco, and P.D. Anderson (2005). Human pharmaceuticals in US surface waters: a human health risk assessment, *Regul. Toxicol. Pharmacol.*, 42(3): 296–312.

Scott, T.M., J.B. Rose, T.M. Jenkins, S.R. Farrah, and J. Lukasik (2002). Microbial source tracking: current methodology and future directions, *Appl. Environ. Microbiol.*, 68: 5796–5803.

Shahamat, M., U. Mai, C. Paszko-Kolva, M. Kessel, and R.R. Colwell (1993). Use of autoradiography to assess viability of *Helicobacter pylori* in water, *Appl. Environ. Microbiol.*, 59: 1231–1235.

She, F.-F., J.-Y. Lin, J.-Y. Liu, C. Huang, and D.-H. Su (2003). Virulence of water-induced coccoid *Helicobacter pylori* and its experimental infection in mice, *World J. Gastroenterol.*, 9: 516–520.

Shin, G.-A., J.-K. Lee, R. Freeman, and G.A. Cangelosi (2008). Inactivation of *Mycobacterium avium* complex by UV irradiation, *Appl. Environ. Microbiol.*, 74: 7067–7069.

Shovlin, M.G., R.S. Yoo, D.R. Crapper-McLachlan, E. Cummings, J.M. Donohue, W.K. Hallman, Z. Khachaturian, J. Orme-Zavaleta, and S. Teefy (1993). *Aluminum in Drinking Water and Alzheimer's Disease: A Resource Guide*. Denver, Colo.: Water Research Foundation.

Shulman, L.M., Y. Manor, D. Sofer, and E. Mendelson (2009). Type 2 polio still in our midst, *Science*, 324: 334.

Simmons, J.E., L.K. Teuschler, C. Gennings, T.F. Speth, S.D. Richardson, R.J. Miltner, M.G. Narotsky, K.D. Schenck, E.S. Hunter III, R.C. Hertzberg, and G. Rice (2004). Component-based and whole-mixture techniques for addressing the toxicity of drinking-water disinfection by-product mixtures, *J. Toxicol. Environ. Health A*, 67:741–754.

Simmons, J.E., S.D. Richardson, L.K. Teuschler, R.J. Miltner, T.F. Speth, K.M. Schenck, E.S. Hunter III, and G. Rice (2008). Research issues underlying the four-lab study: integrated disinfection by-products mixtures research, *J. Toxicol. Environ. Health A*, 71:1125–1132.

Smeds, A., T. Vartianinen, J. Maki-Paakanen, L. Kronberg (1997). Concentrations of Ames mutagenic chlorohydroxyfuranones and related compounds in drinking waters, *Environ. Sci. Technol.*, 31:1033–1039.

Snelling, W.J., J.E. Moore, J.P. McKenna, D.M. Lecky, and J.S.G. Dooley (2006a). Bacterial-protozoa interactions; an update on the role these phenomena play towards human illness, *Microbes Infect.*, 8: 578–587.

Snelling, W.J., M. Matsuda, J.E. Moore, and J.S.G. Dooley (2006b). Under the microscope: *Arcobacter*, *Lett. Appl. Microbiol.*, 42:1: 7–14.

Snyder, S.A., B.J. Vanderford, J. Drewes, E. Dickenson, E.M. Snyder, G.M. Bruce, and R.C. Pleus (2008). *State of Knowledge of Endocrine Disruptors and Pharmaceuticals in Drinking Water.* Denver, Colo.: Water Research Foundation.

Solnick, J.V., and D.B. Schauer (2001). Emergence of diverse *Helicobacter* species in the pathogenesis of gastric and enterohepatic diseases, *Clin. Microbiol. Rev.*, 14:1: 59–97.

Staat, M.A., G. Fairbrother, K.M. Edwards, M. Griffin, P.G. Szilagyi, G.A. Weinberg, C.B. Hall, C.A. Panozzo, D.C. Payne, J.E. Tate, H.A. Clayton, A.L. Fowlkes, M. Wang, A.T. Curns, J. Gentsch, M.M. Cortese, M. Patel, M.A. Widdowson, and U. Parashar (2008). Delayed onset and diminished magnitude of Rotavirus activity–United States, November 2007–May 2008, *Morbidity and Mortality Weekly Rept.*, 57:25: 697–700.

Standridge, J. (2008). *E. coli* as a public health indicator of drinking water quality, *J. AWWA,* 100:2: 65–75.

Stanley, S.L., Jr. (2003). Amoebiasis, *Lancet*, 361: 1025–1034.

Stauffer, W., and J.I. Ravdin (2003). *Entamoeba histolytica*: an update, *Curr. Opin. Infect. Dis.*, 16: 479–485.

Steenland, K., T. Fletcher, and D. Savitz (2008). C8 Science Panel. Status report: Association of perfluorooctanic acid (PFOA) and perfluoroctanesulfonate (PFOS) with lipids among adults in a community with high exposure to (PFOA), http://www.c8sciencepanel.org/

Steenland, K., T. Fletcher, and D. Savitz (2009). C8 Science Panel. Status report: Association of perfluorooctanic acid (PFOA) and perfluoroctanesulfonate (PFOS) with uric acid among adults in a community with high exposure to (PFOA), http://www.c8sciencepanel.org/

Strachan, N.J.C., R.O. Watson, V. Novik, D. Hofreuter, I.D. Ogden, and J.E. Galan (2008). Sexual dimorphism in campylobacteriosis," *Epidemiol. Infect.*, 136: 1492–1495.

Szewzyk, U., R. Szewzyk, W. Manz, and K.-H. Schleifer (2000). Microbiological safety of drinking water, *Ann. Rev. Microbiol.*, 54: 81–127.

Taubes, G. (2008). The bacteria fight back, *Science*, 321: 356–361.

Tenter, A.M., A.R. Heckeroth, and L.M. Weiss (2000). *Toxoplasma gondii*: from animals to humans, *Int. J. Parasitol.*, 30: 1217–1258.

Teuschler, L.K., G.E. Rice, C.R. Wilkes, J.C. Lipscomb, and F.W. Power (2004). A feasibility study of cumulative risk assessment methods for drinking water disinfection by-product mixtures. *J. Toxicol. Environ. Health A*, 67: 755–77.

Thomas, V., T. Bouchez, V. Nicolas, S. Robert, J.F. Loret, and Y. Levi (2004). Amoebae in domestic water systems: resistance to disinfection treatments and implication in *Legionella* persistence, *J. Appl. Microbiol.*, 97: 950–963.

Toppari, J., J.C. Larsen, P. Christiansen, A. Giwercman, P. Grandjean, L.J. Guillette, Jr., B. Jegou, T.K. Jensen, P. Jouannet, N. Keiding, H. Leffers, J.A. McLachlan, O. Meyer, J. Muller, W. Rajpert-De Meyts, T. Scheike, R. Sharpe, J. Sumpter, and N.E. Skakkebaek (1996). Male reproductive health and environmental xenoestrogens, *Environ. Health Persp.*, 104:suppl. 4: 741–803.

Total Coliform Rule/Distribution System (TCRDS) Federal Advisory Committee (2008). Agreement in Principle, Washington, D.C.: U.S. Environmental Protection Agency.

Trubo, R. (2005). Endocrine-disrupting chemicals probed as potential pathways to illness. *Jour. Amer. Med. Assoc.*, 294(3): 291–293.

USDA (U.S. Department of Agriculture) (2009). Nutrient Intakes from Food: Mean Amounts Consumed per Individual, One Day, 2005–2006. U.S. Department of Agriculture, Agricultural Research Service. http://www.ars.usda.gov/Services/docs.htm?docid=17041. Last accessed July 2009.

USEPA (U.S. Environmental Protection Agency) (1979). National Secondary Drinking Water Regulations, *Federal Register*, 44: 42198–42199.

USEPA (1984). Health Assessment Document for 1,1,1,-Trichloroethane (Methyl Chloroform). Washington, D.C.: Office of Health and Environmental Assessment.

USEPA (1985a). National Primary and Secondary Drinking Water Regulations; Synthetic Organic Chemicals, Inorganic Chemicals and Microorganisms. *Federal Register*, 50: 46936–47022.

USEPA (1985b). National Primary Drinking Water Regulations; Volatile Synthetic Organic Chemicals. *Federal Register*, 50: 46880–46901.

USEPA (1986a). Guidelines for Carcinogen Risk Assessment (1986a). Washington, D.C.: Risk Assessment Forum.

USEPA (1986b). National Primary and Secondary Drinking Water Regulations; Fluoride. *Federal Register*, 51: 11396–11412.

USEPA (1987a). National Primary Drinking Water Regulations; Synthetic Organic Chemicals; Monitoring for Unregulated Contaminants. *Federal Register*, 52: 25690–25717.

USEPA (1987b). Water Pollution Controls; National Primary Drinking Water Regulations; Volatile Synthetic Organic Chemicals; Para-Dichlorobenzene. *Federal Register*, 52: 12876–12883.

USEPA (1988a). Drinking Water; Substitution of Contaminants and Drinking Water Priority List of Additional Substances which May Require Regulation under the Safe Drinking Water Act. *Federal Register*, 53: 1892–1902.

USEPA (1988b). Distribution Tables for the National Inorganics and Radionuclides Survey (NIRS) Results. Memorandum from Jon Longtin to Arthur Perler, February 23. USEPA Water Docket.

USEPA (1989a) Drinking Water; National Primary Drinking Water Regulations; Filtration, Disinfection; Turbidity, *Giardia lamblia*, Viruses, *Legionella*, and Heterotrophic Bacteria, Final Rule. *Federal Register*, 54(124): 27486–27541.

USEPA. (1989b) Drinking Water; National Primary Drinking Water Regulations; Total Coliforms (Including Fecal Coliforms and E. Coli); Final Rule. *Federal Register*, 54(124): 27544–27568.

USEPA (1989c). National Primary and Secondary Drinking Water Regulations. *Federal Register*, 54: 22062–22160.

USEPA (1990a). National Primary and Secondary Drinking Water Regulations; Synthetic Organic Chemicals and Inorganic Chemicals. *Federal Register*, 55: 30370–30449.

USEPA (1991a). National Primary and Secondary Drinking Water Regulations; Synthetic Organic Chemicals and Inorganic Chemicals; Monitoring for Unregulated Contaminants; National Primary Drinking Water Regulations Implementation; National Secondary Drinking Water Regulations. *Federal Register*, 56: 3526–3599.

USEPA (1991b). Maximum Contaminant Level Goals and National Primary Drinking Water Regulations for Lead and Copper. *Federal Register*, 56: 26460–26564.

USEPA (1991c). National Primary and Secondary Drinking Water Regulations; Radionuclides. *Federal Register*, 56: 33050–33127.

USEPA (1992a). Consensus Method for Determining Groundwaters Under the Direct Influence of Surface Water Using Microscopic Particulate Analysis (MPA). EPA 910/9-92-029, EPA Region 10, Environmental Services Division, Port Orchard, Wash.

USEPA (1992b). National Primary and Secondary Drinking Water Regulations—Synthetic Organic Chemicals and Inorganic Chemicals; National Primary Drinking Water Regulations Implementation. *Federal Register*, 57: 31776–31849.

USEPA (1993). Draft Drinking Water Health Criteria Document for Bromate. Washington, D.C.: Office of Science and Technology, Office of Water.

USEPA (1994a). National Primary Drinking Water Regulations; Disinfectants and Disinfection Byproducts; Proposed. *Federal Register*, 59: 38668–38829.

USEPA (1994b). Draft Drinking Water Health Criteria Document for Chlorine, Hypochlorous Acid and Hypochlorite Ion. Washington, D.C.: Office of Science and Technology, Office of Water.

USEPA (1994c). Draft Drinking Water Health Criteria Document for Chloramines. Washington, D.C.: Office of Science and Technology, Office of Water.

USEPA (1994d). Final Draft Drinking Water Health Criteria Document for Chlorine Dioxide, Chlorite and Chlorate. Washington, D.C.: Office of Science and Technology, Office of Water.

USEPA (1994e). Final Draft for the Drinking Water Criteria Document on Trihalomethanes. Washington, D.C.: Health and Ecological Criteria Div., Office of Science and Technology.

USEPA (1994f). Draft Drinking Water Health Criteria Document for Chlorinated Acetic Acids/Alcohols/Aldehydes and Ketones. Washington, D.C.: Office of Science and Technology, Office of Water.

USEPA. (1995). Turbidity Criteria Document, Draft (September 1, 1985), Washington, D.C.: Office of Research and Development and Office of Drinking Water.

USEPA. (1996). Microscopic Particulate Analysis (MPA) for Filtration Plant Optimization. EPA 910-R-96-001, EPA Region 10, Office of Environmental Assessment, Seattle, Wash.

USEPA (1997a). Drinking Water Advisory: Consumer Acceptability Advice and Health Effects Analysis on Methyl Tertiary-Butyl Ether (MtBE). Washington, D.C.: Office of Water.

USEPA (1997b). Summary of New Health Effects Data on Drinking Water Disinfectants and Disinfectant Byproducts (D/DBPs) for the Notice of Data Availability (NODA), October 10, 1997. Washington, D.C.: Office of Science and Technology.

USEPA (1998a). Carcinogenic Effects of Benzene: An Update. Washington, D.C.: National Center for Environmental Assessment, Office of Research and Development.

USEPA (1998b). National Primary Drinking Water Regulations: Disinfectants and Disinfection Byproducts; Final Rule. *Federal Register*, 63: 69389–69476.

USEPA (1998c). Health Risk Assessment/Characterization of the Drinking Water Disinfection Byproduct Chlorine Dioxide and the Degradation Byproduct Chlorite. March 13, 1998. Washington, D.C.: Office of Science and Technology, Office of Water.

USEPA (1998d). National Primary Drinking Water Regulations; Disinfectants and Disinfection Byproducts; Notice of Data Availability. *Federal Register*, 63: 15674–15692.

USEPA (1998e). Health Risk Assessment/Characterization of the Drinking Water Disinfection Byproduct Bromate, March 13, 1998. Washington, D.C.: Office of Science and Technology, Office of Water.

USEPA (1999). Extrapolation of the Benzene Inhalation Unit Risk Estimate to the Oral Route of Exposure. Washington, D.C.: NCEA-W-0517.

USEPA (2000a). Methodology for Deriving Ambient Water Quality Criteria for the Protection of Human Health. EPA 822-B-00-004. October, 2000a.

USEPA (2000b). National Primary Drinking Water Regulations; Arsenic and Clarifications to Compliance and New Source Contaminants Monitoring. *Federal Register*, 65, 38887–38983, 2000b.

USEPA (2000c). Toxicological Review of Vinyl Chloride. (CAS No. 75-01-4). In Support of Summary of Information on the Integrated Risk Information System. Washington, D.C.:EPA/635/R00/004.

USEPA (2000d). Radionuclides Notice of Data Availability. Technical Support Document. Washington, D.C.: Office of Ground Water and Drinking Water.

USEPA (2000e). National Primary Drinking Water Regulations; Radionuclides; Final Rule. *Federal Register*, 65(236).

USEPA (2001a). Method 1623–Cryptosporidium and Giardia in Water by Filtration/IMS/FA. EPA/821/R-01-025, Washington, D.C.

USEPA (2001b). Method 1601: Male-specific (F+) and Somatic Coliphage in Water by Two-step Enrichment Procedure. EPA/821/R-01-030, Washington, D.C.

USEPA (2001c). Method 1602: Male-specific (F+) and Somatic Coliphage in Water by Single Agar Layer (SAL) Procedure. EPA/821/R-01-029, Washington, D.C.

USEPA (2001d). National Primary Drinking Water Regulations: Arsenic and Clarifications to Compliance and New Source Contaminants Monitoring. *Federal Register,* 66: 6975–7066.

USEPA (2001e). Occurrence of Unregulated Contaminants in Public Water Systems–A National Summary. Office of Water. EPA 815-P-00-002.

USEPA (2001f). Toxicological Review of Chloroform (CAS No. 67-66-3). In Support of Summary Information on the Integrated Risk Information System (IRIS). EPA/635/R-01/001.

USEPA (2002a). Toxicological Review of 1,1-Dichloroethylene (CAS No. 75-35-4). In Support of Summary of Information on the Integrated Risk Information System. EPA/635/R02/002

USEPA (2002b) Long Term 1 Enhanced Surface Water Treatment Rule, Final Rule. *Federal Register*, 67(9): 1811–1844.

USEPA (2003a). Reregistration Eligibility Decision (RED) for Diuron. List A. Case 0046. Office of Prevention, Pesticide and Toxic Substances. http://www.epa.gov/oppsrrd1/REDs/diuron_red .pdf.

USEPA (2003b). Toxicological Review of Dichloroacetic Acid (CAS No. 79-43-6). In Support of Summary of Information on the Integrated Risk Information System. EPA/635/R03/007.

USEPA (2003c). Drinking Water Advisory: Consumer Acceptability Advice and Health Effects Analysis on Sodium. EPA 822-R-03-006. www.epa.gov/safewater/ccl/pdf/sodium.pdf EPA 822-R-03-006.

USEPA (2004). Drinking Water Health Advisory for Manganese. Document Number: 822-R-04-003.

USEPA (2005a) Microbial Source Tracking Guide Document. Office of Research and Development, EPA/600/R-05/064, Cincinnati, OH.

USEPA (2005b). Guidelines for Carcinogen Risk Assessment, EPA/630/P-03/001B. Risk Assessment Forum. Washington, D.C. 20460.

USEPA. (2006a) National Primary Drinking Water Regulations: Long Term 2 Enhanced Surface Water Treatment Rule; Final Rule. *Federal Register,* 71(3): 654–786.

USEPA. (2006b) National Primary Drinking Water Regulations: Ground Water Rule; Final Rule. *Federal Register*,71(214) 65574–67427.

USEPA (2006c). Drinking Water Standards and Health Advisories. EPA 822-R-06-013. Office of Water, Washington, DC 20460, http://www.epa.gov/waterscience/criteria/drinking/#dw-standards

USEPA (2006d). Triazine Cumulative Risk Assessment. Washington, D.C.: Office of Prevention, Pesticides and Toxic Substances.

USEPA (2006e). Science Advisory Board. Review of EPA's Draft Risk Assessment of Potential Human Health Effects Associated with PFOA and Its Salts. http://www.epa.gov/sab/pdf/sab_06_006.pdf

USEPA (2006f). National Primary Drinking Water Regulations: Stage 2 Disinfectants and Disinfection By-Products Rule; Final Rule. *Federal Register*, 71: 388–493.

USEPA (2007a). Toxicological Review of 1,1,1-Trichloroethane (CAS No. 71-55-46). In Support of Summary of Information on the Integrated Risk Information System. EPA/635/R03/013, Washington, D.C._

USEPA (2007b). Water Contamination Information Tool. Water Security Division. http://www.epa .gov/wcit/.

USEPA (2008a). Ground Water Rule Source Water Monitoring Methods Guidance. Office of Water (4607M), EPA 815-R-07-019, Revised March 2008, Washington, DC.

USEPA (2008b). Health Effects Support Document for Boron. www.epa.gov/safewater/ccl/pdf/boron .pdf. EPA Document Number 822-R-08-002.

USEPA (2008c). Drinking Water Health Advisory for Boron. Document Number 822-R-08-013.

USEPA (2008d). Inhibition of the Sodium-Iodide Symporter by Perchlorate: Evaluation of Lifestage Sensitivity Using Physiologically-Based Pharmacokinetic Modeling. Office of Research and Development, Washington, DC; EPA/600/R-08/106A.

USEPA (2008e). Preliminary Regulatory Determination on Perchlorate in Drinking Water. *Federal Register*, 73: 60262.

USEPA (2008f). Summary and Analysis of the 2008 Gasoline Benzene Pre-Compliance Reports. Office of Transportation and Air Quality. EPA-420-R-08-022.

USEPA (2008g). External Review Draft. Toxicological Review of Tetrachloroethylene (CAS No. 127-18-4). In Support of Summary of Information on the Integrated Risk Information System. EPA/635/R08/011A. Washington, D.C.

USEPA (2008h). Regulatory Determinations Support Document for Selected Contaminants from the Second Drinking Water Contaminant Candidate List (CCL 2). Office of Water (4607M) EPA 815-R-08-012.

USEPA (2008i). Integrated Risk Information System, April. Cincinnati, OH: Office of Research and Development, Office of Health and Environmental Assessment, http://cfpub.epa.gov/ncea/iris/index.cfm, last accessed September 2008.

USEPA (2008j). Health Effects Support Document for Dacthal Degradates: Tetrachloroterephthalic Acid (TPA) and Monomethyl Tetrachloroterephthalic Acid (MTP). Draft, Office of Water. Washington, DC. EPA-822-R-08-005.

USEPA (2008k). Provisional Advisory Levels (PALs) for Hazardous Agents. National Homeland Security Research Center. http://www.epa.gov/nhsrc/news/news121208.html.

USEPA (2009a). Drinking Water Standards and Health Advisories.EPA 822-R-09-011. Office of Water, Washington, DC 20460, http://www.epa.gov/waterscience/criteria/drinking/dwstandards2009.pdf

USEPA (2009b). Science Advisory Board. SAB Advisory on EPA's Draft Third Drinking Water Contaminant Candidate List (CCL 3). January 29, 2009. http://yosemite.epa.gov/sab/sabproduct.nsf/936D6986D4CAF8C28525754E0003B74F/$File/EPA-SAB-09-011-unsigned.pdf.

USEPA (2009c). Drinking Water: Perchlorate Supplemental Request for Comments. *Federal Register,* 74: 41883–41893.

USEPA (2009d). Toxicological Review of Trichloroethylene (CAS No. 79-01-6). External Review Draft. EPA/635/R-09/011A. Washington, DC. October 2009.

USEPA (2010). National Primary Drinking Water Regulations; Announcement of the Results of EPA's Review of Existing Drinking Water Standards and Request for Public Comment and/or Information on Related Issues. *Federal Register,* 75: 15499-15572.

U.S. Geological Survey (2002). Water-Quality Data for Pharmaceuticals, Hormones, and Other Organic Wastewater Contaminants in U.S. Streams, 1999–2000. Iowa City, Iowa. Open-File Report 02-94.

U.S. Geological Survey (2006). Pesticides in the Nation's Streams and Ground Water, 1992–2001—A Summary. National Pesticide Synthesis, http://pubs.usgs.gov/fs/2006/3028/.

Vaerewijck, M.J.M., G. Huys, J.C. Palomino, J. Swings, and F. Portaels (2005). Mycobacteria in drinking water distribution systems: ecology and significance for human health, *FEMS Microbiol. Rev.*, 29: 911–934.

Veeramachaneni, D.N., J.S. Palmer, and G.R. Klinefelter (2007). Chronic exposure to low levels of dibromoacetic acid, a water disinfection by-product, adversely affects reproductive function in male rabbits, *J Androl.* 28: 565–77.

Villanueva, C.M., K.P. Cantor, S. Cordier, J.J. Jaakkola, W.D. King, C.F. Lynch, S. Porru, and M. Kogevinas (2004). Disinfection by-products and bladder cancer: a pooled analysis. *Epidemiology*, 15(3): 357–367.

Villanueva CM, Durand G, Coutté MB, Chevrier C, Cordier S (2005). Atrazine in municipal drinking water and risk of low birth weight, preterm delivery, and small-for-gestational-age status. *Occup. Environ. Med.*, 262(6):400–5.

Villena, I., D. Aubert, P. Gomis, H. Ferte, J.-C. Inglard, H. Denis-Bisiaux, J.-M. Dondon, E. Pisano, N. Ortis, and J.-M. Pinon (2004). Evaluation of a strategy for *Toxoplasma gondii* oocyst detection in water, *Appl. Environ. Microbiol.*,70: 4035–4039.

Wainwright, K.E., M. Lagunas-Solar, M.A. Miller, B.C. Barr, I.A. Gardner, C. Pina, A.C. Melli, A.E. Packham, N. Zeng, T. Truong, and P.A. Conrad (2007). Physical inactivation of *Toxoplasma gondii* oocysts in water, *Appl. Environ. Microbiol.*, 73: 5663–5666.

Walker, E.M., Jr., M.D. Wolfe, M.L. Norton, S.M. Walker, and M.M. Jones (1998). Hereditary hemochromatosis, *Ann. Clin. Lab. Sci.,* 28: 300–312.

Walker, J.E., and D.K. Mitchell (2003). Astrovirus infection in children, *Curr. Opin. Infect. Dis.*,16: 247–253.

Waller S.A., K.Paul, S.E. Peterson, J.E. Hitti (2010). Agricultural-related chemical exposures, season of conception, and risk of gastroschisis in Washington State. *Am J. Obstet. Gynecol.*, 202(3): 241.e1-6.

Ward, M.H., T.M. deKok, P. Levallois, J. Brender, G. Gulis, B.T. Nolan, and J. VanDerslice: International Society for Environmental Epidemiology (2005). Workgroup Report: Drinking-water Nitrate and Health—Recent Findings and Research Needs, *Environ.Health Persp.*, 113(11): 1607–14.

Ward, M.H., J.A. Rusiecki, C.F. Lynch, and K.P. Cantor (2007). Nitrate in public water supplies and the risk of renal cell carcinoma, *Cancer Causes Control,* 18(10): 1141–51.

Ward, M.H., E.F. Heineman, R.S. Markin, and D.D. Weisenburger (2008). Adenocarcinoma of the stomach and esophagus and drinking water and dietary sources of nitrate and nitrite, *International Journal of Occupational and Environmental Health,* 14(3): 193–197.

Ward, R.L., D.I. Bernstein, E.C. Young, J.R. Sherwood, D.R. Knowlton, and G.M. Schiff (1986). Human Rotavirus studies in volunteers: determination of infectious dose and serological response to infection, *J. Infect. Dis.*, 154: 871–880.

Wartenberg, D., D. Reyner, and C.S. Scott (2000). Trichloroethylene and cancer: epidemiologic evidence. *Environ. Health Persp. Suppl.*, 108(S2): 161–176.

Watson, C.L., R.J. Owen, B. Said, S. Lai, J.V. Lee, S. Surman-Lee, and G. Nichols (2004). Detection of *Helicobacter pylori* by PCR but not culture in water and biofilm samples from drinking water distribution systems in England, *J. Appl. Microbiol.*, 97: 690–698.

Webb, S., T. Ternes, M. Gibert, and K. Olejniczak (2003). Indirect human exposure to pharmaceuticals via drinking water, *Toxicol. Lett.*, 142(3): 157–167.

Weber, R., R.T. Bryan, D.A. Schwartz, and R.L. Owen (1994). Human microsporidial infections, *Clin. Microbiol. Rev.*, 7: 426–461.

Welker, M., and H. von Döhren (2006). Cyanobacterial peptides–Nature's own combinatorial biosynthesis, *FEMS Microbiol. Rev.*, 30: 530–563.

Wesley, A., S. Grytdal, and K. Gallagher (2008). Surveillance for acute viral hepatitis–United States, 2006, *Morbidity and Mortality Weekly Rept.*,57(SS-2): 1–24.

White, J.M., J.G. Wingo, L.M. Alligood, G.R. Cooper, J. Gutridge, W. Hydaker, R.T. Benack, J.W. Dening, and F.B. Taylor (1967). Sodium ion in drinking water. I. Properties, analysis, and occurrence, *Jour. Am. Dietetic Assoc.*, 50: 32–36.

Wiegand, C., and S. Pflugmacher (2005). Ecotoxicological effects of selected cyanobacterial secondary metabolites a short review, *Toxicol. Appl. Pharm.*,203: 201–218.

Wilhemi, I., E. Roman, and A. Sanchez-Fauquier (2003). Viruses causing gastroenteritis, *Clin. Microbiol. Infect.*, 9: 247–262.

Wilkes, G., T. Edge, V. Gannon, C. Jokinen, E. Lyautey, D. Medeiros, N. Newmann, N. Ruecker, E. Topp, and D.R. Lapen (2009). Seasonal relationships among indicator bacteria, pathogenic bacteria, *Cryptosporidium* oocysts, *Giardia* cysts, and hydrological indices for surface waters within an agricultural landscape, *Water Research*, 43: 2209–2223.

Winieka-Krusnell, J., K. Wreiber, A. Von Euler, L. Engstrand, and E. Linder (2002). Free-living amoeba promote growth and survival of *Helicobacter pylori*, *Scand. J. Infect. Dis.*, 34: 253–256.

Wolfe, M.S. (1992). Giardiasis, *Clin. Microbiol. Rev.*, 5: 93–100.

Wolfe, G., and L. Kaiser (1996). Final Report, Sodium Bromate: Short Term Reproductive and Developmental Toxicity Study when Administered to Sprague–Dawley Rats in the Drinking Water. Study No. NTP-PEST, 94007. NTP/NIEHS No. NOI-ES-15323.

Woo, Y.T., D. Lai, J.L. McLain, M.K. Manibusan, and V. Dellarco (2002). Use of mechanism-based structure–activity relationships analysis in carcinogenic potential ranking for drinking water disinfection by-products, *Environ Health Perspect.* 110(suppl. 1): 75–87.

World Health Organization (2002). *Guidelines for Drinking Water Quality, (2nd ed.) Addendum: Microbiological Agents in Drinking Water, Legionella,* pp. 40–69. Geneva.

World Health Organization (2003). Acrylamide in Drinking Water. Background document for development of WHO Guidelines for Drinking Water Quality. Geneva.

World Health Organization (2004). Brominated Acetic Acids in Drinking Water. Background document for development of WHO Guidelines for Drinking Water Quality. Geneva.

World Health Organization (2005). Food and Agriculture Organization of the United Nations. Joint FAO/WHO Expert Committee on Food Additives. 64th Meeting. Rome. February 2005. http://www.who.int/ipcs/food/jecfa/summaries/summary_report_64_final.pdf.

World Health Organization (2006). *Health Aspects of Plumbing.* Geneva: World Health Organization and the World Plumbing Council.

World Health Organization (2007). *Legionella and the Prevention of Legionellosis.* Geneva: World Health Organization.

World Health Organization (2008). N-Nitrosodimethylamine in Drinking-water. Background document for development of WHO Guidelines for Drinking Water Quality. WHO/HSE/AMR/08.03/8. Geneva.

Worm, H.C., W.H.M. van der Poel, and G. Brandstatter (2002). Hepatitis E: an overview, *Microbes Infect.*, 4: 657–666.

Xiao, L., K. Alderisio, J. Limor, M. Royer, and A.A. Lal (2000). Identification of species and sources of *Cryptosporidium* oocysts in storm waters with a small-subunit rRNA-based diagnostic and genotyping tool, *Appl. Environ. Microbiol.*, 66: 5492–5498.

Xiao, L., R. Fayer, U. Ryan, and S.J. Upton (2004). *Cryptosporidium* taxonomy: recent advances and implications, *Clin. Microbiol. Rev.*, 17: 72–97.

Yang, W., P. Chen, E.N. Villegas, R.B. Landy, C. Kanetsky, V. Cama, T. Dearen, C.L. Schultz, K.G. Orndorff, G.J. Prelewicz, M.H. Brown, K.R. Young, and L. Xiao (2008). *Cryptosporidium* source tracking in the Potomac River watershed, *Appl. Environ. Microbiol.*, 74: 6495–6504.

Yoder, J.S., and M.J. Beach (2007). Cryptosporidiosis surveillance–United States, 2003–2005 and giardiasis surveillance–United States, 2003–2005, *Morbidity Mortality Weekly Reprt.*, 56(SS-7): 1–10.

Yoder, J., V. Roberts, G.F. Craun, V. Hill, L. Hicks, N.T. Alexander, V. Radke, R.L. Calderon, M.C. Hlavsa, M.J. Beach, and S.L. Roy (2008). Surveillance for waterborne disease and outbreaks associated with drinking water and water not intended for drinking–United States, 2005–2006, *Morbidity and Mortality Weekly Rept.*, 57(SS-9): 39–69.

Yoon, J.W., and C.J. Hovde (2008). All blood, no stool: enterohemorrhagic *Escherichia coli* O157:H7 infection, *J. Vet. Sci.*, 9(3): 219–231.

Yu, V.L., J.F. Plouffe, M.C. Pastoris, J.E. Stout, M. Schousboe, A. Widmer, J. Summersgill, T. File, C.M. Heath, D.L. Paterson, and A. Chereshsky (2002). Distribution of *Legionella* species and serogroups isolated by culture in patients with sporadic community-acquired legionellosis: an international collaborative survey, *J. Infect. Dis.*,186: 127–128.

Zewdie, T., C.M. Smith, M. Hutcheson, and C. Rowan-West (2009). Basis of the Massachusetts Reference Dose and Drinking Water Standard for Perchlorate. *Env. Hlth. Perspct.* doi: 10.1289/ehp.0900635 (available at http://dx.doi.org/) Online 13 July 2009.ehponline.org.

CHAPTER 3
CHEMICAL PRINCIPLES, SOURCE WATER COMPOSITION, AND WATERSHED PROTECTION

James K. Edzwald, Ph.D., B.C.E.E.
Professor Emeritus
Department of Civil and Environmental Engineering
University of Massachusetts
Amherst, Massachusetts, United States

John E. Tobiason, Ph.D., B.C.E.E.
Professor
Department of Civil and Environmental Engineering
University of Massachusetts
Amherst, Massachusetts, United States

INTRODUCTION **3.2**
CHEMICAL PRINCIPLES AND CONCEPTS **3.2**
Water Properties 3.2
Water Dissociation, pH, and pX Notation 3.3
Concentrations 3.4
Ionic Strength, Conductivity, and Dissolved Solids 3.5
Principle of Electroneutrality 3.5
Stoichiometry 3.6
Chemical Equilibrium 3.7
Inorganic Carbon Chemistry 3.17
Alkalinity 3.21
Acidity 3.23
SOURCE WATER COMPOSITION **3.24**
Introduction 3.24
Terminology 3.24
Hydrogeochemical and Biochemical Cycles 3.26
Rainwaters 3.28
Rivers and Streams 3.29
Lakes and Reservoirs 3.34
Groundwaters 3.39
PARTICLES **3.42**
Importance 3.42
Types of Particles: Sources and Materials 3.43
Particle Characteristics 3.45
Measurements of Particle Concentration 3.49
Measurements of Particle Size and Size Distributions 3.52
Measurements of Particle Stability 3.57
NATURAL ORGANIC MATTER **3.58**
Sources 3.58
Types and Fractions 3.59
Collective Measurements of Organic Carbon 3.62
UV_{254} Absorbance as a Surrogate Parameter for DOC and TOC 3.63
Specific UV Absorbance 3.65
Measurements of NOM for Water Utilities 3.66
SOURCE WATER SELECTION AND PROTECTION **3.67**
Selection of Source Water 3.67
Source Protection—General 3.69
Watershed Protection 3.70
Aquifer Protection 3.71
ABBREVIATIONS **3.71**
NOTATION FOR EQUATIONS **3.72**
REFERENCES **3.72**
Water Chemistry References 3.76

INTRODUCTION

Chapters 1 through 5 serve as a foundation for the book. This foundation chapter has three goals. One goal is to summarize basic water chemistry principles. A second goal is to provide the reader with background on the composition of source waters with respect to their bulk water chemistry and on naturally occurring contaminants. Contaminants that enter water supplies through human activity are also summarized. Because protecting the quality of source waters is an important strategy in the drinking water field, principles and some practical material on watershed protection are presented as a third goal.

The chapter is organized in the following way. We start with a summary of chemical principles because chemical reactions affect the composition of source waters as well as treatment processes and the quality of water in distribution systems. Next, we address the composition of surface and groundwaters. Hydrogeochemical and biological geochemical cycles and their effects on the composition of water quality are described. Ranges and typical concentrations for chemical species, particles, and natural organic matter (NOM) are presented for rivers and streams, reservoirs and lakes, and groundwaters. The chapter examines the bulk chemical composition and naturally occurring contaminants in water. It also examines the presence of anthropogenic chemicals that are present in some water sources. Particles and NOM are important contaminants that affect source water quality and many treatment processes, and so they are given extensive coverage. The last section reviews important principles related to the selection and protection of drinking water sources with respect to water quality.

CHEMICAL PRINCIPLES AND CONCEPTS

The composition of source waters and water treatment processes are affected greatly by various chemical reactions such as acid-base (e.g., inorganic carbon chemistry, alkalinity, chlorination) complexation (e.g., Fe and NOM in source waters and coagulation), solubility (e.g., the occurrence of metals in source waters, softening, and coagulation), and oxidation–reduction (redox) (e.g., oxidation of Fe and Mn, disinfection, and corrosion). Kinetics of chemical reactions are not covered here but in Chap. 4. Mass transfer kinetics for describing movement or transport of gases and particles are covered in the chapters dealing with these processes.

Here, these topics are briefly summarized for use in this and other chapters. The reader should refer to primary textbooks on water chemistry such as Stumm and Morgan (1996), Snoeyink and Jenkins (1980), and Benjamin (2002) for extensive theory and applications of the topics covered in this section. Note that these water chemistry references are listed separately in the References section.

Water Properties

Polar Nature. Water is a polar molecule, and although electrically neutral it has a polar nature meaning the molecule has a region of negative charge near the O atoms and positive charge near the H atom. This dipolar nature gives water important properties such as its ability to act as a solvent. We therefore find substances (solutes) dissolved in water to varying degrees, depending on specific chemical properties. For example, simple salts such as Na^+ and Cl^- are found at high concentrations such as in the oceans, while Al is found at low concentrations at typical pH conditions of water supplies because it is insoluble. The polar nature of water leads to H bonding of water molecules and other molecules. It also affects the chemical speciation of dissolved substances.

Physical Properties. The dipolar nature of water also affects important physical properties. Compared to other liquids, water has a high boiling point (100°C) and a high freezing point (0°C). Water density depends on water temperature with a maximum value at 4°C of 1000 kg/m^3. Water density as a function of temperature explains the cause of lake stratification, which is addressed later in this chapter. Water viscosity is another important physical property. Water density and viscosity affect various water treatment processes, as presented in individual chapters. Appendix D contains values as a function of water temperature.

The conductivity of water is affected by the concentration of dissolved salts. The dissolved salt concentration affects corrosion reactions and chemical reactions in general. This is addressed later in this chapter.

Water Dissociation, pH, and pX Notation

Water dissociates and is in equilibrium with H_3O^+ (hydronium ion) and OH^-, as shown in Eq. 3-1. Writing the hydronium ion recognizes that free H^+ does not exist in water, it is bound through H bonding to the water molecule. We should be aware of this fact when we write a shorthand version of Eq. 3-1 as Eq. 3-2.

$$2H_2O \rightleftharpoons H_3O^+ + OH^- \tag{3-1}$$

$$H_2O \rightleftharpoons H^+ + OH^- \tag{3-2}$$

The equilibrium constant for the dissociation of water (K_w) and using Eq. 3-2 describes the following concentration relationships. The activity or active concentrations (a_{H^+} and a_{OH^-}) is the formal and proper way to describe the equilibrium constant relationship—more on activities later in this chapter. For dilute solutions, the molar concentrations (the brackets [ ] indicate molar) are approximately equal to the activities yielding Eq. 3-4.

$$K_w = a_{H^+}\, a_{OH^-} \tag{3-3}$$

$$K_w = [H^+][OH^-] \tag{3-4}$$

At standard temperature of 298 K or 25°C, K_w is 10^{-14}. Because chemical reactions including the dissociation of water are temperature dependent, it is often essential to adjust K_w. Equation 3-5 allows calculation for other water temperatures where T is the absolute temperature. As an example of the significance of water temperature on K_w, consider that we measure pH to control processes in water treatment so the [H^+] is known or fixed for that particular pH value; consequently, the [OH^-] varies with water temperature for that particular pH. Many precipitation reactions such as in coagulation depend on the metal precipitating with OH^-. For example, if you carried out a precipitation reaction at pH 6 in the summer at say 25°C, the [OH^-] is 10^{-8} M, but in winter months at the same pH 6 and say 5°C (where the K_w is 1.86×10^{-15}), the [OH^-] is 0.19×10^{-8} M about five times less OH^- for precipitation. Higher pH conditions are required in the winter for effective precipitation.

$$\text{Log } K_w = -9.971 + \frac{509}{T} - \frac{509581}{T^2} \tag{3-5}$$

pX Notation. In water treatment chemistry, the pX notation is often used to describe concentrations and equilibrium constants on a log (base 10) basis. We are most familiar with pH as minus the base 10 log [H^+]. A summary of other pX notation is given in Table 3-1 for reference. This notation is used here and in several chapters throughout the book.

TABLE 3-1 Guide to pX Notation Used in This Chapter and throughout the Book

pX	Definition	Comments
pC	$-\log [C]$	molar concentration of C
pH	$-\log [H^+]$	molar concentration of H^+
pK	$-\log K$	equilibrium constant
pe°	$1/n \log K$	standard oxidation reduction potential, where K is the equilibrium constant for the half reaction and n is the number of electrons transferred

Concentrations

Moles and Mass. We use various measures of concentration in evaluating water quality and in water treatment process chemistry. A fundamental chemical measure of concentration is moles per liter, which is expressed as M. At low concentrations, it may be expressed as millimolar (mM). One mole of a substance is the gram atomic weight of a substance. We usually describe concentrations of substances in water on a mass basis (e.g., the alum or chlorine dose at a water plant is expressed on a mass concentration basis). To convert from moles to mass concentrations, our conversion factor is the gram atomic weight of the substance. We usually express mass concentrations as g/L, mg/L, or μg/L, depending on the relative concentration.

Equivalents: Charge and Alkalinity Concentrations. We also use equivalent concentrations to describe the charge concentration of ions—more on this under the principle of electroneutrality—and to describe alkalinity. The concept of alkalinity is addressed later in the chapter. Fundamentally, alkalinity is measured in units of eq/L (equivalents per liter) or meq/L (milliequivalents per liter), and it is the concentration of H^+ that can be neutralized or the acid neutralizing capacity of water. In our field, it is reported in terms of an equivalent concentration as $CaCO_3$. $CaCO_3$ has an equivalent weight of 50, so the conversion is 1 meq/L of alkalinity is 50 mg/L as $CaCO_3$.

Other Cases. Some other cases of interest in describing concentrations are for hardness and in reporting nitrogen and phosphorus concentrations. Water hardness is caused by divalent and trivalent cations in water, but because calcium and magnesium are much higher in concentrations compared to others, these are the two cations we attribute to hardness. Total hardness is the sum of Ca^{2+} and Mg^{2+}, both expressed in terms of equivalent concentrations as $CaCO_3$. The equivalent concentration of Ca^{2+} is its mass concentration times the atomic weight of $CaCO_3$ divided by the atomic weight of Ca^{2+}, i.e., 1 mg/L of Ca^{2+} equals 100/40 or 2.5 mg/L as $CaCO_3$. Likewise, 1 mg/L of Mg^{2+} equals 100/24.31 or 4.11 mg/L as $CaCO_3$. Classifications of waters with respect to water hardness are: soft < 75, moderately hard 75 to 150, hard 150 to 300, and very hard > 300 mg/L as $CaCO_3$ (Sawyer et al.,1994).

Phosphorus (P) can occur in water as dissolved organic P or inorganic P. Within the inorganic form, it can exist as polyphosphates and orthophosphate. The phosphorus concentration is usually reported as P. Likewise, nitrogen compounds such as NH_4^+, NH_3, and NO_3^- are reported as N. For example, in chloramination 1 mg/L of NH_3 added to water is reported as N, i.e., 1 mg/L of NH_3 equals 1 mg/L times the atomic weight of N divided by the molecular weight of NH_3 or 1 mg/L times 14/17 or 0.82 mg/L as N.

Gas Phase. Gas phase concentrations are expressed in terms of pressures, partial pressures, or vapor pressures, and they are addressed in the following chapters depending on the application. Units are in atmosphere, Pa (Pascal) or kPa, and bar.

Ionic Strength, Conductivity, and Dissolved Solids

Chemists describe the ion concentration of water in terms of the ionic strength. Ionic strength affects the active concentrations (activities) participating in chemical reactions. It affects acid-base reactions, solubility of solids and gases in water, scaling of solids, corrosion chemistry, and colloidal particle double layer theory and stability. Ionic strength effects are covered in Chap. 8 for colloidal particle stability and in other chapters dealing with specific applications. Here it is defined, and expressions for estimating it from conductivity and dissolved solids measurements are presented.

The ionic strength (I) is defined by Eq. 3-6 where C_i is the ion molar concentration of ion i, and Z_i is the charge of ion i.

$$I = \frac{1}{2}\sum_{1}^{n} C_i Z_i^2 \tag{3-6}$$

A complete water analysis of all ions is required to use Eq. 3-6. Because this is often impractical, correlations have been developed relating the ionic strength to more practical measurements (Snoeyink and Jenkins, 1980). Equation 3-7 is a correlation equation between I and conductivity (κ), and Eq. 3-8 relates I to total dissolved solids (TDS).

$$I = 1.6\times10^{-5}(\kappa) \tag{3-7}$$

$$I = 2.5\times10^{-5}(TDS) \tag{3-8}$$

Values for I, κ, and TDS are given in Table 3-2. Most natural fresh waters have ionic strengths between 5×10^{-4} M and 10^{-2} M, corresponding to estimated conductivities of about 30 to 625 μS/cm and TDS of 20 to 400 mg/L.

Principle of Electroneutrality

Water is electrically neutral, meaning the cations must equal the anions on an equivalent charge basis. This is expressed as Eq. 3-9 where i and j refer to cations and anions.

$$\sum_i C_i Z_i = \sum_j C_j Z_j \tag{3-9}$$

TABLE 3-2 Ionic Strength (I), Conductivity (κ), and Total Dissolved Solids (TDS) for Several Water Types

I (M)	κ (μS/cm)	TDS (mg/L)	Comments
1.2×10^{-5}	0.75	0.5	dimineralized water open to atmospheric CO_2(g) of 380 ppm
5×10^{-4}	31	20	soft groundwater
10^{-3}	63	40	low to moderate hardness surface or groundwater
5×10^{-3}	312	200	high hardness surface or groundwater
10^{-2}	625	400	very hard and brackish water
10^{-1}*	7000	5000	estuarine water
0.68†	~50,000	35,000	seawater at salinity of 35.2 parts per thousand

*Outside of correlation range of data

†Seawater values vary with salinity but are well-known

The principle of electroneutrality has useful applications. These are (1) to determine the accuracy of a reported chemical composition and (2) to determine an anion or cation that was not measured. Example 3-1 shows use of the principle in evaluating the composition of a bottled spring water.

Example 3-1 Principle of Electroneutrality
A bottled water company gives the following chemical composition for their spring water: pH 7.1; calcium, 91 mg/L; magnesium, 4.0 mg/L; sodium, 4.0 mg/L; sulfate, 170 mg/L; bicarbonate, 63 mg/L; fluoride, 0.4 mg/L; and nitrate, 0.26 mg/L. Check on whether the reported composition satisfies the principle of electroneutrality.

Solution Equation 3-9 is applied to solve the problem. The results are presented in the table.

Cations					
Ion	mg/L	Atomic Wgt	C_i (mM)	Z_i	$C_i Z_i$ (meq/L)
Ca^{2+}	91	40	2.275	2	4.550
Mg^{2+}	4	24.31	0.164	2	0.329
Na^+	4	23	0.173	1	0.174
H^+		1	$<10^{-4}$	1	$<10^{-4}$
Sum					**5.053**
Anions					
Ion	mg/L	Atomic Wgt	C_j (mM)	Z_j	$C_j Z_j$ (meq/L)
HCO_3^-	63	61	1.033	1	1.033
SO_4^{2-}	170	96	1.770	2	3.540
F^-	0.4	19	0.021	1	0.021
NO_3^-	0.26	62	0.004	1	0.004
OH^-		17	$<10^{-3}$	1	$<10^{-3}$
Sum					**4.598**

The total cation charge is much higher (5.053 meq/L) compared to the total anion charge (4.598 meq/L). This can mean either that there are errors in the chemical analysis or that one or more anions were not measured and reported. The latter is most likely since all natural waters contain chloride, which is not reported. If we assume that the reported concentrations of cations and anions are accurate, then we can calculate the missing chloride concentration using Eq. 3-9. The missing anion (chloride) concentration is 0.455 meq/L, which is a chloride concentration of 16 mg/L. ▲

Stoichiometry

Stoichiometry is the subject of chemistry that allows us to make calculations of how much of a chemical undergoes reaction and how much of a chemical is produced for known chemical reactions. Chemical reactions actually take place in terms of numbers of ions or molecules or atoms. However, because the numbers are so high the chemist invented the mole concept where 1 mole is the gram atomic weight of the substance. In other words 1 mole of a substance is the gram atomic weight of a substance and it contains 6.023×10^{23} ions or molecules or atoms. A list of atomic weights is given in App. A. In using stoichiometric calculations, one assumes the reactions go to completion. This is true for many types of reactions in water treatment such as dissolving chemicals in water. Example 3-2 illustrates stoichiometric calculations involving the use of alum as a coagulant.

Example 3-2 Alum Dosing and Residuals Produced
In Part a, find the relationship between alum dosing as alum and as Al. Alum dosing in the United States is expressed as dry alum in which alum is $Al_2(SO_4)_3 \cdot 14H_2O$. In many other countries, alum dosing is expressed as Al. Determine the relationship between 1 mg/L of alum and mg/L as Al. In Part b, determine the production of residuals (sludge) from precipitation of $Al(OH)_3(s)$ ((s) refers to a solid phase) from use of alum.

Solution *Part a.* Dissolution of alum ($Al_2(SO_4)_3 \cdot 14H_2O$) follows the stoichiometric equation below. The calculation shown equating 1 mg/L of alum to x mg/L as Al comes from the mole concept expressed through the stoichiometry of the chemical equation.

$$Al_2(SO_4)_3 \cdot 14H_2O \rightarrow 2Al^{3+} + 3SO_4^{2-} + 14H_2O$$

$$1\frac{\text{mg alum}}{\text{L}} \times \frac{\text{mmole alum}}{594\ \text{mg}} \times \frac{2\ \text{mmole Al}}{\text{mmole alum}} \times \frac{27\ \text{mg Al}}{\text{mmole Al}} = 0.0909\ \text{mg/L Al}$$

Thus, 1 mg/L of alum equals 0.0909 mg/L as Al or 1 mg/L of Al equals 11 mg/L of alum.

Part b. The amount of solids (residuals) produced on a dry basis is calculated from the following stoichiometric equation. Assuming all the Al precipitates, which is a good assumption since alum coagulation is often practiced at pH conditions that maximizes the precipitation reaction leaving little residual dissolved Al. The solid ($Al(OH)_3 \cdot 3H_2O(s)$) has three appended waters and this is sometimes considered in sludge calculations—see Chap. 22. Calculating for 1 mg/L alum or 0.0909 mg/L Al follows.

$$Al^{3+} + 3OH^- \rightarrow Al(OH)_3 \cdot 3H_2O(s)$$

$$\frac{0.0909\ \text{mg Al}}{\text{L}} \times \frac{\text{mmole Al}}{27\ \text{mg}} \times \frac{1\ \text{mmole Al(OH)}_3 \cdot 3\text{H}_2\text{O(s)}}{\text{mmole Al}} \times \frac{132\ \text{mg Al(OH)}_3 \cdot 3\text{H}_2\text{O(s)}}{\text{mmole Al(OH)}_3 \cdot 3\text{H}_2\text{O(s)}} = 0.44\ \text{mg/L Al(OH)}_3 \cdot 3\text{H}_2\text{O(s)}$$

Thus, we find that 1 mg/L alum (or 0.0909 mg/L Al) produces 0.44 mg/L of residuals from the precipitation of the metal coagulant. If we did not include the three appended waters to the solid, the sludge generated would be 0.26 mg/L. So we can estimate the solids generated from alum would be about 0.3 to 0.4 mg/L. Additional residuals would come from the particles (turbidity) and organic matter in the raw water that are removed through coagulation. ▲

Chemical Equilibrium

Chemicals in water are either at equilibrium or seeking an equilibrium condition. By equilibrium we mean there is no net change in the concentrations, they have reached a steady state concentration. When we add an acid or base to water, the reactions are extremely fast (<< 1 sec) because they involve only the donation of H^+ (acid) or the acceptance of H^+ (base). Other reactions may take longer (many seconds or minutes or longer) such as precipitation of solids or oxidation-reduction reactions. We use chemical equilibrium principles to describe the speciation of chemicals in water, for example, the fraction of hypochlorous acid (HOCl) versus hypochlorite (OCl^-) in disinfection applications. Other examples of the use of chemical equilibrium include the solubility of metal coagulants

(Al and Fe salts), water softening precipitation reactions for $CaCO_3(s)$ and $Mg(OH)_2(s)$, oxidation of iron and manganese, and corrosion chemistry for copper and lead. In the material that follows, some basic principles are presented that are applied in this and later chapters.

Equilibrium Constant. Consider the following general reaction occurring in water where the capital letters represent chemical species, the lower case letters are the stoichiometric coefficients, and the symbol ($\rightleftharpoons$) depicts a system at equilibrium between the reacting species and the products. In this book an equal sign is often used for an equilibrium reaction rather than the symbol ($\rightleftharpoons$). The reader will know it is an equilibrium reaction from the description of the subject matter.

$$\mathrm{aA} + \mathrm{bB} \rightleftharpoons \mathrm{yY} + \mathrm{zZ} \tag{3-10}$$

The equilibrium constant relationship follows setting the equilibrium composition distribution of reactants and products for standard conditions of 25°C and 1 atm pressure. The equilibrium constant depends on the activity (a_i) of the species (i) raised to the power of the stoichiometric coefficient. The activity is used in formal development because reactions depend on an active or apparent concentration of the species.

$$K = \frac{a_Y^y\, a_Z^z}{a_A^a\, a_B^b} \tag{3-11}$$

The activities for dissolved substances are related to molar concentrations of species [i] by use of activity coefficients (γ_i). The activity coefficients account for non-ideal behavior of the solutions. Expanding Eq. 3-11, we obtain the following expression:

$$K = \frac{a_Y^y a_Z^z}{a_A^a a_B^b} = \frac{\gamma_Y^y [Y]^y \gamma_Z^z [Z]^z}{\gamma_A^a [A]^a \gamma_B^b [B]^b} \tag{3-12}$$

The non-ideal behavior of reactions is accounted for by use of activity coefficients for aqueous systems that are not dilute solutions in terms of the ionic strength or dissolved salt content. For ions, the fundamental model accounting for ionic strength effects and their charge is the Debye-Hückel theory. The reader should consult the water chemistry references for presentation of this model, other models, and details regarding calculation of activity coefficients. Here, the practical Davies equation is presented and can be used for ionic strength (I) up to ~ 0.5 M. In Eq. 3-13, z_i is the charge on the ion and A depends on the dielectric constant (ε) of water and absolute temperature (T) according to Eq. 3-14. A has a value of 0.5 for a water temperature of 298 K (25°C).

$$\log \gamma_i = -A z_i \left(\frac{I^{0.5}}{1 + I^{0.5}} - 0.3 I \right) \tag{3-13}$$

$$A = 1.82 \times 10^{-6} (\varepsilon T)^{-3/2} \tag{3-14}$$

Most water supplies and treatment processes are sufficiently dilute in ionic strength that we can assume ideal solutions. In assuming ideal solutions, we set the activity coefficients to 1, which is a reasonable approximation for ionic strength conditions of about 0.001 M and less. These ionic strength conditions apply to many fresh waters. Some values are presented later in the chapter. For waters or processes at higher ionic strength such as brackish waters, seawater, scaling on membranes, and desalination, we must account for the ionic strength effects on the equilibrium constant reactions.

Concentration Scales for Ideal Systems. The following concentration scales are used for ideal aqueous systems:

- Dissolved substances: molar concentrations
- Water: mole fraction, which is 1 for water (even for seawater it is nearly 1)
- Solid phases: mole fraction of 1 for pure solids, which is assumed
- Gas phase: partial pressure traditionally in atm or may be expressed in bar

In presenting many of the concepts regarding equilibrium chemistry and for making approximate calculations describing the chemistry of fresh waters and water treatment process chemistry, activities are ignored (setting activity coefficients to 1), and hence we consider the waters as ideal solutions. For an ideal solution, Eq. 3-12 is written then as Eq. 3-15. It is noted that in some cases, equilibrium constants are given for specified ionic strength conditions, meaning the effect of non-ideal solutions has been accounted for in determining the equilibrium constant.

$$K = \frac{[\mathrm{Y}]^{\mathrm{y}}[\mathrm{Z}]^{\mathrm{z}}}{[\mathrm{A}]^{\mathrm{a}}[\mathrm{B}]^{\mathrm{b}}} \tag{3-15}$$

Effects of Temperature and Pressure. Equilibrium constants must be adjusted for temperature in order to make equilibrium calculations for the temperature of interest. This is done by using a form of the van't Hoff equation as follows:

$$\log\frac{K_2}{K_1} = \frac{\Delta H_r^o(T_2 - T_1)}{2.303(R)(T_2)(T_1)} \tag{3-16}$$

where K_2 is for the temperature of interest at T_2 in degrees Kelvin, K_1 is for standard temperature of T_1 at 298 K, ΔH_r^o is the standard enthalpy of the reaction, and R is the universal gas constant (8.314 J/mole·K). Values for the standard enthalpy of the reaction may be found in the books listed under water chemistry references or from chemistry handbooks.

Only pressures at 100s of atmospheres and higher affect the equilibrium constants, so pressure effects on the equilibrium constants are ignored in this book.

Reaction Quotient. To determine whether actual measured concentrations are at equilibrium for the reaction of interest, the reaction quotient (Q) is used. The actual measured concentrations are inserted into the right-hand side (RHS) of Eq. 3-17. If Q equals K, then the system is at equilibrium; if Q is > K then the reaction will move to the left, seeking equilibrium; and if Q is < K, then the system will proceed to the right, seeking equilibrium.

$$Q = \frac{[\mathrm{Y}]^{\mathrm{y}}[\mathrm{Z}]^{\mathrm{z}}}{[\mathrm{A}]^{\mathrm{a}}[\mathrm{B}]^{\mathrm{b}}} \tag{3-17}$$

Solubility and Stripping of Gases—Henry's Law. Henry's law describes the equilibrium partitioning of gases with water. In Chap. 6 the stripping of volatile organic compounds (VOCs) is described by the following equation in which the gas phase (i.e., the VOC of interest in the gas phase) is on the RHS of Eq. 3-18; the resulting Henry's law expression is shown by Eq. 3-19. The gas phase concentration is expressed in terms of its partial pressure.

In the stoichiometric description depicted by Eq. 3-18, our interest is in stripping VOCs, so the aqueous phase concentration is shown on the left-hand side (LHS) and leaving solution to the gas phase. Here H has units of atm/molar.

$$A_{(aq)} \rightleftharpoons A_{(g)} \tag{3-18}$$

$$H = \frac{P_A}{[A_{aq}]} \tag{3-19}$$

In other applications we may be interested in the saturation concentration of a gas in water—its solubility. Examples are O_2(aq) and N_2(aq) in dissolved air flotation applications (Chap. 9), O_2(aq) in source waters, and CO_2(aq) in source waters and in treatment processes. For these applications, the stoichiometric equation is written in the reverse direction from Eq. 3-18. Our interest is in dissolving the gas into water. The gas phase is written on the LHS of Eq. 3-20, while the dissolved equilibrium concentration (solubility or saturation concentration) is written on the RHS. Henry's law is shown as Eq. 3-21.

$$A_{(g)} \rightleftharpoons A_{(aq)} \tag{3-20}$$

$$H^* = \frac{[A_{aq}]}{P_A} \tag{3-21}$$

Here, the Henry's law constant is designated as H^*, and it has units of molar/atm. H^* equals $1/H$. Some values are presented in Table 3-3.

TABLE 3-3 Henry's Law Constants (H^*) at 25°C for Selected Gases (Stumm and Morgan,1996)

Gas	Units of molar/atm
Ammonia (NH_3)	57
Carbon dioxide (CO_2)	3.39×10^{-2}
Hydrogen (H_2)[†]	8.04×10^{-4}
Hydrogen sulfide (H_2S)	1.05×10^{-1}
Nitrogen (N_2)	6.61×10^{-4}
Oxygen (O_2)	1.26×10^{-3}
Ozone (O_3)	9.4×10^{-3}

[†]Benjamin (2002)

Solubility of Carbon Dioxide. Dissolved carbon dioxide (CO_2(aq)) is hydrated by water molecules according to Eq. 3-22, forming a diprotic acid, i.e., contains $2H^+$. The reaction lies far to the left, so mostly we have CO_2(aq), but the almost infinite number of water molecules provides for maintenance of equilibrium for the acid, H_2CO_3. To account for both species, they are combined according to Eq. 3-23.

$$CO_2(aq) + H_2O \rightleftharpoons H_2CO_3 \tag{3-22}$$

$$H_2CO_3^* = CO_2(aq) + H_2CO_3 \tag{3-23}$$

The solubility of $CO_2(g)$ is defined then with respect to $H_2CO_3^*$, as shown by Eq. 3-24. Henry's law is applied to give the dissolved concentration of $H_2CO_3^*$ as a function of the partial pressure of $CO_2(g)$, as presented in Eq. 3-25.

$$CO_2(g) + H_2O \rightleftharpoons H_2CO_3^* \tag{3-24}$$

$$H^*_{CO_2} = \frac{[H_2CO_3^*]}{P_{CO_2}} \tag{3-25}$$

The value for the Henry's constant ($H^*_{CO_2}$) at 25°C is 3.388×10^{-2} M/atm. Air at atmospheric pressure of 101.3 kPa (1 atm) contains 380 ppm $CO_2(g)$ by volume or expressed as partial pressure (P_{CO_2}), 3.80×10^{-4} atm. Applying Eq. 3-25 for this condition yields an equilibrium dissolved $H_2CO_3^*$ concentration of 1.29×10^{-5} M or 0.16 mg/L as C. It should be noted that many water systems are not in equilibrium with respect to dissolved inorganic carbon ($H_2CO_3^*$) because of other reactions adding $CO_{2(aq)}$ such as bacterial decomposition or removing $CO_{2(aq)}$ from water such as algae growth. However, if equilibrium is established between the water and air phases, then the concentration of dissolved $H_2CO_3^*$ is fixed at 1.29×10^{-5} M assuming 25°C. For a colder water temperature of say 5°C, the equilibrium concentration of $H_2CO_3^*$ is 2.39×10^{-5} M or 0.29 mg/L as C.

Acid-Base Chemistry. Acids and bases are defined by the Brønsted-Lowrey concept. An acid is a substance that donates a proton (H^+), and the corollary is that a base is a substance that accepts a proton. Monoprotic acids have only one proton to donate. Generically, they are defined as HA (e.g., hypochlorous acid (HOCl)). Some acids are diprotic and have two protons and are generally defined as H_2A (e.g., dissolved $H_2CO_3^*$) and some are triprotic with three protons (H_3A, e.g., H_3PO_4). The strength of an acid has to do with the tendency of the acid to donate the proton to water and it is quantified by acidity constants (K_a). Table 3-4 lists pK_a values for several acids of interest. Acids with high acidity constants (or pK_a of low values) are called strong acids; practically, one can assume complete dissociation when they are added to water. Examples are sulfuric, hydrochloric, and nitric acids. There are several weak acids and bases of interest in natural waters and in water supply and treatment, so principles regarding the equilibrium distribution of acids and their conjugate bases are covered next, and several applications are given.

TABLE 3-4 Acidity Constants for Some Important Acids at 25°C (from Stumm and Morgan, 1996 and Benjamin, 2002)

Acid	Formula	pK_{a1}	pK_{a2}	pK_{a3}	Applications
Sulfuric	H_2SO_4	~ −3	1.99		Coagulation, pH control, membrane cleaning
Phosphoric	H_3PO_4	2.16	7.20	12.3	Speciation in lakes, corrosion control
Arsenic	H_3AsO_4	2.24	6.76	11.6	Oxidized As—in water treatment
Carbonic	$H_2CO_3^*$	6.35	10.33		Many; see inorganic carbon section
Hydrogen Sulfide	H_2S	6.99	12.92		Found in some groundwaters
Hypochlorous	HOCl	7.60			Chlorination
Arsenous	H_3AsO_3	9.23	12.1	13.4	Reduced As found in some groundwaters
Ammonium	NH_4^+	9.25			Chloramination

Acid-Base Equilibria. We begin with monoprotic acids and consider generically the acid HA and its conjugate base A^-. The stoichiometric and equilibrium constant equations follow:

$$HA \rightleftharpoons A^- + H^+ \tag{3-26}$$

$$K_{a1} = \frac{[A^-][H^+]}{[HA]} \tag{3-27}$$

K_{a1} is the acidity constant. Because there is only one proton involved, it may be written simply as K_a. Taking base 10 logs of both sides of Eq. 3-27, invoking the pX notation for $[H^+]$ and K_a, and rearranging, we obtain Eq. 3-28.

$$\log\frac{[A^-]}{[HA]} = pH - pK_a \tag{3-28}$$

Applying Eq. 3-28 to the equilibrium distribution of $HOCl/OCl^-$ (free chlorine), we obtain Eq. 3-29.

$$\log\frac{[OCl^-]}{[HOCl]} = pH - pK_a = \log\frac{[OCl^-]}{[HOCl]} = pH - 7.6 \tag{3-29}$$

If we practice free chlorination at pH 7.6, then chlorine in the form of HOCl equals OCl^-. For lower pH conditions, HOCl is the dominant form of free chlorine, whereas at $pH > pK_a$, OCl^- dominates. This simple acid (HOCl) and base (OCl^-) chemistry is important in chlorination because HOCl is a far more effective disinfectant than OCl^-. In Chap. 17 the chemistry of free chlorine is covered in detail.

Another important application is the use of ammonia in combined chlorination or chloramination (see Chap. 17). Equation 3-30 shows the dependence of ammonium ion (NH_4^+) and dissolved ammonia (NH_3) on pH. It shows that when ammonia is used in chloramination at $pH < 9.25$, ammonium ion dominates.

$$\log\frac{[NH_3]}{[NH_4^+]} = pH - pK_a = \log\frac{[NH_3]}{[NH_4^+]} = pH - 9.25 \tag{3-30}$$

Two diprotic acids of interest are hydrogen sulfide and dissolved inorganic carbon ($H_2CO_3^*$). Because reduced inorganic sulfur can occur in groundwaters, the distribution of $H_2S/HS^-/S^{2-}$ as a function of pH is important to understand. Its occurrence is presented later in this chapter; the removal of H_2S by stripping is covered in Chap. 6, and the removal of reduced sulfur by oxidation is covered in Chap. 7. Inorganic carbon has many important applications presented later in this chapter.

Generically, we describe the diprotic acid systems as $H_2A/HA^-/A^{2-}$. Note that H_2A is an acid, HA^- is both an acid and a base, and A^{2-} is a base. The total molar concentration of A ($C_{T,A}$) containing species must equal the sum of all species, as shown by Eq. 3-31.

$$C_{T,A} = [H_2A] + [HA^-] + [A^{2-}] \tag{3-31}$$

Using this equation and writing the equilibrium constant equations, one can obtain expressions for the fraction of each species (α expressions) as a function of pH, $K_{a,1}$, and $K_{a,2}$. Details are found in any of the water chemistry references.

$$\alpha_0 = \frac{[H_2A]}{C_{T,A}} = \frac{1}{1 + \frac{K_{a,1}}{[H^+]} + \frac{K_{a,1}K_{a,2}}{[H^+]^2}} \tag{3-32}$$

$$\alpha_1 = \frac{[HA^-]}{C_{T,A}} = \frac{1}{1 + \frac{[H^+]}{K_{a,1}} + \frac{K_{a,2}}{[H^+]}} \tag{3-33}$$

$$\alpha_2 = \frac{[A^{2-}]}{C_{T,A}} = \frac{1}{1 + \frac{[H^+]}{K_{a,2}} + \frac{[H^+]^2}{K_{a,1}K_{a,2}}} \tag{3-34}$$

Three triprotic acids are of interest, as indicated in Table 3-4—phosphoric acid (orthophosphate) and oxidized and reduced arsenic. Reduced As (III) occurs in some groundwaters primarily as H_3AsO_3, because of its high pK_a (see Table 3-4). Oxidized As (V) is the form of As subject to removal by water treatment processes. Its speciation ($H_3AsO_4/H_2AsO_4^-/ HAsO_4^{2-}/AsO_4^{3-}$) depends on pH. The speciation of triprotic acids as a function of pH is shown through their alpha expressions, which can be found in any of the water chemistry references. Oxidation and removal of As from water supplies are covered in Chaps. 7 and 12.

Precipitation and Dissolution of Metal Solids. Precipitation and dissolution of metal solids are governed by chemical principles of metal solubility. Some examples of precipitation reactions are: metals undergoing precipitation in coagulation, water softening to remove Ca and Mg, scaling on membranes or pipe walls, precipitation of ferric iron and oxidized manganese in treatment processes and in distribution systems, and precipitation of Ca in lakes at high pH conditions. Dissolution reactions are important in treatment processes and in affecting the chemistry of water supplies. Dissolution of minerals imparts various cations and anions to waters, affecting the chemical composition of source waters; this is discussed later in the chapter.

Metal solubility is classified as follows: simple metal solubility, metal oxide and hydroxide solubility involving hydrolysis of the metal and thus pH effects on solubility, metal carbonate solubility involving effects of pH and inorganic carbon concentration, and metal solubility with other anions or ligands such as sulfate, sulfide, phosphate, and organic matter. A few principles are presented here, and simple solubility of metals is considered. In Chap. 13 there is additional presentation of principles pertaining to water softening and the solubility of calcium carbonate and magnesium hydroxide.

By convention solubility reactions are written with the solid on the LHS of the equation (see Eq. 3-35), and the solubility constant is written as K_{s0}. The subscript 0 (zero) means the metal (Me) does not form a soluble complex with the anion or ligand (L). Sometimes the solubility constant is written as K_{sp}, indicating a solubility product constant.

$$Me_aL_b(s) \rightleftharpoons aMe^{b+} + bL^{a-} \tag{3-35}$$

$$K_{s0} = [Me^{b+}]^a[L^{a-}]^b \tag{3-36}$$

For simple salts, the dissolution and precipitation depend only on the free metal; there is no complexation with OH^- (metal ion hydrolysis) or complexation with other anions. Simple metal solubility is largely restricted to monovalent metals (Na^+, K^+, Ag^+) and can apply to some solubility reactions for Ca^{2+}. A good example of simple solubility for Ca^{2+} is the scaling of gypsum on nanofiltration and reverse osmosis membranes.

Solubility reactions are temperature and ionic strength dependent. In most cases, solids are more soluble as temperature increases. However, an important case where the solid is more insoluble with increasing temperature is the solubility of $CaCO_3$(s). In many references, solubility constants are given for various temperatures. If not available, one can calculate for the temperature of interest using the van't Hoff equation (Eq. 3-16). For high ionic strength (I) applications, activity coefficients must be used to account for the effect of I or by using equilibrium constants determined for the ionic strength of interest.

Example 3-3 Precipitation (Scaling) of Gypsum
Consider scaling due to buildup of Ca^{2+} and SO_4^{2-} on the reject side of nanofiltration or reverse osmosis membranes. The solid formed is gypsum. Because the ionic strength (I) can be high on the reject side, activity coefficients must be calculated and used. In this example, I is varied to examine its effect on what concentrations produce scaling.

Solution Gypsum is $CaSO_4 \cdot 2H_2O$(s) and its equilibrium constant for zero I is $10^{-4.58}$. Writing the equilibrium constant expression with activity coefficients, we have

$$K_{s0} = \gamma_{Ca}[Ca^{2+}]\gamma_{SO_4}[SO_4^{2-}]$$

Solving for the case of the molar concentration of Ca^{2+} equal to SO_4^{2-}, we obtain the following.

$$[Ca^{2+}] = [SO_4^{2-}] = \left(\frac{K_{s0}}{(\gamma_{Ca})(\gamma_{SO_4})}\right)^{0.5}$$

Ionic strength is varied from 0 to 0.5 M. The Davies equation (Eq. 3-13) is used to calculate activity coefficients. The results are shown in the table.

I (M)	γ_{Ca}	γ_{SO_4}	Ca = SO_4 (M)	Ca (mg/L)	SO_4 (mg/L)
0	1	1	5.1×10^{-3}	204	490
0.001	0.867	0.867	5.9×10^{-3}	236	566
0.01	0.667	0.667	7.77×10^{-3}	311	746
0.1	0.380	0.380	1.35×10^{-2}	540	1296
0.5	0.296	0.296	1.73×10^{-2}	692	1660

The results show that the formation of gypsum occurs at Ca^{2+} and SO_4^{2-} concentrations of 100s of mg/L. Note that the activity coefficients (γ_i) decrease as I increases. This means that the background concentration of salt interferes with the precipitation of gypsum so that the concentrations of Ca^{2+} and SO_4^{2-} at which a solid is formed increase with I. The above calculations were made for equal molar concentrations of Ca^{2+} and SO_4^{2-}, but any concentration conditions in which the reaction quotient (Q) exceeds K_{s0} will cause scaling. ▲

Oxidation-Reduction Chemistry. Redox is an abbreviation referring to oxidation-reduction reactions or chemistry. Redox chemistry can often appear complicated because scientists and engineers working in the water field only deal with it occasionally. There are several applications of importance to us, as the following examples illustrate. Reducing conditions (low or no dissolved oxygen [DO]) can exist in reservoir water supplies leading to reduced Fe(II) and Mn(II) at concentrations high enough to cause problems. Likewise, Fe(II) and Mn(II) can occur in groundwaters, and under high reducing conditions we can also find reduced S(II) and the odor of hydrogen sulfide. Another important reduced species that occurs in some groundwater supplies is As(III). Later in this chapter the occurrence and concentrations of some of these chemicals are discussed. In water treatment we make use of oxidation (Chap. 7) processes to transform reduced chemicals to oxidized species. In the oxidized form, these chemicals are more amenable to removal. It is also noted that chemical disinfectants (Chap. 17) are oxidants so knowledge of redox chemistry helps us understand disinfection. Pb(II) and Cu(II) in the consumer's tap water arise from corrosion of metal pipes (Chap. 20) that involve redox chemistry.

Many redox reactions are biologically mediated. Some examples follow: (1) algae in reservoirs use photosynthesis to convert dissolved carbon dioxide to algal biomass, (2) respiration of algal biomass within reservoirs or bacterial decomposition of the organic carbon consuming oxygen and producing dissolved carbon dioxide, (3) microbial growth in filters and in water distribution systems, and (4) use of microorganisms in slow sand filters and riverbank filtration systems to biodegrade organic contaminants.

The reader is referred to the water chemistry references at the end of the chapter for comprehensive coverage of redox chemistry. Some definitions and principles are covered here to serve as a foundation for other chapters.

Definitions. Oxidation reactions involve the loss of electrons (e^-) and, conversely, reduction reactions involve the gain of electrons. Since we cannot have free e^- in water, whenever an oxidation reaction occurs there is also a reduction reaction occurring or vice versa. An oxidizing agent, what we usually call an oxidant, is a substance capable of gaining e^- and, conversely, a reducing agent is a substance capable of losing e^-. In simple terms, redox reactions are e^- transfer reactions in which oxidized substances or oxidants are e^- acceptors, and reduced substances or reducing agents are e^- donors. A useful analogy to redox reactions is acid-base reactions. Recall that acids are proton donors and bases are proton acceptors. We will make further use of this analogy later in this section.

The oxidation state, or number, is a hypothetical number or charge an atom might have when all the electrons are counted according to a set of rules. The assigning of oxidation state by the rules are not so important here; see the water chemistry references on how to determine oxidation states. Roman numerals are used in specifying oxidation states except, of course, when it is zero. For oxidation reactions there is an increase in the oxidation state, while there is a decrease in the oxidation state for substances that are reduced. Some examples of oxidation states follow:

- Reduced Fe and Mn are Fe(II) and Mn(II)
- Oxidized Fe and Mn are Fe(III) and Mn(IV); for permanganate (MnO_4^-) Mn is Mn(VII)
- Dissolved gases in water such as oxygen (O_2(aq)) and chlorine (Cl_2(aq)); the O and Cl atoms have an oxidation state of zero
- The oxidation state of reduced S in H_2S is S(-II), while in SO_4^{2-} it is S(VI)
- Reduced As and oxidized As are As(III) and As(V)

Nernst Equation. An overall oxidation-reduction reaction (or galvanic cell) is shown by Eq. 3-37. Applying the Nernst equation (see the water chemistry references), which describes the electromotive force (EMF) or electrical potential, we obtain Eqs. 3-38a and 3-38b. These are general forms of the Nernst equation where R, T, n, F, and Q are, the universal gas constant (8.314 J mole^{-1} K^{-1}), temperature in K, number of e^- transferred, universal gas constant (8.314 J mole^{-1} K^{-1}), temperature in K, number of e$^-$ transferred, Faraday constant (96.485 kJ V^{-1} mole^{-1}), and reaction quotient, respectively.

$$aA + bB = cC + dD \tag{3-37}$$

$$E = E^o - \frac{RT}{nF}\ln\frac{[C]^c[D]^d}{[A]^a[B]^b} \tag{3-38a}$$

$$E = E^o - \frac{RT}{nF}\ln Q \tag{3-38b}$$

Inserting values for R and F, converting to base 10 log, and setting the temperature to 298 K (25°C), we obtain the following practical form of the Nernst equation:

$$E = E^o - \frac{0.059}{n}\log Q \tag{3-39}$$

At equilibrium E is zero, Q is K, and therefore E^o is related to the equilibrium constant. E^o is a thermodynamic property that is tabulated for half-cell reactions. By definition, the reference half-cell is the hydrogen half-cell and has a value of 0 V (E_H^o). Standard potentials (E^o) for half-cell reactions can be combined to obtain overall cell reactions.

The Nernst equation can be applied to half-cell reactions because we can add the standard hydrogen half-cell reaction with a value of E_H^o of 0 V to any other half-cell reaction to complete the overall reaction since all reactions must be a combination of both oxidation and reduction. Applying the Nernst equation to half-cell reactions without an effect of H^+ (Eq. 3-40), we obtain Eq. 3-41.

$$Ox + ne^- = Red \tag{3-40}$$

$$E = E^o - \frac{0.059}{n}\log\frac{[Red]}{[Ox]} \tag{3-41}$$

A pX notation is used in redox reaction chemistry where pe = – log{e$^-$} and pe° = (1/n) log K (note the positive sign for defining pe° to 1/n log K; see Table 3-1).

The Nernst equation for a half-cell reaction written in terms of p notation follows, where pe equals E/0.059 or 16.95 E and pe° equals E^o/0.059 or 16.95 E^o.

$$pe = pe^o - \frac{1}{n}\log\frac{[Red]}{[Ox]} \tag{3-42}$$

Oxidation-Reduction Potential Measurements. The electric potential (E) or oxidation-reduction potential (ORP) is measured to evaluate the reducing or oxidizing conditions of waters—see *Standard Methods* (APHA, AWWA, and WEF, 2005). It is recognized that there are limitations of measured E values and interpretation of their meaning because E in the Nernst equation relates it to a thermodynamic property (E^o) and other conditions that are not strictly valid such as the presence of a single redox pair (waters contain multiple pairs). Nonetheless, E is used as an *indicator* of the extent of reducing or oxidizing conditions and of which species of a redox pair tends to dominate.

Whether a reduced or oxidized species dominates in concentration depends on the measured E with reference to the standard potential (E^o). It is analogous to acid-base chemistry in which measured pH is compared to the pK_a of the acid-base pair. When the pH is < pK_a, the acid species is high in concentration relative to the base species. For a redox pair, from the *pe* form of the Nernst equation (Eq. 3-42), when the pe is > pe° (or E is > E^o [Eq. 3-41]), there is low electron activity (low $\{e^-\}$) or oxidizing conditions and the oxidized species dominates. Conversely, for high $\{e^-\}$ or reducing conditions, pe < pe° and the reduced species are at a high concentration for the redox pair. Consider, for example, the Fe^{2+}/Fe^{3+} pair, the pe° is 13.03 ($E^o = 0.769$ V). For pe < pe°, we would expect reducing conditions and Fe^{2+} to be at a greater concentration than Fe^{3+}.

Figure 3-1 shows stability limits for water. The figure provides guidelines of redox and pH conditions for waters saturated with oxygen, conditions for transition zones with oxygen at low to anoxic conditions, and conditions for water-logged soils and sediments.

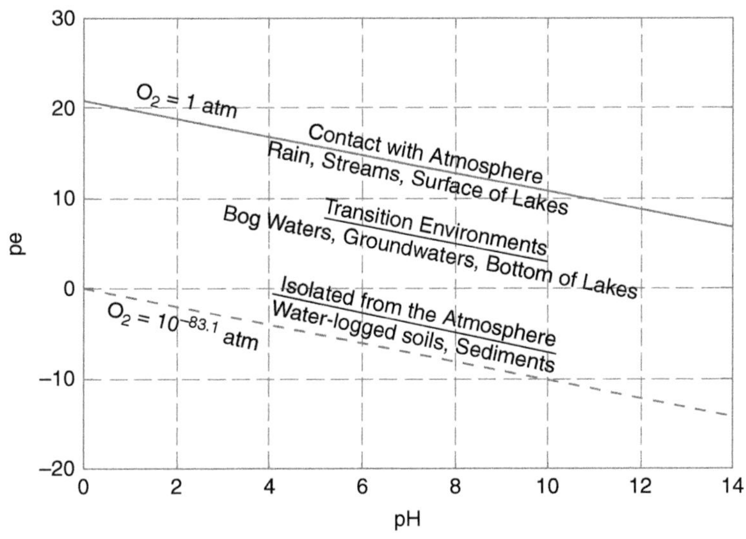

FIGURE 3-1 Redox stability limits for water with respect to pe and pH at 25°C (E [ORP] in volts = pe/16.95).

Inorganic Carbon Chemistry

The total molar inorganic carbon concentration in water is defined by Eq. 3-43.

$$C_{T,C} = [H_2CO_3^*] + [HCO_3^-] + [CO_3^{2-}] \qquad (3\text{-}43)$$

Recall (see previous section on the solubility of carbon dioxide) that $H_2CO_3^*$ is the sum of CO_2(aq) and H_2CO_3, but it is mostly CO_2(aq). Because CO_2(aq) (i.e., $H_2CO_3^*$) is an acid, inorganic carbon is important in many processes in surface waters and groundwaters in which it can affect pH. Several of these processes are summarized in Table 3-5 and include various weathering reactions, algal reactions, and biological degradation reactions. All of these reactions affect the inorganic carbon concentration, pH, and chemical composition of water.

TABLE 3-5 Processes in Surface and Groundwaters Affecting Inorganic Carbon and pH

Process	Inorganic carbon	pH
Weathering of Al-Silicate Minerals	Increases HCO_3^-	Increases
Weathering of Carbonate Minerals	Increases HCO_3^-	Increases
Precipitation of $CaCO_3$(s) (In hard water lakes with algal growth)	Decreases CO_3^{2-}	Decreases
Photosynthesis (algal growth)	Decreases $CO_{2(aq)}$	Increases
Respiration of Algae	Increases $CO_{2(aq)}$	Decreases
Bacterial Degradation of Organic Matter	Increases $CO_{2(aq)}$	Decreases
Anaerobic Degradation of Organic Matter	Increases $CO_{2(aq)}$	Decreases

Table 3-6 summarizes inorganic carbon concentrations for various types of waters. Of course, the system with the lowest concentration is pristine rainwater, which would give an estimate for harvested rainwater systems ignoring interaction with any materials in the collection system (roofs, concrete, etc.). An approximate average concentration for surface waters is 10^{-3} M (12 mg/L). Hard waters contain higher concentrations and soft waters lower concentrations. Groundwaters can contain lower and higher concentrations depending on the minerals in the aquifer and bacterial decay of organic matter in the soils and aquifer.

TABLE 3-6 Dissolved Inorganic Carbon Concentrations* in Various Water Types

Water type	C_T M	C_T mg/L as C
Pristine rainwater	1.52×10^{-5}	0.18
Rivers		
Range	$10^{-4} - 5 \times 10^{-3}$	1.2–60
Average	$\sim 10^{-3}$	12
Groundwaters		
Range	$5 \times 10^{-4} - 8 \times 10^{-3}$	6–96
Seawater	2.3×10^{-3}	28

*Pristine rainwater calculated based on open system in equilibrium with atmospheric CO_2(g) at 380 ppm; other values from Stumm and Morgan (1996).

Inorganic carbon chemistry is important in many water treatment processes including coagulation, water softening, corrosion, oxidation processes, and any processes involving acid-base reactions. Alkalinity is due mainly to inorganic carbon and is discussed in detail later.

Acid-Base Chemistry of Inorganic Carbon. The stoichiometric equations (Eqs. 3-44 and 3-46), equilibrium constant expressions (Eqs. 3-45 and 3-47), and alpha equations (Eqs. 3-48 through 3-50) are given here. General development of alpha equations for diprotic acids was presented earlier. Adjusted values for the equilibrium constants for waters at 10^{-3} M ionic strength (a common condition for water supplies) for K_1 and K_2 are $10^{-6.30}$ and $10^{-10.25}$ at 25°C, respectively, and are given in Eqs. 3-45 and 3-47. A summary of

TABLE 3-7 Inorganic Carbon and Water Equilibrium Constants (K) at 25°C for Zero Ionic Strength (Values from Stumm and Morgan (1996))

Reaction	5°C	10°C	15°C	20°C	25°C	40°C
$CO_2(g) + H_2O = H_2CO_3^*$	1.20	1.27	1.34	1.41	1.47	1.64
$H_2CO_3^* = HCO_3^- + H^+$	6.52	6.46	6.42	6.38	6.35*	6.35
$HCO_3^- = CO_3^{2-} + H^+$	10.56	10.49	10.43	10.38	10.33†	10.22
$H_2O = H^+ + OH^-$	14.73	14.53	14.34	14.16	14.0	13.53

*K_1 for 10^{-3} M ionic strength is $10^{-6.3}$
†K_2 for 10^{-3} M ionic strength is $10^{-10.25}$

equilibrium constants for inorganic carbon and water dissociation is given in Table 3-7 as a function of water temperature and ionic strength approaching zero.

$$H_2CO_3^* \rightleftharpoons HCO_3^- + H^+ \tag{3-44}$$

$$K_1 = 10^{-6.30} = \frac{[HCO_3^-][H^+]}{[H_2CO_3^*]} \tag{3-45}$$

$$HCO_3^- \rightleftharpoons CO_3^{2-} + H^+ \tag{3-46}$$

$$K_2 = 10^{-10.25} = \frac{[CO_3^{2-}][H^+]}{[HCO_3^-]} \tag{3-47}$$

$$\alpha_0 = \frac{[H_2CO_3^*]}{C_{T,C}} = \frac{1}{1+\frac{K_1}{[H^+]}+\frac{K_1K_2}{[H^+]^2}} \tag{3-48}$$

$$\alpha_1 = \frac{[HCO_3^-]}{C_{T,C}} = \frac{1}{1+\frac{[H^+]}{K_1}+\frac{K_2}{[H^+]}} \tag{3-49}$$

$$\alpha_2 = \frac{[CO_3^{2-}]}{C_{T,C}} = \frac{1}{1+\frac{[H^+]}{K_2}+\frac{[H^+]^2}{K_1K_2}} \tag{3-50}$$

Alpha (α) expressions are useful in calculations of inorganic carbon concentrations. Table 3-8 provides a summary of alpha values between pH 2 and 12.

Open System. An open system means there is a gas phase as well as the aqueous phase and any solid phases. Consider the case of an open system for inorganic carbon without any solid carbonate phase. Figure 3-2a shows the distribution of the inorganic species and $C_{T,C}$ as a function of pH. The following are important points that will be useful in understanding the composition of water and water treatment processes. First, for water in equilibrium with air at 101.3 kPa (1 atm) it means that $H_2CO_3^*$ is in equilibrium with $CO_2(g)$ at 380 ppm. The dissolved equilibrium concentration of $H_2CO_3^*$ from Henry's law (Eq. 3-25) is 1.29×10^{-5} M. It is fixed by Henry's law and is independent of pH as shown in Fig. 3-2a. Second, as we increase pH $C_{T,C}$ increases because $H_2CO_3^*$ dissociates forming HCO_3^- and

TABLE 3-8 Inorganic Acid-Base Alphas for K's at 25°C and Adjusted for 10^{-3} M Ionic Strength (pK_1 = 6.30 and pK_2 10.25)

pH	H^+ M	α_0	α_1	α_2
2	1.0×10^{-2}	0.9999	5.01×10^{-5}	2.82×10^{-13}
3	1.0×10^{-3}	0.9995	5.01×10^{-4}	2.82×10^{-11}
4	1.0×10^{-4}	0.9950	4.99×10^{-3}	2.81×10^{-9}
4.5	3.16×10^{-5}	0.984	0.0156	2.78×10^{-8}
5	1.0×10^{-5}	0.952	0.0477	2.69×10^{-7}
5.5	3.16×10^{-6}	0.863	0.137	2.43×10^{-6}
6	1.0×10^{-6}	0.666	0.334	1.88×10^{-5}
6.3	5.01×10^{-7}	0.500	0.500	5.61×10^{-5}
7	1.0×10^{-7}	0.166	0.833	4.68×10^{-4}
7.5	3.16×10^{-8}	0.059	0.939	1.67×10^{-3}
8	1.0×10^{-8}	0.019	0.975	5.48×10^{-3}
8.3	5.01×10^{-9}	0.010	0.979	0.011
9	1.0×10^{-9}	1.88×10^{-3}	0.945	0.053
9.5	3.16×10^{-10}	5.30×10^{-4}	0.849	0.151
10	1.0×10^{-10}	1.30×10^{-4}	0.640	0.360
10.25	5.62×10^{-11}	5.50×10^{-5}	0.500	0.500
11	1.0×10^{-11}	3.0×10^{-6}	0.151	0.849
12	1.0×10^{-12}	3.4×10^{-8}	0.0175	0.983

CO_3^{2-} (see Eqs. 3-44 and 3-46), and through Henry's law, $H_2CO_3^*$ is replaced from the gas phase (air) to keep it constant. Third, HCO_3^- is the dominant form of inorganic carbon for pH conditions slightly greater than 6 to about 10; above pH 10, CO_3^{2-} dominates.

We usually do not have equilibrium with the atmosphere because of various biological and chemical reactions in water supplies that add or remove inorganic carbon. However, open systems such as streams and the surface waters of lakes are seeking equilibrium. Certain water supply systems such as stored water in open distribution reservoirs may be at equilibrium with the atmosphere and the effluent water from gas transfer processes (addition of or stripping of gases using air) is at or near equilibrium.

Closed System. For many applications, inorganic carbon chemistry occurs for closed system (cut off from the atmosphere or, more generally, no gas phase (CO_2(g)). Closed system conditions occur for groundwaters, water in the bottom stratified layer in lakes and reservoirs, water in pressurized transmission mains and distribution pipes, and in many water treatment processes. In some water treatment processes, we assume a closed system even though the water is open to the atmosphere, and thereby assume the gas transfer kinetics are slow for the timescale of our problem; thus we can ignore the gas phase.

$C_{T,C}$ is constant for a closed system because there is no gas exchange. Consider $C_{T,C}$ at a concentration of 10^{-3}M (a common value for many water supplies, see Table 3-6). Figure 3-2b shows the distribution of the inorganic carbon species as a function of pH. The following important points are made. First, $H_2CO_3^*$ is the dominant form of inorganic carbon and approaches 100 percent for pH < 5 (Fig. 3-2b), Second, HCO_3^- dominates between pH 6.3 and pH 10.3. Third, CO_3^{2-} is not an important fraction of inorganic carbon unless the pH is > 10. This figure is invaluable for many applications, as listed above, and for understanding alkalinity, which is covered next.

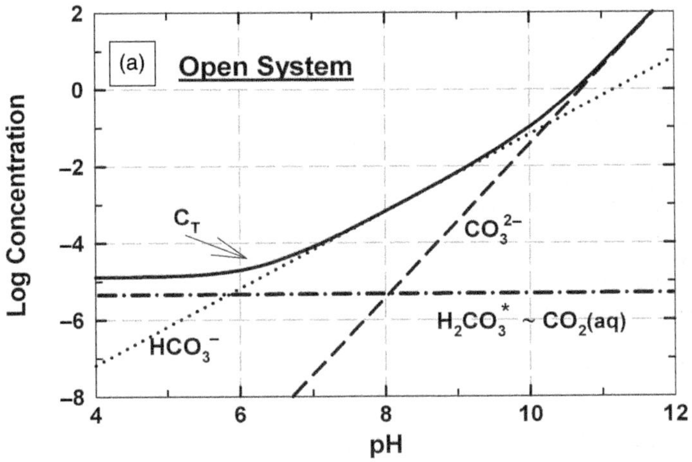

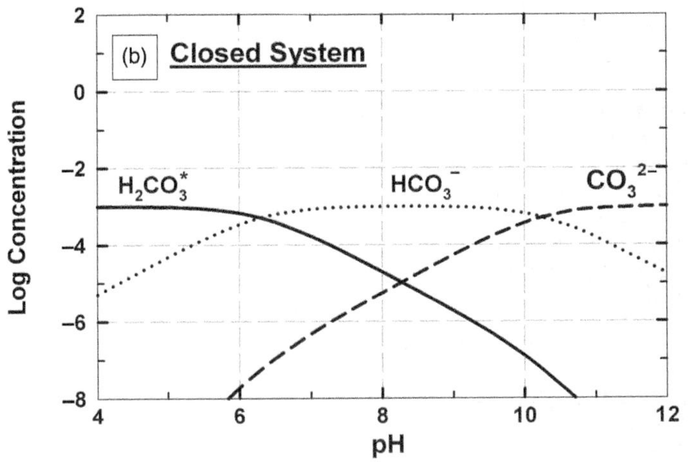

FIGURE 3-2 (*a*) Open system for inorganic carbon as a function of pH (no carbonate solid phase). (*b*) Closed system for inorganic carbon at 10^{-3}M (conditions: 25°C and 10^{-3} M ionic strength).

Alkalinity

Definition, Causes, and Laboratory Measurement. Alkalinity (Alk) is the acid neutralizing capacity (ANC) of a water. The causes of Alk are the presence of conjugate bases in water that can consume acid (H^+) when added to the water. In general, the conjugate bases that can potentially contribute to alkalinity are those from inorganic carbon (HCO_3^- and CO_3^{2-}), orthophosphate ($H_2PO_4^-$, HPO_4^{2-}, and PO_4^{3-}), borate (e.g., in seawater ($B(OH)_4^-$)), and others such as NH_3, silicates, sulfides, and conjugate bases of organic acids. Alk can also be due to OH^- when it is in excess relative to H^+. For most water supplies and water treatment applications, the dominant conjugate bases present are those from inorganic carbon, meaning that the others are relatively low and insignificant. Here and subsequently we describe Alk with respect to inorganic carbon and the net difference between OH^- and H^+.

Alk of water is measured in the laboratory by titrating a water sample with a strong acid to a defined theoretical or operational end point under closed system conditions, i.e., no exchange of the carbon dioxide with the atmosphere. Theoretically, the end point depends on the inorganic carbon concentration ($C_{T,C}$). The greater $C_{T,C}$ is, the lower the pH end point. However, for most water supply and treatment applications, the theoretical end point is near pH 4.5 so this is often set as an operational end point (APHA, AWWA, and WEF, 2005) and is determined by actually measuring the pH during the titration or using a color indicator. The concentration of strong acid required to reach this end point is measured and reported in meq/L. This is the total alkalinity of the sample. Sometimes, different components of Alk are measured such as caustic Alk (due to OH^-) and carbonate Alk (due to CO_3^{2-}), as well as the total Alk (see APHA, AWWA, and WEF, 2005; and water chemistry references). For most water supply and water treatment applications, it is the total Alk that is of interest and that is our focus.

Some people confuse the meaning of Alk, thinking of it as a measure of the resistance to a decrease in pH with strong acid addition for some particular pH condition. This is not the case. The resistance to a pH change at a particular pH is the buffer intensity. Alk is a *capacity* measurement—it is the total concentration of strong acid required to reduce the pH to an end point of say 4.5.

Measurements are made in meq/L of acid required to reach the end point; however, in the water field Alk is reported in terms of an equivalent concentration of $CaCO_3$. The conversion is 1 meq/L equals 50 mg/L as $CaCO_3$.

Basic Equations for Inorganic Carbon and Alkalinity. We can relate the Alk to the concentrations of HCO_3^- and CO_3^{2-} in water and to the net deficiency of protons with regard to OH^- and H^+ (stated another way, the net excess of OH^- relative to H^+). Equation 3-51 is the theoretical description of Alk due to the conjugate bases of inorganic carbon. The millimolar (mM) concentration of HCO_3^- is multiplied by 1 meq/mM because it can neutralize 1 meq of strong acid. Likewise, the mM concentration of CO_3^{2-} is multiplied by 2 meq/mM because it can neutralize 2 meq of strong acid. Thus, the Alk is in units of meq/L and can be converted to mg/L as $CaCO_3$ by multiplying by 50.

$$Alk = [\mathrm{HCO_3^-}] + 2[\mathrm{CO_3^{2-}}] + [\mathrm{OH^-}] - [\mathrm{H^+}] \tag{3-51}$$

Equation 3-51 is rewritten by replacing HCO_3^- with α_1 times $C_{T,C}$ and CO_3^{2-} with α_2 times $C_{T,C}$.

$$Alk = (\alpha_1 + 2\alpha_2)C_{T,C} + [\mathrm{OH^-}] - [\mathrm{H^+}] \tag{3-52}$$

We can learn some interesting and practical things from Eq. 3-52. First, the Alk of water depends simply on the actual pH and $C_{T,C}$ of the water sample when collected. Waters with higher pH (alphas are pH dependent) and higher $C_{T,C}$ have higher Alk. Second, we can rearrange the equation and make $C_{T,C}$ the dependent variable, meaning that its concentration can be calculated from pH and Alk measurements. Finally, using Eq. 3-52 and Fig. 3-2b, we can make the following observations. For waters with pH < 9, CO_3^{2-} and OH^- are low relative to HCO_3^-, so these waters do not contain caustic and carbonate Alk. For waters with pH between about 6 and 9, we can ignore CO_3^{2-}, OH^-, and H^+, yielding the following approximation for Alk.

$$Alk = [\mathrm{HCO_3^-}] = \alpha_1 C_{T,C} \tag{3-53}$$

As stated above, the total inorganic carbon concentration ($C_{T,C}$) and pH of a water sample sets the Alk. $C_{T,C}$ in these equations applies to whatever the inorganic concentrations are and may not necessarily be in equilibrium with atmospheric CO_2(g), i.e., $H_2CO_3^*$ is not in equilibrium with the atmosphere. In other words, Alk is set by pH and $C_{T,C}$ whether the latter is in equilibrium with the atmosphere or not. Often, it is not! Consider closed systems such as groundwaters, the bottom of lakes, and water distribution pipes. Even surface waters and waters within treatment plants, $C_{T,C}$ may be seeking equilibrium, but it is often out of equilibrium with the gas phase.

Classifications. There is no standard classification system for waters with respect to Alk concentrations. In establishing enhanced coagulation requirements for total organic carbon (TOC) removal as a function of alkalinity (see Chap. 8), the U.S. Environmental Protection Agency (USEPA) classifies waters into three Alk groups: low Alk waters, <60 mg/L $CaCO_3$; intermediate or moderate Alk waters, 60 to 120 mg/L $CaCO_3$; and high Alk waters, >120 mg/L $CaCO_3$. Other classifications exist depending on the problem or application. For example, in assessing the sensitivity of water bodies to acidification, five Alk groups of decreasing sensitivity are used: < 2.5 mg/L $CaCO_3$, 2.5 to 5 mg/L $CaCO_3$, 5 to 10 mg/L $CaCO_3$, 10 to 20 mg/L $CaCO_3$, and > 20 mg/L $CaCO_3$.

Example 3-4 Alkalinity of Pristine Rainwater and for a Water Supply
Calculate the Alk of pristine rainwater in equilibrium with atmospheric CO_2(g), and calculate the Alk for a water supply with a pH of 7.5 and $C_{T,C}$ of 10^{-3} M assuming 25°C.

Solution **Pristine rainwater** has a $C_{T,C}$ concentration of 1.52×10^{-5} M (Table 3-6). Its pH is 5.60 and $[HCO_3^-]$ is 2.29×10^{-6} M. Applying Eq. 3-51 and recognizing that OH^- and CO_3^{2-} are small and can be ignored, we have

$$Alk = [HCO_3^-] - [H^+] = 2.29\times10^{-6} - 10^{-5.6} = 2.29\times10^{-6} - 2.51\times10^{-6} \cong 0 \text{ meq/L}$$

Thus, pristine rainwater has no Alk and is susceptible to strong acid inputs decreasing its pH. Because pristine rainwater has no Alk or acid neutralizing capacity, strong acids such as sulfuric and nitric acids, if present in the atmosphere, affect rainwater pH causing acid rain with pH values less than 5.

The water supply has a pH of 7.5 and $C_{T,C}$ of 10^{-3} M. Using the alpha tables (see Table 3-8), we have α_1 of 0.939 and α_2 of 1.67×10^{-3} M. We see that α_2 is low (CO_3^{2-} is low) and can use Eq. 3-53.

$$\begin{aligned} Alk &= [HCO_3^-] = \alpha_1 C_{T,C} = 0.939\times10^{-3}\text{eq/L} \\ &= 50\times(0.939)\text{meq/L} = 47 \text{ mg/L as } CaCO_3 \end{aligned}$$ ▲

Acidity

A converse concept to Alk is the acidity (Acy) of water. Acy is the base neutralizing capacity of water, and it is due to the presence of conjugate acids of weak bases and the net excess of protons (i.e., $[H^+] - [OH^-]$). As done for Alk, we consider only the inorganic carbon weak conjugate acids resulting in Eqs. 3-54 and 3-55 as the basic equations describing Acy.

$$Acy = 2[H_2CO_3^*] + [HCO_3^-] + [H^+] - [OH^-] \tag{3-54}$$

$$Acy = (2\alpha_0 + \alpha_1)C_{T,C} + [H^+] - [OH^-] \tag{3-55}$$

The total Acy when measured in the laboratory is the total concentration of strong base added to a pH end point of about 10.6. The concentration of base added in the titration is measured as meq/L; it is multiplied by 50 and reported in mg/L as $CaCO_3$. It can be shown that by adding Eqs. 3-51 and 3-54, we can obtain Eq. 3-56. This equation shows that if we measure or know the Alk and Acy of water, we can calculate the total inorganic carbon concentration ($C_{T,C}$). Likewise, if we measure and know the Alk and $C_{T,C}$, we can calculate the Acy of a water.

$$Alk + Acy = 2(C_{T,C}) \tag{3-56}$$

Following up on pristine rainwater, which has a $C_{T,C}$ of 1.52×10^{-5} M (Table 3-6) and no Alk (Example 3-4), from Eq. 3-56 its Acy is about 3×10^{-5} eq/L or 1.5 mg/L as $CaCO_3$.

SOURCE WATER COMPOSITION

Introduction

The primary objectives of this section are to (1) describe the dissolved chemical composition of rainwaters, rivers and streams, lakes and reservoirs, and groundwaters; (2) present principles with reference to the roles of hydrogeochemical and biochemical cycles that affect the composition of water and water quality; (3) summarize the presence of anthropogenic chemicals in water supplies; and (4) highlight important dissolved chemicals that affect water quality and treatment. The chemical compositions for specific cases are presented as well as typical concentrations and ranges. Before getting into this material, some terminology is defined and the hydrogeochemical and biochemical cycles are introduced.

Terminology

Naturally Occurring and Anthropogenic Chemicals. Naturally occurring chemicals in water bodies consist of inorganic and organic chemicals derived from hydrogeochemical and biochemical cycles. Major and minor constituents are presented below for rainwaters, rivers, lakes, and groundwaters. Anthropogenic chemicals enter waters from human activities. These chemicals may be inorganic in nature such as acids and metals or they may be organic. Both naturally occurring and anthropogenic organic chemicals are grouped as summarized in Table 3-9. This table is referred to in the material below.

Chemical Phases, Total and Dissolved Concentrations. Chemists classify chemicals according to gas, dissolved, and solid phases. Dissolved chemicals may be ions (cations and anions) or neutral molecules. The sizes of dissolved chemicals are small, about 10 Å or less for simple and low molecular weight substances. Molecular weights and apparent sizes are greater for many organic compounds, especially macromolecules (e.g., fulvic and humic acids with apparent molecular weights of 100s–1000s and apparent sizes of 10–100 Å).

Practically, we must also consider operational definitions of dissolved versus particulate matter used in measuring chemicals. In the water field, particulate matter is separated from a water sample by membrane filtration either in the field or in the laboratory. Often a nominal pore size, such as 0.45 μm, is prescribed for the membrane filter. This means

TABLE 3-9 Organic Compounds in Natural Waters (Sources for Information: Domenico and Schwartz (1990); Koplin et al. (2002))

Name	Naturally occurring	Anthropogenic
Misc., non-volatile	Humic substances, chlorophyll, enzymes, xanthophylls	Tannic acids, dyes
Halogenated Hydrocarbons		Solvents, pesticides (aldrin, DDT, dieldrin), industrial chemicals (vinyl chloride, methyl chloride, PCBs)
Amino acids	Glycine, alanine, aspartic acid	
Phosphorus compounds		Pesticides (diazinon, malathion, parathion)
Organometallic compounds	Metal complexes with humic substances (Fe, Al)	Tetraethyl lead, diethyl mercury, copper phthalocyanine
Carboxylic acids	Acetic acid, benzoic acid, butyric acid, formic acid	Acetic acid, benzoic acid, butyric acid, formic acid, phenoxy acetic pesticides (dinitrocresol, 2,4-dinitrophenol)
Phenols	Phenol, cresol, p-hydroxybenzoic acids	Phenol, cresol, pyrocatechol, napthol, pesticides (dinitrocresol, 2,4-dinitrophenol)
EDCs*		
50–80%		Coprostanol (estrogen); cholesterol (steroid); N-N-diethyltoluamide (mosquito repellant); caffeine (stimulant); tris (2-chloroethyl) phosphate (fire retardant); triclosan (antibiotic); 4-nonylphenol & 4- nonylphenol monoethoxylate (surfactants)
30–45%		Ethanol, 2-butoxy-phosphate & Bisphenol A (plasticizers); 4-octylphenol monoethoxylate & 4- nonylphenol diethoxylate (surfactants); cotinine (nicotine metabolite); 5-methyl-1H-benzotriazole (antioxidant); fluoranthene (PAH); 1,7,-dimethylxanthine (caffeine metabolite)
15–25%		Pyrene (PAH); trimethoprim, erythromycin-H_2O, lincomucin, sulfamethoxazole (antibiotics); 1,4 dichlorobenzene (deodorizer); acetaminophen (analgesic); tetrachloroethylene (solvent); 4-octylphenol diethoxylate (surfactant); estrriol (estrogen); phthalic anhydride (plasticizer); carbaryl (insecticide)
Amines		Diethylamine, dimethylamine, benzidine, pyridine
Ketones		Acetone, 2,-butane, methyl propenyl
Aldehydes		Formaldehyde
Esthers		Dimethrin, omite, vinyl acetate
Ethers	Diethyl ether, diphenyl ether	Tetrahydrofuran, 1,4-dioxane
PAHs		Anthracene, benzol[a]pyrene
Aromatic hydrocarbons		Benzene, ethylbenzene, toluene
Alkanes, alkenes, and alkynes	Methane, propane, propene, 2-hexene, 2-hexyne	Methane, propane, propene, 2-hexene, 2-hexyne

*Endocrine Disrupting Compounds; frequency of detection in U.S. streams by operational definition we are defining the chemicals that we measure after passing through the membrane filter as dissolved matter, when in fact some colloidal particulate matter may pass through the filter and be measured. In some chemical analysis to eliminate colloidal matter, ultrafilters are used. With this background, we classify measurements as follows:

- Total = dissolved + particulate

The dissolved concentration is sometimes referred to as total dissolved without distinguishing the forms or species that compose the chemical. In this section, the dissolved concentration is understood to mean total dissolved, unless otherwise noted. Some examples of total dissolved and speciation follow:

- Total inorganic P or orthophosphate = $H_3PO_4 + H_2PO_4^- + HPO_4^{2-} + PO_4^{3-}$
- Total inorganic N = $NO_2^- + NO_3^- + NH_4^+ + NH_3$
- Total Al = free Al (Al^{3+}) + inorganic complexes (e.g., $Al(OH)^{2+}$, $Al(OH)_2^+$) + organic complexes (Al-humates)

Hydrogeochemical and Biochemical Cycles

All waters contain naturally occurring chemicals that enter water through the hydrogeochemical cycle (Fig. 3-3). This cycle includes effects of the hydrologic cycle and geochemical reactions that include acid-base, mineral dissolution and solubility, and redox reactions. Figure 3-3 is used in the material below when discussing the major constituents in waters,

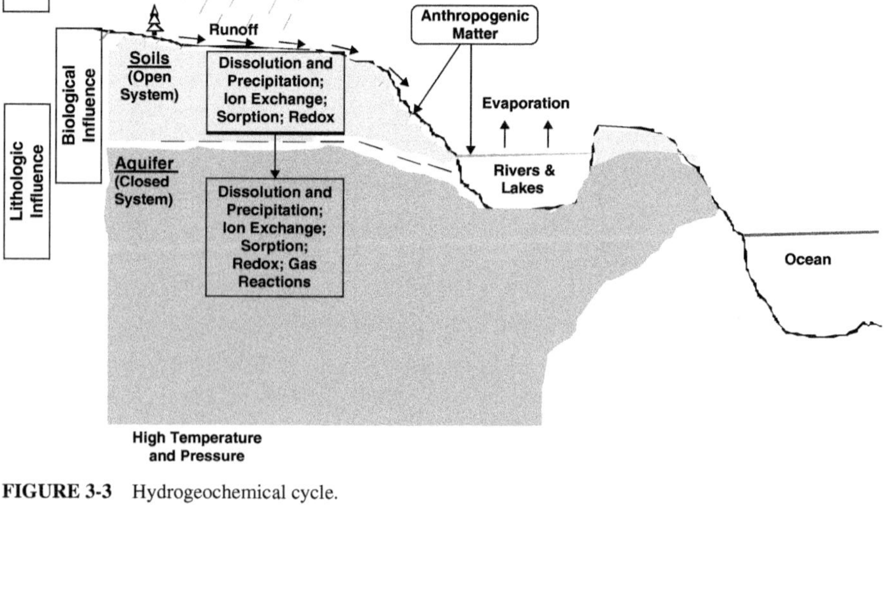

FIGURE 3-3 Hydrogeochemical cycle.

recognizing that local soil and mineral geology play major roles affecting individual source water composition.

Biochemical cycles in water also affect water quality and the biological and chemical composition of water. Figure 3-4a depicts the effects of photosynthesis, algal biomass mineralization, and decomposition effects on major nutrients (C, N, and P) in water. Figure 3-4b depicts the N cycle. The cycling of nutrients involves redox reactions that affect inorganic carbon, organic carbon, P, N, and O_2(aq). Reducing conditions affect the chemistry of Fe, Mn, and S in waters, especially in lakes and groundwaters. The significance of these cycles are incorporated later in the discussion of the water quality of surface waters and groundwaters.

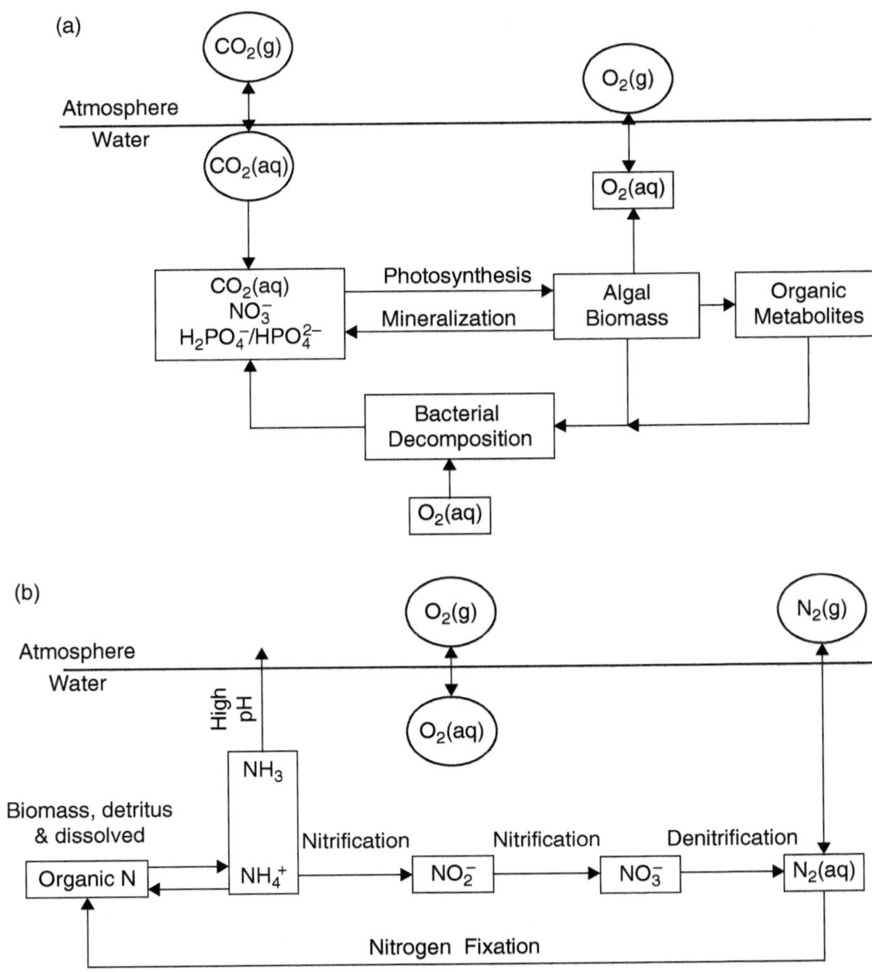

FIGURE 3-4 (*a*) Photosynthesis and decomposition effects on primary nutrients. (*b*) Nitrogen cycle in water.

Rainwaters

We begin with rainwater chemistry. This provides reference conditions for the chemistry of water not impacted by geochemical cycles and biochemical cycles in a watershed and water body. The impacts of naturally occurring and anthropogenic chemicals on rainwater are, however, presented.

The hydrologic cycle provides an important means of purification via evaporation (distillation process). Pristine rainwater chemistry is affected only by $CO_2(g)$ in the atmosphere, which in turn affects the rainwater pH and inorganic carbon concentration (Table 3-10). Pristine rainwater has a pH of 5.6. This acid content comes from $CO_2(aq)$ (i.e., $H_2CO_3^*$), and it is a major factor in affecting natural geochemical weathering in watersheds.

Note that pristine rainwater has no alkalinity (acid neutralizing capacity) and a small amount of acidity (Acy) or base neutralizing capacity (3.06×10^{-2} meq/L or 1.5 mg/L as $CaCO_3$). Consequently, rainwater is affected by airborne contaminants, natural or anthropogenic origin. Naturally occurring contaminants can be sea-salt aerosols (mostly NaCl), dust from urban and farm areas, and particulate matter and gases from volcanic eruption. These contaminants can dissolve into the rainwater. Anthropogenic chemicals can include volatile organic chemicals (VOCs) and acids from fossil fuel combustion, particularly sulfuric and nitric acids. It is important to note that cations in rainwater may be simple salts derived from sea salts or from mineral particles in dust, such as Na^+, K^+, Ca^{2+}, and Mg^{2+}, or they may be base cations, such as Ca^{2+} from carbonate mineral particles in dust or lime dust. Anions in rainwater may be from sea salts (Cl^-, SO_4^{2-}) or dust (SO_4^{2-}), and these do not affect rainwater pH. On the other hand, acid rain is largely due from conversion of sulfur oxides and nitrous oxides in the atmosphere to sulfuric and nitric acids. Therefore, large fractions of the acid anions SO_4^{2-} and NO_3^- are present in acid rain.

TABLE 3-10 Composition of Rainwaters Concentrations (mg/L Except pH and Alk)

Parameter	Pristine rain or demineralized water*	Rainwater Coastal Washington†	Acid rain Penn State, PA†
pH	5.60	5.1	4.15
$H_2CO_3^*$ as C	0.16	NR‡	NR
HCO_3^- as C	0.028	NR	NR
CO_3^{2-} as C	5.2×10^{-7}	NR	NR
Na^+	0	1.81	0.07
K^+	0	0.12	0.05
Ca^{2+}	0	0.08	0.14
Mg^{2+}	0	0.22	0.02
NH_4^+	0	0.04	0.32
Cl^-	0	3.49	0.24
SO_4^{2-}	0	0.73	3.14
NO_3^-	0	0.14	1.93
Acy (mg/L $CaCO_3$)	1.5	NR	NR
Alk (mg/L $CaCO_3$)	~0	<0	<0
Ionic Strength§ (M)	2.4×10^{-6}	1.33×10^{-4}	1.40×10^{-4}

*Equilibrium with atmospheric $CO_2(g)$ at 380 ppm at 25°C.
†from Eby (2004).
‡NR: not reported.
§Calculated from cation and anion concentrations.

Table 3-10 shows the composition of rainwater from Coastal Washington, which is slightly more acid at a pH of 5.1 than pristine rainwater. Its pH is not affected much by fossil fuel combustion since the acid anions are low compared to the acid rainwater (Penn State location) with a pH of 4.15 and relatively high concentrations of the acid anions. The Coastal Washington rainwater chemistry is greatly affected by sea-salt aerosols, as evidenced by the relatively high Na^+ and Cl^- concentrations.

Use of rainwater harvesting is increasing in developing and developed countries (UNEP-IETC, 2002; Han, 2004). It is being considered and used as a potable and non-potable water supply in developing countries. In developed counties, the interest is in non-potable uses (e.g., toilet flushing, watering lawns and gardens, irrigation) and as part of a green engineering strategy of sustainable water use. The chemical quality of the collected water depends on the rainwater chemistry, which varies with location and time. Table 3-10 provides guidance on the concentrations of major inorganic cations and anions that may be present in pristine rainwater, rainwater from a coastal location, and acid rainwater. Generally, rainwater is low in ionic strength, about 10^{-4} M or less (low in conductivity), and even pristine rainwater is acidic. Because rainwater does not contain alkalinity, it is susceptible to reductions in pH with the addition of acids. Furthermore, rainwater has little acidity (base neutralizing capacity), so any addition of bases will increase its pH. Therefore, in rainwater harvesting, the materials used for storage containers and reservoirs may affect the water's chemistry; for example, concrete materials are likely to leach some lime, increasing the pH. Collected rainwater may contain microorganisms that enter the water from wind or from contact with the collection system; consequently, the microbiological quality, particularly with respect to the presence of pathogens, is of paramount interest for any potable use.

Rivers and Streams

The major dissolved inorganic chemicals and nutrients found in surface waters (rivers and streams) are identified and typical concentrations are presented. Ranges in expected concentrations are summarized that reflect their naturally occurring concentrations and effects from human activities. Data are also summarized for alkalinity, hardness, ionic strength, and DOC.

The hydrogeochemical cycle (Fig. 3-3) causes various cations, anions, and silica to be present in surface waters through weathering reactions of rocks (minerals), soils, and sediments. The chemical reactions that take place involve acid-base (dissolved $CO_2(aq)$, i.e., $H_2CO_3^*$ is an acid), solubility, and redox reactions. Here we consider rivers and streams containing oxygen, which is the case for most water supplies. It is realized that deep rivers and run-of-the-river reservoirs may have sufficient depth and organic matter to produce reducing conditions at or near the bottom—these systems are considered under lakes and reservoirs. Figure 3-5 shows oxidized-reduced pairs along a vertical line. The y-axis gives standard potentials adjusted to pH 7—called pe° (W). Recall that pe = 16.95 *E*. ORPs (i.e., *E*) are used as indicators of the oxidizing-reducing conditions of water. Because the presence of many oxidized-reduced pairs and interferences makes precise use of ORPs infeasible, the ORP is used as an indicator.

If we measured pe equal to say about 13.7, which is approximately the pe° (W), then we have high oxidizing conditions in which H_2O can be oxidized to O_2. Moving downward along the vertical line, reducing conditions increase. The vertical line shows conditions in which we have both the oxidized and reduced pairs, i.e., when the measured pe equals the pe° (W). Note that oxidized species exist to the left of the vertical line and would dominate for pe > pe° (W). Considering oxic surface waters, the pe values would be high but less than 13.7, so we would not have oxidation of H_2O. We would expect to have all the oxidized species to the left of the vertical line. Fe exists as ferric iron (Fe(III)) but would be low in

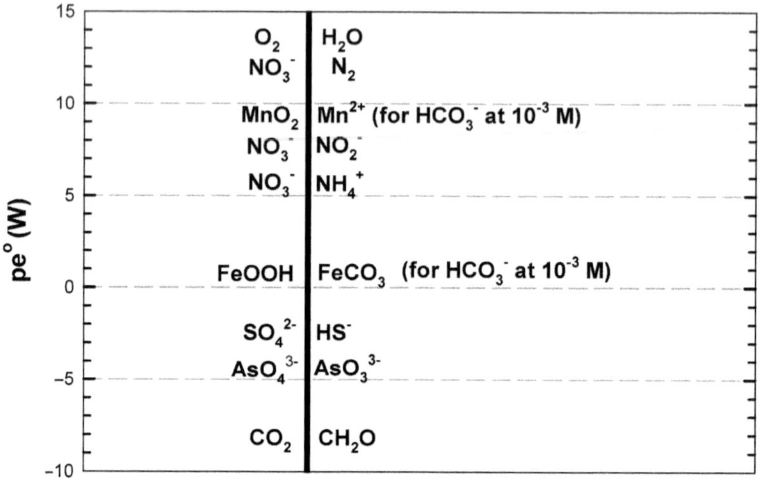

FIGURE 3-5 Sequence of microbially mediated redox processes in natural waters (pe° (W) is pe° adjusted to pH 7).

concentration because of low solubility. Mn exists as oxidized Mn (Mn(IV)), and would also be at low concentration because it has low solubility. S exists in the oxidized form as SO_4^{2-}, and N mainly as NO_3^-.

Upland Rivers and Streams. The chemical composition of upland water supplies is due mainly to the hydrogeochemical cycle and is not greatly affected by human activities (anthropogenic sources of chemicals). Therefore, chemicals are largely naturally occurring. Two examples of weathering reactions from Table 3-11 are used to illustrate related principles. The first example deals with the weathering of granitic rocks, which are representative of mountainous watersheds. A sodium feldspar (Na-Plagioclase), a component of granitic rocks, is used in the example along with its incongruent dissolution by dissolved carbon dioxide ($H_2CO_3^*$), a weak acid in rainwater.

$$\underset{\text{Na-Plagioclase}}{2NaAlSi_3O_8} + 2H_2CO_3^* + 9H_2O \rightleftharpoons \underset{\text{Kaolinite}}{Al_2Si_2O_5(OH)_4} + 2Na^+ + 2HCO_3^- + 4H_4SiO_4 \tag{3-57}$$

The above equation is an acid-base reaction in which $H_2CO_3^*$ is the acid and the feldspar is the base. It is an incongruent dissolution of the rock to a kaolinite (clay mineral) solid phase. From the acid-base reaction, we see that Na^+, HCO_3^-, and Si are imparted to the water. The dissolution is not extensive so we would expect to find fairly low Na^+ and HCO_3^- (i.e., alkalinity) for these waters. A good example of this geochemical weathering and water composition is the Quabbin Reservoir (Chicopee River Basin and a Boston, MA, water supply), which has in part a granitic bedrock geology (see Table 3-12). In general, watersheds with this type of geology have low concentrations of Na^+ and alkalinity or buffering. Furthermore, the solubility of dissolved Si is present as H_4SiO_4 and is independent of pH below 9. Its concentration depends on water temperature and whether it is controlled by amorphous or quartz (SiO_2(s)). Consequently, it is found at concentrations between 2 and 30 mg/L in most water sources, as indicated in the last column of Table 3-12.

TABLE 3-11 Weathering Reactions for Common Primary Minerals

Mineral	Composition	Commonly occur in rock type	Reaction
Olivine	$(Mg,Fe)SiO_4$	Igneous	Oxidation of Fe Congruent dissolution* by acids
Pyroxenes	$(Mg,Fe)SiO_3$ or $Ca(Mg,Fe)Si_2O_6$	Igneous	Oxidation of Fe Congruent dissolution by acids
Amphiboles	$Ca_2(Mg,Fe)_5Si_8O_{22}(OH)_2$ (also some Na and Al)	Igneous Metamorphic	Oxidation of Fe Congruent dissolution by acids
Plagioclase feldspar	$NaAlSi_3O_8$ to $CaAl_2Si_2O_8$	Igneous Metamorphic	Incongruent dissolution† by acids
K-feldspar	$KAlSi_3O_8$	Igneous Metamorphic Sedimentary	Incongruent dissolution by acids
Biotite	$K(Mg,Fe)_3(AlSi_3O_{10})(OH)_2$	Metamorphic Igneous	Incongruent dissolution by acids Oxidation of Fe
Muscovite	$KAl_2(AlSi_3O_{10})(OH)_2$	Metamorphic	Incongruent dissolution by acids
Volcanic glass (not a mineral)	Ca,Mg,Na,K,Al,Fe-silicate	Igneous	Incongruent dissolution by acids Incongruent dissolution by water
Quartz	SiO_2	Igneous Metamorphic Sedimentary	Resistant to dissolution
Calcite	$CaCO_3$	Sedimentary	Congruent dissolution by acids
Dolomite	$CaMg(CO_3)_2$	Sedimentary	Congruent dissolution by acids
Pyrite	FeS_2	Sedimentary	Oxidation of Fe and S
Gypsum	$CaSO_4 \cdot 2H_2O$	Sedimentary	Congruent dissolution by water
Anhydrite	$CaSO_4$	Sedimentary	Congruent dissolution by water
Halite	NaCl	Sedimentary	Congruent dissolution by water

*Congruent dissolution: simple dissolution.
†Incongruent dissolution: mineral breaks down into ions in solution and a solid of different composition.
Source: Berner and Berner (1996), with permission of the authors.

TABLE 3-12 Composition of Major Dissolved Constituents Surface Waters (Concentrations in mg/L Except pH, Alk, Hardness, N, and P)

Parameter	Savannah River*	Columbia River**	Quabbin Reservoir MA (Chicopee R Basin) 2008 mean	CA State Project Water Banks Pumping Station†	Miss R St. Paul*	Missouri R Hermann††	Colo R near Phoenix*	Typical ranges rivers in North America
pH	7.0	NR	6.7	6.9	8.6	7.7	8.3	6.0–8.5
Na^+	12	6.2	5.0	34	10	35	NR	1–100
K^+	NR‡	1.6	0.5	NR	NR	4.9	NR	1–10
Ca^{2+}	4	19	2.0	18	23	55	77	10–100
Mg^{2+}	1	5.1	0.5	12	5	16	29	2–30
Cl^-	9	3.5	8.0	47	21	20	88	1–100
SO_4^{2-}	7	17.1	5.0	30	23	100	250	10–300
HCO_3^- (mg/L as $CaCO_3$)	23	76	~2.6	73	150	175	135	20–200
NO_3^- as N	NR	NR	0.01	NR	NR	NR	NR	<1–5§
Ortho P as P (μg/L)	NR	NR	16	NR	NR	NR	NR	10–100§
SiO_2	NR	10.5	1.7	NR	NR	10	NR	2–30
Hardness (mg/L as $CaCO_3$)	14.1	68.5	7.1	96	78.1	200	311.8	10–350
Alk (mg/L $CaCO_3$)	~23	~76.0	2.6	73	~150	143	~135	5–200
Ionic Strength¶ (M)	0.93×10^{-3}	1.70×10^{-3}	08×10^{-3}	2.5×10^{-3}	3.30×10^{-3}	3.8×10^{-3}	13.4×10^{-3}	10^{-3}–10^{-2}
DOC (mg/L)	3.0	NR	1.8	NR	8.0	3.8	8.3	2–15

*from Benjamin (2002); **from Eby (2004); †median concentrations (from "Water Quality in the State Water Project, 2002 and 2003" State of California, July 2007); ††USGS (Mar 1998); ‡NR: not reported; §low end for undeveloped watersheds, high end for agricultural watersheds; ¶estimated from TDS or specific conductance.

A second and contrasting example is the weathering of calcite and dolomite minerals. Specifically, calcite is used in the example below.

$$\underset{\text{Calcite}}{CaCO_3} + H_2CO_3^* \rightleftharpoons Ca^{2+} + 2\ HCO_3^- \tag{3-58}$$

Calcite is rather soluble in water at pH less than 9 to 10 and can produce fairly high Ca^{2+} and HCO_3^- concentrations. In general, water sources that drain calcite and dolomite watersheds have moderate to high levels of hardness and alkalinity. An example in Table 3-12 is the Missouri River.

Downstream Rivers. Larger rivers as they move to the sea become influenced by anthropogenic chemicals imparted in the drainage from agricultural and urban areas and from industrial and community wastewater discharges. Many major ions such as Ca^{2+}, Mg^{2+}, K^+, and HCO_3^- (see Table 3-12) are generally not affected. Levels of other chemicals increase significantly, especially in rivers in arid regions where irrigation water is used in the watershed. An example is the Colorado River near Phoenix (Table 3-12), which has a high concentration of dissolved solids, as indicated by its high ionic strength. Note the high concentrations of Cl^- and SO_4^{2-} compared to the other surface waters. Na^+ is not reported, but most likely it is high.

Anthropogenic Organic Chemicals. Anthropogenic organic chemicals are listed according to general categories in Table 3-9. These chemicals are normally low in concentration or not detected in upstream and protected water sources. Downstream rivers can contain pesticides and herbicides from agricultural and urban drainage. Other chemicals that may be found are site specific and can be found in major downstream rivers, such as the Mississippi, Ohio, Missouri, and Delaware rivers. A recent problem that is addressed next is the presence of EDCs (endocrine disrupting compounds) in source waters and drinking waters.

A widely quoted study is that of the United States Geological Survey (USGS) performed in 1999 to 2000 (Kolpin et al., 2002). The study sampled 139 streams downstream of urban and agricultural (livestock) drainage areas. The data collected included endocrine disrupting compounds (EDCs) associated with pharmaceuticals and hormones as well as other wastewater organic compounds. Organic contaminants were found in 80 percent of the streams sampled but at low concentrations, generally less than 1 μg/L. The most frequently detected compounds were two steroids (coprostanol and cholesterol), an insect repellant (N,N-diethyltoluamide), caffeine, an antimicrobial disinfectant (triclosan), a fire retardant (tri(2-chloroethyl) phosphate), and a detergent metabolite (4-nonylphenol).

A more recent study of the occurrence of EDCs in drinking water supplies and actual drinking water was reported by Benotti et al. (2009). Source water, finished drinking water, and distribution system tap water were sampled from 2006 through 2007 for 51 pharmaceuticals, other known or potential EDCs, and other organic chemicals. A summary of their findings regarding tap water is given with respect to what was found in the distribution systems of 15 water utilities. No pharmaceuticals or other EDCs were found in two of the systems. The following compounds and median concentrations were found in at least six of the systems at concentrations >1 ng/L: (1) pharmaceuticals: carbamazepine at 6.8 ng/L and phenytoin at 3.6 ng/L; (2) known or potential EDCs: atrazine at 50 ng/L; and (3) other chemicals: TCEP (tris(2-chloroethyl) phosphate) at 150 ng/L, and TCPP (tris(1,3-dichloro-2-propyl) phosphate at 220 ng/L.

Summary. Common inorganic ions found in both upstream and downstream river sources are Na^+, K^+, Ca^{2+}, Mg^{2+}, Cl^-, HCO_3^-, and SO_4^{2-}. Typical ranges in concentration are shown in the last column of Table 3-12. The variations in concentration are due largely to differences in minerals found in rocks, soils, and sediments in the watersheds. Higher concentrations of Cl^-, SO_4^{2-}, and Na^+ are found in downstream rivers influenced by human activities. Rivers that drain areas containing calcite and dolomite minerals have higher concentrations of hardness and alkalinity. Watersheds that drain granitic areas have lower concentrations of all ions and therefore low hardness and alkalinity. The ionic strength of waters varies from about 10^{-3} to 10^{-2} M. High ionic strength waters have high concentrations of dissolved salts or solids. The nutrients of N and P stimulate the growth of algae. This can be a problem in some rivers, but it is more of a problem for lakes and reservoirs and is addressed below. Arsenic (As) usually does not occur except in those areas containing certain minerals and under reducing conditions that occur in groundwaters (see discussion under groundwaters).

Anthropogenic organic chemicals are generally not detected in protected supplies and upstream rivers and streams. Several EDCs and other organic chemicals are found at concentrations generally less than 1 μg/L in downstream rivers that drain urban and agricultural areas. Few EDCs have been detected in drinking waters and at very low concentrations of several ng/L. Atrazine (herbicide) was found at a median concentration of 50 ng/L and two flame retardants were detected: TCEP at 150 ng/L and TCPP at 220 ng/L. See Chap. 2 for information about health effects of EDCs.

All water supplies contain natural organic matter (NOM). Table 3-9 groups NOM by different classes of organic compounds. NOM is the dominant fraction of organic carbon as it occurs at levels of mg/L compared to anthropogenic organic chemicals, which occur at levels of μg/L or much less. For protected water supplies and upstream rivers, NOM accounts for nearly 100 percent of the dissolved organic carbon (DOC). For polluted supplies containing say 10 μg/L of anthropogenic organic chemicals, these organic chemicals would make up much less than 1 percent of the DOC. Extensive information on NOM is presented later in this chapter. NOM is important because it is a precursor to disinfection by-products (DBPs) and affects many treatment processes some of which are designed to remove it.

Lakes and Reservoirs

Classes of Lakes According to Stratification and Mixing. The chemistry of lakes and reservoirs is affected greatly by the mixing and stratification that occurs in most of these water bodies. Stratification occurs in lakes unless they are too shallow and are able to maintain mixing throughout their depth. Stratification occurs because the water's density is temperature dependent. The maximum density of water occurs at 4°C, and the density decreases as water temperature decreases from 4°C to ice formation at 0°C. Likewise, the density decreases as the water temperature increases above 4°C (see App. D).

Whether lakes stratify or not depends on their size, depth, and the extent of wind protection. Shallow lakes do not stratify and are considered as having continuous mixing. Lakes with small surface areas that are protected from the wind can stratify when depths are 3 m or more; while lakes with larger surface areas, stratification can occur at greater depths. Some classes of stratification and mixing are identified here and taken from Kalff (2002).

Cold Polymictic Lakes. These lakes are ice-covered for a large part of the year and free of ice during a summer period. The shallow lakes within this category stratify on sunny summer days and overturn (mix) during the night. Deeper lakes stratify for a period in the summer. Lakes in this category would be found in mountainous areas of the United States and in northern and mountainous areas of Canada.

Warm Polymictic Lakes. These are relatively shallow lakes or deeper lakes with extensive wind exposure. There is no ice cover at any time during the year. Stratification only occurs on a daily basis or for days. Lakes in this category include ponds found in southern warm climates.

Warm Monomictic Lakes. These lakes do not have an ice-cover during the year. They stratify during the warm period of the year and turnover once a year in late fall or winter. They are generally deeper, temperate, and tropical lakes found in southern warm climates.

Dimictic Lakes. These lakes have two periods of stratification and two periods of turnover. Stratification occurs during the winter ice-cover period and in the summer. Overturn occurs in the spring and autumn. These lakes are found extensively in temperate regions of the United States and Canada. Because these lakes are quite common, an extensive description of their mixing and stratification is covered next.

Stratification and Mixing of Dimictic Lakes. The physical aspects of stratification and mixing are explained here. The effects of stratification and mixing on water quality are discussed below in a separate section. Figure 3-6 depicts the four seasons: two stratification and two mixing periods.

During the summer stratification period, lakes stratify into three zones. At the top is the epilimnion, which contains warmer water and is mixed by wind. The epilimnion's surface is in contact with the atmosphere, allowing light penetration and exchange of gases. Below the epilimnion is a transition zone called the metalimnion. This is a temperature gradient region that separates the warm water of the epilimnion from the cold water in the bottom zone. It is often referred to as the thermocline, as labeled in Fig. 3-6. It is a barrier that restricts mixing between the epilimnion and hypolimnion. The hypolimnion contains the cold, dense water in the deeper part of the lake. This zone is separated from the atmosphere and wind, so little mixing occurs.

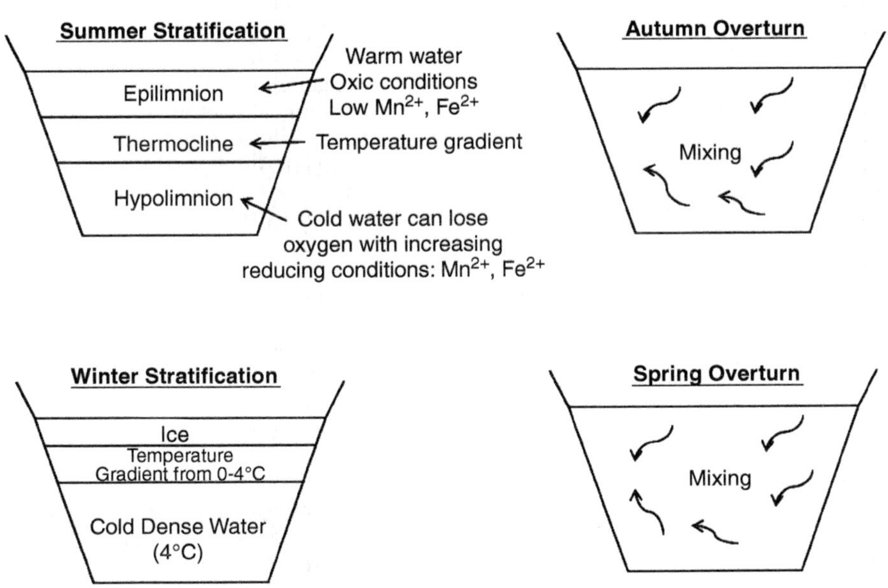

FIGURE 3-6 Lake stratification and seasonal overturn.

During autumn, the air temperatures decrease causing the water to cool in the epilimnion. The density difference between the epilimnion water that is cooled causes the water to sink and mix with the thermocline water, and causes the thermocline to erode. Continued cooling of the air and water along with wind-induced mixing creates an autumn overturn that is marked by circulation and mixing of the lake and what is called turnover. At or near 4°C (can be slightly less or greater), the water temperature becomes approximately constant vertically. This occurs as early as October in northern areas and later in the autumn in southern areas. In the winter, additional cooling produces water that is less dense than the deeper water. At the surface, an ice cover forms, preventing mixing, and a winter stratification period develops. In the spring, the ice cover melts and as the water warms to 4°C, it becomes denser and sinks. With spring winds, the lake is mixed, producing turnover or circulation. This event is called spring overturn, which occurs as late as May in northern regions and as early as March in southern areas.

Water Quality. Algae in lakes affect water quality directly and indirectly. Some direct effects are described here. Algae are particles and therefore affect turbidity. They are low-density particles that may not be removed efficiently in pretreatment sedimentation processes. If this occurs, then some algae may deposit within the pore openings of granular media filters, causing excessive head loss. These algae are referred to as filter-clogging algae. Some filter-clogging algae are listed in Table 3-13. Although only certain algae classes are shown, it is important to note that algae of all classes can clog filters. A common problem for many water utilities in temperate climates is the clogging of filters by diatoms in the spring when this type of algae often dominate over other algae due to the colder water temperatures.

Major problems associated with algae are taste and odor compounds that they produce. A partial list of taste and odor causing algae is found in Table 3-13. All classes of algae produce taste and odor compounds. Common taste and odor compounds are geosmin and 2-methylisoborneol (MIB) that impart earthy and musty odors to waters. These compounds are produced by many of the Cyanobacteria and by the Actinomycetes (bacteria) that are responsible for the decomposition of Cyanobacteria. A good reference on taste and odors identification is Suffet and Rosenfeld (2007). Cyanobacteria also produce microcystins, which can be toxic to animals and humans. Another water quality effect is formation of DBPs from algae. Both the cells and extracellular organic matter imparted into the bulk water serve as DBP precursors (Plummer and Edzwald, 2001). Any algae that pass through a water treatment plant can serve as an organic carbon source for bacteria and biological growth in the water distribution system. For treatment plants that remove particles via clarification and filtration, this should not be a problem; however, it can be a problem for unfiltered supplies.

Within a lake, algae affect water quality including DO and nutrients. Lakes are classified according to their trophic status, as shown in Table 3-14. Oligotrophic lakes are of the highest quality and are usually deeper and characterized by low concentrations of primary nutrients (N and P); low algae concentrations, measured as chlorophyll; and maintain DO in the hypolimnion. At the other end are eutrophic lakes, which are usually shallow compared to oligotrophic lakes and contain high nutrient concentrations, high algae, and anoxic conditions in the hypolimnion. Water supplies should be taken from oligotrophic lakes to avoid water quality problems. Mesotrophic lakes lie in between the other types with regard to nutrient and algae concentrations and characteristically lose oxygen in the hypolimnion in the late summer. When the hypolimnion becomes anoxic in mesotrophic and eutrophic lakes, iron and manganese are soluble in reduced states and affect water treatment.

TABLE 3-13 Common Cyanobacteria and Planktonic Algae in Lakes and Reservoirs

Taste and odor types		
Class	Associated odors	Comments
Cyanobacteria (Blue-Green Algae)		
Anabaena	Earthy, musty	Grassy
Aphanizomenon flos-aquae	Earthy, musty	Grassy
Microcystis aeruginosa	Earthy, musty	
Oscillatoria	Earthy, musty	
Synechocystis minuscula		
Chlorophyceae (Green Algae)		
Chorella vulgaris	Musty	When abundant
Scenedesmus quadricauda	Grassy	When abundant
Selenastrum		
Staurastrum	Grassy	
Bacillariophycaeae (Diatoms)		
Asterionella	Geranium	Fishy when abundant
Cycltella sp.	Geranium	Fishy when abundant
Fragillaria	Geranium	Musty when abundant
Synedra	Grassy	Musty when abundant
Tabellaria	Geranium	Fishy when abundant
Dinophyceae (Pigmented Flagellates)		
Ceratium	Fishy	
Dinobryon	Violet	
Peridinium	Cucumber	Fishy when abundant
Synura	Cucumber	Fishy when abundant
Filter-clogging algae		
Class	Size (μm)	Comments
Cyanobacteria (Blue-Green Algae)		
Anabaena	6–7 (diameter) (colonies much larger)	Single cells and bead-like colonies
Microcystis aeruginosa	5 (colonies much larger)	Forms colonies
Oscillatoria		
Chlorophyceae (Green Algae)		
Closterium		
Palmella		
Spirogyra		
Tribonema		
Bacillariophycaeae (Diatoms)		
Asterionella	2 width 40 length	
Cycltella sp.	6	
Diatoma		
Fragillaria	40–150	Length for species *crotonensis*
Melosira	15–20 diameter 30–40 length	
Synedra	4.5–6 width	

(Continued)

TABLE 3-13 Common Cyanobacteria and Planktonic Algae in Lakes and Reservoirs (*Continued*)

Filter-clogging algae		
Class	Size (μm)	Comments
Tabellaria	100–300	
Dinophyceae (Pigmented Flagellates)		
Dinobryon		
Trachelomonas		

Sources: Palmer (1962), Henderson et al., (2008).

Algal Photosynthesis and Respiration. Equation 3-59 is a stoichiometric equation that shows the relationships between primary nutrients (C, N, and P) and algal biomass from photosynthesis (*P*). The reader is also referred to Fig. 3-4a for the following discussion.

$$106\,CO_2(aq)+16\,NO_3^-+HPO_4^{2-}+18\,H^+ +122\,H_2O+\text{trace elements}$$
$$+\text{solar energy}\underset{\leftarrow R}{\overset{P\rightarrow}{\rightleftharpoons}}C_{106}H_{263}O_{110}N_{16}P+138\,O_2(aq) \tag{3-59}$$

The average dry weight composition of algae, which comes from the classical work of Redfield et al. (1963) is used here. Redfield identified the molar ratios of C:N:P at 106:16:1, which is widely quoted. The photosynthesis equation is a redox equation in which oxidized carbon is reduced to organic carbon. It is also an acid-base equation in which H^+ are removed from the water. There are diurnal effects from the photosynthetic reaction during the day and algal respiration (*R*) during the night. During the day when algae are growing, CO_2(aq) is removed and the pH increases. By the end of the afternoon, water utilities may observe a substantial increase in pH that requires adjustment of coagulation pH conditions within their treatment plants. Overnight, algal respiration increases CO_2(aq) and increases H^+, as indicated in Eq. 3-59. In the morning, water utilities may observe a much lower pH than existed on the prior evening.

The molar relationship between N and P is 16 to 1. For water supplies with higher molar ratios, P is the limiting nutrient. For oligotrophic and mesotrophic lakes, P is usually the limiting nutrient because the available N:P ratio in fresh water lakes is usually greater than needed according to the biomass ratio of 16:1. P is also considered a limiting nutrient because blue-green algae can use N_2(aq) through nitrogen fixation (see Fig. 3-4b).

For stratified lakes in the summer (see Fig. 3-6), photosynthesis exceeds respiration and decay reactions in the euphotic zone (light penetration zone). Therefore, oxic conditions are maintained throughout the epilimnion and into the thermocline. For oligotrophic lakes, oxic conditions are maintained due to low concentrations of organic matter (dead algal cells and other organic matter). For mesotrophic and especially eutrophic lakes, dead algae sink to the bottom and undergo bacterial decomposition consuming DO. This can lead to anoxic

TABLE 3-14 Trophic Status and Water Quality of Lakes (Values Compiled from Kalff (2002); Thomann and Mueller (1987)

Trophic status	Hypolimnetic O_2 (% saturation)	Inorganic N (μg/L)	Total N (μg/L)	Total P (μg/L)	Chlorophyll (μg/L)	Transparency Secchi disc (m)	
						Mean	Min
Oligotrophic	>80	<200	<350	<10	<4	≥6	≥3
Mesotrophic	10–80	200–400	350–650	10–30	4–10	3–6	1.5–3
Eutrophic	<10	300–650	650–1200	>30	>10	1.5–3	0.7–1.5

conditions in the hypolimnion during the summer whereby these reducing conditions can cause Fe and Mn problems.

Iron and Manganese. Under anoxic conditions, Mn and Fe are reduced and are more soluble in reduced states than oxidized states. Mn is reduced first due to its higher standard potential compared to Fe reduction (Fig. 3-5). Therefore, in lakes as DO becomes low and approaches zero from bacterial oxidation of organic matter, reduced Mn appears first. Oxidized Mn (Mn(III) and (IV)) is insoluble and occurs as solids of MnOOH(s) or MnO_2(s) with little soluble Mn present. Upon reduction, the oxidized Mn is reduced to Mn(II), which is soluble at mg/L levels.

Oxidized Fe(III) is quite insoluble and occurs primarily in solid forms such as $Fe(OH)_3$(s) (fresh precipitate) or as a mineral such as goethite (FeOOH(s)). Under reducing conditions that are lower than for Mn (Fe has a lower standard potential), Fe is reduced to Fe(II), which is quite soluble,

Cycling of Nutrients, Fe, and Mn. With reference to stratification and turnover periods for lakes, as depicted in Fig. 3-6, we get cycling of nutrients, Fe, and Mn. For mesotrophic and especially eutrophic lakes, the cycles are as follows. During summer stratification, bacterial decay of algal biomass releases N and P (called mineralization) to the hypolimnion. Fe and Mn are reduced and become soluble. During the fall overturn, mixing occurs and the water is oxygenated. Fe and Mn are oxidized, but these reactions take time. Consequently, for a period of time, reduced Fe and Mn may exist throughout the lake and cause problems for water utilities. The amount of time it takes to oxidize the Fe and Mn depends on the lake water pH (more rapid oxidation occurs at higher pH). These oxidation reactions may take from hours to days depending on the pH. During the late fall and early winter, because Fe and Mn are oxidized and are quite insoluble, low dissolved concentrations are found. During the winter stratification period at the bottom of ice-covered oligotrophic lakes, DO is maintained, and Fe and Mn are oxidized and dissolved concentrations are low. For eutrophic lakes, even though bacterial decomposition is slow due to low water temperatures, there may be sufficient organic matter to produce anoxic reducing conditions and increases in dissolved Fe and Mn. With spring overturn, mixing occurs in the lake and nutrients, Fe, and Mn are recycled. Spring overturn lasts for a short period and then summer stratification is established again.

Groundwaters

Major Ions and Other Chemical Characteristics. Groundwater is an important component of the hydrogeochemical cycle (Fig. 3-3). The chemical composition of groundwater is influenced by the cycle. Figure 3-7 shows a part of the cycle and is used in the following development of principles. The chemical composition of water is affected as the water flows downward through soils and minerals within the unsaturated zone and as it flows within the aquifer (saturated zone) in contact with soils, rocks, and minerals. Many reactions occur such as dissolution (solubility) of the mineral solid phases, acid-base reactions involving $H_2CO_3^*$ and the base minerals, and redox reactions. The chemical composition of a groundwater aquifer reflects the major soils and minerals that the water contacts. A listing of major minerals is found in Table 3-11.

The compositions of major dissolved constituents for several groundwaters associated with different minerals and regions are summarized in Table 3-15. The following points are made:

- Groundwater chemistry varies greatly depending on the mineralogy
- Groundwaters associated with granitic minerals, feldspars, and micas have lower concentrations of base cations, lower alkalinity, lower hardness, and lower ionic strength
- Groundwaters associated with limestone and carbonate mineralogy have higher concentrations of base cations, higher alkalinity, higher hardness, and higher ionic strength

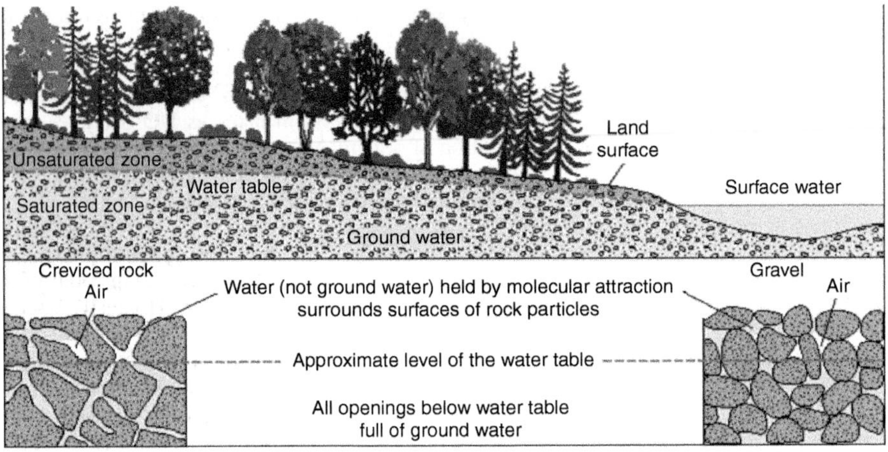

FIGURE 3-7 Depiction of groundwater system. (*Source: USGS, http//pubs.usgs.gov/gip/gw/.*)

Nitrates. Nitrate concentrations are often higher in groundwaters, especially, those that are shallow and occur in agricultural areas. When nitrate reaches 10 mg/L (as N) and greater, it exceeds the primary drinking water standard (Chap. 1) and is considered a health concern (Chap. 2). Nolan et al. (2002) published data showing areas of the United States according to probabilities that nitrate concentrations exceed certain levels.

Redox Effects. As water moves downward through the unsaturated zone, bacterial reactions occur involving decay of organic matter in soils and production of CO_2(aq) (increase in $H_2CO_3^*$). Microbial-mediated reactions produce reducing conditions that affect the oxidation states of Fe, Mn, S, and As. Figure 3-5 shows the order of reduction as follows: Mn, Fe, S, As.

Fe and Mn. As stated previously, Mn is reduced prior to the reduction of Fe. It is common to find reduced Mn in groundwaters at concentrations of 0.1 to 1 mg/L. It is common to find Fe at 1 to 10 mg/L in groundwaters with concentrations as high as 50 mg/L in waters of low pe and low bicarbonate (MWH, 2005).

As and S. Arsenic has both natural (rocks, soils, volcanic activity) and anthropogenic sources. Anthropogenic sources include lumber preservatives, pesticides, insecticides, and various industrial applications. Arsenic at low concentrations poses health problems (Chap. 2). The USEPA drinking water standard is 10 μg/L (Chap. 1). Arsenic in water has two oxidation states: As(V) and As(III), and both forms are toxic (Chap. 2).

Figure 3-5 shows that As is reduced at low pe conditions, lower than Fe reduction. In evaluating the occurrence of As in source waters, the total dissolved As concentration is usually reported. Although As tends to occur in groundwaters in certain areas such as the southwest and some areas in the Rocky Mountains, it can be a problem in many other areas. Table 3-16 shows As occurrence data in groundwaters for community water systems in 17 states. The mean values are below 10 μg/L; however, the standard deviations indicate that many systems exceed this value.

Sulfur can exist in several oxidation states. Common oxidation states for dissolved S in groundwaters are the oxidized form occurring as SO_4^{2-} and the reduced forms occurring primarily as HS^- above pH 7 and as H_2S(aq) below pH 7. H_2S(aq) partitions with the gas

TABLE 3-15 Composition of Major Dissolved Constituents in Groundwaters (SiO_2, Cation, and Anion Information from Eby (2004), Other Parameters Calculated by Authors)

Parameter	Concentrations (mg/L (except Ionic Strength))					
	Rhode Island (granite)	North Carolina (mica, schist)	Maryland (gabbro (igneous basalt))	AL (limestone)	Central PA (carbonate)	New Mexico (gypsum)
Na^+	5.9	6.4	6.2	1.5	8.5	17
K^+	0.8	1	3.2	0.8	6.3	NR*
Ca^{2+}	6.5	17	5.1	46	83	636
Mg^{2+}	2.6	1.7	2.3	4.2	17	43
Cl^-	5	1.1	1.0	3.5	17	24
SO_4^{2-}	0.9	6.9	9.2	4.0	27	1570
HCO_3^- as $CaCO_3$	31	57	30	120	229	117
SiO_2	20	29	39	8.4	NR	29
Hardness mg/L as $CaCO_3$	62.3	50	22	133.0	229	1767
Alk (mg/L $CaCO_3$)	~31	~57		~120	~279	~117
Ionic strength (M)	1.0×10^{-3}	1.2×10^{-3}	1.1×10^{-3}	2.7×10^{-3}	8.0×10^{-3}	7×10^{-2}

Note: pH not reported.
*NR: Not Reported.

TABLE 3-16 Arsenic Occurrence in Groundwater for Community Water Systems (from USEPA (2000))

State	Number of systems	Mean (μg/L)	Standard deviation*
Alaska	304	4.11	7.67
Alabama	263	0.73	0.49
Arizona	279	9.53	13.8
California	1224	4.2	7.57
Indiana	648	0.26	0.66
Kansas	506	2.64	3.83
Michigan	644	5.31	9.00
Minnesota	829	2.77	5.52
Missouri	773	0.77	1.93
Montana	484	1.81	3.53
North Carolina	1735	3.52	1.63
North Dakota	197	4.85	8.13
New Jersey	438	0.92	1.38
New Mexico	559	3.81	6.00
Oregon	316	2.77	5.49
Texas	3105	2.58	4.26
Utah	263	2.89	5.59

*Statistically significant differences at the 5 percent significance level.

phase, creating the characteristic rotten egg odor. Figure 3-5 shows that SO_4^{2-} is reduced at lower pe conditions than ferric reduction.

PARTICLES

This section describes the importance, types, sources, properties, and measurement methods for particles (particulate matter) in drinking water systems. A more in-depth coverage of particles is provided by Gregory (2005).

Importance

Particles in water are significant due to impacts on water quality and treatment process performance. Water quality effects can be aesthetic and or related to human health. The light-scattering properties of particles can cause significant turbidity, i.e., an undesirable cloudy appearance, even if the particles have no health consequences. In contrast, low levels of certain particles may have no visual impact but may be significant to human health, i.e., pathogenic microorganisms such as *Giardia* cysts, *Cryptosporidium* oocysts, *Eschericiia coli* bacteria, or enteric viruses. Particles may also have surface-associated contaminants such as heavy metals and synthetic organics that are of human health concern. Many important aspects of particles in aquatic systems reviewed more than 30 years ago remain relevant (Stumm, 1977; O'Melia, 1980; Kavanaugh and Leckie, 1980).

The amount and nature of particles have significant effects and underlying importance for many treatment processes. For processes that are intended to remove dissolved constituents from water, particulate matter can cause undesired impacts or fouling of processes. Examples include the shielding of microorganisms from disinfectants, fouling of granular activated carbon adsorption and ion exchange media, fouling of reverse osmosis membrane filters, and fouling of heat exchangers. More importantly, many treatment processes are inherently based on the formation, alteration, and removal of particles in water. Coagulation and flocculation involve formation and aggregation of particles; clarification (i.e., dissolved air flotation, sedimentation) and filtration (i.e., media filtration, micro- and ultrafiltration membranes) processes are used for particle removal. Removal of certain dissolved constituents is often based on production of particles, e.g., oxidation and precipitation of iron and manganese, transformation of dissolved NOM to particulate form, and precipitative softening for calcium and magnesium removal. In concentrated form, removed particles are the solids in water treatment residuals, or the sludge, that must be properly managed (see Chap. 22).

Types of Particles: Sources and Materials

Characterization of particles based on their source and materials of composition is useful and important in understanding treatment process performance. Herein, two different sources are defined: raw water particles and particles formed during treatment or distribution. Materials are divided into inorganic and organic categories. Table 3-17 summarizes sources and types of particles.

Raw Water Sources. Most particles in raw surface waters are naturally occurring as a result of watershed and water body processes. Inorganic mineral particles (i.e., aluminosilicate clays, metal oxides and hydroxides, silica oxides) may be present due to soil erosion

TABLE 3-17 Sources and Types of Particles

Category	Type/description/examples	Sources
Inorganic		
Clays	Aluminosilicates (montmorillonite, kaolinite, illite)	Raw water
	Asbestos fibers	
Metal oxides, hydroxides	$Al(OH)_3$ (amorphous) precipitate	Coagulant addition
	$Fe(OH)_3$ (amorphous) precipitate	Coagulant addition Fe(II) oxidation
	MnO_x (i.e., MnO_2) precipitate	Mn(II) oxidation
	$Mg(OH)_2$ precipitate	Softening
	SiO_2 (mineral)	Raw water Ballasted sedimentation
Carbonates	$CaCO_3$ precipitate (calcite)	Raw water, softening
	$Ca,MgCO_3$ preciptate (dolomite)	Raw water, softening
Organic		
Adsorbent	Powdered activated carbon	Added during treatment
Microorganisms	Pathogens (viruses, bacteria, protozoa)	Raw water
	Algae	Raw water
	Bacteria	Bioactive filtration Other process tanks
Detritus	Debris from plants and animals	Raw water
NOM	Macromolecular, colloidal NOM	Raw water
	NOM sorbed to particle surfaces	Raw water Flocs from coagulant addition

during stormwater runoff; levels in treatment plant source water depend on the hydrodynamics of the source water. Run-of-river sources with high velocities maintain higher levels of suspended particles than do lake or reservoir sources with low velocities and long detention times, which allow for sedimentation of most suspended particle mass. Thus, particle levels in the Missouri River, which is one source water for St. Louis, MO, are significant (10s to 100s of mg/L), while the mass of suspended particles is insignificant (order of 1 mg/L) in many Great Lakes intakes and New England reservoir sources. In general, levels of inorganic particles in raw groundwater sources are low due to natural filtration, with transport of the water through the subsurface porous media. As discussed below, exposure of groundwaters to the atmosphere may result in formation of particles that were not present in the source water.

Aquatic biota, from fish to microorganisms as well as aquatic plants, are important sources of naturally occurring organic particles in raw waters. Algae and Cyanobacteria are important types of microorganisms because they can occur at significant concentrations and may cause associated taste and odor problems and, possibly, low levels of toxins (see earlier discussion and Table 3-13). Levels of photosynthetic organisms are dynamic and a result of complex interactions of climate, hydrology, and nutrient cycling. Particle removal processes must be able to respond to periodically elevated levels of microorganisms, and other processes may be needed to address associated dissolved constituents. In general, particulate organic matter includes microorganisms as well as detritus from decaying biomass of both watershed (trees, plants) and water body origin. As noted in the section to follow on NOM, the particulate fraction of NOM is typically small, i.e., 1 to 10 percent of the TOC.

Another important class of raw water microorganisms is human pathogens (viral, bacterial, protozoa). The sources of pathogens include mammals, livestock, and wastewater from humans. Because actual and potential levels of pathogens in source waters are variable on both temporal and spatial scales, treatment processes for removal and inactivation of pathogens are designed to provide desired performance under expected worst-case source water quality. In the United States, federal regulations addressing pathogen occurrence and treatment include the suite of surface water treatment rules (Surface Water Treatment Rule [SWTR], Long Term 1 Enhanced Surface Water Treatment Rule, Long Term 2 Enhanced Surface Water Treatment Rule), the Information Collection Rule, and the Ground Water Rule (see Chap. 1). One goal of source water protection programs is to minimize pathogen input to source waters through best management practices including livestock management, source water selection, and wastewater management.

In addition to pathogens, there are some anthropogenic contaminants that may be present as raw water particles. As noted earlier, contaminants such as heavy metals and hydrophobic synthetic organic chemicals that are released to the environment are often associated with particle surfaces. The occurrence and level of such contaminants are rarely an issue and site specific; investigation should be based on knowledge of potential anthropogenic impacts to source waters. An issue of emerging concern is the potential for the presence of nanoparticles in source waters due to increasing manufacturing and use of nanomaterials (Ju-Nam and Lead, 2008; Wiesner et al., 2008), including nanoparticles made from carbon (i.e., fullerene, carbon nanotubes), silver, gold, titanium, silica, zinc, and many other elements. Little is currently known about the occurrence or levels of anthropogenic nanoparticles in source waters, including their association with other particulate matter. Again, concern and investigation should be based on knowledge of potential inputs to the source water of interest.

Treatment System Sources. An understanding of the types and levels of particles formed (or added) during treatment processes is important for overall process design as well as assessment of treatment performance. Particles formed during treatment must be removed by downstream processes and ultimately become process residuals. Assessment of particle removal across treatment processes may be influenced by lack of knowledge of the source of particles. For example, assessing the potential to remove protozoan pathogens based on levels of protozoan-sized particles in raw and filtered waters ignores the fact that most of the protozoan-sized particles in filtered waters were formed during treatment and as a result have limited relationship to the raw source water particles.

For low-turbidity lake and reservoir source waters, the number and mass of particles formed during treatment are much greater than levels in the raw source water. The most common treatment process source of particles is the formation of amorphous aluminum or iron hydroxide particles ($Al(OH)_3(s)$, $Fe(OH)_3(s)$) by the addition of sparingly soluble salts for coagulation purposes. The primary coagulants also directly precipitate and adsorb or coprecipitate dissolved NOM, creating a mixed inorganic-organic particle. These particles formed during coagulation comprise most of the particle mass to be removed in particle separation processes—raw water particles are a minor component. In some plants such as direct filtration, only synthetic organic polymer coagulants are used, so the precipitated NOM is primarily an organic material.

The precipitative softening process is another significant treatment process source of particles. Calcium hardness is removed by formation of calcite ($CaCO_3(s)$), a carbonate mineral, while magnesium is removed at higher pH by formation of magnesium hydroxide ($Mg(OH)_2(s)$). For many surface or groundwater-based hard water sources, raw water particle levels are low such that particles formed during treatment are dominant. Raw waters may contain naturally occurring calcite or dolomite ($CaMgCO_3(s)$) particles; however, these dense crystalline particles generally settle out within the surface waters where they were formed. This can occur seasonally.

Another important source of particles formed during treatment occurs when reduced iron and manganese are oxidized and precipitate as $Fe(OH)_3(s)$ and $MnO_x(s)$, respectively. This can be a continuous process for groundwaters with consistent levels of reduced Fe or Mn or a periodic process for surface waters that have seasonal occurrence of increased raw water Mn and Fe levels. Knowledge of the form of Fe and Mn in both raw and process waters is essential for the proper design, operation, and troubleshooting of Fe and/or Mn control processes.

Some treatment processes involve the addition of specific particles to the main process flow. One example is the addition of powdered activated carbon (PAC) for adsorption of low levels of organic compounds. PAC addition may occur periodically due to seasonal water quality problems (i.e., taste and odor) or continuously. Mass loadings may be significant relative to other particles, and the PAC typically becomes incorporated within overall plant residuals.

Another process involving the addition of particles is the sand-ballasted high-rate sedimentation process. Dense silica sand particles are added and flocculated with other process particles (i.e., from coagulation and flocculation) to form larger, faster settling floc that are removed via gravity sedimentation. Most of the sand is mechanically recovered from the settled flocs and reused in the process, resulting in little impact on downstream processes or residuals (see Chap. 9).

Particle Characteristics

The interaction of particles and treatment processes is significantly affected by important particle characteristics such as size, density, shape, and surface properties.

Size is perhaps the single most defining characteristic of a particle. Figure 3-8 shows that aquatic particulate matter size can range over five to six orders of magnitude, from millimeter size flocs (aggregates of smaller particles) to colloids on the order of a few nanometers in size. There are a number of generally agreed upon terms used for describing

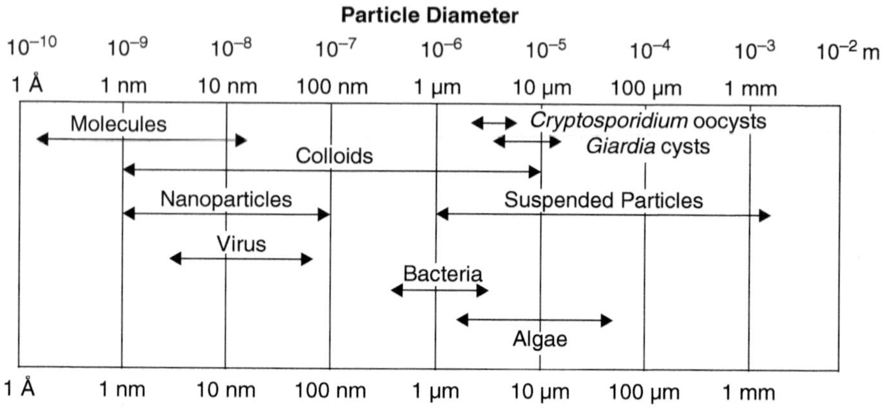

FIGURE 3-8 Size range for particulate matter in water.

size classes of particles, many based on operational definitions using specific methods to separate size classes. Suspended particles are those that are retained on a fibrous laboratory membrane filter, i.e., as defined in *Standard Methods* (APHA, AWWA, and WEF, 2005). These particles are generally in the range of 1 μm to several millimeters. In contrast, colloids are submicrometer particles in the nanometer to micrometer size range, while nanoparticles generally refers to particles with size less than approximately 100 nm; thus, all nanoparticles are also colloids.

The operational distinction between dissolved and particulate matter is based on use of laboratory fiber/membrane filters with pore sizes of a few tenths to several micrometers. Thus, the "dissolved" fraction (the filtrate) often includes colloidal material (and nanoparticles); methods that provide solid/liquid separation at smaller size scales (such as ultrafiltration or centrifugation) are needed to isolate colloidal matter.

Density. Particle density (mass per unit volume) can have important impacts on the fate of particles in natural waters, their removal during treatment, and management of solid residuals. Particle density directly affects the gravitational force on particles, impacting particle removal in natural lakes and reservoirs as well as by many engineered particle separation processes (flotation, sedimentation, media filtration). The mass and volume of solid residuals to be managed is also directly affected by particle density.

Particle density depends on material type and particle (or floc) formation processes. Higher-density particles include crystalline mineral materials such as certain iron oxides (4000–5000 kg/m^3 or 4–5 g/cm^3), silica sand (2650 kg/m^3), and calcium carbonate (2400 kg/m^3); particles of such high density are typically naturally removed from lake and reservoir source waters but may be significant in run-of-river sources. Many source and formed-in-treatment particles have a density that is only a few percent (1–10) greater than water. Examples include microorganisms (including some algae that are buoyant in water) and precipitated amorphous metal hydroxides such as from coagulation using iron or aluminum salts.

Aggregates of smaller particles, or flocs, typically include water-filled pore space as well as solid matter. The net density of flocs is affected by floc porosity and the density of the solid material, or

$$\rho_{\text{net}} = \frac{V_{\text{water}}\,\rho_{\text{water}} + V_{\text{solid}}\,\rho_{\text{solid}}}{V_{\text{total}}} \tag{3-60}$$

where V refers to volume (water, solid, or total floc) and ρ is density. One important floc characteristic is their fractal nature, where flocs become more porous and less dense as they grow in size. This is addressed in other chapters on flocculation (Chap. 8) and clarification (Chap. 9) processes.

Shape. The shape of particles, both naturally occurring and those formed in treatment, can be quite varied and can impact treatment performance. In most mathematical models involving particles and in many measurement techniques for particle sizing, it is common to assume a spherical shape for particles. This assumption can be valid for some biological and inorganic particles (e.g., coccoid bacteria, *Cryptosporidium* oocysts, some algae, some manufactured particles), while it is invalid for others (e.g., rod-shaped bacteria, many algae, flocs). The actual shape of particles can be described qualitatively (e.g., rod, star, oblong) and by some quantitative measures (e.g., aspect ratio of longest to shortest dimension, ratio of perimeter to cross sectional area). Assessment of shape requires analysis of particle images. It may be important to understand particle shape, even if quantitative analysis is not possible, when assessing impacts on removal and process performance (e.g., filter-clogging algae may be filaments with high aspect ratio).

Nature of Particle Surfaces: Stability. An especially important characteristic of particles is the nature of the particle surface or the particle-water interface. The fate and transport of particles in natural waters and engineered water treatment processes is affected greatly by the net force that results when the surfaces of particles interact with each other or with other solid surfaces. If interactions between particles result in repulsive forces that prevent aggregation of particles into larger flocs, the suspension of particles is said to be stable. Similarly, repulsive forces might result during the interaction of a particle with the surface of a sand grain in a media filter or with the synthetic polymer surface of a hollow fiber membrane filter, preventing deposition to that surface. Thus, one objective of coagulation is to destabilize particles, leading to effective flocculation, clarification, and deposition in a media filter.

The net force between interacting surfaces in water is affected in a complex manner by the materials of the surfaces and the chemistry of the water phase. Repulsive forces can result when surfaces have the same electrostatic charge (positive or negative), when portions of surface-associated macromolecules interact unfavorably, and if surfaces are strongly hydrated by water molecules. The distance over which such forces are important is on the order of a few angstroms to tens of nanometers, small compared to many particles but on the order of the size of the smallest colloids. Fundamental aspects of forces between particle surfaces are described in books on colloid and physical chemistry (e.g., Verwey and Overbeek, 1948; Israelachvili, 1992; Hiemenz and Rajagopalan, 1997).

Repulsive forces due to like electrostatic surface charge are common for aquatic particles. Stumm (1992) and Stumm and Morgan (1981) describe three origins for surface charge: (1) isomorphic substitution of elements within a crystal structure, (2) chemical reactions at surfaces (surface complex formation), and (3) sorption of charged macromolecules.

The classic example of isomorphic substitution occurs for clay particles where a permanent net negative "surface" charge results from substitution of Al(III) for Si(IV) in a tetrahedral structure or Fe(II) or Mg(II) substituted for Al(III) in an octahedral layer. This charge is not affected significantly by solution chemistry (pH, etc.), although clay particles can have edges with hydroxyl (–OH) groups that are affected by pH. Interlayer net negative charge can be balanced by sorbed cations, giving rise to the cation exchange capacity of a clay material.

Chemical reactions involving surface-bound atoms (surface complex formation or chemisorption) can also significantly affect surface charge. If ">XO" represents a surface-bound metal atom with a covalently bonded oxygen atom (i.e., $>AlO^-$, $>FeO^-$), often referred to as a surface site or surface functional group, then reversible surface complex formation reactions involving protons (H^+), thus solution pH affected, can control the formation of charged surface groups as illustrated.

$$>XOH + H^+ \leftrightarrow XOH_2^+$$

$$>XOH \leftrightarrow XO^- + H^+$$

As pH decreases, more surface sites are protonated, and the surface becomes more positive while the reverse is true as pH increases. The pH at which the surface has a zero net charge is referred to as the pH of zero point of charge (pH_{ZPC}) or the isoelectric point (pH_{IEP}). The pH of zero net charge depends on particle material as well as solution chemistry. For some metal hydroxides, pH effects on surface charge are significant within the pH range often encountered in water treatment. For example, iron or aluminum hydroxide precipitates may have significant positive charge at pH 5, while at pH 9 or 10 the surface charge is negative. For other oxides, such as silica, the pH of zero charge is low such that bare silica surfaces are negatively charged for pH conditions of water supplies. The effect of pH on surface charge is illustrated schematically in Fig. 3-9.

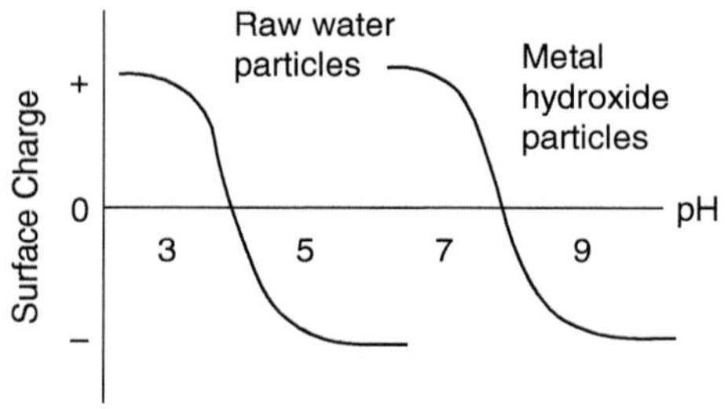

FIGURE 3-9 Illustration of the effect of pH on surface charge.

Surface charge can also be significantly affected by surface complex formation that involves other charged species in solution (Morel, 1983; Stumm 1987, 1992). Complexation of calcium (Ca^{2+}) adds positive surface charge, while a ligand substitution reaction such as phosphate for hydroxide can add negative surface charge. The effective removal of arsenic as As(V) (arsenate) by adsorption to iron oxyhydroxides surfaces is an example of surface complex formation that affects surface charge and results in removal of a trace level contaminant.

Particle surfaces interact with macromolecules such as dissolved NOM and polyelectrolytes (or polymers) added during treatment. These macromolecules often adsorb favorably to surfaces and affect surface charge if they are charged molecules. Thus, in natural waters, most particle surfaces have adsorbed NOM that is negatively charged due to deprotonated carboxyl ($R\text{-}COO^-$) functional groups (protonation to uncharged R-COOH would occur at low pH, i.e., < 3–4).

As described in Chap. 8, the surface charge of particles after coagulation and flocculation is a complex result of the precipitation of metal hydroxides; the precipitation, coprecipitation and adsorption of NOM; and the adsorption of charged species from coagulant addition. However, coagulation typically involves the use of positively charged dissolved species to precipitate and destabilize negatively charged NOM and raw water particles to produce approximately net neutral particles, which can aggregate to form flocs that can be removed effectively by processes such as sedimentation, dissolved air flotation, and filtration.

As noted above, in addition to electrostatic surface charge, surface interaction forces can be affected by the interaction of adsorbed macromolecules and by surface hydration. The interaction of particles with adsorbed macromolecules that have long tails and loops extending into solution can lead to repulsive steric interactions if the macromolecule segments have segment-solvent (water) interactions that are more favorable than between segment-segment interactions. Conversely, high molecular weight (MW) hydrophobic polymers may be used to effectively attach particles together despite other repulsive forces if the particle-molecule interactions are very favorable. The use of high MW synthetic polymers as flocculation or media filtration aids is an example of this. Strongly hydrated surfaces may give rise to repulsive surface interaction forces if the displacement of the water molecules is unfavorable. Repulsive interaction of certain uncharged silica surfaces is one example of this.

Measurements of Particle Concentration

Various analytical measurements are used to quantify the amount of particulate matter in water. Measurements are needed for a variety of purposes including regulatory compliance; process performance assessment; particle detection; and development, calibration, and validation of process theories and models.

Turbidity. The turbidity, or cloudiness, of water is measured as a surrogate for the presence of particles and is a regulated parameter for drinking water quality. As used in the drinking water industry, turbidity is the amount of visible light (400–700 nm wavelength) scattered at a 90 degree angle to the incident light source. Turbidity is measured in nephelometric turbidity units (ntu) using a nephelometer (or turbidimeter). The intensity of scattered light, in ntu, is based on a standard (but arbitrary) concentration of a primary reference colloidal material (formazin) that is used to calibrate a turbidimeter. Various secondary standards comprised of other material are typically used for daily instrument validation and calibration. Turbidity measurements range from the low levels for filtered waters of 0.01 to 0.3 ntu, to typical values in lakes of 0.5 to 10 ntu, to erosion-prone run-of-river levels of 100s to 1000s of ntu.

The turbidity of a sample does not convey any information about fundamental characteristics of particles since the scattered light is the aggregate response for all particles, which is a function of the amount of particles as well as the size, refractive index, and shape of the particles. As shown in Fig. 3-10, for a given mass concentration of particles, measured turbidity is significantly affected by particle size because particle number concentration is inversely proportional to diameter cubed and visible light-scattering efficiency is affected by particle size. Turbidity per unit mass of particles is at a maximum for particles with size similar to the wavelength of visible light (0.4–0.7 μm), decreasing as particles get smaller due to poor scattering and decreasing as particles get larger due to decreased number concentration. Thus, a decrease in turbidity may not always signify a decrease in particle mass concentration if significant size change occurs.

Mass and Volume Concentration. The mass concentration of particles (dry mass/unit volume), for which the total suspended solids (TSS, mg/L) is a common example, is important with respect to solids loading to treatment processes and the management of residuals (sludge). Because finished drinking water has essentially no significant mass concentration of particles, the solids mass in raw water as well as mass added or formed during treatment essentially all becomes residual solids that must be properly managed.

The volume of residuals to be managed is a direct function of the mass concentration and the density of the solids. Total mass concentrations may be the basis for fees for sludge management, while the volume of solids may significantly impact the design and

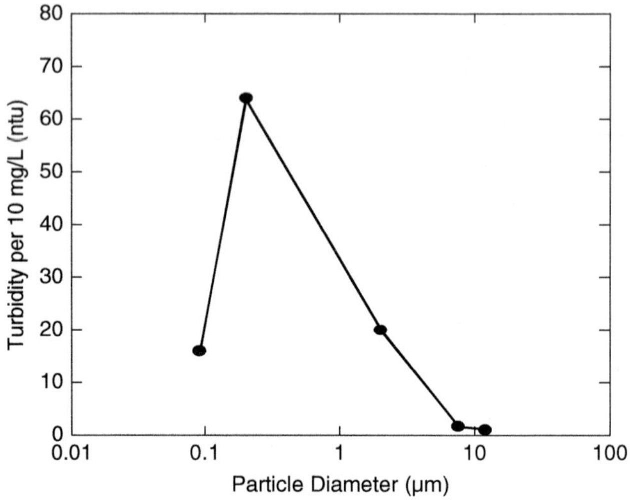

FIGURE 3-10 Effect of particle size on turbidity (polystyrene particles).

operation of residuals management processes. Because volume and mass scale with the cube of diameter, larger particles typically control overall mass or volume in heterodisperse suspensions.

Surface Area Concentration. The surface area associated with particulate matter (i.e., concentration in m^2/L) is important when considering the interaction of dissolved species with particle surfaces, i.e., the adsorption of trace contaminants or of added coagulants. Surface area is directly proportional to mass concentration and inversely proportional to particle size; smaller particles have more surface area per unit mass. Some solids are characterized by their specific surface area (area per unit mass) which can be measured; multiplying this parameter by mass concentration yields surface area concentration.

Number Concentration. Particle number concentration (number of particles per unit volume) is important in performance models for many treatment processes (e.g., rate of aggregation in flocculation, rate of removal in clarification and filtration). Measurement of number concentration requires the counting or detection of individual particles in a known sample volume. Measurement methods include microbiological assays, microscopic observation (such as optical, electron), and various particle counting techniques based on light blockage, light scattering, and electrical resistance, among others.

The 1989 USEPA SWTR (see Chap. 1) focused attention on particle number concentration with its requirements to achieve specified log reduction levels for *Giardia* cysts and enteric viruses. The log reduction achieved by disinfection (or inactivation) and the degree of removal achieved (or credited) for clarification and filtration processes are based on number concentration of microorganisms; 2-log removal is a decrease in number concentration by a factor of 100. The requirements of the SWTR and the availability of light blockage based particle counters for particles larger than 2 μm led to consideration by some of the use of the measured number concentration of cyst-sized particles to assess protozoa removal performance across treatment plants. This practice is not advised for two reasons. The first is that actual concentrations of cysts (or oocysts) in the most contaminated source waters would not be detected by particle counters (i.e., a source water with 10 oocysts/L, highly contaminated, would have only 0.01 oocysts/mL, compared to raw water particle counts of 1000 to 10,000 particles/mL for high-quality, low-turbidity sources). The second reason is that comparison of cyst-sized particles in raw and finished waters is illogical because coagulation processes form high concentrations of cyst-sized particles, the particles most likely to be detected in filtered waters, and they have no relationship to actual cysts in the raw water. On a site-specific basis, monitoring of particle counts in filtered waters can be a very useful and sensitive process-control tool to be used in combination with other performance measures.

Particle characteristics such as size, density, and number concentration can be used to calculate equivalent mass, volume, or surface area concentrations; equations are shown in Table 3-18.

TABLE 3-18 Summary of Measures of Particle Concentration

Measure of concentration*	Equations and comments
Mass (C) [M/L^3]	Measured dry mass/unit volume i.e., Total Suspended Solids (mg/L)
Number (N) [$\#/L^3$]	$N = C/(\rho_p(\pi/6)d_p^3$ (for spheres) $-\ \rho_p$ = particle density $-\ d_p$ = particle diameter
Volume (V) [L^3/L^3]	$V = N(\pi/6)d_p^3$ (for spheres) $V = C/\rho_p$ $V \times 10^6$ = parts per million by volume (ppm_V)
Surface area (S) [L^2/L^3]	$S = N\,\pi\,d_p^2$ (for spheres) $S = (6\,C)/(\rho_p\,d_p)$ (specific surface area of material = S/C)

*[ ], M = mass, L = length, # = number.

Example 3-5 Particle Concentrations
Consider a raw water supply with 3 mg/L of TOC and 1 mg/L of mineral particles. An alum dose of 20 mg/L is added for coagulation. Calculate the volume, number, and surface area concentrations that result from coagulation if the particles have diameters of 50 nm, 5 μm, and 500 μm with assumed densities of 1.5, 1.05, and 1.01 g/cm^3, respectively. Assume that 35 percent of the TOC is converted to particulate form via coagulation and that all of the mineral particles are included. Use the equations in Table 3-18 and assume spherical particles.

Solution Use the result of Example 3-2 and assume that all the added aluminum precipitates as $Al(OH)_3(s)$ without any appended water molecules, i.e., 0.26 mg $Al(OH)_3(s)$/mg of alum. Assume that NOM contains about 50 percent carbon (see upcoming section on Natural Organic Matter), so the amount of NOM precipitated is 0.35×3 mg/L $\times 1/0.5 =$ 2.1 mg/L.

$$\text{Mass of Al(OH)}_3\text{(s) formed} = (20 \text{ mg/L alum})(0.26 \text{ mg Al(OH)}_3\text{(s) per mg of alum}) = 5.2 \text{ mg/L}$$

$$\text{Total mass concentration of particles} = 5.2 + 2.1 \text{ mg/L NOM} + 1 \text{ mg/L mineral} = 8.3 \text{ mg/L}$$

Equations from Table 3-18 are used to determine the following:

Volume concentration = [mass conc.]/[density]
Number concentration = [volume conc.]/[volume per particle]
Surface area concentration = [number conc.] [surface area per particle]

Summary of results is in the table below. Calculations require attention to units.

Particle diameter (μm)	Particle density (g/cm^3)	Volume conc. (ppm by volume)	Number conc. (#/L)	Surface area conc. (cm^2/L)
0.05	1.5	5.5	8.4×10^{13}	6600
5	1.05	7.9	1.2×10^{8}	95
500	1.01	8.2	1.3×10^{2}	0.99

As density decreases (due to aggregation and porosity), volume concentration increases. The results illustrate the sensitivity of number concentration to particle size for a given mass or volume concentration. The very large numbers of 50 nm (0.05 μm) particles aggregate to approximately 120 million 5-μm particles per liter, which in turn can aggregate to only 130 500-μm particles per liter (the kind of particles desired for sedimentation processes). Surface area increases inversely with particle size. ▲

Measurements of Particle Size and Size Distributions

Particle size and concentration are important in fundamental aspects of treatment process performance (see Lawler et al., 1980; Elimelech et al., 1995) and because of the relationship between particle size and specific types of particles (as noted in the example above). Particle size distributions are discussed here, followed by a discussion of particle sizing in general and some specific methods of particle sizing and counting.

Particle Size Distributions. In general, there are particles in a range of different sizes in any sample of source or treated water. Characterization of the concentration of particles in a set of specific, narrow size ranges (or windows, channels, bins, etc.) yields a particle size distribution (PSD). Each size range has an average size (arithmetic or geometric mean of the range, $d_{p,i}$), usually an equivalent spherical size, and the concentration may be expressed on a number, volume, mass, or surface area basis. PSDs can be presented as absolute or relative (fraction of total) differential concentrations versus size or on a cumulative basis (concentration or fraction greater than a certain size); examples of these approaches are shown in Fig. 3-11. A useful approach is to examine absolute, differential PSDs of number concentration (ΔN_i) versus size ($d_{p,i}$) at different locations along a treatment process. A set of representative PSDs for raw, filter influent (flocculated), and filter effluent waters for a direct filtration treatment plant is shown in Fig. 3-12. These PSDs have 16 logarithmically distributed sizes over the 1 to 204 μm size range. The number concentrations are dominated by small particles, with coagulation and flocculation (the filter influent) producing many more, smaller and larger particles as compared to the raw water (3.2×10^4 versus 1.2×10^4 total particles/mL), while the filtered water has very few particles remaining (314 particles/mL).

In order to compare PSDs measured using different particle counting instruments and for research purposes, it can be useful to normalize a PSD by dividing the absolute differential concentration by the width of the size range channel (or window) for that concentration, using a logarithmic basis. Plotting normalized number and volume concentrations ($\Delta N_i/\Delta\log d_{p,i}$ and $\Delta V_i/\Delta\log d_{p,i}$) versus the log of the particle size for samples along a treatment

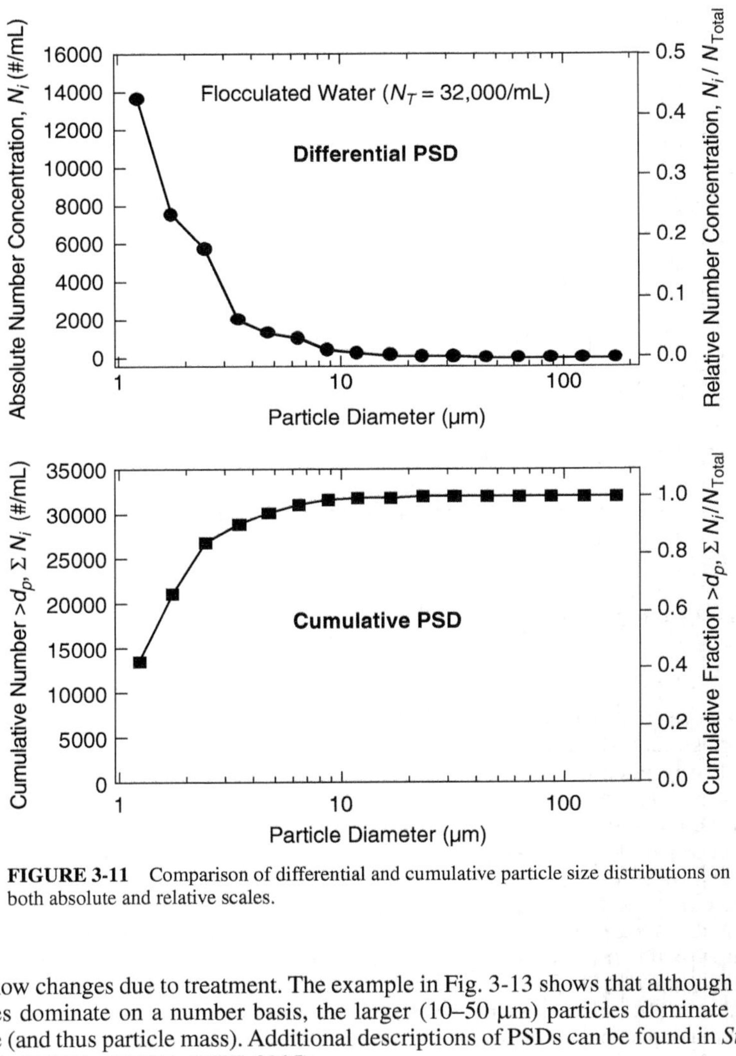

FIGURE 3-11 Comparison of differential and cumulative particle size distributions on both absolute and relative scales.

train show changes due to treatment. The example in Fig. 3-13 shows that although smaller particles dominate on a number basis, the larger (10–50 μm) particles dominate particle volume (and thus particle mass). Additional descriptions of PSDs can be found in *Standard Methods* (APHA, AWWA, WEF, 2005).

It may be convenient or necessary to describe a PSD using a single average size and a total number of particles. Average particle size for a PSD can be calculated using several methods depending on the characteristic of the PSD being emphasized. The most common average size is a number weighted average or mean size for the distribution (d_N). In contrast, the total surface area of a PSD can be captured by determining the surface area weighted average size (d_S), while the total volume of a suspension can be described by a volume weighted average size (d_V). Table 3-19 shows the equations for calculating these characteristic average sizes. Note that the relative order of average size is $d_V > d_S > d_N$.

Particle Sizing and Counting: General. Particles in water are three dimensional and of various shapes. However, it is common to use a single dimension, typically the diameter of a sphere, to characterize particle size. Furthermore, particle size is often

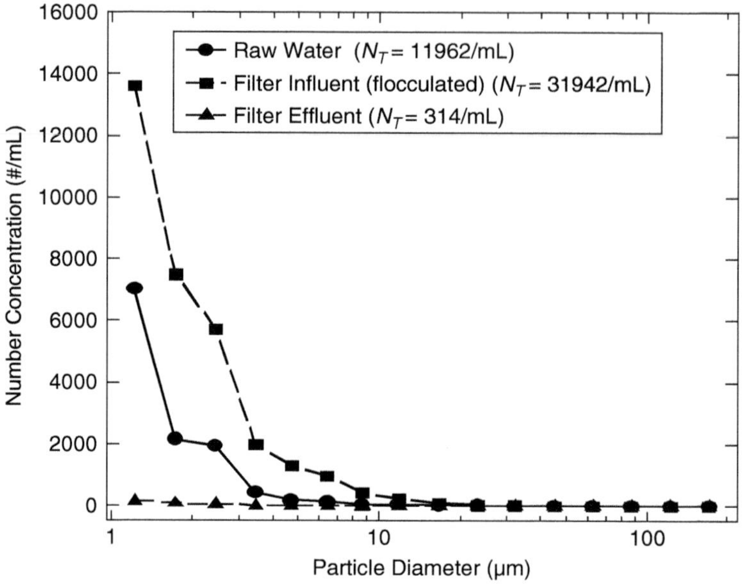

FIGURE 3-12 Representative number based PSDs for raw, flocculated, and filter effluent waters.

inferred, or calculated, based on measured perturbations or phenomena caused by a particle (or particles), such as the scattering or blockage of light, electrical resistance, or motion, in contrast to direct observation via microscopy and image analysis. Thus, particles are most often characterized by a single dimension that is equivalent to a sphere of that diameter that would yield the result that was measured. Particles might also be characterized by major and minor length dimensions that reflect a two-dimensional aspect of the particle shape; such a characteristic is only possible from examination of some type of particle image.

Particle sizing and counting generally require the ability to detect, or sense, an individual particle and to assign the particle to an appropriate size range. Direct microscopic observation using visual light or electron beams produces images of individual particles that can be scaled and assigned to size classes based on major or minor axis length, perimeter, or cross-sectional area. Indirect observation of size involves detecting the magnitude (or timing) of a perturbation or phenomena caused by the particle and relating that magnitude (or timing) to size via calibration using particles, often spheres, of known size. In general, the sizing and counting of suspensions with multiple particle sizes is more easily accomplished for particle sizes greater than approximately 1 μm as compared to submicrometer sized particles. Several books, research reports, and journal articles describing particle counting methods are available (e.g., Lewis et al., 1992; Hargesheimer et al., 1992, 2001; Chowdhury et al., 2000; Broadwell, 2001). A recent book chapter with a focus on nanoparticles provides an excellent summary of particle characterization methods (Hassellöv and Kaegi, 2009).

Particle Sizing and Counting: Light Blockage. The most commonly used particle sizing and counting technology in drinking water treatment applications is the light-blockage-based particle counter. Laser light (visible wavelengths) is directed across a sensing chamber through which a water sample passes at a known flow rate (typically, 50–100 mL/min). A photodiode detects the laser light (photons); when a particle passes through the sensing

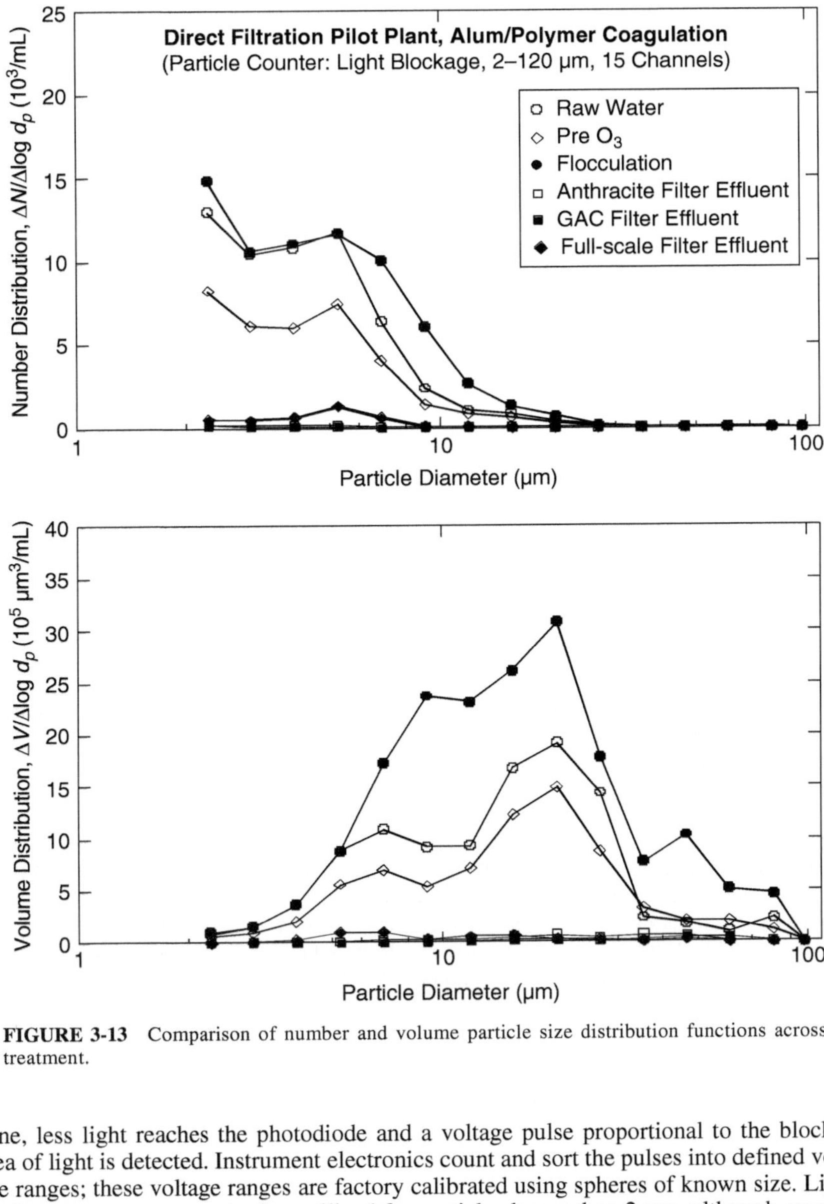

FIGURE 3-13 Comparison of number and volume particle size distribution functions across treatment.

zone, less light reaches the photodiode and a voltage pulse proportional to the blocked area of light is detected. Instrument electronics count and sort the pulses into defined voltage ranges; these voltage ranges are factory calibrated using spheres of known size. Light blockage sensors are most often utilized for particles larger than 2 μm, although sensors with 1 μm detection limits have been utilized. Use of this technology began in the 1970s (e.g., Tate and Trussell, 1978) and saw significantly increased use in the 1980s and 1990s as on-line instruments were developed and concern over *Giardia* cysts and *Cryptosporidium* oocysts increased (e.g., LeChevallier and Norton, 1992). Limitations to light-blockage sensors include maximum concentration limits (avoid coincidence), operation and maintenance (flow control, sensor cleaning, etc.), need for particles that block light (not opaque),

TABLE 3-19 Equations for Calculation of Characteristic Average Particle Sizes

Basis for averaging	Equation for average diameter	
Number	$d_N = \frac{\sum_i N_i d_{p,i}}{\sum_i N_i}$	(3-61)
Surface area	$d_S = \left[\frac{\sum_i N_i d_{p,i}^2}{\sum_i N_i}\right]^{1/2}$	(3-62)
Volume (or mass)	$d_V = \left[\frac{\sum_i N_i d_{p,i}^3}{\sum_i N_i}\right]^{1/3}$	(3-63)

and detection of gas bubbles as particles; some of these issues are addressed by Van Gelder et al. (1999). Despite these limitations, light-blockage particle counters can be effective tools for monitoring the performance of individual filters as well as combined filter effluent because they detect low levels of larger particles that do not create significant turbidity.

Particle Sizing and Counting: Light Scattering. The scattering of monochromatic laser light can be utilized to size and count particles in water. There are both static and dynamic light-scattering techniques, and detection can be at one or more angles relative to the light sources. With respect to natural and treated water samples and continuous use, the light-scattering techniques are often limited by the need for relatively high sample concentration and an inability to effectively resolve broad particle size distributions. Most light-scattering instruments are utilized in research laboratories, although some on-line instruments have been utilized (e.g., Lewis and Manz, 1991). The main advantage of light-scattering methods is the ability to detect submicron size particles that light-blockage methods do not detect. However, the disadvantages noted above regarding the need for relatively high particle concentration and inability to describe broad or multimodal PSDs have resulted in very limited use of light-scattering-based particle counters for water treatment applications.

For many light-scattering approaches as well as for some direct observation approaches, particle sizing is based on detecting the magnitude of the random Brownian motion of a particle in water, relating that to a diffusion coefficient, and then calculating a hydrodynamically equivalent spherical size for that diffusion coefficient using fundamental theory (i.e., the Stokes-Einstein equation for particle diffusion in a liquid).

Particle Sizing and Counting: Electrical Resistance. Another particle sizing and counting technique that is used in research laboratories is based on the electrical resistance caused by the solid particle material as it disrupts an electrical circuit. The electrical circuit involves a high ionic strength solution and electrodes on either side of a small circular orifice in a glass wall. Suction causes fluid flow through the orifice while an electric potential is applied across the electrodes. As a particle passes through the orifice, it causes increased resistance, which leads to a voltage pulse that is proportional to particle solid volume. The pulses can be calibrated to particles of known size. Multiple orifice sizes can be used to measure a broad size range, from approximately 0.5 to several hundred micrometers.

Some researchers have used this technique to study changes in PSDs during treatment and compare results to fundamental theories (i.e., Lawler et al., 1983; Darby and Lawler, 1990; Moran et al., 1993; Nason and Lawler, 2008). The need to dilute the sample with electrolyte and use multiple orifice tubes to characterize a PSD indicates that this technique is not practical for on-line or routine batch sampling at a water treatment plant, despite the method's advantages of a submicron lower size limit and detailed characterization of PSDs.

Measurements of Particle Stability

The stability of particles against aggregation or deposition to surfaces is a critical characteristic with respect to the performance of treatment processes. There are some direct and indirect analytical techniques related to stability that are useful in drinking water treatment.

Electrical Charge of Surfaces. The most common measurements related to particle stability are based on the electrical charge of particle surfaces. Electric potential fields imposed in an electrolyte solution (water sample) that contains charged particles will cause ions and those particles to move; similarly, movement of charged particles (and ions) relative to fixed electrodes can induce an electric current (or potential). These electroosmotic properties are utilized to characterize particles and other surfaces in water. For example, imposing an electric potential across a solution can cause a particle to have an electrophoretic mobility (or velocity) relative to the water. Fundamental theories then relate that mobility to the electric potential at the plane of shear between the particle and the water; this potential is referred to as the zeta potential (see Hunter [1981] and Chap. 8 for schematic of charged ions adjacent to a charged surface in water). Instruments measure electrophoretic mobility and calculate the zeta potential. Often, particles aggregate well when they have no net surface charge or a near zero zeta potential; measurements can help determine at what coagulant dosage this occurs. Streaming current instruments (detectors or monitors) are another useful technique for assessing surface charge. In this method, charged particles adhere to the surfaces of reciprocating (on a piston) and fixed electrodes, inducing an electric field and current that is proportional in magnitude and sign to the charge of the intermittently attached particles. Batch and on-line instruments are available. On-line streaming current detection is typically applied after coagulant chemical addition and rapid mixing and provides an excellent monitor of chemical dosing; measurements can also be utilized for automatic chemical dosing adjustments in certain circumstances (e.g., charge neutralization coagulation with limited or no oxidant addition).

Effectiveness of Destabilization. Most methods for assessing particle stability involve measuring the net impact of chemical addition to destabilize suspensions. The most common method is the use of jar tests in which different doses of coagulant chemicals are added to sample volumes (typically, 0.2–2 L) of raw water, the samples are mixed, and then solid-liquid separation is induced; gravity settling, dissolved air flotation, and filtration may be utilized (see Chap. 8 for description of jar testing). Measurements of turbidity, particles, and NOM are then used to determine the most effective chemical dose for treatment depending on both water quality and cost criteria. Although the absolute levels of particles in treated samples may not be the same as those achieved at full-scale, the relative performance of chemical dosing variation can be assessed and the degree of conversion of dissolved NOM to particulate form that will occur at full-scale can be quantified. Other laboratory-scale techniques to assess particle destabilization include use of small columns of porous filter media as well as direct assessment of particle aggregation via changes in light scattering. These laboratory techniques can be useful tools in assessing chemical dosing impacts as part of pilot-study design as well as ongoing process operation.

NATURAL ORGANIC MATTER

NOM in water can impart color to water, causing aesthetic effects, and react with disinfectants and oxidants, causing formation of DBPs and other by-products that can have health effects. NOM can cause fouling or interfere with some water treatment processes. In addition, NOM can have effects on the quality of water in the distribution system. Several water treatment processes deal with the removal of NOM. Thus, the importance of NOM is manifold. Its many effects are listed in Table 3-20, and the table also lists other chapters in the book where NOM is covered. The reader is directed to the following references for additional information on NOM: Thurman (1985), Suffet and MacCarthy (1988), and Stevenson (1995).

In the remainder of this section, some terminology is presented about sources of NOM, and this is followed by descriptions of the types and fractions of NOM. Next, material is summarized on collective measurement for NOM, the surrogate measurement of UV absorbance at 254 nm (UV_{254}), and characterization of NOM using specific UV absorbance (SUVA).

Sources

NOM is sometimes classified with respect to its origin. Autochthonous organic matter is produced within a water body. In terms of particulate organic carbon (POC), algae are a good example. In terms of dissolved organic carbon (DOC), proteins, amino acids, polysaccharides, and other organic matter produced from algal respiration and decay processes are examples. Allochthonous organic matter is that NOM not formed in the water body

TABLE 3-20 Effects of NOM on Water Quality and Drinking Water Treatment

Effect	Comments (primary chapters of coverage)
Natural color	Aquatic humic matter: Chaps. 3, 8, and 12
Coats inorganic particles in water supplies affecting their stability	Particles: Chaps. 3 and 8
Contributes to alkalinity for waters high in humic matter	Alkalinity is due mostly to inorganic carbon
Source of tastes and odors	Aesthetic parameter: Chaps. 6, 7, and 14
DBP precursors	DBP Formation: Chaps. 8 and 19
Metal complexation—increases metal solubility	For example, Fe is complexed with NOM in water supplies
Reacts with metal coagulants—coagulant demand	Chapter 8
Reacts with metals affecting their precipitation	Chapters 7 to 9, and 13
Reacts with oxidants causing oxidant demand	Chapter 7
Fouling of GAC adsorbers and membranes	Chapters 11 and 14
Reacts with chemical disinfectants—disinfectant demand	Chapter 17
Decreases effectiveness of UV disinfection	Chapter 18
Corrosion	Chapter 20
Carbon source for biofilm growth in distribution systems	Chapter 21

under study but formed elsewhere and transported to the water body. It comes from watershed runoff and from washing out of upstream water bodies, such as bogs, swamps, and reservoirs. The DOC associated with allochthonous origin is composed of aquatic humic matter and many other types of organic compounds. The types of organic compounds are addressed in the next section.

Types and Fractions

Table 3-9 lists various types of organic compounds that make up NOM. The aims of this section are to expand on the chemistry of types of NOM and to discuss in detail aquatic humic substances.

It is convenient to discuss two classes of NOM: aquatic humic matter and non-humic matter. Specific organic compounds can be identified and attributed to non-humic substances. These compounds can be simple plant by-products and metabolites, including tannins, phenols, carbohydrates, sugars and polysaccharides, proteins, amino acids, and fatty acids. Proteins and amino acids are common algal by-products. Non-humic compounds are lower in molecular weight than aquatic humic substances, most are aliphatic, some are acids, some are bases, and some are neutral (no charge from functional groups). Their origin can be allochthonous following rain events, but most are autochthonous. They are also susceptible to further decay.

The NOM that occurs in water bodies and comes from decay reactions of plant and animal matter that take place on the watershed, yielding a residual organic matter referred to as humic matter. Soils with a high organic content or areas containing peat impart aquatic humic matter during runoff events that enters water bodies. Likewise, bogs and swamps are high in aquatic humic matter that can enter downstream water bodies, especially following rain events. Aquatic humic matter refers to both aquatic humic and fulvic acids. Differences between humic and fulvic acids are discussed below, and details can be found in Thurman (1985), Suffet and MacCarthy (1988), and Stevenson (1995). Aquatic humic matter differs from soil humic matter because it is composed of soluble humic matter not retained by soils.

Chemical Properties of NOM Fractions. Measurements for specific organic compounds are rarely done in monitoring the quality of water supplies. Furthermore, aquatic and fulvic acids are a mixture of organic molecules but do have some common properties. There has been considerable work in characterizing NOM into fractions, thereby yielding insight of certain chemical properties. A method developed by Leenheer and Noyes (1984) has been used widely and adapted (e.g., Bose and Reckhow, 1998). With this method, NOM is fractionated into eight fractions: fulvic acid, humic acid, and six non-humic ones. Aquatic humic and fulvic acids are of particular interest and are discussed in detail below. The non-humic fractions are found in all water supplies, and in some cases are the dominant fractions compared to humic substances. Examples of organic substances that compose the non-humic fractions include

- Weak hydrophobic acids: phenols, tannins, intermediate MW alkyl carboxylic acids
- Hydrophobic bases: proteins, aromatic amines, high MW amines
- Hydrophobic neutrals: hydrocarbons, aldehydes, high MW methyl ketones, alkyl alcohols, ethers
- Hydrophilic acids: hydroxyl acids, sugars, low MW alkyl carboxylic acids
- Hydrophilic bases: amino acids, purines, pyrimidines, low MW alkyl amines
- Hydrophilic neutrals: polysaccharides, low MW alkyl alcohols, aldehydes, ketones

The non-humic compounds are not removed as well as humic substances by coagulation, clarification, and filtration, but some removals do occur. Guidelines on removals of NOM are presented later in this section. Some of these compounds may be adsorbed on granular activated carbon (GAC) or oxidized by ozone to varying degrees, aiding removal by adsorption and biodegradation.

Humic and Fulvic Acids. Aquatic humic matter is an important fraction of NOM. In many water supplies it is the dominant fraction of NOM. When the TOC is more than a few mg/L, the presence of humic matter gives water a yellow-brown color. However, for water supplies in which the yellow-brown color is low, aquatic humic matter can still be an important fraction of TOC. Humic and fulvic acids are not specific molecules because they result from the decomposition of organic matter. These acids are recalcitrant in nature and have a range of molecular weights. Humic acid is more aromatic than fulvic acid. Humic acids have higher molecular size and weight, in the 1000s, compared to fulvic acids, which are only several 100. Regarding specific information on the molecular weight of aquatic fulvic acid, Perdue and Ritchie (2003) give an average of about 650 and a fairly narrow range of 540 to 900. Aquatic humic and fulvic acids have similar C, H, and O composition of about 50, 5, and 40 percent, but humic acid has a greater N content of about 2 percent compared to fulvic acid at about 1 percent. You would expect to find a greater amount of aquatic fulvic acid than humic acid in water because it is more soluble in water with its lower molecular weight, and consequently, humic acid is retained in soils to a greater degree.

A model structure for aquatic humic matter is shown in Fig. 3-14. Aquatic humic matter is a generic term referring to either humic or fulvic acids. The structure of aquatic humic substances portrayed in Fig. 3-14 shows that these substances contain acidic carboxyl and phenol-OH groups. These acid groups can donate protons (H^+), creating negatively charged macromolecules. The dissociation of the carboxyl groups occurs at fairly low pH conditions so that for the pH range of natural waters, aquatic humic matter carries a substantial negative charge. Figure 3-15 shows the charge density of fulvic acid as a function of pH.

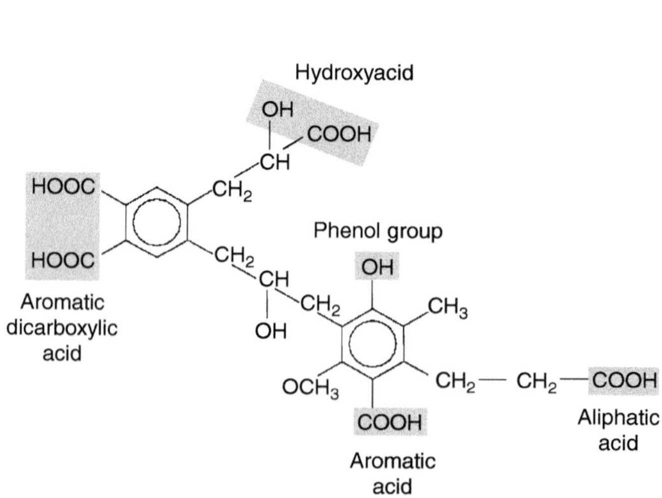

FIGURE 3-14 Hypothetical model structures of aquatic humic matter. [*Source: Thurman (1985) reprinted with permission from Springer, Copyright 1985, Dordrecht, Netherlands*]

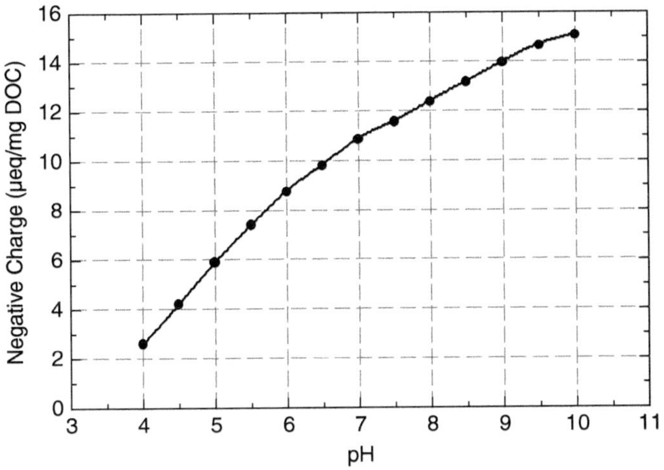

FIGURE 3-15 Aquatic Fulvic Acid Negative Charge Density as a Function of pH. [*Figure developed from data in Van Benschoten (1988).*]

The figure indicates that at about pH 7, the negative charge is approximately 10 μeq/mg of DOC. This negative charge is high compared to the negative charge associated with clay minerals and organic particles, such as algae, protozoan cysts, bacteria, and viruses. This high negative charge affects coagulation processes. In the example that follows, the negative charge associated with NOM and particles are discussed.

Example 3-6 Charge on Particles versus NOM
In this example we compare the negative charge associated with particles and compare it to the charge on aquatic fulvic acid. Clay minerals are used as model particles because they cause mineral turbidity, and in the past they have been viewed as important in controlling coagulation dosing. There are various types of clay minerals and so their negative charge density (analogous to a cation exchange capacity) lies in the range of 0.05 to 0.5 μeq/mg (microequivalents of negative charge per mg of clay). It is reasonable to assume that organic particles such as algae, bacteria, and protozoan cysts have negative charges equal to or less than those of clays. The charge on aquatic fulvic acid depends on pH, as shown in Fig. 3-15. In this example, a negative charge density of 10 μeq/mg (microequivalents of negative charge per mg of DOC or TOC) is used in the calculations.

Calculate the negative charge for a water supply containing clay at a mass concentration of 3 mg/L (say, turbidity of 10 ntu). Compare it to the negative charge of a water containing NOM at a TOC (DOC) of 3 mg/L in which only 50 percent of the NOM is aquatic fulvic acid, and the rest of the NOM is composed of non-humic organic matter without a charge.

Solution Clay particles have charge densities of 0.05 to 0.5 μeq/mg; a mid-range charge density of 0.1 μeq/mg is used for the water supply containing mineral turbidity.

$$\text{Particle negative charge} = 0.1\,\frac{\mu\text{eq}}{\text{mg}} \times 3\,\frac{\text{mg}}{\text{L}} = 0.3\frac{\mu\text{eq}}{\text{L}}$$

For the water supply containing NOM at a TOC of 3 mg/L, but only 50 percent of the TOC is aquatic fulvic acid with a negative charge density of 10 μg/mg TOC:

$$\text{NOM negative charge} = 10\ \frac{\mu\text{eq}}{\text{mg TOC}} \times 0.5 \times 3\ \frac{\text{mg}}{\text{L}} = 15\frac{\mu\text{eq}}{\text{L}}$$

The results of the calculations show that the NOM in a water supply has a negative charge that is about 50 times greater than the charge associated with particles. Obviously, the concentrations of particles and NOM can vary among water supplies, but in fact in many water supplies, the NOM has a far greater negative charge associated with it than particles, and this negative charge must be effectively neutralized by coagulation. In short, for many water supplies, coagulant dosing is controlled by NOM and not turbidity (see Chap. 8). ▲

Collective Measurements of Organic Carbon

TOC is the total concentration of organic carbon in water and is composed of DOC and POC. Measurement of DOC is an operational definition because samples are filtered through 0.45 μm pore size filters (other filters are often used such as glass fiber filters with nominal pore sizes of 1.2–1.5 μm) to obtain this fraction of organic carbon. DOC is defined as the organic carbon measured in what passes the filter. However, it may not be the true dissolved phase of organic carbon since colloidal organic matter smaller than the filter's pore size can pass thorough the filters. In spite of some possible error in measuring DOC, in the water field we accept this operational definition for DOC. POC is defined as the TOC minus the DOC. NOM in the particulate form (POC) can be attributed to viruses, bacteria, algae, other plant matter, and organic detritus. The largest contributor for surface waters is algae, especially in lakes and reservoirs. However, the POC of surface waters is low compared to TOC, usually less than a few percent except for eutrophic lakes. In summary, the DOC composes 90 to 99 percent of the TOC for surface waters used for drinking water. For groundwaters, POC is nearly zero so the DOC is essentially equal to the TOC.

Table 3-21 summarizes ranges and means for TOC concentrations for various types of waterbodies. One can see that rivers and streams can have a wide range in TOC, typically

TABLE 3-21 TOC Concentrations in Various Waterbodies (Based on Authors' Experience and Thurman (1985), Eby (2004), Ogawa and Tanoue (2003); ICR (Information Collection Rule Data from USEPA (2005))

Waterbody	Range (Mean)	Other
Groundwater	0.5–10 (0.7)	Many groundwaters are 2 mg/L or less
Rivers and streams	1.5–20	Typically 5 mg/L or less; higher values downstream of bogs, marshes, and swamps; some can exceed 20 mg/L
Oligotrophic lakes	1.0–3.0 (2.2)	
Mesotrophic lakes	2.0–4.0 (3.0)	
Eutrophic lakes	3.0–30 (12.0)	
Marsh	17.0	Typical
Bog	33.0	Typical
Seawater	0.72–0.96 (DOC)	In upper 100 m for the oceans except the Antarctic Ocean; lower DOC in deep ocean waters
	1–5 (TOC)	Typical range for upper waters and depends on algae concentration and oil spills

from 1.5 to 20 mg/L. Generally, high TOC of greater than 5 to 7 mg/L occurs primarily in these sources when they are influenced by upstream bogs and marshes. In fact, some rivers and streams that are influenced by bogs and swamps can exceed 20 mg/L TOC. For smaller lakes and reservoirs, the TOC concentrations are influenced by the watershed, so a significant fraction of the TOC can have an allochthonous origin. For larger lakes and reservoirs in watersheds in which upstream bogs are not significant, the trophic state of the lake affects the TOC concentration, and a large fraction of the TOC can be autochthonous. As indicated in Table 3-21, oligotrophic lakes (those low in nutrients and algae biomass) have the lowest TOC concentrations. As lakes and reservoirs progress to mesotrophic and eutrophic states, the TOC increases due to the effect of algae. Groundwaters have a low mean concentration of TOC. However, shallow groundwater systems can have higher TOC when under the direct influence of surface waters. Other groundwaters can have high TOC when affected by peat deposits or have other unique characteristics such as some groundwaters in Florida and parts of southern California. Excluding these latter cases, generally the TOC of groundwaters is about 2 mg/L or less.

UV_{254} Absorbance as a Surrogate Parameter for DOC and TOC

Edzwald et al. (1985) showed that UV_{254} can be used as a surrogate parameter for and predictor of DOC, TOC, and trihalomethane (THM) precursors. They also demonstrated its usefulness in monitoring water treatments plants for the removals of TOC and THM precursors. Over the years, UV_{254} has found increasing acceptance and is now widely used for the purposes cited above, and it is also used for setting coagulant dosages for water supplies in which TOC controls coagulation (Edzwald, 1993; Edzwald and Kaminski, 2009) and for monitoring removals of organic matter by GAC and membrane processes. Measuring UV_{254} is simple, rapid (grab sampling or on-line instrumentation), and inexpensive making it a good surrogate parameter. Details on its measurement can be found in *Standard Methods* (APHA, AWWA, and WEF, 2005).

UV light at 254 nm is absorbed by a variety of organic and inorganic molecules, but especially organic compounds with an aromatic structure or compounds that have conjugated C=C double bonds. Aquatic humic matter has these structural features so they absorb more light per unit concentration of DOC (called the specific UV absorbance, or SUVA) than other types of NOM in water supplies, such as hydrophilic acids, hydrophobic bases (e.g., proteins and aromatic amines), and hydrophobic neutrals (e.g., aldehydes); see Edzwald (1993) and Bose and Reckhow (1998). Because the C=C double bonds are sites for donating electrons, oxidants and disinfectants attack and chemically react with these compounds, making UV_{254} a particularly good surrogate and predictor of DBP formation and oxidant demand (ozone and other oxidants and chlorine demand). As stated above, coagulation is controlled by TOC for many water supplies and so UV_{254} can be related to coagulant dosing (Chap. 8). Some water treatment plants operate their coagulant dosing based on achieving a treated water UV_{254} goal (Edzwald and Kaminski, 2009).

Figure 3-16 illustrates the utility of UV_{254} as a surrogate for TOC for two water plants (Edzwald and Kaminski, 2009). The data used to develop the correlations and prediction equations were water treatment plant monthly TOC and UV_{254} measurements on both raw and treated waters. The advantages of pooling the raw and treated water data are that it provides a wide spread in the variables and provides the water plants with the ability to predict raw and treated water TOC concentrations. The two plants control coagulant dosing to achieve UV_{254} (routinely measured as coagulant operating parameter) goals of 0.03 to 0.035 cm^{-1} that correlates with TOC $\leq$ 2 mg/L.

What are raw water UV_{254} levels for water supplies? These can vary from about 0.04 cm^{-1} for water supplies low in TOC and aquatic humic matter to about 0.8 cm^{-1}

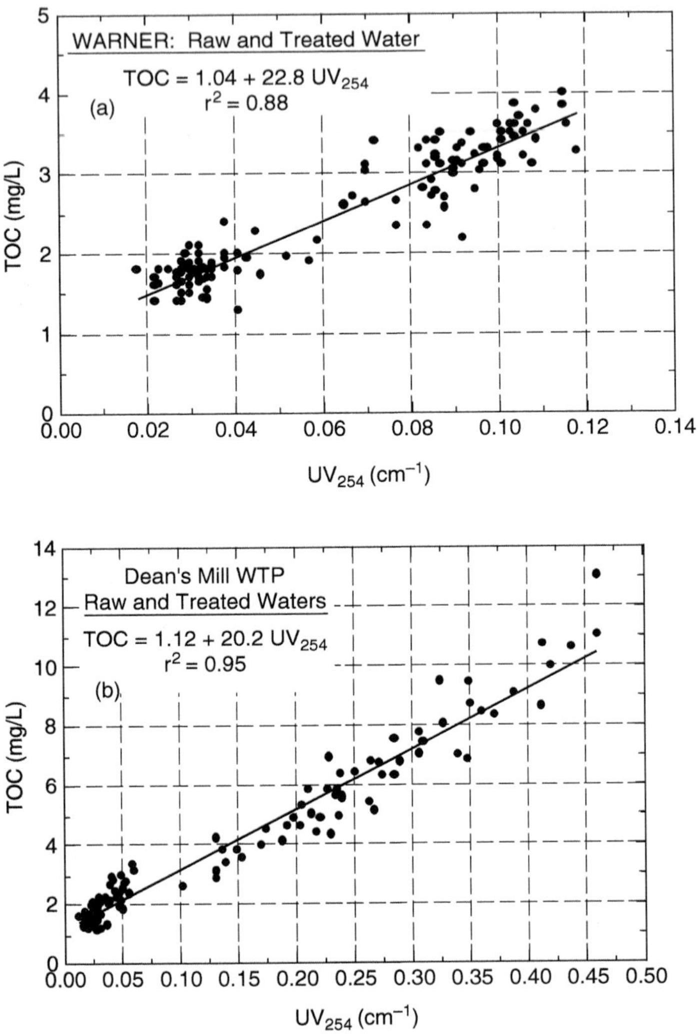

FIGURE 3-16 TOC versus UV_{254} correlation for the Warner and Dean's Mill water plants for the period of 2002–2006. (*Source: Edzwald and Kaminski (2009), reprinted with permission from the Journal of the New England Water Works Association, Copyright 2009, Holliston, MA.*)

for water supplies high in TOC and aquatic humic matter. Many water supplies are in the range of 0.07 to 0.3 cm^{-1}, which corresponds to TOC concentrations of approximately 2 to 7 mg/L, depending on the SUVA of the NOM in the water. Treated water UV_{254} values depend on the extent of TOC removal and the water treatment process. For GAC and nanofiltration and reverse osmosis membranes, low TOC and thus UV_{254} values are achieved (<0.02 cm^{-1}). Edzwald and Kaminski (2009) have shown that full-scale treatment plants operating with good coagulant dosing for particles (turbidity) and TOC removals achieve UV_{254} of 0.023 to 0.033 cm^{-1}, corresponding to treated water TOC of 1.5 to 2 mg/L.

Finally, it is recommended that treatment plants set treated water UV_{254} goals of 0.02 to 0.035 cm^{-1}. These goals apply prior to oxidant or chlorine addition; oxidants and chlorine reduce UV_{254} without removing TOC.

Specific UV Absorbance

SUVA was first developed by Edzwald et al. (1985) to evaluate whether TOC and DBP precursor concentrations could be correlated with UV_{254}. Further development of the utility of SUVA occurred and guidelines were presented relating raw water SUVA values to the NOM composition of water and the ability of coagulants to remove TOC (Edzwald and Van Benschoten, 1990; Edzwald, 1993). SUVA provides a simple way to characterize the nature of NOM and is calculated from measurements of UV_{254} and DOC—samples are filtered. SUVA values are expressed in units of m^{-1} absorbance per mg/L of DOC.

$$\text{SUVA} = \frac{(\text{UV}_{254} \text{ in cm}^{-1})\, 100 \dfrac{\text{m}^{-1}}{\text{cm}^{-1}}}{(\text{DOC in mg/L})} \tag{3-64}$$

As presented above, certain types of NOM absorb UV_{254} light per unit concentration of DOC to a greater degree than other types. Therefore, SUVA is an indicator of the NOM composition of water (Edzwald, 1993; Bose and Reckhow, 1998). SUVA values follow for the following types of NOM: 5 to 7 for humic acid, 3 to 4 for fulvic acid, < 4 for hydrophilic neutrals, about 3 for hydrophilic bases, about 1 for hydrophilic acids, and 1 or less for hydrophobic bases and neutrals. Water supplies contain a mixture of types of NOM, but the SUVA can provide an indication of what types of organic compounds dominate. Guidelines on the composition of NOM for raw waters with respect to three ranges of SUVA are presented in Table 3-22. This table is an updated version of what originally appeared in Edzwald and van Benschoten (1990) and later in Edzwald (1993).

TABLE 3-22 Characterization of NOM and TOC Removals for SUVA Values for Raw Water Supplies

SUVA	Composition	Effects	TOC Removals by coagulation
>4	High fraction of aquatic humic matter High aromatic and hydrophobic character High molecular weight (MW)	High UV absorbance High oxidant and chlorine demand	60–80% Higher end for waters with high TOC
2–4	Mixture of aquatic humic and non-humic matter Mixture of aromatic and aliphatic character Mixture of low to high MW	Medium UV absorbance Medium oxidant and chlorine demand	40–60% Higher end for waters with high TOC
<2	High fraction of non-humic matter High aliphatic and low Hydrophobic character low MW	Low UV absorbance Low oxidant and chlorine demand	<20–40% Higher end for waters with high TOC

Water supplies with high SUVA values of 4 or greater indicate that the NOM is composed mainly of aquatic humic matter. Raw waters with SUVA of 2 to 4 contain a mixture of aquatic humic matter and non-humic NOM. Raw waters with SUVA < 2 contain mainly non-humic material with properties shown in Table 3-22. Table 3-22 also shows expected TOC removals by coagulation indicating greater removals for higher SUVA waters.

Figure 3-17 shows data over a five-year period for a raw water supply. The TOC of this supply varies from about 2.5 to 13 mg/L. The UV_{254} data follow the variation in TOC and can predict it as previously shown in Fig. 3-16b. The SUVA data are also plotted in Fig. 3-17 and vary from 3.2 to 5.6, with a mean value of 4.3. This water supply has bogs in the watershed yielding NOM with a large fraction of aquatic humic matter and high TOC and UV_{254} following heavy rains that flush the upstream bogs.

In water supplies with high SUVA, the NOM has C=C bonds that contain electrons, and as a result, oxidants and disinfectants will attack and react at these chemical groups. Therefore, water supplies with high SUVA will have greater chlorine demand, greater oxidant demand, and greater DBP formation.

Water treatment processes such as coagulation and adsorption will preferentially remove aquatic humic matter or NOM; consequently, the SUVA of treated waters is much lower than that of raw waters. Treated water SUVA values of 2 or less indicate good removal of NOM.

Measurements of NOM for Water Utilities

Water utilities are required to make monthly measurements of TOC for the USEPA Enhanced Coagulation requirements (see Chaps. 1 and 8). It is highly recommended that water utilities collect monthly data from the same TOC samples for raw and finished water DOC and UV_{254}. From these data, utilities should develop correlations between TOC and UV_{254} so that they can use daily raw water and finished water UV_{254} measurements to predict raw water and finished water TOC concentrations; see Fig. 3-16 and discussion above. Additional correlations between DOC and UV_{254} are useful. Using these data, water utilities can develop finished water UV_{254} goals by making correlations to their TOC goals. As stated earlier, recommended guidelines for treated water UV_{254} are 0.02 to 0.035 cm^{-1}. These goals apply prior to oxidant or chlorine addition; oxidants and chlorine reduce UV_{254} without removing TOC. Treatment plants that use oxidants in pretreatment or in intermediate treatment must collect samples to assess the effect of the oxidant on UV_{254}.

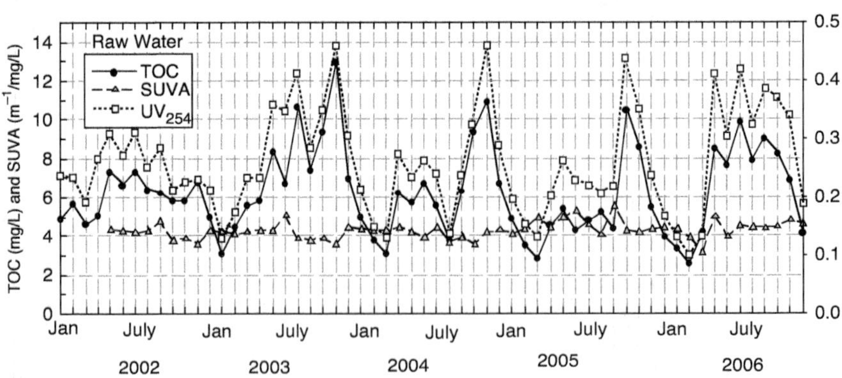

FIGURE 3-17 Raw water TOC, UV_{254}, and SUVA for the Dean's Mill water plant.

Water utilities should also calculate SUVA from their DOC and UV_{254} measurements. Utilities can use raw water SUVA values to determine the fractions of humic matter and non-humic matter in their source water. This was discussed earlier along with the presentation of Table 3-22 and Fig. 3-17. SUVA values also allow utilities to determine expected TOC removals for good coagulation according to the guidelines in Table 3-22. In the previous discussion of SUVA, it was noted that TOC removal processes generally selectively remove aquatic humic fractions, leaving hydrophilic TOC of low molecular weight behind in the finished water. Consequently, the finished water SUVA should be 2 or less and should indicate how well the treatment plant removed the humic matter fraction of TOC. As mentioned above, because oxidants can reduce UV_{254} (without removing DOC), SUVA can be reduced by oxidation. Water utilities should assess the effect of any oxidants on SUVA.

In summary, TOC and DOC should be measured by water utilities on a monthly basis (at minimum) in accordance with the USEPA Enhanced Coagulation requirements. Some utilities may wish to make these measurements more frequently, perhaps weekly basis, in order to capture changes in NOM that may occur in rivers and reservoirs with small watersheds where frequent changes in water quality occur and where seasonal effects and rainfall wash organic matter from the land into supplies. Raw and finished water UV_{254} should be measured on a daily basis and used as a surrogate parameter and predictor of TOC.

There are various other NOM characterization measurements that water utilities can make that provides information on the NOM chemical and physical properties. These are classified as more of a research nature rather than routine water plant operational measures for monitoring and controlling NOM. These include measurements that provide information on the NOM fractions with respect to hydrophobic versus hydrophilic characteristics and acid, base, and neutral fractions (listed earlier in the section, Chemical Properties of NOM Fractions). Other characterization methods are those that give information on molecular weight, charge, elemental composition, and functional groups—the reader is referred to the general references cited at the beginning of the section on NOM and in a new publication by Reckhow (2010) that summarizes various methods. Water utilities can also evaluate the reactivity of NOM to disinfectant dosing and develop models to predict raw water DBP precursors and finished water DBPs from TOC, UV_{254}, and other measurements; see Edzwald et al. (1985) and Chap. 19.

SOURCE WATER SELECTION AND PROTECTION

This section summarizes important principles related to the selection and protection of drinking water sources. Source selection is relevant to development of new sources in areas with increasing water demand. Source protection is important for all sources. The focus is water quality, the topic of this book, not water quantity.

Selection of Source Water

The selection and development of a source of drinking water is often challenging and depends on many factors, including water demand, climate, geography, geology, availability, yield, cost, social and political factors, and water quality. From the perspective of protecting public health and minimizing treatment costs, selection of the best or highest quality source of supply is desirable. This principle of "best available source" was frequently advocated by long-time educator, researcher, and engineer Daniel Okun, as described in an opinion article (Okun, 1991). Okun cited regulatory statements by the U.S. Public Health Service (in 1962) and the USEPA (in 1976) that direct water suppliers to choose the best quality source and to protect that source, subject to economic feasibility constraints.

The best available source concept evolved during the twentieth century from focusing on water quality concerns related to the acute risks to public health from pathogenic microorganisms to focusing on the risks associated with long-term exposure to low levels of chemical contaminants from most, but not all, anthropogenic sources. There is sound logic to avoiding sources with an ever increasing number of measurable contaminants, many of which have unknown health impacts. Selecting well-protected, pristine sources can minimize the types and amounts of possible anthropogenic contaminants and minimize treatment costs along with unknown health impacts.

Along with the best available source concept are the level, density, and location of humans on the watershed. The expanding population increases total amounts of anthropogenic inputs to local, regional, and global hydrologic cycles while decreasing the available land area that could be used to protect surface and groundwaters. In some cases, such as Boston, MA, and New York City, a dense urban population can be provided with drinking water from relatively pristine sources due to favorable climate (available precipitation) and allocation of significant land area for collection, storage, and transmission of the source water, usually from a significant distance. In contrast, large human populations may exist in arid regions and others areas where climate, geography, and human activity make selection of a pristine source technically or economically infeasible. In such cases, the best available source is not pristine, and more extensive treatment technologies must be utilized to provide potable water. Historical treatment process selection reflects water quality impacts of pristine versus non-pristine sources. Protected upland surface waters of pristine quality often require less treatment, i.e., slow sand filtration or conventional coagulation and filtration or in some cases, only disinfection alone. However, treatment for a major downstream run-of-river source may include advanced oxidation and adsorption processes in addition to conventional processes to remove anthropogenic contaminants. More often than not, large, dense urban populations must use non-pristine sources for drinking water due to economic and other factors.

The penultimate non-pristine drinking water source is wastewater generated from human activity, often referred to as water reuse (Chap. 16). In such cases, wastewater is extensively treated prior to being directly or indirectly (via aquifer recharge or discharge to a surface water with non-wastewater derived sources) used as a drinking water source. Examples of near direct reuse of treated wastewater include NEWater in Singapore for industrial uses (as well as for indirect potable uses) and the New Goreangab water reclamation plant in Windhoek, Namibia, which directly augments the potable drinking water supply. An important, long-term example of aquifer recharge with extensively treated wastewater is the Orange County (CA) Water District facility. Discharge of treated wastewaters to surface waters is common such that all "downstream" uses for water supply are indirect reuse. The degree of public concern and attention may be associated with the time between discharge and reuse as well as the relative dilution with more pristine water. Two very different examples include the discharge of treated wastewater from Las Vegas, NV, into Lake Mead from which the city also obtains its raw drinking water source, and the discharge of highly treated low flows of wastewater into the rather pristine upland reservoirs in the supply system for New York City. The major long-term obstacle in the use of wastewater as a drinking water source is the need to remove dissolved constituents, which occur due to use of water to convey waste, from the wastewater. For potable reuse of wastewater, engineers use technology to simulate the natural hydrologic processes of evaporation and condensation of essentially pure water, i.e., desalination. Significant problems are the energy associated with desalination as well as the reject or waste stream containing very high levels of salts.

Not surprisingly, there is growing use of waters of different qualities for different purposes, especially where a sufficient quantity of potable drinking water is not available for all uses. This could be considered a "best suitable source for the purpose" approach. Okun, a proponent of the "best available source," was also an early and frequent proponent of dual water distribution systems, with one system for potable water and one for non-potable water suitable for fire fighting, irrigation, toilet flushing, and similar uses (Okun, 1997). Such an approach can be applied on a community level or for a specific recreational, commercial, or industrial facility.

Source Protection—General

Protection of the quality of natural drinking water supplies is an important and significant objective for most water utilities. Most importantly, source water protection serves as a barrier for protection of public health via minimization of health risks associated with water pollution. In addition, source water protection best management practices (BMPs) are implemented to maintain or improve source water quality such that the cost of drinking water treatment is decreased (Pyke et al., 2003).

Point-Source Control. The federal, state, and local laws and regulations governing the point-source discharge of wastewater and, in certain cases, stormwater runoff are significant for source water protection. The Clean Water Act, originally known as the Federal Water Pollution Control Act, requires the USEPA to regulate and monitor point-source discharges of wastewater via the National Pollutant Discharge Elimination System (NPDES) permit system; this is frequently implemented by state agencies. Required levels of wastewater treatment are based on either technology or water quality, depending on the source of the discharge and the type of environment, usually aquatic, being discharged to. Extensive levels of wastewater treatment may be required based on water quality needs for protection of ecosystems and human health. Traditionally, the assimilative capacity of the receiving aquatic environment with respect to nutrients (N, P), oxygen demand, and pathogens and the required quality for the intended use of the receiving water have been considered in establishing water quality based discharge permit requirements. Thus, a wastewater discharge to a protected drinking water supply reservoir has warranted much more extensive treatment than a discharge to a major river that may or may not be used as a drinking water source.

More recently, concerns have increased with respect to the wide range of low-level contaminants in wastewater that have not been regulated or monitored, so-called contaminants of emerging concern, including pharmaceuticals and personal care products (PPCPs) and EDCs. Despite limited knowledge of the potential health impacts of such contaminants (see Chap. 2), significant research is being directed at assessing and controlling sources, assessing treatment efficacy, and monitoring contaminant levels in the environment. Consequently, activities to eliminate or limit discharges of treated wastewater to drinking water sources are a source water quality protection measure, yet such activities may be infeasible or undesirable from a water quantity perspective. In all cases, reasonable control of contaminant inputs to wastewaters is desired (i.e., non-water based disposal and reuse methods).

Nonpoint Source (Diffuse) Pollution. The control of source water pollution from nonpoint or diffuse sources is often more challenging than the control of point-source inputs. Diffuse contaminants can be input via atmospheric deposition (dry and with precipitation), stormwater runoff and infiltration, subsurface wastewater disposal (i.e., septic systems), and leakage from subsurface storage tanks and pipes. Novotny (2003) provides a comprehensive look at water quality related to diffuse pollution. In contrast to application of treatment technology for point-source discharges (so-called end-of-pipe solutions), control of nonpoint source pollution often involves source control, land use management, and application of ecological engineering principles in order to make sustainable use of natural resources and in turn control contaminant inputs.

Watershed Protection

Watershed protection or management generally refers to activities performed on a topographically and hydrologically defined land area (the watershed) to protect surface water quality. A classic example for drinking water is the topography-based watershed area, or basin, from which surface water runoff can enter a reservoir that was created by impounding a stream or river. Transfer of water between watersheds, or groundwater input and output, can create additional hydrologic-based definitions of a watershed management area. The fact that governmental boundaries (town, county, state, country) often do not coincide with watershed boundaries can present a significant challenge to implementing watershed protection activities. However, this can also present a significant opportunity for citizens to work cooperatively across such boundaries for a common goal. Effective management results when governments create watershed based organizations, such as a water utility, to coordinate watershed activities.

Watershed protection or management plans can take many forms but usually involve identifying possible hazards, identifying hazard mitigation activities, implementing activities, and monitoring the impacts of activities (Novotny, 2003; MA DCR, 2008). Hazards can be identified by routinely conducting a traditional sanitary survey of the watershed to identify locations and sources of contaminants and by careful understanding of watershed geology, topography, vegetation, and land use. Hazards include the potential for natural disaster (hurricanes, flooding) and for pollutant input from human activity. Mitigation activities include land acquisition, control of land use, implementation of best management practices for specific uses (i.e., agricultural), elimination of contaminant sources, and preparation for emergency response. Examples of specific contamination issues include agricultural-based phosphorus inputs (Sharpley et al., 1994) and sources of protozoan pathogens (LeChevallier et al., 1991).

Management and protection of the upland watersheds for the drinking water supplies for New York City (NYC) are very complicated and comprehensive. A National Academy of Sciences (NAS) committee chaired by Dr. Charles R. O'Melia studied the NYC watershed protection plan in light of the potential filtration requirements by the USEPA SWTR (Committee to Review NYC, 2000). NYC is building a coagulation, dissolved air flotation, media filtration, and UV disinfection facility for the Croton watershed source water, while water from the Catskill and Delaware (Cat-Del) sources—the majority of the NYC supply—will be treated via UV and free chlorine disinfection only. Implementation of land acquisition, management of land use (especially agricultural practices), partnership programs with local communities, and extensive treatment of wastewater discharges are utilized by the NYC Department of Environmental Protection (DEP) to maintain the high quality of the Cat-Del source water such that treatment beyond disinfection is not yet required. In order of priority, the NAS report recommended a focus on control of sources of pathogenic microorganisms, NOM precursors to disinfection by-products, phosphorus and turbidity (sediment). The NAS report also recommended that NYC DEP be open to the need for additional treatment and to take a lead role in developing data to document the role of watershed protection activities in reducing public health risk. Specific activities undertaken by NYC to maintain and improve watershed protection are documented in annual reports (e.g., NYC DEP, 2009).

In the United States, watershed-based water quality protection activities have been mandated from a wastewater discharge perspective, for example, under the Section 308 river basin planning portion of the Clean Water Act. Comprehensive consideration of all hydrological activities in a watershed has not typically been undertaken in the United States, although some states have done so (e.g., Massachusetts during approximately 2000 to 2005). In contrast, the United Kingdom has a long history of dividing the country into river basin areas for management of water resources, while the European Union Water Framework Directive of 2000 emphasizes river-basin-based management plans for all member states.

Aquifer Protection

Aquifer protection typically refers to specific activities undertaken to protect the quality and quantity of groundwater sources of drinking water. As described earlier in this chapter, natural groundwater quality is dependent on the geology of the subsurface material from which the water is withdrawn. Specific human activities such as subsurface wastewater disposal, deep-well injection of wastewater, improper underground disposal of wastes, and leaking of wastes from underground tanks and pipes can impact groundwater quality. Natural infiltration of surface water of impaired quality can also impact groundwater quality.

The subsurface water environment is especially challenging to understand and manage, because it is largely unseen and usually spatially non-homogeneous. Understanding the transport and fate of contaminants requires study of groundwater hydrology (Todd and Mays, 2005). Specific activities undertaken to protect and improve groundwater quality include control of risks within the land area near a well (i.e., aquifer protection zones with usage restrictions), treatment to improve the quality of waters infiltrated or pumped into the subsurface, remediation of previously contaminated subsurface material, and hydraulic activities to control the influence of undesirable groundwater quality on water withdrawals (e.g., pumping to prevent saltwater intrusion). Schmoll et al. (2006) cover protection of groundwater from a human health perspective in their book.

ABBREVIATIONS

Acy	acidity
Alk	alkalinity
ANC	acid-neutralizing capacity
BMP	best management practice
DBPs	disinfection by-products
DEP	Department of Environmental Protection
DO	dissolved oxygen
DOC	dissolved organic carbon
EDC	endocrine disrupting compound
EMF	electromotive force
GAC	granular activated carbon
ICR	Information Collection Rule
LT1ESWTR	Long Term 1 Enhanced Surface Water Treatment Rule
LT2ESWTR	Long Term 2 Enhanced Surface Water Treatment Rule
NAS	National Academy of Sciences
NOM	natural organic matter
NPDES	National Pollutant Discharge Elimination System
ntu	nephelometric turbidity units
NYC	New York City
MW	molecular weight
ORP	oxidation reduction potential
PAC	powdered activated carbon
POC	particulate organic carbon
PPCPs	pharmaceuticals and personal care products
PSD	particle size distribution
SUVA	specific ultraviolet absorbance
SWTR	Surface Water Treatment Rule
TDS	total dissolved solids

THM	trihalomethane
TOC	total organic carbon
USEPA	U.S. Environmental Protection Agency
USGS	U.S. Geological Survey
UV_{254}	ultraviolet absorbance at 254 nm
VOC	volatile organic compound
zpc	zero point of charge

NOTATION FOR EQUATIONS

a	activity of the chemical species
A	constant in Davies equation, or gas phase and dissolved gas phase
C	molar or mass concentration
d	diameter
E	electric potential for redox reactions
E^o	standard electric potential for redox reactions
H	Henry's law constant (atm/molar)
H^*	Henry's law constant (molar/atm)
F	Faraday constant (96.485 kJ V^{-1} $mole^{-1}$)
I	ionic strength
K	equilibrium constant
N	number concentration
P	pressure
Q	reaction quotient for a chemical reaction
R	universal gas constant (8.314 J/mole · K).
S	surface area
T	absolute temperature in K (0°C = 273.15 K)
V	volume
Z *or* z	charge of the ion
α	fraction of the chemical species
ΔH_r^o	standard enthalpy of the reaction
ε	dielectric constant
γ	activity coefficient
κ	conductivity
ρ	density

REFERENCES

APHA, AWWA, and WEF (American Public Health Association, American Water Works Association, and Water Environment Federation) (2005) *Standard Methods for the Examination of Water and Wastewater*, 21st ed., APHA: Washington, D.C.

Benotti, M.J., Trenholm, R.A., Vanderford, B.J., Holady, J.C., Stanford, B.D., and Snyder, S.A. (2009) Pharmaceuticals and endocrine disrupting compounds in U.S. drinking water, *Environmental Science and Technology*, 43 (3), 597–603.

Berner, E.K. and Berner, R.A. (1996) *Global Environment: Water, Air, and Geochemical Cycles*, Upper Saddle River, N.J.: Prentice Hall.

Bose, P. and Reckhow, D.A. (1998) Adsorption of organic matter on preformed aluminum hydroxide flocs, *Journal of Environmental Engineering*, 124 (9), 803–811.

Broadwell, M. (2001) *A Practical Guide to Particle Counting for Drinking Water Treatment*, Boca Raton, Fla.: Lewis Publishers.

Chowdhury, Z.K., VanGelder, A., Lawler, D., and Moran, M. (2000) *Particle Count Method for Concentration Standards and Sample Stabilization*, Denver, Colo.: Water Research Foundation.

Committee to Review the New York City Watershed Management Strategy, National Research Council (2000) *Watershed Management for Potable Water Supply: Assessing the New York City Strategy*. Washington, D.C.: National Academy of Sciences.

Darby, J.L. and Lawler, D.F. (1990) Ripening in depth filtration: effect of particle size on removal and head loss, *Environmental Science and Technology*, 24 (7), 1069–1079.

Domenico, P.A. and Schwartz, F.W. (1990) *Physical and Chemical Hydrology*, New York: John Wiley and Sons.

Eby, G.N. (2004) *Principles of Environmental Geochemistry*, Pacific Grove, Calif.: Brooks/Cole.

Edzwald, J.K., Becker, W.C., and Wattier, K.L. (1985) Surrogate parameters for monitoring organic matter and THM precursors, *Journal AWWA*, 77 (4), 122–131.

Edzwald, J.K., and Van Benschoten, J.E. (1990) *Aluminum Coagulation of Natural Organic Matter*, In: *Chemical Water and Wastewater Treatment*, H.H. Hahn and R. Klute, eds., New York: Springer-Verlag.

Edzwald, J.K. (1993) Coagulation in drinking water treatment: particles, organics, and coagulants, *Water Science and Technology*, 27 (11), 21–35.

Edzwald, J.K. and Kaminski, G.S. (2009) A practical method for water plants to select coagulant dosing, *Journal of the New England Water Works Association*, 123 (1), 15–31.

Elimelech, M., Gregory, J., Jia, X., and Williams, R.A. (1995) *Particle Deposition & Aggregation. Measurement, Modelling and Simulation*. Oxford, U.K.: Butterworth-Heinemann, Ltd.

Gregory, J. (2005) *Particles in Water: Properties and Processes,* London: IWA Publishing.

Han, M. (2004) Rainwater harvesting: a new paradigm to meet MDG and sustainability, Presented at the 1st Rainwater Harvesting Workshop, IWA 4th World Water Congress and Exhibition, 21 September 2004, Marrakech, Morocco.

Hargesheimer, E.E., Lewis, C.M., and Yentsch, C.M. (1992) *Evaluation of Particle Counting as a Measure of Treatment Plant Performance*, Denver, Colo.: Awwa Research Foundation and AWWA.

Hargesheimer, E.E., McTigue, N., and Lewis, C. (2001) *Fundamentals of Drinking Water Particle Counting*, Denver, Colo.: Water Research Foundation.

Hassellöv, M. and Kaegi, R. (2009) Analysis and Characterization of Manufactured Nanoparticles in Aquatic Environments. In: *Nanoscience and Nanotechnology: Environmental and Human Health Implications*, J.R. Lead and E. Smith, eds., pp. 211–266.

Henderson, R., Parsons, S.A., and Jefferson, B. (2008) The impact of algal properties and pre-oxidation on separation of algae, *Water Research,* 42, 1827–1845.

Hiemenz, P.C. and Rajagopalan, R. (1997) *Principles of Colloid and Surface Chemistry,* 3rd ed., New York: Marcel Dekker, Inc.

Hunter, R.J. (1981) *Zeta Potential in Colloid Science*, London: Academic Press.

Israelachvili, J. (1992) *Intermolecular and Surface Forces,* 2nd ed., London: Academic Press.

Ju-Nam, Y. and Lead, J.R. (2008) Manufactured nanoparticles: an overview of their chemistry, interactions and potential environmental implications, *Science of the Total Environment*, 400 (1–3), 396–414.

Kalff, J. (2002) *Limnology.* Upper Saddle River, N.J.: Prentice Hall.

Kavanaugh, M.C. and Leckie, J.O., eds. (1980) *Particulates in Water: Characterization, Fate, Effects, and Removal,* Advances in Chemistry Series #189. Washington, D.C.: American Chemical Society.

Kolpin, D.W., Furlong, E.T., Meyer, M.T., Thurman, E.M., Zaugg, S.D., Barber, L.B., and Buxton, H.T. (2002) Pharmaceuticals, hormones, and other organic waste contaminants in U.S. streams, 1999–2000: a national reconnaissance, *Environmental Science and Technology,* 36 (3), 202–1211.

Lawler, D.F., O'Melia, C.R., and Tobiason, J.E. (1980) Integral water treatment plant design: from particle size to plant performance, In: *Particulates in Water: Characterization, Fate, Effects, and Removal,* M.C. Kavanaugh, and J.O. Leckie, eds., Advances in Chemistry Series No. 189. Washington, D.C.: American Chemical Society.

Lawler, D.F., Izurieta, E., and Kao, C-P. (1983) Changes in particle size distributions in batch flocculation, *Journal AWWA*, 75 (12), 604–612.

LeChevallier, M.W., Norton, W.D., and Lee, R.G. (1991) Occurrence of Giardia and Cryptosporidium spp in surface-water supplies, *Applied and Environmental Microbiology*, 57 (9), 2610–2616.

LeChevallier, M.W. and Norton, W.D. (1992) Examining relationships between particle counts and *Giardia, Cryptosporidium,* and turbidity, *Journal AWWA*, 84 (12), 54–60.

Leenheer, J.A. and Noyes, T.I. (1984) A filtration method and column adsorption system for on site concentration and fractionation of organic substances from large volumes of water, *U.S. Geological Survey Water Supply Paper 2230.* Denver, Colo.: U.S. Geological Survey.

Lewis, C.M. and Manz, D.H. (1991) Light-scatter particle counting: improving filtered-water quality, *Journal of Environmental Engineering*, 117 (2), 209–223.

Lewis, C.M., Hargesheimer, E.E., and Yentsch, C.M. (1992) Selecting particle counters for process monitoring, *Journal AWWA*, 84 (12), 46–53.

Massachusetts Department of Conservation and Recreation (2008) *2008 Watershed Protection Plan Update, Volume I, The DCR Watershed System*, MA DCR, Division of Water Supply Protection: Office of Watershed Management.

Moran, M.C., Moran, D.C., Cushing, R.S., and Lawler, D.F. (1993) Particle behavior in deep-bed filtration: Part 1–particle detachment, *Journal AWWA*, 85 (12), 82–93.

Morel, F.M.M. (1983) Reactions on solid surfaces, In: *Principles of Aquatic Chemistry*, New York: Wiley-Interscience, pp. 377–433.

MWH (2005), *Water Treatment Design: Principles and Design*, 2nd ed., J.C. Crittenden, R.R. Trussell, D.W. Hand, D.W. Howe, and G. Tchobanoglous, eds., Hoboken, N.J.: John Wiley & Sons.

Nason, J.A. and Lawler, D.F. (2008) Particle size distribution dynamics during precipitative softening: Constant solution composition. *Water Research*, 42 (14), 3667–3676.

New York City Department of Environmental Protection (2009) *Filtration Avoidance Annual Report For the Period for the Period January 1 through December 31, 2008*, New York.

Nolan, B.T., Hit, K.J., and Ruddy, B.C. (2002) Probability of nitrate contamination of recently recharged groundwaters in the conterminous United States, *Environmental Science and Technology,* 36 (10), 2138–2145.

Novotny, V. (2003) *Water Quality: Diffuse Pollution and Watershed Management*, 2nd ed., New York: John Wiley & Sons.

Ogawa, H. and Tanoue, E. (2003) Dissolved organic matter in oceanic waters, *Journal of Oceanography*, 59, 129–147.

Okun, D.A. (1991) Best available source, *Journal AWWA,* 83 (3), 30–32.

Okun, D.A. (1997) Distributing reclaimed water through dual systems, *Journal AWWA,* 89 (11), 52–64.

O'Melia, C.R. (1980) Aquasols—The behavior of small particles in aquatic systems, *Environmental Science and Technology*, 14 (9), 1052–1060.

Palmer, C.M. (1962) *Algae in Water Supplies*, U.S. Public Health Service, Publication No. 657, Washington, D.C.

Perdue, E.M., and Ritchie, J.D. (2003). Dissolved organic matter in fresh waters, In: *Treatise on Geochemistry; Volume 5: Surface and Ground Water, Weathering, Erosion and Soils*, J.I. Drever, ed., San Diego, Calif.: Elsevier.

Plummer, J.D., and Edzwald, J.K. (2001) Effects of ozone on algae as precursors of trihalomethane and haloacetic acid production, *Environmental Science and Technology*, 35 (18), 3661–3668.

Pyke, G.W., Becker, W.C., Head, R., and O'Melia, C.R. (2003) *Impacts of Major Point and Non-Point Sources on Raw Water Treatability*, Denver, Colo.: Water Research Foundation.

Reckhow, D.A. (2010) Natural organic matter, In: *Emerging Problems: Organic By-products of Potential Health Concern*, I.H. Suffet, A. Bruchet, and D. Khiari, eds., Denver, Colo.: Water Research Foundation.

Redfield, A.C., Ketchum, B.H., and Richards, F.A. (1963) The influence of organisms on the chemical composition of seawater, In The Sea: Ideas and Observations of Progress, In: *The Study of the Seas*, vol. 2: *The Composition of Seawater, Comparative and Descriptive Oceanography,* M.N. Hill, ed., New York: Interscience.

Sawyer, C.N., McCarty, P.L., and Parkin, G.F. (1994) *Chemistry for Environmental Engineering,* 4th ed. New York: McGraw Hill.

Schmoll, O., Howard, G., Chilton, J., and Chorus, I., eds. (2006) *Protecting Groundwater for Health: Managing the Quality of Drinking-water Sources*, London: World Health Organization and IWA Publishing.

Sharpley, A.N., Chapra, S.C., Wedepohl, R., Sims, J.T., Daniel, T.C., and Reddy, K.R. (1994) Managing agricultural phosphorus for protection of surface waters– issues and options, *Journal of Environmental Quality,* 23 (3), 437–451.

Stevenson, F.J., ed. (1995) *Humus Chemistry: Genesis, Composition, Reactions*, 2nd ed., New York: John Wiley and Sons.

Stumm, W. and Morgan, J.J. (1981) *Aquatic Chemistry*, 2nd ed., New York: Wiley-Interscience.

Stumm, W. (1977) Chemical interaction in partical(*sic*) separation, *Environmental Science and Technology*, 11, 1066–1070.

Stumm, W., ed. (1987) *Aquatic Surface Chemistry, Chemical Processes at the Particle-Water Interface*, New York: Wiley-Interscience.

Stumm, W. (1992) *Chemistry of the Solid–Water Interface: Processes at the Mineral–Water and Particle–Water Interface in Natural Systems*, New York: Wiley-Interscience.

Suffet, I.H. and MacCarthy, P. (1988) *Aquatic Humic Substances and Their Influence on the Fate and Treatment of Pollutants*, Advances in Chemistry 219. Washington, D.C.: American Chemical Society.

Suffet, I.H. and Rosenfeld, P.E. (2007) The anatomy of odor wheels for odors of drinking water, wastewater, compost and the urban environment, *Water Science and Technology*, 55 (5), 335–344.

Tate, C.H. and Trussell, R.R. (1978) The use of particle counting in developing plant design criteria, *Journal AWWA,* 70 (12), 691–698.

Thomann, R.V. and Mueller, J.A. (1987) *Principles of Surface Water Quality Modeling and Control*, New York: Harper & Row.

Thurman, E.M. (1985) *Organic Geochemistry of Natural Waters,* Dordrecht, The Netherlands: Martinus Nijhoff.

Todd, D.K. and Mays, L.W. (2005) *Groundwater Hydrology,* 3rd ed., New York: John Wiley and Sons.

UNEP-IETC Urban Environment Series 2 (2002). *Rainwater Harvesting and Utilization, an Environmentally Sound Approach for Sustainable Urban Water Management*, by UNEP-DTIE-IETC/Sumida City Government/People for Promoting Rainwater Utilization.

USEPA (U.S. Environmental Protection Agency) (2000) Arsenic Occurrence in Public Drinking Water Supplies, EPA-815-12-00-023, Office of Water, Washington, D.C.

USEPA (2005) Occurrence Assessment for the Final Stage 2 Disinfectants and Disinfection By-Products Rule, EPA 815-R-05-011, Office of Water, Washington, D.C.

Van Benschoten, J.E. (1988) Speciation and Fate of Aluminum in Water Treatment, PhD Dissertation. University of Massachusetts, Amherst.

Van Gelder, A.M., Chowdhury, Z.K., and Lawler, D.F. (1999) Conscientious particle counting, *Journal AWWA,* 91 (12), 64–76.

Verwey, E.J.W. and Overbeek, J.Th.G. (1948) *Theory of the Stability of Lyophobic Colloids,* New York: Elsevier Publishing.

Wiesner, M.R., Hotze, E.M., Brant, J.A., and Espinasse, B. (2008) Nanomaterials as possible contaminants: the fullerene example, *Water Science and Technology*, 57 (3), 305–310.

Water Chemistry References

Benjamin, M.M. (2002) *Water Chemistry,* New York: McGraw-Hill.

Snoeyink, V.L. and Jenkins, D. (1980) *Water Chemistry*, New York: John Wiley & Sons.

Stumm, W. and Morgan, J.J. (1996) *Aquatic Chemistry*, 3rd ed., New York: Wiley-Interscience.

CHAPTER 4
HYDRAULIC CHARACTERISTICS OF WATER TREATMENT REACTORS AND THEIR EFFECTS ON TREATMENT EFFICIENCY

Desmond F. Lawler, Ph.D., P.E.
Professor
Department of Civil, Architectural and Environmental Engineering
University of Texas
Austin, Texas, United States

INTRODUCTION 4.2
CONTINUOUS FLOW REACTORS: IDEAL AND NON-IDEAL FLOW 4.2
TRACER STUDIES 4.3
The Exit Age and Cumulative Age Distributions 4.4
Discrete Data in Tracer Tests 4.8
Choosing a Step or Pulse Input Tracer Test 4.11
Ideal Plug Flow 4.11
Continuous Flow Stirred Tank Reactor 4.13
MATHEMATICAL MODELS FOR NON-IDEAL FLOW 4.16
CFSTRs-in-Series Model 4.17
Dispersion Model 4.20
Other Models 4.22
COMPUTATIONAL FLUID DYNAMICS 4.24
REACTION RATE EXPRESSIONS 4.26
Reactions in Batch Reactors 4.28
REACTIONS IN CONTINUOUS FLOW SYSTEMS AT STEADY STATE: COMBINING HYDRAULICS AND REACTION KINETICS 4.33
Plug Flow Reactors 4.34
Continuous Flow Stirred Tank Reactors 4.36
Comparison of CFSTRs and PFRs 4.36
Non-ideal Reactors with Common Flow Models 4.38
Using the Exit Age Distribution Directly to Estimate Removal 4.40
REACTORS IN WATER TREATMENT AND THEIR HYDRAULICS CHARACTERISTICS 4.43
SUMMARY 4.47
ABBREVIATIONS 4.47
NOTATION FOR EQUATIONS 4.47
REFERENCES 4.49

INTRODUCTION

The treatment processes at the heart of the production of safe drinking water almost always occur in continuous flow reactors. How water flows through those reactors can have a remarkable influence on the treatment efficiency that is achieved. This fact is well understood in the case of disinfection, where it is embedded in the regulations of the Safe Drinking Water Act (SDWA) in the United States. However, the same principles apply for essentially all of the treatment processes in a drinking water treatment plant. In this chapter, we first explain the common descriptions of the hydraulic characteristics of reactors, beginning with ideal flow reactors and proceeding to non-ideal flow reactors; this hydraulic characterization primarily consists of determining the residence time distribution. Subsequently, the ramifications of the different flow characteristics (or different residence time distributions) for some types of chemical reactions are explained, with reference to specific common water treatment objectives.

CONTINUOUS FLOW REACTORS: IDEAL AND NON-IDEAL FLOW

A typical water treatment plant for a surface water pumps water through a pipe from the source (lake, reservoir, river, or stream) to a series of reactors, usually consisting of a rapid mix tank (where chemicals are added to the water for a variety of purposes), flocculation tank(s), sedimentation or flotation tanks, either granular media or membrane filters, and a clearwell that serves both as a disinfection reactor and a flow equalization basin. In most of these cases, all of the water flows into the reactor in a single stream and likewise flows out of the reactor in a single stream. An exception is the sedimentation or flotation unit, which has a single influent stream but two effluent streams, one for the clarified water (which contains most of the water and few particles) and one for the residuals (which contains a small fraction of the water and most of the particles). Because the residuals stream in such reactors is such a small fraction of the water flow, it can often be ignored in the characterization of the overall water flow. Granular media or (dead-end) membrane filters have a periodic backwash that creates a residuals stream, but while treating water, they have only a single effluent. Therefore, for most of the analysis that follows, we consider all reactors to have a single influent and single effluent stream.

Some reactors are designed to promote mixing while others attempt to prevent mixing as much as possible. In this context, mixing is taken to mean the extent to which the water (and any constituents in the water) that comes into the reactor in some small increment of time (say 1 s or 0.001 s) mingles with water that comes into the reactor at any other time (before or after the increment of focus). At one extreme, absolutely no mixing of one entering packet with any other packet occurs, and every packet travels from the influent to the effluent without any exchange of water molecules with any other packet. A reactor with this type of flow pattern is known as a segregated flow reactor; a *plug flow reactor* is a special type of segregated flow reactor in which all packets have the same residence time within the reactor; there is no mixing up of packets in the direction of flow. A plug flow reactor is often thought of as a conveyor belt. Just like boxes on a conveyor belt do not pass one another or exchange material between one another, the packets of water that enter the reactor sequentially stay in the same order and do not exchange water molecules with one another.

At the other extreme, we can imagine complete and instantaneous mixing of every new packet of water entering a reactor with *all* of the water already in the reactor; this type of reactor is known as a *continuous flow stirred tank reactor* (CFSTR) or as a *continuous flow completely mixed reactor* (CFCMR); although the latter term is perhaps more precise, we follow tradition and refer to such a reactor throughout this book as a CFSTR. Because water is not only entering but also constantly being removed from the reactor, some molecules go essentially directly from the influent to the effluent and spend very little time (and, in the limit, zero time) in the reactor. On the other hand, some molecules of water continually get mixed with new incoming water and escape being in the packets of water leaving the reactor for a very long time, far longer than the theoretical detention time (τ) of the reactor, defined as

$$\tau = \frac{V}{Q} \tag{4-1}$$

Here V is the volume of water in the reactor and Q is the volumetric flow rate of water entering and leaving the reactor. Although the distribution of residence times of different molecules of water in this second type of ideal reactor flow is (probably) not intuitively obvious to the reader, quite clearly it is very different from the plug flow reactor where every water molecule stays in the reactor for the same amount of time. The residence time distribution of a CFSTR is developed subsequently.

Between the two ideal extremes of the plug flow reactor (zero mixing in the direction of flow) and the continuous flow stirred tank reactor (complete and instantaneous mixing of the entire contents of the reactor) is a broad spectrum with some, but incomplete or non-instantaneous, mixing of the influent with the water in the reactor. This spectrum constitutes non-ideal flow. Although the ideal extremes are difficult to accomplish in reality, they are useful concepts and can be approached quite closely in real reactors. Throughout the chapter, the behavior in ideal reactors is used to frame that in non-ideal reactors.

TRACER STUDIES

The primary hydraulic characteristic of interest in most cases is the residence time distribution, that is, a description of how long different molecules of water stay in the reactor. If one could somehow label all of the molecules of water that enter the tank over some small increment of time and then observe the time that each labeled molecule leaves the reactor, the resulting information would yield the desired residence time distribution. One could say, for example, that of all the molecules that came into the reactor in some tiny period of time (say, 1 s in a reactor with a theoretical detention time of 2 h), 5 percent came out of the reactor in less than 6 min, an additional 5 percent came out in the next 8 min, and another 7 percent came out in the following 4 min. The same information could be expressed cumulatively by saying that 5 percent stayed in the reactor less than 6 min, 10 percent stayed in the tank less than 14 min, and 17 percent left the reactor within 18 min. If this description were carried out until 100 percent of the labeled water molecules had been accounted for, we would have the complete residence time distribution.

Because it is not easy to label some water molecules in a way that differentiates them from other water molecules, we add a tracer to the influent water; the tracer must be a measurable and nonreactive soluble constituent that travels through the reactor in exactly the way that water does. By measuring the effluent concentration of this tracer as a function of time, we can learn how water molecules travel through the reactor; in essence, the tracer molecules act like the set of labeled water molecules.

Tracer studies can be performed in two primary ways. In a pulse input tracer test, a known mass of tracer (M) is added to the influent of the reactor all at once; the instant of the addition is defined as time zero ($t = 0$), and the concentration of the tracer in the effluent ($C_p(t)$) is monitored for a long time thereafter, until all of the tracer is accounted for. Alternatively, in a (standard) step input tracer test, a steady concentration of tracer (C_{in}) is added to the influent beginning at time zero and continuously thereafter, theoretically forever but practically until the effluent concentration matches the influent concentration. The concentration of tracer in the effluent ($C_S(t)$) is monitored throughout the time period of tracer addition.

In both types of tracer tests, the chosen tracer is usually a constituent that is not naturally in the water, so that the baseline concentration of the effluent is zero; however, this condition is not required. A nonzero (but steady) baseline concentration could be subtracted from all of the effluent concentration measurements to obtain the concentration that is caused by the tracer. This idea can be extended to obtain the desired residence time information from the depletion of tracer concentration after a step input tracer test is ended, as explained subsequently. First, it is necessary to present how the tracer information is mathematically transformed to yield residence time distribution information. Although tracer tests in real systems always involve discrete data points (concentrations at discrete times), the explanations are given first as if the data were continuous functions; how discrete data are manipulated to yield the same information is explained subsequently.

The Exit Age and Cumulative Age Distributions

Consider first a pulse input tracer test on some arbitrary (i.e., non-ideal) reactor; the mass M of tracer is put into the influent of the reactor instantaneously at time zero, and the tracer concentration in the effluent is monitored over time. At and before time zero, the concentration in the effluent is zero. After sufficient time, all of the tracer will have exited the reactor and the effluent concentration will again be zero; in the meanwhile, the effluent concentration would be nonzero. Most real reactors yield results for $C_p(t)$ that show a relatively smooth rise followed by a relatively smooth and continuous fall in the concentration, but the shape of the $C_p(t)$ curve could be almost anything depending on the specific design of the reactor. To emphasize the arbitrariness of the possible response of the reactor, a double-humped curve for $C_p(t)$ is shown in Fig. 4-1a. To keep the number of data points relatively small in this example, data are shown for only every 2.5 min, but more frequent data points, especially in the first 25 min or so where the effluent concentration is changing dramatically, would yield better information.

If another pulse input tracer test were performed on the same reactor at the same flow rate but twice the mass of tracer were put into the reactor, we would expect to see the effluent concentration at every time to be twice that of the original test. Since our interest is in the distribution of times that water, not tracer, stays in the reactor, it is useful to eliminate the effect of the mass of tracer used in the test. To do so, we multiply the effluent concentration values by the flow rate Q and divide by the mass M of tracer input to the reactor. The product of the effluent concentration (mass/volume) and the flow rate (volume/time) is the instantaneous rate (at time t) at which the mass of tracer is leaving the reactor (i.e., dimensions of mass/time). The subsequent division by the total input mass then yields the instantaneous fractional rate at which the input mass is leaving the reactor, with dimensions of inverse time. This normalization is so useful that we give it its own symbol, $E(t)$; that is,

$$E(t) = \frac{Q}{M} C_p(t) \tag{4-2}$$

The set of all values of $E(t)$ is called the *exit age distribution*. If the value of Q/M for the experiment under consideration is $0.002 \text{ L/mg-min} = 0.002 \text{ m}^3\text{/g-min}$, the resulting $E(t)$ curve is shown in Fig. 4.1b. (The origin of this value of Q/M is explained subsequently.)

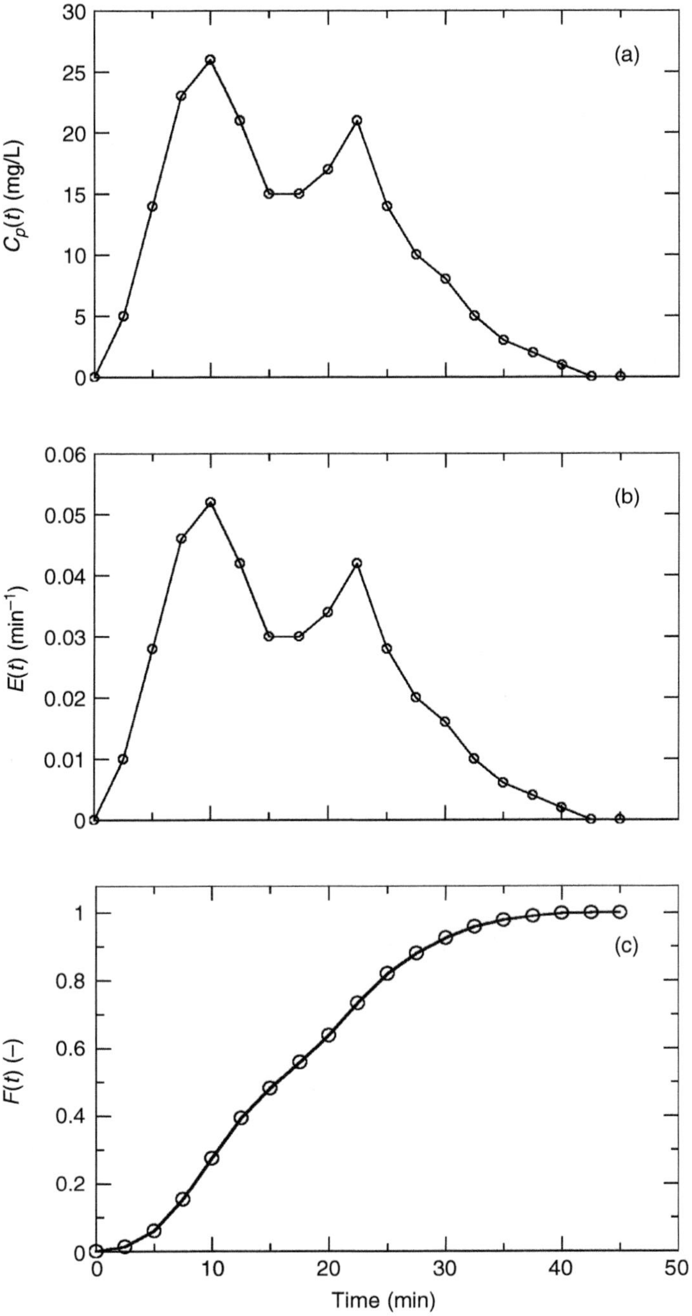

FIGURE 4-1 Pulse input tracer test: (*a*) effluent concentration profile; (*b*) exit age distribution; (*c*) cumulative age distribution.

The exit age distribution describes the distribution of times that *water* stays in the reactor; that is, the fraction of water from any aliquot of influent that stays in the reactor between any time t and $t+dt$ is $E(t)dt$. For example, the value of $E(t)$ in Fig. 4-1b at 15 min is 0.03 min^{-1}, and if we assume that that value is constant for the 1 min from 14.5 to 15.5 min (i.e., $dt \approx \Delta t = 1$ min), then $E(t)dt \approx E(t)\Delta t = (0.03\,\text{min}^{-1})(1\,\text{min}) = 0.03$. That is, 3 percent of the water that enters the reactor at one instant will stay in the reactor between 14.5 and 15.5 min. Likewise, if we take a sample of the effluent at some instant, we can say that 3 percent of the water in that sample stayed in the tank between 14.5 and 15.5 min, or in other words, that 3 percent of the water in that sample came into the reactor between 14.5 and 15.5 min earlier. The exit age distribution describes the distribution of residence times that different molecules of water stay in the reactor.

Often, it is more useful to know what fraction of the water stays in the reactor less than a certain accumulated time (say 15 min) than to know how much stays in for some incrementally small time. That information is contained in the dimensionless *cumulative age distribution* $F(t)$, which is found as the integration of $E(t)$ from time zero to time t; that is,

$$F(t) = \int_0^t E(t)\,dt \tag{4-3}$$

For the experiment under consideration, the values of $F(t)$ are shown in Fig. 4-1c. As can be seen from the example in the figure, $F(t)$ has the characteristics that it starts at zero and rises continuously toward the ultimate value of 1. At $t = 15$ min, the value of $F(t)$ can be seen to be just less than 0.5 (the precise value is 0.483), which means that 48.3 percent of the water that enters this reactor stays in the reactor for less than 15 min.

All of the tracer put into the reactor at time zero in a pulse tracer test eventually exits the reactor; that is, 100 percent of the tracer stays in the reactor for less than infinite time. Mathematically, this fact means that the ultimate value of $F(t)$ is 1 (unity) and that the area under the entire $E(t)$ distribution is also 1. That is,

$$\int_0^\infty E(t) = F(\infty) = 1 \tag{4-4}$$

In real systems, of course, a practical definition of time infinity might be 4 to 15 times the detention time, with the smaller part of this range being acceptable for reactors that are nearly plug flow. The criterion for stopping the monitoring in a pulse input test is that no appreciable mass of tracer remains in the reactor.

A step input tracer test ultimately yields the same information for the exit age distribution $E(t)$ and the cumulative age distribution $F(t)$, but the raw data, and therefore the analysis, are different from the pulse input case. For the standard step input tracer test (in which the baseline concentration of the tracer in the influent water and throughout the reactor is zero), the effluent concentration ($C_s(t)$) always increases with time because water that came into the reactor prior to the initiation of the test (with no tracer) is replaced with water that entered after the start of the test (with concentration C_{in}). An example of the effluent concentration is shown in Fig. 4-2a using an influent concentration of tracer of 10 mg/L; the reactor is the same as used in the example for the pulse input.

A unit volume of effluent with concentration $C_S(t)$ is a mixture of a fraction of water (i.e., $F(t)$) that came into the reactor less than time t ago with concentration C_{in} and the rest ($1 - F(t)$) that came in earlier with concentration zero. That is, the cumulative age distribution $F(t)$ is found by normalizing the effluent concentration by the influent concentration.

$$F(t) = \frac{C_S(t)}{C_{\text{in}}} \tag{4-5}$$

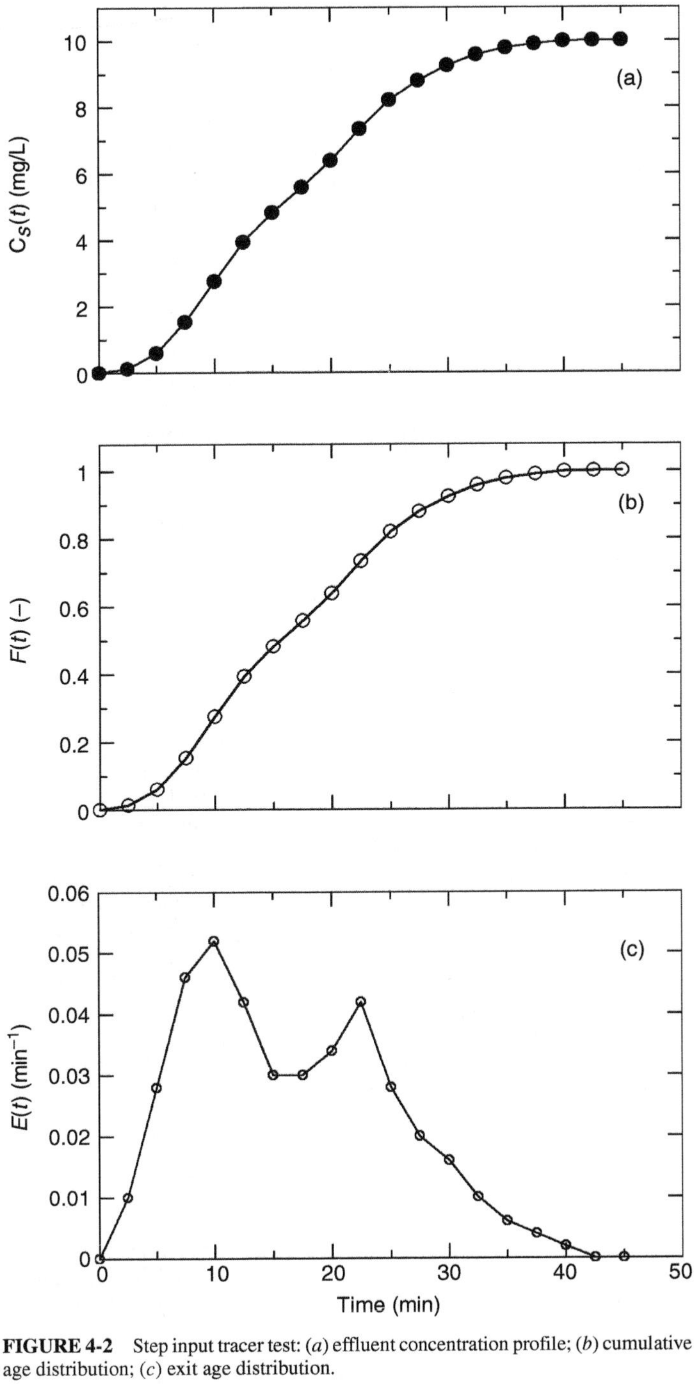

FIGURE 4-2 Step input tracer test: (*a*) effluent concentration profile; (*b*) cumulative age distribution; (*c*) exit age distribution.

The results of this normalization are shown in Fig. 4-2b. As before, $F(t)$ has the characteristics of being zero at time zero, rising continuously over time, and eventually reaching a value of 1.

When the baseline concentration in the influent and throughout the reactor prior to time zero is not zero, a sample of effluent with concentration $C_S(t)$ is a mixture of a fraction ($F(t)$) that came in less than time t ago with concentration C_{in} and the rest $(1 - F(t))$ that came in with concentration C_{baseline}. The result is that the baseline concentration must be subtracted from both the numerator and denominator to obtain the cumulative age distribution. That is,

$$F(t) = \frac{C_S(t) - C_{\text{baseline}}}{C_{\text{in}} - C_{\text{baseline}}} \tag{4-6}$$

Equation 4-6 is completely general and can be used for any (properly performed) step input test, including the possibility that $C_{\text{baseline}} > C_{\text{in}}$. Equation 4-5 is a simplification of Eq. 4-6, where $C_{\text{baseline}} = 0$. At the end of a step tracer test with $C_{\text{baseline}} = 0$, the reactor has the concentration equal to C_{in} throughout the reactor, so it is possible to perform a second test by continuing the effluent monitoring after the influent concentration has been returned to zero (at a time designated as a new time zero). In that case, the new baseline is the old C_{in} and the new C_{in} is zero. Regardless of how a step input tracer test is performed, the criterion for being able to stop the monitoring is that the difference between the influent and effluent concentrations is minimal or immeasurable.

As noted in Eq. 4-3, the cumulative age distribution is the integral of the exit age distribution; conversely, the exit age distribution is the derivative of the cumulative age distribution at any time t. That is,

$$E(t) = \frac{dF(t)}{dt} \tag{4-7}$$

The data for $F(t)$ for Fig. 4-2b were analyzed according to (a discretized version of) Eq. 4-7 to find the $E(t)$ values for this reactor, and the resulting values are plotted in Fig. 4-2c. As noted above, this step input tracer study was performed on the same reactor as the pulse input test shown earlier; as a result, the exit age distributions and cumulative age distributions for the two tests shown in Figs. 4-1 and 4-2 are identical. Both types of tests ultimately yield the identical information, and therefore, the type test that is performed is a matter of choice (perhaps influenced by issues of measurement and precision, discussed subsequently).

Discrete Data in Tracer Tests

The conceptual understanding of the exit age and cumulative age distributions presented above presupposes that these distributions are continuous functions and that the tracer data are likewise continuous. However, the data in real tracer tests are always obtained at discrete times, and the figures given above indicate that samples were taken at discrete times. In this section, we explain how to analyze discrete data from both pulse and step input tracer tests to obtain the desired age distributions. The data for a pulse test used to generate Fig. 4-1 are shown as a spreadsheet in Fig. 4-3. The raw data from the test—the times that samples were taken and the corresponding values of $C_p(t)$—are given in columns A and B, respectively.

In column C, the average concentration in the time interval between adjacent samples is shown in the row corresponding to the end of the time period and is calculated as

$$C_{ave,i} = \frac{C_{i-1} + C_i}{2} \tag{4-8}$$

	A	B	C	D	E	F	G
1	Time	$C_p(t)$	Cave	Cave* Δt	$E(t)$	Eave	$F(t)$
2	(min)	(mg/L)	(mg/L)	(mg-min/L)	(1/min)	(1/min)	(–)
3	0.0	0			0.000		0.000
4	2.5	5	2.5	6.25	0.010	0.005	0.013
5	5.0	14	9.5	23.75	0.028	0.019	0.060
6	7.5	23	18.5	46.25	0.046	0.037	0.153
7	10.0	26	24.5	61.25	0.052	0.049	0.275
8	12.5	21	23.5	58.75	0.042	0.047	0.393
9	15.0	15	18.0	45.00	0.030	0.036	0.483
10	17.5	15	15.0	37.50	0.030	0.030	0.558
11	20.0	17	16.0	40.00	0.034	0.032	0.638
12	22.5	21	19.0	47.50	0.042	0.038	0.733
13	25.0	14	17.5	43.75	0.028	0.035	0.820
14	27.5	10	12.0	30.00	0.020	0.024	0.880
15	30.0	8	9.0	22.50	0.016	0.018	0.925
16	32.5	5	6.5	16.25	0.010	0.013	0.958
17	35.0	3	4.0	10.00	0.006	0.008	0.978
18	37.5	2	2.5	6.25	0.004	0.005	0.990
19	40.0	1	1.5	3.75	0.002	0.003	0.998
20	42.5	0	0.5	1.25	0.000	0.001	1.000
21	45.0	0	0.0	0.00	0.000	0.000	1.000
22							
23			sum(Cave* Δt)	500			

FIGURE 4-3 Spreadsheet for handling pulse input tracer test data.

So, for example, the value of 9.5 mg/L shown in cell C5 is the average of 5 and 14 mg/L shown in cells B4 and B5, respectively. In the subsequent column (col. D), the average concentrations are multiplied by the associated time interval ($\Delta t = t_i - t_{i-1}$); in this example, all of the time intervals are 2.5 min, although the time intervals are often varied to obtain more data when the changes are rapid. The value of 23.75 mg-min/L in cell D5 is obtained as 9.5 mg/L × 2.5 min. The sum of all of the values in column D from the entire test is given in cell D23. This value is found because, when multiplied by the volumetric flow rate, the product is the mass of tracer accounted for in the effluent throughout the tracer test; that is,

$$M_{\text{out}} = Q \sum_{\text{all } i} C_{ave,i} \, \Delta t_i \tag{4-9}$$

This calculation constitutes a quality control on the entire tracer test, because the resulting value can be checked against the known value of the mass input to the reactor. If M_{in} and M_{out} are not in close agreement, the cause should be determined before using the tracer data to calculate the age distributions and ultimately to make treatment decisions. In some cases, one might decide that using M_{out} rather than M_{in} in normalizing from $C_p(t)$ to $E(t)$ is a more accurate approach. In the generation of the exit age distribution in this spreadsheet and in Fig. 4-1, it was assumed that $M_{\text{in}} = M_{\text{out}}$, so that

$$\frac{Q}{M} = \frac{Q}{Q \sum_{\text{all } i} C_{p,\,ave,i} \, \Delta t_i} = \frac{1}{\sum_{\text{all } i} C_{p,\,ave,i} \, \Delta t_i} = \frac{1}{500 \text{ mg-min/L}} = 0.002 \frac{\text{L}}{\text{mg-min}}$$

The values for $E(t)$ shown in column E in Fig. 4-3 are the product of this value and the values of $C_p(t)$ in column B; the resulting units for $E(t)$ are min^{-1}.

The values for $E_{ave,\,i}$ shown in column F are calculated from the values in column E in the same way that the $C_{ave,i}$ values in column C are calculated from column B. Finally, after recognizing that the value of $F(0) = 0$, we calculate the subsequent values of $F(t)$ in column G as

$$F_i = F_{i-1} + (E_{ave,i})(t_i - t_{i-1}) \tag{4-10}$$

So, for example, the value in cell G5 is found as 0.013 + (0.019)(2.5) = 0.060; the value of 0.013 shown in cell G4 is rounded from the precise 0.0125, and the spreadsheet uses this more precise value.

The analysis of discrete data for a step input tracer test is shown in Fig. 4-4, with the raw data of time and measured effluent concentration shown in columns A and B. As indicated by Eq. 4-5 (or Eq. 4-6 with $C_{\text{baseline}} = 0$), the values of $F(t)$ in column C are calculated by dividing the values of $C_S(t)$ in column B by C_{in}, which is 10 mg/L in this case.

Next, the estimate of the derivative of $F(t)$ over a time interval is calculated in column D as follows:

$$\left(\frac{dF}{dt}\right)_i \approx \left(\frac{\Delta F}{\Delta t}\right)_i = \frac{F_i - F_{i-1}}{t_i - t_{i-1}} \tag{4-11}$$

So, for example, the value of 0.019 in cell D5 is calculated as (0.060 − 0.013)/(5.0 − 2.5) min^{-1}. In reality, the values in column D are average values of $E(t)$ over the interval from t_{i-1} to t_i and are shown in the spreadsheet associated with the end of the interval. After recognizing that the initial value of $E(t)$ equals zero, the remaining values of $E(t)$ for the specific values of t are calculated in column E as

$$E(t_i) = 2E_{ave,i} - E_{i-1} \tag{4-12}$$

	A	B	C	D	E
1	Time	$C_S(t)$	$F(t)$	dF/dt	$E(t)$
2	(min)	(mg/L)	(–)	(=Eave)	(1/min)
3	0.0	0.00	0.000		0.000
4	2.5	0.13	0.013	0.005	0.010
5	5.0	0.60	0.060	0.019	0.028
6	7.5	1.53	0.153	0.037	0.046
7	10.0	2.75	0.275	0.049	0.052
8	12.5	3.93	0.393	0.047	0.042
9	15.0	4.83	0.483	0.036	0.030
10	17.5	5.58	0.558	0.030	0.030
11	20.0	6.38	0.638	0.032	0.034
12	22.5	7.33	0.733	0.038	0.042
13	25.0	8.20	0.820	0.035	0.028
14	27.5	8.80	0.880	0.024	0.020
15	30.0	9.25	0.925	0.018	0.016
16	32.5	9.58	0.958	0.013	0.010
17	35.0	9.78	0.978	0.008	0.006
18	37.5	9.90	0.990	0.005	0.004
19	40.0	9.98	0.998	0.003	0.002
20	42.5	10.00	1.000	0.001	0.000
21	45.0	10.00	1.000	0.000	0.000

FIGURE 4-4 Spreadsheet for handling step input tracer test data.

For example, the value of 0.028 in cell E5 is calculated from the values in cells D5 and E4 as 2(0.019) – 0.010. The values in columns B, C, and E for $C_S(t)$, $F(t)$, and $E(t)$, respectively, were plotted against the time values in column A to make the plots shown in Fig. 4-2.

Choosing a Step or Pulse Input Tracer Test

As shown above, both types of tracer tests ultimately yield the same information, so the question of which type of test to perform is a matter of choice. The monitoring of the effluent is essentially the same in both cases, although a step input test has the advantage that it usually requires measurements over a much narrower range of concentrations. The ability to put tracer into the influent either as a pulse or as a step often dictates the choice. For a pulse test, a substantial mass of tracer must be entered in a very short period of time (theoretically, instantly) into the influent in the same way that water enters the reactor; the design of a reactor and its influent structure often makes this impossible. For a step test, the tracer is added continuously to the influent in a small flow rate (relative to the influent flow); it is often easier to accomplish this type of addition with either the existing facility or a slight modification. Many tracer tests in water treatment plants in the United States are undertaken in response to current disinfection regulations that require determining the time (usually designated as T_{10}) at which $F(t) = 0.10$; although this information is available from either type of test, it is far simpler in terms of data handling to find it from a step input test. In some cases, it is convenient to simply turn off the feed of a chemical, such as fluoride which is added continually and is nonreactive,[1] to perform a test in which the concentration decreases from the normal (post addition) value to the natural baseline value; when normal operation is resumed, the effluent concentration will rise and can continue to be monitored as a second tracer test. For all of these reasons, step input tracer tests are performed more often than pulse input tests, although site specific considerations might dictate the opposite choice in some cases. Teefy (1996) delineated many of the issues involved in choosing both the tracer material and the type of tracer test to perform.

Ideal Plug Flow

From the description of a plug flow reactor above, the results of tracer studies on these ideal reactors can be predicted intuitively. In a pulse test, all of the mass that is put into the reactor at time zero travels together from one end to the other, and it all leaves together at the time equal to the theoretical detention time τ. No tracer comes out of the reactor prior to $t = \tau$, nor does any come out afterward; the entire spike of tracer exits the reactor at exactly $t = \tau$. Because M_{in} is not mixed with surrounding fluid, it is difficult to convert to concentration units, but the fact that all of the water that comes in at one time comes out of the reactor at a time τ later leads directly to the exit age distribution shown in Fig. 4-5 (with an assumed theoretical detention time τ of 20 min). The function shown is known as the Dirac delta function and has the following properties: a value of zero everywhere except one time, the value of infinity at that time, and an area under the curve (from time zero to infinity, and therefore under the spike) of 1. In terms of the water age distribution, we can interpret this function in three ways. First, all of the water that enters at one time exits exactly τ later. Second, all of the water that is in a sample of the effluent came into the reactor τ earlier.

[1]Fluoride is reactive with aluminum, so plants that use aluminum-based coagulants cannot use fluoride as a tracer until after filtration. Other commonly considered choices for the tracer include lithium, sodium, and chloride.

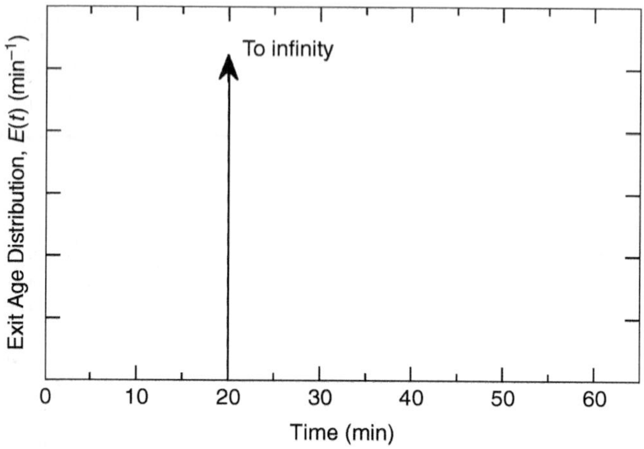

FIGURE 4-5 Exit age distribution for a plug flow reactor with theoretical detention time of 20 min.

Third, the probability that a molecule of water will stay in the tank for a time equal to τ is 1, and the probability that it will stay in the reactor for a time less than or greater than τ is 0.

Translating the exit age distribution to the cumulative age distribution means integrating $E(t)$ from 0 to time t, as indicated in Eq. 4-3. For the plug flow reactor under consideration, the value of $F(t)$ is 0 for all $t < \tau$ and 1 for all $t \geq \tau$. Using the same 20-min detention time indicated in Fig. 4-5, the cumulative age distribution $F(t)$ for the plug flow reactor is shown in Fig. 4-6.

The interpretations of the cumulative age distribution are similar to those of the exit age distribution. First, all of the molecules of water in a sample of the influent will stay in

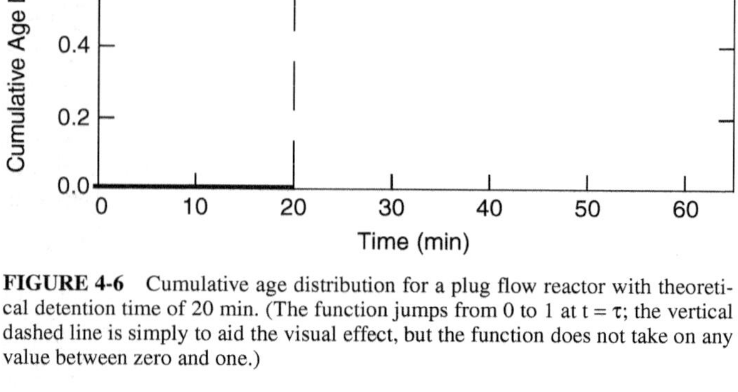

FIGURE 4-6 Cumulative age distribution for a plug flow reactor with theoretical detention time of 20 min. (The function jumps from 0 to 1 at $t = \tau$; the vertical dashed line is simply to aid the visual effect, but the function does not take on any value between zero and one.)

the reactor for exactly τ. Second, all of the molecules of water in a sample of the effluent came into the reactor exactly τ earlier. Finally, the cumulative probability that a molecule of water will stay in the reactor for a time less than τ is 0, and the cumulative probability that a molecule of water will stay in the reactor for a time equal to or greater than τ is 1.

Continuous Flow Stirred Tank Reactor

A drawing depicting a pulse input tracer test being performed on a continuous flow stirred tank reactor (CFSTR) is shown in Fig. 4-7. The flow rate Q must be steady throughout the test, and except for the instant at time zero when the mass M_{in} of tracer is input into the reactor, the influent flow has no tracer in it. Because the contents of the reactor are completely mixed at any instant, the concentration in the effluent at any time is the same as that anywhere in the reactor at that time; that concentration is designated as $C_p(t)$.

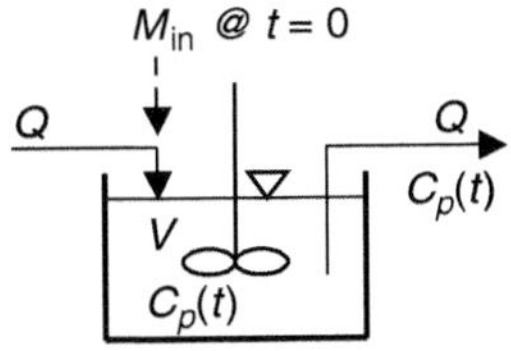

FIGURE 4-7 Schematic diagram for a pulse input tracer test on a CFSTR.

At the initiation of the pulse input test (i.e., at $t = 0^+$), the mass M_{in} is instantaneously mixed throughout the entire volume of the reactor, leading to a concentration throughout the reactor of

$$C_p(0) = C_0 = \frac{M_{in}}{V} \tag{4-13}$$

Because the concentration is the same throughout the reactor at any time (by the definition of a CFSTR), the control volume for a mass balance for the tracer is the entire (water) volume of the reactor. No transfer of mass occurs by dispersion or diffusion across these boundaries, so the mass balance valid for any time after time zero is

$$\begin{bmatrix}\text{Rate of change}\\ \text{of mass of}\\ \text{tracer in the}\\ \text{reactor}\end{bmatrix} = \begin{bmatrix}\text{Rate of advective}\\ \text{input of mass}\\ \text{of tracer}\\ \text{into the reactor}\end{bmatrix} - \begin{bmatrix}\text{Rate of advective}\\ \text{output of mass}\\ \text{of tracer}\\ \text{from the reactor}\end{bmatrix} + \begin{bmatrix}\text{Rate of change}\\ \text{of mass of tracer}\\ \text{in the reactor}\\ \text{due to reactions}\end{bmatrix} \tag{4-14}$$

The first term on the right side is 0 because no tracer is put into the reactor after time zero. Since the tracer is nonreactive, the final term on the right side can be omitted, and the remaining word equation can be translated into symbols as follows:

$$\frac{d(VC_p(t))}{dt} = -QC_p(t) \tag{4-15}$$

Because the volume is constant, it can be taken out of the differential, and both sides of the equation can be divided by V to obtain

$$\frac{dC_p(t)}{dt} = -\frac{Q}{V}C_p(t) = -\frac{1}{\tau}C_p(t) \tag{4-16}$$

When integrated with the initial condition shown in Eq. 4-13, the response of a CFSTR to a pulse input tracer test is found to be

$$C_{p,CFSTR}(t) = C_0 \exp\left(-\frac{t}{\tau}\right) \tag{4-17}$$

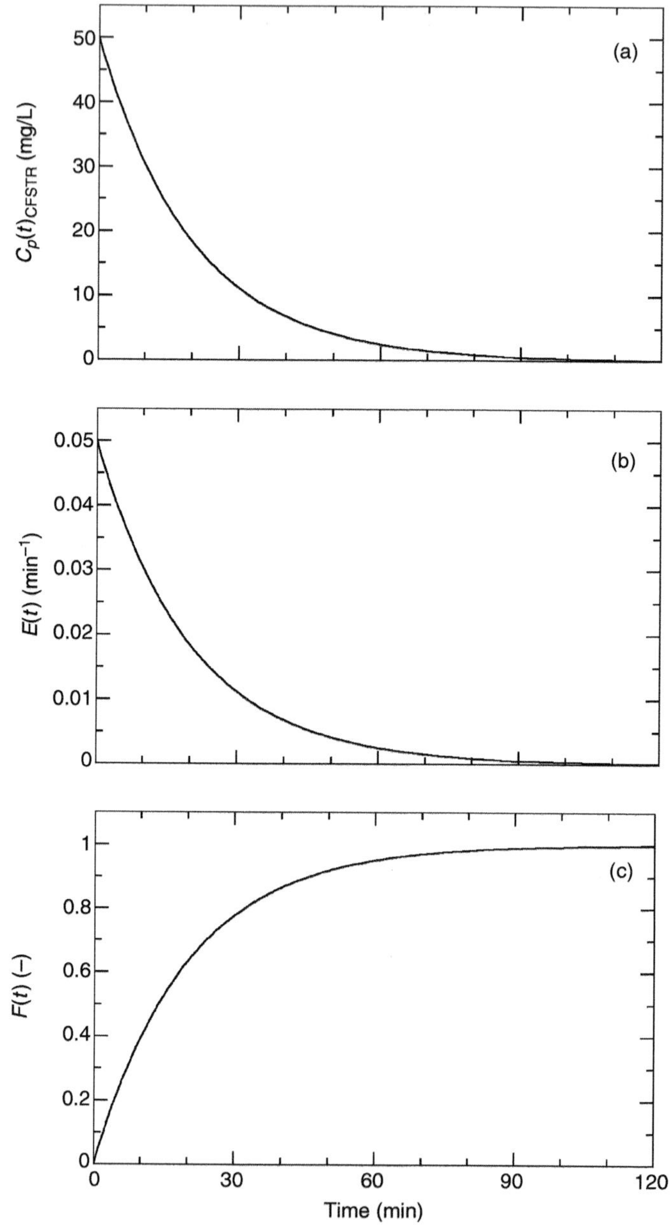

FIGURE 4-8 Pulse input tracer test on a CFSTR: (*a*) effluent concentration profile; (*b*) exit age distribution; (*c*) cumulative age distribution. [Arbitrary choices for this example include a 20 min theoretical detention time (both affects the abscissa of all three plots) and a value of C_o equal to 50 mg/L (both affects Fig. 4-8a only)]

As indicated in Eq. 4-2, the exit age distribution can then be found as

$$E(t)_{CFSTR} = \frac{Q}{M_{\text{in}}} C_{p,CFSTR}(t) = \frac{Q}{M_{\text{in}}} C_o \exp\left(-\frac{t}{\tau}\right) = \frac{Q}{M_{\text{in}}} \frac{M_{\text{in}}}{V} \exp\left(-\frac{t}{\tau}\right) \tag{4-18}$$

or

$$E(t)_{CFSTR} = \frac{1}{\tau} \exp\left(-\frac{t}{\tau}\right) \tag{4-19}$$

The cumulative age distribution $F(t)$ is then found by integration of $E(t)$ to be

$$F(t)_{CFSTR} = \int_0^t E(t)dt = \frac{1}{\tau}\int_0^t \exp\left(-\frac{t}{\tau}\right)dt = \frac{1}{\tau}(-\tau)\exp\left(-\frac{t}{\tau}\right)\Big|_0^t \tag{4-20}$$

$$F(t)_{CFSTR} = 1 - \exp\left(-\frac{t}{\tau}\right) \tag{4-21}$$

Graphs of $C_p(t)$, $E(t)$, and $F(t)$ for a CFSTR with a 20-min detention time are shown in Fig. 4-8; the value of C_0 = 50 mg/L and the detention time were chosen arbitrarily for this example. Since all three functions are mathematically related, all yield essentially the same information. The shape of the $C_p(t)$ and $E(t)$ curves indicate that some tracer or water stays in the reactor for a very short time—the highest point of these functions is at $t = 0$; on the other hand, the exponential decay indicates that some water (or tracer) stays in the tank for extremely long times. The data for $F(t)$ indicate that 10 percent of the water that enters the reactor at one moment exits the reactor within approximately one-tenth of the detention time (i.e., $F(t) = 0.10$ at $t = 0.109\tau$, or just over 2 min in the example shown). Alternatively, we can say that approximately 10 percent of the water in any sample of effluent entered the reactor less than one-tenth of the detention time earlier. At 1 detention time, $F(t) = 0.632$; this means that 63.2 percent of the water stays in the reactor for less than one detention time. On the other hand, some water stays in the tank for a long time; it takes 3 detention times for 95 percent of the water that comes in at one time to leave the reactor, and it takes 4.5 detention times for 99 percent of the water to have left. As shown subsequently, the fact that a fraction of the water escapes after spending little time in the reactor has some significant negative consequences on the extent of reaction (treatment) for most reactive substances, and these consequences are not fully offset by the fact that some water stays in the reactor for a long time.

In a step input tracer test, the tracer is put into the influent of the reactor starting at $t = 0$ and continuing forever, as indicated above. A schematic of such a test being performed on a CFSTR is shown in Fig. 4-9; the picture is the same as that for a pulse test except for the continuous input. The word equation for the mass balance is identical to that for the pulse input, but the translation into symbols now must account for the steady input, as follows:

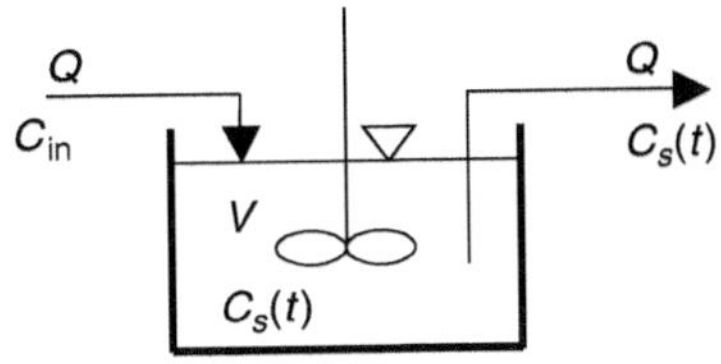

FIGURE 4-9 Schematic diagram for a step input tracer test on a CFSTR.

$$\frac{d(VC_S(t))}{dt} = QC_{\text{in}} - QC_S(t) \tag{4-22}$$

As before, recognizing that the volume V is constant and can be taken out of the differential, then dividing by that volume and substituting the definition of the theoretical detention time, we obtain

$$\frac{d(C_S(t))}{dt} = \frac{1}{\tau}C_{\text{in}} - \frac{1}{\tau}C_S(t) \tag{4-23}$$

The initial condition for this step input situation is that $C_S(0) = C_{\text{baseline}}$; we assume again that this value is zero, the most common case. This initial condition is quite different from that for the pulse input because the amount of tracer put into the reactor in the instant surrounding time zero is different. In a small time period (Δt) after initiating the step input, the small mass of tracer that has been put into the reactor ($QC_{\text{in}}\Delta t$) is diluted into the entire reactor volume. As Δt approaches zero, the concentration in the reactor (and therefore in the effluent) approaches zero (or more generally, C_{baseline}). In the limit, $C_S(0) = 0$.

Integration of Eq. 4-23 with the stated initial condition leads to

$$C_S(t) = C_{\text{in}}\left[1 - \exp\left(-\frac{t}{\tau}\right)\right] \tag{4-24}$$

Invoking Eq. 4-5, we find the cumulative age distribution $F(t)$ as follows:

$$F(t) = \frac{C_S(t)}{C_{\text{in}}} = 1 - \exp\left(-\frac{t}{\tau}\right) \tag{4-25}$$

Finally, taking the derivative of $F(t)$ yields the exit age distribution.

$$E(t) = \frac{dF(t)}{dt} = \frac{1}{\tau}\exp\left(-\frac{t}{\tau}\right) \tag{4-26}$$

The exit age and cumulative age distributions determined from a step input test are identical to those determined from the pulse test. This result is inevitable, since these functions describe how water flows through the tank and that is independent of what kind of tracer test is performed. Hence, the $E(t)$ and $F(t)$ curves shown in Fig. 4-8b and c are valid for the step input tracer test. A graph of $C_S(t)$ for the CFSTR (not shown) would have the same shape as the $F(t)$ curve, except the asymptote would be C_{in} instead of 1.

MATHEMATICAL MODELS FOR NON-IDEAL FLOW

Most real reactors do not meet the ideals of either a plug flow reactor (PFR) (no mixing in the direction of flow) or a CFSTR (complete and instantaneous mixing). Water enters most reactors from a pipe or open channel or directly from a previous reactor through a shared wall. In most cases, the influent structure is designed to distribute that flow across a broad area, either horizontally or vertically or both. That distribution of the flow is accomplished by baffles, multiple small entry ports, or a combination of the two. Similarly, the effluent end of most reactors is designed to remove water from a broad area through submerged or overflow weirs, or again by multiple small holes through a wall or, in the case of granular media filters, the floor. Within the main body of the reactor, that is, between the influent and effluent structures, some reactors (rapid mix tanks, flocculators) have mechanical mixers whereas others (sedimentation tanks or filters) have no intentional mixing.

The efforts to distribute the flow over a broad area at the influent end and withdraw water over a broad area at the effluent end are almost inevitably imperfect. The velocity through an entry or exit port of, say, a concrete wall, will not be evenly distributed across that port because friction at the surface will impose a zero velocity there (the "no-slip" condition) while the velocity at the center of the port will be necessarily greater than the average. Just inside the reactor, the velocity variation will be dissipated to some degree at the influent end by the viscosity of the water, and turbulent eddies will likewise dissipate as they are broken into smaller and smaller sizes. Similarly at the effluent end, some variation in the velocity will occur as the water approaches the effluent weir. In exposed reactors, the wind can cause a higher depth of water above the weir at one end (and therefore more flow) in comparison to the other end of the weir. Excellent design can result in these variations only affecting the flow a short distance into the reactor at both the influent and effluent, but nonideal flow is nevertheless expected.

Although the experimentally determined exit age distribution and the cumulative age distribution are the most accurate descriptors of the flow patterns in a reactor, they require a lot of data in the case of non-ideal flow. It is often more convenient to characterize the flow pattern using a mathematical model with a small number of parameters, especially if one uses the model of the hydraulic characteristics in conjunction with reaction kinetics to predict (or describe) treatment performance. Common models that are useful in water treatment practice are explained below; these models, as well as some other possible models, are also discussed in Levenspiel (1999), Clark (1996), and MWH (2005).

CFSTRs-in-Series Model

A single reactor might be physically or conceptually divided into several reactors in series. Baffles, for example, provide a physical division of a single reactor into somewhat separated compartments. But any reactor might behave, even without baffles, as if the water flowed from one part of the reactor to another with some mixing in each part. If we consider a single reactor with a (total) theoretical detention time τ as if it were divided into N equal compartments and assume that each of those compartments is completely mixed (i.e., acts as a CFSTR), the exit age and cumulative age distributions can be developed mathematically with the following results:

$$E(t) = \frac{1}{\tau}\frac{N^N}{(N-1)!}\left(\frac{t}{\tau}\right)^{N-1}\exp\left(-\frac{Nt}{\tau}\right) \tag{4-27}$$

$$F(t) = 1-\exp\left(-\frac{Nt}{\tau}\right)\sum_{i=1}^{N}\frac{1}{(i-1)!}\left(\frac{Nt}{\tau}\right)^{i-1} \tag{4-28}$$

Graphs showing these distributions for a few selected values of N are shown in Fig. 4-10. For the exit age distribution in Fig. 4-10a, the plot is a dimensionless version of the exit age distribution, obtained by multiplying both sides of Eq. 4-27 by τ. The area under each of these curves is 1, as it is for the usual (dimensional) $E(t)$ curve. When $N = 1$, the reactor is a true CFSTR, and the lines in both parts of Fig. 4-10 are the same as those shown in Fig. 4-8. As N increases from 1 to 10, the distributions gradually become narrower ($E(t)$) or sharper in their rise ($F(t)$). In this model, a value of $N = \infty$ is equivalent to a plug flow reactor, which would be the Dirac delta function on the $E(t)$ curve and the step function on the $F(t)$ curve, as shown in Figs. 4-5 and 4-6, respectively. Focusing on the cumulative age distribution, it is clear that, for values of $F(t) < 0.5$, the time to reach that value increases with increasing value of N, or in other words, with increasing baffling in the reactor. Similarly, for values of $F(t) > 0.7$, the time to reach that value decreases with increasing value of N. That is, a sample of effluent taken from a reactor with a high degree of baffling (i.e., a high value of N) will have a relatively narrow distribution of residence times, meaning that a high fraction of the water stayed in the reactor for a small range of times surrounding the theoretical detention time.

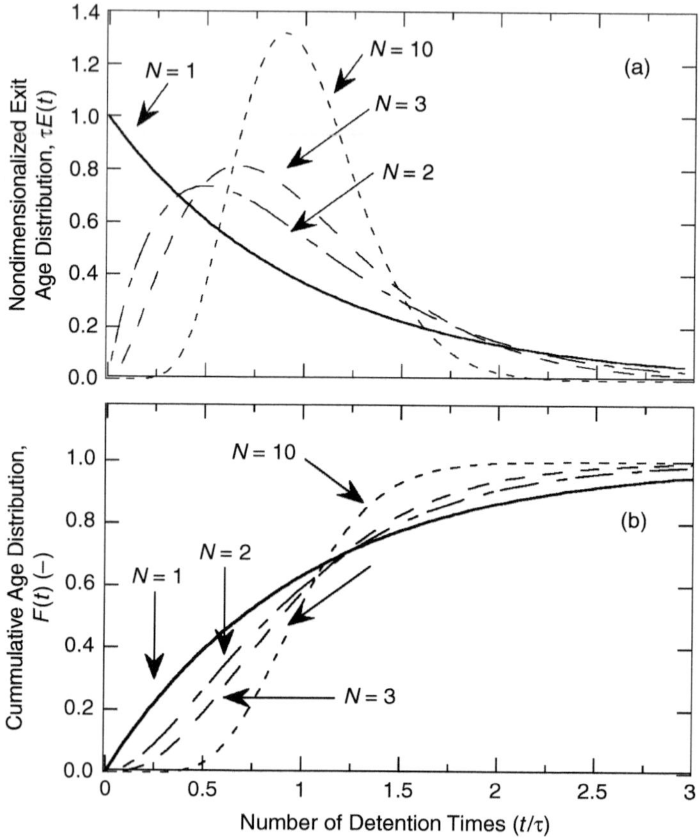

FIGURE 4-10 Age distributions for the CFSTRs in series model (N is the number of equal compartments or CFSTRs that a single real reactor is divided into).

Of particular interest with respect to current regulations are the times associated with $F(t) = 0.10$, that is, the values of T_{10} or T_{10}/τ. For the N values of 1, 2, 3, and 10 shown in Fig. 4-10, the corresponding values of T_{10}/τ (often referred to as the baffling factor) are 0.105, 0.266, 0.367, and 0.622. An open reactor without ostensible mixing or baffling but with inlet and outlet structures that spread the flow over a wide area (e.g., a sedimentation tank) is likely to behave as two or three CFSTRs in series. Rapid mix tanks are likely to come close to behaving as a single ideal CFSTR. Packed bed reactors, such as filters or adsorption columns, act nearly like plug flow reactors; their N value, if modeled with this CFSTRs-in-series model, is likely to be 20 or higher.

Fitting the CFSTRs-in-Series Model to Tracer Data. The equations for the CFSTRs-in-series model have only two parameters: the (total) detention time and the number N of reactors in series. The advantage of mathematical modeling is encapsulated in that fact; instead of having all of the data obtained in a tracer test, one can describe the hydraulic behavior with only two parameters. The model is never (or at least extremely rarely) a perfect fit to the data but, if the fit is good, the slight loss of information by the lack of a perfect fit is often more than made up for by the simplicity of two parameters. We address here a method to find the values of those two parameters to fit the exit age distribution in a way that gives the same average spread. Depending on the task, other methods (such as finding the best fit to the eye or the one that gives the lowest sum of squared residuals) may often be best. But the task of finding equal spread (variance) is a relatively straightforward method that can be achieved in a spreadsheet.

Although the theoretical development of the model uses the theoretical detention time τ, the model should be fit by using the average detention time as determined from the tracer data. After finding the exit age distribution from the tracer test data (as explained above), the average hydraulic detention time $\bar{t}$ can be found as follows:

$$\bar{t} = \frac{\sum_{\text{all } i} t_{i,ave} E_{i,ave} \Delta t_{i,\text{int}}}{\sum_{\text{all } i} E_{i,ave} \Delta t_{i,\text{int}}} \tag{4-29}$$

Here, the subscript *ave* means the average over an interval (int) between adjacent sampling times, as illustrated in Eq. 4-8; to make this calculation requires adding a few columns to the spreadsheet shown in Fig. 4-3. Theoretically, the value of the denominator on the right side of Eq. 4-29 is equal to 1, so the equation would be simplified if that theoretical value were achieved in the tracer test.

The variance of the experimentally determined $E(t)$ distribution must also be calculated, as follows:

$$\sigma_E^2 = \frac{\sum_{\text{all } i} (t_{i,ave})^2 E_{i,ave} \Delta t_{i,\text{int}}}{\sum_{\text{all } i} E_{i,ave} \Delta t_{\text{int}}} - (\bar{t})^2 \tag{4-30}$$

Finally, the value of N, the number of CFSTRs in series, is found as

$$N = \frac{(\bar{t})^2}{\sigma^2} \tag{4-31}$$

For the data in the spreadsheet shown in Fig. 4-4, the following values were found using Eqs. 4-29, 4-30, and 4-31:

$$\bar{t} = 16.63 \text{ min}$$
$$\sigma_E^2 = 74.26 \text{ min}^2$$
$$N = 3.72$$

Although it is mathematically possible to have a noninteger value of N in the model (with some minor changes to Eqs. 4-27 and 4-28), it is usually sufficient, at least when the calculated value of $N \geq 3$, to use only integer values. In this case, the experimentally determined $E(t)$ is compared to the model using both $N = 3$ and $N = 4$ in Fig. 4-11. The model necessarily has only a single peak in the values of $E(t)$ and therefore cannot fit the experimental double-humped curve perfectly. Nevertheless, the fit of the model to the data for either value of N is quite good, especially when viewed as the cumulative age distribution, as shown in Fig. 4-11. In this case, the model with $N = 3$ fits the data for times less than 15 min better than the model with $N = 4$, whereas the reverse is true for the later times. As shown subsequently, the early part of these distributions is far more important than the later parts, at least when concerned with the effectiveness of the reactors in accomplishing a desired removal efficiency or degree of disinfection, so we should choose the value of $N = 3$ as the best fit to these data.

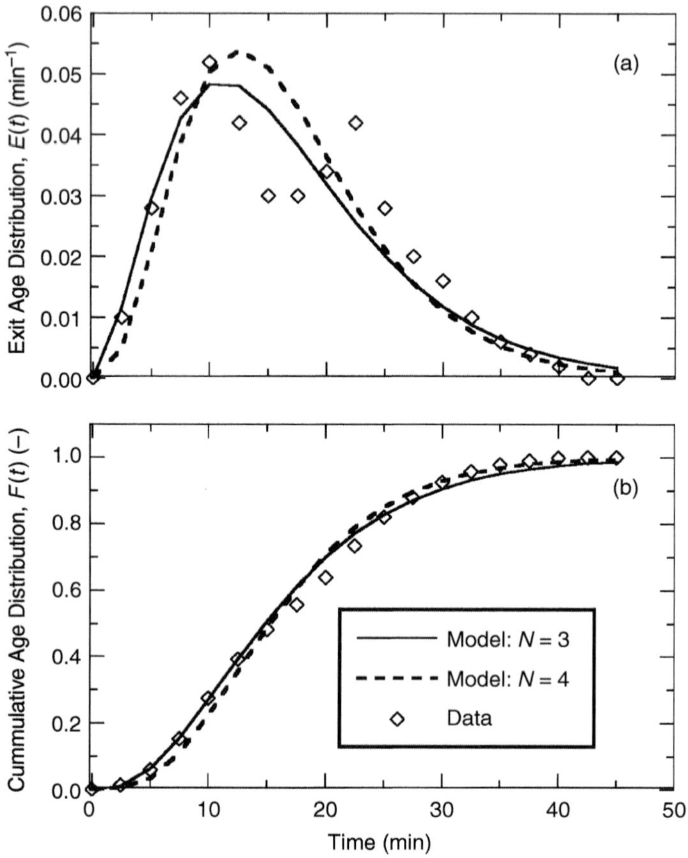

FIGURE 4-11 Fitting the CFSTRs in series model to the data in Fig. 4-1.

Dispersion Model

A second model in common use, especially when the hydraulic behavior of the reactor in question is nearly plug flow, is the dispersion model. Dispersion occurs when small variations in flow velocity occur in the direction of flow; this condition often occurs in turbulent pipe flow and in packed bed reactors. As described above, plug flow occurs when every packet of water stays in the reactor for exactly the same time because every packet has the identical axial flow velocity; when a small and random variation from the average velocity occurs in these one-dimensional situations, a tracer curve shows small variations from the plug flow conditions shown in Figs. 4-5 and 4-6.

In the development of the mathematical model, one develops a mass balance on the tracer in a pulse test over a small differential element of the reactor; water (and tracer) can penetrate the influent and effluent surfaces of that differential element not only by advection (the primary flow), but also by dispersion. Dispersion is mathematically identical to diffusion, although the cause of the dispersion is spatial variation in the flow velocity rather than the molecular motion due to concentration differences that causes diffusion. To solve

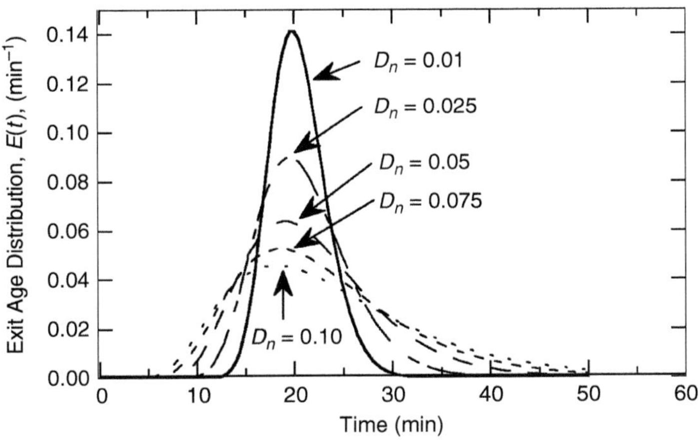

FIGURE 4-12 Sensitivity of the dispersion model to variations in D_n (Figure drawn with theoretical detention time of 20 min).

the differential equation, one needs to describe the boundary conditions at the influent and effluent of the reactor. In this case, one can describe either *open boundaries*, where dispersion can occur across the boundaries, or *closed boundaries*, where no dispersion can occur across the boundaries. Most real reactors in water treatment plants have closed boundaries; that is, the influent and effluent structures (e.g., holes in a wall, weirs of any type, pipe flow into a reactor) do not allow any significant amount of water to disperse back across the boundary against the main flow. Unfortunately, an analytical solution to the differential equation of the mass balance is only available for open boundaries. However, the difference between the numerical solution with closed boundaries and the analytical solution for open boundaries is quite small as long as the dispersion number, the one parameter of this model besides the mean detention time, is low. That fact allows us to use the analytical solution for real reactors when the dispersion is small, that is, when the flow is nearly plug flow.

The model is expressed for the exit age distribution as follows:

$$E(t)=\frac{1}{\tau\sqrt{4\pi D_n\dfrac{t}{\tau}}}\exp\left(-\frac{\left(1-\dfrac{t}{\tau}\right)^2}{4D_n\dfrac{t}{\tau}}\right) \tag{4-32}$$

Here, D_n is the dimensionless dispersion number.[2]

The sensitivity of this model to the dispersion number D_n is shown in Fig. 4-12. The maximum value of D_n in the figure (0.10) should be considered the maximum value for this model as shown here, because the error of using the open boundary solution for the closed boundary situation becomes significant at higher values. The fitting of this model to experimental data is similar to the process used for the CFSTRs-in-series model. One finds

[2]In the development of the model, D_n is defined as the ratio of the dispersion coefficient (with dimensions of length^2 per time) to the product of the flow velocity (length per time) and the reactor length. Many texts show this ratio explicitly in the equation of the model, but since the ratio is used as a single number, we have chosen to use a single symbol here.

the experimental average detention time $\bar{t}$ and the variance σ^2 of the Exit Age Distribution. The value of D_n is then found, by trial and error (including solver routines built into most spreadsheets), as the value that best fits the following equation:

$$\frac{\sigma_E^2}{\bar{t}^2} = \frac{2D_n + 8D_n^2}{(1+2D_n)^2} \tag{4-33}$$

For the low values of D_n under consideration, the denominator of Eq. 4-33 is close to 1.0 and can often be ignored. In that case, the value of D_n can be found using the quadratic equation. The model was not fit to the experimental exit age distribution shown earlier because the best-fit value of D_n would be greater than 0.1, where this model (as shown here) is not valid.

Other Models

A wide variety of other mathematical models have been developed to describe the hydraulic characteristics of a single reactor or a larger portion of a water treatment plant. An infinite number of models can be developed by considering various combinations of ideal reactors either in series or parallel or a combination of the two. No single model can be expected to fit the hydraulic characteristics of all reactors, so one often needs to try several models to find one that fits the data well for a particular reactor, at least over the range of $E(t)$ or $F(t)$ that is of primary interest. A conceptually simple model composed of a small number of reactors in series and/or parallel that gives a reasonable fit to the data is generally of greater value than a far more complex model with a somewhat better fit; good judgment is often required to choose a point in the trade-off between complexity and fit. The value of a model is often to determine what characteristics of a reactor might be changed to make the reactor behave better for its intended purpose; usually, the challenge is to make a reactor more like a plug flow reactor, in delaying the early rise of the $F(t)$ distribution. A simple conceptual model is often easier to understand and interpret than a more complex one, and therefore better in visualizing the physical changes that could or should be made to the reactor.

One example of this type of conceptual model was presented by Rebhun and Argaman (1965) and by Levenspiel (1999). A single reactor with a total volume V is considered to have a dead-space fraction m (in which the water is essentially stagnant so that no exchange of the flowing water with the stagnant water takes place) and an active fraction $1-m$; within the active fraction, the reactor is assumed to behave as two ideal reactors in series: a PFR (with a fraction of the active volume (p) followed by a CFSTR $(1-p)$. At a flow rate Q, the theoretical detention time of the entire reactor is $\tau = V/Q$, but the detention time in the PFR portion is $p(1-m)\tau$ and in the CFSTR portion is $(1-p)(1-m)\tau$. The cumulative age distribution of a reactor with this description can be derived to yield

$$F(t) = 1 - \exp\left\{-\frac{1}{(1-p)(1-m)}\left[\frac{t}{\tau} - p(1-m)\right]\right\} \tag{4-34}$$

To use this model with data for $F(t)$, a plot of $\ln(1 - F(t))$ versus t/τ is made; a perfect fit of this model would be a horizontal line at the ordinate value of 0 for low values of t/τ, followed by a straight line with a negative slope. The breakpoint (value of t/τ) between these two line segments occurs when $t/\tau = p(1 - m)$, and the slope of latter portion is $-1/[(1 - p) \times (1 - m)]$. From that point and slope, the values of the parameters p and m can be determined. Real reactors are unlikely to have the abrupt change from a PFR to a CFSTR, so data from tracer tests plotted this way tend to have some curvature between the two straight-line portions. Although this model has been recommended by the USEPA (1991) as a means of finding the value of T_{10} for compliance with the disinfection regulations, one should ensure that this model is a reasonable fit to the data before applying it for this purpose. If one does the analysis for this model with the data presented in Fig. 4-1c for $F(t)$, for example, the result does not fit the model well at all, as the data past the mean detention time is continuously curved, rather than the straight line proposed by this model.

The double-humped $E(t)$ curve shown in Fig. 4-1b does not fit any simple model well, but a combination of models can be put together to approximate that curve. The most reasonable hypothesis for a double-humped curve is that the flow is divided into two parallel paths through different parts of the reactor, each receiving some fraction of the flow. The curve in Fig. 4-1b can be thought of as the sum of two curves like those shown in Fig. 4-10a for various numbers of CFSTRs in series; that is, we hypothesize that the single reactor with the double-humped exit age distribution is acting like two reactors in parallel, and each of those two is behaving as N-CFSTRs in series described by Eq. 4-27. Each of the two imaginary reactors in parallel has its own flow, detention time, and value of N, but the two fractions of the flow have to add to 1. This conceptual model is depicted in the top part of Fig. 4-13, and the mathematical model can be written as the sum of two fractional exit age distributions.

$$E(t) = f_A\left[\frac{1}{\tau_A}\frac{N_A^{N_A}}{(N_A-1)!}\left(\frac{t}{\tau_A}\right)^{N_A-1}\exp\left(-\frac{N_A t}{\tau_A}\right)\right] + (1-f_A)\left[\frac{1}{\tau_B}\frac{N_B^{N_B}}{(N_B-1)!}\left(\frac{t}{\tau_B}\right)^{N_B-1}\exp\left(-\frac{N_B t}{\tau_B}\right)\right]$$

The five parameters of the model are the detention times (τ_A and τ_B) in each of the two subreactors (A and B) of the overall reactor, the number of CFSTRs in series (N_A and N_B) for each subreactor, and the fraction of flow into Reactor A (f_A). The fraction of the flow that goes into Reactor B (f_B) is shown as $(1 - f_A)$ because the two fractions have to add to 1. The terms inside the brackets on the right side are the exit age distribution of the CFSTRs-in-series model (Eq. 4-27); multiplying each by its respective flow fraction ensures that that the area under the entire $E(t)$ curve will equal 1.

To fit the model, a spreadsheet was used with a graph that showed both the experimentally determined $E(t)$ curve and the model results; the best fit (determined by eye) is shown in the graph in Fig. 4-13. The model matches the experimental curve very well through the first 12.5 min and adequately thereafter. The modeling process (i.e., the determination of the parameter values) was guided by a few facts. First, a characteristic of the curves for the CFSTRs-in-series model is that the peak of the distribution occurs at time, $t = [(N-1)/N]\tau$; this fact suggested that the values of τ_A and τ_B be slightly greater than the times of the two peaks in the experimental curve (10 and 22.5 min). Second, the earlier consideration of the reactor as a single reactor divided into N CFSTRs in series showed that $N = 3$ or $N = 4$ gave a reasonable fit to the early part of the curve. That result suggested that a slightly higher value of N would be useful for reactor A, which was to primarily fit the early part of the overall curve. Reactor B would then require a high value of N so as to give a very small contribution to the overall $E(t)$ value at values of t less than approximately 15 min (see Fig. 4-10). Finally, because both the peak and the spread from Reactor A would be greater than those from Reactor B, the fraction of the flow in Reactor A (i.e., the fraction of the total area of the $E(t)$ curve due to Reactor A) would be greater than that in Reactor B. These facts guided the original selection of values for the five parameters, and these values were then adjusted in small increments to find the final values shown on Fig. 4-13.

It should be noted that more sophisticated methods for finding the parameter values for this model could be used. Also, other models, such as considering the overall reactor to behave like several PFRs in parallel (resulting in a stair-step $F(t)$) could also be fit to the data.

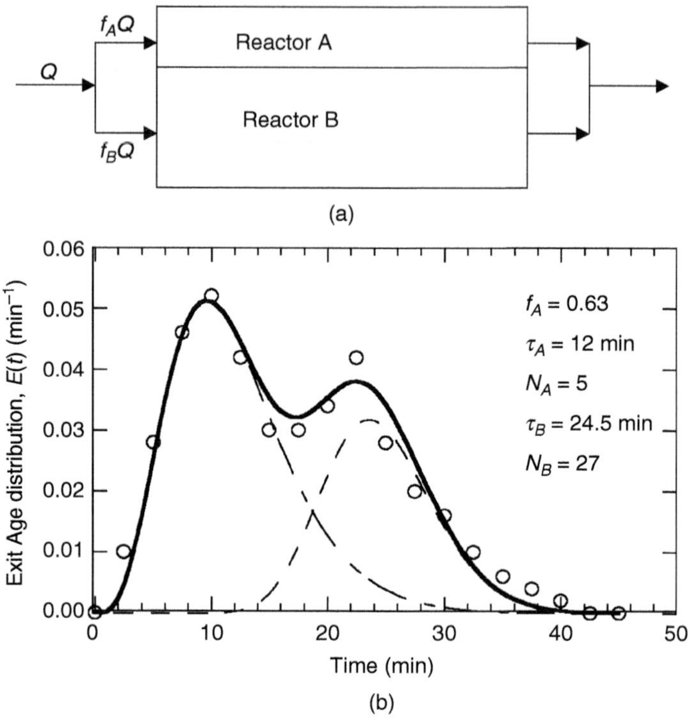

FIGURE 4-13 Fitting the exit age distribution of Fig. 4-1 to a more complex model (*a*) schematic of two sub-reactors in parallel, each behaving as N-CFSTRs in series; (*b*) model fitting.

COMPUTATIONAL FLUID DYNAMICS

In recent years, the field of computational fluid dynamics (CFD) has been developed and used increasingly for evaluating the hydraulic behavior of reactors for water treatment. The fundamental equations of fluid mechanics express the conservation of mass, momentum, and energy; collectively, they are known as the Navier-Stokes equations. Theoretically, these differential equations could be solved numerically for any flow situation, but practically, the computational intensity would be too great to use the full equations due to the short time and small length scales of turbulent motion. For incompressible fluids (such as water), this problem is avoided by using so-called Reynolds averaging, which dramatically reduces the computational needs by solving for mean quantities of the flow and developing models that try to simulate the effects of small-scale fluctuations of turbulence in a small number of averaged parameters. A wide variety of models for these small-scale turbulent effects have been developed, with varying degrees of complexity; many are explained in Garg (1998) and Chen and Jaw (1998). The most commonly used (and simplest) models are those describing the turbulent kinetic energy (k) and the turbulent energy dissipation (ε) and are usually referred to as eddy viscosity, two-equation, or k-ε turbulence models. These k-ε models are semi-empirical models that were validated under specific conditions to simulate the behavior of turbulent structures in the flow. The formulation of these eddy viscosity models includes a number of empirical constants, and various investigators have proposed specific values for these coefficients that have proven useful in a wide variety of situations. When the eddy viscosity models are not able to simulate the flows adequately, more complex models become necessary.

In many variants of the k-ε model described in the literature, the model assumes that turbulence is isotropic. Therefore, the primary shear stresses are well predicted but secondary shear stresses from secondary flow structures and normal stresses are not well predicted. As a result, the simpler k-ε models often have difficulty describing certain 3D flows. Some of these errors are minimized in so-called anisotropic eddy viscosity models, which incorporate additional terms within the two-equation model structure to capture the strong 3D effect flow. A comparison of various turbulence models for the same application is provided by Liu et al. (2007).

Several commercial software packages for performing CFD calculations are available. These products include a menu of the turbulence models from which the user chooses, depending on the problem. These CFD software packages also have user-friendly interfaces to import or re-create the reactor geometry and to set the finite element grid spacing for the discretization of the flow domain. As a result, they have become a versatile tool to analyze a wide variety of problems relevant in drinking water treatment. The algebraic equations expressing the differential versions of the fluid mechanics equations are solved for all of the grid elements in the reactor. Often, it is reasonable to formulate the flow pattern in two dimensions rather than three, at great reduction in the computational complexity (and time). For example, in many disinfection reactors, the length of the flow path from inlet to outlet is far greater than the depth, so that the variation of flow with depth becomes relatively unimportant in comparison to the use of the surface area; in such cases, formulating the CFD problem in the two horizontal dimensions and ignoring the depth yields sufficiently accurate information.

For use in water treatment plants, a CFD model can be used to find the steady-flow velocity vectors at all locations (grid points) within a tank, and then the response to a tracer at the influent can be followed throughout the reactor. Unlike in a real tracer test, the concentrations at all locations of the tank at any time can be calculated, so that graphical software can be used to visualize the flow pattern. In this way, the CFD model produces far more information than a tracer test in identifying specific areas of the reactor that are barely participating in the flow pattern (Templeton et al., 2006).

For an existing plant, the combination of a single tracer test and CFD can be used to determine the effects of variables that had only a certain value in the tracer experiment. For example, any tracer test is run at particular values of flow rate and water temperature, but the plant runs at different flow rates and different temperatures at other times. The user first fits a CFD model to a limited set of experimental results; that is, a set of grid spacing, boundary conditions, and turbulent model in the CFD software is found that yields results in close agreement with the experimental results. Then, the model can be run at different conditions to predict how the reactor would respond. In this way, the CFD model can be used to predict whether the reactor is likely to perform poorly (for example, have an unacceptably low value of T_{10}/τ) under flow conditions that are different than those used in the tracer test.

A common and powerful use of CFD is to aid in design of retrofits (for existing plants) or new facilities. In an existing plant, for example, many designs of new baffles in a clearwell can be tested to arrive at a design that improves the compliance with $C \times T$ regulations for disinfection under various conditions (flow and current storage); examples are shown in Grayman et al. (1996), Hannoun et al. (1998), and Crozes et al. (1998). Grayman et al., for example, showed a dramatic improvement in the T_{10} that could be achieved by the installation of baffling in a circular clearwell that had been originally designed prior to the current disinfection regulations and that had the influent and effluent pipes adjacent to one another. Similarly, Mahmood et al. (2005) showed through CFD modeling that mixing in a distribution system elevated reservoir could be improved by orienting the entrance port vertically rather than horizontally and reducing its diameter to increase the exit velocity; subsequent construction verified these findings. For water quality purposes, mixing is more important than residence time in that situation; the focus in the CFD model in this case was on equalizing the temperature of the water throughout the tank to avoid stratification that would allow warmer water to sit on top of colder water and stagnate for long times.

In the design of new plants, the model cannot be calibrated against tracer results, but the same concept of testing various designs to see the effect on the predicted hydraulic behavior can be undertaken. The results might not be as accurate as a calibrated model in an existing facility would give, but the differences caused by different designs are likely to be well characterized, particularly if a model that has been tested in similar plant conditions (elsewhere) is used.

The use of CFD software is not limited to simulating tracer studies under steady-flow conditions; rather, one can directly calculate the behavior of reactive substances in the flow pattern or study the transient behavior of a reactor under a changed influent flow condition. Several investigators, for example, have applied CFD modeling to UV disinfection modeling; examples include Blatchley and coworkers (Naunovic et al., 2008; Blatchley et al., 2008), Ducoste and coworkers (Zhao et al., 2009; Ducoste et al., 2005), and Munoz et al. (2007). CFD modeling has proven particularly valuable in this situation because the validation of UV reactors has been a difficult experimental (and regulatory) problem (see Chap. 18 for additional information). CFD modeling has also been used for reactive substances including ozone disinfection, including gas transfer and ozone decay (Zhang et al., 2004), and chloramine (Liu and Ducoste, 2006). Finally, CFD has also been applied to modeling particle processes such as flocculation (Prat and Ducoste, 2007) and sedimentation (e.g., Goula et al., 2008); these models are somewhat more complicated inasmuch as particles do not simply follow the fluid flow because of their different density and the effects of Brownian motion and fluid drag.

REACTION RATE EXPRESSIONS

Chemical reactions occur in a wide variety of ways in water treatment, and the rates of those reactions also vary widely. The study of reaction rates is strictly empirical, as the reaction pathways and, particularly, the rate at which those pathways are traversed cannot be determined on purely theoretical grounds. Reaction rate expressions are derived from those experimental investigations and are mathematical formulas that describe the observed kinetics. Once they are developed (usually from batch experiments), these rate expressions can be inserted into appropriate mass balances for reactors with continuous flow and used to predict the behavior of chemical species under various treatment conditions.

Reactions whose kinetics are often of interest in water treatment plants include disinfection, oxidation and reduction reactions, precipitation reactions, and flocculation of particles. Some investigators have also expressed particle removal processes such as granular media filtration, sedimentation, and dissolved air flotation (DAF) in terms of reaction rate expressions. The kinetics of most of these reactions, and the underlying reasons for the form of the kinetic expressions, are discussed in the appropriate later chapters. Here, we present a generic approach for considering the combination of reaction kinetics and reactor hydraulic characteristics that is useful background for those subsequent chapters. The reader may find additional coverage of reaction rates and reactor hydraulics in Weber and DiGiano (1996).

The simplest reaction rate expressions for irreversible reactions (those in which the products cannot be reformed into the reactants) are those of the following form:

$$r_A = -k_n C_A^n \tag{4-35}$$

where C_A is the concentration of species A, k_n is the rate constant, and n is an exponent and is known as the *order* of the reaction. The most common use of this equation is with $n = 1$, a first-order reaction, but it is also used for zero or second or fractional orders (values of n). The original model for disinfection kinetics (see Chap. 17), Chick's law, is an example of such first-order kinetics (Chick, 1908); the expression is

$$r_I = -kC_I \tag{4-36}$$

where C_I is the concentration of viable microorganisms of a particular genus or species (e.g., coliforms) and k is the rate constant.

A slightly more complicated version of this type of rate expression occurs when the concentrations of two species (say, A and B) are known to affect the reaction rate, but each to some power of their respective concentrations. In that case, the form is

$$r_A = -k_n C_A^{\alpha} C_B^{\beta} \tag{4-37}$$

In this case, we say that the reaction is α-order with respect to A, β-order with respect to B, and ($\alpha + \beta$)-order overall. An example in common use in the drinking water field is the Chick-Watson law of disinfection (Watson, 1908), which is written as follows:

$$r_I = -kC_I C_D^n \tag{4-38}$$

where C_D is the concentration of disinfectant present in the water. This reaction is considered first order with respect to the microorganism concentration (C_I) but some fractional order with respect to the disinfectant concentration (n, usually in the range 0.85 to 0.90, although often taken to be 1 for mathematical simplicity). The degradation of many specific organic compounds in the presence of ozone is also commonly reported to be a second-order reaction (see Chap. 7), first order with respect to ozone and first order with respect to the particular organic chemical (von Gunten, 2003a, 2003b).

Other reactions of interest in water treatment engineering have more complex reaction rate expressions. For example, biological treatment of certain contaminants is becoming routine in some areas, and the most common rate expression used in that case is the Monod kinetics:

$$r_S = -\frac{kXS}{K_S + S} \tag{4-39}$$

where S is the contaminant of interest, designated S for substrate; X is the microorganism concentration (considered somewhat differently for suspended versus attached growth systems); and k and K_S are constants for a given system. This rate expression is generally coupled with a second one expressing the net rate of growth of the microorganism population (not shown here), leading to a more complex situation mathematically. Other more complex rate expressions are required for biological systems when competing substrates are available, when toxicity at higher concentrations is possible or when the contaminant degradation occurs by co-metabolism (in which the microorganism gains no advantage by the degradation of the substrate of interest, but performs the degradation fortuitously while oxidizing some other compound).

A final rate expression of great importance in water treatment engineering is the flocculation rate (see Chap. 8). *Flocculation* is the collision and attachment of two particles to form a single particle (a *floc*). When considering a large number of particles of many different sizes as in a water treatment plant, the size distribution is subdivided into a discrete number (N) of size classes with all particles in each class considered to be the same size. Particles of size class k can be formed when two smaller particles (say i and j) attach to each other, with the provision that the volumes of the two smaller particles add to be in the size range of k-size particles. Particles of size class k can also be lost when one such particle collides with and attaches to any size particle and thereby form yet larger particles. To describe flocculation kinetics (i.e., to describe the changes in the particle size distribution), a set of N equations of the following form must be written:

$$r_k = \frac{1}{2}\alpha_{\text{emp}} \sum_{\text{vol}_i + \text{vol}_j = \text{vol}_k} \gamma(i, j)\, n_i n_j - \alpha_{\text{emp}} n_k \sum_{i=1}^{N} \gamma(i, k)\, n_i \tag{4-40}$$

where n_i is the number concentration of particles of size i, $\gamma(i, j)$ is the (theoretical) kinetic coefficient (rate constant) for the collisions between particles of sizes i and j, and α_{emp} is an empirical fitting factor, reflecting the fact that the state of the art of flocculation still requires a fitting factor (between 0 and 1) to match experimental results. The major point to raise in the context of this chapter is that flocculation is a second-order reaction; that is, the rate of gain (first term on the right side) is dependent on both the concentrations of i and j particles (each at first order), and the rate of loss (second term on the right) is dependent on both the concentrations of i and k particles. Because N linked equations must be solved simultaneously, an analytical solution is impossible, and the set of equations resulting from inserting the rate equation into an appropriate mass balance is solved numerically.

Reactions in Batch Reactors

The rate expressions for the various types of reactions described above form one part of a mass balance for a constituent. The overall mass balance for a species i in most situations of interest (those that do not have diffusion or dispersion across the boundary of a control volume) can be expressed by the following word equation:

$$\begin{bmatrix}\text{Rate of change}\\ \text{of mass of}\\ \text{species } i \text{ in the}\\ \text{control volume}\end{bmatrix} = \begin{bmatrix}\text{Rate of advective}\\ \text{input of mass of}\\ \text{species } i \text{ into the}\\ \text{control volume}\end{bmatrix} - \begin{bmatrix}\text{Rate of advective}\\ \text{output of mass of}\\ \text{species } i \text{ from the}\\ \text{control volume}\end{bmatrix} + \begin{bmatrix}\text{Rate of change of}\\ \text{mass of species } i \text{ in}\\ \text{the control volume}\\ \text{due to reactions}\end{bmatrix} \tag{4-41}$$

The dimensions of such a mass balance are mass per time. To translate this word equation into a mathematical equation, one must specify the control volume and the species under consideration. Usually, the species of concern is easily identified as a particular contaminant, reactant, or reaction product. However, in cases of closely related species (e.g., soluble and solid forms of a metal, the different species in a weak acid system such as carbonate, related species with different oxidation states such as arsenic), the choice of species or group of species must be carefully made. The choice of control volume is usually either a differential element with a specified geometry or the entire reactor volume, depending on whether there are spatial variations of concentration in the system of interest.

The most common way to study reaction kinetics is in a well-mixed batch reactor; for that case, the control volume is simply the entire liquid volume in the reactor, because the mixing eliminates spatial variations in the concentration. The defining characteristic of a batch reactor is the absence of any advection (flow) into or out of the reactor, and therefore the first two terms on the right side of the word equation are eliminated. In that case, the equation simplifies to

$$\begin{bmatrix} \text{Rate of change} \\ \text{of mass of} \\ \text{species } i \text{ in the} \\ \text{control volume} \end{bmatrix} = \begin{bmatrix} \text{Rate of change of} \\ \text{mass of species } i \text{ in} \\ \text{the control volume} \\ \text{due to reactions} \end{bmatrix} \tag{4-42}$$

Translating to symbols yields

$$\frac{VdC_i}{dt} = Vr_i \tag{4-43}$$

or, more simply (and with a change of dimensions to mass/(volume-time)),

$$\frac{dC_i}{dt} = r_i \tag{4-44}$$

Equation 4-44 is the (volume-normalized) mass balance for any constituent i in a batch reactor. Those who study kinetics typically perform batch experiments in which they measure C_i as a function of time, and then find an expression for r_i that agrees with the data. For most reactions of interest in water treatment, the rate expressions are known, and the objective is to recognize how that knowledge is best translated into successful design and operation. In most (but not all) situations of interest, the reaction is one in which the constituent is an undesired contaminant, and the reaction rate expression is negative, that is, one in which the constituent is destroyed.

Generally, r_i is a function of the concentration of i, and so Eq. 4-44 is rearranged to separate the variables (concentration and time) for the integration as follows:

$$\int_{C(0)}^{C(t)} \frac{dC_i}{r_i} = \int_0^t dt = t \tag{4-45}$$

The simplest rate expression is that for an irreversible, first-order reaction: $r_i = -k_1 C_i$. Inserting that expression into Eq. 4-45 yields

$$\int_{C_0}^{C(t)} \frac{dC_i}{C_i} = -k_1 \int_0^t dt \tag{4-46}$$

$$\ln \frac{C(t)}{C_0} = -k_1 t \tag{4-47}$$

or

$$C(t) = C_0 \exp(-k_1 t) \tag{4-48}$$

Graphical representations of Eqs. 4-47 and 4-48 are shown in Fig. 4-14. In an experiment, the raw data for $C(t)$ would be like that shown in Fig. 4-14a, and an analysis of the data to check for first order behavior would be done like the plot shown in Fig. 4-14b.

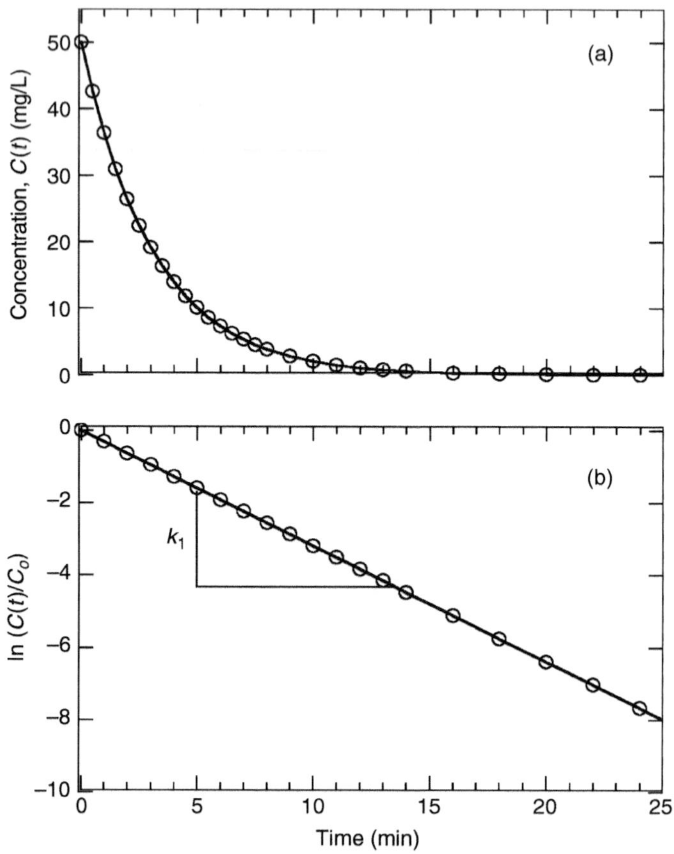

FIGURE 4-14 Graphical representation of first order kinetics in a batch reactor.

Similar analyses for zero-order ($r_i = -k_0$) and second-order ($r_i = -k_2C_i^2$) reactions are performed by inserting these kinetic expressions into Eq. 4-45, with the following results. For zero order,

$$C(t) = C_0 - k_0 t \tag{4-49}$$

For second order,

$$C(t) = \frac{C_0}{1 + k_2 C_0 t} \tag{4-50}$$

or

$$\frac{1}{C(t)} = \frac{1}{C_0} + k_2 t \tag{4-51}$$

For zero-order reactions, Eq. 4-49 indicates that the raw data (concentration versus time) will yield a straight line with intercept equal to the initial concentration C_0 and a negative slope equal to the rate constant k_0. For second-order reactions, a graph of $1/C(t)$ versus time will have an intercept of $1/C_0$ and a positive slope equal to k_2.

The critical fact is that, for first- and second-order reactions, the rate of reaction is dependent on concentration, so that the higher the concentration, the faster the reaction. Unfortunately, our desire in water treatment is to have a low concentration of contaminants where these reaction rates are relatively slow. Engineering a system to account for the kinetics of reaction properly is a critical part of water treatment plant design.

After inserting some of the more complex rate expressions noted in the previous section into Eq. 4-45, integration by analytical methods is possible under some circumstances but not all. For example, we can consider a reaction that is first order with respect to each of two constituents and second order overall (i.e., $\alpha = \beta = 1$ in Eq. 4-37) and further that C_A and C_B are directly linked by stoichiometry, so that the loss of a B molecule is accompanied by the loss of mA molecules (i.e., $m\mathrm{A} + \mathrm{B} \rightarrow$ product). In that case, the concentration of one chemical at any time can be expressed in terms of the concentration of the other (and the initial concentrations); that is, $C_A = C_{A_0} - m(C_{B_0} - C_B)$, where the subscript 0 represents the initial condition. If the two constituents are present originally at their stoichiometric ratio (i.e., $C_{A_0} = mC_{B_0}$), then this ratio of concentrations is maintained at all times (i.e., $C_A(t) = mC_B(t)$) and the mathematical solution is the same as Eq. 4-50 except that k_2 in the integrated expression is replaced by mk. That is,

$$C_B(t) = \frac{C_{B_0}}{1 + mkC_{B_0}t} \tag{4-52}$$

When the two constituents are not present originally at their stoichiometric ratio (i.e., $C_{A_0} \neq mC_{B_0}$), the mass balance for the batch reactor yields

$$\frac{dC_B}{(C_{A_0} - mC_{B_0} + mC_B)C_B} = -k\,dt \tag{4-53}$$

where C_B is a function of time but is written in this shorthand form for convenience. The solution to this differential equation is

$$\frac{C_B}{C_{A_0} - mC_{B_0} + mC_B} = \frac{C_{B_0}}{C_{A_0}} \exp\left[-(C_{A_0} - mC_{B_0})kt\right] \tag{4-54}$$

This expression can be rearranged to yield an explicit expression for C_B as a function of time.

$$C_B(t) = \frac{mC_{B_0} - C_{A_0}}{\left[m - \frac{C_{A_0}}{C_{B_0}} \exp[(C_{A_0} - mC_{B_0})kt]\right]} \tag{4-55}$$

An example of the difference between the two situations described by Eqs. 4-52 and 4-55 is shown in Fig. 4-15 (with $m = 2$). In the case where excess A is present originally (and therefore at all times), the degradation of B is faster because the reaction rate is proportional to C_A. The faster kinetics in this case can be used to advantage in water treatment engineering, as long as A itself is not an undesirable contaminant.

Another reaction of major interest is the Chick-Watson law of disinfection (see Chap. 17) expressed in Eq. 4-38; the rate of disinfection is dependent on both the disinfectant concentration and the concentration of viable microorganisms. In that expression, it is often found that the exponent n on the disinfectant concentration is nearly 1, and for simplicity of the

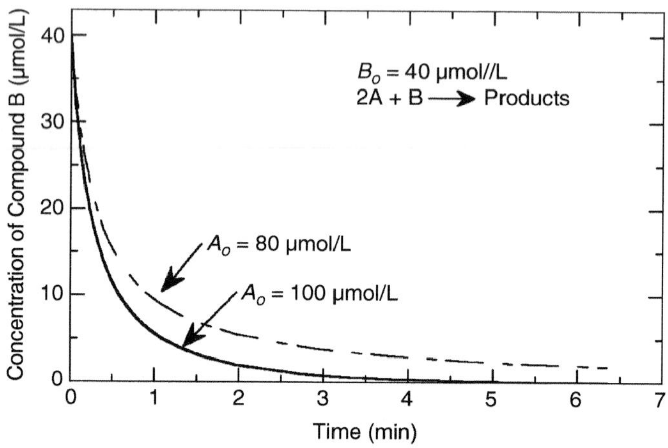

FIGURE 4-15 Progression of a second order bi-molecular reaction: effect of concentration. (In the case shown, $m = 2$; for $C_{A0} = 80$ µmol/L, the two constituents are present in their stoichiometric ratio at all times; for $C_{A0} = 100$ µmol/L, constituent A is always in excess of that ratio.)

mathematics, it is often taken to be 1. For chlorine, for example, the most commonly reported value for n is 0.86 (Berg, 1964), and the error associated with considering that value to be 1.0 is relatively small. The USEPA, in its CT tables for compliance with the Surface Water Treatment Rule, considers $n = 1$ for several disinfectants and target microorganisms but not for chlorine disinfecting Giardia cysts (which is why those tables have separate entries for different chlorine concentrations; see USEPA 1991, guidance documents and tables). With the assumption that $n = 1$, the rate expression is essentially identical to that of the second-order reaction of two constituents (i.e., first order with respect to each). However, the consumption of disinfectant in natural waters is rarely, if ever, linked directly to the degree of disinfection. Two possible scenarios for the concentration of the disinfectant C_D are often considered. In the simpler case, the disinfectant is not considered to decay at all, in which case the disinfection reaction is a first-order reaction with the rate constant being $k_I C_{D_0}$; the solution is similar to that shown in Eq. 4-48.

$$C_I(t) = C_{I_0} \exp(-k_I C_{D_0} t) \tag{4-56}$$

More realistically, the disinfectant or oxidant is consumed by many reactions, and the consumption by disinfection is small in comparison to that from other reactions. As a result, the decay of the disinfectant is considered independent of the disinfection reaction; if that decay is (or is assumed to be) first order (Lawler and Singer, 1993), then the disinfectant concentration varies with time.

$$C_D(t) = C_{D_0} \exp(-k_D t) \tag{4-57}$$

The Chick-Watson law with these assumptions can then be expressed as

$$r_I = -k_I [C_{D_0} \exp(-k_D t)] C_I \tag{4-58}$$

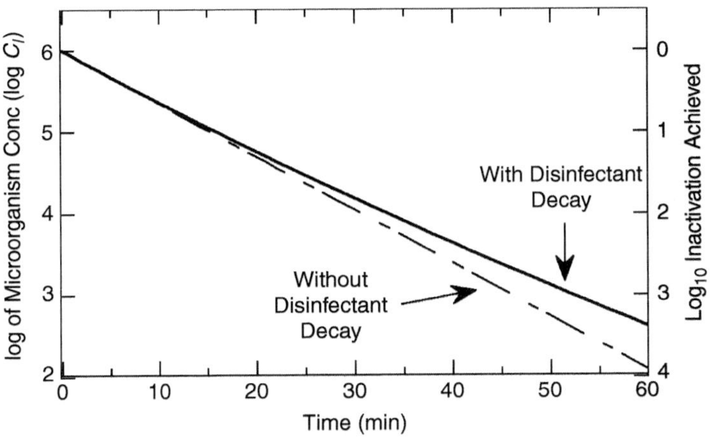

FIGURE 4-16 Disinfection achieved: effect of disinfectant decay during disinfection. (Conditions: $C_{I0} = 10^6$/100 ml; $C_{D0} = 2.5$ mg$_D$/L; $k_D = 0.005$/min; $k_I = 0.06$ L/(mg$_D$-min); disinfectant decayed to 1.85 mg/L in 60 min)

Substituting this expression into Eq. 4-45 and integrating yields

$$C_I(t) = C_{I_0} \exp\left\{-C_{D_0} \frac{k_I}{k_D} [1 - \exp(-k_D t)]\right\} \tag{4-59}$$

The effect of disinfectant decay is illustrated in Fig. 4-16 for conditions that could be considered representative of water treatment conditions; the figure uses Eq. 4-56 for the conditions without disinfectant decay and Eq. 4-59 for conditions with disinfectant decay. With no decay of the disinfectant, the results plot as a straight line on this semilogarithmic plot; base 10 logarithms are used so that the results can be shown (on the right axis) as the (commonly used) log inactivation achieved, defined as $\log_{10} \frac{C_{I_0}}{C_I(t)}$. The slope of that line is $-\frac{C_{D0} k_I}{2.30}$, where 2.30 stems from the conversion from base e to base 10 (i.e., ln 10 = 2.30). Disinfectant decay increases the time required to achieve a specified degree of inactivation; in the example shown, the time to achieve 3-log inactivation increases from 46 min with no disinfectant decay to 52 min with decay (from 2.50 to 1.93 mg/L).

REACTIONS IN CONTINUOUS FLOW SYSTEMS AT STEADY STATE: COMBINING HYDRAULICS AND REACTION KINETICS

In virtually all cases, water treatment plants operate as a series of continuous flow reactors, and so it is critical to understand the interaction of the flow characteristics of a reactor and the kinetics of reactions of interest. In this section, we address that concern by considering reactors that are operating at steady state, that is, reactors in which nothing is changing with time, and we progress from ideal reactors (PFRs and CFSTRs) to non-ideal reactors.

Plug Flow Reactors

As described earlier, a plug flow reactor is one in which water that comes into the reactor at one instant travels from the inlet to the outlet without any mixing with water that came in at any other time. This condition means that all of the water stays in the reactor for exactly the same amount of time, the theoretical detention time τ. To consider the behavior of a reactive substance in a plug flow reactor, we can choose a small differential element as the control volume in the mass balance, as indicated in Fig. 4-17.

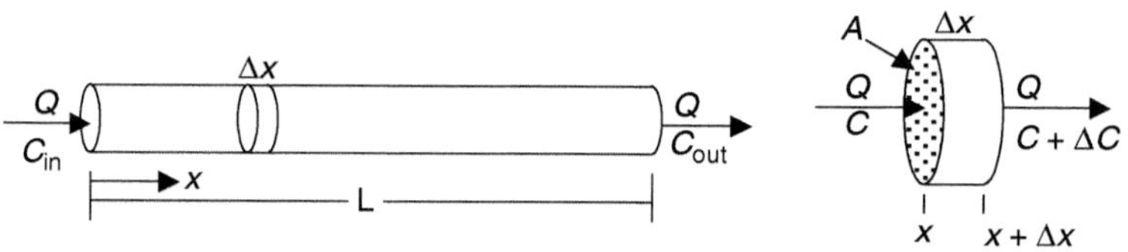

FIGURE 4-17 Schematic of a plug flow reactor, with differential element for the mass balance. (A is the cross-sectional area; x is the distance from the reactor entrance, and L is the length of the reactor.)

Translating the word equation in Eq. 4-41 for the mass balance on the differential element shown on the right side of Fig. 4-17 into symbols yields

$$\frac{\partial(AC\Delta x)}{\partial t} = QC - Q(C+\Delta C) + (A\Delta x)r \tag{4-60}$$

where r is the reaction rate. The stipulation that we are considering steady state conditions means that the left side is zero. Combining the first two terms on the right side and dividing by the cross-sectional area A yields

$$0 = -\frac{Q}{A}\Delta C + r\Delta x \tag{4-61}$$

In the limit as the differential element shrinks in length toward zero, further rearrangement leaves us with

$$\frac{dC}{r} = \frac{A}{Q}dx \tag{4-62}$$

Since r is generally a function of concentration, the integration of the left side must include that functionality, whereas the integration of the right side is straightforward.

$$\int_{C_{in}}^{C_{out}} \frac{dC}{r} = \frac{A}{Q}\int_0^L dx = \frac{AL}{Q} = \frac{V_{pfr}}{Q} = \tau_{pfr} \tag{4-63}$$

Comparing Eq. 4-63 with Eq. 4-45 for batch reactors reveals that they are nearly identical; that is, the amount of reaction that occurs between the influent and effluent of a plug flow reactor with detention time τ is identical to the amount that occurs in a batch reactor in the time $t = \tau$. This identity should perhaps not be surprising: all of the water that comes in to a PFR at one instant travels through the reactor without interacting with water that comes in earlier or later. In essence, a PFR is a moving batch reactor. This result means that all of the expressions from the batch system described above can be directly translated to the PFR case. That translation to the PFR for most of the results shown above for reactions in batch reactors is shown in Table 4-1.

TABLE 4-1 Performance in Plug Flow Reactors

Description	Rate expression	Effluent concentration, C_{out}	Required detention time to achieve C_{out}, τ_{pfr}
Zero order	$r=-k_0$	$C_{in}-k_0\tau_{pfr}$	$\frac{1}{k_0}(C_{in}-C_{out})$
First order	$r=-k_1C$	$C_{in}\exp(-k_1\tau_{pfr})$	$\frac{1}{k_1}\ln\frac{C_{in}}{C_{out}}$
Second order	$r=-k_2C^2$	$\frac{C_{in}}{1+C_{in}k_2\tau_{pfr}}$	$\frac{1}{k_2}\left(\frac{1}{C_{out}}-\frac{1}{C_{in}}\right)=\frac{1}{k_2C_{in}}\left(\frac{C_{in}}{C_{out}}-1\right)$
First order in each of two constituents related by stoichiometry	$r_A=-kC_AC_B$ with $\Delta C_A=m\Delta C_B$	$C_{B,out}=\frac{mC_{B,in}-C_{A,in}}{\left\{m-\frac{C_{A,in}}{C_{B,in}}\exp[(C_{A,in}-mC_{B,in})k\tau_{pfr}]\right\}}$	$\frac{1}{k(C_{A,in}-mC_{B,in})}\ln\left[\left(\frac{C_{B,in}}{C_{A,in}}\right)\frac{C_{B,out}}{C_{A,in}-mC_{B,in}+mC_{B,out}}\right]$
Disinfection Chick-Watson with constant disinfectant concentration $C_{D,out}$	$r_I=-k_IC_{D,out}C_I$	$C_{I,out}=C_{I,in}\exp(-k_IC_{D,out}\tau_{pfr})$	$\frac{1}{k_IC_{D,out}}\ln\frac{C_{I,in}}{C_{I,out}}$
Disinfection Chick-Watson with decaying disinfectant concentration	$r_D=-k_DC_D$ $r_I=-k_IC_DC_I$	$C_{D,out}=C_{D,in}\exp(-k_D\tau_{pfr})$ $C_{I,out}=C_{I,in}\exp\left(-C_{D,in}\frac{k_I}{k_D}(1-\exp(-k_D\tau_{pfr}))\right)$	$-\frac{1}{k_D}\ln\left(1+\frac{k_D}{k_IC_{D,in}}\ln\frac{C_{I,out}}{C_{I,in}}\right)$

Continuous Flow Stirred Tank Reactors

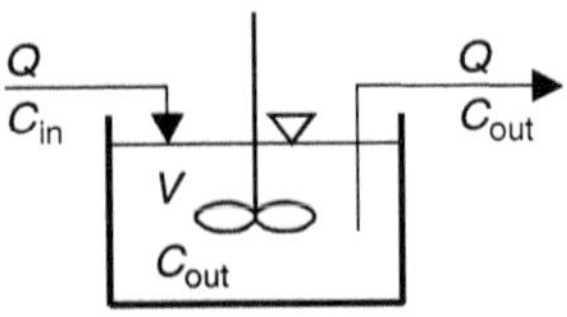

FIGURE 4-18 Schematic diagram of a CFSTR with a reactive substance.

We next consider the extent of reaction that occurs in CFSTRs at steady state. As with the tracer studies considered earlier, the control volume for the mass balance is taken as the entire water volume of the reactors, as depicted in Fig. 4-18. Because the concentration is identical everywhere in the reactor, the concentration of the effluent C_{out} is the same as that in the reactor (or vice versa).

In this case, the translation of the mass balance expressed in words in Eq. 4-41 results in

$$0 = QC_{in} - QC_{out} + Vr_{C_{out}} \tag{4-64}$$

where the zero on the left side reflects that, at steady state, nothing changes with time. Since the concentration is C_{out} everywhere in the reactor, the relevant reaction rate is that associated with this concentration, as indicated by $r_{C_{out}}$. Dividing by V and recognizing that $V/Q = \tau$ leads to

$$C_{out} = C_{in} + r_{C_{out}} \tau_{cfstr} \tag{4-65}$$

or

$$\tau_{cfstr} = -\frac{C_{in} - C_{out}}{r_{C_{out}}} \tag{4-66}$$

The various reaction rate expressions can be substituted into Eqs. 4-65 and 4-66 to predict the performance in CFSTRs. The results of this algebra are collected in Table 4-2 for the same reaction rate expressions considered earlier. For the Chick-Watson law, only the decaying disinfectant equation is shown; the constant disinfectant case would be the same, except that $C_{D,out} = C_{D,in}$.

Comparison of CFSTRs and PFRs

The equations for the performance of the two types of ideal reactors shown in Tables 4-1 and 4-2 are different, and it is worthwhile to explore the meaning of those differences. We limit the discussion to zero-, first-, and second-order reactions. The disinfection expression is similar to a first-order reaction (with the differences from first order being in the direction of second order). Also, the behavior for the reaction that is first order in each of two constituents is essentially identical to the second-order reaction. The differences are most easily explored by determining the ratio of the detention times required for a CFSTR and a PFR to accomplish the same degree of removal (assuming the same influent conditions for both types of reactors). The expressions for these ratios, along with calculated values for a few specific desired removal efficiencies, are different for each reaction order and are collected in Table 4-3.

For zero-order reactions, the flow pattern makes no difference in what is accomplished in a reactor; the extent of reaction that occurs depends only on the time provided and not on the concentration at any point in the reactor. For first- and second-order reactions, however, the results in Table 4-3 are dramatic in showing that the required detention times (or reactor volumes) are much higher for a CFSTR than a PFR. For example, to achieve 99 percent removal of a chemical undergoing a first-order reaction in the two types of reactors, the volume of the CFSTR would have to be 21.5 times that of the PFR. The ratios of required detention times increase with reaction order and also with desired removal efficiency. Plug

TABLE 4-2 Performance in Continuous Flow Stirred Tank Reactors (CFSTRs)

Description	Rate expression	Effluent concentration C_{out}	Required detention time to achieve C_{out} τ_{cfstr}
Zero order	$r=-k_0$	$C_{in}-k_0\tau_{cfstr}$	$\frac{1}{k_0}(C_{in}-C_{out})$
First order	$r=-k_1C$	$\frac{C_{in}}{1+k_1\tau_{cfstr}}$	$\frac{1}{k_1}\left(\frac{C_{in}}{C_{out}}-1\right)$
Second order	$r=-k_2C^2$	$\frac{C_{in}}{1+C_{out}k_2\tau_{cfstr}}$ or $\frac{-1+\sqrt{1+4k_2\tau_{cfstr}C_{in}}}{2k_2\tau_{cfstr}}$	$\frac{1}{k_2C_{out}}\left(\frac{C_{in}}{C_{out}}-1\right)$
First order in each of two constituents related by stoichiometry	$r_A=-kC_AC_B$ with $\Delta C_A=m\Delta C_B$	$C_{B,out}=C_{B,in}(1+mk\tau)-kC_{A,in}\tau-mkC_{B,out}^2\tau$ $=\frac{-1+\sqrt{1+4mk\tau_{cfstr}a}}{2mk\tau_{cfstr}}$ where $a=C_{B,in}(1+mk\tau_{cfstr})+kC_{A,in}\tau_{cfstr}$	$\frac{1}{kC_{A,out}C_{B,out}}\left(\frac{C_{B,in}}{C_{B,out}}-1\right)$ where $C_{A,out}=C_{A,in}-mC_{B,in}+mC_{B,out}$
Disinfection Chick-Watson with decaying disinfectant concentration	$r_D=-k_DC_D$ $r_I=-k_IC_DC_I$	$C_{D,out}=\frac{C_{D,in}}{1+k_D\tau_{cfstr}}$ $C_{I,out}=\frac{C_{I,in}}{1+k_IC_{D,out}\tau_{cfstr}}$	$\frac{1}{k_IC_{D,out}}\left(\frac{C_{I,in}}{C_{I,out}}-1\right)$

TABLE 4-3 Comparison of Required Detention Times in CFSTRs versus PFRs

Reaction order and rate expression	Expression for $\frac{\tau_{CFSTR}}{\tau_{PFR}}$	Value of $\frac{\tau_{CFSTR}}{\tau_{PFR}}$ for 90% removal	Value of $\frac{\tau_{CFSTR}}{\tau_{PFR}}$ for 99% removal
Zero: $r=-k_0$	1	1	1
First: $r=-k_1C$	$\frac{\frac{C_{\text{in}}}{C_{\text{out}}}-1}{\ln\frac{C_{\text{in}}}{C_{\text{out}}}}$	$\frac{9}{\ln(10)}=3.91$	$\frac{99}{\ln(100)}=21.5$
Second: $r=-k_2C^2$	$\frac{C_{\text{in}}}{C_{\text{out}}}$	10	100

flow reactors are simply far more efficient than CFSTRs in their use of space to accomplish constituent removal by many types of reaction. The design of water treatment plant reactors should always account for these differences.

Non-ideal Reactors with Common Flow Models

Most reactors do not exhibit the ideal flow of either a plug flow reactor or a continuous flow stirred tank reactor. As noted earlier in the chapter, several models are in common use to describe the flow characteristics, and some of them can be used in conjunction with reaction kinetic expressions to model the expected conversion accomplished in these reactors under steady state conditions.

The first model to be considered is the CFSTRs-in-series model. If the hydraulic behavior of a single reactor with theoretical detention time τ is adequately modeled as N CFSTRs in series, each with a detention time τ/N, the expressions for the steady state effluent concentration in Table 4-2 can be used successively N times to predict the effluent concentration, as the output concentration of each (imaginary) reactor (or compartment) is the influent concentration of the next one in series. The ease of predicting the performance for reactive species is a major advantage of the CFSTRs-in-series model. For a first-order reaction, the result is particularly straightforward.

$$\frac{C_{\text{out}}}{C_{\text{in}}}=\frac{1}{\left(1+\frac{k_1\tau}{N}\right)^N} \tag{4-67}$$

The sensitivity of the steady state removal accomplished for a constituent undergoing a first-order reaction in reactors whose hydraulic characteristics are described by the CFSTRs-in-series model is shown in Fig. 4-19. The abscissa represents the size (detention time) of the reactor, in dimensionless terms, as the product of the first-order rate constant (with dimensions of inverse time) and the actual detention time. The left ordinate shows the fraction remaining on a logarithmic basis; the value of –2, for example, means that $\frac{C_{\text{out}}}{C_{\text{in}}}=10^{-2}=0.01$; the labeling on the right ordinate expresses the same information in terms of the percent removal, $\left(1-\frac{C_{\text{out}}}{C_{\text{in}}}\right)\times 100\%$. For any value of N, the removal improves with increasing detention time, as would be expected. The ideal CFSTR ($N = 1$) is seen not to be very efficient; to achieve high removal efficiencies requires a high value of the dimensionless detention time. Manipulation of Eq. 4-67 with $N = 1$ shows that the required value of the

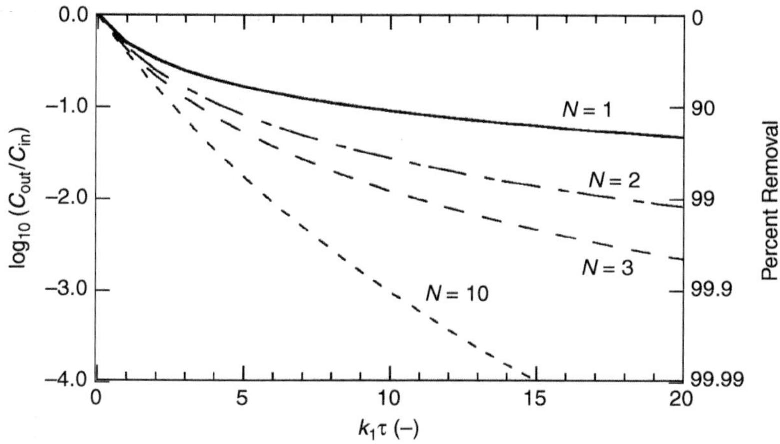

FIGURE 4-19 Steady state performance of a first order reaction according to the CFSTRs-in-series model.

parameter $k_1\tau$ equals the percent removal desired, so a 99 percent removal would require that $k_1\tau = 99$. For a given value of $k_1\tau$, the removal efficiency improves with increasing values of N, that is, as the hydraulic behavior becomes more plug flow-like. The reason why disinfection reactors are built with many baffles to increase the value of N is clear from Fig. 4-19. For the same 99 percent removal that requires $k_1\tau = 99$ for $N = 1$, the values of $k_1\tau$ drop to 18, 11, and 6 for $N = 2$, 3, and 10, respectively. Although not shown, the effect of increasing N on the removal achieved for second-order reactions is even more dramatic than for first-order reactions.

The plug flow with dispersion model can also be solved analytically to predict the steady state performance of a reactor with a first-order reaction. The result is expressed as

$$\frac{C_{out}}{C_{in}} = \frac{4a \exp\left(\frac{1}{2D_n}\right)}{(1+a)^2 \exp\left(\frac{a}{2D_n}\right) - (1-a)^2 \exp\left(-\frac{a}{2D_n}\right)} \tag{4-68a}$$

where D_n is the dimensionless dispersion number (the parameter of the dispersion flow model) and the dimensionless group a is defined as

$$a = (1 + 4k_1\tau D_n)^{1/2} \tag{4-68b}$$

The steady state performance of a first-order reaction according to the plug flow with dispersion model is shown in Fig. 4-20; the axes are identical to those used in Fig. 4-19, and the variation with the dispersion number D_n is indicated. The ideal PFR ($D_n = 0$) is the most efficient reactor possible, and increasing dispersion number reduces the efficiency of removal for the same value of the dimensionless detention time $k_1\tau$. The expected performance is strongly influenced by the dispersion; for example, a reactor designed with a value of $k_1\tau = 10$ will achieve 99.99 percent removal at $D_n = 0.01$ but approximately 99.8 percent removal at $D_n = 0.1$.

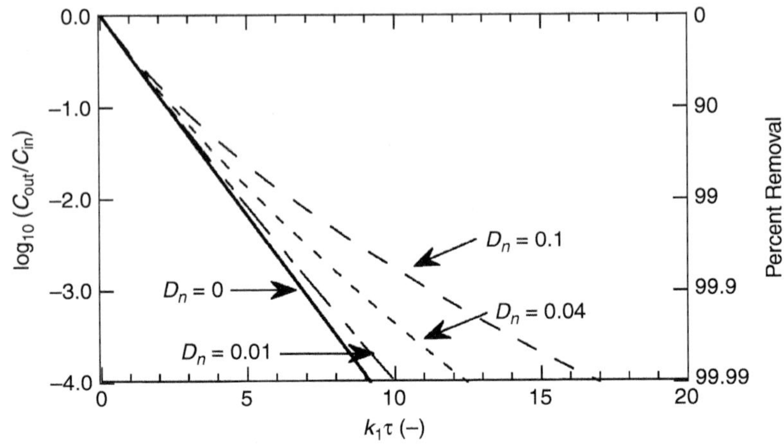

FIGURE 4-20 Steady state performance of a first order reaction according to the PFR with dispersion model.

Using the Exit Age Distribution Directly to Estimate Removal

In discussing the exit age distribution earlier in the chapter, we noted that a value of $E(t)\Delta t$ represented the fraction of the effluent that came into the reactor in an interval of time between $t - [(1/2)\Delta t]$ and $t + [(1/2)\Delta t]$ earlier. The example we used was that, if $E(t)\Delta t$ was 0.03 with $t = 15$ min and $\Delta t = 1$ min, we could say that 3 percent of a sample of effluent came into the reactor between 14.5 and 15.5 min earlier. If we now further hypothesize that this 3 percent of the water traveled together throughout the reactor from influent to effluent (i.e., that this packet was segregated from other parts of the flow), we can calculate the concentration of a reactive constituent in that packet as the concentration in a batch reactor with an initial concentration equal to the influent concentration and a reaction time of 15 min (or, more generally, t). The entire effluent under steady state conditions, then, is a weighted average of the concentrations in all the packets, with the weighting factor being the fraction of the total water in that packet.

$$C_{out} = \frac{\sum_{\text{all } i} C(t_i)E(t_i)\Delta t_i}{\sum_{\text{all } i} E(t_i)\Delta t_i} = \sum_{\text{all } i} C(t_i)E(t_i)\Delta t_i \tag{4-69}$$

The second equality recognizes that the denominator has a value of 1 (i.e., the sum of all the fractional aliquots of water is 1). If $E(t)$ can be expressed as a continuous function, then the equation can be written in integral form as

$$C_{out} = \int_0^\infty C(t)E(t)\,dt \tag{4-70}$$

It is important to realize that these expressions are valid for steady state conditions, despite the variable t in the equations; this time is simply the distribution of times for different packets of water under steady flow. The values of, or expressions for, $C(t)$ to be used in these equations are those for batch reactions (with the substitution of C_{in} for C_0): Eq. 4-48 for a first-order reaction, Eq. 4-49 for zero order, 4-51 for second order, and others shown earlier for more specialized reaction rate expressions.

As indicated, the analysis above assumes that the flow is segregated, and such flow probably never happens in a non-ideal reactor. However, the assumption turns out to be irrelevant for both first-order and zero-order reactions. For such reactions, the value of $C(t)/C_{\text{in}}$ is independent of C_{in}, and that means that mixing of one packet with another along the flow path does not change the fractional removal in the next segment of the reactor. Therefore, the analysis for zero- or first-order reactions does not really rely on the assumption of segregated flow and is accurate under all conditions.[3] For a zero-order reaction, the flow pattern makes no difference in the fractional removal achieved; the only influence in the fraction removed is the average detention time.[4] While an analysis of the type described in Eqs. 4-69 and 4-70 can be used for zero-order reactions, a much simpler analysis based on τ alone leads to the same result. For any other reaction rate expression besides zero or first order, this analysis provides only an estimate (and the most optimistic one possible) of the removal achieved in a reactor; nevertheless, it is a tool that can be used as an estimate of what could occur in a continuous flow reactor under steady state conditions.

We now consider an example of the various equations and methods of estimating the removal achieved in real reactors. For the non-ideal reactor that we have considered throughout the chapter (with the tracer information in Fig. 4-1), the detention time was found to be 16.63 min. If the reactor were a plug flow reactor with that detention time, we would obtain 99 percent removal via a first-order reaction (say, disinfection according to Chick's law) if the reaction rate constant were 0.277 min^{-1}. That is, in a plug flow reactor,

$$\left(\frac{C_{\text{out}}}{C_{\text{in}}}\right)_{PFR} = \exp(-k_1\tau_{pfr}) = \exp[(-0.277\ \text{min}^{-1})(16.63\ \text{min})] = 0.01$$

If this same reaction occurred in a CFSTR with the same detention time, the fraction remaining would be

$$\left(\frac{C_{\text{out}}}{C_{\text{in}}}\right)_{CFSTR} = \frac{1}{1+k_1\tau_{CFSTR}} = \frac{1}{1+(0.277\ \text{min}^{-1})(16.63\ \text{min})} = 0.178$$

or, the fractional removal would be $\eta = (1-0.178) = 0.822$ or 82.2 percent.

For the non-ideal reactor, we have developed three hydraulic models, and for each, we can calculate an expected fraction remaining. Those three models are the N-CFSTRs-in-series model with values of $N = 3$ and $N = 4$ (both with the detention time of 16.63 min) and the two reactors in parallel model depicted in Fig. 4-13. For the CFSTRs-in-series model, we use Eq. 4-67.

For $N = 3$,

$$\left(\frac{C_{\text{out}}}{C_{\text{in}}}\right)_{N=3} = \frac{1}{\left(1+\dfrac{k_1\tau}{3}\right)^3} = \frac{1}{\left[1+\dfrac{(0.277\ \text{min}^{-1})(16.63\ \text{min})}{3}\right]^3} = 0.061$$

or $\eta = 1 - 0.061 = 0.939$.

[3]The reader can show, for example, that a two-reactor system consisting of a PFR with τ_{PFR} and a CFSTR with τ_{CFSTR} in series will give the same fraction remaining regardless of which reactor is first in the series for a first-order reaction but not for a second-order reaction.

[4]Recall that the removal for a zero order reaction is the same in a CFSTR and PFR of identical size, a clear indication that flow pattern makes no difference in the removal for zero order reactions.

For $N = 4$,

$$\left(\frac{C_{\text{out}}}{C_{\text{in}}}\right)_{N=4} = \frac{1}{\left(1+\frac{k_1\tau}{4}\right)^4} = \frac{1}{\left[1+\frac{(0.277\text{ min}^{-1})(16.63\text{ min})}{4}\right]^4} = 0.047$$

or $\eta = 1 - 0.047 = 0.953$.

For the two non-ideal reactors in parallel, each behaving as N-CFSTRs-in-series,

$$\left(\frac{C_{\text{out}}}{C_{\text{in}}}\right) = \frac{f_A}{\left(1+\frac{k_1\tau_A}{N_A}\right)^{N_A}} + \frac{f_B}{\left(1+\frac{k_1\tau_B}{N_B}\right)^{N_B}}$$

$$= \frac{0.63}{\left[1+\frac{(0.277\text{ min}^{-1})(12\text{ min})}{5}\right]^5} + \frac{0.37}{\left[1+\frac{(0.277\text{ min}^{-1})(24.5\text{ min})}{27}\right]^{27}} = 0.0501$$

or $\eta = 1 - 0.050 = 0.950$.

Finally, we use the direct combination of the exit age distribution ($E(t)$) and the integrated reaction rate expression ($\exp(-k_1 t)$) as explained above. A spreadsheet with the appropriate calculations is shown in Fig. 4-21; columns A and C are repeated from the spreadsheet shown in Fig. 4-3 and columns B and D are the average of successive values in

	A	B	C	D	E	F
1	Time	t_ave	*E(t)*	E_ave	exp (–k1*tave)	E_ave*exp (–k1*tave)* Δ*t*
2	(min)	(min)	(1/min)	(1/min)	(–)	(–)
3	0		0			
4	2.5	1.25	0.01	0.005	0.7073	0.0088
5	5	3.75	0.028	0.019	0.3539	0.0168
6	7.5	6.25	0.046	0.037	0.1771	0.0164
7	10	8.75	0.052	0.049	0.0886	0.0109
8	12.5	11.25	0.042	0.047	0.0443	0.0052
9	15	13.75	0.03	0.036	0.0222	0.0020
10	17.5	16.25	0.03	0.03	0.0111	8.32E-04
11	20	18.75	0.034	0.032	0.0056	4.44E-04
12	22.5	21.25	0.042	0.038	0.0028	2.64E-04
13	25	23.75	0.028	0.035	0.0014	1.22E-04
14	27.5	26.25	0.02	0.024	6.95E-04	4.17E-05
15	30	28.75	0.016	0.018	3.48E-04	1.57E-05
16	32.5	31.25	0.01	0.013	1.74E-04	5.66E-06
17	35	33.75	0.006	0.008	8.71E-05	1.74E-06
18	37.5	36.25	0.004	0.005	4.36E-05	5.45E-07
19	40	38.75	0.002	0.003	2.18E-05	1.63E-07
20	42.5	41.25	0	0.001	1.09E-05	2.73E-08
21	45	43.75	0	0	5.46E-06	0
22						
23					sum	0.0618

FIGURE 4-21 Spreadsheet for estimating the conversion in non-ideal flow.

columns A and C, respectively. For the analysis, Eq. 4-69 is rearranged and made explicit for the intervals as follows:

$$\frac{C_{\text{out}}}{C_{\text{in}}} = \sum_{\text{all intervals } i} \exp(-k_1 t_{ave,i}) E_{ave,i}\, \Delta t_i \tag{4-71}$$

In the spreadsheet, the values of $\exp(-k_1 t_{ave,i})$ are calculated in column E; for example, the value in row 4 is calculated as

$$\exp\left[(-0.277 \text{ min}^{-1})(1.25 \text{ min})\right] = 0.7073$$

The values in column F represent the unconverted material that is coming out of the effluent in the fraction of the fluid that stayed in the tank for each time interval. These values are calculated as the product of the values in columns D and E and also Δt_i; in this example, all of the values of Δt_i equal 2.5 min. So, for example, the value in row 4 is calculated as

$$\exp(-k_1 t_{ave,i}) E_{ave,i}\, \Delta t_i = (0.7073)(0.005 \text{ min}^{-1})(2.5 \text{ min}) = 0.0088$$

Finally, the overall fraction remaining is shown as the sum of the values in column F, as 0.0618; that is,

$$\frac{C_{\text{out}}}{C_{\text{in}}} = \sum_{\text{all intervals } i} \exp(-k_1 t_{ave,i}) E_{ave,i}\, \Delta t_i = 0.0618$$

or $\eta = 1 - 0.0618 = 0.9382$.

This last estimate uses the tracer information directly and completely, and so it is the best estimate of the removal that is achieved in the reactor. The other methods use models that encapsulate the tracer information in a small number of parameters, but those models do not fit the original data perfectly and therefore are not quite as reliable as this spreadsheet method in estimating the conversion. Coincidentally, the value calculated for the $N = 3$ CFSTRs in series is essentially identical; this agreement stems primarily from the fact that the model agrees very well with the tracer-derived $E(t)$ curve over the first few time periods, as discussed below.

The calculations in the spreadsheet illustrate a critical point about conversion in continuous flow reactors: the portion of the flow that only stays in the reactor for a few min is primarily responsible for the unreacted material that gets into the effluent. Note, for example, that the sum of the first four values in column F of the spreadsheet in Fig. 4-21 is 0.0529, while the sum of the entire column, as indicated above, is 0.0618; that is, the fraction of the water that stayed in the tank for 10 min or less was responsible for 0.0529/0.0618, or 85.6 percent, of the unreacted material. The $F(t)$ value at $t = 10$ min was reported in Fig. 4-3 as 0.275; that is, the 27.5 percent of the effluent water that stayed in the tank for 10 min or less contained 85.6 percent of the unreacted material. This fact illustrates why baffling to prevent water from going quickly from the influent to the effluent is so critical to obtaining excellent removal in most reactors.

REACTORS IN WATER TREATMENT AND THEIR HYDRAULICS CHARACTERISTICS

Surface water treatment systems are a series of reactors, each designed with a specific purpose (or, in some cases, multiple purposes). Throughout this chapter, we have discussed the possible hydraulic characteristics of reactors and the consequences of those characteristics on the achievement of reaction goals. However, that discussion has been generic in elucidating the principles of exit age distributions and the effects on reactions. In this section, the application of those principles in specific reactors of a water treatment plant is presented. The discussion that follows is summarized in Table 4-4.

A conventional surface water treatment plant has a pipeline from a water source to the plant. This supply pipeline can be, and often is, used as an oxidation or disinfection reactor with the addition of a chemical oxidant at or near a pump near the water source; less frequently, it is used as a particle destabilization reactor by the addition of metal coagulants or polymers. Pipelines leaving the plant are also often used as disinfectant reactors. Pipelines are usually modeled as plug flow reactors, although it would be more accurate

TABLE 4-4 Hydraulic Characteristics of Water Treatment Plant Reactors

Reactor	Assumed hydraulic characteristics in design or process models	Actual hydraulic characteristics in common designs	Possible improvements to common designs
Pipelines	Plug flow	Plug flow with small amount of dispersion so plug flow is good assumption	Improved mixing at point of chemical addition
Rapid mix tanks	CFSTR	CFSTR* or small number of CFSTRs in series	Series of true CFSTRs if order of chemical addition is critical
Flocculation tanks	PFR	2 or 3 CFSTRs in series	Serpentine flow through all but last reactor
Sedimentation	PFR	2 or 3 CFSTRs in series	Baffling to improve plug-flow like conditions
Dissolved air flotation	PFR	Near PFR in contact zone and 2 or 3 CFSTRs in series in separation zone	Addition of tubes or plates in the separation zone to improve hydraulics and performance
Granular media filtration	PFR	Very close to PFR†	None
Membrane filtration	PFR	Very close to PFR	None
Clearwells (traditional)	---	Large dead zones; active area behaves like few CFSTRs in series	Baffling to improve plug-flow-like conditions and eliminate dead zones
Clearwells (designed for disinfection credit)	PFR	Multiple CFSTRs in series	Further baffling if necessary
Elevated storage tanks	---	Susceptible to temperature stratification, leading to large dead zones	Vertical influent nozzles to provide high velocity and induce thorough mixing

* Continuous flow stirred tank reactor.
† Plug flow reactor.

to describe them as plug flow with a small amount of dispersion. Flow in pipelines is usually turbulent, and turbulent flow in a pipe flowing full is characterized by nearly uniform velocity throughout most of the cross section (and a sharply diminishing velocity in a laminar boundary layer as one approaches the pipe's wall). Because of the nearly plug flow characteristics, pipelines are excellent reactors, and should probably be used more often than they traditionally have been as oxidation and/or particle destabilization reactors. The point of addition should be near the pump or other point of hydraulic change (bend, expansion, or reduction in the pipe) to ensure rapid integration of the added chemical into the entire water flow.

The first reactor in most plants is the rapid-mix tank, where one or several chemicals are added for various purposes. Possibilities include any of the oxidants/disinfectants, the particle destabilization chemicals (synthetic polymers, metal coagulants), pH adjustment chemicals (lime and/or soda ash as bases or sulfuric or hydrochloric acid), and precipitation chemicals (lime for softening). Several of these chemicals might need to be added in the same reactor, and the order of addition of some combinations is also critical. The hydraulic design must be performed carefully because of the somewhat conflicting goals of complete and instantaneous mixing required to mix the chemicals with the water and the desire for plug flow conditions for the initiation of the reactions for which those chemicals are added. Many plants include a single completely mixed reactor (CFSTR) for this part of the plant, but the needs for several additions and the importance of the order of addition in several cases mean that a multiple part reactor (i.e., a real set of CFSTRs in series, not just a single reactor modeled that way) is better; a multiple-reactor design is becoming more common. The exit age distribution ($E(t)$) of a single CFSTR suggests that much of the material that enters at any instant leaves the reactor in a small fraction of the detention time; if the reaction required for the added chemical requires contact with other constituents, that rapid exit of material might not be favorable. A growing trend in treatment plant design is to use the pipeline as all or most of the rapid-mix tank, because it combines the benefits of mixing (via the turbulence) with the benefits of plug flow. Klute and Amirtharajah (1991) noted that a major benefit of the intimate mixing of particles and metal coagulants in turbulent pipe flow was to reduce the amount of coagulant required to obtain excellent particle destabilization; apparently, the achievement of particle-coagulant mixing more rapidly than the time required for the coagulant's hydrolysis reactions resulted in a substantial improvement in the efficiency of destabilization.

Flocculation is a second-order reaction in particle number; as indicated earlier in this chapter, achieving conditions that are as close to plug flow as possible is particularly valuable for second-order reactions. On the other hand, maintaining sufficient mixing that the particles stay in suspension so that they can contact one another and not settle to the bottom is also critical. Traditionally, many flocculators have been built as three CFSTRs in series, often with increasing volume and decreasing mixing intensity from the first to the third, so as to avoid breakup of flocs. This design could be improved further by imposing a serpentine flow (much like many disinfection reactors) through the first two compartments to make each act like three or four CFSTRs in series, assuming floc breakup could still be avoided. Adding baffling to existing flocculators to make the hydraulic pattern more plug flow-like would likely improve the flocculation achieved, perhaps substantially. Han and Lawler (1992) suggested that the intensity of mixing (as measured by the velocity gradient G) is less important as a means of bringing particles together than had been previously thought; modern flocculators in sedimentation plants are designed with lower G values (in the range of 15 to 25 s^{-1} and some even lower) than was common in earlier times.

Because flocculation is a second-order reaction (i.e., because high particle concentrations increase the rate of reaction), many designs of flocculators incorporate means of maintaining high particle concentrations. The use of metal coagulants to precipitate new solids is the first way in which this goal is commonly accomplished; especially when the turbidity of the raw water is low, this benefit of metal coagulants is essential. The other means of maintaining high particle concentrations are to either recycle solids from the settled (or floated) residuals in the subsequent gravity separation device or to make it more difficult for particles than for water to leave the flocculator. This latter means is often accomplished in reactors that combine flocculation and sedimentation by maintaining a high concentration of particles in the central section of a circular reactor where flocculation is the dominant process occurring.

Traditional sedimentation tanks are large open reactors with no baffling between the influent structure and the effluent weirs. Both the influent structure and the weirs are designed to spread the flow over the entire surface area, since the surface area (for a given flow rate) is the primary design feature. The traditional (and still widely used) analysis of sedimentation tanks developed by Hazen (1904) assumes plug flow conditions. However, the lack of baffling and the exposure to wind at the surface means that a substantial amount of mixing generally occurs; tracer studies indicate that they behave like a small number (two to four) of CFSTRs in series. Adding baffling can be difficult because of the requirement for mechanisms to remove residuals, but manufacturers now have designs that improve the traditional designs with respect to baffling and hydraulic behavior. While sedimentation is not a reaction in the same sense as those noted earlier in the chapter, the value of nearly plug flow conditions follows the same trend. Sedimentation in water treatment plants is actually "flocculent sedimentation"; that is, flocculation continues simultaneously with sedimentation in these reactors. Improvements in baffling in the sedimentation tanks in a treatment plant in Austin, Texas, for example, reduced the particle load on the subsequent gravity filters so much that the filter runs were increased from 24 to 72 h. Tube or plate settlers not only provide a great increase in the surface area available for settling but also exhibit hydraulic behavior that is much closer to plug flow conditions than traditional sedimentation tanks; retrofitting an existing, traditional open sedimentation tank with tubes or plates can have a dramatic effect on the removal achieved.

Dissolved air flotation units consist of a contact zone and a separation zone. In the contact zone, where the bubbles form and move upward by the balance of gravity and buoyancy (and contact and stick to particles), the hydraulic characteristics are best described as plug flow with dispersion. The drag on the rising bubbles causes water to also flow upward in this part of the reactor; this water flow produces the nearly plug flow conditions. In the separation zone, plug flow is desirable (as in sedimentation), but the flow is often mixed for low-rate DAF systems and the high-rate systems have a complicated flow regime (see Chap. 9).

Granular media filters (either pressure filters or gravity filters) are packed bed reactors with an unbaffled water layer above. As noted earlier in the chapter, packed beds have nearly plug flow hydraulic characteristics. The water layer above the media is mixed to a substantial degree by the momentum of the influent flow which usually comes into the reactor horizontally through a pipe opening; this part of the reactor behaves as one or two CFSTRs in series. Since no reactions are expected to be occurring in that part of a treatment plant and the detention time is low in that water layer, changing the hydraulics of that part of most plants would have little value. Membrane filters also have very good hydraulic characteristics in achieving nearly plug flow conditions, because the head loss is spread evenly over the membrane surface area.

Clearwells are the final reactor in a water treatment plant. In earlier times, clearwells (and elevated storage tanks within distribution systems) were thought of only as storage devices; they were sized to allow for the plant to operate at nearly steady-flow conditions while meeting the fluctuating demand in the distribution system. Now, clearwells are a critical part of the disinfection that is achieved in water plants. As in all disinfection reactors, the importance of avoiding short circuiting (where some of the flow proceeds from the influent to the effluent in a tiny fraction of the detention time) must be emphasized.

Proper baffling to lengthen T_{10} to at least 0.4τ should be a goal in every treatment plant. The prevention of short circuiting usually also prevents the creation of *dead space,* a portion of the reactor where water sits for very long periods of time without participating in the flow pattern. Dead space not only decreases the average residence time of the remaining water but can also directly lead to the deterioration of water quality in the *dead water* (e.g., loss of disinfectant residual), and so it too must be avoided. As in any disinfection reactor (or other reactor in which a high removal efficiency is desired), providing sufficient baffling so that water must take a highly circuitous path from the influent to the effluent creates the best-quality effluent by preventing both short circuiting and dead space.

SUMMARY

The hydraulic characteristics of reactors in existing water treatment plants can be determined by tracer studies. For plants under design, computational fluid dynamics can be a useful tool for modeling the effects of different design details on the expected hydraulic behavior. The results of tracer studies can be converted to the exit age distribution or the cumulative age distribution of a reactor (or series of reactors). These distributions then can be either used directly to predict the effluent quality for known reactions or fit to relatively simple mathematical models of the hydraulic behavior. For most reactions of interest, achieving hydraulic conditions that approach plug flow behavior is of great value in achieving excellent water quality; these conditions are accomplished through careful design of influent and effluent structures, long length to width ratios, and adequate baffling.

ABBREVIATIONS

CFD	computational fluid dynamics
CFCMR	continuous flow completely mixed reactor
CFSTR	continuous flow stirred tank reactor
DAF	dissolved air flotation
PFR	plug flow reactor
SDWA	Safe Drinking Water Act
USEPA	U.S. Environmental Protection Agency
UV	ultraviolet light

NOTATION FOR EQUATIONS

A	cross-sectional area
a	dimensionless group in equation for first-order reaction in a reactor described by the plug flow with dispersion model
$C_{ave,i}$	average concentration in the time interval between t_{i-1} and t_i
C_{baseline}	concentration throughout a reactor and in the influent prior to initiation of a step input tracer test

C_{in}	influent concentration to a continuous flow reactor (generally constant)
C_0	initial concentration in (or exiting) a reactor
$C_p(t)$	effluent concentration at time t in pulse input tracer test
$C_S(t)$	effluent concentration at time t in step input tracer test
D_n	dispersion number, parameter of the plug flow with dispersion model (– –)
$E(t)$	exit age distribution (time^{-1})
$E_{ave,i}$	average value of the exit age distribution in the time interval between t_{i-1} and t_i
$F(t)$	cumulative age distribution (– –)
f_A	fraction of flow in compartment A in non-ideal flow modeled as reactors in parallel
G	velocity gradient in flocculation (s^{-1})
k	rate constant (various dimensions) or
k	turbulent kinetic energy
K_S	half-velocity constant in Monod kinetic expression
L	total length
M or M_{in}	mass of tracer input in pulse test
m	parameter in the PFR-CFSTR dead-space model, the fraction of the reactor that is dead space
m	stoichiometric ratio (one molecule of B reacts with m molecules of A)
M_{out}	mass of tracer accounted for by the effluent in pulse test
N	number of size classes in a discrete particle size distribution
N	parameter of the CFSTRs-in-series model for non-ideal flow; the number of equal CFSTRs that the single reactor is (conceptually) composed of
n	reaction order
n_i	number concentration of particles in size class i
p	parameter in the PFR-CFSTR dead-space model, the fraction of the active volume that is plug flow
Q	volumetric flow rate (L$^3t^{-1}$)
r	reaction rate expression (ML$^{-3}t^{-1}$)
S	substrate concentration
t	time
$t_{ave,i}$	average time in the time interval between t_{i-1} and t_i
$\overline{t}$	average hydraulic detention time calculated from tracer or exit age distribution data
T_{10}	time at which $F(t) = 0.10$
V	volume of water in a reactor
X	microorganism concentration
x	length dimension
α_{emp}	empirical collision efficiency factor in flocculation
ε	turbulent energy dissipation rate

$\gamma(i, j)$	rate constant for collisions between particles of sizes i and j
σ^2	variance of a distribution
τ	theoretical detention time

REFERENCES

Berg, G. (1964). The virus hazard in water supplies. *J. New England Water Works Assoc.* 78, 79–104.

Blatchley, E. R., Shen, C., Scheible, O. K., Robinson, J. P., Ragheb, K., Bergstrom, D. E., and Rokjer, D. (2008) Validation of large-scale, monochromatic UV disinfection systems for drinking water using dyed microspheres. *Water Research* 42, 677–688.

Chen, C., and Jaw, S. (1998). *Fundamentals of Turbulence Modeling,* Washington, DC, Taylor & Francis.

Chick, H. (1908). An investigation of the laws of disinfection. *J. Hygiene* 8, 92–158.

Clark, M. M. (2005). *Transport Modeling for Environmental Engineers and Scientists*. New York: Wiley.

Crozes, G. F., Hagstrom, J. P., Clark, M. M., Ducoste, J., and Burns, C. (1998). *Improving Clearwell Design for CT compliance*, Awwa Research Foundation report. Denver, CO: Water Research Foundation.

Ducoste, J., Linden, K., Rokjer, D., and Liu, D. (2005). Assessment of reduction equivalent fluence bias using computational fluid dynamics. *Environmental Engineering Science* 22, 615–628.

Garg, V. K. (1998). *Applied Computational Fluid Dynamics*, New York, Marcel Dekker.

Goula, A. M., Kostoglou, M., Karapantsios, T. D., and Zouboulis, A. I. (2008). The effect of influent temperature variations in a sedimentation tank for potable water treatment–A computational fluid dynamics study, *Water Research* 42, 3405–3414.

Grayman, W. M., Deininger, R. A., Green, A., Boulos, P. F., Bowcock. R. W., and Godwin, C. C. (1996). Water quality and mixing models for tanks and reservoirs, *Journal AWWA* 88, 60–73.

Han, M. Y., and Lawler, D. F. (1992). The (relative) insignificance of G in flocculation, *Journal AWWA* 84, 79–91.

Hannoun, I. A., Boulos, P. F., and List, E. J. (1998). Using hydraulic modeling to optimize contact time, *Journal AWWA* 90, 77–87.

Hazen, A. (1904). On sedimentation, *Transactions, ASCE* 53, 45–98.

Klute, R., and Amirtharajah, A. (1991). Particle destabilization and flocculation reactions in turbulent pipe flow. In *Mixing in Coagulation and Flocculation,* ed. Amirtharajah, A., Clark, M. M., and Trussell, R. R. (Chap. 6, 217–255), Denver, CO: American Water Works Association.

Lawler, D. F., and Singer, P. C. (1993). Analyzing disinfection kinetics and reactor design: A conceptual approach versus the Surface Water Treatment Rule, *Journal AWWA* 85, 67–76.

Levenspiel, O. (1999). *Chemical Reaction Engineering,* 3d ed., New York, Wiley.

Liu, D., Wu, C., Linden, K., and Ducoste, J. (2007). Numerical Simulation of UV Disinfection Reactors: Evaluation of Alternative Turbulence Models, *Applied Mathematical Modeling* 31, 1753–1769.

Liu, Y., and Ducoste, J.J. (2006). Impact of turbulent mixing on the performance of a CFD chloramine model, *Environmental Engineering Science* 23, 341–356.

Mahmood, F., Pimblett, J. G., Grace, N. O., and Grayman, W. M. (2005). Evaluation of water mixing characteristics in distribution system storage tanks, *Journal AWWA* 97, 74–88.

Munoz, A., Craik, S., and Kresta, S. (2007). Computational fluid dynamics for predicting performance of ultraviolet disinfection–Sensitivity to particle tracking inputs, *Journal of Environmental Engineering Science* 6, 285–301.

MWH. (2005). *Water Treatment Principles and Design,* New York: Wiley.

Naunovic, Z., Lim, S., and Blatchley, E. R., III. (2008). Investigation of microbial inactivation efficiency of a UV disinfection system employing an excimer lamp, *Water Research* 42, 4838–4846.

Prat, O. P., and Ducoste, J. J. (2007). Simulation of flocculation in stirred vessels: Lagrangian versus Eulerian, *Chemical Engineering Research and Design* 85, 207–219.

Rebhun, M., and Argaman, Y. (1965). "Evaluation of Hydraulic Efficiency of Sedimentation Basins, *Journal Sanitary Engineering Division, ASCE* 91, 37–45.

Teefy, S. (1996). *Tracer Studies in Water Treatment Facilities: A Protocol and Case Studies*, Awwa Research Foundation report. Denver, CO: Water Research Foundation.

Templeton, M. R., Hofmann, R., and Andrews, R. C. (2006). Case study comparisons of computational fluid dynamics (CFD) modeling versus tracer testing for determining clearwell residence times in drinking water treatment, *Journal of Environmental Engineering and Science* 5, 529–536.

U.S. Environmental Protection Agency. (1991). Guidance Manual for Compliance with the Filtration and Disinfection Requirements for Public Water Systems Using Surface Water Sources. EPA number 570391001.

Von Gunten, U. (2003a). Ozonation of drinking water: Part I. Oxidation kinetics and product formation. *Water Research* 37, 1443–1467.

Von Gunten, U. (2003b). Ozonation of drinking water: Part II. Disinfection and by-product formation in presence of bromide, iodide or chlorine. *Water Research* 37, 1469–1487.

Watson, H. E. (1908). A note on the variation of the rate of disinfection with change in concentration of the disinfectant, *Journal of Hygiene* 8, 536–542.

Weber, W.J. Jr., and DiGiano, F.A., (1996) Process Dynamics in Environmental Systems, John Wiley & Sons, New York.

Zhang, J., Huck, P. M., and Anderson, W. B. (2004). Optimization of a full-scale ozone disinfection process based on computational fluid dynamics analysis. In *Chemical Water and Wastewater Treatment VIII, Proceedings of the 11th Gothenburg Symposium*, ed. Hahn, H. H., Hoffmann, E., and Odegaard, H. (325–334). London: IWA Publishing.

Zhao, X., Alpert, S. M., and Ducoste, J. J. (2009). Assessing the Impact of Upstream Hydraulics on the Dose Distribution of Ultraviolet Reactors Using Fluorescence Microspheres and Computational Fluid Dynamics, *Environmental Engineering Science* 26, 947–959.

CHAPTER 5
OVERVIEW OF WATER TREATMENT PROCESSES

Doug Elder, P.E.
Senior Water Treatment Process Engineer
Black & Veatch
Kansas City, Missouri, United States

George C. Budd, Ph.D., P.E.
Senior Process Specialist
Black & Veatch
Harborton, Virginia, United States

INTRODUCTION **5.2**
SOURCE WATER QUALITY CONSIDERATIONS (CHAP. 3) **5.2**
Surface Water 5.2
Groundwater 5.3
Water Quality 5.4
CHARACTERISTICS AND GENERAL CAPABILITIES OF UNIT PROCESSES **5.4**
Aeration and Air Stripping (Gas-Liquid Processes; Chap. 6) 5.5
Chemical Oxidation (Chap. 7) 5.5
Coagulation and Flocculation (Chap. 8) 5.7
Sedimentation and Flotation (Chap. 9) 5.9
Granular Media Filtration (Chap. 10) 5.11
Membranes (Chap. 11) 5.13
Ion Exchange and Adsorption of Inorganic Contaminants (Chap. 12) 5.16
Chemical Precipitation (Chap. 13) 5.17
Adsorption of Organic Compounds (Chap. 14) 5.19
Natural Treatment Systems (Chap. 15) 5.21
Chemical Disinfection (Chap. 17) 5.22
UV Technologies (Chap. 18) 5.24
DISTRIBUTION SYSTEM CONSIDERATIONS (CHAPS. 19–21) **5.25**
TREATMENT PROCESS RESIDUALS MANAGEMENT (CHAP. 22) **5.27**
OTHER CONSIDERATIONS **5.28**
Pilot Testing 5.28
Flexibility to Meet More Stringent Future Requirements 5.28
Environmental Considerations 5.28
Use of "Demonstrated Technology" 5.29
TREATMENT PROCESS CONFIGURATIONS **5.29**
High-Quality Reservoir Supply 5.32
Reservoir with Algae and Color/TOC 5.33
River or Reservoir Supply with Low to Moderate Hardness 5.33
River Supply with Hardness 5.34
Groundwater with Iron and Manganese 5.35
Groundwater with Hardness 5.36
ABBREVIATIONS **5.36**
REFERENCES **5.38**

INTRODUCTION

Selection of appropriate unit processes and integration of the processes into an overall water treatment plant when assessing site-specific treatment needs is a complex task. Factors that are typically considered during treatment process evaluation and selection include (1) source water quality, (2) regulatory compliance and contaminant removal requirements, (3) process reliability and flexibility, (4) initial construction and annual operating and maintenance costs, (5) environmental impacts, (6) utility preferences and capabilities, (7) available site space, and (8) residuals handling requirements and site constraints.

The purpose of this chapter is to provide a general description of the various unit processes included in the design of water treatment facilities and their effectiveness and applicability for achieving broad treatment objectives. It is intended to serve as a guide and to direct the reader to more detailed information in the chapters that follow.

SOURCE WATER QUALITY CONSIDERATIONS (CHAP. 3)

Foremost among the considerations for selecting sources of supply for a water system is provision of adequate quantity to meet system demands over a wide range of conditions. Other factors that must be considered are the amount of effort and cost to withdraw and convey the water to the point of use and the water quality characteristics that affect treatment needs and the ultimate acceptability of the potential source. The complexity of these issues varies with location. In some locations, adequate supplies of high-quality water are available in reasonable proximity to the area to be served, whereas others involve vast conveyance systems and multiple jurisdictions.

In recent years, communities with limited available freshwater supplies have begun to look for new types of sources, including brackish and saline waters that can be treated by technologies such as desalination. The potential for reuse of wastewater for either nonpotable or potable applications is another consideration. In the case of potable applications, a distinction is made between indirect potable reuse, where the effluent is returned to a source of supply and subsequently conveyed to a water treatment process, and direct potable reuse, where highly treated effluent is conveyed to a water distribution system for direct consumption (see Chap. 16).

Water sources traditionally have been categorized as either surface or groundwater supplies, both of which have several general characteristics, as discussed below. Details on the compositions of these types of supplies are presented in Chap. 3.

Surface Water

Surface water is open to the atmosphere, which brings it into contact with oxygen, a key regulator of oxidizing conditions in natural waters. Much of the chemistry of surface water is dominated by the presence of oxidized constituents. An exception occurs in the case of impoundments that undergo stratification, with the accompanying oxygen deficit in the lower levels, where the presence of biodegradable substances leads to microbial decay of oxygen residuals. In such impoundments, the level of withdrawal and the time of year dictate the extent to which reducing conditions develop. Strategies such as mixing to limit stratification and various methods of in-lake aeration can be considered to alleviate such conditions.

Potential for exposure to microbial pathogens in surface supplies leads to the use of disinfection as a treatment step in these supplies. In addition, particle-removal processes such as filtration typically are applied as a component of a multiple-barrier approach to provide high levels of safety.

Runoff that enters surface supplies contains natural organic matter (NOM) that reacts with disinfectants to form by-products, some of which are currently regulated. Production of NOM also can result from the activity of algae and other biota within water bodies. The concentration of NOM and the associated potential formation of disinfection by-products (DBPs) depend on the nature of the watershed and can vary seasonally as well. The extent of this variation should be considered when developing treatment strategies.

Point-source waste discharges and non-point-source surface runoff can directly influence the presence of contaminants in surface supplies. Although the volume of flow available in most surface supplies often reduces contaminant concentrations to relatively low levels, concerns have been raised even at extremely low concentrations for new categories of contaminants such as endocrine-disrupting compounds (EDCs) and pharmaceutical and personal care products (PPCPs). While the detrimental impacts of some of these contaminants on aquatic life and wildlife have been documented, the importance of low-level human exposure to EDCs and PPCPs has not yet been determined (see Chap. 2).

Trophic status of impounded supplies can be an important consideration because algal activity can lead to treatment problems that challenge a number of processes. Problems with tastes and odors and with soluble iron and manganese that increase in oxygen-depleted lower strata of impoundments also accompany algal activity. Problems that stem from excess discharges of nutrients often can be resolved by watershed management measures that are also applicable for limiting exposure to a broader range of contaminants.

Storm events can be detrimental to the quality of some surface supplies by increasing turbidity and NOM concentrations, as well as by increasing nutrient loadings and the concentration of microbial contaminants. Facility design and operating concepts should incorporate provisions for dealing with these and other causes of extreme water quality conditions.

Groundwater

Groundwater supplies are confined to substrata, thereby limiting the availability of oxygen, which can lead to the need for different treatment methods than those used with surface supplies. These conditions often lead to treatment to deal with reduced substances such as iron, manganese, hydrogen sulfide, ammonia, and reduced arsenic.

The substrata also can protect the water against contamination by microbial pathogens, resulting in treatment approaches for this category of contaminant that differ from those applied for surface-water supplies. Multiple-barrier approaches that combine disinfection with particle-removal processes are not needed for some groundwater supplies. However, not all groundwater supplies are the same, and some are subject to direct surface-water influence, requiring the use of treatment barriers for pathogens in such cases.

Quality of groundwater varies widely depending on the nature of the geology through which it flows (see Chap. 3). Variations can be regional or across the aquifer layers within a region, resulting in wide differences in required treatment sequences that depend on general characteristics such as hardness and total dissolved solids and the presence of specific natural contaminants such as radon and radionuclides. While total organic carbon (TOC) concentrations in most groundwater supplies are generally low (<2 mg/L), some groundwater supplies contain elevated concentrations of TOC that require treatment strategies to limit the formation of regulated disinfection by-products.

The confinement of groundwater can restrict flow, which can affect the rate of withdrawal from an aquifer and the corresponding yield for a given well. Often a system is supplied from multiple wells, each of which may have its own water quality characteristics and susceptibility to contamination that affects blended quality, thereby requiring management and operating decisions to make best use of these resources.

Groundwater aquifers have long residence times and thus are renewed with less frequency than surface water systems. Therefore, should aquifers become contaminated, the contaminants can occur at much higher concentrations than surface supplies, which places consumers at greater risk of extended exposure to contaminants unless effective treatment measures are provided.

Water Quality

Several general water quality variables must be considered when selecting and configuring the treatment sequence for a given water source regardless of whether it is a surface-water or groundwater supply, including the following:

1. *pH.* This variable has major effects on the chemistry of constituents in water and on treatment-process performance. It is therefore imperative that the pH of the source water is monitored routinely and that the pH profile through the treatment process is controlled. pH also affects corrosion and water quality in the distribution system and thus must be controlled to minimize water quality impacts.
2. *Alkalinity.* Alkalinity is the measure of the acid-neutralizing capacity of a solution. It is defined by the amount of acid required to reduce the pH to a defined end point. For the pH conditions of most supplies, alkalinity is due mainly to the bicarbonate concentration. Alkalinity is an important factor in coagulation and in selecting the corrosion-control methodology to be implemented. Acidity, a related parameter, is a measure of the base-neutralizing capacity. Principles pertaining to alkalinity and acidity are discussed in Chap. 3.
3. *Hardness.* The level of hardness can dictate treatment concepts that should be applied. A high level of hardness tends to favor softening over clarification only and can affect a range of treatment processes. Like alkalinity, it also can affect the choice of corrosion-control methods.
4. *Turbidity.* Turbidity is a measure of the particulate matter in the water. It affects the choice of clarification methods and can dictate whether or not there is a need for pretreatment upstream of some other processes.
5. *Natural organic matter.* NOM presents a concern for disinfection by-product formation, increases coagulant and oxidant demand, and can affect a number of treatment processes. It can be characterized by several surrogate measurements, including TOC concentration and ultraviolet (UV) absorbance (see Chap. 3).
6. *Total dissolved solids (TDS).* This measure of salt and mineral content can affect treatment needs as well as the acceptability of a source of supply.
7. *Dissolved oxygen.* Dissolved oxygen is an important regulator of oxidation-reduction conditions that determine the chemical speciation of a number of constituents in water (see Chap. 3). Oxidizing conditions are produced even when oxygen is present at low concentrations.

In addition to these general parameters, water sources can have unique quality concerns that must be addressed.

CHARACTERISTICS AND GENERAL CAPABILITIES OF UNIT PROCESSES

This section provides an introduction to the treatment processes discussed in the chapters that follow. The objectives are to present a broad overview of the various process components and their typical applications and capabilities and to serve as a "road map" to the contents of the remainder of the book. A more detailed discussion of each unit process is presented in the referenced chapters.

Aeration and Air Stripping (Gas-Liquid Processes; Chap. 6)

Air stripping and aeration are gas-transfer processes that have played important roles in water treatment for many years, including transfer of oxygen for oxidation of constituents such as iron and manganese, removal of certain tastes and odors, and removal of dissolved gases such as hydrogen sulfide and carbon dioxide. Air stripping typically is used to remove volatile organic compounds (VOCs) from contaminated groundwaters that became a concern in the 1980s and newer contaminants such as methyl-*tert*-butyl-ether (MTBE) and radon. Air stripping to remove trihalomethanes (specifically, chloroform) has been used in some facilities, but this approach has proven somewhat limited in scope because most disinfection by-products are not amenable to removal by air stripping.

Aeration is not appropriate for treatment of groundwater containing reduced arsenic and iron because iron is oxidized more quickly than arsenic by oxygen. For effective removal of arsenic, both arsenic and iron need to be oxidized at the same time by a strong oxidant such as free chlorine.

Gas transfer processes are most applicable when there is an imbalance in the activity of the constituent in the gaseous and liquid phases, where activity in the gas phase is generally characterized by the partial pressure, and activity in the liquid phase is characterized by concentration. This imbalance causes constituents to flow from the condition of higher activity in the direction of lower activity to balance the conditions in the two phases. The rate of transfer depends on the extent of the imbalance, the surface area available for transfer, the degree of mixing and turbulence, temperature, and intrinsic diffusion rates of the target constituent. Constituents such as VOCs are readily removed by air stripping. For constituents that can react in solution, the rate and extent of the transfer are also affected by factors such as pH and other reactive constituents in solution, such as chlorine, carbon dioxide, hydrogen sulfide, ammonia, and oxygen. Constituents that have little or no tendency for reaction in solution, such as radon and VOCs, recently have become targets for treatment by air stripping.

The susceptibility of a target constituent to gas transfer depends on its solubility in water relative to its corresponding partial pressure or concentration in the gas phase. Gases that are poorly soluble in water relative to their concentrations in the gas phase can be readily removed. The most common measure of these relationships is *Henry's constant*, which is the ratio of the activity of a constituent in gas to the activity in solution at equilibrium. The higher the Henry's constant of the constituent, the more readily removable it is by gas transfer.

Equipment for bringing gas and water together to produce gas transfer generally is configured to produce a large surface area at the boundary between the two phases and, in some cases, a degree of turbulence, both of which are necessary for accelerating the rate of transfer. This equipment includes spray nozzles, diffusers that control the size of the bubbles for the gaseous phase, and tray aerators and packed-bed towers that spread the surface of the interface across specially designed media. In some cases, treatment of aerator/stripper off-gas may be required to remove contaminants in order meet air quality criteria.

Chemical Oxidation (Chap. 7)

The basic purpose of chemical oxidation is to change the oxidation state of the constituents involved. In the context of water treatment, the intent is to produce changes that improve water quality, but there is also the potential for the formation of undesired reaction products (such as halogenated disinfection by-products when chlorine is used as the oxidant), an outcome that can require wise decision making.

Two different types of chemical oxidation are used commonly in water treatment. The more direct method consists of applying the chemical oxidant to produce the desired treatment outcome in a single step, such as in the oxidation of taste- and odor-causing compounds by ozone. With the second method, chemical oxidation is a component of a multistep sequence whereby the oxidant is used to alter the oxidation state of the target constituent to a form that is more readily removable by subsequent treatment steps. Oxidation of reduced forms of iron and manganese to higher oxidation states that will readily precipitate is one example of this method. Removal of sulfides and arsenic in multistep sequences is another example. Since all chemical disinfectants are oxidants, oxidation will occur to some degree even when the objective is inactivation of microbial contaminants.

Chemical oxidants used most commonly in water treatment include

1. *Chlorine*. Chlorine is used widely in water treatment as a disinfectant. It can oxidize iron and manganese, color, and some taste- and odor-causing compounds. It is especially effective in manganese removal when applied ahead of filtration to maintain an oxidized coating on the granular media. This method of manganese removal, while often inadvertent, is an important consideration if points of chlorine addition are modified to meet other treatment objectives. Chlorine is available commercially in several forms: as a compressed gas, in aqueous liquid form (sodium hypochlorite), and in solid form (calcium hypochlorite). Another alternative that has received increased attention in recent years is on-site electrolytic generation of sodium hypochlorite from salt (see Chap. 17).
2. *Permanganate.* Permanganate can be used for oxidation of iron and manganese and for removal of some taste and odors. Traditionally, it has been available as potassium permanganate, which is delivered as a solid that requires handling and dissolution prior to application. Recently, liquid sodium permanganate has become available and can be applied directly as delivered.
3. *Ozone.* Ozone is a strong oxidant that is effective for reduction of color, control of a broad range of taste- and odor-causing compounds, and conversion of NOM to forms more readily removed by biological filtration. It sometimes can improve coagulation and filtration and reduce the formation of some disinfection by-products, notably trihalomethanes and haloacetic acids. While also effective for oxidation of iron and manganese, use of ozone for control of manganese has proved complex and problematic in some cases because of the formation of colloidal manganese oxide particles. Overdosing of ozone also can result in the formation of permanganate, which can pass through filters and later precipitate manganese oxides in the distribution system. Because ozone is highly reactive and not sufficiently stable for storage, it is produced on site using generation equipment that applies a controlled electrical discharge to interact with oxygen to form ozone. Sources of oxygen for this process include ambient air that has been treated to remove moisture and impurities, commercially manufactured liquid oxygen, and application of equipment to produce a concentrated oxygen stream on site. While use of ambient air was the predominant approach in early ozone-generation systems, newer designs typically use more concentrated oxygen streams that yield improved generation efficiencies and higher ozone concentrations in the feed gas.
4. *Chlorine dioxide.* Chlorine dioxide can oxidize iron and manganese and some taste- and odor-causing compounds, reduce color, and in some cases reduce trihalomethane and haloacetic acid formation potential. It is an alternative to chlorine that reduces production of chlorinated organic by-products, but care must be exercised to manage application rates such that excessive amounts of chlorine dioxide and its inorganic degradation

products (i.e., chlorite and chlorate) are not present in the finished water. Being highly reactive, chlorine dioxide cannot be manufactured off site for delivery to the point of use. It is therefore generated on site by reactions between (1) gaseous chlorine and sodium chlorite, (2) hydrochloric acid, sodium hypochlorite, and sodium chlorite, or (3) sulfuric acid and proprietary blended sodium chlorate–hydrogen peroxide products.

5. *Advanced oxidation.* Advanced oxidation processes produce highly reactive free radicals, most notably the hydroxyl radical. Much of the interest in these processes has been directed toward removal of specific organic contaminants, such as taste- and odor-causing compounds that are resistant to treatment using other oxidants. Interest in removal of a range of other organic contaminants also has emerged. Technologies used to generate these radicals include (a) ozone and hydrogen peroxide, (b) UV radiation and hydrogen peroxide, (c) UV radiation and ozone, and (d) various approaches involving catalysts.
6. *Mixed oxidants.* Mixed oxidants are generated on site from salt solutions and carefully controlled operational settings to produce a hypochlorite-based solution with disinfection and oxidation properties different from hypochlorite alone. Mixed oxidants offer an alternative to traditional forms of chlorination and chloramination for secondary disinfection and residual maintenance in the distribution system.

Treatment results from use of these oxidants are influenced by pH, temperature, oxidant dose, reaction time, and the presence of interfering substances. Characteristics of an oxidant in solution can be affected particularly by pH, whereas increased temperature almost universally increases the rate of reaction. Reaction time and dose are fundamental considerations for providing conditions that affect the extent to which a target constituent will be removed, whereas pretreatment processes and process sequencing may be important considerations for limiting the adverse effects of interfering substances. The susceptibility of various chemical bonds to oxidative attack are also critical considerations and therefore influence selection of chemical oxidants. Each oxidant can lead to different treatment outcomes and effectiveness for a given application.

While chemical oxidants are most commonly used for oxidation of specific constituents, they also can take part in other types of reactions where oxidation is not necessarily the only outcome, such as substitution reactions between chlorine and ammonia to form chloramines. This is a desirable outcome because chlorine is substituting into the structure of ammonia without a change in oxidation state, thereby retaining the capability for disinfection. Other substitution reactions include the formation of halogenated disinfection by-products, an undesired outcome of reactions between oxidants and precursor components of NOM.

Coagulation and Flocculation (Chap. 8)

Coagulation is a fundamental process used at water treatment plants worldwide as a pretreatment step not only for traditional rapid-rate filtration technologies that comprise much of the water treatment infrastructure that still exists but also for many of the newer technologies. In rapid-rate filtration plants it serves as the mechanism for building particles to a size that can be readily removed in the clarification step, as well as for conditioning particles to adhere to the filter media to ensure maximum process efficiency. Its importance to filtration was underscored by the results of a survey by Cleasby et al. (1989) that found coagulation to be more important than physical filter characteristics and loading rates for meeting filtration goals at many of the sites surveyed.

The traditional view of coagulation is that it facilitates agglomeration of small colloidal particles into larger particles of a size that can be physically removed. The mechanisms for achieving this are discussed in detail in Chap. 8. The role of coagulation in removing

TOC has emerged in the last 20 years, and for many surface-water supplies, TOC controls coagulant dosing rather than raw-water turbidity.

Ferric and aluminum salts are the most common primary coagulants. Aluminum salts include alum and various commercial products such as polyaluminum chlorides. Ferric salts include ferric chloride and ferric sulfate and less commonly used formulations.

High-charge-density cationic polymers are sometimes used alone in direct filtration applications or as a dual coagulant with alum or ferric coagulants. High-molecular-weight, long-chained nonionic polymers or anionic polymers of very low charge are used as flocculant aids and filter aids. As flocculant aids, they are capable of bridging particles together to increase floc size and strength, thereby increasing floc settling velocities. As filter aids, they improve attachment of particles to filter grains and previously retained floc particles. Typical points of application of flocculent aids are downstream from the rapid mix or near the midpoint or the end of a conventional flocculation basin. Filter-aid polymers are typically added immediately upstream of the filters.

Discussions of coagulation chemistry emphasize the importance of pH as a critical process condition for coagulation with the common metal salts, which can be especially sensitive to changes in this parameter (see Chap. 8). This is because pH affects the chemical speciation of the dissolved coagulant in rapid mixing and the precipitation of aluminum or ferric hydroxide.

Polyaluminum chlorides, aluminum chlorohydrates, and some other commercial formulations differ from the common metal coagulant salts in that their composition is adjusted during the manufacturing process to maximize the presence of the most effective dissolved aluminum species in rapid mixing. These prehydrolyzed coagulants, which are effective at neutralizing the negative charge of particles, are less acidic than alum and ferric coagulant, so their use in low- to medium-alkalinity waters without the need for addition of supplemental base to control pH may be feasible. In addition to pH, coagulation is strongly affected by a number of other factors, including particle characteristics, concentration and nature of NOM, presence of anions that interact with the metal coagulants, and temperature. There is no single "preferred" coagulant—the selection varies with the treatment conditions and goals.

In addition to these and other direct chemical influences, decisions regarding coagulant application are affected by a number of other secondary effects that can result, such as the production of residual solids, the introduction of contaminants such as manganese that occur as impurities in the coagulant, and changes in pH that affect other aspects of treatment. A more recent concern has emerged regarding the effects that coagulant chemicals have on the finished water chloride-to-sulfate ratio, which may lead to increased lead corrosion in some water systems (see Chap. 20). Corrosion of metal components within a water treatment facility prior to final adjustment of finished water pH can also be affected by the pH during coagulation.

Rapid mixing and flocculation provide the physical mechanisms for dispersion of chemicals and particle contacting following coagulation. Although the physical characteristics of the equipment used in the coagulant dispersion and flocculation processes are similar, the process goals are very different. The objective of mixing is rapid dispersion of the coagulant chemical to achieve relatively uniform distribution. Long detention times are not needed to achieve this goal and actually can be detrimental because dispersion tends to be most effective when it occurs rapidly within a confined space. Older rapid-mix basin configurations based on detention-time criteria are no longer regarded as the best approach. Dispersion-oriented configurations include basins in which mixing is confined to the smallest possible volume as flow passes the mixer blade and a variety of in-pipe configurations that include mechanical mixers, static mixers, and hydraulic-jet dispersion concepts. Other approaches can involve distributed dispersion across weirs and hydraulic jumps. Even simple in-pipe addition can provide better results than some of the older conventional rapid-mix basins (Clark et al., 1994).

Mixing of the coagulant is followed by flocculation to aggregate the chemically conditioned particles into floc that can be removed by downstream processes. The objective of this process step is controlled introduction of mixing energy to induce contact between particles that leads to a progressive increase in floc size. Ideally, this process will result in a uniform floc size with particles in the optimal size range for removal by clarification or filtration. Unlike coagulant dispersion/mixing processes, detention time is important in the flocculation process for developing floc of the desired size. The most common types of equipment for this process are vertical turbines and paddle-type flocculators. Hydraulic flocculators, which rely on the turbulence created as flow is redirected around baffles installed in the floc basin, are another alternative. In some cases, gates that can be adjusted to vary the energy imparted to the process stream are provided to optimize the hydraulic flocculation process. Turbulence within a pipeline or at the entrances to a series of unmixed chambers is also applied in some cases. However, these hydraulic flocculation systems generally offer less operational flexibility and process-control capability over a range of flow rates than conventional mechanically mixed systems.

The design of the flocculation step is generally based on the floc characteristics desired for downstream clarification and filtration. As an example, dissolved air flotation targets smaller floc sizes than conventional sedimentation because the floc particles attach to air bubbles to be floated; excessively large floc size would be detrimental to this process. Other downstream processes are also optimized by using different floc size distributions. For direct filtration, filterable floc size is the target; floc that is too large can cause filter "blinding." Even for conventional settling, there is usually a maximum desirable floc size owing to the lower floc densities and susceptibility to shear that is often encountered with large floc. Therefore, it is essential that particle size be considered in configuring and operating the flocculation process.

The performance of most unit processes within a water treatment plant is highly dependent on effective coagulation, mixing, and flocculation. The pretreatment benefits of coagulation can significantly influence the performance of other unit processes discussed in this book, including chemical oxidation, ion exchange, adsorption of organic compounds, chemical disinfection, and UV irradiation.

Sedimentation and Flotation (Chap. 9)

Sedimentation and flotation processes achieve the objective of separating particles from the process stream. While these processes can be used alone in pretreatment or other applications, they usually follow coagulation or precipitation. In a larger context, these processes fall into the category of clarification technologies. Both processes are gravity processes; in the case of settling, since the density of the particles exceeds the density of water, settling occurs, whereas in flotation, because the density of the bubble-floc aggregate is less than the density of water, the aggregate rises. Principles of settling and rise velocity are similar, and both processes are sized based on hydraulic loading rates—often referred to as the *overflow rate.*

Central to an understanding of sedimentation processes is the concept of surface loading rate or overflow rate (OR), which was developed by Hazen (1904). Its importance was further developed with regard to process effectiveness by Camp (1936). Calculated by dividing the process flow rate by the unit process steeling area, OR corresponds to a vertical particle settling velocity and therefore is commonly expressed in units of m/h (gpm/ft^2 or gpd/ft^2).

Under the OR concept, settling basins with greater settling or surface areas perform better. Recognition of this relationship led to the development of tray, tube, and inclined-plate settling concepts to increase sedimentation rates by incorporating additional settling surfaces within a given volume. The OR concept is commonly applied to all the sedimentation processes and is used as a fundamental basis for characterizing and evaluating these processes.

A number of sedimentation processes are available, including a variety of proprietary configurations. In general, they are divided into the following categories:

1. *Conventional horizontal-flow sedimentation.* This is a traditional approach that consists of simple gravity settling in a basin or tank with no enhancements to accelerate settling. It is simple to operate but requires a larger plan area than other clarification processes.
2. *High-rate gravity settling.* This process involves the use of additional configurations or devices that increase the available effective surface area in the basin. Commercially available inclined-plate and tube settler equipment is the most widely used equipment of this type. Tray sedimentation basins that employ multiple "floors" within a given sedimentation basin area to maximize flow path lengths are another example—these were used in the past but are not used now in new plants.
3. *Solids contact clarifiers.* These basins generally contain mixing and flocculation equipment within a common reaction zone, with settling occurring in another zone of the same basin. Solids are recycled to maintain a high solids inventory in the reaction zone and to provide increased opportunities for contact between particles, thereby increasing the size and settling rates of flocs or precipitates and allowing the use of higher surface loading rates than in conventional gravity settling. One modification of this approach is the IDI Densadeg process, which separates the mixing and settling compartments between different basins; the solids inventory within the reaction zone is maintained through collection and return of settled solids from the settling tank to the mixing/reaction stage. Withdrawal of solids (termed *blowdown*) from the sludge blanket within the solids contact clarifier is necessary to avoid loss of solids from the clarifier into the process stream. Appropriate blowdown rates and duration are critical to maintaining a properly balanced system; excessive rates will remove too many solids from the blanket, whereas an inadequate rate will allow solids to overflow to the process stream.
4. *Floc blanket clarifiers.* Flow into these clarifiers is introduced at the bottom and withdrawn from the surface. The process relies on a solids blanket that is developed at the bottom of the basin to entrain incoming particles. These clarifiers are available in several configurations, but all are equipped with a submerged weir over which the blanket will overflow to be removed by a blowdown process. The appropriate blowdown rate is also critical for this process because excess blowdown will result in larger than necessary waste streams, and inadequate blowdown will result in overflow of solids into the process stream. One application of this approach is the SuperPulsator clarifier, in which a pulsing action is applied to the floc blanket to keep it homogeneous.
5. *Ballasted flocculation.* This process accelerates settling by using polymer to attach floc particles to a concentrated mass of fine sand (microsand). The density of the resulting floc is increased by the sand, thereby allowing use of high overflow rates. Settled solids are collected from the bottom of the settling compartment and pumped to hydrocyclones, where the sand is separated from the lighter floc and returned to the process stream, and the waste solids are directed to the disposal process. The accelerated particle settling features of this process allow use of high loading rates in the settling zone, with corresponding reductions in the system footprint.

Dissolved air flotation (DAF) is an alternative to sedimentation. Available in several configurations, DAF employs a pressurized recycle stream that is saturated with air. The recycle stream discharges at the inlet to the flotation basin, and large quantities of fine bubbles are released. Floc particles from the upstream coagulation/flocculation process attach to the bubbles and are floated to the surface. Conventional DAF systems operate at hydraulic loading rates similar to filter rates. High-rate systems have been introduced in the last 10 years (see Chap. 9).

The sedimentation and DAF processes have site-specific advantages that require recognition of respective benefits and limitations for proper process selection. For example, dissolved air flotation is most effective in treating reservoir supplies, which are low in mineral turbidity (turbidity attributable to silts/clays). In addition, it can have application to supplies with moderate to high color or TOC.

Polymer is used with some of the sedimentation processes, and possible effects of polymer residual on downstream processes are a potential concern depending on the nature and quantity of polymer applied. Ballasted flocculation processes require the application of polymer to attach floc to microsand; polymer usage cannot be avoided for this process, and excessive carryover of polymer has been shown to be detrimental to downstream granular media filter performance if dosing is not regulated properly. Other processes can also benefit from the application of polymer for improved settling. This can be an important consideration for both solids blanket and floc blanket clarifiers, where settling properties within the floc blankets can be critical to effective performance. Gravity settling (both conventional and high rate) can benefit from polymer addition as well, but application generally is less critical for these processes. Dissolved air flotation generally functions well without polymer, although there can be isolated situations where polymer can be beneficial.

Granular Media Filtration (Chap. 10)

Conventional media filtration systems include granular bed, precoat, and slow sand filters. Granular bed filtration, the most commonly used of these processes, consists of a filter box that contains a bed of granular media placed on a support layer with an underdrain below it. Sand, anthracite, granular activated carbon (GAC), and high-density garnet and ilmenite represent common media options, and other commercially produced media have been developed. The dual-media arrangement (typically anthracite over fine sand) is the most common configuration. Efficient removal of particles requires an upstream coagulation step because attachment of flocs to filter grains and previously retained flocs are the mechanisms of floc particle removal within the pores of the filter bed.

Water flows through the filters in the downward direction; filtered water is collected by the underdrain system at the bottom of the filter and conveyed through filter effluent piping. Flow and filter level are usually controlled by valves in the filter effluent piping; however, some types of granular bed filters allow flow to diminish as head loss accumulates to produce a declining-rate mode of operation.

Granular bed filters require periodic backwashing to displace particles that accumulate on the media over the course of a filter run. Older backwashing methods relied on hydraulic scour to clean the filter media during backwash, but newer air-scouring approaches offer significant advantages, including improved media cleaning and reductions in the amount of water required for backwashing.

Precoat filtration consists of applying a fine filter medium to a filter leaf assembly, where it is restrained by a fine-mesh septum. The media are applied initially to build up a cake on the surface of the septum in a precoat step. Naturally occurring diatomaceous earth and commercially produced perlite are the typical sources of this media. Feed water is applied following completion of the precoat step, and it is common to provide a supplemental body feed of the filter material as the run progresses. Precoat filtration differs from granular bed filtration in two respects: (1) particles are removed by straining, and therefore, the process does not require the upstream addition of coagulant for particle conditioning, and (2) the media are disposed of at the end of each filter run. Since coagulant is not added, removal of TOC is not achieved, which limits the process to treating high-quality waters with low TOC.

Slow sand filtration is similar to granular media filtration in that the water passes through a bed of sand, but the process differs mechanistically. Hydraulic loading rates are much lower [typically 0.1–0.3 m/h (0.04–0.12 gpm/ft^2)], and much finer media are used. As with precoat filtration, coagulation is not required because much of the removal takes place through a gelatinous layer that develops on the top of the filter; this layer is commonly referred to as the schmutzdecke. Biological activity that plays a role in the filtration process can develop in this layer and in the upper portion of the sand bed. The filter is cleaned periodically by scraping the schmutzdecke and the top layer of sand when head loss becomes excessive. Required cleaning frequencies can range from several weeks to up to one year, depending on media design and applied water quality. GAC also has been incorporated into slow sand filters to improve removal of organic contaminants, typically with an overlying layer of sand media.

While filtration is important for producing high-clarity water for customers, its primary function is to remove particles that may include microbial pathogens. This is a role that has evolved in the last 20 years following promulgation of the U.S. Environmental Protection Agency's (USEPA's) Surface Water Treatment Rule in 1989. Prior to that time, the regulated limit for filtered water turbidity was 1 nephelometric turbidity unit (ntu). Heightened awareness of the importance of filtration for controlling waterborne pathogens in a multiple-barrier approach in conjunction with disinfection processes emerged in the 1980s with the discovery of pathogens that were resistant to chemical disinfectants. Initially, the focus was on *Giardia*, and the turbidity limit was lowered to 0.5 ntu for granular bed filters to address this concern. Recognition of *Cryptosporidium* as an even more chlorine-resistant waterborne pathogen led to further reduction in the turbidity limit for plants employing coagulation and granular media filtration to 0.3 ntu level or less, and voluntary guidance under the criteria developed for the Partnership for Safe Water set a turbidity goal of less than 0.1 ntu.

These more restrictive requirements have profoundly altered the approach to granular bed filtration in recent years. The need to view it as a component of a larger process sequence has increased, particularly with recognition of the importance of effective coagulation for producing conditioned particles that can be removed efficiently by attachment to filter media. Older, shallow-depth, single-media sand filters that were once common have been displaced by more efficient dual-media filters. Deep-bed monomedia filters are being used in higher-rate filters. Whereas older single-media filters had a smaller sand size and operated at lower hydraulic loading rates, these configurations ignored the importance of attachment rather than straining mechanisms for achieving efficient filter performance. These older configurations were more prone to surface blinding and accompanying turbidity breakthrough that results from shearing of floc even at lower filtration rates.

Hydraulic loading rates, or filter rates, for granular media filters have increased substantially in recent years with the application of newer and more efficient media configurations. While rates as low as 5 m/h (2 gpm/ft^2) were once common, filter rates of 10–15 m/h (4–6 gpm/ft^2) are now routine, and rates of 20 m/h (8 gpm/ft^2) and higher have proved acceptable in some circumstances. Media size and configuration affect the acceptable filter rate because some configurations are more tolerant of higher rates than others. In most cases, larger media applied in deeper beds are more accommodating to higher filter rates than smaller, shallower media.

While granular bed filtration until recently has been the most common approach to media-based filtration, precoat filters and slow sand filters may be preferable in some situations, and membrane filters (discussed in Chap. 11) offer a new option that is increasingly being used in new treatment facilities. These processes differ from granular bed filters in terms of the mechanism of particle removal; precoat and membrane filters remove particles based on exclusion at pore openings. These distinctions in mechanism translate into different criteria for acceptable effluent turbidity. In the case of precoat and slow sand filters, allowable turbidity levels under current regulations are increased from the 0.3-ntu criterion that applies for granular bed filtration to 1 ntu.

Specialized adaptations of granular bed filters are an important consideration in many water treatment systems. One of the most widespread and significant of these is the removal of manganese by adsorption to an oxidized coating that develops at the media surface in the presence of a free chlorine residual; other oxidants are not effective in producing this effect, and changes in chlorine addition procedures or application of an alternative disinfectant or oxidant can alter this mechanism, resulting in manganese breakthrough if appropriate adjustments are not made. Similar issues also may arise when prechlorinated granular bed filters are replaced with other filtration approaches.

Another adaptation of granular bed filters is biological filtration that removes biodegradable organics. Biological activity will develop within any granular media filter when a disinfectant residual is not present at the influent to any granular filter media and particularly when intermediate application of ozone just prior to filtration is practiced. GAC is a common media choice for this application owing to its large surface area, although other media also can work well in this capacity. GAC also can be applied as a filtering layer that can remove taste- and odor-causing compounds.

Granular bed filtration is sometimes practiced without an upstream clarification step. This approach, commonly referred to as direct filtration, follows rapid mixing and flocculation, as required for conditioning the particles and development of a filterable floc. In some cases, the flocculation step is eliminated in a variation referred to as in-line filtration. Use of these approaches is generally restricted to water with low turbidity. Water with low color is also necessary because applicable coagulant dosages for direct and in-line filtration generally are not sufficient to result in significant coagulation and removal of color when high levels are present. While direct filtration was commonly applied in the treatment of low-turbidity source waters during the 1960s and 1970s, it is rarely used in new treatment facilities today due to the desire to provide multiple barriers to contaminants. In addition, direct filtration does not readily accommodate the higher coagulant dosages that may be required to provide for TOC removal to meet regulatory requirements for disinfection by-products control.

Membranes (Chap. 11)

A membrane is a semipermeable barrier that allows certain materials to pass while preventing the passage of others. The two main groups of membrane processes used in water treatment are (1) microfiltration and ultrafiltration (MF/UF) that remove particles and (2) reverse osmosis and nanofiltration (RO/NF) that remove dissolved materials. A generalized comparison of these processes based on the sizes of materials that are removed is presented in Fig. 5-1.

MF/UF, the filtration processes, can be used instead of granular media filtration for treatment of potable water. The use of MF/UF for municipal water treatment is relatively new. The first major water treatment plant in North American to apply MF/UF was placed in service in 1994 in San Jose, California. The success of that facility and others has led to wide acceptance, and MF/UF is now considered commonplace. By 2006, hundreds of North American utilities were using MF/UF plants with a combined capacity exceeding 5678 ML/day (1500 mgd), or about 44 percent of the worldwide installed capacity of such plants (Furukawa, 2008).

The widespread use of MF/UF is attributable to consistent production of low-turbidity filtered water and high microbial pathogen removal rates. Unlike conventional granular

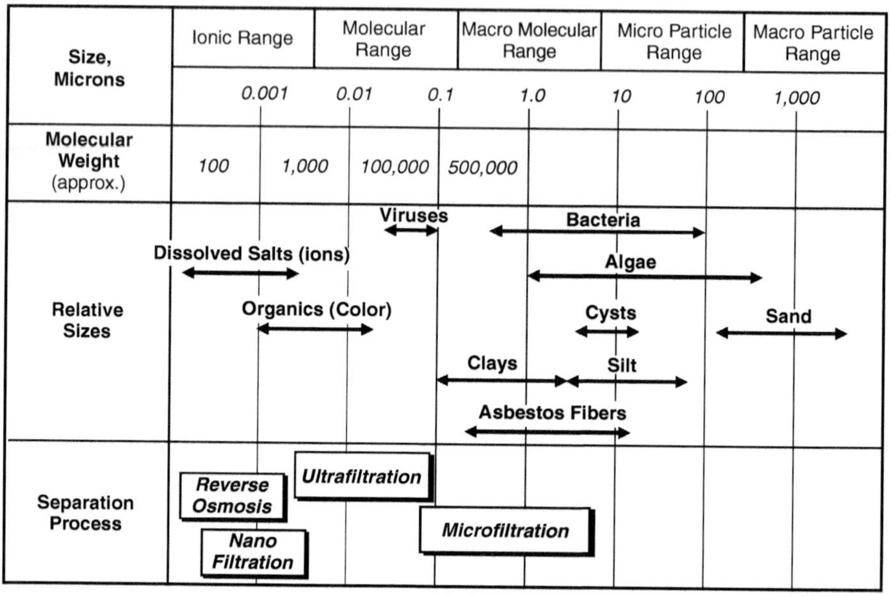

FIGURE 5-1 Membrane removal size ranges.

media filters, MF/UF produces these results without chemical preconditioning of the feed water because the removal mechanism is physical sieving. Speth et al. (2005) summarized a literature review of 122 studies as follows: "MF and UF membranes produce extremely high-quality water regardless of influent turbidity, and there is no apparent difference in turbidity removal between membrane type, manufacturer, or whether a coagulant was used." Mean filtrate turbidity values had a median of 0.06 ntu, and maximum turbidity values had a median of 0.08 ntu. Studies by Jacangelo et al. (1991), Coffey (1993), and others have shown that MF/UF consistently remove *Giardia* cysts and *Cryptosporidium* oocysts at least to the 4-log level, and in some cases removals as high as 6 log to 7 log have been demonstrated. While there is no observed difference between removals of 2- to 15-μm-sized protozoan cysts achieved by MF and UF, UF provides greater removal of viruses, which are one to two orders of magnitude smaller than *Giardia* cysts and *Cryptosporidium* oocysts. Generally, virus removal credits of up to 0.5 log are granted for MF and 2 to 4 log for UF depending on the specific membrane, level of testing sensitivity, and regulatory agency.

MF membranes used for water treatment have pore sizes of 0.1–0.5 μm, and UF membranes have pore sizes of 0.01–0.1 μm. To approximately quantify the removal of dissolved organic material, UF membranes are also rated on the basis of molecular weight cutoff (MWCO); nominal MWCO values of 100,000 to 200,000 Daltons are typical. MF and UF membranes are particle-removal processes that generally do not remove dissolved contaminants unless additional process steps (i.e., coagulation and flocculation) are included. To augment removal of dissolved contaminants, MF/UF is integrated with processes that convert the dissolved constituent into a form that can be removed by the membranes, such as the addition of coagulant or powered activated carbon to adsorb dissolved organics, taste- and odor-causing compounds, and DBP precursors, or the addition of an oxidant to precipitate iron and manganese.

Typical transmembrane pressures for MF/UF are 27.6–242 kPa (4–35 psi). The most common membrane configuration is hollow fibers of polymeric materials arranged in bundles that are mounted inside pressure vessels (typically referred to as *encased membranes*) or directly into atmospheric tanks (typically referred to as *submerged, immersed*, or *vacuum membranes*). The most commonly used membrane materials are polyvinylidene difluoride (PVDF), polysulfone (PS), or polyether sulfone (PES); a few facilities use ceramic or other materials.

Operating cycles for MF/UF membranes consist of filtration, backwashing, maintenance washing or chemically enhanced backwashing, chemical cleaning, and direct integrity testing. Filtration and backwashing (BW) cycles are analogous to those used in granular media filtration, but MF/UF backwash cycles occur more frequently, with backwashing generally conducted two or three times per hour. In some installations, a brief chemical cleaning, sometimes called a *maintenance wash*, is conducted daily or weekly, and a longer, more intensive chemical cleaning may need to be conducted every four to six weeks to remove accumulated material that impedes flow through the membrane. A direct integrity test (typically an air-based test) is conducted at least daily, as required by the Long-Term 2 Enhanced Surface Water Treatment Rule (USEPA, 2006a), to verify that the membranes are intact and achieving the required microbial pathogen log removals. Because of the complexity of the various operating cycles, essentially all MF/UF facilities are automatically controlled.

RO/NF membranes are pressure-driven processes that remove dissolved materials from water. There are a wide range of uses, with the most frequent being desalination of brackish supplies and seawater. Brackish or lower-pressure RO typically is used to control one or more of the following constituents: TDS, chloride, sodium, hardness, nitrate, arsenic, sulfate, and radium, as well as other ions and dissolved organic carbon (DOC), including true color, DBP precursors, and agricultural chemicals. A key difference between RO and NF, which is actually a subclass of RO, is that removal percentages (rejection) for NF are generally lower; NF exhibits more selectivity based on ion charge. For example, for a typical RO membrane, the rejection of divalent sulfate might be about 99 percent (i.e., the sulfate concentration in the permeate flow would be only about 1 percent of the RO feedwater concentration), whereas monovalent chloride rejection might be about 98 percent. In a similar example, a typical NF membrane might yield 85 percent rejection of sulfate but only 40–50 percent rejection of chloride. Both types of membranes would typically yield a rejection of DOC (molecular weight greater than 200 Daltons) of at least 95 percent. Exact values would depend on many variables, including membrane type, temperature, concentration, and ionic strength, but this example illustrates a typical case. The use of NF tends to be limited to controlling color and/or softening and thus is sometimes referred to as *membrane softening*, whereas RO is applied to a wider range of desalination or single-solute applications such as nitrate removal. Neither RO nor NF is effective for removing dissolved gases such as radon and carbon dioxide.

RO/NF also can be used to remove particulate material, for example, to lower turbidity, because RO/NF product water (also known as *permeate*) generally has a turbidity of less than 0.05 ntu; however, this would be a misapplication of the process. To avoid fouling or plugging of RO/NF membranes, the feed water has to be clear, generally with turbidity of less than 0.1 ntu and with a 15-minute silt density index (SDI) of less than 5 and preferably less than 3.

In the past, NF was considered less costly to operate than RO because it operated at lower pressure; however, improvements in RO membranes have reduced the difference in operating pressures when treating comparable feed waters. For a typical low-salinity feed water, operating pressures for an NF system are 620–1040 kPa (90–150 psi), whereas lower-pressure (sometimes mistakenly referred to as *high flux*) RO membrane systems operate at 620–1725 kPa (90–250 psi). Required operating pressures increase as feed-water dissolved solids concentrations increase because the resistance to flow through the membrane (osmotic pressure) increases with TDS concentration. For example, seawater RO plants operate at 5500–6900 kPa (800–1000 psi) while desalinating 32,000–45,000 mg/L of TDS feed water.

The typical configuration of membranes used in municipal-scale RO/NF plants is spiral wound, with flat-sheet membrane, spacer materials, and glues wrapped around a central product water tube into cylindrical elements. These spiral elements are usually 200 mm (8 inches) in diameter by 1000–1525 mm (40–60 inches) long (nominal dimensions). Larger 400- and 460-mm-daimeter (16- and 18-inch-diameter) elements are now available and may become more commonplace. At least one manufacturer provides hollow-fiber RO modules. The most common membrane material is polyamide thin-film composite.

Membrane elements are mounted in pressure vessels arranged in stages, banks, or arrays to meet design goals, such as suitable overall TDS/contaminant rejection, pressure drop, and velocity across the membrane surface. RO/NF is a cross-flow filtration method in which only a portion of the feed water becomes permeate. The remainder leaves the system as concentrate (sometimes called *brine*) that carries away the concentrated material before it precipitates to form scale on the membrane surface. For lower-salinity applications, such as desalination, softening, or removal of color from groundwater, system recovery (the ratio of permeate flow to feed flow) typically is 80–85 percent, whereas for seawater desalination it is typically 45–55 percent.

There are two other types of desalination processes: distillation and electrodialysis, including the related electrodialysis reversal (ED/EDR). While there are a number of distillation methods, distillation is essentially never the low-cost alternative at feed-water concentrations below that of seawater (32,000 mg/L). About half the seawater desalination plants in operation today use RO, and the relative percentage has been increasing in recent years as operating costs for seawater RO systems continue to decline. On the other end of the spectrum, ED/EDR is only cost competitive with RO/NF at low TDS concentrations. It is generally accepted that there is no reason to consider ED/EDR when feed TDS concentrations exceed 1500–2000 mg/L, and at lower TDS concentrations, ED/EDR is typically considered only under special circumstances, such as when achievable recovery for RO/NF would be limited by silica scaling. Even in cases where distillation or ED/EDR might be competitive with RO/NF, a detailed site-specific evaluation is required before an informed decision can be made with respect to the optimal treatment approach.

Ion Exchange and Adsorption of Inorganic Contaminants (Chap. 12)

Ion exchange (IX) has an established place in water treatment practice and has been used for more than 100 years as a means of softening water. Inorganic adsorption media also have been developed recently for removal of certain inorganic constituents. Interest in application of both types of media for removal of a broader range of constituents is emerging, including applications for removal of NOM that can lead to formation of disinfection by-products. Equipment for these processes can include both fixed- and fluidized-bed configurations in columns that provide contact with the process stream. Suspensions of ion exchange media in mixed contact tanks are also used in some specialized applications.

While IX and adsorption processes share similarities in the physical characteristics of the contacting equipment, differences do exist, most notably with the underlying mechanisms that are involved. IX is based on the attraction of ions in solution to oppositely charged sites at the surface of the media. Competition for sites occurs between ions of like charge, an attribute that can be used to achieve regeneration through the application of concentrated salt solutions to displace constituents removed during the in-service cycle; the exchange process is reversed, and the constituents that have been removed are concentrated in a smaller stream for disposal. Many IX media are manufactured as organic resins, but some inorganic IX media also exist, notably natural zeolites that are in widespread use for softening applications.

Inorganic adsorption processes rely on chemical bonding of the constituent to be removed to active sites at the surface of the media. While regeneration is possible for some of these processes through the application of regenerant streams that contain high concentrations of a competing reactant to displace constituents removed during the in-service cycle, regeneration is not always conducted, and the media are often disposed of after only one use.

Applicability of the respective processes depends on the characteristics of the constituent targeted for removal. IX works for charged constituents, but there is a hierarchy of preference that needs to be considered because ions with multiple charges are more favorably attracted to exchange sites than singly charged ions. Other characteristics, such as size of the ions to be removed and chemical affinities for exchange sites, also must be considered during evaluation of alternative IX resins.

Chemical affinity for sites is a key aspect of inorganic adsorption media application because it confines the range of candidate contaminants to those with chemical characteristics that allow bonding to active surface groups. While this limits the range of candidate contaminants, it offers the potential advantage of selectivity for contaminants where this process can be applied.

An illustration of the effect of contaminant characteristic in selecting an applicable approach is provided by comparing the removal of nitrate and arsenate. While arsenate [As(V)] can be removed by IX, it also exhibits an affinity for aluminum and ferric ions that also facilitates removal by inorganic adsorbents such as activated alumina and some ferric media. However, nitrate is not removable by these adsorbents. IX is more effective for nitrate removal. Another illustration of the effect of chemical reactivity on process application is provided by the reduced arsenite [As(III)] form of arsenic, which is much less reactive with metal ions than the more oxidized arsenate ion. Arsenic in the As(III) form must be oxidized to As(V) to facilitate removal by adsorption, thereby demonstrating the effect of chemical speciation on process sequencing.

Softening (removal of calcium and magnesium) is the most widespread and established use of IX. Other potential target constituents for IX and inorganic adsorption include barium, strontium, radium, uranium, manganese, fluoride, nitrate, arsenate, selenate, chromate, sulfate, sulfide, and perchlorate.

Recent advances in IX/adsorbent technologies include specialized inorganic adsorbents for the removal of arsenate and the use of anion-exchange resin with magnetized properties in a mixed suspension (MIEX process) to facilitate its recovery and reuse for NOM removal (see Chap. 12). Also receiving increased attention is the role played by oxidized media coatings on granular media filters as inorganic adsorptive surfaces for the removal of manganese. Since this approach relies on the presence of free chlorine to sustain the process, the coatings can be lost following modifications such as conversion to biofiltration or relocation of points of chlorine addition.

Factors to consider in the application of IX and inorganic adsorption include (1) the potential for fouling of media by suspended solids, precipitates, and biological growth and associated pretreatment requirements to limit fouling, (2) special materials of construction to handle corrosive brine and regenerant solutions, and (3) requirements for disposal of spent regeneration solutions and media. The extraction of contaminants into smaller volumes can require special consideration in some circumstances because the higher concentrations that result may lead to categorization of the resulting waste streams or spent media as hazardous wastes.

Chemical Precipitation (Chap. 13)

Chemical precipitation is commonly used in water treatment facilities to remove hardness and heavy metals such as iron and manganese, as well as regulated contaminants such as radium, uranium, fluoride and, to a lesser extent, natural organic matter. Coagulation with metal salts such as aluminum sulfate or ferric compounds is typically more effective than lime softening for removing natural organic matter. Chemical precipitation also can be used as pretreatment or intermediate treatment in membrane desalination systems to reduce the potential for scaling of the membranes by mineral salts (e.g., strontium, barium, and silica) at high raw-to-product conversion rates, thereby increasing achievable membrane conversion rates and reducing the amount of residual concentrate generated.

Lime softening is by far the most common application of chemical precipitation, with more than 1,000 potable water softening facilities operating throughout the United States (primarily in Florida and the Midwest). A typical softening process train includes either conventional flocculation and sedimentation basins, similar to those used for coagulation and clarification, or, more commonly, upflow solids contact clarifiers, which are more efficient and occupy smaller footprint areas than conventional basins. The precipitation process is followed by pH adjustment (recarbonation) and filtration. Treatment of groundwater supplies with high concentrations of carbon dioxide, which increases required lime or caustic dosages and associated production of chemical solids, will typically also include forced- or induced-draft aeration to reduce carbon dioxide concentrations before softening. Because the high-pH environment of the softening process is not conducive to optimal removal of turbidity and color, when treating variable-quality surface supplies, it is usually preceded by conventional or high-rate sedimentation to reduce the levels of these constituents before softening.

Chemical treatment options for the lime softening process include lime only, lime–soda ash, caustic soda, and variations in the approach to pH adjustment preceding filtration. While water that contains low noncarbonate hardness can be softened adequately with lime alone, water with high noncarbonate hardness may require addition of both lime and soda ash to achieve the desired finished water hardness objectives. Low-alkalinity waters that also contain substantial noncarbonate hardness can be softened with sodium hydroxide (caustic soda) alone or in conjunction with lime, which also produces less sludge solids than conventional lime–soda ash softening. However, since the chemical costs for caustic softening are substantially higher than for lime–soda ash softening at equivalent levels of targeted hardness reduction, caustic softening is rarely employed in potable water treatment.

Water produced by precipitative softening contains high concentrations of caustic alkalinity and high pH and thus has significant scale formation potential. The high pH therefore must be reduced before filtration and distribution to prevent encrustation of the filter media and excessive deposition of calcium carbonate scale in the distribution system. This is typically accomplished using carbon dioxide in either a single- or two-stage process. In the single-stage recarbonation process, carbon dioxide is added to the settled water at the inlet to a small contact basin with a detention time of at least 20 minutes. A two-stage recarbonation process is used where maximum hardness reduction is required or to remove specific constituents such as magnesium hardness, heavy metals, radionuclides, and/or DOC. In the first stage of the two-stage process, the pH of the water is increased to 11.0 or higher to achieve maximum removal of targeted constituents. Carbon dioxide is added to the settled water to lower its pH to the point of minimum calcium carbonate solubility (pH 10.0–10.3). The excess lime initially fed to raise the pH to 11.0 or higher is then allowed to precipitate as calcium carbonate in a second sedimentation stage. Following secondary clarification, additional carbon dioxide is fed to further reduce pH and to adjust the scale formation potential of the finished water. While some utilities practicing lime softening have strived to lower finished water hardness to 70–80 mg/L as $CaCO_3$, a more common current goal is 125–150 mg/L in order to reduce both chemical costs and solids production. Since this trend is expected to continue, single-stage softening/recarbonation probably will become more prevalent because of its lower construction costs. However, the two-stage process has greater flexibility and lower carbon dioxide dosage requirements if high-pH treatment is needed, and it continues to be the preferred precipitative softening process for most large Midwestern utilities that treat surface-water supplies.

While precipitative softening also can be used to remove iron and manganese, it is not typically employed strictly for this purpose. A more common approach is precipitation and filtration following oxidation of reduced iron and manganese using molecular oxygen, potassium permanganate, or chlorine. When the source water contains iron at concentrations exceeding approximately 5–6 mg/L, common practice is to include provisions for sedimentation following aeration and chemical oxidation to reduce the solids loading on the filters and to increase filter efficiencies (see Chap. 7).

Increasingly stringent regulations governing discharge of membrane desalination process concentrate, particularly at inland locations, where surface discharge is not an option, have led to the use of chemical precipitation to increase overall treatment system conversion rates and to reduce the volume of concentrate to be disposed of, thus making membrane desalination of brackish water more economically feasible. While its use for pretreatment to remove scaling constituents such as silica, barium, strontium, and calcium has been evaluated extensively and, in some cases, implemented at full scale, its use as a key step in treating first-stage membrane concentrate is relatively recent. Chemical precipitation of first-stage RO membrane concentrate, followed by additional membrane treatment of the resulting process stream, can help to reduce the discharge of membrane concentrate to zero or near zero. The constituents of concern can be precipitated by conventional slurry-type softening processes or crystallized using a fluidized-bed reactor. However, recent research (Bond and Veerapeneni, 2007) indicates that the fluidized-bed crystallizer is more effective than conventional softening in reducing primary RO concentrate to achieve zero liquid discharge based on the following observations: (1) precipitation can be achieved at lower chemical doses and at lower pH, (2) the fluidized-bed crystallizer produces only about 10 percent of the amount of waste solids (by volume) generated by conventional softening, (3) waste solids produced by the fluidized-bed crystallizer can be dewatered by gravity without the need for additional equipment, such as thickeners, centrifuges, or filter presses, and (4) required footprint area is much less than for conventional precipitative softening processes.

Adsorption of Organic Compounds (Chap. 14)

Adsorption with activated carbon is the predominant method of removing organic contaminants from municipal water supplies. It has been used for decades to remove taste- and odor-causing compounds such as geosmin and 2-methyl isoborneol (MIB). More recently, it has been used to remove synthetic organic chemicals (SOCs), DOC, and DBP precursors. Adsorption with activated carbon is the USEPA-approved best available technology (BAT) for more than 50 organic contaminants regulated under the National Primary Drinking Water Regulations and for many emerging organic contaminants of concern, including EDCs. (Activated carbon is also listed in the Stage 1 Disinfectant/Disinfection By-Products Rule as BAT for TOC removal, although carbon exhaustion rates are typically high when TOC removal is the primary treatment objective. In contrast to oxidation processes, adsorption removes organic contaminants from water rather than chemically transforming them into oxidized by-products that remain in the treated water.

Activated carbon is produced by heating the starting material, typically coal, wood, peat, or coconut shell, to temperatures above 900°C and injecting the material with steam and oxygen or carbon dioxide to activate it. The activation process produces a vast network of internal pores, giving activated carbon an extremely large surface area per unit volume (typically 500–1500 m^2/g). It is the large specific surface area and its chemical properties that make activated carbon a highly effective adsorbent. The affinity for adsorption to activated carbon varies among organic compounds. Hydrophobic compounds of medium to high molecular weight are adsorbed more effectively than low-molecular-weight hydrophilic compounds.

The surface of activated carbon provides a favorable environment for microbial attachment; consequently, under appropriate conditions, contaminant removal by adsorption can be augmented by biodegradation of organic compounds. The surface of activated carbon also provides catalytic destruction of certain chemicals in oxidation-reduction reactions. Examples of this include removal of excess ozone in ozonation off-gas treatment systems and removal of chlorine or chloramines in home point-of-use devices or as pretreatment for dialysis systems in hospitals.

Activated carbon is produced in two forms: granular activated carbon (GAC) and powdered activated carbon (PAC), both of which are produced from the same materials. The physical difference between the two is their particle size. The most commonly used GAC products are U.S. Standard sieve sizes 12×40 (1.68–0.42 mm diameter) and 8×30 (2.38–0.59 mm diameter). In contrast, most PAC particles pass through a 325 (44-μm) sieve and therefore are an order of magnitude smaller than GAC. Equal weights of GAC and PAC of the same material have the same equilibrium adsorption capacity because they have the same surface area. Adsorption equilibrium, however, is reached faster for the smaller-particle PAC because adsorption kinetics increase as particle diameter decreases.

The most important differences between the two forms of activated carbon that affect process selection are the method of their application and the relationship between the concentration of the contaminant being adsorbed and the capacity of the activated carbon for adsorption of the contaminant. GAC is installed in a fixed bed, where it serves as a constant barrier to organic contaminants for as long as it has sufficient adsorption capacity. GAC adsorption capacity is consumed with service time, and once it is exhausted, it is replaced with virgin or regenerated GAC. In contrast, PAC is dosed in the same manner as chemicals used in treatment and can be applied intermittently to coincide with the occurrence of contaminants and at dosages commensurate with contaminant concentrations. Contact time is the design criterion that has the greatest impact on GAC or PAC treatment effectiveness.

GAC is favored over PAC for the removal of organic contaminants in cases where concentrations are too high to be treated with practical doses of PAC and in situations where a constant barrier is desired for sources at risk of unexpected contaminant spikes from runoff, spills, or illegal discharges. GAC can be installed either immediately following coagulation/sedimentation, where it serves as a filter adsorber, or after filtration, where it serves as a postfilter adsorber. Some utilities have elected to replace existing conventional filter media, either partially or completely, with GAC to take advantage of its ability to provide both adsorption and filtration. The bed depth of filter adsorbers is limited by filter headloss considerations, and consequently, postfilter adsorbers are favored where a longer contact time (deeper bed) is needed for effective treatment. Biomass can develop in filter adsorbers and postfilter contactors and thus provide additional treatment by biodegradation. Ozonation ahead of GAC has been used to enhance biodegradation by oxidizing organic compounds into more biologically assimilable forms.

PAC may be more cost-effective than GAC where low to moderate PAC doses are sufficient to achieve treatment goals or where contaminants appear intermittently and therefore do not require continuous treatment. Examples include seasonal occurrences of taste- and odor-causing compounds or seasonal spikes of pesticides from agricultural runoff. PAC is fed as a slurry and can be injected at the raw water intake, at the head of the plant, or in the rapid mix basin. PAC treatment efficiency improves as contact time increases, and available contact time is an important factor in evaluating the cost and effectiveness of PAC treatment. PAC is removed during sedimentation or DAF in conventional treatment, and consequently, PAC use adds to the volume of solids that must be managed at the plant.

Treatment costs for GAC versus PAC can be compared by estimating the amount of each material required annually for treatment and multiplying the projected amount by the unit cost for each. For GAC, carbon use depends on the total bed volume initially installed and the replacement frequency required to meet treatment goals. For PAC, carbon use depends on the PAC dose and its duration.

Air stripping is an alternative to GAC for treatment of groundwater supplies contaminated with VOCs such as trichloroethylene and carbon tetrachloride. Air stripping was discussed briefly earlier in this chapter and is presented in detail in Chap. 6. In situations where the off-gas from air stripping cannot be discharged to the atmosphere without treatment, air stripping and GAC are used in concert; VOCs are removed from the water phase by air stripping, and gas-phase GAC treatment is used to remove them from the stripper off-gas.

Natural Treatment Systems (Chap. 15)

Natural treatment can complement and enhance traditional water treatment methods through improvement of source water quality, thereby reducing the level of treatment required and the associated costs. Natural treatment systems employed most frequently in potable water production (i.e., riverbank filtration and aquifer storage and recovery) have demonstrated the ability to reduce turbidity, pathogens, natural organic matter, and, in some cases, organic contaminants by natural processes during subsurface filtration of surface water. While the concept of using aquifers as natural filtration devices is not new (riverbank filtration has been commonly employed for more than a 100 years in Europe; Kuehn and Mueller 2000), its immergence as a viable treatment technology and implementation in the United States have occurred only more recently.

Riverbank filtration (RBF), sometimes referred to as *bank filtration* or *induced infiltration*, can be defined broadly as the process of surface water flowing from the bank or bed of a river or a lake to extraction wells. During passage through the subsurface sand and gravel layers, the water is subjected to a combination of physical, chemical, and biological processes such as filtration, sorption, dilution, and biodegradation that can substantially improve the quality of the water. The extraction system typically consists of a series of conventional vertical wells or, more recently, one or more high-capacity collector wells equipped with radial horizontal collection laterals. The quality of the filtrate depends on many factors, including initial quality of the source water, degree of groundwater and surface water intermixing, properties of the filter layer, water temperature, and hydraulic detention time in the filter layer (Laszlo et al., 1990). Because of the numerous factors involved, performance of RBF with respect to contaminant-removal efficiencies is highly site-specific and thus difficult to characterize on a general basis. However, ability of RBF to remove NOM and organic micropollutants, such as pesticides and some pharmaceutical products, through adsorption and biodegradation has been well documented. The ability to remove pathogens effectively is also well documented, as evidenced by USEPA's recent inclusion of bank filtration in the Long Term 2 Enhanced Surface Water Treatment Rule (LT2ESWTR) "Microbial Toolbox" for removal/inactivation of *Cryptosporidium*. Under the LT2ESWTR, credits for *Cryptosporidium* removal of up to 1.0 log are available for wells with 15 m (50 ft) or more of setback from a surface-water source and with an average turbidity less than 1 ntu at their discharge.

While contaminant concentrations in surface waters may vary significantly, subsurface passage of the surface water before extraction serves as a barrier against shock loads at the treatment plant influent attributable to chemical spills or routine seasonal runoff. However, because of the residence time required for passage through the alluvium to the point of extraction, this same capability also can result in the presence of these contaminants in the RBF filtrate over longer periods than would be experienced when treating only surface water extracted directly from the source. In addition to suitable subsurface conditions, long-term viability of RBF is also highly dependent on locating the extraction wells where river flows are sufficient to maintain consistent sediment deposition at the river-aquifer interface without excessive scouring.

Aquifer storage, recharge, and recovery (ASR) involves injection of water into the aquifer and subsequent removal to supplement available supplies during periods of high demand. A common approach is to collect and inject runoff into the aquifer during periods of heavy rain or snowmelt and then to extract and treat the water during drought periods or when finished water demands exceed the reliable sustained capacity of the primary raw water supply. ASR systems are designed to promote flow of the water through the aquifer between the points of injection and extraction, which improves its quality through physical, biological, and/or chemical processes similar to those achieved by RBF. However, not all injected water is reclaimed: Typical recovery rates range from 30–80 percent of the water injected (Kline, 2008). Widespread use of ASR has been constrained by obstacles such as permitting requirements, water rights considerations, and concerns regarding endangerment of aquifer quality and capacity.

Chemical Disinfection (Chap. 17)

Disinfection is a fundamental water treatment process for protection against waterborne disease, a long-established public health goal of water treatment. The importance of this aspect of water treatment is underscored by the fact that disinfection generally is accomplished in conjunction with particle-removal processes to provide multiple treatment barriers to waterborne disease. Chemical disinfectants, particularly chlorine, have served as the traditional means of achieving this goal, although other alternatives are becoming available. However, chemical disinfectants are likely to continue to play a significant role, even when these newer alternatives are in use, because maintenance of a protective chemical disinfectant residual in distribution systems remains a common practice in the United States.

Early disinfection practice tended to be confined to the use of free chlorine, with a primary focus on inactivation of bacteria. As new disinfecting agents have emerged, and understanding of microbial contaminants has increased, practitioners now have a choice of several widely available disinfecting agents to address the expanding array of known microbes, which now includes viral and protozoan pathogens as well as bacteria, and possible problems related to other microbes cannot be discounted.

Disinfectants commonly used in water treatment include free chlorine, chloramines, chlorine dioxide, and ozone. The effectiveness of these disinfectants stems from their being reactive in ways that trigger destructive reactions at sites within target organisms. However, their reactivity is not limited to these pathways, and other reactions are inevitable. The most basic are those which result in simple decay of disinfecting residuals even in the absence of other sources of demand. Key factors that affect the rates of decay include pH and temperature. Exposure to sunlight and other sources of light at certain wavelengths can also accelerate decay rates. In some cases, decay reactions are not completed, and undesired products can result, as in the case of chlorine dioxide, which can form chlorite and chlorate as its by-products.

Reactions with other contaminants in water create demand beyond the basic form of decay and generate by-products in the process. NOM and reduced inorganic compounds can produce such results. The availability of these constituents for reaction can be affected by upstream treatment steps that can remove or alter them. Notably, coagulation and other processes can remove NOM to reduce the formation of DBPs, whereas oxidation and pH adjustment can produce changes that can have a similar effect. In other areas of treatment, oxidation of iron and manganese reduces disinfectant demand, and removal of particles can reduce disinfectant demand and improve disinfection effectiveness. Therefore, there are strong relationships between the disinfection process and other components of a treatment train.

Following observations of chloroform in drinking water in Holland (Rook, 1974) and in Cincinnati (Bellar et al., 1974), Symons et al., (1975) showed that the problem of trihalomethanes (THMs) was widespread in drinking water. The list of disinfection by-products has increased since then, and concern over these by-products has become a major consideration in the selection of disinfection strategies. By-product problems are no longer confined to those produced by free chlorine because each disinfecting chemical produces a set of by-products.

Disinfection in the United States is applied to meet two sets of goals for controlling pathogens within a water system. Primary disinfection is used to meet goals for inactivating pathogens at the water treatment plant, thereby providing a barrier to entry into the distribution system, while secondary disinfection is applied to meet the goal of providing a protective disinfectant residual throughout the distribution system as protection against further intrusion. This two-tiered approach tends to be applied in other locations around the world, although there are exceptions where secondary disinfection is not practiced.

In the case of primary disinfection, each disinfectant has a characteristic set of chemical pathways that defines its mechanisms and effectiveness for inactivation of different target microbes. As the array of known microbes expands, several have been identified that are particularly resistant to some chemical disinfectants, thereby making these microbes a focus in developing a disinfection approach. In this regard, a good deal of attention is presently placed on the relative resistance of *Giardia*, *Cryptosporidium*, and some viruses when comparing the effectiveness of alternative disinfecting chemicals. Regulations for inactivation of viruses, *Giardia*, and *Cryptosporidium* are covered in the Surface Water Treatment Rule and the Enhanced Surface Water Treatment Rule (see Chap. 1). The requirements are based on the time of contact T and the concentration C of the disinfectant residual during the period of contacting. Configuration of contacting and the relative level of short-circuiting are also taken into account when establishing credit for contact time T. Regulations apply the product of these two variables ($C \times T$) to determine the level of inactivation credit for a given chemical disinfectant. The underlying basis for this approach is found in disinfection models that are relatively simple in form; more complex models and concepts also exist. The concepts are applied on a relatively conservative basis to provide a safety factor.

Among the chemical disinfectants, free chlorine, ozone, and chlorine dioxide are most frequently applied for primary disinfection. Compliance criteria exist for the use of chloramines, but the required $C \times T$ levels are large, demanding substantial contact times. While not widespread practice, a few utilities with hydraulic retention times sufficient for long periods of disinfectant contact have been able to achieve conditions required for inactivation of viruses and *Giardia* using only chloramines.

In comparing the merits of free chlorine, ozone, and chlorine dioxide as primary disinfectants, it is important to note that, with the exception of *Cryptosporidium*, each can meet the criteria for inactivating target pathogens. Free chlorine is the weakest of these, and while this might appear to be a disadvantage, adequate detention time typically is available within a plant, and the addition of new structures to accommodate this form of disinfection is often not required. Ozone, a much stronger disinfectant that is effective with less contact time, requires specialized contacting facilities for efficient gas transfer and facilities configured for containing and collecting off-gases. The logistics of locating and accommodating the additional structures within a hydraulic profile can be an important consideration for incorporation into a process sequence, especially in existing facilities. Ozone also requires relatively sophisticated generation equipment. The use of chlorine dioxide, also a stronger disinfectant than chlorine, also requires purchase of additional chemicals for on-site generation, and the process must be tightly controlled to avoid issues with excessive residuals and the production of excessive levels of its degradation products, chlorite and chlorate.

With respect to inactivation of *Cryptosporidium*, free chlorine is relatively ineffective. Ozone can be effective, but required $C \times T$ levels are relatively high, particularly in cold water. Chlorine dioxide, while more effective than free chlorine, still requires high levels of $C \times T$ that prove impractical to maintain in many situations.

Each primary disinfectant must be assessed for its effectiveness under a given set of conditions. In many cases, when considering alternatives to free chlorine, more effective disinfection and control of disinfection by-products may offset the greater complexity and higher costs of chlorine dioxide and ozone. Factors beyond disinfection goals should also be considered because each of these disinfectants can provide oxidation functions as well, as discussed in the chemical oxidation section of this chapter. For example, ozone can be used to control tastes and odors, to improve coagulation, and to reduce the formation of total THMs and haloacetic acids (HAA5), whereas chlorine dioxide is effective for removal of manganese and can reduce the formation of total THMs and HAA5 in some situations.

Secondary disinfection operates under a different set of criteria from those which apply to primary disinfection. The most important criterion for secondary disinfection is capability to maintain a disinfectant residual throughout the distribution system. The high reactivity of ozone eliminates it from consideration, and the adverse effects of chlorine dioxide residuals tend to limit its suitability. This leaves free chlorine and chloramines as the secondary disinfectants available for most applications. Even where alternative disinfectants are applied with the intent to limit the use of free chlorine and chloramines for primary disinfection, one of these two disinfectants generally is used as the secondary disinfectant under present rules of practice, and this aspect of disinfection must be integrated into an overall treatment train that will limit disinfection by-product formation appropriately and produce a manageable level of disinfectant demand to allow reasonable use of these two chemicals. Free chlorine is the stronger disinfectant of the two, but chloramines have advantages of longer persistence and lower formation of THMs and HAA5s, although their use can result in other by-products. Chloramines are better at controlling biofilms within the distribution system in most cases. Both are approved for application under federal regulations, and selection is a matter of best fit in a given situation.

UV Technologies (Chap. 18)

Ultraviolet (UV) light, historically used primarily in small groundwater systems and to disinfect wastewater effluent, is rapidly emerging as the preferred primary disinfectant when provisions for inactivation of *Cryptosporidium* are required. There are more than 1000 U.S. facilities that currently use UV for disinfection of public drinking water supplies. UV inactivates pathogens by disrupting their DNA strands, rendering them noninfectious. It has demonstrated the ability to achieve 3.0-log to 5.0-log inactivation of *Cryptosporidium* oocysts and *Giardia* cysts. Benefits of UV disinfection include (1) significantly lower initial and operating costs than for comparable control processes (e.g., ozonation and low-pressure membrane processes), (2) a small footprint, (3) the usual ability to be cost-effectively retrofitted into existing plants, (4) potential reduction in halogenated DBP levels when free chlorine contact times following UV treatment are limited, and (5) high levels of achievable pathogen inactivation. Potential disadvantages include (1) fouling/plating of quartz sleeves housing the UV lamps, (2) inconsistent reliability/accuracy of the UV sensors used to monitor process effectiveness, (3) the need for periodic replacement of lamps, and (4) the potential for lamp breakage and associated exposure to mercury.

UV systems used in potable water facilities typically use medium- or low-pressure high-output lamps installed in stainless steel pipe-type reactor vessels, which facilitates incorporation into existing treatment facilities. The lamps typically are enclosed in high-purity quartz sleeves, which isolate the electrical components from the water, help to maintain optimal operating temperatures, and serve to protect the lamps from liquid-generated forces and thermal shock. Systems typically include online monitors for measuring the

UV transmittance (UVT) of the water being treated. Standby UV reactors are provided to avoid interruption of operation in the event of a breakdown or the need for servicing of online equipment. USEPA's *Ultraviolet Disinfection Guidance Manual for the Long Term 2 Enhanced Surface Water Treatment Rule* (USEPA, 2006b) recommends that UV systems be located after filtration. Drawbacks to placing the reactors upstream from filtration include lower UV transmittance, which results in higher energy requirements to maintain required UV dosages, and the potential for coagulation or softening precipitates to foul lamp sleeves and to shield microorganisms, thereby hindering their inactivation. Also, the UV dosages specified in the LT2ESWTR for inactivation of *Cryptosporidium* apply only to postfilter applications and to certain unfiltered supplies.

The LT2ESWTR dose requirements for inactivation of viruses by UV are based on adenovirus, the most UV resistant virus identified to date. According to the dose tables, UV disinfection is not as cost-effective for inactivation of viruses as it is for inactivation of *Giardia* and *Cryptosporidium.* Therefore, a brief period of free chlorine contact prior to or following UV exposure is usually provided to maintain conditions that will ensure effective inactivation of viruses.

Many water utilities using surface water supplies are taking preemptive measures to prevent exposure of their customers to chlorine-resistant pathogens such as *Cryptosporidium.* One such measure is the installation of UV disinfection. Because of the relatively high benefit-to-cost ratio, many utilities have come to view UV disinfection as a cost-effective additional barrier against outbreaks of waterborne disease, and some are considering its installation regardless of the outcome of source water monitoring mandated by the LT2ESWTR.

Interest in the use of UV and hydrogen peroxide (H_2O_2) as an advanced oxidation process (AOP) for the removal of taste- and odor-causing compounds and organic and inorganic micropollutants is growing. The effectiveness of AOP in removing geosmin and MIB, MTBE, perchlorate, pesticides, and PPCPs has been demonstrated by both pilot- and full-scale testing.

The reaction of UV with H_2O_2 forms hydroxyl radicals, which oxidize contaminants. Advantages of UV/H_2O_2 include a smaller footprint than for other contaminant-control options such as ozonation or carbon adsorption and effective inactivation of *Cryptosporidium* and *Giardia* at all temperatures. Disadvantages include high energy consumption and high operation and maintenance costs when operating in AOP mode and the need for close monitoring of influent contaminant concentrations to maintain proper UV and H_2O_2 dosages. Also, because only about 10–15 percent of the applied hydrogen peroxide dosage is consumed during the oxidation process, the remaining hydrogen peroxide at the UV contactor discharge must be removed through reaction with a quenching agent such as chlorine or calcium thiosulfate. Approximately 2 mg/L of chlorine is required to remove 1 mg/L of hydrogen peroxide residual, which can result in the need to provide additional chlorine storage and feed capability to accommodate the increased chlorine demand.

DISTRIBUTION SYSTEM CONSIDERATIONS (CHAPS. 19–21)

Microbial and chemical changes in distribution system piping have become important considerations in managing the quality of water delivered to the customer's tap. The changes are responses to phenomena that occur both in the bulk water as it flows through the distribution network and at pipe surfaces that serve as habitat for microbes and locations of critical chemical reactions.

An important aspect of distribution system water quality is the practice of maintaining a disinfectant residual in the system. In the United States, this is accomplished by adding either free chlorine or chloramines. Both are reactive chemicals that continue to react as water is conveyed through a system; the extent of these reactions varies with detention time and availability of contaminants with which to react. Some reactions occur with substances in the bulk water, such as NOM. These reactions cause loss of disinfectant residuals and can result in the formation of DBPs. When present, reduced forms of manganese and iron also can react with these residuals, which may result in discolored water at consumer taps.

Reactions with disinfectant residuals are not confined to those which occur in the bulk water because interactions at pipe surfaces can create a high disinfectant demand. These features complicate managing and modeling water quality because they can lead to variations in the rate of dissipation of disinfectant residuals and the formation of DBPs. Important variables that affect the extent of these reactions include the degree of accumulated sediments, effectiveness of corrosion-control practices, and the system's maintenance procedures.

While both free chlorine and chloramines have provided acceptable results in maintaining a disinfectant residual in the distribution system, there are some important differences between them. In the case of chloramines, the presence of ammonia as a component of the chemical structure may require measures to prevent deterioration of microbial quality as a result of biological nitrification. Also, while both chlorine and chloramines are acceptable disinfectants when used alone, mixing chloraminated and chlorinated water should be avoided because their incompatibility can result in water quality problems such as loss of disinfectant residual.

Chemical conditioning of finished water to minimize corrosion in the distribution system is another critical step in water treatment. Goals include protection against damage to distribution infrastructure, development of discolored water, and leaching of regulated contaminants, particularly lead and copper. Indirect effects of corrosion, such as the harboring of microbes and reactions that reduce disinfectant residuals, are often more pronounced at corroded surfaces, and corrosion control can provide improvements in these areas as well. Conditioning methods include pH control, provision of scale-forming conditions, and application of agents that react at the surface of the pipe to form a protective passivating layer.

Consistency of water quality in a distribution system is an important consideration because protective coatings on the pipe surface that prevent corrosion can be compromised as a result of changes in water quality. Several factors can change and create imbalances that accelerate corrosion, including intermixing of sources with different water quality characteristics, such as alkalinity, hardness, pH, chlorine residual, oxygen concentration, conductivity, temperature, and TOC concentration. Recent evaluations indicate that changes in the chloride-to-sulfate ratio, which can occur with changes in coagulant chemicals, can accelerate lead corrosion in some cases. Mixing of water from multiple sources also can affect chemical stability. Another unintended consequence of changes in water quality is the potential for release of absorbed arsenic when iron oxide deposits are released from the walls of old cast iron distribution mains that distribute well water that contains arsenic.

Biological stability is an additional distribution system water quality concern. Biologically stable water can be achieved by maintaining low levels of biodegradable organic material in the water to limit microbial growth in the system. Assimilable organic carbon (AOC) and biodegradable organic carbon (BDOC) measurements have been developed to characterize this biological material. These types of constituents can be removed by biological filtration, which is often applied after ozonation, where levels of biodegradable organics are typically high. Other oxidants, including chlorine and chlorine dioxide, also can form these types of substances because these phenomena are not confined to the effects of ozone oxidation.

Physical conditions and proper maintenance of distribution system components to limit adverse effects have important effects on changes in water quality. Accumulation of sediment and periodic cleaning by flushing or other means can have a significant effect on chlorine demand and availability of sites that harbor bacteria. Where such deposits are problematic, increasing the chlorine residual to offset increased chlorine depletion can result in higher levels of DBP formation.

The integrity of the distribution system is critical to protecting it against intrusion of microbial and chemical agents. Cross-connection programs to maintain integrity traditionally have been a component of water system quality management practices. Other considerations in this area involve procedures for main breaks, cleaning and disinfection practices for new pipes, and methods for protecting storage facilities against intrusion. Adequate system pressure is an important aspect of protecting against intrusion through incidental leaks that can occur in systems, and low-pressure conditions should serve as an alert to possible contamination. More recently, security issues have emerged, and these have resulted in new strategies for monitoring and managing distribution systems.

TREATMENT PROCESS RESIDUALS MANAGEMENT (CHAP. 22)

Water treatment residuals are broadly classified into three categories: solids from clarification (settling or DAF), spent filter backwash solids, and membrane and IX residuals. While historically water treatment residuals were discharged to surface waters, this is rarely permitted (particularly for new treatment facilities), and alternative solids handling methods therefore are being used with increasing frequency. Solids generated through coagulation and lime softening processes have significantly different characteristics but are often thickened and dewatered by many of the same treatment technologies.

Water treatment residuals production is a function of the raw water quality and the quantity of chemicals added during treatment. Solids production varies widely for different water sources and the treatment methods used (e.g., coagulation versus lime softening versus membrane treatment) and often varies seasonally as source water quality changes. Consequently, good historical plant treatment data are imperative to accurately determine solids production rates.

Solids treatment processes can be divided into evaporative and mechanical processes. Evaporative processes are typically used in hot or dry climates that allow rapid evaporation. Since evaporative processes require considerable area, they are difficult to implement at treatment plants with space constraints. Evaporative processes include lagoon dewatering and sand or paved drying beds. Design of evaporative processes is based primarily on local evaporation rates, and approaches therefore vary greatly from region to region. Evaporative processes tend to be less labor-intensive than mechanical processes.

Mechanical processes include both thickening and dewatering technologies. Gravity thickening is typically used for water treatment residuals, which tend to settle well. Mechanical dewatering processes employ belt filter presses, centrifuges, and pressure filters (plate and frame). Both thickening and dewatering system designs are most effective when based on site-specific residuals characteristics, such as settleability and dewaterability, and on pilot-scale testing data. However, in the absence of this information, "rules of thumb" or approximations are often used. Mechanical treatment processes require less space than evaporative processes but are usually more labor-intensive and have higher operating costs.

Regardless of the treatment method employed, final use and ultimate disposal need to be carefully considered when evaluating residuals-handling options. Several options are available, with the most common being disposal of the dewatered cake by landfilling or monofilling (disposal to landfills dedicated solely to water treatment residuals). Beneficial use options include land application, use by soil blenders as a liming/pH adjustment agent, or use as a raw material in a manufacturing process.

OTHER CONSIDERATIONS

Pilot Testing

Pilot-scale testing can provide invaluable information in assessing the applicability of various unit processes to meet site-specific treatment requirements. Side-by-side pilot-scale testing of alternative unit processes or process trains conducted during project predesign phases not only assists in identification of the optimal treatment process configuration and associated design criteria (e.g., process loading rates, filter media configuration, etc.) but also can provide information on chemical requirements and appropriate operating parameters that is useful during startup and initial operation of the full-scale treatment facilities. Pilot testing may be required in some cases to demonstrate the effectiveness of newer, high-rate unit processes in order to obtain state regulatory agency approval.

It is now common practice to incorporate permanent pilot-plant facilities into the design of new or upgraded treatment plants. Pilot-scale testing offers a reliable, low-cost method to evaluate different treatment techniques without the risk of compromising the quality of the water delivered to consumers. Other potential benefits of permanent pilot-plant installations include (1) verification of the ability to operate at increased production rates and associated changes in treatment, (2) identification of optimal treatment conditions as source water quality changes seasonally, (3) evaluation of treatment to address specific source water quality issues such as periodic taste and odor episodes, elevated turbidity, and/or increases in source water iron and manganese concentrations, (4) training of new plant operators, and (5) public tours to increase consumer awareness and confidence in the utility.

Permanent pilot plants typically include all the unit processes that comprise the full-scale treatment process and are commonly sized to treat 0.19–0.63 L/s (3–10 gpm). Since some unit processes do not lend themselves to being scaled down to this capacity range (e.g., DAF and ballasted flocculation/sedimentation), permanent pilot plants have been constructed with some unit processes capable of treating 3.15–6.30 L/s (50–100 gpm) and with downstream processes sized to treat only a small portion of the flow produced by the upstream unit processes.

Flexibility to Meet More Stringent Future Requirements

Considering recent advances in analytical measurements and the current regulatory environment, the ability of a water treatment facility to readily adapt to future changes in regulatory requirements is an important consideration in the selection and design of the unit processes that constitute the overall treatment process. Many utilities that have recently constructed new treatment facilities have specified that the facilities must be able to produce finished water of a quality that not only meets current regulations but also exceeds them by a significant margin, in anticipation of more restrictive future regulations. The ability to produce higher-quality finished water with minimal or no physical modifications often can be incorporated during design and construction of the treatment facility at minimal additional cost by providing flexibility in chemical feed locations and dosing capabilities. Providing greater operational flexibility can enhance the operating staff's capability to cope with unanticipated future water quality conditions.

Environmental Considerations

Recent concerns regarding the potential impacts of climate change on water supply, source water quality, and system demands have resulted in increased attention to many environment-related issues. Water utilities are increasingly dealing with water sustainability concerns, specific areas of system vulnerability, and the impact of their operations on greenhouse gas emissions. Utilities are increasingly monitoring their carbon footprint (an estimate of the amount of carbon dioxide–equivalent gases released into or removed from the environment). In the United Kingdom, assessment of the carbon footprint of any proposed capital improvements is required as part of the preparation of asset-management funding plans (Strutt et al., 2008). Energy use and chemical treatment practices are two key elements contributing to the carbon footprint and associated greenhouse gas emissions.

While raw and finished water pumping constitutes the majority of energy consumption of a water system, continuing increases in energy costs may also influence the selection of unit processes, with those which are less energy-intensive being viewed more favorably during the design of new treatment facilities. The need to treat lower-quality source waters that until recently have remained untapped as potential sources of drinking water, combined with the necessity to meet increasingly stringent regulatory requirements while minimizing the cost of constructing new facilities, has spurred interest in more sophisticated unit processes. However, many of these newer, high-rate processes may be more energy-intensive than older, more conventional ones. AOPs incorporating UV oxidation in conjunction with the addition of hydrogen peroxide have demonstrated the ability to effectively control taste and odors attributable to geosmin and MIB and typically can be incorporated into the treatment facility at significantly lower capital cost than other means of taste and odor control such as ozonation or GAC contactors. However, AOPs also exhibit higher energy demands than alternative processes because of the need to operate the system at high UV dosages. Considering the continuing escalation in energy costs, evaluation of specific energy requirements of alternative unit processes during plant predesign is expected to increase as plant designers and utilities continue to seek methods to reduce overall energy use.

Similar to the need to consider energy consumption when evaluating the various treatment processes, the costs of water treatment chemicals must be considered. Often there are alternatives (e.g., various coagulants) that should be evaluated to minimize their costs.

Use of "Demonstrated Technology"

While plant designers and regulatory agencies sometimes tend to prefer older, "tried and true" technologies when evaluating the requirements for new or expanded treatment facilities, the newer, high-rate unit processes frequently offer substantial benefits with respect to both initial construction and annual operating and maintenance costs. For example, use of the conventional large rectangular flocculation and sedimentation basins with extended hydraulic retention times (4–6 hours) in new treatment facilities has become rare because other unit processes such as inclined-plate sedimentation, ballasted flocculation/sedimentation, and dissolved air flotation continue to demonstrate equal or superior performance. It is also common practice now to consider membrane filtration as an alternative to conventional granular media filtration when evaluating unit process options for new or upgraded/expanded treatment facilities, particularly when there is concern about the presence of microbial pathogens such as *Cryptosporidium* in the source water.

TREATMENT PROCESS CONFIGURATIONS

Information on the applicability of various unit processes for the removal of selected contaminants is presented in Table 5-1. Particulate contaminants are not included in the table; these contaminants are removed by the various filtration processes listed in

TABLE 5-1 Applicability of Unit Treatment Processes for Removal of Contaminants

Contaminant categories	Aeration and stripping	Coagulation, sedimentation or DAF, filtration	Precoat filtration	Lime softening	Chemical oxidation and disinfection	Membrane processes			Ion exchange		Adsorption		
						Nanofiltration	Reverse osmosis	Electrodialysis/ ED reversal	Anion	Cation	Granular activated carbon	Powdered activated carbon	Sorption on inorganic media
Primary contaminants													
Inorganics													
Antimony		BAT				X	BAT	X					X
Arsenic		BAT*		BAT		X	BAT	X	BAT*				BAT*
Barium				BAT			BAT	BAT		BAT			
Beryllium		BAT		BAT		X	BAT	X		BAT			BAT
Cadmium		BAT		BAT		X	BAT	X		BAT			
Chromium(III)		BAT		BAT		X	BAT	X		BAT			
Chromium(VI)		BAT				X	BAT	X	BAT				
Cyanide					BAT		BAT		BAT				X
Fluoride				X			BAT	X	X				BAT
Mercury (inorganic)		BAT		BAT			BAT	X		X	BAT		
Nitrate						X	BAT	BAT	BAT				
Selenium(IV)		BAT		BAT		X	BAT	BAT	X				BAT
Selenium(VI)				BAT		X	BAT	X	BAT				BAT
Thallium							X	X		BAT			BAT
Organic contaminants													
Volatile organics	BAT										BAT		
Synthetic organics						X	X				BAT	X	
Pesticides/herbicides						X	X				BAT	X	

Radionuclides													
Radium 226 + 228				BAT		X	BAT	BAT		BAT			
Uranium		BAT*		BAT		X	BAT	X	BAT	X			X
Secondary contaminants and constituents causing aesthetic problems													
Hardness				X		X	X	X		X			
Iron		XO	XO	X						X			
Manganese		XO	XO	X						X			X
Total dissolved solids							X	X					
Chloride							X	X					
Sulfate						X	X	X	X				
Zinc				X			X	X		X			
Color		X		X	X	X	X		X		X	X	
Taste & odor	X				X						X	X	
Other contaminants/constituents													
Radon	X												
Sulfide	X				X								
Dissolved organic carbon		X		X		X	X		X		X	X	X

X = appropriate process for this contaminant; XO = appropriate when oxidation used in conjunction with this process; BAT = designated as "best available technology" by USEPA.

*Applicable for enhanced coagulation/filtration at pH ~6.

the table and by microfiltration and ultrafiltration. Because site-specific conditions and other factors determine whether a unit process is capable of effectively removing a contaminant, the information presented in the table should be regarded as general guidance. Further information can be obtained from the chapters covering the various unit processes.

The following paragraphs present several treatment process schematics that illustrate how unit processes are typically incorporated into water plants. These are presented for various types of source waters. Since residuals handling and disposal facilities depend on the availability of site space, ultimate disposal requirements, and utility preferences, which are highly site-specific, these facilities are not shown on the process schematics. Also, for simplicity, raw, intermediate, and finished water pumping facilities and some commonly provided chemical feed systems such as fluoride addition are not shown. It is also recognized that other unit process configurations may produce performance results equivalent to the treatment processes illustrated here, depending on site-specific conditions.

High-Quality Reservoir Supply

The membrane filtration treatment process shown in Fig. 5-2 is applicable to treatment of high-quality reservoir sources with little or no iron and manganese, low to moderate levels of turbidity and TOC, low DBP potential, and no history of significant algal blooms and taste and odor episodes. Because membrane filtration does not remove dissolved organic material, provisions for adding coagulant ahead of MF/UF may be needed to remove TOC and DBP precursor compounds. While TOC in some cases can be removed by in-pipe addition of coagulant immediately upstream of MF/UF, the need for flocculation and, potentially, for sedimentation preceding MF is site-specific and must be assessed through bench- and pilot-scale testing.

While a more conventional direct filtration process (i.e., coagulant addition, flocculation, and granular media filtration) also could be considered for this source water, it should be noted that direct filtration does not provide the high levels of microbial pathogen (i.e., *Giardia* cysts and *Cryptosporidium* oocysts) removal that MF/UF membrane filtration processes routinely achieve. Therefore, if direct filtration is specified, it may be necessary to include provisions for postfiltration UV disinfection to achieve overall microbial pathogen removal/inactivation levels comparable with those provided by membrane filtration processes.

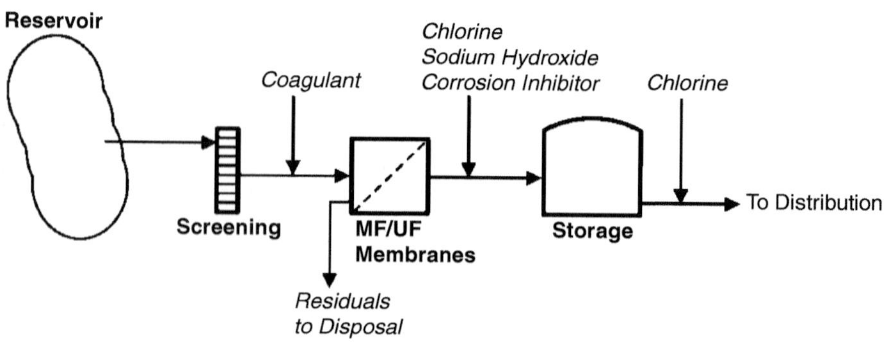

FIGURE 5-2 Membrane filtration of high-quality surface supply.

Reservoir with Algae and Color/TOC

The treatment process train shown in Fig. 5-3 is appropriate for a reservoir supply to address the following water quality conditions:

- Low to moderate levels of non-mineral turbidity (turbidity caused by organic constituents such as algae)
- Moderate to high levels of algae
- Low to high color/TOC
- Provision of high levels of pathogen removal/inactivation
- Iron and manganese may or may not be present.

This approach assumes use of dissolved air flotation to provide effective removal of color and TOC and to ensure acceptable filter run times and productivities when algae concentrations in the source water are high enough to have a negative impact on granular media filter performance. Should manganese concentrations in the source water increase to the extent that treatment is required, the filters could be chemically conditioned to establish a manganese oxide coating on the media, which would ensure highly efficient removal of manganese in the presence of free chlorine residual across the filters.

River or Reservoir Supply with Low to Moderate Hardness

Figure 5-4 presents a typical treatment process train for a reservoir or river supply when reduction of hardness is not required. This process is appropriate to address the following water quality conditions:

- Moderate to high levels of mineral turbidity (turbidity caused by silts/clays)
- Low to moderate levels of color/TOC
- Provision of high levels of pathogen removal/inactivation

This approach assumes use of high-rate clarification with inclined-plate sedimentation basins to reduce the unit process footprint area and construction costs. Provisions for optimization of coagulation pH through the addition of acid or base and for postfiltration adjustment of finished water pH using sodium hydroxide, as shown in Fig. 5-4, are commonly included in the design of new or upgraded treatment facilities. Postfiltration UV

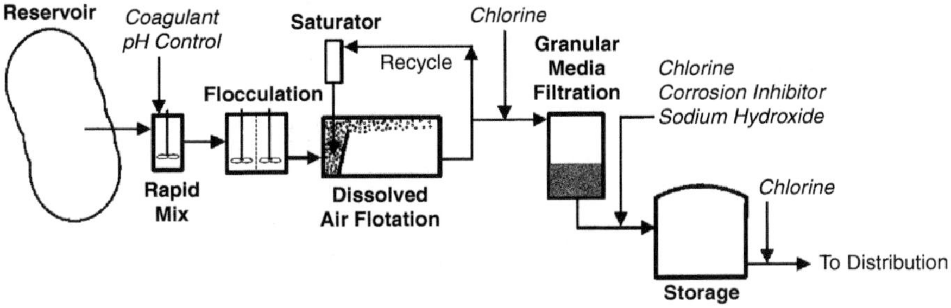

FIGURE 5-3 Dissolved air flotation treatment of reservoir supply.

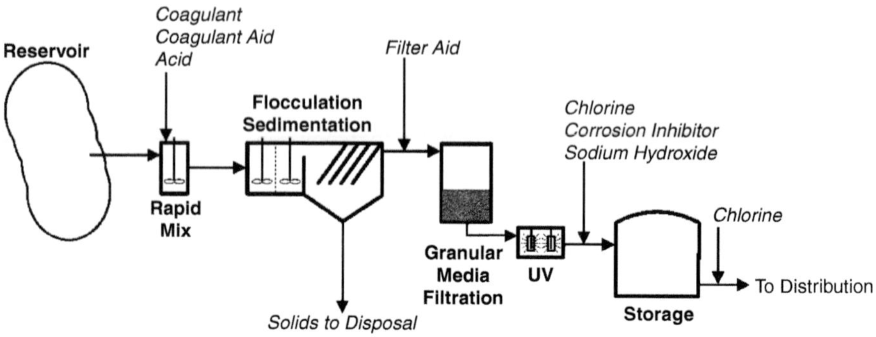

FIGURE 5-4 Conventional treatment of surface supply.

disinfection can be provided if significant concerns regarding source water pathogen levels exist or if inactivation of *Cryptosporidium* oocysts is required to comply with regulatory requirements.

For a reservoir supply having consistently low turbidity but high color and/or episodes of high algae concentrations that would be detrimental to granular filter performance, DAF could be substituted for the sedimentation process shown in Fig. 5-4 because DAF is better able to cope with high color and algae conditions.

River Supply with Hardness

The treatment process shown in Fig. 5-5 is appropriate when both clarification and softening of a river supply with highly variable turbidity and high hardness levels are required, and constitutes the approach currently used by many utilities throughout the Midwestern United States. This treatment process incorporates the following:

- High-rate presedimentation with inclined-plate sedimentation basins to reduce turbidity and color before softening
- Two-stage lime softening/recarbonation with upflow solids contact clarifiers to achieve high lime–soda ash utilization efficiencies and softening performance

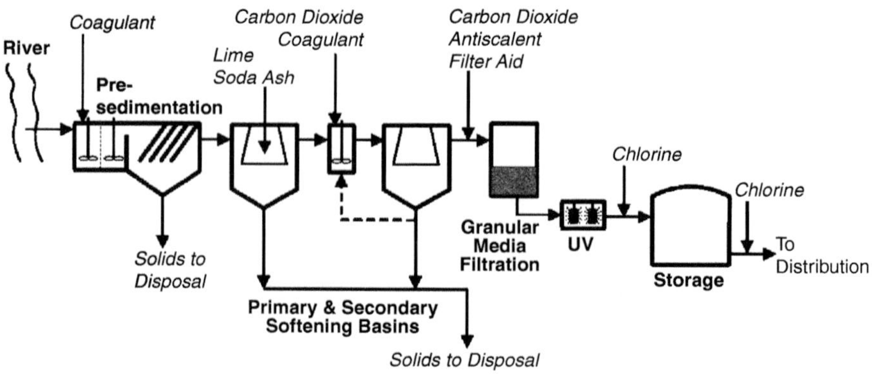

FIGURE 5-5 Clarification/softening of surface supply.

- Provisions for recycling a portion of the solids removed in the secondary softening basins to the secondary basin influent rapid mix to minimize encrustation of mixer impellers and basin influent piping
- Provisions for postfiltration UV disinfection, if needed, to comply with *Cryptosporidium* removal/inactivation requirements

Groundwater with Iron and Manganese

The treatment process shown in Fig. 5-6 is appropriate for a groundwater supply to address the following water quality conditions:

- Iron and/or manganese at concentrations that must be reduced to achieve acceptable finished water quality
- Hydrogen sulfide and/or radon
- Hardness at levels that do not warrant softening treatment

Forced- or induced-draft cascade aeration is provided to initiate the oxidation of iron and manganese and to remove hydrogen sulfide, radon, and other dissolved gases. Because oxidation of manganese to insoluble/filterable form is relatively slow under typical groundwater pH conditions, provisions for maintaining a free chlorine residual across the filters would be included to ensure that effective manganese oxide coatings are maintained on the filter media, thereby ensuring high manganese removal efficiencies.

When removal of hydrogen sulfide or other dissolved gases is not required, aeration may be eliminated, and chemical oxidation of iron and manganese preceding filtration can be employed to reduce construction costs. Use of proprietary high-rate pressure filtration systems for iron and manganese removal is also becoming more common as utilities seek to reduce the complexity of their treatment facilities and to minimize construction costs.

When radon is the only contaminant present at levels that require treatment, aeration and disinfection would be the only unit processes needed. When arsenic is the only contaminant present, aeration is not provided, but chlorine addition for oxidation and a ferric coagulant could be used followed by granular media filtration.

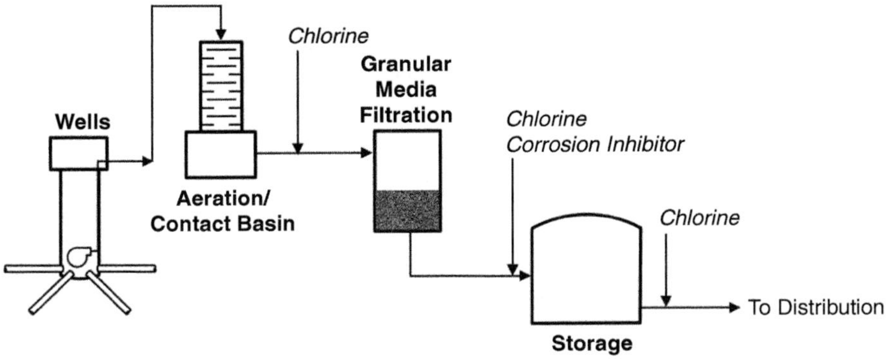

FIGURE 5-6 Iron and manganese removal: groundwater supply.

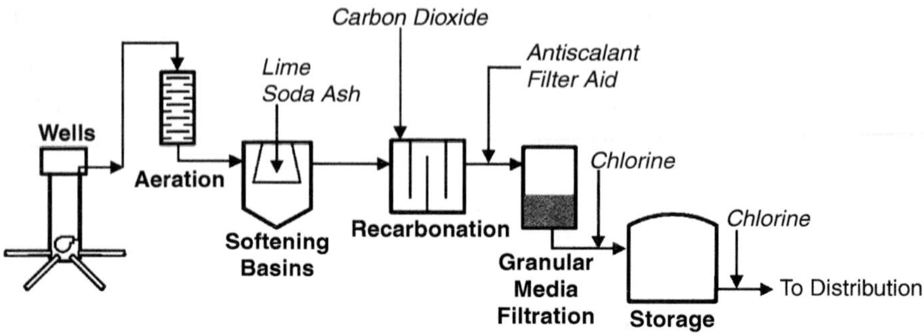

FIGURE 5-7 Lime softening of groundwater supply.

Groundwater with Hardness

Figure 5-7 presents a typical process configuration for a groundwater supply with the following treatment requirements:

- Removal of iron and/or manganese
- Removal of hydrogen sulfide and/or radon
- Reduction of hardness to meet customer expectations

In addition to stripping sulfide and radon, aeration also reduces carbon dioxide concentrations, thereby reducing both the lime dosage and the associated chemical solids production. Following single-stage lime softening, a recarbonation basin with a detention time of at least 20 minutes is typically provided to ensure adequate stabilization of softened water pH before filtration, and provisions for adding an antiscalant (typically sodium hexametaphosphate) at the filter inlet are included to minimize encrustation of filter media by calcium carbonate.

For facilities with capacities of approximately 19 ML/day (5 mgd) or less, IX is often used when softening treatment is required. Because the IX process produces near-zero-hardness water, a portion of the raw water is blended with IX-treated water to achieve the desired finished water hardness. When iron and manganese are present in the source water, provisions for their removal from the IX bypass flow may be required. Treatment of IX feed water to reduce iron and manganese concentrations may also be required to prevent deterioration of the IX resin.

Membrane softening using NF or RO membranes also could be used to treat this source. Pretreatment of membrane feed water typically will be required to ensure that membrane fouling rates are maintained within acceptable ranges.

ABBREVIATIONS

AOC	assimilable organic carbon
AOP	advanced oxidation process
ASR	aquifer storage, recharge, and recovery
As(III)	arsenite
As(V)	arsenate

BAT	best available technology
BDOC	biodegradable organic carbon
BW	backwashing
DAF	dissolved air flotation
DBP(s)	disinfection by-product(s)
DOC	dissolved organic carbon
ED	electrodialysis
EDC(s)	endocrine disruptor(s)
EDR	electrodialysis reversal
GAC	granular activated carbon
HAA5(s)	haloacetic acid(s)
IX	ion exchange
LT2ESWTR	Long Term 2 Enhanced Surface Water Treatment Rule
MF	microfiltration
MIB	2-methyl isoborneol
ML/day	megaliters per day
MWCO	molecular weight cutoff
NF	nanofiltration
NOM	natural organic matter
ntu	nephelometric turbidity unit
OR	overflow rate
PAC	powdered activated carbon
PES	polyether sulfone
PPCP(s)	pharmaceuticals and personal care product(s)
PS	polysulfone
PVDF	polyvinylidene difluoride
RAA	running annual average
RBF	riverbank filtration
RO	reverse osmosis
SDI	silt density index
SOC(s)	synthetic organic chemicals(s)
TDS	total dissolved solids
THMs	trihalomethanes
TOC	total organic carbon
UF	ultrafiltration
USEPA	U.S. Environmental Protection Agency
UV	ultraviolet
UVT	ultraviolet transmittance
VOC(s)	volatile organic chemical(s)
WTP	water treatment plant

REFERENCES

Bellar, T. A., J. J Lichtenberg, and R. C. Kroner (1974). "The Occurrence of Organohalides in Chlorinated Drinking Water," *Journal AWWA*, vol. 66, no. 12, pp. 703–706.

Bond, R., and S. Veerapeneni (2007). *Zero Liquid Discharge for Inland Desalination*. Awwa Research Foundation Report. Denver, CO: Water Research Foundation.

Camp, T. R. (1936). "Study of Rational Design of Settling Tanks," *Sewage Works Journal*, vol. 8, no. 5, p. 742.

Clark, M. M., R. M. Srivastava, J. S. Lang, R. R. Trussell, L. J. McCollum, D. Bailey, J. D. Christie, and G. Stolarik (1994). *Selection and Design of Mixing Processes for Water Treatment*. Awwa Research Foundation Report. Denver, CO: Water Research Foundation.

Cleasby, J. L., A. H. Dharmarajah, G. L. Sindt, and R. E. Baumann (1989). *Design and Operation Guidelines for Optimization of the High-Rate Filtration Process: Plant Survey Results*. Awwa Research Foundation Report. Denver, CO: Water Research Foundation.

Coffey, B. M., M. H. Stewart, K. L. Wattier, and R. T. Wale (1993). "Evaluation of Microfiltration for Metropolitan's Small Domestic Water Systems," in *Proceedings of the AWWA Membrane Technology Conference*. Denver, CO: AWWA.

Furukawa, D. (2008). *A Global Perspective of Low Pressure Membranes*. Fountain Valley, CA: National Water Research Institute.

Hazen, A. (1904). "On Sedimentation," *Trans. ASCE,* paper no. 980, pp. 45–71.

Jacangelo, J. G., J. M. Laine, K. E. Carns, E. W. Cummings, and J. Mallevialle (1991). "Low-Pressure Membrane Filtration for Removing *Giardia* and Microbial Indicators," *Journal AWWA*, vol. 83, no. 9, pp. 97–106.

Keuhn, W., and U. Mueller (2000). "Riverbank Filtration: An Overview," *Journal AWWA*, vol. 92, no. 12, pp. 60–69.

Kline, P. (2008). "What Should I Know about Aquifer Storage and Recovery?" *Opflow*, vol. 34, no. 4, pp. 8–9.

Laszlo, F., Z. Homonnay, and M. Zimonyi (1990). "Impacts of River Training on the Quality of Bank-Filtered Waters," *Water Science and Technology*, vol. 22, no. 5, p. 167.

Rook, J. J. (1974). "Formation of Haloforms during Chlorination of Natural Waters," *Water Treatment and Examination*," vol. 23, no. 2, pp. 234–243.

Speth, T. F., and C. R. Reiss (2005). "Water Quality," in *Microfiltration and Ultrafiltration Membranes for Drinking Water.* AWWA Manual of Water Supply Practice M53. Denver, CO: American Water Works Association.

Strutt, J., S. Wilson, H. Shorney-Darby, A. Shaw, and A. Byers (2008). "Assessing the Carbon Footprint of Water Production," *Journal AWWA*, vol. 100, no. 6, pp. 80–91.

Symons, J. M., T. A. Bellar, J. K. Carswell, J. DeMarco, K. L. Kropp, G. G. Robeck, D. R. Seeger, C. J. Slocum, B. L. Smith, and A. A. Stevens (1975). "National Organics Reconnaissance Survey for Halogenated Organics in Drinking Water," *Journal AWWA*, vol. 67, no. 11, pp. 634–647.

U.S. Environmental Protection Agency (2006a). 40 CFR Parts 9, 141, and 142. National Primary Drinking Water Regulations: Long Term 2 Enhanced Surface Water Treatment Rule; Final Rule. *Fed. Reg.*, vol. 71, no. 3, pp. 654–786.

U.S. Environmental Protection Agency (2006b). *Ultraviolet Disinfection Guidance Manual for the Final Long Term 2 Enhanced Surface Water Treatment Rule*. Washington, DC: USEPA.

CHAPTER 6
GAS–LIQUID PROCESSES: PRINCIPLES AND APPLICATIONS

David W. Hand, Ph.D.
Department of Civil and Environmental Engineering
Michigan Technological University
Houghton, Michigan, United States

David R. Hokanson, Ph.D., P.E., B.C.E.E.
Trussell Technologies, Inc.
Pasadena, California, United States

John C. Crittenden, Ph.D., P.E., D.E.E., N.A.E.
Department of Civil and Environmental Engineering
Georgia Tech. University
Atlanta, Georgia, United States

INTRODUCTION **6.1**
THEORY OF GAS TRANSFER **6.2**
Equilibrium 6.2
Mass Transfer 6.10
PACKED TOWERS **6.14**
Governing Equations 6.17
Design Procedure 6.28
Impact of Dissolved Solids on Tower Performance 6.36
Air Stripping Off-Gas Control Using Adsorption 6.36
DIFFUSED OR BUBBLE AERATION **6.41**
Governing Equations: Air Stripping of VOCs 6.42
Governing Equations: Gas Absorption 6.45
SURFACE AERATION **6.52**
Governing Equations 6.53
SPRAY AERATORS **6.56**
Governing Equations 6.57
NOTATION FOR EQUATIONS **6.61**
REFERENCES **6.63**

INTRODUCTION

Air stripping is defined as a process whereby volatile constituents such as gases and organic compounds are transferred from the water to the air. Absorption is the opposite of air stripping—where constituents such as gases are transferred from the air to the water. Several types of air stripping and aeration systems are used for a variety of water treatment applications. The most common types are diffused air, surface aerator, spray, and packed tower systems. Water treatment applications for these systems include the absorption of reactive

reactive gases for water stabilization and disinfection, precipitation of inorganic contaminants, and air stripping of volatile organic compounds (VOCs) and nuisance-causing dissolved gases. The diffused aeration (or bubble) systems are primarily used for the absorption of reactive gases such as oxygen (O_2), ozone (O_3), and chlorine (Cl_2). For example, ozone is used for disinfection, oxidation of VOCs and pesticides, taste and odor control, disinfection by-products control, as a coagulant aid, and for some other uses. Chlorine is primarily used for disinfection and sometimes as a preoxidant for the oxidation of iron and manganese and for other purposes. Diffused aeration systems have also been used for stripping of odor-causing compounds and VOCs. Surface aeration systems are primarily used for removal of gases and VOCs. The packed tower and spray nozzle systems are primarily used for the removal of NH_3, CO_2, H_2S, and VOCs. The packed tower systems include countercurrent flow, cocurrent flow, and cross-flow configurations. Spray nozzle systems include tower and fountain-type configurations.

Another application is the dissolution of air into recycled water under pressure in packed towers called saturators and subsequent production of air bubbles when the pressurized recycle water is injected into an open tank for water clarification. This application is called dissolved air flotation and is covered in Chap. 9.

A fundamental understanding of the theory of gas transfer is first discussed, followed by a description of the various unit operations, development of the governing equations, and example design calculations.

THEORY OF GAS TRANSFER

Proper design and operation of aeration and air stripping devices require a fundamental understanding of equilibrium partitioning of chemicals between air and water as well as an understanding of the mass transfer rate across the air-water interface is required. Equilibrium is the final state that the system is moving towards. The displacement of the system from equilibrium dictates how much fluid (air) is required for stripping or aeration and defines the driving force that governs mass transfer, i.e., the rate at which chemicals move from one phase to another, which in turn determines the vessel size required for stripping or aeration. Both equilibrium and mass transfer concepts are incorporated into mass balance equations to formulate the governing equations. Consequently, these concepts are reviewed first.

Equilibrium

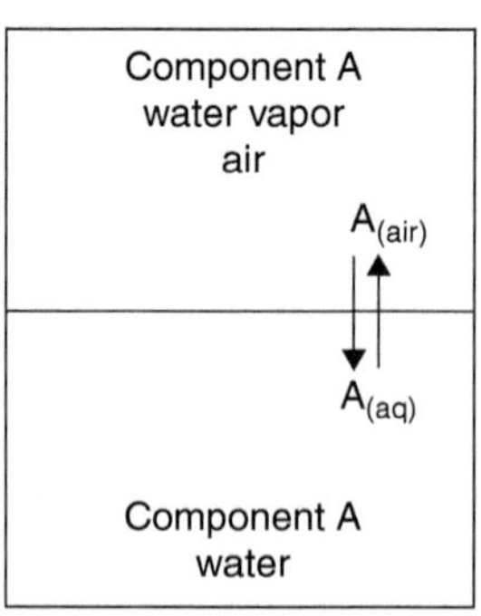

FIGURE 6-1 Schematic of equilibrium conditions for component A in air and water.

For most aeration and air stripping applications in water treatment, equilibrium partitioning of a gas or organic contaminant between air and water can be described by Henry's law. The Henry's law equilibrium description can be derived by considering the closed vessel shown in Fig. 6-1. If the vessel contains both water and air and component A is in equilibrium with both phases at a constant temperature, equilibrium can be described by the following expression:

$$K_{eq} = \frac{a_{air}}{a_{aq}} \tag{6-1}$$

where K_{eq} is the equilibrium constant, a_{air} is the activity of component A in the gas phase, and a_{aq} is the activity of component A in the aqueous phase.

At a pressure of 1 atm, the gas behaves ideally, and Eq. 6-1 reduces to:

$$H = K_{\text{eq}} = \frac{P_A}{\gamma_A A} \tag{6-2}$$

where H is the Henry's law constant (atm-L/mole) of component A, P_A is the pressure A exerts in the gas phase (atm), γ_A is the activity coefficient of A in the aqueous phase, and A is the aqueous-phase molar concentration of A (mole/L).

The presence of air does not affect the Henry's law constant for organic chemicals or gases. For low ionic strength, Henry's law may be written as follows:

$$P_A = H[A] \tag{6-3}$$

When other dissolved organic and inorganic species are present at concentrations less than 0.01 mole/L, the equilibrium partitioning according to Eq. 6-3 is not affected and is generally valid. Further, it has been shown in some cases to be valid for concentrations as high as 0.1 mole/L (Rogers, 1994). The units of H in Eq. 6-3 are atm-L/mole, but H has other units. Figure 6-2 displays the three most commonly used unit measures of H. H is reported in units of atm when the gas-phase concentration of component A is expressed in atm and the aqueous-phase concentration of component A is expressed in terms of mole fraction. H is reported in units of liters of water times atm per mole of gas (L · atm/mole) when the gas-phase concentration of component A is expressed in atm and the aqueous-phase concentration of component A is expressed in terms moles of component A per liter of water. H is also reported in dimensionless units when both the gas and aqueous phases are expressed in the same concentration units. Because the reported units of H vary, it is necessary to convert from one system of units to another. Table 6-1 displays various unit conversions that can be used to perform this conversion. Dimensionless units are convenient for mass balances; consequently, dimensionless units are preferred and are used in this chapter.

Estimating Henry's Constant. When the Henry's law constant for a component of interest is not readily available, it can be estimated if the component's vapor pressure and aqueous solubility are known. There are two situations in which it is possible to estimate the Henry's constant of component A: (1) when component A is perfectly miscible in the aqueous phase and (2) when component A is immiscible in the aqueous phase. When component A is perfectly miscible and the mole fraction of A is equal to 1 ($x_{H_2O} = 0$), the pressure exerted by A is equal to the vapor pressure of pure A at a given temperature. In this case H, can be expressed as:

$$H = P_{v,A} \tag{6-4}$$

in which $P_{v,A}$ is the vapor pressure of pure component A at a given temperature (atm).

When component A is immiscible in the aqueous phase, a third or separate phase of component A is formed within the aqueous phase once the solubility of A is exceeded. If this third phase contains water, then H cannot be determined from vapor pressure and solubility data because the partitioning in the third phase is unknown. However, if the third phase contains only pure component A, then the following expression can be used:

$$H = \frac{P_{v,A}}{C_{s,A}} \tag{6-5}$$

in which $C_{s,A}$ is the aqueous solubility of component A (mg/L). Equations 6-4 and 6-5 are estimation techniques that give H values within ±50 to 100 percent of experimentally reported values and should therefore only be used when measured values of the constants are not available.

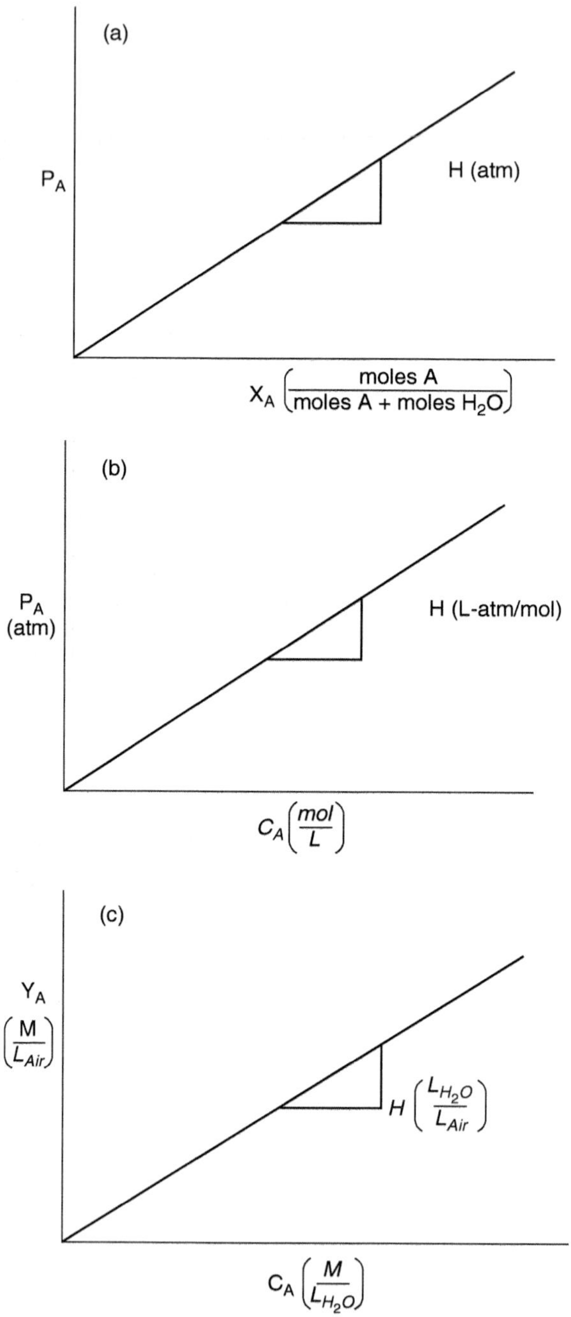

FIGURE 6-2 (*a*) plot of H displayed in atm, (*b*) plot of H displayed in liters of water · atm/mol, (*c*) plot of H displayed in liters of water per liter of air.

TABLE 6-1 Unit Conversions for Henry's Law Constants

$$H\left(L_{H_2O}/L_{Air}\right) = \frac{H\left(\frac{L\cdot atm}{mol}\right)}{RT}$$

$$H\left(\frac{L\cdot atm}{mol}\right) = H(L_{H_2O}/L_{Air}) \times RT$$

$$H(L_{H_2O}/L_{Air}) = \frac{H(atm)}{RT \times 55.6\frac{mol\,H_2O}{L\,H_2O}}$$

$$H\left(\frac{L\cdot atm}{mol}\right) = \frac{H(atm)}{55.6\frac{mol\,H_2O}{L\,H_2O}}$$

$$H(atm) = H\left(\frac{L\cdot atm}{mol}\right) \times 55.6\frac{mol\,H_2O}{L\,H_2O}$$

$$H(atm) = H(L_{H_2O}/L_{Air}) \times RT \times 55.6\frac{mol\,H_2O}{L\,H_2O}$$

$$R = 0.08205\frac{atm\cdot L}{mol\ {}^\circ K} \qquad T = K$$

Effects of Temperature and Solution Properties. Temperature, pressure, ionic strength, surfactants, and solution pH (for ionizable species such as NH_3 and CO_2) can influence the equilibrium partitioning between air and water.

Pressure. The impact of total system pressure on H is negligible because most aeration and stripping devices operate at atmospheric pressure.

Temperature. For the range of temperatures encountered in water treatment, H tends to increase with increasing temperature because the aqueous solubility of the component decreases while its vapor pressure increases. Values of H for several organic compounds at different temperatures are given in Table 6-2. Table 6-3 displays H values for gases at 20°C. Assuming the standard enthalpy change ($\Delta H°$) for the dissolution of a component in water is constant over the temperature range of interest, the change in H with temperature can be estimated using the following van't Hoff-type equation:

$$H_2 = H_1 \times \exp\left(\frac{-\Delta H^\circ}{R}\left(\frac{1}{T_2} - \frac{1}{T_1}\right)\right) \tag{6-6}$$

in which $\Delta H°$ is the standard enthalpy change in water due to the dissolution of a component in water (kcal/kmol), R is the universal gas constant (1.987 kcal/kmol-K), H_1 is a known value of Henry's constant at temperature T_1 (K), H_2 is the calculated Henry's law constant at the desired temperature T_2 (K), and C is a constant.

Equation 6-6 can be simplified to Eq. 6-7, and values of $\Delta H°$ and C for selected compounds are summarized in Table 6-4.

$$\log H = -(\Delta H^\circ/RT) + C \text{ or } H = 10 \exp\,[-(\Delta H^\circ/RT) + C] \tag{6-7}$$

Another common method of expressing the temperature dependence of H is to treat $\Delta H°$, R, and C as fitting parameters A and B using the following equation:

$$H = \exp\left(A - \frac{B}{T}\right) \tag{6-8}$$

Table 6-5 lists values of A and B for several compounds. These are valid for temperatures ranging from 283 to 303 K (Ashworth et al., 1988).

TABLE 6-2 Henry's Law Constants in atm m^3/mole and the H Values in Liters of Water per Liter of Air Are Given in Parentheses for 45 Organic Compounds (Ashworth et al., 1988).

Component	Henry's law constants, H				
	10°C	15°C	20°C	25°C	30°C
Nonane	0.400 (17.2)	0.496 (21.0)	0.332 (13.8)	0.414 (16.9)	0.465 (18.7)
n-Hexane	0.238 (10.3)	0.413 (17.5)	0.883 (36.7)	0.768 (31.4)	1.56 (62.7)
2-Methylpentane	0.697 (30.0)	0.694 (29.4)	0.633 (26.3)	0.825 (33.7)	0.848 (34.1)
Cyclohexane	0.103 (4.44)	0.126 (5.33)	0.140 (5.82)	0.177 (7.24)	0.223 (8.97)
Chlorobenzene	0.00244 (0.105)	0.00281 (0.119)	0.00341 (0.142)	0.00360 (0.147)	0.00473 (0.190)
1,2-Dichlorobenzene	0.00163 (0.0702)	0.00143 (0.0605)	0.00168 (0.0699)	0.00157 (0.0642)	0.00237 (0.0953)
1,3-Dichlorobenzene	0.00221 (0.0952)	0.00231 (0.0978)	0.00294 (0.122)	0.00285 (0.117)	0.00422 (0.170)
1,4-Dichlorobenzene	0.00212 (0.0913)	0.00217 (0.0918)	0.00259 (0.108)	0.00317 (0.130)	0.00389 (0.156)
o-Xylene	0.00285 (0.123)	0.00361 (0.153)	0.00474 (0.197)	0.00487 (0.199)	0.00626 (0.252)
p-Xylene	0.00420 (0.181)	0.00483 (0.204)	0.00645 (0.268)	0.00744 (0.304)	0.00945 (0.380)
m-Xylene	0.00411 (0.177)	0.00496 (0.210)	0.00598 (0.249)	0.00744 (0.304)	0.00887 (0.357)
Propylbenzene	0.00568 (0.245)	0.00731 (0.309)	0.00881 (0.366)	0.0108 (0.442)	0.0137 (0.551)
Ethylbenzene	0.00326 (0.140)	0.00451 (0.191)	0.00601 (0.250)	0.00788 (0.322)	0.0105 (0.422)
Toluene	0.00381 (0.164)	0.00492 (0.210)	0.00555 (0.231)	0.00642 (0.263)	0.00808 (0.325)
Benzene	0.00330 (0.142)	0.00388 (0.164)	0.00452 (0.188)	0.00528 (0.216)	0.00720 (0.290)
Methyl ethylbenzene	0.00351 (0.151)	0.00420 (0.178)	0.00503 (0.209)	0.00558 (0.228)	0.00770 (0.310)
1,1-Dichloroethane	0.00368 (0.158)	0.00454 (0.192)	0.00563 (0.234)	0.00625 (0.256)	0.00776 (0.312)
1,2-Dichloroethane	0.00117 (0.0504)	0.00130 (0.0550)	0.00147 (0.0612)	0.00141 (0.0577)	0.00174 (0.0700)
1,1,1-Trichloroethane	0.00965 (0.416)	0.0115 (0.487)	0.0146 (0.607)	0.0174 (0.712)	0.0211 (0.849)
1,1,2-Trichloroethane	0.000390 (0.0168)	0.000630 (0.0267)	0.000740 (0.0308)	0.000910 (0.0372)	0.00133 (0.0535)
cis-1,2-Dichloroethylene	0.00270 (0.116)	0.00326 (0.138)	0.00360 (0.150)	0.00454 (0.186)	0.00575 (0.231)
trans-1,2-Dichloroethylene	0.000590 (0.0254)	0.00705 (0.298)	0.00857 (0.356)	0.00945 (0.386)	0.0121 (0.469)
Tetrachloroethylene	0.00846 (0.364)	0.0111 (0.467)	0.0141 (0.587)	0.0171 (0.699)	0.0245 (0.985)

Trichloroethylene	0.00538 (0.237)	0.00667 (0.282)	0.00842 (0.350)	0.0102 (0.417)	0.0128 (0.515)
Tetralin	0.000750 (0.0323)	0.00105 (0.0444)	0.00136 (0.0566)	0.00187 (0.0765)	0.00268 (0108)
Decalin	0.0700 (3.015)	0.0837 (3.54)	0.106 (4.41)	0.117 (4.79)	0.199 (8.00)
Vinyl chloride	0.0150 (0.646)	0.0168 (0.711)	0.0217 (0.903)	0.0265 (1.08)	0.028 (1.13)
Chloroethane	0.00759 (0.327)	0.00958 (0.405)	0.0110 (0.458)	0.0121 (0.495)	0.0143 (0.575)
Hexachloroethane	0.00593 (0.255)	0.00559 (0.237)	0.00591 (0.246)	0.00835 (0.342)	0.0103 (0.414)
Carbon tetrachloride	0.0148 (0.637)	0.0191 (0.808)	0.0232 (0.965)	0.0295 (1.21)	0.0378 (1.52)
1,3,5-Trimethylbenzene	0.00403 (0.174)	0.00460 (0.195)	0.00571 (0.238)	0.00673 (0.275)	0.00963 (0.387)
Ethylene dibromide	0.000300 (0.0129)	0.000480 (0.0203)	0.000610 (0.0254)	0.00065(00266)	0.000800 (0.0322)
1,1-Dichloroethylene	0.0154 (0.663)	0.203 (8.59)	0.0218 (0.907)	0.0259 (1.06)	0.0318 (1.28)
Methylene chloride	0.00140 (0.0603)	0.00169 (0.0715)	0.00244 (0.102)	0.00296 (0.121)	0.00361 (0.145)
Chloroform	0.00172 (0.0741)	0.00233 (0.0986)	0.00332 (0.138)	0.00421 (0.172)	0.00554 (0.223)
1,1,2,2-Tetrachloroethane	0.000330 (0.0142)	0.000200 (0.00846)	0.000730 (0.0304)	0.00025(0.0102)	0.000700 (0.0282)
1,2-Dichloropropane	0.00122 (0.0525)	0.00126 (0.0533)	0.00190 (0.0790)	0.00357 (0.146)	0.00286 (0.115)
Dibromochloromethane	0.000380 (0.0164)	0.000450 (0.0190)	0.00103 (0.0428)	0.00118 (0.0483)	0.00152 (0.0611)
1,2,4-Trichlorobenzene	0.00129 (0.0556)	0.00105 (0.0444)	0.00183 (0.0761)	0.00192 (0.0785)	0.00297 (0.119)
2,4-Dimethylphenol	0.00829 (0.357)	0.00674 (0.285)	0.0101 (0.420)	0.00493 (0.202)	0.00375 (0.151)
1,1,2-Trichlorotrifluoroethane	0.154 (6.63)	0.215 (9.10)	0.245 (10.2)	0.319 (13.0)	0.321 (12.9)
Methyl ethyl ketone	0.000280 (0.0121)	0.000390 (0.0165)	0.000190 (0.00790)	0.00013(0.0053)	0.000110 (0.00443)
Methyl isobutyl ketone	0.000660 (0.0284)	0.000370 (0.0157)	0.000290 (0.0121)	0.00039(0.0160)	0.000680 (0.0274)
Methyl cellosolve	0.0441 (1.90)	0.0363 (1.54)	0.116 (4.83)	0.0309 (1.26)	0.0381 (1.53)
Methyl t-butyl ether*	0.5039 (0.0117)	0.7490 (0.0177)	0.9317 (0.0224)	1.194 (0.0292)	1.557 (0.0387)
Trichlorofluoromethane	0.0536 (2.31)	0.0680 (2.88)	0.0804 (3.34)	0.101 (4.13)	0.122 (4.91)

*Fischer, Muller, and Klasmeier (2004).

TABLE 6-3 Henry's Law Constants at 20°C for Gases in Water Compounds (Crittenden et al., 2005)

Compound	H, (L_{H_2O}/L_{Air})
Ammonia	0.000574
Carbon dioxide	0.114
Chlorine	0.442
Chlorine dioxide	0.0408
Hydrogen sulfide	0.389
Methane	28.7
Oxygen	32.5
Ozone	3.77
Radon*	4.08
Sulfur dioxide	0.0287

*Clever (1980).

TABLE 6-4 Temperature Correction Factors for H in atm (Crittenden et al., 2005)

Compound	$\Delta H^{o} \times 10^{-3}$	C
Oxygen	1.45	7.11
Methane	1.54	7.22
Carbon dioxide	2.07	6.73
Hydrogen sulfide	1.85	5.88
Carbon tetrachloride	4.05	10.06
Trichloroethylene	3.41	8.59
Benzene	3.68	8.68
Chloroform	4.00	9.10
Ozone	2.52	8.05
Ammonia	3.75	6.31
Sulfur dioxide	2.40	5.68
Chlorine	1.74	5.75

TABLE 6-5 Parameters for Calculating Henry's Law Constants (in atm m^3/mole) as a Function of Temperature (Ashworth et al., 1988)

Component	A	B	r^2
Nonane	−0.1847	202.1	0.013
n-Hexane	25.25	7530	0.917
2-Methylpentane	2.959	957.2	0.497
Cyclohexane	9.141	3238	0.982
Chlorobenzene	3.469	2689	0.965
1,2-Dichlorobenzene	−1.518	1422	0.464
1,3-Dichlorobenzene	2.882	2564	0.850
1,4-Dichlorobenzene	3.373	2720	0.941
o-Xylene	5.541	3220	0.966
p-Xylene	6.931	3520	0.989
m-Xylene	6.280	3337	0.998
Propylbenzene	7.835	3681	0.997
Ethylbenzene	11.92	4994	0.999
Toluene	5.133	3024	0.982
Benzene	5.534	3194	0.968
Methyl ethylbenzene	5.557	3179	0.968
1,1-Dichloroethane	5.484	3137	0.993
1,2-Dichloroethane	−1.371	1522	0.878
1,1,1-Trichloroethane	7.351	3399	0.998
1,1,2-Trichloroethane	9.320	4843	0.968
cis-1,2-Dichloroethylene	5.164	3143	0.974
trans-1,2-Dichloroethylene	5.333	2964	0.985
Tetrachloroethylene	10.65	4368	0.987
Trichloroethylene	7.845	3702	0.998
Tetralin	11.83	5392	0.996
Decalin	11.85	4125	0.919
Vinyl chloride	6.138	2931	0.970

(Continued)

TABLE 6-5 Parameters for Calculating Henry's Law Constants (in atm m^3/mole) as a Function of Temperature (Ashworth et al., 1988) (*Continued*)

Component	*A*	*B*	r^2
Chloroethane	4.265	2580	0.984
Hexachloroethane	3.744	2550	0.768
Carbon tetrachloride	9.739	3951	0.997
1,3,5-Trimethylbenzene	7.241	3628	0.962
Ethylene dibromide	5.703	3876	0.928
1,1-Dichloroethylene	6.123	2907	0.974
Methylene chloride	8.483	4268	0.988
Chloroform	11.41	5030	0.997
1,1,2,2-Tetrachloroethane	1.726	2810	0.194
1,2-Dichloropropane	9.843	4708	0.820
Dibromochloromethane	14.62	6373	0.914
1,2,4-Trichlorobenzene	7.361	4028	0.819
2,4-Dimethylphenol	−16.34	−3307	0.555
1,1,2-Trichlorotrifluoroethane	9.649	3243	0.932
Methyl ethyl ketone	−26.32	−5214	0.797
Methyl isobutyl ketone	−7.157	160.6	0.002
Methyl cellosolve	−6.050	−873.8	0.023
Trichlorofluoromethane	9.480	3513	0.998

Ionic Strength. In water supplies, gases or VOCs that are high in dissolved solids have higher volatility (or have a higher apparent Henry's law constant) than those with low dissolved solids. This results in a decrease in the solubility of the volatile component, i.e., a "salting-out effect," that can be represented mathematically as an increase in the activity coefficient of component A, γ_A in aqueous solution. γ_A will increase ($\gamma_A > 1$) with increasing ionic strength; this in turn causes the apparent Henry's law constant, H_{app}, to be greater than the thermodynamic value of H. The following equation can be used to calculate H_{app}:

$$H_{\text{app}} = \gamma_A H = \frac{P_A}{[A]} \tag{6-9}$$

in which γ_A is a function of ionic strength and can be calculated using the following empirical equation:

$$\log_{10} \gamma_A = K_s \times I \tag{6-10}$$

where K_s is the Setschenow, or "salting-out," constant (L/mole) and I is the ionic strength of the water (mole/L).

Ionic strength (I), which is discussed in Chap. 3, is defined as follows:

$$I = \frac{1}{2}\sum_i (C_i Z_i^2) \tag{6-11}$$

in which C_i is the molar concentration of ionic species i (mole/L) and Z_i is the charge of species i. The values of K_s need to be determined by experimental methods because there is no general theory for predicting them. Table 6-6 displays the salting-out coefficients for several compounds at 20°C. For most water supplies, the ionic strength is less than 10 mM and the activity coefficient is equal to 1. Significant increases in volatility and the apparent Henry's constant are only observed for high ionic strength waters, such as seawater.

pH. pH does not affect the Henry's constant per se but it does affect the distribution of species between ionized and unionized forms. This, in turn, influences the overall gas-liquid

TABLE 6-6 Setschenow or Salting-Out Coefficients, K_s at (20°C)

Compound	K_s (L·mol^{-1})	Reference
Tetrachloroethylene	0.213	Gossett 1987
Trichloroethylene	0.186	–
1,1,1-Trichloroethane	0.193	–
1,1-Dichloroethane	0.145	–
Chloroform	0.140	–
Dichloromethane	0.107	–
Benzene	0.195	Schwarzenbach et al. 1992
Toluene	0.208	–
Naphthalene	0.220	–

distribution of the compound because only the unionized species are volatile. For example, ammonia is present as ammonium ion at neutral pH and is not strippable. However, at high pH (greater than 10), ammonia is not an ion and may be stripped. To predict the effect of pH on solubility, one must consider the value of the appropriate acidity constant pK_a. If the acid is uncharged, such as HCN or H_2S, and the pH is much less than the pK_a (2 units lower), then equilibrium partitioning can be described using the Henry's law constant for the uncharged species. If the acid is charged, such as NH_4^+, and the pH is much higher than the pK_a (2 units higher), then the compound will be volatile and its equilibrium partitioning can be described using Henry's law constant for the uncharged species. See Chap. 3 for a presentation of acid-base chemistry.

Surfactants. Surfactants can also affect the volatility of compounds. Most natural waters do not have high concentrations of surfactants; consequently, they do not affect the design of most stripping devices. However, surfactants, if present, lower the volatility of compounds by several mechanisms. The most important factor is that they tend to collect at the air-water interface, diminishing the mole fraction of the compound at the interfacial area, thereby lowering its apparent Henry's law constant. In untreated wastewater, for example, the solubility of oxygen can be lowered by 30 to 50 percent due to the presence of surfactants. Another effect for hydrophobic organics is the incorporation of the dissolved organics into micelles in solution (this would occur only above the critical micelle concentration) that, in turn, decreases the concentration of the organic compound at the air-water interface and lowers the compound's volatility (Vane and Giroux, 2000). O'Haver et al. (2004) provide a more detailed discussion of the impact of surfactants on air stripping of VOCs.

Mass Transfer

The driving force for mass transfer between one phase and another derives from the displacement of the system from equilibrium. Figure 6-3 displays two situations where mass transfer is occurring between the air and water. Figure 6-3a displays the situation where mass is being transferred from the water to the air, and Fig. 6-3b displays the situation where mass is being transferred from the air to the water. Because the mechanisms and assumptions for mass transfer are essentially the same for both cases, a detailed explanation of only one is warranted. Consider the case where a volatile contaminant is being stripped from water to air (Fig. 6-3a). The contaminant concentration in the water is high relative to the equilibrium concentration between the air and water. The tendency to achieve equilibrium is sufficient to cause diffusion of the aqueous-phase contaminant molecules from the bulk solution at some concentration, C_b (μg/L), to the air-water interface, where the aqueous phase concentration is C_s (μg/L). Because C_b is larger than C_s, the difference between

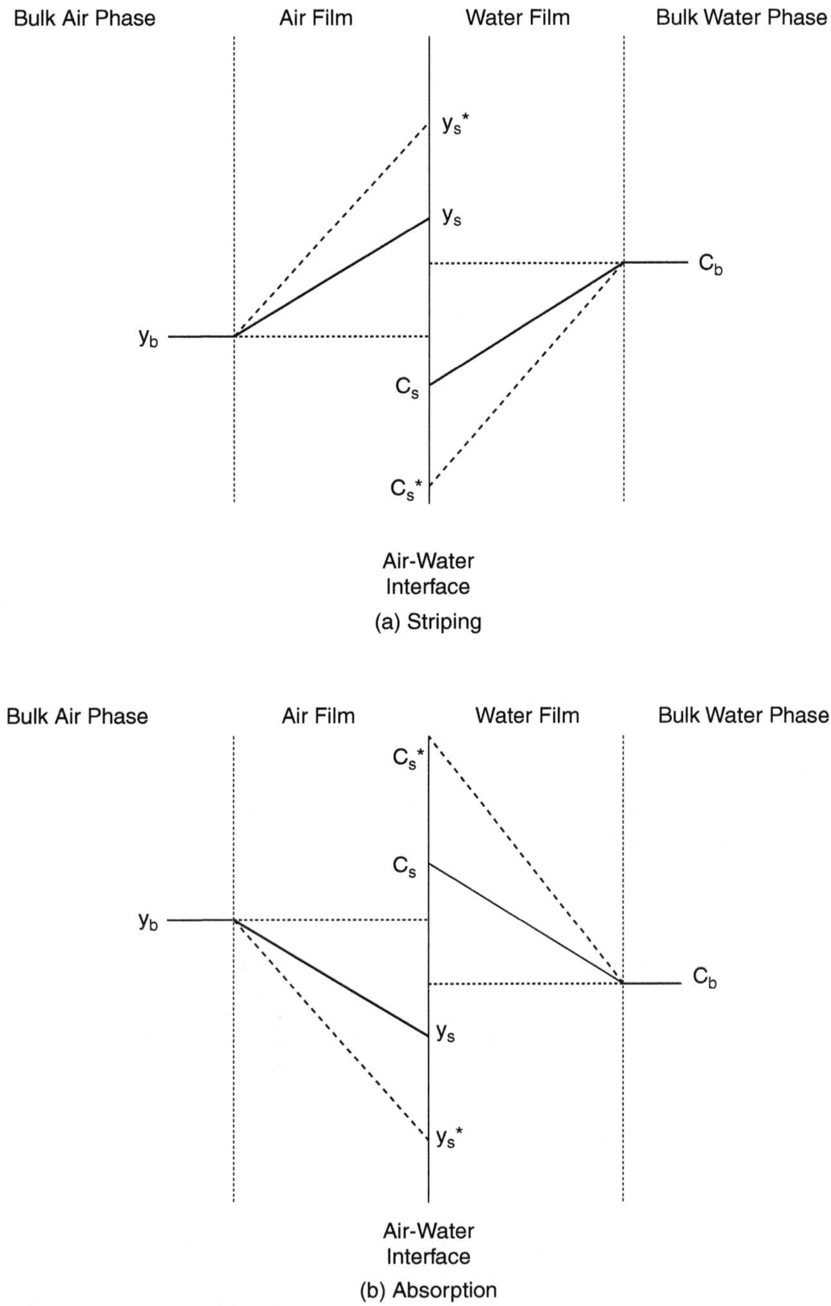

FIGURE 6-3 Diagram describing the equilibrium partitioning of a contaminant between the air and water phases using two film theory.

them provides the aqueous-phase driving force for stripping. Similarly, the contaminant concentration in the air at the air-water interface, y_s, is larger than the contaminant concentration in the bulk air, y_b, and diffusion causes the molecules to migrate from the air-water interface to the bulk air. The difference between y_s and y_b is the driving force for stripping in the gas phase.

Mass Flux. Local equilibrium occurs at the air-water interface because on a local scale of tens of Angstroms random molecular movement causes the contaminant to dissolve in the aqueous phase and volatilize into the air. (On a larger scale, these motions equalize the displacement of the system from equilibrium and cause diffusion.) Accordingly, Henry's law can be used to relate y_s to C_s (Lewis and Whitman, 1924).

$$y_s = HC_s \tag{6-12}$$

Fick's law can be simplified to a linear driving force approximation whereby the flux (the mass transferred per unit of time per unit of interfacial area) across the air-water interface is proportional to the concentration gradient. Mathematically, the flux of A across the air-water interface, J_A, is given by:

$$J_A = k_l(C_b - C_s) = k_g(y_s - y_b) \tag{6-13}$$

in which k_l is the liquid-phase mass transfer coefficient that describes the rate at which contaminant A is transferred from the bulk aqueous phase to the air-water interface (L/t) and k_g is the gas-phase mass transfer coefficient that describes the rate at which contaminant A is transferred from the air-water interface to the bulk gas phase (m/s). Both k_l and k_g are sometimes called local mass transfer coefficients for the liquid and gas phases because they depend upon the conditions at or near the air-water interface in their particular phase. Because the interfacial concentrations y_s and C_s cannot be measured and are unknown, the flux cannot be determined from Eq. 6-13. Consequently, it is necessary to define another flux equation in terms of hypothetical concentrations, which are easy to determine and describe the displacement of the system from equilibrium. If it is hypothesized that all the resistance to mass transfer is on the liquid side, then there is no concentration gradient on the gas side and a hypothetical concentration C_s^* can be defined as shown in Fig. 6-3a.

$$y_b = HC_s^* \tag{6-14}$$

in which C_s^* is the aqueous-phase concentration of A at the air-water interface assuming no concentration gradient in the air phase (μg/L), i.e., equilibrium exists between the bulk gas-phase concentration and the aqueous phase at the interface.

Alternatively, if it is hypothesized that all the resistance to mass transfer is on the gas side, there is no concentration gradient on the liquid side and a hypothetical concentration y_s^* can be defined as shown in Fig. 6-3a.

$$y_s^* = HC_b \tag{6-15}$$

in which y_s^* is the equilibrium gas-phase concentration of A at the air-water interface assuming no concentration gradient in the liquid phase (μg/L); it can be calculated from the bulk liquid-phase concentration.

Mass Flux Based on Water Side. For stripping operations, mass balances are normally written based on the liquid phase. It is convenient to calculate the mass transfer rate using the hypothetical concentration, C_s^*, and an overall mass transfer coefficient, K_L, as shown in the following equation:

$$J_A = K_L(C_b - C_s^*) \tag{6-16}$$

Equating Eq. 6-13 and 6-16 results in:

$$J_A = k_l(C_b - C_s) = k_g(y_s - y_b) = K_L(C_b - C_s^*) \tag{6-17}$$

From Fig. 6-3a, one can obtain the following relationship:

$$(C_b - C_s^*) = (C_b - C_s) + (C_s - C_s^*) \tag{6-18}$$

Substituting $C_s = \frac{y_s}{H}$ and $C_s^* = \frac{y_b}{H}$ into Eq. 6-18 and combining with Eq. 6-17 yields:

$$\frac{J_A}{K_L} = \frac{J_A}{k_l} + \frac{J_A}{Hk_g} \tag{6-19}$$

or

$$\frac{1}{K_L} = \frac{1}{k_l} + \frac{1}{Hk_g} \tag{6-20}$$

Equation 6-20 states simply that the overall resistance to mass transfer is equal to the sum of the mass transfer resistances for the liquid and gas phases. It describes the overall mass transfer rate without the use of the unknown local concentrations at the air-water interface.

Volumetric Mass Transfer Expression. For aeration and stripping process design equations, the rate of mass transfer is often expressed on a volumetric basis rather than an interfacial area basis. The flux term, J_A, is converted to a volumetric basis by multiplying by the specific interfacial area, "a," which is defined as the interfacial area available for mass transfer divided by the system unit volume (m^2/m^3). Equation 6-20 can be expressed in terms of a volumetric mass transfer rate by multiplying J_A by a in Eq. 6-19 and rewriting as:

$$\frac{1}{K_L a} = \frac{1}{k_l a} + \frac{1}{Hk_g a} \tag{6-21}$$

This modification is based on the liquid side controlling the mass transfer and also includes the resistance to mass transfer in the gas phase to provide an exact representation of the mass flux across the air-water interface. For compounds with Henry's law constants less than 0.002 and low solubility, the gas side controls the mass transfer rate. A similar relationship may be developed for an overall mass transfer coefficient on the gas side, which is useful for absorption or scrubbing operations where the constituent is transferred from the air side to the water side. The final equation is given as:

$$\frac{1}{K_G a} = \frac{H}{k_l a} + \frac{1}{k_g a} \tag{6-22}$$

in which K_G is the overall gas-phase mass transfer coefficient (m/s). The mass flux may be obtained from the gas-phase mass balance by multiplying K_G by the hypothetical driving force, $(C_b - C_s^*)$.

Phase Controlling Mass Transfer. Determining which phase controls the mass transfer rate is important in optimizing the design and operation of aeration and air stripping processes. For example, when the mass transfer rate is controlled by the liquid-phase resistance, an increase in the mixing of the air has little impact on removal efficiency. Increasing the interfacial area will increase the removal efficiency. Equation 6-20 states simply that the overall resistance to mass transfer is equal to the sum of the resistance in liquid and gas phases and can be rewritten as:

$$R_T = R_L + R_G \tag{6-23}$$

in which $R_T(1/K_L)$ is the overall resistance to mass transfer (s/m), $R_L(1/k_l)$ is the liquid-phase resistance to mass transfer (s/m), and $R_G(1/k_g)$ is the gas-phase resistance to mass transfer (s/m). Equation 6-20 can also be rearranged into the following equation to evaluate which phase controls the rate of mass transfer:

$$\frac{R_L}{R_T} = \frac{\frac{1}{k_l}}{\frac{1}{k_l} + \frac{1}{Hk_g}} = \frac{1}{1 + \frac{1}{\left(\frac{k_g}{k_l}\right)H}} \tag{6-24}$$

or

$$\frac{R_L}{R_T}(\%) = \frac{100}{1 + \frac{1}{\left(\frac{k_g}{k_l}\right)H}} \tag{6-25}$$

Literature-reported values of the ratio k_g/k_l range from 40 to 200 (Munz and Roberts, 1989) depending upon the type of aeration or stripping system. Assuming k_g/k_l is 100, Eq. 6-25 can be used to determine which phase controls the rate of mass transfer across the interface. Table 6-7 gives the phases that control the rate of mass transfer for a number of compounds with a wide range of Henry's law constants. In general, the liquid phase controls the rate of mass transfer for compounds with H values greater than 0.05 and low liquid solubility. The gas phase controls mass transfer for compounds with H values less than 0.002 and high liquid solubility. For compounds with H values between 0.002 and 0.05 either phase may be preferred, and both the liquid and gas phases affect the rate of mass transfer.

TABLE 6-7 Summary of Controlling Phase for Several Compounds

Compound	H (25°C)	R_L/R_T (%)	Controlling phase
Oxygen	29.29	100	Liquid
CarbonTetrachloride	1.2	99	Liquid
Trichloroethene	0.40	98	Liquid
Chloroform	0.15	94	Liquid
Acetone	0.001	9	Gas

PACKED TOWERS

The primary applications of packed towers in water treatment are the addition of air under pressure in packed towers called saturators in dissolved air flotation (Chap. 9) and stripping of VOCs, carbon dioxide, hydrogen sulfide, and ammonia. An example of a packed tower system is displayed in Fig. 6-4. Water is pumped to the top of the tower and through a liquid distributor, where it then flows by gravity over packing material. At the same time, a blower introduces fresh air into the bottom of the tower; the air flows countercurrent to the water up through the void spaces between the wetted packing material. The packing material provides a large air-water interfacial area, resulting in efficient transfer of the volatile contaminant from the water to the air. The contaminant-free, air-stripped water leaves the bottom of the tower while the air containing the contaminant exits the top of the tower for further treatment or exhaustion to the atmosphere. The packing material consists of individual pieces randomly dumped into the column. Figure 6-5 displays several varieties of commercially available packing material. Table 6-8 lists several packing materials and their physical properties as reported by the manufacturers. The packing materials are made of plastic with a surface tension of 0.033 N/m (0.024 lb/ft).

Cross-flow, cascade, and cocurrent packed tower systems have similar configurations, but the air flow is introduced differently. In the cross-flow system, the air flows across the

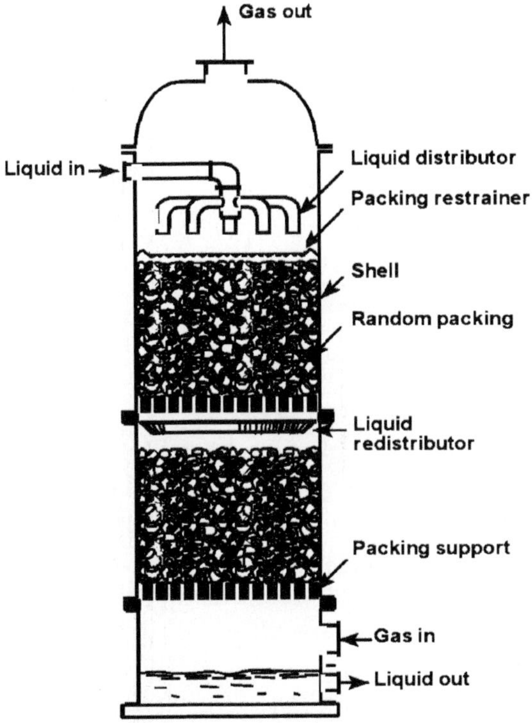

FIGURE 6-4 Schematic of a packed tower.

tower packing at a 90° angle to the direction of the water flow. Methods for estimating the performance of the cross-flow system are available in the literature (Little and Selleck, 1991). In the cascade system, fresh air is introduced at various points along the tower, and fluid flows countercurrent to the water flow (Nirmalakhandan, Jang, and Speece., 1991, 1992). Both the cross-flow and cascade systems provide larger air flow rates at lower gas pressure drops than those typically used in the conventional countercurrent system. The larger air flow rates provide a greater driving force for stripping, leading to more efficient removal of contaminants with low volatility. The cross-flow and cascade systems are recommended for packed tower systems that require high air-to-water ratios to remove semi-volatile and low-volatility contaminants from water as well as larger air treatment devices to remove the VOCs. In the co-current system, both the air and water enter the top of the tower, flow through the packing, and exit at the bottom of the tower. The co-current system is rarely used in water treatment for stripping because it cannot achieve removal efficiencies that are as high as those in the countercurrent process.

Recently, sieve tray columns and low-profile air strippers have been used to remove VOCs from contaminated waters. Sieve tray columns operate as a countercurrent process. The columns are typically less than 3 m (10 ft) high and consist of several perforated trays placed in series along the column. Water enters at the top of the tower and flows horizontally across each tray. Inlet and outlet channels, referred to as downcomers, are placed at the ends of each tray to allow the water to flow from tray to tray. At the same time, fresh air flows up from the bottom of the tower through the tray holes and the layer of water flowing across each tray. Large air flow rates are typically used, causing small bubbles or frothing

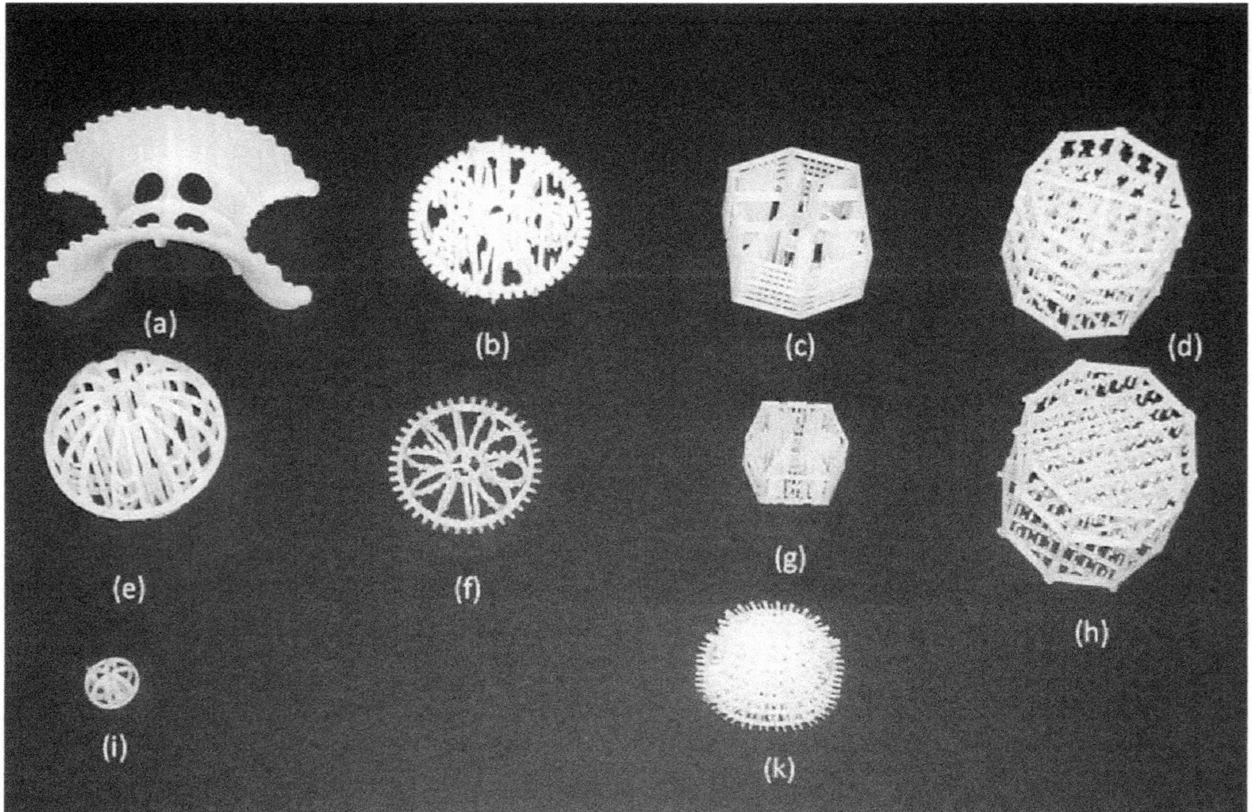

FIGURE 6-5 Random plastic tower packing: (*a*) 3-in. Intalox saddle; (*b*) #3 Type K Tellerette; (*c*) 3.5-in. LANPAC: (*d*) 3.5-in. LANPAC-XL; (*e*) 3.5-in. Tripack; (*f*) #2 Type K Tellerette; (*g*) 2.3-in. LANPAC: (*h*) 4-in. QPAC; (*i*) 1-in. Tripack; (*j*) #2 NUPAC.

TABLE 6-8 Physical Characteristics of Packing Materials

Type of packing	Nominal diameter in	Nominal diameter m	Packing factor ft^{-1}	Packing factor m^{-1}	Specific surface area ft^2/ft^3	Specific surface area m^2/m^3
Plastic Saddles*	3.0	0.0762	16.0	20.0	20.0	66.0
	2.0	0.0508	21.0	69.0	30.0	98.0
Plastic Tripacks*	3.5	0.0889	12.0	15.0	38.0	125.0
	2.0	0.0508	16.0	52.0	48.0	157.0
Plastic Rings*	2.0	0.0508	25.0	82.0	33.0	108.0
	3.5	0.0889	16.0	52.0	26.0	85.0
Plastic Tellerettes†						
No. 2 Type K	2.0	0.0508	11.0	36.0	28.0	92.0
No. 3 Type K	3.0	0.0762	9.0	30.0	22.0	72.0
Q-PAC‡	3.5	0.0889	7.0	20.0	30.0	98.0
NUPAC‡	2.5	0.0635	16.0	53.0	55.0	180.0
	4.5	0.1143	8.0	26.0	40.0	131.0
LANPAC‡	2.3	0.0584	21.0	69.0	68.0	223.0
	3.5	0.0889	14.0	46.0	44.0	144.0
LANPAC-XL‡	3.5	0.0889	10.0	33.0	74.0	242.0

*Jaeger Co., Houston, Texas.
†Ceilcote Co., Cleveland, Ohio.
‡LANTEC Co., Los Angeles, California.

to occur upon air contact with the water. The frothing provides a high air-to-water surface area for mass transfer to occur. Because the water flows horizontally across each tray, the desired removal efficiency can be obtained by increasing the length or width of the trays instead of the height. The advantages of using a low-profile air stripper over a conventional packed tower stripper are that the low-profile air stripper is smaller and more compact and it is easier to provide periodic maintenance. The disadvantage is that for a given removal, the low-profile air stripper requires a significantly higher air flow than the conventional packed tower and the operational costs may be greater. In addition, low-profile air strippers are limited to water flow rates less than 0.0630 m^3/s (1000 gpm) because of manufacturing size constraints.

Governing Equations

Mathematical models that describe countercurrent packed tower processes for water treatment are well established in the literature (e.g., Sherwood and Hollaway, 1940; Treybal, 1980; Kavanaugh and Trussell, 1980; Ball, Jones, and Kavanaugh, 1984; Singley et al., 1981; Umphres et al., 1983; Cummins and Westrick, 1983; McKinnon and Dyksen, 1984; Roberts et al., 1985; Gross and TerMaath, 1985; Roberts and Levy, 1985; Hand et al., 1986; Dzombak, Roy, and Fang, 1993; Hokanson, 1996). These models have been used successfully to describe the packed tower air stripping process and, consequently, to size the towers for a given removal.

Development of Packed Tower Height Expression. Figure 6-6 displays a mathematical representation of the packed tower process shown in Fig. 6-4. The overall assumptions in development of the model are (1) plug-flow conditions prevail in the air and water streams, (2) the incoming air contains no contaminant, (3) the contaminant concentration in the influent water stream is constant, (4) air and water temperatures are constant and equal to the inlet water temperature, and (5) Henry's law describes the chemical equilibrium between the air

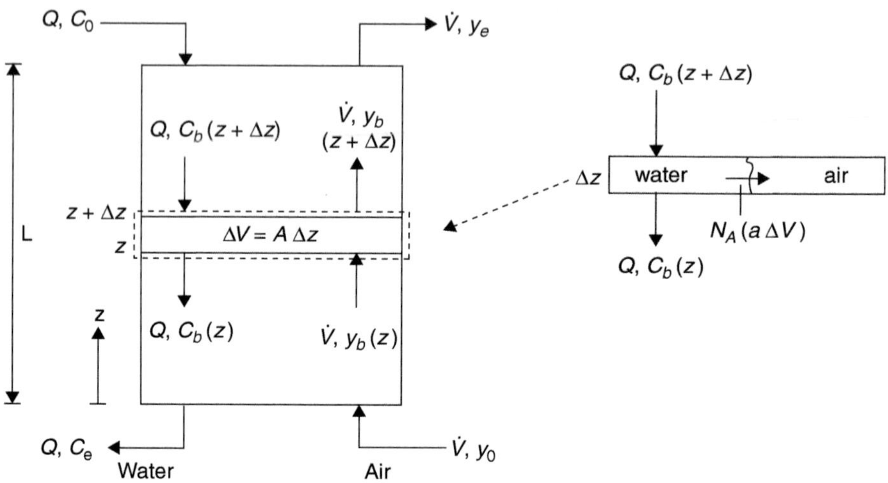

FIGURE 6-6 Schematic of a packed tower used in the development of the design equations.

and water phases. The first step in development of the design equations is to establish the relationship between the air and water phase contaminant concentrations within the packed tower. Figure 6-7 compares the equilibrium line, as defined by Henry's law equation, with the operating line for packed tower air stripping. The operating line relates the bulk air-phase contaminant concentration to the bulk water-phase contaminant concentration and is based on a mass balance around the bottom of the air stripping tower of:

$$y_b(z) = \frac{Q}{\dot{V}}[C_b(z) - C_e] \tag{6-26}$$

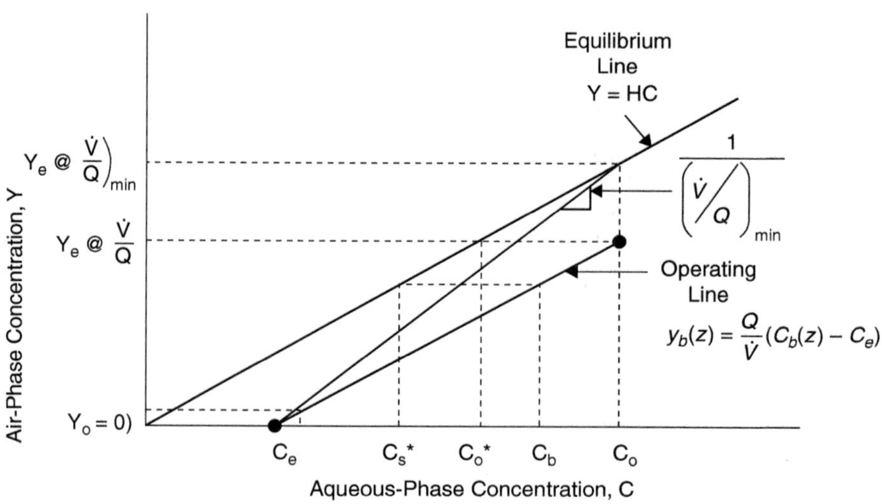

FIGURE 6-7 Operating line diagram for a packed tower.

in which y_b is the bulk air-phase concentration of the contaminant at any point along the packed tower height (μg/L), Q is the water flow rate to the tower (m^3/s), $\dot{V}$ is the volumetric air flow rate to the tower (m^3/s), C_b is the bulk water-phase concentration of the contaminant at any point along the packed tower height (μg/L), and C_e is the bulk water-phase contaminant concentration at the bottom of the tower (μg/L). The driving force for stripping is quantified by comparing the operating line to the equilibrium line. The bulk air-phase contaminant concentration corresponding to a given water-phase contaminant concentration is found by holding C constant and moving vertically until the operating line is intersected. This is the bulk air-phase contaminant concentration. The water-phase contaminant concentration, which would be in equilibrium with the air-phase contaminant concentration, is found by holding y constant until it intersects the equilibrium line. The difference between these two concentrations is the driving force for stripping; it must be positive in order for stripping to occur.

Incorporating the above assumptions into an aqueous-phase mass balance around the differential element circumscribed by the dashed box in Fig. 6-6 yields the following expression for removal of a contaminant within a packed tower:

$$\frac{dC_b}{dz} = \frac{K_L aA}{Q}(C_b - C_s^*) \tag{6-27}$$

in which C_s^* is the water-phase concentration of the contaminant in equilibrium with the bulk air-phase contaminant concentration at any point along the packed tower. The integral form of Eq. 6-27 is expressed as:

$$\int_0^L dz = \frac{Q}{AK_L a} \int_{C_e}^{C_0} \frac{dC_b}{C_b - C_s^*} \tag{6-28}$$

Integration of Eq. 6-28 requires an expression for C_s^* in terms of C_b. From Fig. 6-7, C_s^* is expressed in terms of the bulk air-phase contaminant concentration using Henry's law and substituting into Eq. 6-26 to yield

$$C_s^* = \frac{Q}{\dot{V}H}[C_b - C_e] \tag{6-29}$$

Substitution of Eq. 6-29 into Eq. 6-28 and integrating yields the following design equations for the countercurrent packed tower process:

$$L = \frac{Q}{AK_L a}\left[\frac{C_o - C_e}{C_o - C_e - C_o^*}\right]\log_e\left[\frac{C_o - C_o^*}{C_e}\right] \tag{6-30}$$

$$C_o^*(z = L) = \left(\frac{Q}{\dot{V}H}\right)(C_o - C_e) \tag{6-31}$$

where C_o^* is the aqueous-phase concentration of the contaminant in equilibrium with the exiting air-phase contaminant concentration y_e (μg/L). C_o^* is obtained by rearranging Eq. 6-26 assuming y_b equals y_e and C_b equals C_o and then substituting C_o^* for y_e/H. Equations 6-30 and 6-31 are sometimes combined, and the quantity $\dot{V}H/Q$, defined as the stripping factor (S), is introduced.

Minimum Air-to-Water Ratio and Stripping Factor. The minimum air-to-water ratio required for stripping can be determined by considering the driving force for stripping at the top of the tower. Under the best scenario, the exiting air will be in equilibrium with the incoming water, which implies that C_o^* in Eq. 6-31 is equal to C_o and, upon substitution into Eq. 6-31, the following expression may be obtained:

$$\left(\frac{\dot{V}}{Q}\right)_{\min} = \frac{(C_o - C_e)}{HC_o} \tag{6-32}$$

The minimum air-to-water ratio, $(\frac{\dot{V}}{Q})_{\min}$, is the lowest air-to-water ratio that can be applied for a packed tower and meet a given contaminant's treatment objective, C_e. If the air-to-water ratio applied is less than the minimum air-to-water ratio, it will not be possible to design a packed tower capable of meeting the treatment objective because equilibrium will be established in the tower before the treatment objective is reached. This can be explained using the stripping factor, *S,* which represents an equilibrium capacity parameter. When *S* is greater than 1, there is sufficient capacity in the air to convey all the solute in the entering water, and complete removal by stripping is possible given a sufficiently long column. When $S < 1$, the system performance is limited by equilibrium and the fractional removal is asymptotic to the value of *S* (i.e., $1 - C_e/C_o \rightarrow S$ as $L \rightarrow \infty$, as shown by Eq. 6-32. When $S = 1$, the tower is operating at the minimum air-to-water ratio required for stripping.

Optimum Air-to-Water Ratio. With respect to the selection of the optimum air-to-water ratio, Hand et al. (1986) demonstrated that minimum tower volumes and power requirements are achieved for high-percentage removals using approximately 3.5 times the minimum air-to-water ratio for contaminants with Henry's law constants greater than 0.05. This corresponds to a stripping factor of 3.5.

Packed Tower Height Expression in Terms of Transfer Units. In packed tower air stripping, the tower length is often defined as the product of the height of a transfer unit (*HTU*) and the number of transfer units (*NTU*). The following two equations define *HTU* and *NTU*:

$$HTU = \frac{Q}{AK_L a} \tag{6-33}$$

$$NTU = \left[\frac{C_o - C_e}{C_o - C_e - C_o^*}\right] \ln\left[\frac{C_o - C_o^*}{C_e}\right] \tag{6-34}$$

or incorporating Eq. 6-31 and *R* into Eq. 6-30 yields:

$$NTU = \left[\frac{R}{R-1}\right] \ln\left[\frac{\frac{C_o}{C_e}(R-1)+1}{R}\right] \tag{6-35}$$

Substituting Eq. 6-33 and Eq. 6-34 or 6-35 into Eq. 6-30 results in:

$$L = HTU \times NTU \tag{6-36}$$

The *NTU* is determined by the driving force for stripping and corresponds to the number of stages (or hypothetical completely mixed tanks at equilibrium) required for stripping. Equation 6-35 shows that *NTU* depends upon *R* and VOC removal efficiency; this relationship is displayed in Fig. 6-8. For a given *R*, the removal efficiency increases with increasing *NTU* (or number of hypothetical completely mixed tanks). In addition, for a given removal efficiency, increasing *R* or the air-to-water ratio will decrease the number of *NTU*s required. *HTU* is the height of one stage, as determined by the rate of mass transfer, K_{La}, and the tower area, *A*, which must be known to calculate the packed tower height in Eq. 6-36.

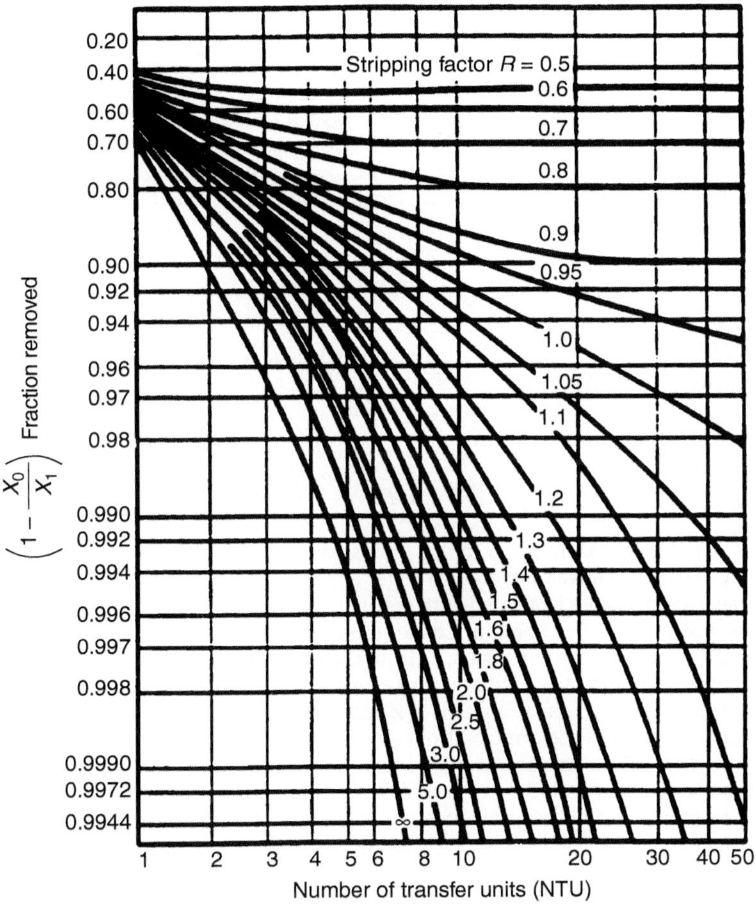

FIGURE 6-8 Dependence of removal efficiency on NTU on and stripping factor. (*Source: Treybal, 1980.*)

Determination of K_La. K_La values for packed towers can determined by performing pilot plant studies or can be obtained from experimental values reported by packing manufacturers and previously reported field studies. When experimental K_La values are not available, they can be estimated from mass transfer correlations. A pilot plant study is the preferred way to determine K_La for a given VOC in contaminated water. Pilot-scale packed towers range in size from 2 to 6 m (6.5 to 20 ft) in height and 0.3 to 0.61 m (1 to 4 ft) in diameter. The column diameter that is used will depend upon the desired packing size. It is generally recommended that ratios of column diameter to packing diameter be greater than 10:1 (>15:1 desired) in order to minimize error due to channeling of the water down the column walls (Treybal, 1980). It is also recommended that the ration of packed column height to diameter be greater than 1:1 in order to provide for proper liquid distribution (Roberts et al., 1985; Treybal, 1980). Pilot plant design guidelines such as these as well as construction and operation of pilot-scale packed tower units are presented elsewhere (Crittenden et al., 2005).

Equation 6-36 is used in conjunction with pilot plant data to determine K_La for a given VOC. The experimental value of K_La can be determined from *HTU*, as shown in Eq. 6-33. The value of K_La is based on VOC removal due to the packed height portion of the tower. However, VOC removal also occurs as the water contacts the air above the packing at the top of the tower and at the bottom of the tower as the water falls into the clearwell below the packing. This removal is sometimes referred to as "end effects" (Umphres et al., 1983). Consequently, an *NTU* correction factor for VOC removal due to end effects is used when determining K_La. The following equation can be used to calculate the *NTU* of the packing (Umphres et al., 1983):

$$NTU_{\text{packing}} = NTU_{\text{measured}} - NTU_{\text{end effects}} \tag{6-37}$$

where NTU_{measured} is the experimentally determined value from pilot data and $NTU_{\text{end effects}}$ is the *NTU* due to the end effects. Combining Eqs. 6-36 and 6-37 yields the following linear expression, which can be used to determine K_La:

$$NTU_{\text{measured}} = \frac{1}{HTU}(Z) + NTU_{\text{end effects}} \tag{6-38}$$

where Z is the distance from the top of the packing to a sample port location along the packed portion of the tower. For a given water and air loading rate, aqueous-phase concentration measurements are evaluated at the influent and effluent and at various sample port locations along the packed column. Equation 6-35 can be used to calculate NTU_{measured} at each sample port location or distance, Z, from the top of the packing, where C_e is assumed to be the concentration measured at each sample port along the packed column. A plot is constructed of NTU_{measured} as a function of Z and should coincide with the linear expression given by Eq. 6-38. K_La is determined from the slope ($1/HTU$) and Eq. 6-33. Experimentally determined K_La values can be correlated as a function of water loading rate for several air-to-water ratios that would be expected during operation of the full-scale column. For a given water loading rate and air-to-water ratio of interest, a correlated K_La value can be determined and used with Eqs. 6-33 through 6-36 to determine the required full-scale packed tower height.

A full-scale packed tower height calculated using a K_La value determined from a pilot study is generally conservative. For a given packing size, K_La values generally increase as the tower diameter increases (Wallman and Cummins, 1986). This is caused by channeling of the water down the inside of the column walls, which is sometimes referred to as "wall effects." The VOC removal rate is less along the walls of the column as compared to the removal within the packing because the air-to-water contact time, surface area, and mixing are less. As the tower diameter increases, the percentage of flow down the walls of the column decreases, minimizing the wall effects. Consequently, the K_La value in a full-scale column will be larger than that in a smaller pilot column.

Table 6-9 displays numerous packed tower field studies that reported experimentally determined K_La values for several VOCs and various contaminated water sources. The K_La values reported in these studies can be used provided the operating conditions (temperature, water, and air loading rate) and packing type and size are the same.

Example 6-1 Determination of K_La from Pilot-Plant Data
An air stripping packed tower pilot plant study was performed on a groundwater contaminated with methyl tert-butyl ether (MTBE). A packed tower with a height of 3.0 m (10 ft) and a diameter of 0.30 m (1 ft) with sample ports distributed down the column every 0.76 m (2.5 ft) was used. The column was operated at an air-to-water

TABLE 6-9 Summary of Several Packed Tower Air Stripping Pilot-Plant Studies Which Determined K_La Values for Several VOCs

Reference	Water matrix	VOCs
Umpheres et al. (1983)	Sacramento-San Joaquin Delta water in Northern California	Chloroform, Dibromochloromethane Bromodichloromethane Bromoform
Ball et al. (1984)	Potomac tidal fresh estuary water mixed with nitrified effluent wastewater	Carbon tetrachloride Tetrachloroethylene Trichloroethylene Chloroform Bromoform
Byers and Morton (1985)	City of Tacoma, WA groundwater	1,1,2,2-Tetrachloroethane trans-1,2-Dichloroethylene Trichloroethylene Tetrachloroethylene
Roberts et al. (1985)	Laboratory Grade Organic Free Water	Oxygen Tetrachloroethylene Freon-12 1,1,1-Trichloroethane Trichloroethylene Carbon tetrachloride
Bilello and Singley (1986)	North Miami Beach, Fl Groundwater & City of Gainsville, Fl groundwater	Chloroform
Wallman and Cummins (1986)	Village of Brewster, NY groundwater	cis-1,2-Dichloroethylene Trichloroethylene Tetrachloroethylene
Lamarche and Droste (1989)	Gloucester, Ottawa, Ont. groundwater	Chloroform Toluene 1,2-Dichloroethane 1,1-Dichloroethane Trichloroethylene Diethyl ether

ratio of 100 for 30 minutes to reach steady-state operation. Water samples were then taken at sample ports and analyzed for MTBE. Given the results tabulated below, determine K_La.

Pilot-Plant Parameters

Parameter	Value
Temperature, °C	15
Influent concentration, μg/L	125
Treatment objective, μg/L	5
Packing type	No. ½ Tripacks
Water flow rate, L/s	0.3
Air flow rate, L/s	300

Measured MTBE Concentration at Each Depth for $\dot{V}/Q = 100$

Depth, m	MTBE concentration, μg/L
0	125
0.76	66.3
1.52	48.5
2.29	34.5
3.05	25.7

Solution

1. Using the Henry's law constant from Table 6-2, calculate the stripping factor for an air-to-water ratio of 100. H from Table 6-2 for MTBE is 0.0177.

$$R = \frac{\dot{V}}{Q} H = 100 \times 0.0177 = 1.77$$

2. For each sample port depth and MTBE concentration, calculate the *NTU* using Eq. 6-35. Sample calculation is made for the sample port at a depth of 0.76 m.

$$NTU = \left[\frac{R}{R-1}\right] \ln\left[\frac{\frac{C_0}{C_e}(R-1)+1}{R}\right]$$

$$NTU_{\text{Depth}=0.762\text{m}} = \left[\frac{1.77}{1.77-1}\right] \ln\left[\frac{\frac{125\ \mu g/L}{66.3\ \mu g/L}(1.77-1)+1}{1.77}\right] = 0.75$$

Depth, m	MTBE concentration, μg/L	NTU
0	125	0
0.76	66.3	0.75
1.52	48.5	1.20
2.29	34.5	1.75
3.05	25.7	2.27

3. Plot the *NTU* as a function of column depth and linearly fit Eq. 6-38 to the data to obtain the *HTU* from the slope.

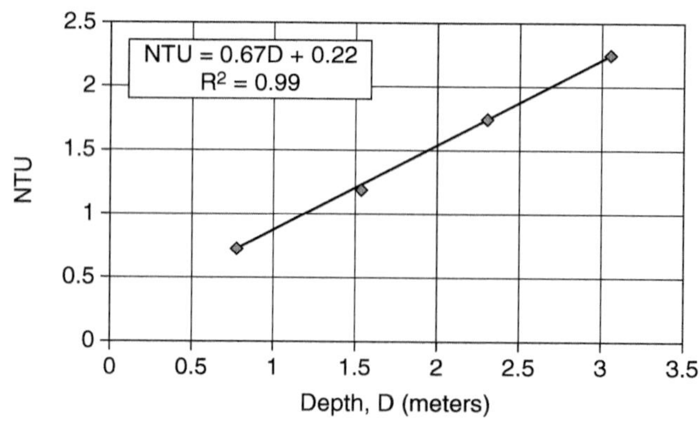

From Eq. 6-38, the *HTU* is determined:

$$Slope = \frac{1}{HTU} = 0.67\,\text{m}^{-1}$$
$$HTU = 1.5\,\text{m}$$

4. Using Eq. 6-33, $K_L a$ is calculated from *HTU* as

$$HTU = \frac{Q}{A \cdot K_L a}$$

$$K_L a = \frac{Q}{A \cdot HTU} = \frac{0.0003\,\text{m}^3/\text{s}}{\left(\frac{\pi}{4}[0.3\text{m}]^2\right)(1.5\,\text{m})} = 0.0028\,\text{s}^{-1} \quad \blacktriangle$$

Estimating $K_L a$. When $K_L a$ values are not readily available, they can be estimated from the following mass transfer correlations (Onda, Takeuchi, and Okumoto, 1968):

$$a_w = a_t\left\{1-\exp\left[-1.45\left(\frac{\sigma_c}{\sigma}\right)^{0.75}\left(\frac{L_m}{a_t\mu_l}\right)^{0.1}\left(\frac{L_m^2 a_t}{\rho_l^2 g}\right)^{-0.05}\left(\frac{L_m^2}{\rho_l a_t \sigma}\right)^{0.2}\right]\right\} \tag{6-39}$$

$$k_l = 0.0051\left(\frac{L_m}{a_w\mu_l}\right)^{2/3}\left(\frac{\mu_l}{\rho_l D_l}\right)^{-0.5}(a_t d_p)^{0.4}\left(\frac{\rho_l}{\mu_l g}\right)^{-1/3} \tag{6-40}$$

$$k_g = 5.23(a_t D_g)\left(\frac{G_m}{a_t\mu_g}\right)^{0.7}\left(\frac{\mu_g}{\rho_g D_g}\right)^{1/3}(a_t d_p)^{-2} \tag{6-41}$$

in which a_t is the total specific surface area of the packing material (m^2/m^3), a_w is the wetted surface area of the packing material, d_p is the nominal diameter of the packing material (m), D_g is the gas-phase diffusivity of the contaminant to be removed (m^2/s), D_l is the liquid-phase diffusivity of the contaminant to be removed (m^2/s), g is the gravitational constant (m/s^2), G_m is the air mass loading rate ($kg/m^2/s$), L_m is the water mass loading rate ($kg/m^2/s$), μ_L is the viscosity of water (kg/m/s), μ_g is the viscosity of air (kg/m/s), ρ_l is the density of water (kg/m^3), ρ_g is the density of air (kg/m^3), σ is the surface tension of water (kg/s^2), and σ_c is the critical surface tension of the packing material (kg/s^2).

Equation 6-40 is valid for water loading rates between 0.8 and 43 $kg/m^2/s$ (1.1 and 63 gpm/ft^2), and Eq. 6-41 is valid for air loading rates between 0.014 and 1.7 $kg/m^2/s$ (2.206 and 267.9 cfm/ft^2). To use these correlations, it is assumed that a_w is equivalent to a. Equations 6-40 and 6-41 were correlated for nominal packing sizes up to 0.05 m (2 in.). The Onda correlation is perhaps the best correlation for predicting $K_L a$ for VOCs in packed tower air stripping applications (Cummins and Westrick, 1983; Roberts et al., 1985; Lamarche and Droste, 1989; Staudinger, Knocke, and Randall, 1990). Although the Onda correlation yields favorable results for laboratory and pilot-plant data, work comparing observed $K_L a$ values for full-scale systems with those obtained by Eqs. 6-39 to 6-41 shows differences of ±30 to 40 percent deviations depending upon the tower operating conditions (Lenzo, Frielinghaus, and Zienkiewicz, 1990; Staudinger, Knocke, and Randall, 1990; Djebbar and Narbaitz, 1995, 1998; Dvorak et al., 1996). Generally, the Onda correlation tends to underpredict $K_L a$ for some large packing material and high gas flow rates (>3.5 $kg/m^2/s$) while overpredicting $K_L a$ for large liquid loadings (>15 $kg/m^2/s$) (Staudinger, Knocke, and Randall, 1990; Djebbar and Narbaitz, 1998, 2002; Dvorak et al., 1996). Djebbar and Narbaitz (1998, 2002) have proposed modifications that improve the predictive capabilities of the Onda correlation. Based

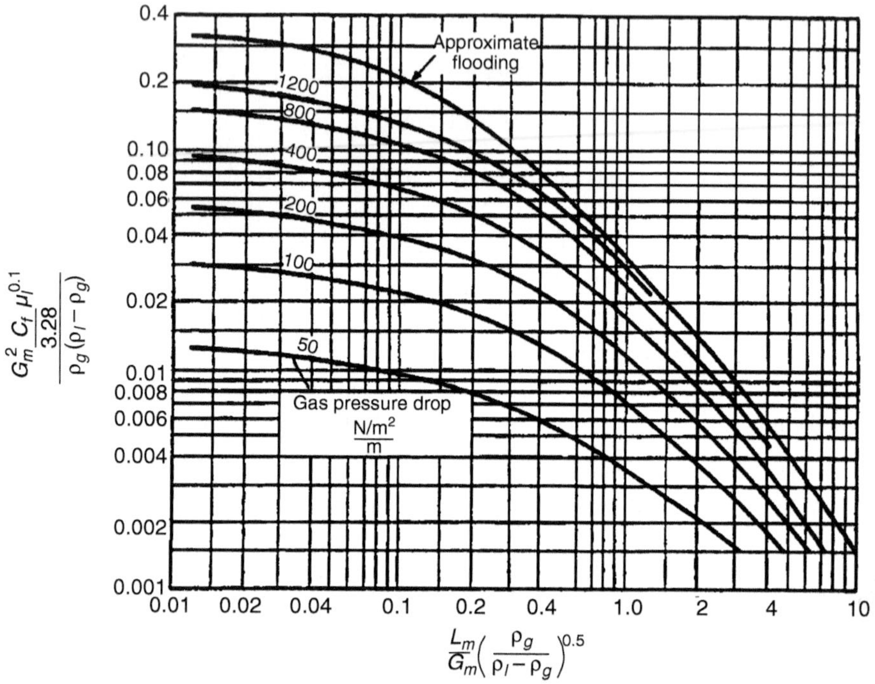

FIGURE 6-9 Flooding and pressure drop in random-packed towers. (*Source: Treybal, 1980.*)

on the literature cited above, it is recommended that a safety factor 0.7 to 0.8 (i.e., K_La/Onda K_La) for packed towers be applied to provide a conservative estimate of the packing height required.

Determination of the Tower Diameter. The tower diameter for a single packed tower is typically designed to provide for a low gas pressure drop (i.e., 50–100 $N/m^2 \cdot m$ (0.06–0.12 in H_2O/ft). Sizing the tower diameter for a low gas pressure drop minimizes the operational power requirements and provides flexibility for the process operation (Hand et al., 1986). The diameter for a single packed tower can be determined using the generalized pressure drop curves shown in Fig. 6-9 combined with the manufacturer's pressure drop properties for a specific packing type. Once the packing factor for the media, the air-to-water ratio, and the air pressure drop per packing height have been specified, it is possible to determine the air loading rate and the tower diameter from Fig. 6-9 or the following set of equations (Cummins and Westrick, 1983):

$$G_m = \sqrt{\frac{M\rho_g(\rho_l - \rho_g)}{C_f(\mu_l)^{0.1}}} \tag{6-42}$$

The empirical parameter M is defined by the following relationship:

$$\log_{10}(M) = a_0 + a_1E + a_2E^2 \tag{6-43}$$

The parameter E is defined as follows:

$$E = -\log_{10}\left[\left(\frac{\dot{V}}{Q}\right)\sqrt{(\rho_g/\rho_l) - (\rho_g/\rho_l)^2}\right] \tag{6-44}$$

The parameters a_0, a_1, and a_2 are defined by the following relationships:

$$a_0 = -6.6599 + 4.3077F - 1.3503F^2 + 0.15931F^3 \tag{6-45}$$

$$a_1 = 3.0945 - 4.3512F + 1.6240F^2 - 0.20855F^3 \tag{6-46}$$

$$a_2 = 1.7611 - 2.3394F + 0.89914F^2 - 0.11597F^3 \tag{6-47}$$

The parameter F that appears in Eqs. 6-41 through 6-43 is defined as:

$$F = \log_{10}(\Delta P/L) \tag{6-48}$$

The pressure drop correlation is valid for $\Delta P/L$ values between 50 and 1200 N/m²/m. Equations 6-42 through 6-48 were developed using a particular set of units. When applying them, it must be assured that values are supplied in the appropriate units: G_m is the air mass loading rate (kg/m²/s); M is a calculated empirical parameter; ρ_g is the air density (kg/m³); ρ_l is the water density (kg/m³); C_f is the packing factor (m^{-1}); μ_l is the water viscosity (kg/m/s); a_0, a_1, a_2 are calculated empirical parameters; E is a calculated empirical parameter; $\frac{\dot{V}}{Q}$ is the air-to-water ratio on a volumetric basis (m³/m³); F is a calculated empirical parameter; and $\Delta P/L$ is the air pressure drop gradient (N/m²/m). The tower diameter can be determined by dividing G_m by the air mass flow rate.

Determination of the tower diameter from Fig. 6-9 is conservative at low gas pressure drops typically used by designers for air treatment of VOCs. If a more accurate value of the pressure drop is desired for a given packing type, the manufacturer's pressure drop curves for a particular packing type can be obtained and used.

The practical operating range for packed towers is between abscissa values of 0.02 and 4.0 in Fig. 6-9. For abscissa values greater than 4, large water loading rates can reduce the water-to-air contact area provided by the packing surface and inhibit proper air flow through the column causing a decrease in VOC removal efficiency. Similarly, high air flow rates (abscissa values less than about 0.02) can cause entrained water in the tower as well as channeling of the air through the tower. For situations where high air flow rates are required for high removal efficiencies (>95 percent), it is important to provide an even air inlet distribution at the bottom of the tower (Thom and Byers, 1993).

Determination of the Operating Power Requirements. The total operating power for a single air stripping packed tower system is the sum of the blower and pump brake power requirements. The blower brake power can be determined from the following relationship (Metcalf and Eddy, 2003):

$$P_{\text{blower}} = \left(\frac{G_m R_g T_{\text{air}}}{1000 n_a \text{Eff}_b}\right)\left[\left(\frac{P_{\text{in}}}{P_{\text{out}}}\right)^{n_a} - 1\right] \tag{6-49}$$

in which Eff_b is the blower net efficiency (expressed as a decimal), which accounts for both the fan and motor on the blower; G_m is the mass flow rate of air (kg/s); n_a is a constant used in determining blower brake power and is equal to 0.283 for air; P_{in} is the inlet air pressure in the packed tower (bottom of tower) (N/m²); P_{out} is the outlet air pressure in the packed tower (top of the tower) and is usually equal to the ambient pressure (N/m²); R_g is the universal gas constant; and T_{air} is the absolute air temperature, which is typically assumed equal to T.

The term P_{in} refers to the pressure at the bottom of the tower, which is the inlet for the air stream. P_{in} is calculated as the sum of the ambient pressure and the pressure drop caused by the packing media, demister, packing support plate, duct work, and tower inlet and outlet. The equation used to find P_{in} is

$$P_{\text{in}} = P_{\text{ambient}} + \left[\left(\frac{\Delta P}{L}\right) L\right] + \Delta P_{\text{losses}} \tag{6-50}$$

ΔP_{losses} is determined by (Hand et al., 1986):

$$\Delta P_{\text{losses}} = \left(\frac{\dot{V}}{A}\right)^2 k_p \tag{6-51}$$

in which $\dot{V}$ is the volumetric air flow rate (m^3/s), A is the tower cross-sectional area (m^2), and k_p is a constant equal to 275 NS^2/m^4 and is used to estimate the air pressure drop for losses other than those caused by the tower packing.

Equation 6-51 represents the air pressure drop through the demister, packing support plate, duct work, and tower inlet and outlet. It is assumed that turbulent flow conditions prevail and most of the losses occur in the tower (i.e., in the packing support plate and the demister).

The pump power requirement can be determined from the following equation:

$$P_{\text{pump}} = \frac{\rho_l Q H g}{1000\text{Eff}_p} \tag{6-52}$$

in which Eff_p is the pump efficiency and H is the vertical distance from the pump to the liquid distributor at the top of the tower (m). Equation 6-52 accounts only for the pressure loss resulting from the head required to pump the water to the top of the tower.

Design Procedure

The procedures used in designing packed towers depend upon the type of design problem. The most common design problems are modifications to existing towers and development of design criteria for new towers. Consequently, the design engineer may use a different approach for each problem. For example, one problem may be to determine ways of improving process efficiency of an existing tower by increasing the air flow rate, inserting a more efficient packing type, or increasing the packed tower height. In another situation, a designer may need to design a new packed tower system using standard vendor column sizes. Problems such as these require many tedious calculations that can easily be performed using commercially available software. Software tools such as ASDC1 (Dzombak, Roy, and Fang, 1993), AIRSTRIP (Haarhoff, 1988), and ASAP (Hokanson, 1996) can be used to evaluate the impact of process variables on process performance. These tools contain the design equations and correlations presented above and graphical user interfaces for ease in software application and provide databases for many commercially available packing types and physical properties of several VOCs that have been encountered in surface and groundwaters.

Presented herein is a step-by-step design procedure for sizing a packed tower and important considerations that should be addressed during the design and operation phases. This procedure is only one of several methods (Kavanaugh and Trussell, 1980; Crittenden et al., 2005) that have been used successfully to design a packed tower. This procedure was presented in detail elsewhere (Hand et al., 1986) and will simply be highlighted herein.

The design parameters for packed tower air stripping are (1) the air-to-water ratio, (2) the gas pressure drop, and (3) the type of packing material. Once the physical properties of the contaminant(s) of interest, the influent concentration(s), treatment objective(s), and water and air properties are known, these design parameters can be selected to give the lowest capital and operation and maintenance costs. As discussed above, the optimum air-to-water ratio for contaminants with Henry's law constants greater than 0.05, which will minimize the tower volume and total power consumption, is around 3.5 times the minimum air-to-water ratio or a stripping factor of about 3.5 (Crittenden et al., 2005). Therefore, the optimum air-to-water ratio can be estimated by multiplying the minimum air-to-water ratio calculated from Eq. 6-32 by 3.5 or using a stripping factor of 3.5. A low gas pressure drop should be chosen to minimize the blower power consumption. In most cases, a gas pressure drop of 50 to 100 $N/m^2/m$ is chosen. In order to determine the optimum gas pressure drop that will result in a least-cost design, a detailed cost analysis should be performed to determine the true costs associated with the tower volume and power requirement. A number of researchers have performed detailed cost analyses, and their results show that using a low gas pressure drop of around 50 to 100 $N/m^2/m$ and an air-to-water ratio of around 3.5 times the minimum yields the lowest total annual treatment cost (Cummins, 1985; Dzombak, Roy, and Fang, 1993). An additional advantage of operating at a low gas pressure drop is that if the blower is sized conservatively, the air flow rate can be increased to meet a greater removal demand without major changes in the process operation. However, the required height using this approach may sometimes be too great for a particular application. In this case, the air-to-water ratio can be increased by increasing the air flow rate to obtain a shorter tower height for a given removal.

The criteria for choosing the type and size of packing depend upon the water flow rate and the desired degree of operational flexibility of the design. For low water flow rates, it is recommended that nominal diameter packing of 0.05 m (2 in.) or less be used to minimize channeling or short-circuiting of the water down the wall of the tower. Minimizing the impact of channeling requires that the ratio of tower diameter to nominal packing diameter be greater than 12 to 15. If a design is needed with a high degree of operational flexibility, a larger, less efficient packing type should be chosen for the initial design of the tower. Tower designs with larger packing sizes will result in larger tower volumes, because the larger packing sizes usually have a lower specific area than the smaller ones. However, if the treatment objective becomes more stringent or other fewer strippable compounds appear in the influent, the less efficient, larger packing can be replaced with smaller, more efficient packing to provide for the needed removal. Building in this type of operational flexibility is much cheaper than adding additional height to the existing tower (if possible) or adding another tower in series. Spending additional capital on providing operational flexibility in the initial design can ultimately reduce future costs and prevent headaches.

***In most water treatment situations, multiple contaminants are present in the water, and the packed tower must be designed to remove all the contaminants to some specified treatment level. For a given tower design, the following equation can be used to determine the effluent concentrations of multiple contaminants for a given tower design:

$$C_e = \frac{C_0(S-1)}{\left\{S\exp\left[\dfrac{LK_La(S-1)}{\left(\frac{Q}{A}\right)S}\right]-1\right\}} \tag{6-53}$$

However, at the design stage, the limiting contaminant that controls the design must first be identified. In general, the contaminant with the lowest Henry's constant is used to determine the required air-to-water ratio and the contaminant with the highest removal efficiency is used to determine the required packing height. The following example problem illustrates a typical design procedure for sizing a packed tower.

Example 6-2 Packed Tower Aeration Calculation for TCA
Design a packed tower to reduce the 1,1,1-trichloroethane (TCA) concentration from 165 to 5.0 μg/L. Listed below are the operating conditions; TCA, air, and water physical and chemical properties; and packing type and packing parameters. For a design water flow of 0.158 m^3/s (2500 gpm), determine the tower diameter, packed tower height, and power consumption requirements.

Operating Conditions

Property	Value
Temperature (T, T_{air}), C	15
Pressure (P, $P_{ambient}$), atm	1

TCA, Water, and Air Characteristics

Property	Value
Molecular weight	133.4
Henry's constant (H), dimensionless	0.487
Influent concentration (C_i), μg/L	165
Effluent concentration (C_e), μg/L	5.0
Molarvolume @ boiling point (V_b), cm³/mole	109
Normal boiling point temperature, °C	74.1
Density of water (ρ_l), kg/m³	999.15
Viscosity of water (μ_l), kg/m · s	1.15×10^{-3}
Density of air (ρ_g), kg/m³	1.22
Viscosity of air (μ_g), kg/ m · s	1.75×10^{-5}
Surface tension of water (σ), N/m	0.0735

Packing Characteristics

Packing type: 3.5-in. nominal diameter LANPAC-XL

Property	Value
Nominal diameter (d_p), m	0.0889
Nominal surface area (a_t), m²/m³	242
Critical surface tension (σ_c), N/m	0.0330
Packing factor (C_f), m⁻¹	33

Constants

Property	Value
Acceleration due to gravity (g), m/s²	9.81
Blower efficiency (Eff$_b$), %	40
Pump efficiency (Eff$_p$), %	85

Solution

1. Calculate the minimum air-to-water ratio ($(\dot{V}/Q)_{\min}$) using Eq. 6-32.

$$(\dot{V}/Q)_{\min} = \frac{C_o - C_e}{HC_o} = \frac{165\frac{\mu g}{L} - 5.0\frac{\mu g}{L}}{0.487 \times 165\frac{\mu g}{L}} = 2.0$$

2. Calculate a reasonable $\dot{V}/Q$ that is some multiple, i.e., assume 3.5 times of $(\dot{V}/Q)_{\min}$.

$$(\dot{V}/Q)_{\text{mult}} = (\dot{V}/Q)_{\min} \times 3.5 = 2.0 \times 3.5 = 7.0$$

3. Calculate the air flow rate, $\dot{V}$.

$$\dot{V} = \frac{\dot{V}}{Q} \times Q = 7.0 \times 0.158\frac{\text{m}^3}{\text{s}} = 1.10\frac{\text{m}^3}{\text{s}}$$

4. Calculate the tower diameter, D.

a. Choose a low value of $\Delta P/L = 75$ N/m²/m.

b. Calculate F, a_0, a_1, a_2, E, M, and G_m using Eqs. 6-42 through 6-48. The tower area can then be determined from either G_m or L_m.

$$F = \log_{10}(\Delta P/L) = \log_{10}\left(75\frac{N}{m^2 \cdot m}\right) = 1.875$$

$$\begin{aligned} a_0 &= -6.6599 + 4.3077F - 1.3503F^2 + 0.15931F^3 \\ &= -6.6599 + 4.3077(1.875) - 1.3503(1.875)^2 + 0.15931(1.875)^3 \\ &= -2.280 \end{aligned}$$

$$\begin{aligned} a_1 &= 3.0945 - 4.3512F + 1.6240F^2 - 0.20855F^3 \\ &= 3.0945 - 4.3512(1.875) + 1.6240(1.875)^2 - 0.20855(1.875)^3 \\ &= -0.2287 \end{aligned}$$

$$\begin{aligned} a_2 &= 1.7611 - 2.3394F + 0.89914F^2 - 0.11597F^3 \\ &= 1.7611 - 2.3394(1.875) + 0.89914(1.875)^2 - 0.11597(1.875)^3 \\ &= -0.2287 \end{aligned}$$

$$\begin{aligned} E &= -\log_{10}\left[\left(\frac{\dot{V}}{Q}\right)\sqrt{(\rho_g/\rho_l) - (\rho_g/\rho_l)^2}\right] \\ &= -\log_{10}\left[15.0\sqrt{\left(\frac{1.22\frac{\text{kg}}{\text{m}^3}}{999.15\frac{\text{kg}}{\text{m}^3}}\right) - \left(\frac{1.22\frac{\text{kg}}{\text{m}^3}}{999.15\frac{\text{kg}}{\text{m}^3}}\right)^2}\right] \\ &= 0.6118 \end{aligned}$$

$$\begin{aligned} \log_{10}(M) &= a_0 + a_1E + a_2E^2 \\ &= -2.280 + -0.7293(0.6118) + (-0.2287)(0.6118)^2 \\ &= -2.8120 \Rightarrow \\ M &= 0.0015424 \end{aligned}$$

c. Calculate air mass loading rate, G_m, using Eq. 6-42.

$$\begin{aligned} G_m &= \sqrt{\frac{M\rho_g(\rho_l - \rho_g)}{C_f(\mu_l)^{0.1}}} \\ &= \sqrt{\frac{1.5424\times10^{-3}\times1.22\frac{\text{kg}}{\text{m}^3}\left(999.15\frac{\text{kg}}{\text{m}^3} - 1.22\frac{\text{kg}}{\text{m}^3}\right)}{33\left(1.15\times10^{-3}\frac{\text{kg}}{\frac{\text{m}}{\text{s}}}\right)^{0.1}}} \\ &= 0.335\frac{\text{kg}}{\frac{\text{m}^2}{\text{s}}} \end{aligned}$$

d. Calculate water mass loading rate, L_m.

$$L_m = \frac{G_m}{\left(\frac{\dot{V}}{Q}\right)\left(\frac{\rho_g}{\rho_l}\right)} = \frac{0.335\frac{\text{kg}}{\frac{\text{m}^2}{\text{s}}}}{\left(7.0\frac{\text{m}^3}{\text{m}^3}\right)\left(\frac{1.22\frac{\text{kg}}{\text{m}^3}}{999.15\frac{\text{kg}}{\text{m}^3}}\right)} = 39.2\frac{\text{kg}}{\frac{\text{m}^2}{\text{s}}}$$

e. Calculate tower cross-sectional area, A.

$$A = \frac{Q\rho_l}{L_m} = \frac{0.1577\frac{\text{m}^3}{s}\times 999.15\frac{\text{kg}}{\text{m}^3}}{39.2\frac{\text{kg}}{\frac{\text{m}^2}{\text{s}}}} = 4.0\ \text{m}^2$$

f. Calculate tower diameter, D.

$$D = \sqrt{\frac{4A}{\pi}} = \sqrt{\frac{4(4.0\ \text{m}^2)}{\pi}} = 2.26\ \text{m}\ (7.4\ \text{ft})$$

Standard packed column sizes are typically available in diameter increments of 0.3048 m (1 ft). Consequently, a tower diameter of 2.44 m (8.0 ft) is chosen for this design. The following design parameters were adjusted for a tower diameter of 2.44 m:

$$A = 4.68\ \text{m}^2 \qquad L_m = 33.74\frac{\text{kg}}{\text{m}^2\cdot\text{s}} \qquad G_m = 0.287\frac{\text{kg}}{\text{m}^2\cdot\text{s}} \qquad \frac{\Delta p}{L} = 64.5\frac{\text{Pa}}{\text{m}}$$

5. Calculate tower length, L.

a. Calculate the TCA aqueous phase concentration in equilibrium with the exiting air, C_0^*, using Eq. 6-31.

$$C_0^* = \left(\frac{1}{\left(\frac{\dot{V}}{Q}\right)\times H}\right)(C_0 - C_e) = \left(\frac{1}{(7.0)\times 0.487}\right)\left(165\frac{\mu\text{g}}{L} - 5.0\frac{\mu\text{g}}{L}\right)$$
$$= 46.93\frac{\mu\text{g}}{L}$$

b. Calculate the specific surface area available for mass transfer, a_w, using Eq. 6-39.

$$a = a_w$$

$$a_w = a_t\left\{1-\exp\left[-1.45\left(\frac{\sigma_c}{\sigma}\right)^{0.75}\left(\frac{L_m}{a_t\mu_l}\right)^{0.1}\left(\frac{L_m^2 a_t}{\rho_l^2 g}\right)^{-0.05}\left(\frac{L_m^2}{\rho_l a_t \sigma}\right)^{0.2}\right]\right\}$$

$$= 125\frac{\text{m}^2}{\text{m}^3}\left\{1-\exp\left[\begin{array}{l} -1.45\left(\frac{0.0330\frac{N}{m}}{0.0742\frac{N}{m}}\right)^{0.75}\left(\frac{33.74\frac{\text{kg}}{\text{m}^2\cdot\text{s}}}{242\frac{\text{m}^2}{\text{m}^3}\times 1.15\times 10^{-3}\frac{\text{kg}}{\text{m}\cdot\text{s}}}\right)^{0.1} \\ \times\left(\frac{\left(33.74\frac{\text{kg}}{\text{m}^2\cdot\text{s}}\right)^2\times 242\frac{\text{m}^2}{\text{m}^3}}{\left(999.15\frac{\text{kg}}{\text{m}^3}\right)^2\times 9.81\frac{\text{m}}{\text{s}^2}}\right)^{-0.05} \\ \times\left(\frac{\left(33.74\frac{\text{kg}}{\text{m}^3\cdot\text{s}}\right)^2}{999.15\frac{\text{kg}}{\text{m}^3}\times 242\frac{\text{m}^2}{\text{m}^3}\times 0.0742\frac{\text{N}}{\text{m}}}\right)^{0.2} \end{array}\right]\right\}$$

$$= 193\frac{\text{m}^2}{\text{m}^3}$$

c. Calculate the TCA diffusivity in water, D_l, using the following correlation (Hayduk and Laudie, 1974):

$$D_l = \frac{13.26\times 10^{-5}}{\mu_w^{1.14}\times V_b^{0.589}} = \frac{13.26\times 10^{-5}}{(1.15cp)^{1.14}\times\left(109\frac{\text{cm}^3}{\text{gmol}}\right)^{0.589}\times\frac{(100\,\text{cm})^2}{1\,\text{m}^2}}$$

$$= 7.13\times 10^{-10}\,\frac{\text{m}^2}{\text{s}}$$

μ_w = viscosity of water (centipoise)
V_b = molar volume, cm^3/gmol

d. Calculate the liquid phase mass transfer coefficient, k_l, using Eq. 6-40.

$$k_l = 0.0051\left(\frac{L_m}{a_w\mu_l}\right)^{2/3}\left(\frac{\mu_l}{\rho_l D_l}\right)^{-0.5}(a_t d_p)^{0.4}\left(\frac{\rho_l}{\mu_l g}\right)^{-1/3}$$

$$= 0.0051\left(\frac{33.74\frac{\text{kg}}{\frac{\text{m}^2}{\text{s}}}}{193\frac{\text{m}^2}{\text{m}^3}\times 1.15\times 10^{-3}\frac{\text{kg}}{\frac{\text{m}}{\text{s}}}}\right)^{2/3}\left(\frac{1.15\times 10^{-3}\frac{\text{kg}}{\frac{\text{m}}{\text{s}}}}{999.15\frac{\text{kg}}{\text{m}^3}\times 7.13\times 10^{-10}\frac{\text{m}^2}{\text{s}}}\right)^{-0.5}$$

$$\times\left(2425\frac{\text{m}^2}{\text{m}^3}\times 0.0889\,\text{m}\right)^{0.4}\left(\frac{999.15\frac{\text{kg}}{\text{m}^3}}{1.15\times 10^{-3}\frac{\text{kg}}{\frac{\text{m}}{\text{s}}}\times 9.81\frac{\text{m}}{\text{s}^2}}\right)^{-1/3}$$

$$= 2.76\times 10^{-4}\,\frac{\text{m}}{\text{s}}$$

e. Calculate the gas-phase diffusivity, D_g, using the following correlation by Hirschfelder, Bird, and Spotz (1949) (Treybal, 1980):

i. To calculate the gas diffusivity, the following parameters must be determined:

$$r_A = 1.18(V_b)^{1/3} = 1.18\times\left(0.109\frac{\text{m}^3}{\text{kgmol}}\right)^{1/3} = 0.564\,\text{nm}\,(\text{TCA})$$

$$\varepsilon_A/K = 1.21(T_b)_A = 1.21\times(74.1+273.15)\text{K} = 420.2\ \text{K}(TCA)$$

in which r_A is the radius of molecule A, nm (TCA); V_b is the molar volume of molecule A at the boiling point (TCA); ε_A is the force constant of molecule A, joule (TCA); K is the Boltzmann's constant, joule/K; T_{bA} is the boiling temperature of molecule A, K (TCA); and r_B is the characteristic radius of the gas medium, nm, and is equal to 0.371 nm for air.

Calculate the molecular separation at collision r_{AB} for TCA in air.

$$r_{AB} = (r_A + r_B)/2 = \frac{0.564\ \text{nm} + 0.371\ \text{nm}}{2} = 0.4676\ \text{nm}\,(\text{TCA in Air})$$

Calculate the energy of molecular attraction between molecule A (TCA) and molecule B (air). The force constant for air, ε_B/K, is 78.6.

$$\frac{\varepsilon_{AB}}{K} = \sqrt{\frac{\varepsilon_A}{K} \times \frac{\varepsilon_B}{K}} = \sqrt{420.2 \times 78.6} = 181.7 (\text{TCA in Air})$$

Calculate the collision function for diffusion of molecule A (TCA) in B (air) (The collision function was presented by Treybal (1980) and correlated by Cummins and Westrick (1983).).

$$EE = \log_{10}(KT/\varepsilon_{AB}) = \log_{10}\left(\frac{283.15\,\text{K}}{185.0}\right) = 0.185$$

$$\begin{aligned} NN &= \begin{pmatrix} -0.14329 - 0.48343(EE) + 0.1939(EE)^2 + 0.13612(EE)^3 \\ -0.20578(EE)^4 + 0.083899(EE)^5 - 0.011491(EE)^6 \end{pmatrix} \\ &= \begin{pmatrix} -0.14329 - 0.48343(0.193) + 0.1939(0.193)^2 \\ +0.13612(0.193)^3 - 0.20578(0.193)^4 + 0.083899(0.193)^5 \\ -0.011491(0.193)^6 \end{pmatrix} \\ &= -0.235 \end{aligned}$$

$$f(KT/\varepsilon_{AB}) = 10^{NN} = 10^{-0.235} = 0.582$$

ii. Calculate gas-phase diffusivity, D_g, as follows:

$$\begin{aligned} D_g &= \left\{ \frac{10^{-4}\left(1.084 - 0.249\sqrt{\frac{1}{M_A} + \frac{1}{MB}}\right)(T^{1.5})\sqrt{\frac{1}{M_A} + \frac{1}{M_B}}}{P_t\,(r_{AB})^2 f\left(\frac{KT}{\varepsilon_{AB}}\right)} \right\} \frac{\text{m}^2}{\text{s}} \\ &= \left\{ \frac{10^{-4}\left(1.084 - 0.249\sqrt{\frac{1}{133.4} + \frac{1}{28.95}}\right)(288.15)^{1.5}\sqrt{\frac{1}{133.4} + \frac{1}{28.95}}}{101325\frac{\text{N}}{\text{m}^2}(0.4676\,\text{nm})^2\,(0.582)} \right\} \frac{\text{m}^2}{\text{s}} \\ &= 8.03 \times 10^{-6}\,\frac{\text{m}^2}{\text{s}} \end{aligned}$$

f. Calculate gas-phase TCA mass transfer coefficient, k_g, using Eq. 6-41.

$$\begin{aligned} k_g &= 5.23(a_t D_g)\left(\frac{G_m}{a_t \mu_g}\right)^{0.7}\left(\frac{\mu_g}{\rho_g D_g}\right)^{1/3}(a_t d_p)^{-2} \\ &= 5.23\left(125\frac{\text{m}^2}{\text{m}^3} \times 8.03 \times 1 - 10^{-6}\frac{\text{m}^2}{\text{s}}\right)\left(\frac{0.287\frac{\text{kg}}{\frac{\text{m}^2}{\text{s}}}}{242\frac{\text{m}^2}{\text{m}^3} \times 1.75 \times 10^{-5}\frac{\text{kg}}{\frac{\text{m}}{\text{s}}}}\right)^{0.7} \\ &\quad \times \left(\frac{1.75 \times 10^{-5}\frac{\text{kg}}{\frac{\text{m}}{\text{s}}}}{1.22\frac{\text{kg}}{\text{m}^3} \times 8.03 \times 10^{-6}\frac{\text{m}^2}{\text{s}}}\right)^{1/3}\left(242\frac{\text{m}^2}{\text{m}^3} \times 0.0889\,\text{m}\right)^{-2} \\ &= 5.09 \times 10^{-4}\,\frac{\text{m}}{\text{s}} \end{aligned}$$

g. Calculate $K_L a$ using Eq. 6-21.

$$\frac{1}{K_L a} = \frac{1}{k_l a_w} + \frac{1}{k_g a_w H}$$
$$= \frac{1}{2.76\times10^{-4}\,\frac{\text{m}}{\text{s}}\times193\frac{\text{m}^2}{\text{m}^3}} + \frac{1}{5.09\times10^{-4}\,\frac{\text{m}}{\text{s}}\times193\frac{\text{m}^2}{\text{m}^3}\times0.487}$$
$$K_L a = 0.0252\,\text{s}^{-1}$$

h. Calculate $K_L a$ assuming a safety factor of 0.75.

$$K_L a = K_L a\times(SF)_{K_L a} = 0.0252\,\text{s}^{-1}\times 0.75$$
$$= 0.0189\,\text{s}^{-1}$$

i. Calculate tower length, L, using Eq. 6-30.

$$L = \frac{Q}{AK_L a}\left[\frac{C_o - C_{TO}}{C_o - C_{TO} - C_0^*}\right]\ln\left[\frac{C_0 - C_0^*}{C_{TO}}\right]$$
$$= \frac{0.1577\frac{\text{m}^3}{s}}{4.68\,\text{m}^2\times0.0189\,\text{s}^{-1}}\left[\frac{165\frac{\mu g}{L} - 5.0\frac{\mu g}{L}}{165\frac{\mu g}{L} - 5.0\frac{\mu g}{L} - 46.93\frac{\mu g}{L}}\right]\ln\left[\frac{165\frac{\mu g}{L} - 46.93\frac{\mu g}{L}}{5.0\frac{\mu g}{L}}\right]$$
$$= 8.0\,\text{m}\ \ (26.0\text{ ft})$$

6. Calculate power requirements.

a. Calculate blower power requirements.

i. Calculate the air mass flow rate, G_m, from the volumetric air flow rate.

$$G_{me} = \dot{V}\times\rho_g = 1.10\frac{\text{m}^3}{\text{s}}\times1.22\frac{\text{kg}}{\text{m}^3} = 1.34\frac{\text{kg}}{\text{s}}$$

ii. Calculate the pressure drop through the demister, the packing support plate, duct work, and inlet and outlet, ΔP_{losses}, using Eq. 6-51.

$$\Delta P_{losses} = \left(\frac{\dot{V}}{A}\right)^2 k_p = \left(\frac{1.1\text{m}^{3/\text{s}}}{4.0\,\text{m}^2}\right)^2 \times 275\,\text{N}\cdot\text{s}^2/\text{m}^4 = 20.80\,\text{N/m}^2$$

k_p = a constant used to estimate gas pressure drop in a tower for losses other than that in the packing, $275\frac{\text{N}\cdot\text{s}^2}{\text{m}^4}$ (Hand et al., 1986).

iii. Calculate the inlet pressure to the packed tower, P_{in}, using Eq. 6-50.

$$P_{\text{in}} = P_{\text{ambient}} + \left[\left(\frac{\Delta P}{L}\right)L\right] + \Delta P_{\text{losses}} = 101325\frac{\text{N}}{\text{m}^2} + \left[\left(64.50\frac{\text{N/m}^2}{\text{m}}\right)9.33\,\text{m}\right]$$
$$+20.80\frac{\text{N}}{\text{m}^2} = 101948\,\frac{\text{N}}{\text{m}^2}$$

iv. Calculate the blower brake power, P_{blower}, using Eq. 6-49.

$$P_{\text{blower}} = \left(\frac{G_{me}R_g T_{\text{air}}}{1000 n_a \text{Eff}_b}\right)\left[\left(\frac{P_{\text{in}}}{P_{\text{out}}}\right)^{n_a} - 1\right]$$
$$= \left(\frac{1.34\frac{\text{kg}}{\text{s}}\times286.7\frac{\text{J}}{\text{kg}\cdot\text{K}}\times288.15\text{K}}{1000\frac{\text{W}}{\text{kW}}\,(0.283)(0.40)}\right)\left[\left(\frac{101948\frac{\text{N}}{\text{m}^2}}{101325\frac{\text{N}}{\text{m}^2}}\right)^{0.283} - 1\right]\left(\frac{1W}{1\frac{\text{J}}{\text{s}}}\right)$$
$$= 1.70\,\text{kW}$$

b. Calculate pump power requirements, P_{pump}, using Eq. 6-52.

$$P_{\text{pump}} = \frac{\rho_l Q L g}{1000 \text{Eff}_p}$$

$$= \left(\frac{999.15\frac{\text{kg}}{\text{m}^3} \times 0.1577\frac{\text{m}^3}{\text{s}} \times 9.33\text{m} \times 9.81\frac{\text{m}}{\text{s}^2}}{1000\frac{\text{W}}{\text{kW}} \times 0.85}\right)\left(\frac{1W}{1\frac{\text{kgm}}{\text{s}^2} \cdot \frac{\text{m}}{\text{s}}}\right)$$

$$= 17.0\text{kW}$$

Eff_p = pump efficiency

c. Calculate total power requirements, P_{total}.

$$P_{\text{Total}} = P_{\text{Blower}} + P_{\text{Pump}} = 1.7\text{kW} + 17.0\text{kW} = 18.7\text{kW}$$ ▲

Impact of Dissolved Solids on Tower Performance

One major concern with packed tower air stripping is the potential for precipitation of inorganic compounds such as calcium, iron, and manganese onto packing media, causing a steady decrease in the tower void volume and a steady increase in tower pressure drop, eventually leading to possible plugging of the tower. This is especially true for some groundwaters and hypolimnetic waters from stratified lakes and reservoirs that contain considerable amounts of dissolved carbon dioxide. It is not uncommon to encounter groundwaters with 30 to 50 mg/L of carbon dioxide. In an air stripping tower, carbon dioxide together with VOCs can be removed from the water. Because carbon dioxide is an acidic gas, its removal increases the water's pH. The reader can find principles of the effects of pH on the speciation of inorganic carbon in Chap. 3. Applying these principles here as the pH increases above 8.3, carbonate (CO_3^{2-}) becomes the dominant species of inorganic carbon. In natural waters containing appreciable amounts of Ca^{2+}, calcium carbonate precipitates when the CO_3^{2-} concentration becomes high enough that the calcium carbonate solubility product (pK_{so} of 8.34) is exceeded.

$$Ca^{2+} + CO_3^{2-} \rightleftarrows CaCO_3(s) \qquad (6\text{-}54)$$

Air contains about 0.038 percent by volume of carbon dioxide. The Henry's law constant for carbon dioxide at 25°C is about 61.1 (1500 mg/L-atm); therefore the equilibrium concentration of carbon dioxide with air is about 0.57 mg/L. Air stripping can reduce the dissolved carbon dioxide concentration to its equilibrium concentration with air. In most situations, theoretical precipitation calculations can be performed to determine the maximum amount and rate of precipitation that will cause plugging of the tower. However, the carbon dioxide concentration is usually reduced to that above its equilibrium concentration with air, and the time taken to plug the air stripping tower is much longer than that predicted from the maximum theoretical precipitation. The actual rate of precipitation can only be determined using pilot-plant testing. The control of precipitation using acid or scale inhibitors can be expensive and, in some cases, it can be a major expense. Consequently, this needs to be carefully considered.

Air Stripping Off-Gas Control Using Adsorption

Air pollution resulting from VOC removal in air stripping processes has prompted many regulatory agencies to set emissions standards. Off-gas treatment technologies such as gas-phase granular activated carbon (GAC) adsorption, thermal incineration, and catalytic incineration are capable of reducing and, in some cases, eliminating the discharge of VOCs from air strippers. Thermal and catalytic incineration is used where the VOC concentrations are high or the VOCs are weakly adsorbing. VOC concentrations are usually dilute in most water treatment applications. Consequently, GAC fixed-bed adsorption is the most widely used process because it is cost effective for treating dilute off-gas streams and many of the common VOC contaminants (e.g., trichloroethene, tetrachloroethene, toluene, xylene isomers) are amenable to adsorption onto GAC (Crittenden et al., 1988). A brief discussion of the important parameters that need to be considered for fixed-bed adsorber design and operation for VOC off-gas control from air strippers is presented below.

The design principles and application of adsorption for water treatment are presented in Chap. 14. The fundamental principles for gas-phase adsorption are similar to those for liquid-phase adsorption, with the exception that the transport fluid is air instead of water. For example, the same equilibrium and mass transfer concepts used in design of liquid-phase systems are used in gas-phase adsorber design. The main differences between these two adsorption processes are the impact of water vapor on VOC adsorption must be considered in gas-phase adsorption; the impact of NOM on the adsorption of organic contaminants is negligible and does not need to be considered; gas-phase adsorbents have properties (e.g., particle size, pore size distribution, hydrophobicity) that are designed to be more amenable to gas-phase adsorption; and the adsorbent reactivation techniques may be different depending upon the type of adsorbent that is used.

A schematic of a typical GAC off-gas control system for an air stripper is displayed in Fig. 6-10. The VOC-laden air stripper off-gas contains water vapor that has a relative humidity (*RH*) of 100 percent. Water vapor can have a large influence on the adsorption capacity of

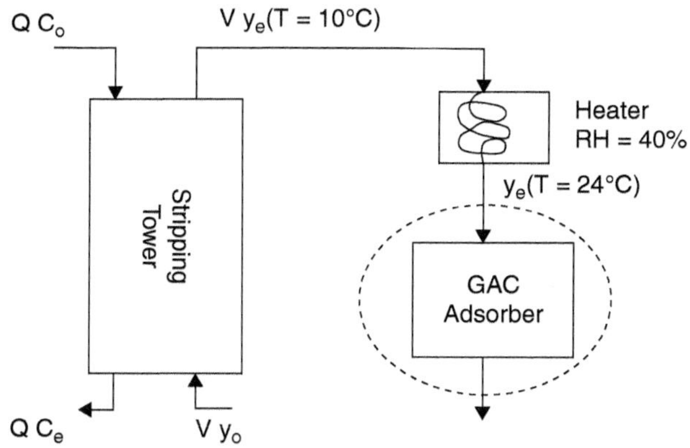

FIGURE 6-10 Schematic of air stripping off-gas control using a GAC adsorber.

adsorbents such as GAC. Figure 6-11 displays water vapor isotherms for several GAC adsorbents. For most of these GAC adsorbents, the adsorption of water is significant when the *RH* is greater than about 40 percent. At high *RH* values, a phenomenon called capillary condensation takes place where the water vapor can condense in the micropores of the adsorbent. The condensed water can then compete with the VOCs for adsorption sites, and the VOC capacity of the adsorbent can be significantly reduced. However, for *RH* values less than about 40 percent, the amount of water adsorbed onto the GAC is small, because capillary condensation of the water vapor is negligible and, consequently, its impact on VOC adsorption capacity can be neglected (Crittenden et al., 1988). For most gas-phase GAC adsorbents, it is recommended the off-gas *RH* be reduced to less than 40 percent prior to the adsorption process. A water vapor isotherm performed on a particular adsorbent of interest should be used to determine the *RH* value that will provide the most cost-effective design. As shown in Fig. 6-11, heating the off-gas is usually the most common method used to reduce the *RH* when GAC is used.

Typical design parameters for gas-phase GAC fixed-bed adsorbers consist of empty bed contact times (EBCTs) from 1.5 to 5.0 s and air loading rates from 0.25 to 0.50 m/s (0.82 to 1.64 ft/s). The GAC used in gas-phase applications is usually a microporous carbon with nominal particle sizes ranging from 4 × 6 to 6 × 16 U.S. standard mesh size. Gas-phase GAC particles are typically larger than those for liquid-phase applications so as to minimize

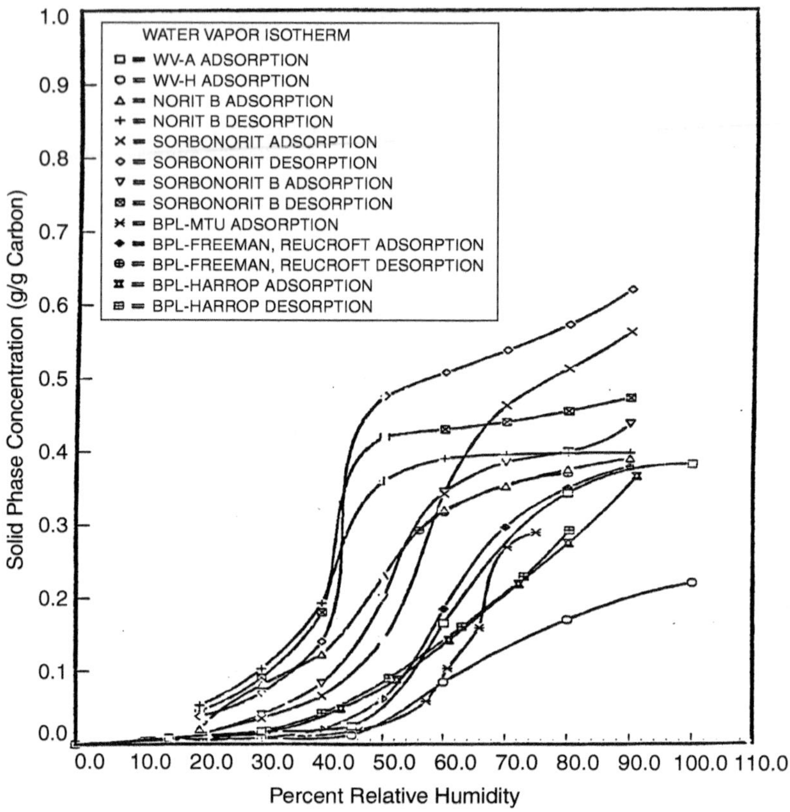

FIGURE 6-11 Water vapor isotherms for various granular activated carbons. (*Source: Tang et al., 1987.*)

the air pressure drop through the fixed-bed. The air pressure drop in gas-phase fixed-bed adsorbers ranges from 0.2 to 5.0 kPa/m of packed bed depth.

In gas-phase adsorption, the mass transfer zone of an organic contaminant is short and occupies a small fraction of the fixed-bed. For moderate to strongly adsorbing VOCs (e.g., trichloroethylene, toluene, tetrachloroethylene), mass transfer zone lengths range from 0.025 to 0.043 m (Crittenden et al., 1988). For a typical packed bed depth of 1.0 m, the mass transfer zone would occupy less than 4.3 percent of the bed. Consequently, for preliminary design calculations, equilibrium models can be used to determine the GAC usage rate for organic contaminants. Based on the preliminary design calculations, pilot studies can be performed and used to evaluate the most cost-effective design and operation.

The Dubinin-Raduskevich (D-R) equation was shown to correlate the isotherms of several gas-phase VOCs (Crittenden et al., 2005). Based on the work of several researchers (Reucroft, Simpson, and Jonas, 1971; Rasmuson, 1984; Crittenden at al., 1988), the following form of the D-R equation can be used to estimate the single solute gas-phase capacity for a VOC:

$$q = \rho_L W = \rho_L W_o \exp\left[\frac{-B\varepsilon^2}{\alpha^2}\right] \tag{6-55}$$

$$\varepsilon = RT \ln\left(\frac{P_s}{P}\right) \tag{6-56}$$

where q is the solid-phase VOC concentration in equilibrium with P (M/M), W is the adsorption space or pore volume occupied by the adsorbate (mL/gm), W_o is the maximum adsorption space (mL/gm), B is the microporosity constant (cm^6/cal^2), ε is the adsorption potential (cal/mole), T is the temperature (K), P_s is the saturation vapor pressure of the VOC at T K (N/m^2), P is the partial pressure of the VOC in the gas (N/m^2), ρ_L is the liquid density of the VOC (kg/m^3), and α is the polarizability of the VOC (m^3/kg) and can be calculated from the Lorentz-Lorentz equation if it is not known.

$$\alpha = \frac{(\eta^2 - 1)M}{(\eta^2 + 2)\rho_l} \tag{6-57}$$

where η is the refractive index of the VOC and M is the molecular weight of the VOC. Equations 6-55 and 6-56 were correlated with isotherm data for several compounds with dipole moments less than 2 debyes (2×10^{-18} esu-cm) (Reucroft, Simpson, and Jonas, 1971; Crittenden et al., 1988). W_o and B depend only on the nature of the adsorbent. When isotherm data is plotted as W versus $(\varepsilon/\alpha)^2$ for a given adsorbent, the data conform to essentially one line for different VOCs and temperatures (Crittenden et al., 2005). Table 6-10 summarizes W_o and B values for three gas-phase adsorbents. The D-R values are based on toluene, which was used as the reference compound. Two important limitations of the D-R equation (Eqs. 6-55 and 6-56) are: it is only valid for relative pressures, P/P_s less than 0.2 because capillary condensation of the VOC occurs, and it is only valid for RH values less than about 50 percent. For relative pressures greater than 0.2, the adsorbent capacity can be calculated by assuming the adsorbent micropore volume is filled with the condensed VOC. For cases where multiple VOCs are present in an air stripper off-gas, equilibrium column model (ECM) calculations can be used to estimate the adsorbent usage rate for each VOC (Cortright, 1986; Crittenden et al., 1987). When the GAC capacity is exhausted, it is usually sent back to the manufacturer for reactivation and reuse.

TABLE 6-10 Typical W_o and B Values for Several Commercial Gas-Phase Adsorbents

Adsorbent type	W_o (cm^3/gm)	B (cm^6/cal^2)	Reference
Calgon BPL (6 × 16 mesh)	0.515	2.99×10^{-5}	Crittenden et al., 1988
Calgon BPL (4 × 6 mesh)	0.460	3.22×10^{-5}	Critenden et al., 1988
CECA GAC–410G	0.503	2.42×10^{-5}	CECA Inc. 1994

Example 6-3 Air Stripper Off-Gas Control Using Adsorption Calculation
Perform preliminary design calculations for a gas-phase adsorber that will remove TCA from the air stripper off-gas in Example 6-2. Determine the GAC usage rate that will be required to meet a nondetectable limit for TCA and the bed life for the following conditions: an EBCT of 2.5 s and a loading rate of 0.5 m/s. A summary of the air stripper off-gas and TCA properties required to perform the calculations is listed below.

Air Stripper Data:

$$\dot{V} = Q \times \frac{\dot{V}}{Q} = 0.158\,\frac{\text{m}^3}{\text{s}} \times 7.0\,\frac{\text{m}^3\text{air}^3/\text{s}}{\text{m}^3\text{water/s}} = 1.10\,\frac{\text{m}^3}{\text{s}}$$

$$C_o(TCA) = 165\,\frac{\mu g}{L} \qquad C_e(TCA) = 5.0\,\frac{\mu g}{L}$$

Off-gas temperature is assumed to be 15°C, RH = 100 percent, the relative humidity entering the adsorber is 40 percent, and air temperature entering the adsorber is 25°C.

Solution

1. Calculate the off-gas TCA concentration leaving the air stripper.

$$y_e\begin{pmatrix}\text{TCA conc.}\\ \text{in off-gas}\\ \text{at } 15^\circ\text{C}\end{pmatrix} = \frac{C_o - C_e}{\dot{V}/Q} = \frac{(165.0-5.0)\,\frac{\mu g}{L}}{7.0} = 22.9\,\frac{\mu g}{L_{\text{Air}}}$$

2. Calculate the TCA gas-phase concentration entering the adsorber.
The TCA concentration in the heated off-gas will be decreased by a small amount due to heating of the off-gas. The ideal gas law at constant pressure of 760 mm Hg can be used to adjust for this change as shown below.

$$y_e(25^\circ\text{C}) = 22.9\,\frac{\mu g}{L_{\text{Air}}}\,(10^\circ\text{C})\left(\frac{273+15}{273+25}\right) = 22.1\,\frac{\mu g}{L_{\text{Air}}}$$

3. Calculate the best possible GAC usage rate assuming no mass transfer resistance. Assume the adsorbent that will be used in the design is Calgon BPL 4 × 6 mesh GAC (D-R Parameters: $B = 3.22 \times 10^{-5}$ cm^6/cal^2, $W_o = 0.460$ cm^3/gm). Performing a mass balance on the GAC adsorber, the following expression can be obtained for calculating the GAC usage rate:

$$\frac{\hat{M}}{\dot{V}t_e} = \frac{y_e}{q_e} = \text{GAC Usage Rate}$$

where $\hat{M}$ is mass of GAC in the fixed-bed, t_e is the time it takes to exhaust the GAC fixed-bed, and q_e is the adsorbed phase concentration on TCA in equilibrium with y_e. q_e can be calculated using Eq. 6-55.

If Eqs. 6-55 and 6-56 are combined and divided by $M \times 10^{-6}$, the following expression can be used to calculate q_e:

$$q_e = \left(\frac{W_o\rho_l}{M\times10^{-6}}\right)\exp\left[\frac{-B}{\alpha^2}\left\{RT\ln\left(\frac{P_s}{P}\right)\right\}^2\right]$$

From *Lange's Handbook of Chemistry* (Speight, 2004), P_s for TCA at 25°C is equal to 78.8 mm Hg.

y_e can be converted in terms of P using the following expression:

$$P(25^\circ\text{C}) = y_eRT$$
$$= \left(22.1\frac{\mu g}{L}\right)\left(0.08205\frac{L-\text{atm}}{\text{mol}-\text{K}}\right)(298\text{K})\left(\frac{\text{mol}}{133.4\,\text{gm}}\right)\left(\frac{\text{gm}}{10^6\,\mu g}\right)\left(\frac{760\,\text{mm Hg}}{1\,\text{atm}}\right)$$
$$= 3.1\times10^{-3}\text{ mm Hg}$$

Equation 6-57 is used to calculate the polarizability of TCA. From the CRC 85th edition, $\eta(\text{TCA}) = 1.4313$ and ρ_l (TCA) = 1.33 gm/cm^3.

$$\alpha(TCE) = \frac{((1.4313)^2-1)33.4\frac{\text{gm}}{\text{mol}}}{((1.4313)^2+2).33\frac{\text{gm}}{\text{cm}^3}} = 25.98\,\frac{\text{cm}^3}{\text{mol}}$$

$$q_e = \left(\frac{\left(0.460 \frac{\text{cm}^3}{\text{gm}}\right)\left(1.33 \frac{\text{gm}}{\text{cm}^3}\right)}{133.4 \times 10^{-6} \frac{\text{gm}}{\mu\,\text{mol}}} \right)$$

$$\times \exp\left[\frac{-3.22\times10^{-5} \frac{\text{cm}^6}{\text{cal}^2}}{(25.98)^2 \frac{\text{cm}^6}{\text{mol}^2}} \left\{ \left(1.987 \frac{\text{cal}}{\text{mol}-\text{K}}\right)(298\,\text{K}) \ln\left(\frac{78.8\,\text{mm Hg}}{3.1\times10^{-3}\,\text{mm Hg}} \right) \right\}^2 \right]$$

$$q_e = 820 \frac{\mu\,\text{mol TCA adsorbed}}{\text{gm GAC}}$$

$$\begin{Bmatrix}\text{GAC}\\ \text{Usage}\\ \text{Rate}\end{Bmatrix} = \frac{y_e}{q_e} = \frac{\left(22.1 \frac{\mu g}{L_{\text{Air}}}\right)\left(\frac{1\,\text{kg}}{1000\,\text{gm}}\right)}{\left(820 \frac{\mu\text{mol}}{\text{gm}}\right)\left(133.4 \frac{\mu g}{\mu\,\text{mol}}\right)\left(\frac{\text{m}^3}{1000\,L}\right)} = 2.02\times10^{-4} \frac{\text{kg GAC}}{\text{m}^3\ \text{Air treated}}$$

or

$$\begin{Bmatrix}\text{GAC}\\ \text{Usage}\\ \text{Rate}\end{Bmatrix} = 4{,}953 \frac{\text{m}^3\ \text{Air treated}}{\text{kg GAC}}$$

4. Calculate the mass of GAC required and the bed life of the GAC adsorber assuming an EBCT = 2.5 s and $v_s = 0.5$ m/s.

$$\begin{Bmatrix}\text{Mass of GAC}\\ \text{in 2.5 s EBCT}\\ \text{Bed}\end{Bmatrix} = \text{EBCT}\dot{V}\rho_F = (2.5\text{s})\left(1.1 \frac{\text{m}^3}{\text{s}}\right)\left(531 \frac{\text{kg}}{\text{m}^3}\right) = 1{,}460\,\text{kg}$$

$$\begin{Bmatrix}\text{Volume of Air}\\ \text{Treated for a}\\ \text{2.5 s EBCT}\end{Bmatrix} = \frac{\begin{pmatrix}\text{Mass of GAC}\\ \text{fo 2.5 s EBCT}\end{pmatrix}}{(\text{TCA Usage Rate})} = \frac{1460\,\text{kg GAC}}{\left(\frac{\text{kg GAC}}{4953\,\text{m}^3\ \text{Air}}\right)} = 7.3\times10^6\,\text{m}^3\ \text{Air}$$

$$\{\text{GAC Bed Life}\} = \frac{(7.3\times10^6\,\text{m}^3\text{Air})}{\left(1.1 \frac{\text{m}^3\ \text{Air}}{\text{s}}\right)\left(\frac{86{,}400\,\text{s}}{d}\right)} = 77\,\text{days} \quad \blacktriangle$$

DIFFUSED OR BUBBLE AERATION

The diffused or bubble aeration process consists of contacting gas bubbles with water for the purposes of transferring gas to the water (e.g., O_3, CO_2, O_2) or removing VOCs from the water by air stripping. The process can be carried out in a clearwell or special rectangular concrete tanks typically 3 to 5 m (9 to 15 ft) in depth. Figure 6-12 displays different types of diffused aeration systems. The most commonly used diffuser system consists of a matrix of perforated tubes (or membranes) or porous plates arranged near the bottom of the tank to provide maximum gas-to-water contact. Various types of diffusers and diffuser system layouts are presented in the U.S. Environmental Protection Agency's technology transfer design manual on fine-pore aeration systems (USEPA, 1989). Jet aerator devices are also used to provide good air-to-water contact (Mandt and Bathija, 1987). These aerators consist of jets that discharge fine gas bubbles and provide enhanced mixing for increased absorption efficiency.

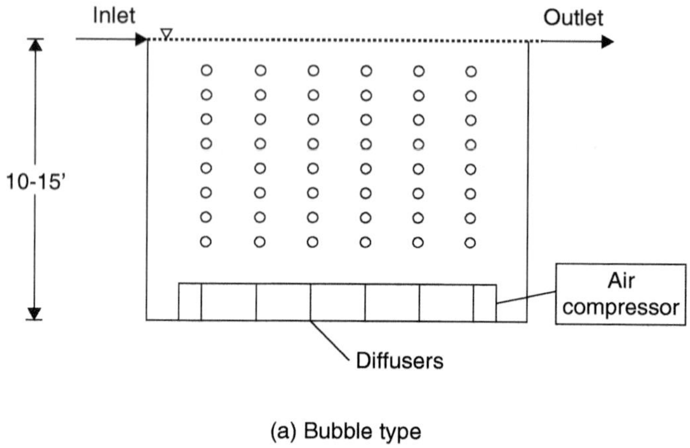

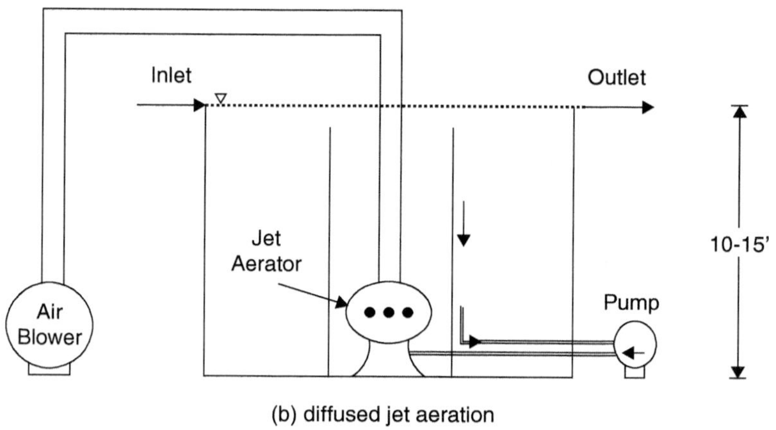

FIGURE 6-12 Schematic of various bubble aeration systems.

Governing Equations: Air Stripping of VOCs

Figure 6-13 shows a schematic of a bubble aeration system for a single tank. Model development for air stripping using bubble aeration has been described in the literature (Matter-Müller, Gujer, and Giger, 1981; Munz and Roberts, 1982; Roberts, Munz, and Dändliker, 1984). The development of the process design equations for bubble aeration incorporates the following assumptions: (1) the liquid phase is completely mixed, (2) the gas phase is plug flow, (3) the process is at steady state, and (4) the inlet VOC gas concentration is zero. A liquid-phase mass balance around the tank shown in Fig. 6-13, incorporating the above assumptions, results in the following relationship:

$$QC_0 - QC_e - \dot{V}(y_o - y_e) = 0 \tag{6-58}$$

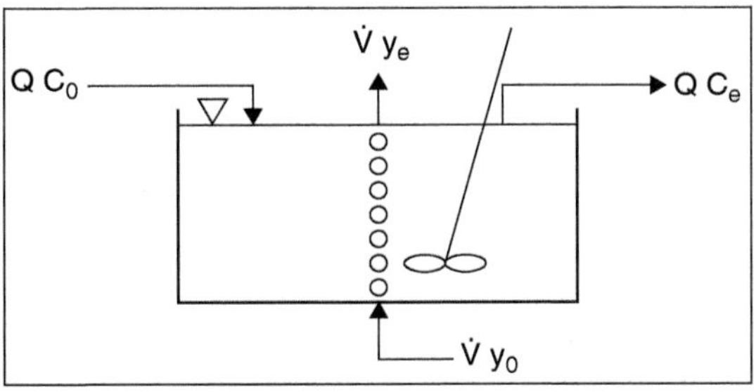

FIGURE 6-13 Schematic of a completely mixed bubble aeration tank for development of the design equations.

Applying the assumptions that the inlet VOC gas concentration is zero (i.e., $y_o = 0$) and the exiting air is in equilibrium with the bulk liquid (i.e., $y_e = HC_e$), Eq. 6-58 can be solved for $(\frac{\dot{V}}{Q})$, which is for the minimum air-to-water ratio for bubble aeration.

$$\left(\frac{\dot{V}}{Q}\right)_{\min} = \frac{C_0 - C_e}{HC_e} \tag{6-59}$$

The minimum air-to-water ratio represents the smallest air-to-water ratio that can be applied to a bubble aeration tank in order to meet the treatment objective, C_e.

Figure 6-14 shows a schematic of a gas bubble within a bubble aeration tank. Eliminating the assumption of equilibrium between the bulk liquid and the exiting air and performing a mass balance around the gas bubble (assuming plug flow conditions prevail in the gas phase) results in an expression for y_e in Eq. 6-58 (Hokanson, 1996; Matter-Müller, Gujer, and Giger, 1981):

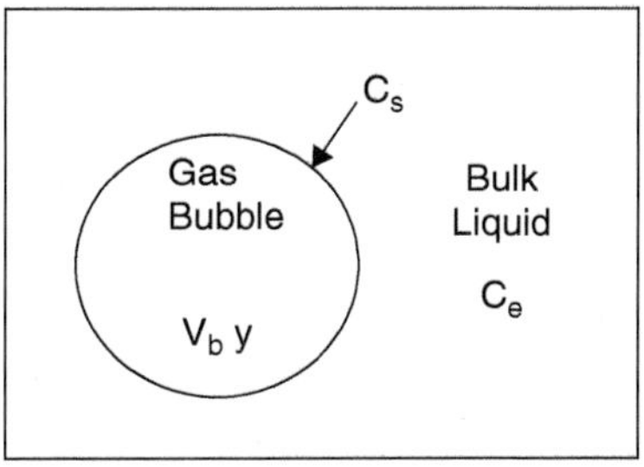

FIGURE 6-14 Schematic of a single bubble inside a bubble aeration tank.

$$y_e = HC_e(1 - e^{-\phi}) \tag{6-60}$$

in which ϕ is equal to the Stanton number, defined by

$$\phi = \frac{K_L a V}{H\dot{V}} \tag{6-61}$$

in which $K_L a$ is the overall mass transfer coefficient for bubble aeration (s^{-1}), V is the volume of a bubble aeration tank (m^3), H is the Henry's law constant (dimensionless), and $\dot{V}$ is the air flow rate (m^3/s).

Substitution of Eq. 6-60 into Eq. 6-58 yields the design equation for bubble aeration:

$$QC_0 - QC_e - \dot{V}HC_e(1 - e^{-\phi}) = 0 \tag{6-62}$$

To design a new bubble aeration tank, Eq. 6-62 can be rearranged and solved for V (note that C_e in Eq. 6-62 represents the treatment objective):

$$V = -\frac{H\dot{V}}{K_L a}\ln\left\{1 - \left[\left(\frac{C_0}{C_e} - 1\right)\left(\frac{1}{\frac{\dot{V}}{Q}H}\right)\right]\right\} \tag{6-63}$$

Recall that the air-to-water ratio, $\dot{V}/Q$ in Eq. 6-63, must be greater than the minimum air-to-water ratio, $(\dot{V}/Q)_{\min}$ defined in Eq. 6-59, for the bubble aeration tank to be able to meet the treatment objective. To estimate the effluent concentrations of various compounds for an existing bubble aeration tank with volume, V, it is necessary to solve Eq. 6-63 for C_e:

$$C_e = \frac{C_0}{1+\frac{\dot{V}}{Q}H(1-e^{-\phi})} \tag{6-64}$$

In either case (new tank or existing tank), the gas-phase effluent concentration from the tank, y_e, can be determined from:

$$y_e = \frac{C_0 - C_e}{\frac{\dot{V}}{Q}} \tag{6-65}$$

Tanks in Series Development. Process performance can often be improved significantly (i.e., lower total volume requirements) by configuring the bubble aeration system as tanks in series. Figure 6-15 shows a schematic of a bubble aeration system operated in a tank in series configuration. Sometimes this is easy to accomplish by placing baffles in the tank. The general case for development of n tanks in series is discussed in Chap. 4. If it is assumed that all the tanks are equally sized and all the assumptions for a single tank system are valid here, the following expression can be used to calculate the effluent concentration for n tanks in series:

$$C_n = \frac{C_{n-1}}{1+\frac{\dot{V}}{Q}H(1-e^{-\phi})} = \frac{C_0}{\left[1+\frac{\dot{V}}{Q}H(1-e^{-\phi})\right]^n} \tag{6-66}$$

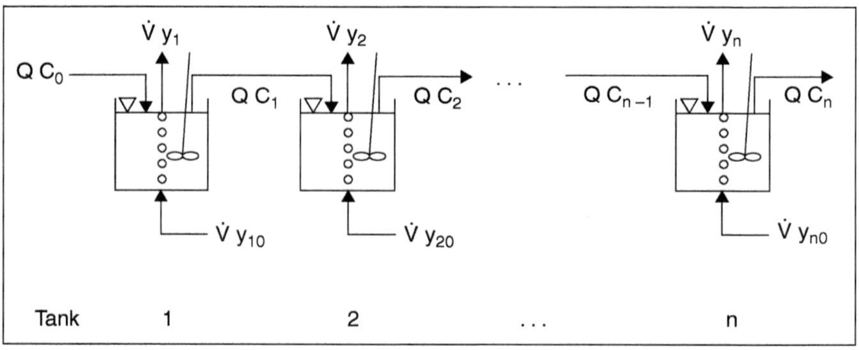

FIGURE 6-15 Schematic of a tanks in series bubble aeration system.

in which C_n is the effluent liquid-phase concentration from tank n. Equation 6-66 is the general expression for the effluent concentration for a bubble aeration system consisting of tanks in series. For the case of designing new tanks in a series systems, both the influent and desired effluent concentrations are known. Consequently, Eq. 6-66 can be rearranged to determine the volume of each individual tank, V_n, using the following equation:

$$V_n = -\frac{H\dot{V}}{K_L a}\ln\left\{1-\left[\left(\frac{C_0}{C_e}\right)^{1/n}-1\right]\left(\frac{1}{\frac{\dot{V}}{Q}H}\right)\right\} \tag{6-67}$$

As for the single-tank case, the air-to-water ratio supplied to each tank must exceed the minimum air-to-water ratio so that the system will be able to meet its treatment objective. The minimum air-to-water ratio for a tanks-in-series configuration can be determined as follows:

$$\left(\frac{\dot{V}}{Q}\right)_{\min} = \left[\left(\frac{C_0}{C_e}\right)^{1/n} - 1\right]\frac{1}{H} \tag{6-68}$$

It is sometimes necessary to determine the gas-phase concentration leaving the tanks. The gas-phase effluent concentration from each tank in series may be calculated from the following:

$$y_n = \frac{C_{n-1} - C_n}{\frac{\dot{V}}{Q}} \tag{6-69}$$

Governing Equations: Gas Absorption

Equations that describe the transfer of compounds from the gas to the water phase are similar to those used to describe stripping in a bubble aeration tank, a process known as absorption. Figure 6-13, the bubble aeration schematic shown previously, is also relevant to this case. The development of the process design equations for absorption using bubble aeration incorporates the same assumptions as for stripping, except that the inlet liquid concentration, rather than the inlet gas phase concentration, is zero. A mass balance around the tank shown in Fig. 6-13 results in the following:

$$-QC_e + \dot{V}y_0 - \dot{V}y_e = 0 \tag{6-70}$$

For sizing a tank, and substituting Eq. 6-60 into Eq. 6-70 and solving it for V, the following expression can be derived:

$$V = -\frac{H \cdot \dot{V}}{K_L a} \ln\left(1 - \left[\frac{1}{H\frac{\dot{V}}{Q}\left(\frac{y_0}{y_e} - 1\right)}\right]\right) \tag{6-71}$$

As for the stripping case, there is a minimum water-to-air ratio that must be exceeded in order for absorption to be viable:

$$\frac{Q}{\dot{V}} \geq \left(\frac{Q}{\dot{V}}\right)_{\min} = H\left(\frac{y_0}{y_e} - 1\right) \tag{6-72}$$

In order to estimate the gas-phase effluent concentrations of various compounds for an existing bubble aeration tank with volume V, it is necessary to solve Eq. 6-70 for C_e, substitute the result into Eq. 6-71, and solve for y_e:

$$y_e = \frac{y_0}{1 + \left[\frac{1}{H\frac{\dot{V}}{Q}(1 - e^{-\phi})}\right]} \tag{6-73}$$

For bubble and surface aeration, it is difficult to determine $K_L a$ for a compound from Eq. 6-21 because the quantities k_l and k_g vary greatly from system to system and are difficult to measure. The general equation for estimating the K_La of a compound in bubble or surface aeration is found by expanding Eq. 6-21 based on oxygen as a reference compound:

$$K_La_i = K_La_{O_2}\left(\frac{D_{l,i}}{D_{l,O_2}}\right)^m\left[1+\frac{1}{H_i\left(\frac{k_g}{k_l}\right)}\right]^{-1} \tag{6-74}$$

where $K_La_{O_2}$ is the overall mass transfer coefficient for oxygen (1/s), $D_{l,i}$ is the liquid diffusivity of component i (m^2/s), and m is the diffusivity exponent (dimensionless). k_g/k_l is the ratio of gas-phase to liquid-phase mass transfer coefficient (dimensionless), which tends to be relatively constant for a given type of aeration system. $D_{l,i}$ can be found from the Hayduk and Laudie (1974) correlation for small molecules in water. D_{l,O_2} is found from a correlation presented in Holmén and Liss (1984):

$$D_{l,O_2} = [10^{(3.15+(-831.0)/T)}](1.0\times10^{-9}) \tag{6-75}$$

where T is the temperature in K. The recommended value for the diffusivity exponent, m, in bubble aeration is 0.6 (Holmén and Liss, 1984). Munz and Roberts (1989) pointed out that a constant value of $k_g/k_l \approx 100$ to 150 is widely used in air-water mass transfer. Therefore, a value of $k_g/k_l = 100$ is recommended as a conservative estimate in bubble aeration. $K_La_{O_2}$ can be found from clean water oxygen transfer test data (Brown and Baillod, 1982; Baillod et al., 1986):

$$K_La_{O_2} = K_La_{20}^*(1.024)^{T-20} \tag{6-76}$$

where T is the temperature in °C, $K_La_{20}^*$ is the true mass transfer coefficient of oxygen at 20°C, and $K_La_{20}^*$ can be determined from the following series of relationships:

$$K_La_{20}^* = \frac{K_La_{20}}{1-\frac{K_La_{20}}{2\phi_O}} \tag{6-77}$$

$$K_La_{20} = \frac{SOTR}{VC_\infty^*} \tag{6-78}$$

$$C_\infty^* = C_s^*\left[\frac{P_b - P_v + \gamma_w d_e}{P_s - P_v}\right] \tag{6-79}$$

$$\phi_O = \frac{M_O\rho_a Q_a}{M_a H_O(P_b + \gamma_w d_e)} \tag{6-80}$$

where ϕ_O is the oxygenation coefficient (1/s), C_∞^* is the dissolved oxygen saturation concentration attained at infinite time (μg/L), SOTR is the standard oxygen transfer rate (kg/s), V is the water volume in the tank (m^3), C_s^* is the dissolved oxygen saturation concentration at 20°C (μg/L), P_b is the barometric pressure (N/m^2), P_s is the barometric pressure under standard conditions (N/m^2), P_v is the vapor pressure of water at 20°C (N/m^2), γ_w is the weight density of water (kg/m^3), d_e is the effective saturation depth (m), which is equal to one third times the water depth, M_O is the molecular weight of oxygen (amu), M_a is the molecular weight of air (amu), Q_a is the volumetric flow rate of air (m^3/s), and H_O is Henry's constant for oxygen (dimensionless). These correlations are valid only for clean waters and should not be used for wastewaters.

The total operating power for bubble aeration is equal to the blower brake power requirement for each tank times the number of tanks times the number of blowers per tank. The blower brake power is calculated from Eq. 6-49. The term P_{in} refers to the pressure at the top of the tank, which represents the sum of the ambient pressure and the head required to raise the water to the inlet of the tank. The equation used to find P_{in} is

$$P_{in,b} = P_{ambient} + \rho_l H_b g \tag{6-81}$$

where the pressure head, H_b, is assumed equal to the water depth in the tank, d_b.

Example 6-4 Bubble Aeration Sample Calculation
A portion of an existing clearwell is to be modified to include a bubble aeration system capable of reducing radon concentration from 6000 pCi/L to less than 300 pCi/L. The total volume of the clearwell is 3750 m^3 (1 Mgal) (depth = 5.0 m, length = 30 m, width = 25 m). The average plant flow rate is 0.22 m^3/s (5 mgd) and the water temperature is 20°C. The chemical and physical properties of radon are summarized below. Tank manufacturer information on clean water oxygen transfer data necessary to determine $K_L a_{O_2}$ is also provided below.

Determine the retention time, air flow rate, and power requirements necessary to meet the treatment objective for a single tank. Also, assume that the clearwell can be modified to perform like three tanks in series. Assume a blower efficiency of 35 percent.

Chemical and physical properties of radon at 20°C.

Property	Value
Molecular Weight (g/mole)	222
D_g (m^2/s)	2.0×10^{-5}
D_l (m^2/s)	1.4×10^{-9}
H (L/L)	4.08

Manufacturer's clean water oxygen transfer test data for a tank with a diffuser/bubble system.

Property	Value
SOTR (kg O_2/day)	1500
Volumetric air flow rate (m^3/hr)	1700
Tank volume (m^3)	500
Tank depth (m)	3

Solution

1. Determine the retention time for a single tank.

$$\tau = \frac{V}{Q} = \frac{3785\text{ m}^3}{\left(0.22\frac{\text{m}^3}{\text{s}}\right)(60\text{ s/min})} = 287\text{ min} = 4.79\text{ hr}$$

2. Find the oxygen mass transfer rate coefficient, $K_L a_{O_2}$.
   a. In the calculation of $K_L a_{O_2}$, the following quantities in addition to the clean water oxygen test data that was given are needed:

      C_s^* Dissolved oxygen surface saturation concentration at 20°C = 9.09 mg/L (see App. D)
      P_b Barometric pressure = 1 atm
      P_s Barometric pressure under standard conditions = 1 atm
      P_v Vapor pressure of water at 20°C = 2.34 kN/m^2 (see App. D
      d_e Effective saturation depth = $\frac{1}{3} \times$ water depth = 1.67 m
      M_O Molecular weight of oxygen (g/mole) = 32.0
      ρ_a Density of air at 20°C = 1240 mg/L
      M_a Molecular weight of air (g/mole) = 28.0
      H_O Henry's constant for oxygen = 50 mg/L/atm

   Using these quantities along with those provided in the problem statement, $K_L a_{O_2}$ is calculated as follows:

b. Calculate C_∞^* using Eq. 6-79.

$$C_\infty^* = C_s^*\left[\frac{P_b - P_v + \gamma_w d_e}{P_s - P_v}\right] = 9.09\frac{\text{mg}}{\text{L}}\left[\frac{1\text{ atm} - \frac{2340\text{ Pa}}{101325.0\frac{\text{Pa}}{\text{atm}}} + \frac{62.3\frac{\text{lb}}{\text{ft}^3}\times 1.67\text{m}\times 3.281\frac{\text{ft}}{\text{m}}}{144\frac{\text{in}^2}{\text{ft}^2}\times 14.696\frac{\text{psi}}{\text{atm}}}}{1\text{ atm} - \frac{2{,}340\text{ Pa}}{101325.0\frac{\text{Pa}}{\text{atm}}}}\right]$$

$$= 10.59\frac{\text{mg}}{\text{L}}$$

c. Calculate K_La_{20} using Eq. 6-78.

$$K_La_{20} = \frac{SOTR}{VC_\infty^*} = \frac{1500\frac{\text{kg}}{\text{d}}\cdot\left(10^6\frac{\text{mg}}{\text{kg}}\right)\left(\frac{1\text{ d}}{86400\text{ s}}\right)}{500\text{ m}^3\cdot 10.59\frac{\text{mg}}{\text{L}}\cdot 1000\frac{\text{L}}{\text{m}^3}} = 3.28\times 10^{-3}\frac{1}{\text{s}}$$

d. Calculate ϕ_O from Eq. 6-80.

$$\phi_O = \frac{M_O\rho_a Q_a}{M_a H_O V_{\text{test}}(P_b + \gamma_w d_e)}$$

$$= \frac{32\times 1240\frac{\text{mg}}{\text{L}}\times\frac{1700\frac{\text{m}^3}{\text{hr}}}{3600\frac{\text{s}}{\text{hr}}}}{28.95\cdot 50\frac{\text{mg}}{\text{L}\cdot\text{atm}}\cdot 500\text{m}^3\left(1\text{ atm} + \frac{62.4\frac{\text{lb}}{\text{ft}^3}\times 1.67\text{ m}\times 3.281\frac{\text{ft}}{\text{m}}}{144\frac{\text{in}^2}{\text{ft}^2}\times 14.696\frac{\text{psi}}{\text{atm}}}\right)}$$

$$= 0.0301\frac{1}{\text{s}}$$

e. Calculate $K_La_{20}^*$ using Eq. 6-77.

$$K_La_{20}^* = \frac{K_La_{20}}{1 - \frac{K_La_{20}}{2\phi_O}} = \frac{3.28\times 10^{-3}\frac{1}{\text{s}}}{1 - \frac{3.28\times 10^{-3}\frac{1}{\text{s}}}{2\left(0.0301\frac{1}{\text{s}}\right)}} = 3.5\times 10^{-3}\frac{1}{\text{s}}$$

f. Calculate $K_La_{O_2}$ using Eq. 6-78.

$$K_La_{O_2} = K_La_{20}^*(1.024)^{T-20} = 3.5\times 10^{-3}\frac{1}{\text{s}}(1.024)^{20-20} = 3.5\times 10^{-3}\frac{1}{\text{s}}$$

3. Calculate the liquid diffusivity of oxygen using the correlation of Holmén and Liss (1984).

$$D_l(\text{O}_2) = 10^{3.15+\frac{-831}{293.15}}\times 10^{-9} = 2.07\times 10^{-9}\frac{\text{m}^2}{\text{s}}$$

4. Calculate the overall mass transfer rate of the radon.

$$K_La_{\text{radon}} = 3.5\times 10^{-3}\frac{1}{\text{s}}\left(\frac{1.4\times 10^{-9}\frac{\text{m}^2}{\text{s}}}{2.07\times 10^{-9}\frac{\text{m}^2}{\text{s}}}\right)^{0.6}\left(1 + \frac{1}{4.08\times 100}\right)^{-1} = 2.76\times 10^{-3}\frac{1}{\text{s}}$$

5. Calculate the air-to-water ratio needed for stripping. First calculate the minimum air-to-water ratio required for stripping using Eq. 6-59: assume a safety factor (SF) of 2.0 to account for the error in estimation of the mass transfer resistance.

$$\left(\frac{\dot{V}}{Q}\right)=\left(\frac{C_0-C_e}{HC_e}\right)_{\min}\times SF=\left(\frac{(6{,}000-300)\frac{\text{pCi}}{\text{L}}}{4.08\times 300\frac{\text{pCi}}{\text{L}}}\right)\times 2.0\approx 10$$

6. Calculate the air flow rate.

$$\dot{V}=\left(\frac{\dot{V}}{Q}\right)\times Q=10\times 0.2191\,\frac{\text{m}^3}{\text{s}}=2.191\,\frac{\text{m}^3}{\text{s}}$$

7. Calculate the Stanton number, ϕ_{radon}, for radon using Eq. 6-61.

$$\phi_{\text{radon}}=\frac{2.76\times 10^{-3}\frac{1}{\text{s}}\times 3785\,\text{m}^3}{2.191\frac{\text{m}^3}{\text{s}}\times 4.08}=1.17$$

8. Calculate the radon effluent concentration, C_e, from the tank using Eq. 6-64.

$$C_{e_{\text{radon}}}=\frac{6{,}000\,\frac{\text{pCi}}{\text{L}}}{1+10\times 4.08\times(1-e^{-1.17})}=206\,\frac{\text{pCi}}{\text{L}}$$

The radon effluent concentration of 206 pCi/L is conservative using a factor of safety of 2.0.

9. Calculate the total blower power required to meet the treatment objective.

a. Calculate the air mass flow rate, G_{me}.

$$G_{me}=2.191\frac{\text{m}^3}{\text{s}}\times 1.24\frac{\text{kg}}{\text{m}^3}=2.72\frac{\text{kg}}{\text{s}}$$

b. Calculate the inlet air pressure to the aeration basin, $P_{\text{in},b}$, using Eq. 6-81.

$$P_{\text{in},b}=101325\,\frac{\text{N}}{\text{m}^2}+998.26\,\frac{\text{kg}}{\text{m}^3}\times 5\,\text{m}\times 9.81\,\frac{\text{m}}{\text{s}^2}=150290\,\frac{\text{N}}{\text{m}^2}$$

c. Calculate the total blower power, P_{blower}, using Eq. 6-49.

$$P_{\text{blower}}=\left[\frac{2.72\frac{\text{kg}}{\text{s}}\times 286.7\frac{\text{J}}{\text{kg}\times\text{K}}\times 293.15\,\text{K}}{1000\frac{\text{W}}{\text{kW}}\times 0.283\times 0.35}\right]\times\left[\left(\frac{150290\frac{\text{N}}{\text{m}^2}}{101325\frac{\text{N}}{\text{m}^2}}\right)^{0.283}-1\right]\times\left(\frac{1\,\text{W}}{1\frac{\text{J}}{\text{s}}}\right)$$

$$=272\,\text{kW}$$

For comparison purposes, assume a portion of the clearwell can be modified to provide a system of three tanks in series with a total retention time of 1 hour.

1. Calculate the total volume if the total retention time is 1 hour.

$$V_n=Q\tau_n=0.2191\,\frac{\text{m}^3}{\text{s}}\times 1\,\text{hr}\times\frac{3600\,\text{s}}{1\,\text{hr}}=789\,\text{m}^3$$

2. Determine the volume per tank, V.

$$V=\frac{V_n}{\#\text{ of tanks}}=\frac{789\,\text{m}^3}{3\text{ tanks}}=266\,\text{m}^3$$

Assume the same tank depth as the single tank and assume that mass transfer coefficients do not change for tanks in series. The removal efficiency after the third tank can be determined as shown in the following calculations.

3. Determine the minimum air-to-water ratio for the system of three tanks in series using Eq. 6-70. Assume a radon effluent concentration of 300 pCi/L and a safety factor of 2.0 for the air-to-water ratio.

$$\left(\frac{\dot{V}}{Q}\right)_{\min} = \left[\left(\frac{C_0}{C_e}\right)^{1/n} - 1\right]\frac{1}{H}\times SF = \left[\left(\frac{6{,}000\frac{\text{pCi}}{\text{L}}}{300\frac{\text{pCi}}{\text{L}}}\right)^{1/3} - 1\right]\frac{1}{4.08}\times 2.0 = 0.840$$

4. Determine the air flow rate for each tank.

$$\dot{V} = \left(\frac{\dot{V}}{Q}\right)_{\min} \times Q = 0.84 \times 0.2191\frac{\text{m}^3}{\text{s}} = 0.184\frac{\text{m}^3}{\text{s}}$$

5. Calculate the Stanton number, ϕ, for radon.

$$\phi_{\text{radon}} = \frac{2.76\times 10^{-3}\frac{1}{\text{s}}\times 266\,\text{m}^3}{0.184\frac{\text{m}^3}{\text{s}}\times 4.08} = 0.98$$

6. Determine the total effluent radon concentration from the tank, C_3.

$$C_{3_{Cl_3CH}} = \frac{6{,}000\,\frac{\text{pCi}}{\text{L}}}{[1+0.84\times 4.08\times(1-e^{-0.98})]^3} = 194\,\frac{\text{pCi}}{\text{L}}$$

The effluent concentration for the three tanks in series design meets the radon treatment objective of 300 pCi/L.

7. Determine the required blower power for the three tanks in series.

a. Find the air mass flow rate, G_{me}.

$$G_{me} = .184\,\frac{\text{m}^3}{\text{s}}\times 1.24\,\frac{\text{kg}}{\text{m}^3} = 0.23\,\frac{\text{kg}}{\text{s}}$$

b. Calculate the inlet air pressure to the aeration basin, $P_{\text{in},b}$.

$$P_{\text{in},b} = 101325\,\frac{\text{N}}{\text{m}^2} + 998.26\,\frac{\text{kg}}{\text{m}^3}\times 5\,\text{m}\times 9.81\,\frac{\text{m}}{\text{s}^2} = 150290\,\frac{\text{N}}{\text{m}^2}$$

c. Calculate the total blower power, P_{blower}.

$$P_{\text{blower}} = \left[\frac{0.23\frac{\text{kg}}{\text{s}}\times 286.7\frac{\text{J}}{\text{kg}\times\text{K}}\times 293.15\,K}{1000\frac{\text{W}}{\text{kW}}\times 0.283\times 0.35}\right]\times\left[\left(\frac{150290\frac{\text{N}}{\text{m}^2}}{101325\frac{\text{N}}{\text{m}^2}}\right)^{0.283} - 1\right]\times\left(\frac{1\,\text{W}}{1\frac{\text{J}}{\text{s}}}\right)$$

$$= 23\,\text{kW per tank}$$

The total blower power required for three tanks in series with a retention time of 1 hour is 69 kW. The power consumption for the single tank is 272 kW, which is significantly larger than the three-tank design. ▲

Example 6-5 Ozone Absorption Problem

A diffused ozone system with a retention time of 15 minutes is set up to apply ozone to surface water at a dosage of 3.0 mg/L. Air is used to generate the ozone at a generation rate of 0.02 mg O_3/mg O_2. The water flow rate is 0.32 m^3/s (7.2 mgd). The temperature of the system (water and air) is 23°C and the pressure is 1 atm. Ozone properties are provided below. The tank manufacturer has provided the clean water oxygen transfer test data needed to estimate $K_L a_{O_2}$. This information is shown below.

Assuming that the decay rate of ozone in the water is negligible, determine the effluent concentration of the ozone in the exiting gas, y_e, and in the water. (Note that one can account for the decay rate by incorporating a first-order reaction term in Eq. 6-58 and solving).

Properties of Ozone

Property	Value
Molecular weight (mg/mmol)	48.0
Henry's constant (–)	3.77
Liquid diffusivity (m^2/s)	1.7×10^{-9}

Manufacturer's Clean Water Oxygen Transfer Test Data for a Tank

Property	Value
SOTR (kg O_2/day)	1500
Volumetric air-flow rate (m^3/hr)	1700
Test tank volume (m^3)	500
Tank depth (m)	3

Solution

1. Calculate the air flow rate, $\dot{V}$.

$$\dot{V} = \frac{Q \cdot \text{dosage}}{\rho_a \cdot (\%\text{O}_3 \text{ in Air})} = \frac{\left(0.32\frac{\text{m}^3}{\text{s}} \times \frac{1000\text{ L}}{\text{m}^3}\right) \cdot 3\frac{\text{mg O}_3}{\text{L}}}{1.19\frac{\text{kg air}}{\text{m}^3\text{ air}} \cdot \left[\left(0.02\frac{\text{mg O}_3}{\text{mg O}_2}\right)\left(0.21\frac{\text{mg O}_2}{\text{mg air}}\right)\right] \cdot \left(10^6\frac{\text{mg air}}{\text{kg air}}\right)} = 0.19\frac{\text{m}^3}{\text{s}}$$

2. Find the volume of the tank, V.

$$V = \tau \cdot Q = (15\text{ min})\left(0.32\frac{\text{m}^3}{\text{s}} \times \frac{60\text{ s}}{\text{min}}\right) = 288\text{ m}^3$$

3. Find the oxygen mass transfer coefficient, $K_L a_{O_2}$.
(See Steps 2a through 2e in previous problem.)

a. Find $K_L a_{O_2}$.

$$K_L a_{O_2} = K_L a_{20}^{*}(1.024)^{T-20} = 0.00408\frac{1}{\text{s}}(1.024)^{23-20} = 0.00438\frac{1}{\text{s}}$$

4. Find the liquid diffusivities of oxygen and ozone.

a. Calculate the liquid diffusivity of oxygen using the correlation of Holmén and Liss (1984).

$$D_l(\text{O}_2) = 10^{\left(3.15 - \frac{831}{296\text{K}}\right)} \times \left(1.0 \times 10^{-9}\frac{\text{m}^2}{\text{s}}\right) = 2.20 \times 10^{-9}\frac{\text{m}^2}{\text{s}}$$

5. Find the overall mass transfer coefficient for ozone, $K_L a_{O_3}$.

$$K_L a_{O_3} = K_L a_{O_2}\left(\frac{D_{l,O_3}}{D_{l,O_2}}\right)^{0.6}\left[1 + \frac{1}{H_i(100)}\right]^{-1}$$

$$= 0.00438\frac{1}{\text{s}}\left(\frac{1.7 \times 10^{-9}}{2.2 \times 10^{-9}}\right)^{0.6}\left[1 + \frac{1}{3.77(100)}\right]^{-1} = 3.7 \times 10^{-3}\frac{1}{\text{s}}$$

6. Calculate y_e.

$$y_e = \frac{y_0}{1 + \left[\frac{1}{H\frac{\dot{V}}{Q}(1 - e^{-\phi})}\right]} = \frac{3\frac{\text{mg}}{\text{L}}}{1 + \left\{1/\left[3.77\left(\frac{0.189}{0.315}\right)\left(1 - e^{-\frac{0.0037\frac{1}{\text{s}} \cdot 284\text{ m}^3}{3.77 \cdot 0.189\frac{\text{m}^3}{\text{s}}}}\right)\right]\right\}} = 1.75\frac{\text{mg}}{\text{L}}$$

7. Calculate C_e assuming no ozone reaction.

$$C_e = \frac{\dot{V}}{Q}(y_0 - y_e) = \left(\frac{0.19}{0.32}\right)(3-1.90)\ \frac{\text{mg}}{\text{L}} = 0.65\ \frac{\text{mg}}{\text{L}} \quad \blacktriangle$$

SURFACE AERATION

Surface aeration has been used primarily for oxygen absorption and the stripping of gases and volatile contaminants when the required removals are less than about 90 percent. Surface aeration devices consist of the brush or turbine types shown in Fig. 6-16. The brush-type aerator consists of several brushes attached to a rotary drum that is half submerged in water in the center of the tank. As the drum rotates it disperses the water into the surrounding air, providing reasonable contact between the air and water for mass transfer to take place. The turbine-type aerator consists of a submerged propeller system located in the center of the tank and surrounded by a draft tube. As the submerged propeller rotates it draws water from outside the draft tube through the inner section and into the air, creating contact between the air and water. These types of systems have been extensively used in the aeration of wastewater, and their design and operation have been well documented (WEF-ASCE, 1991).

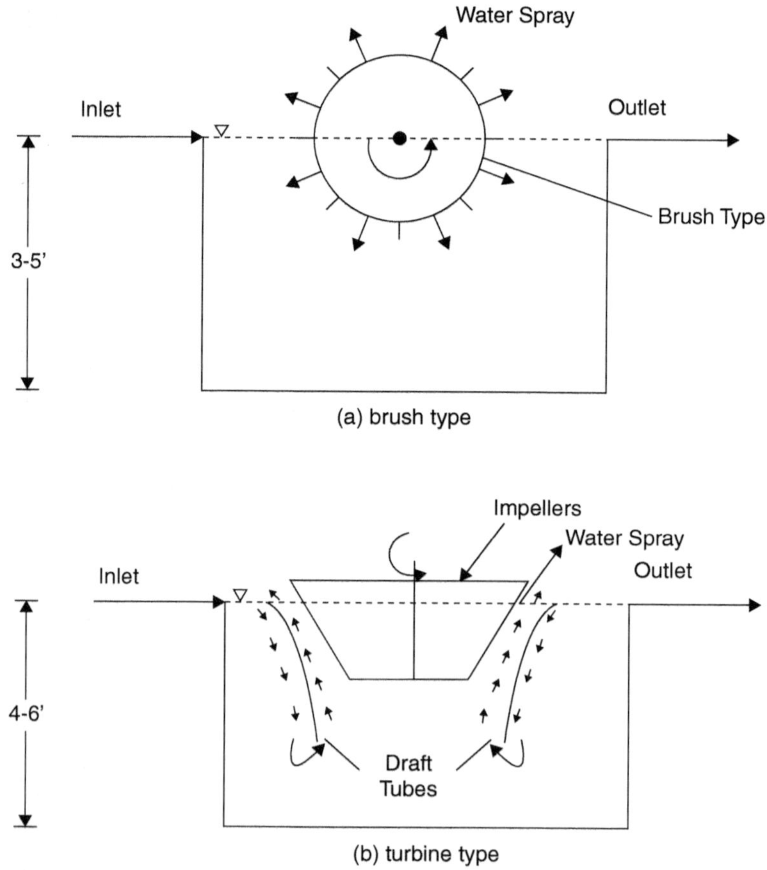

FIGURE 6-16 Schematic of various surface aeration devices.

Governing Equations

Figure 6-17 displays a schematic of a surface aeration system for a single tank. Model development for air stripping using surface aeration has been described in the literature (Matter-Müller, Gujer, and Giger, 1981; Roberts and Dändliker, 1983; Roberts, Munz, and Dändliker, 1984; Roberts and Levy, 1985; Munz and Roberts, 1989). The development of the process design equations incorporate the following assumptions: (1) the tank is completely mixed in the liquid phase, (2) the gas-phase concentration of the contaminant in the tank equals zero, (3) the process is at steady state, and (4) Henry's constant applies. Incorporating these assumptions into an overall mass balance around the tank shown in Fig. 6-17 yields the following expression (Hokanson, 1996):

$$QC_o - QC_e - K_L a C_e V = 0 \tag{6-82}$$

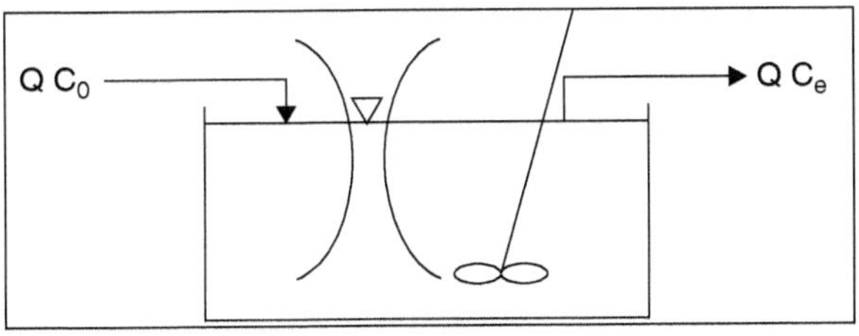

FIGURE 6-17 Schematic of a mechanical surface aerator used for the development of the design equation for a single tank.

Solving Eq. 6-82 for V results in the minimum tank volume required for a single tank to meet the treatment objective, C_e:

$$V = \frac{Q(C_o - C_e)}{K_L a C_e} \tag{6-83}$$

Alternatively, Eq. 6-83 can be expressed in terms of the hydraulic retention time, τ, where τ is equal to V/Q. Substituting this relationship into Eq. 6-83 gives:

$$\tau = \frac{(C_o - C_e)}{K_L a C_e} \tag{6-84}$$

If the removal of a given contaminant is required for an existing tank, Eq. 6-83 or Eq. 6-84 can be solved for C_e to yield:

$$C_e = \frac{C_o}{1 + K_L a\left(\frac{V}{Q}\right)} = \frac{C_o}{1 + K_L a \tau} \tag{6-85}$$

The process performance can be improved significantly with less total volume requirement by configuring the surface aeration system as tanks in series. Figure 6-18 shows a schematic of a surface aeration system operated with a tanks-in-series configuration. From the tanks-in-series development presented in Chap. 4, if it is assumed that all the tanks are equally sized and the assumptions for a single-tank system are also valid here, the effluent concentration for n tanks in series using surface aeration is expressed as:

$$C_n = \frac{C_{n-1}}{1 + K_L a\left(\frac{V}{Q}\right)} = \frac{C_o}{\left[1 + K_L a\left(\frac{V}{Q}\right)\right]^n} \tag{6-86}$$

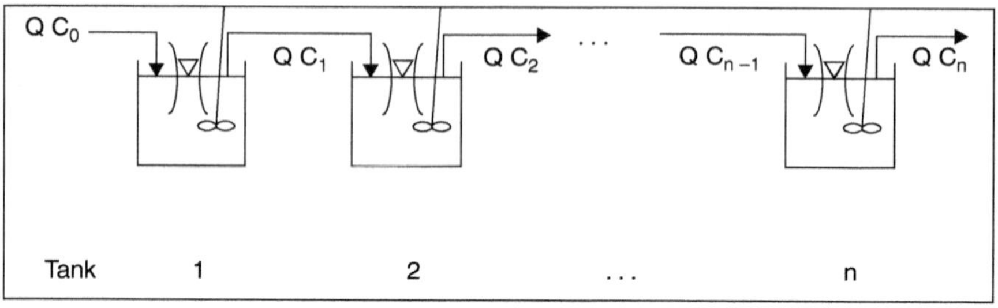

FIGURE 6-18 Schematic of tanks in a series surface aeration system.

in which C_n is the effluent contaminant concentration from tank n and C_{n-1} is the effluent concentration from tank $n-1$.

Equation 6-86 is the general expression for the effluent concentration for a surface aeration system consisting of tanks in series. Solving Eq. 6-86 in terms of the volume for each individual tank yields:

$$V = \frac{Q}{K_L a}\left[\left(\frac{C_o}{C_n}\right)^{1/n} - 1\right] \tag{6-87}$$

Substitution of τQ for V into Eq. 6-87 yields the hydraulic retention time, τ, for each tank in series:

$$\tau = \frac{1}{K_L a}\left[\left(\frac{C_o}{C_n}\right)^{1/n} - 1\right] \tag{6-88}$$

The total volume for all tanks, V_n, can be obtained by multiplying both sides of Eq. 6-88 by n:

$$V_n = V \times n = \frac{nQ}{K_L a}\left[\left(\frac{C_o}{C_n}\right)^{1/n} - 1\right] \tag{6-89}$$

Likewise, the total hydraulic retention time, τ_n, for all tanks can be obtained by multiplying both sides of Eq. 6-89 by n:

$$\tau_n = \tau \times n = \frac{n}{K_L a}\left[\left(\frac{C_o}{C_n}\right)^{1/n} - 1\right] \tag{6-90}$$

Because the quantities V, V_n, τ, and τ_n are not independent, calculation of one of these variables from the above equations allows for determination of the other three parameters because the flow rate, Q, and the number of tanks are specified.

The $K_L a$ for surface aeration can be calculated using Eq. 6-74. A value of 0.5 is recommended for the diffusivity exponent, m, and a value of 40 is recommended for the ratio k_g/k_l (Munz and Roberts, 1989). The overall mass transfer coefficient for oxygen can be estimated using the following correlation (Roberts and Dändliker, 1983):

$$K_L a_{O_2} = 2.9 \times 10^{-5}\left(\frac{P}{V}\right)^{0.95} \tag{6-91}$$

where P/V is the power input per unit tank volume (W/m^3). Equation 6-91 is valid for P/V values between 10 and 200 W/m^3, and it is assumed the reactor is completely mixed. The liquid diffusivity of oxygen can be calculated using the correlation of Hayduk and Laudie (1974).

The total power required for surface aeration is estimated based on the power supplied to the aerators. The following equation can be used to calculate the power requirements:

$$P_{\text{surface}} = \left(\frac{P}{V}\right)\frac{V \times n_{\text{tanks}}}{1000\,\text{Eff}_s} \tag{6-92}$$

in which Eff_s is the surface aerator motor efficiency.

Example 6-6 Surface Aeration
An existing clearwell is to be modified to include a surface aeration system capable of reducing radon concentration from 3000 pCi/L to less than 300 pCi/L. The total volume of the clearwell is 3785 m^3 (1 Mgal) (depth = 5.0 m, length = 30 m, width = 25 m). The average plant flow rate is 0.22 m^3/s (5 mgd) and the water temperature is 20°C. The chemical and physical properties of radon are summarized below.

Determine the power input per unit tank volume and the total power input for aeration of the whole clearwell. Assuming the clearwell can be modified into individual tanks in series, determine the power input per unit tank volume and the total power input for aeration assuming a total retention time of 1.5 hours. Assume the blower efficiency is 30 percent.

Chemical and Physical Properties of Radon at 20°C.

Property	Value
Molecular weight (g/mole)	222
D_g (m^2/s)	2.0×10^{-5}
D_l (m^2/s)	1.4×10^{-9}
H (–)	4.08

Solution

1. Determine the overall mass transfer coefficients for radon by first assuming a power input per unit volume of 35 W/m^3.
   a. Calculate the liquid diffusivity of oxygen using the correlation of Holmén and Liss (1984).

$$D_l(\text{O}_2) = 10^{\left(3.15 - \frac{831}{293^\circ K}\right)} \times \left(1.0 \times 10^{-9}\,\frac{\text{m}^2}{\text{s}}\right) = 2.07 \times 10^{-9}\,\frac{\text{m}^2}{\text{s}}$$

   b. Calculate K_La (O$_2$) using Eq. 6-91,

$$K_La(\text{O}_2) = 2.9 \times 10^{-5} \times \left(\frac{P}{V}\right)^{0.95} = 2.9 \times 10^{-5} \times \left(35\,\frac{\text{W}}{\text{m}^3}\right)^{0.95} = 8.5 \times 10^{-4}\,\text{s}^{-1}$$

   c. Calculate K_La for radon using Eq. 6-90.

$$K_La(\text{radon}) = 8.5 \times 10^{-4}\,\text{s}^{-1}\left(\frac{1.4 \times 10^{-9}\,\frac{\text{m}^2}{2}}{2.07 \times 10^{-9}\,\frac{\text{m}^2}{\text{s}}}\right)^{0.6}\left[1 + \frac{1}{(4.08)(40)}\right]^{-1} = 6.7 \times 10^{-4}\,\text{s}^{-1}$$

2. Calculate the surface aeration removal efficiency using Eq. 6-85.
Volume of clearwell = 3785 m^3
Average flow = 0.22 m^3/s

$$C_e(\text{radon}) = \frac{3000\,\frac{\text{pCi}}{\text{L}}}{1+(6.7\times10^{-4}\,\text{s}^{-1})\left(\frac{3785\,\text{m}^3}{0.22\frac{\text{m}^3}{\text{s}}}\right)} = 239\,\frac{\text{pCi}}{\text{L}}$$

For a *P*/*V* of 35 W/m^3, the radon effluent concentration is less than the treatment objective of 300 pCi/L. The above calculation can be repeated using a slightly lower *P*/*V* value, but the solution is conservative.

3. Calculate the total power requirements for one tank using Eq. 6-92.

$$P = \left(35\frac{\text{W}}{\text{m}^3}\right)\left(\frac{3{,}785\,\text{m}^3}{0.30}\right)\left(\frac{\text{kW}}{1000\,\text{W}}\right) = 442\,\text{kW}$$

4. Calculate the *P*/*V* requirement for the system of three tanks in series using Eq. 6-92 and assuming a total retention time of 1.5 hours (90 minutes).

$$\text{Assume } \frac{P}{V} = 35\frac{\text{W}}{\text{m}^3}$$

$$C_e(\text{radon}) = \frac{3000\,\frac{\text{pCi}}{\text{L}}}{\left[1+6.7\times10^{-4}(30\text{ min})\left(\frac{60\,s}{\text{min}}\right)\right]^3} = 279\,\frac{\text{pCi}}{\text{L}}$$

5. Converting a portion of the clearwell volume into three tanks of equal size in series yields an effluent radon concentration equal to 279 pCi/L. The total power requirement can be calculated as:

$$\left\{\begin{matrix}\text{Total volume of}\\ \text{the 3 tank system}\end{matrix}\right\} = \left(0.22\frac{\text{m}^3}{\text{s}}\right)\left(\frac{3600\text{s}}{\text{hr}}\right)(1.5\text{hr}) = 1188\,\text{m}^3$$

$$\left\{\begin{matrix}\text{Total power requirement}\\ \text{for the 3 tank system}\end{matrix}\right\} = \left(35\frac{\text{W}}{\text{m}^3}\right)\left(\frac{1188\,\text{m}^3}{0.3}\right)\left(\frac{1\text{kW}}{1000\,\text{W}}\right) = 139\,\text{kW}$$

A modification of a small portion of the clearwell to incorporate the three tanks in series system as opposed to aeration of the whole clearwell could reduce the power costs by about 69 percent. ▲

SPRAY AERATORS

Spray aerators have been used in water treatment for many years to oxygenate groundwater in order to remove iron and manganese and for air stripping of gases (i.e., carbon dioxide, hydrogen sulfide) and VOCs. Effective iron oxidation by aeration usually requires at least 1 hour of retention time after aeration. Manganese oxidation by aeration is slow and not practical for waters with pH values below 9.5. Manganese removal usually requires a stronger oxidant. Carbon dioxide and hydrogen sulfide removals have ranged from 50 to 90 percent depending upon the water's pH. Chapter 7 discusses oxidation processes in more detail. VOC removals have been as high as 90 percent depending upon the Henry's law constant.

Spray aerator systems consist of a series of fixed nozzles on a pipe grid. The grids can be placed in towers, commonly known as spray towers or fountains, that spray onto the surface of raw water reservoirs. Pressurized nozzles disperse fine water droplets into the surrounding air, creating a large air-to-water surface for mass transfer. Two types of pressurized spray nozzles, hollow and full cone, are commonly used in water treatment. Full-cone nozzles deliver a uniform spray pattern of droplets. The hollow-cone nozzle delivers a circular spray pattern with most of the droplets concentrated at the circumference. The hollow-cone nozzle is generally preferred over the full-cone type because it provides smaller droplets for better mass transfer, even though it has a larger pressure-drop requirement. Hollow-cone spray droplets are around 5 mm and are prone to plugging. It is recommended that in-line strainers be installed in the spray nozzle manifold to prevent plugging.

The fountain-type spray aerators have been used more widely in water treatment because they can be easily adapted to existing water treatment systems. The design approach and application of the fountain type is presented below. The design approach and application for spray towers is covered elsewhere (Crittenden et al., 2005).

Governing Equations

Figure 6-19 displays a schematic of a single fountain-type spray aerator. A design equation can be derived for estimating the removal of a volatile contaminant by performing a mass balance on a water drop while it is exposed to the air. If it is assumed that all the water drops are the same size and have the same exposure time in the air, the following mass balance can be performed:

$$\left\{\begin{array}{l}\text{Mass lost from}\\ \text{the water drop}\\ \text{per unit time}\end{array}\right\} = \left\{\begin{array}{l}\text{Mass transferred across}\\ \text{the air/water interface}\\ \text{of a water drop}\\ \text{per unit time}\end{array}\right\} \tag{6-93}$$

$$V_d \frac{dC}{dt} = K_L a[C(t) - C_s(t)]V_d \tag{6-94}$$

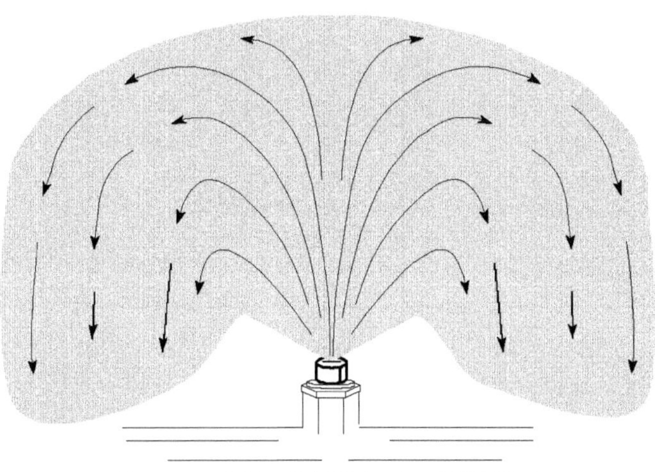

FIGURE 6-19 Schematic of a single fountain spray aerator.

in which V_d is equal to the volume of the drop (m^3). In spray aeration, it is assumed that the gas-phase contaminant concentration in the open air is zero. As a result of this assumption, C_s is also assumed to be zero because it is in equilibrium with the contaminant gas-phase concentration. Rearranging Eq. 6-94 and integrating over the time the drop is exposed to the air yields the following expression for the final contaminant concentration of the water drop after being exposed to the air:

$$C_e = C_o e^{-K_L a t} \tag{6-95}$$

in which C_o is the initial contaminant liquid-phase concentration (μg/L) of the drop before being exposed to the air and t is the time of contact between the water drops and the air and is dependent upon the water drop exiting velocity and trajectory. t is estimated from the following equations:

$$t = \frac{2v_d \sin\alpha}{g} \tag{6-96}$$

where α is the angle of spray measured from the horizontal (degrees), g is the acceleration due to gravity (m/s^2), and v_d is the exit velocity of the water drop from the nozzle (m/s):

$$v_d = C_v\sqrt{2gh} \tag{6-97}$$

where C_v is the velocity coefficient of the nozzle (–) and h is the total head of the nozzle (m). C_v varies from 0.4 to 0.95 and can be obtained from the nozzle manufacturer. The flow rate Q through a given nozzle is estimated from the following equation:

$$Q = C_d A\sqrt{2gh} \tag{6-98}$$

where A is the area of the nozzle opening (m^2) and C_d is the coefficient of discharge from the nozzle (dimensionless).

The overall mass transfer coefficient, K_L, is estimated from the following correlations (Higbie, 1935; Jury, 1967):

$$\text{For } \frac{2D_l t^{1/2}}{d_p} < 0.22 \quad K_L = 2\left(\frac{D_l}{\pi t}\right)^{1/2} \tag{6-99}$$

$$\text{For } \frac{2D_l t^{1/2}}{d_p} > 0.22 \quad K_L = \frac{10 D_l}{d_p} \tag{6-100}$$

where d_p is the sauter mean diameter (SMD) of the water drop (in.) (SMD is obtained by dividing the total volume of the spray by the total surface area), D_l is the contaminant liquid diffusivity (m^2/s), and t is the contact time of the water drop with the air (s).

The interfacial surface area available for mass transfer for the water drop is obtained from the following equation:

$$a = \frac{6}{d_p} \tag{6-101}$$

Example 6-7 Spray Aeration
Well water contains 1.0 mg/L of total sulfide at a pH of 6.2. The well pumps 0.063 m^3/s (1000 gpm) and the pump has the capacity to provide 28 m (91.8 ft) of head. Determine the number of nozzles required and the expected H_2S removal efficiency. It is assumed

that the water has sufficient buffer capacity to maintain the pH at 6.2 during aeration. The following information was obtained from the nozzle manufacturer: SMD = 0.10 cm, $\alpha = 90°$, $C_v = 0.45$, $C_d = 0.1$, nozzle diameter = 1.25 cm.

Solution

1. Determine the number of nozzles required.

$$A_{\text{nozzle}} = \frac{\pi}{4}\left(\frac{1.25\,\text{cm}}{100\,\frac{\text{cm}}{\text{m}}}\right)^2 = 1.23\times10^{-4}\,\text{m}^2$$

The flow rate through one nozzle, Q_{nozzle}, can be calculated using Eq. (6-98):

$$Q_{\text{nozzle}} = C_d A_{\text{nozzle}}\sqrt{2gh} = 0.25(1.23\times10^{-4}\,\text{m}^2)\sqrt{2\left(9.81\frac{\text{m}}{\text{s}^2}\right)(28\,\text{m})} = 7.21\times10^{-4}\,\frac{\text{m}^3}{\text{s}}$$

The number of nozzles can be calculated by dividing the total flow by the flow through each nozzle.

$$\left\{\begin{matrix}\text{Number of}\\ \text{nozzles}\\ \text{required}\end{matrix}\right\} = \frac{Q}{Q_{\text{nozzle}}} = \frac{0.063\,\frac{\text{m}^3}{\text{s}}}{7.21\times10^{-4}\,\frac{\text{m}^3}{\text{s}}} = 88$$

$$K_L = 2\left(\frac{D_l}{\pi t}\right)^{1/2} = 2\left(\frac{1.5\times10^{-5}\,\frac{\text{cm}^2}{\text{s}}}{\pi(2.14\,\text{s})}\right)^{1/2} = 0.0031\frac{\text{cm}}{\text{s}}$$

2. Determine the H_2S removal efficiency.
Equation 6-97 is used to calculate the velocity of the water exiting the nozzle.

$$v_d = C_v\sqrt{2gh} = 0.45\sqrt{2\left(9.81\frac{\text{m}}{\text{s}^2}\right)(28\,\text{m})} = 10.6\,\frac{\text{m}}{\text{s}}$$

The contact time of the water drop with the air can be calculated using Eq. 6-96.

$$t = \frac{2v_d\sin\alpha}{g} = \frac{2(10.6\frac{\text{m}}{\text{s}})\sin(90°)}{9.81\frac{\text{m}}{\text{s}^2}} = 2.16\,\text{s}$$

The overall liquid-phase mass transfer coefficient is calculated using Eq. 6-99.

$$\frac{2D_l t^{1/2}}{d_p} = \frac{2(1.6\times10^{-5}\,\frac{\text{cm}^2}{\text{s}})(2.14\,\text{s})^{1/2}}{0.10\,\text{cm}} = 4.68\times10^{-4}$$

$$4.68\times10^{-4} \ll 0.22$$

The interfacial area for mass transfer is calculated using Eq. 6-101.

$$a = \frac{6}{d_p} = \frac{6}{0.1\,\text{cm}} = 60\,\text{cm}^{-1}$$

Calculate $K_L a$.

$$K_L a = \left(0.0031\frac{\text{cm}}{\text{s}}\right)(60\,\text{cm}^{-1}) = 0.19\,\text{s}^{-1}$$

The H_2S reacts with water to form both HS^- and $S^=$, as shown by the following equations:

$$H_2S + H_2O \text{ Û } H_3O^+ + HS^- \quad K_1 = 1.0\times10^{-6.99}\,\frac{\text{mol}}{\text{L}}\ (25^\circ\text{C})$$

$$HS^- + H_2O \text{Û} H_3O^+ + S^{2-} \quad K_2 = 1.0\times10^{-12.9}\,\frac{\text{mol}}{\text{L}}\ (25^\circ\text{C})$$

On the gas side, the only sulfide species is hydrogen sulfide, H_2S. On the liquid side, sulfide can exist as two species: H_2S or HS^-. Therefore, the following general relationship applies on the liquid side:

$$C_S = C_{\text{H}_2\text{S}} + C_{\text{HS}^-}$$

in which C_S is the total sulfide concentration in the liquid phase (μg/L) and is equal to the sum of $C_{\text{H}_2\text{S}}$ and C_{HS^-}. Conveniently, $C_{\text{H}_2\text{S}}$ can be related to C_S by making use of the ionization fraction, α_0 (see Chap. 3 for details), which is defined as

$$\alpha_0 = \frac{C_{\text{H}_2\text{S}}}{C_S} = \left(1 + \frac{K_{a1}}{[H^+]} + \frac{K_{a1}K_{a2}}{[H^+]^2}\right)^{-1}$$

If Eq. 6-98 is rewritten in terms of total sulfide and integrated using the same assumptions, the following expression to the effluent liquid-phase concentration of hydrogen sulfide can be written as

$$C_{Se} = C_{So} e^{(-\alpha_o K_L a_{\text{H}_2\text{S}} t)}$$

in which C_{Se} is the effluent liquid-phase concentration of total sulfide after spray aeration, C_{Si} is the influent total liquid-phase sulfide concentration, and $K_L a_{\text{H}_2\text{S}}$ is the overall mass transfer coefficient for hydrogen sulfide. The first ionization constant, α_o, is calculated as

$$\alpha_0 = \frac{C_{\text{H}_2\text{S}}}{C_S} = \left(1 + \frac{10^{-6.99}}{[10^{-6.2}]} + \frac{(10^{-6.99})(10^{-18.2})}{[10^{-6.2}]^2}\right)^{-1} = 0.860$$

If the influent H_2S concentration is 1.0 mg/L, then the total initial sulfide concentration is

$$C_{So} = \frac{C_{(\text{H}_2\text{S})_o}}{\alpha_o} = \frac{1.0\,\frac{\text{mg}}{\text{L}}}{0.860} = 1.16\,\frac{\text{mg}}{\text{L}}$$

The effluent total liquid-phase sulfide concentration after stripping is calculated as:

$$C_{Se} = 1.16\,\frac{\text{mg}}{\text{L}} e^{(-0.860\times0.19\text{s}^{-1}\times2.14\text{s})} = 0.818\,\frac{\text{mg}}{\text{L}}$$

The effluent liquid-phase hydrogen sulfide concentration is:

$$C_{e,\text{H}_2\text{S}} = \alpha_o \times C_{Se} = (0.860)\times\left(0.818\,\frac{\text{mg}}{\text{L}}\right) = 0.703\,\frac{\text{mg}}{\text{L}}$$

The H_2S removal due to stripping is calculated as

$$\begin{Bmatrix}\text{H}_2\text{S removal} \\ \text{due to stripping}\end{Bmatrix} = \frac{1.0\,\frac{\text{mg}}{\text{L}} - 0.703\,\frac{\text{mg}}{\text{L}}}{0.1.0\,\frac{\text{mg}}{\text{L}}} \times 100 = 29.7\%$$ ▲

NOTATION FOR EQUATIONS

a	interfacial area available for mass transfer (m^2/m^3)
a_{air}	activity of component A in the gas-phase (dimensionless)
a_{aq}	activity of component A in the aqueous phase (dimensionless)
a_t	total specific surface area of the packing material (m^2/m^3)
a_w	wetted surface area of the packing material (m^2/m^3)
$a_0\ a_1\ a_2$	calculated empirical parameters (dimensionless)
A	cross sectional area (m^2)
B	microporosity constant (cm^6/cal^2)
C_b	bulk aqueous phase concentration (mg/L or mole/L)
C_e	effluent aqueous phase concentration (mg/L or mole/L)
C_f	packing factor (1/m)
C_i	aqueous phase concentration (mg/L or mole/L)
C_s	aqueous phase concentration at the air/water interface (mg/L or mole/L)
$C_{s,A}$	aqueous solubility of component A (mg/L)
C_s^*	aqueous phase concentration of A at the air/water interface assuming no concentration gradient in the gas-phase (mg/L or mole/L)
C_o^*	liquid phase concentration at the air/water interface in equilibrium with the exiting gas-phase concentration at the top of the tower (mg/L or mole/L)
D	tower diameter (m)
d_e	effective depth at saturation (m)
D_g	gas-phase diffusivity (m^2/s)
D_l	liquid-phase diffusivity (m^2/s)
d_p	nominal diameter of the packing material (m)
E	calculated empirical parameter (dimensionless)
Eff_b	blower efficiency (dimensionless)
F	calculated empirical parameter (dimensionless)
g	gravitational constant (9.81 m/s^2)
G_m	air mass loading rate ($kg/m^2{\cdot}s$)
H	Henry's law constant of component A (atm-L/mole)
H_o	Henry's law constant for oxygen (atm·mole/L)
H_{app}	apparent Henry's law constant (atm-L/mole)
HTU	height of transfer unit (dimensionless)
I	ionic strength (mole/L or mg/L)
J_A	mass flux of component A ($kg/m^2{\cdot}s$)
k_l	liquid-phase mass transfer coefficient (m/s)
k_g	gas-phase mass transfer coefficient (m/s)
k_p	constant (275 $N{\cdot}s/m^4$)
K	Boltzmann's constant, (joule/K)

K_{eq}	equilibrium constant (dimensionless)
K_G	overall mass transfer coefficient (m/s)
K_Ga	overall mass transfer coefficient (1/s)
K_L	overall mass transfer coefficient (m/s)
K_La	overall mass transfer coefficient (1/s)
$K_La_{O_2}$	overall mass transfer coefficient for oxygen (1/s)
K_s	Setschenow constant (L/mole)
L	depth of packing in tower (m)
L_m	water mass loading rate (kg/m^2·s)
m	diffusivity exponent (dimensionless).
M	molecular weight of the VOC (gm/mole)
M_a	molecular weight of air (gm/mole)
M_o	molecular weight of oxygen (gm/mole)
n_a	constant in determining blower brake power (dimensionless)
NTU	number of transfer units (dimensionless)
P	partial pressure (N/m^2)
$P_{ambient}$	ambient air pressure (N/m^2)
P_A	pressure A exerts in the gas-phase (atm)
P_b	barometric pressure (atm)
P_{blower}	blower brake power (kW)
P_{in}	inlet air pressure to packed tower (N/m^2)
P_{out}	outlet air pressure from packed tower (N/m^2)
P_{pump}	pump power requirement (kW)
P_s	saturation vapor pressure (N/m^2)
P_s	barometric pressure under standard conditions (1 atm)
$P_{v,A}$	vapor pressure of pure component A (atm)
q	solid-phase VOC concentration (mg/gm or μmole/gm)
Q	water flow rate (m^3/s)
r_A	radius of a molecule A (nm)
r_B	characteristic radius of the gas medium (nm)
r_{AB}	molecular separation at collision (nm)
R	universal gas constant 1.987 kcal/mole-K
R_g	universal gas constant (0.08205 L·atm/mole·K)
R_G	gas-phase mass transfer resistance (dimensionless)
R_L	liquid-phase mass transfer resistance (dimensionless)
R_T	overall resistance due to mass transfer (dimensionless)
S	stripping factor (dimensionless)
T	temperature (°C)
T_b	boiling temperature of molecule (K)
V	volume of bed or tank (m^3)

$\dot{V}$	gas flow rate (m^3/s)
V_b	molar volume of a compound at the boiling point (cm^3/mole)
W	adsorption space or pore volume occupied by the adsorbate (cm^3/gm)
W_o	maximum adsorption space (cm^3/gm)
y_b	bulk gas phase concentration (mg/L or mole/L)
y_s	gas phase concentration at the air/water interface (mg/L or mole/L)
γ_A	activity coefficient of component A in the aqueous phase (dimensionless)
y_s^*	gas-phase concentration at the air/water interface in equilibrium with the wit the bulk aqueous phase concentration (mg/L or mole/L)
z	packing depth (m)
Z_i	charge of species *i* (dimensionless)
α	polarizability of the VOC (m^3/kg)
ΔH^o	standard enthalpy change in water due (kcal/kmole)
ΔP	gas pressure drop (N/m^2)
μ_L	viscosity of water ($N \cdot s/m^2$)
μ_g	viscosity of air ($N \cdot s/m^2$)
ρ_l	density of water (kg/m^3)
ρ_g	density of air (kg/m^3)
σ	surface tension of water (kg/s^2)
σ_c	critical surface tension of the packing material (kg/s^2)
ε_A	force constant of molecule (Joule)
ε	adsorption potential (cal/mole)
ρ_L	liquid density of VOC (gm/L)
η	refractive index of the VOC (dimensionless)
ρ_a	density of air at 20°C
ϕ	Stanton number (dimensionless)

REFERENCES

Ashworth, R.A., G.B. Howe, M.E. Mullins, and T.N. Rogers (1988) "Air–Water Partitioning Coefficients of Organics in Dilute Aqueous Solutions," *Journal of Hazardous Materials*, vol. 18, pp. 25–36.

Baillod, C.R., W.L. Paulson, J.J. McKeown, and H.J. Campbell, Jr. (1986) "Accuracy and Precision of Plant Scale and Shop Clean Water Oxygen Transfer Tests," *Journal WPCF,* vol. 58, no. 4, pp. 25–36.

Ball, W.P., M.D. Jones, and M.C. Kavanaugh (1984) "Mass Transfer of Volatile Organic Compounds in Packed Towers," *Journal WPCF*, vol. 56, no. 127, pp. 405–409.

Bilello, L.J. and J.E. Singley, (1986) "Removing Trihalomethanes by Packed Column and Diffused Aeration," *Journal of AWWA*, vol. 78, no. 2, pp. 62–71.

Brown, L.C. and C.R. Baillod. (1982) "Modeling and Interpreting Oxygen Transfer Data," *Jour. Environ. Eng. Div., Proc. Am. Soc. Civ. Eng.,* vol. 108, no. EE4, pp. 607–616.

Byers, W.D. and C.M. Morton, (1985) "Removing VOCs from Groundwater; Pilot, Scale-up, and Operating Experience," *Environmental Progress*, vol. 4, no. 2, pp.112–118.

CECA (1994) "Bulletin for 410G GAC" Atochem, Tulsa, Oklahoma.

Clever, H.L. (1980) *IUPAC Solubility Data Series: Krypton, Xenon, and Radon*. Pergamon Press.

Cortright, R.D. (1986) "Gas-Phase Adsorption of Volatile Organic Compounds from Air Stripping Off-Gas onto Granulated Activated Carbon," MS Thesis, Houghton, Mich.: Michigan Technological University.

CRC Handbook of Chemistry and Physics, 85th ed. (2004) Lide, D. R., ed., Boca Raton, Fla: CRC Press.

Crittenden, J.C., T.F. Speth, D.W. Hand, P.J. Luft, and B.W. Lykins, Jr. (1987) "The Use of Equilibrium Theory to Evaluate Multicomponent Competition in Fixed Beds," *Jour. Environ. Eng. Div., Am. Soc. Civ. Eng.*, vol. 113, no. 6, pp. 1363–1375.

Crittenden, J.C., R.D. Cortright, B. Rick, S. Tang, and D.L. Perram (1988) "Using GAC to Remove Air Stripper Off-Gas," *Journal AWWA*, vol. 80, no. 5, pp. 73–84.

Crittenden, J.C., R.R. Trussell, D.W. Hand, G. Tchobanoglous, and K. Howe (2005) *Water Treatment Principles and Design*, 2nd ed. New York: John Wiley & Sons, Inc.

Cummins, M.D. and J.J. Westrick (July 1983) Proc. ASCE Environmental Engineering Conference, Boulder, Colo., pp. 442–449.

Cummins, M.D. (1985) "Field Evaluation of Packed Column Air Stripping: Bastrap, Louisiana." Interim Report. Cincinnati, Ohio: U.S. Environmental Protection Agency, Office of Drinking Water, Technical Support Division.

Djebbar, Y. and R.M. Narbaitz (1995) "Mass Transfer Correlations for Air Stripping Towers," *Environmental Progress*, vol. 14, no. 3, pp. 137–145.

Djebbar, Y. and R.M. Narbaitz (1998) "Improved Correlations for VOC Air Stripping Packed Towers," *Water Science & Technology*, vol. 38, no. 4, pp. 295–302.

Djebbar, Y and R.M. Narbaitz (2002) "Neural Network Prediction of Air Stripping K_La," *ASCE Jour. of Environ. Engrg.*, vol. 128, no. 5, pp. 451–460.

Dvorak, B.I., D.F. Lawler, J.R. Fair, and N.E. Handler (1996) "Evaluation of the Onda Correlation for Mass Transfer with Large Random Packings," *Environ. Sci. Tech.*, vol. 30, pp. 945–953.

Dzombak, D.A., S.B. Roy, and H.J. Fang (1993) "Air-Stripper Design and Costing Computer Program," *Journal AWWA*, vol. 85, no. 10, pp. 63–72.

Fischer, A., M. Muller, and J. Klasmeier (2004) "Determination of Henry's Law Constant for Methyl-*Tert*-butyl-Ether (MTBE) at Groundwater Temperatures," *Chemoshpere,* vol. 8, no. 25, pp. 689–694.

Gossett, J.M. (1987) "Measurements of Henry's Law Constants for C1 and C2 Chlorinated Hydrocarbons," *Environ. Sci. Tech.*, vol. 21, no. 2, pp. 202–208.

Gross, R.L. and S.G. TerMaath (1985) "Packed Tower Aeration Strips Trichloroethene from Groundwater," *Environmental Progress*, vol. 4, no. 2, pp. 119–124

Haarhoff, J. 1988. AIRSTRIP. AIRSTRIP Inc., Ames, Iowa.

Hand, D.W., J.C. Crittenden, J.L. Gehin, and B.L. Lykins, Jr. (1986) "Design and Evaluation of an Air-Stripping Tower for Removing VOCs from Groundwater," *Journal AWWA*, vol. 78, no. 9, pp. 87–97.

Hayduk, W. and H. Laudie (1974) "Prediction of Diffusion Coefficients for Non-electrolytes in Dilute Aqueous Solutions," *Jour. AIChE*, vol. 28, no. 611, pp. 611–615.

Higbie, R. (1935) "The Rate of Adsorption of a Pure Gas into a Still Liquid During Short Periods of Exposure," *Trans. Am. Inst. Chem. Engrs.*, vol. 31, no. 365, pp. 356–371.

Hirschfelder, J.O., R.B. Bird, and E.L. Spotz (1949)" The Transport Properties of Gases and Gaseous Mixtures," *Trans. ASME*, 71:921; *Chem. Rev.*, vol. 44, vol. 1, pp. 205–231.

Hokanson, D.R. (1996) "Development of Software Design Tools for Physical Property Estimation, Aeration, and Adsorption," Master's Thesis. Houghton, Mich.: Michigan Technological University.

Holmén, K. and P. Liss (1984) "Models for Air-Water Gas Transfer: An Experimental Investigation," *Tellus*, vol. 36B, pp. 92–100.

Jury, S.H. (1967) "An Improved Version of the Rate Equation for Molecular Diffusion in a Dispersed Phase," *A.I.Ch.E.J.*, vol. 13, pp. 1124–1126.

Kavanaugh, M.C. and R.R. Trussell (1980) "Design of Aeration Towers to Strip Volatile Contaminants from Drinking Water," *Journal AWWA*, vol. 72, no. 12, pp. 684–691.

Lamarche, P. and R.L. Droste (1989) "Air-Stripping Mass Transfer Correlations for Volatile Organics," *Journal AWWA*, vol. 81, no. 1, pp. 78–89.

Lenzo, F.C., T.J. Frielinghaus, and A.W. Zienkiewicz (June 1990) "The Application of the Onda Correlation to Packed Column Air Stripper Design: Theory Versus Reality," Proc. AWWA Annual Conference, Cincinnati, Ohio.

Lewis, W.K. and W.E. Whitman (1924) "Principles of Gas Absorption," *Ind. Engrg. Chem.*, vol. 16, pp. 1215–1220.

Little, J.C. and R.E. Selleck (June 1991) "Evaluating the Performance of Two Plastic Packings in a Crossflow Aeration Tower," *Journal AWWA*, vol. 83, no. 6, pp. 88–95.

Mandt, M.G. and P.R. Bathija (1987) "Jet Fluid Gas/Liquid Contacting and Mixing," *AIChE Symposium Series*, vol. 73, no. 167, p. 15.

Matter-Müller, C., W. Gujer, and W. Giger (1981) "Transfer of Volatile Substances from Water to the Atmosphere," *Water Research*, vol. 15, pp. 1271–1279.

McKinnon, R.J. and J.E. Dyksen (1984) "Removal of Organics from Groundwater Using Aeration Plus Carbon Adsorption," *Journal AWWA*, vol. 76, no. 5, pp. 42–56.

Metcalf and Eddy Inc. (2003) *Wastewater Engineering*, 4th ed. New York: McGraw-Hill.

Munz, C. and P.V. Roberts (1982) "Mass Transfer and Phase Equilibria in a Bubble Column," Proc. AWWA Annual Conference, Miami Beach, Fla., pp. 633–640.

Munz, C. and P.V. Roberts (1989) "Gas- and Liquid-Phase Mass Transfer Resistances of Organic Compounds during Mechanical Surface Aeration," *Water Research*, vol. 23, no. 5, pp. 589–601.

Nirmalakhandan, N., W. Jang, and R.E. Speece (1991) "Evaluation of Cascade Air Stripping–Pilot-Scale and Prototype Studies," *ASCE Jour. of Environ. Engrg.*, vol. 117, no. 6, pp. 788–798.

Nirmalakhandan, N., W. Jang, and R.E. Speece (1992) "Removal of 1,2 Dibromo-3-Chloropropane by Countercurrent Cascade Air Stripping," *ASCE Jour. of Environ. Engrg.*, vol. 118, no. 2, pp. 226–237.

O'Haver, J.H., R. Walk, B. Kitiyanan, J.H. Harwell, and D.A. Sabatini (2004) "Packed Column and Hallow Fiber Air Stripping of a Column Contaminant Surfactant Stream," ASCE *Jour. of Environ. Engrg.*, vol. 130, no. 3, pp. 4–11.

Onda, K., H. Takeuchi, and Y. Okumoto (1968) "Mass Transfer Coefficients between Gas and Liquid Phases in Packed Columns," *Journal Chem. Engrg. Japan*, vol. 1, no. 1, pp. 56–62.

Rasmuson, A.C. (1984) "Adsorption Equilibria on Activated Carbon of Mixtures of Solvent Vapors," *Fundamentals of Adsorption, Proceedings of the Engineering Foundation Conference* (A. Meyers and G. Belfort, eds.). The Engineering Foundation, New York.

Reucroft, P.J., W.H. Simpson, and L.A. Jonas (1971) "Sorption Properties of Activated Carbon," *Journal Phys. Chem.* vol. 75, no. 23, pp. 3526–3534.

Roberts, P.V. and P.G. Dändliker (1983) "Mass Transfer of Volatile Organic Contaminants from Aqueous Solution to the Atmosphere during Surface Aeration," *Environ. Sci. Technol.*, vol. 17, no. 8, pp. 484–489.

Roberts, P.V., C. Munz, and P. Dändliker (1984) "Modeling Volatile Organic Solute Removal by Surface and Bubble Aeration," *Journal WPCF*, vol. 56, no. 2, pp. 157–163.

Roberts, P.V., G.D. Hopkins, C. Munz, and A.H. Riojas (1985) "Evaluating Two-Resistance Models for Air Stripping of Volatile Organic Contaminants in a Countercurrent, Packed Column," *Environ. Sci. Technol.*, vol. 19, no. 6, pp. 164–173.

Roberts, P.V. and J.A. Levy (1985) "Energy Requirements for Air Stripping Trihalomethanes," *Journal AWWA*, vol. 77, no. 4, pp. 138–145.

Rogers, T.N. (1994) "Predicting Environmental Physical Properties from Chemical Structure Using a Modified UNIFAC Model," Ph.D. Dissertation. Houghton, Mich.: Michigan Technological University.

Schwarzenbach, R.P., P.M. Gschwend, and D.M. Imbod, (1992) *Environmental Organic Chemistry*, Second Edition, New York: John Wiley & Sons Inc.

Sherwood, T.K. and F.A. Hollaway (1940) "Performance of Packed Towers–Liquid Film Data for Several Packings," *Trans. Amer. Inst. Chem. Engrs.*, vol. 36, pp. 39–48.

Singley, J. E., A. L. Ervin, M. A. Mangone, J. M. Allan, and H.H. Land, (1981) Trace Organics Removal by Air Stripping. Denver, Colorado: Supplementary Report to Awwa Research Foundation.

Speight, J. (2004) *Lange's Handbook of Chemistry*, 16th ed. New York: McGraw-Hill.

Staudinger, J., W.R. Knocke, and C.W. Randall (1990) "Evaluating the Onda Mass Transfer Correlation for the Design of Packed-Column Air Stripping," *Journal AWWA*, vol. 82, no. 1, pp. 73–79.

Tang, S., (1987) "Description of Adsorption Equilibria for Volatile Organic Chemicals in Air Stripping Off-gas," Master's Thesis. Houghton, Mich.: Michigan Technological University.

Thom, J.E. and W.D. Byers (1993) "Limitations and Practical Use of a Mass Transfer Model for Predicting Air Stripper Performance," *Environmental Progress*, vol. 12, no. 1, pp. 61–66.

Treybal, R.E. (1980) *Mass Transfer Operations,* Chem. Engrg. Series, 3rd ed. New York: McGraw-Hill.

Umphres, M.D., C.H. Tate, M.C. Kavanaugh, and R.R. Trussell (1983) "A Study of THM Removal by Packed Tower Aeration," *Journal AWWA*, vol. 75, no. 8, pp. 414–424.

USEPA (U.S. Environmental Protection Agency) (1989) *Design Manual Fine Pore Aeration Systems.* EPA/625/1-89/023. Cincinnati, Ohio: U.S. Environmental Protection Agency.

Vane, L.M. and E.L. Giroux (2000) "Henry's Law Constants and Micellar Partitioning of Volatile Organic Compounds in Surfactants," *Journal of Chemical Engineering Data*, vol. 45, pp. 38–47.

Wallman, H. and M.D. Cummins (1986) "Design Scale-Up Suitability for Air-Stripping Columns," *Journal of Public Works*, vol. 153, no. 10, pp. 73–78.

WEF-ASCE (Water Environment Federation-American Society of Civil Engineers) (1991) *Design of Municipal Wastewater Treatment Plants*. WEF-ASCE, vol.1, pp. 1–829.

CHAPTER 7
CHEMICAL OXIDATION

Philip C. Singer, Ph.D., P.E., B.C.E.E.
Dan Okun Distinguished Professor of Environmental Engineering
University of North Carolina
Chapel Hill, North Carolina
United States

David A. Reckhow, Ph.D.
Professor of Civil and Environmental Engineering
University of Massachusetts
Amherst, Massachusetts
United States

INTRODUCTION 7.1
PRINCIPLES OF OXIDATION 7.2
Thermodynamic Considerations 7.2
Kinetic Considerations 7.8
Types of Reactions 7.11
OXIDANTS USED IN WATER TREATMENT 7.15
Chlorine 7.15
Chlorine Dioxide 7.21
Ozone and Advanced Oxidation Processes 7.23
Potassium Permanganate 7.29
Potassium Ferrate 7.30
Mixed Oxidants 7.31
APPLICATIONS OF OXIDATION PROCESSES TO WATER TREATMENT 7.31
Control of Iron and Manganese 7.31
Destruction of Tastes and Odors 7.37
Sulfide Removal 7.42
Color Removal 7.42
Oxidation of Organic Micropollutants 7.43
Oxidation as an Aid to Coagulation and Flocculation 7.46
Control of Biological Growths in Treatment Plants 7.47
Summary of Benefits of Oxidation to Water Treatment Processes 7.47
ABBREVIATIONS 7.47
NOTATIONS FOR EQUATIONS 7.48
REFERENCES 7.48

INTRODUCTION

Chemical oxidation processes play several important roles in the treatment of drinking water. Chemical oxidants are used for the oxidation of: reduced inorganic species, such as ferrous iron [Fe(II)], manganous manganese [Mn(II)], and sulfide [S(-II)]; hazardous synthetic organic compounds, such as trichloroethene (TCE) and atrazine; and other emerging contaminants, such as methyl-*tert*-butyl-ether (MTBE), pharmaceutically active compounds and personal care products (PPCPs), endocrine disrupting chemicals (EDCs) and algal toxins. Oxidants can also be used to destroy taste and odor-causing compounds and eliminate color. In addition, in some cases, they may improve the performance of, or reduce the required amount of coagulants. Because many oxidants also have biocidal properties, these oxidants can be used to control nuisance aquatic growths, for example, algae, in pre-treatment basins and may be used as primary disinfectants to meet $C \times T$ (disinfectant concentration times contact time) requirements (see Chap. 17). Oxidants are often added at the head of the treatment plant, prior to or at the rapid mix basin, but they can also be employed after clarification, prior to filtration, after a substantial portion of oxidant-demanding natural organic matter (NOM) has been removed.

The most common chemical oxidants used in water treatment are chlorine, ozone, chlorine dioxide, and permanganate. Ozone is sometimes used in conjunction with hydrogen peroxide or ultraviolet irradiation in what are called advanced oxidation processes (AOPs) to produce radicals that have powerful oxidative properties. Mixed oxidant technologies are also available.

Free chlorine has traditionally been the oxidant (and disinfectant) of choice in the United States, but concerns about the formation of potentially harmful halogenated disinfection by-products (DBPs; see Chap.19) produced by reactions between free chlorine and NOM (exacerbated in some cases by the presence of bromide) have caused many water systems to adopt alternative chemical oxidants (and disinfectants) to lower halogenated DBP formation.

These other oxidants may also react with NOM and bromide to various degrees, depending on the properties of the oxidant, to form oxidation by-products, some of which also have adverse public health effects or result in downstream operational problems in the treatment plant or distribution system.

This chapter reviews thermodynamic and kinetic principles associated with the use of chemical oxidants in general, the types and properties of the chemical oxidants used in water treatment, and specific applications of oxidation processes for the treatment of drinking water. Comparisons among the different oxidant choices are presented where information is available.

PRINCIPLES OF OXIDATION

Thermodynamic Considerations

Thermodynamics establishes the feasibility or constraints of chemical reactions such as oxidation reactions. Chemical kinetics describes the rate of the reactions and pathways or mechanisms by which the reactions occur. Thermodynamic properties of the reactants, such as free energies, enthalpies, and entropies, provide a starting point for characterizing and understanding oxidation reactions. In this section, basic thermodynamic concepts relating to oxidation reactions are presented. Additional background is provided in Chap. 3. For a more comprehensive treatment of the subject, there are many excellent physical chemistry textbooks that can be consulted, for example, Atkins (1994).

Electrochemical Potentials. Oxidation reactions are often viewed as reactions involving the exchange of electrons. While it is sometimes convenient to think of oxidants as electron acceptors and reductants as electron donors, many oxidants actually donate an electron-poor element or chemical group, rather than simply accept a lone electron. Nevertheless, it is useful to treat all oxidation reactions as simple electron transfers for the purpose of balancing equations and performing thermodynamic calculations.

Thermodynamic principles can be used to determine if specific oxidation reactions are possible. This generally involves the calculation of some form of reaction potential. Although in most cases oxidation equilibria lie far to one side or the other, it is sometimes instructive to calculate equilibrium concentrations of the reactants and products.

The first step is to identify the species being reduced and those being oxidized. Appropriate half-reactions and their standard reduction and oxidation potentials (E^o_{red} and E^o_{ox}, respectively) are available in tables of thermodynamic constants. Important reduction and oxidation half-reactions in water treatment are listed, respectively, in Tables 7-1 and 7-2. Table 7-1 lists common oxidants that accept electrons from reducing agents, while Table 7-2 gives illustrative reducing agents that donate electrons and are subsequently oxidized. Oxidation and reduction half-reactions may be combined to get the overall standard reaction potential E^o_{net}.

$$E^o_{net} = E^o_{ox} + E^o_{red} \tag{7-1}$$

Much as an acidity constant describes the tendency of an acid to give up a proton, the reduction and oxidation potentials (E^o_{red} and E^o_{ox}, respectively) describe the tendency of an

TABLE 7-1 Standard Half-Cell Potentials for Chemical Oxidants Used in Water Treatment at 25°C

Oxidant	Reduction half-reaction	E^o_{red}, volts
Ozone	$\frac{1}{2}O_{3(aq)} + H^+ + e^- \rightarrow \frac{1}{2}O_{2(aq)} + \frac{1}{2}H_2O$	2.04
Hydrogen peroxide	$\frac{1}{2}H_2O_2 + H^+ + e^- \rightarrow H_2O$	1.78
Hydroxyl radical	$OH + H^+ + e^- \rightarrow H_2O$	2.59
Permanganate	$\frac{1}{3}MnO_4^- + \frac{4}{3}H^+ + e^- \rightarrow \frac{1}{3}MnO_{2(s)} + \frac{2}{3}H_2O$	1.68
Ferrate	$\frac{1}{3}FeO_4^{2-} + \frac{5}{3}H^+ + e^- \rightarrow \frac{1}{3}Fe(OH)_{3(s)} + \frac{1}{3}H_2O$	2.07
Chlorine dioxide	$ClO_2 + e^- \rightarrow ClO_2^-$	1.15
Hypochlorous acid	$\frac{1}{2}HOCl + \frac{1}{2}H^+ + e^- \rightarrow \frac{1}{2}Cl^- + \frac{1}{2}H_2O$	1.49
Hypochlorite ion	$\frac{1}{2}OCl^- + H^+ + e^- \rightarrow \frac{1}{2}Cl^- + \frac{1}{2}H_2O$	0.90
Hypobromous acid	$\frac{1}{2}HOBr + \frac{1}{2}H^+ + e^- \rightarrow \frac{1}{2}Br^- + \frac{1}{2}H_2O$	1.33
Hypoiodous acid	$\frac{1}{2}HOI + \frac{1}{2}H^+ + e^- \rightarrow \frac{1}{2}I^- + \frac{1}{2}H_2O$	0.98
Monochloramine	$\frac{1}{2}NH_2Cl + H^+ + e^- \rightarrow \frac{1}{2}Cl^- + \frac{1}{2}NH_4^+$	1.40
Dichloramine	$\frac{1}{4}NHCl_2 + \frac{3}{4}H^+ + e^- \rightarrow \frac{1}{2}Cl^- + \frac{1}{4}NH_4^+$	1.34
Oxygen	$\frac{1}{4}O_{2(aq)} + H^+ + e^- \rightarrow \frac{1}{2}H_2O$	1.27

TABLE 7-2 Standard Half-Cell Potentials for Some Oxidation Reactions That Can Occur during Drinking Water Treatment at 25°C

Oxidation half-reaction	E^o_{ox}, volts
$\frac{1}{2}Br^- + \frac{1}{2}H_2O \rightarrow \frac{1}{2}HOBr + \frac{1}{2}H^+ + e^-$	−1.33
$\frac{1}{6}Br^- + \frac{1}{2}H_2O \rightarrow \frac{1}{6}BrO_3^- + H^+ + e^-$	−1.44
$\frac{1}{2}I^- + \frac{1}{2}H_2O \rightarrow \frac{1}{2}HOI + \frac{1}{2}H^+ + e^-$	−0.98
$\frac{1}{6}I^- + \frac{1}{2}H_2O \rightarrow \frac{1}{6}IO_3^- + H^+ + e^-$	−0.26
$\frac{1}{6}Cl^- + \frac{1}{2}H_2O \rightarrow \frac{1}{6}ClO_3^- + H^+ + e^-$	−1.45
$\frac{1}{2}Mn^{2+} + H_2O \rightarrow \frac{1}{2}MnO_2(s) + 2H^+ + e^-$	−1.21
$Fe^{2+} + 3H_2O \rightarrow Fe(OH)_3(s) + 3H^+ + e^-$	−1.01
$\frac{1}{8}NH_4^+ + \frac{3}{8}H_2O \rightarrow \frac{1}{8}NO_3^- + \frac{5}{4}H^+ + e^-$	−0.88
$\frac{1}{2}NO_2^- + \frac{1}{2}H_2O \rightarrow \frac{1}{2}NO_3^- + H^+ + e^-$	−0.84
$\frac{1}{8}H_2S + \frac{1}{2}H_2O \rightarrow \frac{1}{8}SO_4^{-2} + \frac{10}{8}H^+ + e^-$	−0.30
$\frac{1}{2}H_2S \rightarrow \frac{1}{2}S(s) + H^+ + e^-$	−0.14
$\frac{1}{2}SO_3^{2-} + \frac{1}{2}H_2O \rightarrow \frac{1}{2}SO_4^{2-} + H^+ + e^-$	−0.80
$\frac{1}{2}AsO_3^{3-} + \frac{1}{2}H_2O \rightarrow \frac{1}{2}AsO_4^{3-} + H^+ + e^-$	−0.156
$\frac{1}{2}HCOO^- \rightarrow \frac{1}{2}CO_2(g) + \frac{1}{2}H^+ + e^-$	+0.29

oxidant to take up an electron (Table 7-1) or a reductant to give one up (Table 7-2). The standard Gibbs free energy of a reaction (ΔG^o) is related to the standard electrochemical potential by Faraday's constant ($F = 96.5$ kJ/coul) and the number n of electrons transferred.

$$\Delta G^o = -nFE^o_{\text{net}} \tag{7-2}$$

For one-electron transfer reactions such as those shown in Tables 7-1 and 7-2, this becomes

$$\Delta G^o(\text{kJ}) = -96.5E^o_{\text{net}}(\text{V}) \tag{7-3}$$

Classical thermodynamics indicates that reactions with a negative Gibbs free energy (or a positive E^o_{net}) are feasible and will spontaneously proceed in the direction as written (i.e., from left to right), and those with a positive Gibbs free energy (or negative E^o_{net}) are not feasible and will proceed in the reverse direction.

Consider a generic oxidation reaction

$$aA_{\text{ox}} + bB_{\text{red}} \rightarrow aA_{\text{red}} + bB_{\text{ox}} \tag{7-4}$$

where substance A (an oxidant) accepts electrons from substance B (a reductant). The coefficients a and b are stoichiometric coefficients that are required to balance the reaction. In order to determine which substance is being reduced and which is being oxidized, one must calculate and compare oxidation states of the reactant atoms and product atoms (see Chap. 3 and below).

The equilibrium constant K for this reaction defines the relative concentrations of the reactants and products at equilibrium.

$$K = \frac{(A_{\text{red}})^a (B_{\text{ox}})^b}{(A_{\text{ox}})^a (B_{\text{red}})^b} \tag{7-5}$$

The overall standard electrochemical potential for the reaction is then directly related to this equilibrium constant by

$$E^o_{\text{net}} = \frac{RT}{nF} \ln K \tag{7-6}$$

For a one-electron transfer reaction at 25°C, this simplifies to

$$\log K = \frac{1}{0.059} E^o_{\text{net}} \tag{7-7}$$

Oxidation-Reduction Reactions

Oxidation state. An oxidation state is characterized by an oxidation number, which is the charge one would expect for an atom if it were to dissociate from the surrounding molecule or ion (assigning any shared electrons to the more electronegative atom). Oxidation number may be either a positive or negative number, usually an integer between –VII and +VII, although in their elemental forms, for example, S(s) and O_2(aq), atoms have an oxidation number of zero. This concept is useful in balancing chemical equations and performing certain calculations.

Balancing equations. The first step in working with oxidation reactions is to identify the roles of the reacting species. At least one reactant must be the oxidizing agent (i.e., contain an atom or atoms that accept electrons and get chemically reduced), and at least one must be a reducing agent (i.e., contain an atom or atoms that lose electrons and become oxidized). The second step is to balance the gain of electrons by the oxidizing agent with the loss of electrons from the reducing agent. Next, oxygen atoms are balanced by adding water molecules to one side of the reaction or the other as needed, and hydrogen atoms are balanced with H^+ ions. For a more detailed treatment of oxidation reactions, the reader is referred to a general textbook on aquatic chemistry (e.g., Jensen, 2003; Benjamin, 2002; Stumm and Morgan, 1996).

Example 7-1 Oxidation of Reduced Manganese by Ozone
Write a balance stoichiometric reaction for the oxidation of reduced (manganous) manganese by ozone and determine if the reaction is thermodynamically feasible under typical water treatment conditions.

Solution Equation 7-8 shows the species involved in the reaction. The substance being oxidized (losing electrons) is manganese (i.e., the reducing agent), and the one doing the oxidizing (i.e., gaining electrons and being reduced) is ozone.

$$Mn^{2+} + O_3 \rightarrow \text{products} \qquad (7\text{-}8)$$

Next, the products formed need to be evaluated. It might be known from experience that soluble manganous manganese (i.e., Mn^{2+}) can be oxidized in water to the relatively insoluble manganese dioxide. It might also be known that ozone ultimately forms hydroxide and oxygen after it becomes reduced.

$$Mn^{2+} + O_3 \rightarrow MnO_2 + O_2 + OH^- \qquad (7\text{-}9)$$

The next step is to determine the oxidation state of all atoms involved.

$$\overbrace{Mn^{2+}}^{+II} + \overbrace{O_3}^{0} \rightarrow \overbrace{Mn}^{+IV}\overbrace{O_2}^{-II} + \overbrace{O_2}^{0} + \overbrace{O}^{-II}\overbrace{H^-}^{+I} \qquad (7\text{-}10)$$

From this analysis, it is clear that manganese is oxidized from +II to +IV, which involves a loss of two electrons per atom. On the other side of the ledger, the ozone undergoes a gain of two electrons per molecule, as one of the three oxygen atoms goes from an oxidation state of 0 to –II. The two separate half-reactions can be written as single electron transfers, as shown in Tables 7-1 and 7-2. Each half-reaction is balanced by adding water molecules and/or H^+ ions as needed to balance oxygen and hydrogen, respectively. For the oxidation half-reaction:

$$\tfrac{1}{2}\overbrace{Mn^{2+}}^{+II} + H_2O \rightarrow \tfrac{1}{2}\overbrace{Mn}^{+IV}O_2 + 2H^+ + e^- \qquad (7\text{-}11)$$

By convention, when hydroxide appears in a half-reaction, additional H^+ ions are added until all of the hydroxide is converted to water. This is done for the reduction half-reaction as follows:

$$\tfrac{1}{2}\overbrace{O_3}^{0} + H^+ + e^- \rightarrow \tfrac{1}{2}\overbrace{O_2}^{0} + \tfrac{1}{2}H_2\overbrace{O}^{-II} \qquad (7\text{-}12)$$

From this point, it is a simple matter of combining the equations and canceling out terms or portions of terms that appear on both sides. At the same time, the standard electrode potentials from Tables 7-1 and 7-2 can be combined to get the overall standard potential for the net reaction:

$$\tfrac{1}{2}Mn^{2+} + H_2O \rightarrow \tfrac{1}{2}MnO_{2(s)} + 2H^+ + e^- \qquad E^o_{ox} \quad (-1.21\ V)$$
$$\tfrac{1}{2}O_{3(aq)} + H^+ + e^- \rightarrow \tfrac{1}{2}O_{2(aq)} + \tfrac{1}{2}H_2O \qquad E^o_{red} \quad (+2.04\ V)$$
$$\tfrac{1}{2}O_{3(aq)} + \tfrac{1}{2}Mn^{2+} + \tfrac{1}{2}H_2O \rightarrow \tfrac{1}{2}O_{2(aq)} + \tfrac{1}{2}MnO_{2(s)} + H^+ \qquad E^o_{net} \quad (+0.83\ V) \tag{7-13}$$

The balanced reaction indicates that 1 mole of soluble manganous manganese requires 1 mole of ozone to oxidize it to insoluble manganese dioxide. Further, the oxidation reaction produces two protons for each mole of manganous manganese oxidized, resulting in a reduction in pH and the consumption of two equivalents of alkalinity.

It can also be seen that the reaction is thermodynamically feasible and will proceed toward the right (the E^o_{net} is positive). But how far to the right will the reaction proceed? To answer this, Eq. 7- 7 is rearranged to get

$$K = e^{16.95E^o_{net}} \tag{7-14}$$

So for this reaction,

$$K = e^{16.95(0.83)} = 1.29\times10^6 \tag{7-15}$$

Using the concentration quotient (see Eq. 7-5) and the reaction stoichiometry:

$$1.29\times10^6 = \frac{[O_{2(aq)}]^{0.5}[MnO_{2(s)}]^{0.5}[H^+]}{[O_{3(aq)}]^{0.5}[Mn^{2+}]^{0.5}[H_2O]^{0.5}} \tag{7-16}$$

Because the activity of solvents (i.e., water) and solid phases are, by convention, equal to 1, Eq. 7-16 reduces to

$$1.29\times10^6 = \frac{[O_{2(aq)}]^{0.5}[H^+]}{[O_{3(aq)}]^{0.5}[Mn^{2+}]^{0.5}} \tag{7-17}$$

If the pH is 7.0 and a dissolved oxygen concentration of 10 mg/L and an ozone concentration of 0.5 mg/L is achieved in the ozonated water, an equilibrium Mn^{2+} concentration of 1.8×10^{-25} M, or about 10^{-27} mg/L can be calculated. Thermodynamic principles therefore indicate that this reaction essentially goes to completion.

Knowing that the Mn^{2+} should react completely to form manganese dioxide, it might be informative to determine if ozone can oxidize the manganese dioxide to a higher oxidation state, i.e. permanganate. To examine this possibility, the reduction equation for ozone must be combined with the reverse of the permanganate reduction equation (from Table 7-1) to get the net reaction

$$\tfrac{1}{2}O_{3(aq)} + H^+ + e^- \rightarrow \tfrac{1}{2}O_{2(aq)} + \tfrac{1}{2}H_2O$$
$$\tfrac{1}{3}MnO_2 + \tfrac{2}{3}H_2O \rightarrow \tfrac{1}{3}MnO_4^- + \tfrac{4}{3}H^+ + e^-$$
$$\text{Net Reaction: } \tfrac{1}{2}O_{3(aq)} + \tfrac{1}{3}MnO_2 + \tfrac{1}{6}H_2O \rightarrow \tfrac{1}{3}MnO_4^- + \tfrac{1}{3}H^+ + \tfrac{1}{2}O_{2(aq)} \tag{7-18}$$

This allows the net potential to be calculated.

$$E^o_{net} = E^o_{ox} + E^o_{red} = (-1.68\ V) + (+2.04\ V) = +0.36\ V \tag{7-19}$$

Again, this is a thermodynamically favorable reaction. The equilibrium constant is

$$\log K = \frac{1}{0.059} E^o_{\text{net}} = \frac{1}{0.059}(+0.36\ \text{V}) = 6.1$$
$$K = 10^{6.1} \tag{7-20}$$

The equilibrium quotient can now be formulated directly from the balanced equation and the equilibrium constant calculated.

$$K = \frac{[A_{\text{red}}]^a[B_{\text{ox}}]^b}{[A_{\text{ox}}]^a[B_{\text{red}}]^b}$$
$$K = \frac{[\text{MnO}_4^-]^{0.33}[\text{H}^+]^{0.33}[\text{O}_2]^{0.5}}{[\text{O}_3]^{0.5}} = 10^{6.1} \tag{7-21}$$

Note that neither manganese dioxide (MnO_2) nor water (H_2O) appears in this quotient because both are presumed present at unit activity. Manganese dioxide is a solid and is considered to be in a pure, undiluted state as long as it remains in the system. The same may be said for water. As long as the solutes remain dilute, the concentration of water remains constant.

So, under typical conditions where the pH is near neutrality (i.e., $[H^+] = 10^{-7}$), dissolved oxygen is near saturation (i.e., $[O_{2(aq)}] = 3 \times 10^{-4}$M), and the ozone residual is 0.25 mg/L (i.e., $[O_{3(aq)} = 5 \times 10^{-6}$ M), the expected equilibrium permanganate concentration can be calculated:

$$K = \frac{[\text{MnO}_4^-]^{0.33}[10^{-7}]^{0.33}[3\times10^{-4}]^{0.5}}{[5\times10^{-6}]^{0.5}} = 10^{6.1} \tag{7-22}$$

Solving for permanganate,

$$[\text{MnO}_4^-]^{0.33} = 3.5\times10^7$$
$$[\text{MnO}_4^-] = 327 \tag{7-23}$$

Obviously one cannot have 327 moles/L of permanganate. The example illustrates that the reaction will proceed in this direction so that all of the manganese dioxide would be converted to permanganate. Once the manganese dioxide is depleted, the reaction must stop. This simple example explains why permanganate is produced when water containing dissolved manganous manganese is subjected to ozonation during water treatment (see applications below). ▲

Some other examples illustrating the balancing of oxidation/reduction equations that are relevant to water treatment, using the half-reactions in Tables 7-1 and 7-2 and their corresponding oxidation and reduction potentials, include:

1. The reduction of free chlorine by sulfite, which is the common method of dechlorinating water:

$$\tfrac{1}{2}\text{HOCl} + \tfrac{1}{2}\text{H}^+ + e^- \rightarrow \tfrac{1}{2}\text{Cl}^- + \tfrac{1}{2}\text{H}_2\text{O} \qquad (+1.49\ \text{V})$$
$$\tfrac{1}{2}\text{SO}_3^{2-} + \tfrac{1}{2}\text{H}_2\text{O} \rightarrow \tfrac{1}{2}\text{SO}_4^{2-} + \text{H}^+ + e^- \qquad (-0.80\ \text{V})$$
$$\text{Net reaction} \quad \text{HOCl} + \text{SO}_3^{2-} \rightarrow \text{Cl}^- + \text{SO}_4^{2-} + \text{H}^+ \qquad (+0.69\ \text{V}) \tag{7-24}$$

2. The oxidation of ferrous iron by oxygen leading to the formation and precipitation of insoluble ferric hydroxide, which is the net reaction resulting from the aeration of water for iron removal

$$Fe^{2+} + 3H_2O \rightarrow Fe(OH)_3(s) + 3H^+ + e^- \qquad (-1.01\ V)$$

$$\tfrac{1}{4}O_{2(aq)} + H^+ + e^- \rightarrow \tfrac{1}{2}H_2O \qquad (+1.27\ V)$$

$$\text{Net reaction} \quad 4Fe^{2+} + O_{2(aq)} + 10H_2O \rightarrow 4Fe(OH)_3(s) + 8H^+ \qquad (+0.26\ V) \qquad (7\text{-}25)$$

3. The oxidation of sulfide by chlorine, a common method for eliminating odorous sulfide from water

$$\tfrac{1}{2}HOCl + \tfrac{1}{2}H^+ + e^- \rightarrow \tfrac{1}{2}Cl^- + \tfrac{1}{2}H_2O \qquad (+1.49\ V)$$

$$\tfrac{1}{2}H_2S \rightarrow \tfrac{1}{2}S(s) + H^+ + e^- \qquad (-0.30\ V)$$

$$\text{Net reaction} \quad HOCl + H_2S \rightarrow Cl^- + S(s) + H_2O + H^+ \qquad (+1.19\ V) \qquad (7\text{-}26)$$

4. The formation of hypoiodous acid from monochloramine, an important reaction in iodide containing waters associated with the formation of iodinated disinfection byproducts (see the following and Chap. 19)

$$\tfrac{1}{2}NH_2Cl + H^+ + e^- \rightarrow \tfrac{1}{2}Cl^- + \tfrac{1}{2}NH_4^+ \qquad (+1.40\ V)$$

$$\tfrac{1}{2}I^- + \tfrac{1}{2}H_2O \rightarrow \tfrac{1}{2}HOI + \tfrac{1}{2}H^+ + e^- \qquad (-0.98\ V)$$

$$\text{Net reaction} \quad NH_2Cl + I^- + H_2O + H^+ \rightarrow HOI + NH_4^+ + Cl^- \qquad (+0.42\ V) \qquad (7\text{-}27)$$

In all cases, the value of the redox potential for the net reaction is positive, indicating that the reaction proceeds from left to right as written.

As already mentioned, the preceding thermodynamic analysis is useful in establishing which reactions are feasible and the relative concentrations of oxidized and reduced species when the system reaches equilibrium. However, many oxidation reactions are quite slow or, in some cases, kinetically unfavorable, so that the actual concentrations of reactants and products observed during water treatment are far from those predicted by classical thermodynamics. For this reason, kinetic considerations are extremely important in understanding which oxidation reactions occur and the extents to which they occur in water treatment.

Kinetic Considerations

Reaction Kinetics. Thermodynamics indicates whether a reaction will proceed as written. However, it does not indicate whether this reaction will produce significant change within milliseconds, hours, days, or thousands of years. For this, chemical kinetics must be considered. As an example, consider the reaction between hypochlorous acid and bromide ion:

$$HOCl + Br^- = HOBr + Cl^- \qquad (7\text{-}28)$$

In order for a molecule of hypochlorous acid and a molecule of bromide to react to form products, the two molecules must come into contact with each other (contact meaning approach within a certain distance so that bonding forces can take effect). The probability that a single $HOCl{:}Br^-$ molecular encounter will occur within any fixed time period is directly proportional to the number of molecules of each of the reactants in the system. It will also depend on the rate of movement and energy of each of the reactant molecules. As a consequence, the rate of formation of products, for example, HOBr, will be dependent on a number of factors including the concentration of hypochlorous acid and the concentration of bromide in the reacting solution. This is in accordance with the kinetic law of mass action which is expressed mathematically in Eq. 7-29.

$$\frac{d[\mathrm{HOBr}]}{dt} = k_f[\mathrm{HOCl}][\mathrm{Br}^-] \tag{7-29}$$

The reactants and products are expressed in molar units of concentration, and k_f is called the forward reaction rate constant. The units for k_f are liters/mole per unit time. The reaction rate constant is going to be a function of such factors as temperature, the rate of movement of the molecules, and the probability of HOBr formation given that a collision between hypochlorous acid and bromide has already occurred. Because the concentrations of HOCl and Br^- that appear in Eq. 7-29 are raised to the first power, it is said that this rate law is first order in both reactants. The overall order of the reaction is the sum of the individual orders (i.e., second order in this case).

A more careful examination of this reaction reveals that it probably occurs in three separate steps (Kumar and Margerum, 1987) as follows:

$$\mathrm{HOCl} + \mathrm{Br}^- \xrightarrow{k_f} \mathrm{OH}^- + \mathrm{BrCl}$$
$$\mathrm{BrCl} + \mathrm{Br}^- \xrightarrow{fast} \mathrm{Br}_2 + \mathrm{Cl}^-$$
$$\mathrm{Br}_2 + \mathrm{H}_2\mathrm{O} \xrightarrow{fast} \mathrm{HOBr} + \mathrm{Br}^- + \mathrm{H}^+$$

Since the first step is the slow one (called the *rate-limiting step*), it defines the overall rate of the three-step sequential reaction. As a result, the rate law in Eq. 7-29 is still valid even though it is not strictly speaking an elementary reaction. Like many reactions, this one can be catalyzed by the presence of proton-donating acidic solutes in solution (e.g., acetic acid, bicarbonate, dihydrogen phosphate). These are generically represented by HA in the following alternative first step of the reaction sequence. In this case, the value k_3 is a third-order constant, and dependent on the type of proton-donating species in solution.

$$\mathrm{HA} + \mathrm{HOCl} + \mathrm{Br}^- \xrightarrow{k_3} A^- + \mathrm{H}_2\mathrm{O} + \mathrm{BrCl}$$

In a more general sense, Eq. 7-30 is the rate law for any elementary reaction of the type described by Eq. 7-31.

$$-\frac{d[A]}{dt} = k_{\mathrm{fa}}[A]^a[B]^b \tag{7-30}$$

$$aA + bB = cC + dD \tag{7-31}$$

where the capital letters represent chemical species participating in the reaction and the small letters are the stoichiometric coefficients (i.e., the numbers of each molecule or ion required for the balanced reaction). The overall order describes the extent of dependence of the reaction rate on reactant concentrations. For the reaction in Eq. 7-31, the overall order is equal to $(a + b)$. The order with respect to species A is a, and the order with respect to species B is b. Thus, the reaction in Eq. 7-28 is first order in both reactants and second order overall.

Chemical reactions may be either elementary or non-elementary. Elementary reactions are those reactions that occur exactly as they are written, without any intermediate steps. These reactions almost always involve just one or two reactants. The number of molecules or ions involved in elementary reactions is called the *molecularity* of the reaction. Thus, for all elementary reactions, the overall order equals the molecularity, as in Eqs. 7-28 and 7-31.

Nonelementary reactions involve a series of two or more elementary reactions and involve complex reaction pathways. Many environmental reactions are nonelementary in nature. In general, reactions with an overall reaction order greater than two, or reactions with a noninteger reaction order for one or more of the reactants are nonelementary.

Reaction rate constants for various oxidants with similar solutes are often positively correlated. In other words, a compound favored for oxidation by one oxidant is generally favored by others as well. Those that are relatively resistant to oxidation by one will likewise be unreactive with others. A good example is found in the extensive research done on the oxidation of phenolic compounds by Tratnyek and Hoigne (1994). These data highlight the similarities between the chemical structure of a reactant and its reactivity with various oxidants. Chemists have used such relationships to develop quantitative structure-activity relationships (QSARs). The Hammett equations are some of the most widely used QSARs. (See Brezonik, 1994, for more detail on this subject.)

Temperature Dependence. As mentioned previously, the reaction rate constant k is a function of temperature. The Arrhenius equation (Eq. 7-32) is the classic model that describes this relationship.

$$k = Ae^{-E_a/RT} \tag{7-32}$$

where A is called the frequency factor or the pre-exponential factor, E_a is the activation energy, R is the universal gas constant (8.314 J/K-mole), and T is the temperature in K. The values for k and E_a may be found in the literature or determined from experimental measurements.

pH Dependence. The degree of protonation or deprotonation of reacting species is determined by the solution pH. If one form of an oxidant or reduced substance is the sole or preferred reactive form, then changes in pH can profoundly affect the overall rate of reaction. Consider the reaction of aqueous chlorine and phenol. In this case, it has been determined that only the fully protonated, hypochlorous acid (HOCl) and the deprotonated phenate ($C_6H_5O^-$) react to any appreciable degree.

$$\text{HOCl} + \text{C}_6\text{H}_5\text{O}^- \xrightarrow{k_2} \text{products} \tag{7-33}$$

Consequently, the fundamental rate equation should only include these two species. This is not very convenient, as analytical methods to measure the reactants are not species-specific and only tell us how much total free chlorine ($[\text{chlorine}]_{\text{tot}} = [\text{HOCl}] + [\text{OCl}]^-$) and total phenol ($[\text{phenol}]_{\text{tot}} = [\text{C}_6\text{H}_5\text{OH}] + [\text{C}_6\text{H}_5\text{O}]^-$) we have at any given time. Nevertheless, the ratio of a protonated form or a deprotonated form to its total concentration can be readily calculated from the pH and the acidity constant (K_a). Hence, we can reformulate the rate law as follows:

$$\begin{aligned} -\frac{d[\text{phenol}]_{\text{tot}}}{dt} &= k_2[\text{HOCl}][\text{C}_6\text{H}_5\text{O}^-] \\ &= k_2[\text{chlorine}]_{\text{tot}}\frac{[\text{H}^+]}{K_{a,\text{Cl}_2}+[\text{H}^+]}[\text{phenol}]_{\text{tot}}\frac{K_{a,\text{phenol}}}{K_{a,\text{phenol}}+[\text{H}^+]} \end{aligned} \tag{7-34}$$

The apparent or observed rate constant (k_{obs}) is equal to the product of the fundamental second-order rate constant (k_2) times the total analytical concentrations of the reactants times two pH-dependent distribution coefficient, referred to as alpha-values (α).

$$k_{\text{obs}} = k_2[\text{chlorine}]_{\text{tot}}[\text{phenol}]_{\text{tot}}\alpha_{0,\text{Cl}_2}\alpha_{1,\text{phenol}} \tag{7-35}$$

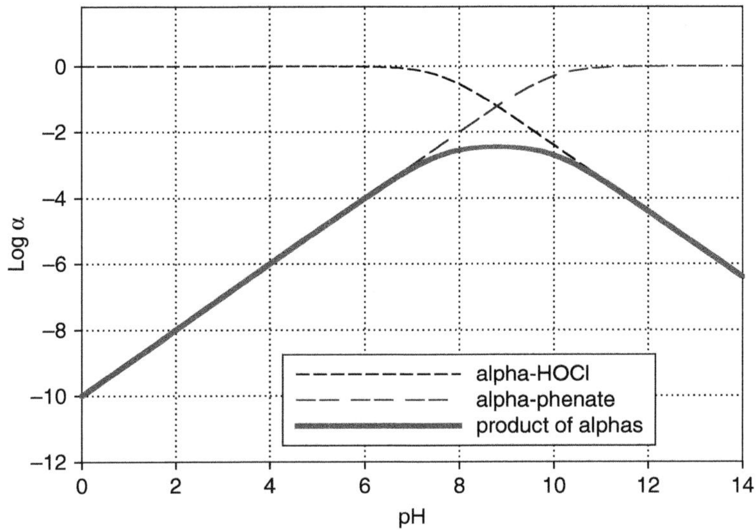

FIGURE 7-1 Illustration of the pH-dependency of the reacting species on the rate constant for the oxidation of phenol by chlorine.

The rate, therefore, takes on a pH dependence that makes the rate increase with pH until a maximum (at a pH halfway between the pK_{a,Cl_2} and the $pK_{a,\ phenol}$) and then declines at higher pH, as shown in Fig. 7-1.

Types of Reactions

Oxidation reactions can generally be categorized as those involving electron transfer and those involving transfer of atoms and groups of atoms. They may also be characterized as reactions involving species with paired electrons (ionic) and those involving unpaired electrons (radical). Aqueous chlorine offers a wide array of ionic reactions (e.g., Morris 1975) that will serve as illustrative examples for this discussion.

Table 7-3 presents a summary of the major types of ionic reactions of interest in drinking water. Chlorine or bromine can be added to olefinic bonds (i.e., carbon-carbon double bonds) forming halohydrins when hypochlorous acid or hypobromous acid, respectively, is added to water. This is an electrophilic reaction where the initial attack is by the halogen atom (on the positive side of the hypohalous acid (HOX) dipole). The most stable configuration places the halogen on the carbon with the most hydrogen atoms (producing the most stable carbonium ion in accordance with Markovnikov's rule). The other carbon becomes a carbonium ion, which subsequently reacts with the HO portion of the HOX species or with water.

Activated ionic substitution can occur with chlorine and bromine and both aromatic and aliphatic compounds. As with the addition reactions, this type of reaction will also lead to the formation of organohalide compounds. Aromatic substitution reactions occur readily when an electron-donating substituent is bound to the ring. Functional groups on the aromatic ring, such as OH and NH_2, can be thought of as creating a partial negative charge on the ortho and para positions (second closest and furthest carbon atoms from the functional group, respectively). The halogen end of the HOX molecule attacks one of these carbons. Next there is a loss of the OH end of the molecule, and displacement of the H atom from

TABLE 7-3 Major Types of Ionic Reactions

Reaction type	Example
1. Addition to an olefinic bond	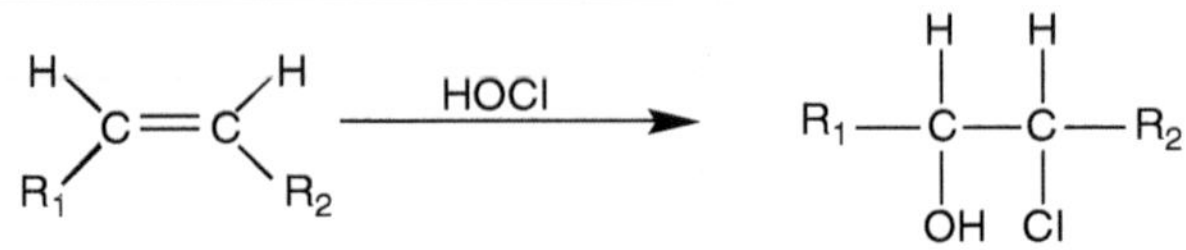
2. Activated aromatic substitution	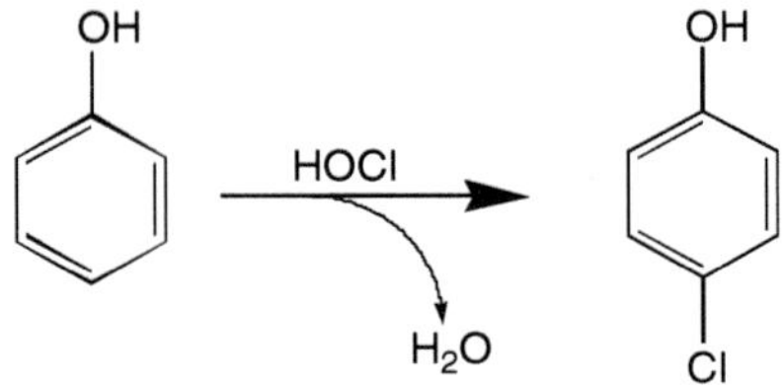
3. Substitution onto nitrogen	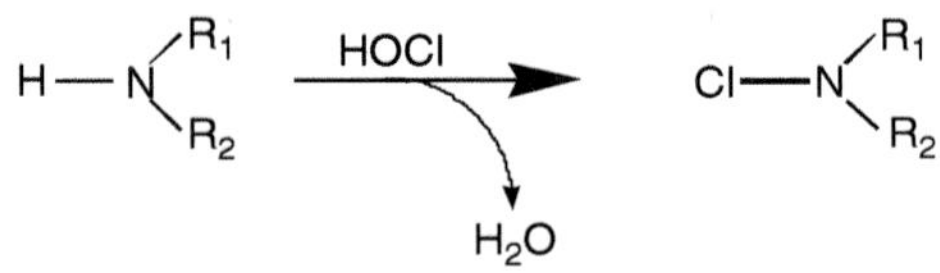
4. Oxidation with oxygen transfer	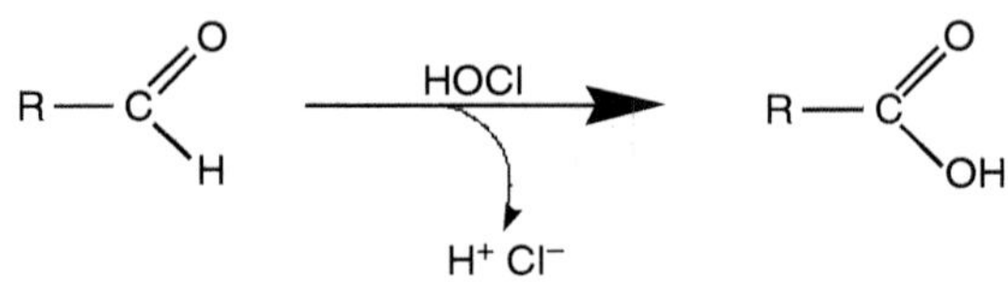
5. Oxidation with electron transfer	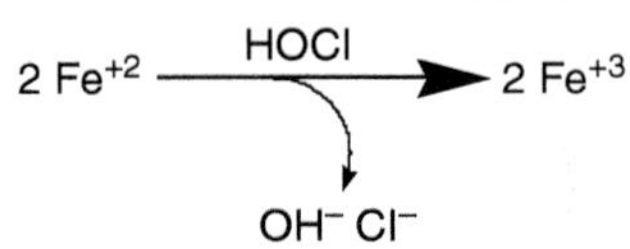

the carbon under attack. Substitution on aliphatic species is also a multi-step reaction as exemplified by the haloform reaction (see next section on pathways).

When substitution (or transfer) of a halogen occurs onto a nitrogen atom, a relatively reactive N-halo organic compound results. These compounds retain some of the oxidizing capabilities of their corresponding hypohalous acid and, consequently, the reactions are not considered to cause an oxidant demand. The rates of substitution reactions with nitrogenous organic compounds generally increase as the basicity of the nitrogen atom increases.

Oxidation reactions involving halogens are characterized by the formation of the corresponding inorganic halide ion and an oxidized (non-halogenated) form of the reacting compound. With organic compounds, it is quite common to observe addition of an oxygen atom. For example, oxidation reactions transform unsaturated hydrocarbons into alcohols, then to aldehydes and ketones, and finally to carboxylic acids. Some oxidations do not result in a net transfer of atoms. For these electron transfer reactions, it is common to form free radical intermediates. When this happens, chain reactions can occur, sometimes leading to the types of reactions in Table 7-4.

There are several types of reactions involving free-radical species that can occur following the addition of oxidants to drinking water. Free radicals are atoms, molecules, or ions

TABLE 7-4 Major Types of Radical Reactions

Reaction type	Example
6. Radical addition reaction	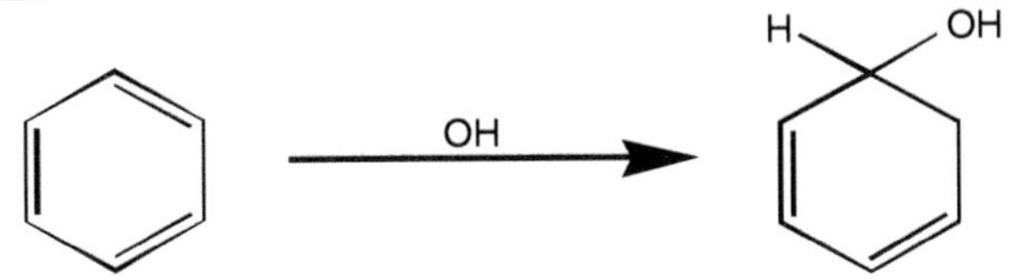
7. Hydrogen abstraction reaction	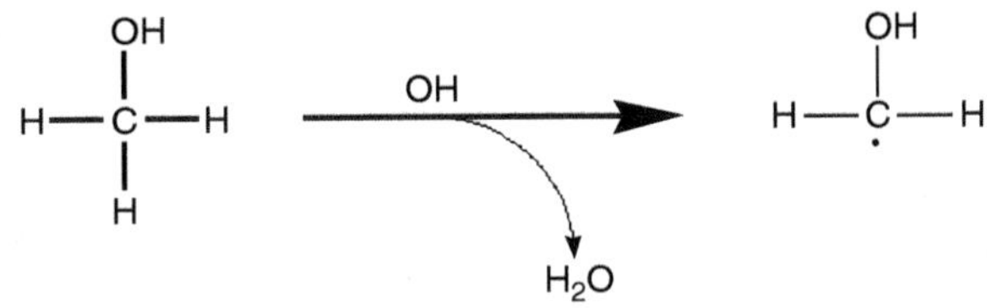
8. Radical oxidation reaction with single-electron transfer	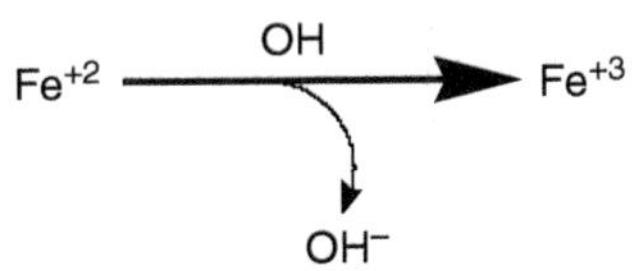

with unpaired electrons that cause them to be highly reactive. These types of reactions are commonly encountered with ozone, chlorine dioxide, and especially *advanced oxidation processes* (*AOPs;* see below). For example, the addition of ozone will always lead to some ozone decomposition and subsequent formation of hydroxyl radicals (OH). These reactive hydroxyl radical species engage in reactions that generally lead to the formation of new free-radical species. The most common types of radical reactions are addition reactions, hydrogen abstractions, and single-electron transfers (see Table 7-4).

Catalysis. Many types of oxidation reactions are strongly affected by the presence of catalysts. Catalysts are agents that alter reaction rates without being formed or consumed in the reaction. They typically participate in some key, rate-limiting step and are regenerated during some later step. Catalysts generally provide an alternative pathway with lower activation energy so that the reaction proceeds faster.

Probably the most important catalytic processes in water treatment involve the participation of acids and bases. Specific acid and specific base catalysis involve H^+ and OH^-, respectively. General acid and base catalysis involves any electron acceptor (e.g., a proton donor) and electron donor (e.g., a proton acceptor), respectively. A good example of general base catalysis is the classic haloform reaction (see the next section).

Another type of catalysis that is important in oxidation processes involves the catalytic initiation of a free-radical chain reaction. Examples include the decomposition of ozone by hydroxide and the decomposition of chlorine by iron (e.g., Brezonik, 1994). In either case, the original oxidant will not react appreciably with recalcitrant compounds such as oxalate. However, in the presence of sufficient catalyst, decomposition is initiated, leading to a series of chain propagation reactions whereby oxalate can be readily converted to carbon dioxide.

Sometimes oxidation reactions are catalyzed by the presence of solid phases in the system which have the ability to adsorb one or more of the reacting species. Manganese dioxide ($MnO_2(s)$) is a common catalyst in the oxidation of dissolved manganous manganese (see below).

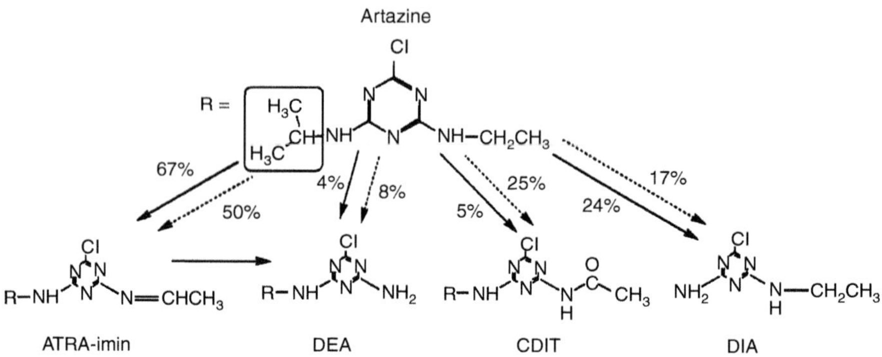

FIGURE 7-2 Ozonation of atrazine. (*Source: Acero et al., 2000; reprinted with permission from the American Chemical Society, Copyright 2000.*)

Reaction Pathways. Oxidation reactions in drinking water can be very complex. They may begin with one of the mechanisms just discussed, but then may be followed by a wide range of non-oxidation processes, such as elimination reactions, hydrolysis reactions, radical chain reactions, and rearrangement reactions.

The ozonation of atrazine, a commonly occurring pesticide found in many raw water sources, provides a good example of the complex reaction pathways associated with ozonation (Acero et al., 2000; see Fig. 7-2). The percentages shown represent the formation of products due to reaction with molecular ozone (O_3; solid lines) or with the hydroxyl radical (·OH; dashed lines). Reaction begins primarily with the attack by ozone on the nitrogen group that is in the α-position to the ethyl group, leading to the formation of an atrazine-imine (ATRA-imin) that slowly hydrolyzes to *diethylatrazine (DEA)* and acetaldehyde. ATRA-imin is also formed by hydroxyl radicals through hydrogen abstraction on the carbon in the α-position to the ethyl-nitrogen. The attack by ozone on the nitrogen group that is in the α–position to the isopropyl group leads to the formation of *diisopropyl atrazine (DIA)*. DIA is also formed by hydrogen abstraction on the isopropyl side of atrazine by the hydroxyl radical, producing an imine, which readily hydrolyzes to DIA and acetone. Another major product is 4-acetamido-2-chloro-6-isopropylamino-s-triazine. At typical ozone doses, atrazine is not mineralized, and although the parent compound atrazine may be eliminated, the triazine ring persists in the daughter products. The lack of mineralization and the formation of daughter products is true for the oxidation of many organic contaminants in water (see below).

Another important reaction is the formation of trihalomethanes (see Chap. 19) which may occur through many different reaction pathways. One of the most widely discussed is the haloform reaction (Fig. 7-3), which involves the stepwise chlorine substitution of the enolate form of a methyl ketone. The rate-limiting step is the loss of a proton leading to the formation of the enol. This classic reaction begins with base-catalyzed halogenations, ultimately leading to a carboxylic acid and chloroform. It is base-catalyzed because the species that reacts with hypochlorous acid is the enol form of the methyl ketone. Although Fig. 7-3 shows the reaction being catalyzed by hydroxide, any strong base will participate. The base (or hydroxide) consumed in the first step is regenerated in the second. The enol undergoes electrophilic substitution forming a mono-halogenated intermediate. The presence of halogens on this carbon speeds subsequent enolization, which leads to complete halogenation of the α-carbon. The resulting intermediate (a trihalogenated acetyl compound) is subject to base-catalyzed hydrolysis, giving a trihalomethane and a carboxylic acid. If hypochlorous acid is the only halogenating species, chloroform is the result.

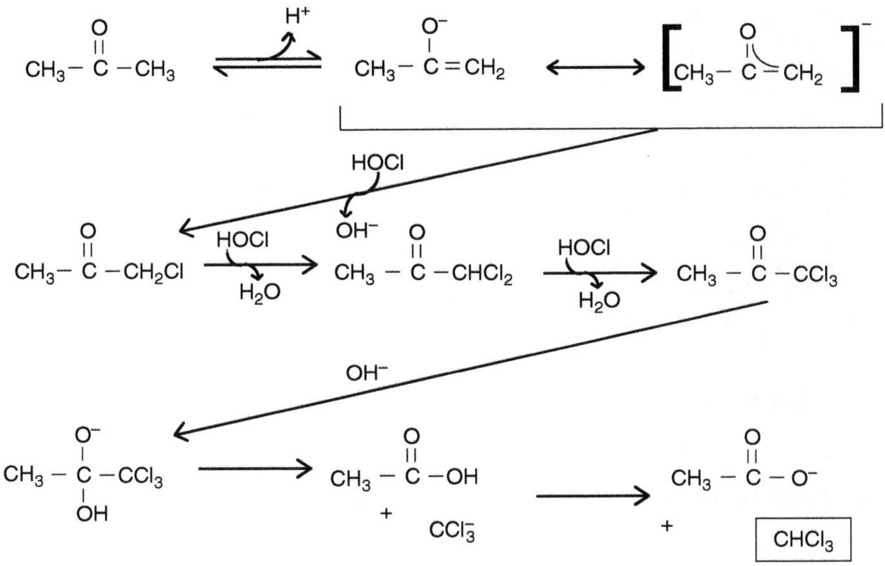

FIGURE 7-3 The haloform reaction.

Many early studies with acetone (or propanone) indicated that the rate-limiting step was the initial enolization. Once the enolate was formed, the molecule quickly proceeded through the entire reaction pathway. Thus, the reaction rate expression often cited in the chemical literature was

$$\frac{-d[\mathrm{CH_3COCH_3}]}{dt} = k[\mathrm{CH_3COCH_3}] \tag{7-36}$$

However, under drinking water treatment conditions (i.e., neutral pH, low chlorine residual), other steps may be rate limiting. This changes the rate law and complicates attempts to characterize it. In addition, a competing pathway exists that leads to the formation of dichloroacetic acid (Reckhow and Singer, 1985). Trichloropropanone may undergo further base-catalyzed chlorine substitution to form pentachloropropanone. This compound will rapidly hydrolyze to form chloroform and dichloroacetic acid. The rate law proposed for the loss of trichloropropanone is

$$\frac{-d[\mathrm{CH_3COCCl_3}]}{dt} = \left\{\frac{k_1k_2[\mathrm{HOCl}][\mathrm{P_T}]}{k_{-1}[\mathrm{P_T}]+k_2[\mathrm{HOCl}]} + k_3[\mathrm{OCl^-}] + k_4[\mathrm{OH^-}]\right\}[\mathrm{CH_3COCCl_3}] \tag{7-37}$$

This complicated rate law reflects catalysis by hypochlorite (OCl^-), hydroxide (OH^-), and phosphate (P_T).

OXIDANTS USED IN WATER TREATMENT

Chlorine

Chlorine is the most widely used oxidant (and disinfectant; see Chap. 17) in water treatment practice. Chlorine is available in gaseous form (as Cl_2), as a concentrated aqueous solution (sodium hypochlorite, NaOCl, i.e., bleach), or as a solid (calcium hypochlorite, $Ca(OCl)_2$), or it can be electrolytically generated on-site.

When chlorine gas is added to water, the chlorine rapidly disproportionates to form hypochlorous acid (HOCl) and the chloride ion (Cl^-).

$$Cl_2 + H_2O \rightarrow HOCl + H^+ + Cl^- \tag{7-38}$$

The equilibrium constant for Eq. 7-38 is 5×10^{-4} at 25°C, indicating that the reaction goes relatively far to the right, as shown. The residual concentration of molecular Cl_2 in solution will represent only a small fraction of the total chlorine concentration, except at very low pH conditions or at high chloride concentrations. For example, even in water at pH 2 with a 10^{-3} M chloride concentration, only 2 percent of the total chlorine will exist as molecular Cl_2. The chloride produced by Eq. 7-38 is essentially inert with respect to its oxidizing and disinfecting properties.

Hypochlorous acid is a weak acid ($pK_a = 7.6$ at 25°C) that can transfer a proton to form the hypochlorite ion (OCl^-).

$$HOCl = H^+ + OCl^- \tag{7-39}$$

The ratio of hypochlorous acid to hypochlorite may be calculated from the expression

$$\log([HOCl]/[OCl^-]) = 7.6 - pH \tag{7-40}$$

The sum of the three species, Cl_2, HOCl, and OCl^-, is commonly referred to as *free available chlorine (FAC),* and the concentrations of the individual species and their sum are usually expressed in units of mg/L as Cl_2.

Equations 7-38 and 7-39 show that the species of chlorine present in water depends on the total concentration of chlorine, pH, and temperature. Figure 7-4 is a diagram of the relative amounts of hypochlorous acid and hypochlorite as a function of pH. At 20°C, hypochlorous acid is the predominant species below pH 7.6, and hypochlorite ion is predominant at pH values greater than 7.6. The concentrations of the two species are equal

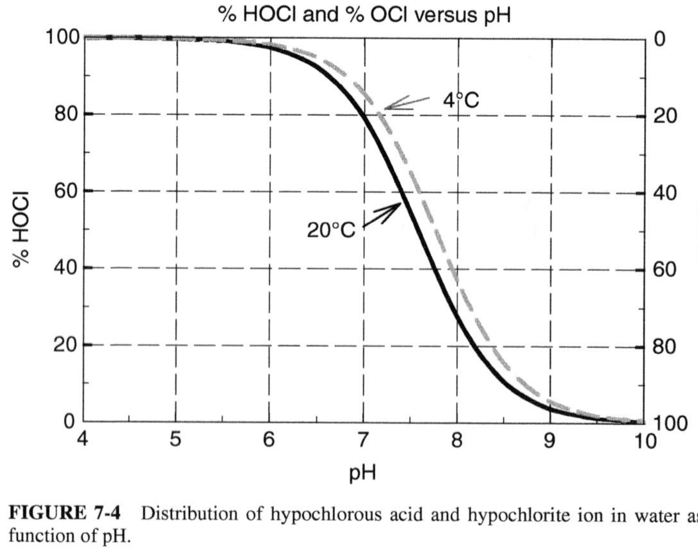

FIGURE 7-4 Distribution of hypochlorous acid and hypochlorite ion in water as a function of pH.

at pH 7.6 (the pK_a value). The distribution shifts somewhat with changing temperature because the equilibrium constants for Eqs. 7-38 and 7-39 are temperature dependent.

If chlorine is added to water as liquid sodium hypochlorite, the following reactions occur.

$$NaOCl \rightarrow Na^+ + OCl^- \tag{7-41}$$

$$OCl^- + H_2O = HOCl + OH^- \tag{7-42}$$

If calcium hypochlorite granules are dissolved in water, they also form the hypochlorite ion

$$Ca(OCl)_2 \rightarrow Ca^{2+} + 2OCl^- \tag{7-43}$$

which, like the hypochlorite formed from the addition of NaOCl, reacts with water to form hypochlorous acid.

$$OCl^- + H_2O = HOCl + OH^- \tag{7-44}$$

Just as in the case of the species formed by the addition of chlorine gas, the relative distribution of HOCl and OCl^- that results from the addition of sodium hypochlorite or calcium hypochlorite will be determined by pH, temperature, and the total chlorine concentration. Again, this distribution will be in accordance with Eq. 7-40 and Fig. 7-4. Regardless of the form in which the chlorine is added to water (Cl_2, NaOCl, or $Ca(OCl)_2$), hypochlorite, hypochlorous acid, and molecular chlorine will be formed as described by the chemical equilibria represented by the previous reactions. The only difference is that chlorine gas produces an acidic reaction which lowers the pH of the solution, whereas sodium hypochlorite and calcium hypochlorite are both bases which will raise the pH of the water. The extent of the pH change will depend on the alkalinity of the water, but is likely to be relatively small because of the low doses of chlorine typically used.

As shown in Table 7-1, hypochlorous acid and the hypochlorite ion are both strong oxidizing agents, but HOCl is the stronger of the two. Hence, oxidation reactions of chlorine, in general, tend to be more effective at low pH values unless the reactant exhibits pH-dependent behavior as in the case of phenol (see Fig. 7-1).

Gaseous chlorine is typically provided for water treatment applications in pressurized tanks so that the chlorine exists as a liquid under pressure. The chlorine is then added to water by reducing the pressure in the tank and releasing the chlorine as a gas. Handling of the chlorine gas is extremely dangerous, and caution must be exercised in working with chlorine feeding equipment. This is discussed further in Chap. 17.

Alternatively, if liquid sodium hypochlorite solution is used, it is usually added from a concentrated NaOCl bulk solution directly to the main plant flow with the help of a liquid metering pump. Solid granules of calcium hypochlorite can be added to the water directly, or a concentrated solution of calcium hypochlorite can be prepared from which the hypochlorite is metered into the main flow as in the case of sodium hypochlorite. Calcium hypochlorite is not widely used at municipal water treatment plants.

Historically, gaseous chlorine in pressurized tanks has been the most common method of applying chlorine in municipal water treatment practice, but more recently, because of concerns about transport and handling of hazardous chemicals, the use of liquid sodium hypochlorite is becoming more widely practiced despite its higher cost. A potential concern involving the use of sodium hypochlorite is its stability. The hypochlorite tends to degrade over time, particularly when it is stored at high temperatures and/or exposed to sunlight (Bolyard et al., 1992, 1993; Adam and Gordon, 1999). One of the degradation products is chlorate (ClO_3^-), a potential health concern (see the discussion of chlorine dioxide below). Common practice is to limit on-site storage of sodium hypochlorite feedstocks to 90 days. Research also indicates the possible presence of bromate, another inorganic disinfection by-product usually associated with ozonation, in sodium hypochlorite feedstocks at levels which could exceed regulatory limits when the hypochlorite is fed at typical doses used in practice (Chlorine Institute, 1999; Weinberg et al., 2003). Appropriate specifications need to be provided when ordering sodium hypochlorite to minimize the presence of bromate.

Free chlorine can also be generated electrolytically on-site from brine (sodium chloride, NaCl) solutions (see Fig. 7-5). The molecular chlorine generated at the anode can

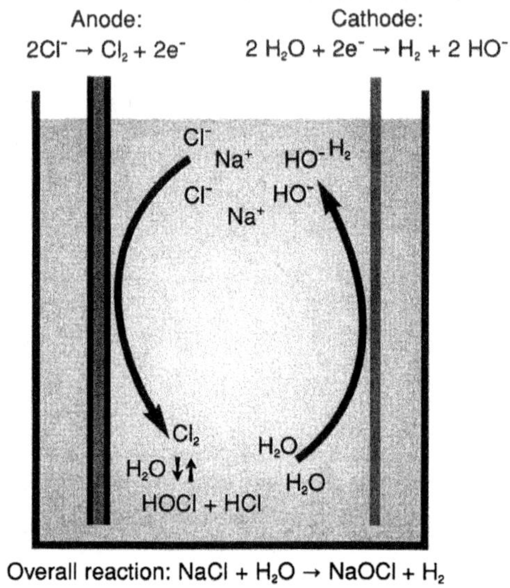

FIGURE 7-5 Electrolysis cell illustrating on-site generation of sodium hypochlorite. (*Source: Boal, 2009; reprinted with permission from the National Environmental Services Center, West Virginia University, Copyright 2009.*)

be dissolved in sodium hydroxide (NaOH), which is also a by-product of the electrolysis reaction, to produce a concentrated sodium hypochlorite solution.

$$Cl_2 + NaOH \rightarrow NaOCl + H^+ + Cl^- \qquad (7\text{-}45)$$

Again, while the use of electrolysis cells to produce chlorine gas on-site tends to be more costly than purchasing chlorine gas directly in pressurized tanks, the hazards associated with the transport and handling of pressurized chlorine containers is avoided. Also, although more expensive than purchasing bulk sodium hypochlorite, electrolytically generated chlorine production can be paced to match the need for chlorine during treatment, thereby eliminating the need for storage of sodium hypochlorite and its associated degradation over time.

Reactions of Chlorine with Organic Compounds. Chlorine reacts with organic material by a combination of oxidation and substitution reactions. For example, chlorine reacts with amino acids to produce non-halogenated oxidation by-products, such as aldehydes and organic acids. Conversely, chlorine reacts with phenol to produce chlorinated phenolic compounds, such as ortho- and parachlorophenol. Formation of the latter contributes to an enhancement of taste and odor in chlorinated water containing phenol (see below). In the case of aromatic compounds, the presence of electron-donating substituents on the aromatic ring facilitates both substitution reactions and oxidative ring cleavage reactions.

Chlorine reacts with NOM (e.g., humic substances) to produce a variety of chlorine-substituted halogenated disinfection by-products, such as chloroform, di- and trichloroacetic acid, dichloroacetonitrile, and chloropicrin. In view of the importance of these reactions to water quality and treatment, the subject of disinfection by-product formation and control is presented separately, in Chap. 19. The rates of reactions between chlorine and NOM or chlorine and trace organic contaminants are dependent on the structure of the organic compounds and on other factors such as pH, temperature, and chlorine residual concentration. See the critical review by Deborde and von Gunten (2008) for a more complete discussion of chlorine reactions and a compilation of rate constants.

Reactions of Chlorine with Ammonia. Ammonia in water exerts a chorine demand in accordance with Eq. 7-46.

$$2NH_3 + 3HOCl \rightarrow N_2(gas) + 3H^+ + 3Cl^- + 3H_2O \tag{7-46}$$

One mole of ammonia consumes 1.5 moles of chlorine, or 1 mg of ammonia as N consumes 7.6 mg of chlorine as Cl_2. Hence if chlorine is used for the oxidation of reduced iron and manganese, sulfide, or organic micropollutants (see below) or for disinfection (see Chap. 17), more than 7.6 mg of chlorine must be applied to overcome the chlorine demand of 1 mg of ammonia as N.

If lesser amounts of chlorine are applied per mg N, inorganic chloramines are produced in accordance with Eqs. 7-47 to 7-49.

$$NH_3 + HOCl = NH_2Cl + H_2O \tag{7-47}$$

$$NH_2Cl + HOCl = NHCl_2 + H_2O \tag{7-48}$$

$$NHCl_2 + HOCl = NCl_3 + H_2O \tag{7-49}$$

The mixture of monochloramine (NH_2Cl), dichloramine ($NHCl_2$), and trichloramine or nitrogen trichloride (NCl_3) that results depends on pH, temperature, and the ratio of chlorine to ammonia. At a chlorine:ammonia molar ratio less than 1:1 or a weight ratio (Cl_2:N) less than 5:1 and a pH of about 8, the principal chloramine species formed is monochloramine. At higher chlorine:ammonia ratios and lower pH values, dichloramine formation becomes important, and at mildly acidic pH values in the range of 6 to 6.5, nitrogen trichloride can be formed. The sum of the concentrations of the three chloramine species is typically referred to as *combined chlorine* and is often expressed in units of mg/L as Cl_2. Combined chlorine is analytically distinguishable from free chlorine (*Standard Methods,* 2005). The sum of the free and combined chlorine concentrations is referred to as *total chlorine*.

Chloramine formation may be done purposefully by adding ammonia to water containing free available chlorine if it is desired to form monochloramine for maintenance of a stable disinfectant residual in the distribution system (see Chap. 17) and for limiting the formation of regulated disinfection by-products (see Chap. 19). Monochloramine, being a weaker oxidant, is a more stable and persistent, albeit weaker, disinfectant species than free chlorine and does not contribute appreciably to the formation of the regulated disinfection by-products. For these reasons, monochloramine is an attractive secondary disinfectant in water distribution systems. Because of its important role in disinfection, chloramine chemistry is discussed in greater detail in Chap. 17. From the standpoint of its oxidation potential, monochloramine is too weak to oxidize reduced iron and manganese and most taste and odor-causing organics, and it is too weak to eliminate natural organic color. While monochloramine will oxidize NOM to some degree and produce some halogenated organic material, it does not generally produce trihalomethanes (see Chap. 19).

Part of the oxidation capacity of the chloramines derives from monochloramine hydrolysis represented by the reverse reaction of Eq. 7-47. The free chlorine liberated, although often in small amounts, may be responsible in part for the formation of some halogenated by-products associated with chloramines. Likewise, the free ammonia liberated may contribute to nitrification problems in water distribution systems.

In general, the principal value of monochloramine in water treatment is not as an oxidant but as a secondary disinfectant (see Chap. 17).

Reactions of Chlorine with Bromide. In drinking waters containing bromide, chlorine will oxidize the bromide to produce hypobromous acid (HOBr).

$$HOCl + Br^- \rightarrow HOBr + Cl^- \tag{7-50}$$

This reaction is very fast, with a second-order rate constant of 2.95×10^3 $M^{-1}s^{-1}$. Hence bromide represents another source of chlorine demand, albeit a minor one, and the presence of bromide contributes somewhat to the depletion of free chlorine.

The resulting hypobromous acid, like hypochlorous acid, is a weak acid ($pK_a = 8.7$ at 25°C) and will dissociate to some degree to form the hyprobromite ion (OBr^-), depending on pH.

$$HOBr \rightarrow OBr^- + H^+ \tag{7-51}$$

As noted in Table 7-1, hypobromous acid is a somewhat weaker oxidant than hypochlorous acid. The formation of hyprobromous acid is important, however, because hypobromous acid is capable of reacting with NOM to produce undesirable brominated disinfection by-products, such as bromoform and dibromoacetic acid. The presence of bromide and its subsequent oxidation to hyprobromous acid is what contributes, along with hypochlorous acid, to the formation of mixed brominated and chlorinated disinfection by-products such as bromodichloromethane and bromochloroacetic acid (see Chap. 19). Because of their relative acidity constants, more hypobromous acid remains in the undissociated HOBr form than hypochlorous acid under most pH conditions encountered in practice. Accordingly, because the undissociated forms of these species are stronger oxidants than their deprotonated counterparts, hypobromous acid behaves as a stronger substituting agent than hypochlorous acid, which results in the formation of greater amounts of halogenated disinfection by-products in bromide-containing waters.

Reactions of Chlorine with Iodide. Chlorine reacts quickly with iodide, leading to the formation of hypoiodous acid (HOI).

$$HOCl + I^- \rightarrow HOI + Cl^- \tag{7-52}$$

The second-order rate constant for the reaction between HOCl and iodide is 4.3×10^8 $M^{-1}s^{-1}$. The resultant HOI is rapidly oxidized to iodate, with OCl^- being the primary oxidant at typical drinking water treatment conditions (Bichsel and von Gunten, 1999).

$$2OCl^- + HOI \rightarrow IO_3^- + 2Cl^- + H^+ \tag{7-53}$$

The reaction is first order with respect to both OCl^- and HOI, and the rate constant is 52 $M^{-1}s^{-1}$ at 25°C. Because the rate constant for the oxidation of HOI by chlorine to produce iodate is so fast, HOI does not persist long enough to react with NOM to produce iodinated disinfection by-products to any appreciable degree. This is not the case with monochloramine, which will rapidly oxidize iodide to HOI (second-order rate constant of $2.4 \times 10^{10} \times [H^+]$ $M^{-2}s^{-1}$), but the subsequent oxidation of HOI by monochloramine is so slow ($<2 \times 10^{-3}$ $M^{-1}s^{-1}$) that HOI persists and will react with NOM to produce iodinated disinfection by-products. This is the reason why higher levels of iodinated disinfection by-products are found in chloraminated waters containing iodide than in chlorinated waters (Weinberg, et al., 2002; Bichsel and von Gunten, 1999).

Chlorine Dioxide

Chlorine dioxide (ClO_2) is unstable at high concentrations and can explode on exposure to heat, light, electrical sparks, or shocks. Accordingly, it is never shipped in bulk, but instead is generated on-site. Aqueous solutions are usually prepared from the gaseous chlorine dioxide generated, as chlorine dioxide is very soluble in water. It does not hydrolyze in water as chlorine does and remains in its molecular form as ClO_2. It is much more volatile than chlorine and can easily be stripped from aqueous solution if not properly handled. Chlorine dioxide is usually generated by reacting sodium chlorite ($NaClO_2$) with either gaseous chlorine (Cl_2) or hypochlorous acid (HOCl) under acidic conditions, in accordance with the following reactions:

$$2NaClO_2 + Cl_2(g) \rightarrow 2ClO_2(g) + 2Na^+ + 2Cl^- \tag{7-54}$$

$$2NaClO_2 + HOCl \rightarrow 2ClO_2(g) + 2Na^+ + Cl^- + OH^- \tag{7-55}$$

In order to drive the reaction to completion and avoid the presence of unreacted chlorite (ClO_2^-) in the product stream, some generators use excess chlorine, although several variations in generator design have been developed which allow for the sodium chlorite to be completely converted to chlorine dioxide without using excess chlorine. Additionally, to overcome the basicity of sodium chlorite and the hydroxide produced by Eq. 7-55, acid is sometimes added along with the hypochlorous acid to maintain the optimal pH for chlorine dioxide generation. Typically, pH values in the range of 3.5 to 5.5 are preferred; more acidic pH values lead to the formation of chlorate (ClO_3^-).

$$NaClO_2 + Cl_2(g) + H_2O \rightarrow ClO_3^- + Na^+ + 2Cl^- + 2H^+ \tag{7-56}$$

Chlorine dioxide may also be prepared by acidification of a sodium chlorite solution.

$$5NaClO_2 + 4H^+ \rightarrow 4ClO_2(g) + 5Na^+ + Cl^- + 2H_2O \tag{7-57}$$

Once generated, chlorine dioxide can be dissolved in water and is stable in the absence of light and elevated temperatures. At high pH values or in the presence of light or at elevated temperatures, chlorine dioxide disproportionates to form both chlorite and chlorate (Bowen and Cheung, 1932), both of which are undesirable in drinking water (see below and Chap. 19):

$$2ClO_2 + 2OH^- \rightarrow ClO_2^- + ClO_3^- + H_2O \tag{7-58}$$

The primary application of chlorine dioxide in the past has been for taste and odor control, although it is also an effective oxidant for reduced iron and manganese and is a good primary disinfectant (see Chap. 17). Hoigne and Bader (1994) and Tratnyek and Hoigne (1994) provide a listing and discussion of the rate constants for reactions between chlorine dioxide and a variety of organic and inorganic compounds, a number of which are of relevance to the treatment of drinking water (see Table 7-5). The rate constants are shown for pH 8 because many of the reactions are pH dependent due to acid/base behavior of the reactant, for example, amines and phenols. For the phenols and amines, the deprotonated forms are generally more reactive than the protonated forms, and the rates tend to increase 10-fold for every unit increase in pH.

One of the principal advantages of chlorine dioxide is that it does not react with ammonia (see the low rate constant in Table 7-5). Hence, much lower doses of chlorine dioxide are required for most oxidative applications compared to chlorine dosage requirements. Another advantage is that chlorine dioxide does not enter into substitution reactions with NOM to the same degree that free chlorine does and, accordingly, does not form trihalomethanes, haloacetic

TABLE 7-5 Second-Order Reaction Rate Constants for Chlorine Dioxide and Several Organic and Inorganic Compounds Found in Water (after Hoigne and Bader, 1994)

Reactant	Rate constant at pH 8, $M^{-1}s^{-1}$
Br^-	<0.05
I^-	1.4×10^3
Ammonia	$<10^{-6}$
Ferrous iron	3×10^3
Manganous manganese	5×10^4
Phenol	5×10^5
Catechol	1.58×10^8
Resorcinol	3×10^6
2-chlorophenol	1.1×10^7
2,4-dichlorophenol	1×10^6
2,4,6-trichlorophenol	5.3×10^6
Pentachlorophenol	1.4×10^3
o-cresol	2.2×10^6
m-cresol	$0,8 \times 10^6$
4-nitrophenol	1.4×10^8
Glycine	$<10^{-5}$
Dimethylamine	<1
Trimethylamine	10^2
Anasole	$<2 \times 10^{-3}$
Naphthalene	<0.3
Phthalic acid	$<10^{-3}$

acids, or most other commonly observed halogenated disinfection by-products that result from chlorination, at least not to any appreciable extent. Richardson et al. (1994) and others have identified a number of disinfection/oxidation by-products of chlorine dioxide treatment, for example, aldehydes, carboxylic acids, and some halogenated compounds, but most of the latter were present at extremely low concentrations. Because it does not tend to form halogenated disinfection by-products to any significant degree, chlorine dioxide has enjoyed renewed interest as a water treatment oxidant. An additional benefit is that chlorine dioxide reacts only very slowly with bromide (see Table 7-5). Hence, brominated by-products, either organic or inorganic (e.g., bromate), are not a concern following treatment with chlorine dioxide.

Chlorine dioxide typically reacts with most reducing agents (e.g., taste and odor compounds, NOM) through a one-electron transfer (see Table 7-1).

$$ClO_2 + e^- \rightarrow ClO_2^- \tag{7-59}$$

As a result, chlorite is considered to be the principal oxidation by-product of chlorine dioxide usage. Most researchers (e.g., Werdehoff and Singer, 1987) have reported that approximately 50 to 70 percent (by mass) of the chlorine dioxide applied during the course of drinking water treatment ends up as chlorite.

Chlorite has been demonstrated to exhibit a number of potential adverse health effects based on studies with laboratory animals (see Chap. 2). Because of this, the Stage 1 and Stage 2 Disinfectants/Disinfection By-Products rules (USEPA, 1998, 2006) establish a maximum contaminant level (MCL) of 1.0 mg/L for chlorite in the distribution system and a maximum residual disinfectant level (MRDL) of 0.8 mg/L for chlorine dioxide at the point of entry to the distribution system. No regulations exist for chlorate at this time. Given the observations that chlorine dioxide is rapidly consumed during the course of water treatment and that up to 70 percent of the applied chlorine dioxide is reduced to chlorite, the practical upper limit for chlorine dioxide doses would be approximately 1.4 to 1.5 mg/L, unless chlorite is removed (see below).

A second concern associated with residual chlorite in the distribution system is that it reacts with free chlorine, producing low levels of chlorine dioxide and/or chlorate.

$$HOCl + 2ClO_2^- \rightarrow 2ClO_2 + Cl^- + OH^- \tag{7-60}$$

$$HOCl + ClO_2^- \rightarrow ClO_3^- + Cl^- + H^+ \tag{7-61}$$

If present in tap water, the chlorine dioxide, being volatile, can be released into the home or office environment when customers open their water taps. This can lead to offensive chlorinous odors, or if new carpeting has recently been installed, the escaping chlorine dioxide can react with organic compounds released from the carpeting to produce other offensive odors (Hoehn et al., 1990).

Chlorite, whether it is present as a result of incomplete oxidation of sodium chlorite in the chlorine dioxide generator (Eqs. 7-54 and 7-55) or by chemical reduction of the chlorine dioxide during the course of treatment (Eq. 7-59), can be removed from water by the application of ferrous iron salts (Iatrou and Knocke, 1992) or reduced sulfur compounds such as bisulfite.

$$4Fe^{2+} + ClO_2^- + 10H_2O \rightarrow 4Fe(OH)_3(s) + Cl^- + 8H^+ \tag{7-62}$$

$$2HSO_3^- + ClO_2^- \rightarrow 2SO_4^{2-} + Cl^- + 2H^+ \tag{7-63}$$

No practical method for the removal of chlorate is available; hence, its formation during chlorine dioxide generation should be minimized.

Ozone and Advanced Oxidation Processes

Ozone is an unstable gas that must be generated on-site. A simplified representation of the chemistry involved in the formation of ozone (O_3) is as follows:

$$O_2 + \text{Energy} \rightarrow O + O \tag{7-64}$$

$$O + O_2 \rightarrow O_3 \tag{7-65}$$

Figure 7-6 shows an illustrative picture of an ozone generator. The energy required to produce nascent or elemental oxygen (O) from molecular oxygen (O_2) is usually supplied by an electric discharge with a peak voltage from 8 to 20 kV depending on the apparatus used. Dry, refrigerated, particle-free air, oxygen, or oxygen-enriched air is passed through a narrow gap between two electrodes and a high energy discharge is generated across the gap between the electrodes. This corona or cold plasma discharge is induced by an alternating current that creates a voltage cycle between the two electrodes. The yield of ozone depends on the voltage, the frequency, the design of the ozone generator, and the type and quality of the feed gas used. Ozone streams containing up to 14 percent ozone by volume can be produced. Current ozone generators are available as low-frequency (50–60 Hz), medium-frequency (400–1000 Hz), and high-frequency (2000–3000 Hz) systems.

Once generated, the ozone-enriched air or oxygen gas is passed through a gas absorption device to transfer ozone into solution. This can be achieved through either a countercurrent multistage bubble contactor (see Fig. 7-7) using gaseous diffusers such as those shown on the left side of Fig. 7-8 or an in-line gas injection system using Venturi-type injectors such as those shown on the right side of Fig. 7-8.

FIGURE 7-6 Illustrative picture of an ozone generator facility. (*Photo courtesy of Jim Kostelecky, Metropolitan Water District of Southern California.*)

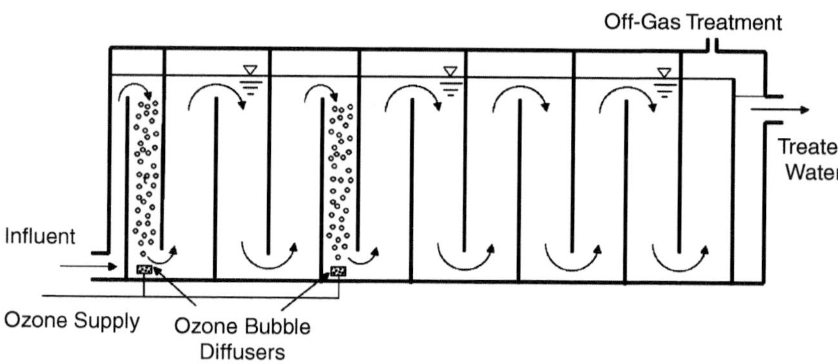

FIGURE 7-7 Schematic of a countercurrent multistage ozone contactor. (*Courtesy of Jaehong Kim, Georgia Tech.*)

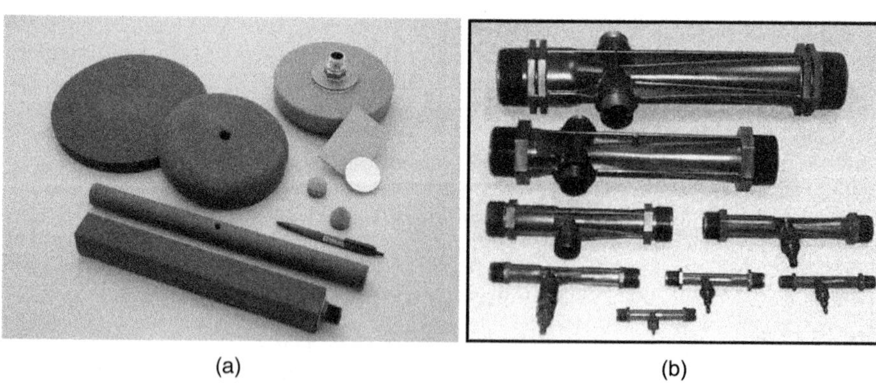

FIGURE 7-8 Illustrative (*a*) gas diffusers, and (*b*) venturi injectors for ozone transfer. (*Reprinted with permission from (a) Refractron Technologies Corp., Newark, NJ, (b) Precision Marine Systems Inc., New Braunfels, Texas.*)

Ozone is unstable in aqueous solution. It is very reactive with a number of common constituents in drinking water (e.g., NOM), and it also undergoes a spontaneous decomposition process, sometimes referred to as *auto-decomposition*. The auto-decomposition of ozone is a complex chain reaction process involving several free-radical species. Decomposition may be initiated by a number of different water constituents, such as hydroxide ion (e.g., high pH values), NOM, and ferrous iron, or it may be initiated by the addition of hydrogen peroxide or by irradiation with ultraviolet light. The reactions shown in Eqs. 7-66 through 7-71 illustrate the auto-decomposition scheme when hydroxide ion is the initiator.

$$OH^- + O_3 \rightarrow HO_2 + O_2^- \tag{7-66}$$

$$HO_2 = H^+ + O_2^- \tag{7-67}$$

$$O_2^- + O_3 \rightarrow O_2 + O_3^- \tag{7-68}$$

$$O_3^- + H^+ \rightarrow HO_3 \tag{7-69}$$

$$HO_3 \rightarrow O_2 + OH \tag{7-70}$$

$$OH + O_3 \rightarrow HO_2 + O_2 \tag{7-71}$$

Because the OH radical reacts rapidly with molecular ozone (Eq. 7-71), the decomposition of ozone is an autocatalytic reaction.

These reactions constitute a chain mechanism because the hydroperoxyl radical (HO_2) produced by Eq. 7-71 generates additional superoxide (O_2^-; Eq. 7-67) that further contributes to ozone decomposition. In pure water, the chain may be very long; for example, hundreds of ozone molecules may be decomposed by a single initiation step. In natural waters, the lifetime of ozone depends on several variables, including pH, temperature, total organic carbon (TOC) concentration, and bicarbonate and carbonate concentrations. Bicarbonate and carbonate increase the lifetime of ozone by reacting with the hydroxyl radical (OH), thereby slowing down the chain reaction mechanism shown in Eqs. 7-66 to 7-71.

$$OH + HCO_3^- \rightarrow OH^- + HCO_3 \tag{7-72}$$

$$OH + CO_3^{2-} \rightarrow OH^- + CO_3^- \tag{7-73}$$

The bicarbonate and carbonate radicals (HCO_3 and CO_3^-, respectively) are relatively unreactive intermediates that cannot propagate the chain. Thus, waters high in bicarbonate and carbonate alkalinity will retain an ozone residual for longer periods of time than low-alkalinity waters. This is especially important when ozone is used as a disinfectant (see Chap. 17). Additionally, the radical-scavenging activity of the carbonate species increases with increasing pH because carbonate is a more effective scavenger than bicarbonate. This partially offsets the more rapid rate of the hydroxide-induced initiation reaction (Eq. 7-66) at higher pH values (higher hydroxide ion concentrations).

The hydroxyl radical, one of the intermediates produced by the decomposition of ozone, is one of the strongest chemical oxidants known (see Table 7-1), and is capable of rapidly reacting with a myriad of organic and inorganic impurities in water (see below). Accordingly, the oxidative properties of ozone depend significantly on the oxidative characteristics of this free-radical species.

Figure 7-9 is an illustrative schematic of the behavior of ozone in aqueous solution based on the fundamental work of Hoigne and his coworkers (e.g., Hoigne and Bader, 1976; Staehelin and Hoigne, 1982). Figure 7-9 shows that ozone reacts by two distinct types of pathways: a direct pathway involving molecular ozone (O_3) and an indirect pathway originating with the decomposition of ozone to produce the hydroxyl radical (OH). Direct reactions involving molecular ozone are selective; ozone reacts rapidly with some species (M_d in the figure), but only slowly with other species. The kinetics of reactions of

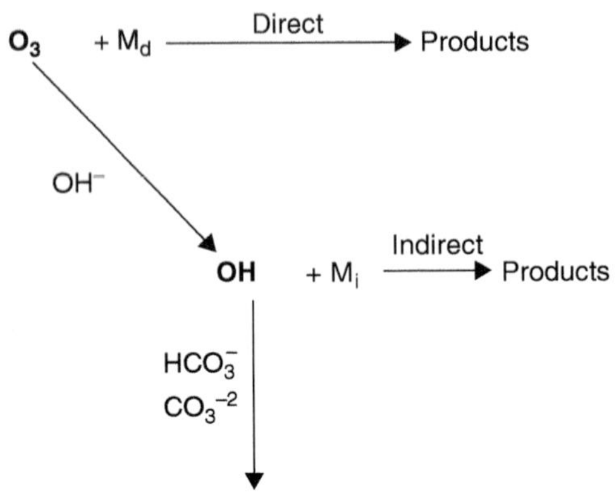

FIGURE 7-9 Illustrative direct and indirect reaction pathways of ozone.

molecular ozone with various impurities in water have been studied extensively. The most comprehensive listings of reaction rate constants are those provided by Hoigne and co-workers (e.g. Hoigne and Bader, 1983a, 1983b; Hoigne et al., 1985) and von Gunten (2003). Illustrative second-order rate constants for molecular ozone with a number of potential drinking water contaminants are shown in Table 7-6. As indicated, some solutes react quickly with molecular ozone whereas others react more slowly. The rate constants vary over ten orders of magnitude. For example, the oxidation of sulfide (S^{-2}) by molecular ozone is

TABLE 7-6 Second-Order Rate Constants for the Oxidation of Selected Drinking Water Contaminants by Molecular Ozone and the Hydroxyl Free Radical (after von Gunten, 2003)

Reactant	Second-order rate constant for O_3, $M^{-1}s^{-1}$	Second-order rate constant for OH, $M^{-1}s^{-1}$
Ammonia, NH_3	20	9.7×10^7
Ferrous iron, Fe(II)	8.2×10^5	3.5×10^8
Manganese, Mn(II)	1.5×10^3	2.6×10^7
Hydrogen sulfide, H_2S	$\sim 3 \times 10^4$	1.5×10^{10}
Sulfide, S^{2-}	3×10^9	9×10^9
Bromide, Br^-	160	1.1×10^9
Geosmin	<10	8.2×10^9
2-methylisoborneol	<10	$\sim 3 \times 10^9$
Microcystin-LR	3.4×10^4	—
Atrazine	6	3×10^9
Alachlor	3.8	7×10^9
Carbofuran	620	7×10^9
Endrin	<0.02	1×10^9
Methoxychlor	270	2×10^{10}
Chloroethene (vinyl chloride)	1.4×10^4	1.2×10^{10}
Cis-1, 2-dichloroethene	540	3.8×10^9
Trichloroethene	17	2.9×10^9
Tetrachloroethene	<0.1	2×10^9
Chlorobenzene	0.75	5.6×10^9
Benzene	2	7.9×10^9
Toluene	14	5.1×10^9
o-Xylene	90	6.7×10^9
MTBE	0.14	1.9×10^9
Chloroform	<0.1	5×10^7
Bromoform	<0.2	1.3×10^8
Diclofenac	$\sim 1 \times 10^6$	7.5×10^9
Carbamazepine	$\sim 3 \times 10^5$	8.8×10^9
Sulfamethoxazole	$\sim 2.5 \times 10^6$	5.5×10^9
17α-ethinylestradiol	$\sim 7 \times 10^9$	9.8×10^9

extremely rapid, with a second-order rate constant of 3×10^9 $M^{-1}s^{-1}$. The second-order rate constants for the oxidation of the solvent vinyl chloride (chloroethene) and the pesticide carbofuran by molecular ozone are also relatively high at 1.4×10^4 and 620 $M^{-1}s^{-1}$, respectively. Conversely, the second-order rate constants for the oxidation of benzene and tetrachloroethene, two common groundwater contaminants, are 2 and <0.1 $M^{-1}s^{-1}$, respectively. The oxidation of trichloroethene and atrazine, two other common contaminants of drinking water, are also relatively low, at 6 and 17 $M^{-1}s^{-1}$. Additionally, ammonia reacts very slowly with molecular ozone in the pH range of interest in drinking water treatment.

The second-order rate constant for the reaction of olefins with molecular ozone decreases by a factor of more than 10 for each additional chlorine substituent. Also, while benzene itself is relatively unreactive toward molecular ozone, derivatives of benzene tend to be much more reactive, especially with the addition of electron-withdrawing substituents on the aromatic ring. Compounds that dissociate in water react with molecular ozone at rates that depend on the individual species present. Amines and amino acids react quickly when the amino group is deprotonated, that is, with increasing pH. The same is true for phenols; phenolate ($C_6H_5O^-$) reacts 10^6 times faster than molecular phenol (C_6H_5OH; Hoigne and Bader, 1981b). Saturated compounds, such as aldehydes, ketones, and organic acids react very slowly with molecular ozone, although deprotonation of the organic acids increases their reactivity. Methyl and other alkyl ethers also react slowly.

Conversely, the OH radical is relatively nonselective in its behavior, reacting rapidly with a large number of species (illustrated by M_i in Fig. 7-9). Table 7-6 presents illustrative second-order rate constants for the hydroxyl radical with a number of potential drinking water contaminants. Whereas the rate constants for molecular ozone range over ten orders of magnitude, the rate constants for the hydroxyl radical range over only three orders of magnitude, from 10^7 to 10^{10} $M^{-1}s^{-1}$. It is for this reason that the hydroxyl radical is often called *nonselective* with respect to its reactivity. Atrazine, benzene, geosmin, trichloroethene, and tetrachloroethene, all relatively inert toward molecular ozone, are oxidized at appreciable rates by the hydroxyl radical. Because of the large magnitude of the rate constants for oxidation reactions involving the hydroxyl radical, the rate-determining step in these reactions is the rate at which the radicals are generated either by ozone decomposition or by other advanced oxidation processes (see below). Hence, in many cases, the oxidation of these solutes is enhanced at elevated pH values because of the more rapid generation of hydroxyl radicals by hydroxide-induced ozone decomposition.

Hydroxyl free radicals can be produced through a number of other pathways in addition to the hydroxide-induced ozone decomposition chain reaction described previously. For example, the addition of both hydrogen peroxide and ozone to water accelerates the decomposition of ozone and enhances production of the hydroxyl radical. The hydrogen peroxide (H_2O_2) dissociates into the hydroperoxide ion (HO_2^-).

$$H_2O_2 = HO_2^- + H^+ \tag{7-74}$$

The hydroperoxide ion then reacts with molecular ozone to produce the superoxide ion (O_2^-) and the hydroxyl radical, plus molecular oxygen.

$$HO_2^- + O_3 \rightarrow OH + O_2^- + O_2 \tag{7-75}$$

The hydroxyl radical and the superoxide ion then participate in the ozone decomposition chain depicted by Eqs. 7-68 to 7-71; this leads to an accelerated production of more hydroxyl radicals. Because reactions 7-74 and 7-75 tend to be appreciably faster than reaction 7-66 under most conditions, the conjunctive use of O_3 and H_2O_2 tends to be a more effective method of generating the highly reactive, nonselective hydroxyl radicals for chemical oxidation reactions.

Another method for generating hydroxyl radicals is by ultraviolet (UV) irradiation of hydrogen peroxide. The UV irradiation provides the energy needed to split the hydrogen peroxide into two hydroxyl radicals.

$$H_2O_2 + UV \rightarrow 2OH \tag{7-76}$$

This process tends to be slower than reactions 7-74 and 7-75 and is therefore a less effective method of generating hydroxyl radicals.

Additionally, ultraviolet irradiation of waters containing dissolved molecular ozone leads directly to the formation of hydrogen peroxide. The hydrogen peroxide then reacts with molecular ozone in the same manner as it does in reactions 7-74 and 7-75, which produces the hydroxyl radical and other species that enter the ozone decomposition chain depicted by Eqs. 7-68 to 7-71.

The preceding reactions, all of which involve the accelerated production of the hydroxyl free radical, are termed advanced oxidation processes. Often, when applying ozone to natural waters containing impurities, it is difficult to distinguish between reactions that are attributable to molecular ozone and those which are attributable to the hydroxyl radical, although there are analytical techniques that can be used in the laboratory to distinguish between the indirect and direct reaction pathways (Hoigne and Bader, 1983a; Elovitz and von Gunten, 1999; Elovitz et al., 2000). It is believed that hydroxyl radical reactions are at the heart of all reactions involving molecular ozone, except perhaps for disinfection reactions, reactions with some solutes that have very high reaction rate constants with molecular ozone (see Table 7-6), and reactions that take place in the presence of high concentrations of hydroxyl radical scavengers.

Formation of Biodegradable Organic Material. When ozone reacts with organic contaminants in water, including NOM, it partially oxidizes them to lower molecular weight and more polar species, including a variety of aldehydes and organic acids. These oxidation by-products, while not believed to be harmful in themselves, tend to be biodegradable and may contribute to biofouling problems in the water distribution system if not properly controlled. Often, ozonation is followed by a biologically active filtration process to remove these biodegradable organic materials (see Chaps. 10 and 14) to produce a biologically stable finished water. Ozonation by-products are discussed in greater detail in Chap. 19.

Reactions of Ozone with Bromide. Ozone is capable of oxidizing bromide to hypobromous acid in the same manner that chlorine does.

$$O_3 + Br^- + H^+ \rightarrow HOBr + O_2 \tag{7-77}$$

This is a fast reaction, with a second-order reaction rate constant of 160 $M^{-1}s^{-1}$ (see Table 7-6). The hypobromous acid can then react with NOM to produce brominated DBPs, such as bromoform, dibromoacetic acid and bromopicrin. This is discussed in greater detail in Chap. 19.

A particular concern associated with the ozonation of bromide-containing waters is the formation of bromate (BrO_3^-). Bromate is a possible human carcinogen and is currently regulated at 10 μg/L, the *practical quantitation limit (PQL)* for bromate, although it is expected that this level will decrease in the future with the development of more sensitive and robust analytical techniques for measuring bromate. Bromate is produced by a number of possible pathways involving molecular ozone and the hydroxyl radical (von Gunten and Oliveras, 1998; see Fig. 7-10).

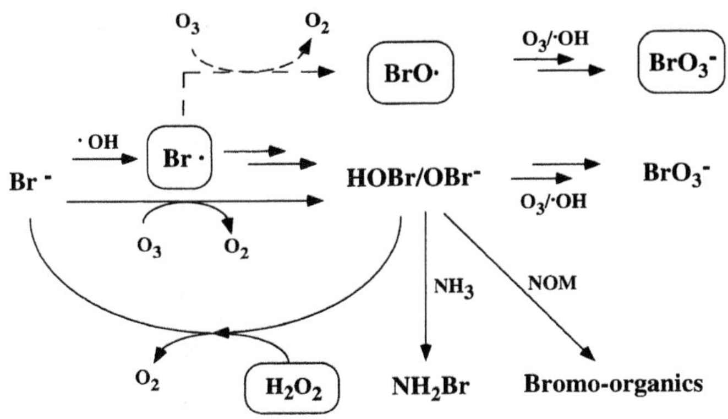

FIGURE 7-10 Pathways illustrating the reaction of ozone and bromide to form bromate and brominated organic compounds. (*Source: Sulzberger et al., 1997, reprinted with permission from Chimia, Copyright 1997.*)

One such pathway proceeds through the formation of hypobromite (OBr^-) and bromite (BrO_2^-) and involves only molecular ozone.

$$O_3 + Br^- \rightarrow OBr^- \tag{7-78}$$

$$OBr^- + O_3 \rightarrow BrO_2^- \tag{7-79}$$

$$BrO_2^- + O_3 \rightarrow BrO_3^- \tag{7-80}$$

In the pH range of most natural waters, a large portion of the OBr^- is protonated as HOBr (the pK_a for hypobromous acid is 8.7 at 25°C), so that the rate of formation of bromite by reaction 7-79 is slowed considerably with decreasing pH.

If the rate of formation of hydroxyl radicals in the water is high, bromate may also be formed through the hypobromite radical (BrO) as follows:

$$HOBr/OBr^- + OH \rightarrow BrO \tag{7-81}$$

$$BrO + O_3 \rightarrow BrO_3^- \tag{7-82}$$

Techniques for controlling bromate formation most often involve ozonation at slightly acidic pH values, multistage ozonation in which the ozone is added at several different application points, and the use of ammonia or chlorine and ammonia together to convert the hypobromous acid produced to bromamine (see Chap. 19).

Once formed, bromate is somewhat difficult to remove, although it can be removed by chemical reduction using reduced sulfur compounds, such as bisulfite (HSO_3^-) or ferrous iron. Granular activated carbon is also capable of adsorbing bromate, albeit to a limited degree, and medium pressure ultraviolet irradiation decomposes bromate to bromide. Bromate control is discussed further in Chap. 19.

Potassium Permanganate

Potassium permanganate ($KMnO_4$) has been used as a water treatment oxidant for decades. It is commercially available in crystalline form and is either fed into solution directly using a dry chemical feeder or a concentrated solution is prepared on-site from which the desired dose is metered into the water. Sodium permanganate is also commercially available.

Permanganate contains manganese in the +VII oxidation state. Under most treatment applications, oxidation by permanganate involves a three-electron transfer, with the permanganate (MnO_4^-) being reduced (see Table 7-1) to insoluble manganese dioxide (MnO_2(s)).

$$MnO_4^- + 4H^+ + 3e^- \rightarrow MnO_2(s) + 2H_2O \tag{7-83}$$

The manganese dioxide produced is a black precipitate which, if not properly removed by a suitable solid-liquid separation process, will create black particulate deposits in the distribution system and in household plumbing fixtures. Most often, removal of manganese dioxide is achieved by conventional clarification and/or filtration processes. Because most operators are fearful of seeing pink water (reflecting unreacted permanganate) coming through their filters, permanganate is commonly added at the head of the treatment plant, as close to the intake as possible. This allows sufficient time for the permanganate to perform its oxidative function and to be reduced completely to solid manganese dioxide prior to filtration.

The kinetics of oxidation reactions involving permanganate tend to be more rapid with increasing pH values. Hence, in some cases, addition of a base prior to filtration may be desirable to hasten the reduction of permanganate.

The manganese dioxide that results from permanganate reduction (Eq. 7-83) has some beneficial attributes. Manganese dioxide is an effective adsorbent for ferrous iron (Fe^{2+}), manganous manganese (Mn^{2+}), radium (Ra^{2+}), and other trace inorganic cationic species. Accordingly, some additional removal of Fe^{2+} and Mn^{2+}occurs as a result of permanganate treatment beyond that achieved simply through oxidation by permanganate. In fact, the adsorptive behavior of MnO_2(s) is the principle underlying the historic manganese greensand process in which the filter media is coated with manganese dioxide which serves as an adsorbent for Fe^{2+}, Mn^{2+}, and Ra^{2+} in the filter influent. The filter backwash water is treated with permanganate, or low doses of permanganate are applied to the filter influent to oxidize adsorbed Fe^{2+} and Mn^{2+}, thereby creating additional adsorption sites.

Solid manganese dioxide is also capable of adsorbing NOM that serves as DBP precursors. This benefit is particularly pronounced in hard waters (Colthurst and Singer, 1982), presumably because of the bridging action of calcium and magnesium which facilitates the adsorption of NOM to the negatively charged MnO_2 surface.

Potassium Ferrate

Potassium ferrate (K_2FeO_4) is a strong oxidant that is relatively new to the water treatment field (Sharma, 2002; Jiang and Lloyd, 2002). Ferrate, which contains iron in the +VI oxidation state, has a standard half-cell potential of 2.07 V (see Table 7-1); this makes it a powerful oxidant. One of the attractive features is that when the Fe(VI) is reduced, it produces Fe(III) which can serve a coagulation function, or an adsorptive function with respect to the removal of arsenic.

$$FeO_4^{2-} + 5H^+ + 3e^- \rightarrow Fe(OH)_3\,(s) + H_2O \tag{7-84}$$

Ferrate is unstable under acidic conditions but is stable in alkaline solutions. It can be produced in the laboratory by oxidizing an Fe(III) solution with concentrated sodium hypochorite under alkaline conditions, or it can be produced electrochemically from metallic iron under strongly alkaline conditions. Recent studies by a number of researchers have shown that ferrate rapidly oxidizes many trace organic contaminants in water, including pharmaceutically active compounds and endocrine disruptors (Lee et al., 2009). Ferrate also has good disinfecting properties.

Mixed Oxidants

Electrolysis of brine has been used since the nineteenth century to produce chlorine on an industrial scale (see Fig. 7-5). This technology has recently been modified and adapted to electrochemically produce a mixture of free chlorine and other powerful oxidants and disinfectants for drinking water applications for small, rural water supplies. Both liquid- and gas-phase generators are available.

The underlying principle is based on fundamental electrochemical theory. At the anode of the electrochemical cell, chloride is oxidized to chlorine, which subsequently hydrolyzes to hypochlorous acid.

$$2Cl^- \rightarrow Cl_2 + 2e^- \tag{7-85}$$

$$Cl_2 + H_2O \rightarrow HOCl + Cl^- + H^+ \tag{7-86}$$

The pH of the anodic stream tends to be on the order of 3 to 5. At the cathode, water is reduced to hydrogen gas, producing a strongly alkaline solution with a pH of about 10 to 11.

$$2H_2O + 2e^- \rightarrow H_2 + 2OH^- \tag{7-87}$$

The anodic and cathodic streams are separated by a semipermeable barrier. While free chlorine is the primary oxidant produced by these mixed oxidant generators, other reactions occur at the anode, purportedly resulting in the formation of ozone, chlorine dioxide, hydrogen peroxide, other short-lived reactive oxidant species, and a number of inorganic by-products, such as chlorite, chlorate, and bromate. However, the composition of the oxidant streams from these mixed oxidant generators has not been fully characterized. More research is needed to characterize these product streams before widespread use of these generators is recommended.

APPLICATIONS OF OXIDATION PROCESSES TO WATER TREATMENT

This section describes typical applications of oxidants and the role they play in overall water treatment practice. These oxidants are most often used for the oxidation of reduced iron and manganese, destruction of taste- and odor-causing organic contaminants, elimination of color, and the oxidation of organic chemicals (micropollutants) of public health concern. Additionally, many of these oxidants act as coagulant aids, and are also employed as part of an overall program for the control of potentially harmful disinfection by-products. The formation and control of oxidation and disinfection by-products is discussed separately in Chap. 19. Many of these oxidants are also powerful disinfectants and therefore serve the dual purposes of oxidation and disinfection. Disinfection is discussed in detail in Chap. 17.

Control of Iron and Manganese

Iron and manganese are relatively soluble under reducing conditions, for example, in groundwaters, stagnant surface waters, and hypolimnetic waters of eutrophic lakes, reservoirs, and impoundments (see Chap. 3 for a discussion of the occurrence of Fe and Mn in source waters). Correspondingly, they are quite insoluble under oxidizing conditions, e.g., in flowing streams and in the epilimnetic waters of lakes or impoundments, at pe values greater than 9 for manganese and at pe values greater than zero for iron (see Fig. 3-5). Accordingly, dissolved iron and manganese are rarely found at appreciable concentrations under oxidizing conditions. This is the basis of hypolimnetic aeration, which introduces oxygen into the bottom waters of impoundments to maintain oxidizing conditions, which keep iron and manganese immobilized in their insoluble oxidized forms. It is also the basis of in-situ control of iron and manganese in groundwaters, where oxygen-enriched water is injected into aquifers to create an oxidized zone around the well point that limits iron and manganese mobility (Braester and Martinell, 1988).

The reduced forms of iron and manganese, ferrous iron [Fe(II)] and manganous manganese [Mn(II)], respectively, may occur as the free metal ions, Fe^{2+} and Mn^{2+}, as is often the case in most groundwaters, or they may be found complexed to various degrees with NOM, as is often the case in surface waters and highly colored groundwaters. During and immediately following lake overturn, that is, when iron- and manganese-rich hypolimnetic water is mixed with the remainder of the lake water, dissolved iron and manganese levels in the upper portions of the lake can increase appreciably. The duration of this increase may be days or weeks, depending on a number of factors, for example, temperature, pH, dissolved oxygen concentration, and the presence of microorganisms capable of catalyzing the oxidation of ferrous iron and manganous manganese to their insoluble oxidized forms. The primary concern with elevated levels of dissolved iron and manganese in water is that, if they are not removed, insoluble ferric hydroxide ($Fe(OH)_3(s)$) and manganese dioxide ($MnO_2(s)$) precipitates are produced when the reduced forms of the metals are oxidized by dissolved oxygen or chlorine. These precipitates cause reddish-orange or black deposits, respectively, to appear on plumbing fixtures and to create stains during laundering operations. As a result, the secondary maximum contaminant levels for iron and manganese in finished drinking water in the United States have been set at 0.3 and 0.05 mg/L, respectively, although many utilities have established finished water goals for manganese of less than 0.02 mg/L.

Dissolved Fe(II) and Mn(II) are usually removed from water by oxidizing them under engineered conditions to their insoluble forms through the addition of an oxidant and then removing the precipitated ferric hydroxide and manganese dioxide by sedimentation and filtration. The oxidants used most often for this purpose are oxygen, chlorine, permanganate, chlorine dioxide, and ozone. Equations 7-88 and 7-89 show the balanced oxidation reactions with dissolved oxygen following aeration.

$$Fe^{2+} + \tfrac{1}{4}O_2 + \tfrac{5}{2}H_2O \rightarrow Fe(OH)_3(s) + 2H^+ \tag{7-88}$$

$$Mn^{2+} + \tfrac{1}{2}O_2 + H_2O \rightarrow MnO_2(s) + 2H^+ \tag{7-89}$$

Table 7-7 shows balanced reactions for the oxidation of reduced iron and manganese with various strong oxidants, and the corresponding oxidant demands. It should be noted that, although manganese dioxide is the principal form of oxidized manganese shown

TABLE 7-7 Oxidant Demands of Reduced Iron and Manganese

Reaction	Oxidant demand
$2Fe^{2+} + HOCl + 5H_2O \rightarrow 2Fe(OH)_3(s) + Cl^- + 5H^+$	0.64 mg of chlorine per mg Fe(II)
$5Fe^{2+} + ClO_2 + 13H_2O \rightarrow 5Fe(OH)_3(s) + Cl^- + 11H^+$	0.24 mg of chlorine dioxide per mg Fe(II)
$3Fe^{2+} + MnO_4^- + 7H_2O \rightarrow 3Fe(OH)_3(s) + MnO_2(s) + 5H^+$	0.71 mg of permanganate per mg Fe(II)
$2Fe^{2+} + O_3 + 5H_2O \rightarrow 2Fe(OH)_3(s) + O_2 + 4H^+$	0.43 mg of ozone per mg Fe(II)
$Mn^{2+} + HOCl + H_2O \rightarrow MnO_2(s) + Cl^- + 3H^+$	1.29 mg of chlorine per mg Mn(II)
$Mn^{2+} + 2ClO_2 + 2H_2O \rightarrow MnO_2(s) + 2ClO_2^- + 4H^+$	2.46 mg of chlorine dioxide per mg Mn(II)
$3Mn^{2+} + 2MnO_4^- + 2H_2O \rightarrow 5MnO_2(s) + 4H^+$	1.45 mg of permanganate per mg Mn(II)
$Mn^{2+} + O_3 + H_2O \rightarrow MnO_2(s) + O_2 + 2H^+$	0.87 mg of ozone per mg Mn(II)

for these oxidants, various forms of solid manganese oxidation products exist, e.g., MnOOH(s), Mn_2O_3(s), Mn_3O_4(s), and MnO_2(s), such that the oxidized insoluble form of manganese is often expressed as MnO_x(s) with values of x ranging from 1.3 to 1.9 (Stumm and Morgan, 1996).

The kinetics of oxidation of Fe(II) by oxygen are relatively rapid at pH values above 7 (Stumm and Morgan, 1996), provided that the ferrous iron is not complexed by organic material. This is often the case for the removal of iron from groundwater, where dissolved organic carbon concentrations tend to be relatively low. The rate expression for the oxygenation of ferrous iron is

$$\frac{-d[\text{Fe(II)}]}{dt} = k[\text{Fe(II)}][\text{OH}^-]^2 p_{\text{O}_2} \tag{7-90}$$

The kinetics are first-order with respect to the concentration of ferrous iron and the partial pressure of oxygen (p_{O_2}) and second-order with respect to the hydroxide ion concentration (Stumm and Morgan, 1996). The latter illustrates the strong pH dependency of Fe(II) oxidation by oxygen and is due to the pathway by which the ferrous iron is oxidized; that is, oxidation proceeds primarily via the oxidation of the zerovalent $Fe(OH)_2$ species. The rate increases tenfold for every 0.1 increase in pH. The rate constant k has a value of 8.0×10^{13} $L^2mole^{-2}atm^{-1}min^{-1}$ at 20°C.

In the case of Mn(II), oxidation by oxygen has been shown to conform to the following kinetic expression:

$$\frac{-d[\text{Mn(II)}]}{dt} = k_0[\text{Mn(II)}] + k[\text{Mn(II)}][\text{MnO}_2(\text{s})] \tag{7-91}$$

where the rate constants k_o and k are dependent on pH and the oxygen concentration. This is an autocatalytic reaction whereby the by-product of the oxidation reaction, manganese dioxide (MnO_2(s)), catalyzes the oxidation of reduced manganese as shown by the second-term on the right (Stumm and Morgan, 1996).

Despite the close chemical relationship between iron and manganese, the physical-chemical conditions for the oxidation of the two metals require different treatment strategies. This is seen by the different oxidation potentials for Fe^{2+} and Mn^{2+} (see Table 7-2), the different kinetic relationships defining their oxidation rates, and the different oxidation states of their oxidized products.

The oxidation of Fe(II) by oxygen is relatively rapid at neutral pH values (pH 7 and above) whereas the initial oxidation step for the chemical oxidation of Mn(II) by oxygen, represented by the first term on the right side of Eq. 7-91, is orders of magnitude slower at neutral pH. Accordingly, whereas simple aeration is commonly used to provide oxygen for the oxidation of ferrous iron in groundwaters, aeration is not an effective means for the chemical oxidation of reduced manganese. If, however, the ferrous iron is complexed by organic material, the rate of oxidation by oxygen can be very slow (Theis and Singer, 1974) and a stronger oxidant, for example, chlorine, permanganate, chlorine dioxide, or ozone, may be needed. Likewise, because the oxidation of manganous manganese by oxygen is relatively slow at neutral pH values, a strong chemical oxidant is often used for the oxidation of manganese. Again, Table 7-7 gives the chemical requirements for the oxidation of reduced iron and manganese by a number of strong chemical oxidants, although higher doses are typically needed to overcome the oxidant demand of dissolved organic material (Knocke et al., 1994).

The rates of oxidation of manganous manganese by permanganate and chlorine dioxide are illustrated, respectively, in Figs. 7-11 and 7-12. The rates of Mn(II) oxidation with both oxidants increase with increasing pH, increase with increasing temperature, and decrease with increasing amounts of NOM. In all cases shown, the half-times for Mn(II) oxidation tend to be less than 60 sec, attesting to rapid oxidation kinetics with these strong

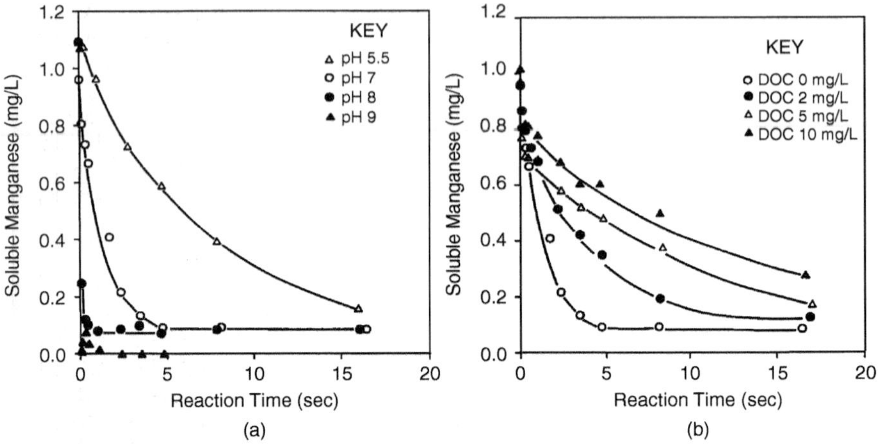

FIGURE 7-11 Rate of Mn(II) oxidation by permanganate under various conditions. (*a*) Effect of pH at 25°C, DOC<1 mg/L; (*b*) Effect of DOC concentration at pH 7, 25°C. (*Source: Knocke et al., 1991a; reprinted with permission from the American Water Works Association, Copyright 1991.*)

oxidants. Gregory and Carlson (1993), however, showed that oxidation kinetics were highly dependent on the concentration of reduced manganese and that, at the low Mn(II) concentrations typically found in raw drinking waters, oxidation was appreciably slower, requiring up to 30 min to lower dissolved manganese concentrations to less than 0.01 mg/L. They found that the oxidation reaction was much faster for chlorine dioxide than for permanganate. In the case of ozone, selection of the proper ozone dose for oxidation of Mn(II) to its insoluble form is more challenging; sufficient ozone must be applied to overcome the oxidant demand of NOM in the water, but too much ozone will convert the Mn(II) to permanganate, as shown by Eq. 7-92 and Example 7-1 at the beginning of this

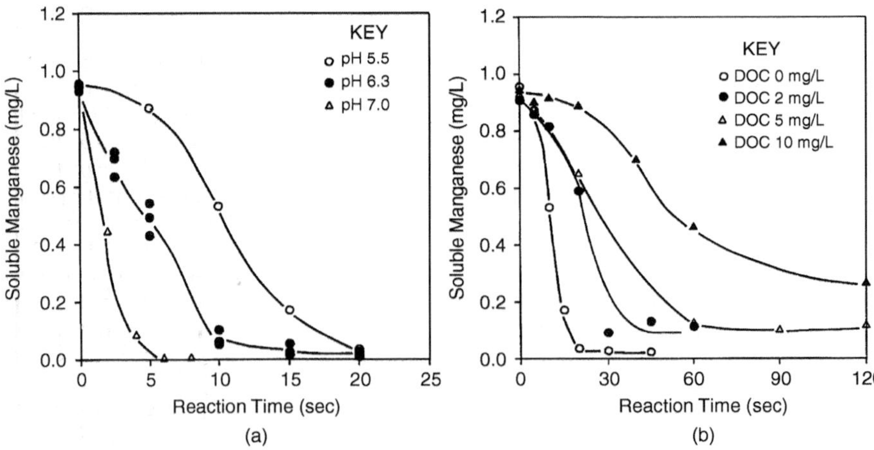

FIGURE 7-12 Rate of Mn(II) oxidation by chlorine dioxide under various conditions. (*a*) Effect of pH at 25°C, DOC<1 mg/L; (*b*) Effect of DOC concentration at pH 7, 25°C. (*Source: Knocke et al., 1991a; reprinted with permission from the American Water Works Association, Copyright 1991.*)

chapter, necessitating additional contact time for the permanganate to be reduced to the insoluble manganese oxides.

$$Mn^{2+} + \tfrac{5}{2}O_3 + \tfrac{3}{2}H_2O \rightarrow MnO_4^- + \tfrac{5}{2}O_2 + 3H^+ \tag{7-92}$$

For all of the strong oxidants, in the presence of NOM, higher doses beyond the stoichiometric amounts shown in Table 7-7 are required due to the oxidant demand of the organics. Figure 7-13a and b illustrates this for chlorine dioxide and permanganate, respectively. The stoichiometric requirements for the oxidation of 0.25 mg/L of Mn(II) are 0.62 mg/L for chlorine dioxide and 0.36 mg/L for permanganate so it is clear that appreciably higher oxidant doses are required to achieve effective manganese oxidation. Furthermore, the rates of Mn(II) oxidation tend to be much slower in the presence of NOM. The data shown in Fig. 7-13 are for a total oxidant contact time of 1 hr. A similar situation occurs with ferrous iron; that is, kinetics are slower in the presence of NOM, and greater amounts of the oxidants are needed beyond the stoichiometric requirements (Knocke et al., 1994). It should also be noted that the oxidized iron and manganese may be colloidal in nature and therefore difficult to remove by conventional solid-liquid separation techniques. This is especially true in the presence of NOM.

An alternative approach to removing manganese from water besides conventional oxidation, clarification and filtration is by adsorption/oxidation in a filter bed. Because the oxidation of Mn(II) is catalyzed by manganese dioxide (see the second term on the right side of Eq. 7-91) and manganese dioxide strongly adsorbs Mn^{2+}, a common practice for manganese oxidation and removal is to allow the filter media to become coated by manganese dioxide and other insoluble oxides of manganese, $MnO_x(s)$. Reduced manganese in the filter influent then adsorbs to these manganese oxide coatings and can be oxidized within the filter bed using chlorine or permanganate in the filter influent or during backwashing. This sequence of reactions can be represented as

$$Mn^{2+} + MnO_x(s) \rightarrow MnO_x(s){\cdot}Mn^{2+} \tag{7-93}$$

$$MnO_x(s){\cdot}Mn^{2+} + HOCl + H_2O \rightarrow 2MnO_x(s) + Cl^- + 3H^+ \tag{7-94}$$

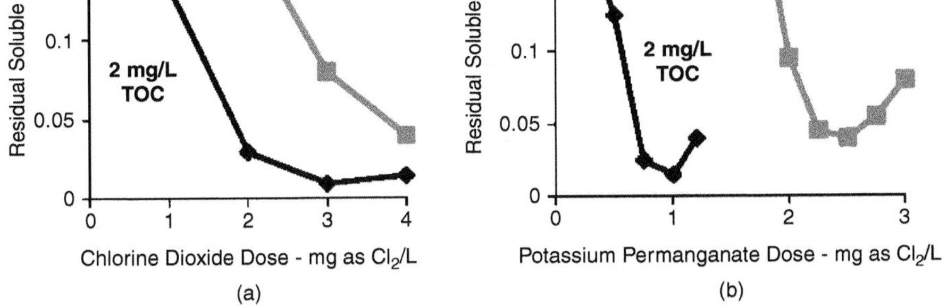

FIGURE 7-13 Effect of TOC on Mn(II) oxidation by (*a*) chlorine dioxide at pH 5.0 and 20°C, and (*b*) permanganate at pH 5.5 and 20°C. (*After Knocke et al., 1987; reprinted with permission from the American Water Works Association, Copyright 1987.*)

Commercially available glauconite (greensand), which has an affinity for the adsorption of Mn(II), can also be used as a filter media in the same manner. Other available proprietary products containing pyrolusite (β-MnO_2) such as Pyrolox® and Filox®, are also commercially available. Such adsorbent media are also effective for the removal of reduced iron and arsenic (Ghurye and Clifford, 2004) and radium (Qureshi and Nelson, 2003).

When employing Mn oxide-coated filter media for Mn removal, care must be exercised with respect to maintaining optimal pH and oxidizing conditions in the filter bed (Knocke et al., 1988; Knocke et al., 1991b). Figure 7-14a shows that Mn removal increases with increasing pH. Tobiason et al. (2008) also demonstrated that both Mn uptake capacity and rate of Mn uptake increased as pH increased from 5.5 to 8.0. To ensure effective removal of manganese, a slightly alkaline pH value (7–8) is recommended.

Figure 7-14b illustrates the importance of the manganese oxide coating, with removal increasing as the extent of the Mn coating increases. The extent of the manganese oxide coating in several cores taken from a filter bed employing adsorption/oxidation for Mn removal is shown in Fig. 7-15. Effective Mn removal in full-scale filter media has been observed even with Mn coating levels as low as 1–2 mg Mn/g media as long as the proper pH is maintained.

Figures 7-16 and 7-17 show that it is important to maintain a free chlorine residual across the filter bed in order to retain manganese in the bed. Many utilities employ an operational target of 0.5 to 1 mg/L of free chlorine in the filter effluent for this purpose. Because MnO_2 is a moderately strong oxidant, it has the potential to undergo reductive dissolution within the filter bed to produce soluble Mn(II) in the absence of a strong oxidant residual, leading to manganese release from the bed. Again, pH is an important variable influencing the reductive dissolution of MnO_2.

An adaption of the oxide-coated media process for manganese removal involves the use of a post-filter, coarse media, manganese-coated adsorptive contactor (Tobiason et al., 2008). Because many utilities have had to (or would like to) eliminate free chlorine application to their conventional filters to reduce DBP production or employ biological filtration for dissolved organic carbon (DOC) removal, their ability to remove soluble Mn is compromised. The post-filter manganese oxide-coated adsorptive contactor concept offers the potential for controlling Mn in a separate bed under conditions that are different from those needed for optimal particle and NOM removal. For example, coagulation for NOM removal with alum is optimal at pH 6 to 6.5 whereas conditions for Mn uptake are better at pH values greater than 7.

Another approach for controlling manganese in water is through biological oxidation and filtration. Manganese biofiltration utilizes naturally occurring manganese-oxidizing bacteria in biofilms attached to the filter media to oxidize Mn(II). The resulting insoluble manganese oxides are retained within the filter bed and are available to adsorb additional Mn(II). Accordingly, the biological process cannot be considered in isolation and is dependent on the chemical behavior of Mn, i.e., the physical chemical factors and at the surface/water interface (e.g., pH, pe, temperature, solution-phase Mn(II) concentration, Mn coating on the filter media).

Microbial oxidation processes for Mn removal in drinking water treatment plants are widely used in Europe (Peitchev and Semov, 1988; Mouchet, 1992; Vandenabeele et al., 1992; Cameron, 1996; Katsoyiannis and Zouboulis, 2004; Pacini et al., 2005; Burger et al., 2008), and the subject is receiving increasing attention in the United States as more utilities move their application point for chlorine further downstream in the treatment train to avoid the formation of DBPs. Further research and development is required before it becomes acceptable for widespread incorporation into drinking water treatment practice in the United States.

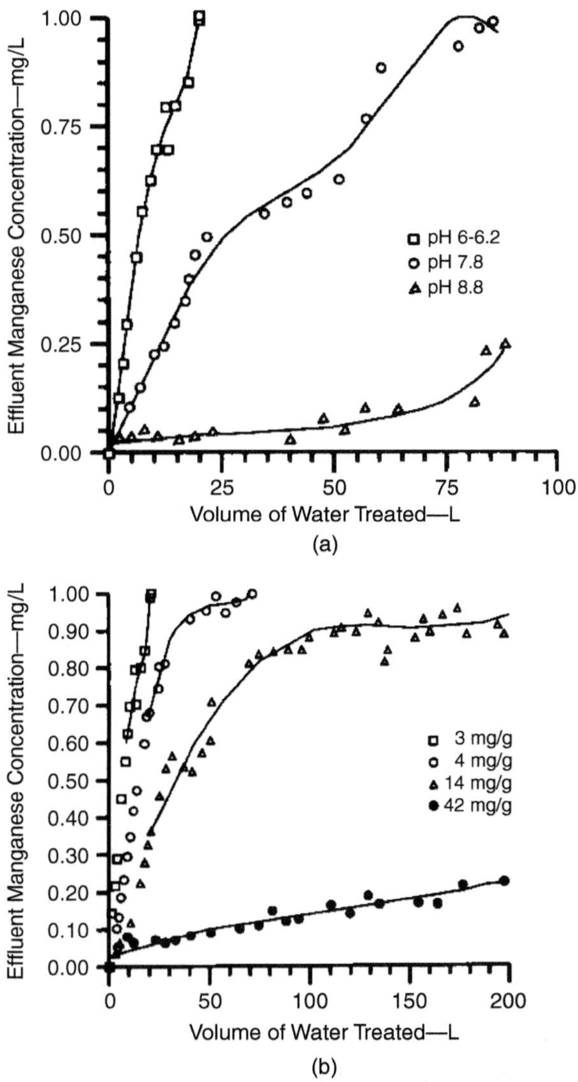

FIGURE 7-14 Impact of (*a*) pH, and (*b*) manganese dioxide coating (pH 6-6.2) on removal of dissolved manganese by manganese dioxide-coated filter media. Influent Mn(II) concentration is 1.0 mg/L. (*Source: Knocke et al., 1991b.; reprinted with permission from the American Water Works Association, Copyright 1991.*)

Destruction of Tastes and Odors

Tastes and odors occur in water from a variety of sources, most notably algae, actinomycetes, organic and inorganic sulfides (e.g., mercaptans and hydrogen sulfide, respectively), and industrial contaminants, such as phenols (Mallevialle and Suffet, 1987). The presence of blue-green algae, in particular, is associated with a variety of specific chemical

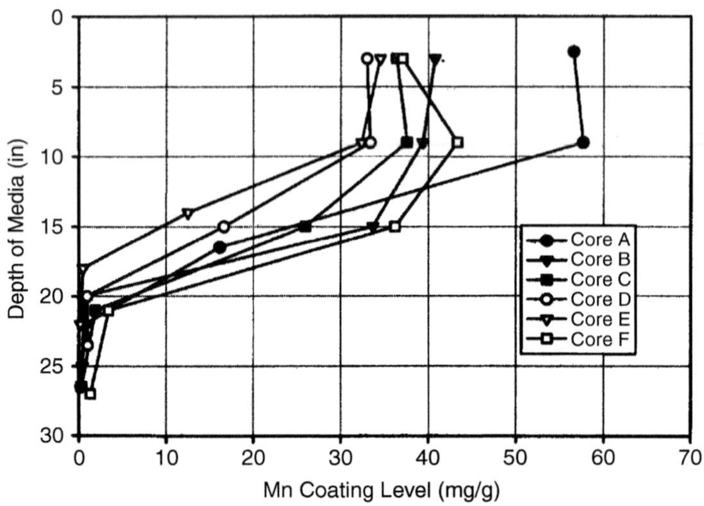

FIGURE 7-15 Manganese coatings on filter media in cores taken from a dual media filter using adsorption/oxidation for manganese removal. (*Source: Tobiason et al., 2008; reprinted with permission from the Water Research Foundation, Copyright 2008.*)

compounds that produce unpleasant odors at ng/L levels, for example, methylisoborneol (MIB) and geosmin, especially in waters drawn from impoundments. Because of the algal origin of these odorous compounds, they tend to occur on a seasonal basis (see Chap. 3 for additional coverage of the occurrence of algal-derived odors in source waters).

Chemical oxidation and activated carbon adsorption (see Chap. 14) are the two most common methods of eliminating taste and odor from drinking water. Free chlorine is an effective chemical oxidant for the destruction of odors associated with reduced sulfur compounds, but is less effective in destroying phenolic compounds and MIB and geosmin

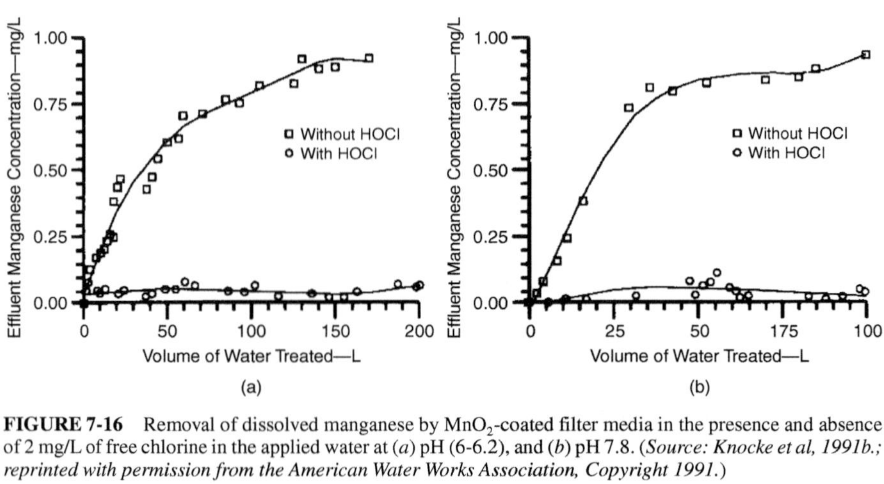

FIGURE 7-16 Removal of dissolved manganese by MnO_2-coated filter media in the presence and absence of 2 mg/L of free chlorine in the applied water at (*a*) pH (6-6.2), and (*b*) pH 7.8. (*Source: Knocke et al, 1991b.; reprinted with permission from the American Water Works Association, Copyright 1991.*)

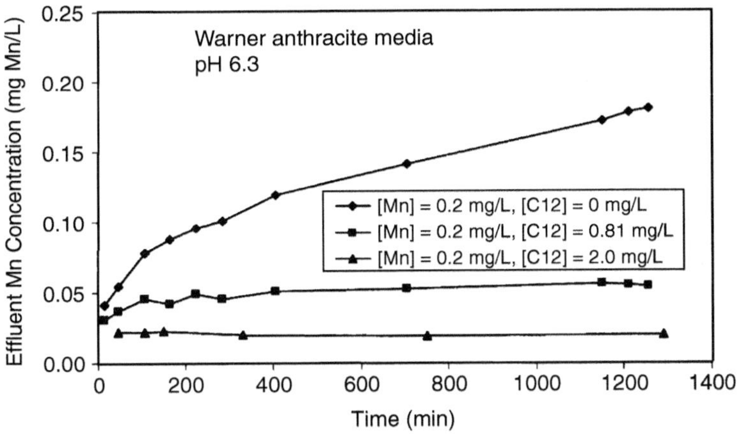

FIGURE 7-17 Effect of free chlorine concentration in the influent water to an anthracite filter coated with MnO_2 on effluent dissolved manganese concentrations. (*Source: Tobiason et al., 2008; reprinted with permission from the Water Research Foundation, Copyright 2008.*)

(Lalezary et al., 1986). In the case of phenols, low doses of chlorine can produce chlorophenols by substitution reactions of the type shown earlier in this chapter. Some of these chlorophenols have stronger odors than the parent phenolic compounds themselves (Burttschell et al., 1959). Hence, the application of chlorine intensifies the odor. In some cases, the chlorinous odor associated with free chlorine itself simply masks the odor associated with the odor-causing contaminant.

Ozone (and AOPs), chlorine dioxide, and permanganate have all been used for oxidizing and eliminating odorous compounds in water, but their success varies and all are limited to some extent for different types of odor-producing compounds and in different types of waters (Glaze et al., 1990). Generally, all are effective for oxidizing phenolic compounds. A good summary of the kinetics of ozonation of a number of important taste and odor compounds is given by Peter and von Gunten (2007).

Lalezary et al., (1986) found that the order for oxidizing MIB and geosmin at low oxidant doses was $ClO_2 > O_3 > Cl_2 > MnO_4^-$. Other researchers (e.g., Glaze et al., 1990) have reported that ozone is the most effective agent for oxidizing MIB and geosmin, although when coupled with hydrogen peroxide or UV irradiation to produce the hydroxyl radical (see the earlier discussion), the resulting AOPs tend to be more effective in destroying MIB and geosmin than ozone alone (Ferguson et al., 1990; Ho et al., 2004; Westerhoff et al., 2006). pH has a significant effect on the ozonation of both compounds (see Fig. 7-18), with geosmin being more rapidly destroyed by ozonation than MIB (Liang et al., 2007). Dissolved organic matter in the water exerts an ozone demand, thereby decreasing the ozone concentration and resulting in lower removals of MIB and geosmin. On the other hand, the formation of OH radicals initiated by the organic matter can accelerate the oxidation of MIB and geosmin by ozone (Glaze et al., 1990; Liang et al., 2007).

Peter and von Gunten (2007) found that the oxidation of MIB and geosmin by molecular ozone is very slow (see Table 7-6). The ability of the OH radical in enhancing the oxidation of MIB and geosmin is shown in Figs. 7-19 and 7-20. Figure 7-19 shows a comparison between ozone and ozone/peroxide in oxidizing MIB and geosmin in Colorado River water. The AOP is clearly superior to molecular ozone by itself under these conditions. Figure 7-20 illustrates that both MIB and geosmin are oxidized rapidly, even in the presence of 2 mg/L

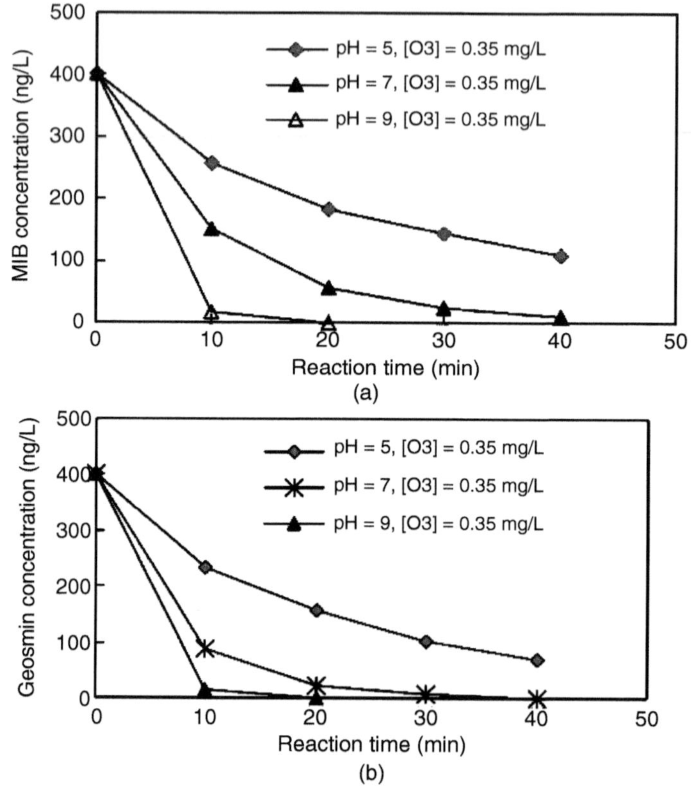

FIGURE 7-18 Effect of pH on the oxidation of (*a*) MIB and (*b*) geosmin by ozone. Conditions: 20°C, TOC = 2.91 mg/L, alkalinity = 120 mg/L as $CaCO_3$. (*Source: Liang et al., 2007; reprinted with permission from Ozone: Science and Engineering, Copyright 2007.*)

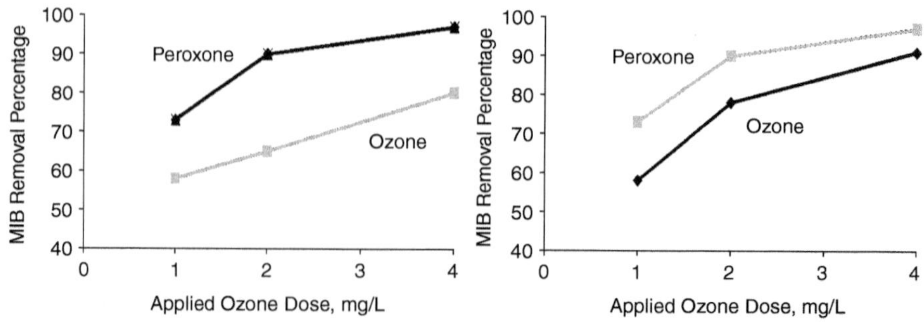

FIGURE 7-19 Removal of MIB and geosmin by ozone and peroxide plus ozone ("Peroxone"). H_2O_2/O_3 ratio = 0.2 mg/mg. Conditions: 12–13°C, pH 8.3–8.4, TOC = 2.6–2.8 mg/L, alkalinity 120–135 mg/L as $CaCO_3$, contact time = 6–12 min. (*After Ferguson et al., 1990; reprinted with permission from the American Water Works Association, Copyright 1990.*)

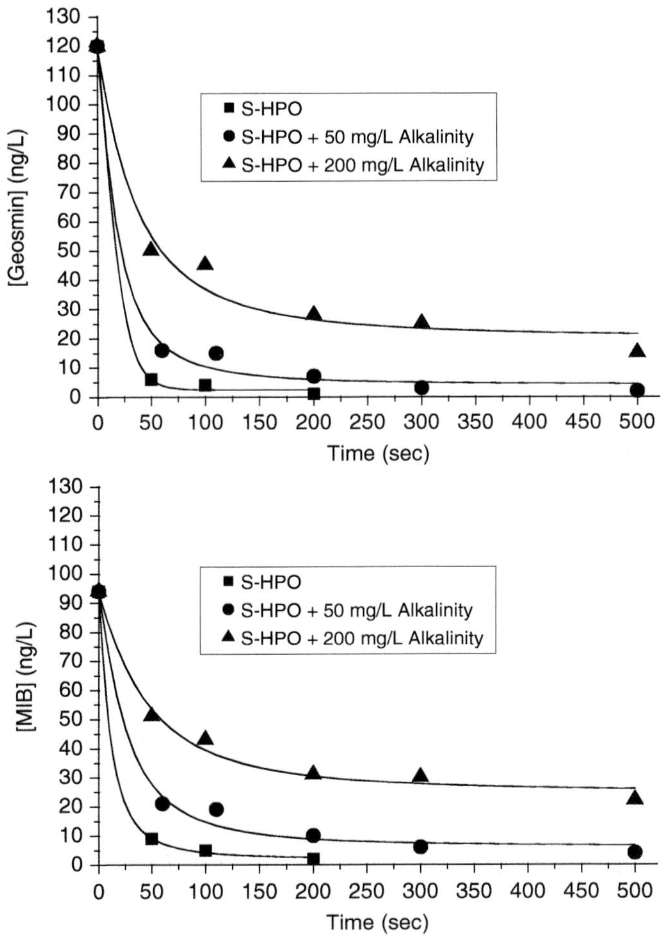

FIGURE 7-20 Oxidation of MIB and geosmin in the presence of dissolved hydrophobic organic material (S-HPO; DOC = 2 mg/L) and carbonate alkalinity. Ozone dose = 2 mg/L, temperature = 20°C, pH = 7.0). (*Source: Ho et al., 2004; reprinted with permission from IWA Publishing, Copyright 2004.*)

of hydrophobic organic carbon (represented here as S-HPO). Carbonate species (alkalinity), because they behave as OH radical scavengers, decrease the rate of oxidation. The relatively rapid rate shown here (seconds vs. minutes) is most likely due to the higher ozone dose used.

It should be noted that aldehydes are often formed from the reactions of ozone with NOM (see Chap. 19). These aldehydes will often impart a fruity odor to the water.

Lastly, as noted previously, the application of chlorine dioxide has been shown, in some cases, to produce noxious (kerosene-like and cat-urine-like) odors. This has been attributed to reactions between chlorite (the principal inorganic by-product of chlorine dioxide), free chlorine (which oxidizes the residual chlorite to chlorine dioxide, which is volatile), and volatile organic chemicals released from new carpeting (Hoehn et al., 1990). When chloramines are used as a secondary disinfectant in place of free chlorine, they are not strong enough to oxidize chlorite to chlorine dioxide, and there is no associated odor problem.

Sulfide Removal

Hydrogen sulfide is a common impurity in groundwater, giving rise to noxious odors. The presence of sulfide arises from the dissolution of sulfidic minerals and the reduction of sulfate under strongly reducing conditions. The most common method of eliminating sulfide from water is by aeration (see Chap. 6). Chemical oxidation can also be used to augment air-stripping or in place of air-stripping, although it tends to be a more expensive approach than air-stripping. The end products of sulfide oxidation are solid elemental sulfur, polysulfides, and sulfate, depending on the oxidant used and solution conditions, such as pH and temperature. Microorganisms can also play an important role in sulfide oxidation, as a number of species are capable of catalyzing sulfide oxidation by oxygen and are commonly found in air-stripping facilities.

Table 7-8 shows the chemical requirements for sulfide oxidation by oxygen and a number of strong oxidants. The stoichiometric requirements are shown for both elemental sulfur and sulfate as end products. Because of the different oxidation states of the sulfur, 4 times as much oxidant is required to convert sulfide to sulfate than to elemental sulfur.

TABLE 7-8 Oxidant Requirements for Sulfide Oxidation

Reaction	Oxidant demand, per mg S(-II)
$2H_2S + O_2 \rightarrow 2S(s) + 2H_2O$	0.5 mg oxygen
$H_2S + 2O_2 \rightarrow SO_4^{2-} + 2H^+$	2 mg oxygen
$H_2S + HOCl \rightarrow S(s) + Cl^- + H_2O + H^+$	2.21 mg chlorine
$H_2S + 4HOCl \rightarrow SO_4^{2-} + 4Cl^- + 6H^+$	8.86 mg chlorine
$H_2S + 2ClO_2 \rightarrow S(s) + 2ClO_2^- + 2H^+$	4.21 mg chlorine dioxide
$H_2S + 8ClO_2 + 4H_2O \rightarrow SO_4^{2-} + 8ClO_2^- + 10H^+$	16.8 mg chlorine dioxide
$3H_2S + 2MnO_4^- + 2H^+ \rightarrow 3S(s) + 2MnO_2(s) + 4H_2O$	2.47 mg permanganate
$3H_2S + 8MnO_4^- + 2H^+ \rightarrow 3SO_4^{2-} + 8MnO_2(s) + H_2O$	9.89 mg permanganate
$H_2S + O_3 \rightarrow S(s) + O_2 + H_2O$	1.50 mg ozone
$H_2S + 4O_3 \rightarrow SO_4^{2-} + 4O_2 + 2H^+$	5.99 mg ozone

The kinetics of sulfide oxidation by oxygen are strongly impacted by pH, as shown in Fig. 7-21. The oxidation reaction by oxygen is relatively slow and elemental sulfur and polysulfides are the primary products, but as noted previously, the reaction is readily catalyzed by sulfide-oxidizing bacteria. Lyn and Taylor (1992) found that the oxidation of sulfide by free chlorine was fast under typical water treatment conditions and that elemental sulfur is a major end product leading to elevated turbidity levels in the chlorinated water, requiring attention to sound solid-liquid separation processes to capture the sulfur precipitate. Sulfide oxidation by ozone and advanced oxidation processes is rapid, as illustrated in Table 7-6; sulfate is the principal end product, necessitating relatively high oxidant doses.

Color Removal

Color in water tends to be associated with the presence of polyaromatic compounds arising from natural vegetative decay processes. These compounds are often referred to as *humic acids* or *humic substances*. They impart a yellowish hue to the water, which can be quantified by the measurement of absorbance at 450 to 465 nm or by comparison to platinum-cobalt standards (*Standard Methods*, 2005). The nature of these substances and the molecular basis of the color varies with the source water. Color can also be correlated to the absorbance of UV radiation at 254 nm, a parameter that is often used to quantify the presence of NOM (see Chap. 3) that contributes to the formation of disinfection by-products (see Chap. 19). The

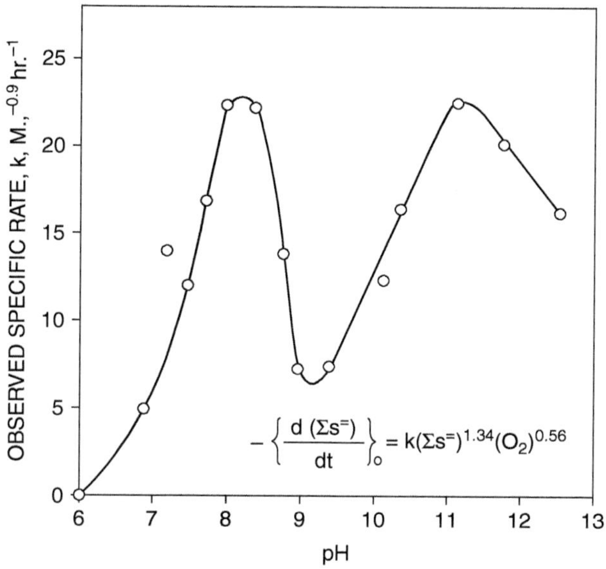

FIGURE 7-21 Effect of pH on the rate constant for the oxidation of sulfide by oxygen. (*Source*: *Chen and Morris, 1972; reprinted with permission from the American Chemical Society, Copyright 1972.*)

humic substances are frequently complexed with trace metals in the water, a phenomenon which contributes to the appearance and intensity of the color.

The most common processes for the removal of natural organic color from water are chemical oxidation and coagulation. In the case of coagulation, the color-causing organics are coagulated and physically removed from solution (see Chap. 8), whereas for chemical oxidation, the color-causing organics are chemically altered but the colorless products remain in solution. The oxidizing agent attacks the chromophoric portion of the molecules that are responsible for the absorption of visible light; these are usually the carbon-carbon double bonds, the groups that bind trace metals, and the polyaromatic structures themselves. Chlorine had been used historically for color removal, but it became apparent in the 1970s that the reactions between chlorine and the color-causing organic materials are the ones that lead to the formation of halogenated disinfection by-products that are of public health concern (see Chap. 19). Ozone and chlorine dioxide are also effective chemical oxidants for color removal, but these oxidants also produce by-products that can be problematic, for example, biodegradable organic material that, if not properly controlled, can lead to microbial growth in the distribution system (see Chap. 21). Accordingly, it may be more desirable to remove the color-causing organic materials by coagulation processes (see Chap. 8) first, thereby lowering the oxidant demand of the water, and then use chemical oxidation to polish any residual color that may remain after clarification and/or filtration.

Oxidation of Organic Micropollutants

Strong chemical oxidants are capable of reacting with a variety of organic micropollutants that may be present in water supplies, such as synthetic organic chemicals (e.g., pesticides), chlorinated solvents, fuel additives, pharmaceutically active compounds and personal care products, endocrine disruptors, and algal toxins. These reactions are highly dependent on the nature of the organic compound, the specific oxidant in question, other constituents in the water that may interfere with or compete with the target contaminant, pH, and temperature. Some organic compounds are relatively easy to oxidize, whereas others are much more resistant to oxidation. Additionally, for any particular oxidant, the products of the reaction vary, depending on the oxidant, the ratio of the concentration of the oxidant to that of the contaminant, the pathway by which the reaction occurs, and general water quality characteristics. Rarely is the contaminant completely mineralized to carbon dioxide and water. In some cases, the oxidation products may be of greater concern than the parent compound.

For example, the oxidation of phenol by low doses of chorine leads to the formation of mono-, di-, and trichlorophenols (Burttschell et al., 1959). At higher doses, the aromatic ring is cleaved to generate halogen-containing products. In the case of ozone, oxidation of phenol at low ozone doses results in the formation of primarily catechol and benzoquinone, before the aromatic ring is cleaved at higher doses to produce a variety of low-molecular-weight carboxylic acids such as cis,cis-muconic acid, maleic acid, and glyoxylic acid (Singer and Gurol, 1983; Ramseier and von Gunten, 2009). With chlorine dioxide, oxidation of phenol produces hydroxophenols, quinones, and several chlorophenols and chloroquinones at low oxidant doses (Wajon et al., 1982; Stevens, 1982). Subsequent ring cleavage at higher chlorine dioxide doses leads to the formation of carboxylic acids, as is the case with ozone. And, of course, chlorination of NOM leads to the formation of halogenated disinfection by-products (see Chap. 19).

As noted earlier, Hoigne and coworkers (1983a, 1983b, 1985) and von Gunten (2003) present rate constants for reactions of ozone with a variety of inorganic and organic compounds in water (e.g., see Table 7-6). The examples in Table 7-6 attest to the selectivity of molecular ozone for various types of contaminants. For example, the rate constants for the oxidation of pesticides span more than five orders of magnitude. Endrin has the lowest reactivity of the compounds shown because of the electron-withdrawing groups of the chloro-substituted double bond in this cyclic compound, whereas the rate constant for carbofuran is much higher because of the presence of a secondary amino group on the aromatic ring.

Phenol and naphthalene are rapidly oxidized by molecular ozone (Hoigne and Bader, 1983a), whereas atrazine, alachlor, benzene, methyl tertiary butyl ether (MTBE), trichloroethene, and tetrachoroethene are only slowly oxidized by molecular ozone (Table 7-6). On the other hand, these same micropollutants are rapidly oxidized by the hydroxyl radical generated from the decomposition of ozone. Hence, conditions that favor the formation of hydroxyl radicals are conducive to the elimination of these harmful contaminants. Such conditions include ozonation at elevated pH values or the use of ozone in conjunction with UV irradiation or hydrogen peroxide addition (AOPs). Conversely, conditions that favor the scavenging of hydroxyl radicals interfere with the effectiveness of these AOPs. This is true for high alkalinity waters that contain high concentrations of bicarbonate and carbonate, both of which are hydroxyl radical scavengers (see the discussion above). Additionally, because these micropollutants are usually present at relatively low concentrations (μg/L) compared to NOM which is present at milligram per liter concentrations, these advanced oxidation processes tend to be less effective in waters that contain high concentrations of NOM, although in some cases the presence of NOM can promote the formation of free radicals and accelerate the oxidation process.

Chlorine dioxide is also a highly selective chemical oxidant. Although little work has been done with pesticides and solvents, Hoigne and Bader (1994) present a summary of reaction rate constants for a variety of organic and inorganic species (see Table 7-5). Their work shows that olefins, aromatic hydrocarbons, primary and secondary amines, aldehydes, ketones, and carbohydrates are not very reactive with chlorine dioxide under water treatment conditions, whereas tertiary amines, thiols, and deprotonated phenols tend to have high reaction rate constants. Again, in waters with high concentrations of NOM, the oxidant demand of the NOM must be overcome before appreciable elimination of the target micropollutants can be achieved.

Given the growing interest in pharmaceutically active compounds, endocrine disruptors, and cyanotoxins, it is worth noting that many of these compounds are also readily oxidized by a number of the strong oxidants discussed in this chapter.

Cyanotoxins. Rositano et al. (2001) reported 100 percent destruction of the cyanotoxins microcystin-LR and -LA and anatoxin-a by ozonation of four treated waters with very different water qualities. The degree of destruction was related to the residual ozone concentration present after 5 min, which was, in turn, related to the water quality. They noted that direct reaction with molecular ozone could be responsible for cyanotoxin destruction.

Rodriguez et al. (2007) studied the oxidation of three cyanotoxins, microcystin-LR, cylindrospermopsin, and anatoxin-a, with ozone, chlorine, chlorine dioxide, and permanganate. Permanganate was found to oxidize anatoxin-a and microcystin-LR effectively, chlorine was found to oxidize cylindrospermopsin and microcystin-LR, and ozone was capable of oxidizing all three toxins with the highest rate constant among the three oxidants examined. Chlorine dioxide was not an effective oxidant for elimination of these cyanotoxins during drinking water treatment. Further, Onstad et al. (2007) found that hydroxyl radicals contributed significantly to toxin oxidation during the ozonation of natural waters. The order of reactivity, anatoxin-a > cylindrospermopsin > microcystin-LR, corresponds to the relative magnitudes of the second-order rate constants for their reaction with ozone and the hydroxyl radical. It was suggested that, because ozone primarily attacks the structural moieties responsible for the toxic effects of these algal toxins, ozone selectively detoxifies these compounds.

Pharmaceuitcals, Personal Care Products, Endocrine Disruptors. Huber et al. (2003) determined second-order rate constants for the reactions of selected pharmaceuticals with ozone and OH radicals. Five of the pharmaceuticals (carbamazepine, diclofenac, 17α-ethinylestradiol, sulfamethoxazole, and roxithromycin) exhibited rate constants $>5 \times 10^4\ M^{-1}s^{-1}$ with molecular ozone at pH 7, indicating that these compounds are completely transformed during ozonation processes. Diazepam, ibuprofen, and iopromide reacted orders of magnitude more slowly. In contrast, values for the rate constants with the OH radical ranged from 3.3 to $9.8 \times 10^9\ M^{-1}s^{-1}$. Compared to other important micropollutants such as MTBE and atrazine, the selected pharmaceuticals reacted about 2 to 3 times faster with OH radicals. It was concluded that ozonation and AOPs are promising processes for an efficient removal of pharmaceuticals in drinking waters.

Westerhoff et al. (2005), in a study of 62 different endocrine disrupting compounds, pharmaceutically active compounds, and personal care products, reported that chlorine and ozone decreased the concentrations of these emerging contaminants by <10 to >90 percent. Ozone oxidized steroids containing phenolic moieties (estradiol, ethinylestradiol, and estrone) more efficiently than those without aromatic or phenolic moieties (androstenedione, progesterone, and testosterone). Six of the compounds studied (tri(2-chloroethyl) phosphate, BHC, chlordane, dieldrin, heptachlor epoxide, and musk ketone) were oxidized <20 percent by chlorine or ozone. Twenty-four of the compounds exhibited better removals with ozone than with chlorine. The remaining compounds were easily oxidized (>80 percent reacted) by chlorine, and were always oxidized at least as efficiently by ozone. Lei and Snyder (2007) used three-dimensional quantitative structure-property relationships as a screening tool to predict the removals of selected endocrine-disrupting compounds and pharmaceuticals and personal care products by ozone and free chlorine.

Broseus et al. (2009) investigated the oxidation of a number of pharmaceuticals, endocrine-disrupting compounds, and pesticides (including caffeine, trimethoprim, carbamazepine, naproxen, gemfibrozil, estrone, estriol, estradiol, 17α-ethinylestradiol, progesterone, cyanazine, and diethylatrazine) and found that ozonation, at ozone doses and contact times typically applied in drinking water treatment, removed over 80 percent of caffeine, pharmaceuticals, and endocrine disruptors. The pesticides examined, some of which are metabolites of atrazine, were found to be the most recalcitrant compounds to oxidize.

In a review of the literature, Ikehata et al. (2008) found that, in most cases, ozonation and AOPs were effective for the degradation of a wide variety of pharmaceuticals, personal care products, endocrine-disrupting compounds, pesticides, and surfactants. Some compounds, such as carbamazepine, 17-estradiol, diclofenac, and pentachlorophenol, were reported to be extremely reactive toward ozone and hydroxyl radicals. On the other hand, it was found that halogenated pesticides, X-ray contrast agents, and surfactants were resistant to oxidative degradation by molecular ozone. UV-assisted advanced oxidation was found to be more effective in the degradation of halogenated organics compared with ozonation due to enhanced generation of hydroxyl radicals and photon-initiated cleavage of carbon-halogen bonds.

In an evaluation of chlorine dioxide for the oxidation of pharmaceutically active compounds, Huber et al. (2005) reported that sulfonamide and macrolide antibiotics, pyrazolone derivatives, and estrogens were rapidly oxidized by ClO_2, but many other PhACs were refractory toward ClO_2. Compared to ozone, ClO_2 reacted more slowly and with fewer compounds, but reacted faster with the compounds investigated than chlorine.

Methyl Tertiary Butyl Ether. Methyl-*tert*-butyl-ether (MTBE), a common oxygenated fuel additive found in reformulated gasoline and an oft-reported groundwater contaminant, was found to be rapidly oxidized by ozone and peroxide, but high doses of ozone and an ozone:peroxide ratio of 1:1 by weight were required (Liang et al., 2001). The products of the reaction were observed to be t-butyl formate, t-butyl alcohol, acetone, and a number of aldehydes, many of which are biodegradable. Acero et al. (2001) showed that MTBE reacts slowly with molecular ozone but rapidly with the hydroxyl radical, and found that the oxidation products reacted only slowly with molecular ozone. They proposed a pathway for the reaction between MTBE and the hydroxyl radical.

It is important to note again that, while the concentration of many of these organic micropollutants may be decreased appreciably by chemical oxidation reactions, the compounds are typically not completely mineralized. The oxidation reactions generate "daughter" products that also may be of public health concern. Hence it is important to understand the pathway by which the contaminants are oxidized and to identify the daughter products generated by the oxidation reactions before these processes are widely implemented for controlling organic micropollutants in drinking water.

Oxidation as an Aid to Coagulation and Flocculation

The application of oxidants to raw waters drawn from surface supplies is known to assist in subsequent coagulation and flocculation processes. While this has been reported to some degree for chlorine and chlorine dioxide, the use of ozone for this purpose has been widely touted (Langlais et al., 1991). The phenomenon appears to be associated with the facts that most particles in raw drinking water are negatively charged due to the adsorption of NOM to the particle surface and that NOM in bulk solution exerts an appreciable coagulant demand. Accordingly, mechanisms cited for the coagulation and flocculation benefits of oxidants include: oxidation of adsorbed organics to more polar forms which desorb from particles, thereby making the particles less stable and more amenable to aggregation; alteration of the configuration of the adsorbed organics so that they more effectively bind coagulants such as Al(III) and Fe(III) at the particle surface; oxidation of organics in bulk solution to form carboxylic acid functional groups that bind metals such as calcium and aluminum, resulting in direct precipitation of the organic material; and oxidation of organics causing them to release bound metals such as Fe(III) which can then assist in particle coagulation by conventional mechanisms (Reckhow et al., 1986). While the actual mechanisms are not entirely clear, there are a number of waters where these phenomena have been observed, although there are many other waters where no such benefits have been found (Reckhow et al., 1986).

Control of Biological Growths in Treatment Plants

One of the many benefits of chlorine is that it controls nuisance aquatic growths in water treatment plants. This includes growths in flocculation and sedimentation basins and on filter media. Maintenance of a free chlorine residual throughout the treatment plant minimizes the potential growth of microorganisms, especially in open basins exposed to sunlight, which encourages algal growth. Ozone, chlorine dioxide, and permanganate also exhibit biocidal attributes, but because of the doses typically employed and the relative chemical instability of these oxidants over time, it is not easy to maintain a residual with these oxidants across sedimentation basins and, in some cases, through filter beds.

In fact, because ozone, AOPs, chlorine dioxide, and permanganate oxidize NOM to produce low-molecular-weight biodegradable organic compounds, it can be anticipated that oxidative treatment will actually increase the potential for microbial growth in pretreatment basins and on filter media in the absence of a disinfectant (oxidant) residual. Accordingly, such growths must be controlled in order to minimize subsequent biofouling problems. This can include the use of free or combined chlorine following the application of these other oxidants, periodic shock treatment of the basins or filters with free chlorine, or engineering and operating the filters in a biological mode (see Chap. 21). Chlorine will also generate biodegradable organic compounds, but often the presence of a chlorine residual will control these biological growths.

Summary of Benefits of Oxidation to Water Treatment Processes

It is clear that chemical oxidants provide a variety of tangible benefits in drinking water treatment practice. Some are easy to define and quantify. Others are more diffuse and tend to be more difficult to delineate and assess. Ultimately, however, chemical oxidation is a multiobjective process that serves many useful purposes and helps to improve the overall quality of drinking water. The specific benefits of the process and of each candidate chemical oxidant need to be evaluated on a case-by-case basis.

ABBREVIATIONS

AOPs	advanced oxidation processes
ATRA-imin	atrazine-imine
CT	product of concentration and contact time
DBPs	disinfection by-products
DEA	diethylatrazine
DIA	diisopropyl atrazine
DOC	dissolved organic carbon
EDCs	endocrine disrupting chemicals
FAC	free available chlorine
HOX	hypohalous acid
MCL	maximum contaminant level
MIB	methylisoborneol
MRDL	maximum residual disinfectant level
MTBE	methyl-*tert*-butyl-ether

NOM	natural organic material
pe	negative log of electron activity
pH	negative log of hydrogen ion activity
PhACs	pharmaceutically active compounds
pK_a	negative log of acidity constant
PPCPs	pharmaceutically active compounds and personal care products
PQL	practical quantitation limit
QSARs	quantitative structure-activity relationships
S-HPO	hydrophobic organic carbon
TCE	trichloroethene
TOC	total organic carbon
USEPA	U.S. Environmental Protection Agency
UV	ultraviolet

NOTATIONS FOR EQUATIONS

α	pH-dependent distribution coefficient
ΔG^o	standard Gibbs free energy change
A	arrhenius frequency factor or pre-exponential factor
E_a	activation energy
E^o_{net}	net standard reaction potential
E^o_{ox}	standard oxidation potential
E^o_{red}	standard reduction potential
F	faraday's constant
K	equilibrium constant
k, k_f, k_{obs}, k_2, k_3, etc.	reaction rate constants
M_d	direct-reacting substrate
M_i	indirect-reacting substrate
n	number of electrons transferred
R	universal gas constant
T	absolute temperature

REFERENCES

Acero, J.L., K. Stemmler, and U. von Gunten (2000) "Degradation Kinetics of Atrazine and Its Degradation Products with Ozone and OH Radicals: A Predictive Tool for Drinking Water Treatment," *Environmental Science and Technology,* 34(4), 591–597.

Acero, J.L., S.B. Haderlein, T.C. Schmidt, M.J.F. Suter, and U. von Gunten (2001) "MTBE Oxidation by Conventional Ozonation and the Combination Ozone/Hydrogen Peroxide: Efficiency of the Processes and Bromate Formation," *Environmental Science and Technology*, 35(21), 4252–4259.

Adam, L.C., and G. Gordon (1999) "Hypochlorite Ion Decomposition: Effects of Temperature, Ionic Strength, and Chloride Ion," *Inorganic Chemistry*, 38(6), 1299–1304.

Atkins, P.W. (1994) *Physical Chemistry*, 5th ed. W. H. Freeman and Co. New York: NY.

Benjamin, M.M. (2002) *Water Chemistry,* McGraw-Hill, New York: NY.

Bichsel, Y., and U. von Gunten (1999) "Oxidation of Iodide and Hypoiodous Acid in the Disinfection of Natural Waters," *Environmental Science and Technology*, 33(2), 4040–4045.

Boal, A.K. (2009) "On-Site Generation of Disinfectants," National Environmental Services Center, West Virginia University, 9(1), 1–4.

Bolyard, M., P.S. Fair, D.P. Hautman (1992) "Occurrence of Chlorate in Hypochlorite Solutions Used for Drinking Water Disinfection," *Environmental Science and Technology*, 26(8), 1663–1665.

Bolyard, M., P.S. Fair, and D.P. Hautman (1993) "Sources of Chlorate Ion in US Drinking Water," *Journal AWWA*, 85(9), 81–88.

Bowen, E., and W.M. Cheung (1932) "The Photodecomposition of Chlorine Dioxide Solutions," *Journal Chemical Society London*, 1200–1208.

Braester, C., and R. Martinell (1988) "The Vyredox and Nitredox Methods of *in situ* Treatment of Groundwater," *Water Science and Technology*, 20(3), 149–163.

Brezonik, P.L. (1994) *Chemical Kinetics and Process Dynamics in Aquatic Systems*. Boca Raton: Lewis Publishers.

Broseus, R., S. Vincent, K. Aboulfadl, A. Daneshvar, S. Sauve, B. Barbeau, and M. Prevost (2009) "Ozone Oxidation of Pharmaceuticals, Endocrine Disruptors and Pesticides during Drinking Water Treatment," *Water Research*, 43, 4707–4717.

Burger, M.S., C.A. Krentz, S.S. Mercer, and G.A. Gagon (2008) "Manganese Removal and Occurrence of Manganese Oxidizing Bacteria in Full-Scale Biofilters," *Journal of Water Supply: Research and Technology–AQUA*, 57, 351–359.

Burttschell, R.H., A.A. Rosen, F.M. Middleton, and M.B. Ettinger (1959) "Chlorine Derivatives of Phenol and Phenol Causing Taste and Odor," *Journal AWWA*, 51(2), 205–214.

Cameron, I. (1996) "Biological Iron and Manganese Removal: An Untapped Potential," *Water, Journal Australian Water and Wastewater Association*, 23(3), 25–28.

Chen, K.Y., and J.C. Morris (1972) "Kinetics of Oxidation of Aqueous Sulfide by O_2," *Environmental Science and Technology*, 6(6), 529–537.

Chlorine Institute (1999) "Bromate in Sodium Hypochlorite Solutions. A Progress Report," Chlorine Institute, Washington, DC.

Colthurst, J.M., and P.C. Singer (1982) "Removing Trihalomethane Precursors by Permanganate Oxidation and Manganese Dioxide Adsorption," *Journal AWWA*, 74(2), 78.

Deborde, M., and U. von Gunten (2008) "Reactions of Chlorine with Inorganic and Organic Compounds during Water Treatment–Kinetics and Mechanisms: A Critical Review," *Water Research*, 42(1–2), 13–51.

Elovitz, M.S., and U. von Gunten. (1999). "Hydroxyl Radical/Ozone Ratios during Ozonation Processes. I. The R_{ct} Concept," *Ozone Science Engineering*, 21(3), 239–260.

Elovitz, M.S., U. von Gunten, and H.-P. Kaiser. (2000). "Hydroxyl Radical/Ozone Ratios during Ozonation Processes. II. The Effect of Temperature, pH, Alkalinity, and DOM Properties," *Ozone Science Engineering*, 22(2), 123–150.

Ferguson, D.W., M.J. McGuire, B. Koch, R.L. Wolfe, and M.E. Aieta (1990) "Comparing PEROXONE and Ozone for Controlling Taste and Odor Compounds, Disinfection By-Products, and Microorganisms," *Journal AWWA*, 82(4), 181–191.

Ghurye, G., and D. Clifford (2004) "As(III) Oxidation Using Chemical and Solid-Phase Oxidants," *Journal AWWA*, 96(1), 84–96.

Glaze, W.H., R. Schep, W. Chauncey, E.C. Ruth, J.S. Zarnoch, E.M. Aieta, C.H. Tate, and M.J. McGuire (1990) "Evaluating Oxidants for the Removal of Model Taste and Odor Compounds from a Municipal Water Supply," *Journal AWWA*, 82(5), 79.

Gregory, D., and K. Carlson (2003) "Effect of Soluble Mn Concentration on Oxidation Kinetics," *Journal AWWA*, 95, 98–108.

Ho, L., J.-P. Croué, and G. Newcombe (2004) "The Effect of Water Quality and NOM Character on the Ozonation of MIB and Geosmin," *Water Science and Technology*, 49(9), 249–255.

Hoehn, R.C., A.M. Dietrich, W.S. Farmer, M.P. Orr, R.G. Ramon, E.M. Aieta, D.W. Wood, and G. Gordon (1990) "Household Odors Associated with the Use of Chlorine Dioxide," *Journal AWWA*, 82(4), 162.

Hoigne, J., and H. Bader (1976) "The Role of Hydroxyl Radical Reactions in Ozonation Processes in Aqueous Solutions," *Water Research*, 10, 377.

Hoigne, J., and H. Bader (1983a) "Rate Constants of Reactions of Ozone with Organic and Inorganic Compounds in Water. I. Non-Dissociating Organic Compounds," *Water Research*, 17(2), 173–184.

Hoigne, J., and H. Bader (1983b) "Rate Constants of Reactions of Ozone with Organic and Inorganic Compounds in Water. II. Dissociating Organic Compounds," *Water Research*, 17(2), 185–194.

Hoigne, J., H. Bader, W.R. Haag, and J. Staehelin (1985) "Rate Constants of Reactions of Ozone with Organic and Inorganic Compounds in Water. III. Inorganic Compounds and Radicals," *Water Research*, 19(8), 993–1004.

Hoigne, J., and H. Bader (1994) "Kinetics of Reactions of Chlorine Dioxide (OClO) in Water. I. Rate Constants for Inorganic and Organic Compounds," *Water Research*, 28(1), 45.

Huber, M.M., S. Canonica, G.-Y. Park, and U. von Gunten (2003) "Oxidation of Pharmaceuticals during Ozonation and Advanced Oxidation Processes," *Environmental Science and Technology*, 37(5), 1016–1024.

Huber, M.M., S. Korhonen, T.A. Ternes, and U. von Gunten (2005) "Oxidation of Pharmaceuticals during Water Treatment with Chlorine Dioxide," *Water Research*, 39(15), 3607–3617.

Iatrou, A., and W.R. Knocke (1992) "Removing Chlorite by the Addition of Ferrous Iron," *Journal AWWA*, 84(11), 63–68.

Ikehata, K., M. Gamal El-Din, and S.A. Snyder (2008) "Ozonation and Advanced Oxidation Treatment of Emerging Organic Pollutants in Water and Wastewater," *Ozone Science and Engineering*, 30(1), 21–26.

Jensen, J.N. (2003) *Aquatic Chemistry*. New York: John Wiley and Sons.

Jiang, J.Q., and B. Lloyd (2002) "Progress in the Development and Use of Ferrate(VI) Salt as an Oxidant and Coagulant for Water and Wastewater Treatment," *Water Research*, 36(6), 1397–1408.

Katsoyiannis, I.A., and A.I. Zouboulis (2004) "Biological Treatment of Mn(II) and Fe(II) Containing Groundwater: Kinetic Considerations and Product Characterization," *Water Research*, 38, 1922–1932.

Knocke, W.R., R.C. Hoehn, and R.L. Sinsabaugh (1987) "Using Alternative Oxidants to Remove Dissolved Manganese From Waters Laden With Organics," *Journal AWWA*, 79(3), 75–79.

Knocke, W.R., J.R. Hamon, and C.P. Thompson (1988) "Soluble Manganese Removal on Oxide-Coated Filter Media," *Journal AWWA*, 80(12), 65–70.

Knocke, W.R., J.E. Van Benschoten, M.J. Kearney, A.W. Soborski, and D.A. Reckhow (1991a) "Kinetics of Manganese and Iron Oxidation by Potassium Permanganate and Chlorine Dioxide," *Journal AWWA*, 83(6), 80–87.

Knocke, W.R., S.C. Occiano, and R. Hungate (1991b) "Removal of Soluble Manganese by Oxide-Coated Filter Media: Sorption Rate and Removal Mechanism Issues," *Journal AWWA*, 83(8), 64–69.

Knocke, W.R., H.L. Shorney, and J.D. Bellamy (1994) "Examining the Reactions between Soluble Iron, DOC, and Alternative Oxidants during Conventional Treatment," *Journal AWWA*, 86(1), 117–127.

Kumar, K., and D.W. Margerum (1987) "Kinetics and Mechanism of General-Acid-Assisted Oxidation of Bromide by Hypochlorite and Hypochlorous Acid," *Inorganic Chemistry*, 26, 2706–2711.

Lalezary, S., M. Pirbazari, and M.J. McGuire (1986) "Oxidation of Five Earthy–Musty Taste and Odor Compounds," *Awwa Research Foundation*, 78(3), 62–69.

Langlais, B., D.A. Reckhow, and D.R. Brink (1991) *Ozone in Water Treatment*, American Water Works Association Research Foundation, Lewis Publishers.

Lee, Y.H., S.G. Zimmerman, A.T. Kieu, and U. von Gunten (2009) "Ferrate (Fe(VI)) for Municipal Wastewater Treatment: A Novel Process for Simultaneous Micropollutant Oxidation and Phosphate Removal," *Environmental Science and Technology*, 43(10), 3831–3838.

Lei, H., and S.A. Snyder (2007) "3D QSPR Models for the Removal of Trace Organic Contaminants by Ozone and Free Chlorine," *Water Research*, 41(18), 4051–4060.

Liang, S., R.S. Yates, D.V. Davis, S.J. Pastor, L.S. Palencia, and J.M. Bruno (2001) "Treatability of MTBE–Contaminated Groundwater by Ozone and Peroxone," *Journal AWWA*, 93(6), 110–120.

Liang, C., D. Wang, J. Chen, L. Zhu, and M. Yang (2007) "Kinetics Analysis on the Ozonation of MIB and Geosmin," *Ozone Science and Engineering*, 29(3), 185–189.

Lyn, T.L., and J.S. Taylor (1992) "Assessing Sulfur Turbidity Formation Following Chlorination of Hydrogen Sulfide in Groundwater," *Journal AWWA*, 84(9), 103–112.

Mallevialle, J., and I.H. Suffet (1987) "Identification and Treatment of Tastes and Odors in Drinking Water," Awwa Research Foundation Report. Water Research Foundation, Denver, CO.

Morris, J.C. (1975) "Formation of Halogenated Organics by Chlorination of Water Supplies," EPA-600/1-75-002. U.S. Environmental Protection Agency, Washington, D.C.

Mouchet, P. 1992. "From Conventional to Biological Removal of Iron and Manganese in France," *Journal AWWA*, 84(4), 158–167.

Onstad, G.D., S. Strauch, J. Meriluoto, G.A. Codd, and U. von Gunten (2007) "Selective Oxidation of Key Functional Groups in Cyanotoxins during Drinking Water Ozonation," *Environmental Science and Technology*, 41, 4397–4404.

Pacini, V.A., A.M. Ingallinella, and G. Sanguinetti (2005) "Removal of Iron and Manganese Using Biological Roughing Up Flow Filtration Technology," *Water Research*, 39, 4463–4475.

Peitchev, T., and V. Semov (1988) "Biotechnology for Manganese Removal from Groundwaters," *Water Science and Technology*, 20, 173–178.

Peter, A., and U. von Gunten (2007) "Oxidation Kinetics of Selected Taste and Odor Compounds during Ozonation of Drinking Water," *Environmental Science and Technology*, 41(2), 626–631.

Qureshi, N., and S. Nelson (2003) "Radium Removal by HMO and Manganese Greensand," *Journal AWWA*, 95(3), 101–108.

Ramseier, M.K., and U. von Gunten (2009) "Mechanisms of Phenol Ozonation–Kinetics of Formation of Primary and Secondary Reaction Products," *Ozone Science Engineering*, 31(3), 201–215.

Reckhow, D.A., and P.C. Singer (1985) "Mechanisms of Organic Halide Formation during Fulvic Acid Chlorination and Implications with Respect to Preozonation." In R.L. Jolley, R.J. Bull, W.P. David, S. Katz, M.H. Roberts, and V.A. Jacobs (eds.), *Water Chlorination: Environmental Impact and Health Effects,* Vol. 5, 1229–1257. Chelsea, MI: Lewis Publishers.

Reckhow, D.A., P.C. Singer, and R.R. Trussell (1986) "Ozone as a Coagulant Aid." In *Ozonation: Recent Advances and Research Needs*, 17–46, American Water Works Association, Denver, CO.

Richardson, S.D., Thruston Jr., A.D., Collette, T.W., Patterson, K.S., Lykins Jr. B.W., Majetich, G., and Y. Zhang. (1994) "Multispectral Identification of Chlorine Dioxide Disinfection By-products in Drinking Water," *Environmental Science and Technology*, 28(4), 592.

Rodrıguez, E., G.D. Onstad, T.P.J. Kull, J.S. Metcalf, J.L. Acero, and U. von Gunten (2007) "Oxidative Elimination of Cyanotoxins: Comparison of Ozone, Chlorine, Chlorine Dioxide and Permanganate," *Water Research*, 41, 3381–3393.

Rositano, J., G. Newcombe, B. Nicholson, and P. Sztajnbok (2001) "Ozonation of NOM and Algal Toxins in Four Treated Waters," *Water Research*, 35(1), 23–32.

Sharma, V.K. (2002) "Potassium Ferrate(VI): An Environmentally Friendly Oxidant," *Advances in Environmental Research*, 6(2), 143–156.

Singer, P.C., and M.D. Gurol (1983) "Dynamics of the Ozonation of Phenol: I. Experimental Observations," *Water Research*, 17, 1163–1171.

Staehelin, J., and J. Hoigne (1982) "Decomposition of Ozone in Water: Rate of Initiation by Hydroxide Ions and Hydrogen Peroxide," *Environmental Science and Technology*, 16(10), 676.

Standard Methods for the Examination of Water and Wastewater (2005), 21st ed. American Public Health Association, American Water Works Association, Water Environment Federation.

Stevens, A.A. (1982) "Reaction Products of Chlorine Dioxide," *Environmental Health Perspectives*, 46, 101.

Stumm, W., and J.J. Morgan (1996) *Aquatic Chemistry*, 3d ed. Wiley Interscience.

Sulzberger, B., S. Canonica, T. Egli, W. Giger, J. Klausen, and U. von Gunten (1997) "Oxidative Transformations of Contaminants in Natural and Technical Systems," *Chimia*, 51(12), 900–907.

Theis, T.L., and P.C. Singer (1974) "Complexation of Iron(II) by Organic Matter and Its Effect on Iron(II) Oxygenation," *Environmental Science and Technology*, 8, 569.

Tobiason, J.E., A.A. Islam, W.R. Knocke, J. Goodwill, P. Hargette, R. Bouchard, and L. Zuravnsky (2008) "Characterization and Performance of Filter Media for Manganese Control," Water Research Foundation, Denver, CO.

Tratnyek, P.G., and J. Hoigne (1994) "Kinetics of Reactions of Chlorine Dioxide (OClO) in Water. II. Quantitative Structure–Activity Relationships for Phenolic Compounds," *Water Research*, 28(1), 57.

U.S. Environmental Protection Agency (USEPA) (1998) "National Primary Drinking Water Regulations: Disinfectants and Disinfection By-products," *Federal Register*, 63(241), 69289.

U.S. Environmental Protection Agency (USEPA) (2006) "National Primary Drinking Water Regulations: Stage 2 Disinfectants and Disinfection By-Products Rule; Final Rule," *Federal Register*, 71(2), 388–493.

Vandenabeele, J., D. deBeer, R. Germonpré, and W. Verstraete (1992) "Manganese Oxidation by Microbial Consortia from Sand Filters," *Microbial Ecology*, 24, 91–108.

von Gunten, U., and Y. Oliveras (1998) "Advanced Oxidation of Bromide-Containing Water: Bromate Formation Mechanisms," *Environmental Science and Technology*, 32(1), 63–70.

von Gunten, U. (2003) "Ozonation of Dinking Water: Part I. Oxidation Kinetics and Product Formation," *Water Research*, 37, 1443–1467.

Wajon, J.E., D.H. Rosenblatt, and E.P. Burrows (1982) "Oxidation of Phenol and Hydroquinone by Chlorine Dioxide," *Environmental Science and Technology*, 16, 396.

Weinberg, H.S., S.W. Krasner, S.D. Richardson, and A.D. Thruston Jr. (2002) "The Occurrence of Disinfection By-Products (DBPs) of Health Concern in Drinking Water: Results of a Nationwide DBP Occurrence Study." EPA/600/R-02/068.

Weinberg, H.S., C.A. Delcomyn, and V. Unnam (2003) "Bromate in Chlorinated Drinking Waters: Occurrence and Implications for Future Regulation," *Environmental Science and Technology*, 37(14), 3104–3110.

Werdehoff, K.S., and P.C. Singer (1987) "Chlorine Dioxide Effects on THMFP, TOXFP, and the Formation of Inorganic By-Products," *Journal AWWA*, 79(9), 107.

Westerhoff, P., Y. Yeomin, S. Snyder, and E. Wert (2005) "Fate of Endocrine-Disruptor, Pharmaceutical, and Personal Care Product Chemicals during Simulated Drinking Water Treatment Processes," *Environmental Science and Technology*, 39(17), 6649–6663.

Westerhoff, P., B. Nalinakumari, and P. Pei (2006) "Kinetics of MIB and Geosmin Oxidation during Ozonation," *Ozone Science and Engineering*, 28(5), 277–286.

CHAPTER 8
COAGULATION AND FLOCCULATION

Raymond D. Letterman, Ph.D., P.E.
Professor
Department of Civil and Environmental Engineering
Syracuse University
Syracuse, New York, United States

Sotira Yiacoumi, Ph.D.
Professor
School of Civil and Environmental Engineering
Georgia Institute of Technology
Atlanta, Georgia, United States

INTRODUCTION **8.2**
Definitions 8.3
CONTAMINANTS **8.3**
Particles 8.3
STABILITY OF PARTICLE SUSPENSIONS **8.6**
Electrostatic Stabilization 8.7
Steric Stabilization 8.12
DESTABILIZATION MECHANISMS **8.14**
Double-Layer Compression 8.14
Surface Charge Neutralization 8.14
Adsorption and Interparticle Bridging 8.15
Application of External Fields 8.15
COAGULANTS **8.15**
Hydrolyzing Metal Salt Coagulants 8.15
Types of Hydrolyzing Metal Salt Coagulants Used in Water Treatment 8.20
Impurities in Hydrolyzing Metal Salt Coagulant Solutions 8.21
Models for Predicting Hydrolyzing Metal Salt Coagulant Dosages 8.22
NOM Removal–Example Jar Test Results 8.30
Acid-Base Chemistry of Hydrolyzing Metal Salt Coagulants 8.33
Residual Aluminum and Iron in Coagulant Treated Water 8.36
Selecting the Type of Coagulant and Estimating the Required Dosage 8.41
Organic Polyelectrolyte Coagulants 8.45
Supplemental (Coagulant Aid) Materials 8.48
THE RAPID MIXING AND FLOCCULATION PROCESSES **8.50**
Purpose of Rapid Mixing 8.50
Purpose of Flocculation 8.50
Transport Mechanisms 8.50
Flocculation Kinetics in Batch and Plug Flow Reactors 8.55
Ideal Continuous-Flow Flocculation Reactors 8.55
Effect of Temperature on Coagulation and Flocculation 8.63
Evaluating Coagulant Performance Using Bench Scale Testing 8.64
Monitoring the Particle Aggregation Process 8.66
Electrokinetic Measurements 8.68
ABBREVIATIONS **8.71**
EQUATION NOTATION **8.71**
REFERENCES **8.72**

INTRODUCTION

Coagulation is a process that increases the tendency of small particles in an aqueous suspension to attach to one another and to surfaces such as the media in a filter bed. It is also used to affect the removal of certain soluble materials by adsorption or precipitation. The coagulation process typically includes promoting the interaction of particles to form larger aggregates. It is an essential component of conventional water treatment systems in which the processes of coagulation, clarification, filtration, and disinfection are combined to treat the water and remove and inactivate microbiological contaminants such as viruses, bacteria, and the cysts and oocysts of pathogenic protozoa. Facilitating the removal of microbiological contaminants is perhaps the key reason for using coagulation; however, other objectives such as the removal of natural organic matter (NOM) to reduce the formation of disinfection by-products (DBPs) and the adsorption of dissolved inorganic contaminants such as arsenic compounds are increasingly important. Coagulation and flocculation are pretreatment processes; they do not actually remove contaminants but rather promote changes that facilitate their efficient removal in subsequent separation processes.

Coagulation has been an important component of granular bed (rapid sand and multimedia) filtration plants in the United States since the 1880s. Alum and iron (III) salts have been used as coagulant chemicals since that time, with alum having the most widespread use. In the 1930s, Baylis (1937) perfected activated silica as a coagulant aid. This material, formed on site, is an anionic polymer or a small, negatively charged colloid. Synthetic organic polymers were introduced in the 1960s, with cationic polymers having the greatest use. Natural starches were used before the synthetic compounds. Polymers have helped change pretreatment and filtration practice, including the use of multimedia filters and filters with deep, uniform grain-size media; high-rate filtration; direct filtration (rapid mixing, flocculation, and filtration, but no clarification); and inline filtration (rapid mixing and filtration only).

Coagulants are also used to enhance the performance of membrane microfiltration and ultrafiltration systems (Wiesner et al., 1989; Judd and Hillis, 2001; Howe and Clark, 2006) and in pretreatment to prolong the bed life of granular activated carbon contactors (Nowack and Cannon, 1996). The development of new coagulant chemicals, advances in floc removal processes and filter design, particularly membrane processes, and challenging particle removal performance standards and goals have stimulated substantial diversity in the design and operation of the coagulation process. Change can be expected to continue into the future.

In evaluating granular bed filtration plants that were producing high-quality filtered water, Cleasby et al. (1989) concluded, "Chemical pretreatment prior to filtration is more critical to success than the physical facilities at the plant." Their report recommends that plant staff use a well-defined coagulant chemical control strategy that considers variable raw-water quality. There is no question that high-rate (rapid sand and multimedia) filtration plants are coagulant-based systems that work only as well as the coagulants that are used.

This chapter begins with introductory sections on definitions and the types of contaminants removed by the coagulation and flocculation processes. The fundamentals of physics and chemistry that explain how particles interact in suspensions are then reviewed. The various types of hydrolyzing metal salt coagulants are described and methods for estimating required dosages and minimizing residual coagulant metal concentrations are explained using some example calculations. Organic polyelectrolyte coagulants and supplemental materials are then discussed. The final sections describe the flocculation process and how it is used with coagulants to produce flocs that can be removed by particle separation processes.

Definitions

Coagulation is a complex process, involving numerous interconnected chemical reactions and mass transfer steps. As practiced in water treatment, the process is essentially four separate and sequential steps: coagulant transformation, uptake of adsorbed species, particle destabilization, and interparticle collisions. Coagulant transformation, adsorbate-coagulant interaction, and particle destabilization typically occur during and immediately after chemical dispersal in rapid mixing; interparticle collisions that cause aggregate (floc) formation begin during rapid mixing but usually occur with greatest effect in the flocculation process. For example, when the aluminum sulfate salt known as alum ($Al_2(SO_4)_3 \cdot 14H_2O$) is used in coagulation, the assortment of chemical species, called *aluminum hydrolysis products,* that causes coagulation is formed during and after the alum is mixed with the water to be treated. Coagulants are sometimes formed (or partially formed) prior to their addition to the rapid-mixing units. Examples include activated silica and synthetic organic polymers and the prehydrolyzed metal salts such as polyaluminum chloride (PACl) and polyiron chloride (PICl).

The terminology of coagulation has not been standardized. However, in most of the water treatment literature, coagulation refers to all the reactions and mechanisms that result in particle aggregation in the water being treated, including coagulant transformations (where applicable), particle destabilization, adsorbate uptake, and physical interparticle contacts. The physical process of producing interparticle contacts and particle aggregates (flocs) is termed *flocculation.*

These definitions of coagulation and flocculation are based on the terminology used by early practitioners such as Camp (1955). However, in the colloid science literature, LaMer (1964) considered only chemical mechanisms in particle destabilization and used the terms *coagulation* and *flocculation* to distinguish between two of them. LaMer defined destabilization by simple salts such as NaCl (a so-called indifferent electrolyte) as coagulation.

The water treatment literature sometimes makes a distinction between the terms *coagulant* and *flocculant.* When this distinction is made, a coagulant is a chemical used to initially destabilize the suspension and is typically added in the rapid-mix process. In most cases a flocculant is used after the addition of a coagulant; its purpose is to enhance floc formation and to increase the strength of the floc structure. It has awkwardly been called a *coagulant aid.* Flocculants are often used to increase filter performance (they may be called *filter aids* in this context) and to increase the efficiency of a sludge dewatering process. In this chapter, no distinction is made between coagulants and flocculants. The term *coagulant* is used exclusively.

CONTAMINANTS

Particles

As discussed in Chap. 3 particles in natural water vary widely in origin, concentration, size, and surface chemistry. Some are derived from land-based or atmospheric sources (e.g., clays and other products of weathering, silts, pathogenic microorganisms, asbestos fibers, and other terrestrial detritus and waste discharge constituents), and some are produced by chemical and biological processes within the water source (e.g., algae, precipitates of $CaCO_3$, $Fe(OH)_3$, MnO_2, and the organic exudates of aquatic organisms). Certain toxic metals and synthetic organic chemicals (SOCs) are associated with solid particles, so coagulation for particle aggregation can be important in the removal of soluble, health-related pollutants.

Particle size may vary by several orders of magnitude, from a few tens of nanometers (e.g., viruses and high molecular weight NOM) to a few hundred micrometers (e.g., zooplankton). The important cysts and oocysts of pathogenic protozoa (e.g., *Giardia* and

Cryptosporidium) are ovoid particles with overall dimensions in the 3- to 12-μm (micrometer) range. All can be effectively removed by properly designed and operated coagulation, floc separation, and filtration facilities.

The smallest particles, those with one dimension less than 1 μm, are usually called *colloidal* and those which are larger than this limit are said to be suspended. The operational definition of dissolved and suspended impurities is frequently established by a 0.45-μm-pore-size membrane filter but colloidal particles can be smaller than this dimension. The effect of gravity on the transport of colloidal particles tends to be negligible compared to the diffusional motion caused by interaction with the fluid (Brownian motion) and, compared to suspended particles, colloidal particles have significantly more external surface area per unit mass.

Measuring Particle Concentration. The principal methods for measuring the performance of particle removal processes in water treatment systems are turbidity and particle counting. Both techniques have limitations, and, consequently, a single method may not provide all the information needed to successfully monitor and control process performance.

Turbidity is measured using an instrument, called a *turbidimeter* or *nephelometer,* that detects the intensity of light scattered at one or more angles to an incident beam of light. Light scattering by particles is a complex process and the angular distribution of scattered light depends on a number of conditions including the wavelength of the incident light and the particle's size, shape, and composition (Sethi et al., 1997). Consequently, it is difficult to correlate the turbidity with the amount, number, or mass concentration of particles in suspensions.

When the turbidity measurement is used for regulatory purposes, it should theoretically be possible to take a given suspension and measure its turbidity at any water treatment facility and obtain an unbiased result that is reasonably close to the average turbidity measured at all other facilities. To achieve reasonable agreement, three factors must be considered: the design of the instruments, the material used to calibrate the instrument, and the technique used to make the measurement. Given this need, turbidimeter design, calibration, and operation criteria have been developed by a number of organizations using the consensus process. However, standardization is a difficult and imperfect process, and it has been shown (Hart et al., 1992) that instruments designed and calibrated using the criteria of these standards can give significantly different responses.

Turbidity measurements were first used to maintain the aesthetic quality of treated water. In 1974, after the passage of the Safe Drinking Water Act, the USEPA lowered the limit for filtered water to 1 ntu (nephelometric turbidity unit) with the explanation that particles causing turbidity can interfere with the disinfection process by enmeshing and, therefore, protecting microbiological contaminants from chemical disinfectants such as chlorine. Today the turbidity measurement is also used to assess filter performance. It is viewed as an important indicator of the extent to which disinfectant-resistant pathogens have been removed by the filtration process. Filtered-water turbidity criteria must be met before the protozoan cyst/oocyst and virus removal credits allowed by the Surface Water Treatment Rule of the 1986 Amendments of the Surface Water Treatment Rule can be applied (see Chap. 2).

Particle-counting instruments are widely used in the drinking water industry, especially for monitoring and controlling clarification and filtration process performance. Plants use them to detect early filter breakthrough and to maintain plant performance at a high level. Online devices that continuously measure particle concentrations in preselected size ranges at various points in the treatment system are especially important. Occasionally batch-sampling devices are used. Two types of particle-counting/sizing sensors are important in water treatment applications: light-blocking (light-obscuration) devices and light-scattering devices (Hargesheimer et al., 1992; Lewis et al., 1992). At the present time instruments with light-blocking-type sensors are more common.

The types of particle-counting instruments used in water treatment plants have limitations (Hargesheimer et al., 1992; Letterman, et al. (2001). Most do not detect any particles smaller than about 1 μm and do detect only a fraction of the particles larger than 1 μm. (Letterman et al., 2001). They should therefore be used in conjunction with turbidimeters that detect smaller particles. Differences in the optical characteristics of the sensors make achieving direct count and size agreement between instruments difficult. There are no industry-wide standards for sensor resolution or for particle-counting and -sizing accuracy and efficiency. For a given particle suspension it is not possible to make similar sensors yield identical particle counts and sizes. Until this is feasible, the regulatory use of particle count measurements will be limited.

The reader may find additional coverage of turbidity and particle measures in Chap. 3.

Natural Organic Material (NOM) and the USEPA's Enhanced Coagulation Requirement. Chapter 3 discusses the characteristics and significance of NOM in water treatment. NOM consists of various types of organic matter that are important precursors in the formation of disinfection and other types of by-products as discussed in Chap. 19. NOM imparts a yellow color to water, especially the humic substances fractions (humic and fulvic acids); causes aesthetic concerns; and can affect water quality in the distribution system. Coagulation and flocculation can be effective ways to remove a significant fraction of the NOM. The dosage required for effective NOM removal is typically greater than that needed for particle (turbidity) removal.

The USEPA's 1998 Stage 1 Disinfection By-Products Rule (Stage 1 DBPR) (USEPA, 1998) requires the use of a NOM removal strategy called *enhanced coagulation* to limit the formation of all disinfection by-products (DBPs). The requirement applies to conventional water treatment facilities that treat surface water or groundwater that is under the influence of surface water. The amount of TOC a plant must remove is based on the raw-water TOC and alkalinity.

Enhanced coagulation ties the TOC removal requirement to the raw water alkalinity to avoid forcing a utility to add high dosages of *hydrolyzing metal salt (HMS)* coagulants to reduce the pH to between 5 and 6, a range where HMS coagulants frequently appear to be most efficient. It also recognizes that higher TOC removal is usually possible when the raw-water TOC concentration is relatively high and the fraction of the NOM that is more readily removed by HMS coagulants is typically greater. The matrix in Table 8-1 gives Stage 1 DBPR's required TOC removal percentages.

TABLE 8-1 Required Percent Removals of Total Organic Carbon by Enhanced Coagulation in the 1998 Stage 1 Disinfection By-Products Rule

Source-water total organic carbon (mg C/L)	Source-water alkalinity (mg/L as $CaCO_3$)		
	0–60	60–120	> 120
> 2.0–4.0	35	25	15
> 4.0–8.0	45	35	25
> 8.0	50	40	30

Example 8-1 Application of the USEPA's Enhanced Coagulation Requirement for TOC

Problem Statement A plant's source water has a TOC of 3.5 mg C/L and an alkalinity of 85 mg/L as $CaCO_3$. Determine the maximum allowable TOC of the combined filter effluent in mg C/L for this water.

Solution According to the table, the required TOC removal is 25 percent. The TOC of the filtered water (after the point where the filter effluents are combined) would have to be less than 2.6 mg C/L, calculated using the relationship 2.6 mg C/L = 3.5 mg C/L × (1 – 0.25). ▲

The Stage 1 DBPR provides alternative compliance criteria for the enhanced coagulation, treatment technique requirement. The six criteria are

1. The system's source water TOC is < 2.0 mg C/L.
2. The system's treated water TOC is < 2.0 mg C/L.
3. The system's source water TOC is < 4.0 mg C/L, the source water alkalinity is > 60 mg/L as $CaCO_3$, and the system is achieving TTHM < 40 μg/L and HAA5 (haloacetic acids) < 30 μg/L.
4. The system's TTHM is < 40 μg/L, HAA5 is < 30 μg/L and only chlorine is used for primary disinfection and maintaining a distribution system residual.
5. The system's source water SUVA prior to any treatment is ≤ 2.0 L/(mg-m).
6. The system's treated water SUVA is ≤ 2.0 L/(mg-m).

The measurements used to test compliance with criteria 1, 2, 5, and 6 are made monthly and a running annual average is calculated quarterly. Compliance with criteria 3 and 4 is based on monthly measurements of TOC and alkalinity and quarterly measurements of TTHMs and HAA5, and the running annual average is calculated quarterly.

The regulatory negotiators who formulated the enhanced coagulation requirement were concerned that coagulant chemical costs might be excessive for utilities that treat water with a high concentration of NOM that is not amenable to removal by coagulants. For these plants the removal requirements of Table 8-1 and the alternative compliance criteria may not be feasible. This is rare, but in that event the Stage 1 DBPR allows them to use a jar test procedure to determine an appropriate, alternative TOC removal requirement (White et al., 1997).

The alternative TOC removal requirement is determined by performing jar tests on at least a quarterly basis for 1 year. In these tests, alum or ferric coagulants are added in 10-mg/L increments until the pH is lowered to a target pH value that varies with the source water alkalinity. For the alkalinity ranges of 0–60, >60–120, >120–240, and >240 mg/L as $CaCO_3$, the target pH values are 5.5, 6.3, 7.0, and 7.5, respectively. When the jar test is complete, the residual TOC concentration is plotted versus the coagulant dosage (in mg coagulant/L) and the alternative TOC percentage is found at the point called the *point of diminishing returns* (*PODR*). The PODR is the coagulant dosage where the slope of the TOC-coagulant dosage plot changes from greater than 0.3 mg C/10 mg coagulant to less than 0.3 mg C/10 mg coagulant. If the plot does not yield a PODR, then the water is considered to be not amenable to enhanced coagulation and the primacy agency may grant the system a waiver from the enhanced coagulation requirement. Details of the jar test procedure are given in the USEPA's Guidance Manual for Enhanced Coagulation and Enhanced Precipitative Softening (USEPA, 1999).

STABILITY OF PARTICLE SUSPENSIONS

In water treatment, the coagulation process is used to increase the rate or kinetics of particle aggregation and floc formation. The objective is to transform a stable suspension, that is, one that is resistant to aggregation (or attachment to a filter grain), into an unstable one. Particles that may have been in lake water for months or years as stable, discrete units can be aggregated in an hour or less following successful destabilization. The design and operation of the coagulation process requires proper control of both the particle destabilization and the subsequent aggregation process.

The suspensions that must be aggregated and separated in water treatment may include particles formed in or added to the treatment system, in addition to the particles in the source water. For example, processes used to remove manganese and iron can produce suspensions of MnO_2 and $Fe(OH)_3$ precipitates, and particles of finely divided activated carbon (powdered activated carbon) are often added before particle separation processes to adsorb organic compounds. In some cases, such as high-quality upland reservoirs and lakes, the quantity of particles formed or added in treatment may greatly exceed that of the particles in the source water.

As particles in a suspension approach one another or as a particle in a flowing fluid approaches a stationary surface such as a filter grain, forces arise that tend to keep the surfaces apart. Also, there are forces that tend to pull the interacting surfaces together. The most well known repulsive force is caused by the interaction of the electrical double layers of the surfaces (electrostatic stabilization). The most important attractive force is called the *London–van der Waals force*. It arises from spontaneous electrical and magnetic polarizations that create a fluctuating electromagnetic field within the particles and in the space between them. These two types of forces, repulsive and attractive, form the basis of the Derjaguin, Landau, Verwey, and Overbeek (DLVO) theory of colloid stability (Derjaguin and Landau, 1941; Verwey and Overbeek, 1948). Other forces include those associated with the hydration of ions at the surfaces (a repulsive force), hydrophobicity of surfaces (an attractive force), and the presence of adsorbed polymers, which can cause either repulsion (steric interaction) or attraction (polymer bridging). Additional forces may arise if external fields, such as magnetic fields, are utilized in aggregation processes. As particles approach one another on a collision course, the fluid between them must move out of the way. The repulsive force caused by this displacement of fluid is called *hydrodynamic retardation*. Details about these forces can be found in the text by Elimelech et al. (1995).

Electrostatic Stabilization

Origins of Surface Charge. Most particles in water, mineral and organic, have electrically charged surfaces, and the sign of the charge is usually negative (Niehof and Loeb, 1972; Hunter and Liss, 1979; Stumm, 1992). As described in Chap. 3, there are three important processes for producing this charge. First, surface groups on the solid may react with water and accept or donate protons. In these reactions, the surface charge on a solid particle depends on the pH in the solution. As the pH increases, the surface charge becomes increasingly negative. Silica is negatively charged in water with a pH above 2; proteins contain a mixture of carboxyl and amino groups and usually have a negative charge at a pH above about 4.

Second, surface groups can react in water with solutes other than protons. These surface complex formation reactions involve specific chemical reactions between chemical groups on the solid surface (e.g., silanol groups) and adsorbable solutes (e.g., calcium and phosphate ions). The surface charge is again related to solution chemistry.

Third, a surface charge may arise because of imperfections within the structure of the particle; this is called *isomorphic replacement* or *substitution*. It is responsible for a substantial part of the charge on many clay minerals. Clays are layered structures and in these structures sheets of silica tetrahedra are typically cross-linked with sheets of alumina octahedra. The silica tetrahedra have an average composition of SiO_2 and may be depicted as shown in Fig. 8-1a: If an Al atom is substituted for a Si atom during the formation of this lattice, a negatively charged framework results (Fig. 8-1b).

Similarly, a divalent cation such as Mg (II) or Fe (II) may substitute for an Al(III) atom in the aluminum oxide octahedral network, also producing a negative charge. The

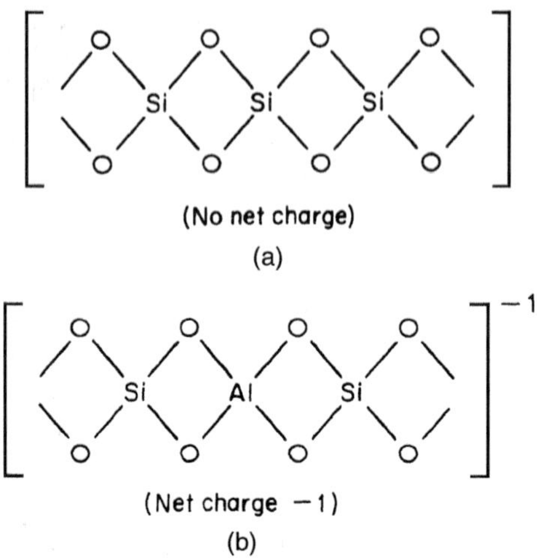

FIGURE 8-1 SiO_2 structure, (*a*) with no net charge and (*b*) with −1 net charge.

sign and magnitude of the charge produced by such isomorphic replacements are independent of the characteristics of the aqueous phase after the crystal is formed.

The Electrical Double Layer. In a colloidal suspension, there can be no net imbalance in the overall electrical charge; the primary charge on the particle must be counter balanced in the aqueous system. Figure 8-2 shows schematically a negatively charged particle with the counterbalancing cloud of ions (the diffuse layer) around the particle. Because the particle is negatively charged, excess ions of opposite charge (positive) accumulate in this interfacial region. Ions of opposite charge accumulating in the interfacial region, together with the primary charge, form an electrical double layer. The diffuse layer results from electrostatic attraction of ions of opposite charge to the particle (*counterions*), electrostatic repulsion of ions of the same charge as the particle (*coions*), and thermal or molecular diffusion that acts against the concentration gradients produced by the electrostatic effects. Schematic representations of the diffuse layer are shown in Figs. 8-2 and 8-3a.

Because of the primary charge, an electrostatic potential (voltage) exists between the surface of the particle and the bulk of the solution. This electrical potential can be pictured as the electric pressure that must be applied to bring a unit charge having the same sign as the primary charge from the bulk of the solution to a given distance from the particle surface, shown schematically in Fig. 8-3b. The potential has a maximum value at the particle surface and decreases with distance from the surface. This decrease is affected by the characteristics of the diffuse layer and, thus, by the type and concentration of ions in the bulk solution.

Dynamic Aspects of the Electrical Double Layer. The interaction of two particles can be evaluated in terms of potential energy (i.e., the amount of energy needed to bring two particles from infinite separation up to a given separation distance). If the potential energy is positive, the overall interaction is unfavorable (repulsive) since energy must be provided to the system. If the potential energy is negative, the net effect is attractive.

When particles are forced to move in a fluid or when the fluid is forced to move past a stationary particle, some of the charge that balances the surface charge moves with the fluid and some of it does not. Thus, the electrical potential at the plane of shear (which is the boundary between the charge that remains with the particle and the charge that does not) is less than the electrostatic potential at the surface of the particle. The exact location of the slipping plane is not known, and, therefore, it is difficult to relate the potential measured at the plane of shear (by electrokinetic techniques such as electrophoresis or streaming current) to the surface potential. Lyklema (1978) locates the slipping plane at the outer border of the Stern layer as shown in Fig. 8-2.

The mathematical treatment of the double layer, shown schematically in Figs. 8-2 and 8-3a, was developed independently by Gouy and Chapman, and the result is called the *Gouy-Chapman model.* Details of the model can be found in several texts (e.g., Elimelech et al., 1995). In this model, electrostatic or coulombic attraction and diffusion are the

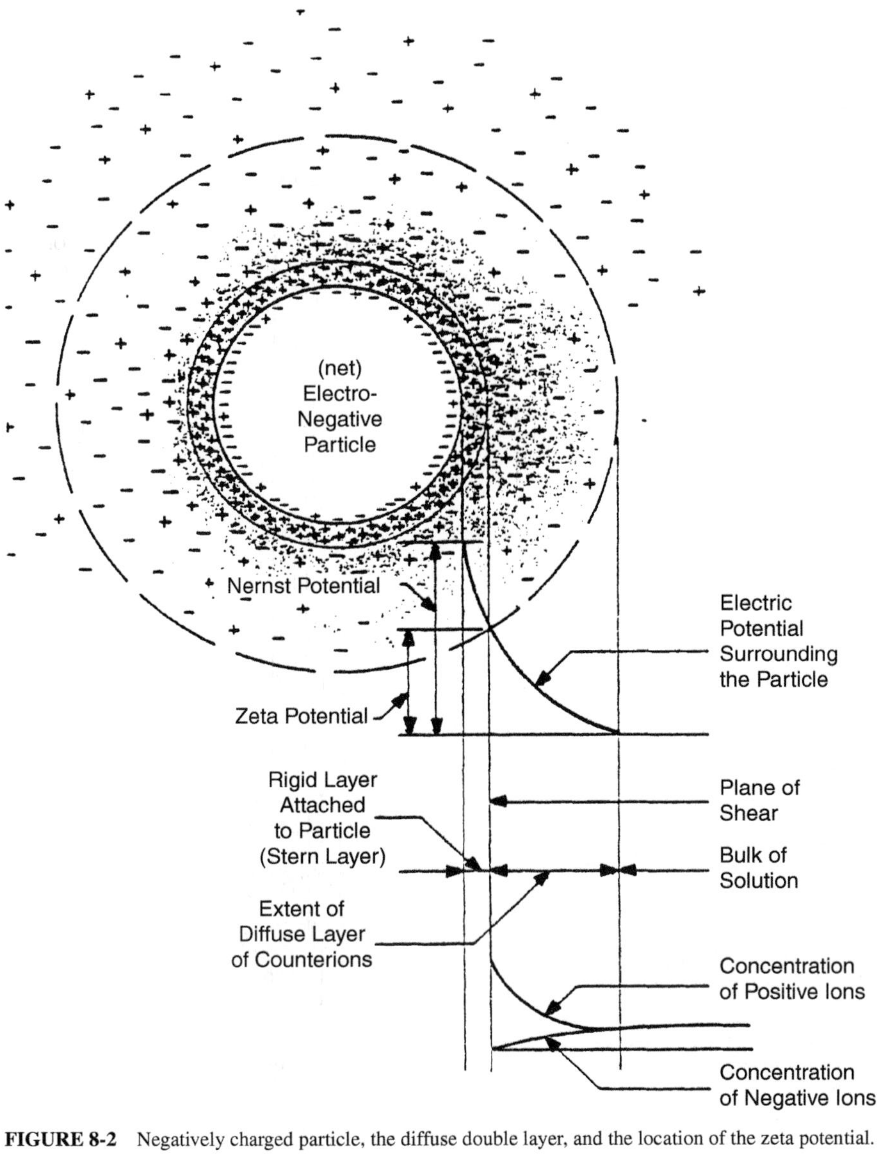

FIGURE 8-2 Negatively charged particle, the diffuse double layer, and the location of the zeta potential.

interacting processes responsible for the formation of the diffuse layer. Ions are treated as point (dimensionless) charges and have no other physical or chemical characteristics.

When two similar colloidal particles approach each other, their diffuse layers begin to overlap and to interact. This comingling of charge creates a repulsive potential energy ψ_R that increases in magnitude as the distance separating the particles decreases. Figure 8-3c and d plots the repulsive energy of interaction as a function of the distance between the interacting surfaces.

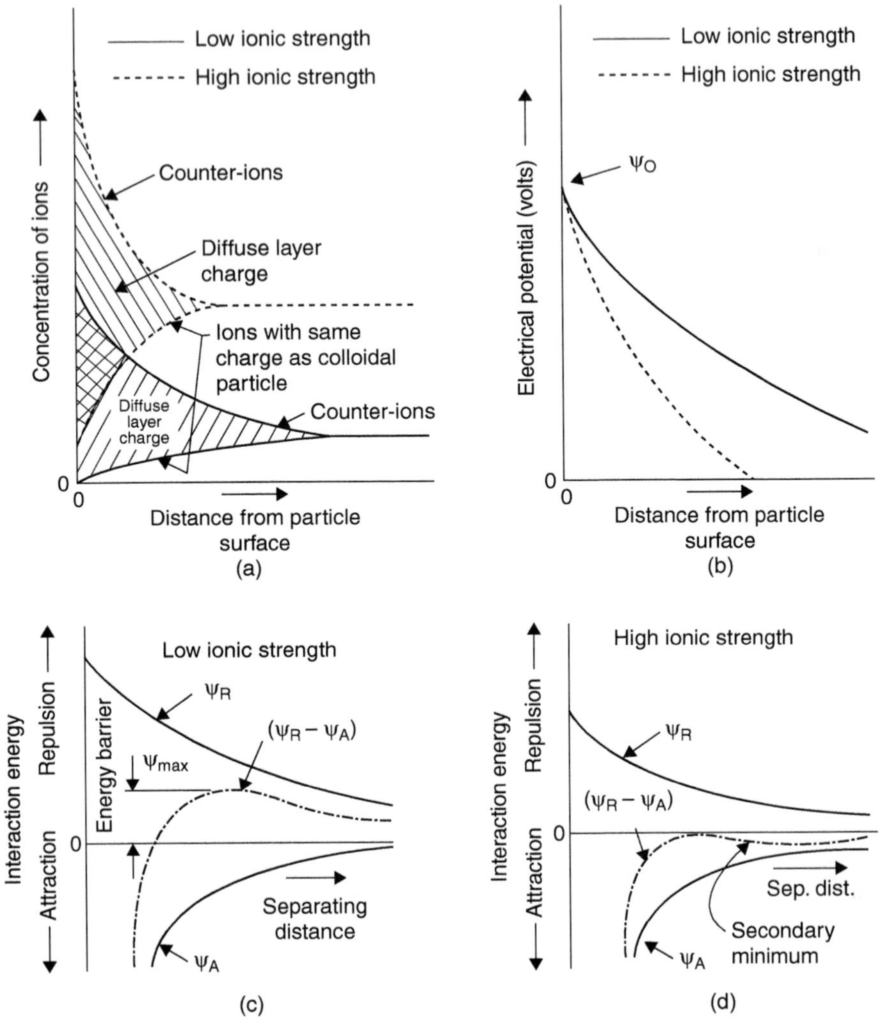

FIGURE 8-3 Schematic representations. (*a*) The diffuse double layer. (*b*) The diffuse layer potential. (*c* and *d*) Two cases of particle-particle interaction energies in electrostatically stabilized colloidal systems.

Attractive forces (van der Waals forces) exist between all types of particles, no matter how dissimilar they may be. These forces arise from interactions among dipoles, either permanent or induced, in the atoms composing the interacting surfaces and the water. The magnitude of the attractive energy of interaction ψ_A decreases with increasing separation distance. Schematic curves of the van der Waals attractive potential energy of interaction are shown in Fig. 8-3c and d. Unlike electrostatic repulsive forces, the van der Waals attractive forces are essentially independent of the composition of the solution; they depend on the kind and number of atoms in the colloidal particles and the continuous phase (water).

The net forces in a colloidal system are a superposition of electrostatic repulsive forces and attractive van der Waals interactions as explained by the DLVO theory. In other words, summation of ψ_R and ψ_A yields the net interaction energy between two colloidal particles. This sum with the proper signs is $(\psi_R - \psi_A)$ and is shown schematically as a function of separation distance in Fig. 8-3c and d. The force acting on the particles is the derivative $d(\psi_R - \psi_A)/ds$ where s is the separation distance. When electrostatic repulsion dominates during particle-particle interactions, the suspension is said to be electrostatically stabilized and to undergo only *slow* coagulation. Electrostatic stabilization is fundamental to the current understanding of coagulation in water treatment. For additional insight into electrostatic stabilization, the mathematical treatment of Derjaquin and Landeau (1941) and Verwey and Overbeek (1948), as well as a summary by Lyklema (1978), should be consulted. Valuable summaries of the methods, mathematics, and meanings of models for electrostatic stabilization are also contained in books by Morel (1983), Stumm and Morgan (1996), and Elimelech et al. (1995).

The DLVO theory, which is based on the Gouy-Chapman model, provides a good qualitative picture of the origins of electrostatic stabilization. It can also describe the influence of ionic strength on particle stability as will be discussed later when double-layer compression is introduced as a destabilization mechanism. It does not provide, however, a quantitative predictive tool for such important characteristics as coagulant requirements and coagulation rates. Its principal drawback lies in its characterization of all ions as point charges. This allows for a physical description of electrostatics, but omits a description of chemical interactions. For example, the sodium ion (Na^+), the dodecylamine ion ($C_{12}H_{25}NH_3^+$), and the proton (H^+) have identical charges but are not identical coagulants. In addition, DLVO theory does not consider adsorption reactions at the surface although it can be used with surface complexation models to combine electrostatics and surface chemistry.

Quantitative models for the surface chemistry of oxides and other types of surfaces with ionizable surface sites are available. Some have been incorporated into computer programs such as MINEQL, allowing self-consistent calculations to be made simultaneously for surface and solution (Davis and Leckie, 1978). The most popular are the surface complexation models, which describe the formation of charge, potential, and the adsorption of ions at the particle-water interface. The fundamental concept on which all surface complexation models are based is that adsorption takes place at defined coordination sites (surface hydroxyl groups present in finite number) and that adsorption reactions can be described quantitatively by mass action expressions (Goldberg et al., 1996). The various models use different assumptions for the electrostatic interaction terms. For example, some use a simple linear relationship between charge and potential at the surface [the *constant capacitance model* of Schindler and Stumm (Stumm, 1992)] and others use more elaborate relationships such as the triple-layer model of Davis and Leckie (1978).

All the surface complexation models reduce to a similar set of simultaneous equations that are typically solved numerically: (1) mass action equations for all surface reactions, (2) mass balance equations for surface hydroxyl groups and other species, (3) equations for calculation of surface charge, and (4) a set of equations that describes the charge and potential relationships of the electrical double layer. Attempts have been made to use these models to predict the effects of solution and suspension variables on flocculation efficiency when aluminum salt coagulants are used (Letterman and Iyer, 1985) and on the rate of coagulation of iron-oxide particles (Liang and Morgan, 1990). Surface complexation models have also been extended to predict the kinetic behavior of particle-water interfaces (Yiacoumi and Tien, 1995), and electrostatic potentials as function of time have been calculated and compared with experimental data (Subramaniam and Yiacoumi, 2001). Such information could be useful in developing models for predicting particle flocculation kinetics.

Secondary Minimum Aggregation. In certain systems the kinetic energy and electrical double-layer characteristics of the interacting particles may be such that aggregation takes place in what is called a *secondary minimum* (see Fig. 8-3d). Under this condition, the particles are held in proximity (at least momentarily) by van der Waals attraction but remain nanometers apart due to the repulsive force associated with the interacting electrical double layers. If the kinetic energy of the interacting particles is increased, the repulsive force may not be great enough to limit the close approach of the particles, and aggregation will tend to occur in the primary minimum, with the surfaces of the particles close together. In this case the separation distance in Fig. 8-3c is reduced until the quantity ($\psi_R - \psi_A$) becomes less than ψ_{max}. Consideration of both primary and secondary minima in particle interactions may explain discrepancies between classical model predictions that consider only primary minimum aggregation and experimental deposition efficiencies of particles in porous media under unfavorable chemical conditions (Hahn and O'Melia, 2004).

Like–Charge Electrostatic Attraction of Colloids. As noted previously the principal drawback of the DLVO theory lies in the characterization of ions as point charges. As a consequence, the electrostatic factor in models of interparticle interaction does not consider interaction between the ions. Work in recent years has been focused on developing molecular modeling techniques that include ion-ion correlations or model ions as separate identities in predicting electrostatic interactions. Based on these studies, the presence of attractive electrostatic interactions between similarly charged surfaces in electrolyte solutions has been identified (Oasawa, 1971). In a study by Gulbrand et al. (1984), the osmotic pressure has been measured and attractive interaction has been observed in a system with two planar walls with divalent counterions. Taboada-Serrano et al. (2006) have studied the interactions between a colloidal particle and a planar surface in systems of symmetric and asymmetric electrolytes and like-charge electrostatic attraction has been observed in various systems. Similarly to the previous discussion of secondary minimum aggregation, like-charge electrostatic attraction of colloids may explain deposition of charged colloidal particles under unfavorable conditions or aggregation behavior that cannot be explained with classical theory. In the review article by Taboada-Serrano et al. (2005), aggregation modeling of colloidal particles is discussed on the basis of classical treatment and new molecular approaches.

Nanoparticles. Nanoparticles have received great interest in recent years. When studying nanoparticle fate and transport in aqueous media, the effect of diminishing particle size on the interaction energies has to be taken into account. For nanoparticles smaller than 20 nm, particle behavior resembles that of a molecular solute and intermolecular forces play a significant role in affecting the aggregation and deposition of these particles. At this scale, it is also important to characterize the heterogeneities of the surfaces with which nanoparticles interact. The transport of nanoparticles at mesoscopic scale is characterized by high diffusion coefficients due to their small size resulting in increased contact opportunities and resulting in fast aggregation even at low attachment efficiencies. Nanoparticles can often spontaneously aggregate to form clusters even in the absence of destabilizing agents (Brant et al., 2005).

Steric Stabilization

Steric stabilization can result from the adsorption of polymers at solid-water interfaces. Large polymers can form adsorbed segments on a solid surface with loops and tails extending into the solution (Lyklema, 1978) as illustrated in Fig. 8-4. Adsorbed polymers can be either stabilizing or destabilizing, depending on the relative amounts of polymer and solid particles, the affinities of the polymer for the solid and for water, electrolyte type and concentration, and other factors. A stabilizing polymer may contain two types of groups, one of which has a high affinity for the solid surface and a second, more hydrophilic group that

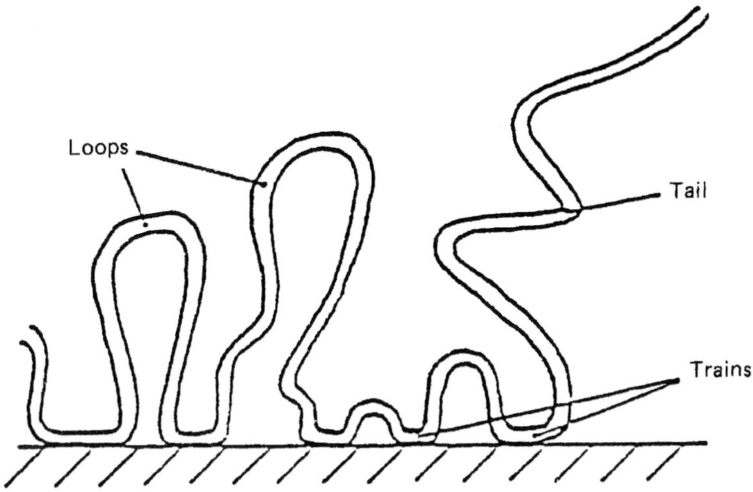

FIGURE 8-4 Illustration of adsorbed polymer configuration with loops, trains, and tails. (*Source: Lyklema, 1978.*)

extends into the solution (Gregory, 1978). The configuration of such an interfacial region is difficult to characterize either theoretically or experimentally. This, in turn, prevents quantitative formulation of the interaction forces between two such interfacial regions during a particle-particle encounter in coagulation. Some useful qualitative descriptions can, however, be made.

Gregory (1978) summarized two processes that can produce a repulsive force when two polymer-coated surfaces interact at close distances. These are illustrated in Fig. 8-5. First, the absorbed layers can each be compressed by the collision; this reduces the volume available for the adsorbed molecules. This reduction in volume restricts the movement of the polymers (a reduction in entropy) and causes repulsion between the particles. Second, and more frequently, the adsorbed layers may interpenetrate on collision, increasing the concentration of polymer segments in the mixed region. If the extended polymer segments are strongly hydrophilic, they can prefer the solvent to reaction with other polymer segments. An overlap or mixing then leads to repulsion. These two processes are separate from and in addition to the effects of polymer adsorption on the charge of the particles and the van der Waals

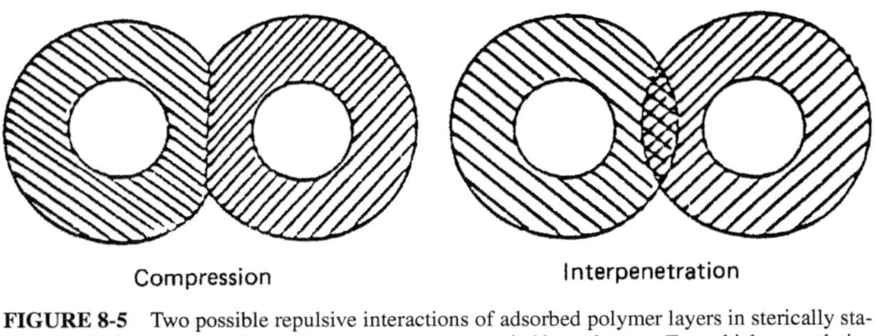

FIGURE 8-5 Two possible repulsive interactions of adsorbed polymer layers in sterically stabilized colloidal systems. (Shaded areas are zones occupied by polymers. Zone thickness relative to particle size is arbitrary.) (*Source: Gregory, 1978.*)

interaction between particles. Charged polymers (polyelectrolytes) can alter particle charge; organic polymers can also reduce the van der Waals attractive interaction energy. Steric stabilization is widely used in the manufacture of industrial colloids such as paints and waxes. Natural organic matter such as humic substances is ubiquitous in water supplies. They are anionic polyelectrolytes, adsorb at interfaces, can be surface active, and may contribute to particle stability by affecting the particle negative charge and by steric effects (O'Melia, 1995).

DESTABILIZATION MECHANISMS

Destabilization is the process in which the particles in a stable suspension are modified to increase their tendency to attach to one another (or to a stationary surface like a filter grain or deposit). The aggregation of particles in a suspension after destabilization requires that they be transported toward one another. In flocculation, particles include those in the source water; those produced in treatment processes, such as oxidation and precipitation; and those added to the treatment system, such as powdered activated carbon. In dissolved air flotation following flocculation the aggregated particles must be transported to air bubbles; see Chap. 9. In filtration they must be transported to the stationary filter surface. Transport processes in suspensions are discussed in this chapter in the section on flocculation. Transport in filtration is discussed in Chap. 10.

Double-Layer Compression

The classical method of colloid destabilization is double-layer compression. To affect double-layer compression, a simple electrolyte such as NaCl is added to the suspension. The ions that are opposite in sign to the net charge on the surface of the particles enter the diffuse layer surrounding the particle. If enough of these counterions are added, the diffuse layer is compressed, reducing the energy required to move two particles of like surface charge into close contact. Destabilization by double-layer compression is not a practical method for water treatment because the salt concentrations required for destabilization may approach that of seawater and, in any case, the rate of particle aggregation would still be relatively slow in all but the most concentrated suspensions. Double-layer compression, however, is an important destabilization mechanism in certain natural systems, for example, estuaries (Edzwald et al., 1974).

Surface Charge Neutralization

Destabilization by surface charge neutralization involves reducing the net surface charge of the particles in the suspension. As the net surface charge is decreased the thickness of the diffuse layer surrounding the particles is reduced and the energy required to move the particles into contact is minimized.

Two processes are used to accomplish surface charge neutralization. In the first, coagulant compounds that carry a charge opposite in sign to the net surface charge of the particles are adsorbed on the particle surface. (In some cases the coagulant is a small particle that deposits on the particle surface.) The coagulants used to accomplish this usually have a strong tendency to adsorb on (attach to) surfaces. Examples include the synthetic and natural organic polyelectrolytes and some of the hydrolysis products formed from hydrolyzing metal salt coagulants. The tendency for these compounds to adsorb is usually attributable to both poor coagulant-solvent interaction and a chemical affinity of the coagulant or chemical groups on the coagulant for the particle surface. Most of the coagulants that are used for charge neutralization can adsorb on the surface to the point that the net surface charge is reversed and, in some cases, increased to the point that the suspension is restabilized.

Adjustment of the chemistry of the solution can be used to destabilize some common types of suspensions by reducing the net surface charge of the particle surfaces. For example, when most of the surface charge is caused by the ionization of surface sites, pH adjustment with acid or base may lead to destabilization. For some surfaces, such as positively charged oxides and hydroxides, the adsorption of simple multivalent anions (such as sulfate and phosphate) or complex polyvalent organic compounds such as humic materials will reduce the positive charge and destabilize the suspension.

Heterocoagulation is a destabilization mechanism that is similar to the process of surface charge neutralization by the adsorption of oppositely charged soluble species. However, in this case, the process involves one particle depositing on another of opposite charge. For example, large particles with a high negative surface charge may contact smaller particles with a relatively low positive charge. Because the particles have opposite surface charge, electrostatic attraction enhances particle-particle interaction. As the stabilizing negative charge of the larger particles is reduced by the deposited positive particles, the suspension of larger particles is destabilized.

Adsorption and Interparticle Bridging

Destabilization by bridging occurs when segments of a high-molecular-weight polymer adsorb on more than one particle, thereby linking the particles together. When a polymer molecule comes into contact with a colloidal particle, some of the reactive groups on the polymer adsorb on the particle surface and other portions extend into the solution. If a second particle with open surface is able to adsorb the extended molecule, then the polymer will have formed an interparticle bridge. The polymer molecule must be long enough to extend beyond the electrical double layer (to minimize double-layer repulsion when the particles approach) and the attaching particle must have available surface. The adsorption of excess polymer may lead to restabilization of the suspension. Ions such as calcium are known to affect the bridging process, apparently by linking sites on interacting polymer chains (Black et al., 1965; Lyklema, 1978; Dentel, 1991).

Application of External Fields

The utilization of external fields in aggregation processes, for example, magnetic fields, introduces an additional force among magnetic (paramagnetic, ferrimagnetic, or ferromagnetic) particles. When the magnetic field is applied, the particles experience an attractive force along the direction of the external field and a repulsive force normal to the external field; consequently, linear aggregates of magnetic particles are formed (Chin et al., 2000). The application of magnetic fields has been proposed to enhance aggregation processes or other separation processes like magnetic seeded filtration (Yavuz et al., 2006; Ying et al., 2000).

COAGULANTS

Hydrolyzing Metal Salt Coagulants

The most widely used coagulants in water treatment are sulfate or chloride salts that contain the metal ions Al^{3+} or Fe^{3+}. In aqueous solutions these small, highly positive ions form such strong bonds with the oxygen atoms of six surrounding water molecules that

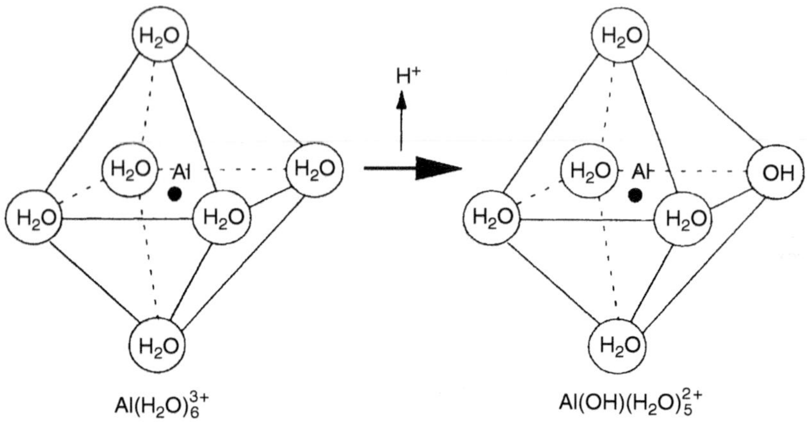

FIGURE 8-6 Deprotonation of the aquo aluminum ion, initial step in aluminum hydrolysis. (*Source: Letterman, 1991.*)

the oxygen-hydrogen atom association in the water molecules is weakened and hydrogen atoms tend to be released to the solution (see Fig. 8-6). This process is known as *hydrolysis* and the resulting aluminum and ferric hydroxide species are called *hydrolysis products*. The water molecules are frequently omitted when writing the chemical formula for these species, but their role is important in determining species behavior. These Fe and Al coagulants are hydrolyzing metal salt (HMS) coagulants.

The chemistry of aluminum and iron hydrolysis reactions and products is complex (see Fig. 8-7). As the pH increases and hydrolysis proceeds and if there is a high total metal ion concentration in the solution, simple mononuclear products can form complex

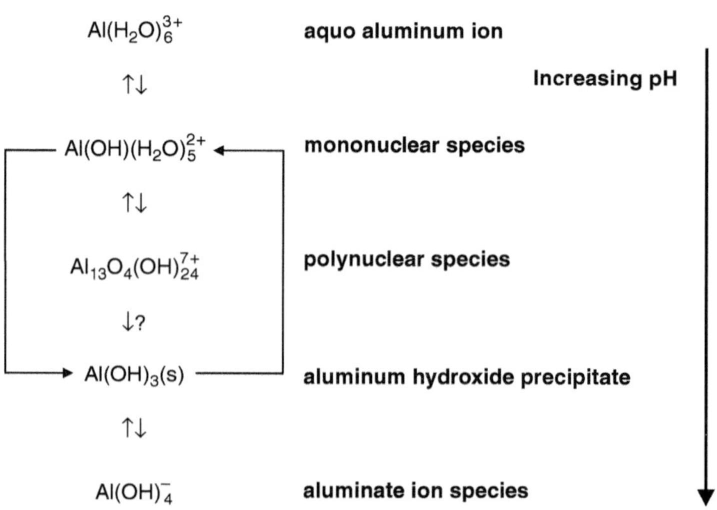

FIGURE 8-7 Schematic diagram of the essential steps in aluminum hydrolysis.

polynuclear species. Eventually microcrystals and the amorphous metal hydroxide precipitate are formed. Under certain conditions (e.g., low total metal ion concentration) the precipitate can form directly from mononuclear species. Dissolution of the metal hydroxide precipitate forms mononuclear species but not polynuclear forms.

The pure amorphous precipitate is a low density, gelatinous material, consisting of agglomerated microcrystals and water. It changes physically with aging, and it can be dehydrated by freezing and thawing. The various hydrolysis products can adsorb (and continue to hydrolyze) on many types of particulate surfaces, and they may form a precipitate with the higher-molecular-weight fraction of NOM. This NOM fraction strongly adsorbs on the metal hydroxide precipitate. The precipitate can be an effective adsorbent for a number of inorganic anions such as fluoride and arsenate. Models for predicting the dosage of coagulants needed for the effective removal of particles and NOM are discussed later in this chapter

The solubility of the metal hydroxide precipitate is one factor that must be considered in maximizing coagulant performance and in minimizing the amounts of residual Al and Fe in treated water. At low pH the dissolution of the metal-hydroxide precipitate produces positively charged, soluble hydrolysis products and the aquo-metal ion (Al^{3+} and Fe^{3+}).

At high pH, the negatively charged, soluble hydrolysis products $Al(OH)_4^-$ and $Fe(OH)_4^-$ are formed. The minimum-solubility pH of aluminum hydroxide precipitate at 25°C is approximately 6.3 (Fig. 8-8); for ferric hydroxide it is about 8 (Fig. 8-9). The theoretical solubility of ferric hydroxide is significantly lower than that of aluminum hydroxide. For example at 25°C with $I = 0.001$, the minimum concentrations of soluble Al and Fe are about 17 μg/L and 0.01 μg/L, respectively. Metal hydroxide solubility should be considered when selecting the type and dosage of a coagulant, especially Al salt products. How to do this and the effect of water temperature on metal hydroxide precipitate solubility are discussed in a later section of the chapter.

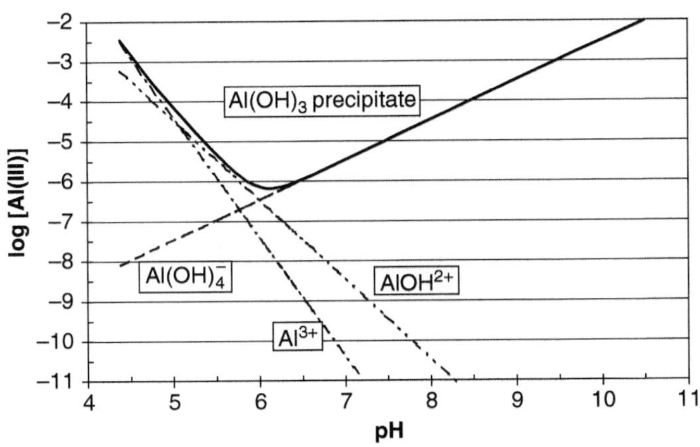

FIGURE 8-8 Solubility diagram for amorphous, freshly precipitated aluminum hydroxide. Water molecules are omitted in species notation, temperature = 25°C.

The solubility diagrams of Figs. 8-8 and 8-9 were plotted using the formation constants ($\beta_{x,y}$) and solubility constants (K_{sp}) listed in Table 8-2. The ionic strength was assumed to be 0.001 M. The formation constants in Table 8-2 are for reactions and mass action expressions of the form

$$x\,M^{3+} + y\,H_2O \rightleftarrows M_x(OH)_y^{(3x-y)+} + y\,H^+ \tag{8-1}$$

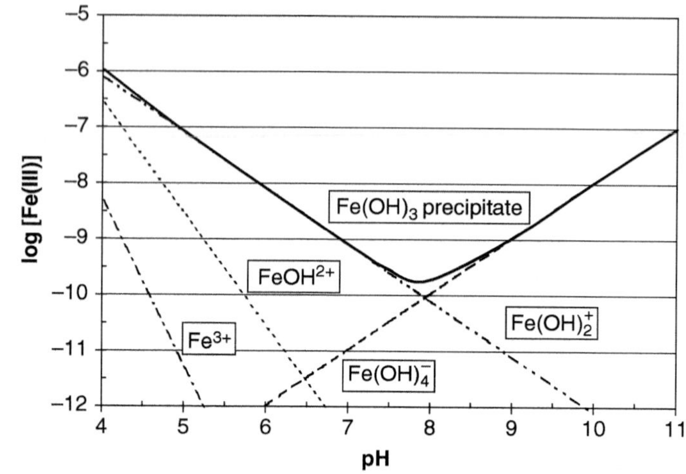

FIGURE 8.9 Solubility diagram for amorphous, freshly precipitated ferric hydroxide. Water molecules are omitted in species notation, temperature = 25°C.

and

$$\beta_{x,y} = \frac{[M_x(OH)_y^{(3x-y)+}][H^+]^y}{[M^{3+}]^x} \tag{8-2}$$

The solubility constants are for reactions and solubility product relationships of the form

$$M(OH)_3\,(s) + 3H^+ \rightleftarrows M^{3+} + 3\,H_2O \tag{8-3}$$

and

$$K_{sp} = \frac{[M^{3+}]}{[H^+]^3} \tag{8-4}$$

TABLE 8-2 Aluminum and Iron Hydrolysis and Solubility Constants Used to Plot Solubility Diagrams for Amorphous Aluminum Hydroxide and Amorphous Ferric Hydroxide Precipitates, $T = 25°C$ and $I = 0.001$

	Amorphous aluminum hydroxide	
Hydrolysis product	Log $\beta_{x,y}$ and Log K_{sp}	Reference
$Al(OH)^{2+}$	−5.07	Ball et al. (1980)
$Al(OH)_4^-$	−23.12	Ball et al. (1980)
$Al(OH)_3(s)$	10.64	Ball et al. (1980)
	Amorphous ferric hydroxide	
$Fe(OH)^{2+}$	−2.2	Stumm and Morgan (1996)
$Fe(OH)_2^+$	−5.7	Stumm and Morgan (1996)
$Fe(OH)_4^-$	−21.6	Stumm and Morgan (1996)
$Fe(OH)_3$ (s)	3.6	Stumm and Morgan (1996)

[For aluminum this is a simplified set of hydrolysis products that have been found to effectively predict filtrate Al concentrations under laboratory and field conditions (Letterman and Driscoll, 1993; Van Benschoten and Edzwald, 1990). A more complete list of hydrolysis products and constants can be found in Sposito (1996).]

When the dissolution reaction is written as

$$M(OH)_3\,(s) \rightleftarrows M^{3+} + 3OH^- \tag{8-5}$$

the corresponding solubility product relationship is

$$K_{so} = [M^{3+}][OH^-]^3 \tag{8-6}$$

The equilibrium constants K_{sp} (Eq. 8-4) and K_{so} are related by

$$K_{so} = K_{sp}(K_w)^3 \tag{8-7}$$

where K_w is the ion product of water ($K_w = 1 \times 10^{-14}$ at 25°C and infinite dilution).

A number of researchers have assumed that the important mononuclear Al hydrolysis products include $Al(OH)^{2+}$, $Al(OH)_2^+$, $Al(OH)_3^o$, and $Al(OH)_4^-$ (Bertsch and Parker, 1996). Others, including Letterman and Driscoll (1993), Hayden and Rubin (1974), and van Benschoten and Edzwald (1990a), have reported that the effect of pH on the solubility of freshly precipitated aluminum hydroxide can be effectively predicted using just the aquo Al species (Al^{3+}) and two mononuclear hydrolysis products ($Al(OH)^{2+}$ and $Al(OH)_4^-$). This assumption was used to plot Fig. 8-8. The term *solubility* in the articles cited means the concentration of aluminium that is not removed by membrane filtration and not the true thermodynamic equilibrium solubility of the precipitate.

The dotted lines plotted in Fig. 8-8 give the concentrations of each Al species when it is in equilibrium with the $Al(OH)_3$ precipitate. The equation of each line was derived by combining Eqs. 8-1 and 8-3 and using the appropriate equilibrium constants corrected for ionic strength. For example, the equation for log $[Al(OH)^{2+}]$ when the ionic strength I = 0.001 M is

$$\log[Al(OH)^{2+}] = \log \beta_{1,1} + \log K_{sp} + 2\log H^+ = 5.57 - 2\,\text{pH} \tag{8-8}$$

Polynuclear hydrolysis products are known to form under certain conditions, typically where hydrolysis is induced by the slow and spatially uniform addition of base to a high-concentration aluminum salt solution. The different types of polynuclear species proposed in the literature range from small units with two aluminum atoms ($Al_2(OH)_2{}^{4+}$) to $Al_{54}(OH)_{144}{}^{18+}$ with 54. A species with 13 aluminum atoms, $Al_{13}O_4(OH)_{24}{}^{7+}$, has been identified with NMR spectroscopy (Bertsch et al., 1986), and a significant number of investigators including Bertsch and Parker (1996) believe this is the predominant species in commercial polyaluminum chloride coagulant solutions. Zhao et al. (2009) used a total Al concentration of 1.5×10^{-4} M and increasing pH to show that a continuum of Al hydrolysis products is formed: monomeric species at pH = 4, small polymeric species ($Al_3 - Al_5$) at pH = 4.8, and aggregation and self-assembly of the small polymeric units to form medium polymeric species ($Al_6 - Al_{10}$) at pH = 5. Above pH 5, the small and medium polymers aggregated into large ($Al_{11} - Al_{21}$) polymeric species. Bertsch and Parker (1996) suggest that polynuclear species are metastable intermediates in solutions that reach or exceed a critical degree of supersaturation with respect to a solid phase such as gibbsite (Al_2O_3) or perhaps amorphous aluminum hydroxide ($Al(OH)_3$).

Bertsch and Parker (1996) have argued that complex formation reactions (such as Eq. 8-1) use electrolyte theory and conventions that do not apply to large, polyvalent, polynuclear species, and consequently, it is not thermodynamically correct to include polynuclear species concentrations in plots such as Figs. 8-8 and 8-9. Also polynuclear species are not formed in the equilibration process when aluminum dissolves from aluminum-hydroxide precipitate, a prerequisite for including them in an equilibrium solubility diagram based essentially on thermodynamic relationships. The next section describes Al and Fe coagulant products that contain polynuclear species.

Types of Hydrolyzing Metal Salt Coagulants Used in Water Treatment

Some of the hydrolyzing metal salt (HMS) coagulant product groups are described in this section. The list is not comprehensive because new types of products are constantly entering the water treatment market in this competitive business.

Simple Metal Salts. The simple HMS coagulants are aluminum sulfate ($Al_2(SO_4)_3 \cdot 14\, H_2O$, alum), ferric sulfate ($Fe_2(SO_4)_3 \cdot X\, H_2O$), and ferric chloride ($FeCl_3 \cdot X\, H_2O$). These are sold as dry crystalline solids and as concentrated aqueous solutions. For example, commercial alum solution is usually about 4.3 percent Al which is equal to about 2 moles Al/L or 8.2 percent Al_2O_3 by weight. Alum is the most widely used hydrolyzing metal salt coagulant.

Prehydrolyzed Metal Salts. When hydrolyzing metal salts are added to and diluted in the water to be treated, the hydrolysis reaction produces hydrogen ions that react with alkalinity species in the solution. If some of this acid is neutralized with base when the coagulant is manufactured, the resulting product is a prehydrolyzed metal salt coagulant solution. The degree to which the hydrogen ions produced by hydrolysis are preneutralized is called the *basicity*. The basicity is given by

$$\text{Basicity (\%)} = B = \left(\frac{100}{3}\right) \times \frac{[\text{OH}]}{[\text{M}]} \tag{8-9}$$

where [OH]/[M] is the weighted average of the molar ratio of the bound hydroxide to metal ion for all the metal hydrolysis products in the undiluted coagulant solution. For example, if a hypothetical product solution contained just one hydrolysis product, the polynuclear species $Al_{13}O_4(OH)_{24}^{7+}$, [OH]/[Al] would be effectively equal to 2.46 = [24 + (2 × 4)]/13 and the basicity according to Eq. 8-9 would be 82 percent = (100/3) × 2.46. For commercial coagulant solutions the basicity varies from ~10 (a low amount of prehydrolysis) to around 83 percent. As the basicity is increased beyond about 75 percent, it becomes increasingly difficult to keep the metal hydroxide precipitate from forming in the product solution during shipping and extended storage. AWWA's Standard B408, Liquid Polyaluminum Chloride (AWWA, 2003), provides a laboratory method for measuring the basicity of prehydrolyzed product solutions.

To avoid forming a precipitate in prehydrolyzed product solutions with higher basicities, they are usually manufactured with chloride as the dominant anion. Multivalent anions such as sulfate, SO_4^{2-}, (Sricharoenchaikit and Letterman, 1987) and high-ionic-strength solutions (de Hek et al., 1978) tend to destabilize positively charged polynuclear and microcrystalline hydrolysis products. Prehydrolyzed metal salt coagulants made with aluminum chloride are called *polyaluminum chloride* or sometimes *polyaluminum hydroxychloride* or simply *PACl*. Some manufacturers refer to the highest basicity products (typically about 83 percent) as aluminum chlorohydrate. These are usually sold with a total metal concentration that is higher than that of the other prehydrolyzed but lower basicity Al products. Use of the term *polyaluminum* (or *polyiron* for Fe salt products) is based on the assumption that the product solution contains significant amounts of relatively stable polynuclear metal hydrolysis products; this tends to be true only for the higher-basicity (60^+ percent) solutions. A number of investigators (Bottero et al. 1987; Xu et al., 2003; Hu et al., 2005; Huang et al., 2006; Chen et al., 2007; Wu et al., 2007, 2008; Lin et al., 2008) appear to believe that the principal hydrolysis product in PACl solutions with basicities greater than about 70 percent is Al_{13}, that is, $Al_{13}O_4(OH)_{24}^{7+}$.

Prehydrolyzed iron solutions exist but are still a relatively uncommon commercial product. The significance of the basicity of iron salt coagulant solutions has been discussed by Tang and Stumm (1987). Wang et al. (2003) observed that, compared to PACl, it is difficult to form stable poly-iron species in a clear solution. In their experiments increasing the basicity without a stabilizer compound tended to form a suspension of small $Fe(OH)_3$ precipitate particles and decreased the performance of the coagulant.

Jiang and Graham (2003) describe the preparation and performance of prehydrolyzed coagulants that are combination of Al and Fe sulfate salts. Gao et al. (2003) report on the preparation and application of a poly-aluminum-silicate-chloride coagulant.

In aged solutions of polynuclear (and microcrystalline) hydrolysis products, most of the metal atoms are interconnected by double-hydroxide bonds. Double-hydroxide bonds are not easily broken. Therefore, when these complex metal hydroxide species are transferred from the solution in which they were formed (e.g., in a commercial coagulant product solution) to one in which they could not be formed and are not stable (the water to be treated), the transition to the new set of species is typically slow. For example, if a concentrated solution of polynuclear species is added to a suspension as a coagulant, it is likely that a significant amount of time (days to months) will be required for a new stable distribution of hydrolysis products to form in the system; that is, the polynuclear species will tend to persist. Consequently, particle or NOM removal diagrams prepared using preformed polynuclear species (Dempsey et al., 1984) may show effective coagulation in areas where the corresponding solubility diagram based on equilibrium conditions (e.g., Fig. 8-8 for $Al(OH)_3$) does not show the formation of destabilizing hydrolysis products.

Metal Salts Plus Strong Acid. Several coagulant manufacturers prepare coagulant solutions that contain the metal salt (e.g., alum) and an amount of strong acid, typically sulfuric acid. The typical acid-supplemented alum product (also called *acidulated alum* or *acid alum*) contains 5 to 20 percent (weight basis) sulfuric acid. Iron salt solutions are available that contain supplemental sulfuric acid. In some cases, especially with iron salt products, the coagulant solution contains up to several weight percent excess acid. A product solution has excess acid when, in the manufacturing process where a metal oxide is dissolved in strong acid, the amount of strong acid added is greater than the stoichiometric amount required by the dissolution reaction. For a given amount of metal ion, Al or Fe, added to the water, products with supplemental or excess strong acid react with more alkalinity and depress the pH to a greater extent than the conventional metal salt solutions.

Metal Salts Plus Additives. Metal salt coagulant solutions are available with special additives including phosphoric acid, sodium silicate, and calcium salts. Alum with phosphoric acid has some of the characteristics of acid-supplemented alum, but $AlPO_4$ precipitate is formed when the solution is added to the water. Metal salt solutions are also sold premixed with polyelectrolyte coagulant compounds such as epichlorohydrin dimethylamine (epiDMA) and polydiallyl dimethylammonium chloride (poly DADMAC). Some prehydrolyzed polyaluminum chloride products are sold with several percent sulfate, and there is evidence that the sulfate can have a positive effect on product performance (Pernitsky and Edzwald, 2003, 2006).

Impurities in Hydrolyzing Metal Salt Coagulant Solutions

Most of the impurities in hydrolyzing metal salt (HMS) coagulants are derived from the raw materials used to make them. For example, alum is usually made by digesting an aluminum source in sulfuric acid. Typical aluminum sources are bauxitic or high-aluminum clays, aluminum trihydrate, and high-purity bauxite. Impurities in the aluminum source tend to appear in the alum product. The most significant contaminant in aluminum salt coagulants is iron. Standard commercial alum solution may contain 1000 mg Fe/L. Small amounts of heavy metals such as chromium and lead can be found in standard-grade alum solution (Lind, 1995).

Iron salt coagulants are manufactured by dissolving various iron sources in sulfuric or hydrochloric acid. In the past they were also made by reprocessing materials such as acidic iron salt solutions from iron mills and foundries, and ferric chloride was sometimes made from reprocessed titanium dioxide liquors. Like alum, the iron salt coagulants typically contain metal contaminants. Manganese (Mn) is often the most significant contaminant; Cu, V, Zn, Pb, and Cd may also be present. The amount varies with the source of the product.

In most cases the low amounts of heavy metal contaminants in HMS coagulants will not have a significant effect on metal concentrations in the treated water. The metals may already be in an insoluble form, or they may be likely to precipitate or adsorb on the floc when the coagulant is added to the water. The metals may, however, increase the heavy metal content of treatment residue. Eyring et al. (2002) have shown that trace contaminants such as lead, chromium, and cadmium in alum will be diluted to less than drinking water standard concentrations in finished water and will not be present in alum sludge at agriculturally significant levels even under worse-case coagulant dosage conditions.

Standards for HMS coagulants (e.g., AWWA B403, Aluminum Sulfate—Liquid, Ground, or Lump [AWWA, 2009a] and NSF/ANSI 60 Drinking Water Treatment Chemicals-Health Effects [NSF, 2009]) may include limits on levels of impurities in products and on maximum dosages. Users of standards should consult with their local regulatory agencies to determine if these limits are requirements.

The amount of residual Al in treated water can be minimized by optimizing filtration to maximize the removal of particulate matter and by maintaining the pH during coagulation and flocculation at a value near the pH of minimum solubility of aluminum hydroxide precipitate (Letterman and Driscoll, 1993). The pH of minimum solubility aluminum hydroxide is between 6 and 7 and depends on the temperature of the water. At the pH of minimum solubility the soluble Al concentration increases with increasing temperature. Frommell et al. (2004) found that the addition of orthophosphate during the rapid mix following alum addition reduced residual aluminum concentrations and decreased the tendency for aluminum to precipitate (or deposit) in the distribution system. Periodically, health concerns about Al in drinking water arise; this issue is addressed in Chap. 2.

A subsequent section of this chapter discusses how to minimize residual aluminum by selecting the appropriate target pH for the coagulation process.

Models for Predicting Hydrolyzing Metal Salt Coagulant Dosages

A number of models have been developed for predicting the Al or Fe salt coagulant dosage required for effective process performance. The models are of two general types. One type calculates the coagulant dosage needed to destabilize a suspension of solid particles, and the other the dosage needed to effectively remove natural organic matter (for background material on NOM; see Chap. 3). With typical surface-water concentrations of NOM and particles the required coagulant dosage (sometimes called the *coagulant demand*) is generally much higher when NOM is present. In waters with both NOM and particles, the NOM concentration is typically the dosage-controlling factor (Semmens and Field, 1980; Edzwald, 1993; White et al., 1997; Shin and O'Melia, 2004; Archer and Singer, 2006a, 2006b).

Particle Suspensions–Destabilization by Surface Charge Neutralization. Models used to predict the dosage needed to destabilize a suspension of solid particles (e.g., mineral particles such as silica and clay) by the process called charge neutralization usually assume that the destabilization process involves the particle surfaces adsorbing coagulant species that have a charge opposite to that of the ionizable sites on the surface of the particles. Typically the pH of the solution determines the speciation of the ionizable sites and, especially with hydrolyzing metal salt coagulants, the charge associated with the coagulant species that adsorb form by reaction on the surface or deposit there after precipitating in the bulk solution.

The amount of surface charge associated with charged sites on particles in a suspension can be calculated with expressions like the following where the quantity 1.66×10^{-24} moles/site is the reciprocal of the Avogadro constant.

$$\begin{aligned} &\text{Surface charge concentration (moles/L)} \\ &\quad = (\text{mass particles/volume}) \times (\text{surface area/mass particles}) \\ &\qquad \times (\text{ionizable sites/surface area}) \times (\text{charge/site}) \\ &\qquad \times (1.66 \times 10^{-24}\ \text{moles/site}) \end{aligned} \tag{8-10}$$

For example, a 100-mg/L colloidal silica suspension with a specific surface area of 0.010 m^2/mg and ionizable surface sites ($\equiv$ SiOH) called silanol groups at a surface density of 5×10^{18} sites/m^2, would have, if every site was ionized to $\equiv$ SiO^-, a negative surface charge concentration of 100 mg/L $\times$ 0.010 m^2/mg $\times$ 5×10^{18} sites/m^2 $\times$ 1.66×10^{-24} moles/site = 8.3×10^{-6} moles of negative surface charge/L. If surface charge neutralization occurred by the quantitative uptake of $Al(OH)^{2+}$, a monomeric Al hydrolysis product with 2 moles of positive charge per mole of aluminum, the Al hydrolysis and sorption reactions might be described by

$$Al^{3+} + \equiv SiO^- + H_2O \rightarrow \equiv SiO^-\ Al(OH)^{2+} + H^+ \tag{8-11}$$

Since each mole of Al in $Al(OH)^{2+}$ has 2 moles of charge the amount of aluminum required for charge neutralization is 0.5 moles of Al/mole charge $\times$ 8.3×10^{-6} moles/L = 4.2×10^{-6} moles Al/L, or 0.12 mg Al/L. For comparison, if the Al that interacts with the particle surface is part of the polynuclear species $Al_{13}O_4(OH)_{24}^{7+}$, which many believe to be the predominant form in prehydrolyzed Al products, the moles of Al per mole of positive charge are 13/7 = 1.86 and the amount of aluminum required for charge neutralization is 1.86 moles of Al/mole of charge $\times$ 8.3×10^{-6} moles of negative surface charge/L = 1.54×10^{-5} moles Al/L, or 0.42 mg Al/L. This hypothetical example illustrates how the amount of hydrolyzing metal salt needed for surface charge neutralization increases as the degree of hydrolysis of the metal ion increases. It also shows that the amount of coagulant needed for charge neutralization is relatively low compared to the dosages used in practice. In practice it is common to use a dosage that is high enough to form a significant amount of precipitate and increase the rate of particle aggregation by increasing the volume concentration of the suspensions. The significance of this factor in the flocculation process is discussed in later sections of this chapter.

While the example assumes a simple one-dimensional binding of Al species to ionizable surface sites, the details of this interaction are poorly understood. Letterman and Iyer (1985) used the chemical equilibrium software MINEQL+ (the most recent version is described by Schecher and McAvoy [2003]) to simultaneously predict the surface site and *aluminum hydrolysis product* (*AHP*) speciation and assumed that the AHPs were quantitatively sorbed on the particle surfaces. Different assumptions were tested when calculating the net surface charge to determine the electrical potentials associated with the double-layer and suspension stability. It was found that the model predicted results gave better agreement with the experimental particle flocculation-separation data when it was assumed that the AHPs produced a three-dimensional layer on the surfaces and that the positive countercharge from the AHPs was arranged in *patches* and was not uniformly distributed over the particle surfaces. More recently, Donose et al. (2007) using atomic force microscopy found that positively charged hydrolyzed aluminum sorbed on negatively charged borosilicate glass was distributed nonuniformly and patchlike. They proposed that this might explain the strong attraction they observed between partially coated silica surfaces. Ye et al. (2007) use a surface patch argument to explain the better coagulation performance they observed for prehydrolyzed Al salts compared to conventional alum.

In some treatment plants, typically direct filtration and microfiltration membrane systems, where large, settleable flocs are not required or even desirable, the target for coagulant addition is frequently the low dosage that yields charge neutralization. As described previously the amount of positive charge required for charge neutralization is proportional to the site density on the particle surface and the total surface area concentration of the suspension. The pH of the solution and the presence of site-binding cations and anions that compete with the hydrolysis products for surface sites and affect the amount of positive charge on the hydrolysis products will also be factors. When destabilization by charge neutralization is used with a dilute suspension, the rate of flocculation is usually much lower than the rate obtained when the voluminous metal hydroxide precipitate is formed.

At coagulant dosages that exceed the charge neutralization requirement, the formation of an unstable (flocculent) precipitate can lead to the relatively rapid formation of visible floc. This process is usually called *sweep flocculation* or *enmeshment*. In most cases the rate of flocculation increases in proportion to the volume concentration of precipitate in the suspension. Floc formation by enmeshment involves both the interaction of particles of metal hydroxide precipitate and contact with contaminant particles. The effect of the amount of metal hydroxide precipitate on the rate of flocculation is discussed in another section of this chapter.

Removal of Natural Organic Matter by HMS Coagulants. Edzwald (1993) listed the possible reactions involved in the coagulation of NOM by Al salt coagulants. It is reasonable to assume that these apply for both Al and Fe salt coagulants.

1. Formation of soluble complexes by the reaction of the coagulant's metal cations with the higher molecular weight fraction of the NOM. Hydrolysis of the metal cations may be a competing reaction.
2. Formation of colloidal metal ion-NOM molecule precipitate particles.
3. Adsorption of NOM species and/or soluble metal-ion–NOM complexes on the amorphous metal hydroxide precipitate.

Others (Dempsey, 1989; Bottero and Bersillon, 1989) have argued that for the common range of pH in water treatment (6 and greater) soluble hydrolysis products binding with soluble NOM is not a significant step in the removal of NOM by HMS coagulants and that the adsorption of the NOM on metal hydroxide precipitate particles is the controlling process in NOM removal.

It is also possible that the initial coagulant-NOM interaction at low dosages is by negatively charged NOM molecules coating and stabilizing (peptizing) small positive metal hydroxide microcrystals (Kodama and Schnitzer, 1980; Bertsch and Parker, 1996). As the dosage of the HMS is increased, the surface area concentration of the microcrystals increases and the amount of NOM adsorbed per unit area of particle surface decreases. Eventually, at a higher HMS dosage, there is not enough negatively charged NOM on the surfaces to stabilize the positively charged particles and the precipitate suspension becomes unstable, forming flocs that consist of the metal hydroxide precipitate and sorbed NOM.

The dosage of HMS required for destabilization and floc formation and, consequently, effective NOM removal has been shown by a number of investigators to be proportional to the initial concentration of NOM in the water (Narkis and Rebhun, 1977; Dempsey, 1987; Edzwald, 1993; Edzwald and Kaminski, 2009; Archer and Singer, 2006a, 2006b). There is evidence that the dosage is also inversely proportional to the initial positive charge on the metal hydroxide precipitate. Differences in initial hydroxide surface charge (in addition to differences in the atomic weights of Al and Fe) may explain why, at a given pH, ferric iron salts generally require a slightly higher amount of coagulant for effective NOM removal than aluminum salts.

Two types of models have been developed for predicting the amount of HMS coagulant needed for effective NOM removal. The first applies the direct proportionality (stoichiometry) that has been observed between the initial amount of NOM and the required coagulant dosage. The second assumes that the NOM adsorbs on the metal hydroxide precipitate and that the amount of NOM removed from solution can be determined using adsorption isotherm relationships. The models are described in the next sections.

A Model That Assumes the Required Dosage Is Proportional to the Initial NOM Concentration. In this model the required dosage m in mg Al/L or mg Fe/L is given by

$$m = R \cdot \mathrm{TOC}_o \tag{8-12}$$

where TOC_o is the raw-water total organic carbon concentration in mg C/L. The magnitude of the coefficient R varies with the type of metal salt, Al or Fe, and with the pH of the solution. It is also possible that water temperature and solution chemistry factors such as the presence of significant amounts of multivalent anions such as sulfate have an effect the magnitude of R but these dependencies, if they are significant, have not yet been quantified. Van Benschoten and Edzwald (1990b), Edzwald (1993), Pernitsky and Edzwald (2006), and Archer and Singer (2006a) have shown that Eq. 8-12 can be used as the basis for a method to estimate coagulant dosage.

Figure 8-10 shows the effect of pH on the magnitude of R for aluminum and iron salt coagulants. The data used to plot Fig. 8-10, were obtained from a number of papers, most involving jar test experiments (see Table 8-3). The determination of R for a given experiment typically involves plotting the amount of residual TOC or UV absorbance (UV_{254}) versus the coagulant dosage expressed as milligrams of metal per liter. The magnitude of R is equal to the *effective dosage* of the coagulant divided by the initial TOC concentration of the water used in the experiment. The corresponding pH (used in plotting the points in Fig. 8-10) is the value measured at the effective dosage. The effective dosage has been determined in a number of ways including (1) estimating the dosage that gives the maximum removal of TOC or UV_{254}, (2) as the dosage that gives a certain percentage removal of TOC or UV_{254}, and (3) as the dosage that corresponds to the *point of diminishing returns* (PODR) as defined by the USEPA in their disinfection by-product precursor removal guidance manual (USEPA, 1999). Van Benschoten and Edzwald (1990b) determined the relationship between

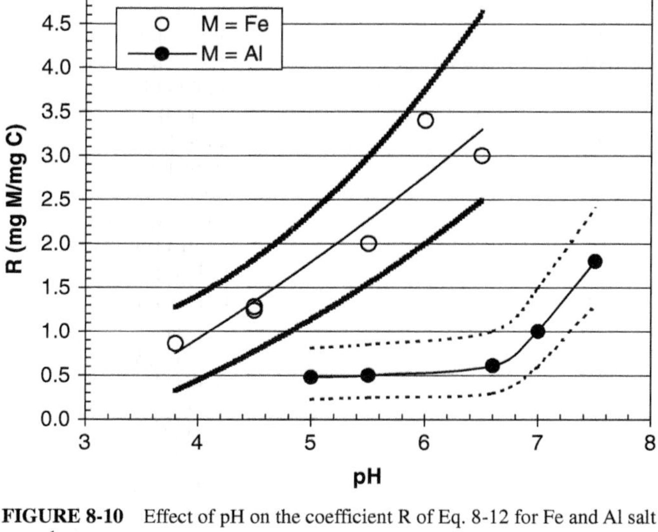

FIGURE 8-10 Effect of pH on the coefficient R of Eq. 8-12 for Fe and Al salt coagulants.

TABLE 8-3 References for the Data Points Plotted in Fig. 8-10 (In some cases the values of mg M/mg C given are averages of a set of measurements at pH values close to the average value listed in the table.)

pH	mg Fe/mg C	Criterion used to determine effective coagulant dosage	Reference
3.8	0.86	90% UV_{254} decrease	Dennett et al. (1996)
4.5	1.24	90% UV_{254} decrease	Dennett et al. (1996)
4.5	1.28	90% UV_{254} decrease	Dennett et al. (1996)
6.0	3.40	90% UV_{254} decrease	Dennett et al. (1996)
6.5	3.00	Maximum removal TOC	Bernhardt et al. (1985)
5.5	2.00	Maximum removal TOC	Lefebvre and Legube (1990)
pH	**mg Al/mg C**	**Criterion used to determine effective coagulant dosage**	**Reference**
5.0	0.5	Maximum removal UV_{254}	Chappell (1996)
5.5	0.5	UV_{254} decrease (1)	Edzwald and Van Benschoten (1990), Edzwald (1993)
6.6	0.6	TOC removal–PODR (2)	Archer and Singer (2006b)
7.0	1.0	UV_{254} decrease	Edzwald and Van Benschoten (1990), Edzwald (1993)
7.5	1.8	Maximum removal UV abs	Chappell (1996)

Notes: (1) mg Al/mg C held constant and pH varied. (2) Point of diminishing returns as prescribed in the USEPA Guidance Manual (USEPA, 1999).

R and pH using jar test experiments and a unique procedure in which the ratio of the Al coagulant dosage to initial NOM concentration was set at given values of about 0.5 or 1 mg Al/mg C and the pH was varied between roughly 3 and 9. The fraction of UV_{254} removed was measured at each pH. The procedure was repeated with three initial NOM concentrations.

A coagulant dosage and pH corresponding to maximum removal of TOC or UV_{254} are only observed when jar test experiments are conducted without pH control; that is, the pH of the coagulant-treated solution decreases as the coagulant dosage increases. In general, when the pH is controlled at a constant value, the TOC concentration and the UV_{254} of filtered- or settled-water supernatant samples decrease continuously as the dosage increases. In constant-pH experiments the effective dosage is usually established by a given percentage removal, for example, 50 percent of TOC or, say, 60 percent UV_{254}.

The estimated range of magnitude of the parameter R is shown in Fig. 8-10 with pairs of dotted lines that bracket the two sets of data points. These give a rough indication of the range of values that might be expected given the different methods typically used to locate the effective coagulant dosage. For example, the dosage that corresponds to 50 percent removal of UV_{254} is typically less than the dosage at 50 percent removal of TOC, and both are less than the dosages that correspond to the maximum removal of TOC and UV_{254}. Locating the effective dosage that maximizes TOC- and UV-adsorbing organics removal can be difficult because maximum and near-maximum removal typically occur within a wide range of coagulant dosages.

According to Fig. 8-10 at pH 6, the magnitude of R in Eq. 8-12 is approximately 0.5 mg Al/mg C for aluminum salts and about 2.5 mg Fe/mg C for ferric iron salts. For both Fe and Al the magnitude of R does not depend to a significant extent on the type of hydrolyzing metal salt coagulant. For example, prehydrolyzed products give essentially the same relationship between R and pH as the conventional salt solutions such as commercial alum, and the acidified products. The utility of these various products derives mostly from how they facilitate achieving a given target pH for a given required coagulant dosage. The acid-base chemistry of hydrolyzing metal salt coagulants and how to use the different types to simultaneously add a required dosage and meet a desired target pH is discussed in a subsequent part of this chapter.

The differences between the magnitude of R at a given pH for Al and Fe salt coagulants is explained only partially by the factor of approximately 2 difference between the equivalent weights of Fe (18.6 mg/meq) and Al (9 mg/meq). For example at pH 6 the magnitude of R expressed as meq/mg C is approximately 2.5/18.6 = 0.13 for Fe and 0.5/9 = 0.056 for Al. This result suggests that the effective charge per mole of hydrolyzed metal is somewhat higher for Al than for Fe at this pH.

Estimating TOC and UV_{254} Removal Using SUVA. The extent of removal of TOC and UV_{254} can be estimated using the SUVA parameter described in Chap. 3. Figure 8-11a and

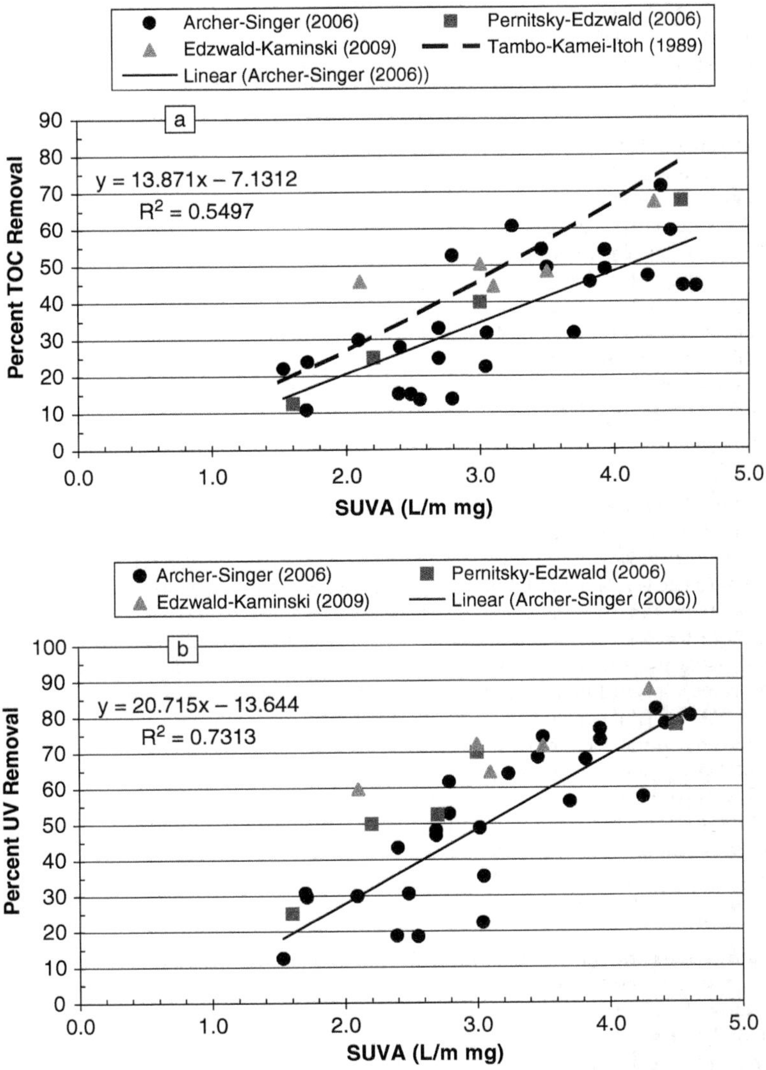

FIGURE 8-11 (*a*) Relationship between TOC removal and source water SUVA: (*b*) Relationship between the percentage removal of UV_{254} and SUVA.

TABLE 8-4 References and Relevant Conditions for the Data Points and Curves Plotted in Fig. 8-11a and b

Reference	Coagulant(s) used	Separation method(s)	Effective dosage criterion
Tambo et al. (1989)	Alum	0.45-μm-pore-size membrane filter (jar test experiments)	Maximum removal of dissolved organic carbon
Archer and Singer (2006)	Alum	Sedimentation (jar test experiments)	Point of diminishing returns (TOC removal)
Pernitsky and Edzwald (2006)	Alum, PACl (several basicities and with and without sulfate)	Sedimentation or dissolved air flotation (jar test)	Maximum removal of TOC and UV_{254}
Edzwald and Kaminski (2009)	Alum and cationic polyelectrolyte	Dual-media filters (full-scale plants–0.8 to 50 MGD)	Filtered water-UV_{254} goal

b shows percent TOC and UV_{254} removal for jar test measurements (Tambo, et al., 1989; Archer and Singer, 2006a; Pernitsky and Edzwald, 2006) and average removals for full-scale plant operation (Edzwald and Kaminski, 2009) in all cases with Al salt coagulants. Table 8-4 summarizes the experimental conditions and methods used in these studies. At least some of the scatter in the data in these figures is attributable to the different criteria used to define an effective dosage of coagulant. The criteria are listed in the table. Edzwald and Kaminski (2009) calculated the average removals obtained in a number of full-scale treatment plants over a period of several years. Each plant's coagulant dosage was determined using a treated-water UV_{254} goal.

The curve in Fig. 8-11a was plotted using an equation that relates the percent organic carbon removal to the SUVA (L/mg-m) of the untreated water.

$$\text{Percent TOC Removal} = 10.8(\text{SUVA})^{1.32} \tag{8-13}$$

Equation 8-13 is based on an empirical relationship developed by Tambo et al. (1989) using data from jar test experiments and regression analysis. Tambo et al. used a quantity they called the *DOC/E ratio* which is equal to 100/SUVA. The substitution DOC/E = 100/SUVA was made to convert the Tambo et al. empirical expression to equation to Eq. 8-13. In the experiments of Tambo et al. each jar test sample was filtered with a 0.45-μm membrane filter before the organic carbon concentration was measured and percent organic carbon removal is the maximum value observed for each source water (i.e., for each DOC/E, or SUVA, value). According to Eq. 8-13 the predicted organic carbon removal is 27 percent when SUVA = 2 L/mg-m and 68 percent when SUVA = 4 L/mg-m.

According to the data and curve plotted in Fig. 8-11a and b, the percent removals of TOC and UV_{254} increase with increasing SUVA and for a given value of SUVA the removal of the UV-absorbing fraction is generally higher than the removal of total organic carbon. On a carbon weight basis UV_{254} is influenced more by the high-molecular-weight fraction of the organic carbon than by the low-molecular-weight fraction. Since the high-molecular-weight fraction is preferentially removed by hydrolyzing metal salt coagulants, its percentage removal tends to be higher.

Models Based on the Adsorption of NOM by the Metal Hydroxide Precipitate. A number of sorption-based models have been developed for predicting NOM removal by metal salt coagulants. Dentel (1988), applying the general approach of the particle coagulation model of Letterman and Iyer (1985), used a surface charge neutralization mechanism. Dentel assumed that fulvic acid can be modeled as a suspension of spherical colloids with high specific surface area and charge. The model uses a patch-site distribution assumption to predict the coagulant metal concentration–pH region of charge neutralization. The effective diffuse-layer charge for a coagulant-treated suspension is the area-weighted average of the diffuse-layer charges of the component patches of uncoated NOM and sorbed aluminum hydroxide.

The agreement between Dentel's predicted region of charge neutralization and NOM removal data from Dempsey et al. (1984) was good at relatively high pH values where the removal of FA, as suggested by Dentel, may occur by adsorption of FA on $Al(OH)_3(s)$. Less satisfactory agreement at lower pH values was probably caused, he explained, by the complexation of Al^{3+} ions by FA. Dentel's study did not provide direct evidence that aluminum hydrolysis products adsorb on FA colloids. Later investigations have used the alternative assumption that NOM removal occurs primarily by adsorption on the metal hydroxide precipitate.

Edwards (1997) and Tseng and Edwards (1999) calculated the fraction of the raw-water dissolved organic carbon $(DOC)_0$ that cannot be adsorbed on metal hydroxide precipitate using the specific UV absorbance (SUVA) and the equation

$$\text{Fraction nonsorbable DOC} = K_1(\text{SUVA})_{\text{raw\_water}} + K_2 \tag{8-14}$$

where the parameters K_1 and K_2 are empirical coefficients that must be determined for each source water by fitting a linear equation to fraction nonsorbable DOC data plotted versus the SUVA.

The amount of adsorbable DOC removed was determined using the Langmuir equation,

$$\frac{x}{M} = \frac{ab[C]_{eq}}{1+b[C]_{eq}} \tag{8-15}$$

where x is the amount of DOC removed (mg DOC/L), b is the sorption constant for sorbable DOC on the metal hydroxide surface (L/mg DOC), and $[C_{eq}]$ is the sorbable DOC (mg/L) remaining in solution at equilibrium. The following equation relates x to the nonsorbable fraction and the raw-water DOC concentration:

$$x = \{1 - \text{SUVA}\ K_1 - K_2\}(\text{DOC})_o - [C_{eq}] \tag{8-16}$$

The magnitude of a in Eq. 8-15, the maximum DOC that can be adsorbed per amount of precipitate (mg DOC/(mM Fe^{3+} or Al^{3+})), is assumed to vary with the final pH of the solution according to the following polynomial equation:

$$a = x_3\text{pH}^3 + x_2\text{pH}^2 + x_1\text{pH} \tag{8-17}$$

Tseng and Edwards (1999) outline how the final pH of Eq. 8-17 can be estimated using the initial alkalinity of the water and assumptions about the acidity of the coagulant solution, the amount of metal ion that precipitates and whether there is exchange of carbon dioxide with the atmosphere.

In Eq. 8-15 the quantity M is the amount of metal hydroxide precipitate in mM/L present at the final pH. Edwards (1997) and Tseng and Edwards (1999) assumed that essentially all the metal ion added forms hydroxide precipitate at pH values between 3 and 8 for ferric salt coagulants and between 5.5 and 8 for aluminum salt coagulants, and consequently in these pH intervals, M is equal to the amount of Fe^{3+} or Al^{3+} coagulant added.

The magnitudes of the six parameters in Eqs. 8-14 to 8-17 (K_1, K_2, x_1, x_2, x_3, *and* b) were determined (Tseng and Edwards, 1999) for each water treatment site using a search algorithm and an objective function that minimized the quantity $\left(\Sigma(\text{DOC}_{\text{measured}} - \text{DOC}_{\text{modeled}})^2\right)$. $\text{DOC}_{\text{modeled}}$ is the nonsorbable DOC plus $[C_{eq}]$, the calculated equilibrium concentration, that is, unadsorbed portion, of the adsorbable DOC. The model's input variables are the raw-water SUVA, the raw-water DOC and TOC, the final pH (measured or calculated), and the coagulant type and dosage. The difference between the total organic carbon (TOC) and the raw-water DOC concentrations was assumed to be a particulate organic carbon fraction that is 100 percent removed by coagulation.

Edwards (1997) and Tseng and Edwards (1999) used USEPA and other data sets for model calibration. Since in most cases the data were from operating water treatment plants, the coagulant dosage was a value that tended to maximize DOC removal. Both papers state that the number of adjustable parameters and the limited range of DOC removals in the calibration data sets limit using their results to test the validity of the assumptions they made about removal mechanisms. In any case, the model-calculated treated-water DOC concentrations were in very good agreement with measured values.

Kastl et al. (2004) developed a modified Langmuir equation-type model in which the number of adjustable parameters is reduced from six (Tseng and Edward, 1999) to five. In contrast to Edwards and Tseng, Kastl et al. assumed that (1) the sorbable DOC fraction consists of humic acid and nonpolar fractions, (2) only the associated form of the humic acid is absorbed during coagulation and the removal of this fraction depends on pH, (3) all compounds (humic and fulvic acids) in the humic acid fraction can be represented by a single humic acid, (4) the nonpolar fraction is adsorbed regardless of pH, and (5) the maximum sorption capacity does not vary with pH. Kastl et al., as Tseng and Edwards (1999), used measured residual DOC concentrations (from the USEPA database and their own jar test results) to calibrate their model for each site. A multiple-parameter optimization method similar to that of Tseng and Edwards (1999) was employed.

Wang and Tang (2006) describe a quantitative model based primarily on Dentel's work (Dentel, 1988) but modified to accommodate the special characteristics of prehydrolyzed metal salt coagulants, particularly PACl. Wang and Tang contend that the aluminum fractions in the prehydrolyzed product solution must be considered in the model calculations. For example, the addition of a given dosage of Al at a given pH using a PACl solution affects the stability of a particle suspension much differently than if the same dosage of Al was added using an alum solution.

The models of Dentel (1988), Edwards (1997), Tseng and Edwards (1999), and Kastl et al. (2004) do not include the effects of variables such as the initial pH and alkalinity, ionic strength. and concentration of inorganic anions (SO_4^{2-}, PO_4^{3-}, etc.) that tend to affect the surface charge of metal hydroxide precipitates and their solubility. The initial alkalinity and pH and the effective acid content of the coagulant solution determine the coagulant dosage–pH relationship. Ionic strength also affects suspension stability by modifying the thickness of the diffuse electrical layer of the particles. Snodgrass et al. (1984) found that high ionic strength increases the particle flocculation rate in systems containing sulfate ions and/or fulvic acid. However, Gu et al. (1994) observed that changes in ionic strength did not significantly affect the adsorption of NOM on oxide surfaces, while sorbed oxyanions (sulfate and phosphate species) decreased DOC uptake significantly.

The effect of inorganic anions such as sulfate on the removal of NOM might be exerted in several ways. Anions might complex with aluminum species in the solution or on the surface of the aluminum hydroxide precipitate, enhancing the NOM removal by accelerating the formation of metal hydroxides. It is also possible that high anion concentrations might hinder NOM removal by competing with NOM for the available metal hydroxide sites. Cheng et al. (2004) studied the effect of phosphate on the removal of Aldrich humic acid (AHA) by alum. They found that at pH 4 the removal of AHA increased with the phosphate concentration, while at high pH (7 to 8) an increasing phosphate concentration decreased the removal of AHA.

NOM Removal–Example Jar Test Results

The jar-test results plotted in Figs. 8-12 to 8-14 are from a study of a small community water supply system in upstate New York. The raw-water sample had a pH of 7.6, an alkalinity of

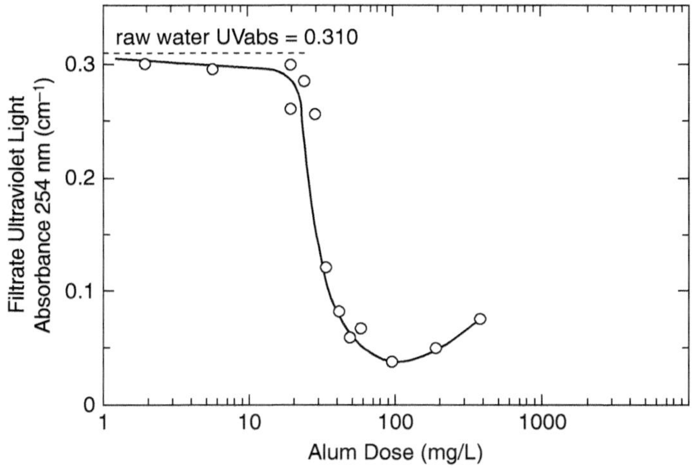

FIGURE 8-12 Jar test results of effect of alum dosage on the UV_{254} of filtrate samples (filtrate samples obtained using 0.45-μm-pore-size membrane filters).

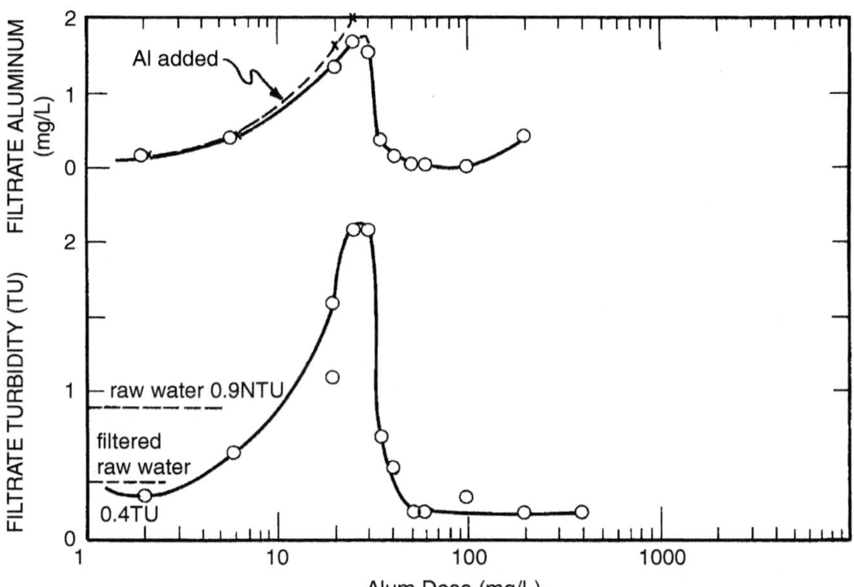

FIGURE 8-13 Jar test results of effect of the alum dosage on filtrate total aluminum concentration and turbidity (dotted line in the upper graph is the amount of Al added; filtrate samples obtained using 0.45-μm-pore-size membrane filters).

1 meq/L (50 mg/L as $CaCO_3$), a total dissolved-solids concentration of 190 mg/L, a turbidity of 0.9 ntu, and a color of 50 CU. The TOC concentration was approximately 6 mg C/L. The alum dose plotted in the figures is reagent-grade aluminum sulfate in milligrams of $Al_2(SO_4)_3 \cdot 18H_2O$ per liter. A dosage of 100 mg/L of reagent-grade alum is 8 mg Al/L. The

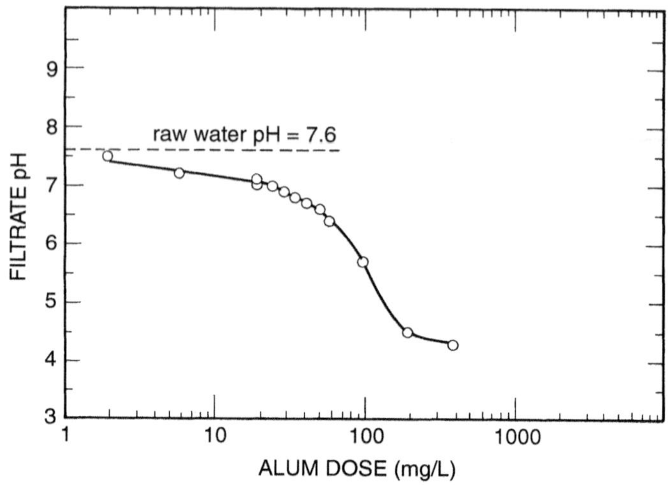

FIGURE 8-14 Jar test results of effect of the alum dosage on the filtrate pH (filtrate samples obtained using 0.45-μm-pore-size membrane filters).

effect of the coagulant concentration on the removal of NOM was determined by filtering an aliquot of supernatant using a 0.45-μm-pore-size membrane filter. Sedimentation was not effective because of the low raw-water turbidity (0.9 ntu) and, therefore, low floc density.

As discussed in Chap. 3, the ratio of the UV absorbance at 254 nm and the dissolved organic carbon concentration (the SUVA) is a useful measure of the fraction of the NOM that can be removed by HMS coagulants. In Fig. 8-12 the UV_{254} of filtrate samples is plotted versus the alum dose. Effective removal of the UV light-absorbing NOM fraction required an alum dose of at least 30 mg $Al_2(SO_4)_3 \cdot 18\ H_2O$/L (2.4 mg Al/L or 0.4 mg Al/mg C). Maximum removal was obtained at approximately 100 mg $Al_2(SO_4)_3 \cdot 18H_2O$/L (8 mg Al/l or 1.3 mg Al/mg C). About 12 percent of the UV-light-absorbing compounds could not be removed by coagulation and filtration at any alum dosage. This high removable fraction (88 percent) is consistent with the high raw-water SUVA of about 5 L/mg-m (See SUVA guidelines in Chap. 3 and Fig. 8-11b).

The filtrate UV_{254} curve in Fig. 8-12 shows a relatively sharp decrease in the NOM concentration as the alum dosage is increased. This is characteristic of NOM that is relatively high in molecular weight and homogeneous in composition. If this system had contained a lower-molecular-weight NOM with a broader molecular-weight distribution (typical of low, <5 mg C/L, NOM concentration waters), the decrease in the residual NOM with increasing coagulant dosage would have been more gradual and the maximum removal of NOM at high coagulant dosages would not have been as great.

Figure 8-13 shows the effect of the alum dose on the filtrate turbidity and aluminum concentration. For alum dosages in the interval 2 to 25 mg/L of $Al_2(SO_4)_3 \cdot 18\ H_2O$/L (0.2 to 2 mg Al/L), most of the aluminum added passed through the membrane filter and was measured in the filtrate. The increasing filtrate turbidity shown in the lower plot (and the high filtrate UV_{254} in Fig. 8-12) suggests that the aluminum was present in a colloidal, light-scattering form. At an alum dose of approximately 30 mg/L (2.4 mg Al/L), the filtrate turbidity and aluminum concentration decreased abruptly, indicating that at this point the suspension of Al-humate precipitate (or possibly aluminum hydroxide microcrystals stabilized by adsorbed negatively charged NOM) had become unstable and filterable flocs had begun to form.

The water pH decreased with increasing alum dose, from about 7 at 10 mg $Al_2(SO_4)_3 \cdot 18H_2O$/L (0.8 mg Al/L) to 4.3 at 200 mg $Al_2(SO_4)_3 \cdot 18H_2O$/L (16.2 mg Al/L) (Fig. 8-14). As aluminum hydrolysis released hydrogen ions, they depleted the limited alkalinity of the system, and the pH was depressed.

According to the filtrate aluminum plot of Fig. 8-13, as the alum dosage was increased beyond about 100 mg $Al_2(SO_4)_3 \cdot 18H_2O$/L (8.1 mg Al/L), the filtrate aluminum concentration increased. This behavior can be attributed, at least in part, to the solubility of aluminum hydroxide at the low pH values indicated at high alum doses in Fig. 8-14. Also, at low pH, colloidal aluminum hydrolysis products may acquire a high positive charge and pass through the membrane filter.

Acid-Base Chemistry of Hydrolyzing Metal Salt Coagulants

When a HMS coagulant solution such as alum or PACl is added in treatment, the hydrolysis reaction produces hydrogen ions (H^+) that decrease the alkalinity and pH of the solution.

$$2H^+ + HCO_3^- \rightleftarrows H_2CO_3(CO_2 + H_2O) + H^+ \tag{8-18}$$

The amount of hydrogen ions formed by hydrolysis or added to the water with the product solution depends on the dosage of the coagulant metal, the basicity of prehydrolyzed products, and the free acid content of products with excess or supplemental acid. When the pH of the coagulant-treated water is near neutral (pH ≈ 5.5 to 8.5), the effect of these products on the pH and alkalinity of the treated water can be determined using a quantity proposed by Letterman et al., (1996) called the *effective acid content* (*EAC*).

The effective acid content of Al and Fe coagulant products including those manufactured with excess acid or supplemented with acid after manufacture can be calculated using the information provided on product data sheets and the following equation:

$$\text{Effective acid content (meq/mg metal)} = \frac{300}{100(AW)} + \frac{A}{EW(M)} \tag{8-19}$$

where A = weight percent of pure acid (H_2SO_4 or HCl) in products with excess acid or products supplemented with acid, EW = equivalent weight of the acid, H_2SO_4 = 49 g/eq, HCl = 36.5 g/eq, AW = atomic weight of the metal (Al = 27 g and Fe = 55.9 g), and M = metal concentration in the product solution (weight percent).

Products have excess acid when, in the manufacturing process where a metal oxide is dissolved in strong acid, the amount of strong acid used is greater than the stoichiometric amount required by the dissolution reaction. Supplemental acid (typically expressed in product data sheets as the weight percent pure acid) is the strong acid added to a product solution, for example, the standard alum solution, to increase its effective acid content.

The effective acid content of prehydrolyzed Al and Fe products can be calculated using the basicity B of the product and the following equation:

$$\text{Effective acid content (meq/mg metal)} = \frac{300 - 3B}{100(AW)} \tag{8-20}$$

Equations 8-19 and 8-20 are based on the assumption that the predominant hydrolysis product present after coagulant addition is the metal hydroxide precipitate, $Al(OH)_3$ or $Fe(OH)_3$, with a molar ratio of hydroxide to metal of 3. A manufacturer would not add supplemental acid or have excess acid in a prehydrolyzed product solution, so it can be assumed that when $B > 0$, $A = 0$, and conversely when $A > 0$, $B = 0$. Examples of calculated coagulant solution effective acid content values are listed in Table 8-5 for four commercial HMS products.

TABLE 8-5 Calculated Effective Acid Content (EAC) of Selected Commercial Coagulant Products (The solution attributes used in the calculations are approximate values from the manufacturer's product data sheets.)

Type of coagulant solution	Basicity (*B*, %)	Excess or supplemental acid (*A*, weight % pure acid)	Al or Fe Conc. in product solution (*M*, weight % metal)	Specific gravity of coagulant solution (ρ_w)	Calculated effective acid content, EAC, (meq/mg metal)
Aluminum sulfate (alum)	0	0	4.3 (Al)	1.335	0.111
Polyaluminum chloride (PACl)	80	0	12.3 (Al)	1.35	0.022
Acid-supplemented alum (acidized alum)	0	10 (H_2SO_4)	2.8 (Al)	1.27	0.184
Ferric sulfate (with 2% excess acid)	0	2 (H_2SO_4)	10 (Fe)	1.45	0.058

The effective acid content of a commercial coagulant solution can be measured in the laboratory by titrating a volume *V* of the solution with strong base (e.g., NaOH) to a pH of 7. If *T* is the milliequivalents of strong base used to reach this end point, then the measured effective acid content of the coagulant solution is given by

$$\text{Measured EAC (meq/mg } M) = \frac{T}{M \times V \times SG \times \rho_w} \tag{8-21}$$

where *M* is the weight percent metal (Al or Fe) in the coagulant solution, *V* and *SG* are the volume in mL and specific gravity of the coagulant solution titrated, respectively, and ρ_w is the mass density of water in g/mL.

The effective acid content of a coagulant product can be used to determine the relationship between the coagulant dosage and the pH after flocculation and floc separation. In one approach to establishing this relationship, the water to be treated is slowly titrated with a coagulant solution of known acidity while the pH is measured. The titration results are plotted as pH versus the coagulant dosage in meq/L. The coagulant dosage in meq/L is equal to the dosage in mg metal/L times the effective acid content of that product in meq/mg metal. If the coagulant dosage is initially recorded as volume of commercial coagulant solution per volume of water, then the conversion to mass concentration units is given by

$$\text{Coagulant dosage(mg metal/L)} = \text{volumetric dosage}(\mu\text{L/L}) \times SG \times \rho_w \, x \times M/100 \tag{8-22}$$

where *SG* is the specific gravity of the coagulant solution, ρ_w is the density of water (1 mg/μL), and *M* is the metal concentration (as weight percent) in the coagulant solution. The specific gravity and metal concentration can be found on coagulant product data sheets. Examples are listed in Table 8-5.

The relationship between the amount of coagulant added and the pH of the solution can vary with time due to relatively slow steps in the hydrolysis and metal hydroxide precipitation reactions. It is also affected by the transport of carbon dioxide between the solution and the atmosphere. Consequently, the relationship obtained from laboratory measurements may not always agree exactly with what is measured in the treatment plant.

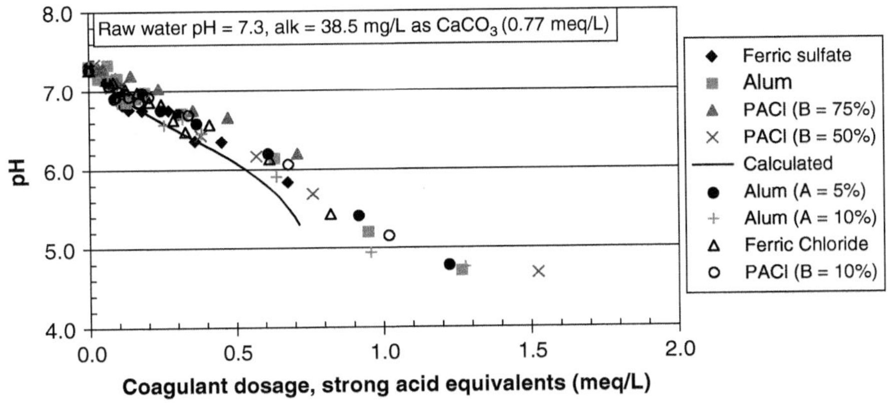

FIGURE 8-15 Titration curves for a diverse assortment of hydrolyzing metal salt coagulant products (coagulant dosages were converted to the equivalent concentration of strong acid using the effective acid content values listed in Table 8-6).

The transport of carbon dioxide across the air-water interface is relatively slow in treatment plants because full-scale tanks have relatively low interfacial surface area per volume of water (Shankar and Letterman, 1996). Consequently, a closed-to-the-atmosphere (no CO_2 transport) assumption is appropriate, especially when mixing is not intense. In small laboratory batch reactors (e.g., 1 or 2 L) with relatively high interfacial surface area per volume of water, CO_2 transport is significantly faster, the loss of inorganic carbon from the solution is higher, and, therefore, measured pH values tend to be higher than the in-plant measured values (and closer to chemical equilibrium model predictions for open systems for CO_2 in the atmosphere and water).

Figures 8-15 and 8-16 illustrate the use of the effective acid content to predict the final pH of water treated with hydrolyzing metal salt coagulants. Aluminium and iron salt commericial products, some prehydrolyzed and some with supplemental acid, were used to titrate 1–L samples of water from the Glenmore Reservoir in upstate New York. These

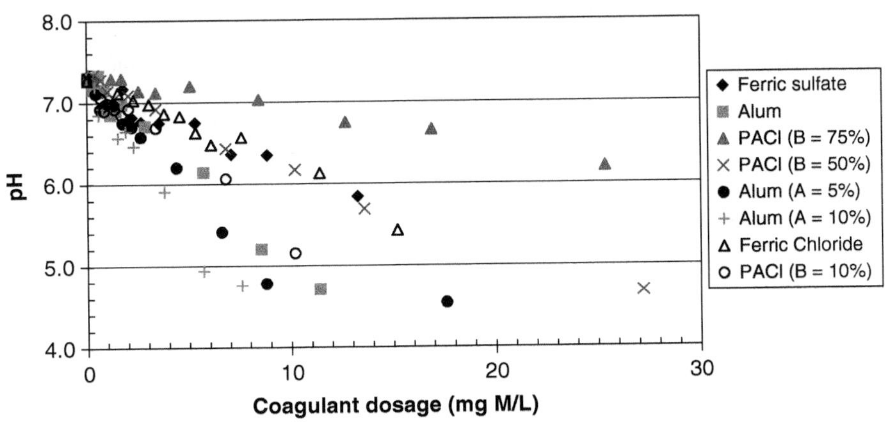

FIGURE 8-16 Titration curves for a diverse assortment of hydrolyzing metal salt coagulant products (coagulant dosages are plotted as milligrams metal (Al or Fe) per liter of jar test solution).

TABLE 8-6 Coagulant Solutions Used to Obtain the Titration Results Plotted in Figs. 8-15 and 8-16 (In the EAC of the third column, M = Fe for iron salt products and Al for aluminum salt products.)

Coagulant product solution	Label in figures	Effective acid content (EAC, meq/mg M)
Acid supplemented alum with $A = 10\%$	Alum ($A = 10\%$)	0.168
Acid supplemented alum with $A = 5\%$	Alum ($A = 5\%$)	0.135
Conventional alum ($A = 0\%$ and $B = 0\%$)	Alum	0.111
Polyaluminum chloride with $B = 10\%$	PACl ($B = 10\%$)	0.100
Polyaluminum chloride with $B = 50\%$	PACl ($B = 50\%$)	0.056
Ferric chloride	Ferric chloride	0.054
Ferric sulfate	Ferric sulfate	0.054
Polyaluminum chloride with $B = 75\%$	PACl ($B = 75\%$)	0.028

water samples had an initial alkalinity and pH of 38.5 mg/L as $CaCO_3$ (0.77 meq/L) and 7.3; the temperature was approximately 20°C. The products used and their EAC values (which range from 0.028 to 0.168 meq/mg metal) are listed in Table 8-6.

Figure 8-15 is a plot of the final pH versus the equivalent strong acid concentration (the product's EAC times its dosage in mg metal/L) for all the products listed in Table 8-6. When the pH is plotted versus the equivalent strong acid concentration of the coagulant (instead of versus the coagulant dosage in mg metal/L as shown in Fig. 8-16), the results tend to follow the same trend line; this illustrates the value of using the EAC to compute the final pH of coagulant treated solutions for all types of coagulant products (and supporting the assumptions used to compute the EAC values in Table 8-6).

The solid curve plotted in Fig. 8-15 is the theoretical pH for the titration of Glenmore Reservoir water with a strong acid for a closed system, that is, constant inorganic carbon concentration, no CO_2 exchange between the water and air. The experimental pH values are slightly above this curve; this suggests that in the small batch reactor used in the titrations some inorganic carbon (carbon dioxide) may have left the solution after the acid from the product solutions and hydrolysis converted bicarbonate ion to carbon dioxide.

Residual Aluminum and Iron in Coagulant Treated Water

When a hydrolyzing metal salt coagulant is added to water buffered by the inorganic carbon system, some of the hydrogen ions produced by hydrolysis of the metal ion (and sometimes added to the coagulant product as supplemental or excess strong acid) react with the carbonate species [in many waters (pH < 9) this is just the bicarbonate, HCO_3^-, ion] and some remain in solution as H^+ and decrease the pH. The difference between the initial alkalinity of the solution and the alkalinity remaining at the target pH determines the amount of coagulant solution needed to reach the traget pH. As the initial alkalinity increases, the capacity to react with hydrogen ions increases and, consequently, the coagulant dosage required to reach a given pH value increases. The next sections illustrate how the amount of metal hydroxide precipitate and soluble metal ion vary as a solution's inorganic carbon species are titrated by H^+ from the addition of a hydrolyzing metal salt coagulant.

Aluminum Hydroxide Precipitate Solubility. Figures 8-17 and 8-18 were plotted using titration data calculated using the chemical equilibrium program MINEQL+ (Schecher and McAvoy, 2003) and the constants listed in Table 8-7. The figures show the theoretical

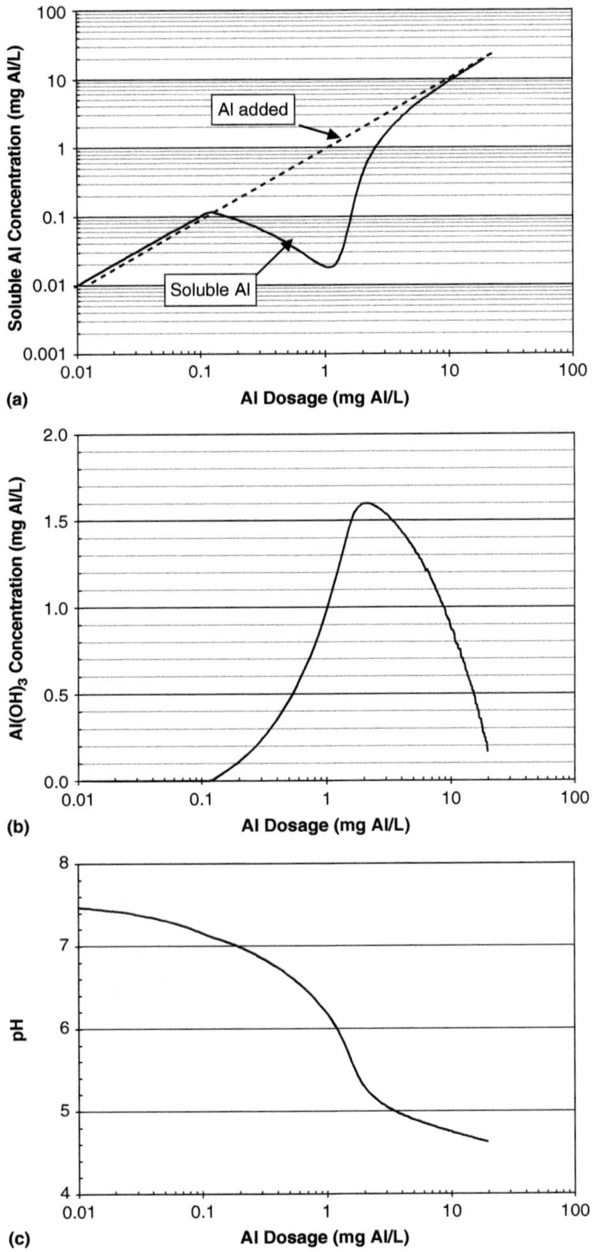

FIGURE 8-17 Titration of a solution with an initial alkalinity of 0.2 meq/L (10 mg/L as $CaCO_3$) with alum (EAC = 0.111 meq/mg Al) using the chemical equilibrium software MINEQL$^+$. Figure (*a*) is the soluble Al concentration, (*b*) is the aluminum hydroxide precipitate concentration and (*c*) is the solution pH as a function of the concentration of Al added in mg Al/L.

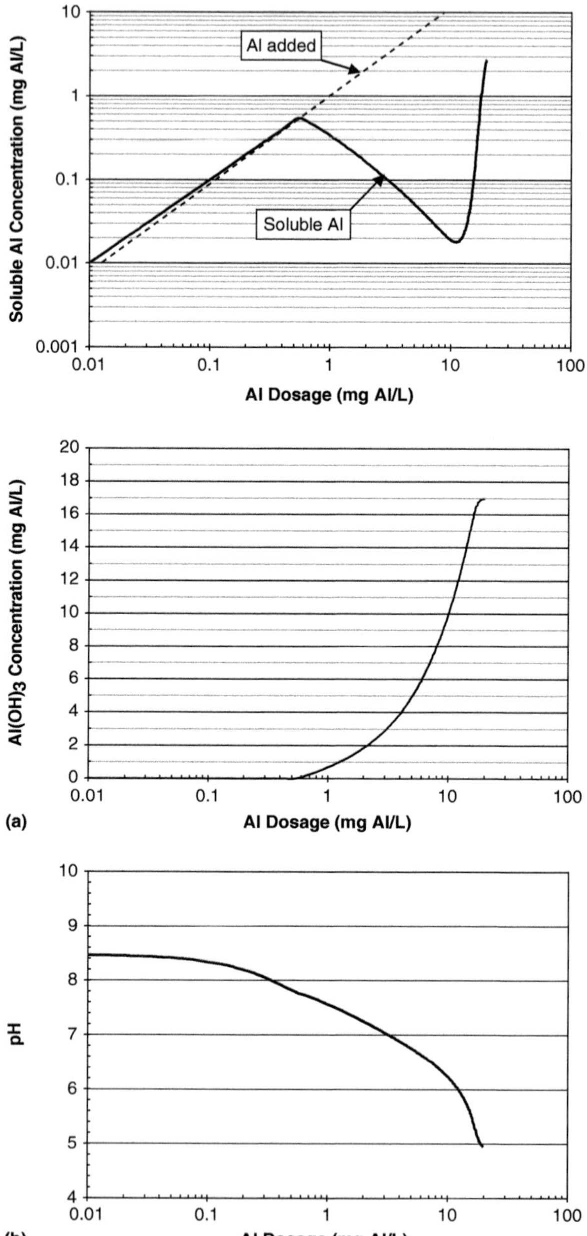

FIGURE 8-18 Titration of a solution with an initial alkalinity of 2 meq/L (100 mg/L as $CaCO_3$) with alum (EAC = 0.111 meq/mg Al) using the chemical equilibrium software MINEQL$^+$. Figure (*a*) is the soluble Al concentration, the middle is the aluminum hydroxide precipitate concentration, and (*b*) is the solution pH as a function of the concentration of Al added in mg Al/L.

TABLE 8-7 Formation Constants, Solubility Constant, and Enthalpy Change Values Used in MINEQL+ (Schecher and McAvoy, 2003) to Plot Figs. 8-17 and 8-18

Hydrolysis product	Type of constant	Magnitude of constant at 25°C	Enthalpy change, ΔH (kcal/mol)*
$Al(OH)^{2+}$	Formation	4.99	11.90
$Al(OH)_4^-$	Formation	23.00	44.06
$Al(OH)_3(s)$	Solubility	−10.60	27.05

*The equilibrium constants are corrected for temperature using the Van't Hoff equation

$$\log K_T = \log K_{25} - \frac{1000 \cdot \Delta H}{2.303 \cdot R}\left[\frac{1}{273+T} - \frac{1}{298}\right]$$

where R is the gas law constant (1.987 cal/deg-mol), T is in degrees Celsius, and K_{25} is the magnitude of the equilibrium constant at 25°C. See Chap. 3 for additional details.

relationship between the Al concentration added (as aluminum sulfate, alum, expressed as mg Al/L) and the pH of the solution and the concentrations of precipitate and soluble aluminum. A constant ionic strength of 0.001 M and a solution temperature of 25°C were used in these calculations. Figure 8-17 is for a low initial alkalinity of 0.2 meq/L (10 mg/L as $CaCO_3$), and Fig. 8-18 is for and a relatively high initital alkalinity of 2 meq/L (100 mg/L as $CaCO_3$). The initial pH value and total dissolved inorganic carbon concentration used in the calculations were 7.5 and 2.14×10^{-4} moles C/L for Fig. 1 and 8.47 and 2.0×10^{-3} moles C/L for Fig. 8-18.

According to Fig. 8-17 for an initial alkalinity of 0.2 meq/L a minimum soluble Al concentration (and the maximum aluminum hydroxide precipitate concentration) is reached at an aluminum dosage of 1.1 mg Al/L. The pH at this point is approximately 6.1 and the soluble Al concentration is about 17 µg/L. For the titration results plotted in Fig. 8-18 (an initial alkalinity of 2 meq/L, or 100 mg/L as $CaCO_3$), the Al dosage at the minimum soluble Al concentration is approximately 11 mg Al/L, essentially 10 times greater than for an initial alkalinity of 0.2 meq/L. The minimum soluble Al concentration and corresponding pH are the same as in Fig. 8-17, 17 µg/L and 6.1, respectively.

The results plotted in Figs. 8-17 and 8-18 show that as alum is added and the pH decreases the concentration of soluble Al initially increases to a relatively high value before the solubility product of aluminum hydroxide is exceeded and the aluminum hydroxide precipitate forms, eventually becoming the predominant form of aluminum. For an initial alkalinity of 0.2 meq/L (Fig. 8-17) up to an aluminum dosage of about 0.1 mg/L (corresponding to a pH of about 7.1), essentially all the Al added becomes soluble Al. For an initial alkalinity of 2 meq/L (Fig. 8-18), the peak soluble Al concentration before the precipitate begins to form is approximately 0.55 mg/L (at a pH of about 7.8). In both examples the pH of minimum Al solubility (pH = 6.1) must be reached to achieve a minimum soluble Al concentration.

The pH of minimum solubility and the minimum soluble Al concentration for aluminum hydroxide precipitate vary with solution temperature (Van Benschoten et al., 1994). Figure 8-19a and b shows these dependancies for a solution with low ionic strength (I = 0.001 M). The plotted points were calculated using MINEQL+ (Schecher and McAvoy, 2003) and the constants listed in Table 8-7. At 25°C the pH of minimum solubility is approximately 6.1 and the total soluble aluminium concentration is 6.7×10^{-7} moles/L, or 17 µg Al/L. At 2°C the pH of minimum solubility is about 6.8 and the soluble Al concentration is approximately 7 µg Al/L.

Pernitsky and Edzwald (2003) measured filtrate aluminum concentrations to compare the solubility of aluminum hydroxide formed from alum with that formed from polyaluminum chloride (PACl) products. High concentrations of different types of products were added to deionized water, the pH was varied, and the concentration of aluminum was measured in samples filtered through a 0.22-μm-pore-size membrane filter. The concentration of filtrate aluminum varied with pH, temperature, and the type of coagulant. The results for alum were similar to the predicted values plotted in Figs. 8-17 and 8-18, but the results for the prehydrolyzed PACl products were not. At a given temperature the pH of minimum solubility and the minimum concentration of filtrate Al increased as the basicity B of the products increased. For example, at 20°C the pH of minimum solubility and filtrate Al concentration were 6.0 and 16 μg/L for alum and 6.7 and 101 μg/L for an aluminum chlorohydrate product with B = 83 percent. The presence of small amounts of sulfate in various products (<2.5 percent) did not affect the filtrate Al concentration. Alum contains 23.5 percent sulfate.

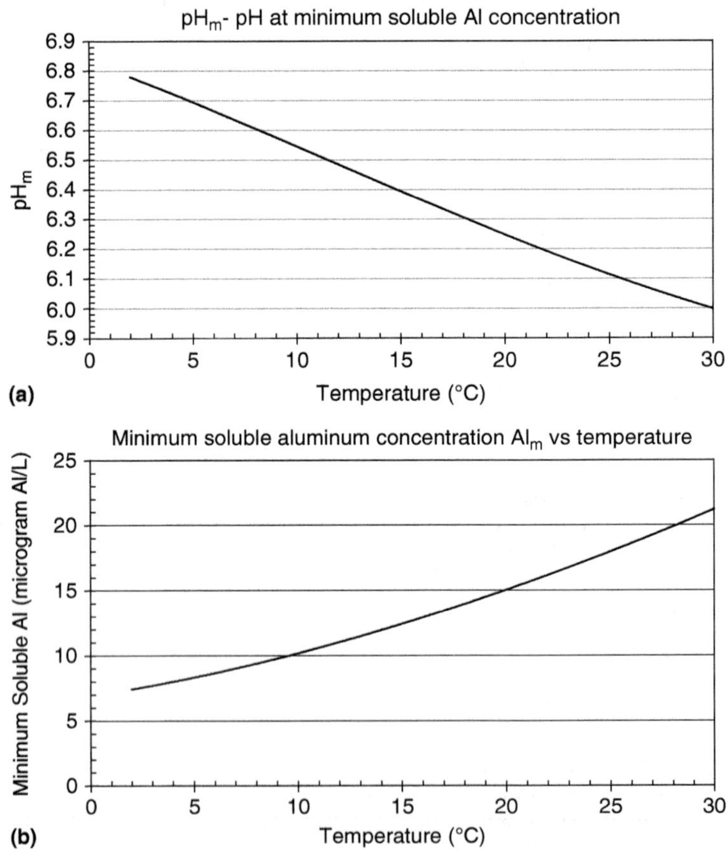

FIGURE 8-19 Effect of solution temperature on the pH of minimum aluminum hydroxide solubility (*a*) and minimum soluble aluminum concentration (*b*).

The trends observed by Pernitsky and Edzwald (2003) are informative but should be interpreted carefully. A 0.22-μm-pore-size membrane filter may not separate the hydrolyzed aluminum into precipitate and truly soluble forms; dispersed microcrystalline aluminum hydroxide may not be removed by membrane filtration under certain conditions, especially in deionized water at low pH with low concentrations of sulfate ion. Solution chemistry (especially multivalent anions such as sulfate and phosphate) can have a significant effect on the speciation of soluble aluminum, and it can affect the stability and physical form (e.g., particle size) of precipitate suspensions (Letterman and Vanderbrook, 1983).

Ferric Hydroxide Precipitate Solubility. As discussed in a previous section the solubility of amorphous ferric hydroxide precipitate is significantly less than that of amorphous aluminum hydroxide. While ferric hydroxide, like aluminum hydroxide, is amphoteric and tends to dissolve at both high and low pH, its overall low solubility leads to theoretically low concentrations of soluble iron (<5 μg/L) across the entire range of pH values (5 to 9) that is of practical importance in the coagulation process. Given ferric hydroxide's overall low solubility, the effect of temperature on it is not a significant concern.

Selecting the Type of Coagulant and Estimating the Required Dosage

The following example illustrates how the calculated or measured effective acid content (EAC) of commercial coagulant product solutions can be used with a dosage-estimating model to evaluate alternative strategies for simultaneously adding the required dosage and achieving a target value of the treated water pH. Selecting and using an appropriate target pH in coagulation is important for minimizing the concentration of residual coagulant metal in the treated water and for maximizing the efficiency of the coagulant in removing particles and natural organic material.

In the approach to coagulant selection and pH control illustrated by this example the type of coagulant, the coagulant solution's EAC and the types and amounts of any separate acids or bases used are chosen to simultaneously accomplish two objectives: (1) add the required Al dosage and (2) achieve a certain final target pH in the coagulation process. This example assumes that the objective is NOM removal with an aluminum salt coagulant.

Similar calculations are used to determine the dosage of an iron salt coagulant. However, because of the low overall solubility of ferric hydroxide, a criterion other than the pH of minimum solubility of the metal hydroxide is used to select the target pH.

Required Al Dosage. As discussed in a previous section, when the principal purpose of coagulation is effective NOM removal, the required aluminum dosage can be assumed to be proportional to the initial TOC concentration in the water, as in Eq. 8-12, where m, the coagulant dosage, is equal to R times TOC_0. The magnitude of R is a function of pH and can be estimated for all types of aluminum salt coagulants using the curve for $M = Al$ in Fig. 8-10.

Target pH. This example applies the strategy described by Edzwald and Kaminski (2009) for choosing alum dosages that they found agreed with full-scale plant data. Their method uses a coagulation process target pH equal to the pH of minimum solubility of the aluminum hydroxide precipitate. As discussed in a preceding section, this pH varies with the water temperature.

Determining the Required Effective Acid Content of the Coagulant Solution. An important determinant in selecting the type of coagulant is the required effective acid content of the coagulant solution, EAC in meq/mg Al, given by

$$\mathrm{EAC} = \frac{S \pm X}{m} \tag{8-23}$$

where m is in mg Al/L and X is the change in the alkalinity (in meq/L) associated with the addition of base or strong acid before or as the coagulant solution is added. When supplemental alkalinity (base) is added along with the coagulant solution, X is positive, and when strong acid is added, X is negative. $S = \mathrm{alk}_0 - \mathrm{alk}$ is the difference between the raw-water alkalinity (alk_0) and the alkalinity at the target pH (alk) after coagulant addition.

The alkalinity of the coagulant-treated solution at the target pH (alk) can be determined using chemical equilibrium calculations (see Chap. 3). Programs such as MINEQL+ (Schecher and McAvoy, 2003) and water chemistry spreadsheets, for example, RTW Model 4.0 (Rothberg and Scuras, 1994), are available for doing this. If it is assumed there is no carbon dioxide transport to or from the solution, S in Eq. 8-23 can be determined easily using a graph called a Deffeyes diagram (Snoeyink and Jenkins, 1980). Najm (2001) has published sets of nomographs that can be used for this purpose.

To calculate the alkalinity of the coagulant-treated solution using equilibrium relationships, first the inorganic carbon concentration (C_T) in the raw water is calculated using alk_0 and the raw(untreated) water pH (see Chap. 3). If no inorganic carbon is added with any supplemental acids or bases and if there is negligible carbon dioxide transport between the solution and the atmosphere, then it can be assumed that C_T is essentially constant during coagulant and any acid or base addition. However, if a supplemental base contains inorganic carbon, for example, sodium bicarbonate ($NaHCO_3$) or sodium carbonate (Na_2CO_3), then this must be included in C_T. Table 8-8 summarizes the alkalinity and inorganic carbon contributions of selected acids, bases, and carbon dioxide. The addition of 1 mg/l carbon

TABLE 8-8 Effect of Selected Acids, Bases, and Carbon Dioxide on the Alkalinity and Inorganic Carbon Concentration

Additive	Additive dosage	Change in alkalinity (meq/L)	Change in inorganic carbon concentration, C_T (moles C/L)
Sodium bicarbonate ($NaHCO_3$)	1 mg/L	1.19×10^{-2}	1.19×10^{-5}
Sodium carbonate (Na_2CO_3)	1 mg/L	1.89×10^{-2}	9.43×10^{-6}
Hydrated lime ($CaOH_2$)	1 mg/L	2.7×10^{-2}	0
Sodium hydroxide ($NaOH$)	1 mg/L	2.5×10^{-2}	0
Sulfuric acid (H_2SO_4)	1 mg/L	-2.04×10^{-2}	0
Hydrochloric acid (HCl)	1 mg/L	-2.82×10^{-2}	0
Carbon dioxide (CO_2)	1 mg/L	0	2.0×10^{-5}

dioxide to reduce the pH does not change the alkalinity but increases the inorganic carbon concentration by 2.0×10^{-5} moles/L.

The calculations that follow use raw-water chemistry for two sites, Vadnais Lake in Minnesota and Hinckley Reservoir in New York. These are 2 of the 27 raw-water samples that Archer and Singer (2006a, 2006b) employed in their study of the effect of the initial TOC concentration and alkalinity on the reduction of organic halide disinfection by-product concentrations by enhanced coagulation. Their sample collection sites were selected to correspond to each of the nine cells of the TOC concentration–alkalinity matrix in the USEPA's enhanced coagulation guidance manual (USEPA, 1999). Vadnais Lake is a raw water in the highest TOC, highest alkalinity category of the USEPA's matrix, and Hinckley Reservoir is in the lowest TOC–lowest alkalinity category. The essential characteristics of these sites are listed in Table 8-9.

Table 8-9 also lists the calculated total inorganic carbon concentration (C_T) for each raw water. Since Archer and Singer conducted their experiments with the water at approximately 15°C, this temperature was used for the equilibrium chemistry calculations. In addition, at 15°C according to Fig. 8-19a, the pH of minimum solubility of aluminum hydroxide is approximately 6.4; therefore, this pH was used as the target pH and, with the listed C_T concentrations, to determine the alkalinity (alk) of each coagulant-treated solution.

Example 8-2 Estimating the Type and Dosage of Aluminum Coagulant and Associated Acids and Bases, if Any, for Effective NOM Removal and Minimization of Residual Soluble Aluminium

Problem Statement Given the source-water chemistries for Vadnais Lake and Hinckley Reservoir listed in Archer and Singer (2006b), determine the type and dosage of aluminium salt coagulant for each of these water sources that will maximize the removal of NOM and minimize the concentration of residual soluble aluminum, and compare these estimates with the dosages and TOC and UV_{254} removals determined by Archer and Singer using jar testing and the USEPA's point of diminishing returns criterion.

TABLE 8-9 Raw-Water Characteristics of Two of the Source Waters Studied by Archer and Singer (2006b) (Both of these supplies are candidates for enhanced coagulation to reduce the concentration of disinfection by-product precursors.)

Raw-water characteristics	Vadnais Lake, St. Paul, MN	Hinckley Reservoir, Utica, NY
Total organic carbon (TOC, mg C/L)	8.1	3.9
SUVA (L/mg–m)	2.4	4.3
Alkalinity (alk_0, meq/L and mg/L as $CaCO_3$)	2.66 (133)	0.2 (10)
pH	8.2	7.6
Calculated inorganic carbon concentration (C_T, moles C/L)	2.69×10^{-3}	2.14×10^{-4}
Calculated alkalinity (meq/L) at the target pH of 6.4 and $T = 15°C$	1.35	0.10

Solution

Vadnais Lake The calculated alkalinity of the coagulant-treated water for Vadnais Lake is 1.35 meq/L, and since $alk_o = 2.66$ meq/L for this water, $S = 2.66 - 1.35 = 1.31$ meq/L. Using $TOC_o = 8.1$ mg C/L, Eq. 8-12, and $R = 0.65$ mg Al/mg C at pH 6.4 from Fig. 8-10, the required Al dosage is $m = 0.65 \times 8.1 = 5.3$ mg Al/L. According to Eq. 8-23 and assuming that no supplemental acids or bases will be added with the coagulant solution (i.e., $X = 0$), the required EAC for the coagulant product according to Eq. 8-23 is EAC = 1.31 meq/L/5.3 mg Al/L = 0.247 meq/mg Al. Using Vadnais Lake's SUVA of 2.4 L/mg-m and Fig. 8-11a and b, the expected removals of TOC and UV_{254} should be about 25 and 37 percent, respectively.

An EAC of 0.247 meq/mg Al is greater than the EAC of conventional alum solution (0.111 meq/mg Al, Table 8-5) so an alum product with supplemental acid is an option. According to Eq. 8-19, with AW = 27 and EW = 49, an alum product with supplemental H_2SO_4 should have an acid to Al ratio (i.e., an A/M value) of 6.67. A product with $A = 15$ percent H_2SO_4 and $M = 2.3\%$ Al ($A/M = 15/2.3 = 6.5 \approx 6.67$) would be appropriate.

Another option is to use conventional alum solution with an aluminum content of 4.3 percent Al and EAC of 0.111 meq/mg Al and independently add a strong acid. For this option Eq. 8-23 is solved for X using $S = 1.31$ meq/L, $m = 5.3$ mg Al/L, and EAC = 0.111 meq/ mg Al. The result is $X = -0.722$ meq/L. According to the acid equivalents listed for H_2SO_4 in Table 8-8, the required dosage is –0.722 meq/L/–2.04 × 10^{-2} meq/mg = 35 mg H_2SO_4/L.

Archer and Singer (2006b) using a jar test procedure with conventional alum and the USEPA's PODR criterion measured an effective Al dosage of 4.5 mg Al/L. Their measured TOC and UV_{254} of 27.8 and 43.3 percent are similar to the values of 25 and 37 percent estimated with Fig. 8-11a and b at an SUVA of 2.4 L/mg-m. The pH at the PODR dosage is not listed in the Archer and Singer paper, but it can be estimated using equilibrium chemistry and was probably about 6.4, which is close to the target pH of 6.4 used in this example calculation.

Hinckley Reservoir (Mohawk Valley Water Authority Supply). The Hinckley Reservoir supply is a low-alkalinity water with a medium concentration of TOC. The calculated alkalinity for the coagulant-treated water based on $C_T = 2.14 \times 10^{-4}$ moles C/L and a target pH = 6.4 is 0.10 meq/L. The change in alkalinity with coagulant addition is therefore $S = 0.2 - 0.1 = 0.1$ meq/L.

With a raw-water TOC of 3.9 mg C/L and $R = 0.65$ mg Al/mg C at pH 6.4 from Fig. 8-10, Eq. 8-12 gives a required Al dosage of $m = 0.65 \times 3.9 = 2.53$ mg Al/L. Using Eq. 8-23 and assuming that no supplemental acid or base will be added with the coagulant solution (i.e., $X = 0$), the required EAC for the coagulant product is EAC = 0.1 meq/L/2.53 mg Al/L = 0.04 meq/mg Al. According to Fig. 8-11a and b and the Hinckley Reservoir's SUVA of 4.3 L/mg-m the expected removals of TOC and UV_{254} are 52 and 75 percent, respectively.

Since the calculated EAC of 0.04 meq/mg Al is significantly less than that of conventional alum solution (0.111 meq/mg Al). this points to a prehydrolyzed aluminum product as an option. According to Eq. 8-19 with $AW = 27$, the prehydrolyzed product should have a basicity of 88 percent. The highest basicity of prehydrolyzed aluminum products is typically about $B = 80$ to 83 percent. If a prehydrolyzed product with $B = 83$ percent is used, since the basicity is slightly less than what is required (83 < 88 percent), the final pH may be slightly lower than the target value of 6.4. The transport of some CO_2 from the solution during coagulant addition may reduce this discrepancy.

An alternative to using a prehydrolyzed product is to add conventional alum solution with an aluminum content of 4.3 percent Al and EAC of 0.111 meq/mg Al and independently add a strong base, for example, NaOH. For this approach Eq. 8-23 is solved for X using $S = 0.1$ meq/L, $m = 2.53$ mg Al/L, and EAC = 0.111 meq/ mg Al. The result is $X = 0.181$ meq/L. According to Table 8-8, the required dosage of NaOH is 0.181 meq/L/2.5 × 10^{-2} meq/mg = 7.2 mg NaOH/L.

Archer and Singer (2006b) measured an effective Al dosage of 0.9 mg Al/L for Hinckley Reservoir, which is significantly less than the 2.53 mg Al/L calculated above. Their measured TOC and UV_{254} removals of 47.0 and 57.3 percent are less than the values of 52 and 75 percent estimated with Fig. 8-11a and b and this site's SUVA of 4.4 L/(mg-m). The pH at the PODR dosage of Archer and Singer is not listed in their paper, but it can be estimated using equilibrium chemistry and was approximately 7.1. This is higher than the pH of minimum aluminum hydroxide solubility (6.4) and near the high end of the pH interval (6 to 7) of highest Al coagulant effectiveness. This suggests that the effective coagulant dosage determined by Archer and Singer using the USEPA's PODR criterion was significantly less than what is needed for maximum TOC and UV_{254} removal. Also, it is likely their dosage and the associated final pH would not minimize the concentration of residual aluminum in the treated water.

The 120 megaliter/d (32 MGD) Mohawk Valley Water Authority treatment plant at Hinckley Reservoir consists of rapid-mix units, contact basins, adsorption clarifiers, and dual-media granular bed filters. Between 1999 and 2010 the average raw-water TOC and UV_{254} were 4.85 mg C/L and 0.179 cm^{-1} and the average raw-water SUVA was 3.7 L/(mg-m). The alum dosage was usually between 1.6 and 1.8 mg Al/L. In the 1999 to 2010 period the average removal of TOC was 62 percent and the average removal of UV_{254} was 79 percent. Both TOC and UV_{254} removal percentages are somewhat higher than the values predicted using their average SUVA with Fig. 8-11a and b and significantly higher than the values obtained by Archer and Singer (2006b) using jar tests and the USEPA's PODR, although as given here, Archer and Singer used a much lower alum dose of 0.9 mg Al/L. ▲

Organic Polyelectrolyte Coagulants

The polyelectrolyte coagulants used in water treatment are high-molecular-weight, synthetic organic compounds called polymers that have a strong tendency to adsorb on the surfaces of most particles in an aqueous suspension. Polyelectrolyte polymers consist of subunits called *monomers*. A polymer's total number and types of monomer units can be varied in manufacture; consequently, a wide variety of polymers can be and has been produced. The fact that polymer chains may be linear, branched, or cross-linked increases their complexity. A review by Bolto and Gregory (2007) covers the types of polyelectrolytes used in water treatment and the mechanisms of their action in removing different types of contaminants.

Types of Polyelectrolytes. The monomer units in a polymer may have positively or negatively charged sites. In some cases, charged sites are formed by ionization reactions, with the overall charge on the molecule a function of the solution's pH and ionic strength. Polymers with a preponderance of negative sites are called anionic; those with predominately positive sites are called cationic. Polymers with no charged sites or a low tendency to develop them in aqueous solution are known as nonionic polymers. (Although not polyelectrolytes by the strictest chemical definition, the term is still used in describing nonionic polymers.) Ampholyte polymers have both positive and negative sites.

Usage. Past estimates have indicated (AWWA, 1982) that over half the water treatment plants in the United States use one or more polyelectrolytes to improve treatment efficiency.

Today this percentage is probably conservative. Cleasby et al. (1989) found that of 23 treatment plants with high-quality filtered water, 20 were using one or more polymeric flocculants and/or filter aids. The Cleasby report states that these additives are essentially required at high-rate (rapid sand) filtration plants.

Types of Polyelectrolytes Used in Water Treatment. Polyelectrolytes used to treat drinking water are often categorized as either primary coagulant polymers or flocculant (coagulant aid) polymers. Primary coagulant polymers are generally cationic and, with few exceptions, have relatively low molecular weight (< 500,000). Primary coagulant polymers seem to enhance the coagulation and deposition (filtration) of negatively charged particles by adsorption and particle surface-charge neutralization. If one adds too much polymer to a suspension, each particle's overall surface charge may become positive. This occurrence, known as restabilization, can adversely affect coagulation and filtration.

Primary coagulant polymers may not be as effective as aluminum and iron salts to treat dilute inorganic suspensions and water with significant amounts of NOM when aggregate removal is by, for example, sedimentation. Enmeshment by the voluminous metal hydroxide precipitates can be used to increase the volume concentration of the suspension and increase the rate of particle aggregation to efficient levels. Enmeshment is not usually a significant factor with polyelectrolytes, and while a dilute suspension can be effectively destabilized by a primary coagulant, the rate of particle aggregation may be too low to produce large, settleable aggregates in a reasonable period of time.

On the other hand, when the pretreatment objective is to produce small, high-density aggregates for direct filtration (i.e., with no intervening aggregate removal step like sedimentation), a primary coagulant used alone is often an effective alternative. One must be careful, however, never to exceed the product dosage limit recognized by the review authority for health protection.

The use of dual coagulants (a hydrolyzing metal salt and a low concentration of cationic polyelectrolyte) is a common procedure. The typical purpose is to reduce the dosage of hydrolyzing metal salt and therefore the amount of treatment residue and the amount of alkalinity consumed in the process. The polyelectrolyte may also increase the strength of the floc, reducing the tendency for flocs to break under adverse flow conditions.

Cationic Polyelectrolytes. The two most widely used cationic polyelectrolytes in water treatment are polydiallyldimethyl ammonium chloride (polyDADMAC) and epichlorohydrin dimethylamine (epiDMA). Both compounds are sold as aqueous solutions, and both are in a class called *quaternary amines*. Each monomer in a quaternary amine contains a nitrogen atom bound to four carbon atoms and carries a positive charge. This positive charge, unlike the charge of an ionizable group, is not affected to a significant extent by a dilute solution's chemistry (especially the pH). According to Bratby (2006) the molecular weight of polyDADMAC is around 2 to 3×10^6 Daltons and for epiDMA it is approximately 8×10^5 Daltons, but manufacturers can prepare specific products with different molecular weights.

Figure 8-20 shows the results of a jar test experiment in which a silica suspension was treated with a cationic polyelectrolyte (polyDADMAC). Turbidity was maximally removed at a polyelectrolyte product solution concentration of 10^{-4} g/L (0.1 mg/L). The product solution in this case contained about 20 percent by mass polyelectrolyte. The range for all the types of products sold as solutions is roughly 10 to 60 percent.

The dosage at maximum turbidity removal in this example is relatively low because, while the silica suspension concentration was relatively high (500 mg/L), the silica particles were rather large in size making the surface area concentration and the concentration of surface charge relatively low. In real source waters with smaller particles and higher

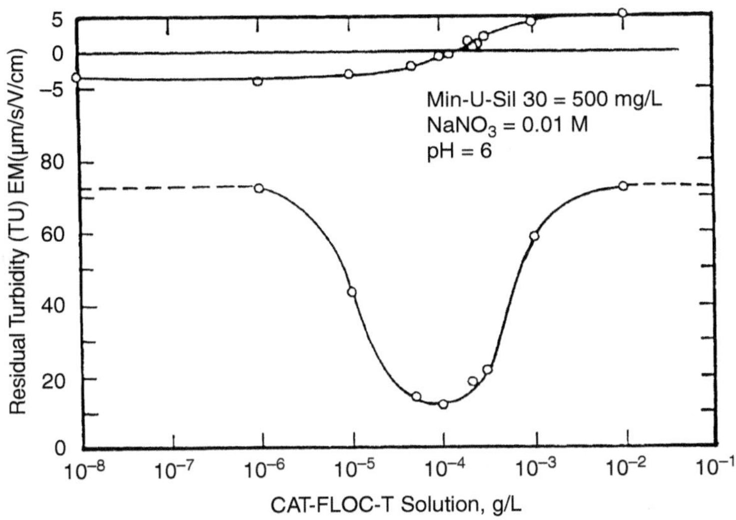

FIGURE 8-20 Effect of polyelectrolyte concentration (PolyDADMAC) on particle electrophoretic mobility and residual turbidity after flocculation and sedimentation.

particle surface area concentrations, polyelectrolyte dosages when used as the primary coagulant are typically in the 1 to 10 mg/L range.

The electrophoretic mobility (EM) at the dosage of maximum effectiveness (minimum residual turbidity) in Fig. 8-20 is approximately zero. Maximum particle removal at the point of zero EM is consistent with a charge neutralization mechanism. Above a product dosage of about 1 mg/L (for this silica particle size and concentration), the adsorption of polyelectrolyte causes the EM to become positive. The high positive charge on the particles causes restabilization of the suspension, with turbidity removal negligible.

Anionic and Nonionic Polyelectrolytes. Flocculant polymers are generally anionic or nonionic and have relatively high molecular weight, perhaps 10 times or more that of the typical primary coagulant polymer but usually less than about 10^7 daltons. They are often added after the flocculation process to increase the size and strength of particle aggregates. Because of their high molecular weight and, therefore, appreciable length, the polymers are able to bridge or interconnect particles in the agglomerates.

Low dosages of high-molecular-weight nonionic polymers (0.005 to 0.05 mg/L) are often applied just before granular bed filtration. This procedure may improve overall effluent quality and reduce the tendency for perturbations, such as an abrupt change in the filtration rate, to have an adverse effect on performance. The ripening period of granular bed filters has been decreased by adding low dosages of nonionic polymers to all or part of the backwash water (Harris, 1972; Yapijakis, 1982).

Most anionic and nonionic polymers are made with the monomer acrylamide and are known as polyacrylamide polymers. Surveys of the water-supply industry (AWWA, 1982; Dentel et al., 1989) indicate that many treatment plants use this type of polymer and, as noted previously, Cleasby et al. (1989) suggest that their use is almost essential to high performance in high-rate filtration. Dentel et al. (1989) have prepared a manual that guides the testing needed to determine a feasible type and cost effective dosage of flocculant polymers used in water treatment.

Polymeric Coagulants from Natural Organic Compounds. Polymeric coagulants made from natural organic compounds are sometimes used for water treatment. This practice is most common in developing countries where these compounds may be cheaper than hydrolyzing metal salt coagulants. An example is the water extract of the macerated seeds of the tropical shade plant *Moringa oleifera* (Jahn, 1988; Sutherland et al., 1989). The extract contains a dimeric cationic protein with a molecular weight of approximately 13 kilodaltons. The treatment performance of *moringa* seed extract has been discussed by Folkard (2002).

Another example of a cationic polymer is chitosan, a partially deacetylated chitin derived from crustacean shells. Commercial chitosan has a molecular weight of about 10^6 daltons (Bratby, 2006) and has a relatively low positive charge at neutral pH. Its charge density is pH dependent (Rinaudo, 2006). Its use in a variety of applications is described by Pariser and Lombardi (1989). Bolto and Gregory (2007) describe a cationic derivative of starch and flocculating agents that is lignin based and an anionic natural polymer lignin sulphonate made from kraft pine lignin, a by-product of papermaking.

Impurities in Polyelectrolyte Products. Polyelectrolyte formulations contain contaminants from the raw materials and the manufacturing process, such as residual monomers and other reactants and reaction by-products that have potential human health significance (Letterman and Pero, 1990). Concern about certain contaminants has limited the use of polyelectrolytes in certain countries and caused some to establish quantitative limits on applications rates. Some states in the United States require plants that use polyacrylamide and epichlorohydrin/dimethylamine (epi/DMA) compounds to notify them of monomer content.

For many years the U.S. Public Health Service and the USEPA evaluated the health significance of polyelectrolyte products and maintained a list of accepted products that many states used in their regulatory/review activities. The list included a maximum dosage for each accepted product. In the 1980s the USEPA sponsored the development and initiation of a new system based on a voluntary, third-party standard prepared through the auspices of NSF International (NSF, 2009). NSF International and other accredited organizations test and certify compliance with NSF/ANSI Standard 60, a service that is paid for by each product's manufacturer. Most of the states and many provinces in Canada require polyelectrolyte products used in their jurisdictions to be certified as in compliance with NSF/ANSI Standard 60.

A large number of polymer products are listed as certified to be in compliance with NSF/ANSI Standard 60, but this does not mean that hundreds of different compounds are being used in potable water treatment. A number of years ago Letterman and Pero (1990) found that fewer than 12 compounds made up the products on the USEPA's list of acceptable products in use at that time. It is reasonable to assume that only polyDADMAC and epi/DMA are widely used.

Supplemental (Coagulant Aid) Materials

Activated Silica. The first application of activated silica was described by Baylis (1937). The material was prepared on-site by acidifying a sodium silicate solution. Acidification formed negatively charged polysilicate compounds which were added in combination with a hydrolyzing metal salt coagulant to enhance flocculation and sedimentation. Activated silica was often used in treating low-temperature, low-turbidity water. O'Melia (1985) has described the mode of action of activated silica and the factors that affect its performance. Its use today is limited.

Clay and Fine Sand. In the process called *ballasted flocculation* a ballasting agent, typically clay or fine silica particles, microsand, in the 50- to 400-μm range (Ghanem et al., 2007), is added before or with the coagulant(s) in the rapid-mix unit. Polyelectrolytes are typically used to bind the mineral particles to the floc. Incorporation of these relatively dense particles in the floc tends to enhance flocculation and floc separation processes by displacing water and increasing floc density and settling properties. Ghanem et al. found that the bulk density of the flocs is directly proportional to the ballasting agent dosage.

Combinations of Coagulants. In a 1982 national survey of treatment plants (AWWA, 1982), a majority of respondents stated that they use cationic polymers in combination with hydrolyzable metal coagulants such as alum. Some benefits of dual coagulant use were discussed earlier.

Results obtained in jar test studies with silica suspensions (Letterman and Sricharoenchaikit, 1982) suggest that under a condition where the positively charged, aluminum hydrolysis products cause destabilization by charge neutralization, a simple trade-off relationship exists between the amounts of Al and cationic polyelectrolyte required to achieve maximum turbidity removal. In experiments where aluminum hydroxide precipitate was formed and enmeshment of the silica particles occurred, the adsorbed polyelectrolyte became covered by the precipitate and the stability of the suspension was controlled by the precipitate coating. A number of companies are marketing coagulant solutions that are a mixture of two active ingredients, for example, alum or prehydrolyzed aluminum salt with a cationic polyelectrolyte.

Hubel and Edzwald (1987) studied the use of combinations of cationic polyelectrolyte and alum for the removal of NOM from low-turbidity, low-alkalinity water. Low-molecular-weight cationic polyelectrolytes used alone were less effective than alum used alone. Without alum the dosage of cationic polyelectrolyte required to achieve charge neutralization and effective removal was proportional to the initial NOM concentration. This dosage was high for all but the lowest initial NOM concentrations and tended to exceed the health-based product dosage limit.

The use of high-molecular-weight polyelectrolytes as flocculants in water that had already been treated with alum improved turbidity removal and reduced the sludge volume but did not improve NOM removal over what was obtained with alum alone (Hubel and Edzwald, 1987). The use of high-charge-density cationic polyelectrolytes as primary coagulants with alum as a supplemental coagulant achieved high NOM removal and reduced the amount of solids produced by the process.

Ozone and Coagulation. In addition to being an effective disinfectant and oxidant, ozone can cause particle destabilization and flocculation. In the Los Angeles Aqueduct treatment plant, preozonation has been shown to improve filtration efficiency and to reduce the need for conventional coagulants and long flocculation periods (Prendiville, 1986; Monk et al., 1985; Becker and O'Melia, 2001). Ozone under certain conditions can reduce the coagulant dosage required for effective NOM removal; see Chap. 7 and the book by MWH (2005). This source recommends that, in general, NOM removal by coagulation be carried out before ozonation.

The mechanism(s) by which ozone causes particle destabilization is not well understood. It is possible that the ozone oxidizes metal ions such as Fe(II) that then form insoluble precipitates. Humic materials, both free and sorbed, may be oxidized and made more polar. It has also been suggested that ozone treatment causes the desorption of stabilizing NOM from the surfaces of mineral colloids (Edwards and Benjamin, 1992; Chandrakanth and Amy, 1996; Paralkar and Edzwald, 1996).

THE RAPID MIXING AND FLOCCULATION PROCESSES

Purpose of Rapid Mixing

Rapid, or flash, mixing is a high-intensity mixing step used before the flocculation process to disperse the coagulant(s) and to initiate the particle aggregation process (Amirtharajah and Mills, 1982). In the case of hydrolyzing metal salts, the primary purpose of the rapid mix is to quickly disperse the salt so that contact between the simpler hydrolysis products and the particles occurs before the metal hydroxide precipitate has formed. Rapid dispersal before precipitation helps ensure that the coagulant is distributed uniformly among the particles. This process is poorly understood, but probably depends on factors such as the concentration of salt in the coagulant feed solution, the coagulant dosage, the concentration and size distribution of the particulate matter, the temperature and ionic constituents of the solution, and the turbulent flow conditions (overall energy input and the flow and kinetic energy spectrum of the turbulent motion) in the rapid-mixing device. Amirtharajah and O'Melia (1990) have reviewed how some of these factors might affect rapid-mix unit performance and a report from the Awwa Research Foundation (now the Water Research Foundation) on mixing in coagulation and flocculation (AwwaRF, 1991) is also useful.

The significance of the rapid-mix step when polyelectrolyte coagulants are used is probably similar to that for hydrolyzing metals except that the reactions involving the coagulant and the water are not as important. Polyelectrolytes rapidly and irreversibly adsorb on the particulate surfaces. Therefore, in the absence of intense mixing at the point of coagulant addition, it is logical to assume that some particles might adsorb more polymer than others. If the overdosed particles became surrounded by other particles that have little or no adsorbed polymer, it is possible that the aggregation process would slow or stop before sufficiently large flocs were formed. Bratby (2006) presents useful practical information about the application of polyelectrolytes in flocculation tanks and before filtration to enhance the performance of these processes.

Rapid mixing is also the start of the flocculation process. When coagulant is added, the particles become destabilized and the high-intensity mixing leads to rapid aggregation. Particle disaggregation may become important as the aggregates grow. Evidence suggests that a steady state size distribution of relatively small aggregates may characterize the suspension leaving the rapid-mix process (AwwaRF, 1991). Furthermore, there is limited evidence that mixing at high intensity for too long can be detrimental to subsequent process performance, possibly because aggregates that are eroded or broken have a reduced tendency to reform with time because of changes in surfaces' chemical or physical properties.

Purpose of Flocculation

The purpose of the flocculation process is to promote the interaction of particles and form aggregates that can be efficiently removed in subsequent separation processes such as sedimentation, flotation, and granular media bed filtration. For efficient flocculation to occur the suspension must be destabilized. This is usually accomplished by the addition of a coagulant.

Transport Mechanisms

A number of mechanisms can cause relative motion and collisions between particles in a destabilized suspension, including Brownian motion, velocity gradients in laminar flow, magnetic attraction, unequal settling velocities, and turbulent transport. All but laminar flow velocity gradients and magnetic attraction are important in typical water treatment systems. Flocculation rate equations have been derived for each of these mechanisms by assuming that the aggregation process is a second-order rate process in which the rate of collision N_{ij} between i- and j-size particles is proportional to the product of the concentrations of the two colliding units n_i and n_j. The general form of these relationships is

$$N_{ij} = \alpha_{ij} k_{ij} n_i n_j \tag{8-24}$$

where k_{ij} is a second-order rate constant that depends on the transport mechanism and a number of factors including particle size, and α_{ij} is a flocculation rate correction factor that is usually called a *collision efficiency factor* in the water treatment literature. It is equal to the inverse of the stability ratio W of the colloid chemistry literature, and its magnitude ranges from 0 (when the suspension is completely stable) to slightly greater than 1 (when an attractive forces affects the motion of interacting particles).

The α_{ij} term in Eq. 8-24 corrects for the simplifying assumption that transport is not influenced by the short-range forces that affect particle motion when two particles move close together. The important short-range forces are double-layer repulsion, van der Waals attraction, and hydrodynamic retardation. The first two were discussed earlier in the chapter; hydrodynamic retardation is caused by the viscous flow of fluid from between the particles as they start to collide. Hydrodynamic retardation and double-layer repulsion tend to slow particle motion and inhibit collisions, and van der Waals attraction tends to promote them. Equations for predicting the magnitude of α_{ij} are usually from studies in which numerical solutions of the complete equations of motion for interacting particles are compared with solutions obtained with the simplified transport equations that do not consider the short-range forces. Han and Lawler (1992) in modeling the flocculation process summarize the various studies regarding particle transport and the significance and incorporation of short-range forces into the models.

Brownian Diffusion. Small particles suspended in a fluid move about in a random way due to continuous collisions with the surrounding water molecules. The intensity of this motion is a function of the thermal energy of the fluid k_B times T, where k_B is Boltzmann's constant and T is the absolute temperature. The process is called *Brownian diffusion* and the particle interaction it causes is Brownian or perikinetic flocculation. The flocculation rate constant (k_{ij} of Eq. 8-24) for perikinetic flocculation in an infinite stagnant fluid, k_p, is given by

$$k_p = \left(\frac{2}{3}\right)\frac{k_B T(d_1 + d_2)^2}{\mu(d_1 d_2)} \tag{8-25}$$

According to Eq. 8-25, the rate of perikinetic flocculation is proportional to the absolute temperature, and, for particles of equal diameter ($d_1 = d_2$), is independent of the particle size. The quantity μ is the dynamic viscosity.

For transport by Brownian diffusion, the flocculation rate correction factor α_{ij} of Eq. 8-24 is α_p. It accounts for the retarding effect of hydrodynamic forces, double-layer repulsion, and other collision inhibiting factors such as steric repulsion and hydration forces on particle-particle interaction. The magnitude of α_p can be estimated with laboratory experiments or by using expressions derived by Kim and Rajagopolan (1982). These expressions include the effects of double-layer repulsion and hydrodynamic forces and use the following parameters: the height of the potential barrier in the net energy of interaction curve (ψ_{max} in Fig. 8-3c, $A/(6k_BT)$, d_1, and d_2. The quantity A is the Hamaker constant.

In systems with significant mixing and convection, Brownian diffusion becomes relatively unimportant in the bulk fluid. However, where small particles contact large floc particles, Brownian diffusion can still control transport of the small particles over a final short distance at the floc surface. The rate of transport to the floc surface is a function of the floc size and the intensity of the turbulent fluid motion as well as the Brownian diffusivity D, where $D = k_B T/(3\pi\mu d)$; d is the diameter of the diffusing particle.

Transport in Laminar Shear. When particles are suspended in a laminar flow field, a particle located at a point with high fluid velocity tends to move faster than one at a point with low velocity. If the particles are close enough together, their different velocities will eventually cause them to come into contact. This process is called orthokinetic flocculation and its flocculation rate constant $k_{ij} = k_o$ is given by,

$$k_o = \left[\frac{(d_1 + d_2)^3}{6}\right]\left(\frac{du}{dz}\right) \tag{8-26}$$

where du/dz is the magnitude of the velocity gradient.

The flocculation rate correction factor in Eq. 8-21 for orthokinetic flocculation ($\alpha_{ij} = \alpha_o$) has been evaluated by van de Ven and Mason (1977). They presented an approximate expression for the factor when double-layer repulsion is negligible ($\psi_{max} \to 0$) and van der Waals attraction and hydrodynamic retardation are the controlling short-range forces. The two interacting particles are assumed to have the same diameter d. The expression is

$$\alpha_o = 0.8\left[\frac{A}{4.5\,\pi\mu\, d^3 (du/dz)}\right]^{0.18} \tag{8-27}$$

According to Eq. 8-27, hydrodynamic retardation reduces the collision rate in orthokinetic encounters even when double-layer repulsion is negligible. The effect is greater for larger particles and higher shear rates. For example, assuming a water temperature of 20°C and using A (Hamaker constant) = 8×10^{-20} J, $\mu = 1 \times 10^{-3}$ kg/(m–s), $d = 2 \times 10^{-6}$m, du/dz = 10 s^{-1}, and $\alpha_o = 0.62$. If du/dz is increased to 100 s^{-1}, α_0 decreases to 0.41. The significance of hydrodynamic retardation is reduced if the particles are porous and intervening fluid can move through the particle structure, rather than just flowing from between the colliding surfaces.

Differential Settling. Flocculation by differential settling occurs when particles have unequal settling velocities and their alignment in the vertical direction makes them tend to collide when one overtakes the other. The driving force for this mechanism is gravity, and the flocculation rate constant $k_{ij} = k_d$ is given by

$$k_d = \frac{\pi g(s-1)}{72\nu}(d_1 + d_2)^3 (d_1 - d_2) \tag{8-28}$$

where s is the specific gravity of the particles and ν is the kinematic viscosity. This expression is based on several assumptions; the particles are spherical and have the same density and their settling velocities are predicted by Stokes' law. According to Eq. 8-28 the rate of flocculation by differential settling is maximized when both particles are large and dense and the difference in their sizes is great. The effect of the short-range attractive forces and hydrodynamic retardation on the rate of flocculation by differential settling has been evaluated by Han and Lawler (1991).

Turbulent Transport. In turbulent flow the fluctuating motion of the fluid forms eddies that vary in size. The largest eddies are of comparable size to the vessel or impeller. The kinetic energy in large-scale eddies is transferred to smaller and smaller eddies. Eventually, below a certain length scale, the energy is dissipated as heat. An eddy size, known as the *Kolmogoroff microscale,* separates the inertial size range, where energy is transferred with very little dissipation, from the viscous subrange, where the energy is dissipated as heat. Within the eddies of turbulent flow are time-varying velocity gradients. These velocity gradients cause relative motion of entrained particles, and this relative motion, as in laminar flow, causes flocculation.

The flocculation rate constant $k_{ij} = k_t$ for collisions between two neutrally buoyant particles (i.e., particles with insignificant settling velocities) in homogeneous and isotropic turbulent fluid motion is given by

$$k_t = \left[\frac{(d_1 + d_2)^3}{6.18}\right]\left(\frac{\varepsilon}{\nu}\right)^{1/2} \tag{8-29}$$

where ε is the local rate of turbulent energy dissipation per unit mass of fluid and ν is the kinematic viscosity. Equation 8-29 was derived by Saffman and Turner (1956) using the assumption that the diameter of the particles is less than the Kolmogoroff microscale of the turbulent motion, $\eta = (\nu^3/\varepsilon)^{1/4}$. According to Eq. 8-29, the rate at which particles collide by turbulent diffusion increases with particle size and with the rate of energy dissipation in the fluid.

Spielman (1978) stated that the magnitude of the flocculation rate correction factor $\alpha_{ij} = \alpha_t$ for the interaction of equal-size particles in turbulent flow can be approximated using Eq. 8-27, the expression derived by van de Ven and Mason (1977) for particle interactions in laminar shear. Spielman substituted the quantity $(\varepsilon/\nu)^{1/2}$ for the laminar velocity gradient du/dz and obtained

$$\alpha_t = 0.8\left[\frac{A}{4.5\,\mu d^3(\varepsilon/\nu)^{1/2}}\right]^{0.18} \tag{8-30}$$

This expression suggests that α_t decreases with increasing particle diameter and increasing mixing intensity.

Delichatsios and Probstein (1975) derived a kinetic model for flocculation in isotropic turbulent flow for two conditions. In the first the radius $(d_1 + d_2)/2$ of the collision sphere is smaller than the Kolmogoroff microscale, and in the second it is greater. For the first condition, the equation for the flocculation rate constant is similar in form to Eq. 8-29 with a constant 1/19.6 instead of 1/6.18. For the second condition, $(d_1 + d_2)/2$ larger than the microscale, the expression is

$$k_{t,d-p} = (0.427)(d_1 + d_2)^{7/3}\varepsilon^{1/3} \tag{8-31}$$

This equation suggests that for larger floc particles the rate of particle interaction is proportional to ε to the 1/3 power instead of the 1/2 power of Eq. 8-29. Also, in contrast to Eq. 8-29, the rate of interaction is independent of the viscosity of the fluid.

G Value Concept. In the 1940's Camp and Stein (1943) used Smoluchowski's (1917) equation for flocculation in uniform laminar shear to derive a widely used flocculation rate equation for turbulent flow. They calculated that for turbulent fluid motion a rms (root-mean-square) velocity gradient can be used in place of the laminar velocity gradient in the laminar shear flocculation rate equation (Eq. 8-26). Their rms velocity gradient, or G value, is given by

$$G = \left(\frac{P}{V\mu}\right)^{1/2} \tag{8-32}$$

where P is the power input to the fluid (through, for example, the blades of a rotating impeller) and V is the volume of water in the vessel. The flocculation rate constant in Camp and Stein's modified Smoluchowski equation for orthokinetic flocculation is

$$k_{o,c-s} = \frac{(d_1 + d_2)^3}{6} G \tag{8-33}$$

Equation 8-33 becomes similar to the more rigorous expression (Eq. 8-29) derived by Saffman and Turner (1956) if it is assumed that the G value is equal to $(\varepsilon/\nu)^{1/2}$ where ε, as noted previously, is the local rate of energy dissipation per unit mass of fluid.

The limitations of the G-value concept have been discussed by a number of authors including Cleasby (1984), McConnachie (1991), Hanson and Cleasby (1990), and Han and Lawler (1992). Clark (1985) has argued that inappropriate assumptions were made in the derivation of the Camp and Stein G value, and therefore, it is not a valid representation of the average velocity gradient in a turbulent fluid. The theoretical validity of the G value of Eq. 8-32 as a design parameter and reactor scale-up tool is questionable because the rate of flocculation is affected not only by the intensity of the fluid motion but also by how the kinetic energy of the fluid is distributed over the size of the eddies in the turbulent flow field. For a given mixing intensity (either overall power input to the fluid or local rate of energy dissipation), the distribution of energy is determined by the size and geometry of the impeller and vessel system.

Use of the G-value concept in flocculation is affected by three other considerations. First, the local rate of energy dissipation in a mechanically mixed vessel varies widely with location in the vessel, from several hundred times the overall energy input rate per unit mass of fluid $P/(V\rho_w)$ near the rotating impeller to a fraction of this overall value away from the impeller (Schwartzberg and Treybal, 1968; McConnachie, 1991). Consequently, the local rate of particle aggregation (and possibly the local rate of floc breakup) will tend to vary widely with location in the vessel, increasing the tendency for floc structure and density to change through breakup and reformation as the flocs are moved around the vessel by the fluid (Clark and Flora, 1991).

Second, some of the energy supplied to the rotating impeller shaft is dissipated directly as heat at the surfaces of the impeller and vessel walls and does not produce turbulent fluid motion. The weighted average (by fluid mass) of ε for all elements of fluid in the vessel will tend to be less than $P/(V\rho_w)$, the overall energy input per unit mass of fluid in the vessel; the efficiency with which the motion of the impeller increases the kinetic energy of the turbulent flow will vary with the geometry of the mixing system.

Third, according to theoretical calculations by Han and Lawler (1992), fluid shear, and therefore the G value, is a controlling factor in flocculation kinetics only when the interacting particles are larger than 1 μm and about the same size. When one particle is large and the other is small (< 1 μm), Brownian diffusion tends to control the relative motion between the particles, and while mixing intensity is still important, the form of the relationship between flocculation rate and the overall energy dissipation rate in the fluid is likely to be different than that predicted by Eq. 8-29.

Our understanding of turbulent flow in mechanically agitated vessels is incomplete but growing rapidly. For example, Liu et al. (2004) have shown how point-to-point calculations using computational fluid dynamics can predict flocculation performance in open-channel flow flocculators more effectively than the mean velocity gradient approach. Unfortunately, sensitive methods for measuring flocculator performance, especially the performance of full-scale units, are not available. Until this capability improves, it seems reasonable to use the Camp and Stein G value for flocculator design and scale-up, along with other empirically derived design best practices related to mixer and reactor configuration.

Flocculation Kinetics in Batch and Plug Flow Reactors

Flocculation in a well-mixed batch reactor is described by Eq. 8-34 for the rate of change in the concentration of k-size particles. The equation assumes that k-size particles are formed by the collision of i- and j-size particles; that is, $k = i + j$. The term dn_k/dt on the left is the rate of change in the concentration of k-size particles in the reactor. The first term on the right is for the rate of formation of k-size particles by the collision of i- and j-size particles and the second term on the right accounts for the loss of k-size particles by their collision with any other particle. This expression only applies to irreversible aggregation; it does not account for the breakup or erosion of aggregates. For continuous particle-size distributions an integral form of this expression can be written (Elimelech et al., 1995).

$$\frac{dn_k}{dt} = \frac{1}{2}\sum_{\substack{i+j\to k \\ i=1}}^{i=k-1} \alpha_{ij}k_{ij}n_i n_j - n_k \sum_{k=1}^{\infty} \alpha_{ik}k_{ik}n_i \tag{8-34}$$

The effective use of Eq. 8-34 is difficult. One of the most significant problems is assigning values to the rate constants k_{ij} and k_{ik} and the flocculation rate correction factors α_{ij} and α_{ik}. These depend on the nature of the particles and on the transport processes that cause the particle interaction (Elimelech et al., 1995). Relationships must be established for how aggregate shape and density vary with size. The simplest assumption is that the particles coalesce like oil droplets, but this is obviously very approximate for the interaction of solid particles.

Most flocculators are mechanically mixed, continuous-flow reactors and the fluid motion is turbulent; particle interaction through laminar velocity gradients and unequal settling velocities tend to be insignificant. The next section describes and applies relationships that predict the performance of completely mixed, continuous-flow flocculation reactors in which turbulent fluid motion is the predominant particle transport mechanism and that make simplifying assumptions about the size distributions of the particles. These relationships are presented as a framework for understanding the flocculation process and not a design methodology.

Ideal Continuous-Flow Flocculation Reactors

Argaman and Kaufman (1970) have shown that the performance of a continuous-flow flocculation reactor can be predicted using relationships based on the completely mixed, tanks-in-series model (see Chap. 4). It was assumed that each completely mixed (CFSTR) compartment in the reactor contains a bimodal particle-size distribution where the smaller particles in the distribution are from the influent and are called *primary particles,* and the larger particles are the aggregates or flocs. The flocs grow by the attachment of primary particles. Since each compartment is assumed to be completely mixed, at steady state the effluent characteristics (a mixture of flocs and primary particles) are the same as the vessel contents.

The following equations are based on a simplified version of Argaman and Kaufman's (1970) model in which floc disruption is ignored. The rate at which primary particles contact the larger floc particles in one of the completely mixed compartments is calculated using Eqs. 8-24 and 8-29. It is assumed that the flocs have a spherical shape and are much larger than the primary particles; that is, $d_2 >> d_1$. By substituting $\Phi = n_2\pi d_2^3/6$, where n_2 and d_2 are the floc number concentration and diameter and Φ is the floc volume fraction (volume of floc per volume of suspension), Eqs. 8-24 and 8-29 yield

$$N_{12} = \alpha_{12}\frac{0.97}{\pi}\Phi\left(\frac{\varepsilon}{\nu}\right)^{1/2} n_1 \tag{8-35}$$

Equation 8-35 is a first-order expression in terms of n_1. A mass balance on primary particles in the completely mixed reactor compartment can be combined with this expression and the assumption that $\alpha_{12} = 1$ to result in

$$\frac{n_1}{n_{10}} = \frac{1}{1 + k\bar{t}} \tag{8-36}$$

where the flocculation rate constant k is

$$k = \frac{0.97}{\pi}\phi G \tag{8-37}$$

$\bar{t}$ is the mean residence time of the fluid in the reactor compartment, and n_1 and n_{10} are the effluent and influent primary particle number concentrations, respectively. It is assumed that ε is equal to $P/(V\rho_w)$, the overall energy input rate per unit mass of fluid where V is the volume of water in the reactor and ρ_w is the density of water.

For destabilization by charge neutralization, the floc volume concentration in the reactor is determined by the volume of primary particles that has been incorporated in the flocs and by the density of the flocs; that is,

$$\phi = V_1 n_{10}\left(1 - \frac{n_1}{n_{10}}\right)\left(\frac{\rho_p - \rho_w}{\rho_f - \rho_w}\right) \tag{8-38}$$

where V_1 is the volume of a primary particle and ρ_f and ρ_p are the floc and primary particle densities, respectively. The quantity $V_1 n_{10}$ is equal to C_o/ρ_p, where C_o is the mass concentration of primary particles in the influent to the first compartment. According to Eqs. 8-36 through 8-38, flocculator performance n_1/n_{10} is a function of n_{10}, ρ_f, ρ_p, the dimensionless product $G \times \bar{t}$, and the amount of primary particles that attach to the flocs in the reactor. These equations are approximate due to the simplifying assumptions that have been made; however, they illustrate the importance of mixing intensity, residence time, and influent particle concentration in affecting the performance of continuous-flow flocculation reactors.

It is frequently noted that raw-water supplies with a relatively high concentration of particulate matter is easier to treat than water with a low concentration. There are also proprietary treatment systems that use a buoyant coarse media in the flocculator (Schulz et al., 1994) or the addition of supplemental suspensions such as silica particles to enhance the rate of flocculation for what would otherwise be dilute and slowly flocculating suspensions. One of the reasons for this observation and the efficacy of the proprietary systems is that, for a given value of $G \times \bar{t}$, as the influent particle concentration increases, the floc volume concentration increases and, consequently, the aggregation rate and the fractional conversion of primary particles to flocs become greater.

The typical continuous-flow flocculator has multiple compartments, and the mean fluid detention time and G value may be different for each compartment. As primary particles contact the flocs in each compartment, the floc volume concentration and the flocculation rate constant increase. The equations listed in Table 8-10 are based on Eqs. 8-36 through 8-38 and were used to calculate values of the quantity n_{1m}/n_{10} as a function of the dimensionless product $G \times \bar{t}$ for three values of the influent particle mass concentration. (n_{10} is the number concentration of primary particles in the influent to the first compartment, n_{1m} is the number concentration in the effluent of the mth compartment, and $\bar{t} = \Sigma\, \bar{t}_i$. For this example it was assumed that the flocculator has three equal volume compartments ($m = 3$) each with the same value of $\bar{t}_i$, and the same mixing intensity ($G_i = G$); the primary particle and floc densities are 2.5 and 1.02 g/cm^3, respectively. The results are plotted in Fig. 8-21.

TABLE 8-10 Equations Used to Calculate the Performance of a Compartmentalized, Completely Mixed, Continuous-Flow Flocculation Reactor*

$$\frac{n_{1m}}{n_{10}} = \prod_{i=1}^{m} \frac{1}{1 + k_i \bar{t}_i} \tag{8-39}$$

$$k_i = \left(\frac{0.97}{\pi}\right) \phi_i G_i \tag{8-40}$$

$$G_i = \left(\frac{\varepsilon}{\nu}\right)_i^{1/2} \tag{8-41}$$

$$\phi_i = \frac{C_o}{\rho_p}\left(1 - \frac{n_{1i}}{n_{10}}\right)\left(\frac{\rho_p - \rho_w}{\rho_f - \rho_w}\right) \tag{8-42}$$

*In these equations the subscript i refers to the ith compartment in a series of m compartments. n_{10} is the number concentration of particles in the influent to the first compartment, and n_{1m} is the concentration in the effluent of the mth compartment. C_0 is the mass concentration of primary particles in the influent to the first compartment. The overall mean fluid residence time for the reactor is $\bar{t} = T = \Sigma \bar{t}_i$. It is assumed that α for primary-particle-floc interaction is equal to 1.

According to Fig. 8-21, flocculator performance increases with increasing $G \times \bar{t}$ and increasing influent particle concentration. For $C_o = 5$ mg/L, as $G \times \bar{t}$ increases, the effect of flocculation on the primary particle concentration in the flocculator effluent is insignificant until $G \times \bar{t}$ is approximately 50,000. From $G \times \bar{t} = 50{,}000$ to 200,000, the quantity n_{1m}/n_{10} decreases from slightly less than 1 to approximately 0.02. For $C_o = 5$ mg/L, $n_{1m}/n_{10} = 0.5$

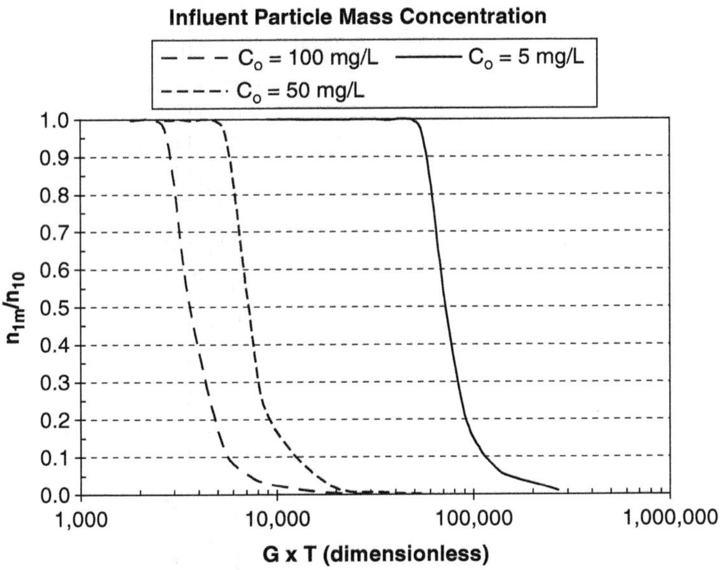

FIGURE 8.21 Effect of the dimensionless parameter G × T (T = $\bar{t}$) on the performance of a flocculation reactor with three equal compartments for several influent primary particle concentrations. The primary particle and floc density are 2.5 and 1.02 g/cm^3 and the G value in each compartment is 25 s^{-1}.

corresponds to $G \times \bar{t} = 72{,}000$. For $C_o = 100$ mg/L, significant primary particle removal begins at $G \times \bar{t} = 2800$ and $n_{1m}/n_{10} = 0.5$ corresponds to $G \times \bar{t} = 3600$. When C_o is equal to 50 mg/L, $n_{1m}/n_{10} = 0.5$ corresponds to $G \times \bar{t} = 7200$. This general relationship between the influent primary-particle concentration and the primary-particle removal efficiency exists because the flocculation rate constant k_i for a given flocculator compartment depends on the magnitude of the floc volume concentration Φ_i in that compartment, and Φ_i depends on the primary-particle concentration in the flocculator influent (see Eqs. 8-40 and 8-42 in Table 8-10). This dependency has been observed in practice (Hudson, 1981).

The equations in Table 8-10 can be used to show that the relationship between flocculator performance n_{1m}/n_{10} and $G \times \bar{t}$ depends on the number m of equal volume compartments in the flocculation reactor. For example, for $C_o = 100$ mg/L and $m = 3$, $n_{1m}/n_{10} = 0.5$ is obtained at $G \times \bar{t} = 3600$. For a single compartment ($m = 1$), this value of $G \times \bar{t}$ decreases to 2200, and for $m = 5$ it increases to 5300. Similar trends are calculated at other values of C_o. In general, a flocculation reactor with a single CFSTR compartment requires the lowest value of $G \times \bar{t}$ to reach a given level of performance. This is because the performance of each compartment depends on the product $[(G \times \bar{t})/m]\Phi_i$ (see Eqs. 8-39, 8-40, and 8-42 in Table 8-10). In the initial compartments of a multiple-compartment flocculator the magnitude of Φ_i is lower than in a single-compartment reactor, and therefore, a larger value of $(G \times \bar{t})/m$ is required to achieve a given level of overall performance.

The curves in Fig. 8-22 were plotted using the equations in Table 8-10 to illustrate the effect of flocculator compartmentalization and the influent primary-particle concentration on flocculator performance for a given value of $G \times \bar{t}$. The primary-particle and floc density were 2.5 and 1.02 g/cm^3 and the G value in each compartment was 25 s^{-1}. The product $G \times \bar{t}$ was constant and equal to 90,000. For the lowest influent primary-particle concentration, $C_o = 10$ mg/L, $m = 5$ yields maximum performance. The fraction of primary particles

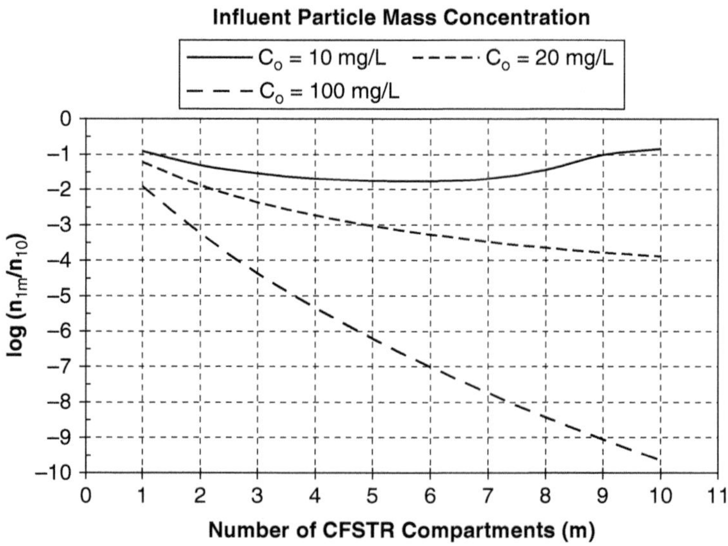

FIGURE 8-22 Effect of influent primary particle concentration and the number of completely mixed compartments on flocculator performance, $G \times T = 90{,}000$, where T is the overall mean fluid detention time. The primary particle and floc density are 2.5 and 1.02 g/cm^3 and the G value in each compartment is 25 s^{-1}.

remaining is approximately 0.12 when $m = 1$ and 0.017 when $m = 5$. For $C_o = 20$ and 100 mg/L and $m < 10$, flocculator performance increases with increasing compartmentalization. For a given number of compartments, the fraction of primary particles remaining in the flocculator effluent decreases with increasing influent primary-particle concentration.

Floc Disaggregation. Floc disaggregation can apparently influence flocculator performance, especially when the mixing intensity is high ($G > 100$ s^{-1}) and the flocs have grown to a significant size. Investigators have observed with batch reactors, using a high intensity and relatively long duration of mixing, that the flocs tend to approach a constant size distribution. One explanation for this behavior is that with time the surfaces of the flocs and primary particles change (chemically and physically), and the floc suspension becomes restabilized. In another explanation it is assumed that a dynamic equilibrium is eventually reached between the rates of particle aggregation and disaggregation. A review of floc strength and breakage by Jarvis et al. (2005) concludes that smaller flocs tend to have greater strength than larger ones and that the use of polymeric flocculation aids tends, in general, to strengthen chemical flocs. Useful models of floc breakage await a better understanding of the internal composition and bonding of the floc structure.

According to a number of investigations conducted in the 1970s (Spielman, 1978; Argaman and Kaufman, 1970; Parker et al., 1972), the principal mechanisms of disaggregation or floc breakup are (1) surface erosion of primary particles from the floc and (2) fracture of the floc to form smaller daughter aggregates.

Argaman and Kaufman (1970) considered both particle aggregation and floc erosion in deriving an equation similar to Eq. 8-29 for predicting the performance of a single-compartment, completely mixed, continuous-flow flocculator. The following first-order rate expression was used to describe the disappearance and formation of primary particles by aggregation and erosion mechanisms:

$$r = -k_a GC + k_b G^2 \tag{8-43}$$

where C is the mass concentration of primary particles, k_a is the agglomeration rate constant, and k_b is the floc breakup (disaggregation) coefficient. For a single-compartment, completely mixed flocculator reactor, Eq. 8-43 yields the following flocculator performance equation:

$$\frac{n_{11}}{n_{10}} = \frac{1 + k_b G^2 \bar{t}}{1 + k_a G \bar{t}} \tag{8-44}$$

The effect of including the erosion term in the rate expression is illustrated by Fig. 8-23, where flocculator performance curves are shown for $k_a = 5 \times 10^{-5}$ and $k_b = 1.0 \times 10^{-7}$ s, and $k_b = 0$ s. With $k_b = 0$ s, the plotted curve is essentially the same as that obtained using Eqs. 8-31, 8-37, and 8-38.

According to Fig. 8-23, including floc disaggregation in the continuous-flow flocculator performance equation yields an optimum mixing intensity, a result that differs from that obtained using Eq. 8-36. Also, when floc erosion is included, constant $G \times \bar{t}$ does not yield constant performance.

Equation 8-44 is appropriate only for completely mixed reactors, where each compartment is populated by a floc suspension of constant size distribution. It is not appropriate to use it (or Eq. 8-36) to derive an expression for flocculation in a plug flow reactor where the floc size distribution is not constant with time.

Floc Size and Density. Experiments have shown that the floc size distribution is a function of the intensity of the turbulence and the structural characteristics of the flocs. A number of investigators (Parker et al., 1972; Mühle and Domasch, 1991; Tambo and François, 1991)

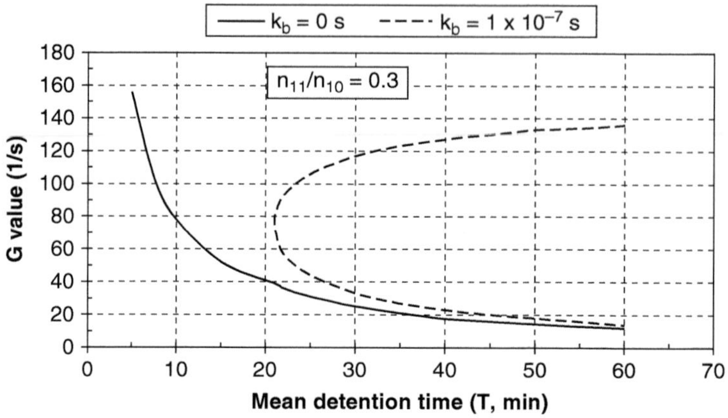

FIGURE 8-23 Flocculator performance curves illustrating the significance of the floc disaggregation term in Eq. 8.43 (G versus T for $n_{11}/n_{10} = 0.3$, $k_b = 1.0 \times 10^{-7}$ s and $k_b = 0$ s).

have suggested the following simple relationship between a characteristic floc size, d_f (e.g., the "maximum" diameter), and G for a steady state condition in which the rate of disaggregation equals the aggregation rate, where the exponent b is a positive integer. According to Eq. 8-45, the steady state floc size decreases as the mixing intensity increases.

$$d_f = \frac{P}{G^b} \tag{8-45}$$

The coefficient P in Eq. 8-45 has been related to the strength of the floc (Parker et al., 1972; Tambo and François, 1991). According to Parker et al., the magnitude of the exponent b depends on the size of the flocs relative to the characteristic size of the turbulent eddies that cause the erosion of primary particles or floc breakage. For flocs that are large relative to the Kolmogoroff microscale of the turbulence, $b = 2$, and for flocs smaller than this scale, $b = 1$. For the floc breakage mechanism, in both turbulence scale regimes, $b = 0.5$. Some experimental investigations (Argaman and Kaufman, 1970; Lagvankar and Gemmell, 1968) have reported the value of the exponent b to be 1.0; that is, the characteristic size of the flocs is inversely proportional to G. Equations similar in form to Eq. 8-45 have been derived using theoretical approaches for floc disruption in turbulent flow (Tambo and François, 1991; Mühle, 1993). The removal of flocs in most separation processes, such as sedimentation, is determined by floc density, and floc density is related to floc size and composition.

It is important to note that the separation process used for floc removal determines what constitutes the best type of floc. For example, with sedimentation the flocs should be relatively large and of high density to maximize their settling velocity. With flotation the flocs should be about 25–50 μm (see Chap. 9) and of low density. The effectiveness of a granular bed filter in a direct filtration system would be increased by relatively small but filterable flocs that can penetrate the filter bed, distributing the deposit over the depth of the bed and minimizing the headloss increase per unit of deposited particles.

Using a computer simulation of the aggregation process, Vold (1963) determined the following inverse relationship between floc buoyant density ($\rho_f - \rho_w$) and diameter, d_f (of a circle with equal projected area):

$$\rho_f - \rho_w = B d_f^{-0.7} \tag{8-46}$$

This relationship, which shows that the floc density tends to decrease as the floc size increases, is in general agreement with experimental results (Lagvankar and Gemmell, 1968). Equations 8-45 and 8-46 considered together suggest that as the mixing intensity used in the flocculator is increased, the steady state, characteristic floc size will decrease (Eq. 8-45) and the floc buoyant density (Eq. 8-46) will increase.

According to Eqs. 8-36 to 8-38, increasing the floc buoyant density tends to have an adverse affect on flocculator performance. For example, for a primary particle density of 1.5 g/cm^3 and with the floc buoyant density increasing from 2×10^{-3} to 1×10^{-2} g/cm^3, the fraction of primary particles remaining in the continuous-flow flocculator effluent increases from 1.6×10^{-5} to 2×10^{-3}. This result is obtained because as the floc buoyant density increases Φ in Eq. 8-38 and the flocculation rate constant k in Eqs. 8-36 and 8-37 decrease. It suggests that the beneficial effect of increasing the mixing intensity on flocculator performance may be offset to some extent by the tendency for higher mixing intensities to produce smaller (see Eq. 8-45), less voluminous, that is, higher-density, flocs (see Eq. 8-46).

Mathematical models of the flocculation process, especially those which use the population balance equation, require assumptions about the physical characteristics of the aggregates formed. These assumptions are important because they shape the calculations used to determine how the aggregate will interact with other particles (Lee et al., 2002) and stationary surfaces such as filter grains, how it will be affected by disruptive forces (Jarvis et al., 2005), and how it will behave in floc separation processes such as sedimentation and flotation (Lee et al., 2000) and be detected in floc measurement instrumentation (Gregory, 2008).

If N spherical primary particles of diameter d_o, density ρ_o, and mass m_o form an aggregate, its mass will be $N \times m_o$, but the determination of its "size" requires assumptions about its geometry, that is, how the primary particles are spatially arranged in the structure of the floc. If the spherical primary particles in the floc are arranged in a linear string the "size", that is, length of the floc, will be $N \times d_o$. However, if the primary particles aggregate by coalescence (like oil droplets) and the aggregate of N primary particles is spherical in shape, then its diameter is given by

$$d = \left[\frac{6Nv_o}{\pi}\right]^{1/3} \tag{8-47}$$

where $v_o = m_o/\rho_o$, the volume of each primary particle. This equation also applies if the primary particles are arranged in a compact spherical shape with constant porosity ε; the quantity in the brackets must be replaced by $6Nv_o/\pi\varepsilon$. This example is an application of Euclidean geometry.

Fractal geometry provides an alternative approach to floc characterization in the flocculation process (Lee et al., 2000; Somasundaran and Runkana, 2003). However, true fractal structure requires a hierarchy of space sizes within the aggregate and uniform fractal dimensionality throughout the floc. Aggregates have a random structure and, logically, varying fractal dimensionality. It might be better to use the term *power law relation* instead of *fractal geometry* for floc particle characterization, but *fractal geometry* is widely used in the literature.

Flocs formed by aggregation processes tend to have porous and irregular structures which change as the aggregates increase in size (Chakraborti et al., 2003). As these irregular objects grow it becomes increasingly difficult to quantify their "size." Fractal geometry has been used to characterize the structural characteristics of particle aggregates (Meakin, 1988; Lee et al., 2000; Somasundaran and Runkana, 2003). As a fractal object, the "size" d of the aggregate is related to its mass ($N \times m_o$) by

$$d \propto (Nm_0)^{1/F} \tag{8-48}$$

where F in the exponent is called the *fractal dimension* (or *mass fractal dimension*). As a limit, if a floc particle is formed by coalescence (the spherical oil droplet described above) or if the primary particles form compact flocs of nearly spherical shape and uniform porosity, then F is equal to 3, the expected magnitude for a Euclidean object. In general, when aggregates of irregular shape are formed from solid primary particles, the magnitude of F can be considerably smaller than 3. Its magnitude decreases as the aggregate structure becomes more open and irregular. Jiang and Logan (1991) observed that aggregates formed by shear (velocity gradient) flocculation had fractal dimensions greater than 2.4 whereas aggregates formed from differential sedimentation (a less energetic mode of interaction) had values in the 1.6 to 2.3 range. According to Elimelech et al. (1995), fractal dimensions in the range 1.6 to 2 are typical for computer model simulation results and model systems.

The exponent S in the relationship between buoyant density and aggregate size is related to the fractal dimension by the equation $F = 3 + S$. According to Eq. 8-48, for the flocs formed by Vold (1966) using computer simulation, $S = -0.7$, and therefore, in this case F is $3 - 0.7 = 2.3$.

$$\rho_f - \rho_w = Bd_f^S \tag{8-49}$$

Floc Density—Significance of Enmeshment by Hydroxide Precipitates. When aluminum and ferric coagulants are used at concentrations in the enmeshment range, the metal hydroxide precipitate can significantly influence the floc volume concentration and buoyant density as well as flocculator performance. This effect is especially significant at low influent primary-particle concentrations. Letterman and Iyer (1985) determined that, for alum and an enmeshment condition, Φ is given by

$$\Phi = \frac{S}{\rho_p} + 120[Al_p] \tag{8-50}$$

where S is the mass concentration of primary particles of density ρ_p that are enmeshed in the precipitate and $[Al_p]$ is the molar concentration of Al in the precipitate. The coefficient, 120 L/mole Al, was determined by measuring the buoyant density of flocs as a function of the ratio of Al precipitate to primary-particle mass.

For aluminum salt coagulants, the buoyant density of the flocs is determined by the ratio of the primary-particle mass concentration and the alum dosage. An empirical relationship, derived by Letterman and Iyer (1985) using floc-density measurements by Lagvankar and Gemmell (1968), is

$$\rho_f - \rho_w = 4.5 \times 10^{-3} \left[1 + \frac{S}{Q}\right]^{0.8} \tag{8-51}$$

where Q is the alum ($Al_2(SO_4)_3 \cdot 18\ H_2O$) dosage. Both S and Q are in mg/L and $\rho_f - \rho_w$ is in g/cm^3. When $S/Q \ll 1$, $\rho_f - \rho_w$ is a minimum and approximately equal to 4.5×10^{-3} g/cm^3, the buoyant density, in these experiments, of the aluminum hydroxide precipitate. For $S/Q \gg 1$, the flocs' buoyant density is roughly proportional to S/Q, the ratio of floc incorporated primary particles and amount of metal hydroxide precipitate. Since the primary particles in Lagvankar and Gemmell's experiments were kaolin clay platelets, Eq. 8-51 pertains only when the primary particles are of this type. There is evidence that the chemical makeup of the solution affects the structure and density of the metal hydroxide precipitate, and this factor must also be considered.

According to Eq. 8-51, the amount of higher-density particulate material that is enmeshed in the lower-density precipitate matrix has a significant effect on the density of the flocs. Decreasing the amount of coagulant relative to the concentration of particulate matter, (increasing S/Q) increases the density of the flocs. Ballasted flocculation uses fine sand added during the rapid mix to increase the density of the flocs (Ghanem et al., 2007). However, as indicated by Eq. 8-50, reducing the amount of coagulant also reduces the floc volume concentration, and according to Eqs. 8-36 to 8-38, this general reduction can decrease the performance of the flocculator. Of course, as the coagulant concentration becomes less than the amount needed for suspension destabilization and NOM removal, the tendency for particles to aggregate and form flocs will become negligible.

The effect of metal hydroxide precipitate on the performance of a flocculation reactor is especially significant when the influent particle concentration is low. This important relationship is illustrated by the following example.

Example 8-3 Flocculator Performance–Significance of the Metal Hydroxide Precipitate and Enmeshment

Problem Statement Example calculations with the flocculator performance equations described previously show that with low influent particle concentrations addition of a hydrolyzing metal salt coagulant and enmeshment in a metal hydroxide precipitate can significantly enhance aggregation efficiency.

Solution This example assumes that the flocculator is a single compartment. Equations 8-36 to 8-38, 8-50, and 8-51 were used with $C_o = S = 5$ mg/L, $G \times \bar{t} = 1 \times 10^5$, ρ_p = 2.5 g/cm^3, and Q = 20 mg/L. For aggregation in the absence of precipitate ($Q = 0$), it was assumed that the flocs have a density of 1.01 g/cm^3, a reasonable density for particles destabilized with a polyelectrolyte, and Eq. 8-38 was used in place of Eq. 8-50 to calculate Φ.

For a low influent particle concentration ($C_o = S = 5$ mg/L), the calculated Φ (Eq. 8-50) for flocs formed in the presence of aluminum hydroxide precipitate is 7.2×10^{-3} L/L and significantly less, 3×10^{-4} L/L, for flocs formed without a precipitate matrix (Eq. 8-38). Consequently, the fraction of primary particles remaining in the flocculator effluent is 0.005 with Al and significantly higher, 0.10, without Al.

When C_o is increased to 100 mg/L and all other parameters are held constant, the fraction of primary particles remaining is approximately 0.005 for both conditions. The higher influent primary-particle concentration yields a floc volume concentration (6×10^{-3} L/L) which does not require augmentation with precipitate to obtain a high level of primary-particle aggregation in a simple, single-compartment flocculator. ▲

Effect of Temperature on Coagulation and Flocculation

The temperature of the water can have a significant effect on coagulation and flocculation (Hanson and Cleasby, 1990; Kang and Cleasby, 1995). In general, the rate of floc formation and the efficiency of primary-particle removal decrease as the temperature decreases. The negative effect is greatest with dilute suspensions. As discussed in a previous section, the solubility of $Al(OH)_3$ decreases (see Fig. 8-19a) and the pH of minimum $Al(OH)_3$ solubility increases (see Fig. 8-19b) with decreasing temperature. The rate of hydrolysis and metal hydroxide precipitation, and the rate of hydrolysis product dissolution or reequilibration (as in the disappearance of a polynuclear species when the product solution is diluted in the water to be treated) decrease with decreasing temperature. At lower temperatures polynuclear species tend to persist for a longer period of time.

Temperature also alters the distribution of kinetic energy over the scale of fluid motion in a turbulent flow field. Kang and Cleasby (1995) concluded that chemical factors are more significant than fluid motion effects, but Morris and Knocke (1984) presented evidence that physical rather than chemical kinetic factors are behind the effect of temperature on coagulation and flocculation performance. In their work they observed a significant effect of temperature on the size distribution of aluminium hydroxide floc.

Evaluating Coagulant Performance Using Bench-Scale Testing

It is difficult to predict a priori how a coagulant will perform in a given application. Consequently, it is common to use bench-scale testing to evaluate the performance of chemical coagulants. The jar test procedure is the oldest and most widely used method.

The bench-scale jar test procedure typically includes the following general steps:

1. Coagulant solution addition during a short period (typically 1 min in duration) of high-intensity mixing (rapid mixing)
2. A period of low-intensity (relatively low impeller speed) mixing for floc formation
3. A floc separation step
4. Sample withdrawal and measurement of residual quantities (e.g., turbidity, UV_{254} (DOC or TOC), particle count, coagulant cation concentration, etc.) and solution acid/base chemistry (e.g., pH and alkalinity)

The USEPA includes a recommended jar test procedure in their guidance manual on enhanced coagulation and enhanced precipitative softening (USEPA, 1999), as does the AWWA in their Manual of Practice, M37, *Operational Control of Coagulation and Filtration Processes,* third edition (AWWA, 2010). AWWA also supplies a video on jar testing (AWWA, 2009b).

Jar testing is typically done using a multiple batch reactor device called a *jar test apparatus* or *gang stirrer*. A number of these are commercially available. In the typical jar test unit, the mixing impellers all turn at the same rotational speed and it is difficult to add the coagulant and rapid mix for a given period of time and then immediately begin the slow mixing step unless the coagulant solution can be added rapidly and simultaneously to all the containers. In an alternative setup an independent rapid-mix unit is used. Each filled jar test container is put in the rapid-mix unit, and the coagulant solution is added. At the end of the rapid-mix period, the container is immediately transferred to the jar test unit and the low-intensity mixing step commences without delay. In any case, a coagulant-treated jar test container should always move without delay between the rapid- and slow-mixing stages. It should never be set aside to wait for a place in the jar test apparatus. If possible, jar tests should be conducted using water that is the same temperature as the water in the full-scale facility.

A disposable-tip microdispenser capable of adding microliter volumes is a convenient and effective way to add coagulant solution in the rapid-mix step. Jar test results tend to be more consistent and reproducible when the same undiluted commercial strength coagulant solution is used in all the jar test containers. If a significant range of coagulant dosages is to be tested, two or more microdispensers of adjustable capacities (e.g., one with a 1- to 10-μL capacity and one with a 10- to 100-μL capacity) may be needed. The use of laboratory-made stock or commercial coagulant solutions of different concentrations obtained by dilution before addition should be avoided especially with PACl and other prehydrolyzed products because the rate of coagulant solution dispersal and the rate of coagulant cation hydrolysis vary with the volume of solution added and the cation concentration.

A number of floc particle separation steps are used in the jar test procedure depending on the type of system to be simulated. The most common method is sedimentation. The jar test container with flocculated suspension is removed from the slow-mixing apparatus and covered (to minimize the effect of air currents in the room), and the floc particles are allowed to settle in a time period that typically ranges from 15 to 60 min. The residual particles are measured in a quantity of supernatant that has been carefully withdrawn from a certain depth in the container. The approximate sedimentation overflow rate (in m^3/m^2-h) for a given settling time S and sampling depth D is equal to D/S. For example, if the sample is withdrawn from a depth of 10 cm (0.1 m) in the jar test container and the settling period is 20 min (0.33 h) the corresponding overflow rate is (0.1 m/0.33 h) = 0.3 m^3/m^2-h (0.12 gpm/ft^2).

Floc particle separation with filter membranes is sometimes used, either alone or after sedimentation, to obtain a water sample for measuring residuals such as DOC (dissolved organic carbon), UV absorbance (UV_{254}), and coagulant cation concentration (Al(III) or Fe(III)). Membrane filtration, compared to, for example, sedimentation, tends to give results that are in better agreement with those from the full-scale facility, but results should be tested for agreement by a comparison with the full-scale system. Membrane filtration is not an appropriate floc separation technique for a plant that uses sedimentation or flotation.

Manufactured membrane filter units with a membrane enclosed within a plastic housing that attaches to a syringe are convenient. The sample is drawn quickly into the syringe and then immediately forced through the filter unit into the sample container. Care should be used in selecting the effective pore size and material of the membrane because some membranes leach significant amounts of organic carbon. Pore sizes that are too small may lead to rapid clogging that limits the amount of sample filtered.

Dissolved air flotation is also used for floc separation. A bench-scale apparatus that resembles a jar test apparatus is commercially available (ECE, 2009). Water supersaturated with air is mixed with the flocculated suspension in the desired proportions in a jar-test-type container that contains the flocculated suspension, and the clarified water is sampled from the bottom of the jar and used to measure the residuals. Jar test results from studies that used a flotation unit for floc separation have been reported in the literature (e.g., Malley and Edzwald, 1991).

Jar tests do not tell us exactly how the coagulant will perform in the full-scale system. One reason for discrepancies is that even with the same apparent chemical conditions, the pH of the treated water of the jar test is significantly higher than the pH in the full-scale plant because there is more postcoagulation dissolved carbon dioxide in the full-scale system compared to the jar test vessels. This CO_2 difference is typically caused by the higher rate of mass transfer of carbon dioxide from the solution to the atmosphere in the jar test compared to the full-scale system. The interfacial area per volume (which is proportional to the depth of the water) in a 1- or 2-L jar test container is much greater than that of the much deeper and many thousands of liters tank of the full-scale treatment system.

The mixing intensity or G value or G (in s^{-1}) used in a jar test should be as close as possible to conditions in the full-scale system. The magnitude of the G in the jar test apparatus (and in the full-scale system if mixing is provided by a single rotating impeller with a vertical shaft) can be estimated using the following relationship:

$$G = K N^{1.5} \tag{8-52}$$

where N is the rotational speed of the impeller in revolutions per second and K (in $s^{0.5}$) is given approximately by

$$K = \left(\frac{D^5 k}{V \nu}\right)^{0.5} \tag{8-53}$$

where D is the diameter of the impeller, ν is the kinematic viscosity of the water, V is the volume of water in the vessel, and k is a dimensionless constant that varies between 0.1 and about 5 depending on the geometry of the impeller. This equation assumes the vessel is shaped or baffled so that fluid rotation and vortex formation is minimal and that the mixing regime is fully turbulent; that is, the impeller Reynolds number $ND^2/\nu > 1 \times 10^4$.

Example 8-4 Estimating the G Value in a Jar-Test-Type Rotating Impeller Mixing Device

Problem Statement Estimate the G value in a 1-L baffled vessel with a single paddle impeller (8 cm long) rotating at 60 rpm. The temperature of the water is 20°C.

Solution According to Eqs. 8-52 and 8-53 the G value for a 1-L baffled vessel (V = 1000 cm^3) with a flat paddle impeller ($k = 2$ and $D = 8$ cm), rotating at 60 rpm (1 rps) in water with a kinematic viscosity of 1.1×10^{-6} m^2/s, is given by

$$K = \left[\frac{(0.08^5)(2)}{(0.001)(1.1\times10^{-6})}\right]^{0.5} = 77.2\,\mathrm{s}^{0.5} \quad \text{and} \quad G = 77.2\,(1)^{1.5} = 77.2\,\mathrm{s}^{-1}$$

The Reynolds number for this case is $DN^2/\nu = 0.08$ m $(1\ \mathrm{s}^{-1})^2/1.1 \times 10^{-6}$ m^2/s $= 7.3 \times 10^4$ which is greater than 1×10^4 so the mixing is fully turbulent and Eq. 8-53 tends to apply. ▲

A more thorough discussion of mixing in the flocculation process can be found in the book by MWH (2005).

Monitoring the Particle Aggregation Process

An alternative way of evaluating the performance of coagulants is to measure the physical characteristics of the particle aggregates as they form or are disrupted in the flocculation process. Gregory (2008) prepared a comprehensive review of the instrumental methods available for monitoring the flocculation process. A focus of the review is the turbidity fluctuation technique developed by Gregory and Nelson (1986).

In the turbidity fluctuation technique (Gregory, 2008) a low flow rate of suspension is continuously drawn from the flocculation reactor through a small transparent tube where the particles are illuminated by a narrow light beam. The transmitted light beam (the classical turbidity of the suspension) is directed to a photodiode tube. The output from the photodiode fluctuates as the particles move through the light beam. The fluctuations vary with the number, concentration, and size of the aggregates. The signal is electronically modified so that it varies in a sensitive way with the state of aggregation of the suspension. As aggregation increases the signal increases allowing the flocculation rate and extent to be monitored in a semiquantitative way. The method can be used to compare the performance of coagulants, and by varying the mixing intensity in the flocculation reactor, it can be used to evaluate aggregate strength, that is, the sensitivity of the particle aggregates to disruption by the fluid motion. A typical experimental setup for the turbidity fluctuation method is shown in Fig. 8-24.

In Fig. 8-25 the turbidity fluctuation instrument's output, called the *flocculation index*, is plotted versus time as filtered tap water is dosed with a ferric salt coagulant. The flocculation reactor impeller rotational speed is set at 100 rpm. A 120- to 180-sec lag period occurs after

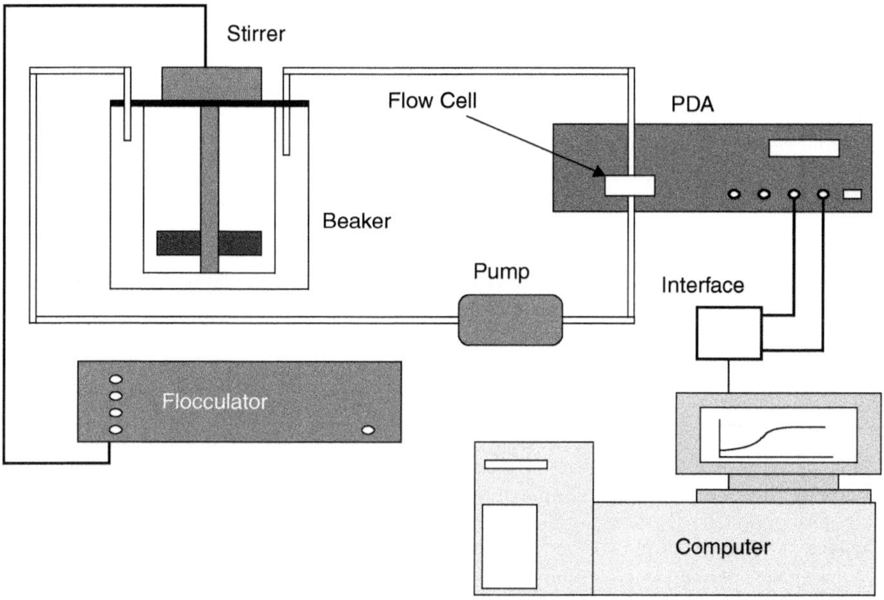

FIGURE 8-24 Typical experimental set-up for the turbidity fluctuation method for monitoring the floc formation and disruption processes. (*Source: Gregory, 2008. Reprinted with permission from Elsevier.*)

coagulant addition because the hydrolysis products are initially too small to scatter sufficient light. At about 250 s the flocculation index begins to increase, and by about 700 s. it reaches a maximum indicating that the ferric hydroxide flocs have grown to a steady state size distribution. At 900 s. the impeller speed is increased to 400 rpm, and the flocculation index decreases and then increases as the aggregates are broken and partially reformed.

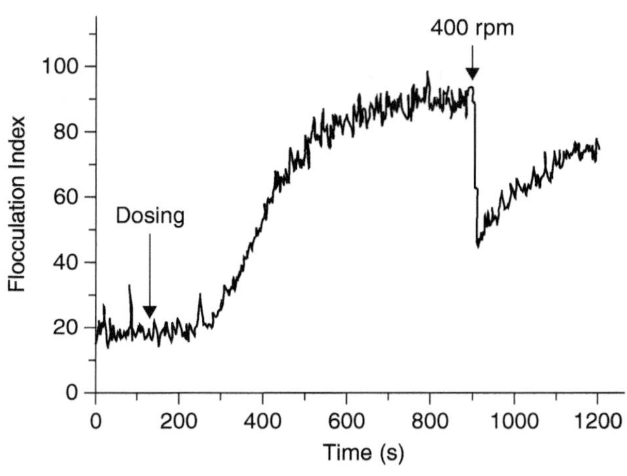

FIGURE 8-25 Example results from the turbidity fluctuation method. (*Source: Gregory, 2008. Reprinted with permission from Elsevier.*)

Electrokinetic Measurements

Electrokinetic measurements are used to monitor the effect of coagulants and changes in solution chemistry on particle surface chemistry and the stability of particle and precipitate suspensions. Two types of electrokinetic measurements, electrophoretic mobility (zeta potential is then calculated and displayed by some instruments) and streaming current are used in water treatment practice to control the addition of coagulants and monitor the conditions that affect coagulant performance. Streaming current is sometimes used to simply trigger an alarm when the coagulant feed stops functioning.

Electrophoretic Mobility Measurements. The engineering literature contains numerous references to the use of electrophoretic mobility (EM) or zeta potential (ZP) measurements as a coagulation process control technique. Many of these papers describe attempts to correlate ranges of EM values or changes in the EM, such as sign reversal, with the efficiency of particle removal by flocculation followed by sedimentation and filtration.

Electrophoresis is an electrokinetic effect and is, therefore, explained by the same fundamental principles as other electrokinetic phenomena such as streaming current (or streaming potential), sedimentation potential, and electro-osmosis. In electrophoresis, particles suspended in a liquid are induced to move by the application of an electric field across the system. This technique has been used by colloid chemists for many years to determine the net electric charge or near-surface (zeta) potential of particles with respect to the bulk of the suspending phase (Hunter, 1981; Anonymous, 2000).

The use of EM measurements as a method to control the application of coagulant chemicals in solid-liquid separation systems has been known for many years. However, while there have been convincing advocates for the use of EM measurements for coagulation process control, inconsistent and difficult-to-interpret results and the time-consuming nature of the EM determination appear to have limited the widespread use of EM measurements in routine treatment plant operation.

A number of different techniques are used to determine particle EM. The most important of these in water treatment applications has been microelectrophoresis. In microelectrophoresis, the suspension is contained in a small-diameter glass or plastic tube that has, in most cases, a round or rectangular cross section. An electric field is applied across the contents of the tube in the axial direction using a stable, constant-voltage power supply and inert (e.g., platinized platinum) electrodes inserted in sealed fluid reservoirs at the end of the tube. When the voltage is applied, the particles tend to migrate in the axial direction. The EM is determined by measuring the average velocity of particle migration and dividing this by the voltage gradient across the cell. Unless special procedures are used, the measured velocity must be corrected for the particle movement caused by electroosmotic flow in the electrophoresis cell.

Streaming Current Measurements. The original *streaming current detector* (SCD) was patented by Gerdes (1966) in 1965 and variants of this device have been used in water treatment plants for coagulant dosage control for a number of years. The electronic output from an SCD has been shown by Dentel and Kingery (1988) to be proportional to the net charge on the surfaces of the particles that have been treated with the coagulant chemical.

The cross section of a widely used type of SCD is shown in Fig. 8-26. The purpose of the sample chamber is to contain the coagulant-treated suspension. Within the chamber is a reciprocating piston that moves vertically inside a cylinder. The movement of the piston causes water to flow inside the annular space between the cylinder and piston and to be pumped through the sample chamber. The streaming current is detected by electrodes attached inside the cylinder.

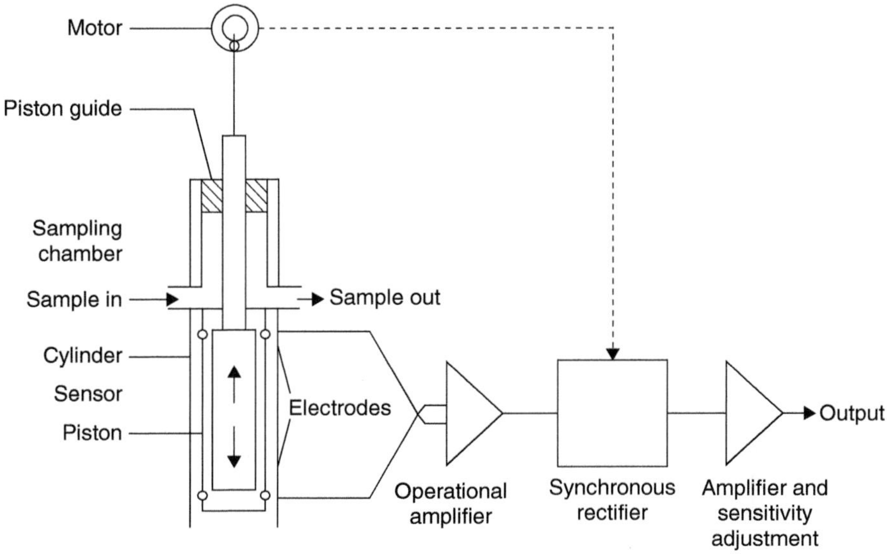

FIGURE 8-26 Cross section of a widely used type of streaming current detector.

The streaming current is determined by the charged particles that attach to the walls of the stationary cylinder and the moving piston. As water flows back and forth cyclically in the annulus, some of the ions in the electrical double layer next to the charged particles are transported with the flow. This displacement of electric charge by the movement of the fluid past the stationary particles creates the sinusoidal current that tends to flow between the electrodes that are in contact with the solution within the cylinder. The magnitude of the current depends on the amount of charge on the attached particles. If the surfaces are not covered by particles, it is also affected to some extent by the charge on the surfaces of the piston and cylinder. The current is amplified, rectified, and time smoothed by circuitry in the instrument. The processed signal is called the *streaming current* (*SC*).

Interpreting EM and SC Measurements. The rate of particle interaction for orthokinetic flocculation is determined (approximately) by the product of the collision efficiency factor α and the volume concentration of the suspension Φ (see Eq. 8-27). Theoretical relationships show that α is related to the EM of the particles in the suspension (Anonymous, 2000). As the EM approaches zero, α should approach a maximum value of approximately 1. (This also applies to the SC if the instrument has been adjusted to give SC $\approx$ 0 at the dosage that yields EM = 0.) As discussed, the magnitude of Φ depends on the volume concentration of the particles and precipitate in the suspension.

EM and SC measurements effectively predict the rate of flocculation when Φ is relatively constant, and the rate is influenced only on the magnitude of α. When the addition of a coagulant, typically a HMS coagulant, affects both α and Φ, the interpretation of EM and SC measurements can be difficult.

The data plotted in Fig. 8-27 were obtained using jar test measurements and a suspension containing 50 mg/L silica and 3×10^{-4} M sulfate. To maintain a constant

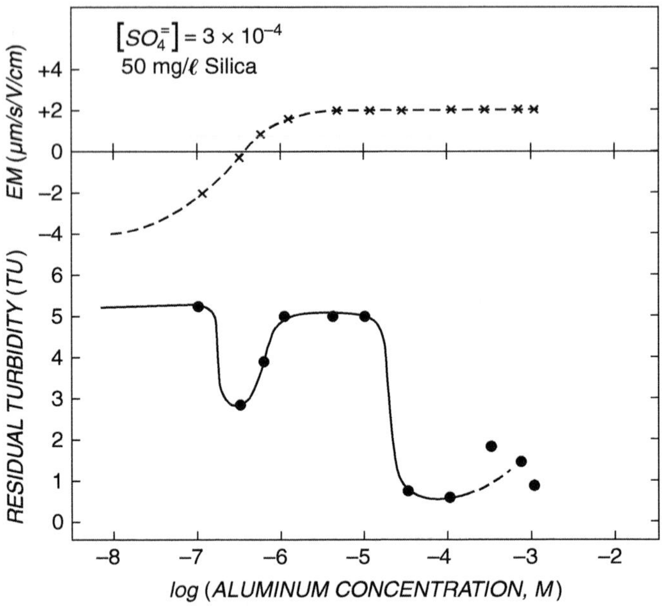

FIGURE 8-27 Jar test data showing the effect of the aluminum nitrate dosage on the electrophoretic mobility and residual turbidity after flocculation and sedimentation for 50 mg/L silica, pH = 6, and $[SO_4^=] = 3 \times 10^{-4}$.

sulfate concentration, the aluminum has been added as $AlCl_3$. This figure illustrates an important aspect of hydrolyzing metal coagulants that can make the application of EM measurements for dosage control difficult. Note that between aluminum concentrations of about 10^{-6} and 10^{-5} (0.027 and 0.27 mg Al/L), the residual turbidity is high suggesting that restabilization has occurred. However, as the aluminum concentration is increased beyond 2×10^{-5} M (0.54 mg Al/L), the residual turbidity decreases to values lower than the minimum observed at the point of charge neutralization (EM = 0). The EM measurement remains at +2 and gives no indication that flocculation efficiency should improve with increasing aluminum concentration beyond the point of charge neutralization.

The results presented in Fig. 8-27 can be explained by a consideration of the effect of the aluminum concentration on the rate Φ of flocculation and on the magnitude of the product α. Apparently under the experimental conditions used to obtain the results plotted in Fig.8-27 and at aluminum concentrations between 10^{-6} (0.027 mg Al/L) and 10^{-5} (0.27 mg Al/L), the magnitude of α and the product $\alpha\Phi$ are not high enough to yield efficient flocculation. However, as the aluminum concentration is increased beyond 10^{-5} (0.27 mg Al/L), the magnitude of Φ increases until the modest value of α is compensated for and efficient flocculation is obtained.

The streaming current measurement has been found to be a useful technique for controlling the HMS coagulant dosage in the removal of NOM. According to Dempsey (1994) the streaming current detector response is sensitive to conditions that lead to effective NOM removal. However, Dentel (1994) has noted that while streaming current appears to have promise, problems may occur with low turbidity, high-NOM-concentration waters, especially at high pH.

ABBREVIATIONS

AHP	aluminum hydrolysis product
ANSI	American National Standards Institute
AWWA	American Water Works Association
CFSTR	continuous-flow stirred tank reactor (tanks in series)
DBP	Disinfection By-Product
DBPR	Disinfection By-Products Rule
DLVO	Derjaguin, Landau, Verwey, Overbeek
DOC	dissolved organic carbon concentration (mg C/L)
EAC	effective acid content (meq/mg M)
EM	electrophoretic mobility
epiDMA	epichlorohydrin dimethylamine
HAA5	haloacetic acid group
HMS	hydrolyzing metal salt
NOM	natural organic matter
NSF	National Sanitation Foundation
ntu	nephelometric turbidity unit
PACl	polyaluminum chloride
PICl	polyiron chloride
PODR	point of diminishing returns
polyDADMAC	polydiallyl dimethylammonium chloride
SC	streaming current
SCD	streaming current detector
SOC	synthetic organic chemical
SUVA	specific ultra-violet light absorbance
TOC	total organic carbon
TTHM	total trihalomethane
USEPA	U.S. Environmental Protection Agency
ZP	zeta potential

EQUATION NOTATION

A	Hamaker constant
B	basicity of hydrolyzing metal salt coagulant solution (%)
C_{eq}	concentration of organic carbon in the solution at equilibrium
C_T	inorganic carbon concentration (moles C/L)
D	sampling depth
d_1, d_2	particle diameters
d_f	floc diameter (circle with equal projected area)

d_o	diameter
du/dz	velocity gradient
F	fractal dimension (or mass fractal dimension)
G	G value or rms velocity gradient
$K_1, K_2, b, a, x_1, x_2, x_3$	constants in organic carbon adsorption equations
k_{ij}, k_o	flocculation rate constants
K_{so}, K_{sp}	solubility constants
K_w	ion product of water
M	aluminum or iron concentration (mg M/L)
M	mass of metal hydroxide precipitate
m_o	mass of particle
n_i, n_j	particle number concentrations
N_{ij}	rate of collision between *i*- and *j*-size particles
P	power input
Q	alum dosage
R	coefficient in Eq. 8-12 (mg M/mg C)
S	mass concentration of clay
s	separation distance between interacting particles
TOC_o	total organic carbon in the source water (mg C/L)
V	volume
$\alpha_t, \alpha_{ij}, \alpha_t$	collision efficiencies
$\beta_{x,y}$	formation constants
ε	energy dissipation rate in the fluid
η	microscale of turbulent fluid motion
μ	dynamic viscosity
ν	kinematic viscosity
ρ_f	floc density
$\rho_f - \rho_w$	floc buoyant density
ρ_o	density of particle
ρ_w	density of water
ψ_A	magnitude of the attractive energy of interaction
ψ_{max}	maximum in potential energy of interaction curve
ψ_R	magnitude of repulsive energy of interaction

REFERENCES

Amirtharajah, A., and Mills, K.M. (1982) Rapid-mix design for mechanisms of alum coagulation, *Journal AWWA*, vol. 74, no. 4, pp. 210–216.

Amirtharajah, A., and O'Melia, C.R. (1990) Coagulation processes: Destabilization, mixing, and flocculation. In *Water Quality and Treatment*, 3d ed., ed. Pontius, F.W., ch. 6. Denver, CO: American Water Works Association, and New York: McGraw-Hill.

Anonymous (2000) *Operational Control of Coagulation and Filtration Processes,* 2d ed., Manual M37. Denver, CO: American Water Works Association.

Archer, A.D., and Singer, P.C. (2006a) SUVA and NOM coagulation using the ICR database, *Journal AWWA*, vol, 98, no. 7, pp. 110–123.

Archer, A.D., and Singer, P.C. (2006b) Effect of SUVA and enhanced coagulation on removal of TOX precursors, *Journal AWWA*, vol. 98, no, 8, pp. 97–107.

Argaman, Y., and Kaufman, W.F. (1970) Turbulence and flocculation, *Jour. Sanitary Eng. Div.*, ASCE, vol. 96, p. 223.

AWWA (American Water Works Association) (1982) Coagulation and Filtration Committee, Survey of polyelectrolyte use in the United States, *Journal AWWA*, vol. 74, pp. 600–607.

AWWA (2010) M37, *Operational Control of Coagulation and Filtration Processes*, 3rd ed. Denver, CO: American Water Works Association.

AWWA (2009a) ANSI/AWWA Standard B403-09, Aluminum Sulfate–Liquid, Ground, or Lump. Denver, CO: American Water Works Association.

AWWA (2009b) *Water Supply Operations: Jar Testing,* Video. ISBN 978-1-58321-610-1, Catalog No. 64083. Denver, CO: American Water Works Association.

AWWA (2003) ANSI/AWWA Standard B408-03, Liquid Polyaluminum Chloride. Denver, CO: American Water Works Association.

Awwa Research Foundation (1991) Mixing. In *Coagulation and Flocculation,* ed. Amirtharajah, A., Clark, M.M., and Trussell, R.R., Awwa Research Foundation report. Denver, CO: Water Research Foundation.

Ball, J.W., Nordstrom, D.K., and Jenne, E.A. (1980) Additional and revised thermochemical data for WATEQ-2 computerized model for trace and major element speciation and mineral equilibrium of natural waters, U.S. Geological Survey Water Resource Investigations. Menlo Park, CA: U.S. Geological Survey.

Baylis, J.R. (1937) Silicates as aids to coagulation. *Jour. Am. Water Works Assoc.*, vol. 29, no. 9, p. 1355.

Becker, W.C., and O'Melia, C.R. (2001) Ozone: Its effect on coagulation and filtration, *Water Science and Technology: Water Supply*, vol. 1, no. 4, pp. 81–88.

Bernhardt, H., Hoyer, O., Schell, H., and Lusse, B. (1985) Reaction mechanisms involved in the influence of algogenic matter on flocculation, *Z. Wass. Abwass. Forsch.*, vol. 18, pp. 18–30.

Bertsch, P.M., and Parker, D.R. (1996) Aqueous polynuclear aluminum species. In *The Environmental Chemistry of Aluminum*, 2d ed., ed. Sposito, G., p. 117. Boca Raton, FL: CRC Press.

Bertsch, P.M., Thomas, G.W., and Barnhisel, R.I. (1986) Characterization of hydroxy aluminum solutions by aluminum-27 nuclear magnetic resonance spectroscopy, *Soil Sci. Soc. Am. J.*, vol. 50, pp. 825–830.

Black, A.P., Birkner, F.B., and Morgan, J.J. (1965) Destabilization of dilute clay suspensions with labeled polymers, *Journal AWWA*, vol. 57, no. 12, pp. 1547–1560.

Bolto, B., and Gregory, J. (2007) Organic polyelectrolytes in water treatment, *Wat. Res.*, vol. 41, pp. 2301–2324.

Bottero, J.Y., Axelos, M., Tchourbar, D., Cases, J.M., Fripiat, J.J., and Fiessinger, F. (1987) Mechanism of formation of aluminum trihydroxide from Keggin Al_{13} polymers, *J. Colloid Interface Sci.*, vol. 117, pp. 47–57.

Bottero, J.Y., and Bersillon, J.L. (1989) Aluminum and iron(III) chemistry: Some implications for organic substance removal. In *Aquatic Humic Substances: Influence on Fate and Treatment of Pollutants*, ed. I. Suffet and P. MacCarthy, pp. 425–442. Washington, DC: American Chemical Society.

Brant, J.A., Lecoanet, H., and Wiesner, M. R. (2005) Aggregation and Deposition Characteristics of Fullerene Nanoparticles in Aqueous Systems, *J. Nanopart. Res.*, vol. 7, pp. 545–553.

Bratby, J. (2006) *Coagulation and Flocculation in Water and Wastewater Treatment,* 2d ed. London: IWA Publishing.

Camp, T.R., and Stein, P.C. (1943) Velocity gradients and internal work in fluid motion, *Jour. of the Society of Civil Engineering*, vol. 30, pp. 219–237.

Camp, T.R. (1955) Flocculation and flocculation basins, *Am. Soc. Civil Eng. Trans.*, vol. 120, pp. 1–16.

Chakraborti, R.K., Gardner, K.H., Atkinson, J.F., and Van Benschoten, J.E. (2003) Changes in fractal dimension during aggregation, *Wat. Res.*, vol. 37, pp. 873–883.

Chandrakanth, M.S., and Amy, G.L. (1996) Effects of ozone on the colloidal stability and aggregation of particles coated with natural organic matter. *Environ. Sci. Technol.*, vol. 30, no. 2, pp. 431–443.

Chappell, R.J. (1996) A method to estimate the performance of metal salt coagulants in removing natural organic material for water treatment, Masters Thesis, Department of Civil and Environmental Engineering, Syracuse University, Syracuse, NY.

Chen, Z., Luan, Z., Fan, J., Zhang, Z., Peng, X., and Fan, B. (2007) Effect of thermal treatment on the formation and transformation of Keggin Al_{13} and Al_{30} species in hydrolytic polymeric aluminum solutions, *Colloids and Surfaces A: Physicochem. Eng. Aspects*, vol. 292, pp. 110–118.

Cheng, W.P., Chi, F.H., and Yu, R.F. (2004) Effect of Phosphate on Removal of Humic Substances by Aluminum Sulfate Coagulant, *Jour. Coll. Inter. Sci.*, vol. 272, pp. 153–157.

Chin, C-J, Yiacoumi, S., Tsouris, C., Relle, S. and Grant, S. B. (2000) Secondary-minimum aggregation of superparamagnetic colloidal particles, *Langmuir*, vol. 16, no. 8, pp. 3641-3650.

Clark, M.M. (1985) Critique of Camp and Stein's rms velocity gradient, *Jour. of Environ. Eng.*, ASCE, vol. 111, no. 6, pp. 741–764.

Clark, M.M., and Flora, J.R.V. (1991) Floc restructuring in varied turbulent mixing, *Jour. of Colloid and Interface Sci.*, vol. 147, no. 2, pp. 407–421.

Cleasby, J.L. (1984) Is velocity gradient a valid turbulent flocculation parameter? *Jour. of Environ. Eng.*, ASCE, vol. 110, no. 5, pp. 875–895.

Cleasby, J.L., Dharmarajah, A.H., Sindt, G.L., and Baumann, E.R. (1989) Design and operations guidelines for optimization of the high-rate filtration process: Plant survey results, Awwa Research Foundation report. Denver, CO: Water Research Foundation.

Davis, J.A., and Leckie, J.O. (1978) Surface ionization and complexation at the oxide/water interface. II. Surface properties of amorphous iron oxyhydroxide and adsorption of metal ions, *Jour. Colloid Interface Sci.*, vol. 67, pp. 90–107.

de Hek, H., Stol, R.J., and DeBruyn, P.L. (1978) Hydrolysis-precipitation studies of aluminum (III) solutions. Part 3. The role of the sulfate ion, *Jour. Colloid Interface Sci.,* vol. 64, pp. 72.–89.

Delichatsios, M.A., and Probstein, R.F. (1975) Coagulation in turbulent flow: Theory and experiment, *Jour. Colloid Interface Sci.*, vol. 51, pp. 394–405.

Dempsey, B.A. (1987) Chemistry of coagulants. In *AWWA Seminar Proceedings, Influence of Coagulation on the Selection, Operation, and Performance of Water Treatment Facilities, Proceedings of AWWA Annual Conference,* Kansas City, MO, June 14, 1987.

Dempsey, B.A. (1989) Reactions between fulvic acid and aluminum: Effects on the coagulation process. In *Aquatic Humic Substances: Influence on Fate and Treatment of Pollutants*, ed. Suffet, I., and MacCarthy, P., pp. 409–424. Washington DC: American Chemical Society.

Dempsey, B.A. (1994) Enhanced coagulation using charge neutralization, *Proceedings of the AWWA Enhanced Coagulation Research Workshop,* Charleston, SC, December 4–6, 1994. Denver, CO: AWWA.

Dempsey, B.A., Ganho, R.M., and O'Melia, C.R. (1984) The coagulation of humic substances by means of aluminum salts, *Journal AWWA*, vol. 76, no. 4, pp. 141–150.

Dennett, K.E., Amirtharajah, A., Moran, T., and Gould, J.P. (1996) Coagulation: Its effect on organic matter, *Journal AWWA*, vol. 88, no. 4, pp. 129–142.

Dentel, S.K. (1988) Application of the precipitation-charge neutralization model of coagulation, *Environ. Sci. Tech.*, vol. 22, pp. 825–832.

Dentel, S.K. (1991) Coagulant control in water treatment, *Critical Reviews in Environmental Control,* vol. 21, no. 1, pp. 41–135. Boca Raton, FL: CRC Press.

Dentel, S.K. (1994) Use of the streaming current detector in enhanced coagulation processes, *Proceedings of the AWWA Enhanced Coagulation Research Workshop,* Charleston, SC, December 4–6, 1994. Denver, CO: AWWA.

Dentel, S.K., and Kingery, K.M. (1988) An evaluation of streaming current detectors, Publication 90536, Awwa Research Foundation report. Denver, CO: Water Research Foundation.

Dentel, S.K., Gucciardi, B.M., Bober, T.A., Shetty, P.V., and Resta, J.J. (1989) Procedures manual for polymer selection in water treatment plants, Publication 90553, Awwa Research Foundation report. Denver, CO: Water Research Foundation.

Derjaguin, B.V., and Landau, L.D. (1941) Theory of the stability of strongly charged lyophobic sols and of the adhesion of strongly charged particles in solutions of electrolytes, *Acta Physico*, URSS, vol. 14, pp. 733–762.

Donose, B.C., Nguyen, A.V., Evans, G.M., and Yan, Y. (2007) Effect of aluminium sulphate on interactions between silica surfaces studied by atomic force microscopy, *Water Research,* vol. 41, pp. 3449–3457.

ECE (2009) EC Engineering, Inc., Edmonton, CAN; available at www.ecengineering.net/html/dbt_daf_batch_tester.html.

Edwards, M. (1997) Predicting DOC removal during enhanced coagulation, *Journal AWWA*, vol. 89, no. 5, pp. 78–89.

Edwards, M., and Benjamin, M.M. (1992) Effect of preozonation on coagulant-NOM interactions, *Journal AWWA*, vol. 84, no. 8, pp. 63–72.

Edzwald, J.K. (1993) Coagulation in drinking water treatment: Particles, organics and coagulants, *Wat. Sci. Tech.*, vol. 27, no. 11, pp. 21–35.

Edzwald, J.K., Upchurch, J.B., and O'Melia, C.R. (1974) Coagulation in estuaries, *Environmental Science and Technology*, vol. 8, pp. 58–63.

Edzwald, J.K., and van Benschoten, J.E. (1990) Aluminum coagulation of natural organic material. In *Chemical Water and Wastewater Treatment*, ed. Hahn, H.H., and Klute, R. Berlin: Springer-Verlag.

Edzwald, J.K, and Kaminski, G.S. (2009) A practical method for water plants to select coagulant dosing, *Jour. New England Water Works Assoc.*, vol. 123, no. 1, pp. 15–31.

Elimelech, M., Gregory, J., Jia, X., and Williams, R. (1995) *Particle Deposition and Aggregation.* Oxford, UK:.Butterworth-Heinemann Ltd.

Eyring, A.M., Weinberg, H.S., and Singer, P.C. (2002) Measurement and effect of tracer element contaminants in alum, *Journal AWWA*, vol. 94, no. 5, pp. 135–146.

Folkard, G. (2002) Development of a naturally derived coagulant for water and wastewater treatment, *Water Supply*, vol. 2, no. 5–6, pp. 89–94.

Frommell, D.M., Feld, C.M., Snoeyink, V.L., Melcher, B., and Feizoulof, C. (2004) Aluminum residual control using orthophophosphate, *Journal AWWA*, vol. 96, no. 9, pp. 99–109.

Gao, B.Y., Yue, Q.Y., Wang, B.J., and Chu, Y.B. (2003) Poly-aluminum-silicate-chloride (PASiC)–A new type of composite inorganic polymer coagulant, *Colloids and Surfaces A: Physicochem. Eng. Aspects*, vol. 229, pp. 121–127.

Gerdes, W.F. (1966) A new instrument—the streaming current detector, *12th Natl. ISA Analysis Instrument Symposium.*

Ghanem, A.V., Young, J.C., and Findley, F.G. (2007) Mechanism of ballasted floc formation, ASCE, *J. Env. Eng.,* vol. 133, no. 3, pp. 271–277.

Goldberg, S., Davis, J.A., and Hem, J.D. (1996) The surface chemistry of aluminum oxides and hydroxides. In *The Environmental Chemistry of Aluminum*, 2d ed., ed. Sposito, G., p. 271. Boca Raton, FL: CRC Press.

Gregory, J. (1978) Effects of polymers on colloid stability. In *The Scientific Basis of Flocculation*, ed. Ives, K.J. The Netherlands: Sijthoff and Noordhoff.

Gregory, J. (2008) Monitoring particle aggregation processes. *Adv. Colloid Interface Sci.*, vol. 147–148, pp. 109–123.

Gregory, J., and Nelson, D.W. (1986) Monitoring of aggregates in flowing suspensions, *Colloids Surf.*, vol. 18, pp. 175–188.

Gu, B., Schmitt, J., Chen, Z., Liang, L., and McCarthy, J.F. (1994) Adsorption and desorption of natural organic matter on iron oxide: Mechanisms and models, *Environ. Sci. Tech.*, vol. 28, pp. 38–46.

Guldbrand, L., Jönnson, B., Wennerström, H. and Linse, P. (1984) Electrical double layer forces. A Monte Carlo study, *J. Chem. Phys.*, vol. 80, pp. 2221.

Hahn, M.W., and O'Melia, C.R., (2004) Deposition and re-entrainment of Brownian particles in porous media under unfavorable chemical conditions: Some concepts and applications, *Environ. Sci. Technol.*, vol. 38, pp. 210–220.

Han, M., and Lawler, D.F. (1991) Interaction of two settling spheres: Settling rates and collision efficiencies, *Journal of Hydraulics Division, Am. Soc. Civil Engrs.*, vol. 17, no. 10, pp. 1269–1289.

Han, M., and Lawler, D.F. (1992) The (relative) insignificance of G in flocculation, *Journal AWWA*, vol. 84, no. 10, pp. 79–91.

Hanson, A.T., and Cleasby, J.L. (1990) The effect of temperature on turbulent flocculation: Fluid dynamics and chemistry, *Journal AWWA*, vol. 82, no. 11, pp. 56–73.

Hargesheimer, E.E., Lewis, C.M., and Yentsch, C.M. (1992) Evaluation of particle counting as a measure of treatment plant performance, Awwa Research Foundation report. Denver, CO: Water Research Foundation.

Harris, W.L. (1972) Use of polyelectrolytes as aids during backwash of filters, *Proceedings of the AWWA Seminar on Polyelectrolytes—Aids to Better Water Quality,* Chicago, IL.

Hart, V., Johnson, C.E., and Letterman, R.D. (1992) An analysis of low-level turbidity measurements, *Journal AWWA*, vol. 84, no. 12, pp. 40–45.

Hayden, P.L., and Rubin, A.J. (1974) Systematic investigation of the hydrolysis and precipitation of aluminum (III). In *Aqueous–Environmental Chemistry of Metals*, ed. A.J. Rubin. Ann Arbor, MI: Ann Arbor Science.

Howe, K.J., and Clark, M.M. (2006) Effect of coagulation pretreatment on membrane filtration performance, *Journal AWWA*, vol. 98, no. 4, pp. 133–146.

Hu, C., Liu, H., and Qu, J. (2005) Preparation and characterization of polyaluminum chloride containing high concentrations of Al_{13} and active chlorine, *Colloids and Surfaces A: Physicochem. Eng. Aspects*, vol. 260, pp. 109–117.

Huang, L., Wang, D., Tang, H., and Wang, S. (2006) Separation and purification of nano-Al_{13} by UF method, *Colloids and Surfaces A: Physicochem. Eng. Aspects*, vol. 275, pp. 200–208.

Hubel, R.E., and Edzwald, J.K. (1987) Removing trihalomethane precursors by coagulation, *Journal AWWA,* vol. 79, no. 7, pp. 98–106.

Hudson, H.E., Jr. (1981) *Water Clarification Processes: Practical Design and Evaluation.* New York: Van Nostrand Reinhold.

Hunter, R.J. (1981) *Zeta Potential in Colloid Science*. London: Academic Press.

Hunter, R.J., and Liss, P.S. (1979) The surface charge of suspended particles in estuarine and coastal waters, *Nature*, vol. 282, pp. 823.

Jahn, S.A.A. (1988) Using Moringa seeds as coagulants in developing countries, *Journal AWWA*, vol. 90, pp. 43–50.

Jarvis, P., Jefferson, B., Gregory, J., and Parsons, S.A. (2005) A review of floc strength and breakage, *Wat. Res.*, vol. 39, pp. 3121–3137.

Jiang, J.-Q., and Graham, N.J.D. (2003) Development of optimal poly-alumino-iron sulphate coagulant, ASCE, *Journal of Environmental Engineering*, vol. 129, no. 8, pp. 699–708.

Jiang, Q., and Logan, B.E. (1991) Fractal dimentions of aggregates determined from steady-state size distribution, *Environ. Sci. Technol*, vol. 25, no. 12, pp. 2031–2038.

Judd, S.J., and Hillis, P. (2001) Optimisation of combined coagulation and microfiltration for water treatment, *Wat. Res.*, vol. 35, no. 12, pp. 2895–2904.

Kang, L.S., and Cleasby, J.L. (1995) Temperature effects on flocculation kinetics using Fe(III) coagulant, *Jour. of Env. Eng*, ASCE, vol. 121, no. 12, pp. 893–910.

Kastl, G., Sathasivan, A., Fisher, I., and van Leeuwen, J. (2004) Modeling DOC removal by enhanced coagulation, *Journal AWWA*, vol. 96, no. 2, pp. 79–89.

Kim, J.S., and Rajagopolan, R. (1982) A comprehensive equation for the rate of adsorption of colloidal particles and for stability ratio, *Colloids and Surfaces,* vol. 4, pp. 17–31.

Kodama, H., and Schnitzer, M. (1980) Effect of fulvic acid on the crystallization of aluminum hydroxides, *Geoderma,* vol. 24, pp. 195–205.

Lagvankar, A.L., and Gemmell, R.S. (1968) A size-density relationship for flocs, *Journal AWWA,* vol. 60, no. 9, pp. 1040.

LaMer, V.K. (1964) Coagulation symposium introduction, *Jour. Colloid Sci.*, vol. 19, pp. 291.

Lee, D.G., Bonner, J.S., Garton, L.S., Ernest, A.N.S., and Autenrieth, R.L. (2000) Modeling coagulation kinetics incorporating fractal theories: A fractal rectilinear approach, *Wat. Res.*, vol. 34, no. 7, pp. 1987–2000.

Lee, D.G., Bonner, J.S., Garton, L.S., Ernest, A.N.S., and Autenrieth, R.L. (2002) Modeling coagulation kinetics incorporating fractal theories: Comparison with observed data, *Wat. Res.*, vol. 36, pp. 1056–1066.

Lefebvre, E., and Legube, B. (1990) Iron (III) coagulation of humic substances extracted from surface waters: Effect of pH and humic substances concentration, *Wat. Res.*, vol. 24, no. 5, pp. 591–606.

Letterman, R.D., Chappell, R., and Mates, B. (1996) Effect of pH and alkalinity on the removal of NOM with Al and Fe salt coagulants, *Proceedings of the AWWA Water Quality Technology Conference,* November 17–21, Boston, MA.

Letterman, R.D., and Driscoll, C.T. (1993) Control of residual aluminum in filtered water, Awwa Research Foundation report. Denver, CO: Water Research Foundation.

Letterman, R.D., and Iyer, D.R. (1985) Modeling the effects of hydrolyzed aluminum and solution chemistry on flocculation kinetics, *Environ. Sci. Tech.*, vol. 19, pp. 673–681.

Letterman, R.D. (1991) *Filtration Strategies to Meet the Surface Water Treatment Rule.* Denver, CO: American Water Works Association.

Letterman, R.D., and Pero, R.W. (1990) Contaminants in polyelectrolyte coagulant products used in water treatment, *Journal AWWA*, vol. 82, no. 11, pp. 87–97.

Letterman, R.D., Ramaswamy, M., Staniec, T., Schulz, C., and Kelkar, U.G. (2001) Development of a count performance evaluation procedure for on-line particle counters used in drinking water treatment, Draft report submitted to Awwa Research Foundation, Denver, Co; available at http://web.syr.edu/~rdletter/particle countingreport.pdf/

Letterman, R.D., and Sricharoenchaikit, P. (1982) Interaction of hydrolyzable metal and cationic polyelectrolyte coagulants, *Jour. of Env. Eng.*, ASCE, vol. 108, no. 5, pp. 883–898.

Lewis, C.M., Hargesheimer, E.E., and Yentsch, C.M. (1992) Selecting counters for process monitoring, *Journal AWWA*, vol. 84, no. 12, pp. 46–53.

Liang, L., and Morgan, J.J. (1990) Chemical aspects of iron oxide coagulation in water: Laboratory studies and implications for natural systems, *Aquatic Sciences–Research Across Boundaries,* vol. 52, pp. 32–55.

Lin, J.-L., Chin, C.-J.M., Huang, C., Pan, J.R., and Wang, D. (2008) Coagulation behavior of Al_{13} aggregates, *Wat. Res.*, vol. 42, pp. 4281–4290.

Lind, C. (1995) A coagulant road map, *Public Works*, March, pp. 36–38.

Liu, J., Crapper, M., and McConnachie, G.L. (2004) An accurate approach to the design of channel hydraulic flocculators, *Wat. Res.*, vol. 38, pp. 875–886.

Lyklema, J. (1978) Surface chemistry of colloids in connection with stability. In *The Scientific Basis of Flocculation*, ed. Ives, K.J. The Netherlnds: Sijthoff and Noordhoff.

Malley, J.P., Jr., and Edzwald, J.K. (1991) Laboratory comparison of dissolved air flotation to conventional water treatment, *Journal AWWA*, vol. 83, no. 9, pp. 56–61.

McConnachie, G.L. (1991) Turbulence intensity of mixing in relation to flocculation, *Jour. of Env. Eng., ASCE,* vol. 117, no. 6, pp. 731–750.

Meakin, P. (1988) Fractal aggregates, *Adv. Colloid Interface Sci.*, vol. 28, pp. 249–331.

Monk, R.D.G., Yoshimura, R.Y., Hoover, M.G., and, Lo, S.H. (1985) Prepurchasing ozone equipment, *Journal AWWA*, vol. 77, no. 8, pp. 49–54.

Morel, F.M.M. (1983) *Principles of Aquatic Chemistry.* New York: Wiley-Interscience.

Morris, J.K., and Knocke, W.R. (1984) Temperature effects on the use of metal-ion coagulants for water treatment, *Journal AWWA*, vol. 76, no. 3, pp. 74–87.

Mühle, K. (1993) Floc stability in laminar and turbulent flow. In *Coagulation and Flocculation, Theory and Applications,* ed. Dobiáš, B., pp. 355–390. New York: Marcel Dekker.

Mühle, K., and Domasch, K. (1991) Stability of particle aggregates in flocculation with polymers, *Chem. Eng. Progress*, vol. 29, pp. 1–8.

MWH (2005) *Water Treatment: Principles and Design,* 2d ed. Hoboken, NJ: John Wiley and Sons, Inc.

Najm, I.N. (2001) User friendly carbonate chemistry charts, *Journal AWWA*, vol. 93, no. 11, pp. 86–93.

Narkis, N., and Rebhun, M. (1977) Stoichiometric relationship between humic and fulvic acids and flocculants, *Journal AWWA*, vol. 69, no. 6, pp. 325–333.

Niehof, R.A., and Loeb, G.I. (1972) The surface charge of particulate matter in sea water, *Limnology Oceanography*, vol. 17, p. 7.

Nowack, K.O., and Cannon, F.S. (1996) Ferric chloride plus GAC for removing TOC, *Proceedings of the AWWA Water Quality Technology Conference,* Boston, MA. November 17–21.

NSF (2009) NSF/ANSI Standard 60, Drinking Water Treatment Chemicals–Health Effects. Ann Arbor, MI: NSF International.

O'Melia, C.R. (1985) Polymeric inorganic flocculants. In *Flocculation, Sedimentation, and Consolidation*, ed. Moudgil, M., and Somasundaran, P. New York: Engineering Foundation.

O'Melia, C.R., (1995) From algae to aquifers: Solid–liquid separation in aquatic systems. ACS Advances in Chemistry Series No. 244, *Aquatic Chemistry: Interfacial and Interspecies Processes*, ed. Huang, C.P., O'Melia, C.R., and Morgan, J.J. Washington, DC: American Chemical Society.

Oosawa, F. (1971) *Polyelectrolytes*. New York: Marcel Dekker.

Paralkar, A., and Edzwald, J.K. (1996) Effect of ozone on EOM and coagulation, *Journal AWWA*, vol. 88, no. 4, pp. 143–154.

Pariser, E.R., and Lombardi, D.P. (1989) *Chitin Sourcebook: A Guide to the Research Literature.* New York: Wiley.

Parker, D.S., Kaufmann, W.J., and Jenkins, D. (1972) Floc breakup in turbulent flocculation processes, *Jour. Sanitary Engineering Div.*, ASCE, vol. 98, pp. 79–99.

Pernitsky, D.J., and Edzwald, J.K. (2003). Solubility of polyaluminum coagulants, *Journal of Water Supply Research and Technology (AQUA)* vol. 52, no. 6, pp. 395–406.

Pernitsky, D.J., and Edzwald, J.K. (2006) Selection of alum and polyaluminum coagulants: Principles and applications, *Jour. Water Supply: Research and Technology-AQUA,* vol. 55, no. 2, pp. 121–141.

Prendiville, P.W. (1986) Ozonation at the 900 cfs Los Angeles water purification plant, *Ozone: Sci. & Eng.*, vol. 8, pp. 77–93.

Rinaudo, M. (2006) Chitin and chitosan: Properties and applications, *Progr. Polymer Sci.*, vol. 31, pp. 603–632.

Rothberg, M.R., and Scuras, S.E. (1994) The Rothberg, Tamburini and Winsor Model for Water Process and Corrosion Chemistry, Version 4.0, distributed by American Water Works Association, Denver, CO.

Saffman, P.G., and Turner, J.S. (1956) On the collision of drops in turbulent clouds, *Jour. Fluid Mech.*, vol. 1, p. 16.

Schecher, W.D., and McAvoy, D.C. (2003) MINEQL+: A Chemical Equilibrium Modeling System, Version 4.6 for Windows, User's Manual, 2d ed. Hallowell, ME: Environmental Research Software.

Schulz, C.R., Singer, P.C., Gandley, R., and Nix, J.E. (1994) Evaluating buoyant coarse media flocculation, *Journal AWWA*, vol. 86, no. 8, pp. 51–63.

Schwartzberg, H.G., and Treybal, R.E. (1968) Fluid and particle motion in turbulent stirred tanks, *Ind. Eng. Chem. Fundamentals*, vol. 7, pp. 1–12.

Semmens, M.J., and Field, T.K. (1980) Coagulation: Experiences in organics removal, *Journal AWWA*, vol. 72, no. 8, pp. 476–483.

Sethi, V., Patniak, P., Biswas, Clark, R.M., and Rice, E.W. (1997) Evaluation of optical detection methods for waterborne suspensions, *Journal AWWA*, vol. 89, no. 2, pp. 98–112.

Shankar, S., and Letterman, R.D. (1996) Modeling pH in water treatment plants–The effect of carbon dioxide transport on pH profiles, poster presented at the AWWA Annual Conf, Toronto, CAN, June 23–27.

Shin, J.Y., and O'Melia, C.R. (2004) Stoichiometry of coagulation revisited, *Proceedings 2004 AWWA Annual Conference,* Orlando, FL.

Smoluchowski, M. (1917) Versuch einer mathematischen Theorie der Koagulationskinetic kolloider Losunger, *Zeitschrift Physicalische Chemie,* vol. 92, p. 129.

Snodgrass, W.J., Clark, M.M., and O'Melia, C.R. (1984) Particle formation and growth in dilute aluminum (III) solution: Characterization of particle size distribution at pH 5.5, *Wat. Res*., vol. 18, no. 4, pp. 479–488.

Snoeyink, V.L., and Jenkins, D. (1980*) Water Chemistry.* New York: John Wiley and Sons.

Somasundaran, P., and Runkana, V. (2003) Modeling flocculation of colloidal mineral suspensions using population balances, *Int. J. Miner. Process*., vol. 72, pp. 33–35.

Spielman, L.A. (1978) Hydrodynamic aspects of flocculation. In *The Scientific Basis of Flocculation*, ed. Ives, K.J. The Netherlands: Sijthoff and Noordhoff.

Sposito, G., ed. (1996) *The Environmental Chemistry of Aluminum.* Boca Raton, FL: CRC Press.

Sricharoenchaikit, P., and Letterman, R.D. (1987) Effect of Al(III) and sulfate ion on flocculation kinetics, *Jour. of Environ. Engineering, ASCE*, vol. 113, no. 5, pp. 1120–1138.

Stumm, W. (1992) *Chemistry of the Solid–Water Interface.* New York: John Wiley and Sons.

Stumm, W., and Morgan, J.J. (1996) *Aquatic Chemistry*, 3d. ed. New York: John Wiley and Sons.

Subramaniam, K., and Yiacoumi, S., (2001) Modeling kinetics of copper uptake by inorganic colloids under high surface coverage conditions, *Colloids and Surfaces A: Physicochemical and Engineering Aspects*, vol. 191, pp. 145–159.

Sutherland, J.P., Folkard, G.K., and Grant, W.D. (1989) Seeds of *Moringa* species as naturally occurring flocculants for water treatment, *Sci. Tech. and Dev.*, vol. 7, pp. 191–197.

Taboada-Serrano, P., Chin, C.J., Yiacoumi, S., and Tsouris, C. (2005) Particle aggregation/phase formation, *Current Opinion in Colloid and Interface Science*, vol. 10, pp. 123–132.

Taboada-Serrano, P., Yiacoumi, S., and Tsouris, C. (2006) Electrostatic surface interactions in mixtures of symmetric and asymmetric electrolytes: A Monte Carlo study, *Journal of Chemical Physics*, vol. 125, pp. 1–11.

Tambo, N., and François, R.J. (1991) Mixing, breakup, and floc characteristics. In *Mixing in Coagulation and Flocculation,* ed. Amirtharajah, A., Clark, M.M., and Trussell, R.R. Awwa Research Foundation report. Denver, CO: Water Research Foundation.

Tambo, N., Kamei, T., and Itoh, H. (1989) Evalution of extent of humic-substance removal by coagulation. In *Aquatic Humic Substances*, Adv. in Chemistry Series 219, ed. Suffett, I.H., and MacCarthy, P., pp. 453–471. Washington, DC: American Chemical Society.

Tang, H.-X., and Stumm, W. (1987) The coagulating behaviors of Fe(III) polymeric species–I, II *Wat. Res.*, vol. 21, no. 1, pp. 115–121, 123–128.

Tseng, T., and Edwards, M. (1999) Predicting full-scale TOC removal, *Journal AWWA*, vol. 91, no. 4, pp. 159–170.

U.S. Environmental Protection Agency (USEPA) (1998) Stage 1 Disinfection By-products Rule. Washington, DC.

USEPA (1999) Enhanced Coagulation and Enhanced Precipitative Softening Guidance Manual, EPA 815-R-99-012, Washington, DC: USEPA Office of Water.

van Benschoten, J.E., and Edzwald, J.K. (1990a) Chemical aspects of coagulation using aluminum salts. I. Hydrolytic reactions of alum and polyaluminum chloride, *Wat. Res.*, vol. 24, no. 12, pp. 1519–1526.

van Benschoten, J.E., and Edzwald, J.K. (1990b) Chemical aspects of coagulation using aluminum salts. II. Coagulation of fulvic acid using alum and polyaluminum chloride, *Wat. Res.*, vol. 24, no. 12, pp. 1527–1535.

van Benschoten, J.E., Jensen, J.N., and Rahman, Md. A. (1994) Effects of temperature and pH on residual aluminum in alkaline-treated waters, *Jour. of Env. Eng.*, vol. 120, no. 3, pp. 543–559.

van de Ven, T.G.M., and Mason, S.G. (1977) The microrheology of colloidal suspensions. VII. Orthokinetic doublet formation of spheres, *Colloid Polymer Science*, vol. 255, pp. 468–479.

Verwey, E.J.W., and Overbeek, J.Th.G., (1948) *Theory of the Stability of Lyophobic Colloids.* Amsterdam: Elsevier.

Vold, M.J. (1963) Computer simulation of floc formation in a colloidal suspension, *Jour. Colloid Sci.*, vol. 18, pp. 684–693.

Wang, D., Luan, Z., and Tang, H. (2003) Differences in coagulation efficiencies between PACl and PICl, *Journal AWWA*, vol. 95, no. 1, pp. 79–86.

Wang, D.S., and Tang, H.X. (2006) Quantitative model of coagulation with inorganic polymer flocculants PACl: Application of the PCNM, *J. ASCE Environ. Eng.*, vol. 132, pp. 434–441.

White, M.C., Thompson, J.D., Harrington, G.W., and Singer, P.C. (1997) Evaluating criteria for enhanced coagulation, *Journal AWWA*, vol. 89, no. 5, pp. 64–77.

Wiesner, M.R., Clark, M.M., and Mallevialle, J. (1989) Membrane filtration of coagulated suspensions, *Jour. of Environ. Engineering*, ASCE, vol. 115, pp. 20–40.

Wu, X., Ge, X., Wang, D., and Tang, H. (2007) Distinct coagulation mechanism and model between alum and high Al_{13}–PACl, *Colloids and Surfaces A: Physicochem. Eng. Aspects*, vol. 305, pp. 89–96.

Wu, X., Wang, D., Ge, X., and Tang, H. (2008) Coagulation of silica microspheres with hydrolyzed Al(III)–Significance of Al_{13} and Al_{13} aggregates, *Colloids and Surfaces A: Physicochem. Eng. Aspects*, vol. 330, pp. 72–79.

Xu, Y., Wang, D., Liu, H., Lu, Y., and Tang, H. (2003) Optimization of the separation and purification of Al_{13}, *Colloids and Surfaces A: Physicochem. Eng. Aspects*, vol. 231, pp. 1–9.

Yapajakis, C. (1982) Polymer in backwash serves dual purpose, *Journal AWWA*, vol. 74, no. 8, pp. 426–428.

Yavuz, C. T., Mayo, J. T., Yu, W. W., Prakash, A., Falkner, J. C., Yean, S., Cong, L., Shipley, H. J., Kan, A., Tomson, M., Natelson, D. and Colvin, V. L. (2006) Low-field magnetic separation of monodisperse Fe_3O_4 nanocrystals, *Science*, vol. 314, no. 5801, pp. 964–967.

Ye, C., Wang, D., Shi, B., Yu, J., Qu, J., Edwards, M., and Tang, H. (2007) Alkalinity effect of coagulation with polyaluminum chlorides: Role of electrostatic patch, *Colloids and Surfaces A: Physico. Eng. Aspects*, vol. 294, pp. 163–173.

Yiacoumi, S., and Tien, C. (1995) *Kinetics of Metal Ion Adsorption from Aqueous Solutions: Models, Algorithms, and Applications*, p. 256. Norwell, MA: Kluwer Academic Publishers.

Ying, T.–Y., Yiacoumi, S., and Tsouris, C. (2000) High-gradient magnetically seeded filtration, *Chemical Engineering Science*, vol. 55, pp. 1101–1113.

Zhao, H., Liu, H., and Qu, J. (2009) Effect of pH on the aluminum salts hydrolysis during coagulation process: Formation and decomposition of polymeric aluminum species, *Jour. Coll. Interface Sci.*, vol. 330, pp. 105–112.

CHAPTER 9
SEDIMENTATION AND FLOTATION

Ross Gregory, MPhil, C.Eng., C.WEM
Consultant
Swindon, Wiltshire, England

James K. Edzwald, Ph.D., B.C.E.E.
Professor Emeritus, Department of Civil and Environmental Engineering
University of Massachusetts
Amherst, Massachusetts, United States

MODERN HISTORY OF SEDIMENTATION **9.2**
Emergence of Modern Sedimentation Methods 9.2
Floc Blanket Sedimentation and Other Innovations 9.2
SEDIMENTATION THEORY **9.3**
Settling of Discrete Particles (Type 1) 9.5
Hindered Settling (Types 3a and 3b) 9.11
Fluidization 9.14
Inclined (Tube and Plate) Settling 9.20
Floc Blanket Clarification 9.23
Hydraulic Characteristics of Sedimentation Tanks 9.26
OPERATIONAL AND DESIGN CONSIDERATIONS FOR SEDIMENTATION **9.26**
Types of Sedimentation Tanks 9.26
Other Factors Influencing Sedimentation Efficiency 9.39
INTRODUCTION TO DISSOLVED AIR FLOTATION **9.45**
History of Flotation 9.45
Types of Flotation Processes 9.46
General Description 9.47
FUNDAMENTALS OF DISSOLVED AIR FLOTATION **9.48**
Air Solubility and Bubbles in the Contact Zone 9.48
Contact Zone Performance Model 9.54
Bubble and Floc-Bubble Aggregate Rise Velocities 9.59
Separation Zone Performance 9.63
OPERATIONAL AND DESIGN CONSIDERATIONS FOR FLOTATION **9.65**
Conventional and High-Rate DAF Systems 9.65
Types of Flotation Tanks 9.68
Air Saturation Systems 9.70
Factors Influencing Dissolved Air Efficiency 9.72
Performance of Dissolved Air Flotation Plants 9.79
APPLICATIONS **9.84**
Factors Influencing Choice of Process 9.84
Emerging Technology 9.88
ABBREVIATIONS **9.89**
NOTATION FOR EQUATIONS **9.89**
REFERENCES **9.92**

Sedimentation and flotation are the principal solid-water separation (clarification) processes used in water treatment mostly to reduce the particle concentration, or load, on granular filters. As a result, filters can be operated more easily and cost-effectively to produce acceptable-quality filtered water. Many sedimentation and flotation processes and variants of them and other clarification processes exist, including types of filtration (e.g., upflow filtration, moving-bed filtration, buoyant-media filtration, horizontal roughing filters, etc.; see Chap. 10 on filtration), and each has advantages and disadvantages. The theory and application of sedimentation are addressed in the next three sections followed by the theory and application of dissolved air flotation in the next three sections. The most appropriate choice of a sedimentation or flotation process for a particular application depends on the water to be treated as well as local circumstances and requirements, and this is addressed in the last section.

MODERN HISTORY OF SEDIMENTATION

Emergence of Modern Sedimentation Methods

The art of sedimentation progressed little until the industrial age and its increased need for water. Storage reservoirs developed into settling reservoirs. Perhaps the largest reservoirs constructed for this purpose were in the United States at Cincinnati, Ohio, where two excavated reservoirs held approximately 1.48×10^6 m^3 (390 million gal) and were designed to be operated by a fill-and-draw method though they never were used in this way (Ellms, 1928). The development of settling basins led to the construction of rectangular masonry settling tanks that assured more even flow distribution and easier sludge removal. With the introduction of coagulation and its production of voluminous sludge, mechanical sludge removal was introduced.

Attempts to make rectangular tanks more cost effective led to the construction of multilayer tanks. Large-diameter, 60 m (200 ft), circular tanks also were constructed at an early stage in the development of modern water treatment. Other applications and industries, such as wastewater treatment, mineral processing, sugar refining, and water softening, required forms of sedimentation with specific characteristics, and various designs of settling tanks particular to certain applications were developed. Subsequently, wider application of successful industrial designs was sought. Out of this emerged circular radial-flow tanks and a variety of proprietary designs of solids-contact units with mechanical equipment for premixing and recirculation.

The inclined plate settler also has industrial origins (Barham et al., 1956), although the theory of inclined settling dates back to experiments using blood in the 1920s and 1930s (Nakamura and Kuroda, 1937; Kinosita, 1949). Closely spaced inclined plate systems for water treatment have their origins in Sweden in the 1950s, resulting from a search for high-rate treatment processes compact enough to be economically housed against winter weather. Inclined tube systems were spawned in the United States in the 1960s. The most recent developments have involved combining inclined settling with ballasting of floc to reduce plant footprint further (de Dianous et al., 1990).

Floc Blanket Sedimentation and Other Innovations

The floc blanket process for water treatment emerged from India in about 1932 as the simple inverted-pyramidal Candy sedimentation tank. Imhoff used a tank of similar shape in 1906 for wastewater treatment (Kalbskopf, 1970). The Spaulding Precipitator (with its

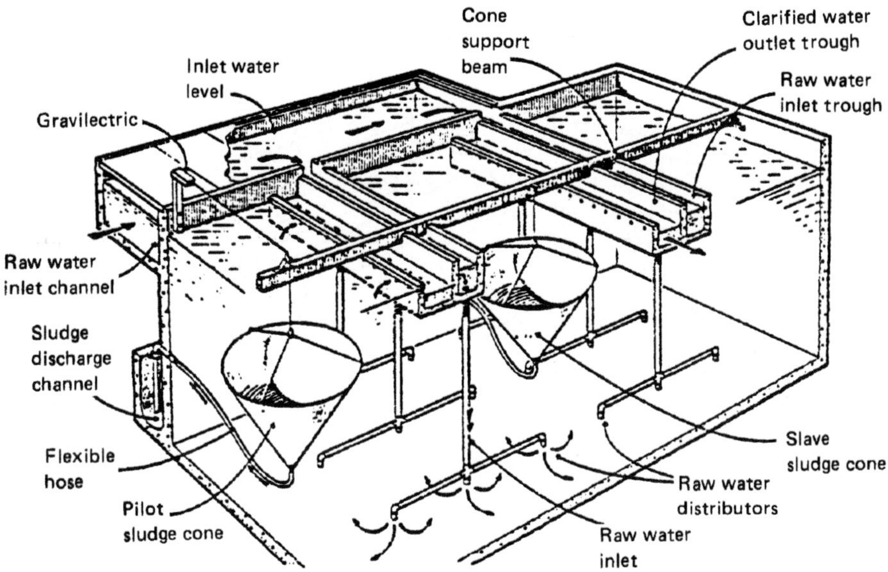

FIGURE 9-1 The flat-bottom clarifier with candelabra flow distribution. (*Source: PWT.*)

annular trough settlement zone surrounding a central mixing and flow distribution chamber) soon followed in 1935 (Hartung, 1951). Other designs that were mainly solids-contact clarifiers (i.e., high concentration of solids in suspension but not usually forming a distinct fluidized bed—further explanation in the sections Sedimentation Theory and Operational and Design Considerations for Sedimentation) rather than true floc blanket tanks (i.e., with a fluidized bed of floc) were also introduced.

The Candy tank can be expensive to construct because of its large sloping sides, so less costly structures for accommodating floc blankets were conceived. The aim was to decrease the hopper component of tanks as much as possible yet provide good flow distribution to produce a stable floc blanket. Development from 1945 progressed from tanks with multiple hoppers or troughs to the present flat-bottom tanks. Efficient flow distribution in flat-bottom tanks is achieved with either candelabra or lateral inlet distribution systems (Figs. 9-1 and 9-2).

An innovation in the 1970s was the inclusion of widely spaced inclined plates in the floc blanket region (Fig. 9-2). Other developments that also have led to increased surface loadings include the use of polyelectrolytes, ballasting of floc with disposable or recycled solids, and improvements in blanket-level control. The principal centers for these developments have been in the United Kingdom, France, and Hungary.

SEDIMENTATION THEORY

The particle water separation processes of interest to water engineers and scientists are difficult to describe by a theoretical analysis, mainly because the particles involved are not regular in shape, density, or size. Consideration of the theory of ideal systems is, however, a useful guide to interpreting observed behavior in more complex cases.

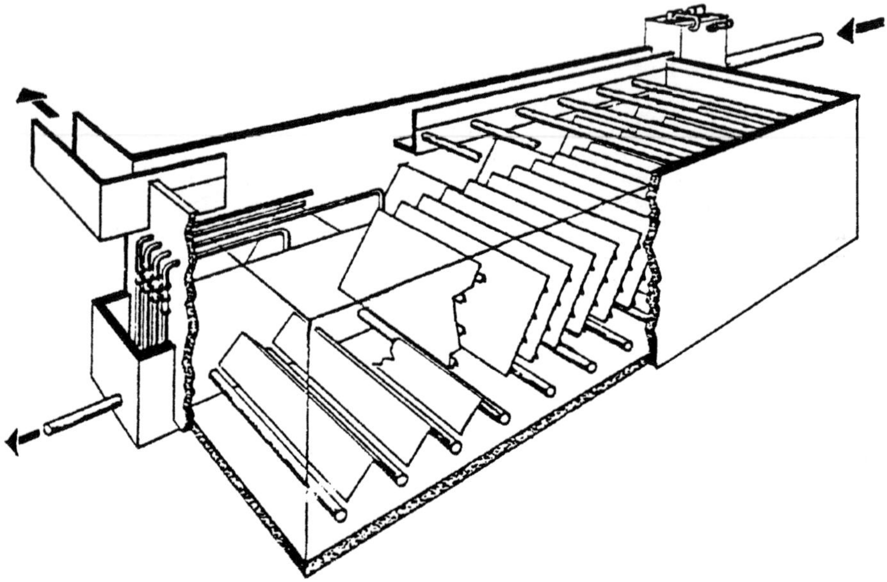

FIGURE 9-2 The Superpulsator flat-bottom clarifier with lateral-flow distribution. (*Courtesy of Infilco Degremont, Inc., Richmond, Va.*)

The various regimes in settling of particles are commonly referred to as Types 1 to 4. The general term *settling* is used to describe all types of particles falling through water under the force of gravity and settling phenomena in which the particles or aggregates are suspended by hydrodynamic forces only. When particles or aggregates rest on one another, the term *subsidence* applies. The following definitions of the settling regimes are commonly used in the United States and are compatible with a comprehensive analysis of hindered settling and flux theory:

Type 1: Settling of discrete particles in low concentration (less than about 1 percent on a volume concentration basis), with flocculation and other interparticle effects being negligible.

Type 2: Settling of particles in low concentration (less than about 8 percent) but with coalescence or flocculation. As coalescence occurs, particle sizes increase and particles settle more rapidly.

Type 3: Hindered or zone settling in which particle concentration (more than about 8 percent) causes interparticle effects, which might include flocculation, to the extent that the rate of settling is a function of solids concentration. Zones of different concentrations may develop from segregation of particles with different settling velocities. Two regimes exist, (a) and (b), with the concentration being less and greater than that at maximum flux, respectively. In the latter case the concentration has reached the point that most particles make regular physical contact with adjacent particles and effectively form a loose structure. As the height of this zone develops, this structure tends to form layers of different concentration, with the lower layers establishing permanent physical contact, until a state of compression is reached in the bottom layer.

Type 4: Compression settling or subsidence develops under the layers of zone settling. The rate of compression is dependent on time and the force caused by the weight of solids above.

Settling of Discrete Particles (Type 1)

Terminal Settling Velocity. When the concentration of particles is very small, each particle settles discretely, as if it is alone, unhindered by the presence of other particles. Starting from rest, the velocity of a single particle settling under gravity in water will increase where the density of the particle is greater than the density of the water. Acceleration continues until the resistance to flow through the water, or drag, equals the effective weight of the particle. Thereafter the settling velocity remains essentially constant. This velocity is called the *terminal settling velocity*, v_t. The terminal settling velocity depends on various factors relating to the particle and the water.

For most theoretical and practical computations of settling velocities, the shape of particles is assumed to be spherical. The size of particles that are not spherical can be expressed in terms of a sphere of equivalent volume. The general equation for the terminal settling velocity of a single particle is derived by equating the forces on the particle. These forces are the drag f_d, buoyancy f_b, and gravity f_g. Hence,

$$f_d = f_g - f_b \tag{9-1}$$

The drag force on a particle traveling in a resistant fluid is (Prandtl and Tietjens, 1957)

$$f_d = \frac{C_D v^2 \rho A}{2} \tag{9-2}$$

where C_D = drag coefficient, v = settling velocity, ρ = mass density of water, and A = projected area of particle in direction of flow. Any consistent dimensionally homogeneous units may be used in Eq. 9-2 and all subsequent rational equations.

At constant, that is, terminal, settling velocity v_t,

$$f_g - f_b = Vg(\rho_p - \rho) \tag{9-3}$$

where V is the effective volume of the particle and ρ_p is the density of the particle. When Eqs. 9-2 and 9-3 are substituted in Eq. 9-1,

$$\frac{C_D v_t^2 \rho A}{2} = Vg(\rho_p - \rho) \tag{9-4a}$$

Rearranging, we get

$$v_t = \left[\frac{2g(\rho_p - \rho)V}{C_D \rho A}\right]^{1/2} \tag{9.4b}$$

When the particle is solid and spherical (d is the diameter of the sphere),

$$v_t = \left[\frac{4g(\rho_p - \rho)d}{3C_D \rho}\right]^{1/2} \tag{9-5}$$

The value of v_t is the difference in velocity between the particle and the water and is essentially independent of horizontal or vertical movement of the water, although in real situations there are secondary forces caused by velocity gradients and other factors. Therefore, the relationship also applies to a dense stationary particle with water flowing upward past it or a buoyant particle with water flowing downward.

Calculation of v_t for a given system is difficult because the drag coefficient C_D depends on the nature of the flow around the particle. This relationship can be described using the

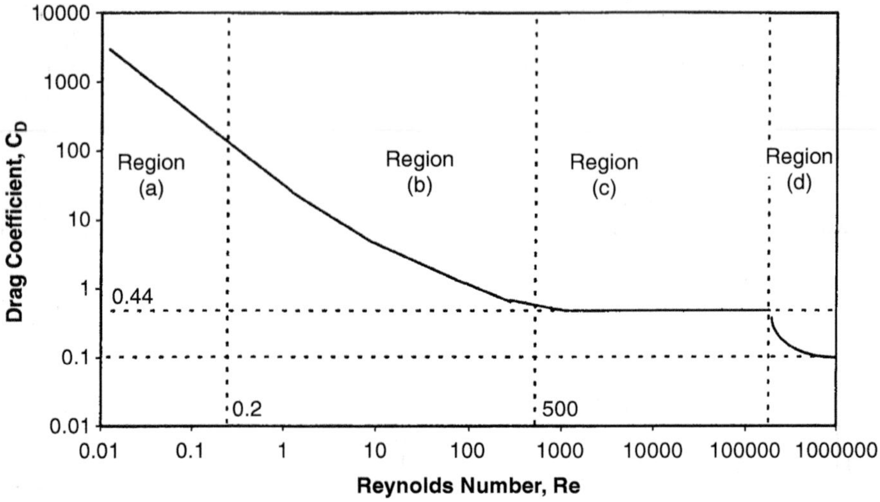

FIGURE 9-3 Variation of drag coefficient, C_D, with Reynolds Number, *Re*, for single-particle sedimentation.

Reynolds number Re (based on particle diameter), as illustrated schematically in Fig. 9-3, where

$$\mathrm{Re} = \frac{\rho v d}{\mu} \tag{9-6}$$

μ is the absolute (dynamic) water viscosity, and v is the velocity of the particle relative to the water.

The value of C_D decreases as the value of Re increases, but at a rate depending on the value of Re. In Region (a) in Fig. 9-3 of small Re value, $10^{-4} < \mathrm{Re} < 0.2$, the laminar flow region, the equation of the relationship approximates to

$$C_D = \frac{24}{\mathrm{Re}} \tag{9-7}$$

This when substituted in Eq. 9-1 gives Stokes' settling velocity equation for laminar flow conditions

$$v_t = \frac{g(\rho_p - \rho)d^2}{18\mu} \tag{9-8}$$

For Region (b), $0.2 < \mathrm{Re} < 500$ to 1000, classically for water treatment, the Fair and Geyer equation is promoted (Fair et al., 1971).

$$C_D = \frac{24}{\mathrm{Re}} + \frac{3}{\sqrt{\mathrm{Re}}} + 0.34 \tag{9-9}$$

Region (c), 500 to $1000 < \mathrm{Re} < 2 \times 10^5$, is one of turbulent flow. The value of C_D is almost constant at 0.44, such that substitution of this into Eq. 9-5 results in Newton's equation

$$v_t = 1.74\left[\frac{(\rho_p - \rho)gd}{\rho}\right]^{1/2} \tag{9-10}$$

In Region (d), $\text{Re} < 2 \times 10^5$, the drag force decreases considerably with the development of turbulent flow at the surface of the particle, such that C_D becomes equal to 0.10. This region is unlikely to be encountered in sedimentation in water treatment.

However, Brown and Lawler (2003) have challenged the acceptance of dividing the relationship for spheres into four zones according to Re value and in particular the use of the Fair and Geyer equation. They noted that the Fair and Geyer equation was the least accurate of those they reviewed. They found, from pooling all published data on settlement of spheres relevant to Regions (a), (b), and (c) and accounting for wall effects, that a single equation can represent the data accurately. They recommended for $\text{Re} < 2 \times 10^5$, the use of

$$C_D = \frac{24}{\text{Re}}(1+0.150Re^{0.681}) + \frac{0.407}{1+8{,}710/\text{Re}} \tag{9-11}$$

Brown and Lawler (2003) also developed equations to determine the terminal settling velocities of spheres. They recommended that for Re < 5000 use of

$$v_{t*} = \frac{d_*^2(22.5+d_*^{2.046})}{0.0258d_*^{4.046}+2.81d_*^{3.046}+18d_*^{2.046}+405} \tag{9-12}$$

where the subscript '$_*$' denotes dimensionlessness, such that in dimensionless terms the Stokes' settling velocity (Eq. 9-8) is given by

$$v_{t*} = \frac{d_*^2}{18} \tag{9-13}$$

For $\text{Re} < 2 \times 10^5$, Brown and Lawler also recommended for general use:

$$v_{t*} = \left[\left(\frac{18}{d_*^2}\right)^{[(0.936d_*+1)/(d_*+1)]0.898} + \left(\frac{0.317}{d_*}\right)^{0.449}\right]^{-1.114} \tag{9-14}$$

Effect of Particle Shape. Equation 9-4b shows how particle shape affects velocity. The effect of a non-spherical shape is to increase the value of C_D at a given value of Re. As a result, the settling velocity of a non-spherical particle is less than that of a sphere having the same volume and density. Sometimes a simple shape factor (Θ) is determined and incorporated into Eq. 9-7; this yields:

$$C_D = \frac{24\Theta}{\text{Re}} \tag{9-15}$$

Typical values found for Θ for rigid particles are sand 2.0, coal 2.25, gypsum 4.0, and graphite flakes 22 (Degremont, 1991). Details on the settling behavior of spheres and non-spherical particles can be found in standard texts (e.g., Richardson and Harker, 2002).

Flocculation. A shape factor value is difficult to determine for floc particles because their size and shape are interlinked with the mechanics of their formation and disruption in any set of flow conditions. When particles flocculate, a loose and irregular structure is formed, which is likely to have a relatively large-value shape factor. Additionally, while the effective particle size increases in flocculation, the effective particle density decreases in accordance with a fractal dimension (Lagvankar and Gemmel, 1968; Tambo and Watanabe, 1979; Bache and Gregory, 2007; also see Chap. 8 on flocculation).

Flocculation is a process of aggregation and attrition. Aggregation can occur by Brownian diffusion, differential settling, and velocity gradients caused by fluid shear. Attrition is caused mainly by excessive velocity gradients (see Chap. 8).

The theory of flocculation detailed in Chap. 8 recognizes the role of particle stability (attractive and repulsive forces between particles), velocity gradient *G*, and time *t*, as well as particle volumetric concentration Φ. For dilute suspensions optimum flocculation conditions are generally considered only in terms of *G* and *t* where

$$Gt = \text{constant} \tag{9-16}$$

Camp (1955) identified optimum *Gt* values between 10^4 and 10^5 for flocculation prior to conventional horizontal-flow settlers. In the case of floc blanket clarifiers (there being no prior flocculators), the value of *G* is usually less than in flocculators and the value of *Gt* is about 20,000 (Gregory, 1979). This tends to be less than that usually considered necessary for flocculation prior to inclined settling or dissolved air flotation, but appropriate for direct (contact) filtration (Bache and Gregory, 2007). Recently this has become considered adequate for high-rate DAF (see the section Operational and Design Considerations for Flotation).

In concentrated suspensions, such as with hindered settling, the greater particle concentration, for example, volumetric concentration Φ, contributes to flocculation by enhancing the probability of particle collisions and increasing the velocity gradient that can be expressed in terms of the head loss across the suspension. Consequently, optimum flocculation conditions for concentrated suspensions may be better represented by (Fair et al., 1971; Ives, 1968; Vostrcil, 1971)

$$Gt\Phi = \text{constant} \tag{9-17}$$

The value of the constant at maximum flux is likely to be about 4000, when Φ is measured as the fractional volume occupied by floc, with little benefit to be gained from a larger value (Gregory, 1979; Vostrcil, 1971).

Measurement of the volumetric concentration of floc particle suspensions is a problem because of variations in particle size, shape, and other factors. A simple settlement test is the easiest method of producing a measurement (for concentrations encountered in floc blanket settling) in a standard way, for example, half-hour settlement in a graduated cylinder (Gregory, 1979). A graduated cylinder (e.g., 100 mL or 1 L) is filled to the top mark with the suspension to be measured. The half-hour settled-solids volume is the volume occupied by the settled suspension measured after 30 min, and it is expressed as a fraction of the total volume of the whole sample.

The process of flocculation continues during conditions intended to allow settlement. Assuming collision between flocculant particles takes place only between particles settling at different Stokes' velocities, then the flocculation rate constant k_d between spherical particles of size d_i and d_j and same specific gravity *s* is given in Chap. 8 and shown next.

$$k_d = \frac{\pi g \rho (s-1)}{72\mu}(d_i + d_j)^3 (d_i - d_j) \tag{9-18}$$

Practical application of Eq. 9-18 and the Stokes' velocity, Eq. 9-8 or similar, is complicated because as floc size increases, its density decreases (Lagvanker and Gemmel, 1968; Tambo and Watanabe, 1979; see Fig. 9-4). The density difference ($\Delta\rho = \rho_f - \rho_w$) between floc and water is the same as the density term in Stokes' equation (Eq. 9-8). The relationship between floc size and density means that although Stokes' velocity is proportional to the square of floc size, the decrease in density reduces the effect of increasing size. It is relevant to note that floc strength also decreases as size increases. These are key characteristics of floc agglomerates, that is, multifractal structures (Bache and Gregory, 2007). A practical implication is that flocculation prior to sedimentation has diminishing return on the size and increase in rate of settlement of the floc that will be produced, notwithstanding the need to ensure the flocculation results in a narrow particle size range and hence settling velocity distribution. Table 9-1 has been prepared to illustrate the interaction between particle size and density at 4°C. It also illustrates the benefit of seeking to increase particle size without loss of particle density.

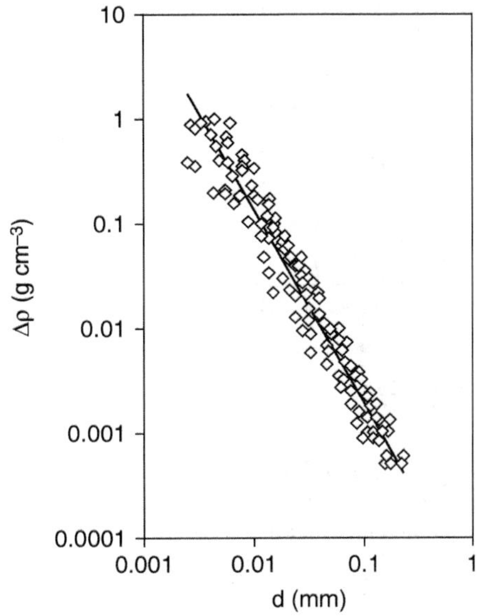

FIGURE 9-4 Example of size-density function for alumino-humic flocs strengthened by polymer addition (density in kg/m^3 is obtained by multiplying g/cm^3 by 1000). (*Source: Bache and Gregory (2007), reprinted with permission from IWA Publishing, Copyright 2007, United Kingdom.*)

Settlement in Tanks. In an ideal upflow settling tank the particles retained are those whose terminal settling velocity exceeds the water upflow velocity.

$$v_t \geq \frac{Q}{A} \tag{9-19}$$

where Q is the inlet flow rate to the tank and A is the horizontal cross-sectional, or plan, area of the tank.

In a horizontal-flow rectangular tank, the settling of a particle has both vertical and horizontal components, as shown in Fig. 9-5.

$$L = \frac{tQ}{HW} \tag{9-20}$$

TABLE 9-1 Stokes' Settling Velocity at 4°C in Water: Values in Shaded Boxes Computed from Newton's Equation because Re > 0.2 for Stokes' Velocity

Particle density, kg/m^3	Terminal settling velocity, m/h				
	Particle size, mm				
	0.1	0.2	0.5	1.0	2.0
2000*	12.5	50.0			
1500	6.25	25.0			
1200	2.50	10.0			
1100	1.25	5.00			
1050	0.63	2.50	13.5		
1020		1.00	5.71		
1010		0.50	2.93	10.46	
1005			1.56	5.50	17.83
1002			0.63	2.30	8.00
1001				1.25	4.26

* Similar to the example for floc ballasted with sand.

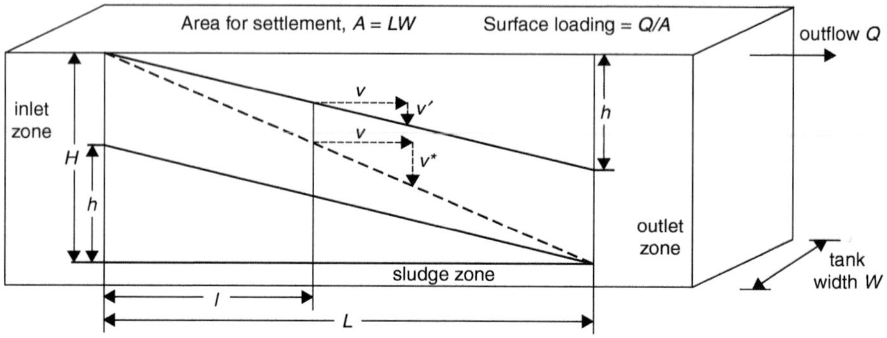

FIGURE 9-5 Horizontal and vertical components of settling velocity.

where L = horizontal distance traveled, t = time of travel, H = depth of water, and W = width of tank. When h = the vertical distance traveled, the settling time for a particle that has entered the tank at a given level h is

$$t = \frac{h}{v} \tag{9-21}$$

Substitution of this in Eq. 9-20 gives the length of tank required for settlement to occur under ideal flow conditions.

$$L = \frac{hQ}{vHW} \tag{9-22a}$$

or

$$v = \frac{hQ}{HLW} \tag{9-22b}$$

If all particles with a settling velocity of v are allowed to settle, then h equals H, and consequently this special case then defines the surface loading or overflow rate v^* of the ideal tank.

$$v^* = \frac{Q}{L^*W} \tag{9-23a}$$

$$v^* = \frac{Q}{A^*} \tag{9-23b}$$

where L^* is the length of tank over which settlement ideally takes place and A^* is the plan area of tank, with horizontal flow, over which the settlement ideally takes place.

All particles with a settling velocity greater than v^* are removed. Particles with a settling velocity $v^/$ less than v^* need a tank of length $L^/$ greater than L^* for total settlement, such that

$$\frac{v^/}{v^*} = \frac{L^*}{L^/} \tag{9-24}$$

This ratio defines the proportion of particles with a settling velocity of $v^/$ that settle in a length L^*.

Equation 9-23b shows that settling efficiency for the ideal condition is independent of depth H and dependent on only the tank plan area. This principle is sometimes referred to as *Hazen's law*. In contrast, detention time t is dependent on water depth H, as given by

$$t = \frac{AH}{Q} \tag{9-25}$$

In reality, depth is important because it can affect flow stability if it is large and scouring if it is small.

Predicting Settling Efficiency (Types 1 and 2). For discrete particle settling (Type 1), variation in particle size and density produce a distribution of settling velocities. The settling-velocity distribution may be determined from data collected from a column-settling test. These data are used to plot a settling-velocity analysis curve, and Eq. 9-26 defines the proportion of particles with settling velocity v smaller than v^*, that will be removed in a given time. If x^* is the proportion of particles having settling velocities less than or equal to v^*, the total proportion of particles which could be removed in settling is defined by (Thirumurthi, 1969)

$$F_t = (1 - x^*) + \int_0^{x^*} \frac{v}{v^*}\, dx \tag{9-26}$$

The settling column test is very rarely used in evaluations for water treatment although may apply to waste suspensions, or possibly preliminary settlement of highly turbid raw waters: It generally applies to concentrated suspensions of unflocculated solids, although it can also be used for flocculated suspensions (Type 2) (Zanoni and Blomquist, 1975).

Hindered Settling (Types 3a and 3b)

The following addresses Type 3 settling relevant to clarification. Types 3 and 4 settling as relevant to thickening are addressed in Chap. 22.

Particle Interaction. When particle volume concentration is greater than about 1 percent, individual particle behavior is influenced, or hindered, by the presence of other particles, and the flow characteristics of the bulk suspension can be affected. With increased particle concentration, the free area between particles is reduced; this causes greater interparticle fluid velocities and alteration of flow patterns around particles. Consequently, the average settling velocity of the particles in a concentrated suspension, especially when particle volume concentration is greater than about 8 percent, is generally less than that of a discrete particle of similar size.

When particles in a suspension are not uniform in size, shape, or density, individual particles will have different settling velocities. Particles with a settling velocity less than the suspension increase the effective viscosity. Smaller particles tend to be dragged down by the motion of larger particles. Flocculation may increase the effective particle size when particles are close together (i.e., flocculation due to differential settling, Eq. 9-18).

Solids Flux. The settling velocity v_s of a suspension depends on the suspension particle concentration. The product of velocity and mass concentration C is the solids mass flux F_M, the mass of solids passing a unit area per unit of time.

$$F_M = v_s C \tag{9-27a}$$

The equivalent relationship holds for the solids volumetric concentration Φ to define solids volume flux F_V.

$$F_V = v_s \Phi \tag{9-27b}$$

The relationship between F_M and C (or F_V and Φ) is shown in Fig. 9-6 and is complex because v_s is affected by concentration. The relationship can be divided into four regions.

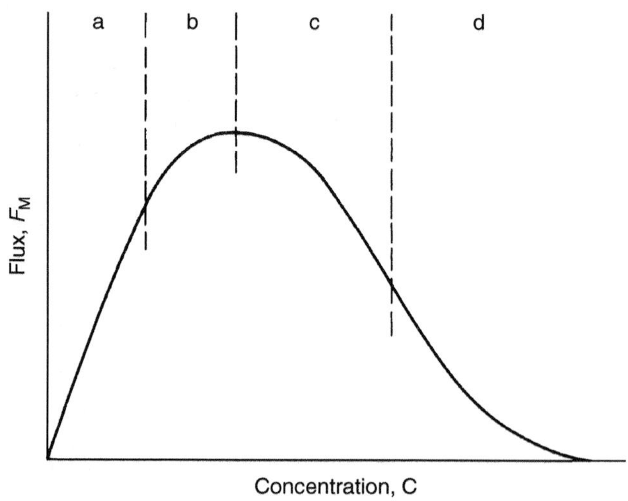

FIGURE 9-6 Typical relationship between flux and concentration for batch settlement.

Region (a), Type 1 and 2 settling. Unhindered settling occurs such that the flux increases in proportion to the concentration. A suspension of particles with different settling velocities has a diffuse interface with the clear water above.

Region (b), Type 3a settling. With increase in concentration, hindered flow settling increasingly takes effect, and ultimately a maximum value of flux is reached. At about maximum flux the diffuse interface of the suspension becomes distinct with the clear water above when all particles become part of the suspension and settle with the same velocity.

Region (c), Type 3b settling. Further increase in concentration reduces flux because of the reduction in settling velocity. In this region the suspension settles homogeneously.

Region (d), Type 4 settling. Associated with the point of inflection in the flux concentration curve, the concentration reaches the point where thickening can be regarded to start leading ultimately to compression settling.

Equations for Hindered Settling. The behavior of suspensions in Regions (b) and (c) has attracted considerable theoretical and empirical analysis and is most important in understanding floc blanket clarification. The simplest and most convenient relationship is represented by the general equation (Gregory, 1979).

$$v_s = v_0 \exp(-q\Phi) \tag{9-28}$$

where q = constant representative of the suspension, v_O = settling velocity of suspension for concentration extrapolated to zero, and Φ = volume concentration of the suspension.

Other empirical relationships have been proposed. The most widely accepted and tested relationship was initially developed for particles larger than 0.1-mm diameter in rigid particle fluidized systems. This relationship has been shown to be applicable to settling and is known as the *Richardson and Zaki equation* (Richardson and Harker, 2002).

$$v_s = v_t E^n \tag{9-29}$$

where E = porosity of the suspension (i.e., volume of fluid per volume of suspension, $E = 1 - \Phi$), n = power value dependent on the Reynolds number of the particle, and v_t = terminal settling velocity of particles in unhindered flow (i.e., absence of effect by presence of other particles).

For rigid particles this equation is valid for porosity from about 0.6 (occurring at around minimum fluidization velocity) to about 0.95. The Reynolds number determines the value of n (Richardson and Harker, 2002). For a suspension with uniform-size spherical particles, n equals 4.8 when Re is less than 0.2. As the value of Re increases, n decreases until Re is greater than 500 when n equals 2.4.

When Eq. 9-29 is used for flocculent suspensions (Gregory, 1979), correction factors must be included to adjust for effective volume to account for particle distortion and compression. If particle volume concentration is measured, say, by the half-hour settlement test, then because such a test is only a relative measurement that provides a measure of the apparent concentration, correction factors are necessary.

$$v_S = v_t k_1 (1 - k_2 \Phi^*)^r \tag{9-30}$$

where k_1, k_2 = constants representative of the system, Φ^* = apparent solids volumetric concentration, and r = power value dependent on the system.

Equation 9-29 can be substituted in Eq. 9-27b for flux with $1 - E$ substituted for Φ (Richardson and Harker, 2002).

$$F_V = v_t E^n (1 - E) \tag{9-31}$$

Differentiating this equation with respect to E gives

$$\frac{dF_V}{dE} = v_t n E^{n-1} - v_t (n+1) E^n \tag{9-32}$$

The flux F_V has a maximum value when dF_V/dE equals zero and E equals E^+ (the porosity at maximum flux). Hence, dividing Eq. 9-32 by $v_t E^{n-1}$ and equating to zero leads to

$$n = \frac{E^+}{1 - E^+} \tag{9-33a}$$

which can be restated as

$$\Phi^+ = \frac{1}{1+n} \tag{9-33b}$$

This means that the porosity E^+ or the volume concentration Φ^+ at maximum flux is an important parameter in describing the settling rates of suspensions. In the case of rigid uniform spheres if n ranges from 2.4 to 4.6, the maximum flux should occur at a volumetric concentration between 0.29 and 0.18. In practice, the range of values generally found for suspensions of aluminum and iron flocs for optimal coagulant dose and coagulation pH, when concentration is measured as the half-hour settled volume, is 0.16 to 0.20 (Gregory, 1979;

Gregory et al., 1996), in which case n ranges from 4.0 to 5.26, reflecting the existence of low Re values.

If Eq. 9-32 is differentiated also, then when $d^2F_V/dE^2 = 0$ for real values of E, a point of inflection will exist, such that

$$E^{++} = \frac{n-1}{n+1} \tag{9-34a}$$

which can be restated as

$$\Phi^{++} = \frac{2}{n+1} \tag{9-34b}$$

This means theoretically that the concentration at the inflection point is twice that at maximum flux. In practice because of the forces on floc as concentration increases, this ratio of 2 is unlikely to apply and the ratio will be less. If the point of inflection is associated with the transition from Type 3 to Type 4 settling, then Eqs. 9-33 and 9-34 provide a basis for characterizing floc with respect to thickening. Types 3 and 4 settling in the context of thickening are considered in Chap. 22.

Prediction of Settling Rate. The hindered settling rate can be predicted for suspensions of rigid and uniform spheres using Eqs. 9-5 and 9-29. For suspensions of non-uniform and flocculent particles, however, settling rate has to be measured. This is most simply done using a settling column; a 1-L measuring cylinder is usually adequate. The procedure is to fill the cylinder to the top measuring mark with the sample and record at frequent intervals the level of the interface between the suspension and the clear-water zone. The interface is only likely to be distinct enough for this purpose if the concentration of the sample is greater than that at maximum flux. The results are plotted to produce the typical settling curve (Fig. 9-7). The slope of the curve over the constant-settling-rate period is the estimate of the Type 3 settling rate for quiescent conditions. If the concentration of the sample was greater than that at the inflection point in the mass flux curve, the transition from Region (c) to (d) in Fig. 9-7, then a period of constant settling rate and the compression point (CP) will not be found as represented by line A.

The compression point signifies the point at which all the suspension has passed into the Type 4 settling or compression regime. Up to that time a zone of solids in the compression regime has been accumulating at the bottom of the suspension with its upper interface moving upward. The compression point thus is where that interface reaches the top of the settling suspension.

Fluidization

When water is moving up through a uniform stationary bed of particles at a low flow rate, the flow behavior is similar to when the flow is down through the bed. When the upward flow of water is great enough to cause a drag force on particles equal to the apparent weight (actual weight less buoyancy) of the particles, the particles rearrange to offer less resistance to flow and bed expansion occurs. This process continues as the water velocity is increased until the bed has assumed the least stable form of packing. If the upward water velocity is increased further, individual particles separate from one another and become freely supported in the water. The bed is then said to be *fluidized*.

For rigid and generally uniform particles, such as filter sand, about 10 percent bed expansion occurs before fluidization commences. The less uniform the size and density of the particles, the less distinct is the point of fluidization. A fluidized bed is characterized

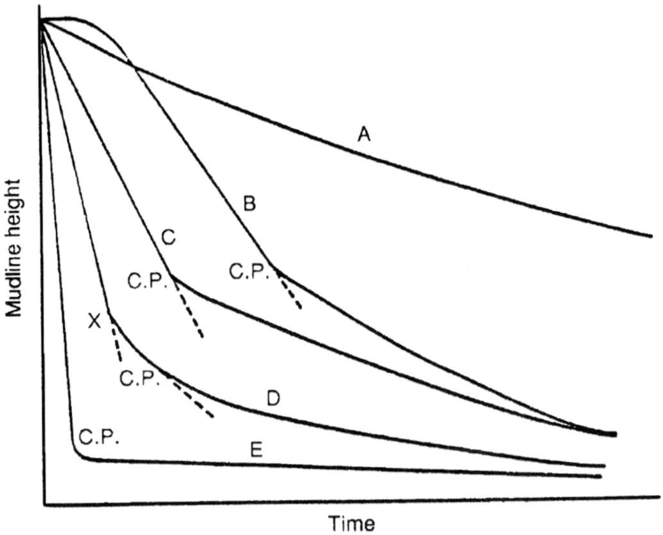

A Concentrated, flocculated or unflocculated pulp
B Intermediate flocculated or unflocculated pulp
C As (B) but showing no induction period
D Dilute, flocculated pulp
E Dilute, unflocculated pulp
C.P. Compression point

FIGURE 9-7 Typical batch-settling curves. (*Source: Pearse, 1977.*)

by regular expansion of the bed as water velocity increases from the minimum fluidization velocity until particles are in unhindered suspension, that is, Type 1 settling.

Fluidization is hydrodynamically similar to hindered, or Type 3, settling. In a fluidized bed particles undergo no net movement and are maintained in suspension by the upward flow of the water. In hindered settling, particles move downward, and in the simple case of batch settling, no net flow of water occurs. The Richardson and Zaki equation (Eq. 9-29) has been found to be applicable to both fluidization and hindered settling (Richardson and Harker, 2002), as have other relationships.

In water treatment, floc blanket clarification is more a fluidized bed than a hindered settling process. Extensive floc blanket data (Gregory, 1979) with Φ^* determined as the half-hour settled-solids volume, such that Φ^+ tended to be in the range 0.16 to 0.20, allowed Eq. 9-30 to be simplified to

$$v_s = v_o(1 - 2.5\Phi^*) \tag{9-35}$$

The data that allowed this simplification were obtained with alum coagulation (optimal coagulant dose and coagulation pH) of a high-alkalinity, organic-rich river water, but the value for k_2 (in Eq. 9-30) for other types of water producing similar-quality floc should be similar. The value of k_2 can be estimated as the ratio of the concentration at the compression point to the half-hour settled concentration. Values predicted for v_o by Eq. 9-35 are less, about one-half to one-third, than those likely to be estimated by Stokes' equation for v_t, assuming spherical particles (Gregory, 1979).

The theory of hindered settling and fluidization of particles of mixed sizes and different densities is more complex and is still being developed. In some situations two or more phases can occur at a given velocity, each phase with a different concentration. This has been observed with floc blankets to the extent that an early but temporary deterioration in performance occurs with increase in upflow (Gregory and Hyde, 1975; Setterfield, 1983, Fig. 9-8). An increase in upflow leads to intermixing of the phases with further increase in

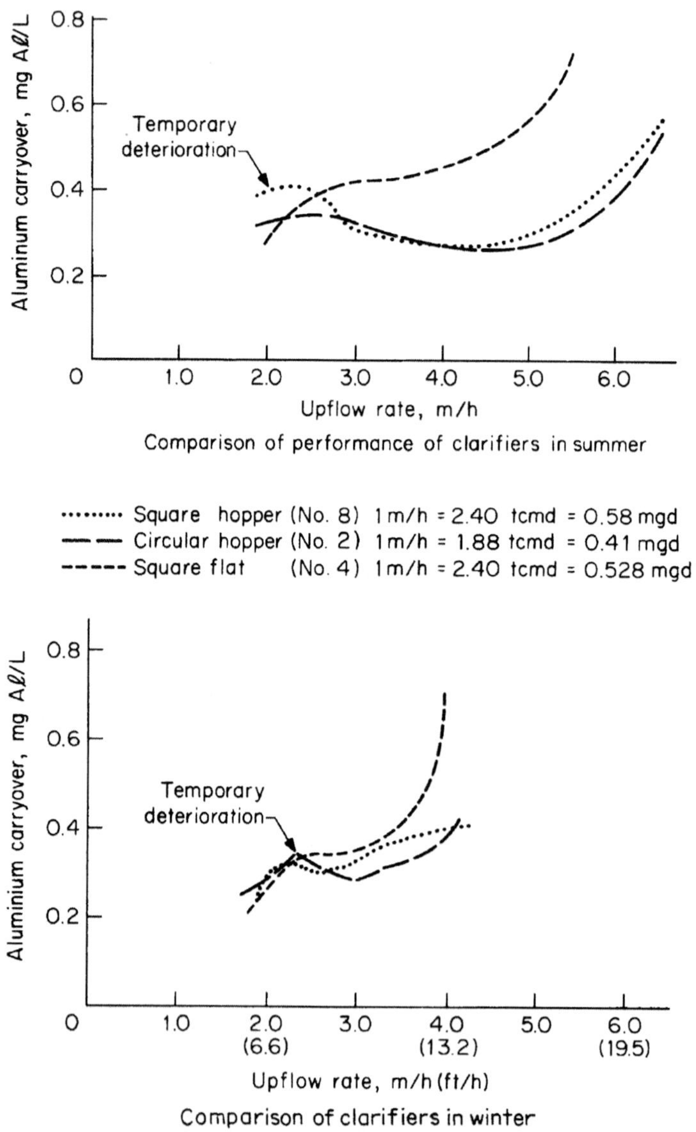

FIGURE 9-8 Floc-blanket performance curves showing "temporary" deterioration in settled water quality. (*Source: Setterfield, 1983.*)

upflow limited by the characteristics of the combined phase. The theory has been used to explain and predict the occurrence of intermixing and segregation in multimedia filter beds during and after backwash (Patwardan and Chi, 1985; Epstein and LeClair, 1985).

Example 9-1 Floc Blanket Sedimentation
Predict the maximum volume flux conditions for floc blanket sedimentation.

Solution For a floc blanket that can be operated over a range of upflow rates, collect samples of blanket at different upflow rates. For these samples measure the half-hour settled volume. Example results are listed here.

Upflow rate, m/h	1.6	1.95	2.5	3.05	3.65	4.2	4.7	5.15
Half-hour floc volume, %	31	29	25	22	19	16	13	10
Blanket flux = upflow rate x Half-hour floc volume, %m/h	49.6	56.6	62.5	67.1	69.4	67.2	61.1	51.5

These results predict that maximum flux occurs at an upflow rate of 3.7 m/h. If flux is plotted against upflow rate and against half-hour floc volume, then the maximum flux is located at 3.4 m/h for a half-hour floc volume of 20 percent, as shown in Fig. 9-9. The above results can be fitted to Eq. 9-35.

$$v_S = v_O(1-2.5\Phi^*)$$

$$3.44 = v_O\ (1 - 2.5 \times 0.2)$$

$$v_0 = 3.44/0.5 = 6.9 \text{ m/h}$$

This means that at the maximum flux the theoretical terminal settling velocity of the blanket is 6.9 m/h. The maximum operating rate for a floc blanket in a stable tank is about 70 percent of this rate (Gregory, 1979), or 4.8 m/h. ▲

Clarifiers as Collectors. Bache and Gregory (2007) regard flocculators, floc blanket clarifiers, DAF, and filters as having much in common, in that in each case the incoming stream of flocs encounters an array of *collectors*. In clarifiers, like flocculators, the collectors are a quasi-stationary distribution of flocs, the collection process coming under the heading of contact flocculation as described in Tambo and Hozumi (1979). The rate of loss of particles from the incoming stream depends on the flux of particles past a collector, the single collector collision efficiency (η_{pc}) and the attachment efficiency (α_{pc}). Schematically, the loss of concentration of incoming particles in passing through an absorbing layer of collectors can be represented by

$$\frac{dn_p}{dt} = -\alpha_{pc}\eta_{pc}kn_pn_c \qquad (9\text{-}36)$$

where n_p is the number of incoming particles/flocs per unit volume, n_c the number of collectors per unit volume, and k a transport coefficient with units (m^3s^{-1}). The mass transfer rate constant depends on the prevailing transport mechanism, for example, diffusion, turbulent transfer, sedimentation, or local stream flow. When particles, or floc, are larger than about 1 μm, k increases with size, in a manner depending on the transfer mechanism. The collision efficiency depends on both relative sizes of particles involved, but tends to follow the trend to be shown for a single collector in the section Fundamentals

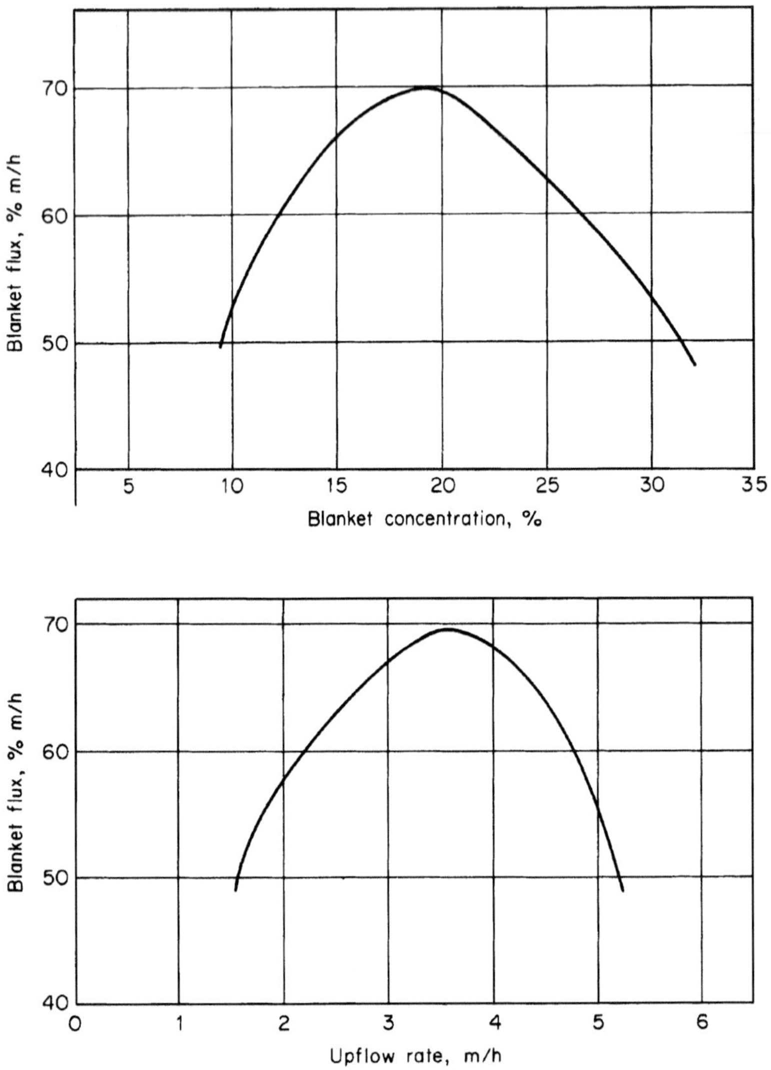

FIGURE 9-9 Relationship between blanket flux, blanket concentration, and upflow rate for Example 9-1.

of Dissolved Air Flotation (see the contact zone efficiency versus particle size figure). Bache and Gregory (2007) identify that removal efficiency increases with particle size, concentration of collectors, and the attachment efficiency. Therefore, clarification (sedimentation and flotation) should be enhanced by practices that lead to an increase of these, including optimization of coagulant dose, coagulation pH, polyelectrolyte dose, choice of chemicals, their preparation, manner of dosing and order of dosing, and flocculation and floc stressing (breakup conditions) before and after entry to the process.

Amirtharajah and O'Melia (1990) considered that when a blanket behaves as a turbulent reactor, the effluent concentration of microflocs can be estimated by (Bache and Gregory, 2007)

$$\frac{n}{n_O} = \exp\left[-\frac{\alpha G\Phi_F t_L}{\pi}\right] \tag{9-37}$$

where n_O is the number concentration of microflocs entering the base of the blanket, α the trapping efficiency, and t_L the water retention time. The equation shows that effluent concentration is reduced by increasing the value of $\alpha G\Phi_F t_L$. It can be shown (Bache and Gregory, 2007) that particle attenuation is steered by the dependence

$$G\Phi_F t_L \propto \Phi_F^{3/2}(1-\Phi_F)^{1/2}(1-k\Phi_F)^{r/2} \tag{9-38}$$

where r is defined by Eq. 9-30. This relationship (which is nominally valid for $0.1 < \Phi_F < 0.25$) highlights the dependence of floc blanket performance on blanket floc concentration (floc volume fraction) Φ_F.

Example 9-2 Floc Volume Fraction

With reference to Eq. 9-38, at what value of Φ_F is $G\Phi_F t_L$ a maximum (and n/n_0 is minimum), when $v_O = 3.5$ m/h, $k = 2.9$, and $r = 1.16$, for which maximum flux occurs when $\Phi_F = 0.16$ (see Bache and Gregory, 2007)?

Solution Calculate the value of each term for $0.9 \geq \Phi_F \geq 0.27$ and determine their products $F(\Phi_F)$.

Φ_F	0.9	0.11	0.13	0.15	0.17	0.19	0.21	0.23	0.25	0.27
$F(\Phi_F)$	0.022	0.029	0.035	0.040	0.046	0.050	0.053	0.055	0.055	0.054

This shows that $F(\Phi_F)$ (i.e., $G\Phi_F t_L$) has a maximum value, not when maximum flux occurs, but when Φ_F is about 0.24. This is probably associated with the transition from Type 3b to Type 4 settling. ▲

The implication from Example 9-2 is that best settled water quality might be produced when upflow rate is greater than at maximum flux, as observed by Su et al., (2004) in laboratory investigations. However, in most cases it would not be cost effective to design for or to operate at such a low flow rate. However, $t_L \propto h/v$, where h is depth of blanket, and $v \propto 1/\mu$. Therefore $\Phi_F \propto 1-(v\mu)^{1/r}$. Also, the velocity gradient in a blanket is proportional to the head loss through the blanket which relates to the mass of floc in suspension such that

$$G \propto \sqrt{\frac{[1-(v\mu)^{1/r}].v\rho}{\mu}} \tag{9-39}$$

Consequently, we obtain

$$G\Phi_F t_L \propto \frac{h}{(v\mu)^{1/2}}[1-(v\mu)^{1/r}]^{3/2}\rho^{1/2} \tag{9-40a}$$

When r is approximately 1 (as per Eq. 9-35).

$$G\Phi_F t_L \propto h\left\{\frac{[1-(v\mu)]^{3/2}}{(v\mu)^{1/2}}\right\}\rho^{1/2} \tag{9-40b}$$

This highlights the dependence of floc blanket performance on blanket depth, upflow velocity, and water viscosity, and hence water temperature and floc density.

Inclined (Tube and Plate) Settling

The efficiency of discrete particle settling in horizontal water flow depends on the area available for settling. Hence, efficiency can be improved by increasing the area. Some tanks have multiple floors to achieve this. A successful alternative is the use of lightweight structures with closely spaced inclined surfaces.

Inclined settling systems (Fig. 9-10) are constructed for use in one of three ways with respect to the direction of water flow relative to the direction of particle settlement: countercurrent, cocurrent, and crossflow. Comprehensive theoretical analyses of the various flow geometries have been made by Yao (1970). Yao's analysis is based on flow conditions in the channels between the inclined surfaces being laminar. In practice the Reynolds number must be less than 800 when calculated using the mean velocity v_θ between and parallel to the inclined surfaces and hydraulic diameter of the channel d_H.

$$d_H = \frac{4A_H}{P} \tag{9-41}$$

where A_H is the cross-sectional area of channel to water flow and P is the perimeter of A_H, so that the equation for the Reynolds number (Eq. 9-6) becomes

$$\mathrm{Re} = \frac{\rho v_\theta d_H}{\mu} \tag{9-42}$$

Countercurrent Settling. With reference to Fig. 9-10, the time t for a particle to settle the vertical distance between two parallel inclined surfaces is

$$t = \frac{w}{v\cos\theta} \tag{9-43}$$

where w is the perpendicular distance between surfaces and θ is the angle of surface inclination from the horizontal. The length L_p of surface needed to provide this time, if the water velocity between the surfaces is v_θ, is

$$L_p = \frac{w(v_\theta - v\sin\theta)}{v\cos\theta} \tag{9-44a}$$

By rearranging this equation, all particles with a settling velocity v and greater are removed if

$$v \geq \frac{v_\theta w}{L_p\cos\theta + w\sin\theta} \tag{9-44b}$$

When many plates or tubes are used,

$$v_\theta = \frac{Q}{Nwb} \tag{9-45}$$

where N is the number of channels made by $N + 1$ plates or tubes and b is the dimension of the surface at right angles to w and Q.

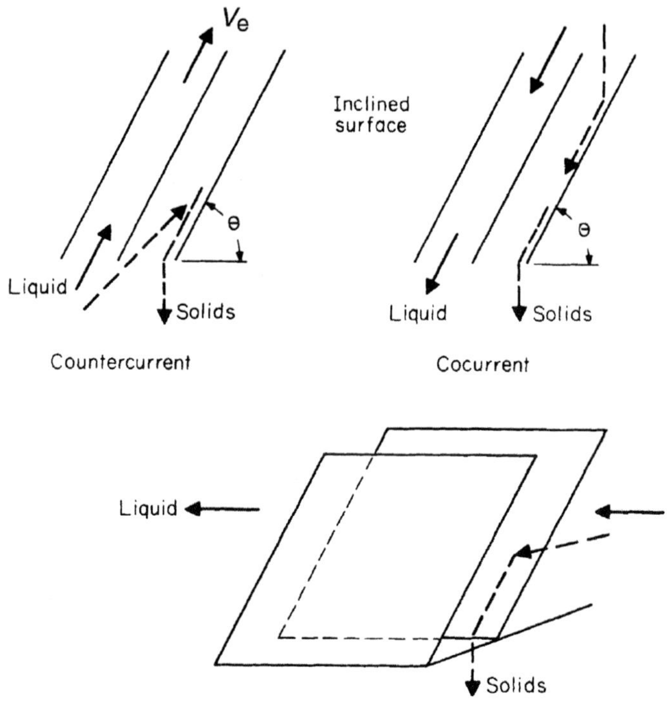

FIGURE 9-10 Basic flow geometries for inclined settling systems.

Cocurrent Settling. In cocurrent settling (Fig. 9-10), the time for a particle to settle the vertical distance between two surfaces is the same as for countercurrent settling. The length of surface needed, however, has to be based on downward and not upward water flow.

$$L_p = \frac{w(v_\theta + v\sin\theta)}{v\cos\theta} \tag{9-46a}$$

Consequently, the condition for removal of particles is given by

$$v \geq \frac{v_\theta w}{L_p\cos\theta - w\sin\theta} \tag{9-46b}$$

Crossflow Settling. The time for a particle to settle the vertical distance between two surfaces in crossflow settling (Fig. 9-10) is again given by Eq. 9-43. The water flow is horizontal and does not interact with the vertical settling velocity of a particle. Hence

$$L_p = \frac{v_\theta w}{v\cos\theta} \tag{9-47a}$$

and all particles with a settling velocity v and greater are removed if

$$v \geq \frac{v_\theta w}{L_p \cos\theta} \tag{9-47b}$$

Other Flow Geometries. The above three analyses apply only for parallel surface systems. In order to simplify the analysis for other geometries, Yao (1970) suggested a parameter S_c defined as

$$S_c = v \frac{(\sin\theta + L_r \cos\theta)}{v_\theta} \tag{9-48}$$

where $L_r = L/w$ is the relative length of the settler.

When v^* is the special case that all particles with velocity v or greater are removed, then for parallel surfaces, S_c is equal to 1.0. But the value for circular tubes is 4/3 and for square conduits is 11/8 (Yao, 1973). Identical values of S_c for different systems may not mean identical behavior.

The design overflow rate is also defined by v^* in Eq. 9-48, and Yao has shown by integration of the differential equation for a particle trajectory that the overflow rate for an inclined settler is given by

$$v^* = \frac{k_3 K v_\theta}{L_r} \tag{9-49}$$

where k_3 is a constant equal to 8.64×10^2 m^3/day/m^2 and

$$K = \frac{S_C\, L_r}{\sin\theta + L_r \cos\theta} \tag{9-50a}$$

For given values of overflow rate and surface spacing and when $\theta = 0$, Eq. 9-50a becomes

$$\frac{S_C}{L} = \text{constant} \tag{9-50b}$$

Equation 9-50b indicates that the larger the value of S_c, the longer the surface length must be to achieve the required theoretical performance. In practice, compromises must be made between theory and the hydrodynamic problems of flow distribution and stability that each geometry poses.

Example 9-3 Inclined Plate Settling

A tank has been fitted with 2.0-m (6.6-ft)-square inclined plates spaced 50 mm (2.0 in.) apart. The angle of inclination of the plates can be altered from 5° to 85°. The inlet to and outlet from the tank can be fitted in any way so that the tank can be used for either countercurrent, cocurrent, or crossflow sedimentation. If no allowances need to be made for hydraulic problems due to flow distribution and so on, then which is the best arrangement to use?

Solution Equation 9-44b for countercurrent flow, Eq. 9-46b for cocurrent flow and Eq. 9-47b for crossflow sedimentation are compared. As an example, the calculation for countercurrent flow at 85° is

$$\frac{v}{v_\theta} = \frac{50}{2000\cos 85 + 50\sin 85} = \frac{50}{174.3 + 49.8} = 0.223$$

The smallest value of v is required. Thus for the range,

Angle (θ)	5	15	30	45	60	75	85
Countercurrent (v/v_θ)	0.025	0.026	0.028	0.035	0.048	0.088	0.223
Cocurrent (v/v_θ)	0.025	0.026	0.029	0.036	0.052	0.106	0.402
Crossflow (v/v_θ)	0.025	0.026	0.029	0.035	0.050	0.096	0.287

From the above, little difference exists between the three settling arrangements for an angle of less than 60°. For angles greater than 60°, countercurrent flow allows settlement of particles with the smallest settling velocity. ▲

Floc Blanket Clarification

A simple floc blanket tank has a vertical parallel-walled upper section with a flat or hopper-shaped base; see Figs. 9-1 and 9-2. Water that has been dosed with an appropriate quantity of a suitable coagulant, and pH adjusted if needed, is fed downward into the base. The resultant expanding upward flow allows flocculation to occur, and large floc particles remain in suspension within the tank. Particles in suspension accumulate slowly at first, but then at an increasing rate due to enhanced flocculation and other effects, eventually reaching a maximum accumulation rate limited by the particle characteristics and the upflow velocity of the water. When this maximum rate is reached, a floc blanket can be said to exist.

As floc particles accumulate, the volume occupied by the suspension in the floc blanket, increases and its upper surface rises. The level of the floc blanket surface is controlled by removing solids from the blanket to keep a zone of clear water or supernatant water between the blanket and the decanting troughs, launders, or weirs.

A floc blanket is thus a fluidized bed of floc particles even though the process can be regarded as a form of hindered settling. However, true hindered settling exists only in the upper section of sludge hoppers used for removing accumulated floc for blanket-level control. Thickening takes place in the lower section of the sludge hoppers. Excess floc removed from the floc blanket becomes a residue stream and may be thickened to form sludge (see Chap. 22).

Mechanism of Clarification. Settling, entrainment, and particle elutriation occur above and at the surface of a blanket. The mechanism of clarification within a floc blanket is more complex, however, and involves flocculation, entrapment, and sedimentation. In practice, the mean retention time of the water within a blanket is in excess of the requirements for floc growth to control the efficiency of the process; that is, the opportunity for the small particles to become parts of larger and more easily retained floc is substantial, and therefore there are other factors that cause particles to pass through a blanket.

Physical removal by a collector process, involving interception and agglomeration, occurs throughout a floc blanket. Probably the most important process is mechanical entrapment and straining in which rising small particles cannot pass through the voids between larger particles that comprise the bulk of the blanket. (The mechanisms are not the same as in filtration through a fixed bed of sand because all the particles are in fluid suspension.) The efficiency of entrapment is affected by the spacing of the larger suspended floc particles, which in turn is related to floc quality (shape, density, etc.) and water velocity. When suspension destabilization, coagulation, is not optimal, then flocculation will be poor and result in a greater number of smaller particles that can pass through the floc blanket (see Chap. 8 for material on coagulation and particle destabilization).

Performance Prediction. Within a floc blanket, the relationship between floc concentration and upflow velocity of the water is represented by Eqs. 9-27 to 9-30 for hindered settling and fluidization. Unsuccessful attempts have been made to establish a simple theory for predicting solids removal (Cretu, 1968; Shogo, 1971). The relationships between settled water quality and floc concentration in the blanket, upflow velocity, and flux (Fig. 9-11) are of practical importance for understanding and controlling plant performance (Gregory, 1979). Floc blanket clarification has been modeled successfully (Hart, 1996; Gregory et al., 1996; Head et al., 1997), although calibration is essential.

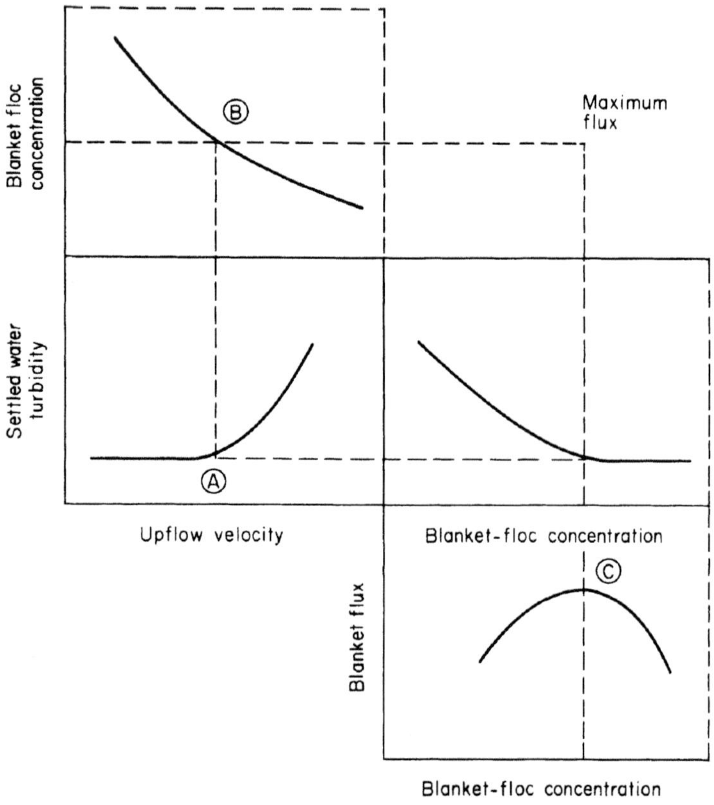

FIGURE 9-11 Typical relationships between settled-water quality and blanket concentration, upflow velocity and blanket flux. (*Source: Gregory, 1979.*)

The modeling by Head and associates has been successfully tested in dynamic simulations of pilot-scale and full-scale plants. The modeling is based on the theories and work of various researchers, including Gould (1967) and Gregory (1979). While the model accommodates the principle that removal rate of primary particles is dependent on blanket concentration, removal is simulated on the basis that the blanket region of a clarifier is a continuous stirred tank reactor (CSTR, see Chap. 4 about CSTR hydraulics and performance). To take into account the possibility of poor coagulation, the model can assume a non-removable fraction of solids.

The relationships in Fig. 9-11 show that settled water quality deteriorates rapidly (point *A*) as the floc concentration (point *B*) is decreased below the concentration at maximum flux (point *C*). Conversely, little improvement in settled water quality is likely to be gained by increasing floc concentration to be greater than that at maximum flux (to the left of points *A* and *B*). This is because for concentrations greater than that at maximum flux, interparticle distances are small enough for entrapment to dominate the clarification process.

As the concentration decreases below that at maximum flux, interparticle distances increase, especially between the larger particles, and the motion of the particles becomes more intense. Some of the larger particles might not survive the higher shear rates that develop. Thus smaller particles may avoid entrapment and escape from the floc blanket. Consequently the maximum flux condition represents possible optimum operating and design conditions. Maximum flux conditions and performance depend on various factors, which account for differences between waters and particle surface and coagulation chemistry and are described later. All this serves to make the point that the blanket floc volume fraction Φ_F has a central role and impact on the behavior of floc blanket clarifiers.

Effects of Upflow Velocity. The surface loading for floc blanket clarification is expressed as the upflow velocity or overflow rate. For some floc blanket systems the performance curve is quartic (fourth order) reflecting an early or premature deterioration in water quality of limited magnitude with increase in upflow velocity (Fig. 9-8). This deterioration is associated with segregation of particles, or zoning, in the blanket at low surface loading (Gregory, 1979) because of the wide range in particle settling velocities. This has been observed not only in the treatment of waters with a high silt content but also with the use of powdered carbon (Setterfield, 1983) and in precipitation softening using iron coagulation (Gregory and Hyde, 1975). As surface loading is increased, remixing occurs at the peak of the "temporary" deterioration as the lower-lying particles are brought into greater expansion.

For floc blanket clarification, the "corner" of the performance curve (point *A* in Fig. 9-11) is associated with the point of maximum flux. The limit to upflow velocity for reliable operation has been expressed by some (Bond, 1965; Tambo and Abe, 1969; Gregory, 1979) in simple terms of terminal velocity. Bond (1965) noted that the blanket surface remained clearly defined up to a velocity of about half of the zero-concentration settling rate v_o, that slight boiling occurred above $0.55v_o$, and that clarification deteriorated noticeably at about $0.65v_o$. Tambo and Abe (1969) found that a floc blanket is stable for velocities less than about $0.7v_o$ and very unstable at velocities greater than $0.8v_o$. Gregory (1979) found that the velocity at maximum flux was about $0.5v_o$, as given by Eq. 9-35, when the maximum flux half-hour settled solids volume is 20 percent. Gregory also observed that for velocities less than at maximum flux the blanket interface was sharp. However, as upflow velocity increased beyond this, the blanket surface became more diffuse to the extent that a blanket was very difficult to sustain for a velocity greater than $0.75v_o$. Su et al., (2004) examined blanket behavior for a synthetic suspension at small laboratory scale and observed that best settled water quality occurred when upflow was less than for maximum flux. If such observation also applies to full scale. then with due regard to the economic utilization of sedimentation and the effectiveness and economics of subsequent filtration, optimal operation lies between the upflow rate that produces best-quality water and the rate that can be safely used without easily resulting in loss of blanket by washout. The optimal rate is most likely to be at a higher than at a lower upflow rate; refer to Example 9-1.

The half-hour settled solids volume at maximum flux has been found to be in the range of 16 to 20 percent when coagulant dose and coagulation pH are selected from jar tests to minimize metal-ion concentration and turbidity. This has been observed for both alum and iron coagulation with and without using polyelectrolyte flocculant aid (Gregory et al., 1996). The actual value depends on the quality of the water as well as the choice of coagulation chemistry, because these govern such characteristics as floc strength, size, and density. The lowest blanket solids volume concentration that a discernible blanket can be found to exist with is about 10–12 percent. If upflow rate is increased to cause further dilution, then the blanket effectively becomes washed out.

Hydraulic Characteristics of Sedimentation Tanks

Reference should be made to Chap. 4 for more detailed information on the nature of the hydraulic characteristics of processes such as sedimentation tanks.

Ideal Flow Conditions. The simplest flow condition is plug flow when all water advances with equal velocity. Conditions only approximate to this when turbulence is small and uniform throughout the water. In laminar, non-turbulent, flow conditions, a uniform velocity gradient exists, with velocity zero at the wall and maximum at the center of the channel through which the water flows, and therefore plug flow cannot exist.

Major departures from plug and laminar flow conditions in sedimentation tanks are associated with currents caused by poor flow distribution and collection, wind, rising bubbles, or density differences caused by temperature or concentration. Currents caused by these factors result in short-circuiting of flow and bulk mixing and reduce the performance of the process predicted by ideal theory. The extent of departure from ideal plug flow performance can be assessed by residence time distribution analysis with the help of tracer studies and modeling with computational fluid dynamics (CFD); see Chap. 4.

Residence Time. The theoretical mean residence time of a process is the volume of the process from the point of dosing or end of the previous process through to the point at which separation efficiency is measured or the outlet of the process, divided by the flow-through rate (see Chap. 4). For horizontal-flow sedimentation, the volume of the entire tank is important in assessing the effect on sedimentation efficiency. For inclined settlers the volume within the inclined surfaces, and for floc blankets the volume of the blanket itself, is most important. Depth can be used to reflect the volume of a floc blanket.

The mean residence time of a process is effectively the length, in the direction of flow, divided by the average flow velocity. Thus, the mean residence time reflects the average velocity or overflow rate and will relate to sedimentation efficiency accordingly.

Flow-through curves (fluid residence time distributions) determined by the use of tracer tests are a graphical depiction of the distribution of fluid element residence times (see Chap. 4). These can be analyzed to produce estimates of efficiency of flow distribution and volumetric utilization of a tank, that is, its hydraulic characteristics. Consequently the extent to which sedimentation efficiency might be improved by modifying flow conditions can be estimated. Clements and Khattab (1968) have shown with model studies that sedimentation efficiency is correlated with the proportion of plug flow. Hydraulic characteristics were determined in Japan for a range of flow rate for five types of circular and rectangular tanks and compared (Kawamura, 1991). The results showed that a rectangular tank with an inlet diffuser wall was generally the superior design for minimizing short-circuiting and maximizing plug flow.

OPERATIONAL AND DESIGN CONSIDERATIONS FOR SEDIMENTATION

Types of Sedimentation Tanks

Horizontal-Flow Tanks

Rectangular Tanks. With rectangular horizontal-flow tanks, the water to be settled flows in one end, and the treated water flows out at the other end. The inlet flow arrangement must provide a flow distribution that maximizes the opportunity for particles to settle. If flocculation has been carried out to maximize floc particle size, then the flow at the inlet should not disrupt the flocs. This requires minimizing the head loss between the distribution channel and the main body of the tank. A certain amount of head loss is necessary, however, to achieve flow distribution. One option is to attach the final stage of flocculation to the head of the sedimentation tank to assist flow distribution.

The length and cross-sectional shape of the tank must not encourage the development of counterproductive circulatory flow patterns and scour. Outlet flow arrangements also must ensure appropriate flow patterns. The principal differences between tanks relate to inlet and outlet arrangements; length, width, and depth ratios; and the method of sludge removal. For horizontal-flow tanks with a small length/width ratio, the end effects dominate efficiency. Inlet and outlet flow distribution substantially affect overall flow patterns and residence time distribution. When the depth is greater than the width, the length/depth ratio is more important than the length/width ratio.

A length/width ratio of 20 or more ideally is needed to approach plug flow (Hamlin and Wahab, 1970; Marske and Boyle, 1973) and maximum efficiency for horizontal-flow and, presumably, inclined settlers, by determination of the reactor dispersion, (see Chap. 4). Such a high-value ratio may not be economically acceptable, and a lower ratio, possibly as low as about 5, may give acceptable efficiency if the flow distribution is good. The length/width ratio can be increased by installing longitudinal baffles or division walls.

Increasing the length/width ratio also has the effect of increasing the value of the *Froude number.*

$$\mathrm{Fr} = \frac{2v^2}{d_H g} \tag{9-51}$$

The value of Fr increases with the length/width ratio because of the increase in velocity v and decrease in hydraulic diameter d_H. Camp (1936) has shown that the increase in value of Fr is associated with improved flow stability.

Poor flow distribution may produce currents or high flow velocities near the bottom of a tank. This may cause scour, or resuspension of particles from the layer of settled sludge. Scour may cause transportation of solids along the bottom of the tank to the outlet end. An adequate tank depth can help to limit scour, and consequently depths less than 2.4 m (8 ft) are rarely encountered (Gemmell, 1971). To avoid scour, the ratio of length to depth or surface area to cross-sectional area must be kept less than 18 (Kalbskopf, 1970). This has implications for limiting the length/width ratio.

Initially, sludge was removed from tanks by simple manual and hydraulic methods. To avoid interruption in operation and to reduce manpower, mechanically aided sludge removal methods were introduced. In the most common method, mechanical scrapers push the sludge to a hopper at the inlet end of the tank. Periodically, the hopper is emptied hydraulically.

If sludge is not removed regularly from horizontal-flow tanks, allowance must be made to the tank depth for sludge accumulation so that the sedimentation efficiency remains unaffected. Sludge can be allowed to accumulate until settled water quality starts to be impaired. The tank floor should slope toward the inlet, because the bulk of solids generally settle closer to the inlet end.

The frequency of sludge removal depends on the rate of sludge accumulation. This can be estimated by mass balance calculations. Sometimes the decomposition of organic matter in the sludge necessitates more frequent sludge removal. Decomposition can be controlled by prechlorination, if this practice is acceptable, otherwise decomposition may produce gas bubbles that disturb the settled sludge, create disruptive flow patterns, and impair settled water quality by releasing reduced manganese (Mn) and iron (Fe).

Frequent sludge removal is best carried out with mechanical sludge scrapers that sweep the sludge to a hopper at the inlet end of the tank. Frequent removal results in easy maintenance of tank volumetric efficiency, and better output efficiency with continuous operation.

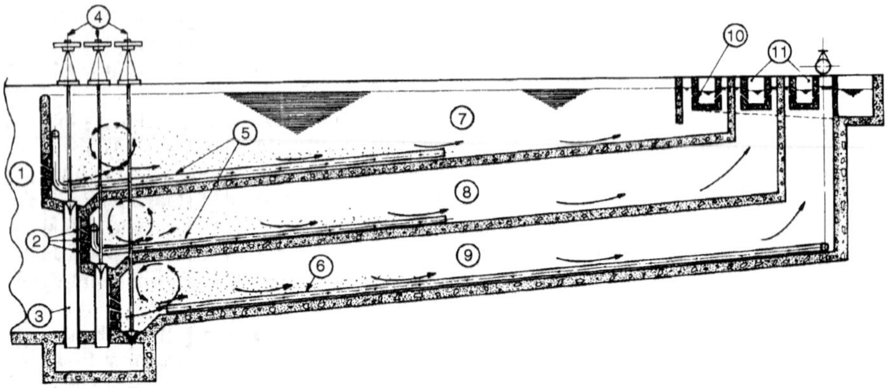

① Flocculation tank
② SPLIT-ROLL inlets
③ Upper floor sludge drain
④ Manually controlled TOP VALVE
⑤ COMBCET: Sludge suction system (optional)
⑥ WATERRINSE: Water flushing system (optional)
⑦ Upper clarifying compartment
⑧ Intermediate clarifying compartment
⑨ Lower clarifying compartment
⑩ DUCK-LIPS: Effluent collectors
⑪ Clarified water collecting channels

FIGURE 9-12 Multistory horizontal tank with parallel flow on three levels. (*Courtesy of OTV, Paris, France, and Kubota Construction Co., Ltd., Tokyo, Japan.*)

Multistory Tanks. Multistory, or tray, tanks are a result of recognizing the importance of settling area to settling efficiency. Two basic flow arrangements are possible with multistory tanks. The trays may be coupled in parallel with flow divided between them (Fig. 9-12) or coupled in series with flow passing from one to the next. A few of the latter reverse-flow tanks exist in the United States with two levels.

Multistory tanks are attractive where land value is high. Difficulties with these tanks include a limited width of construction for unsupported floors, flow distribution, sludge removal, and maintenance of submerged machinery. Successful installations in the United States show that these difficulties can be overcome in a satisfactory manner. However, tanks with reverse flow (180° turn) tend to be the least efficient (Kawamura, 1991).

Circular Tanks. Circular tank flow is usually from a central feedwell radially outward to peripheral weirs (Fig. 9-13). The tank floor is usually slightly conical to a central sludge well. The floor is swept by a sludge scraper that directs the sludge toward the central well. Circular tanks incorporate a central feedwell which is needed to assist flow distribution but also is often designed for flocculation and so incorporate some form of agitation. If this agitation is excessive, it carries through to the outer settling zone and affects sedimentation efficiency (Parker et al., 1996).

Radial outward flow is theoretically attractive because of the progressively decreasing velocity. The circumference allows a substantial outlet weir length and hence a relatively low weir loading. Weirs must be adjustable or installed with great accuracy to avoid differential flow around a tank. Circular tanks are convenient for constructing in either steel or concrete although they might be less efficient in the use of land than rectangular tanks. Sludge removal problems tend to be minimal.

The settling efficiency might be less than expected because of the problem of achieving good flow distribution from a central point to a large area. The principal differences between circular tanks are associated with floor profile and sludge scraping equipment.

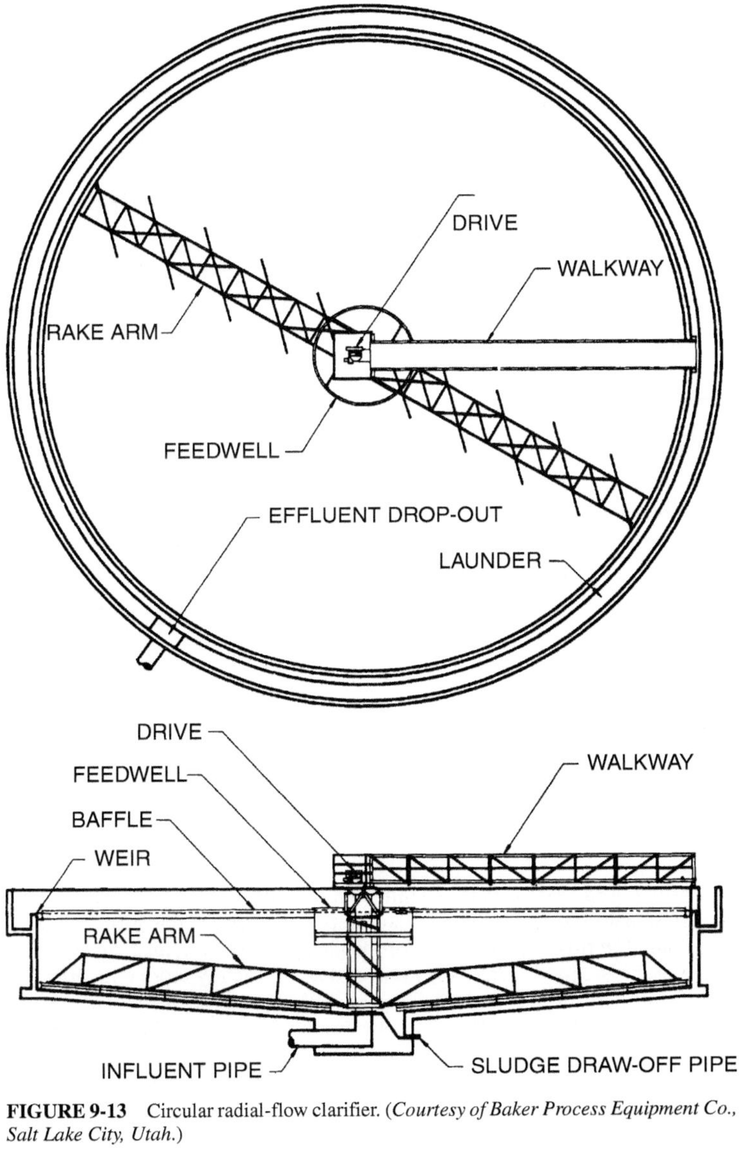

FIGURE 9-13 Circular radial-flow clarifier. (*Courtesy of Baker Process Equipment Co., Salt Lake City, Utah.*)

Inclined (Plate and Tube) Settlers. Individual or prefabricated modules of inclined plate or tube settlers can be constructed of appropriate materials. The advantages of prefabricated modules include efficient use of material, accuracy of separation distances, lightweight construction, and structural rigidity. Inclined surfaces may be contained within a suitably shaped tank for countercurrent, cocurrent, or crossflow sedimentation. Adequate flocculation is a prerequisite for inclined settling if coagulation is carried out. The tank containing the settler system also can incorporate the flocculation stage and preliminary

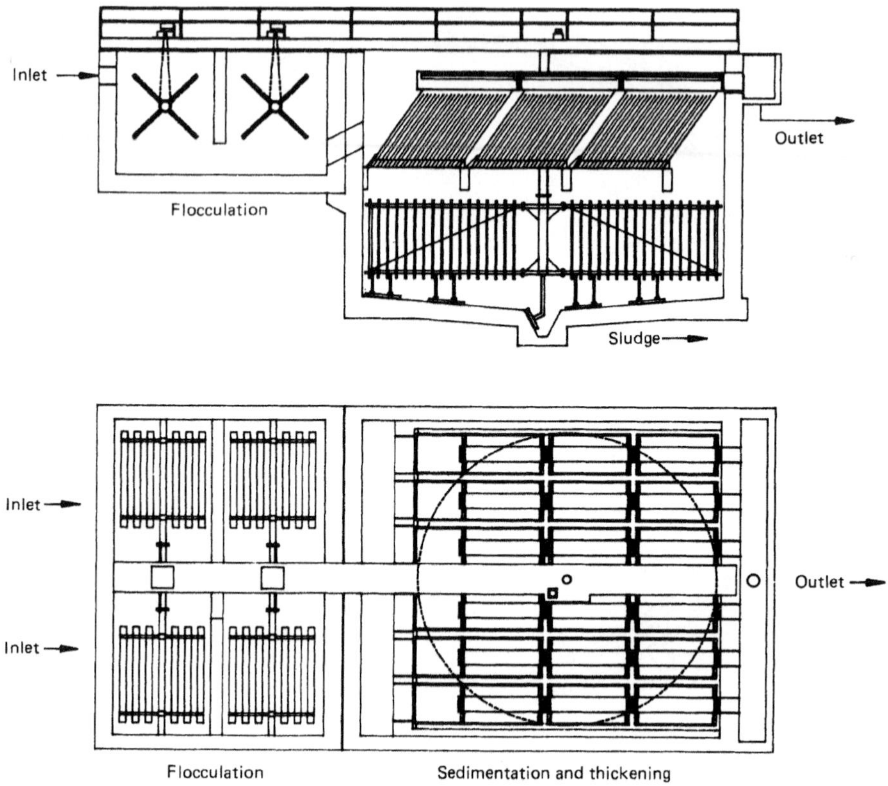

FIGURE 9-14 Inclined-plate settler with preflocculation and combined thickening. (*Courtesy of US Filter-Zimpro Products, Rothschild, Wis.*)

sludge thickening (Fig. 9-14). Particle removal can be enhanced by ballasting the floc. The angle of inclination of the tubes or plates depends on the application, the tendency for self-cleaning, and the flow characteristic of the sludge on the inclined surface. If the angle of inclination of the inclined surfaces is great enough, typically more than 50° to 60° (Yao, 1973), self-cleaning of surfaces occurs. Demir (1995) found for inclined plates fitted at the end of a pilot horizontal-flow settler the optimum angle is about 50°, with this becoming more pronounced as surface loading rate increases. When the angle of inclination is small, the output of the settler must be interrupted periodically for cleaning. This is because the small distance between inclined surfaces allows little space for sludge accumulation. An angle as little as 7° is used when sludge removal is achieved by periodic backflushing, possibly in conjunction with filter backwashing. The typical separation distance between inclined surfaces for unhindered settling is 50 mm (2 in.) with an inclined length of 1 to 2 m (3 to 6 ft).

The main objective in inclined settler development has been to obtain settling efficiencies close to theoretical, for example, as predicted by Eqs. 9-44, 9-46, and 9-47. Considerable attention must be given to providing equal flow distribution to each channel, producing good flow distribution within each channel, and collecting settled sludge while preventing its resuspension.

With inclined settlers, the velocity along the axis of the channels defines the flow regime. In practice, the efficiency is usually related to either the surface loading on the basis of the plan area occupied by the settling system, to the upflow velocity, or to the loading based on the total area available for settlement. Countercurrent settlers are probably more numerous than cocurrent and crossflow settlers, although each type has advantages and advantages.

Countercurrent Settlers. In countercurrent inclined settlers, the suspension is fed below the settling modules and the flow is up the channels formed by the inclined surfaces (Fig. 9-10). Solids settle onto the lower surface in each channel. If the angle of inclination is great enough, the solids move down the surface counter to the flow of the water; otherwise, periodic interruption of flow, possibly with flushing, is necessary for cleaning.

Tube settlers are used mostly in the countercurrent settling mode. Tube modules have been constructed with various configurations (Fig. 9-15), including square tubes between vertical sheets, alternating inclination between adjacent vertical sandwiches, chevron-shaped tubes between vertical sheets, and hexagonal tubes.

Countercurrent modular systems are suitable for installing in existing horizontal flow tanks and some solids contact clarifiers to achieve upgrading and uprating. Closely spaced inclined surface systems are not cost-effective in floc blanket clarifiers although widely spaced (0.3 m [1 ft]) inclined plates are. Tube modules may aid up-rating by acting in part as baffles that improve flow uniformity following poor inlet flow distribution or counteracting the effect of strong wind blowing across the water surface.

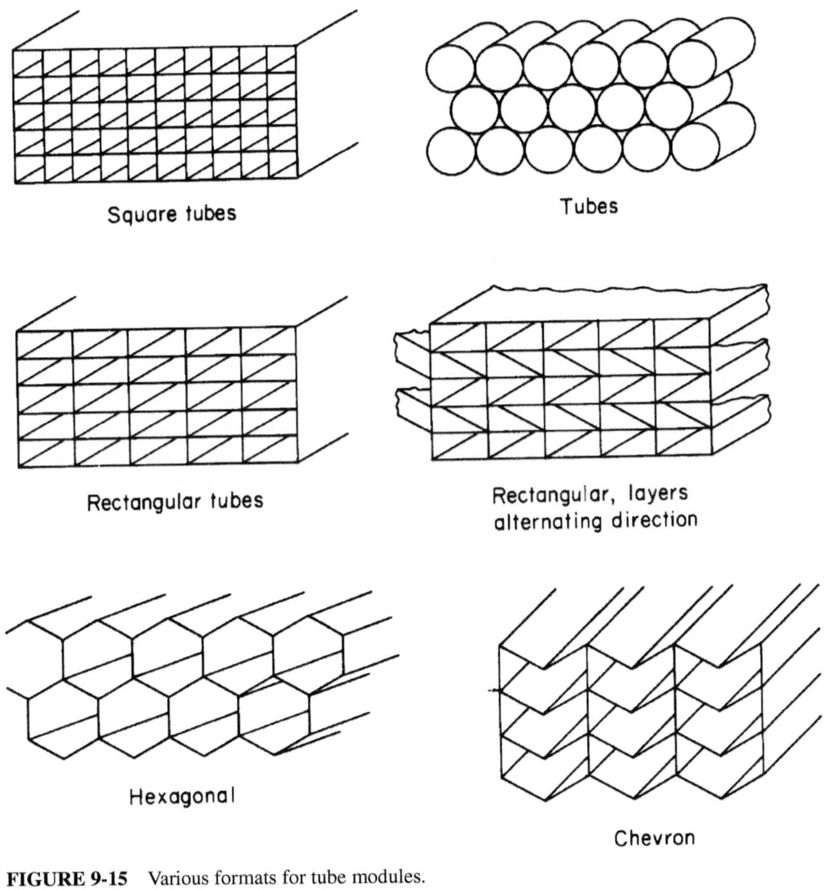

FIGURE 9-15 Various formats for tube modules.

Cocurrent Settlers. In cocurrent sedimentation, the suspension is fed above the inclined surfaces and the flow is down through the channels (Fig. 9-10). Settled solids on the lower surface move down the surface in the same direction as the water above. Special attention must be given to collecting settled water from the lower end of the upper surface of each channel to prevent resuspension of settled solids.

Crossflow Settlers. In crossflow sedimentation, the suspension flows horizontally between the inclined surfaces and the settled solids move downward (Fig. 9-10). In this case, resuspension of settled solids is usually less of a problem than in countercurrent and cocurrent settling. This might not be true in some systems in which the direction of inclination alternates (Fig. 9-16) (Gomella, 1974). Alternating inclination can allow efficient use of tank volume and results in rigidity of modular construction. Development and application of crossflow systems has occurred mainly in Japan.

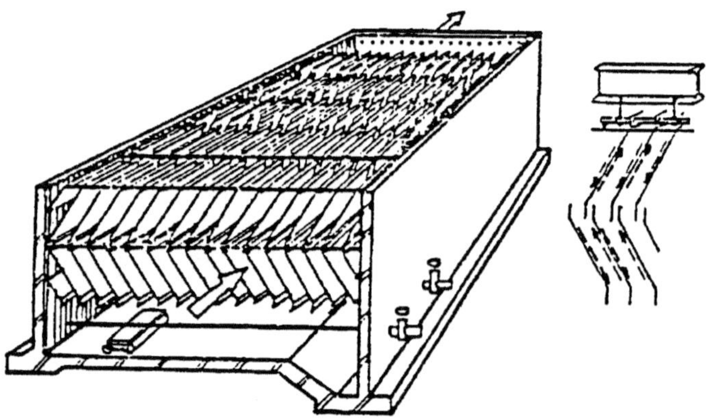

FIGURE 9-16 Alternating cross-flow lamella settler. [*Source: Gomella, C. "Clarification avant filtration; ses progress recents" (Rapport General 1). Int'l Water Supply Assoc. Int'l Conf., 1974.*]

Solids-Contact Clarifiers. Solids-contact clarifiers are generally circular in shape and contain equipment for mixing, flow recirculation, and sludge scraping. There is a wide variety of these tanks and most are of proprietary design. Hartung (1951) has presented a review of eight different designs. Contact clarifiers are of two types: premix, and premix-recirculation like that in Fig. 9-17.

In the simpler premix system, water is fed into a central preliminary mixing zone which is mechanically agitated. This premix zone is contained within a shroud that acts as the inner wall of the outer annular settling zone. Chemicals can be dosed into the premix zone. Water flows from the premix zone under the shroud to the base of the settling zone.

In the premix-recirculation system, water is drawn out of the top of the premix zone and fed to the middle of the settling zone. The recirculation rate can exceed the actual flow of untreated water to the tank such that the excess flow in the settling zone is drawn downward and under the shroud back into the premix zone. This movement recirculates solids that can assist flocculation in the premix zone.

Mechanical equipment associated with solids-contact clarifiers must be adjusted or tuned to the throughput. Excessive stirring motion in the premix zone can be counterproductive, whereas too little stirring can result in poor radial-flow distribution under the shroud as well as poor chemical mixing and flocculation. In solids-contact units, sludge settles to the

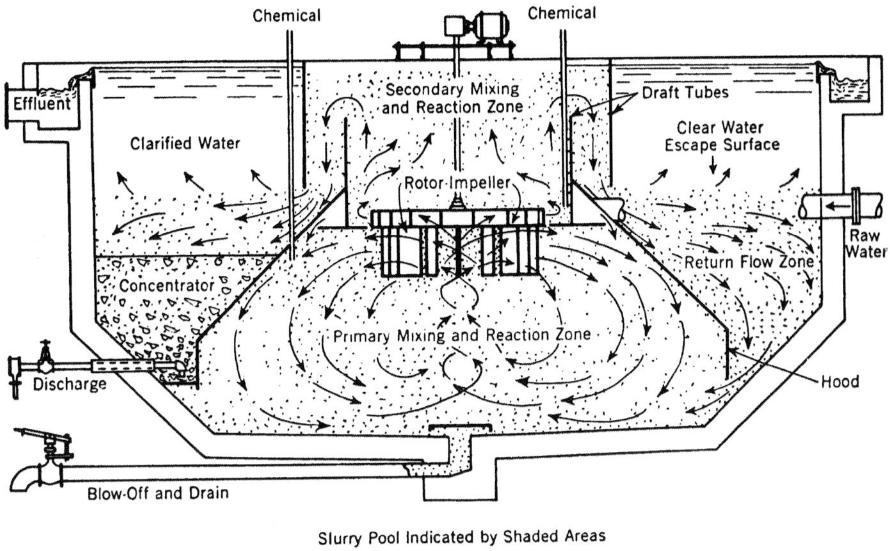

FIGURE 9-17 The Accelator® solids contact clarifier. (*Source: Hartung, 1951.*)

tank floor and is removed with mechanical equipment. Clarifiers with recirculation to keep solids in suspension allow excess solids to accumulate in sludge pockets or concentrators as illustrated in Fig. 9-17. Appropriate operation of these pockets contributes to controlling the concentration of solids in suspension and influences the sedimentation efficiency of the clarifier.

Floc Blanket Clarifiers. Both types of solids-contact tanks, premix and premix-recirculation, can function as floc blanket clarifiers if stable and distinct floc blankets can be established and easily maintained in the settling zone. Only a few designs of solids-contact clarifiers have been developed with this objective. Usually, the volume and concentration of solids in circulation in contact units is not great enough to maintain a blanket in the outer separation zone.

Hopper-Bottomed Tanks. The first-designed floc blanket tanks had a single hopper bottom, square or circular in cross section. In these units coagulant-dosed water was fed down into the apex of the hopper. The hopper shape assists with even flow distribution from a single-point inlet to a large upflow area. The expanding upward flow allows floc growth to occur, large particles to remain in suspension, and a floc blanket to form. The pressure loss through the floc blanket, although relatively small, helps to create homogeneous upward flow.

A single hopper, conical or pyramidal, only occupies 33 percent of available volume relative to its footprint. In addition, it is expensive to construct and its size is limited by constructional constraints. Consequently, alternative forms of hopper tanks have been developed to overcome these drawbacks yet retain the hydraulic advantage of hoppers. These include tanks with multiple hoppers, a wedge or trough, a circular wedge (i.e., premix type of clarifier), and multiple troughs.

As a floc blanket increases in depth, settled water quality improves but with diminishing return (Fig. 9-18). Equation 9-40 has been fitted to the data in Fig. 9-18, assuming constant

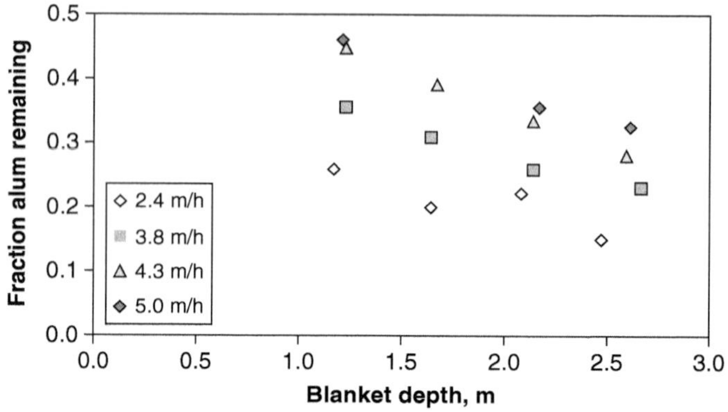

FIGURE 9-18 Example of the effects of upflow velocity and blanket depth on removal of aluminum from water by a floc blanket. (*Based on Fig. 40 in Miller et al., 1965.*)

floc density and an optimum value of $r = 1.62$, on the basis that alum residual is dependent on the collector function $G\Phi t$ resulting in Fig. 9-19. There were problems of flow control with the pilot plant and positive displacement pumps were used to try and overcome these. The outliers to the general trend in Fig. 9-19 are likely to reflect this problem.

Figure 9-19 accounts for the effect of upflow velocity. The blanket depth defines the quantity of solids in suspension and collector time, that is, the value of t in the term $G\Phi t$. If effective depth is defined as the total volume of blanket divided by the area of its upper interface with supernatant, then the effective depth of a hopper is roughly 1/3 the actual depth. This applies similarly for collector time. As a result, flat-bottomed tanks have an actual depth that is much less than that of hopper tanks with the same effective depth (or collector time). Effective depths of blankets are typically in the range 2.5 to 3 m (8 to 10 ft).

The quantity of solids in suspension, established by the blanket depth and concentration, affects sedimentation efficiency because of the effect on flocculation. In addition, the head loss assists flow distribution, ensuring a more stable blanket (delaying the onset of boiling and loss of blanket) and thereby greater blanket concentration.

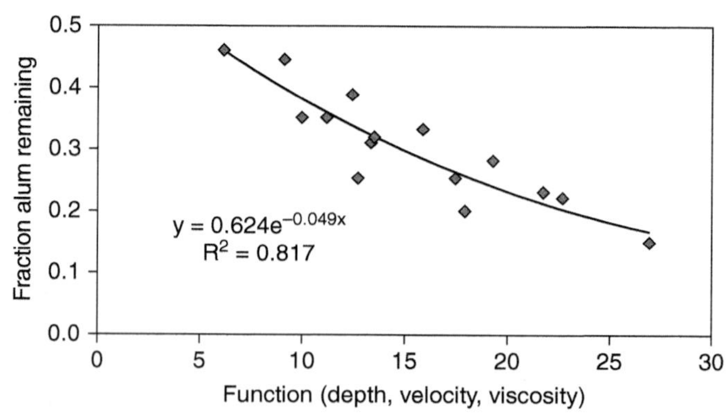

FIGURE 9-19 Figure 9-18 data normalized with collector model, Eq. 9-40.

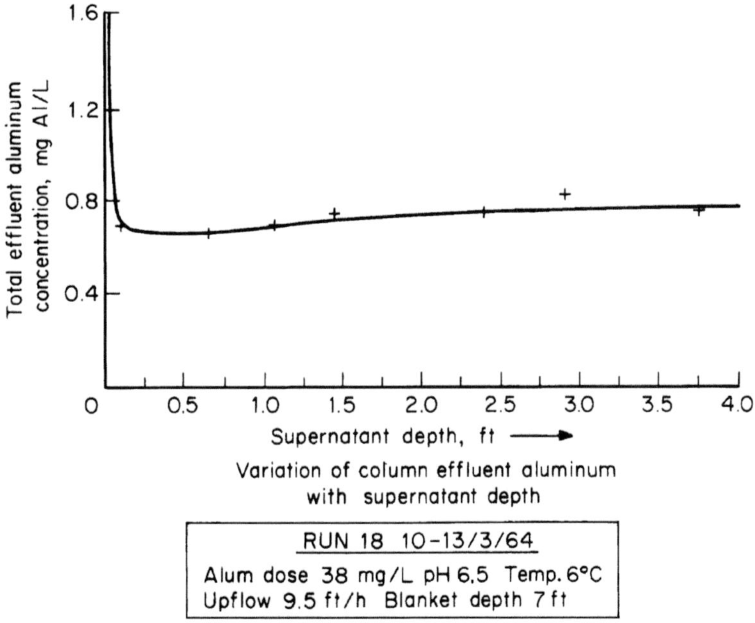

FIGURE 9-20 Change in settled-water quality as depth of supernatant increases above a floc blanket. (*Source: Miller et al., 1966.*)

A very stable floc blanket (e.g., as in a pilot plant) can be operated with little depth of supernatant water, less than 0.3 m (1 ft), without significant solids carryover (Fig. 9-20) (Miller et al., 1966). In practice, the feasibility of this depends on weir spacing and likely disturbance by wind. Generally, blankets are operated in a manner likely to produce a blanket with a diffuse surface, have a tendency to exhibit some unstable boiling, and have poor level control methods. Thus, a supernatant water depth of at least 1 m (3 ft) may be necessary to minimize carryover, especially after increases in upflow velocity. Supernatant water depths of 2 m (6 ft) are commonly provided, but this may be unnecessary if there is good blanket-level control and care is taken when increasing the upflow rate (Hart, 1996).

Control of the blanket surface is easily achieved using a slurry weir or sludge hopper with sills set at an appropriately high level. Sludge hoppers can be emptied frequently by automatically timed valves. The Gravilectric® proprietary suspended sludge hopper system utilizes a strain gauge to initiate drainage of sludge from suspended canvas cones (Fig. 9-1). Another method is to monitor the magnitude of the turbidity in a sample drawn continuously from a suitable point.

Sludge hoppers, pockets, and cones must be sized to allow efficient removal of sludge (Pieronne, 1997). They must be large enough to allow in situ preliminary thickening, even when the sludge removal rate has to be high. High rates might occur following a substantial increase in flow-through rate during periods of high chemical doses. Dynamic simulation modeling by Hart (1996), Gregory et al., (1996), and Head et al., (1997) has shown the importance of correct sizing of sludge hoppers and their operation, especially for handling increases in flow through the clarifiers. If sizing and operation are inadequate, then loss of blanket level control and poor settled water quality will occur when the flow is increased.

Example 9-4 Floc Blanket Clarifier
What area will be needed for removing floc to control the blanket level in a floc blanket clarifier using the conditions for Example 9-1?

Solution Assume the aluminum coagulant dose to the water is 3.2 mg Al/L. The aluminum concentration in the floc blanket at maximum flux (when the floc volume concentration is 20 percent) is 110 mg Al/L. The concentration of aluminum in the blanket is proportional to the floc volume.

From Example 9-1 the conditions are

Upflow rate, m/h	1.6	1.95	2.5	3.05	3.65	4.2	4.7	5.15
Half-hour floc volume, %	31	29	25	22	19	16	13	10

The proportion of the floc blanket tank area needed for removing floc is determined by a mass balance.

$$\text{(Aluminum dose)} \times \text{(total volumetric flow rate to tank)}$$
$$= \text{(blanket aluminum concentration)} \times \text{(volumetric settlement rate into removal area)}$$

But

$$\text{(Total volumetric flow rate to tank)} = \text{(upflow rate)} \times \text{(total tank upflow area)}$$

and

$$\text{(Volumetric settlement rate into removal area)}$$
$$= \text{(upflow rate)} \times \text{(area for removal)}$$

Thus

$$\text{(Area for removal)} \times \text{(blanket aluminum concentration)}$$
$$= \text{(total tank upflow area)} \times \text{(aluminum dose)}$$

This means that the proportion of tank area needed for removing floc is the ratio of aluminum concentration in the blanket to that being dosed. For example, at the upflow rate of 1.95 m/h,

$$\text{Concentration of aluminum in the blanket} = 110 \times 29/20$$

Therefore

$$\text{Proportion of area needed} = 3.2/110 \times 20/29 \times 1/100\% = 2.0\%$$

Hence

Upflow rate, m/h	1.6	1.95	2.5	3.05	3.65	4.2	4.7	5.15
Proportion of area needed, %	1.9	2.0	2.3	2.6	3.1	3.6	4.5	5.8

For a different aluminum dose, the area will need to be accordingly proportionally greater or less. In practice, a greater area will be required to cope with the short-term need to remove excess floc at a high rate to prevent the blanket from reaching the launders when the upflow rate is increased quickly. ▲

Flat-Bottomed Tanks. It is simpler and cheaper to build a floc blanket tank with a flat bottom, and therefore few, if any, tanks are now built with hoppers. In flat-bottomed tanks, good flow distribution is achieved using either multiple downward, inverted-candelabra feed pipes or laterals across the floor (Figs. 9-1 and 9-2).

An inverted-candelabra system can ensure good distribution for a wide range of flows but may obstruct installation of inclined settling systems. The opposite can apply to a lateral distribution system. In the proprietary Pulsator® design, reliability of flow distribution is ensured by periodically pulsing the flow to the laterals.

Inclined Settling with Floc Blankets. Tube modules with the typical spacing of 50 mm (2 in) between inclined surfaces are not cost effective in floc blanket clarifiers (Gregory, 1979). With the blanket surface below the tube modules, the settled water quality is no better than from a stable and efficient tank without modules.

With the blanket surface within the modules, the floc concentration in the blanket increases by about 50 percent, but no commensurate improvement in settled water quality occurs. The failure of closely spaced inclined surfaces to increase hindered settling rates relates to the proximity of the surfaces and a circulatory motion at the blanket surface that counteracts the entrapment mechanism of the blanket (Gregory, 1979).

The problem with closely spaced surfaces diminishes with more widely spaced inclined surfaces. An effective spacing is about 0.3 m (1 ft), but no optimization studies are known to have been published. Large (2.9 m) plates, however, have been shown to be preferable to shorter (1.5 m) plates (Casey et al., 1984). The combined action of suppressing currents and inclined settling with widely spaced surfaces can result in about a 50 percent greater throughput than with a good floc blanket without inclined surfaces. The proprietary Superpulsator® tank is the Pulsator® design with widely spaced inclined surfaces. A further development is the Ultrapulsator® design which although similar to the Superpulsator® design, has short inclined hexagonal tube modules in the supernatant zone.

Sludge Recycling. Many proprietary designs of clarifiers purposely incorporate recycling of sludge. This is relevant for precipitation lime softening, and historically this may account for this design feature, which probably was subsequently promoted as beneficial for coagulation-flocculation. The benefits of producing denser floc, for example, pelletization, have been justified by Tambo and Wang (1993) and Bache and Gregory (2007), and sludge recycling is one way of achieving this. However, so far pelletization has not found application for potable water production probably because of undue dependence on the use of polyelectrolyte. The DensaDeg® process (Degremont, 1991) combines flocculation, inclined tube settling, and sludge settlement in a raked thickener; it also incorporates sludge recycling. The sludge recycling is considered a core feature of the process to enhance particle density and hence increase settlement rate and enhance sludge settling and thickening. The process has found widespread application for uses other than potable water production.

Ballasted Floc Systems. Floc produced by coagulating mineral-bearing water generally settles faster than floc produced by coagulating water containing little mineral turbidity. Consequently, mineral turbidity added purposely to increase floc density can be useful. Bentonite is the usual choice, and fly ash has been considered in eastern Europe. The advantages of ballasting floc also arise with powdered activated carbon, dosed for taste and odor control or pesticide removal (Standen et al., 1995), and when precipitation softening is carried out in association with iron coagulation. Sometimes fine sand has been used as the ballasting agent.

A process based on fine sand ballasting was developed in Hungary in the 1950s and 1960s. In this process, microsand (<200 μm) is recovered and recycled for economy by pumping the sludge through a bank of small hydrocyclones. The basic process was first introduced in the United Kingdom by the proprietary name of *Simtafier* (Webster et al., 1977) and in France by the name of Cyclofloc® (Sibony, 1981). A subsequent development

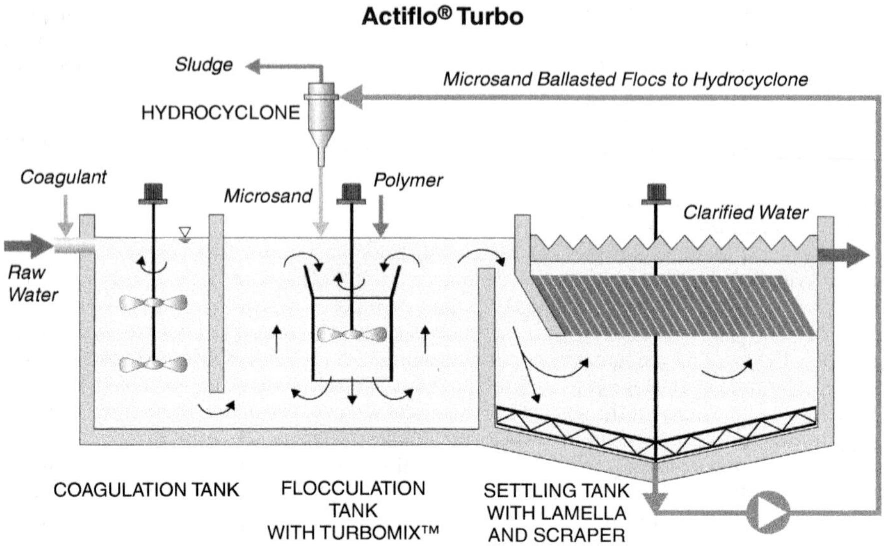

FIGURE 9-21 Actiflo® Turbo: schematic diagram of a typical unit. (*Courtesy of Veolia Water Solutions and Technologies, Saint Maurice, France.*)

is the Fluorapide® system, which combines floc blanket settling and sand ballasting with inclined settling (Sibony, 1981). Another development is the Actiflo® system (de Dianous et al., 1990), which combines sand ballasting with inclined settling and has become widely applied, with Actiflo®Turbo as the most recent development of the process (Fig. 9-21).

Typically in the Actiflo® process, Fig. 9-21, the raw water is first coagulated. The microsand (about 70–100 μm) is then dosed together with the polymer and about 6 min allowed for maturation prior to lamella settling. Typical microsand doses are 3 g/L (range of 3–5 kg/m^3). It is important to ensure the sand size range is within specification for efficient recovery by the hydrocyclones so that sand loss is typically about 2 mg/L. At some plants, part of the polymer dose is dosed to the recycled sand before entering the flocculation tank. Although this does not improve clarified water quality, it can increase filter run lengths. As with ordinary clarification, it is important not to overdose polymer to avoid filter blinding. Desjardins et al. (2002) studied ballasted flocculation in the laboratory by jar tests. The results showed that coagulant dose and coagulation pH have the greatest effects on settled water quality. By contrast, filterability was more sensitive to coagulant and microsand doses. Filterability was also dependent on the ratio of microsand-to-polymer dose and improved when this increased.

The ballasting agent, whether for example, microsand or bentonite, besides enhancing floc density will act also as a collector (see Eq. 9-36). It follows that the ballasting agent must have an optimum dose for a given water quality and choice of chemical treatment. Al Farisi (1988) (Bache et al., 1995) examined the impact of bentonite dose on the size-density function (e.g., Fig. 9-4). As the bentonite dose increased, it was found that the floc density passed through a maximum located in the interval 3–7 mg/L, that is, similar to the doses used in practice for bentonite.

Powdered magnetite, recovered by magnetic drum filters, also has been suggested as a ballasting agent. A process from Australia (Dixon, 1984), with the proprietary name Sirofloc®, is based on recycling magnetite without coagulation, other than for a small dose of cationic polymer. The magnetite is chemically conditioned with sodium hydroxide (changing its surface charge) such that a metal-ion coagulant is not needed. Sirofloc® is almost efficient enough for subsequent filtration to be unnecessary. High-rate (20-m/h) filtration may be needed to remove residual magnetite and manganese that Sirofloc® cannot normally remove (Gregory et al., 1988).

Other Factors Influencing Sedimentation Efficiency

Surface Loading. The surface loading of a sedimentation tank is expressed as the flow rate per unit of surface area of that part of the tank in which settling is meant to happen. The performance of all settling processes is influenced by surface loading. Settled water quality deteriorates when surface loading is increased (Fig. 9-8). The reasons for this deterioration are various, but (particularly for discrete particle settling processes) changes in flow rate may affect flocculation performance as well as sedimentation efficiency.

The point at which the performance curve (Fig. 9-8) crosses the limit of acceptable quality defines the maximum reliable surface loading. Cost effectiveness requires the highest surface loadings possible; however, the efficiency will then be more sensitive to major variations in surface loading and water quality.

The reliable surface loading for a given type of settling process depends on a wide range of factors. For example, when treating a highly colored, low-alkalinity water in winter by floc blanket clarification, the loading might be only 1 m/h (3 ft/h), whereas when treating a minerally turbid, high-alkalinity water in summer with a flocculent aid, the loading could be greater than 7 m/h (23 ft/h).

Size and Shape of Tank

Size and Number. The number of tanks can affect the flexibility of plant operation. When operating close to the limit of acceptable quality, isolation of one tank is likely to impose a reduction in total plant production. No fewer than two tanks should be used for reliability, while three will provide greater operational flexibility. See Baruth (2005) for further guidance on water treatment plant design.

The size of tanks might be limited by construction constraints, and consequently this will dictate the minimum number of tanks. Although factors that affect sedimentation efficiency, such as loading, velocity, and various dimension ratios, can be maintained with any size tank, the performance of extra-large tanks may become unacceptable and limit tank size. For example, tanks with large surface area will tend to be more vulnerable to environmental effects, such as wind-induced circulation.

When considering the number of tanks, the performance of a floc blanket is dependent on the upflow velocity, blanket floc concentration, effective blanket depth, and dosing and delay time conditions. Thus, provided that the size and shape of floc blanket tanks do not affect these factors, any such size or shape might not be expected to affect sedimentation efficiency. The efficiency is affected, however, because differences occur between tanks in their hydrodynamic and hydraulic conditions for flocculation and flow distribution.

The length/width ratio is not relevant to floc blanket clarifiers, although baffles have been shown to be useful (Gregory and Hyde, 1975). With inclined settlers the length of the flow path affects the sedimentation efficiency (Yao, 1973). When lengths are short, end effects may limit efficiency. Additionally, problems of flow distribution may mean that inclined plate settlers with narrow plates are more efficient than those with wide plates.

Depth. According to Hazen's law (Eq. 9-23b), for discrete particle settling, the settling efficiency depends on the tank area and is independent of the depth. However, in practice increasing tank depth promotes settling efficiency and is interrelated with width in horizontal and inclined settlers. A minimum depth may be needed to limit scour as mentioned previously. Depth also defines the spacing between inclined surfaces and, therefore, the number of layers and inclined settler efficiency.

Flow Arrangements

Inlet and Outlet. The purpose of an inlet is to distribute the incoming water uniformly over the cross section of a tank for a wide range of flow rates. The outlet arrangement is as important as the inlet in ensuring good flow distribution. Numerous investigations (Fair and Geyer, 1954; Thirumurthi, 1969) have shown how inlets and outlets in horizontal-flow tanks influence residence time distribution and settling efficiency. Tests with model tanks (Price and Clements, 1974; Kawamura, 1981) have shown that symmetry is desirable and obstacles (such as support walls and piers) can disturb uniformity. Complex arrangements can be as good as simple ones, and a uniformly fed submerged weir can give good results without a baffle. See Baruth (2005) for further guidance on water treatment plant design.

For all types of settlers, flow velocities in approach channels or pipe work that are low enough to allow premature settlement must be avoided. Conversely, flows high enough to disrupt floc must also be avoided.

In floc blanket tanks, the injection velocity from the inlet pipes governs the input energy, which can affect blanket stability and boiling. Boiling causes direct carryover and a reduction in blanket concentration; these result in deterioration of settled water quality.

Outlet weir or launder length and loading affect outlet flow distribution. Therefore, weirs and launders must be kept clean and unobstructed, but the effectiveness of long launders is questionable (Kawamura and Lang, 1986).

Baffling. Baffles are useful in horizontal-flow tanks at the inlet and outlet to assist flow distribution, longitudinally to increase length/width ratio, and as vanes to assist changes in horizontal-flow direction. Kawamura (1981) did numerous model tests with diffuser walls and arrived at a number of design guides relating to the position of walls and the free area in the walls.

In hopper-bottomed floc blanket tanks, baffles might help inlet flow distribution (Gould, 1967; Hale, 1971), but the centering of the inlet should be checked first (Gregory, 1979). A matrix of vertical baffles in the floc blanket is an alternative method for improving tank performance if flow distribution is poor (Gregory and Hyde, 1975).

Vertical baffles have been installed in some solids-contact clarifiers so that the central mixer does not cause the fluid to rotate in the settling zone. Fluid rotation will lift the floc up the outer side of the settling zone. A proprietary tube-module baffle system (Envirotech Corp) has been used to improve flow distribution in radial-flow tanks. Baffles at the water surface can help to counteract wind disturbance.

Particulate and Water Quality

Seasonal Water Quality. Sedimentation efficiency may vary with seasonal changes in temperature, alkalinity, and similar parameters, as well as with changes in the nature of dissolved organic matter (DOM) (e.g., color) and turbidity being coagulated.

Temperature affects efficiency by influencing the rate of chemical reactions, solubilities, viscosity of water, and hence particle settling velocities. Temperature can also be a surrogate for change in other parameters that occur on a similar seasonal basis. Changes in alkalinity, color, turbidity, and orthophosphate concentration for eutrophic waters affect coagulation reactions and the properties and rate of settling of resulting floc particles.

With floc blanket clarification, for example, the reliable upflow velocity in the summer can be more than twice that in the winter (Fig. 9-8) (Gregory, 1979; Setterfield, 1983). This is important when plant size or reliable plant output is defined.

Example 9-5 Floc Blanket Upflow Rate
What will be the floc blanket upflow rate at maximum flux for different temperatures, given the conditions defined in Example 9-1?

Solution The Stokes' settling velocity of particles is inversely proportional to water viscosity, Eq. 9-8.

$$v_{\text{temp 1}} = v_{\text{temp 2}} \times \left(\frac{\mu_{\text{temp 2}}}{\mu_{\text{temp 1}}}\right)$$

The viscosity of water for different temperatures (see App. D) is as follows:

Temperature, °C	5	10	15	20	25
Viscosity, mN-s/m^2	1.52	1.31	1.14	1.00	0.893

If the results given in Example 9-1 were obtained at the temperature of 10°C, then assuming no change in water chemistry and other factors that might affect the floc, the maximum flux upflow rate at 20°C can be estimated, for example, as

$$v_{20°\text{C}} = 3.44 \times 1.31/1.00 = 4.50 \text{ m/h}$$

Hence for the above temperature range,

Maximum flux upflow rate, m/h	2.96	3.44	3.95	4.50	5.05

This means that the upflow rate for reliable operation in the summer could be more than 1.5 times that in the winter for this temperature range. ▲

Raw Water Settlement. The turbidity of raw waters is caused by either or both organic and inorganic matter. Bar screens remove large solids and prevent the entry of larger fish while mesh strainers are intended for removing small fish and organic debris, algae, and other planktonic species. Neither is intended for removing high concentrations of organic or inorganic silt that can occur in some directly abstracted river waters. The settlement characteristics of these solids, and their individual and relative concentrations may vary substantially according to season, for example, between wet and dry seasons. Removal of most of these solids before chemical treatment helps to minimize the chemicals needed and protects from blockages in pipes and channels and from damage to pumps. It is not unusual to have either or both grit channels and preliminary settlement tanks (i.e., without coagulation). Such initial removal of solids may also help with managing solids waste costs. The literature does not provide much help on settlement methods for raw waters entering treatment works. Ideas can be gleaned from methods used to control siltation in canals and impounding reservoirs and sewage works inlet grit removal, with a mind to how accumulating solids can be routinely removed. One of the authors has encountered a site where after grit channels, preliminary horizontal-flow settlement was done in conjunction with marginal coagulation (adequate for destabilization) followed by floc blanket clarification with secondary coagulation (adequate to complete coagulation with respect to residual aluminum, turbidity, and dissolved color).

Coagulation. Particle settling velocity is affected by various particle characteristics, principally size, shape, and density. The choice of coagulant, coagulation pH, and application efficiency affect floc particle characteristics. The choice of coagulant should be dictated by chemical considerations (see Chap. 8). Optimum chemical doses for sedimentation can be predicted reliably by jar tests. Jar tests also reliably predict color and DOC removal and soluble metal-ion concentrations, but not turbidity removal because of the difference in hydrodynamic conditions in the batch jar test and the continuous settling process.

Adequate control of dosing and mixing of the chemicals should be provided. If a change in coagulant dose or pH improves the sedimentation efficiency, the current chemical doses may be incorrect or the mixing or flocculation inadequate. The difference in efficiency between identical tanks supposedly receiving the same water can be because of differences in chemical doses if made separately to each tank, poor chemical mixing before flow splitting, or different flows through the tanks.

Flocculation. The efficiency of horizontal and inclined-flow sedimentation is dependent on prior flocculation if coagulation is carried out. The efficiency of solids-contact units depends on the quality of preliminary mixing. Mixing assists with chemical dispersion, flocculation, solids resuspension, and flow distribution. Because these can conflict with each other, mixing rates should be adjusted when changes are made to the flow-through rate. This is rarely done, however, because it is usually too difficult.

The extent of flocculation is dependent on velocity gradient and time (see Chap. 8). The time delay between chemical dosing and inlet to a floc blanket, to allow preliminary flocculation in connecting pipe work and channels, affects the floc blanket clarification efficiency (Fig. 9-22) (Miller and West, 1968; Yadav and West, 1975). The delay times between dosing the coagulant and dosing a polyelectrolyte and between dosing the polyelectrolyte and the inlet to the tank also affect the efficiency (Fig. 9-23). The reasons for needing such delay times are not entirely understood, but relate to floc development in delivery channels in association with efficiency of mixing, rate of hydrolysis and precipitation, and rate of formation of primary and subsequent floc particles. Useful delay times are likely to be in the range 1/2 to 8 min, similar to those needed for direct filtration. The greatest times tend to be

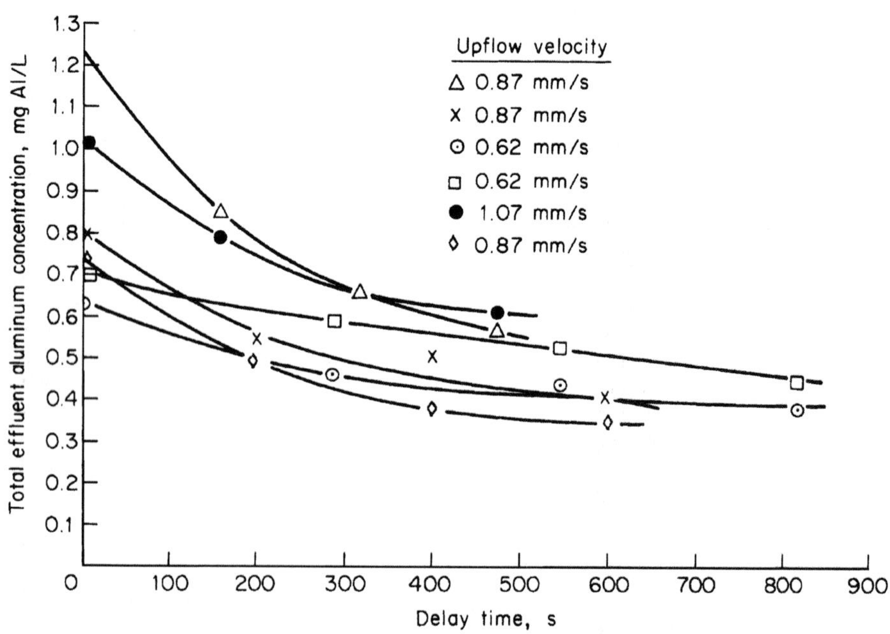

FIGURE 9-22 Change in settled-water quality with increase in delay time between chemical dosing and inlet to a floc blanket. (*Source: Yadav and West, 1975.*)

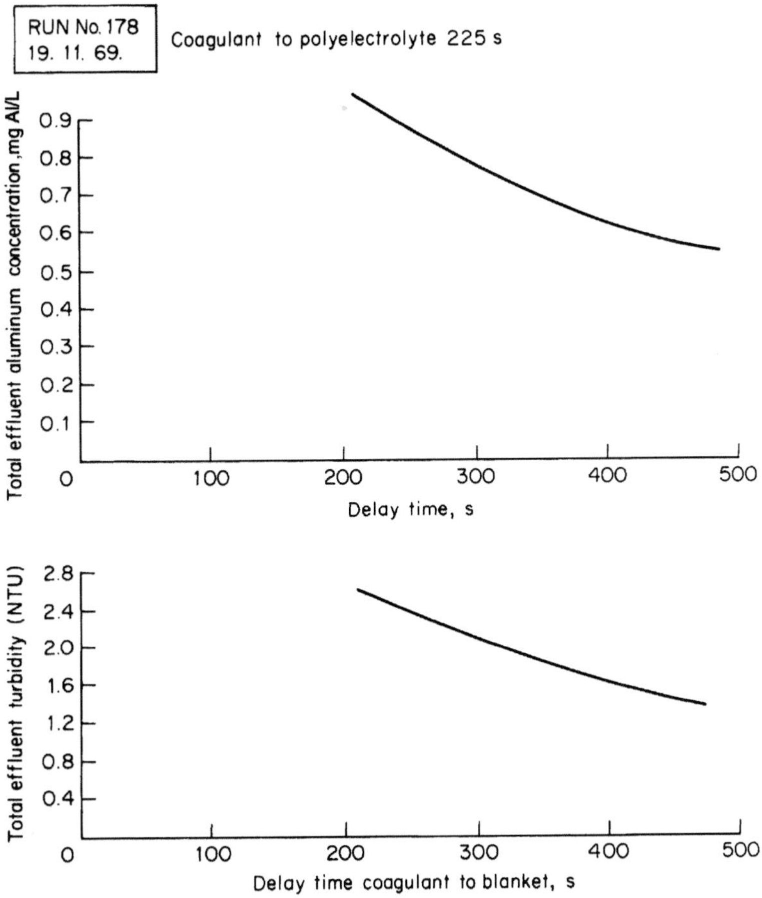

FIGURE 9-23 Change in settled-water quality with increase in delay time between dosing a polyelectrolyte and inlet to a floc blanket. (*Source: Yadav and West, 1975.*)

needed when coagulating very cold water with aluminum sulfate. High-rate flocculation for inclined settling or dissolved air flotation is similarly affected, such that flocculation times of 5 to 10 min may be effective when water is warm but longer times may be needed when water is very cold, especially with aluminum sulfate. Water temperature has less effect on flocculation with polyaluminum chlorides and ferric salts (see Chap. 8).

Delay time is not a substitute for initial rapid chemical dispersion. Many operators have learned that simple improvements to flow conditions and dispersion arrangements at the point of dosing can produce substantial improvements in chemical utilization and sedimentation efficiency.

Polyelectrolytes and Other Agents. Flocculent aids improve sedimentation efficiency by increasing floc strength and size. Jar tests can be used to select suitable polyelectrolytes and determine a trial dosing range. Optimal doses must be carefully determined by full-scale plant tests. An excessive dose producing good settled water may be detrimental to filters. If the apparent optimal dose of polyelectrolyte on the plant is much greater than that indicated by jar tests, then mixing and flocculation conditions must be checked.

Dosing of flocculent aids increases particle settling velocities in floc blanket clarification and may cause in an increase in the blanket-floc concentration if upflow rate isn't increased to match the improvement. A diminishing return also exists in increasing the dose of flocculent aid and is associated with the blanket-floc concentration increasing with increase in flocculent dose and becoming greater than at maximum flux (Fig. 9-24) (Gregory, 1979).

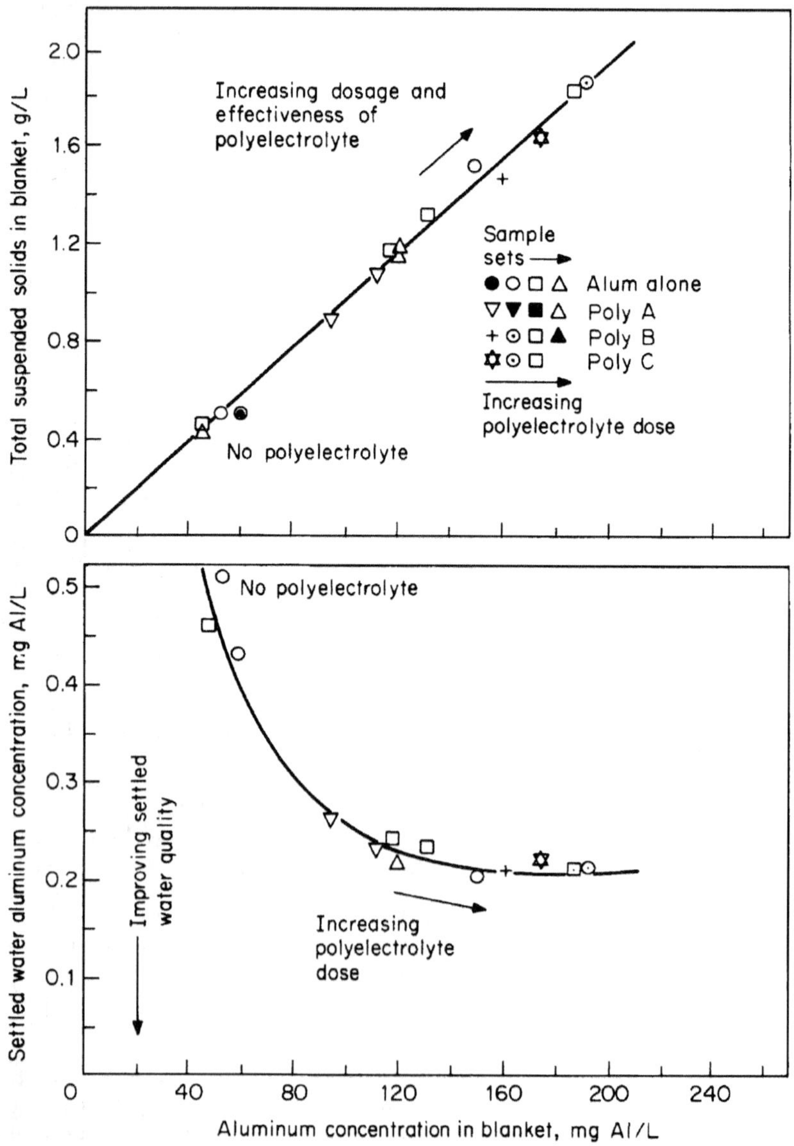

FIGURE 9-24 Effect of polyelectrolyte type and dose on blanket solids concentration and settled water quality. (*Source: Gregory, 1979.*)

Thus the polyelectrolyte dose need be only enough to ensure that the floc concentration is greater than at maximum flux and that the blanket surface is distinct and not diffuse and certainly not so great as to take the floc concentration into the thickening regime. Additives that increase floc density and settling velocity, such as floc ballasting with bentonite or fine sand, have an effect on sedimentation similar to flocculent aids.

Climate and Density Currents. Wind can induce undesirable circulation in horizontal-flow tanks, with crossflow winds reducing sedimentation efficiency the most (Price and Clements, 1974). Therefore, wind effects should be minimized by constructing tanks to align with prevailing winds. Strong winds can disrupt floc blanket stability. This can be counteracted by covering tanks with roofs, placing floating covers between the launders, constructing windbreaks around each tank, aligning the launders across the prevailing wind, installing scum-boards or baffles at the water surface across the prevailing wind, or installing fully submerged baffles in the supernatant water. Wind combined with low temperatures can cause severe ice formation and affect outlet flow. Total enclosure may be needed to prevent ice formation.

High solar radiation can cause rapid diurnal changes in water temperature and density. Rapid density changes because of temperature, solids concentration, or salinity can induce density currents that can cause severe short-circuiting in horizontal tanks (Hudson, 1981) and inclined settlers (White et al., 1974). Hudson (1981) has reviewed methods of minimizing or preventing induced currents. He considered the methods to fall into several categories: (1) use of a surface weir or launder takeoff over a large part of the settling basin, (2) improved inlet arrangement, (3) schemes to provide increased basin drag and friction, and (4) slurry recirculation. Substantial diurnal variation in temperature also can result in release of dissolved gases, especially in biologically active waters. The resulting rising columns of bubbles can disrupt floc blankets.

INTRODUCTION TO DISSOLVED AIR FLOTATION

History of Flotation

Over 2000 years ago, the ancient Greeks used a flotation process to separate the desired minerals from the gangue, the waste material (Gaudin, 1957). Crushed ore was dusted onto a water surface, and mineral particles were retained at the surface by surface tension while the gangue settled. In 1860 Haynes patented a process in which oil was used for the separation of the mineral from the gangue (Kitchener, 1984). The mineral floated with the oil when the mixture was stirred in water.

In 1905 Salman, Picard, and Ballot developed the *froth flotation* process by agitating finely divided ore in water with entrained air. A small amount of oil was added sufficient to bestow good floatability to the sulfide grains (Kitchener, 1984). The air bubbles together with the desired mineral collected as foam at the surface while the gangue settled. The first froth flotation equipment was developed by T. Hoover in 1910 (Kitchener, 1984), and except for size, it was not much different than the equipment used today. Later, in 1914, Callow introduced air bubbles through submerged porous diffusers. This process is called *foam flotation*. The two processes, froth and foam flotation, are generally known as *dispersed air flotation* and are widely used in the mineral industry.

Elmore suggested in 1904 the use of electrolysis to produce gas bubbles for flotation. This process, although not commercially used at that time, has been developed into electrolytic flotation (Bratby, 1976). Elmore also invented the dissolved air (vacuum) flotation process in which air bubbles are produced by applying a vacuum to the liquid, which releases the air in the form of minute bubbles (Kitchener, 1984).

The original patent for the dissolved air pressure flotation process was issued in 1924 to Peterson and Sveen for the recovery of fibers and *white water* in the paper industry (Lundgren, 1976). In pressure dissolved air flotation, the air bubbles are produced by releasing the pressure of a water stream saturated with air above atmospheric pressure. Initially, dissolved air flotation (DAF) was used mainly in applications where the material to be removed, such as fat, oil, fibers, and grease, had a density less than that of water. The first reported use of DAF (based on the ADKA vacuum system) for drinking water treatment was in Scandinavia during the 1920s (Haarhoff, 2008). However, it was not until the 1960s the process also became more acceptable for potable water treatment applications (Haarhoff, 2008). Now it has become accepted as an alternative to sedimentation, not only in the Scandinavian countries and the United Kingdom, but also in the United States and many other countries (Haarhoff, 2008). Flotation is employed mainly for the treatment of nutrient-rich reservoir waters that may have heavy algal blooms and for low-turbidity, low-alkalinity colored waters (Zabel and Melbourne, 1980). These types of waters are difficult to treat by sedimentation, because the floc produced by chemical treatment has a low settling velocity. In the United States, some of the DAF plants are for treating relatively high-quality supplies (low turbidity and low color) that were previously unfiltered or had direct filtration, where the addition of DAF provides an extra pathogen barrier.

Types of Flotation Processes

Flotation is described as a gravity separation process in which gas bubbles attach to solid particles to cause the apparent density of the bubble-solid agglomerates to be less than that of the water, thereby allowing the agglomerate to float to the surface. Different methods of producing gas bubbles give rise to different types of flotation processes. These are *electrolytic flotation, dispersed air flotation,* and *dissolved air flotation.* The basis of electrolytic or electroflotation is the generation of bubbles of hydrogen and oxygen in a dilute aqueous solution by passing a dc current between two electrodes (Barrett, 1975). The application of electrolytic flotation has been restricted mainly to sludge thickening and small wastewater treatment plants in the range 10 to 20 m^3/h (50,000 to 100,000 gpd) and has limited application to drinking water treatment.

Dispersed air flotation involves the formation of bubbles by diffusers, spargers, and other mechanical means. Large bubbles are formed (sizes of a mm) and can be effective in treating suspensions containing large particles such as mineral separation and industrial waste treatment. Dispersed air flotation is generally unsuitable for water treatment as the bubble size is large and because of high turbulence and use of undesirable frothing and surfactant chemicals. However, the French Ozoflot® system applies ozone-rich air as a pretreatment oxidation process for waters high in algae using diffusers to achieve dispersion.

In dissolved air flotation, air bubbles are produced by the reduction in pressure of a water stream saturated with air. The three main types of dissolved air flotation are vacuum flotation (Zabel and Melbourne, 1980), microflotation (Hemming et al., 1977), and pressure flotation (Barrett, 1975). In microflotation the entire volume of water is subjected to increased pressure by passing it down and up a shaft approximately 10 m deep. In the downflow section the water is aerated and the amount of air available for flotation is restricted by the depth of shaft provided. However, pressure flotation in which a proportion of the treated water is recycled under pressure to dissolve the air into the water is the method used in drinking water treatment and is the subject of interest here. A general description is presented next and serves to give the reader a basic description of the DAF process and terminology. This is followed by a section Fundamentals of Dissolved Air Flotation. Then, there is a section Operational and Design Considerations for Flotation. Finally in an applications section we also address emerging technologies.

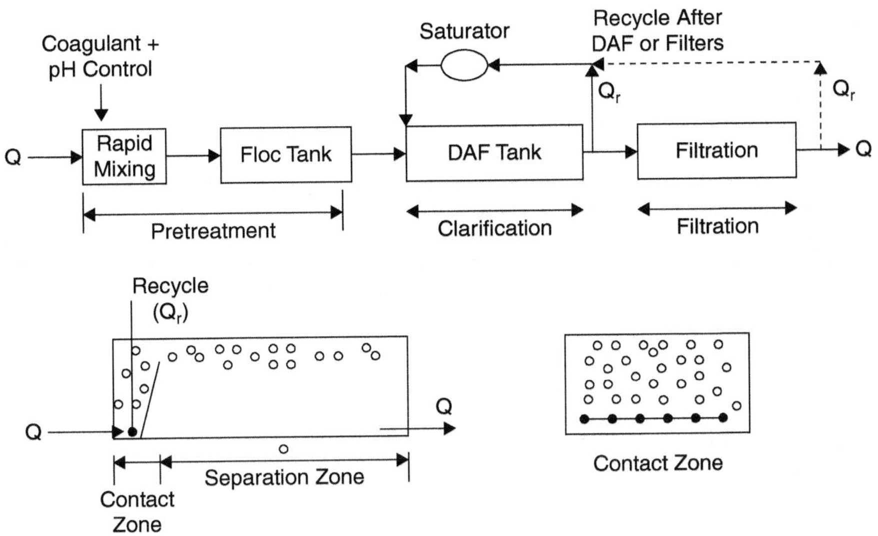

FIGURE 9-25 *Top:* Process schematic of a DAF in a conventional type water plant; *Bottom:* DAF tank showing the contact and separation zones.

General Description

Figure 9-25 is a process schematic diagram for DAF in a conventional type drinking water plant in which it has replaced sedimentation as a clarification process. Coagulation and flocculation are required pretreatment processes. Coagulation is a chemical addition step. Metal coagulants such as alum, polyaluminum chlorides, and ferric salts are commonly used. Coagulation has two purposes: to destabilize particles present in the raw water and to convert dissolved natural organic matter into particles. The latter purpose, over the last 25 years, has become an essential part of water treatment used to remove as much dissolved organic carbon (DOC) as feasible through coagulation and subsequent separation of the formed flocs. Good coagulation chemistry is essential for DAF to work efficiently, as it is for sedimentation and filtration. Its fundamental role in DAF is addressed in the next section on fundamentals of DAF, while the section on operational and design considerations for DAF contains some practical material. Principles of coagulation and flocculation are covered in Chap. 8.

The flocculation step involves mixing of the water to induce collisions among particulate contaminants and precipitated particles formed from metal coagulants so as to yield flocs. The size of particles or flocs in the influent to the DAF tank affects greatly the collisions among flocs and bubbles in the DAF contact zone. Flocculation is used in both sedimentation and DAF plants, but their objectives differ. When settling follows flocculation, the goal is to produce large flocs capable of settling at a fairly rapid rate. Flocs with sizes of 100 μm and greater are required to produce reasonable settling rates. Smaller floc sizes are desired for DAF. This is an important feature of DAF that is examined in the sections on fundamentals and operational and design considerations.

The DAF tank is divided into two sections with different functions as illustrated at the bottom of Fig. 9-25. There is a section at the front end that is baffled, called the *contact zone,* where floc particles are introduced and contacted with air bubbles. Here, collisions occur among bubbles and preformed floc particles. If the floc particles are prepared properly via coagulation with respect to their surface chemistry, then bubbles colliding with the flocs may attach and yield floc-bubble aggregates. This is the principal mechanism for water treatment (Kitchener and Gochin, 1981). Other possible mechanisms for forming floc-bubble aggregates are entrapment of bubbles as flocs form and growth of bubbles from nuclei on flocs.

The water carrying the suspension of bubbles, floc-bubble aggregates, and unattached floc particles flows to the second section of the tank, called the *separation zone*. Here, bubbles not attached to flocs and floc-bubble-aggregates may rise to the surface of the tank. The float layer at the surface of the tank consists of a mixture of bubbles and floc particles attached to bubbles. In drinking water applications, this froth is called the *float*. Over time, this float layer is concentrated producing a sludge that is collected and removed from the tank. Clarified water, often referred to as the *subnatant*, is withdrawn from the bottom of the tank. In a standard type DAF water plant as presented in Fig. 9-25, granular media filtration is placed in a horizontal layout following DAF, and the recycle flow is taken either after DAF or filtration. In some applications of flotation, DAF is placed vertically above the filters and the recycle flow is taken after filtration.

Air bubbles are introduced into the DAF contact zone. First, air is dissolved into the recycle flow by adding air under pressure in a vessel called a *saturator* or *air-dissolving vessel*. The recycle ratio or rate (R) is defined by Eq. 9-52, where Q is the plant throughput flow and Q_r is the recycle flow. Recycle rates are usually in the range of 6 to 12 percent. A rate of 10 percent or a little greater is a typical design value.

$$R = \frac{Q_r}{Q} \tag{9-52}$$

Saturator gauge pressures are usually between 400 and 600 kPa (~60 and 85 psi). The recycle flow is injected through nozzles or special valves at the bottom entrance to the contact zone. Microbubbles are produced with sizes between 10 and 100 μm. These small air bubbles give the water a milky appearance, and so the term *white water* is used to describe the bubble suspension in the DAF tank.

Analogous to sedimentation overflow rates, DAF tank hydraulic loadings are used to describe the rate and size of DAF tanks. There are conventional DAF processes designed at nominal hydraulic loadings of 5 to 15 m/h (2 to 6 gpm/ft^2). More recently, high-rate DAF processes have been developed allowing loadings of 15 to 30 m/h (6 to 12 gpm/ft^2) and greater. Loading rates and the development of high-rate DAF processes are discussed further in subsequent sections.

FUNDAMENTALS OF DISSOLVED AIR FLOTATION

Principles are presented in this section on the following subjects: (1) solubility of air, (2) bubble suspension in the contact zone, (3) performance model for the contact zone, (4) bubble and floc-bubble aggregate rise velocities, and (5) ideal theory for separation zone performance. The material draws on the paper covering the fundamentals of DAF by Edzwald (2007).

Air Solubility and Bubbles in the Contact Zone

Solubility of Air in Water. Our interest is in determining the equilibrium air concentration in water (i.e., solubility) for normal air composition and normal atmospheric pressure of 101.3 kPa (1 atm). The values obtained are used for the air concentrations ($C_{s,\text{air}}$) in the influent recycle water to the saturator and in the influent flocculated water entering the flotation tank; see Fig. 9-26.

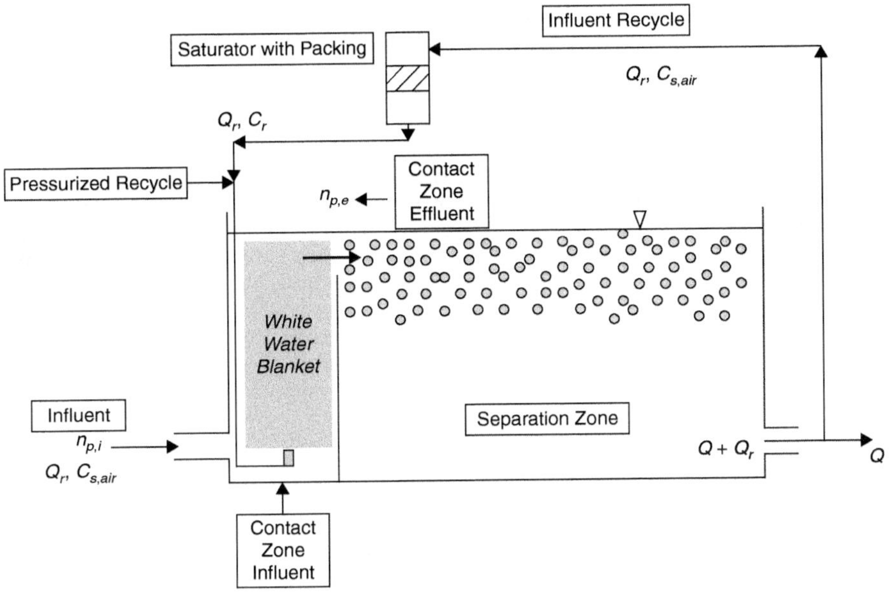

FIGURE 9-26 DAF tank showing the baffled contact zone, the separation zone, the influent recycle flow to the saturator, and the pressurized recycle flow entering the contact zone. (*Source: Edzwald, 2007, reprinted with permission from the* Journal of the New England Water Works Association, *Copyright 2009, Holliston, Mass.*)

Henry's law is used to determine the solubility of a gas in water; Henry's law is discussed in Chap. 3 under chemical concepts and in Chap. 6 on gas-liquid processes. It is important to note here that Henry's law applies to individual gases, and so for air, which is a mixture of gases, one has to calculate solubility for each gas and then add them to obtain the solubility of air in water. Nitrogen and oxygen account for over 99 percent of the composition of air. Thus for practical reasons, nitrogen is set at 79.1 percent (slightly higher than actual to account for other minor gases) and oxygen is set at 20.1 percent. The solubility concentrations of nitrogen ($C_{s,N}$), oxygen ($C_{s,O}$), and air ($C_{s,\text{air}}$) are given in Table 9-2 for various temperatures. As shown, the solubility of air is a little more than 20 mg/L for warm waters increasing to about 30 mg/L for cold water.

TABLE 9-2 Solubility of Oxygen, Nitrogen, and Air in Water for Atmospheric Air (79.1 percent N_2 [accounts for small amounts of other gases] and 20.9 percent O_2)

Temperature (°C)	$C_{s,O}$ (mg/L)	$C_{s,N}$ (mg/L)	$C_{s,\text{air}}$ (mg/L)
5	12.3	19.5	31.8
10	10.9	17.7	28.6
15	9.8	16.1	25.9
20	8.9	14.8	23.7
25	8.1	13.6	21.7
30	7.4	12.6	20.0

Source: Edzwald, 2007, reprinted with permission from the *Journal of the New England Water Works Association*, Copyright 2009, Holliston, Mass.

Solubility of Air in the Pressurized Recycle Water Exiting the Saturator. The composition of air within the saturator differs from atmospheric air. This is because oxygen is more soluble than nitrogen in water so within the saturator a steady state condition is established with saturator air containing more nitrogen than atmospheric air. This saturator air is called *nitrogen-enriched air*. For example, at a typical saturator design pressure of 500 kPa (72.5 psi) and 20°C and for typical water and air loading rates, the saturator air is composed of 86.7 percent nitrogen. Because nitrogen is less soluble than oxygen, the pressurized recycle water contains less dissolved air than it would if the saturator could be operated under conditions of maintaining atmospheric air. Data from Steinbach and Haarhoff (1997) on the composition of saturator nitrogen-enriched air were used to determine the solubility or equilibrium dissolved air concentration (C_r) in the pressurized recycle flow (Q_r) exiting the saturator; see Fig. 9-26.

Figure 9-27 presents C_r as a function of the saturator gauge pressure for temperatures of 5°C (cold water) and 20°C (warm water). Note that C_r is expressed as a function of saturator pressure, but it is the total dissolved air in the recycle water from pressure applied to the

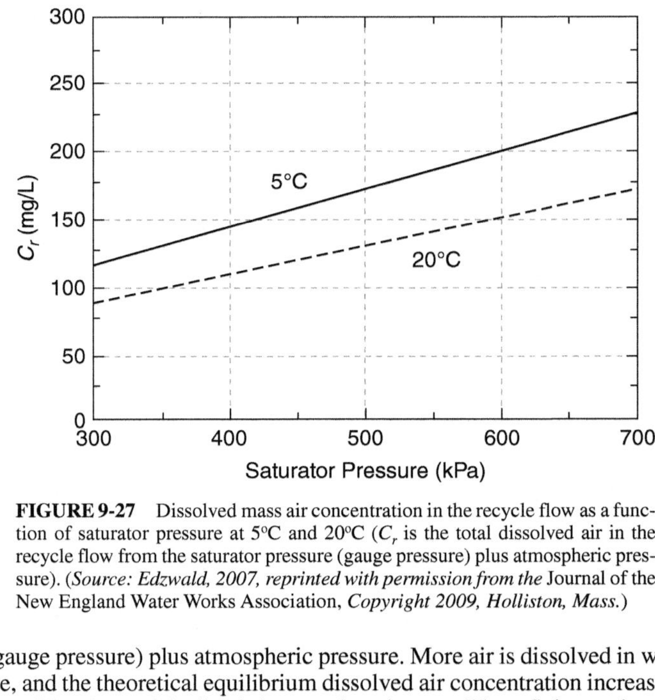

FIGURE 9-27 Dissolved mass air concentration in the recycle flow as a function of saturator pressure at 5°C and 20°C (C_r is the total dissolved air in the recycle flow from the saturator pressure (gauge pressure) plus atmospheric pressure). (*Source: Edzwald, 2007, reprinted with permission from the* Journal of the New England Water Works Association, *Copyright 2009, Holliston, Mass.*)

saturator (gauge pressure) plus atmospheric pressure. More air is dissolved in water at lower temperature, and the theoretical equilibrium dissolved air concentration increases with saturator pressure. The slope of the line is low, so there is a small benefit in varying the saturator pressure to provide more or less delivered air for constant recycle flow. Consequently, the designer and operator should not use the saturator pressure as the primary means to vary the air concentration delivered to the contact zone, but should use the recycle flow.

As pointed out earlier, saturators are designed at gauge pressures in the range of 400 to 600 kPa (~60 to 85 psi). Using a saturator pressure of 500 kPa (72.5 psi) as an example, Fig. 9-27 shows that C_r is about 130 mg/L at 20°C. The theoretical available air concentration that can precipitate in the form of bubbles would be this value (130 mg/L) less the soluble air concentration at atmospheric pressure (24 mg/L), or 106 mg/L. These calculations are for equilibrium conditions, but the actual available air is less than this because saturators do not achieve equilibrium solubility of air in the recycle water. Transfer of air has kinetic mass transport limitations, which is addressed in the next section.

Bubble Suspension in the Contact Zone. The bubble suspension in the contact zone achieves a steady state bubble concentration arising from the continuous input of pressurized recycle flow (Q_r) mixing with the influent water (Q) from the flocculation tank. This bubble suspension is referred to as a *white water* blanket (see the shaded area of Fig. 9-26). There are three measures of bubble concentrations: mass, volume, and number.

Mass Concentration. Equation 9-53, obtained from a mass balance under steady state conditions for the contact zone, gives the mass concentration of air bubbles (C_b) in the contact zone.

$$C_b = \frac{[e(C_r - C_{s,\text{air}})R - k_a]}{1+R} \tag{9-53}$$

The difference between C_r and $C_{s,\text{air}}$ represents the amount of air that will precipitate, that is, released as air bubbles. However, as mentioned, the water leaving the saturators does not reach equilibrium solubility conditions because air transfer into water has kinetic limitations; for principles about gas transfer kinetics, see Chap. 6. There may also be a minor loss in air in delivering the recycle water from the saturator to the point of injection. Thus, a saturator efficiency factor (e) is incorporated into the mass balance. The efficiency may vary from 60 to 70 percent for unpacked saturators to 90 to 95 percent for packed saturators; the latter contain plastic packing to increase the rate of air transfer to the water. Practical information about saturators is presented in the section Operational and Design Considerations for Flotation. The parameter k_a accounts for any air deficit in the incoming water (Q). If the flocculated water is at saturation, then k_a is zero. This is the common situation, but not always. Supplies taken from the bottom of reservoirs that are mesotrophic or eutrophic, may have an oxygen deficit, and therefore an air deficit. One should take water for DAF from a supply that is at saturation or nearly saturated with oxygen; otherwise a significant portion of the air provided will not yield air bubbles but rather will be used to satisfy the dissolved oxygen demand. This may require high recycle rates and so make DAF uneconomical.

Volume Concentration. The density of air bubbles is a fundamental property required for calculating bubble volume concentrations and bubble rise velocities. Moist air densities (ρ_b) for conditions of 100 percent humidity and with the dew point temperature equal to the water temperature are used; see App. D of the book for values. The air bubble volume concentration in the contact zone (Φ_b) is calculated from Eq. 9-54.

$$\Phi_b = \frac{C_b}{\rho_b} \tag{9-54}$$

Number Concentration. The air bubble size (d_b) must be known or assumed to calculate the bubble number concentrations (n_b). Small air bubbles are formed instantaneously after injection of the pressurized recycle flow into the contact zone influent at the bottom of the tank. The bubble sizes formed depend mainly on the saturator pressure and the injection device (nozzle type or needle and gate valves). Nozzles are used most often to inject the recycle water into the contact zone, and they produce a more uniform bubble size distribution than gate and needle valves. The pressure difference across the nozzle is the primary factor affecting bubble size. In theory, smaller-size bubbles are produced as the saturator pressure is increased; however, for typical DAF operation at saturator gauge pressures of 400 to 600 kPa (~60 to 85 psi), the bubble sizes are between 10 and 100 μm with most bubbles about 40 to 80 μm. A good estimate of the mean bubble size for the contact zone is 60 μm (Haarhoff and Edzwald, 2004). For modeling purposes of the contact zone, it is reasonable to assume that the bubble size is maintained throughout the contact zone depth even with the decreasing hydrostatic pressure (Takahashi et al., 1979). However, larger bubbles are found in the separation zone and this is discussed later in the chapter. The number concentration is calculated from Eq. 9-55.

$$n_b = \frac{\Phi_b}{\left(\frac{\pi (d_b)^3}{6}\right)} \tag{9-55}$$

Example 9-6 Bubble Concentrations in the Contact Zone
In this example consider a DAF plant operating at 10 percent recycle with a saturator pressure of 500 kPa (72.5 psi) and 90 percent efficiency. The flocculated water enters the contact zone of the flotation tank with a floc particle concentration ($n_{p,i}$) of 20,000 particles/mL and a floc volume concentration (Φ_p) of 20 ppm. Compute the air mass concentration (C_b), bubble volume concentration (Φ_b), and number concentration (n_b) in the contact zone of the DAF tank, and compare the concentrations of bubbles to floc particles. For these calculations use a water temperature of 20°C; at this temperature ρ_b is 1.19 kg/m³ (see App. D). Assume the flocculated water is saturated with air so k_a is 0.

Solution

1. *Mass of air in DAF tank.* For this calculation, Fig. 9-27 is used to estimate C_r, the equilibrium air concentration in the pressurized recycle water. The value is approximately 130 mg/L. Equation 9-53 is used to calculate the mass concentration of air bubbles in the contact zone using a value for $C_{s,\text{air}}$ from Table 9-2 of 23.7 mg/L.

$$C_b = \frac{[e(C_r - C_{s,\text{air}})R - k]}{1+R} = \frac{[0.90(130-23.7)0.10]}{1+0.10} = 8.7 \text{ mg/L}$$

2. *Bubble volume concentration.* Now Φ_b is calculated from Eq. 9-54.

$$\Phi_b = \frac{C_b}{\rho_{\text{air}}} = \frac{8.7 \text{ mg/L}}{1.19 \text{ kg/m}^3} = 7.31\left(\frac{\text{mg}}{\text{kg}}\right)\left(\frac{\text{m}^3}{\text{L}}\right) = 7.31\left(\frac{\text{mg}}{\text{kg}} \times \frac{\text{kg}}{10^6 \text{mg}}\right)\left(\frac{\text{m}^3}{\text{L}} \times \frac{10^3 \text{L}}{\text{m}^3}\right)$$

$$\Phi_b = \frac{7310 \text{ m}^3 \text{ of air}}{10^6 \text{ m}^3 \text{ of water}} = 7310 \text{ parts per million (ppm)}$$

3. *Bubble number concentration.* n_b is calculated from Eq. 9-55 using a mean bubble diameter of 60 μm (60×10^{-6} m).

$$n_b = \frac{6\Phi_b}{\pi d_b^3} = \left[\frac{6(7310\times10^{-6})}{\pi(60\times10^{-6})^3}\right] = 6.47\times10^{10}\left(\frac{\text{bubbles}}{\text{m}^3}\right)\times\left(\frac{\text{m}^3}{10^3 \text{ L}}\right)$$

$$= 65\times10^6\left(\frac{\text{bubbles}}{\text{L}}\right)$$

4. *Ratios of concentrations.*

$$\frac{n_b}{n_{p,i}} = \frac{65\times10^6}{20\ 000}\left(\frac{\text{bubbles}}{\text{L}}\right)\left(\frac{\text{mL}}{\text{particles}}\right)\times\frac{\text{L}}{10^3 \text{ mL}} = 3.3$$

or about 4 bubbles per floc particle

$$\frac{\Phi_b}{\Phi_p} = \frac{7310 \text{ ppm}}{20 \text{ ppm}} = 365 \text{ bubble volume per floc particle volume} \quad \blacktriangle$$

Discussion of Bubble Concentrations. In Example 9-6 the calculations are instructive, but they were done for specific conditions. A more general analysis is needed to make conclusions regarding bubble concentrations and flotation.

Table 9-3 presents mass, volume, and number concentrations for recycle rates of 6 to 12 percent (typical operating range) at a fixed saturator operating or gauge pressure of 500 kPa (72.5 psi). These are of interest because in DAF plant operation, the saturator pressure is usually kept fairly constant and the recycle rate is varied to change the air bubble concentrations in the contact zone. The following comments are made:

TABLE 9-3 Contact Zone Air Bubble Concentrations Versus Recycle Rate [Conditions: 20°C, saturator pressure of 500 kPa (72.5 psi), C_r for saturator air composition, saturator efficiency of 90 percent, bubble diameter of 60 μm, flocculated water is saturated with air [$k = 0$]].

Recycle rate, R (%)	Mass conc., C_b (mg/L)	Volume conc., Φ_b (ppm)	Number conc., n_b (#/L)
6	5.4	4500	40×10^6
8	7.1	5900	50×10^6
10	8.7	7300	65×10^6
12	10.1	8600	80×10^6

Source: Edzwald, 2007, adapted with permission from the *Journal of the New England Water Works Association*, Copyright 2009, Holliston, Mass.

- For recycle rates of 6 to 12 percent, the air bubble mass concentrations are 5.4 to 10.1 mg/L, the volume concentrations are 4500 to 8600 ppm, and the number concentrations are 40×10^6 to 80×10^6 bubbles per L.
- The contact zone is designed for collisions to occur between bubbles and floc particles. How do the bubble numbers compare to floc particle number concentrations? Field measurements for floc numbers for floc sizes of 1 μm and greater are in the range of 1×10^6 (low end) to 40×10^6 (high end) floc particles per L. For, say, a design recycle rate of 10 percent, the bubble number is 65×10^6 bubbles per L. Therefore, the bubble number far exceeds the floc number for low floc concentrations—65 to 1. For the high floc particle concentrations, there are 1.5 bubbles per floc so here the recycle rate could be increased to provide additional bubbles if needed. One desires a bubble suspension in the contact zone in which bubble numbers exceed floc numbers to ensure good collision opportunities of bubbles with floc particles. Of course, flotation cannot occur unless at least one bubble is attached to a floc particle, so it is essential to have more bubbles than particles.
- The flocs will not float unless the attached bubble volume lowers the density of the floc-bubble aggregate to less than the density of water. One can gain some insight of the potential for this by comparing bubble volumes to floc volumes. Again, for a design recycle rate of 10 percent, the bubble volume concentration is 7300 ppm (from Table 9-3). The physical meaning of the bubble volume concentration is that for every 10^6 m^3 of water flowing through the contact zone, 7300 m^3 of bubble volume is provided. The floc volumes entering the DAF contact zone for most DAF applications are in the range of 5 to 50 ppm.

Using the high end for the floc volume yields a bubble/floc particle volume ratio of about 146. This large bubble volume provides the means to reduce floc densities to less than that of water so flotation can occur.

Contact Zone Performance Model

Mathematical models describing DAF contact zone performance have been developed over the last 25 years. These models have improved our fundamental understanding of DAF and have been instrumental in advancing the use of the technology around the world. There are two approaches to DAF modeling for drinking water treatment. One approach uses flocculation in which floc particles undergo flocculation with bubbles in the contact zone (see Chap. 8 for flocculation modeling) and has been used by several investigators starting with Tambo and coworkers (Tambo and Matsui, 1986; Fukushi et al., 1995; Matsui et al., 1998). The flocculation-type modeling approach was modified by Leppinen (2000) and by Han (2002) using trajectory analysis to model collisions among bubbles and floc particles. The second modeling approach considers the rising bubble suspension in the contact zone as a blanket of rising bubbles collecting floc particles. The bubbles in the blanket act as collectors and are analogous to a granular media filter where filter grains act as collectors (see Chap. 10 for filter modeling). Edzwald and coworkers (Edzwald et al., 1990; Edzwald, 1995) developed this modeling approach and later the model was reviewed and verified (Haarhoff and Edzwald, 2004). Schers and Van Dijk (1992), Baeyens et al. (1995), and Liers et al. (1996) have also used this approach. Here, the Edzwald modeling approach is presented and the variables that affect design and operating performance are identified and discussed.

Model Framework and Assumptions. The DAF tank is divided into two zones with different functions: the contact zone where bubbles collect particles or flocs and the separation zone where free bubbles and bubble-floc aggregates rise to the surface. Our interest now is the contact zone as shown in Fig. 9-26 and from a design view is depicted to reside on the left-hand side of a baffle, which divides it from the separation zone. Additional collection of flocs by particles can continue in the separation zone, but this is not considered in the model.

The following is a summary of the model development and assumptions. Details are found in Haarhoff and Edzwald (2004):

1. In the model, the term *particle* is used in a general sense and refers to either primary non-flocculated particles or to flocs.
2. The model development considers that a blanket (*white water* suspension) of bubbles exists in the contact zone at a dynamic steady state with a high bubble concentration (between ~ 40×10^6 and 80×10^6 bubbles per L; see Table 9-3). This dynamic steady state at a high bubble concentration is maintained by continuous injection of bubbles from the recycle flow and output at the exit of the contact zone plane above the baffle where cross-flow occurs. The bubbles act as collectors of particles analogous to a granular media sand or anthracite filter. In fact, the bubble concentration is so high that the mean separation distance between bubbles is approximately 200 μm, which is about the pore space between filter grains in a filter bed.
3. The basic kinetic equation describing the rate of particle change due to collision and attachment to bubbles is

$$\frac{dn_p}{dt} = -k_c n_p n_b \tag{9-56}$$

where k_c is the rate coefficient and n_p and n_b are the particle and bubble number concentrations, respectively. The rate coefficient depends primarily on how particle collisions occur with bubbles. This is modeled in which the bubbles act as particle collectors using the single collector efficiency (n_T) concept and the particle-bubble attachment efficiency (α_{pb}).

4. A steady state mass balance for particles (n_p) is made across the depth of the contact zone; see Fig. 9-26.
5. Plug flow hydraulic conditions are assumed through the contact zone. This is a reasonable assumption based on tracer studies and the analysis done by Haarhoff and Edzwald (2004), and it is a practical assumption for development of design-based performance models.

Contact Zone Model. The final form of the contact zone performance model is

$$\left(1-\frac{n_{p,e}}{n_{p,i}}\right)=\left[1-\left(\exp\frac{-(3/2)\alpha_{pb}\eta_T\Phi_b v_b t_{cz}}{d_b}\right)\right] \tag{9-57}$$

Equation 9-57 gives the fraction of particles that have collided and attached to bubbles in the contact zone. The equation is instructive in that it identifies variables affecting the contact zone performance. These are

- α_{pb}, the attachment efficiency of particles to bubbles
- η_T, the total collision efficiency accounting for all transport mechanisms
- Φ_b, the bubble volume concentration
- v_b, the bubble rise velocity
- t_{cz}, the detention time of the contact zone
- d_b, the bubble size

In this equation the dimensionless particle transport coefficient (η_T) describes the collision efficiency of a single bubble or collector. Particle trajectory analysis was used to determine expressions for η due to gravitational forces, and the convective-diffusion equation was used to determine a η expression for Brownian diffusion. Four individual η expressions are found. They are for Brownian diffusion (η_D), fluid flow or interception (η_I), differential settling of flocs (η_S), and inertia (η_{IN}). For drinking water applications, the inertia term is small and neglected. Equations 9-58 through 9-60 describe the individual single collector collision efficiencies.

$$\eta_D=6.18\left[\frac{k_bT}{g(\rho_w-\rho_b)}\right]^{2/3}\left[\frac{1}{d_p}\right]^{2/3}\left[\frac{1}{d_b}\right]^2 \tag{9-58}$$

$$\eta_I=\left(\frac{d_p}{d_b}+1\right)^2-\frac{3}{2}\left(\frac{d_p}{d_b}+1\right)+\frac{1}{2}\left(\frac{d_p}{d_b}+1\right)^{-1} \tag{9-59}$$

$$\eta_S=\left[\frac{(\rho_p-\rho_w)}{(\rho_w-\rho_b)}\right]\left[\frac{d_p}{d_b}\right]^2 \tag{9-60}$$

The interception expression (Eq. 9-59) is more complicated than the interception equation for granular media filtration (see Chap. 10). This is because in flotation, particles and bubble sizes are nearly the same size whereas in filtration the filter media collectors are much larger in size than particles applied to the filter yielding a simpler expression for interception. The total single collector collision efficiency (η_T) is the sum of the individual mechanisms.

$$\eta_T = \eta_D + \eta_I + \eta_S \tag{9-61}$$

Discussion of the Contact Zone Model Variables. The contact zone performance model (Eq. 9-57) identifies two types of variables: one type is related to DAF pretreatment; the other type has to do with the design and operation of the DAF tank contact zone. It is instructive to separate the variables as shown in Table 9-4 according to pretreatment variables versus DAF tank variables.

Pretreatment Coagulation Chemistry and Flocculation. The attachment efficiency (α_{pb}) accounts for the fraction of successful collisions between particles and bubbles. If all collisions result in attachment, then α_{pb} is assigned a value of 1. For conditions of

TABLE 9-4 Contact Zone Model Variables

Variable	Dependence	Comments
	Pretreatment variables	
α_{pb} (particle-bubble attachment efficiency)	Interaction forces between particles and bubbles from electrical charge, London-van der Waals forces, hydrophobic force, and hydrodynamic retardation.	Coagulation chemistry including dose and pH critical in maximizing particle attachment to bubbles. α_{pb} approaches 1 with optimum coagulation.
η_T (total single collector collision efficiency)	Brownian diffusion: $d_p^{-2/3}$ Settling and interception: d_p^2	Flocculation increases d_p. Desire flocs of 10s of μm.
	Flotation tank variables	
d_b (mean bubble diameter)	Controlled mainly by pressure difference across the recycle injection device. Effect also from the nozzle type or injection device.	Desire microbubbles; smaller bubbles better performance. Most 40–80 μm, mean of 60 μm.
η_T (total single collector collision efficiency)	Primarily affected by floc and bubble sizes. Floc size : $d_p^{-2/3}$ and $d_p^{\ 2}$ Bubble size: d_b^{-2}	Minimum η_T for floc particles of ≈ 1 μm. Flocculation should produce floc particles of 10s of μm.
Φ_b (bubble volume concentration)	Affected by recycle rate and saturator pressure.	Controlled mainly by the recycle rate. Increasing Φ_b increases collision opportunities (more bubbles or collectors) and increases bubble volume to reduce particle-bubble aggregate density.
v_b (bubble rise velocity)	Affected mainly by the size of bubbles (d^2). Effect of water temperature: decreases with viscosity for colder waters.	Describes motion of bubbles relative to the water flow in the contact zone.
t_{cz} (contact zone detention time)	Determined by the water velocity and depth of the contact zone.	In practice, detention times are in range of 1–2.5 min. Hydraulic behavior approximately plug flow.

low attachment efficiency, α_{pb} has a value << 1. From a theoretical view, α_{pb} depends on the interparticle (particle-bubble) forces: electrostatic repulsion; the van der Waals force, which for dissimilar particles (particles and bubbles) in water is repulsive; the major force of attraction, the hydrophobic force; and hydrodynamic retardation (water layer between the bubble and particle). With no coagulant addition, the particles carry a negative charge and have some hydrophilic character so that bubble attachment is poor. Proper coagulant dose and coagulation pH reduce the particle charge and give particles a more hydrophobic character so that attachment to bubbles can occur. A reasonable estimate of α_{pb} for good coagulation is about 0.5 (Schers and Van Dijk, 1992; Shawwa and Smith, 2000; Haarhoff and Edzwald, 2004). Additional discussion of the importance of coagulation in DAF treatment and some practical points are presented in the section Operational and Design Considerations for Flotation.

The contact zone equation (Eq. 9-57) is used to predict performance as a function of particle or floc size for two cases of coagulation conditions: good coagulation (α_{pb} = 0.5) and poor coagulation (α_{pb} = 0.01). Typical DAF design and operating conditions are set with the contact zone detention time (t_{cz}) at 1.5 min and the bubble volume concentration (Φ_b) at 7300 ppm. The contact zone bubble size is set at 60 μm and the flocs or particles have a density of 1100 kg/m^3—a reasonable value following coagulation for low-density primary organic particles such as viruses, bacteria, protozoa cysts, and algae or flocculated particles for various water supplies containing the aforementioned primary particles, humic substances, and low mineral turbidities. The predictions are presented in Fig. 9-28. Collection of particles on bubbles occurs primarily by Brownian diffusion for small particles (<1 μm) and primarily by interception for large particles (>1 μm). For low-density particles, collection by settling is small relative to interception and can be neglected. For poor coagulation and pH adjustment, the contact zone efficiency is low for all particle sizes smaller than about 100 μm. Hence, good coagulation is essential for good DAF performance. The rest of the discussion focuses on good coagulation conditions (α_{pb} = 0.5).

There is a minimum in the contact zone efficiency for particles with a size of about 1 μm as shown by Fig. 9-28. This is because the single-collector efficiency is low for both

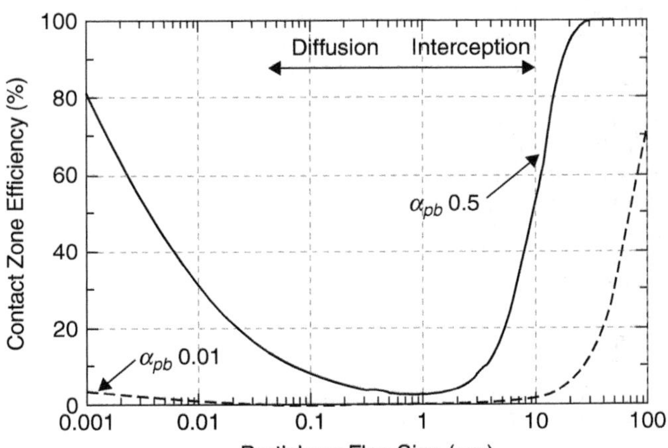

FIGURE 9-28 Contact zone efficiency as a function of particle or floc size for good (α_{pb} = 0.5) and poor (α_{pb} = 0.01) coagulation chemistry (Conditions: detention time 1.5 min, floc density 1100 kg/m^3, bubble size 60 μm, bubble volume 7300 ppm, temp. 20°C).

diffusion and interception; this minimum in collection efficiency also occurs in granular media filtration (see Chap. 10). The contact zone efficiency improves with decreasing particle size (<1 μm) because Brownian diffusion increases as shown by the dependence in Eq. 9-58. However, the addition of coagulant causes flocculation of submicron particles into larger sizes approaching a micron so engineered flocculation is necessary to avoid the minimum in efficiency for 1-μm particles. From Fig. 9-28, we see that as particle or floc size increases, the contact zone efficiency improves greatly through physical interception. In fact, the dependence of η_I (interception) is roughly proportional to the particle size to the second power (d_p^2) (see Table 9-4). If we produce flocs of about 25 μm, the contact zone efficiency is approximately 99 percent. Larger flocs can be produced and floated, but it is not necessary for good collision opportunities between particles with bubbles. A disadvantage of large flocs is multiple bubble attachment will be required to reduce their density sufficiently to achieve high floc-bubble aggregate rise rates for the separation zone. Plants with long flocculation times can produce flocs with a tendency to settle; this is undesirable.

An important finding from the contact zone model predictions is that the desired and optimum size of floc is about 25 to 50 μm. This theory has been incorporated into DAF practice so that flocculation tanks in DAF plants are designed and operated at much shorter detention times to produce small pinpoint flocs. Sedimentation plants require flocs with sizes of 100 μm or greater and thus longer flocculation times (see the section Sedimentation Theory and Table 9-1). Practical aspects dealing with pretreatment flocculation are addressed in the section Operational and Design Considerations for Flotation.

Flotation Tank Variables. Equation 9-57 contains several variables that affect contact zone performance as summarized in Table 9-4. In terms of design and operation, the four important variables are the bubble size (d_b), the total single-collector efficiency (η_T), the bubble concentration described through the bubble volume concentration (Φ_b), and the hydraulic contact zone detention time (t_{cz}). η_T depends strongly on particle and bubble sizes. Flocculation pretreatment affects floc sizes in the contact zone and this was addressed above. Thus, the discussion here is on bubble size (d_b).

The smaller the bubble size, the better the DAF contact zone performance is. This is a primary reason why dissolved air flotation is a more efficient process than dispersed air flotation, where bubble sizes are much larger at about 1 mm. It was presented above that bubbles in the contact zone are mainly between 40 and 80 μm in diameter, with a mean bubble size of about 60 μm. Although smaller bubbles improve performance just like smaller filter grains improve filtration, the bubble size is fixed mainly by the pressure difference across the recycle injection device and by the injection device. Therefore the designer ensures a good bubble size through the saturator design and selection of nozzle, and the operator has little control over bubble size other than to maintain the saturator pressure at the desired level, such as within the range of 400 to 600 kPa (~60 to 85 psi).

Increasing the contact zone detention time (t_{cz}) improves the contact zone performance. The designer sets this time by sizing the contact zone for the design flow. In practice, the contact zone detention time for DAF processes usually exceeds 1 min for design flows. Although the contact zone detention time is not varied directly in plant operation, you can expect shorter (design) times in the summer with greater plant flows and longer times in the winter.

The most important operating and control variable affecting DAF performance is the bubble volume concentration (Φ_b) in the contact zone. The air concentration can be changed by changing the saturator pressure (dissolving more air into the recycle flow) or by changing the recycle rate; see Table 9-3. However, the saturator pressure is not varied very much as explained earlier so the main operational way to change the bubble concentration is to increase or decrease the recycle flow or ratio (R). The plant operator, as needed, can vary the recycle rate from, say, 6 to 12 percent yielding bubble volumes (Φ_b) from about 4500 to 8600 ppm (see Table 9-3). Haarhoff and Edzwald (2004) have shown with the model that contact zone performance is excellent when the bubble volumes exceed about 6000 ppm for a contact zone detention time of at least 1.5 min. Greater bubble volumes yield excellent performance for detention times as short as 1 min. Poor contact zone performance occurs below 1-min detention time for bubble volumes as high as 9000 ppm.

Bubble and Floc-Bubble Aggregate Rise Velocities

Bubble Rise Velocities. A 100-μm bubble is shown in Fig. 9-29a with no attached particles. Bubbles are larger in the separation zone compared to those in the contact zone of DAF tanks. A bubble of 100 μm is a good estimate of the mean size for bubbles in the separation zone (Leppinen and Dalziel, 2004; Haarhoff and Edzwald, 2004). Three forces act on the bubble: the buoyant force which dominates because the density of the air bubble is far less than water, the gravitation force, and a drag force as the bubble rises. The net result is a rise velocity (v_b) for free bubbles (no particles attached) according to Stokes' law as given in Eq. 9-62. Bubble rise velocity depends greatly on bubble diameter (d_b^2), the density difference between water and air, and the water viscosity. The driving force for the rise velocity is the density difference ($\rho_w - \rho_b$). Since water density is about 1000 kg/m^3 and bubble density is about 1 kg/m^3, the driving force value is roughly 1000, which is much greater than the driving force for most cases of floc particle settling.

$$v_b = \frac{g(\rho_w - \rho_b)d_b^2}{18\mu_w} \tag{9-62}$$

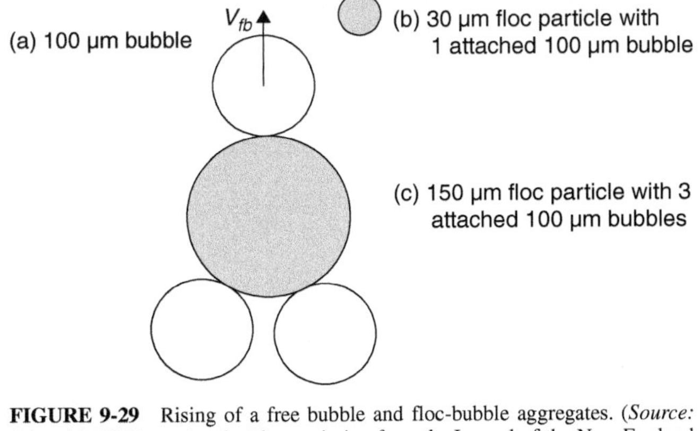

FIGURE 9-29 Rising of a free bubble and floc-bubble aggregates. (*Source: Edzwald, 2007, reprinted with permission from the* Journal of the New England Water Works Association, *Copyright 2009, Holliston, Mass.*)

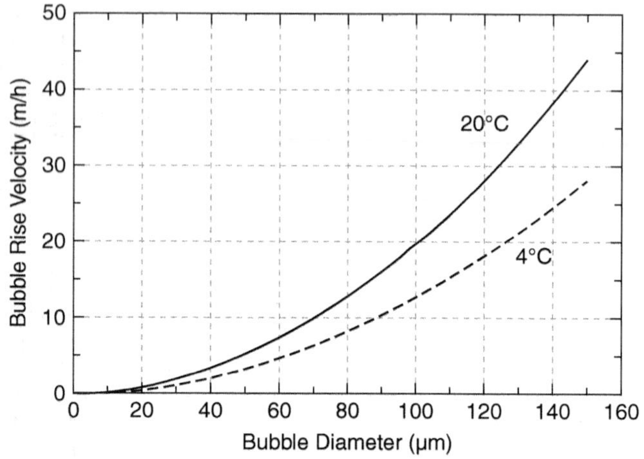

FIGURE 9-30 Bubble rise velocities at 4 and 20°C.

Figure 9-30 presents bubble rise velocities as a function of bubble diameter for cold water (4°C) and warm water (20°C) conditions. The following observations are made:

- Bubbles rise more slowly in cold water than warm water due to the greater water viscosity.
- Both small and large bubbles occur in the separation zone, but using the mean size of 100 μm as a reference case, the bubble rise rate is about 20 m/h at 20°C. This rise rate is far greater than the settling rate of 100-μm flocs, as shown in Example 9-7.

The classic form of Stokes' law is used here with a factor of 1/18. Some people model bubble rise velocities with a factor of 1/12 attributing this to less drag (slipping condition) of the bubble as it rises through the water. Our approach of using 1/18 is conservative in that the bubble rise velocities are greater for slipping conditions.

Floc-Bubble Aggregate Rise Velocities. Figure 9-29a, b, and c shows illustrations of one free 100-μm bubble, one 30-μm particle with one attached 100-μm bubble, and a 150-μm particle with three attached 100-μm bubbles, respectively. To determine the rise velocity of floc-bubble aggregates, one has to determine the aggregate equivalent size and density. For the general case of multiple-bubble attachment, the following three equations are used, where d_{fb} is the aggregate equivalent diameter produced from N bubbles of size d_b attached to a floc particle of size d_f and ρ_{fb}, and v_{fb} are the floc-bubble aggregate density and aggregate rise velocity. K accounts for floc shape and drag force. For spheres, K is 24 (Haarhoff and Edzwald, 2004). Here, it is assumed that K is 24 for flocs smaller than 40 μm and increases gradually to 45 at a floc size at or above 170 μm.

$$d_{fb} = (d_f^3 + Nd_b^3)^{1/3} \tag{9-63}$$

$$\rho_{fb} = \frac{\rho_f d_f^3 + N\rho_b d_b^3}{d_f^3 + Nd_b^3} \tag{9-64}$$

$$v_{fb} = \frac{4g(\rho_w - \rho_{fb})d_{fb}^2}{3K\mu} \tag{9-65}$$

Equation 9-65 is used now to calculate floc-bubble aggregate rise velocities (v_{fb}) for two cases of bubble attachment: 1 bubble attached per floc and half the maximum number of possible bubbles per floc (N_{max}). N_{max} depends on the floc and bubble areas according to Tambo and Matsui (1986), and it is calculated from Eq. 9-66. It is unlikely that the maximum number of bubbles that can possibly attach would do so; thus half that number is assumed for the multiple bubble attachment case. For flocs less than 70 μm, only one bubble is allowed to attach per floc. The number increases gradually to six bubbles per floc for flocs with equivalent sizes of 200 μm.

$$N_{max} = \left(\frac{d_f}{d_b}\right)^2 \tag{9-66}$$

What we learn from the results displayed in Fig. 9-31 (assuming no significant decrease in floc density with increase in size) follows:

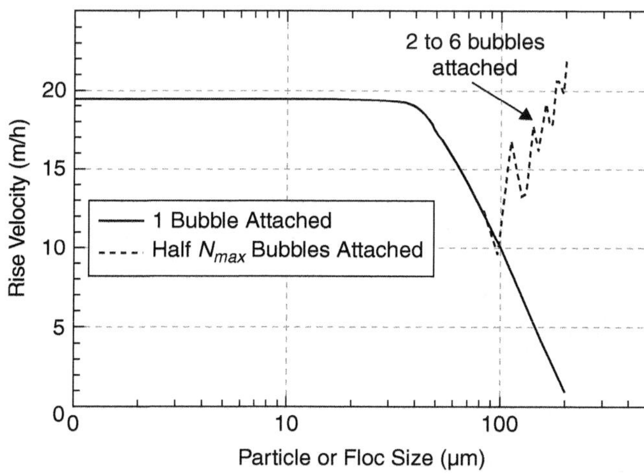

FIGURE 9-31 Rise velocity of floc-bubble aggregates versus particle size for 1 bubble attachment and multiple bubble attachment (1/2 N_{max}) (Conditions: d_b of 100 μm; ρ_f of 1100 kg m^{-3}, 20°C). (*Source: Edzwald, 2007, reprinted with permission from the* Journal of the New England Water Works Association, *Copyright 2009, Holliston, Mass.*)

- Aggregate rise velocities are near their maximum value (reference case for 100-μm free bubbles) of 20 m/h for flocs with sizes of 50 μm or less with just one bubble attached.
- As flocs increase in size, the rise velocity decreases approaching zero for 200-μm flocs with one bubble attached because the driving force is about zero.
- Multiple-bubble attachment is required for flocs >100 μm to achieve rise velocities of 10 to 20 m/h.

In summary, rise rates of free bubbles of 100 μm are about 20 m/h, and floc-bubble aggregates have about the same rise velocity for flocs < 50 μm with one bubble attached per floc. For larger flocs (100–200 μm), multiple-bubble attachment is required to achieve rise velocities of 10 to 20 m/h. What then is the optimum floc size for DAF? For the contact zone, it was shown that the optimum floc size is about 25–50 μm. For the separation zone, we find that maximum rise rates of 20 m/h are achieved with one bubble attached to particles < 50 μm. Overall then, we can say that optimum floc sizes are about 25 to 50 μm.

Example 9-7 Particle Settling Velocity Versus Floc Bubble Rise Velocity
Calculate:

1. The settling velocity of a floc particle of 100 μm with a particle density of 1100 kg/m^3 for summer (20°C) and winter (4°C) water temperatures
2. The rise velocity of a 50-μm floc particle attached to one air bubble of 100 μm for the same two temperature conditions

Solution

1. *Floc settling velocity.* To solve the problem requires water density and absolute viscosity for the two water temperatures. Values are found in App. D and tabulated next.

Temp. (°C)	Density (kg/m^3)	Absolute viscosity (N-s/m^2)
4	998.98	1.570×10^{-3}
20	998.21	1.002×10^{-3}

a. For 20°C, from Stokes' law for settling of particles, we use Eq. 9-8:

$$v_p = \frac{g(\rho_p - \rho_w)d_p^2}{18\mu} = \frac{9.8(1100 - 998.21)(100 \times 10^{-6})^2}{18 \times 1.002 \times 10^{-3}} \frac{\text{m}}{\text{s}} \times \frac{3600\text{s}}{\text{h}} = 2\frac{\text{m}}{\text{h}}$$

b. For 4°C, similar calculations result in $v_p = 1.3$ m/h.

2. *Rise velocity of 100-μm air bubble attached to floc particle of 50 μm.* To solve the problem requires water density and viscosity constants (given above) and the density of air: air densities at 4°C and 20°C are 1.27 and 1.19 kg/m^3, from App. D.

a. For 20°C, in order to compute the rise velocity, the equivalent diameter of the floc particle-bubble aggregate and density must be calculated. Eqs. 9-63 and 9-64 are used.

$$d_{pb} = [d_f^3 + N_{ab}(d_b)^3]^{1/3} = [50^3 + 1(100)^3]^{1/3}\mu\text{m} = 104\ \mu\text{m}$$

$$\rho_{fb} = \left[\frac{\rho_p d_p{}^3 + N_{ab}(\rho_b d_b{}^3)}{d_p{}^3 + N_{ab}(d_b{}^3)}\right] = \left[\frac{1100(50)^3 + 1(1.19)(100)^3}{50^3 + 100^3}\right]\frac{\text{kg}}{\text{m}^3} = 123.3\ \frac{\text{kg}}{\text{m}^3}$$

Note that the equivalent spherical diameter of the floc particle-bubble aggregate is not much larger than the bubble itself of 100 μm. This is because $d_b >> d_f$. From Eq. 9-65, the particle-bubble agglomerate rise velocity is determined.

$$v_{fb} = \frac{g(\rho_w - \rho_{pb})d_{pb}^2}{18\mu} = \frac{9.8(998.21 - 123.3)(104 \times 10^{-6})^2}{18 \times 1.002 \times 10^{-3}} \frac{\text{m}}{\text{s}} \times \frac{3600\text{s}}{\text{h}} = 18.5\frac{\text{m}}{\text{h}}$$

b. For 4°C, the results are

$$\rho_{fb} = 123.4\ \text{kg/m}^3$$
$$d_{fb} = 104\ \mu\text{m}$$
$$v_{fb} = 11.8\ \text{m/h} \quad \blacktriangle$$

Floc bubble aggregate rise velocities are much greater than particle settling velocities and illustrate why DAF tanks can be designed with greater hydraulic loading rates than horizontal-flow settling tanks. The next step is to evaluate these rise velocities with respect to the hydraulics of the separation zone.

Separation Zone Performance

Ideal Separation Zone with Vertical Flow and Conventional Hydraulic Loading Rates. Figure 9-32 depicts a DAF tank with an ideal separation zone analogous to an ideal sedimentation tank; see section Sedimentation Theory. The separation zone is divided into sections: (1) a float layer (sludge) at the top which is periodically withdrawn, (2) a cross-flow section above the baffle dividing the contact and separation zones where flow enters horizontally into the separation zone before moving vertically, (3) the main part of the separation zone with vertical flow and which is considered the clarification section or zone, and (4) an outlet section where water leaves the DAF tank.

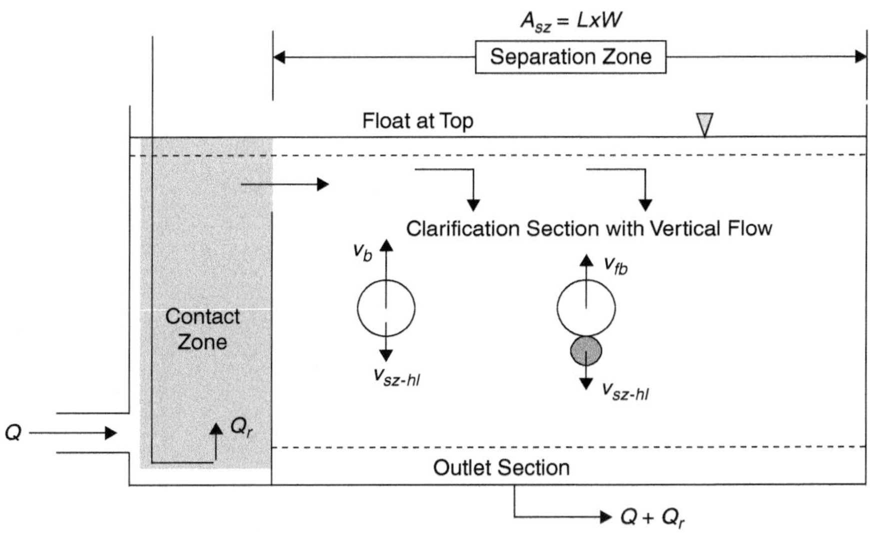

FIGURE 9-32 Idealized DAF tank showing separation zone divided into four sections: float or sludge layer, cross-flow section where flows moves horizontally before moving vertically, vertical flow section for clarification, and the outlet.

This idealized clarification zone with vertical flow is considered for evaluation of separation zone performance, in which we compare the hydraulic loading or upward water velocity through the clarification zone to the rise velocities of bubbles and floc bubble aggregates. The theory is similar to the Hazen concept used in sedimentation tank design in which the sedimentation tank overflow rate is used; see section Sedimentation Theory, Eq. 9-23b. For DAF, it is called the hydraulic loading rather than the overflow rate. In addition to the idealized separation zone described, there are several assumptions and conditions used.

- Vertical plug flow through the main part of the separation zone; the horizontal flow near the top of the tank is ignored in evaluating performance.
- This flow structure is assumed for conventional DAF with nominal hydraulic loading rates of 5 to 15 m/h (2 to 6 gpm/sf).

- The area for clarification called the separation zone area (A_{sz}) is the separation zone surface or footprint area, and for the common rectangular DAF tank is the separation zone length (L) times width (W).
- The flow through the separation zone is the sum of the throughput flow (Q) plus the recycle flow (Q_r).
- The hydraulic loading is ($Q + Q_r$) divided by the separation zone area (A_{sz}), and for vertical flow equals the downward average water velocity (v_w) or $v_{sz\text{-}hl}$.

In evaluating DAF separation zone performance, it is essential to distinguish the separation zone hydraulic loading from the nominal DAF tank hydraulic loading; the latter is defined by Eq. 9-67 as the throughput flow (Q) divided by the gross footprint DAF tank area (A) that includes the contact and separation zones.

$$v_{nom-hl} = \frac{Q}{A} \tag{9-67}$$

To assess the separation zone clarification performance, we must consider the throughput and recycle flows and use the separation zone area. As depicted in Fig. 9-32, free bubbles are removed if their rise velocity is greater than the vertical water velocity or hydraulic loading according to Eq. 9-68. Similarly, floc bubble aggregates are removed if their rise velocity exceeds the hydraulic loading according to Eq. 9-69.

$$v_b \geq v_{sz-hl} = \frac{Q+Q_r}{A_{sz}} \tag{9-68}$$

$$v_{fb} \geq v_{sz-hl} = \frac{Q+Q_r}{A_{sz}} \tag{9-69}$$

In the prior section it was found that free bubbles (reference size of 100 μm for the separation zone) have rise velocities of about 20 m/h, and maximum rise rates of about 20 m/h are achieved for floc-bubble aggregates with one bubble attached to floc particles <50 μm. Therefore, removals of these bubbles and aggregates occur for separation zone hydraulic loadings (see Eqs. 9-68 and 9-69) less than 20 m/h. Example 9-8 shows that DAF processes with nominal conventional loading rates of 5 to 15 m/h have separation zone hydraulic loadings less than 20 m/h and are therefore effective in removing bubbles and floc-bubble aggregates.

Example 9-8 Separation Zone Hydraulic Loading and Area
In this example consider a DAF process with a nominal hydraulic loading of 15 m/h. The recycle rate is 10 percent.

1. Calculate the separation zone hydraulic loading assuming the separation zone footprint area is 90 percent of the gross footprint area, and compare it to bubble and floc bubble aggregate rise velocities.
2. Determine the separation zone area for a design flow of 37,850 m^3/d (10 mgd).

Solution

1. *Separation zone hydraulic loading.* This is calculated simply by recognizing that the flow is 10 percent greater due to the recycle flow and the separation zone area is 90 percent of the gross tank area. Thus, we obtain

$$v_{sz-hl} = v_{nom-hl}\left(\frac{1.1}{0.9}\right) = 15\left(\frac{1.1}{0.9}\right) = 18.3 \text{ m/h}$$

Good performance is expected because bubbles and floc-bubble aggregates with rise velocities of 20 m/h will be removed.

2. *Separation zone area.* This is obtained using the separation zone hydraulic loading of 18.3 m/h and solving for A_{sz} using either Eq. 9-68 or 9-69.
3.

$$v_{sz-hl} = \frac{Q+Q_r}{A_{sz}}$$

Solving for A_{sz}, we get

$$A_{sz} = \frac{Q+Q_r}{v_{sz-hl}} = \frac{(37850+3785)}{18.3} \frac{\text{m}^3/\text{d}}{\text{m}^3/\text{m}^2\ \text{h}} \left(\frac{\text{d}}{24\text{h}}\right) = 94.8\ \text{m}^2$$

If we choose a L/W ratio of 1.3, then the separation zone L is 11 m and W is 8.5 m. ▲

High-Rate DAF Systems. High-rate systems are available from several equipment companies with nominal hydraulic loading rates of 15 to 30 m/h (6 to 12 gpm/sf). According to the theory presented above, these systems would not remove bubbles and floc-bubble aggregates with rise velocities of 20 m/h and less, yet high-rate systems have been built and work. The primary reason the high-rate systems work is that the flow pattern is not the simple vertical flow structure depicted in Fig. 9-32. The flow patterns are more complex and contribute to their ability to remove bubbles and aggregates at these high rates. This is addressed in the next section.

OPERATIONAL AND DESIGN CONSIDERATIONS FOR FLOTATION

Conventional and High-Rate DAF Systems

Conventional rate and high-rate DAF systems are classified by the nominal hydraulic loading rate defined as the throughput plant flow divided by the footprint area of both the contact and separation zones; see Eq. 9-67. Conventional rate systems fall within design hydraulic loadings of 5 and 15 m/h (2 and 6 gpm/sf), whereas high-rate DAF is normally in the range of 15 to 30 m/h but can be greater. Table 9-5 shows a summary of the ranges of values for conventional and high rate DAF systems. The values in Table 9-5 serve as guidance and are used below in discussing various design and operating considerations. It is noted at the outset that there is essentially no difference between conventional rate and high-rate DAF with respect to pretreatment flocculation, recycle and saturator systems, and the method of collecting the float or its sludge characteristics. The main differences between the two systems are DAF contact zone and separation zone loading rates. The DAF separation zone is where free bubbles and floc-bubble aggregates rise to the surface. The separation zone efficiencies for removals of free bubbles and aggregates depend on the separation zone hydraulic loading rate. Note from Table 9-5 that the separation zone loadings for high-rate DAF range typically from 20 to 40 m/h (some greater). However, it is not only the separation zone hydraulic loading rate that affects performance (see below); otherwise the systems would not work. Another difference for some high-rate DAF systems pertains to the basin length/width ratio; some systems have a ratio of 1 or less. Some high-rate systems also use a deeper tank as noted in Table 9-5.

TABLE 9-5 Comparison Between Conventional Rate and High-Rate DAF (Common Range of Values*)

Item	Conventional rate	High rate	Comments
	Pretreatment flocculation		
Mean detention time (min)	10–20	10–15	Some as low as 5 and some at 20–30 min.
Number of stages	2		Some with 3 stages.
Mixing intensity (G) (sec^{-1})	50–100		Some as low as 30 and some as high as 150 sec^{-1}. Propeller or Gate Flocculators used. Some use of tapered flocculation. Some use of hydraulic flocculation.
	DAF tank		
Hydraulic loading rate (m/h)	5–15	15–30	Some high-rate DAF as high as 40 m/h. Based on the throughput flow only and the combined footprint area of the contact and separation zones. Some high-rate systems have plates or tubes placed at the bottom of the separation zone.
Contact zone loading rate (m/h)	100–200	120–300	With about 10% recycle.
Contact zone detention time (min)	1–2.5	1.0–2.0	Some high-rate DAF < 1 min.
Separation zone loading rate (m/h)	6–18	20–40	Some high rate DAF at 40–50 m/h. Based on the throughput flow and 10% recycle.
Basin length/to width ratio	1.0–2.0		Some high-rate DAF as low as 0.5 and as high as 2.5.
Basin depth (m)	2.0–3.5	2.5–4.5	Some high-rate DAF as deep as 5 m.
	Recycle and saturator systems		
Air mass (g/m^3)	6–10	6–11	
Recycle rate %	6–12		10% most typical.
Saturator gauge pressure (kPa)	400–600		Higher pressures for unpacked saturators.
Saturator efficiency (%)	80–95 packed		Higher efficiencies for higher temperatures. Unpacked saturators: 50–70%
Bubble size (μm)	10–100		Depends on saturator pressure and recycle injection device. Common range of 40–80 μm.
	Floated sludge		
Hydraulic removal	0.2–1 % solids		
Chain and Flight or reciprocating skimmer	2–3 % solids		Some as high as 5%.
Beach drum	1–3 % solids		Also called star wheel, sludge roller, and flipper.

*Values are a composite obtained from Enpure Ltd. (UK), ITT WWW Leopold (USA), Hazen and Sawyer P.C. (USA), Infilco Degremont, Inc. (USA), Läckeby Water AB, Purac Division (Sweden), and Insinööritoimisto OY Rictor AB (Finland). In some cases, particular companies design outside the reported values.

During the last 10 years high-rate DAF systems have been developed and introduced into water plant practice at nominal hydraulic loadings of 15 to 30 m/h and greater. Edzwald et al. (1999) demonstrated through pilot studies that DAF can operate at high hydraulic loadings. Their data showed good DAF performance at hydraulic loadings in the range of 20 to 40 m/h with pretreatment flocculation times of only 5 min. Several process equipment companies have developed high-rate DAF systems. These companies include Rictor (Insinööritoimisto OY Rictor AB [Finland]), Purac (Läckeby Water AB, Purac Division, Sweden), Enpure Ltd. (UK) with a system called Enflo-vite® in the United States or DAFRapide® in the United Kingdom, and Leopold with a system called Clari-DAF®. The Rictor Company refers to their process as Turbulent Flotation. The technology is available by license to Infilco Degrémont, and is known by the trade name of AquaDAF®. Figure 9-33 shows a high-rate system available from Enpure.

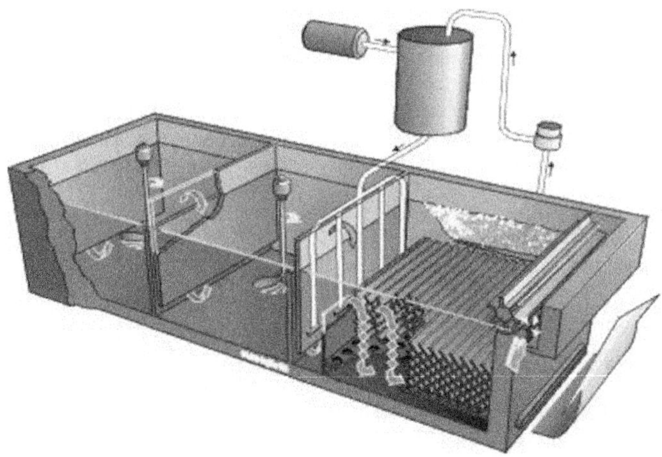

FIGURE 9-33 Schematic of a proprietary high rate DAF. (*Courtesy of Enpure Ltd., UK: Enflo-vite® from Enpure–here fitted with outlet tubes and plate system.*)

How is it that high rate DAF processes can perform efficiently at separation zone hydraulic loadings of, say, 40 m/h if free bubble and floc-bubble aggregate rise velocities are about 20 m/h (see theory on rise velocities)? The answer is that it is not just the hydraulic loading in which one assumes simple vertical flow through the separation zone as depicted in Fig. 9-32. This assumes simple plug flow downward through the separation zone so that the clarification area and separation zone area are the same, but this flow pattern does not occur especially for high-rate DAF. DAF tanks, especially those with higher hydraulic loadings exhibit stratified flow (Lundh et al., 2000) in which the flow moves horizontally near the top of the separation zone to the end wall and then returns below this in a horizontal layer, and then may again reverse direction before proceeding to the exit at the tank bottom. A simple and idealized illustration of this stratified flow is presented in Fig. 9-34. In this illustration the flow pattern effectively triples the clarification separation area. Thus, the clarification loading rate is 1/3 the footprint loading rate so a 40 m/h footprint loading rate is a clarification loading rate (as a first estimate) of 13.3 m/h. Therefore, bubbles and floc-bubble aggregates with rise rates greater than 13.3 m/h are removed. In general then, the stratified flow produces hydraulic flow patterns that effectively increase the clarification area for separation of free bubbles and floc-bubble aggregates and is analogous to increasing the clarification area in settling tanks with plates or tubes but without these physical devices. Some companies have provision for

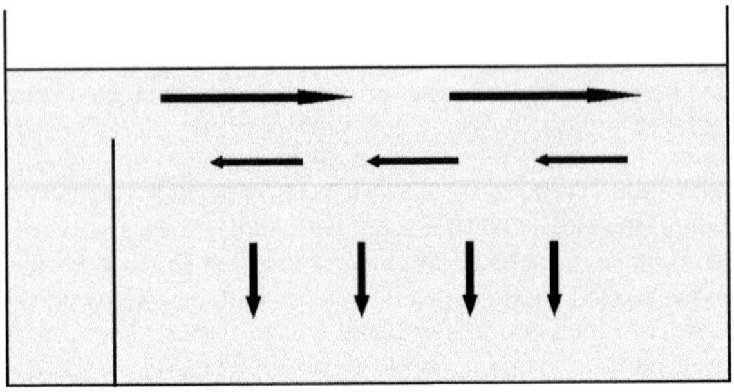

FIGURE 9-34 Idealized stratified flow in the separation zone for high rate DAF.

placing tubes or plates near the bottom of the separation zone (Enflo-vite® or DAFRapide®) as depicted in Fig. 9-33 for industrial applications, but this is not commonly used for drinking water. The plates or tubes improve the flow patterns in the separation zone and would increase the clarification area if at high rates the bubble blanket reaches the tubes or plates. Several papers were presented on these high-rate DAF systems at the international flotation conferences in Helsinki (2000) and in Seoul (2007).

Types of Flotation Tanks

Circular Tanks. Circular tanks are used mainly in small flotation plants treating wastewater or for sludge thickening applications that require no preflocculation prior to flotation. For potable water treatment, a preflocculation stage is typically required prior to flotation to flocculate the impurities present in water to form agglomerates suitable for removal by flotation. For circular tanks the transfer of the flocs to the flotation tank without breakage creates problems. In larger plants the flocculated water must be introduced close to the bottom of the center section of the flotation tank to achieve even distribution. As a result, most large flotation plants for water treatment use rectangular tanks. However, small and large flotation package plants have been built where flocculation and flotation are contained within the same circular tank.

Rectangular Tanks. Rectangular flotation tanks offer advantages in terms of scale-up, simple design, easy introduction of flocculated water, easy float removal, and a relatively small area requirement. Typical design features are summarized in Table 9-5. Many tanks are equipped at the inlet with an inclined baffle (60° to the horizontal) to direct the bubble-floc agglomerates toward the surface and to reduce velocity extremes in the water entering the separation zone in order to ensure minimum disturbance of the float layer accumulating on the water surface. Some facilities are designed with a baffle at 90°, especially those with flotation over filtration. The gap between the top of the baffle and the water surface should be designed to achieve a horizontal water velocity similar to the velocity in the top section of the baffle area. However, the size and nature of design features such as these will continue to be challenged as investigational tools and computational fluid dynamic (CFD) modeling are developed.

Maximum tank size is determined by hydraulic conditions and the design of the sludge removal system. Tanks with surface areas in excess of 80 m^2 (860 ft^2) are in operation with conventional surface loadings. Detention time in the flotation tank depends on tank depth but generally times are 10–20 min for conventional and high-rate DAF (Table 9-5). The flotation tank must be covered because both rain and wind can cause breakup of floated solids and because freezing of the float can cause problems. Treated water should be withdrawn, preferably via a full-width weir, to maintain uniform hydraulic conditions and to minimize changes in water level because of variations in flow through the plant.

Countercurrent Flotation. The basic design of dissolved air flotation has some analogy to horizontal-flow sedimentation, with flocculated water entering one end of a tank where it is mixed with dissolved air, followed by separation in a horizontal dimension. One radical reappraisal of the concept of dissolved air flotation concluded that since flotation is mainly in the vertical dimension, then flow through the flotation tank should be the same. This led to the CoCoDAF® design (CoCoDAFF® when combined with filtration, Fig. 9-35). In this

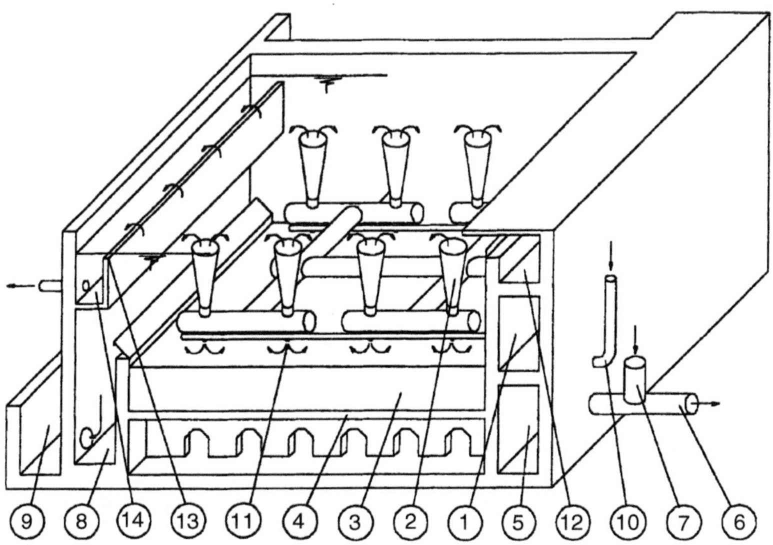

1 - INLET DUCT
2 - INLET DISTRIBUTION CONES
3 - FILTER MEDIA
4 - FILTER FLOOR
5 - OUTLET & UPWASH DUCT
6 - OUTLET PIPE
7 - UPWASH PIPE
8 - WASHOUT BAY
9 - WASHOUT CHANNEL
10 - AIR INLET
11 - AIR DISTRIBUTION NOZZLES
12 - FLUSHING CHANNEL
13 - SCUM WEIR
14 - SCUM CHANNEL

FIGURE 9-35 Typical arrangement of a CoCoDAFF® unit. (*Courtesy of Thames Water and Paterson Candy Ltd, Isleworth, UK.*)

design, the DAF tank bears more resemblance to an upflow floc blanket clarifier in contrast to a normal DAF tank having similarities with a horizontal-flow sedimentation tank. In the CoCoDAF® design, the flocculated water is introduced above the recycle, uniformly across the area of the tank, so that the water flows down through the rising bubble cloud to the treated water outlet, or to the filter bed as in CoCoDAFF®.

Combined Flotation and Filtration. The combination of flotation and filtration was developed in Sweden (Zabel and Melbourne, 1980) in the 1960s–1970s and now is widely used around the world. The system is sometimes called simply flotation over filtration.

A dual media anthracite-sand filter or sand filter is incorporated in the lower section of the flotation tank. The flotation hydraulic loading rate for the plant is, however, limited by the filtration rate that can be achieved. It is also possible to apply the concept to existing filters, of appropriate dimensions, by retrofitting the dissolved air flotation equipment.

Flotation over filtration has major advantages of a compact water plant (small footprint area) and cost savings. The compactness of the system makes it particularly suitable for package plants and is thus a good application for plants with flows less than 4 ML/d (~1 mgd). However, because of the advantage of a small plant footprint, this system has also been adapted in the last 15 years for large water plants that have land constraints. One example is the water plant in Fairfield, Connecticut in the United States that has been in operation since 1997. It is a 190-ML/d plant with a nominal flotation-filtration loading rate of 15 m/h. The performance of this plant is described in Performance of Dissolved Air Flotation Plants, showing that DAF plants with flotation over filtration perform as well as those with separate processes. One of the world's largest DAF plants is being built within the city limits of New York, and this plant is flotation over filtration (DAF/F) (Crossley et al., 2007). The plant is scheduled for commissioning in 2011 or 2012. The flotation and filtration rates will be 12.2 m/h and 15.9 m/h, respectively, treating Croton Reservoir water which has low turbidity (0.6–3.6 ntu), low color (true color of 2–20 cu), low TOC (2.2–4.3 mg/L), and low to moderate alkalinity (40–70 mg/L $CaCO_3$).

Water plants with flotation over filtration have a smaller plant footprint, thereby reducing the land area, which is a significant advantage for large cities with limited land areas. There is also a construction cost savings of having one structure for flotation and filtration compared to conventional rate plants with separate units. There are two drawbacks for such a system. First, flotation and the filters cannot be independently designed and operated. Thus, the flotation unit is limited to conventional loading rates. A second drawback is that one loses the option of placing an ozonation process between DAF and filtration.

The tank depth of a flotation-filtration unit tends to be deeper compared to a separate flotation unit to accommodate the filter media and underdrains. In addition, the flow to the unit has to be stopped periodically to facilitate cleaning of the filter that is backwashed in the normal way by air scour and water wash. This can be accommodated by having several treatment trains, so that with one out of service, the flow into the plant is reduced to maintain constant loadings on the other trains and the coagulant feed adjusts to the reduced flow. However, as with a bank of filters, providing there are at least four treatment units then when one is isolated for washing, or maintenance, the flow to the others can be ramped up so that there is no change in total flow.

Air Saturation Systems

For an efficient packed-column saturator system, the energy used for air compression is about 5 to 10 percent of that used for recycle water pumping. When efficiency is poor, the air compression can account for a much greater proportion of the power cost of producing the saturated recycle. Therefore, optimization of the recycle system design is important in minimizing operating costs.

Various methods are employed for dissolving air under pressure in the recycle stream. These include sparging the air into the water in a pressure vessel, or saturator; trickling the water over a packed bed; spraying the water into an unpacked saturator; entraining the air with eductors; and injecting the air into the suction pipe of the recycle pump (Fig. 9-36) (Zabel and Melbourne, 1980). The packed column is probably the most widely implemented method.

Introducing the air to the recycled water either on the suction side of the recycle pump or through eductors before entering the saturation vessel leads to substantially higher pumping costs compared with using a separate compressed-air supply, although capital cost and footprint might be smaller. Relatively efficient self-aspirating pumps

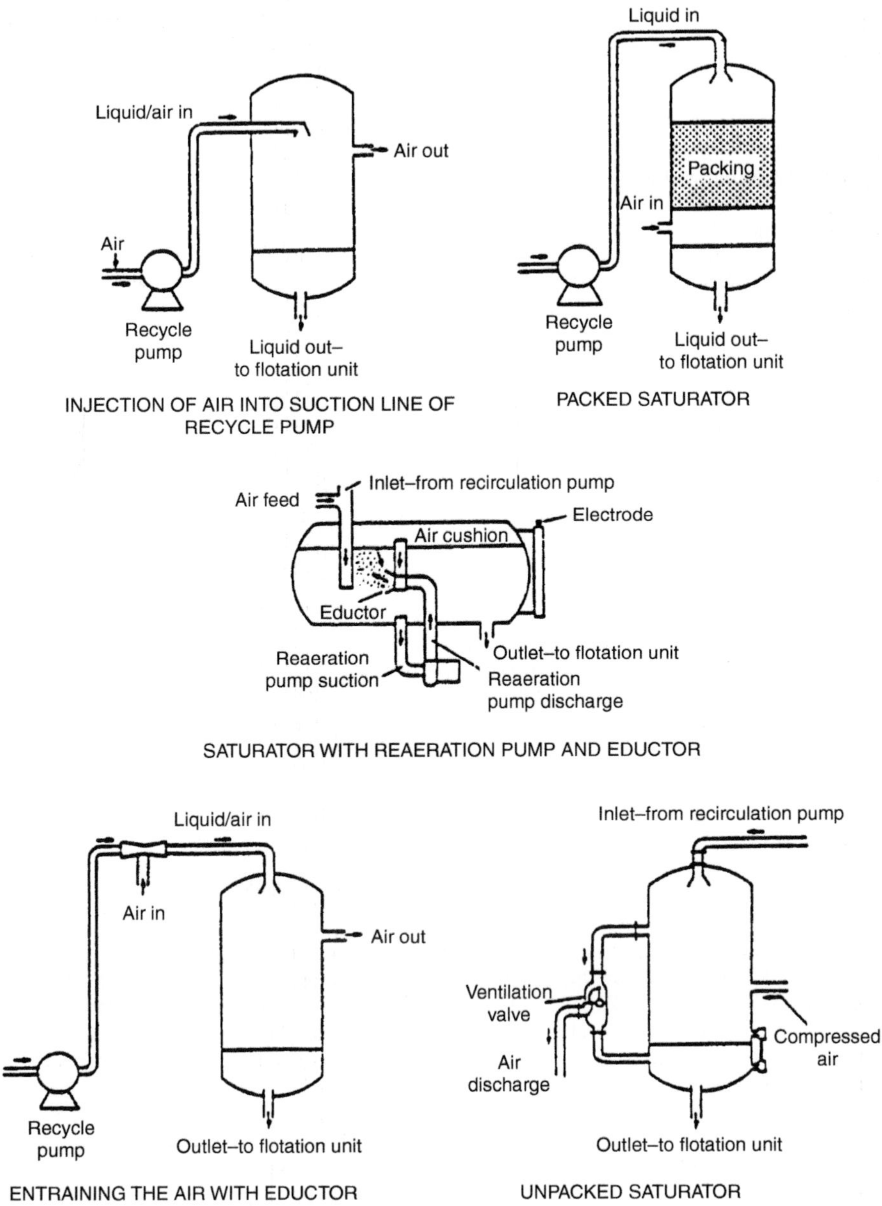

FIGURE 9-36 Different air saturation systems.

have been developed to be suitable for DAF without the need for saturator vessels. Although their operating cost (kW-h/kg air) is greater than for a packed saturator system, the capital cost and footprint are smaller. Consequently, the net present cost of self-aspirating pump systems can be less than for a packed-column system and are attractive where space is limited.

A possible disadvantage of using packing is the danger of blockage caused by biological growth or other precipitates. A problem with iron precipitates has been encountered in a few plants treating drinking water. Fouling of packing can be avoided by recycling filtered water rather than clarified water.

Extensive research has been carried out on optimizing the design of packed saturators (Rees et al., 1980b; Casey and Naoum, 1986). It has been shown that saturators can be operated over a range of hydraulic loadings of 12.5 to 104 m/h (36 to 274 ft/h or 5 to 30 gpm/ft^2) without any decrease in saturation efficiency. A packing depth of 0.8 m (2.6 ft) with 25 mm (1 in.) polypropylene rings is sufficient to achieve greater than 90 percent saturation (Fig. 9-37) (Rees et al., 1980b). This packing depth, however, was substantially higher than the 0.3 m (1 ft) reported elsewhere (Bratby and Marais, 1977). The theory of saturators for flotation has been examined extensively (Haarhoff and Rykaart, 1995; Haarhoff and Steinbach, 1996; Haarhoff and Steinbach, 1998; Valade et al., 2001).

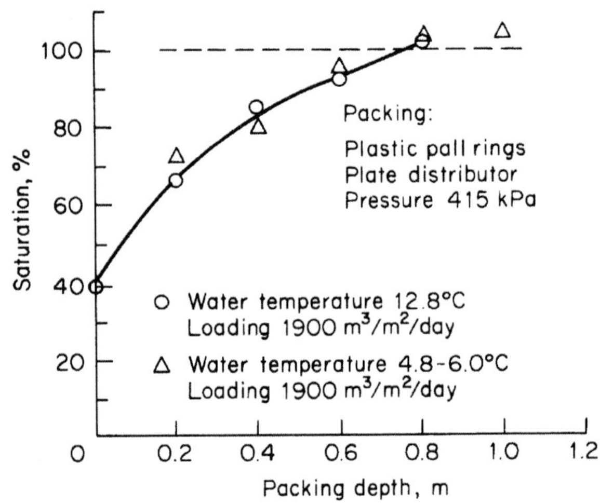

FIGURE 9-37 Effects of packing depth on air dissolution.

In terms of operating costs, an unpacked saturator must be operated at a pressure of about 200 kPa (30 psi) above that of a packed saturator in order to supply the same amount of air to the flotation tank (Zabel and Hyde, 1977). Water saturated with air under pressure is corrosive. If mild steel is used for construction of the saturation vessel, a corrosion-resistant lining should be provided. Stainless steel or plastic pipes should be used for connecting pipe work between the saturator and flotation tank.

Factors Influencing Dissolved Air Efficiency

The reader is referred to the section Fundamentals of Dissolved Air Flotation for presentation of the principles regarding the variables that affect dissolved air flotation performance or efficiency. For the contact zone efficiency (i.e., the efficiency of collisions and attachment of particles to bubbles), the important variables are coagulation, floc size, bubble size, contact zone detention time, and bubble volume concentration; see Table 9-4 for a summary. For the separation zone efficiency (i.e., removal efficiencies of free bubbles and floc-bubble aggregates by rising to the top of separation zone), the fundamental variables affecting rise velocities are bubble size, bubble-floc aggregate size, number of bubbles attached to each floc, floc-bubble aggregate density, and water temperature. Practically, the separation zone efficiency relates rise velocities to separation zone hydraulic loading rates. In the following material, some practical factors influencing DAF efficiency are discussed.

Coagulation. Coagulation is a chemical addition step. Metal coagulants such as alum, polyaluminum chlorides, and ferric salts are commonly used. Some water plants in the United States use two primary coagulants: a metal coagulant and a highly charged organic cationic polymer. Coagulation has two purposes: to destabilize particles present in the raw water and to convert dissolved natural organic matter (NOM) into particles. The latter purpose of coagulation is a technology control strategy to remove NOM, thereby reducing TOC and disinfection by-product precursors. This has become an important coagulation objective in water treatment, not only for supplies with moderate to high concentrations of color and TOC but also for supplies with small amounts.

In the contact zone of the flotation tank, collisions occur between bubbles and floc particles. Coagulation is essential to ensure attachment of bubbles to flocs when collisions occur. Without coagulation, the particles carry a negative charge and are often hydrophilic so that bubble attachment is poor. Good coagulation chemistry depends on coagulant dose and pH. Optimum coagulation conditions are those of coagulant dose and pH which produce flocs with charge near zero and with relatively high hydrophobicity. These optimum coagulation conditions cause high bubble attachment efficiency.

The coagulants that are effective for flotation are the same that are used in sedimentation plants. Coagulant selection depends not on flotation per se, but on raw water quality factors of turbidity, NOM concentration (TOC, DOC, UV [254 nm] absorbance, color) and composition, and alkalinity as well as water temperature. Other considerations in coagulant selection are sludge production and overall costs that consider not only the coagulant cost but also costs from dosing of pH control chemicals and sludge handling and disposal. Local water plant preference and availability of coagulants also affect coagulant selection.

Figure 9-38 illustrates the effect of coagulant dose on flotation efficiency for the removals of DOC and turbidity. Electrophoretic mobility (EPM) data are presented as an indicator of floc charge. Figure 9-38 shows clearly that underdosing of coagulant is associated with negative particle charge and poor particle removal as shown by the small difference in turbidities before and after flotation. Minimum turbidity after flotation is associated with a floc charge of about zero and with good removal of organic matter, both DOC and UV (254 nm).

In practice, measurements of particle charge and hydrophobicity are not needed to determine optimum coagulation conditions. Coagulant dosing may be evaluated using standard sedimentation or flotation jar test apparatus (see Chap. 8) in which the optimum dose is that which produces low turbidity and low NOM measured as DOC or UV (254 nm) as a surrogate of NOM. Favorable pH conditions that produce floc charge at the optimum dose and minimize residual soluble coagulant metal depend on the specific coagulant. Guidance on favorable coagulation pH conditions follows for water supplies containing turbidity and DOC of low or moderate (2–6 mg/L) to high (> 6 mg/L) concentrations. For alum, optimum pH is 6 to 7 depending on water temperature (for warm waters, lower pH near 6 is optimum) for low or moderate levels of DOC whereas for high DOC waters, favorable pH conditions are in the mid 5s to low 6s. For PACls of medium to high basicity (see Chap. 8 for PACl characterization), favorable pH conditions are similar to alum but with less dependence on water temperature. For ferric coagulants, favorable pH conditions are in the mid 5s to low 6s for moderate DOC waters whereas for high DOC waters lower pH of low 5 to about 6 is best. Some water plants with very high DOC (say, > 10 mg/L) actually practice coagulation with alum and ferric salts at lower pH values than given above. Water supplies of moderate to high alkalinity that practice coagulation above pH 7 must add additional coagulant or

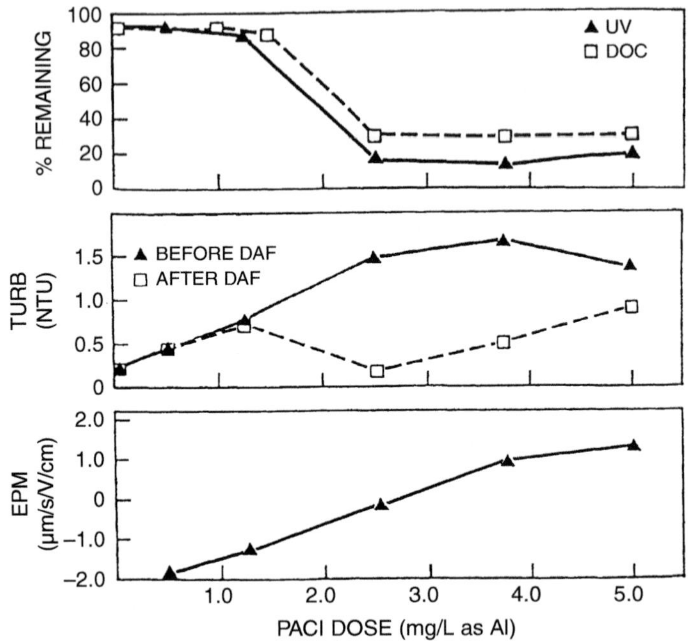

FIGURE 9-38 Effect of polyaluminum chloride dose on the electrophoretic mobility (EPM: indicator of charge) of flocs and flotation performance for turbidity and natural organic matter (Conditions: high basicity PACl at constant coagulation pH of ~5.5, 4°C, 5 mg/L TOC of fulvic acid, and low initial turbidity). (*Source: Edzwald and Malley, 1990.*)

acid to reach the desired pH or use higher doses to make up for the unfavorable pH conditions. Water supplies containing turbidity and low DOC, especially low in aquatic humic matter, can achieve optimum coagulation with smaller chemical doses and over a wider pH range than presented here.

Coagulant dosing for turbidity and NOM is fundamentally the same for flotation and sedimentation processes. Optimum coagulant doses and pH conditions are generally the same. An exception is that lower coagulant doses for DAF than for sedimentation processes may be found in treating high-quality waters of low turbidity and low to moderate levels of NOM. This is because in sedimentation the ability to form large-size flocs for settling depends on the floc particle concentrations. To achieve rapid flocculation and aggregation of flocs into large sizes often requires adding additional coagulant to increase precipitated solids (additional sweep floc), thereby increasing the rate of flocculation (see Chap. 8). Large-size flocs are not required for DAF so additional coagulant dosing for these type water supplies is not necessary.

Another important difference is that flotation does not normally require addition of high-molecular-weight polyelectrolyte flocculants. There are some exceptions where water plant operators have found that small amounts of high-molecular-weight polymers improve retention of the floated sludge.

Chemicals must be thoroughly and rapidly mixed with the raw water. If both a coagulant and a pH adjustment chemical are required, good mixing of the first chemical with the source water should be completed before the addition of the second chemical. This is particularly important when treating soft waters with high DOC. The order of chemical addition can be important if stress conditions of high flow, low water temperature, or poor raw water quality occur. The order of addition can be established through jar tests and plant trials.

Flocculation. Before coagulated particles and NOM can be removed successfully by flotation, flocculation into larger agglomerates (floc) is required. Flocculation time, degree of agitation, and the means of providing agitation affect flotation performance. Flocculation for flotation has a different objective than for sedimentation. Whereas clarification by settling requires flocs with sizes of 100 microns and larger, flotation is effective with smaller floc particles. In DAF, particles with sizes of about 1 μm are not collected efficiently by bubbles so larger flocs are desired but not too large since floc particles larger than about 100 μm require multiple-bubble attachment to achieve high rise rates. Therefore, the theory presented in the section (Fundamentals of Dissolved Air Flotation) predicts optimum floc sizes for DAF are 25 to 50 μm, much smaller flocs than for sedimentation processes.

Flocculation Time. The flocculator usually consists of a tank subdivided by partial baffles into two equal-sized compartments or stages in series each agitated by slow-moving paddles or propeller units; see Table 9-5. Additional compartments can be used, but this is usually not cost effective. The flocculation time required differs with the type of raw water being treated, coagulant being used, and water temperature. Table 9-5 shows common design ranges of flocculation times of 10–20 min for conventional rate DAF and 10–15 min for high-rate DAF systems; however, there has been a dramatic decrease in flocculation times used in practice over the last 20 years, so for new plants flocculation times of about 10 min are often used for both conventional and high-rate DAF systems. Even shorter times such as 5 min may be feasible if demonstrated through pilot studies as was done for the Croton water plant for New York City. In a pilot study done on the Gothenburg (Sweden) water supply, Valade et al. (1996) showed both good DAF and filtration performance with a flocculation time of only 5 min treating water under very cold water temperatures of about 2°C (see Table 9-6).

Degree of Agitation. Besides the flocculation time, the degree of agitation is also important for efficient flocculation. Propeller-type units have become favored since their introduction (Franklin 1997), although gate (paddle) flocculators are still used and were widely used in the past. To avoid excess shear, which prevents adequate floc growth, the tip speed of the paddles should not exceed 0.5 m/s (1.6 ft/s).

The degree of agitation can be expressed by the mean velocity gradient G (see Chap. 8). Tests have indicated that the optimum mean velocity gradient for flotation is about 70 s^{-1} independent of the type of surface water treated (Rees et al., 1979). This compares with

TABLE 9-6 Comparison of the Effect of 5 and 20 min Flocculation on Flotation and Filtered Water Turbidity and Particle Numbers for Cold Water (From work by Valade et al. 1996)

	Flocculation time*			
	5 min		20 min	
Sample	Turbidity (ntu)	Part. #/mL (2–200 μm)	Turbidity (ntu)	Part. #/mL (2–200 μm)
Raw water	1.0	2,040	0.9	4,010
Floc. effl.	6.7	18,480	6.9	14,010
DAF effl.	1.0	1,620	0.8	1,530
Filter effl.	0.01	6	0.04	12

*Conditions: *5 min floc time*: ferric sulfate 5.7 mgFe/L, water temp. 2.1°C, recycle 6%; *20 min floc time*: ferric sulfate 5.1 mgFe/L, water temp. 5.8°C, recycle 9%. Common conditions: three-stage flocculation G of 70 sec^{-1}, DAF loading 8 m/h (3.3 gpm/ft^2), dual-media filters loading, 10 m/h (4.1 gpm/ft^2).

an optimum G value for horizontal sedimentation of between 20 and 100 s^{-1}. Valade et al. (1996) found that flotation performance, using either gate or propeller units for flocculation, was affected little by G in the range 30 to 70 s^{-1}. However, there may need to be a minimum degree of agitation in flocculators to prevent settlement of floc in the flocculators or short-circuiting through the tanks. It is also important to produce relatively small and compact flocs to avoid floc breakup when the air is introduced.

Values of Gt between 40,000 and 60,000 are usually considered necessary for efficient flotation. However, the move toward short flocculation times is showing that Gt values less than 20,000 can be effective (Valade et al., 1996).

Hydraulic Flocculation. An alternative approach to mechanical flocculation is the use of hydraulic flocculation in which the energy required for flocculation is provided by the water flowing through the flocculator, which can be a baffled tank. Tests have shown that one-half the flocculation time with a higher G value (150 s^{-1}) was required for hydraulic flocculation compared with mechanical flocculation (G = 70 s^{-1}) (Fig. 9-39) (Rodman, 1982). The difference in G value required is probably due to the more uniform velocity distribution in the hydraulic flocculators, thus avoiding excess shear and floc breakup. However, from these results the Gt value for efficient flotation is independent of whether hydraulic or mechanical flocculation is employed.

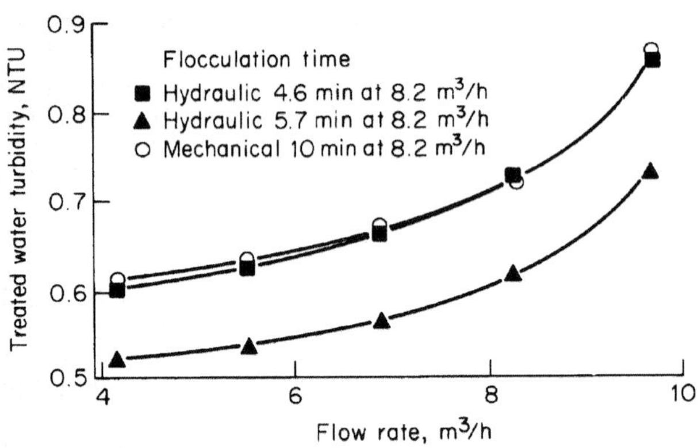

FIGURE 9-39 Comparison of hydraulic and mechanical flocculation for flotation. (*Source: Rodman, 1982.*)

Quantity of Air Required for Flotation. The quantity of air supplied to the flotation tank can be varied by altering the saturator pressure or the amount of recycle or both. Valade and Crossley (1998) concluded that maintaining operating pressure and changing recycle rate is the most energy-efficient way of changing air dose. If a fixed orifice is used for controlling the injection of recycle, an increase in saturator pressure is associated with a small increase in recycle rate. Thus, different nozzle sizes require different combinations of flow and pressure to deliver the same amount of air.

Experiments varying the recycle rate by using different nozzle sizes and saturator pressures have shown that treated water quality is dependent on only the total amount of air supplied, not the pressure and recycle rate employed (Fig. 9-40) (Rees et al., 1979). This confirmed earlier work by Bratby and Marais (1975).

The quantity of air required for treatment of surface waters depends only on the total quantity of water treated and is independent of the suspended solids present, unless the suspended solids concentration is very high (>1000 mg/L). The air/solids ratio required for surface water treatment is approximately 380 mL of air per gram of solids for a

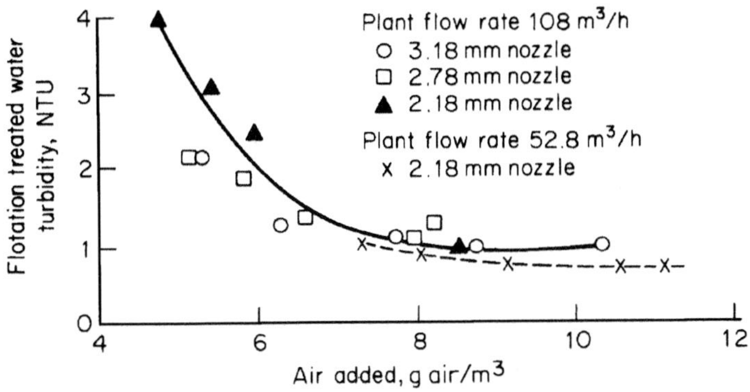

FIGURE 9-40 Effect of quantity of air added and nozzle size on flotation treated water quality. (*Source: Rees et al., 1979.*)

solids concentration in the raw water of 20 mg/L: this corresponds to a bubble volume concentration of 7600 ppm (see the section Fundamentals of Dissolved Air Flotation and Example 9-6). It is much higher than for activated sludge thickening, where 15 to 30 mL of air per gram of solids is required (Maddock, 1977). The large air/solids ratio required for treatment of surface water with low turbidity is probably necessary to ensure adequate collisions and attachment between floc particles and air bubbles before separation.

With a packed saturator, an operating pressure of between 350 and 420 kPa (50 and 60 psi), and a recycle rate of between 7 and 8 percent corresponding to about 8 to 10 g air/m^3 water treated were found adequate for optimum performance (Rees et al., 1979). Subsequent experience has shown this generally to be the case, although with lower efficiency saturators and warm waters the yield of dissolved air may be less.

Fundamentally, volume concentrations of bubbles are more important than mass concentrations. Larger volume concentrations lead to greater collision rates with particles and thereby produce densities of particle-bubble aggregates that are much less than water (see Examples 9-6 and 9-7).

Air-Release Devices. Different types of pressure release devices are employed ranging from proprietary nozzles and needle valves to simple gate valves. To achieve effective air release, the pressure should be reduced suddenly and highly turbulent conditions should exist in the device. The velocity of the recycle stream leaving the pressure-reduction device should be low enough to prevent floc breakup. As water passes through the pressure-release devices at high velocities and as air is released, erosion and cavitation can occur, and so the devices should be made from stainless steel or plastic. For larger plants a number of these devices, usually at a spacing of approximately 0.3 m (1 ft), are used to obtain good mixing and distribution between the flocculated water and the air bubbles. Good mixing of the recycle stream containing the released air bubbles with the flocculated water stream is also essential to facilitate contact between bubbles and floc.

Releasing the pressure of the recycle stream close to the point of injection into the flocculated water is important to minimize coalescence of air bubbles or premature release of air in the recycle prior to injection, which could result in a loss of bubbles available for flotation.

An air injection nozzle was patented and developed (Brit. Pat., 1976) that consists of two orifice plates, to reduce the pressure and to create turbulence, and a shroud section, to decrease the velocity of the stream of recycled water before it is mixed with the flocculated water. The size of the first orifice plate, which is the smaller, controls the amount of recycled water added to the flotation tank. More than 95 percent of the bubbles produced by the nozzle were in the size range 10 to 120 μm, with a mean size of about 40 μm. A comparison of the proprietary nozzle and a needle valve showed that the nozzle produced smaller air bubbles; however, both devices achieved similar flotation treated water quality (Rees et al., 1979).

Because the production and size of air bubbles are important, considerable attention has been given to understanding and improving nozzle design (Van Craenenbroeck et al., 1993; van Puffelen et al., 1995; Heinänan et al., 1995; Rykaart and Haarhoff, 1995; Offringa, 1995; Franklin, 1997). These investigations have demonstrated that orifice diameter, impingement distance, and other features are important in efficient production of bubbles of the preferred mean size and size range. Franklin (1997) found that small improvements in flotation efficiency could be obtained through changes to nozzle design and bubble size. In general, the size and number of bubbles produced is primarily a function of the energy available, which is not otherwise dissipated in turbulence and friction; the higher the nozzle back-pressure the smaller the bubble size (Jackson, 1994; Rykaart and Haarhoff, 1995).

Float Removal. The sludge that accumulates on the flotation tank surface, called *float,* can be removed either continuously or intermittently by flooding or mechanical scraping. Flooding involves raising the water level in the flotation tank sufficiently by closing the treated water outlet or lowering the sludge beach to allow the float and water to flow into the float collection trough. The flooding method has the advantages of low equipment costs and minimal effect on the treated water quality, but at the expense of high water wastage (up to 2 percent of plant throughput) and low sludge solids concentration (less than 0.2 percent) although concentrations of 0.5–1 percent should be possible. Therefore, with this float removal method, one advantage of flotation—the production of sludge with a high solids concentration—is lost.

The most widely used mechanical float removal devices are of two types: (1) part or full-length scrapers usually with rubber or brush blades that travel over the tank surface and push the float over the beach into the collection channel and (2) beach scrapers that consist of a number of rubber blades rotating over the beach.

As float is removed from the beach, float from the remainder of the flotation tank surface flows toward the beach. The beach scrapers, especially if operated continuously, have the advantage of reducing the danger of float breakup during the removal process because the float is minimally disturbed. Beach scrapers are also of simpler construction compared to full-length scrapers. They have the disadvantage, however, of producing relatively thin sludges (1 to 3 percent; see Table 9-5), because thicker sludge will not flow toward the beach. If a part- or full-length scraper is used, selecting the correct frequency of float removal and travel speed of the scraper is important to minimize deterioration in treated water quality because of float breakup. When optimized they can achieve solids concentrations of 2–3 percent.

Effect of Air/Solids Ratio on Float. In water treatment, the aim is to produce water of good quality, and thickness of the float is of secondary importance. In general, the thicker the float, which means the longer the float is allowed to accumulate on the tank surface, the more severe will be the deterioration in treated water quality during the float removal process. The variations in air/solids ratio, produced by varying the amount of air added to the system, has no influence on the float concentration produced.

Influence of Source Water on Float Characteristics. The characteristics of float obtained from the treatment of different source waters vary considerably (Rees et al., 1979). Therefore, the most appropriate float removal system must be selected for the source water being treated. For example, experience has shown that for cases in which a low-alkalinity, highly colored water was treated, the float started to break up after only 30 min of accumulation. The most appropriate device for such an application is a full-length scraper operating at a removal frequency and blade spacing that does not allow the float to remain on the flotation tank surface for longer than 30 min. The optimum scraper speed for this application, in terms of treated water quality and float solids concentration, was 30 m/h (0.028 ft/s), producing a sludge of 1 percent solids concentration. An alternative strategy with such quality float might be to strengthen the floc by dosing a high molecular weight polymer.

Conversely, float produced from turbid river water or stored algal-laden water is very stable. Accumulation of float for more than 24 h does not result in float breakup or deterioration in treated water quality. Beach scrapers and part and full-length scrapers have been used successfully for these applications, producing solids concentrations in excess of 3 percent with little deterioration in treated water quality, provided the float is not allowed to accumulate for too long. These floats were suitable for filter pressing, producing cake solids concentration of between 16 and 23 percent without polyelectrolyte addition.

The float stability is independent of the primary coagulant used for optimum coagulant dose and coagulation pH. The addition of flocculant aid has been found useful when using aluminum sulfate in cold waters, with the flocculant aid also helping in weak floc situations to reduce float breakup and increase filter run length.

For optimum operation in terms of treated water quality and float solids concentration, beach scrapers should be operated continuously, the water level in the flotation tank should be adjusted close to the lower edge of the beach, and a thin, continuous float layer of about 10 mm (0.4 in.) should be maintained on the surface of the flotation tank. This operation produces a float concentration, depending on the source water treated, of 1 and 3 percent with minimum deterioration of treated water quality.

Equipment costs for float removal systems can be as much as 10 to 20 percent of the total plant cost. Therefore, selection of the most appropriate and cost-effective removal system for a particular application is important. For example, although a beach scraper might have the lowest cost because of its relative simplicity, it might not be as effective as a flight scraper for the quality and quantity of float produced.

Performance of Dissolved Air Flotation Plants

The several international flotation conferences held periodically (Ives and Bernhardt, 1995; CIWEM, 1997; Helsinki, 2000; and Seoul, 2007) are good sources of information on DAF performance in treating different types of water supplies. The work reported on at these conferences includes the evaluation of pilot and full-scale plants. Some performance data are presented in the following subsections in which the material is arranged according to either the type of source water or the specific source water contaminants.

Treatment of Turbid Lowland (High-Alkalinity) River Water. Under optimum operating conditions (Rees et al., 1979), flotation reduced source water turbidities of up to 100 ntu to levels rarely exceeding 3 ntu at a design loading of 12 m/h (4.9 gpm/sf). When the source water turbidity exceeded 60 ntu, treated water quality was improved greatly by reducing the loading rate through the plant by about 10 to 20 percent. Color was reduced from 70 to less than 5 color units (CU), and residual coagulant concentrations before filtration were in the range 0.25 to 0.75 mg Al/L (total Al).

A floc blanket sedimentation plant operated at an upflow rate of 2 m/h (0.8 gpm/sf) produced similar treated water quality to that of the flotation plant during low-turbidity periods but better quality (by 1 to 2 ntu) when source water turbidity was greater than 100 ntu. Selection of the correct coagulant dosage and coagulation pH was critical during storm events. Because of the short residence time in the flotation plant, changes in source water quality had to be followed closely to maintain optimum coagulation conditions.

Directly abstracted river water can be treated successfully by flotation. However, if the source water turbidity varies rapidly, has high-turbidity peaks (>100 ntu), and large doses of coagulant are needed, then sedimentation tends to be the more appropriate treatment process. High-turbidity sources are sometimes pumped to bank-side reservoirs or raw water tanks for short-term storage (<3 days); this dampens out short-term quality variation and allows some settlement of solids.

Treatment of Low Turbidity, Low Color Waters. In the United States a large number of DAF plants were built in the 1990s and early 2000s to treat waters of high quality (e.g., New Castle, NY, Fairfield, CT, Greenville, SC, Cambridge, MA, Catskill, NY). In these cases, raw water turbidities are about 1 ntu, color is typically 15 CU or less, and TOC values are 3 mg/L or less. They were designed at DAF loadings of 10–15 m/h (4–6 gpm/ft^2) and produce DAF effluent turbidities below 1 ntu and often in the range 0.2 to 0.5 ntu.

Several U.S. cities have limited land and space so the reduced footprint is a considerable advantage. The design of a 1100 ML/d (290 mgd) plant for New York City allows for 1-min flash mixing and 4.8-min flocculation at full capacity prior to flotation over filtration (DAF/F). The DAF loading rate will be 12.2 m/h (5 gpm/sf) and the filtration rate will be 15.9 m/h (6.5 gpm/sf) with four of the 48 DAF/F units out of service (Crossley et al., 2007). This plant is expected to be commissioned in 2011 or 2012.

Treatment of Waters with Natural Color and Humic Substances. Table 9-7 shows a comparison of the water quality achieved by flotation, sedimentation, and filtration treating a stored water of high color and low alkalinity with ferric coagulation (Rees et al., 1979). The flotation plant was operated at 12 m/h (4.9 gpm/sf) loading rate.

The floc blanket sedimentation plant could only be operated at less than 1 m/h (0.4 gpm/sf), even with the addition of polyelectrolyte. The floc produced by coagulation of these waters is very light and has low settling velocities. The quality of the waters treated by the two processes was quite similar. Initially the residual coagulant concentration of the sedimentation plant treated water was usually lower by about 0.2 mg Fe/L. By increasing the flocculation time from 12 to 16 min, however, the residual coagulant concentration in the flotation treated water was reduced to that of the sedimentation treated water.

Another advantage of flotation was that the plant consistently produced good treated water quality even at temperatures below 4°C (39°F), which occurred frequently during the winter months. At these low temperatures the floc blanket in the sedimentation tank tended to become unstable, resulting in deterioration in treated water quality.

TABLE 9-7 Comparison of Qualities Achieved with Flotation,* Sedimentation,† and Filtration (Colored Low-Alkalinity Water) Following Iron Coagulation (Rees et al., 1979; Zabel and Melbourne, 1980)

Type of water	Turbidity, ntu	Dose, mgFe/L	Color, PtCo	pH	Iron, mg/L	Manganese, mg/L	Aluminum, mg/L
Source	3.2	–	45	6.2	0.70	0.11	0.23
Flotation-treated‡	0.72	8.5	2	4.8	0.58	0.16	0.01
Flotation-filtered	0.19	–	0	9.0	0.01	<0.02	0.01
Sedimentation-treated	0.50	6.0§	0	5.05	0.36	0.14	0.10
Sedimentation-filtered	0.29	–	0	10.5	0.01	<0.02	0.10

*Upflow rate 12 m/h (39 ft/h).

†Upflow rate 1 m/h (3.25 ft/h).

‡Improved flotation-treated water quality similar to that achieved with sedimentation was obtained by increasing the flocculation time from 12 to 16 min.

§Plus 0.8 mg polyelectrolyte/L required to maintain the floc blanket.

Treatment of Waters with Algae. Severe algal problems can be experienced in some storage reservoirs containing nutrient-rich waters. Algae tend to float and are, therefore, difficult to remove by sedimentation. Table 9-8 shows the removal efficiency of flotation and sedimentation for different algal species (Rees et al., 1979). At times the algal counts in the flotation treated water were lower than those in the sedimentation and filtered water. (The results in Table 9-8 refer to use of chlorinated ferrous sulfate. This coagulant is no longer used because of its association with THM production. However, the nature of the results still applies to coagulants currently used.) Efficient coagulation and flocculation are essential for effective algal removal. Only 10 to 20 percent algal removal was obtained when the flotation plant was operated without coagulation.

TABLE 9-8 Comparison of Algae Removal Efficiency of Flotation* and Sedimentation† Using Chlorinated Ferrous Sulfate as Coagulant (Based on Rees et al., 1979; Zabel and Melbourne, 1980)

Algae type	Source water, cells/mL	Sedimentation-treated water‡, cells /mL		Flotation-treated water‡, cells/mL	
			% removal		% removal
Aphanizomenon	179,000	23,000	87.2	2,800	98.4
Microcystis§	102,000	24,000	76.5	2,000	98.0
Stephanodiscus	53,000	21,900	58.7	9,100	82.8
Chlorella	23,000	3,600	84.3	2,200	90.4

*Upflow rate 12 m/h (39 ft/h).
†Upflow rate 2 m/h (6.5 ft/h).
‡Before filtration.
§Aluminum sulfate used as coagulant.

The greater effectiveness of DAF compared to sedimentation in removing algae is because algae have low densities and so attachment of air bubbles will easily cause separation in DAF plants. Figure 9-41 demonstrates this for removing a green alga (*Chlorella*) and a diatom (*Cyclotella*). Log removals of 2–3 log (99–99.9 percent removal) were found for DAF compared to 1–2-log removal by settling. Other reports on algae removal by DAF support these findings as well as removals of blue-green algae, or cyanobacteria (Teixeria and Rosa, 2006).

The percentage removals given are quite high, but were obtained in well-controlled and optimized treatment conditions with raw water algae at high concentrations. In practice, one should not expect such high removals. For guidance purposes, one should expect 90–99 percent removals of algae by DAF unless the algae are at low concentrations, then lower removals are achieved. Settling processes are not as effective (Table 9-8).

Cryptosporidium *and* Giardia *Removal.* Extensive pilot-scale studies (Edzwald et al., 2000; Edzwald et al., 2003) have demonstrated that DAF is very effective in removing *Cryptosporidium* compared to sedimentation. Figure 9-42 illustrates the ability of DAF to produce high-quality clarified water with turbidities of about 0.5 ntu for cold water conditions and at a DAF nominal loading rate of 15 m/h (6 gpm/sf). Plate settling (design hydraulic loading of 0.6 m/h) had higher turbidities of about 1.6 ntu. Likewise as shown in Fig. 9-42, DAF effluent particle counts were much lower than those from sedimentation.

Figure 9-43 presents log removal *Cryptosporidium* data for several pilot runs for summer (17–18°C) and winter conditions (2–5°C), all runs at DAF and dual-media filtration hydraulic loadings of 15 m/h (6 gpm/sf). DAF achieved 2-log removals for summer and

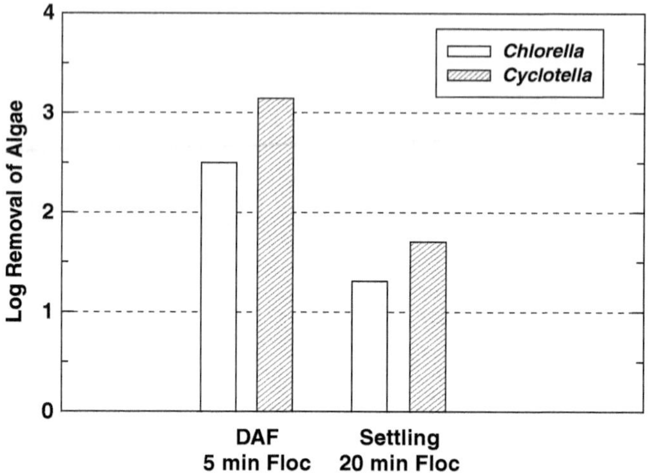

FIGURE 9-41 Bench-scale DAF comparison of log removals of algae by DAF versus sedimentation (4°C, DAF conditions: 5 min flocculation time and 10 min flotation time; sedimentation conditions: 20 min flocculation time and 60 min of settling). (*Source: Edzwald and Wingler, 1990.*)

winter. Plate settling performed well for summer temperatures, but only achieved 1-log removal in the winter. These winter results showing better DAF performance than plate settling are in agreement with the turbidity and particle count data presented in Fig. 9-42. Finally, it is noted that overall *Cryptosporidium* log removals by clarification and dual media filtration were 4–5 for both the DAF and plate settling trains.

The filter is the final barrier and ensures high overall log removals of *Cryptosporidium*, but greater particle and *Cryptosporidium* removals can be achieved by DAF compared to sedimentation. An advantage of DAF plants is filtration serves as a polishing step since most of the pathogens are removed by DAF and leave the system in the floated sludge (Edzwald et al., 2003) as opposed to leaving in waste filter backwash water.

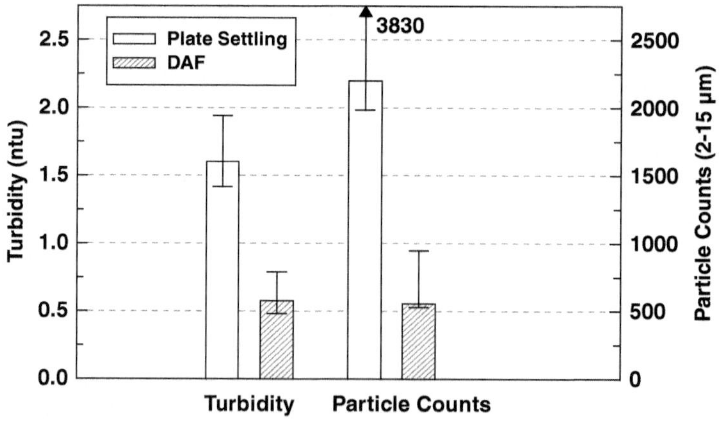

FIGURE 9-42 Pilot-scale comparison of DAF to plate sedimentation for turbidity and particle counts under winter conditions (2–5°C). (*Source: Edzwald et al., 2003.*)

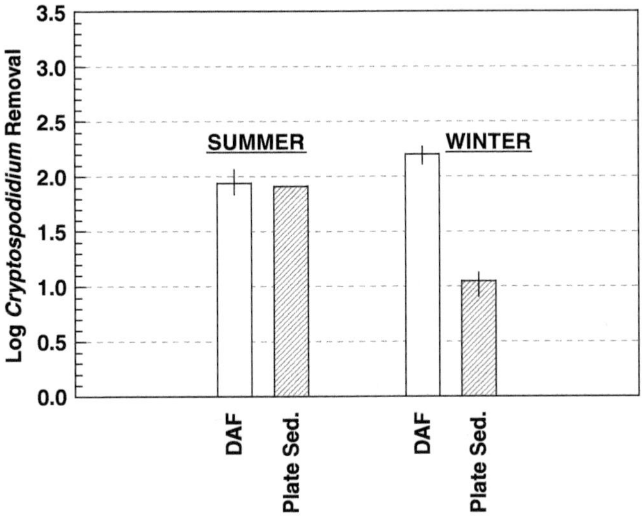

FIGURE 9-43 Pilot-scale comparison of DAF to plate sedimentation for *Cryptosporidium* log removals under summer (17–18°C) and winter conditions (2–5°C). (*Adapted from data of Edzwald et al., 2003, Journal of Water Supply: Research and Technology—AQUA, 52 (4), 243–248, with permission from the copyright holders, IWA Publishing.*)

Edzwald et al. (2000) have also reported on pilot-scale performance for *Giardia* removal. The pilot tests were conducted by spiking *Giardia* into the raw water at high concentrations to represent peak short-term loads for cold-water winter (2–3°C) and spring (13–14°C) water temperatures. It was found that DAF performed better than sedimentation for both seasons. Average log removals for the spring were 2.8 for DAF and 1.5 for sedimentation; for the winter average log removals were lower for DAF with 2.0 but much lower at 0.75 for sedimentation.

Full-Scale Plant Performance for Turbidity, NOM, and Filtration. Tests comparing the performance of rapid gravity sand filters fed with water treated by flotation and by floc blanket clarification showed that these waters have similar turbidities and residual coagulant concentrations. The presence of air bubbles in flotation treated water has no influence on filter performance (Rees et al., 1980a).

An extensive full-scale evaluation of a DAF plant in Fairfield, Connecticut, has been reported by Edzwald and Kaminski (2009). In their study of this 190 ML/d flotation-over-filtration (DAF/F) plant with a nominal design DAF and filter loading of 15 m/h (6 gpm/sf), they evaluated NOM removal [measured by TOC and UV (254 nm)] absorbance) and turbidities following DAF and dual-media filtration. The water supply is a reservoir with low turbidities (<3 ntu), low to moderate NOM (UV (254 nm) of 0.065 to 0.12 cm^{-1} and TOC of 2.2 to 4.0 mg/L. Figure 9-44 presents DAF effluent turbidities and treated water NOM data over a 1-year period. The DAF process performed well with DAF turbidities generally less than 0.4 ntu. The plant performed well in removing NOM producing treated water TOC below 2.0 mg/L and UV (254 nm) generally less than 0.04 cm^{-1}. It was found that over a 5-year period the NOM removals were 44 percent for TOC and 65 percent for UV (254 nm), which are excellent considering the raw water TOC values of 2.2 to 4.0 mg/L. Additional data on filter performance are presented next that were not reported in the paper

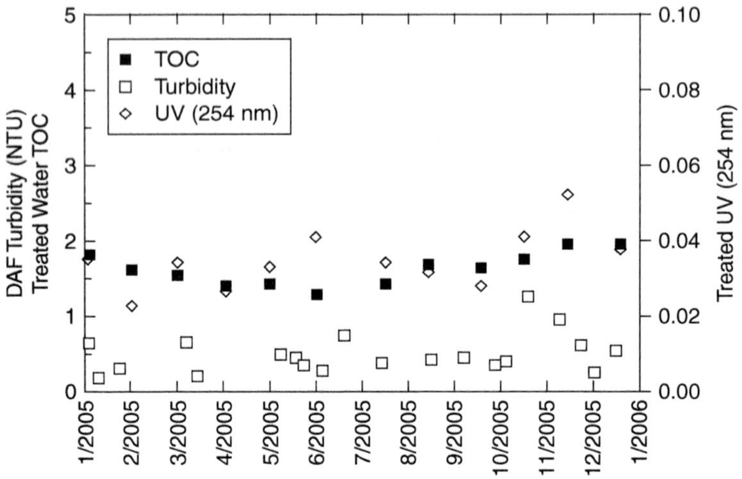

FIGURE 9-44 Full-scale performance for DAF effluent turbidity and plant performance for NOM for a reservoir supply. (*Source: Edzwald and Kaminski, 2009.*)

by Edzwald and Kaminski (2009). For the year, the plant operated at an average filtration rate of 8.1 m/h (3.3 gpm/sf). Excellent performance was achieved with filtered water turbidities not exceeding 0.11 ntu with an average of 0.07 ntu. Good water production was obtained with average unit filter run volume (UFRV) of 640 m^3/m^2/run (15,700 gal/ft^2/run) with values up to 730 m^3/m^2/run (18,000 gal/ft^2/run).

APPLICATIONS

Factors Influencing Choice of Process

A comparison of the various types of dissolved air flotation and sedimentation processes for removing floc prior to normal deep bed filtration is summarized in Table 9-9. The factors influencing choice of process—those principally relating to cost, source water quality, compactness of plant, rapid start-up, and sludge removal—are discussed. Further guidance is provided by Jannssens and Buekens (1993), Valade et al., (2009), and in Chap. 5.

Solids Loading. Solids removal prior to filtration is needed when the solids concentration is too great for filtration to cope with alone. Pretreatment is needed when the coagulant dose required for soft waters exceeds about 1 mg/L Al or Fe or greater than about 2 mg Al/L for hard waters (Gregory, 1991).

The concentration of solids in a floc blanket clarifier using alum coagulation is equivalent to about 100 mg Al/L or 200 mg Fe/L if iron coagulant is used. Floc blanket sedimentation becomes inappropriate for coagulant doses greater than about 7.5 mg Al/L or 15 mg Fe/L because blanket level control requirements become unreasonable (see Example 9-4). Other types of sedimentation then will be more appropriate. In some situations it can be appropriate to have two stages of sedimentation (Chen et al., 2002): first stage with horizontal-flow settlers and primary coagulation to at least achieve particle destabilization, followed by floc blanket sedimentation after a second coagulant dose to achieve sweep floc coagulation.

TABLE 9-9 Comparison of Sedimentation and Dissolved Air Flotation

	Horizontal flow (rectangular)	Radial flow (circular)	Upflow, U.S.-type clarifier	Upflow floc blanket	Floc blanket (widely spaced inclined plates)	Ballasted floc	Inclined plates (unhindered)	Dissolved air flotation
Sedimentation regime	Unhindered	Unhindered	Unhindered or hindered	Hindered	Hindered	Hindered or unhindered	Unhindered	Unhindered (hindered by air)
Theoretical removal of particles with $v_t < v_t^*$	Partial depending on entrance position	Partial depending on entrance position	Partial depending on whether blanket formed	No removal	No removal	No removal	Partial depending on entrance position	Depends on entrance position
Appropriate for heavy silts and lime precipitate as well as light floc	Yes	Yes	Yes if bottom scraped	Flat tanks limited to light floc, hopper tanks can take wider range of settling velocities.		For heavy silts & light floc	Yes	No
Appropriate for eutrophic (algal) waters	No	No	No	No	No	No	No	Yes
Ease of start-up and on/off operation	Easy	Easy	Easy for unhindered	Slow; may take several days to form a new blanket.		Quick but some skill needed	Quick but some skill needed	Quick & easy
Skill level of operation	Low	Low	Medium	Medium	Medium	High	Medium-high	Medium-high
Retention time	High	High	Medium	Medium-low	Low	Low	Low	Low
Short-circuiting	High	High	Medium	Medium-low	Medium-low	Medium-low	Medium-low	Low
Coagulation and flocculation effectiveness	Little difference if adequately designed and properly operated with rapid dispersion of chemicals							
Polymer usage	Low-medium	Low-medium	Low-medium	Low-medium	Low-medium	Medium-high	Medium-high	None-low
Relative capital cost	High-medium	High-medium	Medium	Medium	Medium-low	Medium-low	Medium-low	Medium-low
Relative operating cost	Medium	Medium-low	Medium-low	Low	Low	Medium	Medium	Medium

Floc blanket sedimentation can be used with partial lime softening in conjunction with iron coagulation. For simple precipitation softening, a solids contact clarifier, which is not too complex and is easily maintained, is likely to be the best choice. Consideration should be given to using pellet reactors, however, especially when softening groundwaters. When the raw water bears heavy silts, horizontal and inclined settlers are most appropriate. They will, however, need good arrangements for discharging the sediment with robust sludge scraping equipment. When silt concentrations are very high, sedimentation prior to coagulation may be appropriate. When silt concentrations are low but algae pose a problem, for example, reservoir supplies, and for colored (low alkalinity) waters, flotation is likely to be the best choice for pretreatment. Flotation might also be a better choice for other reasons, as mentioned below.

Costs. The comparison of costs of different sedimentation processes and comparison of sedimentation with flotation must include both capital and operating costs. This is necessary because of the substantial differences in the distribution of costs in the various processes, as summarized in Table 9-9. Floc blanket sedimentation has a relatively high capital cost but low operating cost. In contrast, flotation has a relatively low capital cost but can have a high operating cost because of its energy consumption associated with the saturator; however, smaller flocculation tanks reduce the energy for flocculation mixing.

The chemical cost of basic coagulation should be the same for the various processes. Although generally no advantage occurs from using polyelectrolyte as a flocculation aid with flotation, its use can be cost effective for any form of sedimentation because loading rates can be increased. High loading rates and hence lower capital costs are also possible with ballasted floc sedimentation, but additional operating costs of the ballasting agent, conditioning chemicals, and energy for agent recovery and recirculation are incurred.

The relatively high energy cost of flotation, although still possibly less than the cost of chemicals or labor, is largely incurred in pumping the recycle through the saturator system, to achieve the required level of saturation with air, and through the nozzles, with enough backpressure to prevent premature release of air downstream of the saturators. Most of the remaining energy cost is required for flocculation. The cost of flocculation for discrete-particle settling systems, horizontal flow and inclined sedimentation should be more than for DAF because of the much shorter flocculation times (and smaller flocculator volumes) with DAF. Since a large portion of the operating cost of DAF is in the recycle, it is important that total recycle flow, or pressure, can be reduced for low plant flows. This will need to be done in conjunction with reducing the number of nozzles so as to maintain nozzle backpressure and prevent premature release of air in the recycle. DAF usually produces better clarified water quality than sedimentation; thus filter production will be greater and backwash costs will be lower.

The large differences between processes in fixed and variable costs means that dissolved air flotation is especially attractive for plants with low utilization, less than full flow, or infrequent use, because of its relatively lower capital cost and the reduced variable costs. Other factors besides cost can affect the choice of process, however. Such factors include footprint, rate of start-up, operational flexibility, ease of removing algae or color, coping with high mineral loadings, and producing the best treated water quality.

Low-rate dissolved air flotation (about 8 m/h) is likely to be less expensive than simple floc blanket sedimentation, for a fully utilized plant and if neither or both are covered with a roof or enclosed in a building, only if the sedimentation process cannot be operated at rates greater than about 2 to 3 m/h (0.8 to 1.2 gpm/sf) (Gregory, 1977). For higher-rate, DAF, sedimentation will need to be effective at a greater rate than 2–3 m/h to be financially competitive. As the expected plant utilization decreases, DAF becomes relatively cheaper. The same should apply to ballasted sedimentation also because of its relatively greater operating cost. Whereas, inclined-plate sedimentation (without use of a ballasting agent) should be similar to the other types of unhindered sedimentation and to coarse-bed filtration in being more appropriate for full-utilization applications.

Sometimes the solids concentration in the water, after initial chemical treatment, is low enough to be within the capacity of filtration to operate without prior sedimentation or flotation and still achieve the desired objective. Just as each form of filtration has a limited capacity to remove solids, so does each form of sedimentation and flotation. This limited ability to remove solids efficiently may limit the surface loading and therefore cost effectiveness of the process. However, some of the big new plants in North America are exploiting DAF costs effectively in treating high-quality raw waters: DAF effluent is so good, even at high rates, that high-rate filtration is also possible. This allows very compact plants to be built.

With the typical applications of flotation, solids loading and air requirements are independent of each other. If the influent solids concentration increases, a greater recycle rate or pressure may be required to achieve efficient flotation, however, and result in a greater operating cost. Additionally, even continuous removal of float might prove inadequate. The greater the solids concentration, the more likely horizontal-flow or lamella sedimentation will be selected. The same applies to floc blanket sedimentation when the extreme situation is equivalent to thickening and then presedimentation before coagulation may be justified. Additionally, the volume of discharged sludge becomes so large that the arrangements for thickening become substantial. At some point, the combination of clarification and thickening has to be rationalized.

For certain types of source waters, flotation does produce better water for filtration and therefore better filtrate quality than sedimentation. This may be because of better use of flocculation (this is applicable also to inclined settling), and the mechanism of flotation that makes it especially effective for algae removal. The aeration in flotation also may be attractive because it might increase dissolved oxygen concentration or even cause some desorption or stripping of volatile contaminants. However, higher than typical recycle rates should not be used to satisfy high dissolved oxygen demand of the water being treated: waters with low DO should have an aeration step before chemical treatment or there should be aeration of the raw water reservoir (probably useful anyway to address stratification and its associated problems).

Compactness. Inclined settling, ballasted floc settlement, dissolved air flotation and coarse-bed filtration are regarded as high-rate processes because of the relatively high surface loadings possible. This label also implies they are compact processes and so occupy less area.

Compactness is important where land is at a premium. These processes are also likely to need shallower and smaller tanks (except coarse-bed filtration), which is important in coping with preparing foundations in difficult sites. Both of these points are relevant where a plant has to be housed in a cold climate or for other environmental or strategic reasons. The compactness of the flotation carried out over the filter (DAF/F) has made it especially attractive for package plants, and in the last 15 years many large DAF plants in locations without much land have adapted this strategy, for example, a 190-ML/d plant in Fairfield, Connecticut, and a 1100-ML/d plant for New York City.

Rapid Start-Up. A floc blanket can take several days to form when starting with an empty tank, especially when the solids concentration is low. In contrast, high-rate inclined settlers and dissolved air flotation can produce good quality water within 45 min from start-up. Such rapid start-up is useful where daily continuous operation is not needed. The disadvantage of high-rate processes is they are more sensitive to failure of chemical dosing and flocculation and changes in inlet water quality. Rapid start-up plants are suitable for unstaffed sites using automatic shut-down and call-out alarms as a means of avoiding relying on difficult automatic quality control.

Sludge Removal. The ability to accumulate sludge, prior to removal, requires space and this depends on the process.

1. In horizontal-flow sedimentation, the whole plan area of the floor of the tank is available for sludge accumulation.
2. In dissolved air flotation, the whole plan area of the surface of the flotation tank is available, but thickness of the floated sludge and ease of its removal are more important.
3. In solids contact clarifiers with scrapers, the whole plan area of the floor is available.
4. Floc blanket clarifiers have only a small area for excess floc removal as sludge. However, when it is designed well, this will be adequate to ensure some thickening before discharge.
5. Some inclined settler systems are combined with thickeners.

Some systems, such as dissolved air flotation and those with combined thickeners can produce sludge concentrated enough for dewatering without additional thickening (see Chap. 22). Although systems that produce the more concentrated sludges are likely to incur greater operating costs for the scraping and pumping, this is likely to be more than offset by a saving in subsequent sludge treatment processes.

Emerging Technology

Developments in sedimentation and dissolved air flotation processes, as currently known, have probably reached the point of diminishing returns. Application lags behind knowledge, including making existing plants work better. Developments will occur in part by paying more attention to the chemical engineering of reagent dosing, mixing, and flocculation rather than the clarification process itself. In the case of flotation, more attention also is likely to be given to the transfer of flocculated water to the flotation tank and mixing with air in the contact zone. The use of CFD (Fawcett, 1997; Eades et al., 1997; Ta and Brignall, 1997; Amato and Wicks, 2009) is leading to radical changes in tank design resulting in confidence with high-rate flotation for treating waters with low to medium solids concentrations. More attention also is likely to be given to selecting the best chemistry and process for a particular application and to the design of the chosen process with regard to materials of construction, control equipment, and compactness. A driver for developments in coagulation chemistry and flocculation is the progress taking place in mathematical modeling that helps to highlight how much process performance may benefit from improvements in such factors as floc size, strength, and density, as has been identified in this chapter and in Chap. 8 and by Bache and Gregory (2007).

No clear benefit has been demonstrated for the aeration that water receives (when the water before treatment does not have a significant oxygen demand) in flotation, but no doubt there is some such as stripping of taste and odor compounds. However, an example of combined ozonation and flotation has been reported as the Flottazone® process (Baron et al., 1997). In this process the recycle is saturated with ozone-rich air. This differs from the Ozoflot® process in which the flocculated water passes over porous plate diffusers releasing ozone-rich air. More examples of combined ozonation and flotation might occur should it be found appropriate to use ozonation in the treatment of waters for which flotation is also applicable.

Ballasted floc sedimentation systems, recycling fine sand as ballast, have received more attention during the past decade. Use of buoyant ballast has been suggested for flotation to reduce the need for air (Jarvis et al., 2009). Just as flotation has its niche for application, then there will be a niche for ballasted floc sedimentation, plain sedimentation, and other clarification processes. A wide variety of options has emerged and continues to expand for treatment prior to final filtration. Individual circumstances will dictate the choice. Furthermore, no single process will likely take a major share of new applications because costs of alternatives are generally the same, unless substantial technical advantages exist as has occurred for flotation.

ABBREVIATIONS

CFD	computational fluid dynamics
DAF	dissolved air flotation
DAF/F	dissolved air flotation over filtration
DOC	dissolved organic carbon
DOM	dissolved organic matter
NOM	natural organic matter
PACl	polyaluminum chloride
TOC	total organic carbon
UFRV	unit filter run volume
UV (254 nm)	ultraviolet absorbance at 254 nm

NOTATION FOR EQUATIONS

A	area of particle or tank
A_b	projected area of a bubble
A_H	cross-sectional area of channel to liquid flow
A^*	plan area of tank with horizontal flow for ideal settlement
A_{sz}	separation zone surface area
b	breadth
C	mass concentration
C_b	mass concentration of air or bubbles in contact zone
C_D	drag coefficient
C_{fl}	mass concentration of air or bubbles in flotation tank effluent
C_r	mass concentration of dissolved air in recycle flow
$C_{s.\mathrm{air}}$	air mass concentration at saturation in water for atmospheric pressure
C_{sN}	solubility or equilibrium concentration of nitrogen in water for atmospheric pressure
C_{sO}	solubility or equilibrium concentration of oxygen in water for atmospheric pressure
d	diameter
d_b	diameter of bubble
d_f	diameter of floc
d_{fb}	diameter of the floc-bubble aggregate

d_H	hydraulic diameter
d_p	diameter of particle
e	saturator efficiency
E	porosity
E^+	porosity at maximum flux
F	function of
f_b	buoyancy
f_d	drag force
f_g	external force such as gravity
Fr	Froude number
F_M	solids mass flux
F_t	fraction at time t
F_V	solids volume flux
g	gravitational constant (9.806 m/sec^2)
G	velocity gradient
h	vertical distance $\leq H$
H	height or depth
k_a	air deficit concentration
k_B	Boltzmann's constant
k, k_1, *etc.*	coefficients
K	constant from Yao's inclined settling equation, or factor to account for shape effects on drag
K_H	Henry's law constant
L	horizontal distance or length of tank or DAF separation zone
L_p	length of surface for inclined settlement
L_r	Yao's relative length of settler
L^*	length of tank for ideal settlement
L'	greater than or equal to L^*
m_f	maximum number of attached air bubbles to floc particles
n	power index dependent on *Re*
n_b	bubble number concentration in the contact zone
n_i, n_j	number concentrations
$n_{p,e}$	floc or particle number concentration in the contact zone effluent
$n_{p.i}$	floc or particle number concentration in the contact zone influent
N	number of channels, compartments or stages, or number concentration e.g. of bubbles attached to a floc
$N_{\max}$	maximum number of bubbles that can attach to one floc
p	partial pressure
P	perimeter
q	constant

Q	volumetric flow rate
Q_r	recycle flow rate
r	power index
R	recycle ratio
Re	Reynolds number
s	particle specific gravity = ρ_p/ρ_w
S_c	Yao's inclined flow geometry number
t	detention time
t_{cz}	contact zone hydraulic detention time
$T°$	absolute temperature
v	particle or liquid velocity
v_b	bubble rise velocity
v_d	horizontal velocity in Fig. 7-7
v_{fb}	floc-bubble rise velocity
v_n	velocity relating to h_n and t_n
v_o	settling velocity of suspension for concentration extrapolated to zero
v_p	particle settling velocity
v_s	settling velocity of suspension
v_t	terminal settling velocity
v_θ	velocity at angle θ
v_w	water velocity
$v_{clar\text{-}hl}$	clarification area hydraulic loading
$v_{nom\text{-}hl}$	nominal tank hydraulic loading
$v_{sz\text{-}hl}$	separation zone hydraulic loading
v^*	overflow rate of the ideal settling tank
v'	kess than or equal to v^*
V	volume
w	perpendicular spacing
W	width, e.g., of separation zone
x	mole or mass fraction
x^*	mass fraction of particles with settling velocity v^*
α	collision attachment factor
α_{pb}	particle-bubble attachment efficiency
ε	mean energy dissipation rate
Φ	particle volume concentration
Φ_b	air or bubble volume concentration
Φ^*	apparent volume concentration
Φ^+	particle volume concentration at maximum flux
N_{ij}	collision frequency between particles of concentrations n_i and n_j
η	collector collision efficiency

η_D	collision efficiency due to Brownian diffusion
η_I	collision efficiency due to fluid flow or interception
η_{IN}	collision efficiency due to inertia
η_S	collision efficiency due to differential settling
η_T	total single-collector efficiency
μ	dynamic viscosity
π	geometric constant pi = 3.14
θ	angle of inclination
Θ	shape factor
ρ	density
ρ_b	density of air or gas bubble
ρ_{fb}	floc-bubble density
ρ_p	density of particle or floc
ρ_w	density of water

REFERENCES

al Farisi, A.-L. (1988) Effect of bentonite addition on floc-size density relationships in coloured water, MSc thesis, University of Strathclyde, UK.

Amato, T. and Wicks, J. (2009) The practical application of computational fluid dynamics to dissolved air flotation, water treatment plant operation, design and development, *Jour. water Supply: Research and Technology-Aqua,* vol. 58 , no. 1. pp. 65–73.

Amirtharajah, A., and O'Melia, C.R. (1990) Coagulation processes: Destabilization, mixing and flocculation. In *Water Quality & Treatment*, 4th ed., ch. 6. Denver, CO: American Water Works Association, and New York: McGraw-Hill.

Bache, D.H., Rasool, E., Ali, A., and McGilligan, J.F. (1995) Floc character measurement and role in optimum dosing, *Jour. Water SRT–Aqua*, vol. 44, no. 2, pp. 83–92

Bache, D.H., and Gregory, R. (2007) *Flocs in Water Treatment.* London: IWA Publishing.

Baeyens, J., Mochtar, I.Y., Liers, S., and de Wit, H. (1995) Plugflow dissolved air flotation, *Water Environment Research*, vol. 67, pp. 1027–1035.

Barham, W.L., Matherne, J.L., and Keller, A.G. (1956) *Clarification, Sedimentation and Thickening equipment—A Patent Review.* Bulletin No. 54, *Engng. Expt.* Station: Louisiana State University.

Baron, J., Martin Ionesco, N., and Bacquet, G. (1997). Combining flotation and ozonation—the Flottazone process, *Dissolved Air Flotation*, proc. int. conf., London, CIWEM.

Barrett, F. (1975) Electroflotation-development and application, *Water Pollution Control*, vol. 74, no. 1, pp. 59–62.

Baruth, E.E. (ed.) (2005) *Water Treatment Plant Design*, 4th ed. Denver, CO.: American Water Works Association, and New York: McGraw-Hill.

Bond, A.W. (1965) Water-solids separation in an upflow, *Instn. Engrs. Australia, Civil Engng.Trans.*, vol. 7, October, pp. 141–150.

Bratby, J. and Marais, G.v.R. (1975) Dissolved-air flotation—an evaluation of inter-relationships between process variables and their optimization for design, *Water SA*, vol. 1, no. 2, pp. 57–69.

Bratby, J. (1976) Dissolved-air flotation in water and waste treatment, Ph.D. dissertation. University of Cape Town, South Africa.

Bratby, J., and Marais, G.v.R. (1977) Chap. 5. In *Solid–Liquid Separation Equipment Scale Up*, ed. D.B. Purchas. Croydon, London: Uplands Press.

Brown , P.P., and Lawler, D.F. (2003) Sphere drag and settling velocity revisited, *Jour. Envntl. Engng.*, vol. 129, no. 3, pp. 222–231

Camp, T.R. (1936) A study of the rational design of settling tanks, *Sewage Works Jour.*, vol. 8, pp. 742–758.

Camp, T.R. (1955) Flocculation and flocculation basins, *Trans. Amer. Soc. Civil Engrs.*, vol. 120 (paper 2722), pp. 1–16.

Casey, J.J., O'Donnel, K., and Purcell, P.J. (1984) Uprating sludge blanket clarifiers using inclined plates, *Aqua*, vol. 2, pp. 91–95.

Casey, T.J., and Naoum, I.E. (1986) Air saturators for use in dissolved air flotation processes, *Water Supply*, vol. 4, pp. 69–82.

Chen, L.C., Sung, S.S., Lin, W.W., and Lee, D.J. (2002) Observations of floc blanket characteristics in full-scale floc blanket clarifiers, *Water Science & Technology*, vol. 47, no. 1, pp. 197–204.

Clements, M.S., and Khattab, A.F.M. (1968) Research into time ratio in radial flow sedimentation tanks, *Proc. Inst. Civil Engrs.*, vol. 40, pp. 471–493.

CIWEM (1997) *Dissolved Air Flotation*, proc. int. conf., Chart. Instn. Water & Envtl. Management, London.

Cretu, G. (1968) Contribution to the theory of water treatment using a sludge blanket, (in Romanian), *Hydrotechnia Gospodarirea Apelor, Meterologia*, vol. 13, no. 12, pp. 634–638.

Crossley, I.A., Herzner, J., Bishop, S.L., and Smith, P.D. (2007) Going underground–Constructing New York City's first water treatment plant, a 1,100 Ml/d dissolved air flotation, filtration and UV facility, *Proceedings of the 5th International Conference on Flotation in Water and Wastewater Systems*, Seoul National University, SK.

Degremont (1991) *Water Treatment Handbook,* Vols. I & II, 6th ed. Degremont, Paris, France.

Demir, A. (1995) Determination of settling efficiency and optimum plate angle for plated settling tanks, *Water Research,* vol. 29, no. 2, pp. 611–616.

Desjardins, C., Koudjonou, B., and Desjardins, R. (2002) Laboratory study of ballasted flocculation, *Water Research,* vol. 36, no. 16, pp. 744–754.

de Dianous, F., Pujol, E., and Druoton, J.C. (1990) Industrial application of weighted flocculation: Development of the Actiflo clarification process, *Proceedings 4th Gothenburg Symp.*, Madrid. *Chemical Water and Wastewater Treatment*, ed. H.H. Hahn, and R. Klute , pp. 127–137. Berlin: Springer Verlag.

Dixon, D.R. (1984) Colour and turbidity removal with reusable magnetite particles, VII, *Water Research*, vol. 18, no. 5, pp. 529–534.

Eades, A., Jordan, D., and Scheidler, S. (1997) Counter-current dissolved air flotation filtration COCO-DAFF, *Dissolved Air Flotation,* proc. Conf., London, April 1997, CIWEM.

Edzwald, J.K., and Malley Jr., J.P. (1990). Removal of humic substances and algae by dissolved air flotation, *EPA/600/2-89-032,* U.S. Environmental Protection Agency, OH.

Edzwald, J.K., Malley Jr., J.P. and Yu, C. (1990) A conceptual model for dissolved air flotation in water treatment, *Water Supply,* vol. 8, pp. 141–150.

Edwald, J.K., and Wingler, B.J. (1990) Chemical and physical aspects of dissolved air flotation for the removal of algae, *Jour. Water Supply Research and Technology–Aqua,* vol. 39, no. 2, pp. 24–35.

Edzwald, J.K. (1995) Principles and applications of dissolved air flotation. Flotation processesin water and sludge treatment, *Water Science and Technology,* vol. 31, no.3–4, pp. 1–23.

Edzwald, J.K., Tobiason, J.E., Amato, T., and Maggi, L.J. (1999) Integrating high rate dissolved air flotation technology into plant design, *Journal AWWA,* vol. 91, no. 2, pp. 41–53.

Edzwald, J.K., Tobiason, J.E., Parento, L.M., Kelley, M.B., Kaminski, G.S., Dunn, H.J., and Galant, P.B. (2000) *Giardia* and *Cryptosporidium* removals by clarification and filtration under challenge conditions, *Journal AWWA,* vol. 92, no. 12, pp. 70–84.

Edzwald, J.K., Tobiason, J.E., Udden, C., Kaminski, G.S., Dunn, H.J., Galant, P.B., and Kelley, M.B. (2003) Evaluation of the effect of recycle of waste filter backwash water on plant removals of *Cryptosporidium*, *Jour. Water Supply: Research and Technology–Aqua,* vol. 52, no. 4, pp. 243–258.

Edzwald, J.K. (2007) Fundamentals of dissolved air flotation, *Jour. New England Water Works Assoc.,* vol. 121, no. 3, pp. 89–112.

Edzwald, J.K., and Kaminski, G.S. (2009) A practical method for water plants to select coagulant dosing, *Jour. New England Water Works Assoc.*, vol. 123, no. 1, pp. 15–31.

Ellms, J.W. (1928) *Water Purification.* New York: McGraw-Hill.

Envirotech Corp. Eimco Modular Energy Dissipating Clarifier Feedwells, *Form No. MED 121-10-72-3M*, Envirotech Corporation, Brisbane, CA.

Epstein, N., and LeClair, B.P. (1985) Liquid fluidization of binary particle mixtures–II, *Chem. Engng. Sci.*, vol. 40, no. 8, pp. 1517–1526.

Fair, G.M., and Geyer, J.C. (1954) *Water Supply and Wastewater Disposal.* New York: John Wiley and Sons.

Fair, G.M., Geyer, J.C., and Okun, D.-A. (1971) *Elements of Water Supply and Waste Water Disposal,* 2d ed. New York: John Wiley and Sons.

Fawcett, N.S.J. (1997) The hydraulics of flotation tanks: Computational modelling, *Dissolved Air Flotation*, proc. int. conf., London, CIWEM.

Flotation in Water and Wastewater Systems (2007) 5th International Conference on Flotation, Seoul National University, SK.

Franklin, B. (1997) Ten years experience of dissolved air flotation in Yorkshire Water, *Dissolved Air Flotation,* proc. int. conf., London, CIWEM.

Fukushi, K., Tambo, N., and Matsui, Y. (1995) A kinetic model for dissolved air flotation in water and wastewater treatment, in Flotation Processes in Water Sludge Treatment, *Water Sci. Tech.* vol. 31, no. 3–4, pp. 37–48.

Gaudin, A.M. (1957) *Flotation,* 2d ed. New York: McGraw-Hill.

Gemmell, R.S. (1971) Mixing and sedimentation. In *Water Quality and Treatment,* 3d ed. Denver, CO: American Water Works Association, and New York: McGraw-Hill.

Gomella, C. (1974) Clarification avant filtration ses progres recents, *(Rapport General 1) Intl. Water Supply Assoc. Intl. Conf.*

Gould, B.W. (1967) Low cost clarifier improvement, *Australian Civil Engrg. and Constn.*, vol. 8, August 5, pp. 49–53.

Gregory, R., and Hyde, M. (1975) *The Effects of Baffles in Floc Blanket Clarifiers*, TR7. Swindon, UK: Water Research Centre.

Gregory, R. (1977) A cost comparison between dissolved air flotation and alternative clarification processes, *Papers and Proceedings of the Conference on Flotation for Water and Waste Treatment,* ed. J.D. Melbourne and T.F. Zabel. Swindon, UK: Water Research Centre.

Gregory, R. (1979) *Floc Blanket Clarification*, TR 111. Swindon, UK: Water Research Centre.

Gregory, R., Maloney, R.J., and Stockley, M. (1988) Water treatment using magnetite: A study of a Sirofloc pilot plant, *Jour. Instn. Water & Envir. Management*, vol. 2, no. 5, pp. 532–544.

Gregory, R. (1991) *Controlling Coagulant esiduals: Direct filtration operation and performance,* UM1273. Swindon, UK: Water Research Centre.

Gregory, R., Head, R., and Graham, N.J.D. (1996) Blanket solids concentration in floc blanket clarifiers, *Proc. Gothenburg Symposium*, Edinburgh.

Haarhoff, J., and Rykaart, E.M. (1995) Rational design of packed saturators, *Water Sci. Tech.* vol. 31, no. 3–4, pp. 179–190.

Haarhoff, J., and Steinbach, S. (1996) A model for the prediction of the air composition in pressure saturators, *Water Research*, vol. 30, no. 12, pp. 3074–3082.

Haarhoff, J., and Steinbach, S. (1998) A simplified method for assessing the saturation efficiency at full-scale dissolved air flotation plants, *Water Sci. Tech.*, vol. 38, no. 6, pp. 303–310.

Haarhoff, J., and Edzwald, J.K. (2004) Dissolved air flotation modelling: Insights and shortcomings, *Jour. Water Supply: Research and Technology—Aqua,* vol. 53, no. 3, pp. 127–150.

Haarhoff, J. (2008) Dissolved air flotation: Progress and prospects for drinking water treatment, *Jour. Water Supply: Research and Technology—Aqua*, vol. 57, no. 8, pp. 555–567.

Hale, P.E. (1971) *Floc Blanket Clarification of Water*, Ph.D. dissertation, London University.

Hamlin, M.J., and Abdul Wahab, A.H. (1970) Settling characteristics of sewage in density currents, *Water Research*, vol. 4, pp. 609–626.

Han, M.Y. (2002). Modeling of DAF: The effect of particle and bubble characteristics, *Jour. Water Supply: Research and Technology—Aqua,* vol. 51, no. 1, pp. 27–34.

Hart, J. (1996) Application of process simulation in water treatment, *Association Generale Hygienistes Techniciens Municipaux 76th Congress*, London.

Hartung, H.O. (1951) Committee report: Capacity and loadings of suspended solids contact units, *Journal AWWA*, vol. 43, no. 4, pp. 263–291.

Head, R., Hart, J., and Graham, N. (1997) Simulating the effect of blanket characteristics on the floc blanket clarification process, *Water Sci. Tech.* vol. 36, no. 4, pp. 77–84.

Heinänan, J., Jokela, P., and Ala-Peijari, T. (1995) Use of dissolved air flotation in potable water treatment in Finland. Flotation Processes in Water and Sludge Treatment, *Water Sci. Tech.* vol. 31, no. 3–4, pp. 225–238

Helsinki (2000) *4th International Conference, Flotation in Water and Waste Water treatment,* Helsinski, FIN.

Hemming, M.L., Cottrell, W.R.T., and Oldfelt, S. (1977) Experience in the treatment of domestic sewage by the micro-flotation process, *Papers and Proceedings of the Water Research Centre Conf. on Flotation for Water and Waste Treatment,* ed. J.D. Melbourne and T.F. Zabel, Water Research Centre, Medmenham, UK.

Hudson, H.E. (1981) Density considerations in sedimentation. In *Water Clarification Processes Practical Design and Evaluation.* New York: Van Nostrand Reinhold.

Ives, K.J. (1968) Theory of operation of sludge blanket clarifiers, *Proc. Instn. Civ. Engrs.,* vol. 39, no. 2, pp. 243–260.

Ives, K., and Bernhardt, H.J. (eds.) (1995) Flotation processes in water and sludge treatment, *Water Sci. Tech.*, vol. 31, no. 3–4.

Jackson, M.L. (1994) Energy effects in bubble nucleation, *Ind. Eng. Chem. Res.*, vol. 33, no. 4, pp. 929–933.

Janssens, J., and Buekens, A. (1993) Assessment of process selection for particle removal in surface water treatment, *Jour. Water Supply: Research and Technology—Aqua,* vol. 42, no. 5, pp. 279–288.

Jarvis, P., Buckingham, P., Holden, B., and Jefferson, B. (2009) Low energy ballasted flotation, *Water Research,* vol. 43, no. 14, pp. 3427–3434.

Kalbskopf, K.H. (1970) European practices in sedimentation. In *Water Quality Improvement by Physical and Chemical Processes, Water Resources Symposium No. 3*, ed. E.F. Gloyna and W. W. Eckenfelder. Austin: University of Texas Press.

Kawamura, S. (1981) Hydraulic scale-model simulation of the sedimentation process, *Journal AWWA*, vol. 73, no. 7, pp. 372–379.

Kawamura, S., and Lang, J. (1986) Re-evaluation of launders in rectangular sedimentation basins, *Jour. Water Pollution Control Federation*, vol. 58, pp. 1124–1128.

Kawamura, S. (1991) *Integrated Design of Water Treatment Facilities.* New York: John Wiley & Sons.

Kinosita, K. (1949) Sedimentation in tilted vessels, *Jour. Colloid Science*, vol. 4, pp. 525–536.

Kitchener, J.A., and Gochin, R.J. (1981) The mechanism of dissolved-air flotation for potable water: Basic analysis and a proposal, *Water Research*, vol. 15, pp. 585–590.

Kitchener, J.A. (1984) The froth flotation process: Past, present and future–in brief. In *The Scientific Basis of Flotation,* ed. K.J. Ives, NATO ASI Series. The Hague, Netherlands: Martinus Nijhoff Publishers.

Lagvankar, A.L., and Gemmel, R.S. (1968) A size-density relationship for flocs, *Journal AWWA*, vol. 60, no. 9, pp. 1040–1046.

Leppinen, D.M. (2000) A kinetic model of dissolved air flotation including the effects of interparticle forces, *Jour. Water Supply: Research and Technology—Aqua,* vol. 49, no. 5, pp. 259–268.

Leppinen, D.M., and Dalziel, S.B. (2004) Bubble size distribution in dissolved air flotation tanks, *Jour. Water Supply: Research and Technology—Aqua,* vol. 43, no. 8, pp. 531–543.

Liers, S., Baeyens, J., and Mochtar, I. (1996) Modeling dissolved air flotation, *Water Environment Research*, vol. 68, no. 6, pp. 1061–1075.

Lundgren, H. (1976) Theory and practice of dissolved-air flotation, *Jour. Filtration and Separation*, vol. 13, no. 1, pp. 24–28.

Lundh, M., Jonsson, L., and Dalquist, J. (2000) Experimental studies of the fluid dynamics in the separation zone in dissolved air flotation, *Water Research*, vol. 34, no. 1, pp. 21–30.

Maddock, J.L. (1977) Research experience in the thickening of activated sludge by dissolved-air flotation, *Papers and Proceedings of the Conference on Flotation for Water and Waste Treatment*, ed. J.D. Melbourne and T.F. Zabel. Medmenham, UK: Water Research Centre.

Marske, D.M., and Boyle, J.D. (1973) Chlorine contact chamber design—a field evaluation, *Water and Sewage Works*, vol. 120. January, pp. 70–77.

Matsui, Y., Fukushi, K., and Tambo, N. (1998) Modeling, simulation, and operational parameters of dissolved air flotation, *Jour. Water Supply: Research and Technology—Aqua,* vol. 47, no. 1, pp. 9–20.

Miller, D.G., Robinson, M., and West, J.T., Water Treatment Processes-I, TP43, Water Research Centre, UK, 1965.

Miller, D.G., West, J.T., and Robinson, M. (1966) Floc blanket clarification–1, *Water and Water Engng.*, vol. 70, June, pp. 240–245.

Miller, D.G., and West, J.T. (1968) Pilot plant studies of floc blanket clarification, *Journal AWWA*, vol. 60, no. 2, pp. 154–164.

Nakamura, H., and Kuroda, K. (1937) La cause de l'acceleration de la vitesse de sedimentationdes suspensions dans les recipients inclines, *Keijo Jour. Med.,* vol. 8, pp. 256–296.

Offringa, G. (1995) Dissolved air flotation in Southern Africa, Flotation Processes in Water and Sludge Treatment, *Water Sci. Tech.* vol. 31, no. 3–4, pp. 159–172.

Parker, D., Butler, R., Finger, R., Fisher, R., Fox, W., Kido, W., Merrill, S., Newman, G., Pope, R., Slapper, J., and Wahlberg, E. (1996) Design and operations experience with flocculator-clarifiers in large plants, *Water. Sci. Tech.*, vol. 33, no. 12, pp. 163–170 (also in *WQ International,* Nov/Dec 1996, pp. 32–36.)

Patwardhan, V.S., and Chi, T., (1985) Sedimentation and liquid fluidization of solid particles of different sizes and densities, *Chem. Engng. Sci.*, vol. 40, no. 7, pp. 1051–1060.

Pearse, M.J. (1977) *Gravity Thickening Theories: A Review,* LR261 (MP). Stevenage, UK: Warren Spring Laboratory.

Pieronne, P. (1997) Design of an optimized water treatment plant, in particular sludge extraction, by means of a pilot study, *TSM*, vol. 92, no. 7–8, pp. 17–22.

Prandtl, L., and Tietjens, O.G. (1957) *Applied Hydro and Aeromechanics.* New York: Dover.

Price, G.A., and Clements, M.S. (1974) Some lessons from model and full-scale tests in rectangular sedimentation tanks, *Water Pollution Control*, vol. 73, pp. 102–113.

Rees, A.J., Rodman, D.J., and Zabel, T.F. (1979) *Water Clarification by Flotation–5*, TR 114, Medmenham, UK: Water Research Centre.

Rees, A.J., Rodman, D.J., and Zabel, T.F. (1980a) Dissolved-air flotation for solid–liquid separation, *Jour. Separation Process Technology*, vol. 1, no. 3, pp. 19–23.

Rees, A.J., Rodman, D.J., and Zabel, T.F. (1980b) *Evaluation of Dissolved-Air Flotation Saturator Performance*, TR 143. Medmenham, UK: Water Research Centre.

Richardson, J.E., and Harker, J.H. (2002) *Coulson & Richardson's Chemical Engineering,* Vol. 2, *Particle Technology & Separation Processes,* 5th ed. Woburn, MA: Butterworth-Heineman.

Rodman, D.J. (1982) *Investigation into Hydraulic Flocculation with Special Emphasis on Algal Removal.* Master's thesis. Stevenage, UK: Water Research Centre.

Rykaart, E.M., and Haarhoff, J. (1995) Behaviour of air injection nozzles in dissolved air flotation, in flotation processes in water and sludge treatment, *Water Sci. Tech.*, vol. 31, no. 3–4, pp. 25–36.

Schers, G.J., and van Dijk, J.C. (1992) Dissolved-air flotation: Theory and practice. In *Chemical Water and Wastewater Treatment II*, ed. E. Klute and H.H. Hahn, pp. 223–246. New York: Springer-Verlag.

Seoul (2007), 5th International Conference on Flotation, Flotation 2007, Flotation in Water and Wastewater Systems, Seoul National University, Seoul, South Korea

Setterfield, G.-H. (1983) Water treatment trials at Burham, *Effluent and Water Treat. Jour.,* vol. 23, no. 1, pp. 18–23.

Shawwa, A.R., and Smith, D.W. (2000) Dissolved air flotation model for drinking water treatment, *Canadian Jour. Civil Engng.,* vol. 27, pp. 373–382.

Shogo, T. (1971) Slurry-blanket type suspended solids contact clarifiers. Part 5, *Kogyo Yoshui*, vol. 153, pp. 19–25.

Sibony, J. (1981) Clarification with microsand seeding—a state of the art, *Water Research*, vol. 15, pp. 1281–1290.

Standen, G., Inole, P.J., Shek, K.J., and Irwin, R.A. (1995) Optimization of clarifier performance for pesticide removal, *3rd Int. Conf. Water and Waste Water Treatment,* Harrogate, UK, ed. M. White, BHR Group Conf. Series, Publication No.17, pp. 13–33.

Steinbach, S., and Haarhof, J. (1997) Air precipitation efficiency and its effect on the measurement of saturator efficiency. In *Dissolved Air Flotation.* London: Chartered Institution of Water and Environmental Management.

Su, T., Wu, D.J., and Lee, D.J. (2004) Blanket dynamics in upflow bed, *Water Research,* vol. 38, no. 1, pp. 89–96.

Ta, C.T., and Brignal, W.J. (1997) Application of single phase computational fluid dynamics techniques to dissolved air flotation tank studies. Iin *Dissolved Air Flotation*, proc. int. conf. London: Chartered Institution of Water and Environmental Management.

Takahashi, T., Miyahara, T., and Mochizuki, H. (1979) Fundamental study of bubble formation in dissolved air pressure flotation, *Jour. Chemical Engineering. of Japan,* vol. 12, pp. 275–280.

Tambo, N., and Abe, S. (1969) Behaviours of floc blankets in an upflow clarifier, *Jour. Japan Water Works Assoc.*, no. 415, pp. 7–15.

Tambo, N., and Hozumi, H. (1979) Physical aspect of flocculation process. II. Contact flocculation, *Water Research*, vol. 13, pp. 441–448

Tambo, N., and Watanabe, Y. (1979) Physical characteristics of floc. I. The floc density function and aluminum floc, *Water Research,* vol. 13, pp. 409–419.

Tambo, N., and Matsui, Y. (1986) A kinetic study of dissolved air flotation, pp. 200–203, *World Congress_of Chem. Engr.,* Tokyo.

Tambo, N., and Wang, X.C. (1993) The mechanism of pellet flocculation in a fluidized-bed operation, *Jour. Water SRT–Aqua*, vol. 42, no. 2, pp. 67–76.

Thirumurthi, D.A. (1969) Breakthrough in the tracer studies of sedimentation tanks, *Journal Water Pollution Control Federation*, vol. 41, no. 11 (pt 2), pp. R405–R418.

Teixeria, M.B., and Rosa, M.J. (2006) Integration of dissolved air flotation and nanofiltration for *M. aeruginosa* and associated microcystins removal, *Water Research*, vol. 40, pp. 3612–3620.

Valade, M.T., Edzwald, J.K., Tobiason, J.E., Dahlquist, J., Hedberg, T., and Amato, T. (1996) Pretreatment effects on particle removal by flotation and filtration, *Journal AWWA,* vol. 88, no. 12, pp. 35–47.

Valade, M.T., and Crossley, I.A. (1998) Design of the world's largest DAF recycle system: Efficient and cost effective, *Proc. AWWA Annual Conference*, Dallas, TX.

Valade, M., Nickols, D., Haarhoff, J., Barrett, K., and Dunn, H. (2001) Evaluation of the performance of full-scale packed saturators, *Water Science and Technology*, vol. 43, no. 8, pp. 67–74.

Valade, M.T., Becker, W.C., and Edzwald, J.K. (2009). Treatment selection guidelines for particle and NOM removal, *Jour. Water Supply: Research and Technology—Aqua*, vol. 58, no. 6, pp. 424–432.

Van Craenenbroeck, W., Van den Bogaert, J., and Ceulemans, J. (1993) The use of dissolved air flotation for the removal of algae—the Antwerp experience, *Water Supply*, vol. 11, no. 3/4, pp. 123–133.

van Puffelen, J., Buijs, PJ., Nuhn, P.N.A.M., and Hijnen, W.A.M. (1995) Dissolved air flotation in potable water treatment: the Dutch experience, *Wat. Sci. Tech,* vol. 31, no. 3–4. pp. 149–157.

Vostrcil, J. (1971) The effect of organic flocculants on water treatment and decontamination of water by floc blanket, *Prace a Studie, Sesit 129*, Water Research Inst., Prague.

Webster, J.A., Webster, H.C., and Fairley, J.P. (1977) Aspects of design for efficient plant operation for water treatment processes using coagulants, paper presented to Scottish Section, *Instn. Water Engrs. and Scientists*, UK.

White, M.J.D., Baskerville, R.C., and Day, M.C. (1974) Increasing the capacity of sedimentation tanks by means of sloping plates, paper presented to Inst. Water Pollution Control, East Midlands Branch, U.K, November 1974. Reproduced as *The Application of Inclined Tubes or Plates to Sedimentation Tanks in Waste-Water Treatment, Notes on Water Pollution No.68*, Water Research Centre, UK, March 1975.

Yadav, N.P., and West, J.T. (1975) *The Effect of Delay Time on Floc Blanket Efficiency, TR9.* Medmenham, UK: Water Research Centre.

Yao, K.M. (1970) Theoretical study of high-rate sedimentation, *Journal Water Pollution Control Federation*, vol. 42, no. 2, pp. 218–228.

Yao, K.M. (1973) Design of high-rate settlers, *Proc. Amer. Soc. Civ. Engrs.,* vol. 99, part. EE5, pp. 621–637.

Zabel, T.F., and Hyde, R.A. (1977) Factors influencing dissolved-air flotation as applied to water clarification, *Papers and Proceedings of Water Research Centre Conference on Flotation for Water and Waste Treatment*, ed. J.D. Melbourne and T.F. Zabel. Medmenham, UK: Water Research Centre.

Zabel, T.F., and Melbourne, J.D. (1980) Flotation. In *Developments in Water Treatment,* Vol 1, ed. W.M. Lewis. London: Applied Science Publishers .

Zanoni, A.E., and Blomquist, M.W. (1975) Column settling tests for flocculant suspensions, *Proc. Amer. Soc. Civ. Engrs.,* vol. 101, part. EE3, pp. 309–318.

CHAPTER 10
GRANULAR MEDIA FILTRATION

John E. Tobiason, Ph.D., P.E., B.C.E.E.
Professor
Department of Civil and Environmental Engineering
University of Massachusetts
Amherst, Massachusetts, United States

John L. Cleasby, Ph.D., P.E., B.C.E.E.
Professor Emeritus
Department of Civil and Construction Engineering
Iowa State University
Ames, Iowa, United States

Gary S. Logsdon, D.Sc., P.E., B.C.E.E.
Senior Process Consultant
Fairfield, Ohio, United States

Charles R. O'Melia, Ph.D., P.E.
Professor Emeritus
Department of Geography and Environmental Engineering
Johns Hopkins University
Baltimore, Maryland, United States

OVERVIEW OF PARTICLE FILTRATION PROCESSES **10.2**
Types of Particle Filtration 10.3
Particle Filtration within the Treatment Plant: Configurations and Secondary Roles 10.4
Regulatory Requirements 10.5
Removal of Microorganisms by Particle Filtration Processes 10.7
GRANULAR MEDIA FILTRATION PROCESS DESCRIPTION **10.9**
General 10.9
Filter Media Materials 10.11
Filter Media Configurations 10.15
Rates of Filtration 10.21
Acceptable Filter Run Length and Production per Run 10.23
Underdrain and Support Gravel 10.24
MEDIA FILTRATION THEORY AND MODELING **10.26**
Overview 10.27
Particle Removal Theories 10.29
Head Loss Theories and Models 10.36
RAPID-RATE FILTER PERFORMANCE **10.39**
Effluent Quality During a Filter Run 10.39
The Critical Importance of Adequate Prefilter Coagulation 10.43
Detrimental Effects of Rate Increases and Restarting Dirty Filters 10.45
Head Loss Development During a Filter Run 10.46
The Detrimental Effects of Negative Head 10.49
FLOW CONTROL IN FILTRATION **10.49**
Why Control Is Needed 10.49
Rate Control Systems for Gravity Filters 10.49

Common Elements of Filter Control Systems 10.52
Choice of the Appropriate Control System 10.53
BACKWASHING AND MAINTENANCE OF FILTER MEDIA **10.53**
Alternative Methods of Backwashing 10.53
Backwash Troughs and Wash Water Required 10.57
Hydraulics of Fluidized Beds 10.58
Stratification and Intermixing During Backwashing 10.62
Movement of Gravel During Backwashing 10.64
Underdrain Failures 10.65
Dirty Filter Media and Mineral Deposits 10.66
Backwash of GAC Media 10.68
Waste Backwash Water Management 10.68
Wash Water versus Filtered Water Chemistry 10.69
DIRECT FILTRATION **10.70**
Process Description, Advantages, and Disadvantages 10.70
Chemical Pretreatment for Direct Filtration 10.72
Filter Details for Direct Filtration 10.73
PRESSURE GRANULAR BED FILTERS **10.74**
Description of Pressure Filters 10.74
Comparison of Pressure and Gravity Filtration 10.75
Operation of Pressure Filters 10.75
Rate Control for Pressure Filters 10.75
Applications of Pressure Filters 10.76
SLOW SAND FILTRATION **10.76**
Description and History 10.76
Mechanisms of Filtration and Filter Performance 10.77
Slow Sand Filter Design 10.80
Source Water Quality 10.82
Modifications to Slow Sand Filtration for Enhanced Performance 10.83
Startup and Maintenance Procedures 10.84
Performance Monitoring 10.86
Application to Small Systems 10.86
Summary 10.87
PRECOAT FILTRATION **10.87**
Applications and Performance 10.88
The Filter Element and Septum 10.89
The Filter Vessel 10.89
Filter Media 10.91
Filter Operation 10.92
Theoretical Aspects of Precoat Filtration 10.93
DEDICATION **10.95**
ABBREVIATIONS **10.95**
EQUATION NOTATION **10.96**
REFERENCES **10.97**

OVERVIEW OF PARTICLE FILTRATION PROCESSES

The purpose of this chapter is primarily to present the principles and practice of using porous beds of granular media to remove particulate materials as part of the overall process of drinking water treatment. The precoat filtration process is also discussed. Types and important characteristics of particles in water are described in some detail in Chap. 3 of this book. Particles may be present in the raw water (e.g., clay, silt, microorganisms, detritus) or formed during treatment (e.g., aluminum or iron hydroxide precipitates from coagulation, calcium carbonate precipitates from softening), and their characteristics, such as size, density, and stability, have important effects on the performance of processes used for their removal. Media and precoat filtration are types of solid–liquid separation processes; others include clarification by dissolved air flotation or sedimentation (Chap. 9), processes that are always followed by particle filtration, and low pressure membrane filtration [e.g., microfiltration and ultrafiltration (MF/UF); Chap. 11]. Consistent with this book's objectives, this chapter introduces and reviews important principles of media filtration; the reader should consult separate sources for more in-depth coverage of filtration practice (Logsdon et al., 2002; Logsdon, 2008) and process design (e.g., Kawamura, 2000; AWWA, 2004; MWH, 2005; Hendricks, 2006).

Types of Particle Filtration

The various types of filters used for particle removal in potable water filtration can be classified by several factors, including (1) the dominant overall mechanism for particle removal, (2) the filtration media and apparatus used, (3) the hydraulic loading rate, and (4) the source of head or energy for filtration.

Overall, particle removal mechanisms can be divided into two or three main classifications. *Depth filtration* occurs when particles deposit within the pore spaces of a filter, usually a bed of granular media, due to transport to and attachment on the surface of the granular media or previously deposited particles; the removed particles generally are much smaller in size than characteristic dimensions of the pore spaces through which they are transported. *Straining* or *sieving filtration* occurs when particles are prevented from passing through a pore space because of size exclusion; that is, the particle is too large to pass through the pore opening; removal by straining is the predominant mechanism in low pressure membrane filtration and for microsieves. Straining is also the dominant mechanism in *cake filtration*, wherein the water to be filtered passes through a cake comprised of previously deposited particles; precoat filtration is an example of particle removal by a cake comprised of previously removed particles and very small media or body feed [e.g., diatomaceous earth (DE)], continuously added to the water to be filtered.

In application, most particle filtration processes involve a dominant mechanism, whereas other mechanisms also can be significant. Granular bed, or media, filters are generally dominated by depth filtration, yet straining by constricted pores and cakes of deposited particles also occurs, depending on various fixed and operational parameters of the process. Cake filtration, along with biological activity, is especially important in the slow sand media filtration process. Low pressure membrane filtration processes rely on the straining mechanism as typical dead-end operation (all water passes through the membrane) causes cake filtration to occur on the membrane surface; depth filtration inside a membrane pore is undesired and is one cause of membrane fouling.

Another useful classification of filtration processes is based on the physical material used for particle removal. Thus *granular bed*, or *media filters* are comprised of specified depths of one or more types and sizes of granular media; rapid sand, dual-media, and slow sand filters are some examples. In contrast, *membrane filters* for particle removal are manufactured from a thin layer of material with pores that allow water passage while straining out particles; synthetic polymer hollow fiber membranes and ceramic membranes are two examples (see Chap. 11).

An important design and operational characteristic of particle filtration processes is the *hydraulic loading rate* (HLR), or water flux, defined as the flow rate Q through the filter divided by the surface area of the filter A, or Q/A; typical units for media filtration are meters per hour (m/h = $m^3/m^2/h$) or gallons per minute per square foot (gpm/ft^2), whereas for membrane filtration the typical units are liters per square meter per hour (L/m^2/h or lmh) or gallons per square foot per day (g/ft^2/d or gfd). Rapid-rate media filters operate at much higher HLRs than slow sand filters; in turn, these media filter HLRs are much higher than membrane filter HLRs. The HLR has a significant impact on the capital cost of treatment because the required filtration surface area to produce a desired flow decreases as the design HLR increases.

Filtration processes are also classified according to the source of the energy required to force water to flow through the filter. Historically, the most common source for large treatment plants has been *gravity*, relying on an open surface hydraulic grade line (HGL) upstream of the filter that exceeds the downstream HGL; the granular media are contained in a steel or concrete walled open tank. Flow through the filter also may be caused by applying *pressure* in a closed vessel within which the filter media are housed. The pressure vessel typically is steel for media and precoat filters, whereas synthetic polymer housings are often used for low pressure membrane filters. Filtration flow also can be induced by application of a *vacuum* on the downstream side of the filter, such as used for submerged low pressure membrane filters (see Chap. 11) and some precoat filters.

Thus a filter can be fully described by an appropriate choice of adjectives. For example, a rapid, gravity, dual-media filter describes a deep bed comprised of two media (usually anthracite coal on top of sand) operated at high enough rates to encourage depth removal of particulates within the bed and operated by gravity in an open tank.

Particle Filtration within the Treatment Plant: Configurations and Secondary Roles

It can be argued that particle filtration processes are possibly the most critical component in potable water treatment facilities because of their ability to remove microorganisms, especially pathogens that cause acute human health concerns. The selection and performance of particle filtration processes are affected greatly by raw water quality and the configuration of processes prior to filtration. Particle filtration processes also may have important secondary roles in the overall treatment train.

The number and type of prefiltration treatment processes depends mostly on raw water quality but also can be influenced by such factors as land area, cost, and residuals disposal. Relatively pristine sources with low turbidity and natural organic matter (NOM) may be suitable for direct application to slow sand media filtration without chemical coagulant or oxidant addition prior to filtration. The same or similar source water may be suitable for treatment by *direct filtration* with chemical coagulation and flocculation but without a clarification process prior to filtration; *in-line* or *contact filtration* is similar yet with no or minimal engineered flocculation. Less pristine sources with more NOM, turbidity, and algae are better treated by *conventional filtration* involving chemical coagulation, flocculation, and clarification prior to filtration (and usually media filtration is implied). Useful guidance for selection of treatment processes based on raw water quality has been provided by several authors (e.g., Valade et al., 2009; Janssens and Buekens, 1993). While direct or in-line filtration can be cost-effective by eliminating the clarification process, the inherent reliability of the overall treatment facility is improved by provision of two separate methods for particle removal, that is, clarification and filtration. In the United States, a number of direct filtration facilities were constructed in the 1970s and 1980s, whereas selection of such a process train has been less frequent in the last two decades; this issue is discussed further later in this chapter. The significant increase in selection of low pressure membranes for particle filtration in recent years has resulted in increased consideration of conditions that lead to the use of clarification ahead of the membrane filters; compared to media filtration, less generalization is possible because of the range of filter element geometries and operational modes employed.

Clarification processes used prior to filtration include sedimentation (conventional, plate, tube, ballasted, etc.), dissolved air flotation (DAF), upflow sludge blanket clarifiers, and coarse media roughing filters. These processes occur most often in a tank or vessel that is separate from the filtration process (but often attached via a common wall). An important exception is the application of DAF directly above the granular media filter bed in the same process tank; two examples are the 190 ML/d (50 mgd) plant in Fairfield, Conn., operated by the Aquarion Water Company and the soon-to-be-completed 1100 ML/d (290 mgd) Croton Treatment Plant for New York City (see Chap. 9).

It is important to recognize that particle filtration processes are designed most commonly to only remove particulate matter, although there are important exceptions, as described below. Statements of the performance of particle filtration processes for the removal of raw water constituents that are normally dissolved, such as NOM or manganese, require careful consideration; removal of such constituents by many particle filtration processes requires their transformation to particulate form prior to filtration, typically by coagulation and oxidation processes.

Particle filtration processes also may serve important secondary functions for control of water quality, often involving processes occurring at the surface of the filter media. One important example is the promotion of biological activity on filter media surfaces (via biofilm formation) for the biological transformation of one or more dissolved constituents. A common example is the use of granular activated carbon (GAC) media for particle removal and as support media for biofilms to oxidize biodegradable organic matter (BOM) and/or specific taste- and odor-causing compounds. Prefiltration conditions may be optimized to promote biofiltration; for example use of ozonation to increase BOM levels and no use of chlorine prior to filtration. Another process that can occur at the surface of granular media and be used for treatment is adsorption. GAC media adsorptive capacity most commonly would result in relatively short-term water quality impacts, but long-term continuous impacts may be realized because of combined adsorption and biodegradation (see Chap. 14). An important process used for removal of dissolved (reduced) manganese (Mn^{2+}) is sorption to manganese-oxide coatings on granular filter media (anthracite, sand) with subsequent oxidation and regeneration of the adsorption site by the presence of an oxidant, usually free chlorine, in the water applied to the media filter, sometimes referred to as the *natural greensand effect* or *catalytic oxidation* (see Chap. 7). It can be challenging to design and operate a filtration process for optimal performance with respect to particle removal as well as additional contaminant removal (e.g., Wiesner et al., 1987a); particle removal takes precedence, and sometimes a separate treatment process, often after filtration, is the best option to achieve objectives in addition to particle removal.

Regulatory Requirements

Beginning in 1989, the U.S. Environmental Protection Agency (USEPA) promulgated a number of regulations that require filtration of water sources, set limits on the turbidity of filtered waters, and regulate recycling of certain process residual flows. These regulations have been promulgated in conjunction with requirements for disinfection and for control of disinfection by-products (DBPs). Some pertinent regulations for this chapter are summarized below, and the reader is referred to Chap. 1 for additional coverage of regulations.

Surface Waters. The USEPA 1989 Surface Water Treatment Rule (SWTR), the 1998 Interim Enhanced SWTR (IESWTR), the 2002 Long Term 1 Enhanced SWTR (LT1ESWTR), and the 2006 Long Term 2 Enhanced SWTR (LT2ESWTR) collectively require disinfection of all surface water sources and require filtration of all surface waters unless stringent source water quality criteria are met; these rules also apply to so-called groundwater under the influence (GWUI) of adjacent surface water. In lieu of setting maximum contaminant levels (MCLs) for specific pathogenic microorganisms, the USEPA has used a treatment technique (TT) approach based on required levels of overall removal/inactivation of classes of pathogens (log removals), documented performance of particle separation technologies for pathogen removal, compliance with filtered water turbidity standards, and provision of specified levels of disinfection. As such, certain requirements are technology- and pathogen-specific. Individual states impose regulations that are at least as stringent as the USEPA regulations.

The USEPA regulations require that the combined extent of physical removal and inactivation be at least 3 log (99.9 percent) for *Giardia* cysts and 4 log (99.99 percent) for viruses, independent of raw water quality. Requirements for total removal of *Cryptosporidium* oocysts are 3 log (99.9 percent) unless the results of raw water sampling required by the LT2ESWTR show that greater removal is needed. Guidance documents for the SWTRs indicate assumed pathogen log removal performance for specified treatment technologies that apply when required filtered water turbidity levels are met. The required log inactivation via disinfection is the difference between the total removal/inactivation required and that achieved by particle removal processes.

Assumed log removal credits and required filtered water turbidity for certain specified technologies are shown in Table 10-1. The filtered water turbidity requirement to be met at least 95 percent of the time for conventional and direct filtration processes (both include chemical coagulation; conventional includes clarification and media filtration;

TABLE 10-1 USEPA Log Removal Credits and Turbidity Requirements

Filtration process	Log removal credits: *Giardia*	Virus	*Cryptosporidium*	Combined filter effluent turbidity requirement
Conventional	2.5	2.0	3.0	≤0.3 ntu in 95% of samples each month; never >5 ntu
Direct	2.0	1.0	2.5	≤0.3 ntu in 95% of samples each month; never >5 ntu
Slow sand	2.0	2.0	3.0	≤1.0 ntu in 95% of samples each month; never >5 ntu
Diatomaceous earth	2.0	1.0	3.0	≤1.0 ntu in 95% of samples each month; never >5 ntu

direct filtration has no clarification process) was decreased to 0.3 ntu by the IESWTR from the 0.5 ntu level specified in the 1989 SWTR. The LT2ESWTR allows for an additional *Cryptosporidium* 0.5 log removal credit if a 0.15 ntu level in the combined filter effluent is achieved; the credit increases to 1.0 log if effluent from individual filters is at the 0.15 ntu level. Criteria for slow sand and diatomaceous earth filtration are also included in Table 10-1. In general, some level of inactivation of *Giardia* and viruses is always required, whereas inactivation or an additional filtration method is required only if raw water *Cryptosporidium* levels exceed 1.0 oocysts/L (bin 3 or 4 of the LT2ESWTR). Log removal credits for physical removal and filtered water turbidity requirements for technologies not specified in the regulations, so-called alternative filtration technologies, must be determined by states, often via demonstration studies; low pressure particle filtration membranes generally fall into this category. Because of the relatively large size (4–12 μm) of protozoan pathogens relative to the typically less than 0.2 μm pore size of MF/UF processes, log removal credits for protozoa are typically high (i.e., >4 log) for MF/UF, whereas in the same context log credits for typically sub-100 nm size viruses may not be allowed or are much less than those for protozoa.

The Safe Drinking Water Act (SDWA) Amendments of 1996 directed the USEPA to regulate the practice of recycling of waste or spent filter backwash within the treatment plant; USEPA thus promulgated the Filter Backwash Recycling Rule (FBRR) in 2001 (USEPA, 2001). This rule regulates the practice of recycle of waste filter backwash water, supernatant from sludge thickeners, and liquids from dewatering processes (e.g., centrifugation, belt press) at conventional and direct filtration facilities. These flows must be recycled to the head of the treatment plant in order to be subjected to all the processes in the plant. Utilities have to notify states that they practice recycle and describe their practices; states have the authority to request changes in practice. Recycle is discussed more later in this chapter as well as in Chap. 22.

Groundwaters. The USEPA promulgated the Ground Water Rule (GWR) in 2006 following the mandate in the 1996 SDWA to address disinfection of all water supplies. The approach of the GWR is to identify groundwater sources at risk of fecal contamination and possibly require treatment to provide 4 log inactivation/removal of viruses (other actions can be taken to mitigate risks). The GWR applies to groundwater sources for mixed surface and groundwater systems unless the groundwater is blended with a surface water prior to treatment of the mixed sources; in such a case, the SWTRs apply. Chemical disinfection and membrane filtration are indicated as acceptable technologies, whereas combinations of other inactivation and removal processes are considered alternative technologies subject to approval. In practice, groundwater sources are often treated for removal of inorganic matter (iron, manganese, calcium) and NOM, often via oxidation/precipitation followed by media or membrane filtration. Filtered water turbidity and disinfection requirements for such processes are specified by state agencies because no federal regulation exists.

Removal of Microorganisms by Particle Filtration Processes

The ability of particle filtration processes to remove microorganisms is perhaps the most important reason for their use in the production of potable water. In North America, waterborne disease outbreaks caused by *Giardia lamblia* and *Cryptosporidium parvum*, pathogenic protozoa with high resistance to disinfectants, resulted in federal regulations and numerous pilot- and full-scale studies to evaluate removal of microorganisms by treatment processes. In the preamble language in the *Federal Register* for the SWTR, IESWTR, LT1ESWTR, and LT2ESWTR, the USEPA reviews and cites extensive literature documenting filtration performance. Many of these studies were focused on media filtration process trains involving coagulation, whereas some investigations of the efficacy of slow sand filtration and DE filtration also have been carried out. Some pertinent results for media filtration of protozoan pathogens are discussed below.

Pilot-plant studies (Logsdon et al., 1985; Al-Ani et al., 1986) established the need for attaining effective coagulation and filtered water turbidity in the range of 0.1 to 0.2 ntu for effective removal of *Giardia* cysts. Additional research by Nieminski and Ongerth (1995) on *Giardia* and *Cryptosporidium* confirmed the necessity of attaining low filtered water turbidity and the importance of maintaining proper coagulation chemistry. They evaluated a small (4.9 ML/d or 1.3 mgd) plant for removal of protozoan cysts and concluded that a properly operated treatment plant producing finished water turbidity of 0.1 to 0.2 ntu, using either the direct filtration mode or the conventional treatment mode, could achieve 3.7 to 4.0 log removals for *Giardia* and *Cryptosporidium*. The similarity of results for direct filtration and conventional treatment is different from the outcome reported by Patania et al. (1995), who observed 1.4 to 1.8 log additional removal for *Cryptosporidium* and 0.2 to 1.8 log additional removal for *Giardia* when sedimentation was included in the treatment train, in comparison with in-line filtration treatment employing only coagulation and filtration.

LeChevallier et al. (1991) examined raw and filtered water samples from 66 surface water treatment plants in 14 states and 1 Canadian province. Log removals of *Giardia* and *Cryptosporidium* ranged from less than 0.5 to greater than 3.0. Log removals averaged slightly above 2.0 for each organism. Some of these plants were practicing disinfection (usually chlorination) during clarification, so the *Giardia* results may have been influenced by disinfection. They reported that production of water with turbidity less than 0.5 ntu did not ensure that the treated water was free of cysts or oocysts. Treatment plants evaluated by LeChevallier et al. probably would have removed cysts and oocysts more effectively if they had been attaining lower filtered water turbidity. In pilot-plant testing carried out by Patania et al. (1995), when filtered water turbidity was equal to or less than 0.1 ntu, removals of cysts and oocysts were greater by as much as 1.0 log than removals when filtered water turbidity was between 0.1 and 0.3 ntu.

Payment et al. (2001) reported on biweekly monitoring of a conventional water treatment plant (coagulation, flocculation, sedimentation, and filtration) for over 12 months. Clarification through filtration attained the following before any use of disinfectant: (1) 3.0 log or greater removal of human enteric viruses in 20 of 30 samples, (2) 3.0 log or greater removal of coliphage in 20 of 32 samples, (3) 4.0 log or greater removal of *C. perfringens* in 20 of 33 samples, and (4) 3.0 log or greater removal of *Giardia* cysts in 24 of 32 samples. *Giardia* cysts were detected in the filtered water in only 1 of 32 samples. Removals of *Cryptosporidium* oocysts appeared to be lower than those for *Giardia*, but only 2 of 16 raw water samples contained sufficient densities of *Cryptosporidium* oocysts to permit calculation of 3 log removal based on the detection limit for *Cryptosporidium* in filtered water. During the study, 98 percent of the monthly average turbidity values for individual filters were equal to or less than 0.20 ntu, and 80 percent were equal to or less than 0.10 ntu. These results show that well-run conventional treatment plants present a formidable barrier to the passage of pathogens through effective sedimentation (or other solids separation) followed by media filtration.

In a study of key operational parameters affecting the robustness of media filtration for removal of *Cryptosporidium*, Huck et al. (2002) confirmed the importance of chemical coagulation in maintaining excellent performance; turbidity and particle count breakthrough in the later stages of the filter run also showed concomitant decreases in oocyst removal. Emelko et al. (2005) provide an extensive review of available data for *Cryptosporidium* removal by media filtration.

Slow sand filtration is effective for removal of *Giardia* and *Cryptosporidium*. Pilot-plant studies gave *Giardia* cyst removals of over 3 log at filtration rates up to 0.40 m/h (0.16 gpm/ft^2) (Bellamy et al., 1985a, 1985b). Field testing (Pyper, 1985) at a small slow sand filter (area 37 m^2) operated at 0.08 m/h (0.03 gpm/ft^2) gave 3.7 to 4.0 log *Giardia* removal at temperatures of 7.5 to 21°C, but at temperatures below 1°C, removals ranged from 1.2 to 3 log. Schuler et al. (1988) obtained 3.7 or higher log removals for *Cryptosporidium* in a pilot filter operated at 0.26 m/h (0.11 gpm/ft^2) using 0.27 mm effective size (ES) sand.

DE filtration is effective for removal of pathogens in the size range of *Giardia* cysts and *Cryptosporidium* oocysts. The removal mechanism involved is straining, and when an appropriate grade of DE is selected, the pore structure of the filter cake physically blocks the passage of cysts and oocysts into filtered water. Pilot-plant studies (Lange et al., 1986) using a 0.1 m^2 (1 ft^2) test filter have demonstrated *Giardia* cyst removals ranging from 2 to 4 log, typically at filtration rates of 2.4 to 3.7 m/h (1.0–1.5 gpm/ft^2). Schuler and Ghosh (1990), using three common grades of DE in pilot filtration studies, reported 100 percent removal of *G. muris* and greater than 3 log removals of *Cryptosporidium* oocysts, all at 4.9 m/h (2 gpm/ft^2) and without the use of coagulants. Use of alum or cationic polymer to coat the media did not enhance the cyst removal but aided in the removal of turbidity and coliform bacteria. Principe et al. (1994) reported 3.7 log *Giardia* cyst removals by DE filtration (unspecified grade of DE) at 7.3 m/h (3 gpm/ft^2). Ongerth and Hutton (1997, 2001) reported that greater than 6 log removal of seeded *Cryptosporidium* oocysts was achieved at rates of 2.4 and 4.9 m/h (1 and 2 gpm/ft^2) in bench/pilot-scale studies using several grades of DE media.

Low pressure membrane filtration (MF/UF) has been found to be effective in removing protozoan cysts and oocysts independent of coagulant addition; this is a logical result based on MF/UF pore sizes of less than 0.2 μm. The extent of virus removal depends on membrane pore size and coagulation pretreatment. See Chap. 11 for more information on membrane filtration.

In summary, filtration options for the removal of pathogenic microorganisms include rapid-rate media filtration, slow sand filtration, DE filtration, and membrane filtration. The most widely used filtration process continues to be coagulation and rapid filtration, and generally the process train includes flocculation and clarification by flotation or sedimentation (conventional treatment). For conventional treatment or for direct or in-line filtration, effective removal of microorganisms requires careful control of coagulation chemistry and operation of filters to consistently attain very low filtered water turbidity. Based on pilot-plant and full-scale studies, removal of microorganisms by these processes can be maximized by the following: (1) filtered water turbidity should be 0.10 ntu or lower, (2) the duration of higher turbidity at the beginning of a filter run should be minimized (<1 hour), and (3) filtered water turbidity above 0.2 ntu should be considered turbidity breakthrough, signaling the need to backwash the filters.

GRANULAR MEDIA FILTRATION PROCESS DESCRIPTION

General

Rapid-rate granular media (or bed) filtration usually consists of passage of pretreated water through a granular bed at rates between 5 to 30 m/h (2–12 gpm/ft^2). Flow is typically downward through the bed, although some use of upflow filters is reported in Latin America, Russia, and the Netherlands; upflow roughing filters (very coarse media) are also used. Both gravity and pressure filters are used, although some restrictions are imposed on the use of pressure filters on surface waters or polluted source waters or following lime-soda softening (Great Lakes–Upper Mississippi River Board, 2003).

A schematic of a typical gravity filtration system from one manufacturer is shown in Fig. 10-1, and a generic vertical cross section is shown in Fig. 10-2. One to three layers of specified depths of filter media of selected material and size are supported by an underdrain system that allows for both collection of filter effluent water and application of a reverse-direction water (and often air) backwash. Various methods are used to measure and control filter effluent flow and water level above the filter media. Often piping and valves are placed to provide a filter-to-waste option if filter effluent is not of the desired quality.

Media filtration performance typically is described by measurement of filter effluent quality and head (or pressure) loss as a function of filter operation time; a generic plot is shown in Fig. 10-3. During operation, solids are removed from the filter influent water and accumulate mostly within the voids of the filter bed and to some degree on the surface of the filter bed. Filter effluent turbidity often improves after an initial maximum (the *ripening period*), may be fairly constant for a long period, and then might degrade. The deposition of particles results in an increase in head loss (i.e., clogging head loss) from the initial, or

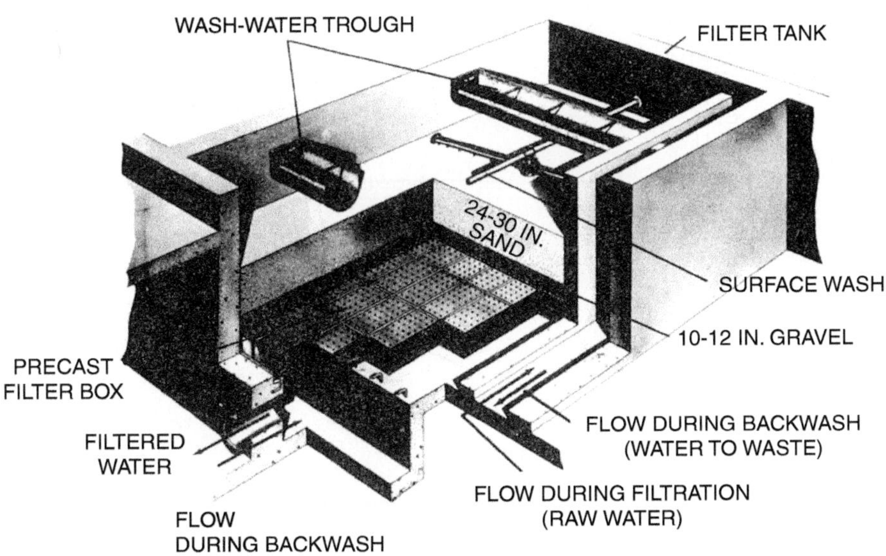

FIGURE 10-1 Typical gravity rapid-rate filter system. (*Courtesy of ITT Leopold.*)

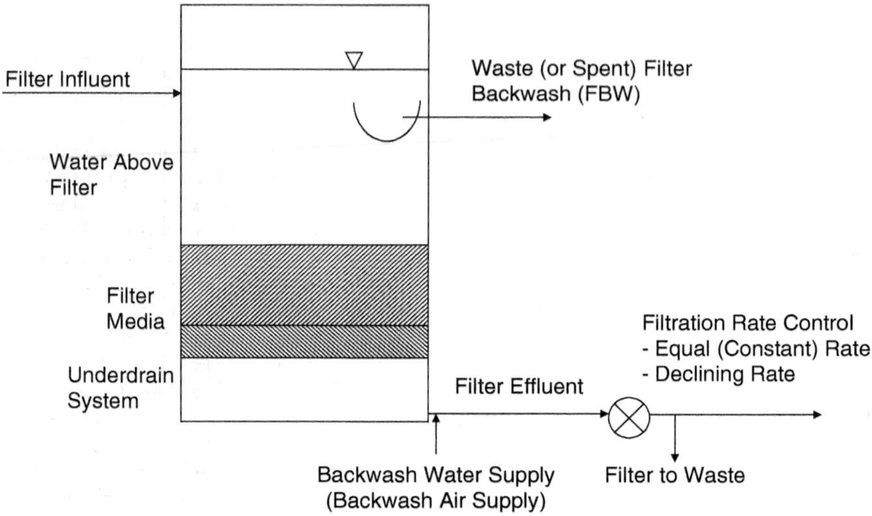

FIGURE 10-2 Schematic of a typical gravity filter cross section.

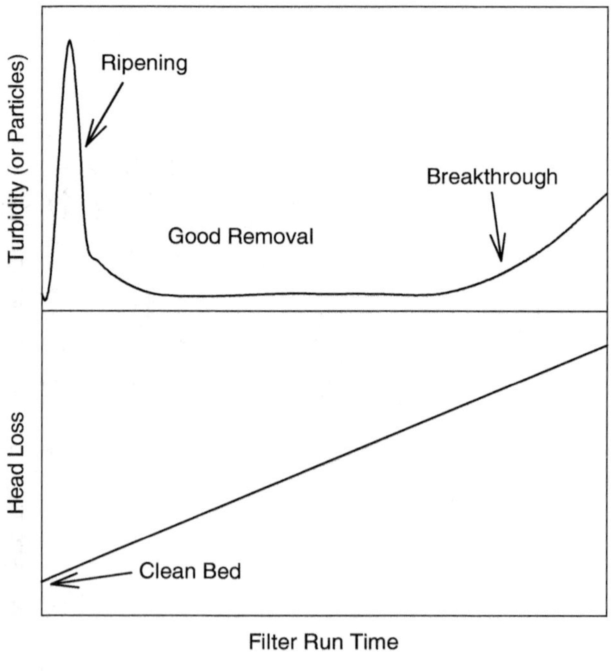

FIGURE 10-3 Generic plot of media filter performance: filtrate quality and head loss.

clean-bed, head loss. If flow remains constant, a linear rate of head loss increase with time often occurs; a decrease in filter flow yields a lower rate of increase in head loss. The total head loss (clean-bed plus clogging) may approach the maximum head loss provided in the plant, sometimes called the *available head loss*.

After a period of operation, the rapid filter is cleaned by backwashing with an upward flow of water, usually assisted by some auxiliary water or air scour system. The operating time between backwashes is referred to as a *filter cycle* or a *filter run*. The head loss at the end of the filter run is called the *terminal head loss*.

The need for backwash is indicated by one of the following three criteria, whichever occurs first: (1) The head loss across the filter increases to the available limit or to a lower established limit, usually 2.4 to 3.0 m (8–10 ft) of water, (2) the filtrate begins to deteriorate in quality or degrades to some allowable limit, or (3) a desired or maximum time limit has been reached (usually 1–4 days).

Filter cycles range from about 12 to 96 hours, although some plants operate with longer cycles; a 24 to 48 hour filter run length is common. Setting an upper time limit for the cycle is desirable because of concern for bacterial growth in the filter and because of concern that compaction of the solids removed in the filter will make backwashing difficult.

In addition to impacts on destabilization of raw water particles to allow aggregation in flocculation and effective clarification and to convert dissolved NOM to particulate NOM, pre-filtration treatment of surface waters by chemical coagulation (see Chap. 8) is essential to achieve effective removal of particles in rapid-rate granular media filters. This is so because of the transport and deposition particle removal mechanism (or depth filtration) predominant in media filtration; particles must be able to attach to the filter media and previously deposited particles. In some cases, filter-aid polymers may be added to the water just ahead of filtration to strengthen the attachment of the particles within the filter media. Groundwaters treated for iron and manganese removal by oxidation, precipitation, and filtration most often do not use other chemical pretreatment.

Filter Media Materials

Types of Media. The most common types of media used in granular bed filters are silica sand and anthracite coal, used alone or in a dual-media combination; an additional layer of garnet or ilmenite sometimes is used in a triple-media configuration. AWWA Standard B100 (AWWA, 2010) provides standard requirements for properties, sampling, testing, placement, and packaging of these filter materials.

Granular activated carbon (GAC) is also used as filter media, sometimes for taste and odor reduction in granular beds that serve both for filtration and adsorption, that is, filter-adsorbers (Graese et al., 1987), as well as for biologically active filters. GAC contactors are also used separately after filtration for adsorption and/or biodegradation of organic compounds (see Chap. 14). AWWA Standard B604 (AWWA, 2006) provides standard requirements for physical properties, sampling, testing, and packaging of GAC.

Garnet and ilmenite are naturally occurring, high-density minerals. Garnet is somewhat of a generic term referring to several different minerals, mostly almandite, andradite, and grossularite, which are silicates of iron, aluminum, and calcium mixtures. Ilmenite is an iron titanium ore that invariably is associated with hematite and magnetite, both iron oxides. Garnet specific gravities range from 3.6 to 4.2 and ilmenite from 4.2 to 4.6.

Other types of granular media include engineered ceramic materials (e.g., Filtralite, Macrolite), naturally occurring and engineered manganese oxide based materials (e.g., greensand), and synthetic polymers (plastic). The engineered ceramic materials include expanded clay aggregates; manufacturing allows for control of both media size and density to produce desired characteristics.

Important Granular Media Properties. A number of properties of filter media are important in affecting filtration performance and also in defining the media. These properties include size, shape, density, and hardness. The porosity of the granular bed formed by the grains is also important.

Grain Size and Size Distribution. Grain size has important effects on particle removal, head loss, and backwashing requirements for a media filter. It is determined by sieve analysis using American Society for Testing and Materials (ASTM) Standard Test Method C136-06, Sieve Analysis of Fine and Coarse Aggregates (ASTM, 2009). A log-probability plot of a typical sieve analysis is presented in Fig. 10-4. Sieve analysis results for most filter materials appear in nearly a linear manner on a log-probability plot.

The size gradation of filter media is commonly described by the effective size (ES) and the uniformity coefficient (UC). The ES is that size for which 10 percent of the grains are smaller by weight. It is read from the sieve analysis curve at the 10 percent passing point on the curve and is often abbreviated by d_{10}. The UC is a measure of the size range of the media. It is the ratio of d_{60} to d_{10}, where d_{60} is the size for which 60 percent of the grains are smaller by weight. In some cases, the lower and upper sizes of the media are specified with some maximum percentage allowance above and below the specified sizes. This can be achieved, for example, by indicating two standard sieves to define the size range; for example, "8 × 20" GAC media are all smaller than the number 8 sieve and larger than the number 20 sieve.

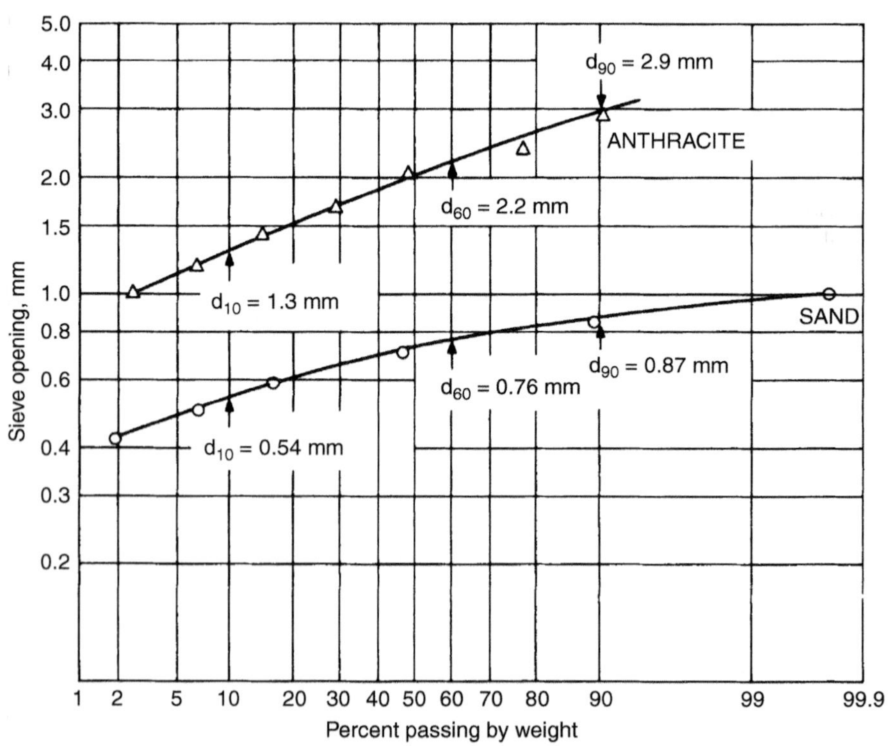

FIGURE 10-4 Typical sieve analysis results for two types of filter media.

Values of d_{10}, d_{60}, and d_{90} can be read from an actual sieve analysis curve such as the one shown in Fig. 10-4. If such a curve is not available, and if a linear log-probability plot is assumed, they can be interrelated by the following equation derived from the geometry of such linear plots for different UCs:

$$d_{90} = d_{10} \times 10^{1.67 \log UC} \tag{10-1}$$

This relationship is useful because the d_{90} size is recommended for calculation of the required backwash rate for a media filter.

Grain Shape and Roundness. The shape and roundness of the filter grains are important because they affect the backwash flow requirements for the filter bed, the fixed bed porosity, the head loss for flow through the filter, the filtration effectiveness, and the ease of sieving.

Different measures of grain shape have evolved in the geologic and chemical engineering literature, leading to considerable confusion in terminology (Krumbein, 1941; McCabe and Smith, 1976). The chemical engineering literature defines the *sphericity* ψ as the ratio of the surface area of an equal-volume sphere (diameter of d_{eq}) to the surface area of the grain (McCabe and Smith, 1976). The equivalent spherical diameter can be determined by a tedious count, weigh, and calculation procedure (Cleasby and Fan, 1981). In the absence of such data, the mean size for any fraction observed from the sieve analysis plot (e.g., Fig. 10-4) can be used as an acceptable approximation of the equivalent spherical diameter.

The chemical engineering definition is used in the following discussion. The sphericity of filter media by this definition can be determined indirectly by measuring pressure drop for flow of water or air through a bed of uniform-sized grains. The Kozeny or Ergun equation for flow through porous media (presented later) is used to calculate ψ after determining all other parameters of the equation (Cleasby and Fan, 1981).

Grain Density or Specific Gravity. Grain density, mass per unit grain volume, is important because it affects the backwash flow requirements for a media filter. Grains of higher density but of the same diameter require higher wash rates to achieve fluidization.

Grain density is determined from the specific gravity following ASTM Standard Test C128-07a, Density, Relative Density (Specific Gravity), and Absorption of Fine Aggregate (ASTM, 2009). This ASTM test uses a displacement technique to determine the specific gravity. Two alternative tests are detailed. The procedure for "bulk specific gravity, saturated surface dry" would be best from a theoretical standpoint for fluidization calculations. However, starting with a reproducible saturated surface dry condition is difficult. Therefore, the "apparent specific gravity" that starts with an oven-dry sample is more reproducible and is an acceptable alternative for fluidization calculations. For porous materials such as anthracite coal or GAC, the sample should be soaked to fill the pores with water before final measurements are made.

Grain Hardness. The hardness of filter grains is important to the durability of the grains during long-term service as a filter. Hardness is usually described by the *MOH hardness number*, which is a scale of hardness based on the ability of various minerals to be scratched by another harder object. A sequence of minerals of specified hardness is listed (Trefethen, 1959).

The two materials of known MOH hardness that can and cannot scratch the filter media are used to estimate the hardness of the filter grain material. This is a rather crude test and difficult to apply to small filter grains. Of the filter media listed earlier, only anthracite coal and GAC have low hardness worthy of concern. Silica sand, garnet, and ilmenite are very hard, and their hardness need not be of concern. A minimum MOH hardness of 2.7 or 3 is often specified for anthracite coal filter media, although measuring fractional values closer than 0.5 is doubtful.

Two standard mechanical abrasion tests are presented in the AWWA Standard B604, Granular Activated Carbon (AWWA, 2006) to evaluate the abrasion resistance of GAC.

TABLE 10-2 Typical Properties of Common Filter Media for Granular-Bed Filters (Cleasby and Woods, 1975; Cleasby and Fan, 1981; Dharmarajah and Cleasby, 1986)

	Silica sand	Anthracite coal	Granular activated carbon	Garnet	Ilmenite
Grain density (ρ_s), kg/m^3	2650	1450–1730	1300–1500[a]	3600–4200	4200–4600
Loose-bed porosity (ε_0)	0.42–0.47	0.50–0.60	0.50	0.45–0.55	b
Sphericity (ψ)	0.7–0.8	0.46–0.60	0.75	0.60	b

[a]For virgin carbon, pores filled with water; density increases when organics are adsorbed.
[b]Not available.

Despite its greater friability than anthracite, the reduction in grain size of GAC because of backwashing and regeneration operations is not reported to be a serious problem (Graese et al., 1987), although some loss of GAC occurs with each regeneration.

Fixed Bed Porosity. Fixed bed porosity is the ratio of void volume to total bed volume, expressed as a decimal fraction or as a percentage. It is important because it affects the backwash flow rate required, the fixed bed head loss, and the solids holding capacity of the filter bed. Fixed bed porosity is affected by the grain sphericity; angular grains (i.e., grains with lower sphericity) have higher fixed bed porosity (Cleasby and Fan, 1981), as is evident in Table 10-2. For the low UC commonly specified for filter media, UC has no effect on porosity; however, for natural materials with high UC, the nesting of small grains within the voids of the large grains can reduce the average bed porosity.

Fixed bed porosity is determined by placing a sample of known mass and density in a transparent tube of known internal diameter. The depth of the filter media in the tube is used to calculate the bed volume. The grain volume is the total mass of media in the column divided by the density. The void volume thus is the bed volume minus the grain volume. The fixed bed porosity is substantially affected by the extent of compaction of the media placed in the column and by the column diameter. The loose bed porosity can be measured in a column of water. If the bed is agitated by inversion and then allowed to settle freely in the water with no compaction, the highest porosity will be obtained, that is, the loose bed porosity. It may be as much as 5 percent greater than the porosity measured after gentle compaction of the bed. Materials of lower sphericity show greater change in porosity between the loose and compacted bed conditions.

The porosity of the granular filter bed is higher near the wall of the filter. This can be important in small pilot-scale filter columns because it causes the average bed porosity to be higher than in full-scale filters and affects the particle removal and head loss behaviors during filtration studies. It is common to make the diameter of such filter columns at least 50 times the grain size of the coarser filter grains to be studied (e.g., d_{90}) to minimize such wall effects.

Sieve Analysis Considerations. The standard procedure for conducting sieve analysis of filter media is detailed in ASTM Standard Test C136-06 (ASTM, 2009). This standard does not specify a sieving time or a specific mechanical apparatus for shaking the nest of sieves. Rather, it specifies that sieving should be continued "for a sufficient period and in such manner that, after completion, not more than 1 weight percent of the residue on any individual sieve will pass that sieve during one minute of hand sieving" and conducted in a manner described in the ASTM standard. With softer materials such as anthracite coal or GAC, abrasion of the material may occur when attempting to meet the 1 percent passing test.

When sieving a 100 g sample of hard materials such as sand on 8-in sieves and using a Ro-Tap type of sieving machine, it is common to require three sieving periods of 5 minutes each to satisfy the 1 percent passing test. The Ro-Tap machine imparts both a rotary shaking and a vertical hammering motion to the nest of sieves. With some other sieving machines, the ASTM requirement will not be achieved even after three 5-minute periods of sieving.

When sieving anthracite coal, the sample should be reduced to 50 g because of its lower density. The Ro-Tap machine should be used and the time fixed at 5 minutes. This will not meet the 1 percent passing test, but prolonged sieving may cause continued degradation of the anthracite, yielding a less accurate result.

Because of sieving and sampling difficulties, and because of the tolerance allowed in manufacture of the sieves themselves (ASTM E11-09 Standard Specification for Wire Cloth and Sieves for Testing Purposes, ASTM, 2009), a reasonable tolerance should be allowed in the specified size when specifying filter media. Otherwise, producers of filter material may not be able to meet the specification, or a premium price will be charged. A tolerance of ±10 percent is suggested. For example, if a sand of 0.5 mm ES is desired, the specification should read "0.45 to 0.55 mm ES." If an anthracite coal of 1.0 mm ES is desired, the specification should read "0.9 to 1.1 mm ES."

Typical Properties of Granular Filter Media. With prior understanding of the importance of various properties of granular filtering materials, Table 10-2 is presented to illustrate typical measured values for some properties. The large difference in grain densities, evident in the table, allows the construction of dual- and triple-media filters with coarse grains of low-density material on top and finer grains of higher-density material beneath; this layering is maintained during fluidized backwashing because of the combination of size and density. Alluvial sands have the highest sphericity, and crushed materials such as anthracite, ilmenite, and some garnet have lower sphericity. Some anthracites contain an excessive amount of flat, elongated, jagged grains resulting in lower sphericity. The loose bed porosity is inversely related to the sphericity; that is, the lower the sphericity, the higher is the loose bed porosity. An approximate empirical relationship between sphericity and loose bed porosity was used in developing a predictive model for fluidization presented later in this chapter (Cleasby and Fan, 1981).

Filter Media Configurations

Some typical configurations of filter media are shown in Fig. 10-5. The most commonly used of these configurations are the conventional sand filter and the dual-media filter, but a substantial number of triple-media filters have been installed. GAC can be used to replace anthracite (or sand), sometimes referred to as *filter-adsorbers*. GAC can be used alone or in dual- or triple-media configurations. The first three configurations in Fig. 10-5 are backwashed with full fluidization and expansion of the bed. Fluidization results in stratification of the finer grains of each media near the top of that layer of media.

The single-media, deep-bed filter using coarse sand, anthracite coal, or GAC (Fig. 10-5, bed 4) differs from the conventional sand filter in two ways. First, because the media in such filters are coarser, a deeper bed is required to achieve comparable removal of particles. Second, because excessive wash rates would be required to fluidize the coarse media, it is often washed without fluidization by the concurrent upflow of air and water. The air-water wash causes mixing of the media, and little or no stratification by size occurs.

The upflow filter is used in some wastewater filtration plants and in a few potable water treatment plants in other countries. It may include a restraining grid to resist uplift, as shown in Fig. 10-5 (bed 5), or it may be operated with a deeper sand layer and to a limited terminal head loss so that the mass of the sand itself acts to resist uplift. The upflow

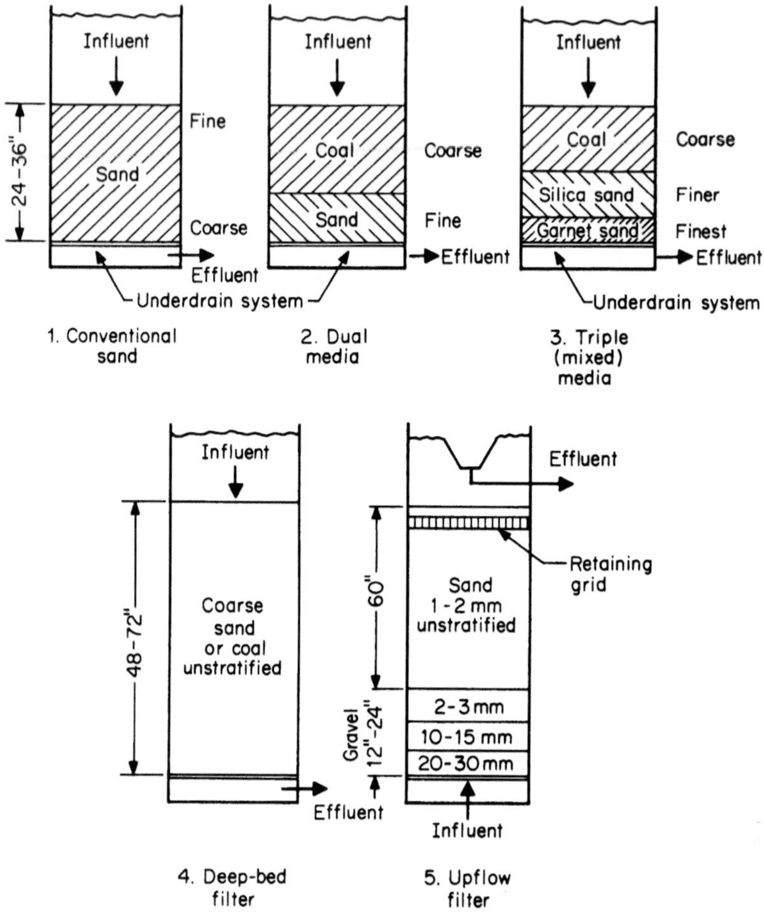

FIGURE 10-5 Schematic diagrams of some typical filter bed configurations. Beds 1, 2, and 3 are washed with fluidization, whereas beds 4 and 5 are washed with air plus water without fluidization.

filter is washed using a forward flow employing air and water together. Adoption of the upflow filter as the only filtration step in potable water treatment is doubtful because of the potential for contamination of the filtered water by both the dirty wash water and the filtered water exiting above the filter bed. Upflow roughing filtration is being used in potable water treatment plants in the United States, where it is used in the pretreatment process in place of sedimentation or other solids separation processes, ahead of downflow filtration. A buoyant crushed plastic media (2–4 mm size) used in upflow mode ahead of a downflow triple-media bed (Benjes et al., 1985) is one common configuration.

Typical grain sizes used in rapid-rate filters are presented in Table 10-3 for various potable water applications. The UC of the filter media is usually specified to be less than 1.65 or 1.7. Use of a lower UC is beneficial for coarser filter media sizes that are to be backwashed with fluidization because this will minimize the d_{90} size and thereby reduce the required backwash flow rate. The lower the specified UC, the more costly will be the

TABLE 10-3 Typical Media Grain Sizes and Bed Depths for Different Applications

	Effective size, mm	Total depth, m
A. Common U.S. Practice after Coagulation and Clarification		
1. Sand alone	0.45–0.55	0.6–0.7
2. Dual media: Add top-layer anthracite, 0.3–0.6 m deep (bottom sand layer, 0.15–0.3 m deep)	0.9–1.1	0.6–0.9
3. Triple media: Add bottom garnet, 0.1 m, to dual media	0.2–0.3	0.7–1.0
4. GAC: Use in place of anthracite in dual media		
5. For direct filtration: dual media as in A(2)		
B. Coarse Single-Media Filters, Simultaneous Air/Water Wash		
1. For coagulated and clarified water	0.9–1.1	0.9–1.2
2. For direct filtration: May add 0.15–0.3 m bottom sand layer	1.4–1.6	1–2
C. U.S. Practice for Fe and Mn Filtration		
1. Dual media similar to A(2) above		
2. Single media, standard	<0.8	0.6–0.9
3. Single media, coarse	1–2	1.5–3

filter media because a greater portion of the raw material falls outside the specified size range. Therefore, the lowest practical UC is about 1.4. Anthracite coal that will meet this UC is available commercially.

There is growing interest in and use of deeper beds of filtering materials, either single- or dual-media, especially for high-rate direct filtration applications. These filter media configurations typically involve 1.0 to 3.0 m (3.2–9.8 ft) of coarse (1.2 to 1.5 mm ES) anthracite alone or above a 0.15 to 0.3 m (0.5 to 1.0 ft) layer of 0.6 to 1.0 mm ES sand and HLRs as high as 33 m/h (13.5 gpm/ft^2). More details and specific examples are described under "Direct Filtration" later in this chapter. One reason for consideration of deep beds is the possibility of having relatively long contact times, which can be advantageous for biologically active filtration, perhaps with GAC media.

For example, Patania et al. (1995) used deep beds (2.0 m or 6.7 ft) of 1.25 mm ES anthracite over 0.25 m (0.83 ft) of 0.6 mm ES sand for in-line filtration pilot studies of cyst removal in Seattle. In 2000, the 454 ML/d (120 mgd) capacity Seattle Tolt plant went into operation, incorporating preozonation, coagulation, and hydraulic flocculation prior to single-media filtration with 1.8 m (6.0 ft) of 1.3 mm ES anthracite with a UC of less than 1.4, which is approved for an HLR of 29 m/h (12 gpm/ft^2) (Tarbuck et al., 2002). In late 2009, the new 1800 ML/d (475 mgd) Seymour-Capilano direct filtration plant began operation for the metropolitan Vancouver, BC, water supply. The configuration for the dual-media filters (24) is 1.7 m (5.6 ft) of 1.4 mm ES anthracite (UC < 1.5) over 0.3 m (1.0 ft) of 0.7 mm ES sand (UC < 1.5) (Phelan, 2010). Pilot studies for proposed plants for Sydney, Australia, included deep-bed single-media and dual-media configurations (Murray and Roddy, 1993). The dual-media filters studied included anthracite 1.0 to 3.0 m (3.2–9.8 ft) in depth with 1.7 to 2.5 mm ES over 0.15 to 0.30 m (0.5–1.0 ft) of crushed sand with 0.65 to 1.0 mm ES. The coarser dual media performed best in water production per cycle and in filtrate quality. One reason expressed for the interest in the deep beds was the possible future conversion of the filters to GAC with empty-bed contact times of at least 7.5 minutes.

In addition to the configurations of filtering materials shown in Fig. 10-5, other approaches are used in some applications. For traveling-bridge, continuous-backwash downflow filters (e.g., the ABW Automatic Backwash Filter), the filter is divided horizontally into lateral cells and uses a shallow layer of fine sand, usually of about 0.3 m (12 in) depth (Medlar, 1974). Another approach involves continuous downward movement of the sand bed while water to be treated flows up through the media; the lowest media is continuously directed to an uplift backwashing system and reapplied to the top of the downflowing media bed (e.g., the DynaSand process). Beds of engineered ceramic media have been used in downflow mode for direct filtration (Saltnes et al., 2002), with special attention to the role of surface roughness for biofilm support (Melin et al., 2000) and particle retention (Emelko et al., 2007).

The Relationship Between Bed Depth and Media Size: The (L/d) Ratio. When considering the use of coarser media, deeper bed depth results when the ratio of bed depth to grain diameter, *L/d,* is held constant or is increased; the goal is to maintain equal filtrate quality when filtering the same influent suspension at the same filtration rate. The concept is supported by the experiments of Ives and Sholji (1965). However, these authors also analyzed the work of other investigators that suggested that the relationship should be L/d^{β}, with β values from 1.5 to 1.67. Fundamental theory (see below) supports higher values of β, and the consideration of a characteristic filter bed depth presented by Lawler and Nason (2006) shows the need for increasing the *L/d* ratio to maintain similar particle removal performance as media size and filtration rate increase.

Nevertheless, the *L/d* ratio is used in practice. For graded beds, the effective size (d_{10}) has been suggested for the diameter term; the MWH (2005) text indicates a range of 1000 to 2000 for the ratio. For dual- and multi-media filters, the sum of the L/d_{10} values for each layer is calculated for the bed. Some typical values for L/d_{10} for common bed configurations from Kawamura (2000) follow: (1) $L/d_{10} \geq 1000$ for ordinary fine-sand and dual-media beds, (2) $L/d_{10} \geq 1250$ for triple-media (anthracite, sand, garnet) beds, (3) $L/d_{10} \geq 1250$ for a deep, single-media bed (1.5 mm > d_{10} > 1.0 mm), and (4) L/d_{10} = 1250 to 1500 for very coarse, deep, single-media beds (2.0 mm > d_e 10 > 1.5 mm).

For a common dual-media filter with 0.6 m (2.0 ft) of anthracite of 1.0 mm ES over 0.3 m (1.0 ft) of sand of 0.5 mm ES, the total L/d_{10} would be 600/1.0 plus 300/0.5 = 1200, meeting Kawamura's criteria. Of course, this is a simplistic concept for a complex process, but it does assist in selecting alternative filter media for pilot-scale evaluations. Kawamura (2000) also notes that for media size greater than 1.5 mm, it is advisable to include a shallow (0.3 m or 1 ft) layer of finer sand below the deep coarse-media bed.

Single-Media Versus Multi-Media Filters. The media used in rapid filters evolved from fine-sand single-media filters about 0.5 to 0.9 m (2–3 ft) deep to dual-media and later to triple-media (mixed) filters of about the same depth. Still later, led by the successful use at Los Angeles, there has been a strong interest in deep-bed [1.2–1.8 m (4–6 ft)], coarse, single-media filters. The rationale for this evolution is discussed in this section.

Early research on rapid sand filters with sand about 0.4 to 0.5 mm ES demonstrated that most of the solids were removed in the top several centimeters (few inches) of the sand, and thus the full bed depth was not well used. The dual-media filter bed, consisting of a layer of coarser anthracite coal on top of a layer of finer silica sand, was developed to encourage better penetration of solids into the bed and thus better use of the filter bed.

The use of dual-media filters is widespread in the United States. Camp (1961) reported using dual-media swimming pool filters beginning about 1940 and later in municipal treatment plant filters. Baylis (1960) described early work in the mid-1930s at the Chicago Experimental Filtration Plant, where a 7.5 cm (3-in) layer of 1.5 mm ES anthracite over a layer of 0.5 mm ES sand greatly reduced the rate of head loss development in treatment of Lake Michigan water.

The benefit of dual-media filters in reducing the rate of head loss development, thus lengthening the filter run, is well proven by a number of later studies (Conley and Pitman, 1960a; Conley, 1961; Tuepker and Buescher, 1968); however, the presumed benefit to the quality of the filtrate is not well demonstrated.

Based on their experiences, Conley and Pitman (1960b) concluded that the alum dosage should be adjusted to achieve low levels of uncoagulated matter in the filtrate (low turbidity) early in the filter run (after 1 hour) and that the filter-aid polymer should be adjusted to the minimum level required to prevent terminal breakthrough of alum floc near the end of the run. The dosage of polymer should not be higher than necessary, however, to prevent excessive head loss development.

Further research comparing dual media with fine-sand media was reported by Robeck et al. (1964). They compared three filter bed designs during filtration of alum-coagulated surface water. These comparisons were made by running filters in parallel so that the benefits of dual media were demonstrated more conclusively. The rate of head loss development for the dual-media filter was about half that of the sand media, but the effluent turbidity was essentially the same prior to breakthrough, which was observed under some weak flocculation conditions.

The evidence clearly demonstrated the lower head loss expected for a dual-media filter compared with a traditional fine-sand filter. For a typical dual-media filter with an anthracite ES that is about double the sand ES, the head loss development rate should be about half the rate of the fine-sand filter when both are operated at the same filtration rate on the same influent water.

The benefits gained by the use of dual-media filters led to the development of the triple-media filter, in which an even finer layer of high-density media (garnet or ilmenite) is added as a bottom layer (Conley, 1965, 1972). The bottom layer of finer material should improve the filtrate quality in some cases, especially at higher filtration rates.

The triple-media filter is sometimes referred to as a *mixed-media* filter because in the original development, the sizes and uniformity coefficients of the three layers were selected to encourage substantial intermixing between the adjacent layers. This was done to come closer to the presumed ideal configuration of "coarse to fine" filtration. The original patents have expired, and other specifications for triple-media filters are being used with various degrees of intermixing.

The initial clean bed head loss will be higher for the triple-media filter because of the added layer of fine garnet or ilmenite. Thus, for a plant with a particular total available filter system head loss, the clogging head loss available to sustain the run is reduced, which may shorten the run compared with the run obtained with a dual-media filter (Robeck, 1966).

Some comparisons of triple-media versus dual-media filters have demonstrated triple-media filters to be superior in filtered water quality. On Lake Superior water at Duluth, a mixed-media (triple-media) filter was superior to a dual-media filter in amphibole fiber removal (Logsdon and Symons, 1977; Peterson et al., 1980). Twenty-nine of 32 samples of filtrate were below or near detection level for the dual-media filter and 18 of 18 for the mixed-media filter. Mixed-media filters were recommended for the plant. Mixed-media filters also were reported to be superior in resisting the detrimental effects of flow disturbances on filtrate quality (Logsdon. 1979).

In contrast, Kirmeyer (1979) reported pilot studies for the Seattle water supply in which two mixed-media and two dual-media filters were compared. No difference in filtrate quality was observed in either turbidity or asbestos fiber content with filtration rates of 13.4 to 24.4 m/h (5.5–10 gpm/ft^2). Some differences in production per unit head loss were observed favoring dual-media filters at lower rates and mixed-media filters at higher rates.

A number of laboratory studies have compared triple-media versus single-media (fine-sand) filters (Diaper and Ives, 1965; Rimer, 1968; Oeben et al., 1968; Mohanka, 1969). All these studies have shown clearly the head loss benefit gained by filtering in the direction of coarse to fine grains. Three of the studies also clearly showed benefits to the filtrate quality for the triple-media filters (Diaper and Ives, 1965; Rimer, 1968; Oeben et al,. 1968).

Cleasby et al. (1992) summarized 12 unpublished pilot- and plant-scale studies conducted for utilities comparing dual- and triple-media filters and deep single-media filters in high-rate filtration of surface waters. The studies supported the superiority of triple-media filters over dual-media filters (with comparable depths of anthracite and sand) in producing the best-quality filtered water; however, higher initial head losses and better solids capture resulted in shorter filter cycles for triple-media filters, other factors being the same. The studies comparing deep bed, coarse, single-media filters with dual- or triple-media filters were inconclusive with regard to quality of filtrate; however, the deep-bed, coarse, single-media filters achieved longer filter cycles and greater water production per cycle than traditionally sized dual- or triple-media filters.

Use of GAC in Rapid Filtration. GAC media are used for both particle removal and adsorption of organic compounds, that is, as a filter-adsorber, as well as in biologically active filters and post-filtration contactors (see Chap. 14 for more details). The principal application of filter-adsorbers to date is to remove taste- and odor-causing compounds where full-scale experience shows successful removal for periods from 1 to 5 years before the GAC must be replaced/regenerated (Graese et al., 1987). High concentrations of competing organic compounds, however, can reduce this duration. Many GAC filter-adsorbers are retrofitted rapid filters in which GAC has replaced part or all of the sand in rapid sand filters or replaced anthracite in dual- or triple-media filters. In addition, selection of GAC media for new filters is not uncommon.

GAC used in retrofitted filters typically is 0.38 to 0.76 m (15–30 in) of 12 × 40 mesh GAC (ES 0.55–0.65 mm) or 8 × 30 mesh GAC (ES 0.80–1.0 mm) placed over 0.15 to 0.30 m (6–12 in) of sand (ES 0.35–0.60 mm) in a dual-media configuration. These GAC materials often have a higher UC (≤2.4) than traditionally used filter materials (≤1.6). The high UC can contribute to more fine grains in the upper layers and shorter filter cycles but results in less size intermixing during backwashing, which is beneficial to the adsorption function. A lower UC can be realized by specifying a narrower mesh range, that is, 8 × 20 or 8 × 16 mesh. Some filters designed for GAC media initially use up to 0.9 to 1.8 m (3–6 ft) of coarser GAC (ES 1.3 mm or greater) with a lower uniformity coefficient (1.4). This provision should result in longer filter cycles and longer periods between the need for GAC replacement/regeneration based on adsorption objectives.

GAC has been used successfully in both single-media and dual-media filters. It is as effective in its filtration function as conventional filter media, either sand or dual-media, provided that an appropriate size media has been selected (Graese et al., 1987; Emelko et al., 2006). Particles removed by the GAC filter and the need for frequent backwashing do, however, impede the adsorption function somewhat compared with postfilter GAC contactors with the same contact time (see Chap. 14).

The layer of fine sand in a dual-media GAC–sand filter may be essential where low filtered water turbidity is the goal. The use of sand and GAC in dual-media beds, however, causes difficulties during replacement/regeneration because it is difficult to remove the GAC media from the filter free of sand or to separate the sand after removal. Sand causes difficulties in regeneration furnaces (Graese et al., 1987).

GAC media are often selected for filters that are designed to be biologically active, that is, where removal of biodegradable organic matter (BOM) is expected; often particle and BOM removal are expected to occur in a single filtration process. Urfer et al. (1997), Liu et al. (2001), and Emelko et al. (2006) review and report on factors such as temperature, media type, and chlorination/chloramination of backwash water that can affect the performance of such filters. Media other than GAC (sand, anthracite) can be effective for BOM removal, although GAC appears to perform better for challenging conditions of lower temperature and oxidants in backwash water. Accumulation of biomass in biofiltration necessitates periodic backwashing, which can, in turn, affect BOM removal. Chlorine residual in filter influent is quickly destroyed by the GAC media and may negatively affect the media (see Chap. 14). In general, chlorination of filter influent does not completely prevent bacterial growth. Since microorganisms detach from biologically active filters, post-filter disinfection is needed.

Because of the adsorption capacity of unused or regenerated activated carbon, there can be rapidly changing impacts of GAC media particle removal filters on certain measures of water quality following filter startup. This is especially true for sorption of NOM, as measured by removal of total organic carbon (TOC) or UV_{254}; short-term (weeks to months) high removal efficiency is not at all representative of long-term (years) performance. Care must be taken in conducting pilot-scale studies with GAC media to recognize this and either use media that has already had its short-term NOM uptake capacity exhausted or operate the pilot study long enough for this to occur. It is also difficult to measure the expected lower-level long-term removal of NOM by biodegradation on GAC filter media during the duration of a typical pilot study.

GAC is softer than anthracite, but the attrition of GAC media in full-scale plants has not been excessive. GAC losses from 1 to 6 percent per year have been reported (Graese et al., 1987), which is not higher than typical losses of anthracite. A modest reduction in grain size of GAC has been measured at some plants. GAC has a lower density than anthracite, posing some concerns about losses during backwashing, a topic discussed later in this chapter.

Rates of Filtration

Fuller (1898) is commonly credited with establishing a rate of filtration of 5 m/h (2 gpm/ft^2) for chemically pretreated surface waters. This filtration rate was considered practically inviolable for the first half of the twentieth century in the United States. Fuller observed that with properly pretreated water, however, higher rates gave practically the same water quality. Of more importance, Fuller acknowledged that without adequate chemical pretreatment, no assurance of acceptable filtered water existed even at filtration rates of 5 m/h (2 gpm/ft^2); much more recently, Bellamy et al. (1993) emphasized that optimal coagulation is the most important factor in achieving excellent performance of granular media filtration.

In the 1950s and 1960s, many plant-scale studies were conducted comparing filter performance at different filtration rates (Brown, 1955; Baylis, 1956; Hudson, 1962). These studies generally were conducted as utilities were considering uprating existing filters or building new plants with filtration rates that were higher than the traditional 5 m/h (2 gpm/ft^2). The results of these and similar studies (Cleasby and Baumann, 1962; Conley and Hsuing, 1969) demonstrated that higher filtration rates resulted in somewhat poorer filtrate quality, as both theory and intuition would predict. However, at the time, the turbidity standard was 1 Jackson turbidity unit, and it was presumed that chlorine disinfection would inactivate pathogens that passed through the filters. Nevertheless, there was a gradual acceptance of filtration rates up to 9.8 m/h (4 gpm/ft^2) with conventional pretreatment and without the use of filter-aid polymers to increase particle retention (King et al., 1975).

Pioneering work at higher filtration rates, assisted by filter-aid polymer, included the following studies: Robeck et al. (1964) compared performance of pilot filters at 5 to 15 m/h (2–6 gpm/ft^2) filtering alum-coagulated surface waters through single-media and dual-media filters. They concluded that with proper coagulation ahead of the filters, the effluent turbidity, coliform bacteria, poliovirus, and powdered carbon removal was as good at 15 m/h (6 gpm/ft^2) as at 10 or 5 m/h (4 or 2 gpm/ft^2). Pretreatment included activated silica when necessary to aid flocculation and a polyelectrolyte as a filter aid (referred to as a "coagulant aid" in the original article); the filter-aid polymer retarded turbidity breakthrough, causing the filter run time to increase from 16 to 22 hours.

Conley and Pitman (1960a) showed the detrimental effect of high filtration rates up to 37 m/h (15 gpm/ft^2) in the treatment of Columbia River water by alum coagulation followed by short-detention flocculation and sedimentation before filtration. A proper dose of nonionic polymer added to the water as it entered the filters, however, resulted in the same filtrate quality from 5 to 85 m/h (2–35 gpm/ft^2).

Over the last 30 years, concerns about *Giardia* and *Cryptosporidium* have emphasized the need to achieve filtered water turbidities at or below 0.10 ntu and have emphasized log reduction of cysts or cyst-sized particles to ensure the absence of protozoan pathogens. Some results of such studies were presented earlier in this chapter, and some are presented below.

When chemical pretreatment was optimized for turbidity and particle removal resulting in filtered water turbidity below 0.10 ntu, Patania et al. (1995) found no difference in *Giardia* cyst and *Cryptosporidium* oocyst removal in pilot studies at Contra Costa at filtration rates of 7 to 15 m/h (3–6 gpm/ft^2) and at Seattle at 12 to 20 m/h (5–8 gpm/ft^2). Conventional pretreatment preceded filtration at Contra Costa, whereas in-line filtration was used at Seattle. Both used deep-bed dual-media filters and dual coagulants (alum plus cationic polymer). Filter-aid polymer also was used in the Seattle pilot plant.

Design and operation guidelines for optimization of high-rate filtration plants were reported by Cleasby et al. (1989, 1992). The reports were based on a survey of 21 surface water treatment plants with consistent operational success in producing filtered water turbidity below 0.2 ntu at filtration rates at or above 10 m/h (4 gpm/ft^2). These plants were characterized by management support of a low turbidity goal, optimal chemical pretreatment, the use of polymeric flocculation and/or filter-aid chemicals, the use of dual- or triple-media filters, continuous monitoring of effluent turbidity for each filter, and good operator training.

The use of unusually high filtration rates was reported at the Contra Costa County Water District plant in California. By precoating the dual-media filters with a small dose of polymer during the backwash operation, Harris (1970) reported successful operation at 24 m/h (10 gpm/ft^2). Harris also reported that the initial period of poorer water quality was eliminated by this precoating operation. The Contra Costa County Water District plant was authorized by the State of California to operate at 24 m/h (10 gpm/ft^2).

Another example of an unusually high design filtration rate is the 33 m/h (13.5 gpm/ft^2) selected for the Los Angeles direct filtration plant. This filtration rate was selected and approved by the state after more than 5 years of extensive pilot-scale studies (Trussell et al., 1980; McBride and Stolarik, 1987). The plant uses ozone, ferric chloride, and cationic polymer in pretreatment and deep-bed (1.8 m or 6 ft) coarse anthracite (1.5 mm ES, 1.4 UC) single-media filters. It has operated successfully at the design rate, producing filtered water below 0.1 ntu on a consistent basis.

Both these treatment plants are large, treating high-quality surface water with skilled management and operation. Thus filtration rates of these magnitudes probably will not become common. With poorer source waters and in smaller plants with less careful surveillance or operation, filtration rates at or below 10 m/h (4 gpm/ft^2) are likely to continue to be the prudent choice. Nevertheless, the success at Los Angeles has stimulated great interest in deep-bed single-media filtration. Other similar plants are now in service, but not at such high filtration rates. For example, the Modesto, Calif, plant (Short et al. 1996) uses the same filter media as in the Los Angeles plant, but the design filtration rate is 15 m/h (6 gpm/ft^2).

Acceptance of filtration rates above 5 m/h (2 gpm/ft^2) varies among state regulatory agencies, however, and plant- or pilot-scale demonstrations are required by some state agencies before acceptance.

The use of higher filtration rates shortens the filter cycle roughly inversely with the rate. This problem is minimized by the use of dual- or triple-media filters or deep beds of coarser media with or without an underlying shallow layer of finer-sand media. As noted earlier, filters with coarser media provide for better penetration of solids into the filter bed and thus better use of the bed pores for solids storage during the filter cycle.

Acceptable Filter Run Length and Production per Run

Higher filtration rates have two conflicting impacts on plant production. First, the clean bed head loss is increased. Thus, for a given amount of available head loss in the plant hydraulic profile, less clogging head loss is available to sustain the filter cycle. This will shorten the filter cycle length if it is terminated based on reaching maximum available head loss. Second, for a given media, the head loss increases at a faster rate per hour at higher filtration rates because solids are captured at a faster rate. However, the water production per unit time also increases so that production per day is increased substantially unless the filter cycles get too short. Therefore, the common importance attached to the length of filter run can be misleading. Rather than emphasizing the effect of filtration rate on length of run, one should emphasize the effect of filtration rate on filtrate quality, net plant production, and production efficiency.

Figure 10-6 illustrates the effect of unit filter run volume on net water production at three filtration rates (Trussell et al., 1980). *Unit filter run volume* (UFRV) is the actual throughput of a filter per unit area during one filter run; it also can be called *gross production per filter run*. Figure 10-6 is based on the assumption that 4 m^3/m^2 (100 gal/ft^2) is used for each backwash operation, and 30 minutes of downtime is required per backwash. Similar figures can be constructed with other assumptions. The following example illustrates preparation of the figure.

- Assume four cycles per day per filter.
- Run time per cycle = 6 hours/cycle – 0.5 hour/wash = 5.5 hours/cycle.
- Assume desired maximum filtration rate = 12.2 $m^3/m^2/h$ (5 gpm/ft^2).
- Unit filter run volume 12.2 × 5.5 = 67.1 m^3/m^2/cycle (1,650 gal/ft^2/cycle).
- Gross volume filtered/m^2/day = 67.1 × 4 cycles/day = 268.4 m^3/m^2/day (6600 gal/ft^2/day).

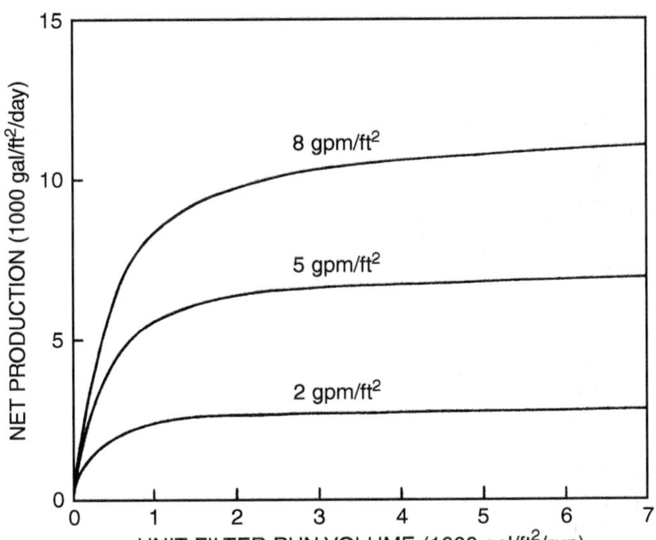

FIGURE 10-6 Net (gross) production versus unit filter run volume assuming 30 minutes of downtime per backwash and a backwash volume of 100 gal/ft^2 per backwash—1 gal equals 3.785 liters. (*Source: Trussell et al., 1980.*)

- Net water production/m^3/m^2/day = 268.4 – backwash volume = 268.4 – 4.1 × 4 cycles/day = 252 m^3/m^2/day (6200 gal/ft^2/day).
- Production efficiency expressed as percent of gross volume filtered = (252/268.4) × 100 = 94 percent.

In general, allowing for backwash volumes of 4 to 8 m^3/m^2/wash (100–200 gal/ft^2/wash) and a downtime of 30 min/wash, production efficiency will remain high if the UFRV exceeds about 200 m^3/m^2/run (5000 gal/ft^2/run). The calculations just presented are the same whether the backwash water is recovered by recycling through the plant or not recovered. Above 200 m^3/m^2/run (5000 gal/ft^2/run) UFRV, the net production increases almost linearly with rate. Thus the key issue when considering the impact of high filtration rates on production is not run length alone but rather the gross production per run (run length × filtration rate) and the filtrate quality. A unit filter run volume of 200 m^3/m^2/run (5000 gal/ft^2/run) means a run length of 41 h at 4.9 m/h (2 gpm/ft^2) or 16.7 h at 12 m/h (5 gpm/ft^2); both rates yield a 98 percent production efficiency, so one is not better than the other by this measure.

In summary, high-rate filtration does have a definite effect on development of head loss and length of filter run that must be anticipated. The benefits of greater net production/m^2/day at higher filtration rates, however, more than justify that consideration because of lower capital cost for new plants or increased production from existing plants.

Underdrain and Support Gravel

The underdrain system serves to support the filter media, collect filtered water (in downflow filters), and distribute backwash water and air scour if air is used. A wide variety of underdrain systems are in use; three major types are described below.

The first and oldest type is the manifold-lateral system, in which perforated pipe laterals are located at frequent intervals along a manifold (Fig. 10-7). Perforations in the laterals are 6 to 13 mm (¼–½ in), located on 8 to 30 cm (3-12 in) spacing.

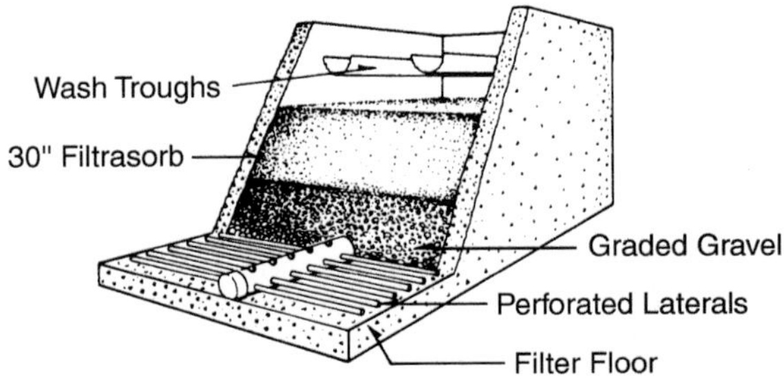

FIGURE 10-7 Rapid gravity filter with manifold and lateral underdrain system. (*Source: After C. P. Hoover, Water Suppy and Treatment, National Lime Association.*)

The second is the fabricated self-supporting underdrain system that is attached (grouted) to the filter floor, as in Fig. 10-1. The older example of this type is a vitrified clay block underdrain; more recently, plastic block underdrains have been used. The plastic block underdrains are capable of delivering backwash air or water or air and water simultaneously. Top openings on these types of underdrains are about 6 mm (¼ in) when a gravel support layer is used. The plastic block underdrains may have a smaller-opening porous cap installed (plastic bead or stainless steel) to eliminate the gravel support layer.

A third type of underdrain is the false-floor underdrain with nozzles (Fig. 10-8). A false-floor slab or a steel plate is located 0.3 to 0.6 m (1–2 ft) above the bottom of the filter, thus providing an underdrain plenum below the false floor. Nozzles to collect the filtrate and distribute the backwash water are located at 13 to 20 cm (5-8 in) centers. Nozzles may have coarse openings of about 6 mm (¼ in), or they may have fine openings, sufficiently small to retain the filter media. The nozzles may be equipped with a stem protruding about 15 to 23 cm (6–9 in) into the underdrain plenum. One or two small holes or slots in the stem serve to distribute air, either alone or in combination with water.

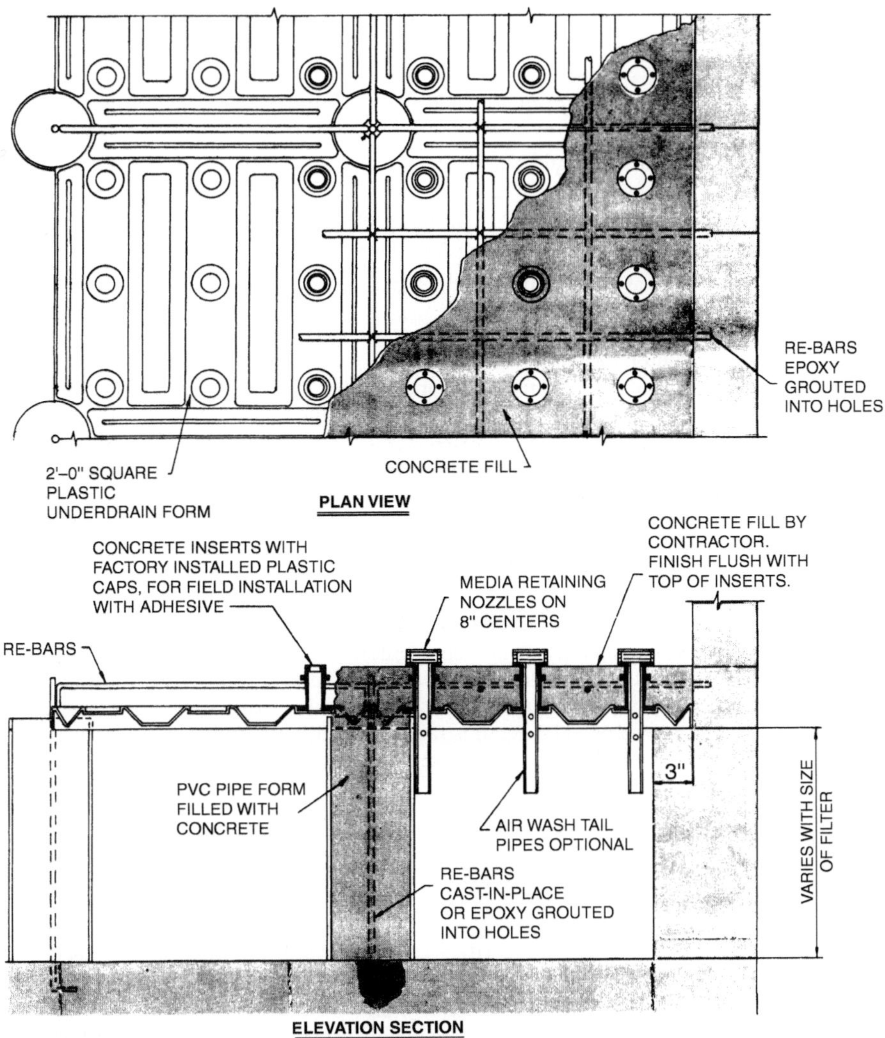

FIGURE 10-8 Nozzle underdrain system consisting of a monolithic cast-in-place concrete slab on a permanent plastic underdrain with nozzles capable of air and water distribution. (*Source: Multicrete II, Courtsey of General Filter Co., Ames, Io.*)

Many filters use underdrain systems with openings larger than the filter media to be supported. As a result, using several layers of graded gravel between the underdrain openings and the filter media is necessary to prevent the media from passing downward into the underdrain system. Details of gravel sizing and layering are available in design texts (e.g., Kawamura, 2000), in standards (AWWA, 2001), and from manufacturers. The conventional gravel system begins with coarse-sized gravel at the bottom with progressively finer-sized gravel layers above up to the filter media. Typical rules of thumb for the gradations of adjacent layers reported by Huisman (1974) in Europe are as follows:

1. Each gravel layer should be as uniform as possible, preferably with the 10 and 90 percent passing diameters not more than a factor of $\sqrt{2}$ apart. In the United States, a sieve size ratio of 2 for the passing and retaining sieves is commonly accepted (AWWA, 2001).
2. Bottom-layer lower grain size limit should be 2 to 3 times the orifice diameter of the underdrain system.
3. Top-layer lower grain size limit should be 4 to 4.5 times the ES of the media to be retained.
4. From layer to layer, the upper grain size limit of the coarser layer should be 4 times the lower grain size limit of the adjacent finer layer.
5. Each layer should be at least 7 cm (3 in) thick or 3 times the upper grain size limit of the layer, whichever is greater.

The U.S. interpretation of these guidelines is that the upper and lower grain size limits refer to the sieve sizes that pass and retain the gravel. No more than 8 percent by dry weight shall be finer than the lower grain size limit, and a minimum of 92 percent shall be finer than the upper grain size limit (AWWA, 2010).

When air scour is to be delivered through the supporting gravel, great danger exists that the gravel will be disrupted during the backwash cycle, especially if air and water are used simultaneously. Two solutions are being used. One is to use a nozzle or capped-block underdrain with small enough openings so that gravel is not required. The other is to use a double-reverse-graded gravel system graded from coarse on the bottom to fine in the middle and back up to coarse on the top. This concept—originally used by Baylis to solve gravel migration problems in large filters in Chicago (AWWA, 1971)—is being used in air and water backwash filters supplied by several companies.

In addition to the systems just described, various proprietary underdrain systems that eliminate the need for supporting graded gravel layers are available. The long-term success of these new systems remains to be demonstrated. Since filter failures are often attributed to underdrain deficiencies, adoption of unproven systems is a matter of concern.

MEDIA FILTRATION THEORY AND MODELING

The earliest recorded instance of the filtration of impure water through sand is found in *Susruta Sanhita*, a medical treatise of the Hindus in India (Baker, 1948). In a more recent era, slow sand filters were introduced in England in 1829, and rapid-rate sand filters came into existence in the latter part of the nineteenth century. Slow sand filters were introduced into the United States in 1872, with the first plant built in Poughkeepsie, NY (Fuller, 1933). Municipal rapid-rate sand filters followed with the construction of a plant at Somerville, NJ, in 1885.

After the introduction of sand filters into the United States, a long period then followed during which attention was focused on refining the filter apparatus. A period of refinement in filter operation followed, after which attention was directed toward proper conditioning of the water prior to the filtration step. Only in the past 50 years or so have significant efforts been made toward determination of the mechanisms of particle removal during packed-bed filtration.

Overview

Much can be learned about filtration of particles in water (aquasols) from past work on filtration of particles in air (aerosols). Observations made by Billings (1966) are particularly helpful (Fig. 10-9). Air filters are often made up of mats of cylindrical fibers. In the experimental results shown in Fig. 10-9, a single cylindrical glass fiber represents an aerosol filter. It receives an airflow of 13.8 cm/s containing polystyrene latex particles at a concentration of approximately 1000/cm^3. As filtration proceeds, latex particles are removed from suspension. At and near the start of the run, the glass fiber dominates particle removal and is the primary filter media. However, by 135 minutes, previously retained latex particles begin to control the removal of latex particles and accumulation of the filter deposit. By the end of the run (420 minutes), most particles removed from suspension have been removed by filter media that are actually previously retained particles; the contribution of the glass fiber to removal is negligible. The fact that the media operative in much of a filtration run are actually particles removed from the fluid and not the media installed initially in the filter is helpful in understanding how packed-bed filtration functions in water (aquasol) as well as air (aerosol) filtration.

Packed-bed filters remove particles from the suspensions that flow through them. As noted previously, these particles may have been present in the raw water supply or have been formed in pretreatment. Particles in raw water supplies vary widely in origin, composition, concentration, and size. Some are constituents of land-based or atmospheric inputs (clays, microorganisms including pathogens, asbestos fibers, etc.), and others are produced

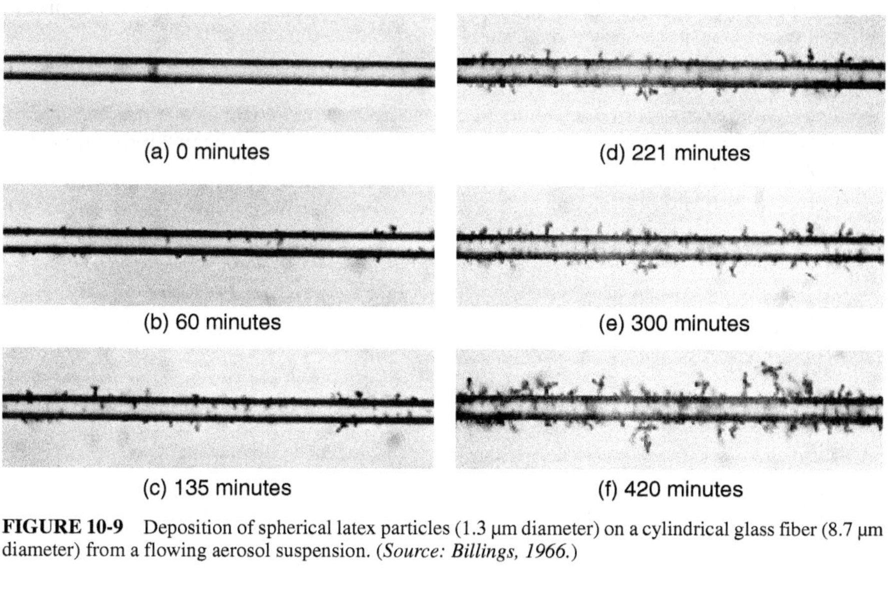

FIGURE 10-9 Deposition of spherical latex particles (1.3 μm diameter) on a cylindrical glass fiber (8.7 μm diameter) from a flowing aerosol suspension. (*Source: Billings, 1966.*)

by chemical or biological processes in the water source (algae, organic detritus, inorganic precipitates of $CaCO_3$, FeOOH, etc.). Soluble natural organic substances (e.g., humic substances) are present in all surface water supplies; these can originate on the land or in the water. Synthetic organic substances and many trace and toxic metals are particle reactive and are associated with particulate materials in the water source. The addition of chemical coagulants aggregates particles present in the water supply, adsorbs and precipitates humic substances, and forms new solid particles [$Al(OH)_3(s)$, etc.]. The particles reaching a filter in a water treatment system reflect all these inputs and processes (O'Melia 1988).

Pretreatment chemistry is a dominant factor affecting particle removal in packed-bed filters. If the solution chemistry of the water to be treated, the surface chemistry of the particles to be removed, and the surface chemistry of the media used to accomplish removal are not properly controlled, effective particle removal will not occur regardless of the bed depth, filtration rate, and media size installed in the filter system. In practice, this is accomplished by proper selection of the types and dosages of pretreatment chemicals. These may include aluminum or iron(III) salts, polymeric inorganic coagulants such as polyaluminum chloride (PACl), polyiron chloride (PICl) and activated silica, synthetic organic polymers and polyelectrolytes, and even ozone (see Chap. 8).

The principal factors affecting head loss development in packed-bed filters are the concentration and *size* of the particles applied to the filters. Media size, filtration rate, and bed depth have important effects, but the physical properties of the particles applied to filters dominate head loss development after the early stages of filtration in these filters. Small particles can produce large head losses; submicron particles can induce rapid and extensive head loss development when treated by conventional packed-bed filters.

The size of the particles in the water being filtered also affects the removal efficiency of a filter bed when it is clean, such as at the start of a filter run. A critical suspended particle size exists, in the region of 1 μm, for which the removal efficiency of clean beds is a minimum. Particles in this size region have the fewest opportunities for contact with the filter media and subsequent removal from suspension. Smaller particles are transported predominately and effectively to the external surfaces of the media by convective diffusion and larger ones by following fluid flow (interception) and by gravity. This 1 μm size is also important in assessing filtrate quality. This is often done by measuring turbidity, a light-scattering property of the particles in the filtrate. As it happens, the turbidity per unit mass of particles in a suspension is a maximum at a size of 1 μm or so; thus the particles that have the smallest chance of removal (based on size) are those that have the largest chance of detection.

It is convenient to describe the removal of a suspended particle in a packed bed of media grains as involving two distinct steps (O'Melia and Stumm, 1967). The first step is transport of the suspended particle from the flowing fluid to the surface of the filter media grain. This is primarily a physical process. The second step is attachment of the particle to the media grain. This is primarily a chemical process.

When filtration is successful in the early stages of a filter run, that is, when the hydrodynamics of particle transport and the chemistry of particle attachment combine to produce effective particle removal in a clean filter bed, these retained particles can serve as filter media during later stages of the run (Fig. 10-9). This produces a ripening of the filter; the efficiency of the removal of particles or turbidity improves with time and can be substantially greater than the "clean" bed in place when the filter run began. Particles removed by the filter become sites (collectors or media) for the deposition of additional material. The important result is that in many and perhaps most of the packed-bed filters used in water and wastewater treatment, the actual filter media operative during most of a filter run are not the sand, coal, or other media installed as specified by the designer. Rather, they are the particles present in the water applied to the filter bed and removed in the filter. Stated another way, the actual filter media active during most of a filter run are formed from particles present in the water supply and altered during pretreatment processes and from new particles formed by chemical precipitation during pretreatment. Hence the "design" of filter media includes the selection of the raw water source and the design of pretreatment facilities that are installed ahead of the filters. The design of filter media is established "upstream" of the filters by the concentration, size, and surface properties of the particles applied to filters and by the solution chemistry of the aqueous phase.

Particle Removal Theories

Theories or models for particle removal by media filters can be classified into two types of approaches: macroscopic (or phenomenological) and microscopic (or fundamental). The former do not explicitly consider the actual characteristics of the media filter (grain size, shape, etc.) or the particles being removed (i.e., size), whereas the latter focuses explicitly on particle- and pore-scale physics and geometry. Both approaches consider the performance of the "clean" filter media (no significant particle deposit) and how that performance changes as particles are deposited within the filter ("dirty" media). Detailed descriptions of these approaches are available in texts by Tien (1989) and Elimelech et al. (1995).

Macroscopic (Phenomenological) Approach. The contributions of Iwasaki and Ives described next are examples of macroscopic approaches. Based on observations, Iwasaki (1937) published an early mathematical model for slow sand filtration. The equations proposed are as follows:

$$\frac{\partial C}{\partial z} = -\lambda C \tag{10-2}$$

$$\frac{\partial C}{\partial t} + \frac{V}{(1-\varepsilon_\sigma)}\frac{\partial C}{\partial z} + \frac{\partial \sigma}{\partial t} = 0 \tag{10-3}$$

$$\lambda = \lambda_o + k_1 \sigma \tag{10-4}$$

where C is the concentration (mass or volume basis) of suspended material at any bed depth or filtration time, z is the bed depth, λ is an empirical coefficient termed the *impediment modulus* or *filter coefficient*, λ_0 is the initial impediment modulus, t is time, σ is the volume of suspended material retained in the bed per unit of filter volume (termed the *specific deposit*), ε_σ is the porosity of the deposited material, V is the approach or superficial filtration velocity, and k_1 is an empirical constant.

Equation 10-2 states that the removal of suspended particles with depth is proportional to the concentration of particles in the water at that depth. A rational basis for this result has been provided by investigators in aerosol filtration (Friedlander, 1958). Equation 10-3 is a volume balance and states (with several assumptions) that the volume of material accumulated within the filter equals the volume removed from suspension. Equation 10-4 expresses Iwasaki's observations that filtration efficiency improved with time for his slow sand filters.

Based on a mass balance across a filter bed, Ives published an extensive series of results building on Eqs. 10-2 to 10-4 proposed by Iwasaki (e.g., Ives, 1960, 1961, 1963). Ives proposed the following relationship to replace Iwasaki's Eq. 10-4:

$$\lambda = \lambda_o + k_2\sigma - \frac{\phi\sigma^2}{\varepsilon - \sigma} \tag{10-5}$$

Here, k_2 and ϕ are empirical coefficients used, respectively, to describe the increase and subsequent decrease in particle removal efficiency with time as the specific deposit increases.

Various other empirical rate expressions have been proposed to relate λ to λ_0 as some function of σ; 10 such expressions have been summarized by Tien (1989). Two examples illustrate the evolution of such equations. Heertjes and Lerk (1967) proposed the following:

$$\frac{\partial \sigma}{\partial t} = k_1(\varepsilon_0 - \sigma)VC \tag{10-6}$$

where ε_0 is the clean bed porosity and k_1 is the empirical attachment coefficient (same as λ_0).

This equation states that the rate of particle deposition at a particular depth in a filter at a given time is proportional to the particle flux ($V \times C$) and the volume remaining available for deposition ($\varepsilon_0 - \sigma$).

Adin and Rebhun (1977) proposed a rate expression that includes a second term to allow for detachment of already deposited particles:

$$\frac{\partial \sigma}{\partial t} = k_1 VC(F - \sigma) - k_2 \sigma J \tag{10-7}$$

where k_2 is the empirical detachment coefficient, F is the theoretical filter capacity (i.e., the amount of deposit that would clog the pores such that no more deposition occurred), and J is the hydraulic gradient, with other terms as defined previously.

The first term is almost identical to the preceding expression, but a new filter capacity term F is proposed rather than the clean bed porosity ε_0. The second term states that the probability of detachment depends on the product of the amount of material deposited already, σ, and the hydraulic gradient J.

The mass-balance equation and one rate expression must be solved simultaneously in order to find C and σ as a function of time of filtration in the cycle and depth in the filter. Various numerical or analytical solutions have been used to solve the equations (Heertjes and Lerk, 1967; Adin and Rebhun, 1977; Saatci and Oulman, 1980; Tien, 1989). One of the analytical solutions (Saatci and Oulman, 1980) was developed based on the close similarity between the Langmuir equation for adsorption with the filtration rate expression.

The use of these phenomenological models depends on the collection of pilot- or full-scale filtration data to generate the appropriate attachment and, in some cases, detachment coefficients for the particular suspension being filtered. This need, plus the complex nature of the solutions, has limited the use of such equations in routine filter design. Nevertheless, such models with the appropriate empirical constants can be used to simulate filtrate quality, specific deposit, and head loss as a function time and depth.

The two rate expressions presented (Eqs. 10-6 and 10-7) draw attention to a long-standing question in filtration research, namely, whether detachment of deposited solids occurs in a filter operating at a constant filtration rate or new influent particles merely bypass previously clogged layers. Evidence indicates that detachment does occur, possibly as a result of impingement of newly arriving particles (Clough and Ives, 1986). This evidence was collected using an industrial endoscope inserted into the filter to magnify, observe, and record the deposition process on videotape. The observations were made while filtering kaolin clay on sand filters but without the use of coagulant. Similar unpublished observations (Ives, 1997) were made while filtering a 100 mg/L kaolin suspension that had been coagulated and flocculated with 25 mg/L of alum [$Al_2(SO_4)_3 \cdot 16H_2O$]. Detachment was observed to occur during constant-rate filtration from the impact of arriving particles on unstable deposits and from hydraulic shear stresses without the influence of arriving particles. Further studies of detachment were presented by Ginn et al. (1992), who indicated that detached aggregates in the filter effluent could distort log-removal results for cyst-sized particles obtained by particle counting. Moran et al. (1993) measured particle size distributions as a function of run time and bed depth; the results supported significant detachment that varied with particle size, showing a maximum for particles of several-micron size. In a controlled laboratory study using several sizes of dyed polystyrene spheres, Kim and Tobiason (2004) provided evidence that for much of a filter run, particles in filter effluents are dominated by detached particles compared with filter influent particles that were never removed.

These types of results, along with rational thinking, make it clear that it is unwise to use measurements of cyst- or oocyst-sized particles in raw water and filter effluent to draw conclusions about the removal of actual protozoan pathogens. Filter influent particles are dominated most often by particles created during treatment, not in the raw water, and filter effluent particles may be detached aggregates of smaller particles. In general, with proper chemical pretreatment, removal performance for actual pathogens is much better than might be inferred erroneously by comparing cyst-sized particles in raw and filtered waters.

Microscopic (Fundamental) Approach. Turning to the microscopic approach, consider a single spherical particle of the filter media. Assume that it is unaffected by its neighbors and fixed in space (Fig. 10-10). This single particle of filter media is termed a *collector*, emphasizing that the ultimate purpose of transporting suspended particles from the bulk flow to the external surfaces of media grains in packed beds is the collection of these particles, thereby accomplishing their removal from the water. The main flow direction is that of gravitational force. A suspended particle following a streamline of the flow may come in contact with the media grain by virtue of its own size (case *A* in Fig. 10-10); this transport process is termed *interception*. If the density of the suspended particle is greater than that of water, the particle will follow a different trajectory because of the influence of the gravitational force field (case *B*). The path of the particle is influenced by the combined effects of the buoyant weight of the particle and the fluid drag on the particle. This transport process is called *sedimentation*. Finally, a particle in suspension is subject to random bombardment by molecules of the suspending medium, resulting in the well-known Brownian movement of the particle (case *C*). The term *convective diffusion* is used to describe mass transport in the flowing fluid by this process.

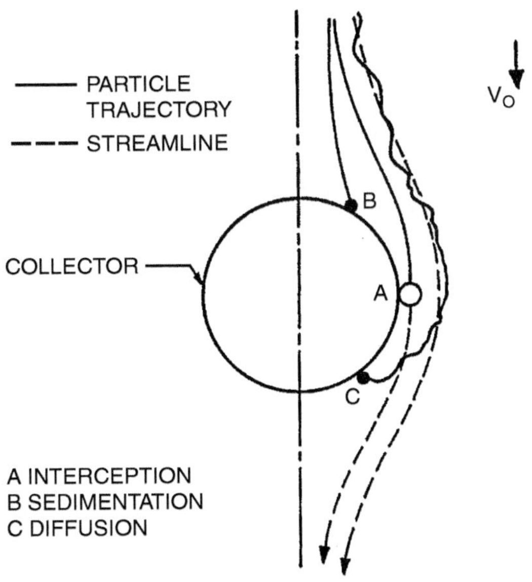

FIGURE 10-10 Basic transport mechanisms in water filtration. (*Source: Reprinted with permission from Yao et al., 1971.*)

Following after Friedlander (1958) and using the *microscopic approach*, Yao et al. (1971) described the temporal and spatial variation of particle concentration around a single spherical media grain (collector) as follows:

$$\frac{\partial C}{\partial t} + V \cdot \nabla C = D_{bm} \nabla^2 C + \left(1 - \frac{\rho}{\rho_p}\right) \frac{mg}{3\pi\mu d_p} \frac{\partial C}{\partial z} \tag{10-8}$$

Again following Friedlander, the following are defined for a single grain of filter media (Yao et al., 1971):

$$\eta_T = \frac{\text{total rate at which particles strike a media grain}}{\text{total rate at which particles approach a media grain}} \tag{10-9}$$

$$\alpha = \frac{\text{total rate at which particles attach to a media grain}}{\text{total rate at which particles strike a media grain}} \tag{10-10}$$

The product $\alpha\eta_T$ is thus the ratio of the flux of particles removed by a single media grain to the flux of particles approaching it. Investigators have had considerably more success describing the transport of particles to media grains (as represented by η_T) than the attachment of these particles to those grains (as described by α, the attachment efficiency factor or sticking probability).

Numerical solutions to Eq. 10-8 have been reported by several authors, notably Yao et al. (1971), Rajagopalan and Tien (1976), and most recently, Tufenkji and Elimelech (2004). The Tufenkji and Elimelech equation follows:

$$\eta_T = 2.4 A_s^{1/3} N_R^{-0.081} N_{Pe}^{-0.715} N_{vdW}^{-0.52} + 0.55 A_s N_R^{1.675} N_A^{0.125} + 0.22 N_R^{-0.24} N_G^{1.11} N_{vdW}^{-0.053} \tag{10-11}$$

The first set of terms on the right-hand side describes transport by convective diffusion, the second by interception, and the third by gravity. The dimensionless groups and parameters in Eq. 10-11 are defined in Table 10-4. Equation 10-11 is based on the Happel sphere-in-cell model of a porous bed and includes the effects of hydrodynamic retardation at small surface separation distances as well as the attractive van der Waals interactions; other repulsive or attractive forces, such as arising from electric double-layer interactions, are not included.

As presented elsewhere (i.e., Yao et al., 1971; Tobiason and O'Melia, 1988), it is instructive to consider the impact of particle size on predicted collision efficiency; Figure 10-11 shows η_T, calculated using Eq. 10-11, as a function of particle size for specified conditions (d_c = 0.5 mm, V = 5 and 20 m/h, T = 20°C, ρ_p = 1050 kg/m^3, ε_0 = 0.4, $H = 1.0 \times 10^{-20}$ J). The plot shows a characteristic minimum in collection efficiency at approximately 1 μm, with collection improving as size decreases due to increased particle diffusivity for submicron particles and collection improving as size increases above 1 μm due to increased interception and sedimentation. Increased velocity decreases collection efficiency, with greater impact for submicron particles. The characteristic minimum in collision efficiency is a universal phenomenon, shown in Chap. 9 of this text for bubble-particle collisions in dissolved air flotation and elsewhere for collisions of aerosols with pine needles and the human nose and lungs. Values of η_T for virus (100 nm), bacteria (1 μm), and cyst (10 μm) sized particles from Fig. 10-11 are 0.0025, 0.00056 and 0.0062, respectively.

The next step in the fundamental approach is to take the results for a single media grain (or collector) as represented by η_T and integrate this over an entire bed of length L. The result is (Yao et al., 1971):

$$\ln \frac{C_L}{C_0} = -\frac{3(1-\varepsilon)}{2d_c} \alpha\eta_T L \tag{10-12}$$

TABLE 10-4 Definitions of Parameters in Eq. 10-11 for Single-Collector Efficiency η_T

Parameter	Definition	Parameter	Definition
N_R	$\frac{d_p}{d_c}$	d_p	Particle diameter
		d_c	Collector (grain diameter)
N_{Pe}	$\frac{Vd_c}{D_\infty}=\frac{3\pi\mu d_c d_p V}{kT}$	D_∞	Particle diffusivity ($kt/3\pi\mu d_p$)
		ε_0	Clean-bed porosity
N_{vDW}	$\frac{H}{kT}$	g	Gravitational acceleration
		H	Hamaker constant
N_A	$\frac{H}{3\pi\mu d_p^2 V}$	k	Boltzmann's constant
		T	Temperature
N_G	$\frac{d_p^2(\rho_p-\rho)g}{18\mu V}$	V	Superficial filtration velocity
		ρ_p, ρ	Density of particle, water
A_S	$\frac{2(1-\gamma^5)}{2-3\gamma+3\gamma^5-2\gamma^6}$	μ	Viscosity of water
γ	$(1-\varepsilon_0)^{(1/3)}$		

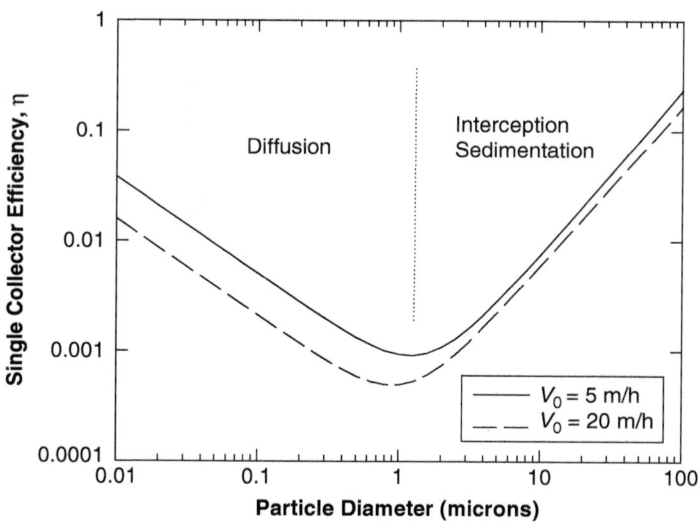

FIGURE 10-11 Calculated clean-bed single-collector removal efficiency as a function of particle size for two filtration velocities according to Eq. 10-11 ($d_c = 0.5$ mm, $V = 5$ and 20 m/h, $T = 20°C$, $\rho_p = 1050$ kg/m^3, $\varepsilon_0 = 0.4$, $H = 1.0 \times 10^{-20}$J).

where C_L and C_0 are the filtrate and influent concentrations of particles, L is the depth of the filter bed, ε is the porosity of the clean bed, and d_c is the diameter of the media (collector) grains. Equation 10-12 can be used to examine the impact of several key variables on particle removal. Figures 10-12 and 10-13 show plots of fraction of influent particles removed (or percent removal) and the negative of the log of the fraction remaining in the filter effluent (the log removal) as a function of particle size for specified other conditions. Figure 10-12 has results for a typical single-media sand filter with a 0.60 m (2.4-ft) depth of 0.5 mm diameter sand operated at a filtration rate of 5 m/h (2 gpm/ft^2) with a porosity of 0.4. Results for two values of the attachment efficiency are shown: $\alpha = 1$ for complete destabilization (good coagulation) and $\alpha = 0.01$ for somewhat stable particles (poor coagulation). The minimum in removal for the approximately 1 μm particle size is again apparent, as is the very significant detrimental impact of poor coagulation. Laboratory studies have demonstrated experimentally the effects of particle size and chemistry that are consistent with the theory (Yao et al., 1971; Fitzpatrick and Spielman, 1973; Tobiason and O'Melia, 1988; Tobiason et al., 1993). One of the most important points from Fig. 10-12 is that if poor removal of submicron particles, such as viruses, is observed, it is not due to their size; most likely surface and bulk chemistry conditions result in the sticking probability α being low for these particles.

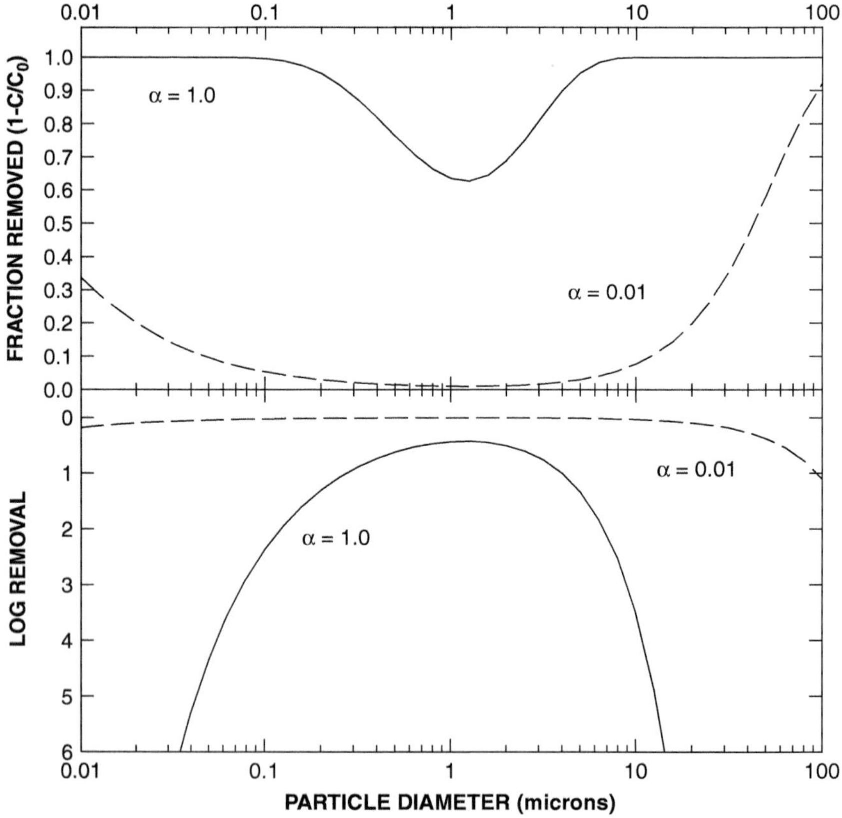

FIGURE 10-12 Effect of particle size on predicted clean-bed removal for typical depth (0.6 m) of typical sand (0.5 mm) media at a low filtration rate (5 m/h) for excellent ($\alpha = 1$) and poor ($\alpha = 0.01$) coagulation (calculated using Eqs. 10-11 and 10-12; other parameters are the same as for Fig. 10-11.)

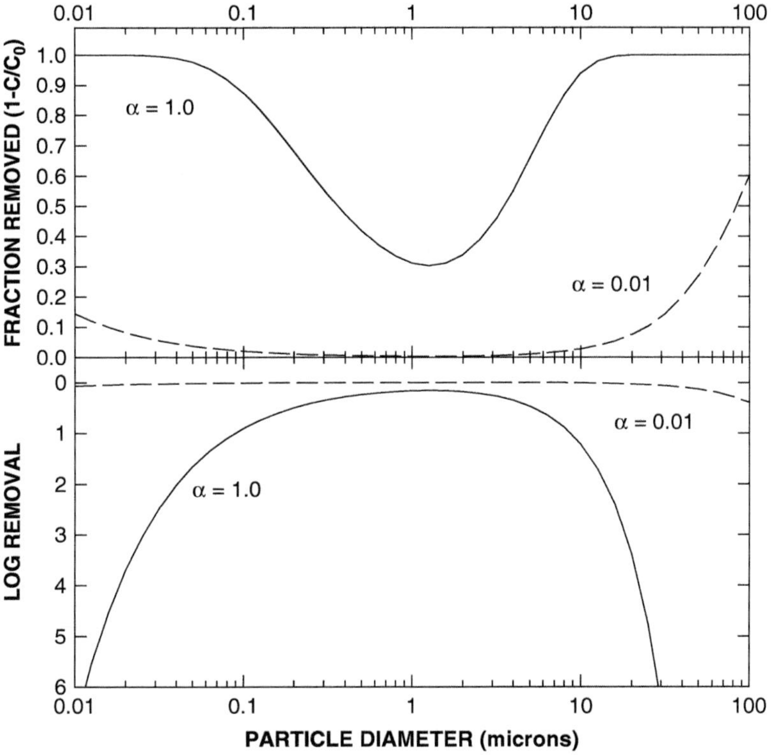

FIGURE 10-13 Effect of particle size on predicted clean-bed removal for deep bed (1.5 m) of coarse (1.0 mm) media at a high filtration rate (15 m/h) for excellent (α = 1) and poor (α = 0.01) coagulation (calculated using Eqs. 10-11 and 10-12; other parameters are the same as for Fig. 10-11.)

Figure 10-13 has results for a deeper filter (1.5 m or 5.9 ft) of coarser media (1 mm diameter) operated at a higher filtration rate (15 m/h or 6 gpm/ft^2). Predicted particle removal is worse than calculated for the shallower bed of finer-grained media operated at a lower rate despite the L/d_{10} ratio of 1500 for the deeper bed versus 1200 for the shallower bed. This result and the variables on the right-hand side of Eq. 10-12 show that maintaining or even increasing somewhat the L/d ratio for higher-rate filters of coarser media does not provide equal particle removal. According to Eq. 10-12, it is the value of $(\eta_T L)/d_c$ that should remain constant (assuming constant sticking probability). The collision efficiency decreases significantly as HLR (or V) and media diameter, d_c, increase; for the interception mechanism of particle removal η_T is approximately proportional to d_c^{-2}, so the value of L/d_c^3 should be maintained, not L/d_c. Lawler and Nason (2006) present the concept of a characteristic removal length and show the need for increasing the L/d_c ratio as media size and filtration velocity increase. The preceding analysis suggests a risk of less than desired particle removal for high-rate, coarse-media filters, especially if no layer of finer sand media is placed below the deep-bed of coarse media; Kawamura (2001) recommends use of 0.3 m (1 ft) of sand if the media effective size exceeds 1.5 mm. Modeling and laboratory results presented by O'Melia and Shin (2001) demonstrated the importance of the layer of smaller-size media below a layer of coarser media in maintaining desired particle removal.

Combined Microscopic/Macroscopic Models for Dirty Media. The fundamental/microscopic approach to filtration theory has had most success in describing clean-bed or initial particle removal. The changes in particle removal and thus within-media particle deposition over time and bed depth have to date proved too complex to model in a rigorous manner on the scale of particles and pores. Deposited particles alter fluid streamlines and provide greatly increased probability for being contacted by a subsequent suspended particle—hence the observed deposition in Fig. 10-9 for aerosols. Some models that have been developed based on a combination of the fundamental approach and the empirical or phenomenological approach are mentioned here.

O'Melia and Ali (1978) presented a model for filter ripening, or improved particle removal caused by previously retained particles. Removal by the original media was modeled with the Rajagopalan and Tien (1976) equation for single-collector efficiency. Deposited particles were accounted for and allowed to act as collectors for additional particles. Chemical effects were accounted for by attachment efficiency factors, and another empirical factor was used to describe the fraction of the deposit that was able to collect other particles. The model was fit to filter-ripening data from laboratory experiments with suspensions of monodisperse polystyrene particles. The model was subsequently incorporated into particle-based models for combined flocculation, sedimentation, and filtration performance (Lawler et al., 1980) and cost-optimal treatment plant design (Wiesner et al., 1987b). Other researchers subsequently have expanded the hybrid microscopic/macroscopic approach to include limits on maximum deposition (ultimate deposit) as well as particle detachment (e.g., Vigneswaran and Tulachan, 1988; Tobiason and Vigneswaran, 1994). While the modeling work, with a special emphasis on suspended particle size and attachment efficiency, has not resulted in significant advances in design, it has allowed for rational interpretation of filter performance, as measured by turbidity and particle counting. For example, observations of faster ripening based on greater than two micron diameter particle counts than that based on turbidity (which is more affected by 0.5 to 1 μm particles) is expected based on fundamental theory at the particle and pore scale. Additional insight regarding the role of particle size on filtration performance is gained by considering the morphology of deposited particles based on fundamental concepts of fractal geometry (Veerapaneni and Wiesner, 1997; Vigneswaran and Tobiason. 1998).

Head Loss Theories and Models

Theories for the loss of head (or pressure) when water flows through a media filter include fundamental and well-established work for clean media and much more empirical (or phenomenological) approaches for head loss development during the filter run. Both are described below.

Clean Bed Head Loss. The head loss (i.e., pressure drop) that occurs when clean water flows through a bed of clean filter media can be calculated from well-known equations. The flow through a clean filter of ordinary grain size (i.e., 0.5–1.0 mm) at ordinary filtration velocities (5–12 m/h or 2–5 gpm/ft^2) would be in the laminar range of flow depicted by the Kozeny equation (Fair et al., 1968) that is dimensionally homogeneous (i.e., any consistent units may be used that are dimensionally homogeneous*):

$$\frac{h}{L}=\frac{k\mu}{\rho g}\frac{(1-\varepsilon)^2}{\varepsilon^3}\left(\frac{a}{v}\right)^2 V \tag{10-13}$$

*Units are not presented for all dimensionally homogenous equations in this chapter.

where h is the head loss in depth of bed L, g is the acceleration of gravity, a/v is the grain surface area per unit of grain volume (i.e., specific surface $S_v = 6/d$ for spheres and $6/\psi d_{eq}$ for irregular grains), d_{eq} is the grain diameter of a sphere of equal volume, μ is the absolute viscosity of the fluid, ρ is the mass density of the fluid, k is the dimensionless Kozeny constant, commonly found to be close to 5 under most filtration conditions (Fair et al., 1968), with the other variables as previously defined.

The Kozeny equation is generally acceptable for most filtration calculations because the Reynolds number (Re) based on superficial velocity is usually less than 3 under these conditions, and Camp (1964) reported strictly laminar flow up to Re of about 6, where Re is

$$\mathrm{Re} = \frac{V d_{eq} \rho}{\mu} \tag{10-14}$$

The Kozeny equation can be derived from the fundamental Darcy-Weisbach equation for flow through circular pipes:

$$h = \frac{f L U^2}{D(2g)} \tag{10-15}$$

where f is the friction factor, a function of pipe Reynolds number and pipe roughness, D is the pipe diameter, and U is the mean flow velocity in the pipe.

The derivation is achieved by considering flow through porous media analogous to flow through a group of capillary tubes of hydraulic radius, r (Fair et al., 1968). The hydraulic radius is approximated by the ratio of the volume of water in the interstices per unit bed volume divided by the grain surface area per unit bed volume. If N is the number of grains per unit bed volume, v is the volume per grain, and a is the surface area per grain, then the bed volume is $Nv/(1-\varepsilon)$, the interstitial volume is $Nv\varepsilon/(1-\varepsilon)$, and the surface area per unit bed volume is the product of N times a, leading to $r = \varepsilon v/(1-\varepsilon)a$. The following additional substitutions are made: $D = 4r$, U = interstitial velocity = V/ε, $f = 64/\mathrm{Re}'$ for laminar flow, and $\mathrm{Re}' = 4(V/\varepsilon)r\rho/\mu$ is the Reynolds number based on interstitial velocity.

For larger filter media, where higher velocities are used in some applications or for velocities approaching fluidization (as in backwashing considerations), the flow may be in the transitional-flow regime, where the Kozeny equation is no longer adequate. Therefore, the Ergun equation (Ergun, 1952a), Eq. 10-16, should be used because it is adequate for the full range of laminar, transitional, and inertial flow through packed beds (Re from 1–2000). The Ergun equation includes a second term for inertial head loss.

$$\frac{h}{L} = \frac{4.17\mu}{\rho g}\frac{(1-\varepsilon)^2}{\varepsilon^3}\left(\frac{a}{v}\right)^2 V + k_2 \frac{(1-\varepsilon)}{\varepsilon^3}\left(\frac{a}{v}\right)\frac{V^2}{g} \tag{10-16}$$

Note that the first term of the Ergun equation is the viscous energy loss that is proportional to V. The second term is the kinetic energy loss that is proportional to V^2. Comparing the Ergun and Kozeny equations, the first term of the Ergun equation (viscous energy loss) is identical to the Kozeny equation except for the numerical constant. The value of the constant in the second term, k_2, was reported originally to be 0.29 for solids of known specific surface (Ergun, 1952a). In a later paper, however, Ergun reported a k_2 value of 0.48 for crushed porous solids (Ergun, 1952b), a value supported by later unpublished studies at Iowa State University. The second term in the equation becomes dominant at higher flow velocities because it is a square function of V. The Kozeny equation, however, is more convenient to use and is quite acceptable up to Re = 6.

The ability to calculate head loss through a clean fixed bed is important in filter design because provision for this head must be made in the head loss provided in the plant;

increased clean-bed head loss for a fixed total available head loss means that there is less available clogging head loss. As is evident from Eqs. 10-13 or 10-16, the head loss for a clean bed depends on the hydraulic loading rate V, grain size, water viscosity, bed porosity, and grain sphericity. Higher HLRs are often accompanied by larger media (and greater bed depth) in part to address head loss issues. In northern climates, water viscosity can change by a factor of 2 from summer to winter, which means that designers must consider the cold water viscosity as a design constraint.

Example 10-1 Head Loss Calculations
Calculate the head loss for the 0.91 m (3 ft) deep bed of the filter sand shown in Fig. 10-4 at a filtration rate of 14.7 m/h (6.0 gpm/ft^2) and a water temperature of 20°C using a grain sphericity of 0.75 and a porosity of 0.42 estimated from Table 10-2. Sphericity and porosity can be assumed constant for the full bed depth.

Solution Because the sand covers a range of size and will be stratified during backwashing, divide the bed into five equal segments, and use the middle sieve opening size for the diameter term in the solution. Solving the Kozeny equation in SI units:

$$\frac{h}{L} = \frac{k\mu}{\rho g}\frac{(1-\varepsilon)^2}{\varepsilon^3}\left(\frac{a}{v}\right)^2 V$$

$$\frac{a}{v} = \frac{6}{\psi d} = \frac{6}{0.75d} \qquad (10\text{-}13)$$

$$\frac{\mu}{\rho} = \nu = 1.004 \times 10^{-6}\ \text{m}^2/\text{s at } 20°\text{C}$$

$g = 9.81$ m/s^2
k = Kozeny's constant, which is typically 5 for common filter media
$V = 14.7$ m/h $= 4.08 \times 10^{-3}$ m/s (0.0134 ft/s)

So

$$\frac{h}{L} = 5\frac{1.004\times10^{-6}}{9.81}\frac{(1-0.42)^2}{0.42^3}\left(\frac{6}{0.75d}\right)^2 4.08\times10^{-3} = \frac{6.067\times10^{-7}}{d^2}$$

From Fig. 10-4, select mid-diameters and calculate h/L for each selected mid-diameter:

Size	Mid-diameter (mm)	$\frac{h}{L}$	Layer depth (m)	h(m)
d_{10}	0.54	2.08	0.18	0.38
d_{30}	0.66	1.39	0.18	0.25
d_{50}	0.73	1.14	0.18	0.21
d_{70}	0.80	0.95	0.18	0.17
d_{90}	0.87	0.80	0.18	0.15
	Average	1.27	Total	1.16

Alternatively, because each layer was the same depth, the average h/L can be used to calculate the head loss for the 0.91 m (3.0 ft) bed of sand, $L = 0.91$ m, average $h/L = 1.27$, $h = 1.27 \times 0.91 = 1.16$ m (3.81 ft) of water. ▲

Head Loss Development During Filter Run. As filtration progresses and solids are deposited within the void spaces of the bed, the porosity of the bed decreases, causing a significant increase in total head loss because of the increasing clogging head loss. Head (or pressure) must be provided in the plant design for both the clean-bed and clogging head loss components. The clogging head loss to be provided is usually based on prior experience for similar waters and treatment processes or on pilot studies because there are no universal fundamental theories or models for clogging head loss. A brief discussion of some empirical modeling work follows.

A general equation for head loss development can be written as

$$h(t,z) = h_0 + fn[\sigma(t,z)] \tag{10-17}$$

where $h(t, z)$ is the total head loss, h_0 is the clean-bed head loss, and the second term is a function of the amount (and character) of specific deposit σ. The amount of specific deposit is directly proportional to the solids loading to the filter (VC) and the particle removal efficiency. The specific deposit varies across the filter depth as well as during run time. In addition, the character of the specific deposit (shape, volume, and strength) is affected by the nature of the deposited particles and physical variables (media size and HLR). Thus fundamental modeling is challenging. Laboratory and field results often show a linear rate of head loss development with time at constant HLR. In such a case, the function in Eq. 10-17 is a constant times run time; the constant is a rate of head loss increase, sometimes expressed in centimeters per hour (inches per hour). Studies show that the value of this constant varies with many factors, including type of raw water, type and dosage of coagulant(s) and filter aid, pretreatment and filtration design, and operating variables.

Results of laboratory studies with model particles have shown dramatic impacts of particle size on head loss development; for the same suspended particle mass concentration, submicron particles cause much greater head loss than larger particles (Habibian and O'Melia, 1975; Tobiason et al., 1993). Researchers have attempted to describe this impact based on greater exposed surface area for smaller particles (O'Melia and Ali, 1978) and on greater deposit volume (more pore clogging) as a result of the fractal nature of the deposit (Veerapaneni and Wiesner, 1997). Implications for treatment application are that optimal pretreatment for filtration, especially direct filtration, may be based on increasing particle size via flocculation to decrease head loss development; a laboratory study (Westerhoff and Tobiason, 1991) and pilot results for in-line versus direct filtration (Glaser and Edzwald, 1979; Edzwald et al., 1987) support this concept.

RAPID-RATE FILTER PERFORMANCE

Effluent Quality During a Filter Run

General Pattern. As shown generically in Fig. 10-3, the effluent quality for an individual filter often varies during the run cycle. Quality typically is poorer at the beginning or ripening period of the filtration run cycle, is relatively stable and excellent for a prolonged period, and may deteriorate near the end of the cycle if the cycle is prolonged for a sufficient period of time. The particles removed in the filter are held in equilibrium with the hydraulic shearing forces that tend to tear them away and carry them deeper into or through the filter. Initially, deposited particles acting as collectors cause improved removal performance with time. As deposits increase, the velocities through the more nearly clogged upper layers of the filter increase, and eventually, these layers become less effective in removal. The burden of removal passes deeper and deeper into the filter (Eliassen, 1941; Stanley, 1955; Ling, 1955). Ultimately, adequate bed depth with removal capacity is not available, so filter effluent quality deteriorates (referred to as *breakthrough*), and the filter must be backwashed. As indicated elsewhere, it is important to note that some filter influent particles pass through the filter throughout the filter cycle and that many filter effluent particles are very likely to be particles that have detached from the media after prior attachment, especially as the filter run progresses (Kim and Tobiason, 2004). In addition, if the filtration rate of a filter that contains deposited solids is increased, the hydraulic shearing forces also increase, causing some particles to detach, and these may appear in the filter effluent. This aspect of filtration is discussed in detail later.

Initial Performance—Filter Ripening. Cleasby and Baumann (1962) presented detailed data on filtered water quality early in the filter run observed during filtration of precipitated iron (Fig. 10-14). The filtrate was observed to deteriorate for a few minutes and then improve over about 30 minutes before reaching the best level of the entire run. Thereafter, as the 16 to 18 hour filter cycle progressed, steady deterioration of filtrate quality occurred at the highest filtration rate of 15 m/h (6 gpm/ft^2) but did not occur at HLRs of 5 and 10 m/h (2 and 4 gpm/ft^2), nor in a constant head loss run, where the filtration rate declined from 15 m/h (6 gpm/ft^2) at the beginning of the run to about 10 m/h (4 gpm/ft^2) at the end of the run.

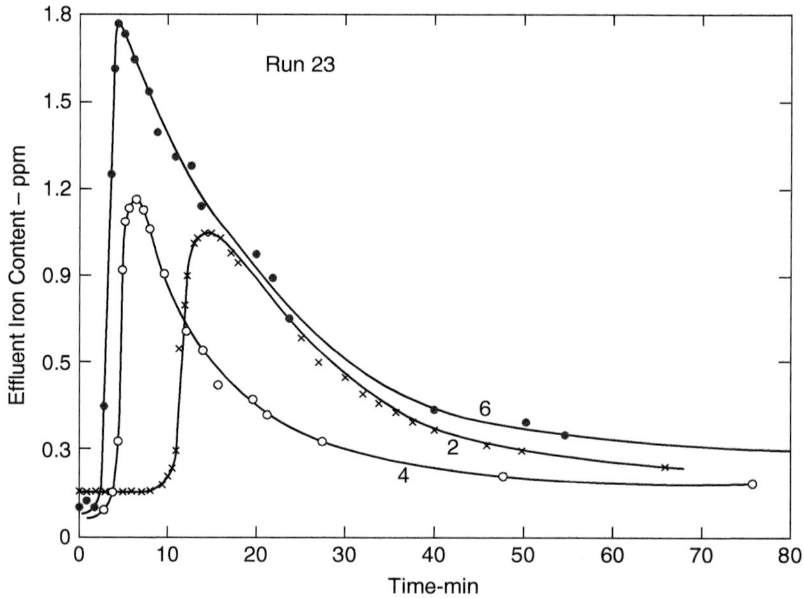

FIGURE 10-14 Initial effluent improvement on filter startup after backwash. Numbers by curves indicate filtration rate in gallons per minute per square foot. (*Source: Cleasby and Bauman, 1962.*)

The initial peaks in Fig. 10-14 occurred at approximately the theoretical hydraulic detention time of the filter plus the detention time of the effluent appurtenances to the point of turbidity monitoring. Based on the work of Amirtharajah and Wetstein (1980), it is useful to consider a generic conceptual plot of filter effluent quality versus volume of water filtered per unit area of filter, as shown in Fig. 10-15. After backwash ends, filter effluent is first the clean backwash supply water from the underdrain system, followed by backwash remnant water from within the pores of the filter media, then backwash remnants above the media bed, and finally, new filter influent water. The actual pattern of filter effluent quality for a specific filter is site-specific and depends on many factors, including details of the backwashing process (i.e., duration, flow rates, and use of air scour), chemistry of the backwash water (i.e., presence of chlorine or phosphate, pH level, and addition of coagulants), method of filter box refilling (i.e., with backwash versus influent), and approach to filter startup (i.e., resting of the filter and initial rate of filtration). It is recommended that operators determine the volume for each component of the filter box (e.g., as in Tobiason et al., 1996) and analyze initial filter effluent, as shown in Fig. 10-15, in order to understand sources of filter effluent particles and thus inform possible improvement strategies. For example, Amburgey et al. (2003) showed that the initial ripening peak could be eliminated by ending the backwash with a period of subfluidization flow to remove within media remnant particles after the media had returned to their filtration-mode packing. Colton et al. (1996) showed that a no-flow resting period after backwash also could decrease the turbidity ripening peak significantly.

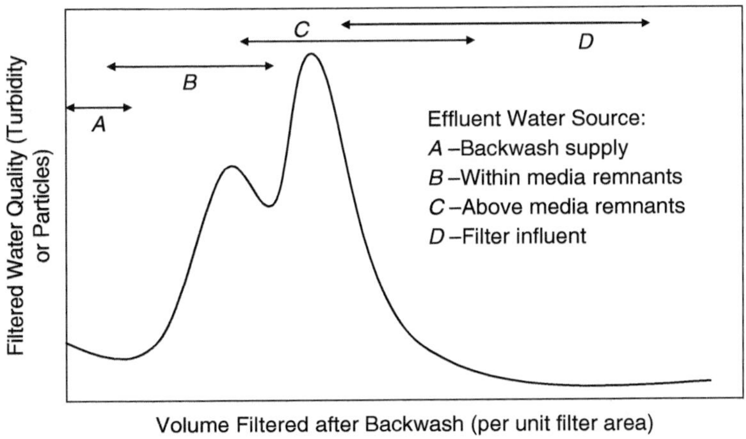

FIGURE 10-15 Filtrate quality during startup after backwash.

The initial filtrate quality degradation period also has been demonstrated based on levels of particles and microorganisms. For example, in a study by Logsdon et al. (1981b), *Giardia muris* (used as a model for the human pathogen *G. lamblia)* was spiked into a low turbidity surface water, coagulated with alum alone or alum and nonionic polymer, flocculated, and filtered through granular media filters. Initial concentrations of cysts in the filtrate were from 10 to 25 times higher than those following the initial improvement period, even though the turbidity improved less than 0.1 ntu during the period. Results from this and other studies were the basis for the USEPA to determine that the poorer initial quality of individual filter effluent was a significant health risk. The 1998 IESWTR established monitoring requirements for individual filter effluent quality during the first 4 hours of operation in addition to the requirement of less than 0.3 ntu in the combined filter effluent for more than 95 percent of measurements; LT1ESWTR and LT2ESWTR maintain these requirements, and the LT2ESWTR identifies the possibility of additional *Cryptosporidium* log removal credit for achieving lower effluent turbidity (see Chap. 1 and earlier in this chapter).

The Practice of Filtering to Waste. In conventional water treatment practice, passage of turbidity during the initial degradation and improvement (i.e., ripening) period is relatively small when averaged over the entire filter run and compared with the total turbidity passed during the run. For many years, little attention was paid to the effect of the initial period on the average filtrate quality on the assumption that disinfection would control pathogenic microorganisms that passed through the filters. Consequently, based on this assumption, the early-twentieth-century practice of filtering to waste at the beginning of the filter cycle was largely abandoned; many plants were constructed without the valves and piping to allow for filtering to waste. However, interest in filtering to waste increased in the latter part of the twentieth century because of increasing outbreaks of giardiasis and cryptosporidiosis caused by protozoan cysts that are very resistant to chlorine disinfection. Provision of piping to be able to conduct a filter-to-waste period is now strongly encouraged because of the low infective dose for these protozoan pathogens (see Chaps. 2 and 18). Analysis of initial filtrate quality, as shown in Fig. 10-15, can be useful in determining the appropriate filtrate volume to waste based on filtrate quality; selection of an arbitrary filter-to-waste fixed time period without such analysis may compromise the quality improvement objective of this practice or be wasteful of excellent quality filtrate.

Steady State Period and End-of-Run Breakthrough. For conditions of proper pretreatment and constant hydraulic loading, it is common to achieve relatively long periods (e.g., 24–72 hours) of very low and constant effluent turbidity (e.g., <0.1 ntu) in media filter effluent. Early work on surface water treatment by Robeck et al. (1964) showed that for strong floc conditions, the filtrate quality was nearly constant after the initial improvement period, whereas for weak floc conditions, terminal turbidity breakthrough was rather abrupt and occurred at low head loss. In other cases, however, the filtrate quality may improve throughout the run, and no terminal breakthrough is observed (Cleasby, 1969b). Effluent turbidity versus run-time results for four pilot filters are shown in Fig. 10-16; specifications for the filter media are shown in Table 10-5 [see Edzwald et al. (1999) for study details]. These data illustrate how HLR and media configuration may affect effluent turbidity. The typical dual-media filter operated at 10 m/h (4 gpm/ft^2) had constant turbidity and no breakthrough over a 24-hour period, whereas three filters of deeper and coarser media operated at 20 m/h (8 gpm/ft^2) HLR had similar steady state effluent turbidity and showed breakthrough in an expected sequence based on media size and configuration (largest single-media first, then finer single-media, then dual-media with a finer lower sand layer).

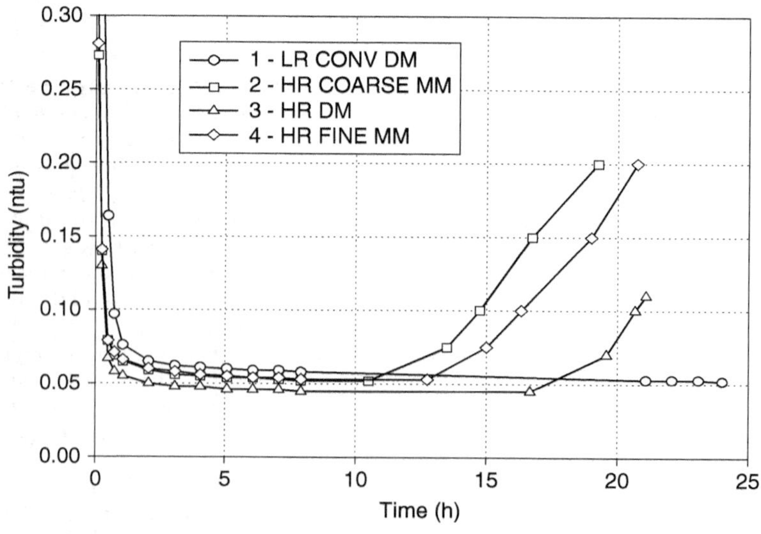

FIGURE 10-16 Turbidity profiles for four pilot filters after alum coagulation; dissolved air flotation prior to filters (see Table 10-5 for filter media descriptions).

TABLE 10-5 Specifications for Pilot Filter Media for Data in Figs. 10-16, 10-20, and 10-21

Filter	Media type(s)	Media depth(s), cm	Effective size, mm	Uniformity coefficient	*L/d* ratios
1. LR CONV DM	Anthracite	60	0.90–1.00	1.4–1.5	630
Conventional dual	Sand	30	0.50–0.55	1.3–1.4	570
media, 10 m/h					1200 (total)
2. HR COARSE MM	Anthracite	200	1.25–1.35	<1.5	1540
Deep-bed, coarser single media, 20 m/h					
3. HR DM	Anthracite	170	1.25–1.35	<1.5	1310
Deep-bed, dual media,	Sand	30	0.55–0.65	1.3–1.4	500
20 m/h					1810 (total)
4. HR FINE MM	Anthracite	200	1.05–1.15	<1.5	1820
Deep-bed, finer single media, 20 m/h					

Source: Edzwald et al., 1999.

The Value of Continuous Turbidity Monitoring and Particle Counting. From the foregoing discussion of effluent quality patterns, the value of providing continuous monitoring of filtrate quality for each individual filter becomes apparent. As noted earlier, it is useful to observe the length of the initial filtrate quality improvement period to guide the duration of filtering to waste. Continuous monitoring is also especially effective for detecting the onset of end-of-run breakthrough. This is especially important because research has shown that along with increased turbidity at breakthrough, even more significant passage of specific particulate contaminants occurs. Robeck et al. (1964) showed increases in filter effluent viruses at breakthrough, whereas Logsdon et al. (1981a) demonstrated this for asbestos fibers. In another study, Logsdon et al. (1985) showed that a threefold end-of-run change in turbidity for an alum-coagulated and settled surface water corresponded to a 30- to 40-fold change in the effluent concentration of *Giardia* cysts. Similarly, Huck et al. (2002) showed significantly decreased log removal of *C. parvum* oocysts at onset of breakthrough and breakthrough conditions.

Important fundamental differences between measurement of turbidity and optical blockage based particle counts are described in Chap. 3. Filter effluents with very low concentrations of greater than 2 μm size particles that can be detected by a particle counter typically have very low turbidity. Because hydraulic shear is more likely to cause detachment of particles of a few microns in size compared with submicron-sized particles, continuous particle count monitoring data frequently show earlier and more significant relative change indicating end-of-run breakthrough compared with turbidity data; a classic example of this is shown in Fig. 10-17, taken from the work of Huck et al. (2002). Early detection of breakthrough is especially significant in light of the associated increases in filter effluent pathogen levels found in the same study. In general, typical online particle counters can detect increases in concentrations of particles greater than 2 μm in size that may not be associated with significant increases in filtrate turbidity and so can provide plant operators with early warning of incipient turbidity breakthrough. This is especially true for low-turbidity waters below 0.1 ntu because the turbidity measurement is reaching its lower limit of detection. The monitoring of particle counts in filtered waters can improve understanding and control of the filtration process and especially the impacts of plant operation on filtered water quality.

The Critical Importance of Adequate Prefilter Coagulation

Because of the particle removal mechanisms in deep-bed filtration, continuous pretreatment that produces destabilized filter influent particles is absolutely essential to producing

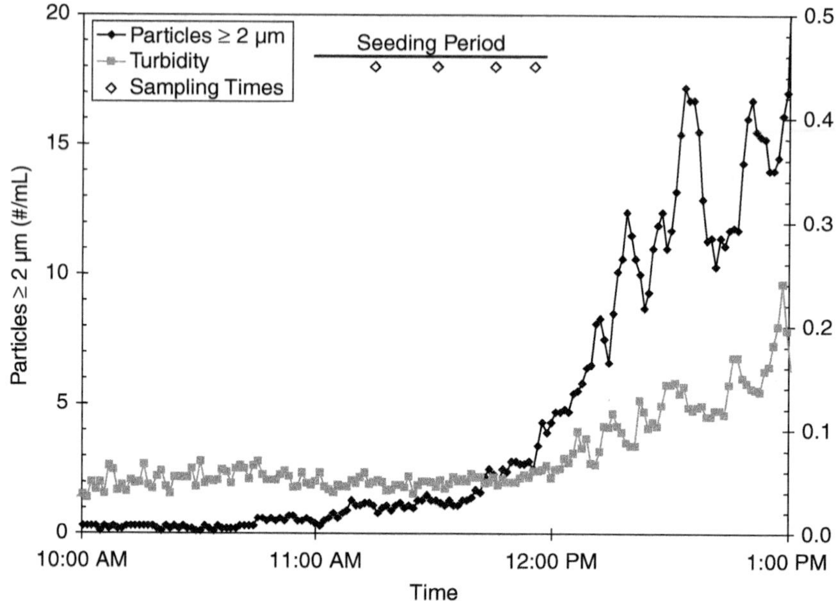

FIGURE 10-17 Turbidity and particle counts during onset of breakthrough for a pilot filter. (*Source: Huck et al., 2002.*)

good-quality effluent from rapid media filters; this is expected based on fundamental filtration theory. Any interruption of good pretreatment (e.g., proper coagulation) results in an almost immediate deterioration of filtrate quality because particle attachment to filter media and previously deposited particles decreases. Long before the development of fundamental theories, Fuller's famous work (1898) recognized this, as evidenced by this quotation from his 1898 report: "In all cases, experience showed that for successful filtration, the coagulation of the water as it enters the filter must be practically complete."

Intermittent causes of inadequate coagulation include failure of coagulant chemical feed pumps and sudden changes in raw water quality without concomitant coagulant dosing changes. The latter case was documented in the Robeck et al. (1964) study at Gaffney, SC. Two filters at 5 and 25 m/h (2 and 5 gpm/ft^2) were operating in parallel and producing about equal turbidity in the filtrate; however, when the quality of the source water suddenly worsened and coagulation was thereby upset, the effluent of both filters deteriorated sharply. The immediate detrimental impact of loss of coagulant dosing on filtrate quality during an in-line direct filtration pilot study was examined by Trussell et al. (1980), as shown in Fig. 10-18. In

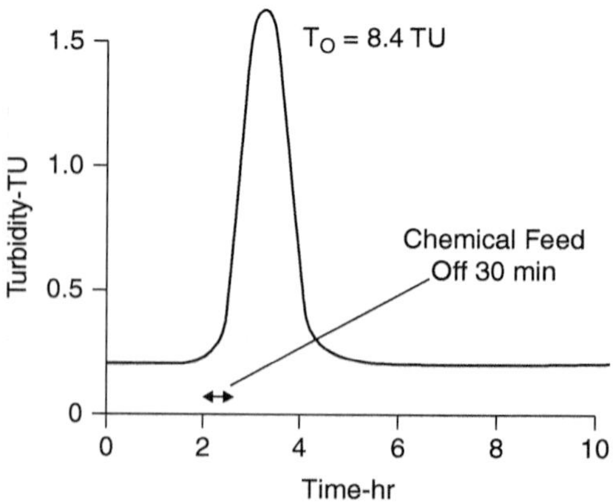

FIGURE 10-18 Influence of loss of chemical feed on filter performance. (*Source: Trussell et al., 1980.*)

this case, the coagulant feed was discontinued for 30 minutes, and filtrate turbidity rose sharply until the chemical feed was resumed.

Most importantly, a number of studies have demonstrated that proper coagulation is the most important factor for effective removal of protozoan pathogens by media filters. Logsdon et al. (1981b) found a 10- to 50-fold increase in effluent *Giardia* cyst concentrations when turbidity increased by only 0.1 to 0.3 ntu following coagulant feed interruption. Patania et al. (1995) showed poor removal of *Cryptosporidium* oocysts and *Giardia* cysts when no coagulant was added during pilot studies of GAC and anthracite/sand dual-media filters. The pilot study by Huck et al. (2002) examined impacts of suboptimal coagulation as well as no coagulation and found dramatic impacts on removal of *Cryptosporidium* oocysts by media filters. Thus all precautions must be taken to ensure that the chemical dosage is appropriate and that the feed is maintained reliably. This is even more critical in direct filtration because of the short detention time ahead of the filters; that is, there is less time to identify and correct a coagulation problem. As noted in Chap. 3, online monitoring of the streaming current after rapid mixing is an excellent method to immediately detect loss of chemical feed or sudden changes in raw water quality. Prefilter treatment processes that affect the size (flocculation) and amount (clarification) of particles in the filter influent also affect filter performance, but mostly with respect to head loss development and the length of a filter run, not the ability of the deep bed to remove particles adequately for some period of time.

Detrimental Effects of Rate Increases and Restarting Dirty Filters

If the rate of filtration is increased on a dirty filter, the equilibrium between the attachment forces holding the solids in the filter and the hydraulic shearing forces tending to dislodge those solids is disturbed. The result is a temporary flushing of solids deeper into the filter and into the filtrate; the effect is more significant if the rate change is sudden and for dirtier filters. The relative effect of rate changes is also affected by the nature of the removed solids, that is, weak floc versus stronger floc, use of filter aids, etc.

Cleasby et al. (1963) showed that the amount of material (iron floc) flushed through the filter was greater for sudden increases in filtration rate than for gradual increases in rate. Also, the amount of material released was greater for large increases than for small increases, but the amount was not affected by the duration of the maximum imposed rate of filtration. The detrimental impact of sudden rate increases on dirty filters also is evident in the work of Tuepker and Buescher (1968) and in the studies of DiBernardo and Cleasby (1980).

Again, the impacts of rate changes on pathogens in filter effluents are of most concern. The studies by Logsdon et al. (1981b) showed that when the filtration rate was increased suddenly from 11 to 27 m/h (4.5–11 gpm/ft^2), the turbidity in the effluent increased 4-fold, whereas effluent concentrations of *G. muris* cysts increased 25-fold.

Significant detrimental filter effluent impacts can occur when starting up a dirty filter after brief idle periods. Trussell et al. (1980) demonstrated this based on particle counts, as shown in Fig. 10-19. In the investigation of the causes of a cryptosporidiosis outbreak in Georgia, Logsdon et al. (1990) reported that restarting dirty filters resulted in filtered water turbidities of 0.5 to 5 ntu that persisted for 1 to 1.5 hours. In contrast, filters operated continuously after backwashing had low turbidities of 0.1 to 0.2 ntu. As a result, it was recommended that dirty filters should not be restarted.

These results demonstrate that filtration rate increases on dirty filters should be avoided or minimized and made very gradually (e.g., over 10 minutes). Routine operation of treatment plants often results in undesired rate increases; overall plant design should seek to minimize the need for this to occur. A common situation is to maintain overall plant production during filter backwashing by increasing the HLR for in-service filters while one filter is taken out of service for backwash. Analysis of effluent turbidity and particle counts versus

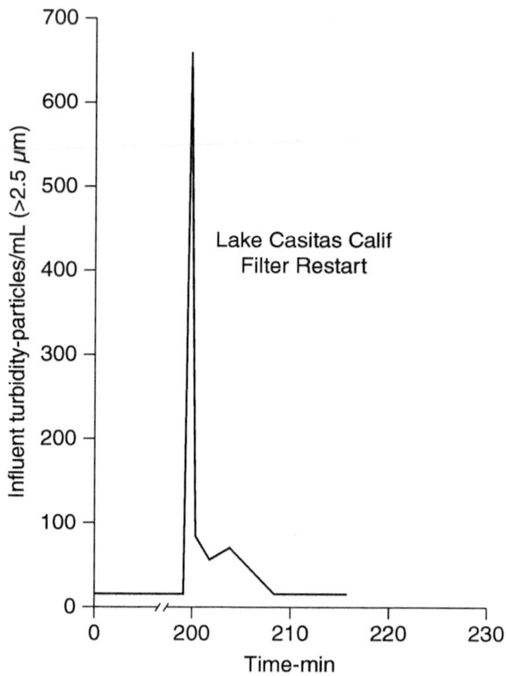

FIGURE 10-19 Influence of stop-start on filter performance. (*Source: Trussell et al., 1980.*)

run time for the in-service filters frequently clearly show when other filters are backwashed by a sudden intermittent increase in these parameters; this is difficult to avoid despite careful rate change control. In some cases, the required filtration rate increase is made more severe if the source of backwash water is a clearwell that must remain at a certain level to meet disinfection requirements. Appropriate provision of storage, a separate backwash source, or allowance for decreased plant production during backwash should be used to minimize required flow rate increases for in-service media filters.

Head Loss Development During a Filter Run

The rate of increase in head loss during the filter run is roughly proportional to the rate of solids capture by the filter. Assuming essentially complete capture of incoming solids, the head loss develops in proportion to the filtration rate V and the influent suspended solids concentration C_0. The rate of head loss development is reduced if the solids capture occurs over a greater depth of the bed rather than in a thin upper layer of the bed. A coarser grain size encourages greater penetration of solids into the bed and possibly different deposit morphology, thereby reducing the rate of head loss development per unit mass of solids captured. As noted earlier, particle size also can affect deposit morphology and density with subsequent impacts on head loss.

The most common head loss pattern encountered in rapid filtration is linear with respect to volume of filtrate or nearly linear, as shown by Cleasby and Baumann (1962) for three constant-rate 5 to 15 m/h (2–6 gpm/ft^2) runs and one declining-rate run for the same filter. This linear head loss would be typical for most alum- or iron-coagulated waters. For

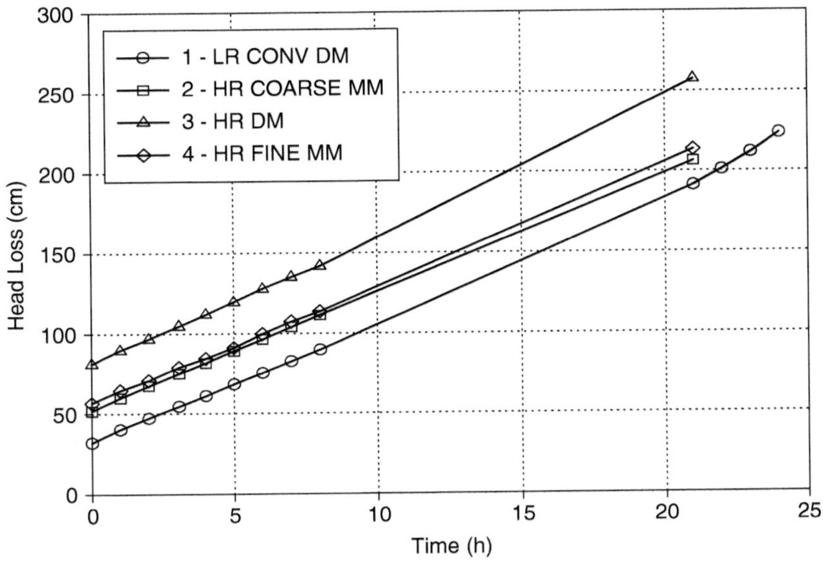

FIGURE 10-20 Head loss development for four pilot filters after alum coagulation; dissolved air flotation prior to filtration (see Table 10-5 for filter media descriptions).

constant-rate filtration, linear head loss increase with filter run time (thus filtrate volume) is typical.

Examples of head loss development for four different pilot filters operated at constant rate are shown in Fig. 10-20; these data correspond to the turbidity data in Fig. 10-16 and media specification in Table 10-5 [see Edzwald et al. (1999) for study details]. The data in Fig. 10-20 reflect the predicted effect of media size and bed depth on clean-bed head loss. The rates of head loss development versus time are almost the same for all four filters despite the two times higher HLR for the three coarser media filters compared with the lower rate for the finer dual-media filter (20 m/h or 8 gpm/ft^2 versus 10 m/h or 4 gpm/ft^2); this means that the higher-rate filters remove twice the amount of suspended solids per unit time (and would have half the head loss rate per unit volume filtered). The reason for this result is the deeper penetration of solids for the coarser-media filters, as shown in the Fig. 10-21 plots of additional (or clogging) head loss development versus run time for different bed depths for the lower-rate, dual-media filter and the higher-rate, coarser-dual-media filter. A much greater fraction of the total head loss occurs in the upper section of the top-layer anthracite for the lower-rate, finer-media filter compared to the higher-rate, coarser-media filter.

An accelerating or exponentially increasing head loss versus unit filtrate volume pattern is observed less commonly. This pattern is caused by partial capture of solids in a surface cake and/or by heavy solids removal within a thin layer of the media at the top (Cleasby and Baumann, 1962). In this case, increasing the filtration rate reduces the exponential tendency by encouraging greater penetration of solids into the bed and increases the production to a given head loss. Quality of filtrate also must be observed, however, if such increases in filtration rate are being considered.

In the absence of plant- or pilot-scale data to assist in predicting development of head loss, experimental values of mass of solids captured per unit filter area per unit increase in head loss sometimes are used for prediction of clogging head loss. This is an admittedly simplistic

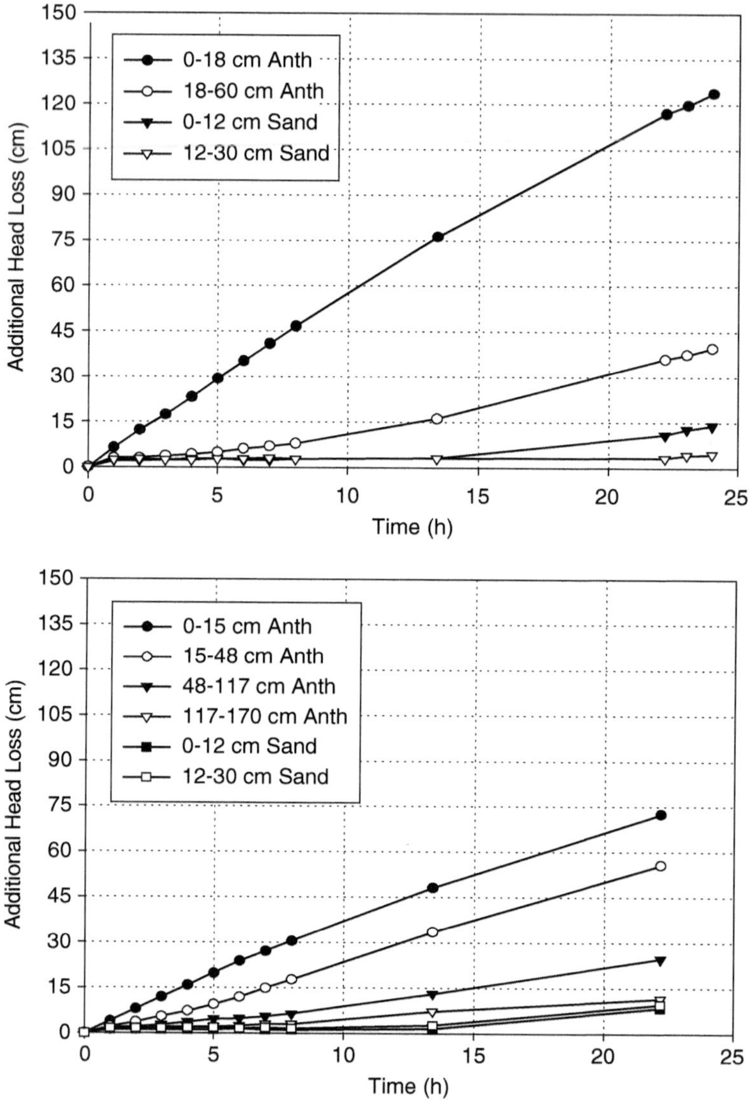

FIGURE 10-21 Clogging head loss development at several bed depths for two different pilot filters [data for filter runs in Figs 10-16 and 10-20 for low-rate (10 m/h) conventional dual media on the top and for high-rate (20 m/h) dual media on the bottom (see Table 10-5 for media descriptions)].

concept. The values used depend on the density of the solids, the ES of the media where flow enters the filter, and the filtration rate. Typical values range from 0.55 to 5.5 kg/m^2/m (0.035–0.35 lb/ft^2/ft) (Montgomery, 1985) and more commonly less than 1.6 kg/m^2/m (0.1 lb/ft^2/ft) for flocculent solids using typical potable water filter media and filtration rates. This approach can be applied only when a linear head loss versus time pattern is expected.

The Detrimental Effects of Negative Head

Some filter arrangements and operating practices can result in pressures below atmospheric pressure (i.e., negative head) in the filter bed during a filter cycle. This can occur in gravity filters when the total head loss at any depth in the filter bed exceeds the static head (i.e., water depth) at that depth; the schematic head profiles in Fig. 10-22 illustrate this concept. Negative head is more likely to occur if gravity filters are operated with low submergence of the media and if the filter effluent exits to the clearwell at an elevation below the top elevation of the filter bed.

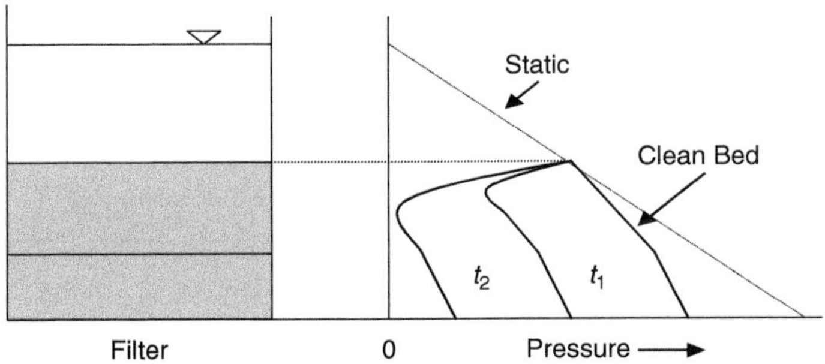

FIGURE 10-22 Head profiles in media filtration; filtration time t_1 is less than time t_2.

Negative head is undesirable because dissolved gases in the influent water may be released in the zone of negative pressure, causing gas bubbles to accumulate during the filter cycle (called *air binding*). Accumulations of gas causes more rapid development of head loss and poorer quality of filtrate because the flow velocity is higher around the gas pockets. Negative head can be avoided completely by terminating the filter run before the total head loss reaches the submergence depth of the media or by causing the effluent to exit at or above the surface of the filter bed.

FLOW CONTROL IN FILTRATION

Why Control Is Needed

The need for some method of filter flow control has been recognized since the early days of potable water filtration. If no attempt is made to control flow, the filters may not share the total plant flow in a reasonably equal manner, or sudden changes in flow rate may occur, both of which could cause filtrate quality to suffer. For these reasons, a number of papers presenting alternative flow control systems have been published (Aultman, 1959; Baylis, 1959b; Hudson, 1959; Cleasby, 1969a, 1981; Arboleda, 1974; Arboleda-Valencia et al., 1985; Cleasby and DiBernardo, 1980; Committee Report, 1984).

Rate Control Systems for Gravity Filters

Filter control systems currently in use can be divided into two categories—mechanical control systems and non-mechanical control systems that achieve control by the inherent hydraulics of the operating filters. Control systems also can be classified as equal-rate

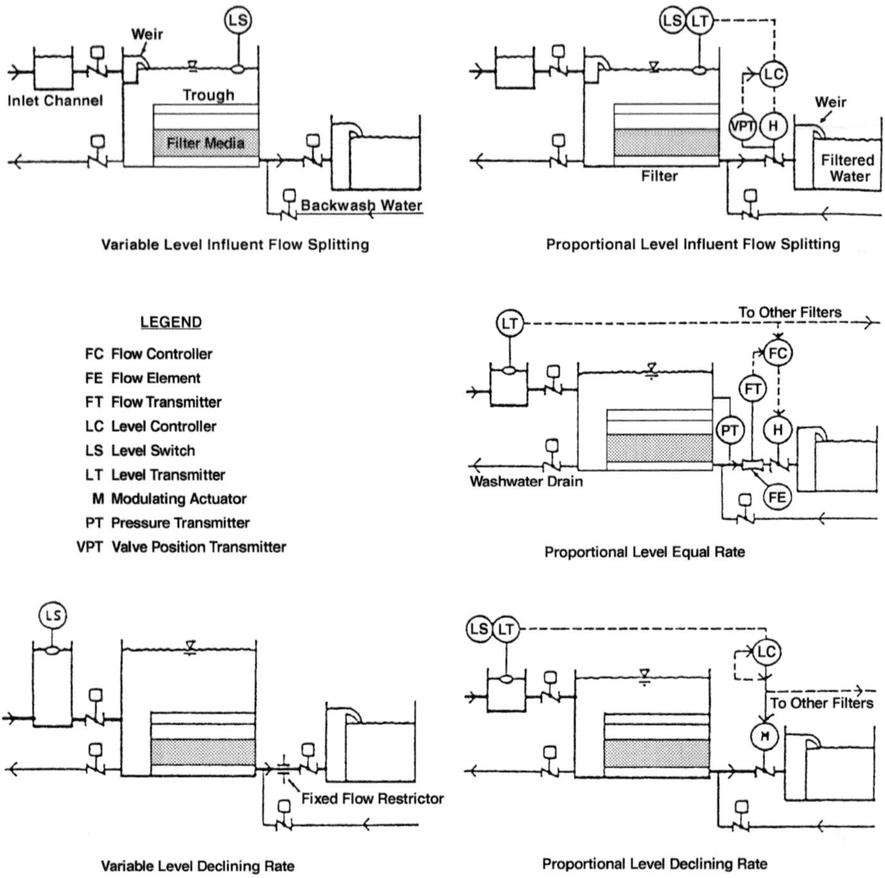

FIGURE 10-23 Schematic diagrams for common gravity filter flow control systems. (*Source: From Browning, 1987, with permission of the Australian Water and Wastewater Association.*)

systems or declining-rate systems, as was done by Browning (1987). Five control systems are shown in Fig. 10-23 using Browning's classification system and nomenclature. They are for *equal-rate systems,* (1) variable-level influent flow splitting, (2) proportional-level influent flow splitting, and (3) proportional-level equal rate, and for *declining-rate systems,* (4) variable-level declining rate and (5) proportional-level declining rate.

It should be noted that none of the systems is called *constant-rate filtration.* True constant-rate filtration can occur only if the total plant flow rate is constant and the number of filters in service is constant. This is seldom the case. Plant flow rates are varied to meet variations in demand. Removing one filter from service for backwashing typically causes the other filters to pick up the load, thereby increasing the rate of filtration on the filters remaining in service. Thus, in most water plants, constant-rate filtration is not possible.

The key features of each of these systems are described briefly in the following paragraphs.

Variable-Level Influent Flow Splitting. This is an equal-rate nonmechanical system in which each filter receives an equal (or near-equal) portion of the total plant flow. This is achieved by splitting the flow by means of an inlet weir box or orifice on each filter inlet above the maximum water level of the filter. The filter effluent discharges to the clearwell at a level above the surface of the filter bed. As solids accumulate in the filter pores, the water level rises in the filter box to provide the head required to drive the flow through the filter. The water level in each filter box is different and depends on the extent to which the filter pores are clogged. When the water level in a filter box rises to some selected maximum level, that filter must be backwashed. If the water level in the filter box is allowed to rise above the inlet weir box or orifice, flow in that filter becomes declining-rate flow.

No instrumentation is required for flow rate or head loss measurement on the individual filters. This is the simplest system for operators because the head loss is clearly evident merely from the water level in the filter box, and there is no control equipment or instrumentation to maintain.

Proportional-Level Influent Flow Splitting. This is an equal-rate mechanical system in which flow control occurs by inlet flow splitting, as in the prior system. A flow splitting weir or orifice at the inlet of each filter is located above the operating water level of the filter. However, each filter requires a level transmitter, a level controller, a modulating valve, and a head loss instrument. The modulating valve acts to control the water level in the associated filter, not as a flow rate controller. The effluent modulating valve for a filter is closed sufficiently to impart enough head loss to maintain the appropriate water level in the clean filter at the start of a filter run. The valve is gradually opened as clogging head loss increases during the filter run to continue to maintain the water level in its designated range.

Head loss is monitored for backwash initiation. No flow-measurement instrumentation is required for the individual filters.

Proportional-Level Equal Rate. This has been the most common equal-rate mechanical control system. It splits the total plant flow equally among the operating filters and maintains the water level in the influent channel within a preselected band. Each filter requires a flowmeter, a flow controller, a modulating control valve, and a head loss instrument. A single level transmitter in the inlet channel, serving a bank of several filters, sends an equal signal to the flow controller in each filter that is proportional to the water level within the band and represents the required flow. The flow controllers adjust the modulating valves to make measured flows match required flows.

When everything is working properly, flow rate through each filter is equal; however, equipment malfunction often results in unequal flow that often goes undetected. Head loss is monitored for backwash initiation.

Variable-Level Declining Rate. This is a declining-rate nonmechanical system. Flow enters the filter below the normal water level in each filter and discharges to the clearwell above the level of the filter bed. Because the inlet to the filters is below the normal water level in each filter, all filters connected by a common inlet channel or pipe operate at approximately the same water level and thus have the same total head loss available to the effluent weir level at any instant. Therefore, the cleanest filter operates at the highest filtration rate, and the dirtiest filter operates at the lowest filtration rate.

As solids accumulate in the filter media, the water level rises in all connected filters to provide the head required to drive the flow through the filter media. Water level is monitored at a single location for backwash initiation. In some plants, backwash is initiated when the filtration rate of the dirtiest filter drops to a preselected level or on a time-interval basis. The filtration rate of each filter declines in a stepwise fashion after each backwash. As a clean (backwashed) filter is returned to service, it assumes the highest rate of filtration.

A fixed flow restrictor (orifice plate or fixed-position valve) is provided for each filter to limit the starting rate on a clean filter. No instrumentation for flow rate or head loss measurement is required on individual filters; however, operating personnel prefer to have some flow rate indication to help them understand the system and for diagnostic purposes. Compared with other control systems, less total filter system head loss must be provided because head consumed in turbulent head losses early in the filter run is reallocated to clogging head loss as the filtration rate declines (DiBernardo and Cleasby, 1980; Cornwell et al,.1991).

Proportional-Level Declining Rate. Browning (1987) suggested one additional flow control system combining the features of declining-rate with proportional-level control from the level in the filter inlet channel. As in the preceding system, submerged filter inlets result in a common water level in all filters. However, in this system, a modulating valve in the effluent of each filter is controlled by the water level in the inlet channel. All valves receive the same set point and are fully closed when water level in the channel is at the bottom of the band, and they are at their maximum opening when the water level is at the top of the band. This system functions like a variable declining-rate system but with filter water level controlled in a preselected bandwidth. Browning claimed that the depth of the filter box could be reduced with this system compared with the previous declining-rate system.

Common Elements of Filter Control Systems

Some common elements of the various rate control systems should be recognized. The total filter system head provided to operate the gravity filter is the vertical distance from the water level (hydraulic grade line) in the filter inlet conduit to the water level at the downstream control point. The downstream control point can be an overflow weir to the clearwell or an upturned elbow in the clearwell or the level in the clearwell if it is higher.

If the downstream control point is located above the top surface of the filter bed, no possibility of having pressures below atmospheric pressure (i.e., negative head) exists anywhere within the filter media or underdrain system because the static water level is above the surface of the media (consider Fig. 10-22). Also, no possibility of accidentally allowing the filter water level to drop below the top surface of the media (i.e., partially dewatering the filter bed) exists. These advantages are not obtained without cost, however, because all the needed head to operate the filter must be placed above the downstream control level, and this means a deeper filter box.

The three equal-rate systems have the common characteristic that the entire filter system operating head provided, which is essentially constant, is consumed throughout the filter run. The head that is not used by the dirty filter bed is wasted either in the modulating control valve of the mechanical systems or in free fall into the filter box for the influent flow splitting system. The declining-rate systems have a lower total filter system head requirement in the plant hydraulic profile because the total head is reallocated as the filter cycle proceeds, as described earlier.

The two non-mechanical systems (i.e., without proportional level control) have the inherent advantage that sudden changes in filtration rate cannot be imposed on the filter. If the total plant flow is increased or a filter is removed from service, the filters can pick up the load only by changing water level to generate the head needed to accommodate the increased flow. This changing of water level takes time so that the change in filtration can occur slowly and smoothly without mechanical devices.

Slow, smooth changes in filtration rate are also possible in properly designed proportional-level mechanical control systems, as well as in variable level influent flow splitting systems. The rate of change of the modulating control valve position should be proportional to the divergence of the measured variable (flow rate or level) from the desired value. The proportional-level control band should be substantial. The drive for the control valve should be an electric motor rather than a pneumatic or hydraulic valve because the latter two drives are prone to sticking and under- or overshooting (i.e., hunting, if repeated) problems as they age. No industry standards for filter control systems are available, so some systems do not have all the attributes just listed.

Choice of the Appropriate Control System

With the foregoing information at hand, some generalizations are possible about the five systems that may affect the choice of control system to meet the desired goals and needs of the utilities. Each system has a place, and a particular system should be used only when conditions are appropriate.

If full automation of the plant is desired, proportional-level systems are favored, although the two non-mechanical systems can be partially automated. If minimizing mechanical equipment and instrumentation is desired, nonmechanical systems are favored. This is especially important when equipment and future repair parts must be imported to a developing country that has scarce foreign exchange. In small, nonautomated plants with four or fewer filters and with unskilled operators, the variable-level influent flow splitting system is ideal because of its simplicity of understanding. In larger, nonautomated plants, the variable-level declining-rate method may be favored because it requires less total filter system head to generate equal filter cycle lengths and because it produces better filtrate quality in some cases (DiBernardo and Cleasby, 1980); however, this is not ensured in all cases (Hilmoe and Cleasby, 1986: Cornwell et al., 1991).

There is legitimate concern about poorer filtered water quality during the ripening phase at the beginning of the filter cycle. This concern is valid for all the control systems discussed earlier, and provision of filter-to-waste is the best available solution, as discussed elsewhere.

The design of variable-level declining-rate filters, including selection of the required total filter system head loss and prevention of excessive starting filtration rate for a clean declining-rate filter, has been presented by Cleasby (1993).

BACKWASHING AND MAINTENANCE OF FILTER MEDIA

Effective backwashing of rapid-rate media filters is essential to long-term successful service. The goal of the backwashing operation is to remove deposited solids and return the media to an acceptably clean condition so that no progressive evidence of the development of problems such as mudballs and filter cracks (discussed later) is seen in the filter bed.

Alternative Methods of Backwashing

The backwashing system is the most frequent cause of filter failure. Therefore, selection of the backwashing system and proper design, construction, and operation of that system are key elements in the success of a treatment plant. Two prominent systems of backwashing currently in use are compared in Table 10-6.

Upflow Wash With Full Fluidization. The traditional backwash system in the United States uses an upflow water wash with full bed fluidization. Backwash water is introduced into the bottom of the bed through the underdrain system. It should be turned on gradually over at least a 30 seconds to avoid disturbing gravel layers or subjecting the underdrain to sudden momentary pressure increases. The filter media gradually assume a fluidized state as the backwash flow rate is increased and the bed expands. The backwash flow is continued with full fluidization until the waste wash water is reasonably clear; a turbidity of about 10 ntu is sufficiently clear. Then the supply valve is shut off. Shutoff is not as crucial as the opening because no danger to the underdrain or gravel exists. A slow shutoff will result in a greater degree of restratification.

TABLE 10-6 Comparison of Two Backwash Alternatives for Granular Bed Filters

	Backwash method	
	With fluidization	Without fluidization
Applications	1. Fine sand 2. Dual media 3. Triple media	Coarse single-media sand or anthracite
Routines used	1. Water wash + surface wash 2. Water wash + air scour Air first, water second No air during overflow	Air scour + water wash Simultaneously during overflow (See text for precautions) Finish with water only wash
Fluidization	Yes, during water wash	No
Bed Expansion	15–30 percent	Negligible
Wash troughs	Usually used	Usually not used
Horizontal water travel to overflow	Up to 0.9 m (3 ft)	Up to 4 m (13 ft)
Vertical height to overflow	0.76–0.91 m (2.5–3 ft)	0.6 m (2 ft)

According to Baylis (1959c), backwash by water fluidization alone is a weak washing method. The reason for this weakness, discussed by Amirtharajah (1978), is attributed to a lack of abrasion occurring between the grains in a fluidized bed. For this reason, backwashing usually is assisted by an auxiliary scour system such as surface wash or air scour. The contrasts between the auxiliary scour systems are presented in Table 10-7 and in the following sections.

Surface Wash Plus Fluidized-Bed Backwash. Surface wash has been used extensively and successfully to improve the effectiveness of fluidized-bed backwashing. Surface wash systems inject jets of water from orifices located about 2.5 to 5 cm (1–2 in) above the surface of the fixed bed. Surface wash jets are operated for 1 to 2 minutes before the upflow wash and usually are continued during most of the upflow wash, during which time they are immersed in the fluidized filter media. Surface wash is terminated 2 or 3 minutes before the end of the upflow wash.

TABLE 10-7 Contrasts Between Backwash Alternatives

	With fluidization			Without fluidization
	Without auxiliary scour	With surface wash auxiliary	With air scour auxiliary	Simultaneous air + water backwash
Wash effectiveness	Weak	Fair	Fair	Good
Solids transport to overflow	Fair	Fair	Fair	Good
Compatible with fine media	Yes	Yes	Yes	No
Compatible with dual and triple media	Yes	Yes	Yes	No
Compatible with graded support gravel	Yes	Yes	Yes	No
Potential for media loss to overflow	Nil	Yes, mainly for coal	Yes, unless used properly	Major, unless used properly

Surface wash is accomplished either with a grid of fixed vertical pipes located above the granular media or with rotary water distribution arms containing orifices or nozzles that supply high-pressure jets of water. Orifice sizes typically are 2 to 3 mm diameter (3/32–1/8 in) and are directed downward 15 to 45 degrees below horizontal. Operating pressures typically are 350 to 520 kPa (50–75 psig). Fixed-nozzle systems deliver flows of 5 to 10 $m^3/m^2/h$ (2–4 gpm/ft^2), and rotary systems deliver 1.2 to 2.4 $m^3/m^2/h$ (0.5–1 gpm/ft^2).

The use of surface wash has a number of advantages and disadvantages (Cleasby et al., 1977). The advantages include the following: (1) It is relatively simple because only a source of high-pressure water is needed in conjunction with a system of distribution nozzles, and (2) it is accessible for maintenance and repair because it is located above the surface of the fixed bed. Some disadvantages of surface wash systems include the following: (1) Rotary-type washers sometimes stick in one position temporarily and do not rotate as intended, (2) if mudballs form in the bed and reach sufficient size and density, they can sink into the fluidized bed and no longer come under the action of the surface wash jets, (3) fixed-nozzle surface wash systems obstruct convenient access to the filter surface for maintenance and repair, and (4) in addition, nozzles are subject to becoming clogged (Logsdon et al., 2002).

Air Scour Assisted Backwash. Air scour systems supply air to the full filter area from orifices located under the filter media. Air scour is used to improve the effectiveness of the backwashing operation. If air scour is used during overflow, there is substantial danger of losing some of the media. Therefore, the system must be designed and operated properly to avoid such loss.

The air scour operating routines are different for finer filter media and for coarse sand media, as shown in Table 10-6. Operating sequences are listed below.

Air Scour Alone Before Water Backwash. This system can be used for fine-sand and dual- and triple-media systems.

- Lower the water level about 15 cm (6 in) below the edge of the backwash overflow.
- Turn on the air scour alone for 1 to 2 minutes.
- Turn off the air scour.
- Turn on water wash at a low rate to expel most of the air from the bed before overflow occurs.
- Increase the water wash rate to fluidize and restratify the bed, and wash until clean. Some additional air will be expelled.

Simultaneous Air Scour and Water Backwash During Rising Level Before Overflow. This system can be used for fine-sand, dual-media, triple-media, and coarse single-media anthracite systems.

- Lower the water level to just above the surface of the filter media.
- Turn on the air scour for 1 to 2 minutes.
- Add low-rate water wash at below half the minimum fluidization velocity as the water level rises.
- Shut off the air scour about 15 cm (6 in) below the overflow level while the water wash continues. Most air will be expelled before overflow.
- After overflow occurs, increase the water wash rate to fluidize and restratify the bed, and wash until clean. Some additional air will be expelled.

Simultaneous Air Scour and Water Backwash During Overflow. This system can be used for coarse (1.0 mm ES or larger) single-media sand or anthracite systems. However, for anthracite, special baffled overflow troughs are essential to prevent loss of anthracite.

- Turn on the air scour.
- Add water wash at below half the minimum fluidization velocity, and wash with simultaneous air plus water for about 10 minutes during overflow.
- Turn off the air scour, and continue the water wash until overflow is clean. The water wash rate is sometimes increased to hasten the cleanup, but usually below fluidization velocity.

For the simultaneous wash routine just outlined, the wash is very effective, even though the bed is never fluidized. A slow transport of the grains occurs, caused by the simultaneous air plus water flow that causes abrasion between the grains. The abrasion plus the high interstitial water velocities results in an effective backwash.

Several precautions are taken to prevent the loss of sand or anthracite when using simultaneous air plus water wash during overflow. The water and airflow rates are varied appropriately for the size of the filter media. For coarse sand, backwash troughs generally are not used, the dirty wash water exits over a horizontal concrete wall or central gullet, and the vertical distance from the surface of the sand to the wash water overflow is at least 0.5 m (1.6 ft). The top edge of the overflow wall(s) is sloped at 45 degrees downward toward the filter bed so that any sand grains that fall on the sloping wall during backwashing roll back into the filter bed (Degremont, 1973). Alternatively, if backwash troughs are used, specially shaped baffles can be located around each trough to prevent the loss of filter media (Dahab and Young, 1977). Such baffled troughs are especially important when using anthracite media.

Air Scour Delivery Systems. Air scour may be introduced to the filter through a pipe system that is completely separate from the backwash water system, or it may be through the use of a common system of nozzles (strainers) that distribute both the air and the water, either sequentially or simultaneously. In either method of distribution, if the air is introduced below graded support gravel, concern exists over movement of the finer gravel by the air, especially by air and water used concurrently by intention or by accident. This concern has led to the use of media-retaining strainers (Fig. 10-8) in some filters that eliminate the need for graded support gravel in the filter.

Backwash Water and Air Scour Flow Rates. Typical flow rates used in backwash practice are summarized in Table 10-8. Hewitt and Amirtharajah (1984) did an experimental study of the particular combinations of air and subfluidization water flow that caused the formation and collapse of air pockets within the bed, a condition they called *collapse-pulsing*. This condition was presumed to create the best abrasion between the grains and the optimal condition for air plus subfluidization water backwashing. An empirical equation relating airflow rate, fluidization velocity, and backwash water flow rate was presented. In a companion paper, Amirtharajah (1984) developed a theoretical equation for collapse-pulsing using concepts from soil mechanics and porous-media hydraulics. The resulting equation is

$$0.45Q_a^{\,2} + 100(V/V_{mf}) = 41.9 \tag{10-18}$$

where Q_a is the airflow rate in standard cubic feet per minute per square foot, and V/V_{mf} is the ratio of superficial water velocity divided by the minimum fluidization velocity based on the d_{60} grain size of the media.

TABLE 10-8 Typical Water and Air Scour Flow Rates for Backwash Systems Employing Air Scour

Filter media	Backwash sequence	Air rate, m/h (scfm/sf)	Water rate,[a] m/h (gpm/sf)
Fine sand, 0.5-mm ES	Air first	37–55 (2–3)	
	Water second		37 (15)
Fine dual and triple media, 1.0-mm-ES anthracite	Air first	55–73 (3–4)	
	Water second		37–49 (15–20)
Coarse dual media, 1.5-mm-ES anthracite	Air first	73–91 (4–5)	
	Air+water on rising level	73–91 (4–5)	24 (10)
	Water third		61 (25)
Coarse sand, 1.0-mm ES	Air + water first simultaneously	55–73 (3–4)	15–17 (6–7)
	Water second		Same or double rate
Coarse sand, 1.0-mm ES	Air + water first simultaneously	110–146 (6–8)	24–29 (10–12)
	Water second		Same or double rate
Coarse anthracite, 1.5-mm ES	Air + water first simultaneously	55–91 (3–5)	20–24 (8–10)
	Water second		Same or double rate

[a]Water rates for dual and triple media vary with water temperature and should fluidize the bed to achieve restratification of the media. See Eq. 10-20.

The empirical and theoretical equations give almost the same results over the range of airflow rates from 37 to 110 m/h (2–6 scfm/ft^2). For a given airflow rate, both the equations predict somewhat higher water flow rates than the typical values presented in Table 10-8. The tabulated values should be used until future research reconciles these differences.

Some of the advantages of auxiliary air scour in contrast to surface wash systems follow (Cleasby et al., 1977): (1) Air scour covers the full area of rectangular filters and is adaptable to any filter dimensions, and (2) it agitates the entire filter depth, and therefore, it can agitate the interfaces in dual- and multiple-media beds and can reach mudballs that have sunk deep into the filter. Some disadvantages of the air scour auxiliary are (1) the need for a separate air blower and piping system, (2) the potential for loss of media, especially if air and water are used simultaneously during overflow, and (3) a greater possibility of moving the supporting gravel if air is delivered through the gravel concurrently with water; special gravel designs are required, as described earlier.

Backwash Troughs and Wash Water Required

If backwash troughs are used, they must be of adequate size and on appropriate spacing to take away the dirty backwash water without surcharging. Spacing is dictated by maximum horizontal travel of the water (see Table 10-6). At the maximum backwash rate, there should be a free fall of the water into the troughs, even at the upstream end. The top edges of the troughs all should be at the same elevation, and all should be level so that the flow is withdrawn evenly over the whole bed.

The volume of wash water required to wash a filter depends on the depth of the filter media and the vertical distance from the fixed-bed surface to the overflow level. The larger the depth of media, the larger is the water volume needed to flush detached particles out of the bed and into space above the filter. The larger the vertical distance from the surface of the media to the wash water overflow level, the larger is the volume needed to wash detached particles out of the filter box. Typically, a wash takes about four displacement volumes for the void volume in the bed plus the water volume above the bed to the overflow level. Typical wash water volumes used per unit area per wash range from 4 to 8 m^3/m^2/wash (100–200 gal/ft^2/wash); greater volumes (up to 12 m^3/m^2/wash or 300 gal/ft^2/wash) may occur for very deep filters, greater than normal freeboard, flotation-over-filtration designs, and other special cases.

In filters with backwash troughs that are washed with full bed fluidization, the expanded media surface should be lower than the bottom of the troughs. If the expanded media rise up between the troughs, the effective vertical flow velocity is increased, and the danger of carrying the filter media into the troughs is increased.

The wash water volume required per wash to clean the filter could be reduced by decreasing the vertical distance from the fixed-bed surface up to the top edge of the troughs. The temptation to do this must be resisted, however, because the danger of loss of filter media would be increased. Loss of filter media is greater when anthracite or GAC is used. In this case, to minimize the loss of media, the vertical distance to the edge of the troughs should be increased above the traditional 0.75 to 0.9 m (2.5–3 ft) to 1.1 to 1.2 m (3.5–4 ft). This, in turn, requires a greater volume of wash water to complete the backwash.

Hydraulics of Fluidized Beds

Head Loss for a Fluidized Bed. The American practice of filter backwashing typically has used backwash flow velocities above the minimum fluidization velocity of the filter media. Therefore, some fluidization fundamentals are essential to proper understanding of this backwashing practice.

Fluidization can best be described as the upward flow of a fluid (gas or liquid) through a granular bed at sufficient velocity to suspend the grains in the fluid. During upward flow, the energy loss (pressure drop) across the fixed bed is a linear function of flow rate at low superficial velocities when flow is laminar. For coarser or heavier grains, it may become an exponential function at higher flow rates if the Reynolds number (Re) enters the transitional regime, Re > 6. As the flow rate is increased further, the resistance equals the gravitational force, and the particles become suspended in the fluid. Any further increase in flow rate causes the bed to expand and accommodate to the increased flow while effectively maintaining a constant pressure drop (equal to the buoyant weight of the media). Two typical curves for real filter media fluidized by water are shown in Fig. 10-24 (Cleasby and Fan, 1981).

The pressure drop Δp after fluidization is equal to the buoyant weight of the grains and can be calculated from the following equation:

$$\Delta p = h\rho g = L(\rho_s - \rho)g(1-\varepsilon) \tag{10-19}$$

in which ρ_s is the mass density of the grains and other terms are as defined previously.

Point of Incipient Fluidization. The point of incipient fluidization, or minimum fluidizing velocity V_{mf}, is the superficial fluid velocity required for the onset of fluidization. It can be defined by the intersection of the fixed-bed and fluidized-bed head loss curves, the points labeled V_{mf} on Fig. 10-24.

The calculation of minimum fluidization velocity is important in determining minimum backwash flow-rate requirements. The rational approach to the calculation is based on the fixed bed head loss being equal to the constant head loss of the fluidized bed at the point of incipient fluidization. Thus the Ergun equation (Eq. 10-16) can be equated to the constant-head loss equation (Eq. 10-19) and solved for the velocity, that is, V_{mf}. The accuracy of the result depends on using realistic values for sphericity ψ and fixed bed porosity ε. Such data

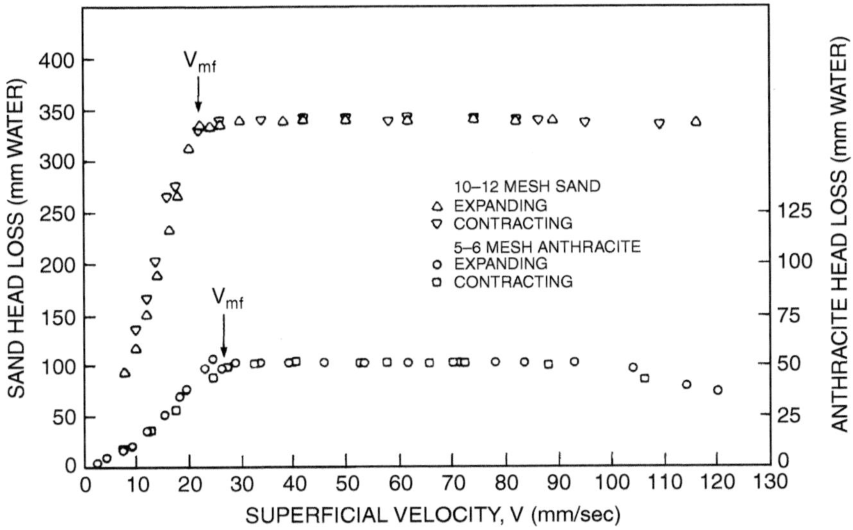

FIGURE 10-24 Head loss versus superficial velocity for 10- to 12-mesh sand at 25°C, $L_0 = 37.9$ cm and $\varepsilon_0 = 0.466$, and for 5- to 6-mesh anthracite at 25°C, $L_0 = 19.8$ cm and $\varepsilon_0 = 0.581$.(*Source: Cleasby and Fan, 1981.*)

for actual media may not be available, making the result uncertain. By using an approximate relation between sphericity and porosity, Wen and Yu (1966) developed a simplified equation for the calculation of V_{mf}. The resulting equation is

$$V_{mf} = \frac{\mu}{\rho d_{eq}}(33.7^2 + 0.0408\text{Ga})^{0.5} - \frac{33.7\mu}{\rho d_{eq}} \qquad (10\text{-}20)$$

where Ga is the Galileo number:

$$\text{Ga} = d_{eq}^3 \frac{\rho(\rho_s - \rho)g}{\mu^2} \qquad (10\text{-}21)$$

For a bed containing a gradation in particle sizes, the minimum fluidization velocity is not the same for all particles. Smaller grains become fluidized at a lower superficial velocity than larger grains. Therefore, a gradual change from the fixed bed to the totally fluidized state occurs. In applying Eq. 10-20 to a filter bed with grains graded in size, calculating V_{mf} for the coarser grains in the bed is necessary to ensure that the entire bed is fluidized. The d_{90} sieve size would be a practical diameter choice in this calculation. Generally, the d_{eq} is not conveniently available, and the d_{90} diameter from the sieve analysis may be used as an acceptable approximation.

Furthermore, the minimum backwash rate selected must be higher than V_{mf} for the d_{90} sieve size (V_{mf90}) to allow free movement of these coarse grains during backwashing. A backwash rate equal to $1.3V_{mf90}$ has been suggested to ensure adequate free movement of the grains (Cleasby and Fan, 1981); however, a rate closer to V_{mf} may be better to avoid movement of graded gravel support layers.

Hemmings and Fitzpatrick (1997) measured minimum fluidization velocities for water only and combined air-water backwashing and reported on a pressure-signal technique to measure the onset of the collapsed pulse condition. The measured result for water only agreed well with results from Eq. 10-20.

For filter media that has a very wide range of grain sizes, the use of V_{mf} for the d_{90} size during backwashing potentially could expand the finest grains so much that they could be lost to overflow. This is never a problem with the uniformity coefficients commonly specified for filter media (UC < 1.7, often <1.5). Calculation of the minimum fluidization velocity is shown in Example 10-2.

Example 10-2 Minimum Fluidization Velocity
Calculate the minimum fluidization velocity for the anthracite shown in Fig. 10-4 at a water temperature of 20°C using an anthracite density of 1600 kg/m^3 estimated from Table 10-2.

Solution Calculate the fluidization velocity V_{mf} for the d_{90} size of the anthracite as suggested in the text. The d_{90} size from Fig. 10-4 is 2.9 mm. Use Eq. 10-20:

$$V_{mf} = \frac{\mu}{\rho d_{eq}}(33.7^2 + 0.0408\text{Ga})^{0.5} - \frac{33.7\mu}{\rho d_{eq}}$$

And the Galileo number (Eq. 10-21) is

$$\text{Ga} = d_{eq}^3 \frac{\rho(\rho_s - \rho)g}{\mu^2}$$

Selecting values of physical water properties from Appendix D of this book and solving in SI units, we get

$$\mu = 1.002 \times 10^{-3}\ \text{N}\cdot\text{s/m}^2$$

$$\rho = 998.21\ \text{kg/m}^3$$

$$g = 9.81\ \text{m/s}^2$$

$$d_{eq}\ (\text{use}\ d_{90}) = 2.9\ \text{mm} = 2.9 \times 10^{-3}\ \text{m}$$

$$\text{Ga} = (2.9\times10^{-3})^3 \frac{998(1600-998)9.81}{(1.002\times10^{-3})^2} = 143{,}170\ (\text{dimensionless})$$

$$V_{mf} = \frac{1.002\times10^{-3}}{998(2.9\times10^{-3})}(33.7^2 + 0.0408\times143{,}170)^{0.5} - \frac{33.7(1.002\times10^{-3})}{998(2.9\times10^{-3})}$$

$$= 0.01725\ \text{m/s}$$

The recommended backwash rate would be up to 30 percent higher, as discussed in the text:

$$\text{Backwash rate} = 1.3 \times 0.01725 = 0.0223\ \text{m/s} = 80.1\ \text{m/h}\ (32.8\ \text{gpm/ft}^2)$$

This is a higher than the normal backwash rate because of the very coarse anthracite grain size, the rather warm water, and the choice of 30 percent factor above V_{mf}. Without this factor applied, the rate would be 61.6 m/h (25.2 gpm/ft^2). ▲

Expansion of Filter Media During Backwashing. When using upflow wash with full fluidization, the filter bed expands about 15 to 30 percent above its fixed bed depth. The degree of expansion is affected by many variables associated with the filter media and the water. Filter media variables include the size and size gradation and grain shape and density. Water variables include viscosity and density. The ability to predict expansion is important, for example, in determining whether the expanded media will rise too high, possibly above the bottom of the troughs.

The following model for predicting expanded-bed porosity during backwashing was developed by extending a Reynolds number versus porosity function that had been used for fixed beds into the expanded-bed region (Dharmarajah and Cleasby, 1986). The modified Reynolds number (Re_1) uses interstitial velocity V/ε for the characteristic velocity and a term approximating the mean hydraulic radius of the flow channel, $\varepsilon/[S_v(1 - \varepsilon)]$, for the characteristic length.

$$Re_1 = \frac{V}{\varepsilon}\frac{\varepsilon}{S_v(1-\varepsilon)}\frac{\rho}{\mu} = \frac{V\rho}{S_v(1-\varepsilon)\mu} \tag{10-22}$$

where S_v is the specific surface of the grains ($6/d$ for spheres and $6/\psi d_{eq}$ for nonspheres). The porosity function used previously to correlate fixed-bed pressure drop data to Re_1 was modified by combining it with the constant-head loss equation for a fluidized bed (Eq. 10-19), resulting in a new dimensionless porosity function for fluidized beds (A_1).

$$A_1 = \frac{\varepsilon^3}{(1-\varepsilon)^2}\frac{\rho(\rho_s-\rho)g}{S_v^3\mu^2} \tag{10-23}$$

Using data for many different sizes and types of filter media, $\log A_1$ was correlated with $\log Re_1$ using a stepwise regression analysis and resulting in the following expansion correlation (Dharmarajah and Cleasby, 1986):

$$\begin{aligned}\log A_1 &= 0.56543 + 1.09348 \log Re_1 + 0.17971(\log Re_1)^2 \\ &\quad - 0.00392(\log Re_1)^4 - 1.5(\log \psi)^2\end{aligned} \tag{10-24}$$

This equation can be used to predict the expanded porosity ε of filter media of any uniform size (i.e., S_v = specific surface) at any desired backwash rate V. Because both A_1 and Re_1 are functions of ε, the solution is trial and error and is best solved numerically. When applying the equation to real filter media with size gradations, the bed must be divided into several segments of approximately uniform size using the sieve analysis data, and the expanded porosity of each size segment must be calculated. The expanded depth of each segment then can be calculated from the following equation that is based on the total grain volume remaining unchanged as the bed expands (Cleasby and Fan, 1981):

$$\frac{L}{L_0} = \frac{1-\varepsilon_0}{1-\varepsilon} \tag{10-25}$$

in which L/L_0 is the ratio of expanded depth L to fixed-bed depth L_0, ε is the expanded porosity, and ε_0 is the fixed loose-bed porosity. Typical values of ε_0 were presented earlier in Table 10-2.

Example 10-3 Filter Bed Expansion
Calculate the expansion for the d_{50} size anthracite shown in Fig. 10-4 at 20°C and at the backwash rate of 80.1 m/h (32.8 gpm/ft^2) calculated in Example 10-2. Use a sphericity of 0.55 and a density of 1600 kg/m^3 estimated from Table 10-2.

Solution in SI Units Use Equation 10-22:

$$Re_1 = \frac{V\rho}{S_v(1-\varepsilon)\mu}$$

Values from Example 10-2 and used here are $V = 0.0223$ m/s, $\rho = 998.21$ kg/m^3, and $\mu = 1.002 \times 10^{-3}$ N · s/m^2. From Appendix D, $\mu/\rho = \nu = 1.004 \times 10^{-6}$ m^2/s.

$$d_{50} = 2.2 \text{ mm or } 2.0 \times 10^{-3} \text{ m (Fig. 10-4)}$$

$$S_v = 6/\psi d = 6/(0.55 \cdot 2.0 \times 10^{-3}) = 5454 \text{ m}^{-1}$$

$$\text{Re}_1 = \frac{0.0223}{5454(1-\varepsilon)1.004\times10^{-6}} = \frac{4.073}{1-\varepsilon} \quad \text{(dimensionless)}$$

where ε is the desired expanded porosity. Then use Eq. 10-23:

$$A_1 = \frac{\varepsilon^3}{(1-\varepsilon)^2}\frac{\rho(\rho_s-\rho)g}{S_v^3\mu^2}$$

$$A_1 = \frac{\varepsilon^3}{(1-\varepsilon)^2}\frac{998(1600-998)9.81}{(5454)^3(1.002\times10^{-3})^2} = \frac{36.184\varepsilon^3}{(1-\varepsilon)^2}$$

Insert these Re_1 and A_1 values into Eq. 10-24 as follows:

$$\log\left[36.184\frac{\varepsilon^3}{(1-\varepsilon)^2}\right] = 0.56543 + 1.09348 \log\left[\frac{4.073}{(1-\varepsilon)}\right] + 0.17971\left[\log\frac{4.073}{(1-\varepsilon)}\right]^2$$

$$-\,0.00392\left[\log\frac{4.073}{(1-\varepsilon)}\right]^4 - 1.5(\log 0.55)^2$$

Note that the only unknown in the preceding equation is ε, but it appears on both sides of the equation. A trial-and-error solution is necessary. The solution to Eq. 10-24 yields $\varepsilon = 0.616$

The expansion of this size anthracite can be calculated from Eq. 10-25, assuming an initial porosity of 0.56 from Table 10-2. Thus

$$\frac{L}{L_0} = \frac{1-\varepsilon_0}{1-\varepsilon}$$

$$\frac{L}{L_0} = \frac{1-0.56}{1-0.616} = 1.15$$

Therefore, this midsize material would be expanded 15 percent. To obtain the expansion of the entire bed, the bed must be divided into about five sizes and the expanded depth of each layer calculated and totaled. To obtain the expansion of dual- or triple-media beds, the expansion of each layer is calculated and totaled without regard to any intermixing at the interfaces between layers. ▲

Stratification and Intermixing During Backwashing

The related phenomena of stratification and intermixing are important issues in filter construction and backwashing. These are discussed below.

Stratification and Skimming. In the case of a single-media type of filter such as a sand filter, during backwash with fluidization, the grains tend to stratify by size, with the finer grains on top and the coarser grains on the bottom. The tendency to stratify at a given backwash rate (above fluidization velocity) is driven by bulk density differences between the fluidized grains of different sizes. Smaller grains expand more and have a lower bulk density (grains plus fluidizing water) and thus rise to the top of the bed. The concept of bulk density is discussed further in the next section on intermixing of adjacent layers. This stratification is upset to varying extents by nonuniform upflow of the backwash water that creates localized regions of above-average upflow velocity. Larger grains are transported upward rapidly into the upper bed, whereas in adjacent regions the sand will be moving downward carrying finer grains down into the bed. These regions of excessive upflow are referred to as *sand boils* or *jet action* in the water filtration literature and as *gulf streaming* in some fluidization literature.

The tendency to stratify is used beneficially during construction of a filter to remove unwanted fine grains from the bed. The filter is washed above fluidization velocity, and the fine grains accumulate at the upper surface. After the backwash is completed, the fine grains are skimmed from the surface to avoid leaving a blinding layer of fines on top.

Better stratification is achieved at lower upflow rates, just barely above the minimum fluidization velocity of the bed. Therefore, in preparing for skimming, the bed first should be fully fluidized, and then the upflow wash rate should be reduced slowly over several minutes to bring to the surface as many of the fine grains as possible. The fluidization and skimming process may be repeated two or three times for maximum effectiveness.

During the installation of dual- or triple-media filter beds, skimming each layer as it is completed is common. The same concepts and procedures described earlier are equally appropriate for each layer. If the very fine grains are not removed from the lower layers, they may rise high in the next upper layer, partially negating the desired benefit of the coarse upper layer (Cleasby et al., 1984a).

Intermixing of Adjacent Layers. Intermixing tends to occur between adjacent layers of dual- and triple-media filters. For example, the upper finer sand grains of a dual-media bed move up into the lower coarser grains of the anthracite bed that lies above. The tendency to intermix between two adjacent layers of filter media can be estimated by comparing the bulk density of the two media calculated independently at any particular backwash flow rate and temperature. The *bulk density* is the mixed density of the grains and fluidizing water calculated as follows (Cleasby and Woods, 1975):

$$\rho_b = (1 - \varepsilon)\,\rho_s + \rho\varepsilon \tag{10-26}$$

in which ρ_b is the bulk density. The first term on the right-hand side is the density contributed by the solid fraction, and the second term is the density contributed by the water fraction.

The tendency to intermix increases with the backwash flow rate because the bulk densities tend to converge at higher flow rates. Intermixing actually begins before the bulk densities become equal because of uneven flow distribution and mixing and circulation patterns that exist in the fluidized layers. Mixing was observed to occur when the bulk densities converged to within 50 to 130 kg/m^3 (3–8 lb/ft^3) (Cleasby and Woods, 1975). Similarly, as the backwash rate is decreased slowly, the bulk densities of the adjacent layers diverge, and intermixing decreases.

The bulk density model would predict that inversion of layers should occur at very high wash rates, and such inversion was observed experimentally (Cleasby and Woods, 1975). Wash rates, however, were far higher than rates used in practice.

The tendency for sand and anthracite to intermix during backwashing is much less than for silica and garnet sand in the usual backwash flow range. This is so because the differences in bulk density are greater for the usual sand and anthracite sizes used in current practice.

Movement of Gravel During Backwashing

Many filters have layers of graded gravel to support the filter media. The conventional gravel layers are graded from coarse on the bottom to fine at the top, according to appropriate size guidelines presented earlier. With this conventional gravel arrangement, some cases of mounding of the gravel caused by lateral movement of the fine gravel have been reported. In severe cases, the finest gravel layer may be completely removed from some areas of the filter bed. This can lead to leakage of the filter media into the underdrain system and ultimately to the need to clean the underdrain and rebuild the filter gravel and media. Mechanisms contributing to movement of gravel have been discussed exhaustively by Baylis (1959a; AWWA, 1971), and those discussions related to fluidized-bed backwashing of sand beds are summarized here.

As backwashing begins, sand grains do not move apart quickly and uniformly throughout the bed. Time is required for the sand to equilibrate at its expanded spacing in the upward flow of wash water. If the backwash is turned on suddenly, it lifts the sand bed bodily above the gravel layer, forming an open space between the sand and gravel. The sand bed then breaks at one or more points, as shown in Fig. 10-25, causing sand boils and subsequent upsetting of the supporting gravel layers. This then requires frequent rebuilding of the graded gravel.

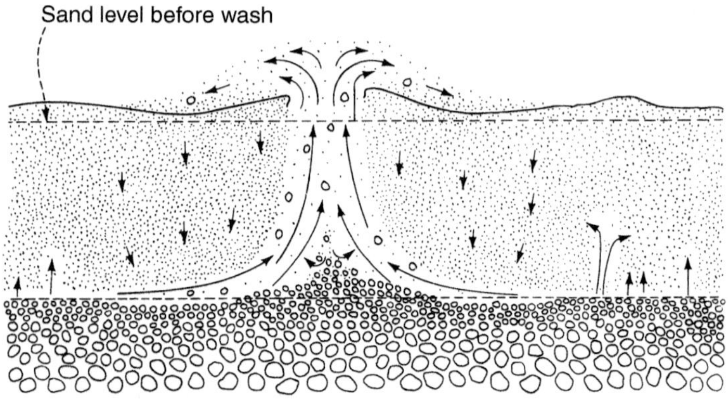

FIGURE 10-25 Sand boil at beginning of wash.

Opening the backwash valve slowly is essential. The time from start to full backwash flow should be at least 30 seconds and perhaps longer and should be restricted by devices built into the plant. The most destructive filter washing blunder is to turn the backwash water on quickly when the bed has been drained. Gross disturbance of the gravel results, and rupture of the filter underdrain system has resulted in many cases.

The upward flow from the gravel is never completely uniform. As a result, in some parts of the expanded sand bed the water and sand travel upward at rates higher than the average backwash velocity, and at other places the sand actually travels downward. This leads to jet action at the sand-gravel junction, as illustrated in Fig. 10-26. In some cases, the jet velocity is so great that it will move grains of small gravel several diameters larger than the sand grains adjacent to the gravel. The problem of gravel movement is greater at wash rates above 37 m/h (15 gpm/ft^2).

The importance of sand in generating the movement of gravel is illustrated by the observation that an upward flow through a graded gravel bed at backwash rates exceeding 61 m/h

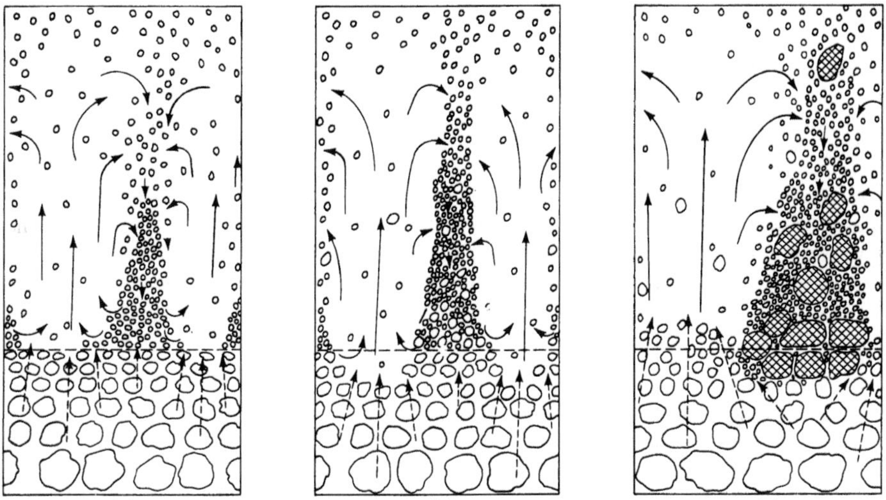

FIGURE 10-26 Jet action at the sand-gravel interface.

(25 gpm/ft^2) will not move any of the gravel unless sand is present on top of the gravel (Baylis, 1959a). The presence of the sand greatly increases the mobility of the gravel.

Movement of the gravel also can occur when air scour and water backwash are used simultaneously. Provisions being used to avoid movement of gravel are the use of double-reverse graded gravel or the use of media retaining underdrains that require no gravel, as discussed earlier.

Three methods are used to determine whether the upper surface of the gravel has moved: (1) push a 6 mm (0.2 in) metal rod gently down to the gravel surface of a newly backwashed filter while the filter is at rest, (2) lower a probe with a flat plate on the end through the expanded filter media during the fluidized-bed backwash, and (3) drain the filter to the gravel, insert a thin-walled pipe of about 2.5 cm (1 in) in diameter into the media, and remove the media in segments until the top layer of gravel is reached in successive probings. Each method has advantages and disadvantages. The first method is quick, but distinguishing the fine gravel from the bottom filter media is difficult. The second method requires a long period of fluidized backwash. The third method is the most positive but also the most time-consuming and labor-intensive. Numerous measurements made on a gridiron pattern will reveal the contour of the top surface of the gravel.

Underdrain Failures

The filter underdrain is a vitally important component of the rapid-rate media filter. It serves to collect the filtered water, support the filter gravel (if used) and filter media, and distribute backwash water.

The most serious failures of rapid-rate filters are usually related to problems with the underdrain. Proper design and construction of the underdrain are essential. Any underdrain system, if designed improperly, can lead to uneven distribution of the backwash water. This, in turn, can lead to problems such as dirty areas of the filter and migration of gravel if gravel is used to support the media.

False-floor underdrains with nozzles and plastic block underdrains with caps have occasionally failed by several modes. If the nozzles or caps have fine openings to retain the fine filter media, the openings may become clogged with debris from construction of the underdrain or from solids in the backwash water or because of mineral deposits. If steel filter vessels are allowed to corrode, the rust can clog backwash nozzles. If the clogging becomes excessive, the backwash flow rate may be diminished, and the pressure drop across the floor may cause excessive uplift during backwashing. The problem can be repaired by removing the filter media and cleaning the nozzles or caps. If not corrected, excessive pressure drop can result in an uplift structural failure of the false floor or filter blocks. An open pressure-relief pipe is often provided that projects an appropriate length above the backwashing water level. Excessive pressure in the underdrain is evident when this pipe starts to flow during the backwash operation.

Some modern plastic nozzles are not strong enough to accept the abuse of construction during or after installation. They may be weakened during construction, and the nozzle heads may break off during future backwashing operations. The result is leakage of media into the underdrain plenum.

The manifold and lateral type of underdrain is not subject to uplift failure. It can fail to function properly, however, if the distribution orifices are reduced in size because of mineral deposits or enlarged in size because of corrosion or erosion.

Vitrified clay block underdrains have proven quite successful if installed properly and if precautions are taken to prevent air from entering the backwash system. The manufacturers' recommendations for installation and grouting must be followed explicitly to obtain a good installation. Careless workmanship can lead to grout in the water-carrying channels. This can lead to maldistribution of wash water and in severe cases to excessive backwash pressures and uplift failure. Some plastic block underdrains eliminate the danger of grout intrusion by the use of O-ring gaskets or other positive exclusion systems.

Dirty Filter Media and Mineral Deposits

So-called dirty filters result from inadequate backwashing of the filter, including the absence or improper operation of auxiliary scour systems. Typical manifestations of inadequate backwashing include filter cracks and mudballs. Inadequate cleaning leaves a thin layer of compressible matter around each media grain. As pressure drop across the filter bed increases during the subsequent filter run, the grains are squeezed together, and cracks form in the surface of the media, usually along the walls first. In severe cases, cracks may develop at other locations as well. The AWWA has published a guidance manual for evaluation of granular media (Nix and Taylor, 2003).

The heavier deposits of solids near the top surface of the media break into pieces during backwashing, resulting in spherical/oblong-shaped accretions referred to as *mudballs*. Mudballs are composed of the filter grains and the solids removed by the filter and range from pea size to 2.5 to 5 cm (1–2 in) or more. The formation of mudballs is accentuated by the use of polymers as coagulant aids or as filter aids that form stronger attachments between the filter grains and the removed solids. If mudballs are small enough and of low enough density, they float on the surface of the fluidized media. If larger or heavier, they may sink into the filter to the bottom or to the coal-sand interface in dual-media filters. Subsurface accumulations can lead to solidified inactive regions of the filter bed that are not remedied in normal backwashing. This, in turn, increases the filtration rate through the remaining active portions of the bed with potential detriment to the quality of the filtrate and shorter filter cycles.

Surface mudballs can be removed manually by (1) lifting them from the surface with a large strainer while the backwash water is running at a low rate, (2) breaking them up with rakes, or (3) breaking them up with a high-pressure hose jet. Subsurface mudballs and agglomerations can be reduced by (1) probing the bed with lances delivering jets of high-pressure water, (2) pumping the media through an ejector, (3) washing the filter and using chemical additions (described below), or (4) digging out the hard spots and removing, cleaning, or replacing the media.

Solid surfaces in water greatly increase the rate of solid-phase formation via precipitation from dissolved species; this often occurs subsequent to solute adsorption and, in some cases, oxidation. Thus strongly attached mineral deposits or coatings can develop on the grains of the filter media, causing them to change size, shape, and density. Calcium carbonate deposits are common in many lime–soda ash softening plants if recarbonation is not sufficient to deliver a stable water to the filters. Such deposits also may occur in surface water plants if lime is added ahead of filtration for corrosion control. The deposition of calcium carbonate can be minimized by adding a low dosage of polyphosphate ahead of the filters. Surface deposition of calcium carbonate is not surprising because one effective softening method involves fluidized beds of granular $CaCO_3(s)$ (or pellets) for calcium removal.

Iron oxides and manganese oxides are common deposits on the filter media of iron- and manganese-removal plants. Aluminum oxide or iron oxide deposits can occur in surface water plants that use alum or iron salts for coagulation. In general, Al, Fe(III), and Mn(IV) are very insoluble, and essentially all precipitation occurs as suspended particles that are removed subsequently by clarification and filtration. Sorption of reduced Fe(II) or Mn(II) to media surfaces and subsequent oxidation can result in strongly attached coatings of insoluble Fe or Mn oxides.

The occurrence (and desirability) of manganese oxide (MnO_x) coatings on standard anthracite and sand filter media is of special importance. It has long been recognized that sorption of Mn(II) by MnO_x surfaces catalyzes oxidation of the sorbed reduced Mn(II) to Mn(IV), thus creating new sites for more Mn(II) uptake (Morgan and Stumm, 1964). Extensive work by Knocke and co-workers rediscovered the importance and mechanisms of Mn(II) removal by MnO_x coatings on anthracite and sand media in filters that are continuously exposed to free chlorine because of the application of prefilter chlorination, the so-called natural greensand effect (e.g., Knocke et al., 1988, 1990, 1991). This process is intentionally used by many utilities for Mn removal (see Chap. 7); MnO_x coatings also can develop unintentionally by removal of low levels of Mn added with iron coagulants. Process changes that eliminate prefilter chlorination, for example, a change to prefilter ozonation and operation of biologically active filters, but do not replace the MnO_x-coated media can result in unintended and undesirable release of Mn from the media (e.g., Gabelich et al., 2006). Results of earlier work by Knocke and a more recent study (Tobiason et al., 2008) show extractable Mn levels of 0.1 to 100 mg Mn/g media depending on Mn loading, years in service, location in the media, and other factors. In general, media size is not significantly affected by typical Mn coating levels, but extensive coatings on smaller media can increase size. These studies also showed appreciable levels of other metals (Al, Fe, Cu, etc.) associated with the extractable Mn coatings.

In general, mineral deposits are subject to later leaching into the filtered water if the pH is changed in a direction to increase solubility. Also, in the case of iron and manganese coatings, the occurrence of reducing conditions, possibly mediated by biofilms, can result in release of the more soluble reduced forms of these metals. In some cases, the increased size of the grains caused by mineral deposits results in an increase in bed depth and can impair filtration efficiency. Thus, cleaning the deposits from the grains or replacing the media may become necessary.

Dirty filters usually can be corrected by proper use of the auxiliary scour devices such as surface washers or air scour. Where auxiliary scour is not successful, use of chemicals may be necessary in some cases to attempt to clean the filter media in place. Various chemicals have been used, including chlorine, copper sulfate, acids, and alkalis (Babbitt et al., 1962). Chlorine may be used where the material to be removed includes living and dead organisms or their metabolites. Copper sulfate is effective in killing algae growing on filter walls or media. Carrying a chlorine residual in the filter influent aids in the control of microorganisms in the filter. Acidifying the water to a pH of about 4 aids in dissolving deposits of calcium carbonate and the oxides of iron, aluminum, and manganese. Sulfuric acid and hydrochloric acid have been used, but care must be taken to prevent high local concentrations to prevent damage to concrete filter walls. Citric acid has been used in the St Louis, MO, softening plant, as well as diluted, food-grade glacial acetic acid at the lime softening plant at Decatur, IL (Mayhugh et al., 1996). These organic acids are less dangerous to handle. Caustic soda has been used at the rate of 5 to 15 kg/m^2 (1–3 lb/ft^2) for removing deposits resulting from the use of alum and for deposits of organic material. Chemical solutions are generally left in contact with the filter media for 1 to 2 days, and the filter then is washed thoroughly and returned to service.

If a filter can be removed from service for an extended period, deposits of calcium carbonate can be removed by feeding at a low flow rate a water acidified with carbon dioxide. The flow rate is selected so that the water leaves the filter with the CO_2 fully consumed at a pH of about 8.

Backwash of GAC Media

The properties of GAC are sufficiently different from conventional anthracite or sand media that some special precautions are required related to backwashing of GAC media filters (Graese et al., 1987). The lower density of GAC means that lower backwash rates are required to fluidize the common size gradations being used. Lower wash rates mean lower hydraulic shear and less effective upflow wash. Mudballs have been a common problem in GAC filters. Therefore, surface wash or auxiliary air scour is essential for GAC backwashing.

The higher UC of gradations of GAC often being used results in a greater percent expansion and better size stratification if the full bed is fluidized. There is greater potential for loss of GAC into the backwash troughs. In retrofitting existing filters with GAC media, raising the backwash troughs may be necessary to reduce the loss of GAC.

GAC is abrasive and corrosive to many metals. Therefore, all metals in contact with GAC should be resistant to abrasion and corrosion. Metals can be coated with corrosion-resistant substances.

Special precautions are important when preparing a new GAC media filter for service. The filter box should be disinfected before installing the GAC. After installation of the GAC, it must be submerged and soaked for at least a day to allow water to penetrate the pores of the GAC. Then backwashing must be done at a very low rate initially to be sure that all air is out of the bed to avoid loss of GAC. When all air is removed, the backwash rate can be increased to wash out undesirable fines from the bed. Release of additional air and fines may be observed during the first week of normal backwashing operation. See Chap. 14 for further information on the backwashing of GAC adsorbers.

Waste Backwash Water Management

Sections of Chap. 22 on management of water treatment plant residuals describe the treatment and recycle of waste (or spent) filter backwash water (FBW). A brief discussion is included here for completeness and because of the importance of the subject. Waste FBW is produced at a high flow rate for a relatively short time period (i.e., 15–30 minutes per backwash), yielding a rather large volume of water with low solids content. The volume is typically 2 to 5 percent of total plant production or, as noted previously, in the range of 4 to 12 m^3/m^2/run (100–300 gal/ft^2/run). Typical management options include discharge to a receiving water (requiring a National Pollutant Discharge Elimination System permit and water quality control), discharge to a sanitary sewer (often not available or feasible), and recycle to the head of the treatment plant, with or without treatment (for direct filtration, solids must be removed from recycled waste FBW). In nearly all cases, flow equalization via storage and controlled discharge or recycle is practiced. As noted in studies conducted prior to the 2001 USEPA Filter Backwash Recycling Rule (FBRR) (USEPA, 2001), recycle of waste FBW is a common practice because of the benefits of recovering the large water volume and the challenges associated with other options.

The impact of the recycle of waste FBW on treatment performance and plant effluent quality needs to be carefully considered; potential and actual effects are quite case-specific. Raw water quality, performance of clarification prior to filtration, backwash water chemistry, waste FBW treatment performance, other recycle streams, and rate of recycle all can affect the impact of recycle.

Concerns about recycling of *Giardia* cysts and *Cryptosporidium* oocysts had significant impact on the language in the 1996 SDWA Amendments that led to the USEPA FBRR. This was logical because when media filters remove cysts or oocysts, their concentration in the untreated waste FBW should be 20 to 50 times higher than in the filter influent water. The significance of this depends *very* much on levels of pathogens in raw water and the performance of prefilter clarification for removal of pathogens. Mass-balance modeling of recycle impacts can show the detrimental impacts of recycle (Cornwell and Lee, 1994), and it also can show that good clarifier performance greatly decreases or eliminates the possibility for waste FBW recycle to negatively affect plant influent (raw plus recycle) pathogen levels (Edzwald et al., 2001, 2003). Pilot studies with spiking of inactivated oocysts showed no negative impact of waste FBW recycle on clarification performance (Cornwell and MacPhee, 2001; Edzwald et al., 2003). A study of well-equalized recycle of untreated waste FBW at several plants for one utility (Tobiason et al., 2003b) showed no detrimental impacts on clarification (DAF or plate settling) or media filtration performance with respect to turbidity; the utility routinely had no detection of protozoan pathogens in its source waters. A companion pilot study showed that high levels of recycle of waste FBW with altered water chemistry (high pH) can have significant negative impacts on clarifier performance and some limited impact on filtration performance (Tobiason et al., 2003a).

As noted in Chap. 22, other water quality concerns associated with waste FBW recycle include organic matter (DBP precursors) and suspended solids and metals. Design of facilities for management of waste FBW and operation of recycle should be based on careful and rational consideration of the risk and impacts on treatment performance. Utilities should regularly monitor recycle stream quality and assess impacts on plant influent quality and process performance.

Wash Water versus Filtered Water Chemistry

The chemistry of the water used to backwash a filter may be significantly different from the chemistry of the filter influent and effluent waters and thus may affect aspects of filter performance and the management of waste backwash water. This is most likely to occur if the wash water source is the plant effluent clearwell and chemicals for disinfection and corrosion control have been added to the media filter effluent; common examples are free chlorine, chloramines, acid or base for pH alteration, a phosphorus-based corrosion inhibitor, and fluoride.

Backwashing with wash water of different chemistry than the filter effluent may have impacts on the stability of particles in backwash remnant water and thus affect particle removal on filter startup; oxidation by disinfectants, surface-charge change due to pH change, and adsorption of corrosion inhibitors may be important. The actual impact is likely to depend on the extent of difference in water chemistry and the relative volume of wash water versus filter influent in the filter on startup; filter to waste may be useful in mitigating potential negative impacts on filtered water quality.

The presence of oxidants such as free chlorine in wash water when not typically present in filter influent or effluent may affect biological activity (Emelko et al., 2006) and may provide minimal regeneration of sorption sites for reduced manganese (Tobiason et al., 2008). Detrimental impacts of chlorine on GAC filter media are also possible.

As noted earlier, the chemistry of wash water may affect the quality of waste FBW and thus the quality of recycle streams. Changes in pH and oxidants relative to filtration conditions may result in dissolution of precipitated particles and/or desorption of material such as NOM. Also, free chlorine could interact with NOM in waste FBW and create additional DBPs that could be recycled. Impacts are site-specific and may or may not be of significance.

DIRECT FILTRATION

Process Description, Advantages, and Disadvantages

Direct filtration is a surface water treatment process that includes addition of coagulant, rapid mixing, flocculation, and filtration. In some cases, the flocculation tank is omitted, and the process is referred to as *in-line* or *contact filtration*. The most significant feature of direct filtration is the elimination of a separate clarification process prior to filtration. There are advantages and disadvantages to selection of direct filtration as well as limitations and special issues.

The advantage of direct filtration over conventional treatment is the potential for reduced costs. Capital and operating costs are lower because of elimination of a clarification process. If the clarification process was sedimentation, then lower capital costs for flocculation can be realized because the detention time for producing a "pinpoint" size filterable floc is much lower than the time required to produce a large settleable floc (e.g., 10–15 minutes versus 30–40 minutes). In some cases, lower coagulant dosages may be used in direct filtration if the comparison is against the use of excess coagulant to produce more solids to be flocculated for sedimentation. In general, coagulant dosing is controlled by levels of NOM and the need to destabilize raw water particles, so coagulant dosing is often the same regardless of the solid separation process. If coagulant doses are lower, then cost savings are realized for both chemicals and sludge treatment and disposal. Overall, the cost savings can be substantial, making direct filtration an attractive alternative.

There are, however, several significant disadvantages to direct filtration. The most significant is the reliance on a single process, media filtration, for all particulate matter removal; thus there is only one physical treatment barrier for pathogens, which is inconsistent with a multiple-barrier approach. Since effective media filter performance requires effective coagulation, and the time between coagulant addition and filter effluent is short, the direct filtration process has more inherent risk of failure than conventional treatment, thus requiring more careful and extensive monitoring and automatic controls. Without a clarification process, the filtration process must respond effectively to all changes in raw water quality; thus there are rational limitations on the types of raw waters amenable to direct filtration (see next section). In order to recycle waste FBW in direct filtration, a high level of removal of solids from the recycle flow must occur because there is no other location for solids removal to occur; recycle stream treatment options are discussed in Chap. 22. Because of less flocculation time and no clarification process, there is less detention time available for controlling seasonal tastes and odors via intermittent adsorbent or oxidant addition; if this is expected, a pretreatment or contact basin may be used to provide the needed contact time with powdered activated carbon or oxidizing chemicals (Conley and Pitman, 1960b).

Appropriate Source Waters for Direct Filtration. Media filtration performance (i.e., particle removal and head loss) typically is limited and challenged by the amount of solids in the filter influent. Thus there are limits on suitable maximum levels of raw water parameters such as turbidity, algae, and NOM (as measured by color, TOC, and UV_{254} absorbance). Turbidity and algae are direct sources of solids, whereas the level and character of source water NOM usually controls coagulant addition and thus the coagulant-produced solids. Oxidation of high levels of reduced iron and manganese also can lead to significant solids production; this is usually an issue for groundwater treatment. The combination of cost-based advantage and process-based water quality limits has resulted in the development of guidelines for appropriate source waters for direct filtration (e.g., Culp, 1977; Letterman, 1987).

The type and dosage of coagulants needed to achieve desired quality of filtrate, as demonstrated by pilot- or full-scale observations, are probably the most important consideration in assessing the feasibility of direct filtration. If alum is the primary coagulant, Hutchison (1976) suggested that 12 mg/L of alum on a continuous basis would yield 16 to 20 hour filter runs at 12 m/h (5 gpm/ft^2). This was considered sufficiently long to limit backwash water volume to 4 percent of the product. Waters with up to 25 color units were considered as suitable candidates for direct filtration with partial substitution of polymer for alum. Hutchison also suggested that diatom levels from 1000 to 2000 areal standard units per milliliter (asu/mL) required use of coarser anthracite and more frequent use of polymer to prevent breakthrough. One areal standard unit (asu) is 20×20 μm, or 400 (μm)^2. Anthracite with 1.5 mm ES was effective for 2500 asu/mL of diatoms and produce 12 hour filter runs at 12 m/h (5 gpm/ft^2).

Wagner and Hudson (1982) suggested the use of jar tests and filtration through standard laboratory filter paper as a screening technique for selecting appropriate waters for direct filtration. Waters requiring more than 15 mg/L of alum (as filter alum) to produce an acceptable quality of filtrate were doubtful candidates for direct filtration. Waters requiring only 6 to 7 mg/L of alum and a small dose of polymer were considered favorable candidates. If the results of this bench-scale screening test are favorable, then pilot-plant tests are needed to determine full-scale design parameters.

An AWWA report on direct filtration (Committee Report, 1980) defined a water meeting the following criteria as a "perfect candidate" for direct filtration: color < 40 color units; turbidity < 5 ntu; algae < 2000 asu/mL; iron < 0.3 mg/L; and manganese < 0.05 mg/L. Cleasby et al. (1984a) considered the AWWA committee guidelines acceptable except for turbidity, which was considered too low. During seasons with low algae, they suggested turbidity limits of 12 ntu when using alum alone or 16 ntu when using cationic polymer alone. During seasons with high algae, they suggested limits of 7 ntu using alum alone and 11 ntu using cationic polymer alone. Best performance was achieved during seasons with low algae with alum dosages between 5 and 10 mg/L. During seasons with high algae, dosages of up to 20 mg/L were attempted with only moderate success and with runs as short as 12 hours at 7.3 m/h (3 gpm/ft^2).

The preceding guidelines were developed prior to the increased attention to removal of DBP precursors (e.g., as TOCs) that arose during the late 1980s and 1990s. Maximizing TOC removal often involves addition of higher levels of precipitating coagulants and thus higher solids levels. Based on pilot and laboratory experiments, Edzwald et al. (1987) recommended TOC < 5 mg/L for direct filtration. More recently, based in part on full-scale plant surveys, Valade et al. (2009) recommended that average TOC and turbidity be less than 3 mg/L and 5 ntu, whereas maximum levels should be less than 5 mg/L and 25 ntu, respectively, for direct filtration.

In addition to the quantitative measures of water quality described earlier, consideration may be given to the probability or risk of the presence of pathogens in the source water. Increased concerns over waterborne disease from chlorine-resistant pathogens may contribute to selection of treatment with both clarification and filtration processes in lieu of direct filtration, especially when combined with the impact of potentially increased coagulant dosing for NOM control. The increasing use of low-pressure membrane filters, with or without prior clarification, instead of media filters, may reflect the potential increased reliability of a sieving removal mechanism independent of chemical dosing; the very frequent backflushing of low-pressure membranes also may be better suited for higher solids loading than direct filtration with media filters (see Chap. 11).

Chemical Pretreatment for Direct Filtration

The selection of the coagulant dosage for direct filtration is best determined by full-length filter cycles using pilot- or full-scale filters. Jar tests are often misleading because the goal in direct filtration is to form small pinpoint flocs that are barely visible but will filter effectively. Thus the usual criteria used to judge jar test observations such as formation of a large floc, or a clear supernatant liquor after settling, are not appropriate for direct filtration. The jar test and filter paper filtration technique (Wagner and Hudson, 1982) is somewhat better, but it provides no information on terminal breakthrough behavior or rate of generation of head loss; in addition, particles that are restabilized by charge due to coagulant overdosing can be retained by the filter paper but would not be removed by media filtration. Substitution of a miniature granular bed filter for the filter paper (Bowers et al., 1982) avoids the latter problem but suffers from the first two weaknesses. Kreissl et al. (1968) demonstrated that a clean filter could be operated at very high rates to simulate the hydraulic shear that would exist in the more clogged layers of the filter late in the filter run. This would predict whether the chemical dosage would be successful through the full length of the filter run. Despite these various techniques of using bench or pilot tests to select chemical dosages, the most common approach involves confirmation of the selected dosage with pilot- or full-scale filter runs of full duration at the design filtration rate.

When using alum or iron salts alone, the optimal dosage of coagulant for direct filtration is the lowest dosage that will achieve filtrate quality goals. An optimal dosage is normally not observed based on filtrate quality alone. Rather, as the dosage is increased, the filtrate quality gets marginally better, but the head loss develops at a higher rate, and early breakthrough is encouraged (Cleasby et al., 1984a).

When using cationic polymer alone as a primary coagulant, a distinct optimal dose may be found that produces the best-quality filtrate. In pilot studies, one method of detecting whether a particular dosage is above or below the optimum is to shut off the polymer feed for a few minutes. If the filtrate improves momentarily, the prior dose is excessive. If the filtrate deteriorates immediately, the prior dose was at or below optimum (Cleasby et al., 1984a).

A number of studies have reported successful use of alum and cationic polymers simultaneously. Typical dosages are 2 to 10 mg/L of alum plus about 0.2 to 2 mg/L of cationic polymer (McBride et al., 1977; Tate et al., 1977; Monscvitz et al., 1978; Cleasby et al., 1984a; Tobiason et al., 1992). These dosages are drawn from pilot- and plant-scale experience.

The impact of flocculation on direct filtration has been the subject of several studies that have been summarized by Cleasby et al. (1984a). These studies can assist in deciding whether flocculation should be provided in direct filtration. Comparing operation with and without flocculation, the provision of flocculation generally was found to improve filtrate quality before breakthrough, shorten the initial improvement period of the filter cycle, and reduce the rate of head loss development, but it also resulted in earlier breakthrough.

As noted earlier in this chapter, results of laboratory studies with model particles have shown that for the same suspended particle mass concentration, submicron particles cause much greater head loss than larger particles (Habibian and O'Melia, 1975; Tobiason et al., 1993) possibly because of greater exposed surface area for smaller particles (O'Melia and Ali, 1978) or greater deposit volume (more pore clogging) because of the fractal nature of the deposit (Veerapaneni and Wiesner, 1997). Implications for treatment application are that optimal pretreatment for filtration, especially direct filtration, may be based on increasing particle size via flocculation to decrease head loss development; a laboratory study (Westerhoff and Tobiason, 1991) and pilot results for in-line versus direct filtration (Glaser and Edzwald, 1979; Edzwald et al., 1987) support this concept.

The benefits or detriments of providing flocculation are best determined on a case-by-case basis by pilot- or full-scale studies. In some cases, flocculation has been found of no benefit (Al-Ani et al., 1986). Nevertheless, current practice usually includes a short flocculation period in direct filtration plants, typically about 10 minutes of high-energy flocculation at a velocity gradient G of up to 100 s^{-1}, although the range of flocculation detention times in existing plants has been very great, ranging from no flocculation to up to 60 minutes of flocculation (Letterman and Logsdon, 1976).

In-line filtration is appropriate for source waters that are consistently very low in turbidity and color. If flocculation were provided for such waters, long flocculation times would be required because of the poor opportunity for collision of particles. For example, flocculation was added to the in-line filtration plant at Las Vegas, but 30 minutes of detention was needed for this excellent source water (Monscvitz et al., 1978). Therefore, source waters are more economically treated by the addition of coagulant and an extended rapid mix of 3 to 5 minutes to achieve only the initial stages of particle aggregation before filtration. Few in-line plants are being built because flocculation adds flexibility to the plant operation. A bypass around the flocculator can be provided for contingencies.

Filter Details for Direct Filtration

Older summaries of direct filtration (Committee Report, 1980; Letterman and Logsdon, 1976) indicated filtration rates of full-scale operating plants of between 2.4 and 15 m/h (1 and 6 gpm/ft^2). Direct filtration pilot studies at three locations showed, however, that effluent turbidity was nearly constant over the range from 5 to 29 m/h (2–12 gpm/ft^2) and in one case as high as 44 m/h (18 gpm/ft^2) (Trussell et al., 1980). In these studies, both alum and cationic polymer were used for chemical pretreatment. Dual-media and deep beds of coarse sand were compared. Some larger direct filtration plants with deep beds of coarser media operated at high HLR have been put into operation; a few examples follow.

A high-rate direct filtration plant is operated by the Los Angeles Department of Water and Power to treat water from the Los Angeles Aqueduct. The plant design was based on extensive pilot studies that led to the choice of deep beds of anthracite (1.8 m, or 6 ft) with an ES of 1.5 mm and UC $<$ 1.4. The design filtration rate is 33 m/h (13.5 gpm/ft^2). Pretreatment includes ozone to assist coagulation with ferric chloride and cationic polymer, and flocculation is provided before filtration. The filters are backwashed by air alone first at 1.2 m/min (4 cfm/ft^2), followed by air plus water at 24 m/h (10 gpm/ft^2). The air is terminated before overflow occurs to prevent loss of filter media, and the water wash is continued alone at a higher rate during overflow (McBride and Stolarik, 1987). Some of the pilot studies were reported by Trussell et al. (1980).

Pilot studies for the then-proposed plants for Sydney, Australia, included deep-bed single-media and dual-media configurations (Murray and Roddy, 1993). The dual-media configuration studied included anthracite 1.0 to 3.0 m (3.2–9.8 ft) in depth with 1.7 to 2.5 mm ES over 0.15 to 0.30 m (0.5–1.0 ft) of crushed sand with 0.65 to 1.0 mm ES. The coarser dual media performed best in production per cycle and in filtrate quality. The Sydney Prospect Water Filtration Plant, the largest direct filtration plant in the world at a capacity of 3000 ML/d (780 mgd), was commissioned in 1996 and has 2 m (7 ft) of 1.7 mm ES sand media.

Patania et al. (1995) used deep beds of dual media (2.0 m, or 6.7 ft) of 1.25 mm-ES anthracite over 0.25 m (0.83 ft) of 0.6 mm-ES sand for in-line filtration pilot studies of cyst removal in Seattle. In 2000, the 454 ML/d (120 mgd) capacity Seattle Tolt Plant went into operation, incorporating preozonation, coagulation (ferric chloride and cationic polymer), and hydraulic flocculation prior to single-media filtration with 1.8 m (6.0 ft) of 1.3 mm ES anthracite with a UC < 1.4 (Tarbuck et al., 2002). The filters are approved for an HLR of 29 m/h (12 gpm/ft^2) and regularly operate at more than 25 m/h (10 gpm/ft^2); the underdrains have nozzles and no gravel.

In late 2009, commissioning was performed for the new 1800 ML/d (475 mgd) Seymour-Capilano Direct Filtration Plant for the metro Vancouver, BC, water supply. The configuration for the 24 dual-media filters is 1.7 m (5.6 ft) of 1.4 mm ES anthracite (UC < 1.5) over 0.3 m (1.0 ft) of 0.7 mm ES sand (UC < 1.5) (Phelan, 2010). The design filter HLR is 15 m/h (6 gpm/ft^2) in winter and 20 m/h (8 gpm/ft^2) in summer; initial winter operation indicates that filter-aid polymer is needed to have filter runs longer than 20 hours.

PRESSURE GRANULAR BED FILTERS

Pressure filters are sometimes used for rapid granular media filtration. The filter media are contained in a steel pressure vessel. The water to be filtered enters the filter under pressure and leaves at slightly reduced pressure because of the head loss caused by the filter media, underdrain, and piping connections.

Description of Pressure Filters

The pressure vessel may be a cylindrical tank with a vertical axis, such as shown in Fig. 10-27, or a horizontal-axis cylindrical tank. The horizontal cylindrical configuration has the disadvantage that the width of the filter bed is not constant from top to bottom, usually being wider at the top. This leaves dead areas along the walls that do not receive adequate fluidization during backwashing and therefore may not be washed effectively.

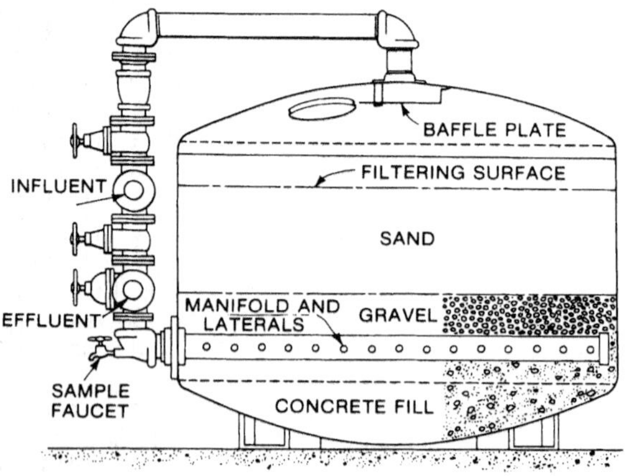

FIGURE 10-27 Cross section of a typical pressure filter.

In some horizontal-axis cylindrical tanks, the filter is divided into multiple (four or five) cells by vertical bulkheads. Cost advantages are gained by this configuration because only a single pressure vessel is needed. One filter cell can be backwashed by the production of the other cells that remain in service. This requires a filtration rate sufficiently high that the total production of the operating cells is sufficient to fluidize the media of the single cell being backwashed.

Comparison of Pressure and Gravity Filtration

While the outward appearances of a pressure and a gravity filter are quite different, the filtration process is the same. The same mechanisms for capturing particles are functioning in both. The same filter media, the same filtration rates, and the same terminal head loss should be used if comparable filtrate quality is desired.

The use of higher filtration rates and high terminal head loss in a pressure filter is tempting because the influent is under pressure, and more potential head loss is available. This temptation should be resisted, however, unless no detriment to the quality of the filtrate can be demonstrated on a case-by-case basis.

One advantage gained by pressure filtration is that water leaves the filter under a positive gauge pressure, and therefore, no negative pressure can ever exist in the filter bed. The potential problems associated with negative pressure, as discussed earlier, are therefore avoided.

Operation of Pressure Filters

Because of the similarities between pressure and gravity filters, the operating principles are identical. For example, appropriate pretreatment is equally important to pressure and gravity filters, patterns of filtrate quality will be the same, the impact of sudden rate increases will be just as detrimental, and the importance of proper backwashing is equally important.

The operation of a pressure filter is similar in most respects to the operation of a gravity filter. Proper backwashing of a pressure filter is more difficult, however, because the filter bed is not conveniently visible to the operator during the backwash operation. The visual observations that can be conveniently made for gravity filters are difficult or impossible with pressure filters. These include (1) the presence of filter cracks before the backwash or mudballs after the backwash, (2) the uniformity of backwash water distribution, (3) the uniformity of rate of cleanup of the wash water over the full filter areas, (4) the proper functioning of the auxiliary scour device such as surface wash or air scour, (5) the elevation and appearance of the top surface of the filter media after the backwash, (6) whether the media are fully fluidized during the backwash (if that is the intended washing method) and the extent of expansion of the filter media, and (7) the extent of loss of the filter media.

Because of these difficulties, failures of pressure filters have resulted in waterborne disease outbreaks partially attributed to the poor condition of the media in the filters (Kirner et al., 1978; Lippy, 1978). Prior concerns about the reliability of pressure filters and other concerns have caused some state regulatory agencies to exclude the use of pressure filters in the treatment of surface waters or other polluted source waters and lime-softened waters (e.g., Great Lakes–Upper Mississippi River Board, 2003).

Rate Control for Pressure Filters

Rate control for pressure filters is as important as it is for gravity filters because the same filtration mechanisms are functioning in both cases. Fewer options are available for pressure filters, however, because the pressure filter operates full of water under pressure. As a result, options involving changing water level are not available, and influent gravity flow splitting is not available. Therefore, the benefits of slow rate changes by allowing water levels to change are not available to pressure filters.

Typically, a bank or set of pressure filters is piped in parallel; thus the total head loss across each filter is identical, and the flow through each filter is inversely proportional to the clogging head loss in the filter. The total filter area of the filter bank is sized appropriately to meet local regulatory requirements based on feed-pump capacity. Individual filter flow controllers and flowmeters are not provided. After the backwash of all the filters in short succession, if one filter is passing more flow than it should, that filter is assumed to clog more quickly than the other filters, and the flow will be reduced because of the clogging resistance. Thus the system is considered self-equalizing during operation. However, if the multiple filters are backwashed at random intervals during a filter cycle, the cleanest filter will have the highest flow and the dirtiest filter the lowest flow. The extent of these differences will not be known if individual flowmeters are not provided.

Despite the common practice described earlier, it is desirable to have a flowmeter on each pressure filter in a bank of several filters. Flowmetering is useful to plant operators for diagnostic purposes to observe if something is wrong with one filter causing the flows not to be equally split. This could happen if one filter is not cleaned properly during backwashing, perhaps because of a clogged underdrain, or one filter flows at an excessive rate due to loss of some or all of the filter media.

Applications of Pressure Filters

Pressure filters tend to be used in small water systems. Many pressure filters are used in industrial water and wastewater filtration applications. They also are used widely in swimming pool filtration.

Advantages of pressure filters over gravity filters include the following: The filtrate, which is under pressure, can be delivered to the point of use without repumping. In some treatment plants, source water (often from a well) can be pumped from the source through the treatment plant and directly to the point of use by the source water pumps. Also, it is somewhat easier to automate the operation of a treatment plant equipped with pressure filters.

Some groundwaters containing iron can be treated by pressure aeration and/or chemical oxidation and then filtered directly by pressure filters. This approach has received considerable application for small communities. Not all iron-bearing waters respond successfully to this method of treatment, however, and prior pilot testing usually is required. Similarly, some groundwaters that require manganese (and possibly iron and organic matter) removal can undergo chemical oxidation and pressure media filtration with direct introduction to the distribution system.

SLOW SAND FILTRATION

Description and History

The slow sand filter is a sand filter operated at low filtration rates (<0.3 m/h, or <0.12 gpm/ft^2), traditionally without the use of coagulation in pretreatment. The sand size is somewhat smaller (~0.15–0.30 mm) than that used in a rapid-rate filter, and this, plus the low filtration rate, results in the solids being removed almost entirely in a thin layer on the top of the sand bed and in the uppermost portion of the sand bed. The layer on top of the sand, composed of dirt and living and dead micro- and macroorganisms from the water (i.e., the *schmutzdecke*, or "dirty skin"), becomes the dominant filter medium as the filter cycle progresses. When the head loss becomes excessive, the filter typically is cleaned by draining the supernatant water below the sand surface and physically removing (scraping) the dirty layer along with 1 to 3 cm (0.5–1 in) of sand. Typical filter cycle duration between cleaning may vary from 1 to 6 months (or longer) depending on source water quality and the filtration rate. Biological activity is a key aspect of the slow sand filtration process, and this must be considered in the design and operation of slow sand filters.

Slow sand filtration was developed in England by James Simpson in the early nineteenth century and continues to be used successfully in the United Kingdom and Europe. In the United States, this treatment process has been used to treat high-quality upland surface waters in New York and New England since the late 1800s and early 1900s, but because it could not treat turbid waters in the Midwest and elsewhere in the United States, the process was not used widely. As a result, rapid-rate filtration came to be the filtration process of choice for most water utilities.

Renewed interest in slow sand filters occurred because numerous waterborne outbreaks of giardiasis occurred in the latter 1970s and early 1980s in communities using unfiltered or poorly filtered water. Slow sand filtration is a simple technology requiring no knowledge of coagulation chemistry and is quite attractive for small installations treating high-quality surface waters. The possibilities for applying this treatment process for small water systems led to research efforts demonstrating the efficacy of the slow sand filter in removal of *G. lamblia* and to its application in numerous communities in New England, the Rocky Mountain Region, and the Pacific Northwest in the United States, as well as in Canada.

Mechanisms of Filtration and Filter Performance

Biological action is the key to removal of microbes, and it influences turbidity removal in slow sand filters. This is one important difference between how these filters and rapid-rate filters work. Design and operation of slow sand filters need to be consistent with the objective of maintaining an active biological population within the sand bed.

The Organisms in a Slow Sand Filter. Haarhoff and Cleasby (1991) undertook an in-depth review of the ecosystem in a slow sand filter, which typically includes bacteria, protozoa such as rhizopods and ciliates, rotifers, copepods, and aquatic worms. They explained that the kinds of organisms found change as water flows deeper into the sand bed. At the surface of the filter bed, a layer of decaying biological matter and bacteria called the *schmutzdecke* develops. This is a German word for "dirty skin." Some straining occurs in the *schmutzdecke*. As the water enters the *schmutzdecke*, biological action breaks down some organic matter, and inert suspended particles may be physically strained out of the water. The water then enters the top layer of sand, where more physical straining and biological action occur, and attachment of particles onto surfaces of the sand grains takes place. Algal mats of filter-clogging species may form at the *schmutzdecke*, and some algal cells may penetrate into the top inch (about 30 mm) into the sand in large numbers.

Duncan (1988) studied the ecology of slow sand filters in facilities owned by Thames Water Utilities and reported that bacteria are most numerous within the top 5 cm (2 in), decreasing at lower depths. In a biologically mature filter, protozoa were most numerous from near the surface to depths of 5 to 10 cm (2–4 in). They feed on bacteria and sometimes also on algae cells. Among the largest organisms found in a slow sand filter bed are round worms, flatworms, and segmented worms, which can be as large as a few millimeters in size. Some of the worms feed on detritus, that is, fragments of organic matter, and tend to be deeper in the bed than bacteria and protozoa.

Understanding that the sand bed is inhabited by living organisms is important to the design and operation of slow sand filters. The organisms need both food and oxygen to metabolize the food they ingest, and having a steady input of food and oxygen to the filter bed helps to stabilize their population. Long-term interruptions of flow can lead to depletion of oxygen in the filter bed and to die-off of organisms that are the key to effective slow sand filtration. Drying a slow sand filter bed has even more serious consequences and could lead to elimination of most or all of the biota in the bed. Therefore, the flow of influent water should be continuous during water production, and filter-bed maintenance activities need to be performed in a prompt manner to minimize downtime and harmful effects to the biota.

The Schmutzdecke. The *schmutzdecke* that develops on the sand bed surface provides significant removal of particulate matter, and an initial improvement period (or ripening period) frequently can be observed at the beginning of each cycle after removal of the *schmutzdecke* along with a thin layer of sand at the end of the previous filter run (Cleasby et al., 1984b; Cullen and Letterman, 1985). The initial improvement period was observed by Cullen and Letterman (1985) to vary from 6 hours to 2 weeks, although most improvement periods have been reported to be less than 2 days (Cleasby et al., 1984b; Cullen and Letterman, 1985). A filter-to-waste period of 2 days was recommended, in which cysts of *G. lamblia* are of concern (Cleasby et al., 1984b), but Bellamy et al. (1985a) found no effect of scraping on *Giardia* cyst removal after 41 weeks of filter operation, when the sand bed of a pilot filter was biologically mature.

In addition to the initial improvement after scraping, a distinct improvement in the performance of a filter with new sand occurs as it matures over several cycles or runs, as indicated by the removal of coliform bacteria, particles, and *Giardia* cysts (Fox et al., 1984; Cleasby et al., 1984b; Bellamy et al., 1985a). In a Midwestern pilot study, after the first four filter cycles, average reduction of turbidity was 97.8 percent or better, removal of cyst-sized particles (7–12 μm) was 96.8 percent or better, removal of total particles (1–60 μm) was 98.1 percent or better, removal of total coliform bacteria was 99.4 percent or better, and removal of chlorophyll *a* was 95 percent or better after the first 2 days (the ripening period) of each run (Cleasby et al., 1984b).

Removal of Microorganisms. Slow sand filtration was effective in removing pathogens and reducing the incidence of waterborne disease in London and other cities in the 1800s even before scientists understood that microbes could cause communicable diseases such as cholera and typhoid fever. Bacteria removal can be as great as 3 log to 4 log (Fox et al., 1984; Cleasby et al., 1984b; Bellamy et al., 1985a; 1985b). Following the identification of viruses as a cause of waterborne illness, researchers at the Metropolitan Water Board in London (now part of Thames Water Utilities) evaluated virus removal and again reported values as high as 3 log to 4 log (Poynter and Slade, 1977). Research sponsored by the USEPA in the 1980s demonstrated that *Giardia* cyst removal also could be in the range of 3 log to 4 log (Bellamy et al., 1985a, 1985b; Pyper, 1985). Similar removals have been attained for *Cryptosporidium* oocysts (Economic and Engineering Services and Thames Water Utilities, 1996). Mature slow sand filters evaluated at the University of Waterloo (Anderson and Huck, 2005) also performed in this range and sometimes better for removal of *Giardia* cysts and *Cryptosporidium* oocysts. These results have been attained with filters having a developed, or mature, biota.

The removal of microorganisms in filtration by predatory action is probably unique to slow sand filtration. The role of ciliate protozoa as bacterial predators was investigated in laboratory studies (Lloyd, 1996) of a pilot filter having a bed of sterilized sand. Removal of bacteria in filtration tests without ciliate protozoa present was about 1 log lower than the 2 log to 3 log removal attained when ciliate protozoa were inoculated into the sand bed. Bichai et al. (2009) presented photomicrographs of *Cryptosporidium* oocysts ingested by rotifers in a review of work on slow sand filters undertaken at the Weesperkarspel plant.

This is further evidence that predation of microbes occurs in slow sand filters and thus is a mechanism by which microbes are removed by slow sand filtration.

Because slow sand filters are biological filters, microorganism removal declines when water is cold and the metabolism rate of the biota facilitating microbial removal slows greatly. *Giardia* cyst removal by a small slow sand filter used for demonstration testing was in the range of 3 log to 4 log when water was 7.5 to 21°C. When water was less than 1°C, cyst removal generally was in the 1 log to 3 log range (Pyper, 1985). The effect of low temperature on filter performance needs to be considered by designers if cold source water will be treated.

Removal of Inorganic Particles. Removal of organic particles in a slow sand filter can be explained in part by predatory action of the larger organisms as they use smaller ones for food, but clay and other mineral particles are not an organic food source, so some other explanation for removal of inorganic turbidity particles is needed.

Experiments in which Horsetooth Reservoir water originating high in the Rocky Mountains was used (Bellamy et al., 1985b) showed that turbidity and total coliform removals were lower in a slow sand filter that had been chlorinated between filter runs to minimize biological activity compared with a control filter operated in the normal manner. On the other hand, a slow sand filter treating water with supplemental sterile nutrients for promotion of biological growth removed both turbidity-causing particles and total coliform bacteria more effectively than the control filter. This work indicated that biological action played an important role in turbidity removal. Huisman and Wood (1974) stated that the action of microbes in a slow sand filter results in the formation of a sticky gelatinous film on surfaces of the *schmutzdecke* and on sand grains that can hold inert matter until it is removed when the filter bed is cleaned.

Additional explanation for removal of inorganic particles by slow sand filtration was found in literature on the activated-sludge wastewater treatment process. Pavoni et al. (1972) reported that some kinds of bacteria release exocellular polymers that can flocculate not only bacteria in the activated-sludge liquor but also clay particles. If higher levels of bacterial growth in a slow sand filter result in release of greater concentrations of exocellular polymers that can make the surfaces of sand grains more "sticky," this could enhance clay particle removal by attachment to the sand. The results of Bellamy et al. (1985b) reinforce the importance of maintaining a healthy and active biological population in the filter bed and strongly suggest a role for exocellular polymers in slow sand filters.

Biological Filtration for Removal of Dissolved Organic Matter. The extent to which dissolved organic matter can be removed in slow sand filters depends on the biodegradability of the dissolved organics. Although readily degradable organic matter can be removed effectively by the biota in a slow sand filter, biota also exist in natural waters, and they use organic matter there. Readily degraded dissolved organic matter may have been used already as a food source before source water is treated by slow sand filtration. Organic matter that is not readily degraded in natural waters and thus is persistent, such as natural color or humic substances, is not likely to be used easily as a food source by the biota in a slow sand filter. Water sources with the potential to have high concentrations of color include marshes, swamps, and lakes having large quantities of vegetation, and removal of color from such waters may be in the range of 25 percent. Rooklidge et al. (2005) studied removal of antimicrobial compounds and reported that only those with high filter media sorption coefficients were removed well by slow sand filtration. Based on these observations, slow sand filters would not be expected to be effective for removal of some of the pharmaceutical compounds and personal-care products that pass through biological wastewater treatment plants and persist in low (e.g., microgram per liter) concentrations in surface waters.

Slow Sand Filter Design

Facility and Filter Bed. Slow sand filters are the least complex type of granular media filter. They consist of a basin with a bed of relatively fine sand placed over support gravel and underdrain piping (Fig. 10-28). The basin can be a concrete filter box or an earthen embankment. In modern designs, the embankment and basin bottom may have an impervious liner to prevent water loss through the bottom of the basin and erosion of the basin slopes. In the usual operating mode, influent water is introduced above the filter bed in a way that does not disturb the sand, but for the initial filling and after the bed has been dewatered, water must be introduced from beneath the filter to drive air out of the bed as it is filled.

Twort et al. (2000) indicated that five valves are needed for a slow sand filter. They are provided for the purposes of filling the filter from the bottom to drive air out of the bed during initial filling and after maintenance procedures, filtering to waste and draining water out of the bed, draining supernatant water down prior to dewatering the bed for maintenance, controlling influent flow, and controlling effluent flow.

At the bottom of the filter box or basin, collector piping is surrounded by support gravel about 0.4 to 0.6 m (1.3–2 ft) deep. Above the gravel layer is filter sand with an initial depth of 0.8 to 1.2 m (2.7–4 ft). Depth of supernatant water over the media is typically about 1 m (3 ft).

The effective size of filter sand usually is in the range of 0.15 to 0.30 mm, with UC < 5 and preferably <3. Smaller media sizes tend to provide more effective particle removal but cause head loss to be higher than larger media. Bellamy et al. (1985b) studied the effect of sand size and reported that total coliform removal was 96 percent for a filter having an effective size of 0.6 mm and 99.4 percent when the effective size was 0.1 mm. The recommendations for sand size, bed depth, UC, and supernatant water depth just given are similar to criteria suggested by Visscher (1990) for filters in developing countries.

Slow sand filters designed for operation in cold regions generally are covered because scraping sand in an ice-covered slow sand filter cannot be done until the ice layer is removed from the water. Providing structures to cover filter beds requires additional capital investment, so some exceptions have been made to this general concept of covering the filters when extended freezing conditions are expected. A second reason for covering slow sand filters to

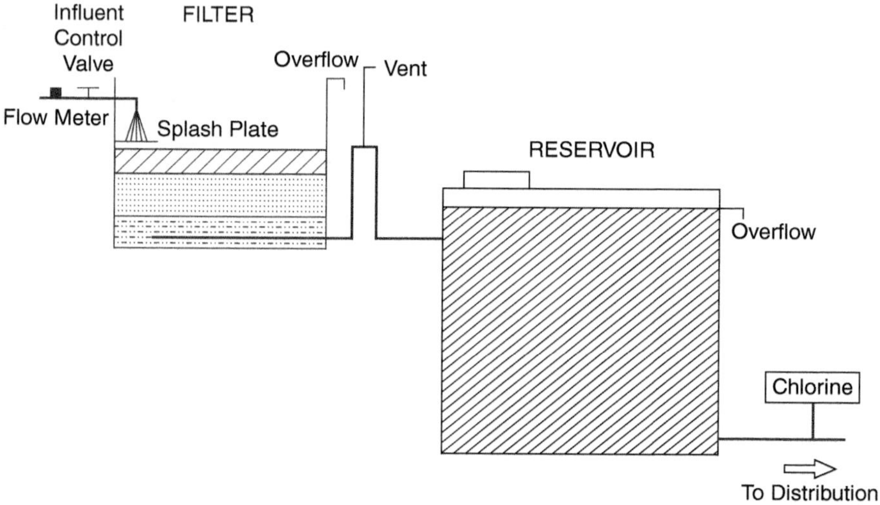

FIGURE 10-28 Slow sand filter influent control. (*Source: Logsdon, 2008.*)

prevent freezing is the possible damage that may be caused if a thick layer of ice forms on the supernatant water and pushes against the filter walls. Note that if filters are covered, it is important that sufficient headroom for maintenance workers be provided between the surface of the sand bed and the bottom of the roof structure covering the filter.

Filtration Rates and Rate Control. The filtration rate may range from 0.1 to 0.3 m/h (0.04–0.12 gpm/ft^2), and sometimes even higher rates are provided. Higher filtration rates result in smaller sizes for filter beds, but removal of microbes and turbidity can decrease when the filtration rate is higher. The low filtration rates require filter bed areas 20 to 50 times larger than filter areas for rapid-rate filtration for production of the same quantity of water per day. For many small installations in rural areas, however, the much larger filter area needed for slow sand filter beds does not present a problem because of the availability of land. Good design practice provides multiple filter beds that are sized so that when one filter is removed from service for cleaning by removal of a thin layer of sand or for replacing sand to restore the initial bed depth, the other filter or filters can produce sufficient water to meet the system demand.

Filter efficiency is influenced by water temperature because the metabolic rate of the biota in a slow sand filter and in the *schmutzdecke* is impeded or decreased by cold water. As a result of the temperature effect, removal of microbiological contaminants has been demonstrated to be reduced somewhat in water when the temperature is near freezing. Operating at a slower filtration rate can help to counteract this decrease in filtration efficiency. In many communities, water demand peaks in warm weather when lawns and gardens are watered, so lower filtration rates can be used in winter. An exception to this could be a community that has a heavy influx of tourists and high water demand during the winter season. If cold water will be encountered, this can be a consideration in selection of the filtration rate used for plant design.

As with rapid-rate filtration, rate control is important and can be accomplished by influent or effluent rate control, as shown in Figs. 10-28 and 10-29. The simplest modes of rate control involve manual operation of valves to adjust the rate of flow to the desired value. Influent rate control can be applied simply by use of an influent water channel and a weir at each filter in plants with multiple filters, with each weir having the same design and elevation for equal flow splitting. A gate or other device is needed to block the weir and stop inflow when maintenance is performed. A splash plate is needed to prevent erosion of the sand as it falls from the weir into the sand filter, and a discharge weir higher than the top of the sand bed is provided to prevent dewatering of the bed during operation. An example of influent rate control is provided in Fig. 10-28. Rate changes will occur gradually as the

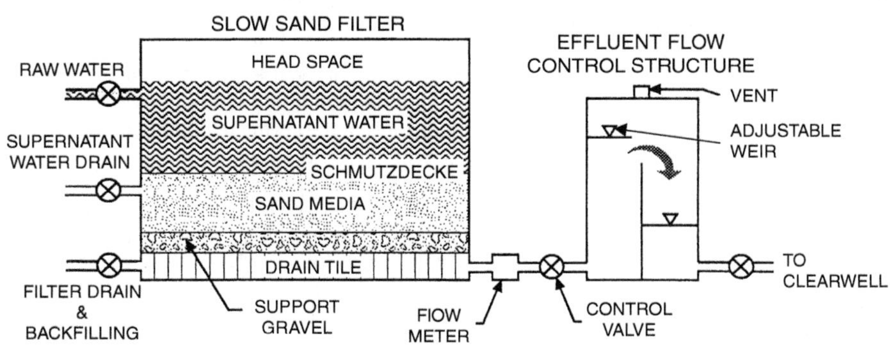

FIGURE 10-29 Slow sand filter and effluent control. (*Source: Logsdon, 2008.*)

water level in the filter box raises or lowers depending on the rate of inflow versus outflow and the available head to cause flow through the filter bed.

For filter effluent rate control (shown in Fig. 10-29), rate changes are made by adjusting an effluent valve, and these changes should be made gradually and as seldom as possible. Guidelines for two water utilities in the United Kingdom limit filtration rate increases to about 10 percent per hour (Ratnayaka, 2009). Particles that are not part of the biofilm in a slow sand filter may be weakly bound and can be flushed out by a rapid increase in filtration rate. A practice to avoid in effluent rate control is to maintain the desired supernatant water level in the filter by frequently opening and closing the effluent valve. Frequent on-off operation is not acceptable, and shutdowns of filter flow are to be avoided except for filter maintenance.

When filtered water storage is sufficient to cover variations in system demand over a day's time, adjustment of filtration rate can be made as infrequently as once per day, which is appropriate for small systems that are operated by part-time personnel who have only an hour or two each day to devote to water treatment operations. Sizing finished water storage so that it is sufficient to account for hourly fluctuations in demand throughout the day enables the filter to operate throughout the day at a production rate similar to the total water demand for that day.

Other Information Sources. With the increased interest in slow sand filtration in the 1980s, the AWWA Research Foundation sponsored development of a design manual for slow sand filters (Hendricks et al., 1991). This is no longer available, but a book by Hendricks (2006) contains a chapter, "Slow Sand Filtration," with about 50 pages of text and photos. The author states that the chapter was adapted, to a large extent, from the slow sand filter design manual for which he was the senior author in 1991.

Another important source of information for slow sand filter design is pilot-plant filtration testing data obtained by treating the source water that will be treated by the full-scale plant that is being planned. This is especially important if no pretreatment is planned to prepare the water for slow sand filtration. Pilot-plant studies are not labor-intensive because with the low filtration rate used for slow sand filtration, changes in filter performance do not occur in short periods of time. One daily check on filter operation for sample collection and data recording is often sufficient, and the testing results can provide guidance for design when a source water has not been treated previously by slow sand filtration.

Source Water Quality

Slow sand filters have not been successful in treating clay-bearing river waters because the clay penetrates too deeply into the filter and cannot be removed in the normal surface-scraping operation. In addition, they are not effective for removing color, typically achieving only 25 percent removal.

A 1984 survey of 27 existing full-scale plants in the United States found that the mean turbidity of the source water was about 2 ntu, with the peak about 10 ntu (Slezak and Sims, 1984). Mean lengths of filter runs varied from 42 days in the spring to 60 days in the winter. Wide variations were reported, with two reports of cycles as long as 1 year.

Performance of seven plants in New York was evaluated by Cullen and Letterman (1985), including three of the plants reported on by Slezak and Sims. In New York, the turbidity of the source waters being used generally was less than 3 ntu. One plant with 6 to 11 ntu source water was not meeting the existing (1984) 1 ntu finished water turbidity MCL. Average run lengths varied from 1 to 6 months.

Short filter runs may occur when algal blooms take place in source water. In a pilot-plant study, Cleasby et al. (1984b) observed a 10 day run when the chlorophyll *a* concentrations were 131 and 132 μg/L in two samples of raw water obtained during the run. Algae and diatoms that have been classified as filter-clogging species are most likely to have an adverse effect on filter run length. Color plates of filter clogging algae can be found in *Standard Methods* (APHA, AWWA, and WEF, 2005).

When slow sand filters are used without pretreatment or posttreatment other than disinfection and corrosion control, source water quality must be good based on how these filters work. Cleasby (1991) has recommended the following guidelines for source water that is ideal for slow sand filters: turbidity < 5 ntu; no heavy seasonal algal blooms occur, and chlorophyll $a < 5$ μg/L; Fe < 0.3 mg/L; and Mn < 0.05 mg/L. His recommendations did not cover synthetic organics or inorganic contaminants such as arsenic. Since the capability of slow sand filtration for removal of microbiological contaminants is excellent, it is most suited for application without supplemental treatment other than disinfection when source waters do not contain contaminants that resist removal by biological filtration.

Modifications to Slow Sand Filtration for Enhanced Performance

When source water quality is beyond the range recommended for slow sand filters used alone, pretreatment can extend the capability of this process so that a wider variety of source waters can be treated. Higher turbidity, color and natural organic matter, and synthetic organic chemicals can be removed when pretreatment or posttreatment processes are added.

For some source waters, roughing filters, infiltration galleries, or infiltration wells in or near streams or rivers can remove sufficient turbidity that the remainder can be dealt with in a slow sand filter. These concepts were reviewed in detail by Cleasby (1991). Although chemical pretreatment is usually not used, an emergency pretreatment practice using chemical coagulation followed by flocculation and sedimentation in channels and basins at Salem's slow sand filter facility on Geren Island was carried out during and following a major flood in Oregon in early 1996. This ad hoc pretreatment attained turbidity reductions ranging from 80 to 90 percent and enabled the filters to continue operating without being plugged over a period of several weeks of abnormally high turbidity (Krueger et al., 1999).

Changing the nature of dissolved organic constituents can increase their biodegradability. This is most readily accomplished by use of ozone for pretreatment ahead of a slow sand filter. Preozonation has been shown to increase the biodegradability of dissolved organics in biologically active rapid-rate filters, and it does the same for slow sand filters (see Chap. 7 for a discussion of ozone). The extra microbiological growth resulting from increased availability of food can result in formation of biofilms on slow sand filter media to the extent that head loss may increase at a faster rate than if ozone were not used. Adapting ozone to an existing slow sand filter might decrease filter run times, whereas for a new filter with preozonation, designers can consider using a slightly larger media size to compensate for the possibility of added head loss caused by more biofilm formation.

Preozonation proved beneficial in a slow sand filtration study performed in the United Kingdom. Reduction of true color by slow sand filtration alone was about 20 percent, whereas reduction by preozonation and slow sand filtration was almost 75 percent (Greaves et al., 1988). In addition, preozonation increased removal of TOC from 8 to 15 percent for slow sand filtration alone to 25 to 35 percent, indicating that ozone was breaking apart organic molecules that were not readily biodegradable and creating substances that were more easily used as food by organisms living in the filter bed.

To enable slow sand filter plants to effectively treat source water containing pesticides and satisfy the European Community limit of 0.1 μg/L for individual pesticides, Thames Water Utilities, Ltd., developed the GAC Sandwich filter (Bauer et al., 1996). A full-scale filter consisting of a layer of 300 mm of filter sand as the bottom layer, 150 mm of Chemviron 400 GAC in the middle, and 450 mm of filter sand on top was operated at filtration rates ranging from 0.1 to 0.3 m/h and demonstrated excellent removal of pesticides.

Treatment with slow sand filtration supplemented by preozonation, a roughing filter for enhanced particle removal, and a GAC filter following the slow sand filter was evaluated at the University of Waterloo in Canada (Anderson and Huck, 2005). Source water was the Grand River, which has a watershed in agricultural and urban areas in Ontario. When the roughing filter and slow sand filter were biologically mature, filtered water turbidity was maintained below 0.3 ntu even when source water turbidity exceeded 25 ntu and the slow sand filter was operated at a rate of 0.4 m/h. TOC removal for the entire process train was 58 to 65 percent compared with about 15 percent for a slow sand filter operated in parallel but with neither pretreatment nor posttreatment. The pretreatment processes combined with slow sand filtration reduced THM formation by 58 to 60 percent. The test results demonstrated that pretreatment compatible with slow sand filtration can extend its capabilities for removal of turbidity and organic contaminants substantially.

Startup and Maintenance Procedures

Placing a New Filter into Service. When a new slow sand filter is placed into service, a period of weeks is needed for the biota to develop within the sand bed and provide effective treatment. With passage of time, the *schmutzdecke* develops, and a variety of organisms grow within the filter bed. Among the approaches taken for evaluation of slow sand filter efficacy are measurement of turbidity, *Escherichia coli*, total coliform bacteria, viruses, and *Giardia* cysts in filter effluent (Haarhoff and Cleasby, 1991). In their review of the slow sand filter literature, Haarhoff and Cleasby reported five studies in which a new filter bed was reported to have matured within a time of 35 to 60 days. In three of the five studies, the monitoring done for filter maturity was for total coliform bacteria or *E. coli*. Colder water would slow the development of biota in the bed and the maturation time, and one study done in Vermont was reported by Haarhoff and Cleasby to have had a 100-day initial operating period before erratic removal results were no longer observed.

Locally obtained sands have been used for the filtering material in some slow sand filters. When this was done without thoroughly washing the local sand, turbidity in filter effluent exceeded that of the filter influent during startup. Tanner and Ongerth (1990) reported that a new slow sand filter with beds of locally obtained sand that initially contained clay fines to the extent of 4 percent by weight produced filtered water turbidity that generally exceeded raw water turbidity from 3 to 15 months of operation after startup. Filtered water turbidity at this plant exceeded the MCL during 6 of the 12 months of Tanner and Ongerth's study. This performance was attributed to incomplete washing of the filter sand before it was placed in the beds. The gradual discharge of fine particles that cause turbidity has the potential to lengthen the time between starting the initial flow of water through the filter and acceptance of the filter as being capable of producing potable water for public consumption.

Cleaning When Terminal Head Loss Is Reached. Slow sand filters are cleaned by removing the *schmutzdecke* along with a small amount of sand, an operation known as *scraping*. The sand that is removed can be cleaned hydraulically and stockpiled for later replacement in the filter. The scraping operation can be repeated until the depth of the filter bed has decreased to about 0.5 to 0.6 m (1.6–2 ft), at which time the design depth should be restored, a procedure referred to as *resanding* the filter. Resanding is described later in this section.

The sand in small plants is usually skimmed by hand using broad shovels. Scraping and resanding are labor-intensive operations, and consideration should be given to reducing the manual labor involved. By the use of efficient motorized buggies or hydraulic transport, and by limiting removal to a 2.5 cm (1-in) depth, 5.4 h/100 m^2 (5 h/1000 ft^2) was considered to be an adequate allowance for labor (Cullen and Letterman, 1985). One small plant reported use of asphalt rakes to scrape the dirty surface into windrows from which it was then shoveled into buckets for transport. Using this technique, the filter could be cleaned using labor at a rate of 1.3 h/100 m^2 (1.2 h/1000 ft^2) when the depth of sand removed from the filter surface was about 0.5 cm (0.2 in), according to Seelaus et al. (1986).

The longer a filter is drained for the scraping operation, the longer will be the initial improvement period during the subsequent run (Cullen and Letterman, 1985). Therefore, cleaning should be accomplished quickly to minimize the time during which the filter has been drained.

Another approach to cleaning a slow sand filter bed is *wet harrowing*. This method of cleaning has been used with success at a large slow sand filter plant in New England (Collins et al., 1991). To accomplish wet harrowing, the supernatant water is drained from the filter down to a depth of about 30 cm (12 in). Then a rubber-tired tractor equipped with a comb-tooth harrow is used to stir the top 30 cm (12 in) of sand while the remaining supernatant water is allowed to drain off to the side of the filter bed. The horizontal flow of water carries away the dirt and *schmutzdecke* stirred up by the harrowing. If the harrowing is not completed by the time the water level drops to 8 cm (3 in) over the bed, harrowing is stopped, and the bed is refilled to about 30 cm (12 in) so that the harrowing can resume. This cleaning method has been reported to extend the time for removing sand from the beds for a thorough cleaning to intervals of 8 to 10 years.

After filter cleaning or scraping, operating the filter and wasting filtered water for a day, or perhaps longer, may be appropriate. A conservative approach would be to test for coliform bacteria in filter effluent and filter to waste until coliform data are acceptable. The time needed to verify that filtered water quality is satisfactory after scraping is a reason for having multiple filter beds.

Replacing Sand When Bed Level Reaches Minimum Depth. Adding sand to the filter bed (resanding) is done when a minimum acceptable bed depth, about 0.5 to 0.6 m (1.6–2 ft), is reached after repeated filter scrapings. To do this, the bed must be dewatered first and then scraped to remove the *schmutzdecke* and accumulated dirt at the top of the sand layer. Next, starting at one wall of a rectangular or square filter bed, a trench parallel to the wall is excavated about 1 m (3 ft) wide, equal to the depth of the layer of sand that needs to be added to restore the filter to its original surface elevation. The excavated sand can be set aside for use at completion of the resanding task. New sand or cleaned, recycled sand that had been scraped off the filter previously is placed in the trench formed by removing the old sand, and then a trench of old sand is excavated adjacent to the previous trench that was filled. The old sand removed from the second trench is placed on top of the new sand that was placed in the first trench. This procedure results in placing sand from the lower part of the filter onto the top portion of the bed in the resanded filter. The pattern is repeated across the entire filter. When the last trench is excavated adjacent to the far wall (the one opposite the wall where excavation was first performed), the sand that was excavated first then is placed in over the new or reused sand that was placed into the last trench.

Use of the trenching procedure is more laborious than merely putting sand on the top of a filter that needs a deeper bed, but a couple of advantages can result from doing the trenching. If the resanding is done in a prompt manner, especially in smaller installations, some of the biota in the sand bed may survive the resanding procedure and be able to serve as a source for organisms in the resanded bed. A second reason for using the trenching procedure instead of just adding more sand to the shallower filter is to prevent development of a less pervious layer of sand in the upper zone of the undisturbed sand. For example, if a filter is repeatedly resanded when the depth decreases to 60 percent of the original depth, the sand just below the 60 percent level may become clogged, resulting in increased clean-bed head loss.

Letterman (1991) has indicated that resanding can require as much as 50 hours of labor per 100 m^2 (53 h per 1000 ft^2). This is a procedure that calls for advanced planning, and for a small system with limited staff, finding extra workers on a temporary basis to help with this task is a way to accomplish it in a timelier manner. Washing used sand again just before resanding would be helpful for keeping dirt and contamination out of the filter bed.

Following resanding, the renovated filter may need to be run to waste for a few days. Letterman (1991) reported that earlier literature had suggested a period of 1 to 2 days for filter-bed ripening. Monitoring filtered water turbidity and coliform bacteria will provide information on the quality of water produced after resanding and will indicate when filter effluent again can be sent to the distribution system.

Recommended Standards for Water Works (Great Lakes–Upper Mississippi River Board, 2003) specifies that slow sand filters should be operated to waste after scraping or resanding until the filter effluent turbidity is consistently less than the standard required for the system. For slow sand filters, the SWTR requires that representative samples of filter effluent should be equal to or less than 1 ntu in 95 percent of the measurements taken each month (USEPA, 1989). Therefore, an appropriate turbidity goal is less than 1 ntu.

When coliform bacteria monitoring is done to assess the condition of a filter bed with regard to ripening, coliform bacteria density in undisinfected filtered water samples during the ripening period can be compared with data for similar samples prior to termination of the filter run for scraping or resanding. Comparable numbers of bacteria in the effluent after filter maintenance or comparable removal percentages would indicate that the filter had returned to its earlier capability for removal of microbiological contaminants.

Performance Monitoring

As a result of the low filtration rate, monitoring of slow sand filter performance can be done on a daily basis, in contrast to the regulatory requirement in the United States that filtered water turbidity must be monitored at 15-minute intervals for rapid-rate granular media filtration. Furthermore, whereas a filter run at a rapid-rate filtration plant sometimes can last less than 24 hours, slow sand filter runs can last for days, weeks, or even months, except for unusual events such as an algae bloom that results in formation of a mat of filter-clogging algae on the slow sand filter surface. Because head loss increases slowly, monitoring head loss once per day is adequate, and this can be done by use of water piezometers rather than by sophisticated instrumentation and electronics. The filtration rate also must be monitored, but if the filter effluent passes through a totalizing water meter that also has a rotating dial that can be used to calculate a rate, this may be sufficient for obtaining needed operating data. Total volume filtered per day for each filter also would be included in daily operating data. In the United States, other regulatory requirements related to disinfectant residual and coliform sampling need to be met.

Application to Small Systems

Slow sand filters are especially well adapted for use by small systems because of their simplicity of operation. These filters are quite suitable for the part-time mode of operation affordable at small systems because coagulation chemistry usually is not involved. Duties for an operator would include checking head loss, checking and adjusting filter production to provide for system demand, and measuring and recording filtered water turbidity and disinfectant residual.

Tanner and Ongerth (1990) reviewed the performance of slow sand filters operated by small water systems in northern Idaho. During their 1-year study, the plants were sampled weekly for measurement of turbidity, total coliforms, fecal coliforms, and heterotrophic plate count bacteria. The authors concluded that if a slow sand filter is designed and operated according to accepted standards, it should provide effective water treatment on a consistent basis. Of special relevance to small systems, they noted that plant operators should be educated about proper operating techniques for slow sand filters. They indicated that drinking water regulatory personnel should provide training to operators and perform sanitary surveys to help ensure that slow sand filters are operated and maintained acceptably.

Green and Benjamin (2000) reported that 23 slow sand filtration plants had been constructed in First Nation communities in British Columbia. Of these, 11 used only slow sand filtration, and 12 included pretreatment consisting of alum coagulation and settling. In some instances, a limestone contactor was used in pretreatment to aid alum coagulation.

Pre-engineered slow sand filtration plants similar in design to the process train studied by Anderson and Huck (2005) at the University of Waterloo, but with the GAC filter placed between the roughing filter and the slow sand filter, have been fabricated and are in use in several Canadian provinces and in some states in the United States. As of 2007, more than two dozen of these full-scale pre-engineered (package) slow sand filtration plants had been installed or were under construction in the United States and Canada (Jobb et al., 2007).

An interesting development for small systems is the use of precast concrete filter boxes (Riesenberg et al., 1995) rather than constructed-in-place concrete filter boxes. The precast boxes were reported to provide better quality control during fabrication at a reduced cost compared with filters constructed in place. This type of construction would lend itself well to a concept of modular slow sand filter units and might decrease the time needed for constructing the plant, as well as overall construction costs.

Summary

Slow sand filtration is a biological process as well as a physical process. Without the biological effects, slow sand filters are not very effective. Therefore, these filters must be managed and operated with the goal of maintaining an active biological system within the bed so that all the benefits of this process can be realized. Because of their simplicity of operation, slow sand filters are well suited for application in small water systems that may rely on part-time operators.

PRECOAT FILTRATION

Precoat filtration is a USEPA-accepted filtration technique for potable water treatment. When the SWTR went into effect in 1989, it provoked a renewed interest in this filtration process. Precoat filters use a thin layer of very fine material such as DE as a filter media. In precoat filtration, the water to be filtered is passed through a uniform layer of the filter media that has been deposited (precoated) on a septum, a permeable material that supports the filter media. The septum is supported by a rigid structure termed a *filter element*. As the water passes through the filter media and septum, suspended particles about 2 μm and larger are captured and removed (AWWA, 1995). The majority of particles removed by the filter are strained (i.e., trapped) at the surface of the layer of filter media, with some being trapped within the layer. As the filter cycle proceeds, additional filter media, called *body feed*, are regularly metered into the influent water flow in proportion to the solids being removed. Without the regular addition of body feed, the head loss across the precoat layer would increase rapidly. Instead, the removed particles intermingle with the body-feed particles so that permeability of the cake is maintained as the thickness of the cake increases gradually. By maintaining permeability of the cake in this way, the length of the filter cycle is extended.

Ultimately, a gradually increasing pressure drop through the filter system reaches a point where continued filtration is impractical. The forward filtration process is stopped,

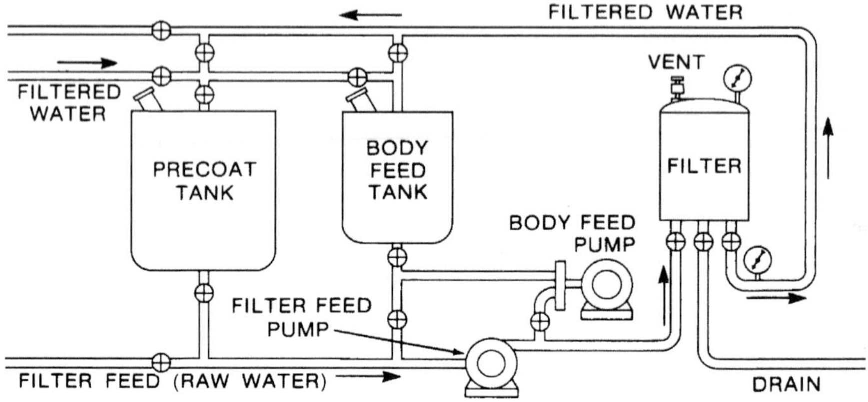

FIGURE 10-30 Typical precoat filtration system. (*Courtesy of Manville Filtration and Minerals Division.*)

the filter media and collected particles are washed off the septum, a new precoat of filter media is applied, and filtration continues. A typical flow schematic diagram is shown in Fig. 10-30.

Precoat filters accomplish particle removal by physical straining. The thickness of the initial layer of precoat filter media is normally about 3 mm (1/8 in), and the water passageways through this layer are so small and numerous that even very fine particles are retained.

Five or six grades of filter media (sometimes called *filter aid*) are used commonly in potable water treatment. They offer a range of performance with respect to clarity and flow characteristics. With an appropriate selection from among these grades, particles as small as 1 μm can be removed by the precoat filter cake. Where colloidal matter or other finely dispersed particles are present, however, filtration alone may not be adequate to produce 1 ntu water, as required by the SWTR based on earlier results obtained by Logsdon et al. (1981b).

Applications and Performance

Precoat filters are used widely in industrial filtration applications and in swimming pool filtration. They also have been used in municipal potable water treatment, primarily in the direct, in-line filtration of high-quality surface waters (turbidity 10 ntu or less and acceptable color) and in the filtration of iron and manganese from groundwaters after appropriate pretreatment to precipitate these contaminants.

Since 1949, more than 170 potable water treatment plants using precoat filtration with DE or other filter media have been designed, constructed, and operated (AWWA, 1995). About 90 percent are surface water supplies, and 10 percent are groundwater supplies. The largest existing plant is the 76 ML/d (20 mgd) San Gabriel, CA, plant.

Where the source water and other conditions are suitable, precoat filtration can offer a number of benefits to the user, including the following: (1) Capital cost savings may be possible because of smaller land and plant building requirements; (2) treatment costs may be slightly less than conventional coagulation/sedimentation/granular media filtration when filterable solids are low (AWWA, 1995; Bryant and Yapijakis, 1977), although sedimentation usually would not be needed for such high-quality source waters; (3) the process is entirely a physical/mechanical operation and can attain high log removals of *Giardia* cysts and *Cryptosporidium* oocysts (as described in the introduction of this chapter) without operator expertise in water chemistry relating to coagulation; (4) the waste residuals are easily dewatered, and in some cases, they may be reclaimed for other uses, including soil conditioning and land reclamation; (5) acceptable finished water clarity is achieved as soon as precoating is complete and filtration starts—a filter-to-waste period is generally not necessary to bring turbidity of the finished water within acceptable limits; and (6) terminal turbidity breakthrough generally is not observed because it is predominantly a surface filtration process.

The disadvantages of precoat filtration are (1) the continued cost of the filter media (body feed) that are usually discarded at the end of each filter cycle and the problems and costs associated with that disposal; (2) less cost-effective for waters that require pretreatment for algae, color, taste and odor problems, or DBP precursor removal, although waters containing only larger plankton such as diatoms sometimes can be treated economically by microstraining prior to the precoat filtration; and (3) proper design, construction, and operation are absolutely essential to prevent dropping the filter cake off the septum or cracking of the filter cake during operation that might result in failure to remove the target particulates.

Precoat filtration has been shown to be very effective in the removal of *Giardia* cysts over a broad range of operating conditions typical of potable water filtration (Logsdon et al., 1981b; Lange et al., 1986). The need for proper operation and maintenance is emphasized, however, and a precoat rate of 1 kg/m^2 (0.2 lb/ft^2) is recommended.

The ability to remove protozoan cysts without chemical coagulation makes precoat filtration attractive to very small communities that may need filtration because of concern about cysts in source waters. Operators skilled in coagulation are less likely to be available in small installations, but operators with mechanical skills are required.

The removal of small (<1–2 μm) particles such as bacteria and viruses depends on the grade of filter aid used and other operating conditions (Lange et al., 1986). Capture of smaller particles can be enhanced by using aluminum or iron coagulants or cationic polymers to coat the filter media (Burns et al., 1970; Brown et al., 1974; Baumann 1975; Lange et al., 1986).

The Filter Element and Septum

The filter element is a vital component of a precoat filter because it supports the septum and the filter cake. The attributes of good filter elements and septa are detailed in the *AWWA Precoat Filtration Manual* (AWWA, 1995). Filter elements may be either flat or tubular (Figs. 10-31 and 10-32). Flat elements, often referred to as *leaves*, may be rectangular or round. Tubular elements are available in several different cross-sectional shapes but generally are round.

Overlaying the filter element is the septum material, consisting of either tightly woven stainless steel wire mesh or a tightly fitted synthetic cloth bag, usually made from monofilament polypropylene weave. The precoat layer forms on and is supported by the septum. For this to occur, the size of the clear openings in the septum must be small enough for the filter media particles to form and maintain stable bridges across the openings. Generally, a clear opening of about 125 μm (0.005 in) or less in one direction is desirable.

The septum must be firmly supported so that it does not yield, flex, or become distorted as the differential pressure drop increases during the filter cycle resulting in temporary cracks in the filter cake.

The Filter Vessel

Two basic types of precoat filter vessels that contain multiple filter elements and septa are available. They are the vacuum filter and the pressure filter.

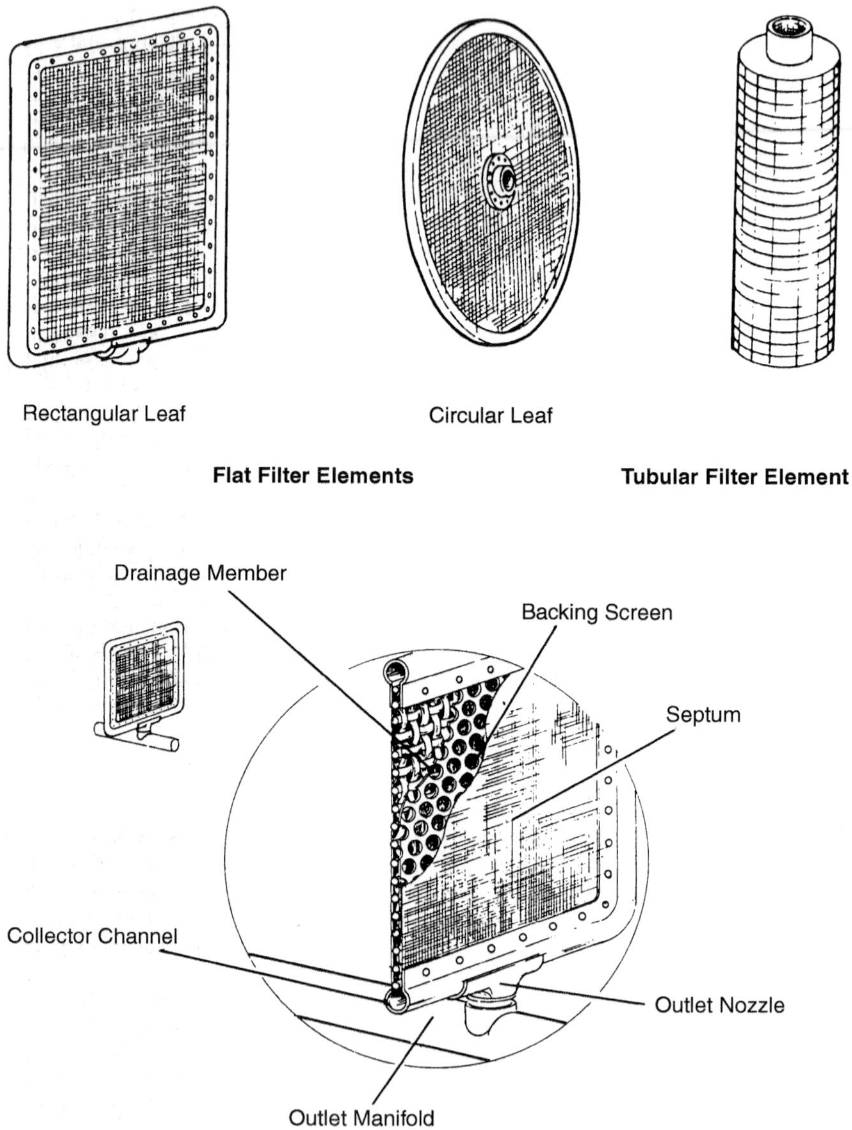

FIGURE 10-31 Types of filter elements for precoat filters and construction details of a flat-leaf element.

Vacuum Filters. The vessel containing vacuum filter elements and their septa is an open tank at atmospheric pressure. A filter discharge pump or a vacuum discharge leg downstream of the filter creates suction, enabling the available atmospheric pressure to move water through the precoat and filter cake as it develops. The open filter tanks permit easy observation of the condition of the septa and elements and general condition of the filter during filtration and cleaning.

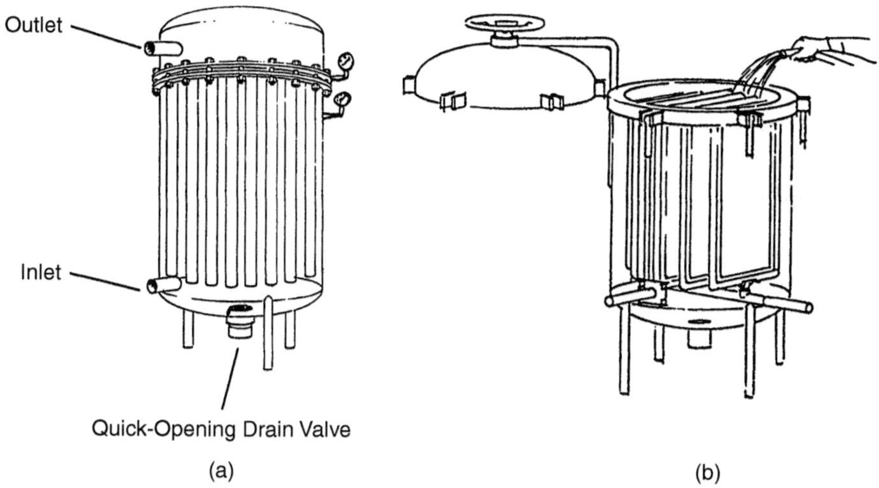

FIGURE 10-32 Typical pressure tubular filter and leaf filter for precoat filtration.

The maximum available differential pressure across the vacuum filter is limited to the net positive suction head of the filter pump or the vacuum leg. This limitation influences the length of filter cycles.

The effect of any entrained air or dissolved gases coming out of solution because of the decrease in pressure also must be considered with vacuum filters. Gas bubbles will have an adverse effect on the filter cake, tending to disrupt the integrity of the filter cake on the septum. One advantage of the vacuum filter is that the condition of the cake can be observed easily at the open water surface; this is not possible in a pressure filter.

Pressure Filters. A filter feed pump or influent gravity flow produces higher than atmospheric pressure on the inlet (upstream) side of a pressure filter, forcing liquid through the filter cake (Fig. 10-32). Large pressure drops across the filter are theoretically possible (limited only by the strength of the filter shell and the filter elements and septa), but the maximum economic differential pressure drop generally is limited to 205 to 275 kPa (30–40 psi). Typically, a higher pressure drop across a pressure filter will yield longer cycles than with vacuum filters, which, in turn, will yield greater removal of suspended solids per pound of filter media. Increased pumping costs for differential pressures much over 205 to 275 kPa (30–40 psi), however, usually offset savings in costs of filter media.

Filter Media

Precoat filters use DE or perlite as a filter media. Both materials are available in various grades to allow creation of a filter cake with the desired pore size. AWWA B101, Standard for Precoat Filter Media, covers quality control, density, permeability, and particle size distribution of precoat filter media (AWWA, 2002). DE (or diatomite) is composed of the fossilized skeletons of microscopic diatoms that grow in fresh or marine waters. Deposits of this material from ancient lakes or oceans are mined and then processed by flux calcining, milling, and air classification into various size grades for various filtration applications. The grades used in potable water filtration have a mean pore size of the cake of from about 5 to 17 μm. A less common material for precoat filtration is perlite, which comes from glassy volcanic rock. It is a siliceous rock containing 2 to 3 percent water. When heated, the rock expands to form a mass of glass bubbles. It is crushed and classified into several size grades. Pilot testing can determine two or three grades of filter media that will produce a desired clarity, simplifying the task of optimizing a final selection when working with a full-scale plant.

Filter Operation

Precoating. Successful precoating requires the uniform application of precoat filter aid to the entire surface of the clean septum. This is accomplished by recirculating a concentrated slurry of clean water and filter media (generally 12 percent or greater) through the filter at 2.4 to 3.7 m/h (1.0–1.5 gpm/ft^2) until most of the media have been deposited on the septum and the recirculating water turbidity is lowered to the desired treated water quality.

Because particles of the media are much smaller than the clear openings in the septum, their retention and the formation of a stable precoat take place as a result of bridging. As the particles crowd together when passing through the openings, they jam and interlock, forming a bridge over the openings (Fig. 10-33). As bridges form, additional particles are caught, and the filter septum is coated. A typical precoating system is shown in Fig. 10-30. The thickness of the precoat layer is typically 1.5 to 3 mm (1/16–1/8 in) or 1 kg/m^2 (0.2 lb/ft^2) of precoat media.

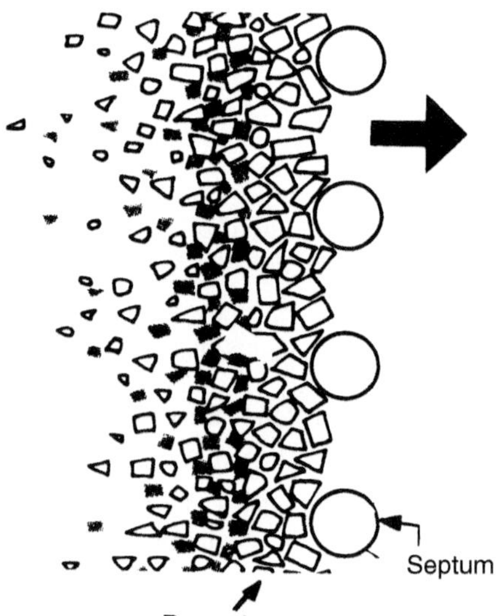

FIGURE 10-33 Illustration of precoat formation, with bridges of precoat between the elements of the septum material. (*Courtesy of Manville Filtration & Minerals Division.*)

Body-Feeding. The amount of body feed to be added to the source water is determined by the nature and amount of solids to be removed. Pilot testing during representative source water quality periods generally will indicate the type and range of solids that will be encountered and the amount and type of body feed needed to provide an incompressible and permeable cake.

Typical body-feed ratios of 1 to 10 mg/L of diatomite for each 1 mg/L of suspended solids are required depending on the type of solids being filtered. Details of selecting the appropriate ratio are presented later. Compressible solids such as alum or iron coagulation solids require the highest ratios. Proper control of the body-feed dosage is the most important factor contributing to economic operation of a precoat filtration plant.

Body-feed equipment can be classified as dry or wet systems. Wet, or slurry, feeders are the most common. The attributes of good precoat and body-feed equipment have been presented in the AWWA Manual M30, *Precoat Filtration* (AWWA, 1995).

Spent Cake Removal. At the end of each filter cycle, the spent filter cake is removed from the filter in preparation for the precoat that will begin the next cycie. If spent solids are not fully removed from the filter vessel, the material could be resuspended and deposited on the filter septa. The resuspended dirty material could foul the septa, although the effect usually would develop gradually so that the operator would become aware of it only after a number of cycles.

The most reliable determination of cleanliness of the septa requires visual observation of the bare septa, followed by observation of the uniformity and completeness of the precoat. When the septa cannot be fully inspected, a higher than normal differential pressure immediately after precoating (at the start of the filtration cycle) would suggest that the septa are becoming fouled.

Techniques for removing the spent cake vary with the different kinds of filter vessels and filter elements. Spent cake removed from a precoat filter is a mixture of filter media and the materials removed from the source water. This waste matter is usually removed from the filter in slurry form. Although some systems may be able to dispose of the slurry into a sanitary sewer system, most plants must dewater the waste material and make separate provisions for the solid and liquid wastes. Methods of handling wastes are covered further in Chap. 22 on water treatment residuals management.

Theoretical Aspects of Precoat Filtration

Understanding the probable effect of flow rate, terminal head loss, and body-feed rate on the operation of the filter is necessary to obtain maximum economy in the design and operation of a precoat filter. The most important of these items affecting filter operation is the prudent use of body feed during a filter run.

Effect of Concentration of Body Feed. The results of a typical series of filter runs conducted at a constant rate of filtration and to a constant terminal pressure drop, but with varying amounts of body feed, are shown in Fig. 10-34. The total volume of filtrate is plotted against the total head loss in feet of water. The initial head loss is caused by the clean precoat layer, septum, and filter element. With no body feed or insufficient amounts of body feed, the head loss will increase more rapidly as the run continues. This is so because suspended solids that are removed on the surface of the precoat layer soon form a compressible, impermeable layer of solids that will not readily permit the passage of water.

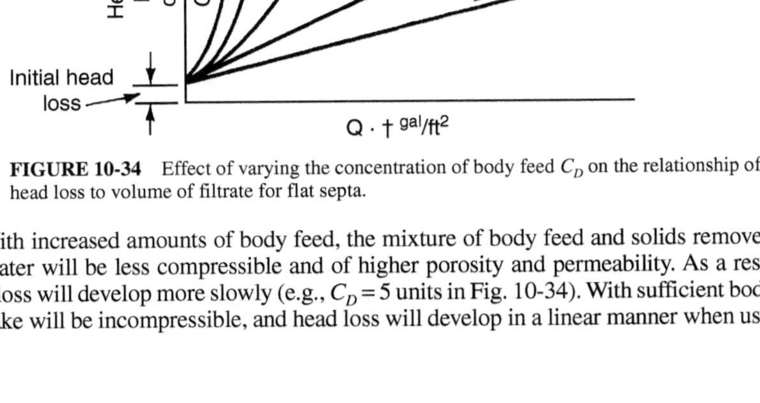

FIGURE 10-34 Effect of varying the concentration of body feed C_D on the relationship of head loss to volume of filtrate for flat septa.

With increased amounts of body feed, the mixture of body feed and solids removed from the water will be less compressible and of higher porosity and permeability. As a result, the head loss will develop more slowly (e.g., $C_D = 5$ units in Fig. 10-34). With sufficient body feed, the cake will be incompressible, and head loss will develop in a linear manner when using flat septa (e.g., $C_D = 10$ units in Fig. 10-34). With still higher rates of body feed, the slope of the straight-line plot will get flatter. The most economical use of body feed generally occurs at the lowest concentration that generates a linear head loss curve when using flat septa.

The linear head loss curves shown in Fig. 10-34 for higher body-feed concentrations only occur with flat septa at constant total flow rate. With tubular septa, as the body-feed layer gets thicker, the area of filtration increases, and the effective filtration rate decreases for the outer layers of the cake. This causes head loss to increase more slowly as the cycle progresses (Dillingham et al., 1967a).

Effect of Filtration Rate. Consider the linear head loss curve for a flat septum at constant flow rate and adequate body feed, as shown in Fig. 10-34. The total head loss at any time is the sum of the initial head loss (caused by precoat layer, filter element, and septum) plus the head loss of the body-feed layer. The flow through the entire filter cake at the usual filtration velocities is laminar. Therefore, the following can be stated with regard to head loss development:

1. Head loss for a particular instant (i.e., cake thickness) is proportional to the filtration rate Q because of laminar flow conditions.
2. Head loss across the body-feed layer is proportional to the thickness of the layer. This, in turn, is proportional to the volume of filtrate $\dot{V}$ = filtration rate Q times running time t at constant-body-feed concentration.
3. Thus, combining 1 and 2 above, the head loss in the body-feed layer at any instant is proportional to Q^2t.

Using these concepts, the rate of head loss development at two different filtration rates Q_1 and Q_2 can be compared:

$$h_1 \propto Q_1^2 t_1 \text{ and } h_2 \propto Q_2^2 t_2 \tag{10-27}$$

Dividing these two expressions for the case of the same terminal head loss, $h_1 = h_2$, we get

$$\frac{t_1}{t_2} = \frac{Q_2^2}{Q_1^2} \tag{10-28}$$

Similarly,

$$V_1' = Q_1 t_1 \quad \text{and} \quad V_2' = Q_2 t_2 \tag{10-29}$$

Dividing these two equations and inserting the previous relation for t_1/t_2, we get

$$\frac{V_1'}{V_2'} = \frac{Q_2}{Q_1} \tag{10-30}$$

According to these expressions, for a given terminal head loss, the length of the filter run is inversely proportional to the square of the filtration rate, and the filtrate volume is inversely proportional to the filtration rate. These relationships are approximate because the impact of the precoat head loss has been ignored as it is usually trivial compared with the body-feed head loss. Equations 10-28 and 10-30 help to explain why the most economical operation of precoat filters usually occurs at fairly low filtration rates, 2.4 m/h (1 gpm/ft^2) being most common, but with 1.2 to 7.3 m/h (0.5–3 gpm/ft^2) used in some plants.

Mathematical Model for Precoat Filtration. A rigorous mathematical model has been presented for head loss development for a precoat filter operating at constant total flow rate. The model describes the head loss for either flat or tubular septa operating with adequate body feed to achieve an incompressible cake (Dillingham et al., 1967a; Weber, 1972). To use the model, empirical filter cake resistance indices must be determined for the precoat and body-feed layers by pilot filter operation. The body-feed resistance index can be correlated empirically with the ratio of source water suspended solids concentration to the body feed concentration (Dillingham et al., 1967b). The model can be used in an optimization program to select the best combination of filtration rate, terminal head loss, and body-feed concentration for best overall economy (Dillingham et al., 1966).

DEDICATION

We dedicate this chapter to the memory of two people whose work as researchers, advisors, mentors, and teachers has greatly affected the understanding and practice of filtration using granular media: Kenneth (Ken) Ives and Appiah (Amit) Amirtharajah. Professor Ives led many fundamental and significant studies of filtration in his nearly 40 years of work at University College London. Professor Amirtharajah, beginning with his M.S. and Ph.D. studies with Professor John Cleasby at Iowa State University and latter in his academic career at Montana State University and Georgia Tech, conducted extensive studies of filter backwashing, initial filter effluent quality, and many other fundamental and applied aspects of media filtration and coagulation.

ABBREVIATIONS

A	cross-sectional area
ASTM	American Society for Testing and Materials
AWWA	American Water Works Association
BOM	biodegradable organic matter
DAF	dissolved air flotation
DBPs	disinfection by-products
DE	diatomaceous earth
ES	effective size
FBRR	Filter Backwash Recycling Rule
GAC	granular activated carbon
gfd	gallons per square foot per day
gpm	gallons per minute
GWR	Ground Water Rule
GWUI	groundwater under the influence
HGL	hydraulic grade line
HLR	hydraulic loading rate
IESWTR	Interim Enhanced Surface Water Treatment Rule
lmh	liters per square meter per hour
L/d	ratio of media depth to media size
LT1ESWTR	Long Term 1 Enhanced Surface Water Treatment Rule

LT2ESWTR	Long Term 2 Enhanced Surface Water Treatment Rule
MCL	maximum contaminant level
MF	microfiltration
ML/d	mega liters per day
NOM	natural organic matter
ntu	nephelometric turbidity units
PAC	powdered activated carbon
PACl	polyaluminum chloride
PIC	polyiron chloride
Q	flow rate
TOC	total organic carbon
UC	uniformity coefficient
UF	ultrafiltration
UFRV	unit filter run volume
USEPA	United States Environmental Protection Agency
SDWA	Safe Drinking Water Act
SWTR	Surface Water Treatment Rule
TT	treatment technique

EQUATION NOTATION

a/v	grain surface area per unit grain volume
A_1	dimensionless porosity function
C	mass or volume concentration of particles
d	filter media grain size (diameter)
d_e	effective size of media = d_{10}
d_{eq}	grain diameter of sphere of equal volume
d_p	particle diameter
d_{10}	media size for which 10% (by weight) of the media grains are smaller
d_{60}	media size for which 60% (by weight) of the media grains are smaller
d_{90}	media size for which 90% (by weight) of the media grains are smaller
D	pipe diameter
D_{bm}	Brownian diffusivity
f	friction factor
F	theoretical filter capacity
g	acceleration due to gravity
Ga	Galileo number
h	head loss
J	hydraulic head loss gradient

k	dimensionless Kozeny constant
k_1, k_2, k_3	empirical constants
L	depth of filter media
L_0	depth of fixed (unexpanded) bed of filter media
Q	filtration flow rate
Q_a	airflow rate
Re	Reynolds number
Re_1	modified Reynolds number
t	time
U	mean flow velocity in pipe
UC	uniformity coefficient
V	approach or superficial velocity
V_{mf}	minimum fluidization velocity
V'	filtration volume
z	bed-depth variable
α	attachment efficiency factor
ε	porosity
ε_0	clean-bed porosity
ε_σ	porosity of the deposited material
η_T	total single-collector efficiency
λ	filter coefficient
μ	absolute viscosity of fluid
ρ	density of fluid
ρ_b	bulk density of media (grains and water-filled pores)
ρ_p	density of particle
ρ_s	density of media grains
σ	specific deposit, volume of deposit per volume of bed
ϕ	empirical coefficient
ψ	grain sphericity

REFERENCES

Adin, A., and M. Rebhun (1977). A Model to Predict Concentration and Head Loss Profiles in Filtration. *Journal AWWA* 69(8), 444–453.

Al-Ani, M. Y., D. W. Hendricks, G. S. Logsdon, and C. P. Hibler (1986). Removing *Giardia* Cysts from Low Turbidity Waters by Rapid Rate Filtration. *Journal AWWA* 78(5), 66–73.

Amburgey, J. E., A. Amirtharajah, B. M. Brouckaert, and N. C. Spivey (2003). An Enhanced Backwashing Technique For Improved Filter Ripening. *Journal AWWA* 95(12), 81–94.

AWWA (American Water Works Association) (1971). *Water Quality and Treatment,* 3rd ed. New York: McGraw-Hill.

AWWA (1995). Manual M30, *Precoat Filtration*. Denver, CO: American Water Works Association.

AWWA (2002). ANSI/AWWA B101, Precoat Filter Media. Denver, CO: American Water Works Association.

AWWA (2004). *Water Treatment Plant Design,* 4th ed., E. Baruth, ed. New York: McGraw-Hill.

AWWA (2006). ANSI/AWWA Standard B604, Granular Activated Carbon. Denver, CO: American Water Works Association.

AWWA (2010). ANSI/AWWA Standard B100, Granular Filter Material. Denver, CO: American Water Works Association.

Amirtharajah, A. (1978). Optimum Backwashing of Sand Filters. *Journal of the Environmental Engineering Division, ASCE* 104(5), 917–932.

Amirtharajah, A. (1984). Fundamentals and Theory of Air Scour. *Journal of the Environmental Engineering Division, ASCE* 110(3), 573–590.

Amirtharajah, A., and Wetstein, D. P. (1980). Initial Degradation of Effluent Quality During Filtration. *Journal AWWA* 72(9), 518–524.

Anderson, W. B., and P. M. Huck (2005). *Sustainable Water Treatment of Grand River Surface Water with Multistage Filtration.* University of Waterloo, Waterloo, Ontario, Canada, Executive Summary of Report submitted to Ontario Ministry of Environment.

Arboleda, J. V. (1974). Hydraulic Control Systems of Constant and Declining Flow Rate in Filtration. *Journal AWWA* 66(2), 87–94.

Arboleda-Valencia, J., R. Giraldo, and H. Snel (1985). Hydraulic Behavior of Declining-Rate Filtration. *Journal AWWA* 77(12), 67–74.

ASTM (2009). *Annual Book of ASTM Standards*, Vol. 04.02: *Concrete and Aggregates.* West Conschohoken, PA: ASTM International.

Aultman, W. W. (1959). Valve Operating Devices and Rate-of-Flow Controllers. *Journal AWWA* 51(11), 1467.

Babbitt, H. E., J. J. Doland, and J. L. Cleasby (1962). *Water Supply Engineering*, 6th ed. New York: McGraw-Hill.

Baker, M. N. (1948). *The Quest for Pure Water*. Denver, CO: American Water Works Association.

Bauer, M. J., J. S. Colbourne, D. M. Foster, N. V. Goodman, and A. J. Rachwal (1996). GAC Enhanced Slow Sand Filtration (GAC Sandwich). In *Advances in Slow Sand and Alternative Biological Filtration*, N. Graham and R. Collins, eds. New York: Wiley.

Baumann, E. R. (1975). Diatomite Filters for Removal of Asbestos Fibers. In *Proc. AWWA Annual Conference*, Minneapolis, MN.

Baylis, J. R. (1956). Seven Years of High-Rate Filtration. *Journal AWWA* 48(5), 585–596.

Baylis, J. R. (1959a). Nature and Effect of Filter Backwashing. *Journal AWWA* 51(1), 131–156.

Baylis, J. R. (1959b). Variable Rate Filtration. *Pure Water*. May.

Baylis, J. R. (1959c). Review of Filter Design and Methods of Washing. *Journal AWWA* 51(11), 1433–1472.

Baylis, J. R. (1960). Discussion of Conley and Pitman. *Journal AWWA* 52(2), 214–218.

Bellamy, W. D., J. L. Cleasby, G. S. Logsdon, and M. J. Allen (1993). Assessing Treatment Plant Performance. *Journal AWWA* 85(12), 34–58.

Bellamy, W. D., D. W. Hendricks, and G. S. Logsdon (1985a). Slow Sand Filtration: Influences of Selected Process Variables. *Journal AWWA* 77(12), 62–66.

Bellamy, W. D., G. P. Silverman, D. W. Hendricks, and G. S. Logsdon (1985b). Removing *Giardia* Cysts with Slow Sand Filtration. *Journal AWWA* 77(2), 52–60.

Benjes, H. H., C. E. Edlund, and P. T. Gilbert (1985). Adsorption Clarifier Applied to Low Turbidity Surface Supplies. In *Proc. AWWA Annual Conference*, Washington, DC.

Bichai, F., Y. Dullemont, M. Tosielle, B. Barbeau, and W. Hijnen (2009). Role of Predation in Transport of Fecal Bacteria and Protozonan (Oo)cysts in Water Treatment. In *Proc. AWWA 2009 Water Quality Technology Conference*, Seattle, WA.

Billings, C. E. (1966). Effects of Particle Accumulation in Aerosol Filtration, unpublished PhD dissertation, California Institute of Technology.

Bowers, D. A., A. E. Bowers, and D. D. Newkirk (1982). Development and Evaluation of a Coagulant Control Test Apparatus for Direct Filtration. In *Proc. AWWA Water Quality Technology Conference*, Nashville, TN.

Brown, W. G. (1955). High-rate filtration experience at Durham, NC. *Journal AWWA* 47(3), 243–247.

Brown, T. S., J. F. Malina, Jr., and B. D. Moore (1974). Virus Removal by Diatomaceous Earth Filtration, Part 2. *Journal AWWA* 66(12), 735–738.

Browning, R. C. (1987). New Method and Revised Nomenclature for Granular Filter Flow Control. In *Proc. Australian Water and Wastewater Association*, Adelaide, Australia, March 23–27, pp. 202–209.

Bryant, A., and C. Yapijakis (1977). Ozonation–Diatomite Filtration Removes Color and Turbidity. *Water and Sewage Works, Part 1*, 124:96; *Part 2*, 124:94.

Burns, D. E., E. R. Baumann, and C. S. Oulman (1970). Particulate Removal on Coated Filter Media. *Journal AWWA* 62(2), 121–126.

Camp, T. R. (1961). Discussion of Conley: Experiences with Anthracite Filters. *Journal AWWA* 53(12), 1478–1483.

Camp, T. R. (1964). Theory of Water Filtration. *Journal of the Sanitation Engineering Division, ASCE* 90(SA4), 1–30.

Cleasby, J. L. (1969a). Filter Rate Control Without Rate Controllers. *Journal AWWA* 61(4), 181–183.

Cleasby, J. L. (1969b). Approaches to a Filterability Index for Granular Filters. *Journal AWWA* 61(8), 372–381.

Cleasby, J. L. (1981). Declining Rate Filtration. *Journal AWWA* 73(9), 484–489.

Cleasby, J. L. (1991). Source Water Quality and Pretreatment Options for Slow Sand Filters. In *Slow Sand Filtration*, G. S. Logsdon, ed. New York: ASCE.

Cleasby, J. L. (1993). Status of Declining Rate Filtration Design. *Water Science and Technology* 27(10), 151–164.

Cleasby, J. L., and E. R. Baumann (1962). Selection of Sand Filtration Rates. *Journal AWWA* 54(5), 579–602.

Cleasby, J. L., and L. DiBernardo (1980). Hydraulic Considerations in Declining-Rate Filtration Plants. *Journal of the Environmental Engineering Division, ASCE* 106(6), 1043–1055.

Cleasby, J. L., and K. S. Fan (1981). Predicting Fluidization and Expansion of Filter Media. *Journal of the Environmental Engineering Division, ASCE* 107(3), 455–471.

Cleasby, J. L., and C. W. Woods (1975). Intermixing of dual and multi-media granular filters. *Journal AWWA* 67(4), 197–203.

Cleasby, J. L., A. Arboleda, D. E. Burns, P. W. Prendiville, and E. S. Savage (1977). Backwashing of Granular Filters. *Journal AWWA* 69(2), 115–126.

Cleasby, J. L., A. H. Dharmarajah, G. L. Sindt, and E. R. Baumann (1989). *Design and Operation Guidelines for Optimization of the High-Rate Filtration Process: Plant Survey Results*. Awwa Research Foundation Report. Denver, CO: Water Research Foundation.

Cleasby, J. L., D. J. Hilmoe, C. J. Dimitracopoulos, and L. M. Diaz-Bossio (1984a). Effective Filtration Methods for Small Water Supplies. Final Report. USEPA Cooperative Agreement CR808837-01-0, NTIS No. PB84-187-905.

Cleasby, J. L., D. J Hilmoe, and C. J. Dimitracopoulos (1984b). Slow Sand and Direct In-Line Filtration of a Surface Water. *Journal AWWA* 76(12), 44–55.

Cleasby, J. L., G. L. Sindt, D. A. Watson, and E. R. Baumann (1992). *Design and Operation Guidelines for Optimization of the High-Rate Filtration Process: Plant Demonstration Studies*. Awwa Research Foundation Report. Denver, CO: Water Research Foundation.

Cleasby, J. L., M. M. Williamson, and E. R. Baumann (1963). Effect of Rate Changes on Filtered Water Quality. *Journal AWWA* 55(7), 869–880.

Clough, G., and K. J. Ives (1986). Deep Bed Filtration Mechanisms Observed with Fiber Optic Endoscopes and CCTV. In *Proceedings 4th World Filtration Congress,* Part II, Ostend, Belgium. Antwerpen, Belgium: Royal Flemish Society of Engineers.

Collins, M. R., T. T. Eighmy, and J. P. Malley, Jr. (1991). Evaluating Modifications to Slow Sand Filters. *Journal AWWA* 83(9), 62–70.

Colton, J. S., Hillis, P., and Fitzpatrick, C. S. B. (1996). Filter Backwash And Start-Up Strategies For Enhanced Particulate Removal. *Water Research* 30(10), 2502–2507.

Committee Report (1980). The Status of Direct Filtration. *Journal AWWA* 72(7), 405–411.

Committee Report (1984). Comparison of Alternative Systems for Controlling Flow Through Filters. *Journal AWWA* 76(1), 91–95.

Conley, W. R. (1961). Experiences with Anthracite Sand Filters. *Journal AWWA* 53(12), 1473–1478.

Conley, W. R. (1965). Integration of the Clarification Process. *Journal AWWA* 57(10), 1333–1345.

Conley, W. R. (1972). High Rate Filtration. *Journal AWWA* 64(3), 203–206.

Conley, W. R., and K. Y. Hsuing (1969). Design and Application of Multimedia Filters. *Journal AWWA* 61(2), 97–101.

Conley, W. R., and R. W. Pitman (1960a). Test Program for Filter Evaluation at Hanford. *Journal AWWA* 52(2), 205–214.

Conley, W. R., and R. W. Pitman (1960b). Innovations in Water Clarification. *Journal AWWA* 52(10), 1319–1325.

Cornwell, D. A., M. M. Bishop, T. R Bishop, and N. E. McTigue (1991). *Full Scale Evaluation of Declining and Constant Rate Filtration.* Awwa Research Foundation Report. Denver, CO: Water Research Foundation.

Cornwell, D. A., and R. Lee (1994). Waste Stream Recycling: Its Effects on Water Quality. *Journal AWWA* 86(11), 50–63.

Cornwell, D. A., and M. J. MacPhee (2001). Effects of Spent Filter Backwash Recycle on *Cryptosporidium* Removal. *Journal AWWA* 93(4):153–162.

Cullen, T. R., and R. D. Letterman (1985). The Effect of Slow Sand Filter Maintenance on Water Quality. *Journal AWWA* 77(12), 48–55.

Culp, R. L. (1977). Direct Filtration. *Journal AWWA* 68(7), 375–378.

Dahab, M. F., and J. C. Young (1977). Unstratified-Bed Filtration of Wastewater. *Journal of the Environmental Engineering Division, ASCE* 103(1), 21–36.

Degremont, G. (1973). *Water Treatment Handbook.* Caxton Hill, Hertford, UK: Stephen Austin & Sons, Ltd.

Dharmarajah, A. H., and J. L. Cleasby (1986). Predicting the Expansion of Filter Media. *Journal AWWA* 78(12), 66–76.

Diaper, E. W. J., and K. J. Ives (1965). Filtration Through Size-Graded Media. *Journal of the Sanitary Engineering Division, ASCE* 91(SA3), 89–114.

DiBernardo, L., and J. L. Cleasby (1980). Declining Rate versus Constant Rate Filtration. *Journal of the Environmental Engineering Division, ASCE* 106(6), 1023–1041.

Dillingham, J. H., J. L. Cleasby, and E. R. Baumann (1966). Optimum Design and Operation of Diatomite Filtration Plants. *Journal AWWA* 58(6), 657–672.

Dillingham, J. H., J. L. Cleasby, and E. R. Baumann (1967a). Diatomite Filtration Equations for Various Septa. *Journal of the Sanitary Engineering Division, ASCE* 93(SA1), 41–55.

Dillingham, J. H., J. L. Cleasby, and E. R. Baumann (1967b). Prediction of Diatomite Filter Cake Resistance. *Journal of the Sanitary Engineering Division, ASCE* 93(SA1), 57–76.

Duncan, A. (1988). The Ecology of Slow Sand Filters. In Slow *Sand Filtration: Recent Developments in Water Treatment Technology*, Nigel J. D. Graham, ed. Chichester, UK: Ellis Horwood, Ltd.

Economic and Engineering Services, Inc., and Thames Water Utilities, Ltd. (1996). Report to the City of Salem, OR.

Edzwald, J. K., W. C. Becker, and S. J. Tambini (1987). Organics, Polymers, and Performance in Direct Filtration. *Journal of the Environmental Engineering Division, ASCE* 113(1), 167–185.

Edzwald, J. K., J. E. Tobiason, T. Amato, and L. J. Maggi (1999). Integrating of High Rate DAF Technology into Plant Design. *Journal AWWA* 91(12), 41–53.

Edzwald, J. K., J. E. Tobiason, M. B. Kelley, G. S. Kaminski, H. J. Dunn, and P. B. Galant (2001). *Impacts of Filter Backwash Recycle on Clarification and Filtration.* Awwa Research Foundation Report. Denver, CO: Water Research Foundation.

Edzwald, J. K., J. E. Tobiason, C. Udden, G. S. Kaminski, H. J. Dunn, and P. B. Galant (2003). Impacts of Recycle of Waste Filter Backwash on the Removal of *Cryptosporidium. Journal of Water Supply: Research and Technology, AQUA* 52(4), 243–258.

Eliassen, R. (1941). Clogging of Rapid Sand Filters. *Journal AWWA* 33(5), 926–942.

Elimelech, M., J. Gregory, X. Jia, and R. A. Williams. (1995). *Particle Deposition and Aggregation, Measurement, Modelling and Simulation.* Oxford, UK: Butterworth-Heinemann, Ltd.

Emelko, M. B., P. M. Huck, and B. M. Coffey (2005). A review of *Cryptosporidium* removal by granular media filtration. *Journal AWWA* 97(12), 101–115.

Emelko, M. B., Huck, P. M., Coffey, B. M., and Smith, E. F. (2006). Effects Of Media, Backwash, And Temperature On Full Scale Biological Filtration. *Journal AWWA* 98(12), 61–73.

Emelko, M. B., D. Scott, and J. Bolton (2007). Using Engineered Media to Minimize Chemical Dependency in Filtration. In *Chemical Water and Wastewater Treatment IX*, H. H. Hahn, E. Hoffman, and H. Ødegaard, eds. *Proceedings of the 12th Gothenburg Symposium, 20–23 May 2007, Ljubljana, Slovenia.* London: IWA Publishing.

Ergun, S. (1952a). Fluid Flow Through Packed Columns. *Chemical Engineering Progress* 48(2), 89.

Ergun, S. (1952b). Determination of geometric surface area of crushed porous solids. *Analytical Chemistry* 24(2), 388–393.

Fair, G. M., J. C. Geyer, and D. A. Okun (1968). *Water and Wastewater Engineering*, Vol. 2. New York: Wiley.

Fitzpatrick, J. A., and L. A. Spielman (1973). Filtration of Aqueous Latex Suspensions Through Beds of Glass Spheres. *Journal of Colloid and Interface Science* 43(2), 350–369.

Fox, K. R., R. J. Miltner, G. S. Logsdon, D. L. Dicks, and L. F. Drolet (1984). Pilot Plant Studies of Slow Rate Filtration. *Journal AWWA* 76(12), 62–68.

Friedlander, S.K. (1958). Theory of Aerosol Filtration. *Industrial and Engineering Chemistry* 50(8), 1161–1164.

Fuller, G. W. (1898). *The Purification of the Ohio River Water at Louisville, Kentucky.* New York: Van Nostrand.

Fuller, G. W. (1933). Progress in Water Filtration. *Journal AWWA* 25, 1556–1576.

Gabelich, C. J., F. W. Gerringer, C. C. Lee, and W. R. Knocke (2006). Sequential Manganese Desorption and Sequestration in Anthracite Coal and Silica Sand Filter Media. *Journal AWWA* 98(5), 116–127.

Ginn, T. M., Jr., A. Amirtharajah, and P. R. Karr (1992). Effects of Particle Detachment in Granular-Media Filtration. *Journal AWWA* 84(2), 66–76.

Glaser, H. T., and J. K. Edzwald (1979). Coagulation and Direct Filtration of Humic Substances with Polyethylenimine. *Environmental Science and Technology* 13(3), 299–305.

Graese, S. L., V. L. Snoeyink, and R. G. Lee (1987). *GAC Filter-Adsorbers.* AWWA Research Foundation Report. Denver, CO: Water Research Foundation.

Great Lakes–Upper Mississippi River Board of State and Provincial Public Health and Environmental Managers (2003). *Recommended Standards for Water Works.* Albany, NY: Health Education Services.

Greaves, G. F., P. G. Grundy, and G. S. Taylor (1988). Ozonation and Slow Sand Filtration for the Treatment of Coloured Upland Waters: Pilot Plant Investigations. In *Slow Sand Filtration: Recent Developments in Water Treatment Technology*, N. J. D. Graham, ed. Chichester, UK: Ellis Horwood, Ltd.

Green, R., and L. Benjamin (2000). Small Systems in British Columbia. *Journal AWWA* 92(5), 10.

Haarhoff, J., and J. L. Cleasby (1991). Biological and Physical Mechanisms in Slow Sand Filtration. In *Slow Sand Filtration*, G. S. Logsdon, ed. New York: American Society of Civil Engineers.

Habibian, M. T., and C. R. O'Melia (1975). Particles, Polymers and Performance in Direct Filtration. *Journal of Environmental Engineering* 101(EE4), 567–583.

Harris, W. L. (1970). High-Rate Filter Efficiency. *Journal AWWA* 62(8), 515.

Heertjes, P. M., and C. E. Lerk (1967). The Functioning of Deep Filters, Part I: The Filtration of Flocculated Suspensions. *Transactions of Instn. of Chemical Engineering* 45(4), T138.

Hemmings, D. G., and C. S. B Fitzpatrick (1997). Pressure signal analysis of combined water and air backwash of rapid gravity filters. *Water Research* 31(2), 356–361.

Hendricks, D., J. M. Barrett., J. Bryck, M. R. Collins, B. A. Janonis, and G. S. Logsdon (1991). *Manual of Design for Slow Sand Filtration.* Awwa Research Foundation Report. Denver, CO: Water Research Foundation.

Hendricks, D. W. (2006). *Water Treatment Unit Processes: Physical and Chemical.* Boca Raton, FL: CRC Press.

Hewitt, S. R., and Amirtharajah, A. (1984). Air Dynamics Through Filter Media During Air Scour. *Journal of the Environmental Engineering Division, ASCE* 110(3), 591–606.

Hilmoe, D. J., and J. L. Cleasby (1986). Comparing Constant Rate and Declining Rate Filtration of a Surface Water. *Journal AWWA* 78(12), 26–34.

Huck, P. M., B. M. Coffey, M. B. Emelko, D. D. Maurizio, R. M. Slawson, W. B. Anderson, J. Van den Oever, I. P. Douglas, and C. R. O'Melia (2002). Effects of Filter Operation on *Cryptosporidium* Removal. *Journal AWWA* 94(6), 97–111.

Hudson, H. E., Jr. (1959). Filter Design—Declining Rate Filtration. *Journal AWWA* 51(11), 1455–1463.

Hudson, H. E., Jr. (1962). High Quality Water Production and Viral Disease. *Journal AWWA* 54(10), 1265–1274.

Huisman, L. (1974). *Rapid Filtration,* Part 1. Delft, The Netherlands: Delft University of Technology.

Huisman, L., and W. E. Wood (1974). *Slow Sand Filtration.* Geneva, Switzerland: World Health Organization.

Hutchison, W. R. (1976). High-Rate Direct Filtration. *Journal AWWA* 68(6), 292–298.

Iwasaki, T. (1937). Some Notes on Sand Filtration. *Journal AWWA* 29(10),1591–1602.

Ives, K. J. (1960). Rational Design of Filters. *Proceedings of the Institute of Civil Engineers* 16, 189–193.

Ives, K. J. (1961). New Concepts in Filtration. *Water and Water Engineering* 65, 307–309, 341–344, and 385–388.

Ives, K. J. (1963). Simplified Rational Analysis of Filter Behavior. *Proceedings of the Institute of Civil Engineers* 25, 345–364.

Ives, K. J. (1997). Personal communication.

Ives, K. J., and I. Sholji (1965). Research on Variables Affecting Filtration. *Journal of Sanitary Engineering Division, ASCE* 91(SA4), 1.

Janssens, J., and A. Buekens (1993). Assessment of Process Selection for Particle Removal in Surface Water Treatment. *Journal of Water Supply: Research and Technology, AQUA* 42(5), 279–288.

Jobb, D. Brian, W. B. Anderson, R. A. LeCraw, and M. R. Collins (2007). Removal of Emerging Contaminants and Pathogens Using Modified Slow Sand Filtration: An Overview. In *Proceedings AWWA Annual Conference, Toronto, Ontario.*

Kau, S. M., and D. F. Lawler (1995). Dynamics of Deep-Bed Filtration: Velocity, Depth, and Media. *Journal of Environmental Engineering, ASCE* 121(12), 850–859.

Kawamura, S. (2000). *Integrated Design of Water Treatment Facilities,* 2nd ed. New York: Wiley.

Kim, J., and J. E. Tobiason (2004). Particles in Filter Effluent: The Roles of Deposition and Detachment. *Environmental Science & Technology* 38(22), 6132–6138.

King, P. H., R. L. Johnson, C. W. Randall, and G. W. Rehberger (1975). High-Rate Water Treatment: The State of the Art. *Journal of the Environmental Engineering Division, ASCE* 101(EE4), 479–498.

Kirmeyer, G. J. (1979). *Seattle Tolt Water Supply Mixed Asbestiform Removal Study*. EPA 600/2-79-125.

Kirner, J. C., J. D. Littler, and L. A. Angelo (1978). A Waterborne Outbreak of Giardiasis in Camus, Wash. *Journal AWWA* 70(1), 35–40.

Knocke, W., K. Hamon, and C. Thompson (1988). Soluble Manganese Removal on Oxide-Coated Filter Media. *Journal AWWA* 80(12), 65–70.

Knocke, W., S. Occiano, and R. Hungate (1990). *Removal of Soluble Manganese from Water by Oxide-Coated Media*. Awwa Research Foundation Report. Denver, CO: Water Research Foundation.

Knocke, W., S. Occiano, and R. Hungate (1991). Removal of Soluble Manganese by Oxide-Coated Filter Media: Sorption Rate and Removal Mechanism Issues. *Journal AWWA* 83(8), 64–69.

Kreissl, J. F., G. G. Robeck, and G. A. Sommerville (1968). Use of Pilot Filters to Predict Optimum Chemical Feeds. *Journal AWWA* 60(3), 299–314.

Krueger, R. J., D. A. Prock, S. D. Sundseth, and G. S. Logsdon (1999). Salem, Oregon, Expands and Improves Its Slow Sand Filtration Facility to Prepare for Water Quality and Regulatory Challenges. In *Proceedings AWWA Annual Conference, Chicago*.

Krumbein, W. C. (1941). Measurement and Geological Significance of Shape and Roundness of Sedimentary Particles. *Journal of Sedimentary Petrology* 11(2), 64.

Lange, K. P., W. D. Bellamy, D. W. Hendricks, and G. S. Logsdon (1986). Diatomaceous Earth Filtration of *Giardia* Cysts and Other Substances. *Journal AWWA* 78(1), 76–84.

Lawler, D. F., and J. A. Nason (2006). Granular Media Filtration: Old Process, New Thoughts. *Water Science and Technology* 53(7), 1–7.

Lawler, D. F., C. R. O'Melia, and J. E. Tobiason (1980). Integral Water Treatment Plant Design: From Particle Size to Plant Performance. In *Particulates in Water: Characterization, Fate, Effects, and Removal*, M. Kavanaugh and J. Leckie, eds. Advances in Chemistry Series No. 189. Washington, DC: American Chemical Society.

LeChevallier, M. W., W. D. Norton, R. G. Lee, and J. B. Rose (1991), *Giardia and Cryptosporidium in Water Supplies*. Denver, CO: AWWA Research Foundation.

Letterman, R. D. (1987). An Overview of Filtration. *Journal AWWA* 79(12), 26–32.

Letterman, R. D. (1991). Operation and Maintenance. In *Slow Sand Filtration*, G. S. Logsdon, ed. New York: American Society of Civil Engineers.

Letterman, R. D., and G. S. Logsdon (1976). Survey of Direct Filtration Practice. In *Proceedings of AWWA Annual Conference*, New Orleans, LA.

Ling, J. T. (1955). A Study of Filtration Through Uniform Sand Filters. *Journal of the Sanitary Engineering Division, ASCE* 81, 1–35.

Lippy, E. C. (1978). Tracing a Giardiasis Outbreak at Berlin, N.H. *Journal AWWA* 70(9), 512–520.

Liu, X. B., P. M. Huck, and R. M. Slawson (2001). Factors Affecting Drinking Water Biofiltration. *Journal AWWA* 93(12), 90–101.

Lloyd, B. J. (1996). The Significance of Protozoal Predation and Adsorption for the Removal of Bacteria by Slow Sand Filtration. In *Advances in Slow Sand and Alternative Biological Filtration*, N. Graham and R. Collins, eds. Chichester, UK: Wiley.

Logsdon, G. S. (1979). *Water Filtration for Asbestos Fiber Removal*. EPA 600/2-79-206.

Logsdon, G. S. (2008). *Water Filtration Practices*. Denver, CO: American Water Works Association.

Logsdon, G. S., and J. M. Symons (1977). Removal of Asbestiform Fibers by Water Filtration. *Journal AWWA* 69(9), 499–506.

Logsdon, G. S., A. F. Hess, M. J. Chipps, and A. J. Rachwal (2002). *Filter Maintenance and Operations Guidance Manual*. Awwa Research Foundation Report. Denver, CO: Water Research Foundation.

Logsdon, G. S., L. Mason, and J. B. Stanley, Jr. (1990). Troubleshooting an Existing Treatment Plant. *Journal of New England Water Works Association* 104(1), 43–56.

Logsdon, G. S., J. M. Symons, and T. J. Sorg (1981a). Monitoring Water Filters for Asbestos Removal. *Journal of the Environmental Engineering Division, ASCE* 107(6), 1297–1315.

Logsdon, G. S., J. M. Symons, R. L. Hoye, Jr., and M. M. Arozarena (1981b). Removal of *Giardia* Cysts and Cyst Models by Filtration. *Journal AWWA* 73(2), 111–118.

Logsdon, G. S., V. C. Thurman, E. S. Frindt, and J. G. Stoecker (1985). Evaluating Sedimentation and Various Filter Media for Removal of *Giardia* Cysts. *Journal AWWA* 77(2), 61–66.

Mayhugh, J. R., Sr., J. A. Smith, D. B. Elder, and G. S. Logsdon (1996). Filter Media Rehabilitation at a Lime-Softening Plant. *Journal AWWA* 88(8), 64–69.

McBride, D. G., and G. F. Stolarik (1987). Pilot to Full-Scale: Ozone and Deep Bed Filtration at the Los Angeles Aqueduct Filtration Plant. AWWA Seminar Proceedings, Coagulation and Filtration: Pilot to Full Scale. No. 20017. 1987 AWWA Annual Conference, Kansas City, MO.

McBride, D. G., R. C. Siemak, C. H. Tate, and R. R. Trussell (1977). Pilot Plant Investigations for Treatment of Owens River Water. In *Proceedings of AWWA Annual Conference, Anaheim, CA.*

McCabe, W. L., and J. C. Smith (1976). *Unit Operations of Chemical Engineering*, 3rd ed. New York: McGraw-Hill.

Medlar, S. (1974). This Filter Cleans Itself. *The American City* 89(6), 63.

Melin, E. S., R. A. Bohne, F. Sjøvold, and H. Ødegaard (2000). Treatment of ozonated water in biofilters containing different media. *Water Science and Technology* 41(4–5), 57–60.

Mohanka, S. S. (1969). Multilayer Filtration. *Journal AWWA* 61(10), 504–511.

Monscvitz, J. T., D. J. Rexing, R. G. Williams, and J. Heckler (1978). Some Practical Experience in Direct Filtration. *Journal AWWA* 70(10), 584–588.

Montgomery, J. M., Consulting Engineers (1985). *Water Treatment Principles and Design*. New York: Wiley.

Moran, M. C., D. C. Moran, R. S. Cushing, and D. F. Lawler (1993). Particle Behavior in Deep-Bed Filtration, Part 1: Particle Detachment. *Journal AWWA* 85(12), 82–93.

Morgan, J., and W. Stumm (1964). Colloidal-Chemical Properties of Manganese Dioxide. *Journal of Colloid Science* 19, 347–359.

Murray, B. A., and S. J. Roddy (1993). Treatment of Sydney's Water Supplies. *Water* 20(2), 17–20, 38.

MWH (2005). *Water Treatment: Principles and Design*, 2nd ed. New York: Wiley.

Nieminski, E. C., and J. E. Ongerth (1995). Removing *Giardia* and *Cryptosporidium* by Conventional Treatment and Direct Filtration. *Journal AWWA* 87(9), 96–106.

Nix, D. K., and J. S. Taylor (2003). *Filter Evaluation Procedures of Granular Media*. Denver, CO: American Water Works Association.

Oeben, R. W., H. P. Haines, and K. J. Ives (1968). Comparison of Normal and Reverse Graded Filtration. *Journal AWWA* 60(4), 429–439.

O'Melia, C. R. (1988). From Filters to Forests: Water Treatment and Supply. In *Pretreatment in Chemical Water and Wastewater Treatment*, H. H. Hahn and R. Klute, eds. Berlin: Springer-Verlag, pp. 3–14.

O'Melia, C. R., and W. Ali (1978). The Role of Retained Particles in Deep Bed Filtration. *Progress in Water Technology* 10, 167–182.

O'Melia, C. R., and J. Y. Shin (2001). Removal of Particles Using Dual Media Filtration: Modeling and Experimental Studies. *Water Science & Technology: Water Supply* 1(4), 73–79.

O'Melia, C. R., and W. Stumm (1967). Theory of Water Filtration. *Journal AWWA* 59(11), 1393–1412.

Ongerth, J. E., and P. E. Hutton (1997). DE Filtration to Remove *Cryptosporidium. Journal AWWA* 89(12), 39–46.

Ongerth, J. E. and P. E. Hutton (2001). Testing of Diatomaceous Earth Filtration for Removal of *Cryptosporidium* Oocysts. *Journal AWWA* 93(12), 54–63.

Patania, N. L., J. G. Jacangelo, L. Cummings, A. Wilczak, K. Riley, and J. Oppenheimer (1995). *Optimization of Filtration for Cyst Removal.* Awwa Research Foundation Report. Denver, CO: Water Research Foundation.

Pavoni, J. L., M. W. Tenny, and W. F. Eichelberger, Jr. (1972). Bacterial Exocellular Polymers and Biological Flocculation. *Journal of WPCF* 44(3), 414.

Payment, P., J. Siemiatycki, L. Richardson, E. Franco, G. Renaud, and M. Prévost (2001). *An Epidemiological Study of the Gastrointestinal Health Effects of Drinking Water.* Awwa Research Foundation Report. Denver, CO: Water Research Foundation.

Peterson, D. L., F. X. Schleppenbach, and T. M. Zaudtke (1980). Studies of Asbestos Removal by Direct Filtration of Lake Superior Water. *Journal AWWA* 72(3), 155–161.

Phelan, T. (2010). Personal communication.

Poynter, S. F. B., and J. S. Slade (1977). The Removal of Viruses by Slow Sand Filtration. *Progress in Water Technology* 9(1), 75–88.

Principe, M. R., D. Mastronardi, D. Bailey, and G. Fulton (1994). New York City's First Water Filtration Plant. In *Proceedings AWWA Annual Conference, New York, NY*, pp. 147–173.

Pyper, G. R. (1985). Slow Sand Filter and Package Treatment Plant Evaluation: Operating Costs and Removal of Bacteria, *Giardia*, and Trihalomethanes. USEPA/600/2-85/052. Cincinnati, OH: U.S. Environmental Protection Agency.

Rajagopalan, R., and C. Tien (1976). Trajectory Analysis of Deep Bed Filtration with the Sphere-in-cell Porous Media Model. *American Institute of Chemical Engineers Journal* 22(3), 523–533.

Ratnayaka, D. (2009). Personal communication.

Riesenberg, F., B. B. Walters, A. Steele, and R. A. Ryder (1995). Slow Sand Filters for a Small Water System. *Journal AWWA* 87(11), 48–56.

Rimer, A. E. (1968). Filtration Through a Trimedia Filter. *Journal of the Sanitary Engineering Division, ASCE* 94(SA3), 521–540.

Robeck, G. G. (1966). Discussion of Conley. *Journal AWWA* 58(1), 94–96.

Robeck, G. G., K. A. Dostal, and R. L. Woodward (1964). Studies of Modification in Water Filtration. *Journal AWWA* 56(2), 198–213.

Rooklidge, S. J., R. J. Miner, T. A. Kassim, and P. O. Nelson (2005). Antimicrobial Contaminant Removal by Multistage Slow Sand Filtration. *Journal AWWA* 97(12), 92–100.

Saatci, A. A., and C. S. Oulman (1980). The Bed Depth Service Time Design Method for Deep Bed Filtration. *Journal AWWA* 72(9), 524–528.

Saltnes, T., B. Eikebrokk, and H. Ødegaard (2002). Coagulation optimization for NOM removal by direct filtration in clay aggregate filters. *Journal of Water Supply: Research and Technology, AQUA* 51(2), 125–134.

Schuler, P. F., and M. M. Ghosh (1990). Diatomaceous Earth Filtration of Cysts and Other Particulates Using Chemical Additives. *Journal AWWA* 82(12), 67–75.

Schuler, P. F., M. M. Ghosh, and S. N. Boutros (1988). Comparing the Removal of *Giardia* and *Cryptosporidium* Using Slow Sand and Diatomaceous Earth Filtration. In *Proceedings AWWA Annual Conference, Orlando, FL.*

Seelaus, T. J., D. W. Hendricks, and B. A. Janonis (1986). Design and Operation of a Slow Sand Filter. *Journal AWWA* 78(12), 35–41.

Short, A. C., M. B. Gilton, and R. E. Henderson (1996). Blending Problems Overcome During the Design and Start Up of the Modesto Domestic Water Project. In *Proceedings AWWA Annual Conference, Water Quality*, Vol. D, Paper 56-7B, p 217–235. Denver, CO: American Water Works Association.

Slezak, L. A., and R. C. Sims (1984). The Application and Effectiveness of Slow Sand Filtration in the United States. *Journal AWWA* 76(12), 38–43.

Stanley, D. R. (1955). Sand Filtration Studies with Radio-Tracers. *Proceedings of the Americal Society of Civil Engineers* 81, 1–23.

Tanner, S. A., and J. E. Ongerth (1990). Evaluating the Performance of Slow Sand Filters in Northern Idaho. *Journal AWWA* 82(12), 51–61.

Tarbuck, B., G. Lindstadt, and D. Hilmoe (2002). Tolt Treatment Facility DBO: The First Year of Operation is a Success! In *Proceedings of the AWWA Annual Conference, New Orleans, LA*.

Tate, C. H., J. S. Lang, and H. L. Hutchinson (1977). Pilot Plant Tests of Direct Filtration. *Journal AWWA* 69(7), 379–384.

Tien, C. (1989). *Granular Filtration of Aerosols and Hydrosols*. Stoneham, MA: Butterworth Publishers.

Tobiason, J. E., and C. R. O'Melia (1988). "Physicochemical Aspects of Particle Removal in Depth Filtration." *Journal AWWA* 80(12), 54–64.

Tobiason, J. E., and B. Vigneswaran (1994). Evaluation of a Modified Model for Deep Bed Filtration. *Water Research* 28(2), 335–342.

Tobiason, J. E., M. Burns, L. Gaffney, and O. D. Schneider (1996). Particles in Filtered Water: Effects of Backwash Remnants and Initial Filtration Rate. In *Proceedings of the AWWA Annual Conference, Toronto, Ontario, Canada*, pp. 445–462.

Tobiason, J. E., J. K. Edzwald, V. Gilani, G. S. Kaminski, H. J. Dunn, and P. B. Galant (2003a). Effects of Waste Filter Backwash Recycle Operation on Clarification and Filtration, *Journal of Water Supply: Research and Technology, AQUA* 52(4), 259–276.

Tobiason, J. E., J. K. Edzwald, B. R. Levesque, G. S. Kaminski, H. J. Dunn, and P. B. Galant (2003b). Full-Scale Assessment of the Impacts of Recycle of Waste Filter Backwash. *Journal AWWA* 95(7), 80–93.

Tobiason, J. E., J. K. Edzwald, O. D. Schneider, M. B. Fox, H. J. and Dunn (1992). Pilot Study of the Effects of Ozone and Peroxone on In-Line Direct Filtration. *Journal AWWA* 84(12), 72–84.

Tobiason, J. E., G. S. Johnson, P. K. Westerhoff, and B. Vigneswaran (1993). Particle Size and Chemical Effects on Contact Filtration Performance. *Journal of Environmental Engineering* 119(3), 520–539.

Tobiason, J. E., W. R. Knocke, J. Goodwill, A. Islam, P. Hargette, R. Bouchard, and L. Zurvansky (2008). *Characterization and Performance of Filter Media for Manganese Control*. Denver, CO: Water Research Foundation.

Trefethen, J. M. (1959). *Geology for Engineers*, 2nd ed. Princeton, NJ: Van Nostrand.

Trussell, R. R., A. R. Trussell, , J. S. Lang, and C. H. Tate (1980). Recent Developments in Filtration System Design. *Journal AWWA* 72(12), 705–710.

Tuepker, J. L., and C. A. Buescher, Jr. (1968). Operation and Maintenance of Rapid Sand Mixed-Media Filters in a Lime Softening Plant. *Journal AWWA* 60(12), 1377.

Tufenkji, N., and M. Elimelech (2004). Correlation Equation for Predicting Single-Collector Efficiency in Physicochemical Filtration in Saturated Porous Media. *Environmental Science and Technology* 38(2), 529–536.

Twort, A. C., D. D. Ratnayaka, and M. J. Brandt (2000). *Water Supply*, 5th ed. London: IWA Publishing.

Urfer, D., P. M. Huck, S. D. J. Booth, and B. M. Coffey (1997). Biological filtration for BOM and particle removal: a critical review. *Journal AWWA* 89(12), 83–98.

U.S. Environmental Protection Agency (1989). 40 CFR Parts 141 and 142, Drinking Water, National Primary Drinking Water Regulations; Filtration, Disinfection; Turbidity, *Giardia lamblia*, Viruses, *Legionella*, and Heterotrophic Bacteria: Final Rule, *Federal Register* 54(124), 27486–27541, June 29.

U.S. Environmental Protection Agency (2001). 40 CFR Parts 9, 141, and 142, National Primary Drinking Water Regulations: Filter Backwash Recycling Rule: Final Rule, *Federal Register* 66(111), 31086–31105.

Valade, M. C., W. C. Becker, and J. K. Edzwald (2009). Treatment Selection Guidelines for Particle and NOM Removal. *Journal of Water Supply: Research and Technology, AQUA* 58(6), 424–432.

Veerapaneni, S., and M. R. Wiesner (1997). Deposit Morphology and Head Loss Development in Porous Media. *Environmental Science and Technology* 31(10), 2738–2744.

Vigneswaran, B., and J. E. Tobiason (1998). Significance of the Volume of Deposit During Deep Bed Filtration. In *Proceedings of the AWWA Annual Conference*, Dallas, TX.

Vigneswaran, S., and R. K. Tulachan (1988). Mathematical modeling of transient behavior of deep bed filtration. *Water Research* 22(9), 1093–1100.

Visscher, J. T. (1990). Slow Sand Filtration: Design, Operation, and Maintenance. *Journal AWWA* 82(6), 67–71.

Wagner, E. G., and H. E. Hudson (1982). Low-Dosage, High-Rate Direct Filtration. *Journal AWWA* 74(5), 256.

Weber, W. J. (1972). *Physicochemical Processes for Water Quality Control*. New York: Wiley-Interscience.

Wen, C. Y., and Y. H. Yu (1966). Mechanics of Fluidization. *Chemical Engineering Progress Symposium Series*, Vol. 62. New York.

Westerhoff, P. K., and J. E. Tobiason (1991). Effects of Flocculation on Direct Filtration. *Journal of New England Water Works Association* 105(2), 95–109.

Wiesner, M. R., J. J. Rook, and F. Fiessenger (1987a). Optimizing the Placement of GAC Filtration Units. *Journal AWWA* 79(12), 39–49.

Wiesner, M. R., C. R. O'Melia, and J. L. Cohon (1987b). Optimal Water Treatment Plant Design. *Journal of Environmental Engrineering* 113(3), 567–584.

Yao, K. M., M. T. Habibian, and C. R. O'Melia, (1971). Water and Waste Water Filtration: Concepts and Applications. *Environmental Science & Technology* 5(11), 1105–1112.

CHAPTER 11
MEMBRANES

Steven J. Duranceau, Ph.D., P.E.
Associate Professor
Department of Civil, Environmental and Construction Engineering
University of Central Florida
Orlando, Florida, United States

James S. Taylor, Ph.D., P.E.
Alex Alexander Professor of Engineering
Department of Civil, Environmental and Construction Engineering
University of Central Florida
Orlando, Florida, United States

SIZE RANGES FOR MEMBRANE PROCESSES **11.4**
CLASSIFICATIONS AND CONFIGURATIONS OF MEMBRANE PROCESSES **11.9**
Classification by Material 11.9
Classification by Geometry 11.12
RO-NF CONFIGURATION **11.18**
Classification by Driving Force 11.21
ED Cell 11.23
MEMBRANE PROPERTIES AND REJECTION CHARACTERISTICS **11.26**
Dissolved Solute–Membrane Interactions 11.27
Influence of Dissolved Solutes on Electrokinetic Properties of Membranes 11.28
Impacts on Membrane Rejection 11.28
Organic Solute Removal 11.29
Pathogen Removal 11.33
Membrane Integrity 11.34
BUBBLE-POINT DIRECT TESTING **11.35**
MASS TRANSPORT AND SEPARATION **11.37**
Mass Transport Considerations in Pressure-Driven Membrane Processes 11.37
Mass Transport and Process Considerations in ED and EDR 11.50
Temperature Effects on Flux 11.55
INTEGRATED MF AND UF PROCESS APPLICATIONS AND PROCESS DESIGN **11.57**
MF and UF Integrated with Either Lime or Coagulation, Flocculation, and Sedimentation (CFS) with or without Powdered Activated Carbon (PAC) 11.58
MF and UF Integrated with Coagulation 11.59
MF and UF Integrated with Coagulation, Flocculation, and Clarification as Pretreatment to NF, RO, and EDR for Surface Water Supplies 11.59
MF and UF Integrated with Oxidation Processes for Removal of Iron and Manganese from Groundwater 11.60
MF and UF Integrated with Coagulant for Arsenic Removal from Groundwater 11.61
MF and UF Integrated with Seawater RO 11.61
MF and UF Process Design 11.62
NF AND RO PROCESS CONCEPTS AND DESIGN CRITERIA **11.65**
RO and NF Processes 11.65
Fouling Indexes 11.67
Pretreatment 11.69
Limiting Salt 11.71
Array Models 11.76
Practical Array Design 11.83
Posttreatment 11.86
RESIDUALS DISPOSAL AND CONCENTRATE MANAGEMENT **11.91**
MF and UF Process Residuals 11.91

Spent Cleaning Solutions 11.92
NF and RO Membrane Concentrate 11.93
PILOT PLANT TESTING **11.94**
MF and UF Piloting 11.95
NF, RO, and ED/EDR Piloting 11.95
ABBREVIATIONS **11.96**
NOTATION FOR EQUATIONS **11.98**
REFERENCES **11.99**

Membranes represent an important set of processes for potable water treatment. Benefits of their use include demonstrated universal treatment capabilities and competitive costs. Membrane systems have experienced significant growth in water quality applications in the last three decades, and are now available in a variety of forms that can differ in size, shape, and operation, and yet each is uniquely able to resolve a particular need or purpose. There are very few drinking water contaminants that cannot be removed economically by synthetic membrane processes. Several applications have been described in books on water treatment (Nalco, 1988; Mallevialle, Odendall, and Wiesner 1996; MWH, 2005). Membrane processes with the greatest immediate application to drinking water treatment are microfiltration (MF), ultrafiltration (UF), nanofiltration (NF), reverse osmosis (RO), and electrodialysis/electrodialysis reversal (ED/EDR). Common terminology for membrane processes appears in Table 11-1. A list of abbreviations and equation symbols along

TABLE 11-1 Membrane Terminology

Term	Definition
Array, train, or pass	Multiple interconnected stages in series by concentrate staging or permeate staging
Asymmetric	A membrane that varies in density or porosity across its structure, and having a varying consistency throughout
Backwash	The intermittent residual stream from a MF or UF membrane system
Brine	Concentrate stream containing a TDS concentration greater than seawater (approximately 35,000 mg/L total dissolved solids)
Brackish water	Water having a TDS concentration ranging from approximately 1000 to 15,000 mg/L
Chemical cleaning or clean in place	The periodic application of a chemical solution to a membrane system to remove accumulated foulants in order to maintain productivity and return the membrane to baseline operation levels
Concentrate, reject, retentate, residual, or blow-down stream	The membrane output stream that contains higher TDS (NF, RO) or higher TSS (MF, UF) than the feed stream
Concentration-polarization	A phenomenon that occurs when dissolved and/or colloidal materials concentrate on or near the membrane surface in the boundary layer
Conventional MF or UF process	A treatment system consisting of a configured group of pressure- or vacuum-driven membranes contained in modules or basins consisting of a pretreatment strainer, air and reverse filtration discharges, filtrate tankage and associated appurtenances, including chemical feed for posttreatment
Conventional RO or NF process	A treatment system consisting of acid and/or scale inhibitor addition for scaling control, cartridge filtration for pretreatment, the RO or NF membrane treatment process, including the processes of aeration, corrosion control stabilization and pH adjustment, and disinfection for posttreatment

(*Continued*)

TABLE 11-1 Membrane Terminology (*Continued*)

Term	Definition
Differential pressure	A pressure drop across a membrane element or module from the feed inlet to concentrate outlet
Feed	Input stream to the membrane array
Flux	The throughput of a membrane process expressed as a mass ($mL^{-2}t^{-1}$) or volume (Lt^{-1}) rate of transfer through a membrane surface
Fouling	Deposition of solid material from the feed stream onto the membrane surface or within a porous membrane structure that restricts the passage of water decreasing productivity
High-recovery array	Array where the concentrate stream from one array becomes feed to a succeeding array to increase recovery
Hollow fiber membrane	A geometry in which the membrane material is configured into a very small diameter and mounted in a housing or pressure vessel, or submerged in a basin
Instantaneous flux	The flux that occurs in a MF or UF membrane when the unit is operating between backwash operations and is generally the highest flux reported for a system, as it does not account for system downtime or filtrate losses during treatment
Lumen	The center or bore of a hollow fiber membrane
Mass transfer coefficient (RO, NF), or specific flux (MF, UF)	Mass or volume unit transfer through a membrane based on driving force expressed in volume per unit area per unit time
Membrane element	A single membrane unit containing a bound group of SW or HFF membranes to provide a nominal surface area
Module or cartridge	The smallest component of a MF or UF membrane unit where the membranes are placed in a housing device with a filtrate outlet structure
Membrane softening	The use of membranes for removing hardness from water most often accomplished using NF membranes
Net driving pressure	The hydraulic pressure differential across the membrane minus the osmotic pressure differential across the membrane that is available to force water through the membrane
Permeate, product (filtrate)	The membrane output stream that contains lower TDS (lower TSS) than the feed stream
Pressure vessel	A single tube with several membrane elements in series
Raw	Input stream to the membrane process
Recovery	The ratio of the permeate flow to the feed flow expressed as a percentage
Rejection	Percentage solute concentration reduction of the permeate stream relative to the feed stream
Scaling	Precipitation of solids in an element due to solute concentration in the feed stream
Solute	Dissolved solids in raw, feed, permeate, and concentrate streams
Solvent	Liquid, usually water, containing dissolved solids
Stage or bank	Parallel pressure vessels
System arrays	Several arrays that produce the required plant flow
Transmembrane pressure	Pressure drop across the membrane barrier

Abbreviation key: TDS, total dissolved solids; TSS, total suspended solids; MF, microfiltration; UF, ultrafiltration; NF, nanofiltration; RO, reverse osmosis.

with their meanings that are used throughout the chapter can be found at the end of this chapter.

The use of synthetic membrane processes for desalination has increased over the past five decades primarily in coastal areas with limited fresh water sources. Between 1970 and 1990 the membrane market was dominated by desalting applications; however, since 1990 the use of MF and UF has increased exponentially, with over 200 operating installations ranging up to 380,000 m^3/d (100 mgd) in size at the turn of the century, primarily a result of increase regulatory focus on *Cryptosporidium* and lower capital and operating costs. Initially, municipal membrane applications were limited to desalting with the first municipal brackish water RO plant commencing operation in October 1971 serving the Key Largo Ocean Reef Club in the Florida Keys. The plant operated at 4200 kPa (600 psi) and produced 225 m^3/d (0.06 mgd) of permeate, unlike today, when RO plants are large as 190,000 m^3/d (50 mgd) can be constructed to operate at half the operating pressure than at the time the Key Largo plant was placed online. The concept of membrane softening was first practiced at Pelican Bay, Florida, in 1977, yet was limited in use until the mid-1980s when spiral-wound polyamide nanofiltration elements became available. Since that time there has been steady growth in the development of NF plants for softening and disinfection by-product precursor removal and continued use of RO and ED/EDR for brackish water desalting. Florida also was host to the introduction of the largest seawater RO plant in the Western Hemisphere at Tampa Bay Water's Big Bend site, when the facility became operational in 2004. The use of ED/EDR also has continued to expand in niche markets, one plant as large as 265,000 m^3/d (70 mgd) placed online in Barcelona Spain in 2009 for bromide removal and disinfection by-product (DBP) control.

SIZE RANGES FOR MEMBRANE PROCESSES

The applicable size ranges for membrane processes are shown in Fig. 11-1. In the broadest sense, a membrane—the common element of all these processes—could be defined as any barrier to the flow of suspended, colloidal, or dissolved species in any solvent. MF and UF membrane technologies are increasingly popular alternatives to conventional treatment for potable water production, and are typically employed to remove particles from water. MF and UF are used to remove turbidity, pathogens, and particles from freshwaters. Both UF and MF processes and coagulation/filtration processes remove particulates from drinking water. However, UF and MF have a significant advantage over conventional processes in that a membrane forms a static pathogen barrier. The pathogens are simply too big to pass through the membrane, so 6-log pathogen removal is not unexpected and small chance remains that pathogen removal will fail.

NF is used to soften freshwaters and remove *disinfection by-product (DBP)* precursors. RO and ED are typically employed to remove dissolved substances from water. Electrodialysis is used to demineralize brackish water and to soften freshwater. RO is primarily used to remove salt from brackish water or seawater, although RO is also capable of very high rejection of *synthetic organic compounds (SOCs)*. Desalting techniques are primarily intended for the removal of *total dissolved salts (TDS)* that generally cannot be removed by conventional treatment processes.

Drinking water contaminants are presented as biological, inorganic, and organic contaminants, as well as radionuclides, particulates, and other groupings. The ability of membranes to remove these contaminants can be inferred from Table 11-2 and Fig. 11-1, both of which show effective size ranges of contaminants that can be partially or completely removed by each membrane process. If sieving or size exclusion is the mechanism of solute rejection, complete removal can be achieved by a membrane free of defects. All membrane processes can reject contaminants such as turbidity or pathogens, but UF and MF are the most cost-effective processes for control of large particles. Smaller contaminants are removed via

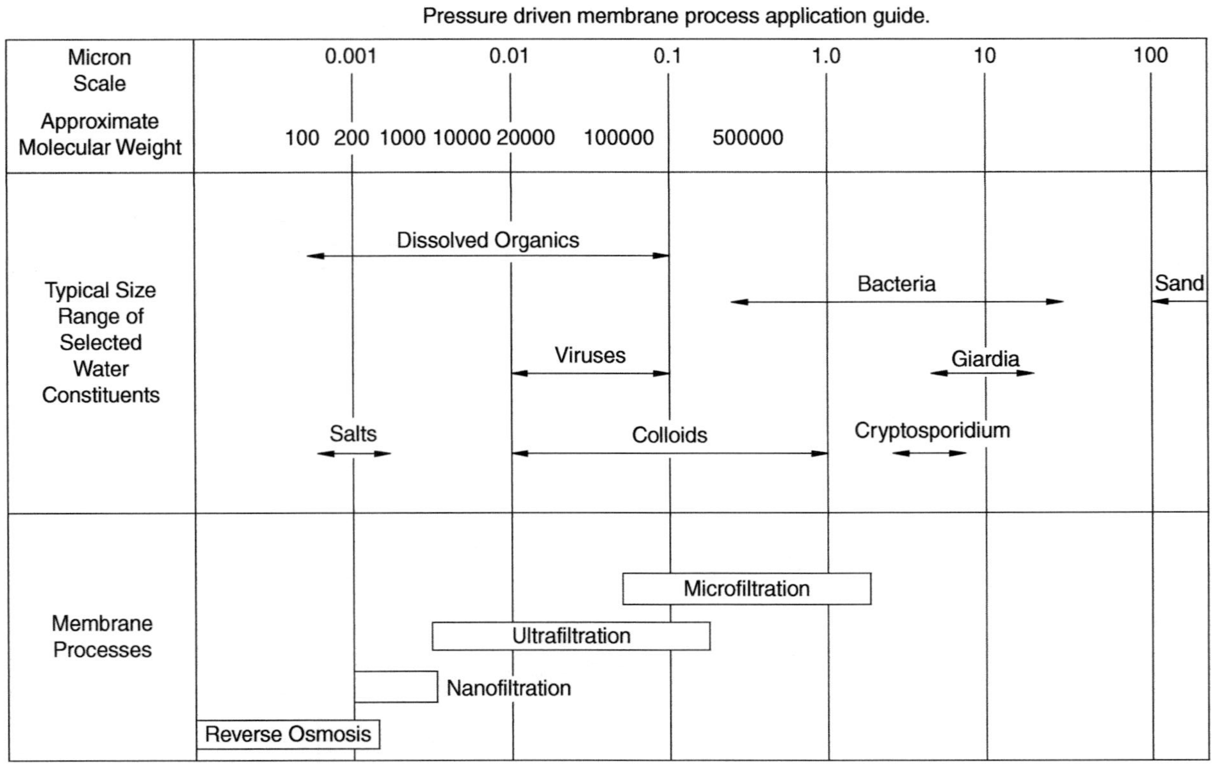

FIGURE 11-1 Size ranges of membrane processes and contaminants.

TABLE 11-2 Potable Water Contaminants Classified by Effective Membrane Size Range and Removal by Application

Particle category	Classification	Regulatory rule	Size range			Membrane process application by name with contaminant size removed				
			Ionic	Macromolecular and colloidal	Fine	MF 0.10 μm	UF 0.01 μm	NF 0.001 μm	RO 0.0001 μm	EDR 0.0001 μm (charged)
Biological										
	Viruses	GWR, SWTR[1]		√		*	√	#	#	N/A
	Bacteria	GWR, SWTR[1]		√		√	√	#	#	N/A
	Helemitis	GWR, SWTR[1]			√	√	√	#	#	N/A
	Algae	N/A			√	√	√	#	#	N/A
	Protozoa	GWR, SWTR[1]			√	√	√	#	#	N/A
	Cysts	GWR, SWTR[1]			√	√	√	#	#	N/A
	Fungi	N/A			√	√	√	#	#	N/A
Inorganic										
	Metals	Phases II &V IOCs, LCR	√			*	*	√	√	√
	Chlorides	Phase II IOCs	√			N/A	N/A	√	√	√
	Fluoride	Phase II IOCs	√			N/A	N/A	√	√	√
	Sulfate	N/A	√			N/A	N/A	√	√	√
	Nitrate	Phase II IOCs	√			N/A	N/A	√	√	√
	Cyanide	Phase V IOCs	√			N/A	N/A	√	√	√
Organic										
	Priority Pollutants	Phases I-V VOCs & SOCs	√	√		*	*	√	√	*
	Surfactants	Phases I-V VOCs & SOCs		√		*	*	√	√	*
	NOM	SWTR[1], DBPR[2]		√		*	*	√	√	*
	THM , HAA and DBP Precursors	DBPR[2]		√		*	*	√	√	*

Radionuclides										
	Radium-226/228	Radionuclide Rule	√			N/A	N/A	√	√	√
Particulates										
	Turbidity	SWTR[1]			√	√	√	√	√	√
	TSS	N/A			√	√	√	*	*	*
Other										
	Color	N/A		√		*	*	√	√	*
	TDS	N/A	√			*	*	√	√	√

√ = very effective; # = very effective; however, process typically not employed for this application due to fouling limitations; * = requires integrating coagulation, flocculation, sedimentation and PAC and oxidation processes, or differing combinations thereof, to be effective.

N/A, not applicable; NOM, natural organic matter; THM, trihalomethanes; HAA, haloacetic acid, DBP, disinfection by-product; TSS, total suspended solids; TDS, total dissolved solids; GWR, Ground Water Rule; SWTR, Surface Water Treatment Rule; LT2ESWTR, Long Term 2 Enhanced Surface Water Treatment Rule; LCR, Lead and Copper Rule; IOC, inorganic compound; VOC, volatile organic compound; SOC, synthetic organic compound; DBPR, Disinfection By-product Rule

[1]SWTR: Includes initial SWTR regulations implemented in 1989 and any revisions to the SWTR including Interim Enhanced Surface Water Treatment Rule (IESWTR), Long Term [1]Enhanced Surface Water Treatment Rule (LT1ESWTR) and Long Term 2 Enhanced Surface Water Treatment Rule (LT2ESWTR)

[2]DBPR: Includes Phase I and Phase II of the DBPR

size exclusion, charge repulsion, or diffusion mechanisms in membrane processes, leaving some residual. Although such a residual may be below the detection level, it is present.

Typically, the cost of membrane treatment increases as the size of the solute removed decreases. The ionic range in Fig. 11-1 encompasses potable water solutes such as sodium, chloride, total hardness, most total dissolved solids, and disinfection by-product precursor matter. The macromolecular range includes large and small colloids, bacteria, viruses, and color. The fine particle range includes larger turbidity producing particles, most total suspended solids, cysts, and larger bacteria. The membrane processes normally used in the ionic range remove macromolecules and fine particles, but because of operational problems, they are not as cost effective as membranes with larger pores.

Contaminants larger than the maximum pore size of the membrane are completely removed by sieving in a diffusion-controlled process. Contaminant rejection by diffusion-controlled membrane processes increases as species charge and size increases. Consequently, satisfactory removal of metals, TDS, biota, radionuclides, and disinfection by-product precursors can be attained. No commercially available membrane effectively removes uncharged species such as hydrogen sulfide (H_2S) and small, uncharged organic contaminants. Other aqueous contaminants should be treatable by membrane processes, although, once again, the cost of this treatment generally increases as the size of the removed contaminant decreases. Many membrane manufacturers specify the molecular weight cutoff (MWC) values for membranes. The MWC represents a nominal molecular weight of a known species that would always be rejected in a fixed percentage, using a specific membrane under specific test conditions.

Although many factors affect solute separation by these processes, a general understanding of drinking water applications can be achieved by associating minimum size of solute rejection with membrane process and regulated contaminants (Taylor et al., 1989). One correct interpretation of Fig. 11-1 is to assume that each membrane process has the capability of rejecting solutes larger than the size shown in the exclusion column. As shown in Table 11-2, regulated drinking water solutes can be simplified to the categories of pathogens, organic solutes, and inorganic solutes. Pathogens can be subdivided into cysts, bacteria, and viruses. Organics can be subdivided into DBPs and SOCs. Inorganic parameters are total dissolved solids, total hardness, and heavy metals, among others.

Electrodialysis (ED) and *electrodialysis reversal (EDR)* processes are capable of removing the smallest contaminant ions to 0.0001 μm, but a charge is required. Consequently ED and EDR are limited to treatment of ionic contaminants and are ineffective for pathogen removal and most organic applications. RO and NF perform both diffusion and sieving. They can remove all pathogens and many organic contaminants by sieving; by diffusion, they can achieve almost total removal of ionic contaminants. RO and NF processes have the broadest span of treatment capabilities. UF can achieve greater than 6-log removal of all pathogens from drinking water. MF can achieve greater than 6-log removal of cysts. Consequently, these processes effectively remove turbidity and microbiological contaminants, making them ideal for treating the majority of drinking water sources in the United States.

Existing regulations have been and will be modified to include more stringent control of chemical and biological toxins. Table 11-2 correlates various drinking water regulations to specific contaminants and the appropriate membrane process application that can satisfy regulation requirements. The SDWA amendments that were modified in 1996 still require the USEPA to create new drinking water regulations. The regulatory changes will continue to create a need for new drinking water technology to meet these challenges. The growth of drinking water regulations for both chemical and biological species has created treatment applications that can be met by membrane processes. Membranes, along with post-disinfection, can be effectively used for total removal of pathogens, high removals of inorganic and organic contaminates, and maintaining the highest possible distribution system integrity. There are very few instances where membranes cannot be utilized to meet or exceed all drinking water regulations.

CLASSIFICATIONS AND CONFIGURATIONS OF MEMBRANE PROCESSES

Some membrane processes rely on pressure as the driving force to transport fluid across the membranes. They can be classified by the types of materials they reject and the mechanisms by which rejection occurs. The progression of microfiltration (MF) to ultrafiltration (UF) to nanofiltration (NF) to reverse osmosis (RO) corresponds to a decreasing minimum size of components rejected by membranes as well as increasing transmembrane pressures required to transport fluid across the membranes and decreasing recoveries. Although a continuum of mechanisms likely contribute to separation in these pressure driven processes, very clear differences in separation methods distinguish reverse osmosis from microfiltration. These mechanisms are discussed later in the text.

While pressure driven processes like RO pass water through the membranes, electrodialysis involves the passage of the solute rather than the solvent through the membrane. As a consequence, both the mechanism of separation and the physical characteristics of membranes in electrodialysis differ substantially from those in pressure driven processes. ED membranes are fundamentally porous sheets of ion exchange resin with a relatively low permeability for water.

Membranes are classified by solute exclusion size, which is sometimes referred to as *pore size.* A reverse osmosis or hyperfiltration membrane rejects solutes as small as 0.0001 μm, which is in the ionic or molecular size range. A nanofiltration membrane rejects solutes as small as 0.001 μm, which is also in the ionic and molecular size range. Solute mass transport in these processes is diffusion controlled. Ultrafiltration and microfiltration membranes have a minimum solute rejection size of 0.01 and 0.10 μm, respectively. These membranes reject colloidal particles, bacteria, and suspended solids by size exclusion and are not diffusion controlled. Pressure drives the transport of water (the solvent) through these membranes. MF and UF require the use of pretreatment processes such as coagulation and/or oxidation or powdered activated carbon (PAC) to remove dissolved organic and inorganic constituents. ED relies on charge for solute separation and pulls ions through ED membranes, so it is unaffected by pore size.

Membranes can be classified by molecular weight cutoffs, solute and solvent permeability, solute and solvent solubility in the membrane film, active film material, active film thickness, surface charge, and active film surface. The molecular weight cutoff is the degree of exclusion of a known solute, as determined for a given set of test conditions in the laboratory. Typical known solutes used for determination of molecular weight cutoff are sodium chloride, magnesium sulfate, dextrose, and some dyes. Solute mass transport through a diffusion-controlled membrane is influenced by solute type and aqueous environment, so attempts to characterize them benefit from the use of additional organic solutes such as aromatic and aliphatic compounds of known molecular weight and structure.

Classification by Material

The development of a variety of membrane materials, both synthetic and modified natural polymers coupled with various manufactured forms, has played an integral part in the development of industrial-scale membrane separation applications. In any such process, several important membrane characteristics must be considered. Membrane selectivity, permeability, mechanical stability, chemical resistance, and thermal stability are among these critical factors, which are highly dependent on the type of material and the process control variables applied during manufacturing (Rautenbach and Albrecht, 1989).

Among the many raw materials used for membrane manufacturing, most basic types involve various forms of modified natural cellulose acetate materials and a variety of synthetic materials. To name a few, these synthetic materials are primarily composed of polyamides, polysulfone, vinyl polymers, polyfuran, polybenzimidazole, polycarbonate, polyolefins, and polyhydantoin. The diverse chemistries of the specific polymers and the associated production kinetics yield the individual attributes that provide for the myriad of membrane selectivity and productivity combinations (Rautenbach and Albrecht, 1989). Common MF and UF membrane materials used today include polyvinylidene fluoride (PVDF), polyethersulfone (PES), polysulfone, and cellulose triacetate. Recently, the use of ceramic materials made by sintering inorganic materials such as aluminum oxide, titanium oxide, or a carbon nanocomposite formulation have been considered for MF applications. Common NF and RO materials include polyaramide and polyamide for spiral-wound configurations, and cellulose triacetate for RO hollow fine fiber membranes.

Porous materials produced by precipitation from a homogeneous polymer solution are termed *phase inversion membranes.* They incorporate both symmetrical (homogeneous) and asymmetrical structures. The basic production process consists of five fundamental steps. A homogeneous polymer solution must first be produced. The polymer film is then cast followed by partial evaporation of the solvent from the polymer film. Immersion of the polymer film in a precipitation solution then allows the solvent to be exchanged for the precipitation agent. Imperfections in the precipitated membrane film are restructured by treatment in a heated bath solution (Bungay et al., 1986). Variations in environmental conditions for each of these steps can generate an assorted range of membrane structures that affect system performance. Typical membrane structural profiles can range from well-defined cavities in the shape of fingers to pores arranged in a dense sponge structure. These configurations result from the membrane structure both at the membrane surface and within the support structure itself (Rautenbach and Albrecht, 1989).

Both symmetric and asymmetric membranes can be produced by the phase inversion process. The difference between these two membrane classifications is the environmental conditions in which the membranes are produced and their resulting structural profiles. In the production of symmetrical membranes, homogeneous conditions for material formation throughout the membrane matrix lead to a uniform polymer structure. Conversely, structural properties vary throughout asymmetric membranes. Production of an asymmetric membrane forms a dense surface layer of submicron thickness that gives the membrane its selectivity properties. This active layer is in turn supported by a porous support structure. The combination yields membrane materials that are both selective and mechanically stable with enhanced productivity. Given the reduced thin active layer, hydraulic losses in head across the membrane are significantly less than those associated with comparable symmetric membranes. Consequently, development of the asymmetric membrane structure is an additional component integral to the commercial success of both RO and UF.

Another classification of phase inversion membranes is the composite membrane. In such a membrane, the materials comprising the active surface are different from those of the support material. These membranes are produced by lamination of the active surface layer onto the support layer, for example, polysulfone. This class of membranes is generally regarded as an improvement in membrane material design in that a specific active surface layer can be matched with a support layer of optimum porosity. This combination enhances membrane productivity while retaining the desired rejection properties offered by the dense active surface layer. Figure 11-2 shows the characteristic layers in a composite membrane, showing three distinct components, the polyamide membrane film (active layer), a polymeric supporting layer, and a fabric-backing material.

The active layer or membrane film is a polymer or combination of polymers forming a composite layer of varying thin films or a single thin-film layer. These polymers are generally in the form of straight chain compounds such as cellulose acetate or aromatic compounds such as a polyamide. Several different interactions occur between

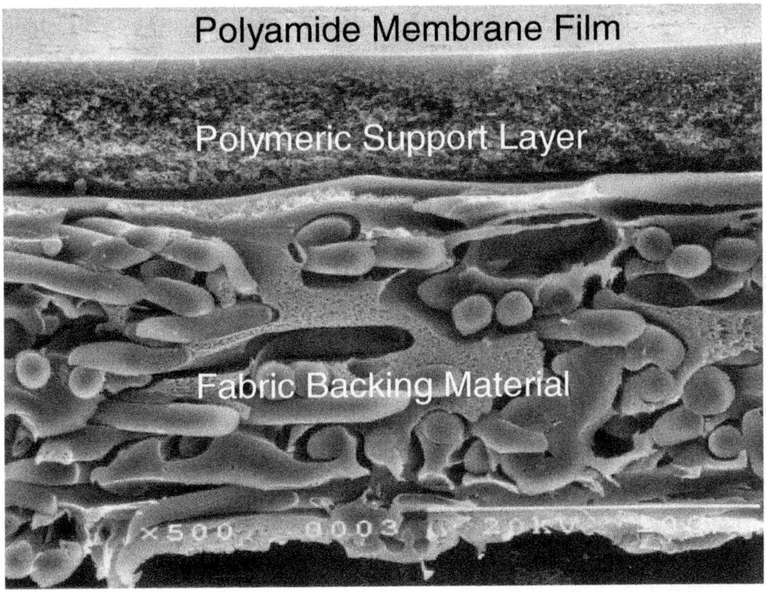

FIGURE 11-2 Composite membrane structure. (*Courtesy of Toray Membrane America, Inc.*)

the polymers that form the active layer of the membrane, and consequently between the membrane and solutes that pass through it. The three types of secondary forces are dipole forces, dispersion forces, and hydrogen bonding forces. Covalent and ionic forces are primary forces, with stronger effects than those of the secondary forces in the active membrane film. Average values of primary and secondary forces in membrane polymers are 400 kJ/mole for covalent and ionic forces, 40 kJ/mole for hydrogen bonding forces, 20 kJ/mole for dipole forces, and 2 kJ/mole for dispersion forces. These membrane forces can interact with corresponding forces associated with solutes, possibly promoted by functional groups on pesticides, for example.

Hyperfiltration (reverse osmosis at 5516 to 8274 kPa [800 to 1,200 psi]) and nanofiltration membrane films have active layers of cellulose compounds, aliphatic or aromatic polyamides, and thin-film composites. Cellulose triacetate is used as the active film for many desalination applications. Cellulose derivatives have good properties for membranes, since their crystalline and hydrophilic properties enhance durability and their capacity to transport water. Cellulose membranes are subject to chemical degradation by hydrolysis and biological degradation by oxidation. They must be operated at ambient temperatures from pH 4.0 to 6.5 with a biocide to avoid degradation.

Polyamide membranes are also effective films. Aromatic polyamides are generally preferred over aliphatic polyamides because of their mechanical, thermal, chemical, and hydrolytic stability and their permaselective properties. The aliphatic polyamides are porous and therefore not permaselective.

The development of the cross-linked fully aromatic polyamide thin-film composite membrane in the late 1970s represented a major advance in membrane technology. Thin-film composites provided thin active films that required much less energy to induce fluid passage than previous materials, making them more economical to use on a large scale. As mentioned previously, thin-film composites layer asymmetric films to form membranes with several different characteristics. Both hydrophilic and hydrophobic films are laid in a composite film by cross-linking different polymers. The thickness of the nonporous layer is typically less than 1 μm. Cross-linking to the porous film provides needed support for the nonporous film. Removal of macromolecules and ionic species is achieved by a nonporous film in a pressure driven membrane process.

Classification by Geometry

MF, UF, NF, and RO membranes are made from different materials and in different configurations. Both the material and configuration of such a membrane affect its mass transport or performance. As the effects of materials and configurations on solute rejection by membranes are largely unknown, the following paragraphs briefly discuss pressure-driven membrane materials and configurations.

MF/UF Configuration. The flow in microfiltration or ultrafiltration processes can run from inside out or outside in through hollow fibers. MF and UF inner and outer diameters vary between manufacturers; however, a typical membrane fiber's inner and outer diameter approximates 90 and 1900 μm, respectively. Figure 11-3a shows a single MF PVDF fiber under magnification, and Fig.11-3b presents the structure and dimensions for a MF PVDF fiber having an inner and outer diameter of 0.7×10^{-4} m (2.5×10^{-3} ft) and 1.3×10^{-3} m (4.2×10^{-3} ft), respectively. Typical MF poresize ratings range from 0.1 to 0.5 μm, whereas commonly used UF membranes range from 150,000 dalton to 2000 dalton molecular-weight cutoff. Although some UF membranes have been produced in spiral-wound (SW) configurations for industrial applications, SW UF configurations have not been applied to any significant degree for drinking water production in the United States.

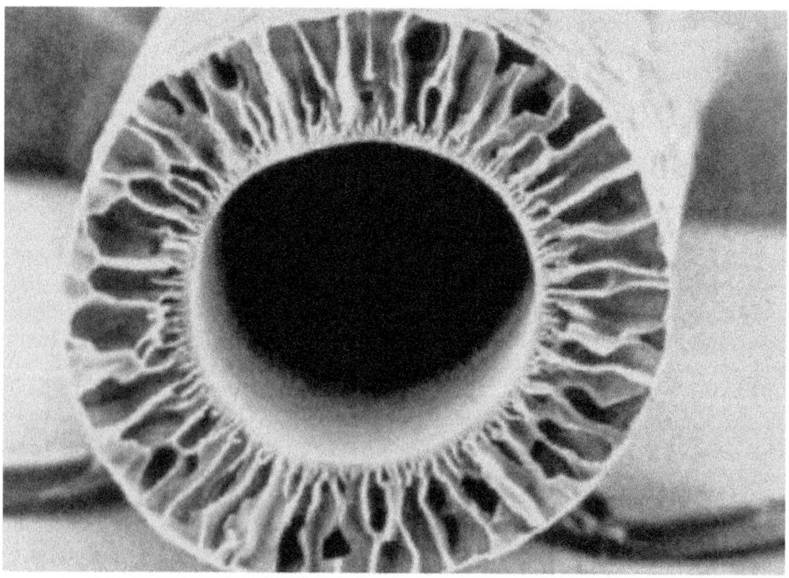

(a)

FIGURE 11-3 (*a*) Microscopic image of an asymmetric MF membrane.

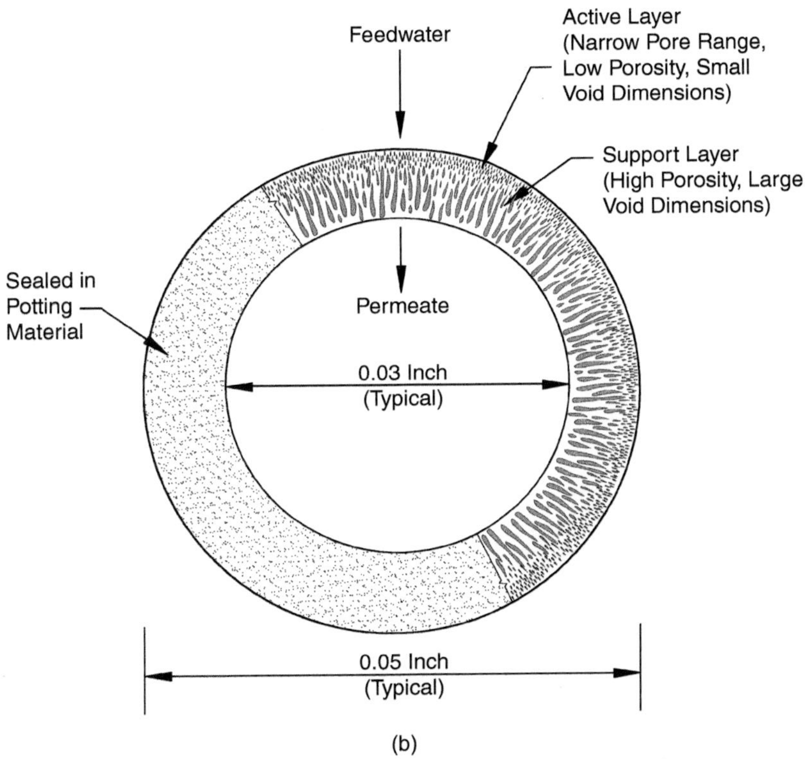

FIGURE 11-3 (*b*) Lumen dimensions of an outside-in asymmetric MF membrane. (*Continued*)

In a conventional UF or MF process, the driving force to produce filtrate can work in two ways; positive pressure moves fluid through the fibers, usually at a rating lower than 241 kPa (35 psi), and negative pressure moves fluid through fibers under vacuum pressure. Combining the two different flow regimes and the two driving forces allows three different processes.

1. Inside-out flow with positive pressure
2. Outside-in flow with positive pressure
3. Outside-in flow with negative pressure

Both inside-out and outside-in flow patterns can be further characterized as either dead end or crossflow operations, as shown in Fig. 11-4. In a dead end operation, the entire feed flow passes through the membrane. The dead end mode of operation is analogous to conventional media filtration in that the retained particles accumulate and form a type of cake layer at the membrane surface. In a crossflow operation, only a portion of the flow passes through the membranes, and only a portion of the retained solutes accumulate at the membrane surface. The remaining flow (retentate) is recycled on the feed side. The crossflow regime also incorporates a tangential flow that shears the cake and minimizes the accumulation of solids on the membrane surface, which can lead to fouling of the material and reduce performance.

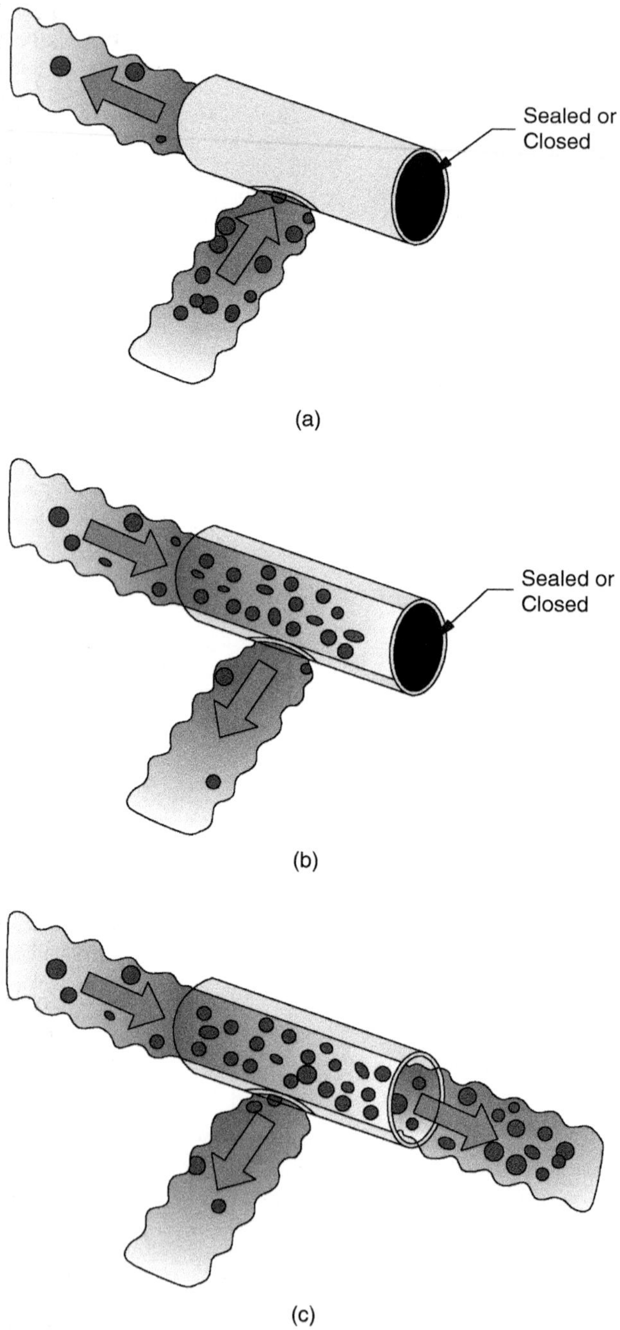

FIGURE 11-4 Individual fiber flow pattern. (*a*) Outside-in dead end filtration, (*b*) Inside-out dead end filtration, and (*c*) Inside-out cross flow filtration.

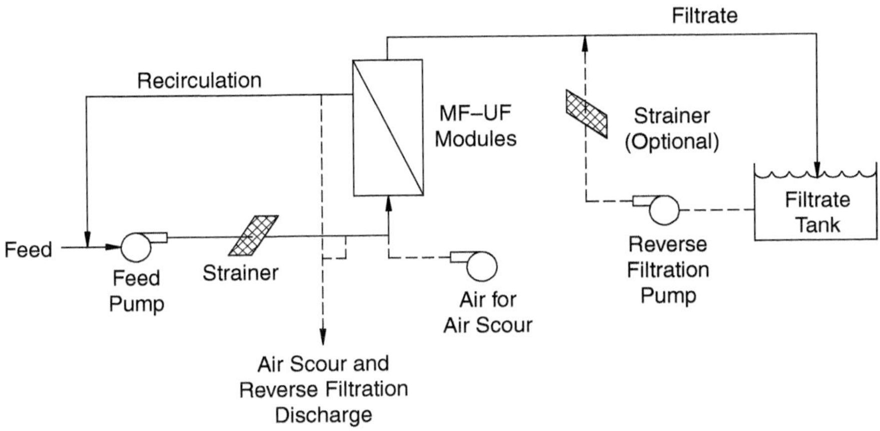

FIGURE 11-5 Typical MF and UF system process flow schematic. (*Courtesy of the American Water Works Association Manual of Water Supply Practices M53 Microfiltration and Ultrafiltration Membranes for Drinking Water.*)

Because the permeate flow produced through a membrane filter is roughly two orders of magnitude lower than loading rates through conventional media filters, MF and UF plants require on the order of 100 times the filter area to produce an equivalent amount of water. However, because of the packing densities achieved through the incorporation of hollow fibers into modules, MF and UF processes require less floor space than typical conventional media filtration processes. Figure 11-5 depicts a typical MF and UF process flow schematic. In Fig. 11-5 the feedwater is pumped to the membrane. In a crossflow mode, the water that does not permeate the membrane is recirculated as concentrate just ahead of the prefilter and blended with additional feedwater. A bleed stream may be employed to control the concentration of the solids in the recirculation loop. In direct filtration during which no crossflow is applied, or dead end filtration, prestrained water is applied directly onto the membrane, and there is 100 percent recovery of this water, except for those times a backwash or system flush is required.

MF processes are typically available in pressure-vessel configurations, with each pressure vessel housing thousands of fibers. UF membrane processes are available in two basic configurations, where the membranes are housed in pressure vessels and vacuum-type systems where the membranes are submerged or immersed in nonpressurized tanks. There are significant differences with regard to the design and layout of these two types of configurations, requiring that the configuration type be determined prior to commencement of facility design. .

Figure 11-6 shows one pressure-vessel module that is generally 10.2 to 45.7 cm (4 to 18 in.) in diameter and from 1.8 to 5.5 m (6 to 18 feet) long, and contain packing densities that provide between 9.3 and 93 m^2 (100 and 1000 ft^2) of filter area. Figure 11-7 illustrates a submerged, or immersed, membrane process where membranes suspended in modules are suspended inside basins containing pretreated feedwater. Transmembrane pressure is developed by using suction on the permeate side of the membranes which limits immersed membrane transmembrane operating pressures to between 41.4 and 48.3 kPa (6 and 7 psi), and pressure on the influent side is limited to the static pressure provided by the water column. Table 11-3 provides the typical operating characteristics of MF and UF facilities.

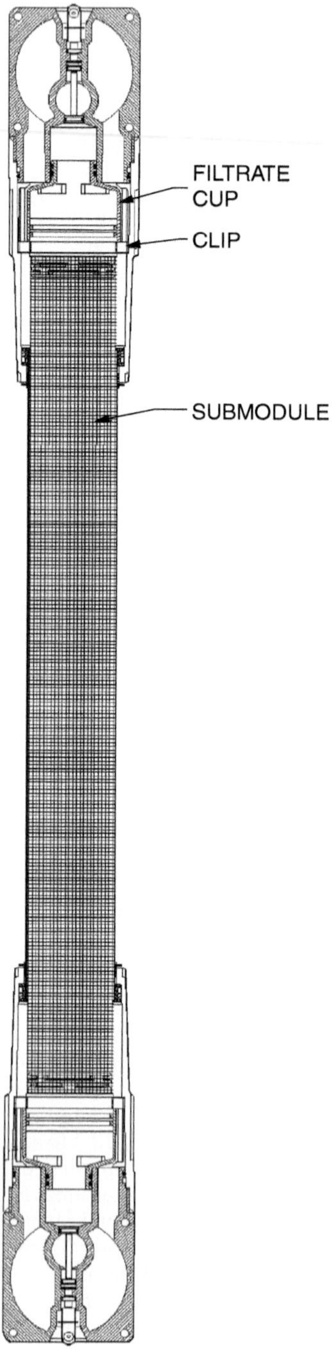

FIGURE 11-6 Representation of membrane fibers housed using the pressure vessel configuration. (*Source: Siemens Water Technology Corp.*)

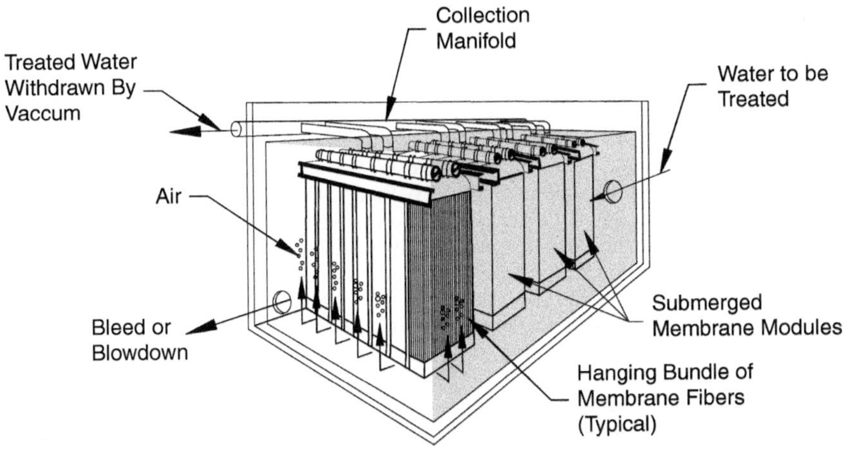

FIGURE 11-7 Representation of a submerged (immersed) UF membrane process. (*Courtesy of GE Water & Process Technologies Zenon Environmental.*)

TABLE 11-3 Typical Operating Characteristics of Membrane Filtration Facilities

Parameter	Units	Range of typical values
Permeate flux		
Pressurized systems	$L/m^2 \cdot h$	30–170
	$Gal/ft^2 \cdot d$	18–100
Submerged systems	$L/m^2 \cdot h$	25–75
	$Gal/ft^2 \cdot d$	15–45
Typical transmembrane pressure		
Pressurized systems	Bar	0.4–1
	Psi	6–15
Submerged systems	Bar	0.2–0.4
	Psi	3–6
Maximum transmembrane pressure		
Pressurized systems	Bar	2
	Psi	30
Submerged systems	Bar	0.5
	Psi	7.4
Recovery	%	>95
Filter run duration	Min	30–90
Backwash duration	Min	1–5
Time between chemical cleaning	Day	5–180
Duration of chemical cleaning	Hour	1–6
Membrane life	Yr	5–10

RO-NF CONFIGURATION

RO and NF membranes for drinking water treatment have either spiral-wound (SW) or hollow fine-fiber (HFF) configurations The SW configuration is the most common for production of drinking water. The HFF configuration was used extensively for desalination of seawater in the Middle and Far East; however, the use of spiral-wound configurations has increased over time. The geometry of a SW membrane is subject to fewer "dead areas" than that of a HFF membrane, it can be cleaned more thoroughly, and it is less subject to fouling. The ratio of surface area to volume is higher for a HFF element than a SW element. The recovery from a hollow fine-fiber element ranges from 10 to 50 percent and is typically higher than that from a SW element. The radial feed-stream velocity along the outside surface of the HFF varies from approximately 0.003 to 0.0003 m/s (0.01 to 0.001 ft/s) which give Reynolds numbers ranging from 100 to 500 for transport through the membrane.

From 1969 until about 2000, E. I. DuPont de Nemours, Inc. (DuPont) produced a polyamide hollow fine-fiber (HFF) membrane that was designated as B-9 and was used for brackish water desalting. DuPont produced a significant number of HFF membrane elements using this type of fiber material. In 1974, DuPont introduced its seawater membrane referred to as B-10, which was used predominantly overseas. In 2000 DuPont ceased the manufacture of B-9 and B-10 HFF membranes. Between 1982 and 2002, the city of Sarasota, Florida, operated a 17,000 m^3/day (4.5 mgd) DuPont HFF B-9 membrane process, which was the largest operating brackish HFF drinking water facility in North America. However, the city converted the HFF process trains to SW thin-film composite membranes after DuPont withdrew from the membrane market. Shortly thereafter, the Key West Aquaduct Authority replaced its DuPont B-10 seawater HFF membranes with Toyobo HFF modules.

Today HFF membranes can be acquired through Toyobo Company Limited, which produces Hollosep® HFF membranes. In the hollow fine-fiber (HFF) RO configuration, feedwater flows radially from the center pipe, passes across the hollow fibers in the bundle, and flows away from the element as concentrated brine flow. A depiction of a double cartridge HFF RO membrane configuration housed in a pressure vessel is in Fig. 11-8. In this configuration, permeate water is recovered from both ends of the RO HFF module. Permeate water is collected with each supporting end that is held in position using compression snaps and an O-ring seal on the face of the open end, so that concentrate water cannot penetrate into the permeate. The compression snap has another function; it is placed in the center of the pressure vessel and forms a narrow space between the pressure vessel and the RO element for the concentrated brine water flow. Brine water flows on the periphery of the RO element, passes through the narrow space between pressure vessel and supporting plate and flows to the brine port of the pressure vessel. The Toyobo HFF RO membranes are made of cellulose triacetate (CTA), that unlike polyamide RO membranes, are chlorine tolerant. This unique chlorine tolerance has demonstrated stable production and easy maintenance by eliminating biofouling in many seawater desalting applications (Burashid, 1994). The Toyobo HFF RO membranes are manufactured using a winding technology in a cross and overlapped arrangement that yields a self-supporting HFF bundle. The winding technology provides uniform flow throughout the entire volume of the bundle, resulting in a lower pressure drop across the bundle that provides additional resistance to accumulation of any fouling matter.

SW thin-film composite RO membranes were introduced in the mid to late 1970s and were designed for brackish- and seawater application. Low-pressure thin-film composite, spiral-wound membranes were first introduced in the early 1980s by Fluid Systems (now Koch Membrane Systems) and by FilmTec Corporation (now a division of Dow Chemical Company). These composite membranes are in common use, and many manufacturers offer similar products.

SW elements are manufactured using flat-sheet membranes, as opposed to bundles of fibers, as shown in Fig. 11-9. A typical SW element consists of envelopes attached to a

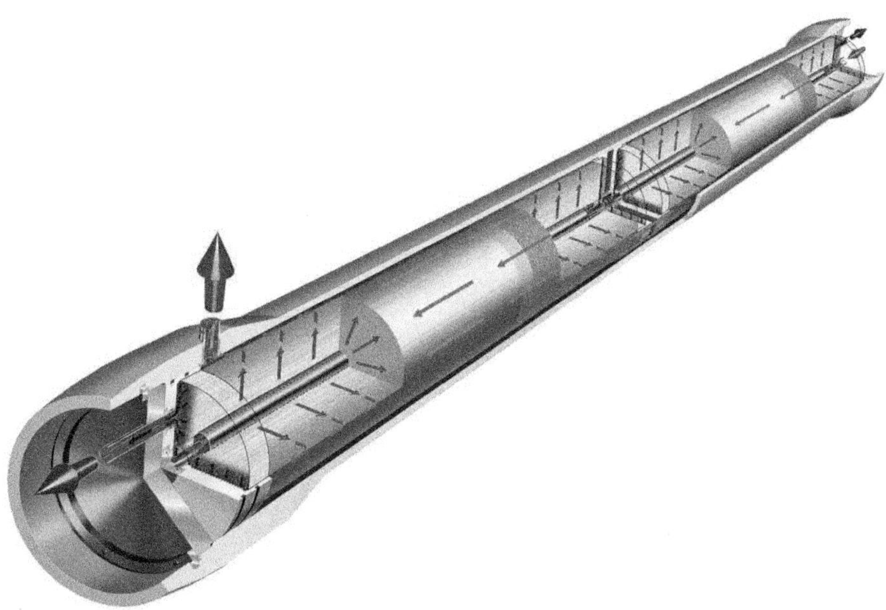

FIGURE 11-8 Depiction of hollow fine-fiber RO bundles housed in a pressure vessel. (*Courtesy of Toyobo Co., Ltd.*)

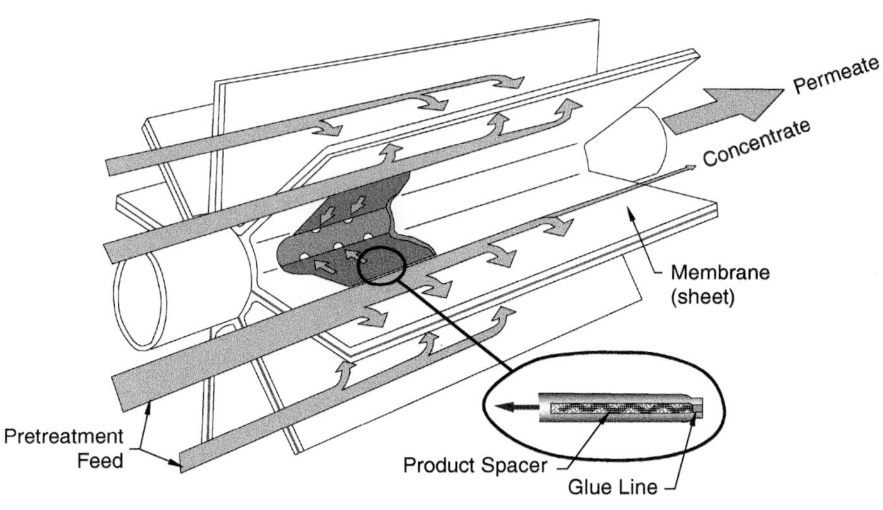

FIGURE 11-9 Diagram of a spiral wound reverse osmosis membrane. (*a*) Open construction, (*b*) Wound construction, (*c*) End construction, and (*d*) Cut-away element view. (*Courtesy of the American Membrane Technology Association, Stuart, FL.*)

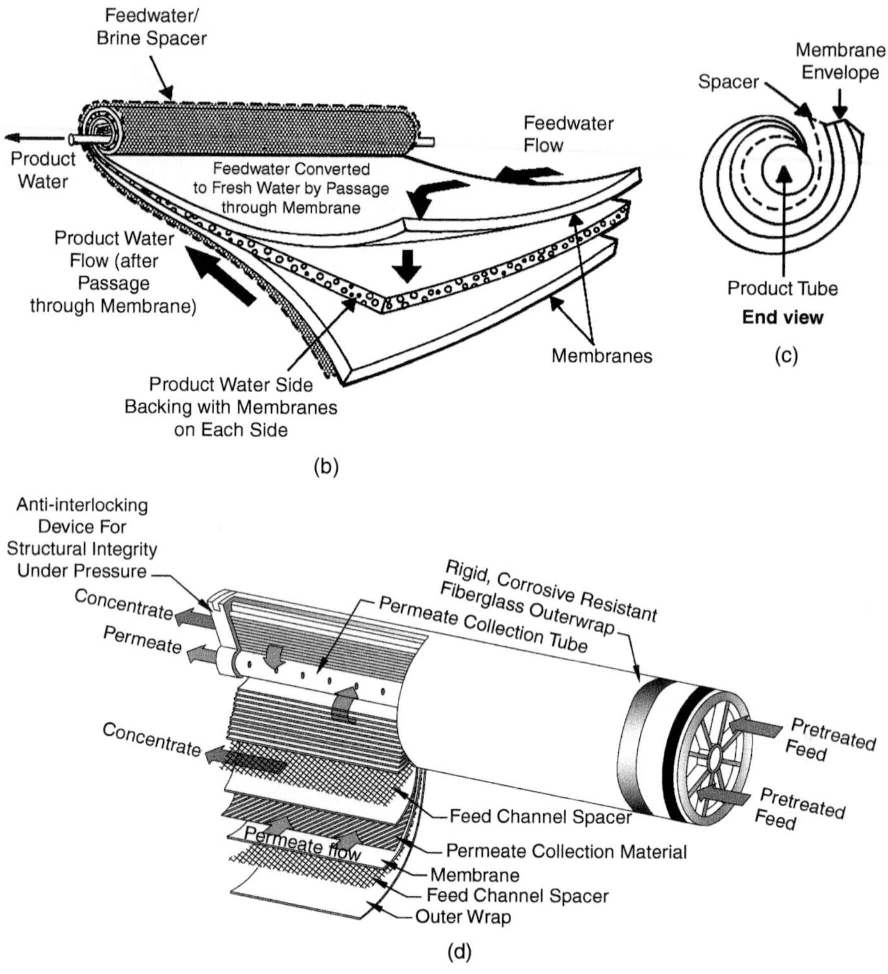

FIGURE 11-9 Diagram of a spiral wound reverse osmosis membrane. (*a*) Open construction, (*b*) Wound construction, (*c*) End construction, and (*d*) Cut-away element view. (*Courtesy of the American Membrane Technology Association, Stuart, FL.*)(*Continued*)

center tube that collects the permeate stream. Designs of SW elements differ among manufacturers; however, the following description is applicable to Filmtec, Toray, Hydranautics, and Koch SW membranes. Typically, in a SW element an envelope is formed by folding one flat sheet over a permeate stream spacer. The sheet itself consists of two layers, a nonporous active membrane film and a porous membrane support. The active layer is on the outside of the fold. The envelope is glued along three open sides and near the fold, completely enclosing the permeate spacer. The glue line on the fold end is a short distance away from the fold, because the fold end is attached to the center collection tube. The glue line at the fold end stops the flow of the feed stream and allows the remaining pressure in the permeate stream to drive it through the membrane into the center collection tube. A feedstream spacer is attached to each envelope prior to establishing the fold end glue line. Several envelopes and feedstream spacers are attached to the center collection tube and wrapped in a spiral around it. An epoxy shell or tape wraps are applied around the envelopes, completing the SW element. An anti-telescoping device is attached both ends of the element to maintain a fixed space between elements and facilitate flow from one element to the next. Multiple elements are housed in a series of cylindrical pressure vessels.

The feed stream enters the end of the SW element in the channel created by the feed stream spacer. The feed stream can flow either in a path parallel to the center collection tube or through the active membrane film and membrane supports into a channel created by the permeate-stream spacers. The permeate stream follows a spiral path into the center collection tube and is taken away as product water in a drinking water application. As with hollow fine-fiber membranes, the feed stream becomes progressively more concentrated as it passes to a succeeding element. As an example, a 100 mm (4 in.) Dow Filmtec NF70 membrane contains four envelopes with approximately 8.3 m^2 (90 ft^2) of surface area in a sheet that measures 0.91 m (3 ft) by 1.14 m (3.75 ft). The total element is 1.01 m (3.33 ft) long, but the feed stream path along the active membrane film is approximately 0.91 m (3 ft).

The amount of membrane surface area offered by SW elements varies between manufacturers, depending on application, element diameter, and spacer thickness. In recent times, both brackish- and seawater elements have been made with increased area. Brackish elements have increased from 34 to 37 m^2 (365 to 400 ft^2) and most recently to 40 or 41 m^2 (430 or 440 ft^2). These changes have mostly been done by changing element construction, not by changing the membrane. A combination of thinner permeate spacers, optimized glue line placement, and material thickness control has led to the improvements. These same improvements are now being applied to seawater membranes as well. Seawater elements in the 1990s were typically around 29 to 30 m^2 (310 to 325 ft^2) and then increased to 34.4 to 35.3 m^2 (370 to 380 ft^2) in the mid- to late 1990s. By the early 2000s, most seawater elements contained 37 m^2 (400 ft^2). Recently, 41 m^2 (440 ft^2) seawater elements have been produced. A 102 mm (4 in.) diameter SW membrane will provide approximately 6.5 m^2 (70 ft^2) of membrane surface area per membrane. In comparison, a 203 mm (8 in.) diameter SW membrane could provide upward of 37.2 m^2 (400 ft^2) of surface area per membrane element. Newer 406 mm (16 in.) diameter RO membrane elements that recently entered the market place possess an active surface area on the order of 160 m^2 (1725 ft^2), more than 4 times that of the standard 203 mm diameter element. The increase in active surface area greatly decreases the number of membrane elements needed for the same quantity of feedwater, reducing the system's footprint and equipment needs and providing capital cost savings for plant operators.

The recovery in a SW element varies from approximately 5 to 15 percent. The maximum feed and concentrate stream flows in a 10.2 cm (4 in.) element are approximately 4.2×10^{-3} m^3/min (16 gpm) and 7.9×10^{-4} m^3/min (3 gpm). Neglecting the effect of the feedstream spacer, the Reynolds number typically ranges from 100 to 1000. The feedstream spacer creates additional turbulence and increases the Reynolds number. The physical configuration of the SW element produces a more turbulent feed stream than that in a HFF element and leaves the membrane more easily accessible to cleaning agents. The highest and lowest feedstream velocities occur at the entrance and exits of the element, respectively. The feed flow is in the laminar region and is most likely to produce chemical or colloidal fouling in the last elements in series. Fouling from particle deposition occurs mainly in the first elements in series.

Classification by Driving Force

Membrane processes can be classified based on the driving force that induces transport of materials across the membranes. Examples of driving forces and corresponding membrane processes are listed in Table 11-4. Although interest in industrial applications for pervaporation is growing, industrial-scale applications of membrane processes for environmental quality control have so far been dominated by the pressure driven processes such as RO and

TABLE 11-4 Categorization of Membrane Processes by Driving Force

Driving force	Examples of membrane processes
Temperature gradient	Thermo-osmosis, distillation (thermal)
Concentration gradient	Dialysis, pervaporation, direct (forward) osmosis
Pressure gradient	RO, NF, UF, MF, piezodialysis
Electrical potential	ED, EDR, electroosmosis

by electrodialysis and electrodialysis reversal (EDR), which employ electrical potential as the driving force.

Electrodialysis. Electrodialysis is a membrane process driven by electric potential and is used for removing charged species (ions) from an aqueous stream using cation- and anion-selective membranes. In this electrochemical separation process ions are transferred through ion exchange membranes by means of a direct current (DC) voltage. In a simple electrolytic cell, negatively charged ions (anions) are drawn toward the positively charged electrode, or anode, and positively charged ions (cations) are drawn toward the negatively charged electrode, or cathode.

The ED/EDR process differs fundamentally from the pressure-driven membrane processes by mechanism and effect. An electrochemical separation process removes ions from a process stream, whereas a pressure-driven separation process removes water from the process stream. Both membrane processes can remove ions, but electrochemical separation processes cannot remove pathogens, as pressure-driven processes can, so they have no role in disinfection. Electrochemical separation processes are often less costly than NF/RO, however, because they achieve higher recoveries through lower salt rejection and mass transport advantages. A basic diagram of a batch electrodialysis process is shown in Fig. 11-10.

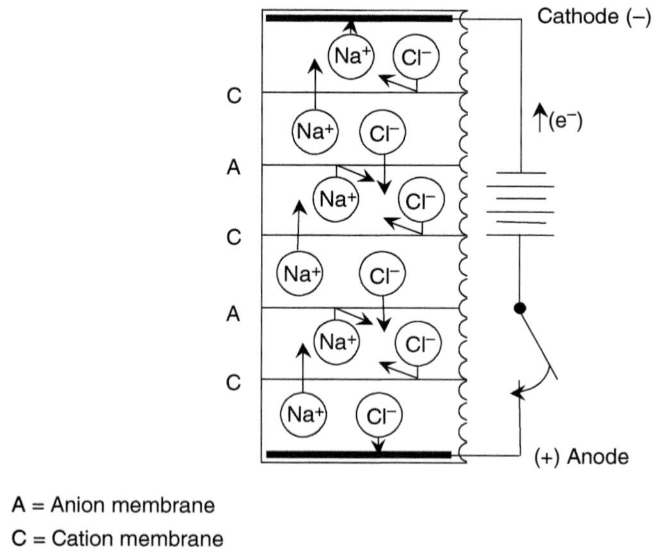

FIGURE 11-10 Diagram of an electrodialysis batch process under applied DC voltage. (*Courtesy of GE Water & Process Technologies Ionics, Inc.*)

ED Cell

The basic ED cell consists of alternating anion-permeable and cation-permeable membranes, which provide a basis for separation of ions under DC voltage. Redox reactions occur during ED. Water and chlorine are oxidized at the cathode, and water is reduced at the anode. These by-products of the ED cell are removed as electrode waste products and treated by aeration and neutralization (reduction), if necessary.

$$H_2O \rightarrow 4H^+ + 4e^- + O_{2(g)} \qquad (11\text{-}1)$$

$$2Cl^- \rightarrow Cl_{2(g)} + 2e^- \qquad (11\text{-}2)$$

$$H_2O + 2e^- \rightarrow 2OH^- + H_{2(g)} \qquad (11\text{-}3)$$

A simplified diagram of a complete cell for NaCl removal is shown in Fig. 11-11. Sodium ions pass through the cation-transfer membrane, and chloride ions pass through the anion-transfer membrane. The sodium and chloride ions are trapped in the brine channel by the alternating ion-exchange membranes, as only cations can pass the cation-permeable membranes and only anions can pass the anion-permeable membranes. The alternating ion-exchange membranes produce a demineralized stream and a concentrate stream, as shown in Fig. 11-11. Water flows across these membranes, not through them, as it does in pressure-driven processes.

As shown in Fig. 11-12 an ion-exchange membrane has a polymeric support structure with fixed sites and representative water-filled passages that reject common ions and pass counter ions through the membrane. In Fig. 11-12 a fixed, negatively charged, sulfonated functional group and a fixed, postively charged, tertiary amine are shown in the anionic and cationic membranes, respectively. These membranes are (1) impermeable to water, although water does participate in the transport process, (2) electrically conductive, and (3) ion selective. The anion membrane is composed of a cast anion exchange resin with a fixed negative charge in sheet form. The fixed negative charge assists cations in passing and

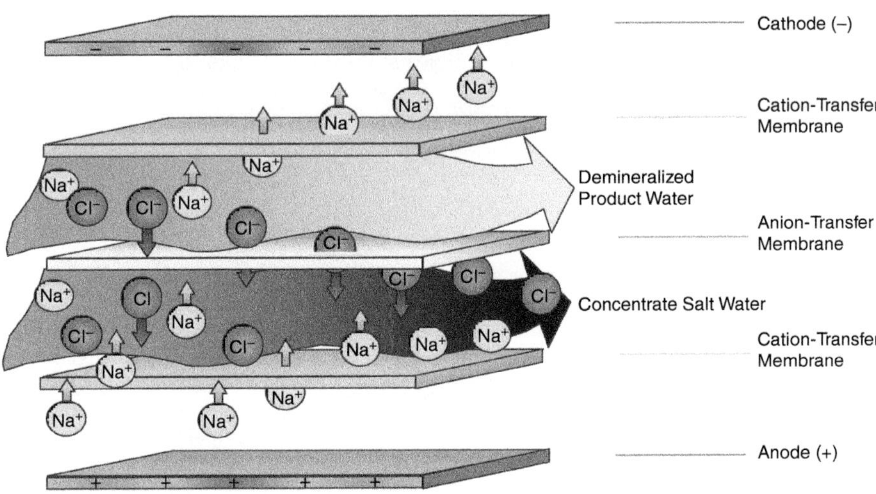

FIGURE 11-11 Simplified diagram of an electrodialytic cell. (*Courtesy of GE Water & Process Technologies Ionics, Inc.*)

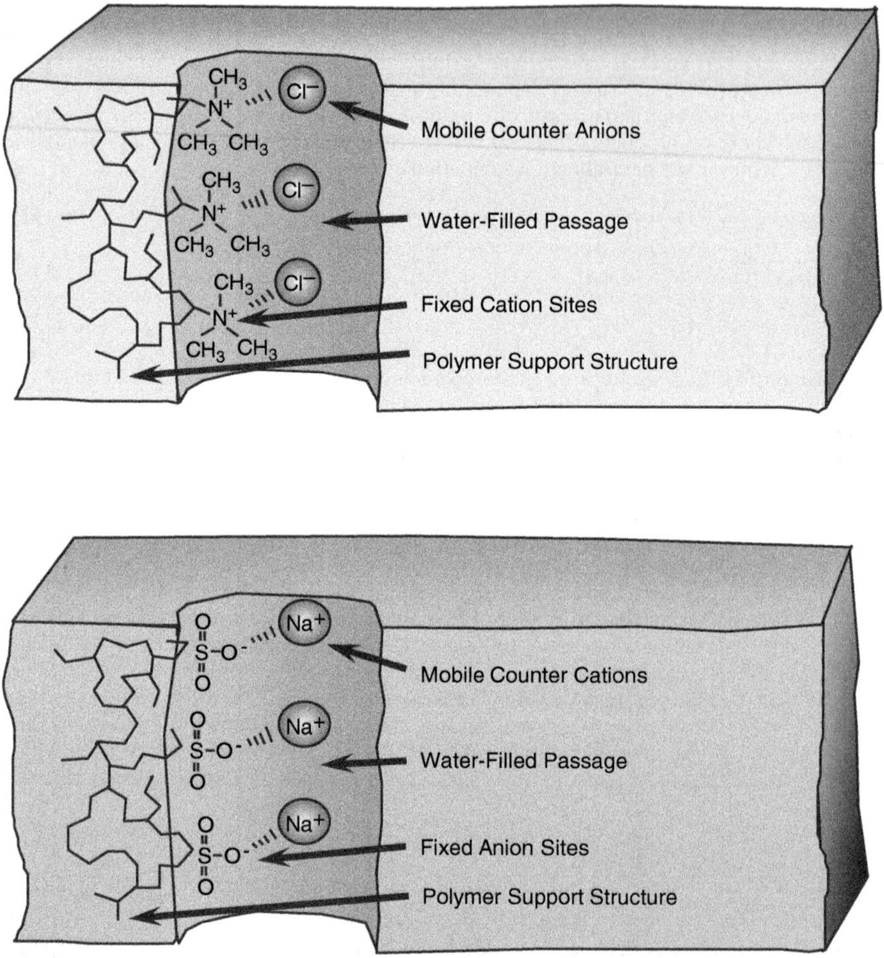

FIGURE 11-12 Ion exchange cast membrane sheets showing a fixed charge and mechanisms of mass transport in the electrodialysis process. (*Courtesy of GE Water & Process Technologies Ionics, Inc.*)

keeps anions from passing through the anion membrane. The cation membrane has a fixed positive charge and utilizes the same mechanism to repel cations and pass anions.

A spacer is placed between the anion-exchange and cation-exchange membranes to form a basic cell pair, as shown in Fig. 11-13. The cell pair consists of an anion-exchange membrane, a concentrating spacer, a cation-exchange membrane, and a demineralizing spacer as shown in Fig. 11-14. The spacers are assembled with cross straps to promote turbulence and reduce polarization at the membrane surfaces. Several hundred cell pairs are grouped into a common arrangement known as a *membrane stack* or *module,* as shown in Fig. 11-15a and are protected by removable stack siding enclosures that must be heat resistant. These streams either pass to succeeding cells for additional treatment or are discharged from the ED cell as concentrate waste and demineralized water (also termed *product*) as shown in

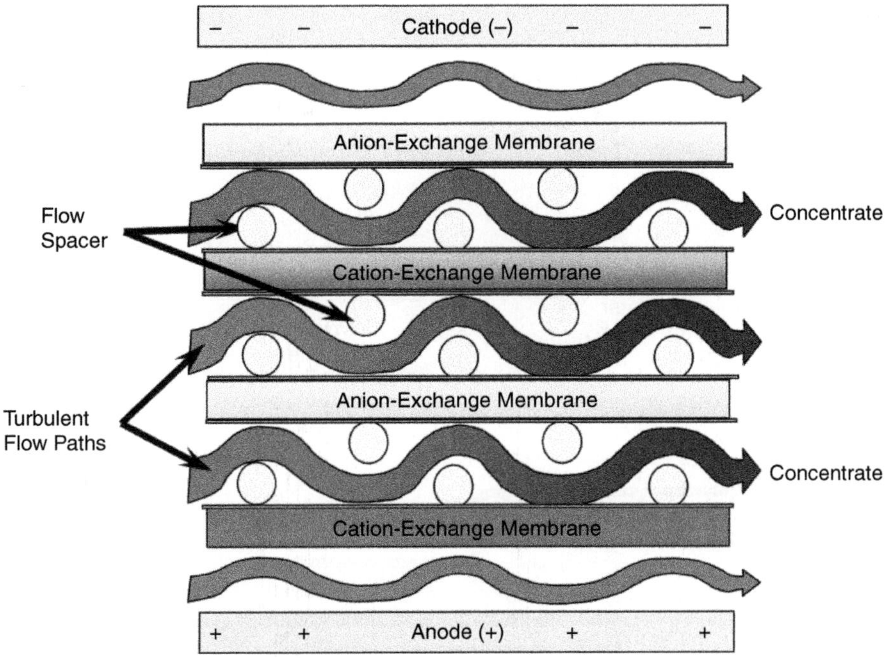

FIGURE 11-13 Basic electrodialysis cell pair. (*Courtesy of GE Water & Process Technologies Ionic, Inc.*)

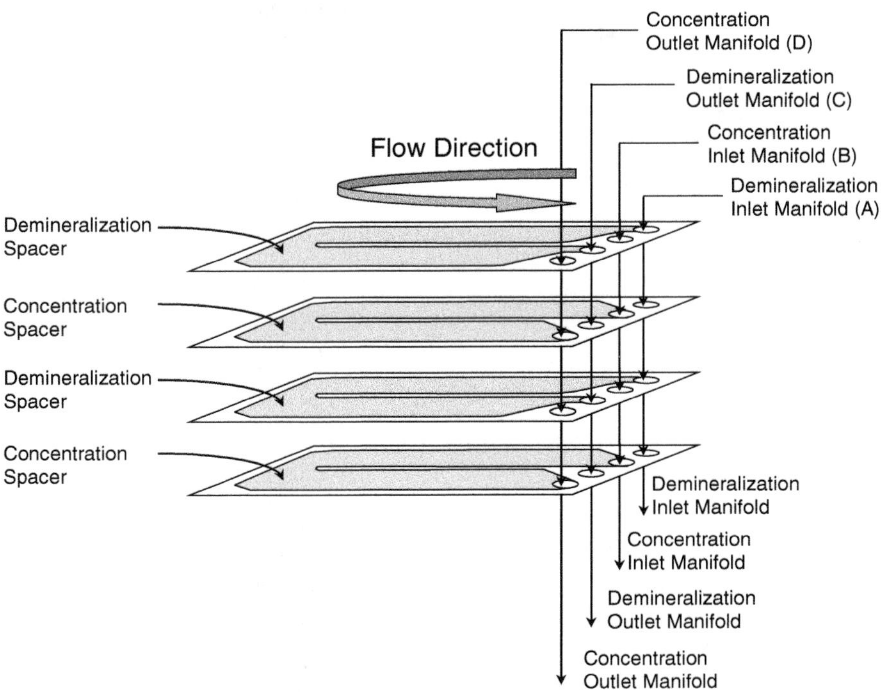

FIGURE 11-14 EDR spacer configuration showing flow direction. (*Courtesy of GE Water & Process Technologies Ionic, Inc.*)

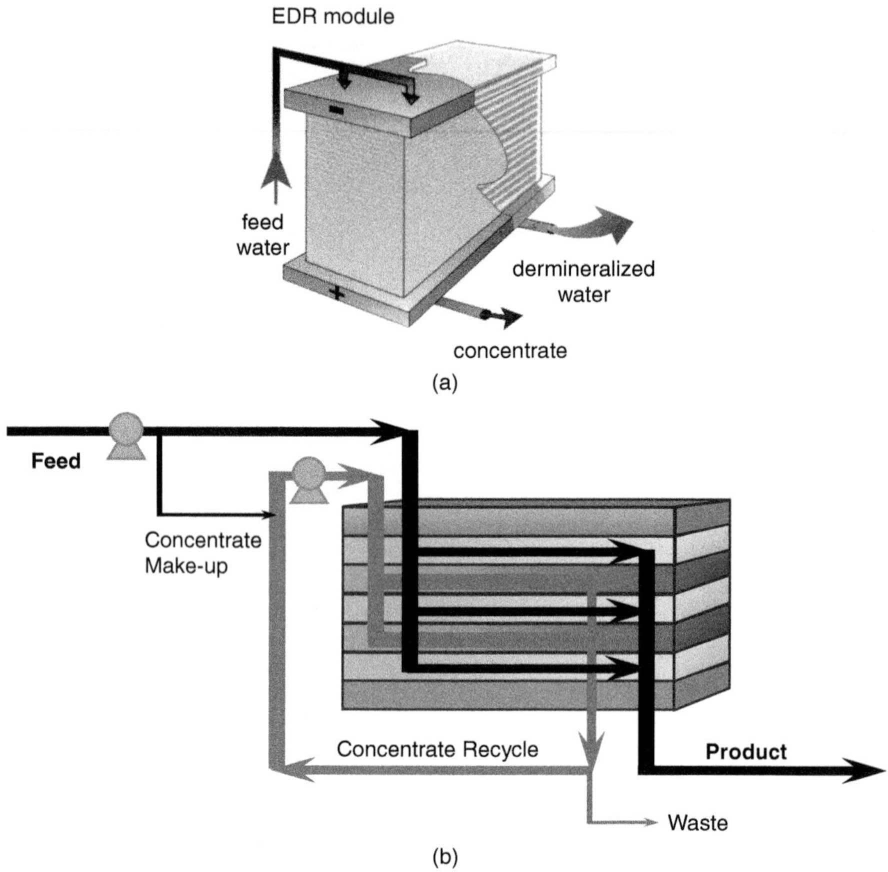

FIGURE 11-15 (*a*) EDR module. (*b*) Flow path. (*Courtesy of GE Water & Process Technologies Ionic, Inc.*)

Fig. 11-15b. The concentrate waste stream does not contain the redox products in the electrode waste streams, although these streams may be combined in some applications.

MEMBRANE PROPERTIES AND REJECTION CHARACTERISTICS

Characterization of membrane properties is desirable because of the potential of relating these characteristics to solute rejection and membrane-fouling studies. This information could be used to modify existing membranes or develop new membranes that would effectively reject inorganic and organic solutes from drinking water sources without fouling the membranes.

Fundamental research for mass transfer, membrane fouling and solute-solvent interactions has been conducted or is in progress using several different techniques. Currently, feasible techniques are capable of measuring membrane surface charge, pore size, thickness, roughness, surface energy, and surface atomic composition. These techniques are presented in Table 11-5.

TABLE 11-5 Techniques to Characterize Membrane Surfaces

Techniques	Application on membrane research	References
Streaming potential	Charge on membrane surface to determine electrostatic interactions.	Elimelech, Chen, and Waypa (1994)
Contact angle	Surface energy to distinguish hydrophilic and hydrophobic membranes	Oldani and Schock (1989)
Attenuated total reflectance-Fourier transform infrared spectroscopy (ATR-FTIR)	Kinetics study on membrane-solute-solvent interface	Ridgway and Flemming (1996)
Secondary ion mass spectrometry (SIMS)	Chemical analysis for sorption and fouling for clean and fouled membrane	Spevack and Deslandes (1996)
X-ray photoelectron spectroscopy (XPS)	Chemical analysis for sorption and fouling for clean and fouled membrane	Jucker and Clark (1994)
Scanning probe microscopy	Topological information to characterize roughness and pore size for clean and fouled membrane	Fritsche, et al. (1992)
Scanning electron microscopy (SEM)	High resolution image to characterize roughness and pore size for clean and fouled membrane	Fritsche, et al. (1992)

Diffusion-controlled solute and solvent permeability are best described by mass transfer coefficients, although percent rejection is sometimes used to describe solute mass transfer for a given environment. The solvent and solute mass transfer coefficients, like molecular weight cutoffs, are measured in the laboratory with standardized test conditions and a test cell. Continuous-flow test cells that operate at less than 1 percent recovery can be used to minimize concentration polarization so that linearly determined mass transfer coefficients can be used to characterize membranes. Such test cells can also be used to determine the changes in solvent mass transfer coefficients over time for different films (membranes) to determine a possible correlation between films and fouling.

Solute and solvent solubility in the membrane film are important characteristics of diffusion-controlled membrane processes. Ideally the active film would have no solubility for the contaminants or foulants and complete solubility for water. The solute mass transfer coefficient increases as the solubility of the contaminant in the film increases. Hence the lower the contaminant solubility in the film, the lower the contaminant concentration in the permeate. If needed, the film material can be modified to effect solubility by the inclusion of functional groups, or a different film material can be substituted.

The thickness and chemical structure of the film are important membrane characteristics and are difficult to measure, although techniques do exist. Plasma etching has been used to measure film thickness. A plasma is introduced to initiate a chemical reaction with the top layer of the film, producing gases. The rate of gas production during etching indicates surface morphology and thickness of a nonporous film. Electron spectroscopy and X-ray photoelectron spectrometry can be used to determine the thickness of the top layer.

Dissolved Solute–Membrane Interactions

Dissolved inorganic and organic solutes markedly influence electrokinetic properties of RO and NF membranes through various solute-membrane interactions. Among such interactions are, adsorption of inorganic and organic solutes on the membrane surface and/or within the membrane pores. These solute-membrane interactions have a significant impact on membrane rejection and fouling behavior. In the following text, the effect of solution composition (i.e., dissolved solutes) on membrane rejection and fouling is documented from recent publications.

Influence of Dissolved Solutes on Electrokinetic Properties of Membranes

The effect of dissolved solutes on electrokinetic properties of RO and NF membranes has been investigated by the streaming potential technique, which estimates charge on membrane surface and in membrane pores (AWWA committee report, 1998). Streaming potential measurements by Childress and Elimelech (1996) demonstrated that typical commercial RO and NF membranes are amphoteric, and the isoelectric point ranges from pH 3 to 5. This pH dependence is attributed to deprotonation of membrane surface functional groups originated from manufacturing processes.

Membrane charge is also greatly influenced by divalent cations. Hong and Elimelech (1997) showed that membrane surface charge becomes less negative with increasing divalent cation concentration. The decrease in the negative charge of the membrane was attributed to charge neutralization and effective screening of the membrane surface charge by divalent cations. Specific adsorption of divalent cations to membrane surface functional groups may also, in part, have been responsible for the decrease in the negative charge of the membrane.

Dissolved organic solutes such as natural organic matter (NOM) readily adsorb to the membrane surface and/or pores, affecting the membrane charge. It has been observed that membrane surface becomes more negatively charged in the presence of humic acids (Childress and Elimelech, 1996; Hong and Elimelech, 1997). The increase in the negative charge is ascribed to adsorption of the humic acids to the membrane surface. The adsorbed humic substances mask inherent membrane charge and dominate the surface charge of the membrane.

Impacts on Membrane Rejection

Charge repulsion is an important rejection mechanism of RO and NF membranes. Thus, changes in electrokinetic properties of the membrane due to solute-membrane interactions have a significant influence on membrane rejection characteristics (Braghetta, 1995). Hong and Elimelech (1997) observed a decline in NOM removal by NF membranes with decreasing pH. This observation can be explained by the reduced electrostatic repulsion between organic matter and membrane surfaces due to the lower membrane surface charge at low pH.

Impacts on Membrane Fouling. The degree of membrane fouling is determined by interplay between several chemical and physical interactions. Electrostatic repulsion, one of the critical interactions, is affected by solution composition (Song and Elimelech, 1995; Zhu and Elimelech, 1997). A recent NF study by Hong and Elimelech (1997) reported that NOM fouling increases with increasing electrolyte concentration and decreasing solution pH. In particular, at fixed solution ionic strength and pH, the presence of divalent cations markedly increases NOM fouling. Divalent cations substantially reduce the electrostatic repulsion between NOM and the membrane surface, resulting in a substantial increase in NOM deposition on the membrane. Jucker and Clark (1994) also recognized the effect of divalent cations on NOM adsorption to membrane surfaces.

Organic Solute Removal

The removal of disinfection by-product (DBP) precursors or *dissolved organic carbon (DOC)* by RO or NF has been studied extensively (Taylor et al., 1986, 1987, 1989a, 1989b; Jones and Taylor, 1992). Nanofiltration membranes have been shown to control trihalomethane formation potential (THMFP) in highly organic (>10 mg/L DOC) potable water sources (Taylor et al., 1986). These efforts have often been necessitated by inadequate efficiency of DBP removal using conventional coagulation and softening treatment processes.

An early investigation using diffusion controlled membrane processes had shown that membrane material would achieve the same solute rejection (Reinhard, 1986). In addition to membrane chemistry, characterization of raw-water organic fractions was found to be an important consideration for appropriate membrane selection (Fouroozi, 1980; Conlon, 1989; Taylor, 1989a, 1989b; Tan and Amy, 1991). In research conducted by Jones and Taylor (1992), over 90 percent of the DOC was removed for both surface water and groundwater supplies by membranes for which a manufacturer reported a molecular weight cutoff (MWC) of 300 to 500 daltons. The THMs and haloacetic acids (HAAs) in the permeate averaged 15 and 4 ug/L, respectively, and represented more than a 90 percent disinfection by-product formation precursors (DBPFP) reduction. However, lower MWC membranes removed little additional DBPFP.

Effective DBPFP removal has also been demonstrated for highly organic potable water sources where limited removal of inorganic contaminants was desired. Research published by Taylor (1989a, 1989b) and Spangenburg et al., (1997) have shown high DBP removal capability while simultaneously meeting treatment objectives for hardness or alkalinity concentrations in the treated permeate. It should also be noted that energy requirements for membranes offering comparable DBPFP removal efficiency have been shown to vary as much as 50 percent. This has been demonstrated in single-stage membrane pilot systems operated in parallel using different film chemistries and membrane manufacturers (Tan and Amy, 1991). Research has also focused on the rejection of specific THM species using membranes. The relative percentage of brominated THMFPs increased in permeates from membrane processes with increasing MWC (Laine et al., 1993). While NF was found to control brominated DBPs, advanced pretreatment was necessary to sustain production when using the bromide-spiked surface water source. Many NF membranes do not remove bromides effectively; hence, higher ratios of brominated DBPs are formed as the Br:NOM ratio increases in the permeate.

High DBPFP reduction has been demonstrated for groundwaters and surface waters alike. However, due to increased fouling potential, surface water is generally more difficult to treat by membrane processes than groundwaters with high organic matter concentrations. Consequently, advanced pretreatment is frequently required to counter significant membrane production losses in these applications. Long-term investigation of THMFP control at the Flagler Beach, Florida, water treatment plant (a groundwater site) and at the Punta Gorda, Florida, water treatment plant (a surface water site) provided for comparison of groundwater and surface water membrane treatment. Although consistent control of permeate THMFP was achieved, severe water flux loss was experienced in the system treating surface water. Membrane cleanings were conducted on 20 occasions in an attempt to sustain a water flux of 0.407 m^3/m^2-day (10 gsfd). The operating system at the groundwater site required prefiltration and acidification in order to sustain a flux of over 0.611 m^3/m^2-day (15 gsfd) with only a semiannual cleaning frequency (Taylor, 1989a). Similar experience was reported for a surface water of high natural color (Kouadio and Madeleine, 2002). In research reported by European investigators, NOM adsorption onto a membrane surface has been verified by XPS analysis (Heimstra and Nederlof, 1997). As a result, an alternative membrane was studied which offered similar organic removal and sustained operation capability (three-month cleaning frequency).

TABLE 11-6 Summary of DBP Precursor Studies with Membrane Processes

Citation	Water source	Pretreatment	Membrane technology	Feed water THMFP (μg/L)	Treated THMFP (μg/L)	Percent THMFP removal (%)
Taylor, Thompson and Carswell (1987)	Ground	Antiscalant,	NF	961	28–32	97
		Prefiltration	NF	961	31–39	96–97
		Screening	UF	961	326–947	2–66
Amy Alleman and Cluff (1990)	Surface	Prefiltration	NF	157–182	55–84	49–70
	Ground	Prefiltration	NF	176–472	6–95	78–98
Parker (1991)	Surface	None	MF	60–630	40–420	20
		Coagulation	MF	70–80	30–40	40–60
Tan and Amy (1991)	Ground	Prefiltration	NF	259	39	85
Duranceau, Taylor, and Mulford (1992)	Ground	pH adjustment, prefiltration	NF	120	6	95
Laine, Clarke, and Mallevialle (1993)	Surface	Prefiltration	UF	40–460	NR	<10
		Prefiltration	NF	40–460	NR	30–90
		UF	NF	40–460	NR	90
Kouadio and Madeleine (2005)	Surface	Media filtration	NF	155	20	87
	Surface	Media filtration	NF	117	19	84
	Surface	Media filtration	NF	170	24	86

NR = not reported.

The referenced DBP investigations are summarized in Table 11-6. In all of these investigations, the removal of naturally occurring DOC or DBPFP by membranes was found to be virtually independent of operating conditions (i.e., flux or recovery); this suggests that natural organics removal is generally sieve controlled. Similar findings using nanofiltration and reverse osmosis to remove trace-level pesticides are reported by Duranceau and Taylor (1990). If the DBPFP are assumed to be uncharged DOC, which is likely, then the removal of uncharged organics such as pesticides and naturally occurring dissolved organic compounds may be controlled by sieving. Removal of sieving controlled organic compounds may be increased if they can be placed in a state of increased size or steric hindrance. Some NOM compounds are diffusion controlled and affected by variations in flux and recovery. However, differences in permeate NOM concentrations due to diffusion-controlled variables (flux and recovery) are in the tenths of mg/L and unlikely to have any effect on DBP MCL compliance.

Suppliers of drinking water are subject to stringent government regulations for potable water quality regarding allowable *synthetic organic compounds (SOCs)* pesticide and herbicide concentrations. In particular, European standards require less than 0.1 μg/L for any one particular pesticide or herbicide and no greater than 0.5 μg/L for total pesticides and herbicides in drinking water. Many investigators have shown that RO and NF are effective techniques for pesticide and herbicide removal. However, specific mechanisms underlying SOC rejection is largely unknown. Some general statements can be made about SOC rejection in membranes. Rejection has been observed to increase as SOC molecular weight and charge increased and membrane polarity increased. In Table 11-7, results and significant findings from published accounts of pesticide and SOC removal are summarized.

TABLE 11-7 Literature Summary for SOC and Pesticide Removal

Membrane	SOC and pesticides	Rejection %	Significant findings	Researchers
CA	Sodium alkyl benzene Sulphonate	99.9	Flat sheet application	Ironside and Sourirajan (1967)
	DDT,	99.9	Flat sheet application	Hindin (1969)
	TDE,	99.5		
	BHC	52.0		
	Lindane	79.0		
TFC (Aromatic polyamide)	Heptachlor	99.5	(1) Application of TFC membrane	Chian, Bruce, and Fang (1975)
	Lidane	99.5	(2) Organic solutes were rejected higher in TFC membrane than in CA membrane	
	DDT	99.5		
	Aalathion	98.0		
	Parathion	98.0		
	Atrazine	72.0		
	Captan	99.0		
RO (Toray PEC100, UOP TFC 4600, FimTec Ft30 Desal DSI)	Acetic acid	11–72	Pilot application	Whittaker and Szaplonczy (1985); Whittaker and Clark (1985)
	Benzene	0–99		
	Propylene oxide	28–70		
	Ethylene dichloride	0–93		
	Formaldehyde	8–95		
	24-D	83–99		
	P-chlorobenzo trifluoride	77–99		
TFC	Alachlor	98.5	(1)Pilot application	Miltner, Fronk, and Speth (1987)
Nylon Amide		84.6	(2)TFC > PA > CA	
CA		71.4		
TFC	Ethylene dibromide (EDB)	35.0	(1)Pilot application	Fronk (1987)
	Alachlor	100	(2)TFC > PA > CA	
	Metolachlor	100		

(*Continued*)

TABLE 11-7 Literature Summary for SOC and Pesticide Removal (*Continued*)

Membrane	SOC and pesticides	Rejection %	Significant findings	Researchers
NF	Ethylene dibromide	0	(1)Pilot application	Duranceau and Taylor (1991)
(NF-70)	Dibromochloropropane	28	(2)SOC rejections were depended on charge and MW	
	Heptachlor	100		
	Methoxychlor	100		
	Chlordane	100		
	Alachlor	100		
NF	Simazine	66–94	(1) Pilot application	Hofman et al. (1993)
(NF-70,	Atrazine	78–99	(2) Pesticide removal	
PVD1,	Diuron	45–92		
PZ,	Bentazone	97–100		
SU-610)	DNOC	38–98		
	Dinoseb	80–100		
NF	Uncharged SOC		(1) Flat sheet and pilot applications	Berg and Gimbel (1997)
(CA-50,	Simazine	0–80	(2) Charged SOC were rejected higher in charged membranes	
BQ-01,	Atrazine	5–90		
Desal 5-DK, NTC-20,	Terbutylazine	12–96		
NTC-60,	Diuron	5–90		
PVD-1,	Metazachlorine	20–95		
NTR-7250)	Charged SOC			
	TCA	61–95		
	Mecoprop	60–90		
RO and NF (20 membranes)	Simazine	14–95	(1)Flat sheet application	Chen and Taylor (1997)
	Atrazine	41–99	(2) PA > CA	
	Cyanazine	33–99	(3) Inorganic and organic solutes didn't affect pesticide rejection.	
	Bentazone	0–99		
	Diuron	15–83		
	DNOC	9–95		
	Pirimicarb	48–97		
	Metamitron	12–98		
	Metribuzin	45–97		
	MCPA	0–99		
	Mecoprop	78–99		
	Vinclozolin	64–95		

More recently, there is concern over the presence of pharmaceuticals, personal care products and endocrine disruptors in water supplies. UF and MF were shown to be ineffective in removing pharmaceuticals, personal care products, and endocrine disruptors from water supplies (Snyder et al., 2008). NF and RO were shown to remove a wide variety of these emerging contaminants, but were not effective on low-molecular-weight organics such as N-nitrosamines, acetaminophen, and phenacetine. Additional studies are ongoing that further evaluate membrane processes for the removal of these organic contaminants.

Pathogen Removal

The removal of *Giardia* cysts by microfiltration and ultrafiltration has been well documented in the literature (Jacangelo et al., 1991; and Coffey, 1993). In these studies, a removal higher than 4 log was reported with ultrafiltration (Fig. 11-16) and microfiltration membranes having provided absolute removal of this protozoan cyst. In these cases, the removals were limited only by the concentration of the cysts in the feedwaters. One study reported that at bench scale, all the tested membranes (three MF and three UF) except one MF membrane (for which the membrane seal ring was found to be defective) removed *Cryptosporidium* and *Giardia* below the detection limit (1 cyst/L) (Jacangelo, Adham, and Laine, 1995). These results were confirmed at pilot scale. Removal effectiveness ranging from 6 to 7 logs was limited only by the influent concentration of the *Cryptosporidium* and *Giardia*. Polymeric MF and UF membranes behave as significant barriers to protozoan cysts as long as they remain intact. However, no process should be regarded as an absolute barrier to pathogens.

Coffey observed a 0.32 mean log removal for heterotrophic plate count (HPC) in a MF process. Permeate contained an average of 182 CFU/mL for a mean of 1480 CFU/mL in the feedwaters. In the same study, no *E. Coli* or *Giardia muris* were found in the filtrate for a feed concentration of 9.8×10^5 to 2.67×10^6 and 2.75×10^4, respectively. The low HPC removal was due to regrowth on the permeate channel and the low HPC in the feed stream. These results are consistent with the results published by Yoo et al., (1995). An intact 0.2 μm MF filtration process appears to provide a significant barrier to *Cryptosporidium* and *Giardia*. A microfiltration pilot plant has been used at the Fishing Creek water supply for Frederick,

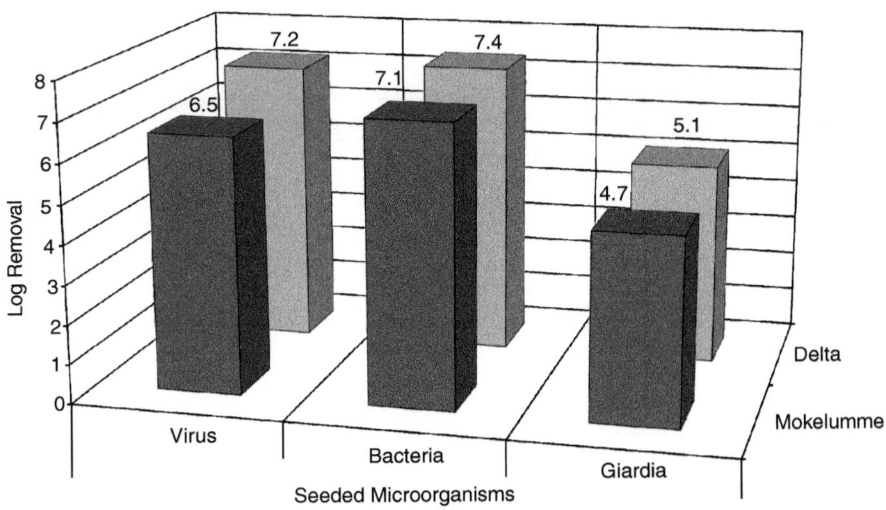

FIGURE 11-16 Removal of seeded microorganisms by UF of Mokelumme and Delta Waters.

Maryland, removing all the total coliforms in the feedwaters, which ranged from 10 to 1000 per mL (Olivieri, 1991). At Manitowoc, Wisconsin, no total coliforms, fecal coliforms, or *E. Coli* were detected in the MF filtrate. The feed contained 0 to 140 total coliform, 0 to 10 fecal coliform per mL, and 0 to 6 *E. Coli* per mL (Kothari, 1997). For the Saratago Water Treatment Plant, a microfiltration process achieved more than 6-log removal of *Giardia* and *Cryptosporidium* (Yoo et al., 1995). Between 1.7 and 2.9 log removal of MS2 virus in a California surface water with a 0.2 μm pore size hollow fiber microfiltration membrane was observed. The range of MS2 virus number in the feedwaters was from 1.3×10^6 to 3×10^7, and in the filtrate was found at 2.2×10^4 to 3.4×10^5 MS2 viruses (Coffey, 1993). Total removal of the MS2 virus was found for these feed concentrations using a hollow fiber UF membrane with a 0.01 μm nominal pore size (Jacangelo et al., 1991). Membrane processes, like any process, will eventually fail, and pathogen passage is more likely to be detected as feed concentration increases. However, membrane processes offer greater pathogen removal than other water treatment processes intended for the removal of suspended or dissolved matter. Typically, UF membranes remove a greater percentage of viruses than MF, greater than 3 logs of viruses, where MF remove less than 2 logs viruses.

Membrane Integrity

Membrane integrity is an essential issue in membrane processes. The high efficacy of MF and UF processes for removal of turbidity and pathogens can be compromised if the membranes are damaged. Membrane integrity can be extended to NF and RO membranes as dissolved contaminants can pass if membrane integrity is compromised. MF and UF integrity breaches can include a severed fiber, a pinhole puncture, or a weakened area due to abrasion. As an example, testing of an intact UF membrane achieved 7.0 log removal of *Bacillus subtillis*, but with a pinhole puncture the log removal was reduced to 5.2, and with two fibers cut, the log removal decreased to less than 4.0 (Côté et al., 2003). NF and RO integrity breaches can include a tear or weakened area of the membrane film, however, like MF and UF, can be compromised by a failed seal or O-ring. Hence, monitoring membrane integrity is important to avoid any water quality degradation.

As shown in Table 11-8, different membrane integrity tests are divided into direct and indirect monitoring methods. Currently available direct integrity test methods represent the

TABLE 11-8 Different Membrane Integrity Tests

Indirect monitoring	Direct monitoring
Particle counting	Air-pressure testing: i. Pressure decay test ii. Diffusive air flow test or water displacement test
Particle monitoring	Bubble-point testing
Turbidity monitoring	Sonic sensors
Conductivity monitoring	Marker or dye-based tests

most sensitive means of detecting integrity breaches, but these tests cannot be conducted on a continuous basis while the membrane filtration system is in operation. Thus, direct integrity testing is implemented at regular intervals and complemented by indirect integrity monitoring, which is generally less sensitive but can be conducted continuously during filtration. Indirect methods do not assess the integrity of the membrane barrier directly. Instead, water quality parameters are used as a surrogate to infer information about membrane integrity based on a comparison of the levels of monitored parameters relative to known baseline conditions of an integral system.

Integrity testing is required for regulatory requirements when MF or UF membranes are used for drinking water production. NF and RO processes are not at this time required to perform direct integrity tests primarily because no established methods are available. For MF and UF indirect integrity monitoring, the *Long Term 2 Enhanced Surface Water Treatment Rule (LT2ESWTR)* requires continuous indirect monitoring; this means measurements are collected at least every 15 minutes on the filtrate of each membrane module (Duranceau, 2004). A direct integrity test is initiated once a problem has been identified. The LT2ESWTR does not specify the use of a particular type of direct integrity test, but rather allows for the utilization of any type of test meeting the requirements for test resolution, sensitivity, and frequency. Specifically, the direct integrity test must be conducted once every 24 hours, detect a breach of size 3 μm or less, and verify the *log reduction value (LRV)* equal to or greater than the credit that had been given to the membrane system initially. The sensitivity of a direct integrity test is expressed in terms of a LRV, which must be equal to or greater than the *Cryptosporidium* removal credit awarded to the system in order to achieve compliance with the LT2ESWTR. However, state or local county regulatory agencies are allowed to make their requirements more stringent than federal regulations, so each membrane facility may have its own specific permit requirements with respect to integrity monitoring.

Air-Pressure Direct Testing Processes. Air-pressure tests are more sensitive than water-quality monitoring and have been reported to be capable of detecting one broken fiber in a membrane module containing 20,000 fibers (MWH, 2005). Air-pressure testing can be subdivided into two general tests, the pressure decay test (PDT) and the diffusive airflow test. The PDT is an online method that determines the membrane's ability to maintain a hold on a specific pressure that is generated by applying air to the membrane module. The rate of pressure decay is measured over time, and if the rate exceeds a specific test value, dependent on manufacturer, the membrane integrity has been breached. Air-pressure testing is accomplished in a series of actions or steps. For a pressurized system, the test requirements are:

1. Isolate each module by closing all their inlets and outlets
2. Drain membrane lumens via the filtrate exhaust line
3. Pressurize the filtrate side to 103 kPa (15 psi) with the feed side open to ambient pressure
4. Turn off the membrane test pressure to the filtrate side and watch for any decay in filtrate pressure
5. Exhaust the pressure to the filtrate exhaust line.

Usually the pressure is monitored after 2 and 4 min. A pressure drop between the readings at 2 and 4 min exceeding 2.8 kPa (0.4 psi) may indicate breaks in one or more fibers. The air-pressure test method cannot be used for air-permeable membranes. For immersed UF processes testing is analogous, except that a vacuum-pressure hold test is performed. The diffusive airflow, or water displacement, test is similar to a PDT, except that the rate of flow of either air or water is measured in lieu of pressure. If the flow of air or water exceeds specific criteria, a breach has occurred.

BUBBLE-POINT DIRECT TESTING

The bubble-point test is used to detect a compromised fiber. The pressure at which the membrane structure is sufficiently deformed to allow air bubbles to pass through the fiber is called the bubble-point pressure. The pressure is measured by applying increasing air pressure to the hollow fiber until bursting occurs. For example, to conduct a bubble-point test, the module is first taken off-line, and then the water outside the fiber is drained. Air pressure of 200 kPa (29.4 psi) is applied on the external side of the fibers. A low-concentration surfactant solution is applied at the surface of the module end to assist in identifying any air bubbles coming through damaged fibers (Adham, 1995). Pressure decays higher than 5 kPa (0.7 psi) in 5 min indicate damage to one or more fibers.

RO and NF Sonic Sensor Direct Testing. Sonic testing is a direct measure of the air-induced vibration within a membrane module during a PDT, as air bubbling through a damaged fiber will make enough noise to be detected by an inline sonic sensor. The sensor located on the module is monitored for hydraulic noise in a certain frequency range, which indicates membrane integrity loss. When membrane integrity is compromised, hydraulic noise increases due to a rise in turbulence in the module. A single broken fiber in a module containing 20,000 fibers can be detected by sonic monitoring (Landsness, 2001).

RO and NF Process Integrity Monitoring. For NF and RO systems there are currently no guidelines or established direct methods for element integrity testing. Typically NF and RO process skids are fitted with sample panels where aqueous samples can be analyzed for conductivity, dissolved solids, color, or hardness and process performance is monitored by profiling each pressure vessel of that process train once or twice daily. If contaminant or salt passage has been detected within a single pressure vessel (PV), then while in operation, the permeate end of the suspect pressure vessel is then probed by inserting a conductivity probe into the permeate collection tube. As the probe moves between elements, conductivity is measured at discrete distances. If an O-ring or membrane has failed, the probe will identify the location or element within the individual PV. The process train is then taken off-line, and the PV is emptied of the elements and either the breached membrane is replaced, an O-ring or stub-tube adapter is replaced, the PV is reloaded with the elements, and the process train is placed back into operation. Desalting plants are monitored online with conductivity instrumentation; however, monitoring of NF plants with conductivity does not offer the sensitivity as perhaps a specific contaminant such as calcium, sulfate, color, or total dissolved solid water-quality parameter monitoring may offer.

Fiber Breakage and Module Repair Considerations. The versatility of membrane filtration with regard to fiber integrity has been documented and shows that in general membrane fibers do not typically break (Coffey, 1993). However, when fibers are found to have been breached, they are typically removed from use by an action referred to as *pinning* where a pin is inserted and glued into the end of a compromised fiber. Removals of fibers from service over time affect the instantaneous flux of the module, effectively increasing the velocity of filtrate through the module in which the fiber or fibers have been pinned. The permeate water quality is thus influenced by compromised fibers. Several different commercially available membrane fibers were recently evaluated for membrane integrity, with an annual fiber failure rate of 1 to 10 per million fibers, with an average annual fiber failure rate of one per 100,000 fibers to be considered representative for state-of-the art membrane facilities (Gijsbertsen-Abrahamse, 2006). It was also found that broken fibers were not randomly distributed through the fiber bundle, rather in proximity of the potting material located in the top and bottom of the fiber bundle or pressure vessel. This finding had previously been documented by mechanical analysis of fiber integrity (Childress, 2005). When a fiber breakage is detected, the module can be repaired by isolating each individually broken fiber and making the necessary repair or replacing the broken fiber.

MASS TRANSPORT AND SEPARATION

For an isothermal process, the total driving force $F_{\text{total},i}$ for the transport of component i across the membrane can be expressed as the sum of contributions from an electrical potential, a concentration gradient (assuming activity coefficients equal to one), and a pressure gradient.

$$F_{\text{total},i} = \frac{R_g T}{\delta_m}\frac{\Delta c_i}{c_i} + \frac{z_i F}{\delta_m}\Delta\phi + \frac{v_i}{\delta_m}\Delta p \tag{11-4}$$

Different driving forces typically come into play in transporting solutes, colloids, larger particles, and water across the membrane. Summing flux values for individual components of a raw water (e.g., Na^+, Cl^-, and H_2O) gives J_i, a weighted sum of a minimum number of driving forces F_k affecting all of the components. This sum is mathematically described by the Onsager relationship (Onsager, 1931) as

$$J_i = \sum_{k=1}^{n} L_{ik} F_k \tag{11-5}$$

Each phenomenological coefficient L_{ik} relates the flux of component i to force F. When one driving force dominates mass transport across the membrane, Eq. 11-4 can be simplified in conjunction with the Onsager relationship to obtain expressions for the transport of water or solutes in a pressure-driven or electrically driven membrane process. Thus, these two expressions are the basis for describing a variety of membrane processes. For example, in electrodialysis, the last term in Eq. 11-4 is negligible and expressions can be derived for the transport of solute across the electrodialysis membrane. This issue will be taken up in a later consideration of principles of separation in electrodialysis. First, however, some simplifications allow descriptions of performance for pressure-driven membrane processes such as microfiltration (MF), ultrafiltration (UF), nanofiltration (NF), and reverse osmosis (RO).

Mass Transport Considerations in Pressure-Driven Membrane Processes

Mathematical Models. If pressure is the only driving force, as in UF and MF, Eqs. 11-4 and 11-5 reduce to a Darcy-type expression for the flux of water across the membrane.

$$J = \frac{\Delta p}{\mu R_m} \tag{11-6}$$

where Δp is the pressure drop across the membrane (the transmembrane pressure drop or TMP), μ is the absolute viscosity (of the water), and R_m is the hydraulic resistance of the clean membrane with dimensions of reciprocal length. In this case, there is a single phenomenological coefficient L_{11}, which by comparison of Eqs. 11-5 and 11-6, is seen to be equal to $(\mu R_m)^{-1}$. This expression is similar in form to the Kedem-Katchalsky equation commonly derived from irreversible thermodynamics to describe solute transport across reverse osmosis membranes. However, in this case, the flux of solute and the buildup of an osmotic pressure across the membrane must be accounted for. If the flux J of permeate is much greater than the flux J_S of solute, then such a derivation leads to the following result:

$$J \approx \left(\frac{L_v}{V_w}\right)(\Delta p - \sigma\Delta\Pi) \tag{11-7}$$

where L_v is a phenomenological coefficient, V_w is the molar volume of water, σ is the reflection coefficient (derived as the ratio of two phenomenological coefficients), and $\Delta\Pi$ is the difference in osmotic pressure across the membrane. Empirically, permeate flux is calculated from Eq. 11-7 as a function of the transmembrane pressure, the osmotic pressure, and two empirical constants corresponding to L_v/V_w and σ. By analogy, Eq. 11-6 may be modified directly to account for the reduction in the net transmembrane pressure due to the effects of osmotic pressure.

$$J = \frac{(\Delta p - \sigma_k \Delta\Pi)}{\mu R_m} \tag{11-8}$$

Thus, permeate flux across a clean membrane is not predicted to occur until the transmembrane pressure Δp exceeds the difference in osmotic pressure across the membrane. The osmotic pressure of a solute is inversely proportional to its molecular weight. Larger macromolecules, colloids, and particles produce little osmotic pressure. As a result, the correction for osmotic pressure is negligible for pressure-driven processes such as MF or UF that reject only these larger species.

Example 11-1 Determination of Hydraulic Resistance Coefficient
The membrane hydraulic resistance coefficient can be calculated from laboratory experiments. The flux of a new membrane can be determined by filtering clean, deionized distilled water for a given water temperature and pressure. One UF membrane's flux was determined by laboratory procedure to be 1700 L/m^2-hour (1000 gal/ft^2-day) at 20°C and 70 kPa (10 psi or 70,000 N/m^2). The viscosity of water at 25°C is about 0.89×10^{-3} kg/m-s (51.7 lb/ft-day) (see Appendix D).

Solution Rearrangement of Eq. 11-6 provides

$$R_m = \frac{\Delta p}{\mu J}$$

$$R_m = \frac{(70\text{ kPa})\left[\dfrac{10^5(\text{kg/sec}^2-\text{m})}{100\text{ kPa}}\right]\left[\dfrac{3600\text{ sec}}{\text{hour}}\right]\left[\dfrac{1000\text{ L}}{\text{m}^3}\right]}{(0.89\times10^{-3}\text{kg/m}-\text{s})\left[\dfrac{1700L}{\text{m}^2-\text{hr}}\right]}$$

$$R_m = 1.7\times10^{11}\text{m}^{-1}(5.1\times10^{10}\text{ft}^{-1}) \quad \blacktriangle$$

Concentration Polarization. Osmotic pressure is one of many phenomena that may reduce permeate flux as the result of the rejection of materials by the membrane. Reductions in permeate flux due to the accumulation of materials on, in, or near the membrane are referred to as *membrane fouling*. As water moves across the membrane, it draws solutes and particles toward the membrane. If these materials do not pass through the membrane, they may begin to accumulate on or near its surface, leading to the formation of additional layers of material through which water must pass. Particles may form cakes on the membrane surface, and macromolecules may form gel layers. Virtually all species achieve higher concentrations near the membrane surface in a flowing concentration boundary layer referred to as the concentration-polarization layer. Concentration-polarization is often a precursor to cake or gel formation. In addition, materials may precipitate or adsorb on or in the membrane leading to reductions in permeate flux that may be difficult to reverse. Elevated concentrations near the membrane resulting from the rejection of and subsequent concentration-polarization of these materials tend to exacerbate precipitative or adsorptive membrane fouling.

Permeate flux decline can be described mathematically by generalizing Eq. 11-8 to the case where resistance to permeate flux is produced by both the membrane and by materials accumulated near, on, and in the membrane. These layers are assumed to act in series to present additional resistance to permeation, designated as R_c and R_{cp}, respectively. These components of resistance vary as a function of the composition of materials rejected and thickness of each layer. The resulting expression for permeate flux is

$$J = \frac{(\Delta p - \sigma_k \Delta \Pi)}{\mu[R_m(t) + R_c(\delta_c(t),\ldots) + R_{cp}(k.J)]} \tag{11-9}$$

where δ_c is the thickness of the cake (or gel) layer and k is the mass transport coefficient (MTC) of the material in the concentration-polarization layer. The resistance terms are all functions of time and are related to the hydrodynamics of the membrane system and the raw-water quality. Adsorption or precipitation of materials within the membrane matrix as well as compaction of the membrane may lead to an increase in the membrane resistance $R_{m(t)}$ over time. The resistance of the cake R_c can be expressed as the product of the specific resistance of the material that forms the cake $\hat{R}_c$ and the cake thickness δ_c. From the Kozeny equation, the specific resistance of an incompressible cake composed of uniform particles can be calculated as

$$\hat{R}_c = \frac{180(1-\varepsilon_c)^2}{d_p^2 \varepsilon_c^3} \tag{11-10}$$

where ε_c is the porosity of the cake, and d_p is the diameter of particles deposited. This expression predicts that resistance to permeation by a deposited cake should increase as the particles comprising the cake decrease in size. The resistance produced by RO and NF membranes is likely to be large in comparison with the resistance of deposited colloidal materials or cakes. However, gel layers of macromolecular materials may produce significant resistance. Cake resistance may also be small in comparison with the resistance of a UF or MF membrane if the particles deposited in the cake are large compared with the effective pore size of the membrane. For feed streams containing large particles, permeate flux may be relatively independent of the concentration of particles. The morphology of the cake appears to be an important variable in particle filtration, and cake porosity appears to vary as a function of the hydrodynamics of the membrane module as well as the size distribution of the particles. By comparison with the resistance produced by cake or gel layers, resistance from the concentration-polarization layer is typically small.

Equations 11-6 through 11-9 at first appear to predict that permeate flux increases indefinitely with increasing transmembrane pressure (TMP). However, it is frequently observed that as pressure increases, a maximum permeate flux is eventually attained and permeate flux becomes pressure-independent. This maximum is usually interpreted as a mass transport-limited flux. Under conditions of mass transport-limited flux, an increase in pressure that would otherwise increase the flow of permeate across the membrane is instantaneously balanced by an increased accumulation of permeate-limiting materials near the membrane. In other words, increases in the resistance terms in the denominator of Eqs. 11-7 through 11-9 offset the increase in Δp in the numerator.

For a given TMP or under conditions of mass transfer-limited permeate flux, the accumulation of materials near the membrane can be envisioned as a balance between advection of materials toward the membrane due to permeation and back-diffusion that occurs as a concentration gradient builds up near the membrane. Assuming, therefore, a constant permeation rate and concentration-polarization layer with axial distance, mass balance on the concentration-polarization layer yields

$$-D\frac{\partial c}{\partial y} = Jc \tag{11-11}$$

where c is the concentration of the species subject to concentration-polarization, D is the diffusivity of this species, and y is the distance with the boundary layer such that $c = c_{\text{mem}}$ at $y = 0$ and $c = c_{\text{bulk}}$ at $y = \delta_{cp}$. Integrating this expression over the thickness δ_{cp} of the concentration-polarization layer results in the following expression for the concentration of solute at the membrane surface c_{mem}:

$$c_{\text{mean}} = c_{\text{bulk}} \exp\left(\frac{\delta_{cp}}{D} J\right) \tag{11-12}$$

For constant operating conditions, the exponential term in Eq. 11-12 can be taken as a constant, referred to as the *polarization factor (PF)*. Thus, concentrations of rejected materials at the membrane surface exceed the local bulk concentration by a factor of PF; sometimes estimated as an exponential function of the recovery r, such that PF = exp(kr). The semiempirical constant k typically takes on values of 0.6 to 0.9 for commercially available RO modules.

When permeate flux is mass transfer limited, Eq. 11-12 can be rearranged to describe the limiting permeate flux J as a function of the bulk concentration c_{bulk}, the limiting concentration at the membrane c_{mem}, the diffusion coefficient for the solute, and the concentration-polarization layer thickness.

$$J = \frac{D}{\delta_{cp}} \ln\left(\frac{c_{\text{mem}}}{c_{\text{bulk}}}\right) \tag{11-13}$$

The ratio of diffusivity to concentration-polarization layer thickness in this "film theory" model defines a mass-transfer coefficient k.

$$k = \frac{D}{\delta_{cp}} \tag{11-14}$$

The thickness of the concentration-polarization layer is a function of the hydrodynamics of the membrane module. For example, for a module with tangential flow across the membrane surface, a higher crossflow velocity tends to decrease the thickness of the concentration-polarization layer if Brownian diffusion is the only mechanism of back transport ($D = D_B$). In this case, the mass transfer coefficient can be calculated directly using correlations for the Sherwood number Sh of the form

$$Sh = \frac{kd_h}{D_B} = A(Re)^{\alpha}(Sc)^{\beta}\left(\frac{d_h}{L}\right)^{\omega} \tag{11-15}$$

where

$$Re = \frac{u_{\text{ave}} d_h}{\nu}$$

$$Sc = \frac{\nu}{D_B}$$

ν = kinematic viscosity

u_{ave} = average crossflow velocity

d_h = hydraulic diameter of the membrane element (e.g., diameter of the hollow fiber)

A, α, β, and ω = adjustable parameters

The Graetz-Leveque correlation, which is valid for laminar flow when the velocity field is fully developed and the concentration boundary layer is not fully developed, is often used to estimate the mass transfer coefficient.

$$Sh = 1.86(Re)^{0.33}(Sc)^{0.33}\left(\frac{d_h}{L}\right)^{0.33} \tag{11-16}$$

The Linton and Sherwood correlation can be used to calculate the mass transfer coefficient in turbulent flow, which may occur in tubular membranes.

$$Sh = 0.023(Re)^{0.83}(Sc)^{0.33} \tag{11-17}$$

Concentration-Polarization and Precipitative Fouling. As one consequence of concentration-polarization in RO and NF membranes, salts may more readily precipitate as a scale on membranes. The average concentrations of scale-forming species rejected by NF or RO membranes in the bulk flow inevitably increase as water permeating through the membrane is removed from the salt-bearing solution. Concentration-polarization further elevates rejected salt concentrations near the membrane and exacerbates the tendency to form a scale.

Principles of solubility and precipitation are covered in Chap. 3. When the reaction quotient (Q) exceeds the solubility constant (K_{s0}), the metal will precipitate and scale membranes. For example, metals such as calcium, magnesium, and iron most commonly precipitate as hydroxide, carbonate, or sulfate scales. In evaluation of whether a metal will precipitate, one must account for the salt concentration (ionic strength) and its effect on activity coefficients. For example, scaling from the precipitation of $CaSO_{4(s)}$ (gypsum) can be a problem. This occurs for any molar concentration combination of calcium and sulfate such that Q exceeds K_{s0} divided by the product of the two activity coefficients as shown in the following equation:

$$Q = [Ca^{2+}][SO_4^{2-}]$$

Precipation Occurs if

$$Q > \frac{K_{s0}}{\gamma Ca \gamma SO_4} \tag{11-18}$$

As water flows across the feed side of the membrane, the dissolved solids are rejected by the membrane and concentrate within the channel. The average (bulk) concentration of the rejected salts (for example, the cation) in the concentrate stream c_r exiting a membrane module increases over the feed concentration c_f as the recovery r and global rejection R increase. This is expressed in equation 11-19.

$$c_r = c_f \frac{1 - r(1-R)}{(1-r)} \tag{11-19}$$

The salt may precipitate when the ratio of the product of ion concentrations in the concentrate (the right-hand side of Eq. 11-19 after substituting $[A]_r$ and $[B]_r$ calculated from the feed concentration) to the solubility product K_{s0} is greater than 1. However, concentration-polarization further increases the ion product at the membrane surface since $c_{mem} = c_{bulk} * PF = c_r * PF$. Thus, taking the case of $x = y = 1$ (calcium sulfate, for example), the theoretical conditions to avoid scale formation are given by

$$(PF)^2\left\{\frac{(1-r)(1-R)}{(1-r)}\right\}^2 (B)_f(A)_f < K_{s0} \tag{11-20}$$

Transport of Colloids and Particles

The preceding discussion clearly suggests that the diffusivity of contaminants in water plays an important role in determining the permeate flux. However, Brownian diffusion alone does not adequately describe the transport of particulate and colloidal species near the membrane. The transport of colloidal and particulate species is of greatest significance in MF and UF systems, which are designed to remove these species. In a flowing suspension, particles may collide with one another and produce random rotary and translational motions that, on the average, result in a net particle migration from regions of high concentration and shear (near the membrane) to regions of lower concentration and shear (Davis, 1993). Building on the work of Eckstein and coworkers (Eckstein et al., 1977), Leighton and Acrivos (1987) proposed the following expression for the shear-induced diffusion coefficient. The equation was reported as valid up to particle volume fractions (the concentration of particle volume per volume of water) of $\Phi = 0.5$.

$$D_{SH} = a_p^2 \omega \hat{D}_{sh}(\Phi) \tag{11-21}$$

where a_p is the particle radius, ω is the shear rate, and D_{sh} is a dimensionless function of Φ estimated for suspensions of rigid spheres as

$$\hat{D}_{sh} = 0.33\Phi^2(1 + 0.5e^{8.8\Phi}) \tag{11-22}$$

Brownian diffusion is more important for smaller species such as solutes and small macromolecules, while shear-induced diffusion is increasingly important for larger particles. Contaminants encountered in water treatment typically span several orders of magnitude in hydrodynamic size, and so require consideration of both Brownian and shear-induced diffusion in estimating concentration-polarization and permeate flux. The sum of Brownian and shear-induced diffusivity D can be rewritten (Sethi and Wiesner, 1997) as

$$D = \frac{kT}{6\Pi\mu_o r_p} + \frac{\tau_{\text{wall}}}{\mu_0 \eta(\Phi)} r_p^2 \hat{D}_{sh}(\Phi) \qquad 11\text{-}23$$

where τ_{wall} is the shear stress at the membrane wall, μ_o is the absolute viscosity of the water at low particle volume fractions, and $\eta(\Phi)$ is the relative viscosity, which varies with volume fraction. Simultaneous consideration of Brownian and shear-induced diffusivity yields a minimum in particle back-transport for particles in the size range of several tenths of a micron and a corresponding minimum in permeate flux. The sum of the Brownian and shear-induced diffusion coefficients exhibits a minimum for particles approximately 0.1 μm in size for flow conditions typical of hollow fiber or SW modules, as shown in Fig. 11-17. Consequently, the relatively low mass transfer coefficients of species in this unfavorable size range are predicted to produce minimum permeate flux, while species either smaller or larger than this intermediate size are predicted to produce higher permeate fluxes. Until recently, only indirect experimental evidence supported this hypothesized minimum in back-transport as interpreted from permeate flux data (Fane, 1984; Wiesner, Clark, and Mallevialle, 1989; Lahoussine-Turcaud et al., 1990). However, one report gave the first direct experimental confirmation of a minimum back-transport based on measurements of particle residence time distributions (Chellam and Wiesner, 1997).

Diffusive transport is significant near the membrane where boundary layers may develop. Differential transport of smaller species in the bulk flow (outside the boundary layers) is generally negligible. However, in laminar crossflow filtration, the transport of larger particles from the bulk into the boundary layer may be affected by interactions between particles and the fluid flow relatively far from the membrane surface. In particular, inertial lift arising from nonlinear interactions of particles with the surrounding flow field may offset the drag force on particles associated with the flow of permeate across the membrane. This effect

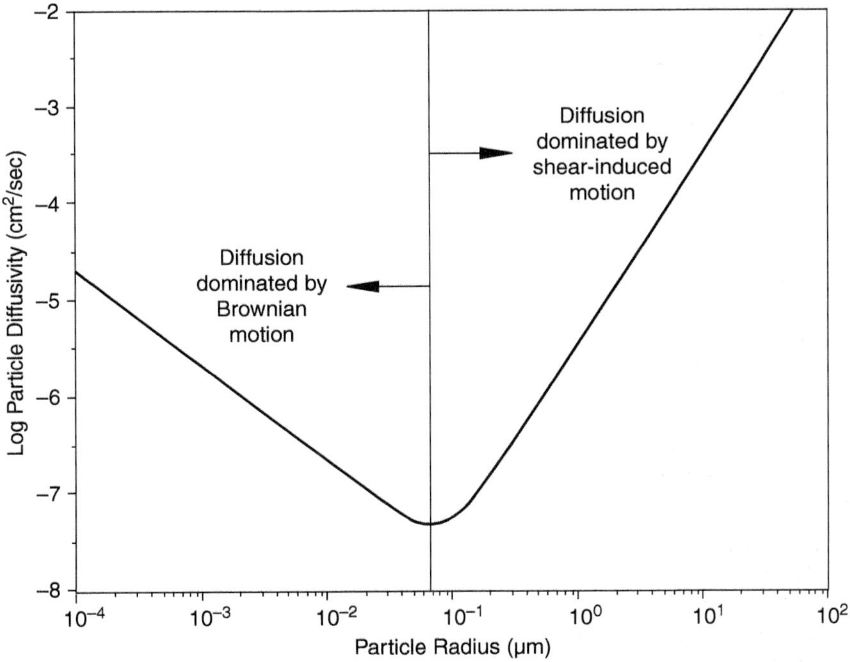

FIGURE 11-17 Brownian and shear-induced diffusivity as a function of particle size for conditions typical of hollow fiber UF membranes.

may occur where the particle Reynolds number becomes important (as it does with large particle sizes or fast crossflow velocities). For fast laminar flows ($Re >> 1$), as is typical in membrane filtration, the maximum value of the lift velocity in a clean channel is given by

$$u_{lo} = \frac{0.036 \rho a_p^3 \gamma_o^2}{\mu_o} \tag{11-24}$$

where ω_o is the shear rate at the membrane (in the absence of a cake layer) and ρ is the fluid density. Particle transport into the boundary layer is favored when the magnitude of the permeation velocity exceeds that of the inertial lift velocity (Belfort, Davis, and Zydney, 1994).

In addition to these processes governing the transport of particulate species to and from boundary layers, rolling or sliding of particles along the membrane surface due to convection must also be considered. In one approach (e.g., Leonard and Vassilieff, 1984; Davis and Birdsell, 1987) particles deposited as a cake and considered as a continuum may, under sufficient shear, begin to flow tangentially along the membrane surface toward the filter exit. An alternative approach (e.g., O'Neill, 1968; Blake et al., 1992; Stamatakis and Tien, 1993) considers the force and torque balances on a single particle on the membrane or cake surface. These balances are used to develop criteria for an estimate of whether the particle will adhere to the surface.

Mechanisms of Separation. Pressure-driven membrane processes may differ from one another substantially in the mechanisms by which they achieve separation, also referred to as rejection. In MF and UF, size exclusion is the primary mechanism of separation.

Rejection of contaminants by RO membranes is largely a function of the relative chemical affinity of the contaminant for the membrane vis-à-vis water. NF and UF membranes often represent intermediate cases that achieve rejection by combinations of mechanisms.

Rejection is usually defined as minus the ratio of the permeate- and feed stream concentration. This *global* rejection R is calculated as

$$R = 1 - \left(\frac{c_p}{c_f}\right) \tag{11-25}$$

where c_p and c_f are the permeate and feed concentrations. Occasionally, as in bench-scale test modules and associated scale-up, analysis focuses on the inherent *local* rejection R_{local} of a given membrane for a specific contaminant. The evaluation must account for local changes in rejection that occur with increases in the bulk concentration of a material as flow proceeds along the membrane (x-coordinate) due to the permeation of water through the membrane. In addition, rejected materials tend to accumulate near the membrane compared with the bulk flow due to concentration-polarization. Thus, the effective rejection of materials by a membrane may be very different from that calculated on the basis of the average feed and permeate concentrations. Local rejection is a function of the concentrations directly on either side of the membrane at a specific location and is defined as

$$R_{\text{local}}(x) = 1 - \left(\frac{c_p(x)}{c_{\text{mem}}(x)}\right) \tag{11-26}$$

where c_{mem} is the concentration at the membrane surface calculated from the bulk concentration at a given location within the module such that $c_{\text{mem}}(x) = (PF)c_{\text{bulk}}(x)$. If a mass balance is performed over a membrane module operating in tangential flow, the following equation is derived relating the global rejection to the apparent rejection:

$$R = 1 - \left(\frac{c_p}{c_f}\right) = 1 - \frac{1-(1-r)^{(1-R_{\text{local}})PF}}{r} \tag{11-27}$$

where r is the single-pass recovery of the module and the local rejection is assumed constant along the membrane. This latter assumption is intimately related to the mechanism of rejection.

Mechanical Sieving at the Membrane Surface. Membranes intended specifically to remove or separate particles and colloidal materials, such as microfiltration and ultrafiltration membranes, reject materials largely by mechanical sieving. However, electrostatic interactions, dispersion forces, and hydrophobic bonding may significantly affect the rejection of materials with dimensions similar to the pore sizes in UF or MF membranes. These considerations tend to be more important for separations involving larger macromolecules such as humic and fulvic acids, where charge, adsorptive affinity to the membrane, and hydrodynamic size are interrelated.

A quantitative analysis of mechanical sieving (or steric rejection) was first presented by Ferry (1936). Assuming pores of cylindrical geometry and spherical particles, he derived an expression for the fraction x of particles passing through a pore of radius a. For particles with radii equal to or greater than the pore radius, perfect rejection is predicted. For particles much smaller than the pore, no significant rejection occurs. Only particles with radii slightly smaller than pore radius show nontrivial rejection behavior. This expression was subsequently modified (Zeman, 1981) to account for the lag velocity of particles with respect to the fluid permeating through the membrane pore. The rejection of particles by a membrane $(1 - p)$ can be estimated using this expression, as a function of the nondimensionalized particle radius $\lambda = a_p/a_{pore}$ as

$$x = \begin{cases} (1-\lambda^2)[2-(1-\lambda)^2]\Gamma & \lambda \le 1 \\ 1 & \lambda > 1 \end{cases} \tag{11-28}$$

where Γ is the lag coefficient, empirically estimated by Zeman and Wales (Zeman, 1981) as

$$\Gamma = e^{-0.7146\lambda^2} \tag{11-29}$$

Rigorously derived, $(1 - p)$ corresponds to the local rejection R_{local} of the membrane. Measurements of apparent rejection can be used to calculate a value p^* that theoretically corresponds to the product of the polarization factor PF and the particle passage ψ. Removal of materials in deposited cake or gel layers may further alter the apparent rejection of the membrane. Further modifications can account for electrostatic repulsion and dispersion forces near the pore wall (Matsuura and Sourirajan, 1983). A more rigorous treatment would incorporate drag on particles transported through membrane pores by advection and diffusion (Deen, 1987) and the diffusion and convection of flexible macromolecules through cylindrical pores of molecular dimensions (Davison and Deen, 198). One approach extends the physical sieving model for particle removal to describe the rejection of macromolecular compounds such as humic materials by substituting each molecule's hydrodynamic radius for the particle radius. Empirical constants in an expression for the molecule's hydrodynamic radius are then manipulated to fit a given data set. For larger molecular weight materials, the hydrodynamic radius can often be correlated with the compound's molecular weight $\overline{M}$. These correlations are typically of the form

$$a_p = Z_1(\overline{M})^{Z_2} \tag{11-30}$$

where Z_1 and Z_2 are empirical constants. The constant Z_2 contains information on molecule geometry. For a perfectly spherical molecule, Z_2 takes on its theoretical minimum value of 1/3. The theoretical maximum value of 1 for Z_2 corresponds to a linear molecule. Z_2 may take on values outside its theoretical range when Z_1 and Z_2 are treated as fitting constants within the context of the sieving model.

Thus, rejection of organic compounds is predicted to increase with molecular weight. Such behavior has been observed in processes removing larger molecular weight fractions of natural organic matter (NOM) by ultrafiltration and nanofiltration membranes (e.g., Taylor et al., 1987, 1988).

While the membrane itself determines the initial removal of materials and serves as an absolute barrier, the rejection characteristics of the membrane system may change over time as materials are deposited on the membrane surface. This concept is exploited in the application of "dynamic" membranes, which formed effective barriers from materials deposited on a porous support that are periodically removed and replenished. Mathematical treatment of this process has borrowed concepts from packed-bed filtration. Suspended colloidal materials transported toward the cake with the permeating water may be removed at the cake surface by simple sieving. Particles entering the cake may be removed by previously deposited particles that form a type of micropacked bed. Removal within the cake may be due to simple interception of mobile colloids, if the streamline they follow brings them in contact with a previously deposited (immobile) colloid. Gravity and Brownian motion may cause particles to deviate from their streamlines, helping to bring particles in contact with the immobile particles composing the cake. The previously deposited particles are termed *collectors* in the packed-bed filtration literature.

Solute Transport and Rejection. The solution diffusion model proposed by Lonsdale, Merten, and Riley (1965) and Merten (1966) describes solute and solvent transport across membranes in terms of the relative affinities of these components for the membrane and their diffusive transport within the membrane "phase." Rejection of a given solute is therefore a function of the degree to which it partitions into the membrane and its diffusivity within the membrane. The driving forces for transport are the differences in chemical potential across the membrane due to differences in concentration and pressure.

A basic diagram representing osmosis and reverse osmosis cells is shown in Fig. 11-18. Osmosis is a special case of diffusion in which the molecules are water and the concentration gradient occurs across a semipermeable membrane. The semipermeable membrane allows the passage of water, but not ions (e.g., Na^+, Ca^{2+}, Cl^-) or larger molecules (e.g., NOM). Diffusion and osmosis are thermodynamically favorable and will continue until equilibrium is reached. Osmosis can be slowed, stopped, or even reversed if sufficient pressure is applied to the membrane from the concentrated side of the membrane. Reverse osmosis occurs when the water is moved across the membrane against the concentration gradient, from lower concentration to higher concentration.

To illustrate, imagine a semipermeable membrane with freshwater on one side and a concentrated aqueous solution on the other side. If normal osmosis takes place, the freshwater

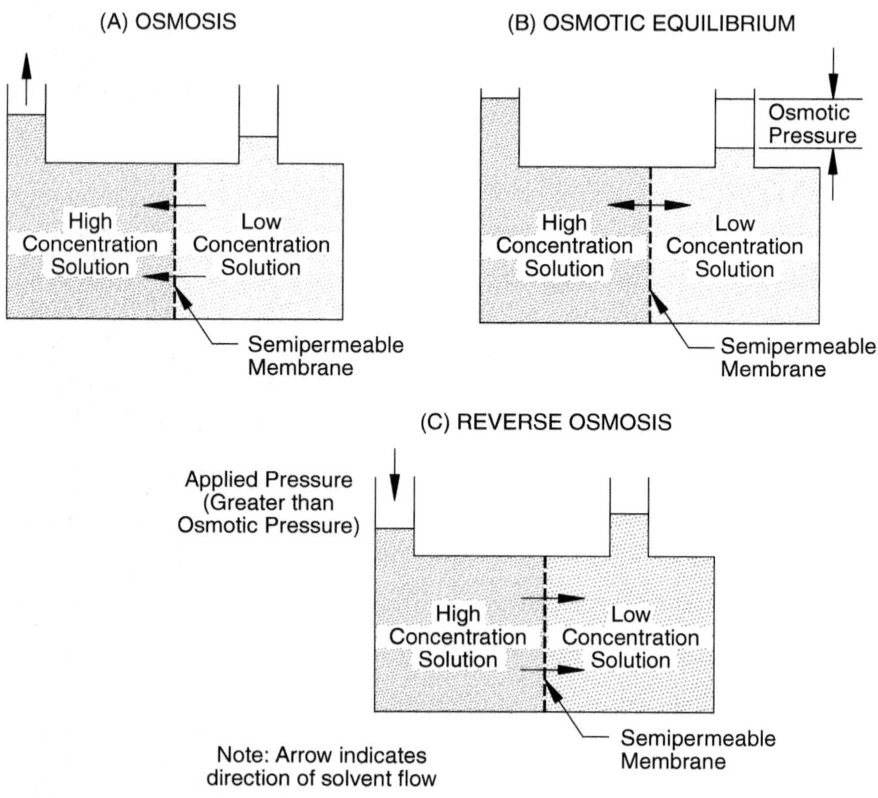

FIGURE 11-18 The principles of osmotic flow.

will cross the membrane to dilute the concentrated solution. In reverse osmosis, pressure is exerted on the side with the concentrated solution to force the water molecules across the membrane to the freshwater side. The transport of water occurs by convection and is dependent on pressure, while the transport of solutes is independent of pressure. Thermodynamically, the osmotic pressure is defined as

$$\Pi = -\frac{R_g T}{V_w} \ln(x_w) \tag{11-31}$$

with the osmotic pressure Π, the molar volume of water V_w, the mole fraction of water x_w, and the ideal gas constant R_g. In dilute solutions, the osmotic pressure can be estimated using van't Hoff's law, which was developed using the ideal gas law.

$$\Pi = -\frac{n_s}{V} R_g T \qquad \text{or} \qquad \Pi = C R_g T \tag{11-32}$$

with the total amount of solutes in solution n_s [moles], total concentration of solutes C (moles/L), and the volume of solvent V. Considering the dissociation of ions in solution, van't Hoff's equation is typically shown as

$$\Pi = i\Phi C R_g T \tag{11-33}$$

with i representing the dissociation constant, which is equal to the number of ions and molecules per mole of solute produced by the dissolution of the solute, and with ϕ representing a correction factor for non-ideal behavior. As a general rule of thumb the osmotic pressure of brackish water can be estimated by estimating that for every 100 mg/L of total dissolved solids that is present in the feedwaters, 7 kPa (1 psi) of osmotic pressure will be present within the membrane feed channel.

$$\Pi = (\text{TDS, mg/L}) \frac{1\text{ psi}}{100\text{ mg/L} - \text{TDS}} \tag{11-34}$$

For a general estimate of the osmotic pressure of seawater, it can be assumed that an NaCl solution of equal total dissolved solids concentration is approximated as (Fritzmann et al., 2007)

$$\Pi = \frac{8\text{ bar}}{\text{wt\%}_{\text{NaCl}}} \tag{11-35}$$

It should be noted that the actual osmotic pressure of seawater has been shown to be approximately 10 percent less than a solution of sodium chloride that is equal to the total dissolved solids concentration due the presence of higher molar mass species present in seawater (AWWA, 1999). This phenomenon is not as apparent in brackish waters as in seawater due differences in composition. Regardless, the permeate water quality is thus a function of the diffusion of salt across the membrane and its associated osmotic pressure gradient, the transmembrane pressure, water recovery, and mass transfer of solute and water with respect to membrane material.

Example 11-2 Determination of Osmotic Pressure by Calculation
The osmotic pressure of a solution can be determined by calculation. Calculate the osmotic pressure of a 7500 mg/L solution of sodium chloride at a temperature of 19°C assuming an osmotic coefficient of 0.95.

Solution Determine the osmotic pressure using van't Hoff's equation Eq. (11-33).

$$\Pi = i\Phi C R_g T$$

$$\Pi = \left(\frac{2 \text{ mols ion}}{\text{mols}}\right)(0.95)\left(\frac{7500\,\text{mg}}{\text{L}}\right)\left(\frac{0.08314\,\text{L}\cdot\text{bar}}{K\cdot\text{mols}}\right)(292K)\left(\frac{1000\text{ mg}}{\text{g}}\right)\left(\frac{58.4\text{ g}}{\text{mols}}\right)$$

$$\Pi = 5.92\text{ bar }(85.9\,\text{psi})$$

Discussion Osmotic coefficients for sodium chloride range from approximately 0.94 to 1.04 for the differing ranges of seawater globally. Since the osmotic pressure is a function of the mole fraction of water in the system, and solutes reduce the concentration of water, the effect of multiple solutes cumulatively reduces the mole fraction of water. Solutes that dissociate will have an additional stoichiometric effect on the mole fraction of water. Reported values for the osmotic pressure of seawater tend to be lower than calculated values because seawater is composed of components having a higher molecular weight than that of sodium chloride.

The concentration of the water passing through the RO membrane decreases as pressure increases. In the case of brackish water systems, a general rule of thumb is that for every 100 mg/L of total dissolved solids that is present in the feedwaters, 7 kPa (1 psi) of osmotic pressure will be present within the membrane feed channel. The permeate water quality is thus a function of diffusion of salt across the membrane and its associated osmotic pressure gradient, the transmembrane pressure, water recovery, and mass transfer of solute and water with respect to membrane material. ▲

Thus far, the focus has been on the transport of water across membranes and the role of materials in water in potentially reducing the permeate flux. However, expressions for the flux of solutes across membranes can also be derived from very general principles beginning with Eqs. 11-4 and 11-5. The ratio of solute to water flux J_s/J yields the permeate concentration c_p, and therefore, the rejection as calculated by Eq. 11-25. Models for solute rejection tend to differ in the degree of mechanistic detail they offer or the specific mechanisms they invoke. However, where solutes can enter the membrane matrix, the underlying assumption is usually that separation of solutes occurs due to the relative rates of transport across the membrane rather than by size exclusion, as is typically assumed in the rejection of macromolecules, colloids, and particles. An important implication of this distinction is that rejection of solutes depends on the feed (or reject) concentration while rejection due to size exclusion is independent of feed concentration. The flux of water is given as

$$J_w = -\frac{D_w c_{w,\text{mem}}}{R_g T}\frac{d\mu_w}{dz} \approx -\frac{D_w c_{w,\text{mem}}}{R_g T}\frac{d\mu_w}{\delta_m} \tag{11-36}$$

Note that the form of Eq. 11-36 gives the permeate flux in proportion to a driving force (chemical potential), concentration, and mobility (as expressed by diffusivity). Substituting the Lewis equation for osmotic pressure, the following equation is derived for water flux:

$$J_w = -K_w \frac{(\Delta p - \Delta\Pi_w)}{\delta_m} \tag{11-37}$$

where the specific hydraulic permeability K_w is a function of the diffusivity D_w, the concentration $C_{w,\text{mem}}$ of the water in the membrane, and the molar volume of water V_w such that $K_w = \frac{D_w C_{w,\text{mem}} V_w}{RT}$. Comparison of this result with Eq. 11-7 shows a similar result to that obtained from irreversible thermodynamics. Indeed, the models are identical, given the underlying interpretations of the phenomenological coefficients.

If water and solute transport are not coupled ($\sigma = 1$), the flux of solute J_s is given as

$$J_s = -D_s k_s \frac{(c_{\text{mem}} - c_p)}{l} \qquad (11\text{-}38)$$

where k_s is the distribution coefficient that describes the relative affinity of the solvent for the membrane. Similar to the result for permeate flux, this equation expresses the solute flux as a function of a driving force (concentration difference across the membrane), mobility (diffusivity), and concentration. Solute flux is predicted to be independent of the transmembrane pressure drop Δp. Therefore, increases in Δp result in an increase in J_w and in local rejection, since

$$R_{\text{local}}(x) = 1 - \left[\frac{\left(\frac{J_s(x)}{J_w(x)}\right)}{c_{\text{mem}}(x)} \right] \qquad (11\text{-}39)$$

This model has been generalized (Sherwood, 1967) to include advective transport through membrane pores as well as diffusion. This modification leads to the following expression for local rejection by a membrane:

$$R_{\text{local}} = \left[1 + \left(\frac{D_s K_d}{K_w} \right) \left(\frac{1}{\Delta p - \Delta \Pi} \right) + \left(\frac{K_{pf}}{K_w} \right) \left(\frac{\Delta p}{\Delta p - \Delta \Pi} \right) \right]^{-1} \qquad (11\text{-}40)$$

where K_{pf} is an additional fitting constant describing pore flow through the membrane. Equation 11-40 predicts increasing local (and therefore global) rejection with increasing pressure, but only to an asymptotically limiting value. In practice, higher operating pressures have been observed to increase global rejection as described by the following empirical equation (Sourirajan, 1970):

$$R = \frac{z_1 \Delta p}{(z_2 \Delta p + 1)} \qquad (11\text{-}41)$$

where z_1 and z_2 are empirically determined constants.

One variation on such a generalized solution-diffusion model, the preferential sorption-capillary flow model, involves a mechanistic interpretation of transport within the membrane matrix based on differential influence on a layer of water by interfacial processes near polymer surfaces. This layer may exist due to (1) repulsion of ions by membrane materials of low dielectric constant, (2) the excessive energy required to strip ions of their hydration spheres as they move from the bulk water to the water-membrane interface, or (3) the hydrophilic nature of the membrane surface, which actively coordinates and orders water near the membrane surface. Transport of the solute across the membrane within this layer is impeded due to the rise in free energies required to move the solute through a progressively more ordered fluid. The solute is assumed to move through the membrane by diffusion, advection, or both (Sourirajan, 1970). Ions with smaller hydrated radii may be able to diffuse through the adsorbed water film or may be convected through pores.

Charges on membrane surfaces can play a significant role in the rejection of ionic solutes. Although these effects are most important in electrically driven membrane separations, they may also play a role in pressure-driven processes. Charged functional groups on membranes attract ions of opposite charge (counterions). This effect in combination with a deficit of like-charged ions (co-ions) in the membrane results in a so-called Donnan potential. Although co-ions are largely unable to enter the membrane, water can pass through under pressure. The accumulation of co-ions in the membrane concentrate is accompanied by an accumulation of counterions as well, due to the need to preserve electroneutrality in the solution. By this mechanism, membrane rejection is predicted to increase with increasing membrane charge and with co-ion valence (Dresner, 1972). Similarly, for membranes that are semi-permeable to some ionic species, the requirement for electroneutrality in the concentrates and permeates may lead to significant modification of rejection characteristics in the presence of mixed solutes. For example, as a NF membrane rejects calcium, higher rejections of monovalent anions such as nitrate or chloride may also occur. Varying the concentration of competing anions in the concentrate can affect anion rejection by NF membranes. For example, increasing sulfate concentration increases sulfate rejection and decreases nitrate and chloride relative to levels in single-solute solutions (Rautenbach, 1990).

Mass Transport and Process Considerations in ED and EDR

Principles of Mass Transport. In electrodialysis, the driving force for separation is a gradient in electrical potential. An ED system applies an electrical potential across charged ion-exchange membranes, and cations or anions are differentially transported across the membrane and removed from the feed flow. A stack of cation- and anion-selective membranes can remove ionic species in applications such as desalination of brackish water.

In the absence of a pressure drop across the membrane, the driving force for transport of ions across an ion selective membrane becomes

$$F_{tot,i} = \frac{R_g T}{\delta_m}\frac{\Delta c_i}{c_i} + \frac{z_i F}{\delta_m}\Delta\phi \tag{11-42}$$

Assuming ion flux proportional to this driving force and to ion concentration and mobility, the following equation is derived for cation transport across a cation-exchange membrane:

$$J_+ = -U_+ c_+ \left(\frac{R_g T}{\delta_m}\frac{\Delta c_+}{c_+} + \frac{z_i F}{\delta_m}\Delta\phi\right) \tag{11-43}$$

The fraction of current I carried by the flux of cations FJ_+ is defined as the cation transport number t_+.

$$t_+ = \frac{FJ_+}{I} \tag{11-44}$$

The flux of cations through a cation exchange membrane under the driving force of an electric potential can then be rewritten as

$$J_+ = \frac{t_+ I}{zF} \tag{11-45}$$

Increases in current are therefore predicted to induce increases in cation transport (or similarly anion transport). The electrical potential E required to produce a given current is calculated form Ohm's law.

$$E = IR_E \tag{11-46}$$

where R_E is the resistance of the total membrane stack. The resistance of the membrane stack can be considered as the sum of n cell pairs, each with a resistance R_{cell} such that

$$R_E = R_{\text{cell}}\, n \tag{11-47}$$

Resistance within each cell pair can be expressed in turn as the sum of resistances due to the anion exchange membrane R_{an}, the cation exchange membrane R_{cat}, the concentrate stream R_{con}, and the dialysate stream R_{dia}.

$$R_{cell} = R_{an} + R_{cat} + R_{con} + R_{dial} \tag{11-48}$$

Unfortunately, the increase in ion flux with current is limited by a theoretical maximum current required to transport all available ions. This maximum is a function of the ion concentration at the membrane surface and the concentration in bulk flows of concentrate and dialysate. Differential transport of cations across the membrane leads to the formation of boundary or concentration-polarization layers near either side of the membrane. On the side of the membrane where cations are depleted (the dialysate), concentrations are lower than the bulk concentration. On the concentrate side of the membrane, cation concentration is greater in the boundary layer than in the bulk flow. The flux of cations through the boundary layers can be calculated as a function of a transport number t_{+bl} for flux due to the applied potential and the concentration gradient. At equilibrium, cation flux through the membrane must equal the flux of cations through the boundary layer.

$$J_+ = \frac{t_+ I}{zF} = \frac{t_{+,bl} I}{zF} - D\frac{dc_+}{dy} \tag{11-49}$$

Integration of Eq. 11-49 with the boundary conditions $c_+ = c_{bulk}$ at $y = \delta_{cp}$ and $c_+ = c_{+,mem}$ at $y = 0$ yields the following equation:

$$c_{+,mem} = c_{+,bulk} \pm \frac{(t_+ - t_{+,bl})\delta_{cp}}{DzF} \tag{11-50}$$

The plus sign applies to calculations of concentrations on the concentrate side of the membrane, and the minus sign applies to calculations of the depletion of cations in the boundary layer near the membrane on the dialysate side. Analogous results are obtained for transport across anion-exchange membranes. The depletion of ions near the membrane leads to an increase in electrical resistance across the dialysate compartment R_{dial}, which increases power consumption and reduces efficiency. Rearranging Eq. 11-50 yields the following expression for the current density in the concentration-polarization layer on the dialysate side of the membrane:

$$I_{density} = \frac{DzF(c_{+,bulk} - c_{+,mem})}{(t_+ - t_{+,bl})\delta_{cp}} \tag{11-51}$$

The limiting case (maximum current) occurs when the current density is high enough to drive the cation concentration (or similarly the anion concentration) to zero on the dialysate side of the membrane. This limiting current density is then given as

$$I_{density,lim} = \frac{DzFc_{+,bulk}}{(t_+ - t_{+,bl})\delta_{cp}} \tag{11-52}$$

In a situation analogous to concentration-polarization in pressure-driven processes, the thickness of the polarization layer is function of the flow across the membrane. The cross flow can therefore be manipulated to a limited extent to optimize performance.

ED and EDR Processes. ED processes must work within polarization and scaling limitations like those in pressure-driven, diffusion-controlled membrane processes, because of ion concentrations at the membrane surfaces. However, polarization problems arise in ED processes due to the lack of ions at the surface. Ions arrive at a membrane surface by electrical transport, diffusion, and convection, and they are transported through the membrane electrically. As ions are transported through the membrane, the demineralized stream-membrane interface becomes depleted of ions, causing an exponential increase in electrical resistance and current density. This process continues until polarization occurs, disassociating water into protons and hydroxide ions. This point is identified as the limiting current density. These problems have been reduced in ED applications by reversing module polarity every 15 to 20 min during plant applications. This process, known as electrodialysis reversal (EDR), has advantages of (1) reducing scaling potential, (2) breaking up fresh scale, (3) reducing microbiological growth on membrane surfaces, (4) reducing membrane cleaning frequency, and (5) cleaning electrodes with acid formed during anodic operation. EDR processes reportedly achieve slightly higher recoveries than NF or RO processes; this can become an economic advantage when removal of unionized solutes such as disinfection by-product formation potential or pathogens is not an issue.

The rate of ion removal in an electrochemical separation process is controlled by the feed-water characteristics, design parameters, and equipment selection. The water quality and temperature of the feed waters determine the system recovery and rate of mass transfer. Ion removal increases as temperature increases and as charge increases. System recovery is typically limited by precipitation of the least soluble salt, as in RO or NF systems. Design parameters include current, voltage, current efficiency, polarization, and finished water quality. As current or current density increases, the removal of ions also increases. Voltage is a function of the current, temperature, and ionic water quality. Current efficiency is determined by the efficiency of salt transfer. A 100 percent efficient process would transfer 1 g equivalent of salt for every 26.8 A hours. Polarization is controlled by limiting the allowable current density to 70 percent of the limiting current density. Finished water quality standards are set by design, within limits of back-diffusion. A high concentration gradient between the demineralized and concentrate streams leads to diffusion of ions from the concentrate to the demineralized stream. This effect, termed *leakage,* is controlled by limiting the concentration ratio of the concentrate stream to the demineralized stream to less than 150.

An example of a three-stage EDR process is shown in Fig. 11-19. The system recovery from this process is 90 percent, and the overall TDS rejection from the feed stream is 75 percent. Contrary to a RO or NF process, the product water is taken only from the final stage, as ions are continually removed from the product stream until the desired TDS concentration is obtained. The feed stream to the EDR process is pretreated with acid and/or an inhibitor, as is a RO or NF feed stream, to prevent scaling at the desired recovery. Recycling of the concentrate stream to the feed stream is not shown in Fig. 11-19, but is often used in an EDR process to (1) reduce equipment requirements for a given recovery, (2) reduce concentrate stream discharge, (3) reduce pretreatment, and (4) increase the transfer of ions to the concentrate stream.

Figure 11-20 presents a process flow diagram for an ED/EDR process. In plants working with maximum recommended polarization levels, the tendency to precipitation of calcium carbonate becomes possible when a bicarbonate ion is greater than 67 percent of anions and calcium accounts for more than 50 percent of cations. The larger relative size of the bicarbonate ion makes it more resistant to movement. This fact implies that, when it is a larger percentage of anions, the high resistance generates polarization of the anionic membrane surface. Hydroxide ions produced as consequence of polarization react with the bicarbonate ion to form carbonate ions, which will precipitate with the existing calcium.

In general EDR can operate at higher water recoveries because the periodic reversal of the DC electrical field removes deposited solids and colloids; the self-cleaning reversal technique breaks up polarization films in the system 3 to 4 times per hour, preventing scale. Moreover, where EDR can operate in the presence of higher levels of dissolved silica as silica is not charged and hence not removed in EDR; RO is limited in water productivity when silica is present due to its concentration on the surface (silica is not ionized under pH 9.5).

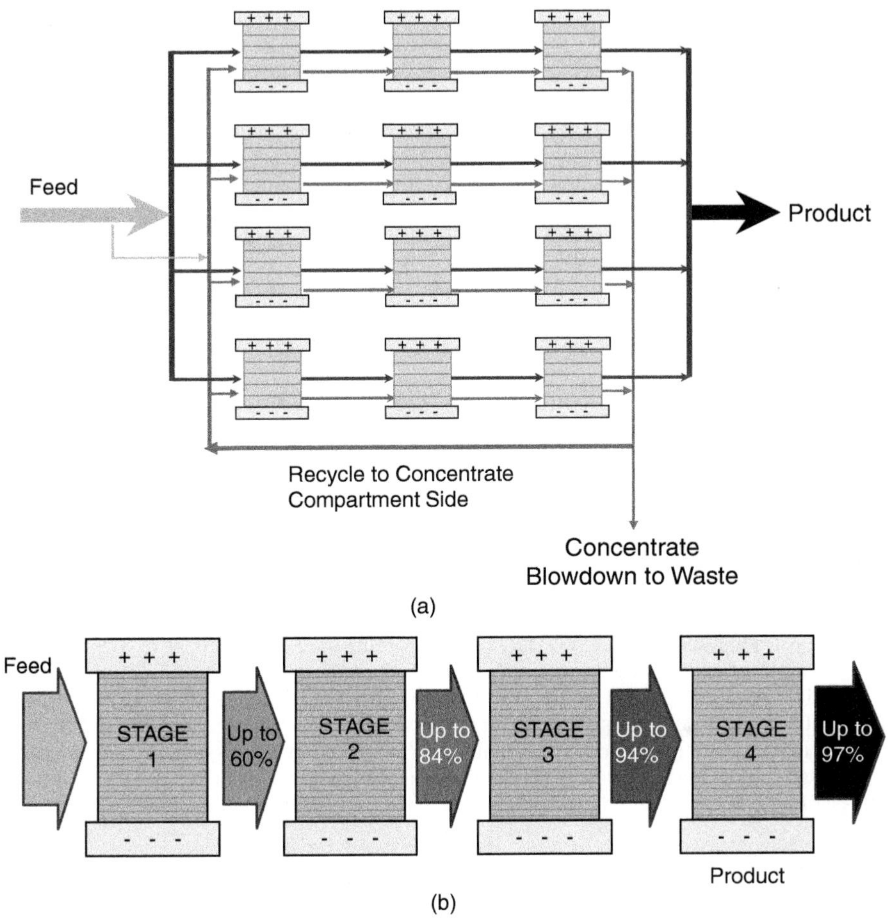

FIGURE 11-19 (*a*) Four line, three-stage EDR process, (*b*) Single line, four-stage EDR train. (*Courtesy of GE Water & Process Technologies Ionic, Inc.*)

One factor of importance that impacts membrane life is cleaning and cleaning intervals. Clean-in-place (CIP) intervals are dependent on site-specific water quality and pretreatment conditions. RO membranes require regular cleaning when employed for surface water applications; monthly cleanings are not uncommon. However, in general EDR CIPs are performed between 800 and 1000 hours of operating time and may take from 30 min to 1 hour (for forward flushing). If EDR stack disassembly is required, then this may take 5 to 6 hours in addition to the cleaning (if CIP unsuccessful in removing a hot spot for example). Caustic cleaning or brine CIPs occur every 3 to 4 months if bacteria are present. Electrode tape is the tape on the electrode that provides protection for the stack to minimize the amount of shorting that takes place. The tape should be confirmed to be greatly heat resistance and caustic resistant. Electrode connecting bars must be monitored, inspected, and maintained, in addition to the confirmation of the presence of electrode gaskets and torque on the stacks must be maintained else the membranes can slip, leading to hotspot formation which can damage the membranes.

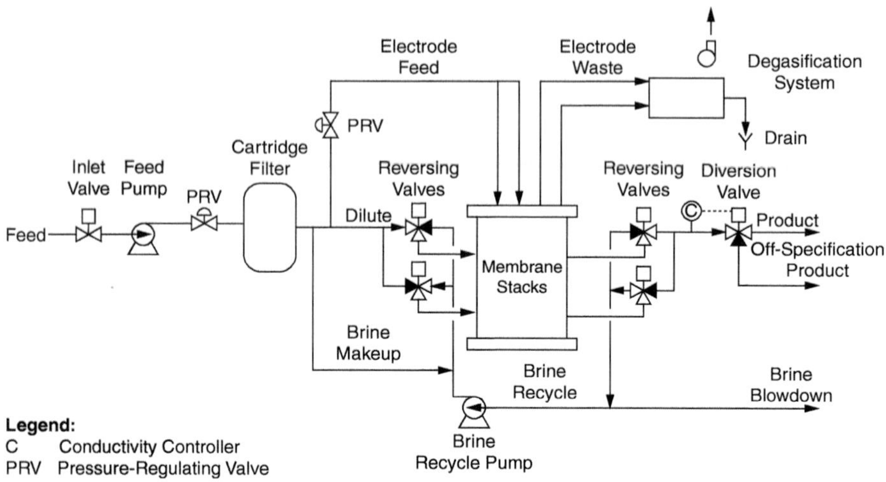

FIGURE 11-20 Process flow diagram for an EDR process. *Note:* When a VFD is connected to the feed pump, the inlet valve and PRV are not required. (*Courtesy of GE Water & Process Technologies Ionic, Inc.*)

Example 11-3 Electrodialysis Power Requirements
An electrodialysis stack consists of 350 cells is to be used to demineralize 378.5 m^3/day (0.1 mgd) of conventionally pretreated brackish surface water for drinking water production. The salt content of the water is 1000 mg/L and the cation or anion content is 0.09 g equivalent weights per liter. Pilot scale studies determined that the current efficiency E_c and efficiency of salt removed E_r was 90 and 50 percent, respectively. The resistance was measured as 5.0 ohms, and the current density/normality ratio was 450. Determine the current required, the area of the membranes, and the power required.

Solution The current I is given by

$$I = \left[\frac{96{,}500\ \text{amp}\cdot\text{sec}}{\text{gm}\cdot\text{eq}\cdot\text{wt}}\right]\left[\frac{378.5\ \text{m}^3}{\text{day}}\right]\left[\frac{1000\ \text{L}}{\text{m}^2}\right]\left[\frac{0.09\ \text{gm}\cdot\text{eq}\cdot\text{wt}}{\text{L}}\right]$$
$$\times\left[\frac{\text{day}}{86400\ \text{sec}}\right]\left[\frac{0.50}{350(0.90)}\right]$$
$$I = 60.4\ \text{amps}$$

Current density/normality ratio is 450, and since normality equals the number of gram-equivalent weight per liter, normality is 0.09.

The current density I_{density} is then equal to

$$I_{\text{density}} = (450)(\text{normality}) = (450)(0.09) = 40.5\ \text{mA/cm}^2$$

The membrane area A is

$$A = [60.4\ \text{amps}]\left(\frac{\text{cm}^2}{40.5\ \text{mA}}\right)\left(\frac{1000\ \text{Amps}}{\text{mA}}\right) = 1490\ \text{cm}^2$$

If the membranes are square in shape, the size is 38.6 cm by 38.6 cm and the power required is

$$P = RI^2 = (5.0\ \Omega)(60.4\ \text{A})^2 = 18{,}240\ \text{W} \quad \blacktriangle$$

Temperature Effects on Flux

The flux of water through a pressure-driven membrane is affected by temperature. The following section presents three techniques to compensate for the effect of temperature on flux.

Theoretical Normalized Flux Equation. The Hagen-Poiseuille equation, Eq. 11-53, measures flow through pores of a membrane as a function of viscosity, pore diameter, porosity, applied pressure, and membrane thickness (Cheryan, 1986). This equation indicates that water flux is inversely proportional to solvent viscosity. The viscosity of water decreases as temperature increases, which increases flux.

$$F_W = \frac{\Delta P}{R_W} = \frac{\varepsilon r^2 \Delta P}{8\delta\mu} = \frac{1}{\mu}\frac{\varepsilon r^2 \Delta P}{8\delta} \tag{11-53}$$

where ε is the porosity, r the pore radius, μ the solvent viscosity, and δ the membrane thickness.

If the porosity, pore radius, viscosity, and thickness are assumed to be constant, Eq. 11-53 can be used to develop a temperature correction factor. The normalized temperature of 25°C is a worldwide reporting standard (PEM, 1982). To derive the normalized flux equation, θ in Eq. 11-54 is determined graphically or by linear regression.

$$\text{TCF} = \frac{F_{T^\circ\text{C}}}{F_{25^\circ\text{C}}} = \theta^{(T-25)} \tag{11-54}$$

where TCF is the temperature correction factor and T the temperature (°C).

Applying the Hagen-Poiseuille equation, Eq. 11-54 can be written as

$$\log \frac{F_{T^\circ\text{C}}}{F_{25^\circ\text{C}}} = \log \frac{\mu_{25^\circ\text{C}}}{\mu_{T^\circ\text{C}}} = (T-25)\log\theta \tag{11-55}$$

Log θ then can be solved by regressing or plotting $\log \frac{\mu_{25^\circ\text{C}}}{\mu_{T^\circ\text{C}}}$ versus $T - 25$. The result for this theoretical derivation is shown in Eq. 11-56. Membrane suppliers have developed temperature correction equations for their products that account for temperature changes of water and membrane. Equation 11-56 accounts only for the viscosity change with temperature.

$$\text{TCF} = \frac{F_{T^\circ\text{C}}}{F_{25^\circ\text{C}}} = 1.026^{(T-25)} \tag{11-56}$$

Temperature Correction Factor Based on Operational Data. The theory developed so far assumes no effect of temperature on the membrane. Another method derives a TCF by developing a statistical relationship of the mass transfer coefficient and temperature using operational data. This technique accounts for temperature effects on both water and the membrane; however, it does assume the membrane is effectively cleaned.

The same equation form as shown in Eq. 11-54 is used.

$$\mathrm{TCF} = \frac{F_{T°\mathrm{C}}}{F_{25°\mathrm{C}}} = \theta^{(T-25)} \tag{11-54}$$

Flux is proportional to the mass transfer coefficient. Taking the logarithm of both sides gives

$$\log \frac{F_{T°\mathrm{C}}}{F_{25°\mathrm{C}}} = \log \frac{\mathrm{MTC}_{T°\mathrm{C}}}{\mathrm{MTC}_{25°\mathrm{C}}} = (T-25)\log\theta \tag{11-57}$$

Again, plotting $\log \frac{\mathrm{MTC}_{T°\mathrm{C}}}{\mathrm{MTC}_{25°\mathrm{C}}}$ of operational data versus $T - 25$ gives $\log \theta$ as the slope, which is determined by linear regression. Manufacturers commonly develop TCFs for their membranes. For example, a *TCF* of 1.04 was developed for a Memtec MF product using operational data from Manitowoc, Wisconsin (Kothari, 1997).

DuPont Equation. The TCF for DuPont B-10 membrane is commonly used for normalized flux and is presented in Eq. 11-58 (PEM, 1982). This figure of 1.03 is essentially equal to the theoretical θ of 1.026.

$$\mathrm{TCF} = \frac{F_{T°\mathrm{C}}}{F_{25°\mathrm{C}}} = 1.03^{(T-25)} \tag{11-58}$$

Example 11-4 Temperature Normalization and Specific Flux
A UF operator determines that the average instantaneous flux was 1.63 m^3/m^2-day (40 gal/sfd) at 89 kPa (13 psi) and 9°C in early spring. One month later the operator finds the flux changed to 1.34 m^3/m^2-day (33 gal/sfd) at a TMP of 69 KPa (10 psi) and 21°C. Determine if a significant change in specific flux has occurred, and what implications the number may have with respect to the operation of the facility.

Solution The flux for the spring measurement is normalized to a standard temperature of 25°C using Eq. 11-56.

$$\mathrm{TCF}_{1.63\mathrm{m}^3/\mathrm{m}^2\cdot\mathrm{day}} = \left(1.63\,\frac{\mathrm{m}^3}{\mathrm{m}^2\cdot\mathrm{day}}\right)(1.026)^{(25°\mathrm{C}-9°\mathrm{C})} = 2.458\,\frac{\mathrm{m}^3}{\mathrm{m}^2\cdot\mathrm{day}}\left(64.2\,\frac{\mathrm{gal}}{\mathrm{sfd}}\right)$$

The specific flux is the flux at standard temperature divided by the TMP.

$$J_{\mathrm{specific}} = \frac{2.458\,\frac{\mathrm{m}^3}{\mathrm{m}^2\cdot\mathrm{day}}}{89\ \mathrm{kPa}} = 0.02792\,\frac{\mathrm{m}^3}{\mathrm{m}^2\cdot\mathrm{day}\cdot\mathrm{kPa}}\cdot\left(4.9\,\frac{\mathrm{gal}}{\mathrm{sfd}\cdot\mathrm{psi}}\right)$$

Accounting for the temperature change one month later,

$$\mathrm{TCF}_{1.34\mathrm{m}^3/\mathrm{m}^2\cdot\mathrm{day}} = \left(1.34\,\frac{\mathrm{m}^3}{\mathrm{m}^2\cdot\mathrm{day}}\right)(1.026)^{(25°\mathrm{C}-21°\mathrm{C})} = 1.485\,\frac{\mathrm{m}^3}{\mathrm{m}^2\cdot\mathrm{day}}\left(37.1\,\frac{\mathrm{gal}}{\mathrm{sfd}}\right)$$

And the specific flux was

$$J_{\mathrm{specific}} = \frac{1.458\,\frac{\mathrm{m}^3}{\mathrm{m}^2\cdot\mathrm{day}}}{69\ \mathrm{kPa}} = 0.02152\,\frac{\mathrm{m}^3}{\mathrm{m}^2\cdot\mathrm{day}\cdot\mathrm{kPa}}\cdot\left(3.7\,\frac{\mathrm{gal}}{\mathrm{sfd}\cdot\mathrm{psi}}\right)$$

The percent reduction can be calculated as

$$\left[\frac{\left[0.02792\,\frac{\text{m}^3}{\text{m}^2\cdot\text{day}\cdot\text{kPa}} - 0.02152\,\frac{\text{m}^3}{\text{m}^2\cdot\text{day}\cdot\text{kPa}}\right]}{0.02792\,\frac{\text{m}^3}{\text{m}^2\cdot\text{day}\cdot\text{kPa}}}\right]\times 100 = 23\%$$

This is a 23 percent reduction, indicating that a significant loss in performance has occurred and the membranes should be chemically cleaned to remove the foulants that have caused the loss in productivity. ▲

INTEGRATED MF AND UF PROCESS APPLICATIONS AND PROCESS DESIGN

Membrane processes are considered for integration into water treatment plants for a variety of reasons. Because MF and UF processes do not effectively remove dissolved organic and inorganic constituents from water, the use of pretreatment processes such as powdered activated carbon (PAC), coagulation, lime addition, and/or oxidation are often integrated into the treatment scheme. It has been recognized that UF is limited in removing disinfection by-product precursors as a stand-alone process (Jacangelo, 1991). MF and UF processes are often integrated with other water treatment processes for application on ground and surface waters, and function as integrated membrane systems. There is a variety of flexibility for integrating MF and UF with other treatment processes (Pressdee et al., 2006). According to a survey conducted by the U.S. Environmental Protection Agency (USEPA, 2001) the principal reasons for integrating MF and UF membranes are

1. Complying with existing and future regulations (with compliance being facilitated by using membrane filtration)
2. Providing a barrier to protozoan cysts and bacteria
3. Improving operating efficiency through automation and ability to treat water of variable quality
4. Using a smaller land area for the plant in comparison with a conventional filtration system
5. Providing an additional barrier against hazards to public health
6. Having lower capital, operation, and maintenance costs in comparison to other technologies
7. Providing a barrier for unfiltered systems that may lose unfiltered status
8. Providing pretreatment before NF or RO
9. Bolstering consumer confidence in water quality

The most common reason for implementing membrane filtration has been to meet or exceed current and identified future regulatory requirements, and for MF and UF particularly the removals of pathogens, organics, and particles. However, certain types of conventional processes should not be considered for integration. For example, the use of polymers (e.g., cationic polymer) that are upstream of membranes can result in excessive fouling. This was observed during pilot studies in Appleton, Wisconsin, when cationic polymer use during upstream lime softening resulting in 10-day cleaning intervals (Roquebart et al., 2002). Also the use of some polymeric coagulants (e.g., chitosan) has been determined to cause significant fouling for hollow fiber UF membranes (Machenback et al., 2002). The following section describes the various types of MF- and UF-integrated membrane systems that are employed to address specific contaminants not associated with simple particle removal and microbial removal.

MF and UF Integrated with Either Lime or Coagulation, Flocculation, and Sedimentation (CFS) with or without Powdered Activated Carbon (PAC)

Many surface water supplies contain elevated levels of TOC (natural organic matter [NOM]) and variable turbidity requiring the use of excessive coagulant dosing to produce sweep floc or addition of polymers to aid floc formation, settling, and filtration. Hard-surface water supplies often use lime softening for pretreatment ahead of filtration. UF has been integrated successfully into lime-softening facilities for surface- and groundwater treatment (Duranceau, 2000; Kweon and Lawler, 2002; Brown et al., 2004). Also, it has been shown that the use of CSF prior to membrane filtration produces higher-quality water for significantly variable turbidity loadings and the UF process can operate more efficiently. Sedimentation is incorporated to reduce the solids loading to the membranes; however, MF and UF processes are able to operate at increased flux rates and reduced cleaning frequencies. thereby reducing overall cost. However, pressurized UF processes operated at instantaneous flux rates over 1.83 m^3/m^2-day (45 gal/ft^2-day) required frequent cleaning events and suffered damage due to excessive backwash regimes when the same water was treated without the use of CSF pretreatment. Figure 11-21 provides a simplified process flow schematic for MF or

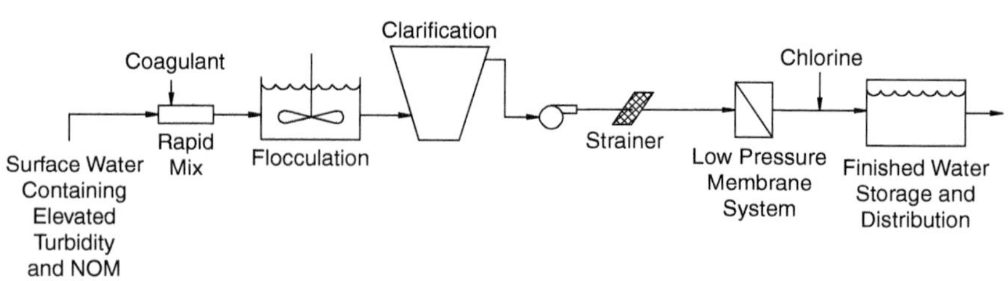

FIGURE 11-21 Process flow diagram for a MF or UF process integrated with CSF pretreatment. (*Courtesy of the American Water Works Association Manual of Water Supply Practices M53 Microfiltration and Ultrafiltration Membranes for Drinking Water.*)

UF membrane processes integrated with conventional CSF treatment. PAC has been used for treatment of organics from water supplies. (Adham, 1991, 1993) PAC has been used to remove pesticide contaminants in the Seine River in Paris, where PAC was added ahead of a cellulose triacetate UF membrane (Anselme, 1991). However, PAC has been reported to decrease filtration run times and productivity, especially at higher operating flux (Huey et al., 1999). Longer backwash intervals improved PAC effectiveness for removal of tastes and odors, and membrane fouling was shown to be independent of PAC dose for a low-turbidity, cold-lake-water supply (Schideman et al., 2001). Preclarification was shown to increase the operating flux in piloted systems in Texas by 19 to 50 percent (Braghetta, Price, and Kolkhorst, 2001). PAC was believed to have reduced the fouling rates due to PAC adsorption of NOM. Presedimentation was shown to only marginally improve the operating flux of a submerged UF system by less than 0.204 m^3/m^2-day (5 gal/ft^2-day), most likely because these immersed systems operate in basins with higher solid loadings than their forward pressure counterpart (Crawford and Bach, 2001).

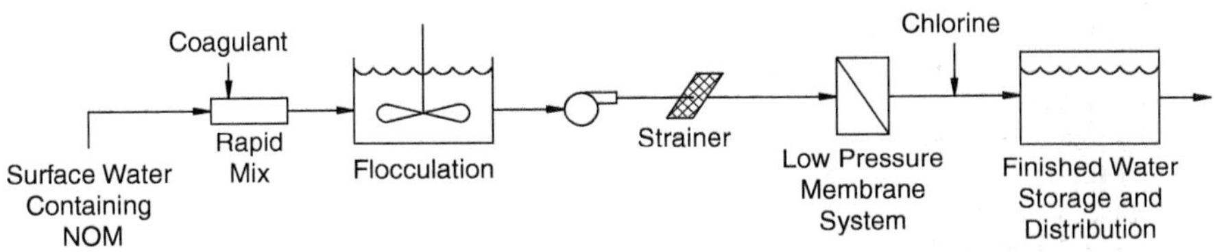

FIGURE 11-22 Process flow diagram for a MF or UF process integrated as direct filtration. (*Courtesy of the American Water Works Association Manual of Water Supply Practices M53 Microfiltration and Ultrafiltration Membranes for Drinking Water.*)

MF and UF Integrated with Coagulation

Water supplies with low turbidity, less than 5 ntu and having no other water quality contaminants (e.g., iron, manganese, low NOM, or other metal foulants) can be fed directly to a MF or UF membrane process, as shown in Fig. 11-22. This is analogous to direct filtration with conventional media filters. Prestrainers are incorporated into the flow stream to serve as a bulk protection strainer and can be automated for backwashing. The applicability of direct membrane filtration is limited by the amount of coagulant the membrane can reliably process without excessive decline in production or clarification will have to be incorporated into the design. Studies have shown that the addition of coagulant prior to MF and UF reduces membrane fouling by organic matter (Lahoussine-Tourcaud et al., 1990; Carroll et al., 2000).

MF and UF Integrated with Coagulation, Flocculation, and Clarification as Pretreatment to NF, RO, and EDR for Surface Water Supplies

As shown in Fig. 11-23, NF, RO, and EDR can be integrated with MF, UF, and CSF for treatment of surface water supplies. Although MF and UF pretreatment of surface water supplies is of benefit for downstream NF and RO processes, fouling is of concern, and most surface water NF or RO membranes will require more frequent cleaning intervals than when NF or RO is used for treatment of ground-, brackish-, or seawater supplies (Taylor et al., 1989a, 2000; Duranceau 2001) even if MF or UF is used in combination with the RO and NF process. This is because aerobic biofouling will impact the NF and RO processes

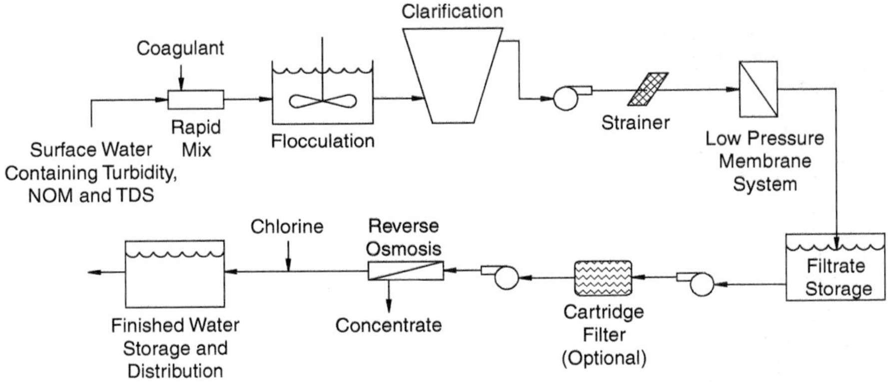

FIGURE 11-23 Process flow diagram for MF or UF process integrated with CSF pretreatment prior to NF, RO or EDR for surface water treatment. (*Courtesy of the American Water Works Association Manual of Water Supply Practices M53 Microfiltration and Ultrafiltration Membranes for Drinking Water.*)

as disinfectant is not available for pretreatment downstream of MF and UF processes else damage will occur in the NF and RO membranes as NF and RO processes are tolerant to only 2000 ppm-hours of chlorine contact. Although NF and RO membranes are somewhat more tolerant to monochloramine on the order of 20,000 ppm-hours of monochloramine contact time, the NF and RO membranes will eventually be damaged. Monochloramine in incremental or low-level treatment can help to reduce biofouling impacts on NF and RO processes. EDR facilities are unique in that they can be disassembled for fouling control if needed unlike the synthetic membrane processes (MF, UF, NF and RO) that are of a fixed membrane construction. EDR does require additional stack maintenance due to the configuration and design of the process (it is an electrically driven process and not pressure driven). Similar chemicals are used in EDR, RO and NF for the removal of organics, scale and bacteria; however, EDR membranes, if severely fouled, can be individually removed and cleaned by disassembling the stack and placing the flat sheet membranes in chemical baths. Severely fouled RO and NF membranes cannot be treated in this same method. Silica when present in sufficient quantities could limit the recovery of RO or NF but does not impact EDR as silica remains uncharged and does not concentrate and potentially scale in the EDR process. Consequently, EDR may have an advantage over RO and NF offering a lower present worth when silica is predominant in brackish surface water (Duranceau, 2006). RO and NF membranes operating as a surface water integrated membrane system will typically require monthly cleaning frequencies, which compares to conventional groundwater RO and NF treatment that experience cleaning frequencies on the order of once or twice per year.

MF and UF Integrated with Oxidation Processes for Removal of Iron and Manganese from Groundwater

There are many options for removing iron and manganese from groundwater. Typically a two-step process is used. First, addition of an oxidant (air, permanganate, or chlorine) reacts with the dissolved Fe and Mn. This is followed by granular media filtration to remove oxidized iron and manganese particulates that remain after the chemical reaction has been completed. Chapter 7 discusses principles of iron and manganese removal; the reader should consult this chapter for more detail. Note that strong oxidants, such as chlorine, can damage membranes that are not oxidant resistant and should not be used without quenching the chlorine or use oxidant-resistant membranes. Figure 11-24 presents a basic process flow diagram for the oxidant integrated

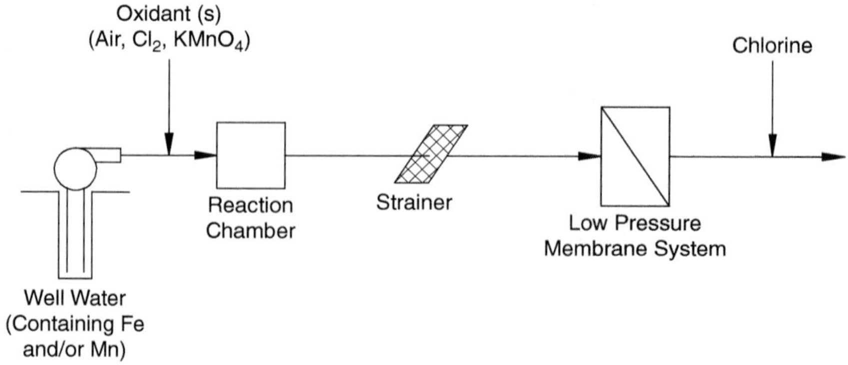

FIGURE 11-24 Process flow diagram for a MF or UF process integrated with oxidation processes for removal of iron and manganese from groundwater. (*Courtesy of the American Water Works Association Manual of Water Supply Practices M53 Microfiltration and Ultrafiltration Membranes for Drinking Water.*)

membrane system. Schneider et al. (2001) demonstrated that chlorine dioxide reduced manganese from 0.094 mg/L in an MF feed stream to 0.001 mg/L in the permeate after a 20 min upstream MF detention time. Permanganate reduced manganese to 0.03 mg/L in MF permeate (from a feed concentration of 0.099 mg/L), also after 20 min of detention time upstream of MF. Of the facilities that use oxidant-resistant membranes, permanganate is the preferred oxidizing chemical. For nonoxidant resistant membranes, designs must be evaluated for compatibility with oxidation pretreatments, as some membranes are not tolerant to oxidizing chemicals (e.g., chlorine and other strong oxidants) and may weaken or change in pore size and permeability.

MF and UF Integrated with Coagulant for Arsenic Removal from Groundwater

Integrated membrane systems (IMSs) using coagulant and MF or UF processes have been shown to be effective at reducing arsenic levels from as high as 100 μg/L to an arsenic concentration of 2 μg/L (Amy et al., 2000, Chwirka et al., 2000). Figure 11-25 presents a process flow diagram of an integrated membrane system that could be used for arsenic removal. Included in the process flow is acidification, which is required to reduce the pH to between 6.8 and 7.0 to maximize arsenic adsorption. Also, because arsenic (V) is better adsorbed and co-precipitated during coagulation than is arsenic (III), an addition of an oxidizing chemical upstream of coagulant may be required if As(III) is the predominant arsenic species. Regarding the coagulant feed, ferric salts are preferred over aluminum salts because they provide better arsenic removal on a molar basis.

MF and UF Integrated with Seawater RO

The integration of MF and UF as membrane pretreatment for seawater desalination is an integrated membrane system and has been demonstrated at the pilot scale (Rosberg, 1997; Van Hoof, 2001; Vial, 2003; Duranceau and Henthorne, 2004). However, operation of an open intake system caused poor UF performance and presented operational challenges because of fouling (Henthorne, 2005). It was found that MF and UF production could be maintained with the use of additional pretreatment using disc filters. Similar results were obtained at another MF and UF facility when disc filters were used for additional pretreatment to resolve membrane failure from sharp and serrated debris present in UF feedwater (Jew et al., 2003). Disc filters were shown to reduce MF and UF fouling rates during pilot-scale testing at the Naval Facilities Engineering Service Center in Port Hueneme, California (Huehmer et al., 2005).

There are many water quality parameters that can influence the choice of pretreatment, such as dissolved organic carbon, silt density index (SDI), turbidity, algae content (that may vary depending on the season and other factors), temperature, oil and grease, and suspended

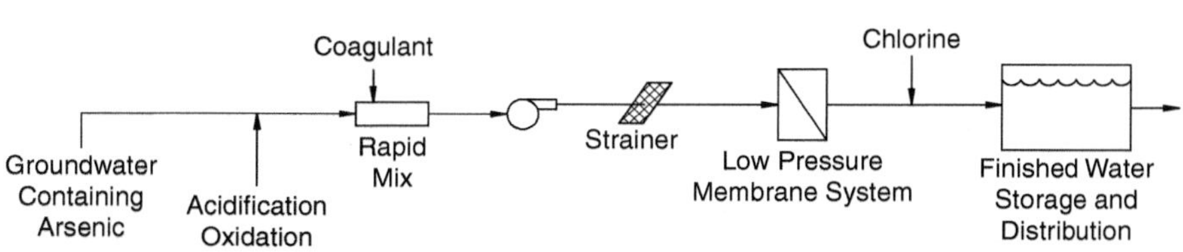

FIGURE 11-25 Process flow diagram for a MF or UF process integrated with coagulant for removal of Arsenic from groundwater. (*Courtesy of the American Water Works Association Manual of Water Supply Practices M53 Microfiltration and Ultrafiltration Membranes for Drinking Water.*)

solids content. As a result, required pretreatment can comprise different technologies, and may include the coupling of an advanced technology such as UF with a conventional process such as coagulation, ballasted sedimentation, dissolved air flotation (DAF), dual media filtration, and the like. Regarding UF, studies have shown that the principal cause of severe fouling in seawater RO is organic matter including algae (Chua et al., 2003; Ma et al. 2007); however, UF pretreatment was found as an effective pretreatment to seawater RO.

MF and UF Process Design

MF and UF membranes are designed primarily as particle removal processes. Additionally, MF and UF design concepts are changing rapidly with this relatively new technology, which presents unique challenges during engineering design. Currently, MF and UF systems are available as modular systems from several manufacturers. MF and UF systems are typically offered as packaged systems that contain the required filtration equipment and appurtenances and vary greatly among manufacturers. Consequently, MF and UF system design should involve direct communication with manufacturers and ancillary supporting system vendors that are familiar with individual products, as many of these systems have proprietary design features that the process design engineer must know. Typically, MF and UF processes require prescreening of the raw water and pumping it under pressure onto a membrane in either a direct or crossflow operation mode. For the direct filtration mode, the transmembrane pressure (TMP) can be calculated as the difference between the pressure of the inlet and the permeate pressure as follows:

$$p_{\text{tmp}} = p_i - p_p \tag{11-59}$$

where p_{tmp} is the transmembrane pressure kPa (psi), p_i the pressure at the inlet to the MF or UF module kPa (psi), and p_p the pressure of permeate kPa (psi).

For the crossflow operation scenario, the average TMP is defined as the average of the difference between the pressure at the inlet and outlet of the process less the permeate pressure.

$$p_{\text{tmp}} = \left[\frac{(p_i + p_o)}{2}\right] - p_p \tag{11-60}$$

where p_o is the pressure at the outlet to the MF or UF module kPa (psi).

When the MF or UF process is operated in the crossflow mode, there is an accompanying pressure drop across the module where Δp is the pressure drop across the module in kPa (psi).

$$\Delta p = p_i - p_o \tag{11-61}$$

Application of transmembrane pressure (TMP) in an MF and UF process will produce a filtrate or permeate, with TMPs typically ranging from 0.2 bar (3 psi) to 1 bar (15 psi). The total production flow of permeate from a MF or UF process can be determined by

$$Q_p = (J_{tm})(S) \tag{11-62}$$

where Q_p is the process permeate flow in L/hr (gal/min), J_{tm} the transmembrane flux in L/hr/m^2 (gal/day/ft^2), and S the total effective membrane surface area.

The flux is directly proportional to TMP, and for a given membrane area, the filtrate or permeate flow rate is also directly proportional to the TMP. The flux is inversely proportional to the water viscosity. Hence, temperature will affect the required driving pressure for membrane processes because of changes in the feedwater viscosity. Therefore, as the water temperature decreases, MF and UF systems must increase the TMP to maintain production. About 50 percent more TMP is needed to maintain a constant flux at 5°C than at 20°C. Assuming a constant TMP and membrane area, the filtrate flow at 5°C would be about 35 percent less than at 20°C. Therefore, designers must account for temperature impacts on sizing of the membrane facilities and must define the desired production rate at a specified temperature. Typical flux rates for MF and UF membranes vary widely, depending on membrane material and specific application, and range from 0.815 m^3/m^2-day to 4.07 m^3/m^2-day (20 to 100 gpd/ft^2).

In MF and UF processes, a frequent backwash (every 15 min to 1 hour or more frequently) removes the cake formed on the membrane surface. Some MF and UF systems incorporate a chemical enhancement routine in backwash events, referred to as chemically enhanced backwashing. However, over time, as the MF and UF membrane ages, backwashing will not completely remove components of the foulant layer, and the process will be unable to achieve its original operating transmembrane pressure.

Figure 11-26 illustrates correlating a backwashing event to operating transmembrane pressure, where at some point in time the MF or UF system will require a chemical clean in place (CIP) to maintain productivity without exceeding some maximum allowable transmembrane pressure (else the membrane become damaged). CIP is typically required when the TMP reaches 207 to 172 kPa (30 to 25 psi). A typical CIP would be implemented in a two-step process that uses (1) caustic chlorine solution to remove organic and microbial foulants and (2) an acid solution to remove inorganic mineral foulants.

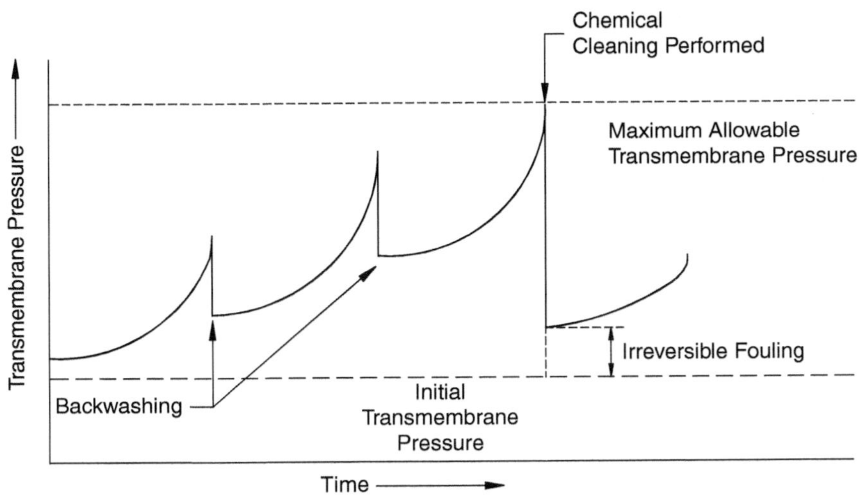

FIGURE 11-26 Transmembrane pressure development over time indicating a chemical cleaning was performed after backwashing illustrating concept of irreversible fouling.

Example 11-5 Ultrafiltration Plant Concept Design

An ultrafiltration treatment plant is to be designed to produce 19,000 m^3/d of treated water. Piloting demonstrates that a UF membrane system can operate at a flux of 100 L/m^2-h at 20°C with a filter run length of 75 min. The selected UF manufacturer offers the following full-scale system:

Item description	Quantity
Number of modules per rack	90
Membrane area per module (m^2)	15
Backwash duration (min)	1
Backwash flux (L/m^2-h)	2810

Develop design criteria for the following:

1. Membrane area required to achieve the design flow
2. Number of membrane racks
3. Total number of UF modules
4. UF permeate flux
5. Water volume produced per filtration cycle
6. Water volume wasted per backwash cycle
7. System recovery

Solution

1. Convert the given design flow rate into L/h, then calculate membrane area.

$$Q = \left(\frac{19{,}000\ \text{m}^3}{\text{day}}\right)\left(\frac{1000L}{\text{m}^3}\right)\left(\frac{\text{day}}{24\ \text{h}}\right) = 791{,}667\frac{\text{L}}{\text{h}}$$

$$\text{Area}_{\text{min}} = \frac{Q}{J} = \frac{791{,}666.67\ \text{L/h}}{100\ \text{L/m}^2\cdot\text{h}} = 7916.67\ \text{m}^2$$

2. Calculate the minimum number of modules, then determine rack quantity.

$$N_{\text{modules,min}} = \frac{7916.67\ \text{m}^2}{(15\text{m}^2/\text{moduls})} = 527.78\ \text{modules} = 528\ \text{modules}$$

$$N_{\text{racks,min}} = \frac{528\ \text{modules}}{(90\,\text{modules/rack})} = 5.86\ \text{racks} = 6\,\text{racks}$$

3. Calculate the total number of UF modules.

$$N_{\text{modules,total}} = (6\ \text{racks})\left(90\ \frac{\text{modules}}{\text{rack}}\right) = 540\ \text{modules}$$

4. Calculate the design membrane area, then the UF permeate flux.

$$\text{Area}_{\text{actual}} = (540\ \text{modules})\left(\frac{15\ \text{m}^2}{\text{modules}}\right) = 8100\ \text{m}^2$$

$$J = \frac{Q}{A_{\text{actual}}} = \frac{791{,}667\ \text{L/h}}{8100\ \text{m}^2} = 97.74\frac{\text{L}}{\text{m}^2\cdot\text{h}}$$

5. Calculate the water produced during one filtration cycle per rack.

$$V_F = \left(\frac{97.748}{\text{m}^2\cdot\text{h}}\right)\left(\frac{15\ \text{m}^2}{\text{modules}}\right)\left(\frac{90\ \text{modules}}{\text{rack}}\right)(75\ \text{min})\left(\frac{1\ \text{h}}{60\ \text{min}}\right) = 164{,}931\frac{\text{L}}{\text{rack}}$$

6. Calculate the water consumed during one backwash cycle per rack.

$$V_{BW} = \left(\frac{280\ \text{L}}{\text{m}^2\cdot\text{h}}\right)\left(\frac{15\ \text{m}^2}{\text{modules}}\right)\left(\frac{90\ \text{modules}}{\text{rack}}\right)(1\ \text{min})\left(\frac{1\ \text{hr}}{60\ \text{min}}\right) = 6300\frac{\text{L}}{\text{rack}}$$

7. Calculate the design water recovery

$$r=\left(\frac{V_F-V_{BW}}{V_F}\right)=\left(\frac{(164.931\ L/\text{rack})-(6300\ L/\text{rack})}{(164.931\ L/\text{rack})}\right)=0.96$$

Thus, recovery is 96 percent. ▲

NF AND RO PROCESS CONCEPTS AND DESIGN CRITERIA

RO and NF Processes

This section introduces priorities for NF and RO design and associated material. A conventional RO or NF treatment system includes pretreatment, membrane process, posttreatment, and concentrated residuals, as shown in Fig. 11-27. As can be seen in Fig. 11.27, the overall process can be rationally separated into advanced pretreatment, conventional pretreatment, membrane process, and posttreatment subsections. Any raw-water stream used as a feed stream to a membrane process must undergo either conventional or advanced pretreatment. Conventional pretreatment includes acid or scale inhibitor addition to prevent precipitation of salts during membrane processing. Advanced pretreatment is required before conventional pretreatment when the raw water contains excessive fouling materials. Membrane treatment is the passage of the pretreated water through an active RO/NF membrane surface with a pore size 0.001–0.0001 μm. Posttreatment includes many unit operations common to drinking water treatment such as aeration, disinfection, and corrosion control.

In a typical membrane process, one stream enters the membrane element and two streams exit. The entering stream is the feed stream. The streams exiting the membrane are referred to as *concentrate* and *permeate streams.* A portion of the concentrate stream is sometimes recycled back to the feed stream to increase crossflow velocity and recovery. The different configurations for softening NF, brackish RO, and seawater RO processes are shown in Fig. 11-28. In multistaged systems the number of pressure vessels decreases in each succeeding stage to maintain sufficient velocity in the feed-concentrate channel and is referred to as *concentrate-staged arrays*. In this type of system, the concentrate stream

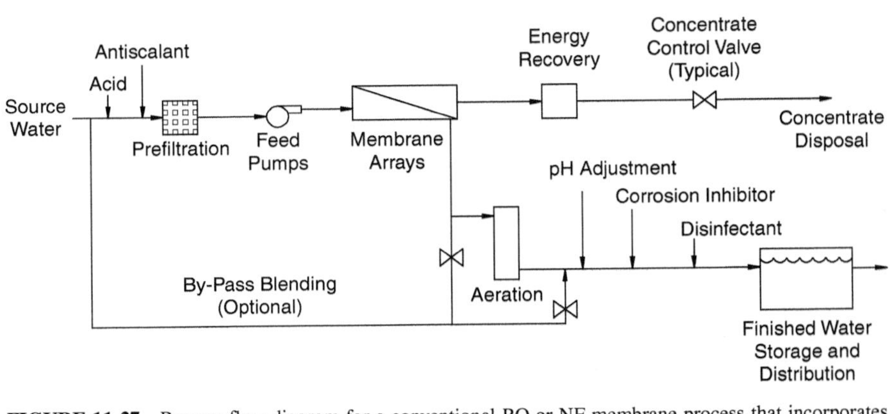

FIGURE 11-27 Process flow diagram for a conventional RO or NF membrane process that incorporates energy recovery and by-pass.

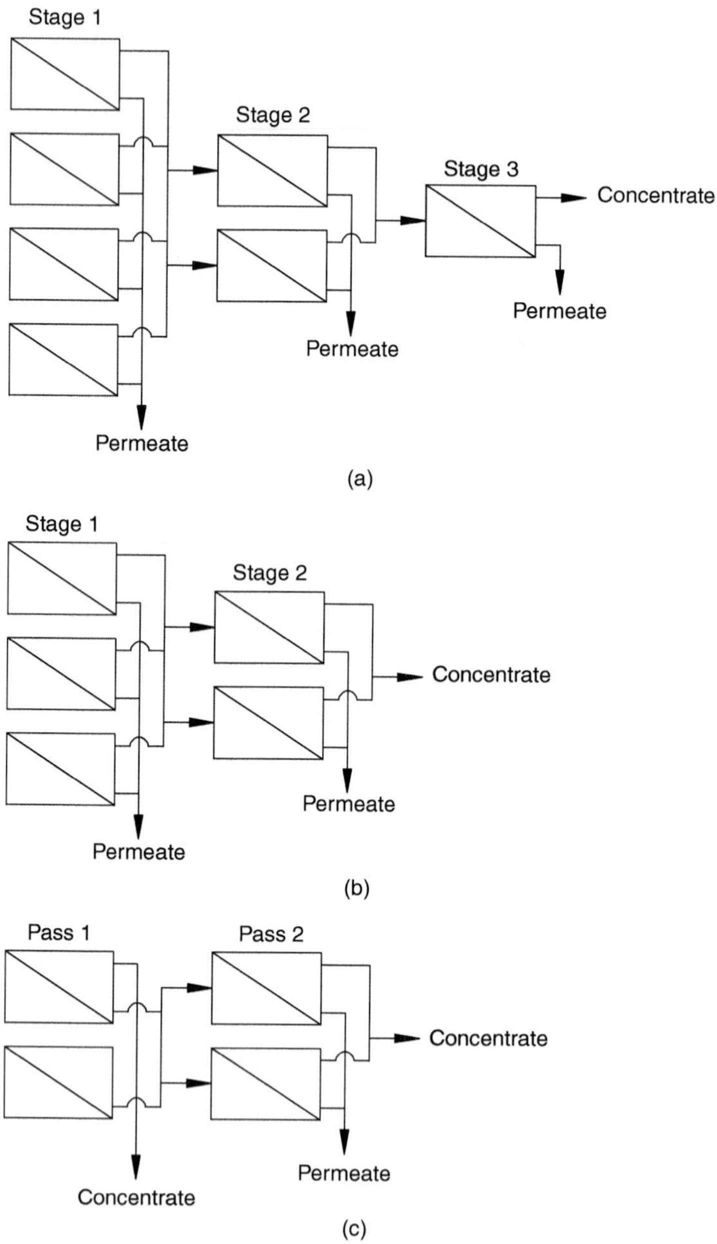

FIGURE 11-28 Configuration of reverse osmosis facilities. (*a*) A concentrate-staged $4 \times 2 \times 1$ NF softening array, (*b*) A concentrate-stage 3×2 RO brackish desalting array, and (*c*) A two-pass permeate-staged RO seawater array.

produced in the first stage serves as the feed stream for the second stage while the permeate produced in all stages is taken for subsequent unit operations or processes. However, for seawater applications, the permeate from the first bank of membrane trains is fed to the second bank of membrane trains and is referred to as *permeate-staged arrays*.

Fouling Indexes

Membrane fouling is a widespread problem limiting the performance and application of synthetic membrane processes and, as such, is an important consideration in the design and operation of a membrane system. Cleaning frequency, pretreatment requirements, operating conditions, cost, and performance are affected by membrane fouling. Fouling indexes are estimates of the fouling and pretreatment requirements for RO or NF systems. The silt density index (SDI) and the *modified fouling index (MFI)* are the most common fouling indexes. SDI and the MFI are defined using the basic resistance model as quantitative indicators of water quality and potential for membrane fouling.

Fouling indexes are determined from simple membrane tests. These values resemble mass transfer coefficients for membranes used to produce drinking water. Although similar data are collected for each index, significant differences separate them. Because of the effects of different filtration equipment, only a Millipore filter apparatus can be used to generate accurate index measurements.

Silt Density Index. The SDI is the most widely used fouling index and is calculated as shown in Eq. 11-63. The SDI is a timed filtration test using three time intervals whereby water flows through a laboratory-grade membrane Millipore filter. The water is passed through a 0.45 μm Millipore filter having a 47 mm internal diameter at 207 kPag (30 psig) to determine the SDI (ASTM, 2001). The time required to complete data collection for such a test varies from 15 min to 2 hours depending on the character of the water. The Millipore test apparatus is used to determine three time intervals for calculation of the SDI. The first two intervals are the times to collect initial and final 500 mL samples. The third time interval—5, 10, or 15 min—is the time between the collection of the initial and final samples. The 15 min interval is used unless the water has such extreme fouling potential that the filter plugs before that period expires. The interval between the initial and final sample collection is decreased until a final 500 mL sample can be collected.

$$\text{SDI} = \frac{100[1-(t_i/t_f)]}{t} \tag{11-63}$$

where t_i is the time to collect initial 500 mL of sample, t_f the time to collect final 500 mL of sample, and t the total running time of the test.

The SDI value is a static measurement of resistance determined by samples taken at the beginning and the end of the test. The SDI does not measure the rate of change of resistance during the test, and it is the least sensitive of the fouling indexes. The SDI is not dynamic, is not measured in a crossflow mode, does not use the same material or pore size as a membrane element, measures only static resistance, and is not reflective of a continuously operated membrane process.

Modified Fouling Index. The MFI is determined using the same equipment and procedure as for the SDI, except that the volume is recorded every 30 sec over a 15 min filtration period (Schippers and Verdouw, 1980). The flow rate is determined from volume and time data, and the inverse of the flow rate is plotted as a function of volume filtered. The development of the MFI is consistent with Darcy's law, in that the thickness of the cake layer

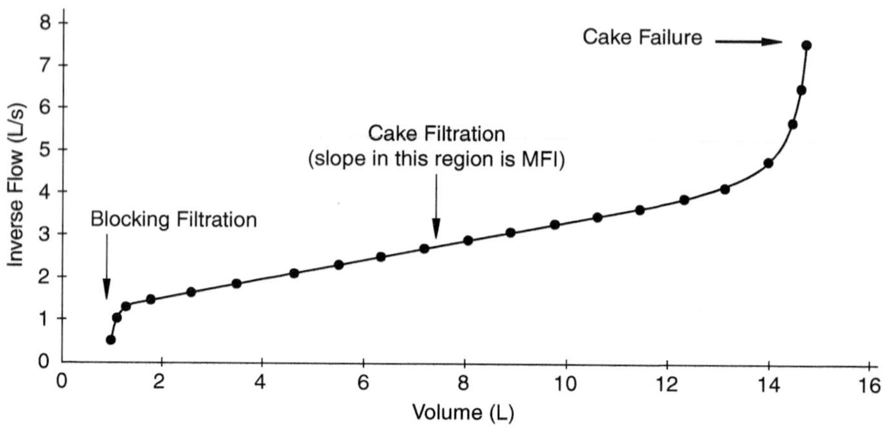

FIGURE 11-29 Typical MFI curve.

formed on the membrane surface is directly proportional to the filtrate volume. The total resistance is the sum of the filter and cake resistance. The MFI is derived in Eqs. 11-64 to 11-66 and is defined graphically as the slope of a curve relating inverse flow to cumulative volume (Fig. 11-29).

$$\frac{dV}{dt} = \frac{\Delta P}{\mu} \frac{A}{(R_f + R_k)} \tag{11-64}$$

$$t = \frac{\mu V R_f}{\Delta P A} + \frac{\mu V^2 \xi}{2 \Delta P A^2} \tag{11-65}$$

$$\frac{1}{Q} = a + MFI \times V \tag{11-66}$$

where R_f = resistance of the filter (L^{-1})
R_k = resistance of the cake (L^{-1})
ζ = measure of fouling potential ($L^2 t$)
Q = average flow (v/t^{-1})
a, b = constants

On the basis of the work of Schippers and Verdouw, Morris (1990) determined an instantaneous MFI by calculating the ratio of flow over volume in 30 sec increments to increase sensitivity. An MFI plot is shown in Fig. 11-29. Typically the formation, buildup, and compaction or failure of the cake can be seen in three distinct regions on a MFI plot. The regions corresponding to blocking filtration and cake filtration represent productive operation, whereas compaction indicates the end of a productive cycle.

Index Guidelines. The SDI and MFI serve as general guidelines for acceptable feedwater quality, and a maximum value of 3 to 4 is typically specified by membrane manufacturers when the SDI and MFI are expressed in units of reciprocal minutes and s/L^2, respectively. Some approximations for index values required prior to conventional membrane treatment are given in Table 11-9. The indexes apply to the raw or advanced pretreated water before conventional pretreatment processes. These numbers are only approximations and do not eliminate the need for pilot plant studies. Pretreatment requirements cannot be determined

TABLE 11-9 Fouling index approximations for RO and NF

Fouling index	Range (s/L^2)	Application
MFI	0–2	Reverse osmosis
	0–10	Nanofiltration
SDI	0–2	Reverse osmosis
	0–3	Nanofiltration
	0–10	Electrodialysis reversal

Source: Morris (1990); Sung (1994); Duranceau (2006).

for most installations without pilot plant studies unless actual plant operating data can be obtained on similar waters. Pilot plant studies have been omitted in the design of some reverse osmosis plants for treatment of raw waters with TOC and SDI values less than 1 unit. EDR can handle feedwaters with SDI_5 values of less than 12, whereas RO requires feed waters to have a 15 min SDI (SDI_{15}) less than 2. Colloidal fouling of EDR has not been experienced below a feedwater 5 min SDI (SDI_5) of 12, while fouling is likely if the SDI_5 is above 16; RO membranes could not operate under similar conditions. NF can process feed waters with a SDI_{15} of 3.0, but suffers from biological and organic fouling which are not predicted by the SDI.

Example 11-6 SDI Problem
A municipal utility is considering using a membrane process as an alternative treatment method for its newly developed well field. The following information is collected using a SDI testing apparatus on an unfiltered raw water sample. Determine the 15 min SDI value.

Test time (min)	Water volume collected (mL)	Time to collect 500 mL (sec)
0	500	9
5	500	11
10	500	12
15	500	14

Solution Using Eq. 11-63,

$$SDI = \frac{100[1-(t_i/t_f)]}{t}$$

$$SDI = \frac{100\left[1-\frac{9\,\text{sec}}{14\,\text{sec}}\right]}{15\,\text{min}}$$

$$SDI = 2.4\ \text{unit}$$

Since the SDI is less than 3 units, this specific water evaluated would be considered acceptable feed water for a reverse osmosis process. ▲

Pretreatment

As mentioned, membrane operations require some feedwater pretreatment upstream of the process. First, however, it is important to realize that pretreatment is specific for individual processes and feedwaters. Needs differ from application to application and site to site.

Control of Fouling. Pretreatment is the first step in controlling membrane fouling, and it can be quite involved. In its simplest form, pretreatment involves microstraining with no chemical addition. However, when a surface water is treated, the pretreatment procedure may be much more involved and include pH adjustment, chlorination, addition of coagulants (e.g., alum, polyelectrolytes), sedimentation, DAF, dechlorination (e.g., addition of sodium bisulphite), adsorption onto activated carbon, addition of complexing agents [e.g., EDTA, sodium hexametaphosphate (SHMP)], and final polishing. Note that when oxidants are used for pretreatment prior to polyamide membranes or other nonoxidant resistant materials, precautions must be made to prevent the oxidant from contacting these types of membranes else irreversible damage will occur. Several important factors must be considered in contemplating pretreatment.

1. Material of membrane construction (asymmetric cellulose or noncellulose membranes, thin-film ether, or amide composite membranes)
2. Module configuration (spiral wrap, hollow fine fiber, tubular)
3. Feedwater quality
4. Recovery ratio
5. Final water quality

A listing of substances potentially harmful to membranes has been compiled (Rautenbach and Albrecht, 1989) and they have been categorized as either those damaging (acids, bases, pH, free chlorine, bacteria, and free oxygen) or those blocking, which include foulants (metal oxides, colloids, inorganic, NOM, and biological substances) and scalents ($CaCO_3$, $CaSO_4$, CaF_2, $BaSO_4$, and silica). Concentrations and/or presence of these components in the feedwaters must be controlled (Taylor and Jacobs, 1996). Biological substances can include bacteria, such as slime forming, iron, and sulfur bacteria, and other microorganisms will impact membrane process performance if not controlled (Duranceau, 2005). Several researchers have identified the characteristics and nature of the foulants using different autopsy procedures and found that humic acids were not as fouling as other organic compounds (Mosqueda-Jimenez, Huck and Basu, 2008) for UF, as an example.

Turbidity levels stipulated by membrane manufacturers are normally attained by conventional clarification techniques such as coagulation followed by sedimentation and sand filtration. In seawaters that are rich in suspended nutrients and for treating surface waters with high TOC, ultrafiltration has been advocated as a viable pretreatment option capable of reducing the suspended solids concentration to acceptable standards to feed finer membrane processes (Strohwald and Jacobs, 1992; Metcalf et al., 1992; Taylor et al., 1989a). The minimum pretreatment processes for RO or NF consist of scale inhibitor and/or acid addition and microfiltration. These pretreatment processes help to control scaling and to protect the membrane elements. Such precautions are required for conventional reverse osmosis or nanofiltration systems.

Advanced Pretreatment. Advanced pretreatment operations precede scaling control and static microfiltration. These processes might include coagulation, dissolved air flotation (DAF), oxidation followed by green-sand filtration, groundwater recharge, continuous microfiltration, and granular activated carbon (GAC) filtration. Any other processes upstream of conventional pretreatment would also be advanced pretreatment by definition. In some pretreatment processes, such as alum coagulation, the feedwater is saturated with a salt, such as alum hydroxide. The solubility of such salts must be accounted for in the feedwater stream to avoid precipitation in the membrane. Several other types of pretreatment may form part of RO or NF membrane systems.

The majority of RO plants operating on surface water and many operating on groundwater have reported fouling problems (Goosen et. al., 2004). Fouling controls productivity and directly impacts cleaning and operating costs in many surface water RO applications (Duranceau 2005, 2006). Membrane fouling is not a well-understood phenomenon, and many different pretreatment processes can be used to control different types of fouling mechanisms. For example, to prevent problems with biological fouling, a bactericide that is not harmful to the membranes might be introduced. To treat for biodegradable foulants, biologically activated carbon would be a feasible option. Microfiltration or ultrafiltration would also be useful for removing some fouling contaminants. Many processes could be combined with membranes for specific applications.

The most serious fouling problems often involve treatment of highly organic surface waters by NF or RO for TOC or DBP control. Such applications may require cleaning frequencies of less than 1 month, a period not typical for conventional RO or NF processes. However membrane applications for organic control are not typically used and new methods will have to be developed with time. One method for controlling fouling is to highly automate the membrane-cleaning process. Currently, conventional RO plants treat nonfouling groundwaters, averaging cleaning frequencies of 6 months or more. Consequently, the cleaning operation is not automated, and it disrupts normal operation. Development of nondamaging membrane cleaning agents and automation of the cleaning process offer potential answers to fouling control.

Impact of Hydrocarbons on RO and EDR. Organic fouling in RO has been studied using model hydrocarbons such as hexane and diesel (Owadally, 2002). A large number of countries that use RO to obtain drinking water also are producers and exporters of hydrocarbons. This makes seawater RO units particularly susceptible to damage from oil spills. Lower molecular weight hydrocarbons are typically present in contaminated seawater feed as it can be safely assumed that organics of higher molecular weight will have already been dealt by passage through the RO pretreatment processes.

Diesel (a likely constituent arising from spillages) and hexane (chosen as a model low-molecular-weight hydrocarbon) were found to be present in oil spills, and the presence of these contaminants in both water-soluble and emulsion form can impact RO performance. Brackish-water membranes and seawater membranes of different structures were tested in saline water under different environmental conditions such as salinity and pressure of feedwaters and duration.

The performance of the RO unit, in terms of salt passage and permeate flux through the membranes, was assessed before and after fouling. It was determined that the polyamide-based membranes exhibit a decrease in permeate flux and an increase in salt passage after a short exposure to the foulants and a catastrophic drop in performance if exposure was prolonged. Although EDR membranes are tolerant to many different chemical conditions, continued exposure to solvents would affect the membrane process (including membranes, spacers, gaskets, and electrode tape). However, EDR would be more tolerant to short-term exposure than RO because of the fact that the anion and cation membranes are hydrophobic, unlike many RO membrane formulations that are hydrophilic.

Scaling Control. Scaling control is essential in RO/NF membrane processes. Control of precipitation or scaling within the membrane element involves identifying the limiting salt and determining an acid and/or inhibitor treatment that will prevent precipitation at the desired recovery.

Limiting Salt

The required dosage of scale inhibitor or acid is determined by the limiting salt. A diffusion-controlled membrane process naturally concentrates salts on the feed side of the membrane. If excessive water is passed through the membrane, this concentration process continues until a salt precipitates and scaling occurs. Scaling reduces membrane productivity and limits recovery to the allowable recovery just before the limiting salt precipitates.

The limiting salt can be determined from the solubility products of potential limiting salts and the actual feed stream water quality, as shown Eq. (11-67).

$$A_nB_m \leftrightarrow nA^{+J} + mB^{-K}$$
$$K_{s0} = (A^{+J})^n(B^{-K})^m \tag{11-67}$$
$$K_c = \left(a\frac{A^{+J}}{x}\right)^n\left(b\frac{B^{-K}}{x}\right)^m$$

where a is the fraction of cation retained and b the fraction of anion retained.

Ionic strength must also be considered in these calculations, because the natural concentration of the feed stream during the membrane process increases the ionic strength which affects solubility and recovery. Calcium carbonate scaling is commonly controlled by sulfuric acid addition; however, sulfate salts are often the limiting salt. Commercially available scale inhibitors can prevent scaling by complexing metal ions and preventing precipitation. Specific scale inhibitor formulas are proprietary, but phosphates, silicates, and other materials commonly used for corrosion control also help with scaling control. Scale inhibitors are not toxic, but they can add significant nutrients for biogrowth in the feed stream and in the concentrate stream after discharge. Equilibrium constants for these scale inhibitors are not available, preventing direct calculations. However, some manufacturers provide computer programs for estimating the required scale inhibitor doses for given recovery, water quality, and membrane parameters. General formulas for the ionic strength and activity coefficient approximations are given in Eqs. 11-68 and 11-69; see Chap. 3 for additional coverage of ionic strength.

$$I = 0.5\sum C_iZ_i^2 \approx (2.5\times10^{-5})(\text{TDS}) \tag{11-68}$$

where I is the ionic strength, C in moles/L, Z the ion charge, and TDS in mg/L.

$$\log\gamma \approx -0.5Z^2\frac{\sqrt{I}}{1+\sqrt{I}} \tag{11-69}$$

where γ is the activity coefficient.

Example 11-7 Limiting Salt

A limiting salt calculation is provided in Table 11-10. Rejection of ions varies by operating conditions, membrane type, molecular weight, and charge. The solubility product has been modified by a and b coefficients, as shown in Eq. 11-67, for consideration of less-than-complete rejection. A 0.90 rejection is assumed for divalent ions. A more exact estimate of the fraction rejected can be obtained from the manufacturer's literature or pilot studies. A raw-water quality is given with the solubility products of limiting salts at 25°C. The limiting salt is identified by determining which salt allows the least recovery or minimizes $1 - x$. The solubility levels of all possible salts are considered here. The first calculations demonstrate supersaturation of both $CaCO_3$ and $BaSO_4$. This conclusion is suggested by the negative decimal fraction for *r;* a positive *r* would indicate the fraction of the feed stream that could be recovered. Although recovery seems impossible, when ionic strength is considered, the allowable recovery rises to 60 percent. $CaCO_3$ precipitation will be controlled by acid addition.

Solution Using Eq. 11-67,

for $$CaCO_3 \rightarrow Ca^{2+} + CO_3^{2-} \qquad K_{s0} = [Ca^{2+}][CO_3^{2-}] = 10^{-8.3}$$

if $$Ca^{2+} = 8\ \text{mg/L} = 0.0002\ \text{M}$$

$$CO_3^{2-} = HCO_3^- \frac{\alpha_2}{\alpha_1} = HCO_3^- \frac{K_2}{H^+} = 0.00005\ M = 3\ \text{mg/L}$$

then $$\left[\frac{0.0002}{X}\right]\left[\frac{0.00005}{X}\right] = 10^{-8.3}$$

$$X = 1.41\ \text{L}$$

$$r = 1 - X = -0.41\ \text{L}$$

Using Eq. 11-67,

for $$BaSO_4 \rightarrow Ba^{2+} + SO_4^{2-} \quad K_{s0} = [Ba^{2+}][SO_4^{2-}] = 10^{-9.7}$$

if $$Ba^{2+} = 0.04\ \text{mg/L} = 3\times10^{-7} M$$

$$SO_4^{2-} = 79\ \text{mg/L} = 0.0008\ M$$

then $$\left(\frac{3\times10^{-7}}{X}\right)\left(\frac{0.0008}{X}\right) = 10^{-9.7}$$

$$X = 1.09\ \text{L}$$

$$r = 1 - X = -0.09\ \text{L}$$

The following calculations demonstrate the effect of ionic strength on recovery for the $BaSO_4$ example, approximated for brevity. Recovery increases because the increasing TDS boosts the ionic strength, which increases the solubility of the limiting salt at equilibrium. Iteration using the adjusted solubility product allows convergence of the recovery calculation, as illustrated in the example. The recovery is observed to increase from 0 to 72 percent when ionic strength is used following iteration.

$$\text{TDS} = \left(\frac{0.8}{0.4}\right)2200 = 4400\ \text{mg/L}$$

Using Eq. 11-68,

$$I = (2.5\times10^{-5})\text{TDS} = 0.11$$

Using Eq. (11-69),

$$\log\gamma = -0.5(2)^2 \frac{\sqrt{0.11}}{1+\sqrt{0.11}} \log\gamma = pk_{s0} - (m)p\gamma - (n)p\gamma = 9.7 - (1)0.50 - (1)0.50 = 87$$

Recalculate x from Eq. (11-67).

$$K_c = \left[a\frac{(A^{+m})}{X}\right]n\left[b\frac{(B^{-n})}{X}\right]m = \left[(0.9)\frac{3\times10^{-7}}{X}\right]\left[(0.9)\frac{0.0008}{X}\right] = 10^{-8.7}$$

$$X = 0.31\ \text{L}$$

$$r = 1 - X = 0.69\ \text{L} \quad \text{or} \quad 69\%$$

Iterate calculations from the beginning but use 0.31 for X in place of 0.4 to determine new TDS concentration in concentrate stream. After two iterations the recovery converges at 72 percent. ▲

TABLE 11-10 Limiting Salt Example

Raw water quality used for example 11-7				Salts to consider	
Parameter	Conc. (mg/L)	Parameter	Conc. (mg/L)	Salt	pK_{s0}@ 25°C
pH	8.0	Cl^-	730	$CaCO_3$	8.3
Na^+	695	NO^{3-}	0	$Ca_3(PO)_4$	6.8
K^+	8	F^-	1.1	CaF_2	10.3
Ca^{+2}	8	SO_4^{-2}	79	$CaSO_4$	4.7
Mg^{+2}	2	$O\text{-}PO_4$	0.7	$BaSO_4$	11.7
Fe^{+2}	0.5	HCO_3^-	631	$SrSO_4$	6.2
Mn^{+2}	0.02	SiO_2	24	SiO_2	2.7
Cu^{+2}	0	TDS	2200		
Ba^{+2}	0.04				

Example 11-8 Acid Addition
Acid is usually added to allow $CaCO_3$ recovery to the limiting salt recovery, as illustrated in the example. The equilibrium constants for the carbonate system, alkalinity, pH, calcium, and recovery are required to make these calculations. In the example, pH 7.18 was required to avoid scaling by $CaCO_3$. Addition of 162 mg/L H_2SO_4 produces this pH. The additional sulfate from acid addition should be considered for calculating the limiting salt recovery.

Solution

1. Determine the pH required on the feed side of membrane.

Given

$$Ca^{2+} = 8 \text{ mg/L} \qquad HCO_3^- = 631 \text{ mg/L} \qquad pH = 8.0$$

$$R = 75\% \qquad I = 0.12 \text{ M}$$

For $$CaCO_3 \rightarrow Ca^{2+} + CO_3^{2-}$$

$$K_{s0} = [Ca^{2+}][CO_3^{2-}] = 10^{-8.3}$$

$$\log\gamma = -0.5(1)^2 \frac{\sqrt{0.12}}{1+\sqrt{0.12}} = -0.13$$

$$p\gamma_{+f-1} = 0.13$$
$$p\gamma_{+f-2} = 0.51$$

$$pK_c = pK_{s0} - (m+n)p\gamma = 8.3 - 1(0.51) - 1(0.51) = 7.31$$

$$K_c = \left(a\frac{(A^{+m})}{x}\right)n\left(b\frac{(B^{-n})}{x}\right)m = \left((0.9)\frac{[Ca^{2+}]}{x}\right)\left((0.9)\frac{[CO_3^{2-}]}{x}\right) = 10^{-7.31}$$

$$Ca^{2+} = \frac{8}{40000} = 0.0002\,M$$

For $$HCO_3^- \rightarrow H^+ + CO_3^{2-}$$

$$K_2 = \frac{[H^+][CO_3^{2-}]}{[HCO_3^-]} = 10^{-10.3}$$

$$K_2 = \frac{[H^+][\gamma_2 CO_3^{2-}]}{[\gamma_1 HCO_3^-]} \; pK_{c2} = pK_2 + p\gamma_1 - p\gamma_2 = 10.3 + 0.13 - 0.51 = 9.92$$

$$CO_3^{2-} = \frac{[K_{c2}][HCO_3^-]}{[H^+]} = \frac{[10^{-9.92}][631/61000]}{[H^+]} = \frac{10^{-11.9}}{[H^+]}$$

Going back to the preceding equation for K_c,

$$K_c = \left(a\frac{(A^{+m})}{x}\right)n\left(b\frac{(B^{-n})}{x}\right)m = \left((0.9)\frac{[Ca^{2+}]}{x}\right)\left((0.9)\frac{[CO_3^{2-}]}{x}\right) = 10^{-7.31}$$

$$K_c = \left((0.9)\frac{[Ca^{2+}]}{0.25}\right)\left((0.9)\frac{[CO_3^{2-}]}{0.25}\right)$$

$$K_c = \left((0.9)\frac{[0.0002]}{0.25}\right)\left((0.9)\frac{\left[\frac{10^{-11.9}}{[H^+]}\right]}{0.25}\right) = \left(\frac{0.9}{0.25}\right)^2 (0.0002)\frac{10^{-11.9}}{[H^+]} = 10^{-7.31}$$

$$[H^+] = 10^{-7.18}$$
$$pH = 7.18$$

The calculations determined the pH that will not scale calcium carbonate at 75 percent recovery. The example assumed 90 percent rejection of calcium and carbonate ions. This calculation is approximate, but it does illustrate the effect of ionic strength and the control of calcium carbonate solubility by adjusting pH. Control of calcium carbonate solubility is required in almost every RO/NF plant that treats groundwater. Once the required pH has been determined for calcium carbonate scaling, the required acid dose can be calculated.

2. Determine the dose of H_2SO_4 required to achieve pH 7.18.

$$K_1 = \frac{[H^+][\gamma HCO_3^-]}{[H_2CO_3]} = 10^{-6.3}$$

$$pK_{c1} = pK_1 - p\gamma$$

$$\log\gamma = -0.5(1)^2\frac{\sqrt{0.12}}{1+\sqrt{0.12}} = -0.13$$

$$p\gamma = 0.13$$

$$pK_{c1} = pK_1 - p\gamma = 6.3 - 0.13 = 6.17$$

$$K_{c1} = \frac{[H^+][HCO_3^-]}{[H_2CO_3]} = 10^{-6.17} = \frac{[10^{-7.18}][(0.9/0.25)(631/61000) - X]}{[0+X]}$$

where X is the H^+ needed to react with HCO_3^-.

$$10^{-6.17}X = 10^{-7.18}X + 10^{-8.61}$$

solve for X and convert to an equivalent amount of sulfuric acid.

$$X = 0.0033\frac{M}{L}H^+ = \frac{(0.0033\times 98000)}{2} = 162 \text{ mg/L } H_2SO_4$$

This is the H_2SO_4 that needs to be added in the last element of the last element of the last stage. Consequently, this concentration is reduced by (0.25)(162) = 40.5 mg/L to determine feed stream H_2SO_4 addition. ▲

Scale Inhibitor Manufacturers' Programs. Scale inhibitors can also be used to determine allowable recovery. Manufacturers provide computer programs to help with estimating recovery. Most scale inhibitors work by complexing or dispersing metal cations, making them unavailable for precipitation. Polyacrylic acid (PAA), one common scale inhibitor, interferes with nucleation and crystal growth. Unfortunately, the equilibrium constants are not available, and the predicted recoveries cannot be checked. However, scale inhibitors certainly work, as they are used worldwide in membrane plants to increase recovery. Many membrane plants combine sulfuric acid and scale inhibitor pretreatment. For example, even though $BaSO_4$ can be 608 percent oversaturated in a feedwater, a computer evaluation will render a recommended inhibitor dose of 4.2 mg/L to control scaling. These manufacturer programs can be used to predict the saturation percentages of selected salts and the doses required to control scaling, but they do not give the exact requirements for a particular membrane. That judgment reflects a design choice, as all scaling could possibly be controlled by scale inhibitor addition, and often requires site-specific investigations to confirm design and operation criteria. Scale inhibitors are typically more costly than acid based on mass, but they cost less than acid in use. Scale inhibitors commonly lose effectiveness if they react preferentially with metals other than the limiting salt metals. High iron concentrations sometimes greatly decrease the effectiveness of certain scale inhibitors, and dispersant formulations are then often recommended. These factors and others point out the need for pilot studies.

Static Microfiltration. Static microfilters typically used for RO/NF pretreatment are sieving filters made of polypropylene with pore diameters of 5 to 20 μm. The pretreatment microfilters have larger pores than the continuous MF membranes. The pressure drop through a microfilter does not usually exceed 34.5 kPa (5 psi) in nanofiltration applications and 68.9 kPa (10 psi) in reverse osmosis applications. Microfiltration upstream of a nanofiltration or reverse osmosis process protects only against foulants or materials in the solid phase. The technology offers no protection against scaling. Also, microfiltration alone cannot remove foulants from a feed stream with a high turbidity or suspended solids concentration. Generally, the size distribution of the foulants removed in this way includes particles with diameters smaller than the microfilter pore sizes. Consequently, a maximum value for the feed-stream fouling indexes is specified prior to microfiltration. A microfilter in a reverse osmosis or nanofiltration process should be thought of as a means of protecting the membrane elements against periodic upsets from failure of solids removal processes or sand entrainment from well pumping.

Array Models

This section describes different models and design techniques used to size arrays of membrane elements. A linear solution diffusion model, a film theory solution diffusion model, and a coupling model are described. Single and multistaged array equations for predicting permeate water quality are developed, and a simple design example is presented.

Linear Solution Diffusion Model. Many different theories and models attempt to describe mass transfer in diffusion-controlled membrane processes. However, a few basic principles or theories are used to develop most of these models: convection, diffusion, film theory, and electroneutrality. These principles or theories could be used to group models

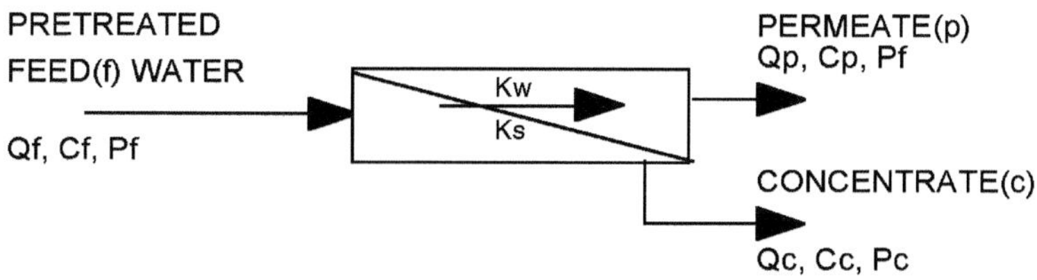

FIGURE 11-30 Basic diagram of mass transport in a membrane.

into linear diffusion models, exponential diffusion models, and coupling models. Most of the modeling efforts have been developed using very small test cells. Also, they have not incorporated product recovery, limiting their practical use. The basic equations used to develop these models are shown in Eqs. 11-70 through 11-74. The standard membrane element configuration is shown in Fig. 11-30.

$$J = k_W(\Delta P - \Delta \Pi) = \frac{Q_p}{A} \tag{11-70}$$

$$J_i = k_i\ \Delta C = \frac{Q_p C_p}{A} \tag{11-71}$$

$$r = \frac{Q_p}{Q_f} \tag{11-72}$$

$$Q_f = Q_c + Q_p \tag{11-73}$$

$$Q_f C_f = Q_c C_c + Q_p C_p \tag{11-74}$$

where J = water flux (L^3/L^2t)
J_i = solute flux (M/L^2t)
k_w = solvent mass transfer coefficient (L^2t/M)
k_i = solute mass transfer coefficient (L/t)
r = recovery
Q_f, Q_p = feed flow rate, permeate flow rate (L/t) (respectively)
C_f, C_p, and C_c = feed concentration, permeate concentration and concentrate concentration (M) (respectively)

If ΔC is defined as the difference of the average feed and brine stream concentrations and the permeate stream concentration, then Eq. 11-75 can be derived from Eqs. 11-70 and 11-74. This model can be described as a linear homogenous solution-diffusion model in that it predicts diffusion-controlled solute flow and pressure-(convection-)controlled solvent flow. Eq. 11-75 can be simplified by including a Z term that incorporates the effects of the mass transfer coefficients, pressure, and recovery into a single term.

$$C_p = \frac{k_i C_f}{k_w(\Delta P - \Delta \Pi)\left(\frac{2-2r}{2-r}\right) + k_i} = Z_i C_f \tag{11-75}$$

where Z is the combined mass transfer term and r the recovery.

This is the simplest model, but it does incorporate the effects of five independent variables on permeate water quality, as shown in Fig. 11-31. If pressure increases and

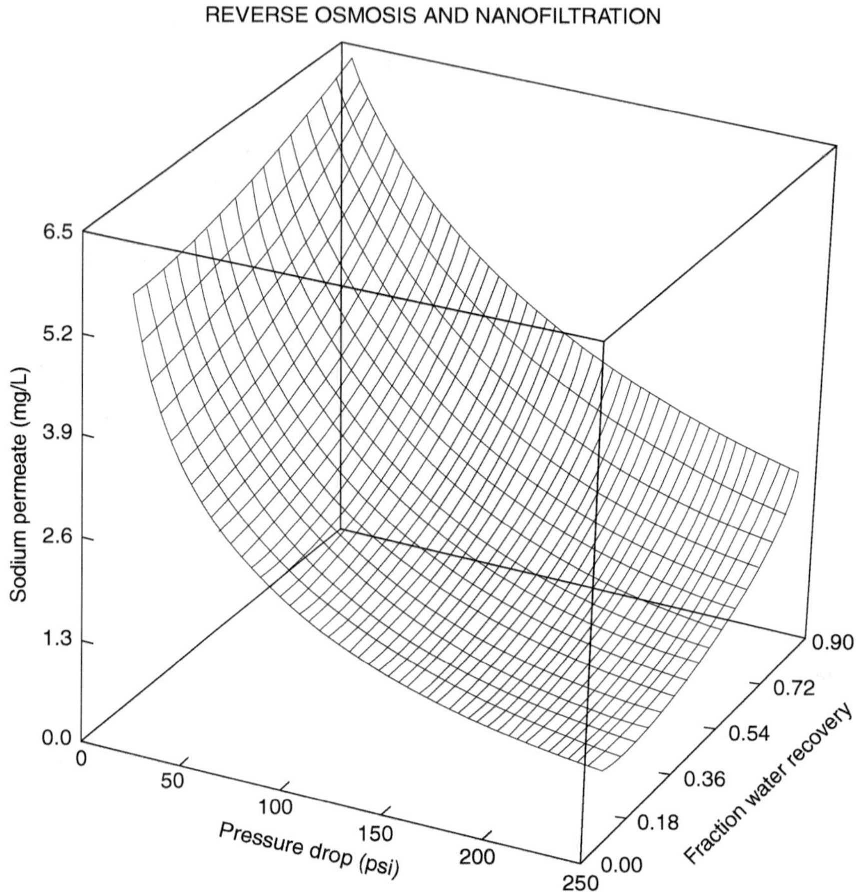

FIGURE 11-31 Sodium permeate concentration for varying pressure and recovery. (*Source: Duranceau, 1990.*)

all other variables are held constant, then the permeate concentration will decrease. If recovery is increased and all other variables are held constant, then the permeate concentration increases. These effects may be hard to realize if an existing membrane array is considered. It is impossible in such an environment to increase pressure without increasing recovery. However, array designs can increase pressure without varying recovery. Decreasing feed stream concentration may identify pretreatment as an option for decreasing the permeate stream concentration. Different membranes may have different mass transfer characteristics. Using a membrane with a lower molecular weight cutoff would probably decrease the permeate concentration, although the solvent and solute MTCs must be considered before such a result can be expected.

Film Theory Model. The linear model for array design can be modified by the incorporation of film theory, which assumes that the solute concentration exponentially increases from the center of the feed stream channel toward the surface of the membrane and diffuses back into the bulk stream. Mathematically, this is shown in Eq. 11-76, and the back diffusion constant is introduced in Eq. 11-77.

$$J_i = -D_s \frac{dC}{dx} + C_i J \tag{11-76}$$

$$\left(\frac{C_s - C_p}{C_b - C_p}\right) = \exp\left(\frac{J}{k_b}\right) \tag{11-77}$$

where x is the path length or film thickness, $k_b = D_s/x$ is the diffusion coefficient from the surface to the bulk.

Incorporation of Eq. 11-77 into Eqs. 11-70 through 11-74 results in Eq. 11-78, which is a development of the homogenous solution diffusion model using concentration-polarization. This model predicts a higher concentration at the membrane surface than in the bulk of the feed stream. Such an effect is documented in the literature (Sung, 1993; Hofman, 1995), and the model in Eq. 11-78 accounts for this phenomenon. The back-diffusion coefficient k_b represents solute diffusion from the membrane surface to bulk in the feed stream which is different than the mass transfer coefficient k_w, which represents solute diffusion through the membrane to the permeate stream.

$$C_p = \frac{C_f k_i e_b^{j/k}}{k_W(\Delta P - \Delta\Pi)\left(\frac{2-2r}{2-r}\right) + k_i \exp\left(\frac{J}{k_b}\right)} \tag{11-78}$$

Solutions are electrically neutral according to the principles laid out in Chap. 3. Therefore, solutes do not pass through a membrane in a charged state but in a coupled state. Such electroneutrality is observed in diffusion-controlled membrane processes when the permeate streams become more concentrated than the feed streams. Relatively high concentrations of sulfates have been observed to increase rejection of calcium and decrease rejection of chlorides. One interpretation of such observations suggests that the strong calcium sulfate couple is retained in the feed stream and the weaker sodium chloride couple is forced to pass through the membrane to maintain equilibrium.

Coupling. The coupling effect on mass transfer through membranes has been modeled using a statistical modification of free energy for single-solute systems as shown in Eq. 11-79 (Rangarajan, 1976; Sung, 1993). The free-energy term is assumed to be different in the bulk solution and at the membrane surface because of a difference of ion concentration. Consequently, the energy required to bring the ions to the surface, shown in Eq. 11-80, is the difference of the free energy in bulk and surface solutions. The $\Delta\Delta G$ value for each ion is determined by experiment.

$$\frac{1}{\Delta G} = -\frac{1}{E}r - \frac{\Delta}{E} \tag{11-79}$$

$$\Delta\Delta G = \Delta G_I - \Delta G_B \tag{11-80}$$

where ΔG is the free energy of coupled ion, E a solution-dependent constant, a a coupled ion radius, Δ a statistical constant, and $\Delta\Delta G$ a difference of coupled ion free energy at interface (I) and bulk (B).

Membrane-specific solute mass transfer coefficients for single-solute systems have been determined by experiment. They are related to free energy as indicated by Eq. 11-81. Once the membrane specific constant $\ln(C^*)$ has been determined for a reference solute (e.g., NaCl), it is possible to determine k_s for any given solute in a single-solute system.

Once k_s is known, the mass transport of any single solute in a diffusion-controlled membrane can be predicted.

$$\ln(k_i) = \ln(C^*) + \sum\left(-\frac{\Delta\Delta G}{RT}\right) \qquad (11\text{-}81)$$

where C^* is the membrane specific constant.

Coupling can be used to model mass transport in a multisolute system. As with the single-solute system, electroneutrality is assumed in all phases (bulk, surface, permeate, film). A coupling model can incorporate either a homogenous or an exponential solution-diffusion model with the free-energy model. The model requires the mass transfer coefficients for water and a reference couple, the back-transfer coefficient and the $\Delta\Delta G$ values for all ions to predict flux, and the concentration of the permeate stream. The mass transfer models shown in Eqs. 11-39 and 11-40 can both be used as coupling models as long as the k_i term is determined using $\Delta\Delta G$ or feed stream solute composition. However, this term is membrane-specific and must be determined for a given product. No model accounts for the effects of the membrane surface on a fundamental basis.

The determination of $\Delta\Delta G$ is based on free energy and is therefore affected by feed stream composition. The determination of free energy for any given reaction is a good method of determining the likelihood that the reaction will proceed, but this method relies on consideration of all the free energies for the major chemical interactions. As the solutes and free energies are not known for many of the constituents in natural waters, a basic free-energy model has not been developed for practical membrane design.

Array Modeling. The design process for membrane element arrays will be illustrated using the linear model for simplicity and to take advantage of a membrane manufacturer's computer program. The mass transfer coefficients for the linear model can be developed from field criteria, and the model can easily be used to predict permeate water quality for given changes in operation. Variations in sodium permeate concentration are shown in Fig. 11-31 for varying pressures and recoveries. Reverse osmosis and nanofiltration processes allow considerable flexibility in permeate water quality. Designers can vary flux and recovery to achieve water quality goals. Anyone involved in the design or operation of a membrane facility should be aware of the coupling phenomena and concentration-polarization. The mass transfer coefficients for the film theory model are more difficult to develop, but they have been shown to increase accuracy of pilot plant data. However, the best design criteria would come from site-specific pilot studies. Pilot studies may not be required for systems to desalinate some brackish waters with low organic concentration and no potential foulants, but many physical, chemical, and biological factors not initially recognized can cause serious operational problems. The only way to avoid these types of unrealized problems in design is to conduct a pilot study.

Modeling a Linear Array. Equations 11-75 and 11-78 are useful tools for determining the effect of pressure gradient (ΔP), feed concentrate (C_f), recovery (r), osmostic pressure ($\Delta\Pi$), k_w, and k_i on permeate stream solute concentration (C_p) without experimentation if the six variables are known. The first four can be accurately assumed by the user. However, k_w and k_i should be measured experimentally for a given membrane and source; they can be taken from the literature if reports describe a given membrane and a similar source. In Eq. 11-82, Z_i states a modified MTC, which represents the effect of solvent mass transfer, solute mass transfer, recovery, and pressure on permeate stream concentration for any single membrane element. In this way, the expression simplifies Eq. 11-75. If Eq. 11-78 were used, the modified MTC would be represented by Z_{cpi}, which would include consideration of back-diffusion.

$$C_{p,i} = \frac{k_{i,i} C_{f,i}}{k_{W,i}\,\Delta P_i\left(\frac{2-2r_i}{2-r_i}\right) + k_{i,i}} = Z_i C_{f,i} \qquad (11\text{-}82)$$

where Z_i is the modified mass transfer coefficient, and $\Delta P_i = \Delta P - \Delta\Pi$ = net average driving pressure including hydraulic losses.

The subscript i in Eq. 11-82 denotes any stage in a multistage membrane array. Membrane systems are configured in arrays that consist of stages. Consequently, Eq. 11-82 can be expanded to describe an entire membrane system. The configuration for a typical two-stage membrane system is shown in Fig. 11-32. The first stage consists of two pressure vessels, each of which typically contains six membrane elements. The second stage consists of one pressure vessel, also typically with six elements. The combination of these two stages is a 2–1 array. The model can be used to predict the system's permeate concentration or the effects of any of the six independent variables on permeate concentration if MTCs, pressures, and recoveries for each stage are known.

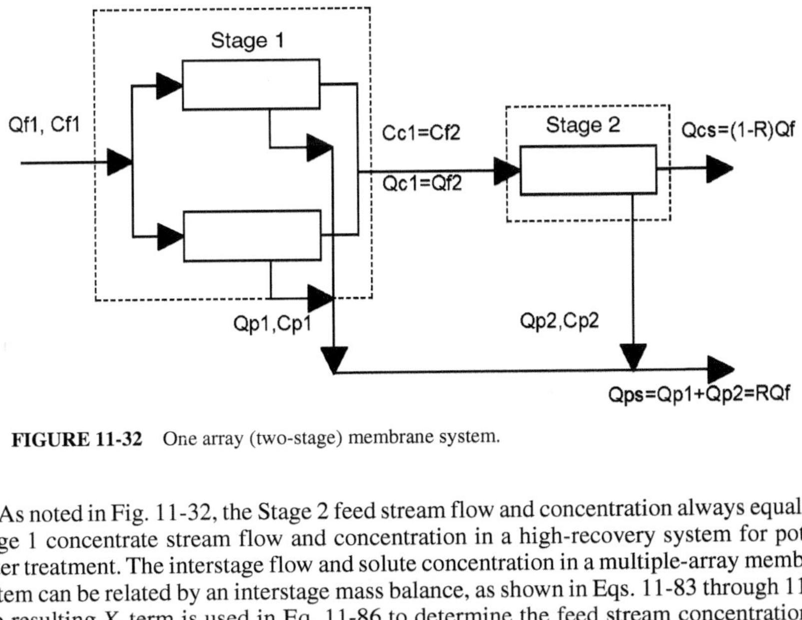

FIGURE 11-32 One array (two-stage) membrane system.

As noted in Fig. 11-32, the Stage 2 feed stream flow and concentration always equals the Stage 1 concentrate stream flow and concentration in a high-recovery system for potable water treatment. The interstage flow and solute concentration in a multiple-array membrane system can be related by an interstage mass balance, as shown in Eqs. 11-83 through 11-86. The resulting X_i term is used in Eq. 11-86 to determine the feed stream concentration for any stage from the initial feed stream concentration. It can be considered a concentration factor of Stage i.

$$C_{C,i} = \frac{Q_{f,i}C_{f,i} - Q_{p,i}C_{p,i}}{Q_{c,i}} = \frac{C_{f,i} - r_i C_{p,i}}{1 - r_i} \tag{11-83}$$

$$C_{c,i} = C_{f,i+1} = C_{f,i}\left(\frac{1 - Z_i r_i}{1 - r_i}\right) \tag{11-84}$$

$$C_{c,i} = X_i C_{f,i} \tag{11-85}$$

$$X_i = \frac{1 - Z_i r_i}{1 - r_i}; \quad \textit{Define } X_0 \equiv 1 \tag{11-86}$$

Consequently an interstage equation for permeate concentration can be developed by interrelating all stages in any diffusion-controlled membrane system. Equation 11-87 interrelates all stages in a simple, flow-weighted concentration equation that represents the total solute mass from the permeate stream in each stage, shown in the numerator of Equation 11-87, and the total volume of solvent from the permeate stream in each stage, shown in the denominator. Equation 11-88 is less complex than an expression incorporating the full Z_i and X_j terms, as shown in Eq. 11-87. This equation can be used to describe membrane system performance with as much versatility as Eq. 11-75 or Eq. 11-78, which describes a single stage or element. The subscripts j and i denote the stage number in a multistage membrane array as shown in Eq. 11-88.

$$C_{p,\text{system}} = \frac{C_f \sum_{i=1}^{n} \left[A_i k_{i,i} \Delta P_i \frac{k_{i,i}}{k_{W,i} \Delta P_i \left(\frac{2-2r_i}{2-r_i}\right) + k_{i,i}} \prod_{j=0}^{i-1} \frac{1 - \frac{k_{i,j} r_j}{k_{W,j} \Delta P_j \left(\frac{2-2r_j}{2-r_j}\right) + k_{i,j}}}{1-r_j} \right]}{\sum_{i=1}^{n} A_i k_{W,i} \Delta P_i} \tag{11-87}$$

$$C_{p,\text{system}} = \frac{C_f \sum_{i=1}^{n} \left(A_i k_{i,i} \Delta P_i Z_i \prod_{j=0}^{i-1} X_j \right)}{\sum_{i=1}^{n} A_i k_{i,i} \Delta P_i} \tag{11-88}$$

In Eq. 11-88, the term $\prod_{j=0}^{i-1} X_j$ is the product of the concentration factors of all stages before Stage i. The concentration factor is therefore mathematically defined as 1 before Stage 1 ($X_0 = 1$) (Duranceau, 1992).

The exponent J/k_b used in the film theory model, as shown in Eq. 11-77, can be approximated by a constant times the ratio of the flow through the membrane to the flow across the membrane, as shown in Eq. 11-89. This expression approximates the compensation for the effects of multisolute solutions in a high-recovery or plant application.

$$\frac{J}{k_b} = k \left(\frac{r_e}{1-r_e} \right) \tag{11-89}$$

where k_b is a statistically derived coefficient, and r_e the average membrane recovery per element.

The system permeate equation can also be modified to incorporate the film theory model, as shown in Eqs. 11-90 and 11-91.

$$C_p = \frac{k_i e^{\frac{J}{k_b}} C_f}{k_W (\Delta P - \Delta \Pi) \left(\frac{2-2r}{2-r} \right) + k_i e^{\frac{J}{k_b}}} = Z_{i,cp} C_f \tag{11-90}$$

$$C_{p,\text{system}} = \frac{C_f \sum_{i=1}^{n} \left(A_i k_{W,i} \Delta P_i Z_{i,cp} \prod_{j=0}^{i-1} X_{j,cp} \right)}{\sum_{i=1}^{n} A_i k_{W,i} \Delta P_i} \tag{11-91}$$

where $Z_{i,cp}$ is the modified film theory feedstream mass transfer coefficient, and $X_{j,cp}$ the modified film theory concentrate stream concentration factor.

Practical Array Design

Membranes are typically incorporated in elements, as mentioned early in the chapter. Unfortunately, the terms in Table 11-1 are not universally accepted, and meanings vary. For example, a membrane *array* can be defined as Stage 1, or as the total of Stages 1, 2, and 3. Care must be taken to ensure that parties discussing membrane systems are using the same terminology.

Example 11-9 Simple Sizing
This example assumes a simplistic 4-2-1 array to illustrate design concepts for membrane configurations. Basic assumptions in this design are 20 cm (8 in.) diameter membrane elements with 32.5 m² (350 ft²) surface areas operating at a flux of 0.6 m³/m²/day (15 gsfd) and a 75 percent recovery. One element produces 19.9 m³/d (5250 gpd). As seven elements are in each pressure vessel, and seven pressure vessels are in one array, an array will produce approximately 946.4 m³/d (250,000 gpd). Four such arrays provide 3785 m³/d (1 mgd) of permeate and 1249 m³/d (0.33 mgd) of concentrate. Note that osmotic pressure, not illustrated, would continuously decrease the recovery in each succeeding element.

Assumptions

1. Membrane element flux = 0.6 m³/m²/day (15 gal/ft²-day)
2. Design for a 4-2-1 array
3. Each pressure vessel can contain 6 to 8 elements
4. Element surface area = 32.5 m² (350 ft²)
5. Recovery = 75 percent

Solution Water 1 element can produce

$$Q_p = JA = \left(0.6\frac{\text{m}^3}{\text{m}^2\text{day}}\right)\left(32.5\frac{\text{m}^2}{\text{element}}\right) = 19.5\frac{\text{m}^3}{\text{day element}}\left(5150\frac{\text{gal}}{\text{day element}}\right)$$

Number of elements on one array

$$\left(7\frac{PV}{\text{array}}\right)\left(7\frac{\text{elements}}{PV}\right) = 49\frac{\text{elements}}{\text{array}}$$

Gallons per day produced per array

$$\left(19.5\frac{\text{m}^3}{\text{day element}}\right)\left(\frac{49\text{ elements}}{\text{array}}\right) = 955.5\frac{\text{m}^3}{\text{array day}}\left(252{,}400\frac{\text{gal}}{\text{array day}}\right)$$

Trains needed to supply 3786 m³/day (1 mgd)

$$\left(3786\frac{\text{m}^3}{\text{day}}\right)\left(\frac{\text{array day}}{955.5\text{ m}^3}\right) = 3.96\text{ arrays}$$

Total number of elements needed

$$(4\text{ arrays})\left(49\frac{\text{elements}}{\text{array}}\right) = 196\text{ elements}$$

Flow in and out of one array

$$Q_f = \frac{Q_p}{r} = \frac{3786\ \text{m}^3/\text{day}}{0.75} = 5048\ \text{m}^3/\text{day}$$

Flow in per array

$$\frac{5048\ \text{m}^3/\text{day}}{4} = 0.88\ \text{m}^3/\text{min}$$

Flow our per array

$$\frac{3786\ \text{m}^3/\text{day}}{4} = 0.66\ \text{m}^3/\text{min} \quad \blacktriangle$$

The film theory model can be used to estimate permeate water quality. In the following example, MTCs were determined from a pilot study or given by the membrane manufacturer. The concentration-polarization factor is also given. The example calculation shows the chloride concentrate in the permeate stream from each stage and the array (Hofman et al., 1993).

Example 11-10 Predicting RO Permeate Concentration
The film theory model can be used to estimate permeate water quality. In this example, mass transfer coefficients were determined from a pilot study or given by the membrane manufacturer. The concentration-polarization factor is also given. The example calculation shows the chloride concentrate in the permeate stream from each stage and the array (Hofman et al., 1993).

This example assumes a simplistic 2-1 array to illustrate design concepts for membrane configurations. Basic assumptions in this design are six elements per pressure vessel with an element surface area (SA) of 30 m^2 (323 ft^2). There are five bars of driving force. The feed concentration C_f is 100 mg/L. For this particular system, pilot studies have shown that the mass transfer coefficient of water k_w is equal to 1.46×10^{-6} m/s-bar (0.2135 gsfd/psi), and the mass transfer coefficient of the solute k_i is equal to 1.87×10^{-6} m/s (0.53 ft/day). The concentration-polarization factor or mass transport coefficient k is equal to 0.75. The recovery of the system r is 0.5, and the average membrane recovery per element r_e is equal to 0.083 per element.

Given

1. 2-1 array with six elements/vessel
2. Element SA = 30 m^2 (323 ft^2)
3. P = five bars
4. $k_w = 1.46 \times 10^{-6}$ m/s-bar (0.2135 gsfd/psi)
5. $k_i = 1.87 \times 10^{-6}$ m/s (0.53 ft/day)
6. $k = 0.75$
7. $r = 0.5$
8. $r_e = 0.083$/element

Solution Using the array film theory model, Eq. 11-91,

$$C_{p,\text{system}} = \frac{C_f \sum_{i=1}^{n} \left(A_i k_{W,i} \Delta P_i Z_{i,cp} \prod_{j=0}^{i-1} X_{j,cp} \right)}{\sum_{i=1}^{n} A_i k_{W,i} \Delta P_i}$$

First calculate the stage permeate flow given as $A_i k_{w,i}\, \Delta P_i$.

$$\sum_{i=1}^{2} A_i k_{W,i} \Delta P_i = (\text{Stage 1})\left(\frac{2\ \text{vessels}}{\text{Stage 1}}\right)\left(\frac{6\ \text{elements}}{\text{vessel}}\right)\left(\frac{30\ \text{m}^2}{\text{element}}\right)\left(\frac{1.46\times10^{-6}\,\text{m}}{s\text{-bar}}\right)(5\ \text{bar})$$

$$+(\text{Stage 2})\left(\frac{1\ \text{vessels}}{\text{Stage 2}}\right)\left(\frac{6\ \text{elements}}{\text{vessel}}\right)\left(\frac{30\ \text{m}^2}{\text{element}}\right)\left(\frac{1.46\times10^{-6}\,\text{m}}{s\text{-bar}}\right)(5\ \text{bar})$$

$$=2.628\times10^{-3}\frac{\text{m}^3}{\text{s}}+1.314\times10^{-3}\frac{\text{m}^3}{\text{s}}=3.942\times10^{-3}\frac{\text{m}^3}{\text{s}}\ (89999\ \text{gal/day})$$

Next, calculate the modified film theory feed stream mass transfer coefficient, $Z_{i,cp}$, using Eq. 11-90,

$$C_p = \frac{k_i e^{\frac{J}{k_b}} C_f}{k_W(\Delta P - \Delta\Pi)\left(\dfrac{2-2r}{2-r}\right) + k_i e^{\frac{J}{k_b}}} = Z_{i,cp} C_f$$

According to Eq. 11-89,

$$\frac{J}{k_b} = k\left(\frac{r_e}{1-r_e}\right)$$

Assume $(\Delta P - \Delta\Pi)$ is equal to the given pressure (five bars).

Plugging in Eq. 11-89 into Eq. 11-90 and solving for $Z_{i,cp}$, you obtain

$$Z_{i,cp} = \frac{k_{S,i} e^{k\left(\frac{r_e}{1-r_e}\right)}}{k_{W,i}\Delta P_i\left(\dfrac{2-2r_i}{2-r_i}\right) + e^{k\left(\frac{r_e}{1-r_e}\right)} k_{S,i}} = Z_1 = Z_2$$

$$Z_1 = Z_2 = \left(\frac{\left(1.87\times10^{-6}\,\frac{\text{m}}{\text{s}}\right) e^{0.75\left(\frac{0.083}{1-0.083}\right)}}{\left(1.46\times10^{-6}\,\frac{\text{m}}{\text{s bar}}\right)(5\,\text{bar})\left(\dfrac{2-2(0.5)}{2-0.5}\right) + \left(1.87\times10^{-6}\,\frac{\text{m}}{\text{s}}\right) e^{0.75\left(\frac{0.083}{1-0.083}\right)}}\right) = 0.29$$

Equation 11-86 can be used to solve for the concentration factor $X_{j,cp}$.

$$\prod_{j=0}^{1} X_j = (1)\left(\frac{1-Z_1 r}{1-r}\right) = (1)\left(\frac{1-(0.29)(0.50)}{1-0.5}\right) = 1.71$$

Plug in the values obtained for stage permeate flow, modified film theory feed stream mass transfer coefficient, and concentration factor, as well as the feed stream concentration into Eq. 11-91.

$$C_{p,\text{system}} = \frac{100\,\frac{\text{mg}}{\text{L}}\left(2.628\times10^{-3}\,\frac{\text{m}^3}{\text{s}}\right)(0.29) + 100\,\frac{\text{mg}}{\text{L}}\left(1.314\times10^{-3}\,\frac{\text{m}^3}{\text{s}}\right)(0.29)(1.71)}{3.942\times10^{-3}\,\frac{\text{m}^3}{\text{s}}}$$

$$= 19.33\frac{\text{mg}}{\text{L}} + 16.53\frac{\text{mg}}{\text{L}} = 35.86\frac{\text{mg}}{\text{L}}\quad \blacktriangle$$

Some comments will clarify this model application. The 5-bar feed stream pressure gradient through each stage has been assumed to include the effects of osmotic pressure and hydraulic losses. These effects can be estimated by assuming that 1 mg/L TDS produces 0.070 kPa (0.01 psi) of osmotic pressure, a 20.0 kPa (2.9 psi) pressure loss across each element, and a 34 kPa (4.93 psi) stage entrance loss. The hydraulic effects on pressure gradient are accounted for during design and typically controlled by high-pressure pumps and flow restrictors. The first stage and second stage Z terms are identical, because stage recovery, pressure gradient, and membranes are identical. The Z_1 term is used to calculate the X_2 term, as the Stage 1 concentrate stream is the Stage 2 feed stream. The denominator of Eq. 11-91 is the permeate flow produced by the two-stage system and is shown as the first calculation in the example application. The final calculation gives the system's permeate chloride concentration, estimated at 35.8 mg/L. The Stage 1 and Stage 2 chloride concentrations are 29.0 and 49.6 mg/L, respectively shown in the calculation of the flow-weighted system concentration. Since recovery, pressure, and membranes are identical from Stage 1 to Stage 2, the second stage permeate stream chloride concentration will always be higher than that in the first stage, because the feed stream chloride concentration is higher in Stage 2 (171 mg/L) than in Stage 1 (100 mg/L).

Membrane Manufacturers' Programs. Manufacturers such as Hydranautics, Toray, Koch (Fluid Systems), Dow-Filmtec, GE-Osmonics, and GE-Ionics provide computer programs for developing RO plant design criteria. These programs do not yield final design specifications, and no manufacturer accepts responsibility for their use by anyone not specifically authorized. These programs are only tools for developing and testing various system configurations, and their output should not be regarded as completed designs. The software can be used to project the performance of a specific membrane over a period of time, and are often used to guarantee performance warrantees. The programs do provide a means of estimating water production and quality for given parameters. Feed flow rate, TDS concentration, feed temperature, recovery, array type (i.e., 4-2-1 array), number of elements per pressure vessel, and element type are typically required program inputs. Other process inputs could include permeate back-pressure or interstage booster pumping which will affect system design and performance. The program outputs typically include pressure, flux, flow, and water quality for all process streams by stage and element for each array. Scaling calculations are provided for calcium sulfate.

Posttreatment

The Need for Posttreatment. Posttreatment consists of several different unit operations for RO and NF membrane systems and are summarized in Table 11-11. A recent overview of the current state of 62 full-scale RO/NF plants conducted for plants greater than 3785 m^3/day (1.0 MGD) of capacity, used for either seawater desalination, brackish-water desalination (including groundwater, surface water, and agricultural runoff), or wastewater reclamation provides an insight into posttreatment practices (Burbano 2007). All of the surveyed facilities reported using at least one posttreatment method for permeate conditioning and corrosion control. These included such methods as caustic addition (31 percent), blending with raw, semitreated, or finished water (29 percent), degasification–decarbonation (25 percent), and addition of corrosion inhibitor (14 percent). Most of the brackish-water RO plants responding to the survey reported using degasification-decarbonation and caustic addition, with the majority blending permeate with groundwater. Permeate disinfection was reported to be used by 85 percent of the surveyed facilities that responded, most of which used chlorine. Other reported disinfection methods included the use of chloramine (24 percent) and ultraviolet irradiation (4 percent).

TABLE 11-11 Typical Posttreatment Processes Based on Source Water

Supply type	Membrane process type(s)	Examples of applicable posttreatment processes
Seawater	RO	Recarbonation Lime addition Calcite bed filtration pH and/or alkalinity adjustment Addition of corrosion inhibitors Primary and secondary disinfection Blending with freshwater supplies
Brackish water (surface)	RO, NF, EDR	pH and/or alkalinity adjustment Addition of corrosion inhibitors Primary and secondary disinfection Blending with freshwater supplies
Brackish water (ground)	RO, NF, EDR	Decarbonation (degasification) Hydrogen sulfide stripping pH and/or alkalinity adjustment Addition of corrosion inhibitors Primary and secondary disinfection Blending with freshwater supplies Bypass blending with raw water supply
Fresh water (ground)	NF, EDR	Decarbonation Hydrogen sulfide stripping pH and/or alkalinity adjustment Addition of corrosion inhibitors Primary and secondary disinfection Blending with fresh water supplies Bypass blending with raw water supply

Desalinated waters are commonly blended with small volumes of more mineral-rich waters to improve their acceptability and particularly to reduce their aggressive attack on materials (WHO, 2004). Blending water should be fully potable; where seawater is used for this purpose, the major ions added are sodium and chloride. This does not contribute to improving hardness or ion balance, and only a small amount, no more than 3 percent, can be added without leading to problems of acceptability. Blended waters from coastal and estuarine areas may be more susceptible to contamination with petroleum hydrocarbons or algal toxins, which could give rise to taste and odor problems. Some groundwaters or surface waters, after suitable treatment, may be employed for blending in higher proportions and may improve hardness and ion balance. As noted by Fritzmann (2007), untreated permeate from sea- or brackish-water RO plants does not conform to the drinking water standards such as WHO Guidelines for Drinking Water Quality. Due to the low TDS values, RO permeate water can be unpalatable, corrosive, and unhealthy. Permeate must be rehardened in order to prevent corrosion of pipes in the distribution network, pH value and carbon dioxide content need to be adjusted for scaling prevention, and permeate water needs further disinfection.

Alkalinity depends mainly on the bicarbonate concentration, and for waters with pH > 9, it also depends on the carbonate, and hydroxide concentrations; see Chap. 3 for a presentation on alkalinity. According to Lahav and Birnhack (2007) for a given pH, the higher the alkalinity, the water has a greater acid-neutralizing capacity. Also, higher-alkalinity water means a higher dissolved inorganic carbon (DIC) concentration. However, too high an alkalinity at higher pH levels may accelerate lead and copper metal release (Duranceau and Henthorne, 2004). Ruggieri and coworkers developed limestone selection criteria for EDR water remineralization (Ruggieri, 2008). Five commercially available limestones were characterized by mineralogical, chemical, and surface methods, including *scanning electron microscopy (SEM)*. The main criteria in selecting a limestone for EDR water remineralization are its compositional purity in terms of both mineralogy and chemistry, and its textural characteristics. The dominant mineral phase in the uncontacted limestones was calcite ($CaCO_3$), and included small amounts of dolomite ($CaMg(CO_3)_2$) and quartz (SiO_2) and trace levels of clay matter. In relation to mineralogical purity, it was recommended that the occurrence other than calcite should be avoided and that the use of SEM was a useful tool in evaluating surface texture influence on suspended particulate matter generation on consumption of the bed (calcite). Recarbonation refers to the process used to introduce inorganic carbon and alkalinity to produce a positive Langelier Saturation Index (LSI) to prevent corrosion in pipes; see Chap. 20 for discussion of corrosion and corrosion indices. Remineralization is a means of increasing the mineral content by addition to those which increase the bicarbonate or carbonate alkalinity of the desalinated water as stated by Withers (2005).

The steps chosen and their sequence depend on the designer's preferences and water-quality goals. The primary posttreatment unit operations are aeration, disinfection, and stabilization. Additional posttreatment operations of concern are removal of hydrogen sulfide, if present, and alkalinity recovery. A systems view of posttreatment can help a designer to realize important goals. The membrane process removes essentially all pathogens and the majority of the DBP precursors, salts, and other solutes in the feed stream. Solute removal eliminates inorganic carbonate alkalinity, but all dissolved gases including carbon dioxide and hydrogen sulfide pass through the membrane (Lovins, 2004). The designer must produce finished water after posttreatment with an appropriate alkalinity profile and disinfection without significant sulfur turbidity. Posttreatment water quality changes by unit operation are provided in Table 11-12 and are based on using the permeate stream water quality illustrated in the previous membrane section. The water quality resulting from each posttreatment unit operation is illustrated in the table.

TABLE 11-12 Posttreatment Water Quality Changes by Unit Operation

	Permeate, mg/L	Alkalinity recovery, mg/L	Aeration/stabilization, mg/L
pH	5.5	7.6	7.6
H_2CO_3	129 mg/L	6 mg/L	0.6 mg/L
HCO_3^-	0.8	122	122
H_2S as S	1	0	0
SO_4^{-2}	1.7	4.7	4.7
Cl^-	19.7	30.1	30.1
Ca^{+2}	4.1	50.9	50.9
TDS	34	79	79
DO	0	0	8.24 (25°C)
Cl_2	0	3	3

The changes in water quality will be discussed in the following sections. The sequence of unit operations assumed here are H_2S removal, alkalinity recovery, and aeration/stabilization. There are different sequences of posttreatment, but this sequence has the advantage of minimizing equipment, as H_2S removal, alkalinity recovery and aeration/stabilization are conducted simultaneously in separate unit operations.

Hydrogen Sulfide Removal. Many of the groundwaters used for feed streams to RO or NF plants contain hydrogen sulfide. Acid addition in the feed stream is common to avoid calcium carbonate scaling in the permeate stream, and permeate pH is commonly between 5.5 and 6.5. While this measure prevents scaling, the acid addition converts dissolved sulfide to gaseous hydrogen sulfide. Neither the conventional pretreatment process (microfiltration, acid, or antiscalant addition) nor the membrane process will remove hydrogen sulfide. Aeration and oxidation are the two primary means of removing hydrogen sulfide; however, the involved chemical reactions are not well defined. An often-neglected problem in the hydrogen sulfide removal processes is the formation of elemental sulfur. Elemental sulfur has been shown not to form during the chlorination of hydrogen sulfide if the reaction pH is less than 3.7; sulfate is the predominant reaction product at this pH. However, both entrained oxygen and chlorine will react with hydrogen sulfide to form elemental sulfur at pH levels above 4.0, as shown here.

$$2H_2S + O_2 \rightarrow 2S + 2H_2O \tag{11-92}$$

$$2H_2S + O_2 + 2H_2O \rightarrow 2H_2SO_4 \tag{11-93}$$

$$H_2S + Cl_2 \rightarrow S + 2HCl \tag{11-94}$$

$$H_2S + 4Cl_2 + 4H_2O \rightarrow H_2SO_4 + 8HCl \tag{11-95}$$

The pK for hydrogen sulfide is 7, and H_2S gas can essentially be *removed* at pHs below 6.5 without turbidity formation in an air-stripping, packed tower process. Aeration using tray aerators is less efficient at hydrogen sulfide removal because there is less available surface area for air stripping, so that residual sulfide remains in the water allowing elemental sulfur to form due to oxygen oxidation. While the air-stripping, packed tower process generally will avoid sulfur formation, the CO_2 is also lost during the volatilization process. Consequently, unless a carbonate salt is added or a significant amount of alkalinity passes through the membrane, there will be no carbonate buffering in the finished water. The final pH can be increased by the addition of sodium hydroxide but the finished water will have very low buffering capacity and will be corrosive. A better application would be to increase the pH to 6.3 prior to packed tower, air stripping with calcium hydroxide or sodium hydroxide to recover 1 to 2 meq/L of alkalinity with minimal elemental sulfur formation. Ultimately, most plants rely heavily on air stripping with packed towers for H_2S removal rather than using chlorination for H_2S destruction.

Alkalinity Recovery. Alkalinity recovery becomes a consideration during scaling control. The addition of acid to the feed stream prior to membrane treatment will prevent scaling, but because of the decrease in pH, alkalinity in the form of bicarbonate (HCO_3^-) will be converted to carbon dioxide, and this will pass through the membrane. Membranes are a closed system, and the carbon dioxide will remain under pressure until exposed to an open system. Exposure to an open system occurs in the removal of hydrogen sulfide using aeration. Unless the pH is increased prior to aeration, the excess CO_2 will leave the system along with hydrogen sulfide and the alkalinity will be lost. In an alkalinity recovery system, the excess dissolved CO_2 is converted to alkalinity in the form of HCO_3^- by adding caustic downstream of permeate prior to air stripping with packed towers. Figure 11-33 depicts an alkalinity recovery process flow diagram that further illustrates this concept. Normally, finished waters with 1 to 3 meq/L of bicarbonate alkalinity are considered highly desirable for corrosion control. The reader is referred to Chap. 3 for principles dealing with inorganic carbon and alkalinity. Table 11-12 illustrates how pH increases and alkalinity and hydrogen sulfide concentrations change following the unit operations discussed.

Aeration and Stabilization. These unit processes can be discussed simultaneously in membrane processes because they will be accomplished simultaneously. Chlorination could also be done here if desired. If calcium and bicarbonate are present, the pH following

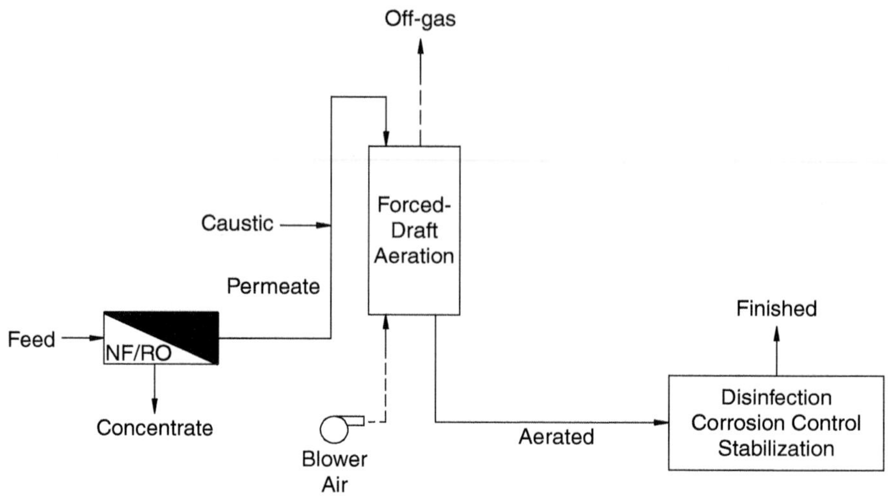

FIGURE 11-33 Alkalinity recovery process flow diagram.

aeration is controlled by $CaCO_3(s)$ buffering and can be estimated by assuming $CaCO_3$ equilibrium. Equation 11-96 shows the calculation made to determine the pH following aeration.

$$pH_s = pK_2 - pK_{s0} + p[Ca^{2+}] + p[HCO_3^-] \tag{11-96}$$

Example 11-11 Stabilization pH Following Aeration
The pH calculation following aeration is determined by equilibrium with $CaCO_3$ and can be determined by the pH_s calculation shown as follows. Table 11-12 gives the permeate stream pH and concentrations of different species following alkalinity recovery. The following equations, along with Eq. 11-96, are referenced and used for calculations:

$$CaCO_3 = Ca^{+2} + CO_3^{-2} \quad K_{s0} = 10^{-8.3} \tag{11-97}$$

$$HCO_3^- = H^+ + CO_3^{-2} \quad K_2 = 10^{-10.3} \tag{11-98}$$

Given

$$Ca^{+2} = 50.9 \text{ mg/L}$$

$$HCO_3^- = 122 \text{ mg/L}$$

Solution Convert the concentration of calcium to mol/L.

$$[Ca^{+2}] = \left(50.9 \frac{\text{mg}}{\text{L}} Ca^{2+}\right)\left(\frac{1 \text{ mmol } Ca^{2+}}{40 \text{ mg } Ca^{2+}}\right) = 1.27 \text{ mM } Ca^{+2}$$

Calculate $p[Ca^{2+}]$.

$$p[Ca^{+2}] = -\log[Ca^{+2}] = -\log[1.27 \times 10^{-3}] = 2.90$$

Convert the concentration of bicarbonate to mol/L.

$$[HCO_3^-] = \left(122\,\frac{mg}{L}\,HCO_3^-\right)\left(\frac{1\ mmol\cdot HCO_3^-}{61\ mg\cdot HCO_3^-}\right) = 2.0\ mM\cdot HCO_3^-$$

Calculate $p[HCO_3^-]$.

$$p[HCO_3^-] = -\log[HCO_3^-] = -\log[2.0\times10^{-3}] = 2.70$$

Using Eq.11-96, calculate the stabilization pH.

$$pH_s = pK_2 - pK_{s0} + p[Ca^{2+}] + p[HCO_3^-]$$

Eq.11-98 gives $pK_2 = 10.3$

Eq.11-97 gives $pK_{s0} = 8.3$

$$p[Ca^{2+}] = 2.90$$

$$p[HCO_3^-] = 2.70$$

Plug in these values into Eq. 11-96 to obtain the stabilization pH.

$$pH_s = pK_2 - pK_{s0} + p[Ca^{2+}] + p[HCO_3^-]$$

$$pH_s = 10.3 + 8.3 + 2.9 + 2.7 = 7.6$$ ▲

RESIDUALS DISPOSAL AND CONCENTRATE MANAGEMENT

Waste disposal is a more significant problem for any membrane plant than for a conventional water treatment plant because concentrate disposal is highly regulated by government agencies. A membrane plant collects significant amounts of waste with a high TDS concentration.

MF and UF Process Residuals

MF and UF process residuals result from the wastes generated during backwashing and chemical cleaning. Chemical-cleaning wastes include residuals generated from either *clean in place (CIP)* or *chemically enhanced backwash (CEB)* operations. Monthly CIP wastes are normally less than one percent of the plant flow, and the volume of chemical-cleaning wastes generated from CEB operations are typically less than 0.5 percent of the plant flow. Chlorine residuals in CIP or CEB wastes may range from 1 to 1000 mg/L as chlorine. The pH of the chemical-cleaning-waste streams may vary from acidic ($pH < 6$) to basic ($pH > 9$) depending on the type of cleaning performed. If citric acid is used in the CIP, the biological oxygen demand of the spent cleaning solution may be high. In addition, if citric acid is used in the process cleaning regimen, then the ability to recirculate combined backwash and CIP solutions is limited if the MF or UF process is integrated into a coagulation process due to chemical reactions between the citric acid and coagulant that will impact process operations and result in loss of operating flux.

MF and UF membranes produce a similar backwash stream of approximately the same quantity of materials as would a conventional water treatment plant. However, MF and UF membrane processes do not contain residual polymers that are sometimes used in conventional treatment facilities as polymers are incompatible with membranes. In many cases considerable coagulant or other pretreatment residuals (such as PAC or potassium permanganate) may not be used in combination with MF and UF. So, the amount of solids removed from a water stream is less than a conventional process. Also, MF and UF processes are backwashed more frequently than a conventional filter. The typical membrane process backwash occurs every 15 to 60 min, with a backwash duration of 1 to 4 min. Although the backwashes are more frequent, the total water lost during MF and UF backwash operations is typically only 2 to 12 percent of the plant flow rate for systems operating at 85 to 98 percent recovery. Backwash flows represent 95 to 98 percent of the total volume of residual wastes from a MF or UF system. The remaining 1 to 5 percent of MF and UF membrane residuals originates from cleaning procedures.

Consequently, the disposal of MF and UF solid residual streams are handled in a similar fashion as a conventional water treatment plant residual stream. The backwash solution may require dechlorination (or pH adjustment) prior to discharge. Dechlorination, if used, is generally designed to match the backwash frequency and typically performed as a batch operation where the backwash water is collected, neutralized, and discharged. Backwash wastes are commonly disposed of by discharging into sewers or surface water. Sewers are used as the predominant method of disposal for cleaning wastes. However, some form of pH adjustment will occur prior to discharge. Some MF and UF facilities may be able to settle solids out of the spent backwash stream and recycle the supernatant to increase the overall system recovery or to further treat the backwash stream using additional unit operations to develop a zero discharge backwash residual operation.

Disposal of MF and UF process residuals is generally regulated by the same guidelines governing water treatment plant residuals. Although the USEPA has not established specific regulations for the disposal of MF and UF water treatment plant residuals, there are federal requirements that apply to membrane residuals as industrial waste streams. Most states have been delegated by the USEPA to administer established regulations that will meet the requirements of the federal regulations. It is important to note that although the recycling of backwash and other recycled flows in conventional treatment plants is regulated by the Filter Backwash Recycling Rule (see Chap. 10), this rule does not apply to membrane residuals recycling unless membrane wastes are mixed with other conventional process wastes prior to recycling.

Spent Cleaning Solutions

Membrane systems require periodic cleaning with chemicals to control fouling and maintain productivity. If a membrane process is suspect of forming mineral scale on the membrane surface, then most often it will be cleaned with an *acid cleaner such as citric acid.* In addition, most membrane processes will be cleaned with a base cleaner, such as sodium hydroxide, to remove organic foulants. Spent chemical-cleaning solutions may be handled and disposed of either separately from the concentrate discharge or, if neutralized, may be blended in with the concentrate prior to disposal. Several waste discharges are generated during membrane cleaning procedures. The concentrated waste cleaning system is the actual spent membrane cleaning chemical; however, permeate (or filtrate) flush water is used to displace the spent cleaning residual remaining in the membrane unit or system and must be disposed of after use. The flush water removed from the concentrate lines of membrane systems must also be accounted for in determining residual quantities and flow.

The various streams typically are routed to a spent cleaning solution scavenger tank for holding and treatment (neutralization) prior to discharge. A survey of 49 NF and RO facilities and 27 MF and UF facilities was conducted to evaluate the disposal methods for spent cleaning solutions (Kenna and Zander, 2001). It was found that the two most common methods for disposal of cleaning solution residuals from NF and RO plants were sewer discharge and mixing with concentrate after the cleaning solution had been neutralized. The most common method for disposal of MF and UF residuals was sewer discharge.

NF and RO Membrane Concentrate

The quality and quantity of concentrate streams from RO/NF/ED plants can be easily estimated using simple recovery and rejection calculations. A 37,855 m^3/d (10 mgd) plant operating at 90 percent recovery will produce a concentrate stream of 4165 m^3/d (1.11 mgd), which is determined by dividing the permeate flow, 37,855 m^3/d (10 mgd), by the recovery, 0.11. The quality of the concentrate stream can be determined by dividing the decimal fraction solute rejection into the feed stream recovery. The same plant producing 37,855 m^3/d (10 mgd) at 90 percent recovery and rejecting 80 percent of 100 mg/L of the feed stream chlorides would produce 20 mg/L of chlorides in the permeate stream and 820 mg/L of chlorides in the concentrate stream. Rejection and recovery are sometimes confused. The calculations are illustrated below. For the purposes of these calculations, r will represent the fraction recovered and R will represent the fraction rejected.

$$Q_c = \frac{Q_p(1-R)}{R} = \frac{37{,}855\ \text{m}^3/\text{day}\ (1.0-0.9)}{0.9} = \frac{3785.5}{0.9} = 4200\,\frac{\text{m}^3}{\text{day}}\ (1.11\ \text{mgd})$$

$$C_p = (1-R)(C_f) = (1-0.8)\left(100\,\frac{\text{mg Cl}^-}{\text{L}}\right) = 20\,\frac{\text{mg Cl}^-}{\text{L}}$$

$$C_p = \frac{Q_f C_f Q_p C_p}{Q_c} = \frac{(1\ L)\left(100\,\frac{\text{mg Cl}^-}{\text{L}}\right) - (0.9\ \text{L})\left(20\,\frac{\text{mg Cl}^-}{\text{L}}\right)}{0.1\ \text{L}} = 820\,\frac{\text{mg Cl}^-}{\text{L}}$$

Table 11-13 lists techniques used for concentrate disposal in the United States by plant size and number (Mickley, 1993). Mickley identified five basic techniques used by 137 plants surveyed in an AWWARF study of concentrate disposal. Other surveys of concentrate disposal methods are available (Kenna and Zander, 2001; Truesdall, Mickley and Hamilton, 1995). Surface discharge is the most common technique, accounting for nearly half of the techniques surveyed. Land application was not planned for any proposed new plant. However, land application of NF concentrates is possible in some locations because of their low TDS concentrations, typically 1000 to 2000 mg/L, relative to RO concentrates. Sewer discharge, usually an option only for very small plants, was

TABLE 11-13 Concentration Disposal Technique Distribution by Plants and Flow

	Plant size (mgd)				
Technique or method	<0.3	0.3–1	1–3	>3	Total
Surface	34	12	9	11	66
Land application*	14	2	0	1	17
Sewer	18	8	3	3	32
Deep-well injection	3	1	5	5	14
Evaporation pond	8	0	0	0	8
Total	77	23	17	20	137

*None planned in future.

the second most common technique for plants under 1100 m^3/d (0.3 mgd). Deep-well injection was more common in Florida than any other states because of differences in regulatory environments.

Concentrate discharge is regulated through programs in the Clean Water Act (CWA) National Pollutant Discharge Elimination System (NPDES) program, Dredge and Fill program and Resource Conservation Recovery Act, and the Safe Drinking Water Act (SDWA) Underground Injection Control (UIC) program. Surface waters fall under NPDES whereas groundwaters fall under the UIC program. Regulatory environments for concentrate disposal are controlled by states or by USEPA, depending on primacy. Although these programs have existed since the 1970s, the regulatory environment is becoming more stringent. Bioassay testing is required now in many states, is common in Florida and California, and can be used to require additional treatment or stop planned disposal of concentrate streams. An overview and summary of regulation and current practices has been documented elsewhere (Mickley, 2001; AWWA 2004).

PILOT PLANT TESTING

Pilot plant testing offers the best method for evaluating the feasibility of a membrane application for a specific water supply, especially since fouling cannot be quantitatively predicted from water-quality measurements alone. Fouling indices do provide an estimate of the potential for fouling, but unlike pilot-scale testing, are not predictive of long-term performance. In many states, pilot testing is required for membrane processes prior to receiving regulatory approvals and applicable permits. There are many reasons for pilot plant testing, some of which include

1. Pilot testing is mandated by state or local primacy regulatory agencies.
2. Collect baseline raw-water-quality profiles that can be used to establish a basis of design.
3. Collect operating membrane process data for cost and performance evaluations.
4. Confirm that the permeate water quality meets the contractual, regulatory, and site-specific needs of the owner and engineer.
5. Provide hands-on training for plant operations personnel.
6. Demonstrate operation protocols and procedures.
7. Allow for continued long-term investigations on process operations.
8. Allow for research and development.

Two of the most critical needs to the design of an UF treatment plant are delineating the quality of the feedwaters going into the system and predicting the desired quality of the water being produced by the system. Additional considerations for design include pretreatment, process feed pumping requirements, process monitoring and flow control, backwash and cleaning cycles, chemical feed equipment, posttreatment, and residuals disposal. These components are necessary to provide an estimate of the cost and allow a cost-benefit evaluation to be conducted. Different membranes can produce different permeate water qualities depending on the feedwater quality. Because most applications are unique, a site-specific understanding is necessary for the proper design of the membrane system. However, the operation of MF and UF processes are different from NF, RO, and ED/EDR membrane processes, requiring that system-specific testing for different manufacturers' membranes and systems be addressed accordingly.

MF and UF Piloting

Important considerations for evaluating MF and UF performance include specific flux, water temperature and associated TMP, backwash and bleed interval impact on productivity and run time, cleaning frequency and interval, and other system-specific operation, such as chemical-enhanced backwashes. Instrument verification and calibration are required for flowmeters, pressure and temperature transmitters, online particle counters, and turbidimeters. Test duration is also critical to obtaining pertinent and applicable information from the pilot operation regarding cost and performance of the projected full-scale facility. If the feedwater supply is subject to seasonal impacts, then the number of applicable testing periods for each season should be conducted. During testing, a minimum of 30 days of run time should be required prior to altering the test conditions or pilot operation set points for any given feedwater. If multiple feedwaters are to be blended or varied during full-scale operating conditions, then worst-case blending scenarios with regard to temperature and water-quality impacts (particularly those related to fouling) must be considered and studied at the pilot scale.

Additional consideration should be given to specific study components of a pilot program, such as challenge tests, integrity testing, and module repair procedures. Fouling impacts on process productivity are best assessed by pilot testing. Evaluation of chemical cleaning is a significant component of MF and UF piloting, and longer run times may be required in order to fully evaluate CIP procedures. Also, if CEB is incorporated into the process operation, then impact on backwash recycle operations with regard to DBP formation potential must be evaluated. If recirculation is to be practiced in the full-scale plant then the pilot-testing program should incorporate the recirculation. Should citric acid be incorporated into the cleaning regimen, then resultant residuals should be disposed of in an acceptable fashion and not be recycled back to the front of the process stream, particularly if coagulant is used as part of the pretreatment process train. Citric acid will interfere with coagulant pretreatment, especially when an iron base is used. They should not be allowed to come into contact with coagulant upstream of the membrane.

It is important that any problems related to scalability, such as membrane packing density, and analogous pretreatment be incorporated into the testing. The MF and UF modules tested must be comparable to anticipated full-scale module configurations. If packing density differences exist between pilot-scale and full-scale systems then inaccuracies in operation evaluations will occur. For MF and UF, the owner and engineer are thus reliant on the manufacturer to provide a pilot system that mimics full-scale operation, and so must be involved in the technical aspects of the pilot test.

NF, RO, and ED/EDR Piloting

The demonstrated benefits of performing bench-, pilot- and demonstration-scale testing of NF, RO, and ED/EDR processes has continued over the years. Brackish groundwater RO and EDR plants have been successfully designed, effectively constructed, and successfully operated over the years. Typically, approximately 2000 hours of run time are required to operate a pilot and obtain quality performance data. Piloting could be considered a method to reduce the risk of unknown potential problems related to operating a membrane process with a specific raw water and does provide hands-on experience for plant operations personnel. Problems related to long-term fouling experienced in nonbrackish ground- and surface water remain with NF, RO, and ED/EDR, and should be assessed with pilot testing. In these cases longer testing intervals should be considered to capture seasonal variations and allow for the development of long-term fouling assessments, particularly if biological and organic fouling are predicted.

Today most RO and NF elements in use are 20.3 cm (8 in.) in diameter, and it is common that piloting be performed using 10.2 cm (4 in.) diameter elements as they tend to mimic full-scale operating conditions (e.g., feed channel hydraulics). However, 6.35 cm (2.5 in.) diameter elements, although available for testing, should not be used to evaluate RO full-scale operating conditions, as these elements do not mimic larger-element manufacture. On the other hand, for 10.2 cm (4 in.) NF pilot studies requiring a third stage, the use of 6.35 cm (2.5 in.) diameter elements are often used because the need to control velocity in the third stage, despite the inherent limitations of the smaller-diameter element. With the advent of 406.4 mm (16 in.) diameter elements entering the market, the use of 10.2 cm (4 in.) membranes for analogous testing conditions would be in question for this application. Demonstration-scale testing (on the order of 1900 m^3/d (0.5 mgd)) using the large-diameter membrane would be recommended for these cases, as these larger diameter elements can be mounted on pilot-scale skids. Instrument verification and calibration of flowmeters, pressure and temperature transmitters, and online pH and conductivity meters, and similar, are required for NF, RO, and EDR pilot facilities.

ABBREVIATIONS

BAT	Best available technology
$CaCO_3$	Calcite
CTA	Cellulose triacetate
CEB	Chemically enhanced backwash
CWA	Clean Water Act
CIP	Clean in place
CFS	Coagulation, flocculation, and sedimentation
m^3/d	Cubic meters per day
DC	Direct current
DBP	Disinfection by-product
DBPFP	Disinfection by-product formation potential
DAF	Dissolved air flotation
DIC	Dissolved inorganic carbon
DOC	Dissolved organic carbon
$CaMg(CO_3)_2$	Dolomite
ED/EDR	Electrodialysis/electrodialysis reversal
EDTA	Ethylenediaminetetraacetic acid
GAC	Granular activated carbon
GWR	GroundWater Rule
$CaSO_{4(s)}$	Gypsum
HAA	Haloacetic acid
HPC	Heterotrophic plate count
HFF	Hollow fine fiber
IDSE	Initial Distribution System Evaluation

IOC	Inorganic compounds
IMS	Integrated membrane system
kJ	Kilojoules
kPa	Kilopascals
LSI	Langelier Saturation Index
LCR	Lead and Copper Rule
LRV	Log reduction value
LT2ESWTR	Long-term 2 Enhanced Surface Water Treatment Rule
MTC	Mass transfer coefficient
MF	Microfiltration
mgd	Million gallons per day
MFI	Modified fouling index
MW	Molecular weight
MWC	Molecular weight cutoff
NF	Nanofiltration
NPDES	National Pollutant Discharge Elimination System
NOM	Natural organic matter
PF	Polarization factor
PAA	Polyacrylic acid
PES	Polyethersulfone
PVDF	Polyvinylidene fluoride
psi	Pounds per square inch
PAC	Powder-activated carbon
PDT	Pressure decay test
PV	Pressure vessel
SiO_2	Quartz
RO	Reverse osmosis
SDWA	Safe Drinking Water Act
SEM	Scanning electron microscopy
SDI	Silt density index
SHMP	Sodium hexametaphosphate
SW	Spiral wound
SWTR	Surface Water Treatment Rule
SOC	Synthetic organic compound
TCR	Total Coliform Rule
TDS	Total dissolved salts
TH	Total hardness
TMP	Transmembrane pressure drop
THMFP	Trihalomethane formation potential
UF	Ultrafiltration

UIC	Underground injection control
VOC	Volatile organic compound
WHO	World Health Organization

NOTATION FOR EQUATIONS

Term	**Symbol**
Absolute viscosity	μ
Activity coefficient	γ
Brownian diffusivity	D_B
Cation transport number	t_+
Concentration of species	c, C
Current	I
Current density	I_{density}
Density	ρ
Diameter	d
Diffusion coefficient from surface to bulk	k_b
Diffusivity	D
Distribution coefficient	k_s
Electric potential	E
Electrical Rresistance	R_E
Fitting constant	K_{pf}
Flow rate	Q
Flux	J
Force	F
Fouling potential	ξ
Fraction of material	x, X
Free energy	G
Global rejection	R
Hydraulic diameter	d_h
Hydraulic resistance	R_m
Ideal gas constant	R_g
Ionic strength	I
Kinematic viscosity	ν
Lag coefficient	Γ
Mass transfer coefficient	k
Mass transfer coefficient of solute	k_i
Mass transfer coefficient of solvent	k_w
Modified film theory mass transfer coefficient	$Z_{i,cp}$

Modified mass transfer coefficient	Z
Molar volume	V
Osmotic pressure	Π
Particle passage	ψ
Phenomenological coefficient	L
Polarization factor	PF
Porosity	ε
Pressure	p, P
Radius	a
Reaction quotient	Q
Recovery	r
Reflection coefficient	σ
Relative viscosity	η
Resistance of the cake	R_k
Resistance of the filter	R_f
Reynolds number	Re
Shear rate	ω
Shear stress	τ
Sherwood number	Sh
Solubility constant	K_{s0}
Specific hydraulic permeability	K_w
Temperature	T
Temperature correction factor	TCF
Temperature correction factor constant	θ
Thickness	δ
Velocity	u

REFERENCES

Adham, S., V.L. Snoeyink, M.M. Clark, and J.L. Bersillon (1991) Ultrafiltration of Groundwater with Powdered Activated Carbon Pretreatment for Organics Removal, *Proceedings of the AWWA Membrane Technology Conference,* Orlando, FL.

Adham, S.S., V.L. Snoeyink, M.M. Clark, C. Anselme, and I. Bausin (1993) Predicting and Verifying Organics Removal by PAC in Large-Scale UF Systems, *Proceedings of the 1993 Membrane Technology Conference,* Baltimore, MD.

Adham, S.S., J.G. Jacangelo, and J.-M. Laine (1995) Assessing Integrity, *Journal AWWA*, vol. 87 no. 3, pp. 62–75.

Amy, G., B.C. Alleman, and C.B. Cluff (1990), Removal of Dissolved Organic Matter by Nanofiltration, *Journal of Environmenal Engineering*, vol. 116, no. 1, pp. 200–205.

Amy, G., M. Edwards, P. Brandhuber, L. McNeill, M. Benjamin, F. Vagliasindi, K. Carlson, and J. Chwirks (2000) *Arsenic Treatability Options and Evaluation of Residuals Management Issues*, American Water Works Association and Water Research Foundation, Denver, CO.

Anselme, C., J.L. Bersillon, and J. Mallevialle (1991) The Use of Powdered Activated Carbon for the Removal of Specific Pollutants in Ultrafiltration Processes, *Proceedings of the AWWA Membrane Processes Conference,* Orlando, FL.

ASTM. (2001) D4189-95 Standard Test Method for Silt Density Index (SDI) of Water, American Society for Testing and Materials, Philadelphia.

Baier, J.H., B.W. Lykins, Jr., C.A. Fronk, and S.J. Kramer (1987) Using Reverse Osmosis to Remove Agricultural Chemicals from Groundwater, *Journal AWWA*, vol. 79, no. 8, pp. 55–60.

Belfort, G. (1984) *Synthetic Membrane Processes, Fundamentals and Water Applications.* New York: Academic Press.

Belfort, G. (1981) Fluid Mechanics in Membrane Filtration Recent Developments, *Journal of Membrane Science*, vol. 40, pp. 123–147.

Belfort, G., D.R. Davis, and A.L. Zydney (1994) The Behavior of Suspensions and Macro-molecular Solution in Crossflow Microfiltration, *Journal of Membrane Science*, vol. 96, pp. 1–58.

Berg, P., and P. Gimbel (1997) Rejection of Trace Organics by Nanofiltration, *Proceedings of the AWWA Membrane Technology Conference,* New Orleans.

El Boukari, M., J.L. Bribess,. and J. Maillols (1990) Application of Raman Spectroscopy to Industrial Membranes. Part I. Polyacrylic Membranes, *Journal of Raman Spectroscopy*, vol. 21, p. 755.

Braghetta, A. (1995) The Influence of Solution Chemistry Operating Conditions on Nanofiltration of Charged and Uncharged Organic Macromolecules, Ph.D. Dissertation, University of North Carolina, Chapel Hill.

Braghetta, A., M. Price, and C. Kolkhorst (2001) Use of Physical and Chemical Pretreatment Ahead of Ultrafiltration for Surface Water Treatment in San Antonio, Texas, *Proceedings of the AWWA Membrane Technology Conference,* San Antonio, TX.

Burashid, K. (1994) Seawater RO operation and maintenance experience: Addur desalination plant operation assessment, *Desalination*, vol. 165, no. 2004, pp. 11–22.

Burbano, A.A., S.S. Adham, and W.R. Pearce (2007) The State of Full-Scale RO/NF Desalination–Results from a Worldwide Survey, *Journal AWWA,* vol. 99, no. 4, p. 116.

Carrol, T., S. King, S.R. Gray, B.A. Bolto, and N.A. Booker (2000) The Fouling of Microfiltration Membranes After Coagulation Treatment, *Water Research*, vol. 31, no. 11, pp. 2861–2868.

Chen, S., and J.S. Taylor (1997) Flat Sheet Testing for Pesticide Removal by Varying RO/NF Membrane, *Proceedings of the AWWA Membrane Technology Conference,* New Orleans.

Chellam, S., and M.R. Wiesner (1997) Particle Back-Transport and Permeate Flux Behavior in Crossflow Membrane Filters, *Proceedings of the AWWA Membrane Technology Conference,* New Orleans.

Cheryan, M. (1988) *Ultrafiltration Handbook.* Lancaster, PA: Technomic Publishing Co.

Chian, E.S.K., W.N. Bruce, and H.H.P. Fang (1975) Removal of Pesticides by Reverse Osmosis, *Journal of Environmental Science and Technology*, vol. 9, no. 1, pp. 52–511.

Childress, A.E., and M. Elimelech (1996) Effect of Solution Chemistry on the Surface Charge of Polymeric Reverse Osmosis and Nanofiltration Membranes, *Journal of Membrane Science*, vol. 119, no. 10, pp. 253–268.

Childress, A.E., P. Le-Clech, J.L. Daugherty, D. Chen, and G.L. Leslie (2005) Mechanical Analysis of Hollow Fiber Membrane Integrity in Water Reuse Applications, *Desalination*, vol. 180, pp. 8–14.

Chua, K.T., M.N.A. Hawlader, and A. Malek (2003) Pretreatment of Seawater: Results of Pilot Trials in Singapore, *Desalination*, vol. 159, pp. 225–243.

Chwirka, J.D., B.M. Thomson, and J.S. Stomp (2000) Removing Arsenic from Groundwater, *Journal AWWA*, vol. 92, no. 3, pp. 79–88.

Clark, M.M. (1998) Committee Report: Membrane Processes, *Journal AWWA*, vol. 90, no. 6, pp. 91–105.

Coffey, B.M. (1993) Evaluation of MF for Metropolitan's Small Domestic Water System, *Proceedings of the AWWA Membrane Technology Conference,* Baltimore, MD.

Conlon, W.J., and S.A. McClellan (1989) Membrane Softening—A Treatment Process Comes of Age. *Journal AWWA*, vol. 81, pp. 47–51.

Coster, H.G.L., K.J. Kim, K. Dahlan, J.R. Smith, and C.J.D. Fell (1992) Characterization of Ultrafiltration Membranes by Impedance Spectroscopy, I: Determination of the Separate Electrical Parameters and Porosity of the Skin and Sublayers, *Journal Membrane Science*, vol. 66, p. 111.

Côté P., I. Sutherland, N. Adams, and J. Cadera (2003) Validation of Membrane Integrity Methods in a Challenge Test with *Bacillus subtillis, Proceedings AWWA Membrane Technology Conference,* Atlanta.

Crawford, S., and D. Bach (2001) Expanding and Upgrading a Conventional Water Treatment Plant–The Membrane Alternative, *Proceedings of the AWWA Membrane Technology Conference,* San Antonio, TX.

Davison, M.G., and W.M. Deen (1988) Hydrodynamic Theory for the Hindered Transport of Flexible Macromolecules in Porous Membranes, *Journal of Membrane Science*, vol. 35, pp. 167–192.

Deen, W.M. (1987) Hindered Transport of Large Molecules in Liquid-Filled Pores, *American Institute of Chemical Engineers Journal*, vol. 33, pp. 1409–1425.

Dresner, L. (1972) Some Remarks on the Integration of the Extended Nernst-Planck Equation in Hyperfiltration of Multicomponent Solution, *DeSalination*, vol. 10, p. 27.

Drew, D.A., J.A. Schonberg, and G. Belfort (1991) Lateral Inertial Migration of a Small Sphere in Fast Laminar Flow through a Membrane Duct, *Chemical Engineer Science*, vol. 46, pp. 3219–3224.

E.I. Dupont de Nemours & Co. (1982) *PEM Products Engineering Manual*, Wilmington, Del.

Duranceau, S.J. (2001) *Membrane Practices for Water Treatment, Trends in Water Series*, American Water Works Association, Denver, CO.

Duranceau, S.J. (2004) Membrane Process Integrity Testing, *Proceedings of the AWWA Annual Conference and Exposition,* Orlando, FL.

Duranceau, S.J. (2005) Membrane Biofouling: Causes and Remedies, *Proceedings AWWA Annual Conference and Exposition,* San Francisco, CA.

Duranceau, S.J. (2006) Chemistry for Membrane Treatment, *Proceedings of the Southeast Desalting Association June 2006 Symposium–Advancing Membrane Treatment Together,* Captiva Island, FL.

Duranceau, S.J. (2006) Membrane Forensics and Troubleshooting Membrane Facilities, *Proceedings of the AMTA Biennial Conference and Exposition,* Anaheim, CA.

Duranceau, S.J., and L.R. Henthorne (2004) Membrane Pretreatment for Seawater Reverse Osmosis Desalination, *Florida Water Resources Journal*, vol. 56, no. 11, pp. 33–37.

Duranceau, S.J., and J.S. Taylor (1990) Investigation and Modeling of Membrane Mass Transfer, *Proceedings of the National Water Improvement Supply Association,*Orlando, FL.

Duranceau, S.J., and J.S. Taylor (1992) SOC Removal in a Membrane Softening Process, *J, AWWA*, vol. 84, pp. 68–78.

Duranceau, S.J., J.A. Manning, H.J. Losch, and F. Kane (2000) Integrating an Immersed Ultrafiltration Process into an Existing Lime Softening Water Treatment Plant, *Proceedings of the Conference on Membranes in Drinking Water and Industrial Water Production, AWWA, International Water Association, European Desalination Society and Japan Water Works Association,* Paris, FR.

Elimelech, M., W. Chen, and J. Waypa (1994) Measuring the Zeta (Electrokinetic)-Potential of Reverse Osmosis Membranes by a Streaming Potential Analyzer, *Desalination*, vol. 95, pp. 269–286.

Fane, A.G. (1984) Ultrafiltration of Suspensions, *Journal of Membrane Science*, vol. 20, pp. 249–2511.

Ferry, J.D. (1936) Statistical Evaluation of Sieve Constants in Ultrafiltration, *Journal of General Physiology*, vol. 20, pp. 95–104.

Fourozzi, J. (1980) Nominal Molecular Weight Distribution of Color, TOC, TTHm Precursors, and Acid Strength in a Highly Organic Portable Water Source, Master's Thesis, University of Central Florida, Orlando.

Fox K.R. (1994) Outbreaks of Waterborne Disease in the United States, *Proceedings of the AWWA Water Quality and Treatment Conference,* Denver, CO.

Fox, K.R., and Lytle, D.A. (1993) The Milwaukee Cryptosporidiosis Outbreak: Investigation and Recommendations, *Proceedings of the AWWA Water Quality and Treatment Conference,* Denver, CO.

Fritsche, A.K., A.R. Arevalo, and A.F. Connolly (1992) The Structure and Morphology of the Skin of Polyethersulfone Ultrafiltration Membranes: A Comparative Atomic Force Microscope and Scanning Electron Microscope Study, *Journal of Applied Polymer Science*, vol. 45, no. 11, p. 1945.

Fritzmann, C., J. Lowenberg, T. Wintgens, and T. Merlin (2007) State-of-the-art Reverse Osmosis, *Desalination*, vol. 216, no. 1–3, pp. 1–76.

Fronk, C.A. (1987) Removal of Low Molecular Weight Organic Contaminants from Drinking Water Using Reverse Osmosis Membranes, *Proceedings of the Annual Conference and Exhibition of the American Water Works Association,* Kansas City, MO.

Fukuzawa, K., et al. (1995) Imaging of Optical and Topographical Distributions by Simultaneous Near-Near-Field Scanning Optical-Atomic Force Microscopy with a Microfabricated Photocantilever, *Journal of Applied Physics*, vol. 78, p. 7376.

Gijsbertsen-Abrahamse, A.J., E.R. Cornelissen, and J.A.M.H. Hofman (2006) Fiber Failure Frequency and Causes of Hollow Fiber Integrity Loss, *Desalination,* vol. 194, pp. 251–258.

Goosen, M.F.A., S.S. Sablani, H. Al-Hinai, S. Al-Obeidani, R. Al-Belushi, and D. Jackson (2004) Fouling of Reverse Osmosis and Ultrafiltration Membranas: A Critical Review, *Separation Science and Technology*, vol. 39, no. 10.

Hayes, E.B., T.D. Matte, T.R. O'Brien, T. W. McKinley, G.S. Logston, J.B. Rose, B.L.P. Ungar, D.M. Word, P.F. Pinskey, M.L. Cummings, M.A. Wilson, E.G. Lung, E.S. Hurwitz, and D.D. Juranek (1981) Large Community Outbreaks of Cryptosporidiosis Due to Contamination of a Filtered Water Supply, *New England Journal of Medicine*, vol. 320, p.1372.

Heimstra, P., and M. Nederlof (1997) Reduction of Color and Hardness of Groundwater with Nanofiltration- Is Pilot-plant Operation Really Important? *Proceedings of the AWWA Membrane Technology Conference,* New Orleans.

Henthorne, L. (2005) *Evaluation of Membrane Pretreatment for Seawater RO Desalination*, U.S. Department of the Interior, Bureau of Reclamation Desalination and Water Purification Research and Development Report No. 106, Agreement No. 01-FC-81-0735.

Heuhmer, R.P., L. Henthorne, and D. Guendert (2005) Increasing MF/UF Reliably in Seawater Desalination Pretreatment Applications Using Enhanced Prefiltration. In K. Howe (ed.), *Membrane Treatment for Drinking Water and Reuse Applications: A Compendium of Peer-Reviewed Papers*, American Water Works Association, Denver, CO.

Hofman, J.A.M.H., Th.H.M. Noij, J.C. Kruithof. and J.C. Schippers (1993) Removal of Pesticides by Nanofiltration, *Proceedings Of the AWWA Membrane Technology Conference,* Reno, NV.

Hong, S., and M. Elimelech (1997) Chemical and Physical Aspects of Natural Organic Matter (NOM) Fouling of Nanofiltration Membranes, *Journal of Membrane Science*, vol. 132, no. 9, pp.159–181.

Huey, B.J., Heckler, R. Joost, G. Croves, and T. Gallier (1999) Combination of Powdered Activated Carbon and Low Pressure Membrane Filtration: A Process Alternative for SRP Water Treatment, *Proceedings of the AWWA Membrane Technology Conference,* Long Beach, CA.

IONICS Incorporated (1997) *Electrodialysis–Electrodialysis Reversal Technology*, Ionics, Incorporated:, Watertown, MA.

Ironside, R., and S. Sourirajan (1967) The Reverse Osmosis Membrane Separation Technique for Water Pollution Control, *Water Research*, vol. 1, no. 2, pp.179–180.

Jacangelo, J.G., S.S. Adham, and J.-M. Laine (1995) *Cryptosporidium, Giardia* and MS2 Virus Removal by MF and UF, *Journal AWWA*, vol. 87, no. 9, pp. 107–121.

Jacangleo J.G., J.-M. Laine, K.E. Carns, E.W. Cummings, and J. Maleville (1991) Low-Pressure Membrane Filtration for Removing Giardia and Microbial Indicators, *Journal AWWA*, vol. 83, no. 9, pp. 97–106.

Jacangelo, J.G., N.L. Patania, J.M. Laine, W. Booe, and J. Mallevialle (1992) *Low Pressure Membrane Filtration for Particle Removal.* AwwaRF report, Water Research Foundation, Denver, CO.

Jew, V., P. Schoenberger, G. Filteau, and S. Alt (2003) Desalination of Pacific Ocean Water Via Microfiltration and Reverse Osmosis, *Proceedings of the AWWA Annual Conference and Exposition,* Anaheim, CA.

Johnson, W., and T. MacCormick (2002) Issues of Operational Integrity in Membrane Drinking Water Plants, *Proceedings of the Membranes in Drinking and Industrial Water Production, 5th Conference,* Mülheim an der Ruhr, Germany: IWW Rheinisch-Westfälisches.

Jones, P.A., and J.S. Taylor (1992) DBP Control by Nanofiltration, Cost and Performance, *Journal AWWA*, vol. 84, pp. 104–116.

Jucker, C., and M.M. Clark (1994) Adsorption of Aquatic Humic Substances on Hydrophobic Ultrafiltration Membranes, *Journal of Membrane Science*, vol. 97, p. 37.

Kenna, E., and A.K. Zander (2001) Survey of Membrane Concentrate Reuse and Disposal. In *Membrane Practices for Water Treatment,* ed. S.J. Duranceau. Denver, CO: American Water Works Association.

Kothari, N., W.A. Lovins, C. Robert, S. Chen, K. Kopp, and J.S. Taylor (1997) Pilot Scale Microfiltration at Manitowoc, *Proceedings of the AWWA Membrane Technology Conference,* New Orleans.

Kouadio, P., and T. Madeleine (2002) Nanofiltration of Colored Surface Water, Quebec's Experience, *Proceedings of the Membranes in Drinking and Industrial Water Production, 5th Conference,* Mülheim an der Ruhr, Germany: IWW Rheinisch-Westfälisches

Kweon, J., and D. Lawler (2001) Evaluating Softening as a Pretreatment for Ultrafiltration, *Proceedings of the AWWA Membrane Technology Conference,* San Antonio, TX.

Lahav, O., and L. Birnhack (2007) Quality Criteria for Desalinated Water Following Post Treatment, *Desalination,* vol. 207, pp. 286–303.

Lahoussine-Turcaud, V., M.R. Wiesner, J. Yves, and J. Mallevialle (1990) Coagulation Pretreatment for Ultrafiltration of a Surface Water, *Journal AWWA*, vol. 82, no. 12, pp. 76–81.

Laine, J.M., J.G. Jacangelo, E.W. Cummings, K.E. Carnes, and J. Mallevialle. Influence of Bromide on Low Pressure Membrane Filtration for Surface Waters. *Journal AWWA*, vol. 85, pp. 87–99.

Landness, L.B. (2001) Accepting MF/UF Technology, Making the Final Cut, *Proceedings of the AWWA Membrane Technology Conference,* San Antonio, TX.

Leighton, D., and A. Acrivos (1987) The Shear-Induced Migration of Particles in Concentrated Suspensions, *Journal of Fluid Mechanics*, vol. 181, pp. 415–4311.

Lonsdale, H.K., U. Merten, and R.L. Riley (1965) Transport Properties of Cellulose Acetate Osmotic Membrane, *Journal of Applied Polymer Science*, vol. 9, p. 1341.

Lovins, W.A., S.J. Duranceau, T. King, and S. Medeiros (2004) Trumping Hydrogen Sulfide, *Water Environment & Technology*, vol. 16, no. 3, pp. 37–41.

Lovins, W., L. Mulford, and J. Taylor (1997) Modeling for Membrane Selection, *Proceedings of the AWWA Membrane Technology Conference,* New Orleans.

Lozier, J., M. Kitis, C. Colvin, J. Kim, B. Mi, and B. Marinas (2003) *Microbial Removal and Integrity Monitoring of High-Pressure Membranes*, AwwaRF report. Denver, CO:Water Research Foundation and U.S. Department of the Interior, Bureau of Reclamation.

Ma, W., Y. Zhao, and L. Wang (2007) The Pretreatment with Enhanced Coagulation and a UF Membrane for Seawater Desalination with Reverse Osmosis, *Desalination*, vol. 203, no. 1–3 pp. 256–259.

Machenbach, I., S. Andersensvei, T. Leiknes, and H. Ødegaard (2002) Coagulation/Submerged Hollow-Fiber Ultrafiltration for NOM Removal, *Proceedings of the Membranes in Drinking and Industrial Water Production, 5th Conference,* Mülheim an der Ruhr, Germany: IWW Rheinisch-Westfälisches.

Mallevialle, J., P.E. Odendall, and M.R. Wiesner (1996) *Water Treatment Membrane Processes*, American Water Works Association, Lyonnaise des Eaux, Water Research Commission of South Africa, McGraw-Hill.

Matsuura, T., Y. Taketani, and S. Sourirajan (1983) Interfacial Parameters Governing Reverse Osmosis for Different Polymer Material—Solution Systems Through Gas and Liquid Chromatography Data, *Journal of Colloid and Interface Science*, vol. 95, no. 1, pp.10–22.

Merten, U (1966) *Desalination by Reverse Osmosis*. Cambridge, MA.: MIT Press.

Metcalf, P.J. (1992) Water Science and Technology, *Proceedings of the IAWPRC,* Cape Town, SA.

Mickley, M., R. Hamilton, and J. Truesdall (1990) *Membrane Concentrate Disposal.* Denver, CO: American Water Works Association.

Miltner, R.J., C.A. Fronk, and T.F. Speth (1987) Removal of Alachlor from Drinking Water. *Proceedings of the National Conference on Environmental Engineering,* Orlando, FL.

Montomery, J.M. (2005) *Water Treatment Principles and Design.* New York: John Wiley & Sons.

Morris, K.M. (1990) Predicting Fouling in Membrane Separation Processes, Master Thesis, University of Central Florida, Orlando.

Mosqueda-Jimenez, D.B., P.M. Huck, and O.D. Basu (2008) Fouling Characteristics of an Ultrafiltration Membrane Used in Drinking Water Treatment, *Desalination,* vol. 230, pp. 79–91.

Nalco Chemical Company (1988) *Nalco Water Handbook.* New York: McGraw-Hill.

Oldani, M., and G. Schock (1981) Characterization of Ultrafiltration Membranes by Infrared Spectroscopy, ESCA, and Contact Angle Measurements, *Journal of Membrane Science*, vol. 3, p. 243.

Olivieri, V.P., G. A. Willingham, J.C. Vickers, and C. McGahey (1991) Continuous Microfiltration of Secondary Effluent, *Proceedings of the Technologies Conference in the Water Industry,* Orlando, FL.

Onsager, L. (1931) Reciprocal Relations in Irreversible Processes. I. *Physical Review,* vol. 37, p. 405.

Owadally, H. (2002) Fouling Effects of Hydrocarbons on Reverse Osmosis Membranes, Postgraduate Colloquium, Department of Mechanical Engineering, University of Glasgow, Glasgow, UK.

Pressdee, J.R., S.Veerapaneni, H.L. Shortney-Darby, and J. Clement (2006) *Integration of Membrane Filtration into Water Treatment Systems*, AwwaRF project 2765. Denver, CO: Water Research Foundation.

Rangarajan, R., E.C. Matsuura, and S. Sourirajan (1976) Free Energy Parameter for Reverse Osmosis Separation of Some Inorganic Ions and Ion Pairs in Aqueous Solutions, *I & EC Process Design and Development*, vol. 15, no. 4, pp. 529–534.

Rautenbach, R., and R. Albrecht (1981) *Membrane Processes*. New York: John Wiley & Sons.

Reinhard, M., N.L. Goodman, P.L. McCarty, and D.G. Argo (1986) Removing Trace Organics Using Cellulose Acetate and Polyamide Membranes, *Journal AWWA,* vol. 78, pp. 163–174.

Ridgway, H. F., and H.-C. Flemming (1996) *Membrane Biofouling, Water Treatment Membrane Processes.* New York: McGraw-Hill.

Rocquebert, V., G. Gozes, B. Alibert, and D. Leaf (2002) Identifying and Resolving Key Design Considerations for Integration of Low Pressure Membrane Filtration and Lime Softening Processes, *Proceedings of the Membranes in Drinking and Industrial Water Production, 5th Conference,* Mülheim an der Ruhr, Germany: IWW Rheinisch-Westfälisches.

Rosberg, R. (1997) Ultrafiltration, A Viable Cost-Saving Pretreatment for Reverse Osmosis, *Desalination*, vol. 110, pp. 107–114

Ruggieri, F., J.L. Fernandez-Turiel, D. Gimeno, F. Valero, J.C. Garcia, and M.E. Medina (2008) Limestone Selection Criteria for EDR Water Remineralization, *Desalination*, vol. 227, pp. 314–326

Schideman, L., M. Kosterman, and L. Rago (2001) Performance and Optimization of Hybrid Membrane-Adsorption Systems–A Case Study for the Racine Water Utility, *Proceedings of the AWWA Membrane Technology Conference,* Atlanta.

Schippers, J.C., and J. Verdouw (1980) The Modified Fouling Index, a Method of Determining the Fouling Characteristics of Water, *Desalination.* vol. 32, pp. 137–148.

"Schneider, C., P. Johns, and R. Huehmer (2001) Removal of Manganese by Microfiltration in a Water Treatment Plant, *Proceedings of the AWWA Membrane Technology Conference,* San Antonio, TX.

Schonberg, J.A., and E.J. Hinch (1988) Inertial Migration of a Sphere in Poiseuille Flow, *Journal of Fluid Mechanics*, vol. 203, pp. 517–524.

Sethi, S., and M.R. Wiesner (1997) Modeling the transient permeate flux in crossflow membrane filtration incorporating multiple transport mechanisms, *Journal of Membrane Science*, vol. 136, nos. 1–2, pp. 191–205.

Sethi, S., G. Crozes, D. Hugaboom, B. Mi, J. Curl, and B. Marinas (2003) Results and Assessment from Full-scale Testing of Integrity Monitoring Methods for MF and UF Processes, *Proceedings of the AWWA Membrane Technology Conference,* Atlanta.

Shane, S.A., B.J. Vanderford, J. Drewes, E. Dickenson, E.M. Snyder, G.M. Bruce, and R.C. Pleus (2008) *State of Knowledge of Endocrine Disruptors and Pharmaceuticals in Drinking Water.* AwwaRF report. Denver, CO: Water Research Foundation.

Shorney, H., W. Vernon, J. Clune, and R. Bond (2001) Performance of MF/UF Membranes with In-Line Ferric-Salt Coagulation for the Removal of Arsenic from a Southwest Surface Water, *Proceedings of the AWWA Annual Conference,* New York.

Song, L., and M. Elimelech (1995) Particle Deposition onto a Permeable Surface in Laminar Flow, *Journal of Colloid and Interface Science*, vol. 173, no. 7, pp. 165–180.

Sourirajan, S. (1970) *Reverse Osmosis.* New York: Academic Press.

Spevack, P., and Y. Deslandes (1996) TOF-SIMS Analysis of Adsorbate–Membrane Interactions, 1. Adsorption of Dehydroabietic Acid on PVDE, *Applied Surface Science*, vol. 99, p. 41.

Spangenburg, C., A. Kalinsky, and E. Akiyoshi (1997) Selection, Evaluation and Optimization of Organic Selective Membranes for Color and DBP Precursor Removal, *Proceedings of the AWWA Membrane Technology Conference,* New Orleans.

Strohwald, N.K.H., and E.P. Jacobs (1992) An Investigation into UF Systems in the Pretreatment of Sea Water for RO Desalination, *Water Science and Technology*, vol. 25, no. 10, pp. 69–78.

Sung, L. (1993) Modeling Mass Transfer in Nanofiltration, Ph.D. Dissertation, University of Central Florida, Orlando.

Taylor, J.S., and S.J. Duranceau (2000) Membrane Case Studies, Past, Present, Future. In_*Membrane Technology in Water and Wastewater Treatment,* ed. P. Hillis. London: Royal Society of Chemistry.

Taylor, J.S., and E.P. Jacobs (1996) Nanofiltration and Reverse Osmosis, in *Membrane Water Treatment Membrane Processes.* New York: McGraw-Hill.

Taylor, J. S., S.J. Duranceau, J. Goigel, and W. Barrett (1989) *Assessment of Potable Water Membrane Application and Research Needs.* AwwaRF report. Denver, CO: Water Research Foundation.

Taylor, J.S., L.A. Mulford, S.J. Duranceau, and W.M. Barrett (1989b) Cost and Performance of a Membrane Pilot Plant, *Journal AWWA*, vol. 81, no. 11, pp. 52–60.

Taylor, J.S., D. Thompson, J.K. Carswell, and J. Keith (1987) Applying Membrane Processes to Groundwater Sources for Trihalomethane Precursor Control. *Journal AWWA,* vol. 79, no. 8, pp. 72–82.

Taylor, J.S., L.A. Mulford, W.M. Barrett, S.J. Duranceau, and D.K. Smith (1989a) *Cost and Performance of Membrane Processes for Organic Control on Small Systems.* Cincinnati, OH: U.S. Environmental Protection Agency Water Engineering Research Laboratory.

Taylor, J.S. (1984) Potable Water Reuse in the U.S.A., *Proceedings of the AWWA Annual Conference and Exposition,* Las Vegas, NV.

Taylor, J.S., D. Thompson, B.R. Snyder, J. Less, and L. Mulford (1986) *Cost and Performance Evaluation of In-Plant Trihalomethane Control Techniques*, EPA 600/2-85-168. Cincinnati, OH: U.S. Environmental Protection Agency.

Taylor, J.S., D. Thompson, and L. Mulford (1988) Comparison of Membrane Processes at Ground and Surface Water Sites, *Proceedings of the Annual Conference AWWA,* Orlando, FL.

Tan, L., and G.L. Amy (1991) Comparing Ozonation and Membrane Separation for Color Removal and Disinfection By-product Control, *Journal AWWA*, vol. 83, pp. 74–711.

Truesdall, J., M. Mickley, and R. Hamilton (1995) Survey of Membrane Drinking Water Plant Disposal Methods, *Desalination*, vol. 102, no. 1–3, pp.93–105.

U.S. Environmental Protection Agency (2001) *Low Pressure Membrane Filtration for Pathogen Removal Application, Implementation and Regulatory Issues*, EPA 815-C-01-001. Washington, DC: U.S. Environmental Protection Agency.

Van Hoof, S.C.J.M., J.G. Minnery, and B. Mack (2001) Dead-end Ultrafiltration as Alternative Pre-treatment to Reverse Osmosis in Seawater Desalination: A Case Study, *Desalination*, vol. 139, pp. 161–181.

Van Hoof, S.C.J.M., L. Boens, A. Nahrstedt, S. Panglisch, and R. Gimbel (2002) Development of a New Integrity Testing System, *Proceedings of the Membranes in Drinking and Industrial Water Production, 5th Conference,* Mülheim an der Ruhr, Germany: IWW Rheinisch-Westfälisches.

Vial, D., G. Doussaur and R. Galindo (2003) Comparison of Three Pilot Studies Using Microza Membranes for Mediterranean Seawater Pre-treatment, *Desalination*, vol. 156, pp. 43–50.

Weber, W.J. (1972) *Physicochemical Process for Water Quality Control.* New York: John Wiley & Sons.

Weigand, R.J., F.W. Altena, and G. Belfort (1985) Lateral Migration of Spherical Particles in laminar Porous Tube Flows: Application to Membrane Filtration, *Physicochem. Hydrodyn.,* vol. 6, pp. 393–413.

Whittaker, H., and R. Clark (1985b) Cleanup of PCB Contaminated Groundwater by Reverse Osmosis, *Proceedings of the 2nd Annual Technical Seminar on Chemical Spills,* Toronto, CAN.

Whittaker, H., and T. Szaplonczay (1985a) Testing of Reverse Osmosis on Chemical Solutions, *Proceedings of the 2nd Annual Technical Seminar on Chemical Spills,* Toronto, CAN.

WHO (2004) *WHO Guidelines for Drinking Water Quality.* Geneva: WHO Press.

Wiesner, M.R, M.M. Clark, and J. Mallevialle (1989) Membrane Filtration of Coagulated Suspensions, *Journal of Environmental Engineering: American Society of Civil Engineers*, vol. 115, no. 1, pp. 20–40.

Withers, A. (2005) Options for Recarbonation, Remineralisation and Disinfection for Desalination Plants, *Desalination*, vol. 179, no. 1–3, pp. 11–24.

Yoo, R.S., D.R. Brown, R.J. Pardini, and G.D. Bentson (1995) Microfiltration: A Case Study, *Journal AWWA*, vol. 38, pp. 411.

Zhu, X., and M. Elimelech (1997) Colloidal Fouling of Reverse Osmosis Membranes: Measurements and Fouling Mechanisms, *Environmental Science and Technology,* vol. 31, pp. 3654–3662.

CHAPTER 12
ION EXCHANGE AND ADSORPTION OF INORGANIC CONTAMINANTS

Dennis Clifford, Ph.D., P.E.. B.C.E.E.
Thomas and Laura Hsu Professor and Director of Environmental Engineering
Department of Civil and Environmental Engineering
University of Houston
Houston, Texas, United States

Thomas J. Sorg, P.E., B.C.E.E.
Research Engineer
U.S. Environmental Protection Agency
Cincinnati, Ohio, United States

Ganesh L. Ghurye, Ph.D., P.E., B.C.E.E.
Engineering Associate
ExxonMobil Upstream Research Company
Houston, Texas, United States

OVERVIEW **12.2**
INTRODUCTION AND THEORY OF ION EXCHANGE **12.2**
Uses of IX in Water Treatment 12.3
Past and Future of IX 12.4
IX Materials and Reactions 12.4
IX Equilibrium 12.11
IX and Adsorption Kinetics 12.17
Column Processes and Calculations 12.19
Design Considerations 12.25
APPLICATIONS OF IX AND ADSORPTION **12.27**
Sodium IX Softening 12.27
Hydrogen IX Softening 12.33
Barium Removal by IX 12.33
Radium Removal by IX 12.34
Nitrate Removal by IX 12.35
Fluoride Removal by Activated Alumina 12.41
Arsenic Removal by Adsorptive Media and IX 12.46
Arsenic Removal by IX 12.55
Natural Color and Organics Removal by Resins 12.61
Uranium Removal by Anion Exchange 12.67
Perchlorate Removal by Anion Exchange 12.73
IX MODELING USING EMCT **12.74**
Exhaustion or Breakthrough Curves and Resin Profile 12.74
Modeling Arsenic Removal on SBA Resins 12.76
Modeling Nitrate Removal: Comparing Conventional Resins with Nitrate-Selective Resins 12.77
Modeling Perchlorate Removal: Comparing Conventional Resins with Nitrate-Selective Resins 12.79
WASTE DISPOSAL **12.82**
SUMMARY **12.82**
ABBREVIATIONS **12.87**
NOTATION FOR EQUATIONS **12.89**
REFERENCES **12.90**

OVERVIEW

In the first part of this chapter, the fundamentals of ion exchange (IX) and adsorption processes are explained, with the goal of demonstrating how these principles influence process design for inorganic contaminant removal. In the second part, ion exchange and adsorption processes that have been proven effective at bench, pilot, and full scale are described for the removal of hardness, barium, radium, nitrate, fluoride, arsenic, dissolved organic carbon (DOC), uranium, and perchlorate. For each contaminant, the factors that influence the choice and design of a process, for example, pH, total dissolved solids (TDS), contaminant speciation, competing ions, resins, adsorbents, foulants, regenerants, and column flow patterns, are discussed. In the third and final part of the chapter, ion exchange modeling using equilibrium multicomponent chromatographic theory (EMCT) with constant separation factors (CSFs) is covered. Summary tables for cation and anion contaminants are included at the end of the chapter to aid the reader in process selection. The basic features, advantages, and disadvantages of the ion exchange and metal oxide/hydroxide packed bed processes covered in this chapter are summarized in Table 12-1.

INTRODUCTION AND THEORY OF ION EXCHANGE

Contaminant cations such as calcium, magnesium, radium, barium, and strontium and anions such as fluoride, nitrate, fulvates, humates, arsenate, selenate, chromate, perchlorate, and anionic complexes of uranium can be removed from water using ion exchange

TABLE 12-1 Advantages and Disadvantages of Packed Bed Inorganic Contaminant-Removal Processes

Ion exchange

Advantages

- Operates on demand.
- Relatively insensitive to flow variations, short contact time required.
- Relatively insensitive to trace level contaminant concentration.
- Essentially zero level of effluent contaminant possible.
- Large variety of "tailored" resins available for specific target ions.
- Beneficial selectivity reversal commonly occurs on regeneration.
- In some applications, spent regenerant may be reused without contaminant removal.

Disadvantages

- Potential for chromatographic effluent peaking when using single beds.
- Variable effluent quality with respect to background ions when using single beds.
- Usually not feasible at high levels of sulfate or total dissolved solids.
- Large volume/mass of regenerant must be used and disposed of.

Metal oxide/hydoxide adsorption

Advantages

- Operates on demand.
- Relatively insensitive to total dissolved solids and sulfate levels.
- Low effluent contaminant level possible.
- Highly selective for fluoride and arsenic.

Disadvantages

- Both acid and base are required for regeneration when regeneration is possible.
- Relatively sensitive to trace level contaminant concentration.
- Media tend to dissolve, producing fine particles.
- Slow adsorption kinetics and relatively long contact time required.
- Significant volume/mass of spent regenerant to neutralize and dispose of.

with resins or adsorption onto granular hydrous metal oxides/hydroxides such as activated alumina (AA), granular ferric oxide (GFO), and granular ferric hydroxide (GFH) or coagulated Fe(II), Fe(III), Al(III), Ti(III), Ti(IV), Zr(III), Zr(IV), and Mn(IV) surfaces. This chapter deals only with the theory and practice of IX with resins and adsorption with granular media including GFH, GFO, and AA. The reader interested in cation and anion adsorption onto hydrous metal oxides in general is referred to Schindler's and Stumm's publications on the solid-water interface (Schindler, 1981; Stumm, 1992) as a starting point.

IX with synthetic resins and adsorption onto hydrous metal oxides/hydroxides are water treatment processes in which a presaturant ion on the solid phase, the *adsorbent*, is exchanged for an unwanted ion in the water. In order to accomplish the exchange reaction, a packed bed of IX resin beads or metal oxide/hydroxide granules is used. Source water is continually passed through the bed in a downflow or upflow mode until the adsorbent is exhausted, as evidenced by the appearance (*breakthrough*) of the unwanted contaminant at an unacceptable concentration in the effluent.

The most useful IX reactions are reversible. In the simplest cases, the exhausted bed is regenerated using an excess of the presaturant ion. Ideally, no permanent structural change takes place during the exhaustion/regeneration cycle. (Resins do swell and shrink, however, and the metal oxides/hydroxides may be slightly dissolved during regeneration.) When the reactions are reversible, the medium can be reused many times before it must be replaced because of irreversible fouling or, in the case of metal oxides/hydroxides, excessive attrition. In a typical water supply application, from 300 to as many as 300,000 bed volumes (BV) of contaminated water may be treated before the medium is exhausted. Regeneration typically requires from 1 to 5 BV of regenerant followed by 2 to 20 BV of rinse water. These wastewaters generally amount to less than 2 percent of the product water; nevertheless, their ultimate disposal is a major consideration in IX and adsorption design practice. Disposal of the spent medium also may present a problem if it contains toxic or radioactive substances such as nitrate, arsenic, perchlorate, uranium, and/or radium.

Uses of IX in Water Treatment

By far the largest application of IX to drinking water treatment is water softening, that is, the removal of calcium, magnesium, and other polyvalent cations, including Ra^{2+}, Ba^{2+}, Sr^{2+}, Fe^{2+}, and Mn^{2+}, in exchange for sodium. The IX softening process is applied to residential, industrial, and municipal water treatment. In residential use, IX can be applied for whole-house [*point-of-entry* (POE)] softening or for softening only the water that enters the hot water heater. For industrial use, generally all the water is passed through the resin to lower the hardness to zero, whereas in municipal water treatment some of the water is bypassed around the softener because some hardness is desirable in the treated water. Radium and barium are ions more preferred by the resin than calcium and magnesium; thus the former are also removed effectively during IX softening. Resins beds containing chloride-form anion exchange resins can be used to remove nitrate, arsenate, perchlorate, chromate, selenate, natural organic matter (NOM)/color, uranium, perchlorate, and other contaminant anions. GFO, GFH, and zirconium and titanium oxides/hydroxides are being used to remove arsenate, whereas AA is being used to remove fluoride. These adsorption processes are especially useful for groundwater treatment in small systems and for residential *point-of-use* (POU) and POE systems.

The choice between IX and metal oxide/hydroxide adsorption, for example, to remove arsenic from water, is largely determined by (1) the background water quality including TDS level, competing ions, alkalinity, and contaminant concentration and (2) the resin or adsorbent medium affinity for the contaminant ion in comparison with competing ions. The affinity sequence determines the run length, chromatographic peaking (if any), and process costs. As mentioned previously, process selection is affected by spent regenerant and spent medium disposal requirements and regenerant reuse possibilities, particularly if hazardous contaminants are involved. Each of these requirements is dealt with in some detail in the upcoming design sections for the specific processes.

Past and Future of IX

Natural zeolites, which are crystalline aluminosilicate minerals, were the first ion exchangers, and they were used to soften water on an industrial scale. Later, zeolites were completely replaced with synthetic resins because of the latter's faster exchange rates, higher capacity, greater resilience, and longer life. Aside from softening, the use of IX and inorganic adsorption for removal of specific contaminants from municipal water supplies has been limited owing to the perceived expense involved in removing what have been seen as minimal health risks resulting from contaminants such as fluoride and nitrate. The production of pure and ultrapure water by IX demineralization (DM) is the largest use of IX resins on an industrial scale. The complete removal of contaminants, which occurs in DM processes, is not necessary for drinking water treatment, however. Furthermore, treatment costs are high compared with membrane processes (reverse osmosis and electrodialysis) for desalting water (see Chap. 11).

Adherence to governmentally mandated maximum contaminant level (MCL) goals for inorganic chemicals (IOCs) will result in more use of IX and adsorption systems for small community water treatment to remove barium, radium, cadmium, chromium, mercury, lead, fluoride, arsenic, nitrate, perchlorate, uranium, and other IOCs.

IX Materials and Reactions

An IX resin consists of a cross-linked polymer matrix to which charged functional groups are attached by covalent bonding. The usual matrix is polystyrene cross-linked for structural stability with 3 to 8 percent divinylbenzene (abbreviated STY-DVB or simply polystyrene). The common functional groups fall into four categories: strongly acidic (e.g., sulfonate, $—SO_3^-$), weakly acidic (e.g., carboxylate, $—COO^-$), strongly basic [e.g., quaternary amine, $—N(CH_3)_3^+$], and weakly basic [e.g., tertiary amine, $—N(CH_3)_2$] (Helfferich, 1962).

A schematic presentation illustrating resin matrix cross-linking and functionality is shown in Fig. 12-1*a*. The schematic depicts a three-dimensional bead (sphere) made up of many polystyrene polymer chains held together by divinylbenzene cross-linking. The negatively charged IX sites ($—SO_3^-$) in a cation exchanger and the positively charged sites [$—N(CH_3)_3^+$] in an anion exchanger are shown in Fig. 12-1*b* fixed to the resin backbone, or *matrix* as it is called. Mobile positively charged counterions (positive charges in Fig. 12-1*a*) are associated by electrostatic attraction with each negative IX site on a cation exchange resin. The resin exchange *capacity* is measured as the number of fixed charge sites per unit volume or weight of resin. *Functionality* is the term used to identify the chemical composition of the fixed charge site, for example, sulfonate ($—SO_3^-$) or quaternary ammonium [$—N(CH_3)_3^+$]. *Porosity*, expressed by the terms *microporous/gel*, *porous*, and *macroporous*, is the resin characterization referring to the degree of openness of the polymer structure. An actual resin bead is much tighter than on the schematic, which is shown as fairly open for purposes of illustration only. The water (40–60% by weight) present in a typical resin bead is not shown. This resin-bound water is an extremely important characteristic of ion exchangers because it strongly influences both the exchange kinetics and thermodynamics.

Strong- and Weak-Acid Cation Exchangers. *Strong-acid cation* (SAC) exchangers operate over a wide pH range because the sulfonate group, being strongly acidic, is ionized

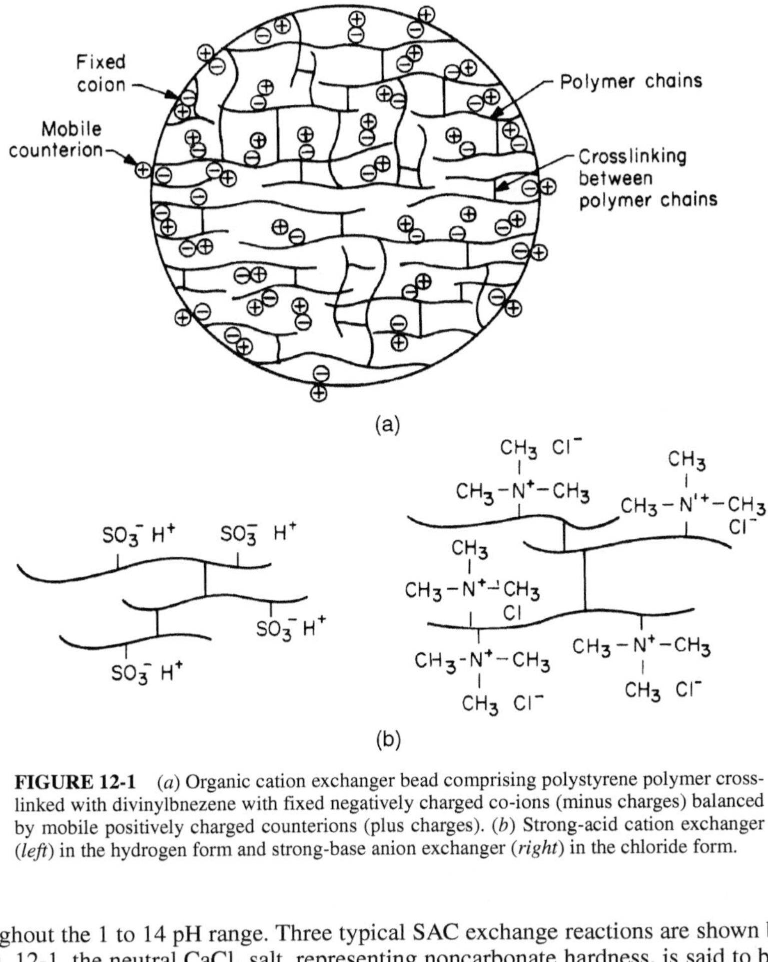

FIGURE 12-1 (*a*) Organic cation exchanger bead comprising polystyrene polymer cross-linked with divinylbnezene with fixed negatively charged co-ions (minus charges) balanced by mobile positively charged counterions (plus charges). (*b*) Strong-acid cation exchanger (*left*) in the hydrogen form and strong-base anion exchanger (*right*) in the chloride form.

throughout the 1 to 14 pH range. Three typical SAC exchange reactions are shown below. In Eq. 12-1, the neutral $CaCl_2$ salt, representing noncarbonate hardness, is said to be *split* by the resin, and hydrogen ions are exchanged for calcium, even though the equilibrium liquid phase is acidic because of HCl production. Equations 12-2 and 12-3 are the standard IX softening reactions in which sodium ions are exchanged for the typical hardness ions: Ca^{2+}, Mg^{2+}, Fe^{2+}, Ba^{2+}, Sr^{2+}, and/or Mn^{2+}, either as noncarbonate hardness (Eq. 12-2) or as carbonate hardness (Eq. 12-3). In all these reactions, R denotes the resin matrix, and the overbar indicates the solid (resin) phase.

$$2\overline{RSO_3^- H^+} + CaCl_2 \leftrightarrow \overline{(RSO_3^-)_2Ca^{2+}} + 2HCl \tag{12-1}$$

$$2\overline{RSO_3^- Na^+} + CaCl_2 \leftrightarrow \overline{(RSO_3^-)_2Ca^{2+}} + 2NaCl \tag{12-2}$$

$$2\overline{RSO_3^- Na^+} + Ca(HCO_3)_2 \leftrightarrow \overline{(RSO_3^-)_2Ca^{2+}} + 2NaHCO_3 \tag{12-3}$$

Regeneration of the spent resin is accomplished using an excess of concentrated (0.5–3 M) HCl or NaCl, and amounts to the reversal of Eqs. 12-1 through 12-3. *Weak-acid cation* (WAC) resins can exchange ions only in the neutral to alkaline pH range because the functional group, typically carboxylate ($pK_a = 4.8$), is not ionized at low pH. Thus WAC resins can be used for carbonate hardness removal (Eq. 12-4) but fail to remove noncarbonate hardness, as is evident in Eq. 12-5.

$$\overline{2RCOOH} + Ca(HCO_3)_2 \rightarrow \overline{(RCOO^-)_2Ca^{2+}} + H_2CO_3 \tag{12-4}$$

$$\overline{2RCOOH} + CaCl_2 \leftarrow \overline{(RCOO^-)_2Ca^{2+}} + 2HCl \tag{12-5}$$

If Eq. 12-5 proceeded to the right, the HCl produced would be so completely ionized that it would protonate, that is, add a hydrogen ion to the resin's weakly acidic carboxylate functional group and prevent exchange of H^+ ions for Ca^{2+} ions. Another way of expressing the fact that Eq. 12-5 does not proceed to the right is to say that WAC resins will not *split neutral salts*; that is, they cannot remove noncarbonate hardness. This is not the case in Eq. 12-4, in which the *basic* salt $Ca(HCO_3)_2$ is *split* because a weak acid, H_2CO_3 ($pK_1 = 6.3$), is produced.

In summary, SAC resins split basic and neutral salts (remove carbonate and noncarbonate hardness), whereas WAC resins split only basic salts (remove only carbonate hardness). Nevertheless, WAC resins have some distinct advantages for softening, namely, TDS reduction, no increase in sodium, and very efficient regeneration resulting from the carboxylate's high affinity for the regenerant H^+ ion.

Strong- and Weak-Base Anion Exchangers. The use of *strong-base anion* (SBA) exchange resins for nitrate removal is a fairly recent application of IX for drinking water treatment (Clifford and Weber, 1978; Guter, 1981), although anion resins have been used in water demineralization for more than 50 years. In anion exchange reactions with SBA resins, the quaternary amine functional group [$—N(CH_3)_3^+$] is so strongly basic that it is ionized, and therefore useful as an ion exchanger, over the pH range of 0 to 13 (Helfferich, 1962). This is shown in Eqs. 12-6 and 12-7, in which nitrate is removed from water by using hydroxide- or chloride-form SBA resins. Note that R_4N^+ is another way to write the quaternary exchange site $—N(CH_3)_3^+$.

$$\overline{R_4N^+OH^-} + NaNO_3 \leftrightarrow \overline{R_4N^+NO_3^-} + NaOH \tag{12-6}$$

$$\overline{R_4N^+Cl^-} + NaNO_3 \leftrightarrow \overline{R_4N^+NO_3^-} + NaCl \tag{12-7}$$

In Eq. 12-6, the NaOH produced is completely ionized, but the quaternary ammonium functional group has such a small affinity for OH^- ions that the reaction proceeds to the right. Eq. 12-7 is a simple ion-exchange reaction without a pH change. Fortunately, all SBA resins have a higher affinity for nitrate than chloride (Clifford and Weber, 1978) and Eq. (12-7) proceeds to the right at near-neutral pH values.

Weak-base anion (WBA) exchange resins are useful only in the acidic pH region, where the primary, secondary, or tertiary amine functional groups (Lewis bases) are protonated, and can act as positively charged exchange sites for anions. In Eq. 12-8, chloride is, in effect, being adsorbed by the WBA resin as hydrochloric acid, and the TDS level of the solution is being reduced. In this case, a positively charged Lewis acid-base adduct (R_3NH^+) is formed that can act as an anion exchange site. As long as the solution in contact with the resin remains acidic (just how acidic depends on the basicity of the R_3N; sometimes pH ≤ 6 is adequate), IX can take place, as is indicated in Eq. 12-9—the exchange of chloride for nitrate by a WBA resin in acidic solution. If the solution is neutral or basic, no adsorption or exchange can take place, as indicated by Eq. 12-10. In all these reactions, R represents either the resin matrix or a functional group, such as $—CH_3$ or $—C_2H_5$, and the overbars represent the resin phase.

$$\overline{R_3N:} + HCl \leftrightarrow \overline{R_3NH^+Cl^-} \tag{12-8}$$

$$\overline{R_3NH^+Cl^-} + HNO_3 \leftrightarrow \overline{R_3NH^+NO_3^-} + HCl \tag{12-9}$$

$$\overline{R^3N:} + NaNO_3 \rightarrow \text{no reaction} \tag{12-10}$$

Although WBA resins are not used commonly for drinking water treatment, such processes are possible as long as the exchange pH is in the 4 to 6 range, which can be costly because this requires neutralization of most of the alkalinity present. Furthermore, activated alumina, when used for fluoride and/or arsenic removal, acts as if it were a WBA exchanger, and the same general rules regarding pH behavior can be applied. Another advantage of weak-base resins in water supply applications is the ease with which they can be regenerated with bases. Even weak bases such as ammonium hydroxide (NH_4OH) and lime ($Ca(OH)_2$) can be used, and regardless of the base used, only a small stoichiometric excess (<20 percent) normally is required for complete regeneration.

Granular Media Adsorption. Packed beds of AA, titanium and zirconium oxides/hydroxides, GFO and GFH, and iron oxides coated onto or incorporated into various media, including anion resins (Cumbal and Sengupta, 2005), diatomaceous earth (NSF International, 2005), and sand (Benjamin et al., 1996), can be used to remove contaminant anions, including fluoride, arsenic, selenium, silica, phosphate, vanadate, and natural organic matter (fulvate) anions from water. A simple schematic of the contaminant adsorption process is shown in Fig. 12-2, where the contaminant (Adsorbate-A) attaches to an

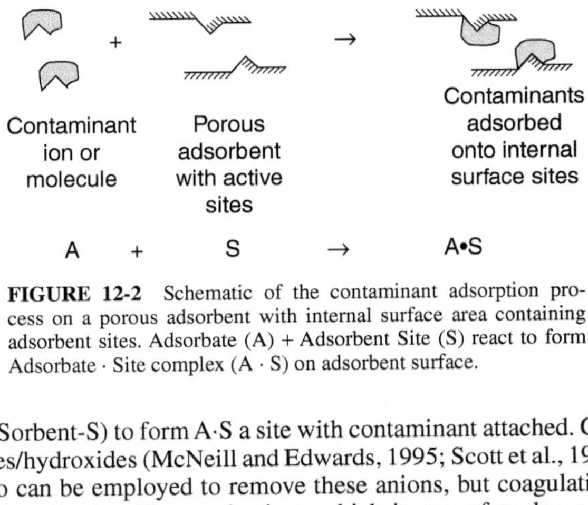

FIGURE 12-2 Schematic of the contaminant adsorption process on a porous adsorbent with internal surface area containing adsorbent sites. Adsorbate (A) + Adsorbent Site (S) react to form Adsorbate · Site complex (A · S) on adsorbent surface.

adsorption site (Sorbent-S) to form A·S a site with contaminant attached. Coagulated Fe(II) and Fe(III) oxides/hydroxides (McNeill and Edwards, 1995; Scott et al., 1985; Lakshmanan et al., 2008) also can be employed to remove these anions, but coagulation processes are not covered in this chapter. The mechanism, which is one of exchange of contaminant anions for surface hydroxides on the alumina or iron oxides/hydroxides, generally is called *sorption* or *adsorption*, although *ligand exchange* is a more appropriate term for the highly specific surface reactions involved (Stumm, 1992). Figure 12-3 depicts the ligand exchange of arsenate for hydroxide on the surface of ferric oxide/hydroxide (FeOOH). Similar reactions can occur on the surfaces of activated alumina (AlOOH) and metal oxides/hydroxides (MeOOH), including zirconium and titanium.

The typical activated aluminas used for fluoride removal in water treatment are 28 × 48 mesh (0.3–0.6 mm diameter) mixtures of amorphous and gamma aluminum oxide (γ-Al_2O_3) prepared by low-temperature (300°C–600°C) dehydration of precipitated $Al(OH)_3$. These highly porous materials have surface areas of 50 to 300 m^2/g. The granular ferric oxides/hydroxides (GFO and GFH) used for arsenate removal are prepared by similar methods and also have surface areas in the 50 to 300 m^2/g range. Using the model of a hydroxylated alumina surface subject to protonation and deprotonation, the following ligand (fluoride) exchange reaction (Eq. 12-11) can be written for fluoride adsorption in acid solution (alumina exhaustion), in which AlO represents the alumina surface and an

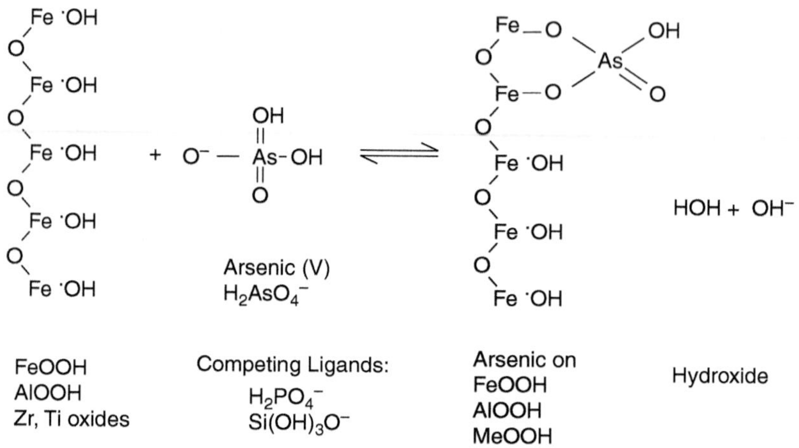

FIGURE 12-3 Schematic of ligand exchange of hydroxide for arsenate on ferric oxide/hydroxide surface. Similar reactions may occur for other metal oxides/hydroxides (MeOOH).

overbar denotes the solid phase. FeOOH surfaces undergo similar reactions with arsenate ($H_2AsO_4^-$) as the ligand (see Fig 12-2).

$$\overline{\text{AlO}-\text{OH}}+\text{H}^+ +\text{F}^- \rightarrow \overline{\text{AlO}-\text{F}}+\text{HOH} \tag{12-11}$$

The equation for fluoride desorption by hydroxide (alumina regeneration) is presented in Eq. 12-12.

$$\overline{\text{AlO}-\text{F}}+\text{OH}^- \rightarrow \overline{\text{AlO}-\text{OH}}+\text{F}^- \tag{12-12}$$

Metal oxide/hydroxide surfaces, including AA, GFH, GFO, and etc., are sensitive to pH, and anions are best adsorbed below the *zero point of charge* (ZPC), where the metal oxide surface has a net positive charge, and excess protons are available to fuel Eq. 12-11. Above the ZPC (pH 8.2), alumina is predominantly a cation exchanger, but its use for cation exchange is relatively rare in water treatment. An exception is encountered in the removal of radium by plain and treated activated alumina (Clifford et al., 1988; Garg and Clifford, 1992).

Ligand exchange, as indicated in Eqs. 12-11 and 12-12, occurs chemically at the internal and external surfaces of activated alumina, GFH, and GFO. A more useful model for process design, however, is one that assumes that the adsorption of fluoride or arsenic onto a metal oxide/hydroxide surface at the optimal pH of 5.5 to 6 is analogous to weak-base anion exchange. For example, the uptake of F^- or $H_2AsO_4^-$ requires the protonation of the metal oxide/hydroxide surface, which is sometimes accomplished by preacidification of the medium with HCl or H_2SO_4 and/or reducing the feed water pH into the 5.5 to 6.0 region. The positive charge caused by excess surface protons then may be viewed as being balanced by exchanging anions, that is, ligands such as hydroxide, fluoride, and arsenate. To reverse the adsorption process and remove the adsorbed fluoride or arsenate, an excess of strong base, for example, NaOH, must be applied. The following series of reactions (Eqs. 12-13 through 12-17) is presented as a model of the adsorption/regeneration cycle that is useful for design purposes. Although less common, these reactions also may be applied to GFH and GFO processes.

The first step in the cycle is acidification, in which neutral (water-washed) alumina (alumina·HOH) is treated with acid, for example, HCl, and protonated (acidic) alumina is formed as follows:

$$\overline{\text{alumina} \cdot \text{HOH}} + \text{HCl} \rightarrow \overline{\text{alumina} \cdot \text{HCl}} + \text{HOH} \qquad (12\text{-}13)$$

When HCl-acidified alumina is contacted with fluoride ions, the chloride ions are strongly displaced, provided that the alumina surface remains acidic (pH 5.5–6). This displacement of chloride by fluoride, analogous to ion exchange, is shown as

$$\overline{\text{alumina} \cdot \text{HCl}} + \text{HF} \rightarrow \overline{\text{alumina} \cdot \text{HF}} + \text{HCl} \qquad (12\text{-}14)$$

To regenerate the fluoride-contaminated adsorbent, a dilute solution of alkali (0.25–0.5 N NaOH) is used. Because alumina is both a cation and an anion exchanger, Na^+ is exchanged for H^+, which immediately combines with OH^- to form HOH in the alkaline regenerant solution. The regeneration reaction of fluoride-spent alumina is

$$\overline{\text{alumina} \cdot \text{HF}} + 2\text{NaOH} \rightarrow \overline{\text{alumina} \cdot \text{NaOH}} + \text{NaF} + \text{HOH} \qquad (12\text{-}15)$$

Recent experiments have suggested that Eq. 12-15 can be carried out using fresh or recycled NaOH from a previous regeneration. This suggestion is based on the field studies of Clifford and colleagues (1998) in which arsenic-spent alumina was regenerated with equally good results using fresh or once-used 1.0 M NaOH. Probably the spent regenerant, fortified with NaOH to maintain its hydroxide concentration at 1.0 M, could have been used many times, but the optimal number of spent-regenerant reuse cycles was not determined in the field study.

To restore the *fluoride-removal capacity*, the basic alumina is acidified by contacting it with an excess of dilute acid, typically 0.5 N HCl or H_2SO_4:

$$\overline{\text{alumina} \cdot \text{NaOH}} + 2\text{HCl} \rightarrow \overline{\text{alumina} \cdot \text{HCl}} + \text{NaCl} + \text{HOH} \qquad (12\text{-}16)$$

The acidic alumina, alumina·HCl, is now ready for another fluoride (or arsenate or selenite) ligand exchange cycle, as summarized by Eq. 12-14. Alternatively, the feed water may be acidified prior to contact with the basic alumina, thereby combining acidification and adsorption into one step, as summarized by Eq. 12-17.

$$\overline{\text{alumina} \cdot \text{NaOH}} + \text{NaF} + 2\text{HCl} \rightarrow \overline{\text{alumina} \cdot \text{HF}} + 2\text{NaCl} + \text{HOH} \qquad (12\text{-}17)$$

The modeling of the alumina/GFH/GFO adsorption-regeneration cycle as being analogous to weak-base anion exchange fails in regard to regeneration efficiency, which is excellent for weak-base resins but quite poor on metal oxides/hydroxides. This is caused by the need for excess acid and base to partially overcome the poor kinetics of the metal oxide/hydroxide granules, which exhibit very low solid-phase diffusion coefficients compared with resins that are well hydrated flexible gels, offering little resistance to the movement of hydrated ions. A further reason for poor regeneration efficiency on metal oxides/hydroxides is that they are amphoteric and react with (consume) excess acid and base to partly dissolve in acidic and basic regenerant solutions. Resins are totally inert in this regard; that is, they are not dissolved by regenerants.

Special-Purpose Resins. Resins are practically without limit in their variety because polymer matrices, functional groups, capacity, and porosity are controllable during manufacture. Thus numerous special-purpose resins have been made for water treatment applications. For example, bacterial growth can be a major problem with anion resins used in some water supply applications because the positively charged resins tend to absorb the negatively charged bacteria that metabolize the adsorbed organic material—negatively charged humate and fulvate anions. To correct this problem, special resins have been developed that contain bacteriostatic long-chain *quaternary amine functional groups* (quats) on the resin surface. These immobilized quats kill bacteria on contact with the resin surface (Janauer et al., 1981).

The strong attraction of polyvalent humate and fulvate anions [components of natural organic matter (NOM)] for anion resins has been used as the basis for removal these compounds and total organic carbon (TOC) from water by using special highly porous resins. Both weak- and strong-base macroporous anion exchangers have been manufactured to remove these large anions from water. The extremely porous resins, originally thought to be necessary for adsorption of the large organic anions, tended to be structurally weak and break down easily. However, it has been shown that both gel and standard macroporous resins, which are well hydrated, highly cross-linked, and physically very strong, can be used to remove NOM (Fu and Symons, 1990). Regeneration of resins used to remove NOM is often a problem because of the strong attraction of the aromatic portion of the anions for the aromatic resin matrix. This problem has been solved at least partially using hydrophilic, well-hydrated, acrylic-matrix SBA resins. More detail on the use of IX resins to remove NOM is given later in this chapter.

Special-purpose SBA resins with large, hydrophobic, widely space functional groups have been developed for nitrate and perchlorate removal applications. These resins are discussed in more detail later in the sections on nitrate and perchlorate removal.

Resins with chelating functional groups such as iminodiacetate (Calmon, 1979), aminophosphonate, and ethyleneamine (Matejka and Zirkova, 1997) have been manufactured that have extremely high affinities for hardness ions and troublesome metals such as Cu^{2+}, Zn^{2+}, Cr^{3+}, Pb^{2+}, and Ni^{2+}. These resins are used in special applications such as trace metal removal and metals recovery operations (Brooks et al., 1991). Table 12-2 summarizes the

TABLE 12-2 Special Ion Exchangers Available Commercially

Type of resin	Functional group	Typical applications
Chelating	Iminodiacetic	Selective removal of heavy metals and transition metals
Chelating	Aminophosphonic	Decalcification of brine and removal of cations of low atomic mass
Chelating	Thiouronium	Selective removal of heavy metals, especially mercury
Chelating	Polyamine	Removal of trace heavy metals and heavy-metal complex anions
Chelating	Amidoxime	Removal of copper and iron from low pH water
Boron-specific	*N*-Methylglucamine	Removal of boron from water
NSS, nitrate-over-sulfate-selective (sulfate/divalent rejecting)	Triethyl and tripropyl quaternary amines	Nitrate removal in high sulfate water
Perchlorate-selective	Tripropyl or ethylhexyl quaternary amines	Perchlorate removal from water, often applied in single-use (nonregenerated resin) applications
Silver-impregnated SAC	Sulfonic	Softening resins with bacteriostatic properties
Iodine-releasing	Quaternary amine in triiodide form—$R_4N^+I_3^-$	Portable disinfection units

Note: Mention of trade names does not indicate endorsement. Similar products are available from alternative suppliers.
Source: www.purolite.com, 2009.

features of some of the special ion exchangers available commercially from a variety of sources (Purolite, 2009).

IX Equilibrium

Selectivity Coefficients and Separation Factors. IX resins do not prefer all ions equally. This variability in preference is often expressed semiquantitatively as a position in a selectivity sequence or quantitatively as a separation factor α_{ij} or a selectivity coefficient K_{ij} for binary exchange. The selectivity, in turn, determines the run length to breakthrough for the contaminant ion; the higher the selectivity, the longer is the run length. Consider, for example, Eq. 12-18, the simple exchange of Cl^- for NO_3^- on an anion exchanger with an equilibrium constant, as expressed numerically in Eq. 12-19 and graphically in Fig. 12-4*a*.

$$\overline{Cl^-} + NO_3 \leftrightarrow \overline{NO_3^-} + Cl^- \tag{12-18}$$

$$K = \frac{\{\overline{NO_3^-}\}\{Cl^-\}}{\{\overline{Cl^-}\}\{NO_3^-\}} \tag{12-19}$$

In Eqs. 12-18 to 12-20, overbars denote the resin phase, and the matrix designation R has been removed for simplicity, K is the thermodynamic equilibrium constant, and braces denote ionic activity. Concentrations are used in practice because they are measured more easily than activities. In this case, Eq. 12-20 based on concentration, the selectivity coefficient $K_{N/Cl}$ describes the exchange. Note that $K_{N/Cl}$ includes activity-coefficient terms that are functions of ionic strength and thus is not a true constant; that is, it varies somewhat with different ionic strengths.

$$K_{N/Cl} = \frac{[\overline{NO_3^-}][Cl^-]}{[\overline{Cl^-}][NO_3^-]} = \frac{q_N C_{Cl}}{q_{Cl} C_N} \tag{12-20}$$

where q_N = resin-phase equivalent concentration (normality) of nitrate, eq/L
C_N = aqueous-phase equivalent concentration (normality), eq/L

and the brackets represent concentration (mol/L).

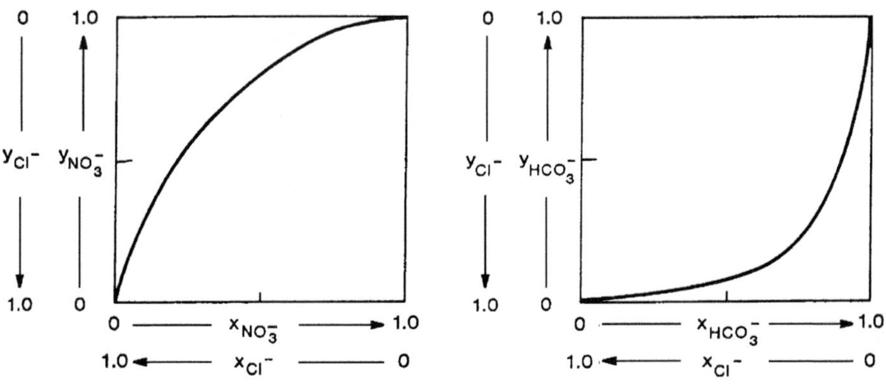

FIGURE 12-4 (*a*) Favorable isotherm for nitrate-chloride exchange according to Reaction 12-18 with constant separation factor $\alpha_{NO_3/Cl} > 1.0$. (*b*) Unfavorable isotherm for bicarbonate-chloride exchange with constant separation factor $\alpha_{HCO_3/Cl} < 1.0$.

The binary separation factor $\alpha_{N/Cl}$, used throughout the literature on separation practice, is a useful description of the exchange equilibria because of its simplicity and intuitive nature:

$$\alpha_{ij} = \frac{\text{distribution of ion } i \text{ between phases}}{\text{distribution of ion } j \text{ between phases}} = \frac{y_i/x_i}{y_j/x_j} \tag{12-21}$$

$$\alpha_{N/Cl} = \frac{y_N/x_N}{y_{Cl}/x_{CL}} = \frac{y_N/x_{Cl}}{x_N/y_{Cl}} = \frac{(q_N/q)(C_{Cl}/C)}{(C_N/C)(q_{Cl}/q)} \tag{12-22}$$

where y_i = equivalent fraction of ion i in resin, q_i/q
y_N = equivalent fraction of nitrate in resin, q_N/q
x_i = equivalent fraction of ion i in water, C_N/C
x_N = equivalent fraction of nitrate in water, C_N/C
q_N = concentration of nitrate on resin, eq/L
q = total exchange capacity of resin, eq/L
C_N = nitrate concentration in water, eq/L
C = total ionic concentration of water, eq/L

Equations 12-20 and 12-22 show that for homovalent exchange, where the exchanging ions have the same charge (i.e., monovalent/monovalent ion and divalent/divalent ion exchange), the separation factor α_{ij} and the selectivity coefficient K_{ij} are equal. This is expressed for nitrate/chloride exchange as

$$K_{N/Cl} = \alpha_{N/Cl} = \frac{q_N C_{Cl}}{C_N q_{Cl}} \tag{12-23}$$

For exchanging ions of unequal charge, that is, heterovalent exchange, the separation factor is not the same as the selectivity coefficient. Consider, for example, the case of sodium IX softening, as represented by Eq. 12-24, the simplified form of Eq. 12-2.

$$\overline{2Na^+} + Ca^{2+} = \overline{Ca^{2+}} + 2Na^+ \tag{12-24}$$

$$K_{Ca/Na} = \frac{q_{Ca} C_{Na}^2}{C_{Ca} q_{Na}^2} \tag{12-25}$$

Using a combination of Eqs. 12-21 and 12-25, we obtain

$$\alpha_{\text{divalent/monovalent}} \text{ or } \alpha_{Ca/Na} = K_{Ca/Na} \frac{q y_{Na}}{C x_{Na}} \tag{12-26}$$

The implication from these equations is that the intuitive separation factor for divalent/monovalent exchange depends inversely on solution concentration C and directly on the distribution ratio (y_{Na}/x_{Na}) between the resin and the water, with constant resin capacity q. The higher the solution concentration C, the lower is the divalent/monovalent separation factor; that is, selectivity tends to reverse in favor of the monovalent ion as ionic strength I (which is a function of C) increases. This reversal of selectivity is discussed in detail below.

Selectivity Sequences for Resins. A selectivity sequence describes the order in which ions are preferred by a particular resin or by a solid porous oxide surface such as AlOOH (activated alumina granules or hydrated aluminum oxide precipitate), FeOOH (hydrous iron oxide), or MnOOH (hydrous manganese oxide). Although special-purpose resins such as chelating resins can have unique selectivity sequences, the commercially available cation and anion resins exhibit similar selectivity sequences. These are presented in Table 12-3,

TABLE 12-3 Relative Affinities of Ions for Resins[a]

Strong-acid cation resins[b]		Strong-base anion resins[c]	
Cation$_i$	$\alpha_{i/Na}$[d]	Anion$_i$	α_{i/Cl^-}
Ra^{2+}	13.0	$UO_2(CO_3)_3^{4-}$	3200
Ba^{2+}	5.8	ClO_4^-[e]	150
Pb^{2+}	5.0	CrO_4^{2-}	100
Sr^{2+}	3.3	SeO_4^{2-}	17
Cu^{2+}	1.9	SO_4^{2-}	9.1
Ca^{2+}	2.6	$HAsO_4^{2-}$	4.5
Zn^{2+}	1.8	HSO_4^-	4.1
Fe^{2+}	1.7	NO_3^-	3.2
Mg^{2+}	1.67	Br^-	2.3
K^+	1.67	SeO_3^{2-}	1.3
Mn^{2+}	1.6	HSO_3^-	1.2
NH_4^+	1.3	NO_2^-	1.1
Na^+	1.0	Cl^-	1.0
H^+	0.67	BrO_3^-	0.9
		HCO_3^-	0.27
		CH_3COO^-	0.14
		F	0.07

[a]Values are approximate separation factors for 0.005 to 0.010 N solution (TDS = 250–500 mg/L as $CaCO_3$).
[b]SAC is polystyrene divinylbenzene matrix with sulfonate functional groups.
[c]SBA resin is polystyrene divinylbenzene matrix with —$N^+(CH_3)_3$ functional groups (i.e., a type 1 resin).
[d]Separation factors are approximate and are based on various literature sources and on experiments performed at the University of Houston.
[e]ClO_4^-/Cl^- separation factor is for polystyrene SBA resins; on polyacrylic SBA resins, the ClO_4^-/Cl^- separation factor is approximately 5.0.

where the most preferred ions, that is, those with the highest separation factors, are listed at the top of the table and the least preferred ions at the bottom. For example, the $\alpha_{Ca/Na}$ value of 1.9 means that, at equal concentrations (eq/L) in the aqueous phase, calcium is preferred by the resin 1.9/1.0 over sodium. Weak-acid cation (WAC) resins with carboxylic functional groups exhibit the same selectivity sequence as SAC resins except that hydrogen is the most preferred cation, and the magnitude of the separation factors differs from that in Table 12-3. Similarly, WBA and SBA resins exhibit the same selectivity sequence except that hydroxide is most preferred by WBA resins, and the WBA separation factors differ in magnitude but have the same trend as those in Table 12-3.

Some general rules govern selectivity sequences. In dilute solution, for example, in the TDS range of natural waters, the resin prefers the ion with the highest charge and lowest degree of hydration.

Selectivity is affected by the nature of the ion. Hydrophobic ions, for example, nitrate, chromate, and perchlorate, prefer hydrophobic resins, that is, highly cross-linked macroporous resins without polar matrices and/or functional groups, whereas hydrophilic ions, for example, bicarbonate and acetate, prefer moderately cross-linked (gel) resins with polar matrices and/or functional groups. Divalent ions, for example, sulfate and calcium, prefer resins with closely spaced exchange sites, where their need for two charges can be satisfied (Clifford and Weber, 1983; Sengupta and Clifford, 1986; Subramonian and Clifford, 1998; Horng and Clifford, 1997).

Selectivity Sequences for Alumina- and Iron-Based Adsorbents. Activated alumina (AlOOH) and GFH/GFO (FeOOH) operated in the acidic to neutral pH range for anion adsorption exhibit selectivity sequences that differ markedly from anion exchange resins. Fortunately, some of the ions, such as fluoride and arsenate, that are least preferred by resins (and therefore are not amenable to removal by resins) are highly preferred by the aluminum and iron oxides/hydroxides. Based on research by the authors and their coworkers, as well as other investigators (Trussell et al., 1980; Sing and Clifford, 1981; Rosenblum and Clifford, 1984; Schmitt and Pietrzyk, 1985), activated alumina operated in the pH range of 5.5 to 8.5 prefers anions in the following order:

$$OH^- > H_2AsO_4^- > Si(OH)_3O^- > F^- > HSeO_3^- > SO_4^{2-} > CrO_4^{2-}$$
$$> HCO_3^- > Cl^- > NO_3^- > Br^- > I^- \qquad (12\text{-}27)$$

NOM including humic and fulvic acid anions are more preferred than sulfate, but because of their widely differing molecular weights and structures and the different pore size distributions of commercial aluminas, no exact position in the selectivity sequence can be assigned. Reliable separation factors for ions in the preceding selectivity sequence such as fluoride, arsenate, silicate, and biselenite are not available in the literature, but this is not particularly detrimental to the design effort because alumina, GFH, and GFO have an extreme preference for these ions. For example, when fluoride or arsenate is removed from water, the presence of the usual competing ions—bicarbonate and chloride—is nearly irrelevant in establishing run length to contaminant ion breakthrough (Singh and Clifford, 1981; Rosenblum and Clifford, 1984; Amy, 2005). Sulfate, however, does offer measurable (up to 20% capacity reduction) competition for adsorption sites. The problem with the extremely preferred ions is that they are difficult to remove from the alumina during regeneration, which necessitates the use of hazardous and potentially destructive (of the medium) regenerants (e.g., NaOH and H_2SO_4).

Iron-based oxide/hydroxide (GFO/GFH) adsorbents are similar in their high affinity for anions such as arsenate, phosphate, and silicate and exhibit the following selectivity sequence, which is based on bench-, pilot-, and full-scale studies (Amy et al., 2005; Lakshmanan et al., 2008):

$$OH^- > H_2AsO_4^- > HSiO_3^- > H_2PO_4^- > VO_3^- > SO_4^{2-} > Cl^-, HCO_3^-$$

Isotherm Plots. The values of α_{ij} and K_{ij} can be determined from a constant-temperature equilibrium plot of resin-phase concentration versus aqueous-phase concentration, that is, the IX isotherm. Favorable and unfavorable isotherms are depicted in Fig. 12-4*a* and *b*, where each curve depicts a constant separation factor $\alpha_{NO_3/Cl}$ for Fig. 12-4*a* and $\alpha_{HCO_3/Cl}$ for Fig. 12-4*b*.

A favorable isotherm (convex to x axis) means that species i (NO_3^- in Fig. 12-4*a*), which is plotted on each axis, is preferred to species j (Cl^- in Fig. 12-4*a*), the exchanging species. An unfavorable isotherm (concave to the x axis) indicates that species i (HCO_3^- in Fig. 12-4*b*) is less preferred than j (Cl^- in Fig. 12-4*b*). During column exhaustion processes, favorable isotherms result in sharp breakthroughs when i is in the feed and j is on the resin, whereas unfavorable isotherms lead to gradual breakthroughs under these conditions. (This is discussed in detail in the section "Column Processes and Calculations.") In viewing these binary isotherms, note that

$$x_i + x_j = 1.0 \qquad (12\text{-}28)$$

$$C_i + C_j = C \qquad (12\text{-}29)$$

$$y_i + y_j = 1.0 \qquad (12\text{-}30)$$

$$q_i + q_j = q \qquad (12\text{-}31)$$

Therefore, the concentration or equivalent fraction of either ion can be obtained directly from the plot, which in Fig. 12-4*a* and *b* is a *unit* isotherm because equivalent fractions (x_i, y_j) rather than concentrations have been plotted in the range 0.0 to 1.0. Figure 12-4*a* represents the favorable isotherm for nitrate-chloride exchange, and Fig. 12-4*b* represents the unfavorable isotherm for bicarbonate-chloride exchange.

For nonconstant separation factors, for example, the divalent/monovalent (Ca^{2+}/Na^+) exchange case described by Eqs. 12-24 and 12-26, a separate isotherm exists for every total solution concentration *C*. As the solution concentration or TDS level decreases, the resin exhibits a greater preference for the divalent ion, as evidenced by a progressively higher and more convex isotherm. The phenomenon can be explained by solution theory: As the solution concentration increases, the aqueous phase becomes more ordered. This results in polyvalent ion activity coefficients that are significantly less than 1.0; that is, the tendency for polyvalent ions to escape from the water into the resin is greatly diminished, leading to a reduction in the height and convexity of the isotherm. This phenomenon of diminishing preference for higher-valent (di- and trivalent) ions with increasing ionic strength *I* of the solution has been labeled *electroselectivity* and eventually can lead to selectivity reversal, whereupon the isotherm becomes concave (Helfferich, 1962). This trend is shown in Fig. 12-5, where the sulfate-chloride isotherm is favorable in ≤0.06 N solution and unfavorable in ≥0.6 N solution.

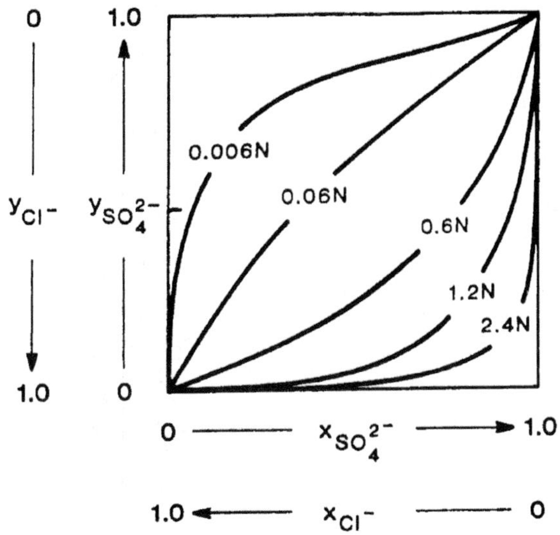

FIGURE 12-5 Electroselectivity of a typical type 1 strong-base anion exchange resin used for divalent-monovalent (SO_4^{2-}/Cl^-) anion exchange.

The exact ionic strength at which electroselectivity reversal occurs depends on the ionic makeup of the solution and highly depends on the resin structure (Boari et al., 1974) and its inherent affinity for polyvalent ions. Electroselectivity reversal is beneficial to the sodium IX softening process in that it causes the divalent hardness ions to be highly preferred in dilute solution ($I \leq 0.020$ M) during resin exhaustion and highly nonpreferred, that is, easily rejected, during regeneration with relatively concentrated (0.50–2.0 M) salt solution.

Example 12-1 Isotherm Data

The following problem briefly describes the experimental technique necessary to obtain isotherm data and illustrates the calculations required to construct a nitrate-chloride isotherm for an SBA exchange resin. By using the isotherm data or the plot, the individual and average separation factors α_{ij} can be calculated. Only minor changes are necessary to apply the technique to weak-base resins or to cation resins. For example, acids (HCl and HNO_3) rather than sodium salts would be used for equilibration of weak-base resins.

To obtain the data for this example, weighted amounts of air-dried chloride-form resin of known exchange capacity were placed into capped bottles containing 100 mL of 0.005 N (5.0 meq/L) $NaNO_3$ and equilibrated by tumbling for 16 hours. Following equilibration, the resins were settled, and the nitrate and chloride concentrations of the supernatant water were determined for each bottle. The nitrate/chloride equilibrium data are listed in Table 12-4. The total capacity q of the resin is 3.63 meq/g. Note that the units of resin capacity used here are milliequivalents per gram rather than equivalents per liter because, for precise laboratory work, a mass rather than a volume of resin must be used.

Outline of Solution

1. Verify that, within the expected limits of experimental error, the total concentration C of the aqueous phase at equilibrium is 0.005 N. Large deviations from this value usually indicate that concentrated salts were absorbed in the resin and leached out during the equilibration procedure. This problem can be avoided by extensive prewashing of the resin with the same normality of salt, in this case 0.005 N NaCl, as is used for equilibration.
2. Calculate the equivalent fractions $x_N = C_N/C$ and $x_{Cl} = C_{Cl}/C$ of nitrate and chloride in the water at equilibrium.
3. Using the known total capacity of the resin q, calculate the milliequivalents of chloride remaining on the resin q_{Cl} at equilibrium by subtracting the milliequivalents of chloride found in the water.
4. Calculate the milliequivalents of nitrate on the resin q_N by assuming that all the nitrate removed from solution is taken up by the resin.
5. Calculate the equivalent fractions $y_N = q_N/q$ and $y_{Cl} = q_{Cl}/q$ of nitrate and chloride in the resin phase at equilibrium.
6. Calculate the separation factor $\alpha_{N/Cl} = (y_N/x_N)/(y_{Cl}/x_{Cl})$, which is equal to the selectivity coefficient K_{ij} for homovalent exchange.
7. Repeat steps 1 through 6 for all equilibrium data points, and plot the isotherm.

Solution The equilibrium data point for 0.2 g of resin is chosen as an example:

1. $C = C_N + C_{Cl} = 1.17 + 3.78 = 4.95$ meq/L (This is within the expected ±5 percent limits of experimental error (5.00 – 4.95)/5.00 = 1.0 percent error).
2. $$x_N = \frac{C_N}{C} = \frac{1.17\ \text{meq/L}}{4.95\ \text{meq/L}} = 0.236$$

$$x_{Cl} = \frac{C_{Cl}}{C} = \frac{3.78\ \text{meq/L}}{4.95\ \text{meq/L}} = 0.764$$

Checking: $x_N + x_{Cl} = 0.236 + 0.764 = 1.00$

3. Calculate chloride remaining on the resin at equilibrium q_{Cl}:

$$q_{Cl} = q_{Cl,\ \text{initial}} - \text{chloride lost to water per gram of resin}$$

$$q_{Cl,\ \text{initial}} = q = 3.63\ \text{meq/g}$$

$$q_{Cl} = 3.63\ \text{meq/g} - 3.78\ \text{meq/L}\left(\tfrac{0.100\ \text{L}}{0.200\ \text{g}}\right) = 1.74\ \text{meq/g}$$

4. Calculate nitrate on resin at equilibrium:

$$q_N = q_{N,\,\text{initial}} + \text{nitrate lost from water per gram of resin}$$

$$q_N = 0 + [(5.00 - 1.17)\ \text{meq/L}]\left(\tfrac{0.100\ \text{L}}{0.200\ \text{g}}\right) = 1.91\ \text{meq/g}$$

Checking: $q_N + q_{Cl} = 1.74 + 1.91 = 3.65$ meq/g (within 5 percent of 3.63)

5. Calculate the resin-phase equivalent fractions y_N and y_{Cl} at equilibrium:

$$y_N = \frac{1.91\ \text{meq/g}}{3.65\ \text{meq/g}} = 0.523$$

$$y_{Cl} = \frac{1.74\ \text{meq/g}}{3.65\ \text{meq/g}} = 0.477$$

6. Calculate the separation factor α_{ij}:

$$\alpha_{ij} = \frac{y_N x_{Cl}}{x_N y_{Cl}} = \frac{0.523 \times 0.764}{0.236 \times 0.477} = 3.55$$

Note: Each data point will have an associated α_{ij} value. These α_{ij} values can be averaged, but it is preferable to plot the isotherm data, construct the best-fit curve, and use the curve at $x_N = 0.5$ to obtain an average α_{ij} value. The bad data points will be evident in the plot and can be ignored. Owing to mathematical sensitivity, resin inhomogeneity, and imprecise experimental data, the calculated α_{ij} values are not constant, as can be seen in Table 12-4. The α_{ij} values at the ends of the isotherm may not be representative.

7. Plot the isotherm of y_N versus x_N. The nitrate versus chloride isotherm plot should appear similar to Fig. 12-4a. ▲

TABLE 12-4 Example 12-1 Data for Plot of Nitrate/Chloride Isotherm

g resin/ 100 mL	C_N, meq/L	C_{Cl}, meq/L	C, meq/L	x_N	x_{Cl}	y_N	y_{Cl}	α_{ij}
0.020	4.24	0.722	4.96	0.854	0.146	0.980	0.020	8.6
0.040	3.56	1.32	4.88	0.730	0.270	0.920	0.091	4.25
0.100	2.18	2.77	4.98	0.440	0.560	0.760	0.240	4.12
0.200	1.17	3.78	4.95	0.236	0.764	0.523	0.477	3.55
0.400	0.53	4.36	4.89	0.108	0.892	0.300	0.700	3.59
1.20	0.185	4.49	4.68	0.040	0.960	0.110	0.890	2.99

Note: The first three columns represent experimental data. The remaining columns were obtained by calculation as described in the example.

IX and Adsorption Kinetics

Pure IX Rates. As is usual with interphase mass transfer involving solid particles, IX kinetics is governed by liquid- and solid-phase resistances to mass transfer (film and particle diffusion processes). It has long been accepted that ion diffusion processes rather than the actual exchange reactions determine IX kinetics. The IX rates depend on solution (ion) concentration, resin particle size, degree of resin cross-linking and hydration, solution-phase turbulence, ion diffusion coefficients in the aqueous and resin phases, and temperature (Kunin, 1958). In early IX kinetic research on the rates of film and particle diffusion measured in batch experiments, Boyd and colleagues (1947) demonstated that for sodium-potassium exchange, film diffusion controlled at 0.001 N solution concentration, whereas particle diffusion controlled at 1.0 N. In the range between 0.001 and 1.0 N solution concentration, where most IX reactions occur, both phases controlled. Thus IX kinetics are difficult to describe mathematically.

The liquid-phase resistance, modeled as the stagnant thin film, can be minimized by providing turbulence around the particle such as that resulting from fluid velocity in packed beds or mechanical mixing in batch operations. The speed of pure IX reactions, that is, reactions not involving (1) WAC resins in the RCOOH form or (2) free-base forms of weak-base resins, can be attributed to the inherently low mass transfer resistance of the resin phase that is caused by its well-hydrated gelular nature. Resins beads typically contain 40 to 60 percent water within their boundaries, and this water can be considered as a continuous extension of the aqueous phase within the flexible polymer network. This pseudocontinuous aqueous phase in conjunction with the flexibility of the resin phase can result in rapid kinetics for "pure IX" reactions, that is, IX of typical inorganic ions using fully hydrated strong resins.

IX and Adsorption Kinetics Compared. Unlike adsorption onto granular activated carbon (GAC), activated alumina, or granular iron oxides/hydroxides requiring on the order of hours to days to reach equilibrium, pure IX using strong-acid or strong-base resins is a rapid process at near-ambient temperature. For example, the half-time to equilibrium for adsorption of arsenate onto granular 28 × 48 mesh (0.29–59 mm diameter) activated alumina was found to be approximately 2 days (Rosenblum and Clifford, 1984), whereas the half-time to equilibrium during the exchange of arsenate for chloride on a strong-base resin was only 5 minutes (Horng, 1983; Horng and Clifford, 1997). This is in general agreement with early observations (Kunin, 1958) that the time for 90 percent attainment of sodium-calcium exchange on strong-acid (sufonic) cation resins (pure IX) was less than 2 minutes.

Rates Involving Tight Resin Forms. In contrast to pure IX, IX with WAC and WBA resins can be slow because of the tight, nonswollen nature of the acid form ($\overline{RCOOH}$) of WAC resins or free-base forms, for example, $\overline{R_3N:}$, of WBA resins. In reactions involving these tight forms, the average solid-phase diffusion coefficients change drastically during the course of the exchange, which is often described using the progressive-shell, shrinking-core model (Helfferich, 1965, 1966) depicted in Fig. 12-6. In these reactions, which are effectively neutralization reactions, either the shell or the core can be the swollen (more hydrated) portion, and a rather sharp line of demarcation exists between the tight and swollen zones. Consider, for example, the practical case of softening with WAC resins in the H^+ form (Eq. 12-4). As the reaction proceeds, the hydrated calcium-form shell comprising $\overline{(RCOO^-)_2Ca^{2+}}$ expands inward and replaces the shrinking, poorly hydrated core of $\overline{RCOOH}$. The entire process is reversed on regeneration with acid, and the tight shell of $\overline{RCOOH}$ thickens as it proceeds inward and replaces the porous, disappearing core of $\overline{(RCOO^-)_2Ca^{2+}}$.

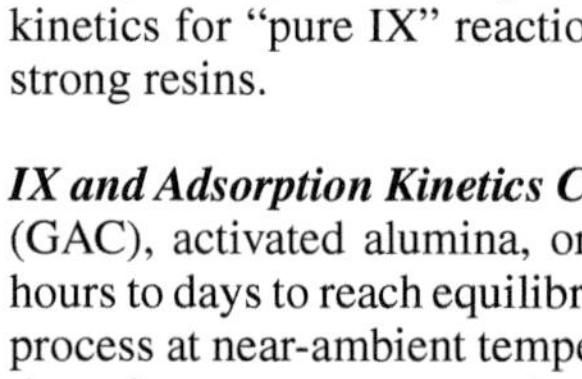

FIGURE 12-6 Progressive shell model of IX with weak resins.

In some cases, pure IX with weak resins is possible, however, and proceeds as rapidly as pure IX with strong resins. For example, the pure exchange of sodium for calcium on a WAC resin (Eq. 12-32) does not involve conversion of the resin RCOOH, in contrast to Eq. 12-4, and would take place in a matter of minutes as with SAC resins (Eq. 12-3).

$$\overline{2RCOO^-Na^+} + Ca(HCO_3)_2 = \overline{(RCOO^-)_2Ca^{2+}} + 2NaHCO_3 \qquad (12\text{-}32)$$

Although weak resins involving $\overline{RCOOH}$ and $\overline{R_3N:}$ may require several hours to attain equilibrium in a typical batch exchange, they still may be used effectively in column processes where the contact time between the water and the resin is only 1 to 5 minutes. There are two reasons for the column advantage: (1) An overwhelming amount of unspent resin capacity (eq/L) is present relative to the number of ions (eq/L) in the water in the column, and (2) the resin is typically exposed to the feed water for periods in excess of 24 hours before it is exhausted. Prior to exhaustion, the overwhelming ratio of resin exchange sites present in the column to exchanging ions present in the column water nearly guarantees that an ion will be removed by the resin before the water carrying the ion exits the column. The actual contaminant removal takes place in the *adsorption* or *ion exchange* or *mass transfer zone* (Fig. 12-7) that characterizes the breakthrough curve of interest.

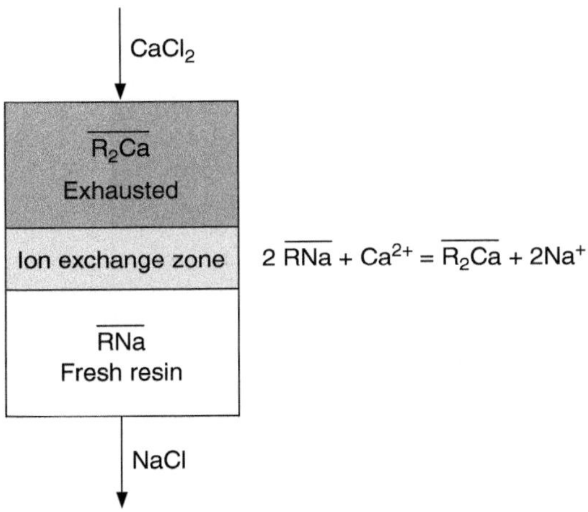

FIGURE 12-7 Resin concentration profile for binary IX of sodium for calcium.

In summary, IX of small inorganic ions (e.g., Na^+, Ca^{2+}, Cl^-, NO_3^-, SO_4^{2-}) using strong resins is fundamentally a fast interphase transfer process because strong resins are well-hydrated gels exhibiting large solid-phase diffusion coefficients and little resistance to mass transfer. This is not the case with weak resins in the acid (RCOOH) or free-base (R_3N:) forms, nor is it true for metal oxides/hydroxides because these media offer considerably more solid-phase diffusion resistance. Irrespective of fast or slow batch kinetics, all these media can be used effectively in column processes for contaminant removal from water because columns exhibit enormous contaminant-removal capacity and are exhausted over a period of many hours to many days. Leakage of contaminants, however, will be much more significant with media that exhibit relatively slow mass transfer rates.

Column Processes and Calculations

Binary IX. IX and adsorption column operations do not result in a fixed percentage of removal of contaminant with time, which would result, for example, in a steady state coagulation process (Lakshmanan et al., 2008). These column processes exhibit a variable degree of contaminant removal and gradual or sharp contaminant breakthroughs similar to (but generally much more complicated than) the breakthrough of turbidity through a granular filter. First, we consider the hypothetical case of pure binary IX before proceeding to the practical drinking water treatment case of multicomponent IX.

If pure calcium chloride solution is softened by continuously passing it through a bed of resin in the sodium form, IX (Eq. 12-2) occurs immediately in the uppermost differential segment of the bed (at its inlet). Here, all the resin is converted to the calcium form in the moving IX zone, where mass transfer between the liquid and solid phases occurs. These processes are depicted in Fig. 12-7.

The resin phase experiences a calcium wavefront that progresses through the column until it reaches the outlet. At this point, no more sodium-form resin exists to take up calcium, and calcium breaks through into the effluent, as shown in Fig. 12-8. In this pure binary IX case, the effluent calcium concentration never can exceed that of the influent; this is, however, generally not true for multicomponent IX, as we show later. The sharpness of the calcium breakthrough curve depends on both equilibrium (i.e., selectivity) and kinetic (i.e., mass transfer) considerations. Imperfect, that is, noninstantaneous interphase mass transfer of sodium and calcium, coupled with flow channeling and axial dispersion, always acts to reduce the sharpness of the breakthrough curve and result in a broadening of the IX zone. This is equivalent to saying that nonequilibrium (non-instantaneous) mass transfer produces a diffuse calcium wave and a somewhat gradual calcium breakthrough.

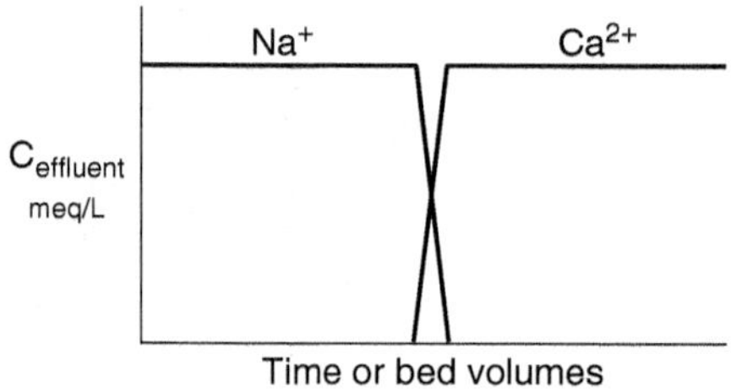

FIGURE 12-8 Effluent concentration histories (breakthrough curves) for softening reaction, that is, exchange of calcium for sodium.

A breakthrough curve can be gradual even if mass transfer is instantaneous and flow channeling and axial dispersion are absent because the first consideration in determination of the shape is the resin's affinity (an equilibrium consideration) for the exchanging ions. Mass transfer is the second consideration. If the exchange isotherm is favorable, as is the case here (i.e., the entering ion, calcium, is preferred to sodium), then a perfectly sharp (squarewave) theoretical breakthrough curve results. If the IX isotherm is unfavorable, as is the case for the reverse reaction of sodium chloride fed to a calcium-form resin, then a gradual breakthrough curve results even for instantaneous (equilibrium) mass transfer. These two basic types of breakthrough curves, sketched in Fig. 12-9, result from the solution of mass-balance equations assuming instantaneous equilibrium and constant adsorbent capacity.

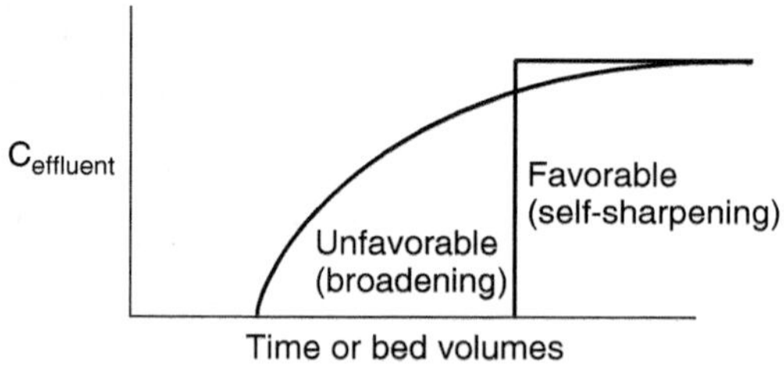

FIGURE 12-9 Theoretical breakthrough curves for equilibrium IX with no mass transfer limitations. An unfavorable isotherm (Fig. 12-4*b*) results in a broadening wavefront (breakthrough), whereas a favorable isotherm (Fig. 12-4*a*) results in a self-sharpening wavefront.

Multicomponent IX. The breakthrough curves encountered in water supply applications are much more complicated than those in Figs. 12-8 and 12-9. The greater complexity is caused by the multicomponent nature of the exchange reactions when treating natural water.

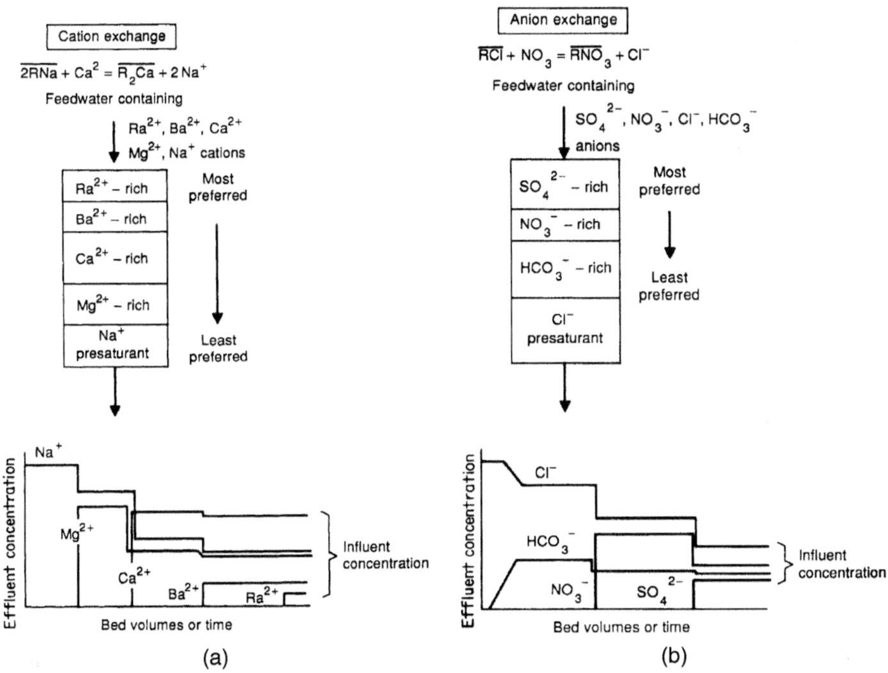

FIGURE 12-10 (*a*) Ideal resin concentration profile (*above*) and breakthrough curves (*below*) for typical softening and radium removal. Note that the column was run far beyond hardness (Mg^{2+}) breakthrough and slightly beyond radium breakthrough. The most preferred ion is Ra^{2+}, followed by $Ba^{2+} > Ca^{2+} > Mg^{2+} > Na^+$. After Ra^{2+} breakthrough, effluent concentration equals influent concentration for all ions. (*b*) Ideal resin concentration profile (*above*) and breakthrough curves (*below*) for nitrate removal by chloride-form anion exchange with strong-base resin. Note that the column was run far beyond nitrate breakthrough and somewhat beyond sulfate breakthrough. The most preferred ion is SO_4^{2-}, followed by $NO_3^- > Cl^- > HCO_3$.

Some ideal resin concentration profiles and breakthrough curves for hardness removal by IX softening and for nitrate removal by chloride-form anion exchange are sketched in Fig. 12-10. The important determinants of the shapes of these breakthrough curves are (1) the feed water composition, (2) the resin capacity, and (3) the resin's affinity for each of the ions as quantified by the separation factor α_{ij} or the selectivity coefficient K_{ij}. The order of elution of ions from the resin, however, is determined solely by the selectivity sequence, which is the ordering of the components i from 1 to n, where 1 is the most preferred and n is the least preferred ion. Finally, before continuing with our discussion of multicomponent IX column behavior, we must remind ourselves of Eq. 12-26, which shows that the α_{ij} values for divalent and higher-valent ions, and thus the order of elution of ions, also will be determined by the total ionic concentration C of the feed water.

In carrying out the cation or anion exchange reactions, ions in addition to the target ion, for example, calcium or nitrate, are removed by the resin. All the ions are concentrated, in order of preference, in bands or zones in the resin column, as shown in the resin concentration profiles of Fig. 12-10. As these resin boundaries (wavefronts) move through the column, the breakthrough curves shown in the figure result. These are based on theory (Helfferich and Klein, 1970) but have been verified in actual breakthrough curves published by Clifford (1982, 1985), Snoeyink and colleagues (1987), and (Guter, 1995).

The following rules can be applied to effluent histories in multicomponent IX (and adsorption) systems theory (Helfferich and Klein, 1970; Clifford, 1982, 1991):

1. Ions higher in the selectivity sequence than the presaturant ion tend to have long runs and sharp breakthroughs; those less preferred than the presaturant ion will always have early, gradual breakthroughs, as typified by HCO_3^- in Fig. 12-10*b*.
2. The most preferred species (radium in the case of softening and sulfate in the case of nitrate removal) are last to exit to column, and their effluent concentrations never exceed their influent concentrations.
3. The species exit the column in reverse preferential order, with the less preferred ions (smaller separation factors with respect to the most preferred species) leaving first.
4. The less preferred species will be concentrated in the column and will, at some time, exit the column in concentrations exceeding their influent concentrations (*chromatographic peaking*). This is a potentially dangerous situation depending on the toxicity of the ion in question. Good examples of chromatographic peaking, that is, effluent concentration greater than influent concentration, are visible in Fig. 12-10. A magnesium peak is shown in Fig. 12-10*a* and bicarbonate and nitrate peaks in Fig. 12-10*b*.
5. When all the breakthrough fronts have exited the column, the entire resin bed is in equilibrium with the feed water. When this happens, the column is exhausted, and the effluent and influent ion concentrations are equal.
6. The effluent concentration of the presaturant ion (Na^+ in Fig. 12-10*a* and Cl^- in Fig. 12-10) decreases in steps as each new ion breaks through because the total ionic concentration of the water C (meq/L) must remain constant during simple IX.

One way to eliminate the troublesome chromatographic peaking of toxic ions such as nitrate and arsenate is by inverting the selectivity sequence so that the toxic contaminant is the ion most preferred by the resin. This requires preparation of special-purpose resins. This has been done in the case of nitrate removal and is discussed later. Potential peaking problems still remain with other inorganic contaminants, notably arsenic [As(V)] (Ghurye et al., 1999) and selenium [Se(IV)] (Clifford, 1991). An alternative means of eliminating or minimizing peaking is to operate several columns in parallel, as will be discussed in the section "Multicolumn Processes" below.

Breakthrough Detection and Run Termination. Clearly, an IX column run must be stopped before a toxic contaminant is dumped during chromatographic peaking. Even without peaking, violation of the MCL will occur at breakthrough when the contaminant feed concentration exceeds the MCL. Detecting and preventing a high effluent concentration of contaminant depends on the sampling and analysis frequency. Generally, continuous on-line analysis of the contaminant, for example, nitrate or arsenate, is too sophisticated for small communities, where most of the inorganic contaminant problems exist. On-line conductivity detection, the standard means of effluent quality determination in IX demineralization processes, is not easily applied to the detection of contaminant breakthrough in single-contaminant removal processes such as radium, barium, nitrate, or arsenate removal. This is so because of the high and continuously varying conductivity of the effluents from cation or anion beds operated on typical water supplies. Conductivity should not be ruled out completely because even though the changes may be small, as the various ions exit the column, a precise measurement may be possible in selected applications.

On-line pH measurement is a proven, reliable technique that sometimes can be applied as a surrogate for contaminant breakthrough. For example, pH change can be used to signal the exhaustion of a weak-acid resin (—RCOOH) used for carbonate hardness removal. When exhausted, the WAC resin ceases to produce acidic carbon dioxide, and the pH quickly rises to that of the feed water. This pH increase is, however, far ahead of the barium or radium breakthrough. Effluent pH change sometimes can be used as an indicator of nitrate breakthrough, as discussed in the section on nitrate removal.

The usual method of terminating an IX column run is to establish the relevant breakthrough curve by sampling and analysis and then to use these breakthrough data to terminate future runs based on the metered volume of throughput with an appropriate safety factor. If a breakthrough detector such as a pH or conductivity probe is applied, the sample line to the instrument can be located upstream (e.g., 15–30 cm, or 6–12 in) of the bed outlet to provide advance warning of breakthrough.

Typical Service Cycle for a Single Column. IX and adsorption columns are operated on similar service cycles consisting of six steps: exhaustion, backwash, regeneration, slow rinse, fast rinse, and return to service. A simple single-column process schematic is shown in Fig. 12-11 and includes an optional bypass for a portion of the feed water. Bypass blending will be a common procedure for drinking water treatment applications because IX resins usually can produce a contaminant-free effluent that is purer than that required by law. Therefore, to minimize treatment costs, part of the contaminated feed water, typically 10 to 50 percent, will be bypassed around the process and blended with the effluent to produce product water approaching some fraction (e.g., 70 percent) of the MCL acceptable to the regulatory agency. An alternative means of providing efficient column utilization when significant contaminant leakage is allowed is to operate several columns in parallel, as discussed in the section "Multicolumn Processes" below.

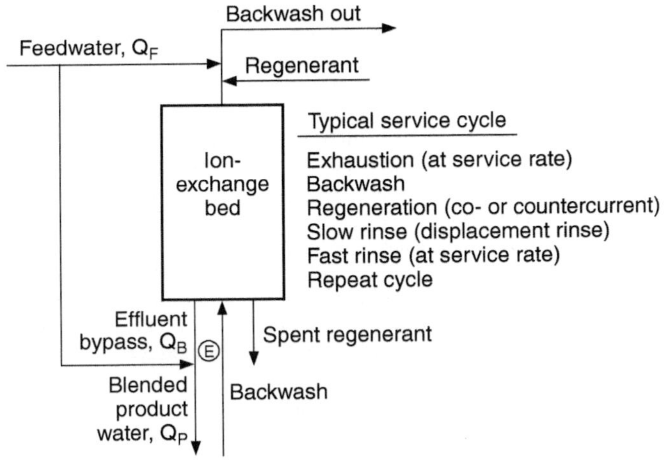

FIGURE 12-11 Schematic and service cycle of a single-column IX process.

Partial Regeneration and Regenerant Reuse. Yet another means of optimizing column utilization and minimizing process costs is to use the technique of partial regeneration. This involves the use of only a fraction, for example, 25 to 50 percent, of the regenerant required for complete, for example, 90 to 100 percent, removal of the contaminant from the exhausted resin. The result is often, but not always, a large leakage of contaminant on the next exhaustion run caused by the relatively high level of contaminant remaining on the resin. This high effluent contaminant concentration (leakage) prior to breakthrough often can be tolerated without exceeding the MCL. Partial regeneration is particularly useful in nitrate removal, as discussed in detail later. Generally, either bypass blending or partial regeneration is used; use of both processes simultaneously is possible but creates significant process control problems.

Reuse of spent regenerant is another means of reducing costs and minimizing waste disposal requirements. In order for a spent regenerant to be reused, either the target contaminant ion must be removed from the regenerant before reusing it, or the resin must have a strong preference for the regenerant ion in favor of the contaminant ion, which accumulates in the recycle brine. The recent literature suggests that spent brine reuse is possible in more applications than previously thought possible. Removing nitrate from the recycle brine by means of biological denitrification was the approach used by me and my colleagues (Clifford and Liu, 1993; Liu and Clifford, 1996a, 1996b) for their nitrate IX process with brine reuse. In our pilot-scale experiments, a denitrified 0.5 M Cl^- brine was reused 38 times without disposal. Hiremath et. al. (2006) developed a similar process for reuse of perchorate-contaminated brine. Clifford and colleagues (2003) determined that spent arsenic-contaminated brine could be reused more than 20 times simply by maintaining the Cl^- concentration at 1.0 M without removing the arsenic. Kim and Symons (1991) showed that DOC anions could be removed from drinking water by SBA exchange with regenerant reuse. No deterioration of DOC removal was noted during nine exhaustion-regeneration cycles with spent brine (a mixture of NaCl and NaOH) reuse when the Cl^- and OH^- levels were maintained at 2.0 and 0.5 M, respectively. Further information on these processes is provided in the "Applications" section below.

Reusing the entire spent-regenerant solution is not necessary. In cases where there is a long tail on the contaminant elution curve, the first few bed volumes of heavily contaminated regenerant are discarded, and only the least-contaminated portions are reused. In this case, a two-step roughing-polishing regeneration can be used. The roughing regeneration is completed with the partially contaminated spent regenerant, and the polishing step is carried out with fresh regenerant. The spent regenerant from the polishing step then is used for the next roughing regeneration.

The regenerant reuse techniques are relatively new to the IX field and are still being proven in full-scale long-term use for water supply applications. Although conserving regenerants and reducing waste discharge volumes, regenerant reuse can result in some significant disadvantages: (1) increased process complexity, (2) increased contaminant leakage, (3) progressive loss of capacity caused by incomplete regeneration and fouling, (4) the need to store and handle spent regenerants, and (5) the buildup (concentration) of trace contaminants as the number of regenerant reuse cycles increases.

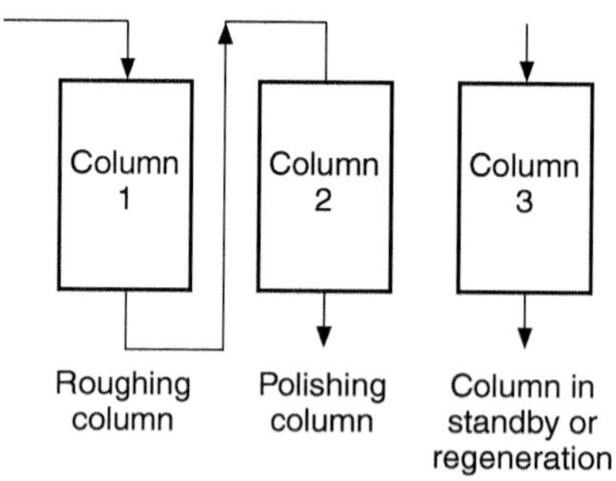

FIGURE 12-12 Two-column roughing-polishing system with third column in standby or regeneration. After exhaustion of column 1, it is taken out of service; partially exhausted column 2 becomes the new roughing column, and freshly regenerated column 3 becomes the polishing column. The cycle is repeated in a merry-go-round pattern.

Multicolumn Processes. IX or adsorption columns can be connected either in series to improve product purity and regenerant usage or in parallel to increase throughput, minimize peaking, and smooth out product water quality variations. If designed properly, multicolumn systems can be operated in parallel, series, or parallel-series modes.

Columns in Series. A series roughing-polishing (lead-lag) sequence is shown in Fig. 12-12. In such a process, a completely exhausted roughing (lead) column is regenerated when the partially exhausted polishing column effluent exceeds the MCL. This unregenerated polishing column becomes the new roughing column, and the old roughing column, now freshly regenerated, becomes the new polishing column. Often three columns are used. While two are in service, the third is being regenerated. A multicolumn system consisting of three or more columns operated in this manner is referred to as a *merry-go-round system*, which should not be confused with a continuously rotating carousel ISEP® system (Calgon Carbon Corp., 2009), which is typically operated as 20 columns in parallel at various stages of exhaustion, rinsing, and regeneration.

Columns in Parallel. In addition to bypass blending (see Fig. 12-11), an alternative means of allowing a predetermined amount of contaminant leakage in the product water is to employ multiple columns in parallel operated at different stages of exhaustion. For example, with three columns operated in parallel, the first one could be run beyond MCL breakthrough, whereas the second and third columns would not have achieved breakthrough. Thus, even after contaminant breakthrough in the first column, the average concentration of the three blended effluents would be below the MCL. Multiple parallel column operation gives a more consistent product water quality and also can prevent, or at least smooth out, chromatographic peaks from serious overruns. During normal operation of multiple parallel column systems, some columns are being exhausted, whereas others are being rinsed and regenerated or are in standby mode. The columns may be stationary (www.basinwater.com, 2009) or moving, as in the *ISEP* system described earlier.

Process Differences: Resins Versus Metal-Oxide Adsorbents. The design of a process for contaminant adsorption and regeneration using hydrous metal oxides (alumina, GFH, GFO, and titanium or zirconium oxides/hydroxides) is similar to that for IX resins but with some significant exceptions. First, contaminant leakage is inherently greater with the adsorptive media; adsorption zones are longer, and breakthrough curves are more gradual because metal oxide/hydroxide adsorption kinetics are much slower than IX. Second, effluent chromatographic peaking of the contaminant (e.g., fluoride, arsenic, or selenium) is rarely seen when using adsorptive media because these contaminants are usually the most preferred ions in the feed water. Finally, complex two-step base-acid regeneration is required to rinse out the excess base and return the metal oxide to a useful form.

Design Considerations

Resin Characteristics. Several hundred different resins are available from U.S. and European manufacturers. Of these, resins based on the polystyrene divinylbenzene matrix see widest use. Representative ranges of properties of these resins are shown in Table 12-5 for the two major categories of resins used in water treatment. IX capacity is expressed in milliequivalents per milliliter (wet-volume capacity) because resins are purchased and installed on a volumetric basis (meq/mL × 21.8 = kgrain $CaCO_3/ft^3$). A wet-volume capacity of 1.0 meq/mL means that the resin contains 6.023×10^{20} exchange sites per milliliter of wet resin, including voids. The dry-weight capacity in milliequivalents per gram of dry resin is more precise and is used often in scientific research.

The *operating capacity* is a measure of the actual performance of a resin under a defined set of conditions, including, for example, feed water composition, service rate, and degree of regeneration. The operating capacity is always less than the advertised exchange capacity because of incomplete regeneration and early contaminant leakage, which cause early run termination. Some example hardness operating capacities during softening are given in Table 12-6, where the operating capacity for softening is seen to be a function of the amount of regenerant used.

Bed Size and Flow Rates. A resin bed depth of 76 cm (30 in) is usually considered the minimum, and beds as deep as 3.7 m (12 ft) are not uncommon. The *empty-bed contact time* (EBCT) chosen determines the volume of resin required and is usually in the range of

TABLE 12-5 Properties of Styrene-Divinylbenzene Gel-Type Strong-Acid Cation and Strong-Base Anion Resins

Parameter	Strong-acid cation resin	Type 1 strong-base anion resin
Screen size, U.S.mesh	−16 + 50	−16 +50
Shipping weight, kg/m^3 (lb/ft^3)	850 (53)	700 (44)
Moisture content, %	45–48	43–49
pH range	0–14	0–14
Maximum operating temp, °C (°F)	140 (280)	OH^- form 60 (140)
Turbidity tolerance, ntu	5	5
Iron tolerance, mg/L as Fe	5	0.1
Chlorine tolerance, mg/L Cl_2	1.0	0.1
Backwash rate, m/h (gpm/ft^2)	12–20 (5–8)	4.9–7.4 (2–3)
Backwash period, minutes	5–15	5–20
Expansion volume, %	50–100	50–75
Regenerant and concentration[a]	NaCl, 3%–12%	NaCl, 3%–12%
Regenerant dose, kg/m^3 (lb/ft^3)	80–320 (5–20)	80–320 (5–20)
Regenerant rate, min/BV(gpm/ft^3)	15 (0.5)	15 (0.5)
Rinse volume, BV (gal/ft^3)	2–8 (15–60)	2–10 (15–75)
Exchange capacity, meq/mL(kgr $CaCO_3$/ft^3)[b]	1.8–2.0 (39–41)	1–1.3 (22–28)
Operating capacity, meq/mL(kgr $CaCO_3$/ft^3)[c]	0.9–1.4 (20–30)	0.4–0.8 (12–16)
Service rate, BV/h (gpm/ft^3)	8–40 (1–5)	8–40 (1–5)

[a]Other regenerants such as H_2SO_4, HCl, and $CaCl_2$ also can be used for SAC resins, whereas NaOH, KOH, and $CaCl_2$ can be used for SBA regeneration.

[b]Kilograins $CaCO_3$/ft^3 are the units commonly reported in the resin manufacturer literature. To convert kgr $CaCO_3$/ft^3 to meq/mL, multiply by 0.0458.

[c]Operating capacity depends on method of regeneration, particularly on the amount of regenerant applied. See Table 9.6 for SAC resins.

TABLE 12-6 Softening Capacity[a] as a Function of Regeneration Level

Regeneration level		Hardness capacity		Regeneration efficiency	
lb NaCl/ft^3 resin	kg NaCl/m^3 resin	kgr $CaCO_3$/ft^3 resin	eq $CaCO_3$/L resin	lb NaCl/kgr $CaCO_3$	eq NaCl/eq $CaCO_3$
4	64	17	0.78	0.24	1.40
6	96	20	0.92	0.30	1.78
8	128	22	1.00	0.36	2.19
10	160	25	1.15	0.40	2.38
15	240	27	1.24	0.56	3.30
20	320	29	1.33	0.69	4.11
Infinite	Infinite	45	2.06	Infinite	Infinite

[a]Operating capacity values are based on the performance of a typical 8 percent cross-linked styrene divinylbenzene SAC resin. Values given are independent of EBCT and bed depth, provided that the minimum criteria [EBCT =1.0 minute; bed depth = 0.76 m (2.5 ft)] are met. More efficient regeneration may be possible using "shallow-shell technology (SST) resins," which are not functionalized to the core (www.purolite.com/SST).

1.5 to 7.5 minutes. The reciprocal of EBCT is the *service flow rate* (SFR) or *exhaustion rate*, and its accepted range is 8 to 40 BV/h (1–5 gpm/ft^3). These relationships are expressed as

$$\text{EBCT} = \frac{V_R}{Q_F} = \text{average water detention time in an empty bed} \tag{12-33}$$

$$\text{SFR} = \frac{1}{\text{EBCT}} = \frac{Q_F}{V_R} \tag{12-34}$$

where Q_F is the feed water volumetric flow rate, L/min (gal/min), and V_R is the resin bed volume including voids, m^3 (ft^3).

Fixed-Bed Columns. IX columns are usually steel pressure vessels constructed so as to provide (1) a good feed and regenerant distribution system, (2) an appropriate bed support, including provision for backwash water distribution, and (3) enough free space above the resin bed to allow for expected bed expansion during backwashing. Additionally, the vessel must be lined so as to avoid corrosion problems resulting from concentrated salt solutions and, in some cases, acids and bases used for regeneration or resin cleaning. There must be minimal dead space below the resin bed where regenerants and cleaning solutions might collect and subsequently bleed into the effluent during the service cycle.

Cocurrent versus Countercurrent Regeneration. Historically, downflow exhaustion followed by downflow (cocurrent) regeneration has been the usual mode of operation for IX columns. However, the recent trend, especially in Europe, is to use upflow (countercurrent) regeneration for the purpose of minimizing the leakage of contaminant ions on subsequent exhaustions of IX demineralizers. Theoretically, countercurrent regeneration is better because it exposes the bottom (exit) of the bed to a continuous flow of fresh regenerant and leaves the resin located near the outlet of the bed in a well-regenerated condition. Our research on nitrate (Clifford et al., 1987) and arsenate removal (Clifford et al., 1999), however, has called into question the conventional wisdom that countercurrent is always better than cocurrent regeneration. It has been found that cocurrent downflow regeneration is superior to countercurrent upflow regeneration for contaminants such as arsenate and nitrate, which are concentrated at the bed outlet at the end of a run. The proposed reason for the superiority of downflow regeneration in these situations is that the contaminant is not forced back through the entire resin bed during regeneration. The forcing of the contaminant back through the bed tends to leave relatively more contaminant in the resin. This is discussed in more detail in the sections on nitrate and arsenic removal.

Spent Brine Reuse. In order for a spent regenerant to be reused, either the target contaminant ion must be removed from the regenerant before reusing it, or the resin must have a strong preference for the regenerant ion in favor of the target ion, which accumulates in the recycle brine. The recent literature suggests that spent brine reuse is possible in more applications than previously thought possible. Brine reuse was discussed earlier in the section "Partial Regeneration and Regenerant Reuse."

APPLICATIONS OF IX AND ADSORPTION

Sodium IX Softening

As already mentioned, softening water by exchanging sodium for calcium and magnesium using SAC resin (see Eq. 12-2) is the major application of ion-exchange technology for drinking water treatment. Prior to the advent of synthetic resins, zeolites, i.e., inorganic crystalline aluminosilicate ion exchangers in the sodium form were utilized as the exchangers. The story of one major application of ion-exchange softening at the Weymouth plant of the Metropolitan Water District of Southern California is well told by Bowers (1980) in *The Quest for Pure Water*. In that application, which included 1.5×10^6 m^3/d (400 mgd) of softening capacity, softening by ion exchange eventually supplanted excess lime-soda ash softening because of better economics, fewer precipitation problems, and the requirement for a high alkalinity level in the product water to reduce corrosion. One advantage of the lime soda ash softening process is that it reduces the TDS level of the water by removing calcium and magnesium bicarbonates as $CaCO_3$(s) and $Mg(OH)_2$(s). The concomitant removal of alkalinity is, however, sometimes detrimental, thus favoring ion-exchange softening, which exchanges only cations leaving the anions intact. Lime-soda ash softening is covered in chapter 13.

As with most IX softening plants, the zeolite media at the Weymouth plant was exchanged for resin in the early 1950s, shortly after polystyrene sulfonate SAC resins were introduced. The SAC softening resins used today are basically the same as those early polystyrene resins. Their main features are high chemical and physical stability (even in the presence of chlorine), uniformity in size and composition, high exchange capacity, rapid exchange kinetics, a high degree of reversibility, and long life. A historical comparison between the life of the zeolites and that of the resins indicated that zeolites could process a maximum of 214,000 BV (1.6×10^6 gal H_2O/ft^3 of zeolite) before replacement, whereas the resins could process up to 2,700,000 BV (20×10^6 gal H_2O/ft^3 of resin) before they needed replacement. The softeners designed and installed for resins at this plant in 1966 were 8.5×17 m (28×56 ft) reinforced-concrete basins filled to a depth of 0.76 m (2.5 ft) and each containing 113 m^3 (4000 ft^3) of resin.

The Weymouth plant used IX softening for over 30 years. Softening ceased in 1975, when the source water hardness was reduced by blending. At that time, the 9-year-old resin in the newest softeners was still good enough to be resold. Other interesting design features of this plant included disposal of waste brine to a wastewater treatment plant through a 32 km long (20 mi) pipe flowing at 0.43 m^3/s (15 ft^3/s) with upflow exhaustion at 14.7 m/h (6.0 gal/min/ft^2)—3.1 minute EBCT in a 0.76 m deep (2.5 ft) bed—followed by downflow regeneration.

Example 12-2 Softening Design Example

This typical design example illustrates how to establish the IX resin volume, column dimensions, and regeneration requirements for a typical water softener.

Design Problem

A groundwater is to be partially softened from 275 down to 150 mg/L of $CaCO_3$ hardness. IX has been selected instead of lime softening because of its simplicity and the ease of cycling on and off to meet the water demand. The well pumping capacity is 3.8×10^3 m^3/d (1 mgd or 700 gpm), and the system must be sized to meet this maximum flow rate. The source water contains only traces of iron; therefore, potential clogging problems because of suspended solids are not significant. In applications where raw water suspended solids would foul the resins, coagulation-filtration or filtration pretreatment with dual- or multimedia filters would be required.

Outline of Solution

1. Select a resin and a regeneration level using the resin manufacturer's literature.
2. Calculate the allowable fraction f_B of bypass source water.
3. Choose the service flow rate (SFR, gpm/ft^3, or EBCT, minutes).
4. Calculate the run length t_H and the bed volumes V_F that can be treated prior to hardness breakthrough.

5. Calculate the volume of resin V_R required.
6. Determine the minimum out-of-service time (in hours) during regeneration.
7. Choose the number of columns in the system.
8. Dimension the columns.
9. Calculate the volume and composition of wastewater.

Calculations
(1) *Selection of resin and resin capacity.* Once the resin and its regeneration level have been specified, the IX operating capacity is fixed based on experimental data of the type found in Table 12-6. The data are for a polystyrene SAC resin subjected to cocurrent regeneration using 10 percent (1.7 N) NaCl. If a regeneration level of 240 kg NaCl/m^3 of resin (15 lb NaCl/ft^3) is chosen, the resulting hardness capacity prior to breakthrough is 1.24 meq/mL of resin. *Note:* Much lower regeneration levels, for example, 80 kg NaCl/m^3 of resin (5 lb NaCl/ft^3 of resin) can be used to minimize salt usage at the cost of shorter runs.

(2) *Calculation of bypass water allowance.* Assume that the water passing through the resin has zero hardness. Actually, hardness leakage during exhaustion will be detectable but usually less than 5 mg/L as $CaCO_3$. The bypass flow is calculated by writing a hardness balance at blending point, point E in Fig. 12-11, where the column effluent is blended with the source water bypass.

Mass balance on hardness at point E:

$$Q_B C_B + Q_F C_E = Q_P C_P \tag{12-35}$$

Balance on flow at point E:

$$Q_B + Q_F = Q_P \tag{12-36}$$

where Q_B is the bypass flow rate, Q_F is the column feed and effluent flow rate, Q_P is the blended product flow rate (i.e., total flow rate), C_B is the concentration of hardness in the bypass raw water (275 mg/L as $CaCO_3$), C_E is the concentration of hardness in the column effluent (assumed to be 0), and C_P is the chosen concentration of hardness in blended product water (150 mg/L as $CaCO_3$).

The solution to these equations is obtained easily in terms of the fraction bypassed f_B:

$$f_B = \frac{Q_B}{Q_P} = \frac{C_P}{C_B} = 0.55 \tag{12-37}$$

The fraction f_F that must be treated by IX is

$$f_F = 1 - f_B = 0.45 \tag{12-38}$$

(3) *Choosing the exhaustion flow rate.* The generally acceptable range of SFR for IX is 8 to 40 BV/h (1–5 gpm/ft^3). Choosing a value of 20 BV/h (2.5 gpm/ft^3) results in an EBCT of 3.0 minutes and an approach velocity v_0 of 15.3 m/h (6.25 gpm/ft^2) if the resin bed is 0.76 m (2.5 ft) deep.

$$\text{EBCT} = \frac{\text{h}}{20\ \text{BVs}} \times \frac{60\ \text{minutes}}{\text{h}} = 3\frac{\text{minutes}}{\text{BV}} \tag{12-39}$$

$$v_0 = \frac{\text{depth}}{\text{detention time}} = \frac{\text{depth}}{\text{EBCT}} = \frac{0.76\text{m}}{3\text{minutes}} \times \frac{60\text{minutes}}{\text{h}} = \frac{15.3\text{m}}{\text{h}} \tag{12-40}$$

$$v_0 = \text{SFR} \times \text{depth} = 2.5\ (\text{gpm/ft}^3) \times 2.5\ (\text{ft}) = 6.25\ \text{gpm/ft}^2 \tag{12-41}$$

(4) *Calculation of run length.* The exhaustion time to hardness breakthrough t_H and the bed volumes to hardness breakthrough BV_H are calculated from a mass balance on hardness, assuming again that the resin effluent contains zero hardness. Expressed in words and equation this mass balance is Equivalents of hardness *removed from the water* during the run = equivalents of hardness *accumulated on the resin* during the run

$$Q_F C_F t_H = V_F C_F = q_H V_R \tag{12-42}$$

where q_H is the hardness capacity of resin at the selected regeneration level (eq/L, kgr/ft^3), V_R is the volume of the resin bed including voids (L), $Q_F t$ is volume of water fed to column V_F during time t.

Then

$$\frac{V_F}{V_R} = BV_H = \frac{q_H}{C_F} \tag{12-43}$$

Based on the hardness capacity in Table 12-6, the bed volumes to hardness breakthrough BV_H following a regeneration at 15 lb NaCl/ft^3 is

$$BV_H = \frac{1.24 \text{ equiv } CaCO_3}{\text{L resin}} \times \frac{1\text{ L } H_2O}{275 \text{ mg } CaCO_3} \times \frac{50{,}000 \text{ mg } CaCO_3}{\text{equiv } CaCO_3} \quad (12\text{-}44$$

$$BV_H = 225 \text{ volumes of } H_2O \text{ treated/volume of resin}$$

The time to hardness breakthrough t_H is related to the bed volumes to breakthrough BV_H and the EBCT:

$$t_H = \text{EBCT} \times BV_H \tag{12-45}$$

$$t_H = 3.0 \frac{\text{minutes}}{\text{BV}} \times 225 \text{ BVs} \times \frac{1\text{ h}}{60 \text{ minutes}} = 11.2 \text{ h} \tag{12-46}$$

If the EBCT is decreased by increasing the flow rate through the bed, that is, SFR, then the run time is proportionately shortened even though the total amount of water treated V_F remains constant.

(5) *Calculation of resin volume V_R.* The most important parameter chosen was the service flow rate (SFR) because it directly specified the necessary resin volume V_R according to the following relationships based on Eq. 12-34:

$$V_R = \frac{Q_F}{SFR} = Q_F(EBCT) \tag{12-47}$$

Numerically, for a column feed flow Q_F of 45 percent (the amount not bypassed) of 3.8×10^3 m^3/d,

$$V_R = \frac{1700 \text{ m}^3}{\text{day}} \times 3 \text{ minutes} \frac{\text{day}}{1440 \text{ minutes}} = 3.5 \text{ m}^3 \text{ or } \frac{(3.28 \text{ ft})^3}{\text{m}^3} = 124 \text{ ft}^3 \tag{12-48}$$

(6 & 7) *Calculation of the number of columns and the minimum out-of-service time for regeneration.* For a reasonable system design, two columns are required—one in operation and one in regeneration or standby. A single-column design with product water storage is possible but provides no margin of safety in case the column has to be serviced. Even with two columns, the out-of-service time t_{OS} for the column being regenerated should not exceed the exhaustion run time to hardness breakthrough t_H.

$$t_{OS} \leq t_{BW} + t_R + t_{SR} + t_{FR} \tag{12-49}$$

where t_{BW} = time for backwashing, 5 to 15 minutes
t_R = time for regeneration, 30 to 60 minutes
t_{SR} = time for slow rinse, 10 to 30 minutes
t_{FR} = time for fast rinse, 5 to 15 minutes

A conservative out-of-service time would be the sum of the maximum times for backwashing, regeneration, and rinsing (i.e., 2 hours). This causes no problem in regard to continuous operation because the exhaustion time is more than 11 hours.

(8) *Calculation of column dimensions.* The resin depth h was specified earlier as 0.76 m (2.5 ft); thus the column height, after we allow for 100 percent resin bed expansion during backwashing, is 1.5 m (5.0 ft). The bed diameter D then is

$$D = \sqrt{\frac{4V_R}{\pi h}} = 2.4 \text{ m (8 ft)} \tag{12-50}$$

The resulting ratio of resin bed depth to column diameter is 2.5:8 or 0.3:1. This is within the acceptable range of 0.2:1 to 2:1 if proper flow distribution is provided. Increasing the resin depth to 1.2 m (4 ft) increases the column height to 2.4 m (8 ft) and reduces its diameter to 1.9 m (6.3 ft). Clearly, a number of depths and diameters are possible. Before specifying these, the designer should check with equipment manufacturers because softening units in this capacity range are available as predesigned packages.

Important: Another alternative would be to use three or more columns with two or more in service and one or more on standby. This offers a more flexible design. For a three-column system with two in service and one on standby, the resin volume of the in-service units would be 3.5 m^3/2 = 1.75 m^3 (62 ft^3) each; that is, the flow would be split between two 1.75 m^3 (62 ft^3) resin beds operating in parallel. Regeneration would be staggered such that only one column would undergo regeneration at any time. *An important advantage of operating columns in parallel with staggered regeneration is that product water quality is less variable than with single-column operation.* This can be a major consideration when the contaminant leakage or peaking is relatively high during a portion of the run; when this happens, combining the high leakage from one column with the low leakage from another produces an average leakage—presumably less than the MCL—over the duration of the run. Also, operating multiple columns in parallel with staggered regeneration is appropriate when nontargeted contaminants are removed for a portion of the run and then are subject to peaking before the target-contaminant run is complete. A good example is the removal of the target contaminant arsenic (which exceeds its MCL in the feed) in the presence of the nontarget contaminant nitrate (which is present but does not exceed its MCL in the feed). Generally, the arsenic run length would be 400 to 1200 BV, whereas nitrate typically would break through before 400 BV and peak at up to three times its feed water concentration. If the nitrate peaking causes the nitrate to exceed its MCL, then averaging the product water from two or more columns in parallel will be necessary to keep nitrate below its MCL while still allowing a long run length for arsenic. Carousel systems generally operate up to 20 columns in parallel, which protects against peaking of all contaminants and provides a consistent (averaged) product water quality.

(9) *Calculation of volume and composition of wastewater.* The spent regenerant solution consists of the regenerant and the slow-rinse (displacement-rinse) volumes. These waste solutions must be accumulated for eventual disposal as detailed in Chap. 22. The actual wastewater volume per regeneration will depend on the size of the resin bed, that is, whether one, two, or three beds are chosen for the design. The following calculations are in terms of bed volumes and can be converted to fluid volume once the column design has been fixed. The chosen regeneration level, 240 kg/m^3 (15 lb NaCl/ft^3) of

resin, is easily converted to bed volumes of regenerant required. Given that a 10% NaCl solution has a specific gravity of 1.07, the regenerant volume applied is

$$\frac{240\,\text{kg NaCl}}{\text{m}^3\,\text{resin}}\times\frac{1.0\,\text{kg solution}}{0.1\,\text{kg NaCl}}\times\frac{1.0\,\text{L solution}}{1.07\,\text{kg solution}}\times\frac{1.0\,\text{m}^3\text{resin}}{1000\,\text{L resin}}$$

$$=2.25\,\frac{\text{L solution}}{\text{L resin}}=2.25\,\text{BVs}$$

$$15\,\frac{\text{lb NaCl}}{\text{ft}^3\ \text{resin}}\times\frac{1\ \text{lb soln}}{0.1\ \text{lb NaCl}}\times\frac{\text{ft}^3\ \text{soln}}{1.07\times 62.4\ \text{lb soln}}=2.25\ \text{BV} \qquad (12\text{-}51)$$

Following the salt addition, a displacement (slow) rinse of 1 to 2 BV is applied. The total regenerant wastewater volume is made up of the spent regenerant (2.25 BV) and the displacement rinse (2.0 BV):

$$\text{Regenerant wastewater} = 2.25 + 2.0 = 4.25\ \text{BV}$$

In this example, the wastewater volume amounts to approximately 1.9 percent [(4.25/225)100] of the treated water or 0.9 percent (1.9×0.45) of the blended product water. Choosing 1.0 L of resin as a convenient bed size (for calculation purposes), we set 1.0 BV = 1.0 L; thus the regenerant wastewater volume from a 1.0 L bed is 4.25 L. For our example bed containing 3.5 m^3 (124 ft^3) of resin, the wastewater volume is 14.9 m^3 [524 ft^3 (3930 gal)].

If the small quantity of ions in the water used to make up the regenerant solution is disregarded, the waste brine concentration (equiv/L) can be calculated as follows:

$$\text{Total ionic concentration of wastewater} = \text{hardness concentration} + \text{excess NaCl concentration} \qquad (12\text{-}52)$$

$$\text{Wastewater hardness concentration} = \frac{\text{equiv hardness removed}}{\text{regenerant wastewater volume}} \qquad (12\text{-}53)$$

From Table 12-6, we see that 1.24 equivalents of hardness are removed from the feed water per liter of resin at a regeneration level of 15 lb NaCl/ft^3 of resin. During steady state operation, this is the amount of hardness removed from the resin with each regeneration. Therefore,

$$\text{Waste hardness concentration} = \frac{1.24\ \text{eq}}{4.25\ \text{L}} = 0.29\,\frac{\text{eq}}{\text{L}} \qquad (12\text{-}54)$$

This amounts to 14,500 mg $CaCO_3$ hardness/L, which can be further broken down into the separate Ca and Mg concentrations using the known ratio of Ca to Mg in the raw water.

The excess NaCl concentration can be calculated from the following relationship where equiv NaCl applied = $3.3 \times$ equiv hardness removed, as seen in Table 12-6:

$$\text{Excess NaCl concentration} = \frac{\text{eq NaCl applied} - \text{eq hardness removed}}{\text{regenerant wastewater volume}}$$

$$\text{Excess NaCl concentration} = \frac{(1.24)3.3 - 1.24}{4.24} = 0.67\,\frac{\text{eq NaCl}}{\text{L}} \qquad (12\text{-}55)$$

This excess NaCl concentration corresponds to 39,300 mg NaCl/L. The total cation composition of the wastewater is 0.96 eq/L made up of 0.29 eq/L hardness and 0.67 eq/L sodium. These, in addition to chloride, are the major constituents of the wastewater. Other minor contaminant cations removed from the source water will be present in the regenerant wastewater. These can include Ba^{2+}, Sr^{2+}, Ra^{2+}, Fe^{2+}, Mn^{2+}, and others. The minor anionic contaminants in the wastewater will be bicarbonate and sulfate from the source water used for regeneration and rinsing. ▲

Brine Disposal from Softening Plants. The usual method for disposing of spent regenerant brine is through metering into a sanitary sewer. In coastal locations, direct discharge into the ocean is a possibility. Other alternatives are properly lined evaporation ponds in arid regions or brine disposal wells in areas where such wells are permitted or already in existence. The uncontrolled discharge (batch dumping) of the spent regenerant brine into surface waters or sanitary sewers is never recommended because of the potential damage to biota from localized high salinity. For the interested reader, the subject of waste disposal is covered in detail in Chap. 22.

Hydrogen IX Softening

Softening water without the addition of sodium is sometimes desirable. In this case, hydrogen can be exchanged for hardness ions using either strong- or weak-base cation exchangers. Hydrogen-form strong-acid resins (Eq. 12-1) are seldom used in this application because of the acidity of the product water, the inefficiency of regeneration with acid, and the problems of handling corrosive acids and of excess acid disposal. Hydrogen-form weak-acid cation (WAC) exchangers are sometimes used for sodium-free softening. Only the removal of temporary hardness is possible, and this proceeds according to Eq. 12-4, resulting in partial softening, alkalinity removal, and TDS reduction. Regeneration is accomplished with strong acids such as HCl or H_2SO_4 and proceeds according to Eq. 12-5 backwards (from right to left).

The partially softened, alkalinity-free column effluent from WAC softening must be stripped of CO_2 and blended with raw water to yield a noncorrosive product water. Alternatively, the pH of the column effluent can be raised by adding NaOH or $Ca(OH)_2$ following CO_2 stripping. This, however, costs more and results in the addition of either sodium or hardness to the product water.

Barium Removal by IX

During all types of IX softening, barium is removed in preference to calcium and magnesium. This is shown graphically in Fig. 12-10*a*, where, in theory, barium breaks through long after hardness (magnesium). This is true even though barium-contaminated groundwater always will contain much higher levels of calcium and magnesium. Snoeyink and colleagues (1987) summarized their considerable research on barium removal using hydrogen- and sodium-form SAC and WAC resins operated to barium breakthrough. They found that the main problem with using SAC resin for barium removal is the difficulty of removing barium from the exhausted resin. Barium accumulates on the resin and reduces the exchange capacity of subsequent runs if sufficient NaCl is not used for regeneration.

When using WAC resins in the hydrogen form for barium removal, the same considerations apply as with WAC softening. Divalent cations are preferentially removed, cation removal is equivalent to alkalinity, partial desalting occurs, CO_2 must be stripped from the column effluent, and bypass blending will be required. The advantage of WAC resins is that barium is removed easily during regeneration with a small excess, typically 20 percent, of HCl or H_2SO_4. In summary, using a hydrogen-form WAC resin for barium or combined hardness and barium removal produces higher-quality product water with less wastewater volume compared with a sodium IX. The WAC process, however, is more complex and more expensive because of chemical costs, the need for acid-resistant materials of construction, wastewater neutralization, and the need to strip CO_2 from the product water.

Radium Removal by IX

Radium-226 and radium-228 are natural groundwater contaminants occurring at ultratrace levels. Their combined MCL is limited to 5.0 pCi/L, which corresponds to 11.1 radium-226 disintegrations/minute/L or 0.185 Bq/L. For radium-226 alone, 5 pCi/L corresponds to 5×10^{-9} mg radium/L. An up-to-date list of radium violations has not been published by U.S. Environmental Protection Agency (USEPA); however, in 1998, approximately 200 community water systems exceeded the radium MCL (USEPA, 2000b), and in 2003, Illinois reported 99 systems in violation of the radium MCL (Illinois EPA, 2003). Cation exchange with either sodium- or hydrogen-form resins is an effective means of radium removal because radium is preferred over all the common cations found in water. As with hardness and barium removal, radium can be removed effectively using sodium-form SAC or hydrogen-form WAC resins.

Radium Removal During Softening. During the normal sodium IX softening process, radium is completely (>95 percent) removed; thus softening is an effective technique for meeting the radium MCL. Pilot studies on a groundwater containing 18 pCi/L total radium and 275 mg/L total hardness in Lemont, IL, resulted in the following conclusions regarding sodium IX softening for radium removal (Subramonian et al., 1990):

1. On the first exhaustion run to radium breakthrough, hardness breakthrough occurred at 300 BV, whereas radium did not break through until 1200 BV.
2. On the second and subsequent exhaustion runs following salt regeneration at 240 kg NaC/m^3 (15 lb NaCl/ft^3) of resin, radium broke through simultaneously with hardness at 300 BV. This long first run to radium breakthrough followed by shorter subsequent runs was not an anomaly but was repeated with other SAC resins. *Note:* Similar performance of softening beds is expected at much lower regeneration levels, for example, 80 kg NaCl/m^3 (5 lb NaCl/ft^3) of resin.
3. When the resin bed was operated in the normal fashion, that is, exhausted downflow and never run beyond hardness breakthrough, no radium was removed during the first three cocurrent regenerations at 240 kg NaC/m^3 (15 lb NaCl/ft^3) of resin. And five exhaustion/regeneration cycles were required to reach a steady state where radium sorption during exhaustion equaled radium desorption during regeneration.
4. WAC resins (containing relatively hydrophilic carboxylate functional groups) had a lower relative affinity for radium than SAC resins. Macroporous SAC resins had the highest affinity for radium but were more difficult to regenerate because of this high affinity.
5. Radium never broke through before hardness in any of the 80 experimental runs with five different SAC and WAC resins. Furthermore, the radium concentration of the effluent never exceeded that of the influent; that is, chromatographic peaking of radium never occurred.
6. Radium was difficult to remove from exhausted resins presumably because it is a large, poorly hydrated ion that seeks the relatively inaccessible hydrophobic regions of the resin phase. Increasing the regeneration time beyond the usual 15 to 30 minutes helped only slightly to remove the radium from the spent resin. In this regard, radium behaved differently from barium (Myers et al., 1985), which was much more efficiently eluted at 30 minutes compared with the 15 minute EBCT for the regenerant.

Dealing with Radium-Contaminated Brines. If one is consistent with existing practices regarding the disposal of radium-contaminated brines, disposal into the local sanitary sewer should be allowed. For example, in many Midwest communities where water is being softened on both a residential and a municipal scale, radium removal is also taking place, and the radium-contaminated brines are being disposed of in the usual fashion, that is, by metering into the local sanitary sewer. To determine the radium concentration in waste brines from softeners, a simple calculation can be made. In a hypothetical situation in which the raw water contains as much as 20 pCi/L of radium and the softening run length is 300 BV, the 5 BV of spent regenerant contain an average of 1200 pCi/L of radium. This is a concentration factor of 60 (i.e., 300/5) and is typical of softeners in general.

If necessary, a spent regenerant brine solution could be decontaminated prior to disposal by passing it through a radium-specific adsorbent such as the Dow complexer or $BaSO_4$-loaded activated alumina (Clifford, 1990). [*Note:* The Dow *radium-selective complexer* (RSC) is not really a complexer but a $BaSO_4$-impregnated SAC resin (Hatch, 1984) that adsorbs radium onto the $BaSO_4$ even in the presence of a high concentration of competing ions such as Ca^{2+}, Mg^{2+}, and Na^+.] This brine decontamination process has been tested on a small municipal scale (Mangelson, 1988) and found to work well. The RSC was loaded to a level of 2700 pCi/cm^3 and was still decontaminating spent brine after 1 year of operation. However, disposal of a radium-containing solid at a level of 2700 pCi/cm^3 (2.7×10^6 pCi/L) is potentially a more serious problem than disposal of the original radium-contaminated brine.

Nitrate Removal by IX

IX of chloride for nitrate is currently the simplest and lowest-cost method for treating nitrate-contaminated drinking water supplies. The process is shown schematically in Fig. 12-11. The anion exchange process for nitrate removal is similar to cation exchange softening except that (1) anions rather than cations are being exchanged, (2) nitrate is a monovalent ion, whereas calcium is divalent, and (3) nitrate, unlike calcium, is not the most preferred common ion involved in the multicomponent IX process used to treat a typical nitrate-contaminated groundwater. The latter two exceptions lead to some significant differences between softening and nitrate removal.

Chloride-form SBA exchange resins are used for nitrate removal according to Eq. 12-7. Excess NaCl at a concentration of 1.5 to 12 percent (0.25–2.0 M) is used for regeneration to produce a reversal of that reaction. The apparently simple process is not without complications, however, as detailed below.

Effects of Water Quality on Nitrate Removal. The source water quality and, in particular, the sulfate content influence the bed volumes that can be treated prior to nitrate breakthrough, which is shown along with that for chloride and bicarbonate in Fig. 12-13. The effect of increasing sulfate concentration on nitrate breakthrough is shown in Fig. 12-14, constructed from data obtained by spiking Glendale, Ariz., water with sodium sulfate (Clifford et al., 1987). As sulfate increased from the natural value of 43 mg/L (0.9 meq/L) to 310 mg/L (6.5 meq/L), that is, an increase of 5.6 meq/L, the experimental run length to nitrate breakthrough decreased 55 percent from 400 to 180 BV. Similar response to increasing sulfate is expected for all conventional type 1 and type 2 strong-base resins, disregarding the special nitrate-selective resins, which are less sensitive to sulfate.

Because all commercially available SBA resins prefer sulfate to nitrate at the TDS and ionic strength levels of typical groundwater, chromatographic peaking of nitrate occurs following its breakthrough. This peaking, when it does occur, means that effluent nitrate concentrations will exceed the source water nitrate level. No easy way is available to

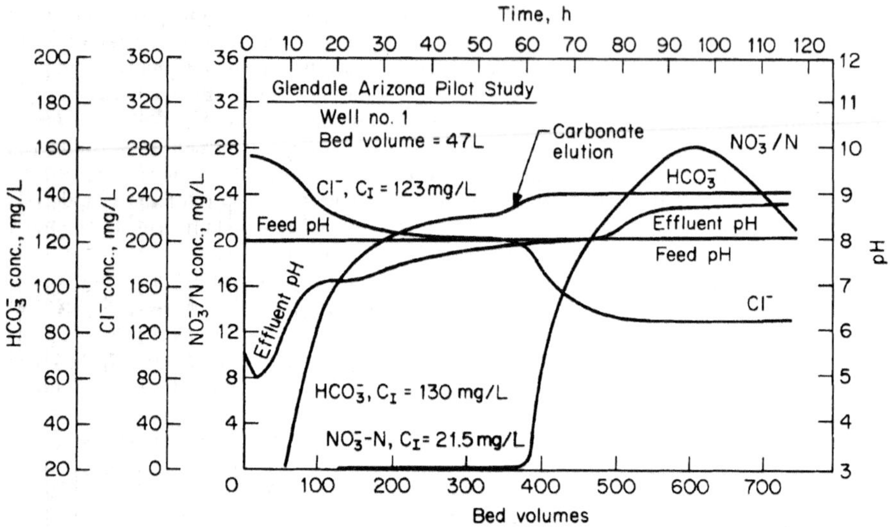

FIGURE 12-13 Breakthrough curves for nitrate, bicarbonate, and chloride following complete regeneration of type 2 gel SBA resin in chloride form. C_I = influent concentration.

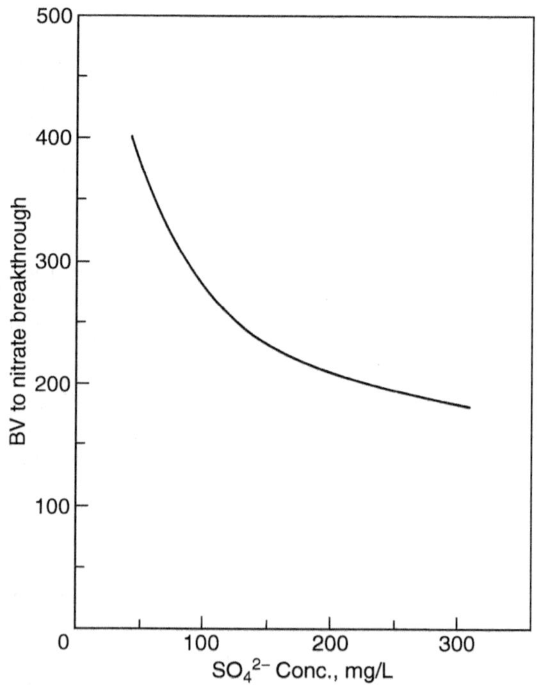

FIGURE 12-14 Effect of sulfate concentration on nitrate breakthrough for source and sulfate-spiked Glendale, Ariz, groundwater.

calculate whether a peak will occur or what the magnitude of the expected nitrate peak will be. Peak magnitude depends primarily on the TDS level, the specific composition of the source water (including its sulfate, nitrate, and alkalinity concentrations), and the type of SBA resin used. Computer estimates of the magnitude of the nitrate peak can be made using the EMCT Windows (Tirupanangadu, 1996) and IX Windows computer programs (Guter, 1998).

As the TDS level of the source water increases, selectivity reversal can occur, with the result that nitrate is preferred to sulfate. In this case, which was reported by Guter in his McFarland, Calif., nitrate-removal pilot studies (Guter, 1982), sulfate broke through prior to nitrate, and no nitrate peaking occurred.

In laboratory and pilot studies (Clifford and Weber, 1978; Clifford et al., 1987), nitrate peaking clearly was observed with both simulated groundwater and actual groundwater. The theoretical nitrate breakthrough curves for the standard type 1 and type 2 SBA resins are depicted in Fig. 12-10*b*, where idealized nitrate peaking is shown. Figure 12-13 depicts the pH, nitrate, and bicarbonate breakthrough curves observed in the Glendale nitrate-removal pilot studies (Clifford et al., 1987). In this groundwater, which is considered typical of nitrate-contaminated groundwater, the effluent nitrate concentration peaked at 1.3 times the feed water value.

Detecting Nitrate Breakthrough. When the usual type 1 and type 2 SBA resins are used to remove nitrate from a typical groundwater containing sulfate and having an approximate ionic strength of 0.010 M or less, chromatographic peaking of nitrate can occur. Even if it does not occur, a single column run still must be terminated prior to the breakthrough of nitrate. If the source water concentration is reasonably constant, this termination can be initiated by a flowmeter signal when a predetermined volume of feed has passed through the column. Terminating a run at a predetermined volume of treated water cannot always be recommended because of the variable nature of nitrate contamination in some types of groundwater. For example, in a 6-month period in McFarland (Guter, 1982), the nitrate content of one well varied from 5 to 25 mg/L NO_3-N. Such extreme variations did not occur during the 15-month pilot study in Glendale (Clifford et al., 1987), where the nitrate content of the study well (which was pumped occasionally to fill a large tank for the pilot study) varied only from 18 to 25 mg/L NO_3-N.

In the Glendale pilot study, a pH change of about 1.0 pH unit accompanied the nitrate breakthrough. This pH wave is visible in Fig. 12-13, where we see that if the run had been terminated when the effluent pH equaled the feed pH, the nitrate peak would have been avoided. The observed pH increase resulted from the simultaneous elution of carbonate and nitrate. The carbonate elution was a fortunate coincidence, and it allowed pH or differential (feed versus effluent) pH to be used to anticipate nitrate breakthrough. Such pH waves are not unique to the Glendale water but are expected when significant (>1.0 meq/L) concentrations of sulfate and bicarbonate are present and when ionic strength is 0.01 M or less. In practice, the pH detector should be located slightly upstream of the column effluent to provide a safety factor. Another way to avoid nitrate peaking is to employ multiple parallel columns as described in the section "Multicolumn Operations."

Choice of Resin for Nitrate Removal. Both laboratory and field studies of nitrate removal by chloride-form anion exchange have shown that no significant performance differences exist among the standard commercially available SBA resins. Both type 1 and type 2 polystyrene divinylbenzene SBA resins have been used successfully. Guter (1982) reported on the performance of a type 1 resin in a full-scale application, and Clifford and colleagues (1987) made extensive use of a type 2 SBA resin during a 15-month pilot-scale study in Glendale. When chromatographic peaking of nitrate must be avoided, however, a special nitrate-selective resin is necessary. Liu and Clifford (1996) also reported that nitrate-selective resins were required with their brine denitrification and reuse process when the feed water contained 200 mg/L sulfate.

Guter (1982) described several such special resins that were nitrate selective with respect to sulfate based on their increased functional group charge separation distance and hydrophobicity. The nitrate-sulfate selectivities of these resins were in accord with the predictions of Clifford and Weber (1978, 1983), who found that (1) the sulfate preference of a resin was reduced significantly by increasing the distance between charged IX sites and (2) the nitrate preference of a resin was improved by increasing matrix and functional group hydrophobicity.

Nitrate-over-sulfate-selective (NSS) resins are available from most resin suppliers. Basically, NSS resins are similar to the standard type 1 resins, which contain trimethyl amine functionality [$RN(CH_3)_3$], but the NSS resins have ethyl, propyl, or butyl groups substituted for the methyl groups to increase the charge-separation distance and resin hydrophobicity. Although the tripropyl and tributyl NSS resins have higher relative nitrate preferences, the less expensive triethyl resins generally have proven to be adequate for applications that require a nitrate-selective resin.

Regeneration of Nitrate-Laden Resin. Regardless of the regeneration method used, all previous studies have demonstrated that sulfate is much easier to elute than nitrate. Typical sulfate and nitrate elution curves are shown in Fig. 12-15. For a complete regeneration, divalent sulfate, the most preferred ion during exhaustion, is the first to be stripped from the resin during regeneration because of the selectivity reversal (nitrate is preferred by the resin in brine) that occurs in high-ionic-strength (0.25–3 M) salt solutions.

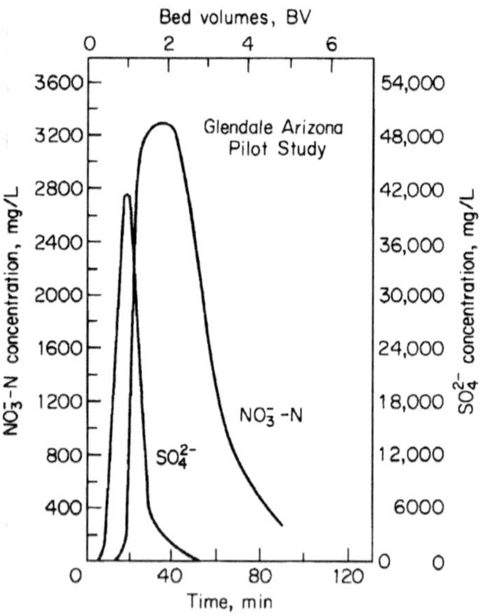

FIGURE 12-15 Elution of sulfate and nitrate during complete regeneration of type 2 gel SBA resin following the exhaustion run shown in Fig. 12-13. Cocurrent downflow regeneration using 12% NaCl (2 N) at an EBCT of 19 minutes—nearly 100 percent recovery of sulfate and nitrate observed.

Regeneration of the nitrate-laden resin was studied extensively in Glendale by using the complete and partial regeneration techniques. For complete regeneration, that is, removal of more than 95 percent of the sorbed nitrate, the more dilute the regenerant, the more efficient it was. For example, 0.5 M NaCl required 4.0 equivalents of chloride per equivalent of resin, whereas 1.7 M NaCl (10% NaCl) required 170 percent of this value. An obvious disadvantage of dilute regenerants is that larger tanks are required when the spent regenerants are stored. In addition, they produce more wastewater and require longer regeneration times.

Complete regeneration is not very efficient in terms of the amount of salt required, but it does result in minimal leakage of nitrate during the subsequent exhaustion run. Because significant nitrate leakage (typically up to 7 mg NO_3-N per liter) is permissible based on the MCL of 10 mg NO_3-N per liter, complete regeneration is not required. Thus nitrate removal is an ideal application for incomplete or partial regeneration, that is, removal of 50 to 60 percent of the sorbed contaminant.

Partial regeneration levels of 0.64 to 2.0 equivalents per equivalent of resin (3–10 lb NaCl per cubic foot of resin) were studied at Glendale with the result that 1.0 eq Cl^- per equivalent of resin was acceptable if 7.0 mg NO_3-N per liter could be tolerated in the effluent prior to the nitrate breakthrough that occurred earlier (350 BV) than in the complete-regeneration case (410 BV). Guter (1982) found that regeneration levels as low as 0.64 eq Cl^- per equivalent of resin (3 lb NaCl per cubic foot of resin) were acceptable in his pilot studies in McFarland. The acceptable level must be determined for each application, and it depends primarily on the feed water quality and the allowable nitrate leakage. Regardless of the level of salt used for partial regeneration, complete mixing of the resin bed is mandatory following regeneration. If mixing is not accomplished, excessive nitrate leakage will occur during the first 80 to 100 BV of the exhaustion cycle. The required mixing can be achieved mechanically or by introducing air into the backwash, but conventional (uniform) backwashing, even for extended periods, will not provide adequate mixing.

In the Glendale pilot study, a comparison of the optimal complete- and partial-regeneration methods indicated that partial regeneration would use 37 percent less salt. Bypass blending could not be used with partial regeneration because the nitrate leakage was 7 mg NO_3-N per liter.

Disposal of Nitrate-Contaminated Brine. Because of its eutrophication potential, nitrate-contaminated brine cannot be disposed of into rivers or lakes, even if it is slowly metered into the receiving water. Furthermore, the high sodium concentration prevents disposal of spent regenerant onto land where its nitrogen content could serve as a fertilizer. To eliminate the high-sodium problem, calcium chloride could be used in place of NaCl, but this would be much more expensive and could result in the precipitation of $CaCO_3(s)$ during regeneration. Potassium chloride (KCl) is also feasible as a regenerant but is currently (2010) at least triple the cost of NaCl.

In one municipal-scale application of nitrate removal by chloride IX (Guter, 1982), the spent regenerant simply was metered into the sanitary sewer for subsequent biological treatment. This is probably possible only in applications where nitrate removal is being carried out in a fraction (e.g., <25 percent) of the wells in the supply system. If all the wells were being treated, the waste salt and varying salinity surely would have a negative effect on the biological treatment process.

In the Glendale studies, direct reuse of the spent brine (without removing the nitrate) was attempted but without success. Electroselectivity reversal does not occur with monovalent ions such as nitrate, and the high affinity of the resin for the nitrate in the brine resulted in excessive nitrate remaining on the resin following regeneration. The nitrate remaining on the resin was driven off subsequently by sulfate in the feed water, causing early nitrate leakage during the next exhaustion.

Biological Denitrification and Reuse of Spent Brine. Removing the nitrate from the spent brine prior to its reuse is possible via biological denitrification. Bench- and pilot-scale studies of this process have been reported by Van der Hoek and colleagues (1988), who found that biological denitrification and reuse were feasible below about 15,000 mg NaCl per liter (1.5% NaCl). Clifford and Liu (1993a, 1993b) developed a *sequencing batch reactor* (SBR) denitrification process for biological denitrification of spent nitrate IX brine. Their bench-scale experiments examined the feasibility of using an SBR to denitrify 0.5 M (3%) NaCl brine containing up to 835 mg NO_3-N per liter. After acclimation, the denitrification rate in 0.5 M NaCl was only 10 percent lower than in the no-salt control reactor. The effect of mass ratio of methanol to nitrate-nitrogen on denitrification rate and residual TOC in the denitrified SBR effluent was studied in the range of 2.2 to 3.2. At the optimal methanol-to-nitrate-nitrogen ratio of 2.7, the time for more than 95 percent denitrification was 8 hours. In one set of runs, actual spent regenerant was reused 15 times; each time it was denitrified in the SBR, filtered, and compensated with NaCl before reuse. Denitrification and reuse of the brine resulted in a 50 percent reduction in NaCl consumption and a 90 percent reduction in spent brine discharge. Their bench-scale research indicated that a denitrifying batch reactor provided simple operation, reliable effluent quality, and compatibility with the inherent batch operation of the nitrate IX process. Later Liu and Clifford (1996) reported on pilot-scale tests of the SBR denitrification brine reuse process (Fig. 12-16) for

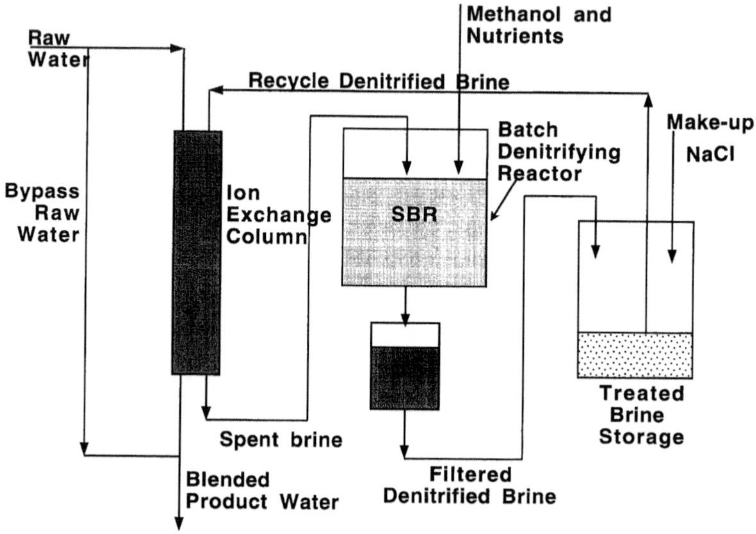

FIGURE 12-16 IX/biological denitrification process with sequencing batch reactor (SBR).

nitrate removal in McFarland, Calif., on a high-sulfate (140–220 mg/L) water. In those tests, which evaluated three different resins and partial and complete regeneration with and without brine reuse, the brine was denitrified and reused successfully up to 38 times. Because of sulfate buildup to 16,000 mg/L in the recycle brine, a nitrate-selective resin (with triethyl or tripropyl amine functional groups) was preferred to minimize nitrate leakage and maximize the number of brine reuse cycles. Even though the biologically denitrified brine was repeatedly in contact with microorganisms, methanol, and high TOC, the product water quality in terms of heterotrophic plate count, volatile organics, TOC, and methanol was not significantly different from product water from a conventional nitrate IX process. The pilot study of the SBR brine denitrification/reuse process verified that NaCl consumption could be reduced up to 75 percent compared with the optimal partial-regeneration process without brine reuse. Depending on the number of reuse cycles, the brine discharge volume could be reduced up to 95 percent compared with operation without brine reuse.

More recently, in a pilot study of nitrate and perchlorate destruction in IX brine using a fluidized-bed reactor (FBR) in California at the West Valley Water District, Lehman and colleagues (2008) demonstrated that IX brine (4.5% NaCl) containing up to 550 mg/L NO_3-N could be completely denitrified in 3 days with a hydraulic retention time (HRT) of 13 hours. The FBR proved to be far more challenging to inoculate and operate than the SBR used in earlier research.

The brine denitrification/reuse process, although more complex than processes using direct disposal of waste regenerants, should be considered where wastewater disposal is a major consideration.

Fluoride Removal by Activated Alumina

Activated alumina, a semicrystalline porous inorganic adsorbent, is an excellent medium for fluoride removal. Alumina is far superior to synthetic organic anion exchange resins because fluoride is one of the ions most preferred by alumina, whereas, with resins, fluoride is the least preferred of the common anions. (Compare the position of fluoride in Eq. 12-27, the activated alumina anion selectivity sequence, with Table 12-3 containing the relative affinity of anions for resins.) The usefulness of alumina as a fluoride adsorbent has been known for more than 70 years, and municipal defluoridation of public water supplies using packed beds of activated alumina has been practiced since the 1930s and 1940s (Fink and Lindsay, 1936; Maier, 1953). More recently, POE and POU fluoride-removal systems using activated alumina have come into common use (Water Quality Research Council, 1987) or at least are being considered for individual-home treatment (Bellen et al., 1986).

The fundamentals of fluoride adsorption onto activated alumina were covered in the section "Granular Media Adsorption" above, where attention was drawn to Eqs. 12-11 through 12-17. The following discussion focuses on the important design considerations for municipal defluoridation. Although pH adjustment is not ordinarily performed prior to POE and POU defluoridation, the discussion also may prove useful to designers of these systems because the important factors, including pH, governing fluoride capacity are explained.

Alumina Defluoridation System Design. A typical fluoride-removal plant using activated alumina consists of two or more adsorption beds operated alternately or simultaneously in parallel or series. The source water pH is adjusted to 5.5 to 6.0 and passed downflow through a 0.9 to 1.5 m deep (3–5 ft) bed of fine, 0.3 to 0.6 mm diameter (28 × 48 mesh) medium that adsorbs the fluoride. The fluoride breakthrough curve (some typical examples are shown in Fig. 12-17) is not sharp compared with the usual IX resin breakthrough curves. The effluent fluoride concentration is continuously increasing, thereby making use of the bypass blending technique shown in Fig. 12-11 difficult to implement. Instead, a product water storage tank sometimes is provided to equalize the column effluent fluoride concentration and to maximize the column run length.

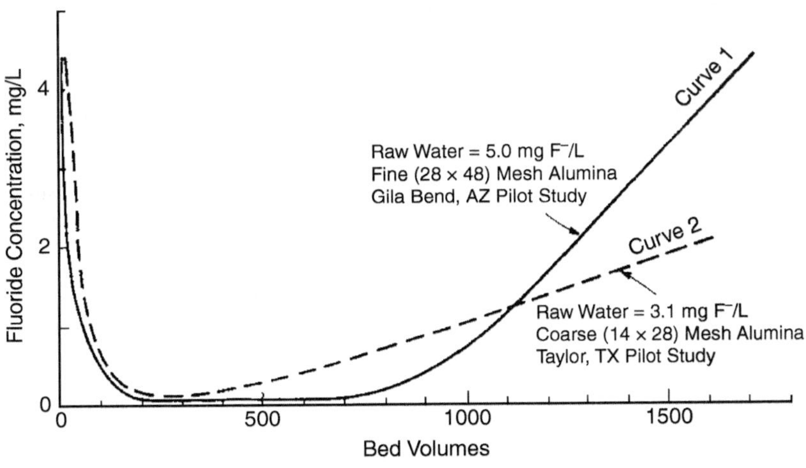

FIGURE 12-17 Typical fluoride breakthrough curves for activated alumina operated at a feed pH of 5.5 and an EBCT of 5 minutes.

A target level of fluoride, usually in the range of 1 to 3 mg/L, must be chosen for the process effluent before the operating procedures and economics can be established. The MCL for fluoride is 4 mg/L, and the secondary maximum contaminant level (SMCL) is 2 mg/L (USEPA, 2009a; see also Chap. 1). However, a recent study by the National Academy of Sciences (2006) concluded that a 4 mg/L fluoride MCL is not sufficiently protective of public health, and the USEPA is under pressure to lower the MCL. When the source water contains 3 to 6 mg/L fluoride and a maximum fluoride effluent (breakthrough) concentration of 1.4 mg/L is chosen, a run length of approximately 1000 to 1300 BV can be expected with a typical alumina. Some typical data from pilot-scale defluoridation processes are presented in Table 12-7 for source water fluoride concentrations of 2 to 5 mg/L, and some detailed design recommendations are given in Table 12-8.

TABLE 12-7 Fluoride Capacity of F-1 Activated Alumina Columns at pH 5.5 to 6.0: Field Results

Feedwater F^- conc., mg/L	Feedwater TDS, mg/L	Alumina particle size[a] (mesh size)	Run length, BV to 1.4 mg F^-/L	Fluoride capacity, g/m^3
2.0[b]	810	0.3–0.6 mm (28 × 48)	2300	3700
3.0[c]	1350	0.6–1.2 mm (14 × 28)	1200	2700
5.0[d]	1210	0.3–0.6 mm (28 × 48)	1150	4600

[a]28 × 48 mesh is "fine" alumina with faster kinetics than14 × 28 "coarse" alumina, which also can be used but yields lower fluoride capacity.
[b]San Ysidro, New Mexico; average of 3 pilot-scale runs (Clifford and Lin, 1991).
[c]Taylor, TX; average of 3 pilot-scale runs.
[d]Gila Bend, Ariz; typical pilot-scale run (Rubel and Woosley, 1979).

Following exhaustion, the medium is backwashed and then subjected to a two-step regeneration with base followed by acid. Because many types of high-fluoride groundwater are located in hot, arid climates, the spent regenerant brines are neutralized and sent to a lined evaporation pond for interim disposal. The ultimate disposal of high-fluoride salt residues is a problem still unsolved.

Factors Influencing the Fluoride Capacity of Alumina. The fluoride capacity of alumina is sensitive to pH, as shown in Fig. 12-18*a* based on laboratory equilibrium data obtained by using minicolumns of granular alumina exposed for up to 30 days to pH-adjusted fluoride solutions in deionized water. The fluoride capacities in Fig. 12-18*a* can be considered the maximum attainable (equilibrium) capacities for similar aluminas because potentially competing anions such as silicate, arsenate, phosphate, selenite, and sulfate were minimized or entirely absent in the test solutions (Singh and Clifford, 1981). Pilot-scale field studies on typical fluoride-contaminated water have resulted in the observation that single columns of fine (28 × 48 mesh) alumina operated to 1.4 mg fluoride per liter breakthrough and a 5 to 10 minute EBCT can attain 25 to 50 percent of the equilibrium capacities in deionized water (Rubel and Woosley, 1979; Clifford and Lin, 1991). See also Fig. 12-18*b* based on previously unpublished field studies in Taylor, Texas. Using coarse-mesh (14 × 28 mesh) alumina under the same conditions yields lower fluoride capacities.

The inability of a pilot- or full-scale continuously operated alumina column to rapidly and closely approach fluoride equilibrium is related primarily to solid-phase mass transfer limitations. These limitations cause the fluoride breakthrough curve to be gradual and the run to be terminated long before the alumina is saturated. This is a classic case in which the process economics can be improved by operating two columns in series in a roughing-polishing

TABLE 12-8 Process Design Criteria for Fluoride Removal by Activated Alumina

Parameter	Typical value or range
Fluoride concentration	3–6 mg/L
Media[a]	Alcoa F-1 activated alumina
Surface area	250 m^2/g
Bulk density	0.85 g/cm^3 (55 lb/ft^3)
Media size[b]	0.29–0.59 mm diameter (28 × 48 mesh)
Media depth	1–2 m (3–6 ft)
Fluoride capacity	3000–5000 g/m^3 (1300–2200 gr/ft^3)
Bed volumes to MCL, 1.4 mg F^-/L^3	1000–1500 (pH 5.5–6.0)
Exhaustion flow rate	EBCT = 5 minutes (1.5 gpm/ft^3)
Exhaustion flow velocity	10–20 m/h (4–8 gpm/ft^2)
Backwash flow rate	20–22 m/h (8–9 gpm/ft^2)
Backwash time	5–10 minutes using raw water
NaOH Regeneration	
Volume of regenerant	5 BV
Regenerant flow rate[d]	EBCT = 15 minutes (0.5 gpm/ft^3)
Regenerant concentration[e]	0.25 N NaOH (1% NaOH)
Total regenerant contact time	75 minutes
Displacement rinse volume	2 BV
Displacement rinse rate	EBCT = 15 minutes (0.5 gpm/ft^3)
H_2SO_4 Neutralization	
Regenerant flow rate[d]	EBCT = 15 minutes (0.5 gpm/ft^3)
Acid concentration[f]	0.4 N H_2SO_4 (2.0% H_2SO_4)
Acid volume[g]	Sufficient to neutralize the bed to pH 5.5, typically 1.5 BV
Displacement rinse volume	2 BV
Displacement rinse rate	EBCT = 15 minutes (0.5 gpm/ft^3)

[a]Aluminum hydroxide activated at 400°C–600°C in a mixture of hot air and combustion gases, 92% Al_2O_3 (Misra, 1986). Mention of trade names does not imply endorsement. Similar activated alumina adsorbents are available from other sources.

[b]Coarse (14 × 28) mesh alumina also has been used successfully.

[c]Capacity and BV to fluoride breakthrough depend on adsorbent particle size, surface area and surface chemistry, EBCT, and feed water quality, including pH and fluoride and sulfate concentrations.

[d]Concurrent (same direction as exhaustion flow) or countercurrent (opposite direction to exhaustion flow) regeneration may be used.

[e]Higher regenerant NaOH concentrations (e.g., 2%–4%) also have been used successfully at shorter contact times. Media losses (dissolution of alumina) increase with increasing NaOH concentration and contact time.

[f]Lower and higher H_2SO_4 concentrations have been used successfully. HCl also can be used, and it may be preferred to H_2SO_4 because chloride does not compete with fluoride for adsorption sites. Media losses (dissolution of alumina) increase with increasing H_2SO_4 concentration and contact time.

[g]Alternatively, the neutralization and acidification steps can be combined. This is accomplished by lowering the initial feed water pH to approximately 2.5 to produce a bed effluent pH slowly dropping from a high initial value (>13) down to a continuous effluent pH in the range of 5.5–6.0.

(lead-lag) sequence, as depicted in Fig. 12-12. The two-column sequence allows the roughing (lead) column to attain a near-saturation fluoride loading prior to regeneration. The obvious disadvantages of such systems are greater complexity and higher capital costs.

In POE and POU alumina systems, a closer approach to equilibrium may be expected because of the intermittent operation of the alumina bed. In these installations, the solid-phase fluoride concentration gradient has a chance to *relax* completely (approaching equilibrium) during the many prolonged off periods experienced by columns in residential applications. This results in a high concentration gradient between the liquid and the surface of the solid and much improved fluoride removal when the column is restarted.

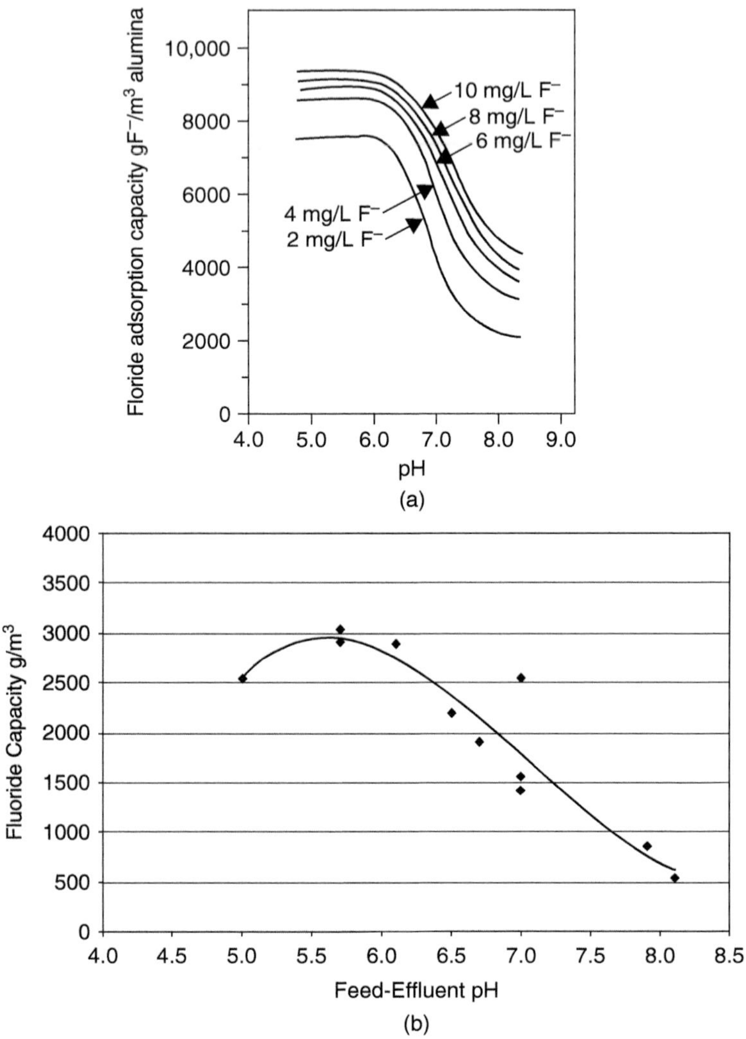

FIGURE 12-18 (*a*) Effects of pH and fluoride concentration on capacity of F-1 activated alumina in equilibrium minicolumn studies using DI water solutions of NaF, pH adjusted with Na_2CO_3 and H_2SO_4. (*b*) Effect of pH on fluoride capacity of 28 × 48 mesh pilot-scale F-1 activated alumina columns operated to 1.4 mg F^- per liter breakthrough. Taylor, Texas, feedwater F^- = 3.0 mg/L, SO_4^{2-} = 380 to 540 mg/L, TDS = 1300 to 1450 mg/L, EBCT = 5 to 10 minutes, pH adjusted with H_2SO_4.

Another reason for the less-than-equilibrium fluoride capacities of actual columns is the effect of competing anions such as silicate and sulfate present in the source water. During a laboratory study (Singh and Clifford, 1981) of the effect of competing ions on fluoride capacity at optimal pH, bicarbonate and chloride ions did not compete measurably with fluoride for adsorption sites. Sulfate, however, did cause a significant reduction in fluoride adsorption. For example, 50 mg of sulfate per liter lowered the fluoride capacity by about 16 percent. The maximum reduction in fluoride capacity because of sulfate was about 33 percent, and it occurred at approximately 500 mg of sulfate per liter. Beyond this, no further fluoride capacity reduction occurred, probably because the alumina contains a high proportion of fluoride-specific sites that are not influenced by sulfate. Fluoride-removal pilot studies in Taylor, Texas, corroborated the sulfate effect because pH adjustment with HCl gave 16 percent higher fluoride capacity than H_2SO_4 at the same adjusted pH (5.7).

Results of fluoride-removal pilot studies in Taylor, Texas, indicated that measurable silicate [$Si(OH)_3O^-$] adsorption onto alumina occurs at about pH 7 and above and can cause a serious reduction in the fluoride capacity of a column operated in the pH range of 7 to 9. This reduction is particularly detrimental during cyclic operation because silicate is only partially removed during a normal (base-acid) regeneration to remove fluoride. The presence of silica in the water had no observable effect on fluoride capacity for columns operated in the optimal (5.5–6.0) pH range, provided that the columns were never operated in the 7 to 9 pH range prior to optimal pH operation. Activated alumina columns that had been operated at pH in the 7 to 8 range adsorbed so much silica, which was not completely removed during regeneration, that subsequent runs at the optimal pH of 5.5 to 6.0 had 30 to 40 percent shorter run lengths, presumably owing to silica retained on the alumina.

Regeneration of Fluoride Spent Alumina Columns. Prior to regeneration to elute fluoride, the exhausted column must be backwashed to remove entrained particles, break up clumps of alumina, and reclassify the medium to eliminate packing and channeling. Following a 5 to 10 minute backwash, regeneration is accomplished in two steps—the first using NaOH and the second using H_2SO_4. Because fluoride leakage is such a serious problem, the most efficient regeneration is accomplished by using countercurrent flow. Nevertheless, cocurrent regeneration is used more commonly because of its simplicity. Rubel and Wossley (1979) recommend a two-stage upflow-then-downflow application of base followed by downflow acid neutralization. A slow rinse follows both the base addition and acid addition steps to displace the regenerant and conserve chemicals. The rinsing process is completed by using a fast rinse applied at the usual exhaustion flow rate. The total amount of wastewater produced amounts to about 4 percent of the blended product water and is made up of 8 BV of backwash, 5 BV of NaOH, 2 BV of NaOH displacement rinse, 1.5 BV of acid, 2 BV of acid displacement rinse, and 30 to 50 BV of fast rinse water. Further details pertaining to regeneration can be found in Table 12-8.

Essentially complete (>95 percent) removal of fluoride from the alumina is possible by using the regeneration techniques described; however, in addition to the complexity of regeneration, its efficiency is poor. For example, an alumina bed employed to remove fluoride from a typical 800 to 1200 mg TDS/L groundwater containing 4 to 6 mg fluoride per liter would contain an average fluoride loading of 4600 g F^- per cubic meter. Removing this fluoride typically would require 5 BV of 1% NaOH (30 gallons of 1% NaOH per cubic foot). This equates to 19 percent base regeneration efficiency, or 5.2 hydroxide ions required for each fluoride ion removed from the alumina. When the amount of acid required (1.5 BV of 0.4 N H_2SO_4) is added into the efficiency calculation, an additional 2.5 hydrogen ions are required for each fluoride ion removed, and the overall efficiency drops to 13 percent, or a total of 7.7 regenerant ions required for each fluoride ion removed from the water. Despite the complexity of the process and the inefficiency of regeneration on an ion-for-ion basis, fluoride removal by activated alumina is probably the lowest-cost alternative compared with other acceptable methods, such as reverse osmosis and electrodialysis (USEPA, 1985). To our knowledge, no study has reported on (1) the influence of NOM adsorption on the alumina capacity for fluoride or (2) the influence of activated alumina dissolution on the aluminum content of the treated water. Because the affinity of alumina for NOM anions (fulvates and humates) is significant, one would expect a decrease in fluoride capacity with increasing NOM concentration. Because aluminum is used widely as the primary coagulant for surface water, there has not been much concern about the effluent aluminum from activated alumina, which is basically low-temperature dehydrated aluminum hydroxide. Nevertheless, aluminum in treated water from activated alumina systems should be studied, especially in low-pH (5.5–6.0) systems. Also, direct reuse of the spent NaOH regenerant should be studied to improve regeneration efficiency.

Arsenic Removal by Adsorptive Media and IX

As with other toxic inorganic contaminants, arsenic is almost exclusively a groundwater problem. Although it can exist in both organic and inorganic forms, only inorganic arsenic in the +III or +V oxidation state has been found to be significant where potable water supplies are concerned (Andeae, 1978).

The MCL for arsenic is 0.010 mg/L or 10 μg/L (USEPA, 2003; see also Chap. 1). To assist small communities in meeting the revised MCL, the USEPA initiated a full-scale demonstration program of commercial-ready arsenic treatment technologies in 2002 (Zeilig, 2005). Over a 5-year period, 50 demonstration projects in 28 states were funded, with the goal to determine the cost and performance of these treatment systems operating under a variety of conditions. The results from the USEPA demonstration projects are summarized in the following sections, along with results from other laboratory and field studies.

The primary arsenate species (As(V)) found in groundwater in the pH range of 6 to 9 are monovalent $H_2AsO_4^-$ and divalent $HAsO_4^{2-}$. These anions result from the dissociation of arsenic acid (H_3AsO_4), which exhibits pK_a values of 2.2, 7.0, and 11.5. Uncharged arsenious acid (H_3AsO_3) is the predominant species of As(III) found in natural water. Only at pH values above its pK_a of 9.2 does the monovalent arsenite anion ($H_2AsO_3^-$) predominate.

Both anion exchange resins and adsorptive media require anionic forms of arsenic to be effective; thus redox potential and pH are important raw water characteristics. As(V) is removed effectively by both processes, whereas uncharged As(III) is not removed by IX, and is poorly removed by most the adsorptive media products. Consequently, if As(III) is the predominant arsenic form, the preoxidation of As(III) to As(V) is generally required. The choice of oxidants among treatment processes depends on many factors, including water quality, wastewater quality and quantity, waste disposal options available at the site, and capital and operating costs.

Oxidation of As(III) to As(V). IX processes require arsenite to be oxidized to arsenate to achieve any removal. For the adsorptive media process to be cost-effective, arsenite also should be oxidized to arsenate. Little if any As(III) is removed by activated alumina (Frank and Clifford, 1984), and a laboratory study of As(III)/As(V) removal by 11 different adsorptive media showed that most of them achieve a significantly higher percent removal of As(V) than of As(III) (Amy et al., 2005).

The importance of oxidizing As(III) to As(V) was clearly demonstrated during a 3500 m^3/d (640 gpm) full-scale test of an iron-based adsorptive media system at Brown City, MI, treating raw water with 15 μg/L As(III) (Chen et al., 2009). It appeared that 10 μg/L arsenic breakthrough would occur at about 30,000 BV; however, the run length was increased to more than 70,000 BV by switching from post- to prechlorination to oxidize As(III) to As(V) ahead of the adsorptive media (Fig. 12-19).

Ghurye and Clifford (2004) conducted a laboratory study of seven oxidants to convert As(III) to As(V): chlorine, permanganate, ozone, monochloramine, chlorine dioxide, a manganese dioxide solid oxidizing media, and 254 nm ultraviolet UV light. The study evaluated the effects of pH (6.3–8.3), temperature (5–25°C), and the interfering reductants Mn^{2+}, Fe^{2+}, S^{2-}, and TOC. Of the seven oxidants investigated, only chlorine and permanganate provided complete oxidation in less than 1 minute under all conditions. Ozone was fast and effective but was significantly affected by TOC. The solid manganese dioxide medium provided complete oxidation except on the test water having low dissolved oxygen

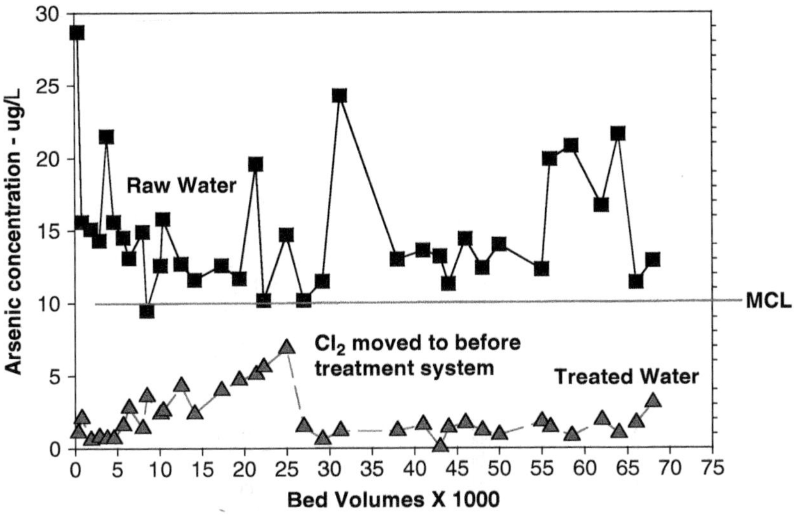

FIGURE 12-19 Effect of moving the point of chlorination on As(III) granular ferric oxide media—Brown City, Mich., demonstration plant.

(0.1 mg/L). Dissolved oxygen (DO) was not studied as an oxidant in this study because DO was known to be unreliable as an oxidant for As(III). For example, Frank and Clifford (1986) showed that pure oxygen sparging could not oxidize As(III) to As(V) in 1 hour, but complete oxidation of As(III) to As(V) occurred during 2 months of ambient temperature storage of synthetic groundwater. Field studies of possible air oxidation of As(III) by Lowry and Lowry (2001) at three test sites with varying water quality found that spray, bubble, and packed-tower aeration were ineffective in oxidizing As(III) to As(V).

Inadvertent, unpredictable microbial-assisted oxidation of As(III) to As(V) has been observed in several University of Houston studies with As(III) (Rosenblum and Clifford, 1984; Frank and Clifford, 1986; Clifford and Lin, 1986, 1991). Lytle and colleagues (2007) also reported the oxidation of As(III) in laboratory-, pilot-, and full-scale adsorption system studies. Analysis of the microbial community from the studies identified the isolates having strong As(III) oxidizing ability. Reduction of As(V) also has been reported by numerous researchers (Tamaki and Frankenberger, 1992; Ahmann et al., 1994; Laverman et al., 1995; Jones et al., 2002; Lloyd et al., 2003; Tadanier et al., 2005; Oremland, 2006). Therefore, As(III) and As(V) should be considered as interconvertible under appropriate conditions.

Adsorptive Media Processes for Arsenic Removal. Until the late 1990s, activated alumina was the dominant adsorptive medium used for arsenic removal, and the process was similar to that already described for fluoride removal. To achieve maximum arsenic removal capacity, the activated alumina process usually was designed with the pH of the raw water being reduced to 5.5 similar to the activated alumina fluoride removal process (Rubel and Hathaway, 1985). Regeneration of the activated alumina also was practiced and usually occurred after 30 to 90 days. Because of the difficulty in removing arsenic from the activated alumina, the regenerant NaOH concentration was 4 percent (1.0 N) compared with the 1 percent NaOH used for fluoride removal. The successful performance of several small activated alumina plants without regeneration was reported by Wang and colleagues (2002).

In the late 1990s, new adsorptive media products tailored for arsenic removal began to be developed and marketed. Most of these new products were iron oxide/hydroxide-based,

TABLE 12-9 List of Adsorptive Media Products in the USEPA Arsenic Demonstration Program

Medium	Composition	Supplier
Bayoxide E33 G	Granular ferric oxide (granular form)	Bayer/Severn Trent
Bayoxide E33 P	Granular ferric oxide (pelletized form)	Bayer/Severn Trent
GFH	Granular ferric hydroxide	GEH Wasserchemie
ARM 200	Iron oxide/hydroxide	Engelhard Corp.
CFH-0818	Iron oxide/hydroxide	Kemira Water Solutions
G2	Iron oxide/hydroxide–coated diatomite	ADI International
AAFS 50	Alumina modified with iron oxide/hydroxide	Alcan Chemical
A/I Complex 2000	Alumina modified with iron oxide/hydroxide	Aquatic Treatment Systems
ArsenXnp	Anion resin modified with iron oxide/hydroxide	Purolite
Isolux	Zirconium oxide/hydroxide	MEI
Adsorbia GTO	Titanium oxide/hydroxide	Dow Chemical

although several titanium oxide/hydroxide and zirconium oxide/hydroxide products also were produced. Other products include iron-modified activated alumina, iron-modified IX resins, and iron-coated diatomite. Table 12-9 lists 11 media products evaluated in the USEPA arsenic demonstration program. Most of these media were developed for one-time exhaustion without regeneration, the exceptions being the iron-modified resin (Cumbal and Sengupta, 2005) and the iron-coated diatomite (ADI G2 Medium), which can be regenerated. Research and full-scale demonstrations have started on the regeneration of other iron-based media, and it is anticipated that routine regeneration of these products will soon occur (Chen and Wang, 2010).

The two most significant factors in media selection are arsenic capacity, which determines the bed volumes to medium exhaustion, and cost. Currently, the cost range of adsorptive media is roughly \$80 to \$500/ft^3. Figure 12-20 shows the water treatment cost (\$/1000 gallons of treated water) as a function of medium cost (\$/ft^3) and arsenic capacity measured as bed volumes of water treated before medium exhaustion. The cost curves shown are for

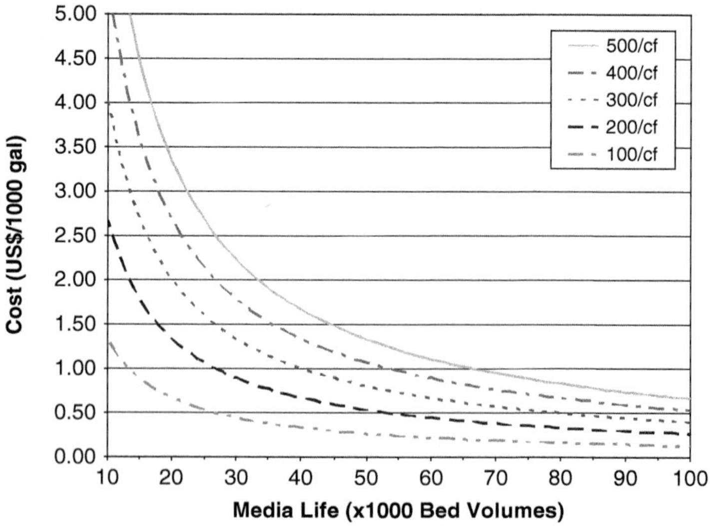

FIGURE 12-20 Effect of adsorptive media life and media cost on treatment cost for arsenic removal.

media only and do not include the costs associated with media replacement and disposal. The relationship is

$$\$/1000 \text{ gal water treated} = 134\ [\text{medium cost } (\$/\text{ft}^3)/\text{BV}]$$

A typical adsorptive media system is a pressurized downflow fixed-bed process. A system can consist of one or more tanks configured in parallel or series. Multibed series systems, although more complex to build and operate than single-bed systems, yield higher arsenic loading on the exhausted medium because of the way in which they are operated. For example, in a two-bed (lead-lag or roughing-polishing) series system, the lead column (the first column to encounter the raw water) typically is exhausted completely when it is regenerated or replaced at the time of MCL breakthrough from the second (lag) column. The partly exhausted lag column then becomes the lead column, and fresh or regenerated medium is used as the lag column. The USEPA arsenic demonstration included both parallel- and series-design systems (Tables 12-10 and 12-11), and performance of the systems listed ranged from treatment runs as low as 3700 BV to greater than 65,000 BV of treated water before arsenic breakthrough at 10 μg/L. Vendor estimates also have been reported above 100,000 BV of treated water. For single parallel columns, as shown in Table 12-10, operating in the pH range of 7.2 to 7.8, longer run lengths are, as expected, associated with lower arsenic, silica, and phosphate concentrations. The data listed in Table 12-11, for the two-columns-in-series configurations, indicate that adding a second column typically increases the run length of the series about 20 to 30 percent compared with a single column. Additional data need to be collected to determine if this trend continues as the lag column is used to replace the lead column on subsequent runs.

Factors Influencing the Arsenic Capacity of Adsorptive Media. Contaminant adsorption onto metal oxide/hydroxide–based media is a ligand exchange process that depends on a number of factors including the type of medium, source water quality, and the fluid-medium contact time [measured as empty-bed contact time (EBCT)]. For metal oxide/hydroxide media and coagulants, arsenic speciation and concentration, pH, and the competing anions phosphate, silicate, and vanadate are considered the most significant source water quality parameters affecting removal capacity (Meng et al., 2000, 2002; Amy et al., 2005; Zeng et al., 2008; Lakshmanan et al., 2008). For all metal oxide/hydroxide media, adsorption capacity decreases significantly as pH increases from 5 to 9. Experimental As(V) capacity determinations on GFH, the first of the iron-based adsorptive media, indicated a sharp decrease in As(V) capacity with increasing pH, as shown in Fig. 12-21 (Driehaus et al., 1998). Laboratory studies by Amy and colleagues (2005) with five iron- and aluminum-based media showed arsenic adsorption capacity increasing as pH dropped from 8 to 7 to 6, even with the test waters containing phosphate, silica, and vanadium. The extent of the pH effect varied with each medium, and their alumina results confirmed the work of Rosenblum and Clifford (1984).

The impact of pH was demonstrated during a full-scale (200 m^3/d, or 37 gpm) test of an iron-modified alumina (AAFS50) at Valley Vista, Ariz., with a pH 7.8 source water containing 40 μg/L As(V) (Valigore et al., 2007). With the two-tanks-in-series treatment system running continuously at a total EBCT of 7.0 minutes, the effluent reached the arsenic MCL of 10 μg/L at 8800 BV at pH 7.8 and at 23,000 BV when the pH was lowered to 6.9.

Because lowering the pH generally increases arsenic capacity, pH adjustment may be cost-effective; thus in the USEPA demonstration program, pH adjustment was initiated at most sites having a pH above 8. Source water pH normally was adjusted into the 7 to 7.6 range using carbon dioxide. The advantage of using CO_2 over sulfuric or hydrochloric acid is that the excess CO_2 can be stripped easily from the process effluent, thereby reducing the potential for any lead or copper problems in the distribution system. Using H_2SO_4 or HCl to adjust the pH would require on-site storage and dilution of these corrosive acids

TABLE 12-10 Performance of Parallel Configuration of Iron Oxide/Hydroxide-Based Arsenic Treatment Systems

Location	Design flow rate, gpm	Source water				Medium	Loading rate, gpm/ft^2	EBCT[a], min/tank	BV to As MCL	BV to SiO_2 TBT	BV to P TBT
		pH	As, μg/L	SiO_2, mg/L	P, μg/L						
Brown City, MI	640	7.3	14	9	<10	E33 G	7.4	4.2	>65,000	<2000	NA
QAC, MD	300	7.8	20	15	17	E33 G	6.9	5.6–6.0	>61,000	<1000	>54,000
Desert Sands, NM, run 2	320	7.7	23	38	NA	E33 P	7.4	3.7	49,500	<1000	NA
Desert Sands, NM, run 1	320	7.7	23	38	NA	E33 G	7.4	4.4	40,600	<500	NA
Lake Isabella, CA	50	7.0	41	43	10	ArsenX	4.0	8.5	33,000	<1000	NA
Wellman, TX	100	7.8	24	46	<10	E33 P	4.0	4.8–6.2	>26,000	<2000	NA
Rollinsford, NH, run 1	100	7.5	37	15	81	E33 G	7.0	3.9	15,000	<1000	NA
Rollinsford, NH, run 2	120	7.8	38	15	81	E33 G	4.8	4.7	12,500	<1000	15,000
Reno, NV, run 1	350	7.1	60	73	150	GFH	4.9	6.5	7200	5000	>9000
Reno, NV, run 2	350	7.2	90	75	112	GFH	4.9	6.5	3700	5800	>8000
Reno, NV, run 2	350	7.2	90	75	112	CFH-0818	4.9	6.5	3700	5800	>8000

[a]EBCT from actual measured flows during performance study.
TBT = total breakthrough.
Source: Data from USEPA Arsenic Demonstration Program.

TABLE 12-11 Performance of Series Configuration Arsenic Treatment Systems: USEPA Arsenic Demonstration Program

		Source water					Loading		BV to as MCL		
Location	Medium	pH	As, µg/L	SiO_2, mg/L	P, µg/L	Media per tank, ft^3	rate gpm/ft^2	EBCT,[a] min/tank	1st tank	2nd tank	% Increase[d]
Wales, ME[b]	GFH	8.6	38	11	32	1.5	13	2.2	8800	11,200	27
Goffstown, NH	E33	7.1	30	25	71	5	5.6	2.9	19,800	25,700	30
Rimrock, AZ	E33	6.9	60	26	10	22	4.2	5.4	39,100	52,100	33
Valley Vista, AZ	AASF-50	7.7	40	19	11	16.7	5.2	3.5	6800	8200	21
Valley Vista, AZ	ARM 200	7.6	40	19	11	22	5.2	4.4	20,200	25,700	27

DFR = design flow rate.
[a]EBCT = empty bed contact time from actual measured flows during performance study.
[b]Pretreatment system included solid-media oxidation tank before three-tank treatment system.
[c]EBCT was 3.1 minutes for first tank and 4.1 minutes for second tank; total system EBCT was 7.2 minutes.
[d]Percent increase is due to addition of second tank.

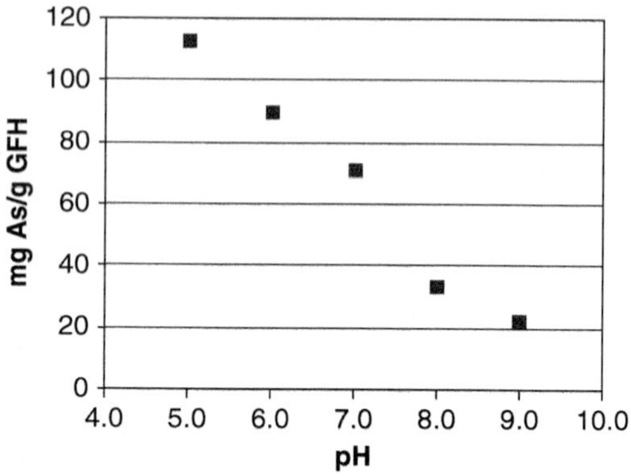

FIGURE 12-21 Arsenate capacity of GFH as a function of pH at arsenic equilibrium concentration of 10 μg As(V) per liter in deionized water. (*Source: Based on the data of Driehaus et al., 1998.*)

and post-pH adjustment using caustic or lime addition. Managers of systems applying pH adjustment must be cognizant of the need to maintain pH control at all times because the loss of pH adjustment can result in the release of arsenic and other anions removed by the adsorptive medium (Min et al., 2007). The degree of desorption (size of the released contaminant spike) depends on the medium and its arsenic loading.

Phosphate, silica, vanadium, chloride, nitrate, fluoride, and TOC have been suggested as possible competing anions that reduce the arsenic capacity of metal oxide/hydroxide media (Amy et al., 2005). More recently, several adsorptive media products have been found to remove uranium and antimony, indicating that these two anions also can compete with arsenic for adsorptive sites (Cumming et al., 2009; Westerhoff et al., 2008). Phosphate, silica, and vanadium are considered to be the most significant competing ions for solid media (Meng et al., 2002; Amy et al., 2005; Xie et al., 2007) and oxide/hydroxide coagulants (Lakshmanan et al., 2008), whereas sulfate, chloride, nitrate, and fluoride do not have any major detrimental effects on arsenic adsorption on iron oxide/hydroxide media and coagulants (Meng et al., 2000; Youngran et al., 2007; Lakshmanan et al., 2008). Sulfate, however, was found to lower As(V) adsorption by 20 percent or more on GFH (Driehaus et al., 1998) and on alumina at pH 6.0 (Rosenblum and Clifford, 1984). Phosphate has been found to be strongly competitive with arsenic on iron oxide/hydroxide media probably because the chemistry of phosphoric acid (H_3PO_4, with pK_a values of 2.1, 7.2, and 12.2) is similar to that of arsenic acid (H_3AsO_4, with pK_a values of 2.2, 7.1, and 12.3); phosphate anions ($H_2PO_4^-$, HPO_4^{2-}) occur naturally in groundwater in microgram per liter (μg/L) concentrations similar to arsenic.

The chemistry of silica is complex, and depending on pH and concentration, silica can occur as soluble silicic acid [$Si(OH)_4$, pK_1 = 9.6] or anionic silicate [$Si(OH)_3O^-$] or as insoluble colloidal silica. Colloidal silica can coat the medium, thereby limiting access to the available sites for arsenic adsorption. Silicate anions compete strongly with arsenate on iron-based media (Xie et al., 2007; Moller and Sylvester, 2007) and coagulants (Lakshmanan et al., 2008). Despite the small fraction of silicate anions present in the neutral pH range, their concentration relative to arsenate is significant because silica generally is present in groundwater at concentrations 1000 times higher than the concentration of arsenic. From laboratory test results, Moller and Sylvester (2007) concluded that at pH 7, the effect of silica as high as 70 mg/L was relatively minor on the arsenic-removal capacity of iron oxide/hydroxide–based (ArsenX^{np}) media because the silica was predominately present as uncharged silicic acid [$Si(OH)_4$]. At pH 7.5 and 30 mg/L of silica, the arsenic-removal capacity was affected significantly presumably because of the higher concentration of silicate. A similar pH dependence of silica interference (i.e., the higher the pH, the greater is the silicate interference) was noted for arsenate sorption onto iron oxide/hydroxide–based coagulants (Lakshmanan et al., 2008).

At a USEPA full-scale demonstration project in the Reno, NV, area with a pH 7.5 raw water containing 67 μg/L arsenic, 75 mg/L SiO_2, and 115 μg/L PO_4-P, the 1900 m^3/d (350 gpm) treatment plant achieved 10 μg/L As breakthrough at only 7000 BV with an iron hydroxide medium (GFH) (Cummings et al., 2008). The source water had 15 μg/L antimony that was removed to less than the Sb MCL (6 μg/L) for approximately 3000 BV. Because of the short treatment runs for both arsenic and antimony, laboratory and field studies using rapid small-scale column tests (RSSCTs) were conducted with eight different media (Westerhoff et al., 2008) in order to indentify higher-capacity media. The results of the laboratory and field RSSCT studies are shown in Table 12-12. Based on the test results,

TABLE 12-12 Summary of Test Results of Rapid Small-Scale Column Tests on Washoe County, NV, Groundwater

Medium	Medium type	Manufacturer	RSSCT results, BV
GFH	Iron-based	GEH Wasserchemie	(L) 11,000 (F) 16,200
CFH-12	Iron-based	Kemira	(F) 12,400
E33	Iron-based	Bayer AG	(F) 8700
ARM200	Iron-based	BASF	(L) 7900
ArsenX	Iron/IX resin-based	Purolite	(L) 7900
Metsorb	Titanium-based	HydroGlobe	(F) 5200
Adsorbsia GTO	Titanium-based	Dow Chemcial	(L) 4500

(L) = lab RSSCT; (F) = field RSSCT.

GFH and Kemira CFH media were chosen for additional full-scale tests. The second GFH run was shorter than the first, with antimony and arsenic breakthroughs at 1200 and 3700 BV, respectively, because of higher 90 μg/L arsenic concentration in the second run. The differences in results between the full-scale system and laboratory RSSCT test results were attributed to the differences in the influent water quality. The laboratory-test water had a consistent arsenic level of 51 μg/L, whereas the full-scale treatment system influent arsenic concentration varied from a low of 35 μg/L to a high of 110 μg/L. The poor performance of the full-scale system and the RSSCT tests for all the media tests was attributed mainly to the high silica in the source water, although the phosphate and high arsenic levels also were believed to have some impact.

Laboratory studies conducted by Amy and colleagues (2005) on the impact of phosphate (125 and 250 μg/L as PO_4^{3-}) showed reductions in arsenic removal capacities with all media products tested. With 125 μg/L of phosphate, the arsenic-reduction capacity for two iron-based media products ranged from 4 to 40 percent depending on the pH. Because phosphate occurs in micrograms per liter (μg/L) rather than milligrams per liter (mg/L), the magnitude of its impact on arsenic is difficult to quantify. Phosphate was removed to some degree at all the USEPA arsenic demonstration program sites, with one site having the phosphate reduced up to 50,000 BV.

The impact of vanadium has not been studied extensively, and therefore, the magnitude of its interference is not fully known. Of the 50 USEPA demonstration projects across the country, 35 sites were found to have vanadium above the detection limit of 0.1 μg/L. The concentrations ranged from 1 to 120 μg/L, with the median value being 5 μg/L. At the demonstration site with the highest level (120 μg/L), vanadium was found to be totally removed for 6000 BV and reached total breakthrough around 15,000 BV. Arsenic (40 μg/L), on the other hand, continued to be reduced to below 10 μg/L at 25,000 BV when the project ended, indicating that arsenic was more preferred than vanadium (Wang et al., 2008). Lakshmanan and colleagues (2008), studying iron oxide/hydroxide coagulants, found that vanadate (50 μg/L as V) did interfere with arsenate removal in the absence of silica and phosphate, but in their presence (SiO_2 at 20 mg/L and PO_4-P at 40 μg/L), the vanadate effect on arsenate removal was not significant.

EBCT can significantly influence arsenic removal, and manufacturers usually recommend an EBCT range for their specific product. EBCT values as low as 1.5 minutes to as high as 10 to 12 minutes have been specified. With 14 × 28 mesh Alcan A-400G alumina at pH 7.5, Simms and Azizian (1997) found that arsenic run length was linearly proportional to EBCT in the range of 3 to 12 minutes and that doubling the EBCT from 6 to 12 minutes increased the run length by 30 percent from 10,700 to 14,000 BV. In arsenic-removal studies at Albuquerque, N.M., Clifford and colleagues (1998) found a similar relationship between EBCT and run length when using a finer 28 × 48 mesh alumina and operating at pH 6.0. At a 5 minute EBCT, the run length was 6400 BV, whereas at a 10 minute EBCT, the run length increased about 37 percent to 8800 BV. At several USEPA arsenic demonstration sites where two or three tanks of media were used in series, arsenic-removal run lengths were found to increase around 30 percent when the EBCT was doubled (see Table 12-11). For example, at the Rimrock, Ariz., site using E33 G medium, the run length to arsenic MCL breakthrough was 391,000 BV from the first tank (5.4 minute EBCT) and 52,100 BV from the second tank (10.8 minute EBCT), an increase of 33 percent (Valigore et al., 2008). A similar increase was found at the Goffstown, N.H., site also using E33 G medium, where at a 2.9 minute EBCT (first tank), the run length was 19,800 BV, and at a 5.8 minute EBCT (two tanks in series), the run length increased 30 percent to 257,000 BV (McCall et al., 2009).

Disposal of Residuals of Arsenic Adsorptive Media. Potential residuals from an adsorptive media process consist of (1) media backwash water, (2) regeneration wastewater, and (3) exhausted (used) media. Some, but not all, adsorptive media require an occasional backwash to fluff the medium and remove particulate material that has accumulated at the top of the medium bed. Backwash water normally includes some suspended solids, with the amount depending on the quality of the source water and the physical properties of the adsorptive medium. The backwash process can produce media fines, which will be removed along with any source water turbidity removed by the filtering action of the medium bed. Source waters having high iron and/or manganese normally will require pretreatment to protect the medium from being fouled, particularly when preoxidation is needed for As(III). When oxidized, untreated sources waters with low levels of iron and/or manganese will have some iron and possibly manganese particulates formed that could be removed in the medium bed. The iron particulates actually can be advantageous because they will absorb some arsenic, which will extend the life of the medium. The potential disadvantage of the iron/arsenic particulates is that they may break through into the product water or end up in the backwash water, creating a disposal problem.

The USEPA arsenic demonstration program collected extensive data on the characteristics of the backwash water where media backwashing was required. The results show that, in most cases, the soluble arsenic (8–56 μg/L, median 16 μg/L) in the backwash water was less than in the raw water used for backwashing (median 20 μg/L), indicating that some arsenic was removed by the medium during the backwashing process. If treated water having low levels of arsenic is used for backwashing, some arsenic desorption from the medium can be expected. Total arsenic levels ranged from 8 to 257 μg/L (median 75 μg/L) in the backwash waste water, with most arsenic the particulate form, likely attached to the iron solids. The total iron in the backwash waste ranged from 0.1 to 18.9 mg/L (median 7.4 mg/L) and was essentially all particulate iron from media fines or precipitated $Fe(OH)_3(s)$ from the raw water or a combination of both. The soluble iron in the backwash waste was less than 0.2 mg/L (median 0.12 mg/L) at all sites and frequently less than the detectable limit of 0.025 mg/L. At sites where the only option for backwash water disposal is recycle, these iron solids can be removed by a combination of settling in a backwash storage tank and poststorage bag filtration. If the arsenic level in the backwash water is low, some states have rules that will allow surface discharge of the wastewater.

The current practice is for exhausted adsorptive media to be removed and replaced with virgin media. Studies have shown that these exhausted media pass the standard federal toxicity characteristic leaching procedure (TCLP) (USEPA, 1992) and thus are classified as nonhazardous waste. The reason is that the arsenic loading is low, and the TCLP test is performed at pH 5, which is near the optimal pH for arsenic adsorption onto most of the adsorptive media products. All the exhausted media from the USEPA arsenic demonstration program projects have passed the TCLP for arsenic, even with an exhausted media loading of 8.3 mg/g. Of the seven other metals included in the TCLP rule, the only detectable measurement was for barium, which ranged from 0.39 to 1.63 mg/L, well below the limit of 100 mg/L. Because these spent media are classified as nonhazardous, they can be disposed of in common sanitary landfills.

Activated alumina, the newly developed iron-modified resins, and the iron-coated diatomite products can be regenerated. The regeneration of other exhausted adsorptive media products, particularly the iron-based products, has not been investigated extensive. However, Chen and colleagues (2008) have shown in laboratory testing that NaOH regeneration of exhausted iron-based media is feasible, and it is anticipated that eventually regeneration will be common. Until regeneration is demonstrated to be practical and cost-effective, adsorptive media simply will be thrown away when exhausted.

Arsenic Removal by IX

If the source water contains less than 500 mg/L TDS, less than 150 mg/L sulfate, and/or multiple contaminants (e.g., nitrate and arsenate), IX may be the arsenic-removal process of choice (Clifford and Lin, 1986; Clifford et al., 1997, 1999). Preoxidation to convert As(III) to As(V) is necessary, but raw water pH adjustment generally is not required. The oxidized and filtered raw water is passed downflow through a 0.8 to 1.6 m deep (2.5–5 ft) bed of chloride-form strong-base anion exchange resin, and the chloride-arsenate IX reaction (Eq. 12-56) takes place in the near-neutral pH range. Regeneration, according to Eq. 12-57, is not difficult because a divalent ion (arsenate) is being replaced by a monovalent ion (chloride) in high-ionic-strength solution, where electroselectivity favors monovalent ion uptake by the resin. Regeneration returns the resin to the chloride form, ready for another exhaustion cycle:

$$2\overline{RCl} + HAsO_4^{2-} = \overline{R_2HAsO_4} + 2Cl^- \tag{12-56}$$

$$\overline{R_2HAsO_4} + 2NaCl = 2\overline{RCl} + Na_2HAsO_4 \tag{12-57}$$

Although chloride-arsenic IX is simple in concept, several issues must be addressed when implementing the process for drinking water treatment. Among the important factors are the following: choice of SBA resin, effect of multiple contaminants such as arsenic and nitrate, arsenic leakage, effect of sulfate concentration, optimal EBCT, regenerant strength (% NaCl), regenerant level (kg NaCl/m^3 resin or lb/ft^3), spent brine reuse, and spent brine treatment. These factors have been examined in field studies in Hanford, Calif. (Clifford and Lin, 1986), McFarland, Calif. (Ghurye et al., 1998), Albuquerque, N.M. (Clifford et al., 1999), and Vale, Ore. (Chen et al., 2008).

Breakthrough Curves in As(V) IX. Prechlorination followed by IX proved to be an effective means of As(V) removal in Hanford. A typical set of breakthrough curves for pH, arsenic, and sulfate is shown in Fig. 12-22. Here, a commercially available type 1 STY-DVB SBA resin (Dowex 11) could effectively treat 4200 BV (16.6 day run length at a 6.6 minute EBCT) of water before the arsenic level reached 0.05 mg/L. However, despite its low concentration, sulfate still was the most preferred anion, and it eventually drove arsenate off the column and caused arsenic to peak at 0.136 mg As(V) per liter (i.e., 160 percent of its feed concentration). (As mentioned earlier, in normal operation, the run would have been stopped at a predetermined volume of throughput, before the peak occurred.)

In contrast with nitrate-chloride IX, a significant pH change did not occur simultaneously with As(V) breakthrough in Fig. 12-22. The effluent pH began at a relatively low value (5.6) and rose to the influent value of 8.7 at about 1300 BV (5 days), whereas As(V) did not reach 0.01 mg/L (the current MCL) until about 3000 BV. In this or a similar situation, a column run simply could be terminated at a predetermined length of, say, 2700 BV, which is far short of the point at which the effluent As(V) concentration reaches its influent value of 0.085 mg/L (5200 BV).

Effect of Process Variables on As(V) Removal BY IX

Choice of Resins for Arsenic Removal. Five different commercially available SBA exchange resins, including polystyrene type 1, were tested for As(V) removal performance in Hanford, Calif. (Clifford and Lin, 1986). After the exchange capacity was taken into account,

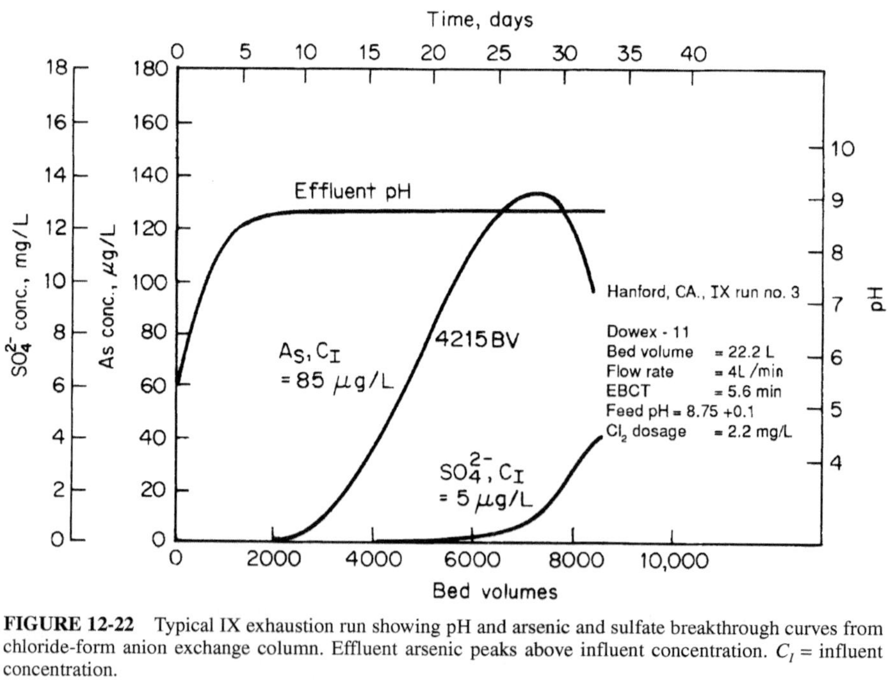

FIGURE 12-22 Typical IX exhaustion run showing pH and arsenic and sulfate breakthrough curves from chloride-form anion exchange column. Effluent arsenic peaks above influent concentration. C_I = influent concentration.

the resins differed very little in their ability to remove As(V). Dowex 11, an improved-porosity polystyrene type 1 resin, and Ionac ASB-2, a polystyrene type 2 resin, seemed to have a slight edge over the others in terms of bed volumes to arsenic breakthrough, so they were studied extensively. In contrast with subsequent field studies in McFarland (Ghurye et al., 1998) and Albuquerque (Clifford et al., 1999), where the run length remained stable after many regenerations, run length deteriorated 3 to 5 percent following each regeneration in the Hanford study. The reason for the decrease was not established, but it was thought to have been caused by the mica fouling observed on the exhausted resin. Most, but not all, of the black coating was removed on NaCl regeneration. Presumably this sort of fouling could be eliminated by multimedia filtration ahead of the anion resin because it is never a good idea to use a resin bed as a particulate filter.

More recently, a STY-DVB macroporous type 2 SBA resin has been used successfully to treat shallow wells near the Republican River in McCook, NB, for the combined removal of arsenic (15 μg/L), nitrate-N (15 mg/L), and uranium (50 μg/L) from a high-sulfate (265 mg/L) water containing TOC (4.0 mg/L). The resin bed also contained 15 percent of an acrylic resin for TOC removal and easier regeneration with NaCl without NaOH. Although the process was successful, the run lengths were quite short at 130 to 150 BV (Boodoo et al., 2008). Resin fouling from TOC was not a problem in McCook. However, the presence of 2.0 mg/L TOC in the Vale, Ore., groundwater seemed to seriously foul the type 1 SBA resin that was used to treat a water containing arsenic (20 μg/L), sulfate (64 mg/L), phosphate-P (300 μg/L), vanadium (50 μg/L), and silica (58 mg/L). Initial run lengths were more than 600 BV but fell to 300 to 400 BV after less than 1 year of operation, apparently owing to TOC fouling of the resin. NaOH was used to clean the resin, which partially restored the capacity, but fouling occurred again during several months of continued operation. Attempts to solve the problem by adding 15 percent acrylic resin to the bed were not successful because run lengths of less than 400 BV were typical.

In addition to TOC, silica can foul anion resins used to treat groundwater to remove arsenic. This apparently occurred in Alamosa, Colo., pilot studies on a pH 8.2 raw water containing 50 μg/L arsenic and more than 100 mg/L silica, most of which was soluble (Brandhuber et al., 2008). Although the water was low in sulfate (<5 mg/L) and projected run lengths were more than 2000 BV, effluent arsenic was erratic, often exceeding the MCL (10 μg/L), and early arsenic breakthrough was apparent. During exhaustion, the pressure drop in the beds exceeded 152 kPa (22 psi), which was attributed to particulate/colloidal silica fouling, which was not removed on backwashing. Installing a particulate filter ahead of the IX unit and regenerating the resin with NaOH and then NaCl appeared to solve the problem of early arsenic leakage. The pilot study was discontinued when a decision was made to install iron coagulation microfiltration for arsenic removal despite the high silica (100 mg/L) concentration.

Combined Arsenic and Nitrate Removal. Combined nitrate and arsenic removal was studied in McFarland, where the chlorinated water contained 13 μg/L arsenic, 10 mg/L NO^3-N, and 40 mg/L sulfate. Here, the triethyl (IRA 996) and tributyl (Ionac SR-6) nitrate-over-sulfate-selective (NSS) SBA resins were compared with two conventional sulfate-selective (CSS) resins. The results of a typical exhaustion cycle are summarized in Fig. 12-23, which presents the arsenic breakthrough curves for all four resins. It is clear from the figure that the conventional SBA resins (IRA 404 and IRA 458) are far superior to the nitrate-selective resins (Ionac SR6, IRA 996) on the basis of run length to arsenic breakthrough. Also evident is the arsenic peaking, which occurs with all the resins, because arsenate is not the most preferred ion by any of the resins. Table 12-13, which gives more details of the runs depicted in Fig. 12-23, shows that because it is the most preferred ion, nitrate does not peak with the nitrate-selective resins but does peak with the conventional sulfate-selective resins.

To summarize, arsenic is removed from water as divalent $HAsO_4^{2-}$, which has poor affinity for monovalent (nitrate) selective resins. Therefore, when removing both nitrate

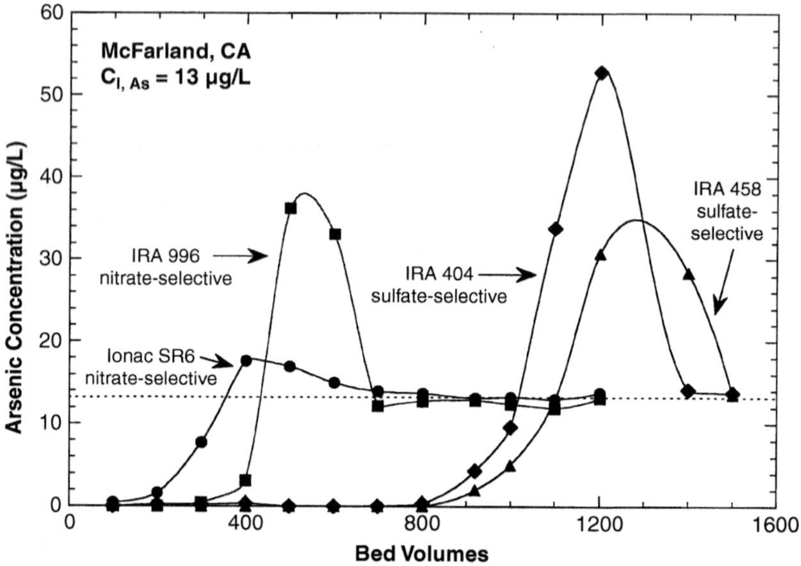

FIGURE 12-23 Comparison of As(V) breakthrough curves for conventional (sulfate-selective) SBA resins compared with nitrate-selective SBA resins.

and arsenic, nitrate-selective resins should be avoided, and the runs should be stopped before nitrate breakthrough to avoid peaking of nitrate and, possibly, arsenate. It is counterproductive to lower the feed pH to produce the monovalent $H_2AsO_4^-$ because $H_2AsO_4^-$ is much lower in the selectivity sequence than is divalent $HAsO_4^{2-}$.

Effect of Arsenic Concentration on Run Length. Because arsenic is a trace species and not the most preferred ion, its concentration does not significantly influence its major breakthrough, according to multicomponent chromatography theory (Helfferich and Klien, 1970). Nevertheless, higher concentration produces higher leakage and shorter run length. To test these predictions, the As(V) concentration was deliberately increased 14-fold from 0.085 to 1.2 mg/L by spiking the feed water during one run at Hanford (Clifford and Lin, 1986). The general trend predicted from theory did occur: The run length to arsenic breakthrough decreased only 46 percent, which was not linearly proportional to the 14-fold

TABLE 12-13 Performance of Nitrate-over-Sulfate-Selective (NSS) and Conventional Sulfate-Selective (CSS) Resins for Combined Arsenic and Nitrate Removal

Parameter	Ionac SR-6	IRA 996	IRA 404	IRA 458
Selective for	Nitrate, NSS	Nitrate, NSS	Sulfate, CSS	Sulfate, CSS
Matrix	STY-DVB	STY-DVB	STY-DVB	Acrylic
Functionality	Tributyl	Triethyl	Trimethyl	Trimethyl
As breakthrough, BV[a]	220	330	850	950
Nitrate breakthough, BV	680	750	560	610
As peaking factor[b]	1.4	3.7	2.7	4.0
Nitrate peaking factor[b]	1.0	0.9	2.5	2.2

[a]Arsenic breakthough to 10 μg/L.
[b]Peaking factor is the ratio of peak concentration to influent concentration.

increase in concentration. The shorter run was primarily a result of greater arsenic leakage at the 14-fold higher influent arsenic concentration.

Arsenic Leakage During Exhaustion. The IX field tests in McFarland, Hanford, and Albuquerque proved that IX could easily meet a 0.010 mg/L arsenic MCL when treating a up to 50 μg/L arsenic-contaminated groundwater containing up to 220 mg/L sulfate. Average arsenic leakages during runs with EBCTs in the range of 1.5 to 3 minutes were 0.2 to 0.8 μg/L for the three SBA resins tested, ASB-2 (polystyrene type 2), IRA 458 (polyacrylic type 1), and IRA 404 (polystyrene type 1). Runs with a 3 minute EBCT produced essentially the same arsenic leakage as runs with a 1.5 minute EBCT. In all cases, cocurrent downflow regenerations with 1 M (6%) NaCl at 320 kg NaCl/m^3 of resin (20 lb/ft^3) were employed. Subsequent experiments with regeneration levels as low as 80 kg NaCl/m^3 of resin (5 lb/ft^3) resulted in similar leakage values.

Arsenic leakage from SBA resin columns increased dramatically in the presence of particulate iron(III), which strongly adsorbed As(V) and was not well filtered by the resin. This fact was discovered during some tests in McFarland using water that was visibly contaminated with particulate Fe(III). Serious arsenic leakage (up to 6 μg/L) tracked the iron contamination in the IX column effluent. The problem was solved by adding a multimedia prefilter ahead of the IX column; the prefilter eliminated the particulate iron in the arsenic-contaminated feed to the SBA column, and arsenic leakage returned to normal.

Effect of Sulfate on Arsenic Run Length. As was observed during nitrate removal, increasing sulfate concentration leads to shorter IX runs because sulfate is preferred to nitrate and arsenate when using the recommended CSS resins. Table 12-14 compares arsenic run lengths for ASB-2 resin using various waters with different TDS values and sulfate concentrations. Even with 100 mg/L sulfate in the feed water, a run length of 540 BV was observed with the spiked McFarland water. The 275 BV run length at 220 mg/L sulfate was considered relatively short for economical full-scale operation, which is why less than 150 mg/L sulfate is suggested as one criterion for selecting IX for arsenic removal.

TABLE 12-14 Arsenic Run Length as a Function of Influent Sulfate Concentration[a]

Location	As, μg/L	TDS, mg/L	Sulfate, mg/L	Run length, BV[b]
Hanford, CA	50	213	5	1600
McFarland, CA, unspiked	13	170	40	1100[c]
Albuquerque, NM	26	328	82	700
McFarland, CA SO_4^{2-} spiked	13	259	100	540
McFarland, CA SO_4^{2-} spiked	13	436	220	275

[a]Arsenic run lengths for ASB-2 type 2 SBA resin regenerated with 320 kg NaCl/m^3 of resin (20 lb/ft^3). When regenerated with 160 kg NaCl/m^3 of resin (10 lb/ft^3), run lengths are decreased by about 25 percent.

[b]Approximate run length based on run termination at 10 μg/L effluent As concentration.

[c]Extrapolated value based on comparison with IRA 404 performance in McFarland, Calif.

Regeneration of Arsenic Spent Resins. During the original Hanford studies in 1984, arsenic recoveries on downflow (cocurrent) regeneration were essentially complete. Regeneration with 3 BV of 1.0 N NaCl (11 lb NaCl/ft^3 of resin) was more than adequate to elute all the adsorbed arsenic, which was even easier to elute than bicarbonate—a non preferred ion. One reason why arsenic elutes so readily is that it is a divalent ion ($HAsO_4^{2-}$) and thus is subject to a selectivity reversal in the high-ionic-strength (>1 M) environment of the regenerant solution. This ease of regeneration is a strong point in favor of IX over adsorptive media for arsenic removal in low-TDS, low-sulfate waters.

In confirmation of similar findings from the nitrate studies in Glendale, Ariz. (Clifford et al., 1987), the Hanford experiments (Clifford and Lin, 1986) showed that dilute regenerants (0.25–0.5 N) were more efficient (use fewer kg NaCl/m^3 of resin) than concentrated ones for eluting arsenic. The greater efficiency of dilute regenerants was further verified in the Albuquerque field study (Clifford et al., 2003), where 0.5 N NaCl outperformed 1.0 and 2.0 N regenerants in downflow regeneration experiments on similar aliquots of exhausted resin. When analyzing the data, however, it appeared that the regenerant flow velocity was an important factor that had not been considered previously. At the low superficial flow velocities (<1.2 m/h) associated with the higher regenerant concentrations, regenerant channeling occurred, which resulted in poor arsenic recoveries. The resulting rule of thumb was that the regenerant *superficial linear velocity* (SLV) should exceed 1.2 m/h with typical U.S. 16 × 50 mesh resins. With SLVs greater than 1.2 m/h, there was not much difference between the performance of 0.5 and 1.0 N regenerants at regeneration levels of 1 to 2 eq Cl^- per equivalent of resin (5–10 lb NaCl/ft^3).

Downflow Versus Upflow Regeneration. It is conventional wisdom that countercurrent upflow regeneration is better than cocurrent downflow to minimize contaminant leakage in cyclic IX operations. Normally this is true, especially in demineralizer operation, where minimal leakage is critical. However, with arsenic IX, it is not necessary to use counterflow (upflow) regeneration to minimize leakage. When upflow was compared with downflow regeneration in the Albuquerque field study (Clifford et al., 1999), conventional downflow regeneration performed somewhat better; that is, it gave lower arsenic leakage during subsequent exhaustions. The reason for the superiority of cocurrent downflow regeneration is that, at exhaustion, most of the arsenic on the resin is concentrated in a thin zone near the column outlet. Based on experimental results, it is preferable to elute the arsenic by the shortest route, that is, cocurrent regeneration, as opposed to flushing it back through the entire resin bed, where some of it apparently remains during countercurrent regeneration.

Reuse of Spent Regenerant. The major finding of the Albuquerque arsenic study was that spent regenerant could be reused without treatment to remove arsenic. In the series of exhaustion/regeneration cycles shown in Fig. 12-24, the ASB-2 column was exhausted

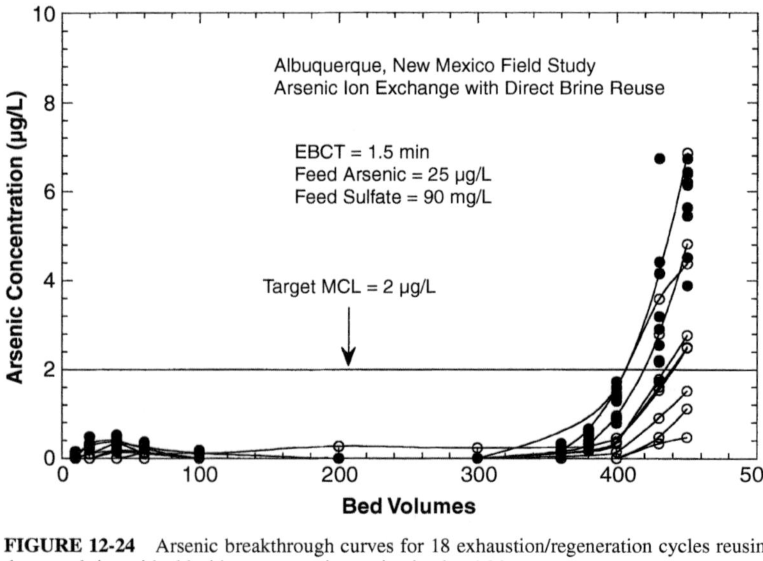

FIGURE 12-24 Arsenic breakthrough curves for 18 exhaustion/regeneration cycles reusing the same brine with chloride concentration maintained at 1 M.

and regenerated 18 times using recycled regenerant that had been compensated each cycle with NaCl to maintain the chloride concentration at 1.0 N. Despite the fact that the arsenic concentration in the reuse brine rose to 19,000 μg/L, sulfate reached 151 g/L (3.1 N), and bicarbonate reached 24.4 g/L (0.39 N), the arsenic leakage during exhaustion was typically less than 0.4 μg/L, and run lengths were consistently in the 400 to 450 BV range. After 18 cycles of reuse, attempts were made to continue reusing the brine without chloride compensation in the hope that the high bicarbonate level in the brine would be sufficient to elute the arsenic. These attempts failed because the arsenic leakage rose to 10 μg/L after only four runs without chloride addition. When chloride addition was resumed with the same recycle brine, the arsenic leakage dropped back to 0.4 μg/L, but the run lengths were a little shorter (380–400 BV). Based on the Albuquerque field-study results, it appears that spent brine can be reused at least 20 times and possibly more. Eventually, the spent brine will have to be wasted, but it should be treated to remove arsenic before disposal. This can be accomplished by iron(III) or aluminum(III) precipitation, as discussed below.

Treatment of Spent IX Regenerant. Arsenic(V) can be precipitated from the waste brine by using iron(III) or aluminum salts such as ferric chloride or alum or lime [$Ca(OH)_2$]. The stoichiometric reactions are

$$3Na_2HAsO_4 + 3H_2O + 2FeCl_3 = Fe(OH)_3(s) + Fe(H_2AsO_4)_3(s) + 6NaCl \qquad (12\text{-}58)$$

$$2Na_2HAsO_4 + NaHCO_3 + 4Ca(OH)_2 = CaCO_3(s) + Ca_3(AsO_4)_2(s) + 3H_2O + 5NaOH \qquad (12\text{-}59)$$

$Al_2(SO_4)_3{\cdot}14H_2O$ may be substituted for $FeCl_3$ in Eq. 12-58 to produce $Al(H_2AsO_4)_3(s)$.

Iron or aluminum is required in greater-than-stoichiometric amounts in practice. A wastewater of pH 9.9 resulting from a 1.0 N NaCl regeneration was treated to remove As(V) in Hanford, Calif. For the wastewater, which contained 90 mg As(V) per liter and 50,000 mg TDS per liter, about 12 times the stoichiometric amount of iron (adjusted final pH 8.5) or aluminum (adjusted final pH 6.5) was required to reduce the As(V) level from 90 mg/L to less than 5 mg/L. To reduce As(V) to less than 1.5 mg/L required a 20-fold stoichiometric dose of either metal salt. The iron floc was heavier and settled better than the alum floc. When subjected to the Extraction Procedure EP test to determine As(V) leachability from the air-dried sludges, both sludges passed the test with about 1.5 mg As(V) per liter in the leachate when the EP MCL was 5 mg As/L (Clifford and Lin, 1986).

Similar Fe(III) precipitation of arsenic experiments were carried out in the Albuquerque field study (Clifford et al., 2003). In these tests, however, the spent brine contained only 3.5 mg/L arsenic initially. The arsenic removals achieved were found to be strongly dependent on pH. When using a molar Fe:As ratio of 20:1, the arsenic removals observed were 98.7 and 99.5 percent at final pH values of 6.4 and 5.5, respectively. TCLP tests will have to be run on precipitated arsenic sludges to verify their suitability for land disposal.

Natural Color and Organics Removal by Resins

Naturally occurring *dissolved organic carbon* (DOC) or NOM consisting primarily of humate and fulvate anions has been recognized for more than 40 years as a common foulant of the anion exchangers used in demineralization processes (McGarvey and Reents, 1954; Frisch and Kuin, 1957; Wilson, 1959). The symptoms of organic fouling are product water quality deterioration, resin darkening, brown-colored regenerant, exchange capacity loss, and increased frequency of regeneration. Early on it was found that occasionally regenerating the fouled SBA resin with a mixture of NaCl and NaOH would reverse most of the fouling and lead to acceptable run lengths and resin life (Wilson, 1959). Despite an early history of problems with DOC in surface waters treated by SBA resins, it was found that DOC could be removed from groundwater and surface water by using resins, especially macroporous anion exchangers in the chloride form. Published research on the IX process for DOC removal is scarce, and much remains to be done to optimize the process. However, the few reported studies to date have suggested that IX is a viable process for DOC and disinfection-by-product (DBP) control in drinking water supplies.

History of DOC Removal by IX. One of the earliest reports on the use of resins to treat drinking water to remove DOC was a 2-year successful pilot study of color removal on the Merrimack River in Massachusetts during the period 1962–1964 (Coogan, 1968). Duolite A-7—a hydrophilic matrix granular weak-base phenol formaldehyde anion resin regenerated with NaOH—was tested extensively. Although A-7 is an anion exchanger at low pH, it had almost no IX capacity at the pH of the influent water (6.4–6.9). Thus the removal of color appeared to be due to adsorption of anion-cation pairs rather than by IX, but this was not proven. The main findings of the study were as follows: (1) With color values of 50 to 100 on the raw water, diatomaceous earth (DE) prefiltration to remove about 10 percent of the presumably larger color bodies was required ahead of the resin bed, (2) EBCTs in the 3.75 to 7.5 minute range were acceptable, with the 3.75 minute time giving only a small decrease in run length, which was 2000 to 2500 BV for 7.5 minutes, (3) typically, 80 to 100 percent color removal was attained at the start of a run, decreasing to 50 to 75 percent the end, (4) NaOH at a level of 64 kg/m^3 (4 lb/ft^3) was adequate to regenerate the resin during more than 200 exhaustion/regeneration cycles in which the resin capacity for DOC remained relatively constant, and (5) iron removals were about 25 percent across the DE filter and 95 percent through the resin bed, which led to the suggestion that soluble iron was being removed by complexation with the adsorbed color bodies. This study and others reported by Abrams (1982) led researchers to try A-7 and similar weak-base resins for adsorption rather than IX of DOC. Later, it was found that SBA resins regenerated with NaOH or a mixture of NaCl and NaOH also could be used in cyclic operation to control organics in surface and groundwaters.

Baker and colleagues (1977) reported that the substitution of the acrylic-matrix gel SBA resin IRA 458 for a conventional polystyrene type 2 resin in a demineralizer system solved the organic fouling problems that had plagued the system for years. The acrylic-matrix resin performed better because it was more efficiently regenerated than the original polystyrene resin. The authors explained that there is minimal van der Waals attraction between the hydrophobic organic molecules and the hydrophilic acrylic resin matrix, which makes regeneration easier and more complete. In 1977, when Baker's paper was published, the acrylic resin had been placed into service at 125 sites worldwide because of its superior regenerability and resistance to organic fouling.

Kolle (1984) described a DOC-removal process using anion exchange for treatment of a highly colored Hannover, West Germany, groundwater. The 2500 m^3/d (15.8 mgd) full-scale process used a macroporous type 1 polystyrene SBA resin (Lewatit MP 500 A) in four single beds operated in parallel with an EBCT of 1.2 minutes. Regeneration was accomplished with 2 BV of a mixture of 1.7 N (10%) NaCl, and 0.5 N (2%) NaOH. A striking feature of the process was the reuse of regenerant seven times with the NaCl and NaOH concentrations replenished after each use. The DOC was not removed from the recycled regenerant, and it built up to 25,000 mg/L prior to disposal, which was triggered when the sulfate concentration increased to the chloride level. This reuse method allowed for a 20,000:1 ratio of product water to spent regenerant.

Typical breakthrough curves for DOC, chloride, and sulfate from the early portion of the run in the Hannover process are shown in Fig. 12-25. After sulfate breakthrough at 250 BV, the DOC removal remained at about 50 percent until 5000 BV—the point of run termination.

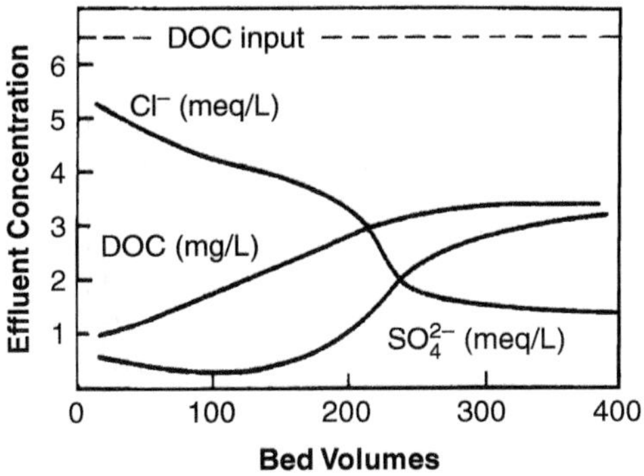

FIGURE 12-25 Typical breakthrough curves for DOC, sulfate, and chloride from a macroporous SBA exchanger removing fulvate and humate anions from Hannover, West Germany, groundwater. Column was operated to 5000 BV before regeneration. Between 400 and 5000 BV, the DOC removal was approximately 50%, with influent sulfate at 130 mg/L. (*Source: Kolle, 1984.*)

Although far from complete, the 50 percent DOC removal was sufficient to increase the bacterial doubling time in the product water from 5 to 12 hours and to drastically reduce the chlorine consumption with its attendant organic halide formation problems. The overall treatment efficiency was reduced only 10 percent because of fouling and resin losses over a period of 2 years of full-scale operation.

Brattebo and colleagues (1987) described a similar DOC-removal process designed to treat a highly colored Norwegian surface water containing 3 to 6 mg/L DOC, primarily in the greater than 10,000-Da molecular weight (MW) range. During their pilot studies, they verified the superiority of macroporous MP 500 A type 1 SBA resin compared with a limited number of other polystyrene resins of macroporous and gel porosity. They also effectively reused the 1.7 N NaCl/0.5 N NaOH regenerant as many as eight times by wasting a fraction of the first eluate containing most of the DOC and making up the remainder with fresh NaCl and NaOH. For water containing up to 4 mg DOC per liter, they recommended the IX method implemented by using a four-column merry-go-round IX system (similar to Fig. 12-12, a three-column merry-go-round system) in which the resting, freshly regenerated fourth column automatically became the final polisher when the effluent DOC from the three columns in series reached a predetermined level. At this time, the lead column was taken out of service and regenerated. They also determined that the EBCT for the series of columns was the most important design parameter, and they suggested a range of 5 to 10 minutes for this surface water, which is considerably longer than the 1.2 minutes used in the full-scale Hannover groundwater system. The difference in optimal EBCT likely was due to differences in the nature (e.g., size, aromaticity, and hydrophobicity) of the various surface and groundwater organics involved.

The same Norwegian research group tested the multibed anion exchange process on a small water supply using a full-scale 3 m^3/h (13 gpm) plant and reported the following conclusions: (1) NaOH was a necessary constituent of the regenerant because with NaCl alone, the DOC recovery was less than 80 percent. (2) reusing up to 75 percent of the spent regenerant did not reduce the DOC adsorption capacity on subsequent runs, (3) mass transfer of DOC was particle diffusion limited such that a reduction in temperature from 15°C to 5°C required a 20 percent increase in EBCT, and (4) for small water supplies with relatively low (<4 mg/L) DOC, the anion exchange process was estimated to be less costly to build and operate than a conventional coagulation process (Brattebo et al., 1987).

Fundamentals of DOC IX. Fu and Symons (1990) conducted research on the fundamentals of DOC IX. Their IX isotherm experiments indicated that within the limits of experimental error, each milliequivalent of DOC in the 1000 to 5000 Da fractions sorbed was balanced by an equal amount of chloride ion released (Fig. 12-26). Overall, their results

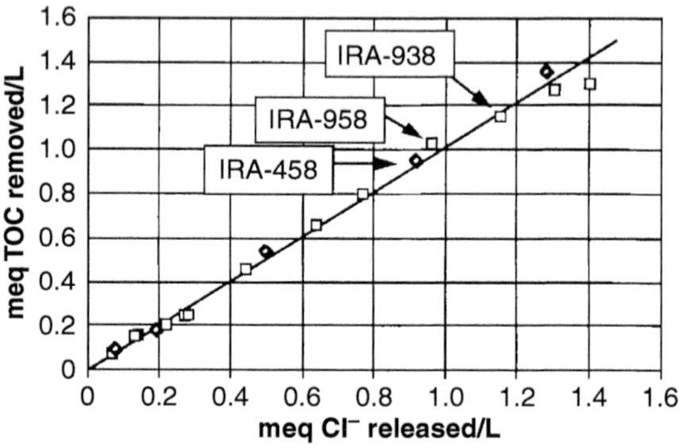

FIGURE 12-26 Proof of 1000 to 5000 Da MW fraction TOC removal by IX (as opposed to adsorption) using chloride-form SBA resin. (*Source: Fu and Symons, 1990.*)

showed that the sorption mechanism for DOC removal was IX, with only a small percentage of the DOC in the less than 1000 Da and greater than 10,000 Da fractions being removed by adsorption. Also using equilibrium isotherm tests, they found that the resin matrix and porosity exhibited significant influences on DOC removal. Generally, for DOC MW of 1000 Da or more, the hydrophilic polyacrylic resins were preferred for DOC removal compared with the relatively hydrophobic polystyrene resins. And, for all MW fractions greater than 1000 Da, macroporous resins performed better than gel types; that is, the greater the resin porosity, the better was the DOC removal. At less than 1000 Da MW, the resin porosity did not influence the DOC removal, which was, however, influenced by the resin matrix. In this low-MW range, hydrophobic polystyrene resins performed somewhat better than the hydrophilic polyacrylic resins. Thus the resin of choice for removing NOM depends on the actual water supply being treated. They also found that the resin's ability to sorb DOC of 1000 Da or greater MW was positively correlated with its tendency to swell in water, that is, its hydrophilicity, which is another way of explaining why acrylic resins were preferred by Baker and colleagues (1977) over polystyrene resins for DOC removal.

Fu and Symons (1990) found that coagulation performed in a similar manner to IX for removing DOC and that neither process was particularly good for the less than 1000 Da fraction. Their column study results suggested that the DOC could be divided into three categories: (1) not removed, (2) less preferred than sulfate, and (3) more preferred than sulfate. Figure 12-27 shows the TOC effluent history and the sulfate breakthrough point during a test with the less than 10,000 Da TOC fraction. With only 6 mg/L sulfate and 18 mg/L chloride in the dialyzed feed water, the 3000-BV run length to sulfate breakthrough was exceptionally long compared with typical drinking waters. Prior to sulfate breakthrough, TOC removal was about 75 percent. At sulfate breakthrough, a TOC peak appeared that was the TOC less preferred than sulfate being driven off the column by the advancing sulfate wavefront. After sulfate breakthrough, the TOC more preferred than sulfate was being removed for a total TOC removal of about 50 to 60 percent until 9000 BV. Then TOC removal dropped gradually to zero at about 12,000 BV. These column results were similar

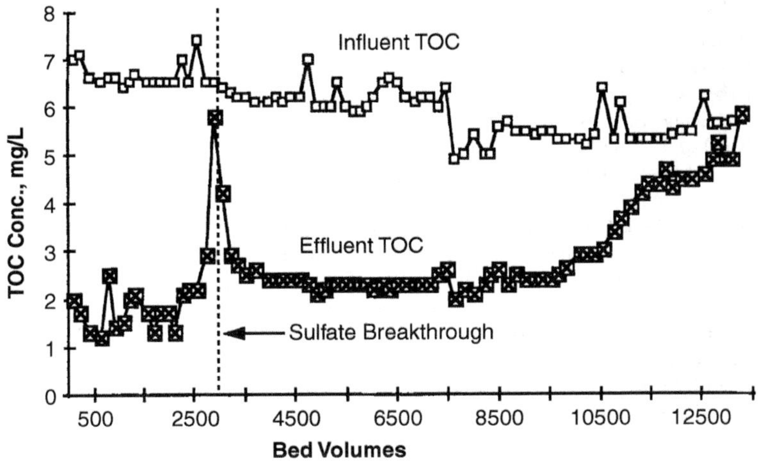

FIGURE 12-27 TOC effluent history for a bench-scale test with IRA 958 (acrylic SBA) resin fed with les than 1000-Da MW TOC fraction in low TDS water. (*Source: Fu and Symons, 1990.*)

to those of Coogan and colleagues (1968) and Kolle and colleagues (1984) in that TOC removal started out at 70 to 80 percent and dropped to a long period of about 50 percent removal following sulfate breakthrough.

Kim and Symons (1991) focused their research on the field performance of the macroporous polyacrylic SBA resin (IRA 958) when treating an alum-coagulated, filtered surface water at a City of Houston water plant. Following conventional treatment, the water fed to their minicolumns was 0.45-μm-cartridge filtered, dechlorinated (Na_2SO_3), and adjusted to pH 6.8. Figure 12-28 illustrates the virgin resin breakthrough curves for the sulfate and TOC (which was DOC for these filtered waters). As with the previously described nitrate- and arsenate-removal processes and all the column experiments with DOC removal, the background sulfate concentration determined the performance of the system. When sulfate broke through at about 500 BV, it caused the TOC that was less preferred than sulfate to peak at about 120 percent of the influent TOC concentration (4.0 mg/L). When the effluent histories of the various TOC fractions were studied, it was found that with the exception of the less than 500-Da fraction, all peaked at essentially the same time, indicating that all fractions, except the less than 500-Da fraction, contained some TOC less preferred than sulfate. The less than 500-Da fraction was only about 25 percent removed throughout the run and was not influenced by sulfate. The TOC effluent histories also indicate that, generally, the higher the MW, the better is the TOC

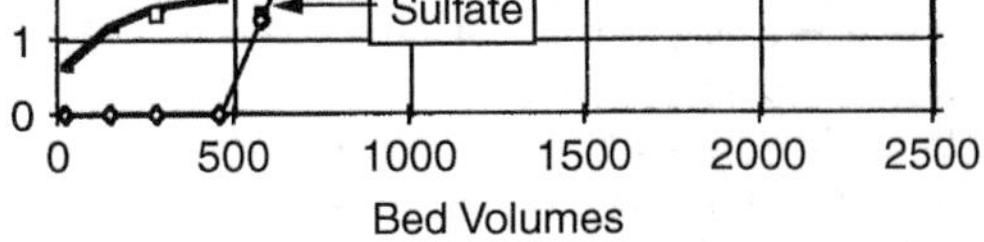

FIGURE 12-28 TOC and sulfate breakthrough curves from virgin macroporous polyacrylic SBA resin (IRA 958) during pilot tests with treated Houston water. pH = 6.8; EBCT = 10 minutes. (*Source: Kim and Symons, 1991.*)

removal before and after the sulfate breakthrough. Running beyond sulfate breakthrough was not very effective with this coagulated water because only about 25 percent overall TOC removal was achieved from 1300 to 2700 BV.

The Kim and Symons (1991) TOC regeneration study determined that the most effective TOC regenerant was a mixture of 2 N NaCl with 0.5 N NaOH, which is in agreement with the 1.7 N NaCl–0.5 N NaOH mixture used by Kolle (1984) and Brattebo and colleagues (1987) in their larger-scale studies. This 2 N NaCl–0.5 N NaOH mixture was used in the regenerant reuse study, which showed that a polyacrylic macroporous resin (IRA 958) could be exhausted and regenerated nine times with the same regenerant that had been compensated with salt and caustic after each regeneration to maintain the original chloride and hydroxide concentrations. Although the TOC removed from the resin during regeneration following the eighth reuse cycle was only 13 percent, the overall performance of the resin for TOC removal up to sulfate breakthrough was essentially unchanged. Kim and Symons suggested that the TOC not removed during regeneration was building up in the deeper pores of the resin, where it did not interfere with the reversible IX, which appeared to be taking place closer to the resin surface.

When Kim and Symons (1991) tested the sequence of IX (IRA 958) followed by GAC (Calgon F-400) for trihalomethane formation potential (THMFP) control, they found that it was much superior to GAC alone. When regenerating the resin at sulfate breakthrough, the IX/GAC process consistently achieved 95 percent THMFP removal to 5000 BV, whereas IX alone achieved only 50 percent removal after 200 BV and GAC alone achieved only 75 percent removal after about 2000 BV.

In summary, experimental evidence shows that NOM in groundwater and surface water can be removed by adsorption onto weak-base resins and is also amenable to removal by using macro- and microporous chloride-form SBA resins with both polystyrene and polyacrylic matrices. With SBA resins, the sulfate concentration of the raw water determines run length to the first DOC breakthrough, which peaks above the influent DOC concentration. After this, DOC removal may be continued if 50 percent and lower removals can be tolerated. Filtration always should be employed ahead of IX to protect against resin fouling. When coagulation precedes IX, the DOC remaining is difficult to remove because the two processes are similar in their DOC-removal profiles.

Magnetic Ion Exchange (MIEX) Treatment Process for DOC Removal. To this point, only packed bed (column-based) IX processes have been described because IX polymers (beads and particles) are almost always used in columns. Equilibrium-stage mixer/settlers and fluidized-bed ion exchangers are not often employed because (1) contaminant loading on the resin is low because resins are in equilibrium with the effluent contaminant concentration as opposed to the influent contaminant concentration for packed beds, and (2) very small resin particles are required in equilibrium-stage contactors for kinetic reasons, and these particles don't settle well because their densities are near that of water. An innovative class of exchangers that has attracted attention in the past decade is MIEX resins, which, although 1/10 the size of resin beads, aggregate and settle quickly in the resin separators (clarifiers or settlers) that must follow the resin contactors (fluidized beds or equilibrium-stage contactors). Although such MIEX resins can be used in both cation and anion exchange applications, magnetic anion resins have attracted the most attention because they have been used successfully at bench-, pilot-, and full-scale applications for DOC removal from water to prevent DBP formation and membrane fouling. One such process using magnetic anion exchangers is the MIEX process developed by Orica, Ltd. (2010), the Australian Water Quality Centre, and the Commonwealth Scientific Industrial Research Organization.

In the dual-stage MIEX process, raw water is contacted with small (20 to 50 μm diameter) magnetic resin particles in a stirred-tank contactor, with a typical resin dose being in the range of 0.5 to 2.5 percent (v/v). Anionic contaminants typically are exchanged for chloride on the resin, and the contents of mixed tank are sent to a settler (typically upflow), where magnetic resin particles aggregate and settle quickly at overflow rates of up to 14.6 m/h (6 gpm/ft^2). Although resin recoveries can exceed 99 percent; there is usually a small carryover of resin, and downstream filtration is necessary. The underflow from the settler is recycled back to the mixer/contactor, with a draw off of 1 to 20 percent of the recycled resin diverted for regeneration. Regeneration typically is performed using a 2 M (12%) NaCl solution, and the regenerated resin is sent to a resin feed tank, where a small amount of fresh resin is added to compensate for the resin lost owing to carryover from the settler. Given the higher solubility of TOC at an alkaline pH, caustic may be added to the NaCl brine solution to efficiently desorb TOC.

Using a MIEX-type process, Boyer and Singer (2006) determined that DOC removal was accompanied by a stoichiometric release of the exchanging chloride ion, verifying the work of Fu and Symons (1990), which showed that DOC removal occurs by anion exchange rather than sorption. Allpike and colleagues (2005) showed that the MIEX process outperformed the more traditional enhanced-coagulation process. MIEX removed more organic carbon than enhanced coagulation, and in contrast to enhanced coagulation, MIEX removed organic carbon across all MW fractions studied and with no preference for aromatic versus aliphatic character. Regarding inorganic contaminants, Seidel (2004) reported that MIEX effectively removed arsenic, hexavalent chromium, and perchlorate from water. Similar to traditional IX, sulfate provided the greatest competition for exchange sites for both TOC- and inorganic contaminant–removal applications. Also, Orica (2010) is promoting the application of MIEX for removing sulfide, nitrate, and arsenate in addition to DOC and color (Pyles, 2009).

Uranium Removal by Anion Exchange

On December 7, 2000, the USEPA set an MCL of 30 μg/L and an MCL goal (MCLG) of 0 pCi/L for uranium owing to health concerns; it is a known kidney chemotoxin and a suspected human carcinogen (Cothern and Lappenbusch, 1983). When the radionuclides rule was published (USEPA, 2000), the USEPA estimated that 500 community water supplies affecting 620,000 persons in the United States would be out of compliance and would have to reduce uranium concentrations to less than the 30 μg/L MCL (see Chaps. 1 and 2 for more information about the regulations and health effects).

Uranium Chemistry and Speciation. Uranium has four oxidation states: III, IV, V, and VI. However, only the IV and VI states are stable. The high affinity of uranium for oxygen results in the formation of uranium's most common oxygen-containing compound, uranyl ion (UO_2^{2+}), which is the most stable state of uranium in aerated aqueous solution under acidic conditions (pH < 5.0). The uranyl ion (UO_2^{2+}) in water readily combines with anionic ligands, including CO_3^{2-}, F^-, Cl^-, NO_3^-, SO_4^{2-}, and HPO_4^{2-}, to form stable complexes. When F^-, $H_2PO_4^-$, or HPO_2^{2-} is present in oxidizing waters, these ligands preferentially complex the UO_2^{2+} at pH below about 8 (Langmuir, 1978).

Carbonate is the most significant uranium ligand in natural water. In the near-neutral pH range, UO_2^{2+} combines with bicarbonate and carbonate anions in aqueous solution to form stable, charged and uncharged uranyl carbonates. Langmuir (1978) has calculated the distribution of typical uranium complexes as a function of pH in aerobic groundwater at a CO_2 partial pressure of 10^{-2} atm. Under mildly acidic conditions of pH 5.0 to 6.5, the principal species is $UO_2CO_3^0$. In the pH range of 6.5 to 7.5, the principal species is $UO_2(CO_3)_2^{2-}$. Above pH 8.0, the predominant species is $UO_2(CO_3)_3^{4-}$. Uranyl hydroxy complexes such as UO_2OH^+ and $(UO_2)_3(OH)_5^+$ are also formed, but generally in small percentages unless at high temperature or for waters with low inorganic carbon (Mo, 1980).

History of IX for Uranium Removal. Both cation and anion exchange resins have been used for uranium removal, but because of their requirements for low feed water pH and their relatively unsatisfactory uranium removal efficiency, cation resins probably are not practical for drinking water treatment (Lee and Bondietti, 1983).

Anion exchange exhibited much better performance than cation exchange for removing uranium from contaminated water. Sorg (1988) summarized the results of USEPA-funded studies for uranium removal by anion exchange, which he suggested would be the cost-effective method for treating small community water supplies. In these bench-, pilot-, and full-scale uranium IX studies, it was apparent that SBA resins exhibited an enormous capacity for the uranyl carbonate complexes—$UO_2(CO_3)_2^{2-}$ and $UO_2(CO_3)_3^{4-}$. For example, during field studies in Colorado and New Mexico, columns receiving 22 to 104 μg/L uranium treated 8000 to 60,000 BV of groundwater with less than 1.0 μg/L uranium in the effluent. In associated studies, two columns containing type 1 polystyrene SBA resins (gel Dowex 21K and macroporous Ionac 641) were operated to 17,400 and 31,300 BV and still removed 95 and 90 percent, respectively, of the total uranium in the feed water (Hathaway, 1983). Uranium-removal field studies by Zhang and Clifford (1994, 1995) have confirmed this earlier work and shown that chloride-form IX is an excellent process for removing uranium from groundwater. In their Chimney Hill, Texas, field study, they reported that a type 1 macroporous polystyrene SBA resin (Ionac A-642) exhibited enormous capacity for uranium removal because of its exceptionally high affinity for $UO_2(CO_3)_3^{4-}$. (Similar resins are available from other suppliers and should perform in a similar manner.) A pilot-scale SBA column was operated continuously for 478 days for a total throughput of 302,000 BV at pH 7.6 to 8.2. The feed water contained 120 μg/L of uranium and 25 pCi/L of radium in a background water quality of 310 mg/L TDS, 150 mg/L alkalinity, 47 mg/L chloride, and less than 1 mg/L sulfate. Even after 300,000 BV, the effluent uranium concentration was less than 6 μg/L, which corresponds to a removal efficiency of 95 percent (Fig. 12-29). Before 260,000 BV, the uranium leakage was less than 1 μg/L, that is, more than 99 percent removal. Compared with typical granular activated carbon (GAC) columns, which showed

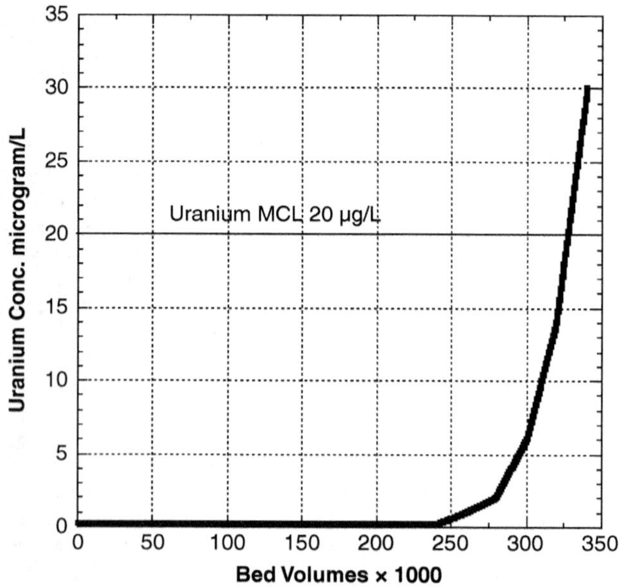

FIGURE 12-29 Effluent uranium levels from pilot-scale IX column during virgin exhaustion of macroporous polystyrene SBA resin (Ionac A-642) at pH 8.0. Chimney Hill Field Study. EBCT = 3 min until 10^5 BV, 2 min thereafter. Feed U = 67-125 μg/L. (*Source: Based on the work of Clifford and Zhang, 1995.*)

a capacity of a few thousand BV of throughput before more than 20 μg/L uranium leakage, the capacity of anion resin for uranium was enormous in this low-sulfate water. After 300,000 BV, the uranium concentration averaged over the whole bed was 30 g/L of resin, which corresponds to 35.4 kg U_3O_8/m^3 of anion resin (2.2 lb U_3O_8/ft^3), or 63 percent of resin sites in uranium form based on a resin capacity of 1.0 meq/mL. Although this ultimate uranium capacity is truly impressive, drinking water treatment columns are not generally operated for 200 to 500 days to uranium breakthrough because of problems with resin fouling and excessive pressure drop. Cyclic run lengths in the range of 30,000 to 50,000 BV are considered to be more appropriate for uranium removal from drinking water.

Effect of pH and Competing Ions on Uranium Removal. Even though SBA resins exhibit enormous capacity for removing uranium from contaminated groundwaters, the removal efficiency and resin capacity are somewhat influenced by the usual parameters, including feed pH and the feed concentrations of uranium, sulfate, chloride, and possibly bicarbonate. The simple IX reactions expected between uranium compounds and SBA resins in the 6.5 to 9.5 pH range, where $UO_2(CO_3)_2^{2-}$ and $UO_2(CO_3)_3^{4-}$ are the predominant forms, can be expressed as follows:

$$2R_4\,N^+Cl^- + UO_2(CO_3)_2^{2-} \leftrightarrow (R_4N^+)_2\,UO_2(CO_3)_2^{2-} + 2Cl^- \tag{12-60}$$

$$4R_4\,N^+Cl^- + UO_2(CO_3)_3^{4-} \leftrightarrow (R_4N^+)_4UO_2(CO_3)_3^{4-} + 4Cl^- \tag{12-61}$$

where R_4N^+ represents a quaternary nitrogen anion exchange site on the resin. Because the resin highly prefers the tetravalent $UO_2(CO_3)_3^{4-}$ complex, there is a tendency for this complex to actually be formed within the resin from neutral and divalent uranyl carbonate complexes. This is similar to the in-resin formation of carbonate from bicarbonate, as described by Horng and Clifford (1997). The in-resin formation of $UO_2(CO_3)_3^{4-}$, as further explained next under the effect of pH, helps to explain why the performance of SBA resin is so good at a feed water pH of 5.8, where the predominant form of uranium in water is the neutral species $UO_2CO_3^0$.

Effect of pH. Based on pH effects on uranium speciation (Langmuir, 1978), pH variations in the 5.8 to 8 range are expected to have a significant impact on resin performance when removing uranium. A substantial decrease in the resin's capacity for uranium is expected at a pH below 7 because the charge on the uranyl carbonate complex decreases with decreasing pH, as does the carbonate concentration in the feed water. However, when pH was decreased by HCl addition from 7.8 to 5.8, uranium removal efficiency was unchanged for 270,000 BV during the Chimney Hill field study (Zhang and Clifford, 1994, 1995). When the pH was decreased further to 4.3, however, serious uranium leakage (about 50 percent of the feed uranium concentration) occurred from the very beginning of a run with virgin chloride-form resin. This unexpectedly excellent performance at pH 5.8 probably resulted from the formation of uranyl tricarbonate complexes within the resin, whereas these complexes were virtually absent from the aqueous phase. At pH 5.8, the dominant uranium species in the aqueous system in the absence of anion resin is the zero-charged complex $UO_2CO_3^0$. It is theorized that the direct formation of charged $UO_2(CO_3)_2^{4-}$ from uncharged $UO_2CO_3^0$ occurs within the ion exchanger because of the resin's high affinity for polyvalent anions and the high stability constant of the tricarbonate uranium complex. The suggested exchange reactions are

$$4R_4N^+Cl^- + 4HCO_3^- \leftrightarrow 4R_4N^+HCO_3^- + 4Cl^- \tag{12-62}$$

$$4R_4N^+HCO_3^- + UO_2CO_3^0 \leftrightarrow (R_4N^+)_4\,UO_2(CO_3)_3^{4-} + 2H_2CO_3\,(aq) \tag{12-63}$$

Reaction 12-62 is the conversion of four resin sites to the bicarbonate form by simple IX with chloride. In Reaction 12-63, the bicarbonate-form ($R_4N^+HCO_3^-$) resin sites are converted to the highly preferred uranyl tricarbonate $(R_4N^+)_4UO_2(CO_3)_3^{4-}$ form with the production of carbonic acid [$H_2CO_3(aq)$]. In the Chimney Hill studies, the resulting lowering of pH by the $H_2CO_3(aq)$ produced was insignificant because of the very low concentration (0.0005 mM) of

$UO_2CO_3^0$ in the feed water compared with the total inorganic carbon ($C_{T,CO3}$) of the carbonate buffer system—3.00 mM in the groundwater.

Adding Reactions 12-62 and 12-63 yields the net IX reaction:

$$4R_4N^+Cl^- + UO_2CO_3^0 + 4HCO_3^- \leftrightarrow (R_4N^+)_4UO_2(CO_3)_3^{4-} + 2H_2CO_3(aq) + 4Cl^- \quad (12\text{-}64)$$

Reaction 12-64 is effectively the uptake of the neutral $UO_2CO_3^0$ molecule from an acidic (pH 5.8) solution by a chloride-form SBA resin.

Effects of Uranium, Sulfate, and Chloride on Run Length. Uranium, sulfate, and chloride concentrations in the feed water have significant effects on uranium breakthrough. Increasing the concentrations of these components results in decreasing run length; however, they affect the runs to different degrees.

Uranium feed concentration was expected to have a big influence on column capacity. However, because of the extremely long runs and the potential regulatory problems associated with uranium spiking of raw water during the Chimney Hill field tests, it was not practical to vary the uranium concentration in the test water to determine its influence on BV to uranium breakthrough. Thus computer predictions of uranium breakthrough were made using the Equilibrium Multicomponent Chromatography Program with Constant Separation Factors (EMCT/CSF) (Horng, 1983; Clifford, 1993; Tirupanangadu, 1996). This program is a user-friendly implementation of the equilibrium multicomponent chromatography theory (EMCT) of Helfferich and Klein (1970), which has been used by Clifford (1982) since the early 1980s for prediction of column effluent concentrations during selected-ion separations.

Using this program, the BV throughputs to uranium breakthrough were calculated at various uranium concentrations in the Chimney Hill water. The results from the computer prediction

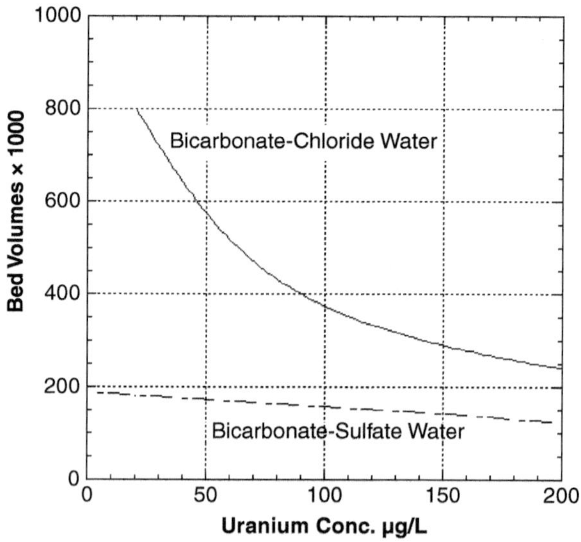

FIGURE 12-30 Calculated (EMCT software) BV throughput from a macroporous polystyrene SBA resin column before uranium breakthrough vs. uranium feed concentration at different background concentrations of chloride, and sulfate. Bicarbonate-chloride water = 183 mg.L HCO_3^-, 49 mg/L Cl^-, 0 mg/L SO_4^{2-}. Bicarbonate-sulfate water = 183 mg/L HCO_3^-, 64 mg/L SO_4^{2-}, 0 mg/L Cl^-. (*Source: Based on the data of Clifford and Zhang, 1995.*)

are presented in the Fig. 12-30 curve labeled "Bicarbonate-Chloride water." This curve predicts that for a sulfate-free water such as the Bicarbonate-Chloride Chimney Hill feed water, an increase in the uranium concentration to 240 μg/L would result in a decreased run length of 203,000 BV. Decreasing the feed water uranium concentration to 20 μg/L would increase the run length to 815,000 BV. This high sensitivity of run length to uranium concentration was due to the fact that (1) uranium was a trace species with an exceptionally high affinity for the resin, (2) uranyl carbonate complexes occupied a significant fraction of the resin sites at exhaustion, and (3) there was no sulfate in the feed water. The Fig 12-30 water labeled "Bicarbonate-Sulfate" is the "Bicarbonate-Chloride" water with sulfate substituted for chloride to illustrate the strong negative influence of sulfate on uranium run length.

Among the common anions in groundwater, sulfate will exhibit the biggest impact on run length. The EMCT/CSF program also was used to predict the influence of sulfate on uranium run length at variable uranium concentration. In this simulation, sulfate replaced chloride in the Chimney Hill water, which was initially free of sulfate. The line labeled "Bicarbonate-sulfate water" in Fig. 12-30 demonstrates that replacing chloride with sulfate greatly affects the SBA resin's capacity for uranium removal. At the natural uranium concentration of 120 μg/L, when the sulfate concentration increases from 0 to 1.32 meq/L (64 mg/L), run length drops 60 percent from the experimental 340,000 BV down to a predicted 135,000 BV. This decrease at higher sulfate concentration is attributed to the fact that compared with chloride, sulfate ion has a high affinity for the SBA resin sites (see Table 12-3). Therefore, water containing significant concentrations of sulfate will produce significantly shorter runs than low-sulfate waters.

Regenerability of Uranium Spent SBA Resins. The Chimney Hill field experiments demonstrated that despite the high affinity of SBA resins for uranyl carbonate complexes, the uranium spent resins are not difficult to regenerate because of the electroselectivity reversal that takes place in the presence of 1 to 6 M chloride regenerant solutions. Furthermore, it is not necessary to completely remove the uranium from the resin during regeneration because the residual uranium does not leak into the effluent on subsequent runs (Zhang and Clifford, 1994). What is necessary for a practical cyclic exhaustion/regeneration process is that a steady state be reached in which the mass of uranium eluted during regeneration is equal to the mass sorbed during exhaustion. Such a steady state can be attained in uranium removal by chloride-form SBA exchange.

Effect of NaCl Concentration and Regeneration Level. Uranium recovery efficiency strongly depends on the regenerant NaCl concentration (Zhang and Clifford, 1994). Increasing NaCl concentration results in increased uranium recovery at a fixed regeneration level. This was established in the Chimney Hill field test, where NaCl concentrations ranging from 0.8 to 4.0 N were applied to columns partially exhausted to 30,000 or 41,000 BV. Although the NaCl concentration varied, the total amount of NaCl applied was maintained constant at 4.0, or 6.0 eq NaCl per equivalent of resin during two sets of experiments. At a regeneration level of 6.0 eq NaCl per equivalent of resin (346 kg NaCl/m^3, or 21.6 lb NaCl/ft^3), the recovery efficiencies for 4.0, 3.0, 2.0, 1.33, 1.0, and 0.8 N NaCl were 91, 86, 75, 67, 54, and 47 percent, respectively, as shown in Fig. 12-31. The uranium recovery did not appear to level off at 4.0 N, which suggests that even higher NaCl concentrations, for example, saturated NaCl (6.15 N at 20°C), might be tried. Increasing the regeneration level from 4 to 6 eq Cl^- per equivalent of resin did improve recovery, but not by very much at any NaCl concentration (see Fig. 12-31). Thus regeneration levels lower than 4 eq Cl^- per equivalent of resin should be considered because this would conserve salt, and as pointed out previously, it is not necessary to remove all or even most of the uranium from the exhausted resin because it does not seriously leak on subsequent exhaustions.

Other observations made during the regeneration experiments in Chimney Hill follow. Uranium was mostly desorbed in the first 1 or 2 BV of regenerant when 2 to 4 N NaCl was used. Addition of NaOH to the NaCl solution greatly reduced the recovery of uranium. When NaOH represented one-third of the regenerant equivalents at 4.0 N regenerant

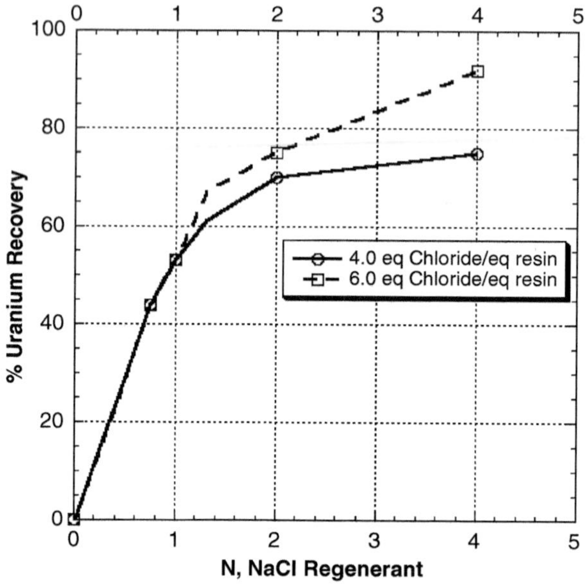

FIGURE 12-31 Effect of equivalents and normality of NaCl on percent uranium recovery during cocurrent regeneration of SBA resin exhausted to 41,000 BV. Regenerant EBCT = 10 min, 5 BV slow rinse. (*Source: Based on the data of Clifford and Zang, 1995.*)

concentration, uranium recovery efficiency dropped from 91 to 15 percent. Reductions in regeneration efficiency also were observed when Na_2CO_3 and $NaHCO_3$ were added to NaCl. Uranium recovery efficiency was not affected by the degree of exhaustion when the anion resin was far away from complete exhaustion. Columns that were exhausted to 500 BV exhibited the same uranium-recovery percentages (75%–78%) during regeneration as did columns exhausted to 30,000 and 40,000 BV, whereas complete exhaustion was approximately 300,000 BV.

Combined Radium and Uranium Removal. In the Chimney Hill field study conducted by Clifford and Zhang (1994), a common NaCl-regenerated SAC resin water softener modified by the addition of approximately 10 percent SBA resin was used for the simultaneous removal of radium (25 pCi/L) and uranium (120 μg/L) from groundwater. Although a high regeneration level (590 kg NaCl/m^3 of resin (37 lb/ft^3) was used in the field tests, typical softener regeneration levels in the range of 160 to 320 kg NaCl/m^3 of resin (10–20 lb/ft^3) are expected to be adequate for a cyclic radium- and uranium-removal process. However, to remove uranium effectively, it is important to keep the regenerant NaCl concentration as high as possible.

Thorough mixing of the SAC and SBA resins was preferred to stratifying the lighter anion resin over the cation resin in the modified softener. Stratified SBA/SAC beds, especially those containing 25 percent or more anion resin, produced greater radium and uranium leakages after cocurrent regeneration than mixed beds. In actual practice, backwashing the softener will tend to stratify the anion resin over the cation resin. This less desirable stratified condition, however, may be adequate for combined radium and uranium removal when the stratified bed contains 10 percent or less anion resin.

As a rule of thumb, an anion bed EBCT of 0.2 minute or more and an anion resin bed depth of greater than 2.5 cm (1 in) are recommended for the stratified SBA resin layer to produce less than 10 μg/L uranium leakage with 120 μg/L uranium in the feed water. Extensive tests with the stratified and mixed beds showed that radium (and possibly uranium) leakage in stratified-bed softeners was attributed to precipitation of $CaCO_3(s)$ during regeneration of the cation resin from carbonate ions taken up and eluted from the anion resin. Minimizing the amount of anion resin and mixing it with the cation resin minimized $CaCO_3(s)$ precipitation and subsequent radium and uranium leakage. Countercurrent regeneration is also expected to eliminate excessive radium and uranium leakage.

Although 10 percent anion resin appeared to be a near-optimal amount, as little as 2.5 percent anion resin removed 50 percent of uranium from the Chimney Hill groundwater at conventional flow rates (EBCT of 3 minutes based on combined cation and anion resin volume). Potassium chloride proved to be superior to NaCl for radium elution from the softener in cyclic operation (Clifford and Zhang, 1994). The use of KCl, although typically triple the price of NaCl, should be considered for the additional health reason that potassium is preferable to sodium as an exchanging ion.

Perchlorate Removal by Anion Exchange

Perchlorate residual from ammonium perchlorate used in rocket fuel and explosives has contaminated U.S. water supplies serving more than 20 million people in 22 states. Perchlorate interferes with the proper operation of the thyroid gland, and a number of health-effect studies have been undertaken to establish the basis of an MCL for perchlorate. Currently, the USEPA has established a health reference level of 15 μg/L, and Massachusetts and California have perchlorate MCLs of 2 and 6 μg/L, respectively. Nitrate, oxidized from ammonium ion, typically is found as a cocontaminant with perchlorate; thus effective treatment technologies often will have to deal with perchlorate and nitrate. Although perchlorate can be removed from drinking water supplies by biological treatment (Kim and Logan, 2001), IX is usually the process of choice because most anion resins have a high to extremely high preference for the hydrophobic perchlorate oxyanion (Tripp and Clifford, 2006). See Chaps. 1 and 2 for additional information on regulations and health effects.

Factors Affecting Perchlorate Resin Selection. Tripp and Clifford (2006) conducted a laboratory study of 15 different SBA resins to relate resin properties to perchlorate/chloride selectivity ($\alpha_{ClO4/Cl}$) and then input the selectivity data into computer programs to aid in process design and selection. They found that the $\alpha_{ClO_4/Cl}$ values varied from 5 to more than 1300, and they sorted the resins into three groups. Group 1 included polyacrylic resins with low perchlorate selectivity ($\alpha_{ClO_4/Cl} = 5\text{–}10$) and a preference order of sulfate > perchlorate > nitrate; group 2 included typical polystyrene resins with high perchlorate selectivity ($\alpha_{ClO_4/Cl} = 80\text{–}170$) and a preference order of perchlorate > sulfate > nitrate; group 3 included nitrate-selective resins and other resins with extremely high perchlorate selectivity ($\alpha_{ClO_4/Cl} = 600\text{–}1300$) and a preference order of perchlorate >> nitrate > sulfate. Group 2 and 3 resins were sensitive to temperature, with $\alpha_{ClO_4/Cl}$ values decreasing about 30 percent for each 20°C increase in temperature. Thus these resins could be regenerated more easily at elevated temperatures in the range of 40 to 60°C (Tripp and Clifford, 2006). As with all contaminants, the background ions, particularly sulfate and nitrate, can reduce the perchlorate run lengths significantly. Thus the concentrations of these background ions must be carefully considered during process selection.

Perchlorate IX Process Considerations. Given these large variations in selectivity among resins, several possible options are available for perchlorate removal depending on the background concentrations of sulfate and nitrate in the water and the availability disposal options for regenerant brine. For water with less than 250 mg/L sulfate and a convenient source for brine disposal, the recommended treatment process would use a polyacrylic gel resin (group 1) with low perchlorate selectivity operated with countercurrent 6% NaCl regeneration at a level of 320 kg NaCl/m^3 or more (≥20 lb NaCl/ft^3) at ambient temperature. The run would be terminated at or before nitrate breakthrough. If spent regenerant disposal is not available, the brine could be treated biologically to remove perchlorate and reused, as demonstrated in the bench- and pilot-scale sequencing batch reactor studies of Liu and colleagues (2007) and Lehman and colleagues (2007). Fluidized-bed reactors for perchlorate destruction in brine were successful, too, but proved to be more challenging to operate (Patel et al., 2008; Lehman et al., 2008). By destroying the perchlorate and nitrate in the spent brine, this available chloride salt can be recycled and used for many more regeneration cycles, which can decrease the amount of salt required for regeneration and the perchlorate-laden brine for disposal significantly.

For perchlorate-contaminated water containing sulfate less than 500 mg/L, NO_3-N less than 5 mg/L, and readily available brine disposal, a polystyrene resin (group 2 with $\alpha_{ClO_4/Cl} = 80–120$) operated with 6 percent NaCl countercurrent regeneration at a level of 320 kg $NaCl/m^3$ or more (≥20 lb $NaCl/ft^3$) at 50 to 60°C might be selected. The process would be operated to sulfate breakthrough, which is often well beyond nitrate breakthrough.

For a similar water (sulfate < 500 mg/L, NO_3-N < 5 mg/L) and no good option for brine disposal, the single use of highly perchlorate-selective resins (group 3, $\alpha_{ClO_4/Cl} > 600$) could be considered. This would require a careful evaluation of the cost of brine disposal versus the cost of highly perchlorate-selective resin. As raw water sulfate concentrations increase, however, the effective run length of the highly perchlorate-selective resin is decreased, diminishing the economic benefits of using this resin type. The highly perchlorate-selective resin typically would be operated as roughing-polishing columns in series with the roughing (lead) column run to nearly complete perchlorate exhaustion. The exhausted resin would be removed for regeneration at a central facility or disposed of and replaced with fresh resin. The one-time use of highly perchlorate-selective resins in waters that also require treatment for nitrate removal is not recommended because such resins cannot be run past nitrate breakthrough. Even though nitrate breakthrough for highly perchlorate-selective resins occurs later than for conventional sulfate-selective resins, the nitrate run length will be only a small fraction of that to perchlorate breakthrough.

Despite potential problems with shortened run length owing to nitrate and sulfate concentrations, the most common solutions to perchlorate contamination problems in California have been single-use IX systems. Aldridge and colleagues (2004b) estimated that for typical California waters contaminated with perchlorate ($ClO_4^- < 20$ μg/L, $SO_4^{2-} < 40$ mg/L, NO_3-N < 9 mg/L, $Cl^- < 30$ mg/L, and $HCO_3^- < 250$ mg/L), single-use systems would be more economical than chloride IX with regeneration. At that time, there were 17 full-scale systems in California with a total capacity of more than 60 mgd. Since then, additional perchlorate-selective resins have come onto the market with even greater effectiveness and economy.

IX MODELING USING EMCT

IX modeling can be a useful tool for researchers and practitioners because it allows for the rapid desktop screening of a number of IX resins and water qualities for a given application. The following discussion shows how IX modeling can help users during resin selection understand the influence (or sometimes noninfluence) of anion resin separation factors (α_{ij} values) and the influence of water quality constituents on contaminant run length.

Exhaustion or Breakthrough Curves and Resin Profile

With respect to the exchange of bicarbonate, nitrate, and sulfate on an SBA exchange resin, as shown in Fig. 12-10*b*, the following chromatographic characteristics that govern IX may be enumerated.

Resin Profile Characteristics

1. For *n* components, there will be *n* zones in the resin phase.
2. Each zone begins and ends with a wavefront and acts as an "affinity cut," with one species disappearing across the wavefront. For example, the sulfate-rich zone shown in Fig. 12-10*b* contains all *n* components. Sulfate, the most preferred anion or the anion with the greatest affinity for the resin, disappears across the wavefront that divides the sulfate- and nitrate-rich zones. Thus there is no sulfate in the nitrate-rich zone. Similarly, there is no sulfate or nitrate in the bicarbonate-rich zone. The chloride zone contains only chloride, the presaturant ion.

Effluent History Characteristics

1. The total effluent concentration of ions (meq/L) remains constant and is equal to the total feed concentration of ions.
2. A decrease in concentration of one ion is always accompanied by an increase in the concentration of another ion in the effluent. For example, in Fig. 12-10*b*, it can be seen that as chloride concentration decreases at the beginning of an exhaustion cycle, the bicarbonate concentration in the effluent increases.
3. The ion with the lowest affinity (bicarbonate in Fig. 12-10*b*) is the first to appear in the effluent (not including chloride, the presaturant ion), followed by the more preferred anions. That is, the order of appearance of ions in solution is the reverse of the selectivity sequence (described again below). The presaturant anion, chloride, is always present in the effluent.
4. For a system with *n* ions, $n - 1$ ions will undergo chromatographic peaking. That is, the effluent concentration will exceed its influent concentration. In Fig. 12-10*b*, bicarbonate peaks first, followed by nitrate. The most preferred anion, sulfate, does not peak, but its effluent concentration rises to equal its feed concentration.

The model used in the IX simulations described in this section is based on the equilibrium multicomponent chromatography theory (EMCT) developed by Helfferich and Klein (1970). Horng (1983) and Tirupanangadu (1996) used EMCT to develop a PC-based simulator, hereafter referred to simply as the *EMCT model*, to simulate IX processes. *Note:* This model is available by writing to D. Clifford at the University of Houston CEE Department (daclifford@uh.edu). Assumptions inherent in the EMCT model are listed below.

1. The resin column is assumed to be of uniform cross section, uniformly packed, and exhibits plug-flow behavior.
2. Instantaneous equilibrium is assumed between the exchanging species; that is, there are no mass transfer limitations. Of course, instantaneous equilibrium is not achieved, but simple mass transfer correction factors may be applied, as will be explained later.
3. Constant binary separation factors α_{ij} are assumed.
4. The resin is uniformly presaturated; that is, the entire resin bed has the same single- or multicomponent anionic composition.
5. The feed water composition and resin capacity are constant.

The mathematic basis for multicomponent equilibrium modeling is described in detail in Helfferich and Klein (1970) and summarized by Clifford (1982).

Selectivity Sequence. The order of preference that an anion has over the resin phase was explained earlier and depends on the ionic strength of the aqueous phase and the chemical structure of the resin. For SBA exchangers in dilute solution (ionic strength range of 0.005–0.015 M), the following order of preference may be obtained from Table 12-3:

$$SO_4^{2-} > HAsO_4^{2-} > NO_3^- > Cl^- > HCO_3^-$$

For specialty resins, this sequence changes. For example, the conventional order of preference of divalent over monovalent ions is reversed, and anions such as nitrate and perchlorate are more preferred over divalent ions such as sulfate. For such nitrate-over-sulfate-selective (NSS) resins, the following order of preference may be used:

$$ClO_4^- > NO_3^- > SO_4^{2-} > HAsO_4^{2-} > Cl^- > HCO_3^-$$

Modeling Arsenic Removal on SBA Resins

Ghurye (2003) undertook an experimental and modeling study to determine the influence of various resin parameters (capacity and anion separation factors) on arsenic run length. Based on EMCT modeling, that study concluded that anion separation factors (including that of arsenic) had little to no influence on arsenic run length. This finding was nonintuitive, but an examination of the resin profile generated by the EMCT model showed that the arsenic-rich zone was extremely small, comprising less than 0.1 percent of total resin sites. Therefore, although an increase in arsenic separation factor increased the fraction of arsenic in the arsenic-rich zone, it did not change the run length to arsenic breakthrough significantly. The conclusion was that the best resin for trace species arsenic removal is the resin with the highest exchange capacity (eq/L). Use of the EMCT model for modeling exhaustion and the resin profiles obtained from EMCT thus provided an insight into a nonintuitive finding and helped to develop a simple rule for resin selection for trace species arsenic removal: Choose a conventional polystyrene, sulfate-selective SBA exchange resin with the highest exchange capacity.

In addition to understanding the relative influence of anion separation factors on the removal of trace species such as arsenic, EMCT modeling also can help to determine the influence of background ions on contaminant run length. For example, while using the NSF International–recommended challenge water quality for arsenic as a baseline, the relative influence of the concentration of commonly co-occurring anions such as bicarbonate, chloride, nitrate, arsenic, and sulfate was modeled using the EMCT model (Fig. 12-32). The

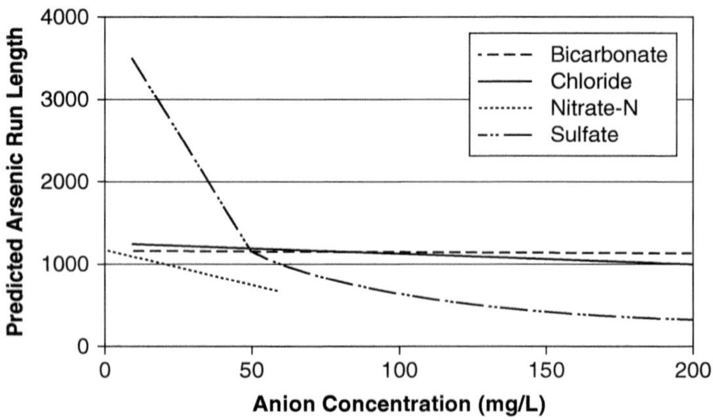

FIGURE 12-32 Influence of water quality on arsenic run lengths using the NSFI challenge water quality as a baseline.

relative impact of increasing anion concentration is greatest for sulfate because sulfate is the common anion most preferred by conventional SBA resins.

Modeling Nitrate Removal: Comparing Conventional Resins with Nitrate-Selective Resins

Based on the work of Clifford and Weber (1978), which showed that sulfate was preferred by resins with closely spaced exchange sites and that nitrate was preferred by hydrophobic resins, NSS resins were developed by Guter (1982), who introduced triethyl, tripropyl, and tributyl functional groups in place of the usual trimethyl group in type 1 resins. These bulkier and more hydrophobic functional groups exhibit much greater affinity for hydrophobic monovalent anions such as nitrate and perchlorate. However, the introduction of bulkier functional groups also reduces resin exchange capacity. For example, type 1 resin capacities range from 0.8 to 1.2 eq/L, and type 2 resins can have exchange capacities as high as 1.4 eq/L, whereas typical tributyl resin capacities are in the 0.8 to 0.85 eq/L range. The following discussion shows how IX modeling can be used to rapidly perform a desktop comparison of these resins for a nitrate-removal application. Note that for comparison, an acrylic resin with high sulfate selectivity but low nitrate selectivity is included in the modeling effort.

For the resins listed in Table 12-15 and using the Glendale water quality (Table 12-16) as a baseline, nitrate run lengths (for three different resins) were modeled as a function of sulfate concentration (Fig. 12-33). Note that caution is required when attempting to determine the influence of concentration simply by increasing the concentration of an anion. For example, Guter (1982) noted that the sulfate selectivity sequence for conventional resins was reversed at an ionic strength of 0.02 M. Clifford and colleagues (1987) experimentally determined selectivity reversal to occur at 0.03 M for nitrate removal on SBA resins in their Glendale study. In the modeling effort shown in Fig. 12-33, the maximum ionic strength modeled was 0.0163 M. Since this was less than the ionic strength reported to cause selectivity reversal by Guter (1982) and Clifford (1987), no attempt was made to keep the total ionic strength of the solution constant by proportionately reducing the concentration of the other anions (such as bicarbonate and chloride). The reason for choosing sulfate concentration as a variable was that the Glendale study (Clifford et al., 1987) found a significant reduction in nitrate run length with increasing sulfate concentration for conventional sulfate-selective (type 1 and type 2 SBA) resins.

TABLE 12-15 Input Parameters for EMCT IX Modeling

	Resin 1 (type 1) nitrate-selective NSS	Resin 2 (type 2) sulfate-selective CSS	Resin 3 (polyacrylic) sulfate-selective CSS
Matrix	STY-DVB	STY-DVB	Polyacrylic
Selectivity	Nitrate/Perchlorate	Sulfate	Sulfate
Functional group	Tributylamine	Dimethylethanol-amine	Trimethylamine
Capacity, meq/mL	0.85	1.40	1.20
Separation factors[a]			
$\alpha_{\text{Bicarbonate/chloride}}$	0.27	0.27	0.27
$\alpha_{\text{Nitrate/chloride}}$	11	3.09	1.79
$\alpha_{\text{Arsenic/chloride}}$		9.36	8.92
$\alpha_{\text{Sulfate/chloride}}$	4	17.9	23.0
$\alpha_{\text{Perchlorate/chloride}}$	Varied (100–3600)	110	5.5

[a]For dilute (0.005 N) or total solution concentration (250 mg/L as $CaCO_3$).

TABLE 12-16 Water Qualities for Model Simulations

Anion	Arsenic removal, NSF Intl. challenge water	Nitrate removal, glendale study[a]	Perchlorate removal, big dalton[b]
Bicarbonate, mg/L	183	124	122
Chloride, mg/L	71	122	30
Nitrate-N, mg/L	2	19.2	6
Arsenic, mg/L	0.05	NA	NA
Sulfate, mg/L	50	42.5	50
Perchlorate, mg/L	NA	NA	0.05

[a]Clifford et al., 1987.
[b]Tripp et al., 2003.

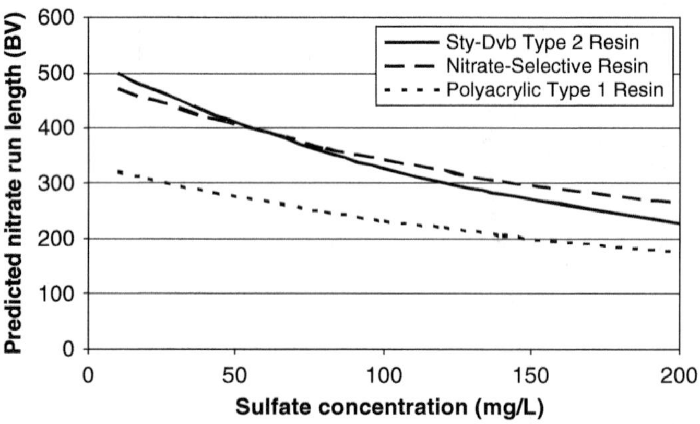

FIGURE 12-33 Influence of sulfate concentration on nitrate run length for three resins (Glendale water).

The polyacrylic resin, by virtue of its hydrophilic nature and low nitrate-chloride separation factor (1.79 versus 11 for a nitrate-selective resin; see Table 12-15) underperforms both the type 2 CSS resin and the NSS resin. The type 2 resin, although possessing a lower nitrate-chloride separation factor (3.09 versus 11 for the nitrate-selective resin), performs nearly the same as the nitrate-selective resin because of the type 2 resin's much higher capacity (1.4 versus 0.85 meq/mL for the nitrate-selective resins).

It has been observed that CSS resins are much more easily regenerated than NSS resins owing to the fact that nitrate is a monovalent ion and is not subject to selectivity reversal at high ionic strength during regeneration. To illustrate how EMCT modeling can be used to model regeneration, an NSS (resin 1) and a CSS (resin 2) resin were used for modeling as follows: At nitrate breakthrough, the NSS resin comprises a single nitrate-rich zone. The fraction of resin sites in this zone was obtained from the resin profile provided by the EMCT model for the NS resin exhausted with Glendale water. However, for a CSS resin at nitrate breakthrough, the resin comprises a sulfate-rich zone and a nitrate-rich zone. As noted earlier, the EMCT model assumes uniform presaturation and cannot accept such multiple zones as an initial condition prior to regeneration. To simplify this problem, a mass balance on all four anions (Glendale water; Table 12-16) was used to determine the size of the sulfate- and nitrate-rich zones. Using this information and the resin profile provided by

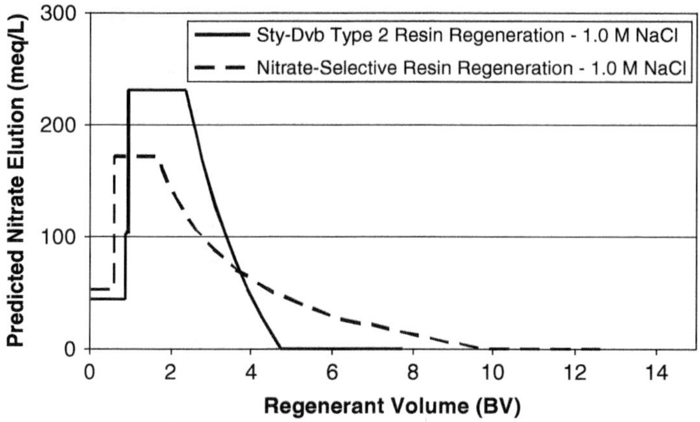

FIGURE 12-34 Nitrate elution (regeneration) from an SS and NS resin with 1.0 M NaCl.

the EMCT model, a uniform presaturation assumption was made by compositing the anion concentration across the entire resin bed. The regeneration results are shown in Fig. 12-34, which shows that about twice as much chloride is required for complete elution of nitrate from the NSS resin than is required for the CSS resin. Note that an earlier modeling study by us (Ghurye and Clifford, 2008) using an IX model that does not require uniform presaturation prior to regeneration showed similar results in terms of salt usage. Thus, with the aid of EMCT modeling, it can be seen that under certain conditions of groundwater quality, it actually may be more advantageous to use a CSS resin despite its lower selectivity for nitrate.

Modeling Perchlorate Removal: Comparing Conventional Resins with Nitrate-Selective Resins

Previous research has shown that hydrophobic resins have much greater capacity for perchlorate removal, but owing to the lack of selectivity reversal during regeneration, hydrophobic resins are extremely difficult to regenerate because of their very high perchlorate selectivity (Tripp et al., 2003). Therefore, perchlorate IX in a single-use application is increasingly the treatment technique of choice (Siedel et al., 2006). Single-use application is favored by the fact that perchlorate separation factors can be very high on hydrophobic resins. For example, Tripp (2003) reported a perchlorate/chloride separation factor of 1300 for a tripropylamine resin. In comparison, the separation factors for bicarbonate, sulfate, and nitrate over chloride were 0.2, 4, and 11, respectively.

Perchlorate is the most preferred anion on hydrophobic resins. Therefore, at breakthrough, the perchlorate zone comprises the entire resin bed, and even small increases in the fraction of resin sites occupied by perchlorate as a result of increasing perchlorate selectivity lead to large increases in perchlorate run length. To illustrate the importance of perchlorate selectivity for single-use applications, the results of an EMCT modeling exercise based on Big Dalton groundwater composition, shown in Table 12-16, are shown in Fig. 12-35. In this exercise, the perchlorate separation factor was varied from 100 (polystyrene type 1 resin) to 3600 (highly perchlorate-selective resin) for varying feed perchlorate concentrations from 10 to 500 μg/L. Unlike the case of arsenic, perchlorate run lengths can be seen to increase substantially with increasing perchlorate separation factor, with run lengths approaching 300,000 BV for a feed perchlorate concentration of 10 μg/L.

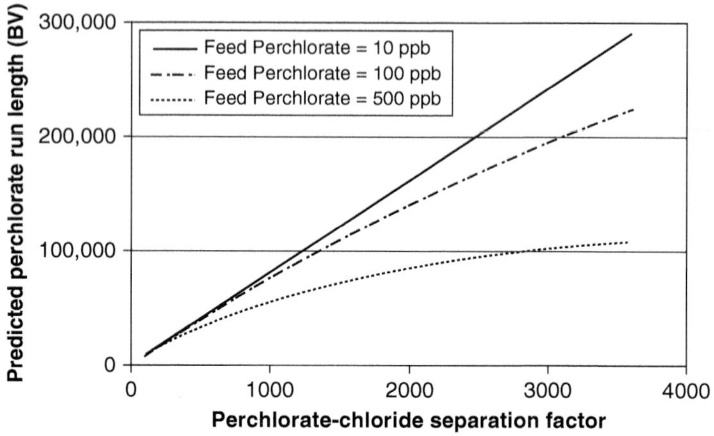

FIGURE 12-35 Influence of perchlorate separation factor on run length for feed perchlorate concentrations in the range of 10 to 500 μg/L (Big Dalton water).

As noted earlier, IX modeling is also a useful tool to determine the impact of various water qualities on contaminant run length. Using Big Dalton groundwater quality as the baseline, hydrophobic perchlorate-selective resin was modeled by varying the concentration of bicarbonate, chloride, or sulfate while holding the perchlorate concentration constant at 0.05 mg/L. The results are shown in Fig. 12-36, where it can be seen that increasing concentrations of all the background ions shortens perchlorate run length, but nitrate has the most significant impact, and bicarbonate has almost no effect.

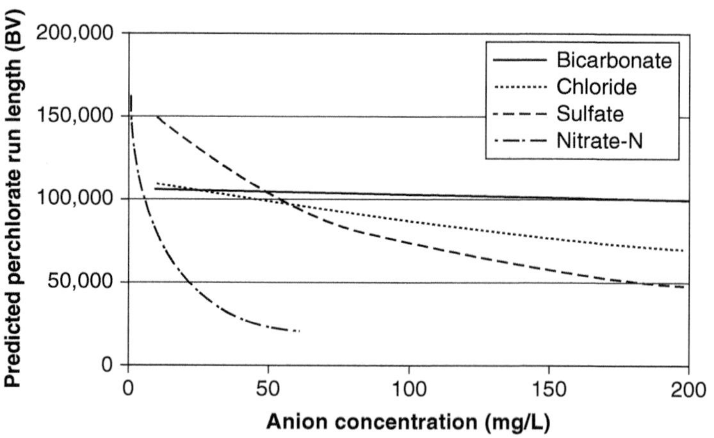

FIGURE 12-36 Modeling the influence of common groundwater anions on perchlorate run length on a hydrophobic resin with perchlorate-chloride separation factor of 1000 (Big Dalton water quality as baseline).

EMCT Correction Factor. It should be noted that contaminant breakthroughs using EMCT model simulations are ideal and appear as square, "self-sharpening" wavefronts for ions that have higher preference for the resin than the presaturant ion (typically chloride). EMCT modeling provides the run length to complete breakthrough without taking into effect mass transfer limitations. Real mass transfer–limited breakthrough curves are

S-shaped, and for anions such as perchlorate, the breakthroughs can last tens of thousands of bed volumes. Further, actual run lengths are decided by the MCL or chosen maximum allowable effluent concentration for a given contaminant, which is generally much lower than the influent concentration.

For example, the breakthrough of nitrate on a type 2 resin is shown in Fig. 12-37. The influent NO_3-N concentration is 24 mg/L, whereas the MCL is 10 mg/L NO_3-N. The EMCT model–predicted run length for nitrate is a sharp wavefront at 356 BV. However, actual breakthrough

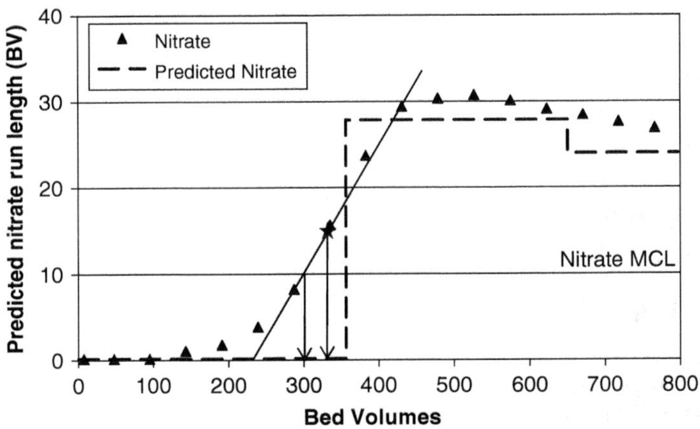

FIGURE 12-37 Use of correction factor for nitrate breakthrough.

is mass transfer–limited and follows an S-shaped curve. Because nitrate is not the most preferred anion, it chromatographically peaks over its influent concentration and then decreases to eventually equal its influent concentration. Major breakthrough for the actual S-shaped curve is estimated as the inflection point of the ascending exhaustion curve (shown by a solid line) and projecting that midpoint on the x axis. The point of major breakthrough thus is estimated as 335 BV, an error of 6 percent. This error may be caused by a number of factors, including errors in estimating separation factors for any of the major ions such as nitrate and sulfate and errors in reported or measured resin capacity, channeling, resin fouling, and fluctuations in both flow rate and anion composition during the course of an exhaustion run. The actual run length termination is user-defined. For the nitrate example shown in Fig. 12-37, the run length may be defined as that point in the exhaustion curve where the effluent concentration first exceeds the nitrate MCL of 10 mg/L as NO_3-N or some fraction (typically 70%–90%) of the MCL. Therefore, the actual run length to 10 mg/L NO_3-N is 300 BV, an error of 16 percent. Thus the EMCT prediction has to be corrected for the user-chosen effluent concentration set point (the MCL or some fraction thereof) that signals run termination. Generally, the greater the difference between the influent concentration of a contaminant and the set-point concentration that terminates a run, the greater is the correction factor that needs to be used to estimate run length from EMCT predictions.

Figure 12-38 shows a similar example but for perchlorate, which generally exhibits prolonged breakthrough on perchlorate-selective resins owing to resin-phase mass transfer limitations for perchlorate in these very hydrophobic resins (Tripp, 2001; Tripp et al., 2004). Note that perchlorate is the most preferred anion on a hydrophobic resin and does not peak chromatographically over its influent but approaches its influent concentration as the resin approaches exhaustion. The EMCT model–predicted run length for an influent perchlorate concentration of 500 μg/L is 69,000 BV, whereas the midpoint of the ascending exhaustion curve is 54,000 BV (22% error), and the run length to 15 ppb (μg/L) effluent perchlorate concentration is 32,500 BV (error of 53%). As noted earlier, the greater the

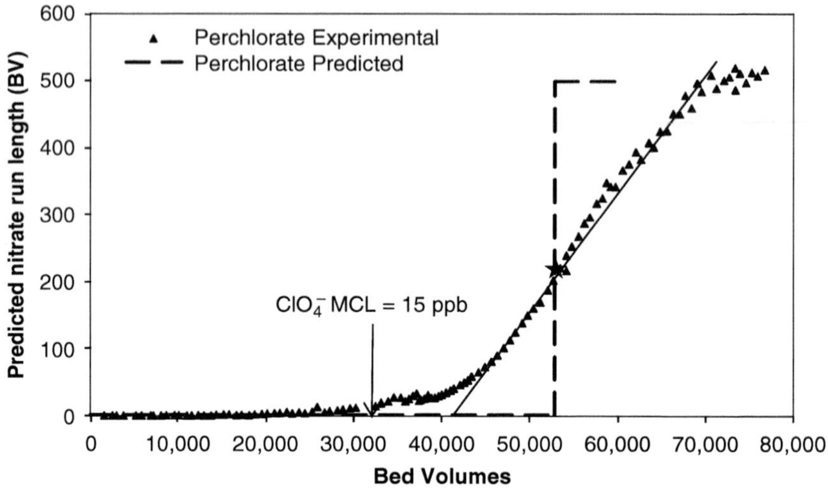

FIGURE 12-38 Use of correction factor for perchlorate breakthrough.

difference between the feed concentration and the run length set point, greater is the correction factor that needs to be used for the EMCT prediction. In practice, the resin bed in the preceding example can be run to complete perchlorate exhaustion by using a two-bed roughing-polishing sequence.

Generally, for arsenic concentrations in the range of 0 to 100 μg/L, nitrate concentrations in the range of 0 to 40 mg/L as NO_3-N and perchlorate concentrations in the range of 0 to 100 μg/L, recommended correction factors are 20, 10, and 35 percent, respectively.

WASTE DISPOSAL

To the extent possible, waste disposal considerations have been covered in the discussions of each contaminant-removal process. The focus has been on the design of processes that minimize regenerant use and on the reuse of spent regenerant to maximize the ratio of product water to spent regenerant for final disposal. Further explanations of how to deal with process residues are given in Chap. 22.

SUMMARY

In the first part of the chapter, the fundamentals of IX and ion adsorption processes were explained with the goal of demonstrating how these principles influence process design for inorganic contaminant removal. In the second part, IX and ion adsorption processes for the removal of hardness, barium, radium, nitrate, fluoride, arsenic, DOC, uranium, and perchlorate were described in detail. In the third part of the chapter, IX modeling using equilibrium multicomponent chromatography theory with constant separation factors (EMCT/CSF) was covered. Because the selection of a process for removal of a given contaminant can be confusing owing to the many variables to be considered, summary Tables 12-17 and 12-18 have been added to aid the reader in process selection. However, the reader should refer back to the discussion of fundamentals and process alternatives to answer questions arising from the tables.

TABLE 12-17 Summary of Processes for Removing Inorganic Cations

Contaminant and its MCL	Usual form at pH 7–8	Removal options	Typical BV to MCL[a]	Pretreatment required	Typical regenerants and % recovery of sorbed contaminant	Effect of TDS on BV	Effect of hardness on BV	Notes
Hardness (no MCL)	Ca^{2+} Mg^{2+}	Na IX softening with SAC resin	200–700	Iron/Mn removal	6%–12% NaCl 90–100% recovery	Very signif. reduction		1–3
		H^+ IX softening with WAC resin	200–500	None	1–2 N HCl or H_2SO_4	Very signif. reduction		4–6
Barium, 1.0 mg/L	Ba^{2+}	Na IX softening with SAC resin	200–700	Iron/Mn removal	1%–12% NaCl 70%–100%	Very signif. reduction	Very signif. reduction	1–3 7, 8
Radium, 5 pCi/L	Ra^{2+}	Na IX softening with SAC resin	200–700	Iron/Mn removal	6%–12% NaCl 0%–100% recovery	Very signif. reduction	Very. signif. reduction	1–3, 9–12, 20
		Ca IX with SAC resin	300–1500	Iron/Mn removal	10%–15% $CaCl_2$ 50%–100% recovery	Slight reduction	Slight reduction	12–16, 20
		Dow radium-selective complexer	20,000–50,000	Iron/Mn removal	Not regenerable	Slight reduction	Slight reduction	12, 17–19
		$BaSO_4$- impregnated alumina	20,000–50,000	Iron/Mn removal	Not regenerable	Slight reduction	Slight reduction	12, 17–19
		Activated alumina	1000–3000	Iron/Mn removal	2% HCl followed by 1% NaOH 70%–100% recovery	None	Slight reduction	12, 20

[a]Generally, run length depends on raw water contaminant concentration, allowable effluent concentration, competing ions, leakage, and the actual resin or adsorbent used.

1. Hardness capacity depends on regeneration level (lb NaCl/ft^3 of resin).
2. Typically, spent regenerant is disposed of to sanitary sewer.
3. Iron and/or Mn removal by oxidation and filtration should be considered if total $Fe_{tot} > 0.3$ mg/L, Mn > 0.05 mg/L.
4. Only carbonate (temporary) hardness can be removed using WAC resins.
5. Carbon dioxide produced in the IX reaction must be removed from product water, and pH adjustment may be necessary.
6. TDS reduction occurs as a result of removing both the hardness and alkalinity.
7. Barium tends to build up on the resin and break through with hardness when insufficient regenerant is used.

(*Continued*)

TABLE 12-17 Summary of Processes for Removing Inorganic Cations (*Continued*)

8. If necessary, $BaSO_4$ can be precipitated from the spent regenerant by adding sulfate.
9. With a virgin resin, radium breaks through long after hardness, but in cyclic operation, they eventually elute simultaneously.
10. Even though radium accumulates on the resin, no leakage of radium occurs before hardness breakthrough.
11. Current disposal practices allow the discharge of radium-contaminated spent regenerant to the sanitary sewer.
12. Radon-222 is continuously generated from the radium-226 on the resin. Radon elution peaks can occur after idle periods.
13. With Ca IX, immediate serious radium leakage occurs if extensive countercurrent regeneration is not used.
14. $CaCl_2$ is much more expensive than NaCl as a regenerant.
15. A process advantage is that sodium is not added to the product water.
16. No softening is achieved in the calcium exchange process, and magnesium is exchanged for calcium.
17. Radon generation is more serious with RSC and $BaSO_4$-alumina because of the large amount of radium on the medium.
18. Disposal of the spent medium is a serious problem because it is considered a low-level radioactive waste.
19. RSC and $BaSO_4$-alumina may have limited application owing to availability and rad-waste disposal problems.
20. Spent regenerant disposal may be a problem if sanitary sewer disposal is not allowed.

TABLE 12-18 Summary of Processes for Removing Inorganic Anions

Contaminant and its MCL	Usual Form at pH 7–8	Removal Options	Typical BV to MCL[a]	Pretreatment Required	Typical Regenerants and % Recovery of Sorbed Contaminant	Effect of TDS on Run BV	Effect of SO_4^{2-} on Run BV	Notes
Fluoride, 4.0 mg/L	F^-	Activated alumina	1000–2500	pH = 5.5–6.0	1% NaOH followed by 2% H_2SO_4 90%–100% recovery	None	Slight reduction	
Nitrate-N, 10 mg/L	NO_3^-	Anion exchange (complete regeneration)	300–600	Usually none	0.25–2.0 N NaCl (1.5%–12% NaCl) 90–100% recovery	Very signif. reduction	Very signif. reduction	1, 2
		Anion exchange (partial regeneration)	200–500	Usually none	1.0–2.0 N NaCl (6.0%–12% NaCl) 50% recovery	Very signif. reduction	Very signif. reduction	1, 3, 4
Arsenic, 0.010 mg/L	$HAsO_4^{2-}$ As(V)	GFO or GFH Adsorption	3000–70,000	pH = 5.5–6.5 Oxidize As(III) to As(V)	GFO and GFH usually not regenerated	None	None	5–8
		Anion exchange	1000–5000	Oxidize and prefilter to remove iron	1.0 N NaCl >95% recovery	Very signif. reduction	Very signif. reduction	1, 5, 9–11
Chromium, 0.10 mg/L	CrO_4^{2-}	Anion exchange	10,000–50,000	None	1.0 N NaCl 60%–90% recovery	Slight reduction	Slight reduction	12–14
DOC	Fulvates	Anion exchange	400–5000	Prefiltration for turbidity	Mixture of 2.0 N NaCl and 0.5 N NaOH	Very signif.	Very signif.	15–17
Uranium, 20 µg/L	$UO_2(CO_3)_3^{4-}$	Anion exchange	30,000–300,000	Prefiltration for turbidity	2–4 N NaCL 60%–90% recovery	Slight reduction	Signif.	18
Perchlorate	ClO_4^-	Anion exchange	400–50,000	Prefiltration for turbidity	0.5–3 N NaCl 0%–90% recovery	Slight reduction	Very. signif. reduction	19, 20

[a]Generally, run length depends on raw water contaminant concentration, allowable effluent concentration, competing ions, leakage, and the actual resin or adsorbent used.

1. Chromatographic peaking of contaminant is possible after breakthrough.
2. No significant leakage of nitrate occurs prior to breakthrough.

(Continued)

TABLE 12-18 Summary of Processes for Removing Inorganic Anions (*Continued*)

3. Continuous significant (>5 mg/L) leakage of nitrate occurs following partial regeneration during all runs
4. Resin must be mixed mechanically following partial regeneration to avoid excessive early nitrate leakage.
5. As(III) should be oxidized to As(V) prior to adsorption and must be oxidized prior to anion exchange.
6. If pH is not lowered into 5.5–6.5 range, adsorption run lengths to arsenic breakthrough will be shorter.
7. Regeneration of GFO or GFH using NaOH and H_2SO_4 is possible and is under further study.
8. Silicate, phosphate, and vanadate anions compete with As(V) for adsorption.
9. IX capacity may decrease following each exhaustion/regeneration cycle owing to organic or particle fouling of resin.
10. Chloride-form anion exchange can be the process of choice for low-sulfate (<50 mg/L) and low-TDS (<500 mg/L) water.
11. As(V) can be coprecipitated from spent regenerant using ferric sulfate or alum.
12. Cr(III) can be precipitated from the spent regenerant after reduction with ferrous sulfate or acidic sodium sulfite.
13. Macroporous resins and polystyrene resins have a higher preference for chromate than gel and acrylic resins.
14. Resins with higher chromate affinity are relatively more difficult to regenerate.
15. DOC has no MCL. Major DOC breakthrough and peaking occur just ahead of sulfate breakthrough. Typical DOC removals are 60%–80% before and 40%–60% after sulfate breakthrough. Highly colored waters should be prefiltered.
16. Direct reuse of spent regenerants for DOC removal by IX is possible if chloride and hydroxide concentrations are maintained. It is not necessary to remove DOC from spent regenerants before reuse.
17. Polyacrylic resins swell more than polystyrene resins in water and consequently are more suited to removal of large organic anions. Furthermore, polyacrylic resins are less subject to organic fouling than polystyrene resins.
18. All SBA resins have an extremely high affinity for uranyl tricarbonate anions. For cyclic operations, shortened run lengths in the range of 30,000–50,000 BV are suggested. Higher regenerant concentrations are more efficient.
19. Polystyrene resins have a much higher perchlorate affinity than polyacrylic resins, which are used in cyclic exhaustion/regeneration processes for perchlorate removal, typically achieving 400–600 BV runs. High salt (NaCl) consumption and perchlorate-contaminated brine disposal are major problems.
20. Single-use, high-perchlorate-selectivity SBA resins, for example, STY-DVB resins with tributyl functionality, may achieve run lengths exceeding 50,000 BV and are reportedly more cost-effective than regenerable polyacrylic resins.
21. STY-DVB weak-base anion resins also may be used for perchlorate removal if pH is lowered to less than 4.5. Posttreatment to strip CO_2 and raise pH using NaOH is necessary. The process is reportedly less costly than NaCl regeneration of SBA resin.

ABBREVIATIONS

A-7	phenol formaldehyde polyamine WBA resin, Dow-Rohm and Haas
AA	activated alumina
AAFS 50	iron(III)-modified activated alumina available from Alcan Corp.
ADI G2	iron(III)-modified diatomaceous earth available from AGI, Inc.
AlOOH	aluminum oxyhydroxide adsorbent solid
Alumina·HCl	activated alumina treated with HCl to protonate the surface
Alumina·HOH	activated alumina in neutral water-washed form
APHA	American Public Health Association
ArsenX	iron(III)-modified anion exchange resin available from Purolite, Inc.
As(III)	arsenic in the +3 oxidation state
As(Tot)	As(III) + As(V)
As(V)	arsenic in the +5 oxidation state
ASB2	Ionac ASB2 STY-DVB type 2 SBA resin available from Beyer-Sybron
BV	bed volume, volume of feed water = volume of media including voids
CSF	constant separation factor
CSS	conventional sulfate selective; also referred to simply as sulfate selective
DBP	disinfection by-product
DM	demineralization
DOC	dissolved organic carbon
Dowex 21K	STY-DVB type 1 gel SBA resin available from Dow Chemical Co.
EBCT	empty-bed contact time, minutes
EMCT	equilibrium multicomponent chromatography theory
EP	extraction procedure test (predecessor to TCLP test)
eq	gram equivalent, Avogadro's number of charges
FBR	fluidized-bed reactor
FeOOH	iron(III) oxyhydroxide adsorbent solid
GAC	granular activated carbon
Gel	standard-porosity IX resin, typically 8 percent DVB cross-linking
GFH	granular ferric hydroxide adsorbent, available from Siemens, Inc.
GFO	granular ferric oxide adsorbent, available from Severn Trent Services
gr	grain, unit of mass; 15.43 gr = 1.00 g
HRT	hydraulic retention time
IOCs	inorganic chemicals
Ionac A 641	STY-DVB type 1 gel SBA resin available from Lanxess
IonacA-642	STY-DVB macroporous type 1 SBA resin available from Lanxess
Ionac SR7	NSS STY-DVB SBA resin with tripropyl functional groups, available from Lanxess

IRA 404	STY-DVB SBA resin available from Dow-Rohm and Haas
IRA 458	polylacrylic SBA resin available from Dow-Rohm and Haas
IRA 958	polyacrylic macroporous SBA resin available from Dow-Rohm and Haas
IRA 996	NSS STY-DVB SBA resin with triethyl groups, available from Dow-Rohm and Haas
IX	ion exchange
kgr	kilograin, 1000 grains
M	molar, g · mol/L
MCL	maximum contaminant level
meq	milliequivalent, 1/1000 of Avogadro's number of charges
mgd	millions of gallons per day flow rate
MnOOH	manganese oxyhydroxide adsorbent solid
MP 500 A	Lewatit macroporous type 1 STY-DVB SBA, Beyer Chemical Co.
MR	macroporous or macroreticular resin
N	normal, g · eq/L
NOM	natural organic matter
NSS	nitrate over sulfate selective; also referred to simply as nitrate selective
pCi	picocuries (10^{-12} Ci)
POE	point of entry
POU	point of use
R_3N:	tertiary weak-base functional group in the free-base form
R_4N^+	quaternary functional group anion exchange site
RCOOH	weak-acid cation resin functional site in hydrogen form
RSSCT	rapid small-scale column test
SBA	strong-base anion
SBR	sequencing batch reactor
SFR	service flow rate or exhaustion rate = 1/EBCT, $L/m^3/min$, gpm/ft^3
SLV	superficial linear velocity, flow rate/area, cm/min, gpm/ft^2
SMCL	secondary maximum contaminant level
STY-DVB	polystyrene polymer cross-linked with divinylbenzene
TCLP	toxicity characteristic leaching procedure
TDS	total dissolved solids
THMFP	trihalomethane formation potential
TOC	total organic carbon in water
USEPA	U.S. Environmental Protection Agency
WAC	weak-acid cation
WBA	weak-base anion
ZPC	zero point of charge on an adsorbent surface

NOTATION FOR EQUATIONS

α_{ij} separation factor of ion i with respect to j, dimensionless

$\alpha_{\mathrm{N/Cl}}$ separation factor of nitrate with respect to chloride, dimensionless

BV_H bed volumes of feed water treatment to hardness breakthrough

C total concentration of ions, eq/L

C_B concentration of contaminant in bypass water, eq/L

C_E concentration of contaminant in column effluent, eq/L

f_F fraction of feed water that must be treated

K_{ij} selectivity coefficient of ion i with respect to j, variable units

n number of exchanging species

Q flow rate, L/min (gpm)

Q_B bypass water flow rate, L/min (gpm)

Q_F column feed flow rate, L/min (gpm)

q_H hardness capacity of resin, eq/L

q_i resin-phase equivalent concentration of ion i, eq/L

Q_P blended product water flow rate, L/min (gpm)

t time, h

t_{BW} time required for backwash, h

t_{FR} time required for fast rinse, h

t_H time to hardness breakthrough, h

t_{os} out-of-service time during backwashing regeneration and rinsing, h

t_R time required for regeneration, h

t_{SR} time required for slow rinse, h

v_0 approach velocity of mobile phase, m/h (gpm/ft^2)

V_F volume of water fed to column prior to contaminant breakthrough, L, gal

V_R volume of resin, m^3 (ft^3)

x_i equivalent fraction of ion i in water, C_i/C, dimensionless

x_{N} equivalent fraction of nitrate in water, C_{N}/C, dimensionless

y_i equivalent fraction of species i in resin, q_i/q, dimensionless

y_{N} equivalent fraction of nitrate in resin, q_{N}/q dimensionless

Q_F column feed flow rate, L/min (gpm)

q_H hardness capacity of resin, eq/L

q_i resin-phase equivalent concentration of ion i, eq/L

Q_P blended product water flow rate, L/min (gpm)

t time, h

t_{BW} time required for backwash, h

t_{FR} time required for fast rinse, h

t_H Time to hardness breakthrough, h

t_{os} out-of-service time during backwashing regeneration and rinsing, h

t_R	time required for regeneration, h
t_{SR}	time required for slow rinse, h
v_0	approach velocity of mobile phase, m/h (gpm/ft^2)
V_F	volume water fed to column before contaminant breakthrough, L (gal)
V_R	volume of resin, m^3 (ft^3)
x_i	equivalent fraction of ion *i* in water, C_i/C, dimensionless
x_N	equivalent fraction of nitrate in water, C_N/C, dimensionless
y_i	equivalent fraction of species *i* in resin, q_i/q, dimensionless
y_N	equivalent fraction of nitrate in resin, q_N/q dimensionless

REFERENCES

Abrams, I. M. (1982). Organic fouling of ion exchange resins. In *Physicochemical Methods for Water and Waste Water Treatment,* L. Pawlowski, ed. Amsterdam: Elsevier.

Aldridge, L., Gillogly, T., Lehman, S. G., D., Roberts, D. and X. Lin (2004a). *Treatability of Perchlorate in Groundwater by Ion-Exchange Technology: Phase II Pilot Plant Study.* Awwa Research Foundation Report. Denver, CO: Water Research Foundation.

Aldridge, L., Gillogly, T., Oppenheimer, J., Lehman, S.G., Witter, K., Burbano, A., Clifford, D.A., and Tripp, A.R (2004b). *Treatment of Perchlorate Using Single-Use Ion-Exchange Resins.* Awwa Research Foundation Report. Denver, CO: Water Research Foundation.

Allpike, B.P., Heitz, A., Loll, C., Kagi, R., Abbt-Braun, G., Frimmel, F.H., Brinkmann, T., Her, N., and Amy, G. (2005). Size exclusion chromatography to characterize DOC removal in drinking water treatment. *Environmental Science and Technology* 39:2334–2342.

Amy, G., Chen, H., Drizo, A., von Guten, U., Brandhuber, P., Hund, R., Chowdhury, Z., Kommineni, S., Sinha, S., Jekel, M., and Banerjee, K. (2005). *Adsorbent Treatment Technologies for Arsenic Removal.* Awwa Research Foundation Report. Denver, CO: Water Research Foundation.

Andreae, M. O. (1978). Distribution and speciation of arsenic in natural waters and some marine algae. *Deep Sea Research* 25(4):391–402.

AST (1995). *Continuous Adsorption Ion Exchange Systems.* Lakeland, FL: Calgon Corporation, Advanced Separation Technologies.

AWWA. (1985). An AWWA survey of inorganic contaminants in water supplies. *Journal AWWA* 77(5):67–72.

Baes, C. F., and Mesmer R. E. (1976). *The Hydrolysis of Cations.* New York: Wiley-Interscience.

Baker, B., Davies, V. R., and Yarnell, P. A. (1977). Use of acrylic strong base anion resins in treatment of organic bearing waters. 38th Annual International Water Conference, Pittsburgh, PA.

Bellen, G. E., Anderson, M., and Gottler, R. A. (1986). *Point-of-Use Treatment of Control Organic and Inorganic Contaminant in Drinking Water.* Cincinnati: USEPA.

Benjamin, M. M., and Sletten, R. S. (1996). Sorption and filtration of metals using iron-oxide-coated sand. *Water Research* 30:2609–2620.

Boari, G., Liberti, L., Merli, C., and Passino, R. (1974). Exchange equilibria on anion resins. *Desalination* 15:145–166.

Boegel, J. V., and Clifford D. A. (1986). *Selenium Oxidation and Removal by Ion Exchange.* Cincinnati, OH: USEPA.

Boodoo, F., Schreiber, G., Satchell, T., Benton, L., Szczesmy, B., Woo, E., and Mielke, D. (2008). Simultaneous ion exchange removal of arsenic, nitrate, uranium and TOC at the City of McCook. In *Proceedings of the AWWA Inorganic Contaminants Workshop.* Denver, CO: American Water Works Association.

Bowers, A. E. (1980). Ion exchange softening. In *The Quest for Pure Water.* Denver, CO: American Water Works Association.

Boyer, T. H., and Singer, P. C. (2006). A pilot-scale evaluation of magnetic ion exchange treatment for removal of natural organic material and inorganic anions. *Water Research* 40:2865–2876.

Boyd, G., Adamson, A. W., and Meyers, L. S., Jr. (1947). The exchange adsorption of ions from aqueous solutions by organic zeolites, part II. *Journal of the American Chemical Society* 69:2836.

Brandhuber, P. (2008). Arsenic removal by ion exchange in high silica groundwater in Alamosa, Colorado. In *Proceedings of the AWWA Annual Conference and Exposition, Chicago.*

Brattebo, H., Odegaard, H., and Halle, O. (1987). Ion exchange for the removal of humic acids in water treatment. *Water Research* 21(9):1045.

Brooks, C. S., Brooks, P. L., Hansen, G., and McCarthy, L. A. (1991). *Metal Recovery from Industrial Waste.* Chelsea, MI: Lewis Publishers.

Calgon Carbon (2009). Continuous Ion Exchange Equipment; available at www.calgoncarbon.com/ion_exchange/equipment.html.

Calmon, C. (1979). Specific ion exchangers. In *Ion Exchange for Pollution Control,* C. Calmon and H. Gold, eds. Boca Raton, FL: CRC Press.

Chen, A., and Wang, L (2010) Regeneration of full-scale adsorptive media system—update. Seventh Annual Workshop on Small Drinking Water Systems, USEPA, Cincinnati, OH.

Chen, A. S. C., Condit, W., Wang, L., and Wang, A. (2008). *Arsenic Removal from Drinking Water by Adsorptive Media U.S. EPA Demonstration Project at Brown City, MI. Final Performance Evaluation Report.* Cincinnati, OH: USEPA.

Clifford, D. A. (1982). Multicomponent ion exchange calculations for selected ion separations, *Industrial and Engineering Chemistry—Fundamentals* 21(2):141–153.

Clifford, D. A. (1990). Removal of radium from drinking water. In *Radon, Uranium, and Radium in Drinking Water*. Chelsea, MI: Lewis Publishers.

Clifford, D. A. (1991). Chromatographic peaking of toxic contaminants during water treatment by ion exchange. In *Proceedings of the International Conference on Ion Exchange.* Tokyo: Japan Association of Ion Exchange.

Clifford, D. A. (1995). Computer prediction of ion exchange, Aqualink column. *Journal AWWA* 85(4):20.

Clifford, D. A., and Lin, C. C. (1986). *Arsenic Removal from Groundwater in Hanford, California—A Summary Report.* Houston, TX: University of Houston.

Clifford, D. A., and Lin, C. C. (1991) *Arsenic(III) and Arsenic(V) Removal from Drinking Water in San Ysidro, New Mexico.* Cincinnati, OH: USEPA.

Clifford, D. A., and Liu, X. (1993a). Biological denitrification of spent regenerant brine using a sequencing batch reactor. *Water Research* 27(9):1477–1484.

Clifford, D. A., and Liu, X. (1993b). Nitrate removal using ion exchange with batch denitrification of spent regenerant brine. *Journal AWWA* 85(4):135–143.

Clifford, D. A., and Weber, W. J., Jr. (1978). *Nitrate Removal from Water Supplies by Ion Exchange.* Cincinnati, OH: USEPA.

Clifford, D. A., and Weber, W. J., Jr. (1983). The determinants of divalent/monovalent selectivity in anion exchangers. *Reactive Polymers* 1:77–81.

Clifford, D. A., and Zhang, Z (1994). Combined uranium and radium removal by ion exchange. *Journal AWWA* 86(4):214–227.

Clifford, D. A., and Zhang, Z. (1995). Removing uranium and radium from groundwater by ion exchange resins. In *Ion Exchange Technology: Recent Advances in Pollution Contro*, A. K. Sengupta, ed. Lancaster, PA: Technomic Publishing Company.

Clifford, D. A., Ghurye, G., and Tripp, A. R. (2003). Arsenic removal from drinking water using ion-exchange with spent brine recycling. *Journal AWWA* 35(6):119–130.

Clifford, D. A., Vijjeswarapu, W., and Subramonian, S. (1988). Evaluating various adsorbents and membranes for removing radium from groundwater. *Journal AWWA* 80(7):94–104.

Clifford, D. A., Lin, C. C., Horng, L. L., and Boegel, J. (1987). *Nitrate Removal from Drinking Water in Glendale, Arizona*. Cincinnati, OH: USEPA.

Clifford, D. A., Ghurye, G., Tripp, A. R., and Tong, J. (1997). *Final Report: Phases 1 and 2 City of Albuquerque Arsenic Study. Field Studies on Arsenic Removal in Albuquerque, New Mexico, Using the University of Houston/USEPA Mobile Drinking Water Treatment Research Facility*. Houston, TX: University of Houston.

Clifford, D. A., Ghurye, G. Tripp, A. R., and Tong, J. (1998). *Final Report: Phase 3 City of Albuquerque Arsenic Study. Field Studies on Arsenic Removal in Albuquerque, New Mexico, Using the University of Houston/USEPA Mobile Drinking Water Treatment Research Facility.* Houston, TX: University of Houston.

Coogan, G. J. (1968). Color removal from surface waters by use of resins-development report. *Journal of the New England Water Works Association* 82(1):1–4.

Cothern, C. R., and Lappenbusch, W. L. (1983). Occurrence of uranium in drinking water in the United States. *Heath Physics* 45(1):89–100.

Cothern, C. R., and Lappenbush, W. L. (1984). Compliance data for the occurrence of radium and gross alpha particle activity in drinking water supplies in the United States. *Health Physics* 46(3):503.

Cumbal, L., and SenGupta, A. K. (2005). Arsenic removal using polymer-supported hydrated iron(III) oxide nanoparticles: Role of donnan membrane effect. *Environmental Science and Technology* 39:6508–6515.

Driehaus, W., Jekel, M., and Hildebrandt, U. (1998). Granular ferric hyroxide: A new adsorbent for the removal of arsenic from natural water. *Journal of Water Supply: Research and Technology—Aqua* 47(1):30–35.

Fink, G. J., and Lindsay, F. K. (1936). Activated alumina for removing fluorides from drinking water. *Industrial and Engineering Chemistry* 28(8):947–948.

Frank, P., and Clifford, D. A. (1986). *Arsenic(III) Oxidation and Removal from Drinking Water.* Cincinnati, OH: USEPA.

Frey, M. M., Edwards, M. A., Amy, G., Owen, D. M., and Chowdhury, Z. K. (1997). *National Compliance Assessment and Costs for the Regulation of Arsenic in Drinking Water.* Denver, CO: American Water Works Association.

Frisch, N. W., and Kunin, R. (1957). Long term operating characteristics of anion exchange resins. *Industrial and Engineering Chemistry* 49(9):1365–1372.

Fu, P. L.-K., and Symons, J. M. (1990). Removing aquatic organic substances by anion exchange resins. *Journal AWWA* 82(10):70–77.

Garg, D., and Clifford, D. A. (1992). *Removing Radium from Water by Plain and Treated Activated Alumina.* Cincinnati, OH: USEPA.

Ghurye, G. L. (2003). Arsenic removal from drinking water using anion exchange and iron coagulation–microfiltration, Ph.D. dissertation, Department of Civil and Environmental Engineering, University of Houston, Houston, TX.

Ghurye, G., and Clifford, D. (2001). *Laboratory Study on the Oxidation of Arsenic III to Arsenic V.* Cincinnati, OH: USEPA.

Ghurye, G., and Clifford, D. (2004). As III oxidation using chemical and solid-phase oxidants. *Journal AWWA* 96(1):84–96.

Ghurye, G. L., and Clifford, D. A. (2008). Modeling ion exchange processes for multicontaminant removal. In *Proceedings of the American Water Works Association Annual Conference.* Denver, CO: American Water Works Association.

Ghurye, G. L., Clifford, D. A. and Tripp, A. R. (1999). Combined arsenic and nitrate removal by ion exchange. *Journal AWWA* 91(10):85–96.

Guter, G. A. (1998). *IX Windows Program.* Bakersfield, CA: Cathedral Peak Software.

Guter, G. A. (1982). *Removal of Nitrate from Contaminated Water Supplies for Public Use.* Cincinnati, OH: USEPA.

Guter, G. A. (1995). Nitrate removal from contaminated groundwater by anion exchange. In *Ion Exchange Technology*, A. K. Sengupta, ed. Lancaster, PA: Technomic Publishing Company.

Hatch, M. J. (1984). *Resin Particulates Capable of Removing Metal Ions from Aqueous Solution.* Canadian Patent No. 117b 799.

Hathaway, S., and Rubel, F., Jr (1987). Removing arsenic from drinking water. *Journal AWWA* 79(8):61–65.

Hathaway, S. W. (1983). Uranium removal process. In *Proceedings of the American Water Works Association Annual Conference.* Denver, CO: American Water Works Association.

Helfferich, F. (1965). Ion exchange kinetics V: Ion exchange accompanied by reactions. *Journal of Physical Chemistry* 69:1178–1187.

Helfferich, F. (1966). Ion exchange kinetics. IN *Ion Exchange: A Series of Advances*, J. A. Marinsky, ed. New York: Marcel Dekker.

Helfferich, F., and Klein, G. (1970). *Multicomponent Chromography: Theory of Interference.* New York: Marcel Dekker.

Helfferich, F. G. (1962). *Ion Exchange.* New York: McGraw-Hill.

Horng, L. L. (1983). Reaction mechanisms and chromatographic behavior of polyprotic acid anions in multicomponent ion exchange, Ph.D. dissertation, Department of Civil and Environmental Engineering, University of Houston, Houston, TX.

Horng, L. L., and Clifford, D. A. (1997). The behavior of polyprotic anions in ion exchange resins. *Reactive and Functional Polymers* 35(1–2):41–54.

Illinois EPA (2003). Radium Compliance Deadline Passes; available at www.epa.state.il.us/environmental-progress/v28/n5/radium-deadline.html.

Janauer, G. E., and Gerba, C. P. (1981). Insoluble polymer contact disinfectants: An alternative approach to water disinfection. In *Chemistry in Water Reuse*, W. J. Cooper, ed. Ann Arbor, MI: Ann Arbor Science.

Kim, K., and Logan, B. E. (2001). Microbial reduction of perchlorate in pure and mixed culture packed-bed bioreactors. *Water Research* 35(13):3071–3076.

Kim, P. H.-S., and Symons, J. M. (1991). Using anion exchange resins to remove THM precursors. *Journal AWWA* 83(12):61–68.

Kolle, W. (1984). *Humic Acid Removal with Macroreticular Ion Exchange Resins at Hannover.* Washington, DC: USEPA.

Kunin, R. (1958). *Ion Exchange Resins*, 2nd ed. New York: Wiley.

Lakshmanan, D., Clifford, D. A., and Samanta, G. (2008). Arsenic removal by coagulation with aluminum, iron, titanium, and zirconium. *Journal AWWA* 100(2):76–87.

Langmuir, D. (1978). Uranium solution–mineral equilibria at low temperatures with applications to sedimentary ore deposits. *Geochemica et Cosmochimica* 42:547–5612.

Lauch, R. P., and Guter, G. A. (1986). Ion exchange for removal of nitrate from well water. *Journal AWWA* 78(5):83–88.

Lee, S. Y., and Bondietti, E. A (1983). Removing uranium from drinking water by metal hydroxides and anion-exchange resins. *Journal AWWA* 75(10):536–540.

Lehman, S. G., Badruzzaman, M., Adham, S., Roberts, D. J., and Clifford, D. A. (2007). Perchlorate and nitrate treatment by ion exchange integrated with biological brine treatment. *Water Research* 42:969–976.

Lehman, S. G., Badruzzaman, M., Adham, S. S., Roberts, D. J., Clifford, D. A., Zuo, G., Patel, A., Rittmann, B. E., Ahn, C., and VanGinkle, S. (2009). *Biological Destruction of Perchlorate and Nitrate in Ion Exchange Concentrate.* AWWA RF Project 3137. Denver, CO: Water Research Foundation.

Liu, X., Roberts, D. J., Hiremath, T., Clifford, D. A., Gilloghly, T., and Lehman, S. (2007). Divalent cation addition (Ca^{2+} or Mg^{2+}) stabilizes biological treatment of perchlorate and nitrate in ion exchange brine. *Environmental Engineering Science* 24(6):1–11.

Lehman S. G., Badruzzaman, M., Adham, S., Roberts D. J., and Clifford, D. A. (2008). Perchlorate and nitrate treatment by ion exchange integrated with biological brine treatment. *Water Research* 42(4–5):969–976.

Liu, C. X., and Clifford, D. A. (1996). Ion exchange with denitrified brine reuse. *Journal AWWA* 88(11):88–99.

Lowry, J. D., and Lowry, S. B. (2001). *Oxidation of Arsenic(III) by Aeration and Storage.* Cincinnati, OH: USEPA.

Lytle, D. A., Chen, A. S. C., Sorg, T. J., Phillips, S. and French, K. (2007). Microbial As(III) oxidation in water treatment plant filters. *Journal AWWA* 99(12):72–86.

Maier, F. J. (1953). Defouridation of municipal water supplies. *Journal AWWA* 45(8):8712.

Mangelson, K. A. (1988). *Radium Removal for a Small Community Water Supply System.* Cincinnati, OH: USEPA.

Matejka, Z., and Zirkova, Z. (1997). The sorption of heavy-metal cations from EDTA complexes on acrylamide resins having oligo(ethyleneamine) moieties. *Reactive and Functional Polymers* 35(1–2):81–88.

McCall, S. E., Chen, A. S. C., and Wang, L. (2009). *Arsenic Removal from Drinking Water by Adsorptive Media U.S. EPA Demonstration Project at Goffstown, NH. Final Performance Evaluation Report.* Cincinnati, OH: USEPA.

McGarvey, F. X., and Reents, C. A. (1954). Get rid of fouling in ion exchangers. *Chemical Engineer* 61:205.

McNeill, L. S., and Edwards, M. (1995). Soluble arsenic removal at water treatment plants. *Journal AWWA* 87(4):105–113.

Meng, X., Bang, S., and Korfiatis, G. P. (2000). Effects of silicate, sulfate, and carbonate on arsenic removal by ferric chloride. *Water Research* 34(4):1255–1261.

Meng, X., Korfiatis, G. P., Bang, S., and Bang, K. W. (2002). Combined effects of anions on arsenic removal by ion hydroxides. *Toxicology Letters* 133(1):103–111.

Min, J. H., Tasser, C., Zhang, J., Haileselassie, H., Boulos, L., and Crozes, G. (2007). *Unintentional pH Variation During Arsenic Removal.* Awwa Research Foundation Report. Denver, CO: Water Research Foundation.

Misra, C. (1986). *Industrial Alumina Chemicals.* Washington, DC: American Chemical Society.

Mo, T. J. (1980). *Chemistry of Uranium in Aqueous Environments.* Washington, DC: USEPA Office of Radiation Programs, Criteria and Standards Division, Radioactive Waste Standard Branch.

Moller, T., and Sylvester, P. (2007). Effect of silica and pH on arsenic uptake by resin/iron hybrid media. *Water Research* 42(11):1760–1766.

Myers, A. G., Snoeyink, V. L., and Snyder, D. W. (1985). Removing barium and radium through calcium cation exchange. *Journal AWWA* 77(5):60–66.

National Research Council (1977). *Arsenic, Drinking Water and Health.* Washington, DC: National Academy of Science.

National Academy of Sciences (2006). *Fluoride in Drinking Water: A Scientific Review of EPA's Standards.* Washington, DC: National Academy Press (www.NAP.edu).

NSF International (2005). *Removal of Arsenic from Drinking Water with ADI Media G2: Environmental Technology Verification Report.* Cincinnati, OH: USEPA.

Patel, A, Zuo, G., Lehman, G. S., Baddruzzaman, M., Clifford, D. A., and Roberts, D. J. (2008). Fluidized bed reactor for the biological treatment of ion exchange brine containing perchlorate and nitrate. *Water Research* 42:4291–98.

Pontius, F. W. (1995). An update of the federal drinking water regulations. *Journal AWWA* 87(2):48–58.

Purolite, Inc. (2009). Chelation Resins; available at www.purolite.com/RelId/606267/ISvars/default/Chelation_Resins.htm.

Purolite, Inc. (2009). Shallow Shell Technology Resins; available at http://www.purolite.com/SST

Pyles, B. (2009). Use of a magnetic ion-exchange resin for TOC removal to enable compliance with the EPA Disinfection Byproduct Rule. NEWWA Spring Regional Joint Conference, April, 1, 2009; available at www.newwa.org/PDF/AprConf09Sess2-900.pdf.

Rohm and Haas (1994). *Amberlite Resin Product Descriptions*. Philadelphia: Rohm and Haas Co. (www.rohmhaas.com/wcm/index.page).

Robertson, F. N. (1975). Hexavalent chromium in the groundwater in Paradise Valley, Arizona. *Groundwater* 13(6):516–527.

Rosenblum, E. R., and Clifford, D. A. (1984). *The Equilibrium Arsenic Capacity of Activated Alumina*. Cincinnati, OH: USEPA.

Rubel, F., Jr., and Hathaway, S. W. (1985). *Pilot Study for the Removal of Arsenic from Drinking Water at the Fallon, Nevada, Naval Air Station*. Cincinnati, OH: USEPA.

Rubel, F., Jr., and Woosley, R. D. (1979). The removal of excess fluoride from drinking water by activated alumina. *Journal AWWA* 71(1):45.

Schindler, P. W. (1981). Surface complexes at oxide-water interfaces. In *Adsorption of Inorganics at Solid–Liquid Interfaces*, M. A. Anderson and A. J. Rubin, eds. Ann Arbor, MI: Ann Arbor Science.

Schmitt, G. L., and Pietrzyk, D. J. (1985). Liquid chromatographic separation of inorganic anions on an alumina column. *Annual Chemistry* 57:2247.

Scott, K. N., Green, J. F., Do, H. D., and Mclean, S. J. (1995). Arsenic removal by coagulation. *Journal AWWA* 87(4):114–126.

Seidel, C., Carr, M., Summers, R. S., Landi, N., Zachman, B., and Frey, M. (2004). The MIEX process: A technology for removing inorganic contaminants. In *Proceedings of the American Water Works Association Annual Conference, Orlando, FL*. Denver, CO: American Water Works Association.

Sengupta, A. K., and Clifford, D. A. (1986). Chromate ion-exchange mechanism for cooling water. *Industrial Engineering and Chemical Fundamentals* 25(2):249–258.

Simms, J., and Azizian, F. (1997). Pilot-plant trials on removal of arsenic from potable water using activated alumina. In *Proceedings of the AWWA Water Quality Technology Conference*. Denver, CO: American Water Works Association.

Singh, G., and Clifford, D. A. (1981). *The Equilibrium Fluoride Capacity of Activated Alumina*. Cincinnati, OH: USEPA.

Snoeyink, V. L., Cairns-Chambers, C., and Pfeffer, J. L. (1987). Strong-acid ion exchange for removing barium, radium, and hardness. *Journal AWWA* 79(8):66–72.

Sorg, T. J. (1988). Methods for removing uranium from drinking water. *Journal AWWA* 80(7):105–111.

Stumm, W. (1992). *Chemistry of the Solid–Water Interface*. New York: Wiley.

Subramonian, S., and Clifford, D. A. (1988). Monovalent/divalent selectivity and the charge separation concept. *Reactive Polymers* 9:195–2012.

Subramonian, S., Clifford, D. A., and Vijjeswarapu, W. (1990). Evaluating ion exchange for the removal of radium from groundwater. *Journal AWWA* 82(5):61–70.

Sybron, I. (1995). *Ionac Resin Prodects Descriptions*. Philadelphia, PA: Sybron, Inc.

Tirupanangadu, M. (1996). A Visual Basic application for multicomponent chromatography in ion exchange columns. M.S. thesis, Environmental Engineering Program, Universtiy of Houston, Houston, TX.

Tong, J. (1997). Development of an iron(III) coagulation-microfiltration process for arsenic removal from groundwater. M.S. thesis, Environmental Engineering Program, University of Houston, Houston, TX.

Tripp, A. R. (2001). Selectivity considerations in modeling the treatment of perchlorate using ion exchange processes. Ph.D. dissertation, Department of Civil and Environmental Engineering, University of Houston, Houston, TX.

Tripp, A. R., and Clifford, D. A. (2004). Selectivity considerations in modeling the treatment of perchlorate using ion-exchange processes. In *Ion Exchange and Solvent Extraction*, Chap. 6, Vol. 16, A. K. SenGupta and Yizhak Markus, eds. New York: Marcel Dekker.

Tripp, A. R., and Clifford, D. A. (2006). Ion exchange for the remediation of perchlorate in drinking water. *Journal AWWA* 98(4):105–114.

Tripp, A. R., Clifford, D. A., Roberts, D. J., Cang, Y., Aldridge, L., Gilloghly, T., and Boulos, L. (2003). *Treatment of Perchlorate in Groundwater by Ion-Exchange Technology.* Awwa Research Foundation Report. Denver, CO: Water Research Foundation.

Trussell, R. R., Trussell, A., Kreft, P., Kavanaugh, M., and Tate, C. (1980). *Selenium Removal from Groundwater Using Activated Alumina.* Cincinnati, OH: USEPA.

USEPA (1976). National Interim Primary Drinking Water Regulations. EPA-570/9-76-003. Washington, DC: USEPA.

USEPA (1979). National Secondary Drinking Water Regulations. EPA-570/9-76-000. Washington, DC: USEPA.

USEPA (1985). 40 CRF Parts 141, 142, and 143, National primary drinking water regulations; fluoride; final rule and proposed rule. *Federal Register* 50:47159-47163 (www.epa.gov/ogwdw000/contaminants/basicinformation/fluoride.html).

USEPA (1991). National primary drinking water regulations: radionuclides. Advanced notice of proposed rulemaking. *Federal Register* 56(138), CFR parts 141 and 142.

USEPA (1992). Toxicity characteristic leaching procedure, method 1311. In *Test Methods for Evaluating Solid Wastes*, Vol 1C: *Laboratory Manual—Physical/Chemical Methods*, SW-846. Washington, DC: USEPA

USEPA (1992). SOCs and IOCs final rule. *Federal Register* 56:3526.

USEPA (1996). *Method 3050B, Acid Digestion of Sediments, Sludges, and Soils.* Washington, DC: USEPA.

USEPA (2000a). National primary drinking water regulations: Radionuclides. Final rule. *Federal Register* 65(236), CFR parts 9, 141 and 142.

USEPA (2000b). *Radionuclides Notice of Data Availability Technical Support Document.* Washington, DC: USEPA Office of Groundwater and Drinking Water.

USEPA (2001). National primary drinking water regulations: Arsenic and clarification to compliance and new source contaminant monitoring. 40 CFR parts 9, 141, and 142. Final rule. *Federal Register* 66(14):6076.

USEPA (2009a). Drinking Water Contaminants: List of Contaminants and Their MCLs; available at www.epa.gov/safewater/contaminants/index.html#inorganic.

USEPA Office of Water (2006). Point-of-use and point-of-entry treatment options for small drinking water systems. EPA 815-R-06-010. Washington, DC: USEPA.

Valigore, J. M., Chen, A. S. C., and Wang, L. (2007a). *Arsenic Removal from Drinking Water by Adsorptive Media. U.S. EPA Demonstration Project at Valley Vista, AZ. Final Performance Evaluation Report.* Cincinnati, OH: USEPA.

Valigore, J. M., Chen, A. S. C., and Wang, L. (2007b). *Arsenic Removal from Drinking Water by Adsorptive Media. U.S. EPA Demonstration Project at Rimrock, AZ. Final Performance Evaluation Report.* Cincinnati, OH: USEPA.

Van der Hoek, J. P., Van der Van, P. J., and Klapwijk, A. (1988). Combined ion exchange/biological denitrification of nitrate removal from groundwater under different process conditions. *Water Research* 22(6):679–684.

Wang, L., Chen, A. S. C., and Wang, A. (2007). *Arsenic Removal from Drinking Water by Adsorptive Media. U.S. EPA Demonstration Project at Alvin, TX. Final Performance Evaluation Report.* Cincinnati, OH: USEPA.

Wilson, A. L. (1959). Organic fouling of strongly basic anion-exchange resins. *Applied Chemistry* 9:352.

Westerhoff, P. K., Benn, T. M., Chen, A. S. C., Wang, L., and Cumming, L. J. (2008). *Assessing Arsenic Removal by Metal (Hydr)Oxide Adsorptive Media Using Rapid Small Scale Column Tests*. Cincinnati, OH: USEPA.

Xie, B., Fan, M., and Van Leeuwen, J. (2007). Modeling of arsenic of arsenic(V) adsorption onto granular ferric hydroxide. *Journal AWWA* 99(11):92–102.

Youngran, J., Fan, M., Van Leeuwen, J., and Belczyk, J. F. (2007). Effect of competing solutes on arsenic(V) adsorption using iron and aluminum oxides. *Journal of Environmental Sciences* 19(8):910–919.

Zeilig, N. (2005). Nationwide demo program gather data on arsenic removal processes. *AWWA Mainstream* (September).

Zeng, H., Arashiro, M., and Giammar, D. E. (2008). Effects of water chemistry and flow rate on arsenate removal by adsorption to an iron oxide-based sorbent. *Water Research* 42(8):4629–4636.

Zhang, Z., and Clifford, D. A. (1994). Exhaustion and regeneration of resins for uranium removal. *Journal AWWA* 86(4):228–241.

CHAPTER 13
PRECIPITATION, COPRECIPITATION, AND PRECIPITATIVE SOFTENING

Stephen J. Randtke, Ph.D., P.E.
Professor
Department of Civil, Environmental & Architectural Engineering
University of Kansas
Lawrence, Kansas, United States

INTRODUCTION **13.1**
PRINCIPLES **13.2**
Precipitation 13.2
Coprecipitation 13.3
Solubility Equilibria 13.5
PRECIPITATIVE SOFTENING **13.12**
Water Hardness 13.12
Chemistry of Precipitative Softening 13.18
Process Variations 13.29
Precipitative Softening Systems 13.52
Coagulation in Precipitative Softening Processes 13.59
Process Monitoring and Control 13.60
Filtration of Lime-Softened Water 13.61
Precipitative Softening Versus Ion-Exchange and Membrane Softening 13.61
Removal of Organic Contaminants by Precipitative Softening Processes 13.63
Residuals 13.70
OTHER APPLICATIONS **13.70**
ACKNOWLEDGMENTS **13.74**
ABBREVIATIONS **13.74**
NOTATION FOR EQUATIONS **13.75**
REFERENCES **13.76**

INTRODUCTION

Precipitation and coprecipitation provide a means of removing certain dissolved contaminants from solution by transforming or transferring them to a solid phase (particles) that can then be removed from solution using processes described in other chapters of this handbook, such as coagulation, flocculation, sedimentation, flotation, and filtration. Contaminants that can be effectively removed in this manner include calcium and magnesium ions (hardness); a portion of the *natural organic matter* (NOM) that reacts with disinfectants to form *disinfection by-products* (DBPs); a limited number of trace organic constituents; and various inorganic contaminants, especially metal ions.

This chapter begins with a discussion of the principles of precipitation and coprecipitation and then describes practical applications of these principles in drinking water treatment. The major focus of this chapter is on precipitative softening, which is employed at approximately 15 percent of water treatment plants in the United States. Other applications of precipitative processes are briefly summarized. Precipitation and coprecipitation are also involved in other processes addressed in this handbook, including coagulation with metal salts, which involves precipitation of aluminum and iron hydroxides and coprecipitation of various dissolved contaminants (Chap. 8); precipitation of iron and manganese after their dissolved forms have been oxidized (Chap. 7); and scale formation, which is important in filters (Chap. 10), membrane processes (Chap. 11), and distribution systems (Chap. 20).

PRINCIPLES

Precipitation

Precipitation occurs when dissolved constituents in a supersaturated solution interact to form a solid. The dissolved constituents and solids may be ionic or nonionic, but further discussion will be limited to those that are ionic, since precipitative processes are used in water treatment applications to remove ionic contaminants. An ionic contaminant can be removed if the concentration of another (oppositely charged) ion can be increased sufficiently to precipitate the contaminant ion out of solution, forming an ionic solid. It is also possible for ionic contaminants present in concentrations below the point of supersaturation to be removed by being incorporated into a precipitate formed by other ions. This process, known as *coprecipitation,* is described below. Treatment processes that remove contaminants by precipitation or coprecipitation are referred to as precipitative processes.

Precipitation of an ionic solid is both a physical and chemical process. It is commonly viewed as occurring in three stages, nucleation, crystal growth, and aging, in which ions in a supersaturated solution join together to form three-dimensional coordination compounds that develop into a solid phase. Coordination is the binding of a central atom (the cation in an ionic solid) by molecules (known as ligands) containing free pairs of electrons. Coordination is a fundamental phenomenon involved in complex formation, precipitation of ionic solids, chemisorption of ions, and coprecipitation. The ions that react to form an ionic solid (or crystal lattice) are referred to as the *lattice ions;* and other ions incorporated into an ionic solid by coprecipitation are referred to as *impurities.*

Nucleation is the coordination of lattice ions to form clusters (nuclei) of sufficiently large size that spontaneous deposition of additional lattice ions (crystal growth) can occur, which is possible only if the solution is supersaturated with respect to the solid being formed. Nucleation can occur through random collisions of lattice ions in solution (homogeneous nucleation) or through adsorption (accumulation at an interface) and nucleus formation on the surfaces of foreign particles (heterogeneous nucleation). Homogeneous and heterogeneous nucleation both require supersaturated solutions, but a higher degree of supersaturation is required for homogeneous nucleation. For this reason, and because foreign particles are present even in relatively clear water supplies, heterogeneous nucleation is expected to occur in precipitative drinking water treatment processes. But, if a high degree of supersaturation is achieved, for example, by adding a relatively large dose of lime or alum, then homogeneous nucleation may also occur.

Nucleation typically occurs instantly in highly supersaturated solutions. In solutions that are only slightly supersaturated, an extended period of time, referred to as an *induction period,* may pass before solids visibly appear. When precipitation is not desired, for example in the concentrate stream in a membrane process, the induction period can be extended, perhaps indefinitely, by adding a chemical that interferes with nucleation or crystallization. Such chemicals, commonly referred to as *antiscalants* or *threshold inhibitors,* can adsorb onto nuclei or small crystals and interfere with adsorption of lattice ions, slowing or halting crystal growth. Solids formed in the presence of an antiscalant will contain antiscalant molecules as impurities; such impurities usually affect the properties of the solids, sometimes dramatically, and may interfere with options for their disposal, use, or reuse.

Crystallization involves (1) transport (diffusion) of lattice ions to the crystal-solution interface, (2) adsorption of lattice ions on the surface, and (3) incorporation of lattice ions into the lattice. Hence, the rate of crystallization can be diffusion controlled or interface controlled. For precipitation of $CaCO_3(s)$ in a lime softening process, the reaction may be diffusion controlled during the initial moments of the reaction, for example, in the rapid mixing unit; but the reaction soon becomes interface controlled with the rate of crystallization being proportional to the degree of supersaturation. This can be expressed mathematically as follows:

$$\frac{d(\mathrm{Ca}^{+2})}{dt} = -kS[(\{\mathrm{Ca}^{2+}\}\times\{\mathrm{CO}_3^{2-}\}) - K_{s0}] \tag{13-1}$$

where k = rate constant
S = available surface area
$\{Ca^{2+}\}$ = calcium ion activity
$\{CO_3^{2-}\}$ = carbonate ion activity
K_{s0} = solubility product constant for $CaCO_3$

Solids recirculation increases the surface area available for crystallization, favoring crystallization over nucleation. This results in the formation of larger particles, which may be of significant benefit if the solids are to be removed by sedimentation or if they will be dewatered prior to disposal or reuse.

Freshly precipitated ionic solids may be noncrystalline (amorphous) or microcrystalline (composed of crystals less than 1 or 2 μm in size), or they may consist of small crystallites in a polymorphous form. As such solids age, they tend to rearrange themselves so as to achieve a more satisfactory (stable) coordination status, forming larger, more well-formed, and less soluble crystals. Smaller, more soluble crystals may dissolve as larger crystals grow. This phenomenon is referred to as *Ostwald ripening* and results in a larger average particle size and a decrease in the total crystal surface area. For aging to occur, the crystals must remain in contact with the solution from which they were precipitated, referred to as the *mother liquor.* Some precipitates, such as calcite crystals composed of $CaCO_3(s)$, age very rapidly, whereas others, such as magnesium, iron, and aluminum hydroxides, age extremely slowly and remain amorphous under water treatment conditions.

Factors influencing contaminant removal by precipitation include the concentration of the other ion forming the solid, the solubility of the solid formed, temperature, and the presence of species able to form complexes with the contaminant ion. For example, to remove a metal ion from solution, the ligand concentration must be high enough to induce precipitation, and the higher the ligand concentration remaining in solution, the lower the metal ion concentration will be at equilibrium. The solubility of many solids increases as water temperature increases, but there are notable exceptions, including $CaCO_3(s)$. Complex formation increases the total solubility of a substance when the concentrations of the complexed species are included in the total.

Coprecipitation

Coprecipitation is the contamination of a precipitate by a substance that would otherwise have remained in solution had the precipitate not formed. Four mechanisms of coprecipitation have been identified (Skoog and West, 1969), each involving adsorption of impurities during crystal growth.

1. *Surface adsorption,* whereby impurities are not incorporated into the crystal lattice but are adsorbed on the external surfaces of the crystals, for example, due to interactions with counter ions in the electrical double layer surrounding crystals having a surface charge
2. *Isomorphic inclusion,* also known as *inclusion* and as *mixed crystal formation,* in which impurities substitute for some of the lattice ions
3. *Nonisomorphic inclusion,* or *solid solution formation,* in which impurities appear to be dissolved in the precipitate and are not part of the crystal lattice
4. *Occlusion,* which occurs when impurities are adsorbed, then entrapped by growing crystals, creating crystal imperfections as they are covered over by lattice ions

Coprecipitation may greatly retard the rate of crystallization and can significantly influence crystal morphology (shape), structure, solubility, and surface properties.

Coprecipitation differs from postprecipitation, surface precipitation, and adsorption. Postprecipitation occurs when a second, more slowly precipitating substance precipitates on the surface of a previously formed precipitate. Postprecipitation is not considered coprecipitation because the solution must be supersaturated with respect to the second precipitate for postprecipitation to occur (Kolthoff, 1932; Skoog and West, 1969). Surface precipitation occurs when a substance adsorbs on a surface and accumulates to the extent that a precipitate forms on the surface (Benjamin, 2002). If a precipitate is formed and is later added to another solution, contaminants may adsorb onto the precipitate, but this is normally considered adsorption (Chap. 14) or ion exchange (Chap. 12) rather than coprecipitation.

If two distinct solid phases are precipitated simultaneously, it is possible for some of the ions associated with one precipitate to become incorporated into the other precipitate. For example, in the excess lime softening process, both $CaCO_3$(s) and $Mg(OH)_2$(s) are formed simultaneously. A small amount of Mg^{2+} would be expected to be incorporated into the rapidly precipitating $CaCO_3$(s). Since the solution is also supersaturated with respect to the Mg^{2+} ion, this would not fall within the classic definition of coprecipitation, but it should nevertheless be considered as coprecipitation.

Factors influencing the removal of contaminants by coprecipitation include

- *The rate of precipitation.* Crystals are purer when they grow more slowly.
- *The amount of precipitate formed.* Coprecipitation involves adsorption, and for crystals of a given size a larger mass of precipitate provides a larger surface on which adsorption can take place.
- *Time.* Crystal purity increases during the aging process and a coprecipitated contaminant can diffuse out of a precipitate over time, though the rate of diffusion may be negligible, e.g., for large molecules removed by occlusion.
- *Crystal size.* Contaminants diffuse more slowly out of larger crystals, but recirculating solids to grow larger crystals is expected to reduce contaminant removal because the crystals will remain in contact with the solution longer and will grow more slowly due to the increase in available surface area. For a given amount of solids, a smaller crystal size corresponds to a larger surface area, which will enhance contaminant removal by surface adsorption.
- *Crystal structure.* Contaminants diffuse more slowly out of well-crystallized solids, whereas a contaminant adsorbed on an amorphous solid is expected to be in equilibrium with the concentration of the contaminant in solution.
- *Crystal surface potential.* As an ionic solid precipitates, it may contain an excess of anions or cations due to differences in the initial lattice ion concentrations in solution,

or it may be positively or negatively charged due to the presence of certain potential determining ions, which typically include H^+ and OH^-. Contaminants having a charge opposite to that of the ion present in excess or to that of the crystal surface are expected to be preferentially adsorbed and coprecipitated.

- *The properties of the contaminant.* Adsorption of a contaminant on a growing crystal depends on the particular contaminant species actually present, its charge and degree of hydration, its ability to coordinate with one of the lattice ions, and its size relative to the lattice ions.
- *Complex formation.* Complexation can decrease coprecipitation by holding the contaminant in solution, and it can increase coprecipitation if the complex adsorbs more strongly to the solid than the uncomplexed contaminant.
- *pH.* The pH of the solution usually affects the charge and speciation of the contaminant, the surface charge of the solid (hence its ability to adsorb ionic impurities), and the behavior of complexing agents, so coprecipitation can be strongly pH dependent.
- *Temperature and ionic strength.* These factors influence the activity of the contaminant ion and can influence its adsorption on charged surfaces as well as the action of other factors listed.

Solubility Equilibria

To determine if a contaminant can potentially be removed by precipitation (or if removal is occurring by precipitation or coprecipitation) and to better understand the factors influencing contaminant removal by precipitation or coprecipitation, it is often helpful to examine solubility equilibria relevant to the contaminant of interest, to other ions present, and to various solids that may be formed under a given set of conditions. The basic concepts and nomenclature pertaining to chemical equilibria and to precipitation and dissolution of metal solids were introduced in Chap. 3.

The dissolution reaction for a sparingly soluble ionic solid may be expressed as

$$A_xB_y(s) \rightleftarrows xA^{y+} + yB^{x-} \tag{13-2}$$

where A and B are the lattice ions and x and y are stoichiometric coefficients. This is similar in form to the dissolution reaction for metal solids shown in Chap. 3 with Eq. 3-35. The solubility product constant (K_{s0}) for the reaction shown in Eq. 13-2 is

$$K_{s0} = [A^{y+}]^x[B^{x-}]^y \tag{13-3}$$

where the square brackets indicate molar concentrations. Table 13-1 provides K_{s0} values for selected sparingly soluble ionic solids containing substances potentially relevant to water treatment or analysis.

In Eq. 13-3, activities must be used instead of molar concentrations to make accurate calculations when ionic strength effects are significant, and a temperature correction must be applied if the actual water temperature differs from the temperature at which the value of K_{s0} was determined. Corrections for ionic strength and temperature were introduced in Chap. 3 and are described in many textbooks, including those by Snoeyink and Jenkins (1980), Stumm and Morgan (1996), Butler (1998), Benjamin (2002), and Jensen (2003). Examples are included in a later section of this chapter. The Davies equation (described in Chap. 3) is valid for ionic strengths up to 0.5 M; ionic strength corrections can be made using Pitzer's specific ion interaction model (Plummer et al., 1988; Pitzer, 1991; Langmuir, 1997; Benjamin, 2002) for ionic strengths up to 1.0 M (or higher if additional terms are included).

TABLE 13-1 Selected Solubility Product Constants

Solid	Formula	K_{s0}†	Reference
Aluminum hydroxide	$Al(OH)_3$	2×10^{-32}	NIST, 2004
Aluminum phosphate	$AlPO_4$	9.84×10^{-21}	CRC Handbook, 2008
Barium arsenate	$Ba_3(AsO_4)_2$	8.0×10^{-51}	Lange's Handbook, 2005
Barium carbonate	$BaCO_3$	2.58×10^{-9}	CRC Handbook, 2008
Barium hydroxide octahydrate	$Ba(OH)_2{\cdot}8H_2O$	2.55×10^{-4}	CRC Handbook, 2008
Barium oxalate	BaC_2O_4	1.6×10^{-7}	Lange's Handbook, 2005
Barium oxalate monohydrate	$BaC_2O_4{\cdot}H_2O$	2.3×10^{-8}	Lange's Handbook, 2005
Barium sulfate	$BaSO_4$	1.08×10^{-10}	CRC Handbook, 2008
Beryllium hydroxide	$Be(OH)_2$	6.92×10^{-22}	CRC Handbook, 2008
Cadmium carbonate	$CdCO_3$	1.0×10^{-12}	CRC Handbook, 2008
Cadmium hydroxide	$Cd(OH)_2$	7.2×10^{-15}	CRC Handbook, 2008
Cadmium oxalate trihydrate	$CdC_2O_4{\cdot}3H_2O$	1.42×10^{-8}	Lange's Handbook, 2005
Cadmium sulfide	CdS	8.0×10^{-27}	Lange's Handbook, 2005
Calcium arsenate	$Ca_3(AsO_4)_2$	6.8×10^{-19}	Lange's Handbook, 2005
Calcium carbonate (calcite)	$CaCO_3$	3.31×10^{-9}	Plummer & Busenberg, 1982
Calcium carbonate (aragonite)	$CaCO_3$	4.61×10^{-9}	Plummer & Busenberg, 1982
Calcium carbonate (vaterite)	$CaCO_3$	1.22×10^{-8}	Plummer & Busenberg, 1982
Calcium carbonatomagnesium	$Ca[Mg(CO_3)_2]$	1×10^{-11}	Lange's Handbook, 2005
Calcium fluoride	CaF_2	3.45×10^{-11}	CRC Handbook, 2008
Calcium hydroxide	$Ca(OH)_2$	5.02×10^{-6}	CRC Handbook, 2008
Calcium oxalate monohydrate	$CaC_2O_4{\cdot}H_2O$	2.32×10^{-9}	CRC Handbook, 2008
Calcium phosphate	$Ca_3(PO_4)_2$	2.07×10^{-33}	CRC Handbook, 2008
Calcium selenate dihydrate	$CaSeO_4{\cdot}2H_2O$	9.5×10^{-4}	NIST, 2004
Calcium sulfate	$CaSO_4$	4.93×10^{-5}	CRC Handbook, 2008
Calcium sulfate dihydrate	$CaSO_4{\cdot}2H_2O$	3.14×10^{-5}	CRC Handbook, 2008
Chromium(III) hydroxide	$Cr(OH)_3$	6×10^{-31}	NIST, 2004
Chromium(II) hydroxide	$Cr(OH)_2$	2×10^{-16}	Lange's Handbook, 2005
Chromium(III) arsenate	$CrAsO_4$	7.7×10^{-21}	Lange's Handbook, 2005
Chromium(III) hydroxide	$Cr(OH)_3$	6.3×10^{-31}	Lange's Handbook, 2005
Cobalt(II) carbonate	$CoCO_3$	6×10^{-12}	NIST, 2004
Cobalt(II) hydroxide	$Co(OH)_2$	5.92×10^{-15}	CRC Handbook, 2008
Cobalt(II) hydroxide (amorphous)	$Co(OH)_2$	1×10^{-15}	NIST, 2004
Cobalt(III) hydroxide	$Co(OH)_3$	4×10^{-45}	NIST, 2004
Copper(I) bromide	$CuBr$	6.27×10^{-9}	CRC Handbook, 2008
Copper(I) chloride	$CuCl$	1.72×10^{-7}	CRC Handbook, 2008
Copper(I) sulfide	Cu_2S	2.5×10^{-48}	Lange's Handbook, 2005
Copper(II) arsenate	$Cu_3(AsO_4)_2$	7.95×10^{-36}	CRC Handbook, 2008
Copper(II) carbonate	$CuCO_3$	6×10^{-12}	NIST, 2004
Copper(II) hydroxide	$Cu(OH)_2$	2×10^{-19}	NIST, 2004
Copper (II) oxalate	CuC_2O_4	4.43×10^{-10}	CRC Handbook, 2008
Copper(II) phosphate	$Cu_3(PO_4)_2$	1.40×10^{-37}	CRC Handbook, 2008
Copper(II) sulfide	CuS	6.3×10^{-36}	Lange's Handbook, 2005
Iron(II) carbonate	$FeCO_3$	3.13×10^{-11}	CRC Handbook, 2008
Iron(II) hydroxide	$Fe(OH)_2$	4.87×10^{-17}	CRC Handbook, 2008
Iron(III) arsenate	$FeAsO_4$	5.7×10^{-21}	Lange's Handbook, 2005
Iron(III) hydroxide	$Fe(OH)_3$	2.79×10^{-38}	CRC Handbook, 2008
Iron(III) phosphate dihydrate	$FePO_4{\cdot}2H_2O$	9.91×10^{-16}	CRC Handbook, 2008
Iron(III) selenite	$Fe_2(SeO_3)_3$	2.0×10^{-31}	Lange's Handbook, 2005
Lead(II) arsenate	$Pb_3(AsO_4)_2$	4.0×10^{-36}	Lange's Handbook, 2005
Lead(II) carbonate	$PbCO_3$	7.40×10^{-14}	CRC Handbook, 2008

(*Continued*)

TABLE 13-1 Selected Solubility Product Constants (*Continued*)

Solid	Formula	K_{s0}†	Reference
Lead(II) fluoride	PbF_2	3.3×10^{-8}	CRC Handbook, 2008
Lead(II) hydroxide	$Pb(OH)_2$	1.43×10^{-20}	CRC Handbook, 2008
Lead(II) oxalate	PbC_2O_4	4.8×10^{-10}	Lange's Handbook, 2005
Lead(II) phosphate	$Pb_3(PO_4)_2$	8.0×10^{-43}	Lange's Handbook, 2005
Lead(II) sulfate	$PbSO_4$	2.53×10^{-8}	CRC Handbook, 2008
Lead(II) sulfide	PbS	8.0×10^{-28}	Lange's Handbook, 2005
Lead(IV) hydroxide	$Pb(OH)_4$	3.2×10^{-66}	Lange's Handbook, 2005
Magnesium ammonium phosphate	$MgNH_4PO_4$	2.5×10^{-13}	Lange's Handbook, 2005
Magnesium arsenate	$Mg_3(AsO_4)_2$	2.1×10^{-20}	Lange's Handbook, 2005
Magnesium carbonate	$MgCO_3$	6.82×10^{-6}	CRC Handbook, 2008
Magnesium carbonate trihydrate	$MgCO_3 \cdot 3H_2O$	2.38×10^{-6}	CRC Handbook, 2008
Magnesium carbonate pentahydrate	$MgCO_3 \cdot 5H_2O$	3.79×10^{-6}	CRC Handbook, 2008
Magnesium fluoride	MgF_2	5.16×10^{-11}	CRC Handbook, 2008
Magnesium hydroxide (brucite)	$Mg(OH)_2$	5.61×10^{-12}	CRC Handbook, 2008
Magnesium hydroxide (amorphous)	$Mg(OH)_2$	3.91×10^{-11}	Loewenthal & Marais, 1976
Magnesium oxalate dihydrate	$MgC_2O_4 \cdot 2H_2O$	4.83×10^{-6}	CRC Handbook, 2008
Magnesium phosphate	$Mg_3(PO_4)_2$	5.2×10^{-24}	NIST, 2004
Manganese(II) arsenate	$Mn_3(AsO_4)_2$	1.9×10^{-29}	Lange's Handbook, 2005
Manganese(II) carbonate	$MnCO_3$	2.24×10^{-11}	CRC Handbook, 2008
Manganese(II) hydroxide	$Mn(OH)_2$	1.9×10^{-13}	Lange's Handbook, 2005
Mercury(I) bromide	Hg_2Br_2	6.40×10^{-23}	CRC Handbook, 2008
Mercury(I) carbonate	Hg_2CO_3	3.6×10^{-17}	CRC Handbook, 2008
Mercury(I) chloride	Hg_2Cl_2	1.43×10^{-18}	CRC Handbook, 2008
Mercury(I) iodide	Hg_2I_2	5.2×10^{-29}	CRC Handbook, 2008
Mercury(I) sulfate	Hg_2SO_4	6.5×10^{-7}	CRC Handbook, 2008
Mercury(II) bromide	$HgBr_2$	6.2×10^{-20}	CRC Handbook, 2008
Mercury(II) hydroxide	$Hg(OH)_2$	3.2×10^{-26}	Lange's Handbook, 2005
Mercury(II) iodide	HgI_2	2.9×10^{-29}	CRC Handbook, 2008
Mercury(II) sulfide (red)	HgS	4×10^{-53}	Lange's Handbook, 2005
Mercury(II) sulfide (black)	HgS	1.6×10^{-52}	Lange's Handbook, 2005
Nickel carbonate	$NiCO_3$	1.42×10^{-7}	CRC Handbook, 2008
Nickel hydroxide	$Ni(OH)_2$	5.48×10^{-16}	CRC Handbook, 2008
Radium sulfate	$RaSO_4$	3.66×10^{-11}	Lange's Handbook, 2005
Silver bromide	AgBr	5.35×10^{-13}	CRC Handbook, 2008
Silver carbonate	Ag_2CO_3	8.46×10^{-12}	CRC Handbook, 2008
Silver chloride	AgCl	1.77×10^{-10}	CRC Handbook, 2008
Silver iodide	AgI	8.52×10^{-17}	CRC Handbook, 2008
Silver sulfide	Ag_2S	6.3×10^{-50}	Lange's Handbook, 2005
Strontium carbonate	$SrCO_3$	5.60×10^{-10}	CRC Handbook, 2008
Strontium oxalate hydrate	$SrC_2O_4 \cdot H_2O$	1.6×10^{-7}	Lange's Handbook, 2005
Strontium sulfate	$SrSO_4$	3.44×10^{-7}	CRC Handbook, 2008
Thallium(III) hydroxide	$Tl(OH)_3$	1.68×10^{-44}	CRC Handbook, 2008
Thorium hydroxide	$Th(OH)_4$	4.0×10^{-45}	Lange's Handbook, 2005
Thorium phosphate	$Th_3(PO_4)_4$	2.5×10^{-79}	Lange's Handbook, 2005
Tin(II) hydroxide	$Sn(OH)_2$	5.45×10^{-27}	CRC Handbook, 2008
Tin(II) sulfide	SnS	1×10^{-25}	Lange's Handbook, 2005
Tin(IV) hydroxide	$Sn(OH)_4$	1×10^{-56}	Lange's Handbook, 2005

(*Continued*)

TABLE 13-1 Selected Solubility Product Constants (*Continued*)

Solid	Formula	K_{s0}†	Reference
Titanium(III) hydroxide	$Ti(OH)_3$	1×10^{-40}	Lange's Handbook, 2005
Titanium(IV) oxide hydroxide	$TiO(OH)_2$	1×10^{-29}	Lange's Handbook, 2005
Vanadium(IV) hydroxide	$VO(OH)_2$	8×10^{-23}	NIST, 2004
Uranyl carbonate	UO_2CO_3	1.8×10^{-12}	Lange's Handbook, 2005
Uranyl hydroxide	$UO_2(OH)_2$	1.1×10^{-22}	Lange's Handbook, 2005
Uranyl phosphate	$(UO_2)_3(PO_4)_2$	2×10^{-47}	Lange's Handbook, 2005
Vanadium(IV) hydroxide	$VO(OH)_2$	5.9×10^{-23}	Lange's Handbook, 2005
Zinc arsenate	$Zn_3(AsO_4)_2$	2.8×10^{-28}	CRC Handbook, 2008
Zinc carbonate	$ZnCO_3$	1.46×10^{-10}	CRC Handbook, 2008
Zinc carbonate monohydrate	$ZnCO_3 \cdot H_2O$	5.42×10^{-11}	CRC Handbook, 2008
Zinc hydroxide	$Zn(OH)_2$	3×10^{-17}	CRC Handbook, 2008
Zinc oxalate dihydrate	$ZnC_2O_4 \cdot 2H_2O$	1.38×10^{-9}	CRC Handbook, 2008
Zinc phosphate	$Zn_3(PO_4)_2$	9.0×10^{-33}	Lange's Handbook, 2005
Zinc sulfide (alpha)	ZnS	1.6×10^{-24}	Lange's Handbook, 2005
Zinc sulfide (beta)	ZnS	1×10^{-21}	Lange's Handbook, 2005

†Values of K_{s0} are at 25°C and zero ionic strength, except those from *Lange's Handbook* (2005), which are for 18 to 25°C. Solubility calculations that fail to consider complexation and other factors can be highly inaccurate. Since the second ionization constant for H_2S is poorly known, it is more useful to examine sulfide solubility as described in the *CRC Handbook* (2008).

If the concentrations of the lattice ions are known, their ion product (Q), also referred to as their *reaction quotient* (Chap. 3), may be calculated as follows (ignoring ionic strength effects):

$$Q = [A^{y+}]^x [B^{x-}]^y \tag{13-4}$$

If $Q > K_{s0}$, the solution is supersaturated with respect to $A_xB_y(s)$ and precipitation can occur, provided that any assumptions made are correct and that all appropriate corrections have been applied. Similarly, if $Q < K_{s0}$, precipitation of $A_xB_y(s)$ is not possible. If $Q = K_{s0}$, the solution is in equilibrium with the solid. These same relationships may also be expressed using the saturation ratio (SR), the saturation index (SI), or the free energy of reaction (ΔG^0_{rxn}), as illustrated in Table 13-2. The Langelier saturation index (LSI) is a special case of the saturation index that applies specifically to precipitation of $CaCO_3(s)$. It is widely used in the field of water supply and treatment and is described in Chap. 20.

TABLE 13-2 Relationships Among Parameters Used to Indicate the Saturation State of a Solution

Parameter	Equivalent expression	For a solution that is		
		Supersaturated	At equilibrium	Undersaturated
Q	See Eq. 13-4	$Q > K_{s0}$	$Q = K_{s0}$	$Q < K_{s0}$
SR	Q/K_{s0}	$SR > 1$	$SR = 1$	$SR < 1$
SI	$\log (Q/K_{s0})$	$SI > 0$	$SI = 0$	$SI < 0$
ΔG^0_{rxn}	$2.3RT \times SI$†	$\Delta G^0_{rxn} > 0$	$\Delta G^0_{rxn} = 0$	$\Delta G^0_{rxn} < 0$

†R is the gas constant (1.987 cal/mol-K), T is absolute temperature in degrees kelvin, and the reaction is written as a dissolution reaction as shown in Eq. 13-2.

It is important to recognize that the solubility of a substance in pure water may differ greatly from its solubility in water containing other dissolved substances. Some handbooks provide information on the solubility of various substances in pure solutions, and such information can be quite helpful when preparing reagents in pure water in the laboratory or when dissolving relatively soluble materials in water to prepare chemical feeding solutions. However, when a sparingly soluble substance is added to water containing dissolved substances, its dissolution may be hindered or promoted by ions already present. For example, if $CaSO_4$(s) is added to water already containing 1,000 mg/L of sodium sulfate, much less $CaSO_4$(s) will dissolve into the water than indicated in Example 3-3. This is expected on the basis of LeChâtelier's principle and is referred to, in the context of precipitation and dissolution, as the *common-ion effect*—the suppression of the solubility of one salt by another when they share a common ion. At equilibrium, the ion product Q will still equal K_{s0} for $CaSO_4$(s), but the Ca^{2+} concentration will be relatively low because dissolution of $CaSO_4$(s) will be limited by the high sulfate concentration.

A contrasting example is the dissolution of $CaCO_3$(s) in water containing a small amount of an acid, for example, ground water containing CO_2. More $CaCO_3$(s) will dissolve into this water than into pure water because the acid will react with carbonate ions, CO_3^{2-}, converting them to bicarbonate ions, HCO_3^-. Thus, additional $CaCO_3$(s) must dissolve before the value of Q equals K_{s0} for $CaCO_3$(s).

In the field of water quality and treatment, questions frequently arise concerning the solubility of a contaminant ion in a water having a given pH, alkalinity, and ionic composition. If the solid limiting the solubility of the contaminant ion is known, the maximum concentration of the contaminant ion can often be estimated as illustrated in Example 13-1.

Example 13-1 Residual Mg^{2+} in Lime-Treated Water
Estimate the residual Mg^{2+} concentration in lime-treated water assuming it is limited by the solubility of magnesium hydroxide, $Mg(OH)_2$(s), and that the pH is 11.0, the temperature is 25°C, and ionic strength effects can be ignored.

Solution

1. Write the appropriate chemical reaction.

$$Mg(OH)_2(s) \rightleftarrows Mg^{2+} + 2OH^-$$

$Mg(OH)_2$(s) produced in a lime softening process will be amorphous; thus, from Table 13-1, K_{s0} for this reaction is 3.91×10^{-11}.

2. Determine the hydroxide ion concentration using the expression for K_w (Chap. 3).

$$K_w = [H^+]\,[OH^-] = 10^{-14} \text{ at } 25°C$$

Since

$$[H^+] = 10^{-pH} = 10^{-11} \text{ mol/L}$$

the concentration of OH^- is

$$[OH^-] = K_w/[H^+] = 10^{-14}/10^{-11} = 10^{-3} \text{ mol/L}$$

3. Establish the solubility product constant expression and solve for the Mg^{2+} concentration.

$$K_{s0} = [Mg^{2+}]\,[OH^-]^2 = 3.91 \times 10^{-11}$$

Rearranging and solving for $[Mg^{2+}]$,

$$[Mg^{2+}] = K_{s0}/[OH^-]^2 = \frac{3.91\times10^{-11}}{(10^{-3})^2} = 3.91 \times 10^{-5} \text{ mol/L (or 0.95 mg/L)}$$

Since hardness ion concentrations are frequently expressed as $CaCO_3$, multiply the concentration in mg/L by the ratio of the equivalent weights of $CaCO_3$ and Mg^{2+}.

$$Mg^{2+} = 0.95 \text{ mg/L} \times \frac{50 \text{ mg } CaCO_3/\text{eq}}{12.15 \text{ mg Mg/eq}} = 3.9 \text{ mg/L as } CaCO_3 \quad \blacktriangle$$

A hypothetical problem, such as Example 13-1, may be straightforward and have an exact solution. Problems encountered in practice may be complex, perhaps more so than those addressing them realize, and simplifying assumptions are often made to facilitate calculations. Such calculations may provide useful quantitative information, but they are subject to a number of limitations and can produce estimates that are inaccurate or even grossly in error. Potential sources of error include

- *Temperature and ionic strength effects.* These effects are commonly ignored when making preliminary estimates, but they can significantly influence contaminant solubility. Failing to apply appropriate corrections may make little difference at low ionic strength or at a temperature close to that at which the reported value of K_{s0} was determined, but can make a great difference when dealing with solutions of higher ionic strength, multivalent contaminants, or solids whose solubility is strongly affected by temperature. The best way to determine whether such effects are significant is to correct for them to see what difference the corrections make. However, most corrections are also approximate, in that they are usually based on correlations of measured values having limited accuracy, and they are sometimes extended, knowingly or unknowingly, beyond the range of ionic strength or temperature for which they were developed.
- *Incorrect assumptions regarding contaminant speciation.* For a particular solid, the value of K_{s0} is valid only for the specific ions forming the solid, but a contaminant may be present in more than one form and many analytical methods measure only the total amount present rather than the individual species. For example, mercury may be present as a free ion (Hg^{1+} or Hg^{2+}); as an inorganic complex, with $HgCl_2^0$ being the dominant dissolved form in most freshwaters (Sawyer et al., 2003); as elemental mercury, for example, if it leaked out of an obsolete pump seal; complexed with natural organic matter; adsorbed on particles; coprecipitated with various solids; or as a component of one or more mercury containing solids. Furthermore, speciation can change during treatment in response to a change in pH, oxidant concentration, temperature, or the concentrations and properties of other dissolved species and solids. If the speciation of the contaminant is not known, the limits of its solubility can be hypothetically estimated or experimentally measured, but they cannot usually be calculated with a reasonable degree of certainty.
- *Complex formation.* Complex formation increases the total concentration of a contaminant that can be held in solution in equilibrium with a particular solid, and it can also increase removal of a contaminant by coprecipitation if the complex is more readily removed than the free ion. Anions commonly present in water supplies that can complex metal ions include carbonate and bicarbonate, hydroxide, sulfate, and as noted above for mercury, chloride and NOM. The solubility of lead in tap water can be significantly influenced by formation of the carbonato-lead(II) complex, $PbCO_3^0$ (Chap. 20). Trivalent aluminum and iron form relatively insoluble hydroxides when used as coagulants (Chap. 8), but at higher pH values their solubility can increase significantly due to tetrahydroxo complex formation, as shown in Figs. 8-8 and 8-9. However, aluminum concentrations as high as those shown in Fig. 8-8 are generally not observed in practice when alum is

added as a coagulant in the lime softening process, presumably because one or more aluminum-containing solids of unknown composition is formed under these conditions. Sulfate interacts with Ca^{2+} to form a sulfato-calcium(II) complex, $CaSO_4^0$, an ion pair that can significantly influence the solubility of $CaSO_4$(s) and $CaCO_3$(s). For example, the solubility of $CaSO_4 \cdot 2H_2O$(s) at zero ionic strength was calculated in Example 3-3 at 694 mg/L (total of calcium and sulfate), but increases to about 1,500 mg/L when the $CaSO_4^0$ complex is taken into consideration. Rules for naming complexes can be found in many water chemistry textbooks, such as those by Snoeyink and Jenkins (1980) and Sawyer et al. (2003).

- *The presence of threshold inhibitors (antiscalants).* These substances may be present naturally (e.g., certain types of natural organic matter) or added intentionally (e.g., polyphosphates and other commercially available antiscalants). In either case, they can retard or prevent precipitation and, if precipitation does occur, they can greatly alter the properties of the precipitate, including its solubility. Calculations based on the assumption that these compounds are absent may not be valid when they are actually present.
- *Incorrect assumptions regarding the solid phase.* If the solid phase actually present differs in composition from the solid assumed to be present, the calculation will be void and will need to be redone using a K_{s0} value corresponding to the correct solid. Solubility can also be significantly influenced by crystal morphology and size, and by the presence of impurities. At least six different crystal forms of $CaCO_3$(s) are known, each having a different solubility. It is also possible for two or more solids to precipitate simultaneously, or for a metastable phase to form. A metastable phase is less thermodynamically stable and more soluble than the phase that will exist at equilibrium, but it may form more rapidly and persist for an indefinite period of time before the more stable phase forms.
- *Failure to reach equilibrium.* Precipitation can occur when $Q > K_{s0}$, but a solution can remain supersaturated for an extended period of time. Once precipitation commences, the system may approach equilibrium very slowly, especially if an amorphous or metastable phase precipitates first and is then slowly transformed into a more stable phase. Precipitation and dissolution involve the physical transfer of matter (ions) from one phase to another, which can also proceed slowly. For example, if water is left standing in a lead service line, it can take hours or days for the lead concentration in solution to reach equilibrium with the lead pipe (Chap. 20), especially if the lead ions must pass through a layer of scale or a biofilm. If other reactions occur simultaneously, such as oxygen consumption and pH changes (commonly associated with both corrosion reactions and biofilm growth), it may take far longer to reach equilibrium.
- *An inaccurate solubility product constant.* Experimentally determined K_{s0} values vary. Some values have been accurately determined and then verified by later investigators, whereas other values may be slightly in error. The most authoritative sources of K_{s0} values, for example, NIST (2004), are those whose authors have critically reviewed the available data and selected one or more values judged to have been accurately determined.

The solubility behavior of most sparingly soluble ionic solids in natural and treated waters is complicated due to numerous competing solubility and acid-base equilibria, complex formation, and other factors including those already noted. Simple calculations can provide much useful information. But, when accurate results and a deeper understanding of contaminant behavior are needed, many investigators find it helpful to conduct empirical studies or to employ a computer-based chemical equilibrium model able to simultaneously address a large number (perhaps hundreds or thousands) of dissolved species and numerous solid phases. Commonly used models include MINEQL+, an updated version of MINEQL (Westall et al., 1976); MINTEQA2 (Allison et al., 1991), a revised version of MINTEQ, which combined MINEQL with WATEQ (Truesdell and Jones, 1973 and 1974); Visual

MINTEQ (Gustafsson, 2009); and PHREEQC, a program based on the work of Parkhurst (1992) and Plummer et al. (1988) that includes redox equilibria and the ability to handle high-ionic-strength environments. Cogley (1998) and Jensen (2003) have summarized the capabilities and use of a number of computer-based chemical equilibrium models. Some suppliers of antiscalants also provide software programs that estimate the dose required to inhibit precipitation of a given salt under a specific set of conditions.

PRECIPITATIVE SOFTENING

Precipitative softening, commonly referred to as *lime softening* (since lime is the chemical most commonly used for this purpose), is primarily used to remove calcium and magnesium hardness from water supplies. It also removes other contaminants commonly found in hard waters including NOM, iron and manganese, and various other inorganic contaminants, especially metal ions. Precipitative softening processes typically include coagulation, flocculation, and sedimentation, in which case they also remove particles, as described in Chaps. 8 and 9. Softening units operated as crystallizers without coagulant addition do not remove particles but can be followed by coagulation and filtration if particle removal is desired.

Water Hardness

Water is considered hard if it forms scale, especially on heating, if it makes soap difficult to lather, if it precipitates soap (forming curdles or soap scum), or if it requires extra detergent for proper cleaning action. Water hardness is caused by dissolved divalent metal cations. In natural waters, Ca^{2+} and Mg^{2+} are the predominant cations causing hardness, since most other divalent cations are typically present at concentrations below 1 mg/L. Trivalent cations can also contribute to water hardness, but their concentrations in drinking water supplies are usually negligible. For analytical purposes, total hardness (TH) is defined as the sum of dissolved Ca^{2+} and Mg^{2+} (APHA, AWWA, and WEF, 2005). Other hardness-causing cations are either ignored or are measured and reported individually.

Units of Expression. When measured separately, Ca^{2+} and Mg^{2+} concentrations are usually reported in units of mg/L (as Ca or Mg) or mg/L as $CaCO_3$. Total hardness, whether determined by direct measurement (titration) or by summing the individually measured concentrations of calcium and magnesium, is usually reported in mg/L as $CaCO_3$. It is not useful or appropriate to report total hardness in mg/L because the atomic weight of Ca differs from that of Mg.

Hardness can also be reported in units of milliequivalents per liter (meq/L) or millimoles per liter (mM). Units of meq/L are most often used when determining the types and concentrations of hardness present; and units of mM are commonly used when calculating chemical dosages. For calculations involving the use of solubility product constants or complex stability constants, the concentrations of Ca^{2+} and Mg^{2+} must be expressed in units of moles per liter (M), corrected for the effects of ionic strength and complex formation when appropriate. For ion exchange softening (Chap. 12), chemists typically prefer units of meq/L for the liquid phase and units such as meq/mL, eq/L, or eq/g for the solid phase. But the units of grains per gallon and kilograins per cubic foot are often used by practitioners.

Example 13-2 Units of Expression
A water sample is analyzed and found to contain 84.2 mg/L of Ca^{2+} (as Ca) and 9.7 mg/L of Mg^{2+} (as Mg). What are the concentrations of calcium, magnesium, and total hardness in mM, meq/L, and mg/L as $CaCO_3$?

Solution

1. Determine the molecular and equivalent weights of Ca, Mg and $CaCO_3$.
From Appendix A, the atomic weights of Ca and Mg are 40.1 and 24.3. These values are rounded to three significant figures because the concentrations were reported to three significant figures. (Hardness concentrations are typically reported only to the nearest mg/L.) Ca^{2+} and Mg^{2+} are divalent and will react with two equivalents of base to precipitate $CaCO_3$ or $Mg(OH)_2(s)$, respectively. Thus, their equivalent weight is their atomic weight divided by 2, or 20.05 for Ca and 12.15 for Mg. The equivalent weight of $CaCO_3$ is its molecular weight (100) divided by 2, or 50 mg/meq (Chap. 3).

2. Determine calcium, magnesium, and total hardness in units of mM.

$$\text{Calcium hardness} = (84.2\ \text{mg/L})/(40.1\ \text{mg/mmol}) = 2.10\ \text{mM}$$

$$\text{Magnesium hardness} = (9.7\ \text{mg/L})/(24.3\ \text{mg/mmol}) = 0.40\ \text{mM}$$

$$\text{Total hardness} = 2.10 + 0.40 = 2.50\ \text{mM}$$

3. Determine calcium, magnesium, and total hardness in units of meq/L.

$$\text{Calcium hardness} = (84.2\ \text{mg/L})/(20.05\ \text{mg/meq}) = 4.20\ \text{meq/L}$$

$$\text{Magnesium hardness} = (9.7\ \text{mg/L})/(12.15\ \text{mg/meq}) = 0.80\ \text{meq/L}$$

$$\text{Total hardness} = 4.20 + 0.80 = 5.00\ \text{meq/L}$$

4. Determine calcium, magnesium, and total hardness in units of mg/L as $CaCO_3$.

$$\text{Calcium hardness} = (4.20\ \text{meq/L}) \times (50\ \text{mg/meq}) = 210\ \text{mg/L as } CaCO_3$$

$$\text{Magnesium hardness} = (0.80\ \text{meq/L}) \times (50\ \text{mg/meq}) = 40\ \text{mg/L as } CaCO_3$$

$$\text{Total hardness} = 210 + 40 = 250\ \text{mg/L as } CaCO_3 \quad \blacktriangle$$

Types of Hardness. Hardness can be classified as calcium or magnesium hardness, with the sum being defined as total hardness as just discussed. The concentration of alkalinity is also important, since lime addition converts carbonate alkalinity (present predominantly in the form of bicarbonate in most natural waters) to carbonate (CO_3^{2-}), which reacts with Ca^{2+} to precipitate $CaCO_3(s)$. Noncarbonate hardness (NCH) is hardness for which a corresponding (equivalent) amount of carbonate alkalinity is not naturally present in the water. NCH cannot be removed by lime addition, but it can be removed by sodium carbonate addition. Thus, when selecting a precipitative softening process (from among the alternative processes) and when determining the chemical dose requirements, there are four types of hardness to consider.

- Calcium carbonate hardness (CCH)
- Magnesium carbonate hardness (MCH)
- Calcium noncarbonate hardness (CNH)
- Magnesium noncarbonate hardness (MNH)

In some circles, primarily those associated with industrial applications, carbonate hardness is referred to as *temporary hardness,* since its concentration can be substantially reduced by boiling (to precipitate $CaCO_3$); and noncarbonate hardness is referred to as *permanent hardness,* since boiling does not reduce its concentration.

Reasons for Removing Hardness. The decision to remove hardness from drinking water is typically made solely for economic and aesthetic reasons. The two most commonly cited reasons are (1) reduced scaling of boilers, water heaters, washing machines, coffee pots, and other appliances that heat water or come in contact with hot water, which can significantly extend their useful life and reduce energy costs, and (2) conservation of soaps and detergents, resulting in cost savings as well as reduced impacts on wastewater treatment facilities and receiving waters. Other commonly cited reasons include improved taste of food and beverages, improved lathering of soap, reduced formation of soap scums and bathtub rings, and improved laundering, resulting in improved appearance and increased life of clothing. The taste of water can be adversely affected by hardness levels that are too high or too low (Durand and Dietrich, 2009). It is rare for hardness to be removed primarily to improve the taste of water, but hardness and alkalinity are often added to demineralized waters to improve their taste and reduce their corrosivity.

Many years ago, hardness was measured using the soap test method (Durfor and Becker, 1964). Modern detergents provide superior lathering and cleaning action compared to the soaps commonly used in the past and they are also much less prone to forming soap scums, but detergent use is still related to water hardness since some detergent products contain chemicals (referred to as *binders*) that complex hardness-causing ions to facilitate proper cleaning action. van der Veen and Graveland (1988) reported that the total cost of centralized softening in Amsterdam was $4 (U.S.) per household per year, while the savings in detergent, soaps, energy, and maintenance were estimated to be $25 to $40 per household per year. Such savings vary geographically and temporally, but many communities in the United States and elsewhere continue to find centralized softening economically attractive.

Questions sometimes arise regarding the potential human health effects associated with hardness removal, but there appears to be no convincing evidence that hardness in drinking water causes or protects against adverse health effects in humans (World Health Organization, 1996). Although Ca and Mg are essential nutrients, water consumption typically accounts for only a small fraction (5–20%) of total dietary intake. For these reasons, the World Health Organization (1996) and regulatory agencies such as the U.S. Environmental Protection Agency (USEPA) decided not to establish health-related guidelines for water hardness. Using soda ash to remove noncarbonate hardness, or using ion exchange to soften water (Chap. 12), increases the sodium concentration in the finished water. Sodium is also an essential nutrient, but if guidance or advisory levels are exceeded, this may raise concerns. Water consumption is typically not a significant source of dietary sodium when compared to the salt intake goal of less than 2,400 mg/day recommended by several government and health agencies (USEPA, 2003), but it may be significant for those on salt-restricted diets (<500 mg/day). See Chap. 2 for additional information on sodium and health effects.

Although there are no regulatory limits on water hardness, precipitative softening offers a convenient and effective means for removing various other constituents for which maximum contaminant levels (MCLs) or secondary maximum contaminant levels (SMCLs) have been established, including barium, radium, iron, manganese, and color. It also removes or inactivates pathogens (qualifying for disinfection credit when used in conjunction with coagulation and filtration), partially removes DBP precursors, reduces total dissolved solids (TDS), and facilitates stabilization of the finished water.

Typical Treatment Objectives. Complete removal of hardness may be appropriate for certain industrial applications but is undesirable for a public water supply. Scaling problems are usually minimal at hardness concentrations below about 80 to 100 mg/L as $CaCO_3$ if the water is properly stabilized. For stabilization (Chap. 20), minimum calcium and alkalinity concentrations of at least 40 mg/L as $CaCO_3$ are generally recommended. Lower hardness levels can reduce consumption of soaps and detergents, but some hardness is needed to help rinse off soap, for example, when showering. In 1968, AWWA adopted water quality goals that included a goal for hardness of 80 mg/L as $CaCO_3$, stating that this represented a balance between scaling and corrosion and that a goal of 90 or 100 mg/L as $CaCO_3$ might be desirable for some water supplies (AWWA, 1968). Utilities that soften typically use this goal as a starting point and then establish local treatment objectives taking into consideration some or all of the following factors:

- Local estimates of the aforementioned economic and aesthetic benefits
- The cost of treatment and residuals handling and disposal, which will depend in part on raw water hardness, the relative amounts of Ca and Mg, and the relative amounts of carbonate and noncarbonate hardness
- The wishes of the consumers (including industrial users) and their willingness to pay for treatment, which can depend greatly on what the consumers are accustomed to
- The silica concentration in the raw water, which can be lowered if magnesium is also removed and which can contribute to scaling of water heaters and boilers
- Whether a higher level of softening (higher chemical dosages) will help meet selected MCLs or SMCLs

Taking these and perhaps other factors into consideration, most utilities that soften produce a finished water hardness between 75 and 150 mg/L as $CaCO_3$, corresponding to moderately hard water based on the classification described below. Treatment objectives are typically relaxed to avoid the use of soda ash or the need to remove magnesium. Soda ash usually costs much more than lime and contributes sodium to the finished water, and the capital cost of equipment needed to store and feed soda ash may not be justified if only a small dosage is needed to achieve an ideal hardness concentration. Many utilities that soften avoid removing magnesium, except for the small amount that coprecipitates with $CaCO_3(s)$. Magnesium removal is often not necessary and, compared to calcium removal, requires a significantly higher lime dose, often requires soda ash addition, produces a greater quantity of residuals that are significantly more difficult to dewater, and requires higher chemical dosages to stabilize the finished water. However, magnesium removal also results in silica removal, which can offer significant benefits to both industrial and residential customers.

Some utilities establish a separate goal for magnesium hardness, especially if magnesium hydroxide or magnesium silicate scales have been found in their customers' water heaters. A typical goal is a maximum magnesium hardness in the range of 40 to 65 mg/L as $CaCO_3$. A maximum finished water pH value is often established as well because the solubility of $Mg(OH)_2(s)$ depends strongly on pH (Example 13-1). In selecting the pH value of finished water, the temperature of the heated water should be taken into consideration because water dissociates more as its temperature increases and, although its pH decreases, its hydroxide ion concentration actually increases; this can cause precipitation of metal hydroxides, carbonates, and silicates. On the basis of magnesium concentrations at the inlet and outlet of a 60°C (140°F) water heater, Tuepker and Hartung (1960) determined the maximum magnesium concentration at which precipitation would not occur at various pH values (Table 13-3). However,

TABLE 13-3 Approximate Maximum Magnesium Concentrations in Lime-Softened Meramec River Water That Will Not Precipitate in 60°C (140°F) Domestic Water Heaters

pH of Water at 25°C	Mg (mg/L as $CaCO_3$)
9.0	65
9.2	57
9.4	51
9.6	45
9.8	40

Source: Tuepker and Hartung (1960)

their results were based on a limited number of experiments using a single source water, and the scale composition and silica concentrations were not reported. Therefore, Table 13-3 should be viewed not as a general guideline applicable to all source waters but as an example of the type of information needed to establish appropriate goals.

Other investigators, including Ryznar et al. (1946), Larson (1951), and Larson et al. (1959), have examined the solubility of $Mg(OH)_2(s)$ at temperatures relevant to domestic water heaters; but, although they recognized the importance of silica in scaling water heaters, they did not consider silica in their calculations. Larson et al. (1959) summarized their recommendations for avoiding $Mg(OH)_2(s)$ precipitation in water heaters and hot water pipes in the form of a magnesium index (MI) based on field data and calculated as follows:

$$\mathrm{MI} = (2 \times \mathrm{pH}) + \log \mathrm{Mg} + (0.02 \times t) - 21.2 \qquad (13\text{-}5)$$

where pH is the measured pH of the treated water at its ambient temperature, Mg is the magnesium hardness in mg/L as $CaCO_3$, and t is the ambient temperature of the treated water in degrees Fahrenheit. This equation was developed for waters having approximately 50 mg/L of alkalinity and a TDS of 200 mg/L. Larson et al. (1959) found that magnesium hydroxide scaling problems were unlikely for MI ≤ 0, isolated for values of 0.1 to 0.3 over an extended period of time, significant for values of 0.4 to 0.7, and numerous after several months for MI ≥ 0.7.

It is difficult to precisely estimate the magnesium and pH limits that will prevent magnesium scaling problems because it may be possible for more than one magnesium containing solid phase to be formed; the relevant equilibrium constants, including solubility product constants and the ionization constants of water (K_w) and carbonic acid, vary with both temperature and ionic strength; the temperature of a heated surface in a water heater will be considerably higher than the operating water temperature; and the change in the pH of water as its temperature increases is a function of alkalinity. The solubility of magnesium-based solids formed in water heaters may not be well understood or easily calculated, but the options for controlling magnesium scales are straightforward: reduce the magnesium concentration, and possibly the silica concentration as well; lower the pH; reduce the water temperature; or add a scale inhibitor.

Classification and Geographic Distribution of Hard and Soft Waters. Water supplies can be classified on the basis of hardness, and the classification shown in Table 13-4 is widely used. The concentration ranges shown in Table 13-4 are arbitrary, and other authors have chosen slightly different ranges, especially when presenting previously published data grouped into other concentration ranges.

TABLE 13-4 Classification of Water Supplies by Degree of Hardness

Hardness (mg/L as $CaCO_3$)	Degree of Hardness
0–75	Soft
75–150	Moderately hard
150–300	Hard
>300	Very hard

Source: Sawyer et al. (2003 and earlier editions), with permission of McGraw Hill, Copyright 2003.

Figure 13-1 shows the average raw water hardness for each of 48 states in the United States, based on a survey of 1315 large cities (Lohr and Love, 1954). On average, raw water supplies in the United States are relatively soft in Hawaii, the Pacific Northwest, the Northeast, and several southeastern states; moderately hard in Alaska, Colorado, and a number of eastern and southeastern states; and relatively hard in Florida and most southwestern and midwestern states. Maps based on other surveys show similar patterns for both raw and finished water hardness (Collins et al., 1934; Lohr and Love, 1954; Durfor and Becker, 1964; Briggs and Ficke, 1977). In Canada, water supplies in provinces near the Pacific or Atlantic Ocean tend to be softer, while those in the central provinces tend to

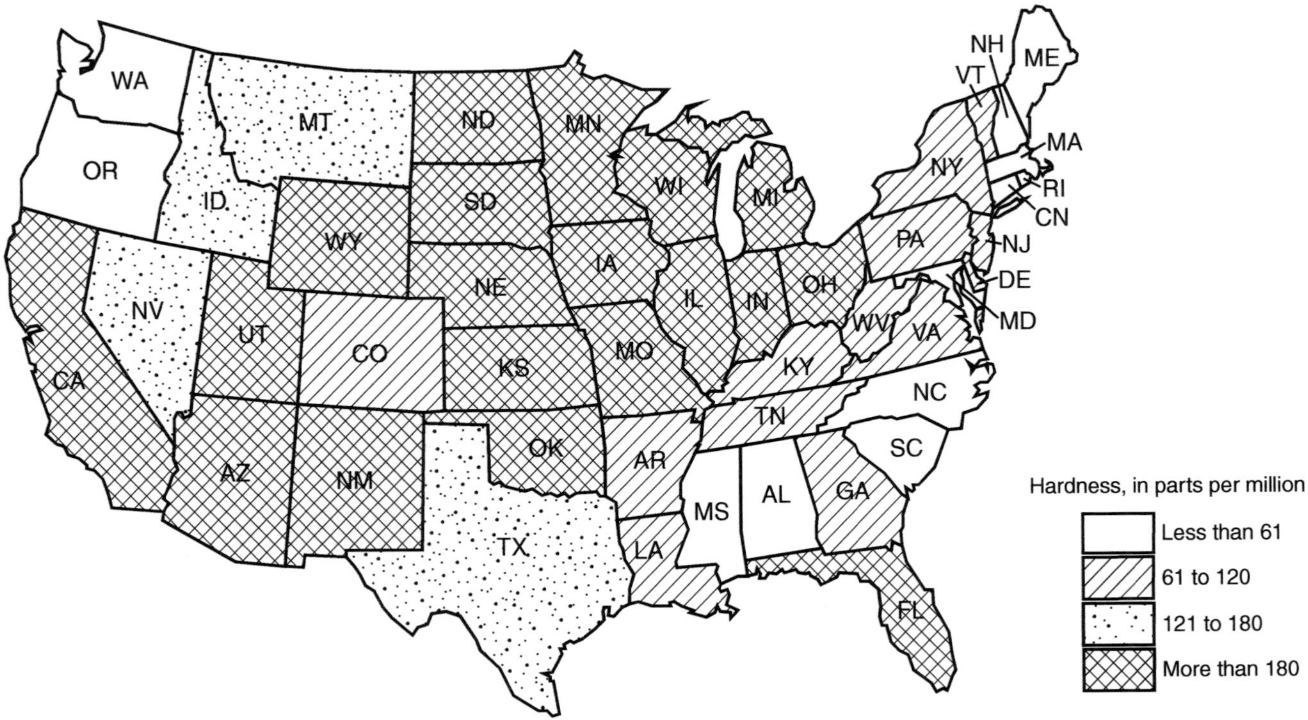

FIGURE 13-1 Average hardness, by state, of raw water from public supplies, drawn from groundwater and surface water sources, for 1,315 of the larger cities in the United States in 1952. (*Adapted from Lohr and Love, 1954.*)

be harder. In some parts of the world, hard surface water supplies are much less common than they are in North America.

Maps such as the one shown in Fig. 13-1 can be helpful for some purposes, such as designing surveys, selecting sampling sites, or identifying regions likely to be impacted by proposed regulations. However, caution should be exercised in drawing conclusions from such maps, as there can be great variations from site to site within a state, especially between surface water and groundwater supplies. The surveys cited focused on larger cities, so the results are biased toward surface water supplies, which Lohr and Love (1954) found to average less than half the hardness concentration of the groundwater supplies surveyed. Although alkalinity tends to vary in patterns similar to those exhibited by hardness, some hard waters contain mostly noncarbonate hardness, which affects the cost of treatment and may influence both the treatment objectives and the choice of a treatment process. Decisions for a specific community or treatment facility should be based on the documented quality of the source water being treated.

Chemistry of Precipitative Softening

In precipitative softening, Ca^{2+} is precipitated as $CaCO_3(s)$, and Mg^{2+} is precipitated as $Mg(OH)_2(s)$. A limited amount of Mg^{2+} may be coprecipitated with (incorporated into or adsorbed onto) $CaCO_3(s)$ (Liao and Randtke, 1985; Russell et al., 2009); thus, some Mg^{2+} may be removed even when the solubility of $Mg(OH)_2(s)$ is not exceeded. The equilibrium reaction equations for precipitation of $CaCO_3(s)$ and $Mg(OH)_2(s)$, written as dissolution reactions, are

$$CaCO_3(s) \rightleftarrows Ca^{2+} + CO_3^{2-} \qquad (13\text{-}6)$$

$$Mg(OH)_2(s) \rightleftarrows Mg^{2+} + 2OH^- \qquad (13\text{-}7)$$

According to LeChâtelier's principle, these reactions can be driven to the left by increasing the concentrations of carbonate (CO_3^{2-}) and hydroxide (OH^-) ions, respectively. This is done by adding a base, usually lime, to the water being treated. Base addition increases the OH^- concentration, which reacts with the carbonate system (described in Chap. 3) to increase the CO_3^{2-} concentration and precipitate $CaCO_3(s)$. A portion of the base is consumed as it reacts with the carbonate system; but if the OH^- concentration increases sufficiently, $Mg(OH)_2(s)$ also precipitates.

Removing Carbonate Hardness Using Lime. Either quicklime or hydrated lime may be used for lime softening. When quicklime, $CaO(s)$, is used, it must first be slaked (combined with water) to form hydrated lime, $Ca(OH)_2(s)$, as follows:

$$CaO(s) + H_2O \rightarrow Ca(OH)_2(s) + \text{heat} \qquad (13\text{-}8)$$

(A note about arrows in chemical equations: When expressing an equilibrium relationship, a double arrow ($\rightleftarrows$) is used; when showing a reaction that is essentially complete or when only the forward reaction is considered, a single, forward arrow ($\rightarrow$) is used.)

Excess water is used in the slaking process, but hydrated lime is only slightly soluble in water, so slaking produces either a lime slurry or a lime paste, depending on the type of slaker used. The slurry or paste is fed by diluting it to produce milk of lime containing up to 180 g/L hydrated lime (AWWA and ASCE, 2005). Quicklime used for water softening must be the fast-slaking type because lime that is incompletely slaked may settle to the bottom of the softening basins and be wasted or cause operational problems. The purity of quicklime varies considerably, ranging from 75 percent to over 90 percent in shipments meeting ANSI/AWWA Standard B202 (AWWA, 2007). Lime purity must be factored in when estimating chemical dosages, and slakers must be designed to handle the impurities that fail to dissolve, referred to as *grit*.

Dissolution of hydrated lime in water is described by the following equilibrium reaction:

$$Ca(OH)_2(s) \rightarrow Ca^{2+} + 2OH^- \tag{13-9}$$

The solubility product constant for $Ca(OH)_2(s)$ is 5.02×10^{-6} at 25°C (CRC Handbook, 2008), so the solubility of $Ca(OH)_2(s)$ in pure water at 25°C is 798 mg/L if complex formation is ignored. Thus, only a small fraction of the lime in milk of lime is dissolved; but the lime should dissolve completely when added to water being treated with the dosages typically associated with lime softening, releasing 2 moles of OH^- ions per mole of lime added. These OH^- ions react with bicarbonate ions associated with carbonate hardness to form carbonate ions.

$$2(OH)^- + 2HCO_3^- \rightarrow 2CO_3^{2-} + 2H_2O \tag{13-10}$$

One mole of carbonate ions can precipitate up to 1 mole of Ca^{2+} ions initially present in the water, and the other mole of carbonate ions can precipitate up to 1 mole of the Ca^{2+} ions that were released into solution as the lime dissolved. Thus, the overall reaction for removal of CCH using quicklime, obtained by summing Eqs. 13-8, 13-9, and 13-10 with the reverse of Eq 13-6 multiplied by 2 is

$$Ca^{2+} + CaO + 2HCO_3^- \rightarrow 2CaCO_3(s) + H_2O \tag{13-11}$$

or, when using hydrated lime,

$$Ca^{2+} + Ca(OH)_2 + 2HCO_3^- \rightarrow 2CaCO_3(s) + 2H_2O \tag{13-12}$$

Precipitation of Mg^{2+} proceeds in similar fashion, with three important differences. First, Mg^{2+} is precipitated as $Mg(OH)_2(s)$, and not as a carbonate salt, with 1 mole of lime producing enough OH^- ions to precipitate up to 1 mole of Mg^{2+}.

$$Mg^{2+} + CaO + H_2O \rightarrow Mg(OH)_2(s) + Ca^{2+} \tag{13-13}$$

Second, as shown in Eq 13-13, precipitation of $Mg(OH)_2(s)$ with lime results in no net removal of hardness, since dissolution of the lime leaves an equivalent concentration of Ca^{2+} in the water, so a second mole of lime must be added to remove these additional Ca^{2+} ions. Assuming that sufficient alkalinity is available, that is, that the Mg^{2+} is present as MCH, the overall reaction for Mg^{2+} removal using quicklime is

$$Mg^{2+} + 2CaO + 2HCO_3^- \rightarrow Mg(OH)_2(s) + 2CaCO_3(s) \tag{13-14}$$

Thus, on a stoichiometric basis, the lime dose required to remove MCH is twice that required to remove CCH. Third, precipitation of $Mg(OH)_2(s)$ requires a relatively high pH, higher than that achieved by stoichiometric addition of lime. Therefore, excess lime must be added to raise the pH to the point that the desired amount of $Mg(OH)_2(s)$ actually precipitates. This process is referred to as excess lime softening.

Removing Noncarbonate Hardness. Lime is unable to remove NCH because it is not possible to precipitate $CaCO_3(s)$ in the absence of carbonate. Carbonate must be added to remove NCH, and the usual chemical of choice for this purpose is sodium carbonate,

Na_2CO_3, commonly known as soda ash. The overall stoichiometric reactions for removal of CNH and MNH are

$$Ca^{2+} + Na_2CO_3 \rightarrow CaCO_3(s) + 2Na^+ \tag{13-15}$$

$$Mg^{2+} + CaO + Na_2CO_3 + H_2O \rightarrow Mg(OH)_2(s) + CaCO_3(s) + 2Na^+ \tag{13-16}$$

Some authors include anions, usually sulfate or chloride, in equations describing these reactions, often referring to these ions as being associated with CNH or MNH. Others show the NCH to be initially present in the form of a salt, such as $CaSO_4(s)$ or $MgCl_2(s)$. These practices stem from the fact that softening was practiced for well over a century before it was first recognized that salts can release dissolved ions into solution, and it took many more years for this to be generally accepted by most chemists. Prior to that time, chemical reactions in water were expressed as reactions between salts. Today it is recognized that anions such as chloride and sulfate do not play a direct role in the softening process, though they may influence softening reactions by contributing to ionic strength or, in the case of sulfate, by forming complexes.

Softening with Sodium Hydroxide. If at least half of the calcium hardness is carbonate hardness, sodium hydroxide (NaOH, also known as caustic soda) can be used to remove both calcium and magnesium hardness. The stoichiometric reactions for removal of Ca^{2+} and Mg^{2+} using NaOH are

$$Ca^{2+} + NaOH + HCO_3^- \rightarrow CaCO_3(s) + H_2O + Na^+ \tag{13-17}$$

$$Mg^{2+} + 2NaOH \rightarrow Mg(OH)_2(s) + 2Na^+ \tag{13-18}$$

As can be seen by comparing Eq. 13-17 with the reaction shown in Eq. 13-11, only half as much carbonate (in the form of HCO_3^- ions) is consumed when using caustic soda instead of lime to remove CCH. This is potentially advantageous when CNH is present, since removing 2 moles of CCH using caustic soda effectively converts 1 mole of CNH to CCH. Also, only half as much $CaCO_3(s)$ is produced; this reduces the cost of residuals disposal. To remove Mg^{2+}, excess NaOH must be added to raise the pH sufficiently to precipitate $Mg(OH)_2(s)$, but this requires conversion of most of the remaining bicarbonate to carbonate, so using caustic soda to remove magnesium is not an attractive option when the magnesium is present as MCH.

Since caustic soda is typically much more expensive than lime or soda ash and since it also increases the sodium concentration in the finished water (which can create a public relations issue), it is rarely used for water softening in the United States. Caustic soda softening is practiced at a number of treatment plants in Europe (see, for example, van Dijk and Wilms, 1991) and elsewhere, especially those employing pellet reactors to soften waters containing NCH. From an operational standpoint, it is easier to inject a solution of NaOH into the inlet of a pellet reactor than to inject lime slurry, and NaOH reacts more rapidly than lime because it is already in solution and does not require time to dissolve. Sodium hydroxide is also easier to store, handle, and feed than lime, especially at smaller treatment plants.

Removing Carbonic Acid. Many hard waters, especially groundwaters and those drawn from the hypolimnion of a lake or reservoir, contain a significant amount of dissolved carbon dioxide (CO_2) along with a much smaller amount of carbonic acid (H_2CO_3). As described in Chap. 3, these species exist in equilibrium with one another. Their sum is commonly designated as $H_2CO_3^*$ (Eq. 3-23), which is referred to as carbonic acid. Carbonic acid reacts instantly with lime, or any other base used to soften water, so carbonic acid must be considered when estimating chemical doses for precipitative softening. For lime, the stoichiometric reaction with $H_2CO_3^*$ is as follows:

$$H_2CO_3^* + CaO(s) \rightarrow CaCO_3(s) + H_2O \tag{13-19}$$

If the concentration of carbonic acid is relatively high, it may be cost-effective to use air stripping (Chap. 6) to reduce its concentration to below 10 mg/L as CO_2 (Horsley et al., 2005).

Recarbonation and Final Stabilization of Lime-Softened Water. During precipitative softening, the water may approach equilibrium with the solid phases being formed, but in most cases it will still be scale-forming; that is, it will have the potential to encrust basins, conduits, valves, filter media, underdrains, clearwells, and wetted surfaces throughout the storage and distribution system. This is especially a problem when adding excess lime to remove Mg^{2+}, since excess lime not removed by adding soda ash can react with residual Ca^{2+} and with CO_2 (adsorbed from the air into the water as it flows over weirs and through open channels) to precipitate additional $CaCO_3(s)$. To minimize or eliminate its scale-forming potential, lime-softened water must be properly stabilized. This is usually accomplished by (1) substantially lowering the pH of the water, prior to filtration, by recarbonation or mineral acid addition; (2) adjusting the pH of the finished water, if necessary, to obtain the desired value of the LSI (Chap. 20); and (3) adding a scale inhibitor, such as phosphoric acid or sodium hexametaphosphate.

Recarbonation is the addition of CO_2 to water from which carbonate has been removed by lime softening or by another process such as membrane softening, reverse osmosis, or ion-exchange demineralization. Recarbonation of lime-softened water restores a portion of the total carbonate concentration lost during the softening process; it also lowers the pH of the treated water, reducing scale formation. The pH of lime-softened water can also be lowered using a strong mineral acid, with sulfuric acid being the usual choice. Addition of a strong mineral acid is sometimes loosely referred to as recarbonation, but it does not change the total carbonate concentration and is more properly referred to as *acid addition* or *pH adjustment*.

Most softening plants practicing recarbonation employ *single-stage recarbonation*, a process in which all of the CO_2 needed to adjust the pH to the desired final value prior to filtration is added at once. However, when excess lime is used to remove Mg^{2+}, the Ca^{2+} associated with the excess lime will remain in the finished water if single-stage recarbonation is employed, since the excess lime in solution will react with CO_2 as follows:

$$2CO_2 + Ca^{2+} + 2OH^- \rightarrow Ca^{2+} + 2HCO_3^- \tag{13-20}$$

If the Ca^{2+} associated with the excess lime must be removed to meet treatment objectives, one option is to remove it by adding soda ash along with the excess lime. However, soda ash is relatively expensive, increases the sodium concentration in the finished water (in essence converting calcium hydroxide into sodium hydroxide), and does not reduce the acid dose required to stabilize the treated water. An alternative is to employ *two-stage recarbonation*. In the first stage, CO_2 is added until the point of minimum $CaCO_3(s)$ solubility is reached, and the additional $CaCO_3(s)$ that is formed is allowed to precipitate and settle. In the second stage, more CO_2 is added to reduce the pH to the desired final value. Since two-stage recarbonation requires an additional recarbonation point and is usually not necessary to meet local treatment objectives, single-stage recarbonation is far more common.

The CO_2 doses required for single-stage and two-stage recarbonation can be calculated on the basis of chemical equilibrium relationships. These relationships and the equations describing them are complex; so, when precise calculations are needed, they are usually made with the aid of a computer program. However, precise estimates of the CO_2 dose are generally not needed in practice, since the dose actually applied by a chemical feeder is only approximately known; some CO_2 is absorbed from the atmosphere; the exact composition and solubility of some of the solids carried over from the softening process may be unknown; and, regardless of the calculated dosage, the CO_2 feed system will be adjusted on the basis of measurements of finished water quality relative to the utility's treatment objectives. Estimates suitable for most purposes can be made using the following stoichiometric relationships:

$$CO_2 + 2OH^- \rightarrow CO_3^{2-} + H_2O \tag{13-21}$$

$$CO_2 + Mg(OH)_2(s) \rightarrow Mg^{2+} + CO_3^{2-} + H_2O \tag{13-22}$$

$$CO_2 + H_2O + CO_3^{2-} \rightarrow 2HCO_3^- \tag{13-23}$$

As these equations indicate, recarbonation converts residual OH^- and CO_3^{2-} ions, whether free or complexed, to HCO_3^- ions. This prevents further precipitation of $Mg(OH)_2(s)$, dissolves $Mg(OH)_2(s)$ particles remaining in the settled water, and diminishes or eliminates the driving force for $CaCO_3(s)$ precipitation. It is commonly believed that recarbonation also dissolves $CaCO_3(s)$ particles carried over from the softening process, but this is expected to occur only if enough CO_2 is added to produce a negative LSI. If a positive LSI is maintained following recarbonation, $CaCO_3(s)$ will have a tendency to precipitate rather than to dissolve; but small or impure $CaCO_3(s)$ crystals, which may have a higher solubility than assumed in calculating the LSI, could potentially dissolve even though the calculated LSI value is positive.

In the first stage of two-stage recarbonation, CO_2 is added to convert OH^- ions to CO_3^{2-} ions, as shown in Eqs. 13-21 and 13-22; then, in the second stage, the remaining CO_3^{2-} ions are converted to HCO_3^- ions, as shown in Eq. 13-23. The carbonate ions formed in the first stage react with the Ca^{2+} ions associated with the excess lime to precipitate additional $CaCO_3(s)$, which is flocculated and then removed by sedimentation. Since the water is already saturated with $CaCO_3(s)$ prior to CO_2 addition, formation of $CaCO_3$ is approximately stoichiometric. Maximizing the carbonate concentration will maximize removal of calcium hardness, and the maximum carbonate concentration will be obtained when the CO_2 dosage increases the acidity from a negative value to 0 mg/L as $CaCO_3$ (Merrill, 1978). Assuming that the concentrations of $H_2CO_3^*$ and H^+ are negligible (an excellent approximation at pH > 10), this corresponds to the point where the concentrations of HCO_3^- and OH^- are equal, or to a pH value of about 10.0 to 10.6 depending on temperature, ionic strength, and alkalinity. This also corresponds to the point where phenolphthalein alkalinity equals one half of the total alkalinity (APHA, AWWA, and WEF, 2005; Schierholz et al., 1976), so alkalinity titrations are sometimes used for process control.

Although these stoichiometric equations ignore important equilibrium relationships involved in recarbonation, CO_2 dose estimates based on stoichiometry are reasonably accurate because inaccuracies introduced by key approximations offset one another. For example, in the first stage of two-stage recarbonation, not all of the hydroxide is converted to carbonate (since the target pH is above 10), but this is offset by the fact that some of the carbonate is converted to bicarbonate as the pH drops. Stoichiometric and equilibrium calculations of CO_2 doses are expected to yield similar results if similar assumptions are made regarding the species present in the settled water.

Following recarbonation, a filter aid is typically added to improve particle capture during filtration. Other chemicals added before or after filtration, including chlorine, ammonia, fluoride, and scale inhibitors, can also influence the pH of the finished water. Therefore, the pH of the finished water should be monitored and adjusted as needed to maintain the desired value. The target value is often selected, in concert with the target values of calcium hardness and alkalinity, to maintain a specified LSI value; but it may also be selected on the basis of empirical observations of scaling and corrosion in the distribution system, tap-water lead levels, or other criteria. The pH of the finished water can be controlled by adjusting the CO_2 dose used for recarbonation or by adding a supplemental dose of acid or base prior to distribution.

Many years ago, CO_2 was often produced on site by the combustion of oil or gas in an external burner, or by gas combustion in an underwater burner. Today, most plants use liquefied CO_2, since its use significantly reduces operation and maintenance problems. Liquefied CO_2 can be vaporized and fed as a gas, using diffusers, or dissolved under pressure into a side stream of water that is then injected back into the settled water prior to filtration. Chemicals and chemical feed systems used for recarbonation, acid addition, and feeding other chemicals used to stabilize water are described in detail by AWWA and ASCE (2005).

Horsley et al. (2005) and Recommended Standards (2007) indicate that a contact time of at least 20 min should be provided for stabilization. The basis for this recommendation is not stated, but it presumably provides time for mixing, for conditioning of the remaining particles at a lower pH value, and in some cases for another opportunity for any remaining particles to settle. Some contend that this detention period is not needed when using side-stream injection because dissolved CO_2 reacts more rapidly than gas-phase CO_2 and because the pH quickly stabilizes following pH addition. However, such arguments do not make sense. Gas-phase CO_2 dissolves rapidly into water, becoming $CO_2(aq)$, and $CO_2(aq)$ reacts with water to form H_2CO_3; but, in this hydration reaction, equilibrium strongly favors $CO_2(aq)$. More than 99 percent of dissolved CO_2 in water is present in the form $CO_2(aq)$ rather than carbonic acid, H_2CO_3 (see Chap. 3 and Snoeyink & Jenkins (1980); and this is true regardless of the manner in which the CO_2 is dissolved into the water. The hydration reaction is relatively slow from a chemist's perspective, but fast enough for the CO_2 in a recarbonation basin to be consumed in a few seconds in cold water. The reaction of H_2CO_3 (the active species in acid-base reactions involving CO_2) with dissolved carbonate and hydroxide ions is virtually instantaneous, so the pH would be expected to stabilize rapidly regardless of the method of CO_2 addition. The important question is how rapidly any remaining particles react with the CO_2, recognizing that heterogeneous reactions (those involving reactants in two different phases) can be quite slow, and whether this affects the filterability of the particles. A much shorter detention period may suffice, especially given the increased use of polymeric filter aids since the criterion of 20 min was first developed; and it was recognized long ago (ASCE, AWWA, and CSSE, 1969) that "some plants are known to be operating [recarbonation basins] with practically no retention time whatsoever." Today it is common to use a recarbonation basin for single-stage recarbonation; to follow the first stage of two-stage recarbonation with flocculation and clarification (or a solids contact basin) rather than a recarbonation basin; and to omit a recarbonation basin following the second stage of two-stage recarbonation.

Chemical Dose Estimates Based on Stoichiometry. Chemical doses for lime softening and recarbonation can be estimated using simple stoichiometric relationships, such as those shown in a number of the preceding equations, or by performing calculations that take equilibrium relationships into consideration. Both approaches have advantages and limitations, and both must be applied with judgment to obtain reasonable results. Chemical dosages can also be determined empirically, at full scale or by conducting jar tests.

Chemical dose estimates based on stoichiometry are easy to understand, can be made very quickly, and require minimal computational resources and ability. However, their accuracy is limited because important equilibrium relationships are either ignored or replaced with simple approximations. For example, the stoichiometry reflected in Eq. 13-11 indicates that 1 mole of lime will convert 1 mole of Ca^{2+} ions to $CaCO_3(s)$, but a portion of the Ca^{2+} will remain in solution in equilibrium with $CaCO_3(s)$, as shown in Eq. 13-6, and some

Ca^{2+} will remain in solution in the form of ion pairs, such as $CaCO_3^0$ and $CaSO_4^0$, or other complexes. To compensate for this, an expected residual Ca^{2+} concentration (or a specific treatment objective) can be assumed and the chemical doses can be adjusted accordingly.

Chemical dose estimates based on stoichiometry are suitable for most purposes, such as setting initial chemical feed rates, sizing chemical handling and storage facilities, estimating solids production, and estimating costs. In practice, operators must adjust chemical doses up or down to achieve specific treatment objectives, but this would be the case even if each chemical dose could be determined very precisely, since there would still be changes in water quality and flow rate, and perhaps in chemical concentration or purity.

The required excess lime dosage, typically about 0.3 to 0.7 mM (Horsley et al., 2005), or 30 to 70 mg/L as $CaCO_3$, is difficult to estimate accurately using stoichiometry alone. One approach is to simply pick a dosage within the typical range and to then adjust it to meet the treatment objective for magnesium. However, the dosage actually required will depend on temperature, ionic strength, and the concentration of magnesium hardness desired in the finished water. The required dosage can be more accurately estimated by introducing simple equilibrium calculations, first to estimate the pH value corresponding to the desired finished-water magnesium concentration, assuming equilibrium with amorphous $Mg(OH)_2(s)$, and then to estimate the lime dose required to reach the calculated pH value. Such calculations are included in Example 13-4.

Estimating Chemical Doses Using Equilibrium Models. Equilibrium models can be used to determine the chemical doses required for lime softening and recarbonation; to evaluate the LSI and calcium carbonate precipitation potential (CCPP) of treated water, as described in Chap. 20; and for other purposes, such as estimating the pH of a water following chemical addition or blending with another water. Equilibrium models used to estimate chemical doses for lime softening and recarbonation incorporate equilibrium constants for dissolution (and precipitation) of $CaCO_3(s)$ and $Mg(OH)_2(s)$, for the carbonate system, and usually for a limited number of metal-ion complexes, along with a means of adjusting each constant to account for the effects of temperature and ionic strength. Equilibrium relationships and other equations needed to construct a basic equilibrium model of the lime softening process are described in the subsection Calcium Carbonate and Magnesium Hydroxide Equilibria.

The equations in simpler models can be solved using only a handheld calculator, as demonstrated in Example 13-5; but a spreadsheet or other computer program is normally used so the calculations can be more easily repeated or refined, and software programs are commercially available for those who lack the time or knowledge to write their own. Prior to the availability of personal computers, many water treatment professionals used Caldwell-Lawrence diagrams to estimate chemical dosages for precipitative softening and water stabilization, and for other purposes, such as determining the pH of blended waters or the pH of a water following chemical addition. These diagrams, prepared using a mainframe computer, provide nomographic representations of $CaCO_3(s)$ and $Mg(OH)_2(s)$ solubility as a function of temperature and ionic strength. Use of these diagrams was described and demonstrated in the previous edition of this handbook (Benefield and Morgan, 1999) and by Caldwell and Lawrence (1953), Merrill and Sanks (1969), Loewenthal and Marais (1976), and Merrill (1978). Since these diagrams are rarely used today, they are not described further in this edition.

When developing or using an equilibrium model, the user should note that although some of the calculations may be more accurate than those based on stoichiometric calculations, their accuracy will be limited by the accuracy of the equilibrium constants used, the corrections applied, and the assumptions regarding the physical and chemical nature of the system. Values of equilibrium constants are based on experimental measurements and are subject to both analytical and procedural errors, so it is important to choose values from reliable sources and to recognize that their accuracy is limited by the accuracy of the data from which they were calculated and by the ability of the investigator to correctly perform the calculations. Furthermore, solubility product constants are usually determined using pure solutions of known composition that are allowed to reach equilibrium before measurements are made. Waters that are lime softened may contain impurities, such as DOC, phosphate, silicic acid, and coagulants, that influence the solubility of $CaCO_3(s)$ and may form complexes not included in the model; and equilibrium may not be attained before the solids are removed. Temperature and ionic strength corrections are based on correlations of measured values. Temperature corrections are often based on the van't Hoff equation (Chap. 3, Eq. 3-16), which in turn is based on the assumption that the enthalpy change does not vary significantly with temperature, which may or may not be the case. Temperature corrections made using correlations based on only a few data points, linear correlations of nonlinear data, or data outside the temperature range of interest may introduce significant error. The Davies equation, commonly used to estimate activity coefficients as a function of ionic strength, was originally developed to estimate the activity coefficients of selected ions to within ±2 percent at 25°C in a 0.1 M solution, so it is presumably less accurate under other conditions. Many models introduce small errors by failing to account for changes in ionic strength that occur as solids precipitate or dissolve, and most models fail to account for the reduction in ionic strength that results from ion-pair formation. Chemical equilibrium models are often based on the assumption of a closed system, but treatment systems are usually not completely closed; thus, gains and losses of CO_2 are often ignored whether or not they are significant.

Calcium Carbonate and Magnesium Hydroxide Equilibria. The solubility product constant for dissolution of $CaCO_3(s)$, shown in Eq. 13-6, can be written as follows when ionic strength effects are considered:

$$K_{s0,\mathrm{CaCO_3}} = \gamma_d[\mathrm{Ca^{2+}}]\gamma_d[\mathrm{CO_3^{2-}}] \tag{13-24}$$

where $K_{s0,\mathrm{CaCO_3}}$ is the solubility product constant for $CaCO_3(s)$, γ_d is the activity coefficient for divalent ions, and the square brackets denote molar concentrations. In very dilute solutions, that is, at very low ionic strengths, activity coefficients have a value of approximately 1 and the activity of an ion is approximately equal to its molar concentration. As ionic strength increases, activity decreases, and activity coefficients must be applied to obtain accurate results. For very precise work involving pure solutions, empirically derived single-ion activity coefficients are commonly used (as described by Butler, 1998). Activity coefficients that vary only with ionic charge are normally used in calculations pertaining to natural (and treated) waters, which contain mixtures of different ions of unlike charge. Using ion-specific activity coefficients would be complicated and would contribute little to the overall accuracy of the calculations given other uncertainties involved in most calculations. Activity coefficients may be estimated using one of several equations in widespread use, including the Davies equation (Davies, 1938) modified as recommended by Davies (1962), and using the Debye-Hückel limiting law to account for the effect of temperature.

$$\log \gamma_z = -Az^2\left(\frac{\sqrt{I}}{1+\sqrt{I}} - 0.3I\right) \tag{13-25}$$

where z is the ionic charge, I is the ionic strength (see Chap. 3), $A = 1.82 \times 10^6\ (\varepsilon T)^{-3/2}$, and ε is the dielectric constant of water (78.38 at 25°C). The value of the dielectric constant varies with temperature and is commonly estimated using the following correlation:

$$\varepsilon = \frac{60{,}594}{T+116} - 68.937 \tag{13-26}$$

where T is absolute temperature (K). Rossum and Merrill (1983) presented this correlation without indicating its source. Compared to accurately determined values of ε from 0 to 100°C (Archer and Wang, 1990; CRC Handbook, 2008), Eq. 13-26 estimates ε with an average error of 0.17 percent and a maximum error of 0.30 percent. The following equation, obtained by fitting an exponential expression to the data of Archer and Wang from 0 to 100°C, estimates ε more accurately, with an average error of only 0.013 percent and a maximum error of 0.025 percent:

$$\varepsilon = 308.67\, e^{-0.0045976(T)} \tag{13-27}$$

where T is absolute temperature (K).

Comparing activity coefficients calculated using Eq. 13-25 (with $A = 0.5$) against 50 values determined experimentally at an ionic strength of 0.1 M and 25°C, Davies (1962) found the average deviation to be 1.6 percent, with three of the deviations being twice that large. Butler (1998) found that activity coefficients estimated using Eq. 13-25 have a possible error of 3 percent when $I = 0.1$ M, and 10 percent when $I = 0.5$ M. Stumm and Morgan (1996) indicated that Eq. 13-25 is appropriate for $I < 0.5$ M, and Butler (1982) stated that it applies from 0 to 50°C.

The value of K_{s0} for $CaCO_3(s)$ depends on the form precipitated and on temperature. Known forms of $CaCO_3(s)$ include calcite, aragonite, vaterite, and several hydrated forms (Randtke et al., 1982). Calcite is the most thermodynamically stable form in most natural freshwaters, and it is the sole or dominant form normally present in lime-softening residuals. Aragonite has been found, typically as a minor component, in some samples of lime-softening residuals, and its formation is favored in seawater under certain conditions. Slack (1980) reported the formation of calcium carbonate hexahydrate ($CaCO_3{\cdot}6H_2O$) at several lime–soda softening plants in England and found that cold temperatures, high phosphate concentrations, and perhaps DOC contributed to its formation. Randtke et al. (1982) observed vaterite formation in laboratory experiments when sodium carbonate was added to Ca^{2+} solutions, but on standing, the solids transformed into calcite and calcite was the only product formed when using quicklime or hydrated lime. Slack found that adding calcite inhibited formation of $CaCO_3{\cdot}6H_2O$, as expected. Since solids are normally recirculated at lime-softening plants and since calcite is more thermodynamically stable than other forms under typical water treatment conditions, calcite formation is strongly favored, fortunately so, since other forms are considerably more soluble than calcite and therefore inhibit the softening process.

The solubility of $CaCO_3(s)$ in water is of interest to investigators in diverse fields and has been extensively studied. Reported values of K_{s0} for $CaCO_3(s)$ vary, but the following equations developed by Plummer and Busenberg (1982) are thought to provide the best available estimates of K_{s0} for calcite, aragonite, and vaterite under conditions relevant to water treatment and distribution (APHA, AWWA, and WEF, 2005):

$$pK_{sc} = 171.9065 + 0.077993T - 2839.319/T - 71.595 \log_{10} T \tag{13-28}$$

$$pK_{sa} = 171.9773 + 0.077993T - 2903.292/T - 71.595 \log_{10} T \tag{13-29}$$

$$pK_{sv} = 172.1295 + 0.077993T - 3074.688/T - 71.595 \log_{10} T \tag{13-30}$$

where $pK = -\log(K)$, K_{sc} = solubility product constant for calcite, K_{sa} = solubility product constant for aragonite, K_{sv} = solubility product constant for vaterite, and T = absolute temperature, K (273.15 + °C). Equations 13-28 through 13-30 are applicable over the temperature range 0 to 90°C (273 to 363K).

The following correlation was widely used in the past to estimate K_{s0}, but although it is still included in some texts today, it should now be considered obsolete:

$$pK_{s0,CaCO_3} = 8.03 + 0.01183t \tag{13-31}$$

where t = temperature in degrees Celsius. This correlation was developed for a temperature range of 0 to 80°C (Loewenthal and Marais, 1976), but it is based on K_{s0} values summarized by Larson and Buswell (1942), which were extrapolated for temperatures above 30°C and which were determined without taking complex formation into consideration.

The solubility product constant expression for dissolution of $Mg(OH)_2(s)$, shown in Eq. 13-7, is as follows when corrected for ion strength effects:

$$K_{s0,Mg(OH)_2} = \gamma_d[Mg^{2+}]\gamma_m^2[OH^-]^2 \tag{13-32}$$

When $Mg(OH)_2(s)$ precipitates in a lime–soda softening process, it forms amorphous or microcrystalline solids that have a much higher solubility than crystalline $Mg(OH)_2(s)$, a mineral known as brucite. Therefore, the values of K_{s0} listed in the chemical and geochemical literature for brucite are not applicable to $Mg(OH)_2(s)$ formed in the lime softening process. Using the data of earlier investigators, Loewenthal and Marais (1976) developed the following equation to describe the solubility of freshly precipitated $Mg(OH)_2(s)$ over a temperature range of 0 to 80°C:

$$pK_{s0,Mg(OH)_2} = 0.0175t + 9.97 \tag{13-33}$$

where t = temperature in degrees Celsius.

Solubility product constants based on molar concentrations (${}^cK_{s0}$) may be written for $CaCO_3(s)$ and $Mg(OH)_2(s)$ by rearranging Eqs. 13-24 and 13-32 as follows:

$${}^cK_{s0,CaCO_3} = \frac{K_{s0,CaCO_3}}{\gamma_d^2} = [Ca^{2+}][CO_3^{2-}] \tag{13-34}$$

$${}^cK_{s0,Mg(OH)_2} = \frac{K_{s0,Mg(OH)_2}}{(\gamma_d\gamma_m^2)} = [Mg^{2+}][OH^-]^2 \tag{13-35}$$

In these expressions and those presented earlier, only free (uncomplexed) Ca^{2+} and Mg^{2+} ions are taken into consideration. To more accurately estimate the total residual concentrations of dissolved Ca and Mg^2, their complexed forms must also be considered. Numerous metal-ligand complexes can potentially form in natural waters. Most can usually be ignored, but some may be present in concentrations that are significant relative to those of the free ions. Ca and Mg complexes commonly included in chemical equilibrium models are listed in Table 13-5, along with expressions for their formation constants and equations that can be used to estimate the values of their formation constants as a function of temperature. The effect of complex ion formation on $CaCO_3(s)$ and $Mg(OH)_2(s)$ solubility can be included in equilibrium calculations by including the complexes in mass balance relationships for total residual calcium and total residual magnesium. For the complexes listed in Table 13-5, these relationships are

$$[Ca]_T = [Ca^{2+}] + [CaOH^+] + [CaHCO_3^+] + [CaCO_3^0] + [CaSO_4^0] \tag{13-36}$$

and

$$[Mg]_T = [Mg^{2+}] + [MgOH^+] + [MgHCO_3^+] + [MgCO_3^0] + [MgSO_4^0] \tag{13-37}$$

At low ionic strength (ignoring activity corrections), Eqs. 13-36 and 13-37 can be reduced to

$$[Ca]_T = \frac{K_{s0,CaCO_3}}{\alpha_2 C_{T,C}}\left(1 + \frac{K_w K_3}{[H^+]} + K_4\alpha_1 C_{T,C} + K_5\alpha_2 C_{T,C} + K_6[SO_4^{2-}]\right) \tag{13-38}$$

TABLE 13-5 Complex Formation Reactions of Ca^{2+} and Mg^{2+} with Selected Ligands

Complex formation reaction	Equilibrium expression	log K (25°C)	ΔH_r^0 (kcal/mole)
$Ca^{2+} + OH^- \leftrightarrow CaOH^+$	$K_3 = [CaOH^+]/([Ca^{2+}][OH^-])$	1.30	2.0
$Ca^{2+} + HCO_3^- \leftrightarrow CaHCO_3^+$	$K_4 = [CaHCO_3^+]/([Ca^{2+}][HCO_3^-])$	1.20	4.6
$Ca^{2+} + CO_3^{2-} \leftrightarrow CaCO_3^0$	$K_5 = [CaCO_3^0]/([Ca^{2+}][CO_3^{2-}])$	3.22	3.6
$Ca^{2+} + SO_4^{2-} \leftrightarrow CaSO_4^0$	$K_6 = [CaSO_4^0]/([Ca^{2+}][SO_4^{2-}])$	2.36	1.7
$Mg^{2+} + OH^- \leftrightarrow MgOH^+$	$K_7 = [MgOH^+]/([Mg^{2+}][OH^-])$	2.58	3.0
$Mg^{2+} + HCO_3^- \leftrightarrow MgHCO_3^+$	$K_8 = [MgHCO_3^+]/([Mg^{2+}][HCO_3^-])$	1.01	1.2
$Mg^{2+} + CO_3^{2-} \leftrightarrow MgCO_3^0$	$K_9 = [MgCO_3^0]/([Mg^{2+}][CO_3^{2-}])$	2.92	2.6
$Mg^{2+} + SO_4^{2-} \leftrightarrow MgSO_4^0$	$K_{10} = [MgSO_4^0]/([Mg^{2+}][SO_4^{2-}])$	2.26	1.4

General equation for estimating log K at absolute temperature T (in degrees K)

$\log K(T) = \log K\ (\text{at } 298.15\text{K}) - [0.73295 \times \Delta H_r^0 \times (298.15 - T)/(298.15)]$

Complex-specific temperature correlations for T in degrees K

$\log K_4 = 1209.12 + 0.31294T - 34{,}765.05/T - 478.782 \log T$

$\log K_5 = -1228.732 - 0.299444T + 35{,}512.75/T + 485.818 \log T$

$\log K_8 = 2.319 - 0.011056T + 2.2812 \times 10^{-5}T^2$

$\log K_9 = 0.991 + 0.00667T$

Note: The equilibrium expressions are written in terms of molar concentrations, but activities must be used when ionic strength effects are significant. Values of log K and ΔH_r^0 are from NIST (2004). The general equation for estimating log K as a function of temperature is based on a rearrangement of the Van't Hoff equation, which is accurate only if ΔH_r^0 does not vary with temperature over the temperature range of interest. The complex-specific temperature correlations for K_4 and K_5 are those found by Plummer and Busenberg (1982). Those for K_8 and K_9 are from Plummer, Jones, and Truesdell (1976) and yield log K values at 25°C that differ slightly from those reported by NIST (2004).

and

$$[\text{Mg}]_T = \frac{K_{s0,\text{Mg(OH)}_2}[\text{H}^+]^2}{(K_w)^2}\left(1 + \frac{K_w K_7}{[\text{H}^+]} + K_8\alpha_1 C_{T,C} + K_9\alpha_2 C_{T,C} + K_{10}[\text{SO}_4^{2-}]\right) \quad (13\text{-}39)$$

where $C_{T,C}$ is the total molar inorganic carbon concentration, K_w is the ion product of water, and K_1 and K_2 are the acid dissociation constants for carbonic acid (see Chap. 3). The value of K_w can be calculated using the following expression:

$$pK_w = \frac{4470.99}{T} + 0.017060(T) - 6.0875 \quad (13\text{-}40)$$

where T = temperature in degrees kelvin (273.15 + °C). Equation 13-40 was developed by Harned and Owen (1958) for temperatures of 273 to 333K (0 to 60°C), but it provides good estimates of K_w at much higher temperatures. To correct for activity, replace the thermodynamic equilibrium constants in Eqs. 13-38 and 13-39 with their corresponding concentration-based constants:

$${}^cK_{s0,\text{CaCO}_3} \text{ and } {}^cK_{s0,\text{Mg(OH)}_2}$$

(as shown in Eqs. 13-34 and 13-35)

$$^cK_w = K_w/(\gamma_m)^2$$

$$^cK_3 = \gamma_d K_3 \qquad ^cK_4 = \gamma_d K_4 \qquad ^cK_5 = (\gamma_d)^2 K_5 \qquad ^cK_6 = (\gamma_d)^2 K_6$$

$$^cK_7 = \gamma_d K_7 \qquad ^c\mathrm{K}_8 = \gamma_d K_8 \qquad ^cK_9 = (\gamma_d)^2 K_9 \qquad ^cK_{10} = (\gamma_d)^2 K_{10}$$

Several of these expressions assume an activity of 1 for uncharged molecules such as the $CaCO_3^0$ ion pair. This is generally considered to be a reasonable assumption for solutions having an ionic strength less than 0.1 M (Snoeyink and Jenkins, 1980). In general, an activity coefficient for an uncharged molecule (γ_0) can be expressed as a simple function of ionic strength up to $I \sim 5$M.

$$\log \gamma_0 = bI \tag{13-41}$$

where b is an empirically determined constant that may be small (or zero) for uncharged ion pairs (Butler, 1998). Note that activity coefficients for uncharged molecules are greater than or equal to 1 (not equal to or less than 1, as they are for ions). Uncharged molecules are salted out of water as ionic strength increases, since ions interact much more strongly with water. Also note that pH equals the negative logarithm of *hydrogen ion activity*, not hydrogen ion concentration. Thus,

$$[\mathrm{H}^+] = \frac{10^{-\mathrm{pH}}}{\gamma_m} = 10^{-\mathrm{pH}-\log(\gamma_m)} \tag{13-42}$$

In many waters, the concentrations of complexed forms of Ca and Mg are small relative to the concentrations of uncomplexed Ca^{2+} and Mg^{2+} and can therefore be ignored. In fact, when the water is cold, the contact time is short, or the solids concentration is low, the errors associated with failure to approach equilibrium are likely to greatly exceed those associated with ignoring complex formation. Complexes are most likely to be important when attempting to accurately model the softening of water high in sulfate or carbonate that is being treated under conditions where equilibrium is approached.

Cadena et al. (1974) concluded that the $CaCO_3^0$ ion pair accounted for a significant amount of the residual calcium remaining after lime softening, but they used a value of K_5 about 17 times higher than values reported by other investigators. Pisigan and Singley (1985) found that the concentration of $CaCO_3^0$ is insignificant in fresh water in the pH range of 6.20 to 9.20. An example by Snoeyink and Jenkins (1980) illustrated that, for a water having $C_{T,C} = 10^{-3}$ at pH 12, the sum of the carbonato, bicarbonato, and hydroxo complexes of Ca^{2+} in equilibrium with $CaCO_3$ is less than 1 mg/L as $CaCO_3$; however, the concentration of free (uncomplexed) Ca^{2+} present at equilibrium was found to be only 0.5 mg/L under these conditions. Thus, in this example, the error introduced by ignoring complexes would be relatively large even though the total concentration of soluble Ca species is quite small.

Process Variations

There are a number of process variations that can be employed to accomplish precipitative softening, several of which are illustrated in Fig. 13-2. Factors considered in process selection include raw water quality, treatment objectives (possibly including the need to remove other contaminants besides those causing hardness), costs (including chemical, solids disposal, and other operating costs), and the preferences of both the consumers and those managing the treatment facilities. Straight lime softening is the process most commonly

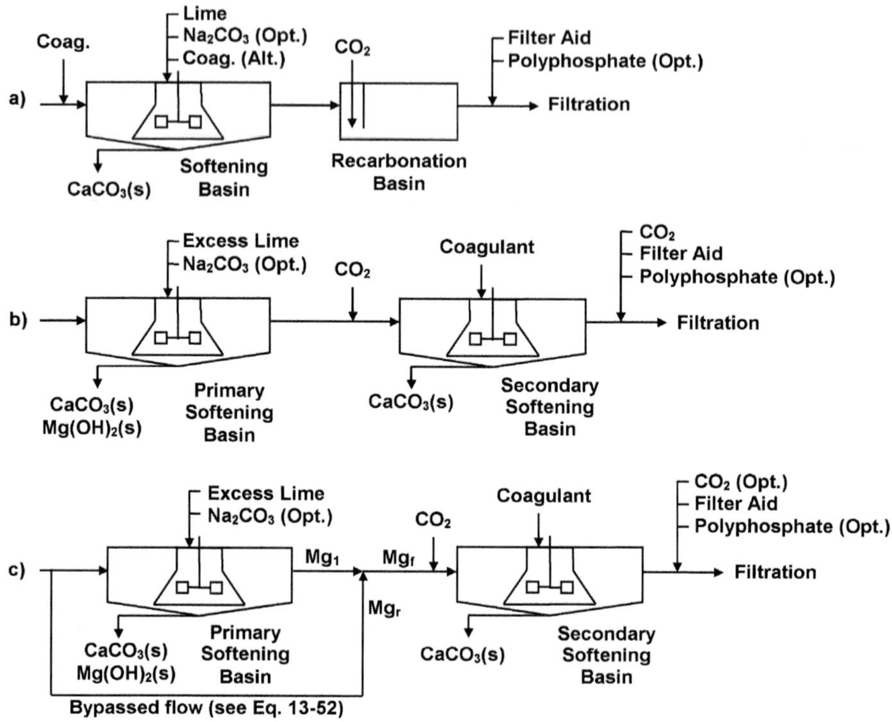

FIGURE 13-2 Common lime-softening process variations. (a) Straight single-stage lime or lime–soda softening. (b) Excess lime or lime–soda softening with two-stage recarbonation (also known as two–stage softening). (c) Split treatment softening.

employed for precipitative softening of public water supplies in North America, but other processes are sometimes found to be advantageous under site-specific conditions.

Straight Lime Softening. *Straight lime softening,* also referred to as *conventional lime softening,* removes carbonic acid and CCH, though a small amount of Mg may be removed by means of coprecipitation. On a stoichiometric basis, the lime dose required is the sum of the doses required to remove carbonic acid and CCH. There are two possible cases.

Case 1. All of the calcium hardness is CCH. In this case, the required lime dose, in units of moles per liter, may be estimated as follows:

$$[CaO] = [H_2CO_3^*] + [CCH] = [H_2CO_3^*] + [Ca^{2+}] \tag{13-43}$$

Case 2. Both CCH and CNH are present. In this case, only the CCH can be removed. Assuming the pH of the raw water is less than 9 and all of the carbonate alkalinity is initially present in the form of HCO_3^-, then the required lime dose may be estimated as follows:

$$[CaO] = [H_2CO_3^*] + [CCH] = [H_2CO_3^*] + ½[HCO_3^-] \tag{13-44}$$

In either case, the lime dosage may be increased to accelerate the reaction or decreased if a higher finished water calcium hardness is acceptable.

Many years ago, waters treated using straight lime or lime–soda softening were sometimes not recarbonated because the water was considered to be reasonably close to equilibrium with respect to precipitation of $CaCO_3(s)$. However, some utilities found it necessary or desirable to lower the pH of the finished water or to add a scale inhibitor, such as sodium hexametaphosphate, to avoid $CaCO_3(s)$ deposition in filters, in the distribution system, and especially in water heaters. Utilities gradually came to recognize the importance of water stabilization and its long-term impacts on the condition of a distribution system, as well as its potential impacts on lead concentrations in tap water. Today most utilities practicing straight lime or lime–soda softening carefully evaluate the stability of the finished water and adjust pH, alkalinity, and hardness as needed to protect the distribution system (Chap. 20) and comply with the Lead and Copper Rule (Chap. 1).

Straight Lime–Soda Softening. *Straight lime–Soda softening,* also referred to as *conventional lime–soda softening,* removes both CCH and CNH, but not magnesium hardness, except for a small amount of magnesium that may be coprecipitated with $CaCO_3(s)$. If all of the carbonate alkalinity is initially present in the form of HCO_3^-, the stoichiometric lime dose is calculated as in Case 2, and the dose of soda ash required to remove CNH is

$$[Na_2CO_3] = [CNH] = [Ca^{2+}] - \tfrac{1}{2}[HCO_3^-] \qquad (13\text{-}45)$$

Example 13-3 Straight Lime and Lime–Soda Softening

A groundwater having a pH of 7.50, an alkalinity of 180 mg/L as $CaCO_3$, and a temperature of 25°C is analyzed and found to have the following major ion concentrations:

$Ca^{2+} = 84.2$ mg/L $\quad Mg^{2+} = 9.7$ mg/L $\quad Na^+ = 23.0$ mg/L

$SO_4^{2-} = 96.1$ mg/L $\quad Cl^- = 14.2$ mg/L

Determine the concentration of each type of hardness and the chemical doses required to soften and recarbonate this water using straight lime and lime–soda softening (Fig. 13-2a). Draw bar diagrams to illustrate the ionic composition of the water after each step of treatment, and ignore ionic strength effects. Assume that a polymer is used as a coagulant. If alum were used instead, an equivalent amount of lime would be needed to neutralize the acidity contributed by the alum, and there would be a corresponding increase in CNH.

Solution

1. Determine the initial HCO_3^- and CO_3^{2-} concentrations.

   Since pH is between 6 and 9, the CO_3^{2-} concentration is negligible, thus $HCO_3^- \approx$ alkalinity = 180 mg/L as $CaCO_3$. Dividing by the equivalent weight of $CaCO_3$,

$$HCO_3^- \approx (180 \text{ mg/L as } CaCO_3)/(50 \text{ mg/meq}) = 3.60 \text{ meq/L}$$

   Since HCO_3^- has one acid-base equivalent per mole,

$$HCO_3^- \approx (3.60 \text{ meq/L})/(1 \text{ meq/mmol}) = 3.60 \text{ mM}$$

2. Determine the $H_2CO_3^*$ concentration.

   At 25°C, $pK_1 = 6.35$ (Table 3-7). Therefore,

$$K_1 = 10^{-6.35} = \frac{[HCO_3^-][H^+]}{[H_2CO_3^*]} = \frac{(3.60\times10^{-3})(10^{-7.50})}{[H_2CO_3^*]}$$

Solving for $H_2CO_3^*$,

$$[H_2CO_3^*] \approx 0.25 \times 10^{-3}\ M = 0.25\ mM$$

Since $H_2CO_3^*$ has two acid-base equivalents per mole,

$$H_2CO_3^* \approx (0.25\ mM)\ (2\ meq/mmol)\ (50\ mg/meq) = 25\ mg/L\ as\ CaCO_3$$

This concentration is high enough to justify evaluating the cost effectiveness of removing CO_2 by air stripping, but it will be removed using lime in this example.

3. Determine the types of hardness present.

A good way to do this is to begin by constructing a bar diagram showing the ionic composition of the water in meq/L. Carbonic acid can also be included in the bar diagram, as a reminder that it must also be considered when estimating chemical dosages. The first step in constructing a bar diagram is to determine the concentration of each constituent in meq/L. The concentrations in mM can also be determined to facilitate calculating dosages in mM.

Constituent	mg/L	Mol. wt.	mM	meq/mmol	meq/L
$H_2CO_3^*$	25†	NA	0.25	2	0.50
Ca^{2+}	84.2	40.1	2.10	2	4.20
Mg^{2+}	9.7	24.3	0.40	2	0.80
Na^+	23.0	23.0	1.00	1	1.00
HCO_3^-	180†	NA	3.60	1	3.60
SO_4^{2-}	96.1	96.1	1.00	2	2.00
Cl^-	14.2	35.45	0.40	1	0.40

†mg/L as $CaCO_3$; units must be converted using equivalent weight.

Construct a bar diagram showing ion concentrations in meq/L. (This bar diagram and those that follow are not drawn to scale.)

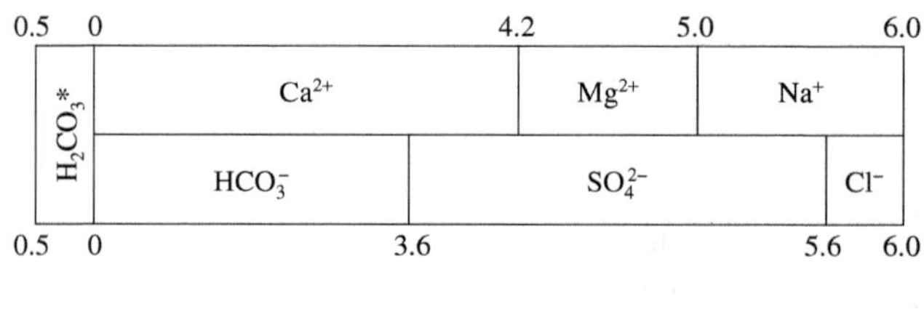

The concentration of each type of hardness can be read from the bar diagram. Recall that carbonate hardness is hardness for which an equivalent amount of HCO_3^- is available, and that removal of calcium hardness with lime requires 2 moles of HCO_3^- per mole of Ca^{2+} (Eqs. 13-11 and 13-12), or one equivalent of HCO_3^- per equivalent Ca^{2+}. Since the concentration of Ca^{2+} in meq/L exceeds the concentration of HCO_3^-, both CCH and CNH are present, whereas all of the Mg^{2+} is NCH. Thus,

$$CCH = 3.60\ meq/L = 1.80\ mM = 180\ mg/L\ as\ CaCO_3$$

$$CNH = 4.2 - 3.6 = 0.6\ meq/L = 0.3\ mM = 30\ mg/L\ as\ CaCO_3$$

$$MCH = 0$$

$$MNH = 5.0 - 4.2 = 0.8\ meq/L = 0.4\ mM = 40\ mg/L\ as\ CaCO_3$$

Sodium, sulfate, chloride, and other ions besides Ca^{2+}, Mg^{2+}, and HCO_3^- are not important with respect to softening, except insofar as they influence ionic strength and form complexes. Thus, they can be omitted from the bar diagram or grouped together as other ions. However, their inclusion in the bar diagram provides a check on the quality of the data, since the sums of the cation and anion concentrations, in meq/L, should be approximately equal (APHA et al., 2005), and checking the balance also helps guard against common errors made when converting concentrations from one set of units to another.

4. Determine the chemical doses for straight lime softening followed by recarbonation.
Straight lime softening will remove only CCH. Using Eq 13-43,

$$[\text{CaO}] = [\text{H}_2\text{CO}_3^*] + [\text{CCH}] = 0.25 \text{ mM} + 1.80 \text{ mM} = 2.05 \text{ mM}$$

Assuming lime purity of 90 percent, the lime dose in mg/L will be

$$(2.05 \text{ mM}) \times (56 \text{ mg/mmol})/(0.90) = 128 \text{ mg/L as 90 percent pure CaO}$$

After softening, before recarbonation, the ionic composition of the water will be as follows, assuming that 40 mg/L of $CaCO_3$ (0.4 mM or 0.8 meq/L) remains in solution and that OH^- is negligible:

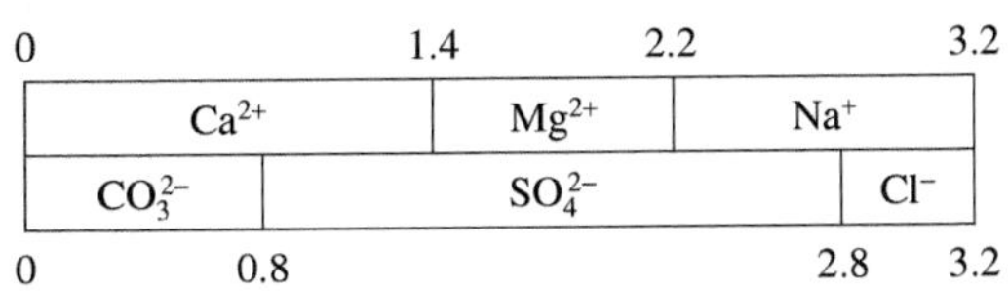

The approximate CO_2 dose required for recarbonation, as shown in Eq. 13-23, equals the residual CO_3^{2-} concentration. Thus,

$$\text{CO}_2 = 0.8 \text{ meq/L} = 0.4 \text{ mM} \times 44 \text{ mg/mmol} = 17.6 \text{ mg/L as CO}_2$$

In practice, the CO_2 dose must be adjusted to achieve the desired treated-water pH value. The treated water will have a calcium hardness of 70 mg/L as $CaCO_3$, a magnesium hardness of 40 mg/L as $CaCO_3$, a total hardness of 110 mg/L, and the following ionic composition:

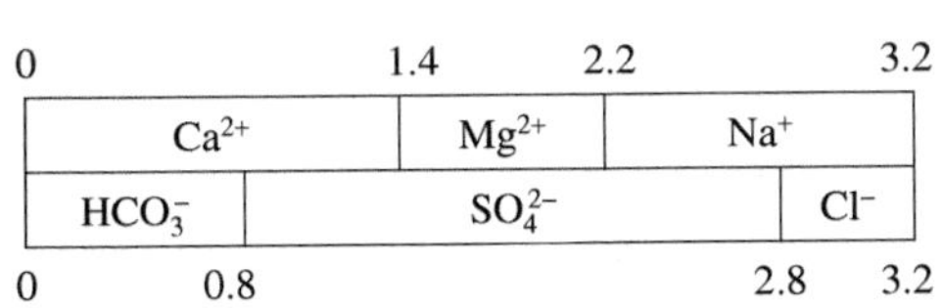

5. Determine the chemical doses for straight lime–soda softening followed by recarbonation.
Soda ash will be added to remove CNH; and the dose, using Eq. 13-45, will be

$$[\text{Na}_2\text{CO}_3] = [\text{CNH}] = 0.6 \text{ meq/L} = 0.3 \text{ mM}$$

Assuming a soda ash purity of 98 percent, the dose in mg/L will be

$$(0.30 \text{ mM}) \times (106 \text{ mg/mmol}) / (0.98) = 32 \text{ mg/L}$$

as 98 percent pure Na_2CO_3.

The doses of lime and CO_2 will remain unchanged from those used for straight lime softening. After softening and recarbonation, assuming that 40 mg/L of $CaCO_3$ remains in solution, the treated water will have a calcium hardness of 40 mg/L as $CaCO_3$, a magnesium hardness of 40 mg/L as $CaCO_3$, a total hardness of 80 mg/L, and the following ionic composition:

0	0.8	1.6	3.2
Ca^{2+} (0–0.8)	Mg^{2+} (0.8–1.6)	Na^+ (1.6–3.2)	
HCO_3^- (0–0.8)	SO_4^{2-} (0.8–2.8)	Cl^- (2.8–3.2)	
0	0.8	2.8	3.2

Soda ash addition removed the CNH, but caused a slight increase in TDS (<2 mg/L), since each Ca^{2+} ion removed was replaced with two sodium ions. However, lime addition removed 2.8 meq/L of Ca^{2+} and HCO_3^- ions, significantly reducing TDS. ▲

Excess Lime or Lime–Soda Softening. Excess lime softening removes both calcium and magnesium hardness. As noted above, excess lime is needed to raise the pH high enough to reduce the magnesium concentration to the desired value. Soda ash is used to remove NCH. Depending on the relative concentrations of alkalinity and hardness in the water being treated, there are two possible cases.

Case 1. All of the hardness is carbonate hardness. In this case, the stoichiometric lime dose, in moles per liter, may be estimated as follows:

$$[CaO] = [H_2CO_3^*] + [CCH] + 2\,[MCH] + \text{Excess} \tag{13-46}$$

In this equation, the excess includes the lime required to convert most of the remaining HCO_3^- to CO_3^{2-} as well as the lime required to increase the OH^- concentration sufficiently to reach the target pH value. When making stoichiometric estimates, it is generally assumed that all of the HCO_3^- is converted to CO_3^{2-} at the target pH value, which is typically about 0.7 to 1.0 pH units higher than the value of pK_2. The reaction of lime with excess HCO_3^-, that is, HCO_3^- not associated with hardness, can be expressed as follows:

$$CaO + 2HCO_3^- \rightarrow CaCO_3(s) + CO_3^{2-} + H_2O \tag{13-47}$$

Assuming that all of the HCO_3^- will be converted to CO_3^{2-}, the stoichiometric lime dosage can also be estimated using the following equation:

$$[CaO] = [H_2CO_3^*] + \tfrac{1}{2}[HCO_3^-] + [MCH] + \text{excess} \tag{13-48}$$

In this equation, the lime required to convert HCO_3^- to CO_3^{2-} is included in the bicarbonate term and not in the excess.

In practice, an excess lime dosage of 0.3 to 0.7 mM is typically required to reduce magnesium hardness to the desired value (Horsley et al., 2005). When estimating lime dosages stoichiometrically, the excess can be estimated on the basis of experience or by choosing a dosage in the typical range. Dosages of 0.50 to 0.65 mM are commonly assumed by those wishing to make a reasonably conservative estimate without performing equilibrium calculations. However, the excess actually required depends on alkalinity, water temperature, ionic strength, the treatment objective for magnesium, and the dissolved silica concentration. (Dissolved silica will be present in the raw water as H_4SiO_4, or silicic acid, which has a pK_1 value of about 9.83 and will consume base as the pH is raised to remove magnesium.) A common approach is to introduce a simple equilibrium calculation, that is, to estimate the pH value corresponding to the solubility of the desired finished water magnesium concentration and to then estimate the lime dose required to achieve the target pH value, as illustrated in Example 13-4.

Case 2. NCH is present. Assuming all of the carbonate alkalinity is initially present in the form of HCO_3^-, the required doses of lime and soda ash may be estimated as follows:

$$[CaO] = [H_2CO_3^*] + [CCH] + 2[MCH] + [MNH] + \text{excess} \tag{13-49}$$

$$[Na_2CO_3] = [NCH] = [Ca^{2+}] + [Mg^{2+}] - \tfrac{1}{2}[HCO_3^-] \tag{13-50}$$

Alternatively, the lime dosage may be estimated as follows:

$$[CaO] = [H_2CO_3^*] + \tfrac{1}{2}[HCO_3^-] + [MCH] + [MNH] + \text{excess} \tag{13-51}$$

Since all of the HCO_3^- is consumed in this case, the excess shown in Eqs. 13-49 and 13-51 includes only the caustic alkalinity needed to raise the pH to the target value.

Soda ash can be added to precipitate the Ca^{2+} associated with the excess lime, but one of three other options is often chosen: (1) allow the Ca^{2+} associated with the excess lime to remain in the finished water, which is the logical option when its removal is not necessary to meet the treatment objectives for calcium hardness and total hardness; (2) remove the Ca^{2+} associated with the excess lime by adding CO_2 in the first stage of two-stage recarbonation; or (3) practice split treatment, treating only a portion of the water with excess lime softening. The first option is most commonly chosen when the entire flow receives excess lime treatment. The second option is generally selected when the goal is to reduce total hardness to a very low level and the utility also desires to avoid increasing the sodium concentration, and to save on chemical costs, as the cost of CO_2 is typically much lower than that of soda ash. The third option is typically chosen when only a fraction of the water requires excess lime softening and the chemical savings associated with split treatment are sufficient to offset its increased capital cost.

Example 13-4 Excess Lime–Soda Softening (Stoichiometric Calculations)
A groundwater having a pH of 7.32, an alkalinity of 200 mg/L as $CaCO_3$, and a temperature of 15°C is analyzed and found to have the following major ion concentrations:

$Ca^{2+} = 68.2$ mg/L $Mg^{2+} = 19.4$ mg/L $Na^+ = 46.0$ mg/L

$SO_4^{2-} = 96.1$ mg/L $Cl^- = 35.5$ mg/L

Determine the concentration of each type of hardness. Estimate the chemical doses required to soften this water to the maximum practical extent (i.e., a calcium hardness of 30 mg/L as $CaCO_3$ and a magnesium hardness of 10 mg/L as $CaCO_3$) using lime–soda softening followed by two-stage recarbonation (Fig. 13-2b). Determine the quantity of softening solids produced, and draw bar diagrams to illustrate the ionic composition of the water after each treatment step. Assume that $Mg(OH)_2$(s) acts as a coagulant in the first stage and a polymer is used in the second stage. Ignore ionic strength effects. Municipal supplies are normally not softened to obtain maximum hardness removal, but may be when it significantly benefits industrial customers or if the softened water is blended with unsoftened water, with the blended water meeting the desired treatment objectives. When only partial removal of Mg^{2+} is desired, some utilities practice split treatment softening, illustrated in Example 13-6.

Solution

1. Estimate the initial HCO_3^- and CO_3^{2-} concentrations.

Since pH is between 6 and 9, the CO_3^{2-} concentration is negligible, thus, $HCO_3^- \approx$ alkalinity = 180 mg/L as $CaCO_3$. Dividing by the equivalent weight of $CaCO_3$,

$$HCO_3^- \approx (200 \text{ mg/L as } CaCO_3)/(50 \text{ mg/meq}) = 4.00 \text{ meq/L}$$

Since HCO_3^- has one acid-base equivalent per mole,

$$HCO_3^- \approx (4.00 \text{ meq/L})/(1 \text{ meq/mmol}) = 4.00 \text{ mM}$$

2. Estimate the initial $H_2CO_3^*$ concentration.

At 15°C, $pK_1 = 6.42$ (Table 3-7). Therefore,

$$K_1 = 10^{-6.42} = \frac{[HCO_3^-][H^+]}{[H_2CO_3^*]} = \frac{(4.00\times10^{-3})(10^{-7.32})}{[H_2CO_3^*]}$$

Solving for $H_2CO_3^*$,

$$[H_2CO_3^*] = 0.50 \times 10^{-3} \text{ M} = 0.50 \text{ mM}$$

Since $H_2CO_3^*$ has two acid-base equivalents per mole,

$$H_2CO_3^* = (0.50 \text{ mM})\ (2 \text{ meq/mmol})\ (50 \text{ mg/meq}) = 50 \text{ mg/L as } CaCO_3$$

This concentration is relatively high, so it will be assumed that air stripping is used to reduce the $H_2CO_3^*$ concentration to 10 mg/L as $CaCO_3$. (Complete removal is not economical.)

3. Determine the types of hardness present.

As shown in Example 13-3, determine the concentration of each constituent in meq/L, then construct a bar diagram. (Note that air stripping to remove CO_2 will convert a small fraction of carbonate hardness to noncarbonate hardness. In this case the change is less than 0.5 percent, so it will be ignored.)

Constituent	mg/L	Mol. wt.	mM	meq/mmol	meq/L
$H_2CO_3^*$	10†	NA	0.10	2	0.20
Ca^{2+}	68.2	40.1	1.70	2	3.40
Mg^{2+}	19.4	24.3	0.80	2	1.60
Na^+	32.2	23.0	1.40	1	1.40
HCO_3^-	200†	NA	4.00	1	4.00
SO_4^{2-}	76.9	96.1	0.80	2	1.60
Cl^-	28.4	35.45	0.80	1	0.80

†mg/L as $CaCO_3$; units must be converted using equivalent weight.

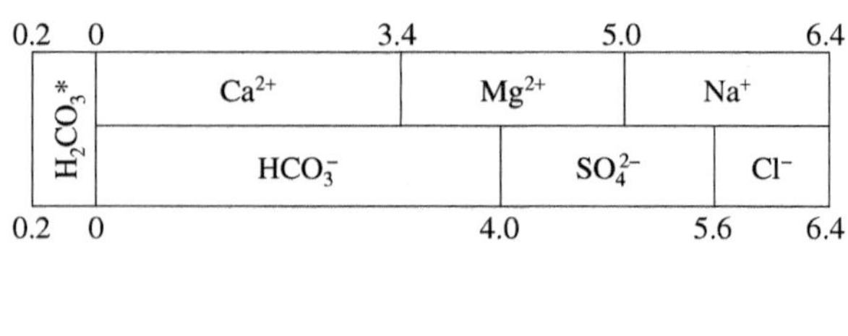

The concentration of each type of hardness can be read from the bar diagram.

$$CCH = 3.40 \text{ meq/L} = 1.70 \text{ mM} = 170 \text{ mg/L as } CaCO_3$$

$$CNH = 0$$

$$MCH = 4.0 - 3.4 = 0.6 \text{ meq/L} = 0.3 \text{ mM} = 30 \text{ mg/L as } CaCO_3$$

$$MNH = 5.0 - 4.0 = 1.0 \text{ meq/L} = 0.5 \text{ mM} = 50 \text{ mg/L as } CaCO_3$$

4. Estimate the pH needed to reduce Mg^{2+} to less than 10 mg/L as $CaCO_3$, or 0.1 mM. Estimate pK_{s0} for $Mg(OH)_2(s)$ at 15°C using Eq. 13-33.

$$pK_{s0} = 0.0175\,(15) + 9.97 = 10.23$$

Determine the pH corresponding to a Mg^{2+} solubility of 0.1 mM (Eq 13-7).

$$[OH^-]^2 = K_{s0}/[Mg^{2+}] = 10^{-10.23}/(0.1 \times 10^{-3}) = 10^{-6.23}$$

$$[OH^-] = 10^{-3.12} \text{ M}$$

To help drive the reaction to completion, and to neutralize silicic acid and any acidic chemicals added, such as alum or chlorine gas, a slightly higher OH^- concentration, 10^{-3} M, will be used. At 15°C, $K_w = 14.34$ (Eq. 13-40 or Table 3-7). Thus,

$$pH = -\log\,(10^{-14.34}/10^{-3.0}) = 11.34$$

5. Estimate chemical doses and solids production for excess lime–soda softening. A good way to do this is to prepare a table summarizing the reactions involved.

Constituent	Conc. (mM)	Equation	Dose (mM) CaO	Dose (mM) Na_2CO_3	Solids (mM) $CaCO_3(s)$	Solids (mM) $Mg(OH)_2(s)$
$H_2CO_3^*$	0.10	13-19	0.10	0.00	0.10	0.00
CCH	1.70	13-11	1.70	0.00	3.40	0.00
CNH	0.00	13-15	0.00	0.00	0.00	0.00
MCH	0.30	13-14	0.60	0.00	0.60	0.30
MNH	0.50	13-16	0.50	0.50	0.50	0.50
Excess	—	†	0.50	†	0.50	0.00
		Subtotal			5.10	0.80
	Solids remaining in solution				−0.30	−0.10
		Total	3.40	0.50	4.80	0.70

†Since all of the HCO_3^- will be consumed by stoichiometric lime addition, the excess lime dose is that needed to achieve the desired pH value of 11.34, which corresponds to an OH^- concentration of 1.00 mM. Since each mole of lime produces 2 moles of OH^-, an excess lime dose of 0.50 mM is needed. An equivalent amount of Na_2CO_3 could be added to remove the Ca^{2+} ions associated with the excess lime. But in this example, the excess Ca^{2+} ions will be removed during the first stage of two-stage recarbonation and the $CaCO_3(s)$ solids produced during recarbonation (0.50 mM) are included in the table.

Alternatively the CaO dose can be estimated using Eq. 13-49.

$$[CaO] = [H_2CO_3^*] + [CCH] + 2[MCH] + [MNH] + \text{excess}$$

$$[CaO] = 0.10 + 1.70 + 2(0.30) + 0.50 + 0.50 = 3.40 \text{ mM}$$

$$(3.40 \text{ mM}) \times (56 \text{ mg/mmol}) = 190 \text{ mg/L}$$

as 100 percent pure CaO. The Na_2CO_3 dose can be estimated using Eq. 13-50.

$$[Na_2CO_3] = [NCH] = 0.50 \text{ mM}$$

$$(0.50 \text{ mM}) \times (106 \text{ mg/mmol}) = 53 \text{ mg/L}$$

as 100 percent pure Na_2CO_3.

Since $CaCO_3$ and $Mg(OH)_2$ have different molecular weights, adding their concentrations together in mM or equivalents is of little practical value. The estimated solids production in mg/L of dry solids is

$$CaCO_3(s) = (4.80 \text{ mM}) \times (100 \text{ mg/mmol}) = 480 \text{ mg/L}$$

$$Mg(OH)_2(s) = (0.70 \text{ mM}) \times (58.3 \text{ mg/mmol}) = 41 \text{ mg/L}$$

$$\text{Total solids (dry weight basis)} = 480 + 41 = 521 \text{ mg/L}$$

Estimates of solids production based on stoichiometric calculations have been found to be reasonably accurate, provided that appropriate adjustments are made when suspended solids, PAC, organic matter, and coagulants also contribute to solids production. See Chap. 22 for additional information regarding the characteristics and handling of lime softening solids.

Prior to recarbonation, the ionic composition of the softened water will be as follows:

<table>
<tr><td>0</td><td>0.6</td><td></td><td>1.6</td><td>1.8</td><td>4.2</td></tr>
<tr><td colspan="2">Ca^{2+} (0–0.6)</td><td>Ca^{2+} (0.6–1.6)</td><td>Mg^{2+} (1.6–1.8)</td><td colspan="2">Na^+ (1.8–4.2)</td></tr>
<tr><td colspan="2">CO_3^{2-} (0–0.6)</td><td>OH^- (0.6–1.6)</td><td>OH^- (1.6–1.8)</td><td>SO_4^{2-} (1.8–3.4)</td><td>Cl^- (3.4–4.2)</td></tr>
<tr><td>0</td><td>0.6</td><td>1.6</td><td>1.8</td><td>3.4</td><td>4.2</td></tr>
</table>

6. Estimate the CO_2 dose required for each stage of two-stage recarbonation.

In the first stage of two-stage recarbonation, the excess lime and $Mg(OH)_2$ remaining in solution will be converted to $CaCO_3(s)$ and $MgCO_3$, respectively. From the bar diagram, the required CO_2 dosage will be (Eqs. 13-21 and 13-22)

$$CO_2 = 1.0 + 0.2 \text{ meq/L} = 1.2 \text{ meq/L}$$

$$(1.2 \text{ meq/L}) \times (22 \text{ mg/meq}) = 26.4 \text{ mg/L as } CO_2$$

A slightly lower CO_2 dose will be needed if an acidic coagulant, such as alum or ferric sulfate, is added to coagulate the $CaCO_3(s)$ produced during the first stage of recarbonation, or if chlorine gas is added to disinfect the water at this point. Following the first stage of recarbonation, the ionic composition will be

<table>
<tr><td>0</td><td>0.6</td><td>0.8</td><td></td><td>3.2</td></tr>
<tr><td>Ca^{2+} (0–0.6)</td><td>Mg^{2+} (0.6–0.8)</td><td colspan="3">Na^+ (0.8–3.2)</td></tr>
<tr><td colspan="2">CO_3^{2-} (0–0.8)</td><td colspan="2">SO_4^{2-} (0.8–2.4)</td><td>Cl^- (2.4–3.2)</td></tr>
<tr><td>0</td><td></td><td>0.8</td><td>2.4</td><td>3.2</td></tr>
</table>

In the second stage of two-stage recarbonation, the residual CO_3^{2-} will be converted to HCO_3^-. From the bar diagram, the required CO_2 dosage will be (Eq. 13-23)

$$CO_2 = (0.8 \text{ meq/L}) \times (22 \text{ mg/meq}) = 17.6 \text{ mg/L as } CO_2$$

The total CO_2 dose for two-stage recarbonation will be

$$\text{Total } CO_2 \text{ required} = 26.4 + 17.6 = 44 \text{ mg/L as } CO_2$$

After the second stage of recarbonation, the ionic composition of the water will be

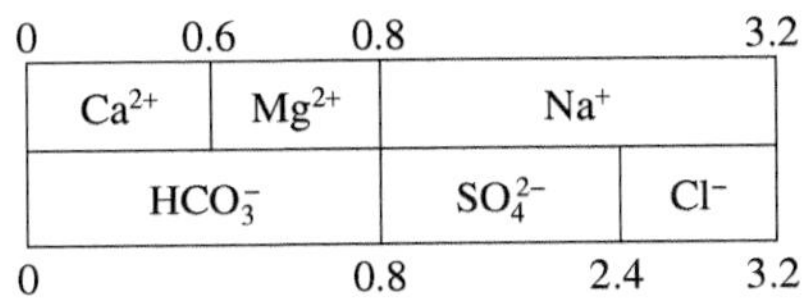

If single-stage recarbonation had been used, the excess lime would have been converted to $Ca(HCO_3)_2$, which would have required an additional 1.0 meq/L of CO_2 (22 mg/L as CO_2), and the finished water would have had the following ionic composition:

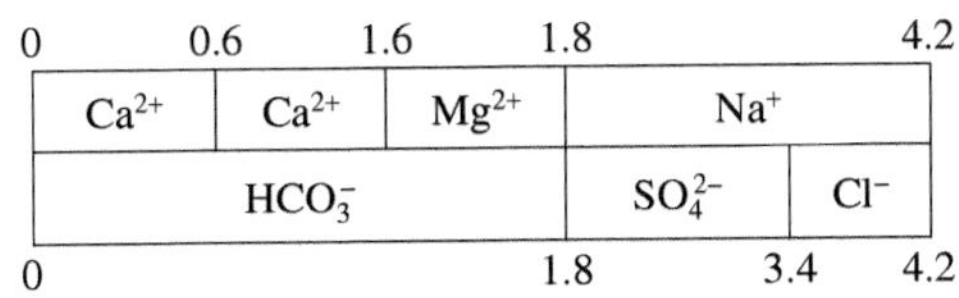

▲

Example 13-5 Excess Lime–Soda Softening (Equilibrium Calculations)
For the groundwater described in Example 13-4, estimate the chemical doses on the basis of chemical equilibrium relationships, taking into consideration the effects of ionic strength, temperature, and complex formation.

Solution

1. Determine the initial ionic strength of the water using Eq. 3-6.
   Using the ion concentrations determined in Step 3 of Example 13-4, converting the units from mM to M, and substituting into Eq 3-6,

$$\begin{aligned} I &= \tfrac{1}{2}[0.0017(2^2)+0.0008(2^2)+0.0014(1^2)+0.004(1^2) \\ &\quad +0.0008(2^2)+0.0008(1^2)] \\ &= 0.0097 \text{ M} \end{aligned}$$

2. Determine the activity coefficients γ_m and γ_d using Eqs. 13-27 and 13-25.

$$\varepsilon = 308.67e^{-0.0045976(273.15+15)} = 82.062$$

$$A = 1.82 \times 10^6(\varepsilon T)^{-3/2} = 1.82 \times 10^6[82.062 \times (273.15+15)]^{-3/2} = 0.501$$

$$\log \gamma_z = -Az^2\left(\frac{\sqrt{I}}{1+\sqrt{I}} - 0.3I\right) = -0.501z^2\left(\frac{\sqrt{0.0097}}{1+\sqrt{0.0097}} - 0.3(0.0097)\right) = -0.0435z^2$$

   Thus, $\log \gamma_m = -0.0435$, so $\gamma_m = 0.905$; and $\log \gamma_d = -0.174$, so $\gamma_d = 0.670$

3. Determine pK_1 and pK_2 for the carbonate system and K_w for water at 15°C (288.15K).

Plummer and Busenberg (1982).

$$pK_1 = 356.3094 + 0.06091964T - 21{,}834.37/T - 126.8339(\log_{10} T) + 1{,}684{,}915/T^2$$

$$pK_2 = 107.8871 + 0.03252849T - 5151.79/T - 38.92561(\log_{10} T) + 563{,}713.9/T^2$$

where T = absolute temperature (kelvin).

At 288.15 K, $pK_1 = 6.419$ and $pK_2 = 10.428$ (consistent with the values shown in Table 3-7).

Using Eq 13-40,

$$pK_w = 4470.99/T + 0.017060T - 6.0875 = 14.345$$

4. Adjust the values of the equilibrium constants determined in Step 3 for ionic strength to obtain concentration-based equilibrium constants.

$$p^cK_1 = -\log \frac{10^{-K_1}}{(\gamma_m)^2} = -\log \frac{10^{-6.419}}{(0.905)^2} = -6.332$$

$$p^cK_2 = -\log \frac{10^{-K_2}}{\gamma_d} = -\log \frac{10^{-10.428}}{(0.670)} = -10.254$$

$$p^cK_w = -\log \frac{10^{-K_w}}{(\gamma_m)^2} = -\log \frac{10^{-14.345}}{(0.905)^2} = -14.258$$

(A note about rounding: Some of the numbers calculated in this example are technically not accurate to the number of decimal places shown, but rounding numbers prematurely, especially logarithmic values used in subsequent calculations, can cause mass, ion, or species balances to be slightly off, leading to confusion as to the correctness of the calculations.)

5. Determine the concentrations of $H_2CO_3^*$, HCO_3^-, CO_3^{2-}, and $C_{T,C}$.

First determine the hydrogen ion concentration, recognizing that pH is a measure of hydrogen ion activity, not its molar concentration. Thus,

$$[H^+] = (10^{-7.32})/0.905 = 10^{-7.277}\ M$$

Then determine the ionization fractions for the carbonate system using Eqs. 3-48, 3-49, and 3-50 with the concentration-based equilibrium constants obtained in step 4.

$$\alpha_0 = \frac{[H_2A]}{C_{T,C}} = \frac{1}{1 + \frac{^cK_1}{[H^+]} + \frac{^cK_1\,^cK_2}{[H^+]^2}} = \frac{1}{1 + \frac{10^{-6.332}}{10^{-7.277}} + \frac{10^{-6.332}\,10^{-10.254}}{(10^{-7.277})^2}} = 0.1018$$

$$\alpha_1 = \frac{[HA^-]}{C_{T,C}} = \frac{1}{1 + \frac{[H^+]}{^cK_1} + \frac{^cK_2}{[H^+]}} = \frac{1}{1 + \frac{10^{-7.277}}{10^{-6.332}} + \frac{10^{-10.254}}{10^{-7.277}}} = 0.8972$$

$$\alpha_2 = \frac{[A^{-2}]}{C_{T,C}} = \frac{1}{1 + \frac{[H^+]}{^cK_2} + \frac{[H^+]^2}{^cK_1\,^cK_2}} = \frac{1}{1 + \frac{10^{-7.277}}{10^{-10.254}} + \frac{(10^{-7.277})^2}{10^{-6.332}\,10^{-10.254}}} = 0.000946$$

Check the sum of the ionization fractions, which should equal 1.

$$\alpha_0 + \alpha_1 + \alpha_2 = 0.1018 + 0.8972 + 0.000946 = 0.999946 \approx 1$$

Solve for $C_{T,C}$ by rearranging Eq. 3-52 and substituting $K_w/[H^+]$ for $[OH^-]$, then, knowing $C_{T,C}$, determine the individual carbonate species from the ionization fractions.

$$C_{T,C} = [\text{Alk} - ({}^cK_w/[H^+]) - [H^+]]/(\alpha_1 + 2\alpha_2)$$

$$= [(200/50{,}000 - (10^{-14.258}/10^{-7.277}) - 10^{-7.277})]/[0.8972 + 2(0.000946)]$$

$$= 0.004449 \text{ M (or 4.449 mM)}$$

$$H_2CO_3^* = \alpha_0 C_{T,C} = 0.1018(4.449) = 0.453 \text{ mM (or 45.3 mg/L as } CaCO_3)$$

$$HCO_3^- = \alpha_1 C_{T,C} = 0.8972(4.449) = 3.992 \text{ mM}$$

$$CO_3^{-2} = \alpha_2 C_{T,C} = 0.000946(4.449) = 0.004 \text{ mM}$$

Check to see that the alkalinity equals 4.00 mM, assuming $[H^+]$ and $[OH^-]$ are negligible.

$$\text{Alk} = HCO_3^- + 2CO_3^{-2} = 3.992 + 2(0.004) = 4.00 \text{ mM}$$

6. Determine $C_{T,C}$ and the concentrations of the individual carbonate species after air stripping.

Air stripping to reduce the concentration of $H_2CO_3^*$ to 10 mg/L as $CaCO_3$ (or 0.100 mM) will reduce $C_{T,C}$, increase pH, and increase CO_3^{-2}. The alkalinity will remain unchanged, but it cannot be assumed that $C_{T,C}$ will be reduced by the exact difference between the initial and final CO_2 concentrations, since speciation will also change as a result of CO_2 loss and the increase in pH. An accurate calculation can be made using a trial-and-error approach, which is typically done by assuming a pH value, then proceeding as in Step 5 until the desired $H_2CO_3^*$ concentration is obtained and the alkalinity check is correct. For pH values between 6 and 9, as expected here, a good first guess can be made by rearranging the expression for K_1 and substituting the known value (after air stripping) of $H_2CO_3^*$ of 0.1 mM (or $10^{-4.00}$ M) and the approximate molar concentration of HCO_3^-.

$$[H^+] \approx \frac{K_1[H_2CO_3^*]}{[HCO_3^-]} = \frac{(10^{-6.332})(10^{-4.00})}{\sim 0.00399} \approx 1.17 \times 10^{-8} \text{ M}$$

$$p[H^+] \approx -\log(1.17 \times 10^{-8}) = 7.932$$

$$pH \approx -\log[(1.17 \times 10^{-8})(0.905)] = 7.975$$

Using this pH value as a starting point, it was found, by trial and error using a spread sheet, that a pH of 7.973 gave the best results, which were as follows (calculated as shown in Step 5):

$$[H^+] = (10^{-7.973})/0.905 = 10^{-7.930} \text{ M}$$

$$\alpha_0 = 0.0245 \qquad \alpha_1 = 0.9709 \qquad \text{and} \qquad \alpha_2 = 0.0046$$

$$\alpha_0 + \alpha_1 + \alpha_2 = 1.000$$

$$C_{T,C} = 4.081 \text{ mM}$$

$$H_2CO_3^* = \alpha_0 C_{T,C} = (0.0245)(4.081) = 0.100 \text{ mM (or 10 mg/L as } CaCO_3)$$

$$HCO_3^- = \alpha_1 C_{T,C} = (0.9709)(4.081) = 3.962 \text{ mM}$$

$$CO_3^{-2} = \alpha_2 C_{T,C} = (0.0046)(4.081) = 0.019 \text{ mM}$$

Check to see that the alkalinity equals 4.00 mM, assuming $[H^+]$ and $[OH^-]$ are negligible.

$$\text{Alk} = HCO_3^- + 2CO_3^{-2} = 3.962 + 2(0.019) = 4.00 \text{ mM}$$

Calculate acidity (for use in Step 8).

$$\text{Acidity} = 2(0.100) + 3.962 = 4.162 \text{ mM}$$

7. Determine the pH value needed to reduce the soluble Mg concentration to 10 mg/L as $CaCO_3$.

This was straightforward in Example 13-4 because only a single Mg species, Mg^{2+}, was considered and ionic strength effects and complexation were ignored. When ionic strength and complex formation are taken into consideration, a computer is typically employed to make iterative calculations since, as the chemical dose is increased to raise the pH, precipitates form and the ionic composition of the water changes, resulting in changes in ionic strength and the concentrations of the species interacting to form complexes. For manual calculations, a good starting point is the approximate composition on the water following lime and soda ash addition, as determined in Step 5 of Example 13-4.

0–0.6	0.6–1.6	1.6–1.8	1.8–3.4	3.4–4.2
Ca^{2+}	Ca^{2+}	Mg^{2+}	Na^+	
CO_3^{2-}	OH^-	OH^-	SO_4^{2-}	Cl^-

(Top scale: 0, 0.6, 1.6, 1.8, 4.2; bottom scale: 0, 0.6, 1.6, 1.8, 3.4, 4.2)

This water has an ionic strength of 0.0124 M. However, some of the ions shown will actually be present in the form of complexes or solids that have not settled, so it is reasonable to assume that there has been no significant change in ionic strength at this point in the process. Accordingly, the values of the parameters calculated that depend only on temperature and ionic strength will be assumed to be unchanged.

The pH needed to reduce Mg to 10 mg/L as $CaCO_3$ (10^{-4} M) can be determined using Eq. 13-39 after determining the values of the constants, corrected for the effects of ionic strength and temperature and assuming an appropriate value of $C_{T,C}$. Using Eq 13-33,

$$pK_{s0,Mg(OH)_2} = 0.0175(15) + 9.97 = 10.233$$

Using Eq 13-35,

$${}^cK_{s0,Mg(OH)_2} = K_{s0,Mg(OH)_2}/(\gamma_d\gamma_m^2) = 10^{-10.233}/[(0.670)(0.905)^2] = 1.067 \times 10^{-10}$$

Using the equations in Table 13-5 to obtain temperature-corrected Mg^{2+} complex formation constants, then correcting them for ionic strength as described following Eq 13-40,

$$K_7 = 10^{2.51} \qquad {}^cK_7 = \gamma_d K_7 = 0.670(10^{2.51}) = 10^{2.33}$$

$$K_8 = 10^{1.03} \qquad {}^cK_8 = \gamma_d K_8 = 0.670(10^{1.03}) = 10^{0.85}$$

$$K_9 = 10^{2.91} \qquad {}^cK_9 = \gamma_d^2 K_9 = 0.670^2(10^{2.91}) = 10^{2.57}$$

$$K_{10} = 10^{2.23} \qquad {}^cK_{10} = \gamma_d^2 K_{10} = 0.670^2(10^{2.23}) = 10^{1.88}$$

The value of $C_{T,C}$ will depend on the doses of lime and soda ash. Since a practical solubility of $CaCO_3(s)$ of 30 mg/L was assumed in the problem statement, it will be assumed that the residual concentration of CO_3^{2-} prior to recarbonation is 30 mg/L as $CaCO_3$ (or 0.3 mM) and that the small amount of HCO_3^- present is included in the assumed CO_3^{2-} concentration; that is, that $C_{T,C} = 0.3$ mM. Other reasonable assumptions can be made, but some of them result in iterative solutions that are best made using a computer-based model. Assuming that $C_{T,C}$ remains unchanged until after $Mg(OH)_2(s)$ has precipitated is not a good assumption because $CaCO_3(s)$ precipitates at the same time and ignoring this fact will cause the concentrations of carbonato and bicarbonato complexes to be significantly overestimated.

Substituting the corrected constants into Eq. 13-39,

$$[Mg]_T = \frac{{}^cK_{s0,Mg(OH)_2}[H^+]^2}{({}^cK_w)^2}(1 + {}^cK_w{}^cK_7/[H^+] + {}^cK_8\alpha_1 C_{T,C} + {}^cK_9\alpha_2 C_{T,C} + {}^cK_{10}[SO_4^{2-}])$$

$$(0.0001) = \frac{(1.067\times10^{-10})[H^+]^2}{(10^{-14.258})^2}[1 + (10^{-14.258})(10^{2.33})/[H^+] + 10^{0.85}\alpha_1(0.0003)$$

$$+ 10^{2.57}\alpha_2(0.0003) + 10^{1.88}(0.0008)]$$

Using a computer to solve for $[H^+]$, recognizing that α_1 and α_2 are functions of $[H^+]$,

$$[H^+] = 10^{-11.35} \qquad pH = -\log[(0.905)(10^{-11.35})] = 11.39$$

$$\alpha_0 = 0.0000 \qquad \alpha_1 = 0.0743 \qquad \text{and} \qquad \alpha_2 = 0.9257$$

and the concentrations of the individual Mg species are

$$Mg^{2+} = 0.0700 \text{ mM}$$

$$MgOH^+ = 0.0185 \text{ mM}$$

$$MgHCO_3^+ = 0.00001 \text{ mM}$$

$$MgCO_3^0 = 0.0072 \text{ mM}$$

$$MgSO_4^0 = 0.0043 \text{ mM}$$

The sum of these species is the target Mg concentration of 0.0001 M (10 mg/L as $CaCO_3$). The most significant Mg complex of those considered, under these conditions, is $MgOH^+$ (18.5% of the total). Although this complex is commonly added in to obtain the total Mg concentration, it is not clear that this approach is correct. In the studies on which Eq. 13-33 is based, the investigators attempted to exclude CO_2 and foreign substances, including sulfate, from the samples. Thus, the concentrations of $MgHCO_3^+$, $MgCO_3^0$, and $CaSO_4^0$ should have been zero or negligible. However, $MgOH^+$ would have been present in these samples, and it is not clear that it was accounted for separately from Mg^{2+}, since $MgOH^+$ was commonly ignored in the time period when the experiments were conducted. If $MgOH^+$ is already included in the calculated Mg^{2+} concentration, then adding it in again will make the pH calculation above slightly inaccurate, though the difference will be small given the other assumptions involved.

8. Determine the doses of lime and soda ash needed to treat this water.

The lime and soda ash doses can be determined by performing chemical balances on the changes in Ca^{2+}, Mg^{2+}, $C_{T,C}$, and acidity during softening, and the calculations can be checked using a balance on alkalinity. The first step is to characterize the carbonate system. Assuming, as before, that $C_{T,C} = 0.0003$ M, and using the pH of 11.39 and the ionizations fractions obtained in Step 7, chemical equilibrium calculations yield

$$H_2CO_3^* = 0.0000\ \text{M} \qquad HCO_3^- = 0.022\ \text{mM} \qquad CO_3^{2-} = 0.278\ \text{mM}$$

$$[H^+] = -\log[(10^{-11.39})/(0.905)] = 10^{-11.35}\ \text{M}$$

$$[OH^-] = {}^cK_w/[H^+] = (10^{-14.258})/(10^{-11.35}) = 1.234\ \text{mM}$$

$$\text{Alk} = 1.812\ \text{mM} \qquad \text{Acidity} = -1.212\ \text{mM} \qquad C_{T,C} = 0.300\ \text{mM}$$

Thus, lime and soda ash addition, coupled with precipitation of $CaCO_3$(s) and $Mg(OH)_2$(s), resulted in the following water quality changes (ignoring complexed OH^- ions):

Loss of acidity = 4.162 – (–1.212) = 5.374 mM
Loss of $C_{T,C}$ = 4.081 – (0.300) = 3.781 mM
Loss of Ca^{2+} = 1.70 – 0.8 = 0.90 mM
Loss of Mg^{2+} = 0.8 – 0.10 = 0.70 mM
Loss of alkalinity = 4.000 – 1.812 = 2.188 mM

The loss of acidity was caused by lime addition, since each mole of lime removes 2 moles (or equivalents) of acidity, and lime was also the source of the OH^- ions removed from the system by precipitation of $Mg(OH)_2$(s). Precipitation of $CaCO_3$(s) and addition of Na_2CO_3 cause no change in acidity. Thus,

$$\text{CaO dose} = 0.5[(5.374 + 2(0.70)] = 3.387\ \text{mM}$$

The loss of Ca^{2+} was the net result of the increase in Ca^{2+} caused by lime addition and loss of Ca^{2+} associated with precipitation of $CaCO_3$(s). Thus,

$$CaCO_3\text{(s) precipitated} = 1.70 - 0.80 + 3.387 = 4.287\ \text{mM}$$

The change in $C_{T,C}$ was the net result of adding soda ash and precipitating $CaCO_3$(s). Thus,

$$Na_2CO_3\ \text{dose} = -(4.081 - 0.300) + 4.287 = 0.506\ \text{mM}$$

These calculations can be confirmed by checking them against the change in alkalinity, recognizing that each mole of lime or soda ash adds 2 moles (equivalents) of alkalinity to the system, while precipitation of a mole of $CaCO_3$(s) or $Mg(OH)_2$(s) removes 2 moles of alkalinity from the system. Thus,

$$\text{Change in alkalinity} = 2[CaO] + 2[Na_2CO_3] - 2[CaCO_3(s)] - 2[Mg(OH)_2(s)]$$

$$= 2(3.387) + 2(0.506) - 2(4.287) - 2(0.7) = -2.188\ \text{mM}$$

This agrees with the value previously calculated; therefore, the calculations are internally consistent.

9. Calculate the CO_2 dose for the first stage of two-stage recarbonation and the resulting pH.

To maximize Ca^{2+} removal in the first stage of two-stage recarbonation (without adding additional Na_2CO_3), the acidity should be increased from a negative value to zero. The acidity of the water leaving excess lime softening is –1.212 mM and each mole of CO_2 ($H_2CO_3^*$) contributes two moles of acidity. Thus,

$$CO_2\ \text{dose} = 1.212/2 = 0.606\ \text{mM (or 26.7 mg/L as } CO_2)$$

Adding CO_2 will not directly affect alkalinity, but will increase $C_{T,C}$ on a 1:1 basis. For each mole of $CaCO_3$(s) precipitated, there will be a 2-mole (equivalent) decrease in alkalinity as well as a 1-mole decrease in $C_{T,C}$. Assuming precipitation of 0.60 mM of $CaCO_3$(s), leaving a residual calcium hardness of 0.30 mM, or 30 mg/L as $CaCO_3$(s), the water leaving the first stage of two-stage recarbonation will have the following characteristics:

Alkalinity = 1.812 − 2(0.60) = 0.612 mM
$C_{T,C}$ = 0.300 + 0.606 − 0.60 = 0.306 mM
Check: $C_{T,C}$ = ½(alkalinity + acidity) = ½(0.612 + 0) = 0.306 mM

The composition of this water is very similar to that shown following the first stage of recarbonation in Step 6 of Example 13-4, the main difference being that HCO_3^- and OH^- are present, whereas only CO_3^{-2} is shown in Example 13-4. Complexes are also present, but are not included in the bar diagram, though they can readily be accounted for in ion, mass, and other balances when using a computer program. The ionic strength of this water is ~ 0.0048 M, or about half the value prior to softening. Therefore, the constants influenced by ionic strength must be recalculated to accurately estimate the pH value of the water leaving this stage of treatment. Using a spreadsheet to make these corrections and to solve, by trial and error, for a pH value that simultaneously satisfies the constraints on alkalinity, acidity, and $C_{T,C}$ shown, the pH value was found to be 10.44, which is in the expected range.

The effluent from the first stage of recarbonation will be unstable, and more CO_2 will be needed to reach the desired final value of pH, LSI, or CCPP. See Chap. 20 for details.

10. Verify that the desired Ca hardness concentration can be obtained.

In this example, it seems intuitively obvious that the desired residual hardness concentration can be obtained; but in more extreme cases the solubility of $CaCO_3$(s) may be sufficiently high, due to the effects of ionic strength, temperature, and complex formation, that the goal cannot be met. This can be checked, as described earlier in this chapter and in Chap. 3, by comparing the ion product Q with the solubility product constant ${}^cK_{s0}$, after making appropriate corrections. The value of pK_{sc} at 15°C (288.15K) can be found using Eq. 13-28.

$$pK_{sc} = 171.9065 + 0.077993(288.15) - 2839.319/(288.15) - 71.595[\log_{10}(288.15)]$$
$$= 8.430$$

Using Eq. 13-34 with a value of γ_d of 0.747 (for I = 0.0048 and T = 288.15 K),

$${}^cK_{s0,CaCO_3} = K_{s0,CaCO_3}/(\gamma_d^2) = (10^{-8.430})/[(0.747)^2] = 10^{-8.177}$$

Following recarbonation, Ca hardness = 0.300 mM and $C_{T,C}$ = 0.306 mM. Using a spreadsheet and assuming I = 0.0048 and T = 288.15K, the following values were obtained:

$$\alpha_0 = 0.0000 \qquad \alpha_1 = 0.4385 \qquad \text{and} \qquad \alpha_2 = 0.5614$$

$$HCO_3^- = 0.134 \text{ mM} \qquad CO_3^{2-} = 0.172 \text{ mM} \qquad \text{and} \qquad OH^- = 0.134 \text{ mM}$$

$$\gamma_m = 0.930 \qquad \text{and} \qquad \gamma_d = 0.747$$

Before determining the ion product Q, the concentrations of the Ca complexes must be deducted from the total Ca concentration, and the concentrations of the carbonato complexes must be deducted from the CO_3^{2-} concentration. The concentrations of the Ca and Mg complexes were determined using the equations in Table 13-5, with the constants corrected for temperature and ionic strength as described earlier, and assuming $Ca_T = 0.30$ mM and $Mg_T = 0.10$ mM, consistent with the treatment objectives. Using a spreadsheet, the concentrations were found to be

$$Ca^{2+} = 0.253 \text{ mM} \qquad Mg^{2+} = 0.0866 \text{ mM}$$

$$CaOH^{+} = 0.00041 \text{ mM} \qquad MgOH^{+} = 0.0025 \text{ mM}$$

$$CaHCO_3^{+} = 0.00024 \text{ mM} \qquad MgHCO_3^{+} = 0.00008 \text{ mM}$$

$$CaCO_3^{0} = 0.028 \text{ mM} \qquad MgCO_3^{0} = 0.0055 \text{ mM}$$

$$CaSO_4^{0} = 0.019 \text{ mM} \qquad MgSO_4^{0} = 0.0053 \text{ mM}$$

(When using a more rigorous approach, these results would be used to adjust the concentrations of other species, such as the carbonate species, which would in turn result in changes in ionic strength, pH, and other parameters. Such an approach is greatly facilitated by using a computer model able to solve many equations simultaneously. Only the first iteration is shown here.)

The CO_3^{2-} concentration, when adjusted for the concentrations of the known carbonato complexes, is

$$CO_3^{2-} = 0.172 - 0.028 - 0.0055 = 0.139 \text{ mM}$$

Knowing the concentrations of Ca^{2+} and CO_3^{2-}, Q can be calculated.

$$Q = (0.253 \text{ mM} \times 0.001 \text{ mol/mmol})(0.139 \text{ mM} \times 0.001 \text{ mol/mmol}) = 10^{-7.454}$$

Since $Q > {}^{c}K_{s0}$, further precipitation of $CaCO_3$(s) can occur, so the goal can be met; and these results suggest that a significantly lower final Ca concentration can be achieved or that the treatment goals could be met using slightly lower doses of CaO and Na_2CO_3.

An alternative and widely used approach is to assume that the softening process produces an effluent Ca^{2+} concentration in equilibrium with $CaCO_3$(s), using Eq. 13-38 to determine the lowest total soluble Ca concentration that can be achieved. This approach results in slightly lower estimated chemical doses, but higher dosages will usually be required in practice due to failure to reach equilibrium, the presence of suspended $CaCO_3$(s) and $Mg(OH)_2$(s) in the settled water, and the presence of other substances that complex hardness ions or otherwise increase the solubility of $CaCO_3$(s) or $Mg(OH)_2$(s). Both approaches, as well as the stoichiometric approach demonstrated in Example 13-4, produce reasonable and similar estimates of the required chemical doses. In practice the operator will in most cases have to use slightly different chemical doses to meet the treatment objectives while minimizing chemical costs. ▲

Two-Stage Softening. *Two-stage softening* consists of excess lime softening followed by a second stage of softening that is actually the first stage of two-stage recarbonation, as shown in Fig. 13-2b. This process variation carries an increased capital cost, relative to single-stage softening followed by single-stage recarbonation, but it offers several advantages including: (1) maximum hardness reduction; (2) reduced CO_2 consumption (since some of the CO_3^{2-} is precipitated as $CaCO_3$(s) instead of being converted to HCO_3^-; (3) greater resistance to treatment upsets, since problems that arise in the first basin can potentially be addressed as the water passes through the second basin; and (4) a stable, high-quality effluent, due to the increased detention time and an additional opportunity to coagulate, flocculate, and settle the water after the first stage of CO_2 addition. Two-stage softening is sometimes referred to as *two-stage treatment,* but the term *two-stage treatment* should be avoided because it may also refer to coagulation followed by softening or softening followed by coagulation. When the source water has a high concentration of organic matter, pretreating the water using coagulation, flocculation, and clarification may remove DBP precursors more effectively than softening alone, especially when using straight lime–soda softening rather than excess lime softening; and it may also enhance the softening process, since organic matter can inhibit $CaCO_3$(s) precipitation.

Split Treatment Softening. This process conserves chemicals and reduces solids production by fully softening only a portion of the flow. It is commonly considered when removing magnesium, since only a portion of the magnesium typically needs to be removed and the chemical dosages required for magnesium removal are relatively high. In this process, the flow is split, as illustrated in Fig. 13-2c, with one portion receiving excess lime or lime–soda softening and the other portion bypassing excess lime softening. The two streams are then recombined and treated using straight lime or lime–soda softening to remove additional calcium hardness. When the two streams are recombined, the excess lime reduces the lime dosage needed to soften the bypassed flow; CO_2 in the bypassed flow helps neutralize the excess lime carried over from the first stage of treatment, and any extra alkalinity (HCO_3^-) present in the bypassed flow can be used to remove CNH from the stream receiving excess lime treatment, thereby reducing or eliminating the need for soda ash. The first-stage softening basin and the bypass line may be sized to handle only a fraction of the flow, thereby reducing capital costs, but they may also be sized to handle the entire flow to increase operational flexibility. The fraction of the flow bypassing the first stage is determined by performing a mass-balance calculation on magnesium hardness at the point where the flows are recombined; this yields

$$x = \frac{\mathrm{Mg}_f - \mathrm{Mg}_1}{\mathrm{Mg}_r - \mathrm{Mg}_1} \tag{13-52}$$

where x = fraction of water bypassing the first stage
Mg_r = magnesium hardness in the raw (unsoftened) water
Mg_1 = magnesium hardness leaving the first-stage (excess lime) softening unit
Mg_f = magnesium hardness in the finished water (treatment objective)

Split treatment is rarely used to soften surface waters because variations in water quality can make it difficult to maintain the proper split ratio and chemical dosages and to reliably meet treatment objectives for hardness, turbidity, color, and taste and odor. Two-stage softening is more commonly recommended when treating surface waters.

The term *split treatment* is also used to describe situations in which only a portion of the flow is softened before being combined with the remaining flow, which may be treated in parallel, using a different process, or which may remain untreated until the flows are recombined. Such an arrangement is common when only partial softening is needed. The CO_2 in the bypassed flow is used to stabilize the softened water (perhaps only partially), and the combined flows are then coagulated and settled prior to filtration. When treating surface waters in this fashion, the entire flow must be properly coagulated at some point prior to filtration to comply with the requirements of the Surface Water Treatment Rule (Chap. 1). In such cases, recarbonation is often unnecessary, but the filtered water should still be properly stabilized before being pumped into the distribution system.

Example 13-6 Split Treatment Softening
For the groundwater described in Example 13-4, determine the chemical doses required for split treatment softening (Figure 13-2c) assuming the following treatment objectives: Mg hardness = 40 mg/L as $CaCO_3$ and Ca hardness = 60 mg/L as $CaCO_3$.

Solution

1. Determine the fraction of flow bypassing the first stage of softening.
Using Eq 13-52,

$$x = \frac{Mg_f - Mg_1}{Mg_r - Mg_1} = \frac{40-10}{80-10} = 0.429 \text{ or } 42.9\%$$

Therefore, 57.1 percent of the flow will be treated using excess lime softening in the first stage of softening. In practice, a higher percentage of the flow is typically treated in the first stage to provide greater operational flexibility. Where air stripping is used to remove CO_2, the bypass stream can also bypass the air stripping process if the utility wishes to use the CO_2 in the bypass stream to recarbonate the water leaving the first stage of softening; but this may not be feasible if air stripping is also used to remove other volatile contaminants from the water.

2. Determine the lime dose to be added to the first stage.
As found in Example 13-4, the lime dose added to the first stage of softening, that is, excess lime softening, will be 3.40 mM (190 mg/L as CaO). However, this dose will be added to only 57 percent of the flow. Based on the total plant flow rate,

$$\text{CaO dose (Stage 1)} = 0.571 \times 3.40 \text{ mM} = 1.94 \text{ mM (or 108 mg/L as CaO)}$$

3. Determine the excess lime leaving the excess lime softening process (Stage 1).
The excess lime dose, from Example 13-4, is 0.50 mM. This lime will be used in Stage 2 to remove $H_2CO_3^*$ and CCH from the bypassed flow. Based on the total plant flow rate, the excess caustic remaining in the effluent from the first stage of softening will be

$$\text{Excess} = 0.571 \times 0.50 \text{ mM} = 0.285 \text{ mM} = 0.571 \text{ meq/L} = 16 \text{ mg/L as CaO}$$

The residual $Mg(OH)_2$ leaving Stage 1 can also be used for softening in Stage 2. It is usually ignored, but will be included in this example to make the ion balance complete.

$$Mg(OH)_2 \text{ (from Stage 1)} = 0.571 \times 0.20 \text{ meq/L} = 0.114 \text{ meq/L}$$

4. Determine the composition of the blended flow entering second-stage softening.
The excess lime will be present in dissolved form, as Ca^{2+} and OH^- ions, at a concentration of 0.285 mM (0.571 meq/L), as found in Step 3. The concentrations of chloride and sulfate will remain unchanged; so will the sodium concentration, since no soda ash was added during the first stage of softening. The initial concentrations of the other constituents, at the blending point, can be determined by performing mass balances:

$$Ca^{2+} \text{ and } OH^- \text{ (from excess lime in Stage 1)} = 0.571 \times 1.0 \text{ meq/L} = 0.571 \text{ meq/L}$$

$$Ca^{2+} + CO_3^{2-} \text{ (not precipitated in Stage 1)} = 0.571 \times 0.6 \text{ meq/L} = 0.343 \text{ meq/L}$$

$$Mg^{2+} \text{ (from bypass)} = 0.429 \times 1.60 = 0.686 \text{ meq/L}$$

$$Mg^{2+} \text{ and } OH^- \text{ (not precipitated in Stage 1)} = 0.114 \text{ meq/L (from Step 3)}$$

$$\text{Total } Mg^{2+} = 0.114 + 0.686 = 0.800 \text{ meq/L (or 40 mg/L as } CaCO_3\text{, as planned)}$$

$$H_2CO_3^* \text{ (from bypass)} = 0.429 \times 0.2 \text{ meq/L} = 0.086 \text{ meq/L}$$

$$HCO_3^- \text{ (from bypass)} = 0.429 \times 4.0 \text{ meq/L} = 1.716 \text{ meq/L}$$

$$Ca^{2+} \text{ (from the bypassed flow)} = 0.429 \times 3.4 \text{ meq/L} = 1.459 \text{ meq/L}$$

Thus, the composition of the blended water entering second-stage softening (with individual constituents not drawn to scale, for clarity) will be

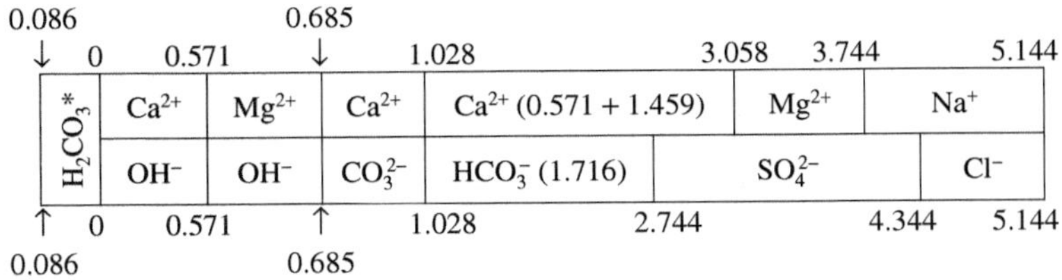

However, the $Ca(OH)_2$ and $Mg(OH)_2$ carried over from Stage 1 will react with $H_2CO_3^*$ and CCH in the bypassed flow. First, allow the $H_2CO_3^*$ to react with excess lime to form $CaCO_3$.

$$Ca(OH)_2 = 0.571 - 0.086 = 0.485 \text{ meq/L}$$

Then use the remaining excess lime to remove CCH.

$$Ca^{2+} = (0.571 + 1.459) - 0.485 = 1.545 \text{ meq/L}$$

$$HCO_3^- = 1.716 - 0.485 = 1.231 \text{ meq/L}$$

Lastly, use the OH^- ions associated with $Mg(OH)_2$ to remove CCH, recognizing that one equivalent of OH^- will react with one equivalent of HCO_3^- to precipitate two equivalents of Ca^{2+}. Thus,

$$Ca^{2+} = 1.545 - 2\,(0.114) = 1.317 \text{ meq/L}$$

$$HCO_3^- = 1.231 - 0.114 = 1.117 \text{ meq/L}$$

Assuming 0.3 mM of $CaCO_3$ (0.6 meq/L) remains in solution, the composition of the water after blending but before adding additional lime will be

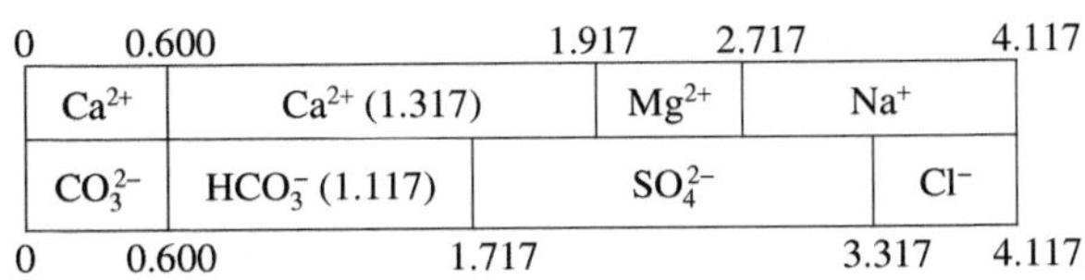

5. Determine the lime and soda ash doses required for the second stage of softening.

The blended water entering the second-stage softening unit contains 1.917 meq/L of Ca^{2+} hardness, or 96 mg/L as $CaCO_3$, so additional lime will be needed to meet the treatment objective of 60 mg/L as $CaCO_3$ (1.20 meq/L). Since the blended water is already saturated with $CaCO_3$(s), the additional dose required to meet the treatment objective is

$$CaO = 1.917 \text{ meq/L} - 1.20 \text{ meq/L} = 0.717 \text{ meq/L} = 0.36 \text{ mM} = 20 \text{ mg/L as CaO}$$

The additional lime can be added to either stage of treatment. Adding it to the first stage of treatment will reduce the number of lime feeders that must be kept in service.

Soda ash is not needed, since the treatment objective can be met without removing CNH. CNH is present in the blended water even though there was none in the raw water because lime was used to remove MNH in the first stage of softening. Depending on water quality, there may or may not be enough alkalinity available for removing Ca^{2+} using only lime in the second stage of softening. If not, soda ash can be added or, if the utility wishes to avoid the use of soda ash, the treatment objective for calcium hardness can be relaxed. Soda ash is usually added to the first stage to simplify chemical feeding; but it can also be added to the second stage; this can help maintain the desired solids concentration in the second-stage softening process.

The total lime dose required for both stages of softening is 1.94 + 0.36 = 2.30 mM, or 32 percent less than required for single-stage excess lime–soda softening in Example 13-4. Furthermore, no soda ash is required, and less CO_2 is required to stabilize the softened water.

6. Determine the CO_2 dose required to stabilize the softened water.

The water leaving the second stage of softening has already been partially stabilized using the bypassed flow. It contains 0.6 meq/L of $CaCO_3$, which can be converted to $Ca(HCO_3)_2$ using a CO_2 dose of 0.6 meq/L, or 13.2 mg/L as CO_2. The finished water will have the following composition:

0 – 1.20	1.20 – 2.00	2.00 – 3.40
Ca^{2+}	Mg^{2+}	Na^+

0 – 1.00	1.00 – 2.60	2.60 – 3.40
HCO_3^-	SO_4^{2-}	Cl^-

▲

Process Variations Facilitating Lime Recovery. Lime softening plants produce large quantities of solids composed largely of $CaCO_3$(s), which can potentially be recovered and recalcined (or "burned," in a calcining furnace) to produce lime and CO_2 which can be reused on site for softening and recarbonation. Since 2 moles of $CaCO_3$(s) are produced for each mole of CCH removed, lime recovery can produce excess lime and CO_2, which can be sold to a neighboring plant. Furthermore, lime recovery greatly reduces the cost of residuals disposal, which can be substantial. Thus, at first glance, lime recovery appears to be an attractive option, and many utilities practicing lime softening have evaluated the feasibility of lime recovery.

For lime recovery to be feasible, the solids sent to the calcining furnace must consist of relatively pure $CaCO_3$(s); otherwise, impurities will build up in the system, adversely affecting finished water quality and making the process inefficient and uneconomical. Lime recovery is not feasible when softening turbid surface waters because $CaCO_3$(s) cannot be economically separated from the solids removed from the raw water, but this limitation can potentially be overcome using an appropriate pretreatment system. For low-turbidity waters, magnesium poses the greatest problem, but iron, silica, and other impurities can also interfere with lime recovery (Burris et al., 1976).

Magnesium hydroxide interferes with lime recovery for two reasons. First, it is light and hydrous, making the lime softening residuals difficult to dewater and the dewatered solids too wet for energy-efficient recalcining. A solids content of 70 percent or higher is recommended for recalcination (Pizzi, 1995). Second, recalcining of $Mg(OH)_2$(s) produces MgO, which unlike CaO is not useful for softening because MgO simply reverts to $Mg(OH)_2$(s) when added to water. Unless magnesium is removed from the system, it will accumulate in the system, making the system increasingly less efficient.

Process variations designed to facilitate lime recovery by removing magnesium from the solids stream sent to the recalcining furnace include (1) three-stage treatment; (2) the Lykken-Estabrook process (Aultman, 1939), which involves using split treatment, discarding the solids from the first stage and calcining solids recovered from the second stage, which is feasible only if a sufficient quantity of $CaCO_3$(s) is produced in the second stage; (3) centrifuging the solids to separate the denser $CaCO_3$(s) from the lighter solids composed primarily of $Mg(OH)_2$(s), aluminum and iron oxides, and organic matter (Sheen and

Lammers, 1944; Burris et al., 1976); and (4) dissolving the $Mg(OH)_2(s)$ with CO_2, as is done in the magnesium-carbonate process (Burris et al., 1976).

Lime recovery was practiced or evaluated at a number of lime softening plants in the past (see, for example, Dye and Tuepker, 1971), but by the mid-1990s, fewer than half a dozen were known to be doing so (Pizzi, 1995). The author is aware of two plants currently recovering lime: Miami-Dade County, Florida, employing a variation of three-stage treatment, and Dayton, Ohio, using the magnesium-carbonate process. Other utilities have stopped recovering lime, primarily because it has not been found to be practical or cost effective under local conditions.

Lime recovery requires a relatively large capital expenditure (for the calcining furnace and ancillary equipment); places an additional burden on the operating staff (who must also be trained to properly run a calcining furnace); requires a relatively pure supply of $CaCO_3(s)$; and can produce poor quality lime if the facilities are not properly designed, constructed, operated, and maintained. Economy of scale is a significant factor, and a lime production rate of at least 45,400 kg/d (50 tons/d) has been cited as a desirable rate for a practical system (Pizzi, 1995). Many lime softening plants are located close enough to a lime supplier that lime recovery saves little in shipping costs. However, if transportation costs and solids disposal costs increase sufficiently, many utilities that soften may decide to take a fresh look at the economics of lime recovery.

Three-Stage Treatment. This process variation, also known as the *Hoover process* and as *two-stage treatment,* has been described by Aultman (1939), by Dye and Tuepker (1971), and in *Hoover's Water Supply and Treatment* (Pizzi, 1995 and earlier editions). Straight lime or lime–soda softening is practiced in the first stage, producing solids consisting primarily of $CaCO_3(s)$, provided that the raw water is relatively free of turbidity. Excess lime softening is practiced in the second stage, producing both $CaCO_3(s)$ and $Mg(OH)_2(s)$, which are discharged to waste. Recarbonation is considered as the third stage. If two-stage recarbonation is used, relatively pure $CaCO_3(s)$ solids are also produced in the first recarbonation stage. The solids from the first stage, combined with solids from the third stage, if applicable, are dewatered and recalcined.

The Miami-Dade County Water & Sewer Department recovers lime using a two-stage lime softening process (Segers, 2009). Straight lime softening is used to remove CCH in the first stage, and the solids are centrifuged to concentrate the $CaCO_3(s)$ to about 70 percent solids, and to remove lighter solids. The concentrated solids are recalcined in one of two calcining furnaces serving a total of three treatment plants. The recalcined lime slakes satisfactorily and contains about 90 to 95 percent available CaO. In the second stage of lime softening, lime and other chemicals (coagulants) are added to increase the removal of color and organic matter (enhanced softening), but little or no magnesium is removed. Single-stage recarbonation is used to reduce the pH of the softened water to the desired value (about 8.9 to 9.1). The high cost of shipping lime to this location contributes significantly to the cost-effectiveness of the lime recovery process.

The Magnesium-Carbonate Process. In this process, also referred to as the *A.P. Black process* in honor of its developer, CO_2 (from the limekiln stack gas) is bubbled through the residual solids to dissolve $Mg(OH)_2(s)$, converting it into a clear solution of magnesium bicarbonate (Black et al., 1971; Thompson et al., 1972; USEPA, 1975). The CO_2 dosage is controlled to avoid dissolving the remaining $CaCO_3(s)$ solids, which are concentrated, dewatered, and fed to the limekiln. The magnesium-carbonate solution is recycled back to the lime softening process, where it serves as a coagulant. Excess magnesium bicarbonate can be discharged downstream or converted to $MgCO_3 \cdot 3H_2O(s)$ and sold.

The city of Dayton, Ohio, has used the magnesium-carbonate process since about 1958 (Zell, 2009). The City's lime recovery facility reclaims $CaCO_3$ from its two large lime softening plants, producing enough lime to supply these plants plus two others and enough CO_2 to run the process and to recarbonate the softened water. Excess lime can be produced by a lime recovery facility because 2 moles of $CaCO_3$(s) are produced for each mole of CCH removed. The city also produced and sold $MgCO_3 \cdot 3H_2O$(s) in the 1970s, but ceased doing so because chemical prices were too low to justify this practice. Since chemical costs have increased recently, the city plans to reevaluate the economic feasibility magnesium-carbonate recovery.

Aeration Softening. *Aeration softening* is a term that has been used to describe the removal of hardness following air stripping, without lime or soda ash addition (Chao and Westerhoff, 2002). Removing CO_2 by air stripping can cause a water supply to become supersaturated with $CaCO_3$(s); this can lead to encrustation of air stripping towers, treatment basins, and filter media, but this also provides a means for removing a limited amount of CCH.

Hot-Process Softening. In hot-process softening the water is preheated, typically to 100°C or higher, in a pressure vessel. Heating the water significantly reduces the solubility of both $CaCO_3$(s) and $Mg(OH)_2$, greatly increasing both the rate and extent of hardness removal. Silica is effectively adsorbed by $Mg(OH)_2$(s), and a magnesium salt can be added if needed to increase silica removal. Heating also dramatically lowers the solubility of dissolved gases, such as carbon dioxide and oxygen, which can then be readily removed using a deaerator. Removing carbon dioxide results in a proportional reduction in the lime dose required for softening. This process is not used to treat public water supplies, but is commonly employed in certain industrial applications, such as pretreating boiler makeup water prior to demineralization. Anthracite filters are normally used to remove the solids carried over from a hot-process softener, since silica sand is relatively soluble in hot alkaline water and would, therefore, release silica back into the treated water.

Precipitative Softening Systems

Precipitative softening may be accomplished using conventional systems, solids contact basins, or pellet reactors. The best system for a particular application depends on a number of factors including raw-water quality, treatment objectives, and space limitations. See Chaps. 8 and 9 for additional information regarding rapid mixing, flocculation, and sedimentation, and Horsley et al. (2005) for additional information regarding the design of lime softening systems.

Conventional Systems. A conventional softening system includes rapid mixing, flocculation, and sedimentation in sequence, or rapid mixing followed by a flocculator-clarifier (see Chap. 9). Such systems closely resemble those used to coagulate and clarify surface water, but there are important design and operational differences in softening systems.

The purpose of rapid mixing in a lime softening system is to dissolve the lime and to disperse other chemicals, such as soda ash or a coagulant, into the water being treated. In softening systems, rapid mixers may require longer mixing times and can use lower mixing speeds than those used for coagulation. It is important to provide enough time for the lime to dissolve, especially in cold, hard water; and the mixing speed only needs to be high enough to keep the hydrated lime particles in suspension as they dissolve (Reh, 1978). A velocity gradient (G value) of 300 sec^{-1} is generally considered adequate, but higher values may be appropriate when a surface water supply is being simultaneously coagulated and softened.

Flocculators in lime softening systems are typically designed to have higher mixing speeds than those used for coagulation in the absence of softening. Higher speeds are needed to keep the flocculating solids in suspension, since they are typically much denser than those produced by coagulation alone. In systems that recirculate solids to promote growth of $CaCO_3(s)$ crystals, individual particles may also be quite large in size. Flocculation time depends not only on flocculation kinetics but also on the amount of time required for the softening reactions to approach equilibrium, which in turn depends on water temperature and on solids recirculation practices. A flocculation time of 30 to 45 min is typical (Horsley et al., 2005), but shorter times may be justified based on local conditions and experience.

Softening solids settle much more rapidly than those typically produced in a coagulation plant, so sedimentation basins in softening plants are typically designed to have significantly shorter contact times and higher surface loading rates, especially when treating groundwater supplies (Recommended Standards, 2007). The rake arms must be designed to withstand the much higher torque associated with high concentrations of dense solids. Larger solids withdrawal pipes are used to handle the larger quantity of solids, to avoid plugging problems, and to facilitate cleaning if plugging does occur. It is recommended that provisions be made to routinely flush the solids withdrawal pipes to prevent solids accumulation. Dissolved air flotation is not used in precipitative softening systems, since softening solids are relatively dense and thus easier to settle than to float, but $CaCO_3(s)$ is floated in various industrial applications.

In coagulation plants, solids recirculation is generally avoided and may even be prohibited, but in lime softening plants it is generally recommended and widely practiced. Indeed, precipitation of $CaCO_3(s)$ can be viewed as a crystallization process, and solids recirculation is perhaps the most obvious way to enhance a crystallization process. Advantages of recirculating solids (back to the rapid mixing basin) include

1. Faster reaction kinetics (since the growth rate of $CaCO_3(s)$ crystals is proportional to the available surface area), resulting in a more stable effluent, improved magnesium removal (Larson et al. 1959; Tuepker and Hartung, 1960), reduced use of chemicals for recarbonation and stabilization, and reduced filter encrustation
2. Elimination of metastable forms of $CaCO_3$, since they are unstable in the presence of calcite and once calcite is present it will remain in the system if solids are recirculated
3. Growth of larger, denser $CaCO_3(s)$ crystals, which reduces the solubility of $CaCO_3(s)$ and produces a more stable effluent and which also improves both the settleability and the dewaterability of the softening solids
4. Promotion of crystal growth over nucleation, resulting in significantly less scale formation on basin walls and mixing equipment

There are added capital and operations and maintenance (O&M) costs associated with the pumps and piping needed to recirculate solids, but the added costs are generally considered minor relative to the benefits provided.

Some conventional lime softening plants operate without solids recirculation. At some plants treating highly turbid, colored, or algae-laden source waters, recirculating settled solids may exacerbate taste and odor problems, especially if some of the settled solids become anoxic. In other cases, solids recirculation has been found to increase settled water turbidity, perhaps due to an unfavorable interaction between the coagulation and softening processes, which often take place concurrently in a lime softening plant. In such cases it is possible that a change in the coagulation process would rectify the problem. Another potential disadvantage of solids recirculation, one not widely recognized, is that it can potentially diminish removal of trace contaminants by coprecipitation, especially if the contaminants diffuse more slowly than the matrix ions, in this case Ca^{2+} and CO_3^{2-}. A softening system is in essence a crystallizer, and in a crystallizer, solids recirculation promotes the growth of crystals that are not only larger but also purer.

Some individuals have expressed concern that recirculating softening solids will reintroduce previously removed pathogens back into the system and might, therefore, give them another chance to escape removal, as has been found to happen in some cases involving recycling of filter backwash water. However, lime softening systems and filtration systems are not analogous in this respect. Solids that accumulate in a lime softening system are discharged from the system at the same rate, on average, that they are produced. Solids recirculation is expected to increase the removal of pathogens by providing more opportunities for them to flocculate with other particles and to be removed by sedimentation and discharged with the solids. Pathogens and fine particles that accumulate on a filter do so because they were not previously removed by sedimentation, so there is a good chance they will again fail to settle out if they are recycled, unless they are recycled back to the rapid-mixing basin and given another chance to be coagulated, flocculated, and settled. (Recycling of filter backwash water, practices, regulations, and possible effects are covered in Chap. 10.)

Solids Contact Basins. Spaulding (1937) and other early investigators noted that lime-softened water could be rapidly stabilized by stirring it with finely divided $CaCO_3$(s) or by passing it through a rapid sand filter. Seeking ways to relieve supersaturation prior to filtration, Spaulding considered recycling softening solids back to the head of the plant, but rejected this option because in his experience it caused increased taste and odor problems. Reasoning that the stabilizing benefit of downward filtration could also be achieved by "upward filtration through suitable material, which might be calcium carbonate of desirable size or light material coated with calcium carbonate," he developed a system he called the *Precipitator* (Fig. 13-3). He experimented with various types of media, including coal and activated carbon, which worked well, but found it unnecessary to add media, since $CaCO_3$(s) crystals could be effectively retained in a funnel-shaped upflow clarifier. Inside the clarifier he placed an inverted cone, providing space for mixing and flocculation in the same basin with the clarifier. Turbulence from the mixer kept the solids in suspension regardless of the flow rate through the system.

In modern upflow clarifiers of various types, flocculated water flows upward through a suspension of solids, being further flocculated and stabilized by what has been termed *sludge*

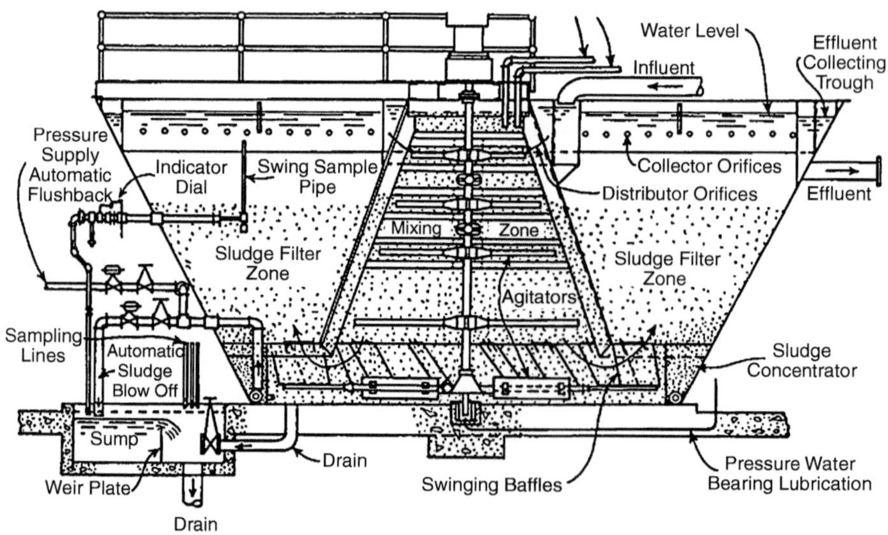

FIGURE 13-3 A solids contact clarifier (the Spaulding Precipitator, marketed by Permutit®, a division of Siemens Water Technologies). (*Source: Hartung, 1951.*)

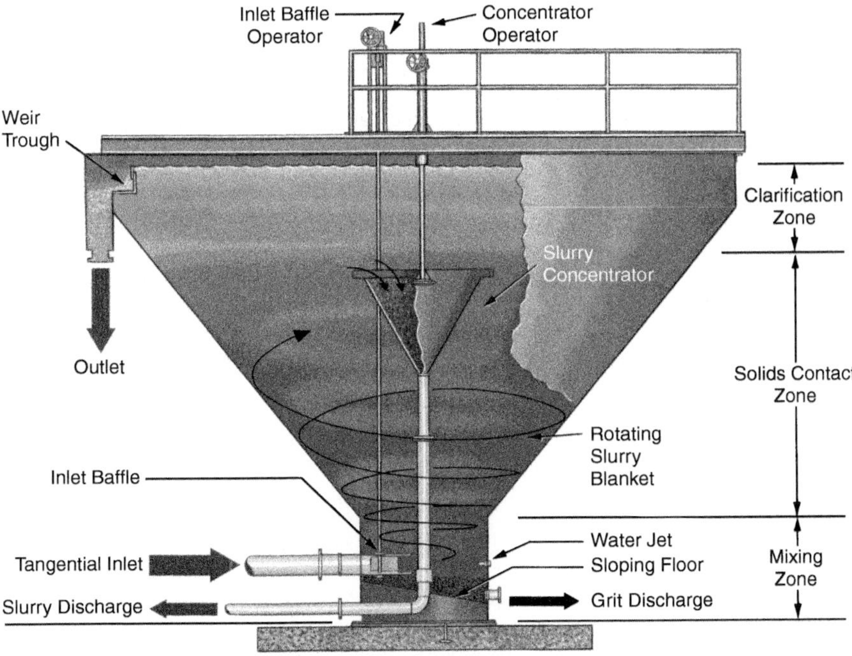

FIGURE 13-4 A helical flow solids contact basin (the ClariCone®; *drawing courtesy of CB&I, The Woodlands, Texas.*)

blanket filtration but is more accurately described as *floc blanket clarification*. Such systems are generically referred to as floc blanket clarifiers. Floc blanket clarifiers were once used in both coagulation and softening applications, but the lime softening market has been dominated for many years by similar but distinctly different basins generically referred to as *solids contact basins, solids contact clarifiers,* or *reactor clarifiers*. Solids contact basins are designed to recirculate a high concentration of solids internally, typically using turbines and draft tubes. They thereby obtain all the benefits of solids recirculation noted above for conventional systems. Examples of solids contact basins are shown in Figs. 13-3, 13-4, and 9-17.

Although some prefer, for the purpose of discussion, to group floc blanket clarifiers with solids contact basins (e.g., Hartung, 1948), the term *solids contact basins* is used here to refer to systems that internally recirculate a high concentration of solids. Solids contact softening basins typically provide a flocculation time of 15 to 30 min (based on the volume under the cone), a sedimentation time of 1 to 2 hours when used to soften ground water, a sedimentation time of 2 to 4 hours when used to treat surface waters, and a rise rate of 0.4 to 0.72 m/hour (1.0 to 1.75 gpm/ft^2), depending on local conditions (Horsley et al., 2005). Helical flow solids contact basins, such as the one shown in Fig. 13-4, share a few similarities with pellet reactors, but they are operated as solids contact basins with the solids kept in suspension hydraulically rather than mechanically.

Solids contact basins are generally more cost effective than conventional systems and are especially attractive when treating waters that do not require turbidity removal, since they can accommodate high loading rates. They generally produce a more stable effluent than a conventional system (Horsley et al., 2005), thereby reducing the chemical costs associated with recarbonation and stabilization, and they are less subject to problems associated with icing and algae growth. Nevertheless, conventional softening basins are still used, especially for treating turbid surface waters and waters that vary considerably in quality. Conventional systems are commonly found in older plants and in larger facilities with basins built using common-wall construction. Floc blanket clarifiers and flocculator-clarifiers, rather than solids contact basins, are sometimes used for the second stage of two-stage softening, especially when relatively few solids are produced and the primary objective is to coagulate and settle the remaining suspended solids prior to filtration.

A question that frequently arises in connection with solids contact basins is whether separate rapid-mixing basins are needed, given that the chemicals can be added directly to the mixing zone within the basin. Rapid mixing basins are generally not needed for the purpose of softening, especially when treating groundwaters or when treating surface waters pretreated or posttreated using coagulation, flocculation, and sedimentation. However, when coagulating and softening surface waters in the same basin, rapid-mixing basins may be helpful or in some cases necessary. Regulatory agencies may require rapid-mixing basins unless the design engineer can demonstrate that adequate mixing and satisfactory process performance can be obtained without them. Operators may find that rapid mixing is not needed, but before turning off a rapid mixing device or otherwise altering coagulation practices, the appropriate regulatory agency should be consulted to confirm that the actions taken will not compromise the plant's compliance with the Surface Water Treatment Rule (summarized in Chap. 1).

Pellet Reactors. *Pellet reactors,* also known as *pelletizers* and as *fluidized bed crystallizers,* are designed to cause deposition of crystalline solids on a fluidized bed of granular media. They can be used to crystallize a variety of solid phases, but are best suited for those that produce hard, well-formed crystals. In the field of water supply and treatment, they are used primarily, if not exclusively, for softening, and are commonly referred to as *pellet softeners*. They are similar in concept to Spaulding's Precipitator and similar in some ways to solids contact basins, but differ in several important ways: (1) they are seeded with fine, dense media, usually quartz or garnet sand, rather than $CaCO_3$(s) or softening solids; (2) the softening chemicals are injected into the water as it enters the bottom of the reactor and are mixed hydraulically rather than mechanically (in a manner favoring crystal growth over nucleation); (3) the solids are withdrawn in the form of relatively large pellets, not as a slurry of finely divided solids; (4) rise rates are typically about 60 to 120 m/hour (25 to 50 gpm/ft^2), much greater than those encountered in conventional softening systems and solids contact basins (thus, pellet softeners have a very small areal footprint); and (5) no coagulant is added to (or formed in) the reactor, so suspended solids are not removed.

The pellet reactor shown in Fig. 13-5 is a modified version of the system developed in 1938 by Dr. E. T. Zentner in Czechoslovakia (Powell, 1954). It consists of a conical tank initially loaded with a bed of fine sand, typically about 0.2 mm in diameter. The sand is often referred to as a catalyst because it accelerates crystallization kinetics by increasing the available surface area, but it is not a true catalyst in a chemical sense. The water and chemicals enter tangentially at the bottom of the cone and mix immediately. The water rises through the reactor in a swirling motion, with an upward velocity sufficient to maintain a fluidized bed. As $CaCO_3$(s) is deposited on the grains, they gradually grow in size, with each grain consisting of a single sand grain coated with $CaCO_3$(s). The larger grains (or pellets) settle toward the bottom of the reactor, where they are drawn off, and fresh seed grains are added to the top of the bed. Some older systems are operated in batch mode, with the entire bed being discharged at the end of a run, but in batch mode performance is reduced at the start and end of a run.

Pellet reactors do not need to have a conical shape. Other types of reactors have been described by Graveland et al., (1983), van der Veen and Graveland (1988), and van Dijk and Wilms (1991). The cylindrical reactor shown in Fig. 13-6 uses nozzles distributed over a

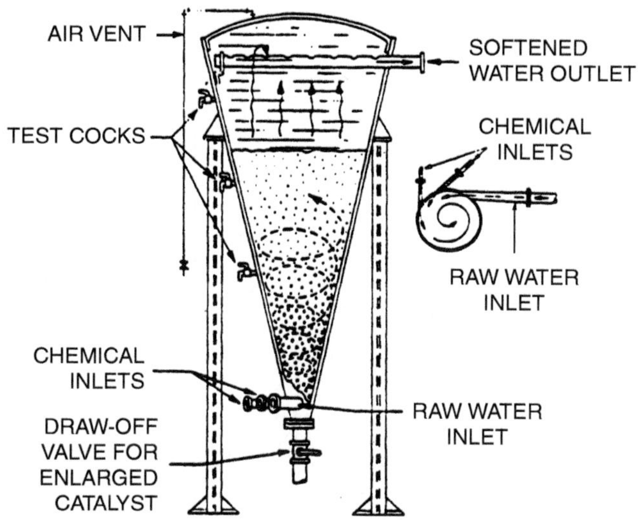

FIGURE 13-5 A conical pellet reactor (the Spiractor™, marketed by Permutit®, a division of Siemens Water Technologies). (*Source: Powell, 1954.*)

flat-bottomed floor to evenly mix chemicals with the inflowing water; this facilitates the use of caustic soda. As a fluidized bed expands, the settling rate of the particles reaches equilibrium with the interstitial velocity of the upflowing water, just as it does when backwashing a rapid sand filter, so the media will be retained provided that the top of the expanded bed is not allowed to approach the effluent weir. In practice, the rise rate is selected to maximize hardness removal and is significantly lower than the rate at which media loss would occur.

Pellet softeners work well when only calcium hardness is removed because $CaCO_3(s)$ crystals are relatively hard and dense. A trace of magnesium may be removed by coprecipitation, but precipitation of $Mg(OH)_2(s)$ must be avoided by keeping the pH sufficiently low. Magnesium hydroxide precipitates as an amorphous or microcrystalline solid that is very light and hydrous, and it does not adhere well to sand or $CaCO_3(s)$. As a result, it passes out of the reactor, creating a high solids loading on the filters (Harms and Robinson, 1992). For similar reasons, coagulants are not added, so the process is not well suited for treating most surface waters. Ferrous iron is reasonably well removed in concentrations up to 3 mg/L, along with the $CaCO_3(s)$, if it is precipitated as $FeCO_3(s)$. However, if dissolved oxygen is present, the iron may be oxidized and precipitated as poorly adhering $Fe(OH)_3(s)$, which may pass through to the filters (Graveland et al., 1983). The process may perform poorly for waters high in color or NOM, which can soften $CaCO_3(s)$, but this problem can potentially be alleviated by using a preoxidation step. For the foregoing reasons, pellet reactors are well suited for treating only certain types of water supplies.

Advantages of pellet reactors include their short hydraulic retention time (small areal footprint), relatively low installation cost, and elimination of waste solids. The pellets are large, typically 0.8 to 1.0 mm in diameter, so they can be drained by gravity to 5 to 10 percent moisture, and in some cases as little as 1 to 2 percent. They are well suited for agricultural and other commercial uses (Graveland et al., 1983), so they can be sold as a product rather than disposed of as waste.

Despite the potential advantages of pellet reactors, they are still rarely used in North America. However, they are used by a number of utilities outside the United States, especially in the Netherlands. Operating data for 16 Dutch plants using pellet reactors were

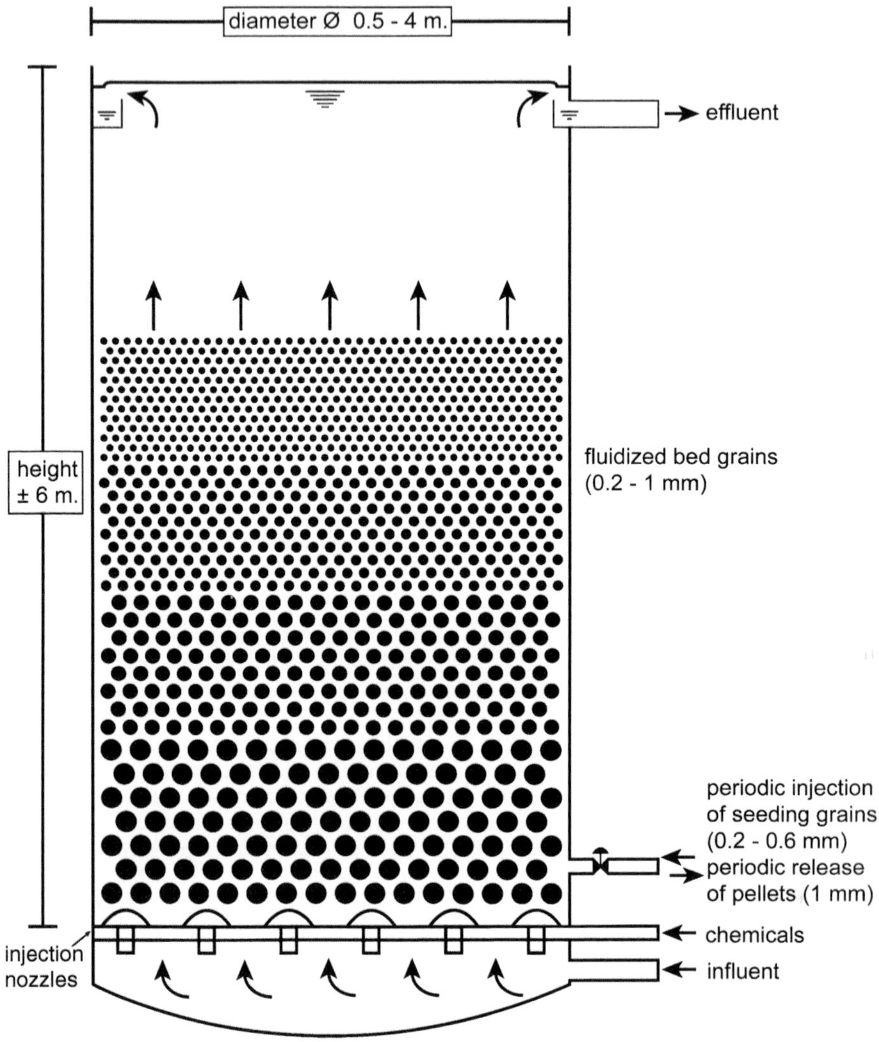

FIGURE 13-6 A cylindrical pellet reactor (the Crystalactor®; *drawing courtesy of DHV BV Water, The Netherlands.*)

reported by van Dijk and Wilms (1991). Interestingly, 11 of them were using caustic soda instead of lime, which cuts solids production in half and conserves alkalinity in waters having CNH (see Softening with Sodium Hydroxide). Under some conditions, including water temperatures below 5°C (van der Veen and Graveland, 1988), use of NaOH in pellet reactors has been found to improve process performance. Use of NaOH is expected to produce higher levels of supersaturation, which could potentially result in the formation of fresh nuclei and solids carryover; but the choice between caustic soda and lime is usually based primarily on local chemical costs, residuals disposal costs, and chemical handling considerations.

Coagulation in Precipitative Softening Processes

Coagulation is an integral part of the precipitative softening process. The solids formed in the softening process must be flocculated and settled, except when using a pellet softener, to avoid overloading the filters. Solids that escape from the process must be properly coagulated so they can be effectively removed by filtration and the finished water turbidity goal can be met. Most precipitative softening plants employ a coagulant dose of about 5 to 15 mg/L of alum or ferric sulfate, 1 to 5 mg/L of polymer, or a proprietary blend of coagulants. Sodium aluminate, activated silica, and ferrous sulfate were commonly used in the past, but are now used infrequently. At most plants a small dose of coagulant (typically a polymer, but sometimes alum or a proprietary coagulant blend) is also added to the filter influent as a filter aid. When using a pellet reactor, a coagulant is typically added after softening, prior to filtration, since coagulants can interfere with crystallization or cause the pellets to be softer and more easily abraded.

Since $Mg(OH)_2(s)$ acts as a coagulant, an additional coagulant is normally not needed when a sufficient amount of magnesium hardness is being removed; but addition of alum or sodium aluminate has been reported to increase removal of magnesium and silica (Larson and Buswell, 1940; Larson, 1951; Tuepker and Hartung, 1960). Since aluminum hydroxide is relatively soluble at the pH values associated with precipitative softening, aluminum evidently forms other solids, such as aluminosilicates, in lime softening applications (e.g., Larson and Buswell, 1940). Thus, when treating waters low in silica, the potential exists for high residual aluminum concentrations and postprecipitation of aluminum hydroxide.

Alum and ferric sulfate are acidic (see Chap. 8), as is gaseous (elemental) chlorine, which may be used as a predisinfectant. When these chemicals are used, a slightly higher lime dose may be required to achieve a given degree of softening. Alternatively, to avoid adding an acid along with the lime (or other caustic chemical), sodium aluminate, which is alkaline rather than acidic, can be used in place of alum or ferric sulfate, and sodium hypochlorite can be used in place of chlorine gas.

Despite the importance of coagulation in precipitative softening, it is not understood nearly as well as it is in plants that coagulate but do not soften (Chap. 8). Surface water sources are often turbid and may contain pathogens, so at plants treating surface water, the coagulant is often added, sometimes with vigorous mixing, prior to or concurrently with the lime, in an effort to effectively coagulate particles present in the raw water. The basis for this practice, which some regulatory agencies may require, appears to be the assumption that coagulation proceeds in the same manner whether or not the water is softened. However, coagulation clearly differs in precipitative softening plants, since

1. At the pH values associated with precipitative softening, positively charged aluminum and iron hydrolysis products are not expected to form, and it is not clear what species actually do form.
2. The softening solids being coagulated are not present in the raw water but form during treatment, and it can take a significant amount of time for stable crystals to form if the softening solids are not recycled. In fact, it can take a significant amount of time for the lime to dissolve, and adding the coagulant prematurely may hinder crystal growth or cause inefficient use of the coagulant.
3. Particles present in the raw water may serve as nuclei for heterogeneous precipitation, as can also occur during coagulation, but a much greater amount of solids is typically precipitated, providing more opportunity for particles in the raw water to be flocculated and settled.

It may be advantageous in some cases to delay coagulant addition. At some plants, the coagulant is added after the lime, and both the lime and a coagulant are commonly added directly into the mixing zone when using solids contact basins. Sometimes one order of addition or point of addition works better than another, but this must be determined empirically, since a theoretical basis for determining the optimum point of coagulant addition is lacking.

Process Monitoring and Control

Parameters typically monitored in the raw and finished water at precipitative softening plants include alkalinity, calcium hardness, total hardness, pH, and temperature. Raw water turbidity is normally monitored at plants treating surface water, and settled- and filtered-water turbidities are usually monitored regardless of the water source. Additional parameters commonly monitored for process control purposes include pH in the mixing zone or flocculation basin; pH, alkalinity, and hardness in the settled water; and pH following recarbonation. Samples are collected manually and analyzed in the plant laboratory at some plants, especially those treating groundwater and having a relatively simple control system. Online instrumentation is used at other plants, especially those with a supervisory control and data acquisition (SCADA) system. Many plants use a combination of online and laboratory methods.

The softening process itself is typically controlled using only a small subset of the parameters listed above, with the choice of parameters and their set points based on the specific process variation employed, treatment objectives, and local preferences that may be strongly influenced by past experience with control parameters and the methods and instrumentation used to monitor them. In developing a control strategy, there are many tools to choose from, but the most important consideration is the operator's need to understand the system and to have the proper tools to control it under changing conditions. The lime or soda ash dosage is usually controlled to maintain a particular value of pH, carbonate alkalinity, calcium hardness, magnesium hardness, or hydroxide alkalinity. At a straight lime or lime–soda softening plant or in the first stage of two-stage recarbonation, removal of calcium hardness may be maximized by maintaining an acidity of 0 mg/L as $CaCO_3$ or a phenolphthalein alkalinity that is half the value of the total alkalinity. The recarbonation process is typically controlled on the basis of measurement of pH or alkalinity following addition of CO_2 or mineral acid. The coagulant dosage is typically selected on the basis of observations of settled water turbidity and the floc characteristics in the mixing zone or flocculator, but it may also be based on jar test results.

Proper control of a precipitative softening process involves more than simply maintaining appropriate chemical dosages. When using quicklime, the slaking process must also be controlled. Incomplete slaking, which can be caused by temperamental lime slakers, poor quality lime, or low water temperatures, can cause degradation of effluent quality and inefficient use of lime. Plugging of chemical feed lines is a common problem when using lime. When using solids contact basins or recycling solids in a conventional system, the solids concentration must be monitored and properly maintained by blowing down excess solids. If the solids concentration is too low, the treated water will be poorly stabilized, settled water turbidity may rise to an unacceptable level, and chemical use may increase. An excessive solids concentration can cause clogging problems and can plug the draft tubes in solids contact basins, causing poor solids recirculation, degradation of effluent quality, and inefficient use of chemicals. Between these extremes, the solids concentration appears to make little difference (Hartung et al., 1951). Solids contact basins are commonly designed to maintain a solids concentration of at least 1 percent by weight in the mixing zone (Recommended Standards, 2007). They are commonly operated to maintain a given volumetric solids concentration in the mixing zone, for example, 10 percent by volume after 10 min of settling, and a higher volumetric concentration in the residuals stream bled to waste.

Settled water turbidity must also be controlled to limit the chemical costs associated with stabilization and to avoid overloading the filters. A high settled water turbidity may indicate that fresh nuclei are forming in the clarifier. This can occur when the softening reactions are not sufficiently complete, which can be caused by a low solids concentration in the mixing zone of a solids contact basin, clogging of a solids return line in a conventional system, or a very low water temperature. A high settled-water turbidity may also indicate that the coagulant dosage needs to be adjusted, but not necessarily that the dosage should be increased. An excessive polymer dosage, for example, can interfere with crystallization of $CaCO_3(s)$, causing pin floc to form, in which case a further increase will only make matters worse. If the operator is in doubt as to whether to lower or raise the coagulant dosage, jar testing is recommended.

Filtration of Lime-Softened Water

Settled water turbidities at lime softening plants are typically higher than those experienced at coagulation plants. A typical settled water turbidity goal for a softening plant is 2 to 5 ntu, but some plants routinely experience values over 10 ntu. Thus, filters at softening plants may receive relatively high solids loadings. However, this generally does not appear to pose a problem, as the solids are typically small, crystalline, and noncompacting; hence, they are amenable to filtration and compatible with relatively high filtration rates. At plants visited by the author, filter run times of 3 to 4 days are common even when the settled water turbidity exceeds 10 ntu, and the run time is typically selected on the basis of a desire to keep the filters clean and not because they reach terminal head loss or breakthrough of turbidity. However, a filter aid must normally be used to achieve compliance with filtered water turbidity requirements.

A question that often arises regarding filtration of lime-softened water is whether the scale inhibitor, most commonly sodium hexametaphosphate, should be added before or after filtration. Filter encrustation is a common problem at lime softening plants, and adding a scale inhibitor prior to filtration helps to avoid this problem. However, scale inhibitors can also act as dispersing agents, stabilizing colloidal particles and causing an increase in filtered water turbidity. Indeed, at several plants that have experimented with the point of addition, a distinct increase in filtered water turbidity has been observed when adding polyphosphate prior to filtration. Perhaps an adjustment to the dosage of filter aid could counteract this effect. In any event, both approaches (pre- and postfilter addition) are currently practiced, and operators should be aware that when changing the point of addition they may be trading one problem for another.

Precipitative Softening Versus Ion-Exchange and Membrane Softening

Hardness can also be effectively removed by ion exchange and membrane softening. As described in Chap. 12, cation exchange resins in the sodium form are used to soften water, and water can also be demineralized using a combination of cation and anion exchange resins in the hydrogen and hydroxide forms, respectively. As described in Chap. 11, nanofiltration (NF) membranes can remove divalent cations, and are sometimes referred to as *softening membranes*, and reverse osmosis (RO) removes not only hardness but other ions as well. Factors influencing process selection include

- *Raw water hardness.* Chemical costs and residuals disposal costs for precipitative and ion exchange softening increase in direct proportion to the amount of hardness removed. For membrane softening, a higher influent hardness may require higher dosages of acid or antiscalants to prevent precipitation of calcium and magnesium salts, and a higher pressure may be required, but recovery is likely to be governed by other factors, as is the cost of residuals disposal.

- *Noncarbonate hardness.* The cost of precipitative softening increases when noncarbonate hardness must be removed, since soda ash is more expensive than lime, but for ion exchange and membrane softening, it makes little or no difference whether carbonate or noncarbonate hardness is being removed.
- *Raw water turbidity.* Precipitative softening is compatible with relatively high levels of turbidity, though presedimentation is typically employed to pretreat highly turbid surface waters. Ion exchange and membrane softening require pretreatment to remove particles unless the source water is very low in turbidity, as is the case for most groundwater supplies, which may still require pretreatment with pressure filtration to remove corrosion products or aquifer materials that pass through well screens.
- *Raw water color and NOM.* Precipitative softening partially removes NOM, especially the fraction contributing to color, but high concentrations of NOM can interfere with the crystallization of $CaCO_3(s)$. Ion-exchange softening resins (cation exchangers) are not adversely affected by color or NOM, but do not remove NOM. NOM tends to foul anion-exchange resins, as well as NF and RO membranes, so pretreatment to remove NOM may be needed when using these materials.
- *Raw water TDS concentration.* High TDS concentrations make calcium and magnesium slightly more soluble (less removable) in precipitative softening processes and can adversely impact ion-exchange softening, since resin selectivity for calcium and magnesium ions decreases as TDS increases. At very high TDS concentrations a cation-exchange resin may even prefer sodium over calcium or magnesium, a phenomenon referred to as *selectivity reversal.* If TDS removal is desired, lime softening can achieve a modest reduction in TDS, as is evident in the bar diagrams in Examples 13-3, 13-4, and 13-6, whereas ion exchange softening increases TDS (since two sodium ions weigh more than the calcium or magnesium ion they are exchanged for). Ion-exchange demineralization and RO can reduce TDS to any desired value.
- *Iron and manganese.* Lime softening effectively removes iron and manganese, but iron and manganese can foul both cation-exchange resins and NF and RO membranes unless pretreatment is provided to remove them.
- *Treatment objectives.* Precipitative, ion-exchange, and membrane softening can all meet typical treatment objectives for municipal water supplies, but only ion-exchange and membrane softening can completely remove hardness. Complete removal of hardness is desirable in certain industrial applications. Lime softening and membrane softening do not increase the sodium concentration, but using soda ash or caustic soda in precipitative softening will increase the sodium concentration, as will ion-exchange softening when using sodium chloride as the regenerant. Softening can influence the corrosivity or scaling potential of the treated water, but regardless of which process is used for softening, the finished water should be properly stabilized as described in Chap. 20.
- *Other constituents requiring treatment.* Precipitative, ion-exchange, and membrane softening are all generally effective in removing divalent cations, but they differ greatly in their ability to remove other contaminants, including silica, fluoride, sulfate, arsenic, chromium, DBP precursors, and other organic and inorganic contaminants. They also differ in their compatibility with other processes that may be needed to remove specific contaminants. See Chap. 5 for process recommendations for contaminant removal.
- *Flow rate and variations in flow rate.* For small systems (households, noncommunity water supplies, small industrial applications, and so forth), ion-exchange softening is the

usual method of choice, but NF and RO systems are often used when other contaminants must also be removed. For mid- to large-size water utilities, lime-softening was routinely selected in the past, but today it is common for utilities to also consider NF (or even RO, for high TDS waters) when building new facilities or rehabilitating existing facilities. Ion-exchange systems have little difficulty handling variable flows, up to their design flow, but both precipitative and membrane softening processes can be adversely affected or difficult to control when the flow rate is highly variable, so additional storage may be required if the demand is highly variable.

- *Process control and operational requirements.* For small systems, especially those that rely on groundwater, ion-exchange and membrane softening are more readily automated and require less operator attention than precipitative softening. For surface water supplies requiring pretreatment, all three options are broadly comparable with respect to process control. Operator training requirements do not vary greatly from one process to another, but different skills are required. A utility may be inclined to select a process that their operators or operators in the area are already familiar with.
- *Options for residuals disposal.* Lime softening produces residual solids, whereas ion-exchange and membrane softening produce residual brines. The cost of residuals disposal is influenced by the type and quantity of waste produced, hauling distances to potential disposal sites, and other factors, as described in Chap. 22. Some disposal options may not be environmentally or socially acceptable. Efforts to control salinity and encourage water reuse, especially in water-short areas, have recently led to limits on brine discharges and in some cases to bans on domestic ion-exchange softeners (e.g., Force, 2009). Efforts are underway to address these challenges.
- *Cost.* Both capital and O&M costs are strongly influenced by many of the factors mentioned here. Process selection is typically based on overall cost, taking non-economic environmental, and social concerns into consideration through public hearings and other appropriate means.

Since precipitative softening removes both particles and organic matter, it is suitable as a pretreatment process for ion-exchange demineralization and membrane processes. It is rarely used ahead of an ion-exchange softening process, except in certain industrial applications where the source water is turbid or colored and complete removal of hardness is desired or, as described by Pizzi (1995), when it is more economical to remove noncarbonate hardness using ion exchange rather than soda ash addition. Kweon and Lawler (2002) found that an increased degree of softening resulted in increased NOM removal and also increased flux (reduced fouling) through an ultrafiltration membrane. Similar results are expected for other types of membranes.

Removal of Organic Contaminants by Precipitative Softening Processes

Precipitative softening can also remove certain organic contaminants, including a fraction of the natural organic matter (NOM) present in every natural water supply. NOM removal is important because NOM reacts with chemical oxidants, exerting an oxidant demand and forming DBPs. Formation of one group of DBPs, the trihalomethanes (THMs), is of particular concern because THM formation is catalyzed by the high-pH environment associated with precipitative softening. NOM can also impart color to the water, transport metal ions, serve as a food source for bacteria, and foul anion-exchange resins and membranes. Additional information regarding regulation of DBPs and other contaminants, the health effects of regulated contaminants, and the characteristics of NOM can be found in Chaps. 1, 3, and 19.

Removal Mechanisms. Organic contaminants may be present in water in either a solid (particulate) or dissolved state. As described in Chap. 3, organic contaminants in water are collectively measured as total organic carbon (TOC), with the dissolved fraction measured as dissolved organic carbon (DOC). Particulate organic carbon (POC) is the difference between TOC and DOC.

In precipitative softening processes, particles (both organic and inorganic) are removed by coagulation, flocculation, and sedimentation, with a subsequent filtration process removing most of the particles that fail to settle. Coagulation is an essential part of precipitative softening, since the solids formed in the process (except in pellet reactors) must be coagulated so they can be effectively removed by sedimentation and filtration. The coagulation mechanisms involved are the same as those described in Chap. 8. For practical purposes, POC can be considered to be completely removed in treatment plants employing precipitative softening, since filtered-water turbidities are typically so low that the POC concentration is negligible. Even when coagulation is done poorly in the softening process itself, coagulant addition prior to filtration allows nearly all of the remaining solids to be captured on granular media filters.

Dissolved organic (and inorganic) contaminants can be removed by precipitative softening only if they are first converted into solid form. The major mechanisms for accomplishing this are precipitation and coprecipitation, which were described earlier in this chapter. Most if not all organic contaminants lack the size and other characteristics to be incorporated into, or dissolved into, the crystal structure of softening solids. Therefore, organic chemicals are expected to be coprecipitated only if they are able to adsorb onto the softening solids as they precipitate (occlusion) or after they precipitate (surface adsorption).

NOM and other organic contaminants can potentially be removed by precipitation (formation of insoluble calcium or magnesium salts), coprecipitation, or adsorption onto previously formed $CaCO_3$ or $Mg(OH)_2(s)$ (Liao and Randtke, 1985; 1986; Russell et al., 2009). Liao and Randtke (1985) argued that coprecipitation is expected to be the dominant mechanism in precipitative softening systems due to the large surface area of the softening solids during the initial moments following chemical addition. They observed that most of the removal of a fulvic acid occurred in less than 1 min and that the Ca^{2+} concentration dropped very rapidly during the initial moments of the reaction. Any organic molecules initially able to be precipitated would also be readily incorporated into the solids by coprecipitation under these conditions, and the opportunity for coprecipitation by adsorption and occlusion would be much greater in the initial moments of the softening reaction than it would be after the solids were fully formed.

Mechanisms that may influence the removal of a limited number of organic contaminants include polymerization and hydrolysis. Some organic chemicals, such as certain polyphenols, can potentially be removed by polymerizing them (chemically or enzymatically) into large insoluble molecules (Liao and Randtke, 1986), and some of the reactions involved may be promoted by a high-pH environment or by the presence of oxidants (e.g., Chrostowski et al., 1983). Similar reactions are involved in the formation of humic substances in the environment. Polymerization may play a role in the removal of organic chemicals from certain industrial wastes, but it is unlikely to be significant in the removal of DOC or trace contaminants from drinking water by precipitative-softening processes because the concentrations of chemicals able to react in this manner are expected to be negligible in drinking water supplies. A limited number of organic chemicals may hydrolyze in the high-pH environment associated with precipitative softening, and Johnson and Randtke (1983) demonstrated that this was the case for a fraction of the total organic halogen produced by prechlorination. This mechanism is expected, however, to play little or no role in the removal of DOC or most organic contaminants from drinking water by precipitative softening.

Factors Influencing TOC Removal. Removal of TOC depends on the fraction that is particulate (POC), since this fraction will be completely removed; on the removability of the dissolved fraction (DOC); and on treatment conditions. TOC removals of 50 percent or more are commonly obtained when softening highly turbid surface waters, which may contain organic-laden clay particles or high concentrations of algal cells, or when softening highly colored waters. Lower TOC removals, in the range of 15 to 30 percent, are commonly observed when treating relatively clear waters containing little POC or color. The DOC in clear surface water supplies is generally less amenable to removal, perhaps due to the source of the DOC, to bleaching by sunlight (fragmentation by free radical reactions), or to microbial activity. DOC in groundwater supplies low in color tends to be poorly removed by precipitative softening processes.

The characteristics of DOC influencing its removal by lime softening are the same as those influencing its removal by coagulation (Chap. 8). Precipitative softening preferentially removes the fraction of DOC that is higher in molecular weight, more hydrophobic, higher in color and UV absorbance, and capable of binding Ca^{2+} ions more strongly (Liao and Randtke, 1985; Randtke, 1988; Thompson et al., 1997; Shorney et al., 1999; Randtke, 1999; Kalscheur et al., 2006; Russell et al, 2009). The ability to bind Ca^{2+} depends on the nature and density of the functional groups involved. Dissociated carboxylic acids (carboxylate ions) are known to have a relatively strong affinity for Ca^{2+} ions; but low-molecular-weight carboxylate ions tend to form relatively soluble complexes that are poorly removed by precipitative softening, as are relatively insoluble compounds that are unable to complex Ca^{2+} or CO_3^{2-} (Liao and Randtke, 1986).

Treatment-related factors influencing DOC removal (or, in some studies, TOC removal) by precipitative softening processes include

1. *Lime dose*. As the lime dose increases, more solids are formed, providing greater surface area for adsorption (coprecipitation) of DOC, so DOC removal typically increases with increasing lime dose, up to a maximum beyond which additional lime has little or no effect on removal. Typical results are shown in Fig. 13-7. As the lime dose increases,

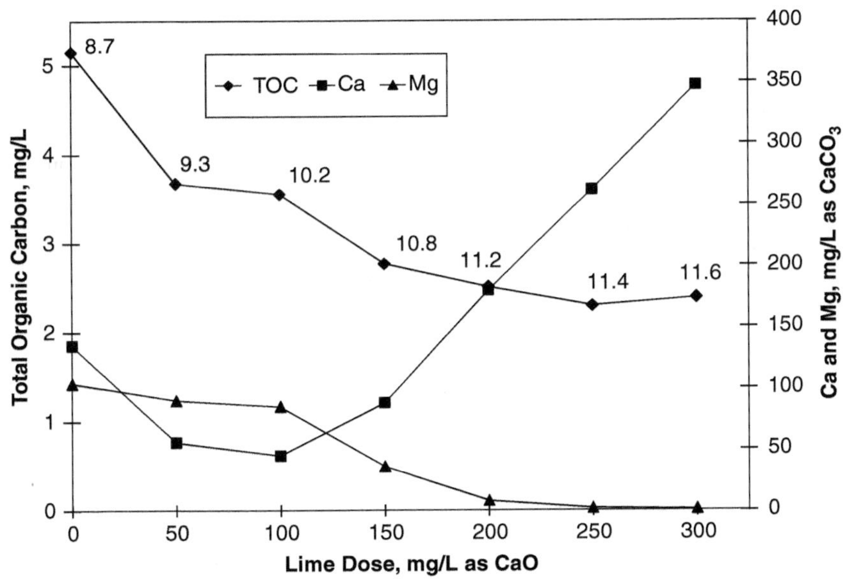

FIGURE 13-7 Removal of TOC, calcium, and magnesium by lime softening. (Data-point labels are treated water pH values, the water source was the Missouri River at Kansas City, Missouri, and no supplemental coagulants were added.) (*Source: Shorney et al., 1999.*)

the solution pH increases, magnesium removal increases, the carbonate concentration decreases, and the calcium concentration eventually begins to increase as well (once most of the carbonate has been precipitated); and these water quality changes also contribute to increased DOC removal, as discussed below. Mustafa and Walker (2006) found that some grades of lime were more effective than others and suggested that this was related to the available Mg content of the lime.

2. *pH.* As pH increases, DOC removal tends to increase. This is attributable to increased dissociation of NOM, enabling it to bind more strongly to Ca^{2+} ions; to increased precipitation of $Mg(OH)_2(s)$ and a reduction in the carbonate concentration, both discussed below; and to an increase in Ca^{2+} ions in vicinity of the $CaCO_3(s)$ surface, which helps to neutralize the surface charge of the solids (or makes them positively charged) and also helps to coprecipitate additional DOC (Liao and Randtke, 1985; Russell et al., 2009).
3. *Magnesium removal.* Many studies have shown that precipitation of $Mg(OH)_2(s)$ helps to remove organic matter (e.g., Liao and Randtke, 1985; Thompson et al., 1997; Shorney et al., 1999; Kweon and Lawler, 2002; Roalson et al., 2003). $Mg(OH)_2(s)$ is precipitated as a noncrystalline solid having a much greater surface area than an equivalent mass of $CaCO_3(s)$, and it is also positively charged, facilitating adsorption of organic anions. (Most NOM is anionic, though individual molecules may be nonionic or cationic.) Incorporating Mg^{2+} into $CaCO_3(s)$ crystals may also improve their ability to remove organic contaminants (Liao and Randtke, 1985; Russell et al., 2009). Russell et al. (2009) found that $CaCO_3(s)$ and $Mg(OH)_2(s)$ were capable of achieving the same maximum removal of NOM but that a much greater mass of $CaCO_3(s)$ was needed to achieve the same degree of removal, and they also suggested that some plants might be able to increase NOM removal by adding magnesium.
4. *Carbonate removal.* As lime is added, CO_3^{2-} ions are formed, from the HCO_3^- ions (alkalinity) naturally present in the water, but they are also consumed as $CaCO_3(s)$ precipitates. Eventually, a point is reached where the supply of carbonate is depleted and the Ca^{2+} concentration begins to increase. This promotes DOC removal because CO_3^{2-} competes with DOC for adsorption sites, while Ca^{2+} can help to remove DOC by both precipitation and coprecipitation (Liao and Randtke, 1985).
5. *Soda ash addition.* Adding soda ash can hinder DOC removal by increasing the carbonate concentration (Randtke et al., 1982; Liao and Randtke, 1985; Russell et al., 2009). Therefore, DOC removal in the primary softening basin can potentially be increased by eliminating use of soda ash (and accepting a higher finished water hardness), by delaying soda ash addition until the second stage of softening, or by substituting two-stage recarbonation for soda ash addition. However, for waters containing high concentrations of noncarbonate hardness, soda ash addition may increase DOC removal by precipitating additional solids and increasing the opportunity for DOC to be coprecipitated.
6. *Solids recirculation.* Solids recirculation should inhibit DOC removal, since it promotes crystal growth over nucleation, which is expected to inhibit both precipitation and coprecipitation of organic contaminants. The $CaCO_3(s)$ lattice ions, Ca^{2+} and CO_3^{2-}, are small and will diffuse to the crystal surfaces more rapidly than the larger organic contaminant molecules; this results in purer crystals and reduced removal of contaminants. In fact, solids recirculation is used in industrial crystallization processes to help make purer crystals. Liao and Randtke (1985) found that that recirculating softening solids reduced DOC removal, and Roalson et al. (2003) found that it had no effect. Mustafa and Walker (2006) observed increased DOC removal as increasing amounts of softening sludge were added during jar testing, but magnesium removal also increased; thus, the increase in DOC removal was attributable to increased precipitation of $Mg(OH)_2(s)$ and it is not clear that the results are applicable to solids recirculation in a full-scale plant.

7. *Mixing.* Mixing intensity is expected to have little or no effect on DOC removal by precipitative softening processes, and Randtke et al. (1982) found this to be the case for removal of a fulvic acid. The reactions responsible for DOC removal occur at a molecular level and are not expected to be significantly influenced by agitation of the bulk solution. However, a minimal amount of mixing is helpful because the opportunity for contaminant molecules to be coprecipitated will be reduced if the contaminants are not present in the immediate vicinity of the softening solids as they are precipitated.
8. *Coagulant addition.* Coagulant addition can enhance DOC removal. However, as noted earlier in this chapter, coagulation is an integral part of precipitative softening, and softening solids must be coagulated to achieve satisfactory settled- and filtered-water turbidities. $Mg(OH)_2(s)$ is an effective coagulant, so it is not necessary to add an additional coagulant when practicing excess lime softening. At most lime-softening plants, alum, ferric sulfate, or a polymer is added as a coagulant. Increasing the coagulant dosage beyond that required to achieve a satisfactory filtered-water turbidity or changing from one coagulant to another is expected to have little or no effect on DOC removal. Shorney et al. (1999) found that a 10- to 15-mg/L dose of alum or ferric sulfate increased TOC removal by an average about 15 percent, relative to softening with no coagulant addition. These dosages are typical of those normally employed in practice. Taylor et al. (1984) found that adding 20 mg/L of alum increased TOC removal by approximately 10 percent when using the magnesium carbonate process, but further alum addition removed no more TOC and resulted in soluble aluminum in the finished water. Other investigators have reported that higher coagulant dosages increased DOC or TOC removal, but in some cases it is not clear that the results are applicable to full-scale systems. For example, Qasim et al. (1992) observed increasing TOC removal with an increasing dose of ferric sulfate, but the filtered-water turbidities were in excess of 2 ntu at all doses below the maximum dose tested, and relevant variables, such as lime dose and magnesium removal, were not reported. There are indications that the timing of coagulant addition influences TOC removal by softening (Randtke, 1999), but until more information is available, the recommended approach is to determine the optimum dose and order of addition empirically, that is, by conducting bench-, pilot-, or full-scale tests on the water being treated. In summary, increasing the coagulant dose or changing coagulants may slightly increase DOC or TOC removal. However, chemical costs, solids production, and possible effects of coagulant overdosing (such as clogging of filter underdrains, postprecipitation of aluminum hydroxide, and increased settled water turbidity) should also be considered.
9. *Preoxidation.* Oxidants such as chlorine, ozone, and chlorine dioxide can increase or decrease the removal of DOC by coagulation, depending on the oxidant dose, characteristics of the organic matter, and water quality (Randtke, 1988; Edwards and Benjamin, 1992; Edwards et al., 1994). Little is known about the effects of preoxidation on removal of DOC by precipitative softening processes, but Johnson and Randtke (1983) found that prechlorination reduced removal of TOC from three different water sources by an average of nearly 10 percent.

Several of these factors were experimentally examined by Liao and Randtke (1985), and the results of one set of experiments are presented in Table 13-6. TOC removal was found to increase with pH, a higher Ca^{2+} concentration, a delay in the addition of soda ash, and addition of magnesium or phosphate. TOC removal was lower in a carbonate-rich environment or when softening solids, produced in prior jars tests under the same conditions, were recycled.

Enhanced Softening. The Stage 1 Disinfection/Disinfectants By-Products Rule (Stage 1 D/DBP Rule) (Chap. 1) requires precipitative-softening plants that treat surface water

TABLE 13-6 Effects of Operational Changes on the Removal of Groundwater Fulvic Acid and Water Hardness by Lime Softening

Process variable	TOC removed, %	Residual hardness, mg/L as $CaCO_3$
pH		
9	28	25
11†	35	16
12	44	12
Chemical addition		
Calcium rich	41	9
Carbonate rich	29	7
Lime, then soda ash‡	53	7
Solids recycling		
1:1 (old:fresh)	31	9
2:1 (old:fresh)	23	8
Additives		
Ca:Mg = 6:1	60	13
Ca:Mg = 6:2	72	14
$Ca:PO_4^{3-}$ = 6:0.1	35	50
$Ca:PO_4^{3-}$ = 6:0.2	43	52

†Standard condition: 3 mM $CaCl_2$ and 6 mM $NaHCO_3$, then 3mM CaO; pH = 11, no solids recycle, no Mg^{2+} or PO_4^{3-} added; TOC = 3 mg/L.
‡3 mM $CaCl_2$ and 3 mM $NaHCO_3$, then 3 mM CaO and, 5 min later, 3 mM Na_2CO_3.
Source: Liao and Randtke(1985)

(or groundwater under the direct influence of surface water) to practice enhanced softening if they use conventional filtration (USEPA, 1998). This requirement can be met by achieving a specified TOC removal depending on raw water TOC and alkalinity (see Table 8-1) and/or by meeting one of the alternative performance criteria for coagulation, based on TOC or specific UV absorbance in the raw or finished water or on finished water DBP concentrations (Chap. 8; USEPA, 1998 and 1999; Pontius and Diamond, 1999); or by meeting one of the two additional alternative performance criteria established for lime softening: (1) softening that reduces the treated water alkalinity to less than 60 mg/L as $CaCO_3$ or (2) softening that achieves a magnesium hardness removal of at least 10 mg/L as $CaCO_3$. For precipitative softening, there is no Step 2 procedure (a jar test procedure for determining the point of diminishing returns in TOC removal as the dosage increases) as there is for coagulation. The rule does not require plants practicing conventional lime softening to change to lime–soda softening or excess lime softening to attempt to remove more TOC, as this would require a plant to operate under conditions for which it was not designed.

If the enhanced softening requirements of the Stage 1 D/DBP Rule are not being met, relatively simple options to change the process to meet them include increasing the lime dose, reducing or eliminating soda ash addition, and changing the coagulant type, dosage, or point of addition. Increasing the lime dose can increase TOC removal in several ways, and, in many cases, a relatively modest increase in the lime dose will result in compliance with one or more of the alternative performance criteria. Reducing or eliminating soda ash addition or changing the coagulation process may result in only a slight increase in TOC removal, but that may be all that is needed. These and other relatively simple options can be evaluated using bench-scale tests, following protocols such as those for enhanced softening described by the USEPA (1999). Shorney et al. (1999) found that, for both softening and coagulation, TOC removals observed in bench-scale tests agreed well with those measured in full-scale plants; and Liao and Randtke (1985) found the bench- and full-scale results for the one softening plant they studied to agree to within 1 percent.

Shorney et al. (1996) found that excess lime softening removed, on average, 18 percent (or 0.5 mg/L) more TOC than conventional lime softening, on the basis of bench-scale testing of five source waters. However, this required an increase in the average lime dose from 88 mg/L as CaO for conventional lime softening to 175 mg/L as CaO for excess lime softening. Most treatment plants are not equipped to handle such a large increase in lime dose or to handle all the additional solids that excess lime softening would produce. Before making such a drastic change, a utility might wish to consider other options, such as changing the raw water source or replacing both precipitative softening and conventional rapid sand filtration with a membrane softening system, especially if these options also offer other site-specific advantages. It is generally best to evaluate the simplest and least expensive options first and to evaluate expensive and more complicated options as part of a utility's long-term planning process.

Softening Versus Coagulation. Softening and coagulation remove similar types of organic contaminants, and when using the chemical dosages typically employed in practice, they generally remove similar amounts of TOC from a given source water. However, coagulation with high dosages of metal salt coagulants has generally been found to remove more TOC than high-pH lime softening (e.g., Taylor et al., 1986; Semmens and Staples, 1986; Shorney, 1998; Shorney et al., 1999), as illustrated in Fig. 13-8. Thus, in plants employing separate coagulation and softening processes, coagulation may be able to remove a small fraction of TOC not removable by precipitative softening. In plants employing both processes, the coagulation process should be placed ahead of the softening process, as this will reduce THM formation (which is base catalyzed) and improve the performance of the

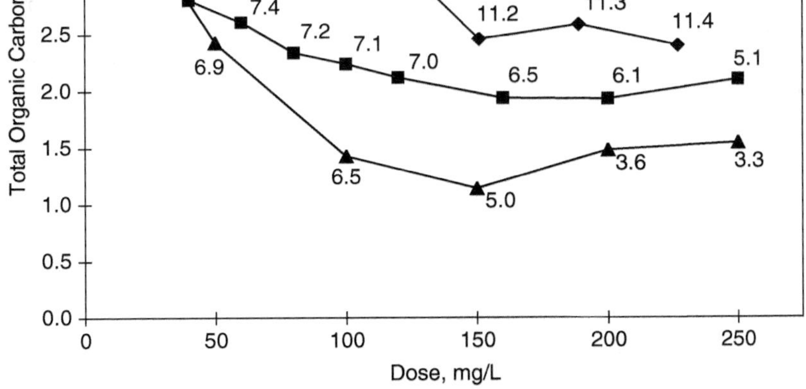

FIGURE 13-8 TOC removal by lime softening and by coagulation with alum and ferric sulfate. (Data-point labels are treated water pH values; the water, drawn from Clinton Lake, Kansas, had an initial alkalinity of 112 mg/L as $CaCO_3$, Ca^{2+} = 100 mg//L as $CaCO_3$, and Mg^{2+} = 29 mg/L as $CaCO_3$.) (*Prepared using data from Shorney, 1998.*)

softening process by removing NOM. NOM inhibits softening by complexing Ca^{2+} and Mg^{2+} and interfering with the formation of $CaCO_3$(s) crystals (Randtke et al., 1982; Liao and Randtke, 1985; Lin et al., 2005; Hsu and Singer, 2009).

Removal of Organic Chemical Contaminants. Precipitative softening processes have the potential to remove organic chemicals able to adsorb onto $CaCO_3$(s) or $Mg(OH)_2$(s). Liao and Randtke (1986) found that softening was generally effective in removing compounds that formed relatively insoluble complexes with Ca^{2+} and higher-molecular-weight or polymeric compounds having acidic oxygen-containing functional groups able to facilitate their adsorption onto $CaCO_3$(s). Most regulated organic contaminants, as well as most other organic chemicals of concern to the drinking water community, adsorb onto $CaCO_3$(s) poorly or not at all, and those able to interact with Ca^{2+} or CO_3^{2-} ions tend to form soluble complexes that are poorly adsorbed.

Organic chemicals can usually be better removed by adsorption on powdered activated carbon (PAC) than by precipitative softening. PAC is compatible with the softening process, but its performance may be influenced by the point of addition. In general, it is better to add the PAC well before or after adding the lime. If lime and PAC are added simultaneously, $CaCO_3$(s) crystals will form on the surface of the PAC and may slow diffusion of the contaminants to the PAC surface. Randtke et al. (2004) found greater removal of atrazine when the PAC was added some time after the lime, presumably because lime softening removed a fraction of the TOC that was competing with atrazine for adsorption sites. On the other hand, Pan et al. (2002) found that the timing of PAC and lime addition had no significant effect on removal of geosmin and 2-methylisoborneol. The optimum point of addition for a given application is best determined by bench-, pilot-, or full-scale testing of the water to be treated, taking into consideration possible chemical addition points and the availability of mixing adequate to keep the PAC in suspension.

Residuals

Precipitative softening processes typically produce residuals composed primarily of $CaCO_3$(s) and, if Mg is removed, $Mg(OH)_2$(s). Residuals composed mostly of $CaCO_3$(s) are dense, stable, and relatively easy to thicken and dewater; but $Mg(OH)_2$(s), coagulants, and organic matter make softening residuals more difficult to thicken and dewater. Stoichiometric estimates of solids production (see Example 13-4) have been found to be reasonably accurate when one accounts for significant contributions of coagulants, suspended solids, organic matter, and PAC. See Chap. 22 and AWWA and ASCE (2005) for additional information regarding the quantities, characteristics, handling properties, and disposal of precipitative softening residuals, as well as those produced by ion-exchange and membrane softening.

OTHER APPLICATIONS

Precipitative softening can remove other contaminants besides calcium, magnesium, and NOM. A variety of trace contaminants can be removed by other precipitative processes, including those coprecipitated with metal hydroxides formed by metal salt coagulation (Chap. 8) or by oxidation of iron and manganese (Chap. 7). Removal of NOM by coagulation is addressed in Chap. 8. Inorganic contaminants that can be removed by precipitative processes include

- *Regulated inorganic chemicals (IOCs).* Coagulation and lime softening can, under certain conditions, effectively remove a number of IOCs from drinking water, including arsenic, barium, cadmium, chromium, lead, mercury, and selenium (Sorg, 1977; Sorg and Logsdon, 1980; AWWA, 1988), and these processes are designated as best available technologies (BATs) for several IOCs. Table 13-7 summarizes IOC removals achieved by coagulation and lime softening, under selected conditions, based on reports by Sorg (1977) and Sorg and Logsdon (1980). Additional information regarding IOCs and their removal can be found in several other chapters in this handbook.

- *Metal ions.* Many metal ions, including those regulated as IOCs, can be effectively removed by precipitation or coprecipitation. The ligands most commonly utilized to precipitate metal ions are hydroxide, carbonate, silicate, and sulfide. Each of these ligands increases in concentration as pH increases, so adding a caustic chemical (usually lime) to increase pH is the most common way to promote precipitation of metal ions. However, due to complex formation, some metal ions become more soluble as pH increases beyond a certain point, as discussed previously. Therefore, an optimum pH for precipitation may exist. Also, some metal ions form insoluble phases with other ligands. For example,

TABLE 13-7 Effectiveness of Chemical Coagulation and Lime Softening for Inorganic Contaminant Removal

Contaminant	Process	Approximate removal, %
Arsenic		
As^{3+}	Oxidation to As^{5+} required for effective removal	—
As^{5+}	Ferric sulfate coagulation, pH 6–8	>90
	Alum coagulation, pH 6–7	>90
	Lime softening, pH > 10.8	>95
	Lime softening, pH < 10.8	30–95
Barium	Lime softening, pH > 10–11	>90
Cadmium	Ferric sulfate coagulation, pH > 8	>90
	Lime softening, pH 8.5–11.3	>95
Chromium†		
Cr^{3+}	Ferric sulfate coagulation, pH 6.5–9.3	>98
	Alum coagulation, pH 6.7–8.5	>90
	Lime softening, pH 10.6–11.3	>98
	Lime softening, pH < 10.6	70–95
Cr^{6+}	*Ferrous* sulfate coagulation, pH 6.5–9.3 (pH may have to be adjusted after coagulant addition to promote reduction to Cr^{3+})	>98
Lead	Ferric sulfate coagulation, pH 6–9	>97
	Alum coagulation, pH 6–9	80–90
	Lime softening, pH 8.5–11.3	>98
Mercury, Hg^{2+}	Ferric sulfate coagulation, pH 7–8	66–97
	Alum coagulation, pH 7–8	38–47
	Lime softening, pH 10.7–11.4	60–80
Selenium, Se^{4+}	Ferric sulfate coagulation, pH 6–7	60–80
	Lime softening, pH 9.5–11.3	20–40
Silver	Ferric sulfate coagulation, pH 7–9	70–80
	Alum coagulation, pH 6–8	70–80
	Lime softening	80–90

Note: Approximate removals are based on a limited number of laboratory experiments and/or pilot-plant or full-scale data. Removal is expected to vary with water quality, pH, contaminant concentration, chemical dosages, temperature, and other treatment conditions.

†Oxidation of Cr^{3+} to Cr^{6+}, e.g., by chlorination, will impair removal.

Source: Based on Sorg (1977) and Sorg and Logsdon (1980).

barium can be readily precipitated using a small amount of sulfate, and phosphate is commonly removed from lake water and wastewater effluents by precipitating aluminum phosphate. Sulfide imparts an offensive odor, exerts a strong oxidant demand, and is more complicated and expensive to use than other alternatives for removing trace metals from drinking water supplies, so its use is normally limited to industrial applications. However, sulfide often limits metal solubility under anaerobic conditions, such as those sometimes found in a groundwater supply or in the hypolimnion of a eutrophic lake.

Most metal ions in drinking water supplies are present only in trace concentrations. Sometimes they must be removed by coprecipitation because precipitation is either not possible or not practical. However, they can often be conveniently removed by a process already being used for another purpose, such as coagulation, precipitative softening, or oxidation of iron and manganese. In some cases, the required degree of removal can be accomplished at no additional cost, although in other cases increased chemical dosages or pH adjustments may be necessary to meet treatment objectives. In addition to the IOCs shown in Table 13-7, other metal ions that can be removed by coprecipitation include beryllium, cobalt, copper, iron, manganese, nickel, zinc, and perhaps antimony and thallium (Jenne, 1968; AWWA, 1988; Faust and Aly, 1998).

- *Radionuclides.* Radium, strontium, and uranium can be removed by coprecipitation. Logsdon (1977) reported that lime softening can remove 80 to 90 percent of radium. Brinck et al. (1978) and Sorg and Logsdon (1980) reported 75 to 96 percent removal of radium by lime softening in pilot- and full-scale systems. Radium removal usually exceeds total hardness removal (Figure 13-9). Brinck et al. noted a positive correlation between radium removal and softening pH, but this may have been due, at least in part, to increased hardness removal at higher pH values. Logsdon (1977), citing Straub (1964), reported that softening followed by filtration achieved 87 to 96 percent removal of strontium. Radium and strontium are well removed by lime softening because they are chemically similar to calcium and magnesium, but radium is poorly removed (less than 25 percent) by coagulation (Logsdon, 1977). Lime softening is generally not economically feasible for removing radionuclides from soft waters, but radium can be removed from soft water by coprecipitating it with barium sulfate.

  Lee and Bondietti (1983) and Sorg (1988) reported that lime dosages of 100 mg/L or greater, when accompanied by magnesium carbonate addition and resulting in a pH of 10.6 or higher, removed 89 to 99 percent of the uranium spiked into pond water. They also reported that coagulation with ferric sulfate or alum at pH 6 or 10 effectively removes uranium, whereas removal is poor at pH 4 or 8. Uranium speciation is strongly influenced by pH, and the data presented by these authors nicely illustrate how greatly this can impact contaminant removal by coprecipitation.

- *Silica.* Silica is not a regulated contaminant. Some water utilities choose to remove silica for the benefit of their industrial customers, since it interferes with a number of industrial water uses. Other utilities remove silica to reduce or eliminate silicate scaling in domestic water heaters. Silica can be removed by excess lime softening, which can precipitate silica as magnesium silicate or coprecipitate it with magnesium hydroxide, and by adsorption (coprecipitation) on freshly precipitated aluminum or iron hydroxide (Powell, 1954; Iler, 1979; NALCO Water Handbook, 1979; Pizzi, 1995). Since silica removal by lime softening is related to the amount of magnesium removed (Fig. 13-10), additional magnesium may need to be added to meet the treatment objective. The effectiveness of silica removal by excess lime softening increases with temperature, so hot-process lime softening is commonly employed for industrial applications. Silica adsorbs more strongly to aluminum and iron hydroxides at lower temperatures and higher pH values, but the pH is usually limited to about 8.5 for aluminum hydroxide and 10 for iron hydroxide to avoid solubilizing aluminum and iron, respectively.

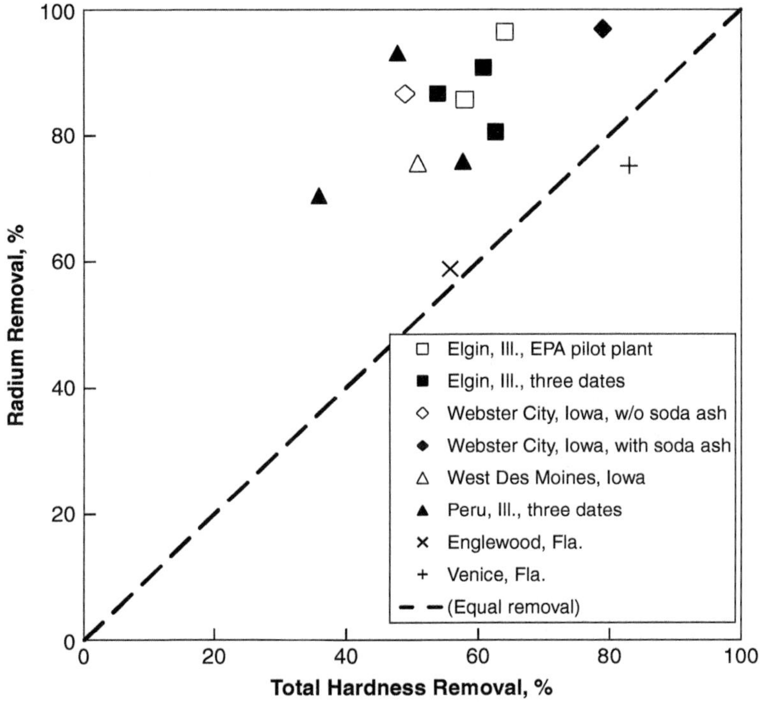

FIGURE 13-9 Radium removal versus total hardness removal. (*Adapted from Logsdon, 1977.*)

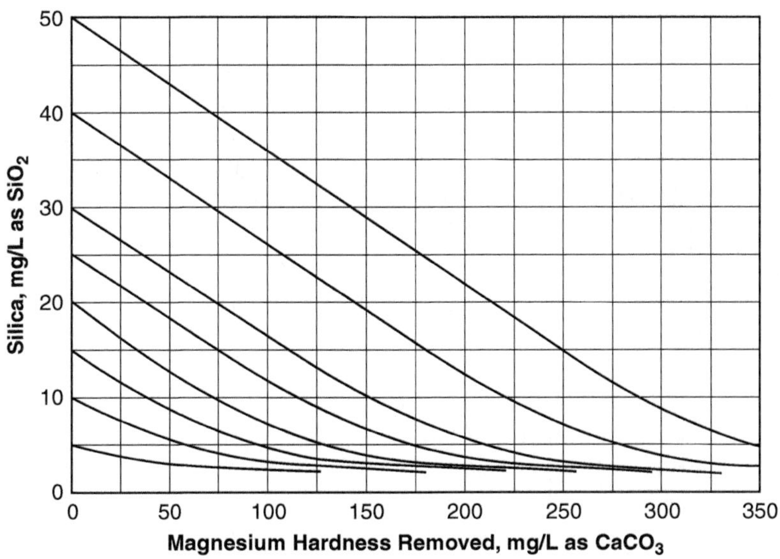

FIGURE 13-10 Silica removal at 70°F as a function of initial silica concentration and the concentration of magnesium hardness removed. (*Courtesy of Sheppard T. Powell Associates, LLC.*)

- *Fluoride.* Fluoride can be coprecipitated with magnesium hydroxide or aluminum hydroxide (Sorg, 1977 and 1978; Sollo et al., 1978). However, these processes remove fluoride inefficiently, so their use is limited to situations where little removal is required or other alternatives are not available. When fluoride must be removed from a public water supply, adsorption on activated alumina (Chap. 12) is often the preferred method.
- *Phosphate.* Phosphate can be removed from the water column of a lake or reservoir using alum, ferric sulfate, or lime. Alum is the preferred chemical because under the anaerobic and slightly acidic conditions normally encountered at the bottom of a eutrophic water body, phosphate is more likely to be released back into the water column if precipitated as iron or calcium phosphate. These same chemicals are commonly used to remove phosphate from municipal and industrial wastewater effluents, and magnesium ammonium phosphate (struvite) precipitates out of anaerobic digester supernatants if the pH is raised slightly, for example, due to off-gassing of CO_2 or if lime is added.
- *Halides.* Chloride, bromide, and iodide can be precipitated with silver, but due to the high cost of silver, this application is currently limited to laboratory applications, including titrimetric analysis of chloride, microcoulometric analysis of total organic halogens, and eliminating chloride interferences in ion chromatography.

ACKNOWLEDGMENTS

The chapters on chemical precipitation and lime softening in the third and fifth editions of this handbook served as helpful starting points for this chapter. The author is grateful to Jim Edzwald for his guidance, patience, and understanding during the preparation of this chapter; to Doug Elder and Mike Horsley for sharing their knowledge of the design and operation of lime-softening systems; to the reviewers (Doug Elder and Nick Dugan) for their many insightful comments and excellent suggestions; and to the many individuals, too numerous to mention individually (though works by many of them are cited herein), who contributed to his understanding of the topics addressed in this chapter.

ABBREVIATIONS

Alk	alkalinity
ANSI	American National Standards Institute
APHA	American Public Health Association
ASCE	American Society of Civil Engineers
AWWA	American Water Works Association
BATs	best available technologies
CCH	calcium carbonate hardness
CCPP	calcium carbonate precipitation potential
CNH	calcium noncarbonate hardness
CRC	Chemical Rubber Company
CSSE	Conference of State Sanitary Engineers
DBP(s)	disinfection by-product(s)
DOC	dissolved organic carbon
gpm/ft^2	gallons per minute per square foot
IOCs	inorganic contaminants

L	liter
LSI	Langelier saturation index
MCH	magnesium carbonate hardness
MCL	maximum contaminant level
MI	magnesium Index
MNH	magnesium noncarbonate hardness
NIST	National Institute of Standards and Technology
NOM	natural organic matter
ntu	nephelometric turbidity units
M	molar (moles per liter)
meq/L	milliequivalents per liter
mg/L	milligrams per liter
mM	millimoles per liter
mm	millimeter
mmol	millimoles
mol. wt.	molecular weight
m/h	meters per hour
NF	nanofiltration
O&M	operation and maintenance
PAC	powdered activated carbon
pH	Potenz hydrogen (Ger.; strength of the hydrogen ion concentration)
POC	particulate organic carbon
RO	reverse osmosis
SI	saturation index
SMCL	secondary maximum contaminant level
SR	saturation ratio
Stage 1	D/DBP Rule
Stage 1	Disinfection/Disinfectants By-Products Rule
TDS	total dissolved solids
TH	total hardness
THM(s)	trihalomethane(s)
TOC	total organic carbon
USEPA	U.S. Environmental Protection Agency
USGS	U.S. Geological Survey
UV	ultraviolet
WEF	Water Environment Federation

NOTATION FOR EQUATIONS

A	constant (in the Davies equation)
b	empirically derived constant (used to calculate activity of uncharged molecules)

$C_{T,C}$	concentration of the sum of the carbonate species in solution
CCH	calcium carbonate hardness
CNH	calcium noncarbonate hardness
I	ionic strength
K	equilibrium constant
k	rate constant
K_{sc}	solubility product constant for calcite
K_{sa}	solubility product constant for aragonite
K_{sv}	solubility product constant for vaterite
K_1	first acid dissociation constant
K_2	second acid dissociation constant
K_{s0}	solubility product constant
${}^cK_{s0}$	concentration-based solubility product constant
MCH	magnesium carbonate hardness
Mg_r	magnesium hardness in the raw (unsoftened) water
Mg_1	magnesium hardness leaving the first-stage softening unit
Mg_f	magnesium hardness in the finished water
MI	magnesium index
MNH	magnesium noncarbonate hardness
Q	ion product or reaction quotient for a chemical reaction
R	universal gas constant (8.314 J/mole-K).
S	available surface area
SI	saturation index
SR	saturation ratio
T	absolute temperature in degrees kelvin (K) (0°C = 273.15 K)
t	temperature in °C or °F (as indicated)
x	fraction of water bypassing the first stage
z	ionic charge
α	fraction of a chemical species in a given form
γ_d	activity coefficient for divalent ions
γ_m	activity coefficient for monovalent ions
γ_z	activity coefficient for ions having an ionic charge of z
γ_0	activity coefficient for uncharged species in solution
ΔG^0_{rxn}	free energy of reaction
ε	dielectric constant of water
{ }	activity
[ ]	molar concentration

REFERENCES

Allison, J.D., D.S. Brown, and K.J. Nono-Gradac (1991) MINTEQA2/PRODEFA2, a Geochemical Assessment Model for Environmental Systems, Version 3.0 User's Manual, Report No. EPA/600/3-91/021, Environmental Research Laboratory, Office of Research and Development,

Athens, GA: USEPA. (This program is maintained by USEPA and is available at www.epa.gov/ceampubl/mmedia/minteq/index.html.)

APHA, AWWA, and WEF (2005) *Standard Methods for the Examination of Water and Wastewater,* 21st ed., eds. A.D. Eaton, L.S. Clesceri, E.W. Rice, and A.E. Greenberg, Washington, DC: American Public Health Association.

Archer, D.G., and P. Wang (1990) The Dielectric Constant of Water and Debye-Hückel Limiting Law Slopes, *J. Phys. Chem. Ref. Data,* 19 (2), 371–411.

Aultman, W.W. (1939) Reclamation and Reuse of Lime in Water Softening, *Journal AWWA,* 31 (4), 640–679.

ASCE, AWWA, and CSSE (1969) *Water Treatment Plant Design,* 1st ed., New York: American Society of Civil Engineers.

AWWA (1968) Quality Goals for Potable Water, *Journal AWWA,* 60 (12), 1317–1322.

AWWA Committee Report (1988) A Review of Solid–Solution Interactions and Implications for the Control of Trace Inorganic Materials in Water Treatment, *Journal AWWA,* 80 (10), 56–64 (Oct.).

AWWA (2007) ANSI/AWWA Standard B202-07: Quicklime and Hydrated Lime, Denver. CO: American Water Works Association.

AWWA and ASCE (2005) *Water Treatment Plant Design,* 4th ed., New York: McGraw-Hill.

Benefield, L.D., and J.M. Morgan (1999) Chemical Precipitation, Ch. 10, in *Water Quality & Treatment,* 5th ed., ed. R.D. Letterman, American Water Works Association, New York: McGraw-Hill.

Benjamin, M. M. (2002) *Water Chemistry*, New York: McGraw-Hill.

Black, A.P., B.S. Shuey, and P.J. Fleming (1971) Recovery of Calcium and Magnesium Values from Lime–Soda Softening Sludges, *Journal AWWA,* 63 (10), 616–622.

Briggs, J.C., and J.F. Ficke (1977) Quality of Rivers of the United States, 1975 Water Year—Based on the National Stream Quality Accounting Network (NASQAN). Open File Report 78-200, U.S. Geological Survey, Reston, VA, 435 pp. (A color map showing surface water hardness concentrations, based on their data, is available at: http://water.usgs.gov/owq/hardness-alkalinity.html#briggs.)

Brinck, W.L., R.J. Schliekelman, D.L. Bennett, C.R. Bell, and I.M. Markwood (1978) Radium-Removal Efficiencies in Water-Treatment Processes, *Journal AWWA,* 70 (1), 31–35.

Burris, M.A., K.W. Cosens, and D.M. Mair (1976) Softening and Coagulation Sludge-Disposal Studies for a Surface Water Supply, *Journal AWWA,* 68 (5), 247–257.

Butler, J.N. (1998) *Ionic Equilibrium: Solubility and pH Calculations*, New York: John Wiley and Sons.

Cadena, F., W.S. Midkiff, and G.A. O'Conner (1974) The Calcium Carbonate Ion-Pair as a Limit to Hardness Removal, *Journal AWWA,* 66 (9), 524.

Caldwell, D.H., and W.B. Lawrence (1953) Water Softening and Conditioning Problems: Solution by Chemical Equilibrium Methods, *Industrial & Engineering Chemistry,* 45 (3), 535–548.

Chao, P.-F., and P. Westerhoff (2002) Assessment and Optimization of Physicochemical Softening Processes, *Journal AWWA,* 94 (3), 109–119.

Chrostowski, P.C., A.M. Dietrich, and I.H. Suffet (1983) Ozone and Oxygen Induced Oxidative Coupling of Aqueous Phenolics, *Water Research,* 17 (11), 1627.

Cogley, D.R. (1998) Automated Computation Methods, Ch. 12, in *Ionic Equilibrium: Solubility and pH Calculations*, Butler, J.N., New York: John Wiley and Sons.

Collins, W.D., W.L. Lamar, and E.W. Lohr (1934) The Industrial Utility of Public Water Supplies in the United States, 1932. U.S. Dept. of the Interior, Geological Survey, Water-Supply Paper 658, Washington, DC: U.S. Government Printing Office, 142 pp.

CRC Handbook of Chemistry and Physics, 89th ed. (2008) ed. D.R. Lide, Boca Raton, FL: CRC Press.

Davies, C.W. (1938) The Extent of Dissociation of Salts in Water. Part VIII. An Equation for the Mean Ionic Activity Coefficient of an Electrolyte in Water, and a Revision of the Dissociation Constants of Some Sulphates, *J. Chem. Soc. (Great Britain),* 141, 2093–2098.

Davies, C.W. (1962) *Ion Association*, Washington, DC: Butterworths.

Durand, M., and A.M. Dietrich (2009) Impact of Calcium and Magnesium on Taste, *Proc. Annual Conference of the American Water Works Association,* San Diego, Calif., June 14–18, Denver, CO: AWWA.

Durfor, C.N., and E. Becker (1964) Public Water Supplies of the 100 Largest Cities in the United States, 1962. Water-Supply Paper 1812, Geological Survey, U.S. Dept. of the Interior, Washington, DC: U.S. Government Printing Office, 372 pp.

Dye, J.F., and J.L. Tuepker (1971) Chemistry of the Lime–Soda Process, in *Water Quality and Treatment,* 3d ed., prepared by the American Water Works Association, New York: McGraw-Hill.

Edwards, M., and M.M. Benjamin (1992) Effect of Preozonation on Coagulant–NOM Interactions, *Journal AWWA,* 84 (8), 63–72.

Edwards, M., M.M. Benjamin, and J.E. Tobiason (1994) Effect of Ozonation on Coagulation of NOM Using Polymer Alone and Polymer/Metal Salt Mixtures, *Journal AWWA,* 86 (1), 105–116.

Faust, S.D., and O.M. Aly (1998) *Chemistry of Water Treatment,* 2nd ed., Chelsea, MI: Ann Arbor Press.

Force, J. (2009) U.S. Salinity Regulations Drive Home Softener Innovation, *Water 21,* pp. 20–22, London: IWA (International Water Association) Publishing.

Graveland, A., J.C. van Dijk, P.J. de Moel, and J.H.C.M. Oomen (1983) Developments in Water Softening by Means of Pellet Reactors, *Journal AWWA,* 75 (12), 619–625.

Gustafsson, J.P. (2009) Visual Minteq Ver. 2.61, Stockholm, Sweden: Kungliga Tekniska Högskolan (www.lwr.kth.se/english/OurSoftware/Vminteq/index.htm).

Harms, W.D., and R.B. Robinson (1992) Softening by Fluidized Bed Crystallizers, *Jour. Environmental Engineering,* 118 (4), 513–529.

Harned, H.S., and B.B. Owen (1958) *The Physical Chemistry of Electrolytic Solutions*, 3d ed., American Chemical Society Monograph Series, New York: Reinhold Publishing.

Hartung, H.O. (1948) Solids-Contact Process Basins, *Journal AWWA,* 40 (10), 1028–1036.

Hartung, H.O. (1951) Committee Report: Capacity and Loadings of Suspended Solids Contact Units. *Journal AWWA,* 43 (4), 263–291.

Horsley, M.B., D.B. Elder, and L.L. Harms (2005) Lime Softening, Ch. 11 in AWWA & ASCE, *Water Treatment Plant Design,* 4th ed., ed. E.E. Baruth, New York: McGraw-Hill.

Hsu, S., and P.C. Singer (2009) Application of Anion Exchange to Control NOM Interference on Lime Softening, *Journal AWWA,* 101 (6), 85–94.

Iler, R.K. (1979) *Chemistry of Silica.* New York: John Wiley and Sons.

Jenne, E.A. (1968) Controls on Mn, Fe, Co, Ni, Cu, and Zn Concentrations in Soils and Water: The Significant Role of Hydrous Mn and Fe Oxides, in: *Trace Inorganics in Water*, ed. R.F. Gould, Advances in Chemistry Series 73, Washington, DC: American Chemical Society, pp. 337–387.

Jensen, J. N. (2003) *A Problem-Solving Approach to Water Chemistry*, Hoboken, NJ: John Wiley and Sons.

Johnson, D.E., and S.J. Randtke (1983) Removing Nonvolatile Organic Chlorine and Its Precursors by Coagulation and Softening, *Journal AWWA,* 75 (5), 249.

Kalscheur, K., C.E. Gerwe, J. Kweon, G.E. Speitel Jr., and D.F. Lawler (2006) Enhanced Softening: Effects of Source Water Quality on NOM Removal and DBP Formation, *Journal AWWA,* 98 (11), 93–106.

Kolthoff, I.M. (1932) Theory of Coprecipitation. The Formation and Properties of Crystalline Precipitates, *Jour. Physical Chemistry,* 36 (3), 860–881.

Kweon, J.H., and D.F. Lawler (2002) Evaluating Precipitative Softening as a Pretreatment for Ultrafiltration of a Natural Water, *Environmental Engineering Science,* 19 (6), 531–544.

Langmuir, D. (1997) *Aqueous Environmental Chemistry,* Saddle River, NJ: Prentice Hall.

Lange's Handbook of Chemistry, 16th ed. (2005) ed. J.G, Speight, New York: McGraw-Hill.

Larson, T.E. (1951) The Ideal Lime-Softened Water, *Journal AWWA,* 43 (8), 649–664.

Larson, T.E., and A.M. Buswell (1940) Theoretical Limits of the Lime–Soda Method of Water Softening, *Industrial and Engineering Chemistry,* 32 (1), 130–132.

Larson, T.E., and A.M. Buswell (1942) Calcium Carbonate Saturation Index and Alkalinity Interpretations, *Journal AWWA*, 34 (11), 1667–1684.

Larson, T.E., R.W. Lane, and C.H. Neff (1959) Stabilization of Magnesium Hydroxide in the Solids-Contact Process, *Journal AWWA*, 51 (12), 1551.

Lee, S.Y., and E.A. Bondietti (1983) Removing Uranium from Drinking Water by Metal Hydroxides and Anion–Exchange Resin, *Journal AWWA*, 75 (10), 536–540.

Liao, M.Y., and S.J. Randtke (1985) Removing Fulvic Acid by Lime Softening, *Journal AWWA*, 77 (8) 78–88.

Liao, M.Y., and S.J. Randtke (1986) Predicting the Removal of Soluble Organic Contaminants by Lime Softening, *Water Research*, 20 (1), 27–35.

Lin, Y.-P., P.C. Singer, and G.R. Aiken (2005) Inhibition of Calcite Precipitation by Natural Organic Material: Kinetics, Mechanism, and Thermodynamics, *Environmental Science & Technology*, 39 (17), 6420–6428.

Loewenthal, R.E., and G.V.R. Marais (1976) *Carbonate Chemistry of Aquatic Systems: Theory and Application*, Ann Arbor, MI: Ann Arbor Science Publishers.

Logsdon, G.S. (1977) Treatment Techniques for the Removal of Radioactive Contaminants from Drinking Water, in Manual of Treatment Techniques for Meeting the Interim Primary Drinking Water Regulations, Report No. 600/8-77-005, USEPA, Cincinnati, OH: Municipal Environmental Research Laboratory. (No document authors are listed. The chapters were written by T.J. Sorg, G.S. Logsdon, and O.T. Love Jr.)

Lohr, E.W., and S.K. Love (1954) The Industrial Utility of Public Water Supplies in the United States, 1952, Part 2, States West of the Mississippi River. U.S. Geological Survey Water-Supply Paper 1300, Washington, DC: U.S. Government Printing Office, 462 pp. (Available at http://pubs.er.usgs.gov/pubs/wsp/wsp1300.)

Merrill, D.T. (1978) Chemical Conditioning for Water Softening and Corrosion Control, in *Water Treatment Plant Design*, ed. R.L. Sanks, Ann Arbor, MI: Ann Arbor Science Publishers.

Merrill, D.T., and R.L. Sanks (1969) Corrosion Control by Deposition of $CaCO_3$ Films: A Practical Approach for Plant Operators, *Journal AWWA*, Part 1, 69 (11), 592–599; Part 2, 69 (12), 634–640; Part 3, 70 (1), 12–18.

Mustafa, B., and H.W. Walker (2006) Lime–Soda Softening Process Modifications for Enhanced NOM Removal, *Jour. Environmental Engineering*, 132 (2), 158–165.

NALCO Water Handbook (1979) ed. F.N. Kemmer, New York: McGraw-Hill.

NIST (National Institute of Standards and Technology) (2004) NIST Critically Selected Stability Constants of Metal Complexes Database, NIST Standard Reference Database 46, ed. R.M. Smith and A.E. Martell, Gaithersburg, MD: National Institute of Standards and Technology (Version 8 for Windows; program developed by R. J. Motekaitis).

Pan, S., S.J. Randtke, F. deNoyelles Jr., and D.W. Graham (2002) Occurrence, Biodegradation, and Control of Geosmin and MIB in Midwestern Water Supplies, *Proceedings of American Water Works Association Annual Conference*, New Orleans, LA, June 16–20, Denver, CO: American Water Works Association.

Parkhurst, D.L. (1992) The User's Guide to PHREEQC—A Computer Program for Speciation, Reaction-Path, Advective Transport, and Inverse Geochemical Calculations, Water Resources Investigations Report 92-4227. Denver. Colo.: U.S. Geological Survey. (Updated versions are available at: http://water.usgs.gov/software/lists/geochemical.)

Pisigan, R.A., and J.E. Singley (1985) Calculating the pH of Calcium Carbonate Saturation, *Journal AWWA*, 77, (10), 83–91.

Pitzer, K.S. (1991) Ion Interaction Approach: Theory and Data Correlation, in *Activity Coefficients in Electrolyte Solutions*, 2nd ed., ed. K.S. Pitzer, Boca Raton, FL.: CRC Press.

Pizzi, N.G. (1995) *Hoover's Water Supply and Treatment, 12th ed.*, Bulletin 211, National Lime Association, Dubuque, IA: Kendall/Hunt Publishing.

Plummer, L.N., and E. Busenberg (1982) The Solubilities of Calcite, Aragonite and Vaterite in CO_2-H_2O Solutions Between 0 and 90°C, and an Evaluation of the Aqueous Model for the System $CaCO_3$-CO_2-H_2O, *Geochim. Cosmochim. Acta* 46, 1011–1040.

Plummer, L.N., D.L. Parkhurst, G.W. Fleming, and S.A. Dunkle (1988) A Computer Program Incorporating Pitzer's Equations for Calculation of Geochemical Reactions in Brines, Water Resources Investigations Report 88-4153, Reston, VA: U.S. Geological Survey. (Updated versions are available at: http://water.usgs.gov/software/lists/geochemical.)

Pontius, F.W., and W.R. Diamond (1999) Complying with the Stage 1 D/DBP Rule, *Journal AWWA,* 91 (3), 16–32.

Powell, S.T. (1954) *Water Conditioning for Industry*, New York: McGraw-Hill.

Qasim, S.R., S.A. Hashsham, and N.I. Ansari (1992) TOC Removal by Coagulation and Softening, *Journal of Environmental Engineering (ASCE*), 118 (3), 432–436.

Randtke, S.J. (1988) Organic Contaminant Removal by Coagulation and Related Process Combinations, *Journal AWWA*, 80 (5), 40–56.

Randtke, S.J. (1999) DBP Precursor Removal by Coagulation and Precipitative Softening, In: Formation and Control of Disinfection By-Products in Drinking Water, ed. P.C. Singer, Denver, CO: American Water Works Association.

Randtke, S.J., C.E. Thiel, M.Y. Liao, and C.N. Yamaya (1982) Removing Soluble Organic Contaminants by Lime-Softening, *Journal AWWA,* 74 (4), 192–202.

Randtke, S.J., A.A. Witt, P.V. Adams, R.E. Pancake, and C.D. Adams (2004) Occurrence and Control of Atrazine in Midwestern Surface Water Supplies, in Advances in Water and Wastewater Treatment, ed. R. Surampalli, Environmental and Water Resources Institute, American Society of Civil Engineers, Reston, VA: American Society of Civil Engineers, pp. 356–394.

Recommended Standards (2007) Recommended Standards for Water Works: Policies for the Review and Approval of Plans and Specifications for Public Water Supplies, A Report of the Water Supply Committee of the Great Lakes–Upper Mississippi River Board of State and Provincial Health and Environmental Managers, Albany: Health Education Services.

Reh, C.W. (1978) Lime–Soda Softening Processes, in *Water Treatment Plant Design*, ed. R.L. Sanks, Ann Arbor, MI: Ann Arbor Science Publishers.

Roalson, S.R., J. Kweon, D.F. Lawler, and G.E. Speitel Jr. (2003) Enhanced Softening: Effects of Lime Dose and Chemical Additions, *Journal AWWA*, 95 (11), 97–109.

Rossum, J.R., and D.T. Merrill (1983) An Evaluation of the Calcium Carbonate Saturation Indexes, *Journal AWWA,* 75 (2), 95–100.

Russell, C.G., D.F. Lawler, and G.E. Speitel Jr. (2009) NOM Coprecipitation with Solids Formed During Softening, *Journal AWWA,* 101 (4), 112–124.

Ryznar, J.W., J. Green, and M.G. Winterstein (1946) Determination of the pH of Saturation of Magnesium Hydroxide, *Jour. Industrial and Engineering Chemistry,* 38 (10), 1057–1061.

Sawyer, C.N., P.L. McCarty, and G.F. Parkin (2003) *Chemistry for Environmental Engineering and Science,* 5th ed., New York: McGraw-Hill.

Schierholz, P.M., J.D. Stevens, and J.L. Cleasby (1976) Optimum Calcium Removal in Lime Softening, *Journal AWWA,* 68 (2), 112–116.

Segers, Allen (Tom), Chief, Water Treatment Plant, Miami–Dade County Water & Sewer Department, Hialeah, FL, personal communication, July 16, 2009.

Semmens, M. J., and A.B. Staples (1986) The Nature of Organics Removal During Treatment of Mississippi River Water, *Jour. AWWA,* 78 (2), 76–81.

Sheen, R.T., and H.B. Lammers (1944) Recovery of Calcium Carbonate or Lime from Water Softening Sludges, *Journal AWWA,* 36 (11), 1145–1169.

Shorney, H.L. (1998) Disinfection By-Product Precursor Removal by Enhanced Coagulation and Softening, Ph.D. Thesis, Dept. of Civil Engineering, University of Kansas, Lawrence.

Shorney, H.L., S.J. Randtke, P.H. Hargette, P.D. Mann, R.C. Hoehn, W.R. Knocke, A.M. Dietrich, and B.W. Long (1996) The Influence of Raw Water Quality on Enhanced Coagulation and Softening for the Removal of NOM and DBP Formation Potential, *Proceedings of the American Water Works Association Annual Conference,* Toronto, CAN, June 23–27, Denver, CO: American Water Works Association.

Shorney, H.L., S.J. Randtke, P.H. Hargette, W.R. Knocke, A.M. Dietrich, R.C. Hoehn, and B.W. Long (1999) Removal of DBP Precursors by Enhanced Coagulation and Lime Softening, Awwa Research Foundation Report No. 90783, Denver, CO: Awwa Research Foundation and the American Water Works Association, 300 pp.

Skoog, D.A., and D.M. West (1969) *Fundamentals of Analytical Chemistry,* 2nd ed., New York: Holt, Rinehart and Winston.

Slack, J.G. (1980) Calcium Carbonate Hexahydrate: Its Properties and Formation in Lime–Soda Softening, *Water Research,* 14 (7), 799–804.

Snoeyink, V.L., and D. Jenkins (1980) *Water Chemistry*, New York: John Wiley and Sons.

Sollo, F.W., T.E. Larson, and H.F. Mueller (1978) Fluoride Removal from Potable Water Supplies, Completion Report No. 136 to the Office of Water Research and Technology, Washington, DC: Dept. of the Interior, 36 pp. (Sept.).

Sorg, T.J. (1977) Treatment Techniques for the Removal of Inorganic Contaminants from Drinking Water, In: Manual of Treatment Techniques for Meeting the Interim Primary Drinking Water Regulations, Report No. 600/8-77-005, USEPA, Cincinnati, Ohio: Municipal Environmental Research Laboratory. (No authors are listed on the cover or title page. The chapters were written by T.J. Sorg, G.S. Logsdon, and O.T. Love Jr.)

Sorg, T.J. (1978) Treatment Technology to Meet the Interim Primary Drinking Water Regulations for Inorganics (Part I), *Journal AWWA,* 70 (2), 105–112.

Sorg, T.J. (1988) Methods for Removing Uranium from Drinking Water, *Journal AWWA,* 80 (7), 105–111.

Sorg, T.J., and G.S. Logsdon (1980) Treatment Technology to Meet the Interim Primary Drinking Water Regulations for Inorganics. Part 5, *Journal AWWA,* 72 (7), 411–422.

Spaulding, C.H. (1937) Conditioning of Water Softening Precipitates, *Journal AWWA,* 29 (11), 1697–1707.

Straub, C.P. (1964) Removal of Radioactivity by Waste-Treatment Processes, Ch. 8, in *Low Level Radioactive Wastes—Treatment, Handling, Disposal,* C.P. Straub, U.S. Atomic Energy Commission, Washington, DC: U.S. Government Printing Office.

Stumm, W., and J.J. Morgan (1996) *Aquatic Chemistry,* 3d ed., New York: John Wiley and Sons.

Taylor J.S., B.R. Snyder, B. Ciliax, C. Ferraro, A. Fisher, P. Muller and D. Thompson (1984) Trihalomethane Precursor Removal by the Magnesium Carbonate Process, U.S. Environmental Protection Agency, Report 600/S2-84/090, Water Engineering Research Laboratory, Cincinnati, OH (Sept.).

Taylor J.S., D. Thompson, B.R. Snyder, J. Less, and L. Mulford (1986) Cost and Performance Evaluation of In-Plant Trihalomethane Control Techniques, USEPA Report 600/S2-85/138, Water Engineering Research Laboratory, Cincinnati, OH (Jan.).

Thompson, C.G., J.E. Singley, and A.P. Black (1972) Magnesium Carbonate—A Recycled Coagulant, *Journal AWWA,* 64 (1), 11–19, and 64 (2), 93–99.

Thompson, J.D., M.C. White, G.W. Harrington, and P.C. Singer (1997) Enhanced Softening: Factors Influencing DBP Precursor Removal, *Journal AWWA,* 89 (6), 94–105.

Truesdell, A.H., and B.F. Jones (1973) WATEQ, A Computer Program for Calculating Chemical Equilibria of Natural Waters, NTIS (National Technical Information Service), Springfield, VA: U.S. Department of Commerce, PB 220464.

Truesdell, A.H., and B.F. Jones (1974) WATEQ, A Computer Program for Calculating Chemical Equilibria of Natural Waters, *J. Res. U.S. Geol. Surv.,* 2 (2), 233–248.

Tuepker, J.L., and H.O. Hartung (1960) Effect of Accumulated Lime-Softening Slurry on Magnesium Reduction, *Journal AWWA,* 52 (1), 106–116.

USEPA (U.S. Environmental Protection Agency) (1975) Magnesium Carbonate–A Recycled Coagulant for Water Treatment, Technology Transfer Capsule Report, EPA-625/2-75-009, Cincinnati, OH.

USEPA (1998) National Drinking Water Regulations: Disinfectants and Disinfection Byproducts; Final Rule, *Federal Register 63,* 241, 69390 (December 16).

USEPA (1999) Enhanced Coagulation and Enhanced Precipitative Softening Guidance Manual, Report EPA 815-R-99-012, Washington, DC: USEPA, Office of Water (May).

USEPA (2003) Drinking Water Advisory: Consumer Acceptability Advice and Health Effects Analysis on Sodium, Office of Water, Health and Ecological Criteria Division, Rept. No. EPA 822-R-03-006 (Feb.).

van der Veen, C., and A. Graveland (1988) Central Softening by Crystallization in a Fluidized-Bed Process, *Journal AWWA,* 80 (6), 51–58.

van Dijk, J.C. and D.A. Wilms (1991) Water Treatment Without Waste Material–Fundamentals and State of the Art of Pellet Softening, *J. Water Supply Research and Technology–Aqua,* 40 (5), 263–280.

Westall, J.C., J.L. Zachary, and F.M.M. Morel (1976) MINEQL, a Computer Program for the Calculation of Chemical Equilibrium Composition of Aqueous Solutions, Technical Note 18, Dept. Civil Engineering, Cambridge: Massachusetts Institute of Technology, 91 pp. (Commercial versions are available at www.mineql.com.)

World Health Organization (1996) Hardness in Drinking Water, in *Guidelines for Drinking-Water Quality,* Vol. 2, 2nd ed., Geneva.

Zell, Shannon, Kiln Supervisor, Division of Water Supply and Treatment, Dayton, OH, personal communication, July 16, 2009.

CHAPTER 14
ADSORPTION OF ORGANIC COMPOUNDS BY ACTIVATED CARBON

R. Scott Summers, Ph.D.
Professor
Civil, Environmental, and Architectural Engineering
University of Colorado
Boulder, Colorado, United States

Detlef R. U. Knappe, Ph.D.
Professor
Department of Civil, Construction, and Environmental Engineering
North Carolina State University
Raleigh, North Carolina, United States

Vernon L. Snoeyink, Ph.D.
Professor Emeritus
Department of Civil and Environmental Engineering
University of Illinois at Urbana-Champaign
Urbana, Illinois, United States

ADSORPTION OVERVIEW ... **14.2**
ADSORBENT CHARACTERISTICS ... **14.4**
Activated Carbon ... 14.4
Other Adsorbents ... 14.7
ADSORPTION THEORY ... **14.8**
Adsorption Equilibrium ... 14.8
Adsorption Kinetics ... 14.28
Desorption and Attenuation ... 14.36
GAC ADSORPTION SYSTEMS ... **14.37**
PERFORMANCE OF GAC SYSTEMS ... **14.39**
Factors Affecting Removal Efficiency ... 14.39
Adsorption Efficiency of Field-Scale Systems ... 14.46
Reactions of Inorganic Compounds with Activated Carbon ... 14.58
GAC PERFORMANCE ESTIMATION ... **14.59**
Isotherm Determination and Application ... 14.59
Bench-Scale Column Tests and Applications ... 14.64
Pilot-Plant Tests ... 14.66
Analysis of GAC Adsorber Data ... 14.67
Modeling of GAC Breakthrough Behavior ... 14.73
POWDERED ACTIVATED CARBON ADSORPTION ... **14.77**
PAC Application ... 14.77
PAC Performance Estimation ... 14.81
Experience with PAC ... 14.83
THERMAL REACTIVATION OF GAC ... **14.85**
Reactivation Principles ... 14.85
Reactivation Furnaces ... 14.86
Reactivation Performance ... 14.86
Reactivation By-Products ... 14.87

ADSORPTION OF ORGANIC MATTER BY OTHER ADSORBENTS 14.87
Synthetic Adsorbent Resins 14.87
Carbonaceous Resins.................... 14.88
Activated Carbon Fibers 14.88
High-Silica Zeolites 14.89
ABBREVIATIONS 14.89
NOTATION FOR EQUATIONS 14.91
REFERENCES 14.91

ADSORPTION OVERVIEW

Adsorption of a substance involves its accumulation at the interface between two phases, such as a gas and a solid or, more germane to this chapter, a liquid and a solid. The molecule that accumulates, or *adsorbs*, at the interface is called the *adsorbate*, and the solid on which adsorption occurs is the *adsorbent*. Adsorbents of interest in water treatment include activated carbon; synthetic and carbonaceous adsorbent resins; ion exchange resins; metal oxides, hydroxides, and carbonates; activated alumina; zeolites; clays; and other solids that are suspended in or in contact with water. Adsorbates of interest in water treatment include both organic and inorganic compounds of natural and anthropogenic origins. The focus of this chapter is the adsorption of organic compounds by activated carbon.

Dissolved organic matter (DOM) in drinking water sources can be divided into several categories. For purposes of this chapter, we will use the categories shown in Fig. 14-1. DOM contains specific organic compounds of both natural and synthetic origin. When these compounds become of health or aesthetic concern to utilities and their consumers, they can be referred to as *micropollutants*. Naturally occurring micropollutants include microbial metabolites that cause taste and odor (e.g., geosmin) or are toxic (e.g., microcystins). There are also anthropogenic micropollutants, which are termed *synthetic organic compounds* (SOCs) for this chapter. SOCs include pesticides, solvents, and pharmaceuticals and personal-care

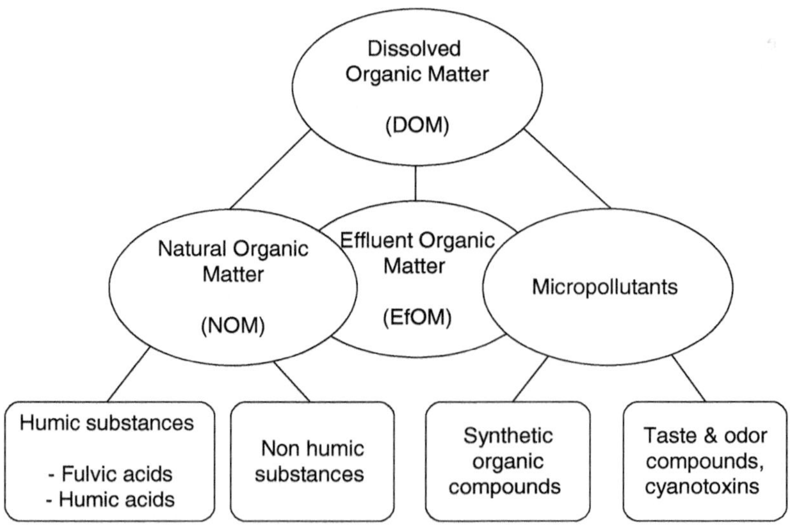

FIGURE 14-1 Categories of dissolved organic matter (DOM).

products (PPCPs). Two other components of DOM are natural organic matter (NOM) and wastewater treatment plant effluent organic matter (EfOM). NOM is a complex mixture of ill-defined organic compounds, such as fulvic and humic acids, hydrophilic acids, and specific compounds such as carbohydrates and proteins, all of natural origin (see Chap. 3 for coverage of NOM). EfOM includes background NOM, compounds from human activity, and metabolites from biological wastewater treatment.

Adsorption plays an important role in the improvement of water quality. Activated carbon, for example, is used most often to adsorb specific organic compounds that cause taste and odor or are of health concern, as well as NOM, which can cause color and can react with chlorine to form disinfection by-products (DBPs). Apart from activated carbon adsorption, a number of other adsorption processes are important in water treatment. The aluminum and ferric hydroxide solids that form during coagulation and the calcium carbonate and magnesium hydroxide solids that form in the lime softening process also adsorb NOM. Pesticides adsorbed on clay particles can be removed by coagulation and filtration. Ion exchange (IX) resins and synthetic and carbonaceous adsorbent resins are available that can be used for efficient removal of selected organic and inorganic compounds, as well as NOM. Removal of NOM by the coagulation process (see Chap. 8) and of organic and inorganic ions by IX resins and activated alumina (see Chap. 12) are also discussed in this book.

Activated carbon can be used as powdered activated carbon (PAC), which is added directly to the water to be treated, typically as a slurry at the rapid mixer, or as granular activated carbon (GAC), which is used in fixed-bed contactors. Historically, PAC has been favored for use with seasonal taste and odor problems because of its low capital cost and flexibility. However, when high PAC doses are required for long periods of time, GAC is an economical alternative to PAC. GAC is used in rapid media filters (filter adsorbers), where both adsorption and particle filtration can occur, in stand-alone units after filtration (postfilter adsorbers), or for groundwater treatment. The use of fixed GAC beds permits higher adsorptive capacities to be achieved and easier process control than is possible with PAC. The higher capital cost for GAC often can be offset by better efficiency, especially when the targeted adsorbate(s) must be removed on a continuous basis. GAC should be seriously considered for water supplies when odorous compounds or organic chemicals of health concern frequently are present, when a barrier is needed against organic compounds from accidental (or deliberate) spills, or in some situations that require DBP precursor removal. While GAC has excellent adsorption capacity for many undesirable substances, it must be removed periodically from the adsorber bed and replaced with fresh or reactivated GAC.

A 1977 study conducted by two committees of the American Water Works Association (AWWA) showed that approximately 25 percent of 645 U.S. utilities, including the 500 largest, added PAC (AWWA Committee Report, 1977). In 1984, 29 percent of the 600 largest utilities reported using PAC (AWWA, 1986), predominantly for odor control. According to the AWWA Water Stats 1996 database, PAC was available for use by 48 percent of the 543 responding surface water systems (AWWA, 1996). The U.S. Environmental Protection Agency (USEPA) Community Water System Survey 2000 (USEPA, 2002) surveyed 1246 community water systems, and PAC was used in 7.6 percent of the 620 surveyed surface water systems. Among surface water utilities serving more than 10,000 customers, more than 17 percent of the responding utilities used PAC (USEPA, 2002).

The number of drinking water plants using GAC increased from 65 in 1977 (AWWA Committee Report, 1977), principally for odor control, to 135 in 1986 (Snoeyink, 1990); in 1996, there were approximately 300 plants treating surface water and several hundred more treating contaminated groundwater (Snoeyink and Summers, 1999). Based on the AWWA Water Stats 1996 database, GAC was being used by 12 percent of the 543 responding surface water systems, with another 4 percent listing it in the planned stage, and 5.3 percent of the 493 responding groundwater systems used GAC (AWWA, 1996).

In the USEPA Community Water System Survey 2000, approximately 14 percent of surface water systems serving more than 3300 people used GAC. Among groundwater systems, 8.6 percent of surveyed utilities serving between 100,000 and 500,000 people employed GAC, but GAC use was less than 2 percent for groundwater systems in other population categories (USEPA, 2002).

ADSORBENT CHARACTERISTICS

Activated Carbon

Production. A wide variety of raw materials can be used to make activated carbon (Hassler, 1974), both PAC and GAC, and the substances used for drinking water treatment carbons predominantly are subbituminous coal, lignite, coconut, and wood. Both the physical and chemical manufacturing processes involve carbonization, or conversion of the raw material to a char, and activation or oxidation to develop the internal pore structure. With physical activation, carbonization or pyrolysis is usually done in the absence of air at temperatures less than 700°C, whereas activation is carried out with oxidizing gases such as steam and carbon dioxide (CO_2) at temperatures of 800 to 900°C. Chemical activation combines carbonization and activation steps by mixing a chemical activating agent, such as phosphoric acid, with the base material and heating the mixture in the absence of oxygen. Patents describing carbonization and activation procedures are discussed by Yehaskel (1978). Both the base material and manufacturing process affect activated carbon characteristics and performance.

Activated carbons are manufactured by either direct activation or a reagglomeration process. Direct activation means that the raw material (e.g., coal) is fed to the furnace as mined, with only some crushing and screening prior to the activation process. The reagglomeration process calls for crushing the raw material, adding some volatile material (such as coal tar pitch), reagglomerating the mixture under high pressure (typically in a pocket briquettor), crushing the material once again, and then activating it.

Physical Characteristics. Activated carbon is a highly porous material with an internal surface area that typically ranges from about 800 to 1500 m^2/g (Bansal et al., 1988). Typically, the manufacturer provides data that include the *BET surface area.* This parameter is determined by measuring the adsorption isotherm for nitrogen (N_2) gas molecules and then analyzing the data using the Brunauer-Emmett-Teller (BET) isotherm equation (Adamson and Gast, 1997) to determine the amount of N_2 to form a complete monolayer of N_2 molecules on the carbon surface. Multiplying the surface area occupied per N_2 molecule (0.162 nm^2 per molecule of N_2) by the number of molecules in the monolayer yields the BET surface area. Not all the BET surface area is accessible to aqueous adsorbates; because N_2 is a small molecule, it can enter pores that are unavailable to larger adsorbates.

The pores that give rise to this large surface area can be envisioned as spaces between irregularly arranged graphite-like platelets (Leon y Leon et al., 1992) or condensed polyaromatic sheets (Bansal et al., 1988) that are the building blocks of activated carbon. To classify pores according to size, the International Union of Pure and Applied Chemistry (IUPAC) differentiates between (1) micropores (<2 nm width), (2) mesopores (2–50 nm width), and (3) macropores (>50 nm width) (Sing et al., 1985). PAC and GAC for water treatment typically exhibit a heterogeneous pore structure, in which micropores, mesopores, and macropores are all present. Activated carbon pore size distributions in the micropores and mesopores are calculated using gas adsorption data, whereas mercury porosimetry data are used to calculate pore size distributions in the macropores.

TABLE 14-1 Physical Characteristics of Representative Activated Carbons and a Carbonaceous Resin

Adsorbent	BET surface area (m^2/g)	Micropore volume[a] (cm^3/g)	Mesopore volume[b] (cm^3/g)
F400 (bituminous coal based)	940	0.340	0.160
HD4000 (lignite based)	525	0.148	0.430
CC-602 (coconut shell based)	1160	0.437	0.060
Picazine (wood based)	1680	0.496	0.655
Ambersorb 563 (carbonaceous resin)	550	0.201	0.318

[a]Micropore volume calculated by density functional theory (DFT) for pores with widths less than 2 nm.
[b]Mesopore volume calculated by Barrett, Joyner, and Halenda (BJH) method for pores with widths ranging from 2 to 50 nm.
Sources: Data from Knappe et al. (2007) and Mezzari (2006).

BET surface areas, micropore volumes, and mesopore volumes for representative activated carbons are shown in Table 14-1. For reference, data for a carbonaceous resin are also shown.

The particle shape of crushed GAC is irregular, but extruded GACs have a smooth, cylindrical shape. Particle shape affects the filtration and backwash properties of GAC beds. Particle size is an important parameter because of its effect on adsorption kinetics and breakthrough, as well as filtration performance. *Particle size distribution* refers to the relative amounts of different-sized particles that are part of a given sample, or lot, of carbon. Commonly used GAC sizes include 12 × 40 and 8 × 30 U.S. Standard Mesh, which range in apparent diameter from 1.68 to 0.42 mm and 2.38 to 0.59 mm, respectively, and have average diameters of about 1.0 and 1.5 mm, respectively. Customized particle size distributions are available to meet site-specific needs. The uniformity coefficient (see Chap. 10) is often quite large, typically on the order of 1.9, to promote stratification during backwashing. Usually, commercially available activated carbons have a small percentage of material smaller than the smallest sieve and larger than the largest sieve, which significantly affects the uniformity coefficient. Extruded carbon particles all have the same diameter, but they vary in length.

The *apparent density*[*] is the mass of nonstratified dry activated carbon per unit volume of activated carbon, including the volume of voids between grains. Typical values for GACs manufactured from bituminous coal, lignite, and coconut shells are 350 to 650 kg/m^3 (22–41 lb/ft^3), whereas typical values for wood based GAC are in the range of 225 to 300 kg/m^3 (14–19 lb/ft^3). Distinguishing between the apparent density and the *bed density, backwashed and drained* (i.e., stratified, free of water) is important for GAC. The former is a characteristic of GAC as shipped. The latter is a characteristic of the GAC as placed in a bed and is about 10 percent less than the apparent density. The bed density is typical of GAC during normal operation unless the GAC becomes destratified during backwashing. The bed density determines how much GAC must be purchased to fill a contactor of a given size and therefore is an important GAC characteristic.

The *particle density wetted in water* is the mass of solid activated carbon plus the mass of water required to fill the internal pores per unit volume of particle. Its value for GAC typically ranges from 1300 to 1500 kg/m^3 (81–94 lb/ft^3), and it determines the extent of fluidization and expansion of a given size particle during backwash.

[*]This definition is based on ASTM Standard D2854-09 (ASTM, 2009). It conflicts with ASTM Standard C128-07a for apparent specific gravity, as used in Chap. 10, which does not include the volume of interparticle voids.

Particle hardness is important because it affects the amount of attrition during backwash, transport, and reactivation. In general, the harder the activated carbon is, the less is the attrition for a given amount of friction or impact between particles. Activated carbon hardness generally is characterized by an experimentally determined hardness or abrasion number using a test such as the American Society for Testing and Materials (ASTM) ball pan hardness test (ASTM Standard D3802), which measures the resistance to particle degradation on agitating a mixture of activated carbon and steel balls (ASTM, 2009). The relationship between the amount of attrition that can be expected when activated carbon is handled in a certain way and the hardness number has not been studied extensively, however. Comparing wood based GAC with an abrasion number of 47 and bituminous coal based GAC with an abrasion number of 75, Grens and Werth (2001) showed that filter performance and changes in GAC bed height were similar during 500 air-water backwash cycles.

The primary difference between PAC and GAC is particle size. As specified by AWWA Standard B600-05 (AWWA, 2005a), not less than 90 percent by mass of PAC shall pass a 44-μm (325 U.S. Standard Mesh) sieve unless the activated carbon is wood based. For wood based PACs, not less than 60 percent by mass shall pass a 44-μm sieve. The particle size distribution is important because the smaller PAC particles adsorb organic compounds more rapidly than large PAC particles (Najm et al., 1990; Adham et al., 1991). The apparent density of PAC ranges from 200 to 750 kg/m^3 (12–47 lb/ft^3) and depends on the base material and the manufacturing process.

Chemical Characteristics. At the edges of the condensed polyaromatic sheets that constitute the building blocks of activated carbons, heteroatoms (i.e., atoms other than carbon) are encountered that define the chemical characteristics of activated carbon surfaces. A typical elemental composition of activated carbon is approximately 88 percent C, 6 to 7 percent O, 1 percent S, 0.5 percent N, and 0.5 percent H, with the remainder being mineral matter (i.e., ash) (Bansal et al., 1988). However, the elemental composition of activated carbons can vary substantially from these average values; for example, the oxygen content can range from as low as 1 percent to as high as 25 percent (Bansal et al., 1988), and the ash content can range from 1 to 20 percent (Jankowska et al., 1991).

Because of its abundance and profound effects on activated carbon hydrophilicity and surface charge, oxygen is generally the most important heteroatom from a standpoint of activated carbon surface chemistry. Oxygen commonly occurs in the form of carboxylic acid groups (–COOH), phenolic hydroxyl groups (–OH), and quinone carbonyl groups (>C=O) (Boehm et al., 1964; Puri, 1970; Mattson and Mark, 1971; Snoeyink and Weber, 1972; Leon y Leon and Radovic, 1992; Boehm, 1994). Activated carbons assume an acidic character when exposed to oxygen at between 200°C and 700°C or to oxidants such as hydrogen peroxide (e.g., Puri, 1970, 1983). The activated carbon acidity is explained primarily by the formation of carboxylic acid and phenolic hydroxyl groups (Leon y Leon and Radovic, 1992). Oxidation of activated carbon surfaces also occurs during the exposure of activated carbon to common oxidants used in water treatment, such as chlorine, permanganate, and ozone.

Adsorption Properties. Both physical and chemical characteristics of activated carbon affect its performance. Pore size distribution is one important characteristic of an activated carbon. The size of adsorbent pores affects the adsorption of organic contaminants in two important ways. First, size exclusion limits the adsorption of contaminants of a given size and shape if pores are too small. Second, the strength of adsorbate-adsorbent interactions increases with decreasing pore size because adsorption potentials between opposing pore walls begin to overlap once the micropore width is less than twice the adsorbate diameter (e.g., Dubinin, 1960; Sing, 1995). Therefore, adsorption of micropollutants preferentially takes place in the smallest pores that are accessible to a given pollutant. As outlined in more detail below, the adsorption of micropollutants and NOM occurs primarily in pores with dimensions that match those of the targeted adsorbate. Thus many micropollutants adsorb in small micropores (<1 nm), whereas the largest percentage of NOM preferentially adsorbs in mesopores and large micropores (>1 nm)

With respect to surface chemistry, hydrophobic adsorbents (i.e., activated carbons with low oxygen content) exhibit larger adsorption capacities for organic micropollutants than hydrophilic adsorbents (i.e., activated carbons with high oxygen content) with similar physical characteristics (Knappe et al., 2003). This trend is principally attributable to enhanced water adsorption on hydrophilic activated carbons (e.g., Kaneko et al., 1989, Li et al., 2002, Quinlivan et al., 2005). More detailed information about surface chemistry effects on the adsorption of micropollutants can be found in Knappe (2006).

A number of surrogate parameters are used to describe the adsorption capacity of activated carbon (Sontheimer et al., 1988). The *iodine number* (ASTM Standard D4607-94; ASTM, 2009) measures the amount of iodine that will adsorb under a specified set of conditions, and it generally correlates well with the surface area available for small molecules. The *molasses number* or *decolorizing index* is related to the ability of activated carbon to adsorb large-molecular-weight color bodies from molasses solution and generally correlates well with the ability of the activated carbon to adsorb other large adsorbates. Other surrogate parameters include the *carbon tetrachloride activity*, the *methylene blue number*, the *acetoxime number*, the *tannin value*, and the *phenol adsorption value*.

Surrogate parameters give some insight into the possible performance of activated carbons in drinking water treatment. BET surface area and iodine number tend to correlate well with adsorption capacity when high solid-phase concentrations can be reached, such as in the treatment of highly contaminated waters or in solvent recovery operations (e.g., Manes, 1998). For the removal of micropollutants from drinking water sources; however, activated carbon performance generally does not correlate well with BET surface area or iodine number (e.g., Quinlivan et al., 2005). In all cases, data from bench- or pilot-scale tests (e.g., isotherm tests, jar tests to evaluate PAC performance, and column tests to evaluate GAC performance) conducted with the specific compound(s) and water of interest are much better indicators of field performance.

Other Adsorbents

Apart from PAC and GAC, a wide range of alternative adsorbents has been tested for their effectiveness to remove organic compounds from water. Among the more frequently studied alternative adsorbents are synthetic adsorbent resins, carbonaceous resins, activated carbon fibers, and hydrophobic zeolites.

Synthetic Adsorbent Resins. Synthetic adsorbent resins are polymeric beads with a large internal surface area that is created during the polymerization process (Neely and Isacoff, 1982). Synthetic adsorbent resins, such as the styrene-divinylbenzene (SDVB) copolymer resin and the phenol-formaldehyde (PF) resin, differ from IX resins because they lack charged or ionizable functional groups. (See Chap. 12 for a discussion of the removal of inorganic substances and NOM by IX resins.) In synthetic adsorbent resins, the degree of cross-linking between the polymers that constitute the matrix can be varied, and thus resins with different pore size distributions can be prepared. In theory, pore size distributions can be tailored to a specific target compound that needs to be removed from water.

Carbonaceous Resins. Carbonaceous resins are prepared by heating macroporous polymer beads in an inert atmosphere. This heat treatment, or pyrolysis, step creates micropores while largely maintaining the macroporous structure of the starting polymer (Neely and Isacoff, 1982).

Thus the pore structure of carbonaceous resins can be tailored in a manner similar to that of synthetic adsorbent resins. Furthermore, pyrolysis conditions can be chosen such that the carbonaceous adsorbents have molecular sieve properties (Neely and Isacoff, 1982). One commonly studied carbonaceous resin is prepared from macroporous SDVB beads, and properties of a representative member (Ambersorb 563) are summarized in Table 14-1.

Activated Carbon Fibers. Activated carbon fibers (ACFs) are manufactured from woven and nonwoven materials such as cross-linked phenolic fibers, polyacrylonitrile, and cellulose (Economy and Lin, 1976; Bahl et al., 1998). ACFs are commonly prepared by pyrolysis and steam activation, and preparation conditions can be controlled such that uniform pore size distributions are obtained (Kasaoka et al., 1989a; Daley et al., 1996). Examination of ACFs by scanning tunneling microscopy showed that both mesopores and micropores are present at the ACF surface; the mesopores were oriented parallel to the fiber axis and did not penetrate the bulk fiber to a depth greater than about 60 nm. Because of their narrower pore size distribution, ACFs are particularly suitable to elucidate pore size effects on organic compound adsorption from aqueous solution (Kasaoka et al., 1989b; Pelekani and Snoeyink, 1999, 2000, 2001; Li et al., 2002; Karanfil et al., 2006).

Zeolites. In water treatment, hydrophilic low-silica zeolites are well known as cation exchangers. These zeolites are aluminosilicates with relatively low SiO_2/Al_2O_3 ratios that give rise to negative charges in the zeolite framework (Dyer, 1984; Townsend, 1984). In contrast, hydrophobic zeolites with high SiO_2/Al_2O_3 ratios show promise for the removal of organic compounds from aqueous solution (e.g., Ellis and Korth, 1993; Kawai et al., 1994; Rossner and Knappe, 2008). Because of their highly ordered lattice structure, zeolites are adsorbents with well-defined pore sizes. For example, ZSM-5 zeolites have straight, elliptical channels with minor and major axis dimensions of 0.53×0.56 nm. The straight channels intersect perpendicularly with sinusoidal, elliptical channels with minor and major axis dimensions of 0.51×0.55 nm. From a drinking water treatment perspective, high-silica zeolites are attractive because pore sizes can be selected that permit the targeted adsorption of smaller organic contaminants while preventing the adsorption of competing substances with larger molecular sizes. High-silica zeolites are marketed in the form of powders and extrudates.

ADSORPTION THEORY

Adsorption Equilibrium

Adsorption of molecules can be represented as a chemical reaction:

$$A + B \rightleftharpoons A{:}B \tag{14-1}$$

where A is the adsorbate, B the adsorbent, and A:B the adsorbed compound. Adsorbates are held on the surface by intermolecular forces such as van der Waal's forces, dipole-dipole interactions, and hydrogen bonds. If the reaction is reversible, as it is for many compounds adsorbed to activated carbon, molecules continue to accumulate on the surface until the rate of the forward reaction (adsorption) equals that of the reverse reaction (desorption). When this condition exists, equilibrium has been reached, and no further accumulation will occur.

Isotherm Equations. One of the most important characteristics of an adsorbent is the quantity of adsorbate that it can accumulate. The constant-temperature equilibrium relationship between the quantity of adsorbate per unit of adsorbent q and its equilibrium aqueous-phase concentration C is called the *adsorption isotherm*. Several equations or models are available that describe this relationship (Sontheimer et al., 1988), but only the commonly used Freundlich model is presented here.

The empirical Freundlich model is useful because it effectively describes adsorption isotherm data for many organic contaminants. The Freundlich isotherm equation has the form

$$q = KC^{1/n} \tag{14-2}$$

and can be linearized as follows:

$$\log q = \log K + \frac{1}{n}\log C \tag{14-3}$$

The parameters q (with units of mass adsorbate/mass adsorbent or mole adsorbate/mass adsorbent) and C (with units of mass/volume or moles/volume) are the equilibrium surface and solution concentrations, respectively. The terms K and $1/n$ are constants for a given system; $1/n$ is unitless, and the units of K are determined by the units of q and C. Although the Freundlich equation was developed to empirically fit adsorption data, the Freundlich equation also can be developed from adsorption theory (e.g., Halsey and Taylor, 1947; Weber and DiGiano, 1996).

The parameter K in the Freundlich equation is related primarily to the adsorption capacity of the adsorbent for the adsorbate, and $1/n$ is a function of the adsorbent heterogeneity. A $1/n$ value of 1 is obtained for homogeneous adsorbents (i.e., adsorbents with a uniform pore size and surface chemistry); $1/n$ values typically are less than 1 for activated carbons, which exhibit a broad distribution of adsorption site energies; much of the heterogeneity results from the variety in pore sizes and shapes in activated carbons. For fixed values of C and $1/n$, the larger the value of K, the larger is the adsorption capacity q. Figure 14-2 depicts an example isotherm; note that both the x- and y-axis scales are logarithmic, in which case

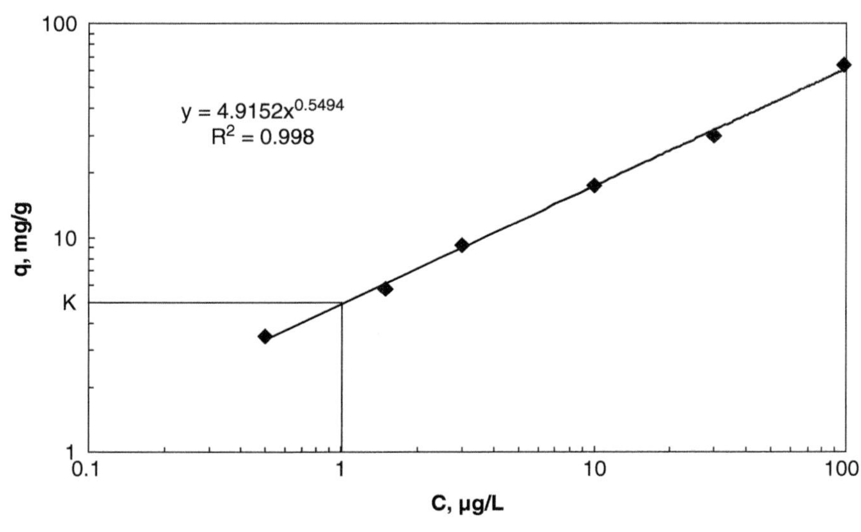

FIGURE 14-2 Example isotherm data and Freundlich isotherm model fit.

the Freundlich model yields a straight line. In Fig. 14-2, equilibrium solid-phase concentrations are presented in units of milligrams of adsorbed compound per gram of activated carbon (mg/g) and equilibrium aqueous-phase concentrations in units of micrograms per liter (μg/L). In this case, the parameter K represents the adsorption capacity (in mg/g) at an equilibrium liquid-phase concentration of 1 μg/L. The parameter $1/n$ represents the slope of the line; for the example isotherm in Fig. 14-2, log q increases by 0.549 for each unit increase in log C.

The Freundlich equation cannot apply to all values of C, however. As C increases, for example, q increases (in accordance with Eq. 14-2) only until all adsorbent pore surfaces are occupied. At saturation, q is a constant, independent of further increases in C, and the Freundlich equation no longer applies. Also, no assurance exists that adsorption data will conform to the Freundlich equation over all concentrations less than saturation, so care must be exercised in extending the equation to concentration ranges that have not been tested.

Tabulations of single-solute isotherm constants are useful when only rough estimates of adsorption capacity are needed to determine whether a more intensive analysis of the adsorption process is warranted. The Freundlich isotherm constants of Speth and Miltner (1990, 1998) are reproduced in Table 14-2 for this purpose. Freundlich isotherm

TABLE 14-2 Freundlich Adsorption Isotherm Parameters for Organic Compounds

Compound	K (mg/g)(L/μg)$^{1/n}$	$1/n$	Source
Cyanazine	102	0.126	*
Trimethoprim (pH 7.8)[a]	98.9	0.148	‡
Metolachlor	98.2	0.125	*
2,4-Dinitrotoluene	96.1	0.157	*
Glyphosate	87.6	0.119	*
Alachlor	81.7	0.257	*
1,1,1-Trichloropropanone	74.4	0.110	†
1,3,5-Trichlorobenzene	63.8	0.324	*
Acifluorfen (pH 6.9)	60.2	0.198	†
Metribuzin	48.7	0.193	*
Hexachlorocyclopentadiene	43.0	0.504	*
2,4,5-Trichlorophenoxy acetic acid	43.0	0.210	*
Pentachlorophenol	42.6	0.339	*
Atrazine	38.7	0.291	*
p-Chlorotoluene	35.9	0.34	*
Dicamba	33.1	0.147	*
Simazine	31.3	0.227	*
Dinoseb	30.4	0.279	*
Chloropicrin	30.2	0.155	†
Sulfamethoxazole (pH 7.8)	29.4	0.234	‡
Picloram	23.4	0.18	*
o-Chlorotoluene	23.2	0.378	*
o-Dichlorobenzene	19.3	0.378	*
Chloral hydrate	18.9	0.051	†
Bromobenzene	17.2	0.364	*
Carbofuran	16.4	0.408	*
Lindane	15.0	0.433	*
p-Xylene	12.6	0.418	*
Styrene	12.2	0.479	*

(*Continued*)

TABLE 14-2 Freundlich Adsorption Isotherm Parameters for Organic Compounds (*Continued*)

Compound	K (mg/g)(L/μg)$^{1/n}$	$1/n$	Source
Trichloroacetic acid (pH 5.6)	11.7	0.216	†
Diquat (pH 6.7)	11.2	0.325	†
Isophorone	9.75	0.271	*
Ethyl benzene	9.27	0.415	*
Chlorobenzene	9.17	0.348	*
Methyl isobutyl ketone	8.85	0.279	†
Aldicarb	8.27	0.402	*
Dibromochloropropane (DBCP)	6.91	0.501	*
2-Methylisoborneol	6.01	0.64	§
m-Dichlorobenzene	5.91	0.63	†
Toluene	5.01	0.429	*
p-Dichlorobenzene	4.97	0.691	*
m-Xylene	4.93	0.614	†
Dalapon	4.92	0.224	*
Methomyl	4.78	0.290	†
Tetrachlorethene (PCE)	4.05	0.516	*
1,1-Dichloropropene	2.67	0.374	*
Methyl ethyl ketone	2.53	0.295	†
Endothall (pH 7.1)	2.28	0.329	†
Trichloroethene (TCE)	2.00	0.482	*
Oxamyl	1.74	0.793	*
Dichloroacetic acid (pH 5.7)	1.63	0.462	†
Benzene	1.26	0.533	*
1,2,3-Trichloropropane	1.08	0.613	*
1,1,1,2-Tetrachlorethane	1.07	0.604	*
Bromoform	0.929	0.665	*
1,3-Dichloropropane	0.897	0.497	*
1,2 Dibromoethane	0.888	0.471	*
Ethylene thiourea	0.716	0.669	†
trans-1,2-Dichloroethene	0.618	0.452	*
Dibromochloromethane	0.585	0.636	*
1,1 Dichloroethene	0.47	0.515	*
Carbon tetrachloride	0.387	0.594	*
1,1,2-Trichloroethane	0.365	0.652	*
1,1,1-Trichloroethane	0.335	0.531	*
1,2-Dichloropropane	0.313	0.597	*
Bromodichloromethane	0.241	0.655	*
Methyl tertiary-butyl ether (MTBE)	0.218	0.479	*
cis-1,2-Dichloroethene	0.202	0.587	*
1,2-Dichloroethane	0.129	0.533	*
Chloroform	0.0925	0.669	*
Dibromomethane	0.0722	0.701	*
1,1-Dichloroethane	0.0646	0.706	*
Methylene chloride	0.00625	0.801	*

[a]When available, pH values are specified for weak acids and bases and ionic compounds.
[*]From Speth and Miltner (1990).
[†]From Speth and Miltner (1998).
[‡]From Rossner (2008).
[§]From Chen et al. (1997).

contants for 2-methylisoborneol (MIB), a taste and odor compound, and sulfamethoxazole and trimethoprim, two antimicrobial compounds, are also shown in Table 14-2. The adsorption isotherm parameters in Table 14-2 were derived from adsorption isotherm experiments conducted with coal based activated carbons with BET surface areas of approximately 1000 m^2/g. Furthermore, adsorbent-adsorbate contact times in these studies were sufficient to reach adsorption equilibrium (5 days for MIB, ≥3 weeks for all other compounds), and the tested aqueous-phase concentrations in were most cases environmentally relevant. Additional isotherm parameters can be found in Sontheimer and colleagues (1988). The parameters in Table 14-2 can be used to judge relative adsorption efficiency. The K values of isotherms that have nearly the same values of $1/n$ show the relative capacity of adsorption. For example, if a GAC column is satisfactorily removing benzene [$K = 1.26(\text{mg/g})(\text{L}/\mu\text{g})^{1/n}$ and $1/n = 0.533$], the removal of compounds with larger values of K and approximately the same concentration likely will be better. (An exception might occur if the organic compounds adsorb to particles that pass through the adsorber.) If the $1/n$ values are much different, however, the capacity of activated carbon for each compound of interest should be calculated at the equilibrium concentration of interest using Eq. 14-2 because the relative adsorbability will depend on the equilibrium concentration.

Adsorption isotherms for representative micropollutants are shown in Fig. 14-3. Among the depicted contaminants, the herbicides metolachlor and atrazine exhibit the largest adsorption capacities, while MTBE exhibits the lowest. At an equilibrium concentration of 1 μg/L, adsorption capacities for MIB and TCE are similar; however, at concentrations that are relevant for MIB (tens of ng/L), the adsorption capacity of activated carbon is considerably lower. The use of isotherm values to estimate GAC adsorber life and PAC usage rate is discussed in the sections "GAC Performance Estimation" and "PAC Adsorption," respectively.

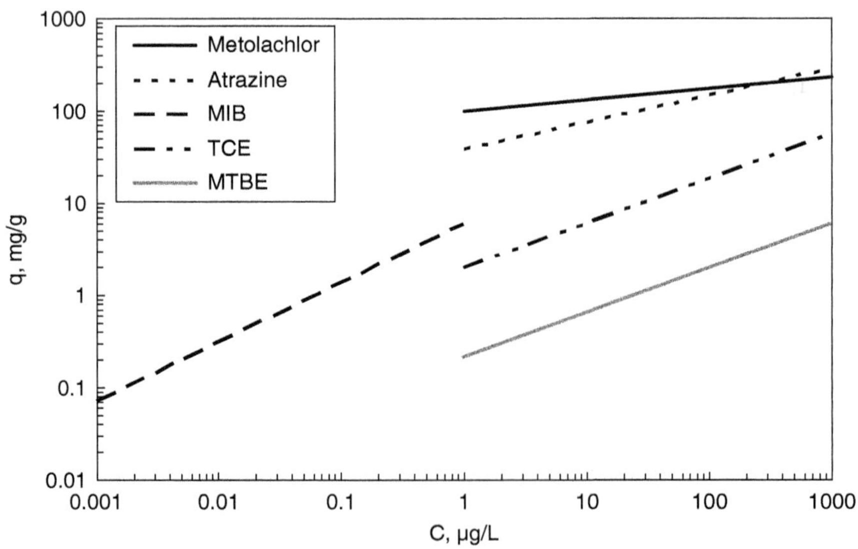

FIGURE 14-3 Adsorption isotherms for representative trace organic contaminants.

Factors Affecting Adsorption Equilibria. Important adsorbent characteristics that affect adsorption uptake at equilibrium include surface area, pore size distribution, and surface chemistry. The maximum amount of adsorption is proportional to the amount of surface area within pores that are accessible to the adsorbate. However, in drinking water treatment applications, organic contaminants typically are present at trace levels well below their aqueous solubility limit, at which the maximum amount of adsorption is observed. At trace levels, adsorption occurs primarily at sites associated with high adsorption energies, that is, in pores whose dimensions closely match those of the targeted micropollutant. Therefore, adsorption uptake by a given activated carbon at a given aqueous-phase micropollutant concentration is related to the volume of pores in a size range that closely matches the size of the targeted adsorbate (e.g., Li et al., 2002; Karanfil and Dastgheib, 2004). Because many micropollutants are small [i.e., total surface areas of about 1–3 nm^2 (Okouchi et al., 1992), which corresponds to spherical diameters of about 0.55–1.0 nm], the presence of small micropores is important for their removal from aqueous solution. For example, the results of Karanfil and Dastgheib (2004) suggest that the adsorption capacity of trichloroethene (TCE) is proportional to the volume of micropores with widths of about 0.5 to 0.8 nm, dimensions that are very close to those of the adsorbate. For emerging contaminants such as pharmaceuticals, endocrine-disrupting chemicals, personal-care products, and flame retardants, Rossner and colleagues (2009) showed that a large micropore volume in the width range of 0.6 to 0.9 nm is important for their effective removal. To achieve effective NOM removal, a large volume of mesopores and micropores larger than 1 nm is required (Ebie et al., 1995; Newcombe et al., 1997; Pelekani and Snoeyink, 1999; Cheng et al., 2005).

The surface chemistry of activated carbon and adsorbate properties also can affect adsorption (e.g., Snoeyink and Weber, 1972; Coughlin and Ezra, 1968; Snoeyink et al., 1974; Pendleton et al., 1997; Karanfil and Kilduff, 1999; Franz et al., 2000; Li et al., 2002, 2005; Quinlivan et al., 2005; Knappe, 2006). These researchers demonstrated that extensive oxidation of carbon surfaces led to large decreases in the amounts of phenol, nitrobenzene, benzene, benzoic acid, benzenesulfonate, 1,2,4-trichlorobenzene, TCE, and MTBE that could be adsorbed. Oxidation of the activated carbon surface with aqueous free and combined chlorine also was found to oxidize the carbon surface and correspondingly to decrease the adsorption capacity for phenol (Snoeyink et al., 1974), MIB (Gillogly et al., 1998a; Lalezary-Craig et al., 1988), and geosmin (Lalezary-Craig et al., 1988). If possible, contact between activated carbon and oxidants such as potassium permanganate and chlorine should be avoided during the treatment of drinking water because the resulting oxygenation of the activated carbon surface decreases micropollutant uptake. In addition, reactions between activated carbon and the oxidizing chemical decrease the oxidant concentration or lead to complete elimination of the oxidant such that the intended oxidation and/or disinfection process is adversely affected. The reaction between activated carbon and ozone produces hydroxyl radicals (Jans and Hoigne, 1998); in the process, the activated carbon surface is oxidized, and ozone decomposition is enhanced. Additional research is needed to better understand micropollutant fate when activated carbon and ozone come into contact at conditions that can be considered typical for drinking water treatment.

The tendency of a molecule to adsorb is a function of its affinity for water compared with its affinity for the adsorbent. Generally, the more hydrophobic the pollutant (or the lower its aqueous solubility), the greater is its tendency to escape from aqueous solution and adsorb on the activated carbon surface, and this tendency is known as *solvent-motivated adsorption* (Weber and DiGiano, 1996). As a molecule becomes larger through the addition of hydrophobic groups such as $-CH_2-$, its solubility decreases, and its extent of adsorption increases as long as the molecule can gain entrance to the pores. When an increase in size causes the molecule to be excluded from some pores, however, adsorption capacity may decrease as solubility decreases. Also, as molecular size increases, the rate of diffusion within the activated carbon particle decreases, especially as molecular size approaches the size of the adsorbent pore.

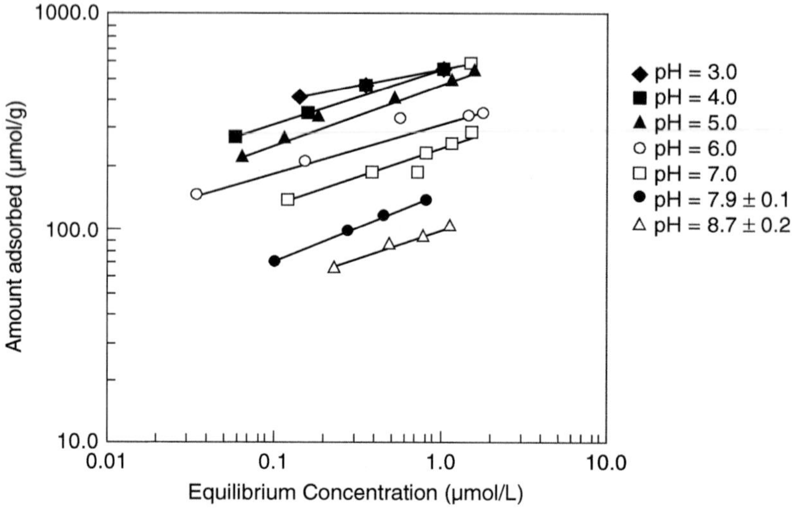

FIGURE 14-4 Effect of solution pH on MCPB adsorbability. (*Source: J. Y. Hu et al. (1998), Adsorptive characteristics of ionogenic aromatic pesticides in water on powdered activated carbon. Water Research 32(9), 2593-2600. Copyright © 1998, with permission from Elsevier.*)

The affinity of weak organic acids or bases for activated carbon is an important function of pH. When the pH is in a range where the molecule is in the neutral form, adsorption capacity is relatively high. When the pH is in a range where the species is ionized, however, the affinity for water (or aqueous solubility) increases, and activated carbon capacity decreases accordingly. Phenol that has been adsorbed on activated carbon at a pH below 8, where phenol is neutral, can be desorbed if the pH is increased to 10 or above, where phenol is anionic (Fox et al., 1973). For the acidic pesticide MCPB (pK_a = 4.8), Hu and colleagues (1998) illustrated that adsorption capacity decreases with increasing solution pH (Fig. 14-4). As MCPB changes from the neutral to the anionic form, its hydrophobicity (as measured by the octanol-water partition coefficient) decreases, and so does its adsorbability. Similar results have been obtained for other acidic pesticides (Hu et al., 1998), as well as for the acidic antibiotic sulfamethoxazole (pK_a = 5.6) (Knappe et al., 2007). For basic compounds that transition from the cationic to the neutral form as pH increases, the reverse trend is observed; that is, for the basic pesticide imazalil (pK_b = 7.5), adsorption capacity increased with increasing solution pH (Hu et al., 1998). A similar trend was observed by Knappe and colleagues (2007) for the basic antimicrobial compound trimethoprim (pK_b = 7).

Competitive Adsorption in Multisolute Systems of Known Composition. Competitive adsorption is important in drinking water treatment because most compounds to be adsorbed exist in solution with other adsorbable compounds. The quantity of activated carbon or other adsorbent required to remove a certain amount of a compound of interest from a mixture of adsorbable compounds is greater than if adsorption occurs without competition because part of the adsorbent's surface is used by the competing substances.

The extent of competition on activated carbon depends on the strength of adsorption of the competing molecules, the concentrations of these molecules, and the type of activated carbon. To illustrate the effect of competition among adsorbates in multisolute systems,

Crittenden and colleagues (1985a) compared single-solute and multisolute adsorption isotherm data. For example, for a trisolute system containing initial bromoform, chloroform, and TCE concentrations in the range of 30 to 40 μM, the bromoform adsorption capacity at an equilibrium concentration of 0.6 μM was approximately 38 percent of the single-solute adsorption capacity. For the more strongly adsorbing TCE, the adsorption capacity in the trisolute system at an equilibrium concentration of 0.6 μM was approximately 80 percent of the single-solute adsorption capacity.

Displacement of previously adsorbed compounds by competition can result in a column effluent concentration of a compound that is greater than the influent concentration, as shown in Fig. 14-5. A dimethylphenol (DMP) concentration about 50 percent greater than the influent resulted when dichlorophenol (DCP) was introduced to the influent of a column saturated with DMP (Thacker et al., 1984). Similar occurrences have been observed in full-scale GAC systems. Effluent concentrations in excess of influent concentrations can be prevented through careful operation and monitoring. Crittenden and colleagues (1980) showed that the magnitude of the displacement decreased when the value of C_{eff}/C_{inf} was lowered at the time the second compound was introduced. Thus a reasonable strategy to prevent the occurrence of an undesirable compound at a concentration greater than the influent is (1) to monitor the column for that compound and (2) to replace the activated carbon before complete saturation at the influent concentration occurs (i.e., before $C_{eff} = C_{inf}$).

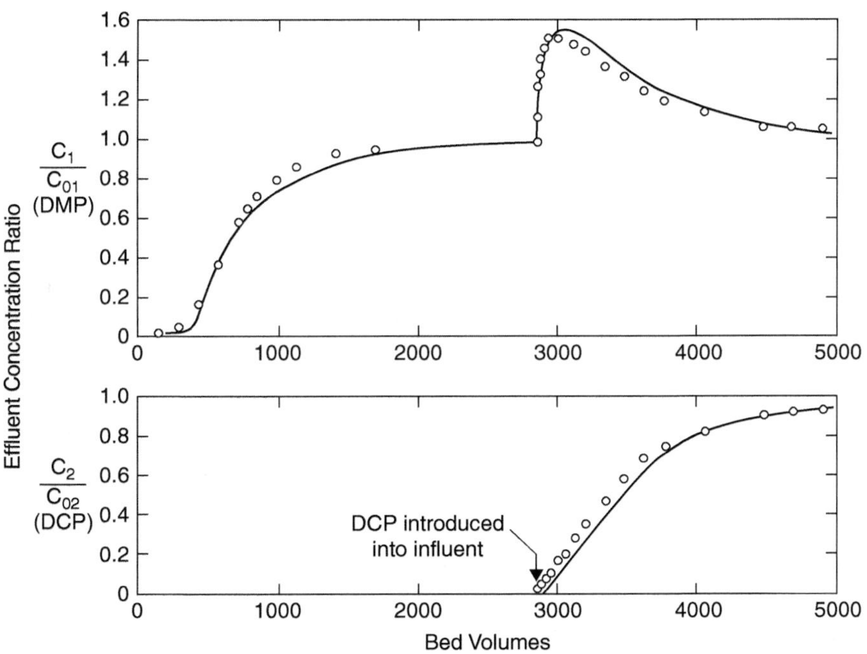

FIGURE 14-5 Breakthrough curves for sequential feed of DMP and DCP to a GAC adsorber (C_{01} = 0.990 mmol/L, C_{02} = 1.02 mmol/L, EBCT = 25.4 seconds). (*Source: W. E. Thacker, J. C. Crittenden, and V. L. Snoeyink, Modeling of adsorber performance: Variable influent concentration and comparison of adsorbents. Journal Water Pollution Control Federation 56:243, with permission. Copyright © 1984, Water Environment Federation, www.wef.org.*)

A number of models have been used to describe competitive adsorption equilibria (e.g., Butler and Ockrent, 1930; Jain and Snoeyink, 1973; Sheindorf et al., 1981). A common model for describing adsorption equilibrium in multiadsorbate systems is the *ideal adsorbed solution theory* (IAST) of Radke and Prausnitz (1972), which has proven to be applicable to a large number of situations. The following two equations, based on the IAST as developed by Radke and Prausnitz (1972) and modified by Crittenden and colleagues (1985a) to include the Freundlich equilibrium expression, describe equilibrium in a two-solute system:

$$C_{2,0} - q_2 C_C - \frac{q_2}{q_1 + q_2}\left(\frac{n_1 q_1 + n_2 q_2}{n_2 K_2}\right)^{n_2} = 0 \tag{14-4}$$

$$C_{1,0} - q_1 C_c - \frac{q_1}{q_1 + q_2}\left(\frac{n_1 q_1 + n_2 q_2}{n_1 K_1}\right)^{n_1} = 0 \tag{14-5}$$

where q_1 and q_2 are the equilibrium solid-phase concentrations of compounds 1 and 2, $q_1 = (C_{1,0} - C_{1,e})/C_c$ and $q_2 = (C_{2,0} - C_{2,e})/C_c$, $C_{1,0}$ and $C_{2,0}$ are the initial aqueous-phase concentrations of compounds 1 and 2, $C_{1,e}$ and $C_{2,e}$ are the equilibrium concentrations of compounds 1 and 2, and K_1 and K_2 are single-solute Freundlich parameters for compounds 1 and 2, $1/n_1$ and $1/n_2$ are single-solute Freundlich exponents for compounds 1 and 2, and C_c is the activated carbon dose.

These equations show the relationship between the initial concentration of each adsorbate, the amount of adsorbed compound per unit mass of activated carbon, and the activated carbon dose. The Freundlich parameters are derived from single-solute tests in organic free water. Extending Eqs. 14-4 and 14-5 to N components yields

$$C_{i,0} - C_c q_i - \left(\frac{q_i}{\sum_{j=1}^{N} q_j}\right)\left(\frac{\sum_{j=1}^{N} n_j q_j}{n_i K_i}\right)^{n_i} = 0 \tag{14-6}$$

where N is the number of components in the solution, $C_{i,0}$ is the initial aqueous-phase concentration of compound i, C_c is the carbon dose, q_i is the equilibrium solid-phase concentration of compound i, and n_i and K_i are the single-solute Freundlich parameters for compound i.

The system of IAST equations can be solved simultaneously to determine the concentrations for each component present in solution. The following example illustrates a spreadsheet solution for a bisolute system in which the single-solute isotherms of two competing adsorbates follow the Freundlich isotherm model. Benjamin (2009) presented a more comprehensive spreadsheet model that solves the IAST equations for up to eight competing adsorbates whose single-solute adsorption isotherms follow either the Freundlich or Langmuir model.

Example 14-1 Adsorption Equilibrium in Multisolute Systems (IAST)
Apply the IAST to predict the adsorption isotherms for trichloroethene (TCE) and bromoform in a system containing 100 μg/L of each compound in ultrapure water. Perform the calculations for activated carbon doses of 10 to 100 mg/L using 10 mg/L dose increments. Plot the resulting isotherms (log q versus log C), and compare them with the respective single-solute isotherms. The single-solute Freundlich isotherm constants

for TCE and bromoform are

$$K_{\text{TCE}} = 1245\ (\mu\text{g/g})(\text{L}/\mu\text{g})^{1/n} \text{ and } 1/n_{\text{TCE}} = 0.4696$$

$$K_{\text{bromoform}} = 436.6\ (\mu\text{g/g})(\text{L}/\mu\text{g})^{1/n} \text{ and } 1/n_{\text{bromoform}} = 0.6889$$

Provided that the initial TCE concentration is kept at 100 μg/L, calculate the bromoform equilibrium liquid-phase and solid-phase concentrations for initial bromoform concentrations of 10 and 1 μg/L. Use the same carbon doses as in part a. On a log-log plot, show the single-solute bromoform isotherm and the isotherms obtained at 100, 10, and 1 μg/L. What do you observe?

Solution

a. Convert initial concentrations to micromolar units:

$$C_{0,\text{TCE}} = 100\ \mu\text{g/L} \times 1\ \mu\text{mol}/131.4\ \mu\text{g} = 0.761\ \mu\text{mol/L}$$

$$C_{0,\text{bromoform}} = 100\ \mu\text{g/L} \times 1\ \mu\text{mol}/252.7\ \mu\text{g} = 0.396\ \mu\text{mol/L}$$

Convert Freundlich K values to micromolar units:

$$\begin{aligned} K_{\text{TCE}} &= 1245\ (\mu\text{g/g})(\text{L}/\mu\text{g})^{1/n} \times 1\ \mu\text{mol}/131.4\ \mu\text{g} \times (131.4\ \mu\text{g}/\mu\text{mol})^{0.4696} \\ &= 93.64\ (\mu\text{mol/g})(\text{L}/\mu\text{mol})^{1/n} \\ K_{\text{bromoform}} &= 436.6\ (\mu\text{g/g})(\text{L}/\mu\text{g})^{1/n} \times 1\ \mu\text{mol}/252.7\ \mu\text{g} \times (252.7\ \mu\text{g}/\mu\text{mol})^{0.6889} \\ &= 78.10\ (\mu\text{mol/g})(\text{L}/\mu\text{mol})^{1/n} \end{aligned}$$

For each carbon dose (10–100 mg/L), set up mass balance Eqs. 14-4 and 14-5; for example, for a carbon dose of 10 mg/L, the following equations result for TCE and bromoform:

$$0.761\ \mu\text{mol/L} - q_1 \times 0.010\ \text{g/L} - \frac{q_1}{q_1+q_2}\left[\frac{(1/0.4696)\times q_1 + (1/0.6889)\times q_2}{(1/0.4696)\times 93.64\ (\mu\text{mol/g})(\text{L}/\mu\text{mol})^{0.4696}}\right]^{1/0.4696} = 0$$

$$0.396\ \mu\text{mol/L} - q_2 \times 0.010\ \text{g/L} - \frac{q_2}{q_1+q_2}\left[\frac{(1/0.4696)\times q_1 + (1/0.6889)\times q_2}{(1/0.6889)\times 78.10\ (\mu\text{mol/g})(\text{L}/\mu\text{mol})^{0.6889}}\right]^{1/0.6889} = 0$$

Using a nonlinear optimization tool (e.g., Solver in Microsoft Excel), vary q_1 (TCE solid-phase concentration) and q_2 (bromoform solid-phase concentration) such that the two equations are satisfied for each carbon dose. A solution can be efficiently obtained by varying q_1 and q_2 for each carbon dose such that the sum of squares of all 20 equations that result for the 10 carbon doses is minimized. The results in Table 14-3 are obtained for a system with initial bromoform and TCE concentrations of 100 μg/L each.

Furthermore, the equilibrium aqueous-phase concentrations of TCE and bromoform can be computed from $C = C_0 - q \times C_c$ (e.g., at a carbon dose of 10 mg/L, the aqueous-phase TCE concentration is 100 μg/L – 6.35 μg/mg × 10 mg/L = 36.5 μg/L). Aqueous-phase concentration results are included in Table 14-3.

Plotting q as a function of C yields the adsorption isotherms shown in Fig. 14-6. This figure illustrates that the presence of the more weakly adsorbing bromoform has only a small effect on TCE adsorption; that is, the predicted TCE adsorption isotherm for the two-solute system is very similar to the single-solute TCE adsorption isotherm. In contrast, the presence of the more strongly adsorbing TCE has a more pronounced effect on the bromoform adsorption isotherm. For example, at an equilibrium aqueous-phase bromoform concentration of 10 μg/L, the adsorption capacity in the two-solute system is only 53 percent of that in the single-solute system (1.14 versus 2.13 mg/g).

TABLE 14-3 IAST Solution for a Bisolute System with Initial Bromoform and TCE Concentrations of 100 mg/L Each

Carbon dose (mg/L)	q_{TCE} (μmol/g)	$q_{bromoform}$ (μmol/g)	q_{TCE} (mg/g)	$q_{bromoform}$ (mg/g)	C_{TCE} (μg/L)	$C_{bromoform}$ (μg/L)
10	48.30	14.08	6.35	3.56	36.54	64.42
20	32.08	11.27	4.22	2.85	15.69	43.02
30	23.31	9.18	3.06	2.32	8.10	30.41
40	18.11	7.65	2.38	1.93	4.79	22.66
50	14.75	6.52	1.94	1.65	3.12	17.62
60	12.41	5.66	1.63	1.43	2.17	14.16
70	10.70	4.99	1.41	1.26	1.59	11.69
80	9.40	4.46	1.23	1.13	1.21	9.85
90	8.38	4.03	1.10	1.02	0.95	8.45
100	7.55	3.67	0.99	0.93	0.76	7.35

b.

$$C_{0,\text{TCE}} = 100\ \mu\text{g/L} \times 1\ \mu\text{mol}/131.4\ \mu\text{g} = 0.761\ \mu\text{mol/L}$$

$$C_{0,\text{bromoform}} = 10\ \mu\text{g/L} \times 1\ \mu\text{mol}/252.7\ \mu\text{g} = 0.0396\ \mu\text{mol/L}$$

or

$$C_{0,\text{bromoform}} = 1\ \mu\text{g/L} \times 1\ \mu\text{mol}/252.7\ \mu\text{g} = 0.00396\ \mu\text{mol/L}$$

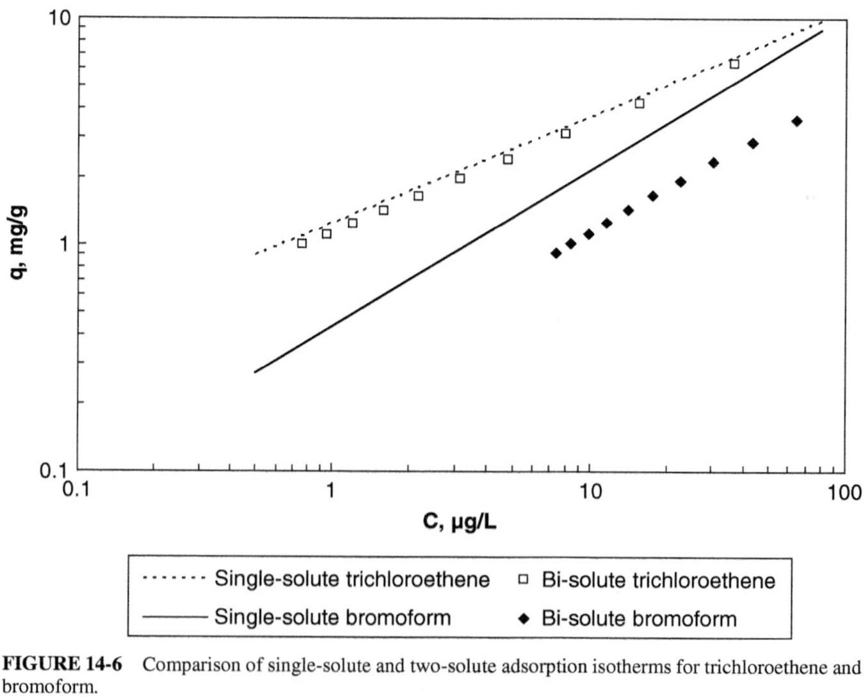

FIGURE 14-6 Comparison of single-solute and two-solute adsorption isotherms for trichloroethene and bromoform.

TABLE 14-4 IAST Solution for a Two-Solute System with Initial Bromoform and TCE Concentrations of 10 and 100 mg/L, respectively

Carbon dose (mg/L)	q_{TCE} (μmol/g)	$q_{bromoform}$ (μmol/g)	q_{TCE} (mg/g)	$q_{bromoform}$ (mg/g)	C_{TCE} (μg/L)	$C_{bromoform}$ (μg/L)
10	49.76	1.40	6.54	0.35	34.62	6.46
20	32.66	1.13	4.29	0.29	14.17	4.30
30	23.57	0.92	3.10	0.23	7.09	3.02
40	18.24	0.77	2.40	0.19	4.11	2.25
50	14.82	0.65	1.95	0.17	2.64	1.74
60	12.45	0.57	1.64	0.14	1.83	1.40
70	10.73	0.50	1.41	0.13	1.33	1.15
80	9.42	0.45	1.24	0.11	1.01	0.97
90	8.39	0.40	1.10	0.10	0.79	0.83
100	7.56	0.37	0.99	0.093	0.63	0.72

For an initial bromoform concentration of 10 μg/L in the presence of 100 μg/L TCE, the results in Table 14-4 are obtained (using the same solution approach as in part a).

Similarly, for an initial bromoform concentration of 1 μg/L in the presence of 100 μg/L TCE, the results in Table 14-5 are obtained.

Using the results shown in Tables 14-3 to 14-5, bromoform isotherms for different initial bromoform concentrations in the two-solute system are compared with the single-solute bromoform isotherm in Fig. 14-7. Fig. 14-7 illustrates that the effect of the competing adsorbate TCE on bromoform uptake becomes more pronounced as the initial bromoform concentration decreases. ▲

NOM Adsorption. Adsorption equilibrium assessments can be made for heterogeneous mixtures of compounds such as NOM. Typically, group parameters such as total organic carbon (TOC), dissolved organic carbon (DOC), and ultraviolet (UV) absorbance at a wavelength of 254 nm (UVA) are used as a measure of the total concentration of substances present. For mixtures such as NOM that have compounds with a wide range of affinities

TABLE 14-5 IAST Solution for a Two-Solute System with Initial Bromoform and TCE Concentrations of 1 and 100 mg/L, respectively

Carbon dose (mg/L)	q_{TCE} (μmol/g)	$q_{bromoform}$ (μmol/g)	q_{TCE} (mg/g)	$q_{bromoform}$ (mg/g)	C_{TCE} (μg/L)	$C_{bromoform}$ (μg/L)
10	49.90	0.14	6.56	0.035	34.43	0.65
20	32.72	0.11	4.30	0.029	14.02	0.43
30	23.59	0.092	3.10	0.023	6.99	0.30
40	18.26	0.077	2.40	0.019	4.05	0.22
50	14.82	0.065	1.95	0.017	2.60	0.17
60	12.46	0.057	1.64	0.014	1.79	0.14
70	10.73	0.050	1.41	0.013	1.31	0.12
80	9.42	0.045	1.24	0.011	0.99	0.097
90	8.39	0.040	1.10	0.010	0.77	0.083
100	7.56	0.037	0.99	0.0093	0.62	0.072

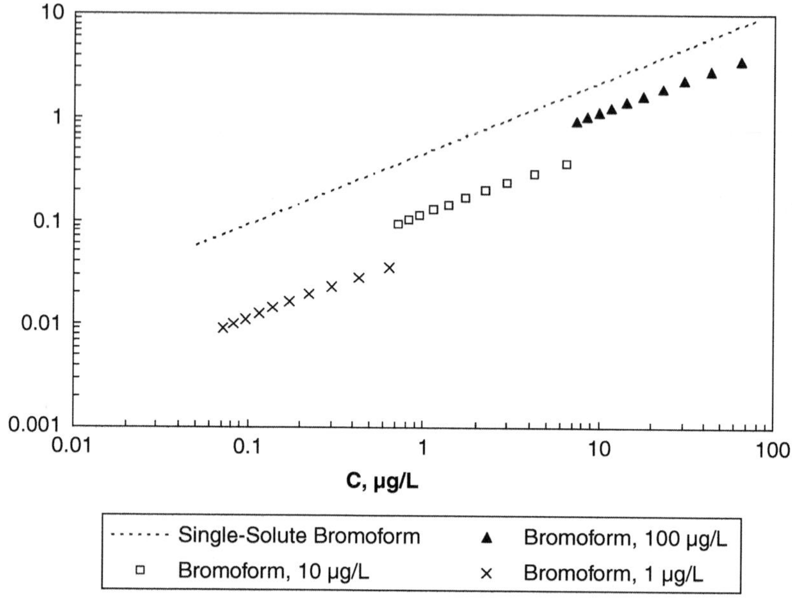

FIGURE 14-7 Comparison of single-solute and two-solute bromoform adsorption isotherms. Two-solute isotherms are in the presence of 100 μg/L trichloroethene.

for an adsorbent, the shape of the isotherm depends on the relative amounts of compounds in the mixture. For example, the isotherms for an isolated groundwater NOM shown in Fig. 14-8*a* display the expected characteristics of a nonadsorbable fraction, about 10 percent of the initial concentration, as well as components with a range of adsorbabilities (Hong, 1996). The strongly adsorbable compounds can be removed with small doses

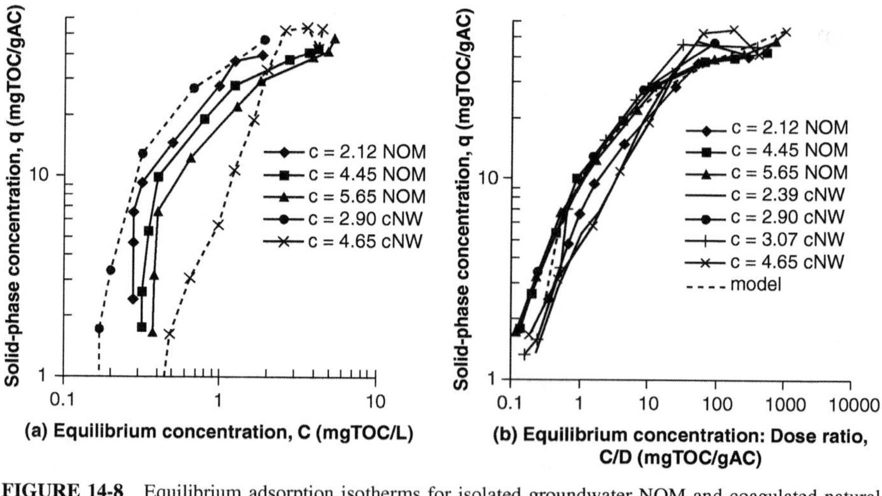

FIGURE 14-8 Equilibrium adsorption isotherms for isolated groundwater NOM and coagulated natural waters from a range of sources using a bituminous-based ground GAC. (*Source: Adapted with permission from Hong, 1996.*)

of adsorbent and yield large values of q. In contrast, the weakly adsorbable compounds can be removed only with large adsorbent doses that yield relatively low values of q. The nonadsorbable compounds produce a vertical isotherm at low C values. This behavior also results in different isotherms at different initial concentrations of the same heterogeneous mixture (Fig. 14-8*a*), in contrast to single-solute isotherms. The IAST model predicts the increase in adsorbability q as a function of decreasing initial concentration. Also shown in Fig. 14-8*a* are isotherms from two coagulated natural waters (cNWs) with initial TOC concentrations of 2.90 and 4.65 mg/L that also display this behavior.

To account for preferential adsorption of the strongly adsorbing compounds in a mixture that causes the impact of initial concentration on equilibrium data, Summers and Roberts (1988a) used an isotherm model that related the solid-phase concentration q to the ratio C/D, where C is the of the equilibrium liquid-phase concentration and D is the adsorbent dose. For isolated humic and fulvic acid fractions of NOM, this approach yielded a good fit between a modified Freundlich model with parameters K_{HS} and $1/n_{\text{HS}}$ and the data.

$$q = K_{\text{HS}}(C/D)^{1/n_{\text{HS}}} \tag{14-7}$$

Using Eq. 14-7, the data in Fig. 14-8a were replotted in Fig. 14-8b, and the impact of initial TOC concentration was mitigated. In addition, data from four cNW isotherms with a range of sources, geographic locations, and initial concentrations (Hong, 1996) are also plotted and show that this approach normalizes the cNW equilibrium data. The modified Freundlich model used by Summers and Roberts (1988a) for isolated NOM fractions does not fit the natural water data in Fig. 14-8b because the natural waters have a wider range of absorbability, including a nonadsorbable fraction. However, a seminatural log equation does fit the data, $q_e = 6.43 \ln (C_e/D) + 9.94$ ($r^2 = 0.91$), as shown in Fig. 14-8b. This model also was applied to the isotherm data from four ozonated natural waters (Hong, 1996), resulting in a slope that was 10 percent lower than that of the nonozonated waters ($r^2 = 0.90$). As will be shown later in this chapter, the relative adsorbability of compounds within a mixture has an important effect on adsorption column performance. The nonadsorbable fraction cannot be removed regardless of the column design, whereas the strongly adsorbable fraction may cause the effluent concentration to slowly approach the influent concentration.

The adsorption of organic acids and bases by GAC can be affected by solution pH (see Fig. 14-4). At environmental pH values, the carboxylic acid and phenolic functional groups in NOM humic substances yield a negative charge that dominates the NOM charactertistics. Several researchers have shown the impact of influent pH on the adsorption of NOM. Unfortunately, some of the work has been done with different initial TOC concentrations, and the increased performance attributed to low pH may be because of the lower initial TOC concentrations (Fig. 14-8). A relationship between the relative adsorption capacity for TOC and pH was developed from batch isotherm data at different doses for 10 different source waters and a bituminous coal based GAC and is shown in Fig. 14-9 (Hooper et al., 1996a). Within the pH range shown, a decrease in the pH of 1 unit yielded a 6 percent increase in adsorption capacity.

The inorganic composition of water also can have an important effect on the extent of NOM adsorption. Divalent cations such as calcium can interact with the NOM, increasing its adsorption capacity through the formation of complexes (Randtke and Jepsen, 1982). Increases in ionic strength have been shown to decrease the size of NOM (Cornel et al., 1986), thus allowing access to more surface area and, in turn, increasing the adsorption (Summers and Roberts, 1988b).

Competitive Adsorption in Natural Waters. Adsorption of organic compounds at trace concentrations from natural waters is an important problem in water purification. Essentially all synthetic organic chemicals that must be removed in water treatment by adsorption must compete with natural or background organic matter for adsorption sites.

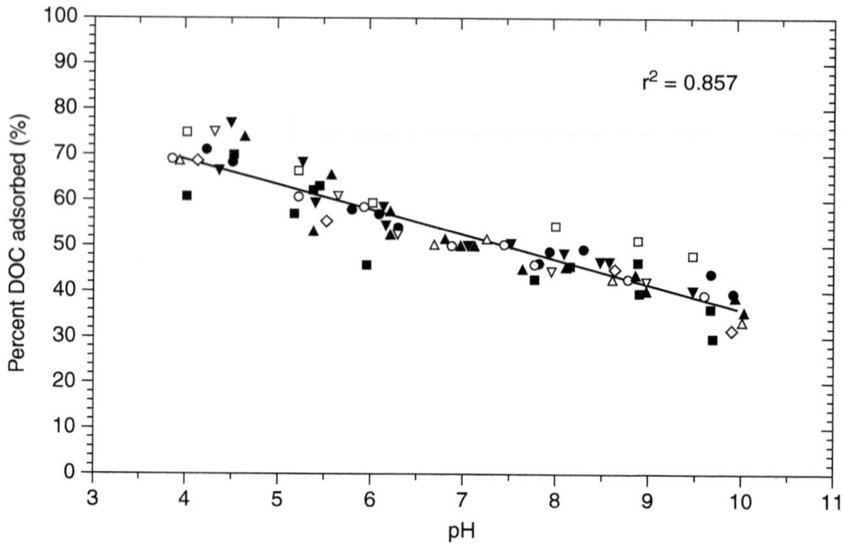

FIGURE 14-9 Impact of pH on percent DOC adsorbed at equilibrium for 10 waters. (*Source: Hooper et al., 1996a.*)

The heterogeneous mixture of compounds in natural waters adsorbs on activated carbon and reduces the number of sites available for the micropollutants either by direct competition for adsorption sites or by pore blockage (Carter et al., 1992; Pelekani and Snoeyink, 1999; Li et al., 2003a). The amount of competition and the adsorption capacity for the micropollutant depend on the nature of the background DOM and its concentration, as well as the characteristics of the activated carbon (Li et al., 2003b). Also important is the concentration of the micropollutant because this concentration affects how much of the compound can adsorb on the carbon. For example, Fig. 14-10 shows that the adsorption capacity of MIB, an important earthy/musty odor compound, has a capacity in natural water that is lower than in distilled water and that this capacity is reduced further as initial concentration decreases (Gillogly et al., 1998b).

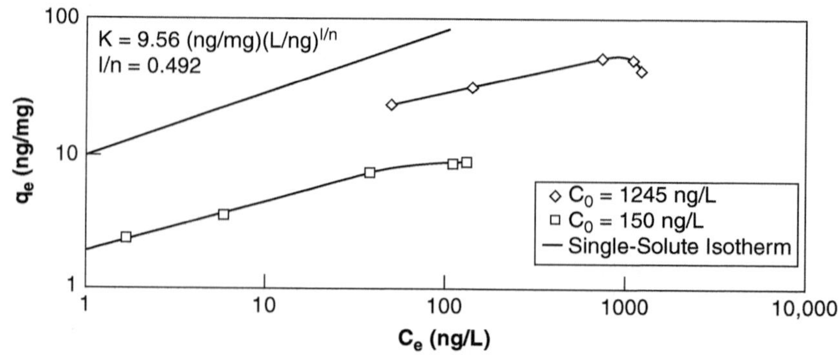

FIGURE 14-10 Effect of initial concentration on MIB adsorption capacity in Lake Michigan water. (*Source: Gillogly et al., 1998b.*)

A procedure is needed to predict adsorption capacity as a function of initial target compound concentration because the adsorption capacity of activated carbon depends in such an important way on initial concentration and because the concentrations of micropollutants vary widely in drinking water sources. The IAST can be used for this purpose, but approaches are needed to represent the competing background organic matter.

In natural waters, the competing organic matter is a complex mixture of many different compounds. Representing each of them, even if they could be identified, would be computationally prohibitive. Several researchers have modeled NOM adsorption by defining several fictive components that represent groups of compounds with similar adsorption characteristics, as expressed by Freundlich K and n values (Sontheimer et al., 1988). Crittenden and colleagues (1985b) used this fictive-component approach to describe the adsorption of a target compound in the presence of NOM. With a single-solute isotherm of the target compound and experimental results from target-compound isotherms in the natural water, parameters for each of the fictive components were found through a best-fit search procedure. These results then were applied to describe the adsorption of other compounds in that water.

The IAST was applied to the problem of micropollutant adsorption in natural waters by Najm and colleagues (1991a) using a procedure that was modified subsequently by Qi and colleagues (1994) and Knappe and colleagues (1998). These researchers assumed that the background organic matter that competed with the micropollutant could be represented as a single compound, called the *equivalent background compound* (EBC). Their approach involved determination of the single-solute isotherm for the micropollutant and an isotherm in natural water for the micropollutant at two different initial concentrations. A search routine was used to find the Freundlich parameters K and $1/n$ and the initial concentration C_0 for the EBC that gave the observed amount of competition. For example, Fig. 14-11

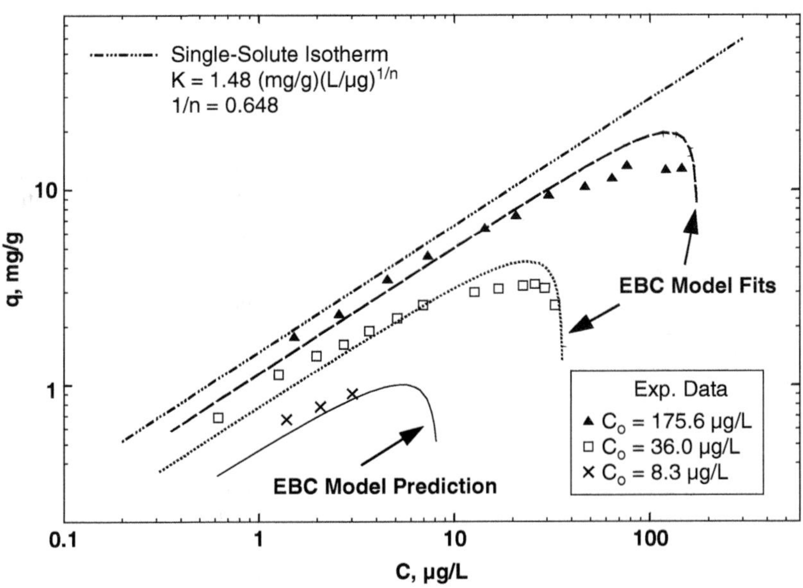

FIGURE 14-11 EBC model results for atrazine isotherms in Illinois groundwater. (*Source: Reprinted with permission from D. R. U. Knappe et al. (1998), Predicting the capacity of powdered activated carbon for trace organic compounds in natural waters. Environmental Science & Technology 32, 1694–1698. Copyright © 1998, American Chemical Society.*)

shows isotherms determined for atrazine in organic-free water and in Illinois groundwater at initial concentrations of 176 and 36 μg/L (Knappe et al., 1998). These data were used to determine the EBC parameters, which are specific for the type of carbon, the type and concentration of the competing background organic matter, and the type of micropollutant. The EBC parameters (K_{EBC}, $1/n_{EBC}$, $C_{0,EBC}$) were used in Eqs. 14-4 and 14-5 together with the initial concentration of the micropollutant and its single-solute Freundlich parameters to calculate the surface coverage of micropollutant as a function of carbon dose C. Given the surface coverage q, the initial concentration C_0, and the carbon dose C_c, the equilibrium concentration of the micropollutant can be calculated from the equation $q = (C_0 - C)/C_c$. This approach was used to determine the predicted isotherm for atrazine at an initial concentration of 8.3 μg/L shown in Fig. 14-11, which compares well with the measured data. At an equilibrium concentration of 1 μg/L, there is a 63 percent reduction in capacity for atrazine as the initial concentration is reduced from 176 to 8.3 μg/L.

An important modification of the EBC model was developed by Knappe and colleagues (1998), who showed that the amount of adsorption on a unit mass of activated carbon was directly proportional to the initial concentration of that micropollutant in an adsorption test if (1) the Freundlich exponents for the micropollutant ($1/n_1$) and the EBC ($1/n_2$) both fall between 0.1 and 1, as is generally the case and (2) the solid-phase concentration of the background organic matter is much greater than that of the micropollutant ($q_2 >> q_1$), which also is often true. When these two approximations hold, the IAST model (Eq. 14-5) simplifies to

$$q_1 = C_{1,0}\left[C_c + \frac{1}{q_2}\left(\frac{n_2\, q_2}{n_1 K_1}\right)^{n_1}\right]^{-1} \tag{14-8}$$

Equation 14-8 shows that the surface loading of micropollutant q_1 is directly proportional to its initial concentration. This equation also can be manipulated to show that the percent removal, or percent remaining, of a compound for any carbon dose, when the preceding two assumptions are valid, is a constant (e.g., Matsui et al., 2002; Qi et al., 2007). The implication of this result is that for low concentrations of micropollutant in a natural water, only one isotherm needs to be determined. This isotherm can be plotted as percent remaining versus carbon dose, as shown in Figs. 14-12 and 14-13 for the same data shown in Figs. 14-10 and 14-11, respectively. In Fig. 14-13, the atrazine data for $C_0 \leq 36$ μg/L plot on 1 percent remaining versus carbon dose line, and the $C_0 = 176$ μg/L line is only slightly above. All the MIB data in Fig. 14-12 plot on one line.

Graham and colleagues (2000) independently derived a different form of Eq. 14-8 and applied it using the EBC approach to the adsorption of MIB and geosmin from four natural waters. For a lignite based PAC, the same EBC isotherm parameters [$K = 1.35$ (μg/mg) (L/μg)$^{1/n}$ and $1/n = 0.20$] were found to be applicable to both MIB and geosmin and, at different initial concentrations (15–51 μg/L), for all four natural waters. For three of the four natural waters, the EBC initial concentration was about 0.45 percent of the TOC initial concentration. Cho and Summers (2007a) extended this approach to six additional waters with the same PAC (lignite based) as Graham and colleagues (2000) and successfully used the same EBC isotherm parameters (K, $1/n$) and EBC concentration function for the lignite based PAC. In addition, they developed EBC parameters for wood- and bituminous-based PACs.

These results allow the following general procedure to be used to determine adsorption capacity for a micropollutant in natural water: (1) determine one adsorption isotherm for the micropollutant in natural water at a sufficiently low initial concentration, (2) plot the data on a log-log plot of percent remaining versus carbon dose, and (3) use this isotherm

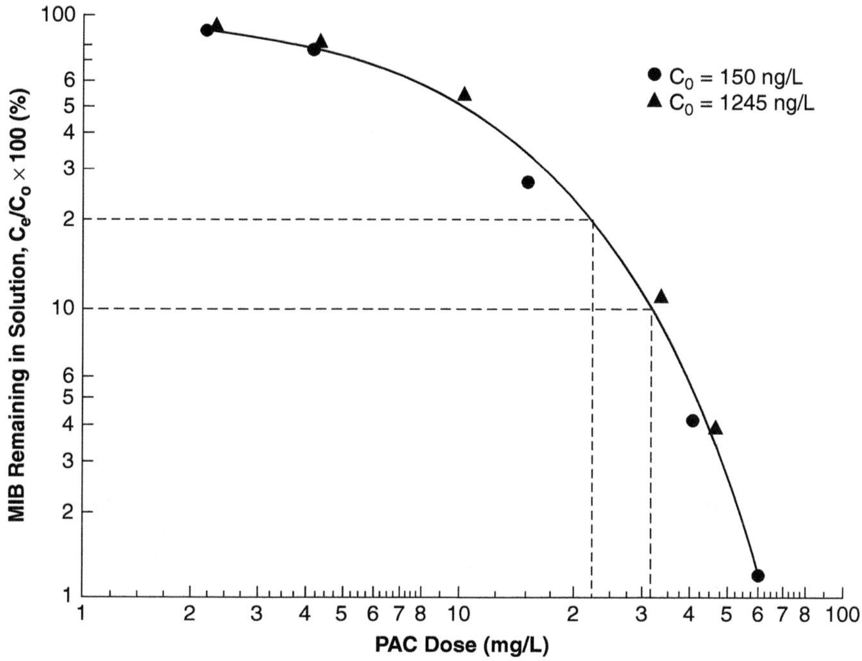

FIGURE 14-12 Percent MIB remaining as a function of PAC dose. (*Source: Gillogly et al., 1998b.*)

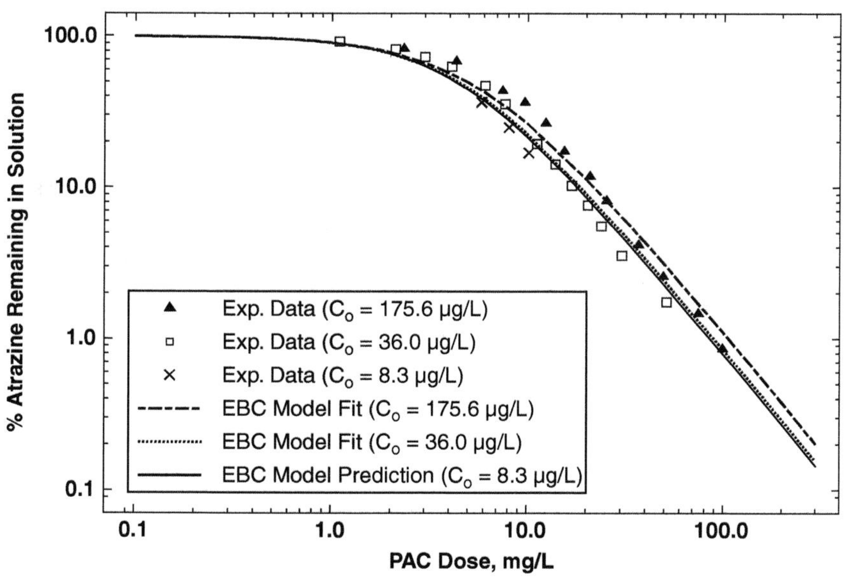

FIGURE 14-13 Removal efficiency as a function of PAC dose for the adsorption of atrazine from Illinois groundwater. (*Source: Reprinted with permission from D. R. U. Knappe et al. (1998), Predicting the capacity of powdered activated carbon for trace organic compounds in natural waters. Environmental Science & Technology 32, 1694–1698. Copyright © 1998, American Chemical Society.*)

plot to determine the carbon dose required for any desired percent removal for any initial concentration that satisfies the assumptions. The validity of this simplified approach depends on the micropollutant concentration being low relative to that of the competing EBC. Research to date has shown that MIB (Gillogly et al., 1998b) and geosmin (Graham et al., 2000) concentrations of less than 1000 ng/L and atrazine concentrations of less than about 50 μg/L (Knappe et al., 1998) will give a satisfactory plot. For emerging contaminants, satisfactory plots were obtained for the antibiotic sulfamethoxazole at concentrations less than 100 μg/L (Rossner et al., 2009) and for the hormone estradiol at concentrations less than 1000 ng/L (Westerhoff et al., 2005). This database needs to be expanded to include other micropollutants.

Preloading in Natural Waters (GAC Fouling). Because water treatment plants typically operate GAC adsorbers on a continuous basis, NOM adsorbs on GAC even during periods when targeted micropollutants are absent in the source water. As a result, the GAC becomes preloaded with or fouled by humic substances (Summers and Roberts, 1988a). As explained in more detail below, both the GAC adsorption capacity for a micropollutant and the rate of micropollutant adsorption are adversely affected by preadsorbed NOM.

Figure 14-14 illustrates the preloading phenomenon by showing typical aqueous-phase concentration profiles of a micropollutant and two fractions of NOM: a small, strongly adsorbing fraction that competes directly for preferred micropollutant adsorption sites and a larger, weakly adsorbing fraction that constricts adsorbent pores and slows adsorption kinetics. The figure shows that only the upper reaches of the GAC bed are exposed to the micropollutant early in an adsorber run, but the entire bed is exposed to the larger, weakly adsorbing NOM compounds. The concentration profile of smaller, strongly adsorbing NOM is similar to the micropollutant but typically proceeds slightly ahead of the micropollutant because of the higher initial concentration of the strongly competing compound (Schideman et al., 2007).

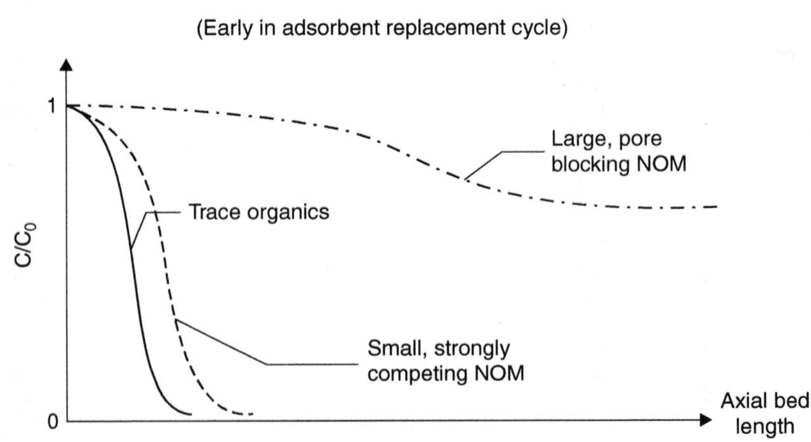

FIGURE 14-14 Conceptual illustration of NOM preloading along length of GAC bed. (*Source: L. Schideman et al. (2007), Application of a three-component competitive adsorption model to evaluate and optimize granular activated carbon systems. Water Research 41(15), 3289–3298. Copyright © 2007, with permission from Elsevier.*)

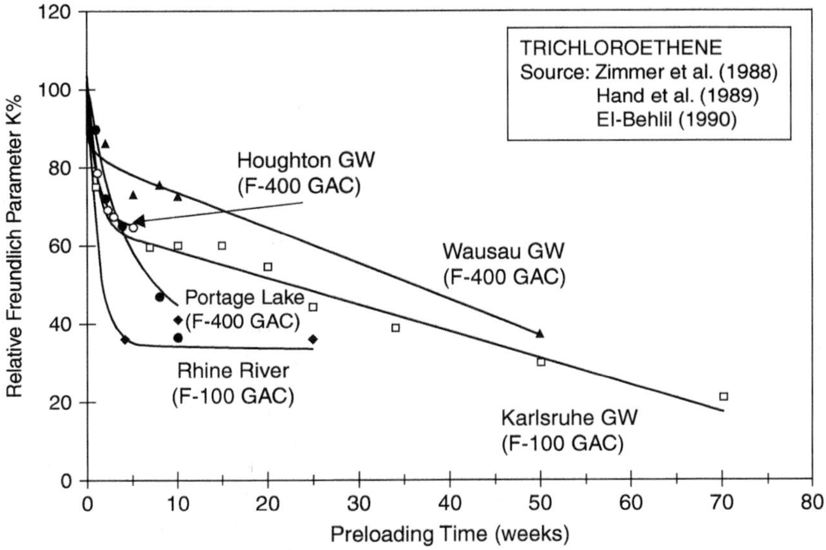

FIGURE 14-15 Effect of NOM preloading time (or GAC service time) on remaining GAC adsorption capacity for TCE in four drinking water sources. (*Source: M. E. Jarvie et al. (2005), Simulating the performance of fixed-bed granular activated carbon adsorbers: Removal of synthetic organic chemicals in the presence of background organic matter. Water Research, 39, 2407–2421. Copyright © 2005, with permission from Elsevier.*)

Two general trends have been observed when studying the effects of NOM preloading on the remaining adsorption capacity for micropollutants: (1) Adsorption capacity, as measured by the Freundlich *K* value, decreased rapidly during the initial weeks of preloading and more gradually thereafter (Fig. 14-15), and (2) differences in adsorption capacity as a function of bed depth diminished as preloading time increased (e.g., Summers et al., 1989; Knappe et al., 1999; Jarvie et al., 2005). With respect to the first observation, the decrease in the adsorption capacity for micropollutants is initially a result of the continuing adsorption of NOM. GAC fouling will continue beyond the point at which the GAC is saturated with NOM, however, because reorientation of preloaded NOM and/or chemical reactions involving preloaded NOM can take place on the GAC surface (Summers et al., 1989, 1990). For example, oxidative polymerization of preloaded NOM may proceed in the presence of dissolved oxygen (Vidic and Suidan, 1991). Regarding the second observation, the remaining adsorption capacity for micropollutants typically varies with GAC bed depth when a GAC bed has been in service for a relatively short time because the adsorbed NOM concentration is highest near the influent end of the GAC bed and lowest near the effluent end. Over time, GAC becomes uniformly saturated with NOM such that the remaining adsorption capacity for micropollutants is similar throughout the entire GAC bed. Because adsorption of humic substances on GAC is governed primarily by size-exclusion effects, and because the largest percentage of adsorption sites is contained in micropores (<2 nm diameter), preferential adsorption of humic substances with lower molecular weights occurs on GAC (McCreary and Snoeyink, 1980; Summers and Roberts, 1988b; Kilduff et al., 1996). Consequently, preloading experiments with different humic acid size fractions showed that the lowest-molecular-weight fraction had the greatest adverse effect on the remaining micropollutant adsorption capacity (Kilduff et al., 1998).

Adsorption Kinetics

Removal of organic compounds by physical adsorption on porous adsorbents involves a number of steps, each of which can affect the overall adsorption kinetics:

1. *Bulk solution transport.* Adsorbates must be transported from bulk solution to the boundary layer of water surrounding the adsorbent particle. Bulk transport to the boundary layer is affected by diffusion, dominant under low levels of mixing, and by turbulence, dominant under high levels of mixing. The latter conditions may be encountered during turbulent flow through a packed bed of GAC or when PAC is being mixed in a rapid-mix unit or flocculator.
2. *External (film) resistance to transport.* Adsorbates must be transported by molecular diffusion through the stationary layer of water (hydrodynamic boundary layer) that surrounds adsorbent particles when water is flowing past them. The distance of transport and thus the time for this step are determined by the bulk flow past the particle. The higher the flow rate, the shorter are the transport distance and time.
3. *Internal (intraparticle) transport.* After passing through the hydrodynamic boundary layer, adsorbates must be transported through the adsorbent's pores to available adsorption sites distributed throughout the particle. Intraparticle transport may occur by molecular diffusion through the solution in the pores (pore diffusion) and by diffusion along the adsorbent surface (surface diffusion) after adsorption takes place.
4. *Adsorption.* After transport to an available site, an adsorption bond is formed between the adsorbate and adsorbent. This step is rapid for physical adsorption (Adamson and Gast, 1997). As a result, one of the preceding diffusion steps will control the rate at which molecules are removed from solution. If adsorption is accompanied by a chemical reaction that changes the nature of the molecule, the chemical reaction may be slower than the diffusion step and thereby control the rate of compound removal.

The transport steps occur in series, so the slowest step, called the *rate-limiting step*, will control the rate of removal. In packed bed GAC adsorbers used for potable water treatment, a combination of film diffusion and intraparticle diffusion typically controls the rate of compound removal. Film diffusion may control initially, and after some adsorbate accumulates within the GAC particle, internal transport may control (Weber and Liu, 1980; Knappe et al., 1999). The mathematical models of the adsorption process therefore usually include both mass transfer steps. In contrast to GAC adsorption systems, film mass transfer typically is rapid in many PAC applications and, therefore, can be neglected (e.g., Najm, 1996)

Both adsorbate molecular size and adsorbent particle size have important effects on the overall adsorption kinetics. Diffusion coefficients are inversely proportional to the molecular size to about the one-third power. Thus humic substances with higher molecular weights diffuse more slowly into activated carbon particles than micropollutants with lower molecular weights. To reach an adsorption site, larger molecules tend to travel a shorter distance into the activated carbon particle because size-exclusion effects limit their access to smaller pores. Thus, despite the smaller diffusion coefficient, large molecules may equilibrate faster than small molecules that must travel farther into the activated carbon particle. Adsorbent particle size is even more important because it determines diffusion path length and thus the time required for transport within the particle to available adsorption sites. If the rate of adsorbate uptake is controlled by intraparticle diffusion, and the effective diffusion coefficient is constant, the time to reach adsorption equilibrium is directly proportional to the diameter of the particle squared. Calculations by Randtke and Snoeyink (1983) for activated carbon illustrate these points. For dimethylphenol (molecular weight 122 g/mol), nearly 8 days are estimated for near-equilibrium

($C_{\text{final}} = 1.01C_{\text{equilibrium}}$) of 2.4 mm diameter activated carbon, but only about 25 minutes would be required for 44 μm diameter activated carbon. For humic acid with a molecular weight of approximately 50,000 g/mol, the 2.4 mm diameter particle is expected to take much longer than a year to equilibrate, but only 2 days are required for the 44 μm diameter particle (assuming the same travel distance as for dimethylphenol). Calculations for fulvic acid with a molecular weight of 10,000 g/mol showed only about 25 percent saturation of 2.4 mm diameter particles after 40 days of contact. Thus the smaller the particle, the faster equilibrium is achieved in both column and complete-mix adsorption systems.

Some conclusions that can be drawn from these observations are that (1) the adsorption capacity of PAC can be used more quickly relative to GAC because of its smaller size, (2) the smallest-size GAC, consistent with other process constraints such as head loss and loss during reactivation, should be chosen for the fastest kinetics, and (3) depending on the compound, all the capacity of large GAC particles in a column may not be used because the time interval between GAC replacements is not sufficient for equilibrium to be achieved.

Estimation of Adsorption Kinetic Parameters. To determine adsorption kinetic parameters for PAC adsorption processes, batch experiments are conducted that measure the time rate of change of the adsorbate concentration in solution (e.g., Qi et al., 1994). For GAC beds, kinetic parameters typically are determined in (1) differential column batch reactors (DCBRs) (Hand et al., 1983) that permit the estimation of intraparticle diffusion coefficients or (2) short-bed adsorber (SBA) tests (Weber and Liu, 1980; Liang and Weber, 1985) that permit the estimation of both film mass transfer and intraparticle diffusion coefficients. A common model that is used to describe adsorption kinetic data is the homogeneous surface diffusion model (HSDM). Spreadsheet solutions of the HSDM for PAC adsorption systems are available (e.g., Hand et al., 1983; Najm, 1996; Zhang et al., 2009), and an example of such a solution is presented below. Numerical solutions to the HSDM, the pore diffusion model, and the pore–surface diffusion model typically are required to mathematically describe the evolution of the adsorbate concentration in the effluent of a GAC adsorber as a function of service time. Models for GAC adsorption systems have been described in detail by Sontheimer and colleagues (1988). Simplified models for fixed-bed adsorption systems have been described by Hand and colleagues (1984).

In the absence of film mass transfer, which typically is not rate limiting in PAC applications, the following equation has been developed as an approximate solution to the HSDM (Najm, 1996):

$$C_0 - C - 3C_c K C^{1/n} \left\{ \begin{aligned} &0.33334 - 0.04903\exp\left(-142.634\frac{tD_s}{R^2}\right) \\ &-0.05399\exp\left(-39.996\frac{tD_s}{R^2}\right) \\ &-0.20240\exp\left(-9.8686\frac{tD_s}{R^2}\right) \end{aligned} \right\} = 0 \qquad (14\text{-}9)$$

This equation illustrates that knowledge of the Freundlich isotherm parameters (K, $1/n$), the surface diffusion coefficient (D_s), the PAC particle radius (R), and the initial target compound concentration (C_0) is required to predict what contaminant concentration remains in solution for a given PAC dose (C_c) and contact time (t). Results obtained with Eq. 14-9 are in good agreement with those of a finite-element model when $tD_s/R^2 > 0.001$. For systems involving the adsorption of micropollutants from waters containing NOM, a pseudo-single-solute approach can be employed (Qi et al., 1994). In this approach, the Freundlich K value is calculated from isotherm data for the target contaminant in the water of interest, whereas the $1/n$ value is taken from the single-solute isotherm. The D_s value then is estimated directly from batch kinetic data obtained in the water of interest. The following example illustrates the application of the HSDM for estimating D_s with the pseudo-single-solute approach.

Example 14-2 Homogeneous Surface Diffusion Model (HSDM): Atrazine Adsorption on PAC

A batch kinetic test was conducted with PAC to measure the rate of atrazine adsorption from a drinking water source. Determine the surface diffusion coefficient for the following data set:

$C_0 = 179\ \mu\text{g/L}$ $\quad R = 6.7\ \mu\text{m}$ $\quad C_c = 2.6\ \text{mg/L}$

Time, min	C/C_0
4.50	0.906
7.17	0.884
10.5	0.864
17.0	0.811
30.0	0.784
50.0	0.729
75.0	0.688
107	0.634
135	0.617
165	0.586
210	0.562
240	0.542
270	0.526
300	0.509
330	0.501
402	0.474
467	0.462

To be able to apply Eq. 14-9 for the determination of D_s, the Freundlich isotherm parameters K and $1/n$ need to be known. From isotherm experiments conducted with the same PAC in the same water, the relationship between the remaining atrazine percentage in solution as a function of PAC dose was determined (Fig. 14-16). This relationship permits calculation of the K value that applies to the kinetic test. The Freundlich $1/n$ value is 0.443.

Solution Using the isotherm data for an initial atrazine concentration of 179 μg/L, the Freundlich K value can be calculated from a point on the line, for which PAC dose and percent atrazine remaining in solution can be easily read from the graph. For example, for a PAC dose of 4 mg/L, Fig. 14-16 shows that approximately 8 percent atrazine remains in solution at equilibrium. The equilibrium solid-phase concentration q for this PAC dose and the Freundlich K value can be calculated from

$$q = \frac{C_0 - C}{m/V} = \frac{C_0 - 0.16 \times C_0}{2.6\ \text{mg/L}} = \frac{(179 - 28.64)\mu\text{g/L}}{2.6\ \text{mg/L}} = 57.2\ \mu\text{g/mg} = 57.2\ \text{mg/g}$$

$$q = K \times C^{1/n} \qquad \text{or} \qquad K = \frac{q}{C^{1/n}} = \frac{57.2\ \text{mg/g}}{(28.64\ \mu\text{g/L})^{0.443}} = 12.9\ (\text{mg/g})(\text{L}/\mu\text{g})^{1/n}$$

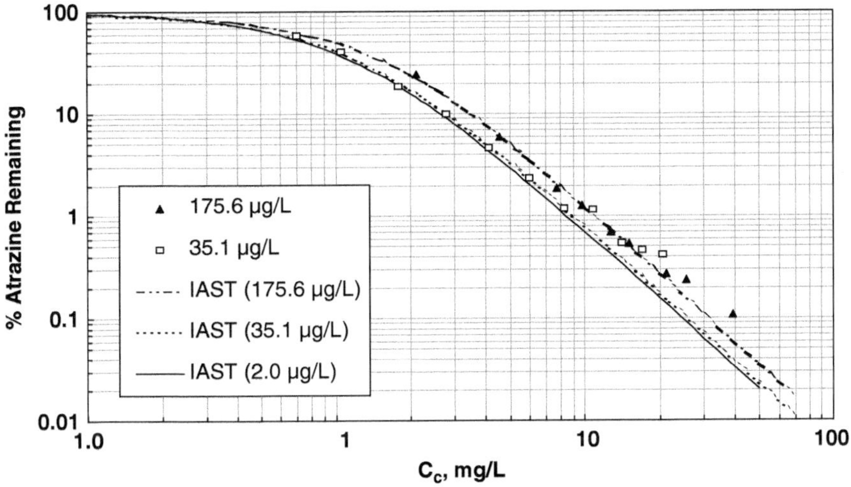

FIGURE 14-16 Atrazine adsorption isotherms in a drinking water source at three initial concentrations.

Approach for Determining D_s from Eq. 14-9
Using a spreadsheet, calculate the result of Eq. 14-9 for each data point by assuming a value for D_s and by using the given/calculated values for C_0, C_c, R, K, $1/n$, and t. Add a column that contains the square or absolute value of the result of Eq. 14-9 for each data point. At the bottom of that column, determine the sum of the squares or absolute values. Then use a nonlinear optimization tool (e.g., Solver in Microsoft Excel) to minimize the value of the sum of squares or absolute values by varying D_s—or better, the log of D_s because D_s is on the order of 10^{-13} cm^2/s. This procedure yields a D_s value of 3.6×10^{-13} cm^2/s as the best estimate for the preceding data set. The resulting D_s value can be used to predict PAC performance in different reactor configurations (i.e., plug flow, completely mixed). ▲

Breakthrough Curves for Packed Bed GAC Reactors. The region of a GAC column in which adsorption is taking place, the *mass transfer zone* (MTZ), is shown in Fig. 14-17*a*. The activated carbon behind the MTZ has reached the adsorption capacity for the given influent concentration C_0, and the solid-phase concentration per unit mass of GAC is q_0. The activated carbon in front of the MTZ has not been exposed to the adsorbate, so both the aqueous- and solid-phase concentrations are zero. Within the MTZ, the solid-phase concentration varies from q_0 to zero. The length of the MTZ L_{MTZ} depends on the rate of internal and external mass transfer and the adsorption capacity. Anything that causes a higher rate of mass transfer, such as a smaller carbon particle size, higher temperature, larger diffusion coefficient of adsorbate, greater strength of adsorption of adsorbate, and larger adsorption capacity (i.e., a larger Freundlich K value), will decrease the length of the MTZ.

In some circumstances, for example, small, strongly adsorbing compounds in long GAC columns, L_{MTZ} will be reduced sufficiently relative to bed length so that it can be assumed to be zero, yielding the ideal plug-flow behavior, as shown in Fig. 14-17*b*. If L_{MTZ} is negligible, analysis of the adsorption process is greatly simplified.

The breakthrough curve is a plot of the column effluent concentration as a function of either the volume treated, the time of treatment, or the throughput in terms of number of bed volumes (BVs) treated (BV = volume of water treated/GAC volume V). The number

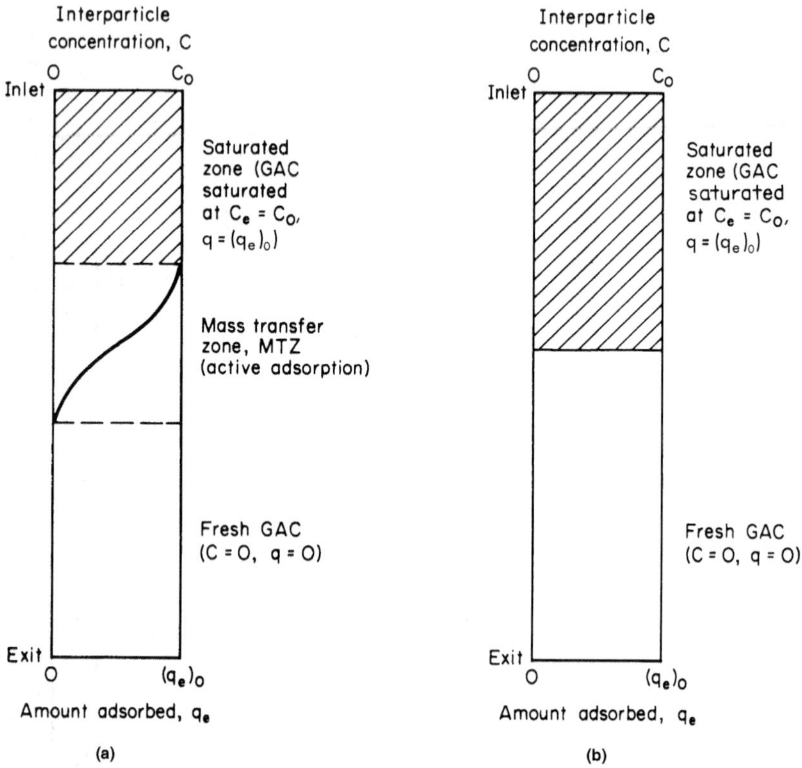

FIGURE 14-17 Adsorption column (*a*) with MTZ and (*b*) without MTZ.

of BVs is a particularly useful parameter because the data from columns of different sizes and with different flow rates are normalized. A breakthrough curve for a single, adsorbable compound is shown in Fig. 14-18. The *breakthrough concentration* C_B for a column is defined as the maximum acceptable effluent concentration. When the effluent concentration reaches this value, the GAC must be replaced. V_B is the volume treated to breakthrough. The shape of the curve is affected by the same factors that affect the length of the MTZ and in the same way. Anything that causes the rate of adsorption to increase will increase the sharpness of the curve. The breakthrough curve will be vertical if $L_{\mathrm{MTZ}} = 0$, as was shown in Fig. 14-17*b*. *Breakthrough capacity* is defined as the mass of adsorbate removed by the adsorber at the maximum acceptable effluent concentration or at breakthrough, $C_{\mathrm{eff}} = C_B$, as illustrated in Fig. 14-18. The degree of column utilization is defined as the ratio of the mass adsorbed at breakthrough relative to the mass adsorbed at complete breakthrough, where $C_{\mathrm{eff}} = C_0$. Values of both parameters increase as the rate of adsorption increases.

The breakthrough curve can be used to determine the activated carbon usage rate (CUR) at breakthrough. The CUR is the mass of activated carbon used per unit volume of water treated:

$$\mathrm{CUR}\left(\frac{\text{mass}}{\text{volume}}\right) = \frac{\text{mass of GAC in column}}{\text{volume treated to breakthrough } V_B} = \text{GAC bed density } \rho/BV_B \quad (14\text{-}10)$$

where BV_B is the number of bed volumes treated at breakthrough V/V_B.

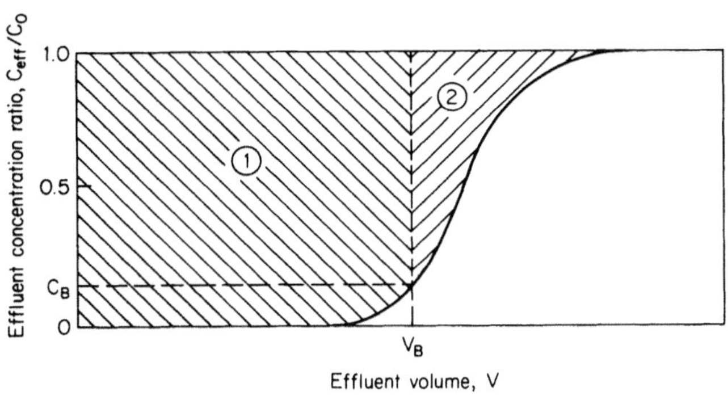

Breakthrough capacity = area (1) x C_0

$$\text{Degree of column utilization (\%)} = \frac{\text{area (1)}}{\text{area (1)} + \text{area (2)}} \times 100$$

FIGURE 14-18 Adsorption column breakthrough curve.

Breakthrough curves of single compounds are affected by the relationship between the L_{MTZ} and the GAC bed depth. Immediate breakthrough of adsorbable compounds occurs if the L_{MTZ} is greater than the GAC bed depth (compare curves *A* and *B* in Fig. 14-19).

For heterogeneous mixtures such as NOM, breakthrough can be strongly affected by the presence of nonadsorbable compounds, biodegradable compounds, and the slow

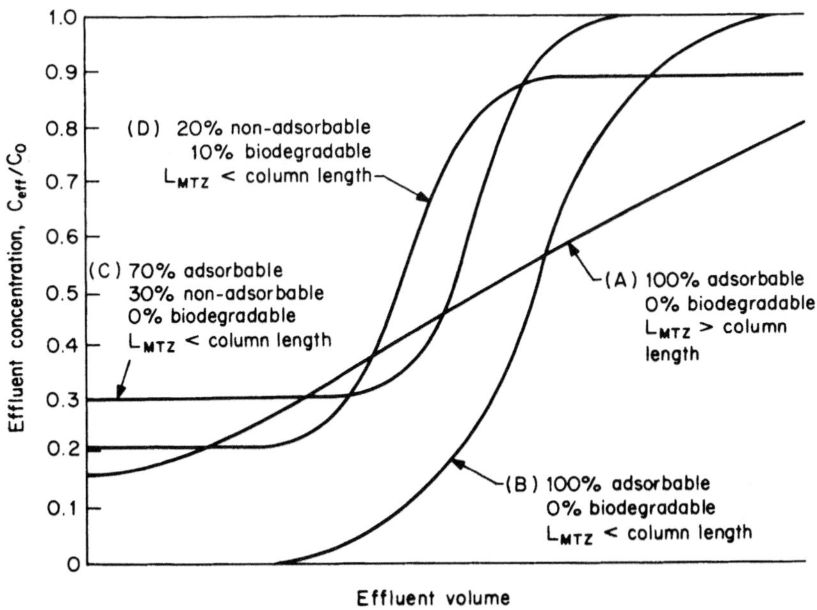

FIGURE 14-19 Effect of biodegradation and the presence of nonadsorbable compounds on the shapes of breakthrough curves.

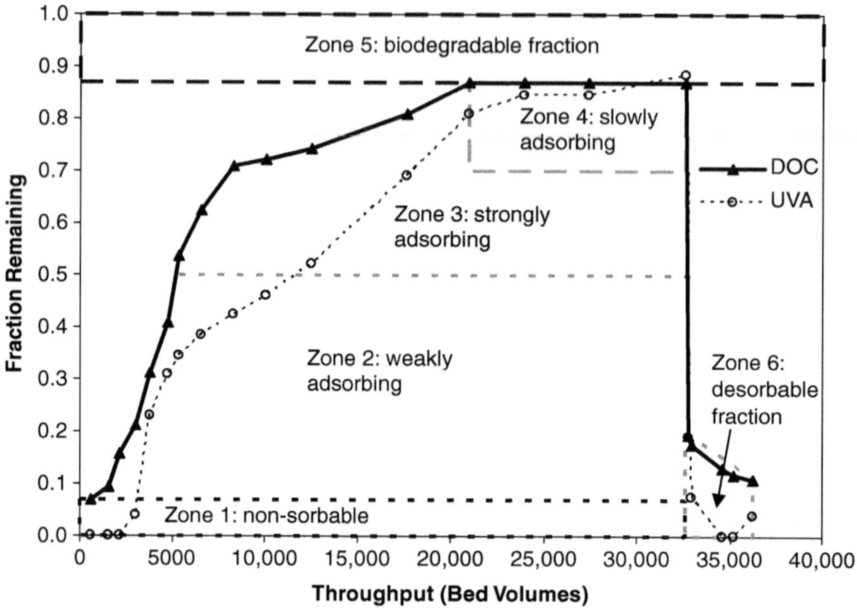

FIGURE 14-20 Schematic of NOM breakthrough.

adsorption of large NOM molecules. Nonadsorbable compounds appear immediately in the column effluent, even when the GAC bed depth is greater than the L_{MTZ} of specific components of the NOM (compare curves *B* and *C* in Fig. 14-19). Removal of adsorbable biodegradable compounds by microbiological degradation in a column results in continual removal, even after the carbon normally would be saturated with adsorbable compounds (see curve *D* in Fig. 14-19). Breakthrough curves shown in Fig. 14-20 also illustrate some of these effects.

Figure 14-20 shows breakthrough of NOM and its different components as measured by DOC. In this case, the MTZ is contained in the bed depth, and zone 1 shows the immediate breakthrough of nonadsorbable compounds. Note that UVA (UV absorbance) does not detect these compounds because most nonadsorbing compounds do not absorb light at the wavelength used. In zone 2, weakly adsorbing compounds break through, and strongly adsorbing compounds are removed until zone 3. A small level of removal continues by slow adsorption of large NOM constituents into small GAC pores in zone 4, as shown by the UVA. Zone 5 shows the long-term biodegradation as measured by the DOC, and zone 6 shows the desorbable fraction, which will be discussed later.

NOM Preloading Effect on Adsorption Kinetic Parameters. For fixed-bed GAC adsorption processes, NOM preloading lowers not only the adsorption capacity of GAC (see above) but also the rate of target contaminant adsorption. Both external film mass transfer and intraparticle diffusion coefficients decrease.

Several studies have shown that the film mass transfer coefficient k_f decreases as preloading (or GAC service) time (e.g., Speth, 1991; Carter and Weber, 1994; Knappe et al., 1999) or NOM surface loading (Schideman et al., 2006a) increases. For TCE adsorption from Huron River water, the k_f value for GAC that was preloaded for 4 weeks was approximately 30 percent of that for fresh GAC (Carter and Weber, 1994), whereas the k_f value describing atrazine adsorption from Seine River water was approximately 50 percent of that obtained with fresh GAC after preloading times of 5 and 20 months (Knappe et al., 1999). A reduction in k_f with increasing preloading time or NOM surface loading may appear surprising given that the k_f value for a given adsorbate should depend only on the adsorbent geometry and the interstitial water velocity in an aqueous system. However, NOM may accumulate near the outer surface of the adsorbent particles as a result of slow adsorption kinetics (Matsui et al., 1994). Consequently, the local water viscosity at the external GAC surface may increase and/or the effective surface area for mass transfer may decrease, which would explain lower k_f values in the presence of preloaded NOM (Carter and Weber, 1994). For the herbicide atrazine, Fig. 14-21 shows the dependence of the external film mass transfer coefficient on NOM surface loading. The effect of NOM preloading on k_f can be important for short GAC beds (e.g., Schideman et al., 2006b) or treatment systems involving the application of PAC in conjunction with membrane filtration processes (e.g., Ding et al., 2006). Breakthrough curve predictions for typical GAC adsorber designs are relatively insensitive to k_f, however (e.g., Knappe et al., 1999).

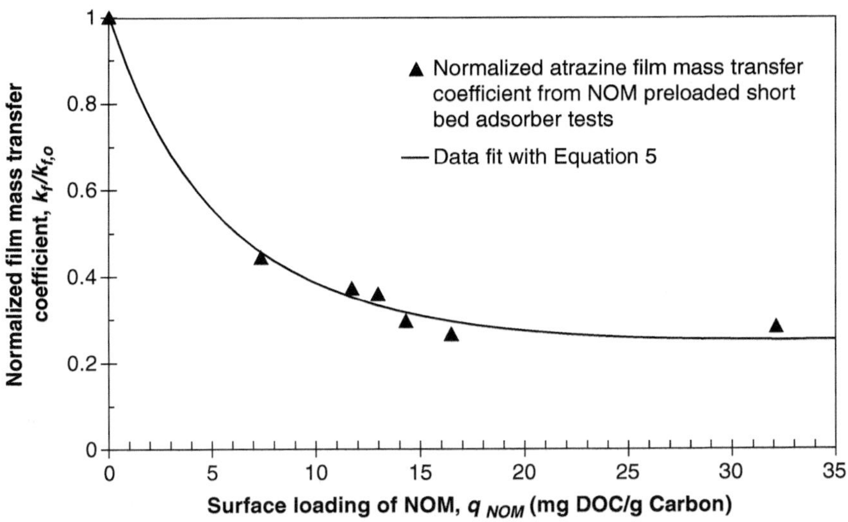

FIGURE 14-21 Dependence of the external film mass transfer coefficient on NOM surface loading. (*Source: Reprinted with permission from L. Schideman et al. (2006a), Three-component competitive adsorption model for fixed-bed and moving-bed granular activated carbon adsorbers: I. Model development. Environmental Science & Technology 40(21), 6805–6811. Copyright © 2006, American Chemical Society.*)

Several researchers have studied the effect of preloaded NOM on internal mass transfer (Zimmer et al., 1987; Hand et al., 1989; Speth, 1991; Carter and Weber, 1994, Li et al., 2003a, 2003b). Internal mass transfer can take place by both surface and pore diffusion. Because the two intraparticle diffusion processes occur in parallel, the faster of the two is rate determining. For TCE adsorption from natural waters, a reduction in the effective surface diffusion coefficient D_s of approximately one order of magnitude was observed after a preloading time of 5 weeks (Zimmer et al., 1987; Hand et al., 1989). The decrease in D_s

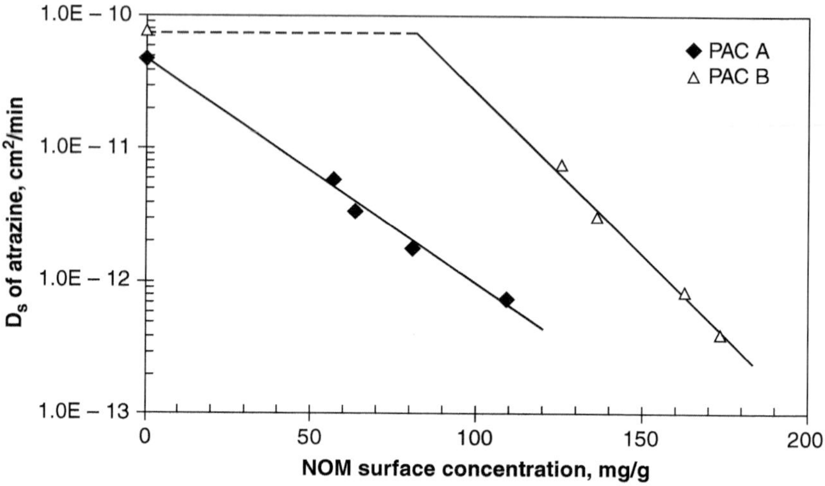

FIGURE 14-22 Effect of NOM surface loading on the magnitude of the surface diffusion coefficient D_s. *(Source: Q. Li et al. (2003b), Pore blockage effect of NOM on atrazine adsorption kinetics of PAC: The roles of PAC pore size distribution and NOM molecular weight. Water Research 37, 4863–4872. Copyright © 2003, with permission from Elsevier.)*

was such that pore diffusion became rate determining, and a pore diffusion model was used to describe the removal rate of TCE (Hand et al., 1989; Weber and Carter, 1994). Relating changes in D_s to NOM surface loading, Li and colleagues (2003b) showed that NOM surface loadings in excess of 100 mg/g can lower D_s for the herbicide atrazine by more than two orders of magnitude relative to the value obtained with fresh GAC (Fig. 14-22). Both activated carbon and NOM characteristics affect the manner in which D_s decreases with NOM surface loading (Li et al., 2003b).

Desorption and Attenuation

Adsorption of many compounds is reversible, which means that they can desorb. Desorption may be caused by displacement by other compounds, as discussed earlier, or by a decrease in influent concentration. An analysis of specific compound desorption at high concentration ranges (>1 mg/L) by Thacker and colleagues (1983) showed that the quantity of adsorbate that can desorb in response to a decrease in influent concentration increased as (1) the diffusion coefficient of the adsorbate increased, (2) the amount of compound adsorbed increased, (3) the strength of adsorption decreased (e.g., as the Freundlich $1/n$ value increased), and (4) the activated carbon particle size decreased. Volatile organic compounds (VOCs) are especially susceptible to displacement because they are weakly adsorbed and diffuse rapidly.

Summers and Roberts (1988a, 1988b) have shown that NOM only partially desorbs, and for the desorbing fraction, the desorption diffusivity is lower than that during adsorption. Corwin and colleagues (2008) found that desorption occurred much more slowly than predicted by the pore–surface diffusion model for target organic compounds at very low concentrations (<1 μg/L) in waters containing adsorbable NOM. To model the results, Freundlich K values of desorption were reduced by 10 to 50 percent compared with the adsorption K values, with more strongly adsorbing compounds showing the largest differences. Longer-term desorption tests indicated a relatively fast decrease in effluent concentration that approached a steady desorption phase that lasted for over 50,000 BVs at about 10 percent of the previously applied influent concentration.

Periodic episodes such as spills and seasonal taste and odor events can occur in surface waters. In these events, the influent concentration increases for a relatively short period, hours to weeks, and then decreases to a baseline concentration that may be very small. A low influent concentration after the episode causes a reversal of the concentration gradient, and desorption occurs. The attenuation of pulses can be an effective way of dampening the exposure to high concentrations, but the total mass passing through the system theoretically should be the same as that in the influent. The peak concentration of the effluent is a function of the factors listed above (Thacker et al., 1983), as well the service life and amount of DOM adsorbed prior to the episode (Matsui et al., 1994; Gillogly et al., 1999, 2006). Hindered desorption has implications in the modeling the micropollutant attenuation by GAC. The actual concentrations are lower initially than those predicted by standard models.

GAC ADSORPTION SYSTEMS

GAC contactors can be classified by the following characteristics: (1) driving force: gravity versus pressure, (2) flow direction: downflow versus upflow, (3) configuration: parallel versus series, and (4) position: filter adsorber versus postfilter adsorber. GAC may be used in pressure or gravity contactors. Pressure filters enclose the GAC and can be operated over a wide range of flow rates because of the wide variations in pressure drop that can be used. In many groundwater systems, pressure filters can take advantage of existing pressure in the system to avoid repumping and the associated costs. Another advantage of these filters is that they can be prefabricated, shipped to the site, and installed quickly. A disadvantage is that the GAC cannot be observed visually or sampled with ease. Gravity contactors are better suited to systems in which wide variations in flow rate are not desirable because of the need to remove turbidity, large pressure drops are undesirable because of their impact on operational costs, there is available hydraulic head in the plant, and visual observation is needed to monitor the condition of the GAC. For many systems, the decision between pressure or gravity contactors is made on the basis of cost. Medium- and large-sized systems normally use gravity contactors, and groundwater systems typically use pressure filters because the flows from a well normally are low.

Water may be applied to GAC either downflow or upflow, and upflow columns may be either packed bed or expanded bed. Downflow columns are the most common and seem best suited for drinking water treatment. McCarty and colleagues (1979) found that carbon fines were produced during packed bed upflow operation and were not produced during downflow operation. The pulsed bed contactor also can be used to decrease carbon usage rate from that of a single contactor. The flow is applied upward through the column; the spent GAC, a fraction of the total amount present, is removed periodically from the bottom of the column, and an equal amount of fresh GAC is applied to the top. Upflow pulsed bed operation followed by membrane filtration can be used to eliminate the problem of fines production (Schideman et al., 2006a, 2006b).

Single-stage contactors are used often for small groundwater systems, but if more than one contactor is required, lower activated carbon usage rates can be achieved by arranging them either in series or in parallel, as shown in Fig. 14-23, possibly yielding a lower-cost system. GAC in a single-stage contactor must be removed about the time the MTZ begins to exit the column. At this point, only a portion of the activated carbon is saturated at the influent concentration, so the activated carbon usage rate may be relatively

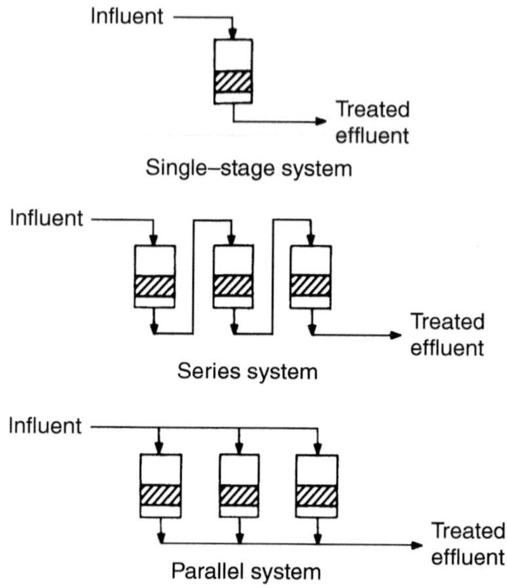

FIGURE 14-23 GAC adsorber systems.

high (see Fig. 14-18). Alternatively, columns may be arranged in series so that the MTZ is contained entirely within the downstream columns after the lead column has been saturated with the influent concentration. When the activated carbon is replaced in the lead column, the flow is redirected so that it goes through the freshest activated carbon last. Thus the activated carbon "moves" countercurrent to the flow of water, and lower activated carbon usage rates are achieved than with single-stage contactors. Series configuration is best used when the effluent criterion is very low compared with the influent concentration (Wiesner et al., 1987). However, for many SOCs at low concentrations, the adsorption capacity is decreased when the GAC is exposed to the background NOM for prolonged times; thus the capacity in the second column in series may be lower than that of the first. The increased cost of plumbing also counters the cost benefit of reduced activated carbon usage rate, especially when more than two columns must be used in series.

When parallel-flow adsorbers are operated in staggered mode, they also can be used to decrease the activated carbon usage rate from that which is possible with a single-stage contactor (Westrick and Cohen, 1976; Roberts and Summers, 1982). Because the effluent from each unit is blended, an individual unit can be operated to effluent concentrations higher than the treated water goal because it is blended with water from another unit that has an effluent below the treatment goal. Only the composite flow must meet the effluent quality goal. Dvorak and Maher (1999) showed that the GAC use rate for NOM removal decreased asymptotically with increasing number of adsorbers in parallel, with the largest decrease occurring when moving from a single adsorber to two adsorbers in parallel.

Denning and Dvorak (2009) used the pore–surface diffusion model to extensively evaluate the performance of lead-lag series operation, blended parallel operation, and bypass operation, which is similar to parallel operation with a column at exhaustion. They found that when high levels of removal were required, $C/C_0 = 0.05$, the lead-lag system was most efficient. For a less stringent criterion, $C/C_0 = 0.25$, the lead-lag series and blended parallel operation yielded similar results. Bypassing was found to be an effective approach in reducing costs.

Other flow arrangements can be used to produce lower activated carbon usage rates. Sontheimer and Hubele (1987) reported the use of pressure filters with two layers of GAC at Pforzheim, Germany. Each of the two layers could be backwashed and replaced independently, and the order of flow through these layers could be reversed. For removing halogenated hydrocarbons from groundwater, a 35 percent lower carbon usage rate (CUR) was reported for this system compared with a single-stage system.

In water treatment plants where filtration is required, for example, surface water treatment, GAC contactors also can be classified by their position in the process train. The filter adsorber employs GAC to remove particles as well as dissolved organic compounds. These contactors may be constructed simply by removing all or a portion of the granular medium from a rapid filter and replacing it with GAC. Alternatively, a new filter box and underdrain system for the GAC may be designed and constructed. Graese and colleagues (1987a) discuss these types of filters in detail. The postfilter adsorber is preceded by a granular media filter and thus has as its only objective the removal of dissolved organic compounds. Backwashing of these adsorbers is unnecessary for particle removal, but if immediately preceded by ozonation and in the presence of extensive biological growth, they may require regular backwashing. In groundwater treatment systems in which precipitates are not formed, filtration typically is not required, and the GAC adsorbers are more like postfilter adsorbers with respect to design and nonbackwashing operation.

PERFORMANCE OF GAC SYSTEMS

Factors Affecting Removal Efficiency

Adsorbate and GAC properties both have important effects on adsorption that have been discussed in earlier sections. Additional factors that must be considered in the design of full-scale systems are presented here.

GAC Particle Size. The effect of particle size on the rate of approach to equilibrium, discussed previously, has a similar effect on rate of adsorption in columns if the rate of adsorption is controlled by intraparticle diffusion. Decreasing particle size will decrease the time required to achieve equilibrium and will decrease the length of the MTZ in a column, with all other factors constant. This is not an important factor if only 50 percent removal is needed, but it becomes important if high levels of removal, for example, greater than 90 percent, are needed. Thus, under high removal requirements, the particle size selected for a contactor should be as small as possible, which will improve fixed-bed adsorption efficiency and minimize the GAC bed size required. For filter adsorbers, the rate of head loss buildup caused by particle removal may limit the GAC particle size that can be used. Small GAC particle sizes lead to a higher initial head loss and rate of head loss buildup. Thus cost of energy and availability of head have an important influence on the GAC size selected for a filter adsorber design. Additionally, if a filter adsorber is constructed by replacing media in an existing rapid filter, turbidity removal efficiency generally increase as the GAC size decreases. If the GAC size is too small, the rate of head loss buildup because of particle accumulation becomes excessive, causing frequent backwashing, low net water production, and uneconomical operation.

The commercial sizes of GAC typically are characterized by a relatively large uniformity coefficient of up to 1.9. This large coefficient causes the bed to restratify more easily after backwashing. Large uniformity coefficients also require greater percentage bed expansion during backwash in order to expand and clean the bottom media (Graese et al., 1987a). Some GAC filters use GAC with a small uniformity coefficient (~1.3) in deep beds to improve depth removal of turbidity and to increase net water production

(Graese et al., 1987b). Mixing of the media in these filters undoubtedly is more than in filters with large uniformity coefficients, so they should not be used in applications where desorption from the mixed GAC will require early GAC replacement.

Filter adsorbers that are used to remove turbidity commonly employ 8 × 30 U.S. Standard Mesh (2.38 × 0.60 mm) or larger activated carbon to promote longer filter runs. Common practice is to use 12 × 40 U.S. Standard Mesh (1.68 × 0.42 mm) or similar activated carbon in postfilter adsorbers. The small particle sizes decrease the L_{MTZ}, and regular backwashing for accumulated particle removal is not required. The option of using custom-sized GAC to obtain a better media design for a particular application also is available.

Contact Time, Bed Depth, Hydraulic Loading Rate. An important GAC adsorber design parameter that affects performance is the contact time, most commonly described by the *empty bed contact time* (EBCT). For a given situation, a critical depth of GAC and a corresponding minimum EBCT exist that must be exceeded to contain the MTZ and minimize or eliminate immediate breakthrough. As the EBCT increases, the run time to a given effluent concentration typically will increase. However, in most cases, the relationship is not linear. Pilot data that show the effect of increasing EBCT on bed life for VOCs have been given by Love and Eilers (1982). In general, they showed that the CUR for *cis*-1,2-dichloroethylene, 1,1,1-trichloroethane, and carbon tetrachloride decreased substantially as the EBCT was increased from 6 to 12 minutes, but little change was noted when the EBCT was increased to 18 minutes. For micropollutants in the presence of background NOM, the CUR of GAC beds with long EBCTs can become larger than that of GAC beds with shorter EBCTs, however. This trend is related to the poorer adsorbability of NOM relative to the micropollutant. NOM will penetrate deeper into the bed and will foul the GAC in the bottom of the adsorber. As a result, the adsorption capacity and adsorption rate of micropollutants are reduced, as discussed earlier in this chapter.

For adsorption of NOM, longer EBCTs have been found to be more efficient in many cases, that is, lower CURs and higher throughput than for shorter EBCTs in the range of 5 to 20 minutes. Summers and colleagues (1997) reported that for three of the four waters examined for TOC and DBP precursor control, no significant change in the BVs treated or CUR occurred when the EBCT was increased from 10 to 15 to 20 minutes. Analysis of the Information Collection Rule (ICR) data by Zachman and Summers (2010) has shown that increasing the EBCT from 10 to 20 minutes increased BVs treated to a given effluent concentration by 10 percent and resulted in a lower CUR. This will be discussed in more detail later in this chapter.

The impact of EBCT on the breakthrough of NOM (measured by UVA) and bisphenol A (BPA) is shown in Fig. 14-24. The mass transfer zones of both the NOM, as measured by UVA, and BPA are contained by the 5.0 minute EBCT. For the first 5000 BVs, the longer EBCTs show better performance for both compounds. This trend continues for the NOM breakthrough; however, the EBCT trend reverses for BPA, reflecting the "fouling" of adsorption sites deeper in the bed with time (Corwin, 2010). Collecting pilot-plant data for a range of EBCT values is important if the lowest activated carbon cost for a given application is to be determined.

Increasing EBCT by increasing the bed depth at a constant hydraulic application rate increases the capital cost of the GAC system (Wiesner et al., 1987; Kornegay, 1979; Lee et al., 1983). Operating costs decrease because of decreasing GAC replacement frequency. Thus the optimal depth, or EBCT, can be calculated by the intersection of the capital and operation and maintenance (O&M) cost curves.

While hydraulic loading rates affect external mass transfer, under normal operating conditions, intraparticle mass transfers controls, rendering hydraulic loading rate impacts negligible (Sontheimer et al., 1988). Cover and Pieroni (1969) review data that show adsorbers with the same EBCT but with different hydraulic loading rates give essentially the same

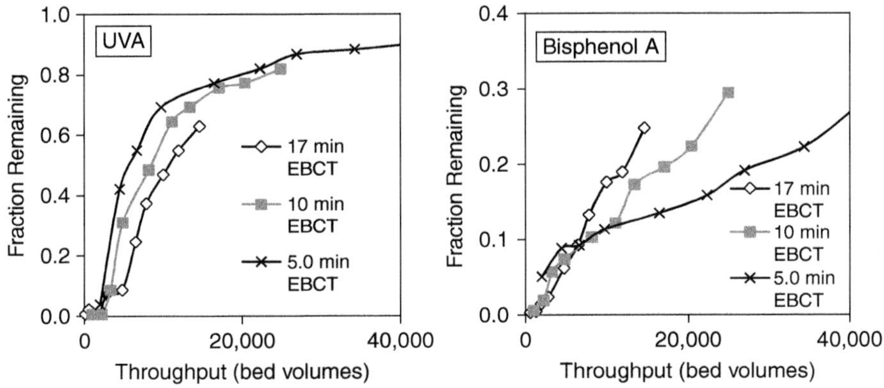

FIGURE 14-24 Impact of EBCT on NOM and bisphenol A breakthrough. (*Source: Adapted with permission from Corwin, 2010.*)

performance in terms of number of BVs processed to breakthrough, provided that the depth is considerably greater than the critical depth L_{MTZ}.

The EBCTs in use range from a few minutes for some filter adsorbers (Graese et al., 1987a) to more than 4 hours for the removal of high concentrations of certain specific contaminants (Snoeyink, 1983). Typical values for filter adsorbers are in the 5 to 10 minute range and for postfilter adsorbers in the range of 15 to 25 minutes. Hydraulic application rates vary from 1 to 30 m/h (0.4–12 gpm/ft^2), with a typical value being 7 to 10 m/h (3–4 gpm/ft^2). For filter adsorbers, selection of hydraulic loading rate and bed depth needs to also include considerations for optimal filtration preformance (see Chap. 10).

Backwashing. Backwashing of GAC filter adsorbers is essential to remove trapped particles and maintain the desired hydraulic properties of the bed. Regular backwashing of postfilter adsorbers is not normally required, although some may be needed to control biological growth and head loss. Backwashing should be minimized, however, because of its possible effect on adsorption efficiency. Mixing of the bed may take place during backwashing, and if so, GAC with adsorbed molecules near the top of the bed may move deeper into the bed, where desorption is possible. Molecules with low K values, such as carbon tetrachloride and other VOCs, may be partially desorbed in this new position, leading to a spreading out of the MTZ and to early breakthrough (Wiesner et al., 1987). Desorption will not occur if the molecules are irreversibly adsorbed or if they are removed by a destructive mechanism such as biodegradation instead of adsorption. The large uniformity coefficient of most commercial activated carbons promotes restratification after backwash. If the underdrain system does not distribute the wash water properly, or if the backwash is not carried out in a manner that aids restratification, substantial mixing of the activated carbon can occur with each backwash (Graese et al., 1987a). Hong and Summers (1994) have shown that backwashing had little impact on the time to 50 percent breakthrough for the TOC and trihalomethane (THM) precursors of four waters. For two of the waters, some increase in TOC after backwashing was detected, but the maximum amount was only 0.2 mg/L, consistent with the TOC being essentially irreversibly adsorbed. Hong and Summers (2006) used models to predict significantly earlier breakthrough of organic contaminants based on complete mixing; however, Corwin and colleagues (2008) did not find a significant impact of mixing using small columns. This discrepancy is likely due to the model assumption of Freundlich and diffusion parameters being constant for adsorption and desorption.

Biological Activity. In a filter or packed column, indigenous microorganisms will accumulate on the surface of the medium, such as sand or GAC, as long as a disinfectant residual is not present in the effluent and if a biodegradable substrate, organic or inorganic, is available. This process is often termed *biofiltration*, and when GAC is the filter medium, the term *biological activated carbon* (BAC) has been used. Chlorine reacts strongly with GAC such that under normal operating conditions, chlorine is not present in the column effluent. Thus all applications of GAC in water treatment have biomass growth and a biological removal component. In the last 10 years, several reviews of biofiltration as used in drinking water treatment have been published (Ufer et al., 1997; Servais et al., 2005; Evans et al., 2010). One review, that by Chowdhury and colleagues (2010), specifically focused on BAC in filter adsorbers with nonozonated influent, the most common use of BAC in the United States. The reviews by Chowdhury and colleagues (2010) and Evans and colleagues (2010) also include utility surveys.

Based on these reviews, some generalizations can be made. GAC seems to provide slightly better biological removal performance than sand or anthracite and adds the potential to attenuate pulses of influent adsorbable compounds. However, there are more similarities than differences between biologically active GAC and inert media with respect to design, operation, and performance. For all biological filters (except in some cases slow sand filters), the following generalizations hold true: Specific organic compounds that are biodegradable, as well as a biodegradable fraction of NOM, can be removed. Increasing EBCT can increase the removal. Lower temperatures can decrease the removal. Preozonation can increase the biodegradable fraction of the NOM. Bioacclimation time can take from days for easily assimilated compounds to months for more resistant compounds, especially at low (submicrogram per liter) concentrations. For GAC, biodegradable compounds may be removed by microbes, without prior adsorption to the GAC, if a microbial community structure capable of degrading such compounds is present. Adsorbable biodegradable compounds may be adsorbed first if the microbial community structure is not developed when the compounds enter the column and then desorbed and degraded as the microbial community structure develops.

Biological oxidation has been shown to be responsible for the removal of specific compounds such as phenol (Chudyk and Snoeyink, 1984), the odor-causing compounds geosmin and MIB (Ho et al., 2007; Chae et al., 2006; Namkung and Rittman, 1987; Silvey and Roach, 1964), the cyanotoxin microcystin-LR (Wang et al., 2007), *p*-nitrophenol and salicylic acid (DeLaat et al., 1985), ammonia (Bablon et al., 1988), trichlorobenzene (Summers et al., 1989), bromate (Kirisits and Snoeyink, 1999; Kirisits et al., 2001), perchlorate (Brown et al., 2002, 2003, 2005), and specific ozonation by-products (Servais et al., 2005). The performance of biofilters for these compounds varies and is a function of influent water quality and filter conditions. Longer EBCTs normally will lead to better removal.

Some portion of the NOM in natural waters can be oxidized biologically by biological filters, as shown in Fig. 14-25 for an anthracite-sand filter and a GAC filter for nonozonated water. The anthracite-sand filter took a few weeks to acclimate but then achieved removals ranging from 5 to 20 percent depending on the time of year. The GAC filter adsorber displayed both adsorption and bioremoval over the first 9 months; thereafter, removal was by slow adsorption and biodegradation. For coagulated waters without preozonation, the biodegradeable NOM fraction is about 10 to 15 percent, of which 50 to 75 percent can be removed in biological filters. This would yield a steady state TOC removal of 5 to 10 percent, which is similar to that shown in Fig. 14-25. For ozonated waters, the biodegradeable NOM fraction increases to about 20 to 40 percent, of which 50 to 75 percent can be removed in biological filters. This would yield a steady state TOC removal of 10 to 30 percent, as reported in Table 14-6 for biofiltration of ozonated and settled Ohio River water.

Under conditions of drinking water treatment, the concentration of biodegradable NOM (0.1–1.0 mg/L) is orders of magnitude greater than that of the specific biodegradable organic compounds (<<0.1 mg/L). Thus most of the biomass accumulation is associated with biodegradable NOM. The dependency of the bioremoval of specific compounds on that of NOM

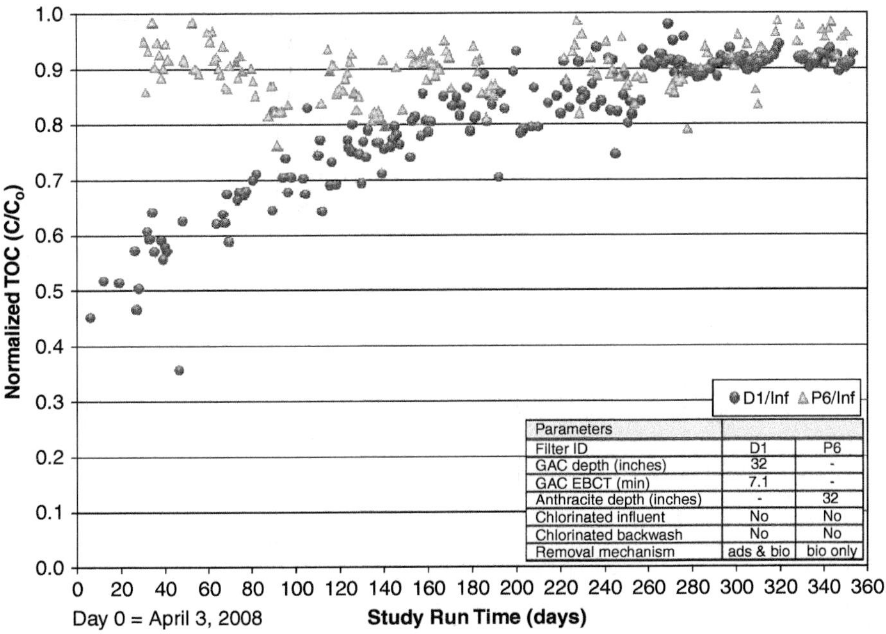

FIGURE 14-25 Normalized effluent TOC from a GAC filter-adsorber (D1) and an anthracite-sand filter (P6) treating nonozonated water at the City of Birmingham's demonstration and pilot-plant facility. (*Source: Chowdhury et al., 2010.*)

and the associated biomass is not understood. The biodegradation of NOM increases with increasing EBCT. This EBCT dependency yields a similar distribution of biomass on the filter beds, high at the top and asymptotically decreasing with depth (Wang et al., 1995; Servais et al., 2005). Under normal operating conditions, the easily degraded NOM fraction, about 50 percent of the total degradable fraction, can be removed in the first 5 minutes of EBCT, with some additional removal occurring in EBCTs up to 30 minutes (Servais et al., 2005; Wang and Summers, 1995). Complete removal of the biodegradable NOM fraction is not possible with the range of EBCTs used in practice (5–30 minutes). Thus, increasing the EBCTs beyond that of a filter adsorber (5–10 minutes) may not yield significant additional removal.

Since the kinetics are that lower at low temperatures, longer EBCTs are need at low temperature to achieve the same results, biological oxidation may vary significantly throughout the year if low temperatures are expected (Servais et al., 2005). TOC removal by biologically

TABLE 14-6 Impact of Media Type on Performance—Ozonated/Settled Ohio River Water

			Removal (%)		
Parameter	Anthracite-sand	Sand	GAC-microporous	GAC-mesoporous	GAC-macroporous
TOC	16	20	29	27	21
AOC-NOX	39	43	51	47	42
THMFP	23	23	40	34	27
TOXFP	28	25	52	44	31

Source: Wang et al. (1995).

active GAC systems seems to be a conservative indicator for DBP precursor removal (Miltner et al., 1996; Wang et al., 1995). Table 14-6 shows the removal of TOC, DBP precursors (as measured by the formation potential for THMs and total organic halide (TOX), and assimilable organic carbon (AOC), a measure of the biodegradable fraction of NOM, by biofiltration of ozonated and settled Ohio River water. All filters had a total depth of 0.76 m (30 in), including a bottom layer of 0.20 m (8 in) of sand, yielding an overall EBCT of 9.2 minutes. The steady state removals reported in the table were taken between 5 and 11 months of operation. The GAC columns outperformed the inert media columns, and the micro- and mesoporous GACs seem to perform best in the biological mode.

Several studies have shown that bacterial counts in the effluent of a biofilter are most often higher than in the influent (Servais et al., 2005). A thorough analysis of this is given by Symons and colleagues (1981). They report data from Beaver Falls, PA, that show that both coliform and standard plate counts were higher in GAC effluent than influent when the water temperature was greater than about 10°C, even though 1 to 2 mg/L of chlorine residual was present in the influent to the bed. Apparently, the GAC reduced the chlorine and allowed the bacteria to regrow. When the water temperature was below about 10°C, no regrowth was noticed. Other data from Philadelphia showed that coliform organisms such as *Citrobacter freundii*, *Enterobacter cloacae*, and *Klebsiella pneumoniae* were found in the GAC filters. In all cases, postdisinfection produced water meeting USEPA regulations.

Another problem may occur if activated carbon particles penetrate the underdrain and provide a habitat for microbes that protects them from being killed by disinfectants (LeChevallier et al., 1984; Servais et al., 2005). These problems can be minimized through proper underdrain design and installation, placement and backwashing of new charges of GAC, and backwashing of beds during operation (Servais et al., 2005).

Macroorganisms can grow in GAC filters if the filters are biologically active. Organisms such as protozoa, rotifers, and nematodes have been reported in biofilters and their effluent, especially at warmer temperatures (Servais et al., 2005). Optimized backwashing conditions have been shown to be effective in controlling this problem.

BAC must be controlled to avoid undesirable effects. Anaerobic conditions may develop, with attendant odor problems, if the system is not kept aerobic. This may happen if large concentrations of ammonia enter the filter (each milligram per liter of NH_3 requires about 3.8 mg/L of dissolved oxygen if it is converted to NO_3^-), if insufficient dissolved oxygen is in the water, or if the bed is allowed to stand idle for a period of time.

Application of chlorine to GAC adsorbers does not prevent biological growth and increases the concentration of adsorbed chlorinated organic compounds. The potential exists for chlorine to make the activated carbon become more friable and break up more easily, especially during backwash, because chlorine destroys some of the activated carbon when it is reduced (Snoeyink and Suidan, 1975). Also, the potential exists for formation of unique organic compounds through the catalytic action of the activated carbon surface, as shown later. Thus application of chlorine to GAC filters is not recommended.

Pretreatment for GAC Systems. Pretreatment can have a significant impact on the performance of GAC systems. The influent concentration of target and competing organic compounds may be lowered, thus increasing the run time to a given target concentration. The species of compounds or NOM composition may be changed, thus changing adsorbability and biodegradability. NOM removal by conventional surface water treatment and by biofiltration is important for specific organic compound adsorption because it reduces the fouling or preloading of GAC. It also decreases the NOM load on the GAC, thus increasing the run time to a target TOC level. In both cases, the CUR and cost of GAC operation are lowered (see Chap. 8 for a discussion of organic compound removal by coagulation).

Kim and colleagues (2007) and Kim (2006) explored the impact of background NOM concentration, methylisoborneol (MIB) concentration, and EBCT on the breakthrough of MIB at the bench scale. Figure 14-26 shows that decreasing background TOC yielded longer

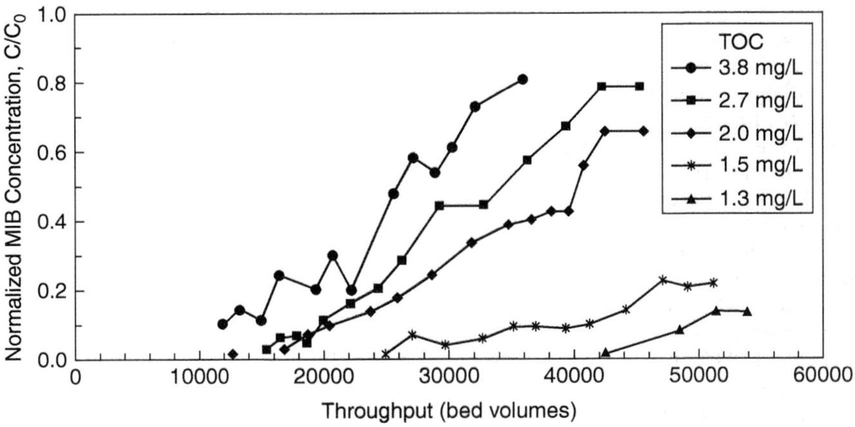

FIGURE 14-26 Impact of background TOC on MIB breakthrough in RSSCTs; MIB C_0 = 100 ng/L, EBCT = 10 minutes, bituminous 12 × 40 mesh GAC. (*Source: Adapted with permission from Kim, 2006.*)

run times to a given effluent MIB concentration. For an influent MIB concentration of 50 ng/L and an EBCT of 15 minutes, the adsorber would last for 6 months to greater than 20 months for this range of influent TOC concentrations and an effluent target of 10 ng/L. These authors also demonstrated that the fractional breakthrough C/C_0 was not affected by the initial concentration when C_0 was at or below 200 ng/L. Similar results have been reported for atrazine (Matsui et al., 2002). After analyzing the TOC and MIB breakthrough results of 35 rapid small-scale column test (RSSCT) runs, Kim and colleagues (2007) found a direct relationship between the throughput in BVs to 70 percent TOC breakthrough and BVs to 10 percent MIB breakthrough for EBCTs of 5, 10, 15, and 20 minutes, influent MIB concentrations ranging from 50 to 200 ng/L, and influent TOC concentrations ranging from 1.3 to 3.8 mg/L.

Summers and colleagues (1994), Hooper and colleagues (1996b), Bond and DiGiano (2004), and Zachman and Summers (2010) have shown that the run time in BVs to a target effluent TOC concentrations is inversely proportional to the influent TOC concentration. They developed models that are discussed later in this chapter.

Bond and DiGiano (2004) and Zachman and Summers (2010) have shown that the run time in BVs to a target effluent TOC concentrations is inversely proportional to the influent pH. The impact of pH on equilibrium capacities was shown in Fig. 14-4 for a specific organic compound and Fig. 14-9 for NOM. These impacts also translate to breakthrough curves; lower pH values yield longer run times for acidic compounds such as humic and fulvic acids that dominate NOM. Several investigators have reported better GAC performance for TOC control after coagulation or after increasing the coagulant dose to achieve enhanced coagulation. Hooper and colleagues (1996a, 1996b, 1996c) have shown that the increase in GAC run time after enhanced coagulation can be attributed to the lower pH and lower initial TOC concentration associated with the coagulated water.

Reactions of ozone with NOM (see Chap. 7) can alter the adsorption performance. For example, ozone can react with humic substances to produce more polar inter mediates that are less adsorbable on GAC (Chen et al., 1987) but usually more biodegradable (Sontheimer and Hubele, 1987). If TOC is more biodegradable, increased removals by microbiological activity in a GAC contactor are expected. However, if biological treatment is not effective, then the weakly adsorbing compounds created by ozonation can have a negative impact on the overall GAC performance. Solarik and colleagues (1996) systematically evaluated the impact of ozonation and biotreatment on subsequent GAC performance for five waters. They showed that ozonation and biotreatment decreased the humic and intermediate-molecular-size DOM fractions, which are the most strongly adsorbing fractions. They found that the early part of the breakthrough was dominated by a relative increase in the weakly adsorbing fraction, which in some cases led to earlier breakthrough, whereas the overall lower influent TOC concentration dominated the latter portion of the breakthrough curve and in some cases led to longer run times.

Chlorine-containing disinfectants (HOCl, ClO_2, or NH_2Cl) may react with both activated carbon and adsorbed compounds. Unusual products not characteristic of solution reactions may be formed when activated carbon is present. For example, the HOCl reaction with adsorbed 2,4-dichlorophenol (2,4-DCP) resulted in a series of hydroxylated polychlorinated biphenyls (PCBs) at HOCl concentrations normally encountered in drinking water treatment practice. A similar product mixture also was obtained when the GAC was first treated with HOCl and then 2,4-DCP was adsorbed. Furthermore, some of these products may desorb from the activated carbon column. Additional data are given on this effect by Voudrias and colleagues (1985). The reaction products that have been found to date have been measured only in laboratory systems. Further research is needed to show whether they also will form in field installations in the presence of humic substances. Because such compounds might form, and because their health effects are unknown, the application of chlorine-containing disinfectants to GAC adsorbers needs to be eliminated, where possible.

Pretreatment to prevent fouling of GAC is also important. Application of water that is supersaturated with salts such as calcium carbonate will lead to blockage of the activated carbon pores and possibly to complete coverage of the particle. Iron and manganese precipitates also may interfere with adsorption.

Adsorption Efficiency of Field-Scale Systems

In this section the performance of field-scale and both pilot- and full-scale GAC adsorbers is discussed for a range of adsorbates. Relationships are evolving that correlate the impact of background DOM on adsorption of micropollutants. However, the results are not yet definitive, and pilot-scale or, at a minimum, bench-scale testing should be carried out with the water in question prior to full-scale design.

TOC and DBP Precursors. Summers and colleagues (1994) and Hooper and colleagues (1996b), as shown in Fig. 14-27, developed a relationship that summarized the impact of the initial TOC concentration TOC_0 on the run time to a 50 percent TOC breakthrough, measured as bed volumes BV_{50}. The relationship can be expressed as

$$BV_{50} = \frac{18{,}000}{TOC_0} \tag{14-11}$$

Twenty-eight case studies of GAC bench-, pilot-, and full-scale contactors from 21 different source waters were evaluated. All systems used bituminous coal based GAC, and the influent pH was between 7 and 8 for the river, lake, and groundwaters examined. The relationship was verified by five GAC runs of the same isolated NOM, diluted to different concentrations (hollow symbols in Fig. 14-27). Part of the variability of the data around the regression line is likely due to differences in the adsorbability of the NOM caused by pretreatment and differences between sources. The type and adsorbability of NOM in source waters may vary widely from location to location; however, after coagulation, most waters have similar specific ultraviolet absorbance (SUVA) values, approximately 2.0 to 2.5 L/mg/m (see Chap. 3 for a discussion of SUVA), such that the influent TOC concentration and pH are the variables that control TOC breakthrough (Zachman et al., 2007a). Figure 14-27 can be used to provide estimates of the range of adsorbability for conventionally treated waters and one specific GAC type. In a later section, models to predict TOC breakthrough are presented.

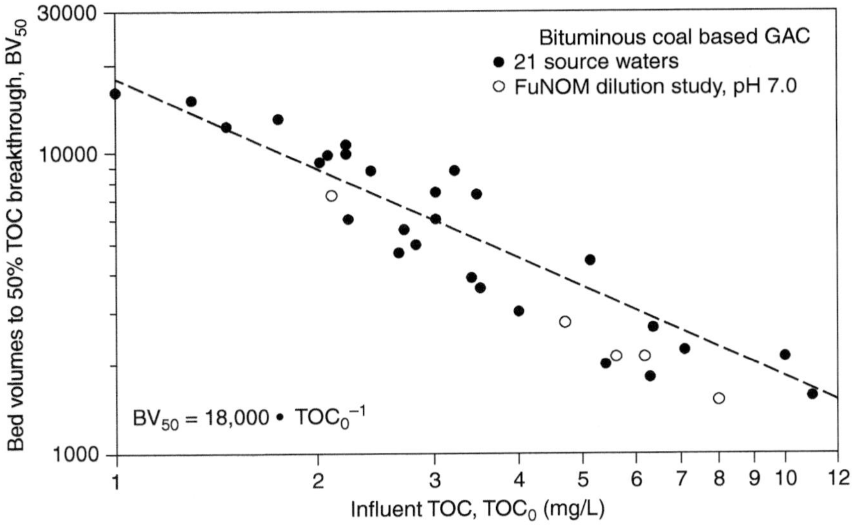

FIGURE 14-27 Correlation between influent TOC concentration and bed volumes to 50 percent TOC breakthrough for 21 source waters and runs with five dilutions of an extracted NOM from a groundwater (FuNOM). (*Source: Hooper et al., 1996b.*)

Roberts and Summers (1982) presented a summary of full-scale plant performance for TOC removal. They evaluated removals from 47 different plants, including some wastewater reclamation plants. The ranges of design conditions are given in Table 14-7. The breakthrough curves reach an effluent concentration plateau about 10 to 25 percent below the influent concentration, indicating biodegradation and that the time to reach this plateau increased as the EBCT increased. These authors found that immediate breakthrough (TOC fraction in the effluent immediately after startup) decreased as the EBCT increased to about 10 minutes, indicating that the MTZ was contained in the bed at this point.

TABLE 14-7 Design Conditions for GAC Adsorbers

Parameter	Median	Range	Typical range
EBCT, minutes	10	3–34	5–24
Depth of bed, m	1	0.2–8	0.5–4
Hydraulic loading, m/h	6	1.9–20	2.6–17
Influent TOC, mg/L	3.5	1–16	2–6

Note: Data from 47 plants were analyzed. Also, m/h × 0.41 = gpm/ft^2.
Source: Roberts and Summers (1982).

Graese and colleagues (1987a) summarized TOC-removal data for a criterion of 50 percent removal and showed that adsorbers with EBCT values of less than 10 minutes had a life of less than 30 days and that service time increased as EBCT increased. To obtain the lowest-cost adsorption system for TOC removal, EBCTs in the range of 10 to 20 minutes should be closely examined.

The performance for GAC removal of DBP precursors usually parallels that for the removal of TOC (Hooper et al., 1996b; Lykins et al., 1988a). Symons and colleagues (1981) have summarized much of the research prior to 1981. Hooper and colleagues (1996b), Solarik and colleagues (1995a,b), and Summers and colleagues (1994) assessed the use of TOC and UVA as indicators of DBP precursor breakthrough for seven source waters and the following DBPs: THMs, haloacetic acids (HAAs), TOX, and chloral hydrate (CH). The precursors were assessed using the uniform formation conditions (UFC) test (Hooper et al., 1996b) (see Chap. 19). In all but one water, TOC was a good, conservative surrogate in that it broke through prior to the DPB precursors. In one case, THM precursors broke through a filter adsorber with an EBCT of 6.3 minutes prior to TOC. Organics measured by UVA tended to break through after the DBP precursors. TOC removal can vary seasonally because the adsorbability and initial concentration may change with season (Solarik et al., 1995b).

The bromide concentration of the water has an important bearing on the use of GAC for DBP control. Symons and colleagues (1981) observed that brominated THMs dominated in GAC adsorber product water. Graveland and colleagues (1981) made a similar observation and noted a shift to the formation of more highly brominated forms of THMs in GAC effluent. The requirement of a chlorine residual forces the DBP formation reactions to be precursor limited. When precursor concentration is low during the first part of the breakthrough curve, the more rapidly forming bromine-substituted THMs dominate relative to chloroform formation (Summers et al., 1993).

An example of field-scale application of GAC that illustrates the advantages of a pilot study over bench-scale studies has been presented by Snoeyink and Knappe (1994). In that study, the authors evaluated the removal of NOM and the control of DBPs by GAC and BAC at EBCTs of 5, 10, 15, and 19 minutes after conventional treatment at Newport News, VA. In addition, the influent to the BAC column was ozonated. The TOC concentration of the influent was 3 mg/L, which gave a distribution system THM of 60 μg/L before ozonation. The duration of the study was 107 days for the GAC columns and 83 days for the BAC columns. Figure 14-28 depicts the trihalomethane formation potential (THMFP) breakthrough curves at different EBCTs. In both cases, EBCT had little impact on the breakthrough curves, with the exception of the GAC breakthrough curve for an EBCT of 5 minutes. (The 5-minute curve was considered unreliable because of the difficulty in maintaining a constant depth of GAC above the first sampling port.) A comparison of GAC and BAC curves in Fig. 14-28 shows that both the GAC and the BAC treatment trains resulted in breakthrough curves that were beginning to plateau at a dimensionless THMFP concentration of about 0.7, an indication that biological removal of organic carbon occurred for both the ozonated and the nonozonated influent. Additional run time is needed to fully establish the effect of ozone on the plateau.

These results highlight some important issues that must be considered when dealing with biologically active filters. It must be determined whether the plateau region meets the treatment objective and whether the plateau region is a function of the ozone dose that was applied to the filter influent and EBCT. If the plateau region meets the treatment objective, long run times can be achieved with BAC filters. Long pilot-test runs are required to establish the plateau value. The life of the filter as a function of ozone dose and EBCT also needs to be examined if the treatment objective is exceeded before the plateau is reached. This long-term treatability data are not available from bench-scale tests.

Taste and Odor. Many types of taste and odor problems are encountered in drinking water, and some of the most frequently occurring ones are attributed to microbial metabolites, for example, MIB and gesomin, for which activated carbon has an excellent history of success. Taste and odor sensitivity to many microbial metabolites can be very low (parts per trillion, i.e., ng/L), which makes them difficult and expensive to quantify directly. In

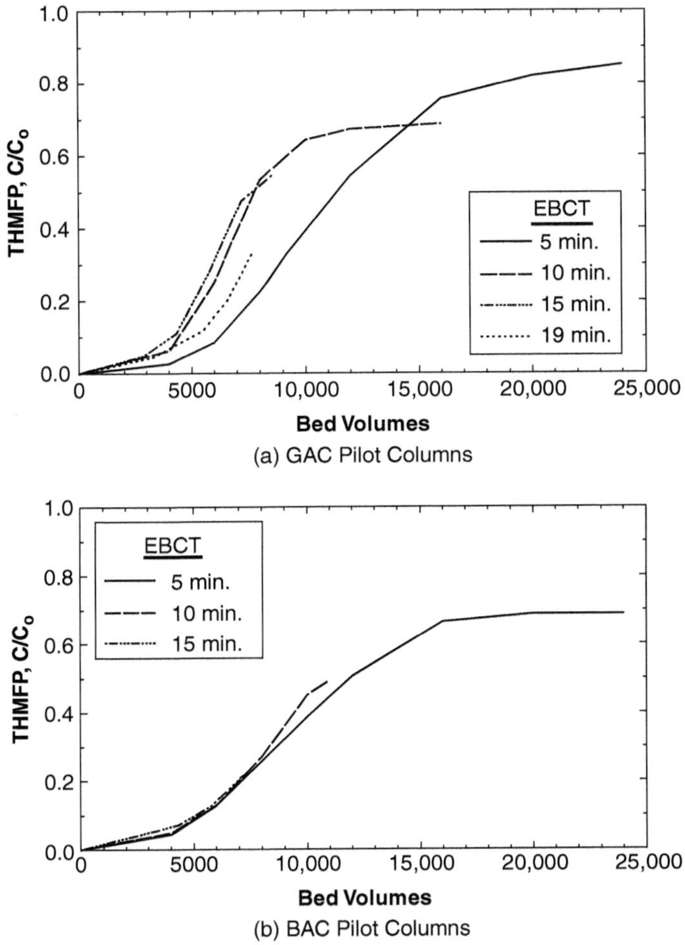

FIGURE 14-28 Distribution system THMFP breakthrough curves: (*a*) GAC pilot columns; (*b*) BAC pilot columns. (*Source: Snoeyink and Knappe, 1994.*)

addition, the specific compound responsible for a given taste and/or odor many not be identified. Thus there are two methods of quantification: human sensory analysis, for example, *threshold odor number* (TON) and flavor profile analysis or direct measurement of the compound, for example, MIB and gesomin.

GAC filter adsorber systems are reported to remove odor effectively from source water. Bed life can range from 1 to 5 years (Graese et al., 1987b) but can be significantly shorter in some situations (Gillogly et al., 1999). The bed life depends on the type, concentration, and frequency of taste and odor compounds; the presence of organics that compete for adsorption sites; and the target effluent concentration, that is, the acceptable level in the treated water. Case-history information is difficult to apply at different utilities because the compounds responsible for the taste and odor are often not known, and the acceptable level of taste and odor varies from community to community. Using pilot-plant data, Gillogly and colleagues (1999) showed that the residual MIB adsorption capacity was a function of EBCT and GAC service time. Also, the MIB-removal percentage was not a function of the influent MIB concentration.

GAC performance at Regina, Saskatchewan, is an interesting example of high-intensity tastes and odors (Gammie and Giesbrecht, 1986; Snoeyink and Knappe, 1994). The water treatment plant has postfilter adsorbers with a bed depth of 3 m (10 ft), which yields an EBCT of 15 minutes. The influent water had a DOC concentration of 2 to 3 mg/L and typical TON values of 5 to 15 with spikes of 40 to 60. The effluent goal was a TON value of 1 with the GAC reactivated once per year. Figure 14-29*a* depicts TON breakthrough

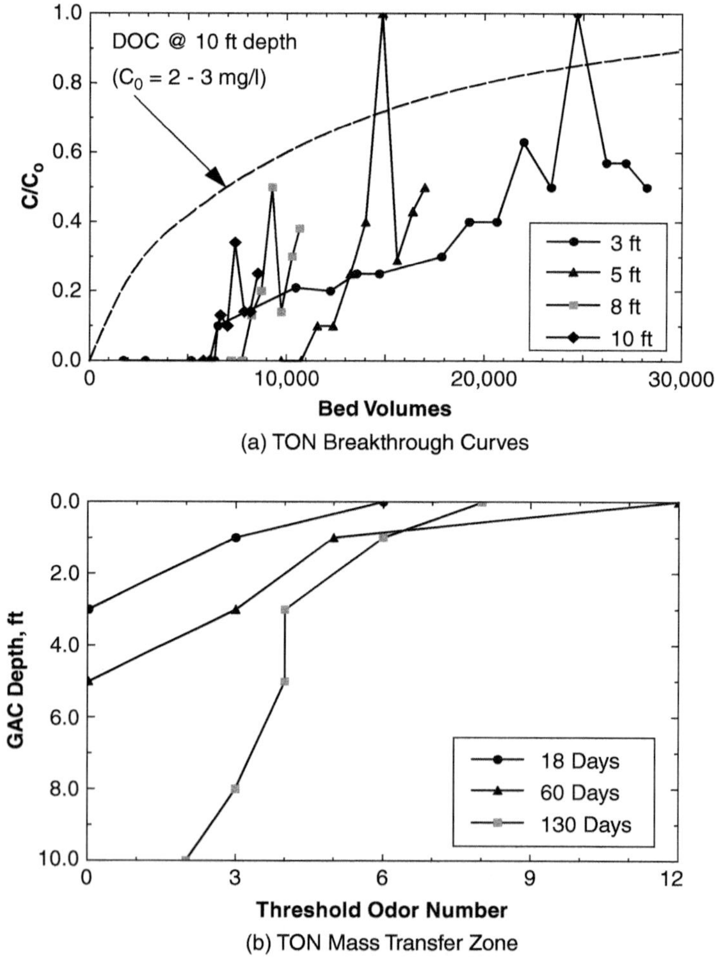

FIGURE 14-29 TON adsorption at Regina: (*a*) TON breakthrough curves; (*b*) TON mass transfer zone. (*Source: Snoeyink and Knappe, 1994.*)

curves at different depths within the GAC adsorber on a throughput basis (the number of BVs treated). The figure shows that the optimal filter depth was 1.5 m (5 ft), and increasing the bed depth to 8 and 10 ft (2.4–3.0 m) reduced the BVs treated before breakthrough, which is indicative of NOM preloading (which reduces the accessible adsorption sites for taste- and odor-causing substances). Decreasing the bed depth to 0.91 m (3 ft) also yielded earlier breakthrough. Figure 14-29*b* shows the progression of the TON mass transfer zone through the GAC bed. The TON value of 3 was reached after 18 days of operation at a GAC depth of 0.3 m (1 ft) (17,000 BVs), after 60 days at a bed depth of 0.91 m (3 ft) (19,000 BVs), and after 130 days at a bed depth of 2.4 m (8 ft) (16,000 BVs). After 130 days of operation (13,000 BVs), the entire GAC bed [3.0 m (10 ft)] had become ineffective for reaching the effluent TON goal.

Some useful observations have been made using data from full-scale GAC systems. Background organic matter, for example, usually breaks through much earlier than the taste- and odor-causing compounds (Love et al., 1973; Robeck, 1975), as was observed in Fig. 14-29*a*. This is also supported by bench-scale work (see Fig. 14-26).

In addition to adsorption, biodegradation has been shown to be responsible for the removal of MIB and geosmin (Silvey and Roach, 1964; Namkung and Rittmann, 1987; Meyer et al., 2005, Summers et al., 2006; Chae et al., 2006). However, Meyer and colleagues (2005) showed that it takes several months for the microorganisms to acclimate to the MIB at environmental concentrations (<500 ng/L). Chae abd colleagues (2006) showed that the biodegradation capacity in filters with inert media is reduced after 2 months without MIB and geosmin in the influent. One hypothesis for an MIB-and geosmin-removal mechanism is that adsorption capacity is needed to prevent these molecules from passing through the GAC adsorber. After the initial accumulation on the GAC, a biofilm develops that is capable of degrading these compounds, and they desorb and diffuse to the biofilm, where they are biologically oxidized. The ability of GAC to remove odor decreases over time because adsorption sites are gradually taken up by NOM and thus are not available for MIB and geosmin. GAC replacement may be required when adsorption capacity has been depleted by these competitors.

Synthetic Organic Compounds. Synthetic organic compounds (SOCs) comprise a range of potentially adsorbable compounds that includes (1) volatile organic chemicals (VOCs), (2) pesticides, and (3) pharmaceutical and personal care products (PPCPs). Many VOCs are associated with solvent contamination of groundwaters and are regulated at about 5 μg/L. Pesticides appear in surface waters affected by runoff from agricultural land and domestic and commercial lawns. There is a wide range of regulated levels. PPCPs are found in surface waters affected by wastewater discharges, and many of them are not regulated. Pesticides and PPCPs often occur at very low concentrations (low μg/L to sub-μg/L and low ppb to ppt levels). In this chapter, these compounds at low concentrations are referred to as *micropollutants*.

VOCs include compounds such as tetrachloroethylene and trichloroethylene that are adsorbed relatively strongly and 1,1,1-trichloroethane, 1,2-dichloroethane, and chloroform that are adsorbed relatively weakly. The isotherm data in Table 14-2 show the relative adsorbability of these compounds. GAC adsorbers can be used to remove VOCs directly from contaminated water (Hess, 1981; Love and Eilers, 1982; Snoeyink, 1983) or from the off-gases from air-stripping towers (Crittenden et al., 1988). The best type of system will depend on the type of VOC to be removed and air emission standards that are in effect.

VOCs are found commonly in contaminated groundwaters, and although this type of water generally is low in TOC (Solarik et al., 1995a; Zimmer et al., 1989), large competitive effects are still noted. For example, Zimmer and colleagues (1989) studied the adsorption of three VOCs, tetrachloroethylene, trichloroethylene, and 1,1,1-trichloroethane, from distilled water and from groundwater in field installations. The full-scale field installations had

TABLE 14-8 GAC Bed Life with and Without Competition from Natural Organic Matter

Compound	Single-solute isotherm constants		Bed volumes to saturation, without competition,	Observed capacity of full-scale adsorbers in practice[a]		
	K	$1/n$	$C_0 = 50$ μg/L[b]	K'	$1/n'$	Bed volumes to breakthrough, $C_0 = \sim 50$ μg/L, $C_{breakthrough} = \sim 5$ μg/L
Tetrachloroethene	219	0.42	620,000	20.4	0.46	51,000
Trichloroethene	78	0.46	197,000	27.4	0.61	44,000
1,1,1-Trichloroethane	23	0.60	38,000	62	1.4	9400

[a]An adsorber isotherm of the form $q = K'C_0^{1/n'}$, where C_0 is the adsorber influent concentration, K' and $1/n'$ are constants, and q is the amount adsorbed in mg/g at the time of adsorber breakthrough, was used to describe the data. These data are based on the performance of several full-scale plants in Germany (Zimmer et al., 1989).

[‡]Units on K and K' are $(mg/g)(L/mg)^{1/n}$. This number is calculated using the assumption that all influent compound is adsorbed until saturation is reached. Thus 52,500 BVs can be processed in 1 year if the EBCT is 10 minutes.

EBCTs ranging from 12 to 15 minutes. The groundwater contained 0.4 to 2 mg/L of DOC and was applied at a rate of 10 to 15 m/h (4–6 gpm/ft^2). As the data in Table 14-8 show, the amount of water containing 50 μg/L of tetrachloroethylene that could be processed in the field was about 8 percent of the distilled water value. The capacity for trichloroethylene and 1,1,1-trichloroethane was reduced to 22 and 24 percent, respectively, of the distilled water value. Even low levels of NOM "fouled," or blocked, significant portions of the GAC surface.

Hess (1981) used pilot-plant data (Love and Eilers, 1982) to predict the effect of influent concentration and type of VOC on GAC life (Fig. 14-30). The figure shows that GAC adsorber life decreases as influent concentration increases. Each water that is treated will

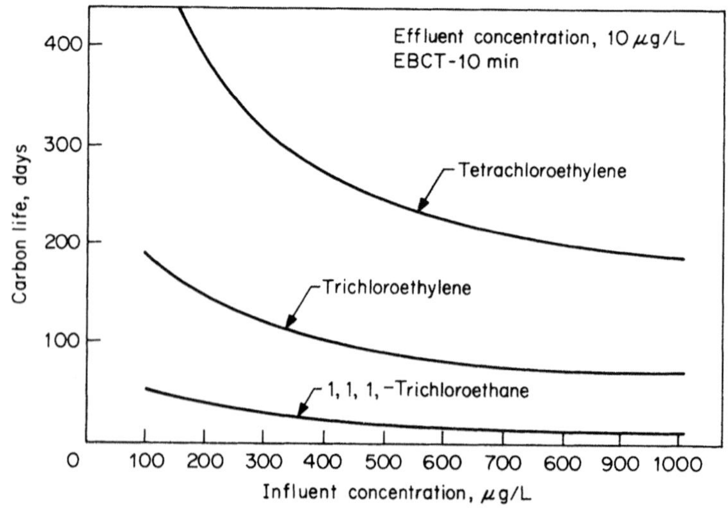

FIGURE 14-30 Effect of contaminant type and influent concentration on carbon life. (*Note:* 100 days = 14,400 BVs if EBCT = 10 minutes). (*Source: Hess, 1981.*)

have different background organic matter, and thus Fig. 14-30 should be taken only as providing estimates of performance. Other data for Mississippi River water at Jefferson Parish, LA, at low VOC concentrations show that a bed life of approximately 200 days (30,000 BVs) can be achieved for trichloroethylene at an influent concentration of 14 μg/L and an effluent concentration of 7 μg/L. Also, 400 days of life (60,000 BVs) can be achieved for tetrachloroethylene for an influent concentration of 22 μg/L and an effluent concentration of 11 μg/L. In each case, the EBCT was 10 minutes (Qi et al., 1992; Snoeyink and Knappe, 1994).

Arbuckle (1980) observed that GAC at the end of an adsorber train adsorbed less SOCs per unit mass than GAC at the front of the adsorber train. Similar effects were observed for VOC adsorption from groundwater (Baldauf and Zimmer, 1986; Zimmer et al., 1989) and for adsorption of chlorinated organic compounds from Rhine River water (Summers et al., 1989). The explanation given for this phenomenon was that large molecules of background organics adsorbed first in the lower reaches of the adsorber and either blocked pores or occupied adsorption sites so that adsorption of micropollutants was decreased. The effect was called the *premature exhaustion* or *preloading effect*, and it results in higher activated carbon usage rates as EBCT is increased beyond a certain value.

Crittenden and colleagues (1988) developed mass transfer models to predict the removal of VOCs in air stripper off-gas by GAC and investigated the regeneration of this GAC with steam. They found that adsorption efficiency decreased as relative humidity increased and that heating of the gas stream to reduce the relative humidity to 40 to 50 percent was beneficial. Air stripping followed by GAC treatment to purify the off-gas was a good alternative to treatment of the water by GAC alone for several VOC treatment applications, even though steam regeneration of spent gas-phase GAC was not feasible. This advantage results because the natural organics that interfere with adsorption from the aqueous phase are not present in the stripper off-gas.

Symons and colleagues (1981) summarized THM adsorption data. Consistent with the Freundlich K values listed in Table 14-2, the brominated THMs are adsorbed much better than chloroform. Graese and colleagues (1987a) presented data from Cincinnati and Miami for an 80 percent chloroform removal criterion that showed that about 5000 to 6000 BVs of water could be processed before breakthrough. If the EBCT was less than 8 to 10 minutes, however, fewer BVs could be processed because of the short length of the MTZ. The performance will vary as the breakthrough criterion and the composition of the water change. In general, the data indicate that it is more efficient to adsorb DBP precursors than to adsorb the DBPs themselves.

Pesticides are common SOCs that require removal, and GAC has long been demonstrated to effective (Robeck et al., 1965). GAC beds were evaluated for the removal of selected pesticides spiked into river water (Robeck et al., 1965). One activated carbon column was exhausted for removal of background organic matter as measured by carbon chloroform extract and chemical oxygen demand (COD). Pesticide-spiked sand-filtered river water was applied to the exhausted activated carbon column, and the effluent from it was applied to a second, fresh column. With concentrations of dieldrin as high as 4.3 μg/L, the effluent from the first column was 0.3 μg/L. Further reduction in the second column reached concentrations as low as 0.05 μg/L and often below the detection limit of 0.01 μg/L.

At Jefferson Parish, La, 18 chlorinated hydrocarbon insecticides at a combined concentration of 18 to 88 ng/L were removed to less than about 5 ng/L for 1 year in an adsorber with an EBCT of 20 minutes (Koffskey and Brodtmann, 1981). Alachlor at concentrations of 13 to 593 ng/L was reduced to the limits of detectability during the same period. Atrazine was present at 30 to 560 ng/L and was removed essentially 100 percent by the postfilter adsorber (24 minute EBCT), but concentrations of 10 to 20 ng/L often were found in the effluent of the filter adsorber (14-minute EBCT). Other analyses at Jefferson Parish showed that phthalates, *n*-alkanes, and substituted benzenes at the nanogram per liter level were not removed by GAC.

TABLE 14-9 Summary of Field-Scale Breakthrough Behavior for a Range of Micropollutants

Compound	C_0(ng/L)	~$BV_{10\%}$	Study
MTBE	100,000	3000	Rossner and Knappe, 2008 DOC_0 = ~2.5 mg/L, EBCT = 9.6 minutes
Atrazine	2000	25,000	Bauldauf and Henkel, 1988 DOC_0 = 2.3 mg/L, EBCT = 12 minutes
Atrazine	4000	21,000	Knappe et al., 1997 DOC_0 = 2.0 mg/L, EBCT = 14 minutes
Bezafibrate	260	38,000	Ternes et al., 2002
Carbamazepine	1030	47,000	DOC_0 = 2.4 mg/L, EBCT = 5 minutes
Clofibric acid	1840	10,000	
Diclofenac	40	42,000	
Carbamazepine	550	12,000	Yu et al., 2009
Naproxen	500	7000	DOC_0 = 4.4 mg/L, EBCT = 2.6 minutes
Nonylphenol	500	15,000	
Atrazine	64	22,000	Chowdhury et al., 2010
DEET	19	22,000	DOC_0 = 2.1 mg/L, EBCT = 7.9 minutes
Deethylatrazine	20	40,000	
Deisopropylatrazine	136	38,000	
Flutolanil	20	>70,000	
Prometon	4	20,000	
Simazine	521	25,000	
Amidotrizoic acid	190	4000	Haist-Gulde and Baldauf, 2006
Bezafibrate	100	110,000	DOC_0 = 1.0 mg/L, EBCT = 12 minutes
Carbamazepine	100	94,000	
Gemfibrozil	100	52,000	
Ibuprofen	100	25,000	
Iopamidol	64	9000	

Table 14-9 provides a summary of field-scale (full or pilot) breakthrough behavior for a range of micropollutants at influent concentrations below 4 μg/L and methyl-*tert*-butyl-ether (MTBE) at 100 μg/L. BVs at 10 percent breakthrough ($BV_{10\%}$) were used to characterize the breakthrough behavior.

Baldauf and Henkel (1988) evaluated atrazine breakthrough in a pilot test using a single-stage adsorber with a total depth of 2.5 m. Samples were taken at different depths corresponding to EBCTs of between 3 and 15 minutes. The influent to the GAC column was a groundwater with a DOC concentration of 2.3 mg/L and an average atrazine concentration of 2.2 μg/L. Figure 14-31*a* shows the breakthrough curves as a function of the volume of water treated, and Fig. 14-31*b* shows the same data as a function of the number of BVs treated. Figure 14-31*a* shows that atrazine reached the 0.1 μg/L maximum contaminant level (MCL) of the European Union (EU) very quickly at EBCTs of 3 and 6 minutes, whereas the longer EBCTs showed a much more gradual breakthrough. Figure 14-31*b* shows similar breakthrough behavior at each EBCT after the data are normalized on the basis of number of BVs treated, but the early breakthrough for EBCTs of 3 and 6 minute is still apparent. The CUR was in the range of 65 to 82 g/m^3 for EBCTs of 3 and 6 minutes, and it decreased to approximately 21 g/m^3 for EBCTs of 9 minutes and longer (Snoeyink and Knappe, 1994).

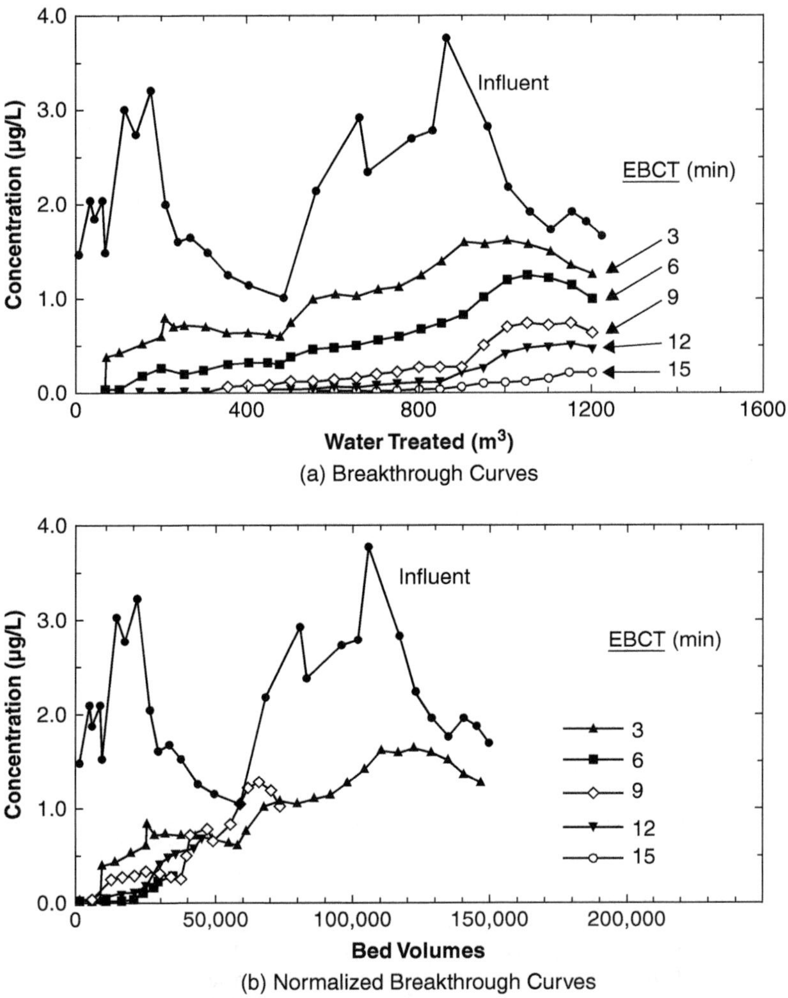

FIGURE 14-31 Atrazine breakthrough curves from groundwater: (*a*) breakthrough curves; (*b*) normalized breakthrough curves. (*Source: Baldauf, G., and Henkel, M. (1988), Entfernung von Pestiziden bei der Trinkwasser-aufbereitung, in 20. Bericht der Arbeitsgemeinschaft Wasserwerke Bodensee-Rhein, Teil 7, 127.*)

At the shorter EBCTs of 3 and 6 minutes, it appears that the MTZ was not fully contained within the bed, leading to the immediate appearance of atrazine in the effluent.

Figure 14-32 shows normalized atrazine breakthrough curves obtained from pilot studies that investigated different EBCTs and NOM preloading times, that is, time of operation before the application of atrazine (Knappe, 1996). Influent atrazine concentrations for the three pilot tests were about 4 µg/L. The data obtained with a GAC that was preloaded for 5 months demonstrated a great decrease in atrazine removal efficiency as the EBCT decreased; whereas an EBCT of 8.6 minutes resulted in a performance similar to that of virgin GAC, an EBCT of 2.3 minutes led to a removal efficiency that was nearly as low as that observed for a GAC that

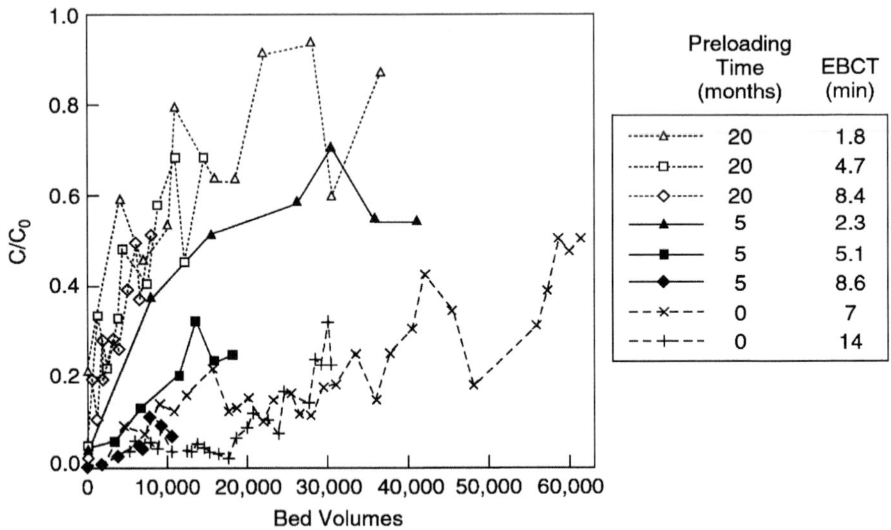

FIGURE 14-32 Effect of EBCT on atrazine removal efficiency at different preloading times. (*Source: Knappe 1996.*)

was preloaded for 1 year and 8 months. In contrast, the GAC that was preloaded for 1 year and 8 months showed a uniformly low atrazine-removal efficiency that was not a function of EBCT. The reduction in atrazine capacity owing to NOM preloading primarily governed the observed breakthrough behavior.

Polynuclear aromatic hydrocarbons are removed effectively by GAC. Tap water spiked with 50 μg/L of naphthalene was passed through a GAC column at Cincinnati, OH (Robeck, 1975). After 7 months of operation at 5 m/h (2 gpm/ft^2), the nonvolatile total organic carbon (NVTOC) front for 50 percent removal had penetrated the first 51 cm (1.7 ft) of the bed, whereas the 50 percent removal point for naphthalene was only about 5 cm (0.2 ft) down the column. River bank filtration followed by activated carbon treatment reduced the concentration of polynuclear aromatic hydrocarbons in water by about 99 percent, whereas rapid sand filtration followed by ozonation or chlorination was not effective, probably because of the absence of GAC (Andelman, 1973).

Figure 14-33 shows the breakthrough of several pesticides and the PPCP N,N-Diethyl-meta-toluamide (DEET) with the influent conditions listed in Table 14-9 (Chowdhury et al., 2010). Four of the five compounds shown in Fig. 14-33 yielded 10 percent breakthrough at BVs ($BV_{10\%}$) from 20,000 to 25,000, and 50 percent breakthrough occurred in the range of 40,000 to 50,000 BVs, whereas that for DOC was less than 10,000 BVs. The herbicide flutolanil was completely removed for more than 1 year.

Table 14-9 summarizes the influent concentration and $BV_{10\%}$ data from this study as well as others, in which influent concentrations were below 4 μg/L. Breakthrough behavior ranged from 7,000 BVs for naproxen in the high-TOC water to more than 70,000 BVs for flutolanil. Three of the studies evaluated atrazine adsorption and yielded similar performance, with $BV_{10\%}$ of about 22,000. The background DOM was similar in all three cases: 2.0 to 2.3 mg/L TOC. The influent atrazine concentrations ranged from 4000 to 64 ng/L, but as discussed in an earlier section, influent concentrations in the sub-μg/L range do not affect the fractional breakthrough behavior when DOM is present. The effect of different influent DOM concentrations is evident in three studies of carbamazepine adsorption. The $BV_{10\%}$ values decreased from 94,000 to 12,000 as the background DOM concentration increased from 1.0 to 4.4 mg/L of TOC.

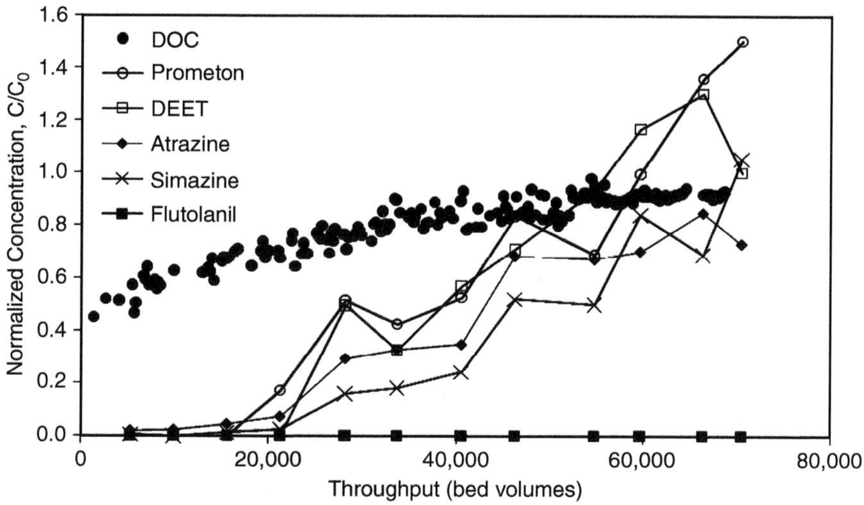

FIGURE 14-33 Breakthrough behavior of pesticides and the personal-care product DEET relative to DOC. (*Source: Adapted with permission from Corwin, 2010.*)

Performance of Filter Adsorbers. Filter adsorbers, which have resulted from the retrofitting of inert media (e.g., sand) beds, have proven to be effective for periodic taste and odor removal. However, their use for removing weakly adsorbed compounds, including THMs and some VOCs, is limited because of the short EBCT typically found in such systems (Graese et al., 1987a). The typical EBCT of operating sand replacement systems is about 9 minutes at operating flow rates but much less at design flow rates. Control of weakly adsorbing organic compounds with such short EBCTs would require replacement of GAC after only a few weeks or months of operation. Replacement at a high frequency would be operationally cumbersome and costly. In all cases, a thorough economic analysis is required to determine whether the costs associated with a higher CUR in a filter adsorber would be justified by the reduced capital costs of the filter adsorber.

Chowdhury and colleagues (1996), Hong and Summers (1994), Summers and colleagues (1994, 1995, 1997), and Wulfeck and Summers (1994) have shown that filter adsorbers operated in parallel may be a cost-effective approach for controlling DBP formation when the influent TOC concentrations are not high. They have shown that the CUR in several waters was not a function of EBCT in the range of 10 to 20 minutes. Based on the relationship shown in Fig. 14-27, a water with an influent TOC concentration of 2.0 mg/L and an EBCT of 10 minutes would lead to an effluent TOC concentration of 1.0 mg/L after 60 days of operation. If the filter adsorber effluents are blended, the replacement interval would be about 4 to 5 months. Regular backwashing of filter adsorbers has been shown to have little impact on the run time to 50 percent TOC breakthrough (Hong and Summers, 1994; Summers et al., 1997).

A review of the available data on the efficiency of GAC for turbidity reduction compared with sand or coal showed that GAC is as good or better (Chowdhury et al., 2010; Graese et al., 1987a). However, when turbidity loadings were high, a more rapid rate of head loss buildup was experienced because of the larger uniformity coefficient of GAC. When the turbidity loading was low, the difference in uniformity coefficient seemed to make little difference in performance.

A sand layer in the filter below the GAC is sometimes retained, often owing to state regulations on filter design. While the efficiency of turbidity removal by GAC indicates that this is not always necessary for good performance, it may be useful if large-diameter GAC is used or if low levels of turbidity are desired. A sand layer below the GAC is a nuisance because the sand mixes with the GAC when GAC is removed. This mixed product interferes with reactivation of the GAC. After several GAC replacement cycles, makeup sand may be required to maintain the required sand depth. Gravel under the GAC can cause similar problems, but gravel is necessary to prevent the GAC from entering many types of underdrains.

Nozzle underdrains can be used that do not require either sand or gravel, and they can be used to provide both air and water for backwashing. These appear to have significant advantages over the types that require gravel. However, a pressure-relief device on the washwater influent line should be provided to prevent high-pressure damage to underdrains when and if the nozzles clog. Monitoring of the plenum pressure during backwashing can give advance warning of nozzle clogging problems.

The backwash system must be designed properly to prevent accumulation of mudballs in the GAC. Mudballs accumulate if the GAC is not cleaned properly. GAC is more difficult to clean because it has a density lower than sand and anthracite. Provision of good surface scour, or air scour, is important for good turbidity control.

Reactions of Inorganic Compounds with Activated Carbon

Activated carbon in water treatment may inadvertently contact oxidants such as oxygen, aqueous chlorine, chlorine dioxide, and permanganate and react with them. Virgin activated carbon has been shown by Prober and colleagues (1975) to react with 10 to 40 mg of aqueous O_2 per gram of carbon over a time span of 1700 hours, and Chudyk and Snoeyink (1981) found that 3 to 4 mg of O_2 per gram of carbon reacted over a time span of 130 hours. Some of this oxygen is converted to surface oxides (Prober et al., 1975).

Free Chlorine-Activated Carbon Reactions. The well-known reactions of HOCl and OCl^- with activated carbon are as follows:

$$HOCl + C^* \rightarrow C^*O + H^+ + Cl^- \tag{14-12}$$

$$OCl^- + C^* \rightarrow C^*O + Cl^- \tag{14-13}$$

where C* and C*O represent the carbon surface and a surface oxide, respectively. These reactions proceed rapidly, with the one in Eq. 14-12 ($pH < 7.5$) being faster than the one in Eq. 14-13 ($pH > 7.5$) (Suidan et al., 1976, 1977a, 1977b). Reactions of free chlorine with activated carbon will result in the production of organic by-products. The TOX on the surface increases as the extent of reaction increases, and some of these compounds may be found in the column effluent if the reaction proceeds for a long time (Dielmann, 1981).

Combined Chlorine-Activated Carbon Reactions. Bauer and Snoeyink (1973) hypothesized that the following reactions were taking place between monochloramine (NH_2Cl) and activated carbon:

$$NH_2Cl + H_2O + C^* \rightarrow NH_3(aq) + C^*O + H^+ + Cl^- \tag{14-14}$$

$$2NH_2Cl + C^*O \rightarrow N_2(g) + 2H^+ + 2Cl^- + H_2O + C^* \tag{14-15}$$

Initially, all the NH_2Cl was converted to NH_3 and Cl^- in accordance with Eq. 14-14, but after a period of reaction, some of the NH_2Cl was converted to $N_2(g)$ and HCl in accordance with Eq. 14-15. The rate of reaction was much slower than the reactions of either free chlorine or dichloramine with activated carbon (Kim and Snoeyink, 1980). Additional studies of this reaction were made by Komorita and Snoeyink (1985), who showed that the rate of reaction was high initially and then reached a plateau value. The higher initial rate is useful in design of some GAC dechlorination systems.

Dichloramine ($NHCl_2$) reacts rapidly with activated carbon according to the following reaction (Bauer and Snoeyink, 1973; Kim et al., 1978):

$$2NHCl_2 + H_2O + C^* \rightarrow N_2(g) + C^*O + 4H^+ + 4Cl^- \tag{14-16}$$

When excess ammonia was present, Kim and colleagues (1978) found evidence of the parallel reaction:

$$NH_4^+ + 3NHCl_2 \rightarrow 2N_2(g) + 7H^+ + 6Cl^- \tag{14-17}$$

Bromate, Chlorine Dioxide, Chlorite, Chlorate, and Perchlorate Reactions with Activated Carbon. Other halogen oxides react with activated carbon. For example, BrO_3^- is converted to Br^- in ultrapure water on fresh activated carbon surfaces (Miller et al., 1995; Kirisits et al., 2000); however, NOM and anions such as chloride, bromide, sulfate, and nitrate can block and/or occupy reducing sites on activated carbon surfaces such that the abiotic bromate reduction capacity of GAC adsorbers is quickly exhausted (Kirisits et al., 2000). ClO_2 reacts rapidly with activated carbon, but the nature of the reaction changes as pH changes. At pH 3.5, Cl^-, ClO_2^-, and ClO_3^- are found in the effluent of a column receiving only ClO_2, but Cl^- is the predominant product. At pH 7.9, the same end products are formed, but ClO_2^- is now the predominant species (Chen et al., 1982). When ClO_2^- solutions at pH 7 were applied to virgin GAC, ClO_2^- readily reacted, presumably in accordance with the reaction (Voudrias et al., 1983)

$$ClO_2^- + C^* \rightarrow C^*O_2 + Cl^- \tag{14-18}$$

The reaction capacity of the fresh carbon was saturated after 80 to 90 mg of ClO_2^- reacted per gram of GAC. The chlorate ion was not reduced by activated carbon but was adsorbed to a slight extent (~0.03 mg ClO_3^- per gram), presumably by an IX mechanism (Dielmann, 1981). Similarly, abiotic perchlorate removal by activated carbon appears to occur via IX only, and calculated ion exchange capacities from fixed-bed experiments are low (~0.17 mg ClO_4^- per gram) (Brown et al., 2002).

Removal of Other Inorganic Ions. Huang (1978) reviewed the removal of inorganic ions by activated carbon. A number of ions can be removed from water by GAC, but the capacity for most substances is quite low. The gold-cyanide complex, for example, is adsorbed on GAC in a widely used method for recovering gold in the mining industry. Huang (1978) reviewed data that show some removal of cadmium(II) at higher pH values and that removal can be increased slightly by complexing before adsorption with chelating agents. The removal of chromium involves adsorption of Cr(III) or Cr(VI), and under some conditions, Cr(VI) is chemically reduced to Cr(III) by the activated carbon. Mercury adsorption is best at low pH. From 0.1 to 0.5 mmol Cu per gram at equilibrium concentrations of 8 mmol/L also have been observed to adsorb. Activated carbon also will catalyze the oxidation of Fe(II) to Fe(III).

GAC PERFORMANCE ESTIMATION

Isotherm Determination and Application

Useful information can be obtained from the adsorption isotherm. Isotherms can be used to show the variation in adsorbability of different types of organic compounds (see Table 14-2), the differences between activated carbons, and the magnitude of competitive effects (see Example 14-1). Because the adsorption capacity of a GAC is a major determinant of its adsorption performance in columns, the isotherm can be used (1) to compare candidate GACs, (2) as a quality-control measure for purchased GACs (some variation from lot to lot of a given manufacturer's GAC is expected), (3) as an input function to mathematical models that can be used to predict performance (discussed later), and (4) to evaluate the residual adsorption capacity of GAC in a contactor that has been in use and thus is partially exhausted. Important limitations to the use of isotherms must be recognized if they are to be used properly, however. The purpose of this section is to demonstrate how isotherms can be used to estimate performance and to note several factors of importance in isotherm determination.

Types of Adsorbates. The determination procedures must take into account the adsorbate characteristics. The simplest procedure can be followed if a single nonvolatile adsorbate is present in an aqueous solution with no background organics that might compete for adsorption sites. This isotherm is not a function of initial concentration, and no precautions need be taken to avoid volatilization. The test is conducted by agitating known volumes of the solution with accurately weighed portions of GAC until equilibrium is achieved. The time required for equilibrium can be estimated using the information in Randtke and Snoeyink (1983) (see discussion in an earlier section of this chapter) and can be determined by measuring solution concentration versus time.

Other adsorbate-related factors must be controlled carefully during isotherm determination. Biological activity in isotherm bottles must be controlled either if the NOM is biodegradable, such as after ozonation of a natural water, or if the targeted micropollutant is biodegradable. Biological growth can be minimized if the isotherm is determined at low temperature, for example, less than 5°C. If the adsorbate is volatile, equilibrium must take place in a headspace-free vessel, and mixing of the system must take place by continual inversion of the test bottle. Control of pH is important for compounds whose adsorption is a function of pH, and other salts in solution can be important for ionic substances or for substances that compete with organic ions for adsorption sites. The reader is referred to Randtke and Snoeyink (1983) for a more detailed discussion of these factors.

When equilibrium is reached, the data are analyzed using the mass balance

$$q = \frac{(C_0 - C)V}{m} \tag{14-19}$$

where q is the amount adsorbed per unit mass of carbon at equilibrium (mol/g or mass/g), C_0 is the initial aqueous-phase concentration (mol/L or mass/L), C is the equilibrium aqueous-phase concentration (mol/L or mass/L), V is the solution volume (L), and m is the mass of carbon (g).

The data can be plotted to determine the Freundlich parameters or other isotherm model parameters as appropriate.

Example 14-3 Determination of Freundlich Isotherm Parameters

The following data were obtained from a single-solute isotherm experiment evaluating the adsorption of methyl *tert*-butyl ether (MTBE) on an activated carbon:

Mass of activated carbon in isotherm bottle (mg)	0	12.6	25.4	63.4	126.7	190.2	253.5	380.1
Aqueous-phase MTBE concentration at equilibrium (μg/L)	864	498	329	131	48.0	25.1	15.3	8.20

Each isotherm bottle was filled headspace-free with 254 mL of water and received the same mass of MTBE. Prepare an isotherm plot and determine the Freundlich isotherm parameters K and $1/n$.

Solution

Calculate solid-phase MTBE concentrations q from the data set given using Eq. 14-19. For the bottle containing 12.6 mg of activated carbon:

$$q = (C_0 - C) \times V/m = (864 - 498)\ \mu\text{g/L} \times 0.254\ \text{L}/12.6\ \text{mg} = 7.38\ \mu\text{g/mg} = 7.38\ \text{mg/g}$$

that is, 7.38 mg of MTBE adsorbs on 1 g of activated carbon at an equilibrium aqueous-phase MTBE concentration of 498 μg/L.

Repeating the calculation for the remaining isotherm bottles yields:

Mass of activated carbon in isotherm bottle (mg)	0	12.6	25.4	63.4	126.7	190.2	253.5	380.1
Aqueous-phase MTBE concentration at equilibrium (μg/L)	864	498	329	131	48.0	25.1	15.3	8.20
Solid-phase MTBE concentration (mg/g)	—	7.38	5.35	2.94	1.64	1.12	0.850	0.572

Plotting the solid-phase MTBE concentrations q as a function of the aqueous-phase MTBE concentrations C yields Fig. 14-34 (note logarithmic scales).

Linear regression of the log-transformed data yields a slope of 0.612 ($1/n$ value) and an intercept of –0.807 (log K value). Thus the K value is $10^{-0.807} = 0.156\ (\text{mg/g})(\text{L}/\mu\text{g})^{1/n}$. The K value suggests that 1 g of the tested activated carbon can adsorb 0.156 mg of MTBE at an equilibrium aqueous-phase MTBE concentration of 1 μg/L. The Freundlich isotherm equation now can be used to estimate solid-phase MTBE concentrations at targeted aqueous-phase MTBE concentrations. For example, if an MTBE concentration of 10 μg/L is targeted, then the tested activated carbon can adsorb

$$q = 0.156\ (\text{mg/g})(\text{L}/\mu\text{g})^{1/n} \times (10\ \mu\text{g/L})^{0.612}$$

$$= 0.638\ \text{mg/g or } 0.638\ \text{mg MTBE per gram of activated carbon}$$ ▲

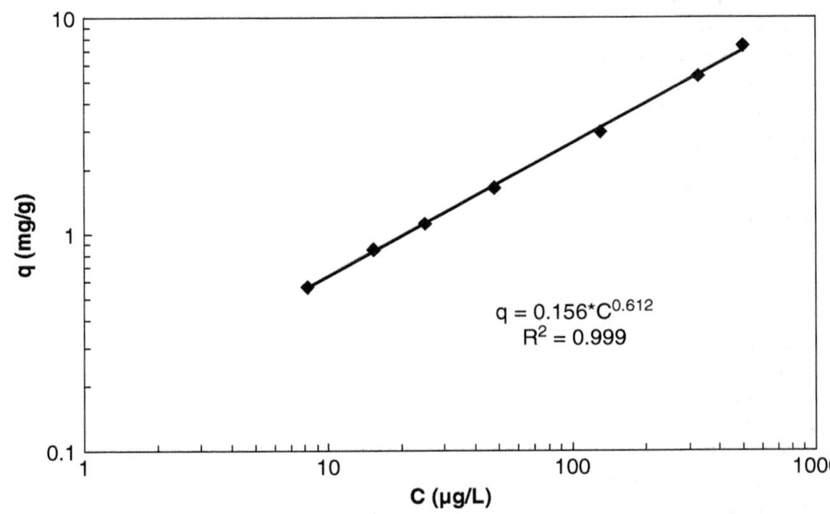

FIGURE 14-34 MTBE adsorption isotherm.

Adsorption from a mixture of adsorbable organic compounds (e.g., NOM) is more complicated because of the effect of the different adsorbabilities of compounds in the mixture. Figure 14-8 shows typical NOM adsorption isotherms. The small fraction of strongly adsorbable compounds is often overlooked when determining an isotherm because of the analytic difficulty of obtaining good data close to C_0 (i.e., for very small doses of activated carbon). The strongly adsorbable fraction is important because it is responsible, at least in part, for the slow approach of effluent concentration to influent concentration in column operation. Similarly, the small, but important, nonadsorbable fraction is often not detected if a sufficiently large dose of carbon is not added to the solution. Initial concentration affects the adsorption of a NOM (see Fig. 14-8). The difference in the isotherms is caused primarily by the different mass concentrations of strongly adsorbable compounds in each of the solutions. The same type of difference is expected if the amount of a single compound that is adsorbed in the presence of different concentrations of background organic matter is measured. The isotherm of the single compound will be a function of the initial concentration of background organic matter if competition occurs between them.

Activated Carbon. The major concerns related to activated carbon are (1) obtaining a representative sample for testing and (2) preparing the sample to obtain a rapid approach to equilibrium without destroying its representative nature. The adsorption capacity of some commercial activated carbons is a function of particle size because of the manufacturing process that is used (Randtke and Snoeyink, 1983); thus, selecting a sample for testing with the same particle size distribution as the bulk activated carbon is important. Activated carbon sizes may separate during shipment, so special equipment or techniques are required to obtain the representative sample. By using equipment such as a sample reducer and a sample splitter, representative samples from large quantities (bags) of GAC can be obtained. The sample reducer reduces sample sizes greater than 4.5 kg (10 lb) to representative samples that are smaller at a 16:1 ratio. Further reduction to a size that can be used for laboratory tests is possible using the sample splitter, or riffle; the particle size distribution and quality of the final sample then are representative of those of the original large quantity of GAC. Using such equipment to establish that a shipment of GAC meets specification is important. A simpler alternative that yields a good approximation of the characteristics of a batch is to sieve the sample and use only the particles in the dominant size range.

The time to reach adsorption equilibrium is an important function of particle size. The adsorption capacity of fresh GAC is not altered when it is crushed to a smaller particle size (Randtke and Snoeyink, 1983). The same is not true for used GAC, however. Crushing used GAC likely gives micropollutants access to pores that were blocked by preloaded NOM prior to crushing (Carter et al., 1992; Knappe et al., 1999). When determining the remaining adsorption capacity of GAC in operating GAC adsorbers, the used GAC therefore should not be crushed. As described earlier in this chapter, long equilibration times are needed to reach adsorption equilibrium when larger GAC grains are tested.

Avoiding significant biological activity should be possible if the sample is not seeded with microorganisms and the test time is 2 days or less. Prefiltration of the solution with a filter membrane having a 0.45-μm pore diameter will remove most bacteria from the sample. Postfiltration of the sample through filter paper having a 0.45-μm pore diameter will be necessary to remove the activated carbon before testing for the concentration of residual organics. Filter paper must be selected that does not interfere with the quantitative analysis of the test compounds either by removing a targeted pollutant or by contributing organic matter.

Isotherms can be used to obtain a rough estimate of activated carbon loading and bed life. These are useful in determining the applicability of GAC. Assuming that (1) all the GAC in an adsorber will reach equilibrium with the adsorber influent concentration and (2) the capacity obtained by extrapolating the isotherm data to the initial concentration is a good value, the bed life, in BVs of water, can be calculated. These assumptions are consistent with the assumption that the length of the MTZ L_{MTZ} is negligible. For example, if q_0 is the mass adsorbed (mg/g) when $C = C_0$, the bed life Y, the volume of water that can be treated per unit BV of carbon, can be calculated by

$$Y = \frac{(q_e)_0 \text{ (mg/g GAC)}}{(C_0 - C_1) \text{ (mg/ L)}} \times \rho_{GAC} \text{ (g/L)} \tag{14-20}$$

where C_0 is the influent concentration, C_1 is the effluent concentration that represents an average for the entire column run, and ρ_{GAC} is the apparent density of the GAC. C_1 is zero for a strongly adsorbed compound that has a sharp breakthrough curve and is the concentration of nonadsorbable compounds when such substances are present. An estimate of the activated carbon usage rate, the rate at which activated carbon is spent as defined in Eq. 14-10, is given by

$$\text{CUR (g/L)} = \frac{(C_0 - C_1) \text{ (mg/L)}}{(q_e)_0 \text{(mg/g)}} \tag{14-21}$$

Example 14-4 GAC Adsorber Bed Life and Carbon Usage Rate
Estimate the bed life and carbon usage rate for a GAC adsorber that is to treat water containing 1 mg/L of toluene from solution. The bed density of the carbon ρ_{GAC} is 500 g/L.

Solution
From Table 14-2, $K = 5.01$ (mg/g)(L/μg)$^{1/n}$ and $1/n = 0.429$ for toluene. Applying the Freundlich equation (Eq. 14-2) gives

$$\begin{aligned}(q_e)_0 &= KC^{1/n} \\ &= 5.01 \text{ (mg/g)(L/}\mu\text{g)}^{0.429}(1000\,\mu\text{g/L})^{0.429} \\ &= 97 \text{ mg/g}\end{aligned}$$

Applying Eq. 14-20, assuming that $C_1 = 0$, gives

$$\begin{aligned}Y &= \frac{(q_e)_0 \text{(mg/g)}}{(C_0 - C_1) \text{(mg/L)}} \times \rho_{GAC} \text{(g/L)} \\ &= \frac{97 \text{ mg/g}}{1 \text{ mg/L}} \times 500 \text{ g/L} \\ &= 48{,}500 \text{ L H}_2\text{O/L GAC} = \text{bed life}\end{aligned}$$

If the EBCT is 15 minutes, 4 BVs per hour or 96 BVs per day can be processed. The bed life in days then is 48,500 BVs/96 BVs per day = 505 days. Applying Eq. 14-21 to obtain CUR gives

$$\begin{aligned}\text{CUR} &= \frac{(C_0 - C_1) \text{ (mg/L)}}{(q_e)_0 \text{ (mg/g)}} \\ &= \frac{1 \text{ mg/L}}{97 \text{ mg/g}} \\ &= 0.0103 \text{ g GAC/L H}_2\text{O}\end{aligned}$$

Note: The bed life will be reduced and the CUR will be increased under practical conditions where other organics are in the water that compete for adsorption sites. ▲

Several limitations exist to the use of isotherm data to estimate the CUR. For example, NOM preloading decreases the adsorption capacity of GAC as GAC service time increases (see Fig. 14-15). Furthermore, it provides no indication of the effect of biological activity. Finally, when competition does occur, its impact in a batch test often is not the same as in a column because the molecules will separate in a column according to their strength of adsorption. The NOM that adsorbs in the upper region of a column will be dominated by the strongly adsorbing fraction of organic matter, especially in the intial portion of a column run, whereas the more weakly adsorbing fraction of NOM will selectively accumulate in the lower reaches of the column. In a batch test, all carbon particles are exposed equally to each fraction of the NOM. Competitive effects of NOM on micropollutants may be much greater in a column for this reason (Zimmer et al., 1989).

One approach to better estimate the adsorption capacity in the field based on isotherm values is to increase the distilled water carbon usage rates based on the adsorbability of the compound. Ford and colleagues (1989) proposed a natural water correction factor be applied to the distilled water CUR. From their data, the following relationship between natural water CURs ($\mathrm{CUR_{NW}}$) taken from full- and pilot-scale adsorbers and those from isotherms in distilled water ($\mathrm{CUR_{DW}}$) can be developed:

$$\mathrm{CUR_{NW}} = 0.7\,\mathrm{CUR_{DW}^{0.5}} \tag{14-22}$$

where the CUR is in pounds per 1000 gallons, and all values of $\mathrm{CUR_{DW}}$ were below 1 lb/1000 gal. While there was a large amount of scatter in the more than 75 data points evaluated, the square-root nature of the relationship shows that the more strongly adsorbing compounds, that is, low CURs, will be more affected by the presence of NOM. For example, a weakly adsorbing compound with a high $\mathrm{CUR_{DW}}$ of 0.1 lb/1000 gal will have a CUR in natural water of 0.22 lb/1000 gal, a 2.2-fold increase, whereas a strongly adsorbing compound with a low $\mathrm{CUR_{DW}}$ of 0.001 lb/1000 gal will have a CUR in natural waters of 0.022 lb/1000 gal, a 22-fold increase.

Bench-Scale Column Tests and Applications

Bench-scale column tests have been developed to obtain data that can be used to estimate the performance of large contactors and are helpful in screening GAC types. Crittenden and colleagues (1986, 1987, 1991) developed a procedure called the rapid small-scale column test (RSSCT) for designing bench-scale columns with ground GAC such that the data obtained could be scaled up to predict the breakthrough behavior of field-scale GAC columns. They used dimensionless parameters that characterize the advection, film mass transfer, intraparticle diffusion, and adsorption processes to develop the relationship between the breakthrough curves of large-scale columns (LCs) with GAC and small-scale columns (SCs) with crushed GAC. Based on the assumption that the internal diffusion coefficient was constant and not a function of GAC particle diameter, they found

$$\frac{\mathrm{EBCT_{SC}}}{\mathrm{EBCT_{LC}}} = \left(\frac{d_{\mathrm{SC}}}{d_{\mathrm{LC}}}\right)^2 = \frac{t_{\mathrm{SC}}}{t_{\mathrm{LC}}} \tag{14-23}$$

where d is the diameter of the GAC particle, and the inverse of the ratio $d_{\mathrm{SC}}/d_{\mathrm{LC}}$ is the scaling factor. The ratio $t_{\mathrm{SC}}/t_{\mathrm{LC}}$ is the time required to conduct a small-column test relative to the time necessary to conduct a large-column test. The constant diffusivity (CD) design equation (Eq. 14-23) predicts that relative to a full-scale column with 1 mm diameter GAC, a column with 0.1 mm diameter GAC (scaling factor of 10) and an EBCT 0.01 times as large as that of a full-scale column will produce the same breakthrough curve. Furthermore, the time required to process a given number of BVs of water through the small column is 0.01 times the time required to process the same number of BVs through the large column.

Subsequent studies by Crittenden and colleagues (1987) showed that the internal diffusion coefficient could not always be assumed to be constant with respect to particle diameter. They developed a second relationship, termed the *proportional diffusivity* (PD) design, based on the assumption that the internal diffusion coefficient varied linearly with particle diameter, which yielded the following equation:

$$\frac{\mathrm{EBCT}_{\mathrm{SC}}}{\mathrm{EBCT}_{\mathrm{LC}}} = \frac{d_{\mathrm{SC}}}{d_{\mathrm{LC}}} = \frac{t_{\mathrm{SC}}}{t_{\mathrm{LC}}} \tag{14-24}$$

For the PD design, the preceding particle sizes (scaling factor of 10) would yield an $\mathrm{EBCT}_{\mathrm{SC}}$ and a run time t_{SC} that is one-tenth as large as that at the full scale. While the time savings is not as much as can be achieved if the diffusion coefficient is constant with respect to particle diameter, factor of 100, it is still a factor of 10 smaller than the run time at full scale.

Particle sizes commonly used range from 0.05 to 0.2 mm, and column diameters range from 5 to 11 mm (Crittenden et al., 1987). Using these particle and column sizes yields RSSCT flow rates of 50 to 150 mL/min for the CD design and 5 to 20 mL/min for the PD design.

For NOM, the PD design approach (Eq. 14-24) has been shown to work in many waters. It is thought that the large-molecular-weight NOM diffuses more quickly in the macropores than in the micropores. When the GAC is ground, the micropore domain dominates because macropore transport is less important in transfering the NOM to the micropores. Thus the overall transport in small GAC particles is controlled by slow diffusion in micropores. Summers and colleagues (1995, 1996) reviewed use of the PD design for nine water sources and four GAC types. The 30 verification runs showed that the RSSCT predicted the pilot- or full-scale breakthrough behavior for TOC, UV absorbance, and precursors of THMs, HAAs, TOX, and CH. An RSSCT design and operation manual was developed for the treatment studies requirement by the USEPA's Information Collection Rule (Summers et al., 1996). Unfortunately, the a priori selection of the design equation, either CD (Eq. 14-23) or PD (Eq. 14-24), is not possible for specific organic compounds such as VOCs, pesticides, and PPCPs. While the scaling for NOM follows the PD approach, the fouling or competition it causes complicates the scale-up design for SOCs. In general, the adsorption capacity for these compounds is greater in the RSSCT than at the full scale (Corwin and Summers, 2010). One approach is to run one pilot-plant column using full-scale GAC with the water to be tested and then adjust the RSSCT design for the target compound. Recently, Corwin and Summers (2010) have proposed using a PD design with a fouling index for adjusting the scaling factor based on the influent concentration of the target compound.

The RSSCT technique is a useful technique because it significantly reduces the time required to obtain typical performance data and can be used effectively to screen GAC types and evaluate pretreatment impacts. However, there are limitations. A test that lasts only a few days cannot account for the changes in the influent concentration expected over a much longer time interval. In addition, the short-term test cannot determine the effect of biological activity in the full-scale column. It also cannot be used to determine the residual capacity of GAC that has been in service for a period of time because grinding this GAC to a smaller size affects the capacity, contrary to virgin carbon.

Pilot-Plant Tests

In most situations, pilot testing is recommended because it provides valuable data that can contribute to the design of efficient GAC systems. Batch equilibrium isotherm or bench-scale column tests or empirical models can provide a preliminary assessment of the technical and economical feasibility of GAC as a solution to a water quality problem. Bench-scale column tests also can be used to screen GAC types and evaluate pretreatment impacts. Long-term pilot testing can capture the day-to-day and season-to-season variation in contaminants, background DOM, and other water quality parameters that can affect adsorption and the type and level of pretreatment. Long-term pilot tests also can capture the potential for biodegradation of the target compound.

A pilot study should (1) show the effectiveness of the carbon through a CUR estimation, (2) permit a determination of the best design parameters to use for the full-scale system, and (3) establish the best operating procedure to use. A good estimate of the cost of the full-scale system then can be made. Although pilot testing is expensive, the excessive cost of operating an improperly designed system because of insufficient information may more than justify the cost of a pilot study to obtain that information. In addition, if the pilot study is carried out with direct utility personnel involvement, it provides an opportunity for them to become comfortable with this new technology.

Pilot testing is not always necessary. Confidence has increased in the use of RSSCTs for predicting the adsorption of NOM and DBP precursors. For small systems, the cost of a pilot study may not be justified by the potential savings that would be made possible by an optimized design. GAC adsorbers for VOC removal from small supplies may be designed on the basis of existing information, possibly supplemented with isotherm or small-scale column-test information, with an appropriate safety factor to compensate for the lack of a pilot test.

The pilot plant should be designed with the same pretreatment that will be used in the full-scale system. It should be operated at the location where the full-scale adsorber is to be constructed so that the water quality for pilot- and full-scale contactors is the same. This may require piloting pretreatment processes as well. A detailed chemical analysis of the water, including the concentrations of target compound(s), for example, SOCs, TOC, DBP precursors, and/or taste- and odor-causing compounds, needs to be conducted. In addition, other water quality parameters that may affect target compound adsorption, such as chlorine residual, pH, ammonia concentration, hardness, alkalinity, and turbidity, need to be measured. Preferably, these analyses will be done at different times so that the range of concentrations expected at the full-scale plant can be estimated accurately. Small-scale column tests on water samples preferably taken at different times, coupled with an analysis of the experience of others, can be used to screen different GAC types. The pilot plant then should be designed and operated to show whether these are optimal or whether they should be modified for best performance.

For postfilter adsorbers, three or four GAC columns in series, such as are shown in Fig. 14-23, provide maximum operating flexibility. Alternatively, one or two long columns with several taps can be used so that samples can be taken at different depths. For filter adsorbers, the GAC and other media, including underdrain media, would be contained in one column. The materials of construction may range from plastics such as clear acrylic and polyethylene to the more expensive stainless steel, polytetrafluoroethylene (PTFE), or glass. Kreft and colleagues (1981) showed no leaching of VOCs or TOC from as well as no adsorption of VOCs and TOC to clear acrylic columns and polyethylene tubing in a pilot plant. They did find that phthalates leached from transparent vinyl plastic tubing, so this material should not be used where phthalates will interfere. The possibility that some SOCs may be adversely affected by some materials of construction cannot be ruled out, however. Stainless steel, PTFE, and glass, while expensive, can be used if avoiding any question of possible organics added by plastic is necessary.

The total depth of the pilot-scale adsorber should be greater than the depth likely to be used in the full-scale plant. For the pilot-scale adsorber, depths in the range of 3 to 4 m (10–13 ft) should be used. Column diameter to average particle diameter should be greater than 30:1, or it should be demonstrated that short-circuiting does not occur if smaller ratios are used (Summers et al., 1996). Hydraulic loading rates of 5 to 12.5 m/h (2–6 gpm/ft^2) are usually used. Within this range, the assumption can be made that at a given EBCT, GAC usage rate is not a function of flow rate. A pilot-plant design and operation manual has been developed for the TOC and DBP precursor treatment studies requirement of the USEPA's Information Collection Rule (Summers et al., 1996).

Analysis of GAC Adsorber Data

The GAC effluent data collected at different EBCT values should be plotted as breakthrough curves to show the effect of EBCT on effluent quality. The effluent concentration C_{eff} or the normalized effluent concentration C_{eff}/C_0 can be used. If there is much variability in C_0, then C_{eff} should be used. One measure of effluent concentration should be plotted against time of operation, volume of water processed (see Fig. 14-35), or throughput in

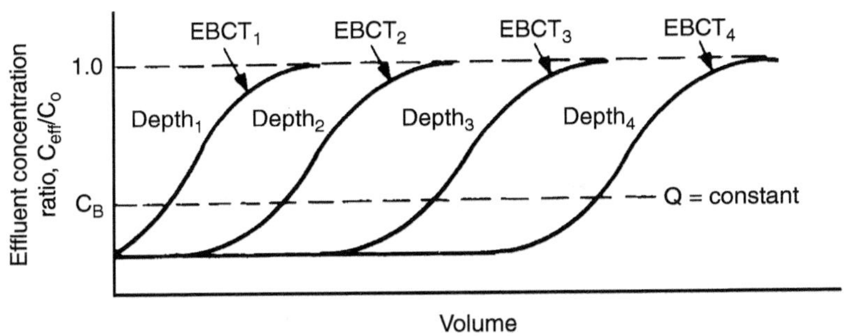

FIGURE 14-35 Breakthrough curves as a function of total volume of water treated.

BVs of water processed (Fig. 14-36). When GACs with different densities are used, the specific throughput in volume treated per mass of GAC is useful because GAC is sold on a weight basis. Because flow rate may vary with time, volume or throughput is the most useful. Because most full-scale plants use time, a double abscissa is suggested, with time

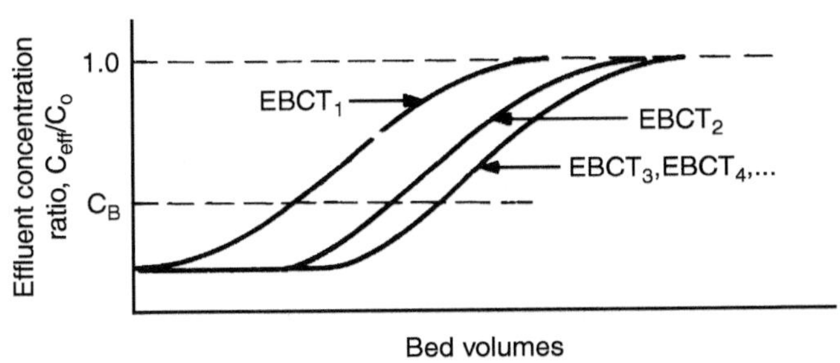

FIGURE 14-36 Breakthrough curves as a function of number of BVs processed.

being one of the independent variables. Using the number of BVs as the abscissa has the important advantage of normalizing the data collected at different depths and at different flow rates. Theoretically, as EBCT values increase, the C_{eff}/C_0 versus BV breakthrough curves eventually will superimpose as shown in Fig. 14-36. However, as has been shown in the discussions for NOM and SOCs, this may not occur because the background DOM affects the adsorption differently at different depths. The CUR, as originally defined in Eq. 14-10, can be calculated from the breakthrough curve for each EBCT:

$$\text{CUR(g/L)} = \frac{\rho_{GAC}\text{(g/L)}}{\text{BVs to breakthrough}} \tag{14-25}$$

where ρ_{GAC} is the apparent density of the GAC.

Parallel-Column Analysis. One of the objectives of pilot- or small-scale column testing should be to determine the CUR if GAC columns are to be used in parallel with staged blending of their effluents. Pilot- and bench-scale column data can be analyzed using the approach of Roberts and Summers (1982), who showed that the fraction, or concentration, of adsorbate remaining in the composite effluent of parallel adsorbers $\bar{f}$ was given by

$$\bar{f} = \frac{1}{n}\sum_{i=1} f_i \tag{14-26}$$

where f_i is the fraction, or concentration, of the adsorbate remaining in the effluent from the ith adsorber, and n is the number of adsorbers in parallel, each of equal capacity. Values of f_i can be determined from a single breakthrough curve such as is shown in Fig. 14-37*a*, assuming that replacement of GAC in each adsorber will take place at equal intervals. Given that θ_n is the number of BVs processed through each adsorber in the parallel system at the time of replacement, the abscissa of the breakthrough curve for the individual contactor from 0 to θ_n is divided into θ_n/n equal increments. For a given value of θ_n, the value of f_i for each increment can be read from the figure, and Eq. 14-26 can be used to calculate the concentration in the blended water. For example, if 10 adsorbers are used in parallel, determination of $\bar{f}$ for several different values of θ_n using Eq. 14-26 leads to the integral curve shown in Fig. 14-37*b*. By operating the 10 contactors in parallel, each adsorber can process 10,000 BVs of throughput if the effluent adsorbate criterion is 50 percent of the influent compared with 5000 BVs if only a single contactor were used or if all 10 contactors were operated in parallel but were replaced at the same time.

Data from an individual GAC pilot contactor, plotted as shown in Figs. 14-36 and 14-37*a*, are needed to develop the integral breakthrough curve shown in Fig. 14-37*b*. The integral curve is developed assuming (1) that each contactor will be the same size, (2) that each contactor will be operated for the same number of BVs θ_n before replacement, and (3) that only one contactor is replaced at one time at intervals of θ_n/n BVs for n contactors in parallel, that is, staggered operation. For a given value of n, f versus θ is calculated by applying Eq. 14-26 for different assumed values of θ_n. (The procedure is illustrated in Fig. 14-37 and related discussion for θ_n = 5000 BVs and an n of 10.) The number of BVs to breakthrough then can be determined from the integral curve. Separate integral curves should be calculated for different numbers of adsorbers. An economic analysis then can be made for each parallel arrangement to determine the best choice.

Example 14-5 GAC Replacement Frequency for GAC Adsorbers Operated in Parallel

An ozone/GAC pilot plant was used to remove THMFP from a shallow reservoir supply after coagulation, sedimentation, and filtration. The GAC columns had a diameter

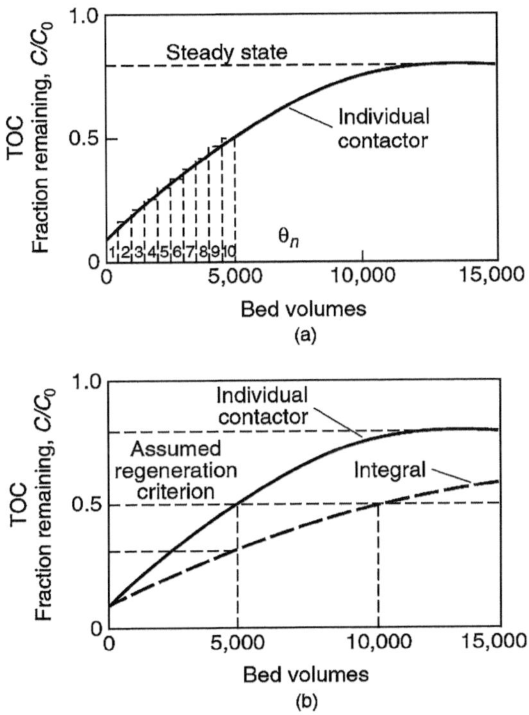

FIGURE 14-37 Integral breakthrough curve approach for characterizing performance of multiple parallel contactors operated in a staged blended effluent mode. (*Source: Roberts and Summers, 1982.*)

of 102 mm (ID or 4 in). Samples were taken from ports located 20 in (port 1), 44 in (port 2), and 64 in (port 3) from the inlet. A flow rate of 0.24 gpm (91 L/min) gave corresponding EBCT values of 4.6, 10.1, and 14.8 minutes. The data (except for points at large volumes processed at ports 2 and 3) are shown in Fig. 14-38. Calculate the GAC replacement frequency if four GAC adsorbers, each with 5-minute EBCT, are to be operated in parallel. The THMFP concentration of the blended GAC effluent is to be 25 μg/L. The C_0 for THMFP is 60 μg/L.

Solution

1. Plot the breakthrough curve (shown in Fig. 14-38) as a function of the BVs of water treated. To convert volume of water processed to BVs, divide the volume processed by the volume of GAC above the port. The volume of GAC above each port is 1.09, 2.39, and 3.48 gal for ports 1, 2, and 3, respectively. For example, the point (vol = 20,000 gal, $C_{\text{eff}}/C_0 = 0.28$) for port 2 in Fig. 14-38 corresponds to

$$\text{BV} = \left(\frac{20{,}000}{2.39} = 8400\right)\left(\frac{C_{\text{eff}}}{C_0} = 0.28\right)$$

in Fig. 14-39.

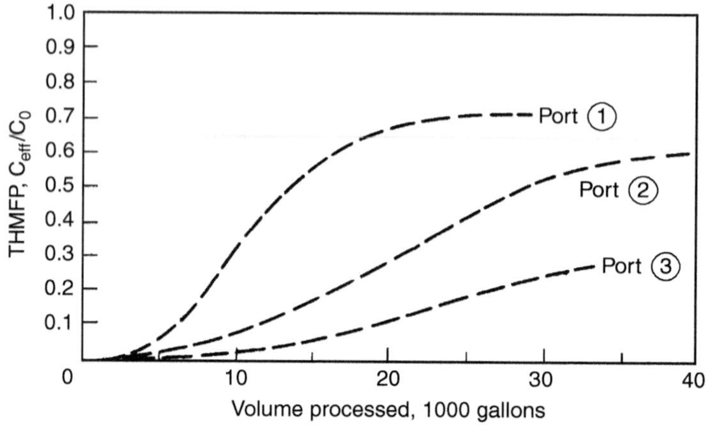

FIGURE 14-38 TTHMFP breakthrough curves as a function of total volume of water treated for example problem.

2. Use Fig. 14-39 to obtain a good estimate of the breakthrough curve expected for a full-scale system. More emphasis is given to data from ports 2 and 3 because the depth of bed for these ports should be much greater than the critical depth. The EBCT for port 3 is also about 15 minutes and should have about the same amount of biological degradation as is expected for the full-scale system. Use the composite of the three curves as the resulting breakthrough curve, which is shown in Fig. 14-40.

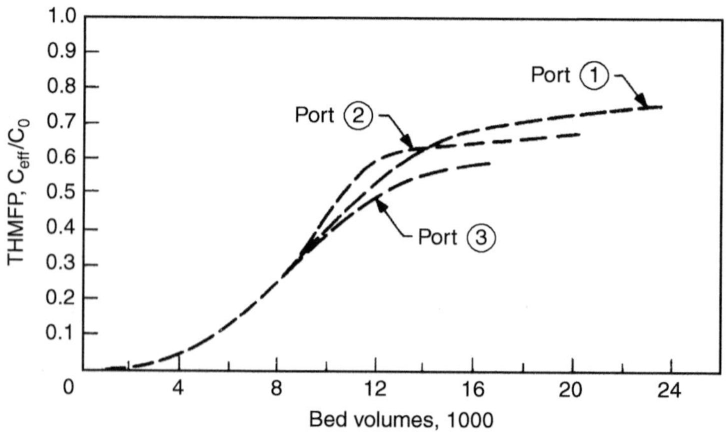

FIGURE 14-39 TTHMFP breakthrough curves as a function of BVs treated for example problem.

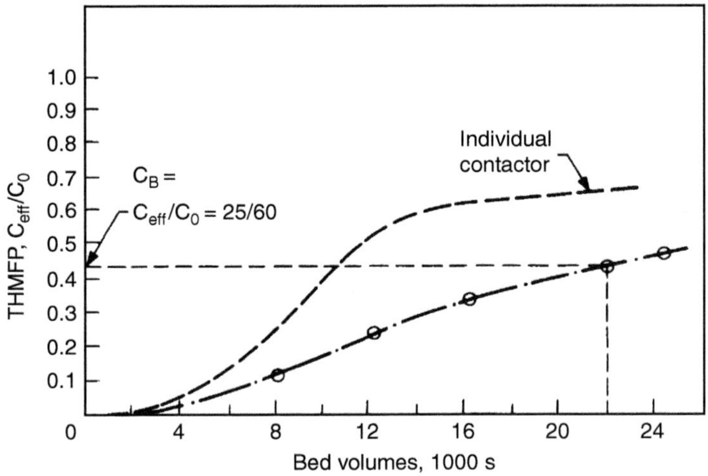

FIGURE 14-40 Integrated TTHMFP breakthrough curve as a function of total volume of water treated for example problem.

3. Determine the expected breakthrough curve when four columns in parallel are blended (integral curve) by applying Eq. 14-26, as discussed above.

Assumed values of θ_4	Corresponding values of f_i (from Fig. 14.39)	$\bar{f}$
8000 BVs	f_1 (2000 BVs) = 0.02	
	f_2 (4000 BVs) = 0.05	
$\left(\frac{8000}{4} = 2000\right)$	f_3 (6000 BVs) = 0.13	
	f_4 (8000 BVs) = 0.23	
	$\Sigma f_i = \overline{0.43}$	$\frac{1}{n}\Sigma f_i = \frac{0.43}{4} = 0.11$
12,000 BVs	f_1 (3000 BVs) = 0.03	
	f_2 (6000 BVs) = 0.13	
$\left(\frac{12{,}000}{4} = 3000\right)$	f_3 (9000 BVs) = 0.35	
	f_4 (12,000 BVs) = 0.52	
	$\Sigma f_i = \overline{1.03}$	$\frac{1.03}{4} = 0.26$
16,000 BVs		0.36
20,000 BVs		0.48

4. Use the integral curve to determine the number of bed volumes that can be processed for a blended effluent concentration, C_B, of 25 μg/L (C_{eff}/C_0 = 25/60 = 0.42). As shown in Fig. 14-40, each column can be operated for 22,000 bed volumes. At this time, the effluent concentration of the single contactor will be 40 μg/L (C_{eff}/C_0 = 0.67), but because it is blended with effluent from the other three columns, which have lower effluent concentrations, the blended water will remain below 25 μg/L.

Given the EBCT of 15 minutes (or 96 BVs/day), each adsorber can be operated for 22,000 BVs/(96 BVs/day) = 230 days before GAC replacement. For a single contactor, the 25 μg/L effluent standard was exceeded at 11,000 BVs, or 115 days, thus blending yields a run time that is twice that of a single contactor. Because carbon must be replaced at equal intervals and the life of each adsorber is 230 days, one of the four adsorbers will be replaced every 230/4 = 59 days. ▲

Series-Column Analysis. The advantage of series operation of columns was discussed earlier. The pilot test should be designed carefully to show the reduction in activated carbon usage rate that can be achieved by such operation.

The simplest case is if only a single compound with no competing adsorbates is present in the water to be treated. The breakthrough curves for each column in a two-column series are shown in Fig. 14-41*a*. Each column is equal in size and thus has the same EBCT. The run can be terminated when the effluent concentration in the second column reaches C_B because moving column 2 to position 1 and placing a fresh GAC column in position 2 should give the same breakthrough curves shown in Fig.14-41*a* between V_1 and V_2. The

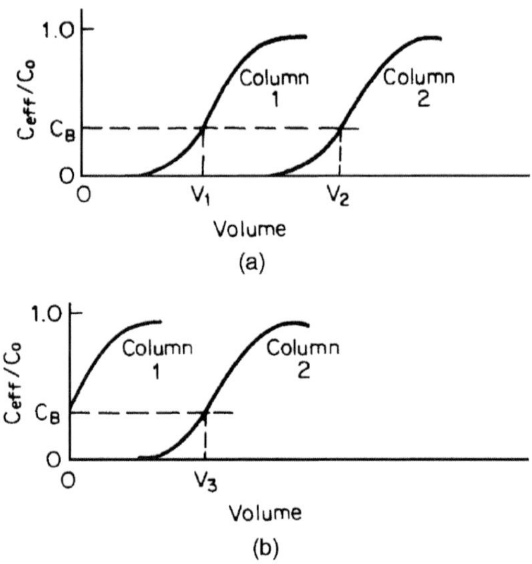

FIGURE 14-41 Breakthrough curves as a function of total volume of water treated for two columns in series. (*a*) Fresh GAC in each column at the start of the run, and (*b*) column 1 was in the second position until C_B was reached and then moved to the first position, and fresh GAC was in column 2.

activated CUR (g/L) then is the mass of GAC in one column (g) divided by the total volume of water processed between replacements, that is, $V_2 - V_1$ (L). The number of BVs processed is ($V_2 - V_1$) divided by the volume of GAC in one column.

The situation is more complex if competing organics are present with the contaminant or if a mixture of compounds is to be removed. The run must begin with fresh GAC in each position and then must be operated until the C_{eff} in the second column reaches C_B. The data should be plotted as shown in Fig. 14-41*a*. Column 2 then must be moved to position 1, a fresh GAC column must be placed in position 2, and operation must continue until C_{eff} for the second column again reaches the breakthrough concentration. These data should be plotted as shown in Fig. 14-41*b*. However, volume V_3 probably will not be the same as volume $V_2 - V_1$ because of the effects of competition. The activated carbon usage rate (g/L) for this case now becomes the mass of GAC in one column (g) divided by V_3 (L).

The plots in Fig. 14-41 show that column 1 is completely saturated when column 2 reaches C_B. If this does not happen, a lower activated carbon usage rate probably can be achieved with more than two columns in series. The pilot test to determine this CUR, assuming competing organics are present, must be run until the column initially in the last position rotates into the first position, followed by continued operation until this sequence of columns reaches breakthrough.

Modeling of GAC Breakthrough Behavior

The modeling of the breakthrough behavior of organic substances in GAC systems can be divided into mathematical models based on the fundamentals of transport and adsorption and empirical models based on performance of existing systems.

Mathematical Models. Several useful mathematical models of the adsorption process are available and have been reviewed (Weber and Smith, 1987; Sontheimer et al., 1988; Worch, 2008). These models have led to a better understanding of the adsorption process to determine high-priority areas for future study and to determine potential cost-effective ways to design and operate activated carbon systems. At this time (2010), the models do not permit accurate long-term predictions of micropollutant removal from a mixture of adsorbable organics. The complex and dynamic effects of NOM preloading reduce GAC effectiveness and complicate micropollutant adsorption modeling to such an extent that often only rough estimates of long-term GAC performance have been possible (Jarvie et al., 2005). The purpose of this section is to summarize what can be done with models and to refer the reader to other sources of information on the subject.

The homogeneous surface diffusion model (HSDM) and modifications of it have been used widely to predict performance of adsorption systems. Crittenden and Weber (1978a, 1978b) developed this model extensively, and it has been employed subsequently to accurately predict adsorption and desorption for single-solute and bisolute systems with similar-size adsorbates (Thacker et al., 1983, 1984). Hand and colleagues (1984) developed a simplified approach to solving the model. A major difficulty in extending the use of the model is the accurate description of the competitive interactions between adsorbates with different characteristics, especially size. The nature of the competition between micropollutants and natural organics also is not well understood and difficult to predict based on laboratory tests.

Some good success with the prediction of TOC adsorption (and other group parameters) from heterogeneous mixtures of organics has been achieved. Lee and colleagues (1981) treated humic substances as a single compound in the HSDM with some success. Summers and Roberts (1984) developed a solution to this same problem using the assumptions that DOC could be treated as a single compound and that it had a linear isotherm. Their model has an analytic solution, and they were able to obtain good agreement between predictions and the performance of operating systems. Sontheimer and colleagues (1988) developed the approach of simulating NOM solutions with many adsorbates by a hypothetical solution with only a few fictive components. The concentration and adsorption properties of the fictive components are assigned to give a composite adsorption isotherm that is the same as the isotherm of the mixture. The HSDM, combined with a competitive adsorption equilibrium model, then can be used to predict the performance of various types of adsorbers for adsorption of the mixture of fictive compounds.

Numerous studies have dealt with the problem of one or more target compounds in the presence of complex mixtures of organics by incorporating the effects of the background organics in the kinetic and equilibrium constants for the target compound(s) (e.g., Weber and Pirbazari, 1982; Smith et al., 1987; Carter and Weber, 1994; Jarvie et al., 2005). The target compound(s) then are modeled as though only they are present in the water. More recently, Schideman and colleagues (2006a, 2006b) provided a solution to the problem of modeling NOM effects on the breakthrough curves of micropollutants in the presence of background organic matter. In this model, background organic matter is represented by (1) a strongly competing compound with similar adsorption characteristics as the target compound and (2) a larger pore-blocking fraction that adversely affects adsorption kinetics, both external mass transfer and intraparticle diffusion, but not the adsorption equilibrium of the target compound. The extent of competition on the adsorption capacity of the target compound was related to the adsorbed concentration of the strongly competing compound at any point in time and space in the GAC adsorber. Similarly, the effects of NOM on film and intraparticle mass transfer of the target compound were related to the adsorbed concentration of the pore-blocking NOM fraction at any point in time and space in the GAC adsorber. By successively running the model first for the pore-blocking fraction, then for the strongly competing fraction, and lastly for the target compound, Schideman and colleagues (2006b) effectively described atrazine breakthrough curves obtained with bench-scale GAC adsorbers.

A good application of mathematical models is to use them to predict small-scale column and pilot-plant results and to determine factors that should be tested in pilot studies.

Empirical Models. Empirical TOC breakthrough models developed by Bond and DiGiano (2004) allow for prediction of the service time in 20 minute EBCT adsorbers with 12 × 40 mesh field-scale bituminous GAC operated to three effluent TOC threshold concentrations: 1.0, 1.5, and 2.0 mg/L. Model input parameters include influent TOC concentration (TOC_0) and pH. Model development was performed using linear regression techniques and included data from 57 RSSCTs run with conventionally treated waters reported in the USEPA's Information Collection Rule (ICR) database (USEPA, 2000). However, they were not developed or verified with field-scale (pilot- or full-scale) data.

Zachman and colleagues (2007a) and Zachman and Summers (2010) extended the approach taken by Bond and DiGiano (2004). Nonparametric (Zachman et al., 2007a) and linear regression (Zachman and Summers, 2010) techniques were used to develop four practical models to predict TOC breakthrough in bituminous GAC adsorbers for two field-scale GAC sizes (8 × 30 and 12 × 40 mesh) and two EBCTs (10 and 20 minutes). Model input parameters include two water quality variables, TOC_0 and pH, that affect performance. The dependent variables for the models were normalized breakthrough time and throughput in bed volumes (BVs) to six fractional ($TOC/TOC_0 = 0.2, 0.3, 0.4, 0.5, 0.6$, and 0.7) effluent and three mass (TOC = 1.0, 1.5, and 2.0 mg/L) effluent concentrations. The model has the general form

$$BV_x = A \times TOC_0^{-\alpha} \times pH^{-\beta} \qquad (14\text{-}27)$$

Model development was performed using small-scale breakthrough data from 35 different source waters in the ICR database with 400 to 500 data points per model. External model validation was performed with small-scale breakthrough data from 14 source waters, and a scalability test was performed to verify the models' ability to predict breakthrough for field-scale GAC adsorbers using at least 100 data points per model.

Table 14-10 shows the model equations and the equations for estimating the model coefficients. The breakthrough predictions are an inverse function of influent TOC and pH, as discussed earlier in this chapter (see Figs. 14-42 and 14-43). Figure 14-42 shows a

TABLE 14.10 Four Models for Predicting TOC Breakthrough

			Normalized effluent concentration model	Mass effluent concentration model
			$BV_x = A - TOC_0^{-1} \times pH^{-1.5}$ $A = f(x)$, where x = 20%, 30%, 40%, 50%, 60%, or 70%	$BV_x' = A' \times TOC_0^{-2} \times pH^{-1.5}$ $A' = f(x')$, where x' = 1.0, 1.5, or 2.0 mg/L
Model 1	12 × 40	10	$A = 196 \times x^2 - 5589 \times x + 252{,}922$	$A' = 1{,}158{,}032 \times x' - 198{,}903$
Model 2	12 × 40	20	$A = 164 \times x^2 - 1938 \times x + 245{,}064$	$A' = 1{,}199{,}245 \times x' - 72{,}818$
Model 3	8 × 30	10	$A = 178 \times x^2 - 6208 \times x + 238{,}321$	$A' = 1{,}125{,}073 \times x' - 429{,}296$
Model 4	8 × 30	20	$A = 202 \times x^2 - 5995 \times x + 261{,}914$	$A' = 1{,}164{,}886 \times x' - 260{,}446$

Source: Zachman and Summers (2010).

sensitivity analysis of model 2 (12 × 40 mesh GAC, 20 minute EBCT) to the influent TOC and pH. The importance of pretreatment to reduce the TOC and pH on extending service time clearly can be seen from this model. Figure 14-43 shows the application of model 2 and model 3 predictions to field-scale data not used in the development of the models. Zachman and colleagues (2007b) used the models to optimize the coagulation and GAC processes for TOC removal as a function of pH.

Kim (2006) and Kim and colleagues (2007) developed predictive models for MIB breakthrough based on the 35 RSSCTs. The three models were designed to predict the BVs treated to target effluent MIB concentration of 10 ng/L (BV_{10}) and 20 ng/L (BV_{20}) and to a 10 percent breakthrough ($BV_{10\%}$). The models were created using multiple linear regression techniques in stepwise manner to determine significant predictors of breakthrough to

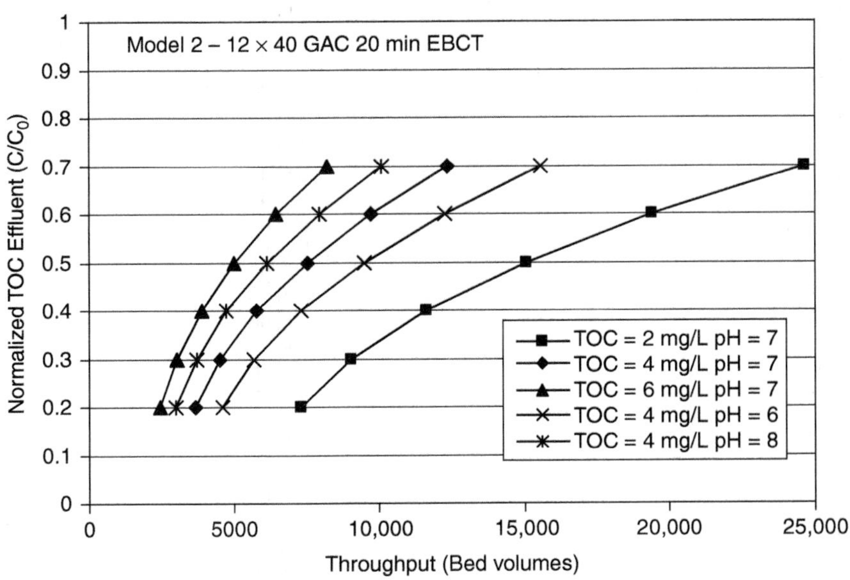

FIGURE 14-42 Model sensitivity analysis showing the impact of influent TOC and pH—Model 2: 12 × 40 mesh GAC, 20 minute EBCT.

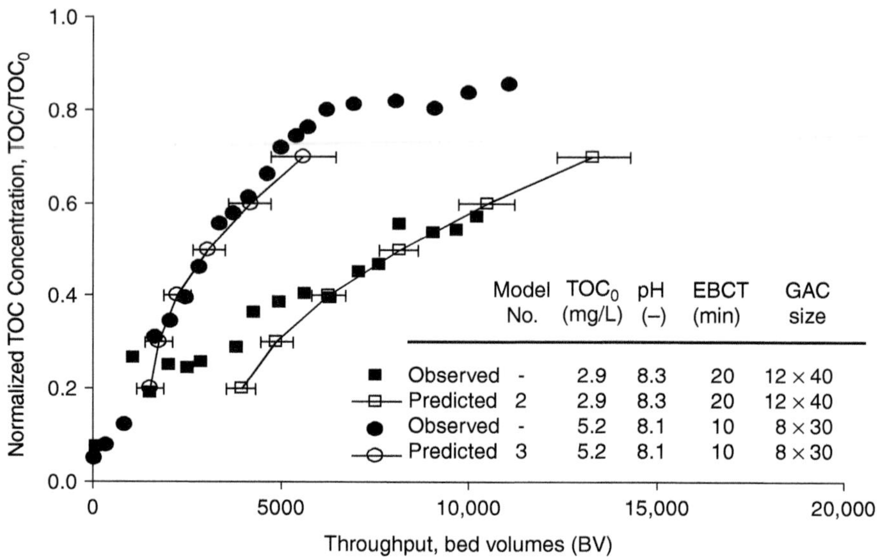

FIGURE 14-43 Comparison of ZS model predictions and external experimental TOC breakthrough data for model 2 (12 × 40 mesh, 20 minute EBCT) and model 3 (8 × 30 mesh, 10 minute EBCT). Error bars represent confidence intervals at 95%. (*Source: Zachman and Summers, 2010.*)

the target effluent conditions. Models were first developed for fresh 12 × 40 mesh GAC at 10 and 20 minute EBCTs and then modified to account for 8 × 20 and 8 × 30 mesh GAC and GAC age. The model input includes influent TOC_0, influent MIB concentration, GAC particle size, and GAC age.

The following equation was found to work for all three models and is shown below for the BV_{10} and the 12- × 40-mesh fresh GAC experimental data.

$$BV_{10}^{12\times40,\ \text{fresh GAC}} = A_1 \times TOC^{x_1} \times MIB^{x_2} \tag{14-28}$$

where $A_1 = 99266$, $x_1 = -0.85$, and $x_2 = -0.2$.

The preceding model was developed using RSSCT breakthrough data from 12 × 40 mesh GAC particle size at EBCTs of 10 and 20 minutes. Also, this model does not take into account GAC age (or NOM preloading effect). Therefore, for more useful and accurate prediction, the model was modified to consider the impact of different GAC sizes and GAC age. To take into account the effect of NOM preloading time into the model, the GAC age coefficient as a function of GAC age α_1 was developed from the relationship between fresh and used GAC BV_{10}.

$$\alpha_1 \approx f\left(\frac{BV_{10}^{\text{used GAC}}}{BV_{10}^{\text{fresh GAC}}}\right) = 1-(0.023\times \text{GAC age}) \tag{14-29}$$

A GAC size coefficient for 8 × 30 mesh GAC ($\alpha_1 = 0.62$) also was calculated using RSSCT breakthrough data. Therefore, the overall MIB model considering the effect of influent

TOC, influent MIB concentration, GAC particle sizes, and GAC age is presented in following equation:

$$BV_{10}^{\text{overall}} = \alpha_1\beta_1(A \times TOC^{x_1} \times MIB^{x_2}) \tag{14-30}$$

α_1 = GAC age coefficient ($\alpha_1 = 1-(0.023\times \text{GAC age})$, if fresh GAC then $\alpha_1 = 1$)
β_1 = GAC size coefficient (if 12×40 $\beta_1 = 1$, if 8×30 then $\beta_1 = 0.62$)

The model has been used successfully to predict pilot-plant breakthrough data at two utilities. Matsui and colleagues (1994), Gillogly and colleagues (1999), and Kim (2006) all have shown that the removal under periodic espisodes was a function of the GAC service time or DOM loading.

POWDERED ACTIVATED CARBON ADSORPTION

The primary advantages of using PAC are the low capital investment costs and the ability to change the PAC dose as the water quality changes. The latter advantage is especially important for systems that do not require an adsorbent for much of the year. The disadvantages, according to Sontheimer (1976), are high operating costs if high PAC doses are required for long periods of time, the inability to regenerate, the low TOC removal, the increased difficulty of residuals management (increased solids loading, residuals are more abrasive), and the difficulty of completely removing the PAC particles from the water.

PAC Application

PAC can be purchased and stored in bags and fed as a powder using dry-feed machines. PAC also can be delivered in bulk and stored dry in silos, from where it is dosed into the water. Alternatively, PAC can be bought in large quantities and fed as a slurry using metering pumps. Storage in bags requires specially designed and operated facilities to avoid generation of nuisance conditions from PAC dust. A survey by Graham and colleagues (1997) in 1995 of 95 utilities using PAC for taste and odor control indicated the following points of PAC addition in conventional plants: presedimentation (16%), rapid mix (49%), flocculation (10%), sedimentation (7%), and filter influent (10%). Twenty-three percent of the plants had the ability to apply PAC at multiple points. Another point of addition that should be considered, although it is not used commonly (7%), is a continuous-flow slurry contactor that precedes the rapid mix. The PAC can be intensely mixed with the water, enabling rapid adsorption onto the small PAC particles, and then incorporated into the floc in the rapid mix for subsequent removal by sedimentation and filtration. Table 14-11 summarizes some of the important advantages and disadvantages to each of these points.

Important criteria for selecting the point of addition include (1) the provision of good mixing, or good contact between the PAC and all the water being treated, (2) sufficient time of contact for adsorption of the contaminant, (3) minimal interference of treatment chemicals with adsorption on PAC, and (4) no degradation of finished water quality. The PAC must be added in a way that ensures its contact with all the flow.

Sufficient time of contact also is necessary, and the time required is an important function of the characteristics and concentration of the molecule to be adsorbed (Meijers and van der Leer, 1983). Figure 14-44 shows the dramatic effect that PAC particle size has

TABLE 14-11 Advantages and Disadvantages of Different Points of PAC Addition

Point of addition	Advantages	Disadvantages
Intake	Generally long contact time, good mixing	Some possibility (slight) of adsorbing compounds that otherwise would be removed by coagulation, thus increasing carbon usage rate
Rapid mix	Good mixing during rapid mix and flocculation; reasonable contact time	Some chance of reduced adsorption rate because of coagulant interference Contact time may be too short to reach equilibrium, thus requiring increased PAC dose
Filter inlet	Sufficient time to reach full adsorption capacity of PAC	PAC may penetrate the filter and deposit in clear well, distribution system, or cause consumer "graywater" complaints; requires good filter aid PAC dose is limited by the rate of head loss buildup and reductions in filter run length
Slurry contactor preceding rapid mix	Excellent mixing; no interference from coagulants	New basin and mixer may have to be installed

on the rate of adsorption of trichlorophenol (TCP) from a groundwater (Adham et al., 1991). The figure shows that 5 μm diameter PAC adsorbs TCP much more rapidly than the 20 μm diameter PAC. Some of the difference in rate in the curves shown in the figure is attributable to the lower capacity of the 20 μm diameter carbon, but the other fractions have the same capacity per unit mass of PAC (Adham et al., 1991). Even faster adsorption

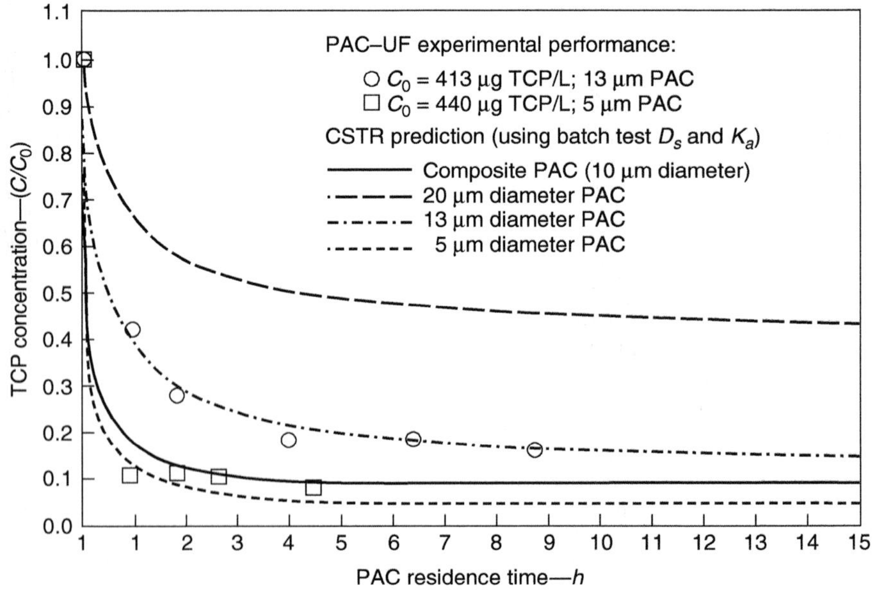

FIGURE 14-44 CSTR model prediction versus PAC-UF performance for the adsorption of TCP from groundwater using all PAC size fractions (PAC dose = 15 mg PAC/L). (*Source: Adham et al., 1991.*)

kinetics have been demonstrated with submicrometer-diameter PAC (Matsui et al., 2008). Adsorption kinetics and equilibrium capacity depend on the type of adsorbate, the type of PAC used, and the competition from background organic matter. If insufficient time is allowed for equilibration, an increased PAC dose must be used to compensate.

Incorporation of PAC into floc particles is one factor that may reduce the rate of adsorption when PAC is added at the rapid mix (Sontheimer, 1976; Gauntlett and Packham, 1973). Gauntlett and Packham (1973) conducted jar tests that showed a lower removal rate of chlorophenol when PAC was added at the same time as alum compared with its addition in the absence of alum. They also found that addition of PAC a short time after alum addition gave a higher rate of removal than when it was added together with alum. They reasoned that PAC added after the alum adhered to the outer surface of the alum floc, thus avoiding rate interference. Najm and colleagues (1989) found little reduction, however, in the rate of TCP adsorption on PAC when PAC particles were incorporated into floc during alum coagulation of a groundwater. Similarly, Graham and Summers (1996) found that alum had only a small impact on the kinetics of TCP adsorption in both organic-free water and water with NOM. Also, Cook and colleagues (2001) showed that MIB and geosmin adsorption rates were not affected by the addition of coagulant in three of four Australian waters. In the fourth water, PAC performance was adversely affected by the presence of alum, however. When alum and PAC were added together at the rapid mix, Ho and Newcombe (2005) showed that increasing alum doses adversely affected MIB adsorption. For alum doses up to 40 mg/L, a PAC dose of 15 mg/L yielded MIB removals of approximately 50 percent; at an alum dose of 100 mg/L, MIB removal was only 20 percent with the same PAC dose. The incorporation of PAC into increasingly larger floc particles was cited as the reason for the deterioration in PAC performance.

Addition at the intake has the advantage of providing extra contact time and generally good mixing in the intake line. Adsorption of compounds that otherwise would be removed by coagulation, flocculation, and sedimentation might increase the required PAC dosage, but this has not been shown to be an important factor in studies to date.

Addition of PAC just before the filter is advantageous because the PAC can be kept in contact with the water longer, thereby better using its capacity. The average PAC residence time is equal to one-half the time between two successive backwashings, assuming that PAC is added continuously to the filter influent. Also, all the PAC could be added at once, just after backwash. The addition must be performed carefully to avoid having PAC escape the filter and penetrate the distribution system, however. The maximum dosage of PAC for this point of addition is limited by the ability of the filter to retain the PAC and by the head loss in the filter, which is expected to increase as PAC dosage increases. Selected polyelectrolytes may be added to retain the PAC, but careful monitoring of the filter to ensure that PAC does not penetrate it is required. The 0.45-μm membrane filter test involving filtration of a fixed quantity of filter effluent is a good method to determine if PAC is penetrating the filter (Greene et al., 1994). Comparison of the color of the filter pad with the color of filter pads that have been used to remove known quantities of PAC from water allows a good estimate to be made of the concentration of PAC that is escaping.

Sontheimer (1976) suggested adding a separate reactor between the sedimentation basin and the filter to increase the time of contact and noted that this procedure would have the advantage of having eliminated competing organics to the maximum extent possible by coagulation-flocculation-sedimentation before PAC addition. A major disadvantage is that the PAC would have to be removed from the water by another coagulation and filtration step, so the cost of this approach would be high.

Careful attention must be paid to the interaction of PAC with water treatment chemicals. Activated carbon is an efficient chemical reducing agent that will chemically reduce substances such as free and combined chlorine, chlorine dioxide, ozone, and permanganate, thereby increasing the demand for these substances and the cost of treatment. Reaction of activated carbon with chlorine will reduce the adsorption capacity of the activated carbon for selected compounds (Snoeyink and Suidan, 1975; McGuire and Suffet, 1984; Gillogly et al., 1998a).

As much as 50 percent of the MIB capacity was lost when 12 mg/L of PAC was reacted with 3 mg/L of free chlorine (Gillogly et al., 1998a). Lalezary-Craig and colleagues (1988) found a reduction in the ability of PAC to adsorb both geosmin and MIB when PAC was applied to water containing free chlorine and monochloramine, and the effect of the monochloramine appeared to be greater than that of chlorine. Furthermore, as chlorine is destroyed by PAC, more chlorine must be added to achieve the desired degree of disinfection.

Addition of PAC to a water that is supersaturated with $CaCO_3$ or other precipitates or an increase in pH to cause supersaturation just after PAC is added, such as in lime softening, may lead to coating of the particle with precipitates and to a corresponding decrease in adsorption efficiency. Greene and colleagues (1994) found that PAC added to a lime-softening basin was much less effective at removing atrazine than PAC added after the softening process. Also, adsorption at high pH is often poorer than at low pH because many organic contaminants are weak acids that ionize at high pH.

Various techniques of applying PAC to improve the ability of PAC to adsorb slow-diffusing compounds such as DBP precursors are available. For example, addition of PAC to solids contact clarifiers has the potential for improved adsorption efficiency because the carbon can be kept in contact with the water for a longer time (Hoehn et al., 1987; Lettinga et al., 1978). Najm and colleagues (1993) found that trichlorophenol reached equilibrium with PAC in a floc blanket reactor; thus lower PAC doses could be used than if PAC were added to the rapid mix followed by flocculation and sedimentation. Similar results were measured and/or predicted for atrazine and DOC adsorption in PAC-floc blanket reactor treatment systems (Campos et al., 2000a, 2000b)

PAC also can be used in conjunction with cross-flow microfiltration (MF) or ultrafiltration (UF) membrane systems. These membranes are effective for removal of particles and microorganisms (see Chap. 11), but pores in the membrane are relatively large, and dissolved organic matter readily passes through the membrane. Removal of organic matter such as pesticides, taste- and odor-causing compounds, and DBP precursors requires pretreatment of the water with an adsorbent. PAC particles are large enough so that PAC can be applied just before the membrane. This system is highly effective for removal of dissolved organic carbon (Adham et al., 1993; Anselme et al., 1997) and atrazine (Campos et al., 1998). Anselme and colleagues (1997) report that several full-scale plants in France are now operating such systems.

A more recent innovation involves the two-stage application of PAC in the floc blanket reactor (FBR)–PAC-UF process (Campos et al., 2000a; Degrémont, 2007). The flow diagram for this process is shown in Fig. 14-45. Fresh PAC is applied continuously just in

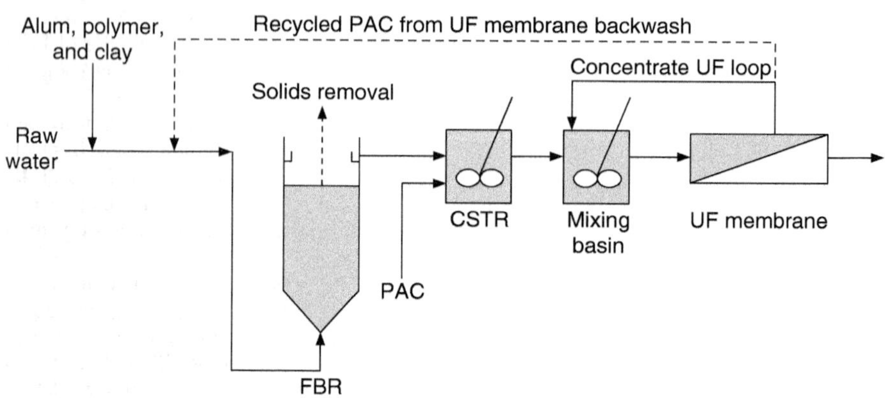

FIGURE 14-45 Schematic of an FBR-PAC-UF system.

advance of the UF process, where it tends to come to equilibrium with the concentration of organic matter in the UF process effluent. As the PAC and other particles accumulate on the membrane, head loss gradually increases until the membrane must be backwashed. The backwash effluent is recycled to the FBR, where the PAC is trapped in the floc. The PAC now tends to equilibrate with the concentration of organic matter in the effluent of the FBR. The solids residence time in the FBR is typically 12 to 36 hours, so there is sufficient time for extensive adsorption to occur. The FBR effluent concentration is much higher than the UF effluent concentration, so the amount of adsorption per unit mass of PAC is increased significantly. The benefit of a countercurrent two-stage process such as this depends on the type of organic matter being removed.

PAC Performance Estimation

Estimation of PAC Dose Using Jar Tests. Jar tests are the simplest way to determine an estimate of the PAC dose required to achieve a desired level of micropollutant removal in conventional surface water treatment plants. Jar tests can be used if water is available with a typical concentration of the micropollutant in question. Alternatively, the micropollutant can be spiked at a typical concentration into the water to which the PAC is to be added. The chemical doses, mixing times and intensities, and settling time should correspond as nearly as possible to those which will be encountered in the full-scale plant. The time of sampling should correspond to the estimated time of contact of the PAC in the plant, keeping in mind that well-coagulated PAC may settle out soon after the PAC enters the sedimentation basin. The percent removal of the micropollutant that was obtained in the jar test can be plotted against PAC dose. The resulting curve is independent of the initial concentration of the micropollutant and can be used to determine the required amount of PAC to be added (see Example 14-6). This dosage will only be an estimate of the full-scale plant dosage because of the many differences between contact in a jar test and in the plant. Therefore, the dosage may have to be modified based on full-scale performance of the PAC. Some bench- or pilot-scale testing also may be necessary to determine the best filter-aid polymer to prevent PAC from penetrating the rapid filter.

Often it is of interest to study the expected performance of PAC for many conditions, various contact times, several types of PAC, and various water quality conditions. Two cases are presented in the following example to illustrate the procedures for estimating performance.

Example 14-6 PAC Usage Rate for Atrazine and MIB Removal

Case 1

Conditions: PAC is to be applied in a way that equilibrium will be achieved, such as to the inlet of a filter. Atrazine concentrations ranging from 3 to 20 μg/L must be lowered to a plant goal of 1 μg/L. The best PAC must be selected.

Approach: An atrazine isotherm should be determined for each PAC to be evaluated. The water sample for the isotherm tests should have the pH, mineral composition, and background organic matter concentration expected in the full-scale plant on a consistent basis. An initial concentration of 50 μg/L or less of atrazine should be used for the test.

Data analysis: The data should be analyzed by plotting the remaining atrazine concentration (as a percentage of the initially added atrazine concentration) versus PAC dose, as shown in Fig. 14-13. One curve will be determined for each carbon. This curve can be used to determine the required PAC dose for each initial concentration expected. For example, if the initial concentration is 20 μg/L, the required percent removal is

$$\frac{(20-1)\,\mu\text{g/L}}{20\,\mu\text{g/L}}\times 100\% = 95\%$$

Assuming that Fig. 14-13 is valid for this water and one of the tested PACs, the required PAC dose to achieve 95 percent removal (5% remaining) is 30 mg/L. Repeating this procedure for each carbon under consideration and knowing the cost of each carbon allow the selection of the most cost-effective carbon to achieve the required 95 percent removal.

The required percent removal will decrease as the influent concentration decreases, given that the desired effluent concentration remains 1 μg/L.

Case 2

Conditions: MIB is present at 20 ng/L, and its concentration must be reduced by 75 percent to the threshold odor level of 5 ng/L. However, several points of addition must be evaluated, each of which has a different contact time, and the adsorption process does not necessarily reach equilibrium.

Approach: A plot of the percentage of adsorbate (in this case MIB) remaining in the water as a function of PAC dose at any contact time is independent of the initial adsorbate concentration if the diffusion coefficient is not a function of test conditions (Gillogly et al., 1998b). Thus the percentage of MIB remaining at each desired contact time can be determined and plotted as a function of PAC dose, as shown in Fig. 14-46. The jar test procedure for MIB adsorption has been described by Gillogly and colleagues (1998b). These tests should be conducted for each desired contact time. Assuming the same PAC and the same water with the same background organic matter as that studied by Gillogly and colleagues (1998b), Fig. 14-46 can be used to determine a PAC dose of 16 mg/L if a contact time of 4 hours is available. The required PAC dose increases to 30 mg/L if the available contact time is only 15 minutes. ▲

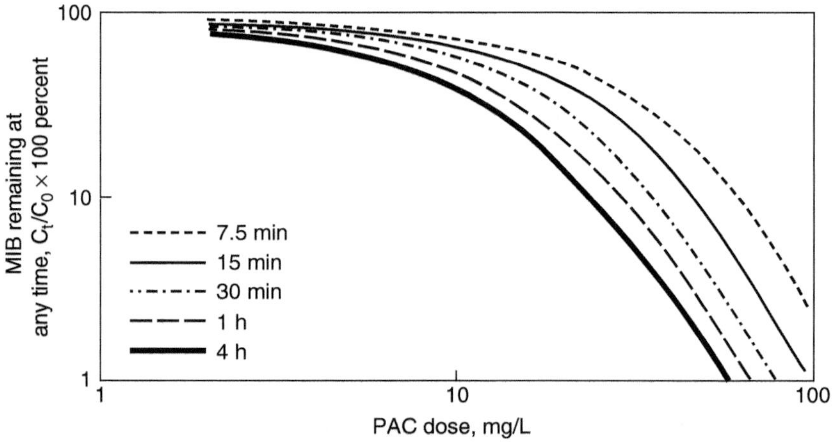

FIGURE 14-46 HSDM fit and simulation of PAC batch kinetic test data for MIB adsorption based on kinetic test plateau values. (*Source: Gillogly et al., 1998b.*)

Estimation of PAC Dose Using Mathematical Models. Apart from the jar-testing approach, mathematical models are useful to predict PAC performance in continuous-flow reactors such as rapid-mix basins, flocculation basins, and intake pipes. For completely mixed reactors such as rapid-mix basins, the continuous stirred tank reactor (CSTR) model is more appropriate, whereas the plug flow reactor (PFR) model is more suitable when PAC is added at the intake and flows with the water through a long pipe. Solutions for the CSTR and PFR models have been presented by Najm (1996).

Experience with PAC

The predominant reason for the use of PAC is taste and odor control, but it is also used for the control of other organic compounds such as pesticides, cyanotoxins, and DBP precursors. PAC doses range from a few to more than 100 mg/L but typically are less than 25 to 50 mg/L (Graham et al., 1997).

Taste- and Odor-Causing Compounds

Numerous studies have evaluated the effectiveness of PAC adsorption processes for taste and odor control. Among earthy-musty odorants, Lalezary and colleagues (1986) showed that activated carbon adsorption is least effective for MIB. Therefore, greater emphasis often is placed on evaluating MIB removal from drinking water. Gillogly and colleagues (1998a) demonstrated that percent MIB removal at a given PAC dose and contact time was independent of the initial MIB concentration. For a contact time of 4 hours, Gillogly and colleagues (1998a) reported MIB removals of approximately 60 to 90 percent with 12 mg/L PAC, the maximum PAC dose that could be applied at the plants participating in the study. Assuming a threshold odor concentration of 5 ng/L for MIB, the results of Gillogly and colleagues (1998a) showed that 13 to 33 mg/L PAC was required to control MIB-related odors in natural water with an initial MIB concentration of 50 ng/L. At an initial MIB concentration of 100 ng/L, PAC doses of 18 to 38 mg/L were required to control MIB-related odors. The PAC dose ranges reflect differences in the effectiveness of five commercially available PACs. A coal based PAC exhibited the largest MIB removal after a contact time of 4 hours.

To lower MIB concentrations in four Australian waters from 50 to 10 ng/L, Cook and colleagues (2001) determined that PAC doses in the range of 39 to 55 mg/L were required when the PAC contact time was 50 minutes. To achieve the same removals for geosmin, PAC doses of 21 to 29 mg/L were necessary. Newcombe and Cook (2002) further showed that mesoporous PACs, such as chemically activated wood based PACs, are more effective for MIB and geosmin removal at short contact times, whereas more microporous PACs, such as steam-activated coal- and coconut shell based PACs, are more effective when longer contact times are available.

Cho and Summers (2007a) evaluated MIB removal with three commercial PAC types at three doses (5, 15, and 30 mg/L) and at four contact times (10, 30, and 60 minutes and 24 hours) for six different waters with TOC concentrations in the range of 2 to 5 mg/L. Doses of 5, 15, and 30 mg/L yielded an average of 30, 50, and 65 percent MIB removal, respectively, with a contact time of 30 minutes and an average of 38, 55, and 75 percent MIB removal, respectively, with a contact time of 60 minutes. During the first 10 minutes of contact, 60 to 75 percent of ultimate MIB removal (24 hours) was achieved. The uptake of MIB was affected by DOC concentration. Higher DOC concentrations yielded lower MIB removals, which were attributed to NOM competition with MIB for PAC adsorption sites. Adding potassium permanganate ($KMnO_4$) with PAC decreased the overall efficiency of MIB removal compared with PAC alone. A simple relationship between percent MIB removal and percent DOC removal was found for all three PACs, suggesting that DOC measurements could serve as a surrogate for more expensive MIB analyses.

For 2t,6c-nonadienal, limited data by Burlingame and colleagues (1992) suggest that PAC can control both the cucumber odor and flavor associated with this compound. PAC doses as large as 24 mg/L were required to control the cucumber flavor in Delaware River water, and similar results were obtained by Krasner and colleagues (2004) for Colorado

River and State Project waters. Krasner and colleagues (2004) further showed that a PAC dose of 25 mg/L yielded at least 50 percent removal for 8 of 12 compounds responsible for fishy, grassy, and swampy odors in Colorado River and State Project waters.

TON removal is used commonly to measure PAC effectiveness. Reports of the effectiveness of PAC for TON removal are complicated because an odor may be caused by the combination of the effects of more than one compound or a compound may mask the odor of another compound, and the odor of the second may appear after the first compound has been adsorbed. Furthermore, odors may form after treatment through chemical or biological reactions in the distribution system (AWWA Research Foundation and Lyonnaise des Eaux, 1987), thus making PAC appear ineffective. A reliable catalog of odors that can be removed well by PAC does not exist probably because of the preceding complicating factors. As analytic capabilities improve and the ability to qualitatively and quantitatively analyze compounds that cause odor advances, however, development of such a catalog may be possible.

Pesticides. Gustafson and colleagues (2003) studied the removal of atrazine and the chloroacetanilide herbicides acetochlor, alachlor, and metolachlor by lignite and coal based PACs. In addition, the removal of the ethanesulfonic acid (ESA) and oxoacetic acid (OA) degradates of the chloroacetanilides was evaluated. For a given water and PAC type, removal of atrazine and the chloroacetanilides was similar, whereas removal of the ESA and OA degradates was lower. For a PAC dose of 20 mg/L and a contact time of 20 minutes, the lignite based PAC performed slightly better than the coal based PAC, whereas the reverse was true for contact times of 60 minutes and longer. For one tested surface water, a lignite based PAC dose of 20 mg/L produced atrazine and chloroacetanilide removals ranging from 66 to 74 percent after a contact time of 20 minutes. In contrast, ESA and OA degradate removals ranged only from 22 to 32 percent. An early study by Robeck and colleagues (1965) showed that PAC effectively removed the pesticides parathion, 2,4,5-T butyl ester, endrin, lindane, and dieldrin from Little Miami River water.

Greene and colleagues (1994) evaluated the effectiveness of PAC in a lime softening plant. Addition of 6 mg/L of lignite based PAC to the stabilization basin (residence time of 3.2 hours) lowered atrazine concentrations from 2.1 to 6.7 μg/L in the basin influent to 0.2 to 3.5 μg/L in the basin effluent. A PAC dose of 9.6 mg/L lowered atrazine concentrations from 2.1 to 8.5 μg/L in the basin influent to 0.09 to 2.3 μg/L in the basin effluent. PAC carryover to the filters produced additional atrazine removal such that atrazine concentrations in the filter effluent were less than 0.3 μg/L at both tested PAC doses. At the same plant, the addition of approximately 10 mg/L of PAC to the stabilization basin lowered alachlor concentrations from approximately 0.5 to 1.1 μg/L in the basin influent to less than 0.25 μg/L in the basin effluent.

Cyanotoxins. Microcystin-LR has been the most widely studied cyanotoxin, and its removal is accomplished most effectively by mesoporous PACs (Donati et al., 1994). In a full-scale lime softening plant, more than 80 percent microcystin-LR removal was achieved with a PAC dose of 30 mg/L (Lambert et al., 1996). Newcombe and Nicholson (2004) calculated PAC doses required to remove 80 percent microcystin-LR and microcystin-LA by a chemically activated wood based PAC; 29 mg/L was required for microcystin-LR, whereas 95 mg/L was required for microcystin-LA after a contact time of 60 minutes. For the removal of the cyanotoxin cylindrospermopsin, Ho and colleagues (2008) determined that both a steam-activated coal based PAC and a chemically activated wood based PAC performed equally well. For a contact time of 30 minutes, 80 percent cylindrospermopsin removal was obtained with a PAC dose of 33 mg/L in the tested Australian water. For saxitoxins, removal of the most toxic variants is accomplished effectively by 30 mg/L of PAC and a contact time of 1 hour (Newcombe and Nicholson, 2004). Little information on the effectiveness of PAC is available on the adsorption of other cyanotoxins such as anatoxin-a and nodularin.

NOM and DBP Precursors. Najm and colleagues (1991b) provided a comprehensive review of the PAC adsorption literature and concluded that relatively high PAC doses are required to achieve substantial removal of TOC, DOC, and/or THM precursors. In some instances, a PAC dose of 50 mg/L yielded negligible TOC and THM precursor removal, whereas in others, the same PAC dose yielded up to 27 percent TOC removal and 49 percent THM precursor removal. Sufficient information about NOM characteristics, PAC characteristics, and PAC contact times was not provided, however, and factors responsible for the differences in TOC and THM precursor removal efficiencies therefore could not be identified.

More recently, Uyak and colleagues (2007) showed that the DOC-removal efficiency of PAC strongly depends on PAC type. For example, for two PACs, a dosage of 20 mg/L yielded approximately 5 percent DOC removal, whereas approximately 25 percent was achieved with another PAC at the same dosage. No information about PAC characteristics was provided that could help to explain the performance differences among the PACs. Cho and Summers (2007b) evaluated three commercial PAC types at three doses (5, 15, and 30 mg/L) and four contact times (10, 30, and 60 minutes and 24 hours) for seven different waters with TOC concentrations in the range of 2 to 5 mg/L. Doses of 5, 15, and 30 mg/L yielded an average of 7, 16, and 23 percent DOC removal, respectively, with a contact time of 30 minutes, with 40 to 50 percent of ultimate DOC removal (24 hours) achieved in the first 10 minutes.

Campos and colleagues (2000a) compared the DOC-removal efficiency of two commercially available PACs with mean diameters of 10 and 6 μm. Using PAC doses of 10 and 20 mg/L and a contact time of 30 minutes, they obtained DOC removals in the range of approximately 10 to 35 percent for the larger PAC and approximately 26 to 43 percent for the smaller PAC. The smaller PAC in this study exhibited higher micropore and mesopore volumes and a larger DOC adsorption capacity than the larger PAC. Thus higher DOC removals obtained with the smaller PAC were in part a result of its higher adsorption capacity and in part a result of faster DOC adsorption rates. The results of Campos and colleagues (2000a) and Cho and Summers (2007b) suggest that typical PAC doses for taste and odor control can yield meaningful DOC-removal percentages.

Emerging Contaminants. Using coal based WPH PAC, Adams and colleagues (2002) studied the adsorption of five sulfonamide antibiotics and the antimicrobial compound trimethoprim (TMP) from Missouri River water. After a contact time of 4 h, sulfonamide removals ranged from 49 to 73 percent, and TMP removal was approximately 55 percent at a PAC dose of 10 mg/L (Adams et al., 2002). Evaluating the removal of pharmaceuticals, endocrine-disrupting chemicals, personal-care products, and flame retardants from four drinking water sources with 5 mg/L of a coal based PAC, Westerhoff and colleagues (2005) obtained average removals ranging from 16 percent for the analgesic ibuprofen to 93 percent for the sunscreen ingredient oxybenzone after a contact time of 4 hours. Most of the tested emerging contaminants exhibited an adsorbability that was similar to or greater than that of the herbicide metolachlor.

THERMAL REACTIVATION OF GAC

Reactivation Principles

The reactivation process can be described as consisting of four basic steps (van Vliet, 1985): (1) drying, including loss of highly volatile adsorbates, at temperatures up to 200°C, (2) vaporization of volatile adsorbates and decomposition of unstable adsorbates to form volatile fragments, at temperatures of 200 to 500°C, (3) pyrolysis of nonvolatile adsorbates and adsorbate fragments to form a carbonaceous residue or char on the activated carbon surface, at temperatures of about 500 to 700°C, and (4) oxidation of the pyrolyzed residue using steam or carbon dioxide as the oxidizing agent, at temperatures above about 700°C.

The drying step eliminates the 40 to 50 percent of the water associated with the spent GAC (Schuliger and MacCrum, 1973). The proportions of adsorbates that are volatilized (steps 1 and 2) or converted to char (step 3) depend on the nature of the adsorbate and the strength of the adsorbate-activated carbon bond (Juntgen, 1976). Oxidation to gasify the char is critical to returning the GAC as closely as possible to virgin conditions. The objective is to remove the char with minimal loss of the activated carbon in a manner that does not alter the structure and adsorption properties of the activated carbon.

Factors affecting reactivation product quality are primarily those which affect the oxidation step. Important factors are the type of oxidizing gas, the time and temperature of oxidation, the amount and type of adsorbate on the carbon, the type and quantity of inorganic substances that accumulate on the carbon, and the type of activated carbon. Reactivation studies indicate a tradeoff between the time required for reactivation and the temperature of reactivation: The higher the temperature, the shorter is the reactivation time (Juhola, 1970). The optimal temperature for reactivation has been shown to depend on the type of activated carbon (Juntgen, 1976).

In the application of activated carbon for purification of surface water and possibly groundwater, the need to remove adsorbed humic substances probably will control the reactivation step, even though the primary purpose in applying the activated carbon is to remove micropollutants. Certain inorganic substances, if present in the water being treated, may form a precipitate on the activated carbon surface or may accumulate on the activated carbon surface because of their association with the organic molecules that are being adsorbed. Knappe and colleagues (1992) and Cannon and colleagues (1993) found that calcium accumulated on many water treatment plant carbons. In agreement with others, they showed that calcium catalyzed the reaction of reactivated gases with GAC and made the recovery of micropores difficult.

Reactivation Furnaces

Common types of furnaces are the rotary-kiln, the multiple-hearth, and the fluidized bed furnace (Sontheimer et al., 1988). The multiple-hearth furnace is used most commonly. Scrubbers are used to remove particles from the off-gases, and afterburners, which are maintained at 750 to 1000°C by burning fuel with excess oxygen, are used to destroy organic compounds that are volatilized but not combusted to CO_2 during the reactivation step.

Reactivation Performance

Loss of mass during reactivation is important because of its impact on the cost of using activated carbon. DeMarco and colleagues (1983) reported total losses (transport plus loss in the furnace) of 16 to 19 percent in an experimental system at Cincinnati, whereas Koffskey and Lykins (1987) observed 9 percent loss (7% in the furnace and 2% during transport) at Jefferson Parish, LA. Proper design of the transport system is very important because of the high losses that can take place in a poorly designed system (Schuliger et al., 1987). Schuliger and colleagues (1987) noted that losses are a function of the transport system design, the type of adsorbate, the loading of adsorbate on the carbon, and the type of furnace.

Along with the loss in mass may come a change in pore size distribution. Pore size distribution has a significant effect on both rate of uptake and capacity for a particular adsorbate. Modifying reactivation conditions to produce the best pore size distribution for a particular adsorbate should be possible, but research to show how this can be done and to show the desired pore size for a given application is needed (Sontheimer et al., 1988). It has been observed that the particle diameter becomes smaller during reactivation, with the disadvantage of creating higher head loss in fixed-bed adsorbers but with the advantage of allowing the rate of uptake of molecules to occur more rapidly. The decrease in particle diameter should be minimized as much as possible because this decrease normally correlates with loss in mass.

Generally, density, molasses number or decolorization index, iodine number, and similar parameters are used to measure the quality of the activated carbon. An improved procedure involves the use of the water to be treated to test the quality of the product with isotherms of RSSCTs. Changes in activated carbon quality produced by reactivation then can be expressed as changes in the length of time to breakthrough and the amount of material adsorbed at breakthrough for a column of a given design.

Kornegay (1979) analyzed the cost of reactivation as a function of furnace size. He showed rapidly increasing unit costs as activated carbon usage became smaller than 2270 to 2720 kg/day (5000–6000 lb/day). Kornegay (1979) noted that on-site thermal reactivation will not be economical if usage is smaller than 227 to 454 kg/day (500–1000 lb/day), whereas Hess (1981) proposed a lower economical limit of about 910 kg/day (2000 lb/day). Transport of spent GAC to off-site thermal reactivation facilities may be appropriate for intermediate values of activated carbon usage rate. The cost of reactivation then would include transportation to and from the site, the cost of makeup activated carbon, and (presumably) profit for the owner of the furnace. This approach to reactivation is common in Europe, where transportation distances are relatively short, the incidence of GAC usage for surface water is high, and the number of on-site furnaces is small. Further, assurance would have to be given that only water treatment plant activated carbon would be returned to a plant. Inclusion of reactivated activated carbon that had been used for municipal or industrial waste treatment would be unacceptable.

Reactivation By-Products

Activated carbon must be regenerated in a way that does not pollute the atmosphere. For this reason, scrubbers and afterburners are used to minimize particulate and gaseous emissions, respectively, from operating facilities. Off-gas control is important in controlling dioxins and furans that are produced during reactivation. Lykins and colleagues (1988b) studied operating reactivation systems at Cincinnati and Jefferson Parish and found these compounds in various effluent streams before treatment of these effluent streams, even though they were not on GAC entering the furnace. A small amount of the dioxins and furans do pass through the stack gas control equipment, but Lykins and colleagues (1988b) found the concentrations to be so low that the estimated cancer risk was negligible.

ADSORPTION OF ORGANIC MATTER BY OTHER ADSORBENTS

Synthetic Adsorbent Resins

The styrene-divinylbenzene (SDVB) resin with no functional groups has received much attention as an adsorbent. The resin was particularly effective for removal of chlorinated pesticides from an industrial wastewater (Kennedy, 1973). Regeneration was readily accomplished with acetone or isopropanol, and the spent regenerant could be reclaimed for reuse. This resin also adsorbs phenols, although its capacity is somewhat less than that of weak-base resins (Kumugai and Kaufman, 1968). The SDVB resin did not remove TOC from Thames River water to an appreciable extent (Gauntlett, 1975). The weak-base phenol-formaldehyde (PF) resin had essentially no capacity for the earthy-musty-odor compound MIB (Chudyk et al., 1979). The SDVB adsorbent resin did adsorb MIB, but its adsorption capacity was lower than those of low-activity activated carbons made from bituminous coal. A general conclusion concerning resins is that they are not applicable as a general adsorbent for drinking water treatment. They are more selective than activated carbon and do not meet the criterion that an adsorbent must be able to remove a wide variety of compounds. They may be applicable in specific situations that require removal of only a particular contaminant.

Carbonaceous Resins

Several studies have evaluated carbonaceous resins as an adsorbent alternative to activated carbon (Hand et al., 1994; Melin, 1999; Knappe et al., 2007; Rossner et al., 2009). Targeted adsorbates in these studies were TCE, MTBE, and emerging contaminants. For MTBE, carbonaceous resins exhibited adsorption capacities that exceeded those of GACs prepared from coconut shells, coal, and wood. Melin (1999) reported that the MTBE adsorption capacity of the carbonaceous resin Ambersorb 572 was not affected by the presence of groundwater NOM, whereas that of a coconut shell based GAC decreased in the presence of the same NOM. Similar results were obtained by Hand and colleagues (1994) when they compared trichloroethene (TCE) adsorption capacities of the carbonaceous resin Ambersorb 563 and a coal based GAC in ultrapure water and groundwater. In the presence of groundwater NOM, the activated carbon adsorption capacity decreased by 35 percent relative to the single-solute adsorption capacity, whereas the carbonaceous resin adsorption capacity was not measurably affected after 10 weeks of NOM exposure (Hand et al., 1994). In contrast, Knappe and colleagues (2007) showed that river water NOM decreased the MTBE adsorption capacity of Ambersorb 563 relative to that obtained in ultrapure water. For the adsorption of an emerging contaminant mixture from lake water, Rossner and colleagues (2009) showed that a coconut shell based GAC produced higher removal percentages at a given dose than the carbonaceous resin Ambersorb 563. The activated carbon had a larger volume of pores in the size range of the targeted contaminants (~0.6–0.9 nm) than the carbonaceous resin, which likely explained the difference in adsorption uptakes.

Activated Carbon Fibers

Compared with GAC and PAC, activated carbon fibers (ACFs) have a relatively uniform and well-defined pore structure. Several research groups have investigated whether the adverse effects of NOM adsorption on micropollutant removal can be minimized through the use of ACFs (Pelekani and Snoeyink, 1999, 2000, 2001; Quinlivan et al., 2005; Karanfil et al., 2006). Pelekani and Snoeyink (1999) investigated the simultaneous adsorption of NOM and atrazine on two ACFs, one containing pores with widths of up to about 0.8 nm and one containing pores with widths of up to about 2 nm. Compared with the single-solute isotherm, atrazine adsorption was reduced by a factor of 10 on the ACF with the narrower pores. Pore blockage likely explained the reduction in atrazine adsorption in the presence of NOM. On the ACF with the larger pores, atrazine adsorption was reduced by a factor of 3 compared with the single-solute isotherm, which was likely a result of direct competition between NOM and atrazine for adsorption sites. The data showed that pore blockage by NOM as a result of size exclusion more severely reduced the micropollutant capacity than direct competition between the micropollutant and NOM for adsorption sites.

Comparing different ACFs, Hopman and colleagues (1995) showed that NOM adsorption on one ACF was minimal. Furthermore, ACF that had been preloaded with NOM was able to adsorb the pesticide chlorotoluron (C_0 = 2 μg/L) from natural water to the same extent as fresh ACF. However, the authors did not state how long this ACF was preloaded, and no information about the physical and chemical characteristics of the ACF were given. To date, the results of Hopman and colleagues (1995) are unique in that the researchers were able to eliminate the adverse effects of NOM preloading on micropollutant removal from a natural water.

High-Silica Zeolites

Several studies have evaluated the use of high-silica zeolites for micropollutant removal from water (Ellis and Korth, 1993; Kawai et al., 1994; Knappe et al., 2007, 2010; Rossner and Knappe, 2008; Rossner et al., 2009). Because of their well-defined pore sizes, it may be possible to select high-silica zeolites that target the removal of specific micropollutants while minimizing access of interfering NOM constituents. For example, silicalite appears to be especially suitable for the adsorptive removal of MTBE and exhibits a larger MTBE adsorption capacity than activated carbons with a considerably larger BET surface area (Knappe et al., 2007). Furthermore, in a packed bed adsorber application, silicalite was immune to NOM preloading effects that markedly decreased the MTBE removal effectiveness of a GAC adsorber that was operated in parallel with the silicalite adsorber (Rossner and Knappe, 2008). In terms of material cost, high-silica zeolites (about $7/lb and up) are more expensive than activated carbons (about $1–2/lb). However, the results of Knappe and colleagues (2007) showed that the higher MTBE adsorption capacity of zeolites compared with activated carbon was sufficient to make up a large part of the cost difference. Apart from niche applications, in which the removal of a specific contaminant, such as MTBE, is targeted, high-silica zeolites cannot compete with the effectiveness of activated carbon. For example, high-silica zeolites with their uniform pores did not provide an effective broad-spectrum barrier against a mixture of 25 emerging contaminants (Rossner et al., 2009).

ABBREVIATIONS

ACF	activated carbon fiber
ANSI	American National Standards Institute
ASTM	American Society for Testing and Materials
AWWA	American Water Works Association
BAC	biological activated carbon
BET	Brunauer-Emmett-Teller
BPA	bisphenol A
BV	bed volume
CD	constant diffusivity
cNW	coagulated natural water

COD	chemical oxygen demand
CUR	carbon usage rate
DBP	disinfection by-product
DCBR	differential column batch reactor
DCP	dichlorophenol
DFT	density functional theory
DMP	dimethylphenol
DOC	dissolved organic carbon
DOM	dissolved organic matter
EBC	equivalent background compound
EBCT	empty bed contact time
EfOM	effluent organic matter
ESA	ethanesulfonic acid
FBR	floc blanket reactor
GAC	granular activated carbon
HAA	haloacetic acid
HSDM	homogeneous surface diffusion model
IAST	ideal adsorbed solution theory
ID	inside diameter
IUPAC	International Union of Pure and Applied Chemistry
LC	large scale
MCPB	4-(2-methyl-4-chlorophenoxy)butyric acid
MIB	2-methylisoborneol
MTBE	methyl *tert*-butyl ether
MTZ	mass transfer zone
NOM	natural organic matter
O&M	operation and maintenance
OA	oxoacetic acid
PAC	powdered activated carbon
PCB	polychlorinated biphenyl
PD	proportional diffusivity
PF	phenol/formaldehyde
PPCP	pharmaceutical and personal-care product
PTFE	polytetrafluoroethylene
RSSCT	rapid small-scale column test
SBA	short-bed adsorber
SC	small scale
SDVB	styrene-divinylbenzene
SOC	synthetic organic compound
SUVA	specific ultraviolet absorbance
TCE	trichloroethylene = trichloroethene
THM	trihalomethane
THMFP	trihalomethane formation potential
TOC	total organic carbon
TON	threshold odor number
TOX	total organic halogens
UF	ultrafiltration
UFC	uniform formation conditions
USEPA	U.S. Environmental Protection Agency
UVA	ultraviolet absorbance at a wavelength of 254 nm
VOC	volatile organic chemical

NOTATION FOR EQUATIONS

C	aqueous-phase concentration
C_0	initial or influent aqueous-phase concentration
C_B	breakthrough concentration
C_c	activated carbon dose
C_{eff}	effluent concentration
d	particle diameter
D_s	surface diffusion coefficient
f	fractional concentration
K	Freundlich isotherm constant
L_{MTZ}	length of mass transfer zone
n	Freundlich isotherm exponent
m	adsorbent mass
q	solid-phase concentration
R	particle radius
t	run time
V	volume
V_B	volume treated to breakthrough
Y	bed life
ρ_{GAC}	bed density
θ_n	number of bed volumes processed through each adsorber in a parallel system at the time of replacement

REFERENCES

Adams, C., Wang, Y., Loftin, K., and Meyer, M. (2002). Removal of antibiotics from surface and distilled water in conventional water treatment processes. *Journal of Environmental Engineering, ASCE* 128(3), 253–260.

Adamson, A. W., and Gast, A. P. (1997). *Physical Chemistry of Surfaces*, 6th ed. New York: Wiley.

Adham, S. S., Snoeyink, V. L., Clark, M. M., and Anselme, C. (1993). Predicting and verifying TOC removal by PAC in pilot-scale UF systems. *Journal AWWA* 85(12) 58–68.

Adham, S. S., Snoeyink, V. L., Clark, M. M., and Bersillon, J. L. (1991). Predicting and verifying organics removal by PAC in an ultrafiltration system. *Journal AWWA* 83(12), 81–91.

Andelman, J. B. (1973). *Proceedings of the 15th Water Quality Conference. University of Illinois Bull.* 70:122.

Anselme, C., Laîné, J. M., Baudin, I. and Chevalier, M. R. (1997). Drinking water production by UF and PAC adsorption. In *Proceedings of the AWWA Specialty Conference on Membrane Technology, New Orleans.*

Arbuckle, W. B. (1980) Premature exhaustion of activated carbon columns. In *Activated Carbon Adsorption,* Vol. 2, J. M. McGuire and I. H. Suffet, eds. Ann Arbor, MI: Ann Arbor Science Publishers.

American Society for Testing and Materials (ASTM) (2009). Refractories; activated carbon; advanced ceramics. In *Annual Book of ASTM Standards*, Vol. 15.01. West Conshohocken, PA: ASTM International.

American Water Works Association (AWWA) (2005a). *ANSI/AWWA Standard B600-05: Powdered Activated Carbon.* Denver, CO: American Water Works Association.

AWWA (2005b). *ANSI/AWWA B604-05: Standard Granular Activated Carbon.* Denver, CO: American Water Works Association.

AWWA (1996). *Water:/Stats 1996 Water Utility Database.* CD-ROM. Denver, CO: American Water Works Association.

AWWA (1986). *1984 Utility Operating Data.* Denver, CO: American Water Works Association.

AWWA Committee Report (1977). Measurement and control of organic contaminants by utilities. *Journal AWWA* 69(5), 267–271.

AWWA Research Foundation and Lyonnaise des Eaux (1987). *Identification and Treatment of Tastes and Odors in Drinking Water,* J. Mallevialle and I. H. Suffet, eds. Denver, CO: Water Research Foundation.

Bablon, G., Ventresque, C., and Ben Aim, R. (1988). Developing a sand-GAC filter to achieve high-rate biological filtration. *Journal AWWA* 80(12), 47–53.

Bahl, O. P., Shen, Z., Lavin, J. G., and Ross, R. A. (1998). Manufacture of carbon fibers. In *Carbon Fibers*, 3rd ed., J. B. Donnet, J. B., Wang, T. K., Rebouillat, S., and Peng, J. C. M., eds. New York: Marcel Dekker.

Baldauf, G., and Henkel, M. (1988). Entfernung von Pestiziden bei der Trinkwasseraufbereitung. *20. Bericht der Arbeitsgemeinschaft Wasserwerke Bodensee–Rhein* 17, 127–142.

Baldauf, G., and Zimmer, G. (1986). Adsorptive Entfernung nichtflüchtiger Halogenkohlenwasserstoffe bei der Wasseraufbereitung. *Vom Wasser* 66, 21–31.

Bansal, R. C., Donnet, J. B., and Stoeckli, F. (1988). *Active Carbon.* New York: Marcel Dekker.

Bauer, R. C., and Snoeyink, V. L. (1973) Reactions of chloramines with active carbon. *Journal Water Pollution Control Federation* 45, 2290–2301.

Benjamin, M. M. (2009). New conceptualization and solution approach for the ideal adsorbed solution theory (IAST). *Environmental Science and Technology* 43, 2530–2536. Spreadsheet model available at http://faculty.washington.edu/markbenj/IAST/.

Boehm, H. P. (1994). Some aspects of the surface chemistry of carbon blacks and other carbons. *Carbon* 32,759–769.

Boehm, H. P., Diehl, E., Heck, W., and Sappok, R. (1964). Surface oxides of carbon. *Angewandte Chemie International Edition* 3, 669–677.

Bond, R. G., and DiGiano, F. A. (2004). Evaluating GAC performance using the ICR database. *Journal AWWA* 96(6), 96–104.

Brown, J. C., Anderson, R. D., Min, J. H., Boulos, L., Prasifka, D., and Juby G. J. G. (2005). Fixed bed biological treatment of perchlorate-contaminated drinking water. *Journal AWWA* 97(9), 70–81.

Brown, J. C., Snoeyink, V. L., and Kirisits, M. J. (2002). Abiotic and biotic perchlorate removal in an activated carbon filter. *Journal AWWA* 94(2), 70–79.

Brown, J. C., Snoeyink, V. L., Raskin, L., and Lin, R. (2003). The sensitivity of fixed-bed biological perchlorate removal to changes in operating conditions and water quality characteristics. *Water Research* 37(1), 206–214.

Burlingame, G. A., Muldowney, J. J., and Maddrey, R. E. (1992). Cucumber flavor in Philadelphia's drinking water. *Journal AWWA* 84(8), 92–97.

Butler, J. A. V., and Ockrent, C. (1930). Studies in electrocapillarity: III. The surface tensions of solutions containing two surface-active solutes. *Journal of Physical Chemistry* 34, 2841–2859.

Campos, C., Schimmoller, L., Mariñas, B.J., Snoeyink, V.L., Baudin, I., and Laîné, J.M. (2000a). Adding PAC to remove DOC. *Journal AWWA* 92(8), 69–83.

Campos, C., Snoeyink, V.L., Mariñas, B.J., Baudin, I., and Laîné, J.M. (2000b). Atrazine removal by powdered activated carbon in floc blanket reactors. *Water Research* 34(16), 4070–4080.

Campos, C., Mariñas, B.J., Snoeyink, V.L., Baudin, I., and Laîné, J.M. (1998). Adsorption of trace organic compounds in CRISTAL processes. *Desalination* 117(1), 265–271.

Cannon, F.S., Snoeyink, V.L., Lee, R.G., Dagois, G., and DeWolfe, J.R. (1993). Effect of calcium in field-spent GACs on pore development during regeneration. *Journal AWWA* 85(3), 76–89.

Carter, M. C., and Weber, W. J., Jr. (1994). Modeling adsorption of TCE by activated carbon preloaded by background organic matter. *Environmental Science and Technology* 28(4), 614–623.

Carter, M. C., Weber, W. J., Jr., and Olmstead, K. P. (1992). Effects of background dissolved organic matter on TCE adsorption by GAC. *Journal AWWA* 84(8), 81–91.

Chae, S., Kim, S. M., Ahn, H., and Summers, R. S. (2006). Biofiltration of MIB and geosmin under varying influent conditions. In *Proceedings of the American Water Works Association Water Quality Technology Conference, Denver, CO.*

Chen, G., Dussert, B. W., and Suffet, I. H. (1997). Evaluation of granular activated carbons for removal of methylisoborneol to below odor threshold concentration in drinking water. *Water Research* 31(5), 1155–1163.

Chen, A. S. C., Snoeyink, V. L., and Fiessinger, F. (1987). Activated alumina adsorption of dissolved organic-compounds before and after ozonation. *Environmental Science and Technology* 21(1), 83–90.

Chen, A. S. C., Larson, R. A., and Snoeyink, V. L. (1982). Reactions of chlorine dioxide with hydrocarbons: Effects of activated carbon. *Environmental Science and Technology* 16, 268–273.

Cheng, W., Dastgheib, S. A., and Karanfil, T. (2005). Adsorption of dissolved natural organic matter by modified activated carbons. *Water Research* 39, 2281–2290.

Cho, H., and Summers, R. S. (2007a). MIB removal by powdered activated carbon: Impact of NOM, kinetics and pretreatment. In *Proceedings of the AWWA 2007 Annual Conference, Toronto, Ontario, Canada.*

Cho, H., and Summers, R. S. (2007b). Powdered activated carbon for controlling total organic carbon: Application and model development. In *Proceedings of the AWWA Water Quality Technology Conference, Charlotte, NC.*

Chowdhury, Z. K., et al. (2010). *GAC Biofilters in Retrofit Applications: An Approach for Cost Effective Regulatory Compliance* (in press). Denver, CO: Water Research Foundation.

Chowdhury, Z.K., Solarik, G., Owen, D., Hooper, D.M., and Summers, R.S. (1996). NOM removal by GAC adsorption: Implications of blending. In *Proceedings of the AWWA 2007 Annual Conference, Toronto, Ontario, Canada.*

Chudyk, W. A., and Snoeyink, V. L. (1984). Bioregeneration of activated carbon saturated with phenol. *Environmental Science and Technology* 18(1), 1–5.

Chudyk, W. A., and Snoeyink, V. L. (1981). *The Removal of Low Levels of Phenol by Activated Carbon in the Presence of Biological Activity.* University of Illinois Water Resources Center Report No. 154.

Chudyk, W. A., Snoeyink, V. L., Beckmann, D., and Temperly, T. J. (1979). Activated carbon versus resin adsorption of 2-methylisoborneol and chloroform. *Journal AWWA* 71(9), 529–538.

Cook, D., Newcombe, G., and Sztajnbok, P. (2001) The application of powdered activated carbon for MIB and geosmin removal: Predicting PAC doses in four raw waters. *Water Research* 35(5), 1325–1333.

Cornel, P. K., Summers, R. S., and Roberts, P. V. (1986). Diffusion of humic acid in dilute aqueous solution. *Journal of Colloid and Interface Science* 110(1), 149–164.

Corwin, C. (2010). Trace Organic Contaminant Removal from Drinking Waters by Granular Activated Carbon: Adsorption, Desorption, and the Effect of Background Organic Matter, Ph.D. dissertation, University of Colorado, Boulder, CO.

Corwin, C., and Summers, R. S. (2010). Scaling trace organic contaminant adsorption by granular activated carbon (in press). *Environmental Science and Technology.*

Corwin, C., Cardenas, J., and Summers, R. S. (2008). Micropollutant control by activated carbon adsorbers: Effect of backwashing and pulse loading. In *Proceedings of the AWWA Annual Conference, Atlanta, GA.*

Coughlin, R. W., and Ezra F. S. (1968). Role of surface acidity in the adsorption of organic pollutants on the surface of carbon. *Environmental Science and Technology* 2(4), 291–297.

Cover, A. E., and Pieroni, L. J. (1969). *Evaluation of the Lliterature on the Use of GAC for Tertiary Waste Treatment.* Report no. twrc-11. Cincinnati, OH: U.S. Department of the Interior, FWPCA.

Crittenden, J. C., and Weber, W. J., Jr. (1978a). Predictive model for the design of fixed-bed adsorbers: Parameter estimation and model development. *Journal of Environmental Engineering, ASCE* 104, 185–197.

Crittenden, J. C., and Weber, W. J., Jr. (1978b). Model for design of multicomponent adsorption systems. *Journal of Environmental Engineering, ASCE* 104, 1175–1195.

Crittenden, J. C., Berrigan, J. K., and Hand, D. W. (1986). Design of rapid small-scale adsorption tests for a constant diffusivity. *Journal Water Pollution Control Federation* 58(4), 312–319.

Crittenden, J. C., Luft, P., and Hand, D. W. (1985b). Prediction of multicomponent adsorption equilibria in background mixtures of unknown composition. *Water Research* 19(12), 1537–1548.

Crittenden, J. C., Berrigan, J. K., Hand, D. W. and, Lykins, B. (1987). Design of rapid fixed-bed adsorption tests for nonconstant diffusivities. *Journal of Environmental Engineering, ASCE* 113(2), 243–259.

Crittenden, J.C., Cortright, R.D., Rick, B., Tang, S.R., and Perram, D. (1988). Using GAC to remove VOCs from air stripper off-gas. *Journal AWWA* 80(5), 73–84.

Crittenden, J.C., Luft, P., Hand, D.W., Oravitz, J.L., Loper, S.W., and Ari, M. (1985a). Prediction of multicomponent adsorption equilibria using ideal adsorbed solution theory. *Environmental Science and Technology* 19(11), 1037–1043.

Crittenden J.C., Reddy, P.S., Arora, H., Trynoski, J., Hand, D.W., Perram, D.L., and Summers, R.S. (1991). Predicting GAC performance with rapid small-scale column tests. *Journal AWWA* 83(1), 77–87.

Crittenden, J.C., Wong, B.W.C., Thacker, W.E., Snoeyink, V.L., and Hinrichs, R.L. (1980). Mathematical model of sequential loading in fixed-bed adsorbers. *Journal Water Pollution Control Federation* 52(11), 2780–2795.

Daley, M. A., Tandon, D., Economy, J., and Hippo, E. J. (1996). Elucidating the porous structure of activated carbon fibers using direct and indirect methods. *Carbon* 34(10), 1191–1200.

Degrémont (2007). *Water Treatment Handbook*, 7th ed. Cachan, France.

DeLaat, J., Bouanga, F., and Dore, M. (1985). Influence of microbiological activity in granular activated carbon filters on the removal of organic compounds. In *Organic Micropollutants in Drinking Water and Health*, H. A. M. de Kruif and H. J. Kool, eds. New York: Elsevier.

DeMarco, J., Miller, R., Davis, D., and Cole, C. (1983). Experiences in operating a full-scale granular activated carbon system with on-site reactivation. In *Treatment of Water by Granular Activated Carbon*, M. J. McGuire and I. H. Suffet, eds. Washington, DC: American Chemical Society.

Denning, P. C., and Dvorak, B. I. (2009). Maximizing sorbent life: Comparison of columns in parallel, lead-lag series, and with bypass blending. *Water Environment Research* 81(2), 206–216.

Dielmann, L. M. J., III (1981). The reaction of aqueous hypochlorite, chlorite and hypochlorous acid with GAC. M.S. thesis, University of Illinois, Urbana, IL.

Ding, L., Mariñas, B.J., Schideman, L.C., Snoeyink, V.L., and Li, Q. (2006). Competitive effects of natural organic matter: Parametrization and verification of the three-component adsorption model COMPSORB. *Environmental Science and Technology* 40(1), 350–356.

Donati, C., Drikas, M., Hayes, R., and Newcombe, G. (1994). Microcystin-LR adsorption by powdered activated carbon. *Water Research* 28(8), 1735–1742.

Dubinin, M. M. (1960). The potential theory of adsorption of gases and vapors for adsorbents with energetically nonuniform surfaces. *Chemistry Reviews* 60, 235–241.

Dvorak, B. I., and Maher, M. K. (1999). GAC contactor design for NOM removal: Implications of EBCT and blending. *Journal of Environmental Engineering, ASCE* 125(2), 161–165.

Dyer, A. (1984). Uses of natural zeolites. *Chemistry and Industry* 7, 241–245.

Ebie, K., Li, F., and Hagishita, T. (1995). Effect of pore size distribution of activated carbon on the adsorption of humic substances and trace organic compounds. *Water Supply* 13(3–4), 65–70.

Economy, J., and Lin, R. Y. (1976). Adsorption characteristics of activated carbon fibers. *Applied Polymer Symposium* 29, 199–211.

Ellis, J., and Korth, W. (1993). Removal of geosmin and methylisoborneol from drinking water by adsorption on ultrastable zeolite-Y. *Water Research* 27, 535–539.

Evans, P. et al. (2010). *GAC Biofilters in Retrofit Applications: An Approach for Cost Effective Regulatory Compliance*. Project 4155 (in press). Denver, CO: Water Research Foundation.

Ford, R., Raczko, R., Phillips, S. L., and Arora, H. (1989). Developing carbon usage rate estimates for synthetic organic chemicals. In *Procceedings of the AWWA Annual Conference, Los Angeles, CA*.

Fox, R. D., Keller, R. T., and Pinamont, C. J. (1973). *Recondition and Reuse of Organically Contaminated Waste Brines*. EPA-R2-73-200. Washington, DC: USEPA.

Franz, M., Arafat, H. A., and Pinto, N. G. (2000). Effect of chemical surface heterogeneity on the adsorption mechanism of dissolved aromatics on activated carbon. *Carbon* 38(13), 1807–1819.

Gammie, L., and Giesbrecht, G. (1986). Full-scale operation of GAC contactors at Regina/Moose Jaw, Saskatchewan. In *Proceedings of the AWWA Annual Conference, Denver, CO*.

Gauntlett, R. B. (1975). *A Comparison Between Ion-Exchange Resins and Activated Carbon for the Removal of Organics from Water*. Technical Report TR 10. Medmenham, U.K.: Water Research Centre.

Gauntlett, R. B., and Packham, R. F. (1973) The use of activated carbon in water treatment. In *Proceedings of the Conference on Activated Carbon in Water Treatment, University of Reading*. Medmenham, U.K.: Water Research Association.

Gillogly, T.E.T., Snoeyink, V.L., Vogel, J.C., Wilson, C.M., and Royal, E.P. (1999). Determining GAC bed life. *Journal AWWA* 91(8), 98–110.

Gillogly, T.E.T., Snoeyink, V.L., Holthouse, A., Wilson, C.M., and Royal, E.P. (1998a). Effect of chlorine on PAC's ability to adsorb MIB. *Journal AWWA* 90(2), 107–114.

Gillogly, T.E.T., Snoeyink, V.L., Elarde, J.R., Wilson, C.M., and Royal, E.P. (1998b). ^{14}C MIB adsorption on PAC in natural water. *Journal AWWA* 90(1), 98–108.

Graese, S. L., Snoeyink, V. L., and Lee, R. G. (1987a). Granular activated carbon filter-adsorber systems. *Journal AWWA* 79(12), 64–74.

Graese, S. L., Snoeyink, V. L., and Lee, R. G. (1987b). *GAC Filter Adsorbers*. AWWA Research Foundation Report. Denver, CO: Water Research Foundation.

Graham, M. R., Summers, R. S., Simpson, M. R., and Macleod, B. W. (2000). Modeling equilibrium adsorption of 2-methylisoborneol and geosmin in natural waters. *Water Research* 34, 2291–2300.

Graham, M., Najm, I., Simpson, M., MacLeod, B., Summers, S., and Cummings, L. (1997). *Optimization of Powdered Activated Carbon Application for Geosmin and MIB Removal*. Awwa Research Foundation Report. Denver, CO: Water Research Foundation.

Graham, M., and Summers, R. S. (1996). The role of floc formation and its presence on adsorption by powdered activated carbon. In *Proceedings of the AWWA Annual Conference, Toronto, Ontario, Canada*.

Graveland, A., Kruithof, J. C., and Nuhn, P. A. N. M. (1981). Production of volatile halogenated compounds by chlorination after carbon filtration. Paper presented at the American Chemical Society Meeting, Atlanta, GA.

Greene, B. G., Snoeyink, V. L., and Pogge, F. W. (1994). *Adsorption of Pesticides by Powdered Activated Carbon*. Denver, CO: Awwa Research Foundation.

Grens, B. K., and Werth, C. J. (2001). Durability of wood-based versus coal-based GAC. *Journal AWWA* 93(4), 175–181.

Gustafson, D.I., Carr, K.H., Carson, D.B., Fuhrman, J.D., Hackett, A.G., Hoogheem, T.J., Snoeyink, V.L., Curry, M., Heijman, B., Chen, S.M., Hertl, P., and van Wesenbeeck, I. (2003). Activated carbon adsorption of chloroacetanilide herbicides and their degradation products from surface water supplies. *Journal of Water Supply: Research and Technology, AQUA* 52(6), 443–454.

Haist-Gulde, B., and Baldauf, G. (2006). Entfernung von Arzneimittelwirkstoffen und iodierten Röntgenkontrastmitteln in Aktivkohlefiltern. In *Organische Spurenstoffe in der Wasserversorgung*. Veröffentlichungen aus dem Technologiezentrum Wasser, Band 30, Karlsruhe, Germany.

Hand, D. W., Herlevich, J. A., Jr., Perram, D. L., and Crittenden, J. C. (1994). Synthetic adsorbent versus GAC for TCE removal. *Journal AWWA* 86(8), 64–72.

Hand, D.W., Herlevich, Jr., J.A., Perram, D.L., and J.C. Crittenden. (1989). Designing fixed-bed adsorbers to remove mixtures of organics. *Journal AWWA* 81(1), 67–77.

Hand, D. W., Crittenden, J. C., and Thacker, W. E. (1984). Simplified models for the design of fixed-bed adsorption systems. *Journal of Environmental Engineering, ASCE* 110(2), 440–456.

Hand, D. W., Crittenden, J. C., and Thacker, W. E. (1983). User-oriented batch reactor solutions to the homogeneous surface diffusion model. *Journal of Environmental Engineering, ASCE* 109(1), 82–101.

Halsey, G., and Taylor, H. S. (1947). The adsorption of hydrogen on tungsten powders. *Journal of Chemical Physics* 15, 624–630.

Hassler, J. W. (1974). *Activated Carbon.* New York: Chemical Publishing Co.

Hess, A. F. (1981). GAC treatment designs and costs for controlling volatile organic compounds in groundwater. Paper presented at the American Chemical Society Meeting., Atlanta, GA.

Ho, L., Slyman, N., Kaeding, U., and Newcombe, G. (2008). Optimizing PAC and chlorination practices for cylindrospermopsin removal. *Journal AWWA* 100(11), 88–96.

Ho, L., Hoefel, D., Bock, F., Saint, C.P., and Newcombe, G. (2007). Biodegradation rates of 2-methylisoborneol (MIB) and geosmin through sand filters and in bioreactors. *Chemosphere* 66(11), 2210–2218.

Ho, L., and Newcombe, G. (2005). Effect of NOM, turbidity and floc size on the PAC adsorption of MIB during alum coagulation. *Water Research* 39, 3668–3674.

Hoehn, R. C., Lavinder, S. R., Hamann, C., Jr., Hoffman, E. R., McElroy, J., and Snyder, E. G. (1987). THM-precursor control with powdered activated carbon in a pulsed-bed, solids contact clarifier. In *Proceedings of the AWWA Annual Conference, Kansas City, MO.*

Hong, S. (1996). Activated carbon adsorption of organic matter: backwashing, desorption and attenuation. Ph.D. disseration, University of Cincinnati.

Hong, S., and Summers, R. S. (2006). Effect of backwashing on activated carbon adsorption using plug flow pore surface diffusion model. *Korean Journal of Chemical Engineering* 23(1), 57–62.

Hong S., and Summers R. S. (1994). Impact of backwashing and desorption on GAC breakthrough of natural organic matter. In *Proceedings of the AWWA Annual Conference, New York.*

Hooper S., Summers, R. S., and Hong, S. (1996a). A systematic evaluation of the role of influent TOC and pH on GAC performance after enhanced coagulation. In *Proceedings of the AWWA Water Quality Technology Conference, Boston, MA.*

Hooper, S. M., Summers, R. S., Solarik, G., and Hong, S. (1996b). GAC performance for DBP control: Effect of influent concentration, seasonal variation, and pretreatment. In *Proceedings of the AWWA Annual Conference, Toronto, Ontario, Canada.*

Hooper, S. M., Summers, R. S., Solarik, G., and Owen, D. (1996c). Improving GAC performance by optimized coagulation. *Journal AWWA* 88(8),107–120.

Hopman, R., Siegers, W. G., and Kruithof, J. C. (1995). Organic micropollutant removal by activated carbon fiber filtration. *Water Supply* 13(3–4), 257–261.

Hu, J.Y., Aizawa, T., Ookubo, Y., Morita, T., and Magara, Y. (1998). Adsorptive characteristics of ionogenic aromatic pesticides in water on powdered activated carbon. *Water Research* 32(9), 2593–2600.

Huang, C. P. (1978). Chemical interactions between inorganics and activated carbon. In *Carbon Adsorption Handbook*, Cheremisinoff, P. and Ellerbusch, F. , eds. Ann Arbor, MI: Ann Arbor Science Publishers.

Jain, J. S., and Snoeyink, V. L. (1973). Adsorption from bisolute systems on active carbon. *Journal Water Pollution Control Federation* 45, 2463–2479.

Jankowska, H., Swiatkowski, A., and Choma, J. (1991). *Active Carbon.* New York: Ellis Horwood.

Jans, U., and Hoigne, J. (1998). Activated carbon and carbon black catalyzed transformation of aqueous ozone into OH-radicals. *Ozone Science and Engineering* 20, 67–90.

Jarvie, M.E., Hand, D.W., Bhuvendralingam, S., Crittenden, J.C., and Hokanson, D.R. (2005). Simulating the performance of fixed-bed granular activated carbon adsorbers: Removal of synthetic organic chemicals in the presence of background organic matter. *Water Research* 39, 2407–2421.

Juntgen, H. (1976). Phenomena of activated carbon regeneration. In *Translation of Reports on Special Problems of Water Technolgy*, Vol. 9: *Adsorption*. Report EPA-600/9-76-030. Cincinnati, OH: USEPA.

Juhola, A. J. (1970). *Optimization of the Regeneration Procedures for Granular Activated Carbon*. USEPA Report No. 17020 DAO. Cincinnati, OH: USEPA.

Kaneko, Y., Abe, M., and Ogino, K. (1989). Adsorption characteristics of organic compounds dissolved in water on surface-improved activated carbon fibers. *Colloids and Surfaces* 37, 211–222.

Karanfil, T., Dastgheib, S. A., and Mauldin, D. (2006). Exploring molecular sieve capabilities of activated carbon fibers to reduce the impact of NOM preloading on trichloroethylene adsorption. *Environmental Science and Technology* 40, 1321–1327.

Karanfil, T., and Dastgheib, S. A. (2004). Trichloroethylene adsorption by fibrous and granular activated carbons: Aqueous phase, gas phase, and water vapor adsorption studies. *Environmental Science and Technology* 38, 5834–5841.

Karanfil, T., and Kilduff, J. E. (1999). Role of granular activated carbon surface chemistry on the adsorption of organic compounds: 1. Priority pollutants. *Environmental Science and Technology* 33(18), 3217–3224.

Kasaoka, S., Sakata Y., Tanaka E., and Naitoh, R. (1989a). Preparation of activated fibrous carbon from phenolic fabric and its molecular-sieve properties. *International Chemical Engineering* 29(1), 101–114.

Kasaoka, S., Sakata, Y., Tanaka, E., and Naitoh, R. (1989b). Design of molecular-sieve carbon: Studies on the adsorption of various dyes in the liquid phase. *International Chemical Engineering* 29(4), 734–742.

Kawai, T., Yanagihara, T., and Tsutsumi, K. (1994). Adsorption characteristics of chloroform on modified zeolites from gaseous phase as well as its aqueous solution. *Colloid & Polymer Science* 272, 1620–1626.

Kennedy, D. C. (1973). Treatment of effluent from manufacture of chlorinated pesticides with a synthetic, polymeric adsorbent, Amberlite XAD-4. *Environmental Science and Technology* 7(2), 138–141.

Kilduff, J. E., Karanfil, T., and Weber, W. J., Jr. (1998). TCE adsorption by GAC preloaded with humic substances. *Journal AWWA* 90(5), 76–89.

Kilduff, J. E., Karanfil, T., Chin, Y. P., and Weber, W. J., Jr. (1996). Adsorption of natural organic polyelectrolytes by activated carbon: A size-exclusion chromatography study. *Environmental Science and Technology* 30(4), 1336–1343.

Kim, S. M., Summers, R. S., and Ahn, H. W. (2007). Predicting MIB breakthrough in granular activated carbon adsorbers. In *Proceedings of the AWWA Annual Conference Toronto, Ontario, Canada.*

Kim, S. M. (2006). Understanding and predicting 2-MIB adsorption by granular activated carbon. Ph.D. disseration, University of Colorado, Boulder, CO.

Kim, S. M., and Summers, R. S. (2006). Control of MIB and geosmin by granular activated carbon. In *Proceedings of the AWWA Annual Conference, San Antonio, TX.*

Kim, B. R., and Snoeyink, V. L. (1980). The monochloramine-GAC reaction in adsorption systems. *Journal AWWA* 72(8), 488–490.

Kim, B. R., Snoeyink, V. L., and Schmitz, R. A. (1978). Removal of dichloramine and ammonia by granular carbon. *Journal Water Pollution Control Federation* 50,122–133.

Kirisits, M.J., Snoeyink, V.L., Inan, H., Chee-Sanford, J.C., Raskin, L., and Brown, J.C. (2001). Water quality factors affecting bromate reduction in biologically active carbon filters. *Water Research* 35(4), 891–900.

Kirisits, M. J., Snoeyink, V. L., and Kruithof, J. P. (2000). The reduction of bromate by granular activated carbon. *Water Research* 34(17), 4250–4260.

Kirisits, M. J., and Snoeyink, V. L. (1999). Reduction of bromate in a BAC filter. *Journal AWWA* 91(8), 74–84.

Knappe, D. R. U. (1996). Predicting the removal of atrazine by powdered and granular activated carbon. Ph.D. dissertation, University of Illinois, Urbana, IL.

Knappe, D. R. U. (2006). Surface chemistry effects in activated carbon adsorption of industrial pollutants. In *Interface Science in Drinking Water Treatment: Theory and Applications*, G. Newcombe and D. Dixon, eds. Oxford, UK: Academic Press.

Knappe, D. R. U., Yuncu, B., and Deng, Q. (2010). *Removal of 2-Methylisoborneol and Geosmin by High-Silica Zeolites and Zeolite-Enhanced Ozonation.* Denver, CO: Water Research Foundation.

Knappe, D. R. U., Li, L., Quinlivan, P. A., and Wagner, T. B. (2003). *Effects of Activated Carbon Surface Chemistry and Pore Structure on the Adsorption of Trichloroethene and Methyl Tertiary-Butyl Ether from Natural Water.* AWWA Research Foundation Report. Denver, CO: Water Research Foundation.

Knappe, D. R. U., Rossner, A., Snyder, S. A., and Strickland, C. (2007). *Alternative Adsorbents for the Removal of Polar Organic Contaminants.* Awwa Research Foundation Report. Denver, CO: Water Research Foundation.

Knappe, D. R. U., Snoeyink, V. L., Dagois, G., and DeWolfe, J. R. (1992). The effect of calcium on the thermal regeneration of granular activated carbon. *Journal AWWA* 84(8), 73–80.

Knappe, D.R.U., Matsui, Y., Snoeyink, V.L., Roche, P., Prados, M.J., and M.M. Bourbigot (1998). Predicting the capacity of powdered activated carbon for trace organic compounds in natural waters. *Environmental Science and Technology* 32, 1694–1698.

Knappe, D.R.U., Snoeyink, V.L., Roche, P., Prados, M.J., and M.M. Bourbigot. (1997). The effect of preloading on rapid small-scale column test predictions of atrazine removal by GAC adsorbers. *Water Research* 31(11), 2899–2909.

Knappe, D.R.U., Snoeyink, V.L., Roche, P., Prados, M.J., and Bourbigot, M.M. (1999). Atrazine removal by preloaded GAC. *Journal AWWA* 91(10), 97–109.

Koffskey, W. E., and Brodtmann, N. V. (1981). *Organic Contaminant Removal in Lower Mississippi Drinking Water by Granular Activated Carbon.* Cincinnati, OH: USEPA.

Koffskey, W. E., and Lykins, B. W., Jr. (1987). Experiences with GAC filtration and on-site reactivation at Jefferson Parish, LA. In *Proceedings of the AWWA Annual Conference,* Kansas City, MO.

Komorita, J. D., and Snoeyink, V. L. (1985). Monochloramine removal from water by activated carbon. *Journal AWWA* 77(1), 62–64.

Kornegay, B. H. (1979). Control of synthetic organic chemicals by activated carbon: Theory, application, and regeneration alternatives. Presented at the Seminar on Control of Organic Chemicals in Drinking Water, USEPA,Covington VA, February 13–14, 1979.

Krasner, S.W., Yates, R.S., Sclimenti, M.J., Pastor, S.J., and Liang, S. (2004). Oxidation and adsorption of fishy/swampy/grassy odors in drinking water. In *Proceedings of the AWWA Annual Conference, Orlando, FL.*

Kreft, P., Trussell, A., Lang, J., Kavanaugh, M., and Trussell, R. (1981). Leaching of organics from a PVC-polyethylene-plexiglass pilot plant. *Journal AWWA* 73(10), 558–560.

Kumagai, J. S., and Kaufman, W. J. (1968). *Removal of Organic Contaminants, Phenol Sorption by Activated Carbon and Selected Macroporous Resins.* SERL Report No. 68-8. Sanitary Engineering Research Laboratory, University of California, Berkeley, CA.

Lalezary-Craig, S., Pirbazari, M., Dale, M.S., Tanaka, T.S., and McGuire, M.J. (1988). Optimizing the removal of geosmin and 2-methylisoborneol by powdered activated carbon. *Journal AWWA* 80(3), 73–80.

Lalezary, S., Pirbazari, M., and McGuire, M. J. (1986). Evaluating activated carbons for removing low concentrations of taste-producing and odor-producing organics. *Journal AWWA* 78(11), 76–82.

Lambert, T. W., Holmes, C. F. B., and Hrudey, S. E. (1996). Adsorption of microcystin-LR by activated carbon and removal in full-scale water treatment. *Water Research* 30(6), 1411–1422.

LeChevallier, M. W., Hassenauer, T. S., Camper, A. K., and McFeters, G. A. (1984). Disinfection of bacteria attached to granular activated carbon. *Applied and Environmental Microbiology* 48(5), 918–923.

Lee, M. C., Snoeyink, V. L., and Crittenden, J. C. (1981). Activated carbon adsorption of humic substances. *Journal AWWA* 73(8), 440–446.

Lee, M. C., Crittenden, J. C., Snoeyink, V. L. and Ari, M. (1983). Design of carbon beds to remove humic substances. *Journal of Environmental Engineering, ASCE* 109(3), 631–645.

Leon y Leon, C. A., and Radovic, L. R. (1992). Interfacial chemistry and electrochemistry of carbon surfaces. In *Chemistry and Physics of Carbon*, Vol. 24, P. A. Thrower, ed. New York: Marcel Dekker.

Leon y Leon, C. A., Solar, J. M., Calemma, V., and Radovic, L. R. (1992). Evidence for the protonation of basal plane sites on carbon. *Carbon* 30(5), 797–811.

Lettinga, G., Beverloo, W. A., and van Lier, W. C. (1978). Use of flocculated powdered activated carbon in water treatment. *Progress in Water Technology* 10, 537–554.

Li, L., Quinlivan, P. A., and Knappe, D. R. U. (2002). Effects of activated carbon surface chemistry and pore structure on the adsorption of organic contaminants from aqueous solution. *Carbon* 40(12), 2085–2100.

Li, Q., Snoeyink, V. L., Mariñas, B. J., and Campos, C. (2003a). Elucidating competitive adsorption mechanisms of atrazine and NOM using model compounds. *Water Research* 37, 773–784.

Li, L., Quinlivan, P. A., and Knappe, D. R. U. (2005). Predicting adsorption isotherms for aqueous organic micropollutants from activated carbon and pollutant properties. *Environmental Science and Technology* 39(9), 3393–3400.

Li, Q., Snoeyink, V. L., Mariñas, B. J., and Campos, C. (2003b). Pore blockage effect of NOM on atrazine adsorption kinetics of PAC: The roles of PAC pore size distribution and NOM molecular weight. *Water Research* 37, 4863–4872.

Liang, S., and Weber, W. J., Jr. (1985). Parameter evaluation for modeling multicomponent mass transfer in fixed-bed adsorbers. *Chemical Engineering Communications* 35, 49–61.

Love, O. T., Jr., and Eilers, R. G. (1982) Treatment of drinking-water containing trichloroethylene and related industrial solvents. *Journal AWWA* 74(8), 413–425.

Love, O. T., Jr., Robeck, G. G., Symons, J. M., and Buelow, R. W. (1973). Experience with activated carbon in the USA. In *Proceedings of the Conference on Activated Carbon in Water Treatment, University of Reading.* Medmenham, UK: Water Research Centre.

Lykins, B. W., Jr., Clark, R. M., and Adams, J. Q. (1988a). Granular activated carbon for controlling THMs. *Journal AWWA* 80(5), 85–92.

Lykins, B. W., Jr., Clark, R. M., and Cleverly, D. H. (1988b). Polychlorinated dioxin and furan discharge during carbon reactivation. *Journal of Environmental Engineering, ASCE* 114(2), 300–316.

Manes, M. (1998). Activated carbon adsorption fundamentals. In *Encyclopedia of Environmental Analysis and Remediation*, Vol. 1, R. A. Meyers, ed. New York: Wiley.

Matsui, Y., Knappe, D. R. U., and Takagi, R. (2002). Pesticide adsorption by granular activated carbon adsorbers: 1. Effect of natural organic matter preloading on removal rates and model simplification. *Environmental Science and Technology* 36(15), 3426–3431.

Matsui, Y. Kamei, T., Kawase, E., Snoeyink, V.L., and Tambo, N. (1994). Adsorption of intermittently loaded pesticides by granular activated carbon. *Journal AWWA* 86(9), 91–102.

Matsui, Y., Murai, K., Sasaki, H., Ohno, K., and Matsushita, T. (2008). Submicron-sized activated carbon particles for the rapid removal of chlorinous and earthy-musty compounds. *Journal of Water Supply: Research and Technology, AQUA* 57(8), 577–583.

Mattson, J. S., and Mark, H. B., Jr. (1971). *Activated Carbon: Surface Chemistry and Adsorption from Solution.* New York: Marcel Dekker.

McCarty, P. L., Argo, D., and Reinhard, M. (1979). Operational experiences with activated carbon adsorbers at Water Factory 21. *Journal AWWA* 71(11), 683–689.

McCreary, J. J., and Snoeyink, V. L. (1980). Characterization and activated carbon adsorption of several humic substances. *Water Research* 14, 151–160.

McGuire, M. J., and Suffet, I. H. (1984). Aqueous chlorine activated carbon interactions. *Journal of Environmental Engineering, ASCE* 110(3), 629–645.

Meijers, J. A. P., and van der Leer, R. C. (1983). The use of powdered activated carbon in conventional and new techniques. In *Activated Carbon in Drinking Water Technology*. AWWA Research Foundation Report. Denver, CO: Water Research Foundation.

Melin, G. (1999). *Evaluation of the Applicability of Synthetic Resin Sorbents for MTBE Removal from Water*. The California MTBE Research Partnership. Fountain Valley, CA: National Water Research Institute.

Meyer, K. J., Summers, R. S., Westerhoff, P., and Metz, D. H. (2005). Biofiltration for geosmin and MIB removal. In *Proceedings of the AWWA Annual Conference, San Francisco, CA*.

Mezzari, I. A. (2006). Predicting the adsorption capacity of activated carbon for organic contaminants from fundamental adsorbent and adsorbate properties. M.S. thesis, North Carolina State University, Raleigh, NC.

Miller, J. A., Snoeyink, V. L., and Harrell, S. (1995). The effect of granular activated carbon surface chemistry on bromate reduction. In *Disinfection By-Products in Water Treatment*, R. Minear and G. Amy, eds. Ann Arbor, MI: Lewis Publishers.

Miltner, R. J., Summers, R. S., Dugan, N. R., Koechling, M. and Moll, D. M. (1996). A comparative evaluation of biological filters. In *Proceedings of the AWWA Water Quality Technology Conference, Boston, MA*.

Najm, I. N. (1996). Mathematical modeling of PAC adsorption processes. *Journal AWWA* 88(10), 79–89.

Najm, I. N., Snoeyink, V. L., and Richard, Y. (1993). Removal of 2,4,6-trichlorophenol and natural organic matter from water supplies using PAC in floc-blanket reactors. *Water Research* 27(4), 551–560.

Najm, I. N., Snoeyink, V. L., and Richard, Y. (1991a). Effect of initial concentration of a SOC in natural water on its adsorption by activated carbon. *Journal AWWA* 83(8), 57–63.

Najm, I. N., Snoeyink, V. L., Lykins, B. W., Jr., and Adam, J. Q. (1991b). Using powdered activated carbon: A critical review. *Journal AWWA* 83(1), 65–76.

Najm, I.N., Snoeyink, V.L., Suidan, M.T., Lee, C.H., and Richard, Y. (1990). Effect of particle-size and background natural organics on the adsorption efficiency of PAC. *Journal AWWA* 82(1), 65–72.

Najm, I. N., Snoeyink, V. L., Suidan, M. T., and Richard, Y. (1989). PAC in floc blanket reactors. In *Proceedings of the AWWA Annual Conference, Los Angeles, CA*.

Namkung, E., and Rittmann, B. E. (1987). Removal of taste-causing and odor-causing compounds by biofilms grown on humic substances. *Journal AWWA* 79(7), 107–112.

Neely, J. W., and Isacoff, E. G. (1982). *Carbonaceous Adsorbents for the Treatment of Ground and Surface Waters*. New York: Marcel Dekker.

Newcombe, G., and Nicholson, B. (2004). Water treatment options for dissolved cyanotoxins. *Journal of Water Supply: Research and Technology, AQUA* 53(4), 227–239.

Newcombe, G., and Cook, D. (2002). Removal of tastes and odours by PAC. *Journal of Water Supply: Research and Technology, AQUA* 51(8), 463–474.

Newcombe, G., Drikas, M., and Hayes, R. (1997). Influence of characterized natural organic material on activated carbon adsorption: II. Effect on pore volume distribution and adsorption of 2-methylisoborneol. *Water Research* 31, 1065–1073.

Okouchi, S., Saegusa, H., and Nojima, O. (1992). Prediction of environmental parameters by adsorbability index: water solubilities of hydrophobic organic pollutants. *Environment International* 18, 249–261.

Pelekani, C., and Snoeyink, V. L. (1999). Competitive adsorption in natural water: Role of activated carbon pore size. *Water Research* 33(5), 1209–1219.

Pelekani, C., and Snoeyink, V. L. (2000). Competitive adsorption between atrazine and methylene blue on activated carbon: The importance of pore size distribution. *Carbon* 38, 1423–1436.

Pelekani, C., and Snoeyink, V. L. (2001). A kinetic and equilibrium study of competitive adsorption between atrazine and Congo red dye on activated carbon: The importance of pore size distribution. *Carbon* 39, 25–37.

Pendleton, P., Wong, S.H., Schumann, R., Levay, G., Denoyel, R., and Rouquerol, J. (1997). Properties of activated carbon controlling 2-methylisoborneol adsorption. *Carbon* 35(8), 1141–1149.

Prober, R., Pyeha, J. J., and Kidon, W. E. (1975). Interaction of activated carbon with dissolved-oxygen. *AIChE Journal* 21, 1200–1204.

Puri, B. R. (1970). Surface complexes on carbons. In *Chemistry and Physics of Carbon*, Vol. 6, P. L. Walker, Jr., ed. New York: Marcel Dekker.

Puri, B. R. (1983). Physicochemical aspects of carbon affecting adsorption from the aqueous phase. In *Advances in Chemistry Series*, Vol. 202, M. J. McGuire and I. H. Suffet, eds. Washington, DC: American Chemical Society.

Qi, S., Adham, S. S. Snoeyink, V. L., and Lykins, B. W., Jr. (1994). Prediction and verification of atrazine adsorption by PAC. *Journal of Environmental Engineering, ASCE* 120(1), 202–218.

Qi, S., Shideman, L., Mariñas, M., Snoeyink, V.L., and Campos, C. (2007). Simplification of the IAST for activated carbon adsorption of trace organic compounds from natural water. *Water Research* 41(2), 440–448.

Qi, S., Snoeyink, V.L., Beck, E.A., Koffskey, W.E., and Lykins, B.W., Jr. (1992). Using isotherms to predict GAC's capacity for synthetic organics. *Journal AWWA* 84(9), 113–120.

Quinlivan, P. A., Li, L., and Knappe, D. R. U. (2005). Effects of activated carbon characteristics on the simultaneous adsorption of aqueous organic micropollutants and natural organic matter. *Water Research* 39(8), 1663–1673.

Radke, C. J., and Prausnitz, J. M. (1972). Thermodynamics of multi-solute adsorption from dilute liquid solutions. *AIChE Journal* 18(4), 761–768.

Randtke, S. J., and Jepsen, C. P. (1982). Effects of salts on activated carbon adsorption of fulvic-acids. *Journal AWWA* 74(2), 84–93.

Randtke, S. J., and Snoeyink, V. L. (1983). Evaluating GAC adsorption capacity. *Journal AWWA* 75(8), 406–413.

Robeck, G. G. (1975). *Evaluation of Activated Carbon*. Water Supply Research Laboratory. Cincinnati, OH: National Environmental Research Center.

Robeck, G. G., Dostal, K. A., Cohen, J. M., and Kreissl, J. F. (1965). Effectiveness of water treatment processes in pesticide removal. *Journal AWWA* 57(2), 181–199.

Roberts, P. V., and Summers, R. S. (1982). Performance of granular activated carbon for total organic carbon removal. *Journal AWWA* 74(2), 113–118.

Rossner, A., Snyder, S. A., and Knappe, D. R. U. (2009). Removal of emerging contaminants of concern by alternative adsorbents. *Water Research* 43(15), 3787–3796.

Rossner, A., and Knappe, D. R. U. (2008). MTBE adsorption kinetics on alternative adsorbents and packed bed adsorber performance. *Water Research* 42(8/9), 2287–2299, 2008.

Rossner, A. (2008). Removal of polar and emerging organic contaminants by alternative adsorbents. Ph.D. dissertation, North Carolina State University, Raleigh, NC.

Schideman, L., Mariñas, B. J., Snoeyink, V. L., and Campos, C. (2006a) Three-component competitive adsorption model for fixed-bed and moving-bed granular activated carbon adsorbers: I. Model development. *Environmental Science and Technology* 40(21), 6805–6811.

Schideman, L., Mariñas, B. J,, Snoeyink, V. L., and Campos, C. (2006b). Three-component competitive adsorption model for fixed-bed and moving-bed granular activated carbon adsorbers: II. Model parameterization and verification, *Environmental Science and Technology* 40(21), 6812–6817.

Schideman, L., Snoeyink, V.L., Mariñas, B.J., Ding, L., and Campos, C. (2007). Application of a three-component competitive adsorption model to evaluate and optimize granular activated carbon systems. *Water Research* 41(15), 3289–3298.

Schuliger, W. G., Riley, G. N., and Wagner, N. J. (1987). Thermal reactivation of GAC: A proven technology. In *Proceedings of the AWWA Annual Conference, Kansas City, MO.*

Schuliger, W. A., and MacCrum, J. M. (1973). GAC reactivation system design and operating conditions. Presented at the AIChE Meeting, Detroit, MI, June 5, 1973.

Servais, P., Prevost, M., Laurent, P., Joret, J.C., Summers R.S., Hambsch, B., and Ventresque, C. (2005). Biodegradable organic matter in drinking water treatment. In *Biodegradable Organic Matter in Drinking Water Treatment and Distribution*, Prevost, M., Joret, J. C., Laurent, P., and Servais, P., eds. Denver, CO: American Water Works Association.

Sheindorf, C., Rebhun, M., and Sheintuch, M. (1981). A Freundlich-type multicomponent isotherm. *Journal of Colloid and Interface Science* 79(1), 136–42.

Silvey, J. K. G., and Roach, A. W. (1964). Studies on microbiotic cycles in surface water. *Journal AWWA* 56(1), 60–71.

Sing, K. S. W. (1995). Physisorption of nitrogen by porous materials. *Journal of Porous Materials* 2, 5–8.

Sing, K.S.W., Everett, D.H., Haul, R.A.W., Moscou, L., Pierotti, R.A., Rouquerol, J., and Siemieniewska, T. (1985). Reporting physisorption data for gas/solid systems with special reference to the determination of surface area and porosity. *Pure and Applied Chemistry* 57, 603–619.

Smith, E. H., Tseng, S. K., and Weber, W. J., Jr. (1987). Modeling the adsorption of target compounds by GAC in the presence of background dissolved organic matter. *Environmental Progress* 6(2), 18–25.

Snoeyink, V. L. (1983). Control stragegy: Adsorption techniques. In *Occurrence and Removal of Volatile Organic Chemicals from Drinking Water*. AWWA Research Foundation/KIWA Report. Denver, CO: Water Research Foundation.

Snoeyink, V. L. (1990). Adsorption of organic compounds. In *Water Quality and Treatment*, 4th ed. New York: McGraw-Hill.

Snoeyink, V. L., and Knappe, D. R. U. (1994). Analysis of pilot and full scale granular activated carbon performance data. In *Proceedings of the AWWA Annual Conference, New York.*

Snoeyink, V. L., and Summer, R. S. (1999). Adsorption of organic compounds. In *Water Quality and Treatment*, 5th ed. New York: McGraw-Hill.

Snoeyink, V. L., and Suidan, M. T. (1975). Dechlorination by activated carbon and other reducing agents. In *Disinfection: Water and Wastewater*, J. D. Johnson, ed. Ann Arbor, MI: Ann Arbor Science Publishers.

Snoeyink, V. L., and Weber, W. J., Jr. (1972). Surface functional groups on carbon and silica. In *Progress in Surface and Membrane Science*, Vol. 5, J. F. Danielli, ed. New York: Academic Press.

Snoeyink, V. L., Lai, H. T., Johnson, J. H., and Young, J. F. (1974). Active carbon: Dechlorination and the adsorption of organic compounds. In *Chemistry of Water Supply, Treatment and Distribution*, A. Rubin, ed. Ann Arbor, MI: Ann Arbor Science Publishers.

Solarik, G., Hooper, S. M., Summers, R. S., Owen, D. M. (1995a). Predicting and characterizing the removal of DBP precursors by GAC. In *Proceedings of the AWWA Annual Conference, Anaheim, CA.*

Solarik G., et al. (1995b). DBP precursor treatment studies at Little Falls, New Jersey, to meet the ICR requirements. In *Proceedings of the AWWA Water Quality Technology Conference, New Orleans, LA.*

Solarik, G., et al. (1996). The impact of ozonation and biotreatment on GAC performance for NOM Removal and DBP control. In *Proceedings of the AWWA Annual Conference, Toronto, Ontario, Canada.*

Sontheimer, H. (1976). The use of powdered activated carbon. In *Translation of Reports on Special Problems of Water Technolgy*, Vol. 9: *Adsorption.* Report EPA-600/9-76-030, Cincinnati, OH: USEPA.

Sontheimer, H., and Hubele, C. (1987). The use of ozone and granular activated carbon in drinking water treatment. In *Treatment of Drinking Water for Organic Contaminants*, P. M. Huck and P. Toft, eds. New York: Pergamon Press.

Sontheimer, H., Crittenden, J. C., and Summers, R. S. (1988). *Activated Carbon for Water Treatment*, 2nd ed. Karlsruhe, Germany: DVGW-Forschungstelle am Engler-Bunte-Institut der Universität Karlsruhe.

Speth, T. F., and Miltner, R. J. (1998). Adsorption capacity of GAC for synthetic organics, part II. *Journal AWWA* 90(4), 171–174.

Speth, T. F. (1991). Evaluating capacities of GAC preloaded with natural water. *Journal of Environmental Engineering, ASCE* 117(1), 66–79.

Speth, T. F., and Miltner, R. J. (1990). Technical note: Adsorption capacity of GAC for synthetic organics. *Journal AWWA* 82(2), 72–75.

Suidan, M. T., Snoeyink, V. L., and Schmitz, R. A. (1976). Performance prediction for the removal of aqueous free chlorine by packed beds of granular activated carbon. In *Water–1976*, Vol. I: *Physical Chemical Wastewater Treatment*. AIChE Symposium Series 73.

Suidan, M. T., Snoeyink, V. L. and Schmitz, R. A. (1977a). Reduction of aqueous HOCl with activated carbon. *Journal of Environmental Engineering, ASCE* 103(4), 677–691.

Suidan, M. T., Snoeyink, V. L., and Schmitz, R. A. (1977b). Reduction of aqueous free chlorine with granular activated carbon pH and temperature effects. *Environmental Science and Technology* 11, 785–789.

Summers, R. S., and Roberts, P. V. (1988a). Activated carbon adsorption of humic substances: I. Heterodisperse mixtures and desorption. *Journal of Colloid and Interface Science* 122(2), 367–381.

Summers, R. S., and Roberts, P. V. (1988b). Activated carbon adsorption of humic substances: II. Size exclusion and electrostatic interactions. *Journal of Colloid and Interface Science* 122(2), 382–397.

Summers, R. S., and Roberts, P. V. (1984). Simulation of DOC removal in activated carbon beds. *Journal of Environmental Engineering, ASCE* 110(1), 73–92.

Summers, R. S., DiCarlo, D., and Palepu, S. (1990). GAC adsorption in the presence of background organic matter: Pretreatment approaches and attenuation of shock loadings. In *Proceedings of the AWWA Annual Conference, Cincinnati, OH*.

Summers, R. S., Hooper, S., and Hong, S. (1996). GAC precursor removal studies. In *ICR Manual for Bench- and Pilot-Scale Treatment Studies*. EPA 814-B-96-003. Cincinnati, OH: USEPA, Office of Water.

Summers, R. S., Benz, M. A., Shukairy, H. M., and Cummings, L. (1993). Effect of separation processes on the formation of brominated THMs. *Journal AWWA* 85(1), 88–95.

Summers, R. S., Chae, S., Kim, S. M., and Ahn, H. W. (2006). Biodegradation of MIB and geosmin in biological sand and BAC filters: Acclimation, steady-state and varying influent conditions. In *Recent Progress in Slow Sand and Alternative Biofiltration Processes*, R. Gimbel, N. Graham, and M. R. Collins, eds. London: IWA Publishing.

Summers, R.S., Haist, B., Koehler, J., Ritz, J., Zimmer, G., and Sontheimer, H. (1989). The influence of background organic matter on GAC adsorption. *Journal AWWA* 81(5), 66–74.

Summers, R. S., Hong, S., Hooper, S., and Solarik, G. (1994). Adsorption of natural organic matter and disinfection by-product precursors. In *Proceedings of the AWWA Annual Conference, New York*.

Summers, R.S., Hooper, S.M., Solarik, G., Owen, D.M., and Hong, S. (1995). Bench-scale evaluation of GAC for NOM control. *Journal AWWA* 87(8), 69–80.

Summers, R. S., et al. (1997). *DBP Precursor Control with GAC Adsorption*. AWWA Research Foundation Report. Denver, CO: Water Research Foundation.

Suzuki, M., Misic, D. M., Koyama, O., and Kawazoe, K. (1978). Study of thermal regeneration of spent activated carbons: Thermogravimetric measurement of various single component organics loaded on activated carbons. *Chemical Engineering Science* 33, 271–279.

Symons, J.M., Stevens, A.A., Clark, R.M., Geldreich, E.E., Love, O.T., Jr., and DeMarco, J. (1981). *Treatment Techniques for Controlling Trihalomethanes in Drinking Water*. EPA-60012-81-156. Cincinnati, OH: USEPA.

Ternes, T.A., Meisenheimer, M., McDowell, D., Sacher, F., Brauch, H.J., Haist-Gulde, B., Preuss, G., Wilme, U., and Zulei-Seibert, N. (2002). Removal of pharmaceuticals during drinking water treatment. *Environmental Science and Technology* 36, 3855–3863

Thacker, W. E., Crittenden, J. C., and Snoeyink, V. L. (1984). Modeling of adsorber performance: Variable influent concentration and comparison of adsorbents. *Journal Water Pollution Control Federation* 56, 243–250.

Thacker, W. E., Snoeyink, V. L., and Crittenden, J. C. (1983). Desorption of compounds during operation of GAC adsorption systems. *Journal AWWA* 75(3), 144–149.

Townsend, R. P. (1984). Ion exchange in zeolites: Basic principles. *Chemistry and Industry* 7, 246–251.

Urfer, D., Huck, P. M., Booth, S. D. J., and Coffey, B. M. (1997). Biological filtration for BOM and particle removal: A critical review. *Journal AWWA* 89(12), 83–98.

USEPA (2002). *Community Water System Survey 2000.* EPA 815-R-02-005B. Washington, DC: USEPA.

USEPA (2000). *ICR Treatment Study Database,* Version 1.0. EPA 815-C-00-003. Washington, DC: Office of Water.

Uyak, V., Yavuz, S., Toroz, I., Ozaydin, S., and Genceli, E.A. (2007). Disinfection by-products precursors removal by enhanced coagulation and PAC adsorption. *Desalination* 216, 334–344.

van Vliet, B. M. (1985). Regeneration principles. In *Proceedings of the Symposium on Design and Operation of Plants for the Recovery of Gold by Activated Carbon.* Johnannesburg, South Africa: South African Institute of Mining and Metallurgy.

Vidic, R. D., and Suidan, M. T. (1991). Role of dissolved oxygen on the adsorptive capacity of activated carbon for synthetic and natural organic matter. *Environmental Science and Technology* 25(9), 1612–1618.

Voudrias, E. A., Larson, R. A., and Snoeyink, V. L. (1985). Effects of activated carbon on the reactions of free chlorine with phenols. *Environmental Science and Technology* 19(5), 441–449.

Voudrias, E.A., Dielmann, L.M.J., Snoeyink, V.L., Larson, R.A., McCreary, J.J., and Chen, A.S.C. (1983). Reactions of chlorite with activated carbon and with vanillic acid and indan adsorbed on activated carbon. *Water Research* 17(9), 1107–1114.

Wang, H.X., Ho, L., Lewis, D.M., Brookes, J.D., and Newcombe, G. (2007). Discriminating and assessing adsorption and biodegradation removal mechanisms during granular activated carbon filtration of microcystin toxins. *Water Research* 41(18), 4262–4270.

Wang, J. Z., and Summers, R. S. (1995). Biomass growth and distribution in drinking water biofilters and its impact on natural organic matter removal. In *Proceedings of the AWWA Annual Conference, Anaheim, CA.*

Wang, J. Z., Summers, R. S., and Miltner, R. J. (1995). Biofiltration performance: 1. Relationship to biomass. *Journal AWWA* 87(12), 55–63.

Weber, W. J., Jr., and DiGiano, F. A. (1996). *Process Dynamics in Environmental Systems.* New York: Wiley.

Weber, W. J., Jr., and Liu, K. T. (1980). Determination of mass transport parameters for fixed-bed adsorbers. *Chemical Engineering Communications* 6, 49–60.

Weber, W. J., Jr., and Pirbazari, M. (1982). Adsorption of toxic and carcinogenic compounds from water. *Journal AWWA* 74(4), 203–209.

Weber, W. J., Jr., and Smith, E. H. (1987). Simulation and design models for adsorption processes. *Environmental Science and Technology* 21(11), 1040–1050.

Westerhoff, P., Yoon, Y., Snyder, S., and Wert, E. (2005). Fate of endocrine-disruptor, pharmaceutical, and personal care product chemicals during simulated drinking water treatment processes. *Environmental Science and Technology* 39(17), 6649–6663.

Westrick, J. J., and Cohen, J. M. (1976). Comparative effects of chemical pretreatment on carbon adsorption. *Journal Water Pollution Control Federation* 48, 323–338.

Wiesner, M. R., Rook, J. J., and Fiessinger, F. (1987). Optimizing the placement of GAC filtration units. *Journal AWWA* 79(12), 39–49.

Worch, E. (2008). Fixed-bed adsorption in drinking water treatment: A critical review on models and parameter estimation. *Journal of Water Supply: Research and Technology, AQUA* 57(3), 171–183.

Wulfeck, W.M., Jr., and Summers R. S. (1994). Control of DBP formation using retrofitted GAC filter-adsorbers and ozonation. In *Proceedings of the AWWA Water Quality Technology Conference, San Francisco, CA.*

Yehaskel, A. (1978). *Activated Carbon Manufacture and Regeneration.* Park Ridge, NJ: Noyes Data Corporation.

Yu, Z., Peldszus, S., and Huck, P. M. (2009). Adsorption of selected pharmaceuticals and an endocrine disrupting compound by granular activated carbon: 2. Model prediction. *Environmental Science and Technology* 43(5), 1474–1479.

Zachman, B. A., and Summers, R. S. (2010). Modeling TOC breakthrough in granular activated carbon adsorbers. *Journal of Environmental Engineering, ASCE* (in press).

Zachman, B. A., Rajagopalan, R., and Summers, R. S. (2007a). Modeling NOM breakthrough in GAC adsorbers using nonparametric regression techniques. *Environmental Engineering Science* 24(9), 1280–1296.

Zachman, B. A., Summers R. S., and Swanson W. (2007b). TOC breakthrough models: Practical application and case studies. In *Proceedings of the AWWA Annual Conference, Toronto, Ontario, Canada.*

Zhang, Q., Crittenden, J., Hristovski, K., Hand, D., and Westerhoff, P. (2009). User-oriented batch reactor solutions to the homogeneous surface diffusion model for different activated carbon dosages. *Water Research* 43(7), 1859–1866.

Zimmer, G., Brauch, H. J., and Sontheimer, H. (1989). Activated carbon adsorption of organic pollutants. In *Aquatic Humic Substances: Influence on Fate and Treatment of Pollutants*, I. Suffet and P. MacCarthy, eds. Advances in Chemistry Series No. 219. Washington, DC: American Chemical Society.

Zimmer, G., Haist, B., and Sontheimer, H. (1987). The influence of preadsorption of organic matter on the adsorption behavior of chlorinated hydrocarbons. In *Proceedings of the AWWA Annual Conference*, Kansas City, MO.

CHAPTER 15
NATURAL TREATMENT SYSTEMS

Saroj K. Sharma, Ph.D.
Senior Lecturer, Urban Water and Sanitation Department
UNESCO–IHE Institute for Water Education
Delft, The Netherlands

Gary Amy, Ph.D.
Director, Water Desalination and Reuse Center,
King Abdullah University of Science and Technology
Thuwal, Saudi Arabia
and
Professor of Water Supply Engineering,
UNESCO-IHE Institute for Water Education
Delft, The Netherlands

INTRODUCTION **15.2**
- Background 15.2
- Managed Aquifer Recharge (MAR) for Water Treatment and Wastewater Reuse 15.2

RIVER (RBF) AND LAKE (LBF) BANK FILTRATION **15.3**
- Removal Mechanisms 15.4
- Factors Affecting Performance of RBF/LBF Systems 15.4
- Attributes and Limitations 15.6

ARTIFICIAL RECHARGE AND RECOVERY (ARR) **15.7**
- Infiltration Rates and Operation 15.9
- Factors Affecting Performance of ARR Systems 15.9
- Attributes and Limitations 15.10

SUBSURFACE GROUNDWATER TREATMENT **15.10**

SOIL AQUIFER TREATMENT (SAT) FOR INDIRECT POTABLE REUSE **15.12**
- SAT Soil Requirements 15.14
- Hydraulic Loading and Operation 15.14
- Water Depth 15.14

WATER QUALITY IMPROVEMENTS IN NATURAL TREATMENT SYSTEMS **15.15**
- Particles and Turbidity 15.15
- Microorganisms 15.16
- Bulk Natural Organic Matter 15.16
- Trace Organic Contaminants 15.17
- Nitrogen and Phosphorus 15.18
- Heavy Metals 15.19
- Comparison of Slow Sand Filtration, Bank Filtration, and Soil Aquifer Treatment 15.19

DESIGN AND OPERATION OF NATURAL WATER TREATMENT SYSTEMS **15.20**
- Type, Number, Spacing, and Distance of Production Wells 15.20
- Hydraulic Loading Rates/Travel Time 15.21
- Redox Conditions 15.21
- Clogging 15.22

SELECTED CASE STUDIES OF NATURAL TREATMENT SYSTEMS **15.23**
- Lake Bank Filtration: Berlin, Germany 15.24
- Riverbank Filtration: Louisville, Ky. 15.25
- Riverbank Filtration: Düsseldorf, Germany 15.25
- Riverbank Filtration and Artificial Recharge: Maribor, Slovenia 15.26
- Dune Infiltration: Leiduin, The Netherlands 15.26

Artificial Recharge: Graz, Austria ... 15.27
Riverbank Filtration and Artificial Recharge: Aurora, Colo. ... 15.27
Soil Aquifer Treatment: Dan Region, Israel ... 15.28
ABBREVIATIONS ... **15.28**
REFERENCES ... **15.29**

INTRODUCTION

Background

Since water demand for the growing world population is increasing rapidly, good quality water sources are becoming scarce, costs of water treatment are increasing, new contaminants are being found in water supply sources, and drinking water quality regulations are becoming more stringent; thus water utilities are looking for comprehensive water supply and treatment technologies to address these emerging challenges. It is well known from the hydrological cycle that the water quality of river or rainwater improves as it passes through the soil layer. Using this natural phenomenon, different natural treatment systems such as bank filtration, aquifer recharge and recovery, subsurface groundwater treatment, and soil aquifer treatment have been applied in many countries throughout the world to improve the quality of water from different sources in the process of drinking water production and, in some cases, wastewater reclamation/reuse. Natural systems have been in use for water and wastewater treatment from time immemorial. There is now renewed interest in these methods because natural treatment systems are robust and, if properly designed and operated, provide a comprehensive, sustainable treatment of multiple contaminants present in water sources.

Natural water treatment systems rely on natural phenomena comprising different physical, chemical, and biological removal mechanisms that take place during soil passage for the improvement in water quality. Furthermore, water treated using natural systems is reported to have relatively high biological stability, thus minimizing the potential of regrowth in the water distribution system. The suitability and performance of such natural treatment systems, however, depend on hydrogeological conditions, source water quality, process conditions applied, and water quality goals to be achieved by the treatment.

This chapter provides an overview of different soil passage–based natural treatment systems that can be applied as the main treatment or as pretreatment for drinking water treatment and wastewater reuse applications. Furthermore, it outlines water quality improvements from and design considerations for these natural treatment systems and finally presents some case studies.

Managed Aquifer Recharge (MAR) for Water Treatment and Wastewater Reuse

MAR is the infiltration or injection of a source water, such as river or lake water, treated wastewater, or urban stormwater, into an aquifer under controlled conditions with the intention of storage and/or treatment of the water. Water can be introduced into the aquifer by a number of methods, including infiltration via basins or galleries or by use of injection wells. The use of MAR has the potential to provide benefits for water resources and environmental management because it can augment quantity as well as improve the quality of water. There are different MAR systems designed to increase the quantity and improve the quality of water. MAR systems for water quality improvement or water treatment include:

1. Bank filtration (river or lake)
2. Artificial recharge and recovery (ARR)

3. Subsurface groundwater treatment
4. Soil aquifer treatment (SAT)

RIVER (RBF) AND LAKE (LBF) BANK FILTRATION

Bank filtration, as an engineered system for water treatment, has been employed dating back to the nineteenth century. During bank filtration, river or lake water is extracted indirectly by drawing it through the subsurface prior to use. Extraction is accomplished by an infiltration gallery or line of wells (horizontal, vertical, or at an angle) located at a short to intermediate distance from the bank of the river or lake. During extraction, the groundwater discharge into the river decreases, and the groundwater table near the waterline may decrease below the river water level. Consequently, surface water enters the aquifer and flows to the gallery or wells. While bank filtration systems generally are not operated to provide storage of surface water underground, they do provide water quality improvements during subsurface transport. To ensure a satisfactory purification, in terms of conventional parameters, of the surface water in the ground, the distance between the river/lake and the extraction well should be such that the travel time exceeds 30 to 60 days (Huisman and Olsthoorn, 1983).

Bank filtration is a traditional, efficient, and well-accepted method of surface water treatment in Europe. For more than 100 years, RBF has been used in Europe for public and industrial water supply along the Rhine, Elbe, and Danube rivers (Grischek et al., 2002a). River and lake BF sites have been operated for over a century to provide drinking water to cities such as Zurich, Dusseldorf, and Berlin. In Dusseldorf, situated on the Rhine River, RBF has been practiced since 1870. Schubert (2002a) reported that the groundwater derived from infiltrating river water provides water supply in densely populated and industrialized regions: up to 50 percent of potable supplies in the Slovak Republic and 45 percent in Hungary. In the Netherlands, the first river bank filtrate was pumped for public drinking water supply around 1879 along the Rhine River at pumping station Nijmegen (KWR, 2004). About 15 percent of the drinking water in the Netherlands is produced from infiltration of surface water (dune and bank filtration) (Hiemstra et al., 2003). RBF systems also have been supplying drinking water to several communities in the United States for nearly half a century (Ray, 2008).

Nowadays, BF is of considerable interest because of its utility as a cost-effective treatment technology. BF enhances water quality by using the adsorption, biodegradation, and natural disinfection that occurs as surface/reclaimed water moves through geological media. Bank filtration is an excellent protection tool to dampen peak concentrations and shock loads resulting, for instance, from chemical spills or failures at industrial or municipal wastewater treatment plants (Schmidt et al., 2008). Underground passage reduces the effects of concentration peaks because of the varying distances covered by the water as it moves from the river/lake to the well. It also dampens temperature fluctuations. In the production well, the withdrawn water is a mixture of water that left the river/lake at different times within a large period. Well-operated BF facilities situated in favorable hydrogeological settings can provide relatively inexpensively high-quality water that needs minimal additional treatment (Grunheid et al., 2005). Because of its ability to reduce dissolved organic carbon (DOC), disinfection by-product (DBP) precursors, and microbiological contaminants such as *Giardia lamblia* and *Cryptosporidium parvum*, there is now increasing interest in RBF in the United States (Ray et al., 2002; Tufenkji et al., 2002; Weiss et al., 2003). The U.S. Environmental Protection Agency (USEPA) Long Term 2 Enhanced Surface Water Treatment (LT2) Rule provides 0.5 and 1.0 log removal credits for travel distances of 7.6 and 15.2 m (25 and 50 ft), respectively, in BF systems (USEPA, 2007).

Removal Mechanisms

Several mechanisms are responsible for improvement of water quality during river/lake BF. During infiltration and travel through the soil and aquifer sediments, surface water is subjected to a combination of physical, chemical, and biological processes such as (1) filtration, (2) solution precipitation, (3) ion exchange, (4) sorption and desorption, (5) complexation, (6) redox reactions, (7) microbial biodegradation, and (8) dilution that significantly improve water quality (Kuehn and Mueller, 2000; Hiscock and Grischek, 2002). Water quality improvements associated with BF include removal of organic matter, removal of suspended solids and odorous compounds, reduction in pathogens, and attenuation of shock loads of chemical contaminants. The top few centimeters of the bank materials act as a screen or filter medium to remove the suspended solids present in the water. Heavy metals, phosphorus, and hydrophobic organic compounds present in the water are removed by adsorption onto certain aquifer materials. In the presence of biomass, the organic matter is further biodegraded (initially under oxic conditions and later under anoxic conditions). The water quality is in most cases improved by dilution of the source surface water with native groundwater (i.e., mixing) if such groundwater is uncontaminated. A schematic of processes affecting water quality improvement during bank filtration is presented in Fig. 15-1. Table 15-1 compares BF with conventional above-ground treatment technology for removal of different contaminants.

Factors Affecting Performance of RBF/LBF Systems

The performance of RBF/LBF for water quality improvements depends on a number of variables: (1) river and lake water quality, (2) characteristic and composition of alluvial aquifer materials, (3) geochemistry, (4) filtration velocity and distance of the well(s) from the river or lake, (5) temperature of the water, (6) pumping rate, (7) soil and sediment characteristics at the river/lake-aquifer interface, and (8) groundwater dilution.

Hydrogeological Conditions. The effectiveness of BF for removing surface water contaminants depends largely on hydrological conditions, including well type and well location with respect to the source water, soil microbiology, characteristics of the bank material and

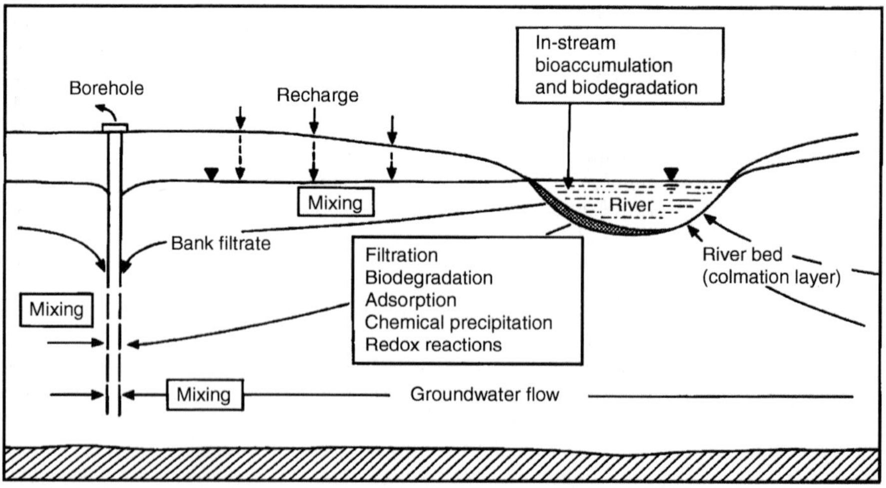

FIGURE 15-1 Schematic of processes affecting water quality during bank filtration. (*Source: Hiscock, K. M., and Grischek, T. (2002). Attenuation of groundwater pollution by bank filtration.* Journal of Hydrology *266(3–4), 139–144, with permission from Elsevier, Ltd., Oxford, UK.*)

TABLE 15-1 Comparison of Bank Filtration and Conventional Treatment Technology

River water contaminant	Bank filtration	Engineering process (excluding biological)
Ammonia	Nitrification (aerobic)	Oxidation (chlorine), air stripping
Nitrate	Denitrification (anoxic)	Ion exchange, reverse osmosis
Particles	Filtration	Coagulation-flocculation/clarification/ filtration, membrane filtration
Microorganisms	Filtration, adsorption, decay (natural die-off)	Coagulation-flocculation/clarification/ filtration, disinfection, membrane filtration
Inorganic micropollutants	Adsorption, ion exchange, precipitation	Coagulation-flocculation, ion exchange, reverse osmosis
Organic micropollutants	Biodegradation, adsorption	Oxidation (ozone, AOP[a]), adsorption
Persistent micropollutants	None	Oxidation (ozone, AOP[a]), adsorption
Phosphorus	Adsorption, precipitation	Chemical precipitation

[a]AOP = advanced oxidation processes.

streambed, and scouring characteristics (Sahoo et al., 2005). In many countries, alluvial aquifers hydraulically connected to a water course are preferred sites for drinking water production (Doussan et al., 1997). Such aquifers are relatively easy to exploit (shallow) and generally highly productive. Alluvium can consist of fluvial (river) deposits and lacustrine (lake) deposits ranging in thickness from a few meters to kilometers. Major deposits usually are found in the lower reaches of river basins forming flood plains.

The topographical relief is usually low, as are natural hydraulic gradients (slope of the groundwater level). The sediments range largely from highly permeable coarse gravel to low-permeable fine-grained silt. Groundwater levels are naturally shallow where the rivers are perennial but may be deeper in arid regions where river flow is intermittent.

The actual biogeochemical interactions that sustain the quality of the pumped bank filtrate depend on numerous factors, including aquifer mineralogy and the extent of the aquifer (Hiscock and Grischek, 2002). The geochemical context of river-aquifer transfers and their evolution thus is a reflection of interactions between biological and physiochemical mechanisms. In addition to influencing the chemistry, the bacterial activity may affect the hydrodynamic parameters of the porous medium (Doussan et al., 1997).

The extension of areas clogged by particulates and the permeability of clogged regions are influenced by river flow dynamics. Schubert (2002a) reported that the effect of clogging increases during flood events in the Rhine River owing to high concentrations of suspended solids and the high gradient between the river level and groundwater table. The interactions that concern the changes of permeability are governed by variations of the concentration of suspended solids in river water, hydraulic gradient from the river to the aquifer, and self-cleaning mechanisms by bed load transport.

In addition, small organisms have been observed seeking food along the silt deposits in the clogged areas. Such activities also influence the permeability of clogged areas through bioclogging, mainly during times of low hydraulic gradient between the river and aquifer. Clogging of river bed sediments can be both beneficial in terms of promoting the biodegradation of contaminants and also undesirable in reducing the hydraulic conductivity of the infiltration zone (Hiscock and Grischek, 2002).

Erosive conditions (i.e., scouring) in the river are advantageous because they limit the formation of a clogging layer. Floods also remove the clogging layer, which might be enriched with heavy metals and adsorbed organic compounds; however, some floods may have a negative effect if the production wells are located near the banks. Even if the wells are not flooded, the destruction of the clogging layer, changes in pore pressure, and higher flow velocities in the aquifer can lead to an increased transport of dissolved compounds, previously adsorbed particles, bacteria, and viruses that potentially could result in contaminant breakthrough. In some settings, moderate floods may be useful in removing the clogging layer (Grischek et al., 2002).

The main hydrogeological factors that should be considered in design of MAR systems are (1) physical and hydraulic boundaries of the aquifer and degree of confinement, (2) hydrogeological properties of the aquifer and overlying formations, especially permeability, (3) hydraulic gradient of the aquifer, (4) depth to aquifer/piezometric surface, (5) groundwater quality, and (6) aquifer mineralogy.

Source Water Quality and Mixing with Native Groundwater. Source water quality is controlled by land use and climatic conditions, and during floods, rivers passing through agricultural watersheds often contain high concentration of farm chemicals. Contaminant concentration peaks often coincide with flow peaks. Ray and colleagues (2002b) observed that cyanazine and atrazine exceed their respective maximum contaminants levels (MCLs) and that elevated concentrations of pesticides and total organic carbon (TOC) occur during many flood occasions. While native groundwater may provide dilution if uncontaminated, landside groundwater may degrade water quality if, for example, contaminated by nitrate. Moreover, reductive dissolution under certain redox conditions may release iron and manganese into the recovered water.

Potential attraction of brackish groundwater from deeper strata is undesirable because of the high salt concentration of the pumped water. Attraction of an undesired quality of native groundwater may need extensive treatment owing to high hardness, iron, manganese, ammonium, nitrate, and/or sulfate.

Distance of the Wells from River Bank and Spacing of Wells. The distance of the production well from the river/lake affects the removal efficiency of contaminants as well as pumping rates. Thus wells located close to the water source generally will have higher capacity but relatively lower quality of bank filtrate. The spacing of wells also affects the drawdown and thus the pumping rates of the wells (Dillon et al., 2002). The retention times for BF sites in Europe are several weeks or even months, designed to eliminate biodegradable dissolved organic carbon (BDOC), pathogens, and degradable trace organic pollutants from the surface water to produce a biostable, high-quality water that can be distributed after little additional treatment without chlorine addition (Hiemstra et al., 2003). In contrast, U.S. practice generally reflects shorter travel distances and times, with more of a focus on *Cryptosporidium* removal.

Pumping Rates and Sediment Permeability. Pumping rates affect the drawdown in the wells and the ratio of the bank filtrate and native groundwater in the extracted water. Higher pumping rates also contribute to clogging of the river/lake bed. It is generally understood that the clogging of riverbed material in the vicinity of a well reduces infiltration rates and decreases the production yield from the well field. When this occurs, streambed infiltration areas extend significantly away from the production center of the well. This causes the water to travel for a longer distance, thus allowing the water to undergo geochemical and biological reactions. The growth of a bacterial biofilm may alter the porosity and permeability of the porous medium significantly, which might result in clogging of the river banks, especially at the river-sediment interface, where the bacterial activity is high (Doussan et al., 1997). However, the process is highly dynamic, with sloughing and degradation of biomass providing a generally steady state biomass profile.

Attributes and Limitations

The main attributes and limitations of BF systems are listed below (Ray et al., 2002a; KWR, 2004; IAH and IHP, 2005).

Attributes

1. BF is a sustainable natural treatment process, avoids or reduces the use of chemicals, and produces biologically stable water.
2. BF improves water quality by removing particles (turbidity and suspended solids), organic pollutants, microorganisms, heavy metals, and nitrogen.
3. BF dampens concentration peaks associated with spills in a river or lake.
4. BF dampens temperature fluctuations.
5. BF replaces or supports other treatment processes by providing a robust barrier for multiple contaminants and reduces the overall cost of water treatment.

Limitations

1. BF is site-specific and is feasible when local hydrogeological conditions, especially permeability, are favorable.
2. There can be leaching of the aquifer materials under reducing conditions (e.g., reductive dissolution when the redox shifts from oxic to anoxic), leading to increased concentrations of iron and manganese in extracted water.
3. One of the main problems is potential clogging of the river/lake bed owing to accumulation of suspended matter that is filtered out when river/lake water enters the aquifer, especially when the system is not sited and designed properly; complex processes in the clogging zone and subsequent aquifer are not fully understood.
4. BF and groundwater recharge may be only a limited barrier for certain contaminants such as persistent organic micropollutants.
5. Influences of surface water and operational conditions on water quality are poorly understood.
6. No methodology or tools are available to facilitate reliable transfer of successful experiences to other locations.
7. Quantitative prediction of the purification potential at a site is limited.

ARTIFICIAL RECHARGE AND RECOVERY (ARR)

ARR is an engineered system designed for intentional treatment, storage, and withdrawal of water in aquifers; storage depends on infiltration versus pumping rates. A schematic showing the main difference between bank filtration and infiltration (ARR) is presented in Fig. 15-2.

ARR methods are employed when the local geological conditions and/or water quality in the river/lake are not suitable for induced infiltration (BF) or when different source waters are available (e.g., stormwater). ARR can be employed without pretreatment of the source, or the source water can be treated to the required level (e.g., filtration) prior to recharge. Artificial recharge is also practiced to control seawater intrusion into coastal aquifers, control land subsidence caused by declining groundwater levels, maintain base flow in some streams, and raise water levels to reduce the cost of groundwater pumping. ARR systems are being used in a number of states in the United States to overcome seasonal imbalances in both potable and reclaimed water projects. The tremendous volumes of storage potentially available in ARR systems means that a greater percentage of the resource, be it raw water or reclaimed water, can be captured for beneficial use. Topper and colleagues (2004) reported that artificial recharge is being used in at least 32 states in the United States and at least 26 countries worldwide.

Artificial recharge can be accomplished by (1) surface spreading, (2) infiltration through the unsaturated zone (vadose zone injection wells), or (3) direct injection into groundwater.

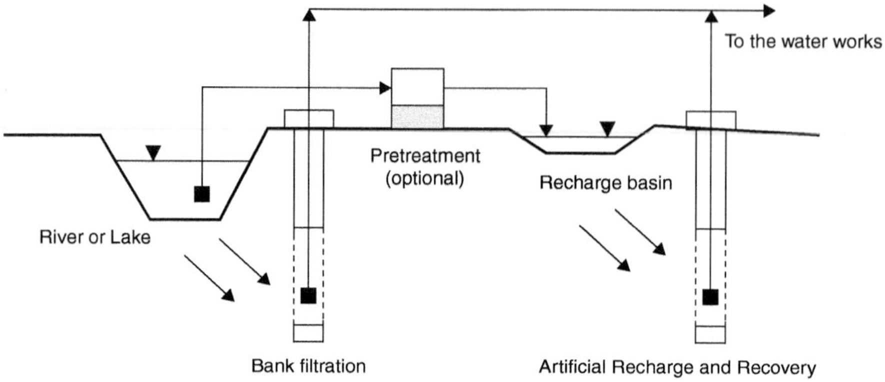

FIGURE 15-2 Schematic of bank filtration and infiltration. (*Source: Adapted from with permission of AWWA, Kuehn, W., and Mueller, U. (2000). Riverbank filtration: An overview. Journal AWWA* ***92****(12), 60–69.*)

ARR systems are also further categorized as (1) aquifer storage and recovery (ASR), which involves injection of water into a well for storage and recovery from the same well, and (2) aquifer storage transfer and recovery (ASTR), which involves injection of water into a well for storage and recovery from a different well, generally to provide additional water treatment (Fig. 15-3). ASTR differs from ASR in that it is more applicable to confined aquifers

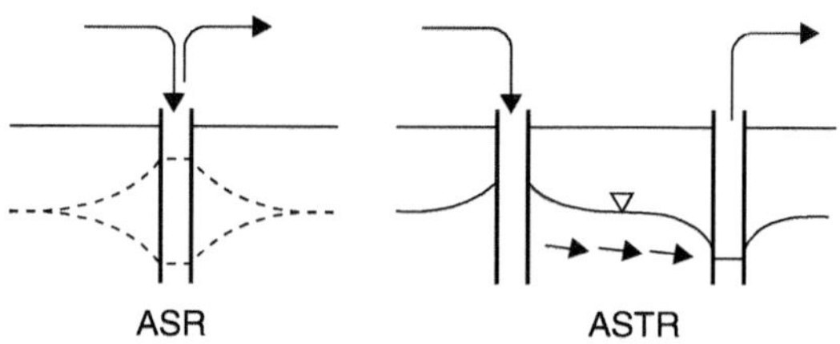

FIGURE 15-3 Schematic of aquifer storage and recovery (ASR) and aquifer storage transfer and recovery (ASTR). (*Source: Dillon, P. (2005). Future management of aquifer recharge. Hydrogeology Journal 13(1), 313–316, with the permission from Springer, Germany.*)

where infiltration is not feasible. Generally, surface infiltration systems can be used to recharge unconfined aquifers only, whereas confined aquifers can be recharged with wells that penetrate confining layers of the aquifer. Recharge by infiltration takes advantage of natural treatment processes as the water moves through the soil/aquifer and therefore is more attractive from a water quality improvement perspective, particularly in relation to the important role of the vadose zone. Furthermore, based on same degree of pretreatment, injection wells are much more vulnerable to clogging than surface infiltration systems because the infiltration rates around the boreholes or injection wells are much higher than in infiltration basins. Clogging is pronounced especially at the borehole interface between the gravel envelope and aquifer, where suspended solids can accumulate and bacterial growth tends to concentrate. A comparison of major engineering factors of three common artificial recharge systems is presented in Table 15-2.

Dune infiltration is a specific method of ARR practiced in coastal zones, in which the valleys between coastal sand dunes are flooded with water from rivers to infiltrate into the underlying sediments and create a recharge mound. The mound can play an important

TABLE 15-2 Comparison of Engineering Factors for Three Major Artificial Recharge and Recovery (ARR) Systems

	Recharge basins	Vadose zone injection wells	Direct injection wells
Aquifer type	Unconfined	Unconfined	Unconfined or confined
Pretreatment requirements	Low technology	Removal of solids	High technology
Estimated major capital costs (US$)	Land and distribution system	$25,000–75,000 per well	$500,000–1,500,000 per well
Capacity	100–20,000 m^3/ha/day	1000–3000 m^3/day per well	2000–6000 m^3/day per well
Maintenance requirements	Drying and scraping	Drying and disinfection	Disinfection and flow reversal
Estimated life cycle	>100 years	5–20 years	25–30 years
Soil aquifer treatment	Vadose zone and saturated zone	Vadose zone and saturated zone	Saturated zone

Source: USEPA, 2004.

role in preventing saline intrusion as well as providing a source of water that is abstracted further inland. This technique has been used for centuries and is highly developed along the coast of the Netherlands, where rivers are the source of water for recharge (Hiemstra et al., 2003; IAH and IHP, 2005).

Recently, an innovative technology using a wick has been used for the artificial recharge of the treated effluent to groundwater. This technology is capable of recharging with a significantly smaller footprint than conventional methods of spreading basins. A *wick* is essentially a column, about 10 m deep and about 2 m in diameter, filled with medium (stone) that acts as a conduit to discharge water to the surrounding aquifer and groundwater. In terms of hydraulic loading/infiltration, a wick can be considered as intermediate between spreading basins and injection wells (Owen et al., 2008).

Infiltration Rates and Operation

The range of infiltration rates for in-channel and off-channel artificial recharge systems is about 0.3 to 3 m/day, including any effects caused by clogging. Systems with year-round recharge and periodic drying and cleaning of the bottom typically have hydraulic loading rates of about 30 to 300 m/yr. If the basin bottoms are not covered by sediment or other clogging material and groundwater levels are sufficiently low to not affect infiltration, infiltration rates are about the same as the vertical hydraulic conductivity of the soil, which may be about 0.3 m/day for sandy loams, 1 m/day for loamy sands, 5 m/day for fine sands, 10 m/day for coarser sands, and 20 to 50 m/day for fine or clean gravel. Sand and gravel mixtures have lower hydraulic conductivities than sand alone (Bouwer and Rice, 1984). To achieve optimal infiltration rates, a number of features need to be considered in the design process, including clogging, water depth, and groundwater level (Bouwer and Rice, 1984). In SAT, clogging is managed by alternating wet (infiltration) and dry (off-line) cycles. In the Netherlands, the residence times of infiltrated water underground ranges from 30 to 400 days (Tielemans, 2007).

Factors Affecting Performance of ARR Systems

Like BF systems, the performance of ARR systems for water quality improvements depend on a number of variables: (1) characteristic and composition of aquifer materials, (2) source and quality of recharge water, (3) degree of pretreatment, (4) method and rates of infiltration, (5) temperature of the water, (6) type of ARR system (ASR or ASTR), and (7) pumping rate.

Attributes and Limitations

The main attributes and limitations of ARR systems are listed below.

Attributes

1. ARR systems can be used for seasonal storage of water in areas where water demands are significantly greater in summer than in winter, or vice versa. With these wells, drinking water treatment plants can be built to meet average rather than maximum daily demands.
2. Recharge water can be treated to the required degree to minimize clogging of the aquifer and to avoid contamination of the aquifer.
3. Infrastructure required for storage is small and cost-effective, and there are minimal water losses compared with, for example, evaporative losses in surface storage.
4. The quality of local groundwater can be improved by recharging with high-quality injected water. Recharge with less saline surface waters or treated effluents improves the quality of saline aquifers.

Limitations

1. Recharge rate may decrease with time owing to precipitation of calcium carbonate, iron oxides, and other compounds in the aquifer, long-term dissolution and collapse of calcarious soils, dispersion and swelling of clay, and air binding.
2. There can be leaching of the aquifer materials when the redox shifts from oxic to anoxic, leading to increased concentrations of iron and manganese in extracted water.
3. There is a potential for contamination of the groundwater by the recharge water if the quality control of the injected water is not adequate.
4. Excessive land area may be needed for spreading basins.
5. Not all recharged water may be recoverable owing to movement beyond the capture zone of the extraction well or mixing with poor-quality groundwater.

SUBSURFACE GROUNDWATER TREATMENT

Subsurface or in situ groundwater treatment involves infiltration of oxygen-rich water into a well and pumping of the groundwater from the same or a different well. This type of treatment is used to remove iron, manganese, ammonium, and arsenic from groundwater. These processes are known as the *vyredox* or *nitredox* methods (van Beek, 1983; Braester and Martinell, 1988; Rott et al., 2002). The method relies on the phenomenon that oxygen present in infiltration water will oxidize soluble iron in the groundwater and form layers of iron hydroxides on the media around the well, which subsequently act as adsorbent for Fe(II) and As(III)/As(V) present when groundwater is pumped from the same or a nearby well. The ratio of the quantity of extracted groundwater to the quantity of infiltrated oxygen-rich water (extraction versus infiltration ratio) is relatively high (it depends on the oxygen concentration of the infiltration water and the type and concentrations of the contaminants intended to be removed or avoided during this treatment process).

Iron oxidation and removal are based on the transformation of the soluble form of iron (Fe^{2+}) to an insoluble form (Fe^{3+}), as shown by the following stoichiometric equations. More extensive coverage of Fe oxidation is provided in Chap. 7.

$$4Fe^{2+} + O_2 + 2H_2O \rightarrow 4Fe^{3+} + 4OH^- \tag{15-1}$$

$$4Fe^{3+} + 4OH^- + 8H_2O \rightarrow 4Fe(OH)_3 + 8H^+ \tag{15-2}$$

$$4Fe^{2+} + O_2 + 10H_2O \rightarrow 4Fe(OH)_3 + 8H^+ \tag{15-3}$$

Equation 15-1 shows that 0.14 mg of oxygen is required to oxidize 1 mg of Fe(II). The removal of Fe(II) by adsorption onto iron oxide surfaces and its subsequent oxidation in the presence of oxygen and regeneration of adsorption sites are given by the following equations:

$$\equiv \text{S–OH} + Fe^{2+} \leftrightarrow \equiv \text{S–OFe}^+ + H^+ \tag{15-4}$$

$$\equiv \text{S-OFe}^+ + \frac{1}{4}O_2 + \frac{3}{2}H_2O \rightarrow \equiv \text{S-OFe}\langle^{OH}_{OH} + H^+ \tag{15-5}$$

where ≡ S represents the adsorption surface (iron hydroxide) (Sharma, 2001).

From these equations, the amount of iron-free water that can be extracted from a given aquifer can be estimated. For example, if the iron concentration in groundwater is 5 mg/L and the oxygen concentration in the recharge water is 7 mg/L, the extraction ratio for iron-free water would be around 10, assuming that all oxygen in the recharge water would be used in oxidation of iron in the aquifer. These equations also show that for a given iron concentration in groundwater, the higher the concentration of oxygen in recharge water, the higher is the ratio of extracted iron-free water to recharge water. However, it should be noted that, in practice, this ratio is affected by flow velocities, reaction times, oxidation kinetics of the contaminant to be removed, oxygen heterogeneity of the aquifer, and other oxygen-consuming processes in the aquifer. Rott and colleagues (2002) reported that the ratio of the delivered water volume to reinfiltrated water volume (efficiency coefficient) ranges between 2 and 12 depending on the aquifer and raw water quality.

Many of the available methods for arsenic removal from water depend on the adsorption of arsenic onto oxides and oxyhydroxides of iron (see Chap. 12). Spent oxide adsorbents normally are not regenerated but are disposed of as stable wastes. Subsurface treatment for arsenic removal relies entirely on these same principles but accomplishes it in the subsurface as an in situ process. When iron is removed in the aquifer, arsenic is removed along with it. By creating renewable iron oxide filters in the aquifer surrounding a production well, arsenic levels in the delivered water are reduced to below the drinking water standard without creation of an arsenic-laden waste stream. Filters are renewed by in situ oxidation of ambient dissolved ferrous iron, which then forms iron oxide and oxyhydroxide-coated geomedia. Oxidation processes are both abiotic and biotically mediated. In aquifers where ambient iron concentrations are insufficient to bring about desired levels of arsenic removal, additional iron is introduced into the aquifer and then oxidized (Sarkar and Rahman, 2001; Miller, 2006; Rott et al., 1996; Rott and Friedle, 1999).

This method of groundwater treatment is attractive because there is no use of chemicals, and the production of sludge is avoided. However, there is potential for contamination of aquifers. Other limitations could be the clogging of the aquifer, excessive bacterial growth around the well, and corrosion of the well screen owing to high oxygen concentration in infiltration water (van Beek, 1983; Rott, 1985).

SOIL AQUIFER TREATMENT (SAT) FOR INDIRECT POTABLE REUSE

SAT is a managed aquifer recharge as well as wastewater treatment technology that, in combination with other available wastewater treatment technologies, can produce effluent of acceptable quality for indirect potable reuse. SAT is similar conceptually to ARR, and together they can be used for wastewater effluents, urban stormwater, or surface water. This technology uses physical, chemical, and biological processes in the soil matrix and aquifer for wastewater treatment. It is a low-cost and appropriate option for wastewater reclamation in developing countries as well as developed countries. It is also considered appropriate for replenishment of underground water to avoid exhaustion of groundwater resources and lowering of groundwater levels. Furthermore, artificial recharge of groundwater basins with SAT effluent contributes to the sustainability of surface water and groundwater resources within the context of integrated water resources management (Asano and Cotruvo, 2004; Rodriguez et al., 2009). A schematic of soil aquifer treatment system is presented in Fig. 15-4, and a comparison of typical SAT zones is presented in Table 15-3.

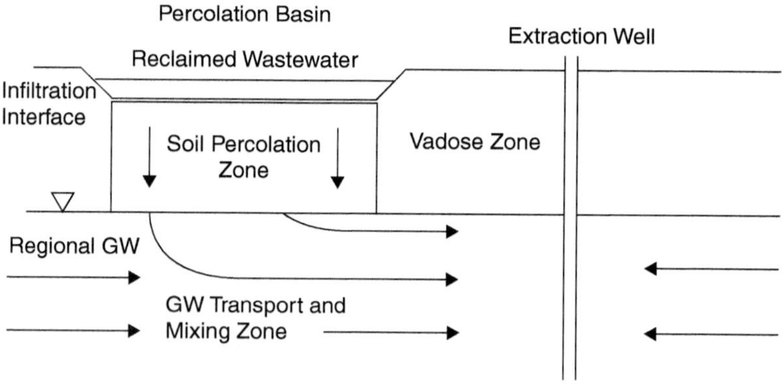

FIGURE 15-4 Schematic of soil aquifer treatment (SAT). (*Source: Amy, G., and Drewes, J. (2007). Soil aquifer treatment (SAT) as a natural and sustainable wastewater reclamation/reuse technology: Fate of wastewater effluent organic matter (EfOM) and trace organic compounds.* Environmental Monitoring and Assessment *129, 19–26, with permission from Springer, Germany.*)

SAT has been practiced in different parts of the world using primary, secondary, and tertiary effluents from wastewater treatment (Fox et al., 2001; Nema et al., 2001). Furthermore, different pretreatment and posttreatment methods have been applied together with SAT to produce water quality suitable for the intended use. The principal design objective for SAT systems is additional wastewater treatment, beyond primary or secondary. Other design objectives can include (1) recharge of stream by interception of shallow groundwater, (2) recovery of water by wells or underdrains, with subsequent reuse, (3) groundwater recharge, and (4) temporary (seasonal) storage of renovated water in an aquifer (Crites et al., 2006). Furthermore, SAT also breaks the pipe-to-pipe connection involved in direct recycling of treated wastewater and serves as a natural psychological barrier that can help to promote public acceptance of (indirect) potable water reuse (Bouwer, 1993). SAT is relatively inexpensive, except in areas of high land costs. The biggest cost is pumping or otherwise collecting water from aquifers. SAT systems are also robust and do not require highly skilled technical personnel for operation (Bouwer, 1992).

TABLE 15-3 Comparison of Typical Soil Aquifer Treatment Zones

Process/parameter	Infiltration interface	Soil percolation	Groundwater transport
Treatment mechanisms	Filtration, biodegradation	Biodegradation, adsorption	Biodegradation, adsorption, dilution
Transport	Saturated	Unsaturated	Saturated
Residence time	Minutes	Hours to days	Months to years
Travel distance	Centimeters	3–30 m	Variable; 0 to several kms
Mixing	No	No	Yes
Oxygen supply	Recharge water	Unsaturated zone	Regional groundwater
Biodegradable carbon availability	Excess	Excess/limiting	Limiting
Redox conditions	Aerobic	Aerobic to anoxic (denitrifying)	Anoxic to anaerobic

Groundwater recharge with reclaimed wastewater is being used increasingly for replenishing aquifers used for municipal water supply. Indirect potable reuse based on groundwater recharge enjoys the following advantages over recycle based on discharge into surface waters (AwwaRF and AWWA, 2006):

1. Groundwater recharge provides additional water quality benefits because of SAT.
2. Seasonal and longer-term storage can be achieved without evaporative losses.
3. Groundwater recharge protects water against recontamination by birds and mammals (with coliforms and parasites) and possibly even by humans.
4. Groundwater recharge keeps sunlight away from water, thereby preventing growth of algae and associated water quality problems such as algae-derived tastes and odors and disinfection by-product formation; to avoid algal growth in a surface reservoir, the wastewater must be treated to remove nitrogen or phosphorus.
5. When reclaimed water is recovered from wells, it is conceived of as groundwater, an aesthetically superior and more publicly acceptable source.

Site selection is important to the success of a SAT system. The important factors in site evaluation and selection are (1) soil depth, (2) soil permeability, (3) depth to groundwater (depth of vadose zone), (4) aquifer thickness (depth from water table to bedrock/confining layer), and (5) groundwater flow direction and velocity. Depending on the local hydrogeological conditions and quality of the effluent available, SAT can be achieved using any of the three commonly employed recharge methods, namely, (1) recharge basins, (2) vadose zone injection well, or (3) direct injection well. The following factors should be considered in the design of SAT systems: (1) pretreatment requirements (degree of wastewater treatment, additional treatment), (2) infiltration (hydraulic loading) rate, (3) land requirement (taking into account wet/dry cycle), (4) number of wells (production capacity per well), (5) spacing between the wells, (6) distance of the wells from infiltration pond or injection well, (7) pumping rate (affects groundwater flow and velocity), (8) share of native groundwater in reclaimed water percentage, (9) water quality obtained from the SAT system, and (10) posttreatment requirements, if any.

Municipal wastewater effluent being recharged via infiltration basins usually is treated to secondary levels (biological treatment) and often chlorinated (in the United States) before being applied to infiltration ponds. The primary objective of water quality improvement is to remove all suspended solids and microorganisms. Removal of nitrogen species through nitrification/denitrification is also a key benefit, as is the reduction in concentration of dissolved organic carbon (DOC) through biological processes. Phosphates and metals also can be removed but are retained in the soil by adsorption. In an indirect potable reuse system, the residual humic substances present in effluent organic matter (EfOM) impart color and serve as a precursor to disinfection by-products (DBPs) if extracted water is chlorinated on recovery, whereas the nitrogen-rich soluble microbial products (SMPs) present in EfOM represent a precursor to nitrogenous DBPs (N-DBPs). In addition to concerns about bulk EfOM, there are various effluent-derived trace organic compounds, including endocrine-disrupting compounds (EDCs), pharmaceutically active compounds (PhACs), and personal-care products (PCPs) that are present in secondary effluents. The presence of these organic micropollutants suggests their persistence through biological wastewater treatment processes (e.g., activated sludge) that embody short-term aerobic biodegradation (Amy and Drewes, 2007).

SAT Soil Requirements

Infiltration basins for SAT systems should be located in soils that are permeable enough to provide high infiltration rates. The soils, however, also should be fine enough to provide good filtration and quality improvement of the effluent as it infiltrates. The best surface soils for SAT systems are in the fine sand, loamy sand, and sandy loam range. Materials deeper in the vadose zone should be granular and preferably coarser than the surface soils. The vadose zones should not contain clay layers or other soils that could restrict the downward movement of water and form perched groundwater mounds (Pescod, 1992).

Hydraulic Loading and Operation

Infiltration basins in SAT systems are intermittently flooded to provide regular drying periods, for restoration of infiltration rates, and for aeration of the soil. This depends on pretreatment: Basins receiving highly treated water have longer flooding cycles. Flooding schedules typically vary from 8 hours dry to 16 hours flooding to 2 weeks dry to 2 weeks flooding. SAT systems therefore should have a number of basins so that some basins can be flooded while others are drying. The total annual infiltration amounts or hydraulic loading rates typically vary from 15 to 100 m/yr depending on soil, climate, quality of sewage effluent, and frequency of basin cleaning.

Intermittent application is critical to the successful operation of all land treatment systems. These drying periods are necessary to restore the infiltration capacity and to renew the biological and chemical treatment capability of soil system. The ratio of wetting to drying in successful SAT systems varies but is always less than 1.0. For those limited cases where primary effluent is used, the ratios generally are less than 0.2 to allow for adequate drying and removal of the applied solids. For secondary effluent, the ratio varies with the treatment objective from 0.1 or less where maximum hydraulic loading or only nitrification is the objective to 0.5 to 1.0 where nitrogen removal is the treatment objective. Typical hydraulic loading rates and wet/dry ratio of selected SAT systems in the United States are presented in Table 15-4. In general, operating conditions depend on a number of environmental factors, including temperature, precipitation, and solar incidence. Therefore, operating conditions must be adjusted to both local site characteristics and weather patterns (Crites et al., 2006; USEPA, 2006).

Water Depth

High hydraulic head produced by deep basin water result in high infiltration rates; however, they also tend to compress clogging layers. Contrary to intuitive expectations, deep basins

TABLE 15-4 Typical Hydraulic Loading Rates, Wet and Dry Periods, and Wet/Dry Ratios for Selected Soil Aquifer Treatment Systems in the United States

Location	Annual loading rate, m/yr	Type of effluent applied	Wet period, days	Dry period, days	Wet/dry ratio
Boulder, CO	30.3–48.5	Secondary	0.1	3	0.03
Ft. Devens, MA	28.8	Primary	2	14	0.14
Hollister, CA	15.2	Primary	1	14	0.07
Phoenix, AZ	60.6	Secondary	9	12	0.75
Vineland, NJ	21.2	Primary	2	10	0.20

Source: Adapted with permission from USEPA (2006).

can produce lower infiltration rates than shallow basins. The rate of turnover of the water in deep basins may be less than in shallow basins, allowing suspended algae to grow during longer exposure to sunlight. With deep basins (e.g., 10 m or more, as in old gravel pits), it can take a long time for all the water to infiltrate into the soil. Water actually may have to be pumped out of the basin to initiate drying. In addition to yielding higher infiltration rates, shallow basins (water depths about 0.5 m or less) have the advantage that drying can start quickly after the inflow of water into the basin has been stopped. In general, for artificial recharge of groundwater or SAT systems, the basin water depths preferably should be less than 0.3 m, and the basins should be hydraulically independent so that each can be flooded, dried, and cleaned according to its best schedule.

WATER QUALITY IMPROVEMENTS IN NATURAL TREATMENT SYSTEMS

In natural treatment systems, subsurface processes during soil passage improve the quality of the water. Results of several experimental and field studies have shown that water quality objectives can be met consistently for the longer term by natural treatment systems. The effects of natural treatment on some of the important water quality parameters are elaborated in the following paragraphs.

Particles and Turbidity

Suspended solids present in the source water used for BF, ARR, or SAT are usually relatively fine, and these solids accumulate at the interface of the water and soil layers. Particle transport in the aquifer is governed by movement of the groundwater coupled with retardation by attachment onto the surfaces and straining and trapping in interstitial pores. Except for medium and coarse uniform sands, soils are effective filters, and suspended solids are essentially completely removed from the water sources after about 1 m of percolation through the soil layer or vadose zone (Bouwer, 1985). Suspended solids typically are less than 2 mg/L in the extracted water as a result of filtration through the soil layer. Based on the studies at the RBF site at Louisville, Ky., Wang (2002) reported that although the turbidity of Ohio River water fluctuated significantly (from 3 to 599 ntu, with an mean of 45 ntu), the turbidity at the collector well remained consistently below 0.1 ntu. Furthermore, it was found that turbidity reduction increased with travel distance, with maximum reduction occurring in the first 0.6 m. In the case of an ARR system in the region of Marchfeld, Austria, the turbidity and suspended solids concentration of the recharge water decreased from 9.9 ntu and 12 mg/L to 0.4 to 0.7 ntu and 1.0 mg/L, respectively, after soil passage (Neudorfer and Weyermayr, 2007).

In BF systems, fine materials present in river/lake water accumulate at the interface and reduce the infiltration rate, consequently causing clogging of the initial infiltration layer. Often these layers are scoured during high floods unless the penetration of the particles is deep in the soil layer. Similarly, infiltration basins for ARR and SAT require regular drying for infiltration recovery and periodic soil cleaning by raking or scraping. For loamy sands and sandy loams, few suspended solids will penetrate into the soil and then usually only for a short distance (a few centimeters, for example). In dune sands and other coarser soils, fine and colloidal suspended solids can penetrate much greater distances.

Microorganisms

Microorganisms may be removed from water during soil passage primarily by straining, inactivation, and attachment to the aquifer grains (in combination with inactivation). The number of infective enteric pathogens decreases with time and eventually decreases to near or below detection by natural processes. Temperature, humidity, pH, the amounts of organic matter in the soil and aquifer material, rainfall, sunlight, and competitive microorganisms all affect the survival of pathogens in water, soil, other unconsolidated material, and aquifer matrices (Schijven et al., 2002).

A summary of microbial removal data obtained from various field studies on riverbank filtration in the Netherlands is presented in Table 15-5. In principle, soil passage is an efficient

TABLE 15-5 Log Removals of Microorganisms During Bank Filtration

	Rhine river at remmerden	Meuse river at zwijndrecht	Meuse river at roostern		
Travel distance (m)	30	25	13	25	150
Travel time (days)	15	63	7	18	43
Total coliforms	5.0	5.0	—	—	—
Fecal streptococi	3.2	3.5	—	—	—
Enteric viruses	4.0	4.0	3.7	—	—
Bacteriophages (type 1)[a]	6.2	5.7	3.9	6.0	—
Bacteriophages (type 2)[b]	—	—	3.8	5.1	7.8

[a]F-specific RNA bacteriophages.
[b]Somatic coliphages.
Source: Adapted with permission from Tufenkji, Ryan, and Elimelech, 2002.

system for removal of microorganisms. Given sufficient flow path distance and time, microbial contaminants are removed or inactivated to levels protective of public health, minimizing risk levels. Under optimal conditions, BF can achieve up to 8 log virus removal over a distance of 30 m in about 25 days (Schijven et al., 2002). Greater removal efficiency may be expected for protozoa and algae (i.e., larger organisms) under the same conditions. Yao and colleagues (1971) stated that a minimum removal in filtration occurs for particles of about 1 μm, and thus one theoretically would expect slightly less removal for bacteria than for viruses (transport to surfaces with a higher diffusion rate than bacteria) during soil passage.

Bulk Natural Organic Matter

Natural systems are effective at removal of bulk natural organic matter (NOM). Adsorption and biodegradation are the dominant mechanisms for removal of organic matter during soil passage. Aerobic, anoxic, and possibly anaerobic conditions promote biodegradation of organic carbon under different electron acceptors. Biodegradation is the primary sustainable removal mechanism for organic compounds during subsurface transport. The concentration of NOM and soluble microbial products (SMPs) that comprise the bulk of the dissolved and particulate organic carbon is reduced during soil passage because high-molecular-weight compounds are hydrolyzed to lower-molecular-weight compounds, and lower-molecular-weight compounds serve as substrates for microorganisms (NRC, 2008).

DOC removal is independent of the concentrations applied but depends on the concentrations of easily biodegradable organic carbon in the source water. It has been reported that during BF, there can be 50 to 90 percent reduction of biodegradable dissolved organic carbon (BDOC) and assimilable organic carbon (AOC), biodegradable portions of NOM, and 26 percent reduction in specific ultraviolet (UV) absorbance (SUVA) (see Chap. 3 for a discussion of SUVA) in UV-absorbing NOM (Weiss et al., 2004). Furthermore, BF can greatly reduce chlorine demand as well as trihalomethane formation potential (THMFP) and haloacetic acid formation potential (HAAFP) in river water; a 50 percent reduction in DBP formation potential was achieved for Ohio River water (Louisville, Ky.) that was abstracted 82 m from the river with a pumping rate ranging from 0.113 to 0.139 m^3/s (Wang, 2002). Additionally, Lehtola and colleagues (2002) reported an average 53 percent removal of AOC along with 60 percent removal of overall TOC and 64 percent removal of microbially available phosphorus (MAP) during soil infiltration.

Trace Organic Contaminants

Removal of trace organic contaminants during soil passage is influenced mainly by their biodegradability and concentration level, redox conditions, adaptation of the biocommunity, travel time, and concentration of bulk organic matter present (to serve as a primary substrate/carbon source). BF and ARR reduce trace organics in water, including pesticides, personal-care products, endocrine-disrupting compounds, and pharmaceuticals to some extent (Massmann et al., 2006; Stuyfzand et al., 2007). Some micropollutants are more degradable under anoxic conditions than under aerobic conditions, and vice versa. Furthermore, the removal capacity of BF for organic micropollutants seems to be quite robust with regard to changes in surface water temperature and discharge (Schmidt et al., 2007).

Studies on trace organic contaminants removal during soil passage in the United States and Europe have revealed the occurrence of some pharmaceuticals, pesticides, and/or industrial chemicals in bank filtrate and treated water from recharge basins. The characteristics common to those compounds that are not attenuated are that they are hydrophilic (polar), and they have structural features that prevent enzymatic attack and render them resistant to biodegradation (Heberer, 2002; NRC, 2008). Compounds such as primidone, carbamazepine, and clofibric acid have been reported to be partly recalcitrant during underground passage. Polar organic molecules, such as complexing agents, pesticides, industrial products like aromatic sulfonates, pharmaceutical compounds, and personal-care products, are of recent concern (Schmidt et al., 2008). Under anoxic conditions, absorbable organic iodine (AOI) and x-ray contrast materials are decreased by bank filtration, whereby the AOI is decreased by 64 percent and the contrast material concentration can be reduced up to 95 percent depending on the compound present (Schittko et al., 2004). Table 15-6 summarizes the removal of some selected pharmaceuticals at the Lake Tegel bank filtration site in Berlin, Germany, which shows that depending on their log K_{ow} values (octanol water partition coefficient, an index of compound hydrophobicity), removal rates for different pharmaceuticals could range from low to very high. Log K_{ow} reflects both hydrophobicity affecting adsorption and compound complexity affecting biodegradation.

TABLE 15-6 Removal of Selected Pharmaceuticals at Lake Tegel Bank Filtration Site

Group	Compound (log K_{ow})	Surface water mean concentration (ng/L)	Mean removal at monitoring well 3303[a]%
	Primidone (0.91)	110	15%
Low removal, 0%–50%	Propyphenazone (2.05)	190	35%
	Carbamazepine (1.58–2.45)	425	35%
Medium removal, 51%–90%	Clofibric acid (1.58–2.45)	50	70%
	Diclofenac (1.13–1.90)	100	70%
High removal, >90%	Bezafibrate (4.2)	40	>97% (<LOD)
	Indometacine (4.27)	15	>95% (<LOD)

LOD = level of detection.
[a]Distance of the well from Lake Tegel = 95 m (312 ft); Travel time = 3.9 months.
Sources: Ameda, 2008; Heberer et al., 2004.

Nitrogen and Phosphorus

Significant biological nitrification and denitrification can occur during soil passage under aerobic and anoxic conditions. The major evolution of nitrogen species during river-to-groundwater transfer takes place in the first few meters of infiltration into the submerged riverbed sediments. This evolution depends on carbon mineralization and bacterial activity. Ammonium adsorption plays an important role in retaining ammonia in the soil long enough for biological conversion, especially at the start of system operations. In aerobic aquifers, ammonium is converted to nitrate by nitrification in the presence of oxygen. During soil passage, there may be a transition from anoxic (nitrate reduction) to anaerobic (iron and manganese reduction) conditions. In the case of SAT systems, during drying cycles, as the soil dries, air/oxygen enters the soil; consequently, oxidation of ammonia to nitrate by nitrifiers also may occur in the vadose zone. Once nitrate reaches deeper into the anoxic zone, heterotrophic denitrification converts nitrogen into nitrogen gas in the presence of organic carbon as an electron donor. Nitrogen removal is a function of detention time, carbon/nitrogen (C:N) ratio, and anoxic conditions. A C:N ratio greater than 3 is necessary to sustain high nitrogen removal efficiency (Fox et al., 2001). Primary effluent usually has a C:N ratio of 3 or higher, whereas for secondary effluent the ratio is about 1:1, and in river/lake water it could be lower. To overcome a low C:N ratio, a longer travel/detention time is required (Crites et al., 2006).

Phosphorus removal during soil passage is mainly by adsorption or chemical precipitation, which is controlled by the interaction of redox potential; pH; native iron, calcium, and aluminum minerals; and iron/phosphorus ratio (Rogers et al., 2005). In calcareous soils and at alkaline pH, phosphate precipitates with calcium to form calcium phosphate. In acid soils, phosphate reacts with iron and aluminum oxides in the soil to form insoluble compounds. Sometimes phosphate is immobilized initially by adsorption to the soil and then slowly reverts to insoluble forms, allowing more adsorption of mobile phosphate. In clean sands with near-neutral pH, phosphate can be relatively mobile (Bouwer, 1985). Ammonium removals of 25 to 98 percent and phosphorus removals of 4 to 84 percent during soil passage have been reported (Rice and Bouwer, 1984; Nema et al., 2001). Because nitrogen removal is driven by biodegradation, it is sustainable; conversely, phosphorus removal is not truly sustainable because adsorption is the dominant mechanism.

Heavy Metals

Soils containing iron, aluminum, calcium, and magnesium have a high affinity for metal adsorption. Metal removal efficiencies vary with the constituent metal. Metals are significantly retained in most soils, but a high pH favors immobilization. Sontheimer (1980) found that heavy metal removal in the lower Rhine River exceeded 90 percent for chromium and arsenic and was greater than 50 percent for cadmium, zinc, lead, copper, and nickel. Heavy metal removal in SAT ranges from 50 to 90 percent. However, metals applied at low concentrations (below drinking water standards) may not be affected by soil passage (Crites et al., 2006). It is noted that longer-term metal removal in the soil layer is not sustainable because the soil minerals have limited adsorption capacities. Furthermore, under reducing conditions or from changes in pH, the already adsorbed metal may desorb or leach and reach deeper into the aquifer or ultimately appear in the extracted well water (Lin et al., 2004).

Comparison of Slow Sand Filtration, Bank Filtration, and Soil Aquifer Treatment

In general, BF and SAT can be considered as the extended version of slow sand filtration (SSF) because all of these systems have a schmutzdecke, where significant removals of microbes and BDOC can take place. BF and SAT, however, have higher removal efficiencies for different contaminants because they have considerable depth and a lower hydraulic loading rate (hence longer travel time) compared with SSF. Table 15-7 presents a comparative assessment of SSF, RBF, and SAT.

TABLE 15-7 Comparative Assessment of Slow Sand Filtration (SSF), Riverbank Filtration (RBF), and Soil Aquifer Treatment (SAT)

Aim	SSF	RBF	SAT
Depth/travel distance	1 m	10 m	>30 m
Residence time	>2 h (~0.1 day)	>10 days	>10 days
Hydraulic loading/infiltration rate	<0.5 m/h (<12 m/day)	<1 m/day	<1 m/day
Schmutzdecke/clogging layer	Yes	Yes	Yes
Flow regime	Saturated	Unsaturated and saturated	Unsaturated and saturated
Redox	Oxic (aerobic)	Oxic (aerobic) and anoxic	Oxic (aerobic) and anoxic
Turbidity	≤1 ntu	≤1 ntu	≤1 ntu
DOC removal	≥15%	≥50%	≥50%
Biostability: BDOC removal	50%	<LOD	<LOD
Trace organic contaminants removal	—	≥50% (except for few persistent PhACs)	≥50% (except for few persistent PhACs)
(Total) Nitrogen achievable	—	≤2 mg/L	≤2 mg/L
Microbial removal (viruses)	≥2 log	≥4 log	≥4 log

LOD = level of detection

DESIGN AND OPERATION OF NATURAL WATER TREATMENT SYSTEMS

The design and operation of natural water treatment systems depend mainly on (1) source water quality, (2) hydrogeological conditions at the site, (3) quantity of water to be abstracted, and (4) intended use of extracted water (water quality requirements). Design parameters for natural treatment systems include (1) number of wells (production capacity per well), (2) spacing between the wells, (3) distance of the wells from the bank. (4) share (percent) of river/lake water and native groundwater, (5) water quality obtained from the system, (6) post-treatment requirements, if any, and (7) capital and operation and maintenance (O&M) costs.

Type, Number, Spacing, and Distance of Production Wells

The numbers of production wells or gallery, their spacing and distance from the water source, and hydrogeological conditions have great impact on productivity. Generally, three types of wells have been used for BF: (1) horizontal collector wells, (2) vertical wells, and (3) angle wells. Horizontal wells have high production capacity; however, there can be more breakthrough of contaminants because the residence time is likely to be shorter. Compared with horizontal collector wells, vertical wells have relatively low production capacity but provide longer residence time. Angle wells are used when the depth/thickness of the aquifer is small (shallow aquifers). An angle well can help to increase the yield by lengthening the screen where water enters the well, potentially allowing more water to flow through the well (Williams, 2008).

The number of wells and their spacing influence the drawdown for a given production capacity. The total number of wells required for a given production capacity and given allowable drawdown in the well can be estimated using groundwater hydraulic models such as MODFLOW (USGS, 2008). The distance of the wells from the river/lake influences the travel time and the dilution ratio. In the Netherlands, the minimum recommended distance of a well from the bank is more than 20 m and preferably more than 100 m. The distance of the abstraction well from the surface water (bank) for 17 BF systems in The Netherlands ranges from 21 to 1500 m, with the resulting travel time ranging from 0.15 to 12 years (Hiemstra et al., 2003). The longer the distance of the wells from the river/lake, the smaller is the percentage of bank filtrate (more native groundwater) in the extracted water, and the longer the travel time, the better is the removal of the contaminants in the source water during soil passage. Therefore, there should be some compromise between the quality and quantity of extracted water by proper design and location of the wells.

The production wells have been constructed in clusters parallel to the bank, perpendicular to the bank, at a certain angle to the bank, or in parabolic configuration in order to optimize the quantity and quality of abstracted water. The best place for abstracting a large quantity of riverbank filtrate is on an island or within a meander, especially if the river has a steeper gradient than the groundwater in the connected aquifer and if the riverbed has high conductivity. The siting of BF schemes is an optimization task that needs a balance between the extracted volume of bank filtrate and the preservation of water quality owing to attenuation and the mixing process during BF (Grischek et al., 2006).

Hydraulic models such as MODFLOW (USGS, 2008) and the NASRI Bank Filtration Simulator (Holzbecher, 2006) can be used to estimate the ratio of bank filtrate and native groundwater in the extracted well water for a given design of the wells for a particular hydrogeological condition. The percentage of river/surface water in the bank filtrates in 17 systems in the Netherlands ranged from 20 to 98 percent, with an average of 63 percent (Hiemstra et al., 2003).

Hydraulic Loading Rates/Travel Time

Most BF sites in the world are located in alluvial sand and gravel aquifers having hydraulic conductivities greater than 10^{-4} m/s. The thickness of exploited aquifers ranges from 5 to 120 m. Different alignments of the abstraction wells with respect to river/lake are possible, which will influence (1) travel time and (2) the ratio of the bank filtrate and native groundwater in the water produced. At most sites in Europe, the distance between riverbank and production wells is more than 50 m, and travel times are more than 50 days. In the United States, the travel times are less than 50 days. This is so because in Europe BF is considered the primary treatment system, whereas in the United States it is considered a pretreatment system only. Table 15-8 summarizes the effects of RBF siting and management on water quality (Grischek et al., 2006).

Surface infiltration systems require permeable soils and vadose zones to introduce the water into the ground and to the aquifer and unconfined and sufficiently transmissive aquifers to provide lateral flow away from the infiltration system without excessive groundwater mounding. Therefore, the hydraulic loading rate applied should match the hydraulic conductivity of the aquifer material. Basin infiltration tests and small-cylinder infiltrometers, double-ring (buffered) as well as single-ring systems, are useful for measurement and comparison of relative infiltration rates or hydraulic conductivity. However, it is noted that the hydraulic conductivity measurements can provide only a good estimate of maximum hydraulic loading rate attainable within the basins. Because of the decrease in infiltration rates with time owing to clogging and the need for regular drying and periodic cleaning of recharge basins or other surface infiltration systems, the actual hydraulic capacities (the amount of water applied per unit area in meters per year) therefore are much smaller than (only a few percent of) the infiltration rates (Bouwer, 2002; USEPA, 2006). Table 15-9 presents the suggested hydraulic rates for surface infiltration systems based on different field measurements.

Redox Conditions

Redox conditions in the aquifer significantly influence the water quality of the product water from natural treatment systems. The reader can find principles of oxidation-reduction chemistry in Chap. 3. Reducing conditions in the aquifer may cause the appearance of

TABLE 15-8 Effects of Design Parameters on Bank Filtration Efficiency

Aim	Long distance between bank and wells	Low abstraction along a line compared to high abstraction at a point	Siting of wells in a meander
Removal of suspended solids and particles	0	0	0
Removal of bacteria, viruses	++	+	0
Elimination of biodegradable compounds	+	+	0
Equilibration of temperature changes	++	+	0
Equilibration of changes in concentration of compounds in water	++	+	0
High proportion of bank filtrate in pumped raw water	– –	0	++

+ = increase/improvement; – = decrease; 0 = no significant effect.
Source: Grischek et al., 2006.

TABLE 15-9 Suggested Hydraulic Loading Rates Based on Different Field Measurements

Field measurement	Annual loading rate (%)
Basin infiltration test	7–10 of minimum measured infiltration rate
Cylinder infiltrometer and air entry permeameter measurements	2–4 of minimum measured infiltration rate
Vertical hydraulic conductivity measurements	4–10 of conductivity of most restricting soil layer

Source: USEPA, 2006.

undesired metals such as iron, manganese, copper, cadmium, zinc, or lead in extracted water. Redox conditions also influence the behavior of a number of organic micropollutants, including pharmaceutically active compounds, halogenated organic compounds, pesticides, and disinfection by-products (Massmann, 2008). The redox conditions along the flow path are determined by the type and quantity of available degradable organic matter and the electron acceptors. Mainly, the oxygen concentration in the source water and the concentration of oxygen-consuming substances present in the water influence the redox conditions. If dissolved oxygen is removed during percolation through the vadose zone, anoxic conditions are likely to develop in the saturated zone because mechanisms for oxygen transport to the saturated zone are mostly due to natural aquifer recharge. Furthermore, mixing of the recharge water into the aquifer may create conditions that are not in thermodynamic equilibrium, consequently driving several redox reactions in the aquifer.

In BF systems, oxygen in the water depletes with the travel distance and ultimately may lead to iron and manganese reduction. For ARR systems, the oxygen concentrations in the aquifer can be increased by aeration and removal of oxygen-consuming contaminants during pretreatment; therefore, aerobic conditions could be more dominant. In SAT systems, a combination of aerobic conditions at the initial depth and anoxic conditions in the deeper layer is beneficial to facilitate organics removal and nitrification of ammonium in the upper layers, followed by denitrification in the subsequent layer. Furthermore, the existence of both aerobic and anoxic conditions in different zones also may favor the removal of specific micropollutants because the removal of trace organics is highly dependent on the redox conditions and concentration of bulk organic matter present (Massmann et al., 2006; Stuyfzand et al., 2007). In the case of ARR and SAT systems, intermittent application of the recharge water (wet/dry cycle) also helps to influence the redox conditions because oxygen can pass deeper in the soil layer during the dry period.

Redox conditions in the aquifer are also important for aquifer integrity in MAR systems. Changes in redox potential in an aquifer may have long-term consequences for aquifer integrity by either dissolution reactions (that dissolve the aquifer media) or precipitation reactions (that plug the aquifer) (NRC, 2008).

Clogging

Clogging of the banks and bottoms of the infiltration facilities is one of the main operational problems associated with natural treatment systems such as BF, ARR, and SAT. Depending on the type of system, clogging may be caused by physical (suspended solids, gas entrapment), chemical (precipitation), and biological (microbial growth) processes or combinations thereof. Furthermore, the suspended solids causing clogging of the banks and basin bottom can be inorganic (e.g., clays, silts, fine sands) or organic (e.g., algae, bacterial flocs, sludge particles, leaf litter).

Clogging is both a desirable and an undesirable process. Clogging may increase the removal efficiency of natural filtration because it is believed to have comparable characteristics to the schmutzdecke in engineered slow sand filtration systems (see Chap. 10)—rich in organic matter with high concentrations of microorganisms. However, it may be detrimental because it leads to a decrease in hydraulic conductivity. The yield of the natural treatment systems therefore can be severely affected by clogging, which is a problem for the sustainability of the drinking water supply (Tufenkji et al., 2002; Schubert, 2004). Clogging rates increase with increasing infiltration rates because of the increased loading rates of suspended solids, nutrients, and organic carbon.

Clogging results in a reduction in the permeability and consequently lower infiltration rates. In the case of ARR and SAT systems, periodic removal of sediments (when the infiltration rate is reduced considerably) through scraping is necessary to restore the infiltration rate and to avoid fine material penetrating too deeply into the infiltration surface. Removing the material by raking or scraping is much better than mixing it with the soil. The latter practice leads to gradual accumulation of clogging materials in the top 10 to 20 cm of the soil, eventually necessitating complete removal of this layer, which can be expensive (Pescod, 1992; NRC, 1994). Furthermore, pretreatment of the source water (e.g., sedimentation, filtration, nutrient removal) considerably reduces the clogging of ARR and SAT systems.

In the case of a BF system, a clogging layer develops with time, thus reducing the infiltration rate. However, the flow and turbidity in rivers vary, and the clogging layer generally is scoured during periods of high floods. Experiences with RBF in the Netherlands indicate that there are serious risks of clogging in the case of a gravel river bed and on sites where sludge is strongly accumulating owing to structurally reduced river flows and hardly any such risks where the river bed is sandy and cleaned periodically by flood waves (Stuyfzand et al., 2004). Furthermore, proper location of intake structure and judicious design of screens and collection well (including type of well and abstraction rate) minimize the clogging effect. Clogging is more common among BF wells near the outer section of a bend. The yield of riverbank filtered water in the inner part of the bend is normally higher because of the movable ground of the riverbed, as well as the natural underground cross-flow owing to the river gradient (Schubert, 2002b).

Schubert (2006) suggested that reducing the pumping rate and in the long term maintaining adequate quality of river water may help to improve the situation in existing RBF plants with severe clogging problems. Based on their study of 11 RBF sites in the United States and Europe, Hubbs and colleagues. (2006) recommended that systems should be designed with the realization that a decline in the capacity in the first few years is possible. However, they suggested that the following measures can be taken to reduce the expected decline in yield:

1. Size the facility to minimize the impact of unsaturated conditions under the riverbed.
2. Design an extraction system that distributes infiltration stresses evenly along the riverbed.
3. Locate the facility to maximize the regenerative impacts of streambed scouring, with preference to locations with movable streambeds, avoiding areas of streambed armoring (areas with rocks or boulders deposited on the streambed and banks, either natural or man-made) and locations where very fine sediment is deposited.

SELECTED CASE STUDIES OF NATURAL TREATMENT SYSTEMS

Table 15-10 presents the salient features of some selected BF systems in the United States and Europe. Details of 17 different BF sites in the Netherlands are presented by Hiemstra and colleagues (2003). Furthermore, BF is now being used increasingly in Asia and Africa

TABLE 15-10 Selected Bank Filtration Systems in the United States and Europe

Site location	Well type (horizontal, H; vertical, V)	Number of wells	Design capacity of well field (m^3/s)	River/lake system
Cincinnati, OH (USA)	V	10	1.750	Great Miami
Columbus, OH (USA)	H	4	1.750	Scioto/Big Walnut
Galesburg, IL (USA)	H	1	0.438	Mississippi
Independence, MO (USA)	H	1	0.656	Missouri
Jacksonville, IL (USA)	H	1	0.350	Illinois
Kalama, WA (USA)	H	1	0.114	Kalama
Kansas City, KS (USA)	H	1	1.750	Missouri
Kennewick, WA (USA)	H	1	0.131	Columbia
Lincoln, NB (USA)	H and V	2 + 44	1.530	Platte
Kearny, NB (USA)	—	12	0.095	Platte
Mt. Carmel, IL (USA)	V	1	0.044	Wabash
Sacramento, CA (USA)	H	1	0.438	Sacramento
Terre Haute, IN (USA)	H	1	0.525	Wabash
Louisville, KY (USA)	H and V	1 + 1	0.875	Ohio
Cedar Rapids, IO (USA)	H and V	4 + 53	1.25	Cedar
Maribor (Slovenia)	V	13	0.750	Drava
Dusseldorf (Germany)	H and V	12 (total)	2.0	Rhine
Tegel, Berlin (Germany)	H and V	8 (total)	1.427	Lake Tegel
Mockritz (Germany)	V	74	1.260	Elbe
Torgau (Germany)	V	42	1.737	Elbe
Csepel Island, Budapest (Hungary)	V and H	280 (total)	3.470	Danube

Sources: Ray et al., 2002; Verstraeten et al., 2002; Eckert et al., 2006; Hunt et al., 2002.

(Ray, 2008; Shamukh and Abdel-Wahab, 2007). Short descriptions of a few selected natural treatment systems are presented in the following paragraphs.

Lake Bank Filtration: Berlin, Germany

About 75 percent of Berlin's drinking water is bank filtrate or artificially recharged groundwater. Lake Tegel, situated in the northwestern part of Berlin, Germany, has an area of about 4 km^2 and average and maximum depths of 7.9 and 16 m, respectively. The lake receives water from its two major tributaries, the Nordgraben and the Tegeler Fließ. This water is influenced by treated domestic wastewater effluent that constitutes about 15 to 40 percent by proportion (Grunheid et al., 2005; Pekdeger, 2006). The Lake Tegel bank filtration field site was established over 100 years ago. The waterworks at Tegel, one of the largest in Berlin, pumps about 50 million m^3 per year from six well fields around the lake and from two on islands. Vertical wells are 30 to 60 m deep with a production capacity of 50 to 150 m^3/h, and one horizontal well is 16 m deep with a production capacity up to 250 m^3/h. The distance of the production wells from the bank ranges from 80 to 100 m. The travel time from the lake through the aquifer to various wells ranges from less than 0.5 to 3.2 months for shallow wells and 1.5 to 4.3 months for deep wells. The bank filtration system at Lake Tegel is characterized by highly transient well operation, whereby hydraulic head differences between lake and groundwater show strong temporal variations.

The DOC of lake water ranges from 6 to 8 mg/L and is reduced to 4.5 to 5.0 mg/L after bank filtration. The BDOC of lake water varies from 1.5 to 3.5 mg/L, with an average of 2.5 mg/L. There is almost complete mineralization of infiltrated BDOC. The nitrate concentration of the source water is reduced significantly during infiltration; however, there is dissolution of iron and manganese in the aquifer. The Fe^{2+} and Mn^{2+} concentrations in the bank filtrate range from 0.1 to 0.5 mg/L and from 0.3 to 0.6 mg/L, respectively. Furthermore, there is some increase in the calcium and bicarbonate concentration of the water owing to dissolution of $CaCO_3$ in the aquifer. The posttreatment of the bank filtrate before supply consists only of aeration followed by rapid sand filtration (for iron and manganese removal), with no chemical disinfection (Grunheid et al., 2005; Ameda, 2008).

Riverbank Filtration: Louisville, Ky.

The Louisville Water Company (Kentucky) has been using the alluvial aquifer along the Ohio River for water supply for many years. Based on extensive studies from 1940 to 1995, a full-scale RBF system of 0.88 m^3/s (20 mgd) capacity was installed in 1999. The RBF facility is a horizontal collector well located on the bank of the Ohio River, about 30 m from the river. It is a 6.1 diameter caisson with seven horizontal laterals. The caisson is about 34 m deep from the ground level. The length of each lateral is 73 m for the four laterals oriented toward the river and 61 m for the remaining laterals. The laterals are 0.30 diameter stainless steel wire wound screens that are installed about 24 m below ground level (Wang, 2002; Hubbs et al., 2006). The average travel time is estimated to be 4 weeks (with a minimum of 2 days), and the percentage of river water in the bank filtrate is estimated to be around 78 percent.

The turbidity of bank filtrate was consistently less than 0.1 ntu. The TOC concentration of the river water ranged from 2.1 to 4.9 mg/L, with an average of 2.9 mg/L. The average DOC concentration was reduced to 2.1 mg/L after 2.75 m of filtration and was reduced further to 1.6 mg/L after some dilution from native groundwater. About 20 percent of the DOC in the river was biodegradable. The BDOC was reduced to less than 0.10 mg/L after 15 m of filtration. The average removals of total coliform and HPC were 3.8 log and 2.2 log, respectively (Wang, 2002).

The bank filtrate from the horizontal well undergoes conventional surface water treatment (coagulation, sedimentation, rapid sand filtration, and disinfection) before distribution. The Louisville Water Company is now expanding and improving the existing RBF facility. A RBF tunnel will extend from a location within the B. E. Payne Water Treatment Plant to the confluence of Harrods Creek and the Ohio River. The tunnel will be 2.41 km (1.5 mi) in length and 3 m (10 ft) in diameter, constructed in bed rock 45.7 m (150 ft) below the ground surface. The tunnel will convey RBF water to a new pump station, where it will be treated and supplied.

Riverbank Filtration: Düsseldorf, Germany

The City of Düsseldorf is situated in the northwest of Germany in the lower Rhine valley. The mean discharge of the Rhine River at Düsseldorf is 2200 m^3/s, whereas the waterworks uses less than 2 m^3/s (46 mgd). The width and dynamics of the Rhine River allow a sustainable application of RBF for drinking water supply. Riverbank filtration has been used as the first step for treating drinking water in the Düsseldorf waterworks since 1870. The raw water is abstracted from a quaternary aquifer with a proportion of 50 to 90 percent of bank filtrate. At present, 600,000 people are supplied with treated bank filtrate by four waterworks using vertical wells and horizontal collector wells. An average water demand of about 165,000 m^3/day

(44 mgd) and a maximum demand of about 210,000 m^3/day (55 mgd) are provided by this system. Depending on the hydraulic situation, the residence time of the bank filtrate in the aquifer varies from 1 week to several months. At the Flehe waterworks site, the production wells are situated at a distance of about 70 m from the riverbank. For the first 80 years of central drinking water supply in Düsseldorf, RBF alone without additional treatment (only disinfection) sufficed to provide safe drinking water (Schubert, 2002a; Eckert et al., 2006).

The concentration of suspended solids in the Rhine River varies between 10 and more than 400 mg/L, with a mean value of less than 40 mg/L. However, the raw water in the production wells is clear, with a turbidity of 0.05 ntu. The average DOC concentration in the well water varies between 1.0 and 1.2 mg/L. The mean reductions for total colony counts and *Escherichia coli* are 3 and 4 log, respectively (Schubert, 2002c).

The existing treatment plant, installed in 1967, consists of a combination of ozonation, biological filtration, and adsorption on granular activated carbon. This posttreatment system was introduced to remove iron, manganese, DOC, and taste and odor levels and meet drinking water quality requirements.

Riverbank Filtration and Artificial Recharge: Maribor, Slovenia

The City of Maribor, Slovenia, has been using bank filtration for control of water quality of local groundwater aquifers. Four wells of 0.075 m^3/s (1.7 mgd) capacity each have been constructed on an island of the Drava River in Maribor to pump the bank filtrate. The island is under natural protection and at the same time represents a recreational area for the city. Some iron and manganese are present in the bank filtrate. This bank filtrate then is treated by applying sedimentation (lamella separators to remove fine sand), cascade aeration, and rapid sand filtration. The treatment plant also has facilities for ozonation, coagulation, and powdered activated carbon dosing for the removal of organic pollutants in bank filtrate, if any. However, at the present time, only lamella settlers, cascade aeration, and rapid filtration are in operation. In the rapid sand filters after cascade aeration, the iron and manganese generated by the bank filtration are eliminated.

The treated bank filtrate then is used for artificial recharge through three wells into the local groundwater aquifer to augment the existing 0.4 m^3/s capacity of the aquifer and at the same time to protect the groundwater used for water supply against the endangering pollution from the city. Finally, groundwater is pumped from 13 wells located 300 m away from the recharge wells and supplied to the city after disinfection. Research is in progress to use and optimize the artificial recharge of the treated bank filtrate to achieve subsurface iron and manganese removal. Thus the water supply system in Maribor is a typical example of a combination of bank filtration and artificial recharge for water treatment and groundwater quality management (Rismal and Kopac, 2008).

Dune Infiltration: Leiduin, the Netherlands

The Leiduin water treatment plant is one of the two plants of Waternet (Amsterdam Water Supply Company) supplying water to the city of Amsterdam, The Netherlands. The raw water from Lekkanaal (a man-made side branch of the Rhine River) is pretreated at Nieuwegein and transported via three transport pipelines of 1200 to 1500 mm (about 4 to 5 ft) diameter for a distance of about 50 km (31 mi) to dune areas near the Leiduin water treatment plant. The total length of the transport pipe is 210 km (130 mi), and the total transport capacity is about 411,000 m^3/day (110 mgd).

The standard pretreatment process employed at the Nieuwegein pretreatment plant is coagulation (with ferric chloride), sedimentation, rapid sand filtration, and pH correction with caustic soda. The typical concentrations of suspended matter in raw water and pretreated water are 37.0 and 0.1 mg/L, respectively, whereas the corresponding DOC concentrations are 3.5 and 2.5 mg/L. After pretreatment, the water is transported to the nearby dune area for infiltration. The system consists of 40 infiltration ditches with a total length of 24.6 km (15 mi) and an average width of 35 m (115 ft). The water is transported to these ditches through open canals with a total length of 10 km (6.2 mi). The dune area used by Waternet is about 36 km^2 (14 mi^2), of which 10 km^2 (4 mi^2) is taken up by actual infiltration areas.

After the average residence time of about 100 days (ranging from 60–400 days), the infiltrated water is abstracted through drains and collected in an open basin. This ARR water then is further treated at the Leiduin water treatment plant by employing cascade aeration, rapid sand filtration, ozonation, pellet softening, two-stage biological activated carbon filtration, and slow sand filtration to achieve drinking water . No final disinfection is applied before supplying the water to the city of Amsterdam. The present production capacity is 192,000 m^3/day (50 mgd) (van der Hoek, 2000; Tielemans, 2007).

Artificial Recharge: Graz, Austria

Artificial groundwater recharge has proved to be a sustainable solution for water supply for the City of Graz, Austria, for many decades. The first groundwater recharge system was operated in the 1920s, with the objective of increasing the capacity of the local groundwater aquifer. Later, with increasing pollution of the Mur River, infiltrating surface water threatened the quality of the local groundwater. The artificial recharge system then was redesigned to control the infiltration of polluted Mur River water into the aquifer. Later, in the 1980s, ARR systems were commissioned in Friesach and Andritz. The waterworks at Andritz maintains the quality of local groundwater using ARR and produces 21,600 to 27,600 m^3/day (5.7–7.3 mgd), providing about 30 percent of the water demand of the City of Graz, with a population of 250,000. Raw surface water from Andritzbach catchments passes through sedimentation tanks, horizontal gravel filters, and cascade aeration before infiltration. Three different systems of lawn infiltration basins, sand basins, and infiltration trenches have been employed for artificial recharge. The sand basins and infiltration trenches operate continuously, while the lawn basins operate intermittently (10 days flooding and 20 days drying). The infiltrated water is later extracted from two horizontal collector wells [each 21.5 m (71 ft) deep with 26 laterals of length 20–30 m (66–98 ft)]. The underground passage ensures the high quality of the drinking water, which does not need any further treatment or disinfection before supply (Tischendorf, 2007).

Riverbank Filtration and Artificial Recharge: Aurora, Colo.

The City of Aurora, Colo., is currently constructing a $754 million Prairie Waters Project as a sustainable water supply solution for the growing population of the city. The Prairie Waters Project is an example a multiple-barrier water treatment approach combining natural systems and conventional water treatment processes. The project uses an innovative natural purification process (combination of bank filtration and artificial recharge) to perform initial treatment of water from the South Platte River. The riverbank filtration along the South Platte River near Brighton extracts water from alluvium and removes nitrate, pathogens, and trace organic chemicals in about 10 days of travel time. Aquifer recharge and recovery then provides additional travel time for additional removal of nutrients and trace organic contaminants.

After the natural purification process, water will be piped 54.7 km (34 mi) south using a 1.5 m (60 in) pipeline to the Aurora Reservoir Water Purification Facility. The new treatment facility will use multiple treatment processes that include chemical softening (to reduce hardness, iron, manganese, and scaling potential), advanced oxidation using ultraviolet (UV) light and hydrogen peroxide (to inactivate pathogens and oxidize remaining trace organics), rapid sand filtration (to remove remaining particles and pathogens), and activated carbon adsorption (to adsorb remaining trace organics and improve taste). Once completed, it will be one of the most advanced water treatment facilities in the United States and will be able to treat 190,000 m^3/day (50 mgd) (Ingvoldstad, 2007).

Soil Aquifer Treatment: Dan Region, Israel

SAT has been practiced continuously on a large-scale in the Dan Region Project in Israel since 1977. It is the largest wastewater reclamation and reuse scheme in Israel and one of the largest in the world. About 330,000 m^3/day (87 mgd) of effluent from the wastewater treatment plant of the Dan Region (Tel Aviv Metropolitan Area) are being recharged to the groundwater aquifer via spreading basins in four recharge zones built in areas of predominantly sandy soils—a northern recharge zone (Soreq) and three southern recharge zones (Yavne 1, 2, and 3). Each recharge zone consists of several recharge basins that are divided into sub-basins. The total area of the infiltration basins is 111 hectares (274 acres). The operation of the recharge basins is intermittent, that is, flooding for 1 to 2 days and drying for 2 to 4 days to maintain high infiltration rates through the upper soil layer and to allow oxygen to penetrate into the soil, thus enhancing and diversifying the soil purification capacity. The average hydraulic loading rate ranges between 0.3 and 0.6 m/day. The sites are mechanically ploughed approximately every 3 months to renew their percolation capacity. The average retention time within the aquifer is 3 to 12 months. An array of 78 recovery wells surrounds the spreading sites at a distance ranging from 500 to 1500 m (3000–4920 ft).

After SAT, the removal of total suspended solids and BOD is virtually complete, 100 percent and more than 98 percent, respectively. The average total suspended solids, BOD, and TOC of SAT effluent are less than 0.1 mg/L, less than 0.5 mg/L, and 2.4 mg/L, respectively. During SAT, complete nitrification and partial denitrification occur, which reduce the total nitrogen concentration from 22.8 to less than 2.1 to 8.76 mg/L. The reclaimed water pumped from recovery wells is conveyed to irrigation areas in the southern coastal plain and the northern Negev by a 120 km long (75 mi) pipeline (Kanarek and Michail, 1996; Ideolovitch et al., 2003).

ABBREVIATIONS

AOC	assimilable organic carbon
AOI	absorbable organic iodine
ARR	artificial recharge and recovery
ASR	aquifer storage and recovery
ASTR	aquifer storage transfer and recovery
BDOC	biodegradable dissolved organic carbon
BOD	biological oxygen demand
BF	bank filtration
C:N	carbon/nitrogen ratio
DBPs	disinfection by-products

DOC	dissolved organic carbon
EDCs	endocrine disrupting compounds
EfOM	effluent organic matter
HPC	heterotrophic plate count
LBF	lake bank filtration
MCLs	maximum contaminants levels
mgd	million gallons per day
NOM	natural organic matter
O&M	operation and maintenance
PCPs	personal-care products
PhACs	pharmaceutically active compounds
RBF	riverbank filtration
SAT	soil aquifer treatment
SMP	soluble microbial product
SSF	slow sand filtration
≡S	adsorption surface
TOC	total organic carbon
THMFP	trihalomethane formation potential

REFERENCES

Ameda, E. (2008). Analysis of the removal of trace organics during bank filtration. M.Sc. thesis MWI 2008-39, UNESCO-IHE, Delft, The Netherlands.

Amy, G., and Drewes, J. (2007). Soil aquifer treatment (SAT) as a natural and sustainable wastewater reclamation/reuse technology: Fate of wastewater effluent organic matter (EfOM) and trace organic compounds. *Environmental Monitoring and Assessment*, 129, 19–26.

Asano, T., and Cotruvo, J. A. (2004). Groundwater recharge with reclaimed municipal wastewater: Health and regulatory considerations. *Water Research* 38(8), 1941–1951.

American Water Works Association Research Foundation (AWWARF) and American Water Works Association (AWWA) (2006). *Advances in Soil Aquifer Treatment for Sustainable Water Reuse.* Boulder, CO: AWWA.

Bouwer, H., and Rice, R. C. (1984). Hydraulic properties of stony vadose zones. *Ground Water* 22(6), 696–705.

Bouwer, H. (1985). Renovation of wastewater with rapid-infiltration land treatment systems. In *Artificial Recharge of Groundwater*, T. Asano, ed. Boston: Butterworths.

Bouwer, H. (1992). Reuse rules. *Civil Engineering* 62(7), 72–75.

Bouwer, H. (1993). Urban and agricultural competition for water, and water reuse. *International Journal of Water Resources Development* 9(1), 13–25.

Bouwer, H. (2002). Artificial recharge of groundwater: Hydrogeology and engineering. *Hydrogeology Journal* 10, 121–142.

Braester, C., and Martinell, R. (1988). The Vyredox and Nitredox methods of in situ treatment of groundwater. *Water Science and Technology* 20(3), 149–163.

Crites, R. W., Middlebrooks, E. J., and Reed, S. C. (2006). *Natural Wastewater Treatment Systems.* Boca Raton, FL: CRC Press.

Dillon, P. J., Miller, M., Fallowfield, H., and Huston, J. (2002). The potential of riverbank filtration for drinking water supplies in relation to microcystin removal in brackish aquifers. *Journal of Hydrology* 266(3–4), 209–221.

Dillon, P. (2005). Future management of aquifer recharge. *Hydrogeology Journal* 13(1), 313–316.

Doussan, C., Poitevin, G., Ledoux, E., and Detay, M. (1997). River bank filtration: Modelling of the changes in water chemistry with emphasis on nitrogen species. *Journal of Contaminant Hydrology* 25(1–2), 129–156.

Eckert, P., and Irmscher, R. (2006). Over 130 years of experience with riverbank filtration in Düsseldorf, Germany. *Journal of Water Supply: Research and Technology- AQUA* 55(4), 283–291.

Fox, P., Nellor, M., Arnold, R., Lansey, K., Bassett, R., Gerba, C., Karpiscak, M., Amy, G., and Richard, M. (2001). *An Investigation on Soil Aquifer Treatment for Sustainable Water Reuse.* AWWA Research Foundation and the American Water Works Association Report. Denver, CO: Water Research Foundation.

Grischek, T., Schoenheinz, D., Worch, E., and Hiscock, K. (2002a). Bank filtration in Europe: An overview of aquifer conditions and hydraulic controls. In *Management of Aquifer Recharge for Sustainability*, P. Dillon, ed. Lisse, The Netherlands: A. A. Balkema.

Grischek, T., Schoenheinz, D., and Ray, C. (2002). Siting and design issues for riverbank filtration schemes. In *Riverbank Filtration: Improving Source-Water Quality*, C. Ray, G. Melin, and R. B. Linsky, eds. Dordrecht, The Netherlands: Kluwer Acadamic Publishers.

Grischek, T., Schoenheinz, D. and Ray, C. (2006). Siting and design issues for riverbank filtration schemes. In *Proceedings of the International Workshop on Riverbank Filtration, Seoul, Korea.*

Grunheid, S., Amy, G., and Jekel, M. (2005). Removal of bulk dissolved organic carbon (DOC) and trace organic compounds by bank filtration and artificial recharge. *Water Research* 39(14), 3219–3228.

Heberer, T. (2002). Occurrence, fate, and removal of pharmaceutical residues in the aquatic environment: A review of recent research data. *Toxicology Letters* 131(1–2), 5–17.

Heberer, T., Mechlinski, A., Fanck, B., Knappe, A., Massmann, G., Pekdeger, A. and Fritz, B. (2004). Field studies on the fate and transport of pharmaceutical residues in bank filtration. *Ground Water Monitoring and Remediation* 24(2), 70–77.

Hiemstra, P., Kolpa, R., van Eekhout, J.M.J.M, van Kessel, T.A.L., Adamse, E.D and van Passesn, J.A.M. (2003). "Natural" recharge of groundwater: Bank infiltration in the Netherlands. *Journal of Water Supply: Research and Technology- AQUA* 52(1), 37–47.

Hiscock, K. M., and Grischek, T. (2002). Attenuation of groundwater pollution by bank filtration. *Journal of Hydrology* 266(3–4), 139–144.

Holzbecher, E. (2006). Calculating the effect of natural attenuation during bank filtration. *Computers and Geosciences* 32(9), 1451–1460.

Hubbs, S. A., Ball, K., and Caldwell, T. (2006). *Riverbank Filtration: An Evaluation of RBF Hydrology and Impact on Yield.* AwwaRF Report 91141F. London: IWA Publishing.

Huisman, L., and Olsthoorn, T. N. (1983). *Artificial Groundwater Recharge.* Boston: Pitman.

Hunt, H., Schubert, J., and Ray, C. (2002). Operation and maintenance considerations. In *Riverbank Filtration: Improving Source-Water Quality.* Dordrecht, The Netherlands: Kluwer Academic Publishers.

International Association of Hydrogeologists (IAH) and International Hydrological Programme (IHP) (2005). *Strategies for Managed Aquifer Recharge in Semi-Arid Areas*, I. Gale, ed. Paris, France: IAH and UNESCO IHP.

Idelovitch, E., Icekson-Tal, N., Avraham, O., and Michail, M. (2003). The long-term performance of soil aquifer treatment (SAT) for effluent reuse. *Water Science and Technology: Water Supply* 3(4), 239–246.

Ingvoldstad, S. (2007). Aurora's Prairie Waters Project: A Sustainable Water Supply Solution. Workshop on Sustaining Colorado's Watershed, October 3, 2007. Available online at www.coloradowater.org/documents/ConferencePowerPoints/ScottIngvoldstad.pdf.

Kiwa Water Research (KWR) (2004). *The Dutch Experience with MARS (Managed Aquifer Recharge and Storage); A Review of Facilities, Techniques and Tools.* KWR 05.001. Amsterdam, The Netherlands: KWR.

Kuehn, W., and Mueller, U. (2000). Riverbank filtration: An overview. *Journal AWWA* 92(12), 60–69.

Kanarek, A., and Michail, M. (1996). Groundwater recharge with municipal effluent: Dan Region reclamation project, Israel. *Water Science and Technology* 34(11), 227–233.

Lehtola, M. J., Miettinen, L. T., Pertti, T. V., and Martikainen, J. (2002). Changes in content of microbially available phosphorus, assimilable organic carbon and microbial growth potential during drinking water treatment processes. *Water Research* 36(15), 3681–3690.

Lin, C., Shacahr, Y., and Banin, A. (2004). Heavy metal retention and partitioning in a large-scale soil-aquifer treatment (SAT) system used for wastewater reclamation. *Chemosphere* 57(9), 1047–1058.

Massmann, G., Greskowiak, J., Dunnbier, U., Zuehlke, S. and Pekdeger, A. (2006). The impact of alternating redox conditions on groundwater chemistry during artificial recharge in Berlin. In *Proceedings of the 5th International Symposium on Managed Aquifer Recharge, ISMAR5, Berlin, Germany.*

Massmann, G., Nogeitzig, A., Taute, T., and Pekdeger, A. (2008). Seasonal and spatial distribution of redox zones during lake bank filtration in Berlin, Germany. *Environmental Geology* 54(1), 53–65.

Miller, G. (2006). *Subsurface Treatment for Arsenic Removal.* Boulder. CO: WERC–New Mexico State University and AWWA Research Foundation.

Nema, P., Ojha, C. S. P., Kumar, A., and Khanna, P. (2001). Technoeconomic evaluation of soil-aquifer treatment using primary effluent at Ahmedabad, India. *Water Research* 35(9), 2179–2190.

Neudorfer, W., and Weyermayr, H. (2007). Securing groundwater use and reestablishing the water balance by artificial recharge of groundwater in the region of Marchfeld, Austria. In *Proceedings of IWA Regional Conference on Groundwater Management in the Danube River Basin and Other Large River Basins, Belgrade, Serbia.*

National Research Council (NRC) (1994). *Ground Water Recharge Using Water of Impaired Quality.* Committee on Ground Water Recharge, National Research Council. Washington, DC: National Academy Press.

NRC (2008). *Prospects for Managed Underground Storage of Recoverable Water.* Committee on Sustainable Underground Storage of Recoverable Water, National Research Council. Washington, DC: National Academy Press.

Owen, M. R., Silbermann, P. T., Boccadoro, J. M., and Reid, A. T. (2008). Aquifer recharge of treated wastewater through wicks: An innovative discharge method. In *Proceedings of AWWA Water Quality and Technology Conference, Cincinnati, OH.*

Pekdeger, A. (2006). *Hydrogeological-Hydrogeochemical Processses during Bank Filtration and Groundwater Recharge Using a Multitracer Approach.* Berlin, Germany: NASRI Project.

Pescod, M. (1992). *Wastewater Treatment and Use in Agriculture.* FAO Irrigation and Drainage Paper 47. Rome, Italy: Food and Agricultural Organization.

Ray, C., Grischek, T., Schubert, J., Wang, J. Z. and Speth, T. F. (2002a). A perspective of riverbank filtration. *Journal AWWA* 94(4), 149–160.

Ray, C., Soong, T. W., Lian, Y. Q., and Roadcap, G. S. (2002b). Effect of flood- induced chemical load on filtrate quality at bank filtrate sites. *Journal of Hydrology* 266(3–4), 235–258.

Ray, C. (2008). Worldwide potential of riverbank filtration. *Clean Technology Environmental Policy* 10(3), 223–225.

Rice, R. C., and Bouwer, H. (1984). Soil aquifer treatment using primary effluent. *Journal WPCF* 56(1), 84–88.

Rismal, M., and Kopac, I. (2008). Artificial groundwater recharge in Slovenia. In *Proceedings of Water Reclamation and Aquifer Recharge: Final Dissemination Workshop, Maribor, Slovenia.*

Rodriguez, C., van Buynder, P., Lugg, R., Blair, P., Devine, B., Cook, A. and Weinstein, P. (2009). Indirect potable reuse: A sustainable water supply alternative. *International Journal of Environmental Research and Public Health* 6, 1174–1209.

Rogers, M., Healy, M. G., and Mulqueen, J. (2005). Organic carbon removal and nitrification of high strength wastewaters using stratified sand filters. *Water Research* 39(14), 3279–3286.

Rott, U. (1985). Physical, chemical and biological aspects of the removal of iron and manganese underground. *Water Supply* 3(2), 143–150.

Rott, U., and Friedle, M. (1999). Subterranean removal of arsenic from groundwater. In *Arsenic Exposure and Health Effects*, C. O. Abernathy, W. R. Chappell, and R. L. Calderon, eds. Oxford, United Kingdom: Elsevier Science.

Rott, U., Meyerhoff, R., and Bauer, T. (1996). In situ treatment of groundwater with increased concentrations of iron, manganese and arsenic. *GWF Wasser-Abwasser* 137, 358–363 (in German).

Rott, U., Meyer, C., and Friedle, M. (2002). Residue-free removal of arsenic, iron, manganese and ammonia from groundwater. *Water Science and Technology: Water Supply* 2(1), 17–24.

Sahoo, G. B., Ray, C., Wang, J. Z., Hubbs, S. A., Song, R., Jasperse, J., and Seymour, D. (2005). Use of artificial neural networks to evaluate the effectiveness of riverbank filtration. *Water Research* 39(12), 2505–2516.

Sarkar, A. R., and Rahman, O. T. (2001). In-situ removal of arsenic: Experiences of DPHE-Danida Pilot Project. In *Technologies for Arsenic Removal from Drinking Water*, M. F. Ahmed, M. A. Ali, and Z. Adeel, eds. Dhaka, Bangladesh: BUET.

Schijven, J., Berger, P., and Miettinen, I. (2002), Removal of pathogens, surrogates, indicators, and toxins using riverbank filtration. In *Riverbank Filtration: Improving Source-Water Quality*, C. Ray, G. Melin, and R. B. Linsky, eds. Dodrecht, The Netherlands: Kluwer Academic Publishers.

Schittko, S., Putshew, A., and Jekel, M. (2004). Bank filtration: A suitable process for the removal of iodinated x-ray contrast media? *Water Science and Technology* 50(5):261–268.

Schubert, J. (2002). Hydraulic aspects of riverbank filtration: Field studies. *Journal of Hydrology* 266(3–4), 145–161.

Schubert, J. (2002b). German experience with riverbank filtration systems. In *Riverbank Filtration Improving Source-Water Quality*, C. Ray, G. Melin, and R. B. Linsky, eds. Dodrecht, The Netherlands: Kluwer Academic Publishers.

Schubert, J. (2002c). Water quality improvements with riverbank filtration at Dusseldorf Waterworks in Germany. In *Riverbank Filtration Improving Source-Water Quality*, C. Ray, G. Melin, and R. B. Linsky, eds. Dodrecht, The Netherlands: Kluwer Academic Publishers.

Schubert, J. (2004). Significance of hydrologic aspects on RBF performance. In *Proceedings of NATO Advanced Research Workshop, Samorin, Slovakia.* Available at www.soulstatic.com/NATORBF/papers/schubert/hydrology.pdf (accessed 27.11.2008).

Schubert, J. (2006). Experience with riverbed clogging along the Rhine River. In *Riverbank Filtration Hydrology*, S. A. Hubbs, ed. Proceedings of the NATO Advanced Research Workshop on Riverbank Filtration Hydrology, Bratislava, Slovakia. NATO Science Series IV: Earth and Environmental Sciences, Vol. 60.

Schmidt, C. K., Lange, F. T., and Brauch, H. J. (2007). Characteristics and evaluation of natural attenuation processes for organic micropollutants removal during riverbank filtration. *Water Science and Technology: Water Supply* 7(3), 1–7.

Schmidt, C. K., Lange, F. T., Brauch, H.-J., and Kühn, W. (2008). Experiences with Riverbank Filtration and Infiltration in Germany. Available at www.tzw.de/pdf/bankfiltration.pdf.

Shamrukh, M., and Abdel-Wahab, A. (2007). Riverbank filtration for sustainable water supply: Application to large-scale facility on the Nile River. *Clean Technology Environmental Policy* 10(4), 351–358.

Sontheimer, H. (1980). Experience with riverbank filtration along the Rhine river. *Journal AWWA* 72(7), 386–390.

Sharma, S. K. (2001). *Adsorptive Iron Removal from Groundwater*. Lisse, The Netherlands: Swets & Zeitlinger B.V.

Stuyfzand, P. J., Juhàsz-Holterman, M. H. A., and de Lange, W. J. (2004). Riverbank filtration in the Netherlands: Well fields, clogging and geochemical reactions. In *Proceedings of NATO Advanced Research Workshop: Clogging in Riverbank Filtration, Bratislava, The Netherlands.*

Stuyfzand, P. J., Segers, W., and van Rooijen, N. (2007). Behaviours of pharmaceuticals and other emerging pollutants in various artificial recharge systems in the Netherlands. in *Proceedings of the 6th International Symposium on Managed Aquifer Recharge, ISMAR6, Phoenix, AZ.*

Tielemans, M. W. M. (2007). Artificial recharge of the groundwater in the Netherlands. In *Proceedings of IWA Regional Conference on Groundwater Management in the Danube River Basin and Other Large River Basins, Belgrade, Serbia.*

Tischendorf, W. (2007). Groundwater management for the city of Graz. In *Proceedings of IWA Regional Conference on Groundwater Management in the Danube River Basin and Other Large River Basins, Belgrade, Serbia.*

Topper, R., Barkmann, P., Bird, D., and Sares, M. (2004). *Artificial Recharge of Ground Water in Colorado: A Statewide Assessment,* CD-ROM. Denver, CO: Colorado Geological Survey.

Tufenkji, N., Ryan, J. N., and Elimelech, M. (2002). The promise of bank filtration. *Environmrntal Science and Technology* 36(21), 422A–428A.

U.S. Environmental Protection Agency (USEPA) (2004). *Guidelines for Water Reuse.* EPA/625/R-04/108. Washington, DC: USEPA.

USEPA (2006). *Process Design Manual: Land Treatment of Wastewater Effluents.* EPA/625/R-06/16. Washington, DC: USEPA.

USEPA (2007). *The Long Term 2 Enhanced Surface Water Treatment Rule (LT2ESWTR): Implementation Guidance.* EPA 816-R-07-006. Washington, DC: USEPA.

U.S. Geological Survey (USGS) (2008) *Three-Dimensional Finite-Difference Ground-Water Model MODFLOW-2005,* Version 1.6.01, Groundwater Software, USGS. Available at http://water.usgs.gov/nrp/gwsoftware/modflow2005/modflow2005.html.

van Beek, C. G. E. M. (1983). *Ondergrondse Ontijzering, een evaluatie van uitgevoered onderzoek. Mededeling 78.* Report 78. Nieuwegein, KIWA, The Netherlands.

van der Hoek, J. P., Hofman, J. A. M. H., and Graveland, A. (2000). Benefits of ozone-activated carbon filtration in integrated treatment processes, including membrane systems. *Journal of Water Supply: Research and Technology, AQUA* 49(6), 341–356.

Verstraeten, I. M., Heberer, T., and Scheytt, T. (2002). Occurrence, characteristics, transport, and fate of pesticides, pharmaceuticals, industrial products, and personal care products at riverbank filtration sites. In *Riverbank Filtration: Improving Source-Water Quality.* C. Ray, G. Melin, and R. B. Linsky, eds. Dordrecht, The Netherlands: Kluwer Academic Publishers.

Wang, J. (2002). Riverbank filtration case study at Louisville, Kentucky. In *Riverbank Filtration Improving Source-Water Quality,* C. Ray, G. Melin, and R. B. Linsky, eds. Dodrecht, The Netherlands: Kluwer Academic Publishers.

Weiss, W.J., Bouwer, E.J., Ball, W.P., O'Melia, C.R., Lechevallier, M.W., Arrora, H. and Speth, T.F. (2003). Riverbank filtration: Fate of DBP precursors and selected microorganisms. *Journal AWWA* 95(10), 68–81.

Weiss, W. J., Bouwer, E. J., Ball, W. P., O'Melia, C. R., Aboytes, R., and Speth, T. F. (2004). Riverbank filtration: Effect of ground passage on NOM character. *Journal of Water Supply: Research and Technology, AQUA* 53(2), 61–83.

Williams, D. E. (2008). *Research and Development for Horizontal/Angle Well Technology.* Desalination and Water Purification Research and Development Program Report No. 151. Denver, CO: U.S. Department of the Interior, Bureau of Reclamation, Technical Service Center.

Yao, K.-M., Habibian, M. T., and O'Melia, C. R. (1971). Water and wastewater filtration: Concepts and applications. *Environmental Science and Technology* 5(11), 1105–1112.

CHAPTER 16
WATER REUSE FOR DRINKING WATER AUGMENTATION

Jörg E. Drewes, Ph.D.
Professor and Director, Advanced Water Technology Center (AQWATEC)
Colorado School of Mines
Golden, Colorado, United States

Stuart J. Khan, Ph.D.
Senior Research Fellow, UNSW Water Research Centre
University of New South Wales
Sydney, N.S.W., Australia

INTRODUCTION TO POTABLE REUSE ... **16.2**
SOURCE WATER CHARACTERISTICS ... **16.5**
Source Control ... 16.6
Contaminants of Concern ... 16.6
Posttreatment Stabilization ... 16.16
SYSTEM RELIABILITY AND HEALTH RISK CONSIDERATIONS ... **16.17**
Treatment Goals from a Health Perspective ... 16.17
Assessment of Unidentified Chemicals ... 16.19
Health Risk Assessment ... 16.19
DESIGN OF POTABLE REUSE SCHEMES ... **16.23**
The Role of the Environmental Buffer ... 16.23
Multiple-Barrier Approach in IPR ... 16.24
Indirect Potable Reuse Through Surface Water Augmentation ... 16.27
Indirect Potable Reuse Through Subsurface Treatment Leading to Groundwater Augmentation ... 16.29
Indirect Potable Reuse Through Direct Injection into a Potable Aquifer ... 16.31
Direct Potable Reuse ... 16.32
MONITORING STRATEGIES FOR PROCESS PERFORMANCE AND COMPLIANCE ... **16.33**
Real-Time Monitoring and Operating Strategies ... 16.33
Surrogate Measures and Indicator Compounds ... 16.34
REGULATIONS AND GUIDELINES FOR DRINKING WATER AUGMENTATION ... **16.36**
United States ... 16.36
Australia ... 16.38
PUBLIC PERCEPTION TO INDIRECT POTABLE REUSE ... **16.39**
ABBREVIATIONS ... **16.42**
REFERENCES ... **16.43**

INTRODUCTION TO POTABLE REUSE

With increasing demands on existing water supplies and limited access to new conventional water resources, some municipalities have begun to intentionally reuse treated municipal wastewater effluents to augment drinking water supplies. The practice of the purposeful addition of highly treated wastewater (i.e., reclaimed or recycled water) via an environmental buffer that is subsequently used to augment a drinking water supply is referred to as *planned or intentional indirect potable reuse* (IPR). Indirect potable reuse can occur through recharge of unconfined or confined aquifers, via surface spreading or direct injection, or by surface water augmentation into a stream or reservoir that serves as a source for drinking. In 2010, approximately 1,350 ML/d (355 mgd) of reclaimed water was used for IPR in the United States, which represents less than 1 percent of all municipal wastewater effluents generated in the country (Table 16-1). However, for municipalities practicing IPR, the contribution of reclaimed water to their drinking water supply can be as high as 30 percent, with some consumers receiving drinking water that originates by more than 50 percent from reclaimed water.

The immediate addition of reclaimed water to a drinking water distribution system is referred to as *direct potable reuse* (DPR). Since 1968, DPR has seen only one significant application worldwide with the commissioning of the direct potable reuse plant in Windhoek, Namibia (21 ML/d). In this case, advanced treated wastewater is fed directly into the distribution system providing up to 35 percent of the city's drinking water supply (du Pisani, 2006). Such direct use of reclaimed water for human consumption is currently not approved for any water system in the United States without the added protection provided by storage in an environment buffer (National Research Council, 1998). While planned IPR receives increasing interest among municipalities worldwide, questions remain regarding how much treatment and monitoring is needed to protect public health when reclaimed water is used for potable purposes. Water quality issues in IPR are associated with pathogens, organic chemicals, residual nutrients, and dissolved solids.

While the traditional maxim for selecting drinking water supplies has been to use the highest-quality source available (Pontius, 2003), in some regions once pristine surface water sources have evolved over time into unintentional indirect potable reuse systems, as wastewater from upstream dischargers has increased to substantial portions of the stream flow (Swayne et al., 1980; WEF/AWWA, 2008). Although planned and incidental or unplanned IPR systems may share similar water quality issues, this chapter provides guidance for systems that are *intentionally* designed and operated to augment drinking water supplies with reclaimed water.

Usually, a planned IPR scheme consists of the following six key components:

1. A *sewage collection system,* which incorporates discharge permits for industries, their monitoring and enforcement as required by USEPA regulations, and additional pollution prevention policies administered by the wastewater agency
2. A *conventional wastewater treatment* train, designed to minimize the presence of organic matter and pathogens, achieving a water quality that is suitable to be discharged to the environment meeting regulatory requirements as defined by the Clean Water Act (CWA)
3. *Advanced water treatment processes* which provide additional barriers to constituents of concern, such as residual organic chemicals, nutrients, dissolved solids, and pathogens
4. An *environmental buffer* or natural system integrated either via surface water or groundwater storage, to provide an opportunity to physically and chemically cut the connection to the source as well as offering time to respond to unforeseen process upsets
5. A *drinking water treatment* plant treating the augmented source water prior to delivery to consumers
6. An *overarching monitoring program,* that assures proper performance of conventional and advanced water treatment unit processes supplying a drinking water quality that is suitable for human consumption at all times

TABLE 16-1 Evolution of Indirect Potable Reuse Schemes and Employed Treatment Technologies

Project location	Type of indirect reuse	Project size ML/d (mgd)	First instalation year	Current status	Treatment technologies: Suspended solids	Organic compounds	Residual nutrients	Residual salts	Pathogens
Montebello Forebay, County Sanitation Districts of Los Angeles County, CA	Groundwater recharge via soil-aquifer treatment	165 (44)	1962	Ongoing	Media filtration	Soil-aquifer treatment	Soil-aquifer treatment	None	Chlorination Soil-aquifer treatment
Upper Occoquan Service Authority, VA	Surface water augmentation	204 (54)	1978	Ongoing	Lime clarification; media filtration	GAC filtration	Ion exchange (optional)	None	Chlorination
Water Factory 21, Orange County, CA	Groundwater recharge via seawater barrier	60 (16)	1976	Terminated 2004	Lime clarification	Reverse osmosis; UV/AOP	Air stripping; reverse osmosis	Reverse osmosis	Chlorination
Hueco Bolson Recharge Project, El Paso Water Utilities, Tx	Groundwater recharge via direct injection	38 (10)	1985	Ongoing	Lime clarification; media filtration	Ozonation; GAC filtration	PAC augmented activated sludge system	None	Ozonation Chlorination
Clayton County Water Authority, GA	Surface water augmentation	66 (17.5)	1985	Ongoing	Land application system	Land application system; wetlands	Land application system; wetlands	None	Chlorination UV
West Basin Water Recycling Plant, CA	Groundwater recharge via direct injection	20 (5.3)	1995	Ongoing	Microfiltration	Reverse osmosis; UV/AOP	Reverse osmosis	Reverse osmosis	Microfiltration Chloramination UV
Gwinnett County, GA	Surface water augmentation	227 (60)	1999	Ongoing	Ultrafiltration	Preozonation; GAC filtration	Chem. P-removal	None	Ultrafiltration Ozone
Scottsdale Water Campus, AZ	Groundwater recharge via direct injection	53 (14)	1999	Ongoing	Media filtration Microfiltration	Reverse osmosis	Reverse osmosis	Reverse osmosis	Microfiltration Chlorination
Toreele Reuse Plant, Wulpen, BEL	Groundwater recharge via infiltration ponds	6.9 (1.8)	2002	Ongoing	Ultrafiltration	Reverse osmosis	Reverse osmosis	Reverse osmosis	Ultrafiltration UV
NEWater, Bedok, Singapore	Surface water augmentation	32 (8.5)	2003	Ongoing	Ultrafiltration	Reverse osmosis	Reverse osmosis	Reverse osmosis	Ultrafiltration UV
NEWater, Seletar, Singapore	Surface water augmentation	24 (6.4)	2003	Ongoing	Ultrafiltration	Reverse osmosis	Reverse osmosis	Reverse osmosis	Ultrafiltration UV

TABLE 16-1 Evolution of Indirect Potable Reuse Schemes and Employed Treatment Technologies (*Continued*)

Project location	Type of indirect reuse	Project size ML/d (mgd)	First instalation year	Current status	Treatment technologies: Suspended solids	Organic compounds	Residual nutrients	Residual salts	Pathogens
Water Replenishment District of Southern California, CA	Groundwater recharge via direct injection	18.9 (5)	2005	Ongoing	Microfiltration	Reverse Osmosis UV	Reverse osmosis	Reverse osmosis	Microfiltration UV
Inland Empire Utility Agency, Chico, CA	Groundwater recharge via soil-aquifer treatment	69 (18)	2007	Ongoing	Media filtration	Soil-aquifer treatment	Soil-aquifer treatment	None	Chlorination
NeWater, Ulu Pandan, Singapore	Surface water augmentation	120 (32)	2007	Ongoing	Ultrafiltration	Reverse osmosis	Reverse osmosis	Reverse osmosis	Ultrafiltration UV
Groundwater Replenishment System, Orange County, CA	Groundwater recharge via direct injection and spreading basins	265 (70)	2008	Ongoing	Ultrafiltration	Reverse osmosis; UV/AOP	Reverse osmosis	Reverse osmosis	Ultrafiltration UV
Western Corridor Project, Southeast Queensland, Au	Surface water augmentation into drinking water reservoir	232 (62)	2008	Ongoing	Ultrafiltration	Reverse osmosis; UV/AOP	Reverse osmosis	Reverse osmosis	Ultrafiltration UV Chlorination
Loudoun County Sanitation Authority, VA	Surface water augmentation	42 (11)	2008	Ongoing	Membrane bioreactor (MF)	GAC	None	None	Microfiltration Chlorination
Arapahoe County/ Cottonwood, CO	Groundwater recharge via spreading operation	34 (9)	2009	Ongoing	Media filtration	Reverse osmosis; UV/AOP	Reverse osmosis	Reverse osmosis	Chlorination
Cloudcroft, NM	Spring water augmentation	0.38 (0.1)	2009	Ongoing	Microfiltration; ultrafiltration	Reverse osmosis; UV-AOP	Reverse osmosis	Reverse osmosis	Chlorination
Prairie Waters Project, Aurora, CO	Groundwater recharge via riverbank filtration	190 (50)	2010	Ongoing	Riverbank Filtration	Riverbank filtration UV/AOP BAC GAC	Riverbank filtration; Artificial recharge and recovery	Precipitive softening	Riverbank filtration UV Chlorination

In the United States, IPR has been practiced for almost 50 years via surface water augmentation, soil treatment leading to groundwater augmentation, or direct injection into a potable aquifer. During this time especially during the last 15 years, IPR reuse schemes in the United States as well as worldwide evolved substantially regarding their capacity and the type and sequence of treatment processes employed (Table 16-1).

For projects favoring direct injection into a potable aquifer, use of *integrated membrane systems* (IMS) incorporating *microfiltration* (MF) or *ultrafiltration* (UF) followed by *reverse osmosis* (RO) have emerged as the industry standard. IPR schemes employing IMS are mostly located in coastal areas where concentrated waste streams (i.e., RO concentrates) may be conveniently discharged to the ocean. In the United States, Singapore, and Australia, utilities have favored IMS, in some cases coupled with subsequent *advanced oxidation processes* (AOP). For inland surface and groundwater augmentation projects, however, IMS are favored less due to the lack of waste stream disposal options. Instead, for these IPR applications various combinations of low-pressure membranes, granular activated carbon (GAC) adsorption/filtration, chemical oxidation (i.e., ozone, UV/AOP), and natural treatment processes have evolved. These practices underscore that multiple options exist for the design of IPR schemes that consider regional conditions but are unified in the goal to lower or eliminate the risk from constituents of concern. Given the nature of the source, public health concerns regarding IPR are foremost related to the presence of pathogens and trace organic chemicals in reclaimed water. Thus, IPR projects must integrate appropriate water treatment processes that are capable of providing effective, reliable, and redundant barriers to pathogens and trace organic chemicals. Fundamental to the design of IPR schemes is the concept of multiple barriers and an environmental buffer to ensure a drinking water quality that is fit for human consumption.

Although these technical components are important for any proposed IPR project, addressing the psychological dimension of IPR (i.e., "toilet to tap," "yuck factor") is essential for a successful project. In several cases over the last 15 years, this aspect of IPR evolved as the determining factor for success or failure of a project and outweighed the technical merit of some proposed projects resulting in their termination prior to completion (e.g., San Diego's Water Recharge Project in 1998; Los Angeles Department of Water and Power's East Valley Groundwater Recharge Projectn 1998; Toowoomba's Water Recycling Project, Australia, in 2006). From the early phase of a proposed project, communication with stakeholders and the general public is the key to success, building confidence that IPR is the best alternative to secure future water supplies and positioning the proposing utility as the trusted source of water quality (Ruetten et al., 2004).

In planning a successful IPR project, the following activities should be carefully considered: (1) source water characterization, (2) appropriate water treatment process selection, (3) quantitative relative risk assessment and development of a risk management strategy, (4) review of institutional and regulatory requirements, (5) communication with stakeholders and the public, and (6) assessment of capital and operational costs.

SOURCE WATER CHARACTERISTICS

Conventional wastewater treatment provides an effluent quality that, in the United States, is ultimately aimed at meeting the requirements of the Clean Water Act (CWA) and therefore in most cases is suitable to be discharged to surface waters. However, conventionally treated effluents remain composed of a wide range of naturally occurring and synthetic, trace organic and inorganic chemicals, residual nutrients, dissolved solids, and residual heavy metals, as well as pathogens. Thus, for potable reuse projects, compliance with established drinking water standards as promulgated by the Safe Drinking Water Act (SDWA) by conventionally treated and reclaimed wastewater does not imply that the reclaimed water is safe for human consumption. In some states and depending on the designated use, surface discharge requirements under the CWA can be more stringent than compliance with SDWA drinking water standards. For example, the state of California's Toxics Rule requires more stringent THM standards (i.e., 5 μg/L instead of 80 μg/L as set by the SDWA) for a reservoir augmentation project discharging to surface water that is designated as a municipal drinking water supply. Thus, for an IPR project a thorough assessment of the source water quality is required that includes a wider range of microbial and chemical constituents than those required by current drinking water standards. Since general water quality characteristics of reclaimed water deviate from those of conventional drinking water sources, water treatment processes employed in IPR schemes must be capable of mitigating and eliminating these differences. Water produced in IPR schemes using advanced treatment processes sometimes requires restabilization to reduce the corrosive nature of the water, which may subsequently lead to deteriorating water quality. The three key elements to maintaining water quality for an IPR system are source control, attenuation of contaminants of concern through tailored treatment, and posttreatment stabilization.

Source Control

Source control is an USEPA regulatory management practice undertaken to minimize the discharge of some pollutants into the sewer. In many cases, wastewater agencies administer additional pollution prevention policies. These programs are conducted with the goal of reducing treatment costs, targeting chemicals of concern that are not primarily removed during conventional wastewater treatment (i.e., heavy metals, trace organic chemicals), and improving the reliability of final water quality. The approach is analogous to watershed management that may normally be exercised for conventional drinking water catchments (e.g., USEPA's Long Term 2 Enhanced Surface Water Treatment Rule; see Chap. 1). Accordingly, effective source control practices require the development and execution of a sewer discharge management plan. Monitoring and compliance assessments of point discharges to the sewer system are the first line of defense in IPR schemes representing the initial component of a multiple-barrier approach.

Contaminants of Concern

Contaminants of concern encompass a wide variety of pathogenic microbial organisms and chemical contaminants, which may be present in conventionally treated municipal effluents. Their presence is generally considered to be of concern due to their potential deleterious impacts on human health.

Pathogens. The control of pathogenic organisms is fundamental to the protection of public health for all potable water supplies as presented in Chap. 2. However, in the case of IPR, where municipal wastewaters make up a major component of source waters, the presence of significant initial concentrations of pathogens can generally be assumed. The diversity and concentrations of pathogens in treated wastewater effluents is highly variable and dependent on numerous locally specific factors. These factors include the patterns of infection within the community and the type of secondary and tertiary as well as disinfection processes employed at the wastewater treatment plant. In terms of their ability to contaminate drinking water supplies and their infectivity, the most significant human pathogens in reclaimed water include a range of bacteria, viruses, and protozoa as summarized in Table 16-2.

TABLE 16-2 Examples of Pathogens Occurring in Reclaimed Water

Pathogen type	Examples	Illness
Bacteria	*Salmonella*	Gastroenteritis, reactive arthritis
	Campylobacter	Gastroenteritis, Guillain–Barré syndrome
	Pathogenic *Escherichia coli*	Gastroenteritis, hemolytic uremic syndrome
	Shigella	Dysentery
	Yersinia	Gastroenteritis, septicemia
	Vibrio cholerae	Cholera
	Atypical *Mycobacteria*	Respiratory illness (hypersensitivity pneumonitis)
	Legionella spp	Respiratory illness (pneumonia, Pontiac fever)
	Staphylococcus aureus	Skin, eye, ear infections, septicemia
	Pseudomonas aeruginosa	Skin, eye, ear infections
Viruses	Enterovirus	Gastroenteritis, respiratory illness, nervous disorders, myocarditis
	Adenovirus	Gastroenteritis, respiratory illness, eye infections
	Rotavirus	Gastroenteritis
	Norovirus	Gastroenteritis
	Hepatitis A	Infectious hepatitis
	Calicivirus	Gastroenteritis
	Astrovirus	Gastroenteritis
	Coronavirus	Gastroenteritis
Protozoa	*Cryptosporidium*	Gastroenteritis
	Giardia	Gastroenteritis
	Entamoeba histolytica	Amoebic dysentery
Helminths	*Taenia* (*T. saginata, T. solium*)	Tapeworm (beef measles), neurocysticercosis
	Ascaris	Roundworm
	Trichuris	Whipworm
	Ancylostoma	Hookworm

Sources: National Resource Management Ministerial Council et al. (2008), Feacham et al. (1983), Geldreich (1990), National Research Council (1996), and Bitton (1999).

Waterborne bacteria, such as *Campylobacter Shigella,* and *Salmonella,* are important human pathogens as described in Chap. 2. However, these species are relatively susceptible to chemical disinfection practices (e.g., chlorination and chloramination; see Chap. 17) and thus can be effectively controlled by wastewater reclamation processes. Accordingly, bacteria are not a primary concern or driver to the implementation of advanced treatment processes in IPR schemes.

Viruses are widely recognized as the microorganisms representing the most significant risks to public health from IPR projects. Although they are generally susceptible to inactivation by chlorine disinfection, their high particle numbers in municipal wastewaters (Costan-Longares et al., 2008) and high infectivity require careful management to maintain adequate disinfection. Viruses responsible for numerous human illnesses, including gastroenteritis and hepatitis A, are known to be commonly present in municipal wastewater effluents. Effective monitoring presents an additional challenge to virus control since few laboratories possess the necessary expertise for proper analysis.

The enteric protozoan parasites *Cryptosporidium* and *Giardia* are notorious agents of waterborne disease and are commonly associated with reported outbreaks (see Chap. 2). *Cryptosporidium* in particular is reported to have been the cause of illnesses (and sometimes deaths) from drinking water supplies with full conventional treatment (coagulation and flocculation, granular media filtration, and chlorine disinfection), usually under challenging or erroneous circumstances as reviewed by Hrudey and Hrudey (2007). In water, protozoa may produce cysts or oocysts that aid in their survival. As a result, some of these organisms, including *Cryptosporidium,* are highly resistant to chlorine disinfection and must generally be controlled by other means, such as ozone oxidation, membrane filtration, soil-aquifer treatment, or riverbank filtration.

Data regarding the reported occurrence of specific pathogens in reclaimed water is limited and more commonly, so-called indicator organisms are analyzed as surrogate measures for microbial water quality (see the discussion in the next section, System Reliability and Health Risk Considerations, and Chap. 2). A study undertaken at three water reclamation plants in Spain reported the occurrence of the pathogens cytopathogenic enteroviruses and viable *Cryptosporidium* oocysts along with a range of potential indicator organisms including total coliforms, *E. coli*, enterococci, spores of sulphite-reducing clostridia, somatic coliphages, RNA F-specific phages, phages that infect *B. fragilis* strain RYC2056, and phages that infect *B. tethaiotaomicron* strain GA-17 (Costan-Longares et al., 2008). The detection limit in both secondary and tertiary effluents was 1 CFU/100 mL for all measured parameters. All three of the tested facilities employed biological nutrient removal for secondary treatment. Tertiary treatment steps including disinfection were variable—Facility 1: coagulation/flocculation → sand filters → UV light → chlorine (5 ppm, 45 min); Facility 2: settling ponds → oxidation ponds → constructed wetlands; Facility 3A: coagulation/flocculation → sedimentation → multilayer filtration → UV, chlorine (0.3 ppm, 1 min); Facility 3B: coagulation/flocculation → sedimentatation → multilayer filtration → chlorine (18 ppm, 90 min). Geometric mean densities of pathogens and indicator organisms, expressed as log (base 10) units, after secondary (prior to disinfection) and tertiary (after disinfection) treatment are illustrated in Fig. 16-1. Mean log reductions observed in this study for pathogens and microbial indicators during tertiary treatment are illustrated in Fig. 16-2. Although tertiary treatment processes were efficient in inactivating microbial indicators, these processes did not provide a pathogen-free water. Therefore, additional barriers for pathogens are needed in IPR applications.

Total Dissolved Solids. Domestic and commercial uses of public water supplies result in an increase of the mineral content in municipal wastewater. In some areas, the use of water softeners may be the major source of an increase in *total dissolved solids* (TDS). The increase in TDS in municipal wastewater effluents results in concentrations that are approximately 150 to 350 mg/L higher than the original potable water supplies (Asano et al., 2007). In a study reported by Drewes and Fox (2001), contributions of major ions in reclaimed water through human activities resulted in an average increase of 75 mg/L sodium, 45 mg/L sulfate, and 75 mg/L chloride, with small changes in hardness and minor ions.

In order to mitigate salinity problems associated with local water reuse activities, especially in inland applications, partial desalination of reclaimed water in IPR schemes may be required. For this reason, groundwater recharge schemes in Orange County, California, incorporated RO to reduce the concentration of dissolved solids to meet the groundwater basin objective for dissolved solids, which was equivalent to the USEPA drinking water secondary standard of 500 mg/L (USEPA, 1991).

Nutrients. Treated municipal effluents may contain numerous forms of nutrients that, if uncontrolled, may lead to excessive eutrophication of drinking water reservoirs in surface water augmentation projects. The most significant growth-limiting nutrients are most commonly nitrogen and phosphorus; however, other important micronutrients include potassium, calcium, magnesium, and sulfur. The risk of eutrophication can be managed by reducing the concentration of phosphorus and nitrogenous compounds through various treatment processes.

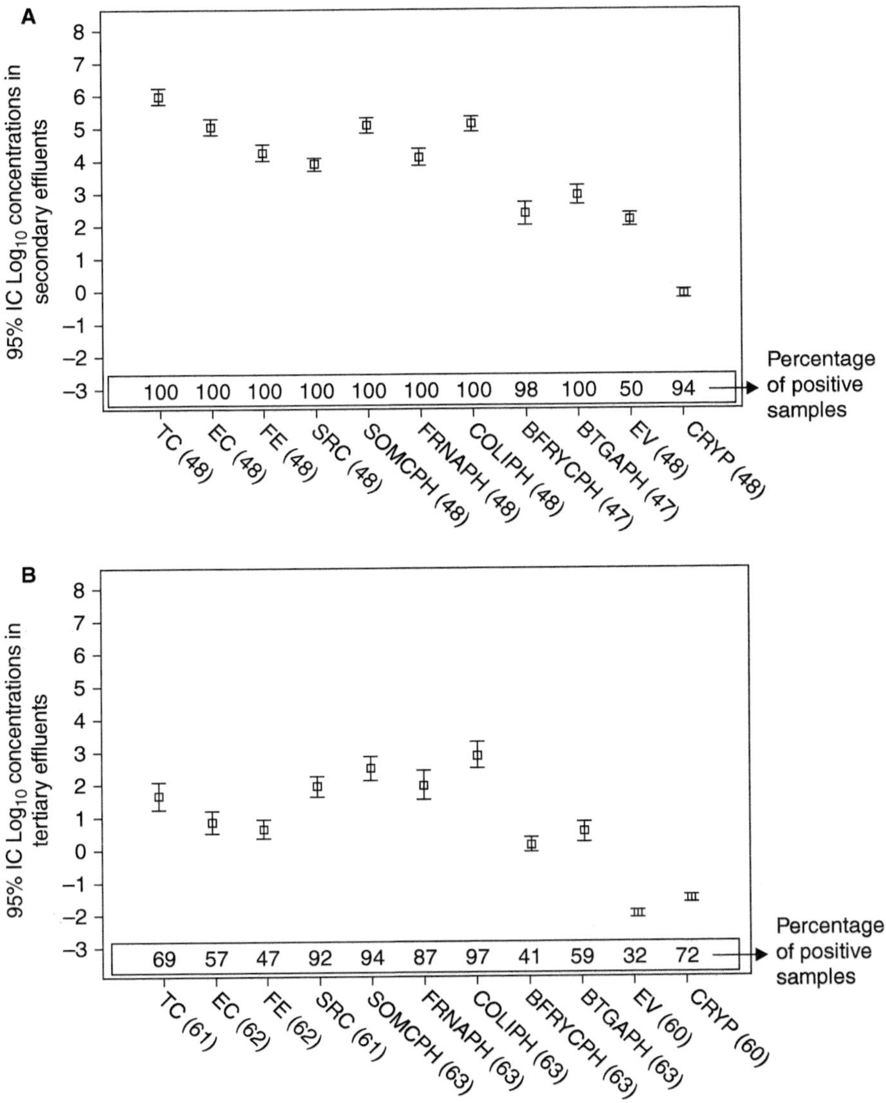

FIGURE 16-1 Microorganism concentrations (mean values and 95 percent confidence levels) in secondary (prior to disinfection) (a) and tertiary effluents (after disinfection) (b). In brackets is the number of analysis performed for every parameter. Indicator microorganisms' concentrations are expressed in PFU/100 mL and pathogens concentrations in PFU/1L. TC: total coliforms, EC: *E. coli*, FE: fecal *enterococci*, SRC: sulphite reducing *clostridia*, SOMCPH: coliphages infecting *E. coli* WG5, FRNAPH: FD RNA specific phages, BFRYCPH: phages that infect *Bacteroides fragilis* strain RYC 2056, BTGA17PH: phages that infect *Bacteroides tethaiotaomicron* strain GA-17, EV: cytopathogenic enteroviruses, CRYP: viable *Cryptosporidium* oocysts. (*Source: adapted with permission from Costan-Longares et al. (2008), copyright 2008, Elsevier.*)

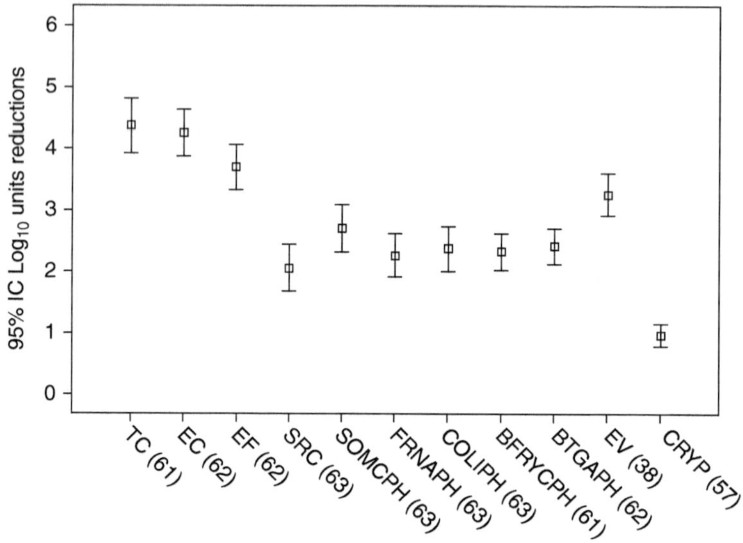

FIGURE 16-2 Microbial inactivation ($\log_{10}$ units reductions) obtained in tertiary treatments studied. (*Source: adapted with permission from Costan-Longares et al. (2008), copyright 2008, Elsevier.*)

In reclaimed water, nitrogen (N) may be present in both ionic and non-ionic forms. The speciation and concentrations will depend on pH, redox conditions, and the efficiency of any BNR process deployed at the wastewater treatment plant. Cationic nitrogenous compounds include ammonium (in equilibrium with ammonia) and amines; anionic forms include nitrite, nitrate, and amino acids. A well-operated *biological nutrient removal* (BNR) process would be expected to produce an effluent with negligible amounts of ammonia, less than 5 mg/L of nitrate and nitrite and less than 1 mg/L of organic N (Table 16-3).

TABLE 16-3 Biological Nutrient Removal (BNR) Treatment Processes and Common Concentration Ranges of Nutrients in Reclaimed Water

BNR treatment processes	Range of NH_4 concentration (mg N/L)	Range of NO_3 concentration (mg N/L)	Range of dissolved organic nitrogen (DON) concentration (mg N/L)	Range of $PO_4^{3-} - P$ concentration (mg P/L)	References
Not nitrifying	15–45	0.1–5	0.3–1.7	1–8	Sedlak, 1991 Krasner et al., 2008
Partially nitrifying	0.1–10	5–20	0.3–1.3	0.4–8	Sedlak, 1991 Krasner et al., 2008
Nitrifying/ denitrifying	<0.1–2	0.5–8	0.2–0.7	0.4–8	Sedlak, 1991 Krasner et al., 2008
Chemical P-removal	<0.1–10	5–20	0.3–1.3	<0.3–1.5	Sedlak, 1991 Krasner et al., 2008
Biological P-removal	<0.1–2	0.5–8	0.2–0.7	<0.1–0.6	Sedlak, 1991 Krasner et al., 2008

Advanced treatment processes that achieve additional nitrogen reduction are membrane separation (e.g., RO, nanofiltration (NF), and electrodialysis), ion-exchange processes, and biologically active filters.

Phosphorus (P) is present in reclaimed water predominantly as organic P and inorganic orthophosphate. Phosphorus may be partially removed biologically in the anaerobic zone of a BNR process. Residual orthophosphate may be removed in the treatment plant by precipitation and filtration using ferric or aluminum salts. Typical phosphorus concentrations in a BNR effluent designed for both nitrogen and phosphorus removal are in the 0.5 to 2.0 mg/L P range, decreasing to <0.3 mg/L if media filters are used (Table 16-3) (Sedlak, 1991).

Dissolved Organic Chemicals. The overall load of organic chemicals in water can be quantified in terms of the *total organic carbon* (TOC) concentration. Dissolved organic chemicals are described after 0.45 μm microfiltration as *dissolved organic carbon* (DOC) concentration. DOC in municipal wastewater effluents is composed of natural organic matter (originating from drinking water), soluble microbial products (generated during the activated sludge process), and small concentrations of a very large number of individual organic chemical contaminants (Namkung and Rittman, 1986; Drewes and Fox, 2000). These contaminants include industrial and domestic chemicals (e.g., pesticides, personal care products, preservatives, surfactants, flame retardants, perfluorochemicals, nanoparticles; see Chap. 2 regarding health effects and Chap. 3 regarding their presence in source waters), and chemicals excreted by humans (e.g., pharmaceutical residues, steroidal hormones), as well as chemicals formed during wastewater and drinking water treatment processes (e.g., disinfection by-products; see Chap. 19). Some representative examples of trace organic chemicals that may be present in treated municipal effluents are summarized in Table 16-4. Chemical contaminants may be present in reclaimed source waters or may be formed as by-products or metabolites via chemical or biological transformation during wastewater collection and treatment. Many of these chemicals may have an adverse effect on human health if present in sufficient quantities and not effectively removed during treatment (see Chap. 2). Potential chronic adverse health effects include carcinogenic, toxicologic, embryonic, and reproductive development impacts caused by interference with the endocrine system.

Anionic surfactants are used in commercial and domestic detergent products. Applications include dishwashing and clothes-washing detergents as well as hair shampoos. Linear alkylbenzene sulphonates (LAS) are the most common. Anionic surfactants are often present in high concentrations in raw wastewater (1–20 mg/L). However, conventional wastewater treatment is effective at eliminating these compounds resulting in much lower concentrations in final effluents. Drewes et al. (2009) studied the fate of LAS in six full-scale wastewater treatment plants and reported average concentrations in raw sewage of almost 6 mg/L that were reduced to approximately 4 μg/L in treated effluents.

Depending on the service area and the extent of the source control program to minimize discharge of chemicals at the source, a wide range of synthetic industrial chemicals are often measurable in influents to a water reclamation plant. Examples include plasticizers and heat stabilizers, biocides, epoxy resins, bleaching chemicals and by-products, solvents, degreasers, dyes, chelating agents, polymers, polyaromatic hydrocarbons, polychlorinated biphenyls, and phthalates. Many of these chemicals are known to be toxic to a diverse range of organisms including humans. Biological processes, however, can significantly reduce these contaminants in reclaimed water to levels usually below drinking water maximum contaminant levels (Trenholm et al., 2008).

Volatile organic compounds (VOCs) are widely used as industrial solvents. Many are constituents of petrochemical products, and a number of halogenated compounds may be formed as by-products of chlorine disinfection. Some VOCs are suspected to be teratogenic or carcinogenic to humans. Because of their high potential to contaminate traditional

TABLE 16-4 Examples of Trace Organic Chemicals That May be Present in Reclaimed Water

Category	Representative examples		
Organic chemicals			
Surfactants	Alkane ethoxy sulphonates (AES)	Linear alkylbenzene sulphonates (LAS)	Secondary alkanesulphonates
Industrial products and by-products	Acrylamide	Dichlorobenzenes	Phthalates
	Alkyl phenols	Ethylenediaminetetraacetic acid (EDTA)	Polyaromatic hydrocarbons
	Alkyltin compounds	Epichlorohydrin	Polychlorinated biphenyls
	Bisphenol A	Hexachloro-butadiene	Styrene
	Chlorinated dioxins	Nitrilotriacetic acid	Trichlorobenzenes
	Chlorobenzene	Perfluorooctanoic acid (PFOA)	Vinyl chloride monomer
		Perfluorooctane sulfonate (PFOS)	
Volatile organics	Benzene	Ethylbenzene	Trichloroethene
	Carbon tetrachloride	Tetrachloroethene	Xylenes
	Dichloroethanes	Toluene	
	Dichloromethane	1,1,1-trichloroethane	
Pesticides or their metabolites	2,4-D	Chlorpyrifos	Heptachlor and epoxide
	Aldicarb	Diazinon	Lindane
	Aldrin/dieldrin	Dichloro-diphenyltrichloroethane (DDT)	Organic mercurials
	Atrazine	Diuron	Pyrethroids
	Chlordane	Endosulfan	Other insecticides, fungicides and herbicides
Algal toxins	Cylindrospermopsin	Nodularin	Saxitoxins
	Microcystins		
Disinfection by-products	Chloral hydrate	Cyanogen	Haloketones
	Chlorate	Formaldehyde	Monochloramine
	Chlorine dioxide	Haloacetic acids	Nitrosamines
	Chlorite	Haloacetonitriles	- Nitrosodimethylamine (NDMA)
	Chlorophenols	Haloaldehydes	Trihalomethanes
	Chloropicrin	Halogenated furanones	

Pharmaceutical residues	Oral contraceptives - Levonorgestrel - Ethinylestradiol Analgesics - Ibuprofen - Paracetamol - Morphine - Naproxen - Ketoprofen Other pharmaceuticals - Methamphetamine - Phenytoin - Carbamazepine - Radiopharmaceuticals	Sedatives - Temazepam Cardiovascular drugs Beta blockers - Atenolol Cholesterol lowering - Simvastatin - Gemfibrozil	H. receptor agonists - Ranitidine Antibiotics - Cephalexin - Cefaclor - Amoxicillin Metronidazole
Estrogenic and androgenic hormones	17β-estradiol	Estrone	Testosterone
Personal care products	Tricosan	DEET	
Inorganic chemicals			
Nanoparticles	Carbon 60	TiO_2	NiO_2

potable water sources and supplies, they are tightly regulated in drinking waters. Many of them are environmentally conservative, so careful control is particularly important for planned IPR schemes.

Pesticides may enter municipal wastewater systems by a variety of means including stormwater influx and illegal direct disposal to sewage systems. Additional routes, of unknown significance, include washing fruit and vegetables prior to household consumption, insect repellents washed from human skin, flea-rinse shampoos for pets, and washing clothes and equipment used for applying pesticides outdoors. Pesticides have been designed and used to have detrimental effects on a wide range of biological species.

Algal toxins, such as microcyctins, nodularins, cylindrospermopsin, and saxitoxins, are all produced by freshwater cyanobacteria (blue-green algae) (Leflaive and Ten-Hage, 2007; Funari and Testai, 2008; Schmidt et al., 2008). Many of these toxins are hepatotoxic and some are neurotoxic. Under suitable conditions, cyanobacteria may grow in untreated or partially treated wastewaters, producing these and other toxins (Barrington and Ghadouani, 2008; Furtado et al., 2009). Numerous algal toxins have been identified as having serious impacts on human and animal health after consumption of contaminated water.

Pharmaceuticals (and their active metabolites) are excreted to sewage by humans as well as direct disposal of unused drugs by households (Shon et al., 2006; Daughton and Ruhoy, 2009). Since pharmaceuticals are designed to instigate biological responses, their inherent biological activity and the diverse range of compounds identified in reclaimed water (and the environment) have been cause for considerable concern during the last decade (Drewes and Shore, 2001; Ternes and Joss, 2006). Specific concerns have not been raised for most classes of drugs, but issues regarding potent endocrine disrupting compounds, aquatic toxicity, and the spread of antibacterial resistance could have significant ecological implications.

Natural steroidal hormones, such as estradiol, estrone, and testosterone, are excreted to wastewater by mammals. During the last two decades, natural steroidal hormones have been linked to a range of endocrinological abnormalities in aquatic species impacted by wastewater effluents (Jobling and Tyler, 2003; Sumpter and Johnson, 2008). Snyder et al. (2008) conducted a comprehensive survey of finished drinking water qualities from 20 different sites in the United States and only three finished water samples exhibited estrogenicity above the method detection limit varying between 0.19 and 0.77 ng/L estradiol equivalents. These impaired finished waters were compared to common dietary items (e.g., beer, soy milk, soy baby formula) that without exception contained far greater estrogenicity than the drinking water samples. Beside these low detects, the authors concluded that there was no evidence of human health risk from consumption of these waters.

Antiseptics, such as triclosan, are commonly used in face washes and anti-gum-disease toothpaste as well as for a wider range of household products including deodorants, antiperspirants, detergents, dishwashing liquids, cosmetics and antimicrobial creams, lotions, and hand soaps. Triclosan has been frequently detected in wastewater effluents (Heidler and Halden, 2008; Trenholm et al., 2008).

Perfluorochemicals (PFCs) and their chemical precursors have been commonly used in many consumer products as discussed in Chap. 2. The recent detection of high levels of PFCs at some sites in the United States has led the USEPA to issue provisional health advisory values for *perfluorooctanoic acid* (PFOA) and *perfluorooctane sulfonate* (PFOS) (USEPA, 2009). These values of 0.4 and 0.2 μg/L for PFOA and PFOS, respectively, were issued out of concern that land-application of biosolids may have leached PFCs into groundwater. PFOS and PFOA have been detected in finished drinking waters in Japan, Spain, and Germany at levels up to 0.6 μg/L (Skutlarek et al., 2006; Takagi et al., 2008; Ericson et al., 2008), though little peer-reviewed data are available for finished U.S. drinking water. Concentrations of 90 to 470 ng/L PFCs have been reported in tertiary treated municipal effluents in California, predominantly PFOA (10–190 ng/L) and PFOS (20–190 ng/L) (Plumlee et al., 2008). Snyder et al. (2010) compared PFC occurrence of source water samples from five different drinking water sites with secondary- and tertiary-treated effluents used at two IPR sites. In this study, the authors report PFOA concentrations varying between nondetect and 36 ng/L for raw drinking water and 15 to 28 ng/L for secondary- and tertiary-treated effluents and PFOS concentrations varying from nondetect to 23 ng/L in raw drinking water and 11 ng/L to 90 ng/L in secondary- and tertiary-treated effluents, respectively. Although, the relatively low affinity of PFCs for environmental solids (Higgins and Luthy, 2006) implies a potential for contamination of groundwater resources through reclaimed water used for irrigation or intentional groundwater recharge efforts, PFC concentrations observed in reclaimed water used for these applications were low and comparable to conventional drinking water supplies using surface waters (Snyder et al., 2010); this suggests a rather low potential for groundwater contamination.

Identification of all organic chemicals present in reclaimed water at a molecular level is neither possible nor practical. Although recent studies attempting to correlate exposure of single trace organic compounds via drinking water to their human health effects suggest no appreciable risk (Schwab et al., 2005; Snyder et al., 2008), these studies are limited to a few individual compounds and fewer data are available for mixtures of compounds. Presented with this uncertainty, the general approach in designing and regulating IPR projects is to ensure that the proposed treatment train includes one or several dedicated processes to significantly reduce the overall concentration of TOC. The technologies used to reduce the TOC content in IPR projects include soil-aquifer treatment and riverbank filtration, high-pressure membranes (RO and NF), granular or powdered activated carbon, biologically activated carbon, ozone, and AOP.

Disinfection By-Products. *Disinfection by-products* (DBPs) are formed by reactions between disinfection agents and other constituents of water; this topic is comprehensively discussed in Chap. 19. However, there are some DBPs that are of particular relevance to IPR applications due to the nature of the highly complex source water and/or the type of water treatment processes employed. High initial concentrations of organic components may lead to excessive production of certain DBPs in the absence of high ammonia concentrations. Similarly, the presence of ammonia in some source waters may lead to the formation of nitrogen-containing DBPs. The vast majority of the compounds of concern originate from chlorine-based disinfectants. Accordingly, excessive formation of well-known DBPs, such as trihalomethanes (THMs) and haloacetic acids (HAAs), must be carefully monitored and controlled. DBPs produced in reclaimed water from five wastewater treatment plants in Southern California employing secondary biological treatment, nitrification (with denitrification at three plants), filtration, and chlorine disinfection exhibited THM levels in the chlorinated effluent of 35 to 86 μg/L, whereas the levels of HAAs were 99 to 262 μg/L (Krasner et al., 2008). Three other classes of DBPs exhibited relatively high concentrations: aldehydes (21–114 μg/L), chloral hydrate (44–76 μg/L), and haloacetonitriles (HAN) (14–33 μg/L) (Krasner et al., 2008).

Ozone treatment (ozonation) of reclaimed water may result in the formation of several groups of DBPs including bromate and carbonyl compounds, such as aldehydes. Bromate is a suspected human carcinogen, and some aldehydes including formaldehyde and acetaldehyde have been classified as probable human carcinogens by the USEPA. Advanced oxidation processes utilize the transient formation of hydroxyl radicals to degrade carbon-carbon and other chemical bonds. The range of by-products formed is a function of the nature of the organic matter present and the relative susceptibility of specific bonds for radical attack. Accordingly, a large number of unidentified low-molecular-weight products is expected to be formed during advanced oxidation of complex solutions.

Ammonia can occur in wastewater effluents, either as a result of incomplete nitrification or by controlled addition for chloramination. The presence of ammonia and suitable organic precursors can lead to the production of N-nitrosamines, such as *N-nitrosodimethylamine* (NDMA) and *N-nitrosodiethylamine* (NDEA). Some N-nitrosamines, including NDMA and NDEA, have been classified as probable human carcinogens. The USEPA has added six of the nine N-nitrosamines, known to occur in reclaimed water, to the unregulated list of drinking water priority contaminants for ongoing monitoring and risk assessment (USEPA, 2007). The California Department of Public Health (CDPH) has set notification levels of 10 ng/L for NDMA and NDEA in drinking water. For reclaimed water used for ground-water recharge, rather than defining concentration limits, CDPH has established performance criteria in terms of achieving a minimum of 1.2 log NDMA and 0.5 log 1,4-dioxane reduction through AOP whether NDMA or 1,4-dioxane are present or not (CDPH, 2008).

Dissolved Inorganic Chemicals. Heavy metals may be present in reclaimed water as a result of municipal and industrial discharges to sewers (Rule et al., 2006; Gasperi et al., 2008a; Gasperi et al., 2008b). Some heavy metals such as cadmium, chromium, and mercury have been associated with human health concerns. However, the occurrence of these contaminants is usually low due to effective source control programs. They are also predominantly adsorbed to wastewater particulates and biosolids (Carletti et al., 2008; Gasperi et al., 2008a) and, as a result, tend to be effectively removed during activated sludge processes (Carletti et al., 2008).

Suspended Solids/Particulate Matter. Suspended solids and other particulate matter include colloidal material, fine particles, and microorganisms, such as protozoan cysts and oocysts, bacteria, and viruses. Removing suspended solids improves the performance and efficiency of downstream processes used to provide additional disinfection and remove organic compounds and dissolved salts. Prior to the early 1990s, conventional water treatment processes, similar to those used in drinking water treatment plants, were used in many IPR projects to reduce the concentration of suspended solids. The treatment steps consisted of flash mixing and flocculation followed by clarification and granular media filtration through sand and/or anthracite. The use of an additional clarifier provided sufficient buffering capacity and treatment residence time to deal with the temporal variations in feed water suspended solids levels. This technology combination was embodied in the Californian Water Recycling Criteria requirements (State of California, 2000) and the Australian-New South Wales Guidelines for Urban and Residential Use of Reclaimed Water (NSW Recycled Water Coordination Committee, 1993).

By the late 1990s, filtration with low-pressure membranes (i.e., MF, UF) had become the preferred technology to control suspended solids in some IPR projects. Membranes require less space, use less energy, and do not generate a solid sludge; this collectively reduces the cost of reclamation significantly (Leslie et al., 1998). In the United States, low-pressure membranes for municipal applications were first used to treat secondary effluent at Orange County's Water Factory 21 in the late 1980s. Subsequent performance tests at this facility demonstrated that the levels of microorganisms, suspended solids, turbidity, and silt density index (SDI) in the filtered water were a factor of 3 lower than conventional processes (Leslie et al., 1999).

Posttreatment Stabilization

The final treatment step in IPR schemes provides, where needed, stabilization of the finished water quality by restoring alkalinity, hardness, and pH as required. The purpose of this stabilization is to reduce the aggressive or corrosive nature of the water toward materials used in treatment plants, conveyance, and storage systems (e.g., concrete, steel or copper, soil). The techniques to stabilize the water include pH correction with caustic or acidic solutions and restoration of alkalinity and mineral hardness through the addition of lime (Withers, 2005).

SYSTEM RELIABILITY AND HEALTH RISK CONSIDERATIONS

Considering the quality of the source, any IPR scheme needs to be designed to reliably supply a finished water quality that is safe for human consumption at all times. System reliability of a reuse project is defined as the probability of adequate performance for a specified period of time under predefined conditions. Several factors affect system reliability: (1) the variability of wastewater characteristics, (2) the inherent variability of biological treatment processes, (3) the inherent variability of advanced water treatment processes, (4) the reliability of mechanical plant components, and (5) the effectiveness of monitoring (Asano et al., 2007).

System reliability requirements may include standby power supplies, readily available replacement equipment, online monitoring of system performance and water quality, redundant process components that are critical for the protection of public health, flexible piping and pumping configurations, and emergency storage or disposal options.

Whether a particular reuse system is able to reliably produce and deliver drinking water that meets health targets can be evaluated through risk assessment. Thus, IPR projects must include an evaluation of the potential health risks and hazards that could compromise the delivery of safe drinking water that is derived from reclaimed water. Health concerns in IPR are primarily related to the insufficient removal of pathogens and trace organic chemicals present in reclaimed water. Although a number of comprehensive studies have been conducted to address the concern about potential health risks of unknown and unidentified chemicals in reclaimed water (Nellor et al., 1984; Lauer and Rogers, 1996; Sloss et al., 1996), there is currently no definitive measure of risk or safety for the use of reclaimed water to augment drinking water supplies.

Treatment Goals from a Health Perspective

The Safe Drinking Water Act gives the USEPA the authority to set drinking water standards (see Chap. 1). In doing so, the USEPA considers a range of factors in determining these standards for specific drinking water contaminants as described in Chaps. 1 and 2. The water quality of any potable reuse project using reclaimed water in the United States has to meet the drinking water standards as promulgated by the USEPA.

Established maximum contaminant level goals (MCLGs) and maximum contaminant levels (MCLs) are generally limited to traditional drinking water contaminants, known to occur in conventional water sources such as surface waters and groundwaters, and DBPs, known to be formed during disinfection of such water sources. These include a range of pesticides and industrial chemicals, but not chemicals that are associated with discharges from municipal wastewater effluents, such as pharmaceutical residues, personal care products, household chemicals, steroidal hormones, or emerging DBPs. Thus, considering the origin of reclaimed water, meeting current drinking water standards does not imply the finished water is fit for human consumption. Although some states have proposed additional water quality requirements for potable reuse, the design of IPR treatment schemes in general should consider processes that are capable of attenuating compounds that are currently unregulated or unidentified. Treatment goals for these compounds for a proposed IPR scheme might consider their toxicological relevance if this information is readily available (Schwab et al., 2005; Snyder et al., 2008) or occurrence levels in current or less impacted drinking water supplies (Benotti et al., 2009; Snyder et al., 2010).

Safe drinking water concentrations of many of these chemicals may potentially be derived on the basis of the same toxicological considerations used for the establishment of current MCLGs and MCLs (as described in Chap. 2). For example, acceptable daily intakes

(ADIs) have been developed for 26 active pharmaceutical ingredients on the basis of various end points from which an appropriate *point of departure* (POD) was determined for the calculation (Schwab et al. 2005). The POD for determining an ADI is commonly the "no observed adverse effects level" (NOAEL) for a given toxic end point. However, for many ADIs, the POD is the lowest dose resulting in an observable effect level (LOEL). Schwab et al. (2005) stated that for pharmaceutical substances, the therapeutic effect usually occurs at a dose considerably below those expected to result in toxicity, and thus a large proportion of the ADIs that they calculated were based on therapeutic effects or minor side effects, such as sensitivity to human intestinal microflora. However, others have reported that for some pharmaceuticals, the most toxic end point is not the therapeutic end point, but rather it is a side effect, such as carcinogenicity (Snyder, 2008).

In response to uncertainties associated with risk posed by these compounds in IPR applications, some scientists and regulators support the adoption of treatment technologies to minimize exposure of humans and aquatic ecosystems (for reservoir augmentation) to wastewater derived chemical contaminants until more data on potential risks are collected. This approach is common practice in the absence of a reliable method that is economically and technically feasible to measure a contaminant at particularly low concentrations. In these cases the USEPA sets treatment techniques for drinking water rather than an MCL. A treatment technique is an enforceable procedure or level of technological performance, which public water systems must follow to ensure control of a contaminant.

Other international agencies have adopted the concept of a 'tolerable level of risk' to assist in setting water quality guidelines or standards. In such cases, health targets are adopted to reflect an accepted tolerable risk level, which could be expressed in a variety of ways including

1. Tolerable risk of infection by a pathogen from drinking water
2. Tolerable increased cancer risk by exposure to carcinogens in water
3. Tolerable amount of disease burden

Health-based targets are used by the World Health Organization (WHO) to assess disease burdens and underpin the development of water safety plans. For chemicals that are known or suspected to cause cancer, the acceptable lifetime exposure is defined in the WHO "Guidelines for Drinking Water Quality" as that which will result in one excess cancer case per 100,000 people (World Health Organization, 2004).

For pathogenic organisms, the WHO has adopted the concept of *disability-adjusted life years* (DALYs) to derive water quality targets (World Health Organization, 2004). DALYs are used to objectively compare water-related hazards and the different health outcomes with which they are associated. The basic principle is to weight each health effect for its severity from 0 (normal health) to 1 (death). This weight is multiplied by the duration of the effect, which is defined as the time in which the disease is apparent. When the outcome is death, the duration is the remaining life expectancy. The product is then multiplied by the number of people affected by a particular outcome. It is then possible to sum the effects of all the possible outcomes due to a particular hazard. Accordingly, the DALY is the sum of years of life lost by premature death and years of healthy life lost in states of less than full health (i.e., years lived with a disability, which are standardized by means of severity weightings). WHO has defined a tolerable level of risk of 1 millionth of a DALY (one microDALY) per person-year (World Health Organization, 2004). This is equivalent to the loss of one DALY per million people per year. This could be interpreted as one person in a million losing 1 year of life or it could be interpreted as a larger number of people suffering less severe impacts. For a pathogen causing watery diarrhea (but rarely causing death), this reference level of risk has been determined to be equivalent to around 1 in 1000 annual risk of diarrhea to an individual, or about 1 in 10 risk of contracting diarrhea over a lifetime.

Assessment of Unidentified Chemicals

Analytical methods involving direct chemical measurements are limited in that they will only identify chemicals that are specifically targeted. This can only ever be a small subset of all the chemicals that may be present. Numerous decades of water quality monitoring have provided a reasonable (though always improving) understanding of which chemicals are likely to be present in drinking waters of conventional sources at significant enough concentrations to present an elevated level of risk. However, we have much less experience in dealing with the diversity of chemicals that may be present in reclaimed water.

Even if all of the organic chemical components in the specific reclaimed water could be identified, there would be scant toxicological data available for most of them and thus little basis for assigning risks. Other important limitations of chemical contaminant monitoring are that the potential additive toxicity of a large number of chemicals, each present at very low concentration, may not be identified unless each of the individual species is analyzed for and determined to be present at a concentration greater than analytical detection limit. Finally, there is some concern that the toxicity of complex mixtures is poorly understood and in some cases may amount to more (or less) than simply additive impacts from the summation of each of the contributing chemical species.

To overcome the limitations associated with the assessment of unidentified chemicals and unknown mixture interactions, many scientists have suggested that direct toxicological testing assessment of reclaimed water may be the only way to ensure safe concentrations of chemical contaminants (National Research Council, 1998). Toxicological testing involves collecting whole water samples and subjecting these to tests for a range of in vitro and/or in vivo toxicological end points, analogous to the toxicological testing described for individual chemicals in Chap. 2. Toxicological end points may include testing for mutagenic activity, carcinogenic activity, hormonal activity such as estrogenicity, or various forms of acute toxicity. Such testing will generally require at least a pilot-scale water treatment plant to be constructed in order to provide relevant samples for testing. Several IPR projects proposed in the past have conducted toxicological testing for their advanced water treatment process trains to demonstrate the absence of adverse human health effects from consumption of the product water (e.g., Denver Potable Water Reuse Demonstration Project, Tampa Water Resource Recovery Project, San Diego's Total Resource Recovery Project, Singapore's Water Reclamation Study). Findings from these studies revealed that exposure to advanced treated water from IPR facilities did not cause any abnormalities or adverse health effects (Western Consortium for Public Health, 1997; National Research Council, 1998; NEWater Expert Panel, 2002).

Health Risk Assessment

Health risk assessments are typically characterized by four distinct steps as described in Chap. 2: hazard identification, dose-response assessment, exposure assessment, and risk characterization. Undertaking a comprehensive health risk assessment for an IPR scheme is a considerable task, and contributions from specialists in the fields of wastewater engineering, advanced water treatment, drinking water management, toxicology, infectious diseases, and statistics may be required. The data obtained will be used for a variety of purposes including projecting final water quality prior to implementation, demonstrating that the proposed scheme can be operated safely, seeking regulatory approval for construction, and identifying *critical control points* (CCPs) for ongoing reliable safe supply of water. A CCP is a point, step, or procedure at which controls can be applied to a hazard such that risks associated with that hazard can be prevented, eliminated, or reduced to acceptable levels.

A health risk assessment might be established to identify and quantify risks in absolute or relative terms. An assessment undertaken to determine risks in absolute terms could involve estimating the exposure of a particular pathogen or chemical contaminant and determining the consequential increased likelihood of identified adverse health outcomes on the basis of the known infectivity of the organism or toxicity of the chemical. Alternatively, a relative risk assessment could compare the degree of exposure to a pathogen or chemical from a new reclaimed water source relative to the exposure from a preexisting water source. Such an approach can be useful for demonstrating reduced or increased risks associated with exposure to a particular contaminant in the absence of quantitative toxicology/or/ infectivity data.

In the past, health risk assessments for water reuse schemes focused on the estimation of chemical and pathogen concentrations expected when a treatment train is operating under normal performance conditions. Although many unit processes employed in treatment facilities provide extremely high quality water under normal conditions, treatment upsets, system failures, and human error can jeopardize the reliability of the treatment system and pose a potential risk to public health that is likely going to exceed the risk associated with normal operation. Thus, few assessments effectively account for the issues of treatment plant reliability and associated water quality variability (Khan, 2009).

For contaminants for which environmental or health impacts are predominantly associated with long-term exposure (e.g. carcinogens), occasional short-term events may be relatively unimportant since the effect is dependent upon the long-term average dose; however, for contaminants with health effects associated with acute exposures (e.g., pathogens or acute chemical toxins), individual events resulting in short-term increases in exposure may be highly significant (Haas and Trussell, 1998). Accordingly, any adequate assessment of health risks from a water reuse scheme needs to include a thorough assessment of potential hazardous events.

The vast majority of observed waterborne disease outbreaks in developed countries during the last few decades have been associated with hazardous events, such as unusual weather patterns, plumbing errors, or treatment failures (Hrudey and Hrudey, 2007; Rizak and Hrudey, 2007). Potential hazardous events can be assessed as a function of their likelihood of occurrence and their consequences if they do occur. Qualitative descriptors for likelihood and consequences are summarized in Tables 16-5 and 16-6, adopted from the "Australian Guidelines for Water Recycling" (EPHC, 2008).

Consequences of hazardous events may be evaluated in terms of change in contaminant concentrations and exposures that may be the result of a specific hazardous event occurring. For example, changes in contaminant exposure could be assessed for a treatment failure situation where an acceptable disinfection dose may not be achieved, and this can lead to increased exposure to viable pathogens.

The likelihood and consequences can then be combined to provide a qualitative estimation of risk, using a suitable risk matrix (Table 16-7). Risks that are judged to be high will generally be the focus of CCPs. Techniques for quantitatively assessing the likelihoods of

TABLE 16-5 Qualitative Measures of Likelihood

Level	Descriptor	Example description
A	Rare	May occur only in exceptional circumstances; may occur once in 100 years
B	Unlikely	Could occur within 20 years or in unusual circumstances
C	Possible	Might occur or should be expected to occur within a 5- to 10-year period
D	Likely	Will probably occur within a 1- to 5-year period
E	Almost certain	Is expected to occur with a probability of multiple occurrences within a year

Source: NRMMC, EPHC & NHMRC (2008), copyright 2008, with permission of EPHC.

TABLE 16-6 Qualitative Measures of Consequence or Impact

Level	Descriptor	Example description
1	Insignificant	Insignificant impact or not detectable
2	Minor	Health—Minor impact for small population Environment—Potentially harmful to local ecosystem with local impacts contained to on-site
3	Moderate	Health—Minor impact for large population Environment—Potentially harmful to regional ecosystem with local impacts primarily contained to on-site
4	Major	Health—Major impact for small population Environment—Potentially lethal to local ecosystem; predominantly local, but potential for off-site impacts
5	Catastrophic	Health—Major impacts for large population Environment—Potentially lethal to regional ecosystem or threatened species; widespread on-site and off-site impacts

Source: NRMMC, EPHC, and NHMRC (2008), copyright 2008, with permission of EPHC.

specific hazardous events could be investigated including the use of historical data, such as weather patterns and frequencies of power failures or plumbing errors. Techniques for assessing the likelihood of treatment failures have also been established using a range of mechanical reliability measures (Shultz and Parr, 1982; Olivieri et al., 1996; Eisenberg et al., 1998; Eisenberg et al., 2001).

Probabilistic Exposure Assessment. Exposure to a contaminant from a water reuse scheme may be determined as a function of its concentration in conventionally treated wastewater and removal by multiple treatment processes, each with variable capability to reduce the concentration of the contaminant (Haas and Trussell, 1998; Tanaka et al., 1998). Water quality of IPR schemes is naturally variable as concentrations of trace chemicals in source waters vary over daily and weekly periods. Furthermore, the performances of advanced water treatment processes, such as RO and AOP, vary depending on a range of factors including hydraulic flow rate, water composition, and temperature. As a result, final concentrations of chemical hazards in water, and hence exposure to them, are variable parameters.

A useful way of dealing with variable parameters in mathematical risk calculations is by describing it, not as a point value but as a distribution of water quality values (Olivieri et al., 1999). A *probability density function* (PDF) is a mathematical function that represents a distribution in terms of the probability or frequency of occurrence of specific values within the distribution.

TABLE 16-7 Qualitative Risk Estimation

	Consequences				
Likelihood	1-Insignificant	2-Minor	3-Moderate	4-Major	5-Catastrophic
A Rare	Low	Low	Low	High	High
B Unlikely	Low	Low	Moderate	High	Very high
C Possible	Low	Moderate	High	Very high	Very high
D Likely	Low	Moderate	High	Very high	Very high
E Almost Certain	Low	Moderate	High	Very high	Very high

Source: NRMMC, EPHC, and NHMRC (2008) copyright 2008, with permission of EPHC.

A range of anticipated exposures requires computing sums, multiplications, and other transformations on multiple PDFs; this is a mathematically challenging task and, in some cases, impossible. For this, probabilistic techniques can provide a powerful alternative approach (Olivieri et al., 1999; Khan and McDonald, 2009). The Monte Carlo model is currently the most widely used method for probabilistic health risk assessment and exposure assessment. In light of their increasing application, the USEPA has developed guidelines for probabilistic risk assessment (USEPA, 1997). The general principle is illustrated in the example provided in Fig. 16-3. In this case, the final contaminant concentration is determined as a function of the concentration of contaminant X in the source water (e.g., tertiary treated effluent) and its subsequent removal during treatment by RO followed by AOP. The source water concentration and the degree of removal by the two advanced treatment processes are all variable parameters and thus they are all represented by PDFs. A Monte Carlo simulation is used to derive the final PDF of the concentration of the target compound X in the final effluent as a function of the three preceding PDFs. This requires randomly sampling each of the three preceding PDFs to derive a random prediction of final effluent concentration according to each random sample. Repeating this process with a sufficient number of random samples (typically 10,000 or more) enables the final effluent PDF to be gradually generated. Treatment performance PDFs for individual unit processes employed in an IPR scheme can be characterized by observed performance, challenge testing, or analysis of fundamental process variables. The outcome of this approach allows the consideration of source water quality variations and an assessment of the final product water quality variability for individual compounds of interest.

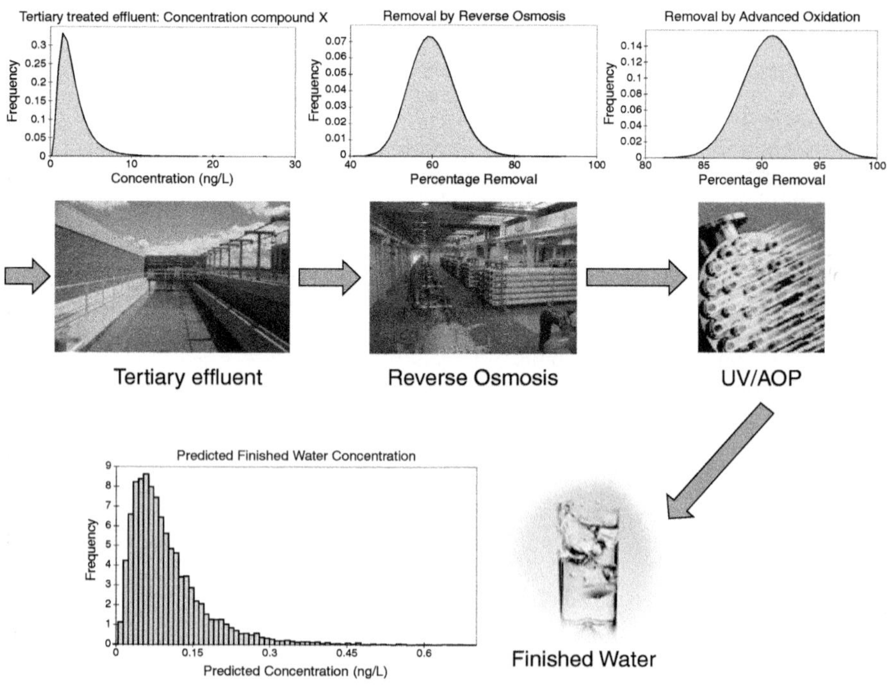

FIGURE 16-3 Illustration of exposure probability functions for compound X for an advanced water treatment train.

DESIGN OF POTABLE REUSE SCHEMES

This section provides information on the type of treatment processes being used in IPR projects to achieve high-quality water and discusses how these technologies reduce specific risk factors and how the efficacy of individual unit processes as well as the overall treatment train can be assured.

The Role of the Environmental Buffer

Potable reuse projects rely on the combination of an environmental buffer with a sequence of unit processes that are employed downstream of conventional wastewater treatment processes to improve water quality prior to introducing reclaimed water into a drinking water supply. IPR can occur through either of the following means or combinations using an environmental buffer:

1. *Surface water augmentation.* Reclaimed water is added to a reservoir, lake, stream, or river that has either the capacity or flow to provide attenuation and dilution.
2. *Groundwater recharge.*
   a. Subsurface treatment leading to groundwater augmentation taking advantage of attenuation processes that occur in the vadose zone and saturated aquifer. These systems utilize spreading, recharge or rapid infiltration basins, vadose zone recharge wells or trenches, and riverbank filtration wells.
   b. Direct injection introduces reclaimed water directly into a potable aquifer. Direct injection requires more stringent water-quality requirements since reclaimed water is introduced into the aquifer with little or no lag time. Direct injection can occur via direct-injection wells, deep vadose zone wells, or aquifer storage and recovery (ASR) wells.

Natural systems can be integrated immediately after the conventional wastewater treatment process, prior to or after advanced water treatment processes. It is noteworthy that the natural system or environmental buffer plays an important element in the design and operation of an overall IPR scheme since it provides an opportunity to physically and chemically cut the connection to the source (being sewage) as well as offering time, if needed, to respond to unforeseen process upsets. Decoupling from the source is achieved by blending with natural waters in the environment and reducing contaminants through natural attenuation processes. Although surface water environments offer opportunities for photolytic decay and some biological transformation of contaminants of concern (Fono et al., 2006), biological and physical attenuation processes are usually more efficient in subsurface environments (Amy and Drewes, 2007; Drewes et al., 2006).

Different requirements exist regarding the minimum retention time of reclaimed water in an environmental buffer prior to extraction. In groundwater recharge projects in California and Washington, regulations have been proposed or adopted, respectively, requiring 6 months of hydraulic retention time in the subsurface for surface spreading operations and for direct injection projects before the water can be recovered as a potable water source, mainly to provide additional protection to pathogens (Washington DOHC, 1997; CDPH, 2008). The Prairie Waters Project in Colorado utilizes a natural subsurface purification system with a hydraulic retention time of approximately 30 days. Since advanced water treatment processes are subsequently employed including several additional barriers for pathogen removal, a shorter retention time was deemed appropriate. Others have defined minimum setbacks (i.e., horizontal separation) between reclaimed water spreading operations and potable wells (e.g., 150 m (500 ft) in Florida; 1520 m (2000 ft) in Washington) (FDEP, 2006; Washington DOHC, 1997).

Multiple-Barrier Approach in IPR

The process of reducing the concentration of any microbial or chemical constituents of concern by applying a series of individual unit processes is referred to as the *multiple-barrier approach* to water treatment. The term *multiple barriers* was introduced in 1970 to describe the series of treatment steps to reduce the concentration of pathogens in a wastewater treatment process, where the receiving water was used as part of the water supply (Velz, 1970). For IPR projects, the concept of multiple barriers has been extended to include organic chemicals. Although the multiple barriers do tend to be relied on to provide cumulative steps toward the achievement of overall treatment goals, there is generally an expectation that they will accommodate a degree of treatment redundancy for pathogens. That is, the protection of public safety will be maintained even if a single treatment barrier fails (National Research Council, 1998). In the case of chemicals, the expectation is that a series of treatment steps will be used to reduce the overall chemical load. Given the wide range of different chemicals present in reclaimed water, multiple barriers should be designed to consider a sequence of diverse processes that are capable of targeting classes of chemicals encompassing different physicochemical properties. The requirement for redundancy normally associated with pathogen removal is not applied to multiple barriers for chemicals. This is because exposure to chemicals is more of a chronic risk, relating to long-term exposure, as compared with the acute risks associated with pathogens, for which even short-term exposure may have significant impacts on human health. From a public health standpoint, adequate and effective disinfection is the most essential process element that requires the highest degree of reliability and need for redundancy. The independence of multiple barriers is a key aspect of system reliability and safety.

Although the actual number of barriers for contaminants of concern might differ among different IPR projects, each scheme is characterized by a combination of *treatment barriers* that are suitable to control the concentrations of contaminants and *preventative measures* that control exposure to certain substances. Treatment barriers are individual unit processes employed at the water reclamation and the drinking water plants. In many cases, these include the use of a natural system, such as riverbank filtration or soil-aquifer treatment, to attenuate organic matter, pathogens, nutrients, and trace organic chemicals. *Preventative measures* include carrying out a source control program, management of the environmental buffer, maintenance of the distribution system, and an overarching monitoring program assuring proper performance of conventional and advanced water treatment unit processes. Sufficient redundancy for the removal of contaminants of concern should be provided to prevent any one system component from becoming absolutely vital to the protection of public health.

Multiple barriers established in IPR schemes are structured in five layers: source control program, conventional wastewater treatment, advanced water treatment, management of the environmental buffer, and drinking water treatment including management of the water distribution system (Fig. 16-4). The performance of each barrier is assured by conducting comprehensive and tailored monitoring programs.

Barrier 1: Source Control. Source control is a regulatory management practice to minimize the discharge of pollutants into the sewer. Best management practices of source control or source protection ensure sustainability and integrated pollution control of the wastewater. Control at the source reduces the treatment costs and improves the reliability of water quality. Effective source control practices involve the following elements:

1. Developing and executing a comprehensive plan for the management of the sewerage systems feeding the sewage treatment plants that supply source waters to the IPR systems

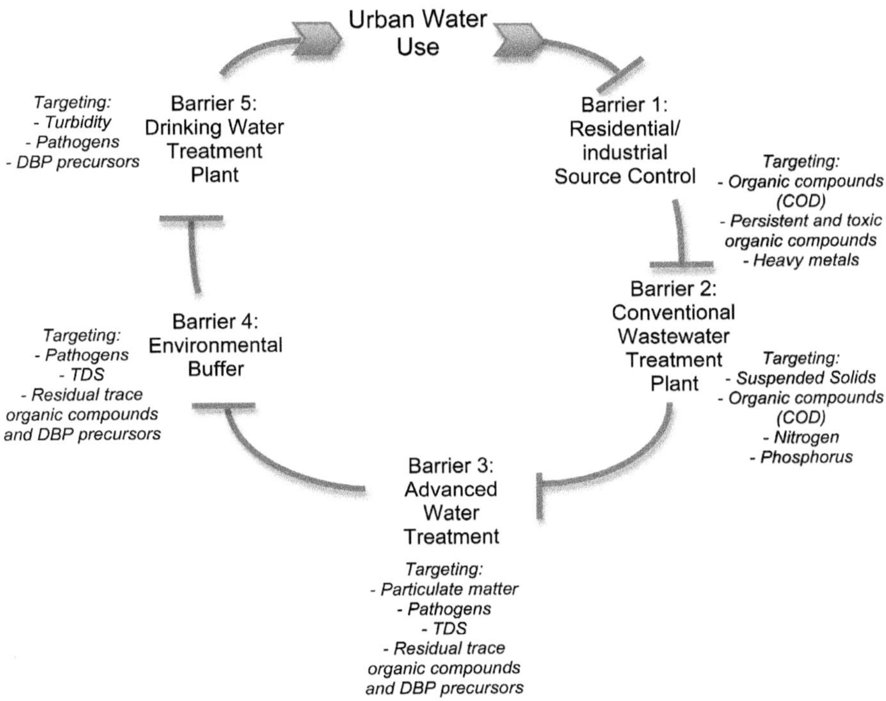

FIGURE 16-4 The multiple-barrier approach in indirect potable reuse.

2. Ensuring protection of reclaimed water from potentially polluting activities
3. Monitoring industrial and commercial discharge permits and compliance assessment
4. Creating awareness within the community of the impact of anthropogenic activities on water quality

Barrier 2: Conventional Wastewater Treatment. Conventional wastewater treatment processes consist predominantly of aerobic and anaerobic biological processes designed to minimize the impact of wastewater effluents discharged to the environment. Processes, such as activated sludge treatment and the use of trickling filters, are used to accelerate aerobic biochemical oxidation of assimilable carbon in order to reduce the occurrence of these processes in the environment which can lead to deoxygenation of waterways. These secondary treatment processes thus degrade easily assimilable organic compounds including proteins, sugars, and other biopolymers. Anaerobic and aerobic conditions are also commonly employed to enhance biological nutrient removal (BNR) by processes, such as nitrification and denitrification and biological phosphorus removal. In order to meet water quality standards, many treatment plants have changed or enhanced their biological treatment to remove nutrients and as a consequence the quality of conventionally treated effluents can vary widely (Table 16-3).

Barrier 3: Advanced Water Treatment. *Advanced water treatment* (AWT) is a term coined to differentiate the treatment processes adopted for IPR from conventional wastewater treatment processes and to indicate a water quality after partial TDS and/or trace chemical

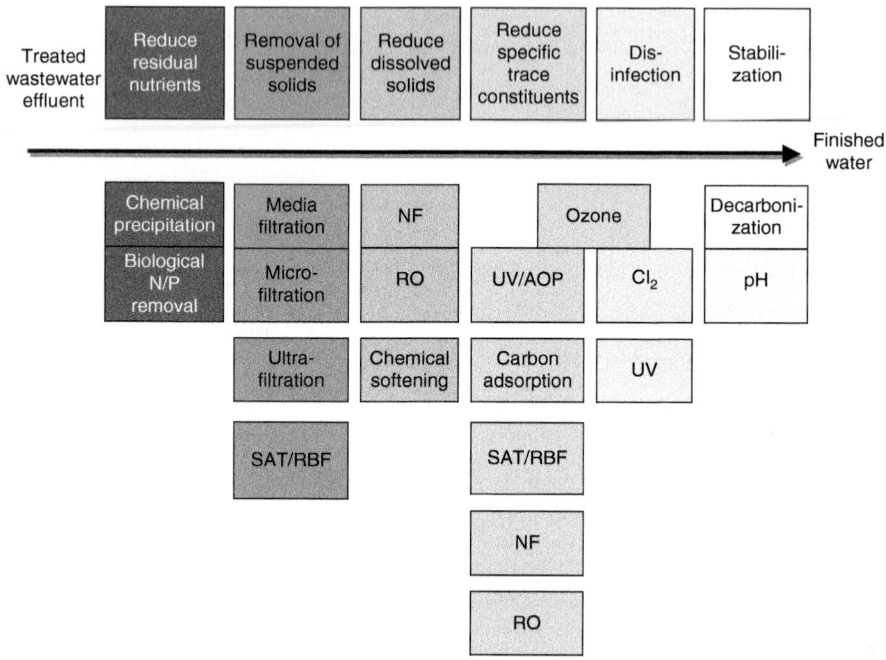

FIGURE 16-5 Treatment objectives of advanced water treatment and commonly employed treatment processes in water reclamation.

removal (Tchobanoglous et al., 2003). AWT processes are employed to meet various IPR treatment objectives, such as increasing the efficacy of disinfection, further decreasing residual nutrient concentrations, removing recalcitrant organic chemicals not metabolized during BNR, lowering the concentration of total dissolved solids, and increasing process redundancy. Various processes and treatment combinations have been employed or proposed to meet these treatment objectives (Fig. 16-5).

Barrier 4: Environmental Buffer. Management of the environmental buffer is another important component of the multibarrier approach; the buffer provides further treatment and/or dilution of reclaimed water, as well as allowing time to commission additional treatment steps or modifications at the reclaimed and drinking water treatment plant if *out-of-specification* reclaimed water is released prior to entering a finished water distribution system.

Attenuation processes in natural systems can result in a loss of identity of water and its wastewater origin. This disconnection from the source is considered a key element in building public trust in IPR.

Barrier 5: Drinking Water Treatment and Distribution System. The final drinking water treatment is mainly intended to address water quality issues associated with the storage, attenuation, blending, and conveyance of water that has been augmented by reclaimed water. In most cases, the treatment process consists of conventional unit processes usually employed for groundwater or surface water sources. A final disinfection residual is provided prior to delivery to the consumer.

Overarching Monitoring Program. In order to assure proper performance of conventional and advanced water treatment unit processes, an overarching monitoring program needs to be tailored to the specific barriers and local conditions of an IPR scheme. Details regarding the design of such a monitoring program are provided in the section Monitoring Strategies for Process Performance and Compliance following.

Indirect Potable Reuse Through Surface Water Augmentation

The practice of using reclaimed water to augment surface water used as a source for drinking is gaining popularity among water purveyors worldwide. The pioneering project of surface water augmentation in the United States was led by the Upper Occoquan Service Authority (UOSA), in Virginia, which began by augmenting surface water supplies in Fairfax County with reclaimed water in 1978 (Table 16-1). Since then, surface water augmentation projects were established in Clayton County (1985) and Gwinnett County (1990) in Georgia. Motivated by a severe drought and population growth, San Diego, California, proposed a surface water augmentation project in the late 1990s, which was terminated due to lack of public and political support in 1998. Ten years later, San Diego's IPR project is being reconsidered. In 2003, Singapore established their first IPR facility via surface water augmentation and has since added four additional facilities with a combined capacity of 467 ML/day as of 2010. Southeast Queensland, Australia, experienced its worst drought in 100 years from 2001 to 2008 with levels of the key drinking water reservoir, Lake Wivenhoe, falling below 17 percent of storage capacity. To counteract this trend, in 2008 the Western Corridor Project, which is designed to augment Lake Wivenhoe with reclaimed water if storage levels again fall below 40 percent of capacity, was established.

Upper Occoquan Service Authority, Virginia. Motivated by population growth, increasing urbanization, and a declining water quality of the Occoquan Reservoir, which is serving as the major raw water supply for people in northern Virginia, the Upper Occoquan Service Authority (UOSA) water reclamation system was established in 1978. The current Millard H. Robbins Jr. Water Reclamation Facility has a capacity of 204 ML/d (54 mgd) and is composed of advanced biological nutrient removal achieving complete nitrification and partial denitrification, lime clarification, two-stage recarbonation, multimedia gravity and pressure filtration, *granular activated carbon* (GAC) adsorption/filtration, ion exchange and breakpoint chlorination (standby), and disinfection (Asano et al., 2007). The BNR combined with the chemical treatment processes provides several-log reduction of viruses, phosphorus removal exceeding 99 percent, and significant reductions in organic and particulate matter, and heavy metals. Subsequent, multimedia filtration produces a reclaimed water quality with average suspended solids of 0.3 mg/L and a turbidity of less than 0.3 ntu. GAC adsorption/filtration serves as a process barrier for disinfection precursors, refractory synthetic organic compounds (SOCs), and other trace organic chemicals (Angelotti et al., 2005). Ion-exchange resins are only employed in standby mode to assure that discharge limits of less than 1.0 mg/L total Kjeldahl nitrogen (TKN) are met and drinking water standards of 10 mg/L nitrate nitrogen in the raw drinking water supply are not exceeded, especially during drought conditions.

Reclaimed water from the UOSA facility is discharged into Bull Run, a tributary of the Occoquan Reservoir. The discharge point is 9.7 km (6 mi) upstream of the headwaters of the reservoir and 32 km (20 mi) upstream of the drinking water supply intake. Reclaimed water currently accounts for about 8 percent of the annual average inflow to the reservoir, but during drought conditions may account for up to 90 percent. Since 2006, the Frederick P. Griffith Water Treatment Plant is treating raw water from the Occoquan Reservoir at a capacity of 454 ML/d (120 mgd) providing high-quality drinking water to the southern portion of Fairfax County. The plant comprises coagulation/flocculation and multimedia filtration coupled with ozone and GAC filters, followed by final chlorination. The finished water quality exhibits average nitrate concentration of 1 mg N/L, TOC concentrations of 1.2 mg/L, and non-detect concentrations for 19 emerging trace organic chemicals (Fairfax Water, 2009).

In the future, the UOSA will be required to meet nutrient loading limits to protect the water quality of Chesapeake Bay. A plant expansion is underway to increase the treatment capacity to 242 ML/d (64 mgd) utilizing membrane bioreactors with chemically enhanced BNR followed by deep-bed GAC filtration (Asano et al., 2007).

Singapore's NEWater Project. The Singapore Water Reclamation Study (NEWater Study) was initiated in 1998 as a joint initiative between the Public Utilities Board (PUB) and the Ministry of the Environment and Water Resources (MEWR) to diversify Singapore's water portfolio and decrease the dependency on imported water from Malaysia. In 2000, a 10 ML/d demonstration plant was constructed at Bedok Water Reclamation Plant to produce *NEWater*. The reclaimed water from this plant was monitored regularly over a period of 2 years, when an expert panel gave an approval in terms of water quality and reliability (NEWater Expert Panel, 2002). By 2003, two *NEWater* Factories at Bedok and Kranji were operational producing 72 ML/d (19 mgd) of high-quality reclaimed water. Approximately 9 ML/d of this water was pumped into raw water reservoirs for blending and mixing (Tortajada 2006). A third plant at Seletar with a capacity of 24 ML/d was commissioned in 2004, and a fourth plant at Ulu Pandan with a capacity of 145 ML/d was brought online in March 2007, bringing the combined capacity of NEWater plants to 240 ML/d (64 mgd). Singapore's largest NEWater plant is the Changi plant that started with a treatment capacity of 68 ML/d (18 mgd) in 2009 and expanded to a total capacity of 227 ML/day (60 mgd) in 2010.

All NEWater plants are treating secondary effluents and subsequently utilize integrated membrane system followed by UV disinfection and chlorination. The largest amount of NEWater (438 ML/d or 116 mgd) is currently used for industrial (semiconductor manufacturing) and commercial use. In 2009 a small amount of NEWater (30 ML/d (8 mgd), or about 1 percent of the daily consumption of the country) began to be blended with raw water in the reservoirs that is subsequently treated for domestic use.

Western Corridor Project, Queensland, Australia. The *Western Corridor Recycled Water Project* (WCRWP) was constructed in South East Queensland, Australia during 2007–2008. The WCRWP has the capacity to produce up to 232 ML/d (61 mgd), which represents approximately 30 percent of the total current water supply needs for South East Queensland (Davies, 2009). The WCRWP utilizes treated effluents from six wastewater treatment plants to produce reclaimed water in three advanced water treatment plants suitable for both industrial and potable uses. The AWT plants, located at Bundamba, Luggage Point, and Gibson Island, incorporate the treatment processes of coagulation/flocculation and settling, MF, RO, AOP (UV/H_2O_2), and chemical stabilization and disinfection.

The scheme is designed to partially supplement potable water supplies by adding highly treated reclaimed water into South East Queensland's largest drinking water reservoir, Lake Wivenhoe. This water is then treated at the potable water treatment plant located at Mt. Crosby before distribution via existing potable water supply infrastructure. Water produced from the WCRWP is currently supplied directly to major industrial users including the Swanbank and Tarong Power Stations which previously drew water from the Lake Wivenhoe system. The concentrate of the IMS is discharged to the Brisbane River and the Pacific Ocean. Prior to discharge, the concentrate is further treated by a nitrification system followed by denitrification filters in order to meet discharge standards to the Brisbane River. After reservoir levels recovered and exceeded 60 percent of storage capacity in 2009, South East Queensland regulators decided that augmentation of Lake Wivenhoe with reclaimed water will only occur once the reservoir storage levels fall below 40 percent of capacity.

Comprehensive validation and verification testing during piloting and start-up of the facilities has been conducted and demonstrated that the reclaimed water quality meets and exceeds the requirements of the "Phase 2 Australian Guidelines for Water Recycling" (NRMMC et al., 2008) as well as the "Australian Drinking Water Guidelines".

Indirect Potable Reuse Through Subsurface Treatment Leading to Groundwater Augmentation

In the United States, drinking water augmentation with reclaimed water was pioneered by the County Sanitation Districts of Los Angeles County (CSDLAC) and the Water Replenishment District of Southern California (WRD) by establishing groundwater recharge spreading operations with reclaimed water in Whittier, California, in the early 1960s. Although the spreading facilities received also untreated surface water and stormwater, subsurface treatment was recognized as being able to provide a barrier to pathogens, to reduce the concentration of nutrients and organic matter, and to attenuate trace organic chemicals present in reclaimed water (Fox et al., 2001).

Since this pioneering project, subsurface treatment for reclaimed water has been employed in various modifications, such as rapid infiltration basins, vadose zone injection wells, infiltration trenches, and riverbank filtration (see Chap. 15).

Recharge basins are often located in, or adjacent to, floodplains, characterized by soils with high permeability. It may be required that excavation is necessary for recharge basins to remove surface soils of low permeability. For vector control and to maintain permeability during operation with reclaimed water, recharge basins are usually operated in wetting and drying cycles. As the recharge basin drains, dissolved oxygen penetrates into the subsurface facilitating biochemical transformation processes and organic material accumulated on the soil surface desiccates allowing for the recovery of infiltration rates. In riverbank filtration, constant scour forces due to stream flow prevent the accumulation of particulate and organic matter in the infiltration layer. All subsurface treatment applications combine the benefits of no need for chemicals, low energy demand, residual free operation, temperature equilibration of the water, and a rather low degree of maintenance.

Different reclaimed water qualities have been applied to recharge operations ranging from primary effluents to highly treated water. Primary effluent has only been applied to demonstration projects and is not currently used for groundwater recharge. CSDLAC in Whittier, California, is applying tertiary-treated effluents to surface spreading operations. The Orange County Water District in Fountain Valley, California, is applying reclaimed water that has been treated with RO and advanced oxidation processes to surface spreading basins. The attenuation processes of subsurface systems include transformations of organic matter, oxidation/reduction (redox) reactions, nitrogen transformations, and inactivation of pathogens. The degree of water quality changes will depend on the feed water quality applied to recharge operations.

The removal of organic matter during infiltration is highly efficient and largely independent from the level of above-ground treatment, since biodegradable organic carbon that is not attenuated during wastewater treatment is readily removed during groundwater recharge (Drewes and Fox, 2000; Rauch and Drewes, 2004; Rauch-Williams and Drewes, 2006). However, organic matter represents an electron donor for microorganisms in the subsurface and reclaimed water after RO/AOP treatment usually lacks sufficient carbon levels to support the growth of microorganisms.

The removal of easily biodegradable organic carbon in the infiltration zone usually results in depletion of oxygen and the creation of anoxic conditions. Although this transition is advantageous regarding denitrification processes, it might also lead to the solubilization of reduced manganese, iron, and arsenic from native aquifer materials. If these interactions occur, an appropriate posttreatment is required after recovery of the recharged groundwater.

Nitrogen removal needs to be carefully managed when not nitrified reclaimed water is applied with nitrogen concentrations in excess of 20 mg N/L representing a nitrogenous oxygen demand in excess of 100 mg/L. Ammonium is usually removed by cation exchange onto soil particles during wetting cycles, followed by nitrification of the adsorbed ammonium during drying cycles. Nitrate is not adsorbed to soils and, providing sufficient carbon is present to create anoxic conditions, can be denitrified during subsequent passage in the subsurface (Fox et al., 2001). Given the high carbon demand for denitrification, reclaimed water with nitrate concentrations in excess of 10 mg NO_3-N/L should not be applied to groundwater recharge.

The combination of filtration and biotransformation processes during subsurface treatment is efficient for the inactivation of pathogens, especially viruses. From an extrapolation of tracer study data derived from a full-scale recharge basin study in the Montebello Forebay, California, a 7-log reduction of bacteriophage should occur within approximately 30 m (100 ft) of subsurface travel (Fox et al., 2001).

Montebello Forebay Spreading Operations, California. Jointly operated by the Water Replenishment District of Southern California (WRD) and CSDLAC, the Montebello Forebay spreading grounds started in 1962 representing the oldest IPR project in the United States. The spreading grounds are used for flood control, storm water conservation, and aquifer recharge using imported surface, storm, and reclaimed water. The San Jose Creek, Whittier Narrows, and Pomona Water Reclamation Plants (owned and operated by the CSDLAC) receive and treat residential and industrial wastewater (more than 90 percent is the former). The treatment processes at the water reclamation plants are composed of primary clarification, BNR including nitrification and denitrification, secondary clarification, tertiary dual-media filtration (anthracite and sand), and a chlorine contact basin. Prior to reuse, sulfur dioxide is added for dechlorination since the water is transferred to the spreading basins via a surface water body. Reclaimed water is diverted to a series of recharge basins along the Rio Hondo and San Gabriel rivers. The spreading operations are characterized by a shallow vadose zone of approximately 10 m (30 ft). Currently, the contribution of reclaimed water to recharge the local groundwater is limited to 35 percent. The reclaimed water blends with the aquifer and remains in the subsurface for approximately 6 months before it is contained in drinking water wells in the north and central portions of the central groundwater basin. For wells that are not subject to the USEPA Groundwater Rule, wellhead treatment is provided using chlorine prior to injection into the distribution system.

Prairie Waters Project, City of Aurora, Colorado. In 2002 and 2003, a historic drought significantly affected the city of Aurora's water supplies and the city was faced with the short-term prospect of rationing water and a long-term unsustainable water resource portfolio in view of significant population growth. Since the city owns transbasin water rights from resources located on the western slope of the Rocky Mountains, which are entitled to reuse, the city decided to recover these resources from the South Platte River downstream of Denver, Colorado. The South Platte River receives treated wastewater effluents and runoff from the Denver/Aurora metropolitan area. Downstream of the Denver urban area, wastewater discharge can make up a significant portion (up to 60 percent annually) of the flow in the South Platte River (Krasner et al., 2008).

The city considered several different options for the recovery and treatment of this water. Because the short timeline to bring a new water supply online, the need for a streamlined permitting process, and the lack of an inland concentrate disposal option called for an environmentally sustainable solution and because of customer preferences, the city settled on a combination of natural purification systems with advanced engineered water treatment processes to establish a project that could be operational by 2010 increasing Aurora's water supply by 20 percent.

The natural purification system, using a combination of riverbank filtration (RBF) wells and artificial recharge and recovery basins, is located along the South Platte River approximately 32 km (20 mi) downstream of Denver. The natural system provides a hydraulic retention time of approximately 30 days in the subsurface and removes particulate matter and pathogens, decreases TOC concentrations from 6–11 mg/L in the river to 3 to 4 mg/L after RBF, achieves nitrification/denitrification with nitrogen concentrations of less than 2 mg N/L, and achieves substantial reduction of trace organic chemicals. Three pump stations and a 56-km (35-mi) pipeline lift the recovered water from the natural purification system and convey it to the Peter Binney Water Purification Facility (BWPF) located at Aurora Reservoir, one of the city's current conventional supplies fed by Rocky Mountain snowmelt. At the BWPF, approximately 33 ML/d (9 mgd) of water recovered from the natural purification system is blended at a ratio of 1:2 with water from Aurora Reservoir, representing a total plant peak capacity of 190 ML/d (50 mgd). The treatment train of the BWPF consists of precipitative softening, followed by UV/AOP, biologically active carbon (BAC) filters, granular activated carbon filtration, and final disinfection. Prior to disinfection, the water is blended at a ratio of 1:2 with conventionally treated water from Aurora Reservoir (i.e., coagulation/flocculation). Partial softening is established to maintain a treatment goal of less than 400 mg/L TDS in the finished water. Advanced oxidation followed by BAC removes any residual trace organic chemicals and pathogens from the natural purification system, followed by GAC serving as a final barrier to trace organic chemicals and DBP precursors.

Indirect Potable Reuse Through Direct Injection into a Potable Aquifer

Direct injection of reclaimed water into a potable aquifer was pioneered in Orange County, California, with the establishment of Water Factory 21 in 1976. This project represented the precursor and benchmark for several subsequent IPR projects around the world, established by the West Basin Municipal Water District, California (1995), the Scottsdale Water Campus, Arizona (1999), the NEWater plants in Singapore (2003), the Groundwater Replenishment System replacing Water Factory 21 (2008), and the Western Corridor Project, Australia (2008). The core treatment processes of these projects are integrated membrane systems, in same cases coupled with subsequent UV advanced oxidation processes. Since all these projects (with the exception of Scottsdale, AZ) are located in coastal settings, the concentrate of the IMS from these projects is disposed via ocean outfalls.

Groundwater Replenishment System, Orange County, California. The *Groundwater Replenishment* (GWR) *System* was completed in 2008 with a capacity of 265 ML/d (70 mgd), representing the largest potable reuse project worldwide. It is a joint project between the *Orange County Water District* (OCWD) and the *Orange County Sanitation District* (*OCSD*). Expansion plans are in preparation to increase the capacity to 380 ML/d (100 mgd) in the near future (Markus, 2009). Approximately half of the reclaimed water provided by this facility is used via surface spreading basins to recharge OCWD's groundwater basin, whereas the other half is used to maintain injection wells of the Talbert Gap Barrier to protect from seawater intrusion from the Pacific Ocean.

The precursor facility, Water Factory 21, operated from 1976 to 2004 reclaiming 60 ML/d (16 mgd) of secondary treated wastewater effluent using lime clarification pretreatment, RO, followed by UV/AOP and final disinfection. The GWR system adopted this approach, and the advanced treatment train consists of submerged microfiltration (MF), RO, advanced oxidation processes using low-pressure, high-output ultraviolet light and hydrogen peroxide. The MF is used to separate suspended and colloidal solids including bacteria and protozoa from the secondary effluent. RO treatment reduces dissolved organic matter, regulated and unregulated trace organic chemicals, and viruses from MF filtrate. The subsequent, ultraviolet light/hydrogen peroxide process provides additional bacterial and viral inactivation and through hydroxyl radical formations can oxidize trace organic compounds. The product water undergoes additional chemical posttreatment using lime prior to groundwater injection and recharge. The finished water quality prior to groundwater injection is characterized by average TOC concentrations of 0.2 mg/L, total nitrogen concentrations of less than 1.7 mg/L, TDS concentrations of less than 34 mg/L, and NDMA concentrations of less than 1.6 ng/L.

Hueco Bolson Recharge Project, El Paso, Texas. The Hueco Bolson Recharge Project uses reclaimed water from the Fred Harvey Water Reclamation Plant and injects it directly into the primary drinking water source for the city of El Paso, Texas. The system has a capacity of 38 ML/d (10 mgd). The core treatment process consists of a two-stage biophysical *powdered activated carbon* (PAC) process. This process, which combines a conventional activated sludge system with the use of PAC, provides most of the removal of organic compounds and all of the removal of nitrogen compounds. Subsequently, high-lime treatment is used for the removal of residual heavy metals, pathogens, and phosphorus. After recarbonation, the water is sand filtered and ozone is used as primary disinfectant. GAC adsorption/filtration serves as a polishing process for removal of residual organic compounds (DBP precursors) and reduction of taste- and odor-producing compounds, and other regulated and unregulated trace organic chemicals. Through 11 injection wells, the reclaimed water is introduced into the Hueco Bolson aquifer, serving as water supply for the El Paso region.

Direct Potable Reuse

The determining factor for direct potable reuse is public confidence in, and reliance on, employed water treatment processes that always produce safe drinking water. Technological concerns are addressed by multiple-barrier systems, providing redundancy and diversity of processes that can remove the wide range of constituents of concern producing a water quality that meets drinking water standards and is safe to drink. In addition, process validation and verification approaches through real-time and frequent monitoring are available today and already embedded in many advanced water treatment plants practicing indirect potable reuse worldwide. What remains challenging is to overcome public and regulator perception to potable reuse. Direct potable reuse, however, offers opportunities to significantly reduce the cost for water collection and water distribution systems that might become more relevant in the near future making direct potable reuse a cost-effective application in the long term.

Windhoek's Goreangab Reclamation Plant, Namibia. Since 1968, Windhoek has pioneered direct potable reuse with the commissioning of the Goreangab Reclamation Plant in Windhoek, Namibia, which still today is the only commercial operation practicing direct potable reuse worldwide. Given the site-specific constraints, Windhoek chose direct potable reuse because it had no other choice. The current plant has a capacity of 21 ML/d (4.4 mgd) providing up to 35 percent of the city's drinking water supply during normal water consumption and up to 50 percent during severe drought conditions (du Pisani, 2006).

The plant design philosophy follows the multiple-barrier concept with established and optional treatment barriers. The treatment train consists of coagulation/flocculation, followed by dissolved air flotation and media filtration. The water is subsequently treated by ozone/hydrogen peroxide followed by biological active granular activated carbon filtration. A final barrier is provided by ultrafiltration prior to final stabilization and disinfection. The finished water quality is subject to ongoing monitoring and testing. Windhoek has managed to overcome negative public perception with positive and proactive marketing providing a viable option of innovative integrated water resource management.

MONITORING STRATEGIES FOR PROCESS PERFORMANCE AND COMPLIANCE

In order to protect public health, careful attention must be given to proper selection of treatment processes that are capable of removing constituents of concern. Performance validation and verification of established treatment processes and compliance water quality monitoring are the most important aspects of overall system reliability. Reliability requirements include provisions for alarms, standby power supplies, readily available replacement equipment, standby treatment units, emergency storage or disposal provisions, flexible piping and pumping facilities, and trained personnel. Pipes or pumps that would allow bypass of critical treatment processes should not be installed. The facility should have procedures in place to operate during power failures, peak loads, equipment failure, treatment plant upsets, and maintenance shutdowns.

Monitoring and operating procedures for indirect potable reuse facilities need to be comprehensive, and several strategies exist to maintain performance integrity of treatment systems.

Real-Time Monitoring and Operating Strategies

The Phase 2 Australian Guidelines for Water Recycling (NRMMC et al., 2008) require a multiple-barrier approach to the management of risks with a strong focus on managing potential hazardous events. In order to illustrate the role of operating strategies and real-time monitoring to manage risks to process upsets, the AWT plants of the Western Corridor project in Southeast Queensland serve as an example. These facilities have established real-time monitoring to create critical control points, which allow the partially treated water stream to be diverted away from the next treatment step. At the beginning of the AWT plant processes, these monitored parameters include nitrate and ammonium concentrations, which indicate if the BNR processes in the wastewater treatment plant are operating appropriately. At the microfiltration step, membrane integrity is tested by automated pressure decay tests up to an hourly frequency, while at the RO step TOC is monitored as a surrogate parameter to detect any leakage of organic molecules through the membranes. At the disinfection/advanced oxidation step, the UV light dose and transmittance is constantly measured to detect lamp failure or opaque water, respectively. These real-time tests are complemented by laboratory tests (both chemical and microbiological) to confirm the data from the real-time sensors, as well as provide information on complex chemicals, such as trace organic chemicals, that cannot be measured in real time. The frequency of the laboratory tests will vary from weekly to annually with the frequency being responsive to the concentrations found, that is, to parameters whose values are rising or enter the recursive concentration range will be monitored more often until the cause of the excursion or violation is identified and rectified.

Simple water quality monitoring techniques, such as continuous turbidity monitoring, while suitable for coarse filtration systems, lack sufficient sensitivity to detect damage in fine filtration systems such as MF or UF membranes (Johnson, 1997). Monitoring techniques that directly assess the integrity of membrane systems include pressure decay tests and diffusive airflow tests, while indirect tests that monitor the quality of the output include particle counting or particle monitoring or microbial challenge tests (see Chap. 11). Challenge tests introduce a known concentration of chemicals or microbial particles into the process and assess the effectiveness of the process by monitoring the final contaminant concentration in the filtrate.

Surrogate Measures and Indicator Compounds

In municipal wastewater effluents, the majority of nontraditional chemical contaminants are typically present at low concentrations (i.e., <300 ng/L) and most analytical methods are optimized to detect only a few compounds at a time. As a result, the effort required to perform a comprehensive analysis of all of the detectable chemical contaminants is unfeasible for all but the most sophisticated analytical laboratories.

Engineered treatment systems that can be used to control wastewater derived chemical contaminants in IPR employ physical, chemical, and biological processes to remove or transform the compounds. Published research on the mechanisms through which treatment processes act indicates that it should be possible to predict the extent of compound removal for compounds (termed *indicators*) exhibiting similar properties provided that those properties determine the behavior of the compound in the treatment process (Snyder et al., 2003; Bellona et al., 2004; Snyder et al., 2006; Deborde and von Gunten, 2008; Dickenson et al., 2009). Furthermore, the removal of specific compounds or families of compounds with closely related properties may be correlated with the removal of other routinely measured compounds or operational parameters (termed *surrogates*) that can be monitored continuously (e.g., conductivity, UV absorbance) (Drewes et al., 2009; Wert et al., 2009).

The most commonly used indicator organisms for pathogens are coliform bacteria, either fecal coliform or *E. coli*. However, since the 1980s, it has been increasingly recognized that coliform bacteria provide a poor indication of the presence of wastewater derived pathogens, such as viruses and protozoa. This is largely because many viruses and protozoa are much more resilient to treatment processes including UV and chemical disinfection as compared to coliform bacteria. Consequently, bacteriophages infecting enteric bacteria and spores of sulphite reducing bacteria have been suggested as more appropriate indicators for viruses and protozoa. Specific bacteriophages that have been recommended include somatic coliphages, F-RNA-specific bacteriophages, and phages infecting *B. fragilis*. Medema and Ashbolt (2006) have proposed that rather than producing a standard and monitoring requirement for all pathogens that could be transmitted through drinking water, the use of a suite of *index pathogens* be adopted.

Although the idea of using indicators (e.g., coliform bacteria) and surrogates (e.g., turbidity) as a means of predicting the behavior of difficult-to-measure contaminants is well established among researchers concerned with waterborne pathogens (see Chap. 2), it has not been adopted widely by regulators concerned about chemical contaminants. Recent studies have proposed a series of indicator compounds and surrogate parameters that were effectively used to assess the performance efficiency for trace organic chemicals of a wide range of advanced water treatment processes commonly employed in IPR schemes (Drewes et al., 2009; Dickenson et al., 2009). An example of chemicals, that have been identified as viable indicator compounds to assess the performance of ozonation processes, is presented in Table 16.8. A road map to adopt a monitoring strategy utilizing surrogates and indicators to assess the performance of unit processes in an IPR scheme has been proposed by Drewes et al. (2009) and is illustrated in Table 16-9. This indicator/surrogate concept has been adopted in the Australian "Guidelines for Water Recycling for Drinking Water Augmentation" (2008).

TABLE 16-8 Treatment Removal Categories for Indicators Compounds of Systems Using Ozone (Conditions: secondary/tertiary treated wastewater, O_3:TOC = 0.6–1 mg/mg, ozone exposure CT_{10} = 4-11[a] mg min/L, pH 7-8, ~20°C, Alkalinity < 300 mg/L as $CaCO_3$)

Good removal			Intermediate removal		Poor removal
90 percent			90–50 percent	50–25 percent	< 25 percent
Hydroxy Aromatic	Nonaromatic C = C	Alkoxy Polyaromatic	Saturated Aliphatic	Nitrosamine	Halogenated Aliphatic
Acetaminophen	Acetyl cedrene	Naproxen[A]	Iopromide[D]	NDMA[D]	Chloroform[D]
Atorvastatin (o-hydroxy)	Carbamazepine[A]	Propranolol[A]	Isobornyl acetate		TCEP
Atorvastatin (p-hydroxy)	Codeine		Meprobamate		TCPP
Benzyl salicylate	Hexylcinnamaldehyde	*Alkoxy Aromatic*:	Methyl dihydrojasmonate		TDCPP
Bisphenol A[A]	Methyl ionine	Gemfibrozil			
Estrone[A]	OTNE	Hydrocodone	*Nitro aromatic*		
Hexyl salicylate	Simvastatin hydroxy		Musk ketone		
Methyl salicylate	Terpineol	*Alkyl Aromatic*:	Musk xylene		
Nonylphenol[A]		Benzophenone			
Oxybenzone	*Deprotonated amine*	Benzyl acetate			
Propylparaben	Atenolol[B]	Bucinal			
Salicyclic acid[B]	Caffeine	DEET			
TriclosanA	Diclofenac[A]	Dilantin			
	EDTA[A]	Dibutyl phthalate		$k_{03,app}$ (M^{-1} s^{-1})	
Amino/Acylamino Aromatic	Erythromycin-H_2O	Diphenhydramine		A > 10^5	
Atorvastatin	Fluoxetine	Galaxolide		B 10^3–10^5	
Sulfamethoxazole[A]	Metoprolol[B]	Ibuprofen[C]		C 10^0–10^3	
Triclocarban	Nicotine	Indolebutyric acid		D < 100	
	Norfluoxetine	Primidone			
	Ofloxacin	Tonalide			
	Paraxanthine				
	Pentoxifylline				
	Trimethoprim[A]				

[a]Determined by the extended integrated CT10 method

Source: Adapted with permission from Dickenson et al. (2009), copyright 2009, American Chemical Society.

TABLE 16-9 Application of the Surrogate/Indicator Framework to Assess Performance of IPR Treatment Trains

	Surrogate parameters	Indicator compounds
Piloting or/and start-up		
Step 1	Define operational conditions for each unit process making up the overall treatment train for proper operation according to technical specification	
Step 2	For each unit process, select surrogate parameter that demonstrate a measurable change under normal operating conditions and quantify this differential $\Delta X_i = \|(X_{i,\text{in}} - X_{i,\text{out}})\|$	Conduct occurrence study to confirm detection ratio (mean concentration/detection limit) of viable indicator compounds is larger than 5 in the feed water of each unit process
Step 3		Conduct challenge or spiking study with select indicator compounds (5–10) during pilot- or start-up to determine the removal differentials under normal operating conditions $\Delta Y_i = (Y_{i,\text{in}} - Y_{i,\text{out}})/Y_{i,\text{in}}$
Step 4	Select viable surrogate and operational parameters for each unit process	Select 3–6 indicator compounds from categories classified as Good removal
Full-scale operation/compliance monitoring		
Step 5	Confirm operational conditions of full-scale operation and removal differential ΔX_i for selected surrogate and operational parameters	
Step 6	Operational monitoring: Monitor differential ΔX_i of select surrogate and operational parameters for each unit process or/and the overall treatment train on a regular basis (daily, weekly)	Verification monitoring: Monitor differential ΔY_i of selected indicator compounds for each unit process or/and the overall treatment train semiannually/annually

Source: Adapted with permission from Drewes et al. (2008), copyright 2008, WaterReuse Foundation.

REGULATIONS AND GUIDELINES FOR DRINKING WATER AUGMENTATION

United States

U.S. Federal Guidelines. In the United States, there are no federal regulations that specifically address potable reuse. The USEPA has published a guidance document on water reuse (USEPA 2004) including guidelines for IPR. However, the document has no regulatory authority and does not make specific recommendations with respect to chemical contaminants. Several states have issued regulations or draft regulations pertaining to potable reuse of municipal wastewater effluents. However, these regulations differ in their guidance on whether potable reuse is an acceptable practice and, if so, what safeguards should be in place (National Research Council, 1998).

The majority of wastewater derived contaminants are not regulated in the United States. Moreover, a comprehensive list of emerging contaminants is not feasible. For instance, *endocrine-disrupting compounds* (EDCs) are a vast group of chemicals that have a toxicological impact and are not simply a list of specific chemical substances. There are currently no federal regulations for pharmaceuticals in drinking water or natural waters. The U.S. Food and Drug Administration (FDA) requires ecological testing and evaluation of a new pharmaceutical only if an environmental concentration in water or soil is expected to exceed 1 mg/L or 100 μg/kg, respectively (FDA, 1998). In light of the recent data on the occurrence of pharmaceutical residues and personal care products in the aquatic environment, these policies may need to be reconsidered. While extensive monitoring programs are underway, toxicological studies conducted at environmentally relevant concentrations are necessary for intelligible regulations to be established. While no federal legislation specifically regulates EDCs and pharmaceutical residues in drinking or wastewater, individual states may regulate these compounds in the absence of any federal mandates. Among the states, California and Florida have made the most progress in establishing uniform approaches for assessing planned IPR and often serve as examples for regulators in other states. Therefore, it is appropriate to consider their regulations in some detail.

State of California. In the late 1980s, the *California Department of Public Health* (CDPH), formerly known as California Department of Health Services (Cal DHS), developed draft criteria for the use of reclaimed municipal wastewater to recharge groundwater basins that are sources of domestic water supply (Crook et al., 2000). The CDPH criteria, which set forth the agency's approach to writing permits for IPR systems, have been updated several times but have not been approved or finalized as of 2010. The CDPH draft groundwater recharge criteria are designed to ensure a groundwater supply that meets all the drinking water standards and other requirements more specific to water derived from wastewater effluent (CDPH, 2008). In formulating the proposed criteria, CDPH considered both acute health effects from microbial pathogens and potential long-term health effects associated with chemical constituents, particularly trace organic chemicals (Geselbracht and Crook, 2000). After receiving the final report prepared by a *science advisory panel* (SAP) submitted to the state in 1987, CDPH selected TOC limits in wastewater effluent prior to recharge as a means and surrogate of ensuring the lowest possible concentration of unregulated wastewater derived organic contaminants (Robeck, 1987). In its summary report, the SAP concluded that concentrations of organic carbon should be removed to "below 1 mg/L by reverse osmosis and essentially all identifiable trace organic compounds of significance should be absent in detectable concentrations."

The current draft criteria (CDPH 2008) couple an even more stringent TOC limit with the fraction of the drinking water supply that is derived from wastewater effluent as a factor in determining system performance requirements (quantified as TOC). This fraction is referred to as the *recycled water contribution* (RWC), which is based on a 60-month running average. The current draft regulations require that subsurface injection projects produce water with a TOC concentration as expressed in Eq. (16-1) prior to injection. Subsurface injection projects are required to treat 100 percent of the reclaimed water by RO and advanced oxidation processes (AOP) to provide sufficient removal of organic matter. For surface spreading operations, TOC must be equal to or less than 0.5 mg/L divided by the RWC in the reclaimed water prior to or after infiltration and prior to blending with the ambient groundwater. Therefore, surface spreading projects can receive credit for TOC removal that occurs within the vadose zone.

$$\mathrm{TOC} \leq \frac{0.5}{\mathrm{RWC}}$$

In recognition of the possible shortcomings of using TOC as a surrogate for wastewater-derived contaminants, the 2008 draft CDPH criteria, while not recommending additional monitoring for specific chemicals, do identify some specific pharmaceuticals, endocrine disruptors, personal care products, and other chemicals with a notification level that should be investigated (CDPH, 2008). The draft criteria provide a provision for CDPH to require monitoring of specific constituents based on (1) a review of the project engineering report;

(2) an inventory of constituents discharged into the wastewater collection system; (3) the affected groundwater basin; (4) a constituent's ability to characterize the presence of pharmaceuticals, endocrine disrupting chemicals, personal care products, and other indicators in municipal wastewater; and (5) the availability of a test method for a constituent. Thus, such non-regulated constituents will be selected for monitoring on a case-by-case basis. The inclusion of compounds for which no notification level had been established was controversial among water professionals because the basis for the selection criteria was unclear and little guidance was presented on analytical methods or detection limits required to quantify these chemicals in water samples.

State of Florida. Florida's water reuse regulations (FDEP 1999, 2006) include sections addressing high-rate infiltration basin systems and soil-absorption field systems, both of which may result in groundwater recharge. Because nearly all groundwater in Florida is classified as G-II, which is defined as groundwater containing 10,000 mg/L or less of TDS and is designated as a potable supply source, any land application system located over G-II groundwater could function as an indirect potable reuse system. If more than 50 percent of the wastewater applied to the system is collected after percolation, the system is classified as an effluent disposal system and not as beneficial reuse. Loading to these systems is limited to 0.23 m/d (0.8 ft/d), and wetting and drying cycles must be used. For systems having higher loading rates or a more direct connection to an aquifer than is normally encountered, the reclaimed water must receive secondary treatment, filtration, and high-level disinfection (no detectable fecal coliforms per 100 mL in at least 75 percent of the samples analyzed over a 30-day period, maximum of 25 fecal coliforms/100 mL in any sample, total-suspended-solid limit of 5.0 mg/L, and a minimum total chlorine residual of 1.0 mg/L after at least 15 min of contact at peak hour flow) and must meet primary and secondary drinking water standards.

The Florida regulations also include criteria directed at planned indirect potable reuse by injection into water supply aquifers and augmentation of surface supplies. The injection regulations pertain to G-I, G-II, and F-I groundwaters, all of which are classified as potable aquifers. Secondary treatment, filtration, and high-level disinfection are required. Reclaimed water must meet G-II groundwater standards prior to injection. G-II groundwater standards are very similar to primary and secondary drinking water standards. For injection into formations of the Floridian and Biscayne aquifers where the concentration of TDS does not exceed 500 mg/L, the regulations are more restrictive and specify that reclaimed water must meet drinking water standards, undergo activated carbon adsorption as an organic removal process, and meet average TOC and TOX limits of 5.0 mg/L and 0.2 mg/L, respectively, in the product water.

Australia

In 2008, Australia was the first country to develop national guidelines for indirect potable reuse with the release of Phase 2 of the Australian Guidelines for Water Recycling (AGWR): Augmentation of Drinking Water Supplies (NRMMC, EPHC, and NHMRC, 2008). Phase 1 of the AGWR (NRMMC & EPHC, 2006) was published in 2006 and addresses non-potable applications of reclaimed water. The Phase 2 Guidelines build on the risk management framework described in Phase 1 and provide additional detail relating to issues specific to IPR.

Management of water resources in Australia is the responsibility of the states, rather than the federal government. Accordingly, the state governments are responsible for approving and regulating the operation of IPR projects, and the federal government has no legislative or regulatory authority in this field. The AGWR are therefore not mandatory regulations in any state, unless the state governments independently choose to adopt them. Some of the state government regulatory agencies have explicitly endorsed the AGWR and will require proponents to adhere to them in order for a proposed water reclamation scheme to be approved or operated. Such requirements are typically implemented through the provision of clauses in plant operating licenses. Other state governments may choose to deviate from the requirements of the AGWR or may refer to them for the development or adaptation of their own state-based guidelines.

The AGWR are notable for the risk management framework that they provide, rather than simply relying on end-product (reclaimed water) quality testing as the basis for managing water recycling schemes. The risk management framework used is based on the framework detailed in the Australian Drinking Water Guidelines (National Water Quality Management Strategy, 2004).

Assessment of the reclaimed water system must be carried out before strategies to prevent and control hazards are planned and implemented. The aim of the assessment is to provide a detailed understanding of the entire reclaimed water supply system, from source to end use. The hazards, sources, and events (including treatment failure) that can compromise reclaimed water quality are to be characterized and preventive measures needed to effectively control hazards and prevent adverse impacts on humans and the environment identified. Examples of potential hazardous events that should be considered are provided in Table 16-10. Hazardous events are assessed in terms of their likelihoods and consequences as described in the earlier section Design of Potable Reuse Schemes.

Chemical guideline values are tabulated in the Phase 2 AGWR along with maximum concentrations of the chemicals that have been reported from international studies of secondary- or tertiary-treated effluents. However, the guidelines note that the table should not be taken as exhaustive and that detailed assessment of individual systems, including surveys of industrial, agricultural, domestic, and urban inputs, needs to be undertaken to identify potential chemical hazards that could affect source water quality. In most cases, this assessment will need to be supported by extensive monitoring of the source water quality.

Where possible, guideline values were acquired from existing guidelines and standards, with the Australian Drinking Water Guidelines being the primary source. Where no existing guideline values could be identified, guidelines were developed from available health, toxicological, or structural information. The list of guideline values is not intended to be regarded as a mandatory set of parameters to be included in monitoring programs. However, key characteristics that need to be considered for system performance verification include

1. Microbial indicator organisms
2. Health-related chemicals
   a. Those identified in the Australian Drinking Water Guidelines
   b. Key organic chemicals of concern (e.g., NDMA)
   c. Indicators or index chemicals for organic chemicals (e.g., contraceptive hormones)
3. Biological activity

The choice of specific parameters must be informed by hazard identification and risk assessment. These, in turn, should be informed by a consideration of source water quality, potential agricultural and industrial inputs, treatment processes, chemicals and by-products, and receiving water quality.

PUBLIC PERCEPTION TO INDIRECT POTABLE REUSE

It is widely recognized that the ultimate success of any indirect potable reuse project is determined by the level of public and key stakeholder acceptance. Communication of risks associated with chemicals present in reclaimed water is an important component of project implementation.

TABLE 16-10 Examples of Potential Hazardous Events

Stormwater catchments	
Chemical use in catchment areas (e.g., use of fertilizers and agricultural pesticides) Sewage overflows and septic system discharges Entry of livestock waste Climatic and seasonal variations (e.g., heavy rainfall, drought) Industrial discharges	Major fires (firefighting chemicals), natural disasters, sabotage Accidental spills or discharge Leaching from existing or historical waste-disposal (e.g. landfill) or mining sites, and contaminated sites and hazardous wastes Road washing
Sewer systems	
Discharges of domestic and household chemicals Discharges of toxic material Infiltration of stormwater Infiltration of saline groundwater to sewer	Trade-waste discharges, including accidental and illegal discharges Infiltration of waste from contaminated sites or waste disposal sites (e.g., landfill)
Water reuse and drinking water treatment systems	
Chemical dosing failures Disinfection malfunctions Equipment malfunctions Failure of alarms and monitoring equipment Formation of disinfection by-products Inadequate Backup for key processes Equipment or unit processes Filter operation and backwash recycling Mixing of treatment chemicals and coagulants	Poor reliability of processes Power failures Sabotage and natural disasters Significant flow variations through water treatment systems Use of unapproved or contaminated water treatment chemicals and materials Failure of staff to respond appropriately to alarms or fluctuations in treatment processes
Receiving waters (reservoirs, rivers, and streams)	
Short-circuiting Forest fires and natural disasters Climatic and seasonal variations (e.g., heavy rainfall, drought) Cyanobacterial blooms Livestock access Inadequate buffer zones and vegetation	Inadequate storage (e.g., during winter or other times of low recycled water usage) Leakage from storage to groundwater Birds and vermin Accidental spillage from public roads Sabotage
Distribution systems	
Cross-connections with lower quality water or storages holding industrial chemicals Inadequate repair and maintenance, inadequate system flushing and reservoir cleaning Inappropriate materials and coatings or material failure	Biofilms, sloughing and resuspension or regrowth Formation of disinfection by-products Pipe bursts or leaks Sabotage and natural disasters
Users of drinking water	
Leaching of metals from piping and fittings Unauthorized plumbing work leading to cross-connections to lower quality water	Inadequate auditing and inspection of internal plumbing systems Use of inappropriate plumbing and construction materials

Source: NRMMC, EPHC, and NHMRC (2008), copyright 2008, with permission of EPHC.

Different communities and different stakeholders within communities will have varying views of any contentious initiative, including IPR. For many communities planning for the future, the role of IPR is untested. Hence, issues that may influence community response to IPR proposals are likely to include those associated with public health, population growth, environmental health, economy and finance, available technology, and emotional factors.

Community views of potable and nonpotable water reuse have been a topic of interest to social scientists since the early 1970s. In the United States, acceptance of potable reuse applications was measured at 48 percent in 1971, but this appeared to have fallen significantly to 26 percent (California) by 1979 and 29 percent (Denver) in 1985 (Bruvold, 1988). By 2002, support in the United States for direct potable reuse was reported at only 18 percent while support for indirect potable reuse was at 40 percent. The difference in support for the two approaches highlights important differences in perceptions of them. Furthermore, the measured degree of support reported for potable reuse schemes tends to be dependent on how the survey questions are stated. For example, questions about "potable water reuse policy" have consistently scored a more positive response than those about "intentions to drink the reclaimed water" (Marks, 2004).

During construction of the Western Corridor Recycled Water Project in Queensland, Australia, the state government commissioned some professional telephone polling to gauge support. The following questions were posed to 800 people (Galaxy Research, 2008):

> In February the first part of the government's water grid is due to come on line. This will include the recycling of wastewater in South East Queensland. Do you support or oppose the inclusion of purified recycled water in the new water grid? Would you support or oppose the inclusion of purified recycled water in the new water grid if it was only to be used as a back-up when dam levels dropped below 40 percent?

In response to these questions, 54 percent of people stated that they supported the addition of reclaimed water to the drinking water supply as a general strategy and an additional 28 percent stated that they supported the scheme as a backup measure (Galaxy Research, 2008). The remaining 18 percent opposed the use of reclaimed water for drinking water supplementation.

To fully understand community attitudes to IPR, it is necessary to consider instinctive and emotional responses that people have to human excrement and sewage issues. It has been illustrated that many people trust their own impressions of water quality (often based on cloudiness of the water) more than they trust medical and scientific evidence (Hartley, 2003). Cognitive factors such as the so-called law of contagion and the law of similarity may explain many of the less rational perceptions that people may have about water reuse (Haddad, 2004). The law of contagion suggests that once water has been in contact with contaminants, it can be psychologically very difficult for people to accept that it has been purified. The law of similarity suggests that the *appearance* of a substance's condition or status is psychologically linked to perceptions of reality. Combined, these factors can create mental barriers to the acceptance of reuse water as a source of pure water. However, the true significance of these cognitive factors is yet to be properly established. It should also be assumed that responses to reuse are also dependent on social and cultural contexts and debate, which are current in a particular social network.

Despite the challenges, a review of research studies has revealed encouraging data about community perceptions and attitudes toward water reuse in the United States (Hartley, 2003; Ruetten et al., 2004). It is anticipated that many of the findings will also apply to other countries. The findings highlight opportunities for effective communication between water reuse organizations and their stakeholders. They include

1. The community is interested in being meaningfully involved in water reuse decisions.
2. The community is interested in finding ways to ensure independent and secure water supplies for their communities.
3. While the community is not well versed in the water cycle, they are generally aware that there are water supply problems in many parts of the country.

4. The community believes that some form of potable reuse is inevitable, given growth and water supply constraints.
5. Information sharing, educational activities, and opportunities for reflection on the concepts of water reuse can increase support.

ABBREVIATIONS

ADI	acceptable daily intake
AOP	advanced oxidation processes
AWT	advanced water treatment
BNR	biological nutrient removal
CCP	critical control point
CDPH	California Department of Public Health
CWA	Clean Water Act
DALYs	disability-adjusted life years
DBPs	disinfection by-products
DOC	dissolved organic carbon
EDC	endocrine-disrupting compound
EfOM	effluent organic matter
GAC	granular activated carbon
IMS	integrated membrane system
IPR	indirect potable reuse
LAS	linear alkylbenzene sulphonates
LOEL	lowest dose resulting in an observable effect
MCL	maximum contaminant level
MCLG	maximum contaminant level goal
MF	microfiltration
mgd	million gallons per day
ML	megaliters
NF	nanofiltration
NOEL	no observed adverse effects level
NOM	natural organic matter
ntu	nephelometric turbidity unit
PAC	powdered activated carbon
PDF	probability density function
PDT	pressure decay test
PFCs	perfluorochemicals
PFOA	perfluorooctanoic acid
PFOS	perfluorooctane sulfonate
POD	point of departure

RBF	riverbank filtration
RO	reverse osmosis
SAT	soil-aquifer treatment
SAP	science advisory panel
SDI	silt density index
SDWA	Safe Drinking Water Act
SSMP	sewer discharge management plan
TDS	total dissolved solids
TOC	total organic carbon
UF	ultrafiltration
USEPA	U.S. Environmental Protection Agency
UV	ultraviolet irradiation
WHO	World Health Organization

REFERENCES

Amy, G., and Drewes, J.E. (2007) Soil-aquifer treatment (SAT) as a natural and sustainable wastewater reclamation/reuse technology: Fate of wastewater effluent organic matter (EfOM) and trace organic compounds. *Environmental Monitoring and Assessment,* 129 (1–3), 19–26.

Angelotti, R.W., Gallagher, T.M., Brooks, M.A., and Kulik, W. (2005) Use of Granular Activated Carbon as a Treatment Technology for Implementing Indirect Potable Reuse. Proceedings 20th Annual WateReuse Symposium. Denver, CO. Alexandria, VA: WateReuse Association.

Asano, T., Leverenz, H.L., Tsuchihashi, R., and Tchobanoglous, G. (2007) *Water Reuse*. New York: McGraw-Hill.

Barrington, D.J., and Ghadouani, A. (2008) Application of hydrogen peroxide for the removal of toxic cyanobacteria and other phytoplankton from wastewater. *Environ. Sci. Technol.*, 42, 8916–8921.

Bellona, C., Drewes, J.E., Xu, P., and Amy, G. (2004) Factors affecting the rejection of organic solutes during NF/RO treatment. *Wat. Res.* 38, 2795–2809.

Benotti, M.J., Trenholm, R., Vanderford, B.J., Holiday, J., Stanford, B.D., and Snyder, S.A. (2009) Pharmaceuticals and endrocrine disrupting compounds in U.S. drinking water. *Environ. Sci. Technol.,* 43 (3), 597–603.

Bruvold, W. H. (1988) Public opinion on water reuse options. *J. Water Poll. Control Fed.*, 60(1), 45–49.

Carletti, G., Fatone, F., Bolzonella, D., and Cecchi, F. (2008) Occurrence and fate of heavy metals in large wastewater treatment plants treating municipal and industrial wastewaters. *Wat. Sci. Technol.*, 57 (9), 1329–1336.

CDPH (2003) *Groundwater Recharge Regulation Draft*. Sacramento, CA: California Department of Public Health.

CDPH (2008) *Groundwater Recharge Regulation Draft*. Sacramento, CA: California Department of Public Health.

Costan-Longares, A., Montemayor, M., Payan, A., Mendez, J., Jofre, J., Mujeriego, R., and Lucena, F. (2008) Microbial indicators and pathogens: Removal, relationships and predictive capabilities in water reclamation facilities. *Wat. Res.,* 42 (17), 4439–4448.

Crook, J., Huiltquist, R., and Sakaji, R. (2000) New and improved draft groundwater recharge criteria in California. Proceedings AWWA 2000 Annual Conference, Denver, CO.

Daughton, C.G., and Ruhoy, I.S. (2009) Environmental footprint of pharmaceuticals: The significance of factors beyond direct excretion to sewers. *Environ. Toxicol. Chem.* (in press).

Davies, K. (2009) Construction completed on Australia's largest recycled water project. *International Desalination and Water Reuse Quarterly,* 19 (1), 17–21.

Deborde, M., and von Gunten, U. (2008) Reactions of chlorine with inorganic and organic compounds during water treatment—Kinetics and mechanisms: A critical review. *Wat. Res.,* 42 (1–2), 13–51.

Dickenson, E., Drewes, J.E., Sedlak, D., Wert, E., and Snyder, S. (2009) Applying surrogates and indicators to assess removal efficiency of trace organic chemicals during chemical oxidation of wastewater. *Environ. Sci. Techn.,* 43 (16), 6242–6247.

Drewes, J.E., and Fox, P. (2000) Effect of drinking water sources on reclaimed water quality in water reuse systems. *Wat. Environ. Res.*, 72 (3), 353–362.

Drewes, J.E., and Fox, P. (2001) Source Water Impact Model (SWIM)—A new planning tool for indirect potable water reuse systems. *Wat. Sci. Technol.,* 43 (10), 267–275.

Drewes, J.E., and Shore, L.S. (2001) Concerns about pharmaceuticals in water reuse, groundwater recharge, and animal waste. In *Pharmaceuticals and Personal Care Products in the Environment,* ed. Ch. Daughton and T.L. Jones-Lepp, pp. 206–228. American Chemical Society Symposium Series No. 791. Washington, DC.

Drewes, J. E., Quanrud, D., Amy, G., and Westerhoff, P. (2006) Character of organic matter in soil-aquifer treatment systems. *J. Environ. Engineer.,* 11, 1447–1458.

Drewes, J.E., Sedlak, D., Snyder, S., and Dickenson, E. (2008). *Development of Indicators and Surrogates for Chemical Contaminant Removal during Wastewater Treatment and Reclamation.* Final Report. Alexandria, VA: WateReuse Foundation.

Drewes, J.E., Dickenson, E., and Snyder, S. (2009). *Contributions of Household Chemicals to Sewage and their Relevance to Municipal Wastewater Systems and the Environment.* Final Report. Alexandria, VA: Water Environment Research Foundation.

Du Pisani, P. (2006) Direct reclamation of potable water at Windhoek's Goreangab reclamation plant. *Desalination,* 188, 79–88.

Eisenberg, D., Olivieri, A., Soller, J., and Gagliardo, P. (1998) Reliability analysis of an advanced water treatment facility. Proceedings of ASCE National Conference on Environmental Engineering. American Society of Civil Engineers.

Eisenberg, D., Soller, J., Sakaji, R., and Olivieri, A. (2001) A methodology to evaluate water and wastewater treatment plant reliability. *Wat. Sci. Technol.,* 43 (10), 91–99.

Ericson, I., Nadal, M., van Bavel, B., Lindström, G., and Domingo, J.L. (2008) Levels of perfluorochemicals in water samples from Catalonia, Spain: Is drinking water a significant contribution to human exposure? *Environ. Sci. Poll. Res.,* 15 (7), 614–619.

Fairfax Water (2009) *Annual Report on Water Quality.* Fairfax, VA.

Florida Department of Environmental Protection (1999) *Reuse of Reclaimed Water and Land Application.* Florida Administrative Code, pp. 62–610. Florida Department of Environmental Protection, Tallahassee, FL.

Florida Department of Environmental Protection (2006) *Reuse of Reclaimed Water and Land Application.* Florida Administrative Code, pp. 62–610. Tallahassee, FL.

Fono, L.J., Kolodziej, E.P., and Sedlak, D.L. (2006) Attenuation of wastewater-derived contaminants in an effluent-dominated river. *Environ. Sci. Technol.*, 40, 7257–7262.

Fox, P., Houston, S., Westerhoff, P., Drewes, J.E., Nellor, M., Yanko, W., Baird, R., Rincon, M., Arnold, R., Lansey, K., Bassett, R., Gerba, C., Karpiscak, M., Amy, G., and Reinhard, M. (2001) *Soil-Aquifer Treatment for Sustainable Water Reuse.* Awwa Research Foundation report. Denver, CO: Water Research Foundation.

Funari, E., and Testai, E. (2008) Human health risk assessment related to cyanotoxins exposure. *Crit. Rev. Tox.,* 38 (2), 97–125.

Furtado, A., Calijuri, M.D., Lorenzi, A.S., Honda, R.Y., Genuario, D.B. and Fiore, M.F. (2009) Morphological and molecular characterization of cyanobacteria from a Brazilian facultative wastewater stabilization pond and evaluation of microcystin production. *Hydrobiologia*, 627, 195–209.

Galaxy Research (2008) *Queensland Water Poll.* 26–27 November 2008, Glebe, New South Wales, AU.

Gasperi, J., Garnaud, S., Rocher, V., and Moilleron, R. (2008a) Priority pollutants in wastewater and combined sewer overflow. *Sci. Total Environ.*, 407 (1), 263–272.

Gasperi, J., Kafi-Benyahia, M., Lorgeoux, C., Moilleron, R., Gromaire, M. C. and Chebbo, G. (2008b) Wastewater quality and pollutant loads in combined sewers during dry weather periods. *Urban Water Journal,* 5 (4), 305–314.

Haas, C.N., and Trussell, R.R. (1998) Frameworks for assessing reliability of multiple, independent barriers in potable water reuse. *Wat. Sci. Technol.,* 38 (6), 1–8.

Haddad, B.M. (2004) *Research Needs Assessment Workshop: Human Reactions to Water Reuse.* Alexandria, VA: WateReuse Foundation.

Hartley, T.W. (2003) *Water Reuse: Understanding Public Perception and Participation.* Alexandria, VA: Water Environment Research Foundation.

Heidler, J., and Halden, R.U. (2008) Meta-analysis of mass balances examining chemical fate during wastewater treatment. *Environ. Sci. Technol.,* 42 (17), 6324–6332.

Higgins, C.P., and Luthy, R.G. (2006) Sorption of perfluorinated surfactants on sediments. *Environ. Sci. Technol.*, 40 (23), 7251–7256.

Hrudey, S.E., and Hrudey, E.J. (2007) Published Case Studies of Waterborne Disease Outbreaks–Evidence of a Recurrent Threat. *Water Environ. Res.*, 79 (3), 233–245.

Jobling, S., and Tyler, C.R. (2003) Endocrine disruption in wild freshwater fish. *Pure Appl. Chem.*, 75 (11–12), 2219–2234.

Khan, S.J. (2009) *Chemical Quantitative Exposure Assessment for Water Recycling Schemes.* Canberra, AU: National Water Commission.

Khan, S.J., and McDonald, J.A. (2009) Quantifying human exposure to contaminants for multiple-barrier water reuse systems. *Water Science & Technology* (i press).

Krasner, S., Westerhoff, P., Chen, B., Amy, G., Nam, S., Chowdhury, Z., Sinha, S., and Rittmann, B. (2008). *Contribution of Wastewater to DBP Formation.* Awwa Research Foundation report. Denver, CO: Water Research Foundation.

Leflaive, J., and Ten-Hage, L. (2007) Algal and cyanobacterial secondary metabolites in freshwaters: A comparison of allelopathic compounds and toxins. *Freshwater Biology,* 52 (2), 199–214.

Leslie, G.L., Mills, W.R., Wehner, M.P., Rigby, M.G., Dunivin, W.R., and Sudak, R.G. (1998) *Treatment costs for membrane processes in water reuse applications: A sensitivity analysis.* In American Desalting Association Biennial Conference & Exposition, Williamsburg, VA.

Leslie, G.L., Dawes, T.M., Snow, T.S., Mills, W.R., and Macintyre, D. (1999) Meeting the Demand for Potable Water in Orange County in the 21st Century: The Role of Membrane Processes. Proc. AWWA Membrane Technology Conference, Long Beach, CA.

Marks, J. (2004) Advancing community acceptance of reclaimed water. *Water,* 31 (5), 46–51.

Markus, M.R. (2009) Groundwater Replenishment and Water Reuse. The Water Report. *Water Rights, Water Quality and Water Solutions in the West,* 59, 1–9.

Medema, G., and Ashbolt, N. (2006) *QMRA: Its Value for Risk Management.* Report from MicroRisk—European Commission Fifth Framework Programme.

Moody, C.A., and J.A. Field (1999) Determination of perfluorocarboxylates in groundwater impacted by fire-fighting activity. *Environ. Sci. Techn.,* 33 (16), 2800–2806.

Moody, C.A., Hebert, G.N., Strauss, S.H., and Field, J.A. (2003) Occurrence and persistence of perfluorooctanesulfonate and other perfluorinated surfactants in groundwater at a fire-training area at Wurtsmith Air Force Base, Michigan. *J. Environ. Monitoring,* 5, 341–345.

Namkung, E., and Rittman, B.E. (1986) Soluble microbial products (SMP) formation kinetics by Biofilms. *Wat. Res.,* 20 (6), 795–806.

National Research Council (1998) *Issues in Potable Reuse: The Viability of Augmenting Drinking Water Supplies with Reclaimed Water*. Washington, DC: National Academies Press.

National Resource Management Ministerial Council and Environment Protection & Heritage Council (2006) *National Guidelines for Water Recycling: Managing Health & Environmental Risks (Phase 1)*. Government of Australia, Canberra, AU.

National Resource Management Ministerial Council, Environment Protection and Heritage Council and National Health and Medical Research Council (2008) *National Guidelines for Water Recycling: Managing Health and Environmental Risks (Phase 2)-Augmentation of Drinking Water Supplies*. Government of Australia, Canberra, AU.

National Water Quality Management Strategy (2004) *Australian Drinking Water Guidelines, National Health and Medical Research Council*. Natural Resource Management Ministerial Council. Government of Australia, Canberra, AU.

Nellor, M., Baird, R.B., and Smith, J.R. (1984) *Summary of Health Effects Study: Final Report*. County Sanitation Districts of Los Angeles County, Whittier, CA.

NEWater Expert Panel (2002) *Singapore Water Reclamation Study: Expert Panel Review and Findings*. Singapore.

Olivieri, A., Eisenberg, D., Soller, J., Eisenberg, J., Cooper, R., Tchobanoglous, G., Trussell, R., and Gagliardo, P. (1999) Estimation of pathogen removal in an advanced water treatment facility using Monte Carlo simulation. *Wat. Sci. Technol.*, 40 (4–5), 223–234.

Olivieri, A. W., Eisenberg, D. M., Cooper, R. C., Tchobanoglous, G., and Gagliardo, P. (1996) Recycled water—A source of potable water: City of San Diego health effects study. *Wat. Sci. Technol.*, 33 (10–11), 285–296.

Plumlee, M.H., Larabee, J., and Reinhard, M. (2008) Perfluorochemicals in water reuse. *Chemosphere*, 72, 1541–1547.

Pontius, F.W. (2003). *Drinking Water Regulation and Health*. New York: John Wiley.

Rauch, T., and Drewes, J.E. (2004). Assessing the removal potential of soil-aquifer treatment systems for bulk organic matter. *Wat. Sci. Technol.*, 50 (2), 245–253.

Rauch-Williams, T., and Drewes, J.E. (2006). Using soil biomass as an indicator for the biological removal of effluent-derived organic carbon during soil infiltration. *Wat. Res.*, 40, 961–968.

Rizak, S., and Hrudey, S.E. (2007) Achieving safe drinking water—Risk management based on experience and reality. *Environ. Rev.*, 15, 169–174.

Robeck, G.G., Cantor, K.P., and Christman, R.F. (1987) *Report of the Scientific Advisory Panel on Groundwater Recharge with Reclaimed Wastewater.* Prepared for State of California, State Water Resources Control Board, Department of Water Resources, Department of Health Services, Sacramento, CA.

Ruetten, J., Birkhoff, J., Darr, K., Drewes, J.E., Flatt. J., Kelso, D., McDaniel, M., and Noble, D. (2004) *Best Practices for Developing Indirect Potable Reuse Projects: Phase 1 Report*. Alexandria, VA: WateReuse Foundation.

Rule, K. L., Comber, S.D.W., Ross, D., Thornton, A., Makropoulos, C.K., and Rautiu, R. (2006) Diffuse sources of heavy metals entering an urban wastewater catchment. *Chemosphere*, 63 (1), 64–72.

Schmidt, W., Bornmann, K., Imhof, L., Mankiewicz, J., and Izydorczyk, K. (2008) Assessing drinking water treatment systems for safety against cyanotoxin breakthrough using maximum tolerable values. *Environ. Tox.*, 23 (3), 337–345.

Schultz, M.M., Barofsky, D.F., and Field, J.A. (2006) Quantitative determination of fluorinated alkyl substances by large–volume-injection liquid chromatography tandem mass spectrometry—Characterization of municipal wastewaters. *Environ. Sci. & Techn.*, 40 (1), 289–295.

Schwab, B.W., Hayes, E.P., Fiori, J.M., Mastrocco, F.J., Roden, N.M., Cragin, D., Meyerhoff, R.D., D'Aco, V.J., and Anderson, P.D. (2005) Human pharmaceuticals in US surface waters: A human health risk assessment. *Regul. Toxicol. Pharm.*, 42 (3), 296–312.

Sedlak, R. (1991) *Phosphorus and Nitrogen Removal from Municipal Wastewater: Principles and Practice*, 2d ed. New York: Lewis Publishers.

Shon, H.K., Vigneswaran, S., and Snyder, S.A. (2006) Effluent organic matter (EfOM) in wastewater: Constituents, effects, and treatment. *Crit. Rev. Environ. Sci. Tech.*, 36 (4), 327–374.

Shultz, D., and Parr, V. (1982) *Evaluation and Documentation of Mechanical Reliability of Conventional Wastewater Treatment Plant Components.* EPA/600/2-82-044, U.S. Environmental Protection Agency, Washington, DC.

Skutlarek, D., Exner, M., and Färber, H. (2006) Perfluorinated surfactants in surface and drinking waters. *Environ. Sci. Poll. Res.*, 13 (5), 299–307.

Sloss, E.M., McCaffrey, D.F., Fricker, R.D., Geschwind, S.A., and Ritz, B.R. (1999) *Groundwater Recharge with Reclaimed Water. Birth Outcomes in Los Angeles County, 1982–1993.* Prepared for the Water Replenishment District of Southern California, Rand, Santa Monica, CA.

Snyder, S.A. (2008) Occurrence, treatment, and toxicological relevance of EDCs and pharmaceuticals in water. *Ozone: Science and Engineering,* 30, 65–69.

Snyder, S.A., Westerhoff, P., Yoon, Y., and Sedlak D.L. (2003) Pharmaceuticals, personal care products and endocrine disrupters in water: Implications for water treatment. *Environ. Eng. Sci.*, 20, 449–469

Snyder, S.A., Wert, E., Rexing, D., Zegers, R., and Droury, D. (2006) Ozone oxidation of endocrine disruptors and pharmaceuticals in surface water and wastewater. *Ozone Sci. Eng.*, 28, 445–460.

Snyder, S.A., Trenholm, R., Snyder, E.M., Bruce, G.M., Pleus, R.C., and Hemming, J.D. (2008) *Toxicological relevance of EDCs and Pharmaceuticals in Drinking Water.* Awwa Research Foundation report. Denver, CO: Water Research Foundation.

Snyder, S.A., Benotti, M., Drewes, J.E., Dickenson, E., and Rosario, F. (2010) *Comparisons of Chemical Compositions of Reclaimed and Conventional Waters.* Final Report. Alexandria, VA: WateReuse Foundation.

State of California (2000) *California Code of Regulations, Title 22, Division 4, Environmental Health,* Chap. 3 Recycling Criteria. Sacramento, CA.

Sumpter, J.P., and Johnson, A.C. (2008) 10th Anniversary Perspective: Reflections on endocrine disruption in the aquatic environment: from known knowns to unknown unknowns (and many things in between). *J. Environ. Monitoring,* 10, 1476–1485.

Swayne, M.D., Boone, G.H., Bauer, D., and Lee, J.S. (1980) *Wastewater in Receiving Waters at Water Supply Abstraction Points.* Final report. Aug 77–Jan 79. Cincinnati: Sponsor: Municipal Environmental Research Laboratory.

Takagi, S., Adachi, F., Miyano, K., Koizumi, Y., Tanaka, H., and Mimura, M. (2008) Perfluorooctanesulfonate and perfluorooctanoate in raw and treated tap water from Osaka, Japan. *Chemosphere,* 72 (10), 1409–1412.

Tanaka, H., Asano, T., Schroeder, E.D., and Tchobanoglous, G. (1998) Estimating the safety of wastewater reclamation and reuse using enteric virus monitoring data. *Wat. Environ. Res.,* 70 (1), 39–51.

Tang, C.Y.Y., Fu, Q.S., Robertson, A.P., Criddle, C.S., and Leckie, J.O. (2006) Use of reverse osmosis membranes to remove perfluorooctane sulfonate (PFOS) from semiconductor wastewater. *Environ. Sci. Technol.*, 40 (23), 7343–7349.

Tang, C.Y., Fu, Q.S., Criddle, C.S., and Leckie, J.O. (2007) Effect of flux (transmembrane pressure) and membrane properties on fouling and rejection of reverse osmosis and nanofiltration membranes treating perfluorooctane sulfonate containing wastewater. *Environ. Sci. Techn.,* 41 (6), 2008–2014.

Tchobanoglous, G., Burton, F.L., and Stensel, H.D. (2003) *Wastewater Engineering: Treatment and Reuse*. Boston: McGraw-Hill.

Ternes, T., and Joss, A. (eds.) (2006) *Human Pharmaceuticals, Hormones and Fragrances: The Challenge of Micropollutants in Urban Water Management.* London: IWA Publishing.

Tortajada, C. (2006) Water management in Singapore International. *J. Wat. Res. Develop.,* 22 (2), 227–240.

Trenholm, B., Vanderford, B.J., Drewes, J.E., and Snyder, S.A. (2008) Determination of household chemicals using gas chromatography and liquid chromatography with tandem mass spectroscopy. *J. Chromatog. A.,* 1190, 253–262.

USEPA (1997) *Guiding Principles for Monte Carlo Analysis.* Washington, DC.

USEPA (1991) National Secondary Drinking Water Regulations. *Federal Register,* 40, 143. Washington, DC.

USEPA (2007) *Unregulated Contaminant Monitoring Regulation (UCMR) for Public Water Systems Revision. Federal Register* 72, 2, 367–398. Washington, DC.

USEPA (2008) *Drinking Water Contaminant Candidate List 3—Draft. Federal Register* 73 (35), 9627–9664. Washington, DC.

USEPA (2009) *Provisional Health Advisories for Perfluorooctanoic Acid (PFOA) and Perfluorooctane Sulfonate (PFOS).* Washington, DC.

Van den Berg, M., Birnbaum, L.S., Denison, M., De Vito, M., Farland, W., Feeley, M., Fiedler, H., Hakansson, H., Hanberg, A., Haws, L., Rose, M., Safe, S., Schrenk, D., Tohyama, C., Tritscher, A., Tuomisto, J., Tysklind, M., Walker, N., and Peterson, R.E. (2006) The 2005 World Health Organization reevaluation of human and mammalian toxic equivalency factors for dioxins and dioxin-like compounds. *Toxicol Sci.,* 93 (2), 223–241.

Velz, C. (1970). *Applied Stream Sanitation.* New York: John Wiley.

Washington Department of Health, Washington Department of Ecology (1997) *Water Reclamation and Reuse Standards.* Publication #97-23. Olympia, WA.

WEF/AWWA (2008). *Using Reclaimed Water to Augment Potable Water Resources.* 2d ed. Alexandria, VA: WEF, and Denver, CO: American Water Works Association.

Wert, E., Rosario-Ortiz, F., and Snyder, S. (2009). Using UV absorbance and color to assess the oxidation of pharmaceuticals during the ozonation of wastewater. *Environ. Sci. Techn.,* 43 (13), 4858–4863.

Western Consortium for Public Health (1997) *The City of San Diego—Total Resource Recovery Project*, San Diego, CA.

Withers, A. (2005) Options for recarbonation, remineralisation and disinfection for desalination plants. *Desalination*, 179 (1–3), 11–24.

World Health Organization (2004) *Guidelines for Drinking-water Quality.* Geneva, Sui.

CHAPTER 17
CHEMICAL DISINFECTION

Charles N. Haas, Ph.D.
LD Betz Professor of Environmental Engineering
Drexel University
Philadelphia, Pennsylvania, United States

INTRODUCTION **17.1**
HISTORY OF DISINFECTION **17.3**
Chlorine 17.3
Chlorine Dioxide 17.3
Ozone 17.3
UV Radiation 17.4
Other Agents 17.4
REGULATORY ISSUES FOR DISINFECTION **17.4**
DISINFECTANTS AND THEORY OF DISINFECTION **17.5**
Basic Chemistry 17.5
Disinfectant Demand Reactions 17.10
ASSESSMENT OF MICROBIAL QUALITY (INDICATORS) **17.19**
Pathogens of Concern 17.21
DISINFECTION KINETICS **17.21**
Chick's Law and Elaborations 17.21
Hom's Law 17.24
Power Law Models 17.26
Models Accounting for Disinfectant Decay 17.26
Impact of Temperature and Other Water Chemistry Factors 17.27
MODE OF ACTION OF DISINFECTANTS **17.27**
Chlorine 17.27
Chlorine Dioxide 17.28
Ozone 17.28
UV Radiation 17.29
Influence of Physical Factors on Disinfection Efficiency 17.30
Influence of Physiological Factors on Disinfection Efficiency 17.30
DISINFECTANT RESIDUALS FOR POSTTREATMENT PROTECTION **17.30**
DESIGN AND APPLICATION OF TECHNOLOGIES **17.31**
Chlorination 17.31
Chlorine Dioxide 17.34
Ozonation 17.36
UV Radiation 17.37
From Kinetics to Process Design 17.37
RELATIVE COMPARISONS **17.41**
ABBREVIATIONS **17.42**
NOTATION FOR EQUATIONS **17.43**
REFERENCES **17.43**

INTRODUCTION

Disinfection is a process designed for the deliberate reduction of the number of pathogenic microorganisms. While other water treatment processes, such as filtration or coagulation-flocculation-sedimentation, may achieve pathogen reduction, this is not generally their primary goal. A variety of chemical or physical agents may be used to carry out disinfection.

The concept of disinfection preceded the recognition of bacteria as the causative agent of disease. Averill (1832), for example, proposed chlorine disinfection of human wastes as a prophylaxis against epidemics in 1832. Chemical addition during water treatment for disinfection became accepted only after litigation on its efficacy (Race, 1918). The prophylactic benefits of water disinfection soon became apparent, particularly with respect to the reduction of typhoid and cholera.

Although significant progress is being made in controlling the classic water-borne diseases, newly recognized agents have added to the challenge. These include viruses (Melnick et al., 1978; Mosley, 1966); certain bacteria, for example, *Campylobacter* (Palmer et al., 1983), *Yersinia* (Brennhovd et al., 1992; Reasoner, 1991) or *Mycobacterium* (Geldreich, 1971; Iivanainen et al., 1993; Reasoner, 1991); and protozoans such as *Giardia* (Brown et al., 1992; Lechevallier et al., 1991; Miller, 1992; Reasoner, 1991; Renton et al., 1996; Rose et al., 1991) and *Cryptosporidium* (Bridgman et al., 1995; Centers for Disease Control and Prevention, 1995; Gallaher et al., 1989; Goldstein et al., 1996; Hayes et al., 1989; Lechevallier et al., 1991; Leland et al., 1993; MacKenzie et al., 1994; Miller, 1992; Reasoner, 1991; Richardson et al., 1991; Rose et al., 1991; Rush et al., 1990; Smith, 1992). Occasional outbreaks of hepatitis associated with drinking water have also occurred (Nasser, 1994; Rosenberg et al., 1980). In addition, new water-borne pathogens are continually being identified, *E. coli* O157:H7 and noroviruses being two examples. There has been a progression of revolutions in disinfection challenges over the history of drinking water disinfection, and what is sure is that there will continue to be more challenges (McGuire, 2006).

Many water utilities have moved to the use of multiple disinfectants (rather than simply chlorination), switched to combined residual chlorination, or abandoned chlorine/chloramines entirely. This is shown by comparing surveys taken in 1998 and 2007 by the AWWA Disinfection Committee (Table 17-1) (AWWA Disinfection Systems Committee, 2008; AWWA Water Quality Division Disinfection Systems Committee, 2000). This chapter covers the use of chlorine, as well as the major alternative chemical agents, for the purpose of disinfection. Specifically, history, chemistry, kinetics, and design implications will be presented. Chapter 18 of this book covers the application of ultraviolet light for disinfection.

TABLE 17-1 Water Utility Disinfection Practices According to 1998 and 2007 AWWA Surveys

Chemical	Percentage of utilities using	
	1998	2007
Chlorine gas	70	61
Chloramines	11	30
Sodium hypochlorite	22	31
Onsite generation of hypochlorite	2	8
Calcium hypochlorite	4	8
Chlorine dioxide	4	8
Ozone	2	9
UV	0	2

Notes:

1. Numbers sum to more than 100% because of multiple disinfectant use at a given facility.
2. Since this survey, there has been a reduction in the use of chlorine gas and an increased use of sodium hypochlorite, but quantitative survey data are not available.

Source: AWWA Disinfection Systems Committee, 2008; AWWA Water Quality Division Disinfection Systems Committee, 2000.

HISTORY OF DISINFECTION

Chlorine

Chlorine gas was first prepared by Scheele in 1774, but chlorine was not regarded as a chemical element until 1808 (Belohlav and McBee, 1966). Early uses of chlorine include the use of Javelle water (chlorine gas dissolved in an alkaline potassium solution) in France for waste treatment in 1825 (Baker, 1926) and its use as a prophylactic agent during the European cholera epidemic of 1831 (Belohlav and McBee, 1966).

Disinfection of water by chlorine was first done in 1908 at Bubbly Creek (Chicago) and by the Jersey City Water Company. Within 2 years, chlorine was introduced as a disinfectant at New York City (Croton), Montreal, Milwaukee, Cleveland, Nashville, Baltimore, and Cincinnati, as well as at other smaller cities. Frequently, dramatic reductions in typhoid accompanied the introduction of this process (Hooker, 1913). By 1918, over 1000 cities, treating more than 1.1×10^7 m^3/d (3 billion gpd) of water, were employing chlorine as a disinfectant (Race, 1918).

Chloramination, the addition of both chlorine and ammonia, either sequentially, or simultaneously, was first employed in 1917 at Ottawa, Ontario, and Denver, Colorado. Both of these early applications employed prereaction of the two chemicals prior to their addition to the full flow of water. Sometime later, preammoniation (the addition of ammonia prior to chlorine) was developed. In both cases, the process was advocated for its ability to prolong the stability of residual disinfectant during distribution and to diminish the disinfectant's propensity to produce chlorophenolic taste and odor substances. Shortages of ammonia during World War II and recognition of the superiority of free chlorine as a disinfectant reduced the popularity of the chloramination process. Concerns for organic by-products of chlorination, for example, trihalomethanes and haloacetic acids, however, have increased the popularity of chloramination (Wolfe et al., 1984), although chloramines have been found to produce other classes of by-products (Mitch et al., 2003).

Chlorine Dioxide

Chlorine dioxide was first produced from the reaction of potassium chlorate and hydrochloric acid by Davy in 1811 (Miller et al., 1978). However, not until the industrial scale preparation of sodium chlorite, from which chlorine dioxide can more readily be generated, did its widespread use occur (Rapson, 1966).

Chlorine dioxide has been used widely as a bleaching agent in pulp and paper manufacture (Rapson, 1966). Despite early investigations on the use of chlorine dioxide as an oxidant and disinfectant (Aston and Synan, 1948), however, its ascendancy in both water and wastewater treatment has been slow, with no evidence of its use even in the early 1970s. (Morris, 1971).

In 1998, 8.1 percent of medium and large utilities reported using chlorine dioxide as a disinfectant (perhaps intermittently or in combination with other disinfectants), up from 4.5 percent in 1986 (AWWA Water Quality Division Disinfection Systems Committee, 2000). The increase in use has occurred because of the potency of ClO_2 as a disinfectant combined with the lack of formation of organo-chlorine disinfection by-products produced by pure chlorine dioxide.

Ozone

Ozone was discovered in 1783 by Van Marum, and named by Schonbein in 1840. In 1857, the first electric discharge ozone generation device was constructed by Siemens, with the first commercial application of this device occurring in 1893 (Water Pollution Control Federation, 1984).

Ozone was first applied as a potable water disinfectant in 1893 at Oudshoorn, Netherlands. In 1906, Nice, France, installed ozone as a treatment process, and this plant represents the oldest ozonation installation in continual operation (Rice et al., 1981). In the United States, ozone was first employed for taste and odor control at the New York City Jerome Park Reservoir in 1906. In 1998, 5.6 percent of medium or large utilities reported using ozone as a disinfectant, up from 0.37 percent in 1989 (AWWA Water Quality Division Disinfection Systems Committee, 2000).

UV Radiation

The biocidal effects of ultraviolet radiation (UV) have been known since it was established that short-wavelength UV was responsible for microbial decay often associated with sunlight (Downes and Blount, 1877). By the early 1940s, design guidelines for UV disinfection were proposed (Huff et al., 1965). UV has been accepted for treating potable water on passenger ships (Huff et al., 1965). Until recently, however, it has met with little enthusiasm in public water supply applications because of the lack of a residual following application. With findings that UV is highly effective against protozoan pathogens, a strong interest in the use of UV as a primary disinfectant in water treatment has occurred (Hijnen et al., 2006). This is seen in the recent (2007) AWWA Survey (Table 17-1).

Other Agents

A variety of other agents may be used to effect inactivation of microorganisms. These include heat, extremes in pH, metals (silver, copper), surfactants, permanganate, and electron beam irradiation. Heat is useful only in emergencies, as in boil-water orders, and is uneconomical. An alkaline pH (during high lime softening) may provide some microbial inactivation, but is not usually sufficient as a sole disinfectant. Potassium permanganate has been reported to achieve some disinfecting effects; however, its efficacy has not been well characterized. High-energy electrons for disinfection of wastewaters and sludges have also been studied (Farooq et al., 1993); however, their feasibility in drinking water is uncertain. In this chapter, therefore, primary consideration is given to chlorine compounds, including chlorine dioxide. In water treatment (as discussed in Chaps. 7 and 18), combinations of ultraviolet radiation and oxidizing agents (typically ozone or hydrogen peroxide) are used in advanced oxidation processes. While the free radicals generated in such processes may be effective at reducing concentrations of certain organic compounds, these free radicals are not highly effective disinfectants, however the individual processes (UV radiation, ozonation) used in such processes by themselves may provide inactivation. More detailed discussions of ozone and UV appear in Chaps. 7 and 18, respectively.

REGULATORY ISSUES FOR DISINFECTION

Amendments to the Safe Drinking Water Act require that all surface water suppliers in the United States filter and/or disinfect to protect the health of their customers. The filtration and disinfection treatment requirements for public water systems using surface water sources or groundwater under the direct influence of surface water are included in what is called the Surface Water Treatment Rule (USEPA, 1991). See Chap. 1 for additional information.

The Surface Water Treatment Rule (SWTR) requires that all surface water treatment facilities (except those meeting specific exemption criteria) provide filtration and disinfection that achieves at least (1) a 99.9 percent (3-log) removal-inactivation of *Giardia lamblia* cysts and (2) a 99.99 percent (4-log) removal-inactivation of enteric viruses. The SWTR credits a treatment plant having effective conventional treatment including filtration with achieving 2.5-log removal of *Giardia* and a 2-log removal of viruses. Disinfection is required for the remainder of the removal-inactivation.

The amount of disinfection credit to be awarded is determined with the $C \times T_{10}$ concept is defined as the residual disinfectant concentration (C, mg/L) multiplied by the contact time (T_{10}, min) for the fastest 10 percent of the flow; T_{10} can be estimated from the mean hydraulic residence time between the point of disinfectant application and the point of residual measurement and a baffling factor (varying between 0.1 and 1) describing the hydraulic characteristics of the contactor. The SWTR Guidance Manual (USEPA, 1991) provides tables of $C \times T_{10}$ values for several disinfectants achieving particular degrees of inactivation of *Giardia* and viruses. A large safety factor was incorporated into the $C \times T_{10}$ values included in the Guidance Manual tables. In addition to relying on the $C \times T_{10}$ tables to calculate disinfection credit, the SWTR allows utilities to demonstrate the ability of their disinfection systems to inactivate microorganisms by conducting pilot-scale studies, which may be prohibitively expensive for smaller operations. The SWTR was revised to take into account knowledge developed since the mid 1980s, including the need to achieve inactivation of *Cryptosporidium*. This resulted in the Long Term 2 Enhanced Surface Water Treatment Rule (LT2ESWTR) in 2006. A more complete discussion of regulatory requirements is included in Chap. 1.

Along with disinfection requirements, explicit regulations have been imposed on disinfection by-products since 1974, first with respect to trihalomethanes and more recently with respect to haloacetic acids, bromate, and other possible by-products. The combination of the requirement to achieve disinfection along with the requirement to minimize disinfection by-products has led to an increasing spectrum of options being considered. For more information on either by-product or disinfection regulations, Chaps. 1 and 19 should be consulted.

DISINFECTANTS AND THEORY OF DISINFECTION

Basic Chemistry

Chlorine and Chlorine Compounds. Chlorine may be used as a disinfectant in the form of compressed gas under pressure which is dissolved in water at the point of application, solutions of sodium hypochlorite, or solid calcium hypochlorite. The three forms are chemically equivalent because of the rapid equilibrium that exists between dissolved molecular gas and the dissociation products of hypochlorite compounds.

Molecular chlorine (Cl_2) is a dense gas that, when subject to pressures in excess of its vapor pressure, condenses into a liquid with the release of heat and with a reduction in specific volume of approximately 450-fold. Hence, commercial shipments of chlorine are made in pressurized tanks to reduce shipment volume. When chlorine is to be dispensed as a gas, supplying thermal energy to vaporize the compressed liquid chlorine is necessary. Due to safety and security concerns, the prevalence of use of gaseous chlorine has been decreasing.

The relative amount of chlorine present in chlorine gas, or hypochlorite salts, is expressed in terms of available chlorine. The concentration of hypochlorite (or any other oxidizing disinfectant) may be expressed as available chlorine by determining the electrochemical equivalent amount of Cl_2 to that compound. By Eq. 17-1, 1 mole of molecular chlorine is capable of reacting with two electrons to form inert chloride.

$$Cl_2 + 2e^- = 2Cl^- \tag{17-1}$$

From Eq. 17-2, 1 mole of hypochlorite (OCl^-) may react with two electrons to form chloride.

$$OCl^- + 2e^- + 2H^+ = Cl^- + H_2O \tag{17-2}$$

Hence, 1 mole of hypochlorite is electrochemically equivalent to 1 mole of molecular chlorine and may be said to contain 70.91 g of available chlorine (identical to the molecular weight of Cl_2).

Calcium hypochlorite ($Ca(OCl)_2$) and sodium hypochlorite (NaOCl) contain 2 moles and 1 mole of hypochlorite per mole of chemical, respectively, and, as a result, 141.8 g and 70.91 g of available chlorine per mole, respectively. The molecular weights of $Ca(OCl)_2$ and NaOCl are, respectively, 143 and 74.5, so that pure preparations of the two compounds contain 99.2 and 95.8 weight percent of available chlorine; hence, they are effective means of supplying chlorine for disinfection purposes. Calcium hypochlorite is available commercially as a dry solid. In this form, it is subject to a loss in strength of approximately 0.013 percent per day (Laubusch, 1963). Calcium hypochlorite is also available in a tablet form, for use in automatic feed equipment at low-flow treatment plants. Sodium hypochlorite is available in 1 to 16 weight percent solutions. Higher concentration solutions are not practical because chemical stability rapidly diminishes with increasing strength. At 20°C to 25°C, the half-life of sodium hypochlorite solutions varies between 60 and 1700 days, respectively, for solutions of 18 and 3 percent available chlorine (Baker, 1969; Laubusch, 1963).

It is noted that the loss of strength in sodium hypochlorite solutions may also result in the formation of by-products that may be undesirable. Thermodynamically, the autodecompositon of hypochlorite to chlorate is highly favored by the following overall process (Bolyard et al., 1992):

$$3ClO^- \rightarrow 2Cl^- + ClO_3^- \tag{17-3}$$

Measurements of sodium hypochlorite disinfectant solutions at water utilities have revealed that the mass concentration of chlorate is from 1.7 to 220 percent of the mass concentration of free available chlorine (Bolyard et al., 1993; Bolyard et al., 1992). The concentration of chlorate present in these stock solutions is kinetically controlled and may be related to the solution strength, age, temperature, pH, and presence of metal catalysts (Gordon et al., 1995; Gordon et al., 1993).

When a chlorine-containing compound is added to water containing insignificant quantities of Kjeldahl nitrogen, organic material, and other chlorine-demanding substances, a rapid equilibrium is established among the various chemical species in solution. The term *free available chlorine* is used to refer to the sum of the concentrations of molecular chlorine (Cl_2), hypochlorous acid (HOCl), and hypochlorite ion (OCl^-), each expressed as available chlorine.

The dissolution of gaseous chlorine to form dissolved molecular chlorine is expressible as a phase equilibrium, and is called *Henry's law.*

$$Cl_2(g) = Cl_2(aq)\ H^*\ (\text{mol/L-atm}) = [Cl_2(aq)]/P_{Cl2} \tag{17-4}$$

In Eq. 17-4, quantities within square brackets represent molar concentrations, P_{Cl2} is the gas phase partial pressure of chlorine in atmospheres, and H^* is the Henry's law constant, estimated from the following equation (Downs and Adams, 1973):

$$H^* = 4.805 \times 10^{-6} \exp(2818.48/T)\ (\text{mol/L} - \text{atm}) \tag{17-5}$$

where T is the temperature in Kelvins (K = °C + 273).

Dissolved aqueous chlorine reacts with water to form hypochlorous acid, chloride ions, and protons (Downs and Adams, 1973) as indicated by Eq. 17-6.

$$Cl_2(aq) + H_2O = H^+ + HOCl + Cl^-$$

$$K_H = \frac{[H^+][HOCl][Cl^-]}{[Cl_2(aq)]} = 2.581\exp\left(-\frac{2581.93}{T}\right)(\text{mol}^2/\text{L}^2) \tag{17-6}$$

This reaction typically reaches completion in 100 msec (Aieta and Roberts, 1985) and involves elementary reactions between dissolved molecular chlorine and hydroxyl ions. The extent of chlorine hydrolysis, or disproportionation (because the oxidation state of chlorine changes from 0 on the left to +*I* and –*I* on the right), as described by Eq. 17-6, decreases with decreasing pH and increasing salinity; hence, the solubility of gaseous chlorine can be increased by the addition of alkali or by the use of fresh, rather than brackish, water.

Hypochlorous acid is a weak acid and may dissociate according to the following reaction:

$$HOCl = OCl^- + H^+$$

$$K_a = [OCl^-]\,[H^+]/[HOCl] \tag{17-7}$$

The pK_a of hypochlorous acid at 20°C is approximately 7.6. Morris (1966) has provided a correlating equation for *Ka* as a function of temperature.

$$\ln(K_a) = 23.184 - 0.0583\,T - 6908/T \tag{17-8}$$

Figure 17-1 illustrates the effect of pH on the distribution of free chlorine between OCl^- and HOCl. At a pH < the pK_a, HOCl is the dominant species, while for pH > pK_a, OCl^- is the dominant species.

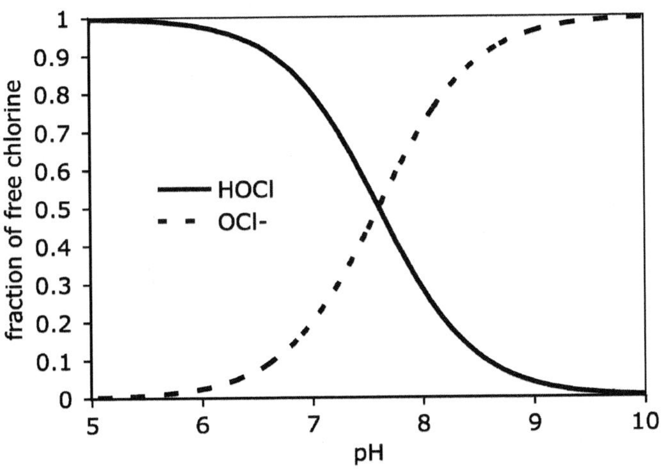

FIGURE 17-1 Effect of pH on relative amount of hypochlorous acid and hypochlorite ion at 20°C.

One practical consequence of the reactions described by Eqs. 17-4 to 17-8 is that the chlorine vapor pressure over a solution depends on solution pH, decreasing as pH increases (because of the increased formation of nonvolatile hypochlorite acid). Therefore, the addition of an alkaline material such as lime or sodium bicarbonate will reduce the volatility of chlorine from accidental spills or leaks and minimize danger to exposed personnel.

The acid-base properties of gaseous chlorine, or the hypochlorite salts, will also result in a loss or gain, respectively, of alkalinity, and also a reduction or increase, respectively, in pH. For each mole of free chlorine (i.e., 1 mole of Cl_2 or of NaOCl or 1/2 mole of $Ca(OCl)_2$), there will be a change of one equivalent of alkalinity (increase for sodium and calcium hypochlorite and decrease for chlorine gas).

Example 17-1 The solution produced by a gas chlorinator contains 3500 mg/L available chlorine at a pH of 3. What is the equilibrium vapor pressure of this solution at 20°C (given that the value of the hydrolysis constant, K_H, is 1.49×10^4 at this temperature)?

Solution

1. The pH is sufficiently low so that the dissociation of hypochlorous acid to form hypochlorite can be ignored (Fig. 17-1). Therefore, a balance over chlorine species yields

$$[Cl_2] + [HOCl] = (3500 \times 10^{-3})/71$$

The factor of 71 reflects the fact that one mole of either dissolved chlorine or hypochlorous acid contains 71 g of available chlorine.

2. The hydrolysis equilibrium constant can be used to develop an additional equation.

$$1.49 \times 10^4 = [H^+][Cl^-][HOCl]/[Cl_2]$$

or, because the pH is given,

$$1.49 \times 10^7 = [Cl^-][HOCl]/[Cl_2]$$

3. Because chlorine gas was used to generate the dissolved free chlorine, the disproportionation reaction requires that for each mole of HOCl produced 1 mole of Cl^- must have been produced. If the initial concentration of chloride (in the feed water to the chlorinator) was minimal, then a third equation results.

$$[Cl^-] = [HOCl]$$

4. These three equations can be manipulated to produce a quadratic equation in the unknown $[Cl_2]^{1/2}$:

$$[Cl_2] + 3860[Cl_2]^{1/2} - 0.049 = 0$$

The single positive root is the only physically meaningful one; hence,

$$[Cl_2]^{1/2} = 1.27 \times 10^{-5}$$

or

$$[Cl_2] = 1.6 \times 10^{-10}$$

5. The Henry's law constant can be computed from Eq. 17-5 as 0.072 moles/L-atm, and therefore the partial pressure of chlorine gas is found.

$$P_{Cl} = 1.6 \times 10^{-10}/0.072 = 2.23 \times 10^{-9} \text{ atm}$$
$$= (2.23 \text{ ppb})$$

The OSHA permissible exposure limit (PEL) is reported as 1 ppm (ACGIH, 1994). Therefore, this level is of no apparent health concern to the workers. ▲

Chlorine Dioxide. Chlorine dioxide (ClO_2) is a neutral compound of chlorine in the +IV oxidation state. It has a boiling point of 11°C at atmospheric pressure. The liquid is denser than water and the gas is denser than air (Noack and Doeff, 1979). Unlike Cl_2, chlorine dioxide does not hydrolyze in water and remains as a dissolved gas.

Chemically, chlorine dioxide is a stable free radical that, at high concentrations, reacts violently with reducing agents. It is explosive, with the lower explosive limit in air variously reported as 10 percent (Masschelein, 1979b) or 39 percent (Noack and Doeff, 1979). As a result, virtually all applications of chlorine dioxide require the synthesis of the gaseous compound in a dilute stream (either gaseous or liquid) on location as needed.

The solubility of gaseous chlorine dioxide in water may be described by Henry's law, and a fit of the available solubility data (Battino, 1984) results in the following relationship for the Henry's law constant (in units of atm^{-1}):

$$\begin{aligned}\ln(H) &= \text{mole fraction dissolved } ClO_2(aq)/P_{ClO2} \\ &= 58.84621 + (47.9133/T) - 11.0593\ln(T)\end{aligned} \tag{17-9}$$

Under alkaline conditions, the following disproportionation into chlorite (ClO_2^-) and chlorate (ClO_3^-) occurs (Gordon et al., 1972):

$$2ClO_2 + 2OH^- = H_2O + ClO_3^- + ClO_2^- \tag{17-10}$$

In the absence of catalysis by carbonate, the reaction (Eq. 17-10) is governed by parallel first- and second-order kinetics (Gordon et al., 1972; Granstrom and Lee, 1957). The half-life of aqueous chlorine dioxide solutions decreases substantially with increasing concentration and with pH > 9. Even at neutral pH values, however, in the absence of carbonate at room temperature, the half-life of chlorine dioxide solutions of 0.01, 0.001, and 0.0001 mol/L is 0.5, 4, and 14 h, respectively. Hence, the storage of stock solutions of chlorine dioxide for even a few hours is impractical. For research purposes, acidified solutions in laboratory pure water may be held under refrigeration without appreciable loss for substantially longer (possibly up to months).

The simple disproportionation reaction to chlorate and chlorite is insufficient to explain the decay of chlorine dioxide in water free of extraneous reductants. Equation 17-10 predicts that the molar ratio of chlorate to chlorite formed should be 1:1. Medir and Giralt (1982), however, found that the molar ratio of chlorate:chlorite:chloride:oxygen produced was 5:3:1:.75, and that the addition of chloride enhanced the rate of decomposition and resulted in the predicted 1:1 molar ratio of chlorite:chlorate. Thus, the oxidation of chloride by chlorate and the possible formation of intermediate free chlorine may be of significance in the decay of chlorine dioxide in demand-free systems (Gordon et al., 1972), conceivably in some circumstances in water distribution.

The concentration of chlorine dioxide in solution is generally expressed in terms of g/L as chlorine by multiplying the molarity of chlorine dioxide by the number of electrons transferred per mole of chlorine dioxide reacted and then multiplying this by 35.5 g Cl_2 per electron mole. Conventionally, the five-electron reduction (Eq. 17-11) is used to carry out this conversion.

$$ClO_2 + 5e^- + 4H^+ = Cl^- + 2H_2O \tag{17-11}$$

Note, however, that the typical reaction of chlorine dioxide in water, being reduced to chlorite, is a one-electron reduction as follows:

$$ClO_2 + e^- \Leftrightarrow ClO_2^- \tag{17-12}$$

Hence, according to Eq. 17-11, 1 mole of chlorine dioxide contains 67.5 g of mass, and is equivalent to 177.5 (= 5 × 35.5) g Cl_2. Therefore, 1 g of chlorine dioxide contains 2.63 g as chlorine. In examining any study on chlorine dioxide, due care with regard to units of expression of disinfectant concentration is warranted.

Ozone. Ozone is a colorless gas, produced from the action of electric fields on oxygen. It is highly unstable in the gas phase; in clean vessels at room temperature the half-life in air is 20 to 100 hours (Manley and Niegowski, 1967). Hence it must be produced on a continuous basis. Details of the chemistry and production of ozone are discussed in Chap. 7.

Disinfectant Demand Reactions

Chlorine

Reactions with Ammonia. In the presence of certain dissolved constituents in water, each of the disinfectants may react and transform to less biocidal chemical forms. In the case of chlorine, these principally involve reactions with ammonia and amino nitrogen compounds. This typically occurs in drinking water treatment when ammonia is added along with, immediately before, or after the addition of chlorine or hypochlorites. In the presence of ammonia nitrogen ion, free chlorine reacts in a stepwise manner to form chloramines (Jafvert and Valentine, 1992). This process is depicted in Eqs. 17-13 through 17-15.

$$NH_3 + HOCl = NH_2Cl + H_2O \qquad (17\text{-}13)$$

$$NH_2Cl + HOCl = NHCl_2 + H_2O \qquad (17\text{-}14)$$

$$NHCl_2 + HOCl = NCl_3 + H_2O \qquad (17\text{-}15)$$

These compounds, monochloramine (NH_2Cl), dichloramine ($NHCl_2$), and trichloramine (NCl_3), each contribute to the total (or combined) chlorine residual in a water. The terms *total available chlorine* or *total oxidants* refer, respectively, to the sum of free chlorine compounds and reactive chloramines, or the total oxidating agents. Under normal conditions of water treatment, if any excess ammonia is present, at equilibrium the amount of free chlorine is much less than 1 percent of total residual chlorine. Each chlorine atom associated with a chloramine molecule is capable of undergoing a two-electron reduction to chloride; hence, each mole of monochloramine contains 71 g available chlorine; each mole of dichloramine contains 2 × 71 or 142 g; and each mole of trichloramine contains 3 × 71, or 223, g of available chlorine. Inasmuch as the molecular weights of mono-, di-, and trichloramine are 51.6, 86, and 110.5, respectively, the chloramines contain, respectively, 1.38, 1.65, and 2.02 g of available chlorine per gram. The efficiency of the various combined chlorine forms as disinfectants differs, however, and thus, the concentration of available chlorine does not completely characterize process performance. On an approximate basis, for example, for coliforms, the biocidal potency of $HOCl{:}OCl{:}NH_2Cl{:}NHCl_2$ is approximately 1:0.0125:0.005:0.0166, and for viruses and cysts, the combined chlorine forms are considerably less effective (Chang, 1971). As the reaction in Eq. 17-13 indicates, the formation of monochloramine is accompanied by the loss of a proton, because chlorination reduces the affinity of the nitrogen moiety for protons (Weil and Morris, 1949a).

The significance of chlorine speciation on disinfection efficiency was graphically demonstrated by Weber as shown in Fig. 17-2 (Weber et al., 1940). As the dose of chlorine is increased, the total chlorine residual (i.e., remaining in the system after 30 min) increases until it reaches a dose of approximately 50 mg/L, whereupon residual chlorine decreases to a very low value, and subsequently increases linearly with dose indefinitely. The "hump-and-dip" behavior is paralleled by the sensitivity of microorganisms to the available chlorine residual indicated by the time required for 99 percent inactivation of test bacteria. At the three points indicated, the total available chlorine is approximately identical at 22 to 24 mg/L yet a

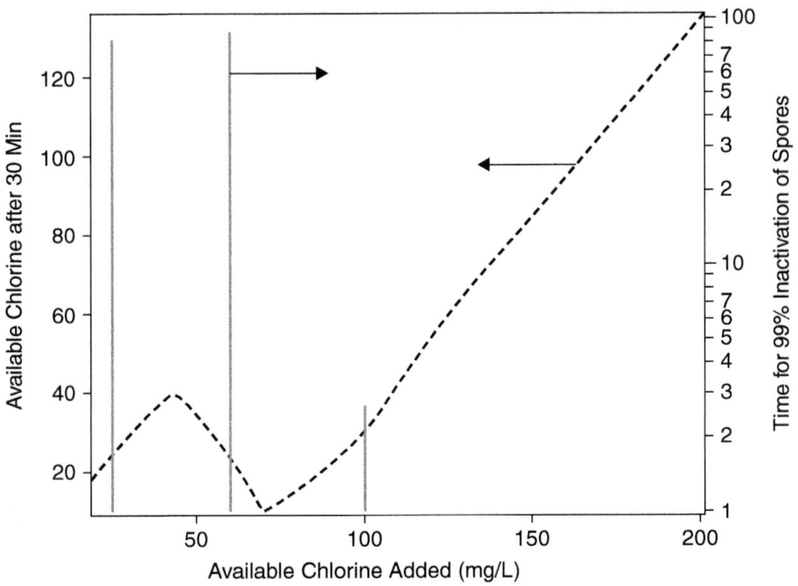

FIGURE 17-2 Effect of increased chlorine dosage on residual chlorine and germicidal efficiency; pH 7.0, 20°C, NH_3 10 mg/L. (*Adapted from Weber et al.,* Journal AWWA, *Vol. 32, No. 11 (November 1940) by permission.*)

32-fold difference in microbial sensitivity occurred. It should be noted that there is substantial impact of the breakpoint curve on the formation of disinfection by-products. which is discussed in Chap. 19.

The explanation for this behavior is the breakpoint reaction between free chlorine and ammonia (Fig. 17-3). At doses below the "hump" in the chlorine residual curve (zone 1), only combined chlorine is detectable. At doses between the "hump" and the "dip" in the curve, an oxidative destruction of combined residual chlorine accompanied by the loss of nitrogen occurs (zone 2) (Taras, 1950). One possible reaction during breakpoint is

$$2NH_3 + 3HOCl = N_2 + 3H^+ + 3Cl^- + 3H_2O \qquad (17\text{-}16)$$

This reaction also may be used as a means to remove ammonia nitrogen from water or wastewaters (Pressley et al., 1972). Finally, after the ammonia nitrogen has been completely oxidized, the residual remaining consists almost exclusively of free chlorine (zone 3). The minimum in the chlorine residual versus dose curve (in this case Cl_2:NH_4^+ -N weight ratio of 7.6/1) is called the *breakpoint* and denotes the amount of chlorine that must be added to a water before a stable free residual can be obtained. If it is desired to maintain a free residual following disinfection, the dose must be at least as large as that required to achieve breakpoint.

In their investigations of the chlorination of drinking water, Griffin and Chamberlin (1941a and b) observed that

1. The classical "hump-and-dip" curve is only seen at pH between 6.5 and 8.5.
2. The molar ratio between chlorine and ammonia nitrogen dose at the breakpoint under ideal conditions is 2:1 corresponding to a mass dose ratio (Cl_2:NH_4^+ -N) of 10:1.
3. In practice, mass dose ratios of 15:1 may be needed to reach breakpoint.

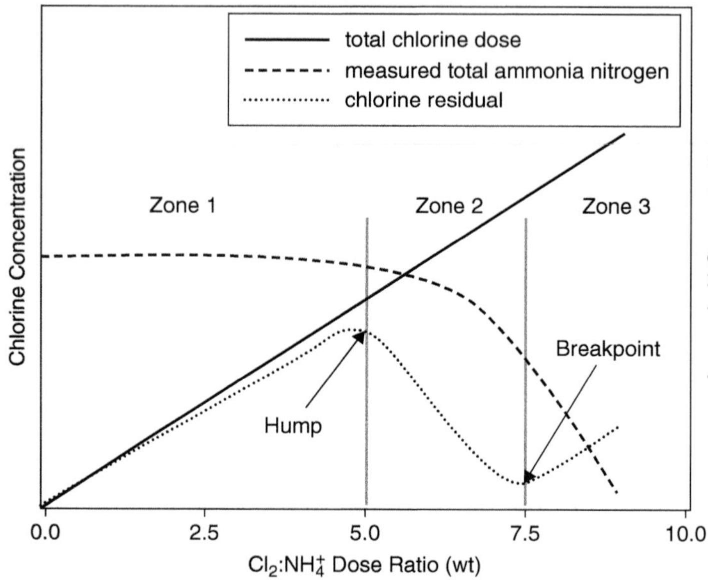

FIGURE 17-3 Schematic idealization of breakpoint curve. (*Modified after White, 1978*)

More recent investigators have reported breakpoint in waters with lower chlorine demand that may occur at mass dose ratios as low as 7.6:1.

The breakpoint reaction may also affect the pH of a water. If sodium hypochlorite is used as the source of active chlorine, as breakpoint occurs, the pH decreases caused by an apparent release of protons during the breakpoint process (Eq. 17-16). If gaseous chlorine is used, this effect is reinforced by the release of protons by hydrolysis of gaseous chlorine according to the reactions in Eqs. 17-6 and 17-7 (Mc Kee, 1960).

The oxidation of ammonia nitrogen by chlorine to gaseous nitrogen at the breakpoint would theoretically require 1.5 mole of chlorine (Cl_2) per mole of nitrogen oxidized according to the reaction in Eq. 17-16. The observed stoichiometric molar ratio between chlorine added and ammonia nitrogen consumed at breakpoint is typically about 1.6-1.7:1, suggesting that more oxidized nitrogen compounds are produced at breakpoint than N_2 gas. Experimental evidence (Saunier and Selleck, 1979) indicates that the principal additional oxidized product may be nitrate formed via the reaction in Eq. 17-17.

$$NH_4^+ + 4HOCl = NO_3^- + 4Cl^- + 6H^+ + H_2O \qquad (17\text{-}17)$$

Depending on the relative amount of nitrate formed in comparison to nitrogen at breakpoint, between 1.5 and 4.0 mole of available chlorine may be required, which is consistent with the available data.

Below the breakpoint, inorganic chloramines decompose by direct reactions with several compounds. For example, monochloramine may react with bromide ions to form monobromamine (Trofe, 1980). If trichloramine is formed, as would be the case for applied chlorine dosages in excess of that required for breakpoint, it may decompose either directly to form nitrogen gas and hypochlorous acid or by reaction with ammonia to form monochloramine and dichloramine (Saguinsin and Morris, 1975). In distilled water, the half-life of monochloramine was approximately 100 hours (Kinman and Layton, 1976). Even in this simple circumstance, however, the decomposition products have not been completely characterized. Valentine et al. (1986) found that the decomposition of pure solutions of monochloramine produces an unidentified product that absorbs UV light at 243 nm and is capable of being oxidized or reduced.

Where the pH is below 9.0 (so that the dissociation of ammonium ion is negligible), the amount of combined chlorine in dichloramine relative to monochloramine after the reactions in Eqs. 17-13 and 17-14 have attained equilibrium, is given by the following relationship (McKee, 1960)

$$A = -1 + \left(\frac{BZ}{1 - \sqrt{1 - BZ(2 - Z)}} \right) \tag{17-18}$$

where A is the ratio of available chlorine in the form of dichloramine to available chlorine in the form of monochloramine, Z is the ratio of moles of chlorine (as Cl_2) added per mole of ammonia nitrogen present, and B is defined by Eq. 17-19.

$$B = 1 - 4\,K_{eq}[H^+] \tag{17-19}$$

The equilibrium constant in Eq. 17-19 refers to the direct interconversion between dichloramine and monochloramine as follows:

$$\begin{aligned} H^+ + 2NH_2Cl &= NH_4^+ + NHCl_2 \\ K_{eq} &= [NH_4^+][NHCl_2]/[H^+][NH_2Cl]^2 \end{aligned} \tag{17-20}$$

At 25°C, K_{eq} has a value of 6.7×10^5 L/mol (Gray et al., 1978). From these relationships, determination of the equilibrium ratio of dichloramine to monochloramine as a function of pH and applied chlorine dose ratio is possible (assuming no dissipative reactions other than those involving the inorganic chloramines). As pH decreases and the Cl:N dose ratio increases, the relative amount of dichloramine also increases (Fig. 17-4). As the Cl:N molar dose ratio increases, the relative amount of dichloramine also increases. As the Cl:N molar dose ratio increases beyond unity, the amount of dichloramine relative to monochloramine rapidly increases as well. For the conversion from dichloramine to trichloramine, the equilibrium constant given at 0.5M ionic strength and 25°C indicates that the amount of trichloramine to be found in equilibrium with di- and monochloramine at molar dosage ratios of up to 2.0 is negligible (Gray et al., 1978). This agrees with experimental measurement of the individual combined chlorine species as a function of approach to breakpoint (White, 1999).

These findings, coupled with the routine observation of the breakpoint at molar dosages at or below 2:1 (weight ratios Cl_2:N below 10:1), indicate that trichloramine is not an important species in the breakpoint reaction. Rather, the breakpoint reaction leading to oxidation of ammonia nitrogen and reduction of combined chlorine is initiated with the formation of dichloramine.

The kinetics of formation of chloramine species has been investigated by various researchers since initial attempts by Weil and Morris (1949b). The formation of monochloramine is a first-order rate process with respect to both hypochlorous acid and unionized ammonia. However, determining whether this or a process involving hypochlorite ions reacting with ammonium cations is the actual mechanism of reaction is not possible solely through kinetic arguments. If the neutral species are selected as the reactants, then the rate of formation of monochloramine (r) may be described by (Morris and Isaac, 1983).

$$r \text{ (mol/L-s)} = 6.6 \times 10^8 \exp(-1510/T)\,[HOCl][NH_3] \tag{17-21}$$

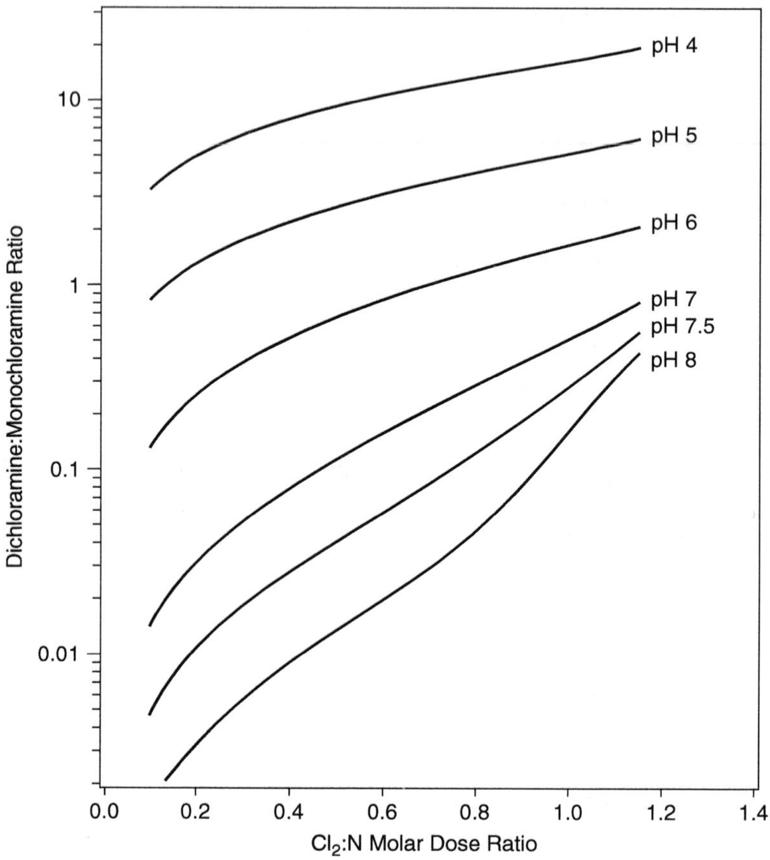

FIGURE 17-4 Effect of pH and Cl_2:NH_4^+ molar ratio on dichloramine-to-monochloramine ratio (25°C).

Because hypochlorous acid dissociates into hypochlorite with a *pKa* of approximately 7.6 and ammonia is able to associate with a proton to form the ammonium cation, with the *pKa* for the latter of approximately 9.3, for a constant chlorine:nitrogen dose ratio, the maximum rate of monochloramine formation occurs at a pH where the product HOCl × NH_3 is maximized, which is at the midpoint of the two *pK* values, or 8.4. At this optimum pH and the usual temperatures encountered in practice, the formation of monochloramine attains equilibrium in seconds to 1 min; however, at either a higher or lower pH, the speed of the reaction slows.

A number of the other reactions in the chlorine-ammonia system may be kinetically limited. Recently, these have been reviewed, and Table 17-2 is a compilation of the known reaction kinetics involving chlorine, ammonia, and intermediate species.

The reaction of NH_2Cl with HOCl to form $NHCl_2$ is catalyzed by a number of acidic species that may be present in water (Valentine and Jafvert, 1988). Possibly, a number of the other reactions in Table 17-2 can also be catalyzed in a similar manner; however, insufficient data are available to evaluate this possibility.

TABLE 17-2 Summary of Chlorine Reaction Kinetics

Reaction	Forward rate expression	Reverse rate expression
$NH_3 + HOCl \Leftrightarrow NH_2Cl + H_2O$	$6.6 \cdot 10^8 \exp\left(-\frac{1510}{T}\right)$	$1.38 \cdot 10^8 \exp\left(-\frac{8800}{T}\right)$
$NH_2Cl + HOCl \Leftrightarrow NHCl_2 + H_2O$	$3 \cdot 10^5 \exp\left(-\frac{2010}{T}\right)$	7.6×10^{-7} L/mol-s†
$NHCl_2 + HOCl \Leftrightarrow NCl_3 + H_20$	$2 \cdot 10^5 \exp\left(-\frac{3420}{T}\right)$	$5.1 \cdot 10^3 \exp\left(-\frac{5530}{T}\right)$
$2NH_2Cl \Leftrightarrow NHCl_2 + NH_3$	$80 \exp\left(-\frac{2160}{T}\right)$	24.0 L/mol-s(*)

Notes: Rates are in units of L/mol-s.
Concentrations are in mol/L.
Reactions are elementary and water is at unit activity
†This is the rate constant at 25°C.
Source: (Morris and Isaac, 1983).

When free chlorine is contacted with a water containing ammonia, the initial velocity of monochloramine formation is substantially greater than the velocity of the subsequent formation of dichloramine. Hence, relative to equilibrium levels, an initial accumulation of monochloramine will occur if large dose ratios (>5:1 Cl_2:N on a mass basis) are employed, until the dichloramine formation process can be driven (Palin, 1983).

The kinetic evolution of the chlorine-ammonia speciation process in batch systems is described by a series of coupled ordinary differential equations. While these are highly nonlinear, various authors have applied numerical integration techniques for their solution and, below the breakpoint, have found reasonable concordance between model predictions and experimental measurements (Haag and Lietzke, 1980; Isaac et al., 1985; Saunier and Selleck, 1979; Valentine and Jafvert, 1988).

The breakpoint process involves a complex series of elementary reactions, of which Eqs. (17-16) and (17-17) are the net results. Saunier and Selleck (1979) proposed that hydroxylamine (NH_2OH) and NOH may be intermediates in this reaction. Sufficient evaluation of their proposed kinetic scheme for the breakpoint process, however, has not yet been achieved to justify its use for design applications.

Example 17-2 A water supply is to be postammoniated. If the water has a pH of 7.0, a free chlorine residual of 1.0 mg/L, and a temperature of 25°C, how much ammonia should be added such that the ratio of dichloramine to monochloramine is 0.1 (assume that, on the addition of ammonia, none of the residual dissipates)?

Solution

1. From Eqs. 17-18 and 17-19, the following is determined:

$$B = 1 - 4\,K_{eq}\,(10^{-7}) = 1 - 4\,(6.7 \times 10^5)(10^{-7})$$
$$= 0.732$$

From Eq. 17-16, noting that the problem condition specifies A = 0.1, the following equationis to be solved:

$$0.1 = -1 + \frac{0.732\,Z}{1 - \sqrt{1 - 0.732(2 - Z)Z}}$$

This can be rearranged into a quadratic equation

$$-0.289Z^2 + 0.134Z = 0$$

2. The single nonzero root gives $Z = 0.463$, which is molar ratio of chlorine (as Cl^2) to ammonia nitrogen. Because chlorine has a molecular weight of 71, 1 mg/L of free chlorine has a molarity of 1.43×10^{-5} (i.e., 1/71,000). Therefore, 3.09×10^{-5} molarity of ammonia is required, or (multiplying by the atomic weight of nitrogen, 14) a concentration of 0.43 mg/L as N of ammonia must be added. ▲

Reactions with Organic Matter. Morris (1967) has determined that organic amines react with free chlorine to form organic monochloramines. The rate laws for these reactions follow patterns similar to the inorganic monochloramine formation process, except that the rate constants are generally less. In addition, the rate constants for this process correlate to the relative basicity of the amine reactant. Organic chloramines may also be formed by the direct reaction between monochloramine and the organic amine, and this is apparently the most significant mechanism of organic N-chloramine formation at higher concentrations such as might exist at the point of application of chlorine to a water (Isaac and Morris, 1980). Some solutions of amino acids and some proteins yield breakpoint curves resembling those of ammonia solutions (Baker, 1947; Wright, 1936). However, with the chlorination of organic amines, formation of other intermediates, for example, the toxic cyanogen chloride, can occur (Shang et al., 2000).

Free chlorine reacts with organic constituents to produce chlorinated organic by-products. Murphy et al. (1975) noted that phenols, amines, aldehydes, ketones, and pyrrole groups are readily susceptible to chlorination. Granstrom and Lee (1957) found that phenol could be chlorinated by free chlorine to form chlorophenols of various degrees of substitution. The kinetics of this process depend on both phenolate ions and hypochlorous acid. If excess ammonia was present, however, the formation of chlorophenols was substantially inhibited.

DeLaat et al. (1982) determined that polyhydric phenols are substantially more reactive than simple ketones in the production of chloroform and that the rates of these processes are first order with respect to the phenol concentration and the free chlorine concentration. More significantly, the reactivity of these compounds was observed to be greater than the reactivity of ammonia with hypochlorous acid. Therefore, even if subbreakpoint chlorination is practiced, some chloroform (typically in small amounts) may be formed rapidly prior to the conversion of free to combined chlorine. Chapter 19 presents additional discussion of the formation of trihalomethanes and other disinfection by-products that can arise from reactions with natural organic matter.

The reactivity of the chlorine species with compounds responsible for some tastes and odors depends on the predominant form of chlorine present. In field tests, Krasner et al. (1986) determined that free chlorine, but not combined chlorine, could remove tastes and odors associated with organic sulfur compounds. However, many taste and odor compounds are not amenable to removal by chlorine oxidation.

Reactions with Other Inorganic Compounds. The rates of reaction between free chlorine residuals and other inorganic compounds likely to be present in water are summarized in Table 17-3 (Wojtowicz, 1979). These reactions are generally first order in each of the oxidizing agent (hypochlorous acid or hypochlorite anion) and the reducing agent.

Nitrites present in partially nitrified waters react with free chlorine via a complex, pH-dependent mechanism (Cachaza, 1976). While combined chlorine residuals were generally thought to be unreactive with nitrite, Valentine (1985) has found that the rate of decay of monochloramine in the presence of nitrite was far greater than would be predicted on the basis of the reaction of the equilibrium free chlorine; this implicates a direct reaction between NH_2Cl and NO_2–N.

TABLE 17-3 Summary of Kinetics of HOCl and OCl⁻ Reduction by Miscellaneous Reducing Agents

Oxidizing agent	Reducing agent	Oxidation product	Log *k*, L/mol-s, 25°C
OCl–	IO–	IO4–	–5.04
OCl–	OCl	ClO2–	–7.63
OCl–	ClO2–	ClO3–	–5.48
OCl–	SO3	SO4–2	3.93
HOCl	NO2–	NO3–	0.82
HOCl	HCOO–	H2CO	–1.38
HOCl	Br–	BrO–	3.47
HOCl	OCN–	HCO3–, N2	–0.55
HOCl	HC2O4–	CO2	1.20
HOCl	I–	IO–	8.52

Source: After Wojtowicz, 1979.

Overall Chlorine Demand Kinetics. Chlorine demand is defined as the difference between the applied chlorine dose and the chlorine residual measured after a particular reaction or contact time (Fig. 17-3). The rate of exertion of chlorine demand in complex aqueous solutions has been the subject of numerous studies. The most systematic work has been that of Taras (1950), who chlorinated pure solutions of various organic compounds and found that chlorine demand kinetics could be described by a power function (Eq. 17-22).

$$D = kt^n \tag{17-22}$$

where t is the time in hours, D is the chlorine demand, and k and n are empirical constants. In subsequent work (Feben and Taras, 1950; Feben and Taras, 1951), it was found that chlorine demand exertion of waters blended with wastewater could be correlated to Eq. 17-22 with the value of n correlated to the 1 hour chlorine demand.

Haas and Karra (1984c) developed Eq. 17-23 to describe chlorine demand exertion kinetics.

$$D = C_0\{1-[x\exp(-k_1t)+(1-x)\exp(-k_2t)]\} \tag{17-23}$$

In Eq. 17-23, x is an empirical parameter, typically 0.4 to 0.6, and k_1 and k_2 are rate constants, typically 1.0 min^{-1} and 0.003 min^{-1}, respectively, and C_0 is the chlorine dosage in mg/L.

Dugan et al. (1995) found that a Monod (Langmuir Hinshelwood) kinetic model could describe free chlorine decay in drinking water in the absence of ammonia, although the mechanism for this process was not presented. It describes chlorine decay as a reaction with *total organic carbon* (*TOC*) in water according to the following differential equation:

$$\frac{dC}{dt} = \frac{k(\text{TOC})C}{K(\text{TOC})+C} \tag{17-24}$$

where TOC (assumed constant) is in mg/L as carbon, C is the free chlorine concentration in mg Cl^2/L, and both k and K are functions of TOC. Equation 17-24 can be integrated to the following implicit equation for chlorine concentration at time t (C_t):

$$C_t = K(\text{TOC})\ln\left(\frac{C_0}{C_t}\right) - k(\text{TOC})t + C_0 \tag{17-25}$$

Furthermore, k and K were correlated with the initial chlorine dose (C_0) and TOC concentration. By conducting tests on a variety of waters, the constants were found to be given by the following equations (it should be noted that the pH and temperature were fixed at 8 and 20°C and the waters were of relatively low ionic strength, so that the applicability of these relationships under other conditions is unclear).

$$K=-0.85\left(\frac{C_0}{\text{TOC}}\right)$$
$$k=0.03-0.006\left(\frac{C_0}{\text{TOC}}\right) \qquad (17\text{-}26)$$

Chlorine Dioxide. The chlorine dioxide demand in natural waters appears to be associated as much with constituents in the water (e.g., TOC) as with the water itself. This results in interconversions to chlorite and chloride, as outlined (Eqs. (17-10) and (17-11)). At mg/L concentrations, ammonia nitrogen, peptone, urea, and glucose have insignificant chlorine dioxide demand in 1 hour (Ingols and Ridenour, 1948; Sikorowska, 1961). However, a variety of inorganic and biological materials will react (Werderhoff and Singer, 1987).

Masschelein (1979a) concluded that only the following organic-ClO_2 reactions are of significance to water applications:

1. Oxidation of tertiary amines to secondary amines and aldehydes
2. Oxidation of ketones, aldehydes, and (to a lesser extent) alcohols to acids
3. Oxidation of phenols
4. Oxidation of sulfhydryl-containing amino acids

Wajon et al. (1982) found a reaction stoichiometry of 2 mol of chlorine dioxide consumed per mole of phenol (or hydroquinone) consumed. Products formed included chlorophenols, aliphatic organic acids, benzoquinone, and (in the case of phenol) hydroquinone. The mechanism appeared to include the possible formation of hypochlorous acid as an active intermediate, and the rate of this process was found to be base catalysed and first order in each of the reactants.

In general, chlorine dioxide itself has been found to produce fewer organic by-products with naturally occurring dissolved organic material although some nonpurgeable organic halogenated compounds are formed (Rav-Acha, 1984; Richardson et al., 1994). Therefore, unless chlorine dioxide is generated in a manner in which highly pure ClO_2 is produced, chlorine may be present as an impurity, and then, the reactions of such a stream may also include those discussed regarding chlorine reactions. The inorganic by-products of chlorine dioxide decay consist of chloride (typically trace), chlorate (typically 30 percent of the applied dose), and chlorite (typically 70 percent of the applied dose); specific ratios may depend on the precise application conditions (Noack and Doeff, 1981; Werderhoff and Singer, 1987). A more detailed discussion on by-products can be found in Chap. 19.

Ozone. On addition to water, ozone reacts with hydroxide ions to form hydroxyl radicals and organic radicals. These radicals cause increased decomposition of ozone, and also are responsible for nonselective (compared to the direct ozone reaction) oxidation of a variety of organic materials. Carbonate, and possibly other ions, may act as radical scavengers and slow this process (Hoigne and Bader, 1975; Hoigne and Bader, 1976).

Gurol and Singer (1982) determined that ozone decomposition kinetics in various aqueous solutions are second order in ozone concentration and base promoted. Some systematic difference between various buffer systems employed does occur, with borate giving higher decomposition rates than phosphate and with phosphate at higher ionic strength giving lower decomposition rates than phosphate at lower ionic strength (1 M vs. 0.1 M). This effect was suggested as being caused by phosphate being a radical scavenger (and radical decomposition being important at higher pH values). As with any rate process, ozone decay is more rapid at higher temperatures.

As a result of these decomposition processes, the half-life of ozone in water, even in the absence of other reactive constituents, is quite short, on the order of seconds to minutes. Water chemistry may exert a strong influence on the rate and extent of ozone demand in a given application. Reactions of ozone in aqueous solution are discussed further in Chap. 7. Formation of disinfection by-products with ozone is discussed in Chap. 19.

Demand of UV. For UV disinfection, the applied dose may be described in terms of the emitted lamp power in the germicidal range per unit volume of fluid under irradiation, for example, W/m^3. This can also be expressed as an integral over the disinfection reactor volume of the surface intensity (e.g., in W/m^2) (Severin et al., 1983b; Severin et al., 1984b).

With UV-light disinfection systems, the equivalent of demand results from dissolved and suspended materials, such as proteins, humic material, and iron compounds that absorb radiation and thus shield microorganisms. It has become increasingly practicable to use intensity monitoring within the reactor itself to monitor and correct for such effects (Larason and Ohno, 2006; Sommer et al., 2008).

Details of UV performance characteristics and modeling are discussed in Chap. 18.

ASSESSMENT OF MICROBIAL QUALITY (INDICATORS)

The microbial quality of a source water, or the efficacy of a treatment system for removing microorganisms, can be assessed either by direct monitoring of pathogens or by the use of an indicator system. Because pathogens are a highly diverse group, generally requiring a highly specialized (and often insensitive and expensive) analytical technique for each pathogen, the use of indicator organisms is a more widespread and practical strategy.

An indicator group of organisms is used to assess either source water contamination or degree of treatment; however, often the same indicator group is used to assess both properties. This poses severe constraints on the group of indicator organisms chosen. Criteria from Bonde (1966) for an ideal indicator remain valid, that is, that an ideal indicator must

1. Be present whenever the pathogens concerned are present
2. Be present only when the presence of pathogens is an imminent danger (i.e., they must not be able to proliferate to any greater extent in the aqueous environment)
3. Occur in much greater numbers than pathogens
4. Be more resistant to disinfectants and to the aqueous environment than pathogens
5. Grow readily on relatively simple media
6. Yield characteristic and simple reactions enabling, as far as possible, an unambiguous identification of the group
7. Be randomly distributed in the sample to be examined, or it should be possible to obtain a uniform distribution by simple homogenization procedures
8. Grow widely independent of other organisms present when inoculated in artificial media (i.e., the indicator bacteria should not be seriously inhibited in their growth by the presence of other bacteria)

The use of coliforms as indicator organisms stems from the pioneering work of Phelps (1909). The basic rationale was that coliforms and enteric bacterial pathogens originate from a common source, namely, human fecal contamination. Subsequent work by Butterfield and others (Butterfield and Wattie, 1946; Butterfield et al., 1943; Kabler, 1951; Wattie and Butterfield, 1944) confirmed that these organisms were at least as resistant to free or combined chlorine as enteric bacterial pathogens.

The coliform group is a heterogeneous conglomerate of microorganisms, including forms native to mammalian gastrointestinal tracts, as well as a number of exclusively soil forms. The common *fermentation tube* (FT) and *membrane filter* (MF) *procedures* are subtly different in the organisms they enumerate. Classically, total coliforms have been defined as gram-negative, non spore-forming bacteria that ferment lactose at 35°C to 37°C, with the production of acid and gas" (APHA et al., 1989). The FT procedure, however, ignores anaerogenic and lactose-negative coliforms, and the MF procedure ignores nonlactose fermenting strains (Clark and Pagel, 1977).

Furthermore, interferences can selectively reduce coliforms as measured by one or the other method. Allen et al. (1977), for example, found that high concentrations (> 500 to 1000/mL) of standard plate count (SPC) organisms appeared to reduce the recovery of coliforms by the MF technique when compared to the FT technique.

The fecal coliform group of organisms is that subset of coliforms that are capable of growing at elevated temperature (44.5°C). The original rationale for development of this test was to provide a more selective indicator group, excluding mesophillic coliforms indigenous, primarily, to soils. Total coliforms, however, continue to be the basic U.S. microbiological standard for drinking water because their absence ensures the absence of fecal coliforms and so is a conservative standard. However, there are continuing developments in the direct measurement of *E. coli* as an indicator group.

While coliforms, either fecal or total, may be reasonably good indicators of fecal contamination of a water supply, as early as 1922 (Anonymous, 1922) reservations were expressed about the relative resistance of coliforms to chlorine vis a vis pathogenic bacteria, and so the adequacy of the coliform test as an indicator of disinfection efficiency. It has been found that coliforms are more sensitive to disinfection by one or more forms of chlorine than various human enteric viruses (Grabow et al., 1983; Kelly and Sanderson, 1958) and the protozoan pathogens *Naegleria* (Rubin et al., 1983), *Giardia* (Jarroll et al., 1981; Korich et al., 1990; Leahy, 1985; Rice et al., 1982), and *Cryptosporidium* (Korich et al., 1990). In addition, viruses (Scarpino et al., 1977) and protozoan cysts (Leahy, 1985) have been found to be more resistant to ClO_2 inactivation than coliforms. Farooq (1976) has determined that coliforms are more resistant to ozone than viruses. Human enteric viruses have been isolated in full-scale water treatment plants practicing conventional treatment and meeting turbidity and coliform standards in the presence of free residual chlorine (Payment et al., 1985; Rose et al., 1986).

As a result of the problems with the coliform group of organisms, a number of workers have investigated alternative indicator systems with greater resistance to disinfectants than coliforms. Among the most successful of these are the acid-fast bacteria and yeasts studied by Engelbrecht and associates (Engelbrecht et al., 1979; Engelbrecht et al., 1977; Haas et al., 1983a; Haas et al., 1983b; Haas et al., 1985a; Haas et al., 1985b). In addition, work using endotoxins (Haas et al., 1983c), *Clostridia* (Cabelli, 1977; Payment and Franco, 1993), and bacteriophage (Abad et al., 1994; Grabow, 1968; Grabow and et al., 1983; Payment and Franco, 1993) has been carried out. In addition, to some degree, heterotrophic plate count (HPC) organisms may provide a conservative indicator of treatment efficiency. Indigenous aerobic spores have also been shown to be useful particularly in assessing performance of processes in which removal occurs via physical mechanisms (Rice et al., 1996). However, quantitative correlations between removal/inactivation of indicator organisms and pathogens in water treatment, including disinfection, remain elusive (Costan-Longares et al., 2008).

Therefore in U.S. practice, design, operation and regulation of water treatment remains based on total coliform, supplemented with the direct pathogen removal requirements associated with the enhanced surface water treatment rule.

Pathogens of Concern

A variety of pathogenic organisms capable of transmission by the fecal-oral route may be found in raw wastewaters. Water-borne outbreaks of shigellosis, salmonellosis, and various viral agents have been reported; in many cases these have been associated with sewage-contaminated water supplies (Blostein, 1991; Drenchen and Bert, 1994; Herwaldt et al., 1992; Levine et al., 1990; Reeve et al., 1989). Among the bacteria, *Salmonella, Shigella,* and *Vibrio cholerae* organisms are the classical agents of concern (Mosley, 1966). In more recent times, concern has expanded to other agents that have been found in wastewater—viruses and protozoa.

Among the viruses, enteroviruses (ECHO virus, Coxsackievirus), rotavirus, reovirus, adenovirus, and parvovirus have been isolated from wastewater (Melnick et al., 1978). New viruses that are suspected of water-borne transmission continue to be identified. Among the more important of these newly identified agents may be the Norwalk virus (Norovirus) and Calicivirus.

Over the past 20 years, significant concerns have increased over the risk from pathogenic protozoa in drinking water, particularly *Giardia* and *Cryptosporidium* (Gallaher et al., 1989; Goldstein et al., 1996; Lechevallier et al., 1991; Leland et al., 1993; Richardson et al., 1991; Rose et al., 1991; Smith, 1992). Under the USEPA Candidate Contaminant Rule, there is the periodic reassessment of additional microorganisms to consider. This process is discussed in Chap. 1.

DISINFECTION KINETICS

The information needed for the design of a disinfection system includes knowledge of the rate of inactivation of the target, or indicator, organism(s) by the disinfectant. In particular, the effect of disinfectant concentration on the rate of this process determines the most efficient combination of contact time and dose to employ.

Chick's Law and Elaborations

The major precepts of disinfection kinetics were enunciated by Chick (1908), who recognized the close similarity of microbial inactivation by chemical disinfectants to chemical reactions.. A good overview of the principles of kinetic modeling of disinfection has been presented by Gyurek and Finch (1998). Disinfection is modeled as a bimolecular chemical reaction, with the reactants being the microorganism and the disinfectant, and can be characterized by a rate law as are chemical reactions.

$$r = -kN \qquad (17\text{-}27)$$

In Eq. 17-27, r is the inactivation rate (organisms killed/volume-time) and N is the concentration of viable organisms. In a batch system, this results in an exponential decay in organisms, because the rate of inactivation equals dN/dt, assuming that the rate constant k is actually constant (e.g., the disinfectant concentration, temperature, and pH are constant).

Watson (1908) proposed Eq. 17-28 to relate the rate constant of inactivation k to the disinfectant concentration C.

$$k = k'C^n \tag{17-28}$$

In Eq. 17-28, n is termed the coefficient of dilution, and k' is presumed independent of disinfectant concentration, and, by virtue of Eq. 17-27, microorganism concentration. From the Chick-Watson law, when C, n, and k' are constant (i.e., no demand, constant concentration), Eq. (17-28), the rate law, may be integrated so that in a thoroughly mixed batch system,

$$\ln (N/N_0) = k'C^n t \tag{17-29}$$

In Eq. 17-29, N and N_0 are, respectively, the concentrations of viable microorganisms at time t and at time zero. When disinfectant composition changes with time or when another configuration than a batch (or plug flow; see Chap. 4 for an explanation that batch and plug flow reactors yield the same performance) system is used, the appropriate rate laws characterizing disinfectant transformation (Haas and Karra, 1984b) along with the applicable mass balances must be used to obtain the relationship between microbial inactivation and concentration and time.

Inactivation of microorganisms in batch experiments, even when disinfectant concentration is kept constant, does not always follow the exponential decay pattern predicted by Eq. 17-29. Indeed, two common types of deviations area noted (Fig. 17-5). In addition to the

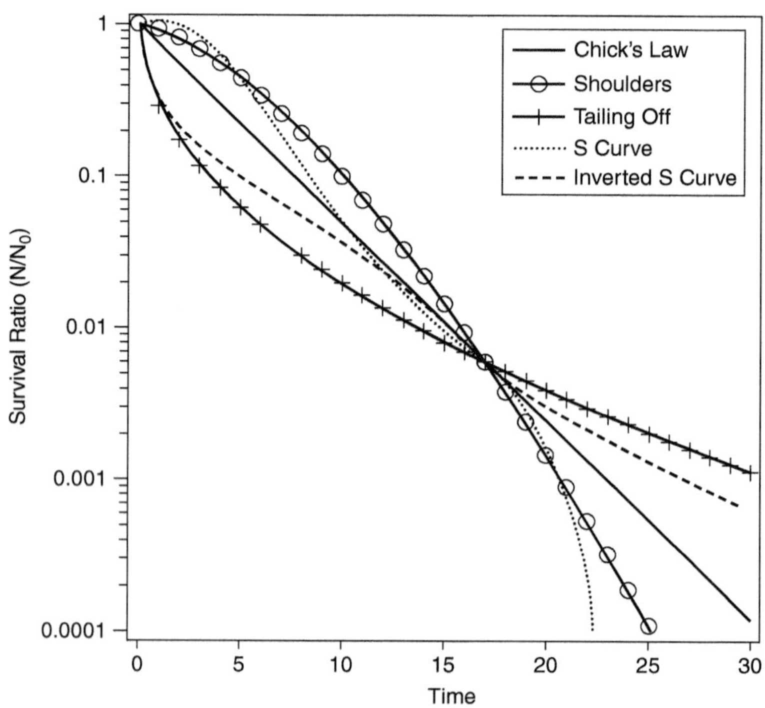

FIGURE 17-5 Chick's law and its deviations.

linear Chick's law decay, the presence of "shoulders," or time lags, until the onset of disinfection is often observed. Also, some microorganisms and disinfectants exhibit a "tailing" wherein the rate of inactivation progressively decreases. In some cases, a combination of both of these behaviors is seen.

Even if deviations from Chick-Watson behavior are observed, plotting combinations of disinfectant concentration and time to produce a fixed percent inactivation is generally possible. Such plots tend to follow the relationship C_{nt} = constant, where the constant is a function of the type of organism, pH, temperature, type of disinfectant, and extent of inactivation. Such plots are linear on a log-log scale. If the value of n is greater than 1, a proportionate change in disinfectant concentration produces a greater effect than a proportionate change in time. In many cases (Hoff, 1986), the Chick-Watson law n value is close to 1.0, and hence a fixed value of the product of concentration and time ($C \times T$) results in a fixed degree of inactivation (at a given temperature, pH, etc.).

In the chemical disinfection of a water, the disinfectant residual may change with time, and, particularly when organic or ammonia nitrogen is present, during the initial moments of contact the chemical form(s) of halogens such as chlorine undergo rapid transformations from the free to the combined forms. Because C would thus not be a constant, typically disinfection results obtained in batch systems exhibit "tailing," the degree of which may depend on the demand and the concentration of reactive constituents (such as ammonia and organic nitrogen compounds) in the system (Jafvert and Valentine, 1992; Olivieri et al., 1971). Determination of the disinfectant residual (and its chemical forms) is more critical than the disinfectant dose in these systems.

For chlorine disinfection, for example, by knowing the rate laws for inactivation by individual separate species and the dynamics of chlorine species interconversions, as described previously, an overall model for chlorine inactivation can be formulated (Haas and Karra, 1984b). In doing this computation, the individual rates are usually assumed to be additive (Fair et al., 1948), although this assumption has not yet been experimentally verified.

"Shoulder" inactivation curves may be explained by a multitarget model (assuming that multiple locations within an organism need to be deactivated before death occurs) (Hiatt, 1964), by a series event model (assuming that a location with an organism must be subject to multiple sequential reactions before death occurs) (Severin et al., 1984a), or by a diffusional model (assuming that transport of disinfectant serves to retard the rate of inactivation) (Haas, 1980). Tailing inactivation curves may be explained either by a vitalistic hypothesis in which individuals in a population are nonidentical, and their inherent resistance is distributed in a permanent (time-independent) manner or by a mechanistic concept (Cerf, 1977). In the latter case, four particular mechanisms have been advanced leading to tailing.

1. Conversion to resistant form during inactivation (hardening)
2. Existence of genetic variants of differing sensitivity
3. Protection of a subpopulation or variations in received dose of disinfectant
4. Clumping of a subpopulation

The hardening process and resultant tailing have received wide attention following discoveries of apparent hardening in the formaldehyde inactivation of poliovirus prepared for the Salk vaccine (Nathanson and Langmuir, 1963). Gard (1960) proposed an empirical rate law for this behavior, which has been used by Selleck et al. (1978) in the analysis of wastewater chlorination kinetics. Tailing behavior has been found for viral and coliform inactivation by ozone (Katzenelson et al., 1974) and for coliform inactivation by free chlorine (Haas and Morrison, 1981; Olivieri et al., 1971).

Hom's Law

Hom (1972) developed a flexible but highly empirical kinetic formulation for the inactivation rate based on modifying Eqs. 17-27 and 17-28 to the following form:

$$r = -k''mNt^{m-1}C^n \tag{17-30}$$

The parameter m quantifies nonlinearity from the Chick-Watson behavior with time (if $m = 1$, Eq. 17-30 becomes identical to the combination of Eqs. 17-28 and 17-29. Eq. 17-30 is difficult to use as a rate model since it contains time as an explicit variable. A formulation leading to the classical Hom integrated relationship can be written as (Haas and Joffe 1994)

$$r = -mN(k''C^n)\left[\ln\left(\frac{N}{N_0}\right)\right]^{(1-1/m)} \tag{17-31}$$

On integration, if C is constant, this results in the following relationship:

$$\ln\left(\frac{N}{N_0}\right) = -k''C^n t^m \tag{17-32}$$

Depending upon the value of m, both shoulders and tailing may be depicted by Eq. 17-32. In early work, Fair et al. (1948) used a model of the form of Eq. 17-32 with $m = 2$ to analyze *E. coli* inactivation by free and combined chlorine.

Example 17-3 A certain water supply has experienced regrowth of HPC organisms during distribution. To maintain adequate system water quality, a decision has been made to keep the concentration of HPC organisms below 10/mL at the entry point to the distribution system (i.e., following disinfection). Disinfection using free residual chlorine is practiced. As part of the laboratory investigation to develop design criteria for this system, the inactivation of the HPC organisms is determined in batch reactors (beakers). The pH and temperature are held constant at the expected final water conditions. Using water with an initial HPC of 1000/mL, the following data are taken:

Cl_2 residual, mg/L	Contact time, min	HPC remaining, #/mL
0.5	10	40
0.5	20	4
0.5	30	1
1.0	5	35
1.0	10	4
1.0	15	1
1.5	2	98
1.5	5	10
1.5	10	1

From this information, determine the best fit using the Hom inactivation model, and compute the necessary chlorine residual that will achieve HPC < 10/mL (from an initial concentration of 1000/mL) at a contact time of 10 minutes.

Solution

1. This problem can be solved using a maximum likelihood technique to fit the Hom model to the data. Regression methods, however, can be used in two different

ways—multiple linear regression, and nonlinear regression (Haas and Heller, 1989; Haas et al., 1988). In a batch system, without chlorine demand, the Hom model is that of Eq. 17-32.

Equation 17-32 can be rearranged as

$$\ln\left[-\ln\left(\frac{N}{N_0}\right)\right]=\ln(k'')+n\cdot\ln(C)+m\cdot\ln(t)$$

2. A multiple linear regression using $\ln\left[-\ln\left(\frac{N}{N_0}\right)\right]$ as the dependent variable and ln (C) and ln (t) as the independent variables produces an intercept (equal to ln (k'')) and slopes equal to n and m. This computation can be handled on common spreadsheet programs, as well as statistical packages. Transformation of the data given produces the following values of the dependent and independent variables:

$\ln\left[-\ln\left(\frac{N}{N_0}\right)\right]$	ln(t)	ln(C)
1.169	2.303	–0.693
1.709	2.996	–0.693
1.933	3.401	–0.693
1.210	1.609	0
1.709	2.303	0
1.933	2.708	0
0.843	0.693	0.405
1.527	1.609	0.405
1.933	2.303	0.405

The results are

$$k''=1.11$$
$$n=0.68$$
$$m=0.70$$
$$\text{Correlation coefficient}=0.997$$

3. The final estimation equation then becomes

$$\ln\left(\frac{N}{N_0}\right)=-1.11C^{0.68}t^{0.7}$$

4. Inserting the design specifications gives the following results:

$$\ln\left(\frac{10}{1000}\right)=-1.11C^{0.68}t^{0.7}$$
$$C^{0.68}=0.83$$
$$C=0.77\,\frac{\text{mg}}{\text{L}}$$

Hence, if a 10 minute contact time is accepted as a worst-case condition and good contactor hydraulics are assumed, the maximum chlorine residual required to achieve the design inactivation is 0.77 mg/L. From this information, and the chlorine demand of the water, the capacity of a chlorine feed system can be computed. ▲

Power Law Models

Another class of models can be obtained by assuming that inactivation is other than first order in surviving microbial concentrations. Depending on the order chosen, either tailing or shoulders can be produced. For example, Roy et al. (1979, 1981, 1981), using continuous flow stirred tank reactor (CFSTR; see Chap. 4 for the hydraulic characterization of CFSTRs) studies on inactivation of Poliovirus I with ozone in demand-free systems, developed the following rate law:

$$r = -kCN^{0.69} \tag{17-33}$$

A similar model was used by Benarde et al. (1967) to analyze *E. coli* inactivation by chlorine dioxide and for analysis of various organisms in ozone contacting (Zhou and Smith, 1994; Zhou and Smith, 1995).

The model of Roy, which is power law in N, and the Hom model can be combined into a generalized formulation for the reaction rate and for the survival ratio (under demand-free conditions) as shown in Eq. 17-34. This was first developed by Anotai (1996) and written in more convenient form by Gyurek and Finch (1998).

$$\begin{gathered} r = -kmN^{x}t^{m-1}C^{n} \\ \ln\left(\frac{N}{N_0}\right) = \left(\frac{1}{x-1}\right)\ln[1 + N_0^{x-1}(x-1)kC^{n}t^{m}] \end{gathered} \tag{17-34}$$

This relationship, having four parameters (x, k, n, m) is highly flexible. In addition to describing inactivation curves with shoulders and tailing off, it is capable of describing inactivation curves with inflection points, such as the S curve and inverted S curve in Fig. 17-5. In contrast to the Chick, Chick-Watson, and Hom models, the Hom power law (and power law) models predict inactivation to be a function of initial microorganism concentration (N_0) as long as $x \neq 0$. This dependency has been observed experimentally (Haas and Kaymak, 2003; Haas and Kaymak, 2008).

Models Accounting for Disinfectant Decay

If the disinfectant concentration is not constant, then the rate laws for inactivation must be combined with those for disinfectant demand. For example, in the case of the Hom model with first-order decay in a batch system, the following approach has been developed (Haas and Joffe, 1994). The time course of disinfectant concentration is given by

$$C = C_0 \exp(-k^* t) \tag{17-35}$$

where C_0 is the initial concentration and k^* is the demand constant. This equation is substituted into the Hom rate expression and then into a batch mass balance to yield

$$\frac{dN}{dt} = -mN[k(C_0 \exp(-k^* t))^{n}]^{1/m}\left[-\ln\left(\frac{N}{N_0}\right)\right]^{(1-1/m)} \tag{17-36}$$

This equation has an analytical solution in terms of an incomplete gamma function, which under conditions generally present in disinfection can be approximated as (Haas and Joffe, 1994) by

$$\ln\left(\frac{N}{N_0}\right) = -kC_0^{n}t^{m}\left[\frac{1-\exp\left(\dfrac{-nk^*t}{m}\right)}{\dfrac{nk^*t}{m}}\right]^{m} \tag{17-37}$$

For other disinfection decay rate laws, the solution may be obtained by numerical integration (Finch et al., 1993).

Impact of Temperature and Other Water Chemistry Factors

Disinfection, like all other rate processes, is temperature dependent. This dependency may be quantified by the Arrhenius relationship.

$$k = k_0 \exp\left(-E_a/RT\right) \tag{17-38}$$

where k is a rate constant characterizing the reaction (such as the Chick-Watson k' value), T is the absolute temperature, R is the ideal gas constant, k_0 is called the *frequency factor*, and E_a, with units of energy/mole, is the activation energy. E_a is always positive, and as it increases, the effect of temperature becomes more pronounced. The values of E_a and k_0 may be determined from rates of inactivation obtained as a function of temperature. As E_a increases, the effect of temperature on the rate increases. For example, E_a of 10 kcal/mol doubles the rate between 10°C and 20°C. In contrast, activation energies for breaking hydrogen bonds are 3 to 7 kcal/mol (Bailey and Ollis, 1986). Activation energies less than this range suggest physical (e.g., transport) limitations rather than chemical reactions.

It should be noted that, like pH, other water chemistry factors can influence the kinetics of disinfection, although the mechanism for many of these effects are unknown. This has been shown to be the case for chlorine and chloramines (Haas et al., 1996) and ozone (Finch et al., 2001). While generic kinetics (e.g., the $C \times T$ tables in Tables 17-7 and 17-8) have been published, these generally utilize laboratory buffered water, and a prudent designer particularly for a large system would consider doing testing to verify inactivation rates under water quality conditions particular to the situation.

MODE OF ACTION OF DISINFECTANTS

Chlorine

Since Nissen (1890), free chlorine at low pH has been known to be more biocidal than free chlorine at high pH. Holwerda (1928) proposed that hypochlorous acid was the specific agent responsible for inactivation, and thus the pH effect. Fair et al. (1948) determined that the pH dependency of free chlorine potency correlated quantitatively with the dissociation constant of hypochlorous acid. Chang (1944) determined that, at low pH, the association of chlorine molecules with cysts of *Entamoeba hystolytica* was greater than at high pH. Friberg (1957; Friberg and Hammarstrom, 1956), and Haas and Engelbrecht (1980a), using radioactive free chlorine, found that similar results with respect to bacteria and that the microbial binding of chlorine could be described by typical chemical isotherms. With respect to viruses, this association of chlorine parallels the biocidal efficacy of hypochlorous acid, hypochlorite, and monochloramine (Dennis et al., 1979a; Dennis et al., 1979b; Olivieri et al., 1980).

Once taken into the environment of the living organism, chlorine may enter into a number of reactions with critical components causing inactivation. In bacteria, respiratory, transport, and nucleic acid activity are all adversely affected (Haas and Engelbrecht, 1980a and b; Venkobachar et al., 1975; Venkobachar et al., 1977). In bacteriophage f2, the mode of inactivation appears to be disruption of the viral nucleic acid (Dennis et al., 1979a). With poliovirus, however, the protein coat, and not the nucleic acid, appears to be the critical site for inactivation by free chlorine (Fujioka et al., 1985; Tenno et al., 1980).

The rate of inactivation of bacteria by monochloramine is greater than could be attributed to the equilibrium-free chlorine present in solution. This argues strongly for a direct inactivation reaction of combined chlorine (Haas and Karra, 1984a). Although organic chloramines are generally measured as combined or total chlorine by conventional methods, they are of substantially lower effectiveness as disinfectants than inorganic chloramines (Feng, 1966; Wolfe and Olson, 1985).

In general, the rate of inactivation of microorganisms by various disinfectants increases with increasing temperature. This may be characterized by an activation energy, a Q_{10} (factor of increase for every 10°C temperature increase) or a temperature multiplier (the latter two methods being an approximation of an Arrhenius approach).

Surprisingly, Scarpino et al. (1972) reported that viruses were more sensitive to free chlorine at high pH than at low pH. A variety of subsequent authors confirmed these findings with viruses and with bacteria (Berg et al., 1989; Haas, 1981; Haas et al., 1986, 1990). Hypochlorite can form neutral ion pairs with sodium, potassium, and lithium, and (particularly at ionic strengths approaching 0.1 M) these can increase disinfection efficiency by free chlorine at high pH (above 8.0) (Haas et al., 1986). Calcium enhancement of chlorine inactivation of coliforms has also been reported (Haas and Anotai, 1996).

Chlorine Dioxide

The dependence of inactivation efficiency on pH is weaker for chlorine dioxide than for chlorine and more inconsistent. Benarde et al. (1965), working with *E. coli,* and Scarpino et al. (1979), working with Poliovirus I, found that the degree of inactivation by chlorine dioxide increases as pH increases. However, for amoebic cysts, as pH increases, the efficiency of inactivation by chlorine dioxide decreases (Chen et al., 1985). Barbeau et al. (2005) found that the efficiency of chlorine dioxide against bacteria spores and bacteriophage was also greater at pH 8.5 than at pH 6.5, and it was hypothesized that some inactivation processes are catalyzed by the presence of hydroxide ion.

The physiological mode of inactivation of bacteria by chlorine dioxide has been attributed to a disruption of cell envelope permeability (Berg et al., 1986). In the case of viruses, chlorine dioxide preferentially inactivated capsid functions, rather than nucleic acids (Noss et al., 1985; Olivieri et al., 1985).

Benarde et al. (1967) computed the activation energy for the inactivation of *E. coli* by chlorine dioxide at pH 6.5 as 12 kcal/M. An identical number was computed for the disinfection of Poliovirus Type I by chlorine dioxide at pH 7 (Scarpino et al., 1979).

Ozone

Understanding of the mode of inactivation of microorganisms by ozone remains hindered by difficulties in measuring low concentrations of dissolved ozone. The effect of pH on ozone inactivation of microorganisms appears to be predominantly associated with changing the stability of residual ozone, although additional work is needed. Farooq (1976) found little effect of pH on the ability of dissolved ozone residuals to inactivate acid-fast bacteria. Roy (1979) found a slight diminution of the virucidal efficacy of ozone residuals as pH decreased; however, Vaughn, as cited in Hoff (Hoff, 1986; Hoff and Akin, 1986), noted the opposite effect. The principal action of ozone as a disinfectant occurs from the dissolved ozone, rather than physical contact with ozone gas bubbles (Dahi and Lund, 1980; Farooq, 1976). The disparate conclusions of the early studies may have been due to difficulties with measuring ozone residuals. In the USEPA *ct* tables, the efficacy of ozone residuals are regarded as being independent of pH.

Bacterial cells lacking certain DNA polymerase gene activity were found to be more sensitive to inactivation by ozone than wild-type strains, strongly implicating physicochemical damage to DNA as a mechanism of inactivation by ozone (Hamelin and Chung, 1978). For poliovirus, the primary mode of inactivation by ozone also appears to be nucleic acid damage (Roy et al., 1981).

Energies for ozone inactivation of *Giardia* and *Naegleria* cysts were reported by Wickramanayake et al. (1985). At pH 7, the activation energies were 9.7 and 16.7 kcal/M. For Poliovirus Type I, Roy et al. (1981) estimated an activation energy of 3.6 kcal/M at pH 7.2. If the latter value is correct, its low value suggests that ozone inactivates a virus by a diffusion process, rather than a reaction controlled one.

UV Radiation

The application of UV radiation is covered in detail in Chap. 18. In this section, broad aspects of UV are summarized in a parallel fashion to chemical disinfectants.

The mode of inactivation of microorganisms by UV radiation is quite well characterized. Specific deleterious changes in nucleic acid arise on exposure to UV radiation (Jagger, 1967). These may be repaired by light-activated as well as dark repair enzymes in vegetative microorganisms. The phenomenon of photoreactivation of UV disinfected microorganisms has been demonstrated in municipal effluents (Scheible and Bassell, 1981). However, the operation of these repair processes in microorganisms discharged to actual distribution systems or receiving waters is not clear (Bohrerova and Linden, 2007). Pathogenic protozoan cysts and oocysts do not appear capable of UV photoreactivation.

Severin (1983a) showed that the series event model for inactivation (in which it is assumed that a sequential set of processes must occur to achieve kill) describes the kinetics of UV disinfection quite well. Kinetic parameters for inactivation are shown in Table 17-4.

TABLE 17-4 Kinetics of UV Inactivation

	k (cm^2/(mW-s))	Number of events	Activation energy (kcal/mole)
E. coli	1.538	9	0.554
Candida parapsilosis	0.891	15	0.562
f2 virus	0.0724	1	1.023

Source: Severin et al., 1983a.

Activation energies for UV inactivation using this model were lower than for chemical disinfection (indicating the relative insensitivity to temperature) (Severin et al., 1983a). These are considerably less than the activation energies for chemical disinfectants in accord with the apparent (purely physical) mechanism of UV inactivation. This also indicates, as a practical matter, that the effect of temperature on UV performance is much less than for chemical agents such as ozone or chlorine.

The pH dependency of UV inactivation has not been characterized in controlled systems. However, since the mechanism of UV inactivation appears to be purely photochemical, it is not anticipated that pH would dramatically alter the efficiency of UV disinfection. However, insofar as pH may affect the light absorption characteristics of humic materials, an indirect effect of pH (by changing the extent of demand) on inactivation efficiency may exist.

Influence of Physical Factors on Disinfection Efficiency

The apparent increase in microbial resistance by clumping has already been discussed. Microorganisms can also be partially protected against the action of disinfectants by adsorption to or enmeshment in nonviable solid particles present in water. Stagg et al. (1978) and Hejkal et al. (1979) found that fecal material protected poliovirus against inactivation by combined chlorine. Boardman and Sproul (1977) found that kaolinite, alum flocs, and lime sludge increased the resistance of Poliovirus Type 1 to free chlorine in demand-free systems. For chlorine dioxide, bentonite turbidity protects poliovirus against the action of chlorine dioxide (Brigano and et al., 1978). Although there was some protection from ozone inactivation afforded to coliform bacteria and viruses by fecal matter and by cell debris, at ozone doses usually employed in disinfection, it was still possible to achieve more than 99.9 percent inactivation within 30 sec (Sproul 1978).

The particular conditions wherein particles can protect viruses from disinfection have been recently reviewed (Templeton et al., 2008).

Influence of Physiological Factors on Disinfection Efficiency

The physiological state of microorganisms, especially vegetative bacteria, may influence their susceptibility to disinfectants. Milbauer and Grossowicz (1959) found that coliforms grown under minimal conditions were more resistant than cells grown under enriched conditions. Similarly, Berg et al. (1985) found that chemostat-grown cells produced at high growth rates were more sensitive to disinfectants than cells harvested from low growth rates. A simple subculturing of aquatic strains of *Flavobacterium* has been found to increase sensitivity to chlorine disinfection (Wolfe and Olson, 1985).

Postexposure conditions can also influence apparent microbial response to disinfectants (Milbauer and Grosswicz, 1959). In three New England water treatment plants and distribution systems sampled for total coliforms using both standard MF techniques and media to recover sublethally injured organisms (m-T7 agar medium) from 8 to 38 times as many coliforms were recovered on the latter medium than on the former medium (McFeters et al., 1986). This sublethal injury does not adversely affect pathogenicity to mice (Singh et al., 1986).

With viruses, the phenomenon of multiplicity reactivation can occur when individual viruses, inactivated by different specific events, are combined in a single host cell to produce a competent and infectious unit. This has been demonstrated to occur in enteric viruses inactivated by chlorine (Young and Sharp, 1979).

The survivors from disinfection can exhibit inheritable increased resistance to subsequent exposure. This was first demonstrated for poliovirus exposed to chlorine (Bates et al., 1978). However, demonstration in bacteria has not been consistent (Haas and Morrison, 1981; Leyval, 1984). Subculturing of survivors from iodine disinfection of *Burkholderia* (*Pseudomonas*) *cepacia* have been found to have increased resistance to iodination, although expression of this increased resistance could be modulated by environmental conditions (Pyle et al., 1994).

DISINFECTANT RESIDUALS FOR POSTTREATMENT PROTECTION

One factor that is important in evaluating the relative merits of alternative disinfectants is their ability to maintain microbial quality in a water distribution system. With respect to chlorine, it has been suggested that free chlorine residuals may serve to protect the distribution system against regrowth, or at least to serve as a sentinel for the presence of contamination (Snead et al., 1980). However, other studies have noted the lack of correlation between distribution system water quality and the form or concentration of chlorine residual (Haas et al., 1983a). Similarly, Lechevallier (1990) has reported that microbial slimes grown in tap water may be more sensitive to inactivation by combined chlorine than by free chlorine transported by the overlying water. It must be recognized that, regardless of the disinfectant chosen, the water distribution system can never be regarded as biologically sterile. As shown by Means et al. (1986a), shifts in the dominant form of disinfectant (e.g., from free chlorine to monochloramine) can result in shifts in the taxonomic distribution of microorganisms that inhabit the distribution system.

It must be noted that there are differences between the distribution system in the United States and many distribution systems in Europe (Haas, 1999; Hydes, 1999; Trussell, 1999). In the United States, most utilities strive to maintain minimum chlorine residuals in the distribution system (and there is a strong incentive to do so under regulatory requirements), with minimum residuals generally exceeding 0.2 mg/L total chlorine. In a number of countries in Europe and elsewhere the philosophical approach to maintaining distribution system water quality relies on nutrient control (principally, degradable organic matter) rather than on disinfectant residuals.

While chlorine dioxide has somewhat greater decay than chlorine in distribution, it has been found to be as or more effective than free chlorine (Gagnon et al., 2004). With respect to both ozone and UV, their absence of a residual may necessitate the addition of a second disinfectant if a residual in the distribution system is desired. Further discussion on this point is in Chap. 21.

DESIGN AND APPLICATION OF TECHNOLOGIES

With the increasing and conflicting objectives that must be met by disinfection processes, water utilities are moving toward the use of multiple disinfectants. In general, disinfectants may be applied at three locations within treatment. When applied prior to coagulation, it is generally termed preoxidation or predisinfection. While some inactivation may occur due to predisinfection, generally this is minor due to substantial disinfectant demand. Primary disinfection occurs when disinfectant is applied after clarification by sedimentation or dissolved air flotation, either before or after filtration, and with contact time prior to distribution. The majority of inactivation by disinfection processes is expected to occur due to this application. If another application occurs following primary disinfection (or if ammonia is added to convert a free residual to a combined residual), this is termed *secondary disinfection*. When secondary disinfection is used, this disinfectant is applied just before entry into the distribution system. Generally the objective of secondary disinfection is to allow the maintenance of an active disinfectant residual into the distribution system.

Chlorination

As noted previously, chlorine may be obtained for disinfection in three forms, gaseous Cl_2, liquid NaOCl, and solid $Ca(OCl)_2$, as well as generated on-site. Generally, on a per-unit mass basis of active chlorine, the least expensive form with large use rates is liquified chlorine gas. The use of liquified chlorine gas carries with it certain risks associated with accidental leakage of the gas. As a result, a number of utilities have elected to use the somewhat more expensive sodium hypochlorite, NaOCl, as a source of disinfectant.

The intrinsic hazards associated with gaseous chlorine at wastewater (and drinking water) disinfection facilities has long been recognized (White, 1999). It is increasingly recognized that storage of large amounts of gaseous chlorine could present a potential target for malicious activity such as by terrorists (Copeland, 2007). As a result, particularly since 2001, water and wastewater utilities have switched from gaseous chlorine to sodium hypochlorite (Jones et al., 2007; U.S. Government Accountability Office, 2007).

The use of sodium hypochlorite solutions as sources of chlorine for water disinfection is not without its own risks. As a strong oxidant, the solutions must be stored and transported in chemically resistant tanks and pipeline systems, and organic matter must be prevented from entering the tanks. Tanks must be vented since the solutions will decompose on a constant basis. Finally, sodium hypochlorite solutions are corrosive and present potential worker safety hazards in the event of a spill or tank breach.

On addition to water, chlorine gas reduces the pH and alkalinity, while sodium hypochlorite raises the pH and alkalinity. In poorly buffered water, when gaseous chlorine is used, the addition of a pH-buffering chemical may be necessary to control the distribution system water aggressiveness.

Chlorine and hypochlorites have been produced from the electrolysis of brines and saline solutions since the early twentieth century (Rideal, 1908). This remains an attractive option for remote treatment plants near an inexpensive source of brine. The basic principle is the use of a dc electric field to effect the oxidation of chloride ion with the simultaneous and physically separated reduction usually of water to gaseous hydrogen.

In practice, it is necessary to operate electrolytic chlorine generating units at voltages as high as 3.85 volts in order to provide reasonable rates of generation. At these high voltages, however, additional oxidations such as formation of chlorate, ohmic heating, and incomplete separation of hydrogen from oxidized products with subsequent dissipative reaction combine to produce system inefficiencies. Current on-site hypochlorite generators have salt efficiencies of 3 kg NaCl/kg (3 lb/lb) free chlorine and power efficiencies of 0.91 kw-hr/kg (2 kw-h/lb) free chlorine (Garibi, 2007).

Chlorine can be applied at a variety of points within treatment. Table 17-5, from the 1989 AWWA Committee Survey illustrates the frequency with which chlorine is applied at various locations. The most recent survey does not contain data on details of points of application (AWWA Disinfection Systems Committee, 2008); however, it is likely that chlorination before sedimentation has decreased in prevalence.

TABLE 17-5 Points of Application of Chlorine or Hypochlorite

Point of application	Frequency of plants ($N = 268$)
Before coagulation	18.66%
After coagulation	5.97%
After sedimentation and before filtration	31.34%
After filtration	47.01%
Within the distribution system	15.67%

Notes: Percentages sum to more than 100% due to multiple points of application at a given facility.
Source: Haas et al., 1992.

Source Water Chlorination/Preoxidation. Prechlorination, the addition of chlorine at an early point within a treatment plant, is designed to minimize operational problems associated with biological slime formation on filters, pipes, and tanks, and also to release potential taste and odor problems from such slimes. In addition, prechlorination can be used for the oxidation of hydrogen sulfide or reduced iron and manganese. Probably the most common point of addition of chlorine for prechlorination is the rapid-mix basin (where coagulant is added). However, with present concerns for minimizing the formation of chlorine by-products, the use of prechlorination is being supplanted by the use of other chemical oxidants (e.g., chlorine dixoide, permanganate) for the control of biological fouling, odor, or amount of iron or manganese.

Postchlorination. Postchlorination, or *terminal disinfection*, is the primary application for microbial reduction. It has been most common to add chlorine for these purposes either immediately before the clearwell or immediately before the granular media filter. In the latter case, the filter itself provides a contact time for disinfection.

In general, the use of specific contact chambers subsequent to the addition of chlorine to a water has been uncommon. Instead, the clearwell, or finished water reservoir, serves the dual function of providing contact to ensure adequate time for microbial inactivation prior to distribution. The distribution system itself, from the entry point until the first consumers tap, provides additional contact time. Many times, however, the water treatment plant itself is the first point of water consumption. Table 17-6 summarizes distributions of first contact times.

TABLE 17-6 Residual and Contact Time to First Customer: 1989 AWWA Disinfection Committee Survey of Utilities (N = 178)

	Contact time to first customer (min)						
Residual (mg/L)	0	1–9	10–29	30–60	75–240	>240	Total
0–.35	4.49%	0.56%	0.56%	0.56%	1.12%	0.56%	7.87%
.4–.95	10.1%	3.93%	3.93%	3.93%	3.37%	2.81%	28.09%
1–1.5	14.6%	5.62%	2.25%	4.49%	1.69%	5.62%	34.27%
1.6–2	3.93%	1.69%	1.69%	3.93%	3.37%	1.69%	16.29%
>2	6.74%	0.56%	0.56%	0.56%	2.81%	2.25%	13.48%
Total	39.9%	12.4%	8.99%	13.5%	12.4%	12.9%	100%

Source: Haas et al., 1992.

The hydraulic characteristics of most finished water reservoirs, however, are not compatible with the ideal characteristics of chlorine contact chambers. The latter are most desirably plug flow, whereas the former most usually have a large degree of dispersion. Chapter 4 has a detailed discussion of ideal and non-ideal reactors (contact chambers) and how they may be characterized.

Superchlorination/Dechlorination. In the process of superchlorination/dechlorination, which has generally been employed when it is desired to treat a poor-quality water (with high ammonia-nitrogen concentrations, or perhaps with severe taste and odor problems), chlorine is added beyond the breakpoint. This process has become infrequent due to concerns with by-product formation. In this process, the ammonia-nitrogen present is oxidized. Generally, the residual chlorine obtained at this point is higher than may be desired for distribution. The chlorine residual may be decreased by the application of a dechlorinating agent (sulfur compounds or activated carbon).

Chloramination. Chloramination, the simultaneous application of chlorine and ammonia, or the application of ammonia prior to or after the application of chlorine, resulting in a stable combined residual, has been a long-standing practice at many utilities, with Denver using it since 1914 (Trussell and Kreft, 1984). Typical dose ratios are 5 to 7:1 on a Cl_2 to ammonia-nitrogen weight basis. With increasing concerns for disinfection by-products, the use of chloramination has been increasing as well.

A major concern that has been identified in chloramination arises during the transition from free chlorination to chlorammoniation (Means et al., 1986a). Before and after data on distribution system water quality were collected at the Metropolitan Water District of Southern California when the distribution system was changed from free chlorine residual to combined (monochloramine) chlorine residual. There was no effect on coliform counts; however, the plate counts on m-R2A medium increased dramatically (with some increase also observed using pour plates/TGEA medium). In one of the reservoirs, a precipitous drop in chlorine residual occurred following the switchover, associated with nitrification in the reservoir and growth of microorganisms. This was postulated to occur as a reaction between nitrites and monochloramine. The occurrence of postchloramination nitrification may also result in problems with pH control and corrosion control.

Influence of Relative Point of Addition of Chlorine and Ammonia. Details of chlorammoniation practice can dramatically influence process performance. Options as to pre- versus postammoniation, or pH, in particular, must be considered.

Prereacted chloramine residuals are more effective bactericides (*E. coli*, demand free batch tests) at pH 6 than at pH 8, and at high Cl_2:N ratios (5:1) rather than low ratios (down to 2:1). Concurrent addition of ammonia and chlorine was as effective as preammoniation (and at pH 6 was nearly as effective as free residual chlorination). Both simultaneous addition and preammoniation were more effective than prereaction of the chlorine, except at pH 8, where all three modes behaved in a similar manner (Ward et al., 1984).

In pilot plant studies, Means et al. (1986b) found that concurrent and sequential methods (chlorine at rapid mix and ammonia at end of flocculation tank) gave better performance at removing m-SPC bacteria than preammoniation (but poorer than free chlorination). Concurrent addition gave about as low a total trihalomethane value as preammoniation, and lower than the use of free chlorine alone.

In the presence of concentrations of organic nitrogen similar to ammonia, prereacted chloramines may give better performance than dynamically formed chloramines from preammoniation due to the favorable competition for chlorine by many organic nitrogen compounds (and their low biocidal potency) relative to inorganic nitrogen (Wolfe and Olson, 1985). However, such waters will likely produce high concentrations of N-Nitrosodimethylamine (NDMA), an emerging disinfection by-product (Mitch et al., 2003).

Contact Tank Hydraulics Characterizing and Improving. Any mixing—deviations from ideal plug flow behavior—reduces the disinfection efficiency of chlorine contactors. In serpentine contactors, the degree of mixing (i.e., the dispersion) can be estimated from the geometry of the tank, in particular the length-to-width ratio (Stover et al., 1986).

When a finished water reservoir is used as a contact tank, severe mixing, and other deviations from ideal flow (due for example to density, and temperature differences and wind) may occur (Boulos et al., 1996; Grayman et al., 1996). Various types of baffles may be used to counteract these tendencies (Grayman et al., 1996). In addition, both hydraulic scale models and computational fluid dynamic (CFD) models may be useful in assessing potential improvements in flow patterns from alternative baffling arrangements (Boulos et al., 1996; Grayman et al., 1996; Stefan et al., 1990). Similarly, chlorine contact chambers may be upgraded in terms of hydraulic performance by installation of baffles (Bishop et al., 1993).

Chlorine Dioxide

It is necessary to generate chlorine dioxide on a continuous basis for use as a disinfectant. Although a few European potable water treatment plants have been reported to use the acid-chlorite generation process (Miller et al., 1978), the most common synthesis route for disinfectant ClO_2 generation is the chlorine-chlorite process.

Chemistry of Generation. For practical purposes in water treatment, chlorine dioxide is generated exclusively from chlorite in as much as the reductive processes using chlorate as a starting material are capital intensive and competitive only at larger capacities (Masschelein, 1979b).

In the acid-chlorite process, sodium chlorite and hydrochloric acid react according to Eq. 17-39.

$$5NaClO_2 + 4HCl = 4ClO_2 + 5NaCl + 2H_2O \tag{17-39}$$

The resulting chlorine dioxide may be evolved as a gas or removed in solution. Mechanistically, this process occurs by a series of coupled reactions, some of which may involve the in situ formation of chlorine, the catalysis by chloride, and the oxidation of chlorite by chlorine (Gordon et al., 1972; Masschelein, 1979a; Noack and Doeff, 1979). In addition, the yield of the reaction as well as the rate of the process is improved by low pH values in which both gaseous chlorine and chlorous acid formation are favored. Under these favorable conditions, the reaction proceeds in the order of minutes; however, to achieve these conditions, excess hydrochloric acid is required.

During the acid-chlorite reaction, the following side reactions result in chlorine production:

$$5ClO_2^- + 5H^+ = 3ClO_3^- + Cl_2 + 3H^+ + H_20$$

$$4ClO_2^- + 4H^+ = 2Cl_2 + 3O_2 + 2H_2O$$

$$4HClO_2 = 2ClO_2 + HClO_3 + HCl + H_2O$$

if equal weights of the reactants are added, than close to 100 percent of the conversion of chlorite may occur; this means a final pH below 0.5 (Masschelein, 1979b).

Alternatively, chlorine dioxide may be produced by the oxidation of chlorite with chlorine gas according to the reaction in Eq. 17-43.

$$2NaClO_2 + Cl_2 = NaCl + 2ClO_2 \tag{17-43}$$

As in the previous case, decreases in pH accelerate the rate of this process, as do excess amounts of chlorine gas. However, if chlorine gas is used in stoichiometric excess, the resultant product may contain a mixture of unconsumed chlorine as well as chlorine dioxide. The excess unreacted chlorine may result in the formation of undesirable disinfection by-products such as the trihalomethanes.

The rate of the direct reaction between dissolved Cl_2 and chlorite has been measured (Aieta and Roberts, 1985), with a forward second-order rate constant given by the following:

$$k_f = 1.31\times10^{11}\exp\left(\frac{-4800}{T}\right)\ \mathrm{L/M\text{-}s} \tag{17-44}$$

In the chlorine-chlorite process, sodium chlorite is supplied as either a solid powder or a concentrated solution. A solution of chlorine gas in water is produced by a chlorinator-ejector system, of similar design to that used in chlorination. The chlorine-water solution and a solution of sodium chlorite are simultaneously fed into a reactor vessel packed with Raschig rings to promote mixing (Miller et al., 1978). From Eq. 17-42, 1 mole of chlorine is required for moles of sodium chlorite, or 0.78 parts of Cl_2 per part of $NaClO_2$ by weight; however, for this reaction to proceed to completion, it is necessary to reduce the pH below that provided by the typically acidic chlorine-water solution produced by an ejector. At 1:1 feed ratios by weight, only 60 percent of the chlorite typically reacts (Miller et al., 1978). To provide greater yields, several options exist. First, it is possible to produce chlorine-water solutions in excess of 3500 mg/L using pressurized injection of gas. In this case, however, there will be an excess of unreacted chlorine in the product solution, and the resultant disinfectant will consist of a mixture of chlorine and chlorine dioxide. The second option consists of acid addition to the chlorine and chlorite solution. For example, the addition of 0.1 mole of HCl per mole of chlorite enabled the production of a disinfectant solution of 95 percent purity in terms of chlorine dioxide, and achieved a 90 percent conversion of chlorite to chlorine dioxide (Jordan 1980). A third process, developed by CIFEC (Paris, FR), involves recirculation of the chlorinator-ejector discharge water back to the ejector inlet to produce a strong (5000–6000 mg/L) chlorine solution, typically at pH below 3.0, and in this manner to increase the efficiency of chlorite conversion (Miller et al., 1978). It has been reported that this last option is capable of producing 95 to 99 percent pure solutions of chlorine dioxide. The intricacies of chlorine dioxide reactions and by-products necessitate careful process monitoring during operation of the generator (Lauer et al., 1986).

Example 17-4 A water utility has chlorination capacity of 907 tonne/day (1000 tons/day) and is considering a switchover to chlorine dioxide to be generated using the chlorine-chlorite process. If the existing chlorination equipment is to be used, what is the maximum production capacity of chlorine dioxide, and how much sodium chlorite must be used under these conditions? Assume ideal stoichiometry and no excess chlorite or chlorine requirements.

Solution From Eq. 17-43, 2 moles of sodium chlorite ($NaClO_2$) react with 1 mole of chlorine to produce 2 moles of chlorine dioxide. Because chlorine has a molecular weight of 70, the current chlorinators have a capacity of 454,000/70 = 6486 moles/day of chlorine. Therefore, 12,972 moles/day of sodium chlorite are required, and the result would be an equal number of moles of chlorine dioxide. The molecular weights are 67 for ClO_2 and 89 for NaOCl.

Therefore, the sodium chlorite required is 12,971(89) = 1.15×10^6 g/day (2540 lb/day). The chlorine dioxide produced would be 12,971(67) = 0.87×10^6 g/day (1914 lb/day). ▲

Application of Chlorine Dioxide. The use of chlorine dioxide is limited by two factors. First, the maximum residual which does not cause adverse taste and odor problems is 0.4–0.5 mg/l as ClO_2 (Masschelein, 1979b). Second, the chlorite produced by the reduction of chlorine dioxide has been found to cause certain types of anemia. Health effects from disinfectants and disinfection by-products are reviewed in Chap. 2. Therefore the maximum chlorine dioxide residual is limited by a maximum residual dsinfectant level (MRDL) of 0.8 mg/L as ClO_2 (Chap. 1).

Insofar as chlorine dioxide is produced free of chlorine (in the acid-chlorite process, or in optimized chlorine-chlorite processes), the reactions with organic material to produce chlorinated by-products appear less significant than with chlorine (Aieta and Berg, 1986; Lykins and Griese 1986). Pure chlorine dioxide would not be expected to produce THMs or HAAs. Additional discussion of by-products can be found in Chap. 19.

Ozonation

The generation and application of ozone is presented in Chap. 7.

The role of hydraulics in ozone contactors has two influences on performance. It influences ozone transfer efficiency, and by virtue of the influence of dispersion on performance, it influences the effectiveness with which a given combination of residual and time can be effective. The complex interactions involved are illustrated in several papers (Lev and Regli, 1992; Martin et al., 1992; Roustan et al., 1991). The designer is faced with a trade-off in which multiple stages, each approximating a completely mixed system, are used to transfer ozone. As the number of stages increase, the overall system hydraulics and performance improves (see Chap. 4), and also potentially the transfer efficiency may improve. The improvement in residence time distribution and transfer efficiency may be sufficient to reduce the size of reactor required, thus offsetting the increase in complexity of the internal reactor design.

UV Radiation

The application and design of UV systems is considered in Chap. 18.

The role of hydraulics in UV disinfection has two effects on performance. In addition to a negative influence of mixing, due to dispersion as noted, the movement of microorganisms relative to lamp surfaces may dramatically influence the integrated exposure to radiation during contacting. The effect of turbulence to decrease the potential for stratification (defined as travel of a given microorganism at a fixed distance from lamp surfaces) is desirable to increase UV disinfection efficiency (Haas and Sakellaropoulous, 1979).

There is an interaction between the residence time distribution and the distribution of intensities of exposure which microorganisms experience in transiting a UV disinfection unit. The use of CFD models may assist in providing a more precise prediction of the effect of flow rate and water quality parameters (absorbancy) on the inactivation expected in full-scale UV units (Chiu et al., 1999; Lyn et al., 1999).

From Kinetics to Process Design

In general, microbial inactivation kinetics has been determined in batch systems. Real contactors, however, are continuous and may have non-ideal flow patterns with mixing and short-circuiting. If the residence time distribution of a continuous flow reactor is known, for example, from tracer experiments, then it is possible to develop reasonable estimates of inactivation efficiency in these systems (Haas, 1988; Haas et al., 1995; Lawler and Singer, 1993; Stover et al., 1986; Trussell and Chao, 1977). There are several design approaches, the use of a $C \times T$ concept (as embodied in regulations), the integration of residence time and kinetics (termed the *integrated disinfection design framework* (IDDF)), and the direct solution of the coupled fluid transport and kinetic relations using CFD (see Chap. 4 for discussion of CFD).

The residence time distribution is determined from a pulse or step tracer experiment. In a pulse experiment, a virtually instantaneous slug of tracer is introduced, while in a step experiment a virtually instantaneous step change in concentration is effected. Experimental aspects of these experiments in disinfection systems have been described by a number of authors (Bishop et al., 1993; Boulos et al., 1996; Teefy and Singer, 1990). From this experiment, descriptive characteristics of the residence time distribution may be obtained, such as the mean residence time and dispersion; in addition the cumulative residence time distribution (F curve) and its density function (E curve) is also directly obtained. The E curve, written as a function of time $E(t)$, is the probability density function that gives the fraction of the fluid elements leaving the reactor that were in the reactor a period of time between t and $t + dt$. The tracer tests, tracer curves, and data analysis are discussed in detail in Chap. 4.

Two commonly used analytical forms for residence time distribution (though they may not always be appropriate) are the axial dispersion and the tanks-in-series models (Haas et al., 1997; Nauman and Buffham, 1983). These are given respectively by

Tanks in series model

$$E(t) = \frac{1}{t\,\Gamma(N_{\text{CFSTR}})}\left(\frac{N_{\text{CFSTR}}t}{\bar{t}}\right)^{N_{\text{CFSTR}}} \exp\left(-\frac{N_{\text{CFSTR}}t}{\bar{t}}\right) \tag{17-45}$$

Axial dispersion model

$$E(t) = \sqrt{\frac{\bar{t}}{2\pi t^3 \nu}} \exp\left[-\frac{(t-\bar{t})^2}{2\bar{t}t\nu}\right] \tag{17-46}$$

where N_{CFSTR} is the number of equal volume tanks in series and ν is the dimensionless variance, which is related to the Peclet number by Eq. 17-47. The Peclet number (*Pe*) is defined as $Pe = v_z D/L$ where v_z is the axial velocity, D is the dispersion coefficient, and L is the flow path length. The reciprocal of the *Pe* number is referred to as the *dimensionless dispersion number;* see Chap. 4.

$$\nu = \frac{2}{Pe} - \frac{2}{Pe^2}(1 - \exp(-Pe)) \tag{17-47}$$

The axial dispersion model also has the integrated solution that gives F, the cumulative fraction of fluid that spends a residence time $\leq t$ in the system as

$$F(t) = \Phi\left[\left(\frac{t}{\bar{t}} - 1\right)\sqrt{\frac{\bar{t}}{t\nu}}\right] + \exp\left(\frac{2}{\nu}\right)\Phi\left[-\left(\frac{t}{\bar{t}} + 1\right)\sqrt{\frac{\bar{t}}{t\nu}}\right] \tag{17-48}$$

where $\Phi(z)$ is the standard normal probability integral (i.e., area under the normal distribution from negative infinity to z).

Use of the Integrated Disinfection Design Framework (IDDF). Using the E curve and the batch kinetic rate expression, the composition of the effluent can be obtained. The approach assumes a completely segregated system (Danckwerts, 1958; Levenspiel, 1972) and estimates the survival ratio in a disinfection reactor by the following equation:

$$\ln\left(\frac{N}{N_0}\right)_{\text{continuous}} = \int_0^{\infty} S(t)E(t)\,dt \tag{17-49}$$

where $\ln(N/N_0)$continuous = predicted continuous survival ratio
$S(t)$ = predicted batch survival at time t, given influent concentration as initial condition (N/N_0)
$E(t)$ = normalized density function for the residence time distribution

The complete segregation model is one extreme of micromixedness behavior. In this model, all fluid (and particles) entering remain in the vicinity of the material with which they entered until the last possible time (given by the residence time distribution) before their exit. This (complete segregation model) underlies the approach used by USEPA in the $C \times T$ tables, and therefore there is a systematic bias to greater predicted inactivation than other micromixedness reactor properties (Craik. 2005; Haas, 1988).

Application of this method is illustrated by the following example.

Example 17-5 Batch studies of inactivation of *Giardia* in water by chlorine have shown the applicability of the Hom model with first-order decay. The kinetic parameters are $k^* = 0.03$, $k = 0.1$, $m = 1.2$, and $n = 0.9$. A contactor is to be designed with a mean residence time of 30 minute and a Peclet number of 50. If a chlorine dose of 2 mg/L is to be used, what is the anticipated survival ratio?

Solution

1. First, from Eq. 17-47, the dimensionless variance is computed as

$$v = \frac{2}{50} - \frac{2}{50^2}(1 - e^{-50}) = 0.0392$$

2. Now, the trapezoidal rule (Chapra and Canale, 1988) can be used to evaluate the integral as shown in the following table:

Time (min)	$S(t)$	$E(t)$	$S(t) \times E(t)$	Trapezoid area
0	1	0	0	
5	0.3000	8.189E–24†	2.457E–24	6.141E–24
10	0.0750	1.435E–08	1.076E–09	2.690E–09
15	0.0192	3.228E–04	6.206E–06	1.552E–05
20	0.0053	0.0147	7.808E–05	2.107E–04
25	0.0016	0.0577	9.234E–05	4.261E–04
30	5.308E–04	0.0672	3.565E–05	3.200E–04
35	1.934E–04	0.0393	7.609E–06	1.081E–04
40	7.713E–05	0.0151	1.162E–06	2.193E–05
45	3.349E–05	0.0044	1.461E–07	3.271E–06
50	1.574E–05	0.0010	1.637E–08	4.062E–07
55	7.954E–06	2.158E–04	1.716E–09	4.522E–08
60	4.299E–06	4.035E–05	1.735E–10	4.724E–09

†E in the columns means 10; e.g., E-24 means 10^{-24}.

Time intervals are selected on the basis that the finer the time spacing, the more precise is the evaluation of the integral. More sophisticated integration approaches, such as Simpson's rule, or higher-order methods may also be chosen to improve precision (Chapra and Canale, 1988). The second and third columns are the computed values of the survival (S) and residence time density function (E) at the particular time (t). The fourth column is the product of S and E. The final column represents the contribution of the rectangle to the area. If t_0 and t_1 represent the prior row and the current row, then the values in the final column are given by

$$\left[\frac{S(t_0)E(t_0) + S(t_1)E(t_1)}{2}\right](t_1 - t_0)$$

The sum of the final column is 0.0011061, indicating that there is just under three logs removal, that is, 0.11 percent survival. Note that times beyond 60 minute were not examined, since at these higher times, as shown in the right-hand column the contribution to the integral is very small. ▲

The $C \times T$ Approach in Regulation. Under the Surface Water Treatment Rule, the adequacy of disinfection is judged using the product of the final residual concentration of disinfectant (in mg/L) and the contact time in minutes ($C \times T$). This is meant to be a simpler computational method than the IDDF described previously. The contact time is evaluated as that which is exceeded by 90 percent of the fluid (this is designated as the t_{10}, denoting that 10 percent of the fluid has a smaller residence time in the system; see Chap. 4 for further information. The t_{10} for disinfection systems may be evaluated from tracer studies or by the use of default multipliers of the theoretical hydraulic residence time (V/Q). Critical values of the $C \times T$ to achieve varying levels of disinfection of *Giardia* and viruses, incorporating a margin of safety, have been published under USEPA guidance documents (Malcolm Pirnie and HDR Engineering, 1991; United States Environmental Protection Agency, 1991). For ozone, chlorine dioxide, and chloramines, they are indicated in Table 17-7. In the case of

TABLE 17-7 CT Values for 99.9 Percent Reduction of *Giardia lamblia* with Ozone, Chlorine Dioxide, and Chloramines

		Temperature (°C)					
Disinfectant disinfectant	pH	1	5	10	15	20	25
Ozone	6–9	2.9	1.9	1.4	0.95	0.72	0.48
Chlorine dioxide	6–9	63	26	23	19	15	11
Chloramines	6–9	3800	2200	1850	1500	1100	750

Source: Malcolm Pirnie and HDR Engineering, 1991.

free chlorine the $C \times T$ values are functions of temperature, pH, and concentration (under the regulations, measured at the end of the contact chamber). Space does not permit a recapitulation of the full $C \times T$ tables for free chlorine; the values at a few conditions are given in Table 17-8.

TABLE 17-8 CT Values for 99.9 Percent Reduction of *Giardia lamblia* with Free Chlorine Under Selected Conditions

Temperature (°C) pH	5 7	5 9	20 7
C (mg/L)			
0.4	139	279	52
0.6	143	291	54
0.8	146	301	55
1	149	312	56
1.2	152	320	57
1.4	155	329	58
1.6	158	337	59
1.8	162	345	61
2	165	353	62
2.2	169	361	63
2.4	172	368	65
2.6	175	375	66
2.8	178	382	67
3	182	389	68

Source: Malcolm Pirnie and HDR Engineering, 1991.

The use of the $C \times T$ approach is illustrated in the following example.

Example 17-6 The disinfection contactor described in Example 17-5 is to be used to achieve 3-log inactivation of *Giardia.* Compute the required residual necessary to achieve this result from the $C \times T$ tables if the pH is 7 and the temperature is 20°C.

Solution

1. First, the t_{10} must be computed. Since the contactor is an axial dispersion contactor, Eq. 17-48 can be used. We therefore need to solve the following equation for the time at which 10 percent of the fluid has exited:

$$0.1 = \Phi\left[\left(\frac{t_{10}}{30} - 1\right)\sqrt{\frac{30}{t_{10}(0.0392)}}\right] + \exp\left(\frac{2}{0.0392}\right)\Phi - \left[\left(\frac{t_{10}}{30} + 1\right)\sqrt{\frac{30}{t_{10}(0.0392)}}\right]$$

   This equation must be solved by trial and error; however, this can readily be done on a spreadsheet such as Microsoft Excel (in which the cumulative normal distribution function is available). Doing this, it is found that $t_{10} = 25.3$ minute.

2. We now need to examine the last column of Table 17-8 to determine what chlorine residual would satisfy the requirement at this t_{10}. The following table illustrates the computation:

C	CT	$t = (CT/C)$
2.4	65	27.08
2.6	66	25.38
2.8	67	23.93

3. Finally, an inverse interpolation is needed to obtain the C that yields the desired t. Therefore, the final residual chlorine should exceed 2.61 mg/L. ▲

Computational Fluid Dynamic Design. Computational fluid dynamic (CFD) methods directly simulate the motion of liquids and potentially the reactions of materials within the fluids. They are computer-intensive methods and are increasingly being used for process design, particularly for the design of large systems. A more detailed explanation of CFD techniques is given in Chap. 4.

The simplest application of CFD techniques in disinfection systems is to simulate the flow patterns in a contactor (not considering reactions) (Hannoun et al., 1998; Henry and Freeman, 1995). This can be used either to estimate a t_{10} (for application in CT-based design) or to generate the residence time distribution for use in an IDDF approach. This avoids the need to perform tracer tests and can be used before a system is constructed.

The decay rate of disinfectant can be incorporated into a CFD model, and then an average integrated ct can be computed by mathematical simulation. This results in a direct integration of residual and time that can be compared with ct tables. The method has been illustrated by Do-Quang et al. (2000).

If experiments are done to elucidate the inactivation kinetics (e.g., Hom, Chick-Watson, etc.) as well as the disinfectant decay rates, then both of these phenomena can be incorporated into a CFD simulation to directly estimate the inactivation ratio of a microorganism in a contactor. This has been illustrated both in the case of homogenous chlorine contactors (Greene et al., 2004; Greene et al., 2006) and with multiphase ozone contactors (where both the gas and liquid phases must be simulated along with the two phase mass transfer) (Bartrand, 2007; Bartrand et al., 2009).

RELATIVE COMPARISONS

The dominant disinfection technology in the United States remains free residual chlorination. However, chloramination, chlorination with partial dechlorination, and chlorine

TABLE 17-9 Applicability of Alternative Disinfection Techniques

Consideration	Cl_2[†]	O_3	ClO_2	UV
Equipment reliability	Good	Fair to good	Good	Fair to good
Relative complexity of technology	Simple	Complex	Moderate	Moderate
Safety concerns	Yes	Moderate	Yes	Moderate
Bactericidal	Good	Good	Good	Good
Virucidal	Moderate	Good	Moderate	Moderate to good
Efficacy against prototozoa	Fair	Moderate	Fair	Good
By-products of possible health concern	Yes	Some	Some	None known
Persistent residual	Long	None	Moderate	None
Reacts with ammonia	Yes	No	No	No
pH dependent	Yes	Slight	Slight	No
Process control	Well developed	Developed, more complex	Developing	Developing
Intensiveness of operations and maintainance	Low	High	Moderate	High

[†]Includes hypochlorites as well as chlorine gas.

dioxide treatment are viable alternative primary disinfectants. In addition, ozone, or possibly UV, when supplemented with a chemical which can produce a lasting residual in the distribution system have application.

Table 17-9 summarizes important aspects of the major technologies and their technical pros and cons.

ABBREVIATIONS

CFD	computational fluid dynamics
CFSTR	continuous flow stirred tank reactor
$C \times T$	product of contact time and residual
HAA	haleoacetic acids
HPC	heterotrophic plate count
IDDF	integrated disinfection design framework
LT2ESWTR	Long Term 2 Enhanced Surface Water Treatment Rule
MRDL	maximum residual disinfectant level
NDMA	N-nitrosodimethylamine
PEL	permissible exposure limit
SWTR	Surface Water Treatment Rule
OSHA	Occupational Safety and Health Administration
THM	trihalomethane
TOC	total organic compounds
USEPA	U.S. Environmental Protection Agency
UV	ultraviolet

NOTATION FOR EQUATIONS

C	concentration, M/L^3
C_0	initial concentration, M/L^3
CT	concentration-time product, MT/L^3
$E(t)$	residence time distribution density function, $1/T$
E_a	activation energy
$F(t)$	residence time distribution cumulative function, dimensionless
H	Henry's law constant, L^2/T^2
k'	Chick-Watson rate constant, $(M/L^3)^{-n}/T$
k''	Hom rate constant, $(M/L^3)^{-n}T^{-m}$
n	Chick-Watson coefficient of dilution, dimensionless
N	microorganism concentration, $\#/L^3$
N_{CFSTR}	number of CFSTRs in tanks in series model, dimensionless
Pe	Peclet number, dimensionless
T	temperature
t	time, T
t_{10}	time of passage for fastest 10% of fluid, T
$\bar{t}$	mean residence time, T
P	pressure
r	reaction rate, M/L^3T, or for microbial inactivation, $\#/L^3T$
R	ideal gas law constant
$\Gamma(x)$	gamma function
ν	dispersion number, dimensionless
$\Phi(t)$	cumulative standard normal integral, dimensionless

REFERENCES

Abad, F. X., Pinto, R. M., Diez, J. M., and Bosch, A. (1994). Disinfection of Human Enteric Viruses in Water by Copper and Silver in Combination with Low Levels of Chlorine. *Applied and Environmental Microbiology*, 60(7), 2377–2383.

ACGIH. (1994). *Threshold Limit Values for Chemical Substances and Physical Agents and Biological Exposure Indices*, American Conference of Governmental Industrial Hygienists, Cincinnati, OH.

Aieta, E., and Berg, J. (1986). A Review of Chlorine Dioxide in Drinking Water Treatment. *Journal AWWA*, 78(6), 62.

Aieta, E., and Roberts, P. (1985). The Chemistry of Oxo-Chlorine Compounds Relavent to Chlorine Dioxide Generation. In *Water Chlorination: Environmental Impact and Health Effects,* ed. R. Jolley. Lewis Publishers.

Allen, M. (1977). The Impact of Excessive Bacterial Populations in Coliform Methodology. Proc. American Society for Microbiology Annual Conference, New Orleans.

Anonymous. (1922). Is Chlorination Effective Against all Waterborne Disease? *Journal of the American Medical Association*, 78, 283.

Anotai, J. (1996). Effect of Calcium Ion on Chemistry and Disinfection Efficiency of Free Chlorine at pH 10, Doctoral Dissertation, Drexel University, Philadelphia, PA.

APHA, AWWA, and WPCF (American Public Health Association, American Water Works Association, and Water Pollution Control Federation). (1989). *Standard Methods for the Examination of Water and Wastewater*. Washington, DC.

Aston, R., and Synan, J. (1948). Chlorine Dioxide as a Bactericide in Waterworks Operation. *Journal of the New England Water Works Association*, 62, 80.

Averill, C. (1832). Facts Regarding the Disinfecting Powers of Chlorine: Letter to Hon. J. I. Degraff, Mayor of the City of Schenectady NY. Schenectady, NY: SS Riggs Printer.

AWWA Disinfection Systems Committee. (2008). Committee Report: Disinfection Survey, Part 1—Recent Changes, Current Practices and Water Quality. *Journal AWWA*, 100(10), 76–90.

AWWA Water Quality Division Disinfection Systems Committee. (2000). Committee Report: Disinfection at Large and Medium-Size Systems. *Journal AWWA*, 92(5), 32–43.

Bailey, J. E., and Ollis, D. F. (1986). *Biochemical Engineering Fundamentals.* New York: McGraw-Hill.

Baker, J. (1926). Use of Chlorine in the Treatment of Sewage. *Surveyor*, 69, 241.

Baker, R. (1947). Studies on the Reaction Between Sodium Hypochlorite and Proteins. I. Physicochemical Study of the Course of the Reaction. *Biochemical Journal*, 41, 337.

Baker, R. J. (1969). Characteristics of Chlorine Compounds. *Journal of the Water Pollution Control Federation*, 41, 482.

Barbeau, B., Huffman, D., Mysore, C., Desjardins, R., Clement, B., and Prevost, M. (2005). Examination of Discrete and Confounding Effects of Water Quality Parameters During the Inactivation of ms2 Phages and *Bacillus subtilis* Spores with Chlorine Dioxide. *Journal of Environmental Engineering and Science*, 4(2), 139–51.

Bartrand, T. A. (2007). High resolution experimental studies and numerical analysis of fine bubble ozone disinfection contactors, Doctoral Dissertation, Drexel University, Philadelphia.

Bartrand, T. A., Farouk, B., and Haas, C. N. (2009). Countercurrent gas/liquid flow and mixing: Implications for water disinfection. *International Journal of Multiphase Flow*, 35(2), 171–184.

Bates, R., Sutherland, S., and Shaffer, P. (1978). Development of Resistant Poliovirus by Repetitive Sublethal Exposure to Chlorine. In *Water Chlorination: Environmental Impact and Health Effects*, ed. R. Jolley. Stoneham, MA: Butterworths.

Battino, R. (1984). *Chlorine Dioxide Solubilities*. New York: Pergamon.

Belohlav, L. R., and McBee, E. T. (1966). Discovery and Early Work. Chlorine: Its Manufacture, Properties and Uses: ACS Monograph No. 154, ed. J. Sconce. New York:.Van Nostrand Reinhold.

Benarde, M. A., Israel, B. M., Olivieri, V. P., and Granstrom, M. L. (1965). Efficiency of Chlorine Dioxide as a Bactericide. *Applied Microbiology*, 13(5), 776.

Benarde, M. A., Snow, W. B., Olivieri, V. P., and Davidson, B. (1967). Kinetics and Mechanism of Bacterial Disinfection by Chlorine Dioxide. *Applied Microbiology*, 15(2), 257.

Berg, G., Sanjaghsaz, H., and Wangwongwatana, S. (1989). Potentiation of the Virucidal Effectiveness of Free Chlorine by Substances in Drinking Water. *Applied and Environmental Microbiology*, 55(2), 390–393.

Berg, J. D., et al. (1985). Disinfection Resistance of *Legionella pneumophila, Escherichia coli* Grown in Continuous and Batch Culture. Water Chlorination: In *Environmental Impact and Health Effects*, ed. R. Jolley. Chelsea, MI: Lewis Publishers.

Berg, J. D., Roberts, P. V., and Matin, A. (1986). Effect of chlorine dioxide on selected membrane functions of *Escherichia coli*. *Journal of Applied Microbiology*, 60(3), 213–220.

Bishop, M., Morgan, J., Cornwell, B., and Jamison, D. (1993). Improving the Disinfection Detention Time of a Water Plant Clearwell. *Journal AWWA*, 85(3), 68–75.

Blostein, J. (1991). Shigellosis from Swimming in a Park Pond in Michigan. *Public Health Reports*, 106(3), 317–21.

Boardman, G. D., and Sproul, O. J. (1977). Adsorption as a Protective Mechanism for Poliovirus. Proc. American Water Works Association, 97th Annual Conference, Anaheim, CA.

Bohrerova, Z., and Linden, K. G. (2007). Standardizing photoreactivation: Comparison of DNA photorepair rate in Escherichia coli using four different fluorescent lamps. *Water Research*, 41(12), 2832–2838.

Bolyard, M., Fair, P. S., and Hautman, D. (1993). Sources of Chlorate Ion in US Drinking Water. *Journal AWWA*, 85(9), 81–88.

Bolyard, M., Fair, P. S., and Hautman, D. P. (1992). Occurrence of Chlorate in Hypochlorite Solutions Used for Drinking Water Disinfection. *Environmental Science & Technology*, 26(8), 1663–1665.

Bonde, G. (1966). Bacteriological Methods for Estimation of Water Pollution. *Health Laboratory Science*, 3(2), 124.

Boulos, P., Grayman, W., Bowcock, R., Clapp, J., Rossman, L., Clark, R., Deininger, R., and Dhingra, A. (1996). Hydraulic Mixing and Free Chlorine Residual in Reservoirs. *Journal AWWA*, 88(7), 48–59.

Brennhovd, O., Kapperud, G., and Langeland, G. (1992). Survey of Thermotolerant *Campylobacter* spp and *Yersinia* spp in 3 Surface Water Sources in Norway. *International Journal of Food Microbiology*, 15(3–4), 327–338.

Bridgman, S. A., Robertson, R. M. P., and Hunter, P. R. (1995). Outbreak of cryptosporidiosis associated with a disinfected ground water supply. *Epidemiology and infection*, 115(3), 555–566.

Brigano, F. A. O., et al. Effect of Particulates on Inactivation of Enteroviruses in Water by Chlorine Dioxide. Proc. American Society for Microbiology Annual Confeerence.

Brown, T. J., Hastie, J. C., Kelly, P. J., Vanduivenboden, R., Ainge, J., Jones, N., Walker, N. K., Till, D. G., Sillars, H., and Lemmon, F. (1992). Presence and Distribution of Giardia cysts in New-Zealand Waters. *New Zealand Journal of Marine and Freshwater Research*, 26(2), 279–282.

Butterfield, C. T., and Wattie, E. (1946). Influence of pH and Temperature on the Survival of Coliforms and Enteric Pathogens When Exposed to Chloramine. *U.S. Public Health Service Reports*.

Butterfield, C. T., Wattie, E., Megregian, S., and Chambers, C. W. (1943). Influence of pH and Temperature on the Survival of Coliforms and Enteric Pathogens When Exposed to Free Chlorine. *U.S. Public Health Reports*, 58, 1837.

Cabelli, V. (1977). *Clostridium perfringens* as a Water Quality Indicator. In *Bacterial Indicators/Health Hazards Associated with Water*, ed. A. Hoadley and B. Dutka. Philadelphia: ASTM International.

Cachaza, J. (1976). Kinetics of Oxidation of Nitrite by Hypochlorite in Aqueous Basic Solution. *Canadian Journal of Chemistry*, 54, 3401.

Centers for Disease Control and Prevention. (1995). Assessing the Public Health Threat Associated with Waterborne Cryptosporidiosis: Report of a Workshop. *Morbidity and Mortality Weekly Reports*, 44(RR6), 1–19.

Cerf, O. (1977). Tailing of Survival Curves of Bacterial Spores. *Applied Bacteriology*, 42, 1.

Chang, S. (1944). Destruction of Microorganisms. *Journal AWWA*, 36(11), 1192.

Chang, S. (1971). Modern Concept of Disinfection. *ASCE Journal of the Sanitary Engineering Division*, 97, 689.

Chapra, S. C., and Canale, R. P. (1988). *Numerical Methods for Engineers*. New York: McGraw-Hill.

Chen, Y., Sproul, O., and Rubin, A. (1985). Inactivation of *Naegleria gruberi* Cysts by Chlorine Dioxide. *Water Research*, 19(6), 783.

Chick, H. (1908). An Investigation of the Laws of Disinfection. *Journal of Hygiene*, 8, 92–157.

Chiu, K., Lyn, D. A., Savoye, P., and Blatchley, E. R., III. (1999). Integrated UV Disinfection Model Based on Particle Tracking. *Journal of Environmental Engineering*, 125(1), 7–16.

Clark, J., and Pagel, J. (1977). Pollution Indicator Bacteria Associated with Municipal Raw and Drinking Water Supplies. *Canadian Journal of Microbiology*, 23, 465.

Copeland, C. (2007). Terrorism and Security Issues Facing the Water Infrastructure Sector, Congressional Research Service, Washington, DC.

Costan-Longares, A., Montemayor, M., Payan, A., Mendez, J., Jofre, J., Mujeriego, R., and Lucena, F. (2008). Microbial indicators and pathogens: Removal, relationships and predictive capabilities in water reclamation facilities. *Water Research*, 42(17), 4439–4448.

Craik, S. A. (2005). Effect of Micro-Mixing Conditions on Predictions of *Cryptosporidium* Inactivation in an Ozone Contactor. *Ozone: Science & Engineering: The Journal of the International Ozone Association*, 27(6), 487–494.

Dahi, E., and Lund, E. (1980). Steady State Disinfection of Water by Ozone and Sonozone. *Ozone Science and Engineering*, 13.

Danckwerts, P. V. (1958). The Effect of Incomplete Mixing on Homogeneous Reactions. *Chemical Engineering Science*, 8(2), 93–102.

De Laat, J. (1982). Chloration de Composes Organizes: Demande On Chlore et Reactivite Vis a Vis de la Furmation des Trihalomethe Incidence de l'Azote Ammoniacal. *Water Research*, 16, 143–147.

Dennis, W. H., Olivieri, V. P., and Kruse, C. W. (1979a). Mechanism of Disinfection: Incorporation of ^{36}Cl into f2 Virus. *Water Research*, 13, 363.

Dennis, W. H., Olivieri, V. P., and Kruse, C. W. (1979b). The Reaction of Nucleotides with Aqueous Hypochlorous Acid. *Water Research*, 13, 357.

Do-Quang, Z., Roustan, M., and Duguet, J.-P. (2000). Mathematical Modeling of Theoretical Cryptosporidium Inactivation in Full-Scale Ozonation Reactors. *Ozone: Science & Engineering*, 22(1), 99–111.

Downes, A., and Blount, T. (1877). Research on the Effect of Light Upon Bacteria and Other Organisms. *Proceedings of the Royal Society of London*, 26, 488.

Downs, A., and Adams, C. (1973). *The Chemistry of Chlorine, Bromine, Iodine and Astatine*, Pergamon, Oxford.

Drenchen, A., and Bert, M. (1994). A Gastroenteritis Illness Outbreak Associated with Swimming in a Campground Lake. *Journal of Environmental Hhealth*, 57(2), 7–10.

Dugan, N., Summers, R., Miltner, R., and Shukairy, H. (1995). An Alternative Approach to Predicting Chlorine Residual Decay, AWWA Water Quality Technology Conference, New Orleans, pp. 1317–1337.

Engelbrecht, R. S., Haas, C. N., Shular, J. A., Dunn, D. L., Roy, D., Lalchandani, A., Severin, B. F., and Farooq, S. (1979). Acid-Fast Bacteria and Yeasts as Indicators of Disinfection Efficiency, EPA-600/2-79-091, U.S. Environmental Protection Agency, Washington, DC.

Engelbrecht, R. S., Severin, B. F., Masarik, M. T., Farooq, S., Lee, S. H., Haas, C. N., and Lalchandani, A. (1977). New Microbial Indicators of Disinfection Efficiency, EPA-600/2-77-052, U.S. Environmental Protection Agency, Washington, DC.

Fair, G. M., Morris, J. C., Chang, S. L., Weil, I., and Burden, R. P. (1948). The Behavior of Chlorine as a Water Disinfectant. *Journal AWWA*, 40, 1051–1061.

Farooq, S. (1976). Kinetics of Inactivation of Yeasts and Acid-Fast Organisms with Ozone, PhD Dissertation, University of Illinois at Urbana-Champaign.

Farooq, S., Kurucz, C. N., Waite, T. D., and Cooper, W. J. (1993). Disinfection of Wastewaters: High Energy Electron vs. Gamma Irradiation. *Water Research*, 27(7), 1177–1184.

Feben, D., and Taras, M. J. (1950). Chlorine Demand Constants of Detroit's Water Supply. *Journal AWWA*, 42, 453–461.

Feben, D., and Taras, M. J. (1951). Studies on Chlorine Demand Constants. *Journal AWWA*, 43, 922–932.

Feng, T. (1966). Behavior of Organic Chloramines in Disinfection. *Journal of the Water Pollution Control Federation*, 38, 614.

Finch, G. R., Black, E. K., Labatiuk, C. W., Gyurek, L., and Belosevic, M. (1993). Comparison of *Giardia lamblia* and *Giardia muris* Cyst Inactivation by Ozone. *Applied and Environmental Microbiology*, 59(11), 3674–3680.

Finch, G. R., Haas, C. N., Oppenheimer, J. A., Gordon, G., and Trussell, R. R. (2001). Design Criteria for Inactivation of *Cryptosporidium* by Ozone in Drinking Water. *Ozone Science and Engineering*, 23(4), 259–84.

Friberg, L. (1957). Further Qualitative Studies on the Reaction of Chlorine with Bacteria in Water Disinfection. *Acta Pathologica et Microbiologica Scandanavica*, 40, 67.

Friberg, L., and Hammarstrom, E. (1956). The Action of Free Available Chlorine on Bacteria and Bacterial Viruses. *Acta Pathologica et Microbiologica Scandanavica*, 38, 127.

Fujioka, R., Tenno, K., and Loh, P. (1985). Mechanism of Chloramine Inactivation of Poliovirus: A Concern for Regulators. Water Chlorination. In *Environmental Impact and Health Effects*, ed. R. Jolley. Chelsea, MI: Lewis Publishers.

Gagnon, G. A., O'Leary, K. C., Volk, C. J., Chauret, C., Stover, L., and Andrews, R. C. (2004). Comparative Analysis of Chlorine Dioxide, Free Chlorine and Chloramines on Bacterial Water Quality in Model Distribution Systems. *Journal of Environmental Engineering*, 130(11), 1269–1279.

Gallaher, M. M., Herndon, J. L., Nims, L. J., Sterling, C. R., Grabowski, D. J., and Hull, H. F. (1989). Cryptosporidiosis and Surface Water. *American Journal of Public Health*, 79(1), 39–42.

Gard, S. (1960). Theoretical Considerations in the Inactivation of Viruses by Chemical Means. *Annalys of the New York Academy of Sciences*, 83, 638.

Garibi, A. (2007). Applying the Make or Buy Decision Process to Sodium Hypochlorite in Water Treatment Plants. *Journal AWWA*, 99(3), 118–124.

Geldreich, E. (1971). Waterborne Pathogens. In *Water Pollution Microbiology*, ed. R. Mitchell. New York: Wiley-Interscience.

Goldstein, S., Juranek, D., Ravenholt, O., Hightower, A., Martin, D., Mesnick, J., Griffiths, S., Bryant, A., Reich, R., and Herwaldt, B. (1996). Cryptosporidiosis: An Outbreak Associated with Drinking Water Despite State of the Art Water Treatment. *Annals of Internal Medicine*, 124(5), 459–68.

Gordon, G., Adam, L. C., and Bubnis, B. P. (1995). Minimizing Chlorate Ion Formation. *Journal AWWA*, 87(6), 97–106.

Gordon, G., Adam, L. C., Bubnis, B. P., Hoyt, B., Gillette, S. J., and Wilczak, A. (1993). Controlling the Formation of Chlorate Ion in Liquid Hypochlorite Feedstocks. *Journal AWWA*, 85(9), 89–97.

Gordon, G., Kieffer, R. G., and Rosenblatt, D. H. (1972). The Chemistry of Chlorine Dioxide. *Progress in Inorganic Chemistry*, 15, 201.

Grabow, W. (1968). The Virology of Waste Water Treatment. *Water Research*, 2, 675.

Grabow, W. O. K., and et al. (1983). Inactivation of Hepatitis a Virus and Indicator Organisms in Water by Free Chlorine Residuals. *Applied and Environmental Microbiology*, 46, 619.

Granstrom, M. L., and Lee, G. F. (1957). Rates and Mechanisms of Reactions Involving Oxychlorine Compounds. *Public Works*, 88(12), 90–92.

Gray, E., Margerum, D., and Huffman, R. (1978). Chloramine Equilibria and the Kinetics of Disproportionation in Aqueous Solution. In *Organometals and Metalloids, Occurrence and Fate in the Environment,* ed. F. E. Brenchman and J. M. Bellama. Washington, DC: American Chemical Society, 264–277.

Grayman, W., Deininger, R., Green, A., Boulos, P., Bowcock, R., and Godwin, C. (1996). Water Quality and Mixing Models for Tanks and Reservoirs. *Journal AWWA*, 88(7), 60–73.

Greene, D. J., Farouk, B., and Haas, C. N. (2004). CFD Design Approach for Chlorine Disinfection Processes. *Journal AWWA*, 96(8), 138–50.

Greene, D. J., Haas, C. N., and Farouk, B. (2006). Computational Fluid Dynamics Analysis of the Effects of Reactor Configuration on Disinfection Efficiency. *Water Environment Research*, 78(9), 909–19.

Griffin, A., and Chamberlin, N. (1941a). Relation of Ammonia Nitrogen to Breakpoint Chlorination. *American Journal of Public Health*, 31, 803.

Griffin, A., and Chamberlin, N. (1941b). Some Chemical Aspects of Breakpoint Chlorination. *Journal of the New England Water Works Association*, 55, 3.

Gurol, M., and Singer, P. (1982). Kinetics of Ozone Decomposition: A Dynamic Approach. *Environmental Science and Technology*, 16(7), 377.

Gyurek, L. L., and Finch, G. R. (1998). Modeling Water Treatment Chemical Disinfection Kinetics. *Journal of Environmental Engineering*, 124(9), 783–93.

Haag, W. R., and Lietzke, M. H. (1980). A kinetic model for predicting the concentrations of active halogen species in chlorinated saline cooling waters. *Water Chlorination: Environ. Impact Health Eff*, 3, 415–426.

Haas, C. N. (1980). A Mechanistic Kinetic Model for Chlorine Disinfection. *Environmental Science and Technology*, 14, 339–340.

Haas, C. N. (1981). Sodium Alterations of Chlorine Equilibria: Quantitative Description. *Environmental Science and Technology*, 15, 1243–1244.

Haas, C. N. (1988). Micromixing and Dispersion in Chlorine Contact Chambers. *Environmental Technology Letters*, 9(1), 35–44.

Haas, C. N. (1999). Benefits of Using a Disinfectant Residual. *Journal AWWA*, 91(1), 65–69.

Haas, C. N., and Anotai, J. The Effect of Calcium Ion on the Disinfection Efficiency of Free Chlorine at pH 10. In *Disinfecting Wastewater for Discharge and Reuse.* Water Environment Federation. Portland, OR, 3.13 – 3.24.

Haas, C. N., and Engelbrecht, R. S. (1980a). Chlorine Dynamics During Inactivation of Coliforms, Acid-Fast Bacteria and Yeasts. *Water Research*, 14, 1749–1757.

Haas, C. N., and Engelbrecht, R. S. (1980b). Physiological Alterations of Vegetative Microorganisms Resulting From Aqueous Chlorination. *Journal of the Water Pollution Control Federation*, 52, 1976–1989.

Haas, C. N., et al. (1992). Survey of Water Utility Disinfection Practices. *Journal AWWA*, 84(9), 121–128.

Haas, C. N., Heath, M. S., Jacangelo, J., Joffe, J., Anmangandla, U., Hornberger, J. C., and Glicker, J. (1995). *Development and Validation of Rational Design Methods of Disinfection*, American Water Works Association Research Foundation.

Haas, C. N., and Heller, B. (1989). Statistics of Microbial Disinfection. *Water Science and Technology*, 21(3), 197–201.

Haas, C. N., and Joffe, J. (1994). Disinfection Under Dynamic Conditions: Modification of Hom's Model for Decay. *Environmental Science and Technology*, 28(7), 1367–1369.

Haas, C. N., Joffe, J., Anmangandla, U., Jacangelo, J., and Heath, M. (1996). The Effect of Water Quality on Disinfection Kinetics. *Journal AWWA*, 88(3), 95–103.

Haas, C. N., Joffe, J., Heath, M., and Jacangelo., J. (1997). Continuous Flow Residence Time Distribution Function Characterization. *Journal of Environmental Engineering*, 123(2), 107–114.

Haas, C. N., Karalius, M. G., Brncich, D. M., and Zapkin., M. A. (1986). Alteration of Chemical and Disinfectant Properties of Hypochlorite by Sodium Potassium and Lithium. *Environmental Science and Technology*, 20, 822–826

Haas, C. N., and Karra, S. B. (1984a). Kinetics of Microbial Inactivation By Chlorine. I. Review of Results in Demand-Free systems. *Water Research*, 18, 1443–1449.

Haas, C. N., and Karra, S. B. (1984b). Kinetics of Microbial Inactivation By Chlorine. II. Kinetics in the Presence of Chlorine Demand. *Water Research*, 18, 1451–1454.

Haas, C. N., and Karra, S. B. (1984c). Kinetics of Wastewater Chlorine Demand Exertion. *Journal of the Water Pollution Control Federation*, 56, 170–3.

Haas, C. N., and Kaymak, B. (2003). Effect of Initial Microbial Density on Inactivation of Giardia muris by Ozone. *Water Research*, 37, 2980–2988.

Haas, C. N., and Kaymak, B. (2008). Effect of Initial Microbial Density on Inactivation of Escherichia coli by Monochloramine. *Journal of Environmental Engineering and Science*, 7, 237–245.

Haas, C. N., Meyer, M. A., and Paller, M. S. (1983a). Microbial Alterations in Water Distribution Systems and Their Relationship to Physical–Chemical Characteristics. *Journal AWWA*, 475–481.

Haas, C. N., Meyer, M. A., and Paller, M. S. (1983b). The Ecology of Acid-Fast Organisms in Water Supply Treatment and Distribution Systems. *Journal AWWA*, 75, 139–44.

Haas, C. N., Meyer, M. A., Paller, M. S., and Zapkin, M. A. (1983c). The Utility of Endotoxins as a Surrogate Indicator in Potable Water Microbiology. *Water Research*, 17, 803–807.

Haas, C. N., and Morrison, E. C. (1981). Repeated Exposure of Escherichia coli to Free Chlorine: Production of Strains Possessing Altered Sensitivity. *Water Air and Soil Pollution*, 16, 233–242.

Haas, C. N., and Sakellaropoulous, G. P. Rational (1978). Analysis of Ultraviolet Disinfection Reactors, ASCE Environmental Engineering Specialty Conference, San Francisco.

Haas, C. N., Severin, B. F., Roy, D., Engelbrecht, R. S., and Lalchandani, A. (1985a). Removal of New Indicators by Coagulation–Flocculation and Sand Filtration. *Journal AWWA*, 77, 67–71.

Haas, C. N., Severin, B. F., Roy, D., Engelbrecht, R. S., Lalchandani, A., and Farooq, S. (1985b). Field Observations on the Occurrence of New Indicators of Disinfection Efficiency. *Water Research*, 19, 323–9.

Haas, C. N., Sheerin, J. G., and Lue-Hing, C. (1988). Effects of Ceasing Disinfection on a Receiving Water. *Journal of the Water Pollution Control Federation*, 60, 667–673.

Haas, C. N., Trivedi, C. D., and O'Donnell., J. R. (1990). Further Studies of Hypochlorite Ion Pair Chemistry and Disinfection Efficiency. In *Water Chlorination Environmental Impact and Health Effects,* ed. R. Jolley et al. Boca Raton, FL: Lewis Publishers, pp. 729–740.

Hamelin, C., and Chung, Y. (1978). Role of the POL, REC and DNA Gene Products in the Repair of Lesions Produced in *E. col i*by Ozone. *Studia* (Berlin), 68, 229.

Hannoun, I. A., Boulos, P. F., and List, E. J. (1998). Using Hydraulic Modeling to Optimize Contact Time. *Journal AWWA*, 90(8), 77–87.

Hayes, E. B., Matte, T. D., O'Brien, T. R., McKinley, T. W., Logsdon, G. S., Rose, J. B., Ungar, B. L., Word, D. M., Pinsky, P. F., and Cummings, M. L. (1989). Large Community Outbreak of Cryptosporidiosis Due to Contamination of a Filtered Public Water Supply. *New England Journal of Medicine*, 320(21), 1372–1376.

Hejkal, T. W., Wellings, F. M., LaRock, P. A., and Lewis, A. L. (1979). Survival of Poliovirus Within Organic Solids During Chlorination. *Applied and Environmental Microbiology*, 38, 114.

Henry, D. J., and Freeman, E. M. (1995). Finite Element Analysis and T_{10} Optimization of Ozone Contactors. *Ozone: Science & Engineering*, 17(6), 587–606.

Herwaldt, B., Craun, G. F., Stokes, S. L., and Juranek, D. D. (1992). Outbreaks of Waterborne Disease in the United States: 1989–90. *Journal AWWA,* 84(4), 129–135.

Hiatt, C. (1964). Kinetics of the Inactivation of Viruses. *Bacteriological Reviews*, 28, 150.

Hijnen, W. A. M., Beerendonk, E. F., and Medema, G. J. (2006). Inactivation Credit of UV Radiation for Viruses, Bacteria and Protozoan (Oo)cysts in Water: A Review. *Water Research*, 40(1), 3–22.

Hoff, J. (1986). Inactivation of Microbial Agents by Chemical Disinfectants, EPA/600/2-86/067, U.S. Environmental Protection Agency, Washington, DC.

Hoff, J., and Akin, E. (1986). Microbial Resistance to Disinfectants: Mechanisms and Significance. *Environmental Health Significance*, 69, 7–13.

Hoigne, J., and Bader, H. (1975). Ozonation of Water: Role of Hydroxyl Radicals as Oxidizing Intermediates. *Science*, 190, 782.

Hoigne, J., and Bader, H. (1976). The Role of Hydroxyl Radical Reactions in Ozonation Processes in Aqueous Solutions. *Water Research*, 10, 377.

Holwerda, K. (1928). *Mededeelingen van den Dienst der Volksgezondheid in Ned-Indie*, 17, 251.

Hom, L. W. (1972). Kinetics of Chlorine Disinfection of an Ecosystem. *Journal of the Sanitary Engineering Division, ASCE*, 98(SA1), 183–194.

Hooker, A. (1913). *Chloride of Lime in Sanitation*, New York: Wiley.

Huff, C. B., Smith, H. F., Boring, W. D., and Clarke, N. A. (1965). Study of Ultraviolet Disinfection of Water and Factors in Treatment Efficiency. *Public Health Reports*, 80(8), 695–705.

Hydes, O. (1999). European Regulations on Residual Disinfection. *Journal AWWA*, 91(1), 70–74.

Iivanainen, E. K., Martikainen, P. J., Vaananen, P. K., and Katila, M. L. (1993). Environmental Factors Affecting the Occurrence of Mycobacteria in Brook Waters. *Applied and Environmental Microbiology*, 59, 398–404.

Ingols, R., and Ridenour, G. (1948). Chemical Properties of Chlorine Dioxide in Water Treatment. *Journal AWWA*, 40(11), 1207.

Isaac, R., and Morris, J. (1980). Rates of Transfer of Active Chlorine Between Nitrogenous Substances. In *Water Chlorination: Environmental Impact and Health Effects,* ed. R. Jolley. Stoneham, MA: Butterworth.

Isaac, R. A., Wajon, J. E., and Morris, J. C. (1985). Subbreakpoint Modeling of the HOBr-NH3-OrgN Reactions. *Water Chlorination: Environmental Impact and Health Effects*, 5, 985–998.

Jafvert, C. T., and Valentine, R. L. (1992). Reaction Scheme for the Chlorination of Ammoniacal Water. *Environmental Science and Technology*, 26(3), 577–586.

Jagger, J. (1967). *Introduction to Research in Ultraviolet Photobiology.* Englewood Cliffs, NJ: Prentice Hall.

Jarroll, E. L., Bingham, A. K., and Meyer, E. A. (1981). Effect of Chlorine on *Giardia lamblia* Cyst Viability. *Applied and Environmental Microbiology*, 41(2), 483–487.

Jones, G., Stephens, H., and Ott, G. (2007). Hypochlorite Conversion Provides Improved Disinfection Safety and Reliability in Portland, OR. *Water Practice*, 1, 1–14.

Jordan, R. (1980). Improved Method Generates More Chlorine Dioxide. *Water and Sewage Works*, 127(10), 44.

Kabler, P. (1951). Relative Resistance of Coliform Organisms and Enteric Pathogens in the Disinfection of Water with Chlorine. *Journal AWWA*, 43(7), 553.

Katzenelson, E., Kletter, B., and Shuval, H. (1974). Inactivation Kinetics of Viruses and Bacteria in Water by Use of Ozone. *Journal AWWA*, 66(12), 725.

Kelly, S., and Sanderson, W. (1958). The Effect of Chlorine in Water on Enteric Viruses. *American Journal of Public Health*, 48, 1323.

Kinman, R. N., and Layton, R. F. (1976). New Method for Water Disinfection. *Journal AWWA*, 68(6), 298–302.

Korich, D. G., Mead, J. R., Madore, M. S., Sinclair, N. A., and Sterling, C. R. (1990). Effects of Ozone, Chlorine Dioxide, Chlorine, and Monochloramine on *Cryptosporidium parvum* Oocyst Viability. *Applied and Environmental Microbiology*, 56(5), 1423–1428.

Krasner, S. Free Chlorine Versus Monochloramine in Controlling Off Tastes and Odors in Drinking Water, AWWA Annual Conference, Denver, CO.

Larason, T., and Ohno, Y. (2006). Calibration and Characterization of UV Sensors for Water Disinfection. *Metrologia,* 43(2), S151.

Laubusch, E. (1963). Sulfur Dioxide. *Public Works*, 94(8), 117.

Lauer, W. C., et al. (1986). Experience with Chlorine Dioxide at Denver's Reuse Plant. *Journal AWWA*, 78(6), 79.

Lawler, D. F., and Singer, P. C. (1993). Analyzing Disinfection Kinetics and Reactor Design: A Conceptual Approach Versus the SWTR. *Journal AWWA*, 85(11), 67–76.

Leahy, J. (1985). Inactivation of *Giardia muris* Cysts by Chlorine and Chlorine Dioxide, MS Thesis, Ohio State University.

LeChevallier, M. W., Lowrey, C. D., and Lee, R. G. (1990). Disinfecting Biofilms in a Model Distribution System. *Journal AWWA*, 82(6), 87–99.

Lechevallier, M. W., Norton, W. D., and Lee, R. G. (1991). *Giardia* and *Cryptosporidium* spp. in Filtered Drinking Water Supplies. *Applied and Environmental Microbiology*, 57, 2617–2621.

Leland, D., McAnulty, J., Keene, W., and Stevens, G. (1993). A Cryptosporidiosis Outbreak in a Filtered-Water Supply. *Journal AWWA*, 85(6), 34–42.

Lev, O., and Regli, S. (1992). Evaluation of Ozone Disinfection Systems–Characteristic Concentration-C. *ASCE Journal of Environmental Engineering*, 118(4), 477–494.

Levenspiel, O. (1972). *Chemical Reaction Engineering*, New York: John Wiley.

Levine, W. C., Stephenson, W. T., and Craun, G. F. (1990). Waterborne Disease Outbreaks, 1986–1988. *Mmwr Cdc Surveill Summ*, 39(1), 1–13.

Leyval, C., Arz, C., Block, J.C. & Rizet, M., (1984). *Escherichia coli* resistance to chlorine after successive chlorinations *Environmental Technology Letters*, 5(8), 359.

Lykins, B., and Griese, M. (1986). Using Chlorine Dioxide for Trihalomethane Control. *Journal AWWA*, 78(6), 88.

Lyn, D. A., Chiu, K., and Blatchley, E. R., III. (1999). Numerical Modeling of Flow and Disinfection in UV Disinfection Channels. *Journal of Environmental Engineering*, 125(1), 17–26.

Mac Kenzie, W. R., Hoxie, N. J., Proctor, M. E., Gradus, M. S., Blair, K. A., Peterson, D. E., Kazmierczak, J. J., Fox, K. R., Addias, D. G., Rose, J. B., and Davis, J. P. (1994). Massive Waterborne Outbreak of *Cryptosporidium* Infection Associated with a Filtered Public Water Supply, Milwaukee, Wisconsin, March and April 1993. *New England Journal of Medicine*, 331(3), 161–167.

Malcolm Pirnie, and HDR Engineering. (1991). *Guidance Manual for Compliance with the Filtration and Disinfection Requirements for Public Water Systems Using Surface Water Sources*, American Water Works Association, Denver CO.

Manley, T. C., and Niegowski, S. (1967). Kirk Othmer Encyclopedia of Chemical Technology, New York.

Martin, N., Benezet-Toulze, M., Laplace, C., Faivre, M., and Langlais, B. (1992). Design and Efficiency of Ozone Contactors for Disinfection. *Ozone Science and Engineering*, 14(5), 391–405.

Masschelein, W. (1979a). *Chlorine Dioxide*. Ann Arbor, MI: Ann Arbor Science.

Masschelein, W. (1979b). Use of Chlorine Dioxide for the Treatment of Drinking Water. Oxidation Techniques in Drinking Water Treatment, EPA 570/9-79-020, USEPA, Cincinnati, OH.

Mc Kee, J. (1960). Chemical and Colicidal Effects of Halogens in Sewage. *Journal of the Water Pollution Control Federation*, 32, 795.

McFeters, G. A., Kippin, J. S., and LeChevallier, M. W. (1986). Injured Coliforms in Drinking Water. *Applied and Environmental Microbiology*, 51, 1–5.

McGuire, M. J. (2006). Eight Revolutions in the History of US Drinking Water Disinfection. *Journal AWWA*, 98(3), 123–50.

Means, E., Scott, K., Lee, M., and Wolfe, R. (1986). Bacteriological Impact of a Changeover from Chlorine to Chloramine Disinfection in a Water Distribution System, American Water Works Association Annual Conference, Denver, CO.

Means, E., Tanaka, T., Otsuka, D., and McGuire, M. (1986b). Effect of Chlorine and Ammonia Application Points on Bactericidal Efficiency. *Journal AWWA*, 78(1), 62.

Medir, M., and Giralt, F. (1982). Stability of Chlorine DIoxide in Aqueous Solution. *Water Research*, 16, 1379.

Melnick, J. L., Gerba, C. P., and Wallis, C. (1978). Viruses in Water. *Bulletin of the World Health Organization*, 56(4), 499–508.

Milbauer, R., and Grossowicz, N. (1959). Effect of Growth Conditions on Chlorine Sensitivity of *Escherichia coli*. *Applied Microbiology*, 7, 71–74.

Milbauer, R., and Grosswicz, N. (1959). Reactivation of Chlorine-Inactivated *Escherichia coli*. *Applied Microbiology*, 7, 67.

Miller, D. G. (1992). A Survey of Cryptosporidium Oocysts in Surface and Groundwaters in the UK. *Journal of the Institution of Water and Environmental Management*, 6, 697–703.

Miller, G. W., Rice, R. G., Robson, C. M., Scullin, R. L., Kuehn, W., and Wolf, H. (1978). An Assessment of Ozone and Chlorine Dioxide for Treatment of Municipal Water Supplies, EPA 600/8-78-018, U.S. Environmental Protection Agency. Washington DC.

Mitch, W. A., Sharp, J. O., Trussell, R. R., Valentine, R. L., Alvarez-Cohen, L., and Sedlak, D. L. (2003). N-Nitrosodimethylamine (NDMA) as a Drinking Water Contaminant: A Review. *Environmental Engineering Science*, 20(5), 389–404.

Morris, J. (1966). The Acid Ionization Constant of HOCl from 5°C to 35°C. *Journal of Physical Chemistry*, 70, 3798.

Morris, J. (1967). Kinetics of Reactions Between Aqueous Chlorine and Nitrogen Compounds. In *Principles and Applications of Water Chemistry,* ed. S. Faust. New York: John Wiley.

Morris, J. C. (1971). Chlorination and Disinfection–State of the Art. *Journal AWWA*, 63(12), 769.

Morris, J. C., and Isaac, R. A. (1983). A Critical Review of Kinetics and Thermodynamic Constants for Aqueous Chlorine-Ammonia Systems. In *Water Chlorination: Environmental Impact and Health Effects,* ed. R. L. Jolley et al. Ann Arbor, MI: Ann Arbor Science Publishers, pp. 49–63.

Mosley, J. (1966). Transmission of Viral Diseases by Drinking Water. In *Transmission of Viruses by the Water Route,* ed. G. Berg. New York: John.Wiley.

Murphy, K. L., Zaloum, R., and Fulford, D. (1975). Effect of Chlorination Practice on Soluble Organics. *Water Research*, 9, 389.

Nasser, A. M. (1994). Prevalence and Fate of Hepatitis A in Water. *Critical Reviews In Environmental Science And Technology*, 24(4), 281.

Nathanson, N., and Langmuir, A. D. (1963). The Cutter Incident: Poliomyelitis Following Formaldehyde-Inactivated Poliovirus Vaccination in the United States During the Spring of 1955: The Relationship of Poliomyelitis to Cutter Vaccine. *American Journal of Epidemiology*, 78, 29–60.

Nauman, E. B., and Buffham, B. A. (1983). *Mixing in Continuous Flow Systems.* New York: John Wiley.

Nissen, F. (1890). Ueber die desinficirende Eigenschaft des Chlorkalks. *Zeitsch. fur Hygiene*, 8, 62–77.

Noack, M., and Doeff, R. (1979). Chlorine Dioxide, Chlorous Acid and Chlorites. Kirk-Othmer In *Encyclopedia of Chemical Technology,* ed. H. Mark. New York: John Wiley.

Noack, M., and Doeff, R. (1981). Reactions of Chlorine, ClO_2 and Mixtures Therof with Humic Acid. In *Water Chlorination, Environmental Impact and Health Effects,* ed. R. Jolley et al. Ann Arbor, MI: Ann Arbor Science.

Noss, C., Dennis, W., and Olivieri, V. (1985). Reactivity of Chlorine Dioxide with Nucleic Acids and Proteins. In *Water Chlorination: Environmental Impact and Health Effects,* ed. R. Jolley. Chelsea, MI: Lewis Publishers.

Olivieri, V. P., Dennis, W. H., Snead, M. C., Richfield, D. T., and Kruse, C. W. (1980). Reaction of Chlorine and Chloramines with Nucleic Acids Under Disinfection Conditions. In *Water Chlorination: Environmental Impact and Health Effects,* ed. R. Jolley. Stoneham, MA: Butterworths.

Olivieri, V. P., Donovan, T. K., and Kawata, K. (1971). Inactivation of Virus in Sewage. *ASCE Journal of the Sanitary Engineering Division*, 97(5), 661.

Olivieri, V. P., Hauchman, F. S., Noss, C. I., and Vasl, R. (1985). Mode of Action of Chlorine Dioxide on Selected Viruses. In *Water Chlorination: Environmental Impact and Health Effects,* ed. R. Jolley. Chelsea, MI: Lewis Publishers.

Palin, A. T. (1983). *Chemistry and Control of Modern Chlorination*, La Motte Chemical Products Co., Chestertown, MD.

Palmer, S. R., Gully, P. R., White, J. M., Pearson, A. D., Suckling, W. G., Jones, D. M., Rawes, J. C., and Penner, J. L. (1983). Waterborne Outbreak of *Campylobacter* Gastroenteritis. *Lancet*, 287–90.

Payment, P., and Franco, E. (1993). *Clostridum perfringens* and Somatic Coliphages as Indicators of the Efficiency of Drinking Water Treatment for Viruses and Protozan Cysts. *Applied and Environmental Microbiology*, 59(8), 2418–2424.

Payment, P., Trudel, M., and Plante, R. (1985). Elimination of Viruses and Indicator Bacteria at Each Step of Treatment During Preparation of Drinking Water at Seven Water Treatment Plants. *Applied Environmental Microbiology*, 49, 1418.

Phelps, E. (1909). The Disinfection of Sewage and Sewage Filter Effluents. *USGS Water Supply Paper*, 229.

Pressley, T., Dolloff, F., and Roan, S. (1972). Ammonia Nitrogen Removal by Breakpoint Chlorination. *Environmental Science and Technology*, 6, 622.

Pyle, B. H., Watters, S. K., and McFeters, G. A. (1994). Physiological Aspects of Disinfection Resistance in *Pseudomonas cepacia. Journal of Applied Microbiology*, 76(2), 142–148.

Race, J. (1918). *Chlorination of Water.* New York: John Wiley.

Rapson, W. H. (1966). From Laboratory Curiosity to Heavy Chemical. *Chemistry Canada*, 18(1), 2531.

Rav-Acha, C. (1984). The Reactions of Chlorine Dioxide with Aquatic Organic Materials and Their Health Effects. *Water Research*, 18, 1329.

Reasoner, D. J. (1991). Pathogens in Drinking Water: Are There Any New Ones? Proc. Water Quality Technology Conference, American Water Works Association, Orlando, FL.

Reeve, G. D. L., Martin, J., Pappas, R. E., Thompson, and Greene, K. D. (1989). An Outbreak of Shigellosis Associated with the Consumption of Raw Oysters,. *New England Journal of Medicine*, 321(4), 224–227.

Renton, J. I., Moorehead, W., and Ross, A. (1996). Longitudinal Studies of *Giardia* Contamination in Two Community Drinking Water Supplies: Cyst Levels, Parasite Viability and Health Impact. *Applied and Environmental Microbiology*, 62(1), 47–54.

Rice, E. W., Fox, K. R., Miltner, R. J., Lytle, D. A., and Johnson, C. H. (1996). Evaluating Plant Performance with Endospores. *Journal AWWA*, 88(9), 122–130.

Rice, E. W., Hoff, J. C., and Schaefer, F. W., III. (1982). Inactivation of *Giardia* Cysts by Chlorine. *Applied and Environmental Microbiology*, 43(1), 250–251.

Rice, R. G., Robson, C., Miller, G., and Hill, A. (1981). Uses of Ozone in Drinking Water Treatment. *Journal AWWA*, 73(1), 44.

Richardson, A. J., Frankenberg, R. A., Buck, A. C., Selkon, J. B., Colbourne, J. S., Parsons, J. W., and Mayon, W. R. (1991). An Outbreak of Waterborne Cryptosporidiosis in Swindon and Oxfordshire. *Epidemiology and Infection*, 107(3), 485–495.

Richardson, S. D., Thruston, A. D., Collette, T. W., Patterson, K. S., Lykins, B. W., Majetich, G., and Zhang, Y. (1994). Multispectral Identification of Chlorine Dioxide Disinfection By-products in Drinking Water. *Environmental Science & Technology*, 28(4), 592–599.

Rideal, S. (1908). Application of Electrolytic Chlorine to Sewage Purification and Deodorization in the Dry Chlorine Process. *Transactions of the Faraday Society*, 4, 179.

Rose, J. B., Gerba, C. P., and Jakubowski, W. (1991). Survey of Potable Water Supplies for *Cryptosporidium* and *Giardia. Environmental Science and Technology*, 25, 1393–1400.

Rose, J. B., Gerba, C. P., Singh, S. N., Toranzos, G. A., and Keswick, B. (1986). Isolating Viruses from Finished Water. *Journal AWWA*, 78(1), 56.

Rosenberg, M. L., Koplan, J. P., and Pollard, R. A. (1980). The Risk of Acquiring Hepatitis from Sewage Contaminated Water. *American Journal of Epidemiology*, 112, 17.

Roustan, M., Stambolieva, Z., Duget, J., Wable, O., and Mallevialle, J. (1991). Influence of Hydrodynamics on *Giardia* Cyst Inactivation by Ozone. Study by Kinetics and by 'CT' Approach. *Ozone Science and Engineering*, 13(4), 451–462.

Roy, D. (1979). Inactivation of Enteroviruses by Ozone, PhD Thesis, University of Illinois at Urbana-Champaign, Urbana, IL.

Roy, D., Engelbrecht, R. S., and Chian, E. S. K. (1981). Kinetics of Enteroviral Inactivation by Ozone. *Journal of the Environmental Engineering Division, ASCE*, 107(EE5), 887–899.

Roy, D., et al. (1981). Mechanism for Enteroviral Inactivation by Ozone. *Applied and Environmental Microbiology*, 41(3), 718.

Rubin, A. J., Engel, J. P., and Sproul, O. J. (1983). Disinfection of Amoebic Cysts in Water with Free Chlorine. *Journal of the Water Pollution Control Federation*, 55(9), 1174.

Rush, B. A., Chapman, P. A., and Ineson, R. W. (1990). A Probable Waterborne Outbreak of Cryptosporidiosis in the Sheffield Area. *Journal of Medical Microbiology*, 32(4), 239–242.

Saguinsin, J. L. S., and Morris, J. C. (1975). The Chemistry of Aqueous Nitrogen Trichloride. In *Disinfection: Water and Wastewater,* ed. J. Johnson. Ann Arbor, MI: Ann Arbor Science, 277–279.

Saunier, B. M., and Selleck, R. E. (1979). The Kinetics of Breakpoint Chlorination in Continuous Flow Systems. *Journal AWWA*, 71, 164–172.

Scarpino, P. V., Berg, G., Chang, S. L., Dahling, D., and Lucas, M. (1972). A Comparative Study of the Inactivation of Viruses in Water by Chlorine. *Water Research*, 6, 959.

Scarpino, P. V., Cronier, S., Zink, M. L., and Brigano, F. A. O. (1977) Effect of Particulates on Disinfection of Enteroviruses and Coliform Bacteria in Water by Chlorine Dioxide, AWWA Water Quality Technology Conference, Kansas City, MO.

Scarpino, P. V., Cronier, S., Zink, M. L., and Brigano, F. A. O. (1979). Effect of Particulates on Disinfection of Enteroviruses and Coliform Bacteria in Water by Chlorine Dioxide, EPA-600/2-79-054, USEPA.

Scheible, O., and Bassell, C. (1981). Ultraviolet Disinfection of a Secondary Wastewater Treatment Plant Effluent, EPA-600/1-81-139, USEPA.

Selleck, R., Sollins, H., and White, G. (1978). Kinetics of Bacterial Deactivation with Chlorine. *ASCE Journal of the Environmental Engineering Division*, 104, 1197.

Severin, B. F., Suidan, M. T., and Engelbrecht, R. S. (1983a). Effects of Temperature on Ultraviolet Light Disinfection. *Environmental Science and Technology*, 17, 717.

Severin, B. F., Suidan, M. T., and Engelbrecht, R. S. (1983b). Kinetic Modeling of UV Disinfection of Water. *Water Research*, 17(11), 1669–1676.

Severin, B. F., Suidan, M. T., and Engelbrecht, R. S. (1984a). Series-Event Kinetic Model for Chemical Disinfection. *Journal of Environmental Engineering*, 110(2), 430–439.

Severin, B. F., Suidan, M. T., Rittmann, B. E., and Engelbrecht, R. S. (1984b). Inactivation Kinetics in a Flow-through UV Reactor. *Journal of the Water Pollution Control Federation*, 56(2), 164–169.

Shang, C., Gong, W.-L., and Blatchley, E. R. (2000). Breakpoint Chemistry and Volatile Byproduct Formation Resulting from Chlorination of Model Organic-N Compounds. *Environmental Science & Technology*, 34(9), 1721–1728.

Sikorowska, C. (1961). Influence of Pollutions on Chlorine Dioxide Demand of Water. *Gig. Woda Tech. Sanita.*, 35(12), 4645.

Singh, A., Yeager, R., and McFeters, G. (1986). Assessment of *in vivo* Revival, Growth and Pathogenicity of *Escherichia coli* Strains after Copper and Chlorine Induced Injury. *Applied and Environmental Microbiology*, 52(4), 832.

Smith, H. V. (1992). *Cryptosporidium* and Water–A Review. *Journal of the Institution of Water and Environmental Management*, 6(4), 443–451.

Snead, M. C., Olivieri, V. P., Kruse, C. W., and Kawata, K. (1980). Benefits of Maintaining a Chlorine Residual in Water Supply Systems, EPA-600/2-80-010, USEPA.

Sommer, R., Cabaj, A., Hirschmann, G., and Haider, T. (2008). Disinfection of Drinking Water by UV Irradiation: Basic Principles—Specific Requirements—International Implementations. *Ozone: Science & Engineering: The Journal of the International Ozone Association*, 30(1), 43–48.

Sproul, O. J. (1978). Effect of Particulate Matter on Viral Inactivation by Ozone. Proc. of American Water Works Association Annual Conference, Atlantic City, NJ.

Stagg, C., et al. (1978). Chlorination of Solids Associated Coliphages. *Progress in Water Technology*, 10(12), 3817.

Stefan, H. G., Johnson, T. R., McConnell, H. L., Anderson, C. T., and Martenson, D. R. (1990). Hydraulic Modeling of Mixing in Wastewater Dechlorination Basin. *Journal of Environmental Engineering ASCE*, 116(3), 524–541.

Stover, E. L., Haas, C. N., Rakness, K. L., and Scheible, O. K. (1986). *Design Manual: Municipal Wastewater Disinfection*, U.S. Environmental Protection Agency, Cincinnati, OH.

Taras, M. J. (1950). Preliminary Studies on the Chlorine Demand of Specific Chemical Compounds. *Journal AWWA*, 462–473.

Teefy, S. M., and Singer, P. C. (1990). Performance and Analysis of Tracer Tests to Determine Compliance of a Disinfection Scheme with the SWTR. *Journal AWWA*, 82(12), 88–98.

Templeton, M. R., Andrews, R. C., and Hofmann, R. (2008). Particle-Associated Viruses in Water: Impacts on Disinfection Processes. *Critical Reviews in Environmental Science and Technology*, 38(3), 137–164.

Tenno, K., Fujioka, R., and Loh, P. (1980). The Mechanism of Poliovirus Inactivation by Hypochlorous Acid. In *Water Chlorination: Environmental Impact and Health Effects,* ed. R. Jolley. Stoneham, MA: Butterworths.

Trofe, T. W. (1980). Kinetics of Monochloramine Decomposition in the Presence of Bromine. *Environmental Science and Technology*, 14, 544.

Trussell, R., and Kreft, P. (1984). Engineering Considerations of Chloramine Application. *AWWA Seminar Proceedings: Chloramination for THM Control*, Dallas, TX.

Trussell, R. R. (1999). Safeguarding Distribution System Integrity. *Journal AWWA*, 91(1), 46–54.

Trussell, R. R., and Chao, J.-L. (1977). Rational Design of Chlorine Contact Facilities. *Journal of the Water Pollution Control Federation*, 49(7), 659–667.

U. S. Environmental Protection Agency. (1991). Guidance Manual for Compliance with the Filtration and Disinfection Requirements for Public Water Systems Using Surface Water Sources, USEPA, Washington DC. http://www.epa.gov/safewater/mdbp/guidsws.pdf.

U.S. Government Accountability Office. (2007). Securing Wastewater Facilities, Washington, DC.

Valentine, R. (1985). Disappearence of Monochloramine in the Presence of Nitrite. In *Water Chlorination: Environmental Impact and Health Effects,* ed. R. Jolley. Chelsea, MI: Lewis Publishers.

Valentine, R., Brandt, K. L., and Jafvert, C. T. (1986). A Spectrophotometric Study of the Formation of an Unidentified Monochloramine Decomposition Product. *Water Research*, 20(8), 1067.

Valentine, R. L., and Jafvert, C. T. (1988). General Acid Catalysis of Monochloramine Disproportionation. *Environmental Science and Technology*, 22, 691.

Venkobachar, C., Iyengar, L., and Rao, A. (1975). Mechanism of Disinfection. *Water Research*, 9, 119.

Venkobachar, C., Iyengar, L., and Rao, A. (1977). Mechanism of Disinfection: Effect of Chlorine on Cell Membrane Functions. *Water Research*, 11, 727.

Wajon, J. (1982). Oxidation of Phenol and Hydroquinone by Chlorine Dioxide. *Environmental Science and Technology*, 16, 396.

Ward, N., Wolfe, R., and Olson, B. (1984). Disinfectant Activity of Inorganic Chloramines with Pure Culture Bacteria: Effect of pH, Application Technique and Chlorine to Nitrogen Ratio. *Applied and Environmental Microbiology*, 48, 508.

Water Pollution Control Federation. (1984). *Wastewater Disinfection, Manual of Practice FD-10*, Water Pollution Control Federation, Alexandria, VA.

Watson, H. E. (1908). A Note on the Variation of the Rate of Disinfection with Change in the Concentration of the Disinfectant. *Journal of Hygiene*, 8, 536–542.

Wattie, E., and Butterfield, C. (1944). Relative Resistance of *Escherichia coli* and *Eberthella typhosa* to Chlorine and Chloramines. *U,S. Public Health Service Reports*, 59, 1661.

Weber, G. R., Bender, R., and Levine, M. (1940). Effect of Ammonia on the Germicidal Efficiency of Chlorine in Neutral Solutions. *Journal of the American Water Works Assocation*, 32(11), 1904.

Weil, I., and Morris, J. (1949a). Equilibrium Studies on N-Chloro Compounds. *Journal of the American Chemical Society*, 71, 3.

Weil, I., and Morris, J. C. (1949b). Kinetic Studies on the Chloramines. I. The Rates of Formation of Monochloramine, N-Chloromethylamine and N-Chlorodimethylamine. *Journal of the American Chemical Society*, 71, 1664–1671.

Werderhoff, K., and Singer, P. (1987). Chlorine Dioxide Effects on THMFP, TOXFP and the Formation of Inorganic By-products. *Journal AWWA*, 79, 107–113.

White, G. C. (1978). *Disinfection of Wastewater and Water for Reuse.* New York: Van Nostrand Reinhold.

White, G. C. (1999). *Handbook of Chlorination and Alternative Disinfectants*, 4th ed. New York: Wiley-Interscience.

Wickramanayake, G. B., Rubin, A. J., and Sproul, O. J. (1985). Effects of Ozone and Storage Temperature on *Giardia* Cysts. *Journal AWWA*, 77(8), 74–77.

Wojtowicz, J. A. (1979). Chlorine Monoxide, Hypochlorous Acid, and Hypochlorites. In *Kirk-Othmer Encyclopedia of Chemical Technology,* ed. H. F. Mark et al. New York: John Wiley, 580–611.

Wolfe, R., and Olson, B. (1985). Inability of Laboratory Models to Accurately Predict Field Performance of Disinfectants. In *Water Chlorination: Environmental Impact and Health Effects,* ed. R. Jolley. Chelsea, MI: Lewis Publishers.

Wolfe, R., Ward, N., and Olson, B. (1984). Inorganic Chloramines as Drinking Water Disinfectants: A Review. *Journal AWWA*, 76(5), 74.

Wright, N. (1936). The Action of Hypochlorites on Amino Acids and Proteins: The Effects of Acidity and Alkalinity. *Biochemical Journal*, 30, 1661.

Young, D., and Sharp, D. (1979). Partial Reactivation of Chlorine Treated Enterovirus. *Applied and Environmental Microbiology*, 37(4), 766.

Zhou, H., and Smith, D. (1994). Kinetics of Ozone Disinfection in Completely Mixed System. *Journal of Environmental Engineering*, 120(4), 841–858.

Zhou, H., and Smith, D. W. (1995). Evaluation of Parameter Estimation Methods for Ozone Disinfection Kinetics. *Water Research*, 29(2), 679–686.

CHAPTER 18
ULTRAVIOLET LIGHT PROCESSES

Karl G. Linden, Ph.D.
Professor, Department of Civil, Environmental, and Architectural Engineering
University of Colorado at Boulder
Boulder, Colorado, United States

Erik J. Rosenfeldt, Ph.D., P.E.*
Assistant Professor, Department of Civil and Environmental Engineering
University of Massachusetts
Amherst, Massachusetts, United States

INTRODUCTION TO ULTRAVIOLET LIGHT PROCESSES **18.1**
FUNDAMENTALS OF UV LIGHT **18.4**
Electromagnetic Spectrum and UV Classification 18.4
UV Lamp Types 18.4
UV Dose Theoretical Considerations 18.6
Full-Scale UV Reactor Configurations 18.10
UV DISINFECTION **18.10**
Fundamentals of UV Disinfection 18.11
UV Disinfection Practice 18.13
Effects of Water Quality and By-Products 18.14
Integration of UV into a Water Treatment Plant 18.15
Examples of UV Installations 18.17
UV PHOTOLYSIS **18.17**
Photochemistry Basics 18.18
UV Photolysis for Treatment of Chemical Pollutants 18.21
UV ADVANCED OXIDATION PROCESSES (AOPs) **18.24**
The UV/H_2O_2 AOP Mechanism 18.24
UV/H_2O_2 Models 18.28
UV/H_2O_2 AOP in Practice: Case Studies 18.30
Alternative UV-Based AOPs 18.33
ABBREVIATIONS **18.34**
NOTATION FOR EQUATIONS **18.36**
REFERENCES **18.36**

INTRODUCTION TO ULTRAVIOLET LIGHT PROCESSES

Ultraviolet (UV) light has long been an important tool in biological and chemical research. UV research has developed into the fields of photobiology and photochemistry and has increased our understanding of biological and chemical processes in the environment. The

*Presently with Hazen and Sawyer, Fairfax, Virginia, United States.

effects of UV light on the survival and mutation of microorganisms have been studied for over a century. The first recorded discovery of the bactericidal effects of sunlight was made in England by Downes and Blunt (1877). In 1901, Peter Cooper Hewitt invented the mercury vapor arc lamp, which, when coupled with a quartz sleeve in 1906, allowed the production of UV lamps and stimulated research into the bactericidal properties of UV light. The first recorded use of UV light for disinfection of water was in Marseilles, France, in 1910, where disinfection of 36 m^3/h (9500 gph) of pathogen-spiked water was achieved in a few seconds using a Westinghouse Cooper Hewitt mercury lamp in fused quartz. The use of UV light for the disinfection of water on ships began in 1916, a practice that continues to this day. Baker (1948) reported, based on personal contacts, that UV disinfection facilities in the United States, supplied by the R.U.V. Co., Inc., of New York, were installed in Henderson, Ky., operating from 1916 to 1923; Berea, Ohio, operating from 1923 to 1936; Horton, Kan., beginning in 1923 (end date unknown); and Perrysburg, Ohio, operating from 1928 to 1939. These facilities were abandoned owing to the cost of operation compared with other water disinfection techniques, problems with electrical supply reliability, and the belief among state officials that chlorine was a more effective technology. The R.U.V. Co. also noted that the small size of quartz lamps limited the applications for municipal systems. The use of UV light for disinfection of water gradually grew out of favor in the 1920s owing to the increasing use of chlorination, the difficulties with reliability and maintenance of UV systems, and the comparatively higher costs associated with the UV technology available at that time.

Despite the decline of UV light for applications in water disinfection, an intense interest in the effects of UV light on biological and chemical processes was sparked. In 1929, Gates reported the inactivation efficacy of individual wavelengths of monochromatic UV light, developing the first germicidal action spectrum (Gates, 1929). This work and the work of subsequent researchers demonstrated the similarity between the absorbance spectrum of DNA and the germicidal action spectrum, suggesting that the target of action of UV light leading to inactivation was the nucleic acids. The wavelengths that were determined to be germicidally effective were found to be those in the far UV range of 200 to 300 nm. Within this range, the most effective germicidal wavelengths were between 250 and 270 nm. Although not as germicidally efficient as the far UV range, near UV light (300–390 nm) also was found to produce essentially all the biological effects produced by far UV light, as well as other sublethal physiologic effects that may interfere with proper cellular functions and contribute to eventual inactivation (Jagger, 1983).

Although chlorination was a proven effective disinfectant, concerns in the 1970s surrounding the toxicity of chlorine residuals to the aquatic biota in receiving waters and the discovery of the potential carcinogenic and mutagenic properties of chlorinated organic compounds formed during disinfection of water and wastewater led to the investigation of alternative disinfectants. Advances in ballast and electronics technology in the 1970s and 1980s, along with the desire to find new disinfection technologies, helped the resurgence of UV light treatment as a viable technology for disinfection. Since that time, UV disinfection has grown into a major commercial industry and increasingly has been the subject of academic and industrial research.

UV light has developed a large market share in wastewater disinfection applications, where it has been shown to be competitive with chlorination in terms of disinfection efficacy and economics (Darby et al., 1995). The use of UV light for wastewater disinfection took root for a variety of reasons: Its small footprint, lack of a need for a large contact basin, elimination of the need for dechlorination before discharge into a natural water body, and ease of use. However, until the late 1990s, about 20 years after adoption for use in wastewater in North America, UV light had not found widespread acceptance for disinfection of potable water (Wolfe, 1990; USEPA, 1996). It was at this time that Bolton and colleagues (1998) and Clancy and colleagues (1998, 2000) discovered that UV treatment was effective in inactivating the chlorine-resistant protozoan pathogen *Cryptosporidium*. This discovery led to a rapid rise in interest for implementation of UV technologies as a primary disinfectant for drinking water, especially in the wake of recent regulations.

Regulation (see Chap. 1 for coverage of regulations) of the water industry in the form of the Ground Water Rule (GWR), the Enhanced Surface Water Treatment Rule (ESWTR), the Information Collection Rule (ICR), the Disinfection/Disinfectant By-Products Rule (D/DBPR), the Uniform Fire Protection Code (UFP), and the Long Term 2 Enhanced Surface Water Treatment Rule (LT2ESWTR) has motivated the water industry to examine disinfection as it is currently practiced with increasing scrutiny. Among the challenges facing this industry are control of disinfection by-products, inactivation or removal of persistent pathogens such as protozoan cysts and some viruses, and limiting bacterial regrowth in distribution systems (USEPA, 2006b). One response to these challenges has been the investigation of alternative disinfectants. UV light is one of a number of disinfection alternatives available. Depending on the type of UV technology applied, UV can meet the needs of the water industry regarding disinfection by-product formation control and inactivation of persistent pathogens. Because UV light does not provide a residual disinfectant concentration for control of regrowth, a chemical disinfectant may be needed when biofilm formation in the distribution system is of concern or regulations demand that a residual be supplied. A summary of the main differences between UV light and chlorine disinfection processes is presented in Table 18-1. The advantages and disadvantages of using UV for disinfection of water are summarized in Table 18-2.

TABLE 18-1 Comparison of UV and Chlorine Disinfection

Parameter	UV Light	Chlorine
Action	Physical	Chemical
Mechanism	DNA damage	Oxidation
Resulting cell structure	Intact	Damaged
Reactivation	Photo/dark	Resuscitation
Design dose	Fixed/semivariable	Adjustable
Calculated dose	$I \times T$	$C \times T$
Exposure time	Seconds	Minutes
Residual	None	Varies with chlorine demand
Demand	Absorbance	Organics/inorganics
	Attenuation	Sunlight

TABLE 18-2 UV Treatment for the Disinfection of Drinking Water Supplies

Advantages

⇒ Effective against bacteria, viruses, and protozoa
⇒ Germicidal effectiveness for standard bacterial and viral indicator organisms is representative of actual pathogen inactivation
⇒ No known toxic by-products formed
⇒ Low operation and maintenance costs
⇒ Small footprint
⇒ Short contact times

Disadvantages

⇒ No residual disinfectant
⇒ Limited effectiveness against adenoviruses
⇒ Complexity measuring germicidally effective UV dose
⇒ Potential for bacterial reactivation
⇒ Lamp fouling

Now that UV light is an accepted technology for drinking water disinfection, the use of UV treatment for applications involving chemical contaminants in water is also growing. UV processes can be effective for degrading organic pollutants via direct UV photolysis or via generation of hydroxyl radicals in advanced oxidation processes (AOPs) such as UV combined with hydrogen peroxide. Therefore, this chapter covers applications of UV light for water treatment both for the disinfection of pathogenic microorganisms and for the destruction of chemical contaminants. The scope of the chapter includes the fundamentals of UV light applications, the science behind UV disinfection processes and inactivation of pathogens, practical aspects of UV disinfection, the use of UV for destruction of contaminants via both photolysis and oxidation pathways, and theory and practice of UV-based oxidation processes.

FUNDAMENTALS OF UV LIGHT

Electromagnetic Spectrum and UV Classification

The electromagnetic spectrum covers a very wide range of energies from radio waves through visible and UV light down through gamma rays (Fig. 18-1). The UV part of that

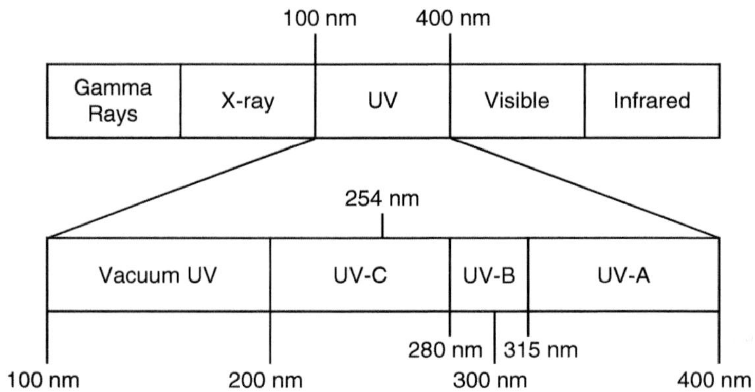

FIGURE 18-1 The location of the UV wavelength region within the electromagnetic spectrum. (*Source: USEPA, 2006a.*)

spectrum extends from 100 to 400 nm. The UV spectrum is divided into four ranges: vacuum UV (100–200 nm), UV-C (200–280 nm), UV-B (280–315 nm), and UV-A (315–400 nm). UV light below 200 nm exists only in a vacuum because these wavelengths are readily absorbed by chemical constituents (principally O_2) in air. The UV-C range is considered to be the most germicidally active UV range because these wavelengths are strongly absorbed by nucleic acids. The UV-B portion of the spectrum can be thought of as the sun-burning wavelengths because it contains wavelengths from sunlight that reach the earth's surface (wavelengths > 300 nm) and can damage exposed skin. The UV-A range is often dubbed the sun-tanning portion of the UV spectrum.

UV Lamp Types

UV light typically is delivered to water using UV lamps, which are constructed out of quartz to allow transmission of UV light. Several different types of UV lamps are available commercially. The most common are the continuous-wave mercury vapor–based lamps including low-pressure (LP) and medium-pressure (MP) UV lamps, but nonmercury UV lamps are also being developed, including pulsed UV sources, excimer sources, and UV light-emitting diodes.

Mercury-Based Lamps. The first UV disinfection systems to be developed commercially employed LP mercury vapor lamps. Initially, these lamps were suspended above a thin liquid film to achieve disinfection. Current designs allow for lamps to be submerged in the water, configured parallel or perpendicular (horizontal or vertical) to the flow path. Approximately 85 percent of the UV light emitted by LP lamps is monochromatic at 253.7 nm. LP UV systems can be divided into low output (LPLO) and high output (LPHO). LPLO lamps typically are used for low flow applications, where the number of lamps can be kept to a minimum and also where UV transmittance and turbidity are not limiting factors in the water quality. LPHO lamps contain solid spots of a mercury amalgam attached to the inside of the lamp's quartz tube. These amalgams are a mercury alloy with an element such as gallium or indium. This mercury amalgam helps to control the mercury vapor pressure, allowing a characteristic LP output at 253.7 nm to be emitted at a higher intensity than is possible without the amalgam. LP UV light from either LPLO or LPHO sources has been proven effective for the inactivation of bacteria, viruses, and protozoa (Hijnen et al., 2006).

Medium-pressure (MP) mercury vapor lamps have grown in use as UV disinfection is applied to systems with large flows and more diverse water quality. MP UV lamps have a polychromatic emission with a germicidal efficiency of approximately 10 to 20 percent. Thus MP lamps are less germicidally efficient than LP lamps; however, because the UV output is much greater, they emit approximately 10 to 50 times the germicidal UV output of LPLO lamps and 4 to 10 times the germicidal output of LPHO lamps. The higher UV output translates into a relatively smaller footprint for a disinfection system, significantly fewer lamps for the equivalent effect, and potentially lower operation and maintenance costs. Most MP UV systems have contact times on the order of 0.5 to 5 seconds. MP UV systems, inherently containing the same wavelength as LP UV along with other wavelengths across the germicidal range, are likewise effective for inactivating bacteria, viruses, and protozoa. Table 18-3 provides a summary of the principal characteristics of typical LP and MP mercury vapor lamps. The emission spectra of LP and MP UV lamps are presented in Fig. 18-2.

TABLE 18-3 Characteristics of Typical Low and Medium-Pressure Mercury Vapor Lamps

Characteristic	Low pressure	Medium pressure
Emission	Monochromatic (85% at 253.7 nm)	Polychromatic (185–1367 nm)
Mercury vapor pressure (torr)	10^{-3} to 10^{-2}	10^2 to 10^4
Operating temperature (°C)	40 to 60	500 to 800
Arc length (cm)	40 to 150	5 to 150
Lifetime (hours)	8000 to 12,000	4000 to 8000
Light intensity (relative)	Low/Medium	High

Note: 100 torr = 13.3 kPa.

The use of mercury-based UV sources in water treatment has led to an investigation of the possibility of mercury release owing to lamp breakage. This issue is addressed in detail in the *UV Disinfection Guidance Manual* (USEPA 2006a). UV lamps can break within a UV disinfection system if overheated owing to loss of water, overtightening during installation,

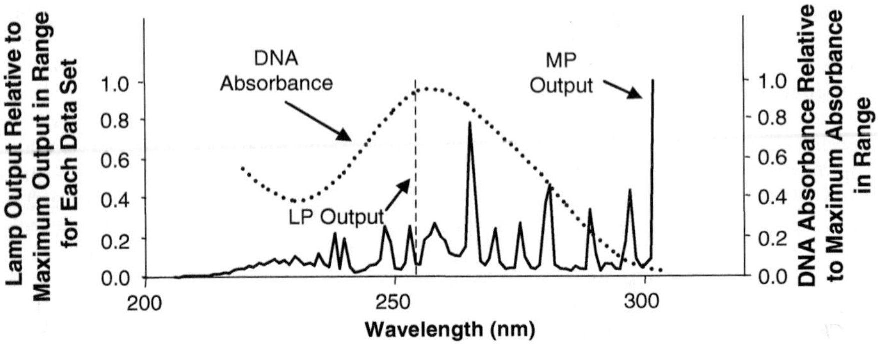

FIGURE 18-2 The absorbance spectra of DNA overlayed with the UV lamp output from LP and MP UV sources. (*Courtesy of Bolton Photosciences, Inc.*)

if some foreign debris strikes the lamp envelope, malfunction of the lamp wiper mechanism, or some extreme pressure event such as water hammer. Proper engineering of the UV system would avoid these types of occurrences, but design considerations for possible mercury containment should be reviewed. UV lamps also may break off-line during lamp handling and bulb changes, and in this case, operator safety precautions and procedures should be followed.

Emerging Lamps. Because of a general concern for use of mercury in the water environment, a number of new mercury-free UV lamps have been developed in recent years. The most prevalent nonmercury UV lamp is the pulsed-UV xenon arc lamp, or flashlamp (Dunn et al., 1995; Bank et al., 1990; Bohrerova et al., 2008). These lamps are characterized as high-intensity broadband sources, with an emission spectrum from the far UV through the infrared. They have seen commercial use in packaging disinfection and UV curing; however, they are not very efficient in their germicidal UV output compared with mercury-based lamps. Commercialization of these lamps into UV disinfection systems has not yet been successful. Surface-discharge lamps also emit pulsed UV light but operate on the principle of plasma emitting from a dielectric (Schaefer et al., 2007) and have improved efficiencies over flashlamps. Excimer lamps are UV sources that can be tuned to emit specific wavelengths (Oppenländer, 2007; Naunovic et al., 2008). UV light-emitting diodes have been developed recently and now can produce emissions down to below 250 nm. However, these UV LEDs are low intensity, are currently expensive to produce, and do not yet have the lifetime needed for municipal applications (Gaska, 2007; Bettles et al., 2007; Chatterley and Linden, 2010). All these sources, by their characteristic wavelength emission in the germicidal UV range, are effective for inactivation of pathogenic microorganisms.

UV Dose Theoretical Considerations

The accurate measurement of UV dose (or UV fluence) both at the bench scale and in field-scale systems is essential to the development of reproducible data regarding the UV dose response of microorganisms and the degradation of contaminants. Because UV light cannot be seen or captured to measure off-site, special methods have been developed to measure UV fluence.

Measurement of UV Energy. A number of techniques have been developed to measure UV irradiance or UV fluence, including physical, chemical, and biological approaches.

Physical Techniques. Physical methods are techniques involving radiometers equipped with detectors calibrated to narrow- or broadband spectral responses. Radiometer detectors can accurately measure only the irradiance incident and near-normal to the plane of the detector; therefore, these systems are applied principally to laboratory-scale UV exposure experiments, where the UV light can be collimated effectively. Radiometers measure the incident intensity on the surface of the detector; therefore, corrections for absorption in and the depth of the liquid samples must be made to provide a measure of the average irradiance in a batch sample. Radiometry has been employed routinely in full-scale reactors, where UV sensors monitor the constancy of the UV output and indicate when the irradiance has decreased owing to lamp fouling, water quality changes, or aging of the lamps.

Chemical Techniques. Chemical actinometry is the measurement of the quantity of UV light exposure by the change produced in a chemical. Some common actinometers in use for measurement of UV irradiance include potassium ferrioxalate, potassium iodide/iodate, and uridine. Each has its benefits and challenges for use with monochromatic and polychromatic UV sources (Jin et al., 2006). Ideal actinometers for measuring polychromatic germicidal UV light would have the following characteristics (von Sonntag and Schuchmann, 1992): (1) They would absorb only germicidal UV light, (2) they would have an absorbance spectrum similar to that of DNA, (3) they would have constant quantum yield throughout germicidal wavelength range, (4) they would be easy to quantify using standard laboratory equipment, and (5) they would be nontoxic in the environment. With any chemical actinometer, care must be taken when evaluating it for use with water sources of a diverse chemical nature to ensure that nonphotochemically induced degradation of the actinometer is not a factor. Similarly, actinometry with polychromatic and/or high-intensity sources must validate that the change produced in the chemical actinometer is solely attributable to the direct photon absorption process and not the result of a photochemically mediated secondary reaction.

Biological Techniques. Biological measurement of fluence has centered on the bioassay or biodosimetry technique for assessing the UV dose in a flow-through UV disinfection system. In the bioassay approach, a UV dose-response curve is established for a given organism under controlled conditions where the UV dose can be determined accurately. A culture of the same organism then is released into a more complex system and the log inactivation determined. This log inactivation value then is used along with the initial UV dose-response curve to estimate what the UV dose must have been to achieve the resulting inactivation response. Because it is a biologically based test, the method provides a measure of the effective germicidal UV dose—the UV dose necessary for a desired germicidal effect. This method provides a single average UV dose value [called the *reduction equivalent dose* (RED)] to the UV disinfection system. Sommer and Cabaj (1993) identified four criteria for selection of a bioassay test organism, including (1) steady inactivation over a wide UV dose range, (2) simple and sufficient culturability, (3) nonpathogenic, and (4) temporal stability with regard to UV susceptibility. In addition, techniques used in culturing the organisms may play a role in the subsequent UV susceptibility. Organisms that have been used for bioassays include *Bacillus subtilis* spores (Qualls and Johnson, 1983) and the MS2 coliphage (Braunstein et al., 1996). This method is described further in the section on calculation of UV dose at the field scale.

Calculating the Average UV Irradiance. The attenuation of UV light in water can be described using an adaptation of the Beer-Lambert law, derived as follows. *Spectrophotometric absorbance* is defined as

$$A = \log\left(\frac{I_0}{I}\right) \tag{18-1}$$

or as the familiar Beer-Lambert law:

$$A = \frac{\varepsilon}{C} \tag{18-2}$$

where A = absorbance, or optical density of solution, as read from the spectrophotometric output
I_0 = irradiance entering the solution in the spectrophotometer cuvette
I = irradiance exiting the solution in the spectrophotometer cuvette
ε = molar absorption coefficient (M^{-1} cm^{-1})
l = solution path length (cm) (see next equation)
C = molar concentration of dissolved constituent (M or mol/L)

When analyzing a complex soluble-particulate mixture such as water, the molar concentration value C is ambiguous, which makes it difficult to determine a value for the molar absorption coefficient ε. However, both the solution absorbance A and the path length l are well defined, such that the absorption coefficient is

$$a_{10} = \frac{A}{l} = \varepsilon c \tag{18-3}$$

Rewriting Eq. 18-1 in terms of this absorption coefficient, we get

$$I = I_0 10^{-a_{10} l} \tag{18-4}$$

or

$$I = I_0 e^{-\alpha_e l} \tag{18-5}$$

where $\alpha_e = a_{10}$ ln(10).

Equation 18-5 was used by Morowitz (1950) to model extinction of light in a stirred sample and UV attenuation coefficients in water. The percent UV transmittance (UVT) of a water is calculated using Eq. 18-6:

$$\%\text{UVT} = 100 \times 10^{-A}$$

Calculation of UV Dose: Bench Scale. Bench-scale UV testing typically is carried out in an apparatus referred to as a *collimated beam* (Fig. 18-3). When considering the calculation of UV

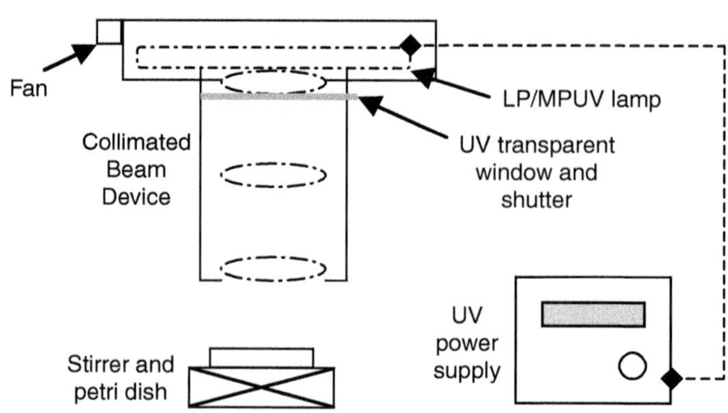

FIGURE 18-3 Collimated-beam apparatus for bench-scale UV exposure experiments. (*Reprinted with permission from* ASCE Journal of Environmental Engineering *[Bolton and Linden, 2003].*)

dose as the product of irradiance and exposure time ($D = I \times T$), both the irradiance and the exposure time are complex entities that need further definition and evaluation. The UV irradiance is a function of two primary factors: the nature of the UV source (e.g., mono/polychromatic, pulsed/continuous wave, low/medium pressure, lamp age, and geometry/spacing/configuration) and the nature of the path to the target (e.g., water quality). The exposure time is a simple measurement on the bench scale using a timer to monitor the time but is much more complex at the reactor scale, as discussed below. Bolton and Linden (2003) detail the specific procedures for proper calculation of the average UV irradiance in the water exposed to UV light in a collimated-beam apparatus. For LP UV systems, the UV dose is calculated as the product of average irradiance in the water at 253.7 nm and the exposure time in seconds. For MP UV sources, because the UV output is polychromatic and each wavelength output has a unique absorbance and germicidal effectiveness depending on the water quality and the target organism of interest, measuring the germicidal UV irradiance is more complex. Methods based on chemical actinometry, mathematical models, and a bioassay have been evaluated for polychromatic sources (Jin et al., 2006).

Calculation of UV Dose: Reactor Scale. The *UV Disinfection Guidance Manual* (USEPA, 2006a) provides detailed procedures for determining and validating the UV fluence (UV dose) in a UV disinfection system. The cornerstone of this method is the bioassay or biodosimetry, as described earlier. Figure 18-4 describes the biodosimetry test procedure.

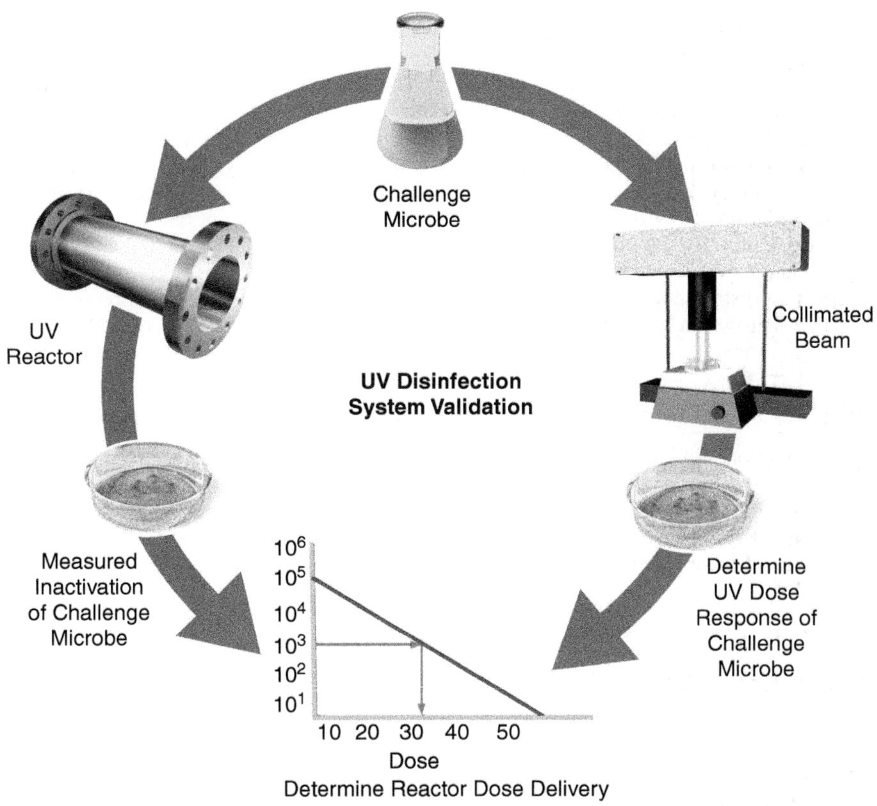

FIGURE 18-4 The biodosimetry test procedure. (*Courtesy of Trojan Technologies.*)

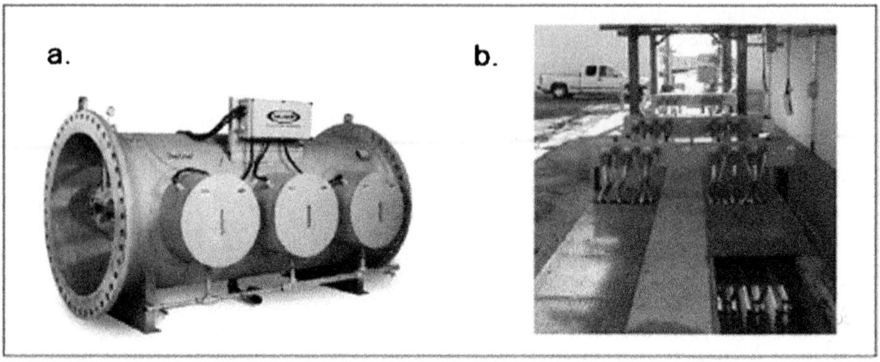

Source: (a) Courtesy of Calgon Carbon Corporation and (b) Courtesy of WEDECO UV Systems

FIGURE 18-5 Examples of UV reactors: (*a*) closed channel; (*b*) open channel. (*Source: USEPA, 2006a.*)

The biodosimetry method ascribes a single UV dose value (the RED) to a UV disinfection reactor under a given set of flow, water quality, and lamp output conditions. Although this may be helpful for assessing the general trends of a UV system and setting operating parameters, the actual behavior of the UV system is not well characterized using this method. Mathematical models have been proposed to characterize the light distribution and flow conditions of a UV reactor system (Jacob and Dranoff, 1970; Chiu et al., 1999; Bolton, 2001; Lyn and Blatchley, 2005; Ducoste et al., 2005). By using irradiance distribution models coupled with computational fluid dynamics analyses of the system hydraulics, a clearer picture of the range of delivered doses provided in a UV reactor can be obtained. Based on modeling efforts, it has been determined that UV disinfection systems, much like other disinfection processes, do not provide a single uniform UV dose during water treatment, but a distribution of UV doses is delivered that depends on UV reactor characteristics (i.e., lamp orientation, baffles, up and down stream piping). Recently, an experimental method employing chemical actinometry embedded in or on microspheres has been used to verify the UV dose distribution concept and validate mathematical models that have predicted this UV dose distribution (e.g., Bohrerova et al., 2005; Blatchley et al., 2008).

Full-Scale UV Reactor Configurations

UV disinfection and oxidation reactors for drinking water treatment are most often pressurized vessels in-line with the existing piping system of the treatment plant. UV lamps in these systems may be oriented parallel or perpendicular to the flow. In some cases, an open-channel system may be desired, which is more common in wastewater UV disinfection systems. Examples of closed-vessel pressured and open-channel systems are presented in Fig. 18-5.

UV DISINFECTION

Disinfection of water can be achieved using a photochemical-based process such as UV light, as described below, or chemical methods such as chlorine or ozone, which are described in Chap. 17.

Fundamentals of UV Disinfection

UV light is a nonchemical means of disinfecting water. Essentially, UV light transmitted into a water column is absorbed by the nucleic acids of a microorganism. The energy from the photons absorbed by the organism causes damage to the genome, rendering the microbes unable to replicate. An organism that cannot replicate cannot cause an infection. Because a UV-inactivated organism is not completely destroyed, the UV damage may undergo some repair, which in cases of insufficient UV exposure may lead to reactivation.

Microbial Inactivation Mechanisms. UV light can damage the nucleic acids of a target pathogen as the photon energy is absorbed by the biological target. The absorbance spectrum of the combined nucleic acids of DNA is presented in Fig. 18-2. The primary mechanism of UV inactivation is the formation of dimers on adjacent pyrimidine bases on a DNA strand. These damage sites typically occur between adjacent thymine bases. The formation of the dimer interferes with replication of the cellular DNA (or RNA), rendering the organisms unable to replicate. An organism that cannot replicate cannot cause an infection, although it still may be metabolically active. An example of this type of nucleic acid damage is illustrated in Fig. 18-6. Other cellular damage that can occur from UV light absorption includes single- and double-strand DNA breaks, DNA-DNA cross-links, protein-DNA cross-links, pyrimidine hydrates, and pyrimidine(6-4) pyrimidine photoproducts. A review of the biological effects of UV light treatment can be found in Harm (1980).

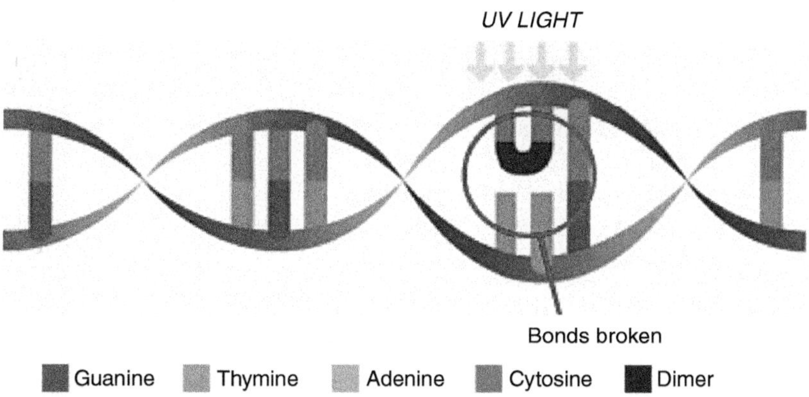

FIGURE 18-6 DNA double helix with thymine dimer formed from UV exposure. (*Courtesy of International Water-Guard Industries, Inc.*, www.water.aero/disinfection.html.)

Repair of UV-Induced Damage. UV light absorption causes damage mainly to the nucleic acids but does not destroy the integrity of the microorganism; therefore, some types of DNA damage can be repaired by the organism, and its ability to replicate can be restored. DNA repair generally is divided into two types: light repair and dark repair. Light repair, also called *photoreactivation*, uses an enzyme called *photolyase* that attaches to nucleic acid dimers and, when exposed to light in the UVA and visible spectra (available in sunlight or daylight lamps), repairs the damage and restores the nucleic acid integrity. Dark repair occurs without any light energy and typically includes excision repair, recombinatorial repair, and inducible error-prone repair. A review of nucleic acid repair processes is covered by Harm (1980). Repair typically is minimized or negligible when an adequate UV dose is delivered. A UV dose of 40 mJ/cm^2 typically is enough to eliminate the possibility of a repair process occurring that would affect the inactivation of a pathogen (Hoyer, 1998). For drinking water applications, it is not likely that the UV-disinfected water will be exposed to visible light for periods of time that would lead to light repair, and a secondary disinfectant can serve to minimize repair and recovery of UV-inactivated microorganisms (Mofidi and Linden, 2004).

UV Doses Required for Pathogen Inactivation. The UV dose is a product of the delivered UV irradiance and the exposure time, or $I \times T$, similar to the $C \times T$ concept, and the UV doses required for some pathogens are published in the LT2ESWTR (USEPA, 2006b). Different types of pathogens require different levels of UV dose to achieve adequate inactivation. As a benchmark, the UV doses needed to achieve 4 log inactivation (99.99 percent) for various pathogenic microorganisms are presented in Fig. 18-7 (Hijnen et al., 2006; USEPA 2006b). In general, the most UV-sensitive organisms are the protozoan pathogens, including *Cryptosporidium* and *Giardia.* Almost as sensitive are the vegetative bacteria, such as *Escherichia coli* and *Salmonella.* Viruses tend to be the least sensitive to UV, but most viruses still are easily inactivated by UV light, such as polio and hepatitis viruses. Spores are known to be more resistant than their vegetative bacterial form. Adenoviruses are thought to be the most resistant pathogens to UV light, requiring a UV dose of 186 mJ/cm^2 for 4 log inactivation, according to the LT2ESWTR (USEPA, 2006b; Yates et al., 2006). However, this UV dose requirement is based on LP UV light treatment, but use of MP UV has been shown to reduce the required UV dose by two- to fourfold (Malley et al., 2004; Linden et al., 2007). Interestingly, both LP and MP UV light cause identical levels of DNA damage to adenoviruses (Eischeid et al., 2009). However, after assaying the viruses in cell culture, MP UV light is more effective than LP UV light, likely owing to the suppression of any viral repair activity by the MP polychromatic light emission. The UV inactivation of adenoviruses is an area of continued research interest. A review of the UV doses required for inactivation of various pathogens is provided in Hijnen and colleagues (2006) and Chevrefil and colleagues (2006).

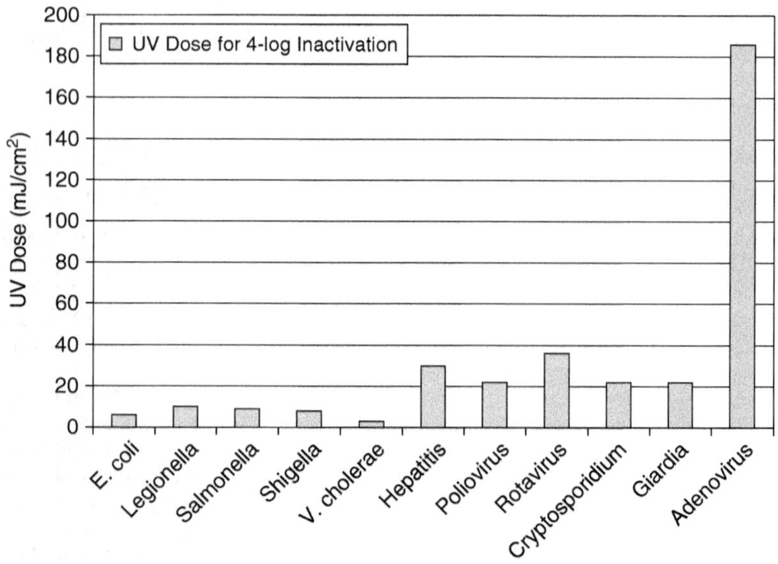

FIGURE 18-7 UV doses required for 4 log inactivation of various waterborne pathogens. (*Source: Hijnen et al., 2006, and USEPA, 2006b.*)

Characterizing UV Reactor Performance. As mentioned earlier, UV disinfection reactor performance is commonly represented by a single UV dose value (the RED) generated from a bioassay validation test (Mackey et al., 2002). However, this representation ignores the complexities of UV dose delivery in a reactor. Rather than a single UV dose provided in a disinfection reactor, UV systems actually deliver a distribution of UV doses depending on the path that a microorganism travels through the reactor, subject to the specific hydraulics and the UV irradiance field within that specific UV system. This complexity was first quantified by Chiu and colleagues (1999) and has since been studied using advanced mathematical techniques such as computational fluid dynamics (CFD) coupled with light irradiance distribution (LID) modeling (e.g., Lyn and Blatchley, 2005; Ducoste et al., 2005).

CFD Analysis and Light Distribution Modeling. The use of CFD has modernized the design and analysis of UV disinfection systems. With CFD, coupled with LID, the actual performance of a UV reactor system can be detailed. The output of these models provides specific information on the fraction of the flow treated by the UV system that received a given UV dose. By analyzing these types of data, information on design optimization and expected disinfection performance can be estimated. The dose distribution of a UV reactor as predicted by mathematical models has been validated experimentally using a technique called *microsphere actinometry* (Bohrerova et al., 2005; Blatchley et al., 2008). In this technique, small particles embedded or tethered with a UV-sensitive chemical are injected into the reactor and captured, each recording a unique UV dose as it travels through the UV system. These spheres are analyzed for their UV exposure history via their chemical change. This emerging method of reactor characterization provides detailed information on UV reactor behavior, which, when coupled with the bioassay information, provides a robust means for validating UV reactor performance.

UV Disinfection Practice

The effectiveness of chlorine disinfection is followed by monitoring the chlorine residual with a simple chemical test on a chlorinated water sample. Because UV light cannot be measured using a method parallel to this chemical analysis, UV systems have been held to high standards regarding validation of performance and development of tools to monitor operation.

Validation of UV Disinfection Systems. A table of UV doses required for various levels of inactivation of viruses, *Cryptosporidium*, and *Giardia* is published in LT2ESWTR (USEPA, 2006b). These UV doses have to be delivered within a UV disinfection system, and that system must be validated and then monitored to ensure proper disinfection. The *UV Disinfection Guidance Manual* (USEPA, 2006a) provides detailed guidance on how UV reactors should be validated. Each UV disinfection system is designed to achieve a given level of disinfection performance within an envelope of operating conditions. The validation process is a means to verify that the claims made on reactor performance by vendors, specifically in the areas of UV dose delivery and monitoring, are achieved during operation. A typical validation protocol includes (1) documentation of the UV system as received from the vendor, along with its components, (2) measurement of the UV dose delivery using a bioassay testing system, (3) correlation of the bioassay-measured UV performance with monitoring devices, and (4) assessment of the uncertainty and bias in the data analysis. A validation takes place over a range of water quality UV transmittance values, flow rates, and UV lamp intensities. The validation process generates a validation factor, much like a safety factor in design, that accounts for the data uncertainties. This factor is used to calculate the validated UV dose required to achieve a given level of disinfection credit. For example, if the UV reactor is designed to achieve 3 log inactivation of *Cryptosporidium*, the required UV dose for that level of disinfection is 12 mJ/cm^2, as indicated in LT2ESWTR. The UV disinfection system must demonstrate a validated UV dose equal to the design UV dose multiplied by the validation factor. Therefore, if the validation factor is determined to be 1.9, the validated UV dose must equal 12×1.9, or 23 mJ/cm^2. More details on how the validation factor is calculated can be found in the *UV Disinfection Guidance Manual* (USEPA 2006a).

Monitoring the UV Dose. During the validation process, data must be collected for numerous monitoring tools, typically including a UV sensor, a UV transmittance monitor, and a flowmeter. The UV sensor is a detector that is embedded in the shell of the UV reactor such that it can monitor the UV irradiance. This sensor reading then is tied to a given level of reactor performance during the validation process. For a given set of flow, UVT, and lamp output, the UV system is validated to provide the desired target (validated) UV dose, and a sensor reading is recorded. These sensor readings are documented for validated conditions over a range of flow, UVT, and UV lamp outputs under which the system is expected to operate. This sensor reading then can be used in a monitoring program during UV system operation to indicate whether the UV disinfection process is within validated conditions. In more complex monitoring scenarios, some UV reactor systems are designed to maintain validated UV dose levels by modulating UV lamp intensity using sensor readings, UVT, and flow information in real time.

Effects of Water Quality and By-Products

The transmission of UV light can be affected by various water quality constituents, both dissolved and suspended. These constituents can absorb and/or scatter UV light or shield microorganisms from lethal UV exposure. The water quality also can influence the rate of fouling onto the quartz envelope surrounding the UV lamp.

UV Absorbers, Foulants, and Particles. UV absorbers of concern are constituents that absorb germicidal UV light and lower the UV irradiance within a disinfection system. Typically, compounds containing aromatic groups are strong UV absorbers, including natural organic matter (humic and fulvic substances) and some organic chemicals (e.g., those with phenolic structures). Other important absorbers are some metals (e.g., ferric iron) and nitrate. The UV absorbance typically is monitored at a wavelength of 254 nm, although a scan of the complete germicidal wavelength range from 200 to 300 nm is more revealing for applications involving polychromatic UV lamps.

Upstream water treatment processes can have an effect on UV absorbance via addition of chemicals such as chlorine, ferric iron, and permanganate. Because iron has a very high molar UV absorbance, even very low levels (<0.01 mg/L) can result in large impacts to UV systems. Residual ozone also significantly absorbs UV light. Proper upstream treatment can serve to improve the UV transmission and thus reduce the size of the UV system required. For instance, enhanced coagulation, activated carbon treatment, and ozonation all reduce the UV absorbance, improving UV disinfection performance.

Polyvalent cations such as iron, calcium, and magnesium deposited onto the quartz sleeve cause fouling and reduced UV transmittance from the lamp into the water. Therefore, water quality constituents such as alkalinity, hardness, pH, and temperature influence lamp sleeve fouling rates. The mechanism of fouling is precipitation of calcium carbonate owing to higher quartz sleeve temperatures, which decrease the solubility of calcium at the water–quartz interface. MP lamps run hotter than LP lamps and are more susceptible to this type of fouling. Other processes that contribute to fouling are gravity settling or impaction, organic fouling, and photochemical reactions (Lin et al., 1999a, 1999b; Wait et al., 2007). The rate of fouling can vary from the time scale of hours to months. Fouling is easily managed using a regular cleaning regime tailored to the fouling rate of the system or through the use of mechanical and/or chemical quartz sleeve wipers, common to many UV systems.

In addition to dissolved constituents, particles also can interfere with UV disinfection. Organic and metallic colloids and solids that form turbidity can absorb and/or scatter UV light. Typically, below a turbidity of 10 ntu, the presence of particles does not significantly affect the operation of UV disinfection systems in terms of delivery of UV light to water in a reactor (Christensen and Linden, 2003; Passantino and Malley, 2004). Elevated levels of turbidity present in water would require higher levels of UV energy to achieve disinfection, similar to the situation with higher UV absorbance (Clancy et al., 2000). If particles shield a pathogen from UV light via particle-microbe association, then UV disinfection may be compromised (Mamane-Gravetz and Linden, 2006; Templeton et al., 2008). Particle size analysis could be used to monitor the presence of protecting particles, which would indicate upstream process (e.g., filtration) performance concerns.

UV-Induced Disinfection By-Products. Another factor in the rapidly increasing interest in UV disinfection of water is the fact that application of UV disinfection does not negatively affect the formation of halogenated disinfection by-products (DBPs) or other by-products such as bromate (Bellamy et al., 2004). Although UV light itself does not contribute to the formation of regulated DBPs such as trihalomethanes (THMs) and haloacetic acids (HAAs), UV light may influence DBP precursors at high UV doses, outside the range of disinfection practice. Small changes in natural organic matter (NOM) structure have been reported (Magnuson et al., 2002), but no effects on THMs and HAAs were noted (Liu et al., 2002). The one area of potential concern for UV-induced by-products is in the photolysis of nitrate to nitrite. Nitrate may be present in groundwaters and some surface waters and absorbs UV light below 240 nm, in the range of output for MP UV lamp sources (but not LP UV). The formation of nitrite is complex (Mack and Bolton, 1999) and may be influenced by the presence of organic matter and pH, but when nitrate is below the regulated maximum contaminant limit (MCL), there is minimal chance that nitrite formation would occur at levels above the USEPA-regulated limits, although the European Union (EU) limits, which are more strict (0.5 mg/L as NO_2^-) may be approached (Sharpless and Linden, 2001). The photolysis of nitrate also may induce the formation of nitrogenous DBPs such as chloropicrin, but not much is yet known about the formation mechanisms, and little formation is expected for UV doses at disinfection levels (Reckhow et al., 2010).

Integration of UV into a Water Treatment Plant

The placement of UV disinfection in a water treatment plant can vary and often will depend on the source water quality. In surface water plants, owing to UV transmittance and particle considerations, it is often best placed following filtration and ideally after a polishing step such as enhanced coagulation or activated carbon. In this location, the UV transmittance is maximized, optimizing UV system sizing and minimizing any issues with particles in the water. UV disinfection also can be employed at other locations within a surface water treatment plant, on the raw water, upstream of filtration, or after chlorination. Upstream of filtration, the water quality will be poorer, and therefore, the UV system required to disinfect the water may be larger. UV employed following chlorination can cause some chlorine photolysis, leading to minor dechlorination (Ormeci et al., 2005), or may under some conditions of low pH form oxidizing radicals (Watts and Linden, 2007). Ideally, UV can be used as a primary disinfectant on filtered water for the inactivation of protozoa, bacteria, and viruses, followed by application of chlorine as a postdisinfectant for maintaining a distribution system residual. For groundwaters low in NOM without reduced iron, UV can be used without pretreatment. With reduced iron, UV should be used following oxidation and removal of iron. For unfiltered water treatment systems, UV disinfection can be employed effectively for waters meeting turbidity standards and those low in NOM. UV is an ideal disinfectant as a primary barrier for inactivation of protozoan pathogens, such as *Cryptosporidium* and *Giardia*, in unfiltered systems, as well as being effective against bacteria and most viruses.

Small Systems and Groundwater Applications. UV disinfection can be an important component for proper treatment of water produced for small community systems (Malley et al., 2001). In cases where there is insufficient contact time for proper chlorination, UV disinfection can serve as a primary disinfectant for inactivation of water borne pathogens. UV is also often used as the primary disinfectant in noncommunity supplies (i.e., households) that draw water from protected groundwater, surface water, or spring sources. When UV is used as a primary disinfectant without sufficient free chlorine application, the required UV doses will need to be high enough to meet the requirements for virus inactivation.

UV Inactivation of Viruses. Presently, UV disinfection systems cannot receive much credit for virus inactivation at UV doses typically used in disinfection practices. This is so because the U.S. Environmental Protection Agency (USEPA) regulates the UV dose required for all viruses based on the results reported for the LP UV inactivation of adenoviruses. As noted in Fig. 18-7, the UV dose needed for 4 log inactivation of other pathogenic viruses (e.g., hepatitis, polio, and rotavirus) is below the commonly applied UV dose of 40 mJ/cm^2 (Snicer et al., 2000), whereas that for adenoviruses is almost five times higher. The fact that water utilities interested in receiving credit for virus inactivation have to apply UV doses approximately five times higher than is typical is a major issue in practice and is due solely to the UV dose regulations being based on adenoviruses. This issue, unless overturned based on new research findings on adenovirus susceptibility to UV light, will virtually require water utilities to achieve virus disinfection credit using high doses of UV light or free chlorine or other chemical disinfectants.

Multi-Barrier Approach. UV disinfection is an important component to the multiple-barrier concept in water treatment. Water treatment processes should establish a series of barriers to stop the passage of pathogens and harmful chemical pollutants to the greatest extent practicable. These protection barriers should be at multiple levels and include source water protection as well as treatment processes. In addition to multiple layers of treatment processes, incorporating processes with differing mechanisms of action is essential. UV disinfection, a photochemical process, directly complements processes such as filtration and chlorination by providing a unique mode of action not found in other treatment processes. The application of UV in combination with free chlorine is an approach that will be increasingly necessary for water utilities to achieve disinfection of both *Cryptosporidium* and viruses (Ballester and Malley, 2004).

Unfiltered Water Supply Systems. Because of its effectiveness for inactivation of *Cryptosporidium*, UV disinfection is an essential part of the water treatment process for unfiltered systems. Since filtration is a primary barrier in the treatment of protozoan pathogens, the water systems granted waivers from filtration are required to implement additional measures to achieve disinfection or removal of these pathogens. These unfiltered supplies represent some of the largest UV disinfection systems in North America and employ multiple disinfectant barriers, including combinations of ozone, UV light, chlorine, and chloramines. These multiple barriers are in place to cover inactivation of all pathogens in the absence of a conventional water treatment train.

Examples of UV Installations

Although UV disinfection has been used extensively for wastewater disinfection and has been used in small communities and thousands of household water treatment systems in the United States, it was not until the promulgation of LT2ESWTR that the design and construction of UV systems into water treatment plants in North America accelerated. Given these new regulations that now require stricter limits on DBPs and inactivation of *Cryptosporidium* for some vulnerable water supplies, it is estimated that over 3000 UV disinfection facilities will be constructed in the United States (USEPA, 2005) to meet new drinking water regulations. A number of prominent UV installations already exist, and some of these are described below.

Fort Benton, Montana. The UV disinfection system in Ft. Benton is one of the oldest and longest-operating UV systems, serving 30,000 residents. Installed in 1987, this system uses a medium-pressure UV disinfection system to disinfect 8.3 million L/day (2.2 mgd) of water from the Platte River, which is collected through a horizontal collection well approximately 7.6 m (25 ft) under the riverbed. Other than bank filtration, there is no other treatment process employed before UV disinfection. The water is postchlorinated and distributed via pipe network to the town of Ft. Benton.

Seattle Public Utilities. The City of Seattle uses a multiple-barrier system consisting of a watershed-protected source water that is unfiltered and goes through an initial ozonation process before being disinfected by a medium-pressure UV system followed by lime addition and postchlorination. This combination provides 3 log *Cryptosporidium*, 4 log *Giardia*, and 5 log virus inactivation. The system serves 1.3 million people and was installed in 2004. It treats up to 680 million L/day (180 mgd), with the UV process being the primary barrier against bacteria and protozoa, whereas ozone and chlorine are used as a dual virus barrier.

Thornton, Colorado. The Wes Brown Water Treatment Plant is a 190 million L/day (50 mgd) facility receiving source water from the South Platte River. The treatment processes include ultrafiltration membranes for filtration coupled with UV disinfection as the primary disinfectant for *Cryptosporidium* and *Giardia* and chlorine as the main disinfectant for viruses.

New York City. The unfiltered Catskill–Delaware (CAT–DEL) source water serving the City of New York will be treated with UV disinfection using 56 low-pressure, high-output UV reactors. The UV system will treat 4.9 million L/day (1300 mgd). The city also uses postchlorination.

UV PHOTOLYSIS

A *photochemical reaction* typically is defined as a chemical reaction caused by the absorption of UV or visible electromagnetic radiation (light), with UV photolysis involving the study of photochemical transformation reactions induced by light within the UV range (200–400 nm) (Barltrop and Coyle, 1978). In contrast to UV for disinfection, where UV energy generally is visualized as a field of energy through which target microbes must pass for a definitive amount of time to achieve an effective inactivation UV dose, in UV photolysis, UV energy is visualized as discrete packets of energy, known as *photons*, that are absorbed by molecules within the water matrix and may be capable of causing photochemical change.

Photochemistry Basics

Laws of Photochemistry. *UV photolysis* is defined by two main principles. The first, known as the *Grotthuss-Draper law*, states that only energy absorbed by a molecule can bring about chemical change. The second, known as the *Stark-Einstein law*, indicates that photochemical degradation of a molecule corresponds to the absorption of a single photon (this law holds for UV light sources, with exceptions possible for extremely intense light sources, such as lasers) (Coyle, 1986). The two laws serve as the basis for photochemical study, giving rise to concepts such as the molar absorption coefficient and quantum yield.

A third principle of interest requires that the energy of an absorbed photon must be equal to or greater than the weakest bond in the molecule to induce photochemical transformation (Schwarzenbach et al., 2003). UV light (200–400 nm) absorbed by molecules is often greater than the energy required for bond dissociation. U_λ, the photon energy at wavelength λ ($J \cdot Es^{-1}$), is calculated using Eq. 18-7. Photon energy is inversely related to λ, indicating that as the wavelength gets shorter, the available photon energy is greater:

$$U_\lambda = \frac{N \cdot h \cdot c}{\lambda} = \frac{0.1196}{\lambda}\left(\frac{\text{joule}}{\text{einstein}}\right) \tag{18-7}$$

where N = Avogadro's number (6.02×10^{23} photons $\cdot\, Es^{-1}$), h = Plank's constant (6.6256×10^{-34} $J \cdot s \cdot \text{photon}^{-1}$), c is the speed of light (2.9979×10^8 $m \cdot s^{-1}$), and λ is the wavelength (m).

Interactions of Light and Matter. When a photon of UV energy is absorbed by a chemical entity, the result is the excitation of electrons within the chemical structure, promoting an electron from the highest occupied molecular orbital (HOMO) to the lowest unoccupied molecular orbital (LUMO). Figure 18-8, adapted from Oppenländer (2003), displays a generalized

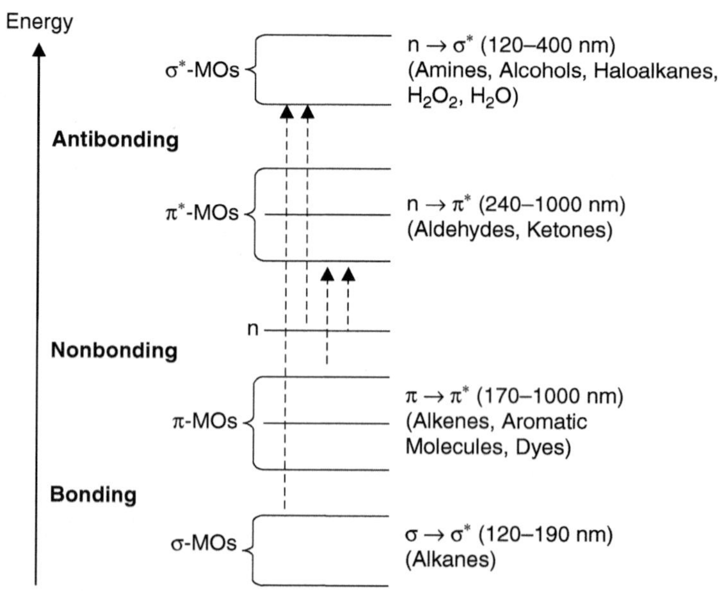

FIGURE 18-8 Molecular orbital diagram describing possible excitations caused by the absorption of a photon of UV light. (Also included are the wavelengths that have the energy to cause the electron jumps and compounds with these types of electrons.) (*Source: Oppenlander, 2003; reprinted with permission from Wiley-VCH. Copyright © 2003.*)

molecular orbital diagram and shows possible electronic transitions in organic molecules, as well as associated absorption regions and examples of common chromophores.

On excitation, the molecule can follow several paths to return to the ground energy state. The Jablonski diagram (Fig. 18-9) is a convenient way to visualize the possible paths between the various electron states of a molecule. To understand Fig. 18-9, the nomenclature must be explained. S_0 indicates the unexcited ground electron state for the molecule, with S_1 and S_2 representing the first or second excited singlet states, respectively, and T_1 indicates the excited triplet state. P_1 and P_2 are photoproducts of the reaction.

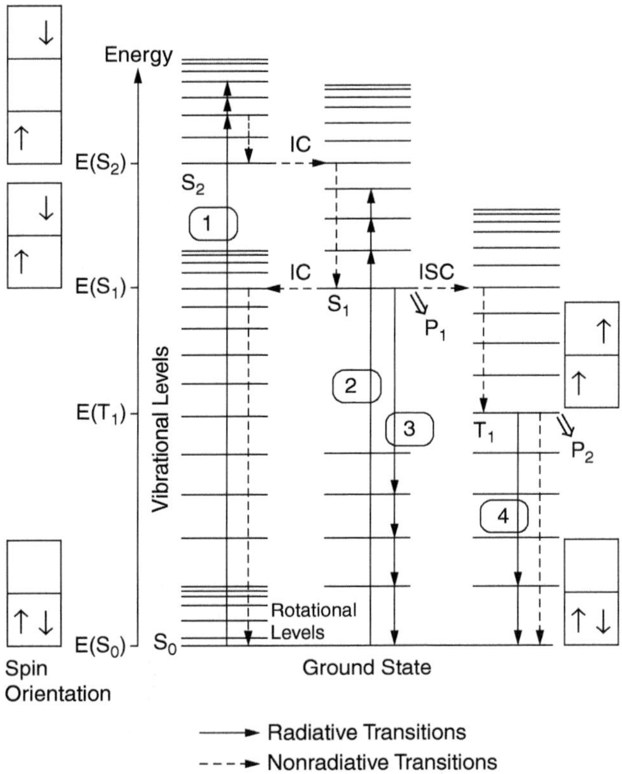

FIGURE 18-9 Modified Jablonski diagram for visualization of radiative, nonradiative, and photoproduct forming processes. (*Source: Oppenlander, 2003; reprinted with permission from Wiley-VCH. Copyright © 2003.*)

Depending on the photon energy, the molecule is excited from the S_0 state to either the S_1 or S_2 state. After excitation, several processes may occur, including internal conversion (IC) from the S_2 state and/or relaxation of vibrationally excited levels of the S_1 state. All subsequent processes occur from the S_1 excited state. S_1 can deactivate via IC and by vibrational relaxation to S_0, transforming the excitation energy into vibrational energy (i.e., heat), or the S_1 deactivates by fluorescent emission. Both the S_2 and S_1 states are present for only very short lifetimes of 10^{-13} to 10^{-12} and 10^{-12} to 10^{-6} second, respectively. Because of its short lifetime, the only chemical reactions that can occur from the S_1 state involve rearrangement of the molecular structure or bond cleavage processes leading to reaction product P_1. If special interactions occur, the relatively long-lived triplet state (T_1) can occur via intersystem crossing (ISC) and vibrational deactivation. Once at the T_1 state, the molecule can deactivate to the S_0 state via phosphorescence emission, a much lower energy and slower energy release than fluorescence. Nearly all intermolecular photochemical reactions occur from the T_1 state, mainly owing to the comparatively long lifetime of this energy state.

Molar Absorption Coefficient. The pattern of wavelength specific UV energy absorption is defined by the molar absorption coefficient ε, which is a measure of the amount of UV energy transferred from an electromagnetic field to the molecular entity divided by the absorption path length and concentration. This absorption of UV energy is chemical-specific and depends heavily on moieties within the chemical. These absorptive moieties are known as *chromophores*, and this absorption is a necessary starting point for photochemical reactions. As an example, Figure 18-10 displays the molar absorption coefficients for several xenoestrogens, as well as for hydrogen peroxide, from 200 to 350 nm. The molar absorption coefficient ε for a particular compound can be determined by using Eq. 18-3 by measuring a_{10} for a solution containing only the compound of interest at a known concentration.

Quantum Yield. The amount of photoproducts (P_1, P_2, etc.) formed or the initial compound degraded via photochemical reactions as a function of the number of photons of energy absorbed is known as the *quantum yield* **Φ**. The quantum yield is a measure of the molecular photon efficiency of a photochemical reaction. It is defined as the number of moles of reactant removed per einstein (mole of photons) absorbed by the chemical (Bolton, 2001). There are no simple rules to predict reaction quantum yields from chemical structure, so **Φ** values need to be determined experimentally (Schwarzenbach et al., 2003).

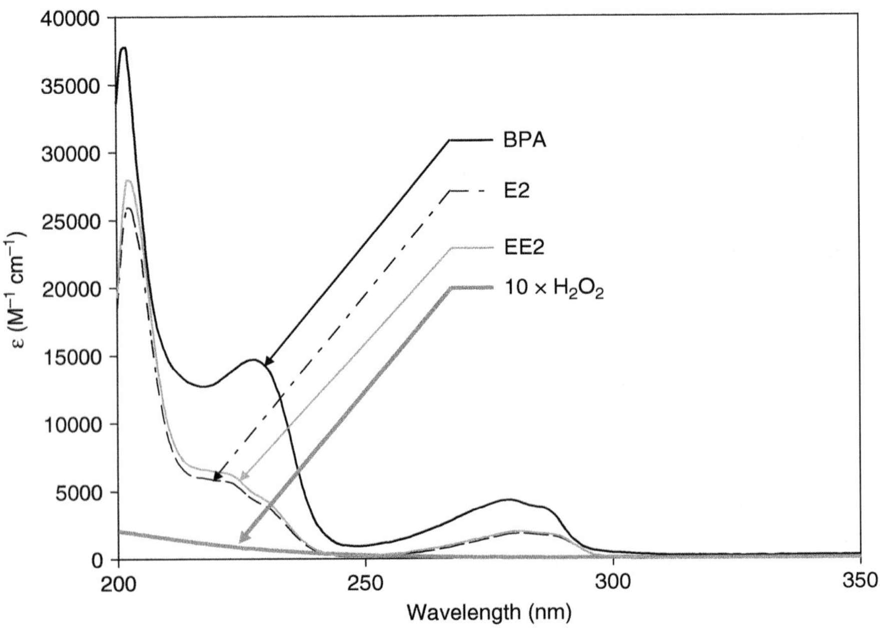

FIGURE 18-10 Molar absorption spectra for bisphenol A (BPA), 17-β-estradiol (E2), 17-α-ethinyl estradiol (EE2), and hydrogen peroxide (H_2O_2) (×10).

Additionally, the wavelength dependence of Φ must be considered when using polychromatic light sources. According to Zepp (1982), quantum yield values can be assumed to be wavelength independent, at least over the wavelength range of a given absorption band, corresponding to one mode of excitation. If multiple light absorption maxima are displayed, quantum yields may have to be determined for various wavelengths to accurately predict the transformation rate of a given compound.

UV Photolysis for Treatment of Chemical Pollutants

Monochromatic UV Photolysis. In determining the efficiency of photolysis for treatment of chemical pollutants, two major chemical-specific parameters are of importance. As mentioned previously, the molar absorption coefficient (ε, in $M^{-1} \cdot cm^{-1}$) identifies how much UV energy at a particular wavelength a compound will absorb. In addition, the quantum yield (Φ in $mol \cdot Es^{-1}$) indicates the efficiency of the absorbed energy for causing a photochemical degradation reaction. While UV photolysis can display zero-order or pseudo-first-order reaction kinetics depending on the concentration of the pollutant and the nature of the water matrix (Bolton, 2001), most photolysis situations of interest in natural or engineered water systems are described by first-order decay kinetics, with a fluence-based rate constant, as described in Eq. 18-8. The complete derivation of the relationship can be found in Bolton and Stefan (2002). Equation 18-8 has been modified slightly from the final equation found in Bolton and Stefan (2002) to reflect a fluence-based rate constant with units of cm^2/mJ rather than m^2/J, as in the referenced text:

$$k_1' = \frac{\Phi_{\lambda,c}\,\varepsilon_{\lambda,c}\ln(10)}{U_\lambda} \tag{18-8}$$

where for micropollutant C, k_1' is the pseudo-first-order rate constant with units of $cm^2 \cdot mJ^{-1}$, $\Phi_{\lambda,C}$ is the wavelength-specific quantum yield ($mol \cdot Es^{-1}$), $\varepsilon_{\lambda,C}$ is the wavelength-specific molar absorption coefficient ($M^{-1} \cdot cm^{-1}$), and U_λ is the wavelength-specific photon energy ($J \cdot Es^{-1}$), as calculated in Eq. 18-7.

Table 18-4 provides several approximate molar absorption coefficients and quantum yield values for some inorganic and organic compounds at 254 nm, as reported in the

TABLE 18-4 Molar Absorption Coefficients and Quantum Yield Values for Several Compounds at 254 nm

Chemical	ε_{254} ($M^{-1} \cdot cm^{-1}$)	Φ (mol Es^{-1})	Reference
N-Nitrosodimethylamine (NDMA)	1,400	0.3	Sharpless and Linden, 2003
	1,470	0.32	Bolton and Stefan, 2002
	1,650	0.31 (pH 2–8)	Lee et al., 2005
Hydrogen peroxide (H_2O_2)	19.6	1.0	Baxendale and Wilson, 1957
Hypochlorous acid (HOCl)	54	1.4	Watts and Linden, 2007
Monochloramine (NH_2Cl)	400	0.45	Watts and Linden, 2007
Atrazine	3,150	0.05	Beltran et al., 1993
17β-Estradiol	223	0.043	Rosenfeldt and Linden, 2004
Naproxen	5,000	0.0093	Pereira et al., 2007
Ciprofloxacin	11,000	0.0103	Pereira et al., 2007
Carbamazepin	8,000	0.0006	Pereira et al., 2007
Ketoprofin	15,000	0.2364	Pereira et al., 2007

literature. Example 18-1 illustrates the potential photochemical treatment of a relatively UV-labile compound, *N*-nitrosodimethylamine (NDMA).

Exampe 18-1 NDMA Photolysis
Using the parameters from Table 18-4, what is the low-pressure (monochromatic, 254 nm) UV fluence required for 50 and 90 percent (1 log) NDMA photodegradation, assuming pseudo-first-order kinetics?

Solution Assuming first-order, fluence-based kinetics, the UV fluence required for removal of a fraction of contaminant M can be described with the following equation:

$$\ln\left(\frac{M_0}{M}\right) = k_1' \cdot \text{UV fluence}$$

Rearranging this equation to find the UV fluence required for degradation and substitution in from Eq. 18-8 for k_1' yields:

$$\text{UV fluence} = \frac{\ln\left(\frac{M_0}{M}\right)}{k_1'} = \frac{\ln\left(\frac{M_0}{M}\right) U_\lambda}{\ln(10)\varepsilon_\lambda \Phi_\lambda}$$

To calculate UV_{254}, use Eq. 18-7:

$$UV_{254} = \frac{hc}{\lambda} = \frac{0.1196}{254\times10^{-9}}\left(\frac{\text{J}}{\text{Es}}\right) = 4.71\times10^5 \text{ J/Es}$$

From Table 18-4, there are several values for ε_{254} (1400, 1470, and 1650 $M^{-1} \cdot cm^{-1}$), but a consistent value and Φ_{254} (~0.3 mol · Es^{-1}) for NDMA. Using an average value of these three published ε_{254} constants and calculating UV fluence for 50 percent NDMA photolysis, we get

$$\text{UV fluence} = \frac{\ln(2)\times4.71\times10^5}{\ln(10)\times1507\times0.3} = 313.6 \text{ mJ}\cdot\text{cm}^{-2}$$

However, because of the variability in reported ε_{254} values observed in Table 18-4, the range of calculated UV fluence values for 50 percent photolysis of NDMA is 286 to 337 mJ · cm^{-2}. Likewise, when calculated using the average ε_{254} value, one obtains a fluence value of 1041 mJ · cm^{-2} for 90 percent NDMA photolysis. ▲

When considering fluence-based kinetics, Eq. 18-8 indicates that UV photolysis rates for a monochromatic light source are constant, irrespective of water quality (assuming that no side reactions leading to the decay or regeneration of *C* are taking place). This observation has been substantiated for NDMA kinetics in phosphate buffer and experimental waters containing 2, 7, or 12 mg/L organic carbon (Stefan et al., 2002). No significant change was found in measured fluence-based degradation. However, this does not indicate that the amount of energy required for photodegradation is independent of water quality. As described earlier, the UV fluence is a function of the average irradiance (which is affected by the amount of energy absorbed by the water) and the exposure time. Since full-scale UV reactors typically are designed with a fairly constant exposure time, the amount of energy absorbed by the water is the parameter often adjusted to increase or decrease UV fluence. The average irradiance is determined by the amount of energy input to the system (from the lamps) and the UV absorbance of the solution. In other words, as absorbance increases, so must the energy input from the lamps in order to maintain the same average irradiance.

Polychromatic UV Photolysis. For polychromatic UV lamp sources, such as MP mercury lamps, understanding of the photolysis mechanism is complicated by the presence of multiple emission wavelengths throughout the UV range. These wavelengths are each absorbed to a different degree by the contaminant, reflecting its molar absorption coefficient, and the possibility of different quantum yield values for various wavelengths throughout the UV range exists.

The discussion in Sharpless and Linden (2003) describes many of these complexities in the context of photolysis of NDMA. Starting with Eqs. 18-9 and 18-10, modified from Sharpless and Linden (2003), it is observed that many of the parameters describing both lamp emission and photon absorbance are wavelength-dependent:

$$-\frac{d[C]}{dt} = k'_d[C] = \left(\sum_{\lambda} k_{s,\lambda}\Phi_{\lambda,c}[C]\right) \tag{18-9}$$

where k'_d is the first-order rate constant for direct photolysis of C (s^{-1}), $k_{s,\lambda}$ is the specific rate of light absorption by compound C ($Es \cdot mol^{-1} \cdot s^{-1}$), $\Phi_{\lambda,c}$ is the wavelength-specific quantum yield of compound C ($mol \cdot Es^{-1}$), $E^0_{p,\lambda}$ is the incident photon irradiance (10^{-3} $Es \cdot cm^{-2} \cdot s^{-1}$), $\varepsilon_{\lambda,c}$ is the molar absorption coefficient of C ($M^{-1} \cdot cm^{-1}$), a_λ is the solution absorbance coefficient (cm^{-1}), and l is the solution path length (cm). All the parameters in Eq. 18-10 are measurable.

$$k_{s,\lambda} = \frac{E^0_{p,\lambda}\varepsilon_\lambda(1-10^{-a_\lambda l})}{a_\lambda l} \tag{18-10}$$

Assuming that the quantum yield is path length independent for a range of wavelengths and that accurate measurements of the lamp emission spectrum and the wavelength-specific molar absorption coefficient are available, Eq. 18-9 can be rewritten in the form of Eq. 18-11:

$$-\frac{d[C]}{dt} = \left(\sum_{\lambda_1 \to \lambda_2} k_{s,\lambda}\Phi_\lambda + \sum_{\lambda_2 \to \lambda_3} k_{s,\lambda}\Phi_\lambda + \cdots\right)[C] \tag{18-11}$$

The wavelength independence of quantum yield, at least for a range of wavelengths (i.e., from λ_1 to λ_2 or λ_2 to λ_3), has been observed for many compounds, and the general rule of thumb is that the quantum yield value will be constant over an observed UV absorbance band, as defined as the area from one UV absorbance minimum to the next UV absorbance minimum (i.e., in Fig. 18-10, an absorption band can be observed for BPA ranging from approximately 250 to 290 nm).

In the case of NDMA (Sharpless and Linden, 2003), the quantum yield is found to be constant over the entire UV wavelength range covered by the polychromatic MP UV lamp, and as a result, the observed quantum yield for NDMA is 0.3 $mol \cdot Es^{-1}$ for both the monochromatic LP source and the MP sources. For atrazine photolysis, however, there are different observations (Sharpless et al., 2003). Atrazine exhibits two distinct UV absorption bands in the 200- to 400-nm range, one ranging below 250 nm and one ranging from approximately 250 to 290 nm. Using bandpass filters, Sharpless and colleagues found the quantum yield at 228 nm to be 0.066 $mol \cdot Es^{-1}$, whereas at 270 nm the quantum yield was 0.038 $mol \cdot Es^{-1}$. By using these constant quantum yield values for each absorbance band, the authors were able to accurately predict atrazine photolysis using the entire MP UV lamp spectrum. A photolysis study of several endocrine-disrupting compounds (EDCs) (Rosenfeldt and Linden, 2004) yielded very different observed quantum yield values for LP

UV and overall MP UV for BPA and two other hormone EDCs, 17-α-ethinyl estradiol and 17-β-estradiol), also indicating the possibility of different quantum yield values associated with each of the absorbance maxima in the 200- to 300-nm range.

If quantum yield values are measured for each of the absorption bands within the UV emission range of a particular lamp source, photolysis rates can be predicted for polychromatic lamp sources as accurately as those for monochromatic lamps through the use of Eqs. 18-10 and 18-11.

UV ADVANCED OXIDATION PROCESSES (AOPs)

The addition of hydrogen peroxide (H_2O_2) to a UV process yields an *advanced oxidation process* (AOP), where hydroxyl radicals (·OH) are formed via the homolytic cleavage of the central HO–OH bond to initiate a propagation cycle, with the net reaction leading to water and molecular oxygen. AOPs are described by a complex chain-reaction mechanism involving the formation and rapid consumption of intermediate radical species, such as the hydroxyl radical (·OH). Short descriptions of UV/H_2O_2 and other AOPs are given in Chap. 7.

UV/H_2O_2 AOP has been shown to be effective for degrading many organic and inorganic contaminants in water applications. Table 18-5 displays references for studies examining the efficiency of UV/H_2O_2 AOP for degrading many organic contaminants of concern in water. The list was compiled with the aid of several reviews, including Ikehata and El-Din (2006), Ikehata and colleagues (2006), Gultekin and Ince (2007), and Oppenländer (2003). The extent of this list indicates the potential of the UV/H_2O_2 AOP for degrading many organic contaminants, including emerging contaminants of concern such as pesticides, pharmaceuticals, and EDCs.

One of the major limitations of UV AOP is the inefficient absorption of UV energy by H_2O_2 ($\varepsilon = 19.6\ M^{-1} \cdot cm^{-1}$ at 254 nm). The quantum yield for H_2O_2 removal associated with the photolytic decomposition of H_2O_2 to form ·OH is relatively high (0.5 mol · Es^{-1}), so photons that are absorbed by the molecule efficiently initiate AOP chemistry. Additionally, the ·OH radical is a relatively unselective oxidant that reacts equally well with many organic and inorganic species in water.

Advantages associated with the UV/H_2O_2 AOP for water treatment over other AOP technologies include short reactor detention time leading to small reactor footprints and the fact that UV disinfection of pathogens in the water readily occurs at the elevated UV doses applied for AOP conditions, so two treatment objectives are met at once. In addition, UV/H_2O_2 AOP provides two mechanisms for transformation of organic chemicals—direct photolysis and hydroxyl radical oxidation. Disadvantages of the process include high energy requirements associated with operating the reactors at high UV doses and the need to store and inject H_2O_2 into the water before the UV reactors at relatively high concentrations (2–10 mg · L^{-1} to initiate effective AOP conditions). Another drawback of the process is based on the fact that typically only a small portion of the added H_2O_2 will degrade to ·OH radicals during the process, leaving 80 to 90 percent of the injected H_2O_2 in the outlet stream. Processes including chemical quenching (with free chlorine, bisulfite, or enzymatic catalysts) and adsorption through granular activated carbon (GAC) or biologically active media filters are needed to remove the excess H_2O_2 before postdisinfection processes.

The UV/H_2O_2 AOP Mechanism

A mechanism for the UV/H_2O_2 AOP, known as the *Haber-Weiss mechanism*, is outlined in Lunak and Sedlak (1992) and described below. The initiation reaction to form hydroxyl

TABLE 18-5 References Describing UV/H_2O_2 AOP Oxidation of Aqueous Compounds

Compound	Reference	Compound	Reference
Groundwater pollutants		**Pharmaceuticals**	
1,4-Dioxane	Stefan and Bolton, 1998	Amoxicillin	Arslan-Alaton and Dogruel, 2004; Andreozzi et al., 2005
Benzene	Kang and Le, 1997	Bezafibrate	Huber et al., 2003
MTBE	Stefan et al., 2000	Carbamazepine	Vogna et al., 2004a; Huber et al., 2003
PCE	Kang and Le, 1997	Clofibric acid	Andreozzi et al., 2003b
TCE	Kang and Le, 1997	Diazepam	Huber et al., 2003
Toluene	Kang and Le, 1997	Diclofenac	Vogna et al., 2004b; Andreozzi et al., 2004
		17-α-ethinyl estradiol	Rosenfeldt and Linden, 2004; Huber et al., 2003
Taste and odor		17-β-estradiol	Onda et al., 2002; Rosenfeldt and Linden, 2004; Shishida et al., 2000
Geosmin	Rosenfeldt and Linden, 2005		
MIB	Rosenfeldt and Linden, 2005	Ibuprofen	Huber et al., 2003
		Iopentol	Sprehe et al., 2001
Pesticides		Iopromide	Sprehe et al., 2001
2,4-Dichlorophenoxyacetic acid	Alfano et al., 2001; Benitez et al., 2004b; Scheuer et al., 1995	Ofloxacin	Andreozzi et al., 2004
Atrazine	Beltran et al., 1993; Hessler et al., 1993; Prados et al., 1995;	Paracetamol	Andreozzi et al., 2003a; Vogna et al., 2002
		Propranolol	Andreozzi et al., 2004
Bendiocarb	Aaron and Oturan, 2001	Sulfamethoxazole	Andreozzi et al., 2004; Huber et al., 2003
Bentazone	Beltran-Heredia et al., 1996	**Endocrine disruptors**	
Butachlor	Benitez et al., 2004a	Bisphenol-A	Rosenfeldt and Linden, 2004
Carbetamide	Mansour et al., 1992	Nonylphenol	Chen et al., 2007
Carbofuran	Benitez et al.., 1995b		
Chlorfenvinphos	Kowalska et al., 2004	**Phthalates**	Okamoto et al., 2006
Cyanazine	Benitez et al., 1995a	DPP	Zhao et al., 2004
Cyanuric acid (did not work)	Minero et al., 1997; Goutailler et al., 2001	DMP	Chiou et al., 2006
Diazinon	Doong and Chang, 1998	DBP	Xu et al., 2007
Dichlorvos	Nitoi et al., 2001	DEP	

(*Continued*)

TABLE 18-5 References Describing UV/H_2O_2 AOP Oxidation of Aqueous Compounds (*Continued*)

Compound	Reference	Compound	Reference
Pharmaceuticals (cont.)		**Other**	
Fenitrothion	Kowalska et al., 2004	Acetone	Stefan et al., 1996; Stefan and Bolton, 1999
Fenuron	Acero et al., 2002	Azo dyes	Galindo and Kalt, 1999
HPN (2104-64-5)	Doong and Chang, 1998	Chlorophenols	Hirvonen et al., 2000
Linuron	Barlas, 2000	EDTA	Sorensen et al., 1996, 1998
Malathion	Doong and Chang, 1998	Hospital wastewater	Sprehe et al., 2001
MCPA (94-74-6)	Benitez et al., 2004b	Humic acids	Wang et al., 2001
Methamidofos	Doong and Chang, 1998	Metol	Andreozzi et al., 2000
Metazachlor	Hessler et al, 1993	N-Nitrosodimethylamine	Sharpless and Linden, 2003
Metolachlor	Benitez et al., 2004a	PAHs	Miller and Olejnik, 2001
Metoxuron	Mansour et al., 1992	Remazol Black-B	Ince et al., 1997
Monolinuron	Barlas, 2000	Sulfonates	Sorensen and Frimmel, 1996
Parathion	Chen et al, 1998		
Pentachlorophenol (PCP)	Trapido et al., 1997		
Phorate	Doong and Chang, 1998		
Propachlor	Benitez et al., 2004a		
Simazine	Beltran et al., 2000		

radical is governed by the absorption of UV energy by H_2O_2, as well as the quantum yield of the reaction (Eq. 18-12):

$$H_2O_2 + h\nu \rightarrow 2 \cdot OH \qquad (18\text{-}12)$$

As with any photochemical reaction, Eq. 18-12 is governed by the absorption of UV energy by H_2O_2, as well as by the quantum yield for the process. The molar absorption coefficient of H_2O_2 is displayed in Fig. 18-10 and indicates that H_2O_2 does not greatly absorb UV light throughout the 200- to 300-nm range. The molar absorption coefficient for H_2O_2 at 254 nm has been reported as 19.6 $M^{-1} \cdot cm^{-1}$, which is exceptionally low for a primary absorber (Glaze et al., 1987). The energy that is absorbed by H_2O_2 initiates the reaction with a relatively high ·OH radical quantum yield. According to the stoichiometry, the photolysis of 1 mol of H_2O_2 should produce 2 mol of hydroxyl radicals with a calculated maximum quantum yield of 1 mol H_2O_2 degraded per einstein or 2 mol ·OH formed per einstein, but in aqueous solution, this is not the case. In water, the so-called cage effect has been postulated to explain the fact that only one hydroxyl radical is formed per einstein absorbed by H_2O_2 (Oppenländer, 2003). Equation 18-13 describes this cage effect, in which only 50 percent of the ·OH radicals seem to be able to escape from the aqueous solvent cage.

$$(H_2O_2)aq \underset{50\%}{\overset{h\nu}{\rightleftarrows}} \begin{matrix}(HO\cdot + \cdot OH)aq \\ \text{“Solvent Cage”}\end{matrix} \xrightarrow[50\%]{} \cdot OH_{\text{free}} \qquad (18\text{-}13)$$

Once initiated, a chain radical mechanism described by Eqs. 18-14 and 18-15 involving conversion of hydroxyl radical to $HO_2\cdot$ while consuming H_2O_2 occurs, propagating the radical reaction (Lunak and Sedlak 1992).

$$H_2O_2 + \cdot OH \rightarrow H_2O + HO_2\cdot \qquad (18\text{-}14)$$

$$HO_2\cdot + H_2O_2 \rightarrow H_2O + O_2 + \cdot OH \qquad (18\text{-}15)$$

Equation 18-15 displays a slow rate constant (3.0 $M^{-1} \cdot s^{-1}$) (Koppenol et al., 1978) when compared with Eq. 18-14 (2.7×10^7 $M^{-1} \cdot s^{-1}$) (Buxton et al., 1988) and is therefore justifiably neglected. Termination reactions outlined in Eqs. 18-16 through 18-18 act to lower the apparent quantum yield.

$$2 \cdot OH \rightarrow H_2O_2 \qquad (18\text{-}16)$$

$$HO\cdot + HO_2\cdot \rightarrow H_2O + O_2 \qquad (18\text{-}17)$$

$$2HO_2\cdot \rightarrow H_2O_2 + O_2 \qquad (18\text{-}18)$$

Of these reactions, Eq. 18-16 is entirely negligible owing to the extremely low steady state concentration of ·OH radicals. Noticeably missing from the reaction mechanism outlined are the reactions with organic and inorganic species found in actual waters. These reactions typically are regarded as termination reactions but can in some cases propagate the ·OH radical mechanism (NOM) or initiate new radical reaction mechanisms (e.g., carbonate species or free chlorine). The hydroxyl radical is a nonselective oxidant that displays high reaction rates with organic and inorganic species present in natural waters. Table 18-6 displays some common species present in natural water, as well as some micropollutants of concern and their rate constants with hydroxyl radical. The rates are exceptionally fast, and the oxidation reactions are nonselective, producing a variety of oxidation products.

TABLE 18-6 Second-Order Rate Constants Between Hydroxyl Radical and Common Water Constituents or Micropollutants of Concern

Pollutant or constituent	·OH radical rate constant ($M^{-1} \cdot s^{-1}$)	Reference
DOM (TOC)	2.5×10^4 L · mg · $C^{-1} \cdot s^{-1}$	Larson and Zepp, 1988
DOM (TOC)	2.23×10^8 L · mol · $C^{-1} \cdot s^{-1}$ (1.9×10^4 L · mg · $C^{-1} \cdot s^{-1}$)	Westerhoff et al., 2007
DOM isolates	$1.39–4.53 \times 10^8$ L · mol · $C^{-1} \cdot s^{-1}$	Westerhoff et al., 2007
Suwanee River DOM	1.33×10^4 L · mg · $C^{-1} \cdot s^{-1}$	Westerhoff et al., 2007
Hydrogen peroxide (H_2O_2)	2.7×10^7	Buxton et al., 1988
Bicarbonate ion (HCO_3^-)	8.5×10^6	Hoigne et al., 1985
Carbonate ion (CO_3^{-2})	3.9×10^8	Hoigne et al., 1985
Hypochlorous acid (HOCl)	8.5×10^4	Watts and Linden, 2007
Hypochlorite (OCl^-)	8.0×10^9	Nowell and Hoigne, 1992a
Monochloramine (NH_2Cl)	5.2×10^8	Poskrebyshev et al., 2003
N-Nitrosodimethylamine (NDMA)	3.3×10^9	Wink et al., 1991
Trichloroethyline (TCE)	2.6×10^8	Buxton et al., 1988
Tetrachloroethyline (PCE)	4.2×10^8	Buxton et al., 1988
Methyl-*tert*-butyl-ether (MTBE)	1.9×10^9	Cater et al., 2000
Atrazine	3×10^9	Acero et al., 2000
Methylisoborneol (MIB)	8.2×10^9	Glaze et al., 1990
Geosmin	1.4×10^{10}	Glaze et al., 1990
Sulfamethoxazole	5.5×10^9, 5.0×10^9	Huber et al., 2003; Boreen et al., 2004
Sulfamethazine	3.7×10^9	Boreen et al., 2004
Sulfadiazine	5.8×10^9	Boreen et al., 2004
17-β-Estradiol (E2)	1.41×10^{10}	Rosenfeldt and Linden, 2004
17-α-Ethinyl estradiol (EE2)	1.08×10^{10}	Rosenfeldt and Linden, 2004
Bisphenol A (BPA)	1.02×10^{10}	Rosenfeldt and Linden, 2004
Nitrobenzene	3.9×10^9	Buxton et al., 1988
p-Chlorobenzoic acid (pCBA)	5.0×10^9	Elovitz and von Gunten, 1999
Methanol	9.7×10^8	Buxton et al., 1988
t-Butanol (tBuOH)	5.0×10^8	Hoigne and Bader, 1975
1,2-Dibromo-3-chloropropane (DBCP)	1.5×10^8	Glaze et al., 1995

UV/H_2O_2 Models

Several approaches for modeling UV/H_2O_2 AOPs exist. At the basic level, these include a simplified oxidation model, called the *steady state ·OH radical model* and outlined in Glaze and colleagues (1995), as well as a complex set of differential equations, described in Crittenden and colleagues (1999) and in Cater and colleagues (2000). All models are based on use of Eqs. 18-12 through 18-18, as well as other termination reactions.

Steady State ·OH Radical Model. The model presented by Glaze and colleagues (1995) is used to model micropollutant oxidation in water matrices. The model is based on the simplifying assumption that the hydroxyl radical is extremely reactive and is essentially consumed as quickly as it is produced. Using a set of equations describing production of hydroxyl radical (dominated by the initiation Eq. 18-12 in most UV/H_2O_2 applications but including the effects of promotion reactions) and consumption of hydroxyl radical (termination and promotion reactions) and the key assumption that hydroxyl radical concentration will achieve steady state (consumption and production are equal), the general Eq. 18-19 can be generated:

$$[\cdot\mathrm{OH}]_{ss} = \frac{\text{rate of hydroxyl radical fomation}}{\text{rate of hydroxyl radical consumption}} \tag{18-19}$$

More specifically, the rate of hydroxyl radical formation can be described by Eq. 18-20:

$$\frac{d[\cdot\mathrm{OH}]}{dt} = E'^{0}_{p,\lambda}\frac{(1-10^{-a_\lambda \cdot l})}{a_\lambda \cdot l}\varepsilon_{\lambda,\mathrm{H_2O_2}}[\mathrm{H_2O_2}]\,\Phi_{\lambda,\mathrm{H_2O_2}\rightarrow\cdot\mathrm{OH}} \tag{18-20}$$

Additionally, the rate of hydroxyl radical consumption can be described with Eq. 18-21:

$$\frac{d[\cdot\mathrm{OH}]}{dt} = k_{\mathrm{H_2O_2},\cdot\mathrm{OH}}[\cdot\mathrm{OH}][\mathrm{H_2O_2}] + k_{\cdot\mathrm{OH},\cdot\mathrm{OH}}[\cdot\mathrm{OH}]^2 + k_{\mathrm{HO_2}\cdot,\cdot\mathrm{OH}}[\cdot\mathrm{OH}][\mathrm{HO_2}\cdot] + \sum k_{\mathrm{S},\cdot\mathrm{OH}}[\cdot\mathrm{OH}][\mathrm{S}] \tag{18-21}$$

where $k_{\mathrm{S,OH}}$ is a rate constant between the hydroxyl radical and some background scavenger S (including, but not limited to, carbonate and bicarbonate species, NOM, etc.). If one assumes steady state, $d[\cdot\mathrm{OH}]/dt = 0$, the rate of hydroxyl radical formation can be set equal to the rate of hydroxyl radical consumption, resulting in Eq. 18-22:

$$[\cdot\mathrm{OH}]_{\mathrm{SS}} = \frac{E'^{0}_{p,\lambda}\dfrac{(1-10^{-a_\lambda \cdot l})}{a_\lambda \cdot l}\varepsilon_{\lambda,\mathrm{H_2O_2}}[\mathrm{H_2O_2}]\,\Phi_{\lambda,\mathrm{H_2O_2}\rightarrow\mathrm{OH}} + k_{\mathrm{HO_2}\cdot,\mathrm{H_2O_2}}[\mathrm{HO_2}\cdot][\mathrm{H_2O_2}]}{k_{\mathrm{H_2O_2},\cdot\mathrm{OH}}[\cdot\mathrm{OH}][\mathrm{H_2O_2}] + k_{\cdot\mathrm{OH},\cdot\mathrm{OH}}[\cdot\mathrm{OH}]^2 + k_{\mathrm{HO_2}\cdot,\cdot\mathrm{OH}}[\cdot\mathrm{OH}][\mathrm{HO_2}\cdot] + \sum k_{\mathrm{S},\cdot\mathrm{OH}}[\cdot\mathrm{OH}][\mathrm{S}]} \tag{18-22}$$

Further assumptions are often used to negate the contributions in formation and consumption of the propagation and lesser termination reactions, so the equation is typically used in its simplified form (Eq. 18-23):

$$[\cdot\mathrm{OH}]_{\mathrm{SS}} = \frac{E'^{0}_{p,\lambda}\dfrac{(1-10^{-a_\lambda \cdot l})}{a_\lambda \cdot l}\varepsilon_{\lambda,\mathrm{H_2O_2}}[\mathrm{H_2O_2}]\,\Phi_{\lambda,\mathrm{H_2O_2}\rightarrow\cdot\mathrm{OH}}}{k_{\mathrm{H_2O_2},\cdot\mathrm{OH}}[\cdot\mathrm{OH}][\mathrm{H_2O_2}] + \sum k_{\mathrm{S},\cdot\mathrm{OH}}[\cdot\mathrm{OH}][\mathrm{S}]} \tag{18-23}$$

AdOx Kinetic Model for UV/H_2O_2 AOP. The AdOx model simultaneously solves a series of 44 differential equations describing elementary chemical and photochemical reactions showing initiation, the chain reaction mechanism, and termination reactions and using reported reaction rates and photochemical parameters to predict contaminant destruction without the assumption of steady state hydroxyl radical concentrations (Crittenden et al., 1999). Figure 18-11 [from Crittenden et al. (1999)] directly compares the AdOx and steady state ·OH radical models for degradation of 1,2-dibromo-3-chloropropane (DBCP) in laboratory water supplemented with inorganic carbon.

The steady state ·OH radical model (Glaze et al., 1995) consistently underpredicts the observed DBCP degradation rate constants, whereas the AdOx model consistently provides reasonably accurate predictions for the system. The more accurate predictions are credited directly to the fact that the AdOx model does not employ the pseudo–steady state or constant-pH assumptions (Crittenden et al., 1999) that are critical to the steady state ·OH radical model.

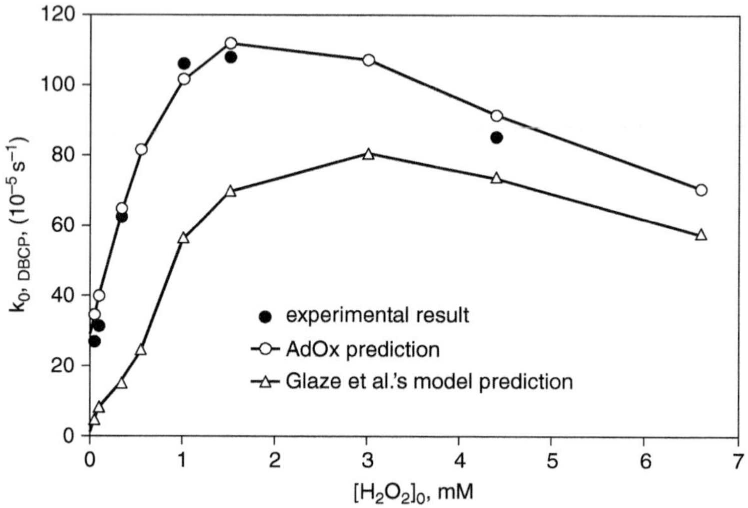

FIGURE 18-11 Effect of initial H_2O_2 on the pseudo-first-order rate constant for the oxidation of DBCP with UV/H_2O_2 AOP. (*Source: Reprinted with permission from* Water Research *[Crittenden et al., 1999]*.)

While these approaches perform decently in approximating chemical destruction for the UV/H_2O_2 AOPs, they rely on published rate constants to determine the amount of ·OH radical–mediated oxidation. Problems arise because of simplifications, such as using a universal rate constant for dissolved organic carbon as a surrogate for NOM, which are used in these models for more complex, natural water systems. Research is currently under way to address these issues, including better predictors for describing the reaction between NOM and hydroxyl radicals.

UV/H_2O_2 AOP in Practice: Case Studies

Although there are many facilities using UV technology for mitigation of pathogenic microbial contaminants in wastewater and drinking water treatment processes, the use of UV processes for chemical contaminants is less common. However, these processes are capable of photolyzing or oxidizing a wide range of contaminants and have advantages over traditional oxidation techniques in areas of safety and reactor footprint size. For these reasons, UV-based AOP is being used or research is being performed to examine the feasibility of using this technology for indirect potable water reuse and drinking water applications. The following discussion will relate some of the full-scale uses of UV technology for the treatment of contaminants in each of these water matrices.

Indirect Potable Water Reuse Applications. The issue of potable water reuse is gaining acceptance throughout the entire United States and the rest of world, as it has already done in more arid parts of the country. Since the supply of pristine water sources is dwindling, utilities reach for new sources, including reusing water that has been treated previously, used, and passed through a wastewater treatment facility. The idea of reusing wastewater as a drinking water source is not a pleasant thought for the general public, but as demand for water increases and supplies shrink, this concept will become a reality throughout the United States. Orange County, CA, has implemented UV/H_2O_2 AOP as part of an advanced water purification plant designed to provide water for indirect potable reuse and is the subject of the following case study. For more information regarding water reuse, refer to Chap. 16.

Reuse Case Study: UV/H_2O_2 for Removal of NDMA in the Orange County Groundwater Replenishment System (OCWD and OCSD 2008A, 2008B). Water demands in northern and central Orange County, CA, are expected to increase nearly 20 percent by 2020. To combat this increased demand of southern California's scarce water supplies, a joint effort by the Orange County Water District (OCWD) and the Orange County Sanitation District (OCSD) created the Groundwater Replenishment System (GWRS). The GWRS is the largest potable reuse water purification project of its kind in the world and will help to increase Orange County's water independence by providing a locally controlled, drought-proof supply of safe, high-quality water.

The GWRS uses an influent of filtered secondary effluent from the neighboring OCSD treatment plant and treats it to a level exceeding all U.S. and California drinking water quality standards. The 380 million L/day (100 mgd) system consists of a three-step process known as the *Advanced Water Purification Facility* (AWPF)—microfiltration, reverse osmosis, and UV light with hydrogen peroxide. Roughly half the water from the GWRS is injected into Orange County's seawater barrier, an underground pressure ridge of water formed by injection wells along the Pacific coast. The remaining water is piped to recharge lakes in Anaheim, CA, where it will take the natural path of rainwater as it filters through clay, sand, and rock to the deep aquifers of the groundwater basin.

While microfiltration (MF) and reverse osmosis (RO) provide a physical barrier that rejects many organic compounds, *N*-nitrosodimethylamine (NDMA) is not rejected by MF and only partially removed by RO. NDMA is present as a by-product formed during the upstream wastewater treatment processes or as an industrial chemical in the production of rocket fuel and must be removed to below 10 ng/L before discharge because it is a known carcinogen. UV light is a proven technology for degrading NDMA and was included in the three-step process primarily to provide a barrier ensuring NDMA destruction. In addition, 3 mg/L H_2O_2 is added to the UV system, creating AOP conditions capable of oxidizing trace organic constituents that may be present in the water, such as 1,4-dioxane or other organic compounds.

The UV/H_2O_2 AOP system employed in the AWPF consists of three horizontally stacked chambers, each chamber made up of two reactors. Each reactor contains seventy-two 756-W low-pressure, high-output (LPHO) mercury amalgam lamps capable of emitting UVC energy at 254 nm. Each reactor consumes 18.5 kW running at 100 percent power, and power is adjusted based on water temperature, flow rate, UVT_{254}, lamp age, and fouling.

With an inlet total organic carbon (TOC) of less than 0.2 mg/L and an average UVT of 98.6 percent, the AWPF is an ideal situation for maximizing the efficiency of the UV/H_2O_2 AOP. Studies at the facility have shown that the system is capable of removing greater than 1.2 log NDMA and inactivating greater than 5 log MS2 with only a fraction of the total system in operation. The AWPF went on-line in January 2008, providing clean water for replenishing the diminishing groundwater basin and diverting wastewater from ocean outflow.

Drinking Water. While UV processes have been used in drinking water treatment for years throughout the world, for pathogen inactivation in drinking water (see the section on UV disinfection), using the technology for chemical oxidation has been slow to gain acceptance. This stems from the much higher UV doses required for oxidation compared with disinfection. In addition, the unselective nature of ·OH radical oxidation leads to process inefficiencies and a greater possibility of unidentified oxidation by-products.

Recently, two water treatment plants have designed processes to use UV/H_2O_2 technology to meet chemical oxidation needs within their water treatment trains. These operations decided to install UV/H_2O_2 AOP to address targeted chemical and biological contaminants and had specific concerns related to employing other oxidation technologies.

Drinking Water UV AOP Case Study: PWN (Kruithof et al., 2004; Trojan UV, 2005a). In October 2004, the PWN Water Supply Company North Holland (PWN) officially opened its water treatment plant in Andjik, Netherlands. This was the first large-scale water treatment plant in the world designed to combine UV disinfection with UV oxidation. The plant treats more than 95 million L/day (25 mgd) of water, providing drinking water to half a million people and was designed to eliminate chemical contaminants such as pesticides, in addition to providing primary disinfection, reducing the amount of chlorine added to the water.

The system was designed in association with Trojan Technologies of Canada, with a footprint of only 49 m^2, significantly smaller than the requirements of many other oxidation technologies and thus reducing capital costs associated with the project. The system is designed to deliver a UV fluence of 540 $mJ \cdot cm^{-2}$ in combination with the addition of 6 mg/L H_2O_2. At this combination, an 80 percent degradation of pesticides, EDCs, and pharmaceuticals can be achieved, primary metabolite formation was insignificant, and no bromate was formed. Bromate is of concern because ozonation technologies create bromate as a by-product, a suspected human carcinogen. Following the UV/H_2O_2 AOP in the treatment train is a GAC filter designed to remove any residual H_2O_2, assimilable organic carbon (AOC), and nitrate from the AOP-treated drinking water.

Drinking Water Case Study: Cornwall, Ontario, Canada (Cornwall, Ont., 2009; Trojan UV, 2005b, 2008). The City of Cornwall (Ontario, Canada) uses Lake St. Lawrence (St. Lawrence River) as a source water that is low in turbidity and color and uses conventional treatment with GAC adsorbers/filters. The plant is designed to treat a peak flow of up to 100 million L/day and provides drinking water for 50,000 people. In 2006, four UV units were installed after filtration to upgrade disinfection and control of seasonal algal taste and odor.

The UV system operates in two modes. In disinfection mode, which is used year round for primary disinfection, the system is designed to provide a RED of at least 40 $mJ \cdot cm^{-2}$. This RED is achieved by operating a minimal number of lamps in each reactor (approximately half the lamps in a reactor are powered in disinfection mode) at low power of approximately 30 percent maximum. During late summer and early fall, hydrogen peroxide is injected downstream of the filter beds and upstream of the UV reactor, and the UV dose is increased to well over 100 $mJ \cdot cm^{-2}$ to create AOP conditions. The UV dose is increased by powering on the remaining lamps in each reactor and by increasing operating power to each lamp. Peroxide is added to a concentration of at least 1 mg/L and is increased as necessary to degrade methylisoborneol (MIB) and geosmin in the water to levels below taste and odor thresholds. In addition, because the UV power is increased, UV disinfection occurs at levels greater than or equal to year-round disinfection mode. The AOP conditions used in the Cornwall system also have been shown to be effective in eliminating other algal products, including surrogates for algal toxins, at levels comparable to MIB and geosmin.

Depending on water quality, the increase in UV dose to treat taste- and odor-causing compounds requires three to six times more energy than disinfection mode, and the power-up is achieved by energizing additional lamps and reactors, which are not used during disinfection mode. In addition to the increased power usage during AOP mode, residual hydrogen peroxide must be quenched after the UV reactor. Quenching is performed by adding free chlorine at a 2.1:1 ratio and is required to ensure that residual peroxide does not interfere with the addition of residual chlorine prior to distribution.

If the system were operated continually in AOP mode, concerns regarding energy requirements and cost of hydrogen peroxide and additional chlorine would surface rapidly. However, the ability to use AOP conditions in a seasonal manner to address particular contaminants in a targeted manner and still use year-round UV primary disinfection offers a high level of water quality protection at minimized additional cost to the utility.

Alternative UV-Based AOPs

In addition to UV/H_2O_2 AOP, other UV-based AOPs have been shown to be effective for degrading organic pollutants in water. Each of these technologies works similarly to the UV/H_2O_2 AOP in that a photoinitiated reaction is used to start a radical chain reaction in which the highly reactive ·OH radical is formed and used for oxidation. Brief descriptions of two homogeneous UV-based AOPs (UV/O_3 and UV/Cl_2) and one heterogeneous AOP (UV/TiO_2) are included below to serve as an introduction to these technologies.

***UV/O_3*.** While ozone alone is a powerful oxidant, and the addition of hydrogen peroxide in the presence of ozone is known to create ·OH radicals (see Chap. 7 for a further discussion), the photoinitiated decay of the ozone molecule also will result in hydroxyl radical–mediated advanced oxidation conditions.

Ozone is an efficient UV absorber (ε_{254} = 3300 for O_3 versus 19.6 for H_2O_2) and has a quantum yield of 0.61 mol O_3 degraded per einstein absorbed (Wittmann et al., 2002). It must be recognized, though, that the primary photoproduct of O_3 photolysis is hydrogen peroxide (Glaze et al., 1987). This formed peroxide then either reacts directly with ozone to form hydroxyl radicals (see Chap. 7) or undergoes additional UV photolysis to form hydroxyl radicals via the Haber-Weiss mechanism described earlier. As such, the UV/O_3 AOP essentially be thought of as a method for generating in situ hydrogen peroxide via photolytic degradation of O_3 (Oppenländer, 2003).

***UV/Cl_2*.** The photoinitiated decay of free chlorine species (HOCl and OCl^-, depending on pH) has been shown to produce ·OH radicals. When free chlorine species absorb a photon (<511 nm), the known photoproducts are a hydroxyl radical (·OH) and chlorine radical (·Cl) (Molina et al., 1980; Nowell and Hoigne, 1992b; Oliver and Carey, 1977; Vogt and Schindler, 1991). Watts and Linden (2007) determined that the quantum yields for degradation of HOCl and OCl^- to ·OH and ·Cl radicals under UV conditions at 254 nm in laboratory water were both greater than 1. These reported yields, corroborated by Feng and colleagues (2007), are most likely a measure of HOCl/OCl^- degradation through a chain reaction with photoproducts in addition to direct photolysis rather than the true quantum yields of direct photolytic conversion.

Watts and colleagues (2007) directly compared oxidation of an ·OH radical probe compound (nitrobenzene, which displays no significant reaction with chlorine radicals) (Nowell and Hoigne, 1992b) created via UV/H_2O_2 AOP and UV/HOCl (or UV/OCl^-) with some interesting results. The work found that at equimolar oxidant concentration in laboratory water, nitrobenzene degradation rates were approximately three times faster using the LP UV/HOCl than using LP UV/H_2O_2 AOP. However, pH adjustment had a significant impact on nitrobenzene degradation rates in the UV/Cl_2 system, whereas it had no effect on the UV/H_2O_2 system. This observation was explained by the fact that formed ·OH radicals will readily react with chlorite ion (k_{OH,OCl^-} = 8.0 × 10^9 M^{-1} · s^{-1}), whereas it reacts much more slowly with hypochlorous acid ($k_{OH,HOCl}$ = 8.0 × 10^4 M^{-1} · s^{-1}), thus providing competition for hydroxyl radicals with the nitrobenzene as solution pH is increased above 6.0.

Similarly, results from the experiments performed in natural waters indicated that the UV/Cl_2 process still oxidized nitrobenzene at a greater rate than the UV/H_2O_2 process; however, the differences were decreased dramatically as water quality deteriorated. An operating cost analysis considering only the cost of electricity to run the UV lamps and the cost of added oxidant indicated that UV/Cl_2 was more efficient in a pH 5 postfiltration water (neglecting the cost associated with pH adjustment). However, at pH 7, UV/H_2O_2 and UV/Cl_2 operating costs were roughly equivalent, depending on oxidant concentration, and at pH 9, UV/H_2O_2 was more efficient than UV/Cl_2. Also of concern with the UV/Cl_2 process are the unknown products associated with chlorine radical chemistry in a solution with typical levels of NOM.

Heterogeneous Photocatalysis: UV/TiO$_2$. Carey and colleagues (1976) first reported the photodegradation of several polychlorinated biphenyls (PCBs) in water in the presence of titanium dioxide (TiO_2). Since then, many applications of the process have been reported in the literature (Legrini et al., 1993). The mechanism of the UV/TiO_2 process has been described using the bandgap model (Fox, 1983; Bard, 1982; Zafiriou et al., 1984; Childs and Ollis, 1980). The model states that excitation of a TiO_2 molecule by UV energy causes an electron-depleted valence-band hole (h^+), as well as a reactive electron, according to Eq. 18-24:

$$TiO_2 \xrightarrow{h\nu} TiO_2(e^- + h^+) \tag{18-24}$$

After excitation, an electron can transfer from an adsorbed organic substrate (creating a radical organic molecule), or from an adsorbed water molecule into the electron hole (h^+). The second transfer reaction yields an adsorbed hydroxyl radical molecule and has been identified to be of greater importance in oxidative degradation processes owing to the high concentration of water molecules adsorbed to the particle surface.

In addition, the liberated electron (e^-) can react with oxygen in solution to form superoxide molecule, initiating a chain reaction mechanism that includes formation of the hydroxyl radical. It also has been found that the addition of hydrogen peroxide enhances the rate of photodegradation, most likely by surface-catalyzed dismutation of H_2O_2 (Legrini et al., 1993). Despite these multiple mechanisms of action, the UV/TiO_2 process is not an efficient AOP because of the low quantum yield for the generation of $\cdot OH$ radicals (Sun and Bolton, 1996).

Additionally, the fact that the reaction occurs on the TiO_2 surface leads to several limitations—the compounds to be degraded must be adsorbed to the dispersed TiO_2 particles or a solid TiO_2 surface. Hydroxyl radicals formed by the second mechanism (e^-) are constrained by the fact that the chain reaction mechanism is most likely initiated a very short distance from the particle surface. Another persistent process found in application of photocatalysis in drinking water has been the potential for irreversible fouling of the TiO_2 surfaces owing to natural organic material.

The particle nature of the process adds to inefficient UV treatment. Because the surfaces are not UV transparent, reactor geometry and particle light screening are both important factors of concern. TiO_2 molecules are excited by UV-C, UV-B, and UV-A energy, indicating that light sources other than UV-C lamps can be used to create AOP conditions. Even solar energy has been shown to be effective at creating AOP conditions via the UV/TiO_2 AOP (Bahnemann, 2004).

ABBREVIATIONS

AOC	assimilable organic carbon
AOPs	advanced oxidation processes
AWPF	advanced water purification facility
CAT-DEL	Catskill-Delaware
CFD	computational fluid dynamics
D	UV dose ($mJ \cdot cm^{-2}$)
D/DBPR	Disinfection/Disinfectant By-Products Rule
DBCP	1,2-dibromo-3-chloropropane

DBPs	disinfection by-products
DNA	deoxyribonucleic acid
EDCs	endocrine-disrupting compounds
ESWTR	Enhanced Surface Water Treatment Rule
EU	European Union
GWRS	Groundwater Replenishment System
GWR	Ground Water Rule
h^+	electron hole
HAAs	haloacetic acids
HOMO	highest occupied molecular orbital
I	irradiance ($mW \cdot cm^{-2}$)
IC	internal conversion
ICR	Information Collection Rule
ISC	intersystem crossing
LEDs	light-emitting diodes
LID	light irradiance distribution
LP	low pressure
LPHO	low pressure, high output
LPLO	low pressure, low output
LT2ESWTR	Long Term 2 Enhanced Surface Water Treatment Rule
LUMO	lowest unoccupied molecular orbital
MCL	maximum contaminant level
MF	microfiltration
mgd	million gallons per day
MP	medium pressure
NDMA	*N*-nitrosodimethylamine
NOM	natural organic matter
ntu	nephelometric turbidity unit
OCSD	Orange County Sanitation District
OCWD	Orange County Water District
P_1, P_2	photoproducts
PWN	PWN Water Supply Company North Holland
RED	reduction equivalent dose
RNA	ribonucleic acid
RO	reverse osmosis
S_0	unexcited ground electron state
S_1, S_2	first or second excited singlet state
T_1	excited triplet state
THMs	trihalomethanes
TOC	total organic carbon

UFP	Uniform Fire Protection Code
USEPA	U.S. Environmental Protection Agency
UV	ultraviolet
UV-C	germicidal ultraviolet light of wavelengths 200 to 280 nm
UV-B	ultraviolet light of wavelengths 280 to 315 nm
UV-A	ultraviolet light of wavelengths 315 to 400 nm
UVT	ultraviolet transmittance

NOTATION FOR EQUATIONS

A	absorbance, or optical density of solution
a_{10}	log base 10 absorption coefficient (cm^{-1})
a_λ	solution absorbance coefficient (cm^{-1})
c, C, or $[C]$	molar concentration of dissolved constituent (M or mol $\cdot$ L^{-1})
$E^0_{p,\lambda}$	incident photon irradiance (10^{-3} Es $\cdot$ cm^{-2} $\cdot$ s^{-1})
h	Planck's constant (6.6256×10^{-34} J $\cdot$ s $\cdot$ $photon^{-1}$)
I_0	irradiance entering the solution in the spectrophotometer cuvette
I	irradiance exiting the solution in the spectrophotometer cuvette
k'_d	first-order rate constant for direct photolysis (s^{-1})
k'_1	pseudo-first-order rate constant ($m^2 \cdot J^{-1}$)
$k_{s,\lambda}$	specific rate of light absorption (Es $\cdot$ mol^{-1} $\cdot$ s^{-1})
$k_{x,\cdot OH}$	second-order rate constant between compound x and ·OH
l	solution path length (cm)
N	Avogadro's number (6.02×10^{23} photons $\cdot$ Es^{-1})
U_λ	wavelength-specific photon energy (J $\cdot$ Es^{-1})
%UVT	percent UV transmittance
α_e	natural log transform of a_{10}
ε	molar absorption coefficient ($M^{-1} \cdot cm^{-1}$)
$\varepsilon_{\lambda,c}$	wavelength-specific molar absorption coefficient for c ($M^{-1} \cdot cm^{-1}$)
λ	wavelength (m or nm)
$\Phi_{\lambda,c}$	wavelength-specific quantum yield for c (mol $\cdot$ Es^{-1})
$\Phi_{\lambda,H_2O_2 \rightarrow \cdot OH}$	wavelength-specific quantum yield for $H_2O_2 \rightarrow \cdot OH$ (0.5 mol $\cdot$ Es^{-1})
ν	wavelength frequency (s^{-1})

REFERENCES

Aaron, J. J., and M. A. Oturan (2001). New photochemical and electrochemical methods for the degradation of pesticides in aqueous media: Environmental applications. *Turkish Journal of Chemistry* 25, 4, pp. 509–520.

Acero, J. L., K. Stemmler, and U. von Gunten (2000). Degradation kinetics of atrazine and its degradation products with ozone and OH radicals: A predictive tool for drinking water treatment. *Environmental Science and Technology* 34, 4, pp. 591–597.

Acero, J. L., F. J. Benitez, M. González, and R. Benitez (2002). Kinetics of fenuron decomposition by single-chemical oxidants and combined systems. *Industrial and Engineering Chemical Research* 41, 17, pp. 4225–4232.

Alfano, O. M., R. J. Brandi, and A. E. Cassano (2001). Degradation kinetics of 2,4-D in water employing hydrogen peroxide and UV radiation. *Chemical Engineering Journal* 82, 1-3, pp. 209–218.

Andreozzi, R., V. Caprio, A. Insola, and R. Marotta (2000). The oxidation of metol (*N*-methyl-*p*-aminophenol) in aqueous solution by UV/H_2O_2 photolysis. *Water Research* 34, 2, pp. 463–472.

Andreozzi, R., M. Canterino, R. Marotta, and N. Paxeus (2005). Antibiotic removal from wastewaters: The ozonation of amoxicillin. *Journal of Hazardous Materials* 122, 3, pp. 243–250.

Andreozzi, R., V. Caprio, R. Marotta, and A. Radovnikovic (2003a). Ozonation and H_2O_2/UV treatment of clofibric acid in water: A kinetic investigation. *Journal of Hazardous Materials* 103, 3, pp. 233–246.

Andreozzi, R., V. Caprio, R. Marotta, and D. Vogna (2003b). Paracetamol oxidation from aqueous solutions by means of ozonation and H_2O_2/UV system. *Water Research* 37, 5, pp. 993–1004.

Andreozzi, R., L. Campanella, B. Fraysse, J. Garric, A. Gonnella, R. Lo Giudice, R. Marotta, G. Pinto, and A. Pollio. (2004). Effects of advanced oxidation processes (AOPs) on the toxicity of a mixture of pharmaceuticals. *Water Science and Technology* 50, 5, pp. 23–28.

Arslan-Alaton, I., and S. Dogruel (2004). Pre-treatment of penicillin formulation effluent by advanced oxidation processes. *Journal of Hazardous Materials* 112, 1–2, pp. 105–113.

Bahnemann, D. (2004). Photocatalytic water treatment: Solar energy applications. *Solar Energy* 77, 5, pp. 445–459.

Baker, M.N. (1949). *The Quest for Pure Water: The History of Water Purification from the Earliest Records to the Twentieth Century*. Denver, CO: American Water Works Association, Inc.

Ballester, N. A., and J. P. Malley, Jr. (2004). Sequential disinfection of adenovirus type 2 with uv-chlorine-chloramine. *Journal AWWA* 96, 10, pp. 97–103.

Bank, H. L., J. John, M. K. Schmehl, and R. J. Dratch (1990). Bactericidal effectiveness of modulated UV light. *Applied and Environmental Microbiology* 56, 12, pp. 3888–3889.

Bard, A. J. (1982). Design of semiconductor photo-electrochemical systems for solar-energy conversion. *Journal of Physical Chemistry* 86, 2, pp. 172–177.

Barlas, H. (2000). Treatment of chlorinated organic materials containing wastewater by oxidation processes. *Fresenius Environmental Bulletin* 9, 9–10, pp. 590–596.

Barltrop, J. A., and J. D. Coyle (1978). *Principles of Photochemistry.* Chichester, UK: Wiley.

Baxendale, J. H., and J. A. Wilson (1957). The photolysis of hydrogen peroxide at high light intensities. *Transactions of the Faraday Society* 53, pp. 344–356.

Bellamy, W. P., P. W. Wobma, J. P. Malley, Jr., and D. A. Reckhow (2004). *UV Disinfection and Disinfection By-Product Characteristics of Unfiltered Water.* Awwa Research Foundation Report. Denver, CO: Water Research Foundation.

Beltrán-Heredia, J., F.J. Benitez, T. Gonzalez, J.L. Acero, and B. Rodriguez. (1996). Photolytic decomposition of bentazone. *Journal of Chemical Technology and Biotechnology* 66, 2, pp. 206–212.

Beltrán, J., G. Ovejero, and B. Acedo (1993). Oxidation of atrazine in water by ultraviolet-radiation combined with hydrogen-peroxide. *Water Research* 27, 6, pp. 1013–1021.

Beltrán, J., J.F. García-Araya, J. Rivas, P.M. Álvarez, and E. Rodriguez. (2000). Kinetics of simazine advanced oxidation in water. *Journal of Environmental Science and Health, Part B: Pesticides, Food Contaminants, and Agricultural Wastes* 35, 4, pp. 439–454.

Benitez, F. J., J. Beltrán-Heredia, T. Gonzalez, and J. L. Acero (1995a). Advanced oxidation processes in the degradation of cyanazine. *Ozone Science and Engineering* 17, 3, pp. 237–258.

Benitez, F. J., J. Beltrán-Heredia, T. Gonzalez, and F. Real (1995b). Photooxidation of carbofuran by a polychromatic UV irradiation without and with hydrogen peroxide. *Industrial and Engineering Chemical Research* 34, 11, pp. 4099–4105.

Benitez, F. J., J. L. Acero, F. J. Real, and C. Maya (2004a). Modeling of photooxidation of acetamide herbicides in natural waters by UV radiation and the combinations UV/H_2O_2 and UV/O_3. *Journal of Chemical Technology and Biotechnology* 79, 9, pp. 987–997.

Benitez, F. J., J. L. Acero, F. J. Real, and S. Roman (2004b). Oxidation of MCPA and 2,4-D by UV radiation, ozone, and the combinations UV/H_2O_2 andO_3/H_2O_2. *Journal of Environmental Science and Health, Part B: Pesticides, Food Contaminants, and Agricultural Wastes* 39, 3, pp. 393–409.

Bettles, T., S. Schujman, J. A. Smart, W. Liu, and L. Schowalter. (2007). UV light emitting diodes: Their applications and benefits. In *Proceedings of the International Ultraviolet Association World Congress, Los Angeles, CA*.

Blatchley III E.R., C. Shen, O.K. Scheible, J.P. Robinson, K. Ragheb, D.E. Bergstrom, and D. Rokjer (2008). Validation of large-scale, monochromatic UV disinfection systems for drinking water using dyed microspheres. *Water Research* 42, 3, pp. 677–688.

Bohrerova Z., G. Bohrer, S. Mohanraj, J. Ducoste, and K.G. Linden. (2005). Experimental measurements of fluence distribution in a UV reactor using fluorescent dyed microspheres. *Environmental Science and Technology* 29, 22, pp. 8925–8930.

Bohrerova, Z., H. Shemer, B. Lantis, C. Impellitteri, and K.G. Linden. (2008). Comparative disinfection efficiency of pulsed and continuous-wave UV irradiation technologies. *Water Research* 42, 12, pp. 2975–2982.

Bolton, J.R., B. Dussert, Z. Bukhari, T. Hargy, and J.L. Clancy. (1998). Inactivation of *Cryptosporidium parvum* by medium-pressure ultraviolet light in finished drinking water. In *Proceedings of the 1998 AWWA Annual Conference, Dallas, TX.*

Bolton, J. R. (2000). Calculation of ultraviolet fluence rate distributions in an annular reactor: significance of refraction and reflection. *Water Research* 34, 13, pp. 3315-3324.

Bolton, J. R., and K. G. Linden (2003). Standardization of methods for fluence (UV dose) determination in bench-scale UV experiments. *ASCE: Journal of Environmental Engineering* 129, 3. pp 209–215.

Bolton, J. R. (2001). *Ultraviolet Applications Handbook*, 2nd ed. Edmonton, Alberta, Canada: Bolton Photosciences.

Bolton, J. R., and M. I. Stefan (2002). Fundamental photochemical approach to the concepts of fluence (UV dose) and electrical energy efficiency in photochemical degradation reactions. *Research on Chemical Intermediates* 28, 7–9, pp. 857–870.

Boreen, A. L., W. A. Arnold, and K. McNeill (2004). Photochemical fate of sulfa drugs in the aquatic environment: Sulfa drugs containing five-membered heterocyclic groups. *Environmental Science and Technology* 38, 14, pp. 3933–3940.

Braunstein, J., F. J. Loge, G. Tchobanoglous, and J. L Darby (1996). Ultraviolet disinfection of filtered activated sludge effluent for reuse applications. *Water Environment Research* 68, 2, pp. 152–161.

Buxton, G., C. Greenstock, W. Helman, and A. Ross (1988). Critical review of rate constants for reactions of hydrated electrons, hydrogen atoms and hydroxyl radicals ($\cdot OH/O^-$) in aqueous solution. *Journal of Physical and Chemical Reference Data* 17, 2, pp. 513–886.

Carey, J. H., J. Lawrence, and H. M. Tosine (1976). Photodechlorination of PCBs in the presence of titanium dioxide in aqueous suspensions. *Bulletin of Environmental Contamination and Toxicology* 16, 6, pp. 697–701.

Cater, S. R., M. I. Stefan, J. R. Bolton, and A. Safarzadeh-Amiri (2000). UV/H_2O_2 treatment of methyl *tert*-butyl ether in contaminated waters. *Environmental Science and Technology* 34, 4, pp. 659–662.

Chatterley, C.A. and K.G. Linden (2010) Demonstration and evaluation of germicidal UV-LEDs for point-of-use water disinfection. *Journal of Water and Health*, 8(3) 479-486.

Chen, P.J., E.J. Rosenfeldt, S.W. Kullman, D.E. Hinton, and K.G. Linden. (2007). Biological assessments of a mixture of endocrine disruptors at environmentally relevant concentrations in water following UV/H_2O_2 oxidation. *Science of the Total Environment* 376, 1–3, pp. 18–26.

Chen, T. F., R. A. Doong, and W. G. Lei (1998). Photocatalytic degradation of parathion in aqueous TiO_2 dispersion: The effect of hydrogen peroxide and light intensity. *Water Science and Technology* 37, 8, pp. 187–194.

Chevrefil, G., E. Caron, H. Wright, G. Sakamoto, P. Payment, B. Barbeau, and B. Cairns. (2006). UV dose required to achieve incremental log inactivation of bacteria, protozoa and viruses. *IUVA News* 8, 1, pp. 38–45.

Childs, L. P., and D. F. Ollis (1980). Is photocatalysis catalytic. *Journal of Catalysis* 66, 2, pp. 383–390.

Chiou, C.S., Y.H. Chen, C.T. Chang, C.Y. Chang, J.L. Shie, and Y.S. Li. (2006). Photochemical mineralization of di-*n*-butyl phthalate with H_2O_2/Fe^{3+}. *Journal of Hazardous Materials B* 135, 1–3, pp. 344–349.

Chiu K., D. A. Lyn, P. Savoye, and E. R. Blatchley (1999). Integrated UV disinfection model based on particle tracking, *Journal of Environmental Engineering, ASCE* 125, 1, pp. 7–16.

Christensen, J. A., and K. G. Linden (2003). How particles affect UV light in the UV disinfection of unfiltered drinking water. *Journal AWWA* 95, 4, pp. 179–189.

Clancy, J. L., T. M. Hargy, M. M. Marshall, and J. E. Dyksen (1998). UV light inactivation of *Cryptosporidium* oocysts. *Journal AWWA* 90, 9, pp. 92–102.

Clancy, J.L., Z. Bukhari, T.M. Hargy, J.R. Bolton, B.W. Dussert, and M.M. Marshall. (2000). Using UV to inactivate *Cryptosporidium. Journal AWWA* 92, 9, pp. 97–105.

Cornwall, ON (2009). Personal communication.

Coyle, J. D. (1986). *Introduction to Organic Photochemistry.* Chichester, UK: Wiley.

Crittenden, J. C., S. M. Hu, D. W. Hand, and S. A. Green (1999). A kinetic model for H_2O_2/UV process in a completely mixed batch reactor. *Water Research* 33, 10, pp. 2315–2328.

Darby, J., M. Heath, J. Jacangelo, F. Loge, P. Swaim, and G. Tchobanoglous. (1995). Comparison of UV irradiation to chlorination: Guidance for achieving optimal UV performance. Final report to Water Environment Research Foundation.

Doong, R. A., and W. H. Chang (1998). Photoassisted iron compound catalytic degradation of organophosphorous pesticides with hydrogen peroxide. *Chemosphere* 37, 13, pp. 2563–2572.

Downes, A., and T. P. Blunt (1877). Researches on the effect of light upon bacteria and other organisms. *Proceedings of the Royal Society of London* 26, pp. 488–500.

Ducoste, J., D. Liu, and K. G. Linden (2005). Alternative approaches to modeling dose distribution and microbial inactivation in ultraviolet reactors: Lagrangian vs Eulerian. *ASCE Journal of Environmental Engineering* 131, 10, pp. 1393–1403.

Dunn, J., T. Ott, and W. Clark (1995). Pulsed-light treatment of food and packaging. *Food Technology* 49, 9, pp. 95–98.

Eischeid, A. C., J. Meyer, and K. G. Linden (2009). UV disinfection of adenoviruses: Molecular indications of DNA damage efficiency. *Applied and Environmental Microbiology* 75, 1, pp. 23–28.

Elovitz, M., and U. von Gunten (1999). Hydroxyl radical ozone ratios during ozonation processes: I. The R-ct concept. *Ozone Science and Engineering* 21, 3, pp. 239–260.

Feng, Y., D. W. Smith, and J. R. Bolton (2007). Photolysis of chlorine species (HOCl and OCl^-) with 254 nm ultraviolet light. *Journal of Environmental Engineering and Science* 6, 3, pp. 277–284.

Fox M. A. (1983). Organic heterogeneous photocatalysis: Chemical conversions sensitized by irradiated semiconductors. *Accounts of Chemical Research* 16, 9, pp. 314–321.

Galindo, C., and A. Kalt (1999). UV/H_2O_2 oxidation of azodyes in aquous media: Evidence of a structure-degradability relationship. *Dyes and Pigments* 42, 3, pp. 199–297.

Gaska, R. (2007). Deep ultraviolet light emitting diodes for water monitoring and disinfection. In *Proceedings of the International Ultraviolet Association Conference, Los Angeles, CA.*

Gates, F. L. (1929). A study of the bactericidal action of ultraviolet light: I. The reaction to monochromatic radiation. *Journal of General Physiology* 13, 2, pp. 231–248.

Glaze, W. H., J. W. Kang, and D. H. Chapin (1987). The chemistry of water treatment processes involving ozone, hydrogen peroxide and ultraviolet radiation. *Ozone Science and Engineering* 9, pp. 335–352.

Glaze, W. H., Y. Lay, and J. W. Kang (1995). Advanced oxidation processes: A kinetic model for the oxidation of 1,2-dibromo-3-chloropropane in water by the combination of hydrogen peroxide and UV radiation. *Industrial and Engineering Chemical Research* 34, 7, pp. 2314–2323.

Glaze, W.H., R. Schep, W. Chauncey, E.C. Ruth, J.J. Zarnoch, E.M. Aieta, C.H. Tate, and M.J. McGuire. (1990). Evaluating oxidants for the removal of model taste and odor compounds from municipal water supply. *Journal AWWA* 82, 5, pp. 79–84.

Goutailler, G., J.C. Valette, C. Guillard, O. Païssé, and R. Faure. (2001). Photocatalysed degradation of cyromazine in aqueous titanium dioxide suspensions: Comparison with photolysis. *Journal of Photochemistry and Photobiology A: Chemistry* 141, 1, pp. 79–84.

Gultekin, I., and N. H. Ince (2007). Synthetic endocrine disruptors in the environment and water remediation by advanced oxidation processes. *Journal of Environmental Management* 85, 4, pp. 816–832.

Harm, W. (1980). *Biological Effects of Ultraviolet Radiation.* New York: Cambridge University Press.

Hessler, D. P., V. Gorenflo, and F. H. Frimmel (1993). Degradation of aqueous atrazine and metazachlor solutions by UV and UV/H_2O_2: Influence of pH and herbicide concentration. *Acta Hydrochimica et Hydrobiologica* 21, 4, pp. 209–214.

Hijnen W. A. M., E. F. Beerendonk, and G. J. Medema (2006). Inactivation credit of UV radiation for viruses, bacteria and protozoan (oo)cysts in water: A review. *Water Research* 40, 1, pp. 3–22.

Hirvonen, A., M. Trapido, J. Hentunen, and J. Tarhanen (2000). Formation of hydroxylated and dimeric intermediates during oxidation of chlorinated phenols in aqueous solution. *Chemosphere* 41, 8, pp. 1211–1218.

Hoigne, J., and H. Bader (1975). Ozonation in water: Role of hydroxyl radicals as oxidizing intermediates. *Science* 190, 4216, pp. 782–784.

Hoigne, J., H. Bader, W. Haag, and J. Staehelin (1985). Rate constants of reactions of ozone with organic and inorganic compounds in water: III. Inorganic compounds and radicals. *Water Research* 19, 7, pp. 903–1004.

Hoyer, O. (1998). Testing performance and monitoring of UV systems for drinking water disinfection. *Water Supply* 16, 1–2, pp. 419–424.

Huber, M. M., S. Canonica, G. Y. Park, and U. von Gunten (2003). Oxidation of pharmaceuticals during ozonation and advanced oxidation processes. *Environmental Science and Technology* 37, 5, pp. 1016–1024.

Ikehata, K., and M. G. El-Din (2006). Aqueous pesticide degradation by hydrogen peroxide/ultraviolet irradiation and Fenton-type advanced oxidation processes: A review. *Journal of Environmental Engineering and Science* 5, 2, pp. 81–135.

Ikehata K., N. J. Naghashkar, and M. G. Ei-Din (2006). Degradation of aqueous pharmaceuticals by ozonation and advanced oxidation processes: A review. *Ozone Science and Engineering* 28, 6, pp. 353–414.

Ince, N. H., M. I. Stefan, and J. R. Bolton (1997). UV/H_2O_2 degradation and toxicity reduction of textile azo dyes: Remazol Black-B, a case study. *Journal of Advanced Oxidation Technology* 2, 3, pp. 442–448.

Jacob, S. M., and J. S. Dranoff (1970). Light intensity profiles in a perfectly mixed photoreactor. *AICHE Journal* 16, 3, pp. 359–363.

Jagger, J. (1983). Physiological effects of near-ultraviolet radiation on bacteria. In *Photochemical and Photobiological Reviews*, Vol. 7, K. C. Smith, ed.. New York: Plenum Press.

Jin, S., A. A. Mofidi, and K. G. Linden (2006). Polychromatic UV fluence measurements using chemical actinometry, biodosimetry, and mathematical techniques. *ASCE Journal of Environmental Engineering* 132, 8, pp. 831–841.

Kang, J. W., and K. H. Le (1997). A kinetic model of the hydrogen peroxide UV process for the treatment of hazardous waste chemicals. *Environmental Engineering Science* 14, 3, pp. 183–192.

Koppenol, W. H., J. Butler, and J. W. Van Leeuwen (1978). Haber-Weiss cycle. *Photochemistry and Photobiology* 28, 4–5, pp. 655–660.

Kowalska, E., M. Janczarek, J. Hupka, and M. Grynkiewicz (2004). H_2O_2/UV enhanced degradation of pesticides in wastewater. *Water Science and Technology* 49, 4, pp. 261–266.

Kruithof, J., P. Kamp, M. Belosevic, and M. Stefan (2004). UV/H_2O_2 retrofit of PWN's water treatment plant Andijk for primary disinfection and organic contaminant control. Second IWA Leading-Edge Conference on Water and Wastewater Treatment Technologies, Prague, June 1–4, 2004.

Larson, R., and R. Zepp (1988). Reaction of the carbonate radical with aniline derivatives. *Environmental Toxicology and Chemistry* 7, 4, pp. 265–274.

Lee, C., W. Choi, and J. Yoon (2005). UV photolytic mechanism of *N*-nitrosodimethylamine in water: Roles of dissolved oxygen and solution pH. *Environmental Science and Technology* 39, 24, pp. 9702–9709.

Legrini, O., E. Oliveros, and A. M. Braun (1993). Photochemical processes for water treatment. *Chemical Reviews* 93, 2, pp. 671–698.

Lin, L. S., C. T. Johnston, and E. R. Blatchley (1999a). Inorganic fouling at quartz:water interfaces in ultraviolet photoreactors: I. Chemical characterization. *Water Research* 33, 15, pp. 3321–3329.

Lin, L. S., C. T. Johnston, and E. R. Blatchley (1999b). Inorganic fouling at quartz:water interfaces in ultraviolet photoreactors: II. Temporal and spatial distributions. *Water Research* 33, 15, pp. 3330–3338.

Linden, K. G., J. Thurston, R. Schaefer, and J. P. Malley, Jr. (2007). Enhanced UV inactivation of adenoviruses under polychromatic UV lamps. *Applied and Environmental Microbiology* 73, 23, 7571–7574.

Liu, W., S.A. Andrews, J.R. Bolton, K.G. Linden, C.M. Sharpless, and M. Stefan. (2002). Comparison of disinfection by-product (DBP) formation from different UV technologies at bench scale. *Water Science and Technology: Water Supply* 2, 5–6, pp. 515–521.

Lunak, S., and P. Sedlak (1992). Photoinitiated reactions of hydrogen peroxide in the liquid phase. *Journal of Photochemistry and Photobiology A: Chemistry* 68, 1, pp. 1–33.

Lyn, D. A., and E. R. Blatchley (2005). Numerical computational fluid dynamics-based models of ultraviolet disinfection channels. *Journal of Environmental Engineering, ASCE* 131, 6, pp. 838–849.

Mack, J., and J. R. Bolton (1999). Photochemistry of nitrite and nitrate in aqueous solution: A review. *Journal of Photochemistry and Photobiology A: Chemistry* 128, 1–3, pp. 1–13.

Mackey, E. D., T. Hargy, J. P. Malley, Jr., and R. S. Cushing (2002). MS-2 bioassays and *Cryptosporidium* challenges: Comparing and contrasting UV reactor validation techniques. *Journal AWWA* 94, 2, pp. 62–69.

Magnuson, M.L., C.A. Kelty, C.M. Sharpless, K.G. Linden, W. Fromme, D. Metz, and R. Kashinkunti. (2002). Effect of UV irradiation on organic matter extracted from treated Ohio River water studied through the use of electrospray mass spectrometry. *Environmental Science and Technology* 36, 23, pp. 5252–5260.

Malley, J.P., Jr., N.A. Ballester, A.M. Margolin, K.G. Linden, and A. Mofidi. (2004). *Inactivation of Pathogens with Innovative UV Technologies.* Awwa Research Foundation Report. Denver, CO: Water Research Foundation.

Malley, J.P., Jr., B.A. Petri, G.L. Hunter, D. Moran, M. Nadeau, and J. Leach. (2001). *Full-Scale Implementation of UV in Groundwater Disinfection Systems.* Awwa Research Foundation Report. Denver, CO: Water Research Foundation.

Mamane-Gravetz, H., and K. G. Linden (2006). Impact of particle aggregated microbes on UV disinfection: I. Discrepancy between spore-clay aggregates and suspended spores. *ASCE Journal of Environmental Engineering* 132, 6, pp. 596–606.

Mansour, M., P. Schmitt, and A. Mamouni (1992). Elimination of metoxuron and carbetamide in the presence of oxygen species in aqueous solutions. *Science of the Total Environment* 123, pp. 183–193.

Miller, J. S., and D. Olejnik (2001). Photolysis of polycyclic aromatic hydrocarbons in water. *Water Research* 35, 1, pp. 233–243.

Minero, C., V. Maurino, and E. Pelizzetti (1997). Heterogeneous photocatalytic transformations of *S*-triazine derivatives. *Research on Chemical Intermediates* 23, 4, pp. 291–310.

Mofidi, A. A., and K. G. Linden (2004). Disinfection effectiveness of ultraviolet light for heterotrophic bacteria leaving biologically active filters. *Journal of Water Supply: Research & Technology, AQUA* 53, 8, pp. 553–566.

Molina, M. J., T. Ishiwata, and L. T. Molina (1980). Production of OH radical from photolysis of HOCl at 307–309 nm. *Journal of Physical Chemistry* 84, 8, pp. 821–826.

Morowitz, H. J. (1950). Absorption effects in volume irradiation of microorganisms. *Science* 111, 2879, pp. 229–230.

Naunovic, Z., S. Lim, and E. R. Blatchley (2008). Investigation of microbial inactivation efficiency of a UV disinfection system employing an excimer lamp. *Water Research* 42, 19, pp. 4838–4846.

Nitoi, I., C. Cosma, and A. Ballo (2001). Some considerations on the advanced oxidation procedures applied in toxical compounds degradation of the organophosphoric pesticides in wastewaters type. *Revista de Chimie* 52, 5, pp. 235–239.

Nowell, L. H. and J. Hoigne (1992a). Photolysis of aqueous chlorine at sunlight and ultraviolet wavelengths: I. Degradation rates. *Water Research* 26, 5, pp. 593–598.

Nowell, L. H., and J. Hoigne (1992b). Photolysis of aqueous chlorine at sunlight and ultraviolet wavelengths: II. Hydroxyl radical production. *Water Research* 26, 5, pp. 599–605.

OCWD and OCSD: Orange County Water District and Orange County Sanitation District (2008a). *Groundwater Replenishment System Fact Sheet.* Fountain Valley, CA.

OCWD and OCSD: Orange County Water District and Orange County Sanitation District (2008b). *Groundwater Replenishment System: Major Milestones 1997–2008.* Fountain Valley, CA.

Okamoto, Y., T. Hayashi, C. Toda, K. Ueda, K. Hashizume, K. Itoh, J. Nishikawa, T. Nishihara, and N. Kojima. (2006). Formation of estrogenic products from environmental phthalate esters under light exposure. *Chemosphere* 64, 7, pp. 1785–1792.

Oliver, B. G., and J. H. Carey (1977). Photochemical production of chlorinated organics in aqueous solutions containing chlorine. *Environmental Science and Technology* 11, 9, pp. 893–895.

Onda, K., S. Y. Yang, A. Miya, and T. Tanaka (2002). Evaluation of estrogen-like activity on sewage treatment processes using recombinant yeast. *Water Science and Technology* 46, 11–12, pp. 367–373.

Oppenländer, T. (2003). *Photochemical Purification of Water and Air—Advanced Oxidation Processes (AOPs): Principles, Reaction Mechanisms, Reactor Concepts.* Amsterdam: Wiley-VCH Verlag.

Oppenländer, T. (2007). Mercury-free sources of VUV/UV radiation: Application of modern excimer lamps (excilamps) for water and air treatment. *Journal of Environmental Engineering and Science* 6, 3, pp. 253–264.

Ormeci, B., J. Ducoste, and K. G. Linden (2005). UV disinfection of chlorinated water: Impact on chlorine concentration and UV dose delivery. *Journal of Water Supply: Research & Technology, AQUA* 54, 3, pp. 189–199.

Passantino, J.P. Malley Jr., M. Knudson, R. Ward and J. Kim. (2004). Effect of low turbidity and algae on UV disinfection performance, *Journal AWWA,* vol. 96, no.6, pp. 128–137.

Pereira, V. J., K. G. Linden, and H. S. Weinberg (2007). Evaluation of UV irradiation for photolytic and oxidative degradation of pharmaceutical compounds in water. *Water Research* 41, 19, pp. 4413–4423.

Poskrebyshev, G. A., R. E. Huie, and P. Neta (2003). Radiolytic reactions of monochloramine in aqueous solutions. *Journal of Physical Chemistry A* 107, 38, pp. 7423–7428.

Prados, M., H. Paillard, and P. Roche (1995). Hydroxyl radical oxidation processes for the removal of triazine from natural-water. *Ozone Science and Engineering* 17, 2, pp. 183–194.

Qualls, R. G., and J. D. Johnson (1983). Bioassay and dose measurement in UV disinfection. *Applied and Environmental Microbiology* 45, 3, pp. 872–877.

Reckhow, D., Linden, K., Kim, J., Shemer, H., and G. Makdissy (2010). Effect of UV treatment on DBP formation. *Journal AWWA*. 2010, 102(6), 100–113.

Rosenfeldt, E. J., and K. G. Linden (2004). Degradation of endocrine disrupting chemicals bisphenol A, ethinyl estradiol, and estradiol during UV photolysis and advanced oxidation processes. *Environmental Science and Technology* 38, 20, pp. 5476–5483.

Rosenfeldt, E. J., B. Melcher, and K. G. Linden (2005). UV and UV/H_2O_2 treatment of methylisoborneol (MIB) and geosmin in water. *Journal of Water Supply Research and Technology, AQUA* 54, 7, pp. 423–434.

Schaefer, R., M. Grapperhaus, I. Schaefer, and K. Linden (2007). Pulsed UV lamp performance and comparison with UV mercury lamps. *Journal of Environmental Engineering and Science* 6, 3, pp. 303–320.

Scheuer, C., B. Wimmer, H. Bischof, L. Nguyen, J. Maguhn, P. Spitzauer, A. Kettrup, and D. Wabner. (1995). Oxidative decomposition of organic water pollutants with UV-activated hydrogen peroxide: Determination of anionic products by ion chromatography. *Journal of Chromatography A* 706, 1–2, pp. 253–258.

Schwarzenbach, R. P., P. M. Gschwend, and D. M. Imboden (2003). *Environmental Organic Chemistry,* 2nd ed. New York: Wiley.

Sharpless, C. M., and K. G. Linden (2003). Experimental and model comparisons of low- and medium-pressure Hg lamps for the direct and H_2O_2 assisted UV photodegradation of *N*-nitrosodimethylamine in simulated drinking water. *Environmental Science and Technology* 37, 9, pp.1933–1940.

Sharpless, C. M., and K. G. Linden (2001). UV photolysis of nitrate: Quantum yields, effects of natural organic matter and dissolved CO_2, and implications for UV water disinfection. *Environmental Science and Technology* 35, 14, pp. 2949–2955.

Sharpless, C. M., D. A. Seibold, and K. G. Linden (2003). Nitrate photosensitized degradation of atrazine during UV water treatment. *Aquatic Science* 65, 4, pp. 359–366.

Shishida, K., S. Echigo, K. Kosaka, M. Tabasaki, T. Matsuda, H. Takigami, H. Yamada, Y. Shimizu, and S. Matsui. (2000). Evaluation of advanced sewage treatment processes for reuse of wastewater using bioassays. *Environmental Technology* 21, 5, pp. 553–560.

Snicer, G. A., J. P. Malley, Jr., A .B. Margolin, and A. A. Hogan (2000). *UV Inactivation of Viruses in Natural Waters*. Awwa Research Foundation Report. Denver, CO: Water Research Foundation.

Sommer, R., and A. Cabaj (1993). Evaluation of the efficiency of a UV plant for drinking water disinfection. *Water Science and Technology* 27, 3–4, p. 357.

Sorensen, M., and M. H. Frimmel (1996). Photochemical degradation of hydrophilic xenobiotics in the UV/H_2O_2 process: Influence of bicarbonate on the degradation rate of EDTA, 2-amino-1-naphthalenesulfonate, diphenyl-4-sulfonate, and 4,4'-diamonostilbene-2,2'-disulfonate. *Acta Hydrochimica et Hydrobiologica* 24, 4, pp. 185–188.

Sorensen, M., S. Zureell, and F. H. Frimmel (1998). Degradation pathway of the photochemical oxidation of ethylenediaminetetraacetate (EDTA) in the UV/H_2O_2 process. *Acta Hydrochimica et Hydrobiologica* 26, 2, pp. 109–115.

Sprehe, M., S. U. Geissen, and A. Vogelpohl (2001). Photochemical oxidation of iodized x-ray contrast media (XRC) in hospital wastewater. *Water Science and Technology* 44, 5, pp. 317–323.

Stefan, M. I., K. Atasi, K. G. Linden, and M. Siddiqui (2002). Impact of water quality on the kinetic parameters of NDMA photodegradation. In *Proceedings of the American Water Works Association Water Quality and Technology Conference, Seattle, WA*.

Stefan, M. I., and J. R. Bolton (1998). Mechanism of the degradation of 1,4-dioxane in disulte aqueious solution using the UV/hydrogen peroxide process. *Environmental Science and Technology* 32, 11, pp. 1588–1595.

Stefan, M. I., and J. R. Bolton (1999). Reinvestigation of the acetone degradation mechanism in dilute aqueous solution by the UV/H_2O_2 process. *Environmental Science and Technology* 33, 6, pp. 870–873.

Stefan, M. I., A. R. Hoy, and J. R. Bolton (1996). Kinetics and mechanism of the degradation and mineralization of acetone in dilute aqueous solution sensitized by the UV phololysis of hydrogen peroxide. *Environmental Science and Technology* 30, 7, pp. 2382–2390.

Stefan, M. I., J. Mack, and J. R. Bolton (2000). Degradation pathways during the treatment of methyl *tert*-butyl ether by the UV/H_2O_2 process. *Environmental Science and Technology* 34, 4, pp. 650–658.

Sun, L., and J. R. Bolton (1996). Determination of the quantum yield for the photochemical generation of hydroxyl radicals in TiO_2 suspensions. *Journal of Physical Chemistry* 100, 10, pp. 4127–4134.

Templeton, M. R., R. C. Andrews, and R. Hofmann (2008). Particle-associated viruses in water: Impacts on disinfection processes. *Critical Reviews in Environmental Science and Technology* 38, 3, pp. 137–164.

Trapido, M., A. Hirvonen, Y. Veressinina, J. Hentunen, and R. Munter. (1997). Ozonation, ozone/UV and UV/H_2O_2 degradation of chlorophenols. *Ozone Science and Engineering* 19, 1, pp. 75–96.

Trojan UV (2005a). *Environmental Contaminant Treatment Case Studies: Site Example: Treating Micropollutants and Disinfecting with UV in Drinking Water—PWN Water Supply Company North Holland, the Netherlands.* London, Ontario, Canada: Trojan Technologies, Inc.

Trojan UV (2005b). *Environmental Contaminant Treatment: Trojan's UV-Oxidation Solutions for Seasonal Taste & Odor.* London, Ontario, Canada: Trojan Technologies, Inc.

Trojan UV (2008). *Case Studies: Taste and Odor Treatment at Cornwall, Ontario with UV-Oxidation.* London, Ontario, Canada: Trojan Technologies, Inc.

USEPA (1996). *Ultraviolet Light Disinfection Technology in Drinking Water Application: An Overview.* EPA 811-R-96-002. Washington, DC: U.S. Environmental Protection Agency.

USEPA (2005). *Economic Analysis for the Long Term 2 Enhanced Surface Water Treatment Rule*. EPA 815-R-06-001. Washington, DC: U.S. Environmental Protection Agency.

USEPA (2006a). UV Disinfection Guidance Manual for the Final Long Term 2 Enhanced Surface Water Treatment Rule. United States Environmental Protection Agency, Office of Water, EPA-815-R-06-007, November 2006; available at www.epa.gov/ogwdw000/disinfection/lt2/pdfs/guide_lt2_uvguidance.pdf.

USEPA (2006b). Long Term 2 Enhanced Surface Water treatment Rule. Available at www.epa.gov/ogwdw000/disinfection/lt2/regulations.html#prepub.

Vogna, D., R. Marotta, A. Napolitano, and M. d'Ischia (2002). Advanced oxidation chemistry of paracetamol: UV/H_2O_2-induced hydroxylation/degradation pathways and ^{15}N-aided inventory of nitrogenous breakdown products. *Journal of Organic Chemistry* 67, 17, pp. 6143–6151.

Vogna, D., R. Marotta, A. Napolitano, R. Andreozzi, and M. d'Ischia. (2004a). Advanced oxidation of the pharmaceutical drug diclofenac with UV/H_2O_2 and ozone. *Water Research* 38, 2, pp. 414– 422.

Vogna, D., R. Marotta, R. Andreozzi, A. Napolitano, and M. d'Ischia. (2004b). Kinetic and chemical assessment of the UV/H_2O_2 treatment of antiepileptic drug carbamazepine. *Chemosphere* 54, 4, pp. 497–505.

Vogt, R., and R. N. Schindler (1991). Product channels in the photolysis of HOCl. *Journal of Photochemistry and Photobiology A: Chemistry* 66, 2, pp. 133–140.

von Sonntag C., and H. P. Schuchmann (1992). UV disinfection of drinking water and by-product formation–Some basic considerations. *Journal Water Supply Research and Technology, AQUA* 41, 2, pp. 67–74.

Wait, I. W., C. T. Johnston, and E. R. Blatchley (2007). The influence of oxidation reduction potential and water treatment processes on quartz lamp sleeve fouling in ultraviolet disinfection reactors. *Water Research* 41, 11, pp. 2427–2436.

Wang, G. S., C. H. Liao, and F. J. Wu (2001). Photodegradation of humic acids in the presence of hydrogen peroxide. *Chemosphere* 42, 4, pp.379–387.

Watts, M. J., E. J. Rosenfeldt, and K. G. Linden (2007). Comparative OH radical oxidation using UV–Cl_2 and UV–H_2O_2 processes. *Journal of Water Supply: Research and Technology, AQUA* 56, 8, pp. 469–477.

Watts, M. J., and K. G. Linden (2007). Chlorine photolysis and subsequent ·OH radical production during UV treatment of chlorinated water. *Water Research* 41, 13, pp. 2871–2878.

Westerhoff, P., S. P. Mezyk, W. J. Cooper, and D. Minakata (2007). Electron pulse radiolysis determination of hydroxyl radical rate constants with Suwannee River fulvic acid and other dissolved organic matter isolates. *Environmental Science and Technology* 41, 13, pp. 4640–4646.

Wink, D.A., R.W. Nims, M.F. Desrosiers, P.C. Ford, and L.K. Keefer. (1991). A kinetic investigation of intermediates formed during the Fenton reagent mediated degradation of *N*-nitrosodimethylamine: Evidence for an oxidative pathway not involving hydroxyl radical. *Chemical Research in Toxicology* 4, 5, pp. 510–512.

Wittmann, G., I. Horvath, and A. Dombi (2002). UV-induced decomposition of ozone and hydrogen peroxide in the aqueous phase at pH 2–7. *Ozone Science and Engineering* 24, 4, pp. 281–291.

Wolfe, R. L. (1990). Ultraviolet disinfection of potable water. *Environmental Science and Technology* 24, 6, pp. 768–772.

Xu, B., N.Y. Gao, X.F. Sun, S.J. Xia, M. Rui, M.O. Simonnot, C. Causserand, and J.F. Zhao. (2007). Photochemical degradation of diethyl phthalate with UV/H_2O_2. *Journal of Hazardous Materials B* 139, 1, pp. 132–139.

Yates, M., J. P. Malley, Jr., P. Rochelle, and R. Hoffman (2006) Impact of adenovirus resistance on UV disinfection requirements: A report on the state of adenovirus science. *Journal AWWA* 98, 6, pp. 93–106.

Zafiriou, O. C., J. Joussotdubien, R. D. Zepp, and R. G. Zika (1984). Photochemistry of natural waters. *Environmental Science and Technology* 18, 12, pp. A358–A371.

Zepp, R. G. (1982). Experimental approaches to environmental photochemistry. In *The Handbook of Environmental Chemistry*, Vol. 2, Part B, O. Tutziner, ed. Berlin: Springer.

Zhao, X., G. Yang, Y. Wang, and X. Gao (2004). Photochemical degradation of dimethyl phthalate by Fenton reagent. *Journal of Photochemistry and Photobiology A: Chemistry* 161, 2–3, pp. 215–220.

CHAPTER 19
FORMATION AND CONTROL OF DISINFECTION BY-PRODUCTS

David A. Reckhow, Ph.D.
Professor of Civil and Environmental Engineering
University of Massachusetts
Amherst, Massachusetts, United States

Philip C. Singer, Ph.D., P.E., B.C.E.E.
Dan Okun Distinguished Professor of Environmental Engineering
University of North Carolina
Chapel Hill, North Carolina, United States

INTRODUCTION **19.1**
FORMATION OF DISINFECTION (AND OXIDATION) BY-PRODUCTS **19.2**
General Considerations 19.2
Identity of Disinfection By-Products 19.3
Factors Influencing By-Product Formation 19.16
Disinfection By-Product Formation Models 19.27
CONTROL OF OXIDATION/ DISINFECTION BY-PRODUCTS **19.27**
Removing Organic Precursors 19.27
Modifying Disinfection 19.36
Removing Disinfection By-Products after Formation 19.39
DISINFECTION BY-PRODUCTS IN THE DISTRIBUTION SYSTEM **19.43**
Stability of Disinfection By-Products 19.43
Monitoring for Regulatory Compliance 19.45
Spatial Temporal Variability of DBP Levels 19.46
ABBREVIATIONS **19.46**
REFERENCES **19.48**

INTRODUCTION

Disinfection (and oxidation) by-products (DBPs) are chemical compounds produced as an unintended consequence of disinfection or oxidation processes in drinking water treatment. Most of the compounds of greatest concern contain chlorine and bromine atoms; however, the formation of iodinated compounds also has been noted (Weinberg et al., 2002). Some of these compounds have been found to be carcinogenic or to cause adverse reproductive or developmental effects in animal studies (see Chap. 2). Others have been shown to be mutagenic and hepatotoxic. As a result, the U.S. Environmental Protection Agency (USEPA) promulgated rules in 1979, 1998, and 2006 regulating DBP concentrations in finished drinking waters (see Chap. 1).

This chapter focuses on a group of chemical contaminants (DBPs in this case), rather than on treatment process technology. As a result, we will refer the reader to specific chapters on treatment processes as they are brought into the discussion of DBPs. In some cases we will provide supplemental information on the performance of treatment processes as they pertain specifically to the control of DBPs. The chapter opens with a summary of the types of by-products formed from each of the commonly used disinfectants, followed by a discussion of factors affecting their formation. Next, the chapter turns to DBP control, including removal of precursors, use of modified disinfection practices, and removal of DBPs after formation. The chapter concludes with a discussion of DBP concentrations and losses in distribution systems.

FORMATION OF DISINFECTION (AND OXIDATION) BY-PRODUCTS

General Considerations

Water treatment oxidants/disinfectants derive their effectiveness from their general chemical reactivity. The same attributes that give disinfectants the ability to react with cell membranes, nuclear materials, and cellular proteins also lead to reactions with abiotic dissolved organic matter and extracellular biomolecules. Except for the occasional source with high ammonia or sulfide concentrations, most of the oxidant/disinfectant demand in raw and treated drinking water can be attributed to reactions with such dissolved organic molecules in water.

Most organic matter in surface and groundwater is of natural origin. Some of this natural organic matter (NOM) is highly reactive with a wide range of oxidants. The reaction products include reduced forms of the oxidants (e.g., chloride, hydroxide, and chlorite when using chlorine, ozone, and chlorine dioxide, respectively) and oxidized forms of the organic or inorganic reactants (e.g., bromate) (Fig. 19-1).

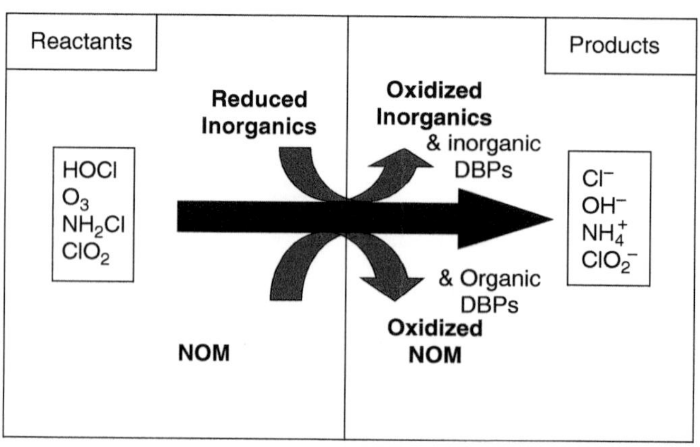

FIGURE 19-1 Schematic illustration of reactions of various oxidants with natural organic material (NOM) and reduced inorganic substances.

The sites of disinfectant (oxidant) attack on NOM are often carbon-carbon double bonds and reduced heteroatoms (e.g., N and S). The organic by-products formed are more highly oxidized, often containing more oxygen atoms. As the extent of the reaction increases, the organic matter becomes more fragmented, and the specific by-products are simpler in structure. General oxidation by-products include the C_1–C_3 acids, diacids, aldehydes, ketones, and ketoacids (e.g., Griffini and Iozzelli, 1996). Specific examples include oxalic acid, pyruvic acid, and formaldehyde.

Several of the disinfectants are capable of producing by-products that have halogen atoms (i.e., chlorine, bromine, and iodine) incorporated into their structure. Aqueous chlorine and bromine do this to the greatest extent, followed by chloramines and ozone. In the case of ozone, high concentrations of bromide are required for substantial bromine incorporation. The organic halide by-products can be measured collectively by the total organic halide analytical method (abbreviated TOX, or more accurately, dissolved organic halide, DOX; for more on this method, see APHA et al., 2005). Because NOM contains very low levels of TOX, this analysis presents an opportunity to easily measure a large and diverse group of compounds that are indisputably DBPs. It also targets a subset of the total DBPs (i.e., just the halogenated ones) that are viewed as the compounds of greatest concern, allowing the calculation of halogen mass balances (e.g., see Singer and Chang, 1989; Shukairy et al., 2002).

Aqueous chlorine, chloramines, and ozone are all capable of oxidizing naturally occurring bromide to form active bromine (i.e., hypobromous acid (HOBr) or bromamines; see Chap. 7). The latter will react with NOM to form brominated organic compounds (e.g., bromoform and dibromoacetic acid) and, in the presence of free chlorine, mixed bromochloro-organics. The same is true with respect to the formation of iodinated DBPs in the presence of iodide, although iodinated DBPs tend to be found only in chloraminated waters (see section on chloramine by-products). These halogenated by-products all contribute to the TOX concentration of the water. Furthermore, it is possible to measure halogen-specific TOX (e.g., TOCl, TOBr, and TOI) by replacing microcoulometric detection with ion chromatography (e.g., Hua and Reckhow, 2007).

Identity of Disinfection By-Products

Since the discovery of trihalomethanes (THMs) in chlorinated drinking water in the early 1970s (Rook, 1974; Bellar et al., 1974), hundreds of specific compounds have been identified as DBPs. Many of the major groups are summarized in Table 19-1. More detailed listings of individual compounds can be found in the review by Richardson (1998).

Each of the four disinfectants presented in the table has its own unique chemistry. For example, ozone is the only disinfectant that produces measurable quantities of bromate. Nevertheless, many by-product classes and specific compounds are common to two or more of the major disinfectants. This is illustrated by the simple aliphatic carboxylic acids (e.g., acetic acid), which are universal by-products regardless of the disinfectant/oxidant. Itoh and Matsuoka (1996) found that all oxidants produce carbonyls (e.g., formaldehyde), with ozone and chlorine dioxide producing the most and chlorine and inorganic chloramines only slightly behind. Some halogenated compounds, such as dihaloacetic acids, may be produced by all four disinfectants, but the amounts produced range over several orders of magnitude, depending on the disinfectant, the disinfectant dose, and the bromide level. For this reason, an attempt has been made in the table to classify by-product abundance based on an order-of-magnitude scale (very high >100 μg/L; high = 10–100 μg/L; medium = 1–10 μg/L; low = 0.01–1 μg/L; very low < 0.01 μg/L) as assessed for an average drinking water under typical treatment conditions.

TABLE 19-1 Chemical By-Products of the Four Major Disinfectants

By-product class	Examples	Chlorine	Chloramines	Chlorine dioxide	Ozone
Compounds with O–X Bonds					
Oxychlorines	Chlorate, Chlorite			V.High[34,35]	
Oxybromines	Bromate, hypobromate				Med[51–53,58]
Compounds with C—X Bonds					
	Chloroform, bromodichloromethane, chlorodibromomethane	High	Low		
Trihalomethanes	Bromoform	Med	Low		Med[49,52]
	Dichloroiodomethane		Low		
Other haloalkanes	1,2-Dibromoethane, 1,2-dibromopropane	Low[14,15]			
Halohydrins	3-Bromo-2-methyl-2-butanol, 9-chloro-10-hydroxyl methyl stearate	(NOM[20])	(Models[3,23])	(Models[37])	Med[62]
Haloacids	Dichloroacetic acid, trichloroacetic acid	High	Med[17,18]	(NOM[29])	
	Monochloroacetic acid, bromochloroacetic acid, bromodichloroacetic acid,	Med	Low		
	monobromoacetic acid, dibromoacetic acid, tribromoacetic acid,	Low	Low		Med[50]
	diiodoacetic acid,		Low		
	6,6-dichlorohexanoic acid		(NOM[17,18])		
Haloacids (unsaturated)	3,3-Dichloropropenoic acid	(NOM[15])	(NOM[17,18])		
Halodiacids	2,2-Dichlorobutanedioic acid, 2,3-dichlorobutenedioic acid	(NOM[15,16])	(NOM[17,18])		
Halohydroxy-acids	3,3,3-Trichloro-2-hydroxypropanoic acid, 2-chloro-4-hydroxybutanoic acid, 2,3-dichloro-3,3-dihydroxy propanoic acid, 4-chloro-4-hydroxypentenoic acid	(NOM[15])	(NOM[17,18])		
Haloketones	1,1,1-Trichloropropanone	Med[1,2,3]			
	Chloropropanone		(NOM[17,18])		
	Bromopropanone				Unkn[60]
	1,1,3,3-Tetrachloropropranone			Unkn[28]	
	1,1,1-Trichloro-2-butanone, pentachloro-3-buten-2-one	(NOM[4])			
Haloaldehydes	Chloral	Med			
	Chloroacetaldehyde, dichloroacetaldehyde		(NOM[17,18])		
	Dichloropropanal, 3-chlorobutanal, 2,3,3-trichloropropenal	(NOM[4])	(NOM[17,18])		

(*Continued*)

TABLE 19-1 Chemical By-Products of the Four Major Disinfectants (*Continued*)

By-product class	Examples	Chlorine	Chloramines	Chlorine dioxide	Ozone
Haloketoacids	2,3-Dichloro-4-oxopentenoic acid, 2,5-dichloro-4-bromo-3-oxopentanoic acid	(NOM[15])			
Halonitriles	Dichloroacetonitrile, trichloroacetonitrile, dibromoacetonitrile	Med[1,2,5]			
Cyanogen aalides	Cyanogen chloride	Low	Med		
	Cyanogen bromide		Low		
C-Chloro amines			(Models[3,21])		
Halophenols		(NOM[1,6])			
Chloroaromatic acids	5-Chloro-2-methoxybenzoic acid, dichloromethoxybenzoic acid	(NOM[7,8])			
Halothiophenes	Tetrachlorothiophene	(NOM[1])			
Chlorinated PAHs		Unkn[9]			
MX and related compounds	MX, EMX, red-MX, ox-EMX	Low[11,12,13]	(NOM[17,18,27])		
	2,2,4-Trichlorocyclopentene-1,3-dione	Low[10]			
Halo-nitromethanes	Chloropicrin	Med			
	Bromopicrin	Low			Med
<u>*Compounds with N—X Bonds*</u>					
N-Chloro-amino acids	*N*-Chloroglycine	(Models)			
N-Chloro-amines		(Models[26])	(Models[22,24,25])		
<u>*Compounds without Halogens*</u>					
Aliphatic	Formic acid, acetic acid, butyric acid, pentanoic acid	High		High[28,29,38,40]	High[43,58]
Monoacids	Hexadecanoic acid				Low[60]
Aliphatic diacids (saturated)	Oxalic acid	High		High[32,40]	V.High[58]
	Succinic acid, glutaric acid, adipic acid			(NOM[29])	Unkn[47]
Aliphatic diacids (unsaturated)	Butenedioic acid	(NOM[15,16])			
	2-*tert*-Butylmaleic acid, 2-ethy-3-methylmaleic acid			Unkn[28,32]	
Aromatic acids	Benzoic acid, 3,5-dimethylbenzoic acid	(NOM[15,16])		Unkn[28]	Unkn[44-47]
	p-Benzoquinone			(Models[33])	
Other aromatics	Hydroxy-PAHs			(Models[36])	
	3-Ethyl styrene, 4-ethyl styrene			Unkn[28]	
	Naphthalene, 1-methylnaphthalene			Unkn[28]	
	Formaldehyde, acetaldehyde, propanal		(Models[1])	Unkn[30]	High
Aldehydes	Glyoxal, methylglyoxal				High
	Benzaldehyde, ethylbenzaldehyde				Unkn[47,57]
	Acetone, propyl ethyl ketone				Med[41,54]

(*Continued*)

TABLE 19-1 Chemical By-Products of the Four Major Disinfectants (*Continued*)

By-product class	Examples	Chlorine	Chloramines	Chlorine dioxide	Ozone
	Dioxopentane, 1,2-dioxobutane				Unkn[41]
Ketones	Acetophenone, 4-phenyl-2-butanone				Unkn[47]
	2-Hexenal, 6-methyl-5-hepten-2-one				Low[60]
	2,3,4-Trimethylcyclopent-2-en-1-one, 2,6,6,-trimethyl-2-cyclohexene-1,4-dione			Unkn[28]	
	Pyruvic acid, glyoxalic acid, ketomalonic acid				High
Ketoacids	Oxobutanoic acid, 4-oxo-2-butenoic acid				Unkn[41,47]
	Ketosuccinic acid, ketoglutaric acid				Unkn[41,61]
	Dioxopropanoic acid, dioxopentanoic acid				Unkn[41]
Hydroxyacids	Hydroxymalonic acid				(NOM[44])
Hydroxy-carbonyls	Hydroxyacetaldehyde				(Models[59])
Furans	Methylfurancarboxylic acid			(NOM[29])	
Epoxides			(Models[23])	(Models[31,39])	(Models[42,55])
Organic Peroxides					(Models)
Nitriles			(Models[1,19])		
Nitrosamines	Nitrosodimethylamine	V.low	Low		
Nitramines	Dimethylnitramine	(Models[63])			
Hydrazines	1,1-Dimethylhydrazine		(Models)		
Miscellaneous	5-Methoxy-α-pyrone				(NOM[44])

Notes:

1. Data marked "NOM" and "Models" are from studies using solutions of natural organic matter extracts and model compounds, respectively.

2. All other data are from treated drinking waters and unaltered natural waters. Concentrations are classified as follows: V. high (very high): >100 µg/L; High: 10–100 µg/L; Med (medium): 1–10 µg/L; Low: 0.01–1 µg/L; V. low (very low): <0.01 µg/L; Unkn (not quantified).

Sources:

1. Le Cloirec & Martin, 1985
2. Brass et al., 1977
3. Minisci & Galli, 1965
4. Smeds et al., 1990
5. Kanniganti, 1990
6. Shank & Whittaker, 1988
7. Backlund et al., 1988
8. Kronberg & Vartiainen, 1988
9. Burttschell et al., 1959
10. Fielding & Horth, 1986
11. Franzén & Kronberg, 1994
12. Peters et al., 1994
13. Franzén & Kronberg, 1994
14. de Leer et al., 1985
15. Oliver, 1983
16. Kanniganti et al., 1992
17. Kanniganti, 1990
18. Kanniganti et al., 1992
19. Hausler,
20. Havlicek et al. 1979
21. Neale, 1964
22. Scully 1986
23. Carlson & Caple, 1977
24. Jensen & Johnson, 1989
25. Crochet & Kovacic, 1973
26. Kringstad et al. 1985
27. Backlund et al., 1988
28. Richardson et al., 1994
29. Colclough, 1981
30. Stevens et al., 1978
31. Legube et al., 1981
32. Masschelein, 1979
33. Wajon et al., 1982
34. Steinbertg, 1986
35. Werdehoff & Singer, 1987
36. Luikkonen et al.,
37. Ghanbari et al., 1983
38. Somsen, 1960
39. Carlson & Caple, 1977
40. Griffini & Iozzelli, 1996
41. Le Lacheur et al., 1993
42. Carlson & Caple, 1977
43. Lawrence, 1977
44. Benga, 1980
45. Paramisigamani et al., 1983
46. Killips et al., 1985
47. Glaze, 1986
48. Edwards, 1990
49. Cooper et al., 1986
50. Daniel et al., 1989]
51. Haag & Hoigne, 1983
52. Glaze et al., 1993
53. Krasner et al., 1993
54. Fawell et al., 1984
55. Chen et al., 1979
56. Chappell et al., 1981
57. Lawrence et al., 1980
58. Griffini & Iozzelli, 1996
59. Le Lacheur & Glaze, 1996
60. Richardson et al., 1996
61. Hwang et al., 1996
62. Cavanagh et al. 1992
63. Mitch, 2007

Many worthwhile studies have been conducted with solutions of isolated NOM (e.g., aquatic fulvic acids). Since these studies are often conducted under extreme conditions (i.e., high TOC, high chlorine dose, and sometimes high or low pH) designed to maximize DBP formation, no attempt has been made to render a judgment on the likely concentration level expected in tap water based on such studies. Also useful, but further removed from practice, are the studies using model compounds (designated "Models" in Table 19-1). Entries that are not labeled with "NOM" or "Model" refer to expected occurrence levels in typical finished drinking water.

Chlorination By-Products. The chlorination by-products include a wide range of halogenated and nonhalogenated organic compounds. Regulatory agencies have focused on the halogenated compounds, especially the THMs and haloacetic acids (HAAs) (see Chap. 1 on regulations). These are small, highly substituted end products of the reaction of chlorine with organic matter. In waters with low bromide levels, the fully chlorine-substituted forms predominate (e.g., chloroform and di- and trichloroacetic acid). Waters with high levels of bromide are likely to contain elevated levels of the bromine-containing analogues (e.g., bromoform and dibromoacetic acid) following chlorination. Waters with moderate levels of bromide will contain the mixed bromo/chloro analogues (e.g., bromodichloromethane and bromodichloroacetic acid).

Nearly all DBP studies in the 1970s focused on the THMs. Given their volatility, chemical stability, and high halogen-carbon ratio, this class of compounds could be easily analyzed with minimal analytical equipment and expertise. Consequently, the THMs were the first by-products to be found in finished drinking waters (Rook, 1974; Bellar et al., 1974), the first to be the subject of an established analytical method, and the first to be included in a large survey of public water supplies (Symons et al., 1975). In a matter of just a few years, it was recognized that the THMs were always present whenever chlorine was used as a disinfectant.

The discovery of HAAs in chlorinated waters (Miller and Uden, 1983; Christman et al., 1983) and subsequent occurrence studies trailed the THMs by several years. One early survey (Krasner et al., 1989) showed the HAAs, like the THMs, to be ubiquitous in chlorinated waters, although present at somewhat lower levels. More recent data have supported this finding. However, the lack of available standards for all the HAAs in these earlier studies may have resulted in underestimation of their concentrations (Cowman and Singer, 1996).

Many other halogenated by-products have been widely reported in chlorinated drinking waters. The most intensively studied of these nonregulated compounds are the halopropanones, the haloacetonitriles, chloropicrin, and chloral hydrate. This group owes its large industry-wide database to the analytical method that it shares with the THMs (i.e., like the THMs, they are all volatile neutral compounds that respond well to analysis by gas chromatography with electron-capture detection). The haloacetonitriles are thought to be largely derived from the chlorination of amino acids and proteinaceous material (Bieber and Trehy, 1983). Nitrogenous structures in humic substances also will form haloacetonitriles via cyano acid intermediates (Backlund et al., 1988). The halopropanones (e.g., 1,1,1-trichloropropanone) and chloral hydrate (a haloaldehyde) are halogenated analogues of some common ozonation by-products. They are commonly found at elevated concentrations in waters that had been previously ozonated (Reckhow et al., 1986; McKnight and Reckhow, 1992).

From July 1997 to December 1998, a comprehensive set of data on DBP occurrence in U.S. drinking waters was collected as part of the Information Collection Rule (ICR). All large drinking water utilities (defined as those serving populations > 100,000) provided DBP occurrence information and corresponding water quality characteristics and treatment conditions for six quarters. A total of 299 utilities participated, encompassing 500 water treatment plants. The following DBPs were measured: THMs, HAAs, haloacetonitriles (HANs), haloketones such as the chloropropanones (CPs), chloral hydrate (CH), chloropicrin

(CP), cyanogen chloride (CNCl), and TOX. Bromate, chlorite, chlorate, and aldehydes, which are oxidation by-products of ozone and chlorine dioxide (see Chap. 7 and below), also were measured in the utilities employing these oxidants. The ICR data set was generated with a high degree of quality assurance and was used to finalize the stage 1 and stage 2 DBP rules (see Chap. 1). A description of the ICR activity and its results is provided in an ICR data analysis report (McGuire et al., 2002). Figures 19-2 to 19-4 illustrate the occurrence findings from the ICR database. It should be noted that the majority of the ICR utilities had raw water sources with relatively low bromide concentrations, a characteristic shared by many high-quality surface waters, so the distribution of DBPs was skewed toward the chlorine-containing species.

About 50 percent of the TOX produced on chlorination can be attributed to the major by-products discussed earlier (Singer and Chang, 1989). The remainder is largely unknown and has been the subject of research for the past 20 years. A similar balance on mutagenic activity (Ames test) reveals that more than 50 percent of the activity can be accounted for among the known by-products. Much of the identified mutagenicity is found in a single compound given the abbreviated name MX (Backlund et al., 1988; Meier et al., 1988), a chlorinated furanone that readily undergoes ring opening. It is produced in small quantities along with several related compounds by free chlorine and chloramines. Relatively little is known about its occurrence because it is difficult to measure. Early studies showed finished water concentrations as high as 60 ng/L (Kronberg and Vartiainen, 1988). Wright and colleagues (2002) conducted a focused study of MX in U.S. finished waters and found concentrations up to 80 ng/L. Weinberg et al (2002) documented a median value of 20 ng/L and a maximum of 310 ng/L.

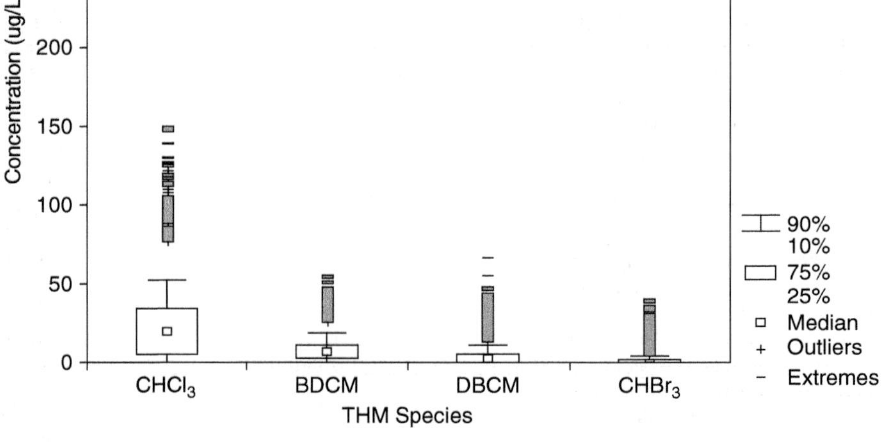

FIGURE 19-2 Distribution of ICR results for individual trihalomethane species. (*Source: McGuire et al., 2002, Awwa Research Foundation.*)

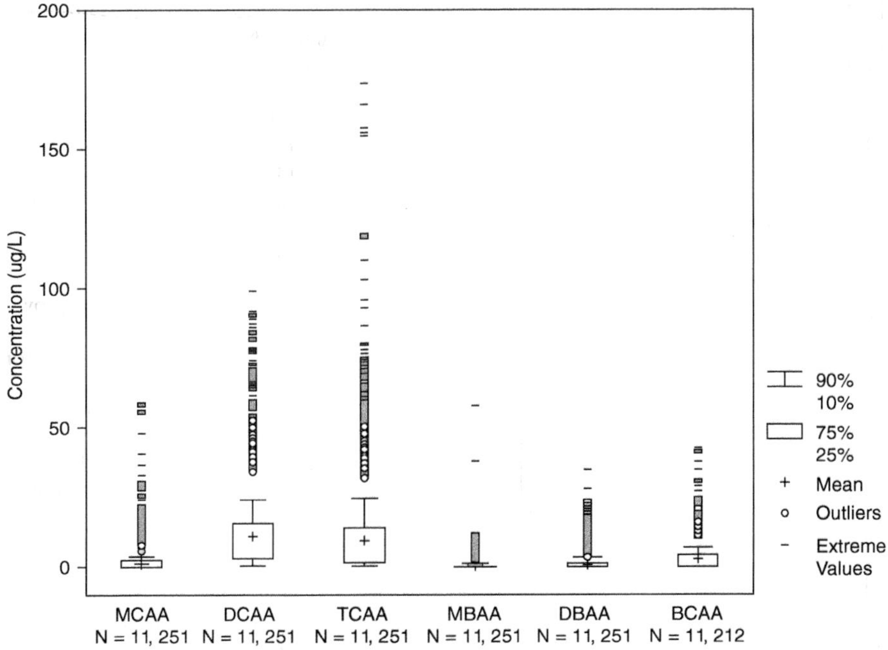

FIGURE 19-3 Distribution of ICR results for individual haloacetic acid species. Note that only six of the nine bromine- and chlorine-containing HAAs were measured by all the participating utilities. (*Source: McGuire et al., 2002, Awwa Research Foundation.*)

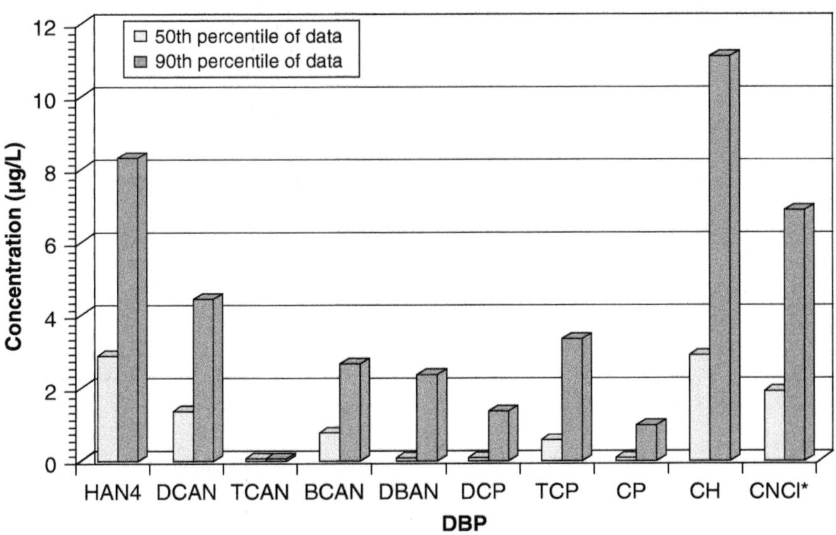

FIGURE 19-4 DBP concentrations in ICR distribution systems. (*Source: McGuire et al., 2002, Awwa Research Foundation.*)

Because TOX formation represents only a small fraction of the total chlorine consumed (typically < 10 percent), it can be concluded that most of the organic chlorination by-products do not contain chlorine and that most of the chlorine consumed leads to the formation of chloride. A number of these nonhalogenated by-products have been identified (Table 19-1). Most are aliphatic mono- and diacids and benzenepolycarboxylic acids. Because these compounds probably are not of health concern, they have not been studied extensively. It should be noted, however, that many of these compounds are readily biodegradable and therefore contribute to the biodegradable dissolved organic carbon (BDOC) content, sometimes measured as assimilable organic carbon (AOC). Hence, in the absence of a disinfectant residual in the distribution system, the presence of these compounds can encourage the growth of biofilms.

Chloramine By-Products. Although monochloramine and dichloramine are less reactive than free chlorine with NOM and most model compounds, inorganic chloramines can form some of the DBPs commonly associated with free chlorine. Identifiable by-products include dichloroacetic acid (DCAA), cyanogen chloride, and small amounts of chloroform and trichloroacetic acid (TCAA; see Table 19-1). While it's not clear that TCAA and the THMs are true by-products of chloramines (i.e., they may be formed owing to the presence of a small free chlorine residual), it does seem that DCAA and cyanogen chloride are true by-products (Singer et al., 1999). Cyanogen chloride concentrations generally are higher in systems using chloramines, and this is due, at least in part, to the greater stability of this compound in the presence of chloramines compared with free chlorine. Both mechanistic and occurrence studies have shown that haloacetonitriles also can be formed as true by-products of inorganic chloramines (e.g., Young et al., 1995; Weinberg et al., 2002). Backlund and colleagues (1988) found that monochloramine forms MX from humic materials, although the amount measured was less than 25 percent of that formed during chlorination.

Weinberg and colleagues (2002) found iodoacetic acid, bromoiodoacetic acid, (*E*)-3,3-bromoiodopropenoic acid, (*Z*)-3,3-bromoiodopropenoic acid, and (*E*)-2-iodo-3-methylbutenedioic acid in a drinking water system using chloramines with no free chlorine contact time. When there is a substantial free chlorine contact time preceding ammonia addition, lower levels of iodinated organics are produced (e.g., von Gunten et al., 2006; Hua et al., 2007). This is attributed to the rapid oxidation of reactive iodine to the nonreactive iodate anion by pretreatment with free chlorine (see Chap. 7). Inorganic chloramines are not capable of this rapid oxidation, so precursor organics have a greater exposure to reactive iodine and thereby produce more iodinated organic by-products (Fig. 19-5).

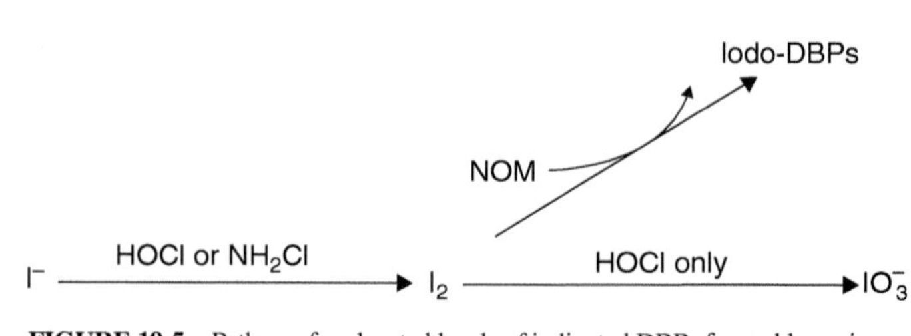

FIGURE 19-5 Pathway for elevated levels of iodinated DBPs from chloramines.

Jensen and colleagues (1985) found that at high monochloramine-carbon ratios (~10 mg Cl_2/mg C), about 5 percent of the oxidant demand goes toward TOX formation. This is identical to the results for chlorine at an equally high chlorine-carbon ratio. Most of the chlorinated organic by-products remained tied up in high-molecular weight compounds. This contrasts with free chlorine because monochloramine is a much weaker oxidant than free chlorine and is much less likely to create the smaller fragments that can be analyzed by gas chromatography.

Monochloramine is known to transfer active chlorine to the nitrogen of amines and amino acids, forming organic chloramines (Scully, 1986). This reaction also occurs with free chlorine. Model compound studies have shown that monochloramine also can add chlorine to activated aliphatic carbon-carbon double bonds (Johnson and Jensen, 1986). The adjacent carbon may become substituted with an amine group or with oxygen. Other types of reactions involve the simple addition of amine or chloramine to unsaturated organic molecules. Under certain conditions, chlorine substitution onto activated aromatic rings has been observed. For example, monochloramine will slowly form chlorophenols from phenol (Burttschell et al., 1959). Inorganic chloramines also will add chlorine to phloroacetophenone, a highly activated aromatic compound, to produce chloroform (Topudurti and Haas, 1991). However, the rate of reaction for this compound is low, and the molar yield (i.e., moles DBP formed per mole precursor) is only 3 percent compared with 400 percent for free chlorine.

In addition to chlorine transfer, inorganic chloramines also undergo addition reactions with many types of organic molecules. This results in the formation of new organic amines and organic chloramines where the nitrogen atom originates from the inorganic disinfectant. Studies designed to determine if this is a major pathway for DBP formation in drinking water have produced mixed results. Hirose and colleagues (1988) found that this was not the case when looking at cyanogen chloride (CNCl) formation from the chloramination of the amino acid leucine. Their findings are supported by Krasner and colleagues (1989). On the other hand, chloramination of formaldehyde does produce CNCl, a clear indication of the ammoniation reaction (Pedersen et al., 1999). Using ^{15}N-labeled monochloramine, Young and colleagues (1995) studied this question with natural waters and dichloroacetonitrile (DCAN) formation. They concluded that most of the nitrogen in DCAN came from the inorganic chloramines and not from naturally occurring nitrogen in the NOM.

One possible result of ammoniation reactions is the formation of carcinogenic nitrosamines from organic amines. Among this group, it is nitrosodimethylamine (NDMA) that has been observed most commonly and studied most widely. NDMA levels in finished drinking water are rarely above the low nanogram per liter level, and it is observed more often in systems using chloramines, especially those affected by wastewater. Although cationic polymers and anion exchange resins used in water treatment can harbor NDMA precursors (e.g., Najm and Trussell, 2001), it now seems clear that a significant amount of organic precursors to NDMA formation originate from municipal and domestic wastewater. Schreiber and Mitch (2006) have formulated a mechanism for NDMA formation from dimethylamine precursors (Fig. 19-6). This explains many of the observations surrounding NDMA formation in drinking water, including the elevated concentrations under conditions that favor dichloramine formation (e.g., high Cl_2:N ratios). Since some NDMA formation has been observed at plants using free chlorine, there may be other important mechanisms not involving inorganic chloramines.

When used in water treatment, chloramines are formed in situ, that is, within the plant. This usually involves addition of ammonia after or at the location of addition of chlorine. While the reaction between chlorine and ammonia is rapid at near-neutral pH (Weil and Morris, 1949), yields of the intended product (i.e., monochloramine and dichloramine) typically are less than 100 percent, especially at high Cl_2:N ratios. Nitrogen species of elevated oxidation state have long been postulated as necessary intermediates in the formation of the final breakpoint products (i.e., nitrogen gas and nitrate). These include N(–I) species (e.g., hydroxylamine, NH_2OH) and N(+I) species (e.g., nitroxyl, HNO, and hyponitrous acid, $H_2N_2O_2$) (Chapin, 1931; Wei and Morris, 1974; Saunier and Selleck, 1979). There is also strong evidence for at least one related N-Cl species (Leung and Valentine 1994a, 1994b).

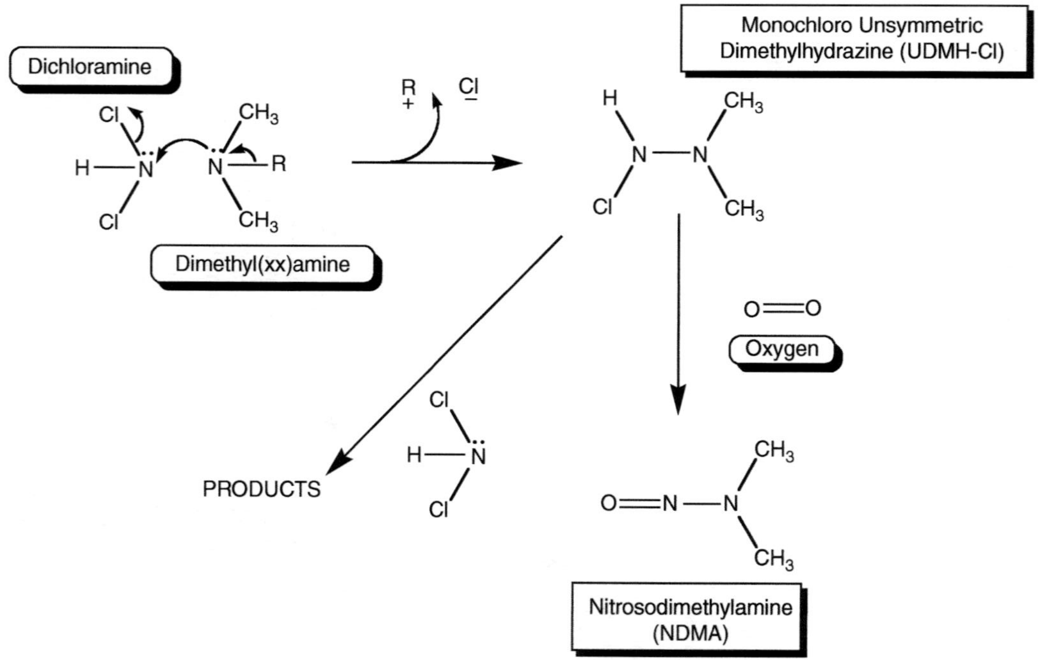

FIGURE 19-6 Proposed pathway for NDMA formation during chloramination. (*Source: Based on work by Schreiber and Mitch, 2006.*)

Little is known about the reactivity of these compounds or their occurrence in drinking water systems.

Although dismissed by Saunier and Selleck (1979), Shank and Whittaker (1988) asserted that small amounts of the known carcinogen hydrazine ($H_2N–NH_2$) could be formed by reaction of ammonia with monochloramine when drinking waters are chloraminated. Models based on known hydrazine formation rates predict that 10 ng/L concentrations could form in chloraminated waters at pH above 9 (Najm et al., 2006). However, little data exist on actual hydrazine concentrations in drinking water distribution systems.

Chlorine Dioxide By-Products. Chlorine dioxide undergoes a wide variety of oxidation reactions with organic matter to form oxidized organics such as aldehydes, ketones, and acids, as well as inorganic chlorite, ClO_2^- (see Table 19-1 and Chap. 7). The concentration of the resulting chlorite accounts for 50 to 70 percent of the chlorine dioxide consumed (Rav Acha et al., 1984; Werdehoff and Singer, 1987). Chlorite also may be formed, along with chlorate (ClO_3^-), by the disproportionation of chlorine dioxide (see Chap. 7). All three of the oxidized chlorine species (chlorine dioxide, chlorite, and chlorate) are considered to have adverse health effects, and their presence in finished water above regulated levels is a source of concern; see Chaps. 1 and 2 for coverage of regulations and health effects.

Chlorine dioxide also can undergo a limited number of chlorine substitution reactions. For example, reaction with dimethoxyphenylethanol, a lignin model compound, produces many initial products, including a ring-chlorinated derivative (Svenson et al., 2002). Trihalomethanes, however, have not been detected as reaction products when water containing NOM is treated with chlorine dioxide. As with the other chlorine-containing oxidants, chlorine addition/substitution products are favored at low oxidant-carbon ratios, and oxidation reactions are favored at high ratios. Studies using drinking waters and NOM have shown that small amounts of TOX form on treatment with typical levels of chlorine dioxide (e.g., Hua and Reckhow, 2007). Aside from direct reactions between NOM and molecular ClO_2, as noted earlier, this also may be due to the formation of HOCl when chlorine dioxide reacts with NOM and subsequent reaction with other NOM compounds (Werdehoff and Singer, 1987). The relatively small amount of HOCl formed in this manner probably leads to sparsely halogenated macromolecular TOX, which would account for the lack of identifiable organohalide by-products.

Ozonation By-Products. Ozonation can lead to the formation of brominated by-products when applied to waters with moderate to high bromide levels. This is a direct result of ozone's ability to oxidize bromide to hypobromous acid and related species (see Chap. 7). Some of this oxidized bromine continues to react to form bromate ion. Much of the remaining hypobromous acid reacts with NOM to form brominated organic compounds. These by-products encompass the same general classes reported for the halogenated by-products of chlorine [i.e., THMs (bromoform), HAAs (dibromoacetic acid), HANs (dibromoacetonitrile), and halonitromethanes (bromopicrin)]. It has been estimated that 7 percent of the raw water bromide becomes incorporated as TOX (or total organic bromide, TOBr) following ozonation under conditions typical of drinking water treatment (Song et al., 1997). The extent of TOBr formation from HOBr versus further oxidation of HOBr to form bromate depends markedly on pH, with elevated pH values (above 6.5–7) favoring bromate formation.

In contrast to the situation with bromide, ozonation will not lead to the formation of iodinated DBPs when water containing iodide is ozonated. This is so because ozonation rapidly oxidizes the hypoiodous acid formed to iodate (see Chap. 7).

Most ozonation by-products are not halogenated, and the majority of these are similar to the general oxidation products reported for other disinfectants. For example, field studies with ozone and chlorine dioxide showed that both produced about the same level of low-molecular weight carboxylic acids (Griffini and Iozzelli, 1996). Nevertheless, there

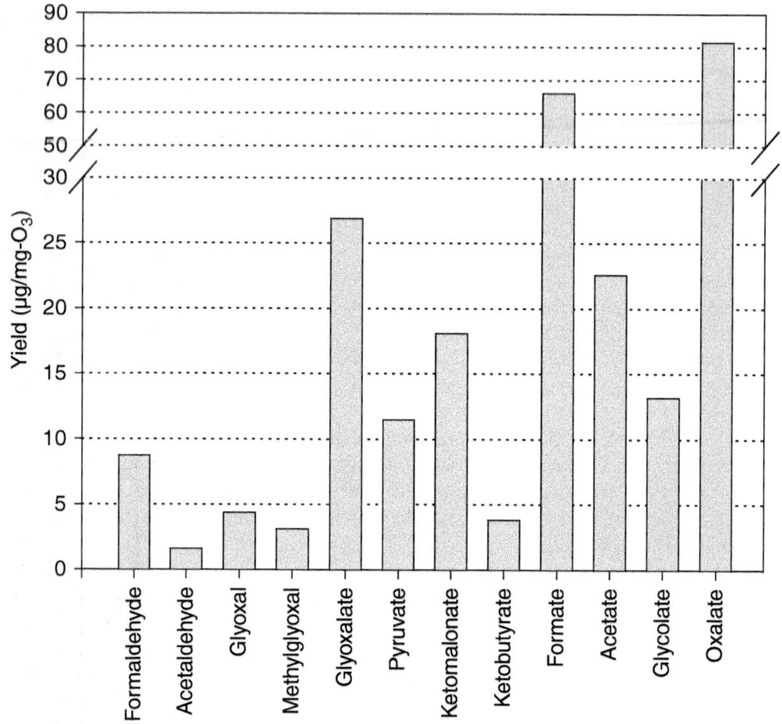

FIGURE 19-7 Yields of major ozonation by-products normalized per milligram of ozone applied, as reported in the literature. (*Source: Reckhow, 1999, Awwa Research Foundation.*)

are a number of studies that suggest ozone produces higher levels of simple aldehydes and ketoacids (or aldoacids) than the other major disinfectants. Figure 19-7 presents some dose-specific yield data for these major by-products. High levels of aldehyde and ketone by-products are characteristic of ozonation and generally are attributed to the Creigee ozonolysis mechanism (Fig. 19-8).

Other oxidation by-products attributed to ozone include hydroxyacids (e.g., pyruvic acid), aromatic acids (e.g,. benzoic acid), and hydroxyaromatics (e.g., catechol) (see Table 19-1). Organic peroxides and epoxides are also expected ozonation by-products, although their detection in treated drinking waters has proved to be a challenge.

Because of the strong correlation between ozonation by-product formation and biodegradable organic matter (BOM) concentrations, attempts have been made to attribute this BOM to specific compounds. Krasner and colleagues (1996) have shown that as much as 40 percent of the assimilable organic carbon (AOC) and 20 percent of the BDOC can be assigned to known major ozonation by-products. However, these known by-products still represent less than 5 percent of the overall DOC concentration.

UV By-Products. Given the recent growth of ultraviolet (UV) disinfection technology in drinking water treatment (see Chaps. 17 and 18), questions have arisen as to how UV treatment might affect finished water DBP levels. Many also have asked whether UV irradiation produces unique DBPs. While it is widely acknowledged that THMs and HAAs do not

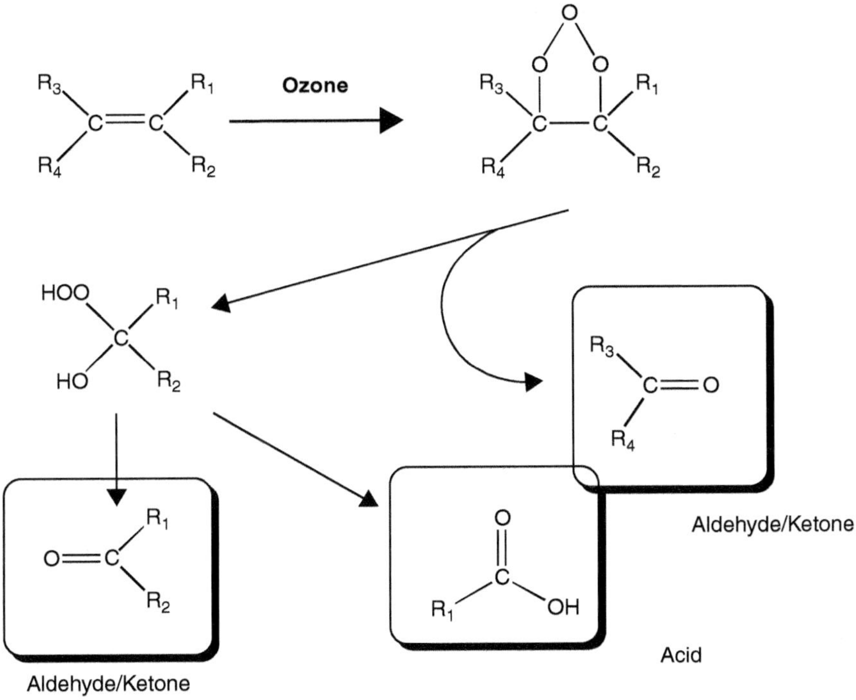

FIGURE 19-8 Creigee ozonolysis mechanism, which results in the formation of acids, aldehydes, and ketones. (*Source: Reckhow, 1999, Awwa Research Foundation.*)

form from UV irradiation, several researchers have shown that low levels of aldehydes and carboxylic acids can form from UV irradiation (Malley et al., 1995; Peldszus et al., 2000; Thomson et al., 2002; Liu et al., 2002).

Most studies also have shown that UV irradiation at doses typical of drinking water disinfection (i.e., ≤40 mJ/cm^2) does not significantly affect THM and HAA formation from subsequent chlorination and chloramination (Malley et al., 1995; Liu et al., 2002; Reckhow et al., 2010). Nevertheless, there are indications that UV treatment can alter the structure and reactivity of NOM in measurable ways (e.g., the formation of aldehydes and acids, as mentioned previously). Others have studied the overall effect of UV irradiation on organic matter from treated surface waters using high-performance size exclusion chromatography (Frimmel, 1998; Frimmel et al., 2005) and electrospray mass spectrometry (Magnuson et al., 2002). The results suggested that the molecular size distribution of NOM shifts toward smaller molecules after UV irradiation at levels typical of drinking water disinfection. Changes in NOM size and functional group content would be expected to have an impact on its reactivity with chlorine. Most studies to date on UV/Cl_2 have focused on the two major regulated groups, the THMs and the HAAs. Far less has been done with other known DBPs. However, Reckhow and colleagues (2010) have shown recently that medium-pressure UV can cause elevated chloropicrin concentrations during subsequent chlorination. These authors have attributed this effect to formation of nitroorganic precursors by photonitration reactions.

Factors Influencing By-Product Formation

Time. Reaction time is among the most important factors determining DBP concentrations under conditions where a disinfectant residual persists. Because the major halogenated DBPs (e.g., THMs and HAAs) are chemically stable, their concentrations typically increase with reaction time for as long as a disinfectant residual exists (Fig. 19-9). However, there are cases where HAA concentrations drop to near zero after long residence times in drinking water distribution systems. This phenomenon is generally attributed to biodegradation, which does not appear to occur with the THMs (the latter appear to be biodegradable only under anoxic conditions).

Laboratory tests have shown that HAAs form somewhat more rapidly than THMs. Studies also have shown the brominated analogues to form more rapidly than the purely chlorinated compounds. This causes the HAA-THM or the THM-chloroform ratio to be high in the early stages of the reaction and drop slowly with reaction time. These observations are supported by data collected from full-scale treatment plants and distribution systems.

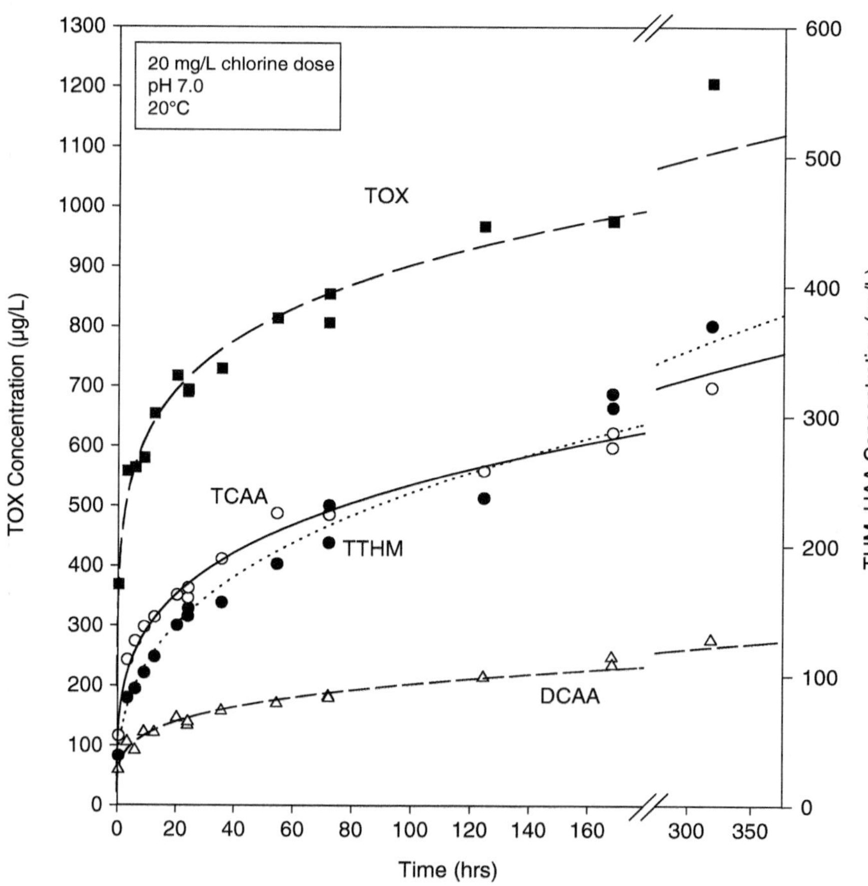

FIGURE 19-9 Effect of reaction time on the major chlorination by-products. (*Data from Reckhow and Singer, 1984.*)

In contrast to the case for chlorine, ozonation by-products form quickly and show little increase with time. This is due to the rapid dissipation of ozone residuals. Once dissolved ozone is depleted, by-product formation can continue only by means of hydrolysis reactions, which represent a relatively minor contribution to the total formation of these by-products.

In contrast to the THMs and HAAs, many other halogenated DBPs are chemically unstable and are subject to hydrolysis or further oxidation. For these compounds, the concentration reaches a maximum value, after which the concentration declines with time (see discussion on pH effects below). Some DBPs, such as dichloroacetonitrile, decompose slowly and reach a maximum concentration after reaction times on the order of days. Others (e.g., 1,1-dichloropropanone) are more reactive and decrease to undetectable levels within minutes to hours though hydrolysis or halogenation pathways.

Disinfectant Dose and Residual. Disinfectant dose has a variable impact on DBP formation. Small changes in the dose used for residual disinfection often have minor effects on DBP formation. This is so because these systems have an excess of disinfectant, and therefore, the DBP formation reaction is precursor limited.

When the disinfectant residual drops below about 0.3 mg/L, DBP formation becomes disinfectant limited. Under these circumstances, changes in disinfectant dose have a large effect. Figure 19-10 shows that when 3 and 5 mg/L of chlorine is added to Connecticut River water, THMs are produced in direct proportion to that dose. However, when an excess is added (>5 mg/L), a residual persists, and the extent of THM formation with increasing dose levels off. The other two waters in this figure show only the latter behavior because they had lower TOC values and their chlorine demand was less than the minimum dose tested (3 mg/L).

As a general rule, disinfectant dose plays a greater role in DBP formation during primary disinfection than during secondary disinfection. This is so because primary disinfectants usually are added in amounts well below the long-term demand. Therefore, the disinfectant is the limiting reactant, not the organic precursors. Figure 19-10 shows a near-linear

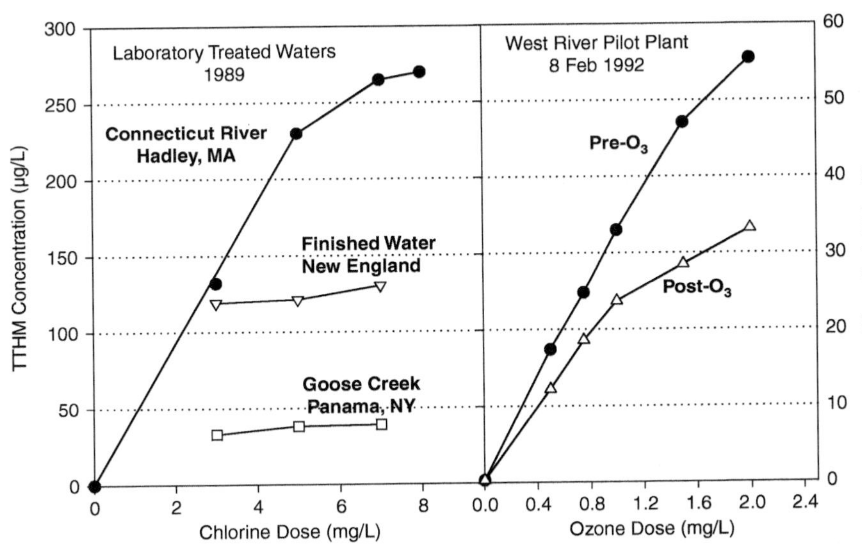

FIGURE 19-10 Observations on effect of disinfectant dose on DBP formation (chlorination conditions: pH 7, 20°C, 3-day reaction time). (*Figures redrawn from Coombs, 1990, and Reckhow et al., 1993.*)

relationship between ozone dose and glyoxalic acid formation. When the ozone is applied after coagulation and filtration (postozone in the figure), glyoxalic acid formation starts to plateau. This is a reflection of the removal of ozone-demanding organics and the approach to precursor-limiting conditions.

The relationship between disinfectant dose and DBP formation can be illustrated with a simple kinetic model. Figure 19-11 shows the results of a kinetic simulation for the simple second-order reaction (Eq. 19-1):

$$A + B \xrightarrow{k} C \tag{19-1}$$

where A represents the NOM precursor material, B is the disinfectant, C is the particular DBP, and k is the second-order rate constant. If the initial precursor level (A_0) is held constant, and a rate constant and reaction time t is arbitrarily chosen, the extent of DBP formation can be calculated as a function of the disinfectant dose (B_0). The three curves in Fig. 19-11 are the result of this calculation for three different combinations of k and t. As

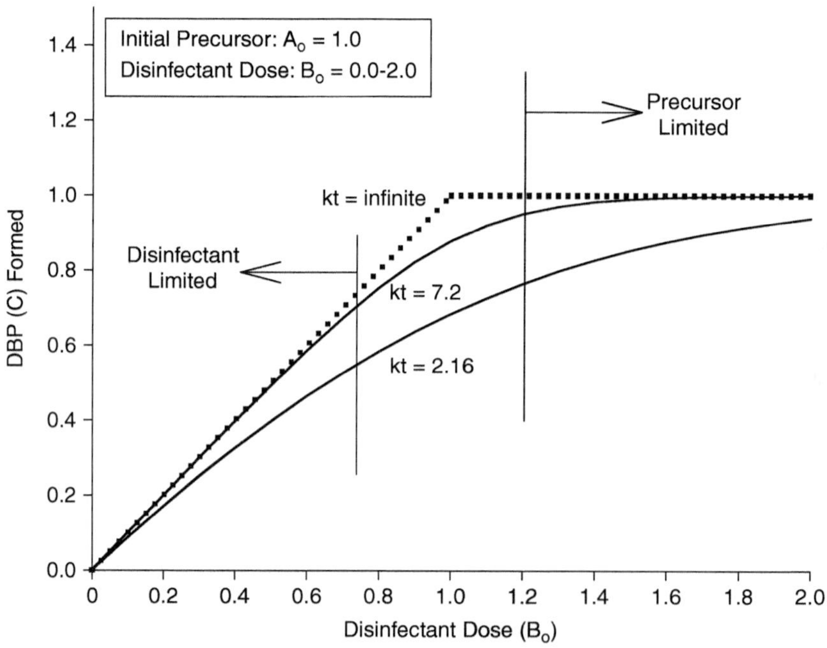

FIGURE 19-11 Theoretical second-order kinetic plot showing effect of reactant dose on product formation.

the product of k and t gets large, the DBP formation curve approaches two straight lines (one where DBP formation increases linearly with dose and one where DBP formation is independent of dose). The upward-=sloping line corresponds to the disinfectant-limiting case. The horizontal line corresponds to the precursor-limiting case. This model is consistent with the experimental observations in Fig. 19-10.

When using chloramines, the chlorine-nitrogen ratio is also an important consideration. As the Cl_2:N ratio increases from 2 to 5 (on a weight basis), the presence of transient free chlorine residuals increases substantially. For example, at a Cl_2:N ratio of 2 mg/mg and a chlorine dose of 2 mg/L (pH 7, 10°C), the measureable free chlorine residual should disappear (i.e., drop below 10 μg/L) at about 8 seconds after initial mixing, whereas it takes about 4 minutes at a Cl_2:N ratio of 5 mg/mg (calculation based on data from Morris and Isaac, 1983). This results in a shift in the DBP character (yields and types) from that typical of pure chloramines to that of free chlorine. Model compound studies (Merlet, 1986) have shown that highly reactive precursors (e.g., resorcinol) will experience this shift at lower Cl_2:N ratios than the slower-reacting compounds (e.g., acetone). It is likely that the same could be said for waters with highly reactive NOM versus less reactive waters. This observation regarding transient free chlorine residuals attests to the importance of mixing when ammonia is added to convert a free chlorine residual to a combined residual.

pH. The overall reaction between chlorine and NOM is relatively insensitive to pH over the range of typical water treatment practice. However, the formation of TOX and specific halogenated by-products is strongly influenced by pH (e.g., Fleischacker and Randtke, 1983; Reckhow and Singer, 1984). Nearly all the by-products (e.g., trihaloacetic acids, halopropanones, haloacetonitriles, etc.) decrease in concentration with increasing pH, but one important exception is the THMs. Although pH can influence chlorination reactions in many ways, it is probably base-catalyzed hydrolysis mechanisms that have the greatest overall effect. Many DBPs (e.g., 1,1,1-trichloropropanone) are decomposed by base-catalyzed hydrolysis. These compounds are less prevalent in waters with high pH, and they tend to decrease in concentration at long residence times. The THMs increase at high pH because many hydrolysis reactions actually lead to THM formation. Other by-products are themselves unaffected by base hydrolysis (e.g., the HAAs), but their formation pathways may be altered at high pH, resulting in lesser formation. It is important to note that the dihaloacetic acids have a different pH dependency than the trihaloacetic acids. Their pathway of formation is different, and dihaloacetic acid formation tends to be relatively independent of pH, whereas trihaloacetic acid formation, as indicated earlier, decreases with increasing pH.

Chloramines exhibit some unique pH behavior that ultimately affects DBP formation from these disinfectants. pH values below 7 tend to favor the formation of dichloramine over monochloramine, especially at chlorine-nitrogen values near the breakpoint (Palin, 1975). This may affect DBP formation because monochloramine and dichloramine react differently with precursor materials. For example, the NDMA formation pathway in Fig. 19-6 suggests that conversion of monochloramine to dichloramine should increase the formation of this DBP. Also, chloramine decay is acid catalyzed, so the total residual drops rapidly at pH values of 6 and below. If not compensated with higher doses, the resulting loss of residual could lead to suspension of DBP-forming reactions and growth of DBP-degrading organisms.

None of the major ozonation by-products are subject to alkaline hydrolysis. Instead, pH plays a role by altering the rate of decomposition of ozone to hydroxyl radicals (see Chap. 7). As pH increases, ozone decomposition accelerates, and this is thought to be responsible for a decrease in the classical by-products of ozonation (e.g., aldehydes) (Schechter and Singer, 1995). However, the chemistry of ozone and hydroxyl radical reactions with NOM is not well understood, and there may be cases where some carbonyl by-products increase with pH (e.g., Itoh and Matsuoka, 1996).

Bromate is formed in ozonated waters from a series of reactions between ozone and/or hydroxyl radicals and naturally occurring bromide (see Chap. 7). A key intermediate in the formation of bromate is hypobromite ion (OBr^-). At lower pH, more of the hypobromite becomes protonated, forming hypobromous acid (HOBr), which is less reactive with respect to bromate formation. This causes the overall concentration of bromate to decrease as pH is decreased. On the other hand, lower pH favors the formation of TOBr during ozonation (Song, 1997). This is probably due to suppressed decomposition of ozone, changes in the speciation of the oxidized bromine favoring HOBr, and less base-catalyzed hydrolysis of brominated by-products.

Temperature. The rate of formation of DBPs generally increases with increasing temperature. Laboratory and full-scale plant data suggest that chloroform formation is more sensitive to temperature than DCAA formation. The relationship is not as clear for TCAA formation. At high temperatures, accelerated depletion of residual chlorine slows DBP formation even though the rate constants for DBP-producing reactions might increase. This may be especially true for TCAA because its formation is more sensitive to chlorine residual than is chloroform or DCAA formation. This is another example that suggests that dihaloacetic acid chemistry is quite different from trihaloacetic acid chemistry. Nevertheless, because more chlorine typically is added to compensate for the more rapid depletion of residual chlorine at high temperatures, overall DBP formation increases with increasing temperature.

High temperatures also may accelerate DBP degradation processes. Biodegradation of HAAs (see below) would be expected to proceed more quickly at high temperature. Reactive DBPs, such as the haloacetonitriles and halopropanones, would undergo abiotic reactions with bases or active chlorine at a faster rate as temperature increases. For this reason, increases in temperature actually may cause a decrease in the concentration of certain DBPs.

Bromide and Iodide. Bromide is readily oxidized by aqueous chlorine and ozone to form hypobromous acid (see Chap. 7). The hypobromous acid reacts with NOM to form brominated DBPs. In the presence of residual free chlorine, mixed chloro-bromo DBPs are formed. As bromide levels increase, formation of the more heavily brominated DBPs increases. Figure 19-12 shows how increasing bromide concentration causes shifts in the THM speciation from chloroform to progressively more brominated members of the group.

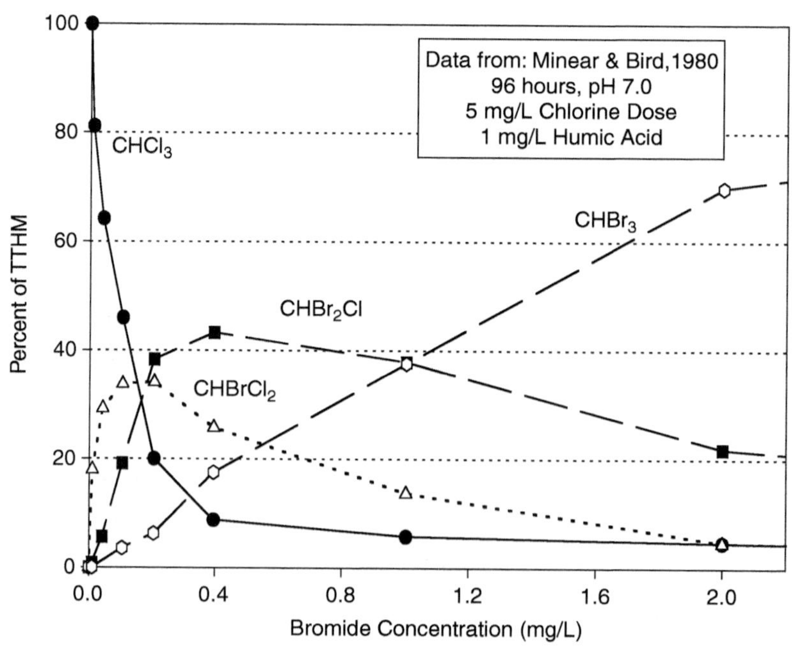

FIGURE 19-12 The effect of bromide concentration on THM speciation. (*Redrawn from Minear and Bird, 1980.*)

A similar effect has been observed for the HAAs, halopicrins, halopropanones, haloacetaldehydes, and haloacetonitriles (e.g., Trehy and Bieber, 1981; Pourmoghaddas and Dressman, 1993; Xie and Reckhow, 1993; Cowman and Singer, 1996). Although Fig. 19-12 extends to high bromide levels rarely seen in practice, Fig. 19-13 presents some actual occurrence data on bromine content of THMs. Note that both indicate a bromide concentration of just over 100 μg/L as the point where the THMs become more heavily populated with bromine atoms. Figure 19-13 also suggests that this characteristic bromide level drops

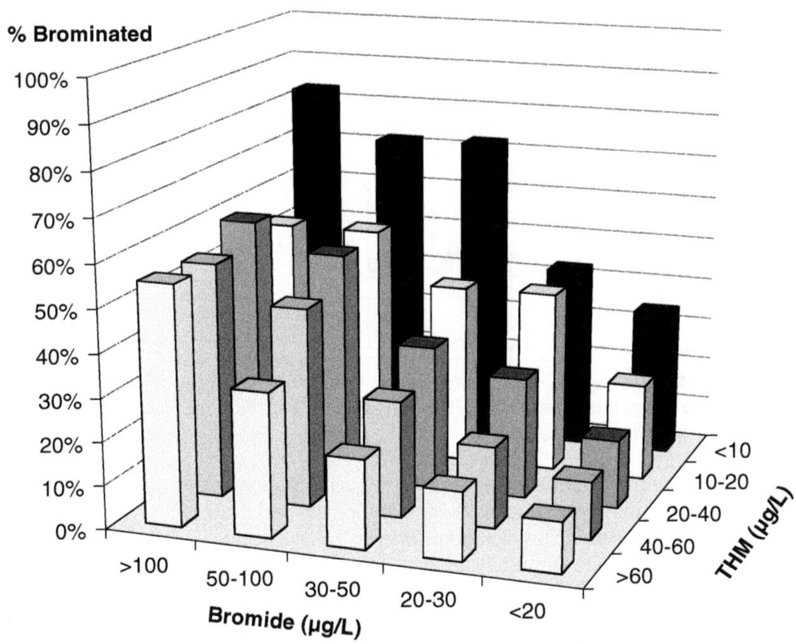

FIGURE 19-13 Dependence of bromine incorporation among trihalomethane species on bromide concentration in ICR waters. (*Source: McGuire et al., 2002, Awwa Research Foundation.*)

considerably as the THM level, and therefore the precursor level, drops. Most laboratory studies also have shown that the molar formation of THMs increases slightly with bromide concentration. Since the brominated forms are heavier than their chlorinated analogues, the mass-based THM level (e.g., micrograms per liter) increases even more sharply with increasing bromide level. The formation of nonhalogenated DBPs is probably insensitive to varying bromide levels. One study of ozonation by-products showed only minor effects of bromide on aldehyde concentrations (Schechter and Singer, 1995).

Bromine (and iodine) incorporation among species in different DBP classes occurs in a predictable fashion. Cowman and Singer (1996) presented a simple competitive model that results in a binomial distribution function describing the relative concentrations of the different bromine-containing species. This model has been shown to accurately reflect actual full-scale THM concentrations (Fig. 9-14). The x axis in this figure represents the mole ratio of bromine to total halogen averaged over all THM molecules in any particular sample [sometimes called the *bromine substitution factor* (BSF)]. This is related to the *bromine incorporation factor* (BIF), which is the average number of bromine atoms per molecule. Plant-scale data from the ICR database show that the extent of bromine

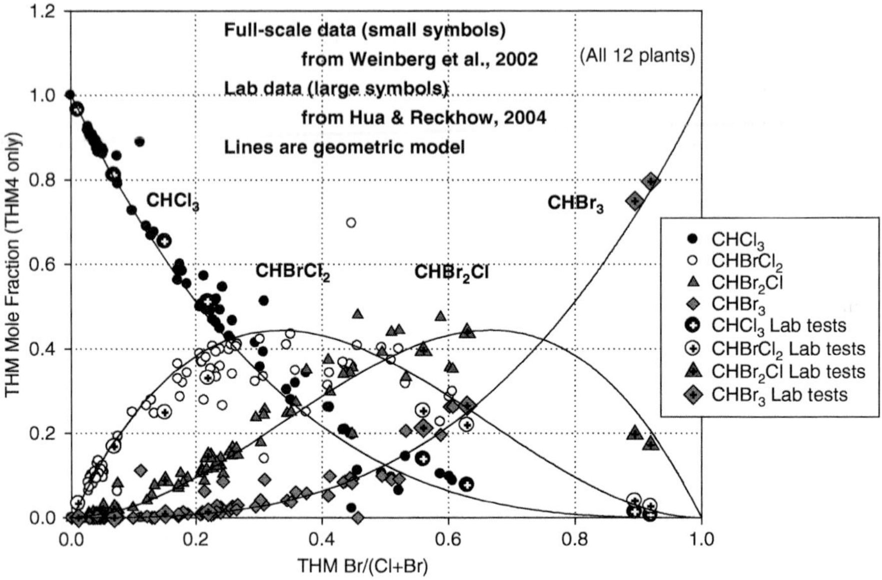

FIGURE 19-14 Relationship between THM species and degree of bromine incorporation. (The *x* axis is the molar ratio of bromine incorporated in the THM species to the total THM concentration.) (*The geometric model is after Cowman and Singer, 1996.*)

incorporation among the dihaloacetic and trihaloacetic acid species closely matches bromine incorporation among the THM species (Obolensky and Singer, 2005). In fact, this parallelism has allowed researchers to predict overall HAA formation (all nine bromine- and chlorine-containing HAAs) from measured values of the four THMs and the five regulated HAAs (Roberts et al., 2002).

The impact of iodide on DBP speciation is qualitatively similar to that of bromide. As inorganic iodide levels increase, the formation of iodinated DBPs is greater, as is the incorporation of iodide into the THMs. Because iodine is more volatile than chlorine or bromine, continental waters (i.e., surface waters and groundwaters that have not been affected by salt water) tend to be enriched in iodide compared with seawater. Nevertheless, iodide levels are thought to be roughly correlated with bromide levels. As a result, Weinberg and colleagues (2002) showed about 10 percent molar ratio of I:Br in THM species.

Organic Precursor Material. It is well established that halogenated DBP formation is strongly correlated with the DOC concentration of the water being chlorinated. For this reason, many calculations regarding DBP formation are based on specific yields normalized to a fixed organic carbon level (e.g., micrograms of chloroform per milligram of DOC). While the relationship between TOC or DOC and DBP formation may be quite strong for selected waters, the correlation becomes much weaker when considering waters of different character, originating from different ecoregions (i.e., geographic areas differing in hydrology, landforms, soil types, and vegetation), different climatic conditions, and different times of the year. The reasons for this can be seen in model compound studies. Common biological molecules such as carbohydrates and sugars are relatively poor DBP precursors, whereas tannins and lignins produce large amounts of measurable by-products on a per-carbon basis. Waters collected from different locations and at different times of the year will have distinctly different mixtures of these biomolecules. As a result, the specific DBP yields can vary substantially between natural waters.

Chlorination studies with NOM extracts and whole waters have suggested that the activated aromatic content of the organic material is an important determinant in its tendency to form major chlorination by-products (Reckhow et al., 1990). This probably explains why UV absorbance is such a good predictor of a water's tendency to form THMs and HAAs because UV absorbance by NOM generally is attributed to activated aromatic chromophores. Accordingly, UV absorbance is a better predictor of chlorine reactivity and halogenated DBP formation potential than overall DOC. See Chap. 3 for coverage of UV absorbance as a surrogate parameter for TOC/DOC and DBP precursors.

Lignin is a major constituent in terrestrial vascular plants. It is quite resistant to biodegradation yet reactive with oxidants owing to a high density of activated aromatic rings. This combination of high terrestrial production, persistence, and high level of reactivity makes lignin an extremely important source of aquatic NOM and DBP precursors.

Perdue and Ritchie (2004) have summarized five literature sources and found a mean of 1 percent of total DOC attributable to phenolic lignin monomers. This is probably an underestimate because direct analysis of commercially derived lignin results in low recoveries of the monomeric compounds (Kim and Reckhow, 2009). If recoveries with commercial lignin and aquatic NOM are assumed to be comparable, aquatic fulvic acid could be in the range of 15 to 50 percent lignin derived.

Terpenoids are hydrocarbons produced in large quantities by a wide range of plants, both terrestrial and aquatic. There are indications that these compounds, which exist in various linear, branched, and cyclic forms, may account for a major fraction of the NOM in fresh waters (Leenheer et al., 2003; Lam et al., 2007). In particular, they may be dominant in waters of low aromaticity. These compounds are not likely to be as reactive as the activated aromatic structures in lignin. However, they certainly will react with oxidants by virtue of their abundant olefinic bonds. Given their presumed abundance and likely reactivity, it seems that the terpeniods could play a major role in oxidant demand and by-product formation in many types of waters.

General NOM reactivity often is expressed as some measure of the tendency to form major organic by-products. For chlorination, this means formation of THMs and HAAs, among others, and it is commonly done by means of an empirical laboratory chlorination test. To assess DBP formation in a valid reproducible fashion requires the use of a standard set of chlorination conditions. Standardized conditions facilitate the comparison of data from different investigators and allow the evaluation of precursor characteristics of different types of raw water sources at different times of the year. Since the 1970s, researchers have used a wide range of chlorination protocols, some calling for fixed chlorine doses and high residuals, some using fixed chlorine dose–TOC ratios, and some aiming for fixed chlorine residuals that are more representative of real systems. The former frequently has been called a *formation potential* (FP) *test*, and it is intended to maintain a relatively constant $C \times T$ (concentration times time) product in the face of changing chlorine demands. While each has its own advantages, the lack of standardization has hindered direct comparisons.

In an effort to solve this problem, many researchers are now using a single low-dose precursor test protocol called the *uniform formation conditions* (UFC) *test*. Summers and colleagues (1996), based on a meeting with several DBP researchers, developed this test, in which a sufficient amount of chlorine is applied to the water to produce a 1 mg/L free chlorine residual after 24 hours of contact at pH 8.0 and 20°C This protocol is especially useful for assessing the effectiveness of various types of treatments on DBP precursor removal while using conditions that are typical of many systems. Because the chlorine doses are typical of those used in practice (on the order of 1.5–2 mg Cl_2 per milligram of C) (Obolensky et al., 2007), the distribution among bromine- and chlorine-containing DBP species is not distorted by use of an excessive chlorine dose. A related protocol using a 7-day reaction time at 25°C, pH 7.0, and a target chlorine residual of 4 mg/L also has been widely endorsed (APHA et al., 2005).

The preceding tests are distinguishable from the *simulated distribution system* (SDS) *test*, in which treated water is held under the specific conditions (pH, temperature, and chlorine residual) of distribution (e.g., APHA et al., 2005). The SDS test is useful to predict the extent of THM and HAA formation in a given water at various distribution system water ages. Comparisons between SDS tests and actual distribution system measurements made as part of the ICR effort were found to be good (McGuire et al., 2002). Although useful for predictive purposes in a specific system, the SDS test cannot be used in a more generic sense to predict DBP formation at conditions other than those used in the actual test.

Several researchers have examined THM precursor levels across a broad geographic distribution of sites. One of the largest of these studies encompassed 133 data points from five literature sources (Chapra et al., 1997). They concluded that the THM formation potential (THMFP) of this broad spectrum of natural waters increases with TOC in a nearly linear fashion. Reckhow and colleagues (2007) have reanalyzed these data, correcting for differences in chlorination conditions, and they found almost no relationship between THMFP and TOC. Not surprisingly, laboratory tests have shown that THM formation from dilutions of a single sample of NOM is constant when expressed on a per milligram of carbon basis. This is only true when the conditions are held reasonably constant, including the profile of chlorine residual versus time. For this reason, a normalized THM yield, defined as the specific THM formation under a UFC protocol (specific THM–UFC), has been used by Reckhow and colleagues (2007) as the characteristic measure of THM yield. This value is the concentration of THMs formed from reaction under the uniform formation conditions divided by the initial TOC of the water. Analysis of the preceding data by Chapra eand colleagues (1997) show a regressed value of 21.5 μg/mg of C at 1 mg/L TOC, which increases only slightly with TOC (Eq. 19-2). The yield is expected to change for different conditions of temperature, pH, and contact time, as discussed below.

$$\text{Specific THM}-\text{UFC} \equiv \frac{\text{THM}-\text{UFC}}{\text{TOC}} = 21.5(\text{TOC})^{0.095} \tag{19-2}$$

There are many other good databases that can be incorporated into a statistical analysis of THM formation. Reckhow and colleagues (2007) have compiled 27 data sets from the open literature and critically assessed the THM formation potential under the standardized specific conditions cited earlier in connection with Eq. 19-2. Within these 27 papers and reports were hundreds of single water assessments. A visually direct way of presenting the data is through a cumulative frequency plot, as found in Fig. 19-15. This reinforces the belief that groundwaters generally have lower specific formation potentials than surface waters, albeit not that much lower. In addition, one can extract median values and population percentiles from these data. Note that the median THM yield for surface waters is about 24 μg/mg of C under the standardized conditions.

Figure 19-15 illustrates clearly that even when normalized to the organic carbon level and subject to the same set of reaction conditions, there still remains a substantial range in THM formation from one NOM sample to the next. The same has been observed for dihaloacetic acid and trihaloacetic acid formation (Reckhow et al., 2007). Some of this variability can be captured with the specific UV absorbance, as will be shown below.

In an effort to better understand the extent of DBP formation from NOM (DOC) in different types of waters and at different times of the year, many investigators have resorted to methods of fractionating the DOC into different subgroups. The most common and one of the simplest of these fractionation methods uses a series of macroreticular resins with different degrees of hydrophobicity and is based on an extraction procedure by Thurman and Malcolm (1981). Most often the organic material is fractionated into hydrophobic acids, transphilic acids, and hydrophilic constituents. Chlorination of the different fractions tends to show that, in most waters, the THM and HAA yield on a per milligram

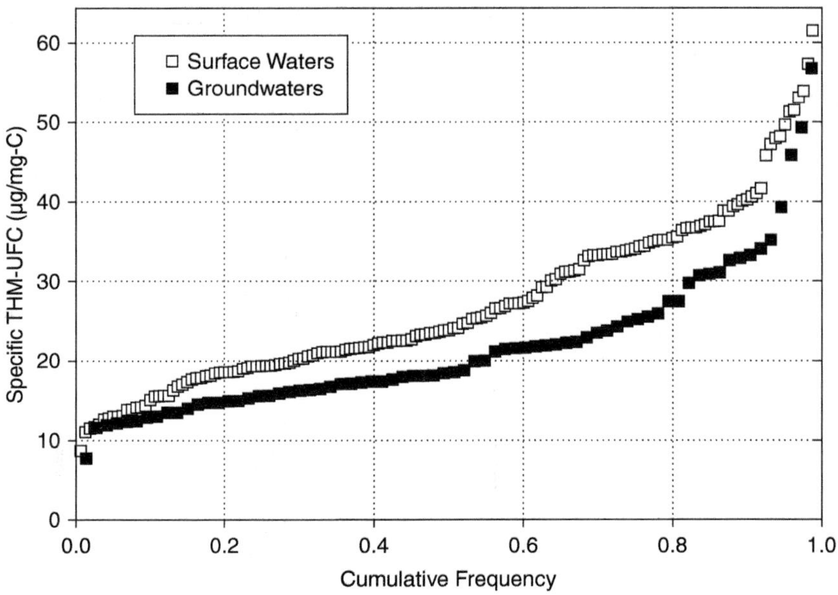

FIGURE 19-15 Specific precursor content in North American waters. (*Source: Reckhow et al., 2007, Awwa Research Foundation.*)

of C basis is greatest for the hydrophobic acid fraction (e.g., Liang and Singer, 2003). More recent studies have focused on fractionating dissolved organic nitrogen (DON) and assessing the formation of nitrogenous DBPs from these fractions (Leenheer et al., 2007; Dotson et al., 2009).

Another common, simple metric for characterizing DOC in waters with respect to its DBP formation potential is UV absorbance, with units of cm^{-1}. As indicated earlier, UV absorbance measurements reflect the activated aromatic content of organic carbon, and it is these activated aromatic structures that are dominant contributors to THM and HAA formation. While UV absorbance reflects the bulk concentration of precursors in a water (Fig. 19-16), the nature and reactivity of the precursors can be assessed by normalizing the UV absorbance, dividing it by the DOC concentration, and multiplying that number by 100 cm/m. The resulting specific UV absorbance (SUVA, in units of L/mg · m) is a good indicator of the specific DBP formation potential of a water, as shown in Fig. 19-17 for three independent data sets. See Chap. 3 for additional coverage of the use of UV absorbance as a surrogate parameter for DOC and DBP precursors and coverage of SUVA.

Seasonal Effects. DBP concentrations are strongly influenced by season. The underlying factors are temperature and seasonally induced changes in water quality (NOM concentrations and characteristics, bromide, pH). In general, DBP concentrations are highest in summer. This is at least partly attributed to high temperatures, which accelerate DBP-forming reactions. Furthermore, disinfectant demand reactions are faster, requiring higher disinfectant doses to maintain target residuals. Competing with this are the facts that chemical degradation of selected DBPs (see below) is faster at high temperatures, and biodegradation may play a larger role in warmer waters.

With some sources, water quality changes also may be important. For example, in many watersheds, heavy spring rains cause allochthonous (terrestrial-derived) NOM, which has

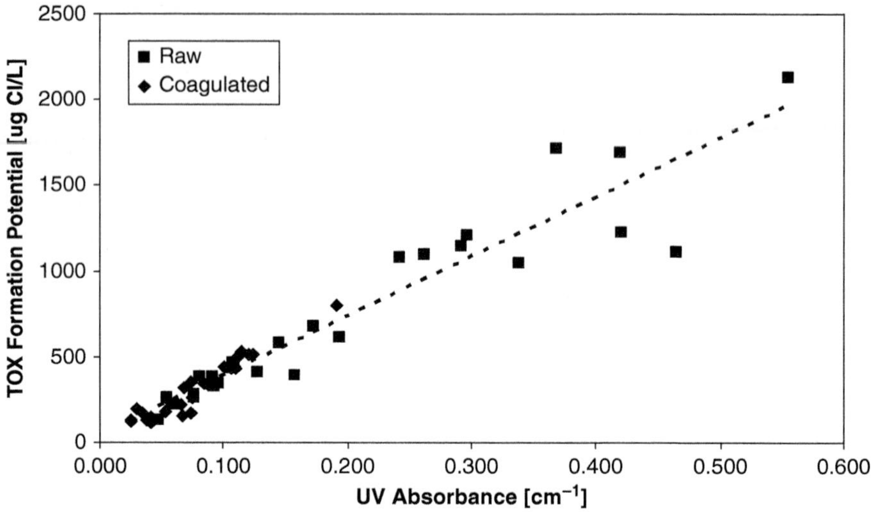

FIGURE 19-16 Relationship between total organic halide formation potential and UV absorbance for 27 different raw and coagulated waters from across the United States. (*Source: Archer and Singer, 2006.*)

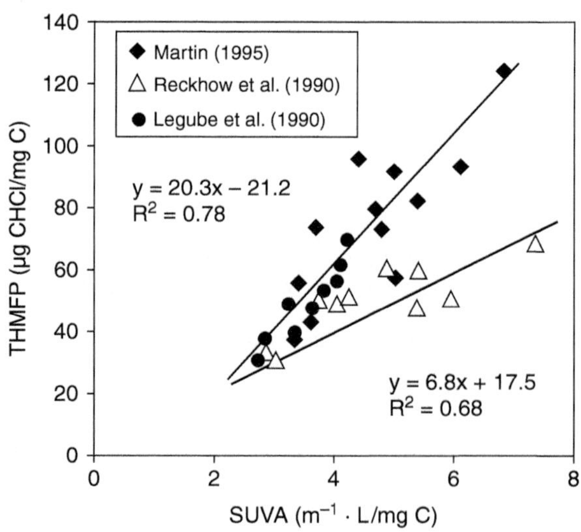

FIGURE 19-17 Correlations for normalized THM formation potential versus SUVA. (*Source: Croué et al., 1999.*)

a high lignin content and is rich in hydrophobic organic acids, to run off into the water source. In the summer, a larger portion of the NOM is autochthonous and hydrophilic in nature, arising from algal activity. In the fall, decomposed leaf litter washes into source waters. Hence the seasonal DBP signatures resulting from the application of disinfectants are expected to be different.

Likewise, the San Francisco Bay Delta, which provides raw drinking water for a large portion of the population in central and southern California, experiences significant seasonal variations in bromide concentration. During the spring season, high freshwater flows from the Sierra Nevada Mountains keep salt water, with its concomitant high bromide content, out of the delta so that chlorinated DBPs comprise most of the DBP content of the treated drinking water. In contrast, during the summer and early fall when fresh water flows are low, more salt water intrusion into the delta occurs, and bromine-containing species are the dominant DBPs formed.

Disinfection By-Product Formation Models

Since the very earliest work on DBPs, researchers and practitioners have proposed mathematical models to describe the formation of these compounds at different contact times and as a function of treatment conditions (e.g., disinfectant dose, pH, temperature) and treated water quality (e.g., DOC, UV absorbance, bromide). Fully empirical models composed of multiplicative power terms have proven to be useful and robust tools (e.g, Amy et al., 1987). This approach has been used in formulating a more general water treatment model that has been valuable in assessing central tendencies on a regional and national basis (Harrington et al., 1992; Solarik et al., 2000; Swanson et al., 2002). Unfortunately, the complex and diverse nature of naturally occurring DBP precursors requires that these DBP models be partially recalibrated before they can be used on specific systems. Models based on simplified chemical rate laws have been proposed in an effort to improve model performance (e.g., McClellan et al., 2000). These are more computationally intensive, and their development is an ongoing area of research.

CONTROL OF OXIDATION/DISINFECTION BY-PRODUCTS

There are three general approaches to controlling DBP concentrations: (1) minimize DBP formation by reducing the concentration of organic precursor material at the point of disinfection, (2) minimize DBP formation by reducing the disinfectant dose or contact time, changing the nature of the disinfectant or optimizing the conditions of disinfection, and (3) remove DBPs after their formation. Most efforts have focused on approach 1, as represented by optimizing the coagulation process (see Chap. 8) for removal of both TOC and particles (turbidity), and approach 2, reflected by changing from free chlorine to ozone or UV for primary disinfection and from free chlorine to chloramination for secondary disinfection. The third approach (DBP removal) is most appropriate for control of biodegradable by-products, especially those produced by ozonation, and possibly for removal of highly volatile DBPs such as chloroform.

Removing Organic Precursors

The objective of this particular strategy is to minimize the amount of organic precursors at the point of disinfection. This can be done by either reducing the precursor content of the raw water, improving precursor removal through the plant, shifting the point of disinfection to a later stage of treatment after precursors have been removed, or some combination of the three. It has been suggested that watershed management practices that help to reduce primary productivity in impoundments also will result in reduced THM precursor levels (Karimi and Singer, 1991; Chapra et al., 1997). Reductions in organic precursor and bromide levels also may be achieved through careful source water selection, blending, or storage. Once they have entered the plant, DBP precursors may be removed or rendered less active by coagulation, adsorption, anion exchange, oxidation, biodegradation, or membrane separation. Each of these is discussed in the text that follows, with special focus on DBP control. For a broader and more comprehensive discussion of each process, the reader should consult the appropriate chapters earlier in this book.

Source Control. Control of DBP precursors at the source is an option that has received some attention, but with little practical experience to date. Cooke and Carlson (1989) prepared an early summary of established methods for reservoir and watershed management, including the likely impacts on DOC levels. Their ideas were based on a general understanding of allochthonous and autochthonous sources of NOM. Techniques intended to minimize terrestrial inputs (e.g., buffer strips) would control the former, and techniques intended to reduce algal and macrophyte contributions (e.g., application of copper sulfate) would control the latter. If DOC could be reduced, so would the levels of DBP precursors.

Twenty years later, we still know very little about the long-term impacts of watershed management practices on DBP precursor levels. However, we have made progress on one of the most fundamental questions—the relative importance of terrestrial versus aquatic sources of precursors. Water quality investigations and watershed modeling studies across the United States have revealed that some supplies are dominated by terrestrial or allochthonous sources, whereas others are largely algal or autochthonous in nature (e.g., Palmstrom et al., 1988; Canale et al., 1997; Stepczuk et al., 1998; Fuji et al., 1998; Speiran, 2000; Waldron and Bent, 2001; Nguyen et al., 2002; Garvey and Tobiason, 2003; Reckhow et al., 2007).

Aquifer storage and recovery (ASR) has been used as a method of source water control (see Chap. 15). By storing high-quality treated water in the subsurface when raw water quality is good, this stored water later can be withdrawn when the raw water quality is poor, with only a minor degree of additional treatment. Similar raw water storage concepts can be considered in cases where the raw water is affected by high bromide concentrations. Storage of raw (or treated) water during times of the year when bromide levels are low allows utilities to use the stored water when bromide levels are elevated. ASR also may result in lower DOC and reduced DBP precursor levels as a result of long storage times in contact with aquifer minerals and microorganisms.

Oxidation. Pretreatment with a nonhalogenating oxidant can result in a net destruction of precursor sites and thereby reduce the subsequent formation of chlorinated DBPs after final disinfection with chlorine or chloramines. This phenomenon has been most widely studied with ozone, although the effect also has been demonstrated with chlorine dioxide, permanganate, and several advanced oxidation processes. From model compound studies, it has been shown that preozonation most often results in intermediates with lower DBP formation potential, although the reverse does sometimes occur. Since NOM is a broad mix of molecules with different structures and reactivities, it is not surprising that most types of NOM seem to show a downward trend in DBP formation with oxidation, much like the majority of model compound studies. Because of its selectivity for olefinic bonds, molecular ozone appears to be more effective than the less selective hydroxyl radicals. Oxidative precursor destruction is minor for most waters at typical oxidant doses used in water treatment. For example, Reckhow and colleagues (1986) showed that ozone doses of 0.25 to 0.5 mg of O_3 per milligram of C can result in THM precursor destruction of about 5 to 25 percent depending on the concentration of bicarbonate. The role of bicarbonate is to suppress ozone decomposition (by serving as a hydroxyl radical scavenger), thereby enhancing the persistence of molecular ozone and possibly encouraging the formation of bicarbonate radicals.

The tendency of ozone to destroy DBP precursors is greater for systems where chlorine is added at acidic pH. The effect of ozone may disappear entirely or even reverse for systems that chlorinate at pH above 8 (Reckhow et al., 1986). Impacts of ozonation on subsequent trihaloacetic acid formation (e.g., TCAA) generally follow its impact on THMs. In contrast, precursors of dihaloacetic acids (e.g., DCAA) are far less affected by preoxidation.

Chlorine dioxide has been shown to produce similar removals of THM precursors as ozone at the same dose levels (Fig. 19-18).

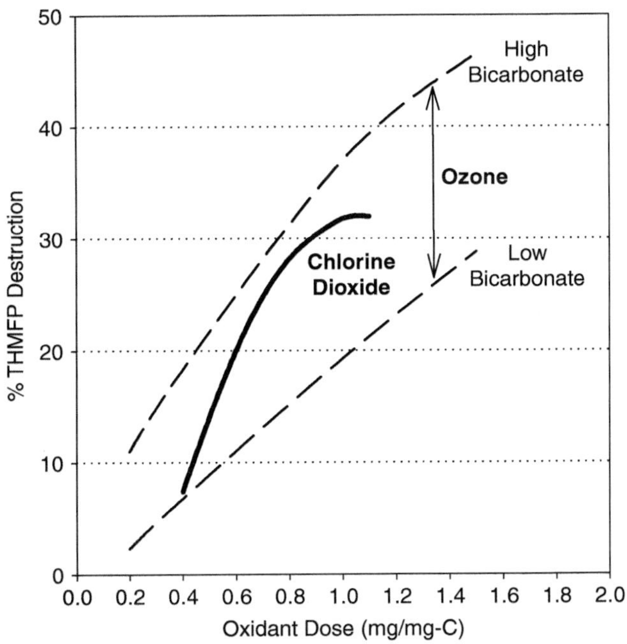

FIGURE 19-18 Comparison between ozone and chlorine dioxide in the destruction of THM precursors. (*Data from Reckhow, 1984; Werdehoff and Singer, 1987.*)

Permanganate also appears to be capable of ooxidizing THM precursors to a limited degree, although effective removal has been observed in hard waters owing to calcium-assisted adsorption of NOM on the manganese oxides produced (Colthurst and Singer, 1982).

Ozonation also has been shown to most effectively destroy the fastest-reacting DBP precursors. This has the effect of making ozonation appear more effective in systems where the chlorine contact time is short. Figure 19-19 shows the relative destruction of TCAA precursors by ozonation when the results are evaluated at different chlorine contact times. Precursor destruction in this case is as high as 75 percent when the chlorine is allowed to react for only 30 minutes. Precursor destruction was only 35 percent when the chlorine contact time was 1 week.

Advanced oxidation processes (AOPs) may be less effective for precursor destruction than simple ozonation. Frimmel and colleagues (2000) examined the use of moderate ozone doses (~1 mg/mg C) with waters of low TOC (1.2–2.3 mg/L) and found that increasing

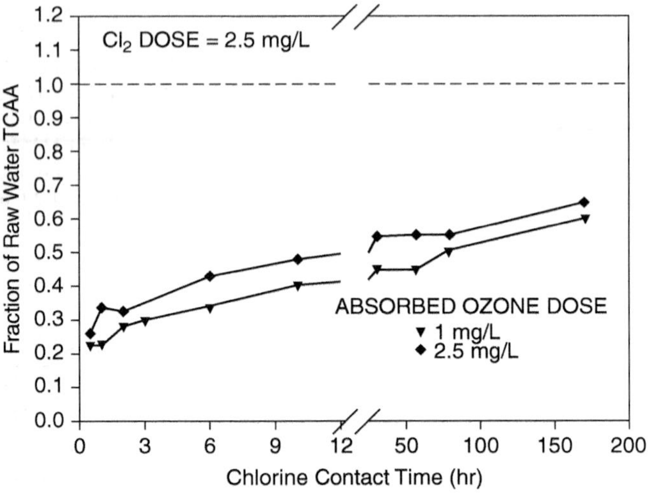

FIGURE 19-19 Destruction of TCAA precursor by ozone, as determined at various chlorine contact times. (*Source: Data from Reckhow et al., 1993.*)

levels of peroxide caused a decrease in the destruction of THM precursors (Fig. 19-20). In related work, Kleiser and Frimmel (2000) explored the impacts of peroxide and UV irradiation on Ruhr River water (2.3 mg/L DOC). They found that UV treatment caused a small increase in THM precursor content even though the UV absorbance was partly destroyed (Fig. 19-21). Parallel experiments with ozone alone showed a monotonic decrease in THM precursor with increasing ozone dose.

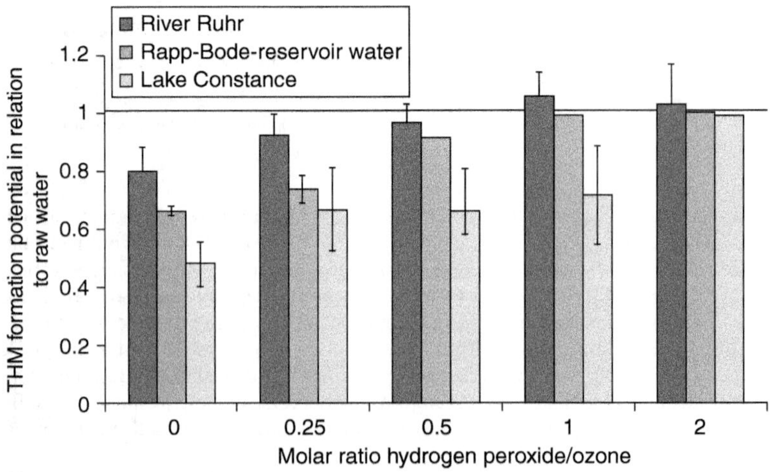

FIGURE 19-20 Impact of peroxide on destruction of THM precursors by moderate doses of ozone. (*Source: Reprinted with permission from Frimmel et al., 2000. Copyright © 2000 American Chemical Society.*)

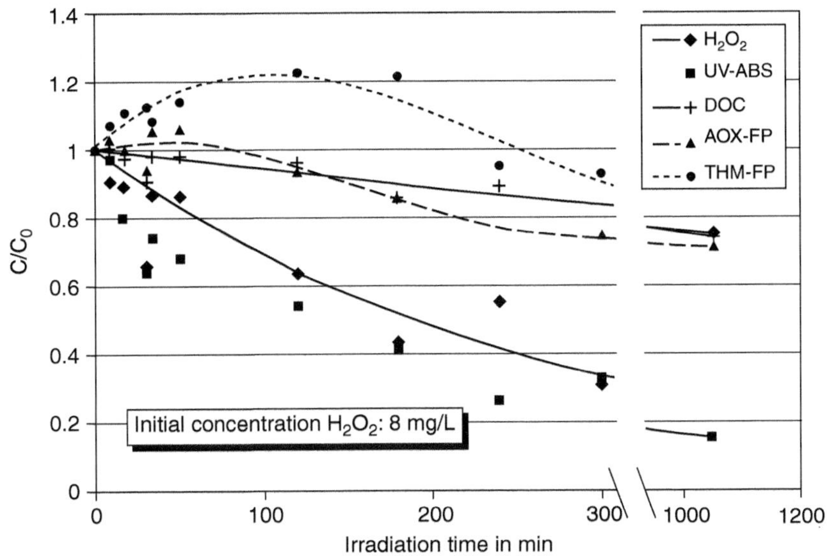

FIGURE 19-21 Impact of UV dose on THM precursor control from combined UV-peroxide treatment. (*Source: Kleiser and Frimmel, 2000.*)

Use of ozone and UV light may present some unique disadvantages, especially with respect to halonitromethane formation. The highest levels of halonitromethanes (e.g., chloropicrin) have been reported for plants with preozonation followed by chlorine/chloramines. It has been proposed that nitration reactions may be partly responsible (e.g., Choi and Richardson, 2004), although there also may be some activation of preexisting N organics. Thibaud and colleagues (1986) have shown that *m*-nitrophenol can produce halonitromethanes at 5 to 50 percent yield depending on pH. UV light from medium-pressure lamps has been shown to enhance halonitromethane formation substantially, probably due to photonitration reactions resulting in the incorporation of ambient nitrate into the NOM (Reckhow et al., 2009).

Coagulation and Clarification. The reader is also referrred to Chap. 8 for additional coverage of this subject. Coagulation with alum or ferric salts generally is quite effective for removal of NOM (e.g., as measured by DOC) and its constituent DBP precursors (Kavanaugh, 1978; Young and Singer, 1979; Babcock and Singer, 1979; Edzwald et al., 1985). In fact, there is a tendency for the humic and fulvic fractions of NOM to be removed more completely during coagulation and subsequent settling or filtration than the more hydrophilic NOM (Liang and Singer, 2003; Archer and Singer, 2006). Since these humic materials bear a higher density of THM, trihaloacetic acid, and TOX precursor sites, this leads to slightly higher removals of THM, trihaloacetic acid, and TOX precursors on a percentage basis than bulk DOC (e.g., see Reckhow et al., 2005; Archer and Singer, 2006) (Fig. 19-22).

There are indications that some DBP precursors are more closely aligned with the hydrophilic fraction of organic carbon. One example is the 1,1,1-trichloropropanone precursors, which are probably associated with aliphatic ketones and related compounds. Figure 19-23 shows that these precursors are less amenable to removal by alum coagulation than the bulk organic carbon.

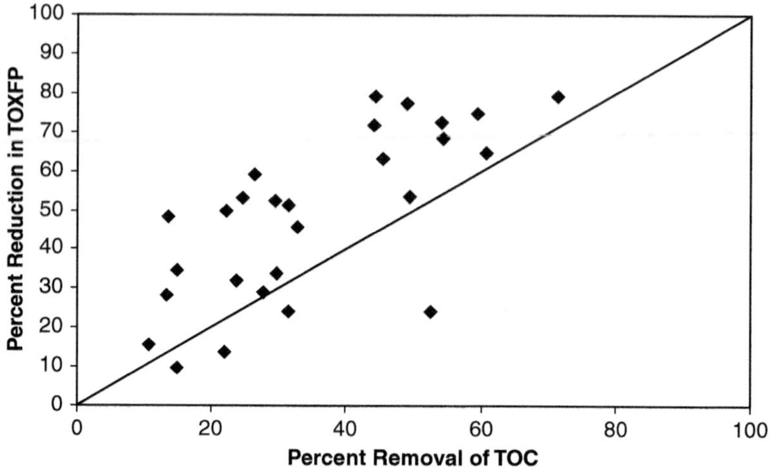

FIGURE 19-22 Comparison of the reduction in total organic halide formation potential and removal of TOC by coagulation for 27 raw waters across the United States. (*Source: Archer and Singer, 2006.*)

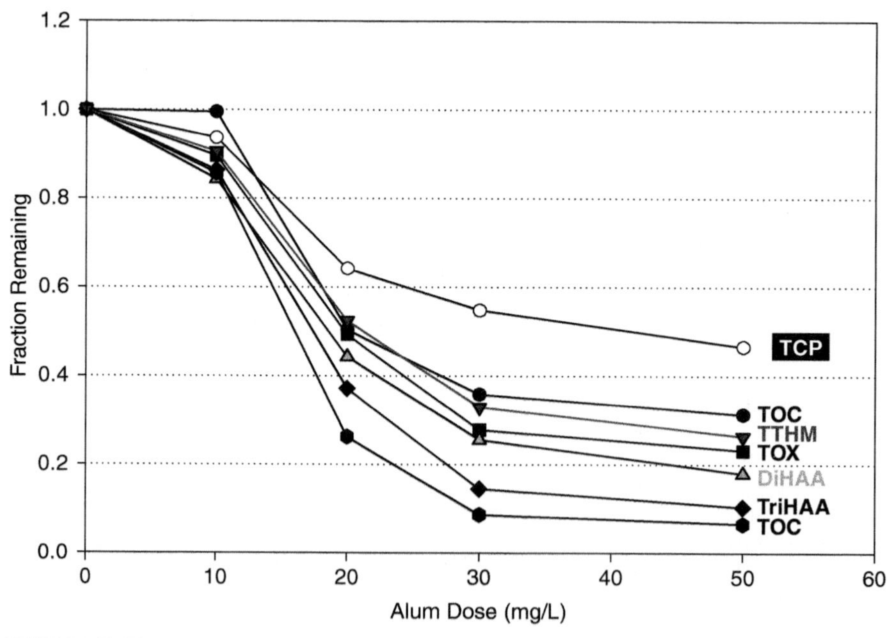

FIGURE 19-23 Removal of NOM and DBP precursors by alum coagulation, settling, and filtration. (*Source: Reckhow and Singer, 1990.*)

The effectiveness of coagulation for the removal of NOM and DBP precursors is the basis for including "enhanced" coagulation as a requirement of the Stage 1 Disinfectants/Disinfection By-Products Rule (see Chaps. 1 and 8). In addition to promulgating specific maximum contaminant levels for THMs and HAAs, the Stage 1 Rule requires a specified degree of TOC removal prior to disinfection. The objective of this requirement is to limit the formation of unidentified DBPs that may have an adverse public health effect. The TOC removals specified depend on the TOC concentration and alkalinity of the raw water (see Chap. 8). In general, raw waters with high SUVA values (and high DBP formation potentials) are more amenable to TOC removal by coagulation than waters with low SUVA values. This is in recognition of the general consensus that raw waters with higher TOC concentrations tend to contain higher concentrations of hydrophobic organic carbon that is more amenable to TOC removal by coagulation than waters with low TOC concentrations.

Ion Exchange. Because much of the NOM that harbors DBP precursor structures contains carboxylic acid functional groups and therefore is anionic in nature, anion exchange has been proposed as a promising technology for precursor control. Most recently, this interest has been directed toward the use of magnetic ion exchange resins (MIEX) of low particle size. Early batch tests suggested that MIEX could remove both hydrophilic and hydrophobic organic acids from water (Boyer and Singer, 2005) and thereby offer some advantages over coagulation, which removes primarily the hydrophobic fractions. Additional studies (Johnson and Singer, 2004; Boyer and Singer, 2006) also showed that MIEX could remove bromide in waters with low levels of competing anions (e.g., bicarbonate and sulfate). These studies also demonstrated THMFP, HAAFP, and DOC removals of 70 percent or more (Fig. 19-24). There are concerns, however, that iron and hydrophobic

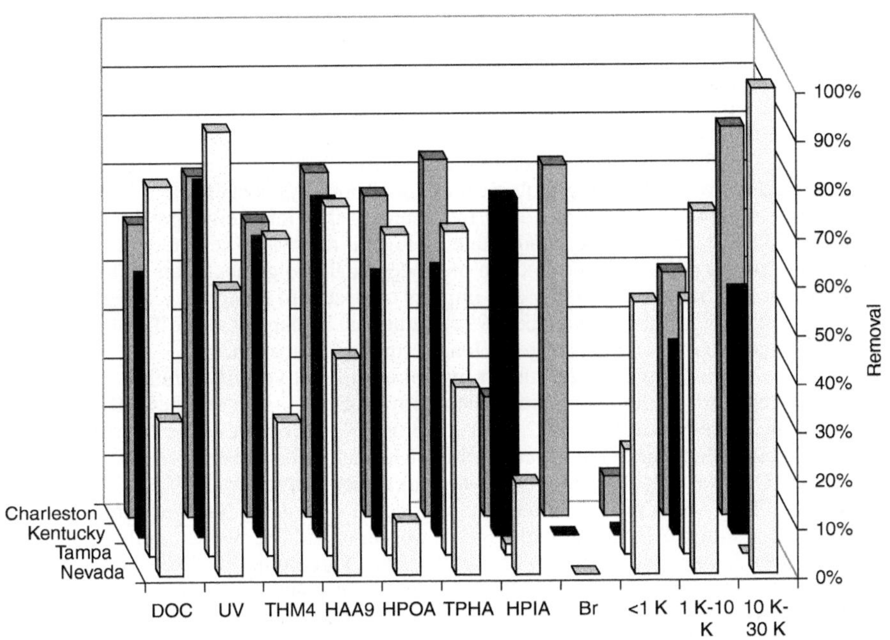

FIGURE 19-24 Summary of optimal MIEX treatment from four pilot-plant studies. (*Source: Singer et al. 2007.*)

organic matter in some waters may foul the MIEX resin and limit its reuse (Mergen et al., 2008). Ion exchange processes are discussed more extensively in Chap. 12.

Adsorption. Use of granular activated carbon (GAC) for adsorption of NOM has been studied widely, including its impacts on the removal of constituent DBP precursors. Because biological activity occurs so readily in GAC beds, it is sometimes difficult to separate removal owing to adsorption from loss to microbial degradation. Nevertheless, fresh GAC medium is effective at removing DBP precursors through abiotic adsorption. Important variables are the empty-bed contact time, GAC surface area, pore size distribution of the GAC, and temperature. In general, THMFP breakthrough parallels DOC breakthrough. Sometimes it lags the DOC breakthrough, making DOC a good conservative surrogate (e.g., Summers et al., 1995). Nevertheless, it should be recognized that use of GAC in a strictly adsorptive mode to control DBP precursors is relatively expensive for most systems, although it has been shown that biodegradation of DBP precursors in biologically active GAC beds can remove on the order of 20 percent of these precursors.

Powdered activated carbon (PAC) also can be used for the removal of DBP precursors. However, at typical water treatment doses and contact times, a limited degree of precursor removal can be expected. Processes that increase the standing concentration of PAC in the contactor or its residence time, such as contact clarifiers containing suspended PAC, are expected to be more effective. For greater coverage of adsorption, the reader should consult Chap. 14.

Biological Filtration. Biological filtration generally refers to the use of granular media filters in a way that encourages growth of attached biomass and results in high levels of biodegradation in the filter bed. Biological filtration has been practiced for more than a century in the form of slow sand filtration. More recently, rapid media filters have been operated for biological activity, especially when following ozonation. Important parameters that affect the performance of biological filters include pretreatment, contact time in the filter bed, oxidant/disinfectant residuals, media surface properties, and temperature.

There are numerous reports in the literature documenting the effectiveness of biological filtration for removing DOC and DBP precursors (e.g., Wang et al., 1995; Urfer et al., 1997; Niquette et al., 1999). These types of studies usually are designed to isolate the impacts of biological activity, separating biological effects from physicochemical adsorption. For GAC processes, this is sometimes done by assuming that long-term (postbreakthrough) removal of DOC represents the steady state biological removal. The extent of this steady state removal will depend on the fraction of biodegradable organics in the influent water, as determined by raw water quality and impacts of upstream treatment (e.g., ozonation). As an example, Snoeyink and Knappe (1994) found that 30 percent of the THM precursors were removed on GAC columns through long-term biodegradation.

In general, removal of THM precursors by biologically active GAC parallels the removal of DOC. While studied less commonly, the removal of precursors to other DBPs appears to reflect the gross removal of DOC, with some notable differences. These are illustrated here using pilot-plant studies conducted for the Massachusetts Water Resources Authority (MWRA). The raw water is a low-DOC, low-alkalinity surface water. Three trains were examined, all including preozonation, with one going directly to filtration, one with dissolved air flotation (DAF) and granular media filtration, and one with an upflow roughing filter (contact adsorption clarifier) with GAC media. Three types of filters were studied: an anthracite single-medium filter operated intermittently and two GAC/sand filters, one operated intermittently and one operated continuously. The intention was to limit biological activity in some of the filters by "starving" them via intermittent operation. The acclimated dual-media filters resulted in substantially lower precursor levels than the nonacclimated single-medium filter in the direct filtration train. Table 19-2 shows average precursor

TABLE 19-2 Average Precursor Removal from Wachusett Pilot-Plant Studies

Treatment process or change in process	TTHM	DCAA	TCAA	Choral hydrate
Percent removal across process				
DAF alone	13	27	38	20
DAF + anthracite bed filtration	37	40	65	27
Anthracite bed filtration	37	40	58	30
Acclimated anthracite bed filtration	49	54	70	67
Acclimated GAC bed filtration	62	65	76	78
Additional percent removal across process				
Acclimated anthracite bed filtration	12	14	9	38
Anthracite → GAC	13	9	6	11

Source: From Reckhow et al. (1992).

removal at various stages of the different process trains. Removals shown are calculated as the difference between the concentration in the ozone contactor effluent and the indicated process effluent divided by the concentration in the ozone contactor effluent.

Table 19-2 shows that removals by DAF always were less than removals by treatment trains that included some form of filtration. Some of this difference was due to removal of particulate organic carbon that escaped DAF, and some was due to biodegradation in the filters. As expected, pretreatment with DAF did not ultimately result in better filtered water quality (compare "DAF + Anthracite Bed Filtration" with "Anthracite Bed Filtration"), although it certainly would reduce head-loss buildup in the filters. The acclimated anthracite filters showed better removal of all four classes of precursors than the intermittently operated (i.e., nonacclimated) filters. This difference is shown in the second to last row in Table 19-2. Note that the difference is greatest for the chloral hydrate precursors, a group that is expected to be dominated by aliphatic aldehydes, which are readily biodegradable. The acclimated GAC media seems to have supported a higher level of removal for all the precursors compared with the acclimated anthracite filters. This difference is shown in the bottom row of Table 19-2. Elevated biological activity with GAC media relative to sand or anthracite has long been noted (e.g., Wang et al., 1995) and has been attributed to differences in the pore size distribution or surface chemistry of the GAC.

While ozonation and biofiltration ultimately will result in lower overall DBP levels in the finished water, they also will result in a shift toward a higher fraction of brominated species. This is especially noticeable in waters that contain high levels of bromide because ozone will oxidize some of the bromide to hypobromous acid and bromate. The hypobromous acid will react with NOM in the water to form low levels of brominated DBPs, even before the addition of any halogen-containing disinfectants. Second, improved removal of DOC by the GAC will result in higher ratios of bromide to DOC in the finished water. This has the unintended consequence of more heavily populating precursor carbon with bromine atoms rather than chlorine atoms. The end result is a higher level of bromine incorporation (e.g., Shukairy and Summers, 1996). This same phenomenon exists for most processes that target DOC removal, for example, coagulation, GAC adsorption, anion exchange, nanofiltration (see below).

Membrane Separation. Although low-pressure membranes (microfiltration and ultrafiltration) generally are not effective at removing dissolved organic substances, high-pressure processes can be quite effective. Nanofiltration (NF) membranes typically reject 50 to 95 percent of the THM precursors depending on the molecular weight cutoff of the membrane (Chellam, 2003). Since NF is capable of removing a large fraction of the organic

DBP precursors, but not bromide, NF-treated water can be expected to have high Br:DOC ratios. This leads to higher levels of bromine incorporation during subsequent chlorination (e.g., Chellam, 2003; Ates et al., 2009), similar to that observed for biofiltration. Reverse-osmosis (RO) processes are capable of removing essentially all the THM and HAA precursors, as well as rejecting some of the bromide. The performance, design, and operational considerations for NF and RO are covered in detail in Chap. 11.

Modifying Disinfection

Various elements of the disinfection step sometimes can be changed to help control DBP formation. One obvious approach is to reduce the disinfectant dose or contact time to the extent possible within the constraints of good disinfection practice. For example, primary and secondary disinfectant doses may be reduced in some plants while relying on booster disinfection out in the distribution system to take care of areas susceptible to low residuals (see Carrico and Singer, 2009). Additionally, disinfectant addition may be delayed until after a substantial portion of the DBP precursors have been removed, for example, by coagulation, anion exchange, adsorption, or membrane filtration. pH adjustment also can be used to favorably affect DBP formation. Base addition for corrosion control might be delayed until after disinfection with chlorine to help depress formation of THMs (see earlier section on the effects of pH). A widely used approach to control the formation of currently regulated DBPs is to substitute an alternative disinfectant for chlorine. Common examples include the use of chloramines as a secondary disinfectant in place of free chlorine or the use of ozonation, UV irradiation, or chlorine dioxide as a primary disinfectant in place of chlorination. Implementation of any of these options must be done in such a manner that disinfection is not compromised.

Moving the Point of Chlorination. It is well established that halogenated DBP formation is, by a first approximation, proportional to the TOC concentration at the point of chlorination. This is why, in waters with high DBP precursor concentrations, shifting the point of chlorination from before coagulation to after clarification yields substantial reductions in DBP formation. This was first demonstrated in full scale by Young and Singer (1979). These authors monitored the Durham, N.C., water treatment plant over the time period when they switched from addition of chlorine prior to rapid mix to addition at a point between sedimentation and dual-media filtration. The drop in THMs after the switch (early January 1977) was quite clear and was based on a series of bench-scale chlorination tests using raw and coagulated water (Fig. 19-25). The Stage 1 Disinfectants/Disinfection By-Products Rule requirement that certain levels of TOC removal be achieved prior to final disinfection is based on recognition that chlorinating after effective coagulation and clarification substantially reduces the formation of THMs and HAAs and most (if not all) other halogenated DBPs.

Furthermore, a 40 percent reduction in TOC across conventional treatment often translates to a 50 percent or more reduction in some DBPs when this switch is made. This is partly attributed to preferential removal of precursor organics by coagulation and the accompanying shift in the nature of the NOM at the point of chlorination to less reactive compounds (see Fig. 19-23).

Delaying the addition of chlorine until after filtration may serve to further reduce finished-water DBP concentrations. Organic precursors deposited on filter media are subject to reaction with chlorine in the water applied to the filters. The end result is that these retained organic precursors are converted to DBPs at a high yield (Corbin et al., 2003). However, the need to control manganese by means of catalytic removal on filter media may preclude this option in some plants (see Chap. 7).

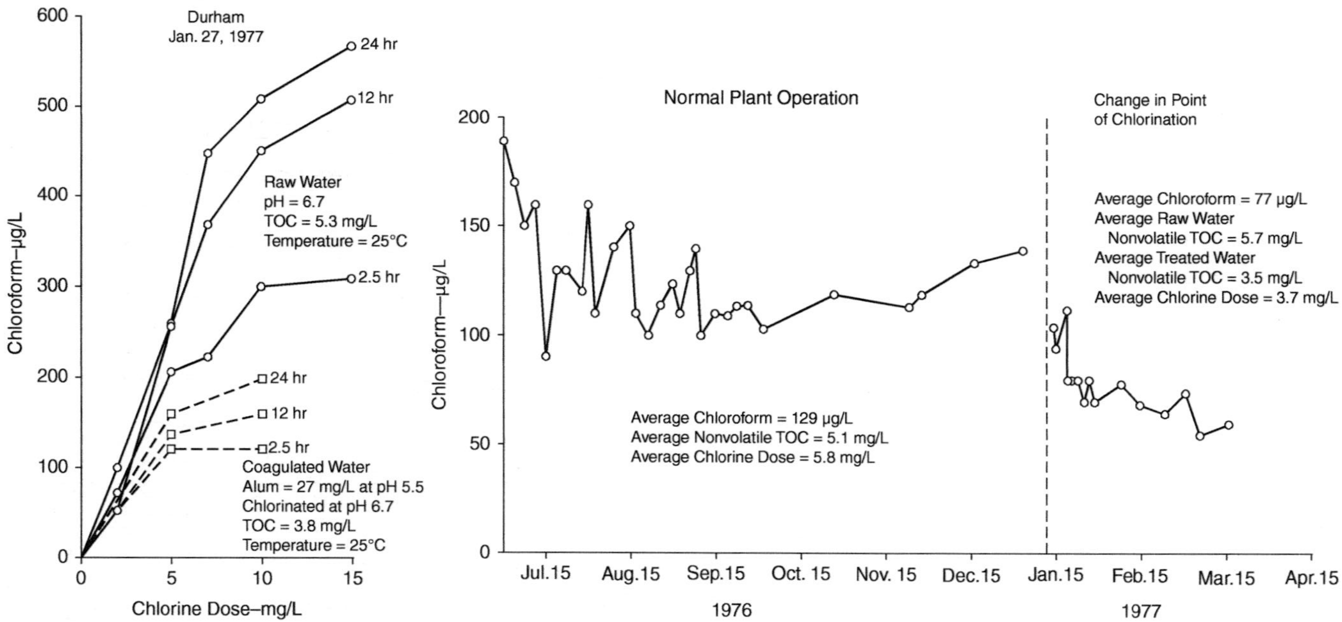

FIGURE 19-25 Impact of moving the point of chlorination from before to after coagulation and sedimentation. (*Source: Young and Singer, 1979.*)

Alternative Primary Disinfectants. A number of utilities have adopted the use of ozone, UV irradiation, or chlorine dioxide as alternative primary disinfectants in place of free chlorine. These agents, although more expensive, generally are better disinfectants than free chlorine (see Chap. 17). As noted earlier and in Chap. 7, however, while these alternative disinfectants do not form THMs and HAAs to any appreciable degree, ozone is associated with the formation of bromate in bromide-containing waters, and chlorine dioxide produces chlorite as a by-product. Both bromate and chlorite are public health concerns and are regulated in finished drinking water.

Alternative Secondary Disinfectants: Chloramines. Research has shown that inorganic chloramines result in little or no formation of trihalogenated species, that is, THMs and the trihaloacetic acids (Cowman and Singer, 1996, Zhang et al., 2000). In contrast, monochloramine produces dichloroacetic acid, but at much lower concentrations than free chlorine. The addition of preformed monochloramine, however, is not commonly practiced. Still, because chloramines are almost always applied after a period of free chlorine contact (or use of an alternative primary disinfectant), they have the effect of greatly reducing the formation of THMs and HAAs (e.g., Speitel, 1999). Figures 19-26 and 19-27 demonstrate the impact of ammonia addition to chlorinated finished water from three water treatment plants in North Carolina (Singer et al., 1998). Ammonia was added at a 1:4 weight ratio relative to chlorine, and waters were held at pH 8.0 and 20°C for the times indicated. As shown, little additional THM or HAA6 formation is apparent.

As a result of the Stage 2 Disinfectants/Disinfection By-Products Rule and the shift to a locational running annual average instead of the historical system-wide running annual average as a basis for regulation and the fact that chloramines essentially stop the continuing formation of THMs and HAAs, many utilities have switched or are planning to switch to the use of chloramines as a secondary disinfectant to comply with the Stage 2 Rule.

Changing the Chemistry of Disinfection. A good example of how knowledge of the chemistry of DBP formation can be used to effect DBP control is the case of bromate. This

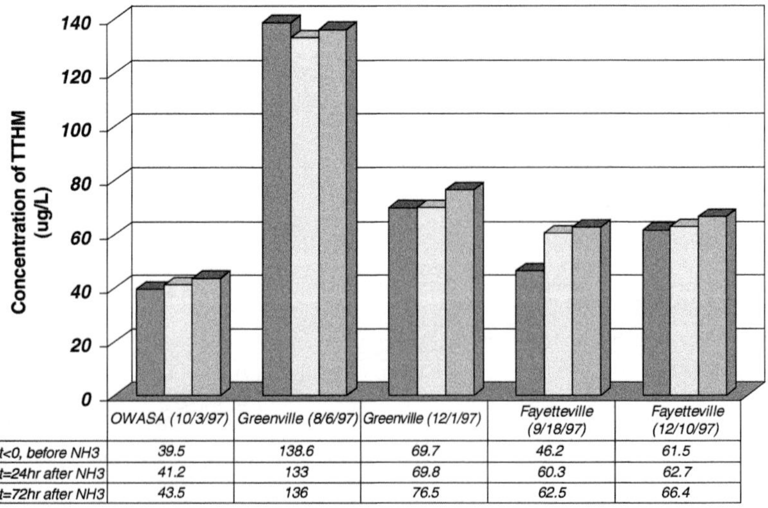

	OWASA (10/3/97)	Greenville (8/6/97)	Greenville (12/1/97)	Fayetteville (9/18/97)	Fayetteville (12/10/97)
TTHM ug/L @ t<0, before NH3	39.5	138.6	69.7	46.2	61.5
TTHM ug/L @ t=24hr after NH3	41.2	133	69.8	60.3	62.7
TTHM ug/L @ t=72hr after NH3	43.5	136	76.5	62.5	66.4

FIGURE 19-26 Impact on THM formation from ammonia addition to chlorinated finished water from three water treatment plants in North Carolina. (*Source: Singer et al., 1998.*)

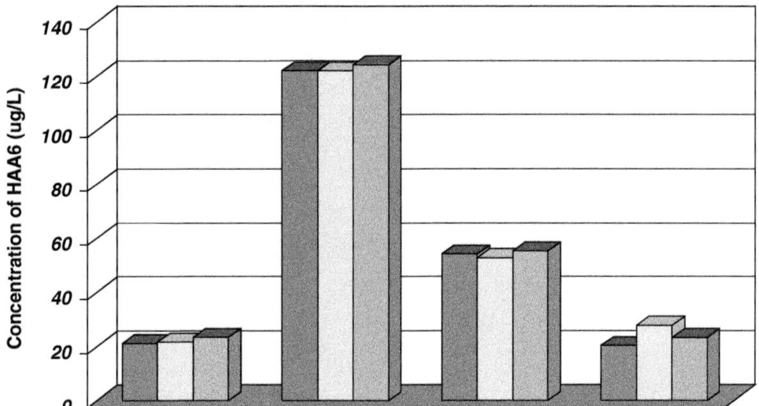

	OWASA (10/3/97)	Greenville (8/6/97)	Greenville (12/1/97)	Fayetteville (9/18/97)
HAA6 ug/L @ t<0, before NH3	21.2	121.9	54.1	20.3
HAA6 ug/L @ t=24hr after NH3	21.8	121.9	54.8	28
HAA6 ug/L @ t=72hr after NH3	23.5	124.2	55.3	23.4

FIGURE 19-27 Impact on HAA formation from ammonia addition to chlorinated finished water from three water treatment plants in North Carolina. (*Source: Singer et al., 1998.*)

inorganic DBP is produced almost exclusively from the ozonation of inorganic bromide ion. Von Gunten and Hoigne (1994) have shown that multiple pathways are important in bromate formation in drinking water, involving some combination of molecular ozone and the hydroxyl radical (see discussion in Chap. 7 and Fig. 19-28).

With this knowledge, it is easy to see that there are at least three feasible strategies for interrupting the formation of bromate: (1) prechloramination: depress bromide (Br^-) concentrations by adding chlorine to produce $HOBr/OBr^-$ and ammonia to convert the $HOBr/OBr^-$ formed to bromamines prior to ozonation; (2) acidification: depress hypobromite (OBr^-) concentration by lowering the pH and forming HOBr, which is less reactive than OBr^- with respect to bromate formation; and (3) preammoniation: add ammonia to convert hypobromous acid and hypobromite formed from the oxidation of bromide by ozone to bromamines. All three of these have been tested and found to be effective to varying degrees, with prechloramination attracting the most attention (e.g., Buffle et al., 2004; Wert et al., 2007).

Removing Disinfection By-Products after Formation

In general, DBPs are hydrophilic and have low molecular weights, characteristics that make them difficult to remove by most physicochemical processes. For example, the removal of THMs in chlorinated water requires either air stripping or granular activated carbon (GAC) adsorption with frequent regeneration. On the other hand, some DBPs are biodegradable and therefore may be removed by biologically active filtration. The most common example of this is the removal of aldehydes, acids, and ketoacids produced by ozonation. There is also evidence that HAAs produced from chlorination can be removed in this way. These postdisinfection options are discussed below.

Adsorption. Granular active carbon is the adsorbent of choice for removing trace levels of organic compounds, especially synthetic organic contaminants, from drinking water.

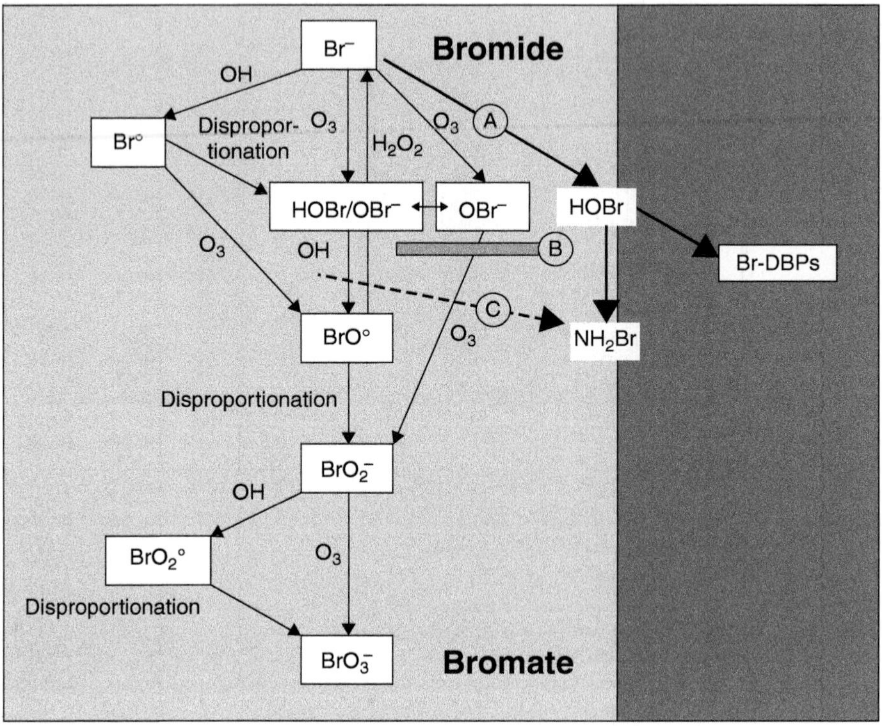

FIGURE 19-28 Bromate formation pathways and strategies for bromate control.

Batch isotherm tests show that the capacity and affinity of GAC for low-molecular-weight polar organic compounds is limited, rendering GAC adsorption an expensive solution for removal of many DBPs. While chloropicrin and 1,1,1-trichloropropanone are strongly adsorbed, chloroform and the other THM species are not. The HAAs are also poorly removed largely because they are almost completely ionized at pH conditions typical of drinking water treatment.

Pilot- and full-scale adsorber studies support the conclusions from bench-scale testing. Graese and colleagues (1987) showed that only 5000 to 6000 bed volumes (BV) of water could be processed before reaching 20 percent breakthrough (80 percent chloroform removal), corresponding to 50 to 100 days of service (Fig. 19-29). Throughput and bed life diminish as the empty-bed contact time (EBCT) drops below 8 to 10 minutes owing to changes in the length of the mass transfer zone. Nevertheless, Savitz and colleagues (2005) have shown that several point-of-use devices containing GAC have the ability to remove substantial amounts of THMs and HAAs for extended periods of time.

It has been suggested that bromate can be removed by GAC (Kirisits et al., 2000). Such removal probably occurs through reaction of bromate with reduced sites on the carbon, and as a result, competition exists with NOM and some inorganics for those sites. There are indications that preozonation may help to improve bromate removal via this mechanism because ozone can render NOM more hydrophilic and less readily adsorbable.

Biological Filtration. While biodegradation has been used for enhancing removal of NOM and DBP precursors in filters, it also may be used to remove preformed DBPs. The

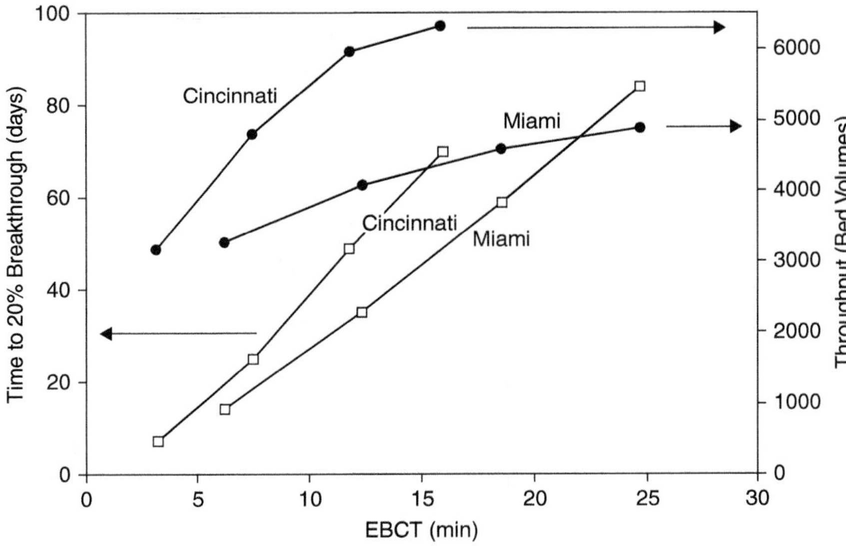

FIGURE 19-29 Field data on chloroform removal by GAC. (*Source: Adapted from Graese et al., 1987.*)

most widely studied cases involve ozonation by-products such as the low-molecular-weight aldehydes and ketones. Reckhow and colleagues (1992) conducted biofiltration pilot studies in a Connecticut utility where they challenged the filters with high levels of ozonation by-products. They found that common by-products such as glyoxalic acid are removed almost completely at conventional filtration rates (Fig. 19-30). They also found that backwashing the filters with water containing a chlorine residual could be detrimental to the activity of the attached growth; thus filters 2 and 3 were less effective than filters 1 and 4. Similar findings have been reported by DiGiano and colleagues (2001) for removal of formaldehyde, glyoxal, and methyl glyoxal at EBCTs of 10 minutes.

Removal of chlorination by-products via biofiltration is practiced less commonly. Wu and Xie (2005) demonstrated that all HAAs could be removed effectively by GAC/sand filters at high temperatures. However, at lower temperatures (e.g., <10°C), removal of TCAA was poor (<50 percent), except at EBCTs of 10 minutes or more. Kim and Kang (2008) confirmed that TCAA was not as well removed as DCAA by biological filtration. They also confirmed that the THMs are not degraded in biologically active filters.

Based on the limited research on HAA degradation in filters, and by analogy with the larger body of work on ozonation by-products, it seems that the following factors are likely to encourage biodegradation of a broad range of oxidation by-products: (1) absence of a disinfectant residual in the filter bed during normal operation, (2) little or no disinfectant residual in the backwash water, (3) extended EBCTs, that is, 10 minutes or more, (4) warm temperatures, (5) prior acclimation by exposure to the DBPs that are to be removed (this is often accompanied by partial exhaustion of adsorption capacity if GAC is used), and (6) media with high external surface area for good microbial attachment.

Biological filtration also may be effective for control of bromate. Pilot-plant studies have shown removals in the range of 40 to 60 percent (Kirisits and Snoeyink, 1999). Bromate seems to serve as a terminal electron acceptor, and as a result, this process may require that most of the dissolved oxygen be depleted prior to filtration. In like manner, bromate removal has been seen in anoxic groundwaters (Butler et al., 2004).

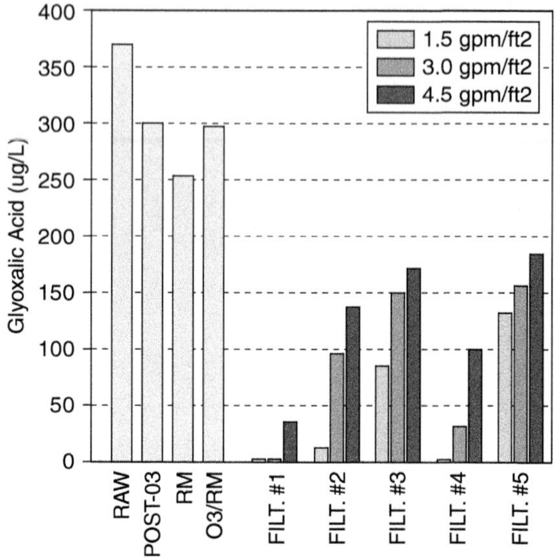

FIGURE 19-30 Impact of filtration rate and backwash on removal of glyoxalic acid by biologically active filters. (Post-O_3, RM and O_3/RM refer to raw water samples treated with ozone, alum, and combined ozone and alum, respectively.) (*Source: Reckhow et al., 1992.*)

Aeration. THMs are among the most volatile of the chlorination by-products. Accordingly, packed tower aeration has been studied by Umphres and colleagues (1983) for removal of THMs. As the level of bromine incorporation increases, the rate of loss by aeration decreases owing to the lower volatility of the heavier bromine-containing species (Fig. 19-31). Aeration is discussed further in Chap. 6.

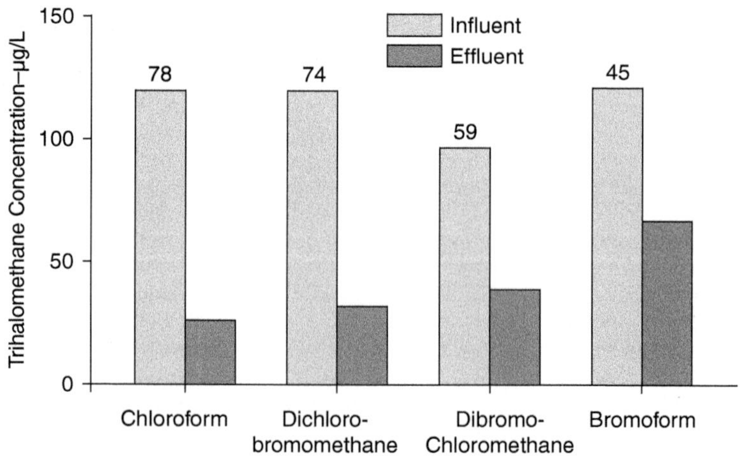

FIGURE 19-31 Comparison of THM removals from pilot-plant studies on packed tower aeration. (*Source: Umphres et al., 1983.*)

Membrane Separation. While high-pressure membrane processes are quite effective at removing DBP precursors, preformed DBPs usually are much smaller in molecular size and therefore less well removed. Reverse osmosis (RO) has exhibited THM rejections as high as 90 percent (Reinhard et al., 1986). Membrane composition is likely to be more important than molecular weight cutoff ratings. For these small-molecular-weight compounds, rejection depends more on specific chemical properties of both the compound and the membrane and on the affinity between the solute and the membrane. In particular, charge effects may be an important characteristic in determining rejection. For example, Chalatip and colleagues (2009) found that high-density negatively charged NF membranes are capable of removing 90 percent or more of the anionic HAAs. On the other hand, uncharged DBPs such as NDMA are poorly removed, even by RO (Steinle-Darling et al., 2007). Partial fouling of membranes may result in improved removal of these difficult solutes. The performance, design, and operational considerations for NF and RO are covered in detail in Chap. 11.

DISINFECTION BY-PRODUCTS IN THE DISTRIBUTION SYSTEM

Stability of Disinfection By-Products

The THMs generally are considered to be stable end products of the reaction between NOM and chlorine. Although the HAAs are also considered to be stable end products of chlorination, they are known to degrade abiotically by decarboxylation or by reductive dechlorination. The rate and extent of their degradation tend to be proportional to their bromine content (Zhang and Minear, 2002), with tribromoacetic acid being the least stable of the bromine- and chlorine-containing trihaloacetic acids. The trihaloacetic acids primarily exchange a bromine atom for a hydrogen atom, forming the corresponding dihaloacetic acids. There also may be some conversion of dibromoacetic acid to bromoacetic acid and bromochloroacetic acid to chloroacetic acid. Loss of bromine atoms is substantially faster than loss of chlorine atoms. Zero valent iron associated with pipe walls can readily dehalogenate HAAs (Hozalski et al., 2001), with chloroacetic acid being the final product of di- and trichloroacetic acid degradation. Furthermore, linear free-energy calculations by Zhang and Minear (2002) suggest that all iodinated trihaloacetic acids except dichloroiodoacetic acid are also likely to undergo decarboxylation in drinking water distribution systems.

Reports of the biological degradation of HAAs in drinking water distribution systems are well known. Starting with the work of Williams and colleagues (1994), numerous researchers have observed a decrease in HAA concentrations in regions of the distribution system that have no chlorine residual (e.g., Speight and Singer, 2005), especially when the water temperature is high. This loss is particularly noticeable for the dihaloacetic acids. Figure 19-32 illustrates, for a utility in the southeastern United States using combined chlorine as a secondary disinfectant, the loss of HAAs in the system at high residence times when the chlorine residual is depleted. The THM concentration at this same location is essentially unchanged. Nitrification also was observed at this same location, indicating the presence of ammonia-oxidizing bacteria. Figure 19-33 shows a similar situation for a utility in California using combined chlorine. Again, HAA concentrations are decreased at locations where the combined chlorine residual is low. [Note that the water ages (residence times) shown are based on hydraulic modeling, and therefore, the hierarchical order of residence times may not be accurate.] The speciation of the HAAs for this utility is shown in Figure 19-34, which illustrates that the decrease in total HAA concentration is due to loss of the dihaloacetic acid species.

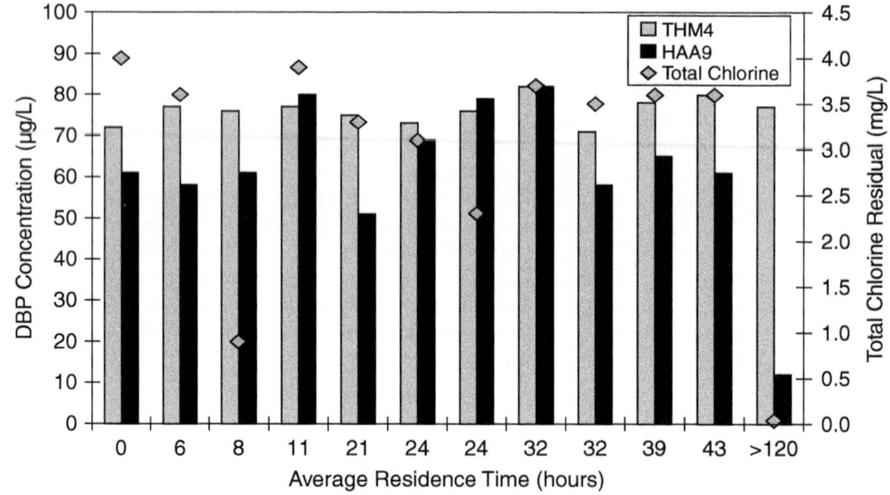

FIGURE 19-32 Trihalomethane and haloacetic acid concentrations and total chlorine residual at various water ages for a utility in the southeastern United States. (*Source: Baribeau et al., 2006, Awwa Research Foundation.*)

The fact that HAA levels tend to decrease at high water ages in the distribution system means that unlike THM concentrations, which are highest at distribution system locations with the highest residence times, HAA concentrations often tend to peak elsewhere in the system. For many systems, different patterns are observed at different times of the year because biodegradation is limited at colder temperatures, and it is also easier to maintain chlorine residual in remote parts of the system when the temperature is colder. This

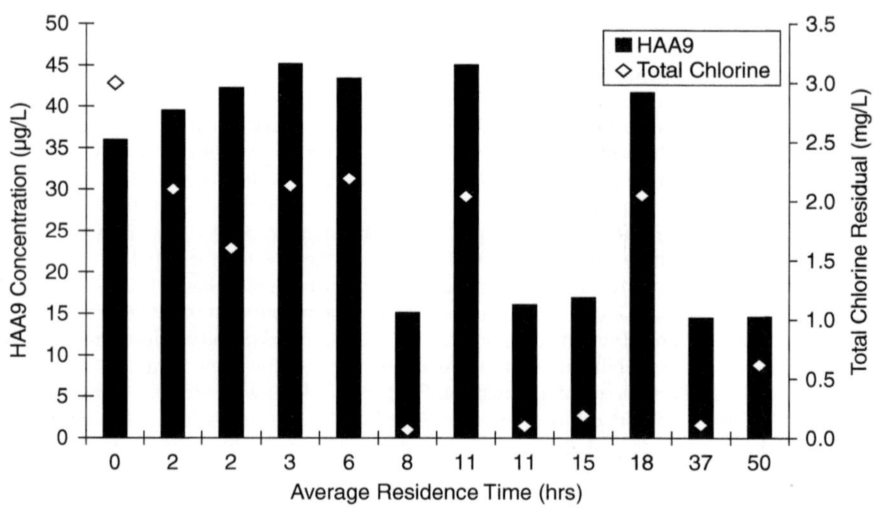

FIGURE 19-33 Haloacetic acid concentrations and total chlorine residual at various water ages for a utility in California. (*Source: Baribeau et al., 2006, Awwa Research Foundation.*)

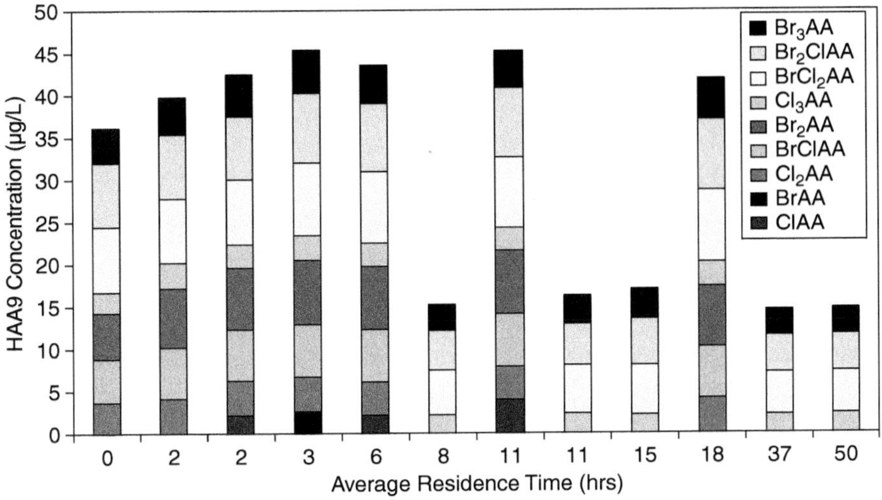

FIGURE 19-34 HAA speciation at various water ages for a utility in California. (*Source: Baribeau et al., 2006, Awwa Research Foundation.*)

has regulatory implications with respect to choosing locations for compliance monitoring where THM and HAA concentrations are at a maximum.

Many of the other halogenated DBPs normally formed during chlorination are chemically unstable. Once in the distribution system, they will continue to react with constituents of the water or materials attached to pipe walls to form secondary by-products. Examples of reactive DBPs, as noted earlier in this chapter, include the haloacetonitriles, the haloketones, and the cyanogen halides. It is quite likely that many other metastable compounds have eluded detection in finished drinking water because of their innate chemical instability.

Monitoring for Regulatory Compliance

Historically, DBPs have been regulated based on annual average levels in the distribution system. For larger systems, the Stage 1 Disinfectants/Disinfection By-Products Rule requires that THMs and HAAs be measured quarterly at four locations in the system for each water treatment plant and that the *system-wide* running annual average (the average of the last four quarters) be below the maximum contaminant levels of 0.08 mg/L for the THMs and below 0.06 mg/L for the HAAs that are regulated—five of the nine bromine and chlorine-containing HAAs. The four locations were designated as three with average distribution system residence times and one remote location. For small systems, compliance is based on monitoring once a year, when the levels are expected to be highest, and at only one location, again where the levels are expected to be highest. The Stage 2 Rule keeps the same maximum contaminant levels but bases compliance on a *locational* running annual average, meaning that compliance must be achieved at all sampling locations. Moreover, the four sampling locations must be at points in the system where THM and HAA levels are expected to be highest.

As noted earlier, because of their stability, THM concentrations in chlorinated systems are expected to be highest where the contact time between free chlorine and the water is greatest, that is, at locations with the highest water age. For HAAs, this is not always the case, particularly in warmer waters, where biodegradation occurs at high water ages. For chloraminated systems, because monochloramine does not produce THMs and produces only limited amounts of the dihaloacetic acids, THM and HAA levels are expected to be relatively uniform throughout the distribution system, except at high water ages, where biodegradation may affect HAA levels. The fact that THMs and HAAs do not increase appreciably in the system when monochloramine is the residual disinfectant is one of the reasons why many utilities have converted or are planning to convert to a combined chlorine residual for compliance with the Stage 2 Rule.

Spatial Temporal Variability of DBP Levels

Until recently, the temporal and spatial variability in DBP concentrations in distribution systems have not been fully appreciated. While it has always been understood that DBP levels change seasonally because of changes in temperature (kinetics of DBP formation) and the nature and amount of precursors in the raw water and that DBP levels change with increasing residence time in chlorinated systems, short-term (day-to-day and hour-to-hour) variations generally have not been recognized.

Singer (2001) showed that THM concentrations at various locations in chlorinated systems exhibit appreciable diurnal variations that match the diurnal variations in residual chlorine and water age. At times of the day when water demand is low and storage tanks in the system are filling, water age at sampling locations influenced by these tanks is low, and THM concentrations tend to be at a minimum. At other times of the day, when water demand is high and water levels in storage tanks are being drawn down, water age is higher, chlorine residuals are lower, and THM levels are correspondingly higher. Such diurnal variations are not seen in chloraminated systems because DBP levels do not increase appreciably with water age. Of course, as noted earlier, such variations can be seen if biodegradation of HAAs occurs with increased water age.

Short-term variations in DBP levels also have been reported at the point of entry to the distribution system, particularly for THMs (Periera et al., 2004; Obolensky et al., 2008). THM levels were observed to change by a factor of up to 2 over a relatively short period of time. These variations were unexpected and were traced to operational variations in the treatment plants associated with chlorine addition, ammonia addition, and pH adjustment. The findings suggest that water utilities should examine their operations routinely to make sure that their chemical feed systems are designed and operated appropriately.

Accordingly, in view of this short-term variability in DBP levels in the distribution system, regulators and utility personnel may wish to review the current reliance on quarterly averaging to assess DBP exposure and to determine MCL compliance.

ABBREVIATIONS

AOC	assimilable organic carbon
BCAA	bromochloroacetic acid (also BrClAA)
BCAN	bromochloroacetonitrile
BDCM	bromodichloromethane
BDOC	biodegradable dissolved organic carbon
BIF	bromine incorporation factor
BOM	biodegradable organic matter

BSF	bromine substitution factor
CH	chloral hydrate
CP	chloropicrin
DAF	dissolved air flotation
DBAA	dibromoacetic acid (also Br_2AA)
DBAN	dibromoacetonitrile
DBCM	dibromochloromethane
DBPs	disinfection by-products
DCAA	dichloroacetic acid (also Cl_2AA)
DCAN	dichloroacetonitrile
DCP	dichloropropanone
D/DBP	disinfectants/disinfection by-products
DiHAA	dihaloacetic acids (DCAA + BCAA + DBAA)
DOC	dissolved organic carbon
DON	dissolved organic nitrogen
DOX	dissolved organic halogen
EBCT	empty-bed contact time
FP	formation potential
GAC	granular activated carbon
HAAFP	haloacetic acid formation potential
HAA9	nine haloacetic acids (MCAA + MBAA + DiHAA + TriHAA)
HAAs	haloacetic acids
HAN4	four haloacetontriles (DCAN + TCAN + BCAN + DBAN)
HANs	haloacetontriles
HPIA	hydrophilic acids
HPOA	hydrophobic acids
ICR	Information Collection Rule
MBAA	monobromoacetic acid (also BrAA)
MCAA	monochloroacetic acid (also ClAA)
MCL	maximum contaminant level
MIEX	magnetic ion exchange resin
NDMA	nitrosodimethylamine
NF	nanofiltration
NOM	natural organic matter
MW	molecular weight
MWRA	Massachusetts Water Resources Authority
PAC	powdered activated carbon
RM	rapid mix
RO	reverse osmosis
SDS	simulated distribution system

SUVA	specific UV absorbance
TBAA	tribromoacetic acid (also Br_3AA)
TCAA	trichloroacetic acid (also Cl_3AA)
TCAN	trichloroacetonitrile
TCP	trichloropropanone
THMFP	trihalomethane formation potential
THM4	four trihalomethanes (chloroform, BDCM, DBCM, and bromoform)
THMs	trihalomethanes
TPHA	transphilic acids
TriHAA	trihaloacetic acids (TCAA + bromodichloroacetic acid + dibromochloroacetic acid + tribromoacetic acid)
TOBr	total organic bromine
TOC	total organic carbon
TOCl	total organic chlorine
TOI	total organic iodine
TOX	total organic halogen
UFC	uniform formation conditions
USEPA	U.S. Environmental Protection Agency
USGS	U.S. Geological Survey
UV	ultraviolet
UV_{254}	ultraviolet absorbance at 254 nm

REFERENCES

Aieta, E. M., Reagan, K. M., Lang, J. S., McReynolds, L., Kang, J. W., and Glaze, W. H. (1988). Advanced Oxidation Processes for Treating Groundwater Contaminated with TCE and PCE: Pilot-Scale Evaluations. *Journal AWWA* 80(5):64–72.

Amy, G. L., Chadik, P. A., and Chowdhury, Z. K. (1987). Developing Models for Predicting Trihalomethane Formation Potential and Kinetics. *Journal AWWA* 79(7):89–97.

APHA, AWWA, WEF (2005). *Standard Methods for the Examination of Water and Wastewater*, 21st ed., A. D. Eaton, L. S. Clesceri, E. W. Rice and A. E. Greenberg, eds. Washington, DC: APHA.

Archer, A. D., and Singer, P. C. (2006a). An Evaluation of the Relationship Between SUVA and NOM Coagulation Using the ICR Database. *Journal AWWA* 98(7):110–123.

Archer, A. D., and Singer, P. C. (2006b). Impact of SUVA and Enhanced Coagulation on the Removal of Total Organic Halide Precursors. *Journal AWWA* 98(8):97–107.

Babcock, D. B., and Singer, P. C. (1979). Chlorination and Coagulation of Humic and Fulvic Acids. *Journal AWWA* 71(3):149–152.

Backlund, P., Kronberg, L., and Tikkanen, L. (1988). Formation of Ames Mutagenicity and of the Strong Bacterial Mutagen 3-Chloro-4-(Dichloromethyl)-5-Hydroxy-2-(5H)-Furanone and Other Halogenated Compounds During Disinfection of Humic Water. *Chemosphere* 17(7):1329–1336.

Baribeau, H., L Boulos, H. Haileselassie, G. Crozes, P.C. Singer, C. Nichols, S.A. Schlesinger, R.W. Gullick, S L. Williams, R.L. Williams, L. Fountleroy, S.A. Andrews, and E. Moffat. (2006). *Formation and Decay of Disinfection By-Products in the Distribution System.* Awwa Research Foundation Report. Denver, CO: Water Research Foundation.

Bellar, T. A., Lichetenberg, J. J., and Kroner, R. C. (1974). The Occurrence of Organohalides in Chlorinated Drinking Water. *Journal AWWA* 66(12):703–706.

Benga, J. (1980). Oxidation of Humic Acid by Ozone or Chlorine Dioxide. Ph.D. dissertation, Miami University, Ohio.

Bieber, T. I., and Trehy, M. L. (1983). Dihaloacetonitriles in Chlorinated Natural Waters. In *Water Chlorination: Environmental Impact and Health Effects*, Vol. 4, Book 1: *Chemistry and Water Treatment. Proceedings of the Fourth Conference on Water Chlorination*, R. L. Jolley, Robert L. Jolley, William A. Brungs, Joseph A. Cotruvo, Robert B. Cumming, Jack S. Mattice, and Vivian A. Jacobs. eds. Pacific Grove, CA., Ann Arbor Science Publ., Ann Arbor, MI

Boyer, T. H., and Singer, P. C. (2005). Bench-Scale Testing of a Magnetic Ion Exchange Resin for Removal of Disinfection By-Product Precursors. *Water Research* 39(7):1265–1276.

Boyer, T. H., and Singer, P. C. (2006). A Pilot-Scale Evaluation of Magnetic Ion Exchange Treatment for Removal of Natural Organic Material and Inorganic Anions. *Water Research* 40(15):2865–2876.

Brass, H. J., Feige, M. A., Halloran, T., Mello, J. W., Munch, D., and Thomas, R. F. (1977). The National Organic Monitoring Survey: Samplings and Analyses for Purgeable Organic Compounds. In *Drinking Water Quality Enhancement Through Source Protection*, R. B. Pojasek, ed. Ann Arbor, MI: Ann Arbor Science Publishers.

Brezonik, P. L. (1994). *Chemical Kinetics and Process Dynamics in Aquatic Systems.* Boca Raton, FL: Lewis Publishers.

Buffle, M. O., Galli, S., and von Gunten, U. (2004). Enhanced Bromate Control During Ozonation: The Chlorine–Ammonia Process. *Environmental Science and Technology* 38(19):5187–5195.

Burttschell, R. H., Rosen, A. A., Middleton, F. M., and Ettinger, M. B. (1959). Chlorine Derivatives of Phenol and Phenol Causing Taste and Odor. *Journal AWWA* 51(2):205–214.

Butler, R., Cartmell, E., Godley, A., and Lytton, L. (2004). Remediation of Bromate Contaminated Groundwater. In *Bringing Groundwater Quality Research to the Watershed Scale*, N. R. Thomson, ed. Wallingford, UK: IAHS Press.

Canale, R. P., Chapra, S. C., Amy, G. L., and Edwards, M. A. (1997). Trihalomethane Precursor Model for Lake Youngs, Washington. *Journal of Water Resources Planning and Management, ASCE* 123(5):259–265.

Carlson, R., and Caple, R. (1977). *Chemical Biological Implications of Using Chlorine and Ozone for Disinfection.* EPA-600/3-77-066 (Ecological Research Series). Washington, DC: USEPA.

Carrico, B., and Singer, P. C. (2009). Impact of Booster Chlorination on Chlorine Decay and THM Production: A Simulated Analysis. *Journal of Environmental Engineering* 135(10):928–935.

Cavanagh, J. E., Weinberg, H. S., Gold, A., Sangalah, R., Marbury, D., Glaze, W. H., Collette, T. W., Richardson, S. D., and Thruston, A. D. J. (1992). Ozonation By-products: Identification of Bromohydrins from the Ozonation of Natural Waters with Enhanced Bromide Levels. *Environmental Science and Technology* 26(8):1658–1662.

Chapin, R. M. (1931). The Influence of pH on the Formation and Decomposition of the Chloro Derivatives of Ammonia. *Journal of the American Chemical Society* 53:912–920.

Chapra, S. C., Canale, R. P., and Amy, G. L. (1997). Empirical Models for Disinfection by-Products in Lakes and Reservoirs. *Journal of Environmental Engineering* 123(7):714–715.

Chellam, S. (2000). Effects of Nanofiltration on Trihalomethane and Haloacetic Acid Precursor Removal and Speciation in Waters Containing Low Concentrations of Bromide Ion. *Environmental Science and Technology* 34(9):1813–1820.

Chen, P. N., Junk, G. A., and Svec, H. J. (1979). Reactions of Organic Pollutants: I. Ozonation of Acenaphthylene and Acenaphthene. *Environmental Science and Technology* 13(4):451–454.

Christman, R. F., Norwood, D. L., Millington, D. S., Johnson, J. D., and Stevens, A. A. (1983). Identity and Yields of Major Halogenated Products of Aquatic Fulvic-Acid Chlorination. *Environmental Science and Technology* 17(10):625–628.

Colclough, C. A. (1981). Organic Reaction Products of Chlorine Dioxide and Natural Aquatic Fulvic Acids. M.S. thesis, University of North Carolina at Chapel Hill.

Colthurst, J. M., and Singer, P. C. (1982). Removing Trihalomethane Precursors by Permanganate Oxidation and Manganese Dioxide Adsorption. *Journal AWWA* 74(2):78–83.

Cooke, G. D., and Carlson, R. E. (1989). *Reservoir Management for Water Quality and THM Precursor Control.* Awwa Research Foundation Report. Denver, CO: Water Research Foundation.

Coombs, K. K. (1990). Development of a Rapid Trihalomethane Formation Potential Test. M.S. thesis, University of Massachusetts, Amherst.

Cooper, W. J., Amy, G. L. Moore, C. A., and Zika, R. G. (1986). Bromoform Formation in Ozonated Groundwater Containing Bromide and Humic Substances. *Ozone: Science & Engineering* 8(1):63–76.

Corbin, M. J., Reckhow, D. A., Tobiason, J. E., Dunn, H. J., and Kaminski, G. S. (2003). Controlling DBPs by Delaying Chlorine Addition Unitl After Filtration. *Journal of the New England Water Works Association* 117(4):243–251.

Cowman, G. A., and Singer, P. C. (1996). Effect of Bromide Ion on Haloacetic Acid Speciation Resulting from Chlorination and Chloramination of Aquatic Humic Substances. *Environmental Science and Technology* 30(1):16–24.

Crochet, R. A., and Kovacic, P. (1973). Conversion of *o*-Hydroxyaldehydes and Ketones into *o*-Hydroxyanilides by Monochloramine. *Journal of the Chemical Society: Chemical Communications* 1973(19):716–717.

Croué, J.-P., Debroux, J.-F., Amy, G. L., Aiken, G. R., and Leenheer, J. A. (1999). Natural Organic Matter: Structural Characteristics and Reactive Properties. In *Formation and Control of Disinfection By-Products in Drinking Water*, P. C. Singer, ed. Denver, CO: AWWA.

Croué, J.-P., and Reckhow, D. A. (1989). The Destruction of Chlorination By-products with Sulfite. *Environmental Science and Technology* 23(11):1412–1419.

Daniel, P. A., Meyerhofer, P. F., Lanier, M., and Marchand, J. (1989). Impact of Ozonation on Formation of Brominated Organics. In *Ozone in Water Treatment: Proceedings of the 9th Ozone World Congress*, L. J. Bollyky, ed. Zurich, Switzerland: International Ozone Association.

de Leer, E. W. B., Damste, J. S. S., Erkelens, C., and de Galan, L. (1985). Identification of Intermediates Leading to Chloroform and C-4 Diacids in the Chlorination of Humic Acid. *Environmental Science and Technology* 19(6):512–522.

DiGiano, F. A., Singer, P. C., Parameswar, C., and LeCourt, T. D. (2001). Biodegradation Kinetics of Ozonated Natural Organic Matter and Aldehydes. *Journal AWWA* 93(8):92–104.

Dotson, A., Westerhoff, P., and Krasner, S. (2009). Nitrogen Enriched DOM Isolates and Their Affinity to Form Emerging DBPs. *Water Science and Technology* 60(1):135–143.

Dowd, M. T. (1994). Assessment of THM Formation with MIOX. M.S. thesis, Department of Environmental Science and Engineering, University of North Carolina at Chapel Hill.

Edwards, M. (1990). Transformation of Natural Organic Matter, Effect of Organic Matter–Coagulation Interactions, and Ozone-Induced Particle Destabilization. Ph.D. disssertation, University of Washington at Seattle.

Edzwald, J. K., Becker, W. C., and Wattier, K. L. (1985). Surrogate parameters for monitoring organic matter and THM precursors. *Journal AWWA* 77(4):122–131.

Fawell, J. K., and Watts, C. D. (1984). The Nature and Significance of Organic By-products of Ozonation: A Review. In *Seminar on Ozone in UK Water Treatment*. London: Institution of Water Engineers and Scientists.

Fielding, M., and Horth, H. (1986). Formation of Mutagens and Chemicals During Water Treatment Chlorination. *Water Supply* 4:103–126.

Fleischacker, S. J., and Randtke, S. J. (1983). Formation of Organic Chlorine in Public Water Supplies. *Journal AWWA* 75(3):132–138.

Franzén, R., and Kronberg, L. (1994). Determination of Chlorinated 5-Methyl-5-Hydroxyfuranones in Drinking Water in Chlorinated Humic Water and in Pulp Bleaching Liquor. *Environmental Science and Technology* 28(12):2222–2227.

Frimmel, F. H. (1998). Impact of Light on the Properties of Aquatic Natural Organic Matter. *Environment International* 24(5–6):559–571.

Frimmel, F. H., Hesse, S., and Kleiser, G. (2000). Technology-Related Characterization of Hydrophilic Disinfection By-Products in Aqueous Samples. In *Natural Organic Matter and Disinfection By-Products: Characterization and Control in Drinking Water*, S. W. Krasner, S. E. Barrett, and G. L. Amy, eds. Washington, DC: ACS.

Frimmel, F. H., Vercammen, K., and Schmitt, D. (2005). Influence of UV-Oxidation on the Metal Complexing Properties of NOM. In *Use of Humic Substances to Remediate Polluted Environments: From Theory to Practice*, Irina V. Perminova, Kirk Hatfield, and Norbert Hertkorn. eds. Ansterdam: Springer.

Fuji, R., Ranalli, A. J., Aiken, G. R., and Bergamaschi, B. A. (1998). Dissolved Organic Carbon Concentrations and Compositions, and Trihalomethane Formation Potentials in Waters from Agricultural Peat Soils, Sacramentao–San Joaquin Delta, California: Implications for Drinking-Water Quality. Water Resources Investigations Report 98-4147, Sacramento, CA, U.S. Geological Survey.

Garvey, E. A., and Tobiason, J. E. (2003). Assessment of Natural Organic Matter in Quabbin Reservoir. *Journal of Water Supply Research and Technology, AQUA* 52(1):19–36.

Ghanbari, H. A., Wheeler, W. B., and Kirk, J. R. (1983). Reactions of Chlorine and Chlorine Dioxide with Free Fatty Acids, Fatty Acid Esters, and Triglycerides. In *Water Chlorination: Environmental Impact and Health Effects*, Vol. 4, Book 1: *Chemistry and Water Treatment. Proceedings of the Fourth Conference on Water Chlorination*, Robert L. Jolley, William A. Brungs, Joseph A. Cotruvo, Robert B. Cumming, Jack S. Mattice, and Vivian A. Jacobs. eds. Pacific Grove, CA., Ann Arbor Science Publ., Ann Arbor, MI

Glaze, W. H. (1986). Reaction Products of Ozone: A Review. *Environmental Health Perspectives* 69(11):151–157.

Glaze, W. H., Schep, R., Chauncey, W., Ruth, E. C., Zarnoch, J. J., Aieta, E. M., Tate, C. H., and McGuire, M. J. (1990). Evaluating Oxidants for the Removal of Model Taste and Odor Compounds from a Municipal Water Supply. *Journal AWWA* 82(5):79.

Glaze, W. H., Weinberg, H. S., and Cavanagh, J. E. (1993). Evaluating the Formation of Brominated DBPs During Ozonation. *Journal AWWA* 85(1):96–103.

Glezer, V., Harris, B., Tal, N., Iosefzon, B., and Lev, O. (1999). Hydrolysis of Haloacetonitriles: Linear Free Enerby Relationship, Kinetics and Products. *Water Research* 33(8):1938–1948.

Gordon, G. (1998). *Electrochemical Mixed Oxidant Treatment: Chemical Detail of Electrolyzed Salt Brine Technology*. Cincinnati, OH: USEPA.

Graese, S. L., Snoeyink, V. L., and Lee, R. G. (1987). Granular Activated Carbon Filter-Adsorber Systems. *Journal AWWA* 79(12):64–74.

Griffini, O., and Iozzelli, P. (1996). The Influence of H_2O_2 in Ozonation Treatment: Experience of the Water Supply Service of Florence, Italy. *Ozone: Science and Engineering* 18(2):17–126.

Haag, W. R., and Hoigné, J. (1983). Ozonation of Bromide-Containing Water: Kinetics of Formation of Hypobromous Acid and Bromate. *Environmental Science and Technology* 17(5):261–267

Harrington, G. W., Chowdhury, Z. K., Owen, D. M. (1992). Developing a Computer Model to Simulate DBP Formation During Water Treatment. *Journal AWWA* 84(11):78–87.

Hausler, C. R., and Hausler, M. L. (1930). Research on Chloramines: I. Orthochlorobenzalchlorimine and Anisalchlorimine. *Journal of the American Chemical Society* 52(5):2050–2054.

Havlicek, S. C., Reuter, J. H., Ingols, R. S., Lupton, J. D., Ghosal, M., Ralls, J. W., El-Barbary, I., Strattan, L. W., Cotruvo, J. H., and Trichilo, C. (1979). Reaction of Aquatic Humic Material with Chlorine: Isolation and Characterization of Some New Chlorinated Organics. In *Abstracts of Papers, 177th National Meeting of the American Chemical Society*. Washington, DC: American Chemical Society.

Hirose, Y., Maeda, N., Dhya, T., Nojima, K., and Kanno, S. (1988). Formation of Cyanogen Chloride by Reaction of Amino Acids and Free Chlorine in the Presence of Ammonium Ion. *Chemosphere* 17(5):865–873.

Hoehn, R. C., Dietrich, A. M., Farmer, W. S., Orr, M. P., Lee, R. G., Aieta, E. M., Wood, D. W., and Gordon, G. (1990). Household Odors Associated with the Use of Chlorine Dioxide. *Journal AWWA* 82(4):166–172.

Hoigne, J., and Bader, H. (1976). The Role of Hydroxyl Radical Reactions in Ozonation Processes in Aqueous Solutions. *Water Research* 10(5):377–386.

Hoigne, J., and Bader, H. (1994). Kinetics of Reactions of Chlorine Dioxide (OClO) in Water: I. Rate Constants for Inorganic and Organic Compounds. *Water Research* 28(1):45–55.

Hoigne, J., and Bader, H. (1983a). Rate Constants of Reactions of Ozone with Organic and Inorganic Compounds in Water: I. Non-Dissociating Organic Compounds. *Water Research* 17(2):173–183.

Hoigne, J., and Bader, H. (1983b). Rate Constants of Reactions of Ozone with Organic and Inorganic Compounds in Water: II. Dissociating Organic Compounds. *Water Research* 17(2):185–194.

Hoigne, J., Bader, H., Haag, W. R., and Staehelin, J. (1985). Rate Constants of Reactions of Ozone with Organic and Inorganic Compounds in Water: III. Inorganic Compounds and Radicals. *Water Research* 19(8):993–1004.

Hua, G. H., and Reckhow, D. A. (2007). Comparison of Disinfection By-product Formation from Chlorine and Alternative Disinfectants. *Water Research* 41(8):1667–1678.

Hwang, C. J., Sclimenti, M. J., Trinh, L., and Krasner, S. W. (1996). Consolidated Method for the Determination of Aldehydes and Oxo-Acids in Ozonated Drinking Water. In *Proceedings of the 1996 AWWA Water Quality Technology Conference, Boston, MA*.

Itoh, S., and Matsuoka, Y. (1996). Contributions of Disinfection By-products to Activity Inducing Chromosomal Aberrations of Drinking Water. *Water Research* 30(6):1403–1410.

Jensen, J. N., and Johnson, J. D. (1989). Specificity of the DPD and Amperometric Titration Methods for Free Available Chlorine: A Review. *Journal AWWA* 81(12):59–64.

Jensen, J. N., Johnson, J. D., Staubin, J., and Christman, R. F. (1985). Effect of Monochloramine on Isolated Fulvic-Acid. *Organic Geochemistry* 8(1):71–76.

Johnson, J. D., and Jensen, J. N. (1986). THM and TOX Formation: Routes, Rates, and Precursors. *Journal AWWA* 78(4):156–162.

Johnson, C. J., and Singer, P. C. (2004). Impact of a Magnetic Ion Exchange Resin on Ozone Demand and Bromate Formation During Drinking Water Treatment. *Water Research* 38(17):3738–3750.

Kanniganti, R. (1990). Characterization and Gas Chromatography/Mass Spectrometry Analysis of Mutagenic Extracts of Aqueous Monochloraminated Fulvic Acid. MS Thesis, University of North Carolina at Chapel Hill.

Kanniganti, R., Johnson, J. D., Ball, L. M., and Charles, M. J. (1992). Identification of Compounds in Mutagenic Extracts of Aqueous Monochloraminated Fulvic Acid. *Environmental Science and Technology* 26(10):1998–2004.

Karimi, A. A., and Singer, P. C. (1991). Trihalomethane Formation in Open Reservoirs. *Journal AWWA* 83(3):84–88.

Kim, J., and Reckhow, D. A. (2009). Role or Lignin in DBP Formation. Unpublished manuscript.

Kirisits, M. J., Snoeyink, V. L., and Kruithof, J. P. (2000). The Reduction of Bromate by Granular Activated Carbon. *Water Research* 34(17):4250–4260.

Leiser, G., and Frimmel, F. H. (2000). Removal of Precursors for Disinfection By-Products (DBPs): Differences Between Ozone- and OH-Radical-Induced Oxidation. *Science of the Total Environment* 256(1):1–9.

Krasner, S. W., Chinn, R., Pastor, S. J., Sclimenti, M. J., Richardson, S. D., Thruston, A. D., Jr., and Weinberg, H. S. (2002). Relationships Between the Different Classes of DBPs: Formation, Speciation and Control. In *Proceedings of the 2002 AWWA Water Quality Technology Conference, Seattle, WA*.

Krasner, S. W., Croue, J. P., Buffle, J., and Perdue, E. M. (1996). Three Approaches for Characterizing NOM. *Journal AWWA* 88(6):66–79.

Krasner, S. W., Glaze, W. H., Weinberg, H. S., Daniel, P. A., and Najm, I. N. (1993). Formation and Control of Bromate During Ozonation of Waters Containing Bromide. *Journal AWWA* 85(1):73–81.

Krasner, S. W., McGuire, M. J., Jacangelo, J. G., Patania, N. L., Reagan, K. M., and Aieta, E. M. (1989). The Occurrence of Disinfection By-products in US Drinking Water. *Journal AWWA* 81(8):41–53.

Kronberg, L., Vartiainen, T. (1988). Ames Mutagenicity and Concentration on the Strong Mutagen 3-Chloro-4-(Dichloromethyl)-5-Hydroxy-2(5H)-Furanone and of Its Geometric Isomer E-2-Chloro-3-(Dichloromethyl)-4-Oxo-Butenoic Acid in Chlorine-Treated Tap Waters. *Mutation Research* 206(2):177–182.

Lam, B., Baer, A., Alaee, M., Lefebvre, B., Moser, A., Williams, A., and Simpson, A. J. (2007). Major Structural Components in Freshwater Dissolved Organic Matter. *Environmental Science and Technology* 41(24):8240–8247.

Langlais, B., Reckhow, D. A., and Brink, D. R. (1991). *Ozone in Water Treatment: Application and Engineering.* Chelsea, MI: Lewis Publishers.

Lawrence, J. (1977). The Oxidation of Some Haloform Precursors with Ozone. In *Proceedings of the 3rd Ozone World Congress.* International Ozone Institute, Paris.

Lawrence, J., Tosine, H., Onuska, F. I., and Comba, M. E. (1980). The Ozonation of Natural Waters: Product Identification. *Ozone: Science and Engineering* 2(1):55–64.

Le Cloirec, C., and Martin, G. (1985) Evolution of Amino Acids in Water Treatment Plants and the Effect of Chlorination on Amino Acids. In *Water Chlorination: Environmental Impact and Health Effects*, R. L. Jolley, R. J. Bull, Jolley, Robert L., Bull, Richard J., David, William P., Katz, Sidney, Roberts, Morris H. Jr., and Jacobs, Vivian A. eds. Chelsea, MI: Lewis Publishers.

Le Lacheur, R. M., and Glaze, W. H. (1996). Reactions of Ozone and Hydroxyl Radicals with Serine. *Environmental Science and Technology* 30(4):1072–1080.

Le Lacheur, R. M., Sonnenberg, L. B., Singer, P. C., Christman, R. F., and Charles, M. J. (1993). Identification of Carbonyl Compounds in Enviromental Samples. *Environmental Science and Technology* 27(13):2745–2753.

Leenheer, J., Dotson, A., and Westerhoff, P. (2007). Dissolved organic nitrogen fractionation. *Annals of Environmental Science* 1:45–56.

Leenheer, J. A., Nanny, M. A., and Mcintyre, C. (2003). Terpenoids as Major Precursors of Dissolved Organic Matter in Landfill Leachates, Surface Water, and Groundwater. *Environmental Science and Technology* 37(11):2323–2331.

Legube, B., Langlais, B., Sohm, B., and Dore, M. (1981). Identification of Ozonation Products of Aromatic Hydrocarbon Micropollutants: Effect on Chlorination and Biological Filtration. *Ozone: Science and Engineering* 3(1):33–48.

Leung, S. W., and Valentine, R. L. (1994a). An Unidentified Chloramine Decomposition Product: I. Chemistry and Characteristics. *Water Research* 28(6):1475–1483.

Leung, S. W., Valentine, R. L. (1994b). An Unidentified Chloramine Decomposition Product: II. A Proposed Formation Mechanism. *Water Research* 28(6):1485–1495.

Liang, L., and Singer, P. C. (2003). Factors Influencing the Formation and Relative Distribution of Haloacetic Acids and Trihalomethanes in Drinking Water. *Environmental Science and Technology* 37(13):2920–2928.

Liu, W., Andrews, S. A., Bolton, J. R., Linden, K. G., Sharpless, C., Stefan, M. (2002). Comparison of DBP formation from different UV technologies at bench scale. *Water Science and Technology: Water Supply* 2(5–6):515–521.

Liukkonen, R. J., Lin, S., Oyler, A. R., Lukasewycz, M. T., Cox, D. A., Yu, Z.-j., and Carlson, R. M. (1983). Product Distribution and Relative Rates of Reaction of Aqueous Chlorine and Chlorine Dioxide with Polynuclear Aromatic Hydrocarbons. In *Water Chlorination: Environmental Impact and Health Effects*, Vol. 4, Book 1: *Chemistry and Water Treatment. Proceedings of the Fourth Conference on Water Chlorination*, R. L. Jolley, W. A. Brungs, Jolley, Robert L., Brungs, William A., Cotruvo, Joseph A., Cumming, Robert B., Mattice,

Jack S., and Jacobs, Vivian A. eds. Pacific Grove, CA. , Ann Arbor Science Publ., Ann Arbor, MI.

MacNeill, A. L. (1994). Mechanistic Modeling of Trihalomethanes and Other Neutral Chlorination Byproducts. M.S. thesis, University of Massachusetts, Amherst.

Magnuson, M. L., Kelty, C. A., Sharpless, C. M., Linden, K. G., Fromme, W., Metz, D. H., and Kashinkunti, R. (2002). Effect of UV Irradiation on Organic Matter Extracted From Treated Ohio River Water Studied Through the Use of Electrospray Mass Spectrometry. *Environmental Science and Technology* 36(23):5252–5260.

Mallevialle, J., and Suffet, I. H. (1987). *Identification and Treatment of Tastes and Odors in Drinking Water.* Awwa Research Foundation Report. Denver, CO: Water Research Foundation.

Malley, J. P., Jr., Shaw, J. P., and Ropp, J. R. (1995). *Evaluation of By-products Produced by Treatments in Ground Waters with Ultraviolet Irradiation.* Awwa Research Foundation Report. Denver, CO: Water Research Foundation.

Masschelein, W. J. (1979). *Chlorine Dioxide: Chemistry and Environmental Impact of Oxychlorine Compounds.* Ann Arbor, MI: Ann Arbor Science Publishers.

McClellan, J. N., Reckhow, D. A., Tobiason, J. E., Edzwald, J. K., and Smith, D. B. (2000). A Comprehensive Kinetic Model for Chlorine Decay and Chlorination Byproduct Formation. In *Natural Organic Matter and Disinfection By-Products: Characterization and Control in Drinking Water*, S. E. Barrett, S. W. Krasner, G. L. Amy, eds. ACS Symposium Series 761. Washington, DC: American Chemical Society.

McGuire, M. J., McLain, J. L., and Obolensky, A. (2002). *Information Collection Rule Data Analysis.* Awwa Research Foundation Report. Denver, CO: Water Research Foundation.

McKnight, A., and Reckhow, D. A. (1992). Reactions of Ozonation By-products with Chlorine and Chloramines. In *1992 Annual Conference Proceedings: American Water Works Association, Vancouver, British Columbia, Canada.*

Meier, J. R. (1988). Genotoxic Activities of Organic Chemicals in Drinking Water. *Mutation Research* 196(3):211–245.

Merlet, N. (1986). Contribution a l'Etude du Mecanisme de Formation des Trihalomethanes et des Composes Organohalogenes non Volatils lors de la Chloration de Molecules Modeles. PhD Dissertation, University of Poitiers, Poitiers, France.

Miller, J. W., and Uden, P. C. (1983). Characterization of Nonvolatile Aqueous Chlorination Products of Humic Substances. *Environmental Science and Technology* 17(3):150–157.

Minear, R. A., and Bird, J. C. (1980). Trihalomethanes: Impact of Bromide Ion Concentration on Yield, Species Distribution, Rate of Formation and Influence of Other Variables. In *Water Chlorination: Environmental Impact and Health Effects*, Vol. 3: *Proceedings of the Third Conference on Water Chlorination: Environmental Impact and Health Effects*, R. L. Jolley, W. A. Brungs, R. B. Cumming, eds. Colorado Springs, CO., Ann Arbor Science Publ., Ann Arbor, MI

Minisci, F., and Galli, R. (1965). A New Highly Selective Type of Aromatic Substitution: Homolytic Amination of Phenolic Ethers. *Tetrahedron Letters* 6(8):433–436.

Mitch, W. (2007). Nitrosamine, Nitrile and Nitramine Formation Relevant to Nitrification Control. In *Proceedings of AWWA 2007 Water Quality Technology Conference, Charlotte, NC.*

Morris, J. C. (1975). *Formation of Halogenated Organics by Chlorination of Water Supplies.* EPA-600/1-75-002. Washington, DC: USEPA.

Morris, J. C., and Isaac, R. A. (1983). A Critical Review of Kinetic and Thermodynamic Constants for the Aqueous Chlorine: Ammonia System. In *Water Chlorination: Environmental Impact and Health Effects*, Vol. 4, Book 1, Robert L. Jolley, William A. Brungs, Joseph A. Cotruvo, Robert B. Cumming, Jack S. Mattice, and Vivian A. Jacobs eds. Ann Arbor, MI: Science Publishers.

Najm, I., Brown, N. P., Guo, Y. C., Hwang, C. J., and Barrett, S. E. (2006). *Formation of Hydrazine as a Chloramination By-product.* Awwa Research Foundation Report 91122. Denver, CO: Water Research Foundation.

Najm, I., and Trussell, R. R. (2001). NDMA Formation in Water and Wastewater. *Journal AWWA* 93(2):92–99.

Neale, R. (1964). The Chemistry of Ion Radicals: The Free-Radical Addition of *N*-Chloramines to Olefinic and Acetylenic Hydrocarbons. *Journal of the American Chemical Society* 86:5340–5342.

Nguyen, M. L., Baker, L. A., and Westerhoff, P. (2002). DOC and DBP Precursors in Western US Watersheds and Reservoirs. *Journal AWWA* 94(5):98–112.

Niquette, P., Prevost, M., Merlet, N., and LaFrance, P. (1999). Influence of Factors Affecting the Removal of Chlorine and Chlorination By-product Precursors in Organic Filters. *Water Research* 33(10):2329–2344.

Obolensky, A, Singer, P. C., and Shukairy, H. M. (2007). Information Collection Rule Data Evaluation and Analysis to Support Impacts on Disinfection By-product Formation. *Journal of Environmental Engineering* 133(1):53–63.

Obolensky, A., Speight, V., and Singer, P. C. (2008). Short-Term Variations in DBP Concentrations at the Point of Entry to the Distribution System. In *Proceedings of the 2008 AWWA Water Quality Technology Conference, Cincinnati, OH.*

Oliver, B. G. (1983). Dihaloacetonitriles in Drinking Water: Algae and Fulvic Acid as Precursors. *Environmental Science and Technology* 17(2):80–83.

Onstad, G. D., and Weinberg, H. S. (2005). Evaluation of the Stability and Analysis of Halogenated Furanones in Disinfected Drinking Waters. *Analytica Chimica Acta* 534(2):281–292.

Palin, A. T. (1975). Water Disinfection: Chemical Aspects and Analytical Control. In *Disinfection: Water and Wastewater*, J. D. Johnson, ed. Ann Arbor, MI: Ann Arbor Science.

Palmstrom, N. S., Carlson, R. E., and Cooke, G. D. (1988). Potential Links Between Eutrophication and the Formation of Carcinogens in Drinking Water. *Lake and Reservoir Management* 4(2):1–15.

Pankow, J. F. (1991). *Aquatic Chemistry Concepts*. Chelsea, MI: Lewis Publishers.

Paramisigamani, V., Malaiyandi, M., Benoit, F. M., Helleur, R., and Ramaswamy, S. (1983). Identification of Ozonated and/or Chlorinated Residues of Fulvic Acids. In *Proceedings of the 6th Ozone World Congress*. Vienna, VA: International Ozone Association.

Pedersen, E. J., Urbansky, E. T., Marinas, B. J., and Margerum, D. W. (1999). Formation of Cyanogen Chloride from the Reaction of Monochloramine with Formaldehyde. *Environmental Science and Technology* 33(23):4239–4249.

Peldszus, S., Souza, R., Bolton, J. R., Dussert, B. W., Smith, F., and Andrews, S. A. (2000). Impact of Medium Pressure UV-Disinfection on the Formation of Low Molecular Weight Organic By-products and Nitrite, and the Reduction of Bromate. In *Proceedings of the 2000 AWWA Water Quality Technology Conference, Salt Lake City, UT.*

Perdue, E. M., and Ritchie, J. D. (2004). Dissolved Organic Matter in Fresh Waters. In *Treatise on Geochemistry*, Vol. 5: *Surface and Ground Water, Weathering, Erosion and Soils*, J. I. Drever, ed., Elsevier, Amsterdam.

Peters, R. J. B., de Leer, E. W. B., and Versteegh, J. F. M. (1994). Identification of Halogenated Compounds Produced by Chlorination of Humic Acid in the Presence of Bromide. *Journal of Chromatography A* 686(2):253–261.

Plewa, M. J., Wagner, E. D., Richardson, S. D., Thruston, A. D., Woo, Y. T., and McKague, A. B. (2004). Chemical and Biological Characterization of Newly Discovered Iodoacid Drinking Water Disinfection By-products. *Environmental Science and Technology* 38(18):4713–4722.

Pourmoghaddas, H., and Dressman, R. C. (1993). Determination of Nine Haloacetic Acids in Finished Drinking Water. In *Proceedings Water Quality Technology Conference,* AWWA, Toronto, Ontario, Canada.

Reckhow, D. A. (1984). Organic Halide Formation and the Use of Preozonation and Alum Coagulation to Control Organic Halide Precursors. Ph.D. dissertation, University of North Carolina at Chapel Hill.

Reckhow, D. A. (1999). Control of Disinfection By-product Formation Using Ozone. In *Formation and Control of Disinfection By-Products in Drinking Water*, P. C. Singer, ed. Denver, CO: American Water Works Association.

Reckhow, D. A., Legube, B., and Singer, P. C. (1986). The Ozonation of Organic Halide Precursors: Effect of Bicarbonate. *Water Research* 20(8):987–998.

Reckhow, D. A., Linden, K. G., Kim, J., Shemer, H., and Makdissy, G. (2010). "Effect of UV treatment on DBP formation." *Journal AWWA*, 102(6): 100–113.

Reckhow, D. A., and Pierce, P. D. (1992). *A Simple Spectrophotometric Method for the Determination of THMs in Drinking Water*. Awwa Research Foundation Report. Denver, CO: Water Research Foundation.

Reckhow, D. A., Platt, T. L., MacNeill, A., and McClellan, J. N. (2001). Formation and Degradation of DCAN in Drinking Waters. *Journal of Water Supply Research and Technology, AQUA* 50(1):1–13.

Reckhow, D. A., Rees, P. L., Nusselin, K., et al. (2007). *Long-Term Variability of BDOM and NOM as Precursors in Watershed Sources*. Awwa Research Foundation Report. Denver, CO: Water Research Foundation.

Reckhow, D. A., and Singer, P. C. (1985). Mechanisms of Organic Halide Formation During Fulvic Acid Chlorination and Implications with Respect to Preozonation. In *Water Chlorination: Environmental Impact and Health Effects*, Vol. 5, R. L. Jolley, R. L. Jolley, R. J. Bull, W. P. David, S. Katz, M. H. J. Roberts, and V. A. Jacobs, eds. Chelsea, MI, Lewis Publishers.

Reckhow, D. A., and Singer, P. C. (1990). Chlorination By-products in Drinking Waters: From Formation Potentials to Finished Water Concentrations. *Journal AWWA* 82(4):173–180.

Reckhow, D. A., Singer, P. C., and Malcolm, R. L. (1990). Chlorination of Humic Materials: By-product Formation and Chemical Interpretations. *Environmental Science and Technology* 24(11):1655–1664.

Reckhow, D. A., Singer, P. C. and Trussell, R. R. (1986). Ozone as a Coagulant Aid. In *Ozonation: Recent Advances and Research Needs*. Denver, CO: American Water Works Association.

Reckhow, D. A., Tobiason, J. E., Switzenbaum, M. S., McEnroe, R., Xie, Y., Zhu, Q.-w., Zhou, X., McLaughlin, P., and Dunn, H. J. (1992). Control of Disinfection Byproducts and AOC by Pre-Ozonation and Biologically-Active In-Line Direct Filtration. In *1992 Annual Conference Proceedings of the American Water Works Association, Vancouver, British Columbia, Canada.*

Reckhow, D. A., Xie, Y., McEnroe, R., Byrnes, P., Tobiason, J. E., and Switzenbaum, M. S. (1993). The Use of Chemical Surrogates for Assimilable Organic Carbon. In *1993 Annual Conference Proceedings of the American Water Works Association, San Antonio, TX.*

Reckhow, D. A., Zhu, Q.-W., Weiner, J. M., and MacNeill, A. L. (1993). Effect of Ozonation on the Kinetics of DBP Formation. In *Proceedings of the 10th World Ozone Congress of the International Ozone Association, San Francisco, CA.*

Reinhard, M., Goodman, N. L., McCarty, P. L., and Argo, D. G. (1986). Removing Trace Organics by Reverse-Osmosis Using Cellulose-Acetate and Polyamide Membranes. *Journal AWWA* 78(4):163–174.

Richardson, S. D. (1998). Drinking Water Disinfection By-products. In *Encyclopedia of Environmental Analysis and Remediation*, R. A. Meyers, ed. New York: Wiley.

Richardson, S. D., Thruston, A. D., Collette, T. W., Patterson, K. S., Lykins, B. W., Majetich, G., and Zhang, Y. (1994). Multispectral Identification of Chlorine Dioxide Disinfection By-products in Drinking Water. *Environmental Science and Technology* 28(4):592–599.

Roberts, M., Singer, P. C., and Obolensky, A. (2002). Comparisons Between Total Haloacetic Acid and Total Trihalomethane Concentrations in Finished Drinking Water: An Analysis of ICR Data. *Journal AWWA* 94(1):103–114.

Rook, J. J. (1974). Formation of Haloforms During Chorination of Natural Waters. *Water Treatment and Examination* 23:234–243.

Saunier, B. M., and Selleck, R. E. (1979). The Kinetics of Breakpoint Chlorination in Continuous Flow Systems. *Journal AWWA* 71(3):164–172.

Savitz, D.A., P.C. Singer, K.E. Hartmann, A.H. Herring, H.S. Weinberg, C. Makarushka, C. Hoffman, R. Chan, R. Maclehose. (2005). *Drinking Water Disinfection By-Products and Pregnancy Loss.* Awwa Research Foundation Report 91088F. Denver, CO: Water Research Foundation.

Schechter, D. S., and Singer, P. C. (1995). Formation of Aldehydes During Ozonation. *Ozone: Science & Engineering* 17(1):53–69.

Schreiber, I. M., and Mitch, W. A. (2006). Nitrosamine Formation Pathway Revisited: The Importance of Chloramine Speciation and Dissolved Oxygen. *Environmental Science and Technology* 40(19):6007–6014.

Scully, F. E., Jr. (1986). *N*-Chloro Compounds: Occurrence and Potential Interference in Residual Analysis. In *Proceedings of the 1986 Water Quality Technology Conference: Advances in Water Analysis and Treatment.* Denver, CO: American Water Works Association.

Shank, R. C., and Whittaker, C. (1988). *Formation of Genotoxic Hydrazine by the Chloramination of Drinking Water.* Technical Completion Report, Project No. W-690. Berkeley, CA: University of California Water Resources Center;

Shukairy, H. M., and Summers, R. S. (1996). DBP Speciation and Kinetics as Affected by Ozonation and Biotreatment. In *Disinfection By-products in Water Treatment*, R. Minear and G. L. Amy, eds. Boca Raton, FL: Lewis Publishers

Shukairy, H. M., Blank, V., and McLain, J. L. (2002). TOX in Finished Water and Distribution Systems: Relationships with Other DBPs. In *Information Collection Rule Data Analysis*, M. J. McGuire, J. L. McLain, and A. Obolensky, eds. Denver, CO: American Water Works Association Research Foundation.

Singer, P. C., Shi, J., and Weinberg, H. S. (1998). Evaluation of pH Adjustment and Chloramination for the Control of Haloacetic Acid Formation in North Carolina Drinking Water. In *Proceedings of the 1998 Water Quality Technology Conference.* Denver, CO: AWWA.

Singer, P. C. 2001. Variability and Assessment of Disinfection By-product Concentrations in Water Distribution Systems. In *Microbial Pathogens and Disinfection By-products in Drinking Water: Health Effects and Management of Risks*, G. F. Craun, F. S. Hauchman, and D. E. Robinson, eds. Washington, DC: International Life Sciences Institute.

Singer, P. C., and Chang, S. D. (1989). Correlations Between Trihalomethanes and Total Organic Halides Formed During Water Treatment. *Journal AWWA* 81(8):61–65.

Singer, P. C., and Gurol, M. D. (1983). Dynamics of the Ozonation of Phenol: I. Experimental Observations. *Water Research* 17(9):1163–1171.

Singer, P. C., Schneider, M., Edwards-Brandt, J., and Budd, G. C. (2007). Magnetic Ion Exchange for the Removal of Disinfection By-product Precursors: Pilot Plant Findings. *Journal AWWA* 99(4):128–139.

Smeds, A., Holmbom, B., and Tikkanen, L. (1990). Formation and degradation of mutagens in kraft pulp mill water systems *Pulp and Paper Research Journal* 5 (3):142-147

Snoeyink, V. L., and Jenkins, D. (1980). *Water Chemistry.* New York: Wiley.

Solarik, G., Summers, R. S., Sohn, J., Swanson, W. J., Chowdhury, Z. K., and Amy, G. L. (2000). Extensions and Verification of the Water Treatment Plant Model for Disinfection By-product Formation. In *Natural Organic Matter and Disinfection By-Products: Characterization and Control in Drinking Water*, S. E. Barrett, S. W. Krasner, and G. L. Amy, eds. Washington, DC: American Chemical Society.

Somsen, R. A. (1960). Oxidation of Some Simple Organic Molecules with Aqueous Chlorine Dioxide Solutions: II. Reaction Products. *Journal of the Technical Association of the Pulp and Paper Industry* 43(2):157–160.

Song, R., Westerhoff, P., Minear, R., and Amy, G. L. (1997). Bromate Minimization During Ozonation. *Journal AWWA* 89(6):69–78.

Speight, V. L., and Singer, P. C. (2005). Association Between Residual Chlorine and Reduction in Haloacetic Acid Concentrations in Distribution Systems. *Journal AWWA* 97(2):82–91.

Speiran, G. K. (2000). Dissolved Organic Carbon and Disinfection By-product Precursors in Waters of the Chickahominy River Basin, Virginia, and Implications for Public Supply. *Water Resources Investigations.* Report 00-4175. Richmond, VA: U.S. Geological Survey.

Speitel, G. E., Jr. (1999). Control of Disinfection By-product Formation Using Chloramines. In *Formation and Control of Disinfection By-products in Drinking Water*, P. C. Singer, ed. Denver, CO: American Water Works Association.

Staehelin, J., and Hoigne, J. (1982). Decomposition of Ozone in Water: Rate of Initiation by Hydroxide Ions and Hydrogen Peroxide. *Environmental Science and Technology* 16(10): 676–681.

Steinbergs, C. Z. (1986). Removal of By-products of Chlorine and Chlorine Dioxide at a Hemodialysis Center. *Journal AWWA* 78(6):94–98.

Stepczuk, C., Owens, E. M., Effler, S., W., Effler, S., W., Auer, M. T., and Bloomfield, J. A. (1998). A Modeling Analysis of THM Precursors for a Eutrophic Reservoir. *Journal of Lake and Reservoir Management* 14(2–3):367–378.

Stevens, A. A., Slocum, C. J., Seeger, D. R., and Robeck, G. G. (1978). Chlorination of Organics in Drinking Water. In *Water Chlorination: Environmental Impact and Health Effects*, Vol. 1: *Proceedings of the Conference on the Environmental Impact of Water Chlorination*, R. L. Jolley, ed. Oak Ridge, TN., Ann Arbor Science Publ, Ann Arbor, MI.

Stevens, A. A. (1982). Reaction Products of Chlorine Dioxide. *Environmental Health Perspectives* 46(12):101–110.

Stumm, W., and Morgan, J. J. (1996). *Aquatic Chemistry*, 3rd ed., New York: Wiley Interscience.

Summers, R. S., Hooper, S. M., Solarik, G., Owen, D. M., and Hong, S. H. (1995). Bench-Scale Evaluation of GAC for NOM Control. *Journal AWWA* 87(8):69–80.

Svenson, D., Kadla, J. F., Chang, H. M., and Jameel, H. (2002). Effect of pH on the Mechanism of OClO Center Dot Oxidation of Aromatic Compounds. *Canadian Journal of Chemistry* (*Revue Canadienne De Chimiei*) 80(7):761–766.

Swanson, W. J., Chowdhury Z. K., Aburto L., and Summers R. S. (2002). Disinfection Practices and Pathogen Inactivation in ICR Surface Water Plants. In *Information Collection Rule Data Analysis*, M. J. McGuire, J. L. McLain, and A. Obolensky, eds. Awwa Research Foundation Report. Denver, CO: Water Research Foundation.

Symons, J. M., Bellar, T. A., Carswell, J. K., DeMarco, J., Kropp, K. L., Robeck, G., G., Seeger, D. R., Slocum, C. J., Smith, B. L., and Stevens, A. A. (1975). National Organics Reconnaissance Survey for Halogenated Organics. *Journal AWWA* 67(11):634–647.

Tratnyek, P. G., and Hoigne, J. (1994). Kinetics of Reactions of Chlorine Dioxide (OClO) in Water: II. Quantitative Structure-Activity Relationships for Phenolic Compounds. *Water Research* 28(1):57–66.

Thibaud, H., Merlet, N., and Dore, M. (1986). Formation de Chloropicrine lors de la Chloration de Quelques Composes Organiques Nitres en Solution Aqueuse. Incidence d'une Preozonation. *Environmental Technology Letters* 7(3):163–176.

Thomson, J., Roddick, F. A., and Drikas, M. (2002). UV Photooxidation Facilitating Biological Treatment for the Removal of NOM from Drinking Water. *Journal of Water Supply Research and Technology, AQUA* 51(6):297–306.

Thurman, E. M., and Malcolm, R. L. (1981). Preparative Isolation of Aquatic Humic Substances. *Environmental Science and Technology* 15(4):463–466.

Topudurti, K. V., and Haas, C. N. (1991). THM Formation by the Transfer of Active Chlorine from Monochloramine to Phloroacetophenone. *Journal AWWA* 83(5):62–66.

Trehy, M. L., and Bieber, T. I. (1981). Detection, Identification, and Quantitative Analysis of Dihaloacetonitriles in Chlorinated Natural Waters *Advances in the Identification and Analysis of Organic Pollutants in Water*, Vol. 2. Ann Arbor, MI: Ann Arbor Science.

Umphres, M. D., Tate, C. H., Kavanaugh, M. C., and Trussell, R. R. (1983). Trihalomethane Removal by Packed-Tower Aeration. *Journal AWWA* 75(8):414–418.

Urfer, D., Huck, P. M., Booth, S. D. J., and Coffey, B. M. (1997). Biological Filtration for BOM and Particle Removal: A Critical Review. *Journal AWWA* 89(12):83–98.

Venczel, L. V., Arrowood, M., Hurd, M., and Sobsey, M. D. (1997). Inactivation of *Cryptosporidium parvum* Oocysts and *Clostridium perfringens* Spores by a Mixed-Oxidant Disinfectant and by Free Chlorine. *Applied and Environmental Microbiology* 63(4):1598–1601.

Vongunten, U., and Holgne, J. (1994). Bromate Formation During Ozonation of Bromide-Containing Waters: Interaction of Ozone and Hydroxyl Radical Reactions. *Environmental Science and Technology* 28(7):1234–1242.

Wajon, J. E., Rosenblatt, D. H., and Burrows, E. P. (1982). Oxidation of Phenol and Hydroquinone by Chlorine Dioxide. *Environmental Science and Technology* 16(7):396–402.

Wang, J. Z., Summers, R. S., and Miltner, R. J. (1995). Biofiltration Performance: 1. Relationship to Biomass. *Journal AWWA* 87(12):55–63.

Wei, I. W., and Morris, J. C. (1974). Dynamics of Breakpoint Chlorination. In *Chemistry of Water Supply, Treatment, and Distribution*, A. J. Rubin, ed. Ann Arbor, MI: Ann Arbor Science.

Weinberg, H. S., Krasner, S. W., Richardson, S. D., and Thruston, A. D., Jr. (2002). *The Occurrence of Disinfection By-products (DBPs) of Health Concern in Drinking Water: Results of a Nationwide DBP Occurrence Study.* EPA/600/R-02/068. Cincinnati, OH: USEPA.

Werdehoff, K. S., and Singer, P. C. (1987). Chlorine Dioxide Effects on THMFP, TOXFP, and the Formation of Inorganic By-products. *Journal AWWA* 79(9):107–113.

Wert, E. C., Neemann, J. J., Johnson, D., Rexing, D., and Zegers, R. (2007). Pilot-Scale and Full-Scale Evaluation of the Chlorine-Ammonia Process for Bromate Control During Ozonation. *Ozone: Science and Engineering* 29(5):363–372.

Williams, S. L., Rindfleisch, D. F., and Williams, R. L. (1995). Deadend on Haloacetic Acids (HAA). In *Proceedings of the 1994 Water Quality Technology Conference.* Denver, CO: American Water Works Association.

Wright, J. M., Schwartz, J., Vartiainen, T., Maki-Paakkanen, J., Altshul, L., Harrington, J. J., and Dockery, D. W. (2002). 3-Chloro-4-(Dichloromethyl)-5-Hydroxy-2(5H)-Furanone (MX) and Mutagenic Activity in Massachusetts Drinking Water. *Environmental Health Perspectives* 110(2):157–164.

Xie, Y., and Reckhow, D. A. (1996). Hydrolysis and Dehalogenation of Trihaloacetaldehydes. In *Disinfection By-products in Water Treatment: The Chemistry of Their Formation and Control*, R. A. Minear and G. L. Amy, eds. Boca Raton, FL: Lewis Publishers.

Xie, Y., and Reckhow, D. A. (1993). Stability of Cyanogen Chloride in the Presence of Sulfite and Chlorine. In *Proceedings of the 1993 Water Quality Tecnology Conference*. Denver, CO: American Water Works Association.

Young, J. S., and Singer, P. C. (1979). Chloroform Formation in Public Water Supplies: A Case Study. *Journal AWWA* 71(2):87–95.

Young, M. S., Mauro, D. M., Uden, P. C., and Reckhow, D. A. (1995). *The Formation of Nitriles and Related Halogenated Disinfection By-products in Chlorinated and Chloraminated Water: Applications of Microscale Analytical Procedures*. Chicago: American Chemical Society.

Zhang, X., Echigo, S., Minear, R. A., and Plewa, M. J. (2000). *Natural Organic Matter and Disinfection By-products: Characterization and Control in Drinking Water*, S. W. Krasner, S. E. Barrett, and G. L. Amy, eds. Washington, DC: ACS.

Zhang, X., and Minear, R. A. (2002). Decomposition of Trihaloacetic Acids and Formation of the Corresponding Trihalomethanes in Drinking Water. *Water Research* 36(14):3665–3673.

CHAPTER 20
INTERNAL CORROSION AND DEPOSITION CONTROL

Michael R. Schock
Chemist
Darren A. Lytle
Environmental Engineer

U.S. Environmental Protection Agency
Water Supply and Water Resources Division
Cincinnati, Ohio, United States

INTRODUCTION 20.2
CORROSION, PASSIVATION, AND IMMUNITY 20.3
Electrochemical Reactions 20.3
The Nernst Equation 20.4
Types of Corrosion 20.5
Pitting Corrosion 20.7
Corrosion Products on Pipe Surfaces 20.11
Corrosion Kinetics 20.12
Solubility Diagrams 20.13
Pourbaix or Potential-pH Diagrams 20.18
PHYSICAL FACTORS AFFECTING CORROSION AND METALS RELEASE 20.20
Physical Characteristics 20.21
Manufacturing-Induced Characteristics 20.24
CHEMICAL FACTORS AFFECTING CORROSION 20.25
Alkalinity/Dissolved Inorganic Carbon (DIC) 20.25
Ammonia 20.25
Buffer Intensity (β), Buffer Capacity, Buffer Index 20.26
Chloride and Sulfate 20.29
Copper 20.30
Disinfectant Residual 20.30
Dissolved Oxygen 20.32
Hardness and Calcium Carbonate Scaling 20.32
Hydrogen Sulfide (H_2S) 20.35
Magnesium 20.35
Metallic Precipitates 20.35
Natural Color and Organic Matter 20.38
Orthophosphate 20.38
pH 20.38
Polyphosphates 20.39
Silicate 20.39
Total Dissolved Solids (TDS) 20.39
CORROSION OF SPECIFIC MATERIALS 20.40
Cast and Ductile Iron (unlined) 20.40
Iron Release Models 20.42
Iron Corrosion Control Strategies 20.42
Iron Sequestration 20.45
Copper 20.46
Brasses and Bronzes 20.53
Lead and Lead-Containing Solder 20.54
Galvanized Steel 20.63
Asbestos-Cement (A-C) Pipe, Cement-Mortar Linings, and Concrete Pipe 20.64
DIRECT METHODS FOR THE ASSESSMENT OF CORROSION 20.66
Physical Inspection 20.66
Corrosion Rate Measurements 20.66
Pipe Rig Systems 20.67
Immersion Testing in the Laboratory 20.68

Chemical Characterization of Deposits.. 20.69

CORROSION CONTROL ALTERNATIVES **20.73**

Materials Selection.......................... 20.73

Engineering Considerations......... 20.74

Water Chemistry Modification 20.75

Secondary Effects of Treatment for Corrosion Control 20.75

Linings, Coatings, and Paints....... 20.76

WATER SAMPLING FOR CORROSION CONTROL.................. **20.77**

Defining the Problem.................... 20.77

Statistical Considerations............. 20.79

Chemical and Physical Considerations.............................. 20.80

ACKNOWLEDGMENTS **20.81**

DISCLAIMER.. **20.82**

ABBREVIATIONS **20.82**

NOTATION FOR EQUATIONS **20.82**

REFERENCES **20.83**

INTRODUCTION

Corrosion is one of the most important problems in the drinking water industry. It can affect public health, public acceptance of a water supply, and the cost of providing safe water (Edwards, 2004; WHO, 2006). Maintaining the integrity of drinking water distribution system materials and distribution system water quality are important objectives of drinking water suppliers. Corrosion, microbiological regrowth, disinfectant demand, water flow restrictions due to excessive pipe scale buildup, discolored or turbid water resulting from metal release from pipe materials or scale, and metallic tasting water are examples of specific distribution system issues.

Deterioration of materials resulting from corrosion can necessitate huge yearly expenditures of resources for repairs, replacement, and system maintenance. A recent assessment by the U.S. Environmental Protection Agency (USEPA) of drinking water infrastructure needs in the United States estimated that a 20-year capital investment of $335 billion will be needed for costs for repairs and replacement of transmission lines and storage, treatment equipment, and other health-related projects for the 70,000 community and not-for-profit, non-community water systems (USEPA, 2009a). Many times the problem is not given the attention it needs until expensive changes or repairs are required. Often overlooked, plumbers and plumbing code development and enforcement play vital roles in risk identification and management in drinking water (WHO, 2006).

Corrosion tends to increase the concentrations of many metals in tap water, although in the United States only two of them (lead and copper) are regulated as corrosion by-products at the consumers' taps. The occurrence of the toxic metals lead and cadmium is attributable almost entirely to leaching caused by corrosion of plumbing materials. Three metals—copper, iron, and zinc—cause staining of fixtures, or metallic taste, or both. Low levels of tin and antimony can be caused by the corrosion of lead-free solders (Herrera et al., 1982; Subramanian et al., 1991; Subramanian et al., 1994). Nickel has sometimes been mentioned as a potential contaminant from the plating of decorative plumbing fixtures (Kimbrough, 2009). The promulgation of the Lead and Copper Rule by USEPA (1991), which has undergone several revisions, has created an awareness and emphasis on corrosion control in distribution systems, as well as domestic, public, and institutional plumbing systems (USEPA, 2009b).

Corrosion products attached to pipe surfaces or accumulated as sediments in the distribution system can shield microorganisms from disinfectants (see Chap. 21). These organisms can reproduce and cause problems such as bad tastes, odors, slimes, and additional corrosion. Corrosion control within the distribution system, therefore, is an important factor in secondary disinfection and maintenance of high water quality (Camper, 1996; Kiéné et al., 1996; Olson, 1996; Rompré et. al., 1996; Schreppel et al., 1997). Corrosion deposits on pipes also provide a major reservoir for a broad variety of elements, some of which could have considerable health consequences if ingested (Reiber and Dostal, 2000; Schock, 2005; Schock et. al., 2008a).

The term *corrosion* is also commonly applied to the dissolution and carbonation reactions of cement-based materials. An increase in pH often occurs as a result of this reaction, which can be detrimental to disinfection and the aesthetic quality of the water, as well as reducing the effectiveness of phosphate corrosion inhibitor chemicals intended to control the corrosion of metals. The release of asbestos fibers is of regulatory concern, and in extreme cases, the chemical attack on the pipe by the water may cause a reduction of structural integrity and, ultimately, failure.

Even when a water system passes all regulatory requirements, the release of corrosion byproducts by miles of distribution system and domestic piping and the application of corrosion inhibitor chemicals containing metals such as zinc can be significant sources of metal loading of wastewater treatment plants. This contamination source can affect compliance with discharge or sludge disposal limits. Phosphate-based corrosion inhibitors can increase nutrient loading to wastewater plants and can cause violations of wastewater or other discharge regulations or contribute to water quality problems in ecosystems receiving the water.

Long-term corrosion-related failure of piping can cause many indirect health consequences, ranging from subtle effects such as intrusion from pressure transients to catastrophic failures and massive potential contamination. Edwards (2004) describes in detail many of these potential impacts.

Each material that can corrode has a body of literature devoted to it. Detail on the form of corrosion for each metal or piping material, possible secondary trace contaminants that can be released into the water, and specific corrosion inhibition practices that might be effective, can be found in a comprehensive text (AwwaRF-TZW, 1996) and numerous specific water treatment journal articles. In general, plastic plumbing materials are more corrosion resistant, but they have their own degradation mechanisms and other potential problems, such as brittleness, taste, odor, biofilms support, and both organic and inorganic leachates.

The purpose of this chapter is threefold: (1) to provide an introduction and overview of the chemistry and treatment concepts involved in corrosion and deposition phenomena in drinking waters, (2) to provide a guide to more specialized literature that describes the intricacies of material corrosion in more depth, and (3) to provide general guidance on preventing, diagnosing, and mitigating corrosion and deposition effects in drinking water distribution, storage, and premise plumbing systems.

CORROSION, PASSIVATION, AND IMMUNITY

Electrochemical Reactions

Metal species can be released into water either from the simple dissolution of, or desorption or scouring from, existing scale materials. For corrosion of any metal to occur, all of the components of an electrochemical cell must be present (AwwaRF-TZW, 1996; MWH, 2005). These include an *anode*, a *cathode*, a *connection* between the anode and cathode for electron transport, and an *electrolyte solution* that conducts ions between the anode and cathode. The anode and cathode are sites on the metal that have different electrical potential. If any one of these components is absent, a corrosion cell does not exist and corrosion does not occur.

Oxidation and dissolution of the metal take place at the anode. The electrons generated by the anodic reaction migrate to the cathode, where they are discharged to a suitable electron acceptor, such as oxygen. The positive ions generated at the anode tend to migrate to the cathode, and the negative ions generated at the cathode tend to migrate to the anode. Migration occurs to maintain electroneutrality.

At the phase boundary of a metal in an electrolyte solution an electrical potential difference exists between the solution and the metal surface. This potential is the result of the tendency of the metal to reach chemical equilibrium with the electrolyte solution. This oxidation reaction, representing a loss of electrons by the metal, can be written as

$$Me \rightleftharpoons Me^{z+} + ze^- \tag{20-1}$$

Equation 20-1 indicates that the metal corrodes, or dissolves, as the reaction goes to the right. This reaction will proceed until the metal is in equilibrium with the electrolyte containing ions of this metal.

Corrosion results from the flow of electric current between electrodes (anodic and cathodic areas) on the metal surface. These areas may be microscopic and in very close proximity in the case of uniform corrosion. Alternatively, they may be large and somewhat remote from one another, causing non-uniform corrosion or pitting, with or without *tuberculation*. Two examples of this situation are tuberculation of iron and galvanized steel piping, and *pinhole leak* formation in copper pipes. The former can greatly impede water flow, whereas the latter causes leakage and loss of integrity of premise plumbing or service lines. The type of corrosion can be controlled by various factors that include but are not limited to the type of metal, characteristics of the water, flow patterns, impurities in or on the metal surface, sediment accumulations and bacterial slimes.

The current that results from the oxidation of the metal is called the *anodic current*. In the companion reaction the electrons are removed by reaction with oxidants via the *cathodic current* (the reaction, Eq. 20-1, going to the left). In drinking water, the cathodic reaction is with a combination of oxygen, disinfectant (usually, but not always, free chlorine or monochloramine), or occasionally some other strong oxidizing agent. The situation of true equilibrium is only rarely the case. In almost all forms of pipe corrosion, the metal goes into solution at the anodic areas. As the metal dissolves, a movement of electrons occurs and the metal develops an electric potential.

The Nernst Equation

The Nernst equation is a relationship that allows the driving force of the reaction to be computed from the difference in free-energy levels of corrosion cell components (Pankow, 1991; Stumm and Morgan, 1996; Langmuir, 1997; Butler and Cogley, 1998). The free-energy difference under such conditions depends on the electrochemical potential, which, in turn, is a function of the type of metal and the solid- and aqueous-phase reaction products. Electrons (electricity) will then flow from certain areas of a metal surface to other areas through the metal. A metal may go into solution as an ion, or may react in water with another element or molecule to form a soluble aqueous complex, ion pair, or insoluble compound. The equilibrium potential of a single electrode can be calculated by using the Nernst equation, which is described in Chap. 3. In drinking water systems, the oxidation half-cell reaction of a metal, such as iron, zinc, copper, or lead, is coupled with the reduction of some oxidizing agent, such as dissolved oxygen or chlorine species. Example half-cell reactions are the following:

$$O_2 + 2H_2O + 4e^- \rightleftharpoons 4OH^- \tag{20-2}$$

$$HOCl + H^+ + e^- \rightleftharpoons (1/2)\ Cl_2\ (aq) + H_2O \tag{20-3}$$

$$(1/2)\ Cl_2\ (aq) + e^- \rightleftharpoons Cl^- \tag{20-4}$$

Equations 20-3 and 20-4 can be combined to yield the net reaction.

$$HOCl + H^+ + 2e^- \rightleftharpoons Cl^- + H_2O \qquad (20\text{-}5)$$

which represents a significant oxidizing half-cell reaction for metals in drinking water.

If Eq. 20-2 is written in the Nernst form (Eq. 20-6) and the ionic strength of the water is low enough that it can be assumed to have unit activity ($\{H_2O\} = 1$), then

$$E_{O_2/OH^-} = E^\circ_{O_2/OH^-} - \frac{0.0591}{4} \log \frac{\{OH^-\}^4}{\{O_2\}} \qquad (20\text{-}6)$$

The oxidation potential for this reaction clearly depends on pH, because the $[OH^-]$ is raised to the fourth power in the numerator. Similarly, the oxidation potential of the hypochlorous acid reaction is directly related to pH.

$$E_{HOCl/Cl^-} = E^\circ_{HOCl/Cl^-} - \frac{0.0591}{2} \log \frac{\{Cl^-\}}{\{HOCl\}\{H^+\}} \qquad (20\text{-}7)$$

because of the $[H^+]$ term in the denominator.

Reactions such as Eqs. 20-2 and 20-5 can be combined with metal oxidation half-cells (Eq. 20-1) to show overall corrosion reactions likely to occur in drinking water. Examples are

$$2Pb\ (\text{metal}) + O_2 + 2H_2O \rightleftharpoons 2Pb^{2+} + 4OH^- \qquad (20\text{-}8)$$

$$2Fe\ (\text{metal}) + O_2 + 2H_2O \rightleftharpoons 2Fe^{2+} + 4OH^- \qquad (20\text{-}9)$$

$$Pb\ (\text{metal}) + HOCl + H^+ \rightleftharpoons Pb^{2+} + Cl^- + H_2O \qquad (20\text{-}10)$$

$$Cu\ (\text{metal}) + OCl^- + 2H^+ \rightleftharpoons Cu^{2+} + Cl^- + H_2O \qquad (20\text{-}11)$$

As an example, the overall Nernst expression for Eq. 20-11, at 25°C, is

$$E_{Cu^{2+}/OCl^-} = E^\circ_{Cu^{2+}/OCl^-} - \frac{0.0591}{2} \log \frac{\{Cu^{2+}\}\{Cl^-\}}{\{OCl^-\}\{H^+\}^2} \qquad (20\text{-}12)$$

Note that in Eq. 20-12, the activities are for the free aqueous species, not the total concentrations. The overall potential of the reaction will depend on several factors: the ionic strength, the temperature, the degree of hydroxide complexation of the metal, the presence of complexing agents for the metal, side reactions of any ligands with other species, and the limits on the free metal ion caused by solubility, such as the formation of corrosion product solids. The amount of the driving force and the rate of the corrosion reactions can also be impacted by barriers that slow or limit the diffusion of oxidants to the surfaces of the materials, where "fresh" metal is available to oxidize.

Although metal levels and oxidant depletion in drinking water may ultimately be controlled by near-equilibrium solubility mechanisms, particularly with prolonged stagnation and low flow situations, corrosion cell activity is rarely entirely stopped. Thus, the control of the electrochemical cell reaction rate is usually the most critical factor in corrosion control.

Types of Corrosion

Many different types of corrosion exist (Pourbaix, 1973; Snoeyink and Wagner, 1996; MWH, 2005), and only some of the most significant to the corrosion of drinking water systems are covered here.

The kind of attack depends on the material, the construction of the system, the scale and oxide film formation, and the hydraulic conditions. Corrosion forms range from uniform to intense localized attack. Specialized corrosion texts should be consulted for more comprehensive discussions.

Uniform Corrosion. Uniform corrosion of a metal surface can proceed via sites that are anodic one instant and cathodic the next. When these sites are microscopic, close to one another, and the potential differences between the sites are small, uniform corrosion is favored (Snoeyink and Wagner, 1996). Reduction of oxygen occurs at the film surface, and transport of ions to and away from the oxide film takes place. The overall rate of corrosion is controlled by the presence and properties of the film or by transport of reaction products, especially hydroxide ion, away from the film-solution interface. This duplex film model was originally described by other researchers, notably by Ives and Rawson (1962a) for copper.

Plumbosolvency and *cuprosolvency* are terms that should specifically apply to the release of soluble lead and copper, respectively, resulting from uniform corrosion of piping, although in common use the terms sometimes generate confusion by being incorrectly applied to total metal release.

Galvanic Corrosion. Galvanic corrosion occurs when two different types of metals or alloys physically contact each other. One of the metals serves as the *anode*, with its corrosion rate accelerated, while the other serves as the *cathode*, with its corrosion rate reduced (AwwaRF-TZW, 1996; MWH, 2005). Available metals and alloys can be arranged in order of their tendency to be anodic, called the *galvanic series*. This order, in a potable water environment, depends on the temperature and solution chemistry (particularly the anionc components), as well as the simple relative ordering of the standard electrode potentials of the oxidation or reduction half-cells or simple metal-ion redox reaction couplings. Hence, tables developed for seawater would not always be the same as those for demineralized water or for any given finished drinking water. Therefore, any such published galvanic series should only be used as approximate guidance. Depending on generalizations about the range of chemical characteristics expected for the water in contact with the metals, different galvanic series can be measured or computed. For example, Larson (1975) described an empirical series (without giving solution composition specifics) as the following, going from the *anodic* end (corroded) to the *cathodic* end (noble): zinc, aluminum 2S, cadmium, aluminum 17ST, steel or iron, cast iron, lead-tin solders, lead, brasses, copper, bronzes, monel, silver solder, and silver.

The rate of galvanic corrosion is increased by greater differences in potential between the two metals. When cathodically controlled, the rate is increased by large areas of cathode relative to the area of the anode, and it is generally increased by closeness of the two metals and by increased mineralization or conductivity of the water. The relative size of the cathode relative to the anode may be of particular concern in the corrosion of lead-tin soldered joints in copper pipe, or when piping of two different types are interconnected, such as replacement of lengths of lead pipe with copper pipe (Britton and Richards, 1981; Gregory, 1990; Reiber, 1991; Singley, 1994).

Research has also suggested that the galvanic corrosion of copper alloys, lead, and leaded solder, when coupled with copper, creates local areas of low pH, increasing the solubility and rate of release of the metal or metals that are at the anodic sites (Dudi and Edwards, 2005; Edwards and Triantafyllidou, 2007; Nguyen et al., 2008, 2009). Further research is needed to refine the mechanisms, particularly with respect to possible different roles of free chlorine versus monochloramine in driving the reactions, buffering systems controlling pH changes, and the role of specific solids and complexes in accelerating or mitigating the effect.

Galvanic corrosion is often a great source of difficulty where brass, bronze, or copper is in direct contact with aluminum, galvanized iron, or iron. An example of a combination of deposition corrosion and galvanic corrosion occurs when copper can attack spots on galvanized pipe, thereby causing copper-zinc galvanic cells (Kenworthy, 1943; Fox et al., 1986). In waters of low conductivity, the influence of the galvanic cell is intense, but only in the immediate vicinity of the contact between the metals (MWH, 2005). In waters of high conductivity, however, the effect of the galvanic cell is spread widely across the surface of the less-noble (anodic) metal.

Pitting Corrosion

Pitting is a type of non-uniform corrosion that forms pits or holes in the pipe surface (Fig. 20-1). It takes relatively little metal loss or overall corrosion to cause a pinhole in a pipe wall, and failure can occur in as little as a few weeks. Pitting failures have little or no correlation to high copper release. Pitting can begin or concentrate at a point of surface imperfections, scratches, or surface deposits.

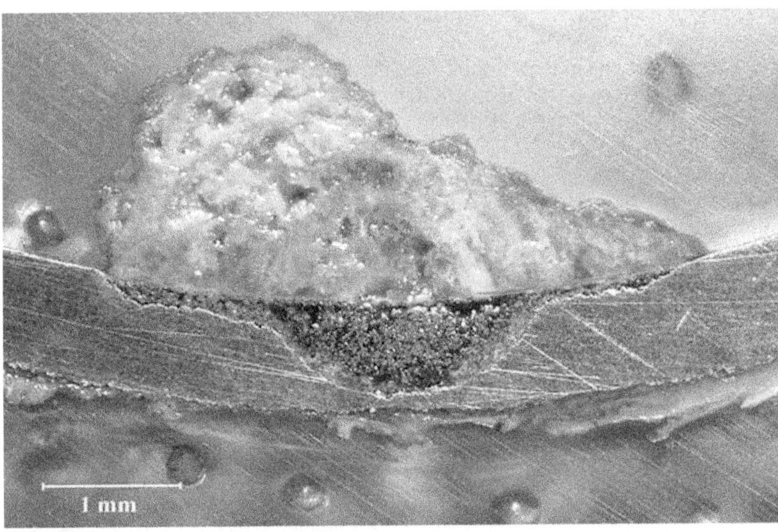

FIGURE 20-1 Pit formed in copper pipe, showing build-up of cupric hydroxycarbonate, hydroxysulfate and hydroxychloride solids in cap, and filling of cuprous oxide and cuprous chloride crystals inside the pit. (*USEPA Photo.*)

Dealloying or Selective Leaching. *Dealloying*, or *selective leaching*, is the preferential removal of one or more metals from an alloy in a corrosive medium, such as the removal of zinc from brass (dezincification), or the removal of pockets of lead from brass (Schock and Neff, 1988; Lytle and Schock, 1996; Oliphant and Schock, 1996; Schock et al., 1996). This type of corrosion weakens the metals and can lead to pipe failure in severe cases. The stability of brasses and bronzes in natural waters depends in a complex manner on the water chemistry and the formation of protective films. Dezincification is common in brasses containing 20 percent or more zinc and is rare in brasses containing less than 15 percent zinc (Oliphant and Schock, 1996). The occurrence of plug-type dezincification and dezincification at threaded joints suggests that debris and crevices may initiate oxygen concentration cells and result in dezincification.

Lead occurs in lead-tin solder as a disseminated phase, acting principally as a diluent for the tin that actually does the binding with the copper (Parent et al., 1988). Factors governing the removal of lead from the soldered joint are presumably similar to those affecting the leaching of lead from brass. Soldered joints in service for a long time frequently show a depletion of lead from the exposed solder.

Selective leaching also applies to the dissolution of asbestos-cement (A-C) pipe, or the deterioration of cement mortar linings of iron water mains (see the section Corrosion of Specific Materials).

Tuberculation. Tuberculation occurs when pitting corrosion products build up at the anode next to the pit (Smith and McEnaney, 1979; Snoeyink and Wagner, 1996; Sarin et al., 2001; Sarin et al., 2004a; 2004b). Tuberculation can occur in many types of metallic pipes. In iron or steel pipes, the tubercles are made up of various iron oxides and oxyhydroxides. These tubercles are usually rust colored and soft on the outside and are both harder and darker toward the inside. Sometimes they are considerably layered with iron minerals indicative of anoxic environments, such as sulfide minerals (Singley et al., 1985; Benjamin et al., 1996).

Tuberculation of iron has often been attributed to poorly buffered waters, where pH can get extremely high under localized surface conditions. Hence, many consultants and researchers have recommended a balance of minimum hardness and alkalinity to provide a water that is resistant to iron corrosion and tuberculation (Stumm, 1956, 1960; Larson and Skold, 1957, 1960; Weber and Stumm, 1963; Merrill et al.,1978; Smith and McEnaney, 1979).

Microbiologically Influenced Corrosion. Microbiologically influenced corrosion (MIC) pertains to the general class of corrosion resulting from a reaction between the pipe material and organisms such as bacteria, their metabolic by-products, or both (Little and Wagner, 1997). Algae and fungi may also influence corrosion by producing changes in the pH, dissolved oxygen level, or other chemistry changes to the microenvironment at the metal surface or under corrosion deposits in treatment plant or water storage scenarios. MIC may be an important factor in the taste and odor problems that develop in a water system, as well as in the degradation of the piping materials. Microbial activity and its impacts potentially extend from the finished water entry point, all the way through distribution system and storage, and on into the premise plumbing, finally reaching all of the way to consumers' taps. Thus, residential impacts have been poorly studied, yet they can be very significant in some circumstances.

Biofilms in pipes are often characterized by stratification, with different families of organisms with different functions existing in the different zones and conditions (Little and Wagner, 1997). Controlling such growths is complicated because they can take refuge in many protected areas, such as mechanical crevices or accumulations of corrosion products. Bacteria can exist within porous scale layers or under tubercles, where neither chlorine nor oxygen can destroy them. Mechanical cleaning may be necessary in some systems before control can be accomplished by residual disinfectants. Preventative methods include good corrosion control to avoid tuberculation and accumulating corrosion by-products, avoiding dead ends, and preventing stagnant water in the system by flushing or bleeding.

Microbial activity can strongly influence the mineralogy of iron corrosion deposits (Allen et al., 1980; Lazaroff et al., 1982; LeChevallier et al., 1993; Camper, 1996; Postma and Jakobsen, 1996; Brown et al., 1997; Lewandowski et al., 1997; Appenzeller et al., 2002; Mikutta et al., 2008; von Blanckenburg et al., 2008). Some researchers have linked reduced microbial activity in distribution mains with the application of effective corrosion control treatment in distribution system mains, as well as controlling limiting nutrients. One proposed mechanism for the effectiveness of corrosion control toward bacterial regrowth and biofilms by corrosion inhibitors is the prevention or saturation of sorption sites at the pipe surfaces, allowing less *natural organic matter* (NOM) sorption and attachment (LeChevallier et al., 1993; Camper, 1996). Thus, there would be a decreased attached nutrient base for the colonizing organisms if potential organic material attachment sites were already occupied by phosphate or other corrosion inhibitor material. Because microbial activity can affect pH, metal solubility, and the oxidation-reduction potential of the surrounding microenvironment, microbial activity in pipe or storage tank scales and deposits could also impact the stability of manganese or other redox sensitive metals in deposits and the mobility of trace metals associated with the formation or decomposition of those coatings.

The ways in which bacteria can increase corrosion rates are numerous. One pathway is the formation of oxygen concentration cells that produce lowered pH and localized corrosion and pitting. Lee et al. (1980), for example, showed increased localized corrosion when cast iron was exposed to a water with extensive biological activity, compared with the same water under sterile conditions. Associations have been found between nitrification activity and copper corrosion problems (Marani, 1992; Murphy et al., 1997a–c). Cantor has reported several cases of particularly copper pitting corrosion in which microbes are believed to play a pivotal role (Cantor et al., 2006; Cantor, 2009).

Nitrification, depending on the extent of completion of the reaction, can produce 2 moles of hydrogen ion, for every mole of ammonia consumed (Metcalf and Eddy Inc., 2003; Schock, 2005). Accordance with the sequential microbially-mediated reactions is shown in Eqs. 20-13 and 20-14.

$$NH_4^+ + \left(\frac{3}{2}\right)O_2 \rightleftharpoons NO_2^- + 2H^+ + H_2O \tag{20-13}$$

$$NO_2^- + \left(\frac{1}{2}\right)O_2 \rightleftharpoons NO_3^- \tag{20-14}$$

Recently, a comprehensive review of nitrification and drinking water corrosion has been published (Zhang et al., 2009a). Additional recent research has provided direct evidence for long-suspected increases in lead and copper release caused by lower pH conditions resulting from nitrification (Zhang et. al., 2008a and b, 2009b). Microbes may also enhance corrosion or solubility because extracellular material and metabolites of microorganisms may include polymers that are good complexing agents (Garrels and Christ, 1965; Geesey et al., 1986, 1988; Geesey and Bremer, 1991), or they may influence the redox chemistry of the scales at the surface (Allen et al., 1980; Lee et al., 1980; LeChevallier et al., 1993; Lewandowski et al., 1997). Conversely, microbes can also dramatically reduce corrosion and leaching of metals such as galvanized iron, if they remove agents that would reduce dissolved oxygen (Zhang et al., 2008a and b, 2009b).

Iron bacteria derive energy from oxidation of ferrous iron to ferric iron. Nuisance conditions often result because the ferric iron precipitates in the gelatinous sheaths of the microbial deposits, and these can be sloughed off and be the cause of red-water complaints. Sontheimer et al. (1981) noted that they can also interfere with the development of passivating scales and, thus, that the rate of corrosion is higher in their presence than in their absence.

The sulfate-reducing bacteria may be one of the most important organisms causing MIC in many water systems (Sontheimer et al., 1981). Sulfate can act as a reducing agent at low redox potentials.

$$SO_4^{2-} + 8H^+ + 8e^- \rightleftharpoons S^{2-} + 4H_2O \tag{20-15}$$

and the sulfate-reducing organisms facilitate this reaction. Sulfate reducers have been found in the interior of tubercles, and thus may be responsible for maintaining corrosion at these locations (Ainsworth, 1980). Sulfate reducers are often present regardless of whether the supply is aerated (Little and Wagner, 1997). Some observations that higher levels of sulfate cause more corrosion (Larson, 1975) may possibly be related to microbial activity (Snoeyink and Wagner, 1996; Lytle et al., 2005a). Recent research has found sulfide to be a contributor to copper pitting that is particularly destructive and difficult to control (Jacobs et al., 1998), and once started, the microbial mediation may be very difficult or impossible to stop. The high-chloride, low-pH environments of pits can lead to continued propagation within pits, even if the microbes are killed by biocides (Little and Wagner, 1997).

Erosion Corrosion. Erosion corrosion involves multiple mechanisms of attack, which can include cavitation and concentration cells formed by differential velocity. High velocities disrupt the formation of protective films, such as metal oxides, hydroxycarbonates, and carbonates, that serve as protective barriers against corrosive attack. When cavitation occurs, it can also remove the metal of the pipe itself. Erosion corrosion can be identified by grooves, waves, rounded holes, and valleys it causes on the pipe walls.

Crevice Corrosion. Crevice corrosion is a subset of concentration cell corrosion, in which changes in acidity, oxygen depletion, and dissolved ions and the absence of an inhibitor cause irregular localized corrosion. As the name implies, this corrosion occurs in crevices at gaskets, lap joints, rivets, and surface deposits.

Concentration Cell Corrosion. Concentration cell corrosion is usually deduced by inference. It occurs when differences in the total or the type of mineralization of the environment exist. Corrosion potential is a function of the concentration of aqueous solution species that are involved in the reaction, as well as of the characteristics of the metal. Differences in acidity (pH), metal ion concentration, anion concentration, or dissolved oxygen cause differences in the solution potential of the same metal. Differences in temperature can also induce differences in the solution potential of the same metal.

When concentration cell corrosion is caused by dissolved oxygen, it is often referred to as *differential oxygenation corrosion* as discussed by Snoeyink and Wagner (1996). Common areas for differential oxygenation corrosion are between two metal surfaces, for example, under rivets, under washers, under debris, or in crevices.

Oxygen concentration cells develop at metal-water interfaces exposed to air, such as in a full water tower, accelerating corrosion a short distance below the surface. The dissolved oxygen (DO) concentration is replaced by diffusion from air and remains high at and near the surface, but does not replenish as rapidly at lower depths because of the distance. Therefore, the corrosion takes place at a level slightly below the water contact surface.

Graphitization. Graphitization is a form of corrosion of cast iron in highly mineralized water or waters with a low pH that results in the removal of the iron-silicon metal alloy making up one of the phases of the cast-iron microstructure. A black, spongy-appearing, but hard mass of graphite remains.

Stray Current Corrosion. Stray current corrosion is a type of localized corrosion usually caused by the grounding of home appliances or electrical circuits to the water pipes. It is well accepted that stray currents can cause failures on the outside of pipes, but the actual contribution of stray current to internal corrosion remains controversial. There has been some evidence presented by a number of researchers that both ac and dc current can affect corrosion rates and metal levels in the water (Williams, 1986; Horton and Behnke, 1989; Horton, 1991; Bell et al., 1995a and b; Goidanich et al., 2010). However, in some studies there was a concurrent substantial water temperature increase as a result of the amount of current passed, and in other studies electrochemical models did not seem to fit the observed polarization measurement data, leaving interpretation of mechanisms still inadequate. In municipal distribution systems, increased corrosion of steel-reinforced concrete pressure pipe has occasionally been noted and has been believed to be related to stray currents.

Corrosion Products on Pipe Surfaces

Metal surfaces may be protected by their being either immune or by rendering them passive. If a metal is protected by *immunity*, the metal is thermodynamically stable, and is therefore incorrodible (Pourbaix, 1973). In practice this occurs for plumbing materials used for water supplies only for copper in deaerated water without sulfides. *Passivation* occurs when the metal is not stable, but becomes protected by a stable film. The protection from a film can vary from nearly perfect to highly imperfect, depending on the extent to which the film effectively shields the metal from contact with the solution (Pourbaix, 1973). In pure water, the reaction product would consist of a metal oxide or hydroxide. However, in drinking waters, anions and other cations are present that can precipitate under these local surface chemistry conditions and contribute to passivation.

Several studies have shown that the pH at the surface of pipe can be significantly different from that in the bulk solution (Snoeyink and Wagner, 1996), although many studies reporting extremely high pH at the surface have neglected to adequately include consideration of buffering by the carbonate system in the water and the possible role of solids such as $CaCO_3(s)$ and $Mg(OH)_2(s)$ (Watkins and Davies, 1987; Dexter and Lin, 1992; Lewandowski et al., 1997). The deposits that form on pipe surfaces may be: (1) a mixture of corrosion products that depend both on the type of metal that is corroding and the composition of the water solution [e.g., $FeCO_3(s)$, $Fe_3O_4(s)$, $FeOOH(s)$, $Pb_3(CO_3)_2(OH)_2(s)$, $Zn_5(CO_3)_2(OH)_6(s)$], (2) precipitates that form because of pH changes that accompany corrosion [e.g., $CaCO_3(s)$], (3) precipitates that form because the water entering the system is supersaturated (e.g., $CaCO_3(s)$, $SiO_2(s)$, $Al(OH)_3(s)$, $FeOOH(s)$, $Fe(OH)_3(s)$, $MnO_2(s)$, etc.), and (4) precipitates or coatings that form by reaction of components of inhibitors, such as silicates or phosphates with the pipe materials (e.g., lead or iron).

The nature of the scales or deposits that form on metals is important because of the effect that these scales have on the corrosion rate, through a combination of physical and chemical factors such as (but not only): chemical composition and crystal structure; porosity to water, oxidants, complexing ligands, and corrosion product ions; semiconductor properties; solubility; magnetic properties; particle size; and surface area.

A long time, from many months to years, may be required for the corrosion rate of iron, steel, and copper to stabilize because of the complex nature of the scales. If a scale considerably reduces the rate of corrosion, it is said to be a relatively *protective* scale; if it only reduces the rate of corrosion slightly, it is termed *non-protective*.

Changes in water treatment or source water chemistry over time can produce successive layers of new solid phases, remove or change the nature of previously existing deposits, or both. Scales of similar chemical composition can have a significantly different impact on corrosion and metal protection because properties such as uniformity, adherability, and permeability to oxidants can vary depending on such factors as trace impurities, presence of certain organics, temperature of deposition, length of time of formation, and so forth.

Scales that form on pipes may have deleterious effects in addition to the beneficial effect of protecting the metal from rapid corrosion or limiting the levels of toxic metals (such as lead) in solution. Aluminum scales have had well-documented hydraulically detrimental impacts (Snoeyink et al., 2003), as have many unwanted secondary deposits formed by corrosion control approaches for lead and copper (Grigg, 2010). In theory, water quality should be controlled so that the scale is protective but as thin as possible, because as the bulk of scale increases, the capacity of the main to carry water is reduced. In practice, it is difficult to maintain uniform chemistry conditions throughout a distribution system to achieve the ideal pipe surface coating. The formation of uneven deposits such as tubercles increases the roughness of the pipe surface, reducing the ability of the mains to carry water, and may provide shelters for the growth of microorganisms.

Understanding that there may be significantly different reactions occurring between the water constituents and the surfaces of new pipes compared with old pipes is important to the interpretation of field and laboratory data from corrosion control studies. Conceptually, this is illustrated in Fig. 20-2 for lead pipe. On the new surface Fig. 20-2a, the full suite of anodic and cathodic reactions can occur, with oxidation of lead followed by the beginning of the development of a passivating film. Once the film is sufficiently developed (Fig. 20-2b), in this case after many decades, the oxidants in the water no longer can directly contact the metal of the pipe material itself. Therefore, the oxidation step will not occur, and metal release will become a strong function of the physical adherence or the solubility of the surface deposit. Thus, corrosion inhibitor chemicals that stifle reactions occurring at cathodic surface sites may appear much better in tests using new metal surfaces than they may operate when applied to distribution system pipes covered by thick scales or corrosion deposits. With well-developed surface deposits present, the solubility and surface chemistry of the existing scales is much more important in developing water treatment targets than predictions based on the pure corrosion chemistry of the metal.

FIGURE 20-2 Different reactions dominate the corrosion and metal release of lead pipe, depending on the surface condition. (*a*) Fresh lead surface, where a full suite of anodic and cathodic reactions take place. (*b*) Exhumed 90-year-old lead service line coated with a thick mixture of $PbCO_3$ and $Pb_3(CO_3)_2(OH)_2$, where the surface reaction is dominated by simple Pb(II) solubility chemistry and ion diffusion. (*Photo by M. DeSantis, Pegasus Technical Services.*)

Corrosion Kinetics

In addition to the understanding of the overall *thermodynamic* driving forces for corrosion and corrosion control approaches, an examination of important dynamic electrochemical reactions taking place is also important to understand many drinking water corrosion-related phenomena. A three-step process is involved in governing the rate of corrosion of pipe: (1) transport of dissolved reactants to the metal surface, (2) electron transfer at the surface, and (3) transport of dissolved products from the reaction site (James M. Montgomery Consulting Engineers, 1985). When either or both of the transport steps are the slowest, *rate-limiting step*, the corrosion reaction is said to be under *transport control*. When the transfer of electrons at the metal surface is rate limiting, the reaction is said to be under *activation control*. The formation of solid natural protective scales that inhibit transport are often an important factor in transport control.

Corrosion has often been described in terms of numerous tiny galvanic cells on the surface of the corroding metal. Such localized anodes and cathodes as those described are not fixed on the surface, but are statistically distributed on the exposed metal over space and time. The electrochemical potential of the surface is determined by the mixed contributions to potential of both the cathodic and anodic reactions, averaged over time and over the surface area. When the electrode is at its equilibrium, the rates of reaction in both the cathodic and anodic half-cells are equal. A sophisticated treatment of the electrokinetics of corrosion reactions based on *mixed-potential theory* is given by MWH (2005). Mixed-potential theory is the most powerful conceptual tool now available for understanding the short-term dynamics of the corrosion process.

Given the thermodynamic basis for corrosion described above, and the body of knowledge about kinetic factors that affect the rate of corrosion of metals, several properties of the water passing through a pipe or device that influence the rate of corrosion can be identified. Some of the water-related properties are (1) concentration of dissolved oxygen, (2) pH, (3) temperature, (4) water velocity, (5) concentration and type of disinfectant residual, (6) chloride and sulfate ion concentrations, (7) concentration of dissolved inorganic carbon (DIC) and calcium, and (8) the concentration and type of remaining NOM.

These properties interrelate, and their effect depends on the plumbing material as well as the overall water quality. References specific to the type of plumbing situation (pipe, soldered joint, galvanic connection, faucet, or flow-control device) and the material of interest should be consulted for the most appropriate information on corrosion rate control.

Solubility Diagrams

The solubility of passivating films on the pipe surface is the most important factor in determining whether a given water quality can meet many drinking water regulations that are based on health effects of ingested metals. In the case of distribution mains, where solubility equilibrium is rarely attained, asbestos fibers are frequently physically released into the water because of the dissolution of the cement matrix originally holding them. Reducing the solubility of that cement matrix, or depositing an insoluble coating over the fibers, would be two pathways to mitigate the problem. Solubility is also a governing factor in the deposition of often unwanted interior coatings of iron, manganese, and aluminum present in the water entering the distribution system.

Focusing more on premise plumbing where cyclical stagnation periods are the norm, solubility is a major factor in lowering maximum soluble metals levels that can be controlled by modifying water chemistry at the treatment plant. Solubility reflects an ideal equilibrium condition, and should not necessarily be expected to correspond to tap water levels (Britton and Richards, 1981; Schock, 1990). Other factors are important, such as the physical location of the plumbing materials relative to the sample collected, the release of particulates from the deposits on the pipes, the rate of the chemical reactions that affect the mobilization of the metals relative to the standing time of the water before collection, and many other variables. However, trends in the response of tap water copper levels to changes in key water chemistry variables followed the predictions of solubility models in principle, in a survey of over 2500 U.S. utilities of all sizes (Edwards et al., 1999). More recent studies on lead pipes that revealed the extensive presence of PbO_2 deposits in waters of low pH and frequently high alkalinity (Schock et al., 2001, 2005a) have now explained several of the apparent discrepancies in lead release trends noted in the earlier Edwards et al. (1999) work, which was limited to consideration of Pb(II) equilibria.

To display solubility relationships in a relatively simple two-dimensional manner, the total solubility of a constituent is plotted as a function of a master variable (Stumm and Morgan, 1996; Butler and Cogley, 1998). Other solution parameters can affect solubility such as the ionic strength, or the concentration of dissolved species (ligands) that can form coordination compounds or complexes with the metal. The solubility diagram does not necessarily give the information needed to minimize the rate of corrosion because rate is affected by kinetic parameters that depend on the relative rates of the oxidation, dissolution, diffusion, and precipitation reactions. It also cannot predict the ability of a pipe coating to adhere to the pipe surface, or the permeability of a coating to oxidants from the water solution or pitting agents. It does, however, give important information for estimating trends in water quality impacts on compliance with of maximum contaminant levels (MCLs) or action levels (ALs) for drinking water, the potential for precipitating passivation films on pipe surfaces. Solubility diagrams are also useful for predicting the onset of scaling phenomena that can constrain effective corrosion control treatment, such as precipitation of calcium carbonate, octacalcium phosphate, and aluminum or ferric hydroxide.

To construct this type of diagram, an aqueous mass balance equation must be written for the metal (or other constituent of interest), with the total solubility (S_T) as the unknown. The mass balance expression should include the concentration of the uncomplexed species (free metal ion), along with all complexes to be included in the model. For Pb(II), the simple relationship for the free ion, plus only the hydroxide and carbonate complexes (the simplest system for potable waters), is

$$\begin{aligned} S_{T,\mathrm{Pb(II)},\mathrm{CO_3}} &= [\mathrm{Pb^{2+}}] + [\mathrm{Pb(OH)_2^{\circ}}] + [\mathrm{Pb(OH)_3^{-}}] + [\mathrm{Pb(OH)_4^{2-}}] \\ &\quad + 2[\mathrm{Pb_2(OH)^{3+}}] + 3[\mathrm{Pb_3(OH)_4^{2+}}] + 4[\mathrm{Pb_4(OH)_4^{4+}}] + 6[\mathrm{Pb_6(OH)_8^{4+}}] \\ &\quad + [\mathrm{PbHCO_3^{+}}] + [\mathrm{PbCO_3^{\circ}}] + [\mathrm{Pb(CO_3)_2^{2-}}] \end{aligned} \tag{20-16}$$

If the concentration of a complex is going to be relatively negligible compared with the total solubility across the pH range of interest, it can be excluded from the model for simplicity. Frequently, only four or five species are significant in a system. Substituting in the aqueous species and conditional formation constants, the full expression for computation becomes (Schock, 1980, 1981)

$$\begin{aligned} S_{T,\mathrm{Pb(II)},\mathrm{CO_3}} &= [\mathrm{Pb^{2+}}] + \frac{\beta'_{1,1}[\mathrm{Pb^{2+}}]}{[\mathrm{H^+}]} + \frac{\beta'_{1,2}[\mathrm{Pb^{2+}}]}{[\mathrm{H^+}]^2} + \frac{\beta'_{1,3}[\mathrm{Pb^{2+}}]}{[\mathrm{H^+}]^3} + \frac{\beta'_{1,4}[\mathrm{Pb^{2+}}]}{[\mathrm{H^+}]^4} \\ &\quad + 2\frac{\beta'_{2,1}[\mathrm{Pb^{2+}}]^2}{[\mathrm{H^+}]} + 3\frac{\beta'_{3,4}[\mathrm{Pb^{2+}}]^3}{[\mathrm{H^+}]^4} + 4\frac{\beta'_{4,4}[\mathrm{Pb^{2+}}]^4}{[\mathrm{H^+}]^4} + 6\frac{\beta'_{6,8}[\mathrm{Pb^{2+}}]^6}{[\mathrm{H^+}]^8} \\ &\quad + \beta'_{1,1,1}[\mathrm{Pb^{2+}}][\mathrm{H^+}][\mathrm{CO_3^{2-}}] + \beta'_{1,0,1}[\mathrm{Pb^{2+}}][\mathrm{CO_3^{2-}}] \\ &\quad + \beta'_{1,0,2}[\mathrm{Pb^{2+}}][\mathrm{CO_3^{2-}}]^2 \end{aligned} \tag{20-17}$$

In these equations, $\beta'_{m,h,c}$ represents the formation constant for the complex having a stoichiometry of m metal ions, h hydrogen ions, and c carbonate ions, corrected for ionic strength and temperature (Stumm and Morgan, 1996). To obtain the predicted solubility, the solubility constant expression for each solid of interest is rearranged and solved to isolate the free metal species concentration, here $[\mathrm{Pb^{2+}}]$, and is substituted into each term of the mass balance equation (e.g., Eq. 20-16). Sometimes, an iterative technique is used to simultaneously solve several related mass balances, depending on exactly the type of diagrams desired and assumptions about the availability of components involved in the precipitation and dissolution reactions in the system.

A diagram is then constructed for each solid, and the curves are superimposed. The points of minimum solubility are then connected, giving the final diagram. This procedure is discussed in more detail by Schock (1980, 1981) for Pb(II), Schock et al. (1995) for copper, and Snoeyink and Jenkins (1980) for Fe(II).

Additional equivalent terms for sulfate, chloride, orthophosphate, and any other relevant complexes should be appended to this expression when the concentrations of the given Pb(II) species would be significant with respect to $S_{T,\mathrm{Pb(II),CO_3}}$. As an example, the contribution to Pb(II) solubility from aqueous orthophosphate species is represented by

$$S_{T,\mathrm{Pb(II),PO_4}} = [\mathrm{PbHPO_4^\circ}] + [\mathrm{PbH_2PO_4^+}] \tag{20-18}$$

So, the total lead(II) solubility becomes $S_{T,\mathrm{Pb(II)}} = S_{T,\mathrm{Pb(II),CO_3}} + S_{T,\mathrm{Pb(II),PO_4}}$, or

$$\begin{aligned} S_{T,\mathrm{Pb(II)}} = {} & [\mathrm{Pb^{2+}}] + [\mathrm{Pb(OH)_2^\circ}] + [\mathrm{Pb(OH)_3^-}] + [\mathrm{Pb(OH)_4^{2-}}] + 2[\mathrm{Pb_2(OH)^{3+}}] \\ & + 3[\mathrm{Pb_3(OH)_4^{2+}}] + 4[\mathrm{Pb_4(OH)_4^{4+}}] + 6[\mathrm{Pb_6(OH)_8^{4+}}] + [\mathrm{PbHCO_3^+}] \\ & + [\mathrm{PbCO_3^\circ}] + [\mathrm{Pb(CO_3)_2^{2-}}] + [\mathrm{PbH_2PO_4^+}] + [\mathrm{PbHPO_4^\circ}] \end{aligned} \tag{20-19}$$

Because these diagrams are two-dimensional (solubility on the *y*-axis, pH on the *x*-axis), there are often additional variables that have an important impact on solubility (such as temperature, ionic strength, carbonate concentration or alkalinity, and orthophosphate concentration), these diagrams must display solubility and species concentration lines for fixed concentrations or values of the other variables. For instance, Fig. 20-3 displays a solubility diagram for lead in the system containing only carbonate species and water, showing the important aqueous species and the stability domains of two lead solids. In this figure, the DIC concentration was fixed at 2.5×10^{-4} mol/L (3 mg C/L), with a temperature of 25°C and an assumed ionic strength of 0.005 M. Figure 20-4 shows the same basic system, but with a total DIC concentration of 2.5×10^{-3} mol/L (30 mg C/L). In both diagrams, the implicit assumption was also made that the solution redox potential was not high enough to cause the formation of Pb(IV) aqueous or solid species. Additionally, when metal solubility becomes high, care must be taken to ensure that computational constraints on the system, such as the fixed ionic strength and the total ligand concentration (mass balance), are not violated by the presence of high concentrations of dissolved metal species.

Note that if the same complex formation constants were used, but a less soluble constant were used for $PbCO_3$(s) (cerussite) solubility, the simple carbonate solid would be predicted to be stable over a wider pH range. Likewise, a less soluble constant for $Pb_3(CO_3)_2(OH)_2$ (hydrocerussite) would expand its stability field relative to that of cerussite. Analogous information can be obtained through the careful construction and study of solubility diagrams for other metals, such as copper and zinc. Similarly, different assumptions about the formation constants for the Pb(II) complexes or the inclusion of mixed Pb(II) hydroxide carbonate complexes would alter the shapes of the diagrams, and therefore the predicted lead concentrations at different pH values.

When constructing solubility diagrams for any metal, the selection of solid and aqueous species must truly represent the system to be modeled, or very erroneous conclusions can result. For example, the aragonite form of calcium carbonate, rather than the calcite form, is frequently found in deposits formed in systems having galvanized pipe. In addition, the ferric iron deposits formed in mains are frequently a relatively soluble hydroxide or oxyhydroxide form [$Fe(OH)_3$ or FeOOH], rather than an ordered form such as hematite (Fe_2O_3). Manganese also tends to form a variety of solid phases with varying degrees of valence state

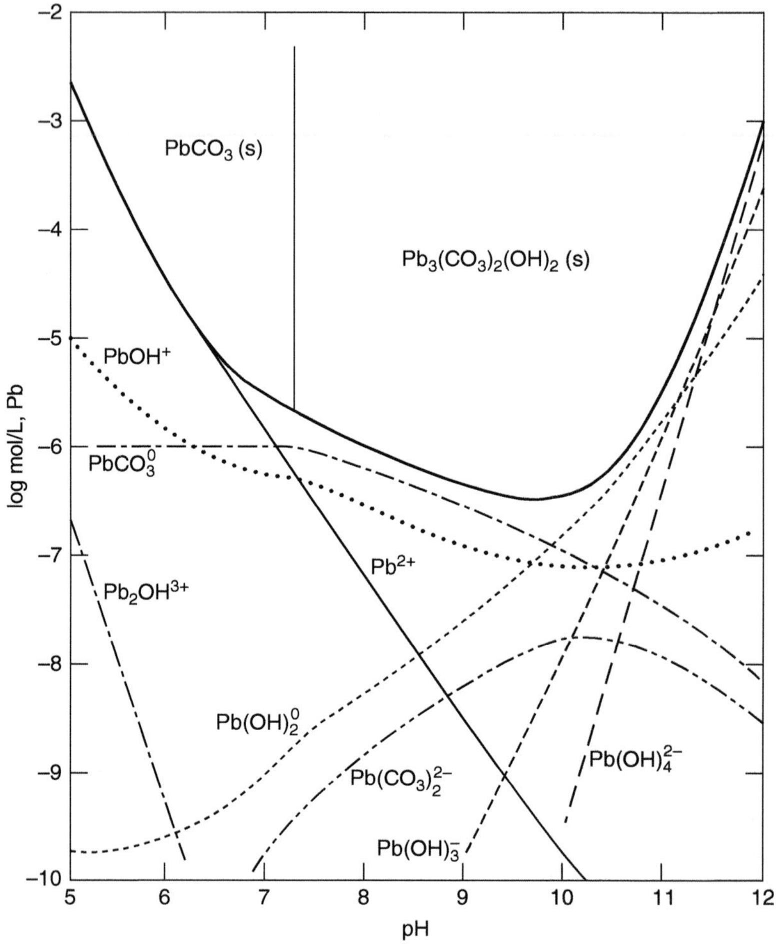

FIGURE 20-3 Solubility diagram showing dissolved lead (II) species in equilibrium with lead (II) solids in a pure system containing 3 mg C/L DIC (2.5×10^{-4} mol/L) at 25° and $I = 0.005$ mol/L. (*Source: Data from Schock et al., 1996.*)

mixing, crystal disorder, degrees of hydration, and so on. To go along with this concept, the equilibrium constants selected must be appropriate to the nature of the solid phases of interest and accurate to give realistic concentration estimates. Knowledge of changes in the equilibrium constants with temperature is essential, especially when projections of depositional tendency have to be made into hot or very cold water piping systems. Critical evaluation of data appearing in handbooks and published papers is necessary to avoid using incorrect values, and occasionally review articles or major works by rigorous researchers can be consulted for reliable values.

Sometimes improvements can be made in predictions by using metastable species in the calculations, although it is not thermodynamically rigorous to do so. Metastable solids have been found to govern copper levels in drinking water via what has been

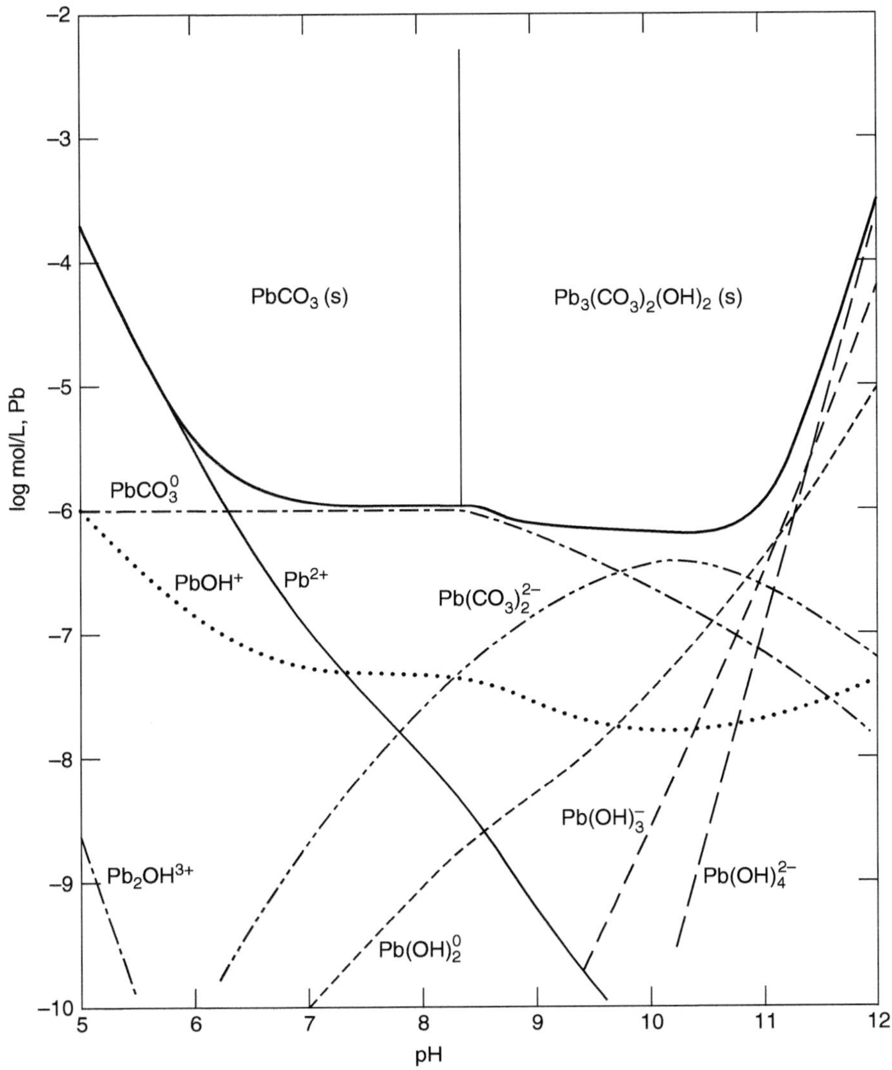

FIGURE 20-4 A solubility diagram showing dissolved lead (II) species in equilibrium with lead (II) solids in a pure system containing 30 mg C/L DIC (2.5×10^{-3} mol/L) at 25° and $I = 0.005$ mol/L. (*Source: Data from Schock et al., 1996.*)

termed the *aging process* (Schock and Lytle, 1994; Schock et al., 1995, 2000; Edwards et al., 1996, 2000).

Stumm and Morgan have discussed the formation of precipitates and how this is related in a conceptual way to solubility diagrams, as illustrated by Fig. 20-5 (Stumm and Morgan, 1996). They define an *active* form of a compound as one that is a very fine crystalline precipitate with a disordered lattice. It is generally the type of precipitate formed incipiently from strongly oversaturated solutions. Such an active precipitate may persist in metastable

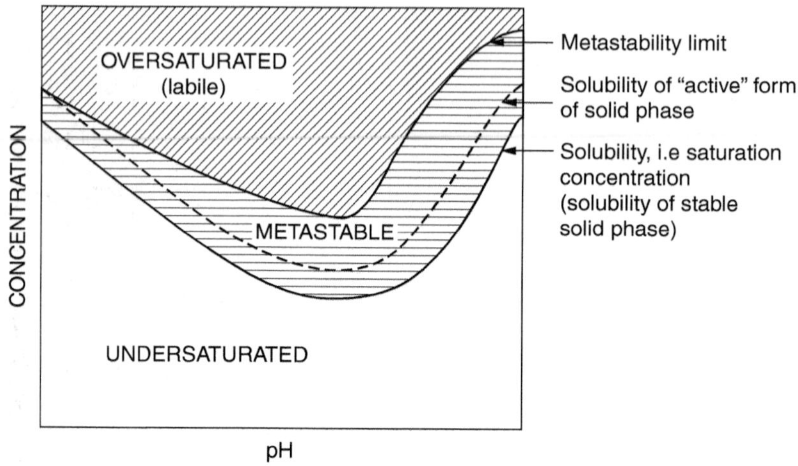

FIGURE 20-5 Solubility and saturation. A schematic solubility diagram showing concentration ranges versus pH for supersaturated, metastable, saturated, and undersaturated solutions. (*Source: Aquatic Chemistry, W. Stumm and J.J. Morgan. Copyright © 1980, John Wiley & Sons, Inc. Reprinted by permission of John Wiley & Sons, Inc.*)

equilibrium with the solution and may convert (age) slowly into a more stable *inactive* form. Measurements of the solubility of active forms of solids give solubility products that are higher than those of the inactive forms. Laitinen and Harris (1975) cite a definition of aging that includes all irreversible structural changes that occur in a precipitate after it has formed, such as recrystallization of primary particles; cementing of primary particles by recrystallization in an agglomerated state; thermal aging, transformation of a metastable modification into a more stable form; and processes involving changes in composition. Langmuir has described a similar natural analogous process for particle size effects on iron mineral transformation (Langmuir, 1971). Examples of changing solubility related to various physical properties abound in the literature. Noteworthy for copper have been both controlled bench solubility experiments (Scaife, 1957; Schindler, 1967; Patterson et al., 1991; Marani et al., 1995) and practical observations in water treatment or natural systems (Ma et al., 2006). For lead, the transformation in a phosphate-treated system from $Pb_5(PO_4)_3OH$ (hydroxypyromorphite) to $Pb_5(PO_4)_3Cl$ (chloropyromorphite) may also be an aging phenomenon (Harmon, 1999).

Hydroxides and sulfides often occur in amorphous and several crystalline modifications. Initially formed amorphous precipitates or active forms of unstable crystalline modifications tend to evolve toward forming a more stable modification, though it may be a nonhomogenous solid. Similar phenomena can occur with basic carbonates, such as a transition from one form to another with changes in pH or DIC over time, or the alteration of existing solid phases by prolonged interaction with sorbed silicates species (Leggett, 1978).

Pourbaix or Potential-pH Diagrams

Potential-pH diagrams, which are also called *E_h-pH* or *Pourbaix diagrams,* are designed to predict what aqueous species or corrosion by-product solid phases are thermodynamically stable under different conditions of electrochemical potential and pH. Potential-pH diagrams have been popularized by Pourbaix and his coworkers in the corrosion field to define areas of passivation, immunity, and active corrosion (Pourbaix, 1966, 1973; Obrecht and Pourbaix, 1967), and by numerous authors in natural water and geochemistry (Garrels and Christ, 1965; Stumm and Morgan, 1996; Langmuir, 1997) for mineral stability predictions. An equivalent version of the diagram uses the concept of electron activity, *pE*, which is analogous to the concept of pH (Snoeyink and Jenkins, 1980; Stumm and Morgan, 1996). Most aquatic chemistry textbooks give detailed instructions on the computation of potential-pH diagrams, and many computer software packages include them as options.

A potential-pH diagram is related to the solubility versus pH plots discussed earlier. This relationship is shown schematically in Fig. 20-6. If a conventional two-dimensional solubility diagram is considered as a vertical plane, a potential-pH diagram may be thought of as a slice that is perpendicular to the solubility plane, which cuts through that plane at a single level concentration of the metal. In actuality, potential-pH diagrams are usually computed in terms of thermodynamic activities rather than concentrations, but the difference is usually not important for practical purposes. The activities of all aqueous and solid species must be fixed, while the electrical potential of the solution and the pH become the master variables.

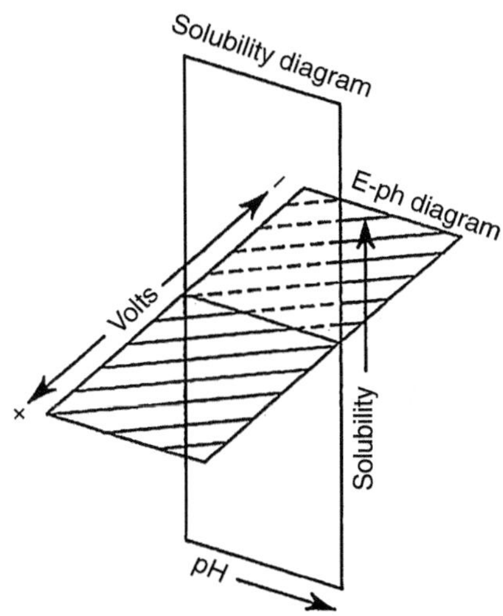

FIGURE 20-6 Schematic relationship between potential pH and conventional solubility versus pH diagrams.

Potential-pH diagrams are particularly useful to study speciation in systems that could contain species of several possible valence states within the range of redox potential normally encompassed by drinking water, such as manganese, iron, arsenic, lead, and copper. Obtaining an accurate estimate of the redox potential of the drinking water is usually an important limitation in using potential-pH diagrams, and the actual corrosion potential at the surface of the pipe is important. The latter differs from the bulk-solution redox potential (MWH, 2005).

Figure 20-7 is a potential-pH diagram for lead in water, illustrating the simultaneous incorporation of two weak polyprotic acids: carbonic and phosphoric. The aqueous species in the diagram are heavily influenced by the carbonate concentration in the water, but the orthophosphate phases are less soluble than the lead(II) carbonates, so they are predicted to provide the actual passivation and plumbosolvency control.

Many references can be consulted for details of potential-pH diagrams for other metals of interest, bearing in mind that the metal concentration, and concentrations of the background constituents in the water, the forms of solid and aqueous species considered in the diagrams, and the stability (or metastability) of the solids must all be relevant for the system under consideration. Because of the metastibility and aging issues noted previously, environmental time frames are often different from time frames implicit in potential-pH diagrams for geological systems, so diagrams are not automatically applicable. Equilibrium constants and other key thermodynamic data are not even available for some of the important aqueous and solid species involved in potable water system corrosion. Note that because the total activity is fixed across the entire pH range, the phase relationships predicted by the potential-pH diagram may deviate somewhat from those predicted in the previously described solubility diagrams.

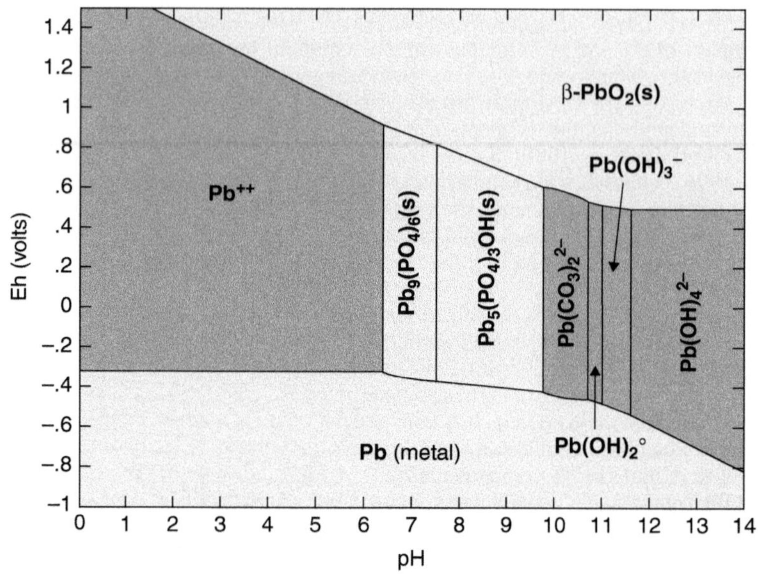

FIGURE 20-7 Potential-pH diagram assuming the activities of Pb, PO_4, and DIC to be 0.015, 2.3, and 18 mg/L, respectively. Stability fields of solids are in a lighter shade.

Bockris and Reddy (1973) provide several good suggestions and cautions for the use of potential-pH diagrams.

- Potential-pH diagrams can be used to give yes or no answers on whether a particular corrosion process is thermodynamically possible or not.
- The diagrams provide a compact pictorial summary of the electron-transfer, proton-transfer, and electron-and-proton-transfer reactions that are thermodynamically favored when a metal is immersed in a particular solution.
- When a potential-pH diagram indicates that a particular metal is immune to corrosion, it is so provided the pH in the close vicinity of the surface is what it is assumed to be.
- Even if a potential-pH diagram indicates that a particular corrosion process can spontaneously take place, it does not mean that significant corrosion will be observed. The corrosion reaction rate and the rate of passivation (if it occurs) cannot be predicted thermodynamically.

PHYSICAL FACTORS AFFECTING CORROSION AND METALS RELEASE

The characteristics of drinking water that affect the occurrence and rate of corrosion can be classified as physical, chemical, and biological. In most cases, corrosion is caused or increased by a complex interaction among several factors. Some of the more common characteristics in each group are discussed in this section to familiarize the reader with their potential effects. Controlling corrosion may require changing more than one of these because of their interrelationships.

Very few statements regarding corrosion resistance can be made that do not have noteworthy exceptions (NACE, 1984). The corrosion of metals and alloys in potable water systems depends on both the environmental factors (solution composition and flow) and the composition of the plumbing or fitting material. Concern about the consequences of corrosion varies with the material. Sometimes, it is toxicity from trace metal dissolution and contamination of the drinking water, but for materials such as iron, steel, and mortar linings, the concern is more one of aesthetic and material degradation or, ultimately, pipe failure.

Physical Characteristics

Flow velocity and temperature are the two main physical characteristics of water that affect corrosion.

Velocity. High flow velocities can sometimes aid in the formation of protective coatings by transporting the protective material to the surfaces at a higher rate. High flow velocities, however, are sometimes associated with erosion corrosion or impingement attack in copper pipes, in which the protective wall coating or the pipe material itself is mechanically removed. High-velocity waters combined with other corrosion-causing water constituents can rapidly deteriorate pipe materials with minimal metal release to the water because of dilution. For example, the combination of high velocity with low-pH and high-DIC concentration is extremely aggressive toward copper. High flow velocity can also increase the rate at which oxidant species come in contact with pipe surfaces, increasing corrosion. It is hard to quantify *high*, but a common guideline for 1.3 cm ID (11/2 in) copper is about 1.2 m/s (4 ft/s) (AwwaRF, 1990), and plumbing handbooks should be consulted for the materials of interest for initial guidelines.

A water that behaves satisfactorily at medium to high velocities may still cause incipient or slow corrosion with serious red water problems at low velocities, because the slow water movement allows for the development of reducing conditions that facilitates rapid dissolution of iron from the metal surface.

High water velocity can also mobilize loosely adherent scale material and cause elevated metal release levels and erratic exposure to consumers. Britton and Richards (1981) pointed out that there are several reasons why the pickup of lead into the water is not a simple function of sampling flow rate. Differences in flow rates and velocity at the pipe-water interface can severely complicate assessments of metal exposures and comparisons to regulatory standards.

Temperature. The influence of temperature is also often misunderstood. Confusion can often be avoided by considering basic equilibrium in water chemistry and by remembering that temperature effects are complex and depend on both the water chemistry and the type of plumbing material present in the system. Multiple phenomena often operate simultaneously, such as changes in solubility of solids; changes in the formation of complex ions; changes in diffusion rates of dissolved gases and solution species; changes in composition or physical properties of the metals, alloys, or solids; and changes in water properties. Thus, generalizations are often stated that are overly simplistic and often misapplied.

The electrode potential (the driving force for any corrosion cell) is proportional to the absolute temperature, and therefore, theory predicts that the thermodynamic driving force for corrosion rate will increase with temperature. This relationship is observed in some controlled laboratory experiments over short time periods, but in practice is less obvious unless wide differences occur (i.e., hot versus cold water systems). In distribution system transmission and service lines, temperature fluctuations are somewhat limited over short time frames, so whatever effect they have is obscured by other factors. Seasonal changes in temperature are often accompanied by significant changes in one or more major chemistry parameters.

Household and building plumbing systems present some additional complications. The relatively high thermal conductivity of the metallic piping materials normally used causes the water standing inside pipes or passing through long stretches of pipe to rapidly equilibrate to the temperature of the air surrounding the pipe. Water temperature is governed by many site-specific construction characteristics, such as (but not constrained to) the locations of the pipes being sampled relative to heating or air-conditioning ductwork, plumbing routed through slab flooring, existence of crawl spaces, presence or absence of pipe or floor/wall insulation, or spaces above ceilings or below floors. Thus, interpreting the metal release and corrosion mechanisms is difficult. For example, cold water pipes running next to a heating duct could actually produce the counterintuitive situation of higher water temperature in the winter in those homes. Fluctuations in temperature from many causes also makes it difficult to define a single temperature as "representative" of sampling or normal usage conditions. Attempts to apply simple correlations with water temperature leaving the treatment plant are tempting, but could be very misleading.

Hot water is often observed to be more corrosive than cold water (AWWA, 1971). Water showing few corrosive characteristics in the distribution system can cause severe damage to copper or galvanized iron hot-water heaters at elevated temperatures. In some hot water systems, however, the opposite is true, as is shown by the example in Fig. 20-8, where the net effect is the facilitation of a better passivating duplex scale of Cu_2O and CuO on the copper tube. Lead monitoring data also have shown in some cases that lead orthophosphates sometimes form more quickly in waters with higher temperatures and, therefore, the highest tap lead levels occur in the winter in some water systems. Water systems should collect appropriate water samples for corrosion and metal release throughout the year and

FIGURE 20-8 Temperature induced change in mineralogy of copper pipe scale between hot and cold water having free chlorine, high pH and low alkalinity. The net result was better passivation and lower metal release at elevated temperature, where CuO formation was accelerated. (*D. Lytle, USEPA.*)

determine their own trends. Water chemistry characteristics play a critical role, and blind generalizations on temperature effects should not be made.

Larson has pointed out that the effect of temperature on pH is seldom recognized (AWWA, 1971). For pure water, as temperature increases the water dissociates more (the pK_w goes down), causing increases in both the H^+ concentration and the OH^- concentration. Drinking waters all contain carbonate species at some concentration, so the degree of influence of temperature on pH is also a function of the alkalinity and DIC of the water, as a result of changes in the dissociation constants for carbonic acid with temperature. Increasing concentrations of bicarbonate increasingly buffer or reduce this effect of temperature on pH. The net pH decrease from heating tends to be less than the net decrease in overall $CaCO_3$ solubility, causing increased scaling tendency at higher temperatures for waters with higher alkalinities. For waters of low alkalinity (less than approximately 50 mg $CaCO_3$/L), the higher temperatures decrease the pH at a faster rate than the rate of decrease in solubility of $CaCO_3$. Consider a water at 15°C having the following characteristics: pH = 8.71; Ca = 17.3 mg/L; Na = 17 mg/L; Cl = 35 mg/L; and total alkalinity = 30.6 mg $CaCO_3$/L. The water is at saturation equilibrium with calcite. Therefore, it has a *saturation index* (or *Langelier Saturation Index* (LSI)) of 0.0. At this temperature, the solubility constant for the simple dissolution of calcite

$$Ca^{2+} + CO_3^{2-} \rightleftharpoons CaCO_3(s) \tag{20-20}$$

is equal to $10^{-8.43}$. If that water is warmed to 55°C, the saturation index drops to –0.02, even though the solubility constant decreases to $10^{-8.71}$. This phenomenon occurs because the pH also decreases, to 8.16.

Figure 20-9 gives an estimation of how the pH of waters of different alkalinities changes when warmed to a temperature of 55°C from 15°C. This figure is based on calculations

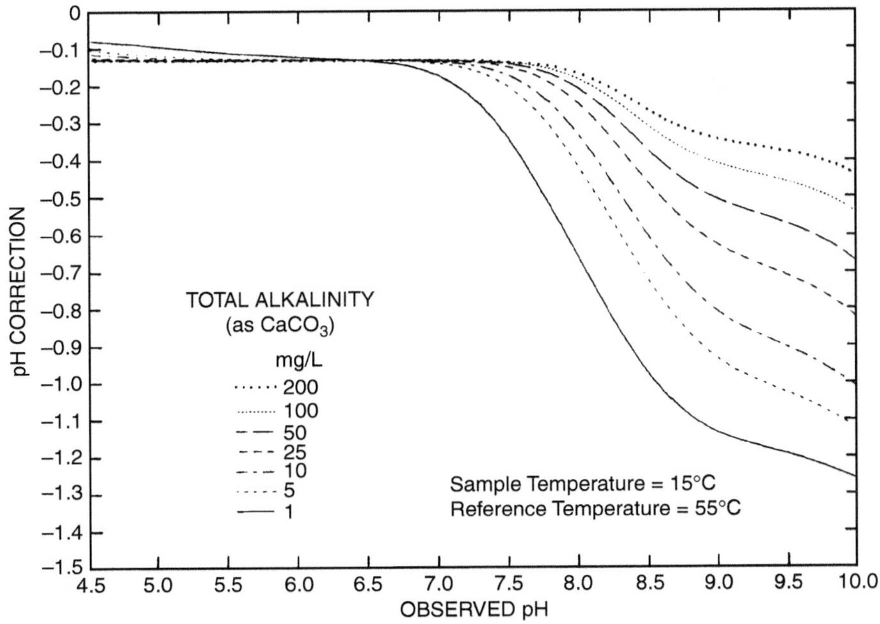

FIGURE 20-9 Change in pH for waters of different alkalinities caused by warming a water originally at 15°C to 55°C under closed-system conditions, assuming ionic strength of 0.001. The pH predicted for 55°C is obtained by *adding* the pH correction to the original (observed) pH at 15°C (e.g., $pH_{55} = pH_{15} + pH_{corr}$).

that assume the following: (1) an ionic strength of 0.001 M, (2) non-carbonate species contributing to alkalinity are negligible, (3) no change in inorganic carbon content during the temperature change, (4) carbonate and bicarbonate complexation are negligible, and (5) no other redox or hydrolysis reactions that significantly affect the proton balance of the system. More details on a general computational approach and some additional figures for computing the pH change resulting from intrinsic changes to water and carbonic acid dissociation have been published (AwwaRF, 1990).

An additional effect of temperature is that an increase can change the entire nature of the corrosion. For example, a water that exhibits pitting at cold temperatures may cause uniform corrosion when hot (Singley et al., 1984; AWWA, 1986). Another example in which the nature of the corrosion is changed as a result of changes in temperature involves the zinc-iron couple (Trussell and Wagner, 1996). Normally, the anodic zinc is sacrificed, or corroded, to prevent iron corrosion. In some waters, the normal potential of the zinc-iron couple may be reversed at temperatures above 46°C. The zinc becomes cathodic to the iron, and the corrosion rate of galvanized iron is much higher than would normally be anticipated, a factor that causes problems in many hot-water heaters or piping systems.

By changing solubility (as well as pH), temperature changes can influence the precipitation of different solid phases or transform the identities of corrosion products. These changes may result in either more or less protection for the pipe surface, depending on the materials and water qualities involved. When attempting to do solubility or other chemical speciation calculations in drinking waters, the effects of temperature on both the solubility constants for the potential passivating solids and the effects on the formation constants of the aqueous complexes of the metals of interest must be considered. Adjusting solubility constants but not complex formation constants, or only adjusting constants for some of the important water or solid species, are frequent errors observed in many published papers and reports. Unfortunately, analytical polynomial functions for changes in equilibrium constants with temperature, or various thermodynamic constants to enable other methods of approximations of adjustments to the equilibrium constants for temperature, are often not available for the aqueous species, solid phases, or both, in important corrosion and metal release situations for drinking water.

Manufacturing-Induced Characteristics

For several types of common plumbing materials, the manufacturing processes used may play an important role in determining the types of corrosion that will occur, and the durability of the pipe or fixtures. Pitting of galvanized pipe has been associated with several characteristics of its manufacture, such as thin, improper, or uneven galvanizing coating, poor seam welds, and rough interior finish (Fox et al., 1983; Trussell and Wagner, 1996; MWH, 2005). Impurities in the galvanizing dip solution, such as lead and cadmium, could cause concern because of their potential for leaching into the water.

In England, rapid pitting failures of copper pipe were found to be associated with a residual carbon film on their interior surfaces (Cruse and Pomeroy, 1974; Ferguson et al., 1996; MWH, 2005). Many copper corrosion researchers believe that the early history of the pipe and initial installation in residential and building plumbing has a profound impact on its susceptibility to localized corrosion later (Lane, 1993). Debris from manufacturing, transport and storage, and patchy or incomplete oxidation coatings from air and moisture exposure may set-up conditions favorable to differential oxygenation corrosion, depending on the initial exposure of the installed piping. If plumbing is not flushed thoroughly and housing is not occupied soon after construction, anoxic conditions favorable to microbiological growth and, later, to differential oxygenation corrosion, will likely result, and the corrosion cells may be hard to stifle once normal water usage is started.

CHEMICAL FACTORS AFFECTING CORROSION

Dissolved substances in water have an important effect on both corrosion and corrosion control. Many water chemistry factors may affect the rate of corrosion, metal release or dissolution, but not the equilibrium concentration. Therefore, in addition to understanding the way the chemistry factors act and interrelate, it is also important to understand the relationship between metal release and stagnation time in a water in a given situation, in order to interpret water sampling data, or to estimate impacts of various chemical parameters. This section provides a brief overview to point out some of the most important chemical factors affecting corrosion.

Alkalinity/Dissolved Inorganic Carbon (DIC)

Alkalinity is a measure of the ability of a water to neutralize acids and bases (Weber and Stumm, 1963; Loewenthal and Marais, 1976; Snoeyink and Jenkins, 1980; Butler, 1982; Pankow, 1991; Morel and Hering, 1993; Stumm and Morgan, 1996; Butler and Cogley, 1998). In most potable waters, total alkalinity is described by the relationship:

$$\text{TALK} = 2[CO_3^{2-}] + [HCO_3^-] + [OH^-] - [H^+] \qquad (20\text{-}21)$$

where [ ] indicates concentration in mol/L and total alkalinity is in equivalents/L (eq/L). Alkalinity and DIC are discussed in detail in Chap. 3. In some waters, where the total inorganic carbon content is low (such as below approximately, 5 mg/L), other weak acids and bases such as phosphoric acid or dissociated sodium or calcium hypochlorite can contribute to alkalinity if present in high concentration, and complicate the computation of DIC from pH and alkalinity results. Because the pH of the equivalence point for the alkalinity titration changes with DIC concentration, traditional analytical methods to fixed pH endpoints (such as pH 4.5) are inherently inaccurate, except for waters with the coincidentally correct DIC (Schock and George, 1991). For waters with low DIC, this can cause appreciable errors in estimating metal solubility and even chemical dosages for optimal corrosion control.

The bicarbonate and carbonate species affect many important reactions in corrosion chemistry, including the ability of a water to form a protective metallic carbonate-based scale or passivating film, such as $CaCO_3$, $FeCO_3$, $Cu_2CO_3(OH)_2$, $Zn_5(CO_3)_2(OH)_6$, or $Pb_3(CO_3)_2(OH)_2$. They also affect the concentration of calcium ions that can be present, which, in turn, affects the dissolution of calcium from cement-lined or from asbestos-cement (A-C) pipe. The formation of strong soluble complexes with metals such as lead, copper, and zinc (Schock, 1980, 1981; Schock and Buelow, 1981; Schock and Gardels, 1983; Schock et al., 1995, 1996; Edwards et al., 1996; Ferguson et al., 1996; Trussell and Wagner, 1996) can accelerate corrosion or cause high levels of metal pickup given the right pH/alkalinity or pH-DIC conditions.

Ammonia

Ammonia forms strong soluble complexes with many metals, particularly copper and probably lead. Thus, ammonia may interfere with the formation of passivating films or increase the corrosion rate. Source water ammonia reacts with chlorine applied for disinfection and, therefore, results in inadvertent chloramination (see Chap. 17). Water systems that have substantial ammonia in their source waters and do not remove it need to take into account optimizing a chloramine residual and conducting oversight to guard against nitrite and nitrate production resulting from nitrifying biofilms and to monitor for possible increases in corrosivity (Lytle et al., 2008).

Buffer Intensity (β), Buffer Capacity, Buffer Index

The ability of a water to provide buffering against a pH increase or decrease caused by a corrosion process or water treatment chemical addition is closely related to the alkalinity, DIC concentration, and pH of the water (Weber and Stumm, 1963; Loewenthal and Marais, 1976; Butler, 1982; Stumm and Morgan, 1996; Butler and Cogley, 1998; Urbansky and Schock, 2000c). The *buffer intensity* of a water is defined as $\beta = (d\text{TALK}/d\text{pH})_{\text{DIC}}$, which is essentially the inverse of the slope of the alkalinity titration curve. The basic equation for buffer intensity, β_{tot}, in terms of *total alkalinity* (TALK) is the following, for a system containing only the carbonic weak acid system in addition to water.

$$\beta_{\text{tot,Alk}} = 2.303\left[\left(\frac{[\text{H}^+]\text{TALK}}{[\text{H}^+]+K_2'}\right)\times\left(\frac{[\text{H}^+]}{[\text{H}^+]+K_1'}+\frac{K_2'}{[\text{H}^+ + K_2']}\right)+[\text{H}^+]-\frac{K_w'}{[\text{H}^+]}\right] \qquad (20\text{-}22)$$

TALK is expressed in eq/L units, $[\text{H}^+]$ is the hydrogen ion concentration in mol/L, K_w' is the dissociation constant for water, and K_1' and K_2' are the first and second dissociation constants for carbonic acid, respectively, all corrected for temperature and ionic strength. Similarly, buffer intensity can be written in terms of DIC (moles/L) rather than alkalinity. The resulting equation is

$$\beta_{\text{tot,DIC}} = 2.303\times\text{DIC}\times K_1'[\text{H}^+]\left[\frac{K_1'K_2'+[\text{H}^+]^2}{(K_1'K_2'+K_1'[\text{H}^+]+[\text{H}^+]^2)^2}\right]+2.303\times\left[\frac{K_w'}{[\text{H}^+]}+[\text{H}^+]\right] \qquad (20\text{-}23)$$

The units of β are normally mol/L per pH unit. If other weak acids or bases such as orthophosphate or silicate are present or if the ammonia concentration is high, additional terms must be added to Eqs. 20-22 and 20-23. These equations relate to what is often termed a *homogeneous buffer system*, in which all buffering components are aqueous species.

Figure 20-10 shows the buffer intensity for several different concentrations of DIC. Note the minimum point near pH 8, which corresponds to the point where pH is equal to $\frac{1}{2}(-\log K_1' + -\log K_2')$ for carbonic acid. The buffering at the pH extremes is from either hydrogen ion (low pH) or hydroxide ion (high pH). Figure 20-11 shows the effect on buffer intensity of the presence of 20 mg SiO_2/L silicate, and 5 mg PO_4/L orthophosphate, in a water having a DIC concentration of only 4.8 mg C/L. Clearly, the carbonate system provides almost all of the buffering, except when the pH is over about 9 and the silicate anion (present at a high total SiO_2 concentration) contributes sufficiently. This demonstrates that corrosion inhibitor chemicals of all types at normal drinking water dosages are a negligible component of pH buffering, except for waters of extremely low DIC (<2 mg/L). When a water body is open to the atmosphere, the exchange of carbon dioxide gas will affect the DIC concentration, and consequently, both the buffer intensity and capacity.

The difference between alkalinity and buffering intensity is often misunderstood. For example, consider four waters all with an identical total alkalinity of 20 mg/L as $CaCO_3$, at pH 6.0, 7.0, 8.0, and 9.0. The DIC concentrations would be 15, 5.8, 4.9, and 4.4 mg C/L, respectively, and the corresponding values for buffer intensity (β) would be 6.3, 0.2, 0.3, and 0.7 (mmol/L)/pH. The buffer intensity and DIC are considerably different for each water corresponding to the trend in Figs. 20-10 and 20-11. Thus, finished waters in the pH 8 to 8.5 range will have much poorer pH stability than if the water with the same DIC concentration was distributed at pH 7 or 9. Prevention of localized sites of high pH (such as at cathodic corrosion cell sites) by better buffering will improve the evenness of formation of corrosion-reducing scales and localized pitting, especially for ferrous materials (Lewandowski et al., 1997). Clement et al. (1998) have confirmed the practical application of these buffer relationships in the Concord, New Hampshire, distribution system, dramatically

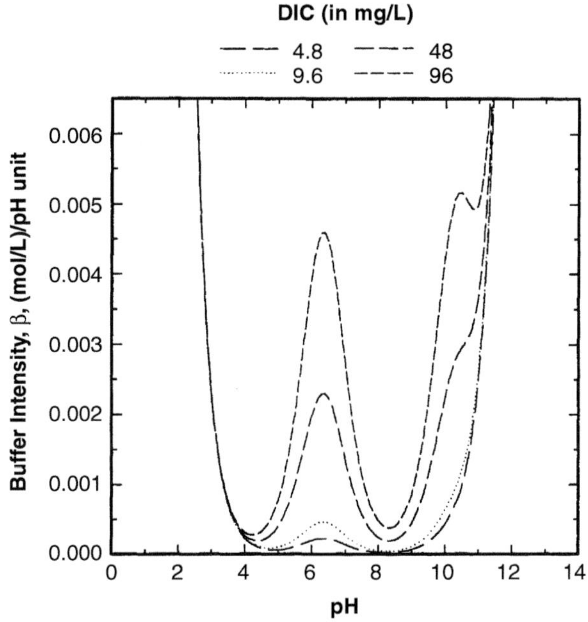

FIGURE 20-10 Effect of DIC on buffer intensity, for 25°C, $I = 0$.

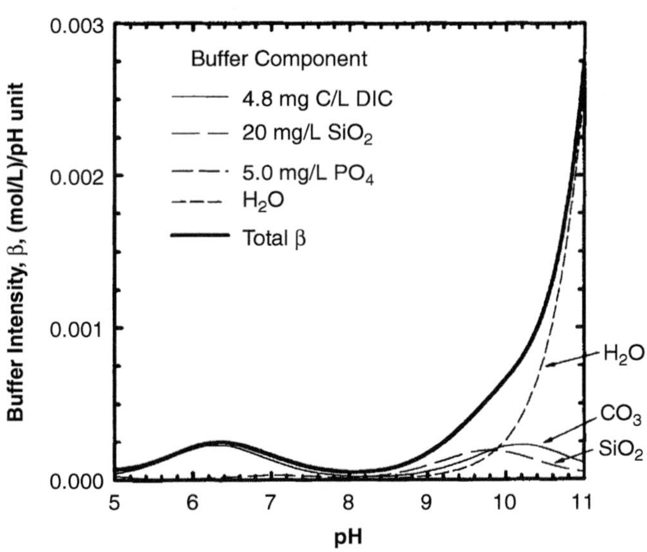

FIGURE 20-11 Combined components of buffer intensity, for DIC, silicate, and orthophosphate at high dosages and low DIC. Computed for 25°C, $I = 0.005$.

improving corrosion control and distribution system pH stability by elevating the pH above the problematic 8 to 8.5 range to 9. Additionally, it improved the overall lead and copper control characteristics of the water (see Fig. 20-12).

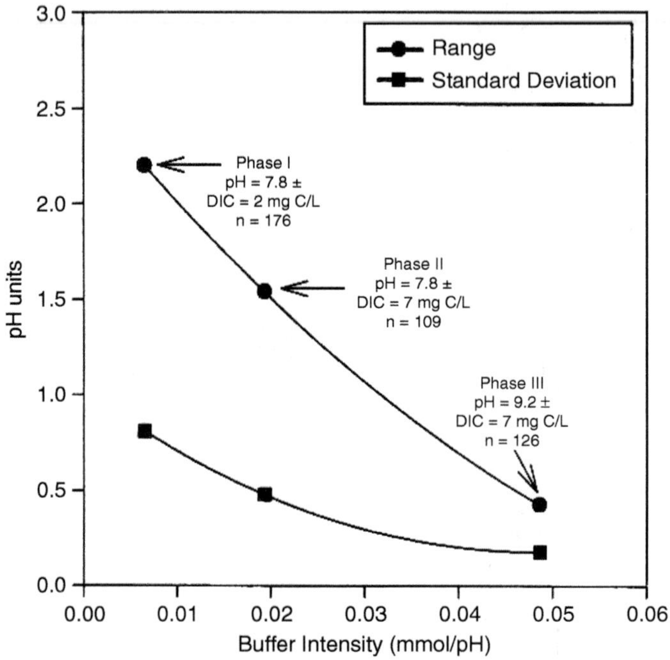

FIGURE 20-12 Improvement in distribution system pH stability in Concord, New Hampshire, treatment study. (*Source: Clement and Schock, 1998, Clement et al., 1998.*)

Buffering can also be provided by solid materials that can consume $[H^+]$ or $[OH^-]$ directly or indirectly. This defines a *heterogeneous buffer system.* Common minerals that provide buffering in natural water systems or scales at the water-pipe interface include aluminosilicates like kaolinite (a clay mineral), magnesium hydroxide, and calcium carbonate (Watkins and Davies, 1987; Dexter and Lin, 1992; Stumm and Morgan, 1996; Lewandowski et al., 1997). Although the aluminosilicates provide relatively strong resistance to pH change, the reaction rates are quite slow in comparison to those of carbonates, oxides, and hydroxides, so they may not be as significant in short time frames of water passage through pipes. In the restricted surface environments of pipe scales, essentially any of the metallic carbonate and hydroxycarbonate minerals could serve as buffers to some extent. Even oxides and oxyhydroxides can play a role in buffering, and the buffering of these scale solids is an important part of the lag time to completely achieve the new target equilibrium after treatment changes involving pH and alkalinity adjustment.

An expression for the buffer intensity at constant DIC for a solution in equilibrium with calcium carbonate (Butler, 1982; Butler and Cogley, 1998) is

$$\beta_{CaCO_3}^{DIC} = \beta_{tot,DIC} + 4.61[Ca^{2+}]\left(\frac{[HCO_3^-]+2[CO_2]}{DIC}\right) \tag{20-24}$$

The derivation of this expression assumed that ion pairing is negligible. To use this equation, the calcium concentration is input presuming all calcium is present as free Ca^{2+} ion. Analyzed concentrations of $[HCO_3^-]$ and $[CO_2]$ must also be input, or computed values derived from a combination of pH and total alkalinity or DIC for the proper temperature and ionic strength. In principle, similar or analogous expressions can be derived for other oxide, carbonate, or hydroxide solids of relevance to corrosion and passivating scales on drinking water pipes.

Chloride and Sulfate

Chloride (Cl^-) and sulfate (SO_4^{2-}) may cause increased corrosion of metallic pipe by reacting with the metals in solution and causing them to stay soluble or interfering with the formation of normal oxide, hydroxide, or hydroxycarbonate film. They also contribute to increased conductivity of the water. Chloride is about 3 times as active as sulfate in this effect for iron materials, but recent research into copper corrosion has shown the action of chloride to be complicated and not always detrimental for copper (Edwards et al., 1994a; Meyer, 1994; Rehring, 1994; Duthil et al., 1996). Pitting of copper tubing is often strongly related to the concentrations of Cl^- and SO_4^{2-} relative to HCO_3^-, where the high salt contributions can contribute to the acidification in pits and enhanced conductivity of the water (Cruse and Pomeroy, 1974; Ferguson et al., 1996). Recently, it has been shown that sulfate is not necessary to form pits in copper tubing at high pH and low alkalinity (Lytle and Schock, 2008).

With lead-containing materials, increases in chloride and sulfate have been implicated in several cases of increased lead contamination, relating to changes in coagulant from alum to ferric chloride (Edwards and Triantafyllidou, 2007; Triantafyllidou et al., 2007; Stone et al., 2008).

Based on iron corrosion rate measurements in a large number of midwestern waters, Larson and Skold (1957, 1975) suggested that the ratio, sometimes called the *Larson ratio* (LR),

$$LR = \frac{[Cl^-] + 2[SO_4^{2-}]}{[HCO_3^-]} \tag{20-25}$$

should ideally be less than approximately 0.2 to 0.3 when expressed using concentrations as mol/L. Originally, the ratio was simply chloride over bicarbonate, but the originators felt that sulfate ion would be similarly aggressive. Other researchers suggest a higher target for the ratio, with values up to 1 as the maximum recommended value (van den Hoven and van Eekern, 1988). Some engineers and researchers express the ratio as the inverse, bicarbonate to chloride plus sulfate. In that case, the ratio should be larger than 3 to 5. Larson's ratio is also logical considering the results of other research in which the chloride ion was noted for its role in breaking down passivating films on many ferrous metals and alloys and is one of the main causes of pitting of stainless steels (Singley et al., 1985). Research in France, principally by LeRoy (1996), reported that some fundamental chemistry mechanisms support the concept of Larson's ratio.

There were limitations to Larson's original research, which did not allow the determination to be made if there were individual trends amongst the examined waters that might be directly caused by differences in pH, hardness, or alkalinity. By doing controlled laboratory investigations, LeRoy (1996) has shown that at constant pH and bicarbonate concentration, necessary because ferrous iron forms ion pairs with bicarbonate ion, the oxidation of ferrous iron is enhanced in the order $Cl^- > SO_4^{2-} > NO_3^-$. This order parallels the strength of the different aqueous complexes with free ferrous ion. The mechanism was proposed that bicarbonate plays a role in mitigating the corrosion reaction, because calcium bicarbonate ion pairs ($CaHCO_3^+$) fix free oxygen radicals formed during the iron oxidation reaction. Calcium carbonate may then deposit at the high-pH cathodic sites, whereas various ferrous or ferric iron solids may form at anodic sites. This model is consistent with the frequently reported anecdotal observations that the corrosion of iron is less significant in harder and relatively well-buffered waters.

Copper

The presence of dissolved copper can cause rapid corrosion of galvanized steel and iron pipe. Free copper may occur in recirculating water systems, such as the hot-water systems commonly found in industry facilities, hotels, and apartments that have consecutive sections of copper and galvanized steel pipes in the loops. This free copper adsorbs or reacts with the galvanized steel piping, setting up small galvanic cells, which produce rapid pitting failure of the galvanized steel piping. Similar corrosion has occurred in unprotected galvanized steel pipes in consumers' homes as a result of the addition of even trace amounts of copper, either from copper sulfate as an algaecide in distribution reservoirs, from flexible copper tubing connectors to hot-water heaters, or from mixed copper and galvanized plumbing in the same water line (Fox et al., 1986; Trussell and Wagner, 1996; MWH, 2005).

Disinfectant Residual

Secondary disinfectants can affect corrosivity, metal release, or both, either by their impact on the *oxidation-reduction potential* (ORP) of the bulk water, corrosion potential of the surface of the metal, or direct reaction mechanisms. The latter has been infrequently studied. The choice and amount of secondary disinfectant can play the dominant role in bulk water ORP and must be considered when determining metal release potential and treatment change impacts. For example, James et al. (2004) compared five common chemical oxidants at pH 7 and DIC of 10 mg/L and noted considerably different impacts on ORP (Fig. 20-13). Chlorine dioxide caused the greatest increase in ORP at low doses, followed by free chlorine. All oxidants tend to have diminishing effects as concentrations rise above several mg/L, presumably due in great part to the intrinsic redox buffering ability of water.

Knowing how system ORP has been impacted historically by oxidative treatments provides one of the most critical keys to determining if there are any likely adverse impacts and metal release episodes that would result from disinfection-related or other treatment changes. ORP also dramatically affects manganese speciation and solubility (Garrels and Christ, 1965; Stumm and Morgan, 1996; Langmuir, 1997), which can govern the formation and persistence of surface coatings on pipes that promote particulate release and can reduce the effectiveness of corrosion inhibitors (Fig. 20-14). In this example, three common manganese oxides were considered, along with manganese(II) carbonate (rhodochrosite). Assumptions for the calculations are given in the caption. A reference box is shown for the approximate ORP and pH ranges commonly found in drinking waters.

Gaseous chlorine can lower the pH of the water by reacting with the water to form hypochlorous acid, hydrogen ion, and chloride ion (particularly as the aqueous chlorine species decomposes). In poorly buffered waters, chlorine can increase corrosivity through a reduction in pH and, in general, by increasing corrosion rates and ORP, though there are notable exceptions to the latter. Conversely, the addition of chlorine as sodium hypochlorite, a base, raises the pH. In waters with low alkalinity or carbonate content, the tendency of chlorine addition to affect pH is greater because such waters have limited buffering at typical finished water pH values. Chlorine residual interactions with iron, copper, and lead are complex and are described in the section, Corrosion of Specific Materials.

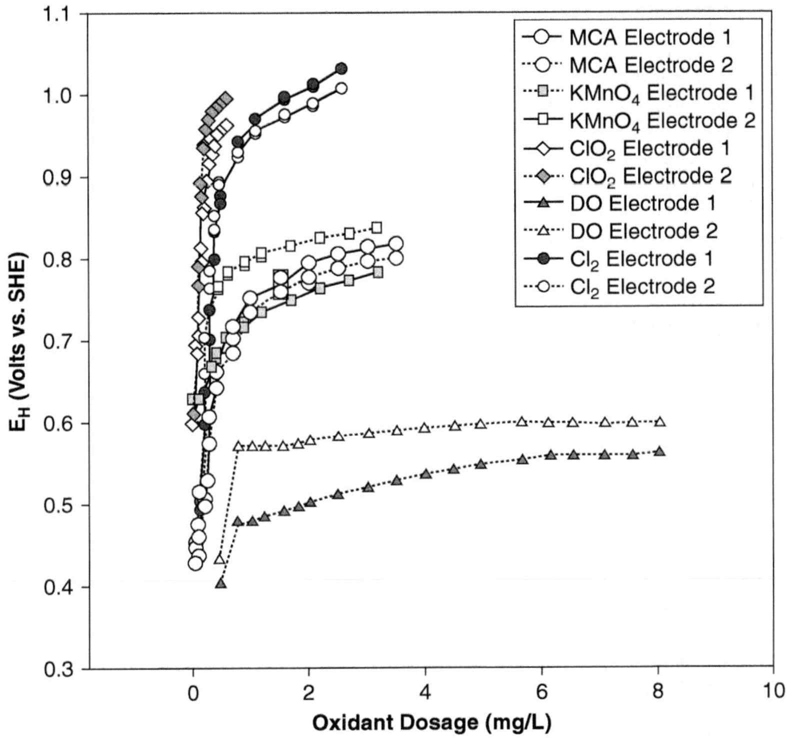

FIGURE 20-13 Effect on ORP of different dosages of strong oxidants into a water at 5 mg C/L DIC, 23°C, pH 8. MCA = monochloramine. (*Source: James et al., 2004.*)

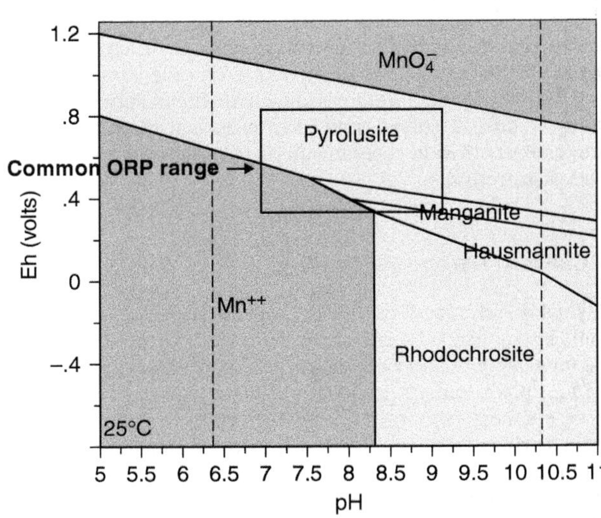

FIGURE 20-14 Potential-pH relationships for several manganese minerals at 25°C, I = 0, DIC = 18 mg C/L, and Mn species activities = 0.1 mg/L, Pyrolusite = $Mn^{4+}O_2(s)$; Manganite = $Mn^{3+}O(OH)$ (s); Hausmannite = $MN^{2+}Mn_2^{3+}O_4$ (s); and Rhodochrosite = $MnCO_3$ (s). Box shows approximate typical ORP-pH range for municipal drinking waters in the United States.

Whether chloramines and accompanying residual free ammonia mechanistically increase or modify corrosivity of most metals, has not been frequently studied in the detail that it deserves. Several investigations have shown differences in corrosion or metal release behavior by some metals, particularly lead and copper, when chloramine is the secondary disinfectant (James M. Montgomery Consulting Engineers, 1982, 1983; Edwards and Dudi, 2004; Eisnor and Gagnon, 2004; Edwards and Triantafyllidou, 2007). The effect could be direct or indirect, and mechanisms need considerable further study.

Dissolved Oxygen

Oxygen is often the agent that accepts the electrons given up by the corroding metal, allowing corrosion reactions to continue. For example, oxygen reacts with any ferrous iron ions and converts them to ferric iron. Ferrous iron ions, Fe^{2+}, are much more soluble in water than ferric iron, which readily forms a gelatinous orange ferric hydroxide floc when dissolved ferrous iron encounters oxygen or chlorine. Ferric iron accumulates at the point of corrosion, forming a tubercle, or settles out at some point in the pipe and interferes with flow. Similarly, oxygen reacts with metallic copper to form cuprous ions (Cu^{+}) or cupric ions (Cu^{2+}), depending on the amount of oxygen present (Pourbaix, 1966; Werner et al., 1994; Schock et al., 1995; Ferguson et al., 1996). Cuprous hydroxide and oxide solids are much less soluble than cupric forms over nearly all of the pH range covered by drinking waters.

When oxygen is present in water, red water may occur if soluble ferrous iron is exposed to a sudden high oxygen concentration either in the distribution system or at the tap (AwwaRF, 2002). However, it must be borne in mind that municipal drinking water systems in the United States typically maintain (or attempt to) a secondary disinfectant residual. Hence, dissolved oxygen is usually a minor factor in corrosion, metal speciation, and metal release. Reiber (1989), for example, showed the relative importance of chlorine relative to oxygen in the case of copper corrosion.

High dissolved oxygen levels may actually provide corrosion benefits in some circumstances, by facilitating the production of different and more protective mineral oxide and hydroxide solid corrosion products than in environments with limited oxygen availability. The many subtleties of the interactions of oxygen with pipe surfaces, and the effects on the physical properties and mineralogy of scales, still needs substantial research. On the basis of current knowledge, dissolved oxygen from advanced oxidation processes would not be expected to cause corrosion problems in distribution systems that have historically practiced chemical disinfection.

Hardness and Calcium Carbonate Scaling

Hardness can play many different roles that can sometimes reduce corrosivity toward many plumbing materials but, conversely, can create secondary constraints on optimizing treatment by forming unwanted calcium carbonate scales in various parts of the water transmission system (e.g., pipes, valves, meters) and in consumer or industrial facilities and processes (e.g., boilers, heat exchangers, hot water tanks, etc.). Cloudy and turbid water can also result from detrimental interactions with other treatment process residuals, causing sediment buildup in the distribution system.

Hardness is caused predominantly by the presence of calcium and magnesium ions, and is most frequently expressed as the equivalent quantity of $CaCO_3$. Hard waters are generally less corrosive than soft waters toward unlined iron pipe if sufficient calcium ions and alkalinity are present at the given pH to participate in the development of calcareous scales, such as a mixed iron-calcium carbonate deposits (Stumm, 1956; Sontheimer et al., 1981; Singley et al., 1985;

Trussell, 1985). High levels of calcium also help stabilize calcium silicate phases in cement-mortar linings and A-C pipes and promote the formation of calcium carbonate to block the release of free lime (calcium hydroxide) (LeRoy, 1996). In spite of common beliefs, significant calcium carbonate films only rarely form on lead, galvanized, or copper cold-water pipes, so they are not primarily the causes of corrosion inhibition in these cases. Exceptions are sometimes found in systems that practice lime or lime-soda softening and put supersaturated (with $CaCO_3$) water into the distribution system, and an indirect beneficial mechanism to reduce copper corrosion has also been proposed (Elfström Broo et al., 1997).

Early research by Stumm (1956, 1960), as well as Larson (1957, 1971, 1975) and many others, showed good associations of deposition of calcium-containing carbonate scales with smooth inner pipe walls, and freedom from tuberculation in unlined iron pipe. Calcium may also assist in the heterogeneous corrosion buffering reactions at the pipe surface (Watkins and Davies, 1987; Dexter and Lin, 1992; MWH, 2005), and by coordinating with polyphosphate functional groups, it may somewhat mitigate otherwise aggressive properties toward lead, copper, and cement materials.

The prediction of the threshold for calcium carbonate formation is important for several reasons. First, it can create a constraint on the ability to adjust pH, by forming a detrimental deposit. Second, for some materials such as cement linings, it is a consideration for adjusting water chemistry to arrest deterioration. Third, it can be used to monitor the consistency of treatment processes such as lime softening or neutralization with limestone or dolomitic materials.

The Langelier saturation index (Langelier, 1936) is the most widely used and misused index in the water treatment and distribution field. The index is based on the effect of pH on the equilibrium solubility of $CaCO_3$. The pH at which a water is saturated with $CaCO_3$ is known as the pH of saturation, or pH_s. At pH_s, a $CaCO_3$ scale should be neither deposited nor dissolved. The LSI is defined by the following equation:

$$LSI = pH - pH_s \tag{20-26}$$

The results of the equation are interpreted as follows:

LSI > 0. Water is supersaturated and tends to precipitate a scale layer of $CaCO_3$.

LSI = 0. Water is saturated (in equilibrium) with $CaCO_3$; a scale layer of $CaCO_3$ is neither precipitated nor dissolved.

LSI < 0. Water is undersaturated, tends to dissolve solid $CaCO_3$.

In 1994, recognition that the LSI was being frequently misused as a corrosion index and that inappropriate treatment approaches were being adopted and used in different parts of the United States, the USEPA repealed the section of the 1980 amendment to the National Interim Primary Drinking Water Regulation that required all community water supply systems to determine either the LSI or an even more approximate version, the aggressiveness index (AI), and report these values to the state regulatory agencies (USEPA, 1980, 1994). Although the LSI tends to predict if $CaCO_3$(s) will precipitate or dissolve, it does not predict how much $CaCO_3$(s) will precipitate or whether its structure will provide resistance to corrosion (Larson, 1975).

Even when a calcium carbonate saturation or precipitation index is of value for the prediction of scaling phenomena, many computational issues have been noted, depending on how the particular index is derived and needs to be solved (Rossum and Merrill, 1983; Schock, 1999a). Additionally, there are critical chemistry issues with the computations and attempts to interpret and apply the resulting index value to precipitation and scaling (Schock, 1984; Snoeyink and Kuch, 1985; Snoeyink and Wagner, 1996).

1. Corrections in the computations are often not made for complexation and ion pairing of Ca^{2+} and HCO_3^-, although this is possible if needed analytical data are available and

if the algorithms are set up in the computation scheme. This problem is most severe in hard waters and ones containing high sulfate or DIC concentrations (or both).

2. In the presence of polyphosphates, the essential equations defining pH_s are invalid. The LSI (and related) equations will *always overestimate* calcium carbonate saturation unless correction factors are added to account for complexation and poisoning of crystal growth. Little information exists that is useful to quantify such effects for different water chemistries.
3. Crystal growth poisoning and increases in the solubility of calcium carbonate polymorphs and solid solutions (such as high magnesian calcites) are commonly caused by high concentrations of sulfate, orthophosphate, magnesium, NOM, and some trace metals (zinc has been widely associated with this phenomenon). This phenomenon *essentially invalidates* the numbers obtained for LSI and related indexes.
4. In the past, the crystalline form of $CaCO_3(s)$ has usually been assumed to be calcite. The presence of a more soluble form of $CaCO_3(s)$, aragonite, has also been observed in drinking water plumbing deposits.
5. $CaCO_3$ can also be deposited from water with a negative LSI if there is localized high pH, generated by cathodic reactions at the pipe wall. This is most commonly observed in poorly buffered, low-DIC waters. Inconsistent deposition can lead to spotty surface coverage and can aggravate localized corrosion attack.
6. Maintaining a positive LSI has occasionally led to excessive deposition of $CaCO_3(s)$, near the point of chemical adjustment, in valves and meters and other locations prone to pressure drops or other physical clogging conditions, and to significant decreases in the capacity of distribution systems to carry water.

Although the LSI is thermodynamically related to the driving force for the amount and rate of deposition (Trussell et al., 1977), sometimes evaluating the theoretical approximate mass of calcium carbonate that could precipitate on a pipe surface is more useful. This quantity is called the *calcium carbonate precipitation potential* (CCPP). Procedures have been given in several sources for these calculations (Loewenthal and Marais, 1976; Merrill et al., 1978; Rossum and Merrill, 1983; MWH, 2005), and it is most conveniently calculated by a variety of computer programs or spreadsheet-based applications (APHA, AWWA, and WEF, 1995; Schott, 1998; RTW, 2008). Geochemical-modeling computer codes capable of mass-transfer and reaction path modeling are also excellent tools for this kind of computation (Parkhurst et al., 1980; Schecher and McAvoy, 1998; Appelo and Postma, 1999; Bethke and Yeakel, 2007).

The relative state of saturation with respect to calcium carbonate has also been determined empirically by physical means for many years, through the use of what is commonly called the *marble test*. Though often considered crude and obsolete, this physical testing approach is often advantageous. When there are complex forming ligands in the water, such as polyphosphates, or there are other water constituents (such as phosphates, sulfate, magnesium, NOM, and certain trace metals) that tend to alter or inhibit the nucleation, crystal growth, rate of precipitation, solubility, or crystal structure of calcium carbonate that could form, the numerical methods are largely inapplicable. Thus, the empirical test is the only accurate indicator of scaling potential. Several papers have described various methodologies for this marble test (Hoover, 1938; Dye, 1964; Balzer, 1980; Notheis et al., 1995). This test simply looks at the change in pH, alkalinity, or both, when a test water is equilibrated with some calcium carbonate powder for a number of hours. A water that does not change is apparently at equilibrium with calcium carbonate, and is considered to be *stable*. Notheis (1995) has described a recent case in which the marble test enabled diagnosis and solution of a corrosion problem that was not apparent from the numerical indices. This kind of test is particularly useful for water systems applying small amounts of polyphosphate to inhibit postprecipitation of calcium carbonate, to ensure the lowest dosage necessary to solve the problem and not aggravate lead or copper release.

A more elaborate electrochemical approach to determine the calcium carbonate saturation state has been used by chemical limnologists and oceanographers for many years. This device is called a *carbonate saturometer,* and it operates on a similar principle to the marble test (Weyl, 1961; Ben-Yaakov and Kaplan, 1971; Ingle et al., 1973; Plath et al., 1980; Königsberger and Gamsjäger, 1990). The instrument is a modification of a pH meter, with a glass cell reaction system setup so that the pH change when calcium carbonate is introduced generates a potential difference from a reference cell at equilibrium with the water. This potential difference can then be mathematically related directly to the amount of over or undersaturation of the water. The saturometry approach could have considerable merits if applied to drinking water treatment in several areas. Conceivably, the saturometer could be designed with minerals other than calcium carbonate, and the status of a water relative to the formation of passivating films on metallic pipe could be directly estimated.

Hydrogen Sulfide (H_2S)

Hydrogen sulfide accelerates corrosion by reacting with metal ions to form nonprotective insoluble sulfides (Singley et al., 1985; AWWA, 1986). It attacks iron, steel, copper, and galvanized piping to form *black water,* even when oxygen is absent. Attack by H_2S is often complex, and the results may be evident immediately or they may not become apparent for months and then suddenly become severe. Hydrogen sulfide has been found to promote a particularly devastating kind of pitting in copper piping (Jacobs and Edwards, 1997). Hydrogen sulfide also has been implicated in the attack on A-C pipe in some waters, possibly through microbial reactions (LeRoy et al., 1996).

Magnesium

Magnesium, and possibly some other trace metals (such as zinc), are known to inhibit the formation of the calcite form of $CaCO_3$. Instead of calcite, the aragonite form or some magnesium calcites may be deposited, which are more soluble. With these conditions, most calcium carbonate precipitation or saturation indices will give erroneous predictions. Whether these $CaCO_3$ forms make any difference in protection against corrosion is not known.

Metallic Precipitates

Soluble iron, zinc, and to some extent, manganese can play a role in reducing the corrosion rates of A-C pipe (LeRoy et al., 1996). Through a reaction that is not yet fully understood, these metallic compounds may combine with the pipe's cement matrix to form a protective coating on the surface of the pipe. Waters that contain natural amounts of iron have been shown to reduce the rate of A-C pipe corrosion and bind asbestos fibers to the surface. When zinc is added to water in the form of zinc chloride or zinc phosphate, protection from corrosion has been demonstrated by the formation of a hard surface coating of basic zinc carbonate or a zinc silicate.

Aluminum may be widespread as a component of films on distribution system piping that can act as diffusion barriers to reduce corrosion or metal release. Many U.S. water systems have aluminum residuals leaving the treatment plants at over 0.05 mg/L. Because soluble aluminum can take a long time in distribution systems to reach equilibrium with a precipitated Al solid and is highly dependent on the total Al concentration, pH, competing ligands, such as NOM and phosphates, and water temperature, it is likely that many U.S. drinking waters treated with alum are supersaturated with aluminum (MWH, 2005).

Even though it has not been systematically studied, several investigations have found aluminum films to significantly reduce lead leaching (Lauer and Lohman, 1994; AwwaRF, 1998; Atassi et al., 2004; Tesfai et al., 2006) and to adversely affect the hydraulic efficiency of distribution mains through depositions of solids (Kriewall et al., 1996; Snoeyink et al., 2003; Schock et al., 2008a). Aluminum was found on copper pipes even at low influent concentrations (Schock et al., 1995), suggesting also that it can be widespread. Aluminum residual and pipe coatings can consume inhibitor chemical and exacerbate the release of particulate-bound Pb from the scales (Kvech, 2001; Cantor, 2006, 2009; Cantor et al., 2006). In natural aquatic systems and groundwater, aluminum readily combines to form aluminosilicate minerals of low solubility, also suggesting that they may be common in distribution systems. Destabilization of aluminum films has been found to aggravate lead levels in the water in one system in Wales (Fuge and Perkins, 1991; Fuge et al., 1992).

Two examples of these foreign deposits on lead service lines are shown in Figs. 20-15 and 20-16, though similar deposits have been observed to form on iron, asbestos-cement,

FIGURE 20-15 Lead service line specimen coated with complex material mainly consisting of aluminum, iron, and manganese mixed with phosphate, causing problems with erratic lead release and ineffectivness of orthophosphate treatment. (*USEPA photo.*)

and various types of plastic pipes. Manganese and iron coatings occur often on lead and other types of pipes and can be responsible for greatly prolonging the time it takes to build passivating films on the pipes with corrosion inhibitors, and they tend to also increase the risk of sporadic spikes of lead from particulate release. Manganese buildups throughout water distribution mains and storage are likely far more common than have been reported. Legacy deposits from decades of suboptimal manganese removal, amounting to literally tons of material in a typical distribution system, can cause discoloration from physical dislodgement (roadwork, main repairs, and flow changes) or changes in water chemistry (Fig. 20-17). Even with treatment to meet a 0.1- or 0.05-mg/L target, manganese solids can still accumulate in all types of piping, including PVC (Cerrato et al., 2006).

FIGURE 20-16 Lead service line with mainly manganese oxyhydroxide coating on top of normal lead carbonate, resulting in erratic lead release and prolonging the time necessary for passivation and conversion to hydrocerussite in the lead scale. (*USEPA Photo.*)

FIGURE 20-17 Effluent from a pigging operation showing slurry of accumulated manganese deposits causing hydraulic impairment and black water episodes.

Beyond just the presence of hydraulically or physically detrimental aspects of the surficial coatings, it has now been demonstrated by many researchers (and references cited by them) that radionuclides, trace elements, and rare earth elements can all accumulate within or on top of the corrosion and scale deposits (Valentine and Stearns, 1994; Reiber and Dostal, 2000; Lytle et al., 2004a; Schock, 2005; Copeland et al., 2007; Morris and Lytle,

2007; Schock et al., 2008a; Friedman et al., 2009). Therefore, water discoloration episodes are likely to be accompanied by the release of some quantity of these accumulated contaminants and should not be considered a purely *aesthetic* annoyance. Even of more concern, such contaminants could be released invisibly into clear, drinkable water, depending on the particular chemistry changes.

Natural Color and Organic Matter

The presence of natural organic matter (NOM), manifested sometimes as color, may affect corrosion in several ways (see Chap. 3 for coverage of the chemistry of NOM). Some natural organics may react with the metal surface and provide a protective film and reduce corrosion, especially over a long period of time (Campbell, 1971). Others have been shown to react with the corrosion products to increase corrosion, such as with lead (Korshin et al., 1996, 1999, 2000, 2005). NOM is regarded as one of the major challenges to plumbosolvency control using orthophosphate in the United Kingdom (Colling et al., 1987; Hayes et al., 2008). Organics may complex calcium ions and keep them from forming a protective $CaCO_3$ coating. In some cases, the organics may become food for organisms growing in the distribution system or at pipe surfaces. This can increase the corrosion rate in instances when those organisms attack the surface, as discussed in the section on biological characteristics. Which of these instances will occur for any specific water has been impossible to determine, so using natural color and organic matter as corrosion-control methods are not recommended (Singley et al., 1984; AWWA, 1986).

Orthophosphate

Orthophosphate usually forms insoluble passivating films on the pipe, reacting with the metal of the pipe itself (particularly with lead, iron, and galvanized steel) in restricted pH and dosage ranges. Orthophosphate formulations that contain zinc can decrease the rate of dezincification of brass (Oliphant and Schock, 1996), and can deposit a protective zinc coating (probably basic zinc carbonate or zinc silicate) on the surface of cement or A-C pipe, given the proper chemical conditions (Schock and Buelow, 1981; LeRoy et al., 1996). Available reports have not led to a consensus, but the zinc formation may be advantageous by providing cathodic inhibition in some situations with iron, galvanized, or steel pipe, particularly if a fresh surface is exposed to the water. Research has generally shown that zinc is unnecessary in the formulation for the control of lead from pipes (Schock and Buelow, 1981; LeRoy et al., 1996; Lytle and Schock, 1996; Lytle et al., 1996). Detailed well-controlled experiments have not been published so far that are definitive with respect to the desirability of zinc to mitigate lead release from leaded brasses or soldered joints. Specific effects of orthophosphate on the control of lead, copper, and iron corrosion are covered in the section Corrosion of Specific Materials. Orthophosphate dosing using blended phosphates raises the complication of the specific chemical form and complexation or sequestration ability of the polyphosphate component. Therefore, although most studies show some benefit to higher ratios of orthophosphate to polyphosphate, it is not always a benefit if the polyphosphate component is a strong complexing agent and stable against reversion. The background water chemistry, particularly iron, calcium, and magnesium concentrations, also plays a major role in blended phosphate effectiveness.

pH

pH is a measure of the activity of hydrogen ions, H^+, present in water. In most potable waters, the activity of the hydrogen ion is nearly equal to its molar concentration. Because H^+ is one of the major substances that accept the electrons given up by a metal when it corrodes, pH is an important factor to measure. The pH also affects greatly the formation or solubility of protective films for both metallic and cementitious materials. The effects of pH on different pipe materials are covered in the section *Corrosion of Specific Materials*.

Polyphosphates

Polyphosphates have been frequently used to successfully control tuberculation and restore hydraulic efficiency to transmission mains. Polyphosphates can sometimes cause the type of corrosion to change from pitting or concentration cell corrosion to a more uniform type, which causes fewer leaks and aesthetic complaints. Pipe walls are usually thick enough that some increase in dissolution rate is not of practical significance. A major role for polyphosphates is the sequestration and mitigation of discolored water from source water manganese and iron, as well as reducing some scaling from hard or lime-softened water. The impacts of polyphosphates on different pipe materials are covered in the section *Corrosion of Specific Materials*.

Silicate

Silicates can form protective films that reduce or inhibit corrosion by providing a barrier between the water and the pipe wall. Research has not clearly established the equivalence of natural silicate species to manufactured ones for the express purpose of corrosion control and metal release mitigation, because there are possibly polymeric forms of silicate that exist in commercial products that impart a greater reactivity toward surface deposits, facilitating passivation. Some applications of silicate used in combination with chlorination to sequester iron in groundwater while providing corrosion control have been reviewed (Schock et al., 2005b). Silicates often do not perform well in pilot-testing or coupon-testing situations for one or both of two reasons: one, because such tests are often performed with fresh metallic surfaces and fail to properly simulate the environments on an existing corroded distribution system, building and residential piping systems; and two, because the passivating film forming reactions tend to be slow relative to carbonates, phosphates, and oxides. Some insight into this issue can be gleaned from research on the observed interference of naturally occurring silica on arsenic removal using ferric-iron-based media. Logically, silicate dosages for corrosion control must exceed natural background levels to be effective.

Total Dissolved Solids (TDS)

A high *total dissolved solids* (TDS) concentration is usually associated with high concentrations of ions that increase the conductivity of the water. This increased conductivity in turn increases the water's ability to complete the electrochemical circuit and conduct a corrosive current. The dissolved solids may also affect the formation of protective films, depending on their particular nature. If sulfate and chloride are major anionic contributors to high TDS, it is likely to show increased corrosivity toward iron-based materials. If the high TDS is mainly composed of bicarbonate and hardness ions, the water may tend to be non-corrosive toward iron and cementitious materials, but highly corrosive toward copper.

CORROSION OF SPECIFIC MATERIALS

Cast and Ductile Iron (unlined)

Iron-based metals, including cast iron, steel, ductile iron, and galvanized and cement-lined iron, are common pipe materials used to distribute drinking water from the water treatment plant to the consumer's tap. Corrosion of iron materials in drinking water distribution systems can negatively impact water quality and hydrodynamics and jeopardize structural pipe integrity. The majority of distribution piping installed in the United States from the late 1800s through the late 1960s, was one form of cast and then ductile iron, and they are still the predominant materials (MWH, 2005). Iron corrosion is generally non-uniform in nature, and is characterized by localized pitting corrosion and thick tuberculated growth of corrosion deposits and scale, which can grow to 50 mm (2 in.) or more in thickness. Severe pitting and tuberculation can lead to significant energy loss through water-flow restrictions, intrusion from breaches in the pipe wall, and eventual pipe failure. Iron corrosion by-products are often layered in structure and consist of reduced, oxidized, and a mixture of these iron phases. The reduced forms can exert a significant oxidant demand, and all forms have a strong affinity to adsorb organic matter, trace metals, arsenic, radionuclides, and a variety of ions. The nature of the deposits makes them a suitable living environment for microorganisms, and they support biofilms by shielding the microorganisms from disinfectants, such as free chlorine, and providing a carbon nutrient source through sorbed organic carbon or mineral carbonate.

Iron release from distribution system sources can lead to red water as a result of two mechanisms: physical action and dissolution of iron scales. Physical action refers to the dislodgement of material from the pipe wall or scale by shearing action, and it is typically associated with hydraulic changes (e.g., changes in flow rate or direction). Dissolution of iron scales can result from chemical processes such as reduction, complexation, and protonation. Certain bacteria having the capability to reduce ferric iron can also induce dissolution. *Iron release* is a term often used to describe the transport of iron, either soluble or particulate, from a corroded iron drinking water pipe wall or scale to the bulk water. Iron corrosion differs from iron release in that corrosion refers to the specific process where electrons are transferred from the iron metal to an electron acceptor such as oxygen.

Iron Corrosion. Innumerable books and papers have described details of reactions and pathways of the oxidation, corrosion, and release of iron drinking water distribution system materials. Although strong oxidants such as free chlorine are present in drinking water at significant concentrations, they are generally not considered as the primary cathodic reactant in iron corrosion cells. Because strong oxidants react rapidly with the thick iron deposits that contain oxidizable materials (e.g., microorganisms and organic matter) and reduced iron phases, they probably do not penetrate the scale too far beyond the scale-water interface. Oxygen is less reactive and has a greater oxidation capacity; therefore, it is able to penetrate farther into the scale.

Oxygenation kinetics are affected by many factors including (1) anions in water that can complex ferric (Weiss, 1935) or ferrous iron (Schenk and Weber, 1968; Tamura et al., 1976a; Sung and Morgan, 1980; Millero, 1985, 1989, 1990a, 1990b; Millero et al., 1987a and b; King, 1998), (2) temperature (Faust and Aly, 1983), (3) ionic strength (Faust and Aly, 1983), and (4) the presence of a variety of Fe(III) [hydr]oxides acting as catalysts (Tamura et al., 1976a and b; 1980; Tamura, 2008). Tests have shown that the corrosion rate of steel is increased by free chlorine concentrations greater than 0.4 mg/L (Singley et al., 1984; AWWA, 1986). Lytle et al. (2004b) has demonstrated the improvement in iron particle behavior and scale solid characteristics when substantial dissolved oxygen is present.

The *siderite model* has been widely cited as a model for the corrosion and release of iron in drinking water systems, which has often been at least qualitatively successful (Sontheimer et al., 1981; Benjamin et al., 1996). The model is based on the premise that protective scales of ferrous solids such as $FeCO_3$ or *green rust* (intermediate corrosion deposits between ferrous and ferric hydroxides, typically containing anions such as sulfate and chloride) are critical in protecting iron pipe surfaces from corrosion attack and iron release. The model presumes that the primary mode of iron corrosion is by the cathodic reduction of oxygen, which is controlled by diffusion and reactions between the diffusing scales. The rate of iron corrosion appeared to be dependent on the buffer capacity, as highly buffered waters tended to form more protective scales that contained a relatively large fraction of ferrous iron (presumed to be siderite) relative to scales formed in poorly buffered waters. The dependence is described by two factors: (1) the reduction of oxygen at the cathode produces OH^-, increasing pH at the pipe wall and (2) the rate of oxidation of Fe(II) to Fe(III) increasing dramatically with increasing pH. Therefore, when the buffer intensity is high, hydroxide ions formed at the cathode can be neutralized by HCO_3^-, the pH at the wall does not change much, and the rate of oxidation of Fe(II) is slowed. This allows the formation of siderite or green rust and, consequently a more protective scale. The model also postulated that any material that interferes with the oxidation of Fe(II), such as certain organic materials, can enhance the formation of protective scales.

Kuch (1983) observed that under non-flowing conditions, oxygen was lost and high concentrations of ferrous iron emerged in the overlaying water. He hypothesized that the alternative oxidant was the ferric outer layer (FeOOH, for example) of corrosion scales, and this produced the conditions wherein both the anodic and cathodic reactions produced Fe^{2+}. Depending on pH, either the released ferrous iron would precipitate as one of several potential ferrous minerals or the dissolved form would diffuse, out into the water. The resumption of water flow would bring fresh oxidant into contact with the dissolved ferrous iron, forming relatively insoluble ferric iron.

Some revisions of the *siderite model* have been proposed, to better explain the variables behind iron release into the water (Beckett et al., 1998; Sarin et. al., 1999, 2000, 2001, 2004a–c).

Discolored Water and Turbidity. Discolored drinking water often results from the presence of colloidal iron (generally defined as having a diameter <1 μm), and particulate (>1 μm) iron is commonly referred to as *yellow* or *red water*, although the color has been reported to range from light yellow to dark red, brown, or black, depending on the water chemistry. Suspended iron in the drinking water distribution system can originate from distribution system materials as well as the source water. Relatively soluble Fe(II) is the dominant oxidation state of iron in reducing environments (e.g., many groundwaters, anaerobic reservoirs, dead ends in distribution networks, and regions within pipe scale). On exposure to oxygen and/or disinfectant during water treatment and distribution, Fe(II) is oxidized to the relatively insoluble Fe(III) oxidation state, which is responsible for discolored water. Iron released from distribution system pipe corrosion scale, accumulated deposits, or sediments, is generated through some combination of physical, chemical, and microbiological pathways.

Given that iron corrosion provides sources of ferrous iron in the scales and distribution system sediments, several investigations have systematically isolated the various influences of water chemistry variables on the chemical and physical properties, aggregation, and transport of the resulting iron particles. Critical factors include the type and amount of oxidant (Eisnor and Gagnon, 2004; Lytle et al., 2004b), background pH, alkalinity, and major ions (Lytle and Snoeyink, 2003a and b; Lytle et al., 2003), and presence of different phosphates (Lytle and Snoeyink, 2002; Lytle et al., 2003). Silicates are also known to affect the formation and chemical properties of iron particulates in natural systems.

Several Awwa Research Foundation (now Water Research Foundation) projects have looked at various aspects of control of water quality from iron pipes (Singley et al., 1985; Benjamin et al., 1996; Snoeyink and Wagner, 1996; Kirmeyer et al., 1999; Clement et al., 2002). Source water iron can be removed by cation exchange, oxidation using air/oxygen, chlorine, or potassium permanganate followed by sedimentation and filtration. Combinations of oxidation and filtration are incorporated into many available packaged removal systems.

Prevention of conditions that favor the formation of discolored water is the best solution, and this may involve both chemical water treatment and a vigorous flushing program to remove sediments and loose materials. The flushing program should be in place before any significant water treatment change to minimize adverse impacts. Aesthetic impacts of discoloration can often be mitigated by sequestration.

Iron Release Models

Some of the best descriptions of the process of iron corrosion and release in water come from work conducted in the early to mid-twentieth century (Shipley et al., 1925; Baylis, 1926a and b, 1953; Smith and McEnaney, 1979; Tuovinen et al., 1980). It was recognized then, as it is today, that corrosion of iron is typically non-uniform and is characterized as pitting and tuberculation (see section Physical Factors Affecting Corrosion and Metals Release).

Recently, the structure and chemical nature of iron corrosion scales has been studied extensively using a variety of surface and solids analysis techniques (Sarin et al., 1999, 2001, 2004a). The results have shown that iron corrosion scales are complex and typically consist of layers of ferrous, ferric, and mixed (ferrous and ferric) oxides and oxy-hydroxides. An inner layer (closest to the pipe wall) consists of relatively porous ferric oxy-hydroxide (goethite). Ferrous corrosion scales such as $Fe(OH)_2$ and $FeCO_3$ are also thought to be present; however, they are difficult to identify since they are sensitive to oxygen introduced during handling and analysis. The inner layer serves as a large source of soluble Fe^{2+}. The porous layer is covered by a dense *shell-like* layer consisting of relatively insoluble ferric phases such as α-FeOOH and β-FeOOH, and mixed oxides such as Fe_3O_4. Inside the tubercles, iron sulfide minerals, sulfur, microbe-rich layers, calcium carbonate, and other unique minerals, all of which could play some role in iron release, have been reported to be present in various microenvironments (Tuovinen et al., 1980; Singley et al., 1985; Benjamin et al., 1996; Lytle et al., 2005a). Some research investigations have uncovered an unusual morphology to the formation of ferrotubes or chimneys resulting from a complex suite of inorganic and conceivably microbial reactions with iron that would be plausible in a distribution system piping environment (Smith and McEnaney, 1979; Stone and Goldstein, 2004). Some extracted unlined iron pipes display features similar to those described, as noted in Fig. 20-18, though the specific relationships among water chemistry parameters and tuberculation morphology are not understood.

Iron Corrosion Control Strategies

Carbonate, pH, and Buffering. Unlike the cases with lead and copper, calcium hardness appears to play an important role in reducing iron corrosion. The role of calcium as a corrosion inhibitor has been investigated by a number of researchers. The importance of $CaCO_3$ in undersaturated oxygenated waters for the formation of dense protective iron scales has been studied internationally for a long time (Tillmans et al., 1927, 1929; Stumm, 1956; Larson and Skold, 1957; Shams El Din, 1986; Sander et al., 1996). More resistant passive $CaCO_3$ layers were easier to form in alkaline conditions.

FIGURE 20-18 Close up of tubercles on an unlined cast iron pipe from a northeastern U.S. water system, showing actual chimney structures as described by Smith and McEnaney (1979) and Stone and Goldstein (2004). (*Photo by M. DeSantis, Pegasus Technical Services.*)

Pisigan and Singley (1987) examined the influence of buffer capacity, chlorine residual, and flow rate on the corrosion rate and pitting index of mild steel at water utilities in the state of Florida. In addition, total dissolved solids, Ca^{2+}, Mg^{2+}, Na^+, Cl^-, SO_4^{2-}, alkalinity, and pH were monitored. They found that an increase in buffer capacity at a constant alkalinity of 100 mg/L as $CaCO_3$ over a pH range of 6.0 to 9.0 decreased the corrosion rate of mild steel. However, when the pH was increased by increasing the alkalinity, the corrosion rate was not lowered. This effect was attributed to the increased ionic strength and conductivity contributed by the additional alkalinity. If a water contains an appreciable amount of calcium, it may also act in conjunction with the pH and bicarbonate concentration to buffer the pH rise from corrosion reactions by the instantaneous formation of calcium carbonate solid, forming a heterogeneous buffer. Benjamin et al. (1996) also summarized additional research on buffering impacts.

Orthophosphate Inhibitors. Relatively little information has been reported on the use of orthophosphate for iron corrosion control and red water reduction in drinking water distribution systems. Orthophosphate has been suggested to reduce the corrosion rate of steel and iron release rate (AwwaRF-TZW, 1996). One report found that the corrosion rate of mild steel and scale accumulation was reduced by 2/3 following the addition of 5 mg/L orthophosphate whereas 1 mg/L had no effect (AwwaRF, 1990). The formation and precipitation of sparingly-soluble metal phosphate salts such as calcium phosphate phases may form protective diffusion barriers (Nancollas, 1983). Various experimental approaches have shown that the main component of films formed on iron and steel in phosphate solutions was vivianite, $Fe_3(PO_4)_2{\cdot}8H_2O$ (Melendres et al., 1989; Smart et al., 1990). The pitting of carbon steel by chloride was reported to be reduced by phosphate addition (Ergun and Turan, 1991).

Polyphosphates. Polyphosphates consist of linear chains of PO_4^{3-} held together by a —P—O—P— linkage (Van Wazer and Callis, 1958; Boffardi, 1993). Chained polyphosphate compounds can be described by the general formula $M_{n+2}P_nO_{3n+1}$, where M is a metal such as Na and the average chain length is defined by the $Na_2O{:}P_2O_5$ ratio.

The effectiveness of polyphosphates for corrosion control of iron and steel, as well as other metals including lead and zinc, has been reported to be dependent on the presence of divalent cations, such as calcium and zinc (Hatch and Rice, 1940, 1945a; Boffardi and Sherbondy, 1991; Rangel et al., 1992; Kowata and Takahashi, 1996), DO concentration (Uhlig et al., 1955), and flow or turbulence (Hatch and Rice, 1945a; Boffardi, 1993). Several interpretations of mechanisms have been developed to describe the inhibitory action of polyphosphates. In most cases, the theories were based on relatively extreme experimental conditions (e.g., high polyphosphate concentrations and temperature) applicable to industrial applications rather than drinking water conditions. Further, differentiation is not always made between the control of corrosion symptoms (e.g., red water or reduction in tuberculation) versus actual decreases in the speed and amount of iron corrosion and pipe degradation.

A sufficient amount of calcium has been suggested for good iron corrosion control, even with polyphosphate applications (Hatch, 1952). The ratio of calcium to polyphosphate is the most critical factor, with a weight ratio of 1:5 calcium:sodium polyphosphate (Na_2O:P_2O_5 = 1.1:1) recommended. Adequate water flow is necessary to transport the colloidal material to the metal surface. In the case of old deposits, polyphosphates may adsorb to metal salts or oxides. If applied too rapidly, they can destabilize loose deposits of iron oxide or calcium carbonate, resulting in red water complaints, while tightly held tubercles and corrosion products are not likely affected. Polyphosphates are most protective over a pH range of 5 to 7 (Hatch and Rice, 1945a and b; Lamb and Eliassen, 1954).

Koudelka (1982) employed X-ray photoelectron spectroscopy (XPS) analysis of scale on iron surfaces along with anodic polarization analysis to conclude that an iron oxyhydroxide and iron orthophosphate (ferric or ferrous), rather than an iron polyphosphate compound, provides corrosion inhibition. Under aggressive conditions, polyphosphate was believed to inhibit film growth, but may provide some protection when soluble ferrous or ferric polyphosphate complex ions near the anode revert and lead to the precipitation of iron orthophosphate. Calcium orthophosphate was thought to enhance the reversion of polyphosphate to orthophosphate. Positively-charged colloidal calcium-polyphosphate particles are attracted to the cathode where reversion and subsequent precipitation of calcium orthophosphate takes place.

Blended Phosphates. Blended phosphates are mixtures of orthophosphate and polyphosphates. The polyphosphate to orthophosphate ratio is likely to be important since sequestration and passivation needs must be met. It has also been reported that ortho- and polyphosphate blends containing small amounts of orthophosphate actually increase corrosion rates (Benjamin et al., 1996). Currently there is no simple guidance or formula for the selection of a product for a given case, as product and dosage choices still are largely based on experience, as well as trial and error.

Silicates. Water-soluble sodium silicates are defined by the ratio of SiO_2 to Na_2O. Viscosity and molecular weight increase with increasing ratio (Iler, 1979). N-type sodium silicate refers to a 3.22 SiO_2:Na_2O, contains 28.7 percent SiO_2. This is the form most commonly applied for lead and copper corrosion control and has been effective at controlling red water problems. The role of silicate for iron, lead, and copper corrosion control is uncertain, although some have suggested that a thin silicate coating is formed that acts as a protective diffusion barrier (Stericker, 1945); there is not a strong body of scientific evidence to support this claim. Since sodium silicates are basic, they do increase the pH, which is generally beneficial with regard to reducing iron release and lead and copper solubility. Regardless of the exact mechanism of corrosion inhibition provided by silicates, several studies had documented improvements in lead, copper, and iron release (LaRosa-Thompson et al., 1997). Silicate species are known to sorb readily onto ferric oxides, and therefore, there will likely be long-term transformations of weak and soluble ferric oxyhydroxides in the scales into stronger and more adherent iron silicates. Unfortunately, while the anecdotal evidence from most utilities using high dosages of sodium silicate is that there is a consistent improvement in color and turbidity of the water, studies have not been done to examine the mineralogy of the transformation over time and attempt to evaluate quantitatively benefits of silicate dosing for iron corrosion and water quality.

Anion Effects. Both corrosion rates and iron uptake (into the water from the pipe) have been determined to be increased sharply as the concentration of chloride or sulfate is increased in solution (Larson and Skold, 1957; Larson, 1975; Singley et al., 1985; Benjamin et al., 1996; Lytle et al., 2003, 2005b). Research concerning the Larson ratio suggests that the molar ratio of chloride plus sulfate to bicarbonate should be below approximately 0.2 to 0.3 to prevent enhanced corrosion of unlined iron.

Iron Sequestration

A symptom-based alternative to source water iron removal and actual iron corrosion control involves the addition of a sequestering chemical such as sodium silicate or polyphosphate chemicals, which prevent the visible precipitation of iron. The term *sequester* (or *chelate*) technically refers to chemical complexation, although it is most often used in the water supply industry to describe any mechanism that prevents visible precipitation.

Polyphosphate chemicals have been suggested to sequester iron through complexation of Fe(II) and Fe(III), inhibition of Fe(III) oxidation rate, and dispersion of iron oxyhydroxides. Polyphosphates complex many metals and, therefore, are used as sequestering agents to prevent iron and manganese precipitation in the distribution system and calcium deposition in filters. The strength of the complex is dependent on the water chemistry and the polyphosphate compound. Polyphosphates are also strong colloidal dispersants, which makes them effective for reducing metal precipitation and floc growth.

The use of polyphosphates for sequestration of source water iron has been practiced with success. Polyphosphates should be added prior to chlorination (Henry, 1950; Illig, 1960, 1987; Klueh and Robinson, 1988). The results of Klueh and Robinson (1988) suggested that nearly all of the sequestered iron was in the colloidal or polymeric form. They found that calcium inhibited treatment effectiveness. Without calcium, polyphosphate at a dose of 1 mg PO_4/L was able to sequester 2 mg Fe/L. When 100 mg Ca^{2+} as $CaCO_3$/L was present, a 2.5 to 5 mg PO_4/L was necessary. Harwood (1989) reported that the Fe(III) capacity of four polyphosphates was 2.66 ± 0.41 mg Fe/mg P, while orthophosphate had a capacity of 0.2 mg Fe/mg P. The capacity of orthophosphate to sequester Fe(II) was only 0.31 ± 0.04 mg Fe/mg P. As a negative secondary effect, polyphosphates have been shown to directly increase the solubility of metals such as lead, which raises concerns about their use (Holm and Smothers, 1990; Holm and Schock, 1991a and b; Cantor et al., 2000).

Silicates have also been used to successfully sequester iron for decades (Dart and Foley, 1970, 1972; Griffith, 1972; Robinson et al., 1987; Armstrong and Zhou, 1988; Schock et al., 2005b). The work of several researchers supports the conclusion that silicates control red water by a colloidal dispersion mechanism rather than a silica-iron complexation mechanism (Hazel, 1945; Henry, 1950; Robinson et al., 1981, 1987). In practice, sodium silicate is added at the same time or shortly after the point of oxidant (typically chlorine) addition when used to control iron precipitation from groundwater to avoid significant precipitation or increase in turbidity. After iron oxidation, silica adsorbs to and disperses the colloids. Robinson et al. (1987) found that 2 mg Fe/L was stabilized by a silicate dosage of 12 mg SiO_2/L for 10 days. Calcium, however, was reported to negatively impact the ability of silicate to disperse iron colloids. They found that 10 mg/L of calcium caused iron precipitation to occur in only 3 days, and iron precipitated after less than 1 day when 100 mg/L of calcium was present. Although sequestration can be effective at reducing the water color associated with suspended iron particles, removal of iron from the source water is the only way to truly eliminate the particles.

Copper

Uniform Corrosion. The performance of copper in potable water systems depends on whether relatively thin and adherent films of corrosion products, cuprous oxide (cuprite, Cu_2O), cupric oxide (tenorite, CuO), or basic copper carbonate (malachite, $Cu_2(OH)_2CO_3$), can be formed. Copper tends to be less prone to the effects of concentration cell or differential aeration corrosion than galvanized steel, steel, and iron pipes. The influence of copper ion concentration on the potential of copper in solution is marked. When solution velocities vary over a copper surface, the parts washed by the solution with the higher rate of movement become anodes and not cathodes, as would be the case with iron (Butler and Ison, 1966). Impingement attack of copper by high water velocities is one of the most common problems of copper pipe. High DIC concentration aggravates impingement attack, and it may manifest itself at velocities and turbulent flow areas that otherwise would not seem to be prone to damage under conventional design specifications for building and household plumbing systems.

In the most ideal case, uniform corrosion of copper is believed to be inhibited by the formation of a *duplex film,* characterized by a thin compact adherent Cu_2O (cuprite) semiconductor layer attached to the copper metal (MWH, 2005). On top of that layer is a more porous, usually CuO (tenorite) or $Cu_2CO_3(OH)_2$ (malachite), depending on water chemistry conditions. When the cuprite underlayer is not able to form uniformly, both non-uniform corrosion and high copper release episodes can result.

Normally, the rate of uniform corrosion of copper in potable water systems is not rapid enough to cause failure of the tubing. However, consumer complaints of *blue water*, *green water*, and staining can result. When such copper dissolution is sufficient to cause complaints, frequently a fine dispersion of copper corrosion products discolors the water. When inspected, the inside surface of the copper tube under such circumstances is characterized by a loosely adhering powdery scale and beneath it, or in areas where no scale is present, by general dissolution of corrosion of the copper (Ferguson et al., 1996; Reiber et al., 1997). Related to green/blue water is green/blue staining, as even a few parts per million of copper in water can react with soap scum and cause green staining of plumbing fixtures. Several cases of blue water have resulted from what is apparently a dislodging of a slime created by microorganisms (Reiber et al., 1997).

The most important chemical variables in general uniform copper corrosion and copper release are pH, DIC, redox potential (ORP), and plumbing age. The ORP is important mostly in defining the conditions in which the oxidizing potential (pE, E_h) is high enough to form cupric ions. At lower redox potentials, either copper is immune to corrosion or highly insoluble cuprous oxide or hydroxide solids form that produce extremely low copper levels in the water (Schock et al., 1995; Ferguson et al., 1996). Many untreated groundwaters of neutral pH and high DIC fall into this category.

The corrosion rate and release rate of copper into the water are sensitive to the level of oxidizing agents in the system, especially free chlorine (Atlas and Zajicek, 1982; Pisigan and Singley, 1987; Schock et al., 1994, 1995; Werner et al., 1994; Ferguson et al., 1996; Lytle and Schock, 2000). Even low levels (such as 0.2 mg/L) affect the oxidation and corrosion rate. Reiber (1989) has shown that the chlorine effect is much more important than dissolved oxygen in normal drinking water situations. Several studies have been made of the influence of oxygen on the oxidation rate of Cu^+ to Cu^{2+} in natural and seawater, which is of some applicability to drinking water situations (Millero et al., 1987a; Millero, 1989, 1990a and b; Eary and Schramke, 1990). In the presence of oxidants, copper levels in standing water in plumbing systems can increase for as many as 48 to 72 hours, so overnight standing samples do not necessarily represent building systems or nearly worst-case scenarios (Schock et al., 1995). Conversely, when oxidants are depleted, copper levels may decrease (Werner et al., 1994; Schock et al., 1995; Ferguson et al., 1996). When oxidative processes are introduced, such as for iron or manganese removal, or disinfection is initiated after low-ORP conditions, high corrosion of copper can result.

Recent research has shown that copper levels in disinfected or oxic drinking water are controlled for many years by the formation of metastable solids, apparently cupric hydroxide or cupric oxide of high surface area and solubility, which ages to either the basic cupric carbonate mineral $Cu_2(OH)_2CO_3$ (malachite) or a more ordered CuO (tenorite), depending on pH (Schock et al., 1994, 1995, 2000; Hidmi and Edwards, 1999; Edwards et al., 2000; Lagos et al., 2001; Turek et al., 2006; Schock and Sandvig, 2009). The formation rate of malachite, which would lead to low copper levels in many waters of moderate DIC, appears to be slow, and in high DIC waters, rapid carbonate complexation disrupts the uniform formation of the desired uniform cuprous oxide protective underlayer. Thus, dissolved copper release for most of the sites targeted for regulatory monitoring and for newer sites is best described by looking at control by cupric hydroxide or cupric oxide solids of high surface area and small particle size. High concentrations of bicarbonate and carbonate ions at neutral to alkaline pH values enhance significantly the release of copper through aqueous complexation (e.g., $CuHCO_3^+$, $CuCO_3^\circ$, $Cu(CO_3)_2^{2-}$) (Schock et al., 1994, 1995; Ferguson et al., 1996; Edwards et al., 2000; Schock and Sandvig, 2009). Electrochemical analyses and copper release experiments of Edwards et al. (1994) also indicate bicarbonate is highly aggressive toward copper, especially below pH 8.5 (Edwards et al., 1994a).

Figure 20-19a shows the strong effect of both carbonate complexation and pH on copper solubility for relatively new plumbing systems. Thus, hard, high alkalinity (DIC) groundwaters that are protective of iron and cement-based piping are particularly aggressive toward copper materials. This is a fact that is often not appreciated, and treating these kinds of water chemistries to reduce copper release is often difficult. Conventional pH adjustment is generally impractical or inadequate, as buffering is extreme and hardness places strong limits on the pH achievable without precipitating calcium carbonate. The neutral pH conditions of these waters generally make direct aeration impractical as a mechanism for reducing DIC (Lytle et al., 1998a). Thus, conventional lime or lime-soda softening and dosing of approximately 3 mg/L or more orthophosphate are the most feasible treatment solutions for high copper release from new plumbing. A combination of hardness and DIC

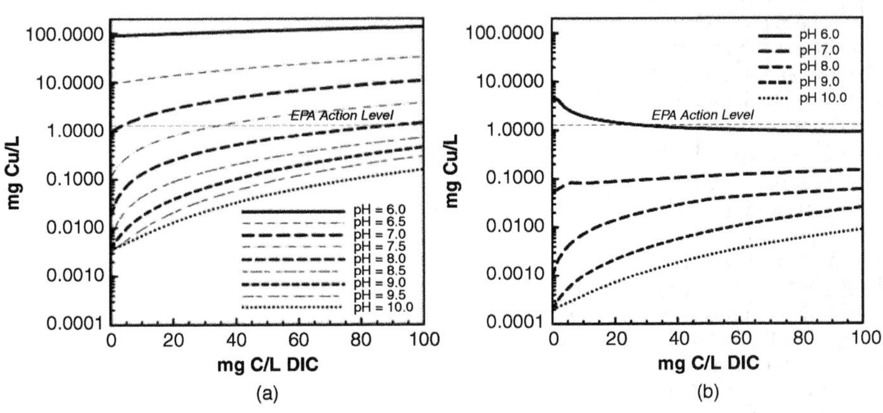

FIGURE 20-19 Solubility diagrams for Cu(II), assuming conditions for new pipe (*a*) with cupric hydroxide scale, or (*b*) assuming a malachite scale.

reduction (to approximately 25 mg C/L) using membranes or ion exchange, followed by pH adjustment, should also be effective.

Figure 20-19b shows the solubility of copper(II) after the development of an aged malachite or tenorite surface scale, such as may happen after 10 to 20 years of use. This aging phenomenon, described in detail by many studies (Schock et al., 1994, 1995, 2000; Edwards et al., 2000; Lagos et al., 2001; Schock and Sandvig, 2006, 2009; Turek et al., 2006) makes field data interpretation complicated. It further illustrates the contradictory nature of treatments needed to reduce acute copper exposure to consumers or high copper contamination into wastewater, as opposed to long-term copper passivation (Schock and Sandvig, 2009).

Soft waters, particularly those with an acidic pH and typified by low alkalinities, dissolve high levels of copper rapidly. These systems can usually be treated in a straightforward manner by any of several relatively simple approaches, such as pH adjustment or pH/DIC adjustment (if added buffering for pH stability is needed). For smaller water systems, it is extremely advantageous to use mechanically simple systems, such as aeration (Whipple, 1913; Spencer, 1996; Lytle et al., 1998a–c; Schock et al., 2002) and limestone contactors (Letterman et al., 1987, 1991; Letterman and Kathari, 1996; Spencer, 1996; Jalil et al., 2002).

Generally, cations (calcium, magnesium, sodium, and potassium) exert little or no effect on the rate of corrosion or copper release, though some researchers have suggested calcium can play a beneficial role at levels and conditions substantially below calcium carbonate saturation (Ives and Rawson, 1962b; Elfström Broo et al., 1997). Anions (chloride, sulfate, bicarbonate, orthophosphate), however, do exert considerable influence on the rate of corrosion, the levels of dissolved copper in the water, and on the nature and stability of corrosion product solids on the pipe surfaces (Reiber et al., 1988; Meyer, 1994; Rehring, 1994; Schock et al., 1995; Ferguson et al., 1996). Traditionally, chloride has been considered to be a very aggressive ion toward copper. Some research suggests that chloride modifies the nature of the surface film formation (Adeloju and Hughes, 1986) and that it may not be as aggressive in the context of uniform corrosion as previously believed (Edwards et al., 1994a; Meyer, 1994; Rehring, 1994). At pH $\geq$ 7, studies by Edwards et al. (1994) indicate that chloride has beneficial long-term effects toward copper corrosion and release (Edwards et al., 1994a).

Sulfate plays a complex role in copper corrosion, sometimes producing lower copper levels at slightly alkaline to somewhat acidic pH conditions, where any of several basic cupric sulfate solids may form a passivating film. Several studies, including electrochemical analyses of different types, have produced data indicating sulfate is very aggressive toward copper, particularly for aged surfaces above pH 7 (Edwards et al., 1994a; Meyer, 1994; Rehring, 1994). At higher pH, some research has suggested that these metastable cupric sulfate solids inhibit the beneficial precipitation of cupric hydroxide and subsequent transformation to tenorite (Schock et al., 1995). Marani (1992) has summarized considerable research that has investigated the conditions favorable to the formation of basic cupric sulfate solids under water chemistry conditions appropriate for drinking waters.

The role of orthophosphate in the corrosion inhibition of copper has been a source of confusion, largely because the roles of carbonate complexation and scale aging have usually been ignored. Additionally, the role of aging of the cupric (hydr)oxide and malachite scales must be considered, when evaluating whether orthophosphate would be beneficial or detrimental (Schock and Sandvig, 2006, 2009). Under some conditions, orthophosphate has appeared to promote higher copper levels (Werner et al., 1994). Electrochemical studies by Reiber (1989) suggested that orthophosphate might provide some corrosion reduction by changing the fundamental form of the anodic reaction. Extensive laboratory and field research by the USEPA and others has led to a practical understanding of the efficacy of orthophosphate treatment for cuprosolvency control, even if some mechanistic details remain to be determined (Schock et al., 1995; Lytle and Schock, 1996, 2006; Schock and Clement, 1998; Edwards et al., 2001; Edwards and McNeill, 2002; Qiu and Dvorak, 2004; Schock and Sandvig, 2006, 2009). Although the exact stoichiometry of the solid may differ, orthophosphate effects on new plumbing can be closely modeled up to a pH of approximately 8 to 8.5 by presuming the formation of a passivating film composed of the solid $Cu_3(PO_4)_2 \cdot 2H_2O$. In hard waters, formation of a calcium orthophosphate deposit on the pipes is also possible, although not clearly proven. With fresh plumbing surfaces, a thin phosphate film is formed on the surface of copper pipe, inhibiting oxidation by chlorine and oxygen and subsequent growth of normal hydroxide, oxide, or hydroxycarbonate passivating solids. These results were generally supportive of the observations of Reiber (1989) about possible interference and altering of normal anodic site reactions and the solubility models predicting lower solubility of the phosphate solids at these pH conditions. At higher pH of 8 to 9, higher copper levels were observed in the presence of the orthophosphate, consistent with the inhibition of growth of the less soluble and protective cupric oxide. These observations help explain contradictory conclusions about the effectiveness of orthophosphate in full-scale applications with different water qualities. The experimental evidence also suggests that the orthophosphate may be less effective in inhibiting copper corrosion if an oxidized carbonate, hydroxycarbonate, or oxide scale already exists on the pipe surface under many water chemistry conditions. Figure 20-20 shows trends of copper solubility with different dosages of orthophosphate at different pH values and DIC concentrations. In practice, the amount of cuprosolvency

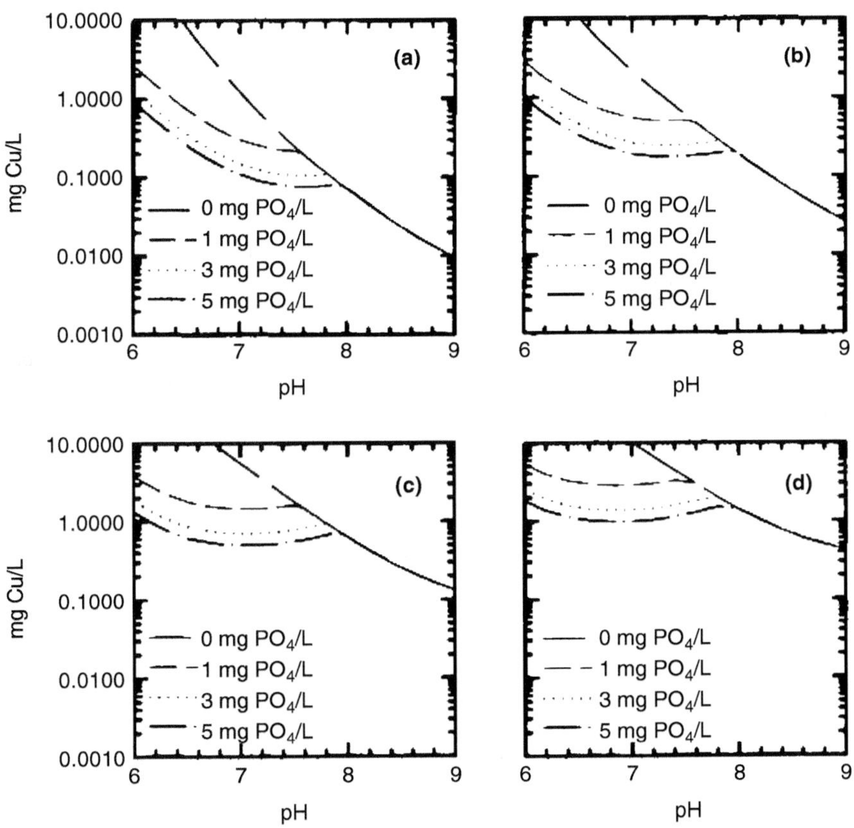

FIGURE 20-20 Effect of orthophosphate on solubility of new copper assuming equilibrium with either $Cu(OH)_2$ or $Cu_3(PO_4)_2 \cdot 2H_2O$ at different DIC and orthophosphate concentrations. Computed for 25°C: (*a*) DIC = 4.8 mg C/L, $I = 0.005$; (*b*) DIC = 14.4 mg C/L, $I = 0.005$; (*c*) DIC = 48 mg C/L, $I = 0.01$; (*d*) DIC = 96 mg C/L, $I = 0.02$. (*Source: Schock, Lytle, and Clement, U.S. EPA, EPA 600/R-95/085, Cincinnati, Ohio, 1995b.*)

reduction in new plumbing is slightly less than predicted by the current solubility model, so dosages should be increased slightly over the predictions. Long- versus short-term impacts have been recently discussed (Schock and Sandvig, 2009).

For water systems with high DIC and new copper plumbing, determining the correct orthophosphate dosage to ensure compliance with the Lead and Copper Rule requirements were shown to be possible using a straightforward jar test approach (Lytle and Schock, 2006). For these waters, a generalization that has proven successful for DIC levels of 60 to 75 mg C/L is a residual of about 3 to 4 mg PO_4/L orthophosphate. Despite the recommendation by many distributors, blended phosphates have often been unsuccessful at sufficiently reducing copper release from new plumbing to a 1.3 mg/L level for overnight standing samples in high-alkalinity groundwaters.

In the presence of oxidants, such as chlorine or dissolved oxygen, complexing ligands such as ammonia accelerate the corrosion and dissolution of copper. Excess ammonia in the presence of chloramines was found to be associated with increased copper levels in one study (AwwaRF, 1990), but the results were not conclusive and more research needs to be done. Natural organic matter has also been found to increase copper corrosion and copper release under many circumstances (Kristiansen, 1982; Rehring, 1994; Rehring and Edwards, 1994, 1995).

Dissolved silica and silicate anion have not been studied in great detail with respect to corrosion effects on copper. Several studies have shown that high dosages of silicate at pH near 8 are effective in reducing copper release, though a mechanism has not been clearly defined (Lytle et al., 1996; LaRosa-Thompson et al., 1997). In one study, malachite was replaced by an amorphous cupric silicate of low solubility (Schock et al., 2005b, 2006).

The presence of dissolved sulfide, or sulfide produced by sulfate reducing bacteria, is especially troublesome to copper (Jacobs and Edwards, 1997). Copper sulfide scales have been shown experimentally to increase copper corrosion rates by 1 to 2 orders of magnitude at pH 6.5 and 9.2 and copper release by 500 to 5000 percent, respectively. Subsequent removal of sulfide from the water by deaeration, chlorination, and superchlorination all failed to restore normal corrosion rates to the pipe containing sulfide scales. Experiments suggested that more than a year might be needed for the corrosion rate to reach levels below those observed in the presence of sulfide.

Pitting Corrosion. Pitting corrosion of copper in water has been studied by many researchers (Lucey, 1967; Kasul and Heldt, 1993; Edwards et al., 1994a and b; Duthil et al., 1996; Marshall and Edwards, 2005), which has led to a number of theories to describe copper pit initiation and growth in drinking water systems through chemical reactions. Pit initiation has also been associated in various studies with microbial activity, material imperfections, carbonaceous manufacturing residues on the pipe surface, stray currents, soldering flux, and others. Despite past research efforts and given its complexity, pitting corrosion of copper in water remains poorly understood and difficult to predict and control. Most likely, pitting has multiple causes and no single all-encompassing theory will be applicable.

Pitting attack is most common in new installations, with 80 to 90 percent of the reported failures occurring within the first 2 to 3 years (Trussell, 1985). In extreme cases, pits can break through within only a few months. Pitting occurs in all types of copper pipe and tubing. If unfavorable water quality conditions occur before the protective coat has formed, then serious pitting attack may occur. These conditions often result from leaving water stand in unoccupied housing for months, after pressure and leak-testing the plumbing systems. Incomplete flushing of flux residue and particulate debris set up conditions favorable to differential oxygenation corrosion. Disinfectant residuals are usually lost within a day or two at most, and microbial growth may occur over long periods of water stagnation. Microbes residing in sheltered areas can promote and accelerate reactions within incipient pits, and they become very hard to disinfect. Once started, pits of this type cannot usually be stopped without strong physical cleaning processes.

Localized copper corrosion in water has been categorized into at least three types of water based on chemistry and physical features: cold water (type I) (Campbell, 1950; Cruse and Pomeroy, 1974; Kasul and Heldt, 1993), hot water (type II) (Mattsson and Fredriksson, 1968), and soft water (type III) (Page et al., 1974). Soft water copper pits are described as being relatively wide and shallow and consist of an exterior layer of bronchantite and/or malachite over a layer of crystalline red-brown cuprite layer and the corroding copper surface. The voluminous corrosion deposits of the exterior layer are released to the water and can potentially even cause water blockage. Waters having low conductivity, low alkalinity, and relatively high pH are typically associated with soft water pitting.

Lucey (1967) is generally credited with developing the modern framework for the structure and propagation of copper corrosion pits and proposed the *membrane cell theory* of copper pitting corrosion that identified the chemical reactions involved in this type of corrosion. The work was based on field observations, bench-scale studies, and electrochemical approaches. Lucey stated that a porous membrane of cuprous oxide (cuprite) is an essential feature of a copper corrosion pit and concluded that pitting arises as a result of a deposited cuprous chloride layer between the corroding copper and the membrane. In the model, the anodic reaction occurs between the cuprite membrane and pipe wall where copper metal is oxidized into cuprous ions. Some of cuprous ions react with oxygen to form cuprite, which precipitates between the membrane and the pipe wall. The remaining ions pass through the membrane and are further oxidized into cupric ions. Some of these cupric ions precipitate as insoluble copper salts (i.e., malachite or bronchantite), and the remaining ions are cathodically reduced to the cuprous state. This cathodic reaction differs from conventional pitting theory in that the cathode's external cathodic surface is not necessary and no large-scale flow of electrons through the metal occurs. In the past decade, many researchers have continued to investigate and expand on the work of Lucey. Some have suggested modifications to his theory, and others have proposed alternative pitting theories. For example, Sosa et al. (1999) revisited the work of Lucey on cold water pitting by repeating experimental conditions using an electrochemical approach. They showed that currents carried through the bulk metal accounted for more than 80 percent of the weight loss in the pit. This observation contradicts the membrane cell theory that places the cathode above the anode in location.

Water chemistry is clearly important with respect to pitting. In addition to relatively high pH, high free chlorine, and low alkalinity (i.e., having low dissolved inorganic carbon or bicarbonate concentrations) specific to soft water attack, sulfate, and chloride are typically associated in one way or another to all forms of pitting. The relative importance of chloride versus sulfate in copper pitting corrosion is debatable. A number of studies have been conducted to evaluate the role of chloride in copper pitting corrosion (Akkaya and Ambrose, 1985; Drogowska et al., 1987; Nishikata et al., 1990). Historically chloride has been thought to be the species responsible for pitting in most theories (Lucey, 1967; Cornwell et al., 1973; Shalaby et al., 1989). Edwards et al. (1994b) concluded that sulfate and nitrate play important roles in pit initiation and development, and they believed these ions may be more critical to copper pitting than chloride. They found that chloride increased copper corrosion rates short term; however, with time, chloride actually produced protective surfaces at pH 7.6. Sulfate initially was not aggressive, but corrosion rates increased above those of chloride over time as scale formed. They noted that brochantite, $Cu_4(OH)_6(SO_4)$, was always present over soft water pits and hypothesized that its formation may be the key to hot water pitting. Thermodynamic calculations showed that brochantite formation was favored over malachite and atacamite in waters having high sulfate to chloride or sulfate to bicarbonate ratios. Research by Duthil et al. (1996) also reported on the role of chloride and sulfate on the pitting of copper, based on pit germination rates using an electrochemical approach. They showed both chloride and sulfate can be responsible for copper pitting when present alone in water. When both anions are present, chloride can have either an accelerating or inhibiting effect depending on the relative concentrations of both sulfate and chloride.

Schmitt et al. (2001) used an electrochemical noise technique to investigate the effect of chlorides, sulfates, and bicarbonate anions on copper pitting corrosion. They concluded that sulfate typically has an activating effect, chloride has a passivating effect, and bicarbonate acts as buffering agent. However, the combination of all three anions in a narrow range of concentrations leads to pitting corrosion.

Although there are a few differences in the details, many of these observations and interpretations are corroborated by Taxèn (1996), who especially observed differences in pitting when Cu(I) was involved more than Cu(II). He also found sulfate and carbonate were most aggressive toward copper, especially in the bulk solution. In contrast, Lytle and Schock (2008) presented experimental data from sulfate-free systems, showing that a combination of free chlorine, low DIC (5 mg/L), and high pH (9.0) alone were sufficient to cause pitting corrosion within several months. Modest amounts of carbonate appeared to reduce the initiation of pits, presumably through buffering against localized acidic conditions believed to be present in soft water pitting. High redox potential could be a factor in pit propagation.

A number of other works have proposed theories on how levels of chloride, sulfate, and/or alkalinity in water impact the nature of copper minerals that are often associated with localized copper corrosion and affect the protective nature of the cuprous oxide layer (Feng et al., 1996a and b; Mankowski et al., 1997). Proper identification of these solids is particularly useful in understanding pitting mechanisms and the conditions within the pit. Unfortunately, many past studies did not apply appropriate analysis approaches, and their descriptions of the nature of these solids have largely been speculative. Secondly, appropriate thermodynamic analyses and evaluations of redox sensitive relationships were rarely computed or performed, even semiquantitatively, to validate the interpretations proposed.

Marshall and Edwards (2005) investigated the role of aluminum on copper pitting corrosion in drinking water. They demonstrated that water containing aluminum, high chlorine residual, and relatively high pH caused pinhole leaks in a laboratory setting. Specifically, a chlorine residual goal of 1.5 mg/L at pH 9.2 and 2 mg/L of aluminum hydroxide solids produced conditions that caused pinhole leaks. Also the water had a low alkalinity and significant levels of chloride and sulfate, putting it in the broad category of soft water pitting corrosion. Pitting did not occur in the complete absence of aluminum in tests of a few weeks duration, but the authors observed pitting tendencies in the complete absence of aluminum if Cl_2 was maintained at high levels for an extended period of time (Rushing and Edwards, 2004; Marshall and Edwards, 2005).

Historically, thin carbonaceous films, derived from carbonization of the lubricant used in manufacture, have resulted in pitting of copper tubes, though it is not clear that this mechanism has operated for decades after changes in manufacturing processes. Films of manganese oxides, formed by slow deposition from soft moorland waters, can also cause localized attack. A review of the literature on pitting corrosion and an evaluation of reported field experience indicate that the presence of certain types of NOM apparently reduce the corrosion of copper in certain waters by promoting the formation of more protective scales, and also by reducing the tendency for pitting attack (Edwards et al., 1994b). This evaluation recommended that careful consideration be given to an increased Cu pitting incidence resulting from the removal of NOM for the purpose of controlling DBPs. Another study found similar effects of NOM. Initially, NOM promoted pitting in certain narrow water quality and NOM concentration ranges, but over long periods of time NOM reduced the tendency toward pitting at higher NOM levels (Korshin et al., 1996). Generally, the study concluded that NOM interacts with the copper through sorption, and it tends to increase the rate of copper leaching and the dispersion of the inorganic corrosion by-product scale materials.

Many instances of microbially mediated copper corrosion have been reported, though the mechanisms and differentiation from inorganic causes are not always clear (Geesey et al.,

1986, 1988; Iwaoka et al., 1986; Jolley et al., 1988; Bremer and Geesey, 1991a and b; Geesey and Bremer, 1991; Fischer et al., 1992; Arens et al., 1996; Camper, 1996; Murphy et al., 1997a–c; Reiber et al., 1997; Cantor et al., 2006; Cantor, 2009).

Brasses and Bronzes

Although many studies on factors involved in the corrosion and dissolution of brasses and bronzes have been conducted, a clear understanding of the role of water chemistry in these processes has not emerged (Schock and Neff, 1988). Even now, little has been published on normal corrosion scales which cover various brass components within a water system and which might give evidence for passivation against lead and copper release or of aggravation of metal release by galvanic reactions (Frenkel and Korshin, 1999; Korshin et al., 2000). Although lead contamination from brass has been well known for over 20 years, some recent studies implicate the brass corrosion as an important source of several other metals to tap water and wastewater metal loadings (Kimbrough, 2001, 2007, 2009).

Most studies on dezincification problems have mainly focused on materials longevity and economic implications. Several aspects of brass dezincification in water systems have been studied (Oliphant, 1978). There are also several useful reviews of important performance issues (Oliphant and Schock, 1996; Sadayappan, 2005), including investigations of the new generation of truly non-leaded brasses (Sandvig et al., 2007).

Corrosion causes the failure of mechanical devices such as valves and faucets made from brass or bronze, but it also contributes lead, copper, and zinc to the water. Metal leaching from the tap and other brass or bronze plumbing fittings can confound the interpretation of field studies for which the intent is to correlate metal levels at the tap to human exposure, corrosion rates, or the effectiveness of corrosion control treatment programs.

In the drinking water corrosion field, the contribution of brass fixtures to first draw lead levels in tap water monitoring programs is now well accepted (Lee et al., 1989; Dodrill and Edwards, 1995; Oliphant and Schock, 1996; Schock et al., 1996; Reiber et al., 1997; Kimbrough, 2001, 2007; Sandvig et al., 2008; Kimbrough, 2009). The effects of many water quality properties on lead release from brass have been reviewed, particularly, the interrelationships of lead content, pH, DIC concentration, and orthophosphate concentration (Schock and Neff, 1988; Lytle and Schock, 1996, 2000; Oliphant and Schock, 1996). Important variables in the performance of brasses and bronzes in potable waters include the pH, flow, temperature, lead content, surface area of lead and brass, alkalinity (or DIC), chloride concentration, orthophosphate concentration, and the presence of natural inhibitors. Several studies have implicated changes in chloride content or the ratio of chloride to sulfate resulting primarily from coagulant changes, as an aggravating cause of elevated lead release from brasses, particularly in the context of in-place galvanic connections (Dudi et al., 2005; Edwards and Triantafyllidou, 2007; Triantafyllidou and Edwards, 2007). Lead, zinc, and copper leaching vary greatly with the types of alloys used and the manufacturing and fabrication processes (Lytle and Schock, 1996, 2000; Oliphant and Schock, 1996; Dudi et al., 2005).

In the United States, the leaching of lead, cadmium, zinc, and other metals from brass and solders is regulated in an indirect way. The Safe Drinking Water Act (SDWA) Amendments of 1996 limit the content of Pb in solder and flux to be used in contact with drinking water to 0.2 percent (USEPA, Sec. 1417(d), 1996). Because the definition of *lead free* for pipes, well pumps, plumbing fixtures, and fittings under the original statute was a content of 8 percent or less, it was ineffective in controlling contamination from this source, as virtually all fixtures and fittings implicated in high levels of metal leaching contain much less than 8 percent Pb (Lytle and Schock, 1996; Oliphant and Schock, 1996; Kimbrough, 2001, 2007). Thus, the SDWA Amendments of 1996 incorporated a performance standard in Sec. 1417(e) requiring fixtures and fittings intended to dispense water for human consumption to be certified under a voluntary standard established by the act (NSF/ANSI, 2007a). As of this writing, there is current research activity and marketing of several new lead-free brasses (Maas and Patch, 2004), which should help water systems develop specifications for materials that will reduce the amount of lead in contact with drinking water in the future. This activity is largely due to the reduction of leachable lead permitted under ANSI/NSF 61, including a new Annex G, which will take effect in 2012 (NSF/ANSI, 2007a and b).

Lead and Lead-Containing Solder

Regulatory requirements for the permissible lead levels in the drinking water of many countries are so low that the rate of lead corrosion is of no practical concern. Even only slightly aggressive water standing for a brief time in contact with lead pipe or lead-containing materials will pick up lead levels that are considered a public health risk; see Chap. 2 for health effects.

Particulate Lead. Lead release is a significant problem whether released as dissolved lead species, associated with particulate material like dislodged pipe deposit solids, or sorbed on other particulate matter present in the water. The relative contribution of dissolved lead to particulate matter almost certainly varies with the water chemistry, the configuration of the sampling site, and the history of the plumbing materials containing the lead. Numerous studies have investigated different aspects of the occurrence of particulate lead in drinking water (Harrison and Laxen, 1980; Sheiham and Jackson, 1981; Walker, 1983; de Mora and Harrison, 1984; Colling et al., 1987; de Mora et al., 1987; AwwaRF, 1990; Hulsmann, 1990; Breach et al., 1991; Fuge et al., 1992; AwwaRF-TZW, 1996; Reiber et al., 1997; AwwaRF, 1998; Edwards and Dudi, 2004; Cantor, 2006, 2009; Triantafyllidou et al., 2007). Aggravation of solder corrosion by galvanic stimulation has been implicated as an important contributor to elevated lead levels after disinfectant or coagulant changes (Edwards and Dudi, 2004; Edwards and Triantafyllidou, 2007; Nguyen et al., 2008; Stone et al., 2008). Comparisons among studies of the significance of lead particle release are difficult, because the analytical schemes (such as filtration) developed to differentiate dissolved from particulate lead species are very prone to both negative and positive biases. Regardless of its prevalence, when it occurs, the release of particulate lead makes achieving regulatory standard objectives for lead levels very difficult. Utilities should make every effort to minimize all types of lead release by the selected corrosion control treatment approach as much as possible.

Divalent Lead. Historically and currently in most drinking water utilities, passivation of lead usually results from the formation of a surface film composed of a Pb(II) hydroxycarbonate or orthophosphate solid. These protective minerals include $Pb_3(CO_3)_2(OH)_2$ (hydrocerussite), $Pb_{10}(CO_3)_6(OH)_6O$ (plumbonacrite), $Pb_5(PO_4)_3(OH,Cl,F)$, $Pb_9(PO_4)_6$, or some combination of these forms. The soluble oxide PbO (litharge or massicot) is sometimes a component of the lowest level of a pipe scale, or may exist as a transitory breakdown product of some other solid phase, such as β-PbO_2 (plattnerite) or hydrocerussite. Though present historically on many lead pipes and though it forms a physically adherent passivating film, the solid $PbCO_3$ (cerussite) is not desirable, because its equilibrium solubility is dangerously high from a health standpoint. The formation of these passivating compounds depends primarily on pH, the concentrations of DIC, and orthophosphate. These interrelationships are complicated and occasionally misunderstood (Schock, 1980, 1981, 1989; Sheiham and Jackson, 1981; Schock and Gardels, 1983; Trussell, 1985; LeRoy, 1993; LeRoy and Cordonnier, 1994; Dodrill and Edwards, 1995; Schock et al., 1996). The carbonate containing films are usually off-white, pale bluish-grey, or light brown, are usually slightly chalky when dry, and are frequently mistaken for coatings of calcium carbonate.

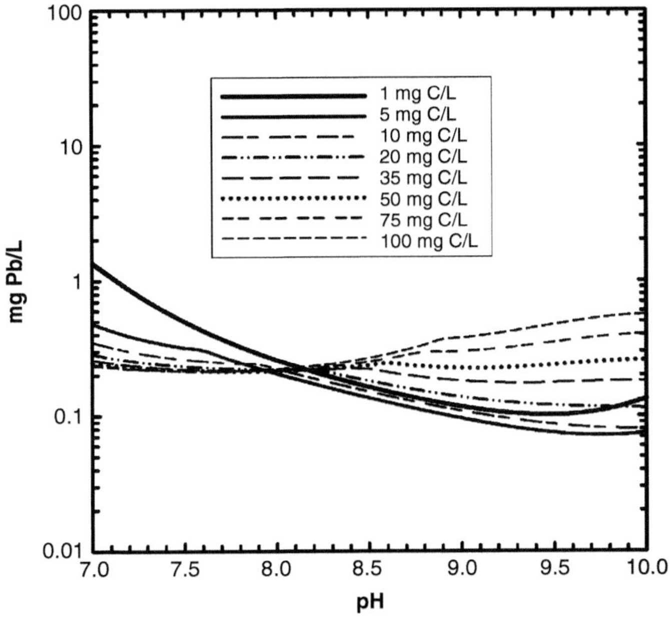

FIGURE 20-21 Effect of DIC on the solubility of Pb(II), $I = 0.01$, 25°C, assuming both hydrocerussite and cerussite can form.

The Pb(II) solubility diagram shown as Fig. 20-21 shows four critically important aspects of plumbosolvency behavior. First, as the DIC level is increased, the domain of lead carbonate stability is extended to a higher pH limit (approximately pH 8.5 instead of approximately pH 7) and the solubility of lead is decreased within the lead carbonate stability domain. Second, the aqueous carbonate complexes compete successfully for lead primarily at the expense of $Pb(OH)_2^\circ$ and $PbOH^+$. Third, lead solubility is enhanced in part of the region where basic lead carbonate is the controlling solid, primarily around pH 8.5, brought about by carbonate complexation. Fourth, for the pH range of approximately 7.2 to 8.2, the lead solubility is nearly independent of the DIC concentration, particularly when DIC is in the range of approximately 15 to 100 mg C/L. This clearly explains why many corrosion control field studies and many corrosion control optimization studies performed by large- and medium-sized utilities required by the Lead and Copper Rule failed to find a significant relationship between lead release and alkalinity. Low lead levels at modest to high DIC were also found to be attributable to the discovery that many of these moderate- to high-alkalinity water systems are likely protected by PbO_2 deposits, and not the Pb(II) carbonates that have been commonly assumed to be the protective solids (Schock et al., 2001, 2005a).

The divalent lead solubility model appears to be accurate in predicting plumbosolvency trends on the basis of many years of experience with carbonate passivation to meet the Lead and Copper Rule. However, it seems to be inaccurate for cases with either dissimilar material coating the pipe surface or for pHs above approximately 9.7. This is likely caused by some combination of inaccurate formation constants for key hydroxide, carbonate, or hydroxycarbonate complexes that dominate at high pH, uncertainty in hydrocerussite formation kinetics with respect to pH and DIC, and uncertainty of unusual plumbonacrite solid phase solubility (Rego and Schock, 2007).

Few systematic studies have been performed that have adequately identified or isolated the effect of various cations, such as calcium, magnesium, sodium, and zinc, on lead corrosion in potable waters (LeRoy, 1993; Schock et al., 1996). Existing data and a consideration of corrosion thermodynamics suggest that in the absence of a uniform diffusion barrier of calcium carbonate or other solids, the significance of the cations will be minor. One study indicated an increase of lead release in the presence of higher concentrations of calcium (LeRoy, 1993). It has been hypothesized that the greatest (and perhaps only) effect would be on new surfaces, where metals could function as inhibitors of cathodic site reactions (Schock and Clement, 1998), which would rarely be the case with lead piping in drinking water plumbing installations. Calcium is also a common substituting cation in the pyromorphite ($Pb_5(PO_4)_3(Cl,OH)$) mineral structure, and therefore, it could be incorporated into protective films in orthophosphate-treated systems.

As noted in the previous section on metallic deposits on pipe surfaces, lead leaching has been shown to be reduced by some of these kinds of surficial deposits. A mixed postprecipitate, probably composed of some combination of calcium, iron, and aluminum, was noted to cause turbidity when orthophosphate application was begun in the Washington, D.C., water system (Tesfai et al., 2006). This precipitation event coincided with rapid reduction in lead release, and therefore, is also likely to be responsible for augmenting the formation of simple Pb(II) orthophosphate solids from PbO_2 breakdown that would occur over time.

Generally, studies have also shown the effects of sulfate, chloride, and nitrate on plumbosolvency in potable waters to be negligible (Schock et al., 1996). However, these anions could have an assortment of kinetic effects on the rate of lead release, or effects on the physical structure of the scale minerals. Sudden increases in chloride levels or ratios of chloride to sulfate resulting primarily from changes in types of coagulant have been implicated in several episodes of increased lead release and failures to meet the lead Action Level in the United States (Edwards and Triantafyllidou, 2007; Triantafyllidou et al., 2007). Metz et al. (2006) showed conclusively that when the key water chemistry variables were accurately controlled, neither sodium fluoride nor fluosilicic acid added at a normal utility dosage (1 mg/L as fluoride) as either sodium fluoride or fluosilicic acid, had any adverse impact on lead levels, as predicted by lead complexation models (Urbansky and Schock, 2000a and b).

Organic acids apparently increase lead solubility, and some complexation of dissolved lead by organic ligands has been determined. In the United Kingdom, NOM is considered to be a major aggravating factor for plumbosolvency control (Colling et al., 1987; Hayes et al., 1997, 2006, 2008). Some organic materials, however, have been found to coat pipe walls and reduce corrosion. The effect of the organic material on lead release or passivating film stability may depend on the chemical and physical nature of the NOM in the particular water supply; this will make a broad generalization about the effect of various natural organic substances impossible.

Orthophosphate has been shown in field and laboratory tests to greatly reduce lead solubility through the formation of Pb(II) orthophosphate films. However, the effectiveness depends on the proper control of pH and DIC concentration, and a sufficient orthophosphate dosage and residual in the distribution system through the premise plumbing (Gregory and Jackson, 1984; Schock, 1989; Colling et al., 1992; Schock et al., 1996; Schock and Clement, 1998; Hayes et al., 2008; Cardew, 2009). The conventional wisdom of using orthophosphate versus pH adjustment to reduce plumbosolvency is summarized for a water with DIC of 5 mg/L in Fig. 20-22.

From solubility considerations alone, considerably higher dosages of orthophosphate are needed in waters with higher carbonate contents, as is illustrated in Fig. 20-23. The dosages of orthophosphate applied throughout the United Kingdom that have been convincingly shown to be highly effective for plumbosolvency control are generally 2 to 4 times the dosages commonly applied in the United States (Hayes et al., 2008; Cardew, 2009). Cardew

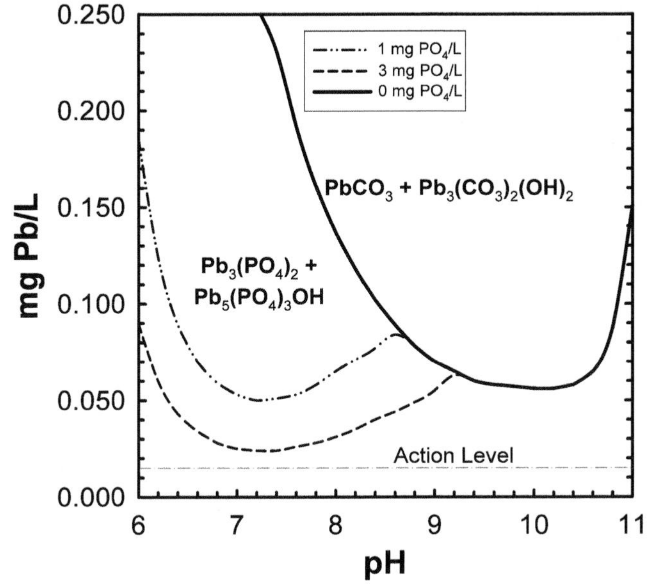

FIGURE 20-22 Solubility of Pb(II) with and without orthophosphate, assuming DIC = 5 mg C/L, 25°C, I = 0.005.

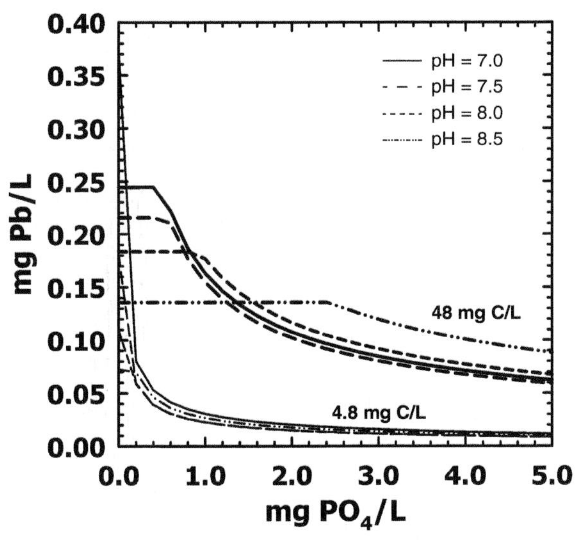

FIGURE 20-23 Computed solubility for $Pb_5(PO_4)_3OH$ (hydroxypyromorphite) at I = 0.01 and 25°C showing strong dependence of the solubility on both orthophosphate and carbonate concentrations. Note the distinct minimum dose thresholds for the high DIC condition, and higher dosages needed for optimal plumbosolvency reduction.

(2009) clearly points out the success of long-term application of high orthophosphate dosages to mitigate both particulate Pb release and plumbosolvency in difficult waters. The low lead levels resulting from the higher orthophosphate dosage from the Cardew (2009) and Hayes et al. (2008) studies argue that the lead release in most U.S. water systems is not truly optimized as intended by the Lead and Copper Rule.

In the United States, several water systems typically having low DIC levels, have reported difficulty with achieving good lead release control at pH over 8 using phosphate. Though the mechanism of the phenomenon is not fully understood, the phenomenon has been observed in USEPA laboratory experiments with low DIC waters and approximately 1 mg PO_4/L orthophosphate dosage (Fig. 20-24) (Schock, 1989; Schock et al., 1996, 2008b). Whether this difficult pH range remains with higher levels of carbonate or orthophosphate has not been tested.

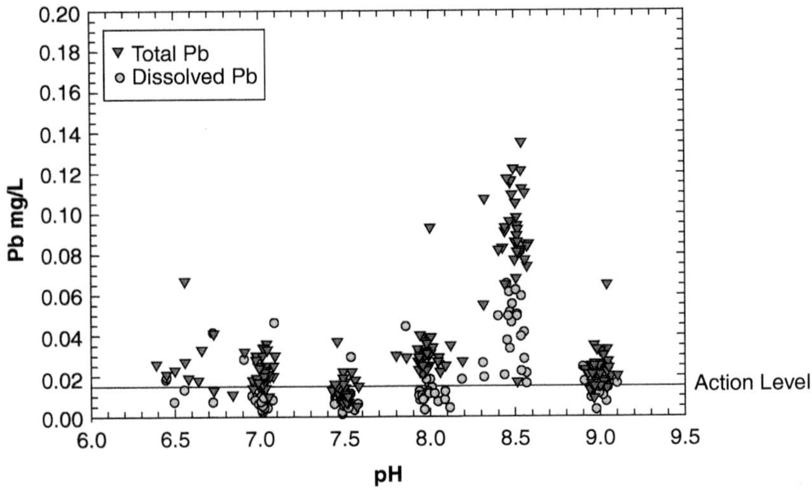

FIGURE 20-24 Soluble and total Pb release versus pH for test systems with new pipe, DIC = 5 mg C/L, orthophosphate = 1.0 mg PO_4/L. (*Source: Schock et al., 2008.*)

The rate of formation of lead orthophosphate passivating films seems to be slower than the rate of carbonate or hydroxycarbonate film formation. There is field and mineralogical evidence that they undergo an aging process similar to the cupric hydroxycarbonate and oxide films. Slow conversion from the hydroxide form of pyromorphite to the chloride form, with accompanying reduction in solubility, is inferred from pipe analysis data from several water systems in the United States. Considerable time must be allowed for the reactions to take place. Some studies have shown that many months to several years are needed to reduce the rate of lead release down to essentially constant levels (Lyons et al., 1995; Cook, 1997). The speed and amount of reduction appear to be proportional to the applied dosage of orthophosphate (Wagner, 1989; Colling et al., 1992; Schock et al., 1996; Schock and Clement, 1998). No systematic and quantitative information exists on the rate of film formation on new versus used pipe.

Most polyphosphate treatment chemicals are strong complexing agents for lead and calcium, and would be expected either to directly enhance lead solubility or to interfere with the formation of calcareous films. In both theory and practice, polyphosphate chemicals have been shown to be detrimental to lead control when all the important factors have been isolated in the tests (Schock, 1989; Holm and Smothers, 1990; Holm and Schock, 1991a and b; Schock et al., 1996; Cantor et al., 2000). Some of the apparently successful applications of polyphosphate-lead corrosion control may actually be caused by pH adjustment or the reversion of a fraction of the polyphosphate to a protective orthophosphate form.

Silicate may also reduce lead solubility or release, especially at levels exceeding 10 to 20 mg SiO_2/L (Lytle et al., 1996; Schock et al., 1995b, 2006). Little is known about the nature of films formed, except that reaction rates may be slow, similar to the case of orthophosphate, and that the use of fresh surfaces (such as commonly employed in pipe loop, coupon, and similar studies) may give a false prediction of poor performance.

Research has also shown that the rate of oxidation of lead is fast and that stagnation curves for lead showing the concentration increases in pipes with time can often be modeled accurately with a diffusion-based model (Kuch and Wagner, 1983), indicating that levels tend to approach equilibrium conditions within about 8 to 12 hours (Schock and Gardels, 1983; AwwaRF, 1990; Schock et al., 1996).

A study in Portland, Oregon, observed that the use of chloramines for disinfection had the effect of decreasing the rate of lead corrosion compared to the use of free chlorine although the final pH values of the two alternatively treated waters were approximately equal (Trussell, 1985). This could be attributed to a lower redox potential in the system with combined chlorine, and lead corrosion being controlled by Pb(II) chemistry in both cases. The addition of ammonia to chlorinated water to generate chloramines, however, tended to increase the concentration of lead by-products in solution. The role of ammonia in possible complexation of metal ions and its effect on corrosion rates are poorly understood, but the effect (if any) must be greater at high pH where Pb-NH_3 complexation is favored. Several recent research studies have shown detrimental impacts of chloramines on the release of Pb(II) from pipe, brass, and soldered joints under some circumstances (Edwards and Dudi, 2004; Zhang et al., 2008a and b).

The leaching of lead from soldered joints, unlike lead pipe dissolution, has a strong galvanic corrosion component. Lead is essentially a diluent in the solder and does not appear to affect how the solder alloys with the copper pipe when the joint is made (Parent et al., 1988). Lead release from solder is somewhat variable, over time and across different sites in a distribution system of similar general age and construction. This is likely caused by differences in exposed surface areas and their locations relative to water samples being collected. Particles of solder can also be eroded or otherwise released, creating the occurrence of random spikes of high lead levels in some samples (Lytle et al., 1993). Lyon and Lenihan (1977) found that the quality of workmanship and the presence of excess flux on the pipe interior enhanced lead release. Though degreasing reduced the lead release from this source, higher levels of copper dissolution were then seen (Lyon and Lenihan, 1977). Several other researchers have investigated lead release and the electrochemistry of soldered joints (Gregory, 1990; Reiber, 1991; Singley, 1994; Dudi and Edwards, 2005; Edwards and Triantafyllidou, 2007; Nguyen et al., 2008, 2009), and mineralogical evidence has recently been presented confirming low pH conditions in the proximity of galvanic connections between lead pipe and solder, brass, or copper (DeSantis et al., 2009). Two studies of lead release from soldered joints, in particular, have attempted to assess the contribution of the different plumbing components to lead levels at the tap (Lee et al., 1989; Sandvig et al., 2008).

Oliphant (1983) did several electrochemical studies of solder corrosion and concluded that even if the exposed area of solder were small, considerable contamination would be possible for years. He suggested that lead leaching from solder is not affected by conductivity, carbonate hardness, or pyro and orthophosphate concentration (Oliphant, 1983; Oliphant and Schock, 1996). Lead leaching was increased by decreasing pH, increasing chloride, and increasing nitrate levels. Leaching rates were decreased by increasing sulfate and silicate concentrations. The effectiveness of sulfate in reducing the rate of Pb release was related to the ratio of sulfate to chloride, with a 2:1 ratio allowing the formation of good crystalline corrosion product layers. Numerous studies of municipal corrosion control treatment in the United States since the promulgation of the Lead and Copper Rule have affirmed that proper control of pH, DIC, orthophosphate, and sometimes blended phosphate levels have been able to achieve considerable reduction in lead release from soldered joints in building and domestic plumbing systems.

Tetravalent Lead. Tetravalent lead, which usually occurs in drinking water pipes as the solid polymorph β-PbO_2 (plattnerite), has been an integral feature of known potential-pH and Pourbaix diagram solubility relationships in potable waters, dating back at least to the original work of Pourbaix (1966), if not earlier. In this regard, lead parallels iron (in contrast to copper) in that the highly insoluble valence state is the most oxidized form. The well-publicized incident of the high lead release caused by conversion from free chlorine disinfection to chloramine that surfaced in 2004 was the first noted instance of this logical reductive dissolution phenomenon by many water systems and regulators (Renner, 2004; Schock and Giani, 2004), though it should not have been surprising. In laboratory corrosion experiments in 1983, Schock and Gardels (1983) observed the formation and accumulation of PbO_2 in the apparatus, away from the pipe sections, where high-ORP conditions were maintained. PbO_2 was a major scale component in an EPA field study on partial lead service line replacement that revealed surprisingly low lead concentrations in service lines in a water supply with high pH and alkalinity (Wysock et al., 1991), and its presence in several pipe scales was further documented in the 1996 edition of the AwwaRF-TZW *Internal Corrosion of Water Distribution Systems* manual (Schock et al., 1996). The low lead levels throughout most of the Cincinnati distribution system were found to be the result of almost pure PbO_2 passivating films (Schock et al., 2001).

Results of solids analysis of pipe scales from the Washington, D.C., Water and Sewer Authority distribution system confirmed the reductive dissolution pathway for the breakdown of PbO_2 (Schock et al., 2001), and many research projects have now subsequently explored various aspects of the kinetics of PbO_2 formation and breakdown (Lytle and Schock, 2005; Switzer et al., 2006; Rajasekharan et al., 2007; Lin and Valentine, 2008a and b; Liu et al., 2008; Lytle et al., 2009b). A representation of the phenomenon of reductive dissolution of PbO_2 with respect to Pb-E_h-pH relationships is shown in Fig. 20-25. The reversibility of this reaction (in the absence of phosphate) over time frames of only weeks has also been shown (Giani et al., 2005; Lytle and Schock, 2005). Tetravalent lead deposits

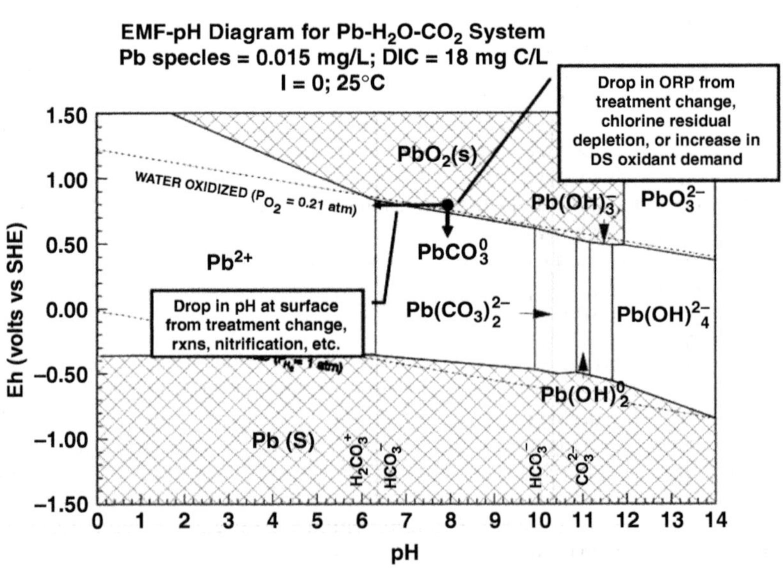

FIGURE 20-25 Potential-pH diagram for 0.015 mg/L Pb and 18 mg C/L showing conceptually the two major pathways by which Pb(II) can be resolubilized to high release levels from a protective low-solubility passivating PbO_2 scale.

have turned out to be relatively common, and they can be categorized generally into three modes (Schock et al., 2005a): (1) complete, relatively uniform PbO_2 scale, often much thinner than normal lead carbonate and hydroxycarbonate deposits; (2) distinctly layered scale, PbO_2 occurring as an upper water contact layer on top of typical Pb(II) carbonate or hydroxycarbonate deposit; and (3) in patches or comingled with divalent lead and other scale minerals. As of this writing, the presence of PbO_2 has been confirmed in 1/3 of the 43 water systems from which USEPA has received over 230 individual lead and lead-lined service line specimens from the United States and one Canadian province.

Persistently high concentrations of chlorine have been observed to alter divalent lead carbonate or hydroxycarbonate under drinking water conditions to form PbO_2 in a matter of weeks to months in recent USEPA field research (Fig. 20-26), which supports the hypothesis that the conversion reaction of Pb(II) (hydroxyl)carbonates to PbO_2 is faster at

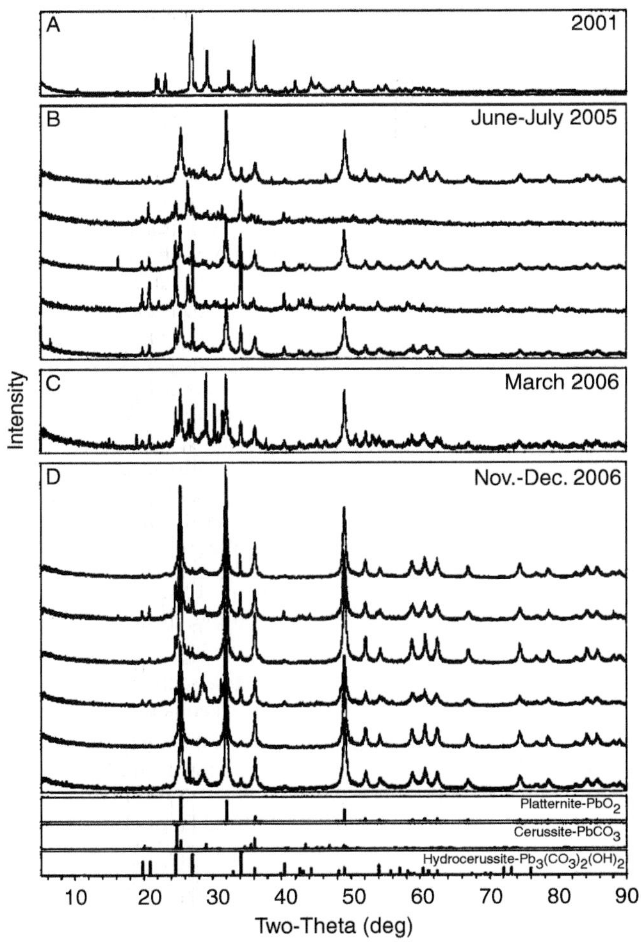

FIGURE 20-26 X-ray diffraction patterns of lead pipe scale samples showing growth of β-PbO_2 (platternite) over time, following pH adjustment to over 9, implemented in early 2005 for a soft, low-alkalinity surface water system in the United States. (*Source: USEPA.*)

higher pH. This further suggests that some water systems that presume they have achieved plumbosolvency control through carbonate passivation are actually achieving it through PbO_2 formation. The implications for the impacts of future changes become radically different in these cases.

Notably, no tetravalent lead compounds other than α-PbO_2 (scrutinyite) or plattnerite have been reported thus far in drinking water pipe scales. For water systems having large numbers of lead service lines, stable and abundant PbO_2 scales appear to be both theoretically and practically of comparable or superior effectiveness to orthophosphate dosing for the control of lead release. Ephemeral or inexplicable episodes of slightly elevated lead levels during Lead and Copper Rule (LCR) monitoring, without apparent correlation to pH fluctuations, temperature changes, alkalinity changes, DIC changes, or other major water quality changes, may actually be caused by cycles of deposition and dissolution of PbO_2 scales.

Little specific information is available that would allow solubility modeling for Pb(IV) of the type successfully applied to copper, divalent lead solubility, or iron corrosion control. Data from Pourbaix (1966) and limited thermodynamic data tabulations only hypothesize the existence of two hydroxide complexes that are only relevant at pH well above those normally encountered by drinking water systems. Carbonate, bicarbonate, chloride, sulfate, and orthophosphate complexes for Pb^{4+} have not been reported, but they would be hard to measure because of the low PbO_2 solubility and extreme ORP conditions. Mixed oxidation state oxides (Pb_3O_4 and Pb_2O_3) have been observed in some artificial conditions (Pourbaix, 1966), but are very uncommon in drinking water pipes. Existing solubility constant and Gibbs free energy of formation values (ΔG_f°) result in solubility predictions that are considerably lower than are observed in practice. This may also cause the pH/E_h stability field for PbO_2 in the diagrams to be slightly overestimated. Largely unpublished sequential sampling data from at least three water systems and bench-scale experimental data, all point to upper-bounds equilibrium plumbosolvency in the range of 0.001 to 0.01 mg/L in the pH range of 7 to 10. It is not currently possible to define a precise relationship to major water quality variables, other than the approximation with ORP. Orthophosphate has now been shown to inhibit the oxidation of Pb(II) to PbO_2, though a threshold pH and phosphate concentration have not been established (Lytle and Schock, 2005; Lytle et al., 2009a).

There is enough occurrence information to be able to offer drinking water utilities some guidance for determining relative likelihood of PbO_2 deposits in lead service lines or pigtails, and thus, the potential vulnerability for process or disinfection changes that would lower ORP (such as changing from free chlorine to chloramination).

- History of very low 90th percentile Pb levels in water systems having large numbers of lead pipes and a pH of below approximately 8.5, without the use of phosphate or silicate corrosion inhibitors
- Recent or historical use of high free chlorine dosages, such as 2 to 3 mg/L or more, to control biofilms or coliform or to ensure residual to the end of an old distribution system
- Systems with low distribution system chlorine demand that maintain near mg/L level free chlorine residuals, such as systems that are non-corrosive toward iron, have few unlined iron pipes, or have removed oxidant demand (including NOM) through optimized coagulation, filtration, GAC, or greensand (or similar Mn and Fe oxidative removal processes)
- Systems which have elevated pH (over 9) to optimize carbonate passivation and which concurrently notice stable free chlorine residuals that approach or exceed 1 mg/L
- Systems that supplement free chlorine with chlorine dioxide, particularly if a ClO_2 residual tends to persist into areas with lead pipes

There are several water chemistry, water treatment, and distribution system characteristics that make the formation and presence of tetravalent lead compounds (such as PbO_2)

unlikely. These include (1) absence of lead piping, (2) low or no free chlorine residual in large areas of the distribution network, (3) long-time use of monochloramine as secondary disinfectant, (4) somewhat frequent red or discolored water complaints, (5) long-term use of phosphate corrosion inhibitors, with higher phosphate residuals making PbO_2 formation progressively less likely, and (6) unchlorinated groundwater systems.

The best way to maintain the stability of tetravalent lead deposits is through optimal corrosion control and DBP precursor removal (i.e., NOM). The removal of the precursor material reduces DBP formation, reduces nutrient material (often a food source for biofilms), and thus reduces disinfectant demand translating into reduced chlorine dosage. Coupling with iron corrosion control is also important. This reduced disinfectant demand reduces needed chlorine dosage and potential for DBP formation. It also reduces microbial habitat, so less disinfectant is needed. A threshold ORP and pH must be maintained by a chlorine residual that is sufficiently high. It is hard to accurately and precisely determine those values, as there are utility-specific components to the amount of oxidant demand observed, but the trends follow those found in the E_h-pH diagrams. Additionally, complete removal of iron and manganese at the treatment plant is very desirable to reduce chlorine demand out in the distribution network.

Galvanized Steel

Galvanized steel pipe was the dominant material used to transport domestic drinking water for most of the twentieth century, only being replaced in the last several decades by copper and plastic. Galvanized plumbing is still the material choice in many developing countries (Trussell and Wagner, 1996). Galvanized pipe consists of a base steel layer, underlying a sequence of layers that are zones of iron-zinc alloy that successively approach pure zinc at the interior surface of the pipe (Trussell and Wagner, 1996). One of the complexities of this material is that once the zinc layer is corroded away, the pipe behaves as if it were black iron pipe. Therefore, the corrosion potential of the water and optimal corrosion inhibition strategies should change, depending on how much of the zinc layers are left.

Uniform corrosion, pitting, metal release, and tuberculation can all be important in galvanized pipe corrosion (Wagner, 1989; Trussell and Wagner, 1996; MWH, 2005). Mechanical failure because of uniform corrosion is rare but, when the zinc corrosion rate is high, failures are often induced by the tuberculation that follows rapid depletion of the zinc layer. When the pipe is new, the uniform corrosion depends strongly on pH. The nature of the film formed on the pipe changes in response to many chemical factors, and galvanized pipe corrosion seems to be sensitive to the type of scale produced. Some scales are voluminous, chalky, and relatively unprotective, even if the rate of metal release is not extremely high. Poor film protection can accelerate pitting or tuberculation. Field experience indicates that the effect of corrosion scales on the corrosion of galvanized pipe does not appear to be strictly proportional to zinc solubility and release. Moderate levels of DIC and high buffer intensity seem to produce good passivating films in cold water (MWH, 2005).

Studies have shown increases in the corrosion rate of zinc with increasing carbonate hardness, even in the presence of orthophosphate (Trussell and Wagner, 1996). This increase may be caused by the formation of aqueous carbonate complexes [$ZnHCO_3^+$, $ZnCO_3^\circ$, $Zn(CO_3)_2^{2-}$] increasing zinc solubility (Schock and Buelow, 1981; Schock, 1985; Trussell and Wagner, 1996).

The calcium carbonate seen in deposits is often aragonite (Schock and Neff, 1982; Trussell and Wagner, 1996), and its role in galvanized pipe protection is not clear. Calcium carbonate, along with basic zinc carbonate [$Zn_5(CO_3)_2(OH)_6$, hydrozincite], and forms of zinc hydroxide and zinc oxide, appear to be the most significant components of natural scale layers.

The effects of orthophosphate and silicate, and the tendency for galvanized pipe to fail through non-uniform corrosion have previously been reviewed in some detail (Trussell and Wagner, 1996; MWH, 2005). Orthophosphate effectiveness is a function of the pH, alkalinity, and the orthophosphate level in solution (Schock and Neff, 1982; Schock, 1985). Silicate has been shown to reduce galvanized pipe corrosion, but most studies have focused on hot water systems. Zinc silicate compounds are generally of low solubility; however, and theoretically should provide some protection in the slightly alkaline pH range.

Pitting failure of galvanized pipe is more common in hot than cold water systems. Pipe failures have been caused by poor pipe coating quality, the presence of dissolved copper, and the reversal of electrochemical potential of zinc and iron at high temperature (Trussell and Wagner, 1996; MWH, 2005).

There have been numerous anecdotal reports of drinking water contamination by impurities of lead and cadmium in the galvanizing layer. Usually, tap water samples have been involved, making it difficult or impossible to separate the probable contribution of brass faucets and shutoff valves.

Asbestos-Cement (A-C) Pipe, Cement-Mortar Linings, and Concrete Pipe

A-C pipes and cement-mortar linings (without seal coat or from in situ rehabilitation) behave similarly in drinking waters. The principal difference is the additional possibility of the release of asbestos fibers into the water by A-C pipe deterioration. Deterioration of cementitious materials causes substantial water quality degradation. Leaching of free lime from the cement causes considerable increase in pH of the water, especially in long lines or low flow areas and with poor to moderately buffered waters. Because the equilibrium pH for calcium hydroxide solubility is approximately 12.4 (note that calcium hydroxide is a common high-pH analytical buffer solution), there is a considerable driving force for pH increase. This pH increase, in turn, can cause reduced effectiveness of disinfectants and phosphate corrosion inhibitors, unwanted precipitation of a variety of minerals (causing cloudy or turbid water), and poor taste. In extreme cases of attack by aggressive water, the attack can result in reduced pipe strength and a roughened pipe wall that can cause increased head loss.

The cement matrix of A-C pipe is a complicated combination of compounds and phases (Schock et al., 1981; LeRoy, 1996; LeRoy et al., 1996). Several different types of cement can be used in the production of cement linings (LeRoy et al., 1996; Guo et al., 1998). The main components of the Portland cement used most frequently are tricalcium silicate, dicalcium silicate, and tricalcium aluminate, together with smaller amounts of iron and magnesium compounds. Free lime ($Ca(OH)_2$) is also present, but it is limited by product standards to be less than or equal to 1.0 percent by weight in U.S. commercial type II autoclaved A-C pipe (LeRoy et al., 1996).

Groundwaters having high mineral content but low pH values, show accelerated rates of pipe leaching (LeRoy et al., 1996). In soft waters with low mineral content the mechanism of attack is believed to proceed as follows: First, the calcium hydroxide is removed, and if the water still possesses sufficient acid content, calcium carbonate is either dissolved or prevented from forming at the water-cement boundary. Because alkaline conditions cannot be maintained within the cement after the initial calcium hydroxide is removed, attack begins on the hydrated calcium silicates. These convert to calcium hydroxide, and the cycle of reactions can continue until sufficient calcium carbonate is formed to fill voids in the cement matrix and prevent lime leaching, the water is neutralized to the point at which the cycle stops, or no cementitious material remains to bind the aggregates together (LeRoy et al., 1996).

Waters with low pH (less than approximately 7.5 or 8, unless they contained high calcium, alkalinity, and silicate levels), and high sulfate concentrations are particularly destructive to A-C pipe, and probably to all concrete or cement-mortar-lined pipes. Strong sequestering or complexing agents, such as polyphosphate chemicals, have been shown to attack the pipe by enhancing calcium, aluminum, iron, and magnesium leaching from the cement matrix (LeRoy et al., 1996). They tend to prevent the formation of protective coatings by metals such as zinc, manganese, iron, and calcium.

No simple chemical index exists that can directly predict the behavior and service life of A-C pipe. In addition to the use of CCPP or saturation indices to suggest chemistries where a porosity sealing process is unlikely to occur, Trussell and Morgan (2006) have proposed the use of the saturation index for calcium silicate (wollastonite) as providing several advantages over calcium carbonate. With A-C pipes, the primary drawback to calcium carbonate saturation treatment is that in the absence of the formation of an actual surface coating or the formation of protective coatings by naturally occurring substances such as iron and manganese, fibers are left exposed at the surface of the pipe where they are vulnerable to erosion.

Coatings containing iron have been found on many A-C pipe specimens exposed to a variety of aggressive to non-aggressive water. Iron coatings sometimes have a granular, porous structure that does not necessarily retard calcium leaching from the cement matrix. The iron frequently helps prevent the exposure of asbestos fibers at the pipe surface; however, the iron coatings may also become a habitat for the growth of biofilms or the retention of microorganisms.

Mn(IV) oxide can provide a similar protection to that provided by iron. Silica may also be beneficial as an agent that can help maintain the hardness of the pipe (Schock et al., 1981; LeRoy et al., 1996). In some cases, silicate dosing could enhance protection by the formation and adsorption of iron colloids on to the pipe surface.

Orthophosphate, sulfate, and chloride salts of zinc have been used in laboratory experiments to prevent the softening of A-C pipe specimens to slow calcium leaching (Schock and Buelow, 1981). These results should also apply to the preservation of any uncoated cementitious materials in a water supply system. Though often overlooked and misunderstood, the required dosage of zinc depends at least on the pH and DIC concentration of the system (Schock and Buelow, 1981). The utility of zinc addition for extending the potential lifetime of newly installed cement linings or A-C pipe lines or for increasing the service life of existing mains has not been established in the short field tests performed thus far, usually because of inadequacies in the maintenance of appropriate pH, alkalinity, and zinc concentrations. In the laboratory experiments, zinc, and not the anion of the compound, has been found to be the active agent (with the exception of the bicarbonate and carbonate ion role); this was also observed in a field investigation (Mah and Boatman, 1978). The reaction mechanisms to form the final long-term coating observed in tests have not been precisely delineated. The experiments suggest the hypothesis that zinc first reacts with the water to form a zinc hydroxycarbonate precipitate, which delivers zinc to the surface of the pipe (Schock et al., 1981; LeRoy et al., 1996). The zinc solid then reacts with dissolved silica released from the pipe surface to convert some or all of the coating to a harder zinc silicate solid phase. In parallel, dissolved zinc probably also will interact and sorb on the surface of the pipe to some degree.

The choice of zinc salt can be based on economics and the presence of materials benefiting from phosphate addition (e.g., unlined iron, lead, brass, copper). Considerations of wastewater discharge regulations could rule out phosphate-containing formulations. More research needs to be done to determine if the orthophosphate inhibits the calcium carbonate protection mechanism of the cement matrix or if the calcium orthophosphate solids help stabilize the cement material.

Polyphosphate use should be avoided or only done cautiously with A-C or cement-lined pipes (LeRoy et al., 1996). In an extreme case, aggressive water from a membrane desalinization plant and containing polyphosphate was introduced into a water system in the Dutch Antilles with a mix of iron and cement-lined pipes. The water leached such high concentrations of aluminum unknown to the hospitals that it caused serious illnesses and several deaths of kidney dialysis patients (Berend and Trouwborst, 1999).

A potential for the leaching of metals such as barium, cadmium, and chromium from some cements has been reported (Guo et al., 1998). Care must be taken especially when cement is produced in kilns cofired with hazardous waste-derived fuel. It is not clear if there is a real problem in the United States, as manufacturing standards that will alleviate the potential for this kind of problem may be in place or developed soon.

DIRECT METHODS FOR THE ASSESSMENT OF CORROSION

Physical Inspection

Water treatment plant operators and engineering staff should be aware of the appearance of the interior of the distribution piping and how their water quality is affecting the materials in their system. Physical inspection is a useful inspection tool to a utility, and it can also be done routinely at a low cost. Both macroscopic (human eye) and microscopic observations of scale on the inside of the pipe are valuable tools for diagnosing the type and extent of corrosion, and digital photography provides a flexible and convenient means to maintain good historical documentation of pipe condition. Macroscopic studies can be used to determine the amount and type of tuberculation, postprecipitation accumulation (Fe, Mn, Al, and others), pitting, corrosion at joints, and possible crevice corrosion. A record of historical observations of pipes in different parts of a distribution system and exposed to different water treatments over time could yield insights into slow corrosion or passivation reactions.

Utility personnel should try to obtain pipe sections from the distribution or customer plumbing systems whenever possible, such as when old lines and equipment are replaced. This is important even when there are no active complaints, because the pipe scale information may be important to determine if future treatment changes that are contemplated for regulatory or other reasons would be feasible. When lines are tapped, the cookies generated can be an excellent record and history of water quality impacts on the scaling or corrosion of mains. An examination of the pipe wall can yield valuable information about the type and extent of corrosion and corrosion-product formation (such as pitting or tubercles), though it may not indicate the most probable cause. Pipe sections of small diameter may be conveniently examined by sawing them lengthwise.

Examination under a microscope can yield even more information, such as hairline cracks and local corrosion too small to be seen by the unaided eye, mineralogical layering of the scale, scale porosity, or associations of particular solid phases in the scales. Such an examination may provide additional clues to the underlying cause of corrosion by relating the type of corrosion to the metallurgical structure of the pipe.

Photographs of specimens should be taken for comparison with future visual examinations. High-magnification photographs should also be taken, if possible.

Corrosion Rate Measurements

Corrosion rate measurements are another method frequently used to identify and monitor corrosion. Corrosion rate measurements rarely correlate well with metal release into the water. The mechanisms involved in the attack on the metal are not necessarily the same reactions as those that result from the interaction of the water with a developed scale on the pipe surface. The corrosion rate of a material is commonly expressed in mils (0.001/in) penetration per year (mpy). Common methods used to measure corrosion rates include (1) *weight-loss methods* (coupon testing and loop studies) and (2) *electrochemical methods*.

Weight-loss methods measure corrosion over a period of time. Electrochemical methods measure either instantaneous corrosion rates or rates over a period of time depending on the method used.

Coupon Weight-Loss Method. Four important criteria for corrosion tests are (1) the metal sample must be representative of the metal piping, (2) the quality of the water to which the pipe sample is exposed should be the same as that transported in the plumbing system, (3) the flow velocity and stagnation times should be representative of those in the full-scale system, and (4) the duration of the test must allow for development of the pipe scales that have an important effect on corrosion rate and on the quality of the water passing through the pipes (Reiber et al., 1996).

Coupons are often employed in *planned interval tests,* which are often useful and desirable if the test duration is to be 12 months or longer. The test duration can then be modified on the basis of the results obtained (Thompson, 1970). Planned interval tests can also be used with removable pipe sections for scale examination, not just weight-loss coupon studies.

Coupon preparation, handling, and examination are relatively time consuming and exacting operations that usually require a trained laboratory analyst or corrosion specialist to provide reproducible results (Kuch et al., 1985; Reiber et al., 1988, 1996). Several ways have been developed to improve the processing of coupons for corrosion rate determination, and the statistical validity of the results (Reiber et al., 1988; USEPA, 1993). Contamination during the manufacture of the coupons has also been identified in some studies, and careful examination, cleaning and data evaluation may need to be used to overcome that problem (Lytle et al., 1992).

Electrochemical Rate Measurements. These methods are based on the electrochemical nature of corrosion of metals in water, and they have been widely discussed in the drinking water corrosion literature (Silverman, 1995; Reiber et al., 1996). The primary types of techniques try to determine corrosion rates through the measurement of the electrical resistance, linear polarization, corrosion current, or galvanic current of test specimens. Although the techniques have been used widely, often successfully, there are limitations inherent to all of them, and the user must be cautious that fundamental mathematical or chemistry assumptions behind the analytical and data analysis procedures are not violated or the results biased by the technique itself (Reiber, 1989; Vitins et al., 1994; Silverman, 1995; Reiber et al., 1996, 1997). Frequently, the experiments and studies focus only on electrochemical parameters and neglect corroboration and simultaneous acquisition of metal release and water chemistry data, to ensure that results and interpretations aren't biased or misleading when applied to their ultimate purpose. In addition to corrosion rate information, some of the techniques can yield valuable information on corrosion mechanisms, such as whether a given variable can act as an anodic site or cathodic site inhibitor and what concentrations lead to protection or adverse behavior. It is also important to realize that only sometimes do electrochemical measurement techniques correspond well to metal release to solution and metal solubility.

Pipe Rig Systems

Another method for determining water quality effects on materials in the distribution system, metal release, or both, is the use of sections of pipe (AwwaRF, 1990, 1994; AwwaRF-TZW, 1996; Eisnor and Gagnon, 2003; Cantor, 2009). The pipe sections can be used to measure the form and extent of corrosion and the effect of corrosion control methods. Pipe loop sections can be used also to determine the effects of different water qualities on a specific pipe material. The advantage is that actual pipe is used as the corrosion specimen. The loop may be made, or a rig made from long or short sections of pipe. For direct monitoring of dissolved constituent levels or changes in background water parameters, such as chlorine residual depletion, or pH changes, longer pipe sections are required (Reiber et al., 1988; AwwaRF, 1990, 1994; AwwaRF-TZW, 1996; Cantor, 2009). Pipe rig systems can also be useful as part of an overall holistic corrosion control optimization strategy, incorporating water quality, scale development, and corrosion treatment monitoring. The effectiveness of this integrated approach has been shown for several water systems (Cantor, 2009).

Excavating and removing existing pipe invariably causes the disturbance of the deposits in the pipe. Special precautions, which have rarely been studied, may be needed to improve the probability of success and restabilization time. Such precautions may include continual contact with water until reinstallation, attachment to cushioned rigid frameworks for transport without damaging vibration, and maintenance of original shape to prevent scale cracking. Hence, a long period of stabilization (many months to years) may be necessary to reestablish equilibrium in the scales and regain consistency of metal release before introducing the test treatment changes to the designated loops. Systematic comparisons between the length of restabilization time necessary for exhumed pipes versus the development of representative characteristic scales on new pipe surfaces have not been reported. For studies involving comparisons among potential treatment changes, stabilization and comparability likely need to be determined among all loops to be used for the study. Substantial disagreement with in situ performance may be a clue that the operating mechanisms are not comparable between the experimental pipe rig conditions and actual distribution system conditions. All of the physical and chemical factors involved with exhumation of pipes and their reinstallation into pipe rig systems are not completely understood. Whereas in many water systems, consultants and researchers have successfully employed exhumed copper, unlined iron, A-C, and lead pipes with Pb(II)-based scales for experimental treatment studies, one area where there has been inadequate investigation is whether the same considerations apply to lead pipes with nearly complete PbO_2 scales on them. Anecdotal evidence from some largely unsuccessful attempts, including one published study (Wysock et al., 1991, 1995), suggests that there may be additional complications with these kinds of scales.

The question of how to make proper statistical comparisons is a difficult problem in using pipe loop data for treatment evaluation or other interpretations of water quality effects. This has been addressed in different ways in many practical studies, but some major considerations and suggested approaches have been given in some recent publications (AwwaRF, 1994; Wysock et al., 1995; Cantor, 2009).

Immersion Testing in the Laboratory

Immersion tests are sometimes useful to make an *initial* appraisal of the effectiveness of inhibitors (Kuch et al., 1985; Reiber et al., 1996), but it is only truly valid for plumbing materials that are being currently installed and have new surfaces. Immersion tests use metal coupons in batch jars of water to evaluate inhibitors, dosages, and pH control. The procedures of ASTM Standard 631-72 should be utilized for immersion tests.

Limitations of all immersion tests are that accurate representation of flow and water chemistry conditions in the field situation are sacrificed for testing simplicity and efficiency. Important parameters like pH, DIC concentration, DO, chlorine, and chloramine residuals may frequently be difficult or impossible to control, leading to incorrect results. Thus, results of immersion tests might not prove to be accurate simulations when extrapolated to practice at the treatment plant and in the whole distribution system. They should only be used as a general screening tool, and results contrary to expectations from chemistry theory or experience of utilities with similar pipe materials and water quality should be critically examined for possible experimental problems or errors.

Chemical Characterization of Deposits

Valuable information about probable corrosion causes or inhibition mechanisms can be found by chemically analyzing the corrosion by-product material on the pipes. Analytical approaches to characterization may be broadly separated into three categories: elemental/functional group characterization, optical/visual characterization, and compound-based characterization. Only several of the most common methods proven widely useful in solving practical drinking water corrosion and scaling problems are described in this section.

Scraping off a portion of the corrosion by-products, dissolving the material in acid, and qualitatively analyzing the solution for the presence of suspected metals or compounds can sometimes suffice to indicate the type or cause of corrosion. These analyses are relatively quick and inexpensive. If a utility does not have its own laboratory, samples of the pipe sections can be sent to an outside laboratory for analysis. The numerical results of these analyses cannot be quantitatively related to the amount of corrosion occurring because only a portion of the pipe is being analyzed. However, such analyses can give the utility a good overview of the type of corrosion that is taking place. Whenever possible, analytical techniques should be specified that can determine the abundance of anionic constituents (carbonate, phosphate, sulfate, chloride, silicate) as well as major metals, toxic and regulated trace metals, and other constituents (such as oxygen, silica, and others). Assumptions based on physical appearance and only data on metal composition can often be misleading (Schock and Smothers, 1989). For complex scales, especially when silicates or organic materials may be a significant component of the deposits, a laboratory should be consulted to advise on the most appropriate digestion technique and analytical methods for the deposit constituents. If samples come from distribution systems fed by water sources likely to have radionuclides based on geologic setting, those should be included in the analysis, along with radon if radium is present. Some useful guidance is given by ASTM procedures (ASTM, 2003a and c).

Comparing sampling data of scales from various locations within the distribution system can isolate sections of pipe that may be corroding. Increases in levels of metals such as iron or zinc, for instance, indicate potential corrosion occurring in sections of iron and galvanized iron pipe, respectively. The presence of cadmium, a minute contaminant in the zinc alloy used for galvanized pipe, also indicates the probable corrosion of a galvanized iron pipe. The presence of zinc, copper, lead, and nickel are often indicative of corrosion of brass plumbing devices. Corrosion of cement-lined or A-C pipe is generally accompanied by an increase in pH, aluminum, and calcium throughout the system, sometimes in conjunction with an elevated asbestos fiber count.

Elemental Spectroscopy after Decomposition. The most common analytical tools for the analysis of total recoverable metal concentrations in environmental samples are flame or graphite furnace atomic absorption spectroscopy (F/ or GF/AAS) and inductively coupled plasma-atomic emission spectrometry (ICP-AES) or mass spectrometry (ICP-MS). The ICP methods add a greater degree of selectivity and sensitivity in elemental analysis.

For total recoverable elemental content determination, solid samples must be prepared for analysis by extraction of the solid-phase metals into solution. This is accomplished through digestion or fusion. Table 20-1 is a list of a few method references for preparation procedures. It should be stressed that while all of the references below are considered acceptable methods, methods should be chosen on the basis of the target constituents and the nature of the information the data user seeks. Because different preparation methods are not necessarily equivalent, the method used must remain a consideration when evaluating data, and preparation and analytical methods should be referenced in any forum in which the data is presented.

It is imperative that digestion efficiency and analytical performance be monitored through the use of appropriate standard reference materials, which can be obtained through

TABLE 20-1 Common Standard Procedures for Accurate Analyses of Solid Materials Related to Pipe Scales and Distribution System Deposits

Method reference	Brief method description
EPA SW846 3050B	Strong hot plate, block, or microwave acid digestion. Applicable to soils, sediments, and sludges. Will recover most elements that could become environmentally available. It is not a total digestion technique. There are two versions, depending on the analysis method that will follow.
EPA SW846 3052	Microwave-assisted acid digestion. Applicable to difficult matrices (siliceous and other complex matrices). It is a total digestion technique. It is not appropriate for regulatory applications that require use of leachate preparations.
Standard Methods for the Analysis of Water and Wastewater, Method 3030D (E, F, G, H, and I) 21st edition	3030D is an introduction to five methods (E through I; K is not for solid samples). All five methods are variations of hot plate acid digestions, each using a different combination of acids. In addition to water; these are applicable to solid samples and liquid sludges. At the end of 3030D, four other references that may be more suitable for solid sample digestions are recommended.
USGS	Very strong hot plate acid digestion. Applicable to most rock types, soils, sediments, and ashed botanical samples. The sections describing Interferences list examples of specific minerals for which complete digestion of certain metals is not achieved.
Method codes: T20 (for ICP-MS) E011/T01 (for ICP-AES) M020/T19 (REEs, ICP-MS)	J.E. Taggart Jr. (ed.), *Analytical Methods for Chemical Analysis of Geologic and Other Materials*; Open File Report 02-223; U.S. Department of the Interior, U.S. Geological Survey: Denver, CO, 2002; available online at URL:http://pubs.usgs.gov/of/2002/ofr-02-0223/OFR-02-0223.pdf.
USGS	Analysis-specific decomposition as part of instrumental procedure.
Method codes: N021/T25 (S) N011/T10 (C) C011/T08 (Total CO_3)	J.E. Taggart Jr. (ed.), *Analytical Methods for Chemical Analysis of Geologic and Other Materials*; Open File Report 02-223; U.S. Department of the Interior, U.S. Geological Survey: Denver, CO, 2002; available online at URL:http://pubs.usgs.gov/of/2002/ofr-02-0223/OFR-02-0223.pdf.
ASTM D 4698–92	Fusion and digestion procedures are described. Both are applicable to suspended and bottom sediments of natural waters. Both recover >95% of the metals listed in each procedure.

the National Institute of Standards and Technology (Gaithersburg, MD), the U.S. Geological Survey (Denver, CO), or international organizations with equivalent programs. It is not always necessary to achieve complete digestion for all purposes, but the objective of the analytical work needs to be carefully considered and methods chosen accordingly.

X-Ray Fluorescence. For surficial deposits, *X-ray fluorescence spectrometry* (XRF) is helpful for elemental analyses (Rose, 1983; Schock and Smothers, 1989; Miller et al., 1995; ASTM, 2003c; Sandvig et al., 2008; Cantor, 2009). The sample is ground finely into a powder and is either compressed into a flat discoidal pellet under high pressure or fused with a compatible flux into a glass bead (Buhrke et al., 1998; SPEX CertiPrep Inc., 2005). X-ray fluorescence can be done quickly in comparison to wet chemical analysis (including flame or flameless atomic absorption spectrophotometry), because of the relative simplicity of sample preparation and processing. Quantitative analysis can be performed with proper sample preparation precautions, calibration techniques with corrections for matrix chemistry and interelemental effects, and suitable mathematical processing of the raw intensity data. For most major constituents and for most metals down to the order of high ppm levels by weight, the accuracy is comparable to dissolution and analysis techniques. The drawback to XRF is that it usually requires a large amount of scale relative to the amounts typically found in lead and copper pipes.

Electron Microscopy Methods. The scanning electron microscope (SEM) has been widely applied to the evaluation of A-C pipe deterioration (Schock and Buelow, 1981; Schock et al., 1981; LeRoy et al., 1996). Studies have also attempted to directly relate the morphology of corrosion product crystals of lead carbonate, basic lead carbonate, or a lead phosphate to the protectiveness against corrosion (Colling et al., 1987, 1992; Breach et al., 1991).

On the microscale, *energy-dispersive X-ray analysis* (EDXA) is useful for elemental identification and qualitative analysis, and *wavelength-dispersive X-ray analysis* (WDXA) is also possible with some electron microscopes. WDXA is generally better for quantitative analysis and lighter elements in the scale. Theoretically, electron diffraction using a transmission electron microscope should be capable of identifying specific solid phases, analogous to X-ray diffractometry. However, the difficulty in applying the technique and isolating sufficient particles in the scale to be representative usually make it impractical for this application.

For all electron microscopic analysis methods, observing representative sections of the pipe is important because the same pipe might vary in color and texture. Often, many small areas on several sections of the same pipe must be analyzed.

Compound Analysis and Visual Techniques. The field of analysis of pipe scales and inhibitor films can make use of a variety of sophisticated instrumental analysis techniques, in addition to the traditional *wet chemical procedures* (ASTM, 2003a). Often, no single technique will provide all the answers in film identification and interpretation of corrosion and corrosion control data. Judicious use of a combination of methods, however, can give invaluable insight into the problems and solutions. A plethora of advanced instrumental analysis techniques exist, which have been or can be applied to scale analysis. For example, techniques such as secondary ion mass spectrometry (SIMS) have been applied to the elemental analysis of pipe scales in Scotland (Davidson et al., 2004). Infrared spectroscopy has been employed extensively by many investigators (Hogan and Jackson, 1981; Sheiham and Jackson, 1981; Boireau et al., 1997). A full description of the many surface and compound analysis techniques available to modern laboratories and their relative merits is beyond the scope of this section. Overviews are available of several useful scale analysis approaches and their applicability to different materials have been given my many authors (Schock and Smothers, 1989; Miller et al., 1995; Scheetz et al., 1997; Davidson et al., 2004; Rego and Schock, 2007; Borch et al., 2008; Cantor, 2009).

Mounting a pipe sample with scale in place using an epoxy resin can be employed to get a cross-sectional analysis, which can reveal detailed information, layer by layer. Sometimes this can reveal important structural features of the scale. Techniques such as WDXA, EDXA and laser-ablation ICP spectrometry can be used to do the scans or elemental maps. The major challenge to such an approach is physically obtaining a flat, contamination-free surface for analysis.

Optical and Photomicroscopy. Optical microscopic techniques have been used widely for the evaluation of A-C pipe corrosion and have been summarized and described (Schock et al., 1981; LeRoy et al., 1996). They have also been employed in the examination of particulates in drinking water (Walker, 1983) and in the examination of films on iron pipe (Singley et al., 1985; Benjamin et al., 1996). Photomicrographs have been extremely useful in examining the structure of the alloy and zinc layers on galvanized pipe (Trussell and Wagner, 1996).

X-Ray Diffraction. *X-ray diffraction* (XRD) is a procedure with which actual crystalline compounds can be identified in the scales, or aggregates of particulates. The fundamental principle behind the method is that when a highly collimated beam of X-rays of a particular wavelength strikes a finely ground, randomly oriented sample of material, the X-rays are diffracted in a pattern that is characteristic of the crystalline structure of the compound (Gould, 1980; Schock and Smothers, 1989; ASTM, 2003b). Mixtures of compounds give a pattern that is a sum of those compounds present. The analyzed patterns are compared to a reference library of known patterns for various solids, minerals, metals, and alloys. Particularly when used in conjunction with other analytical techniques to reduce the amount of pattern-searching necessary (by identifying the combinations of elements present or probable class of compounds—phosphates, silicates, oxides, and so forth—XRD provides a tremendous insight into pipe corrosion and corrosion inhibitor effects. For very crystalline materials, XRD can often detect down to 2 to 5 percent by weight, but it often fails to detect or identify solids that frequently occur as amorphous material, such as iron and manganese oxyhydroxides and aluminum phosphates. Microcrystallinity causes peak broadening which can hinder identification. Because XRD relies on crystallography, there are some minerals that have overlapping significant analytical peaks, making identification in mixtures of certain concentration combinations difficult (Schock et al., 2005a). An example is the misinterpretation of a combination of $PbCO_3$ and tetragonal PbO as appearing to be a trace to minor amount of β-PbO_2 (plattnerite). Sometimes solution chemical context is needed to accurately confirm identifications.

Figure 20-27 shows the easy differentiation between a dominantly β-PbO_2 (plattnerite) pipe scale from Newport, Rhode Island, and a dominantly $Pb_3(CO_3)_2(OH)_2$ (hydrocerussite) pipe scale from the Boston, Massachusetts area.

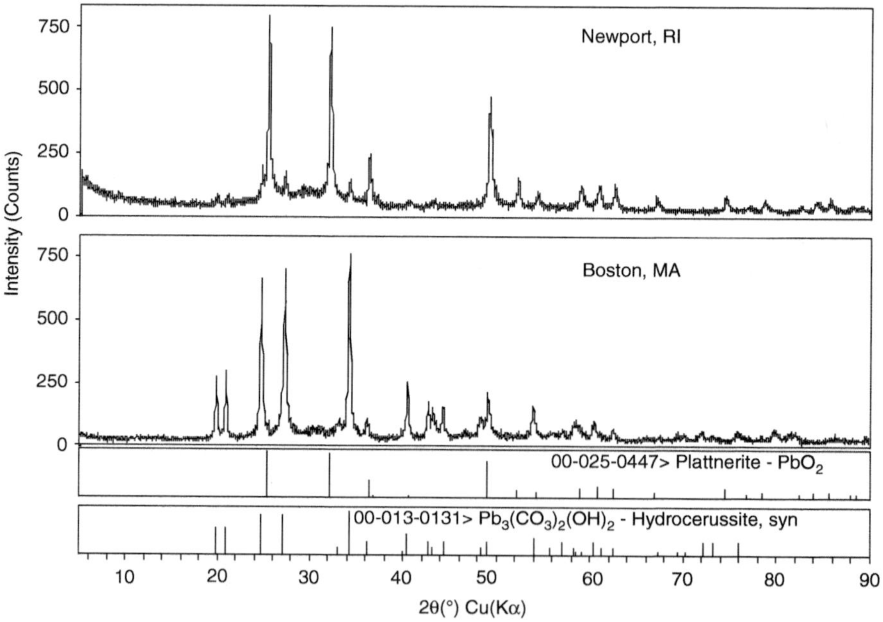

FIGURE 20-27 Comparison of powder X-ray diffraction patterns for a city with predominantly β-PbO_2 scales with a trace of $Pb_3(CO_3)_2(OH)_2$ (Newport, *top*) with the XRD pattern for a scale where $Pb_3(CO_3)_2(OH)_2$ is the dominant solid phase (Boston, *bottom*).

Several studies have reported many analyses of lead pipe by X-ray diffractometry (Elzenga and Graveland, 1981; van den Hoven, 1987; Schock and Smothers, 1989; van den Hoven and van Eekeren, 1990; Schock et al., 1996, 2001, 2005a; Lytle and Schock, 2005; Rego and Schock, 2007), and corroboration of copper solubility models and analysis of pitting phenomena by scale analyses using XRD have also been described (Schock et al., 1995; Lytle and Schock, 2008).

Synchrotron Methods (XANES, XAFS). Recently, powerful compound identification methods have been developed, that have excellent capability for valence state determination, compound determination, and even quantitative analysis. *X-ray absorption fine structure* (XAFS) refers to the interaction of how X-rays are absorbed by an atom at energies near and above the core-level binding energies of that particular atom. XAFS spectra are especially sensitive to the formal oxidation state, coordination chemistry, and the interatomic distances, coordination number, and species of the atoms in the surrounding proximity of the selected element of interest. As a result, XAFS provides a practical and simple way to determine the chemical state and local atomic structure for a selected atomic species. There is significant chemical data in the spectral region near the absorption edge, particularly the formal valence and coordination environment. The valence is identifiable by the position of the maximum edge energy by taking the derivative in this region, which is referred to as *X-ray absorption near-edge spectrocopy* (XANES). Employing a statistical spectral analysis methodology (linear combination fitting) facilitates a quantitative analysis capability with appropriate reference standards (Scheckel et al., 2005).

While environmental applications are voluminous, these analytical methods have only been infrequently applied to drinking water (Frenkel and Korshin, 1999; Schock and Giani, 2004; Schock et al., 2005a; Gerke et al., 2009). They are likely to achieve more use in the future, however, because of common ambiguities in the interpretation of X-ray diffraction, IR, and Raman spectra in pipe scales.

CORROSION CONTROL ALTERNATIVES

The complete elimination of corrosion is difficult if not impossible. Technologies generally exist to reduce or inhibit corrosion, though the determination of feasibility may depend mostly on economics or risk/aesthetic trade-offs. Corrosion depends on both the specific water quality and pipe material in a system, and a particular method may be successful in one system and not in another or even show different results in different parts of a distribution system of a single utility or water supplier. The most common ways of achieving corrosion control are to (1) properly select system materials and adequate system design, (2) modify water quality, (3) use inhibitors (may require pH or DIC adjustment as well), (4) provide cathodic protection (mainly for external corrosion), and (5) use corrosion-resistant linings, coatings, and paints.

Materials Selection

In many cases, corrosion can be reduced by properly selecting system materials and having a good engineering design (Singley et al., 1984; AWWA, 1986). When selecting materials for replacing old lines, putting new lines in service, or rehabilitating existing pipes (e.g., cement lining), the utility should select a material that will not corrode in the water it contacts. Putting a freshly cleaned cast-iron line or a freshly cleaned and cement-relined pipe back into service in a soft and poorly buffered water can cause serious water quality problems. It also undermines the rehabilitation investment just made. To protect sizable investments in rehabilitation, the utility should always evaluate the cost on the basis of integrating distribution system corrosion control adjustments with the cleaning, relining, or replacement. A realistic assessment of the impact of the water quality and the resources available to appropriately modify treatment to protect the newly installed materials in the future is often neglected, but is a wise strategy for long-term economy.

First and foremost, the water chemistry must be examined to determine which material or materials are appropriate. For example, many high-alkalinity waters with high sulfate and neutral to slightly acidic pH is incompatible with copper pipe. Many soft waters are basically incompatible with galvanized pipe, although it provides excellent performance in many hard water situations that are problematic for copper pipe.

The drinking water utility may not have control over the selection and installation of the materials for household plumbing, but should not overlook the possibility of working to influence plumbing codes to be most compatible with the water treatment options available.

Engineering Considerations

Some of the important design considerations for plumbing systems of all sizes include (1) avoiding dead ends and stagnant areas, (2) providing adequate drainage where needed, (3) selecting an appropriate flow velocity, (4) selecting an appropriate metal thickness, (5) eliminating shielded areas, (6) reducing mechanical stresses, (7) avoiding uneven heat distribution, (8) avoiding sharp turns and elbows, (9) providing adequate insulation, (10) choosing a proper shape and geometry for the system, (11) providing easy access to the structure for periodic inspection, maintenance, and replacement of damaged parts, and (12) eliminating grounding of electrical circuits to the system (Singley et al., 1984; AWWA, 1986).

Many plumbing codes are outdated and allow undesirable situations to exist. Such codes may even create problems, such as with grounding (Bell et al., 1995a amd b) and in the allowance or requirement of materials which could be readily corroded by a given water, but for which corrosion control by central water treatment could be difficult. Water utilities seeking to specify low-lead or genuinely lead-free materials for distribution system devices (pressure reducers, control valves, meters, etc.) may not be able to find them currently permitted by applicable state or local codes and industry standards.

Water systems need to be actively engaged with the responsible government agencies and plumbing code bodies to modify outdated codes, to tailor them to reduce treatment costs and protect the health of consumers, and to enforce voluntary standards in the sale and distribution of plumbing materials, such as NSF/ANSI 61, Secs. 8 and 9, as well as the enforcement of the lead solder ban. Surveys since 2001 in Pennsylvania have uncovered a rising number of occurrences of sale of prohibited lead solders (Goss, 2008), and it is widely known that plumbing inspectors rarely examine household and building plumbing devices to determine certification with passing the metal leaching requirements of NSF/ANSI 61.

On a larger scale, a poorly designed distribution system can produce areas of locally poor water quality, loss of disinfection, turbidity, taste, and odor problems. Renovation and rerouting of pipe lines may be more effective in solving water quality problems than treatment in some cases. Consolidating multiple distribution entry points can enable much better consistency and control of water quality. Many references review and discuss construction and design issues that relate to corrosion control (Singley et al., 1984; James M. Montgomery Consulting Engineers, 1985; AWWA, 1986; Medlar and Kim, 1993a and b; Schock et al., 2002; MWH, 2005).

Water Chemistry Modification

Water chemistry adjustment most commonly consists of the feeding single or combinations of solid, liquid, or gaseous chemicals to achieve desired target levels of parameters, such as pH, alkalinity, corrosion inhibitor residual, or buffer intensity, in the finished water and throughout the distribution system. Details of how to properly set-up and operate such processes are beyond the scope of this chapter.

Numerous corrosion control studies have identified mixing zones, or areas of a distribution system where water quality continually shifts, sometimes back and forth, as particularly problematic. Stable passivating films cannot form in these areas, as they are constantly being deposited and redissolved by the water quality changes. Employing any water chemistry adjustment requires a commitment to continually maintaining a consistent operation and prior investigation to ensure the absence of unintended detrimental side effects.

Two treatment approaches that have found growing successful application for the control of corrosion and metal release and that do not consist of the feeding of chemicals are *aeration* and *limestone contactors*. To be more precise, limestone contactors belong to a family of processes that can better be termed to be *neutralization*, as any of several similar natural and solid media can be used. In the United States, the concept of neutralization using filter beds consisting of or amended with limestone has been widely presented since the 1930s. Often mistakenly considered primarily a technology only for small water systems, some use of neutralization beds for acidic waters has been applied to a diverse size array of water systems (Cox, 1933; Letterman et al., 1987; Benjamin et al., 1992; Smith and Bager, 1995; Mackintosh et al., 1998; Spencer, 1998; Kettunen and Keskitalo, 2000; de Souza et al., 2002; Jalil et al., 2002; Rooklidge and Ketchum, 2002; Hammarstrom et al., 2003; Withers, 2005). An additional important application is to provide substantial pH and carbonate adjustment posttreatment for demineralizing processes, such as nanofiltration or reverse osmosis. Supplementing the product water with either orthophosphate or additional pH adjustment to ensure low corrosivity may be necessary. Nonetheless, it is an effective and relatively low-maintenance type of process that can provide a simplified consistent background that is amenable to minor chemical additions. Several good sources are readily available to the design engineer to provide computational support for sizing, final water quality, and contact time considerations (Letterman et al., 1987, 1991; Letterman, 1995; Letterman and Kathari, 1996; Jalil et al., 2002; Robinson and Carmical, 2005).

Aeration offers an excellent option for systems with low source water pH and an abundance of dissolved CO_2 in the water, though it is historically associated more with VOC or radon removal than with its usefulness for corrosion control (Lytle et al., 1998a–c). Typically, pH can be raised to the range of 7.8 to 8.2 without scaling, and DIC can be held at a level compatible with plumbosolvency and cuprosolvency control. For water systems with source water radon, it offers a particularly useful dual treatment option (Spencer, 1998; Schock et al., 2002), and the final pH achievable is automatically in a nearly ideal range for combining with orthophosphate addition. Aeration systems most commonly come in four configurations, all of which can be useful depending on the amount of CO_2 removal needed: Venturi injector systems, tray systems, packed tower systems, and diffuse bubble systems. The effectiveness of radon removal can concurrently be ensured and conveniently monitored by controlling the final pH (Spencer and Brown, 1997). In field application, particularly for smaller water systems, diffuse bubble aeration systems are relatively inexpensive to operate and more robust with respect to pH and DIC consistency than chemical feeds (Lytle et al., 1998a–c; Schock et al., 2002).

Secondary Effects of Treatment for Corrosion Control

A utility must meet a broad spectrum of regulatory and aesthetic water quality goals, aside from those associated with corrosion and corrosion by-products. Therefore, the potential effects of different corrosion control strategies must enter into the determination of the best course to follow. There are both positive and negative effects of corrosion control with respect to other regulations (Schock, 1999b).

Adjustment of pH may affect the ability of coagulation to remove turbidity or organic matter effectively, as can the adjustment of alkalinity or the addition of phosphate. The pH will also affect the formation of disinfection by-products. Consideration must be given to the impact of changes in water quality on industrial processes, building heating and cooling systems, wastewater loadings of metals (such as copper and zinc) and nutrients (such as phosphates), and on algal or plant growth in open storage reservoirs.

Certain treatment processes cause mineral imbalances in the water that can exacerbate corrosion control, without substantial posttreatment. Examples are nanofiltration, reverse osmosis, ion exchange, ozonation, sequestration, enhanced coagulation, enhanced softening optimized for NOM removal (insufficient hardness, and insufficient DIC for buffering), and possibly oxidative processes for iron and manganese removal.

On the other hand, there are particular benefits to a holistic corrosion control approach for the entire distribution system. Good corrosion control throughout the distribution system is necessary to optimize many aspects of distribution system water quality and performance, even aspects targeted by different specific regulations (Schock, 1999b). Examples are improved (1) aesthetic characteristics, (2) longevity of infrastructure, (3) disinfection and biofilm control through reduced chlorine demand and NOM sorption, (4) DBP formation through reduced chlorine demand, and (5) reduction of noxious constituents like Cu, Fe, Zn, and Pb in wastewater and sludges.

These multiple objectives can be achieved with proper attention to the broad view of treatment optimization throughout the whole distribution system. Although not necessarily simple and inexpensive, achieving good lead and copper corrosion control can be made chemically compatible with protecting the other distribution system piping materials. Further, this can be done while also meeting both other regulatory and aesthetic water quality requirements throughout the distribution system.

Linings, Coatings, and Paints

Another technique to keep corrosive water away from the pipe wall is to line the wall with a protective coating. These linings are usually mechanically applied, either when the pipe is manufactured or in the field before it is installed. Some linings can be applied even after the pipe is in service, though this method is much more expensive. The most common pipe linings are coal-tar enamels, epoxy paint, cement mortar, and polyethylene. The use of coatings must be carefully monitored, because they can be the source of several water quality problems, such as support of microbiological growth, taste and odor, and solvent leaching, and only coatings meeting ANSI/NSF and AWWA standards should be used. Although coal-tar based products have been widely used in the past for contact with drinking water, concern exists about their use because of the presence of polynuclear aromatic hydrocarbons and other hazardous compounds in coal tar and the potential for their migration in water.

The process of cement-mortar lining is historically the most common rehabilitation approach in the United States, but it can have important detrimental water quality impacts (LeRoy et al., 1996; Water Research Foundation, 2010). Like calcium carbonate films, the cement linings are far from inert. To protect the investment in time, money, and materials, as well as to avoid water quality degradation, a utility must critically examine alternative materials and not automatically select cement. In Europe, the use of in situ epoxy and polyurethane coatings is becoming far more widespread. A recent study has compared some of the water quality impacts of different lining materials (Water Research

Foundation, 2010). For some applications, in-situ lining products have been developed, consisting of collapsed tubing that can be inserted through even small-diameter pipes, and then expanded with heat and pressure to seal the pipe interior surface against water contact.

WATER SAMPLING FOR CORROSION CONTROL

The effects of corrosion may not be evident without monitoring. The effects can be expensive, and in the case of corrosion by-products such as lead, copper, and cadmium, they can be injurious to the health of segments of the population (see Chap. 2).

The first concern for a utility is to meet regulatory sampling requirements, in terms of the number of sampling sites, the location and frequency of the samples, and the use of approved and generally accepted analytical methods. Beyond these requirements, monitoring programs may address other questions. Some recent reviews discuss several aspects of investigating and monitoring for corrosion control and system studies (AwwaRF-DVGW Forschungsstelle, 1985; Schock et al., 1988; Schock and George, 1991; AwwaRF, 1994; Wysock et al., 1995; AwwaRF-TZW, 1996; AwwaRF, 1998).

Defining the Problem

Many factors are responsible for variability in metal concentrations that can limit the accurate assessment of exposure levels, treatment performance, and regulatory compliance. The goal of a sampling program must be to control as many of the analytical, chemical, and physical factors as possible, so that proper exposure assessments and decisions on the existence of water quality problems or the performance of a treatment program can be made.

The variability represents the actual range of exposures that occur routinely in the population. A monitoring system designed to accurately reflect exposure to lead or any other metal in drinking water must simultaneously capture that diversity while providing adequately reliable information to critically evaluate exposure. A monitoring program for corrosion control effectiveness must address similar concerns, although the criteria for sample size based on the desired levels of confidence in the mean values could be different. A study done in France obtained both good exposure estimates and good correspondence of lead levels to predictions of solubility models by a program of proportional sampling at many household sites (Anjou Recherche, 1994).

Another important piece of information that would be helpful in diagnosing and solving corrosion or metal release problems is determining the extent to which release of particulate corrosion by-products contributes to the observed metal levels (Cantor, 2006).

Human exposure evaluations require sampling considerations that have been frequently overlooked in health-effects studies. The sampling and monitoring structure under the U.S. drinking water regulations were never intended to represent human exposure estimates. To guard against a worst-case scenario, sampling locations and sampling collection protocols that are likely to coincide with occurrences of elevated by-product levels in drinking water to which consumers are likely to be exposed must be established. Several studies have investigated these factors (Schock et al., 1988; Schock, 1990; AwwaRF, 1994; Wysock et al., 1995; Reiber et al., 1996; AwwaRF, 1998), and research continues in that area.

Careful thought must be given to the selection of sites within the target groups. For example, when lead is of concern, sites can be lead interior plumbing or service lines, the most recent construction before the lead solder ban, or faucet replacements when lead-containing brass is used (remembering that lead-free was defined as no more than 8 percent). Aerators should be checked for the presence of trapped particulate material.

For copper, the age of the scale plays a major role in copper solubility. Hence, the most dangerous copper exposures would be in the newest construction, not covered as targeted monitoring sites under the Lead and Copper Rule.

The nature of exposure varies depending on the configuration of the plumbing system involved (e.g., single-family dwelling, apartment building, office building, school, etc.), as well as water usage pattern, contact time of the water with the plumbing, and water chemistry. Different sampling strategies are essential, and they must be carefully oriented toward the layout of the customer's system. At this time, the best sampling program design which adequately addresses this objective is proportional sampling, reported in studies from Canada, the Netherlands, and France (Meranger and Khan, 1983; van den Hoven, 1987; Anjou Recherche, 1994). The age of the plumbing materials, fittings, and devices in which the sampled water is in contact is important information, especially for copper and brasses (see the section, Corrosion of Specific Materials).

To determine whether the contamination is from a particular part of the plumbing system (e.g., faucets, soldered joints, parts of a service line, or pig-tail connector), sequential sampling can be conducted via the collection of small sample volumes, one immediately following the other. The isolation of a specific area of plumbing in question may also be obtained by selectively wasting an appropriate volume of water to get to the water parcel in contact with the area of interest, although it creates the complication that the water contacts other materials during flow to the sampling tap (Schock and Neff, 1988; AwwaRF, 1990; Wysock et al., 1995; Edwards et al., 2004; Sandvig et al., 2008). Guidelines were developed by USEPA, which are applicable to many other corrosion problems as well, to isolate locations of contamination in building (USEPA, 1988) or school (USEPA, 1989) drinking water systems. Table 20-2

TABLE 20-2 Distance-Volume Relationships for Some Common Residential Piping, Useful for Planning, Sampling, and Diagnosing Lead Sources

Material	Type	Nominal size (in)	ID (in)	ID (mm)	mL/ft	mL/m	ft/L	m/L
Copper tube	K	1/2	0.527	13.4	43	141	23.3	7.1
Copper tube	L	1/2	0.545	13.8	46	151	21.8	6.6
Copper tube	M	1/2	0.569	14.5	50	164	20.0	6.1
Copper tube	K	1/4	0.745	18.9	86	281	11.7	3.6
Copper tube	L	3/4	0.785	19.9	95	312	10.5	3.2
Copper tube	M	3/4	0.811	20.6	102	333	9.8	3.0
Copper tube	K	1	0.995	25.3	153	502	6.5	2.0
Copper tube	L	1	1.025	26.0	162	532	6.2	1.9
Copper tube	M	1	1.055	26.8	172	564	5.8	1.8
Galvanized steel†	Sched 40	1/2	0.622	15.8	60	196	16.7	5.1
Galvanized steel†	Sched 40	3/4	0.824	20.9	105	344	9.5	2.9
Galvanized steel†	Sched 40	1	1.049	26.6	170	558	5.9	1.8
Lead	¼-in wall	1/2	0.500	12.7	39	127	25.9	7.9
Lead	¼-in wall	5/8	0.625	15.9	60	198	16.6	5.1
Lead	¼-in wall	3/4	0.750	19.1	87	285	11.5	3.5
PVC, CPVC‡	Sched 80	1/2	0.546	13.9	46	151	21.7	6.6
PVC, CPVC‡	Sched 80	3/4	0.742	18.8	85	279	11.8	3.6
PVC, CPVC‡	Sched 80	1	0.957	24.3	141	464	7.1	2.2
Polyethylene (HDPE)‡	200 psi	1	1.023	26.0	162	530	6.2	1.9

†For old galvanized steel pipe, corrosion usually reduces the ID and volume considerably.
‡For plastic pipe, wall thickness can vary slightly, changing exact ID.

provides some translations of water volume to length of plumbing that is helpful in designing accurate sampling programs.

For schools, there are particularly difficult sampling issues, because of complex usage patterns and the considerable variability in age and configuration of plumbing from room to room, and often poor recording of the origin and model of the plumbing products installed. Multiple lead-containing devices may be located within one sample volume of an end-point tap, and they may have been installed over a long time period. There is essentially no such thing as a representative sampling site for most schools, necessitating sampling at virtually every location where drinking water could be obtained, to ensure adequate protection against lead contamination. Additionally, drinking water could be obtained from non-intuitive locations, such as showers or janitor closets, when large containers of drinking water are filled for activities such as sporting events.

A final aspect to water quality sampling is ensuring the complete chemical characterization of the water. Many direct and indirect reactions govern the corrosivity of the water and metal release, and the lack of complete water characterization causes several problems that are potentially more expensive than the cost of the analyses themselves. This fact has been lamented by Davis (1988) for groundwater studies, and his principles are equally applicable to drinking water problems. Background analyses of parameters that often are not covered directly in the regulations are important in defining the applicability of certain treatment processes and the selection of appropriate treatment. Because of the cost, it is tempting to wait to undertake these kinds of assessments until a specific regulation has been promulgated and a deadline set for initial monitoring. However, selection of appropriate treatment combinations for a water system requires as much information as possible about the raw water to be treated, to anticipate secondary impacts. A few hundred (or even in extreme cases, a few thousand) dollars in analytical costs at the beginning of a project are likely to pay huge dividends in terms of tens to hundreds of thousands of dollars or more saved in the long run by avoiding costly treatment choice mistakes and enabling a good integration of current and future treatments at one time. This benefit needs to be emphasized. Recommendations for analyses to be considered in corrosion and treatment evaluations are given in Table 20-3.

A corollary of obtaining sufficient analytical information for water characterization is that the data must be sufficiently accurate for the purposes. Even seemingly simple measurements, such as pH, alkalinity, and carbon dioxide concentrations are fraught with opportunities for analytical error, particularly with the training available for and instrumentation commonly employed by smaller water systems and many of their consultants. Errors of more than 0.5 pH units are common, in the experience of the authors. Poor data quality can mislead design engineers and regulators in determining and selecting the most economical and appropriate treatment scenarios for the water system. Thus, considerable emphasis should be put on determining and ensuring the quality of the data used for making expensive optimization or design decisions. If necessary, engineers should consult with aquatic chemists to help spot data that might be of dubious quality on the basis of known chemical interrelationships.

Statistical Considerations

The number of sites to be monitored for exposure evaluation or corrosion control must be related to the level of constituent or corrosion by-product variability over the proposed range of sites, and the relationship of that variability to the population mean (mean of all sites) (Schock et al., 1988; AwwaRF, 1990; Wysock et al., 1991, 1992, 1995; AwwaRF, 1994; Cantor, 2009). A good before-sampling program must be developed to obtain a baseline of data. The baseline is used to get a handle on the required number and frequency

TABLE 20-3 Recommended Analyses for a Thorough Corrosion Monitoring Program

General parameters for all investigations	
In situ measurements	pH, temperature, CO_2 if low-pH groundwater
Dissolved gases, oxidants	Oxygen, hydrogen sulfide,[†] free chlorine residual, total chlorine residuals (if ammonia present or used)
Parameters required to calculate $CaCO_3$-based indices[†]	Calcium, total hardness (or magnesium), alkalinity (or DIC), total dissolved solids (or conductivity)[‡]
Parameters for A-C pipe	Add to general parameters: Fiber count, iron, zinc, silica, polyphosphate, aluminum, manganese
Background parameters for metal pipe	Add to general parameters:
Iron or steel pipe	Iron, chloride, sulfate
Lead pipe or lead-based solder	Lead, copper, chloride, sulfate
Copper pipe	Copper, lead, sulfate, chloride
Galvanized-iron pipe	Zinc, iron, cadmium, lead, chloride
Brass (faucets and valves)	Zinc, copper, lead, sulfate
All metal pipes	Add to background parameters:
Corrosion inhibitor constituents (all metal pipes)	Orthophosphate, polyphosphate,[§] silica
Biofilm-related inorganics	Nitrate, nitrite, ammonia (if ammonia is present in source water or if chloramination is used)

[†]It only needs to be analyzed when suspected.
[‡]If a complete water analysis is done, TDS and conductivity can be neglected because ionic strength can be directly computed.
[§]For most potable waters; derived from analyzing total phosphate and subtracting orthophosphate.
[¶]When no inhibiting agents are present.

of sampling for the rest of the program and to enable valid comparisons to be made of water characteristics before and after changes in the system (e.g., treatment changes, design changes, source water changes). A multistage sampling design can sometimes be used to obtain information more economically (AwwaRF, 1990; Schock, 1990; Cantor, 2009).

One problem that arises is the practical issue of the proper statistical tests and the confidence levels for establishing whether two or more treatments or water chemistry conditions are showing equivalent effects on corrosion rates or metal release. Automatically using common statistical criteria, such as 90 or 95 percent confidence limits, may lead to misleading conclusions because of the nature of metal release (and possibly other corrosion test) data. A small episode of high variability or periodic episodes of widely different metal release may bias the statistical calculations of parametric or even nonparametric tests to indicate there is no difference in effect between or among test conditions (Wysock et al., 1991, 1992, 1995; AwwaRF, 1994). Simple examination of the data may show very consistently different behavior, except for the periodic excursions from what is clearly the most characteristic behavior. Therefore, purely statistical testing must be tempered with careful interpretation.

Chemical and Physical Considerations

Many important decisions are likely to be made on the basis of the sampling and chemical analyses performed by a utility or contract laboratory. Therefore, care must be taken during the sampling and analysis to obtain the best data. Handling of samples for pH, alkalinity, and CO_2 analyses often requires special precautions (Schock and Schock, 1982). Samples should be collected without adding air and with minimal agitation, as air tends to remove CO_2 and also affects the oxygen content in the sample. To collect a sample without additional air and to minimize the exchange of volatile gases (such as CO_2 in waters that are frequently out of equilibrium with the atmosphere), fill the sample container to the top so that a convex dome is formed at the opening and no bubbles are present. If possible, the sample bottle should be filled below the surface of the water using tubing, so that the water is not contaminated by the faucet material. Cap the sample bottle as soon as possible. For general purposes, high-density linear polyethylene bottles are very suitable for metals and most other water quality constituents. For pH measurements off-site, the bottles should be glass or a material ensured to be impermeable, especially toward carbon dioxide, and to a lesser extent, oxygen. Most plastic bottle manufacturers can provide tables of gas permeabilities for the plastics used in their products, which will be helpful for the selection of the proper bottles for the purpose.

When available, a direct analysis of DIC concentration (or TIC concentration, if unfiltered) is generally more accurate than computing DIC from pH and alkalinity measurements, unless particular care is taken in the analyses and sophisticated equivalence point and pH stabilization point techniques are employed (Schock and George, 1991; Schock and Lytle, 1994). Glass, or other material impermeable to CO_2, must be used for the container, and the cap must allow no air space.

To determine if metal release is primarily particulate or dissolved, frequently filtration is used. This is a deceptively difficult procedure, as there are numerous opportunities for erroneous data and biases that can significantly affect conclusions. Losses of dissolved metals to container walls, filters, and filtration apparatus will cause an overestimation of the contribution of particulates to the total metal release levels observed. Sorption losses are especially acute on glass materials and with cellulose acetate membranes, so polycarbonate, Teflon, or stainless steel apparatus and filter supports are preferable. Polycarbonate filter membranes have generally been shown to cause less bias than membranes made of most other materials, as have nylon syringe filters. Some cleaning and filter apparatus preparation precautions have been described in some studies (Schock and Gardels, 1983; Schock et al., 1995). Before using data from sample filtration, a careful laboratory and field test evaluation should be conducted of all materials used and all steps in the procedure to determine if there is a potential for losses of dissolved constituents or contamination from the materials.

When corrosion control studies obtain good baseline data and have a comprehensive, well-designed monitoring program in place throughout their corrosion control effort, the data are invaluable to other utilities and corrosion scientists. They can then work with a larger body of drinking water chemistry and treatment knowledge to implement future corrosion control and public health protection strategies much more effectively and efficiently.

ACKNOWLEDGMENTS

The authors gratefully acknowledge the extensive contributions of all of the writers of the AWWA publication *Corrosion Control For Operators*, the AwwaRF/DVGW-Forschungsstelle/TZW manual *Internal Corrosion of Water Distribution Systems*, and the late Dr. T. E. Larson, who originated this chapter and whose works guided much of the evolution of the contents. The authors are also indebted to the late Dr. Heinrich Sontheimer for inspiration for much of the copper research, Dr. Vernon Snoeyink for inspiration on the iron research, and to Dr. Marc Edwards for many interesting ideas, editorial comments, and discussion on lead, copper, corrosion, and historical perspectives. Michael K. DeSantis of

Pegasus Technical Services provided several illustrations and photographs, as well as many useful presentation ideas. John Consolvo of the Philadelphia Water Department contributed review and information for the solids analysis sections. Many improvements were brought about by the editorial and technical reviews from Dr. James Edzwald and France Lemieux (Health Canada). Robert N. Hyland of Pegasus Technical Services assisted greatly in the editorial review, equations, and layout of the chapter.

DISCLAIMER

The U.S. Environmental Protection Agency through its Office of Research and Development funded and managed the production of this chapter. It has not been subject to agency review and, therefore, does not necessarily reflect the views of the agency. Mention of trade names of commercial products and companies does not constitute endorsement or recommendation for use.

ABBREVIATIONS

A-C	asbestos-cement (pipe)
ANSI	American National Standards Institute
ASTM	American Society for Testing of Materials
AwwaRF	Awwa Research Foundation
CCPP	calcium carbonate precipitation potential
DIC	dissolved inorganic carbon, equivalent to $C_{T,C}$ in Chap. 3
DO	dissolved oxygen
LR	Larson ratio
LSI	Langelier saturation index
MCL	maximum contaminant level
NOM	natural organic matter
NSF	NSF International
ORP	oxidation-reduction potential
TALK	total alkalinity
TDS	total dissolved solids
USEPA	U.S. Environmental Protection Agency
WHO	World Health Organization

NOTATION FOR EQUATIONS

{ }	braces denote the *activity* of an aqueous species
[ ]	brackets denote the concentration of an aqueous species
β	buffer intensity, (mmol/L)/pH unit

E potential generated by an electrochemical reaction

$S_{T,X}$ total solubilitiy for species X

REFERENCES

Adeloju, S.B., and H.C. Hughes (1986) The Corrosion of Copper Pipes in High Chloride–Low Carbonate Water Mains, *Corrosion Science*, 26(10): 851–870.

Ainsworth, R.G. (1980) *The Introduction of New Water Supplies into Old Distribution Systems*, Technical Report, Report No. TR-143, Water Research Centre.

Akkaya, M., and J.R. Ambrose (1985) The Effect of Ammonium Chloride and Fluid Velocity on the Corrosion Behavior of Copper in Sodium Bicarbonate Solutions, *NACE Corrosion*, 41(12): 707–714.

Allen, M.J., R.H. Taylor, and E.E. Geldreich (1980) The Occurrence of Microorganisms in Water Main Encrustations, *Journal AWWA*, 72(11): 614–625.

Anjou Recherche (1994) *Etude des Phenomenes de Solubilisation du Plomb par l'Eau Distribuee et des Moyens Pour les Limiter*, Rapport de Synthese.

APHA, AWWA, and WEF (American Public Health Association, American Water Works Association, Water Environment Federation) (1995) *Standard Methods for the Examination of Water and Wastewater*, 19th ed. Washington, DC: American Public Health Association, and Denver, CO: American Water Works Association.

Appelo, C.A.J., and D. Postma (1999) A Consistent Model for Surface Complexation on Birnessite (δ-MnO_2) and Its Application to a Column Experiment, *Geochimica et Cosmochimica Acta*, 63(19–20): 3039–3048.

Appenzeller, B.M.R., Y.B. Duval, F. Thomas, and J.-C. Block (2002) Influence of Phosphate on Bacterial Adhesion onto Iron Oxyhydroxide in Drinking Water, *Environmental Science and Technology*, 36(4): 646–652.

Arens, P.L., G.-J. Tuschewitzki, H. Follner, H. Jacobi, and S. Leuner (1996) Experiments for Stimulating Microbiologically Induced Corrosion of Copper Pipes in a Cold-Water Plumbing System, *Werkstoffe und Korrosion*, 47 (2): 96–102.

Armstrong, R.D., and S. Zhou (1988) The Corrosion Inhibition of Iron by Silicate Related Materials, *Corrosion Science*, 28 (12): 1177–1181.

ASTM (2003a) Standard Practices for Examination of Water-Formed Deposits by Chemical Microscopy. vol. 11.02, D1245-1284, American Society for Testing and Materials, Conshohocken, PA.

ASTM (2003b) Standard Practices for Identification of Crystalline Compounds in Water-Formed Deposits by X-Ray Diffraction, Vol. 11.02, D934-980, American Society for Testing and Materials, Conshohocken, PA.

ASTM (2003c) Standard Practices for Sampling Water-Formed Deposits, Vol. 11.02, D887-882, American Society for Testing and Materials, Conshohocken, PA.

Atassi, A., C. Feizoulof, B. Melcher, V.L. Snoeyink, and P. Sarin (2004) *Corrosion Control Optimization Using Lead Pipe Loops*. Proc. AWWA Water Quality Technology Conference San Antonio, TX, November 14–18, 2004.

Atlas, D., J. Coombs, and O.T. Zajicek (1982) The Corrosion of Copper by Chlorinated Drinking Waters, *Water Res.*, 16: 693–698.

AWWA (American Water Works Association) (1971) *Water Quality and Treatment*; 3d ed. Denver, CO: American Water Works Association.

AWWA (1986) *Corrosion Control for Operators*. Denver, CO: American Water Works Association.

AwwaRF–DVGW-Forschungsstelle. (Awwa Research Foundation/DVGW-Forschungsstell) (1985) *Internal Corrosion of Water Distribution Systems*; Awwa Research Foundation/DVGW-Forschungsstelle report. Denver, CO: Water Research Foundation.

AwwaRF (1990) *Lead Control Strategies*, Awwa Research Foundation report. Denver, CO: Water Research Foundation.

AwwaRF. (1994) *Development of a Pipe Loop Protocol for Lead Control*. Awwa Research Foundation report. Denver, CO: Water Research Foundation.

AwwaRF–TZW (1996) *Internal Corrosion of Water Distribution Systems*; 2d ed.; Awwa Research Foundation/DVGW-TZW report. Denver, CO: Water Research Foundation.

AwwaRF (1998) *Water Quality Changes in the Distribution System Following the Implementation of Corrosion Control*. Awwa Research Foundation report. Denver, CO: Water Research Foundation.

AwwaRF (2002) *Development of Red Water Control Strategies*. Awwa Research Foundation report No. 90883. Denver, CO: Water Research Foundation.

Balzer, W. (1980) Calcium Carbonate Saturometry by Alkalinity Difference Measurement, *Oceanologica Acta*, *3* (2): 237–243.

Baylis, J.R. (1926a) Factors Other Than Dissolved Oxygen Influencing the Corrosion of Iron Pipes, *Industrial and Engineering Chemistry*, 18(4): 370–380.

Baylis, J.R. (1926b) Prevention of Corrosion and "Red Water," *Journal AWWA*, 15: 598.

Baylis, J.R. (1953) Cast-Iron Pipe Coatings and Corrosion, *Journal AWWA*, 45(8): 807–831.

Beckett, M.A., V.L. Snoeyink, K. Jim, P. Sarin, D. Lytle, W.M. Kriven, and J. Clement (1998) A Pipe Loop System for Evaluating Iron Uptake in Disrtibution Systems. AWWA Water Quality and Technology Conference, San Diego, CA, November 1–5.

Bell, G.E.C., M.J. Schiff, and S.J. Duranceau (1995a) *Review of the Effects of Electrical Grounding on Water Piping*. Corrosion/95, Orlando, FL, pp. 603/601–603/616.

Bell, G.E.C., M.J. Schiff, R.L. Bianchetti, and S.J. Duranceau (1995b) *Grounding Can Affect Water Quality*. Proc. AWWA Water Quality Technology Conference, New Orleans, LA, November 12–16, 1995, pp. 1789–1797.

Ben-Yaakov, S., and I.R. Kaplan (1971) Deep-Sea In Situ Calcium Carbonate Saturometry, *Journal of Geophysical Research*, 76(3): 722–730.

Benjamin, L., R.W. Green, and A. Smith (1992) Pilot Testing a Limestone Contactor in British Colombia, *Journal AWWA*, 84(5): 70–79.

Benjamin, M.M., H. Sontheimer, and P. Leroy (1996) Corrosion of Iron and Steel, in *Internal Corrosion of Water Distribution Systems*; pp. 29–70. Awwa Research Foundation/TZW report. Denver, CO: Water Research Foundation.

Berend, K., and T. Trouwborst (1999) Cemented Drinking Water Distribution Pipes as a Cause of Acute Aluminum Intoxication of Dialysis Patients, *Journal AWWA*, 91(7): 91–100.

Bethke, C.M., and S. Yeakel (2007) *The Geochemist's Workbench® Release 7.0 Reference Manual*. Golden, CO: Rockware, Inc.

Bockris, J.O.M., and A.K.N. Reddy (1973) *Modern Electrochemistry*, Vol. 2. New York: Plenum Press.

Boffardi, B.P., and A.M. Sherbondy (1991) Control of Lead Corrosion by Chemical Treatment, *NACE Corrosion*, 27(12): 966–975.

Boffardi, B.P. (1993) The Chemistry of Polyphosphate, *MP* (8): 50–53.

Boireau, A., G. Randon, and J. Cavard (1997) Positive Action of Nanofiltration on Materials in Contact with Drinking Water, *Journal of Water Supply: Research and Technology–Aqua*, 46(4): 210–217.

Borch, T., A.K. Camper, J.A. Biederman, P.W. Butterfield, R. Gerlach, and J.E. Amonette (2008) Evaluation of characterization techniques for iron pipe corrosion products and iron oxide thin films, *Journal of Environmental Engineering–ASCE*, 134(10): 835–844.

Breach, R., S. Crymble, and M.J. Porter (1991) *Systematic Approach to Minimizing Lead Levels at Consumers' Taps*. Proc. AWWA Annual Conference, Philadelphia, PA, June 23–27.

Bremer, P.J., and G.G. Geesey (1991a) An Evaluation of Biofilm Development Utilizing Non-Destructive Attenuated Total Reflectance Fourier Transform Infrared Spectroscopy, *Biofouling*, 3: 89–100.

Bremer, P.J., and G.G. Geesey (1991b) Laboratory-Based Model of Microbiologically Induced Corrosion of Copper, *Applied and Environmental Microbiology*, 57(7): 1956–1962.

Britton, A., and W.N. Richards (1981) Factors Influencing Plumbosolvency in Scotland, *Journal of the Institute of Water Engineers and Scientists*, 35(5): 349–364.

Brown, D.A., B.L. Sherriff, and J.A. Sawicki (1997) Microbial Transformation of Magetite to Hematite, *Geochimica et Cosmochimica Acta*, 61(16): 3341–3348.

Buhrke, V.E., R. Jenkins, and K.K. Smith (1998) *A Practical Guide for the Preparation of Specimens for X-ray Fluorescence and X-ray Diffraction Analysis.* New York: Wiley-VCH.

Butler, G., and H.C.K. Ison (1966) *Corrosion and Its Prevention in Waters.* London: Leonard Hill.

Butler, J.N. (1982) *Carbon Dioxide Equilibria and Their Applications.* Reading, MA: Addison-Wesley Publishing Co.

Butler, J.N., and D.R. Cogley (1998) *Ionic Equilibrium Solubility and pH Calculations.* New York: John Wiley and Sons.

Campbell, H.S. (1950) Pitting Corrosion in Copper Water Pipes Caused by Films of Carbonaceous Material Produced During Manufacture, *Journal of the Institute of Metals*, *77*, 345.

Campbell, H.S. (1971) Corrosion, Water Composition and Water Treatment, *Wat. Treat. Exam.*, 20(1): 11–34.

Camper, A.K. (1996) *Interaction between Pipe Materials, Corrosion Inhibitors, Disinfection, Organics, and Distribution System Biofilms.* Proc. National Conference on Integrating Corrosion Control and Other Water Quality Goals, Cambridge, MA, May 19–21.

Cantor, A.F., D. Denig-Chakroff, R.R. Vela, M.G. Oleinik, and D.L. Lynch (2000) Use of Polyphosphate in Corrosion Control, *Journal AWWA*, 92(2): 95–102.

Cantor, A.F. (2006) Diagnosing Corrosion Problems Through Differentiation of Metal Fractions, *Journal AWWA*, 98(1): 117–126.

Cantor, A.F., J.B. Bushman, M.S. Glodoski, E. Kiefer, R. Bersch, and H. Wallenkamp (2006) Copper Pipe Failure by Microbiologically Influenced Corrosion, *Materials Performance*, 62(6): 38–41.

Cantor, A.F. (2009) *Water Distribution System Monitoring: A Practical Approach for Evaluating Drinking Water Quality.* Boca Raton, FL: CRC Press.

Cardew, P.T. (2009) Measuring the Benefit of Orthophosphate Treatment on Lead in Drinking Water, *Journal of Water and Health,* 7(1): 123–131.

Cerrato, J.M., L.P. Reyes, C.N. Alvarado, and A.M. Dietrich (2006) Effect of PVC and Iron Materials on Mn(II) Deposition in Drinking Water Distribution Systems, *Water Research*, 40(14): 2720–2726.

Clement, J., M. Hayes, P. Sarin, W.M. Kriven, J. Bebee, K. Jim, M. Beckett, V.L. Snoeyink, G.J. Kirmeyer, and G. Pierson (2002) *Development of Red Water Control Strategies*; Awwa Research Foundation report. Denver, CO: Water Research Foundation.

Clement, J.A., W.J. Daly, H.J. Shorney, and A.J. Capuzzi (1998) An Innovative Approach to Understanding and Improving Distribution System Water Quality. Proc. AWWA Water Quality Technology Conference, San Diego, CA, November 1–4.

Colling, J.H., P.A.E. Whincup, and C.R. Hayes (1987) The Measurement of Plumbosolvency Propensity to Guide the Control of Lead in Tapwaters, *Journal of the Institute of Water and Environment Management*, 1(3): 263–269.

Colling, J.H., B.T. Croll, P.A.E. Whincup, and C. Harward (1992) Plumbosolvency Effects and Control in Hard Waters, *Journal of the Institute of Water and Environment Management*, 6(6): 259–268.

Cook, J.B. (1997) Achieving Optimum Corrosion Control for Lead in Charleston, S.C.: A Case Study, *Journal of the New England Water Works Association*, 111(2): 168–179.

Copeland, R.C., D.A. Lytle, and D.D. Dionysiou (2007) Desorption of Arsenic from Drinking Water Distribution System Solids, *Environmental Monitoring and Assessment*, 127: 523–535.

Cornwell, F.J., G. Wildsmith, and P.T. Gilbert (1973) Pitting Corrosion in Copper Tubes in Cold Water Service, *Br. Corros. J.*, 8(9): 202–209.

Cox, C.R. (1933) The Removal of Aggressive Carbon Dioxide by Contact Beds of Limestone or Marble, *Journal AWWA*, 25(11): 1505–1522.

Cruse, H., and R.D. Pomeroy (1974) Corrosion of Copper Pipes, *Journal AWWA*, 66(8): 479–483.

Dart, F.J., and P.D. Foley (1970) Preventing Iron Deposition with Sodium Silicate, *Journal AWWA*, 62(10): 663–668.

Dart, F.J., and P.D. Foley (1972) Silicate as Fe, Mn Deposition Preventative in Distribution Systems, *Journal AWWA*, 64(4): 244–249.

Davidson, C.M., N.J. Peters, A. Britton, L. Brady, P.H.E. Gardiner, and B.D. Lewis (2004) Surface Analysis and Depth Profiling of Corrosion Products Formed in Lead Pipes Used to Supply Low Alkalintiy Drinking Water, *Water Science and Technology*, 49(2): 49–54.

Davis, S.N. (1988) Where Are the Rest of the Analyses? *Ground Water*, 26(1): 2–5.

de Mora, S.J., and R.M. Harrison (1984) Lead in Tap Water: Contamination and Chemistry, *Chemistry in Britain*, (10), 901–904.

de Mora, S.J., R.M. Harrison, and S.J. Wilson (1987) The Effect of Water Treatment on the Speciation and Concentration of Lead in Domestic Tap Water Derived From a Soft Upland Source, *Water Resources*, 21(1): 83–94.

de Souza, P.F., G.J. du Plessis, and G.S. Mackintosh (2002) *An Evaluation of the Suitability of the Limestone Based Sidestream Stabilisation Process for Stabilisation of Waters of the Lesotho Highlands Scheme*. Biennial Conference of the Water Institute of Southern Africa, Durban, FSA, May 19–23.

DeSantis, M.K., M.M. Welch, and M.R. Schock (2009) *Mineralogical Evidence of Galvanic Corrosion in Domestic Drinking Water Pipes*. Proc. AWWA Water Quality Technology Conference, Seattle, WA, November 15–19.

Dexter, S.C., and S.H. Lin (1992) Calculation of Seawater pH at Polarized Metal Surfaces in the Presence of Surface Films, *NACE Corrosion*, 48(1): 50–60.

Dodrill, D.M., and M. Edwards (1995) Corrosion Control on the Basis of Utility Experience, *Journal AWWA*, 87(7): 74–85.

Drogowska, M., L. Brossard, and H. Menard (1987) Anodic Copper Dissolution in the Presence of Cl^- Ions at pH 12, *NACE Corrosion*, 43(9): 549–552.

Dudi, A., and M. Edwards (2005) *Galvanic Corrosion of Lead Bearing Plumbing Devices*. Proc. AWWA Annual Conference and Exposition, San Francisco, CA, June 12–16.

Dudi, A., M. Schock, N. Murray, and M. Edwards (2005) Lead Leaching from Inine Brass Devices: A Critical Evaluation of the Existing Standard, *Journal AWWA*, 97(8): 66–78.

Duthil, J.-P., G. Mankowski, and A. Giusti (1996) The Synergetic Effect of Chloride and Sulphate on Pitting Corrosion of Copper, *Corrosion Science*, 38(10): 1839–1849.

Dye, J.F. (1964) Review of Anticorrosion Water Treatment, *Journal AWWA*, 56(4): 457–465.

Eary, L.E., and J.A. Schramke (1990) Rates of Inorganic Oxidation Reactions Involving Dissolved Oxygen. In *Chemical Modeling of Aqueous Systems II*, ed. D.C, Melchior and R.L. Bassett, pp. 379–396. Washington, DC: American Chemical Society.

Edwards, M., J. Rehring, and T. Meyer (1994a) Inorganic Anions and Copper Pitting, *NACE Corrosion*, 50(5): 366–372.

Edwards, M., J.F. Ferguson, and S.H. Reiber (1994b) On the Pitting Corrosion of Copper, *Journal AWWA*, 86(7): 74–90.

Edwards, M., T.E. Meyer, and M.R. Schock (1996) Alkalinity, pH and Copper Corrosion By-Product Release, *Journal AWWA*, 88(3): 81–94.

Edwards, M., S. Jacobs, and D. Dodrill (1999) Desktop Guidance in Mitigation of Pb and Cu Corrosion By-Products, *Journal AWWA*, 91(5): 66–77.

Edwards, M., K. Powers, M.R. Schock, and L. Hidmi (2000) *Role of Pipe Aging in Copper Corrosion By-Product Release*. Workshop on Pipe Material Selection for Drinking Water Systems, Sustainable Drinking Water Distribution Management, Goteborg, SWED.

Edwards, M., L.S. McNeill, T.R. Holm, and M.C. Lawrence (2001) *Role of Phosphate Inhibitors in Mitigating Lead and Copper Corrosion*, Report No. 90823. Denver, CO: Water Reseaerch Foundation.

Edwards, M., and L.S. McNeill (2002) Effect of Phosphate Inhibitors on Lead Release, *Journal AWWA*, 94(1): 79–90.

Edwards, M. (2004) Controlling Corrosion in Drinking Water Distribution Systems: A Grand Challenge for the 21st Century, *Water Science and Technology*, 49(2): 1–8.

Edwards, M., and A. Dudi (2004) Role of Chlorine and Chloramine in Corrosion of Lead-Bearing Plumbing Materials, *Journal AWWA*, 96(10): 69–81.

Edwards, M., R. Giani, J. Wujek, and C. Chung (2004) *Use of Lead Profiles to Determine Source of Action Level Exceedences from Residential Homes in Washington, DC*. Sunday Workshop, AWWA Water Quality Technology Conference, San Antonio, TX.

Edwards, M., and S. Triantafyllidou (2007) Chloride-to-Sulfate Mass Ratio and Lead Leaching to Water, *Journal AWWA*, 99(7): 96–109.

Eisnor, J.D., and G.A. Gagnon (2003) A Framework for the Implementation and Design of Pilot-Scale Distribution Systems, *Journal of Water Supply, Research, and Technology—AQUA*, 52(7): 501–519.

Eisnor, J.D., and G.A. Gagnon (2004) Impact of Secondary Disinfection on Corrosion in a Model Water Distribution System, *Journal of Water Supply, Research, and Technology—AQUA*, 53: 441–452.

Elfström Broo, A., B. Berghult, and T. Hedberg (1997) Copper Corrosion in Drinking Water Distribution Systems–The Influence of Water Quality, *Corrosion Science*, 39(6): 1119–1132.

Elzenga, C.H., and A. Graveland (1981) Proposal of the European Communities for a Directive Relating to the Quality of Water for Human Consumption, Actual Metal Levels in Drinking Water and Some Possibilities for Reduction by Central Water Conditioning by Waterworks, *KIWA and Gemeentewaterleidingen*, Amsterdam, NETH.

Ergun, M., and A.Y. Turan (191) Pitting Potential and Protection Potential of Carbon Steel for Chloride Ion and the Effectiveness of Different Inhibiting Anions, *Corrosion Science*, 32(10): 1137–1142.

Faust, S.D., and O.M. Aly (1983) *Chemistry of Water Treatment*. Woburn, MA: Butterworth Publishers.

Feng, Y., W.-K. Teo, K.-S. Siow, K.-L. Tan, and A.-K. Hsieh (1996a) The Corrosion Behaviour of Copper in Neutral Tap Water. Part I. Corrosion Mechanisms, *Corrosion Science*, 38(3): 369–385.

Feng, Y., W.-K. Teo, K.-S. Siow, K.-L. Tan, and A.-K. Hsieh (1996b) The Corrosion Behaviour of Copper in Neutral Tap Water. Part II. Determination of Corrosion Rates, *Corrosion Science*, 38(3): 387–395.

Ferguson, J.L., O. von Franqué, and M.R. Schock (1996) Corrosion of Copper in Potable Water Systems. In *Internal Corrosion of Water Distribution Systems*, 2d ed., Awwa Research Foundation/DVGW-TZW, Denver, CO, pp. 231–268.

Fischer, W., H.H. Paradies, D. Wagner, and I. Hänßel (1992) Copper Deterioration in a Water Distribution System of a County Hospital in Germany Caused by Microbially Induced Corrosion-I. Description of the Problem, *Werkstoffe und Korrosion*, 43: 56–62.

Fox, K.P., C. Tate, and E. Bowers (1983) The Interior Surface of Galvanized Steel Pipe: A Potential Factor in Corrosion Resistence, *Journal AWWA*, 75(2): 84–86.

Fox, K.P., C.H. Tate, G.P. Treweek, R.R. Trussell, A.E. Bowers, M.J. McGuire, and D.D. Newkirk (1986) *Copper Induced Corrosion of Galvanized Steel Pipe*. EPA Report, EPA/600/2-86/056. Cincinnati, OH: USEPA Drinking Water Research Division.

Frenkel, A.I., and G.V. Korshin (1999) EXAFS Studies of the Chemical State of Lead and Copper in Corroison Products Formed on the Brass Surface in Potable Water, *Journal of Synchrotron Radiation*, 6: 653–655.

Friedman, M.J., A.S. Hill, S.H. Reiber, R.L. Valentine, and G.V. Korshin (2009) *Assessment of Inorganics Accumulation in Drinking Water System Scales and Sediments*. Denver, CO: Water Research Foundation.

Fuge, R., and W. Perkins. (1991) Aluminium and Heavy Metals in Potable Waters of the North Ceredigion Area, Mid-Wales, *Environmental Geochemistry and Health*, 13(2): 56–65.

Fuge, R., N.J.G. Pearce, and W.T. Perkins (1992) Unusual Sources of Aluminum and Heavy Metals in Potable Waters, *Environmental Geochemistry and Health*, 14(1): 15–18.

Garrels, R.M., and C.L. Christ (1965) *Solutions, Minerals and Equilibria.* New York: Harper and Row.

Geesey, G.G., M.W. Mittelman, T. Iwaoka, and P.R. Griffiths (1986) Role of Bacterial Exopolymers in the Deterioration of Metallic Copper Surfaces, *Materials Performance*, 1986: 37–40.

Geesey, G.G., L. Jang, J.G. Jolley, M.R. Hankins, T. Iwaoka, and P.R. Griffiths (1988) Binding of Metal Ions by Extracellular Polymers of Biofilm Bacteria, *Water Science Technology*, 20(11–12): 161–165.

Geesey, G.G., and P.J. Bremer (1991) *Evaluation of Copper Corrosion Under Bacterial Biofilms.* NACE Corrosion/91, Cincinnati, OH, pp. 1–7.

Gerke, T.L., K.G. Scheckel, and M.R. Schock (2009) Identification and Distribution of Vanadinite ($Pb_5(V^{5+}O_4)_3Cl$) in Lead Pipe Corrosion By-Products, *Environmental Science & Technology*, 43(12): 4412–4418.

Giani, R., M. Donnelly, and T. Ngantcha (2005) *The Effects of Changing Between Chloramine and Chlroine Disinfectants on Lead Leaching*. Proc. AWWA Water Quality Technology Conference, Quebec City, CAN, November 6–10.

Goidanich, S., L. Lazzari, and M. Ormelles (2010) AC Corrosion. Part 1. Effects on Overpotentials of Anodic and Cathodic Processes, *Corrosion Science*, 52(2): 491–497.

Goss, C. (2008) *Pennsylvania Department of Environmental Protection Lead Ban Surveillance Project 2008*, Pennsylvania Department of Environmental Protection, Bureau of Water Standards and Facility Regulation: 2008; available online at www.depweb.state.pa.us/watersupply/lib/watersupply/pb_ban_rpt_2008.pdf.

Gould, R.W. (1980) *The Application of X-Ray Diffraction to the Identificaiton of Corrosion Products*. Proc. AWWA Water Quality Technology Conference, Miami Beach, FL., December 7–10, pp. 235–244.

Gregory, R., and P.J. Jackson (1984) *Central Water Treatment to Reduce Lead Solubility*. Proc. AWWA Annual Conference, Dallas, TX, June 10–14.

Gregory, R. (1990) Galvanic Corrosion of Lead Solder in Copper Pipework, *Journal of the Institute of Water and Environment Management*, 4(2): 112–118.

Griffith, E.J. (1972) *The Chemistries and Applications of Phosphates, Silicates, and Sequestering Agents in Water Treatment*. 14th Water Quality Conference, University of Illinois, pp. 115–125.

Grigg, N.S. (2010) *Secondary Impacts of Corrosion Control on Distribution System and Treatment Plant Equipment*, Water Research Foundation, Denver, CO.

Guo, Q., P.J. Toomuluri, and J.O. Eckert Jr. (1998) Leachability of Regulated Metals from Cement-Mortar Linings, *Journal AWWA*, 90(3): 62–73.

Hammarstrom, J.M., P.L. Sibrell, and H.E. Belkin (2003) Characterization of Limestone Reacted with Acid-Mine Drainage in a Pulsed Limestone Bed Treatment System at the Friendship Hill National Historical Site, Pennsylvania, U.S., *Applied Geochemistry*, 18(11): 1705–1721.

Harmon, S.M. (1999) *The Effects of Polyphosphate Sequestration on Plumbosolvency as Experienced by the Water Utility of Hopkinton, MA*. Proc. AWWA Water Quality Technology Conference, Tampa, FL, October 31–November 3.

Harrison, R.M., and D.P.H. Laxen (1980) Physicochemical Speciation of Lead in Drinking Water, *Nature*, 286(5775): 791–793.

Harwood, J. (1989) *Sequestration Characteristics of Iron Control Chemicals Used in Drinking Water Treatment Plants*, Final Report, Tennessee Technological University.

Hatch, G.B., and O. Rice (1940) Corrosion Control with Threshold Treatment, *Ind. Eng. Chem.*, 32: 1572–1579.

Hatch, G.B., and O. Rice. (1945a) Corrosion Control with Threshold Treatment, *Ind. Eng. Chem.*, 37(8): 710–715.

Hatch, G.B., and O. Rice. (1945b) Corrosion Control with Threshold Treatment, *Ind. Eng. Chem.*, 37: 752–759.

Hatch, G.B. (1952) Protective Film Formation with Phosphate Glasses, *Ind. Eng. Chem.*, 44: 1775–1780.

Hayes, C.R., A.J. Bates, A.H. Goodman, J.P. Vinson, and T.P. Sadler (1997) Meeting Standards for Lead in Drinking Water, *Journal of the Chartered Institute of Water Engineering Management*, 11(8): 257–263.

Hayes, C.R., A.J. Bates, L. Jones, A.D. Cuthill, D. Van Der Leer, and N.P. Weatherill (2006) Optimisation of Plumbosolvency Control Using a Computational Model, *Water and Environment Journal*, 20(4): 256–264.

Hayes, C.R., S. Incledion, and M. Balch (2008) Experience in Wales (UK) of the Optimization of Ortho-phosphate Dosing for Controlling Lead in Drinking Water, *Journal of Water and Health*, 6(2): 177–185.

Hazel, F. (1945) The Effect of Sodium Silicates on Iron Oxide Surfaces, *Journal of Physical Chemistry*, 49(6): 520–553.

Henry, C.R. (1950) Prevention of the Settlement of Iron, *Journal AWWA*, 42(9): 887–896.

Herrera, C.E., J.F. Ferguson, and M.M. Benjamin (1982) Evaluating the Potential for Contaminating Drinking Water From the Corrosion of Tin-Anitmony Solder, *Journal AWWA*, 74(7): 368–375.

Hidmi, L., and M. Edwards (1999) Role of Temperature and pH in $Cu(OH)_2$ Solubility, *Environmental Science and Technology*, 33(15): 2607–2610.

Hogan, I., and P.J. Jackson (1981) *The Examination of Lead Pipe Deposits by Infra Red Spectroscopy. Preliminary Report*, Report No. 81-S, Water Research Centre.

Holm, T.R., and S.H. Smothers (1990) Characterizing the Lead-Complexing Properties of Polyphosphate Water Treatment Products by Competing-Ligand Spectrophotometry Using 4-(2-Pyridylazo) Resorcinol, *International Journal of Environmental Analytical Chemistry*, 41: 71–82.

Holm, T.R., and M.R. Schock (1991a) Polyphosphate Debate (Reply to Comment). *Journal AWWA*, 83(12): 10–12.

Holm, T.R., and M.R. Schock (1991b) Potential Effects of Polyphosphate Products on Lead Solubility in Plumbing Systems, *Journal AWWA*, 83(7): 74–82.

Hoover, C.P. (1938) Practical Application of the Langelier Method, *Journal AWWA*, 30(11): 1802–1807.

Horton, A.M., and R.E. Behnke (1989) *AC-DC Effects of Electrical Grounding on Water Pipe*. Proc. AWWA Annual Conference, Los Angeles, CA, June 18–22.

Horton, A.M. (1991) *Corrosion Effects of Electrical Grounding on Water Pipe*. Corrosion/91, Cincinnati, OH, pp. 519/511–519/520.

Hulsmann, A.D. (1990) Particulate Lead in Water Supplies, *Journal of the Institute of Water and Environment Management*, 4(2): 19–25.

Iler, R.K. (1979) *The Chemistry of Silica.* New York: Wiley-Interscience.

Illig, G.L. (1960) Use of Sodium Hexametaphosphate in Manganese Stabilization, *Journal AWWA*, 52(7): 867–873.

Ingle, S.E., C.H. Culberson, J.E. Hawley, and R.M. Pytkowicz (1973) The Solubility of Calcite in Seawater at Atmospheric Pressure and 35% Salinity, *Marine Chemistry*, 1: 295–307.

Ives, D.J.G., and A.E. Rawson (1962a) Copper Corrosion III. Electrochemical Theory of General Corrosion, *Journal of the Electrochemical Society*, 109(6): 458–462.

Ives, D.J.G., and A.E. Rawson (1962b) Copper Corrosion IV. The Effects of Saline Additions, *Journal of the Electrochemical Society*, 109(6): 462–466.

Iwaoka, T., P.R. Griffiths, J.T. Kitasako, and G.G. Geesey (1986) Copper-Coated Cylindrical Internal Reflection Elements for Investigating Interfacial Phenomena, *Applied Spectroscopy*, 4(7): 1062–1065.

Jacobs, S., S. Reiber, and M. Edwards (1998) Sulfide-Induced Copper Corrosion, *Journal AWWA*, 90(7): 62–73.

Jacobs, S.A., and M. Edwards (1997) *Sulfide-Induced Copper Corrosion*. Proc. AWWA Water Quality Technology Conference, Denver, CO, November 9–12.

Jalil, A., R.B. Robinson, R.D. Letterman, and G.S. Mackintosh (2009) *Small Public Water System Technology Guide Volume II: Limestone Contactor*, University of New Hampshire Water Treatment Technology Assistance Center, University of New Hampshire and University of Tennessee–Knoxville 2002; available online at www.unh.edu/erg/wttac/WTTAC_Water_Tech_Guide_Vol2/default.html (accessed May 12, 2009).

James, C.N., R.C. Copeland, and D.A. Lytle (2004) *Relationships between Oxidation-Reduction Potential, Oxidant, and pH in Drinking Water*. Proc. AWWA Water Quality Technology Conference, San Antonio, TX, November 14–18.

James M. Montgomery Consulting Engineers (1982) *Internal Corrosion Mitigation Study Final Report*. Portland, OR: Bureau of Water Works.

James M. Montgomery Consulting Engineers (1983) *Internal Corrosion Mitigation Study Addendum Report*. Portland, OR: Bureau of Water Works.

James M. Montgomery Consulting Engineers (1985) *Water Treatment Principles & Design*. New York: John Wiley and Sons.

Jolley, J.G., G.G. Geesey, M.R. Hankins, R.B. Wright, and P.L. Wichlacz (1988) Auger Electron Spectroscopy and X-ray Photoelectron Spectroscopy of the Biocorrosion of Copper by Gum Arabic, Bacterial Culture Supernatant and *Pseudomonas atlantica* Exopolymer, *Surface and Interface Analysis*, 11: 371–376.

Kasul, D.B., and L.A. Heldt (1993) *Characterization of Pitting Corrosion of Copper Pipe Carrying Municipal Water*. NACE Corrosion/93, Houston, TX, pp. 512/511–512/516.

Kenworthy, L. (1943) The Problem of Copper and Galvanized Iron in the Same Water System, *Jour. Inst. Metals*, 69,67–90.

Kettunen, R., and P. Keskitalo (2000) Combination of Membrane Technology and Limestone Filtration to Control Drinking Water Quality, *Desalination*, 131(1–3): 271–283.

Kiéné, L., W. Lu, and Y. Lévy (1996) *Relative Importance of Phenomena Responsible of the Chlorine Consumption in Drinking Water Distribution Systems*. Proc. AWWA Water Quality Technology Conference, Boston, MA, November 10–13.

Kimbrough, D.E. (2001) Brass Corrosion and the LCR Monitoring Program, *Journal AWWA*, 93(2): 81–91.

Kimbrough, D.E. (2007) Brass Corrosion as a Source of Lead and Copper in Traditional and All-plastic Distribution Systems, *Journal AWWA*, 98(8): 70–76.

Kimbrough, D.E. (2009) Source Identification of Copper, Lead, Nickel, and Zinc Loading in Wastewater Reclamation Plant Influents from Corrosion of Brass in Plumbing Fixtures, *Environmental Pollution*, 157(4): 1310–1316.

King, D.W. (1998) Role of Carbonate Speciation on the Oxidation Rate of Fe(II) in Aquatic Systems, *Environmental Science and Technology*, 32: 2997–3003.

Kirmeyer, G.J., G. Pierson, J.A. Clement, A. Sandvig, V.L. Snoeyink, W.M. Kriven, and A. Camper (1999) *Distribution System Water Quality Changes Following Corrosion Control Strategies*, AwwaRF report 90764. Denver, CO: Water Research Foundation.

Klueh, K.G., and R.B. Robinson (1988) Sequestration of Iron in Groundwater by Polyphosphates, *Journal of the Environmental Engineering Division, ASCE*, 114(5): 1192–1199.

Köningsberger, E., and H. Gamsjäger (1990) Solid–Solute Phase Equilibria in Aqueous Solution. III. A New Application of an Old Chemical Potentiometer, *Marine Chemistry*, 30: 317–327.

Korshin, G.V., S.A.L. Perry, and J.F. Ferguson (1996) Influence of NOM on Copper Corrosion, *Journal AWWA*, 88(7): 36–47.

Korshin, G.V., J.F. Ferguson, A.N. Lancaster, and H. Wu (1999) *Corrosion and Metal Release for Lead-Containing Materials: Influence of NOM*, Subject Area: AwwaRF Report No. 90759. Denver, CO: Water Research Foundation.

Korshin, G.V., J.F. Ferguson, and A.N. Lancaster (2000) Influence of Natural Organic Matter on the Corrosion of Leaded Brass in Potable Water, *Corrosion Science*, 42: 53–66.

Korshin, G.V., J.F. Ferguson, and A.N. Lancaster (2005) Influence of Natural Organic Matter on the Morphology of Corroding Lead Surfaces and Behavior of Lead-Containing Particles, *Water Research*, 39(5): 811–818.

Koudelka, M., J. Sanchez, and J. Augustynski (1982) On the Nature of Surface Films Formed on Iron in Agressive and Inhibiting Polyphosphate Solutions, *Journal of the Electrochemical Society*, 129(6): 1186–1191.

Kowata, K., and K. Takahashi (1996) *Interaction of Corrosion Inhibitors with Corroded Steel.* Corrosion/96, Houston, TX.

Kriewall, D., R. Harding, E. Naisch, and L. Schantz (1996) The Impact of Aluminum Residual on Transmission Main Capacity, *Public Works*, 127(12): 28.

Kristiansen, H. (1982) Corrosion of Copper by Soft Water with Different Content of Humic Substances and Various Temperatures, *Vatten*, 38, 181–188.

Kuch, A., and I. Wagner (1983) Mass Transfer Model to Describe Lead Concentrations in Drinking Water, *Water Research,* 17(10): 1303.

Kuch, A., V.L. Snoeyink, R.A. Ryder, and R.R. Trussell (1985) Experimental and Investigation Techniques. In *Internal Corrosion of Water Distribution Systems*; 2d ed., Chap. 9, pp. 657–700. Awwa Research Foundation/DVGW-Forschungsstelle report. Denver, CO: Water Research Foundation.

Kvech, S. and Edwards, M. (2001) Role of Aluminosilicate Deposits in Lead and Copper Corrosion, *Journal AWWA,* 93(11): 104–112.

Lagos, G.E., C.A. Cuadrado, and M. Victoria-Letelier (2001) Aging of Copper Pipes by Drinking Water, *Journal AWWA,* 93(11): 94–103.

Laitinen, H.A., and W.E. Harris (1975) *Chemical Analysis*. New York: McGraw-Hill.

Lamb, J.C., and R. Eliassen (1954) Mechanism of Corrosion Inhibition by Sodium Metaphosphate Glass, *Journal AWWA*, 46(5): 445–460.

Lane, R.W. (1993) *Control of Scale and Corrosion in Building Water Systems*. New York: McGraw-Hill.

Langelier, W.F. (1936) The Analytical Control of Anti-Corrosion Water Treatment, *Journal AWWA*, 28(10): 1500.

Langmuir, D. (1971) Particle Size Effect on the Reaction Goethite = Hematite + Water, *American Journal of Science*, 271: 147–156.

Langmuir, D. (1997) *Aqueous Environmental Geochemistry*. Upper Saddle River, NJ: Prentice Hall.

LaRosa-Thompson, J., B.E. Scheetz, M.R. Schock, D.A. Lytle, and P.J. Delaney (1997) *Sodium Silicate Corrosion Inhibitors: Issues of Effectiveness and Mechanism*. Proc. AWWA Water Quality Technology Conference, Denver, CO, November 9–13.

Larson, T.E., and R.V. Skold (1957) Corrosion and Tuberculation of Cast Iron, *Journal AWWA*, 49(10): 1294–1302.

Larson, T.E. (1975) *Corrosion by Domestic Waters*, Bulletin, Report No. 59, Illinois State Water Survey, Urbana, ILL.

Lauer, W.C., and S.R. Lohman (1994) *Non-Calcium Carbonate Protective Film Lowers Lead Values*. Proc. AWWA Water Quality Technology Conference, San Francisco, CA, November 6–10.

Lazaroff, N., W. Sigal, and A. Wasserman (1982) Iron Oxidaton and Precipitation of Ferric Hydroxysulfates by Resting *Thiobacillus ferrooxidans* Cells, *Applied and Environmental Microbiology*, 43(4): 924–938.

LeChevallier, M.W., C.D. Lowry, R.G. Lee, and D.L. Gibbon (1993) Examining the Relationship Between Iron Corrosion and the Disinfection of Biofilm Bacteria, *Journal AWWA*, 85(7): 111–123.

Lee, R.G., W.C. Becker, and D.W. Collins (1989) Lead at the Tap: Sources and Control, *Journal AWWA*, 81(7): 52–62.

Lee, S.H., J.T. O'Connor, and S.K. Banerji (1980) Biologically Mediated Corrosion and Its Effects on Water Quality in Distribution Systems, *Journal AWWA*, 72(11): 636–645.

Leggett, G.E. (1978) Interaction of Monomeric Silicic Acid with Copper and Zinc and Chemical Changes of the Precipitates with Aging, *Soil Sci. Soc. Am. J.*, 42, 262–268.

LeRoy, P. (1993) Lead in Drinking Water–Origins; Solubility; Treatment, *Journal of Water Suply: Research and Technology–Aqua*, 42(4): 233–238.

LeRoy, P., and J. Cordonnier (1994) Réduction de la Solubilité du Plomb par Décarbonation Partielle, *Journal Français d'Hydrologie*, 25(1): 81–96.

LeRoy, P. (1996) *Corrosion and Degradation of Pipes—Mechanisms and Remedies*. Union of African Water Supply Conference, Yaoundi, Cameroon.

LeRoy, P., M.R. Schock, H. Holtschulte, and I. Wagner (1996) Cement-Based Materials., In *Internal Corrosion of Water Distribution Systems,* 2d ed., Awwa Research Foundation/DVGW-Forschungsstelle report. Denver, CO: Water Research Foundation, Chap. 7.

Letterman, R.D., C.T. Driscoll Jr., M. Haddad, and H.A. Hsu (1987) *Limestone Bed Contactors for Control of Corrosion at Small Water Utilities.* Research and Development Report, EPA/600/S2-86/099, Water Engineering Research Laboratory, Cincinnati, OH.

Letterman, R.D., M. Haddad, and C.T. Driscoll Jr. Limestone Contactors (1991) Steady State Design Relationships, *Journal of the Environmental Engineering Division, ASCE*, 117(3): 339–358.

Letterman, R.D. (1995) *Calcium Carbonate Dissolution Rate in Limestone Contactors.* Research and Development Report, EPA/600/SR-95/068. Cincinnati, OH: Risk Reduction Engineering Laboratory.

Letterman, R.D., and S. Kathari (1996) A Computer Program for the Design of Limestone Contactors, *Journal of the New England Water Works Association*, 110(1): 42–47.

Lewandowski, Z., W. Dickinson, and W. Lee (1997) Electrochemical Interactions of Biofilms with Metal Surfaces, *Water Science Technology*, 36(1): 295–302.

Lin, Y.-P., and R.L. Valentine (2008a) The Release of Lead from the Reduction of Lead Oxide (PbO_2) by Natural Organic Matter, *Environmental Science and Technology*, 42(3): 760–765.

Lin, Y.-P., and R.L (2008b) Valentine. Release of Pb(II) from Monochloramine-Mediated Reduction of Lead Oxide (PbO_2), *Environmental Science and Technology*, 42(24): 9137–9143.

Little, B., and P. Wagner (1997) Myths Related to Microbiologically Influenced Corrosion, *Materials Performance*, 36(6): 40–44.

Liu, H., G.V. Korshin, and J.F. Ferguson (2008) Investigation of the Kinetics and Mechanisms of the Oxidation of Cerussite and Hydrocerussite by Chlorine, *Environ. Sci. Technol.*, 42(9): 3241–3247.

Loewenthal, R.E., and G.v.R. Marais (1976) *Carbonate Chemistry of Aquatic Systems.* Ann Arbor, MI: Ann Arbor Science Publishers.

Lucey, V.F. (1967) Mechanisms of Pitting Corrosion of Copper in Supply Waters, *Br. Corros. Jour.*, 2: 175–187.

Lyon, T.D.B., and J.M.A. Lenihan (1977) Corrosion in Solder Jointed Copper Tubes Resulting in Lead Contamination of Drinking Water, *British Corrosion Journal*, 12(1): 41–45.

Lyons, J.J., J. Pontes, and P. Karalekas (1995) *Optimizing Corrosion Control for Lead and Copper Using Phosphoric Acid and Sodium Hydroxide.* Proc. AWWA Water Quality Technology Conference, New Orleans, LA, November 12–16, pp. 2457–2473.

Lytle, D.A., M.R. Schock, and S. Tackett (1992) *Metal Corrosion Coupon Study Contamination, Design, and Interpretation Problems.* Proc. AWWA Water Quality Technology Conference, Toronto, CAN, November 15–19.

Lytle, D.A., M.R. Schock, N.R. Dues, and P.J. Clark (1993) Investigating the Preferential Dissolution of Lead from Solder Particulates, *Journal AWWA*, 85(7): 104–110.

Lytle, D.A., and M.R. Schock (1996) *Stagnation Time, Composition, pH and Orthophosphate Effects on Metal Leaching from Brass.* EPA/600/R-96/103, Office of Research and Development, Washington, DC; available online at www.epa.gov/nrmrl/pubs/600r96103/600R96103.pdf.

Lytle, D.A., M.R. Schock, and T.J. Sorg (1996) Controlling Lead Corrosion in the Drinking Water of a Building by Orthophosphate and Silicate Treatment, *Journal of the New England Water Works Association*, 110(3): 202–216.

Lytle, D.A., M.R. Schock, J.A. Clement, and C.M. Spencer (1998a) Using Aeration for Corrosion Control, *Journal AWWA*, 90(3): 74–88.

Lytle, D.A., M.R. Schock, J.A. Clement, and C.M. Spencer (1998b) Using Aeration for Corrosion Control–Erratum, *Journal AWWA*, 90(5): 4.

Lytle, D.A., M.R. Schock, J.A. Clement, and C.M. Spencer (1998c) Using Aeration for Corrosion Control–Erratum, *Journal AWWA*, 90(9): 4.

Lytle, D.A., and M.R. Schock (2000) Impact of Stagnation Time on Metal Dissolution from Plumbing Materials, *Journal of Water Supply: Research and Technology—AQUA*, 49(5): 243–257.

Lytle, D.A., and V.L. Snoeyink (2002) Effect of Ortho- and Polyphosphates on the Properties of Iron Particles and Suspensions, *Journal AWWA*, 94(10): 87–99.

Lytle, D.A., and V.L. Snoeyink (2003a) The Effect of Dissolved Inorganic Carbon on the Properties of Iron Colloidal Suspensions, *Journal of Water Supply: Research, and Technology–AQUA*, 52(3): 165–180.

Lytle, D.A., and V.L. Snoeyink (2003b) The Effect of pH and Dissolved Inorganic Carbon on the Properties of Iron Colloidal Suspensions, *Journal of Water Supply, Research, and Technology–AQUA*, 52(3): 165–180.

Lytle, D.A., P. Sarin, and V.L. Snoeyink (2003) *The Effect of Chloride and Orthophosphate on the Release of Iron from a Drinking Water Distribution System Cast Iron Pipe*. Proc. AWWA Water Quality Technology Conference, Philadelphia, PA, November 2–6.

Lytle, D.A., T.J. Sorg, and C. Frietch (2004a) Accumulation of Arsenic in Drinking Water Distribution Systems, *Environmental Science and Technology*, 38(20): 5365–5372.

Lytle, D.A., M.L. Magnuson, and V.L. Snoeyink (2004b) The Effect of Oxidant on the Properties of Fe(III) Particles and Suspensions Formed from the Oxidation of Fe(II), *Journal AWWA*, 96(8): 112.

Lytle, D.A., and M.R. Schock (2005) The Formation of Pb(IV) Oxides in Chlorinated Water, *Journal AWWA*, 97(11): 102–114.

Lytle, D.A., T.L. Gerke, and J.B. Maynard (2005a) Effect of Bacterial Sulfate Reduction on Iron-Corrosion Scales, *Journal AWWA*, 97(10): 109–120.

Lytle, D.A., P. Sarin, and V.L. Snoeyink (2005b) The Effect of Chloride and Orthophosphate on the Release of Iron from a Cast Iron Pipe Section, *Journal of Water Supply: Research, and Technology–AQUA*, 54(5): 267–281.

Lytle, D.A., and M.R. Schock (2006) *The Solubility and Surface Chemistry of Freshly Precipitated Copper Solids*. Proc. AWWA Annual Conference and Exposition, San Antonio, TX, June 11–15.

Lytle, D.A., and M.R. Schock (2008) Pitting of Copper in High pH and Low Alkalinity Waters, *Journal AWWA*, 100(3): 115–128.

Lytle, D.A., C. Muhlen, and B. Almassalkhi (2008) *Elevated Natural Source Water Ammonia and Nitrification in the Distribution System*. Proc. AWWA Water Quality Technology Conference, Cincinnati, OH, November 16–19.

Lytle, D.A., M.R. Schock, and K. Scheckel (2009a) The Inhibition of Pb(IV) Oxide Formation in Chlorinated Water by Orthophosphate, *Environmental Science & Technology*, 43(17): 6624–6631.

Lytle, D.A., C. White, M.N. Nadagouda, and A. Worrall (2009b) Crystal, Morphological Phase Transformation of Pb(II) to Pb(IV) in Chlorinated Water, *Journal of Hazardous Materials*, 165(1–3): 1234–1238.

Ma, Y., E. Lombi, I.W. Oliver, A.L. Nolan, and M.J. McLaughlin (2006) Long-Term Aging of Copper Added to Soils, *Environ. Sci. Technol.*, 40(20): 6310–6317.

Maas, R.P., and S.C. Patch (2004) Update on Research Findings and Regulatory/Legal Activities Regarding Tapwater Lead Exposure from Traditional Leaded-Brass and "No-Lead" Type Plumbing Parts, *Journal of the New England Water Works Association*, 118(2): 99–111.

Mackintosh, G.S., G.J. du Plessis, H.A. De Villiers, and R.E. Loewenthal (1998) *Sidestream Stabilization of Aggressive, Corrosive Waters Using Limestone.* WISA 1998 Biennial Conference and Exhibition, Cape Town, FSA, May 4–7.

Mah, M., and E.S. Boatman (1978) Scanning and Transmission Electron Microscopy of New and Used Asbestos-Cement Pipe Utilized in the Distribuiton of Water Supplies, *Scanning Electron Microscopy*, 1: 85.

Mankowski, G., J.P. Duthil, and A. Giusti (1997) The Pit Morphology on Copper in Chloride- and Sulphate-Containing Solutions, *Corrosion Science*, 39(1): 27–42.

Marani, D. (1992) "Precipitation and Conversion of Basic Cupric Sulfate to Cupric Hydrous Oxides," Ph.D. Dissertation, Illinois Institute of Technology.

Marani, D., J.W. Patterson, and P.R. Anderson (1995) Alkaline Precipitation and Aging of Cu(II) in the Presence of Sulfate, *Water Research*, 29(5): 1317–1326.

Marshall, B., and M. Edwards (2005) *Copper Pinhole Leak Development in the Presence of $Al(OH)_3$ and Chlorine.* Proc. AWWA Annual Conference, San Francisco, CA, June 12–16.

Mattsson, E., and A.-M. Fredriksson (1968) Pitting Corrosion in Copper Tubes—Cause of Corrosion and Counter-Measures, *Br. Corros. J.*, 3(9): 246–257.

Medlar, S.J., and A.J. Kim (1993a) *Corrosion Control in Small Water Systems*, Technical Paper, Camp Dresser and McKee.

Medlar, S.J., and A.J. Kim (1993b) *Corrosion Control in Small Water Systems.* Proc. NEWWA Symposium on Solutions to Controlling Lead and Copper Corrosion II, Nashua, NH, November 2–3.

Melendres, C.A., N. Camillone III, and T. Tipton (1989) Laser Raman Spectroelectrochemical Studies of Anodic Corrosion and Film Formation on Iron in Phosphate Solutions, *Electrochimica Acta*, 34 (2): 281–286.

Meranger, J.C., and T.R. Khan (1983) Lake Water Acidity and the Quality of Pumped Cottage Water in Selected Areas of Northern Ontario, *International Journal of Environmental Analytical Chemistry*, 15: 185–212.

Merrill, D.T. Jr., R.L. Sanks, and C. Spring (1978) *Corrosion Control by Deposition of $CaCO_3$ Films*, American Water Works Association, Denver, CO.

Metcalf and Eddy Inc. (2003) *Wastewater Engineering, Treatment and Reuse*, 4th ed., New York: McGraw-Hill.

Metz, D.H., M.R. Schock, and D.D. Dionysiou (2006) *The Effect of Fluoride Additives on Lead Solubility and Corrosion.* Proc. AWWA Annual Conference and Exposition, San Antonio, TX, June 11–15.

Meyer, T.E. (1994) "The Effect of Inorganic Anions on Copper Corrosion," Master of Science in Environmental Engineering, University of Colorado at Boulder.

Mikutta, C., R. Mikutta, S. Bonneville, F. Wagner, A. Voegelin, I. Christl, and R. Kretzschmar (2008) Synthetic Coprecipitates of Exopolysaccharides and Ferrihydrite. Part I: Characterization, *Geochimica et Cosmochimica Acta*, 72(4): 1111–1127.

Miller, N.T., P. Delaney, M.R. Schock, and D.A. Lytle (1995) *Analytical Issues in Corrosion and Deposition Studies.* Proc. AWWA Water Quality Technology Conference, New Orleans, LA, November 12–16, pp. 1799–1827.

Millero, F.J. (1985) The Effect of Ionic Interactions on the Oxidation of Metals in Natural Waters, *Geochimica et Cosmochimica Acta*, 49: 547–553.

Millero, F.J., M. Izaguirre, and V.K. Sharma (1987a) The Effect of Ionic Interaction on the Rates of Oxidation in Natural Waters, *Marine Chemistry,* 22: 179–191.

Millero, F.J., S. Sotolongo, and M. Izaguirre (1987b) The Oxidation Kinetics of Fe(II) in Seawater, *Geochimica et Cosmochimica Acta*, 51: 793–801.

Millero, F.J. (1989) Effect of Ionic Interactions on the Oxidation of Fe(II) and Cu(I) in Natural Waters, *Marine Chemistry*, 28: 1–18.

Millero, F.J. (1990a) Effect of Ionic Interactions on the Oxidation Rates of Metals in Natural Waters, in *Chemical Modeling of Aqueous Systems II,* pp. 447–460. Washington, DC: American Chemical Society.

Millero, F.J. (1990b) Marine Solution Chemistry and Ionic Interactions, *Marine Chemistry*, 30: 205–229.

Morel, F.M.M., and J.G. Hering (1993) *Principles and Applications of Aquatic Chemistry*. New York: John Wiley and Sons.

Morris, E.A., and D.A. Lytle (2007) *Accumulation of Radionuclides in Drinking Water Distribution Systems*. Proc. AWWA Water Quality Technology Conference, Charlotte, NC, November 4–8.

Murphy, B., J.T. O'Connor, and T.L. O'Connor (1997a) Willmar, Minnesota Battles Copper Corrosion. Part 1, *Public Works*, 128(11): 65–68.

Murphy, B., J.T. O'Connor, and T.L. O'Connor (1997b) Willmar, Minnesota Battles Copper Corrosion. Part 2, *Public Works*, 128(12): 44–47.

Murphy, B., J.T. O'Connor, and T.L. O'Connor (1997c) Willmar, Minnesota Battles Copper Corrosion. Part 3, *Public Works*, 128(13): 35–39.

MWH (2005) Internal Corrosion of Water Conduits. In *Water Treatment Principles and Design,* 2d ed., eds., J.C. Crittenden, R.R. Trussell, D.W. Hand, K.J. Howe, and G. Tchobanoglous, Chap. 21, p. 1948. New York: John Wiley and Sons.

NACE (1984) *Corrosion Basics, An Introduction.* Houston, TX: National Association of Corrosion Engineers.

Nancollas, G.H. (1983) Phosphate Precipitation in Corrosion Protection: Reaction Mechanisms, *Corrosion*, 39(3): 77.

Nguyen, C., M. Edwards, K. Stone, and B. Clark (2008) *Mechanistic Effects of Chloride-to-Sulfate Ratio on Lead Corrosion*. Proc. AWWA Annual Conference and Exposition, Atlanta, GA, June 8–12.

Nguyen, C., M. Edwards, K. Stone, and B. Clark (2009) *The Effect of Partial Lead Service Line Replacements on Lead Leaching*. Proc. AWWA Annual Conference and Exposition, San Diego, CA, June 14–18.

Nishikata, A., M. Itagaki, T. Tsuru, and S. Haruyama. (1990) Passivation and Its Stability on Copper in Alkaline Solutions Containing Carbonate and Chloride Ions, *Corrosion Science*, 31: 287–292.

Notheis, M.J., D.A. Rein, A. Struble, D. Yahraus, and W. Harter (1995) *Marble Test Solves Corrosion Mystery*. Proc. AWWA Water Quality Technology Conference, New Orleans, LA, Nov. 12–16, pp. 1687–1693.

NSF/ANSI (2007a) Drinking Water Treatment Components–Health Effects. NSF/ANSI 61-2007a. Ann Arbor, MI: NSF International.

NSF/ANSI (2007b) Revisions to the Evaluation of Lead. NSF/ANSI 61-2007a. Ann Arbor, MI: NSF International.

Obrecht, M.F., and M. Pourbaix (1967) Corrosion of Metals in Potable Water Systems, *Journal AWWA*, 59(8): 977–992.

Oliphant, R. (1978) *Dezincification by Potable Water of Domestic Plumbing Fittings: Measurement and Control,* Technical Report, Report No. TR88, Water Research Centre, UK.

Oliphant, R.J. (1983) *Summary Report on the Contamination of Potable Water by Lead from Soldered Joints*, External, Report No. 125E, Water Research Centre, UK.

Oliphant, R.J. and M.R. Schock (1996) Copper Alloys and Solder. In *Internal Corrosion of Water Distribution Systems*, 2d ed., Awwa Research Foundation/DVGW-TZW report, Chap. 6, pp. 269–312. Denver, CO: Water Research Foundation.

Olson, S.C. (1996) *Phosphate Based Corrosion Inhibitors Effects on Distribution System Regrowth.* Proc. National Conference on Integrating Corrosion Control and Other Water Quality Goals, Cambridge, MA, May 19–21.

Page, G.G., P.C.A. Bailey, and G.A. Wright (1974) Mechanisation of New Type of Copper Corrosion in Water, *Australasian Corros. Eng.*, 18: 13–19.

Pankow, J.F. (1991) *Aquatic Chemistry Concepts.* Chelsea, MI: Lewis Publishers.

Parent, J.O.G., D.D.L. Chung, and I.M. Bernstein (1988) Effects of Intermetallic Formation at the Interface Between Copper and Lead–Tin Solder, *Jour. Met Sci.*: 2564–2571.

Parkhurst, D.L., D.C. Thorstenson, and L.N. Plummer (1980) *PHREEQE–A Computer Program for Geochemical Calculations*, Water-Resources Investigations, Report No. 80-96, U.S. Geological Survey.

Patterson, J.W., R.E. Boice, and D. Marani (1991) Alkaline Precipitation and Aging of Copper from Dilute Cupric Nitrate Solution, *Environmental Science and Technology*, 25(10): 1780–1787.

Pisigan, R.A., and E. Singley (1987) Influence of Buffer Capacity, Chlorine Residual, and Flow Rate on Corrosion of Mild Steel and Copper, *Journal AWWA*, 79(2): 62–70.

Plath, D.C., K.S. Johnson, and R.M. Pytokowicz (1980) The Solubility of Calcite—Probably Containing Magnesium—In Seawater, *Marine Chemistry*, 10: 9–29.

Postma, D., and R. Jakobsen (1996) Redox Zonation: Equilibrium Constraints on the Fe(III)/SO_4-Reduction Interface, *Geochimica et Cosmochimica Acta*, 60(17): 3169–3175.

Pourbaix, M. (1996) *Atlas of Electrochemical Equilibria in Aqueous Solution.* Oxford, UK: Pergamon Press.

Pourbaix, M. (1973) *Lectures on Electrochemical Corrosion.* New York: Plenum Press.

Qiu, J., and B.I. Dvorak (2004) *The Impact of Phosphate Treatment on Copper Corrosion in Two Nebraska Public Water Supply Systems.* Proc. AWWA Water Quality and Technology Conference, San Antonio, TX, November 14–18, p. 13.

Rajasekharan, V.V., B.N. Clark, S. Boonsalee, and J.A. Switzer (2007) Electrochemistry of Free Chlorine and Monochloramine and Its Relevance to the Presence of Pb in Drinking Water, *Environmental Science and Technology*, 41(12): 4252–4257.

Rangel, C.M., D. Damborenea, A.I. De Se, and M.H. Simplicio (1992) Zinc and Polyphosphates as Corrosion Inhibitors for Zinc in Near Neutral Waters, *British Corrosion Journal*, 27(3): 207–211.

Rego, C.A., and M.R. Schock (2007) *Discovery of Unforeseen Lead Level Optimization Issues for High pH and Low DIC Conditions.* Proc. AWWA Water Quality Technology Conference, Charlotte, NC, November 4–8.

Rehring, J.P. (1994) "The Effects of Inorganic Anions, Natural Organic Matter, and Water Treatment Processes on Copper Corrosion," Master of Science, University of Colorado at Boulder.

Rehring, J.P., and M. Edwards. (1994) *The Effects of NOM and Coagulation on Copper Corrosion.* Proc. ASCE National Conference on Environmental Engineering, Boulder, CO, July 11–13, pp. 26–33.

Rehring, J.P., and M. Edwards (1995) Copper Corrosion in Potable Water Systems: Impacts of Natural Organic Matter and Water Treatment Processes, *NACE Corrosion*, 52(4): 307–317.

Reiber, S., J.F. Ferguson, and M.M. Benjamin (1988) An Improved Method for Corrosion-Rate Measurement by Weight Loss, *Journal AWWA*, 80(11): 41–46.

Reiber, S. (1989) Copper Plumbing Surfaces: An Electrochemical Study, *Journal AWWA*, 81(7): 114–122.

Reiber, S. (1991) Galvanic Stimulation of Corrosion on Lead-Tin Solder-Sweated Joints, *Journal AWWA*, 83(7): 83–91.

Reiber, S., R.A. Ryder, and I. Wagner (1996) Corrosion Assessment Technologies. In *Internal Corrosion of Water Distribution Systems*; 2d ed., Awwa Research Foundation/TZW report, Chap. 9, pp. 445–486. Denver, CO: Water Research Foundation.

Reiber, S., S. Poulsam, S.A.L. Perry, M.A. Edwards, S. Patel, and D.M. Dodrill (1997) *A General Framework for Corrosion Control Based on Utility Experience*, Awwa Research Foundation report. Denver, CO: Water Research Foundation.

Reiber, S., and G. Dostal (2000) Well Water Disinfection Sparks Surprises, *Opflow*, 26(3): 1.

Renner, R. (2004) Plumbing the Depths of D.C.'s Drinking Water Crisis, *Environmental Science & Technology*, 38(12): 224A–227A.

Robinson, R.B., T. Demirel, and E.R. Baumann (1981) Identity and Character of Iron Precipitates, *Journal of the Environmental Engineering Division, ASCE*, 107(6): 1211–122.

Robinson, R.B., R.A. Minear, and J.M. Holden (1987) Effects of Several Ions on Iron Treatment by Sodium Silicate and Hypochlorite, *Journal AWWA*, 79(7): 116–125.

Robinson, R.B., and A.J. Carmical (2005) A Web-Based Educational Module on Limestone contactors Technology for Drinking Water Professionals, *Computer Applications in Engineering Education*, 13(4): 240–249.

Rompré, A., M. Prévost, J. Coallier, T. Braekman, P. Servais, and P. LaFrance (1996) *Impacts of Corrosion Control Strategies on Biofilm Growth in Drinking Water Distribution System.* Proc. AWWA Water Quality Technology Conference, Boston, MA, November 10–13.

Rooklidge, S.J., J. Ketchum, and H. Lloyd (2002) Corrosion Control Enhancement from a Dolomite-amended Slow Sand Filter, *Water Research*, 36(11): 2689–2694.

Rose, M. (1983) X-Ray Fluorescence Anlaysis of Waterborne Scales, *American Laboratory*, 15: 46.

Rossum, J.R., and D.T. Merrill Jr. (1983) An Evaluation of the Calcium Carbonate Saturation Indices, *Journal AWWA*, 75(2): 95.

RTW, Inc. (2008) *The RTW Model for Corrosion Control & Process Chemistry.* Rothberg, Tamburini and Winsor, Inc. (Tetra Tech RTW); available online at www.rtwengineering.com/products/model.html (accessed June 3, 2009).

Rushing, J.C., and M. Edward (2004). Effect of Aluminium Solids and Chlorine on Cold Water Pitting of Copper, *Corrosion Science*, 46(12): 3069–3088.

Sadayappan, K. (2005) Mitigating Lead in Copper Alloys for Drinking Water Applications, *Foundry Management and Technology*, 39(3): 59–62.

Sander, A., B. Berghult, A.F. Broo, E.L. Johansson, and T. Hedberg (1996) Iron Corrosion in Drinking Water Distribution Systems—The Effect of pH, Calcium and Hydrogen Carbonate, *Corrosion Science*, 38(3): 443–455.

Sandvig, A., G. Boyd, G. Kirmeyer, M. Edwards, S. Triantafyllidou, and B. Murphy (2007) *Performance and Metal Release of Non-Leaded Brass Meters, Components and Fittings*, Awwa Research Foundation Report No. 91174. Denver, CO: Water Research Foundation.

Sandvig, A., P. Kwan, G. Kirmeyer, B. Maynard, D. Mast, R.R. Trussell, S. Trussell, A. Cantor, and A. Prescott (2008) *Contribution of Service Line and Plumbing Fixtures to Lead and Copper Rule Compliance Issues*, Awwa Research Foundation Report No. 90721. Denver, CO: Water Research Foundation.

Sarin, P., K. Jim, D. Lytle, M.A. Beckett, W.M. Kriven, and V.L. Snoeyink (1999) *Systematic Study of Corrosion Scales and Their Effect on Release of Iron Particles in Drinking Water.* AWWA/USEPA Particle Measurement and Characterization in Drinking Water Treatment Symposium, Proc. Nashville, TEN, March 28–30.

Sarin, P., J. Bebee, M.A. Beckett, K.K. Jim, D.A. Lytle, J.A. Clement, W.M. Kriven, and V.L. Snoeyink (2000) *Mechanism of Release of Iron from Corroded Iron/Steel Pipes in Water Distribution Systems.* Proc. AWWA Annual Conference, Denver, CO, June 11–15.

Sarin, P., V.L. Snoeyink, J. Bebee, W.M. Kriven, and J.A. Clement (2001) Physico-chemical Characteristics of Corrosion Scales in Old Iron Pipes, *Water Research*, 35(12): 2961–2969.

Sarin, P., V.L. Snoeyink, D.A. Lytle, and W.M. Kriven (2004a) Iron Corrosion Scales: Model for Scale Growth, Iron Release, and Colored Water Formation, *Journal of Environmental Engineering*, 130(4): 364–373.

Sarin, P., V.L. Snoeyink, J.A. Clement, and W.M. Krive (2004b) Iron Release from Corroded Unlined Cast Iron Pipe, *Journal AWWA*, 95(11): 85–96.

Sarin, P., V.L. Snoeyink, J. Bebee, K.K. Jim, M.A. Beckett, W.M. Kriven, and J.A. Clement (2004c) Iron Release from Corroded Iron Pipes in Drinking Water Distribution Systems: Effect of Dissolved Oxygen, *Water Research*, 38(5): 1259–1269.

Scaife, J.F. (1957) The Solubility of Malachite, *Canadian Journal of Chemistry*, 35: 1332–1340.

Schecher, W.D., and D.C. McAvoy (1998) *MINEQL+: A Chemical Equilibrium Modeling System, Version 4.0 for Windows User's Manual*. Hallowell, ME: Environmental Research Software.

Scheckel, K.G., J.A. Ryan, D. Allen, and N.V. Lescano (2005) Determining Speciation of Pb in Phosphate-Amended Soils: Method Limitations, *Science of the Total Environment*, 350(1–3): 261–272.

Scheetz, B.E., J. LaRosa-Thompson, and P.J. Delaney (1997) *XPS Characterization of Films Formed on Distribution Systems Using Additives to Control Pb/Cu Levels in Drinking Water*. Proc. AWWA Water Quality Technology Conference, Denver, CO, November 9–12.

Schenk, J.E., and W.J. Weber Jr. (1968) Chemical Interactions of Dissolved Silica with Iron (II) and Iron (III), *Journal AWWA*, 76(2): 199–212.

Schindler, P.W. (1967) Heterogeneous Equilibria Involving Oxides, Hydroxides, Carbonates and Hydroxide Carbonates. In *Equilibrium Concepts in Natural Water Systems*, Advances in Chemistry Series No. 67. Washington, DC: American Chemical Society.

Schmitt, G., E. Slavcheva, and P. Plagemann (2001) ECN–Measurements at Copper in Artificial Tap Water–Investigation of Anion-Effects, *Materials and Corrosion*, 52: 439–444.

Schock, M.R. (1980) Response of Lead Solubility to Dissolved Carbonate in Drinking Water, *Journal AWWA*, 72(12): 695–704.

Schock, M.R. (1981) Response of Lead Solubility to Dissolved Carbonate in Drinking Water, *Journal AWWA*, 73(3): 36.

Schock, M.R., and R.W. Buelow (1981) Behavior of Asbestos-Cement Pipe under Various Water Quality Conditions. Part 2. Theoretical Considerations, *Journal AWWA*, 73(11): 609.

Schock, M.R., R.W. Buelow, and P.J. Clark (1981) *Evaluation and Control of Asbestos-Cement Pipe Corrosion*. Toronto, CAN: Corrosion/81.

Schock, M.R., and C.H. Neff (1982) *Chemical Aspects of Internal Corrosion: Theory, Prediciton and Monitoring*. Proc. AWWA Water Quality Technology Conference, Nashville, TEN, December 5–8, pp. 361–390.

Schock, M.R., and S.C. Schock (1982) Effect of Container Type on pH and Alkalinity Stability, *Water Research*, 16: 1455.

Schock, M.R., and M.C. Gardels (1983) Plumbosolvency Reduction by High pH and Low Carbonate–Solubility Relationships, *Journal AWWA*, 75(2): 87–91.

Schock, M.R. (1984) Temperature and Ionic Strength Corrections to the Rangelier Index (Revisited), *Journal AWWA*, 76(8): 72.

Schock, M.R. (1985) *Treatment or Water Quality Adjustment to Attain MCLs in Metallic Potable Water Plumbing Systems*. Plumbing Materials and Drinking Water Quality. Proceedings of a Seminar, Cincinnati, OH, pp. 82–103.

Schock, M.R., and C.H. Neff (1988) Trace Metal Contamination from Brass Fittings, *Journal AWWA*, 80(11): 47–56.

Schock, M.R., R. Levin, and D.D. Cox (1988) *The Significance of Sources of Temporal Variability of Lead in Corrosion Evaluation and Monitoring Program Design*. Proc. AWWA Water Quality Technology Conference, St. Louis, MO, November 13–17.

Schock, M.R. (1989) Understanding Corrosion Control Strategies for Lead, *Journal AWWA*, 81(7): 88–100.

Schock, M.R., and K.W. Smothers (1989) *X-Ray, Microscope, and Wet Chemical Techniqes: A Complementary Team for Deposit Analysis*. Proc. AWWA Water Quality Technology Conference, Philadelphia, PA, November 12–16.

Schock, M.R. (1990) Causes of Temporal Variability of Lead in Domestic Plumbing Systems, *Environmental Monitoring and Assessment*, 15: 59–82.

Schock, M.R., and G.K. George (1991) *Comparison of Methods for Determination of Dissolved Inorganic Carbonate (DIC)*. Proc. AWWA Water Quality Technology Conference, Orlando, FL, November 10–14.

Schock, M.R., and D.A. Lytle (1994) *The Importance of Stringent Control of DIC and pH in Laboratory Corrosion Studies: Theory and Practice*. Proc. AWWA Water Quality Technology Conference, San Francisco, CA, November 6–10, pp. 1665–1692.

Schock, M.R., D.A. Lytle, and J.A. Clement (1994) *Modeling Issues of Copper Solubility in Drinking Water*. Proc. ASCE National Conference on Environmental Engineering, pp. 17–25. Boulder, CO, July 11–13.

Schock, M.R., D.A. Lytle, and J.A. Clement (1995) *Effect of pH, DIC, Orthophosphate and Sulfate on Drinking Water Cuprosolvency.* EPA/600/R-95/085, Office of Research and Development, Cincinnati, OH; available online at www.epa.gov/NRMRL/pubs/600r95085/600r95085.pdf.

Schock, M.R., I. Wagner, and R. Oliphant (1996) The Corrosion and Solubility of Lead in Drinking Water. In *Internal Corrosion of Water Distribution Systems*, 2d ed., AWWA Research Foundation/TZW report, Chap. 4, pp. 131–230. Denver, CO: Water Research Foundation.

Schock, M.R., and J.A. Clement (1998) Control of Lead and Copper with Non-Zinc Orthophosphate, *Journal of the New England Water Works Association*, 112(1): 20–42.

Schock, M.R. (1998a) Internal Corrosion and Deposition Control. In *Water Quality and Treatment: A Handbook of Community Water Supplies*; 5th ed., ed. R.D. Letterman, Chap. 17. New York: McGraw-Hill, and Denver, CO: American Water Works Association.

Schock, M.R. (1998b) Reasons for Corrosion Control Other than the Lead and Copper Rule, *Journal of the New England Water Works Association*, 113(2): 128–150.

Schock, M.R., M. Edwards, K. Powers, L. Hidmi, and D.A. Lytle (2000) *The Chemistry of New Copper Plumbing*. Proc. AWWA Water Quality Technology Conference, Salt Lake City, UT, November 5–9.

Schock, M.R., S.M. Harmon, J. Swertfeger, and R. Lohmann (2001) *Tetravalent Lead: A Hitherto Unrecognized Control of Tap Water Lead Contamination*. Proc. AWWA Water Quality Technology Conference, Nashville, TEN, November 11–15.

Schock, M.R., T.R. Lovejoy, J. Holldber, and J. Lowry (2002) California's First Aeration Plants for Corrosion Control, *Journal AWWA*, 94(3): 88–100.

Schock, M.R., and R. Giani (2004) *Oxidant/Disinfectant Chemistry and Impacts on Lead Corrosion*. Sunday Workshop, AWWA Water Quality Technology Conference, San Antonio, TX.

Schock, M.R. (2005) Distribution Systems as Reservoirs and Reactors for Inorganic Contaminants. In *Distribution System Water Quality Challenges in the 21st Century*, 6th ed., Chap. 6. Denver, CO: American Water Works Association.

Schock, M.R., K.G. Scheckel, M. DeSantis, and T.L. Gerke (2005a) *Mode of Occurrence, Treatment and Monitoring Significance of Tetravalent Lead*. Proc. AWWA Water Quality Technology Conference, Quebec City, CAN, November 6–10.

Schock, M.R., D.A. Lytle, A.M. Sandvig, J.A. Clement, and S.M. Harmon (2005b) Replacing Polyphosphate with Silicate to Solve Problems with Lead, Copper and Source Water Iron, *Journal AWWA*, 97(11): 84–93.

Schock, M.R., and A.M. Sandvig (2006) *Long-Term Impacts of Orthophosphate Treatment on Copper Levels*. Proc. AWWA Annual Conference and Exposition, San Antonio, TX, June 11–15.

Schock, M.R., D.A. Lytle, A.M. Sandvig, J.A. Clement, and S.M. Harmon (2006) Erratum: Replacing Polyphosphate with Silicate to Solve Problems with Lead, Copper and Source Water Iron, *Journal AWWA*, 98(4): 178–179.

Schock, M.R., R.N. Hyland, and M.M. Welch. (2008a) Occurrence of Contaminant Accumulation in Lead Pipe Scales from Domestic Drinking-Water Distribution Systems, *Environmental Science and Technology*, 42(12): 4285–4291.

Schock, M.R., M.K. DeSantis, D.H. Metz, M.M. Welch, R.N. Hyland, and M.N. Nadagouda (2008b) *Revisiting the pH Effect on the Orthophosphate Control of Plumbosolvency*. Proc. AWWA Annual Conference, Atlanta, GA, June 8–12.

Schock, M.R., and A.M. Sandvig (2009) Long-Term Impacts of Orthophosphate Treatment on Copper Levels, *Journal AWWA*, 101(7): 71–82.

Schott, G.J. (1998) *WATERPRO Corrosion Control Treatment Process Evaluation Program.* Proc. AWWA Water Quality Technology Conference, San Diego, Calif., November 1–4.

Schreppel, C.K., D.W. Frederickson, and T.A. Geiss (1997) *The Positive Effects of Corrosion Control on Lead Levels and Biofilms.* Proc. AWWA Water Quality Technology Conference, Denver, CO, November 9–12.

Shalaby, A.M., F.M. Al-Kharafi, and V.K. Gouda (1989) A Morphological Study of Pitting Corrosion of Copper in Soft Tap Water, *Corrosion–NACE,* 45(7): 536–547.

Shams El Din, A.M (1986) The Problem of "Red Waters": A New Approach to Its Solution, *Desalination*, 60: 75–88.

Sheiham, I., and P.J. Jackson (1981) Scientific Basis for Control of Lead in Drinking Water by Water Treatment, *Journal of the Institute of Water Engineers and Scientists*, 35(6): 491–515.

Shipley, J.W., I.R. McHaffie, and N.D. Clare (1925) Corrosion of Iron in Absence of Oxygen, *Industrial and Engineering Chemistry*, 17(4): 381–385.

Silverman, D.C. (1995) *Measuring Corrosion Rates in Drinking Water by Linear Polarization—Assumptions and Watchouts*. Proc. AWWA Water Quality Technology Conference, pp. 2435–2455. New Orleans, LA, November 12–16.

Singley, J.E., B.A. Beaudet, and P.H. Markey (1984) *Corrosion Manual for Internal Corrosion of Water Distribution Systems,* EPA/570/9-84-001. Washington, DC: USEPA Office of Ground Water and Drinking Water.

Singley, J.E., H. Sontheimer, A. Kuch, W. Kölle, and A. Ahmadi (1985) Corrosion of Iron and Steel. In *Internal Corrosion of Water Distribution Systems*; Awwa Research Foundation/DVGW Forschungsstelle report, Chap. 2, pp. 33–125. Denver, CO: Water Research Foundation.

Singley, J.E.(1994) Electrochemical Nature of Lead Contamination, *Journal AWWA*, 86(7): 91–96.

Smart, N.R., P.M. Scott, and R.P.M. Procter (1990) Repassivation Kinetics and Stress Corrosion of Mild Steel in Phosphate Solutions, *Corrosion Science*, 30(8–9): 877–901.

Smith, D.C., and B. McEnaney (1979) The Influence of Dissolved Oxygen Concentration on the Corrosion of Grey Cast Iron in Water at 50°C, *Corrosion Science*, 19(6): 379–394.

Smith, P.G., and A. Bager (1995) The Use of Limestone Bed Filtration for the Treatment of Ferruginous Groundwater, *Journal of the Chartered Institute of Water Envrionment Management*, 9(4): 192–198.

Snoeyink, V.L., and D. Jenkins (1980) *Water Chemistry.* New York: John Wiley and Sons.

Snoeyink, V.L., and A. Kuch (1985) Principles of Metallic Corrosion in Water Distribution Systems. In *Internal Corrosion of Water Distribution Systems*; Awwa Research Foundation/DVGW Forschungsstelle report, Chap. 1, pp. 1–32. Denver, CO: Water Research Foundation.

Snoeyink, V.L., and I. Wagner (1996) Principles of Corrosion in Water Distribution Systems. In *Internal Corrosion of Water Distribution Systems*; 2d ed., Awwa Research Foundation/DVGW-TZWreport , Chap. 1, pp. 1–27. Denver, CO: Water Research Foundation.

Snoeyink, V.L., M.R. Schock, P. Sarin, L. Wang, A.S.-C. Chen, and S.M. Harmon (2003) Aluminium-Containing Scales in Water Distribution Systems: Prevalence and Composition, *Journal of Water Supply: Research and Technology–Aqua,* 52(7): 455–474.

Sontheimer, H., W. Kolle, and V.L. Snoeyink (1981) The Siderite Model of the Formation of Corrosion-Resistant Scales, *Journal AWWA*, 73(11): 572–579.

Sosa, M., S. Patel, and M. Edwards (1999) Concentration Cells and Pitting Corrosion, *NACE Corrosion*, 55(11): 1069.

Spencer, C.M. (1996) *Aeration and Limestone Contact for Radon Removal and Corrosion Control.* Proc. National Conference on Integrating Corrosion Control and Other Water Quality Goals, Cambridge, MA, May 19–21.

Spencer, C.M., and W.E. Brown (1997) *pH Monitoring to Determine Aeration Effectiveness for Carbon Dioxide and Radon Removal.* Proc. AWWA Water Quality Technology Conference, Denver, CO.

Spencer, C.M. (1998) Aeration and Limestone Contact for Radon Removal and Corrosion Control, *Journal of the New England Water Works Association*, 112(1): 60–69.

SPEX CertiPrep Inc. (2005) Obenauf, R. H., Bostwick, R., Fithian, W., DeSteafano, M., and Torres, A. (eds.). *SPEX CertiPrep Handbook of Sample Preparation and Handling,* 8th ed., SPEX CertiPrep Inc., Metuchen, NJ.

Stericker, W. (1945) Protection of Small Water Systems from Corrosion, *Ind. Eng. Chem.*, 37: 716.

Stone, D.A., and R.E. Goldstein (2004) Tubular Precipitation and Redox Gradients on a Bubbling Template. *Proceedings of the National Academy of Sciences of the United States of America*, 101(32): 11537–11541.

Stone, K., M. Edwards, C. Nguyen, and B. Clark (2008) *Effect of Coagulant and Other Treatment Changes on Lead Leaching*. Proc. AWWA Annual Conference, Atlanta, GA.

Stumm, W. (1956) Calcium Carbonate Deposition at Iron Surfaces, *Journal AWWA*, 48(3): 300.

Stumm, W. (1960) Investigation on the Corrosive Behavior of Waters, *Journal of the Sanitary Engineering Division, ASCE*, 96 (SA6): 27–46.

Stumm, W., and J.J. Morgan (1996) *Aquatic Chemistry,* 3d ed. New York: John Wiley and Sons.

Subramanian, K.S., J.W. Connor, and J.C. Meranger (1991) Leaching of Antimony, Cadmium, Copper, Lead Silver, Tin and Zinc From Copper Piping with Non-Lead-Based Soldered Joints, *J. Environ. Sci. Health*, A26(6): 911–929.

Subramanian, K.S., V.S. Sastri, and J.W. Connor (1994) Drinking Water Quality: Impact of Non-Lead-Based Plumbing Solders, *Toxicological and Environmental Chemistry*, 44(1): 11–20.

Sung, W., and J.J. Morgan (1980) Kinetics and Product of Ferrous Iron Oxygenation in Aqueous Systems, *Environmental Science and Technology*, 14(5): 561–568.

Switzer, J.A., V.V. Rajasekharan, S. Boonsalee, E.A. Kulp, and E.W. Bohannan (2006) Evidence that Monochloramine Disinfectant Could Lead to Elevated Pb Levels in Drinking Water, *Environmental Science and Technology*, 40(10): 3384–3387.

Tamura, H., K. Goto, and M. Nagayama (1976a) Effect of Anions on the Oygenation of Ferrous Iron in Neutral Solutions, *J. Inorg. Nucl. Chem.*, 38: 113–117.

Tamura, H., K. Goto, and M. Nagayama (1976b) The Effect of Ferric Hydroxide on the Oxygenation of Ferrous Ions in Neutral Solutions, *Corrosion Science*, 16(4): 197–207.

Tamura, H., S. Kawamura, and M. Hagayama (1980) Acceleration of the Oxidation of Fe^{2+} Ions by Fe(III)-Oxyhydroxides, *Corrosion Science*, 20: 963–971.

Tamura, H. (2008) The Role of Rusts in Corrosion and Corrosion Protection of Iron and Steel, *Corrosion Science*, 50(7): 1872–1883.

Taxèn, C. (1996) *Pitting Corrosion of Copper: An Equilibrium–Mass Transport Study*, KI Rapport, Report No. 1996:8 E, Swedish Corrosion Institute.

Tesfai, F., P. Constant, S. Reiber, R. Giani, and M. Donnelly (2006) *Precipitate Formation in the Distribution System Following Addition of Orthophosphate*. Proc. AWWA Water Quality Technology Conference, Denver, CO, November 5–9.

Thompson, D.H. (1970) General Tests and Principles. In *Handbook on Corrosion Testing and Evaluation.* New York: John Wiley and Sons.

Tillmans, J., P. Hirsch, and W. Weintraud (1927) Die Korrosion von Eisen unter Wasserleitungswasser, *gwf–Das Gas- und Wasserfach*, 35(70): 845–849, 877–884, 919–925.

Tillmans, J., P. Hirsch, and K. Schilling (1929) Die selbsttätige Schutzschichtbuilding in eisernen Rohren, *gwf –Das Gas- und Wasserfach*, 37(72): 49–53, 78–82, 689–690.

Triantafyllidou, S., and M. Edwards (2007) Critical Evaluation of the NSF 61 Section 9 Test Water for Lead, *Journal AWWA*, 98(9): 133–143.

Triantafyllidou, S., J. Parks, and M. Edwards (2007) Lead Particles in Potable Water, *Journal AWWA*, 99(8): 107–117.

Trussell, R.R., L.L. Russell, and J.F. Thomas (1977) *The Langelier Index*. Proc. AWWA Water Quality Technology Conference, Kansas City, MO.

Trussell, R.R. (1985) Corrosion. In *Water Treatment Principles and Design*, pp. 392–434. New York: John Wiley and Sons.

Trussell, R.R., and I. Wagner (1996) Corrosion of Galvanized Pipe. In *Internal Corrosion of Water Distribution Systems*; Awwa Research Foundation/TZW report, Chap. 3, pp. 71–129. Denver, CO: Water Research Foundation.

Trussell, R.R., and J.J. Morgan (2006) *A Saturation Index for Cement Surfaces Exposed to Water*. Proc. AWWA Water Quality Technology Conference, Denver, CO, November 5–9.

Tuovinen, O.H., K.S. Button, A. Vuorinen, L. Carlson, D.M. Mair, and L.A. Yut (1980) Bacterial, Chemical, and Mineralogical Characteristics of Tubercles in Distribution System Pipelines, *Journal AWWA*, 72(11): 626–635.

Turek, N.F., L. Kasten, D.A. Lytle, and M.N. Goltz (2006) *Investigation of Copper Contamination and Corrosion Scale Mineralogy in Aging Drinking Water Distribution Systems*. Proc. AWWA Annual Conference and Exposition, San Antonio, TX, June 11–15.

Uhlig, H.H., D.N. Triadis, and M. Stern (1955) Effect of Oxygen, Chlorides, and Calcium Ion on Corrosion Inhibition of Iron by Polyphophates, *Jour. of the Electrochem. Soc.*, 102(2): 59–66.

Urbansky, E.T., and M.R. Schock (2000a) Can Fluoridation Affect Lead(II) in Potable Water? Hexafluorosilicate and Fluoride Equilibria in Aqueous Solution, *International Journal of Environmental Studies–Section B*, 57, 597–637.

Urbansky, E.T., and M.R. Schock (2000b) *Can Fluoridation Affect Water Lead(II) Levels and Lead(II) Neurotoxicity?* Proc. AWWA Annual Conference, Denver, CO.

Urbansky, E.T., and M.R. Schock (2000c) Understanding, Deriving, and Computing Buffer Capacity, *Journal of Chemical Education, 77*(12): 1640–1644.

USEPA (1980) Interim Primary Drinking Water Regulations; Amendments. 40 CFR Part 141, *Federal Register*, 45168:57331, August 27.

USEPA (1988) *Suggested Sampling Procedures to Determine Lead in Drinking Water in Buildings other than Single Family Homes,* EPA-570/B-88/015, Office of Water, Washington, DC.

USEPA (1989) *Lead in School Drinking Water.* EPA-570/9-89-001, Office of Water, Washington, DC.

USEPA (1993) *Seminar Publication: Control of Lead and Copper in Drinking Water.* EPA/625/R-93-001, Office of Research and Development, Washington, DC.

USEPA (1994) Analytical Methods for Regulated Drinking Water Contaminants. 40 CFR Parts 141 and 143, *Federal Register*, 59:232:62456, December 5.

USEPA (1996) Safe Drinking Water Act Amendments of 1996. Public Law 104-182,Sec. 118:Aug. 6.

USEPA (2009a) *2007 Drinking Water Infrastructure Needs Survey and Assessment.* EPA 816-R-09-001, Office of Ground Water and Drinking Water, Washington, DC; available online at www.epa.gov/OGWDW/needssurvey/index.html.

USEPA (2009b) Subpart I—Control of Lead and Copper. 40 CFR Part 141, July 1; available online at frwebgate.access.gpo.gov/cgi-bin/get-cfr.cgi?TITLE=40andPART=141andSUBPART=Iand TYPE=PDFandYEAR=2009 (accessed December 15, 2009).

Valentine, R.L., and S.W. Stearns (1994) Radon Release from Water Distributon System Deposits, *Environmental Science and Technology*, 28(3): 534–537.

van den Hoven, T.J.J. (1987) A New Method to Determine and Control Lead Levels in Tap Water, *Journal of Water Supply: Research and Technology–Aqua*, 6: 315–322.

van den Hoven, T.J.J., and M.W.M. van Eekern (1988) *Optimal Composition of Drinking Water*, KIWA Report No. 100, Kiwa, NV.

van den Hoven, T.J.J., and M.W.M. van Eekeren (1990) *10. Corrosion and Corrosion Control in Drinking Water Systems in the Netherlands*. Corrosion and Corrosion Control in Drinking Water Systems; Proceedings from a Corrosion Workshop and Seminar, Oslo, NOR, pp. 43–46.

Van Wazer, J.R., and C.F. Callis (1958) Metal Complexing by Phosphates, *Chem. Rev.*, 58: 1011.

Vitins, A., M.C. Gardels, D.A. Lytle, M.R. Schock, and T.J. Sorg (1994) Investigation of the Electrochemical Properties of Copper in Potable Water, *Latvijas Kimijas Zurnals*, 1: 66–77.

von Blanckenburg, F., M. Mamberti, R. Schoenberg, B.S. Kamber, and G.E. Webb (2008) The Iron Isotope Composition of Microbial Carbonate, *Chemical Geology*, 249(1–2): 113–128.

Wagner, I. (1989) *Effect of Inhibitors on Corrosion Rate and Metal Uptake in Drinking Water Systems*. Proc. AWWA Seminar on Internal Corrosion Control Developments and Research Needs, Los Angeles, CA, June 18, pp. 1–17.

Walker, A.P. (1983) The Microscopy of Consumer Complaints, *Journal of the Institute of Water Engineers and Scientists*, 37(3): 200–214.

Water Research Foundation (2010) *Impacts of Lining Materials on Water Quality*, Report No. 4036 (in press), Denver, CO: Water Research Foundation, and London: United Kingdom Drinking Water Inspectorate.

Watkins, K.G., and D.E. Davies (1987) Cathodic Protection of 6351 Aluminum Alloy in Sea Water: Protection Potential and pH Effects, *British Corrosion Journal*, 22(3): 157–161.

Weber Jr., W.J., and W. Stumm (1963) Mechanism of Hydrogen Ion Buffering in Natural Waters, *Journal AWWA*, 55(12): 1553–1578.

Weiss, J. (1935) Elektronenubergangsprozesse im Mechanismus von Oxydations und Reduktionsreaktionen in Losungen, *Naturwissenschaften*, 23: 64.

Werner, W., H.-J. Groß, and H. Sontheimer (1994) Corrosion of Copper Pipes in Drinking Water Installations, *Translation of: gwf-Wasser/Abwasser*, 135(2): 1–15.

Weyl, P.K. (1961) The Carbonate Saturometer, *Journal of Geology*, 69, 32–43.

Whipple, G.C. (1913) Decarbonation as a Means of Removing the Corrosive Properties of Public Water Supplies, *Journal of the New England Water Works Association*, 27: 193.

WHO (2006) *Health Aspects of Plumbing*. Geneva, SWITZ: World Health Organization and World Plumbing Council.

Williams, J.F. (1986) Corrosion of Metals Under the Influence of Alternating Current, *Materials Protection*, 5(2): 52–53.

Withers, A. (2005) Options for Recarbonation, Remineralisation and Disinfection for Desalination Plants, *Desalination*, 179(1–3): 11–24.

Wysock, B.M., M.R. Schock, and J.A. Eastman (1991) *A Study of the Effect of Municipal Ion-Exchange Softening on the Corrosion of Lead, Copper and Iron in Water Systems*. Proc. AWWA Annual Conference, Philadelphia, PA, June 23–27, pp. 853–881.

Wysock, B.M., A.M. Sandvig, M.R. Schock, C.P. Frebis, and B. Prokop (1992) *Statistical Procedures for Corrosion Studies*. Proc. AWWA Water Quality Technology Conference, Toronto, CAN, November 15–19, pp. 799–837.

Wysock, B.M., A.M. Sandvig, M.R. Schock, C.P. Frebis, and B. Prokop (1995) Statistical Procedures for Corrosion Studies, *Journal AWWA*, 87(5): 99–112.

Zhang, Y., S. Triantafyllidou, and M. Edwards (2008a) Effect of Nitrification and GAC Filtration on Copper and Lead Leaching in Home Plumbing Systems, *Journal of Environmental Engineering*, 134(7): 521–530.

Zhang, Y., A. Griffin, and M. Edwards (2008b) Nitrification in Premise Plumbing: Role of Phosphate, pH and Pipe Corrosion, *Environmental Science and Technology*, 42(12): 4280–4284.

Zhang, Y., N. Love, and M. Edwards (2009a) Nitrification in Drinking Water Systems, *Critical Reviews in Environmental Science and Technology*, 39(3): 153–208.

Zhang, Y., A. Griffin, M. Rahman, A. Camper, H. Baribeau, and M. Edwards (2009b) Lead Contamination of Potable Water due to Nitrification, *Environmental Science & Technology*, 43(6): 1890–1895.

CHAPTER 21
MICROBIOLOGICAL QUALITY CONTROL IN DISTRIBUTION SYSTEMS

Mark W. LeChevallier, Ph.D.
Director, Innovation & Environmental Stewardship
American Water
Voorhees, New Jersey, United States

Marie-Claude Besner, Ph.D.
Postdoctoral Fellow
École Polytechnique de Montréal,
Montreal, Quebec, Canada

Melinda Friedman, P.E.
President
Confluence Engineering Group, LLC
Seattle, Washington, United States

Vanessa L. Speight, Ph.D., P.E.
Associate
Malcolm Pirnie, Inc.
Arlington, Virginia, United States

MICROBIAL RISKS FROM DISTRIBUTION SYSTEM CONTAMINATION 21.2
Waterborne Disease Outbreaks 21.2
Epidemiological Studies of Distribution System Risk 21.4
MICROBES IN DISTRIBUTION SYSTEMS 21.7
Differentiation between Suspended and Biofilm Microbes 21.7
Pathogen Occurrence in Distribution Systems 21.9
Identification of Bacteria in Distribution Systems 21.9
Heterotrophic Bacteria 21.10
Coliform Bacteria 21.13
Stressed and Viable but Nonculturable Bacteria 21.14
Legionella 21.14
Aeromonas 21.16
Mycobacterium 21.17
Free-Living Protozoa 21.18
Fungi 21.18
Actinomyces and Other Related Organisms 21.19
Nitrifying Bacteria 21.19
Iron-, Manganese-, and Sulfur-Using Bacteria 21.20
FACTORS CONTRIBUTING TO MICROBIAL OCCURRENCES IN DISTRIBUTION SYSTEMS 21.21
Treatment Processes 21.21
Microbial Growth in Distribution Systems 21.22
Recontamination of the Distribution System 21.28

MONITORING DISTRIBUTION SYSTEMS **21.34**
Regulatory Requirements 21.35
Measurement of Water Quality Parameters 21.36
Theoretical Approaches to Sampling Plan Design 21.38
Sampling Related to Water System Security 21.41
Integrated Distribution System Management 21.43
ENGINEERING AND DESIGN OF DISTRIBUTION SYSTEMS **21.43**
Use of Distribution System Models 21.44
Water Quality Impacts Associated with Distribution System Design Practices 21.46
CONTROLLING MICROBIAL OCCURRENCES IN DISTRIBUTION SYSTEMS **21.47**
Treatment Approaches 21.48
Operations and Maintenance Approaches 21.50
Prevention of Distribution System Contamination 21.55
FINAL COMMENTS **21.65**
ABBREVIATIONS **21.66**
REFERENCES **21.67**

MICROBIAL RISKS FROM DISTRIBUTION SYSTEM CONTAMINATION

Waterborne Disease Outbreaks

Since the mid-1980s, the number of waterborne outbreaks in the United States has declined (Fig. 21-1), largely attributed to the promulgation of more stringent drinking water standards, including the Surface Water Treatment Rule (SWTR), the Total Coliform Rule (TCR), and other regulations; see Chap. 1 for a discussion of these and other regulations. In addition, many water utilities have made voluntary improvements, such as the Partnership for Safe Water Program, to reduce the risk of waterborne cryptosporidiosis. The Partnership program involves a comprehensive evaluation of treatment practices with a focus on achieving filtered

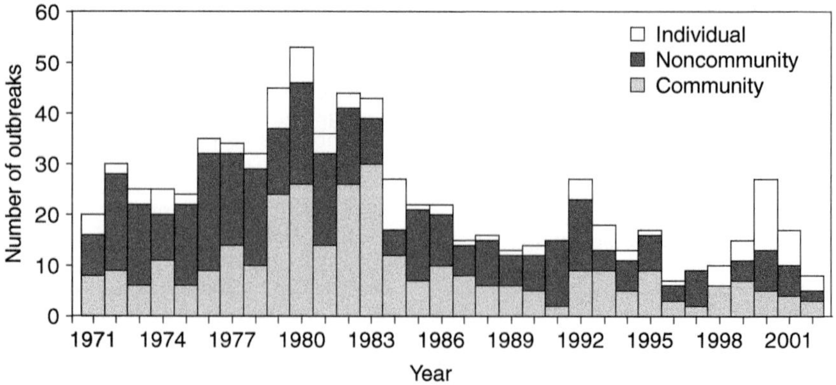

*Excludes outbreaks of Legionnaires disease.

FIGURE 21-1 Number of drinking water disease outbreaks in the United States, 1971–2002. (*Source: Adapted from Blackburn et al., 2004.*)

drinking water turbidities of less than 0.1 ntu. The number of reported community- and non-community-reported outbreaks decreased sharply in the 1984–1986 reporting period, largely in anticipation of promulgation of the 1989 SWTR and TCR regulations. In contrast, the increase in outbreaks reported during 1999–2001 was attributable primarily to individual home owner systems, comprising 40 percent of the waterborne outbreaks during the period (see Fig. 21-1). Among the seven outbreaks associated with community water systems in 2001–2002, four (57.1 percent) were related to problems in the water distribution system. Investigations conducted in the late 1990s suggested that a substantial proportion of waterborne disease outbreaks, both microbial and chemical, was attributed to problems within distribution systems (Craun and Calderon, 2001).

In 2001, there were significant changes in the reporting of outbreaks by the U.S. Centers for Disease Control and Prevention (CDC) and the U.S. Environmental Protection Agency (USEPA) with the inclusion of *Legionella* outbreaks (Blackburn et al., 2004). Additionally, in 2003, there were changes to the classification of system deficiencies that resulted in the outbreaks (Liang et al., 2006). Importantly, the number of deficiency classes increased from 5 to 13 categories with a total of 23 subcategories and distinguished between contamination under the control of the water utility and events outside the jurisdiction of the utility. Examples of contamination outside the jurisdiction of the utility include plumbing deficiencies, deficiencies in the building/home-specific treatment systems, contamination at the point of use, and contamination during the bottling or shipping of water. Outbreaks of *Legionella* typically are outside the jurisdiction of the utility (Liang et al., 2006).

The result of this reclassification was a more detailed understanding of waterborne disease outbreaks, particularly those associated with distribution systems. Prior to 2003, all outbreaks associated with any piping, even those internal to buildings or homes, were categorized as "distribution system." Since 2003, it has been clear that the majority of "distribution system" outbreaks have been due to problems in premise plumbing and other problems outside the jurisdiction of the utility, and relatively few have been due to contamination of community water system pipelines. Since its categorization in 2001 as a waterborne disease agent, *Legionella* has emerged as the most commonly identified pathogen (Fig. 21-2).

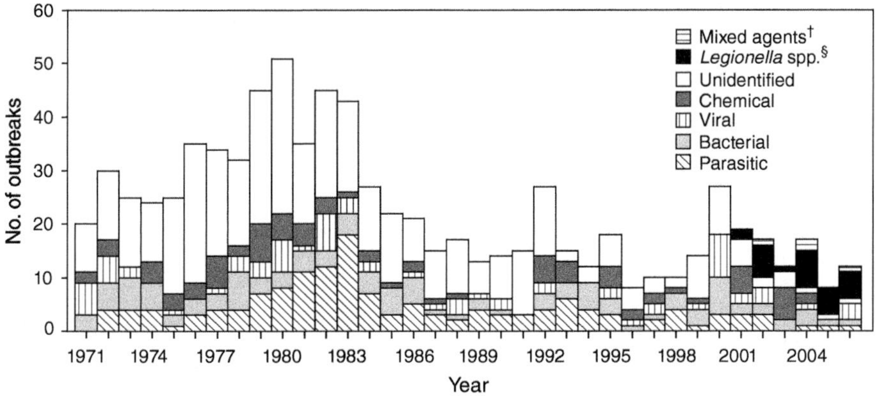

†Beginning in 2003, mixed agents of more than one etiologic agent type were included in the surveillance system. However, the first observation is a previously unreported outbreak in 2002.
§Beginning in 2001, Legionnaires' disease was added to the surveillance system, and Legionella species were classified separately in this figure.

FIGURE 21-2 Number of waterborne-disease outbreaks associated with drinking water by year and etiologic agent—United States, 1971–2006. (*Source: Yoder et al., 2008.*)

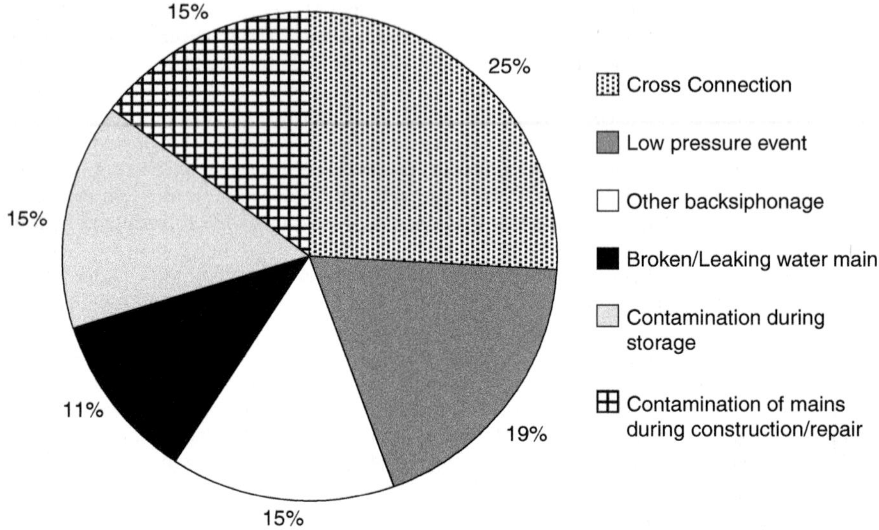

FIGURE 21-3 Reclassification of 27 waterborne disease outbreaks caused by distribution system deficiencies for community and noncommunity systems, 1974–2004. *Legionella* is not included in these outbreaks. (*Source: USEPA, 2007b.*)

The USEPA and the CDC are in the process of reclassifying outbreaks from 1971 through 2004 to better understand the underlying deficiency that caused the outbreak. This process is somewhat subjective because it requires interpretation of historical accounts, and the results do not include outbreaks owing to *Legionella.* Cross-connections or backsiphonage (where low pressures in the distribution system draw in contaminants) accounted for 40 percent of the reported outbreaks (Fig. 21-3). Contamination of mains during construction or broken or leaking mains accounted for 26 percent of the outbreaks. Low-pressure events were associated with 19 percent of the outbreaks, whereas contamination of storage tanks was the cause of 15 percent of the reported deficiencies. Attention to these routes of contamination was the focus of the research agreement formed as part of the 2008 revisions to the TCR (USEPA, 2008b). The importance of cross-connections and backsiphonage is supported in a separate compilation by the USEPA of backflow events that revealed a total of 459 incidents resulting in 12,093 illnesses from 1970 to 2001 (USEPA, 2002a). For the period 1981–1998, the study found that only 97 of 309 incidents were reported to public health authorities, suggesting that the magnitude of the public health concern due to cross-connections is an underestimated.

Epidemiological Studies of Distribution System Risk

Epidemiological studies also point to potential risks in distribution systems. Payment and colleagues (1991, 1997) conducted two epidemiological studies that examined the health of people who drank tap water compared with people receiving water treated by reverse osmosis. For the 1991 study, reverse osmosis units were installed in 299 households (1206 persons), and another 307 households (1202 persons) were followed as controls with no device installed. Both groups were monitored for a 15-month period for episodes of highly credible gastrointestinal illness

(HCGI). The water source for the study area was a river that was contaminated by human sewage discharges, including combined sewer overflows. The community had a single water treatment plant with predisinfection, alum coagulation, flocculation, rapid sand filtration, ozonation, and final disinfection with chlorine or chlorine dioxide. The overall incidence of highly credible gastroenteritis was 0.66 episode/person per year and was highest in children 5 years of age and younger. The authors concluded that approximately 35 percent of the self-reported gastrointestinal illnesses were attributed to tap water consumption. Although not designed to evaluate the source of contamination, the study suggested that contamination could come from the distribution system because those living closest to the treatment plant (or point of rechlorination) had some of the lowest illness rates.

The 1997 study was designed to examine a number of hypotheses as to the possible source of illnesses; however, it should be noted that between the time of the first and second studies, the water treatment plant was upgraded significantly with higher disinfection doses and better filtration. The study included groups receiving (1) regular tap water, (2) tap water from a continuously purged tap, (3) bottled plant effluent water, or (4) bottled plant effluent water purified by reverse osmosis. The difference in gastroenteritis rates between groups 2 and 3 was assumed to be due to pathogens breaking through the treatment process. The average turbidity of the finished water was 0.1 ntu and never exceeded 0.5 ntu. The study found no statistical difference in illness rates between groups 2 and 3, suggesting that no observable pathogens were in the treated effluents, although the authors acknowledged the possibility that "microbursts" would not be observed. The difference in gastroenteritis rates between groups 1 and 2 versus group 3 was assumed to be due to changes in water quality that occurred between the time the water left the treatment plant and the time it reached the household. The water ingested by group 1 represented tap water that had gone through the distribution system and also had residence time in the household plumbing. The water ingested by group 2 represented tap water quality in the distribution system without any significant residence time in the household plumbing. The 1997 study by Payment and colleagues attributed 14 to 40 percent of the gastrointestinal illness to the consumption of tap water and concluded that the distribution system played a role in waterborne disease because the rates of HCGI were similar for group 3 (ingested purified bottled water) and group 4 (ingested bottled water from the treatment plant), but groups 1 and 2 (ingested water from the distribution system) had higher HCGI rates than group 4. Interestingly, there appeared to be no correlation between the relatively short residence time of the water in the distribution system (which varied from 0.3 to 34 hours) and the incidence of HCGI in a family. Furthermore, microbiological testing of the water in the distribution system did not reveal any bacterial indicators of contamination, but these water samples were not tested for viruses or protozoa. Contrary to their expectation, the investigators observed higher HCGI rates in families that ingested water from the continuously purged taps compared with families with regular tap water that may be subject to bacterial regrowth in household pipes. The investigators suggested that the shorter residence time for water from the continuously purged taps may have transported pathogens in the distribution system to the household sooner than regular tap water and that there may have been inadequate contact time with residual chlorine in the distribution system to inactivate pathogens introduced by intrusion.

Subsequent studies by Kirmeyer and colleagues (2001) and Besner (2007) and Besner et al. (2010) examined the susceptibility of the Laval system to intrusion of contaminants during low-pressure events. Transient pressure modeling conducted by Kirmeyer and colleagues (2001) suggested that the distribution system was particularly prone to negative pressures, with more than 90 percent of the nodes within the system drawing negative pressures under certain modeling scenarios (e.g., power outages). During the 1997 study by Payment and colleagues, the system reported some pipe breaks, particularly during the fall and winter, when temperature changes placed added stresses on the distribution system. Although the system employed state-of-the-art treatment, the distribution network maintained low disinfectant residuals, particularly at the ends of the system. Low disinfectant residuals and a vulnerability of the distribution system to pressure transients could account for the viral-like etiology of the illnesses observed.

A double-blinded, randomized trial completed in Melbourne, Australia (Hellard et al., 2001), examined the contribution of drinking water from an unfiltered surface water supply with a protected forested watershed to the rates of gastroenteritis in the distribution system. The rates of HCGI ranged from 0.79 episodes/person per year for those with functional treatment units to 0.82 episodes/person per year with the placebo device (a sham unit that looked the same but provided no additional treatment). The study concluded that the water was not a source of measurable gastrointestinal disease (the ratio of illness rates between the group drinking treated water and those drinking normal tap water was 0.99, with a 95 percent confidence interval of 0.85–1.15; $p = .85$). There was no difference between the groups despite the rather poor water quality in the distribution system. Total coliform bacteria were detected in 18.9 percent of 1167 routine 100-mL distribution system water samples, but fecal coliform bacteria were not detected. Distribution system samples were positive for *Aeromonas* spp. (50 percent of 68 weekly samples), *Campylobacter* (one occasion), and *Giardia* (two positive samples by reverse-transcriptase polymerase chain reaction (RT-PCR). Although the study was not designed specifically to examine the risks from the distribution system, the results call into question traditional measures of water quality (e.g., total coliforms and *Aeromonas* spp.) with respect to predicting the risks of gastrointestinal illness.

A study conducted in Davenport, IO, was designed to determine the incidence of gastrointestinal illness associated with consumption of drinking water meeting all U.S. federal and state treatment guidelines (LeChevallier et al., 2003; Wade et al., 2004; Colford et al., 2005). The municipal water system used a single source (the Mississippi River) and was treated at a single plant with conventional treatment consisting of coagulation, flocculation, sedimentation, filtration (dual filters with granular activated carbon and sand), and chlorination. A total of 456 households with 1296 participants were randomized into two groups. One group received a household water treatment device with a 1-μm absolute ceramic filter and ultraviolet (UV) light with 35,000 to 38,000 $\mu W \cdot s/cm^2$ output. The other group received a placebo (sham) device that was identical to the active device but had an empty filter chamber and a UV light that was shielded to block transmission of radiation but still generated the same light and heat as the active unit. Each study household had an active device for 6 months and a sham device for 6 months and was blinded to the status of their device during the study. Overall, the rate of HCGI for households with the sham device was 2.12 and 2.20 episodes/person per year for households with the active device. Multivariate analyses showed no effect of the household water treatment device on illness rates during the 12-month study period. The overall conclusion was that fewer than 11 percent of the gastrointestinal illnesses observed in this community were due to drinking water.

During the Davenport study in April and May 2001, the Mississippi River experienced a 1 in 300-year flood that swamped upstream sewage treatment plants and elevated raw water levels of *Giardia* cysts, total and fecal coliforms, and male-specific coliphage. Average turbidity exceeded 40 ntu over a 6-week period and peaked at 197 ntu. However, treatment plant performance remained good, with individual filter and combined filter effluent turbidities below 0.1 ntu. The flood period was associated with significantly increased rates of gastrointestinal illness for people (especially in children) with direct contact with flood water and flood cleanup activities; however, drinking water quality was not a factor in the increased illness rate (Wade et al., 2004).

A study conducted in Wales and northwest England in 2001–2002 found a very strong association (a probability $p < .001$) between self-reported diarrhea and reported low water pressure at the home tap based on a postal survey of 423 subjects (Hunter et al., 2005).

This study was part of a larger case-control study of risk factors associated with sporadic cryptosporidiosis. The study populations were drawn from two large regions that include both heavily industrialized areas and rural areas and about 240 water treatment plants. The overall microbiological water quality for the utilities in these regions was described to be excellent, with less than 0.05 percent of water samples positive for *Escherichia coli* during this study period. Four risk factors for diarrhea in the control group remained significant: feeding a child under 5 years old, contact with another person who had diarrhea, loss of water pressure at home, and how often the subject ate yogurt. The first three risk factors had a positive association with diarrhea [odds ratios (ORs) of 2.5, 7.0, and 12.5, respectively, after adjusting for the effects of the other variables in the model]. The investigators hypothesized that most of the reported episodes of pressure loss were due to main breaks in which contamination entered the distribution system. The investigators concluded that up to 15 percent of gastrointestinal illness may be associated with consumption of drinking water that was contaminated from main breaks or other pressure-loss events.

Nygård and colleagues (2007) examined 626 households that lived downstream from main breaks or planned maintenance that resulted in loss of pressure in seven distribution systems and 549 control families that lived outside the affected area. A total of 88 low-pressure episodes were examined, varying from 2 to 24 events per system. Only one system chlorinated the affected areas, but flushing was done in 87 of the cases. No boil-water notice was provided in any of the systems, and none maintained a disinfectant residual in the distribution system. The interview questionnaires were administered 1 week after the low-pressure event. Gastrointestinal illnesses occurred in 22.7 percent of the affected households compared with 8 percent of the controls, with an attributed fraction among exposed households of 37 percent. Households with higher average daily water consumption (>2 glasses per person per day) had higher [relative risk (RR) of 4.9, with a 95 percent confidence interval (CI) of 1.6–15.2) compared with a lower average daily water consumption (≤2 glasses per person per day). The authors concluded that breaks and maintenance work in the water distribution systems increased the risk of gastrointestinal illness among exposed households and that the public health burden caused by contamination of drinking water within the distribution network may be larger than previously anticipated.

MICROBES IN DISTRIBUTION SYSTEMS

Differentiation between Suspended and Biofilm Microbes

The microbiology of distribution systems essentially consists of two different environments—microorganisms suspended in the bulk water column and those in biofilms attached to the surfaces of pipes, sediments, and other materials. Microorganisms in the bulk water column originate from either the source water, bacterial growth within the treatment process (e.g., within the treatment filters), biofilms within the distribution system, or recontamination of the water from cross-connections, intrusion, pipe breaks, or other external sources. The presence of a disinfectant residual will have an important selective effect for suspended microorganisms in treated systems. Organisms that are most resistant to disinfectants include those with spores, cysts, and waxy or thick peptidoglycan cells walls. In general, gram-positive bacteria and acid-fast or partially acid-fast bacteria are more resistant to disinfection than gram-negative bacteria (Norton and LeChevallier, 2000).

Biofilms can form by attachment of microbes from the bulk water or can originate from organisms present on pipe surfaces when the system was installed. A biofilm is a collection of organic and inorganic, living and dead material collected on a surface (Characklis and Marshall, 1990). It may be a complete film or, more commonly in water systems,

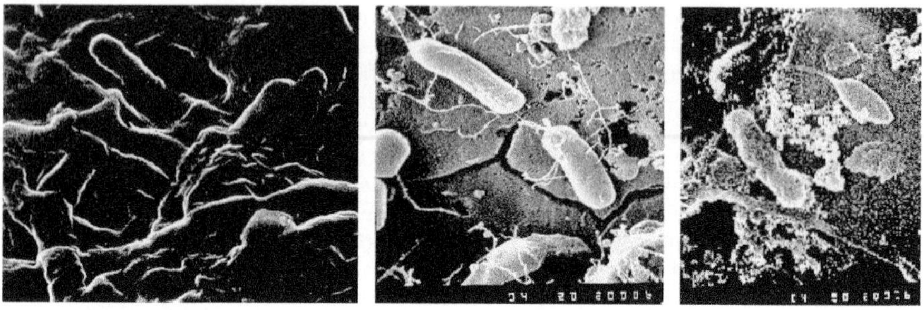

FIGURE 21-4 Examples of biofilm bacteria. (*Source: LeChevallier, 2005.*)

a small patch on a pipe surface. Microorganisms in biofilms can include bacteria [including coccoid (round), rod-shaped, filamentous, and appendaged bacteria] (Fig. 21-4), fungi, and higher organisms, such as nematodes, larvae, and crustaceans. The nature of the surface material can have a strong influence on the composition and activity of the attached microbial community.

Studies have shown that biofilms developed more quickly on iron pipe surfaces than on plastic polyvinyl chloride (PVC) pipes despite the fact that adequate corrosion control was applied, that the water was biologically treated to reduce assimilable organic carbon (AOC) levels, and that chlorine residuals were consistently maintained (Haas et al., 1983a; Camper, 1996). Iron pipes supported a more diverse microbial population than did PVC pipes (Norton and LeChevallier, 2000). Part of the reason that certain bacteria associate with certain pipe types is because materials may leach compounds that support bacterial growth. For example, pipe gaskets and elastic sealants (containing polyamide and silicone) can be a source of nutrients for bacterial proliferation. Colbourne and colleagues (1984) reported that *Legionella* were associated with certain rubber gaskets. Organisms associated with joint packing materials include populations of *Pseudomonas aeruginosa*, *Chromobacter* spp., *Enterobacter aerogenes*, and *Klebsiella pneumoniae* (Schoenen, 1986). Coating compounds for storage reservoirs and standpipes can contribute organic polymers and solvents that may support regrowth of heterotrophic bacteria (Schoenen, 1986; Thofern et al., 1987). Liner materials may contain bitumen, chlorinated rubber, epoxy resin, or tar–epoxy resin combinations that can support bacterial regrowth (Schoenen, 1986). Kilb and colleagues (2003) identified the source of coliforms in a number of nonchlorinated distribution systems in Germany as originating from biofilms grown on rubber-coated valves and concluded that usable carbon compounds leached from the coating material. PVC pipes and coating materials may leach stabilizers that can result in bacterial growth. Studies performed in the United Kingdom reported that coliform isolations were four times higher when samples were collected from plastic taps than from metallic faucets (Geldreich and LeChevallier, 1999). The purpose of these studies was not to indicate that certain pipe materials are preferred over another but to demonstrate the importance of considering the type of materials that come into contact with potable water. Although procedures are available to evaluate the growth-stimulation potential of different materials (Bellen et al., 1993), these tests are not applied in the United States. In parts of the distribution system where there is a large surface-to-volume ratio (such as smaller-diameter pipes), biofilm bacteria will have a greater impact on bulk water quality. The greater surface area of small pipes also increases reaction rates that deplete chlorine residuals. Because smaller-diameter pipes are found in buildings and homes, premise plumbing is particularly prone to growth of biofilm bacteria and the resulting water quality problems.

Pathogen Occurrence in Distribution Systems

Excluding opportunistic pathogens that grow in distribution system biofilms (e.g., *Legionella*, *Mycobacterium*, amoebae, etc.), pathogenic microbes in drinking water networks can be considered transitory (USEPA, 2002b), resulting from failures of the treatment process or subsequent pipeline contamination. Contamination events, however, can lead to transitory accumulation of pathogens within distribution system biofilms. For example, researchers have shown that viruses and parasites such as *Cryptosporidium* can be trapped in biofilms after a contamination event (Quignon et al., 1997; Piriou et al., 2000; Lehtola et al., 2007). Bacterial pathogens such as *Helicobacter pylori* (Mackay et al., 1998; Gião et al., 2008), enterotoxigenic *E. coli* (Szewzyk et al., 1994), *Salmonella typhimurium* (Armon et al., 1997), and *Campylobacter* spp. (Buswell et al., 1998) can persist within biofilms formed in experimental laboratory systems. How long these microbes persist in actual full-scale distribution systems is unknown, but it is thought that microbial competition, predation, and detachment from the biofilm eventually dissipate their numbers. For example, *E. coli* O157:H7 did not persist in biofilms after 20 days, but cells of *S. typhimurium* were detected by fluorescent antibody methods throughout the 60-day experiment (Camper et al., 1998). After this prolonged exposure, no *S. typhimurium* were recovered by culturing methods, but the organism was recovered after incubation in a dilute nonselective broth, indicating that low levels of viable cells still were present. Wolyniak et al., (2009) reported that the number of oocysts adsorbed to a biofilm surface was related to the thickness of the biofilm, with increased adsorption during the summer when biofilms were thicker than in the winter.

These findings are most important for rehabilitation of distribution systems after intentional or unintentional contamination events or after prolonged depressurization (e.g., following flooding or natural disaster), and extensive flushing and disinfection is recommended—even after samples from the water column are clean. A study of decontamination protocols for distribution system infrastructure (Welter et al., 2008) showed that the presence of biofilm and pipe tuberculation/scale tended to increase contaminant attachment to pipe surfaces. Decontamination protocols varied depending on the nature of the contaminant material, but flushing and chlorination may not be sufficient if the pipe is heavily tuberculated. In these cases, mechanical cleaning of the pipe may be necessary to remove both the tubercles and the contaminants.

Identification of Bacteria in Distribution Systems

Identification of microorganisms typically relies on culture methods that do not detect all microbes that may exist in water, and typically such methods recover only a fraction (10 percent or less) of viable organisms. In addition, most culture methods detect only relatively rapidly growing heterotrophic bacteria, whereas slowly growing organisms and fastidious or autotrophic organisms generally are not detected. Diagnostic kits are unreliable for many heterotrophic bacteria because the methodology often requires the analyst to perform a Gram stain, which is difficult because of the slow growth and acid-fast or partially acid-fast nature of bacteria surviving in disinfected drinking water.

Fatty acid profiling has been used to identify organisms from drinking water (Briganti and Wacker, 1995; Norton and LeChevallier, 2000), but the organisms were cultured prior to identification, and therefore, the limitations associated with culturing are still present. Additionally, for identification, the lipid profile must match an established profile in a database; these databases are predominated by clinical isolates. The use of fatty acid profiles has been further developed by Smith and colleagues (2000), who used biofilm samples without prior culturing to demonstrate that predominantly gram-negative bacteria were present, but no further identification was accomplished. A similar approach was taken by Keinanen et al. (2004), who compared profiles from two drinking water systems and showed that they differed, but again, no identifications were obtained.

Molecular methods offer the promise of a more complete determination of the microbiology of drinking water. DNA extraction and polymerase chain reaction (PCR) amplification and analysis of RNA sequences can be used to identify waterborne microbes (Amann et al., 1990; Amann et al., 1995; Liu and Stahl, 2006). These procedures can be combined with quantitative real-time PCR, fluorescence in situ hybridization, or flow cytometry to provide quantitative assessments of bacterial populations. However, careful quality assurance is necessary to ensure complete extraction and recovery of environmental DNA. In an early study in which fluorescence in situ hybridization (FISH) was used along with a redox dye to assess activity, probes specific for the domains Eucarya and Bacteria, the α, β, γ, and δ subclasses of Proteobacteria, the flavobacteria-cytophaga group and Legionellaceae were used. The research demonstrated that the β subclass of the Proteobacteria predominated (Kalmbach et al., 1997), although no further characterization of the individuals in this subclass was done. Another investigation used terminal restriction fragment length polymorphisms (T-RFLPs) to identify members of a biofilm consortium over a 3-year time period (Martiny et al., 2003). In this study, several organisms were identified (*Pseudomonas*, *Sphingomonas*, *Aquabacterium*, *Nitrospira*, *Planctomyces*, and *Acidobacterium*), but for the majority of peaks, no sequence match could be made. This group was unsuccessful in applying FISH probes to visualize the identified species. Other methods for developing microbial fingerprints include denaturing gradient gel electrophoresis (DGGE), PCR–amplicon-length heterogeneity (ALH), single-strand-conformation polymorphism (SSCP), and DNA microarray (Liu and Stahl, 2006).

The point of this section is to illustrate that any discussion of the types of microbes present in distribution systems should be taken with a bit of caution because all methods have some drawbacks and give an imperfect picture of the ecology of drinking water networks.

Heterotrophic Bacteria

Heterotrophic bacteria (a broad classification that takes into account all bacteria that use organic carbon) are commonly found in distribution systems because they readily form biofilms in such systems. They are measured by using heterotrophic plate counts (HPCs). As noted earlier, the composition of the HPC population changes as the water goes through treatment. Phylogenetic analysis of rRNA genes amplified from bulk water and biofilms samples were significantly different and included isolates of *Acidobacterium*, *Nitrospirae*, *Planctomyces*, and *Verrucomicrobium* (Martiny et al., 2005). Identification of bacteria using fatty acid analysis (Norton and LeChevallier, 2000) showed that chlorination resulted in a rapid shift from predominately gram-negative bacteria (97 percent) in the raw water to mostly gram-positive organisms (98 percent) in the chlorinated water (Table 21-1). Bacteria in the raw water were diverse, with *Acinetobacter* spp., *Pseudomonas* spp., and *Klebsiella* spp. predominant among the 20 genera identified. Ozonation of the raw water reduced the microbial diversity to 13 genera, dominated by *Pseudomonas* spp. and *Rhodococcus* spp. However, following biologically active granular activated carbon filtration, 19 genera were identified in the filter effluent, the majority of which (63 percent) matched isolates observed in the raw water. The predominant genera were *Pseudomonas* spp. and *Sphingomonas* spp., which are known to grow attached to the carbon fines of the filter while using natural organic compounds found in the aquatic environment. Final chlorination of the filtered water resulted in a shift to *Nocardia* spp. as the water entered the pipe system. *Nocardia* spp. possess characteristic fatty acids that are closely related to *Rhodococcus*, *Mycobacterium*, and

TABLE 21-1 Microbial Populations Isolated from the Water Column During Treatment

Bacterial identification	% Population raw water	% Population ozone contactor	% Population filter effluent[*]	% Population distribution system influent
Gram negative:				
Acidovorax spp.	2		4	7
Acinetobacter spp.	29	6		
Alcaligenes spp.	12	2	1	
Alteromonas spp.	2			
Comamonas spp.	1		3	
Enterobacter spp.	2		5	
Flavobacterium spp.	2		5	
Hydrogenophaga spp.	8	3	1	
Klebsiella spp.	10	1	3	
Methylobacterium spp.	1		2	
Pseudomonas spp.	14	53	22	
Rhodobacter spp.	2	1		
Sphingomonas spp.	2	2	19	
Stenotrophomonas spp.	2	1	2	
Xanthobacter spp.	3			
Others[†]	2	1	5	
Gram positive:				
Bacillus spp.				7
Nocardia spp.	1	3	7	53
Rhodococcus spp.		16	4	
Staphylococcus spp.	1	1		
Others[*]	1	1	1	
Unidentified:	3	9	16	33

[*]Biologically active GAC filter.
[†]Includes organisms isolated from only one site at a frequency of 1:100 isolates were identified from each site.
Source: Adapted from Norton and LeChevallier, 2000.

Corynebacterium. Its partially acid-fast cell wall and possession of the catalase enzyme, which breaks down hydrogen peroxide, are important factors that enable the organism to survive disinfection. Other gram-positive bacteria found in chlorinated drinking water include *Bacillus* spp. and *Staphylococcus* spp. *Bacillus* spp. form environmentally resistant spores that can withstand prolonged contact with chlorine. Some strains of *Bacillus* and *Staphylococcus aureus* can produce toxins when contaminated water is used in food preparation (LeChevallier and Seidler, 1980).

Treated drinking water will include a mixture of gram-negative and gram-positive HPC bacteria. In the absence of a disinfectant residual, gram-negative bacteria typically will out grow gram-positive bacteria and dominate the bacterial population. These organisms typically include *Pseudomonas*, *Acinetobacter*, *Flavobacterium*, and *Sphingomonas* spp. Figure 21-5 shows the microbial populations grown on PVC and iron pipe surfaces. *Stenotrophomonas* spp. dominated on plastic pipe, but *Nocardia*, *Acidovorax*, *Xanthobacter*, and *Pseudomonas* spp. were most dominant on iron pipe (Norton and LeChevallier, 2000). For the most part, these organisms have limited public health significance, except for *P. aeruginosa,* which is a possible opportunistic pathogen in drinking water and in the biofilms of water systems. The organism has been known to colonize point-of-use carbon

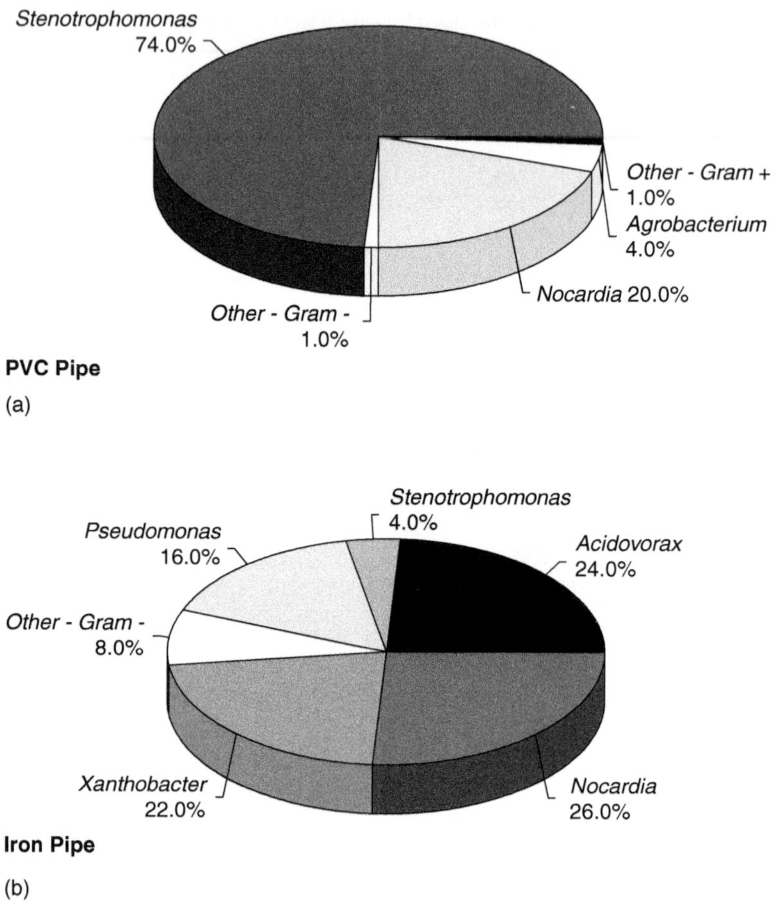

FIGURE 21-5 Microbial populations isolated from PVC (*a*) and iron pipe (*b*) surfaces. (*Source: Adapted from Norton and LeChevallier, 2000.*)

filters in drinking water systems (de Victoria and Galvan, 2001; Chaidez and Gerba, 2004). *P. aeruginosa* is of concern in bathing waters, especially in swimming pools and spas, where skin infections may result owing to exposure. In the case of drinking water, there are a few studies that suggest a relationship between the presence of this organism in the water and disease. In one hospital setting, 5 of 17 patients with a *Pseudomonas* infection carried a genotype also detected in the tap water (Trautmann et al., 2001). In another outbreak of pediatric *P. aeruginosa* urinary tract infections, two isolates had genotypes similar to those in the water. The outbreak was resolved when the taps in the unit were changed (Ferroni et al., 1998). In another instance, *P. aeruginosa* was used as an indicator of long storage and the potential for the presence of bacteria causing gastroenteritis in drinking water from a neighborhood in Mexico (de Victoria and Galvan, 2001).

With regard to the public health significance of HPC bacteria generally in drinking water, a workgroup convened by the World Health Organization (WHO) addressed this issue (Bartram et al., 2003). The WHO concluded that HPC bacteria were not associated with any adverse health effect: "In the absence of fecal contamination, there is no direct relationship between HPC values in ingested water and human health effects in the population at large" (Bartram et al., 2003). HPC levels can be high in foodstuffs, but there is no evidence of a health effect in the absence of pathogen contamination. One of the difficulties in interpreting the significance of HPC data is that test methods involve a wide variety of conditions that lead to a wide range of quantitative and qualitative results. Nonetheless, the USEPA recommends that HPC levels be maintained at or below 500 colony-forming units (CFU)/mL so as to not interfere with the detection of total coliform bacteria (USEPA, 1989).

Coliform Bacteria

Total coliform bacteria (a subset of gram-negative bacteria) are used primarily as a measure of water treatment effectiveness and occasionally can be found in distribution systems. The origins of total coliform bacteria include untreated surface and groundwater, vegetation, soils, insects, and animal and human fecal material. Typical coliform bacteria found in drinking water systems include *K. pneumoniae*, *Enterobacter aerogenes*, *E. cloacae*, and *Citrobacter freundii*. Other typical species and genera are shown in Table 21-2, although

TABLE 21-2 Coliform Isolates Typically Found in 111 Drinking Water Systems from Various Distribution Systems in Six States (United States) and Ontario Province (Canada)

Citrobacter	Enterobacter	Escherichia	Klebsiella
C. freundii	*E. aerogenes*	*E. coli*	*K. pneumonia*
C. diversus	*E. agglomerans*		*K. oxytoca*
	E. cloacae		*K. rhinoscleromatis*
			K. ozaena

Sources: From Geldreich et al., 1977; LeChevallier, Seidler, and Evans, 1980; Olson and Hanami, 1980; Herson and Victoreen, 1980; Reilly and Kippin, 1981; Clark et al., 1983; Staley, 1983.

the range of genera and species that can be recovered by various total coliform methods is expansive. Although most coliforms are not pathogenic, they can indicate the potential presence of fecal pathogens and thus may be a measure of public health risk. Indeed, the presence of coliforms in the distribution system usually is interpreted to indicate an external contamination event, such as injured organism passage through treatment barriers or introduction via water line breaks, cross-connections, or uncovered or poorly maintained finished water storage facilities (Geldreich et al., 1992; Clark et al., 1996). Biofilms within distribution systems can support the growth and release of coliforms, even when physical integrity (i.e., breaches in the treatment plant or distribution system) and disinfectant residual have been maintained (Characklis, 1988; Haudidier et al., 1988; Smith et al., 1990) such that their presence may not necessarily indicate a recent external contamination event.

Thermotolerant coliforms (capable of growth at 44.5°C), also termed *fecal coliforms*, have a higher association with fecal pollution than total coliforms. *E. coli* is considered to be even more directly related to fecal pollution because it is commonly found in the intestinal track of warm-blooded animals. Although most fecal coliform and *E. coli* strains are not pathogenic, some strains are invasive for intestinal cells and can produce heat-labile or heat-stable toxins (AWWA, 1999). *E. coli* and most of the thermotolerant coliforms do not grow in biofilms, although they most likely can be trapped and retained within biofilms (see below).

Stressed and Viable but Nonculturable Bacteria

Although most of the attention on stressed organisms has been on indicator bacteria (especially total and fecal coliforms), all organisms can become stressed. Stressed organisms are present under ordinary circumstances in treated drinking water and may be associated with partial or inadequate disinfection or the presence of metal ions or other toxic substances. These and other factors, including extremes of temperature and pH and solar radiation, may result in significant underestimations of the number of viable bacteria. In addition, laboratory manipulations following sample collection also may produce injury or act as a secondary stress to the organisms (McFeters et al., 1982; McFeters, 1990). These include excessive sample storage time, prolonged holding time (>30 minutes) of diluted samples before inoculation into growth media and of inoculated samples before incubation at the proper temperature, incorrect media formulations, incomplete mixing of sample with concentrated media, and exposure to untempered liquefied agar media. Excessive numbers of nonindicator bacteria also interfere with detection of indicators by causing injury (LeChevallier and McFeters, 1985b).

Coliforms also may enter the distribution system as injured organisms that pass through treatment barriers or by introduction into water line breaks, cross-connections, uncovered finished water storage, or low water pressure [<140 kPa (20 psi)]. For these reasons, more attention should be paid to the search for injured coliforms in an effort to provide a more sensitive measurement of water quality (LeChevallier and McFeters, 1985a; McFeters, 1990; Bucklin et al., 1991). To increase the sensitivity of detection of injured bacteria, the utility may process high-volume samples (≥500 mL) and use methods designed to detect stressed coliform bacteria (LeChevallier and McFeters, 1985a; McFeters, 1990) or HPC bacteria (Reasoner and Geldreich, 1985). For chloraminated water, inclusion of sodium sulfite in the media can reverse the oxidation of bacterial enzymes and increase microbial recoveries significantly (Watters et al., 1989).

The *viable but nonculturable* (VBNC) *response* is a temporary state during which organisms are not culturable using conventional nonselective culture conditions. The VBNC state has been reported for a wide variety of bacteria (Table 21-3). Whereas entrance into the VBNC state is easily demonstrated, proving resuscitation has been difficult for most bacterial species. In some cases, a simple temperature upshift allows for cellular resuscitation, whereas in other cases (such as *Legionella pneumophila*), where induction into the VBNC state was caused by nutrient depletion, resuscitated occurred when co-incubated with an *Acanthamoeba* host (Steinert et al., 1997; Oliver, 2005). Other mechanisms for the loss of culturability are likely to exist for other bacteria, but they are frequently unknown or not rigorously examined.

Organisms not well adapted to survival in a low-nutrient drinking water environment might persist in the VBNC state. Juhna and colleagues (2007) placed coupons made of cast iron, polyvinyl chloride, and stainless steel into 15 locations at six distribution networks in France and Latvia and examined them after 1 to 6 months of exposure to the drinking water. *E. coli* was detected in 56 percent of the coupons using a peptide nucleic acid (PNA) 15-mer probe in the fluorescence in situ hybridization (PNA FISH), but no *E. coli* was detected using culture or enzymatic methods (detection of β-D-glucuronidase). PCR analyses confirmed the presence of *E. coli* in samples that were negative according to culture-based and enzymatic methods. The viability of *E. coli* cells in the samples was demonstrated by cell elongation after resuscitation in low-nutrient medium supplemented with pipemidic acid, suggesting that the cells were present in an active but nonculturable state but unable to grow on agar medium.

Legionella

Legionella is regulated through the Surface Water Treatment Rule, with the maximum contaminant level goal (MCLG) set at zero (USEPA, 1999b). However, there is little evidence

TABLE 21-3 Bacteria Described to Enter the Viable but Not Culturable State

Aeromonas salmonicida	*Lactobacillus plantarum*	*Serratia marcescens*
Agrobacterium tumefaciens	*Lactococcus lactis*	*Shigella dysenteriae*
Alcaligenes eutrophus	*Legionella pneumophila*	*S. flexneri*
Aquaspirillum sp.	*Listeria monocytogenes*	*S. sonnei*
Burkholderia cepacia	*Micrococcus flavus*	*Sinorhizobium meliloti*
B. pseudomallei	*M. luteus*	*Streptococcus faecalis*
Campylobacter coli	*M. varians*	*Tenacibaculum sp.*
C. jejuni	*Mycobacterium tuberculosis*	*Vibrio anguillarum*
C. lari	*M. smegmatis*	*V. campbellii*
Cytophaga allerginae	*Pasteurella piscida*	*V. cholerae*
Enterobacter aerogenes	*Pseudomonas aeruginosa*	*V. fischeri*
E. cloacae	*P. fluorescens*	*V. harveyi*
Enterococcus faecalis	*P. putida*	*V. mimicus*
E. hirae	*P. syringae*	*V. natriegens*
E. faecium	*Ralstonia solanacearum*	*V. parahaemolyticus*
Escherichia coli (including EHEC)	*Rhizobium leguminosarum*	*V. proteolytica*
Francisella tularensis	*R. meliloti*	*V. shiloi*
Helicobacter pylori	*Rhodococcus rhodochrous*	*V. vulnificus* (types 1 and 2)
Klebsiella aerogenes	*Salmonella enteritidis*	*Xanthomonas campestris*
K. pneumoniae	*S. typhi*	
K. planticola	*S. typhimurium*	

Source: Adapted from Oliver, 2005.

that filtration and disinfection of surface water prevent the growth of *Legionella* spp. in distribution systems or premise plumbing. In 2001–2002, six drinking water outbreaks were attributed to *Legionella* spp. (19.4 percent of the total), caused illness in 80 persons, and resulted in 41 hospitalizations. All these outbreaks occurred in large buildings or institutional settings and were related to multiplication of *Legionella* spp. in the respective plumbing systems. In 2003–2004, *Legionella* was responsible for 8 of the 17 (61.5 percent) outbreaks with known etiologies, and in 2005–2006, *Legionella* was responsible for 10 of the 20 (50 percent) drinking water outbreaks (see Fig. 21-2). In fact, all 12 fatalities associated with drinking water from 2001 to 2006 were due to *Legionella*.

Chloramines appear to be more effective than free chlorine in reducing the risks from *Legionella*. Kool et al., (1999) examined 32 hospital-acquired (*nosocomial*) outbreaks of legionnaires' disease from 1979 to 1997 where drinking water was implicated and tabulated the characteristics of the hospital (i.e., size, transplant program) and the primary disinfectant treatment, disinfectant residual, water source, community size, and pH of the water. The researchers found that the odds of a nosocomial *Legionella* outbreak were 10.2 (95 percent CI 1.4–460) times higher in systems that maintained free chlorine than in those using a chloramine residual. Heffelfinger and colleagues (2003) reported that 25 percent (38) of 152 hospitals surveyed had reported cases or outbreaks of hospital-acquired legionnaires' disease during the period 1989–1998. However, hospitals supplied with drinking water disinfected with monochloramine were less likely (OR = 0.20, 95 percent CI 0.07–0.56) to have hospital-acquired legionnaires' disease than other hospitals. Cunliffe (1990) reported that suspensions of *L. pneumophila* were more sensitive to monochloramine disinfection, with a 99 percent level of inactivation when exposed to 1.0 mg of monochloramine per liter for 15 minutes compared with the 37-minute contact time required for *E. coli* inactivation under similar conditions. Donlan and colleagues (2002) reported that monochloramine was significantly more effective than free chlorine at eradicating laboratory-grown biofilms of *L. pneumophila.*

Legionella spp. have been shown to proliferate in biofilms in institutional and domestic plumbing and can be found in water heaters, shower heads, and cooling towers (Pryor et al., 2004; Thomas et al., 2006). There is evidence that the organism must be taken up by protozoa to proliferate (Murga et al., 2001). Epidemiological studies have linked water contaminated with both *Legionella* and protozoa to outbreaks of legionellosis (Moore et al., 2006). Flannery and colleagues (2006) showed a 93 percent reduction in the occurrence of *Legionella* spp. in building plumbing systems in San Francisco after the utility converted from free chlorine to chloramines. Amoebae at sampled sites were associated with *Legionella* spp. colonization predominately when chlorine was used for residual disinfection. *Legionella* spp. were cultured from 61 (36 percent) of 169 samples in which amoebae were present versus 291 (24 percent) of 1236 samples without amoebae ($p = .01$). After conversion to monochloramine, *Legionella* were found in 1 (1 percent) of 78 samples containing amoebae and 8 (1 percent) of 866 samples without amoebae ($p = .75$). The prevalence of amoebae decreased from 169 (12 percent) of 1405 samples when chlorine was the residual disinfectant to 78 (8 percent) of 944 samples collected after conversion to monochloramine ($p = .006$). *Legionella* occurrence in Pinellas County, FL, was reduced when the system converted from chlorine to monochloramine disinfection (Moore et al., 2006). Water samples were collected from 96 buildings (public buildings and individual homes) for a 4-month period when chlorine was the primary disinfectant and from the same sampling sites for a 4-month period after monochloramine was introduced into the municipal water system. When free chlorine was used, 20 percent of the buildings were colonized with *Legionella* in at least one sampling site. *Legionella* colonization was reduced by 69 percent within a month after chloramination. Monochloramine appeared to be more effective in reducing *Legionella* in hotels and single-family homes than in county government buildings perhaps because of more consistent water usage.

Aeromonas

Aeromonas spp. are gram-negative bacteria comprised of 19 recognized DNA groups, are found in fresh and salt waters, and cause a wide variety of human infections, including septicemia, wound infections, meningitis, pneumonia, respiratory infections, hemolytic-uremic syndrome, and gastroenteritis (Carnahan and Altwegg, 1996; Alavandi et al., 1999; Moyer, 2006). The ability of these microorganisms to grow at low temperatures and low levels of organic carbon [total organic carbon, AOC, or biodegradable dissolved organic carbon (BDOC)] is an important factor in their occurrence in drinking water supplies. Through the Unregulated Contaminant Monitoring Rule, USEPA examined the occurrence of *Aeromonas* spp. in 308 drinking water systems and found detectable concentrations in 2.6 percent of 5060 samples and in 13.6 percent of the systems. In a 16-month study conducted on the presence of *A. hydrophila* in drinking water in Indiana, 7.7 percent of the biofilm samples were positive for *A. hydrophila* (Chauret et al., 2001). The health significance of detecting aeromonads in drinking water is not well understood. Chopra (2008) examined 242 *Aeromonas* isolates obtained from water and clinical samples and concluded that both bacterial factors and host responses were responsible for virulence. Chopra (2008) identified both water and clinical isolates with a specific pulsotype based on pulse-field gel electrophoresis that possessed the same set of virulence factors. The European Community has set standards for aeromonads in drinking water leaving the treatment plant (<20 CFU/200 mL) and in the distribution system (<200 CFU/100 mL). Canada has an *Aeromonas* maximum contaminant level (MCL) of 0 CFU/100 mL for bottled drinking water (Moyer, 2006). *Aeromonas* are effectively controlled by disinfection practices used by public drinking water supplies to low levels that do not represent a threat to public health (USEPA, 2006a).

Mycobacterium

The genus *Mycobacterium* comprises both the pathogenic species that are transmitted by human or animal reservoirs (*M. tuberculosis* and *M. leprae*) and the so-called environmental nontuberculous mycobacteria (NTM) that are associated with soil and water. Although some of the NTM are increasingly reported as opportunistic pathogens, the primary risks in drinking water typically are associated with the *M. avium* complex (MAC), comprised of *M. avium* and *M. intracellulare* (Pedley et al., 2004). Studies have detected MAC organisms in drinking water distribution systems with concentrations ranging between 0.08 and 45,000 CFU/mL (LeChevallier, 2006). *M. avium* are resistant to disinfectants, especially free chlorine (Taylor et al., 2000). It is postulated that they may in fact be selected for in distribution systems as a result of their resistance to chlorine (Briganti and Wacker, 1995). However, there is also evidence that MAC organisms are susceptible to chlorine dioxide and chloramine (Vaerewijck et al., 2005). In fact, additions of compounds that cause oxidative stress have been shown to stimulate the formation of biofilms by mycobacteria (Geier et al., 2008).

Falkinham et al., (2001) examined eight well-characterized drinking water systems and reported that 20 percent of the water isolates and 64 percent of the biofilm isolates were identified as *M. avium* or *M. intracellulare*. Additionally, 8 percent of the water isolates were identified as *M. kansasii*. Most of these isolates were detected in raw water samples, with MAC organisms detected in five of six surface water sites ranging from 6 to 35 percent of the organisms isolated. MAC organisms were not detected in any plant or well effluent sample but were detected occasionally at low levels (<1 CFU/mL) in drinking water samples. However, *M. avium* and *M. intracellulare* were recovered frequently from drinking water biofilm samples, indicating that *M. avium* levels were growing in the distribution system. Increases in *M. avium* levels in drinking water were correlated with levels of AOC ($r^2 = 0.65$, $p = .029$) and BDOC ($r^2 = 0.64$, $p = .031$) (Falkinham et al., 2001; LeChevallier, 2004).

MAC organisms were the second most common opportunistic infection among children with human immunodeficiency virus (HIV) infection in the United States (second to *Pneumocystis pneumoniae*) prior to the availability of highly active antiretroviral treatment (HAART), but the incidence of MAC infections has decreased from 1.3 to 1.8 episodes/100 person-years to 0.14 to 0.2 episodes/100 person-years with the availability of HAART for HIV-infected individuals (Mofenson et al., 2009). MAC organisms are ubiquitous in the environment and presumably are acquired by routine exposures through inhalation, ingestion, or inoculation. Evidence for environmental transmission, especially in immunocompromised individuals, includes the following (Pedley et al., 2004):

- The frequency of gastrointestinal colonization increases as the stage of HIV advances.
- Isolation of MAC organisms from the gut occurs more often tha from the respiratory tract.
- Gastrointestinal symptoms (e.g., nausea, vomiting, and diarrhea) are common with MAC infections.

Because MAC organisms are ubiquitous in the environment, the CDC does not recommend any actions to avoid exposure from water, soil, or air. Instead, prophylactic HAART is recommended when levels of immunity (CD4 white blood cells) decline in HIV-infected individuals (Mofenson et al., 2009).

Free-Living Protozoa

Naegleria spp. and *Acanthamoeba* can be pathogenic, but these infections are usually associated with recreational rather than drinking waters. However, cysts also have been isolated from drinking water distribution systems in France (Jacquemin, Simitzia, and Chaneoau, 1981). *Acanthamoeba* is an opportunistic pathogen affecting contact lens wearers (USEPA, 2003a). A study in Arizona detected *Naegleria fowleri* DNA by PCR in 11 (7.7 percent) of 143 wells, and of 185 total samples, 30 (16.2 percent) tested positive for *N. fowleri* (Blair et al., 2008).

Of the genera of protozoa present in distribution systems, *Acanthamoeba*, *Hartmanella*, and *Naegleria* are known to feed on bacteria and biofilms by grazing. Transient association with amoebae has been reported for a number of different bacteria, including *L. pneumophila* (Abu Kwaik, 1996; Bozue and Johnson, 1996), *M. avium* (Adékambi et al., 2006), and other *Mycobacterium* spp. such as *M. leprae* (Adékambi et al., 2004; Yu et al., 2007). *Acanthamoeba* also can support intracellular growth of *Francisella*, the agent of tularaemia (Abd et al., 2003), the enteric pathogens *Listeria monocytogenes* (Ly and Müller, 1990) and *Helicobacter pylori* (Winiecka-Krusnell et al., 2002), *E. coli* O157, and the respiratory pathogen *Chlamydia pneumoniae* (Molmeret et al., 2005). Since most of these bacteria are human pathogens, amoebae have been suggested to represent environmental reservoirs for infection (Molmeret et al., 2005). However, some *Legionella*-like amoebic pathogens (LLAPs) appear to occur exclusively as endosymbionts of protozoa (Adeleke et al., 1996, Birtles et al., 1996, La Scola et al., 2004).

Specific PCR-mediated amplification of bacterial rRNA genes in free-living amoebae, comparative sequence analysis, and fluorescence in situ hybridization led to the identification of five novel evolutionary lineages of bacterial endosymbionts of free-living amoebae. These lineages have been found to be affiliated with (1) Alphaproteobacteriae, (2) Betaproteobacteriae, (3) Gamaproteobacteriae, (4) Bacteroides, and (5) Chlamydiales (Molmeret et al., 2005). Interestingly, co-occurrence of phylogenetically different endosymbionts in a single amoeba isolate has never been observed, and significant differences regarding the host range have been shown for the different endosymbionts. The observation by Flannery and colleagues (2006) that *Legionella* spp. were predominately associated with amoebae when chlorine was used for residual disinfection—36 percent of 169 samples in which amoebae were present) versus 1 (1 percent) of 78 samples containing amoebae when chloramines were used—suggested that the colonization of amoebae by *Legionella* spp. can be influenced by the type of disinfectant used. Understanding the relationship between opportunistic pathogens and free-living amoebae is clearly important in understanding the ecology of distribution system and the factors that contribute to human health risks.

Fungi

Although many fungi have been found in drinking water systems, their levels are typically low (<10 organisms/100 mL), and the organisms have not been directly associated with disease (Kelley et al., 2003). The origin of fungi in drinking water systems has not been well characterized, but it is assumed that they come from environmental sources, including surface water and groundwater, soils, and vegetation. Fungi occurrences in distributed water are more frequent during summer water temperature conditions. *Aspergillus fumigatus* was the predominant species detected in the distribution system water supplies in Finland (Niemi et al., 1982). A variety of fungi (*Cephalosporium* spp., *Verticillium* spp., *Trichodorma sporulosum*, *Nectria veridescens*, *Phoma* spp., and *Phialophora* spp.) were identified from water service mains in England (Bays et al., 1970; Dott and Waschko-Dransmann, 1981).

Nagy and Olson (1982) compared the distribution of fungal isolates in two distribution systems, one chlorinated and the other unchlorinated. In the unchlorinated system, *Penicillium* and *Acremonium* represented approximately 50 percent of 538 colonies identified among 14 genera, whereas *Sporocybe* and *Penicillium* accounted for 56 percent of 923 fungal strains distributed within 19 genera that occurred in the chlorinated supply. The mean densities of fungi from the unchlorinated and chlorinated systems were 18 and 34 organisms/100 mL, respectively.

The conidia of *A. fumigatus*, *A. niger*, and *P. oxalicum* isolated from distribution systems of three small water supplies in Pennsylvania showed a greater resistance to chlorine inactivation than yeast (Rosenzweig et al., 1983). The resistance of yeast to free available chlorine is primarily a result of the thick and rigid cell wall, which presents a greater permeability barrier to chlorine. Specific species identified in distribution waters include *Candida parapsilosis*, *C. famata*, *Cryptococcus laurentis*, *C. albidus*, *Rhodotorula glutinis*, *R. minuta*, and *R. rubra* (Hinzelin and Block, 1985; Engelbrecht and Haas, 1977). Densities of yeasts reported in finished drinking water average 1.5 organisms per liter.

Outside specialized research studies, potable water supplies are not tested routinely for fungi. Most of the focus on these organisms in water supplies has been related to their degradation of gasket and joint materials and taste and odor complaints by consumers (Burman, 1965; Bays et al., 1970; Niemi et al., 1982). Water supplies with fungal densities of 10 to 100 organisms/100 mL are frequently responsible for customer complaints of bad taste and odor (Burman, 1965).

Actinomyces and Other Related Organisms

These microorganisms create some of the objectionable tastes and odors reported in water supplies (see Chap. 3). *Actinomyces* in source waters of temperate climates may be reduced by approximately 10-fold with storage in holding basins prior to other treatment measures. Some drinking water treatment process configurations that include slow sand filtration or granulated activated carbon (GAC) filter adsorbers may support an increase in density and species that enter the distribution network. *Nocardia* spp. were predominant in finished water from water treatment trains consisting of aeration and filtration or aeration, sand filtration, ozonation, and activated carbon adsorption. *Micromonosporium* made up the greatest percentages in finished water from treatment trains consisting of flocculation, settling, and slow sand filtration or aeration and granular bed filtration. Some growth also may be expected on PVC-coated walls in finished water reservoirs (Dott and Waschko-Dransmann, 1981) and in the distribution lines where organic material accumulates in sediments (Burman, 1965; Bays et al., 1970). In addition to their presence in cold tap water, thermophilic *Actinomyces* and mesophilic fungi were found in several hot-water samples in three municipalities in Finland (Niemi et al., 1982). A 2004 study of 450 waterworks revealed that 10 percent had odor or taste problems related to the growth of fungi/*Actinomyces* in their distributed water (Miettinen et al., 2007). *Thermoactinomyces vulgaris* was the predominant actinomycetes found in 11 of 15 water distribution systems examined. In two studies of treated water supplies in England, chlorination alone was not effective in eliminating *Streptomyces*. Median densities were 2 streptomycetes per 100 mL of distributed water. Taste and odor complaints involving *Actinomyces* often were from waters that had *Streptomyces* or *Nocardia* spp. counts of more than 10 organisms per 100 mL (Bays et al., 1970).

Nitrifying Bacteria

Nitrification is a potential problem for utilities that use chloramines as a disinfectant residual, especially where there is an excess level of free ammonia. Nitrification is a microbial process by which ammonia is sequentially oxidized to nitrite and nitrate. The process is accomplished principally by two groups of chemolithotrophic bacteria (bacteria that obtain carbon and energy from inorganic sources), the ammonia-oxidizing bacteria (AOB) and the nitrite-oxidizing bacteria (NOB). Various groups of heterotrophic bacteria and fungi also can carry out nitrification, although at a slower rate than chemolithotrophic organisms (Verstraete and Alexander, 1973). In the first step of nitrification, ammonia is oxidized to nitrite:

$$NH_4 + \frac{3}{2}O_2 \rightarrow NO_2^- + H_2O + 2H^+ \qquad (21\text{-}1)$$

In the second step of the process, nitrite is oxidized to nitrate:

$$NO_2^- + H_2O \rightarrow NO_3^- + 2H^+ \qquad (21\text{-}2)$$

Nitrosomonas is the most frequently identified genus associated with the ammonia oxidation step, although other genera, including *Nitrosolobus*, *Nitrosococcus*, *Nitrosovibrio*, and *Nitrosospira* also can oxidize ammonia autotrophically (Watson et al., 1989). Recent development of molecular methods for the characterization of ammonia-oxidizing bacteria have shown that *Nitrosomonas oligotropha* is the commonly identified ammonia oxidizer in full- and pilot-scale chloraminated distribution systems (Regan et al., 2002, 2003). The molecular analysis also suggests that the currently identified species of ammonia-oxidizing bacteria are not representative of the sequences observed in environmental samples, indicating that many species are yet uncharacterized. Recently identified are the ammonia-oxidizing archaea (AOA), a novel microbial group that also catalyzes the aerobic autotrophic oxidation of ammonia to nitrite (Francis et al., 2005; Konneke et al., 2005), and these organisms have been detected in treatment processes and distribution systems where they were responsible for the removal of ammonia (van der Wielen et al., 2009).

Nitrobacter is the most frequently identified genus associated with nitrite oxidation, although other genera, including *Nitrospina*, *Nitrococcus*, and *Nitrospira*, also can oxidize nitrite autotrophically (Watson et al., 1989). Molecular analyses have identified *Nitrospira* and *Nitrobacter* as ubiquitous nitrite oxidizers in full- and pilot-scale chloraminated systems (Regan et al., 2002, 2003).

Iron-, Manganese-, and Sulfur-Using Bacteria

Other autotrophs (organisms that produce complex organic compounds from simple inorganic molecules using energy from inorganic chemical reactions) that occur in drinking water supplies include those which use iron, manganese, or sulfur as sources of energy. Iron bacteria derive energy by oxidizing dissolved ferrous iron, resulting in an insoluble ferric oxide that appears as a brown gelatinous slime. Iron bacteria can grow at levels as low as 0.1 μg/g of iron but require at least 0.3 mg/L of dissolved oxygen. Common genera of iron-oxidizing bacteria include *Gallionella*, *Sphaerotilus*, *Crenothrix*, *Sidelocapsa*, and *Leptothrix*.

Deposits of manganese can collect in distribution or plumbing systems as black sediments, and manganese-oxidizing bacteria often will be present in such sediments. A number of microorganisms have been reported to oxidize manganese, including isolates belonging to the genera *Pseudomonas*, *Comomonas*, *Cytophaga*, *Flexibacter*, *Arthrobacter*, *Bacillus*, *Corynebacterium*, *Rhodococcus*, *Leptothrix*, and *Metallogenium*, among others. *Metallogenium* can form star-shaped manganese oxide minerals (called *metallogenium*) through the action of manganese oxide precipitation along its surface. The USEPA (1979) has set a recommended limit of 0.05 mg/L for manganese for aesthetic reasons (i.e., taste, discoloration, and staining of laundry or plumbing), but a much lower limit is recommended to prevent chronic problems (see Chap. 7).

Sulfur is a compound that can be either oxidized or reduced by bacteria in drinking water systems. Sulfur-oxidizing bacteria can produce gelatinous, filamentous growths that can be white, yellow, or gray-brown. Sulfur-oxidizing bacteria include species of *Thiobacillus*, *Thiothrix*, and other sulfur-oxidizing bacteria within the e-Proteobacteria class. Sulfur-reducing bacteria live in oxygen-deficient environments such as deep wells, distribution system biofilms, water softeners, and water heaters and usually flourish at the higher temperatures in the hot-water system. Sulfur-reducing bacteria produce hydrogen sulfide, which produces an objectionable "rotten egg" smell and is corrosive to most metals. More than 130 species of sulfur-reducing bacteria have been described, and they comprise a phylogenetically diverse assemblage of organisms consisting of members of at least four bacterial phyla and one archaeal phylum. Most sulfate-reducing bacteria also can use other oxidized sulfur compounds such as sulfite and thiosulfate or elemental sulfur. In drinking water, *Desulfobulbus* and *Desulfovibrio* are the genera most commonly associated with sulfate-reducing bacteria in drinking water.

FACTORS CONTRIBUTING TO MICROBIAL OCCURRENCES IN DISTRIBUTION SYSTEMS

To ensure delivery of a high-quality potable water supply to each consumer, managers of public water supply systems must be continually vigilant for any occurrence of microbial degradation that may occur as a result of inadequate treatment or treatment upsets, microbial growth, or direct intrusions of contamination into the distribution network. This job is complicated by the very nature of a distribution system—a network of mains, fire hydrants, valves, auxiliary pumping, chlorination substations, storage reservoirs, standpipes, and service lines. After a microbial contamination event, any of these component parts may serve as a habitat suitable for colonization by certain microorganisms in the surviving flora. The persistence and potential growth of organisms in the pipe network are influenced by a variety of environmental conditions that include physical and chemical characteristics of the water, system age, type of pipe materials, and the availability of sites suitable for colonization, often located in slow-flow sections, dead ends, and areas of pipe corrosion activity.

Treatment Processes

Water supplies using a single barrier (disinfection) for surface water treatment will not prevent a variety of organisms (algae, protozoan, and multicellular worms and insect larvae) from entering the distribution system (Allen et al., 1980). While many of these organisms are not killed immediately by disinfectant concentrations and contact times ($C \times T$ values) that control coliforms and viruses (USEPA, 1999a), they die eventually because of the lack of sunlight (algae) or adverse habitat (multicellular worms and insect larvae). Disinfection is also less effective on a variety of environmental organisms that include spore-forming organisms (*Clostridium*), fungi, yeast, and protozoan cysts. All these more resistant organisms can be found in the pipe environment (NRC, 2006).

Clarification and filtration are important treatment barriers for protozoan cysts (*Entamoeba*, *Giardia*, and *Cryptosporidium*) and are more effective than the usual disinfectant concentration and contact times applied in processing raw water (Logsdon, 1987). Improperly operated filtration systems have been responsible for releasing concentrated numbers of entrapped cysts (*Giardia* and *Cryptosporidium*) as a result of improper filter backwashing procedures or filter bypasses and channelization within the filter bed (Amirtharajah and Wetgstein, 1980). Filter sand may become infested with nematodes from stream or lake-bottom sediments that shed into the process water and pass into the distribution system. Nematodes, rotifers, and protozoa have been observed in treatment plant effluent (van Lieverloo, 1997; Schreiber et al., 1997). While nematodes are not pathogenic, they can shelter viable bacteria ingested from source water or filter media beds and, if these food-chain organisms are not digested quickly, provide a passage for escaping survivors to reach the distribution system by offering the ingested bacteria a protection against disinfection (Locas et al., 2007; Ding et al., 1995).

Application of powdered carbon or granular activated carbon (GAC) adsorption/filtration or biologically activated carbon treatment (BAC) introduces other opportunities for bacteria to enter the microbial community of processed water. Highly adsorptive activated carbon (in either granular or powdered form) is often used in specific situations for removal of organic matter, including taste- and odor-causing substances, but the granular carbon must be changed periodically when the carbon bed reaches near saturation for expected removal rates. In contrast, the biologically activated carbon (BAC) process depends on the establishment of a permanent biofilm for degradation of conversion products created by ozonation of recalcitrant organics in process water, as well as biodegradable organic matter already in the water. Significantly higher heterotrophic bacterial densities were reported in distributed water from a full-scale treatment train using GAC (Symons et al., 1981) in comparison with a similar full-scale treatment train that did not employ GAC (Haas et al., 1983b).

Several coliform species (*Klebsiella*, *Enterobacter*, and *Citrobacter*) have been found to colonize GAC filters, grow during warm-water periods, and discharge into the process effluent (Stewart et al., 1990; Camper et al., 1987). Activated carbon particles also have been detected in finished water from several water plants using powdered activated carbon (PAC) or GAC treatment. Over 17 percent of finished water samples examined from nine water treatment facilities contained activated carbon particle fines colonized with coliform bacteria (Ridgway and Olson, 1981). These observations confirm that activated carbon fines provide a transport mechanism by which microorganisms penetrate treatment barriers and reach the distribution system. Other mechanisms that could be involved in transport of bacteria into the distribution system include release of aggregates or clumps of organisms from colonization sites in GAC/BAC filtration and by passage with unsettled coagulants. The particulate matter that embeds bacteria offers the microorganisms protection against disinfection (Gauthier et al., 1999a; Morin et al., 1999; Stringfellow et al., 1993).

The use of new and improved (molecular-based) microbial detection methods is suggesting that pathogenic microorganisms may be present in treated water following conventional treatment. Monitoring studies for the detection of specific microorganisms in treated drinking water meeting prescribed water quality regulations (absence of indicators) have shown that viable enteroviruses could be detected in treated water. Using an integrated cell culture/nested PCR approach followed by restriction enzyme analysis, Vivier and colleagues (2004) detected coxsackie B viruses in 11 and 16 percent of water samples from two surface water treatment plants (of a total of 172 samples). Infectious *Cryptosporidium* was detected in 1.4 percent of 1690 100 L finished water samples collected from 22 of 82 surface water treatment plants using cell culture–polymerase chain reaction (Aboytes et al., 2004). Water quality and treatment plant characteristics were no different for the positive sites than for the sites that had no oocyst detections. Consequently, the use of molecular techniques for the detection of pathogenic organisms is gradually confirming that the acceptable water quality indicators may not necessarily reflect the virus and parasitical content of treated water.

Microbial Growth in Distribution Systems

Microbial growth can be defined as the increase in bacterial numbers in the distribution system owing to cell reproduction. Significant growth always occurs at the expense of an organic or inorganic substrate. Most microbial growth is thought to occur in biofilms on distribution pipe surfaces. Key factors in the establishment of microbial colonization within a distribution system include water temperature, the ineffectiveness of disinfectant residuals, the availability of nutrients, corrosion and sediment accumulation, hydraulic effects, and nitrification (NRC, 2006; Besner et al., 2002; LeChevallier et al., 1996; LeChevallier, 1990).

Impact of Treatment. The types of processes used to treat source water are likely to affect the biological stability of drinking water supplies (biologically stable water does not promote the growth of microorganisms). Given the natural percolation of water through the soil environment, groundwaters typically are stable with little change in bacterial levels. However, the presence of methane, ferrous iron, reduced sulfur compounds, hydrogen gas, manganese, ammonia, and nitrite can serve as either carbon or energy sources that promote the growth of certain microbes. In some cases, excess ammonia levels have been related to serious bacterial growth problems (Wolfe et al., 1988; Rittmann and Snoeyink, 1984).

For surface waters, some treatment processes (including the type of coagulant, clarification process, filter media, and disinfection regime) can alter the biological stability of treated water (Volk and LeChevallier, 2000). Essential nutritive substances, including those naturally occurring and human-made, containing phosphorus, nitrogen, trace metals, and carbon are introduced in varying concentrations from source waters. These organics are discharged through municipal wastewater effluents, industrial wastes, and agricultural activities. While some of the nutrient content in dissolved organic carbon may be removed (20–80 percent) through conventional treatment, more attention to treatment refinements is needed to further reduce trace organic residuals. Applying ozone coupled with GAC or other equivalent biological treatment processes (e.g., for improved disinfection by-product precursor control) also can minimize the available organic materials and thereby provide fewer opportunities for microbial biofilm development and coliform growth. For water utilities with a relatively clean surface water source that rely only on disinfection treatment, a seasonal threat of organic contributions from humic substances, natural lignins, algal blooms, and recirculating bottom sediments during lake destratification always will exist. These organic materials pass into distribution pipe networks, where they stimulate growth of a wide range of aquatic bacteria. Unfiltered surface water supplies were found to be particularly susceptible to coliform growth by LeChevallier and colleagues (1996). Application of nanofiltration membranes can be effective for removal of many inorganic and large-molecular-weight carbon molecules, but small-molecular-weight substances may serve as nutrients for coliform growth downstream in the distribution system (Escobar and Randall, 1999).

Temperature. Water temperature affects all processes involved in microbiological water quality—microbial growth rate, disinfection efficiency, decay of disinfectant residual, corrosion rates, and distribution system hydraulics (increased water velocity from increased consumer demand) (LeChevallier, 1990). Water temperature above 10°C accelerates the growth of adapted organisms with slow generation times. On average, the occurrences of coliform bacteria in distributed water are significantly higher when water temperatures are greater than 15°C (Besner et al., 2001; Volk and LeChevallier, 2000; LeChevallier et al., 1996; Volk and Joret, 1994) and during summer months when water temperatures are at their highest. However, the minimum temperature at which coliform activity is observed may vary from system to system. Strains of coliform bacteria that are better adapted to grow at lower temperatures (psychrophiles) may be found in systems experiencing cold water.

Disinfectant Type and Residual. The maintenance of a disinfectant residual in distribution systems is recommended to control microbial regrowth and to act as a barrier against any contamination resulting from the loss of integrity of the system. Such a management approach is widely accepted in North America and the United Kingdom and is endorsed by the World Health Organization (Ainsworth, 2004; Bartram et al., 2003). However, in several European countries, distribution systems are operated with low or no secondary disinfectant mostly because of concerns about the formation of disinfection by-products and low tolerance of chlorine taste and odor. For example, some utilities located in The Netherlands and Germany are successfully producing and distributing hygienically safe and biologically stable drinking water without a disinfectant residual (van der Kooij et al., 1999; Fokken et al., 1998). However, in order to adopt a low, or no, secondary disinfectant residual strategy, water suppliers apply extensive treatment to limit microbial regrowth and care to prevent distribution system contamination, which could be difficult for all water systems considering the age and vulnerability of some distribution infrastructures.

It is known that free chlorine is less efficient for the inactivation of biofilm bacteria than for bacteria present in bulk water (Parent et al., 1996; Van der Wende and Characklis, 1990). However, relatively low chlorine concentrations (0.05–0.5 mg/L) have, in some cases, been successful in controlling the presence of coliform bacteria in distribution systems (Besner et al., 2001; Volk and LeChevallier, 2000; Volk and Joret, 1994), especially when combined with a good biological stability of the treated water (Gatel et al., 2000). By contrast, it has been reported that even high free chlorine doses (between 0.6 and 4.0 mg/L) were ineffective in controlling coliform occurrences in some cases (Schreppel and Geiss, 1996; LeChevallier et al., 1987; Wierenga, 1985). Even though increasing chlorine residuals have helped to control coliform occurrence in some cases, high chlorine may not always be an acceptable solution owing to the formation of potentially carcinogenic disinfection by-products. LeChevallier and colleagues (1996) recommended maintaining at the far ends of the distribution system at least 0.2 mg/L of free chlorine or 0.5 mg/L of chloramines to limit coliform occurrences.

Although free chlorine is a more effective disinfectant than chloramines, research has indicated that in some cases chloramines may penetrate and control coliform bacteria in biofilms better than free chlorine (LeChevallier, 1991; LeChevallier et al., 1990). The penetration of free chlorine into a biofilm has been shown to be limited by its fast reaction rate (DeBeer et al., 1994), with the consumption of the free chlorine residual before it can react with the bacterial components of the film (Chen and Stewart, 1996). Because chloramines react more slowly, they can diffuse into the biofilm and eventually inactivate attached bacteria. Stewart and colleagues (2000) showed that free chlorine did not penetrate alginate beads containing bacterial cells, but chloramines did penetrate into the alginate material and reduced bacterial levels nearly 1 million-fold over a 60-minute interval (2.5 mg/L chloramines, pH 8.9).

A case illustrating the effectiveness of a chloramine residual for controlling coliform occurrences attributed to biofilm regrowth in distribution pipelines was reported by Friedman and colleagues (2009). As shown in Fig. 21-6, the system experienced coliform occurrences even when free chlorine residuals averaged between 2 and 2.5 mg/L in the distribution system (about June 1991 through May 1993). When m-T7 medium, a technique that recovers injured bacteria (LeChevallier et al., 1983), was used, coliform occurrence rates ranged between 10 and 40 percent even during months when coliforms were not recovered on the standard m-Endo medium. Conversion of the disinfectant to chloramines in June 1993 resulted in dramatic decreases in coliform occurrences measured by both m-Endo and m-T7 media, and the bacteria were not detected in the finished drinking water for the 3 years following the change (results are shown until April 1994 in Fig. 21-6) (Norton and LeChevallier, 1997).

The ability of chloramines to penetrate biofilms also has led to a better control of *Legionella* in water systems (Flannery et al., 2006; Pryor et al., 2004). In San Francisco, a prospective study was conducted on 53 buildings to measure the impact of a conversion from chlorination to chloramination on the prevalence of *Legionella* in hot water systems.

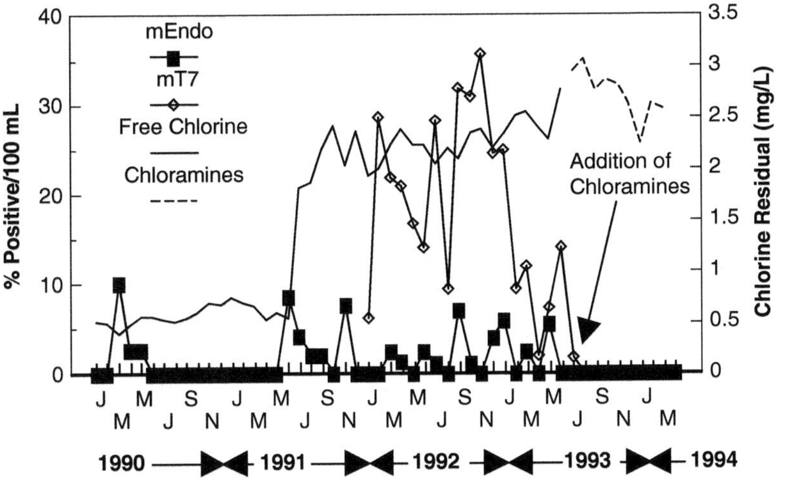

FIGURE 21-6 Coliform occurrence in a system before and after conversion from free chlorine to chloramines (coliform bacteria were enumerated using either m-T7 or m-Endo medium). (*Source: LeChevallier et al., 1996.*)

A striking decrease in the colonization of hot water systems from 60 percent before conversion to 4 percent after conversion led the authors to conclude that chloramination of municipal water systems decreases the risk for legionnaires' disease (Flannery et al., 2006). However, another study found higher levels of mycobacteria after chloramination (Pryor et al., 2004).

Because chloramines are less reactive, the residual can be carried and maintained at the ends of the system, which is another important advantage of chloramination, particularly if the system does not have a large network of storage tanks where stagnation and nitrification may become a problem. However, chloramines generally are less efficient for controlling a sudden pulse of contamination (Snead et al., 1980; Besner et al., 2008a) and also may have other inherent problems (nitrification, corrosion) and should be implemented after considering water quality and other potential unintended consequences.

Bacterial Nutrients. In drinking water, the limiting nutrient for the growth of heterotrophic bacteria is usually the biodegradable fraction of dissolved organic carbon (DOC), expressed in terms of AOC or BDOC concentrations. AOC is determined using a bioassay (Van der Kooij, 1990, 1992) and measures the microbial response to biodegradable materials in water, whereas BDOC is the difference in the concentration of DOC before and after bacterial growth in a sample and measures the amount of nutrient readily available for bacterial growth (Joret and Levi, 1986).

Studies have shown that AOC (LeChevallier et al., 1987) and BDOC (Servais et al., 1992, 1995) generally decrease with increased water residence time in the distribution system, which would suggest carbon use by bacteria and subsequent growth. Thus, decreasing the AOC and BDOC content of the water leaving the plant will help to limit bacterial regrowth in distribution pipes. Recommendations regarding the threshold concentration of biodegradable organic matter below which water can be considered biologically stable have been offered. van der Kooij and colleagues (1989) proposed an AOC value of 10 μg/L, and Servais and colleagues (1995) suggested a BDOC value of 0.15 mg of carbon per liter. Volk and LeChevallier (2000) determined that AOC levels in 94 North American distribution systems ranged from 20 to 214 μg/L, with a median of 100 μg/L, and that levels of BDOC from 30 North American water systems ranged from 0 to 1.7 mg/L, with a median level of 0.38 mg/L. Coliform occurrences in distribution systems have been associated with AOC concentrations greater than 50 μg/L (LeChevallier et al., 1991) and 100 μg/L (Volk and LeChevallier, 2000) and BDOC consumption in the distribution system higher than 0.15 mg of carbon per liter (Volk and Joret, 1994). Levels of AOC alone do not dictate the occurrence of coliform bacteria in drinking water but are only one factor. High water temperature, high AOC/BDOC levels, and low disinfectant residuals especially may result in microbial growth problems (Volk and LeChevallier, 2000).

Because heterotrophic bacteria require carbon, nitrogen, and phosphorus in a ratio of approximately 100:10:1 (C:N:P), organic carbon is often a growth-limiting nutrient. However, limitation of bacterial growth because of the availability of phosphorus also has been demonstrated in some distribution systems (Lehtola et al., 2002). In such cases, the use of phosphate-based corrosion inhibitors could lead to a decrease in biological stability of the water and result in regrowth (Miettinen et al., 1997). For nitrogen, it is generally assumed that it is not a growth-limiting factor. Excess ammonia resulting from the use of chloramination may result in the growth of nitrifying bacteria and in water quality problems related to nitrification (e.g., accelerated loss of chlorine residuals, increased nitrite levels, and stimulated growth of HPC bacteria). Rittmann and Snoeyink (1984) found that ammonia concentrations in groundwater supplies frequently were high enough to cause biological instability.

Corrosion, Pipe Materials, and Sediment Accumulation. Water mains, storage reservoirs, standpipes, joint connections, valves, service lines, and metering devices have the potential to be sites suitable for microbial habitation. No pipe material is immune from potential microbial colonization once suitable attachment sites are established. Given sufficient time, aggressive waters or microbial activity will initiate corrosion of metal pipe surfaces, and water characteristics may change the surface structure of asbestos/cement mains. The nature of the water chemistry and continuous movement of water under high-velocity conditions helps to prevent the buildup of chemical and microbial species that contribute to corrosion and pitting. Therefore, not all pipe sections may show evidence of deterioration even after years of active service.

Corrosion of pipe surfaces provides not only a habitat for bacterial proliferation but also is a source of substrate and protection from chlorine disinfectant residuals. In drinking water systems, the occurrence of coliform bacteria in corrosion tubercles on iron pipes has been reported by a number of investigators (Clement and Snoeyink, 1998; Emde et al., 1992 ; LeChevallier et al., 1987). Laboratory studies showed that the densities of HPC and coliform group bacteria were 10 times higher when grown on mild steel coupons than on noncorroded polycarbonate surfaces (Camper, 1996). The increased surface area owing to tuberculation of the pipe walls, the concentration of organic substances within the tubercles, and the secretion of organic compounds by iron-using bacteria have been postulated as reasons why iron corrosion stimulates bacteria growth.

LeChevallier and colleagues (1990) showed that the disinfection of biofilm on galvanized, copper, or PVC pipes was effective at 1 mg/L of free chlorine or monochloramine, but disinfection of organisms on iron pipes was ineffective even at free chlorine residuals as high as 5 mg/L for several weeks. Follow-up studies showed that a combination of the corrosion rate, the ratio of the molar concentration of chloride and sulfate to bicarbonate (known as the *Larson index*), the chloramine residual, and the level of corrosion inhibitor could account for 75 percent of the variation in biofilm disinfection rates for microorganisms grown on iron pipes (LeChevallier et al., 1993). It is generally agreed that unlined iron pipes induce the highest level of attached bacterial densities (Laurent et al., 2005), although site-specific conditions (nature and quantity of organic material present) may influence biofilm growth on various materials (Clement et al., 2003). The accumulation of fixed bacteria is also more rapid on iron than plastic surfaces [PVC and medium-density polyethylene (MDPE)] (Norton and LeChevallier, 2000; Kerr et al., 1999). Therefore, the choice of pipe material and the accumulation of corrosion products can have a dramatic impact on the ability to control the effects of biofilms in drinking water systems. Besides, biofilm bacteria also may play an important role in pipe corrosion. Iron-using bacteria (such as *Gallionella*) and sulfur-using bacteria (such as *Thiobacillus*) are the bacteria most often associated with pipe corrosion.

Coliform colonization of the pipe network also may occur in areas where porous sediments accumulate (Lu et al., 1997; Oliver and Harbour, 1995). These sediments may originate from the incomplete removal of particles from raw water, release of fines from filters, precipitation of metal oxides or calcium carbonate, external contamination in pipes and reservoirs, postflocculation, biological growth, and corrosion and then settle in low-flow and dead-end areas. Such sites are attractive to bacterial colonization because these deposits adsorb trace nutrients from the passing water and provide numerous surface areas for bacterial attachment against the flow of water. Although usually mostly mineral in nature, pipe and reservoir deposits include variable fractions of organic matter (Carrière et al., 2005; Gauthier et al., 1999b). The mineral fraction acts mainly as a support for bacteria, whereas the organic matter concentration drives the number of microorganisms (Gauthier et al., 1996). Sediments are problematic because they can shield microorganisms from disinfectants, exert a high disinfectant demand, and be the source of aesthetic problems when rapid changes in velocity, flow direction, or turbulence take place.

Residence Time and Hydraulic Effects. Various changes in water quality such as bacterial regrowth, loss of disinfectant residual, nitrification, formation of disinfection byproducts, taste and odor problems, and water discoloration may result from high water age. Storage facilities and dead-end mains where water velocities and turnover can be very low are locations that are more susceptible to water quality degradations.

Long water retention time may result from the mixing characteristics prevailing in storage tanks (short circuiting, stagnation zones), their operation, and location in the distribution system. Water is often not drawn from the bottom of distribution storage tanks, so accumulations of sediments may occur. These sediments can contribute significant amounts of assimilable organic complexes that support bacterial colonization of the distribution network. Loss of disinfectant residuals has been observed in storage tanks (Gauthier et al., 2000; Rossman et al., 1995). Subsequent microbiological proliferation such as coliform occurrences (Gauthier et al., 2000) and nitrification episodes (Baribeau et al., 2000; Burlingame et al., 1995) have been linked directly to tank operation in distribution systems. Dead-end mains may be characterized by long water residence times, which may result in lower chlorine residuals, higher turbidity, and higher HPC and iron concentrations (Barbeau et al., 1999), although microbial levels are not always higher systematically. This could be due to settling of bacteria during stagnant or low-flow conditions in the water main (Carter et al., 1997) or to the limited amount of nutrients supplied to the biofilm because of low water usage or their consumption by upstream biofilm (Lu et al., 1997).

Abrupt changes in flow conditions may result in biofilm being detached from the pipe wall and may lead to microbial water quality variations. Increase in shear stress can lead to a significant increase in effluent suspended bacteria concentration (Choi and Morgenroth, 2003). Bacterial levels in an experimental system were found to increase 10-fold when flows were started and stopped (Opheim et al., 1988). Reversal of water flows within the distribution system also can shear biofilms, and water hammer can dislodge tubercles from pipe surfaces (Lehtola et al., 2006). Clumps of bacteria were observed (using epifluorescence microscopy and on-line particle counters) following flow changes in a 1.3-km experimental drinking water distribution system (Maier et al., 1999).

Nitrification. Nitrification is a microbial process by which reduced nitrogen compounds (primarily ammonia) are sequentially oxidized to nitrite and nitrate. Nitrification can result in rapid loss of chloramine residual and considerable microbial growth, among other water quality transformations, all of which can be problematic in potable water distribution systems that use chloramines for residual (secondary) disinfection. A nitrification episode typically begins with steadily decreasing total chlorine residuals, increased heterotrophic bacteria counts (HPCs), free ammonia decreasing to near zero milligrams per liter, and increasing nitrite levels up and above 0.05 mg/L of nitrogen. The rapid increase of AOB in the bulk water (or attached to sediments, in biofilms, on pipe walls, or on reservoirs liners) results eventually in complete or almost-complete loss of total chlorine residual, conversion of free ammonia to nitrite and nitrate, and often a decrease in pH (AWWA, 2006b).

There are numerous conditions that can contribute to nitrification in distribution systems. These include

- Inadequate chloramine dose, resulting in insufficient inactivation of nitrifying bacteria
- Improper chlorine-to-ammonia ratio, resulting in excess ammonia entering the distribution system
- Excessive chloramine demand, resulting in liberation of free ammonia and reduced total chlorine residuals
- Unstable chloramine residual, resulting in autodecomposition and liberation of free ammonia
- Increased temperature associated with seasonal variations or extended water age in storage facilities or low-flow areas of the distribution system
- Unstable finished water pH and alkalinity, which affect the stability of the monochloramine residual and AOB inactivation
- Other distribution system characteristics, such as biofilm and scale/sediment buildup that can exert a disinfectant demand
- Blending chlorinated and chloraminated sources, resulting in nonoptimal chlorine-to-ammonia ratios

The AWWA Manual M56, *Fundamentals and Control of Nitrification in Chloraminated Drinking Water Distribution Systems* (AWWA, 2006b), provides an in-depth discussion of nitrification causes and control strategies.

Recontamination of the Distribution System

Deteriorated and badly maintained infrastructure used for distribution of drinking water certainly may lead to intrusion of pathogenic microorganisms into the distribution system. Loss of a distribution system's integrity may result in waterborne disease outbreaks, as in the case of Gideon, Mo., where contamination originating from a poorly maintained storage tank resulted in the death of 7 people and 600 others being sick owing to a *Salmonella* outbreak (Clark et al., 1996). However, intrusion of pathogenic microorganisms may take place even in relatively well-maintained systems through four direct routes of entry that include (1) cross-connections, (2) transitory contamination events through water main leakage points and other orifices, (3) covered and uncovered storage facilities, and (4) water main installation and repair sites (Kirmeyer et al., 2001). External contamination events may result in the introduction of microorganisms (that can colonize the pipe system), nutrients, and sediments and decreased residual disinfectant concentrations.

Cross-Connections. A *cross-connection* is defined as a connection between the potable drinking water supply and a nonpotable, undesirable, or contaminated source. Cross-connections pose a threat to distribution system integrity if they are not protected or if the device used for protection fails. Contaminated water may be introduced into distribution pipes from the backflow of water through the cross-connections owing to a differential in pressure between the connected systems, especially during low-pressure events in the distribution system (backsiphonage) or increased pressure from a nonpotable source (backpressure).

In the United States, the USEPA has compiled data from 459 backflow incidents and estimated 12,093 illnesses owing to those incidents between 1970 and 2001 (USEPA, 2002a). Two outbreaks were reported in Finland with cross-connections involving sewage and seawater, both showing evidence of the presence of *E. coli* (Lahti and Hiisvirta, 1995). In the Netherlands, cross-connection incidents between the drinking water supply system and water supply for nonpotable use was one of the reasons leading to the termination of projects where a dual water supply was installed in some new housing estates (Oesterholt et al., 2007).

Cross-connections can result in either chemical or biological contamination of the drinking water supply. Some common types include irrigation, fire systems, garden/washdown hoses, and boilers. Different degrees of hazard may be associated with cross-connections depending on the type of fluid to which the pipe is connected. For example, Duranceau and colleagues (1998) showed that fire systems may be less of a risk in terms of microbial contamination. Total coliforms were mostly absent from the 84 wet-pipe fire sprinkler systems investigated. The situation could be quite different in a case where the plumbing fixture was mistakenly connected to sewage.

Biological contaminants introduced through cross-connections are reported most often as sewage or nonspecific microbes, but pathogens such as *Giardia* and some strains of *E. coli* have been associated with cross-connection contamination events (USEPA, 2002a). Because of the pressure drop induced (and some other site-specific favorable conditions), some maintenance activities (meter repair, main shutdown for valve replacement, and main break) may result in backflow of contaminated water into distribution systems.

Most cross-connections generally take place inside the private plumbing system of the customer. A survey compiled from 719 water utilities showed that 83 percent of documented cross-connections were in private buildings, whereas 16 percent were in areas under the direct control of the water utility (Lee et al., 2003). Although more than half (55.5 percent) of the cross-connections identified by utilities were connected to non-health hazards, 39 percent were connected to health hazards and 5.5 percent to sewage. In the United States, although the program requirements may vary widely between states, 50 states have a requirement for some control of cross-connections and/or backflow prevention (USEPA, 2002a).

Intrusion of Contaminants. Transitory contamination events are related to the intrusion of contaminants into drinking water distribution systems resulting from the occurrence of low or negative pressures in water mains. In comparison with backflow events through cross-connections, *intrusion events* refer to the flow of nonpotable water into mains through pathways such as leakage points, submerged air valves, faulty seals, or other openings (LeChevallier et al., 2002). Pressure transients (also referred to as *surges* or *water hammers*) are caused by abrupt changes in the velocity of water, and multiple factors may lead to their occurrence (Table 21-4).

In a distribution system, intrusion events will result from the combination of three conditions: (1) occurrence of low or negative pressure, (2) presence of a pathway for intrusion, and (3) availability of a contaminant in the soil or water surrounding a pipe. Intrusion events can take place even if the pressure in the pipe remains positive, provided that the external

TABLE 21-4 Causes of Transient Pressure Events in Distribution Systems

Cause	Example
System operation	Controlled pump startup and shutdown
	Booster pump startup and shutdown
	Opening or closing a valve
	Malfunctioning air release/vacuum valve
	Malfunctioning pressure-release valve
	Altitude valve closure
	Air-valve slam
	Check-valve slam
	Opening or closing a fire hydrant
	Surge tank draining
	Feed tank draining
Service interruptions	Pump trip due to power failure
	Break in a pipeline
	Closure of an overhead storage tank
Sudden change in demand	Flushing operations
	Fire fighting
	Tank filling

Sources: Adapted from Friedman et al., 2004; Kirmeyer et al., 2001.

pressure on the pipe is greater than the internal water pressure. Leaking pipes located in areas of high water table would be locations with a high potential for intrusion when negative-pressure events take place. Although no data are available on the actual measurement of intruded material owing to the occurrence of low/negative-pressure transients in full-scale distribution systems, the following observations are discussed next.

Low- and Negative-Pressure Transients Occur in Full-Scale Distribution Systems. One of the first occurrences of low-pressure transients to be reported was by Walski and Lutes (1994). These were found to be caused by pump and valve operation. More recently, the measurement of transient pressures has been facilitated with the use of high-speed pressure-transient data loggers. This type of equipment allows the recording of several pressure measurements per second, thus helping in the detection of short transient events (that can last only a few seconds) that likely would go unnoticed with standard pressure recorders. Since 2000, several Water Research Foundation (formerly Awwa Research Foundation or AwwaRF) studies have been published on the investigation of low/negative-pressure transients in distribution systems (Kirmeyer et al., 2001; Friedman et al., 2004; LeChevallier et al., 2003; Fleming et al., 2006). Transient null (zero) or negative-pressure events have been reported by Gullick and colleagues (2004), Hooper and colleagues (2006), and Besner (2007) and Besner et al. (2010). In these studies, the main cause of negative pressure was identified as a sudden shutdown of pumps (at a treatment plant or pump station) owing to either unintentional (power outages) or intentional (pump stoppage/startup tests) circumstances or as a pressure drop at the point of entry in the system.

Microbial Contamination Present in the Vicinity of Some Intrusion Pathways. Possible pathways for pathogen intrusion include the ones related to pipeline structural integrity: pipe fracture/leaks, deflections at flexible couplings, leaking joints, and deteriorating seals. Leakage rates corresponding to 10 percent of the water produced are not uncommon in distribution systems (with some systems having even higher values). Consequently, leak-detection programs usually are recommended to minimize the risk from transitory events as well as for other economical and operational considerations. Analyses of soil and water samples collected at sites immediately exterior to drinking water pipelines during pipe repairs have shown various levels of microbial contamination. Karim and colleagues (2003) detected indicator microorganisms (total and fecal coliforms, *C. perfringens* spores, *B. subtilis* spores, and coliphage) and enteric viruses (detected by cell culture and RT-PCR) in more than 50 percent of the 65 samples examined from eight distribution systems. Fifty-six percent of the 32 sites investigated were positive by either cell culture or RT-PCR for one of the three viruses tested (enterovirus, Norwalk virus, and hepatitis A virus). A similar study by Besner et al. (2008b) showed lower evidence of fecal pollution and enteric viruses in the soil/water surrounding water mains in two Canadian distribution systems. In this case, *E. coli* was detected in only 1 soil sample out of 17, and no culturable enteric viruses were detected. The distance between the sewer and drinking water pipes, the conditions of soil saturation, and the geologic structure of the area are factors that can influence the level of fecal contamination detected. Leaking sewer mains certainly can contaminate the soil surrounding drinking water pipes because microorganisms can migrate significant distances in soil through macropores, worm holes, cracks, and fractures (Abu-Ashour et al., 1994). Transport mechanisms in soil involve advection and dispersion, which are influenced by the effects of filtration, adsorption, desorption, growth, decay, and sedimentation.

Submerged air-vacuum valves can be another pathway for microbial intrusion. These valves are specifically designed to exhaust large quantities of air when a pipe is being filled or to let air inside the main to prevent vacuum conditions from developing. Accordingly, the orifice may reach up to 600 mm (24 in) in some cases, providing a large entry point for contaminated water into the system if the orifice is completely flooded. In the United Kingdom, McMath and Casey (2000) measured subatmospheric pressures at one distribution system point downstream of a pumping station during surges. The authors observed that this point (an air-valve vault) was flooded with dirty water during wet weather, thus explaining the high coliform failure rate in the corresponding distribution system area. Flooded air-valve vaults may not be that uncommon. Besner (2007) and Besner et al. (2010) were able to collect water samples from 67 percent of 45 air-valve vaults in a distribution system, and 22 percent of these vaults had completely submerged air valves. In a survey of 26 water utilities, 12 utilities indicated that the number of flooded air-valve vaults in their system could vary between 0 to 80 percent of total vaults (Kirmeyer et al., 2001). The accumulated water found in air-valve vaults also may be contaminated with various microorganisms. Besner et al. (2010) found that the frequency and level of detection of various fecal indicator microorganisms (*E. coli*, *C. perfringens*, coliphages) were higher in the water from flooded air-valve vaults than in the soil and water from pipe trenches of a distribution system, although no enteric viruses were detected either by cell culture or by RT-PCR. The level of fecal contamination in the various sources investigated (soil and water from trenches and water from air-valve vaults) was more typical of that found in river water rather than in wastewater.

The evidence that intrusion of contaminated water may take place under negative-pressure transients has been demonstrated under pilot-system simulation conditions (Boyd et al., 2004a, 2004b). The use of a chemical tracer (cesium) method provided strong laboratory evidence that intrusion can occur during transient pressure events and that part of the intruded material remains in the pipe. Other information related to intrusion volumes has been obtained through the use of transient/surge analysis. Calculation of intrusion volumes under different scenarios such as power loss, main break, and fire flow have been performed, but the results were not field verified and served as a comparison purpose only (Kirmeyer et al., 2001; LeChevallier et al., 2003). It should be noted that there are many uncertainties involved in the modeling of intrusion volumes, and this is still a developing science.

Storage Facilities. The presence of open finished water reservoirs has been associated with microbial water quality problems (such as increased coliform occurrences) in distribution systems (LeChevallier et al., 1996). Such uncovered storage facilities are subject to microbial contamination from fecal material from birds and small mammals that come in contact with the water, air contaminants, and surface water runoff. The health concern with bird excrement is that this wildlife may be infected with *Salmonella* and pathogenic protozoa. Within the wildlife population in every area (as is true for any community of people), a persistent pool of infected individuals exists that shed pathogenic organisms in their fecal excretions. Sea gulls are scavengers and often are found at landfill locations and waste-discharge sites searching for food that is often contaminated with a variety of pathogens. At night, birds often return inland to aquatic areas, such as source water impoundments and open finished water reservoirs, to roost, thereby introducing pathogens through their fecal excrement. Increased protozoa concentrations have been measured following open reservoir storage, although the organisms appeared to be nonviable (LeChevallier et al., 1997). Air pollution contaminants and surface water runoff can contribute dirt, decaying leaves, lawn fertilizers, and accidental spills to a water supply that is not covered. Such materials increase productivity of the water by providing support to food-chain organisms and nitrogen-phosphate requirements for algal blooms. This degrades the treated water quality.

For covered storage facilities, appurtenances such as access hatches, overflow pipes, vents, roofs, and poorly constructed sidewall joints are potential portals of entry for airborne microorganisms. Air movement in or out of the vents as a result of water movement in the structure could lead to occasional contamination. During air transfer, the covered reservoir is exposed to fallout of dust and air pollution contaminants from the inflowing air. Air vents with defective screen protection also may allow birds and rodents to gain access to the tank. Bird or rodent excrement around the vents may enter the water supply and become transported into the distribution system before dilution is able to dissipate and residual disinfection inactivate the associated organisms. Tears or rips in floating covers can lead to the loss of chlorine residual (Harvey and Kopansky, 2007) or to contamination of treated water with accumulated rain water, potentially contaminated by bird droppings (Kirmeyer et al., 1999). Inadequate maintenance of a municipal water storage tank was identified as the cause of a *Salmonella* outbreak reported in Gideon, MO, in 1993 (Angulo et al., 1997; Clark et al., 1996). Pigeons roosting in elevated storage tanks were believed to be the source of *S. typhimurium* that contaminated the water supply.

Main Installation and Repairs. Water main repairs have been associated with waterborne disease outbreaks. In Cabool, Mo., record low temperatures in December 1989 caused 43 outdoor service meters to freeze and two major distribution lines to break (Geldreich et al., 1992). The meters had to be replaced (some were found to be submerged under water) and the pipes repaired. It is believed that contaminated runoff from an undersized and deteriorating sewer collection system found to be overflowing through manholes could have entered the system while the repairs were conducted. An outbreak of hemorragic *E. coli* serotype O157:H7 resulted from this contamination and caused 243 known cases of diarrhea and 4 deaths in the population of 2090 inhabitants. In Sweden, a waterborne disease outbreak affecting 10,000 people was due to a pipeline containing stagnant raw water that was brought into use without first being flushed (Andersson and Bohan, 2001). In France, the servicing of a large transmission main [600 mm (24 in)] after several weeks of construction work corresponded with an increase in diarrhea among the population, although the number of affected people could not be evaluated (Deshayes et al., 2001). Approximately 60,000 people were without drinking water for a 15-day period because fecal coliforms and sulfite-reducing clostridial spores were detected in the water.

Corrosion, unstable soil, faulting, land subsidence, extreme low temperatures, and other physical stresses often cause line breaks and necessitate repair or replacement of pipe sections. To avoid possible bacteriologic contamination of the water supply during these construction projects, a rigorous protective protocol must be followed. Microbial contamination arising from pipe repair activities can take place prior to, during, and after construction/repair work. Some potential contamination sources are listed in Table 21-5.

TABLE 21-5 Potential Sources of Contamination During Water Main Repairs

Phase	Potential contamination source	Relative level of risk	
		Common	Higher risk
Preinstallation exposure	Soil/sediment entry	X	
	Animal/human wastes		X
	Dead animals		X
	Unsanitary human contact	X	
	Incidental dirty water (runoff, sprays)	X	
	Trash/garbage	X	
Repair activities	Soil/sediment entry	X	
	Animal wastes		X
	Unsanitary human contact	X	
	Water from broken water pipe enters trench	X	
	Groundwater enters trench	X	
	Sewage water enters trench		X
	Agricultural runoff enters trench		X
	Rainfall fills trench	X	
Postrepair conditions	Transitory pressure variations	X	
	Backflow/cross-connection to service line	X	
	Cross-connection to system (hydrant)	X	
	Contaminated potable water supply		X

Source: Pierson et al., 2001.

Preinstallation contamination sources may deposit significant fecal material in the interior of pipe sections awaiting installation. Once introduced, this fecal material may become lodged in pipe fittings and valves. Thus such sites become protected habitats from which coliforms and any associated pathogens in the contaminated material may be released into the bulk flow of water supply. A case of contamination prior to installation has been reported by Haas and colleagues (1998). Although swabs of an internal surface of a 305-mm (12-in) pipe stored on the curbed side of the street and exposed to the local environment were negative for the presence of total coliforms, 75 percent of the swabs were positive for noncoliform bacterial growth on membrane filters, and 100 percent were positive for HPCs. The presence of noncoliform growth can be used as an indicator for contamination (as done by the Philadelphia Water Department), reflecting some impact of pipe storage conditions prior to installation. In Halifax, Nova Scotia, a new pipeline was found to contain pieces of wood used during construction work (Martin et al., 1982). Coliform bacteria from the wood were transported downstream, resulting in persistent positive bacteriologic results. Since the wood debris was difficult to remove, the problem was resolved by the addition of more than 5 mg/L of lime to the process water, which elevated the water pH to 9.1 in the distribution system. At this pH, *Klebsiella* were either inactivated or entrapped in the resulting carbonate scale, and the problem was eliminated.

During repair activities, the main points of concern are unsanitary construction practices and contamination from soil and trench water exposure [from a survey of 250 water utilities conducted by Haas and colleagues (1998)]. Soil and water found in main repair trenches may be contaminated with various levels of fecal indicators and pathogenic microorganisms (Karim et al., 2003; Besner et al., 2008b). Factors that may affect the potential for contamination during repair work include shutdown time, nature of the excavation, method of repair, storage or cleanliness of materials, and size of the job. As such, to minimize the potential contamination, main repairs should be conducted *under pressure*, where the water main is not isolated or completely shut down by closing valves. However, this may not always be feasible, and many water utilities report that repairs of water mains under pressure are not conducted (Kirmeyer et al., 2001). Water quality changes also may take place in areas adjacent to construction or repair sites. Valves that are closed to isolate the section of main to be repaired may result in changes in flow direction and temporary dead ends in the network area still providing water to customers. In such areas, loss of chlorine residuals can take place (Besner et al., 2008b). When flushing of the repaired main is conducted, upstream distribution system areas also may see increases in turbidity, especially if the utility does not conduct a unidirectional flushing program. Increase in the water velocity may result in the resuspension of accumulated sediments.

After the repair is completed, flushing of the pipe, disinfection, and water quality testing are the recommended actions for contamination control (Pierson et al., 2001). Besner et al. (2008b) investigated 16 planned water main repairs to evaluate the risk of microbial contamination associated with this activity in two chlorinated distribution systems. Overall, 424 water samples were collected from customers' homes and flushed hydrants, and 4 percent of these were found positive for total coliforms. All but one positive sample were collected during the flushing step, suggesting that adequate pipe flushing was effective in removing potential contamination from the distribution system. In this study, *E. coli* was detected on one occasion in the water flushed from a hydrant. In the Netherlands, results of water quality testing after invasive distribution system operations provided by three water companies showed that 0.57 percent of 16,047 water samples collected after operations contained *E. coli* or thermotolerant coliforms (van Lieverloo et al., 2006). A study in Norway investigated the link between pipe breaks and maintenance work (with presumed loss of water pressure) in distribution systems and risk of gastrointestinal illness among water recipients (Nygård et al., 2007). Following 88 low-pressure episodes taking place in seven distribution systems, exposed and unexposed households were selected and interviewed. During a 1-week period after the episode, 12.7 percent of the exposed households reported HCGI compared with 8 percent in the unexposed households showing an increased risk of HCGI for households exposed to pressure-loss events. Unfortunately, little water quality data were available to correlate with the results obtained. Water samples were available for only 18 episodes, and one sample was found positive for *E. coli*. No information regarding the level of residual disinfectant used in the distribution systems was given.

MONITORING DISTRIBUTION SYSTEMS

Water quality in distribution systems depends on a series of complex reactions that vary in both space and time, making it difficult to accurately assess the true status of water quality. To maximize the value of the investment in monitoring, a carefully designed sampling program should be implemented.

Water utilities in the United States collect samples from distribution systems to meet the requirements of several different regulations, including the Total Coliform Rule (TCR) and the Disinfection By-Product Rule (DBPR). Yet utilities are given little guidance from regulatory agencies about the collection of samples from the distribution system, particularly regarding the time and location of sample collection. For example, the requiremenrs for the selection of sampling sites for coliform monitoring vary widely from state to state

(Narisimhan et al., 2004). In addition to regulatory sampling, utilities often conduct operational monitoring and special sampling for a variety of different goals.

Regulatory Requirements

The USEPA regulates water quality in the distribution system for U.S. water utilities under several different rules. Utilities must meet regulatory requirements for disinfectant residual, disinfection by-products, lead and copper, and total coliform bacteria. The USEPA in 2007 and 2008 convened a Federal Advisory Committee to revise the total coliform rule (TCR), including provisions related to monitoring (USEPA, 2008b). Water utilities perform a significant amount of distribution system water quality monitoring to meet these regulatory requirements, as summarized in Table 21-6.

TABLE 21-6 Water Quality Parameters and Associated U.S. Federal Regulations for Water Distribution Systems

Parameter	Sample location	Regulatory limit	Reference	Comments
Disinfectant residual*	Entry point to distribution system	Minimum 0.2 mg/L on a continuous basis	Surface Water Treatment Rule (SWTR)	Only applies to systems using surface water supplies or groundwater supplies under the influence of surface water. In U.S., *Legionella* is also regulated by a treatment technique
Disinfectant residual	Distribution system	Maximum residual disinfectant level (MRDL) for chlorine of 4.0 mg/L; MRDL for chloramine 4.0 mg/L, running annual average	D/DBP Rule Stage 1	Applies to all public water systems that add a primary or secondary disinfectant to the water
Disinfectant residual or heterotrophic bacteria count	Throughout distribution system	Detectable level of disinfectant residual or heterotrophic bacteria count of 500 or less CFU/mL in 95% of samples collected each month for any two consecutive months	SWTR	Only applies to systems using surface water supplies
Total trihalomethanes (TTHMs)	Throughout distribution system	80 μg/L, locational running annual average based on quarterly samples	D/DBP Rule Stage 1	
Haloacetic acids (HAA5)	Throughout distribution system	60 μg/L, locational running annual average based on quarterly samples	D/DBP Rule Stage 1	

(Continued)

TABLE 21-6 Water Quality Parameters and Associated U.S. Federal Regulations for Water Distribution Systems (*Continued*)

Parameter	Sample location	Regulatory limit	Reference	Comments
Total coliform bacteriaw	Throughout distribution system	5% positive for systems collecting ≥40 samples per month. >1 positive for systems collecting <40 samples per month	Total Coliform Rule	Number of samples determined by population served (generally 1 sample per 1,000 people served per month).
Lead and copper	At customer taps	Action levels: Lead 0.015 mg/L at 90% Copper 1.3 mg/L at 90%	Lead and Copper Rule	Number of samples determined by population served

*Disinfectant residual may be regulated for some systems using groundwater supplies under the forthcoming Groundwater Rule.

***Source*:** Speight et al., 2009.

In addition to regulatory sampling, utilities often conduct additional sampling for specific projects. These projects may include assessment of the impact of operational adjustments and determination of compliance with future regulations. Water quality and pressure monitoring are used to help evaluate and optimize distribution system operational issues such as pressure maintenance, water age, nitrification, blending, storage tank operations, and emergency response. Water quality monitoring is also an important tool in planning and conducting maintenance activities. For example, color and turbidity data may be used to prioritize areas to be flushed and cleaned. Water quality testing can be viewed as a nonintrusive method for inspecting water mains and can be used to identify an underground leak by distinguishing between natural groundwater and treated water (Speight et al., 2009).

Measurement of Water Quality Parameters

The value of water quality data is influenced by the practices and procedures used to collect representative samples. Sample collection methods include grab sampling from taps or dedicated sampling devices and on-line continuous sample collection and analysis. Standard sample collection procedures are described in various AWWA publications (AWWA, 1985, 1990). Additional guidance can be found in American Society for Testing and Materials (ASTM) Standard D3370 for grab, composite, and continuous sampling and ASTM Standard D4840 for sampling chain-of-custody forms as well as in *Standard Methods for the Examination of Water and Wastewater* (APHA et al., 2005).

General Sampling Techniques. Water quality samples must be collected with care to minimize contamination and ensure meaningful results. Each test parameter that will be evaluated in the laboratory has a unique collection and storage requirement. Essential elements for the collection of good distribution system samples and associated field data include overall cleanliness of the sampling site, proper preparation of the sample collection container, adherence to aseptic sampling techniques, and proper transportation of samples (Kirmeyer et al., 2000).

Grab samples can be collected from sampling taps in a secure building, vault, or finished water storage facility. Ideally, the tap would be dedicated for water quality sampling only. Sample taps in a finished water storage facility can provide sample collection from several fixed depths and eliminate the time and safety concerns that may be involved in sampling through the access hatches. Dedicated sampling stations offer many advantages over the more traditional faucet or dedicated sampling tap. Burlingame and O'Donnell (1994) give examples of these advantages, which include the ability to sample water flowing in the mains without any potential impact from or interference with a sample site's internal plumbing, accessibility to areas where access to public buildings or residential sites is limited, as well as the ability to be disassembled, cleaned, and sanitized at regular intervals. Disadvantages of using dedicated sampling stations include a susceptibility to vandalism and the need to plan for the safe disposal and dechlorination of water flushed from these devices (Speight et al., 2009).

Burlingame and Choi (1998) explain why sampling from hydrants may not yield samples representative of the distribution system: "Hydrants are designed to be fully opened or closed. There is a drainage channel and port in a hydrant that allows water trapped in the barrel to drain out when the hydrant is shut down. If the hydrant is not open all the way, this drain hole is not closed all the way and water from the drain area can be siphoned back into the water flowing through the barrel."

Sampling in storage tanks can be performed using a depth-sampling device, also commonly used for lake sampling, that has been cleaned and disinfected. Such sampling devices are designed to be engaged and closed at predetermined depths within the storage facility, enabling sampling at varying depths from a single entry point. To ensure that a consistent sampling depth is maintained between sampling periods and among sampling teams, a calibrated cord should be employed when lowering the sampler to a prescribed depth (Kirmeyer et al., 1999). Alternatively, a sampling pump can be used to collect a grab sample at an access hatch (Speight et al., 2009).

On-Line Monitoring. On-line monitoring equipment offers utilities an opportunity to collect continuous data from strategic points in the distribution system. While chlorine residual is the most commonly monitored parameter in the distribution system, on-line monitors are also used for pH, turbidity, color, conductivity, oxidation reduction potential (ORP), temperature, total organic carbon (TOC), UV-254, and system pressure. The primary limitations of on-line monitoring at this time are technology and cost, but advances may make on-line monitoring for a wider range of parameters more readily usable in the future. General reviews of on-line monitoring technology are available in AWWA (2006a), USEPA (2005a), ASCE (2004), and AwwaRF (2002).

Prior to 2001, on-line monitoring in distribution systems generally was limited to chlorine monitors in some larger utilities. With the increased water security concerns following the terrorist attacks of 2001, the water industry sought an available technology that could detect a wider range of potential contaminants. The primary solution has been the installation of an array of available monitors that, in tandem, hopefully would respond to a broad range of contaminants. A panel or suite of monitors typically will consist of some or all of the following monitors: pH, conductivity, temperature, ORP, turbidity, dissolved oxygen (DO), pressure, chlorine, and total organic carbon (TOC). Recently, UV spectrophotometers have been installed in some systems as a surrogate for TOC. An automated sampler that is activated when one or more of the monitors trigger an alarm is also frequently included (Grayman et al., 2007).

It is generally recognized that chlorine and TOC are the two most important components in detecting and maybe identifying the broadest range of contaminants. An approximate cost for all the instruments just described is in the $40,000 to $50,000 range (2009 dollars). However, depending on the specific location and availability of shelter, electricity, sewer, and related items, the fully installed cost could increase to $250,000 or even more (Grayman et al., 2007).

Other advanced on-line monitors that have been used (less commonly) to monitor water in distribution systems include gas chromatographs, fluorometers, photoionization units, and radiation monitors. In general, these instruments are relatively expensive and have been used very sparingly in distribution systems with installation primarily in laboratory settings. There is also experimental use of biological monitors in distribution systems. These instruments monitor the response of living organisms such as fish, mussels, algae, daphnia, and bacteria to the water and set off an alarm when an unusual response is seen. Although used quite widely as part of early-warning systems in rivers, their adaptation to distribution systems has been slow owing to cost, maintenance, and the need to dechlorinate the water. Various grab sample and manual tests have supplemented on-line monitoring primarily as part of a screening or confirmation process. These tests include toxicity tests, PCR, and immunoassays (Grayman et al., 2007).

There are many emerging technologies within the monitoring area. These include on-line toxicity tests, ultrafiltration for contaminant concentration and separation, automatic pathogen detection (APD) systems, microchip technology, cell and tissue biosensors, and multiangle light-scattering measurements. In most cases, these technologies are being developed by commercial interests. Actual implementation for these technologies as affordable and dependable products is likely to be 2 to 10 years in the future (Grayman et al., 2007).

Theoretical Approaches to Sampling Plan Design

While the practical aspects of sampling for water quality in distribution systems, such as access to suitable sample sites, often dictate the sampling plan design, there are elements of theoretical sampling theory that can be applied to strengthen confidence in the data that are collected. Randomization is an important part of a statistically based sampling methodology. When samples are selected without using randomization, bias can be introduced into the results. For water quality in distribution systems, samples collected from locations receiving fresh water from the treatment plant may not reflect the water quality in remote areas of the system, possibly giving a false impression that water quality is acceptable everywhere. The degree of bias introduced by the use of nonrandom sampling depends on the variability in the system under investigation. If the parameters being sampled are relatively constant over space and time, the amount of bias from nonrandom sampling may be negligible. Logistically, it is difficult for water utilities to sample at all locations throughout a distribution system, so the goal of true random sampling may not be achievable. However, utilities should strive to design sampling programs that incorporate some elements of sampling design theory to maximize the effectiveness of their investment in sample collection and analysis.

The critical first step in developing a sampling protocol for a water distribution system is to define the goal of the sampling event. Often there will be multiple sampling goals. For quality-control purposes, the sampling goal may be to ensure that a certain percentage of the samples meets a predetermined water quality level (e.g., free chlorine concentration above 0.2 mg/L). For operational purposes, a sampling program may be designed to identify areas of the system where water quality problems are occurring (e.g., in cast iron pipes or in a specific pressure zone). The goal of the sampling program dictates the total number of samples to be collected and the allocation of samples across space and time.

Number of Samples. Sampling is used to estimate information about a population, which in this case is the population of all water quality samples in a distribution system, by collecting a small number of data points. Information from the small number of samples then is used to estimate properties about the population. The total number of samples n required to meet a sampling goal can be calculated for a given sampling program based on the desired precision of the estimate and historical information about the system to be sampled.

A commonly used measure of precision for an estimate obtained from samples is the margin of error d. The precision of an estimate is linked to the variability in the parameters being sampled. An upper bound on the number of samples that would be required to provide a given margin of error at a desired confidence level can be determined using simple random sampling variance estimator (Kish, 1995):

$$V = \frac{S^2}{n} = \left(\frac{d^2}{t^2}\right) \tag{21-3}$$

where V is the variance of the estimate, t is the Student's t statistic for the desired confidence level (equal to 1.960 for the 95 percent confidence level, two-tailed), S is the standard deviation, and n is the sample size. Historical data or data from other similar studies can be used to estimate S for the system and parameter to be sampled. Equation 21-3 does not depend on the size of the population, which in this case is the total of all potential 100-mL samples of water quality throughout the distribution system, because it is assumed that the population is much greater in size than the number of samples. For cases where the population is closer in size to the sample, a finite population correction factor would be added (Kish, 1995).

Selection of the appropriate margin of error for a sampling program also must consider the accuracy of the analytical method. If a test method has a large analytical error, a larger sample size will be required to estimate parameters accurately from the data set.

Example 21-1 Determine Sample Size

A water utility would like to measure the incidence of low chlorine residual throughout the distribution system using the average chlorine concentration. An analysis of the results for October for chlorine residual samples has yielded the following:

- Number of October samples = 200.
- Average chlorine concentration $\mu = 1.04$ mg/L.
- Standard deviation of chlorine concentration $S = 0.20$ mg/L.

Rearranging Eq. 21-4 for three different margin-of-error values and three different confidence levels to assess the impact of those parameters on the number of samples, we obtain Eq. 21-4 and the results in the following table:

$$n = \frac{S^2 t^2}{d^2} \tag{21.4}$$

Number of Samples n Based on $S = 0.20$

Confidence level	Margin of error d (mg/L)		
(t statistic, two-sided)	0.05	0.1	0.15
99% (2.57)	106	26	12
95% (1.96)	61	15	7
80% (1.28)	26	7	3

Therefore, the utility could take 26 samples to provide that the estimate of chlorine residual would yield a margin of error of 0.1 mg/L with a confidence of 99 percent. This sample size is quite small and reflects the low variability in the historical data, with a standard deviation of 0.2 mg/L. Consider the difference if the utility had analyzed 1 year of historical chlorine residual data and found a standard deviation of 0.42 mg/L for that data set. Recalculating the number of samples would yield the following:

Number of Samples n Based on $S = 0.42$

Confidence level (t statistic, two-sided)	Margin of error d (mg/L) 0.05	0.1	0.15
99% (2.57)	466	117	52
95% (1.96)	271	68	30
80% (1.28)	116	29	13

For the second case, the increase in variability in the historical data set is reflected in an increase in the number of samples required to achieve a given statistical precision. The utility now would need to take 117 samples instead of 26. ▲

To further refine the sampling design, the system to be sampled can be divided into categories, or strata, to allow for analysis of both the overall water quality and the water quality within each stratum. This type of design can be helpful in understanding differences in water quality behavior and in identifying underlying factors contributing to problems. For the best results, the stratification variables used to categorize samples should be correlated with the outcome measurement and should address the sampling goal.

Example 21-2 Allocation of Samples to Strata
The primary goal of a special sampling program is to determine the extent of low chlorine residuals within the entire distribution system, defined as free chlorine ≤ 0.1 mg/L. For this sampling study, a secondary goal is to evaluate if low chlorine residual is related to pipe material.

- Total number of samples = 60.
- Select stratification variable of pipe material. Pipe material within the distribution system is as follows:
  - 10 percent cast iron
  - 70 percent cement-lined ductile iron
  - 20 percent PVC

To meet the primary goal of a good overall system estimate, choose proportional allocation of the samples across the pipe materials. Do not consider time variations in this example. Therefore, the sampling program design has

- 6 samples in cast iron pipe
- 42 samples in cement-lined ductile iron pipe
- 12 samples in PVC pipe

However, to meet the secondary goal of evaluating if chlorine residual is related to pipe material, a proportional allocation as shown above does not provide many samples in the cast iron pipes, which are suspected of having a higher incidence of low chlorine residual than the other pipe materials. Therefore, to meet the secondary goal, a balanced allocation would provide a better distribution of samples while still achieving the overall goal of estimating the system-wide low-chlorine incidence. For a balanced allocation, the sampling program design would allocate one-third of the samples to each of the three pipe materials as follows:

- 20 samples in cast iron pipe
- 20 samples in cement-lined ductile iron pipe
- 20 samples in PVC pipe ▲

Sampling Frequency. The frequency of sampling for a given sampling program is directly related to the sampling goal(s). The theoretical number of samples required for a given sampling program is calculated without explicit regard for the frequency of sampling. However, implicit in the calculation is the concept that the sampling frequency should be compatible with the frequency of occurrence of the phenomena under investigation. For example, to investigate seasonal shifts in water quality parameters, samples should be taken at least quarterly. Water quality sampling in distribution systems poses some challenges with regard to sampling frequency in that a number of phenomena are occurring, many at different time scales. Seasonal changes in water temperature and organic content are contrasted with daily changes in treatment such as chlorine residual. Further, hydraulic transient events occur at a time scale on the order of seconds. For short-duration events on the order of a few hours, it has been shown for a simple distribution system that daily grab samples are unlikely to capture the contamination (Grayman et al., 2007a).

Practical considerations often dictate sampling frequency. Most water utilities do not operate laboratories 24 hours per day, so sampling schedules need to reflect the schedule and timing of different laboratory tests. Utilities using contract laboratories need to collect and ship samples before a shipping deadline. Most water utilities conduct their sampling during the day-shift hours, except in emergencies. However, the increasing availability of on-line or continuous monitors for water quality parameters is shifting the way that water utilities conduct their sampling. These on-line monitors can provide a continuous stream of water quality data, often linked directly to the utility information system for remote monitoring. Most water quality regulations specify sampling frequency for compliance measurements.

Optimization. In general terms, the design of sampling programs is highly suited to optimization. If the sampling goal can be expressed in terms of an objective function, the number of samples, location of sampling sites, and sampling frequency all could be optimized to meet the objective function. Several optimization software packages are available commercially, and many distribution system modeling software programs also include optimization routines.

The application of optimization to the problem of sampling program design for water quality in distribution systems is not a straightforward task. There is a wide range of different, competing objectives that are important to consider, such as minimizing the number of samples, minimizing the installation and maintenance costs of monitoring equipment, minimizing the response time to a contamination event, and minimizing the extent of contamination (Berry et al., 2005). In addition, it is difficult to quantify the health impact of contaminant occurrence because consumer water usage is poorly characterized in terms of both water consumption patterns and dose-response for health outcomes. Consequently, optimization routines require the use of surrogate measures such as average annual water consumption to model health impacts.

Sampling Related to Water System Security

The water security field has evolved over the past decade, primarily in response to a small number of actual or potential water quality–related events. Various industrial spills or unregulated discharges around the world have led to contamination of surface water and groundwater sources, resulting in the development of sophisticated source water monitoring technologies. Following the events of September 11, 2001, the interest in water security in the United States shifted to the potential for intentional contamination of water supplies, with primary concern on impacts within the distribution system. As a result, there has been extensive research and development related to technology and procedures that could be used to prevent, detect, analyze, and mitigate intentional contamination events. Advances in several areas of water security have potential applications beyond security concerns for water quality monitoring. Real-time, on-line monitors have come to be considered as the most expeditious mechanism for detecting the presence of a contaminant in a water distribution system.

In the most extensive project to date in this area, the USEPA has developed and is applying the Threat Ensemble Vulnerability Assessment (TEVA) program (Murray et al., 2004; USEPA, 2009). TEVA software analyzes the vulnerabilities of drinking water distribution systems, measures public health and economic impacts, and designs and evaluates threat mitigation and response strategies. Using a hydraulic model of the distribution system as a basis, TEVA applies a probabilistic framework for assessing the vulnerability of the system to a variety of contamination attacks and the potential for a water utility to decrease this vulnerability through a set of mitigation strategies (Grayman et al., 2007). This procedure is being applied to several large and medium-sized water utilities across the United States (Murray et al., 2008).

One of the TEVA programs was developed to assist the user to design sensor networks as part of a contaminant warning system. The TEVA sensor placement optimization tool (TEVA-SPOT) program models the impacts from intentional contamination events to drinking water distribution systems and uses the results to optimally locate on-line monitors in the distribution system. The optimization algorithm uses objective functions specific to security applications to solve the sensor placement problem. For example, the sensors might be located to provide notification of a contamination event with a given time period or prior to exposure of a given population of consumers. A dual-use adaptation of TEVA-SPOT is under development (Murray et al., 2008). This extension to the current TEVA-SPOT program would provide a dual-use sensor placement tool that allows the user to identify ideal locations for placing water quality monitors for both security goals and operational and/or regulatory sampling goals. The challenge in using this tool successfully will be to develop measurable metrics for a utility's current operational and regulatory sampling, which often do not have easily quantifiable outcomes.

On-line monitors, supervisory control and data acquisition (SCADA), and other sources of real-time or near-real-time data provide the potential for rapid detection of an unusual water quality event in a distribution system. These events could be related to intentional contamination or to an operational problem. Event-detection software is designed to analyze data to determine whether the data indicate an actual contamination event or just a normal variation in the data. In order to be useful, the software should minimize both false-positive (incorrectly detecting that an event is occurring) and false-negative results (missing the detection of an event).

The field of signal analysis and event detection is quite advanced and provides a basis for recent research related to water quality event detection (USEPA, 2005b). Several event-detection products have become available recently. CANARY is a research-oriented water quality event-detection algorithm development tool being developed by Sandia National Laboratories with partial support from the USEPA TEVA program (Hart et al., 2007). CANARY analyzes time-series data, most often water quality data from on-line monitors, using a variety of mathematical algorithms for detecting events. Several proprietary event-detection products are being developed to analyze on-line water quality data and include vector analysis of a suite of parameter measurements to recognize normal data patterns and triggers for levels that deviate from normal (Kroll, 2007). However, event detection in water distribution systems is in the early stages of application, and considerable research remains to be done before on-line monitoring data can be used reliably to detect all types of distribution system water quality events (Grayman et al., 2007).

Integrated Distribution System Management

With the increasing use of electronic databases, geographic information systems (GIS), SCADA, hydraulic models, and on-line water quality monitoring, water utilities are beginning to realize the value of integrated distribution system management. The concept is based on the desire to link data regarding the distribution system from all available sources to provide near-real-time information on the status of the distribution system. Many of the security-related projects are pursuing this type of data integration, but the concept has broad applications in distribution system water quality and operations. The implementation of integrated distribution system management systems varies widely depending on the available databases within the utility and the desired uses for different utility departments. Figure 21-7 presents a data-flow schematic with one data process mapped. For the data process shown, the GIS and SCADA provide information for the hydraulic model of the distribution system, which is a tool used by engineering, operations, and field staff. Many of the different databases shown in the figure are currently implemented at water utilities in the United States. However, the integration of these databases generally requires a central data-processing application. Integrated distribution system management also supports asset management applications (Speight, 2005). The development of sophisticated data-integration systems is in early stages, but the current trends indicate that this concept will be adopted widely by water utilities in the future.

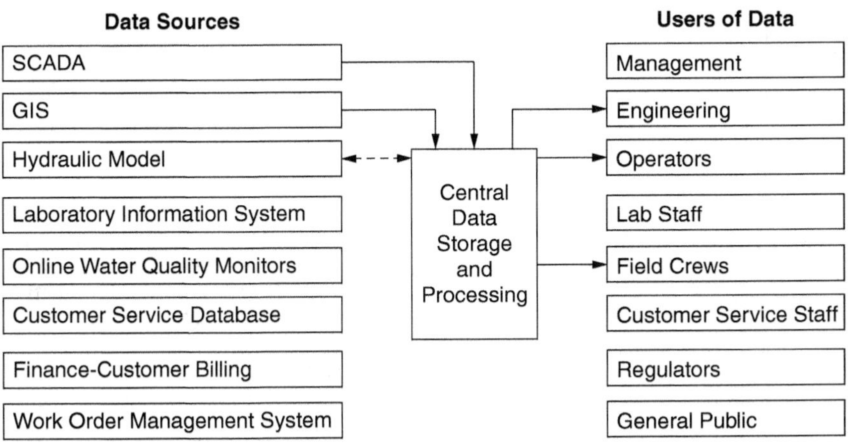

FIGURE 21-7 Example schematic of data-flow processes for integrated distribution system management. (*Source: Adapted from Speight, 2005.*)

ENGINEERING AND DESIGN OF DISTRIBUTION SYSTEMS

While operational procedures have a great impact on distribution system water quality, it is equally important to engineer and construct distribution systems using best practices to ensure system integrity and promote water movement to reduce adverse water quality impacts. A number of engineering practices, standards, and tools are available to water utilities to assist in controlling distribution system water quality.

Use of Distribution System Models

Hydraulic and water quality modeling tools provide a way to link hydraulic and water quality parameters. These models can provide valuable insight into the behavior of flows and water quality constituents and are an important component of a holistic approach to managing distribution system water quality.

Hydraulic models have been in use for several decades and can provide reliable simulations of flows and pressures when calibrated appropriately to real-world conditions (Ormsbee and Lingireddy, 1997). Figure 21-8 shows the input data required for hydraulic and water quality models using chlorine modeling as an example (Speight, 2008). A number of publications are available that outline the development, calibration, and application of distribution system models, including Speight and colleagues (2010), Walski and colleagues (2003), USEPA (2005c), Cesario (1995), Clark and Grayman (1998), AWWA (2004), and Boulos and colleagues (2006). Several commercial modeling software packages are available. Most of the commercial packages are compatible with EPANET, which is a free software developed by the USEPA that is available for download (USEPA, 2008a).

Grayman and colleagues (2007) list the following five areas where hydraulic modeling can be used for distribution system water quality management:

1. *Fate and transport of contaminants.* A hydraulic/water quality model can be used to simulate the movement and behavior of a contaminant in the distribution system and to determine its impacts over time. When a range of different contaminants are studied, models can be used to assess alternative schemes for reducing vulnerability of distribution systems to contamination.

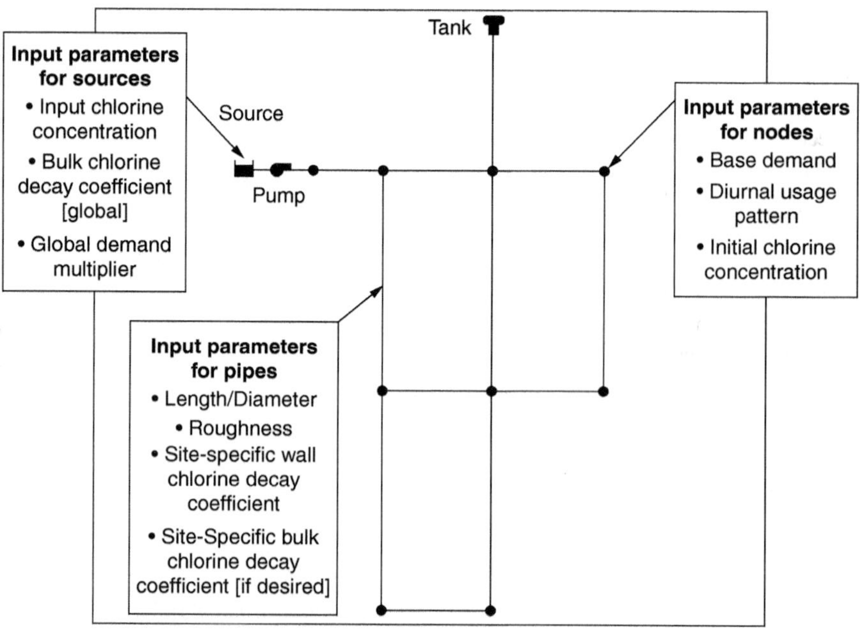

FIGURE 21-8 Summary of model input parameters. (*Source: Speight, 2008.*)

2. *Design of monitoring networks*. On-line continuous monitors have been a major focus as a means of detecting a contaminant in the distribution system. Optimization, simulation, and heuristic methods that combine hydraulic/water quality models with search techniques can be used to gauge the effectiveness of a particular set of monitors in the distribution system.
3. *Contamination source determination*. Hydraulic/water quality models can be used in a forensic mode to determine the likely source of a contaminant that has been observed in a distribution system. The models could be applied in real time to determine and characterize a source or could be used following an event to understand its origin. This form of analysis is referred to as *contaminant source determination, backtracing*, or an *inverse solution*.
4. *Real-time assessment, emergency response, and remediation*. When a contamination event occurs and is detected, a rapid response is needed to minimize the impact of the event. A properly formulated and current hydraulic/water quality model of the distribution system can be used to assess and predict the movement of contaminants through the distribution system, determine the source of the contamination (if not known), provide information on how the movement of the contaminant can be controlled, and assist in the development of a cleanup strategy.
5. *Historical reconstruction of waterborne outbreaks*. There have been many cases of contamination of water systems that have resulted in waterborne outbreaks associated with consumption of the water. Distribution system hydraulic/water quality models have served as key tools in the quantitative reconstruction of a number of major waterborne events.

Accurate modeling of hydraulic and water quality behavior in distribution systems is heavily dependent on the accuracy of the model input data. For many water utilities with old pipes, collecting and verifying pipe diameter and material and location of the installation are still a large focus of effort. A key challenge in hydraulic modeling is the determination of customer water usage at all points in the distribution system over time. Where customer meters exist, they are generally read on a monthly or quarterly basis and do not provide real-time data. Innovations in automated meter reading (AMR) may improve the ability to collect real-time customer usage data, but the data management challenges are significant (Speight, 2008).

Water quality modeling is an emerging field, but it is used increasingly to assist in operational decisions and planning studies. Water quality modeling applications include water age analysis, source tracing, and modeling of water quality constituents such as chlorine. Water quality models generally require a greater degree of detail within the model than for modeling applications that focus strictly on hydraulics, such as master planning modeling. An extended-period simulation (EPS) is required for water quality modeling to calculate constituent reactions over time. The development of "all pipes" models that include even the small-diameter water mains in a system is often driven by the desire to include the flow pathways to the point of consumption for more accurate water quality analysis. While skeletonized models, which include only larger-diameter mains, are not preferred for water quality applications, they can be used if the precision of the water quality predictions is not important. Water trace and water age analyses can be conducted without further input data beyond the base hydraulic information. However, modeling of water quality parameters requires reaction kinetic coefficients and additional field data for calibration (Speight et al., 2010).

Chlorine is the most commonly modeled parameter for water quality. A relatively simple first-order decay model for chlorine has been shown to provide good simulations of field data (Vasconcelos et al., 1997). Modeling of disinfection by-products has been less successful using deterministic models (Speight et al., 2000). Probabilistic models of water quality hold some promise in that they allow the use of uncertain mechanistic formulations with appropriate accounting of uncertainty; however, these probabilistic models are computationally intensive and currently remain in the research arena (Speight et al., 2009).

Modeling of microbial contaminants within the distribution system is limited by the lack of knowledge about biofilm processes, attachment or detachment of microbes to particles, growth mechanisms in low-nutrient environments, concentrations of microbes that enter the system through different pathways (e.g., low-pressure transient versus cross-connection with a sewer), and occurrence of pathogens. Particle transport and deposition modeling is especially challenging for distribution systems because of the highly variable nature of flow, which, in turn, is caused by the highly variable nature of individual water usage (Speight, 2008).

Water Quality Impacts Associated with Distribution System Design Practices

Several standards exist for design and construction of distribution system components. In the United States, designs for water systems typically are reviewed and approved by state regulatory agencies. *Ten States Standards* (Great Lakes Upper Mississippi River Board 2007) summarizes a number of best practices for sizing mains, storage tanks, and other distribution system facilities in a general sense. But many water utilities are moving toward a site-specific assessment that includes a water quality evaluation to appropriately size water mains and storage facilities. The use of hydraulic models is valuable in assessing the impact on water age for different design alternatives. For example, while head loss may be minimized by selecting a large-diameter water main, a water age analysis may reveal that the large-diameter main increases water age and the potential for water quality degradation.

The design standards in the United States require the provision of fire flows in most cases. The quantities of water required for fire flow typically are much higher than normal production levels, and therefore, concerns for adequate fire flow dominate capacity calculations. Research is ongoing in Europe to investigate the potential to size water mains for domestic usage only and therefore reduce water quality concerns, including accumulation of sediment (Blokker et al., 2007). The implementation of such a design philosophy would require the cooperation of fire departments and state regulators in the United States.

As users change and distribution systems evolve, areas of the system may contain oversized mains designed to serve customers that are no longer present. In addition, large mains are often built in anticipation of future development that does not occur. The best practice for main sizing includes an assessment of current and future uses and considerations for water quality impacts. Construction of looped pipe systems that eliminate dead ends is also considered to be a water quality management tool. Many water utilities are instituting development standards that require developers to construct looped systems. However, it is not possible in all situations to construct a fully looped system, and dead-end mains are still prevalent in distribution systems across the United States. As with all water main sizing, water utilities need to consider the impact of construction of loops and connectors on water quality. In some cases, if the loop connection piping requires a long main with few users connected, the construction of the loop may produce an undesired effect by increasing water age.

The continuous maintenance of positive pressure at all locations within the distribution system is likely the most important daily operational requirement for a water system. However, maintenance of pressure is not only an operational practice. The distribution system must be designed properly to minimize the occurrence of low pressures. Maintenance of pressure on a regular basis requires the correct sizing and operation of pumps, tanks, and water mains. Hydraulic models can be used to evaluate the ability of distribution system facilities to provide adequate pressure under different flow conditions. Field data collected from pressure gauges at key points in the system such as pump stations, pressure-reducing valves, and areas with high ground elevation can be used to further understand the ability of the system design to maintain appropriate pressure.

AWWA (2005c) has developed standard procedures that are used, with variations, by most of the water supply industry for disinfecting new and repaired water mains. These recommendations address six general areas of concern: (1) protection of new pipe sections at the construction site, (2) restriction on the use of joint packing materials, (3) preliminary flushing of pipe sections, (4) pipe disinfection, (5) final flushing, and (6) bacteriological testing for pipe disinfections.

Pipe sections, fittings, and valves stockpiled in yard areas or at the construction site should be protected from soil, seepages from water or sewer-line leaks, storm water runoff, and habitation by pets and wildlife. Each of these sources could deposit contamination in the interior of pipe sections awaiting installation. Septic tank drain fields, subsurface water in areas of poor drainage or high water table, and seasonal or flash flooding also may introduce significant contamination into unprotected pipe sections. Contamination from these sources may become lodged in pipe fittings and valves and be released into the bulk flow of water supply after the main is placed into service. Commonsense protective measures include end covers for these pipe materials, drainage of standing water from trenches before pipeline assembly, and flushing of new construction or line repairs to remove all visible signs of debris and soil.

Water use in a community varies continuously as a reflection of the activities of the general public and local industries. Water treatment operations cannot easily gear production to frequent and sudden changes in water demand from all consumers. For this reason, storage facilities are an essential element of a distribution system. Storage tanks supplement flows during periods of fluctuating demand on the system, equalize operational water pressures, augment supply from production facilities that must be pumped at a uniform rate, and provide a backup source of water in case of system outages. Design and construction standards specify that adequate storage capacity be provided for fire emergencies.

Care must be taken to prevent potential contamination of the high-quality water entering storage facilities. Water volumes in large reservoirs mix and interchange slowly with fresh water from the distribution system, and under certain circumstances, old water with degraded water quality can be distributed to customers. Temperature differences between stored and fresh water result in stratification of water in the tank. Old water may be discharged from the tank as a consequence of a major water loss in the system caused by a major main break or intensive system flushing. Storage tanks also can be a deposition zone for sediments and associated microbes. Abrupt changes in water flow resuspend the sediment deposits, moving the contamination into the distributed flow of water.

CONTROLLING MICROBIAL OCCURRENCES IN DISTRIBUTION SYSTEMS

It is often difficult to determine the location of microbial infiltration given the vastness of many distribution systems, and therefore, determining the cause of total coliform positive results can be extremely challenging. Increasingly, utilities are focusing on assessments of best management practices and distribution system integrity rather than on monitoring to ensure the public health integrity of distribution systems.

Treatment Approaches

The utility needs to verify that any cause of microbial occurrence in treated water is not the result of a deficiency in treatment. Deficiencies associated with nutrient loading, excessive disinfectant demand owing to inadequate corrosion control, inadequate removal/inactivation of microbes, or needed modifications to secondary disinfection can contribute to microbial loading or growth in the distribution system. The utility should be able to adequately demonstrate to regulatory agencies that the system is meeting all *best available treatment* (BAT) requirements. In addition, a sanitary survey by an approved state agency can provide an independent evaluation of treatment plant operations. For the treatment plant, the continuity of disinfection and maintenance of low turbidity levels are critical items to document. Surface water plants should ensure that they are meeting the turbidity and disinfection requirements outlined in the SWTR and subsequent revisions. Groundwater systems should ensure that they are not under the direct influence of surface water and have an active wellhead protection program. Monitoring of treatment plant effluent needs to be complete enough to document the absence of total coliforms and *E. coli*. To increase a treatment plant effluent bacterial detection database, the utility may process high-volume samples (Hargy et al., 2008; Friedman et al., 2009) and examine the samples for the presence of injured coliform bacteria (LeChevallier and McFeters, 1985a; McFeters, 1990) or HPC bacteria using sensitive media such as R2A agar.

Nutrient Loading. Microorganisms in the distribution system can thrive if provided with organic and inorganic nutrients that promote growth. Carbon typically is the growth-limiting nutrient in North American drinking water systems (Camper et al., 2000). However, as summarized in the USEPA White Paper, "Effect of Treatment on Nutrient Availability" (USEPA, 2002c), the presence or absence of nitrogen, phosphorus, and metals can enhance or limit growth in drinking water distribution systems. For example, a few studies have determined that phosphorus levels could limit microbial growth in the distribution system. Investigators (Sang et al., 2003; Miettinen et al., 1997) indicate that in some regions, such as Finland, China, Norway, or Japan, phosphorus can be a limiting nutrient in drinking water supply sources. Carbon becomes the limiting nutrient, however, when nitrogen is added.

Some systems have investigated source waters to determine if excessive nutrient levels may be linked to widespread recurring coliform problems. Indiana-American Water Company (Muncie, Ind.) determined that 53 percent of samples collected had assimilable organic carbon (AOC) levels greater than 100 μg/L, with 16 percent having levels greater than 200 μg/L. AOC levels greater than 100 μg/L have been shown to enhance coliform regrowth in certain conditions (Norton and LeChevallier, 1997). In response to historical coliform problems, one of several mitigation actions taken by the Muncie system in 1997 included changing the anthracite in the multimedia filters to granular activated carbon (GAC). Research (Volk and LeChevallier, 2002) has indicated that GAC reduces AOC levels 25 percent better than conventional anthracite/sand filters. Following the implementation of GAC adsorption/filtration in 1997, coliform rates over an 8-year period were the lowest they had been over the past 25 years. A complete case study of Muncie's coliform control program can be found in Friedman and colleagues (2009b).

Corrosion Control and Finished Water Quality Stability. Corrosion control can be an important yet often overlooked aspect of microbial water quality control. The accumulation rate of corrosion products (measured from 22 unlined cast iron mains collected from 9 different utilities) can be as much as 3555 kg/km (6.3 tons/mi) of 200-mm (8-in) pipe (Friedman et al., 2010). This voluminous amount of scale, which is partly comprised of biomass, represents a huge potential source of inorganic and microbial water quality degradation products. Key recommendations for maximizing corrosion control benefits and minimizing corrosion scale/biomass release largely center around providing a stable finished water quality with regard to pH, alkalinity, and oxidation-reduction potential (ORP). Enhanced stability of pH and alkalinity often can be achieved through careful application of corrosion control chemicals at the treatment plant. Ideally, the water should be well enough buffered to avoid pH changes of ±0.2 pH units. Wide shifts in pH can affect corrosion scale stability owing to changes in ORP, sorption/desorption reactions, and precipitation/dissolution reactions. Other methods of enhancing corrosion control with regard to ORP include (Friedman et al., 2010) the following:

- Avoid uncontrolled blending of groundwater and surface water sources because these types of water supplies typically have widely differing ORPs. Switching back and forth affects corrosion scale stability and could exacerbate corrosion scale/biomass release.
- Avoid uncontrolled blending of chlorinated and chloraminated sources. The ORP of waters with free chlorine residuals typically will be higher than that of waters with chloramine residuals. Uncontrolled blending or switching back and forth can cause scale destabilization and exacerbate corrosion scale/biomass release.

Secondary Disinfection. Maintenance of a disinfectant residual throughout the distribution system is considered (in the United States) to be a critical part of assessing and enhancing the integrity of the distribution system. With regard to microbial water quality, the maintenance of a disinfectant residual can contribute to inactivation of microorganisms in bulk water as well as provide some control of biofilms. The USEPA (1999a) has identified free chlorine, chloramines, and chlorine dioxide as being appropriate for secondary disinfection. While none of these options is ideal for all systems, each has characteristics that may meet a specific system's needs for secondary disinfection. Selection of the most appropriate secondary disinfectant must be made on a system-by-system basis, with consideration given to the system's concerns regarding inactivation requirements, DBP formation potential, water quality, distribution system condition, and treatment experience and capabilities.

Systems should investigate the stability of disinfectant residual levels at locations that coincide with positive coliform samples. However, it should be noted that high disinfectant residuals will not always be associated with an absence of positive coliform samples, especially if the cause is biofilms. For example, Pinellas County Utilities (PCU) investigated total chlorine residual levels associated with positive coliform samples collected from November 2002 through December 2004. The average total chlorine concentration for coliform positive samples was 2.89 mg/L (Friedman et al., 2009). In contrast, Metro Vancouver found that 82 percent of samples had less than 0.1 mg/L of free chlorine and that 90 percent of samples contained less than 0.2 mg/L of free chlorine. Metro Vancouver and member utilities installed booster chlorination at five locations to provide a higher disinfectant residual in response to frequent coliform positive samples. A unidirectional flushing program also was initiated in several member municipality distribution systems. Figure 21-9 shows the number of communities served by Metro Vancouver that experienced more than 10 percent positive coliform samples in a given month prior to and after implementing booster chlorination in 1997. The figure shows a significant decline after implementing booster chlorination (Friedman et al., 2007).

Utilities may address microbial distribution system problems by increasing disinfectant residual throughout the distribution system or at problematic locations or by changing the disinfectant used to provide a distribution system residual. Table 21-7 provides a summary of secondary disinfection changes made by several utilities in response to coliform occurrences (Friedman et al., 2009).

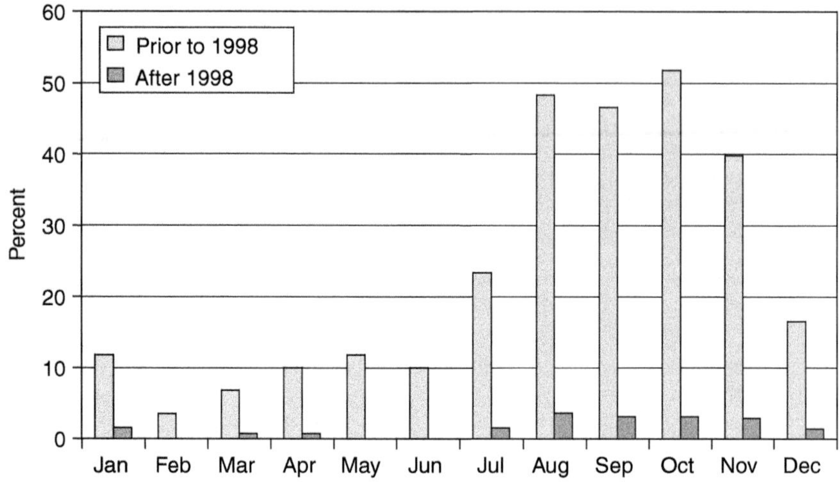

FIGURE 21-9 Metro Vancouver: Percentage of communities with more than 10% positive monthly coliform samples (booster chlorination implemented in 1997). (*Source: Friedman et al., 2007.*)

Operations and Maintenance Approaches

Operations and maintenance procedures can be implemented and/or optimized to address microbial problems in the distribution system. For example, the Greenville Water System (Greenville, S.C.) implemented the following changes in operation and maintenance practices to address microbial issues (Friedman et al., 2007): (1) comprehensive flushing program, (2) tank cleaning and inspection program, (3) criteria for return to service after pipe replacement and repair, and (4) goal for total chloramine residual throughout the distribution system of 2.0 to 2.5 mg/L. Figure 21-10 presents the percentage of positive monthly coliform samples between January 1991 and October 2005 collected by Greenville Water System. This figure shows significant reductions in the percentage of positive coliform

TABLE 21-7 Summary Secondary Disinfectant Strategies for Responding to Total Coliform Occurrences

Case study utility	Secondary disinfectant strategy used
Pinellas County Utilities	Average total chlorine concentration for positive coliform samples was 2.89 mg/L.
Virginia American	During a period of time associated with positive coliform samples, fewer than 3% of distribution system chlorine residual samples were below 1.0 mg/L free chlorine.
Indiana American	Determined that free chlorine levels in the distribution system associated with positive coliform samples were 0.5 to 2.0 mg/L.
MWRA	Found that 65% of samples had total chlorine levels of 0.2 mg/L and fewer than 1% had total chlorine residuals of 1.0 mg/L.
Metro Vancouver	82% of samples had less than 0.1 mg/L free chlorine and 90% of samples contained less than 0.2 mg/L free chlorine.

Source: Friedman et al., 2009.

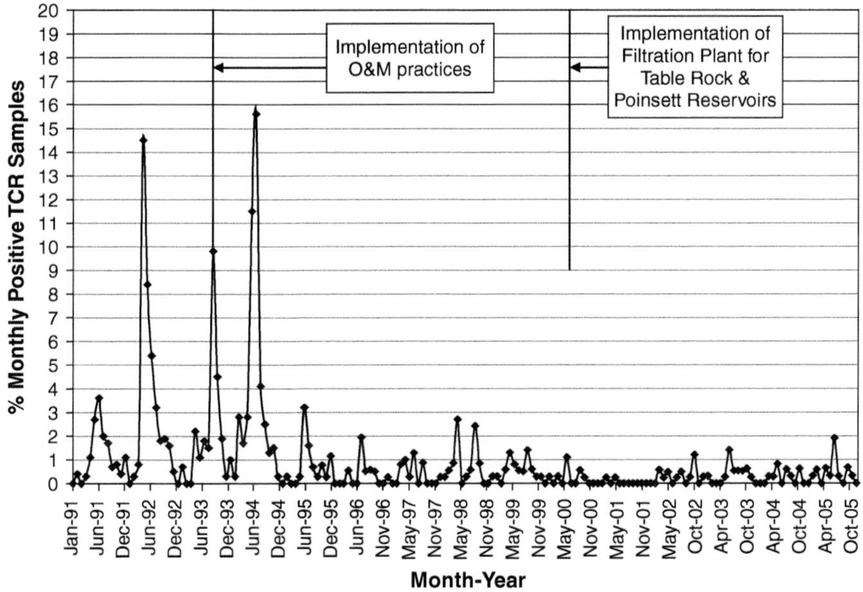

FIGURE 21-10 Greenville water system TCR percent positive, 1991–2005. (*Source: Friedman et al., 2007.*)

samples after implementation of the operations and maintenance (O&M) practices just described. Additionally, the percentage of positive coliform samples declined further after Greenville Water System implemented a new filtration plant.

Maintaining Disinfectant Residuals. Approaches for increasing disinfectant residuals include (1) reducing water age in storage facilities and pipe network, (2) reducing disinfectant demand in the source water, (3) reducing disinfectant demand associated with sediment and scale buildup on pipe walls, and (4) increasing disinfectant dose.

The effectiveness of each approach will vary from system to system and may well vary from location to location within a specific system. Recommendations associated with reducing water age through storage facilities and reducing disinfectant demand in the source water and demand associated with sediment and scale buildup are discussed elsewhere in this chapter.

Reducing Water Age in Pipe Networks. Prudent capital planning necessitates installation of facilities that have excess capacity for water storage and distribution. It is not uncommon to size pipelines for water demands that will occur 20 years or more into the future (AWWA and EES, 2002a). *Ten State Standards* (Great Lakes Upper Mississippi River Board,1997) specifies a minimum pipe size of 150 mm (6 in) at all locations for providing fire protection. Table 21-8 shows the volumetric effect of increased pipe diameter

TABLE 21-8 Pipe Diameter versus Pipe Volume per Mile or Kilometer

Pipe diameter	2 in (50 mm)	4 in (100 mm)	6 in (150 mm)	8 in (200 mm)	10 in (250 mm)	12 in (300 mm)	18 in (450 mm)
Gal/mi	86	3,466	7,755	13,786	21,540	31,019	69,792
L/km	202	8,154	29,362	52,197	81,556	117,445	262,249

Source: AWWA and EES, 2002.

on a per-kilometer and per-mile basis. Thus, for every kilometer (or mile) of 100-mm (4-in) pipe that is replaced with 200-mm (8-in) pipe, the effective volume of the distribution system increases by greater than 37,850 L (10,000 gal).

Excess capacity within piping networks can lead to significantly increased water age, causing loss of disinfectant residual, increased microbial growth, and other water quality problems. It is important to note that there is no *ideal* water age that is appropriate for all systems. Variables such as the degree of source water treatment, disinfectant type and dose, water temperature, and distribution system materials all will have an impact on the degree of water quality degradation that might occur as a result of water age.

Water age–related problems can be analyzed using a range of techniques from simple manual calculations through computer models (Brandt et al., 2006). The most common tools used for assessing water age in piping networks include use of existing water quality and operational information, use of tracer studies, and use of calibrated hydraulic models (AWWA and EES, 2002a). Existing water quality data such as temperature, disinfectant residual, and trihalomethanes (THMs) can provide valuable insight into locations that are experiencing water age problems. Alternatively, in systems with multiple sources with varying water quality characteristics (such as differences in water hardness or conductivity), these natural constituents can be used as the tracer. Finally, during transitional periods in system operation, such as changeovers from chlorination to chloramination, the resulting constituents can be traced. Other utilities have taken advantage of a fluoride system shutdown, use of alternative water sources, or a switch in coagulant to trace water through the distribution system (AWWA and EES, 2002a). Brandt and colleagues (2006) describe the selection of appropriate trace chemical, duration and nature of dose, sampling frequency, sampling method, travel time and network complexity, and tracer studies of storage tanks.

Mathematical models that represent the hydraulic behavior of the movement of water have been used to estimate water age in distribution systems (Clark and Grayman, 1998). As described in the preceding section, water quality models can be used in conjunction with hydraulic models to predict concentrations of chlorine, DBPs, and other constituents in a distribution system (Speight et al., 2009). The effect of interaction of the flowing water with the pipe wall can be estimated using relationships proposed by Rossman and colleagues (1994). For each nonconservative water quality parameter, rates and mechanisms of reaction (decay and growth) must be identified and measured for a particular system.

Once problematic areas of the distribution system are identified and confirmed through water quality monitoring, the utility can evaluate potential water age mitigation strategies. The most appropriate strategy will be associated with the likely contributing causes of excess water age. Example strategies applicable to piping networks are summarized in Table 21-9.

Controlling retention time will affect other operational or service-level requirements of the network (such as fire flow or pressure requirements). Potential impacts should be considered prior to the implementation of water age-reduction strategies.

Increasing Disinfectant Dose. Typically, one of the first responses to positive coliform results is to increase disinfectant residuals by increasing the dose. For systems that maintain minimal disinfectant levels (generally considered to be 0.2 mg/L for free chlorine or 0.5 mg/L for chloramines in all parts of the system), higher residuals may help to penetrate biofilms or sediments to achieve better levels of microbial inactivation.

Free chlorinated systems should attempt to maintain at least 0.5 mg/L residuals at dead-end sites. According to AWWA Manual M56, *Fundamentals and Control of Nitrification in Drinking Water Distribution Systems* (AWWA, 2006b), chloraminating systems should maintain a minimum chloramine concentration of 2.0 mg/L leaving the plant. Systems should maintain a minimum chloramine concentration of 0.5 mg/L at all monitoring points in the distribution system. Many utilities set internal goals of 1.5 mg/L chloramine or

TABLE 21-9 Methods for Controlling Water Age

Method	Rationale
Altering the valving of networks	By changing valve arrangements and hydraulic boundaries, travel times can be reduced and water rerouted to increase velocities in low-flowing pipes to maximize velocities.
Installing time-varying valves	Time and other control valves can be used to control flows.
Manual flushing	Periodic flushing to remove sediment and reduce water age in dead ends and low-velocity sections of pipe.
Automated flushing	Automatic programmed flushing to remove sediment and reduce water age in dead ends and low-velocity sections of pipe.
Abandoning mains	Remove surplus capacity from the system.
Downsizing mains	Reduce system capacity to increase velocities and reduce retention time.
Adjusting pump schedules	Optimize pumping regime to match supply and demand, minimize energy requirements and retention times (linked to optimizing system storage).

Source: Adapted from Brandt et al., 2006.

higher for all distribution system sites. LeChevallier and colleagues (1996) recommend that chloraminated systems set goals to maintain a minimum of 1.0 to 2.0 mg/L at dead-end locations.

Disinfectant dosages can be increased at the treatment plant or through booster disinfection at specific locations within the distribution system. The advantage of this approach is that most systems can make these changes relatively quickly. In some cases, the utility should get approval from the state regulatory agency before increasing residual levels, but in all cases, it is good to be in close contact with the regulatory authorities.

The disadvantages of increasing disinfectant residuals include (1) the potential for increased disinfectant by-products, (2) increased customer complaints about chlorinous taste and odor, (3) increased chemical costs, and (4) most important, it may not have any impact on coliform occurrence. It is well recognized that increased disinfection residuals alone may not suffice to eliminate coliform occurrences.

Flushing. Flushing is recommended as one of many tools for controlling biofilm growth, removing accumulated sediment, restoring disinfectant residual, improving hydraulic capacity, and improving taste and odor. Most loose deposits are readily removed from smooth pipes [such as PVC pipe or new ductile iron (DI) pipe] and slightly tuberculated pipe at velocities between 0.6 and 1.2 m/s (2 and 4 fps) (Friedman et al., 2003). With respect to tuberculated mains, higher flushing velocities are needed, but they tend to produce a diminishing increase in particle-removal performance for severely tuberculated mains. If the flushing objective is to scour and clean the pipe wall, high-velocity flushing [i.e. ≥1.5 m/s (5 fps)] generally is needed. However, flushing at excessively high velocities can damage pipe linings and result in decreased pressures upstream. Additionally, water may be wasted without obtaining a water quality benefit.

Unidirectional flushing (UDF) is the recommended method of removing the hydraulically mobile (and in some cases loosely adhered) solids. Unidirectional flushing involves the purposeful and sequential isolation of specific water mains, control of overflow direction and velocity, and introduction of only clean water (i.e., previously flushed or close to the source of supply or a clean water front) into the main to be flushed. UDF protocols and recommendations regarding flushing duration, velocity, and other programmatic elements are described in *Establishing Site-Specific Flushing Velocities* (Friedman et al., 2003).

Many utilities still conduct conventional flushing in which hydrants are randomly opened to temporarily increase flow within a specific area of the distribution system or to respond to customer complaints related to taste, odor, or discolored water. Utilities typically choose conventional flushing over UDF because it is less time-consuming and requires significantly less planning. The disadvantages of conventional flushing, however, include stirring up of mobilized solids and sediments in adjacent areas of the distribution system and failure to remove the mobilized solids from the distribution system. Discolored water often reaches the customer tap after conventional flushing events, and this practice actually can increase coliform episodes rather than reducing them. It also may result in customer exposure to elevated levels of regulated inorganic contaminants such as lead, barium, arsenic, and other trace elements (USEPA, 2006b; Friedman et al., 2010).

Storage Facility Operations Practices. Excessive water age in many storage facilities can be an important factor related to water quality deterioration and disinfectant residual loss. The excess water age is caused by (1) underutilization (i.e., water is not cycled through the facility) and (2) short circuiting within the reservoir. Poor mixing (including stratification owing to temperature differences) can exacerbate the water quality problems by creating zones within the storage facility where water age significantly exceeds the average water age throughout the facility (AWWA and EES, 2002b). Distribution systems with storage facilities where water cascades from one facility to another (such as pumping up through a series of pressure zones) can result in exceedingly long water age in the most distant tanks and reservoirs.

Operationally, water age in storage facilities is managed by routine turnover of the stored water and fluctuation of the water levels in storage facilities. Kirmeyer and colleagues (1999) recommended a 3- to 5-day complete water turnover as a starting point but cautioned that each storage facility should be evaluated individually and given its own turnover goal. Water storage management for water quality must take into account influent water quality, environmental conditions, retention of fire flow, and demand management, as well as factors specific to the design and operation of the tank, such as velocity of influent water, operational level changes, and tank design. Consequently, water level fluctuations in a distribution system are managed as an integrated operation within pressure zones, demand-service areas, and the system as a whole rather than on an individual-tank basis. Available guidelines for water turnover rates are summarized in Table 21-10.

The Philadelphia Water Department estimated mean residence time and turnover rate in several standpipes by measuring fluoride residual and water levels. Mean residence time of water in the standpipes was determined to be 50 percent longer than expected because old water reentered the standpipes from the distribution system. One major conclusion from this work was that for water to get out into the distribution system and away from the standpipes, standpipe drawdown needs to correspond to peak demands or precede peak demands (Burlingame et al., 1995).

Mixing processes within a storage facility should be controlled to minimize water age (Grayman et al., 2000). When mixing does not occur throughout the storage facility, stagnant zones can form where water age will exceed the overall average water age in the facility. Therefore, mixed flow is preferable to plug flow in distribution system storage. Mixing can be encouraged through the development of a turbulent jet. In order to ensure turbulent jet flow, the following relationship between inflow Q in gallons per minute and inlet diameter d in feet must hold (Grayman et al., 2000):

$$Q/d > 11.5 \quad \text{at } 20^\circ\text{C} \tag{21-5}$$

$$Q/d > 17.3 \quad \text{at } 5^\circ\text{C} \tag{21-6}$$

TABLE 21-10 Guidelines on Water Turnover Rate for Distribution Storage

Source	Guideline	Comments
Georgia Environmental Protection Division	Daily turnover goal equals 50% of storage facility volume; minimum desired turnover equals 30% of storage facility volume	As part of this project, state regulators were interviewed by telephone
Virginia Department of Health, Water Supply Engineering Division, Richmond, VA	Complete turnover recommended every 72 hours	As part of this project, state regulators were interviewed by telephone
Ohio EPA	Required daily turnover of 20%; recommended daily turnover of 25%	Code of state regulations; turnover should occur in one continuous period rather than periodic water level drops throughout the day
Baur and Eisenbart, 1988	Maximum 5- to 7-day turnover	German source, guideline for reservoirs with cement-based internal surface
Braid, 1994	50% reduction of water depth during a 24-hour cycle	Scottish source
Houlmann, 1992	Maximum 1- to 3-day turnover	Swiss source

Source: Kirmeyer et al., 1999.

Temperature differences between the inflow and the ambient water temperature within the storage facility can cause the water to form stratified layers that do not mix together. Stratification is more common in tall tanks such as standpipes and tanks with large-diameter inlets. This can be avoided by increasing the inflow rate.

Mahmood and colleagues (2005) provide an evaluation of water mixing characteristics in distribution system storage tanks. For example, these authors found that standpipes in the Virginia Beach, VA, distribution system had an average fill time of approximately 3 hours and an inflow rate of approximately 126 L/s (2000 gpm) with an inlet pipe diameter of 600 mm (24 in). This resulted in inadequate momentum to mix the bulk water within the 3-hour fill time. Computational fluid dynamics modeling showed that reducing the inlet diameter to 300 mm (12 in) would result in complete mixing within 1 hour of filling. For standpipes and elevated tanks, Mahmood and colleagues (2005) recommend reducing the inlet pipe diameter so that an inlet momentum of approximately 1.7 to 2.6×10^{-1} m^4/s^2 (20–30 ft^4/s^2) can be achieved. The authors concluded that momentum, pipe location and orientation and temperature differences will control the effectiveness of mixing within storage facilities. Mixing systems also are described in Kirmeyer and colleagues (1999).

Prevention of Distribution System Contamination

Direct microbial and chemical contamination of the distribution system is possible where there are physical breaches in distribution system integrity. *Physical integrity* has been defined as the ability of the distribution system to act as a physical barrier that prevents external contamination from affecting the quality of the internal drinking water supply (National Academy of Sciences, 2006). Examples of physical integrity breaches include unprotected cross-connections; leaks in water main, seals, or gaskets; submerged air-vacuum valves; and open access hatches, roof holes, or unscreened vents associated with storage facilities. A description of physical barriers and potential associated contaminant sources is provided in Table 21-11.

TABLE 21-11 Infrastructure Components, What They Protect Against, and Common Materials

Component	External contamination the barrier protects against	Materials used
Pipe	Soil, groundwater, sewer exfiltration, surface runoff, human activity, animals, insects, and other life forms	Asbestos cement, reinforced concrete, steel, lined and unlined cast iron, lined and unlined ductile iron, PVC, polyethylene and HDPE, galvanized iron, copper, polybutylene
Pipe wrap and coatings	Supporting role in that it preserves pipe integrity	Polyethylene, bitumastic, cement-mortar
Pipe lining	Supporting role in that it preserves pipe integrity	Epoxy, urethanes, asphalt, coal tar, cement-mortar, plastic, insects
Service lines	Soil, groundwater, sewer exfiltration, surface runoff, human activity, animals, insects, and other life forms	Galvanized steel or iron, lead, copper, chlorinated PVC, cross-linked polyethylene, polyethylene, polybutylene, PVC, brass, cast iron
Premise plumbing	Air contamination, human activity, sewage, and industrial nonpotable water	Copper, lead, galvanized steel or iron, iron, steel, chlorinated PVC, PVC, cross-linked polyethylene, polyethylene, polybutylene
Fittings and appurtenances (meters, valves, hydrants, ferrules)	Soil, groundwater, sewer exfiltration, surface unoff, human activity, animals, insects, and other life forms	Brass, rubber, plastic
Storage facility walls, roof, cover, vent hatch	Air contamination, rain, algae, surface runoff, human activity, animals, birds, and insects	Concrete, steel, asphaltic, epoxy, plastics
Backflow-prevention devices	Nonpotable water	Brass, plastic
Liquids	Not applicable	Oils, greases, lubricants
Gaskets and joints	Soil, groundwater, sewer exfiltration, surface runoff, human activity, animals, insects, and other life forms	Rubber, leadite, asphaltic, plastic

Source: National Academy of Sciences, 2006.

Numerous AWWA standards, manuals of practice, white papers, research reports, and committee reports address the causes and mitigation strategies associated with most of the physical integrity components, including main repairs, storage facility maintenance practices, and cross-connection control. Table 21-12 summarizes methods for preventing loss of physical integrity, as presented in National Academy of Sciences (2006). Table 21-13 lists some of the relevant AWWA standards that pertain to distribution system water quality (for a complete listing of standards, go to the AWWA Web site at www.awwa.org).

Preventing Intrusion through Water Main Leaks or Other Appurtenances. Friedman and colleagues (2004) and Flemming and colleagues (2006) provide detailed guidance on measuring and controlling pressure transients and intrusion. Guidelines for use of high-speed data loggers, conventional pen-and-chart recorders, and distribution system surge models are described. A multidisciplinary approach is needed for adequate transient and intrusion control, as shown in Fig. 21-11.

TABLE 21-12 Summary of Approaches for Preventing Loss of Physical Integrity

Component	Mechanism of integrity loss	Prevention by
Pipe	Permeation	Standards on pipe applications, local assessments of soil and groundwater for contamination
	Structural failure (leak or break)	Better design and installation, early leak detection with rehabilitation and repair, optimized scheduling of pipe renewals, optimized placement of valves for effective shutoffs and isolations
	Improper installation	Standards and certification for installation, followed by inspection
	Unsanitary activity	Strict requirements and inspections during repair, rehab, installation
Fitting and appurtenance	Structural failure	Improved materials quality as well as quantity in the operating components of valves and hydrants, periodic valve exercising followed by maintenance or replacement as needed
	Improper installation	Strict requirements on installation and design
	Unsanitary activity	Strict requirements during repair, rehab, installation, and inspection
Storage facility wall, roof, cover, vent, hatch	Structural failure (crack, hole)	Better design and installation, early leak detection with rehabilitation and repair
	Absence of proper installation	Inspection and better design with inspection Strict requirements on installation and design
	Unsanitary activity	Strict requirements during repair, rehab, installation, and inspection
Backflow	Absence of proper installation	Inspection and certification Strict requirements on installation and design
	Operational failure	Annual testing and maintenance

Source: National Academy of Sciences, 2006.

TABLE 21-13 Examples of Relevant AWWA Standards Pertaining to Distribution System Water Quality

Standard	Title
G430	Security Practices for Operation and Management
G200	Distribution Systems Operation and Management
C652	Disinfection of Water-Storage Facilities
C651	Disinfecting Water Mains
C652	Disinfection of Water Storage Facilities
C605	Underground Installation of PVC Pressure Pipe and Fittings for Water
C604	Installation of Steel Water Pipe—4 in. (100 mm) and Larger
C602	Cement-Mortar Lining of Water Pipelines in Place—4 in. (100 mm) and Larger
C600	Installation of Ductile-Iron Water Mains and Their Appurtenances
C511	Reduced-Pressure Principle Backflow Prevention Assembly
C510	Double Check Valve Backflow Prevention Assembly
D102	Coating Steel Water Tanks

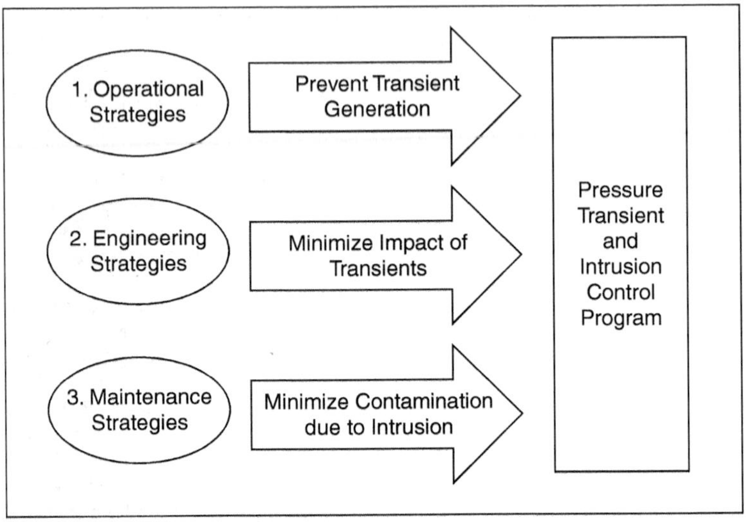

FIGURE 21-11 Relationships between control strategies and utility functions. (*Source: Adapted from Friedman et al., 2004.*)

Small systems may have characteristics that contribute to susceptibility to low or negative pressures, including a lack of surge-control devices (Flemming et al., 2006). The smaller systems [<37.9 million L/d (10 mgd)] examined also were found to have fewer floating storage facilities (e.g., storage facilities that provide pressure by gravity, or "float" on the hydraulic grade line), more supply inputs per mile of main, and operation at lower steady state pressures, all of which are potential contributing factors to increased low- or negative-pressure development. Flemming and colleagues (2006) identified the following additional system characteristics that were found to increase susceptibility to low- and negative-pressure surges with individual systems:

- Located near a pump station with downstream velocity greater than 0.9 m/s (3 fps)
- Located greater than 1 mile away from elevated storage
- Elevations greater than 12 to 15 m (40–50 ft) above surroundings
- Located near a dead end
- Located near a hydrant on a major main

Operational Transient Control Strategies. Operational transient control strategies generally focus on minimizing activities that cause rapid changes in water velocity, thereby minimizing the generation of pressure transients. These activities include (Friedman et al., 2004)

- *Rapid imposition of demands.* Rapid imposition of fire flows, high-demand customers, or due to tanker truck filling should be avoided through increased valve opening time.
- *Pump shutdown time.* Increasing pump shutdown time should be implemented by ramping down speeds for pumps with variable-speed drives and by slowly closing a pump control valve for constant-speed pumps.
- *Pump sequencing.* For multiple-pump pumping stations, the largest pump should be shut down first, and the smallest pump should be shut down last. By staggering the shutdown sequences, the magnitude of the change in flow at any one time is reduced.

- *Valve operations.* The magnitude of transients initiated by valve operations can be reduced by slowing the rate of opening and closing, as described earlier.

Figure 21-12 documents the pressure transients that were generated during a pump startup and shutdown at a water treatment plant (LeChevallier et al., 2002; Friedman et al., 2004). Measurements were made using high-speed pressure data loggers placed at three

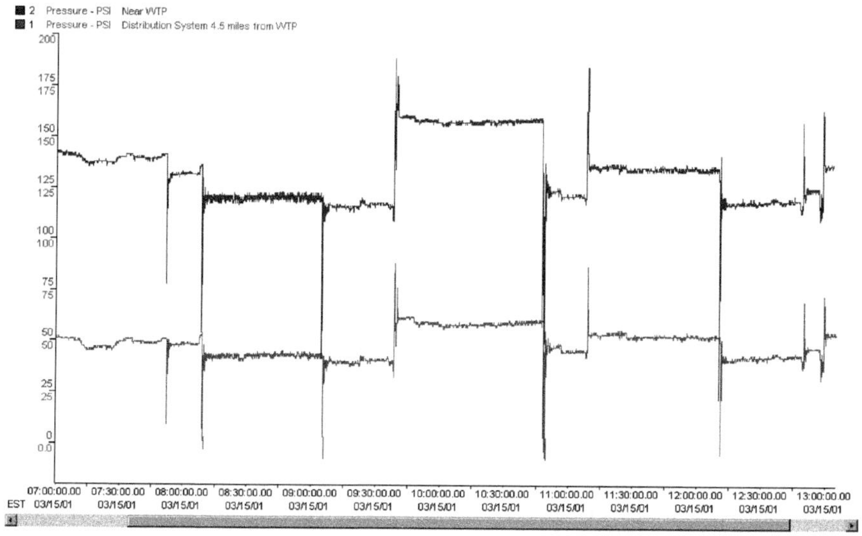

(a)

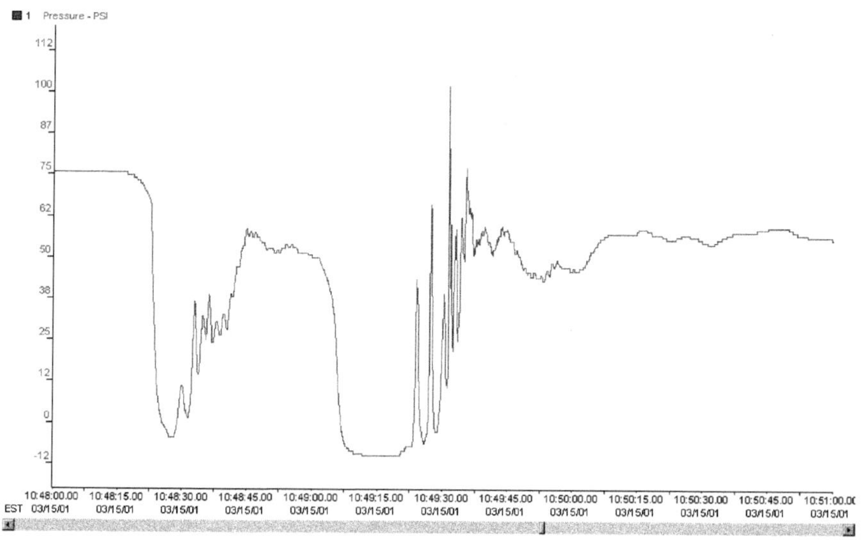

(b)

FIGURE 21-12 Pressure transients caused by pump shutdown and startup at a treatment plant (*a*) and magnification of one of the transients (*b*) (1 psi = 6.894 kPa). (*Source: LeChevallier et al., 2002, and Friedman et al., 2004.*)

different locations within the system, as described below. During this test, the main service pumps were shut down at the treatment plant clearwell and restarted with all flow going through one venturi meter. Two high-speed electronic monitors were installed at high elevation points on a 760-mm (30-in) main [one was ~4 km (2.5 mi) from the plant, and the other was about 7.2 km (4.5 mi) from the plant], and a third monitor was located about 24 m (80 ft) from the treatment plant's high-service pumps. Pressure readings both near the treatment plant and within the distribution system showed large pressure fluctuations (Fig. 21-12a). While the static pressures near the plant ranged between 125 and 150 psi (860–1030 kPa), the pressure transients caused by the pump shutdowns resulted in pressures as low as 18 psi (124 kPa) in the plant effluent. However, several miles away in the distribution system, these fluctuations resulted in pressures as low as minus 10 psi (~70 kPa) lasting for 16 seconds (Fig. 21-12b). The valve closure speed for the main service pumps was 20 seconds. It was felt that this valve closure speed may have been too fast, thus contributing to the pressure transient. A second test was conducted with the valve closure speed slowed to 30 seconds, but negative pressures resulted from this second test as well.

Engineering Transient Control Strategies. Proper engineering control strategies will help to minimize the impacts of transients should they be generated. Such engineering strategies include (1) installation of surge-control tanks, (2) installation of other surge-control devices, (3) evaluating the impacts of increasing steady state pressures in vulnerable sections of the distribution system, (4) providing overhead storage, (5) looping dead ends, and (6) encasement of new mains.

Additionally, air relief valves and similar appurtenances should be designed to have above-grade venting (designed to be tamperproof to avoid deliberate contamination of the system). All below-grade vacuum or air relief valves should be retrofitted to above-grade venting or modified in a way to prevent flooding of the vault (e.g., drainage or pump) (LeChevallier et al., 2002). Both the generation and propagation of pressure transients can be reduced as a result of these design elements (Friedman et al., 2004; Flemming et al., 2006).

Maintenance Control Strategies. Maintenance strategies can help to minimize the likelihood of intrusion should a surge occur and propagate through the distribution systems. Such maintenance strategies include (1) leak detection, (2) main repairs, (3) inspections of valve and meter boxes, (4) cross-connection control, and (5) hydrostatic testing of new mains.

Efforts to reduce distribution system pipeline leakage are beneficial not only from a water conservation standpoint but also minimize the potential for microbial intrusion into potable water supplies. Leaks are not simply a loss of revenue for a water utility, but the leak is a potential pathway for contamination. The public health benefits of leak control should be recognized and encouraged. Repair of leaking sewer lines similarly should be a top priority not only to minimize the occurrence of pathogens near drinking water pipelines but also to reduce these sources of contamination being transported to groundwater supplies and receiving streams, particularly under wet weather conditions (LeChevallier et al., 2002).

Inspection of valve and meter boxes, especially air-vacuum valve boxes that are below grade, is critical for controlling the potential for intrusion. The standing water in flooded air-vacuum valve boxes, which can be backsiphoned into the distribution system during transient pressure events, is highly unsuitable for consumption, as demonstrated in Fig. 21-13.

Preventing Contamination Through Cross-Connections. Prevention of cross-connections is paramount to maintaining the safety of water delivered to the customer. According to USEPA (2002a), contamination owing to backflow incidents may not be detected or reported for several reasons:

- Bacterial contamination tends to be transient and highly localized (ABPA, 1999).
- Water system operators monitor routinely for coliform bacteria, but most often that is the only microbial monitoring conducted (USEPA, 2002a). While these bacteria are important

FIGURE 21-13 Photographs of flooded air-release and air-vacuum vaults (the vault on the right contains a dead mouse). (*Source: LeChevallier et al., 2002.*)

indicators of distribution system problems, some microbial contaminants may go undetected. The limited nature of biological monitoring, especially in smaller systems (as infrequent as once per year), makes it unlikely that contamination will be detected in a timely manner.

- Most backflow incidents are detected and reported to the local authority only if customers detect an irregularity in their water supply. Not all contamination that produces illness and disease can be detected by taste, color, or odor (Hoxie et al., 1997).
- Even if an irregularity is detected, it may not be reported by the consumer.
- When water system operators suspect backflow incidents, they have a disincentive to document and report them because of concerns about legal liability and loss of consumer confidence, as noted by an USEPA Office of the Inspector General Report (USEPA, 1995). (Fortunately, these same concerns provide the utility with an incentive to protect the distribution system.)
- The difference between epidemic and endemic transmission is obscured by limitations in recognizing when an outbreak occurs (Frost et al., 1996). A study of waterborne cryptosporidiosis estimates that out of every 10,000 infections by *Cryptosporidium*, only 3 would be reported and concludes that surveillance for detected cases of a reportable illness may underestimate rates of infection and morbidity substantially (Perz et al., 1998).
- Some contaminants that enter the distribution system through cross-connections and backflow may not be reportable.
- The incidents of reduced pressure and some cross-connections are often transient in nature. Pressure changes may not be detected by conventional pressure monitoring equipment. Reduced pressures also may affect only a portion of the distribution system, a specific pressure zone, or only piping beyond the service connection.

Common backflow control devices include air gaps, reduced pressure zone backflow preventers, double check valves, vacuum breakers, and complete isolation. Numerous organizations provide detailed description of approved backflow prevention assemblies as well as guidance for implementing a cross-connection control program. As summarized by the National Academy of Sciences (2006), examples include the University of Southern California (USC), the American Society of Sanitary Engineering (ASSE), Underwriters Laboratories (UL), the International Association of Plumbing and Mechanical Officials, Factory Mutual, and the Canadian Standards Association. The three most commonly used guidance manuals are the *USC Manual of Cross Connection Control* (USC, 1993), AWWA's *Recommended Practice for Backflow Prevention and Cross-Connection Control Manual M14* (AWWA, 2004b), and USEPA's *Cross Connection Control Manual* (USEPA, 2003b). Table 21-14 describes many of the critical elements of cross-connection control program and summarizes the number of states that require each program element (USEPA, 2002a).

TABLE 21-14 Summary of Cross-Connection Control Program Elements and State Requirements

Requirement	Number of states with requirement
Does the state have a requirement for the control of cross-connections and/or backflow prevention?	50
Is it specified in the requirement that the system must implement or develop a cross-connection control and/or a backflow-prevention program?	32
Does the state require authority to implement a local ordinance or rule for cross-connection control and/or backflow prevention?	33
Must the authority cover testing of backflow-prevention assemblies?	27
Must the authority cover the use of only licensed or certified backflow assembly testers?	16
Must the authority cover the entry of the premises for the sake of inspecting the premises?	14
Must the authority cover the entry of the premises for the sake of inspecting and/or installing backflow-prevention assemblies?	15
Does the state require training, licensing, or certification of backflow-prevention assembly testers?	26
Does the state require training, licensing, or certification of backflow-prevention assembly and/or device installers?	6
Does the state require training, licensing, or certification of backflow-prevention assembly and/or device repairers?	10
Does the state require training, licensing, or certification of cross-connection control inspectors?	19
Does the state require inspection of backflow-prevention devices and/or testing of backflow-prevention assemblies?	37
Does the state require the system to include recordkeeping as part of cross-connection control?	34
Does the requirement include recondkeeping of hazard assessment surveys?	11
Does the state require the system to notify the public following the occurrence of a backflow event?	3
Does the state require the local rule or ordinance to allow the system to take enforcement action against customers that do not comply with the cross-connection control and backflow-prevention requirements?	23
Does the state conduct periodic reviews of cross-connection control programs?	3
Does the state regulation or plumbing code require public education regarding cross-connection control and/or backflow prevention?	7

Source: USEPA, 2002a.

Considerable variability exists in state statutes, regulations, and policies related to cross-connection control and backflow prevention. In some cases where states do not require programs, some water systems within the state have implemented comprehensive and active programs in the absence of a state requirement to do so (USEPA, 2002a).

Preventing Contamination Associated with the Installation and Repair of Water Mains. Exposure of piping materials to contaminants can begin at the point of manufacture. Subsequent handling and storage of piping also present opportunities for exposure. The distribution system white paper entitled, "New or Repaired Water Mains" (AWWA and EES, 2002c), describes the potential for contamination associated with main break and repair and also discusses contamination prevention and mitigation strategies. Excerpts from AWWA and EES (2002c) are provided below.

Design Practices. Water mains should be designed to provide adequate separation from potential sources of contamination such as pipes carrying sewage, storm water, or reclaimed wastewater. The selection of adequate separation is based on factors such as pipe material and joint type, soil conditions, and space for repair. The *Ten State Standards* (Great Lakes Upper Mississippi River Board, 2007) recommend the following minimum separation distances: (1) 3 m (10 ft) horizontal separation between water mains and sanitary sewer force mains or sewers installed in parallel and (2) 0.460 m (18 in) vertical separation for water main crossing above or below a sewer or force main.

In practice, separation distances vary based on site-specific conditions. Best management practices for pipeline design are outlined in Pierson and colleagues (2001). These include (but are not limited to) (1) adequate numbers and types of hydrants, blow-offs, and valves, (2) the ability of valves to provide for complete isolation, (3) minimization of the number and length of tie-ins, (4) minimization of the distance between flushing point and valve that terminates the new main, and (5) simple installation configurations to enable unidirectional flow. Design practices should facilitate the effective and sanitary repair or installation and should facilitate contamination removal through flushing.

Condition of Piping Materials. The sanitary handling and storage of materials constitute perhaps the most critical step in protecting against microbial contamination during water main construction (Burlingame and Neukrug, 1993). The AWWA C600-99 Standard provides guidelines on pipe storage and handling procedures but not on sanitary protection of materials and tools used in main work. AWWA C600-99 was replaced by AWWA C600-05 (AWWA, 2005a). In addition to storing materials at the site prior to construction, many utilities stockpile pipe materials for future use at their utility yard. Protection from contamination is equally important for stockpiled pipes. The use of pipe caps and wraps can secure new mains and prevent contamination prior to installation. The literature cites examples where pipe caps are recommended during the curing of cement mortar linings both at the factory (North American Society for Trenchless Technology, 1999) and in situ (CSIRO, 2002). In these instances, pipe end caps prevent the circulation of air that otherwise could accelerate drying and lead to cracking of the cement lining.

Construction and Repair Practices. AWWA Standards C600 (AWWA, 2005a), C602 (AWWA, 2006c), C604 (AWWA, 2007), C605 (AWWA, 2005b), and C606 (AWWA, 2006d) address pipe installation procedures, as well as guidelines on inspection, trench construction, pipe installation, joint assembly, flushing, and pressure and leakage testing. Best management practices during the construction or repair of water mains are provided by Pierson and colleagues (2001). Topics of interest include (1) maintenance of pipe caps, plugs, or other protective coverings until pipes are joined, (2) provision positive-flow shut-off, (3) protection of existing mains and service connections with watertight caps or covers, (4) completion of all pipe and fitting joints in a trench before work is stopped, (5) use of recommended packing and sealing materials, (6) gasket cleaning, (7) disinfection of fittings, joints, valves, and exposed existing connections with swab or spray technique, and

(8) disinfection of hand tools, tapping machines, and other equipment that come in contact with pipes and fittings.

Main breaks are a pervasive problem for many utilities. The causes for breaks are varied; however, according to Deb and colleagues (2000), one of the primary causes of main breaks in otherwise structurally sound water mains is improper installation. Broken mains may become depressurized owing to water loss or service shutdown. According to AWWA Manual M20, *Water Chlorination Principles and Practices* (AWWA, 1973), contractors and workers should be thoroughly familiar with all pertinent state and local requirements governing installations of mains. The manual recommends the use of watertight plugs at all times when construction is not actually in progress.

Trenchless technologies have gained increasing popularity over the past several years because they reduce disruption and can be cost efficient (Pierson et al., 2001). The authors discuss contamination control methods for trenchless rehabilitation and replacement techniques, including pit excavation, pit dewatering, storing and handling of new liners, coupling methods, extraction of bentonite slurry, lubricant use, flushing, and chlorination. The AwwaRF manual, *Demonstration of Innovative Water Main Renewal Techniques* (Deb et al., 1999), demonstrates and evaluates various trenchless technologies and identifies conditions under which each technology can be best applied.

Trench Work. As described previously, trench water has a high potential for contaminating water mains during installation and repair. Dewatering the trench to a level below the pipe invert may be necessary to avoid contamination. Pierson and colleagues (2001) suggest using submersible pumps and fittings used for clean water work only to avoid cross-contamination from sewage water applications.

Preparing for Service. The elements for contamination control associated with preparing a new, rehabilitated, or repaired mains for service include hydrostatic testing, flushing and cleaning, disinfection, and water quality testing. In North America, disinfection typically is performed in accordance with AWWA Standard C651, which was updated in 2005 (AWWA, 2005c). Based on a survey of 250 utilities, Haas and colleagues (1998) found that 75 percent of respondents reference AWWA Standard C651 in their construction documents. Haas and colleagues (1998) document the results of actual field evaluations to test the adequacy of AWWA Standard C651 for disinfecting water mains. The researchers concluded that the AWWA standard provides adequate disinfection: The AWWA-recommended disinfection dose of 25 mg/L for a 24-hour contact time provides more than a 4 log (99.99 percent) inactivation of HPC organisms. It also was found that approximately 10 mg/L of free chlorine reduces HPC bacteria to less than 100 CFU/mL.

AWWA Standard C600 (AWWA, 2005a) states that the interior of all pipes, fittings, and other appurtenances shall be kept free from dirt or foreign matter at all times. While the current ANSI/NSF 61 standard does not provide specific requirements for the sanitary conditions of piping materials that are intended for use in drinking water systems, other manuals of practice, such as AWWA Manual M20, *Water Chlorination/Chloramination Principles and Practices* (AWWA, 2006), recommend precautionary measures such as the use of watertight plugs at all times when construction is not actually in progress.

Several AWWA standards provide recommendations for pressure and leakage testing (Table 21-13). Problems that can occur if the new or repaired pipe section is not pressure tested adequately include (1) deflection of flexible couplings greater than that acceptable for the type of coupling may lead to leaks and subsequent contaminant intrusions, (2) pipe joints are designed to seal under high internal pressure conditions but may leak if the external pressure is higher than the internal pressure (e.g., a low-pressure pipe where there is a high groundwater level), and (3) adapters may leak if not installed properly.

Haas and colleagues (1998) report another potential source of contamination from stagnant water created by closed valves directly adjacent to the area of main construction or repair. Distribution system water that lies stagnant for extended periods in dead-end sections around the area of construction or repair can have no disinfectant residual and low dissolved oxygen, conditions that may promote the growth of bacteria. (AWWA and EES, 2002c).

Preventing Contamination Associated with Storage Facilities. Storage facilities are most susceptible to external contamination owing to the absence or failure of an essential component, such as a cover, vent, hatch, and so on. Even when covered, storage facilities can suffer from algal growth on the tops of floating covers that can gain entry into the tank through rips and tears or missing hatches. Algae a can lso be airborne or carried by birds and gain entry into storage tanks through open hatches and vents. Algae increase the chlorine demand of the stored water, reduce its oxygen content on their degradation, affect taste and odor, and in some cases release toxin by-products (National Academy of Science, 2006).

Tank inspections provide information used to identify and evaluate current and potential water quality problems. Both interior and exterior inspections are employed to ensure the tank's physical integrity, security, and high water quality. Inspection type and frequency are driven by many factors specific to each storage facility, including its type (i.e., standpipe, ground tank, etc.), vandalism potential, age, condition, cleaning program or maintenance history, water quality history, funding, staffing, and other utility criteria (AWWA and EES, 2002b). Depending on the nature of the water supply chemistry, detailed inspections should be made every 3 to 5 years and consist of tanks needing to be drained, sediment removed, and appropriate rustproofing applied to the metal surfaces (such as where the water level rises and falls more frequently). These inspections are in addition to daily or weekly inspections for vandalism, security, and water quality purposes (such as identifying missing vents, open hatches, leaks, etc.) (Kirmeyer et al., 1999).

Storage facility maintenance activities include cleaning, painting, and repair to structures to maintain serviceability. Based on a utility survey conducted by Kirmeyer and colleagues (1999), it appears that most utilities that have regular tank cleaning programs employ a cleaning interval of 2 to 5 years. This survey also showed that most tanks are painted (exterior coating) on an interval of 10 to 15 years. Three finished water steel elevated spheroids at the City of Brookfield Water Utility in Brookfield, Wis., were the subject of a field study conducted to document the underwater cleaning process and its water quality impacts. The time since last cleaning was 15 years for one tank and 7 years for the other two tanks. The tank with the longest cleaning interval contained the most accumulated sediment of 710 mm (28 in) maximum depth compared with 100 to 300 mm (4–12 in) in the other two tanks and the highest HPC bacteria levels before cleaning of 1300/mL compared with 640 and 80/mL in the other two tanks. As a result of underwater cleaning, HPC bacteria and turbidity levels were reduced significantly (Kirmeyer et al., 1999).

FINAL COMMENTS

Maintenance of the microbial quality within the distribution requires a multidisciplinary approach that integrates the biological, physicochemical, engineering, material sciences, and operations to ensure a safe, high-quality product at the consumer's tap. The availability of advanced tools, including molecular assays, computer models, geographic information system (GIS) interfaces, and on-line monitors and information management systems, is changing and deepening our understanding of the microbial ecology of the distribution system network. Although not under the jurisdiction or control of water utilities but nonetheless important to consumers' perceptions of water quality, increasing attention is being focused on the ecology of premise plumbing. It would not be surprising to find that the ecology of plumbing systems is as different from that of the rest of the distribution system as treated water is different from raw water. Over the next decade, this will be a challenging and active area of research.

Advances in water treatment technology and increased regulation over the past three decades have resulted in marked improvements in drinking water quality and reductions in drinking water–related waterborne disease. Yet there are still challenges with risks emanating within the distribution system. The transient nature of contamination events, the complicated hydraulics of the pipe network, and the limited monitoring capabilities of even the most progressive water utilities mean that detecting and controlling distribution system microbial risks will require greater investments in research for on-line and remote monitoring. The concept that collecting a few grab samples can reveal any meaningful microbial information must be abandoned and replaced with a new scientifically based program that links online process control data for disinfectant residuals, pressure management, and infrastructure integrity into a real-time hydrologic model so that the distribution system operator can monitor and control the distribution system like the treatment operator does with a SCADA system.

All this means that the system operator for the twenty-first century will need to continue to be highly skilled, well-trained, competent, and knowledgeable water professional. Their constant vigilance is the essential glue that makes all the pieces fit together to produce safe drinking water day after day and year after year.

ABBREVIATIONS

ADP	automatic pathogen detection
AMR	automated meter reading
ANSI	American National Standards Institute
AOA	ammonia-oxidizing archaea
AOB	ammonia-oxidizing bacteria
AOC	assimilable organic carbon
ASSE	American Society of Sanitary Engineering
BAC	biologically activated carbon
BAT	best available treatment
BDOC	biodegradable dissolved organic carbon
CDC	U.S. Centers for Disease Control and Prevention
CFU	colony-forming units
DBP	disinfection by-product
DGGE	denaturing gradient gel electrophoresis
EPS	extended-period simulation
FISH	fluorescent in situ hybridization
GAC	granular activated carbon
GC-MS	gas chromatography–mass spectroscopy
GIS	geographic information systems
HAART	highly active antiretroviral treatment
HCGI	highly credible gastrointestinal illness
HIV	human immunodeficiency virus
HPC	heterotrophic plate count
LLAP	*Legionella*-like amoeba pathogens

MAC	*Mycobacterium avium* complex
MCLG	maximum contaminant level goal
MDPE	medium-density polyethylene
NOB	nitrite-oxidizing bacteria
NTM	nontuberculous mycobacteria
ntu	nephelometric turbidity unit
ORP	oxidation-reduction potential
PCR	polymerase chain reaction
PCR-ALH	polymerase chain reaction–amplicon length heterogeneity
PCU	Pinellas County Utilities
PNA	peptide nucleic acid
PVC	polyvinyl chloride
qPCR	quantitative polymerase chain reaction
RT-PCR	reverse-transcriptase polymerase chain reaction
SCADA	supervisory control and data acquisition
SSCP	single-strand-conformation polymorphism
SWTR	Surface Water Treatment Rule
TCR	Total Coliform Rule
TEVA	threat ensemble vulnerability assessment
THM	trihalomethane
TOC	total organic carbon
T-RFLP	terminal restriction fragment length polymorphisms
UDF	unidirectional flushing
USC	University of Southern California
USEPA	U.S. Environmental Protection Agency
UV	ultraviolet light
VNBC	viable but nonculturable
WHO	World Health Organization

REFERENCES

Abd, H., T. Johansson, I. Golovliov, G. Sandström, and M. Forsman (2003). "Survival and Growth of *Francisella tularensis* in *Acanthamoeba castellanii*." *Applied and Environmental Microbiology* 69:600–606.

Aboytes, R., G.D. Di Giovanni, F.A. Abrams, C. Rheinecker, W. McElroy, N. Shaw, and M.W. LeChevallier (2004). "Detection of Infectious *Cryptosporidium* in Filtered Drinking Water." *Journal AWWA* 96(9):88–98.

ABPA (1999). "1999 Survey of State and Public Water System Cross-Connection Control Programs." American Backflow Prevention Association, Bryan, TX.

Abu Kwaik, Y. (1996). "The Phagosome Containing *Legionella pneumophila* Within the Protozoan *Hartmanella vermiformis* Is Surrounded by the Rough Endoplasmic Reticulum." *Applied and Environmental Microbiology* 62:2022–2028.

Abu-Ashour, J., D.M. Joy, H. Lee, H.R. Whiteley, and S. Zelin (1994). "Transport of Microorganisms Through Soil." *Water, Air and Soil Pollution* 75(1–2):141–158.

Adékambi, T., M. Reynaud-Gaubert, G. Greub, M.-J. Gevaudan, B. La Scola, D. Raoult, and M. Drancourt (2004). "Amoebal Coculture of '*Mycobacterium massiliense*' sp. nov. from the Sputum of a Patient with Hemoptoic Pneumonia." *Journal Clinical Microbiology* 42:5493–5501.

Adékambi, T., S. Ben Salah, M. Khlif, D. Raoult, and M. Drancourt (2006). "Survival of Environmental Mycobacteria in *Acanthamoeba polyphaga*." *Applied and Environmental Microbiology* 72:5974–5981.

Adeleke, A., J. Pruckler, R. Benson, T. Rowbotham, M. Halablab, and B. Fields (1996). "*Legionella*-like Amebal Pathogens: Phylogenetic Status and Possible Role in Respiratory Disease." *Journal of Emerging Infectious Disease* 2:225–230.

Ainsworth, R. (2004). *Safe Piped Water: Managing Microbial Water Quality in Piped Distribution Systems*. London: WHO and IWA Publishing.

Alavandi, S. V., M. S. Subashini, and S. Ananthan (1999). "Occurrence of Haemolytic and Cytotoxic *Aeromonas* Species in Domestic Water Supplies in Chennai." *Indian Journal of Medical Research* 110:50.

Allen, M. J., R. H. Taylor, and E. E. Geldreich (1980). "The Occurrence of Microorganisms in Water Main Encrustations." *Journal AWWA* 72(11):614–625.

Amann, R. I., B. J. Binder, R. J. Olson, S. W. Chisholm, R. Devereux, and D. A. Stahl (1990). "Combination of 16S rRNA-Targeted Oligonucleotide Probes with Flow Cytometry for Analyzing Mixed Microbial Populations." *Applied and Environmental Microbiology* 56:1919–1925.

Amann, R. I., W. Ludwig, and K.-H. Schleifer (1995). "Phylogenetic Identification and In Situ Detection of Individual Microbial Cells without Cultivation." *Microbiology Reviews* 59:143–169.

Amirtharajah, A., and D. P. Wetgstein (1980). "Initial Degradation of Effluent Quality during Fitration." *Journal AWWA* 72(9):518.

Andersson, Y., and P. Bohan (2001). "Disease Surveillance and Waterborne Outbreaks." In L. Fewtrell and J. Bartram (eds.), *Water Quality: Guidelines, Standards and Health*. London: WHO and IWA Publishing.

Angulo, F.J., S. Tippen, D.J. Sharp, B.J. Payne, C. Collier, J.E. Hill, T.J. Barrett, R.M. Clark, E.E. Geldreich, H.D. Donnell JR., and D.L. Swerdlow (1997). "A Community Waterborne Outbreak of Salmonellosis and the Effectiveness of a Boil Water Order." *American Journal of Public Health* 87(4):580–584.

APHA, AWWA, and WEF (American Public Health Association, American Water Works Association, and Water Environment Federation) (2005). *Standard Methods for the Examination of Water and Wastewater*, 21st ed., A. D. Eaton, L. S. Clesceri, E. W. Rice, and A. E. Greenberg, eds. Washington, DC: American Public Health Association. Denver, CO: American Water Works Association.

Armon R., J. Starosvetzky, T. Arbel, and M. Green (1997) "Survival of *Legionella pneumophila* and *Salmonella typhimurium* in Biofilm Systems." *Water Science and Technology* 35(11–12): 293–300.

ASCE (2004). "Interim Voluntary Guidelines for Designing an Online Contaminant Monitoring System." Available at www.asce.org/static/1/wise.cfm, American Society of Civil Engineers.

AWWA (American Water Works Association) (1985). *Maintaining Distribution System Water Quality*. Denver, CO: American Water Works Association.

AWWA (1990). *Pocket Guide to Water Sampling: 1. Microbiological Contamination*. Denver, CO: American Water Works Association.

AWWA (1999). M48, *Waterborne Pathogens: Manual of Water Supply Practices*, 1st ed. Denver, CO: American Water Works Association.

AWWA (2004a). M32, *Computer Modeling of Water Distribution Systems*, 2nd ed. Denver, CO: American Water Works Association.

AWWA (2004b). M14, *Recommended Practice for Backflow Prevention and Cross-Connection Control*, 3rd ed. Denver, CO: American Water Works Association.

AWWA (2004c). M7, *Problem Organisms in Water: Identification and Treatment.* Denver, CO: American Water Works Association.

AWWA (2005a). AWWA C600-05 Standard, *Installation of Ductile-Iron Water Mains and Their Appurtenances.* Denver, CO: American Water Works Association.

AWWA (2005b). AWWA Standard, *Underground Installation of Polyvinyl Chloride (PVC) Pressure Pipe and Fittings for Water* (ANSI/AWWA C605-05). Denver, CO: American Water Works Association.

AWWA (2005c). AWWA C651-05, *Standard for Disinfecting Water Mains.* Denver, CO: American Water Works Association.

AWWA (2006). M20, *Water Chlorination Principals and Practices.* Denver, CO: American Water Works Association.

AWWA (2006a). *Contamination Monitoring Technologies.* Workshop Participant Manual. Denver, CO: American Water Works Association.

AWWA (2006b). M56, *Fundamentals and Control of Nitrification in Chloraminated Drinking Water Distribution Systems,* 1st ed. Denver, CO: American Water Works Association.

AWWA (2006c). AWWA C602-06, *Standard for Cement Mortar Lining of Water Pipelines in Place 4 Inch (100 mm) and Larger.* Denver, CO: American Water Works Association.

AWWA (2006d). AWWA C606-06, *Standard for Grooved and Shouldered Joints.* Denver, CO: American Water Works Association.

AWWA (2007). AWWA C604-07, *Standard for Installation of Steel Water Pipe 4 in. (100 mm) and Larger.* Denver, CO: American Water Works Association.

AWWA and EES (2002a). *Effects of Water Age on Distribution System Water Quality.* Available online at http://epa.gov/safewater/disinfection/tcr/pdfs/whitepaper_tcr_waterdistribution.pdf. Accessed on January 8, 2009.

AWWA and EES (2002b). *Finished Water Storage Facilities.* Available online at http://epa.gov/safewater/disinfection/tcr/pdfs/whitepaper_tcr_storage.pdf. Accessed on January 8, 2009.

AWWA and EES (2002c). *New or Repaired Water Mains.* Available online at http://epa.gov/safewater/disinfection/tcr/pdfs/whitepaper_tcr_watermains.pdf. Accessed on January 8, 2009.

AwwaRF (2002). *Online Monitoring for Drinking Water Utilities,* E. Hargesheimer, O. Conio, and J. Popovicova (eds.). Awwa Research Foundation–CRS Proaqua Report. Denver, CO: Water Research Foundation.

Barbeau, B., K. Julienne, V. Gauthier, R. Millette, and M. Prévost (1999). "Dead-End Flushing of a Distribution System: Short and Long-Term Impacts on Water Quality." In *Proceedings of AWWA Water Quality Technical Conference, Tampa, FL.*

Baribeau, H., C.A. Kinner, J.R. Stephen, R. De Leon, P.A. Rochelle and D.L. Clark. (2000). "Microbial Population Characterization of Suspended and Fixed Biomass in Drinking Water Reservoirs." In *Proceedings of AWWA Water Quality Technical Conference, Salt Lake City, UT.*

Bartram, J., J. Cotruvo, M. Exner, C. Fricker, and A. Glasmacher. (2003). *Heterotrophic Plate Counts and Drinking-Water Safety.* London: IWA Publishing.

Baur, A., and K. Eisenbart (1988). "Einfluß der Standzeit in Wasserbehältern auf die Wasserqualität." *GW F- Wasser/Abwasser* 129(2S):109–115.

Bays, L. R., N. P. Burman, and W. M. Lavis (1970). "Taste and Odour in Water Supplies in Great Britain: A Survey of the Present Position and Problems for the Future." *Journal Society Water Treatment Examination* 19:136.

Bellen, G. E., S. H. Abrishami, P. M. Colucci, and C. Tremel (1993). *Methods for Assessing the Biological Growth Support Potential of Water Contact Materials.* Denver, CO: Water Research Foundation.

Berry, W., L. Fleischer, W. Hart, C.A. Phillips, and J-P. Watson (2005). "Sensor Placement in Municipal Water Networks." *Journal of Water Resources Planning and Management, ASCE* 131(3): 237–243.

Besner, M-C., V. Gauthier, B. Barbeau, R. Millette, R. Chapleau, and M. Prévost (2001). "Understanding Distribution System Water Quality." *Journal AWWA* 93(7):101–114.

Besner, M. C., V. Gauthier, P. Servais, and A. Camper (2002). "Explaining the Occurrence of Coliforms in Distribution Systems." *Journal AWWA* 94(8):95–109.

Besner, M. C. (2007). "Risk Evaluation of Drinking Water Distribution System Contamination Due to Operation and Maintenance Activities." Ph.D. thesis, Ecole Polytechnique de Montreal, Canada.

Besner, M.C., J. Lavoie, C. Morissette, P. Payment, and M. Prévost (2008b). "Effect of Water Main Repairs on Water Quality." *Journal AWWA* 100(7):95–109.

Besner, M.-C., P. Servais, and M. Prévost (2008a). "Efficacy of Disinfectant Residual on Microbial Intrusion: A Review of Experiments." *Journal AWWA* 100(10):116–130.

Besner, M.C., Broseus, R., Lavoie, J., Di Giovanni, G., Payment, P. and Prévost, M. (2010). "Pressure monitoring and characterization of external sources of contamination at the site of the Payment drinking water epidemiological studies". *Environmental Science and Technology*, 44. 269-277.

Birtles, R. J., T. J. Rowbotham, D. Raoult, and T. G. Harrison (1996). "Phylogenetic Diversity of Intra-amoebal Legionellae as Revealed by 16S rRNA Gene Sequence Comparison." *Microbiology* 142(3):525–3530.

Blackburn, B. G., G. F. Craun, J. S. Yoder, V. Hill, R. L. Calderon, N. Chen, S. H. Lee, D. A. Levy, and M. J. Beach (2004). "Surveillance for Waterborne-Disease Outbreaks Associated with Drinking Water—United States, 2001–2002." *Morbidity Mortality Weekly Report* 53(8):23–45.

Blair, B., P. Dakar, K. R. Bright, F. Marciano-Cabral, and C. P. Gerba (2008). "*Naeglaria Fowleri* in Well Water." *Journal of Emerging Infectious Disease* 14(9):1499–1500.

Blokker, E. J. M., J. H. G. Vreeburg, P. Schaap, and P. Horst (2007). "Self-Cleaning Networks Put to the Test." In *Proceedings of the World Environmental and Water Resources Congress, EWRI, ASCE, Tampa, FL.*

Boulos P. F., K. E. Lansey, and B. W. Karney (2006). *Comprehensive Water Distribution Systems Analysis Handbook for Engineers and Planners*, 2nd ed. Boulder, CO: MWH.

Boyd, G.R., H. Wang, M.D. Britton, D.C. Howie, D.J. Wood, J.E. Funk, and M.J. Friedman (2004a). "Intrusion Within a Simulated Water Distribution System Due to Hydraulic Transients: I. Description of Test Rig and Chemical Tracer Method." *Journal of Environmental Engineering* 130(7):774–777.

Boyd, G.R., H. Wang, M.D. Britton, D.C. Howie, D.J. Wood, J.E. Funk, and M.J. Friedman (2004b). "Intrusion Within a Simulated Water Distribution System Due to Hydraulic Transients: II. Volumetric Method and Comparison of Results." *Journal of Environmental Engineering* 130(7):778–783.

Bozue, J. A., and W. Johnson (1996). "Interaction of *Legionella pneumophila* with *Acanthamoeba castellanii*: Uptake by Coiling Phagocytosis and Inhibition of Phagosome-Lysosome Fusion." *Infection and Immunity* 64:668–673.

Braid, P. L (1994). "Upgrading Existing Drinking Water Assets to Meet Standards." Presented at Institution of Water and Environmental Management Joint Symposium, National Scientific Section and Scottish Branch, Hamilton, Scotland, March 9.

Brandt, M., J. Clement, J. Powell, R. Casey, D. Holt, N. Harris, and C.T. Ta (2006). *Managing Distribution Retention Time to Improve Water Quality, Phase 1*" Denver, CO: Water Research Foundation.

Briganti, L. A., and S. C. Wacker (1995). *Fatty Acid Profiling and the Identification of Environmental Bacteria for Drinking Water Utilities.* Denver, CO: American Water Works Association.

Bucklin, K. E., G. A. McFeters, and A. Amirtharajah (1991). "Penetration of Coliforms Through Municipal Drinking Water Filters." *Water Research* 25:1013.

Burlingame, G. A., and H. M Neukrug (1993). "Developing Proper Sanitation Requirements and Procedures for Water Main Disinfection." In *Proceedings of the AWWA Annual Conference, Denver, CO.*

Burlingame, G. A., and L. O'Donnell (1994). "Coliform Sampling at Routine and Alternate Taps: Problems and Solutions" In *Proceedings of the AWWA Water Quality Technology Conference, Denver, CO.*

Burlingame, G. A., and J. J. Choi (1998). "Philadelphia's Guidelines for Obtaining Representative Samples from Throughout Drinking Water Systems." In *Proceedings of the AWWA Water Quality Technology Conference, Denver, CO.*

Burlingame, G. A., G. Korntreger, and C. Lahann (1995). "The Configuration of Standpipes in Distribution Affects Operations and Water Quality." *Journal of the New England Water Works Association* 109(4):281–289.

Burman, N. P (1965). "Taste and Odour Due to Stagnation and Local Warming in Long Lengths of Piping." *Journal Society for Water Treatment Examination* 14(1965):125–131.

Buswell C.M., Y. M. Herlihy, L. M. Lawrence, J. T. M. McGuiggan, P. D. Marsh, C. W. Keevil, and S. A. Leach (1998). "Extended Survival and Persistence of *Campylobacter* spp. in Water and Aquatic Biofilms and Their Detection by Immunofluorescent-Antibody and -rRNA Staining." *Applied and Environmental Microbiology* 64(2):733–741.

Camper, A. K., S. C. Broadaway, M. W. LeChevallier, and G. A. McFeters (1987)."Operational Variables and the Release of Colonized Granular Activated Carbon Particles in Drinking Water." *Journal AWWA* 79(5):70–74.

Camper, A. K (1996). *Factors Limiting Microbial Growth the Distribution System: Laboratory and Pilot-Scale Experiments.* Denver, CO: Water Research Foundation and American Water Works Association.

Camper, A. K., M. Warnecke, W. L. Jones, and G. A. McFeters (1998). *Pathogens in Model Distribution Systems.* Denver, CO: Water Research Foundation and American Water Works Association.

Camper, A. K., P. Butterfield, B. Ellis, W. L. Jones. W. B. Anderson, P. M. Huck, R. Slawson, C. Volk, N. Welch, and M. LeChevallier (2000). *Investigation of the Biological Stability of Water in Treatment Plants and Distribution Systems.* Denver, CO: Water Research Foundation.

Carnahan, A. M., and M. Altwegg (1996). "Taxonomy." In S. Joseph (ed.), *The Genus Aeromonas.*" New York: Wiley.

Carrière, A., V. Gauthier, R. Desjardins, and B. Barbeau (2005). "Evaluation of Loose Deposits in Distribution Systems through Unidirectional Flushing." *Journal AWWA* 97(9):82–92.

Carter, J. T., Y. Lee, and S. G. Buchberger (1997). "Correlations Between Travel Time and Water Quality in a Deadend Loop." In *Proceedings of the AWWA Water Quality Technical Conference, Denver, CO.*

Cesario, L (1995). *Modeling, Analysis and Design of Water Distribution Systems.* Denver, CO: American Water Works Association.

Chaidez, C., and C. Gerba (2004). "Comparison of the Microbiological Quality of Point-of-Use (POU)–Treated Water and Tap Water." *International Journal of Environmental Health Research* 14:253–261.

Characklis, W. G (1988). *Bacterial Regrowth in Distribution Systems.* Denver, CO: Water Research Foundation and American Water Works Association.

Characklis, W. G., and K. C. Marshall (1990). *Biofilms.* New York: Wiley.

Chauret, C., C. Volk, R. Creason, J. Jarosh, J. Robinson, and C. Warnes (2001). "Detection of *Aeromonas hydrophila* in a Drinking-Water Distribution System: A Field and Pilot Study." *Canadian Journal of Microbiology* 47:782.

Chen, X., and P. S. Stewart (1996). "Chlorine Penetration into Artificial Biofilm Is Limited by a Reaction-Diffusion Interaction." *Environmental Science and Technology* 30(6):2078–2083.

Choi, Y. C., and E. Morgenroth (2003). "Monitoring Biofilm Detachment under Dynamic Changes in Shear Stress Using Laser-Based Particle Size Analysis and Mass Fractionation." *Water Science and Technology* 47(5):69–76.

Chopra, A. K (2008). *Characterization of Waterborne Aeromonas Species for Their Virulence Potential.* Denver, CO: Water Research Foundation. London: Department for Environment, Food, and Rural Affairs.

Clark, J. A.,C. A. Burger, and L. E. Sabatinos (1983). "Characterization of Indicator Bacteria in Municipal Raw Water, Drinking Water, and New Main Water Samples." *Canadian Journal of Microbiology* 28:1002–1013.

Clark, R. M., E. E. Geldreich, K. R. Fox, E. W. Rice, C. H. Johnson, J. A. Goodrich, J. A. Barnick, and F. Abdesaken (1996). "Tracking a *Salmonella* Serovar *typhimurium* Outbreak in Gideon, Missouri: Role of Contaminant Propagation Modeling." *Journal of Water Supply Research and Technology, AQUA* 45(4):171–183.

Clark, R. M., and W. M. Grayman (1998). *Modeling Water Quality in Drinking Water Distribution Systems.* Denver, CO: American Water Works Association.

Clement, J. A., and V. Snoeyink (1998). "Analyses and Interpretation of the Physical, Chemical and Biological Characteristics of Distribution System Pipe Scales." In *Proceedings of AWWA Water Quality Technical Conference, San Diego, CA.*

Clement, J.A., Spencer, C., Capuzzi, A.J., Camper, A., Van Andel, K. and A. Sandvig (2003). *Influence of Distribution System Infrastructure on Bacterial Regrowth.* Denver, CO: Water Research Foundation.

Colbourne, J.S., D.J. Pratt, M.G. Smith, S.P. Fisher-Hoch, and D. Harper (1984). "Water Fittings as Sources of *Legionella pneumophila* in a Hospital Plumbing System." *Lancet* 323(8370):210–213.

Colford, J. M., Jr., T. J. Wade, S. K. Sandhu1, C. C. Wright, S. Lee, S. Shaw, K. Fox, S. Burns, A. Benker, M. A. Brookhart, M. van der Laan, and D. A. Levy (2005). "A Randomized, Controlled Trial of In-Home Drinking Water Intervention to Reduce Gastrointestinal Illness." *American Journal of Epidemiology* 161(5):472–482.

Craun, G. F., and R. L. Calderon (2001). "Waterborne Disease Outbreaks Caused by Distribution System Deficiencies." *Journal AWWA* 93(9):64–75.

CSIRO (2002). "Spray Lining." Avalable at www.dbce.csiro.au/research/urbanwater/pipes/tech-spray2.cfm.

Cunliffe, D. A (1990). "Inactivation of *Legionella pneumophila* by Monochloramine." *Journal of Applied Bacteriology* 68(5):453–459.

DeBeer, D., R. Srinivasan, and P. S. Stewart (1994). "Direct Measurement of Chlorine Penetration into Biofilms During Disinfection." *Applied and Environmental Microbiology* 60(12):4339–4344.

De Victoria, J., and M. Galvan (2001). "*Pseudomonas aeruginosa* as an Indicator of Health Risk in Water for Human Consumption." *Water Science and Technology* 43:49–52.

Deb, A. K., Y. J. Hasit, and C. Norris (1999). *Demonstration of Innovative Water Main Renewal Techniques.* Denver, CO: Water Research Foundation and American Water Works Association.

Deb, A. K., F. M. Grablutz, Y. J. Hasit, and K. A. Momberger (2000). *Guidance for Management of Distribution System Operation and Maintenance.* Denver, CO: Water Research Foundation.

Deshayes, F., M. Schmitt, M. Ledrans, C. Gourier-Frery, and H. De Valk (2001). "Pollution du Réseau d'Eau Potable à Strasbourg et Survenue Concomitante de Gastro-Entérites—Mai 2000." *Bulletin Épidémiologique Hebdomadaire* 2:1–8.

Ding, G., N. Sugiura, Y. Inamori, and R. Sudo (1995). "Effect of Disinfection on the Survival of *Escherichia coli* Associated with *Nematoda* in Drinking Water." *Water Supply* 13(3–4):101–106.

Donlan, R., R. Murga, J. Carpenter, E. Brown, R. Besser, and B. Fields (2002). "Monochloramine Disinfection of Biofilm-Associated *Legionella pneumophila* in a Potable Water Model System." In R. Marre, Y. Abu Kwaik, C. Bartlett, N. P. Cianciotto, B. S. Fields, M. Frosch, J. Hacker, and P. C. Luck (eds.), *Legionella.* Washington, DC: American Society for Microbiology.

Dott, W., and D. Waschko-Dransmann (1981). "Occurrence and Significance of Actinomycetes in Drinking Water." *Zbl. Bakt. Hyg. I., Abt. Orig. B,* 173:217–232.

Duranceau, S. J., J. V. Foster, and J. Poole (1998). *Impact of Wet-Pipe Fire Sprinkler Systems on Drinking Water Quality.* Denver, CO: Water Research Foundation.

Emde, K. M., D. W. Smith, and R. Facey (1992). "Initial Investigation of Microbially Influenced Corrosion (MIC) in a Low Temperature Water Distribution System." *Water Research* 26(2):169–175.

Engelbrecht, R. S., and C. N. Haas (1977). "Acid-Fast Bacteria and Yeast as Disinfection Indicators: Enumeration Methodology." In *Proceedings of the AWWA Water Quality Technical Conference, Kansas City, MO.*

Escobar, I. C., and A. A. Randall (1999). "Influence of NF on Distribution System Biostability." *Journal AWWA* 91(6):76–89.

Falkinham, J. O., III, C. D. Norton, and M. W. LeChevallier (2001). "Factors Influencing Numbers of *Mycobacterium avium*, *Mycobacterium intracellulare*, and Other Mycobacteria in Drinking Water Distribution Systems." *Applied Environmental Microbiology* 67(3):1225–1231.

Ferroni, A., L. Nguyen, B. Pron, G. Quesne, M. C. Brusset, and P. Berche (1998). "Outbreak of Nosocomial Urinary Tract Infection Due to *Pseudomonas aeruginosa* in a Pediatric Surgical Unit Associated with Tap-Water Contamination." *Journal of Hospital Infection* 39:301–307.

Flannery, B., L.B. Gelling, D.J. Vugia, J.M. Weintraub, J.J. Salerno, M.J. Conroy, V.A. Stevens, C.E. Rose, M.R. Moore, B.S. Fields, and R.E. Besser (2006). "Reducing *Legionella* Colonization of Water Systems with Monochloramine." *Emerging Infectious Diseases* 12(4):588–596.

Flemming, K. K., R. W. Gullick, J. P. Dugandzic, and M. W. LeChevallier (2006). *Susceptibility of Distribution Systems to Negative Pressure Transients*. Denver, CO: AwwaRF.

Fokken, B., F. Karrenbrock, and I. Hübner (1998). "Raw Water Protection and High Standard in Pipe Network Enables Cologne Municipal Water Supply to Cease Disinfection." *Water Supply* 16(1–2):551–554.

Francis, C. A., K. J. Roberts, J. M. Beman, A. E. Santoro and B. B. Oakley (2005). "Ubiquity and Diversity of Ammonia-Oxidizing Archaea in Water Columns and Sediments of the Ocean." *Proceedings of the National Academy of Sciences of the United States of America* 102(41):14683–14688.

Friedman, M. J., K. Martel, A. Hill, D. Holt, S. Smith, T. Ta, C. Sherwin, D. Hiltebrand, P. Pommerenk, Z. Hinedi, and A. Camper (2003). *Establishing Site Specific Flushing Velocities*. Denver, CO: Water Research Foundation and American Water Works Association.

Friedman, M., L. Radder, S. Harrison, D. et al (2004). *Verification and Control of Pressure Transients and Intrusion in Distribution Systems.* Awwa Research Foundation Report. Denver, CO: Water Research Foundation.

Friedman, M., A. Hanson, and K. Dewis (2007). "Case Studies of Utility Approaches to Managing Coliform Occurrences." In *Proceedings from the Distribution System Research Symposium, Reno, NV.* Denver, CO: American Water Works Association.

Friedman, M., A. Hill, R. Valentine, G. Korshin, S. Reiber (2009). *Inorganics Accumulation in Drinking Water Scales and Sediments*. Denver, CO: Water Research Foundation.

Friedman, M., M. LeChevallier, J. Rosen, G. Gagnon, M.C. Besner, L. Truelstrup-Hansen, T. Hargy, A. Hanson, K. Dewis, and G. Kirmeyer (2009). *Strategies for Managing and Responding to Total Coliform and E. coli in Drinking Water Distribution Systems.* Denver, CO: Water Research Foundation.

Frost, F. J., G. F. Craun, and R. L. Calderon (1996). "Waterborne Disease Surveillance." *Journal AWWA* 88:66–75.

Gatel, D., P. Servais, J.C. Block, P. Bonne, and J. Cavard (2000). "Microbiological Water Quality Management in the Paris Suburbs Distribution System." *Journal of Water Spply Research and Technology, AQUA* 49(5):231–241.

Gauthier, V., B. Gérard, J.M. Portal, J.C. Block, and D. Gatel (1999b). "Organic Matter as Loose Deposits in a Drinking Water Distribution System." *Water Research* 33(4):1014–1026.

Gauthier, V., S. Redercher, and J.-C. Block (1999). "Chlorine Inactivation of Sphingomonas Cells Attached to Goethite Particles in Drinking Water." *Applied and Environmental Microbiology* 65(1):355–357.

Gauthier, V., C. Rosin, L. Mathieu, J.M. Portal, J.C. Block, P. Chaix, and D. Gatel (1996). "Characterization of the Loose Deposits in Drinking Water Distribution Systems." In *Proceedings of AWWA Water Quality Technical Conference, Boston, MA.*

Gauthier, V., M.C. Besner, B. Barbeau, R. Millette, and M. Prévost (2000). "Storage Tank Management to Improve Drinking Water Quality: Case Study." *Journal of Water Resources Planning and Management* 126(4): 221–228.

Geier,H., S. Mostowy, G. A. Cangelosi, M. A. Behr, and T. E. Ford (2008). "Autoinducer-2 Triggers the Oxidative Stress Response in *Mycobacterium avium*, Leading to Biofilm Formation." *Applied and Environmental Microbiology* 74(6):1798–1804.

Geldreich, E. E., H. D. Nash, and D. Spino (1977). "Characterizing Bacterial Populations in Treated Water Supplies: A Progress Report." In *Proceedings of the AWWA Water Quality Technology Conference, Kansas City, MO.*

Geldreich, E. E., K. R. Fox, J. A. Goodrich, et al (1992). "Searching for a Water Supply Connection in the Cabool, Missouri: Disease Outbreak of *Escherichia coli* 0157:H7." *Water Research* 26(8):1127–1137.

Geldreich, E. E., and M. W. LeChevallier (1999). "Microbial Water Quality in Distribution Systems." In R. D. Letterman (ed.), *Water Quality and Treatment*, 5th ed., New York: McGraw-Hill. Denver, CO: American Water Works Association.

Gião, M. S. N. F. Azevedo, S. A. Wilks, M. J. Vieira, and C. W. Keevil (2008). "Persistence of *Helicobacter pylori* in Heterotrophic Drinking-Water Biofilms." *Applied and Environmental Microbiology* 74(19):5898–5904.

Grayman W., N. Khanal, and V. Speight (2007). "Future Opportunities for Synergies Between Security Research and TCR Issues." Report to the American Water Works Association, WITAF Project 268, Denver, CO.

Grayman, W., L.A. Rossman, C. Arnold, R.A. Deininger, C. Smith, J.F. Smith and R. Schnipke (2000). *Water Quality Modeling of Distribution System Storage Facilities*. Denver, CO: Water Research Foundation and American Water Works Association.

Grayman, W., V. Speight, and J. Uber (2007a). "Use of Distribution System Modeling in Designing Microbial Monitoring Program." In *Proceedings of the World Environmental and Water Resources Congress, EWRI, ASCE, Tampa, FL.*

Great Lakes Upper Mississippi River Board (GLUMRB) (2007). *Recommended Standards for Water Works*. Albany, NY: Health Research, Inc.

Gullick, R. W., M. W. LeChevallier, R. C. Svindland, and M. J. Friedman (2004). "Occurrence of Transient Low and Negative Pressures in Distribution Systems." *Journal AWWA* 96(11):52–66.

Haas, C. N., M. A. Meyer, and M. E. Paller (1983a) "The Ecology of Acid-fast Organisms in Water Supply, Treatment, and Distribution Systems." *Journal AWWA* 75:39–144.

Haas, C. N., M. A. Meyer, and M. S. Paller (1983b). "Microbial Dynamics in GAC Filtration of Potable Water." *Journal of Environmental Engineering* 109: 956–961.

Haas, C.N., R.B. Chitluru, M. Gupta, W.O. Pipes, and G.A. Burlingame (1998). *Development of Disinfection Guidelines for the Installation and Replacement of Water Mains*. AWWA Researcg Foundation Report. Denver, CO: Water Research Foundation.

Hargy, T., J. Rosen, J. Sobrinho, M. LeChevallier, and M. Friedman (2008). "Field Evaluation of High Volume Sampling Method for Total Coliform and *E. coli*." In *Proceedings of the AWWA Water Quality Technical Conference, Cincinnati, OH.*

Hart, D., S. McKenna, K. Klise, V. Cruz and M. Wilson (2007). "CANARY: A Water Quality Event Detection Algorithm Development Tool." In *Proceedings of the World Environmental and Water Resources Congress, EWRI, ASCE, Tampa, FL.*

Harvey, A., and R. Kopansky (2007). "Maintaining Distribution Water Quality during a Major Reservoir Repair Effort." In *Proceedings of the AWWA Water Quality Technology Conference, Charlotte, NC.*

Haudidier, K., J. L. Paquin, T. Francais, P. Hartemann, G. Grapin, F. Colin, M. J. Jourdain, J. C. Block, J. Cheron, O. Pascal, Y. Levi, and J. Miazga (1988). "Biofilm Growth in Drinking Water Networks: A Preliminary Industrial Pilot Plant Experiment." *Water Science and Technology* 20:109–115.

Heffelfinger, J. D., J. L. Kool, S. K. Fridkin, V. J. Fraser, J. C. Carpenter, J. Hageman, J. Carpenter, and C. G. Whitney (2003). "Risk of Hospital-Acquired Legionnaires' Disease in Cities Using Monochloramine versus Other Water Disinfectants." *Infection Control and Hospital Epidemiology* 24(8):569–574.

Hellard, M. E., M. I. Sinclair, A. B. Forbes, and C. K. Fairley (2001). "A Randomized, Blinded, Controlled Trial Investigating the Gastrointestinal Health Effects of Drinking Water Quality." *Environmental Health Perspective* 109:773–778.

Herson, D. S., and H. Victoreen (1980). "Identification of Coliform Antagonists." In *Proceedings of the AWWA Water Quality Technical Conference, Miami, FL.*

Hinzelin, F., and J. C. Block (1985). "Yeasts and Filamentous Fungi in Drinking Water." *Environmental Technology Letters* 6:101–106.

Hooper, S. M., C. L. Moe, J. G. Uber, and K. A. Nilsson (2006). "Assessment of Microbiological Water Quality after Low Pressure Events in a Distribution System." In *Proceedings of the 8th Annual Water Distribution Systems Analysis Symposium, Cincinnati, OH.*

Houlmann, N (1992). "Reservoirs d'eau, Conception et Entretien vus sous l'angle des Directives de la. SSIGE." *Gas, Wasser, Abwasser* 72(5):320–323.

Hoxie, N.J., J.P. Davis, J.M. Vergeront, R.D. Nashold, and K.A. Blair (1997). "Cryptosporidiosis-Associated Mortality Following a Massive Waterborne Outbreak in Milwaukee, Wisconsin." *American Journal of Public Health* 87(12):2032–2035.

Hunter, P. R., R. M. Chalmers, S. Hughes, and Q. Syed (2005). "Self-Reported Diarrhea in a Control Group: a Strong Association with Reporting of Low-Pressure Events in Tap Water." *Clinical Infectious Diseases* 40(4):e32–e34.

Jacquemin, J. L., A. M. Simitzia, and N. Chaneoau (1981). "Free-Living Amoebae in Fresh Water: A Study of the Water Supply of the Town of Poiters." *Bulletin de la Societe de Pathologie Exotique et de ses Filiales (Paris)* 74 :521–534.

Joret, J. C., and Y. Levi (1986). "Méthode Rapide d'Évaluation du Carbone Éliminable des Eaux par Voie Biologique." *Tribune du Cebedeau* 510(39):3–9.

Juhna,T., D. Birzniece, S. Larsson, D. Zulenkovs, A. Sharipo, N.F. Azevedo, F. Ménard-Szczebara, S. Castagnet, C. Féliers, and C.W. Keevil (2007). "Detection of *Escherichia coli* in Biofilms from Pipe Samples and Coupons in Drinking Water Distribution Networks." *Applied and Environmental Microbiology* 73(22):7456–7464.

Kalmbach, S., W. Manz, and U. Szewzyk (1997). "Isolation of New Bacterial Species from Drinking Water Biofilms and Proof of their in situ Dominance with Highly Specific 16S rRNA Probes." *Applied and Environmental Microbiology* 63(11):4164–4170.

Karim, M. R., M. Abbaszadegan, and M. LeChevallier (2003). "Potential for Pathogen Intrusion during Pressure Transients." *Journal AWWA* 95(5):134–146.

Keinanen, M. M., P. J. Martikainen, and M. H. Kontro (2004). "Microbial Community Structure and Biomass in Developing Drinking Water Biofilms." *Canadian Journal of Microbiology* 50(3):183–191.

Kelley, J., G. Kinsey, R. Paterson, D. Brayford, R. Pitchers, and H. Rossmoore (2003). *Identification and Control of Fungi in Distribution Systems.* Denver, CO: Water Research Foundation and American Water Works Association.

Kerr, C. J., K. S. Osborn, G. D. Robson, and P. S. Handley (1999). "The Relationship Between Pipe Material and Biofilm Formation in a Laboratory Model System." *Journal of Applied Microbiology Symposium Supplement* 85:29S–38S.

Kilb, B., B. Lange, G. Schaule, H.C. Flemming, and J. Wingender (2003). "Contamination of Drinking Water by Coliforms from Biofilms Grown on Rubber-Coated Valves." *International Journal of Hygiene and Environmental Health* 206:563–573.

Kirmeyer, G.J., L. Kirby, B.M. Murphy, P.F. Noran, K.D. Martel, T.W. Lund, J.L. Anderson, and R. Medhurst (1999). *Maintaining and Operating Finished Water Storage Facilities.* Awwa Research Foundation Report. Denver, CO: Water Research Foundation and American Water Works Association.

Kirmeyer, G.J., M. Friedman, J. Clement, A. Sandvig, P.F. Noran, K. Martel, D. Smith, M. LeChevallier, C. Volk, E. Antoun, D. Hiltebrand, J. Dykesen, and R. Cushing (2000). *Guidance Manual to Maintain Distribution System Water Quality.* Denver, CO: Water Research Foundation and American Water Works Association.

Kirmeyer, G.J., M. Friedman, K. Martel, D. Howie, M. LeChevallier, M. Abbaszadegan, M. Karim, J. Funk, and J. Harbour (2001). *Pathogen Intrusion into the Distribution System.* Awwa Research Foundation Report. Denver, CO: Water Research Foundation.

Kish, L (1995). *Survey Sampling.* New York: Wiley.

Konneke, M., A. E. Bernhard, J. R. de la Torre, C. B. Walker, J. B. Waterbury and D. A. Stahl (2005). "Isolation of an Autotrophic Ammonia-oxidizing Marine Archaeon." *Nature* 437(7058): 543–546.

Kool, J. L., J. C. Carpenter, and B. S. Fields (1999). "Effect of Monochloramine Disinfection of Municipal Drinking Water on Risk of Nosocomial Legionnaires' Disease." *Lancet* 353(9149):272–277.

Kroll, D (2007). "Intelligent Monitoring for Contamination Events in the Drinking Water Supply." In *Proceedings of the 2007 IEEE Conference on Technologies for Homeland Security: Enhancing Critical Infrastructure Dependability, Woburn, MA.* Available at www.hachhst.com/uploadedFiles/Hach's_Water_Distribution_Monitoring_System/How_It_Works/2007%20IEEE.pdf.

La Scola, B., R. J. Birtles, G. Greub, T. J. Harrison, R. M. Ratcliff, and D. Raoult (2004). "*Legionella drancourtii* sp. nov., A Strictly Intracellular Amoebal Pathogen." *International Journal of Systematic Evolutionary Microbiology* 54:699–703.

Lahti, K., and L. Hiisvirta (1995). "Causes of Waterborne Outbreaks in Community Water Systems in Finland: 1980–1992." *Water Science and Technology* 31(5–6):33–36.

Laurent, P., M.C. Besner, P. Servais, V. Gauthier, M. Prévost, and A. Camper (2005). "Water Quality in Drinking Water Distribution Systems." In M. Prévost, P. Laurent, P. Servais, and J. C. Joret (eds.), *Biodegradable Organic Matter in Drinking Water Treatment and Distribution.* Denver, CO: American Water Works Association.

LeChevallier, M. W., and R. J. Seidler (1980). "*Staphylococcus aureus* in Rural Drinking Water." *Applied and Environmental Microbiology* 39:739–742.

LeChevallier, M. W., R. J. Seidler, and T. M. Evans (1980). "Enumeration and Characterization of Standard Plate Count Bacteria by Chlorinated and Raw Water Supplies." *Applied and Environmental Microbiology* 40:922–930.

LeChevallier, M. W., S. C. Cameron, and G. A. McFeters (1983). "New Medium for Improved Recovery of Coliform Bacteria from Drinking Water." *Applied and Environmental Microbiology* 45:484–492.

LeChevallier, M. W., and G. A. McFeters (1985a). "Enumerating Injured Coliforms in Drinking Water." *Journal AWWA* 77:81–87.

LeChevallier, M. W., and G. A. McFeters (1985b). "Interactions Between Heterotrophic Plate Count Bacteria and Coliform Organisms." *Applied and Environmental Microbiology* 49:1338–1341.

LeChevallier, M. W., T. M. Babcock, and R. G. Lee (1987). "Examination and Characterization of Distribution System Biofilms." *Applied and Environmental Microbiology* 53(12):2714–2724.

LeChevallier, M. W (1990). "Coliform Regrowth in Drinking Water: A Review." *Journal AWWA* 82(11):74–86.

LeChevallier, M. W., C. D. Lowry, and R. G. Lee (1990). "Disinfecting Biofilms in a Model Distribution System." *Journal AWWA* 82(7):87–99.

LeChevallier, M. W (1991). "Biocides and the Current Status of Biofouling Control in Water Systems." In *Proceedings of an International Workshop on Industrial Biofouling and Biocorrosion.* New York: Springer-Verlag.

LeChevallier, M. W., W. Schulz, and R. G. Lee (1991). "Bacterial Nutrients in Drinking Water." *Applied and Environmental Microbiology* 57(3):857–862.

LeChevallier, M. W., C. D. Lowry, R. G. Lee, and D. L. Gibbon (1993). "Examining the Relationship Between Iron Corrosion and the Disinfection of Biofilm Bacteria." *Journal AWWA* 85(7): 111–123.

LeChevallier, M. W., N. J. Welch, and D. B. Smith (1996). "Full-Scale Studies of Factors Related to Coliform Regrowth in Drinking Water." *Applied and Environmental Microbiology* 62(7): 2201–2211.

LeChevallier, M. W., W. D. Norton, and T. B. Atherholt (1997). "Protozoa in Open Reservoirs." *Journal AWWA* 89(9):84–96.

LeChevallier, M. W., R. Gullick, and M. Karim (2002). "The Potential for Health Risks from Intrusion of Contaminants into the Distribution System from Pressure Transients." Available at http://epa.gov/safewater/disinfection/tcr/pdfs/whitepaper_tcr_intrusion.pdf. Accessed on January 8, 2009.

LeChevallier, M.W., Gullick, R. W., Karim, M. R., Friedman, M. F., and Funk, J. E (2003). "The Potential for Health Risks from Intrusion of Contaminants into the Distribution System from Pressure Transients." *Journal of Water Health* 1(1):3–14.

LeChevallier, M.W., M. Karim, R. Aboytes, R. Gullick, J. Weihe, B. Earnhardt, J. Mohr, J. Starcevich, J. Case, J. S. Rosen, J. Sobrinho, J. L. Clancy, R. M. McCuin, J. E. Funk, and D. J. Wood (2003). *Profiling Water Quality Parameters: From Source Water To The Household Tap*. Awwa Research Foundation Report. Denver, CO: Water Research Foundation and American Water Works Association.

LeChevallier, M. W (2004). "Control, Treatment, and Disinfection of *Mycobacterium avium* Complex in Drinking Water." In S. Pedley, J. Bartram, G. Rees, A. Dufour, J. Cotruvo (eds.), *Pathogenic Mycobacteria in Water.* Geneva: World Health Organization.

LeChevallier, M. W (2005). "Microbial Water Quality Within the Distribution System." In M. J. MacPhee (ed.), *Distribution System Water Quality Challenges in the 21st Century: A Strategic Guide*. Denver, CO: American Water Works Association.

LeChevallier, M. W (2006) "Mycobacterium Avium Complex." In: *Waterborne Pathogens; Manual of Water Supply Practices—M48*, 2nd ed. Denver, CO: American Water Works Association.

Lee, J.J., P. Schwartz, P. Sylvester, L. Crane, J. Haw, H. Chang, and H.J. Kwon (2003). *Impacts of Cross-Connections in North American Water Supplies*. Awwa Research Foundation Report. Denver, CO: Water Research Foundation.

Lehtola, M. J., I. T. Miettinen, T. Vartiainen, and P. J. Martikainen (2002). "Changes in Content of Microbially Available Phosphorus, Assimilable Organic Carbon and Microbial Growth Potential during Drinking Water Treatment Processes." *Water Research* 36(15):3681–3690.

Lehtola M. J., M. Laxander, I. T. Miettinen, A. Hirvonen, T. Vartiainen, and P. J. Martikainen (2006). "The Effects of Changing Water Flow Velocity on the Formation of Biofilms and Water Quality in Pilot Distribution System Consisting of Copper or Polyethylene Pipes." *Water Research* 40(11):2151–2160.

Lehtola, M.J., E. Torvinen, J. Kusnetsov, T. Pitkänen, L. Maunula, C.H. von Bonsdorff, P.J. Martikainen, S.A. Wilks, C.W. Keevil, and I.T. Miettinen (2007). "Survival of *Mycobacterium avium, Legionella pneumophila, Escherichia coli*, and Caliciviruses in Drinking Water-Associated Biofilms Grown under High-Shear Turbulent Flow." *Applied and Environmental Microbiology* 73(9):2854–2859.

Liang JL, Dziuban EJ, Craun GF, Hill V, Moore MR, Gelting RJ, Calderon RL, Beach MJ, Roy SL (2006). "Surveillance for Waterborne Disease and Outbreaks Associated with Drinking Water and Water Not Intended for Drinking—United States, 2003–2004." *Morbidity and Mortality Weekly Report, Surveillance Summary* 55(12):31–65.

Liu, W. T., and D. A. Stahl (2006). "Molecular Approaches for the Measurement of Density, Diversity, and Phylogeny." In C.J. Hurst, Knudsen, G.R., McInerney, M.J., Stetzenbach, L.D., and Water, M. V (eds.), *Manual of Environmental Microbiology*, 3rd ed. Washington, DC: American Society for Microbiology.

Locas, A., B. Barbeau, and V. Gauthier (2007). "Nematodes as a Source of Total Coliforms in a Distribution System." *Canadian Journal of Microbiology* 53(5):580–585.

Logsdon, G. S (1987). "Comparison of Some Filtration Processes Appropriate for *Giardia* Cyst Removal." In *Proceedings of the Calgary Giardia Conference, Calgary, Canada.*

Lu, L., M. Fang, X. Wang, N. Patni, A. Ashendorff, and M. Principe (1997). "In Situ Water Main Biological Study in New York City." In *Proceedings of AWWA Water Quality Technical Conference, Denver, CO.*

Ly, T. M. C., and H. E. Müller (1990). "Ingested *Listeria monocytogenes* Survive and Multiply in Protozoa." *Journal of Medical. Microbiology* 33:51–54.

Mackay W. G., L. T. Gribbon, M. R. Barer, and D. C. Reid (1998). "Biofilms in Drinking Water Systems: A Possible Reservoir for *Helicobacter pylori*." *Water Science and Technology* 38(12):181–185.

Mahmood, F., J. G. Pimblett, N. O. Grace, and W. Grayman (2005). "Evaluation of Water Mixing Characteristics in Distribution System Storage Tanks." *Journal AWWA* 97(3):74–88.

Maier, S.H., C.A. Woodward, A. Delanoue, D.M. Holt, S.M. McMath, and R.S. Powell (1999). "Relationships Between Total Chlorine, Particle Counts and Biofilm Detachment in a Drinking Water Distribution System." *Special Publications of the Royal Society of Chemistry* 242:231–238.

Martin, R.S., W.H. Gates, R.S. Tobin, D. Grantham, R. Sumarah, P. Wolfe, and P. Forestall (1982). "Factors Affecting Coliform Bacteria Growth in Distribution Systems." *Journal AWWA* 74(1):34–37.

Martiny, A. C., T. M. Jorgensen, H. J. Albrechtsen, E. Arvin, and S. Molin (2003). "Long-Term Succession of Structure and Diversity of a Biofilm Formed in a Model Drinking Water Distribution System." *Applied and Environmental Microbiology* 69(11):6899–6907.

Martiny, A.C., T.M. Jorgensen, H.J. Albrechtsen, E. Arvin, and S. Molin (2005). "Identification of Bacteria in Biofilm and Bulk Water Samples from a Nonchlorinated Model Drinking Water Distribution System: Detection of a Large Nitrite-Oxidizing Population Associated with *Nitrospira* spp." *Applied and Environmental Microbiology* 71(12):8611–8617.

McFeters, G. A., S. C. Cameron, and M. W. LeChevallier (1982). "Influence of Diluents, Media and Membrane Filters on the Detection of Injured Waterborne Coliform Bacteria." A*pplied and Environmental Microbiology* 43:97–103.

McFeters, G. A (1990). "Enumeration, Occurrence and Significance of Injured Indicator Bacteria in Drinking Water." In G. A. McFeters (ed.), *Drinking Water Microbiology*. New York: Springer-Verlag.

McMath, S. M., and R. J. Casey (2000). "The Use of Network Modeling Techniques to Resolve Water Quality Issues in Distribution Systems." In *Proceedings of AWWA Water Quality Technical Conference, Salt Lake City, UT.*

Miettinen, I.T., T. Vartiainen, and P.J. Martikainen (1997). "Phosphorus and Bacterial Growth in Drinking Water." *Applied and Environmental Microbiology* 63(8):3242–3245.

Miettinen, I. T., K. Malaska, L. Korhonen, U. Lignell, P. Kärkkäinen, H. Rintala, E. Kauhanen, and A. Nevalainen (2007). "Occurrence of Fungi and *Actinomyces* in Finnish Drinking Waters." In *Proceedings of the World Environmental and Water Resources Congress*. Avail;able at http://dx.doi.org/10.1061/40927(243)483.

Mofenson, L.M., M.T. Brady, S.P. Danner, K.L. Dominguez, R. Hazra, E. Handelsman, P. Havens, S. Nesheim, J.S. Read, L. Serchuck, and R. Van Dyke (2009). "Guidelines for the Prevention and Treatment of Opportunistic Infections Among HIV-Exposed and HIV-Infected Children." *Morbidity and Mortality Weekly Report* 58:1–166.

Molmeret, M., M. Horn, M. Wagner, M. Santic, and Y. Abu Kwaik (2005). "Amoebae as Training Grounds for Intracellular Bacterial Pathogens," *Applied and Environmental Microbiology* 71(1):20–28.

Moore, M. R., M. Pryor, B. Fields, C. Lucas, M. Phelan, and R. E. Besser (2006). "Introduction of Monochloramine into a Municipal Water System: Impact on Colonization of Buildings by *Legionella* spp." *Applied and Environmental Microbiology* 72(1):378–383.

Morin, P., V. Gauthier, S. Saby, and J.-C. Block (1999) "Bacterial Resistance to Chlorine through Attachment to Particles and Pipes Surfaces in Drinking Water Distribution Systems." In W.C. Keevil, A. Godfree, D. Holt, and C.S. Dow (eds.), *Biofilms in the Aquatic Environment*, Vol. 242. London: Special Publication of the Royal Society of Chemistry.

Moyer, N. P (2006). "Aeromonas." In *Waterborne Pathogens*, AWWA Manual M48, 2nd ed. Denver, CO: American Water Works Association.

Murga, R., T. S. Forster, E. Brown, J. M. Pruckler, B. S. Fields, and R. M. Donlan (2001). "Role of Biofilms in the Survival of *Legionella pneumophila* in a Model Potable Water System." *Microbiology* 147:3121–3126.

Murray, R., R. Janke, and J. Uber (2004). "The Threat Ensemble Vulnerability Assessment (TEVA) Program for Drinking Water Distribution System Security." In *Proceedings of the World Water and Environmental Resources Congress, EWRI, ASCE, Reston, VA.*

Murray, R., T. Baranowski, W. E. Hart and R. Janke (2008). "Risk Reduction and Sensor Network Design." In *Proceedings of the Water Distribution System Analysis Conference, Kruger National Park, South Africa.*

Nagy, L. A. and B. H. Olson (1982). "The Occurrence of Filamentous Fungi in Drinking Water Distribution Systems." *Canadian Journal Microbiology* 28:667–671.

Narasimhan R., J. Brereton, M. Abbaszadegan, A. Alum, and P. Ghatpande (2004). Sample Collection Procedures and Locations for Bacterial Compliance Monitoring. Denver, CO: Awwa Research Foundation.

National Academy of Sciences (2006). *Drinking Water Distribution Systems: Assessing and Reducing Risks.* Washington, DC: National Academies Press.

Niemi, R. M., S. Knuth, and K. Lundstrom (1982). "Actinomycetes and Fungi in Surface Waters and in Potable Water." *Applied and Environmental Microbiology* 43:378–388.

North American Society for Trenchless Technology (1999). *Water Pipeline Rehabilitation Method Fact Sheet: Cement Mortar Lining.* Available at www.nastt.org/water/cclnr.pdf.

Norton, C. D., and M. W. LeChevallier (1997). "Chloramination: Its Effect on Distribution System Water Quality." *Journal AWWA* 89(7):66–77.

Norton, C. D., and M. W. LeChevallier (2000). "A Pilot Study of Bacteriological Population Changes through Potable Treatment and Distribution." *Applied and Environmental Microbiology* 66(1):268–276.

National Research Council of the National Academy (NRC) (2006). *Drinking Water Distribution Systems: Assessing and Reducing Risks.* Washington, DC: National Academy Press.

Nygård, K., E. Wahl, T. Krogh, O.A. Tveit, E. Bohleng, A. Tverdal and P. Aavitsland (2007). "Breaks and Maintenance Work in the Water Distribution Systems and Gastrointestinal Illness: A Cohort Study." *International Journal of Epidemiology* 36(4):873–880.

Oesterholt, F., G. Martijnse, G. Medema, and D. Van der Kooij (2007). "Health Risk Assessment of Non-Potable Domestic Water Supplies in the Netherlands." *Journal of Water Supply: Research and Technology, AQUA* 56(3):171–179.

Oliver, E. D., and J. D. Harbour (1995). "Speciation of Total Coliforms Isolated from Sediments in Brown Water Released During a Flushing Program Undertaken by the Seattle Water Department." In *Proceedings of AWWA Water Quality Technical Conference, New Orleans, LA.*

Oliver, J. D (2005). "The Viable but Nonculturable State in Bacteria." *Journal of Microbiology* 43:93–100.

Olson, B. H., and L. Hanami (1980). "Seasonal Variation of Bacterial Populations in Water Distribution Systems," In. *Proceedings of AWWA Water Quality Technical Conference, Miami Beach, FL.*

Opheim, D., J. G. Grochowski, and D. Smith (1988). "Isolation of Coliforms from Water Main Tubercles." In *Proceedings of the Annual Meeting of the American Society for Microbiology. p. 245.* American Society for Microbiology, Washington, D.C.

Ormsbee, L. E., and S. Lingireddy (1997). "Calibrating Hydraulic Network Models." *Journal of American Water Works Association* 89(2):42–50.

Parent, A., S. Fass, M.L. Dincher, D. Reasoner, D. Gatel, and J.C. Block (1996). "Control of Coliform Growth in Drinking Water Distribution Systems." *Journal of the Chartered Institution of Water and Environmental Management* 10(6):442–445.

Payment, P., L. Richardson, J. Siemiatycki, R. Dewar, M. Edwardes, and E. Franco (1991). "Randomized Trial to Evaluate the Risk of Gastrointestinal Disease Due to Consumption of Drinking Water Meeting Microbiological Standards." *American Journal of Public Health* 81(6):703–708.

Payment, P., J. Siemiatycki, L. Richardson, G. Renaud, E. Franco, and M. Prevost (1997). "A Prospective Epidemiological Study of Gastrointestinal Health Effects Due to the Consumption of Drinking Water." *International Journal of Environmental Health Research* 7:5–31.

Pedley, S., J. Bartram, G. Rees, A. Dufour, J. A. Contruvo (eds.) (2004). *Pathogenic Mycobacteria in Water.* London: World Health Organization and International Water Association Publishing.

Perz, J. F., F. K. Ennever, and S. M. LeBlancq (1998). "*Cryptosporidium* in Tap Water." *American Journal of Epidemiology* 147(3):289–301.

Pierson, G., K. Martel, A. Hill, G. Burlingame, and A. Godfree (2001). *Pratices to Prevent Microbiological Contamination of Water Mains*. Denver, CO: Water Research Foundation.

Piriou, P, K. Helmi, M. Jousset, N. Castel, E. Guillot, and L. Kiene (2000). "Impact of Biofilm on *C. parvum* Persistence in Distribution Systems." In *Proceedings International Distribution System Research Symposium.* Denver, CO: American Water Works Association.

Pryor, M., S. Springthorpe, S. Riffard, T. Brooks, Y. Huo, G. Davis, and S. A. Satter (2004). "Investigation of Opportunistic Pathogens in Municipal Drinking Water Under Different Supply and Treatment Regimes." *Water Science and Technology* 50(1):83–90.

Quignon, F., M. Sardin, L. Kiene, and L. Schwartzbrod (1997). "Poliovirus-1 Inactivation and Interaction with Biofilm: A Pilot-Scale Study." *Applied and Environmental Microbiology* 63(3):978–982.

Reasoner, D. J., and E. E. Geldreich (1985). "A New Medium for the Enumeration and Subculture of Bacteria from Potable Water." *Applied and Environmental Microbiology* 49(1):1–7.

Reilly, J. K., and J. Kippin (1981). *Interrelationship of Bacterial Counts with Other Finished Water Quality Parameters Within Distribution Systems.* EPA-0600/52-81-035. Cincinnati, OH: USEPA.

Regan, J. M., G. W. Harrington, and D. R. Noguera (2002). "Ammonia- and Nitrite-Oxidizing Bacterial Communities in a Pilot-Scale Chloraminated Drinking Water Distribution System." *Applied and Environmental Microbiology* 68(1):73–81.

Regan, J.M., G.W. Harrington, H. Baribeau, R. De Leon, and D.R. Noguera (2003). "Diversity of Nitrifying Bacteria in Full-Scale Chloraminated Distribution Systems." *Water Research* 37(1):197–205.

Ridgway, H. F., and B. H. Olson (1981). "Scanning Electron Microscope Evidence for Bacteria Colonization of a Drinking Water Distribution System." *Applied and Environmental Microbiology* 41:274–287.

Rittmann, B. E., and V. L. Snoeyink (1984). "Achieving Biologically Stable Drinking Water." *Journal AWWA* 76(10):106–114.

Rosenzweig, W. D., H. A. Minnigh, and W. O. Pipes (1983). "Chlorine Demand and Inactivation of Fungal Propagules." *Applied and Environmental Microbiology* 45:182–186.

Rossman, L. A., R. M. Clark, and W. M. Grayman (1994). "Modeling Chlorine Residuals in Drinking-Water Distribution Systems." *Journal of Environmental Engineering ASCE* 120(4):803–820.

Rossman, L. A., J. G. Uber, and W. M. Grayman (1995). "Modeling Disinfectant Residuals in Drinking Water Storage Tanks." *Journal of Environmental Engineering* 121(10):752–755.

Sang, J., X. Zhang, and Z. Wang (2003). Role of phosphorus on controlling bacterial regrowth in drinking water. In *Proceedings of the Water Quality Technology Conference*. Denver, CO: AWWA.

Schoenen, D (1986). "Microbial Growth Due to Materials Used in Drinking Water Systems." In H. J. Rehm and G. Reed (eds), *Biotechnology*, Vol. 8. Weinheim, Germany: VCH Verlagsgesellschaft.

Schreiber, H., D. Schoenen, and W. Traunspurger (1997). "Invertebrate Colonization of Granular Activated Carbon Filters." *Water Research* 31(4):743–748.

Schreppel, C. K., and A. J. Geiss (1996). "Total Coliform Occurrence and Corrosion Control." In *Proceedings of the National Conference on Integrating Corrosion Control and Other Water Quality Goals,* Cambridge:MA.

Servais, P., G. Billen, P. Laurent, Y. Lévi, and G. Randon (1992). "Studies of BDOC and Bacterial Dynamics in the Drinking Water Distribution System of the Northern Parisian Suburbs." *Revue des Sciences de l'Eau* 5:69–89.

Servais, P., P. Laurent, and G. Randon (1995). "Comparison of the Bacterial Dynamics in Various French Distribution Systems." *Journal of Water Supply: Research and Technology, AQUA* 44(1):10–17.

Smith, D. B., A. F. Hess, and S. A. Hubbs (1990). "Survey of Distribution System Coliform Occurrences in the United States." In *Proceedings of the AWWA Water Quality Technology Conference, Denver, CO.*

Smith, C. A., C. Phiefer, S. J. Macnaughton, A. Peacock, R. S. Burkhalter, R. Kirkegaard, and D. C. White (2000). "Quantitative Lipid Biomarker Detection of Unculturable Microbes and Chlorine Exposure in Water Distribution System Biofilms." *Water Research* 34(10):2683–2688.

Snead, M. C., V. P. Olivieri, K. Kawata, and C. W. Krusé (1980). "The Effectiveness of Chlorine Residuals in Inactivation of Bacteria and Viruses Introduced by Post Treatment Contamination." *Water Research* 14:403–408.

Speight, V.L., J.R. Nuckols, L. Rossman, A.M. Miles and P.C. Singer (2000). "Disinfection By-Product Exposure Assessment Using Distribution System Modeling." In *Proceedings of the AWWA Water Quality Technology Conference, Salt Lake City, UT.*

Speight V (2005). "GIS Tools for Data Mining and Master Planning." Workshop presentation, AWWA Water Quality Technology Conference, Quebec City, Quebec, Canada.

Speight V (2008). "Distribution Systems: The Next Frontier." *The Bridge, National Academy of Engineering* 38(3):31–37.

Speight V.L., J. Uber, W.M. Grayman, K. Martel and M. Friedman (2009). *Probabilistic Modeling Framework for Assessing Water Quality Sampling Programs.* Awwa Research Foundation Report. Denver, CO: Water Research Foundation.

Speight V., N. Khanal, D. Savic, Z. Kapelan, P. Jonkergouw and M. Agbodo (2010). *Guidelines for Developing, Calibrating and Using Hydraulic Models.* Denver, CO: Water Research Foundation.

Staley, J. T (1983). *Identification of Unknown Bacteria from Drinking Water.* Project CR-807570010. Washington, DC: USEPA.

Steinert, M., L. Emody, R. Amann, and J Hacker (1997). "Resuscitation of Viable but Nonculturable *Legionella pneumophila* Philadelphia JR32 by *Acanthamoeba castellanii*." *Applied and Environmental Microbiology* 63:2047–2053.

Stewart, M. H., R. L. Wolf, and E. G. Means (1990). "Assessment of the Bacteriological Activity Associated with Granular Activated Carbon Treatment of Drinking Water." *Applied and Environmental Microbiology* 56:3822–3829.

Stewart, P. S., G. A. McFeters, and C. T. Huang (2000). "Biofilm Control by Antimicrobial Agents." In J. D. Bryers (ed.), *Biofilms.* New York: Wiley.

Stringfellow, W. T., K. Mallon, and F. A. DiGiano (1993). "Enumerating and Disinfecting Bacteria Associated with Particles Released from GAC Filter-Adsorbers." *Journal AWWA* 85(9):70–80.

Symons, J.M., A.A. Stevens, R.M. Clark, E..E. Geldreich, O.T. Love Jr., and J. DeMarco (1981). Treatment Techniques for Controlling Trihalomethanes in Drinking Water. EPA-600/2-81-156. Cincinnati, OH: USEPA.

Szewzyk U., W. Manz, R. Amann, K. H. Schleifer, T. A. Stenström (1994). "Growth and in Situ Detection of a Pathogenic *Escherichia coli* in Biofilms of a Heterotrophic Water-Bacterium by use of 16S- and 23S-rRNA-Directed Fluorescent Oligonucleotide Probes." *FEMS Microbiology Ecology* 13:169–176.

Taylor, R. H., J. O. Falkinham, III, C. D. Norton, and M. W. LeChevallier (2000). "Chlorine, Chloramine, Chlorine Dioxide, and Ozone Susceptibility of *Mycobacterium avium*." *Applied and Environmental Microbiology* 66(4):1702–1705.

Thofern, E., D. Schoenen, and G. J. Tuschewitzki (1987). "Microbial Surface Colonization and Disinfection Problems." *Off Gesundh.-wes.* 49:14–20.

Thomas, V., K. Herrera-Rimann, D. S. Blanc, and G. Greub (2006). "Biodiversity of Amoebae and Amoeba-Resisting Bacteria in a Hospital Water Network." *Applied and Environmental Microbiology* 72(4):2428–2438.

Trautman, M., T. Michalsky, H. Wiedeck, V. Radlosavljevic, and M. Ruhnke (2001). "Tap Water Colonization with *Pseudomonas aeruginosa* in a Surgical Intensive Care Unit (ICU) and Relation to *Pseudomonas* Infections of ICU Patients." *Infection Control in Hospital Epidemiology* 22:49–52.

University of Southern California (USC) (1993). *Manual of Cross-Connection Control*, 9th ed. Los Angeles: USC Foundation for Cross-Connection Control and Hydraulic Research.

USEPA (1979). "National Secondary Drinking Water Regulations: Final Rule." *Federal Register* 44(140):42195–42202.

USEPA (1989). "Drinking Water: National Primary Drinking Water Regulations: Total Coliforms (Including Fecal Coliforms and *E. coli*): Final Rule." *Federal Register* 54:27544–27568.

USEPA (1995). *Survey Report on the Cross-Connection Control Program.* EIHWG4-01-5400070. Washington, DC: Office of Inspector General.

USEPA (1999a). *Alternative Disinfectants and Oxidants Guidance Manual.* EPA 815-R-99-014. Washington, DC: USEPA.

USEPA (1999b). *Legionella: Human Health Criteria Document.* EPA 822-R-99-001. Washington, DC: EPA Office of Science and Technology, USEPA.

USEPA (2001). "National Primary Drinking Water Regulations." In *Current Drinking Water Standards.* Washington, DC: USEPA. Available at: www.epa.gov/safewater/mcl.html.

USEPA (2002a). "Potential Contamination Due to Cross-Connections and Backflow and the Associated Health Risks." Available at http://epa.gov/safewater/disinfection/tcr/pdfs/issuepaper_tcr _crossconnection-backflow.pdf. Accessed on January 8, 2009.

USEPA (2002b). "Health Risks from Microbial Growth and Biofilms in the Distribution System." Available at http://epa.gov/safewater/disinfection/tcr/pdfs/whitepaper_tcr_biofilms.pdf. Accessed on January 8, 2009.

USEPA (2002c). "Effects of Treatment on Nutrient Availability." Available at http://epa.gov/safewater/disinfection/tcr/pdfs/issuepaper_tcr_treatment-nutrients.pdf. Accessed on September 25, 2009.

USEPA (2003a). *Health Effects Support Document for Acanthamoeba.* EPA 822R03012. Washington, DC: Office of Science and Technology, USEPA.

USEPA (2003b). *Cross-Connection Control Manual.* Washington, DC: USEPA.

USEPA (2005a). "Technologies and Techniques for Early Warning Systems to Monitor and Evaluate Drinking Water Quality: A State-of-the-Art Review." Available at www.epa.gov/nhsrc/pubs/reportEWS120105.pdf.

USEPA (2005b). "Overview of Event Detection Systems for WaterSentinel." Draft, version 1.0. Available at www.epa.gov/safewater/watersecurity/pubs/watersentinel_event_detection.pdf.

USEPA (2005c). *Water Distribution System Analysis: Field Studies, Modeling, and Management: A Reference Guide for Utilities.* EPA 600/R-06/028. Washington, DC: Office of Research and Development, USEPA. Available at www.epa.gov/nrmrl/pubs/600r06028/600r06028.pdf.

USEPA (2006a). *Aeromonas: Human Health Criteria Document.* Washington, DC: Office of Science and Technology, USEPA.

USEPA (2006b). "Inorganic Contaminant Accumulation in Potable Water Distribution Systems." Available at http://epa.gov/safewater/disinfection/tcr/pdfs/issuepaper_tcr_inorganiccontaminantaccumulation.pdf. Accessed on January 8, 2009.

USEPA (2007a). "Effect of Treatment on Nutrient Availability." Available at http://epa.gov/safewater/disinfection/tcr/regulation_revisions.html#issuepapers.

USEPA (2007b). "WBDO Surveillence_Roy.pdf." Available at: www.epa.gov/safewater/disinfection/tcr/pdfs/presentations/presentations_tcrdsac_october2007.pdf.

USEPA (2008a) EPANET. Available at www.epa.gov/nrmrl/wswrd/dw/epanet.html.

USEPA (2008b). *Total Coliform Rule/Distribution System (TCRDS).* Washington, DC: Federal Advisory Committee Agreement in Principle, USEPA. Available at www.epa.gov/safewater/disinfection/tcr/.

USEPA (2009). "Threat Ensemble Vulnerability Assessment Research Program." Available at/www.epa.gov/NHSRC/water/teva.html.

Vaerewijck, M. J. M., G. Huys, J. P. Palomino, J. Swings, and F. Portaels (2005). "Mycobacteria in Drinking Water Distribution Systems: Ecology and Significance for Human Health." *FEMS Microbiology Review* 29:911–934.

van der Kooij, D., W. A. M. Hijnen, and J. C. Kruithof (1989). "The Effects of Ozonation, Biological Filtration and Distribution on the Concentration of Easily Assimilable Organic Carbon (AOC) in Drinking Water." *Ozone: Science & Engineering* 11(3):297–311.

van der Kooij, D (1990). "Assimilable Organic Carbon (AOC) in Drinking Water." In G. A. McFeters (ed.), *Drinking Water Microbiology.* New York: Springer-Verlag.

van der Kooij, D (1992) "Assimilable Organic Carbon as an Indicator of Bacterial Regrowth." *Journal AWWA* 84(2):57–65.

van der Kooij, D., J. H. M. Van Lieverloo, J. Schellart, and P. Hiemstra (1999). "Maintaining Quality Without a Disinfectant Residual." *Journal AWWA* 91(1):55–64.

Van der Wende, E., and W. G. Characklis (1990). "Biofilms in Potable Water Distribution System." In *Drinking Water Microbiology.* New York: Springer-Verlag.

van der Wielen, P. W. J. J., S. Voost, and D. van der Kooij (2009). "Ammonia-Oxidizing Bacteria and Archaea in Groundwater Treatment and Drinking Water Distribution Systems." *Applied and Environmental Microbiology* 75(14):4687–4695.

Van Lieverloo, H (1997) "How to Control Invertebrates in Distribution Systems: By Starvation or by Flushing?" In *Proceedings of AWWA Water Quality Technical Conference, Denver, CO.*

Van Lieverloo, J.H.M., E.J.M. Blokker, G. Medema, B. Hambsch, R. Pitchers, G. Stanfield, M. Stanger, P. Agutter, R. Lake, J.F. Loret, and E. Soyeux (2006). "Contamination During Distribution" (preliminary Web version at www.microrisk.com). Microrisk Research Project.

Vasconcelos, J. J., L. A. Rossman, W. M. Grayman, P. F. Boulos and R. M. Clark (1997). "Kinetics of Chlorine Decay." *Journal AWWA* 89(7):54–65.

Verstraete, W., and M. Alexander (1973). "Heterotrophic Nitrification in Samples of Natural Ecosystems." *Environmental Science and Technology* 7:39–42.

Vivier, J. C., M. M. Ehlers, and W. O. K. Grabow (2004). "Detection of Enteroviruses in Treated Drinking Water." *Water Research* 38(11):2699–2705.

Volk, C., and J. C. Joret (1994). "Paramètres Prédictifs de l'Apparition des Coliformes dans les Réseaux de Distribution d'Eau d'Alimentation." *Revue des Sciences de l'Eau* 7:131–152.

Volk, C. J., and M. W. LeChevallier (2000). "Assessing Biodegradable Organic Matter." *Journal AWWA* 92(5):64–76.

Volk, C. J., and M. W. LeChevallier (2002). "Effects of Conventional Treatment on AOC and BDOC Levels." *Journal AWWA* 94(6):112–123.

Wade, T. J., S. K. Sandhu, D. Levy, S. Lee, M. W. LeChevallier, L. Katz, and J. M. Colford, Jr. (2004). "Did a Severe Flood in the Midwest Cause an Increase in the Incidence of Gastrointestinal Symptoms?" *American Journal of Epidemiology* 159(4):398–405.

Walski, T. M., and T. L. Lutes (1994). "Hydraulic Transients Cause Low-Pressure Problems." *Journal AWWA* 86(12):24–32.

Walski, T.M., D.V. Chase, D.A. Savic, W.M. Grayman, S. Beckwith, and E. Koelle (2003). *Advanced Water Distribution Modeling and Management.* Watertown, CT: Haestad Press.

Watson, S.W., E. E. Bock, H. Harms, H.P. Koops, and A.B. Hooper (1989). "Nitrifying Bacteria." In J. T. Saley, M. P. Bryant, N. Pfennig, and J. G. Holt (eds.), *Bergey's Manual of Systematic Bacteriology,* Vol. 3. Baltimore: Williams & Wilkins.

Watters, S. K., B. H. Pyle, M. W. LeChevallier, and G. A. McFeters (1989). "Enumeration of *Enterobacter cloacae* After Chloramine Exposure." *Applied and Environmental Microbiology* 55:3226–3228.

Welter G., M. LeChevallier, S. Spangler, J. Cotruvo, and R. Moser (2008). *Guidance for Decontamination of Water System Infrastructure.* Awwa Research Foundation Report 2981. Denver, CO: Water Research Foundation.

Wierenga, J. T (1985). "Recovery of Coliforms in the Presence of a Free Chlorine Residual." *Journal AWWA* 77(11):83–88.

Winiecka-Krusnell, J., K. Wreiber, A. von Euler, L. Engstrand, and E. Linder (2002). "Free-Living Amoebae Promote Growth and Survival of *Helicobacter pylori.*" *Scandinavian. Journal of Infectious Disease* 34:253–256.

Wolfe, R. L., E. G. Means, M. K. Davis, and S. Barrett (1988). "Biological Nitrification in Covered Reservoirs Containing Chloraminated Water." *Journal AWWA* 80(9):109–114.

Wolyniak, E. A., B. R. Hargraves, and K. L. Jellison (2009). "Retention and Release of *Cryptosporidium parvum* Oocysts by Experimental Biofilms Composed of a Natural Stream Community." *Applied and Environmental Microbiology* 75(13):4624-4626.

Yoder J, V. Roberts, G.F. Craun, V. Hill, L.A. Hicks, N.T. Alexander, V. Radke, R.L. Calderon, M.C. Hlavsa, M.J. Beach, and S.L. Roy (2008). "Surveillance for Waterborne Disease and Outbreaks Associated with Drinking Water and Water not Intended for Drinking—United States, 2005–2006." *MMWR Surveillance Summary* 57(9):39–62.

Yu, H. S., H. J. Jeong, Y.-C. Hong, S.-Y. Seol, D.-I. Chung, and H.-H. Kong (2007). "Natural Occurrence of *Mycobacterium* as an Endosymbiont of *Acanthamoeba* Isolated from a Contact Lens Storage Case." *Korean Journal of Parasitology* 45:11–18.

CHAPTER 22
WATER TREATMENT PLANT RESIDUALS MANAGEMENT

David A. Cornwell, Ph.D., P.E., B.C.E.E.
Damon K. Roth, P.E.
Environmental Engineering & Technology, Inc.
Newport News, Virginia
United States

INTRODUCTION **22.2**
COAGULATION AND LIME-SOFTENING RESIDUALS **22.4**
Quantities 22.4
Characteristics 22.10
THICKENING **22.16**
NON-MECHANICAL DEWATERING **22.21**
Sand Drying Beds 22.21
Solar Drying Beds 22.24
Dewatering Lagoons 22.25
Freeze-Thaw Beds 22.25
MECHANICAL DEWATERING **22.29**
Vacuum Filters 22.30
Belt Filter Presses 22.31
Centrifuges 22.32
Filter Presses 22.35
SPENT FILTER BACKWASH TREATMENT **22.37**
RECYCLE **22.40**
MEMBRANE RESIDUALS **22.43**
Low-Pressure Membranes 22.43
Desalting Membranes 22.47
ION-EXCHANGE AND INORGANIC ADSORPTION PROCESS RESIDUALS **22.54**
Softening 22.54
Fluoride Removal by Activated Alumina 22.57
Removal of Nitrate and Perchlorate Using Anion Exchange 22.57
Disposal of Brines 22.61
RESIDUALS CONTAINING ARSENIC **22.62**
Regulations 22.63
Residuals Quantities and Characteristics 22.64
Treatment of Arsenic-Containing Liquid Residuals 22.65
Handling of Solids from Arsenic Removal 22.66
RESIDUALS CONTAINING RADIOACTIVITY **22.66**
Characterization 22.67
Regulations 22.68
State TENORM Regulations 22.68
Ultimate Disposal Options 22.69
Mixed Waste 22.69
ULTIMATE DISPOSAL AND UTILIZATION OF SOLIDS **22.70**
ABBREVIATIONS **22.75**
NOTATION FOR EQUATIONS **22.76**
REFERENCES **22.77**

INTRODUCTION

Increasingly, the water treatment process itself and the residuals produced by it are interwoven and considered together rather than looked on as two separate systems. The quality and characteristics of these residuals are completely interrelated to the main treatment process and two different treatment processes achieving the same finished water goals can produce much different residuals. From economic and environmental standpoints, the residuals produced by different treatment processes can have significantly different treatment costs and disposal or reuse options for safe handling. Historically, despite these strong interrelationships between the treatment processes and their residuals, water treatment plant residuals management has historically been treated as a stand-alone management problem. With increasing costs and environmental concerns and desires for beneficial reuse, it has become increasingly common when evaluating or designing the main water process to consider the residuals quality and characteristics and view residuals management as part of the overall process to be optimized when determining the most economical manner in which to meet a specific set of finished water quality goals. This water process and residuals interrelationship is becoming more critical as removal of specific pollutants such as arsenic and radionuclides has emerged.

The original term used to describe water treatment plant wastes was *sludge*. In fact, sludge is just the solid component of many different types of waste streams. The term *residuals* is now used to describe the range of process side-streams—liquid, solid, or gaseous—that can be produced by water treatment processes. These water treatment plant wastes are referred to as residuals and are outlined in Table 22-1. The names of individual waste streams are generally a function of how the residual is produced.

There are five general types of water treatment processes that produce residuals. The first is produced at those plants that coagulate and oxidize a surface water to remove

TABLE 22-1 Major Water Treatment Plant Residuals

Solid/liquid residuals
1. Alum sludges
2. Iron sludges
3. Polymeric sludges
4. Softening sludges
5. Spent filter backwash water
6. Spent GAC or discharge from carbon systems
7. Slow sand filter wastes
8. Wastes from iron and manganese removal plants
9. Spent precoat filter media
10. Spent membrane backwash
Liquid phase residuals
11. Ion-exchange regenerant brine
12. Spent regenerant from activated alumina
13. Membrane concentrate or brine
Gas phase wastes
14. Air stripping off-gasses
Specific residuals
15. Arsenic residuals
16. Radionuclide residuals
17. Perchlorate residuals

particles (both organic and inorganic) and dissolved contaminants such as color, organic carbon, iron, manganese, and occasionally trace metals. These coagulation plants produce two major residuals, *clarifier* (sedimentation or dissolved air flotation (DAF)) *sludge* and *spent filter backwash water* (SFBW).

The coagulation process itself generates most of the solid residuals. Generally a metal salt (aluminum or iron) is added as the primary coagulant. In addition to the coagulant other solids producing chemicals such as powdered activated carbon (PAC), polymer, clay, lime, or activated silica may be used. These added chemicals all produce solids. They are usually removed, along with the solids in the raw water, in a clarifier. These residuals are referred to as *clarifier sludge* or *clarifier blowdown.* They are more specifically referred to by the type of coagulant used. For example, alum sludge is residuals produced from the use of an aluminum-based coagulant and iron sludge is residuals produced by the use of an iron-based coagulant. When dissolved air flotation is used as the clarification step, the residuals are referred to as *float*. The second major residual produced at coagulation plants is from the batch process of backwashing the filters as described in detail in Chap. 10. Note that the residuals are not called backwash water but rather spent filter backwash water. Backwash water is used to clean the filters. The solids collected on the filters are those remaining after sedimentation or caused by the addition of a filter aid or formed by oxidation of perhaps iron or manganese. In a direct filtration process, these are the only solids produced. The volume of SFBW produced is a function of the amount of water used for backwashing. In areas with good raw water quality, the clarification step is occasionally omitted and the solids are removed by filtration only. This process, commonly known as *direct filtration,* is usually used for water with low turbidity and one that requires low levels of coagulant.

The residuals can be further treated resulting in additional residuals streams. Much of the terminology is somewhat obvious. A thickener treating clarifier sludge or clarifier sludge plus SFBW produces *thickened sludge* (which could be thickened alum or iron sludge) and thickener supernatant. A thickener that only treats SFBW will produce SFBW settled solids. A dewatering device will produce a *sludge cake* (also called *dewatered solids*) as well as a liquid stream. The liquid stream is referred to by the type of dewatering used, such as filtrate, decant, and centrate.

The second type of treatment plants are those that practice lime-soda ash softening for the removal of calcium and magnesium as described in Chap. 13. These plants may also remove trace metals, radioactivity, and particles if surface water is being treated. The residuals produced are the same types as in coagulation and include clarifier sludge (*lime sludge*), spent filter backwash water, thickened sludge, and so on.

Ion-exchange processes are the third type of treatment processes that produce residuals. As their name implies, these processes are used to remove cation or anion contaminants such as calcium and magnesium, arsenic, nitrate, and barium as described in Chap. 12. These processes produce a *brine* residual as well as *spent rinse water*. There are also adsorption processes that produce similar types of residuals. Some adsorption processes use throw-away media such that the residual produced is the *spent adsorption material*. Adsorption media may also need to be backwashed so *spent backwash water* is produced.

The fourth general category of treatment is when membranes are used to remove particulates or dissolved solids as described in Chap. 11. Membranes can be used to remove a variety of contaminants. The size of contaminant removed depends on the type of membrane selected and its associated pore size. As membrane systems are increasingly used for water utility applications, the management of their residuals has become a growing challenge. Membrane technology uses a driving force (e.g., electrical, pressure/vacuum, etc.) to separate contaminants from the water. Pressure-driven membranes include *microfiltration* (MF), *ultrafiltration* (UF), *nanofiltration* (NF), and *reverse osmosis* (RO). Electrically driven membranes include *electrodialysis* (ED) and its variant *electrodialysis reversal* (EDR). Whereas MF and UF membranes are designed for particle removal and use low pressure, NF, RO, and ED/EDR are designed for desalination (and softening).

MF and UF processes are low-pressure membranes that primarily remove particles. The particles buildup on the membranes and are backwashed off to clean the membrane. This membrane backwash residual contains the particles that were removed from the source water. Occasionally a coagulant or PAC is added to the raw water prior to the membrane and that will also be in the membrane backwash. These residuals are referred to as spent *membrane backwash*

Desalting membranes such as RO and NF primarily remove dissolved ions. These membranes produce a fairly continuous residual that contains the concentrated ions that the membrane rejects. The residuals are referred to as *concentrate* and occasionally *brine.*

Additional residuals generated by membrane processes are spent chemical solutions that are used to perform deep cleaning of the membranes to remove foulants via *clean-in-place* (CIP) processes; these are labeled *CIP waste.*

Finally, *gaseous residuals* are produced by air stripping processes that release gases to the atmosphere. These releases are primarily volatile organic compounds and radon.

There can also be a wide variety of what are referred to as *minor waste streams* produced at water treatment including analysis waters, sample line and sink drainage, leakage from pipes and pump seals, pump packing water, floor and laboratory drains, roof drains, granular activated carbon (GAC) transport water, leakage from chemical tanks, and process overflows.

There are several additional residuals that are referred to by the fact that they result from the removal of a specific contaminant. The contaminant presence in the residuals describes them, such as arsenic residuals and radionuclide residuals. These residuals have significant regulatory issues associated with their proper management.

COAGULATION AND LIME-SOFTENING RESIDUALS

Quantities

The quantity of solids or sludge dry weight generated from water treatment plants depends on the raw water quality and dosage of treatment chemicals used.

One of the most difficult tasks facing the water utility or engineer in planning and designing a residuals treatment process is determining the amount of residual to be handled. The residual quantity is usually determined as an annual average for a given design year and is a function of flow demand projections. Sometimes overlooked, but important, is information on seasonal or monthly variations. It is not unusual for order of magnitude differences in residuals production to exist for different months of the year.

The amount of alum (or iron) sludge generated can be calculated fairly closely by considering the reactions of alum or iron in the coagulation process. Using an empirical relation to account for the sludge contribution from turbidity improves the estimate and the contribution from other sources can be added as required.

When alum is added to water as aluminum sulfate under dosing and pH conditions such that the solubility of Al is exceeded, an amorphous (am) Al-hydroxide solid is produced. Assuming all of the added Al ends up in the solid, the reaction is typically represented by the simplified equation (see Chap. 8 for details of alum coagulation).

$$Al_2(SO_4)_3 \cdot 14H_2O + 6HCO_3^- = 2Al(OH)_3 \cdot 3H_2O(am) + 6CO_2 + 11H_2O + 3SO_4^{2-} \quad (22\text{-}1)$$

If inadequate alkalinity is present, lime or sodium hydroxide is normally added to maintain the proper pH for precipitation of the aluminum hydroxide. There are three waters of hydration associated with the aluminum hydroxide. These three waters of hydration satisfy the covalent bonding number of 6 for aluminum. Not including the waters of hydration in the reaction underestimates the amount of solids that are produced. This chemically bound water increases the sludge quantity, increases the sludge volume, and also makes it more difficult to dewater since the chemically bound water cannot be removed by normal mechanical methods. The resulting aluminum hydroxide species has a molecular weight of 132, and therefore, 1 mg/L of alum added to water will produce approximately 0.44 mg/L of inorganic aluminum solids. Suspended solids present in the raw water produce an equivalent weight of sludge solids since they are non-reactive. It can be assumed that other additives such as polymer (expressed on a dry weight basis) and powdered activated carbon produce sludge on a one-to-one basis. The amount of sludge produced in an alum coagulation plant for the removal of suspended solids is then

$$S = Q\,(0.44\text{Al} + \text{DOC}_r + SS + A) \tag{22-2}$$

where
S = sludge produced (kg/day)
Q = plant flow (MLD)
Al = alum dose as 17.1 percent Al_2O_3 (mg/L)
SS = raw water suspended solids (mg/L)
DOC_r = dissolved organic carbon removed
A = additional chemicals added such as polymer, clay, or PAC (mg/L)

If iron is used as the coagulant, then the equation becomes

$$S = Q\,(2.9\text{Fe} + \text{DOC}_r + SS + A) \tag{22-3}$$

where the iron dose is expressed as mg/L of Fe^{3+} added or produced via Fe^{2+} oxidation (note that significant Fe^{2+} in the raw water will also produce sludge at a factor of 2.9 if it is oxidized). It should not be interpreted from the above two equations that iron produces several times the amount of sludge that alum produces. The units for the coagulant are significantly different for the two equations. In reality 1 mole of coagulating equivalent of iron produces about 20 to 25 percent more dry weight sludge than 1 mole of aluminum. Iron can be purchased as ferric chloride or ferric sulfate. Ferric chloride ($FeCl_3$) dosing is usually reported as equivalent dry weight of chemical without waters of hydration (although this should be confirmed with the manufacturer) and thus has a molecular weight of 162.3. This relationship results in 1.0 mg of solids produced for each milligram of ferric chloride added. Ferric sulfate is reported by different manufacturers with different waters of hydration, and the individual products need to be referenced.

These equations can be used to track yearly or even daily variation changes in sludge mass produced. The difficulty in applying the relationships is that most plants do not routinely analyze raw water suspended solids concentrations. The logical correlation is to equate a turbidity unit to a suspended solids unit. Unfortunately the relationship is generally not 1 to 1.

$$SS\ (\text{mg/L}) = b \cdot \text{Tu} \tag{22-4}$$

where Tu is raw water turbidity.

The value of b for waters low in color, predominately turbidity removal plants can vary from 0.7 to 2.2. It may vary seasonally for the same raw water supply. A utility can therefore either continually measure suspended solids or develop a correlation between turbidity and suspended solids. Figure 22-1 shows one such correlation for a low-color raw water source. Ideally, this correlation should be done weekly until information is learned as to seasonal

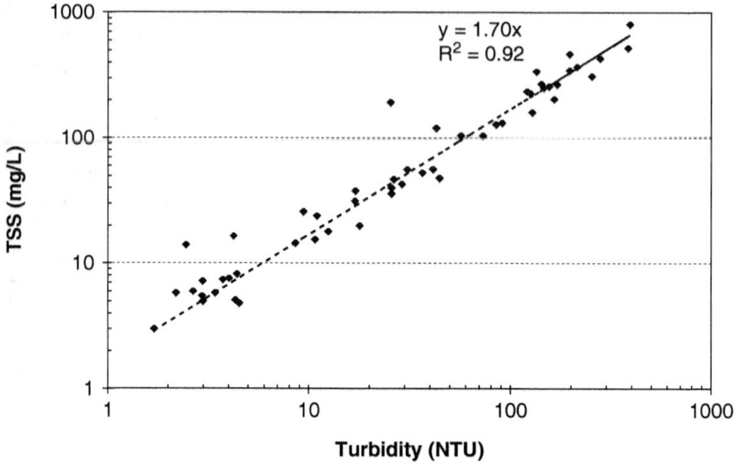

FIGURE 22-1 Suspended solids versus turbidity. (*Courtesy of EPCOR, Inc., Edmonton, Alberta.*)

variations in the suspended solids/turbidity relationship. Afterward a monthly correlation may be sufficient.

Another complication exists for raw water sources that contain a significant amount of natural color. Color can be a large contributor to sludge production. For some waters, color is filterable on a 0.45-μm filter and therefore is measured in a suspended solids analysis. Values of b for colored raw waters can be as high as 20, but unless turbidity and color vary together, a correlation between SS and Tu will not exist.

Figure 22-2 compares calculated and measured solids production done by the city of Philadelphia. The city used iron as the coagulant during this time period. A correlation was developed that showed the ratio of suspended solids to turbidity was 1.4. The calculated quantities were within 5 percent of the measured quantities. Through careful calibration and measurements, a complete solids mass balance can be prepared, as was done by the city of Philadelphia and is shown in Fig. 22-3.

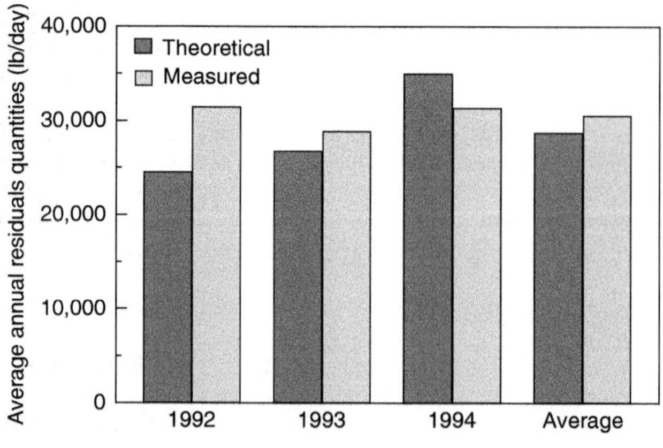

FIGURE 22-2 Baxter WTP theoretical versus measured residuals quantities. Note: multiply by 0.454 to convert to kg/day. (*Source: EE&T, 1996.*)

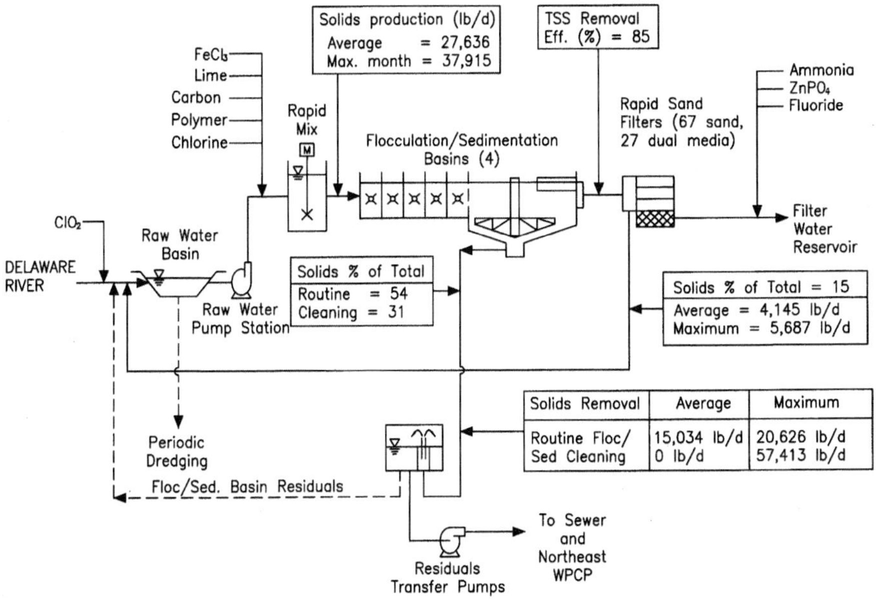

FIGURE 22-3 Baxter WTP baseline residuals distribution. Note: multiply by 0.454 to convert to kg/day. (*Source: EE&T, 1996.*)

Example 22-1 Calculating Sludge Solids

The residuals treatment processes at a water treatment plant are to be sized to accommodate the equivalent of the 90th percentile of the historical daily solids production. The available historical data are summarized in the following table. Although the data shown are monthly values, normally daily data for several years would be utilized. Additionally, after conducting a correlation study between suspended solids and turbidity, it was determined the b value for this raw water to be 1.4 mg/L-ntu.

Solution

			Monthly average	
Month	Plant production (ML)	Turbidity (ntu)	$FeCl_3$ dose (mg/L)	PAC dose (mg·L)
January	1550	4	25	—
February	1540	5	24	—
March	1860	4	25	—
April	1800	15	30	—
May	2015	20	35	3
June	2250	12	25	3
July	2635	10	22	3
August	2790	14	28	3
September	2250	20	35	3
October	1860	15	32	—
November	1500	12	30	—
December	1395	6	25	—

The solids production for each month can be found using Eq. 22-3. The data presented are expressed in mg/L as $FeCl_3$, so it is necessary to convert to Fe for the August data.

$$\frac{\text{MW Fe}}{\text{MWFeCl}_2} = \frac{56 \text{ g/mole Fe}}{161 \text{ g/mole FeCl}_3}(28 \text{ mg/L as FeCl}_3) = 9.74 \text{ mg/L as Fe}$$

Continuing to use the August data as an example, the solids production calculated using Eq. 22-3 is

$$\begin{aligned} S &= 0.001\frac{\text{L}\cdot\text{tonne}}{\text{ML}\cdot\text{mg}}(2790 \text{ ML}) \\ &\times\left[1.4\frac{\text{mg}}{\text{L}\cdot\text{ntu}}(14 \text{ ntu}) + 2.9\frac{\text{mg}}{\text{mgFe}}(9.74 \text{ mg/L as Fe}) + 3 \text{ mg/L FAC}\right] \\ &= 141.9 \text{ tonne} \end{aligned}$$

With this approach, we can determine the average solids production for each month, as shown in the following table.

Month	Plant production (ML)	Monthly average			Solids production (tonne)
		Turbidity (ntu)	$FeCl_3$ dose (mg/L)	PAC dose (mg/L)	
January	1550	4	25	—	47.8
February	1540	5	24	—	48.1
March	1860	4	25	—	57.3
April	1800	15	30	—	92.3
May	2015	20	35	3	133.6
June	2250	12	25	3	101.3
July	2635	10	22	3	103.3
August	2790	14	28	3	141.9
September	2250	20	35	3	149.2
October	1860	15	32	—	99.1
November	1500	12	30	—	70.6
December	1395	6	25	—	46.9

A statistical analysis will indicate that the 90th percentile of the monthly average solids production is 141 tonnes. Converting to kilograms and dividing by 30 days per month provides a daily solids production of 4700 kg/day. This can be compared to the average monthly solids production of 88.9 tonnes, equivalent to a daily average of approximately 2969 kg/day. This difference between the 90th percentile solids production and the average solids production illustrates the problem of sizing facilities based only on average data. Furthermore note that in the example, the turbidity and coagulant dose are paired to the flow for that month. If unpaired data are used (e.g., calculating solids production by calculating lb/MG and multiplying that by the design plant flow), bias can be introduced because peak solids production may or may not occur during periods of peak plant flow. ▲

Polyaluminum chloride (PACl) is the third major coagulant used. Care needs to be used when converting PACl dose to solids production since each purchased chemical has different strengths and utilities report these doses differently. Some utilities report PACl dose as a neat solution, some as Al_2O_3, and some as PACl product. A typical manufactured PACl liquid contains about 30 to 35 percent PACl and around 10 percent Al_2O_3. As an example, one manufactured product contains 33.3 percent PACl and 10.3 percent Al_2O_3. In this case, 1 mg of PACl will produce 0.8 mg of solids, assuming that the coagulation process is operated at a dose and pH that promotes precipitation of aluminum hydroxide.

Through similar theoretical considerations a general equation has been developed for plants that use a softening process with or without the use of alum, iron, or polymer (Cornwell, 2006). The equation is

$$S = Q\,[2.0\text{CaCH} + 2.6\text{MgCH} + \text{CaNCH} + 1.6\text{MgNCH} + 0.44\text{Al} + 2.9\text{Fe} + SS + A] \qquad (22\text{-}5)$$

where S = sludge production (kg/day)
CaCH = calcium carbonate hardness removed as $CaCO_3$ (mg/L)
MgCH = magnesium carbonate hardness removed as $CaCO_3$ (mg/L)
CaNCH = calcium noncarbonate hardness removed as $CaCO_3$ (mg/L)
MgNCH = magnesium noncarbonate hardness removed as $CaCO_3$ (mg/L)
Fe = iron dose as Fe (mg/L)
Al = alum dose as 17.1 percent Al_2O_3 (mg/L)
Q = plant flow (MLD)
SS = raw water suspended solids (mg/L)
A = other additives (mg/L)

If the source of the water to be softened contains turbidity (suspended solids) or DOC, then those are added to the equation the same as for coagulant solids. If the lime dose is known by operating records or separate calculation, then the lime sludge production can be estimated by the lime added (as $CaCO_3$) plus the calcium removed (as $CaCO_3$) plus 0.6 times the Mg removed (as $CaCO_3$).

These equations or prediction procedures allow the estimation of the dry weight of sludge produced. They do not predict the volume of sludge that is produced.

Volumes and suspended solids concentrations of sludges leaving the sedimentation basins or clarifiers are a function of raw water quality, treatment process, and the sludge removal method. Sludges which are allowed to build up in basins and are being cleaned only periodically by manual procedures, tend to compact and thicken in the bottom of the basins. There is often a stratification of solids with the heavier particles settling to the bottom and the hydroxide or lighter particles at the top. However, the actual concentration produced will depend largely on the amount of water used to flush the solids out of the basin during cleaning. With increasing finished water quality standards, there is a trend to remove the solids as quickly as possible, generally with continuous collection equipment. In this case the solids concentrations are lower since compaction height and time has been less. Solids concentrations for sludges produced with alum or iron coagulants and for low to moderate turbidity raw waters are about 0.1 to 1.0 percent. The higher the ratio of coagulant to raw water solids, the lower the solids concentration is, and the higher the sludge volume. Coagulant sludges from highly turbid raw waters may be in the 2 to 4 percent range and occasionally higher. Sludge volumes tend to be 0.1 to 3 percent of the raw water flow, with one survey (Cornwell and Susan, 1979) finding an average of 0.6 percent, but this is highly dependent on the removal method. The type of clarification process used also impacts the solids concentration and volumes. For example some clarification processes like ballasted sand sedimentation can produce weak solids concentrations in the 0.1 percent range, whereas a process like DAF can produce solids concentrations of up to 3 percent or greater (see Chap. 9). Softening sludges will concentrate higher, usually as a function of the $CaCO_3$:$Mg(OH)_2$ ratio and the type of clarifier. Conventional sedimentation basins in softening plants may only produce solids concentrations of 2 to 4 percent, whereas sludge blanket clarifiers can produce solids concentrations of up to 15 percent. Sludge volumes correspondingly vary considerably, from less than 0.5 up to 5 percent of the water plant flow.

Spent filter backwash water is characterized by its large water volume and low solids concentration. Filters can be backwashed at rates of 35 to 70 m/h (15 to 30 gpm/ft^2) depending upon the media size and water temperature, and the backwash time may be 15 to 20 min. Backwash water volumes are typically in the range of 3 to 10 percent of plant production. Accurate plant records often exist on the amount of backwash water used. The percentage of plant production used for backwashing can be computed from the ratios of the unit run volumes. For example, a filter producing water at 9.65 m/h (4 gpm/ft^2) with a 24-hour run time has a unit production of 235 m^3/run/m^2 (5760 gal/run/ft^2). If it is backwashed at 48.9 m/h (20 gpm/ft^2) for 20 min, the unit volume of backwash water is 16.3 m^3/run/m^2 (400 gal/run/ft^2), for a ratio of about 7 percent backwash water compared to production water. Typically operators will try to maintain backwash water use to around 4 percent of plant influent flow. Spent filter backwash water contains 10 to 15 percent of the total solids production depending on the efficiency of the clarification process and have suspended solids concentrations of 30 to 300 mg/L depending on the applied turbidity to the filters and the ratio of backwash water to production. Table 22-2 shows example SFBW suspended solids concentrations reported by Cornwell et al. (2001).

TABLE 22-2 Untreated Composite Spent Filter Backwash Water (SFBW) Characteristics

Treatment plant	pH	Turbidity (ntu)	Total suspended solids (mg/L)	Particle counts >2 μm (counts/mL)
Manheim WTP–Waterloo, ON	7.90	82	208	340,000
Sturgeon Point WTP–Derby, NY	7.65	20	96	83,000
Baxter WTP–Philadelphia, PA	7.54	24	67	51,000
Val Vista WTP–Phoenix, AZ	7.73	93	190	230,000
Kanawha Valley WTP–Charleston, WV	7.80	164	618	304,000
Williams WTP–Durham, NC	7.15	47	133	322,000
Utah Valley Water Purification Plant–Orem, UT	7.73	200	309	$>2.5 \times 10^6$

Source: Cornwell et al., 2001.

Characteristics

The terms *solid/liquid residuals* or *sludges* are used to describe free-flowing liquids that are predominantly water all the way up to mixtures that are predominantly solids and behave like a soil texture. Therefore, whenever referring to the physical properties of sludge, it is important to know the suspended solids concentration of the solid/liquid mixture in order to assess the physical state. Cornwell et al. (1992) used the Atterberg limit test to classify sludges. The Atterberg test describes quantitatively the effect of varying the water content on the consistency of fine-ground soils. The test consists of measuring five limits; however, the liquid and the plastic limits are the most applicable to sludges. The plastic limit identifies the solids concentration at which a sludge transitions from a semisolid to a plastic (the plastic state ranges from soft butter to stiff putty) stage. The liquid limit is the solids concentration below which the sludge exhibits viscous behavior; the consistency could be described as ranging from soft butter to a pea-soup-type slurry. Coagulant sludges that were tested had liquid limits in the 15 to 20 percent solids concentrations range. Solids concentrations below but near this range result in a material that still has free water associated with it but may not be flowable. Generally, a coagulant sludge is still free flowing up to an 8 to 10 percent solids concentration. The plastic limit for coagulant sludges was found to be anywhere from 40 to 60 percent solids concentration.

Knocke and Wakeland (1983) divided the physical properties of sludge into their macro and micro properties. Macro properties are such parameters as specific resistance, settling rates, and solids concentrations. The indices described could be considered macro properties. Micro properties include particle size distribution and density. Vandermeyden et al. (1997) studied the micro and macro properties of about 80 water plant sludges. Coagulant sludges had a median particle diameter of 0.005 mm with a range of approximately 0.001 to 0.03 mm as shown in Fig. 22-4. Lime sludge had a similar range but the median diameter was 0.012 mm. They also measured the specific gravity of the solid material in the sludge mixtures shown in Fig. 22-5. The coagulant residuals had an average specific gravity of 2.32

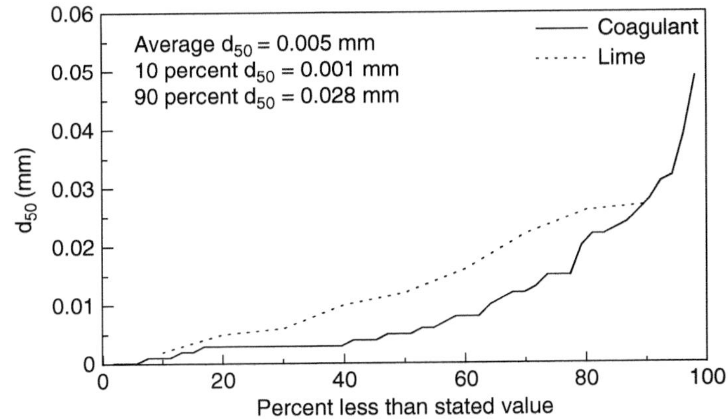

FIGURE 22-4 Average particle diameter for coagulant and lime residuals. (*Source: Vandermeyden et al., 1997.*)

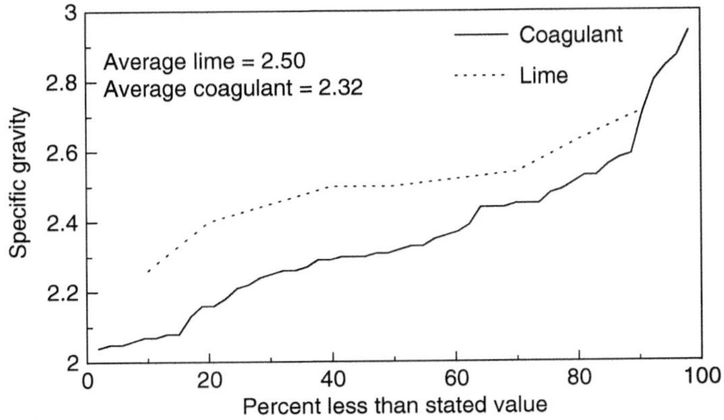

FIGURE 22-5 Specific gravity distribution for coagulant and lime residuals. (*Source: Vandermeyden et al., 1997.*)

and the lime residuals averaged 2.50. Note that this specific gravity is of the solid material. Wet densities tend to be close to water, around 1.05.

Vandermeyden et al. (1997) also evaluated drainage properties of the 80 residuals using the *capillary suction time* (CST), *specific resistance* (SR), and *time to filter* (TTF) tests.

The CST test is a fast and relatively simple test that is performed to determine the rate of free water release from a residual sample. The test is especially useful for comparing the drainage characteristics of different residuals and for optimizing polymer conditioning of residuals. The test consists of measuring the time in seconds for free water to travel 1 cm when a 5 to 7 mL sample of residuals is placed in a special cylinder which is located on a Whatman No. 17 chromatography paper. As the free water drains from the residuals through the chromatography paper, it passes by an electronic sensor that activates a timer. The timer stops when the free water reaches a second electronic sensor, 1 cm away. The time, in seconds, recorded by the instrument is the CST time.

The resistance to fluid flow exerted by a cake of unit weight of dry solids per unit area is defined as the SR. In order to evaluate SR, a sample of residuals is subjected to a vacuum using a Buchner funnel apparatus. Typically, 100-mL portions of residuals are added to the Buchner funnel which is lined with a paper filter and a vacuum is applied to the filter apparatus. The volume of filtrate generated at various times is recorded. This procedure is continued until enough water has been drawn out to produce cracking in the cake on the filter paper and subsequent loss of vacuum.

A simplification of the SR test is the TTF test. This test is set up with the same Buchner funnel apparatus as the SR test, but is much simpler to run. The only data collected is the amount of time it takes for one half of the sample volume to filter. The result is expressed in seconds. Details of the procedures for all these tests are in Vandermeyden et al. (1997).

Vesilind (1988) proposed that the water released from the sludge material is a function of solids concentration and viscosity. It has long been recognized that solids concentration has an effect on CST. The sludge concentration is directly proportional to the filterability constant. The filterability constant can be determined as follows:

$$\chi = \Phi\left[\frac{\mu SS}{\text{CST}}\right] \tag{22-6}$$

where χ = filterability constant (kg^2/s^{2}m^4)
Φ = dimensionless instrument constant
μ = viscosity (centipoise)
SS = solids concentration (mg/L)

The purpose in utilizing such a model is to make the filterability constant a fundamental measure of sludge dewatering rate. Due to the cumbersome nature of the SR test it is often preferable to use an easily determined value such as CST that allows calculation of a measure of dewatering rate that is independent of solids concentration.

If the filterability index and specific resistance are both measures of dewatering rate, then they should plot linearly. Figure 22-6 shows a plot of the inverse of the filterability constant and SR. As predicted by Vesilind, a strong correlation exists between the filterability index and the specific resistance.

The CST and the TTF tests are both direct measurements of the rate of water release from the sludge, and therefore one would expect a relationship to exist between the two tests. A plot of such a relationship is shown in Fig. 22-7. The strong correlation suggests that either test can be used successfully to evaluate sludge drainage characteristics.

Table 22-3 shows the summary statistics for each of the measured drainage tests (Vandermeyden et al., 1997). Interestingly, the general trends from all the tests were similar. For example, lime sludges had the lowest CST, TTF, and SR. The tests would predict the ease of dewatering order to be lime, ferric, alum, and PACl.

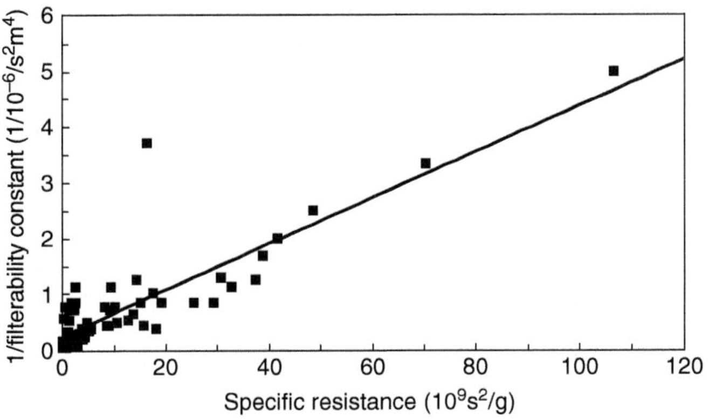

FIGURE 22-6 Inverse filterability constant versus specific resistance. (*Source: Vandermeyden et al., 1997.*)

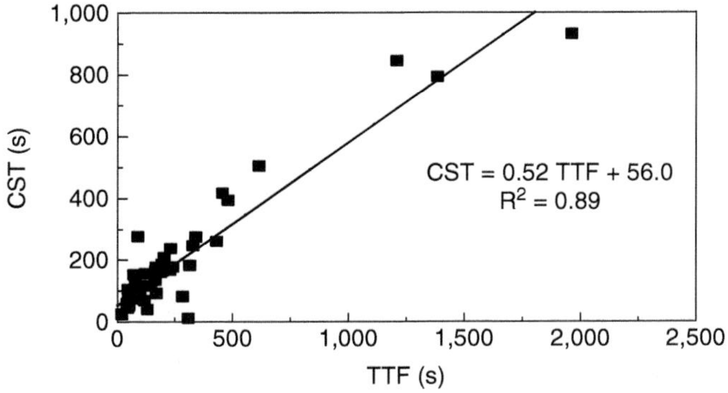

FIGURE 22-7 CST versus TTF for coagulant residuals. (*Source: Vandermeyden et al., 1997.*)

TABLE 22-3 Summary Statistics for Drainage Parameters

	CST (s)			TTF (s)			SR (10^9 s^2/g)		
Sludge type	n	mean	s^2	n	mean	s^2	n	mean	s^2
Alum	38	194.1	195.4	38	319.5	412.6	38	16.1	21.7
Ferric	9	103.0	64.5	9	104.7	79.5	9	6.5	8.3
PACl	5	289.8	258.8	5	410.9	562.5	5	14.1	11.2
Lime	9	70.0	34.5	9	34.3	20.4	9	0.55	0.84

Note: s^2 is the variance.
Source: Vandermeyden et al., 1997.

Two additional physical factors are important when dealing with a dry sludge that is to be disposed of in a landfill or monofill or utilized as backfill or a soil substitute. Those factors are the compaction density and the shear strength. Compaction tests are used to determine the achievable dry density for a particular sludge. This value is useful when estimating volumes required for a landfill or the truck volumes required to haul a certain weight of sludge. Generally, water plant sludges have been found to exhibit increasing density with increasing solids concentrations (Cornwell et al., 1992). At high solids concentrations, 80 percent dry weight densities for coagulant sludges are approximately 1000 kg/m^3 (60 lb/ft^3); at 30 percent solids concentrations, the dry unit weights are about 284 kg/m^3 (20 lb/ft^3). Wet unit weight, the weight to be hauled, is around 1065 kg/m^3 (75 lb/ft^3) at 20 percent solids concentration increasing to 1278 to 1,420 kg/m^3 (90 to 100 lb/ft^3) at 60 percent solids concentration for alum and iron sludges.

The shear strength of sludges relates to the overall ability of the sludge to support itself and external loadings, such as vehicle traffic or earth-moving equipment. Measuring shear strengths of water plant sludges is complicated by the fact that the shear strength varies with sludge age and with disturbance. Figure 22-8 shows an example of shear strength as a function of solids concentrations for three different water plant sludges (Cornwell et al., 1992).

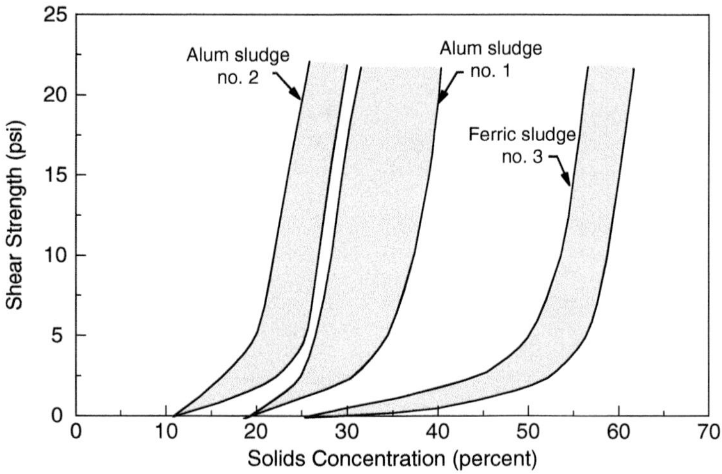

FIGURE 22-8 Shear strength versus solids concentration, cone penetration, and triaxial compression tests (Note: 1 psi equals 6.895 kPa). (*Source: Cornwell et al., 1992.*)

Chemical characterization of solid/liquid water plant residuals is primarily concerned with determining total metal concentrations, leachable metals, and nutrient levels. Several publications (Cornwell et al., 2006; AWWA, 1996; Cornwell et al., 1992; Cornwell and Koppers, 1990) have presented values for total metal concentrations of coagulant sludges; one such example is shown in Table 22-4. The source of the metals can be the raw water as well as the coagulant itself. Metals that are often found in coagulant sludges include aluminum, arsenic, occasionally cadmium, chromium, copper, iron, lead, manganese, nickel, and zinc.

Extraction tests are designated by USEPA to be one of the methods to determine if a waste is hazardous as per Subtitle C of Resource Conservation and Recovery Act. The *toxicity characteristic leach procedure* (TCLP) is used to determine if a waste is toxic, and therefore classified as hazardous. The presence in the extract from the TCLP test of any one

TABLE 22-4 Example Total Metal Analysis for Coagulant Sludges

Metal	Alum sludge 1 (mg/kg dry weight)	Alum sludge 2 (mg/kg dry weight)	Alum sludge 3 (mg/kg dry weight)
Aluminum	107,000	123,000	28,600
Arsenic	25.0	32.0	9.2
Barium	30	<30	230
Cadmium	1	1	2
Chromium	120	130	50
Copper	168	16	52
Iron	48,500	15,200	79,500
Lead	11	9	40
Manganese	1,180	233	4,800
Mercury	0.1	<0.1	0.2
Nickel	24	23	131
Selenium	<2	<2	<2
Silver	<2	<2	<2
Zinc	91.7	393	781

Source: Cornwell et al., 1992.

of a number of contaminants above a specified regulatory level constitutes failure of the test and results in the waste being classified as hazardous. There are no reported coagulant or lime sludges that have failed the TCLP test. In fact, it is rare to even find detectable levels of the regulated contaminants in the extract from a TCLP test on a coagulant or lime sludge.

Leaching tests in a lysimeter were subjected to rainfall for an equivalent time period of about 12 years, and the leachate was analyzed for metal concentrations. The actual test was conducted over a 6-month period. Figure 22-9 shows an example of the leaching of arsenic from the three sludges and compares the arsenic in the leachate to in-stream water quality standards and to the drinking water MCL. Only arsenic, copper, iron, manganese, and zinc leached from all three sludges. A slight degree of leaching was exhibited by the ferric sludge only, and cadmium leached from the ferric sludge only. Selenium leached from two sludges, but only during the first week. Aluminum did not leach from any of the sludges,

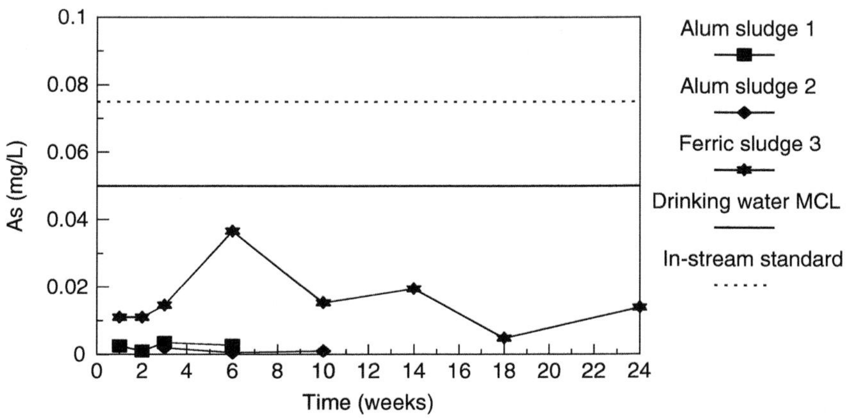

FIGURE 22-9 Arsenic leaching in relation to drinking water and in-stream water quality for an alum sludge. (*Source: Cornwell et al., 1992.*)

and none of the primary drinking water MCLs for metals monitored were exceeded in any of the leachate samples analyzed.

Water plant coagulant sludges are generally considered to be low in nutrient content, particularly nitrogen and phosphorus. However, some residuals can have as much as 3 percent total nitrogen content which can have some fertilizer value. In fact, in some cases nitrogen can be the limiting constituent in determining the maximum amount of residuals that can be applied to a specific land (EE&T, 1992).

THICKENING

After removal from a clarifier or sedimentation basin most water plant sludges can be further thickened in a gravity concentration tank. Thickening can be economically attractive in that it reduces the sludge volume and results in a more concentrated sludge for further treatment in the dewatering process. Some dewatering systems will perform more efficiently with higher solids concentrations. Thickening tanks can also serve as equalization facilities to provide a uniform feed to the dewatering step.

Although there are a few types of thickeners available on the market, the water industry almost exclusively uses gravitational thickening.

Sludge thickening is performed primarily for reduction in the volume of sludge. The relationship between the volume of sludge and the solids concentration is expressed as

$$V = \frac{M}{\rho s P} \tag{22-7}$$

where V = volume of sludge (m^3)
M = mass of dry solids (kg)
ρ = density of water, 10^3 kg/m^3 (at 5°C)
s = specific gravity of the sludge
P = percent solids expressed as a decimal (weight/weight)

An approximation for determining volume reduction on the basis of percent solids is expressed as

$$\frac{V_2}{V_1} = \frac{P_1}{P_2} \tag{22-8}$$

This is a quick and useful equation since the specific gravity of the sludge is not always known, but generally does not change within the limits of most thickening operations. Therefore, for thickening a 1 percent solids concentration sludge to 10 percent solids concentration, a volume reduction of approximately 90 percent, is achieved.

Example 22-2 Sludge Volumes
Alum sludge is removed from a sedimentation basin at a 0.3 percent solids concentration. The plant produces 150 kg of dry solids per day. What is the volume of sludge that is produced? What would be the volume of thickened sludge produced if a thickener were used to increase the solids concentration to 2 percent?

Solution
To determine the volume from Eq. 22-7, the specific gravity of the sludge must be known. For coagulant sludges below approximately 10 percent solids, the specific gravity

of the sludge is essentially 1.0. This range encompasses the performance of most gravity thickeners. Therefore, the specific gravity of 1.0 can be used in Eq. 22-7, and Eq. 22-8 applies.

$$V = \frac{M}{rsP} = \frac{150 \text{ kg/day}}{10^3 \text{kg/m}^3 (1.0)(0.003)} = 50 \text{ m}^3/\text{day}$$

Assuming there are essentially no solids in the overflow, the volume after thickening would be

$$V_2 = \frac{P_1}{P_2}(V_1) = \frac{0.3}{2.0}(50 \text{ m}^3/\text{day}) = 7.5 \text{ m}^3/\text{day} \quad \blacktriangle$$

Gravity sludge thickeners are generally circular settling basins with either a scraper mechanism in the bottom (see Fig. 22-10) or equipped with sludge hoppers (Fig. 22-11). They may be operated as continuous flow or as batch fill-and-draw thickeners. For continuous flow thickeners, the sludge normally enters the thickener near the center of the basin and is distributed radially. The decanted water exits the thickener over a peripheral weir or trough, and the thickened sludge is drawn off the basin. For tanks equipped with a scraper mechanism, the scraper is located at the thickener bottom and rotates slowly. This movement directs the sludge to the draw-off pipe near the bottom, center of the basin. The slow rotation of the scraper mechanism also prevents bridging of the sludge solids. The basin's bottom is sloped to the center to facilitate collection of the thickened sludge.

Batch fill and draw thickening tanks are often equipped with bottom hoppers as shown in Fig. 22-11. In these tanks, sludge flows into the tank, usually from a batch removal of sludge from the sedimentation basin, until the thickening tank is full. The sludge is allowed to quiescently settle, and a telescoping decant pipe is used to remove supernatant. The decant pipe may be continually lowered as the solids settle until the desired solids concentration is reached or the sludge will not thicken further. The thickened sludge is then pumped out of the bottom hoppers to further treatment or disposal. Occasionally a mixing system is added in the sludge layer to allow for efficient pumping of the thickened sludge.

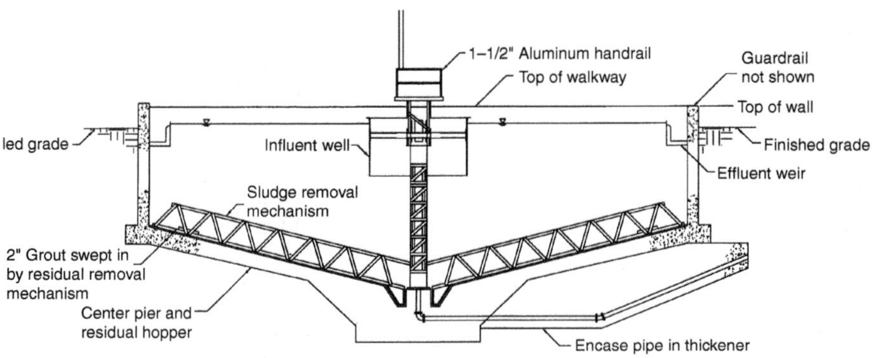

FIGURE 22-10 Continuous flow gravity thickener schematic.

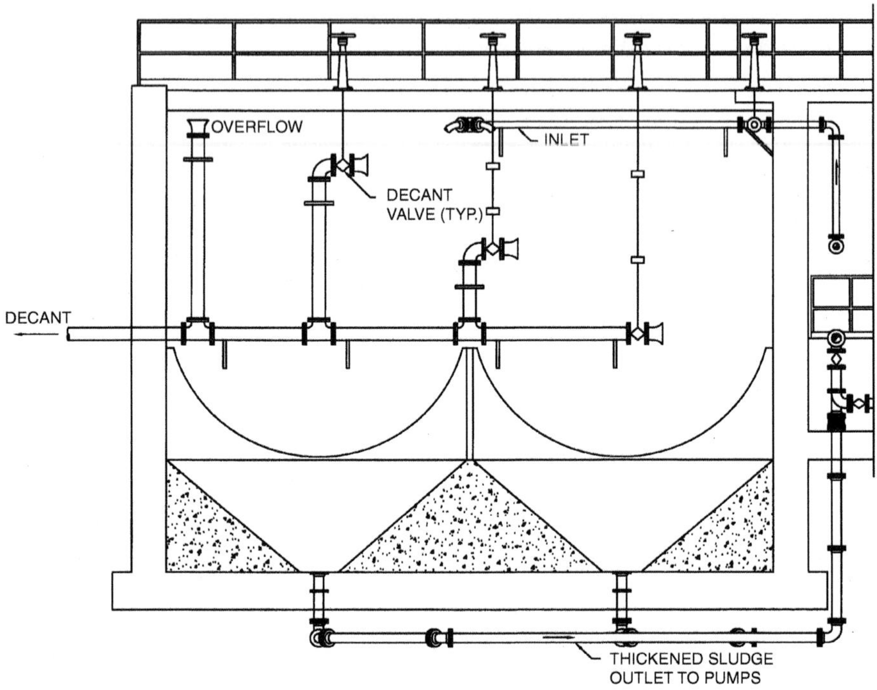

FIGURE 22-11 Batch-thickening tank schematic.

Design of batch or continuous flow thickeners is usually accomplished on the basis of previous experience of similar full-scale installations or on laboratory or pilot settling tests. Thickener design can be based on one of two parameters. If the supernatant is required to meet a specific quality goal (e.g., a discharge permit requirement), then the settling velocity and allowed overflow rate needs to be determined similarly as is done for a main treatment plant clarifier in order to design for the allowed clarified turbidity. A second design parameter can be to meet a specific thickened sludge solids concentration. In the former situation, the thickener is designed on the basis of hydraulic loading (e.g., m/h (gpm/ft^2)), whereas in the second case the thickener is designed on the basis of solids loading (e.g., kg/day-m^2 (lb/ft^2-hr)). When both parameters are important, they must each be determined and the most limiting design condition used.

If the goal is to meet a specific overflow turbidity, then a standard jar test procedure (see Chap. 8) can be used with or without polymer settling aids. However, as the solids concentration of the influent sludge increases, the thickener design may be constrained by the amount of solids the thickener can accommodate. When the influent solids loading exceeds the rate at which the sludge blanket compresses, then the depth of the sludge blanket will increase. This can lead to excessive solids carryover in the thickener supernatant and low solids concentrations for the thickener underflow. Pilot testing is typically required to determine the limiting solids loading rate for a particular sludge.

Depending on the sludge, either solids loading or hydraulic loading will limit the thickener design. Figure 22-12 shows the relationship between the solids concentration of the influent sludge and the hydraulic and solids loading rates. After the limiting loading conditions have been determined by testing, this figure can be used to determine whether

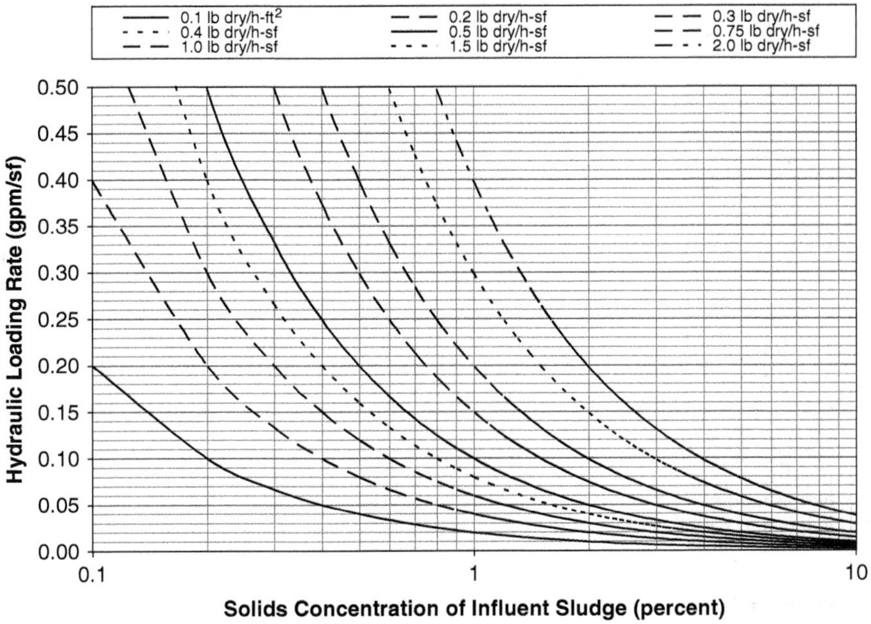

FIGURE 22-12 Equivalent solids and hydraulic loading rates by influent sludge solids concentration. Note: multiply gpm/ft^2 by 2.44 to convert to m/h; multiply lb dry/hr-ft^2 by 4.88 to convert to kg dry/h-m^2.

the hydraulic loading rate or the solids loading rate is the limiting design factor, given the influent sludge design condition. For a given target solids loading rate and hydraulic loading rate, if the intersection between the influent sludge solids concentration and target hydraulic loading rate is to the left of the target solids loading rate curve, the thickener design is limited by the hydraulic loading rate. If the intersection is to the right of the curve, the target solids loading rate is the limiting factor.

As mentioned above, determining the solids loading rate that will provide a specific underflow solids concentration is more involved and more difficult to do on a small scale than testing to determine acceptable hydraulic loading rates. The common settling test used in the laboratory is conducted in a transparent cylinder filled with sludge and mixed to evenly distribute the solids. At time zero the mixing is stopped and the solids are allowed to settle. Water plant sludges from clarifiers and sedimentation basins will generally settle as a blanket with a well-defined interface. By recording the height of the interface with time, a plot such as Fig. 22-13 can be created. The free settling velocity is then determined as the slope of the straight-line portion of the plot.

Pilot scale thickening studies require careful on-site considerations in order to maintain representative sludge flows and solids concentrations to the pilot thickener. However, done properly they are a much better design tool than the laboratory methods as they provide direct data on loading rates (liquid or solids) and overflow turbidities and underflow solids concentrations.

Example 22-3 Sizing of a Thickener

A thickener is to be sized to thicken underflow solids from four sedimentation basins from an initial average solids concentration of 0.2 to 2 percent. Each sedimentation basin is blown down twice per hour, and each blow down lasts for 5 min. The thickener

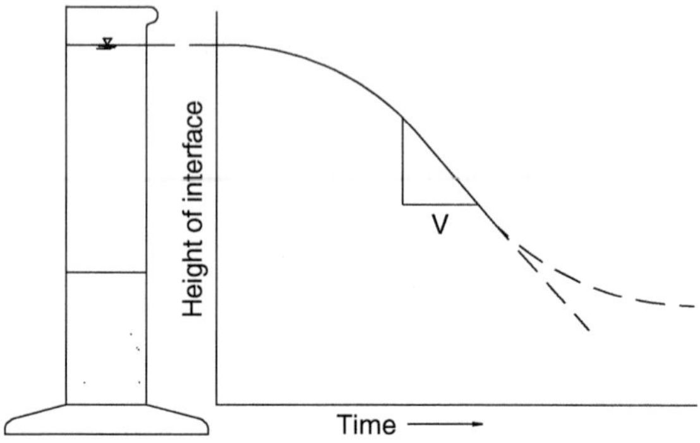

FIGURE 22-13 Thickening test in a cylinder with resulting interface height-versus-time curve.

supernatant is discharged to a nearby creek and must meet a discharge limit of 10 mg/L suspended solids. The plant produces 2720 kg (6000 lb) of solids per day.

Solution

To properly size this thickener, two types of pilot study are required. The first study would determine the effective solids settling velocity to achieve a supernatant suspended solids of 10 mg/L. This testing could be done using techniques described in AWWA M37 (1992). The second type of pilot testing required could include on-site thickener column piloting. For this example, testing determined that the supernatant discharge limit can be met provided the hydraulic loading rate is maintained at or below 0.61 m/h (0.25 gpm/ft^2). Testing also indicated that that the maximum solids flux for this system is 0.98 kg/h-m^2 (0.2 lb/h-ft^2).

Whether the design is liquid or solids limited can be determined by comparing the influent solids concentration and design loading rates to Fig. 22-12. From this figure it can be seen that the intersection of the influent solids concentration (0.2 percent) and design hydraulic loading rate (0.61 m/h (0.25 gpm/ft^2)) lies above the curve equivalent to the design solids loading rate 0.98 kg/h-m^2 (0.2 lb/h-ft^2); thus, this thickener will be solids limited. Therefore, the maximum solids flux rate will be the factor used for design.

To determine the surface area required for thickening, it is first necessary to determine the influent solids loading rate. Assuming the sedimentation basin sludge collector system completely removes all of the solids present in the basin, the loading rate is

$$\frac{2720\text{ kg/day}}{4\text{ sed. basins}}\left(\frac{1\text{h}-\text{sed. basin}}{2\text{ blowdowns}}\right)\left(\frac{1\text{blowdown}}{5\text{ min}}\right)\left(\frac{60\text{ min}}{1\text{ hr}}\right)\left(\frac{1\text{ day}}{24\text{ h}}\right)=170\text{ kg/h}$$

Note that this solids loading rate is greater than the average solids loading rate given by dividing the daily solids production by 24 h/day. This method accounts for the fact that solids are not continuously removed from the sedimentation basins, but are instead removed in periodic bursts. Therefore, the *instantaneous* loading rate must be used rather than the *average* loading rate. At a larger facility with more sedimentation basins, the instantaneous loading rate may approach the average loading rate because it can be assumed that at least one sedimentation basin is being blown down at any given time.

Once the appropriate solids loading rate has been determined, it is a simple matter of applying the maximum solids flux rate to calculate the required surface area for the thickener.

$$\frac{170 \text{ kg/h}}{0.98 \text{ kg/h-m}^2} = 174 \text{ m}^2 \quad \blacktriangle$$

NON-MECHANICAL DEWATERING

Non-mechanical dewatering, as the name implies, is the dewatering of water treatment plant residuals through means that do not require the use of mechanical appurtenances such as centrifuges or filter presses. Often used in locations where land is available and monetary constraints prohibit the use of mechanical dewatering, non-mechanical dewatering can be both economical and efficient for the dewatering of water treatment plant residuals.

A variety of means are employed to accomplish non-mechanical dewatering. The most basic of these is separation of solids and free water through sedimentation followed by natural air drying of the residuals. A second method allows free water to be percolated through sand and into an underdrain system, whereas additional solids concentration increases are achieved through evaporation. In northern climates, a third system is utilized whereby water treatment plant residuals are subjected to freezing and thawing which results in a dramatic reduction in residuals volume and a corresponding increase in solids concentration.

Sand Drying Beds

Sand drying beds were initially developed for dewatering municipal wastewater biosolids but have since also been used to dewater residuals from water treatment plants. Drainage (percolation), decanting, and evaporation are the primary mechanisms for dewatering residuals in sand drying beds and are utilized in a two-step process until the desired cake concentration is achieved. Following residuals application, free water is allowed to drain from the residuals into a sand bottom from which it is transported via an underdrain system consisting of a series of lateral collection pipes. This process continues until the sand is clogged with fine particles or until all the free water has been drained, which may require several days. Secondary free water removal by decanting can take place once a supernatant layer has formed. Decanting can also be utilized to remove rainwater that would otherwise hinder the overall drying process. Water remaining after initial drainage and decanting is removed by evaporation over a period of time necessary to achieve the desired final solids concentration. Manufacturers have developed variations of this standard design, such as replacing the sand with a porous, rigid material or adding vacuum to the underdrain area to speed the water release process.

Layout and construction of sand drying beds is site specific; topography, available land, and operational constraints must all be considered. Topography plays a key role in how beds are laid out on a site, and operational constraints, such as residuals pumping distance, must also be considered when siting a bed location. Materials used in construction are typically cast in-place concrete or concrete block when the beds are constructed at grade or earthen sides with a liner when the beds are constructed below grade.

Underdrain systems for sand drying beds are used to collect water that has percolated down through the sand and gravel and transmit it to a point of discharge. When the plant has the capability to decant, the flow from the decant mechanism is often tied to the underdrains so that a combined effluent is produced. Underdrains are typically constructed of vitrified clay or plastic piping and a host of underdrain configurations exist, but the most common is collecting drainage with laterals and conveying the flow to a header pipe. Figure 22-14 shows a section view of a typical sand drying bed.

Several sand drying beds are typically used at a given site, which offers some advantages from an operations point of view. Chief among these is the ability to rotate bed use, so that as one sand drying bed is loaded and the residuals begin to dry, another bed can be cleaned and readied for a new application of residuals. Once they have been initially loaded with residuals, the sand beds should not be loaded with additional residuals until the original solids have fully dewatered and been removed; otherwise, the partially dewatered residuals from the initial loading will blind the sand layer and prevent the drainage of free water from the newly applied residuals. For this reason, some form of residuals storage is required to allow a sufficient volume of residuals to accumulate so that a bed can be loaded to capacity during the initial loading. For plants that practice manual cleaning of sedimentation basins, this storage can be achieved within the sedimentation basins themselves. Plants that have continuous sludge removal equipment will likely need a separate sludge storage tank to hold the sludge prior to loading the sand drying beds.

Although manual cleaning of sand drying beds may be feasible for the smallest water treatment plants, mechanical equipment is generally required to efficiently clean sand drying beds of any significant size. Concrete support runners are essential for mechanical cleaning, serving to protect the underdrain system and prevent excessive sand removal and/or sand compaction (Vandermeyden et al., 1997). Front-end loaders and Vac-haul trucks have been used successfully by utilities operating sand drying beds.

The time required to evaporate the water remaining after percolation and decanting is the controlling factor in determining the bed size required. Therefore maximizing the removal of water prior to evaporation will also maximize the bed yield. The solids that remain after drainage and decanting are referred to as the drained solids. The value of the drained solids concentration is dependent on the initial solids concentration, the use of conditioning agents such as polymer, the applied depth, and the efficiency of the water removal system. A series of pilot tests can be conducted to determine the combination of loading depth, initial solids concentration, and polymer use that maximizes the drained solids concentration. Vandermeyden et al., (1997) developed the following equation to determine the bed yield once the drained solids concentration is known:

$$Y = (0.833)\frac{SS_f SS_d E}{SS_f - SS_d} \tag{22-9}$$

where Y = sand bed yield (kg/m^2/yr)
SS_f = desired final solids concentration (percent)
SS_d = drained solids concentration (percent)
E = net pan evaporation (cm/mo)

Alternatively, by determining the depth of evaporation required to achieve the required evaporative drying, it is possible to use numerical methods to model non-mechanical dewatering performance using historical plant data, which in turn can be used to determine the

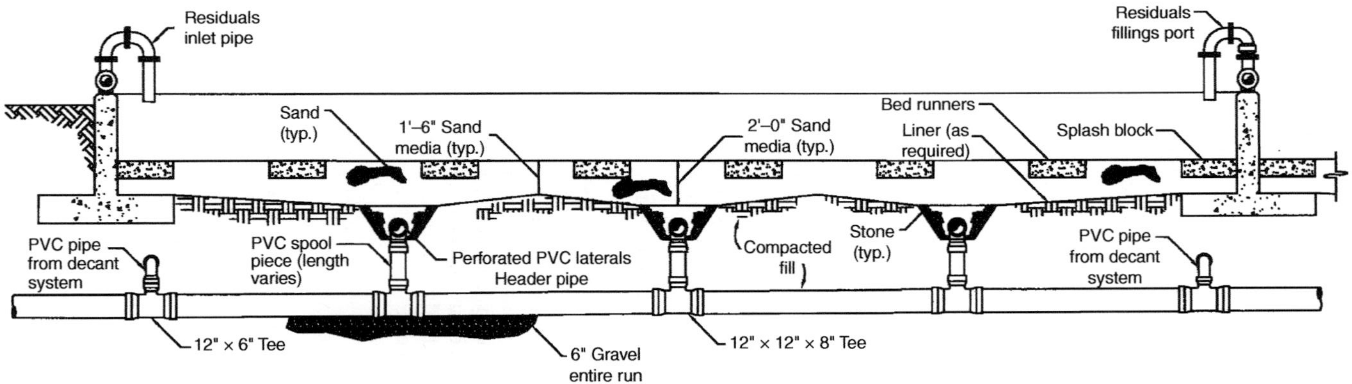

FIGURE 22-14 Typical sand drying bed section.

required bed area. This approach is based on the following equation from Cornwell and Vandermeyden (1999):

$$\Delta D_e = D_d - \left(\frac{D_d SS_d}{SS_f} \right) \tag{22-10}$$

where ΔD_e = required depth change due to evaporation (cm)
D_d = drained depth (cm)

$$D_d = D_i \frac{SS_i}{SS_d}$$

where D_i = initial depth of residuals at loading (cm)
SS_i = initial solids concentration (percent)

Using the required depth change due to evaporation, historical evaporation data can be utilized to determine the time required for a sand bed to reach the desired solids concentration, depending on the time of year during which the bed was initially loaded. Coupling this model to solids production projections based on historical plant data and can allow sand drying bed performance to be numerically simulated. For best performance, sand drying bed sizing should be determined on the basis of monthly, or even daily, solids production and evaporation numbers rather than annual averages in order to account for production and climatic variables (Vandermeyden et al., 1997).

Vandermeyden also found that when using net pan evaporation to predict the evaporation rate that the evaporation was underestimated. In three different field trials, a 25 percent dry cake solids concentration was reached in about 2/3 of the time predicted by using pan evaporation data. Thus, under the test conditions studied, using pan evaporation would result in a conservative sizing of the sand drying beds. At higher solids concentrations (>30 percent); however, pan evaporation overestimated the actual field evaporation. The point of crossover was speculated to be when the free water was evaporated and the chemically bound water remains. The bound water would be more difficult to evaporate, and the loss of bound water would be unlikely to correspond to volume reduction of the cake solids matrix, as is assumed by the model.

Solar Drying Beds

Solar drying beds are similar to sand drying beds in terms of shape and operation; however, they are constructed with sealed bottoms and have sometimes been referred to as *paved drying beds*. These beds have little or no provisions for water to be removed through drainage; all residuals thickening and drying is accomplished through decant of free water and evaporation. A principal advantage of this type of drying bed is low maintenance costs and ease of cleaning. No sand replacement costs are associated with this type of drying bed, and since the bottoms of these beds are sealed, neither initial underdrain costs nor underdrain repair costs are incurred. Also, because the entire solar bed bottom is often paved or concrete, cleaning with front-end loaders can be done quickly and efficiently. Because solar beds rely primarily on evaporation, they typically have lower solids loading rates than sand drying beds. Most solar beds are located in the southern and southwestern parts of the country where evaporation rates are high.

Dewatering Lagoons

Dewatering lagoons are similar to sand drying beds except that they operate at much higher loadings. The dewatering lagoons are equipped with a decant structure and underdrains, similar to a sand bed. For a dewatering lagoon, the lagoon is filled over a long time and then allowed to dry for a long period while another lagoon is filled. Dewatering lagoons can have an advantage over sand drying beds in reducing peaks, since the loading is often spread over several months. Because dewatering lagoons use a much higher loading rate, the drainage volume would generally be lower than a sand drying bed. The main difficulty in sizing a dewatering lagoon is in predicting the drained solids concentration after the loading is complete. Plugging of the sand media on the bottom of the dewatering lagoon with multiple loadings is difficult to predict and would require a carefully planned pilot test with dewatering columns, or even pilot-scale dewatering lagoons would be necessary to accurately size and design the system. The bottom of the lagoon would have a higher solids concentration than the top of the lagoon and a net average solids concentration must be estimated. During the evaporation phase, the bottom layers often do not dry out. Some utilities have found that tilling the sludge during the evaporative cycle helps to expose all of the residuals to drying.

A lagoon can be sized by selecting a desired fill cycle and estimating the drained solids concentration. The lagoon area can be calculated for varying drained solids concentrations as

$$\text{Dewatering lagoon area} = \frac{(\text{kg sludge})}{(\text{Depth in cm})\ (SS_d)(10)} \tag{22-11}$$

Freeze-Thaw Beds

It has long been recognized that when coagulation residuals are subjected to freezing, the resulting volume reduction and increased solids concentration is appreciable. Typically, the volume reduction is well over 70 percent and solids concentrations may reach as high as 80 percent when freeze-thaw is followed by evaporation. Freeze-thaw followed by evaporation dramatically converts the residuals from a fine particle suspension to granular particles. The granular particles often resemble coffee grounds in both size and appearance, and they do not break apart even after vigorous agitation. This change is irreversible, and the resultant solids drain almost without resistance if subsequently rewetted (Vesilind and Martel, 1990). Thus, if the frozen residuals are placed on a porous medium, the melt water drains away easily on thawing (Martel and Diener, 1991). As noted by Parker et al. (1998), improvement in dewaterability of the residuals is influenced by the rate of freezing. At slow freezing rates, the interface between the ice and the melt water advances as a planar interface that causes the residual flocs to migrate ahead of the ice interface and subsequently aggregate into larger flocs before becoming entrained by the ice. At faster freezing rates, the ice-melt interface is composed of dendrites that entrap and fragment the flocs. This results in smaller flocs following thawing, and this reduces the dewaterability of the residuals.

Traditionally, freeze-thaw beds have relied on natural freezing and were thus operated most effectively in northern climates, although mechanical freeze-thaw conditioning may be cost effective in regions where electricity rates are low and residuals disposal costs are moderate to high (Parker and Collins, 2000). Under climatic conditions the rate of freezing will generally be slow, which enhances the dewaterability of the resultant residuals. Generally, the range of effective operation for non-mechanical *freeze-thaw* beds begins at approximately 40° North latitude and extends northward.

Some water treatment plants in cold climates already take advantage of this process by modifying the operation of their lagoons or drying beds. One technique is to decant a lagoon down to the residuals interface and allow it to freeze over the winter months. Martel and Diener (1991) report that this technique is not always successful because the residuals do not freeze to the bottom. Another technique is to pump a shallow layer (20 to 45 cm) of residuals from a storage lagoon into drying beds or ponds that are then allowed to freeze in the winter. This technique works well because the residuals usually freeze completely, but it requires a considerable amount of land and storage volume.

Combination sand drying beds and *freeze-thaw* beds can also be utilized. In this case the design must consider the evaporative condition for the drying bed cycle and the freezing and thawing conditions for the freeze-thaw cycle.

The area required for a freeze-thaw bed is generally determined by the depth of residuals that can be frozen. For some climates with long freezing periods, the depth that can be thawed can be the controlling depth. Martel (1989) developed equations to allow the calculation of these two depths. The freezing depth can be found by

$$D(z) = \frac{t(f)(T_f - T)}{\rho_f F\left(\frac{1}{h} - \frac{d(z)}{2K}\right)} \tag{22-12}$$

where $D(z)$ = total depth of sludge that can be frozen (m)
T_f = freezing point temperature = 0°C
T = average ambient temperature (0°C)
$t(f)$ = freezing time (hours)
ρ_f = density of frozen sludge = 917 kg/m^3
F = latent heat of fusion = 93 W·h/kg
h = convection coefficient = 7.5 W/m^2°C
$d(z)$ = thickness of the sludge layer (m)
K = conductivity coefficient = 2.21 W/m°C

Since many of these parameters are known or assumed constants, the equation was reduced by Vandermeyden (1997) to

$$D(z) = \frac{-t(f)T}{11{,}371 + 19{,}294\, d(z)} \tag{22-13}$$

This equation would be used when the design calls for multiple layers of sludge to be frozen, each layer of thickness $d(z)$. In this case the utility personnel would apply the layer to the bed, and as soon as one layer had frozen then another would be applied. In Eq. 22-13, $t(f)$ and T can be found (for U.S. locations) from NOAA through records that are generally on file at the National Climatic Data Center in Asheville, N.C.

For the case of a one-time bed loading, $D(z)$ would be set equal to $d(z)$. Solving this equation requires use of the quadratic rule which results in the following expression (Vandermeyden et al., 1997):

$$D(z) = \frac{-11{,}371 + \sqrt{1.3\times10^8 - 7.7\times10^4\, t(f)T}}{3.9\times10^4} \tag{22-14}$$

Example 22-4 Sand Drying Beds

The feasibility of non-mechanical dewatering via sand drying beds is being investigated for a treatment plant that produces thickened sludge at a solids concentration of 2 percent. Pilot testing has indicated that the optimum loading rate for the beds is 18.48 kg of dry solids per square meter. At this loading rate, the beds achieve a drained solids concentration of 12 percent once the free water has percolated through the bed bottom. The drying beds will be designed to further dewater the solids to a solids concentration of 20 percent, at which point the solids will be removed from the beds and transported off-site.

On the basis of the historical solids production and evaporation data shown in the following table, it is possible to numerically simulate bed utilization, which in turn can provide the number of beds that are required.

Week	Weekly solids production (metric tons)	Monthly average net evaporation (cm/week)
0	0.0	1.482
1	7.3	1.482
2	10.9	1.482
3	9.1	1.482
4	13.6	1.778
5	11.8	1.778
6	20.9	1.778
7	16.3	1.778
8	9.1	0.889
9	6.4	0.889
10	5.4	0.889

Solution

First, it is necessary to determine the initial depth of loading, based the initial solids concentration and solids loading rate.

$$D_i = L/SS_i = \left(18.48\frac{\text{kg}}{\text{m}^2}\right) \Big/ \left(20{,}000\frac{\text{mg}}{\text{L}}\right)\left(\frac{1\text{ kg}}{10^6\text{mg}}\right)\left(\frac{1000\text{ L}}{\text{m}^2}\right) = 0.924\text{ m}$$

From this, we can determine the drained depth of the solids using the initial and drained solids concentrations.

$$D_d = D_i\frac{SS_i}{SS_d} = 0.924\text{ m}\left(\frac{2\text{ percent solids}}{12\text{ percent solids}}\right) = 0.154\text{ m}$$

As the drained solids continue to lose water to evaporation, the depth of the solids will decrease proportionally to the evaporation rate. The amount of depth change required to reach the final solids concentration of 20 percent is given by Eq. 22-10.

$$\Delta D_e = D_d - D_d\frac{SS_d}{SS_f} = 0.154\text{ m} - 0.154\text{ m}\left(\frac{12\text{ percent solids}}{20\text{ percent solids}}\right) = 0.062\text{ m}$$

Thus, after a bed is loaded and the solids have drained, 0.062 m of water must be lost through evaporation to dewater the sludge to the target solids concentration. A numerical simulation of this process using the data presented previously is shown in the following table. For the purposes of this simulation, it is assumed that each sand drying bed will be 15.25 m wide by 38.1 m long. At the optimum solids loading rate of 18.48 kg/m^2, it will take 536,474 L of sludge at a solids concentration of 2 percent to load the drying beds.

Week	Weekly solids production (metric tons)	Monthly average net evaporation (cm/week)	Weekly sludge production (liters)	EQ volume (liters)	Bed 1 (m)	Bed 2 (m)	Bed 3 (m)	Bed 4 (m)	Bed 5 (m)	Bed 6 (m)	Bed 7 (m)	Bed 8 (m)
0	0.0	1.482	0	0	—	—	—	—	—	—	—	—
1	7.3	1.482	365,000	365,000	—	—	—	—	—	—	—	—
2	10.9	1.482	545,000	373,526	0.062	—	—	—	—	—	—	—
3	9.1	1.482	455,000	292,051	0.047	0.062	—	—	—	—	—	—
4	15.8	1.778	790,000	9,102	0.029	0.044	0.062	0.062	—	—	—	—
5	17.5	1.778	875,000	347,628	0.011	0.026	0.044	0.044	0.062	—	—	—
6	20.9	1.778	1,045,000	319,679	—	0.008	0.026	0.026	0.044	0.062	0.062	—
7	16.3	1.778	815,000	61,730	0.062	—	0.008	0.008	0.026	0.044	0.044	0.062
8	9.1	0.889	455,000	516,730	0.053	—	—	0.000	0.017	0.035	0.035	0.053
9	6.4	0.889	320,000	300,256	0.044	0.062	—	—	0.008	0.026	0.026	0.044
10	5.4	0.889	270,000	33,782	0.035	0.053	0.062	—	—	0.017	0.017	0.035

In order to ensure that sufficient sludge is available to load an entire bed at once, it is necessary to provide an equalization volume to store the sludge prior to bed loading. The volume of sludge produced weekly can be calculated assuming a unit weight of 1 kg/L.

As shown in the preceding table, once sufficient sludge has accumulated in the sludge storage tank, 536,474 L of sludge are removed from storage and loaded into an empty bed. At the time the bed is loaded and the drained depth has been achieved (this occurs through drainage/decant of the free water in the sludge, which typically takes less than a day), the remaining sludge depth must decrease by 0.062 m to achieve the target dewatered solids concentration. This is simulated in the numerical model by showing an active bed depth of 0.062 m in one of the eight-bed columns representing the bed that was loaded. This active bed volume is iteratively reduced by the weekly evaporation (as taken from the monthly average net evaporation for the treatment plant locale) until the entire depth of 0.062 m is lost to evaporation. This simulation assumes that the beds will be available for loading 1 week after the final solids concentration has been achieved, as can be seen in Week 7 where Bed 1 is utilized for a second loading. ▲

Based on the 10 weeks of data presented in this example, eight drying beds will be sufficient to accommodate the sludge production from this treatment plant. Therefore, at least 4,645 m^2 of sand drying bed area will be required. In practice, it is recommended that at least 1 year's worth of data be simulated to determine seasonal effects on dewatering performance. A spot check must be done on the starting month to make sure that the assumption of a bed being available is corrected. For more detail on these calculation procedures, see Vandermeyden et al. (1997).

MECHANICAL DEWATERING

Several mechanical devices are available to dewater water treatment plant residuals. However, only five general types are in common use—centrifuges, plate and frame filter presses, diaphragm filter presses, belt filter presses, and vacuum filters. Vacuum filters have only found use on lime sludges whereas the other four are used on coagulant as well as lime sludges. Centrifuges, belt presses, and vacuum filters are all considered low-pressure systems, whereas the other two types of filter presses can both operate at higher pressures and thus produce a higher-solids-concentration cake. For coagulant sludges, centrifuges and belt presses produce a sludge cake in the 15 to 25 percent solids concentration range, although some centrifuge applications may be capable of producing sludge cake up to 30 percent solids at higher polymer loading rates/lower solids loading rates. Diaphragm and plate and frame presses can produce a cake of 30 to 45 percent solids concentration. All of the devices tend to produce cakes in the 55 to 65 percent range on lime sludges produced by softening groundwater, although vacuum filters can produce cakes in the 40 percent range. The choice of equipment can be first narrowed down by the required final cake concentrations, which for coagulant sludges especially, separate the low-pressure devices from the high-pressure devices. However, even when a high cake solids concentration is required, some utilities have found it desirable to use a lower pressure device followed by air drying in order to produce the desired final cake solids concentration since the low-pressure devices tend to be more economical and easier to operate. Ultimately pilot studies comparing performance of the different devices on a site-specific situation are necessary to select the best equipment. Scale up from pilot results to full scale then becomes an important design issue.

Vacuum Filters

Many types of vacuum filters exist. Each is subject to the same limitation; that is, the maximum theoretical pressure differential that can be applied is atmospheric, 103 kPa (14.7 psi). In practice, a differential pressure of about 70 kPa (10 psi) is achieved.

The equipment itself consists of a horizontal cylindrical drum that rotates partially submerged in a vat of sludge that, in order to assist dewatering is usually conditioned by either a coagulant or a body feed such as fly ash (Fig. 22-15). The drum surface is covered by a

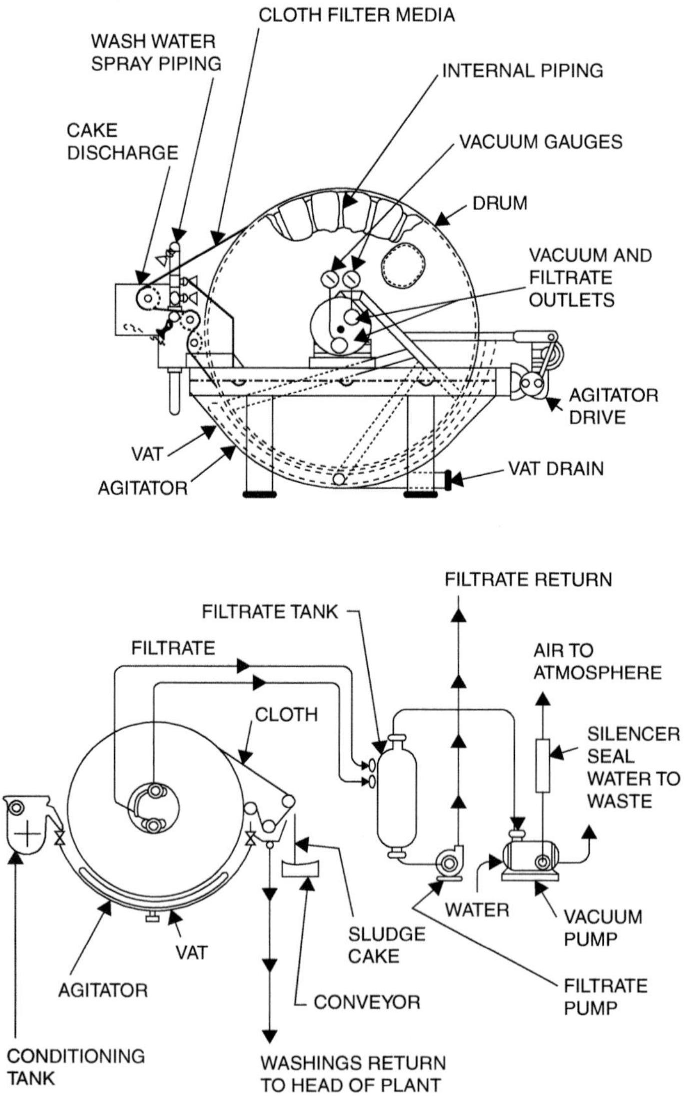

FIGURE 22-15 Typical vacuum filter schematic. (*Source: Innocenti, 1988.*)

filtering medium that is fine enough to retain a thin cake of sludge solids as it is formed. The filtering medium usually consists of a fabric mesh. The drum surface is divided into sections around its circumference. Each section is sealed from its adjacent section and the ends of the drum. A vacuum is applied to the appropriate zone and subsequently to each section of the drum. From 10 to 40 percent of the drum surface is submerged in a vat containing the sludge slurry. The submerged area is the cake-forming zone. When the vacuum is applied to this zone, it causes filtrate to pass through, leaving a cake formed on the cloth. The next zone, the cake-drying zone, represents from 40 to 60 percent of the drum surface. In this zone, moisture is removed from the cake under vacuum. The zone terminates at the point where the vacuum is shut off. Finally, the sludge cake enters the cake discharge zone where it is removed from the medium. This is accomplished by the filter cloth belt leaving the drum surface and passing over a small diameter discharge roll that facilitates cake discharge. No vacuum is applied to this zone. Additional information on vacuum filters can be found in *Water Treatment Residuals Engineering* (Cornwell, 2006).

Belt Filter Presses

Belt filter presses use a combination of gravity draining and mechanical pressure to dewater sludges. A typical belt filter press consists of a chemical conditioning stage, a gravity drainage stage, and a compression dewatering stage (see Fig. 22-16).

The dewatering process starts after the feed sludge has been properly conditioned, usually with polymer. The slurry enters the gravity drainage stage, where it is evenly distributed onto a moving porous belt. Readily drainable water passes through the belt as the slurry travels over the full length of the dewatering stage. Typically, 1 or 2 min are necessary to allow for the filtrate separation in the drainage stage.

Following gravity drainage, the partially dewatered sludge enters the compression dewatering stage. Here, the sludge is sandwiched between two porous cloth media belts which travel in an S-shape path over numerous rollers. Both belts operate under a specific tension which induces dewatering pressure onto the sludge. The S-shape path the sludge follows creates shear forces which assist in the dewatering process. The compressive and shear forces working on the sludge increase over the length of this dewatering stage. The final sludge cake is removed from the belts by blades.

Proper sludge conditioning is considered critical for obtaining acceptable dewatering results. A typical sludge-conditioning unit consists of chemical conditioner storage, metering pumps, mixing equipment (chemical and chemical/sludge), controls, and process piping.

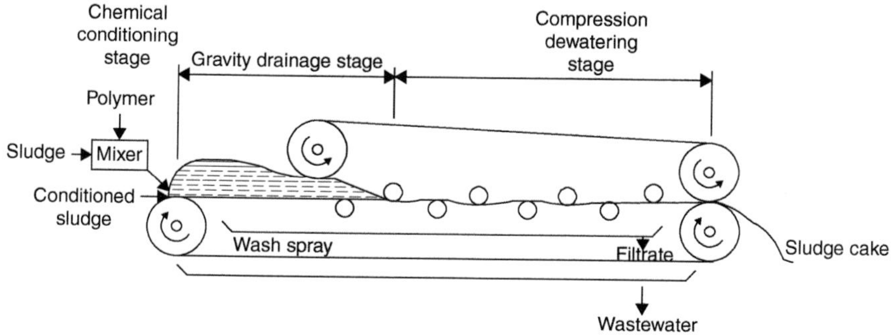

FIGURE 22-16 The three basic stages of belt press. (*Source: Innocenti, 1988.*)

In general, polymer is used for chemical conditioning. To achieve proper sludge conditioning, the polymer is first diluted to between 0.25 and 0.50 percent by weight before it is applied to the feed sludge. Next, the sludge and the polymer are thoroughly mixed. The required mixing time depends on sludge characteristics and the type of polymer used.

The design and selection of a belt filter press is often based on the throughput of the machine; that is, the rate the sludge can be dewatered by the press. The throughput capacity can be limited either by the water in the sludge (hydraulically) or can be solids limited. A belt filter press having a particular type of belt at a particular width has a maximum loading capacity for a particular sludge. Generally, the solids loading is considered the most critical factor and the throughput is expressed in terms of solids loading. The loading units are usually similar to a yield except expressed as belt width, mass/width/time.

Centrifuges

Centrifugation of sludge is basically a shallow depth settling process enhanced by applying centrifugal force. Centrifugation enhances settlement of the solids. In conventional settling tanks, the solids are acted on by the force of acceleration due to gravity (g). In centrifugation, the applied force is $r\omega^2$, where r is the distance of the particle from the axis of rotation and ω is the rotational speed. In modern centrifuges $r\omega^2$ may be 1500 to 4000 times the value of g.

The comparison of $r\omega^2$ to g has led to efforts by many to develop equations for centrifugation by substituting $r\omega^2$ for g. However, this substitution relates to discrete particles only and does not account for hindered settling and the effect of scrolling (moving the solids out of the bowl). These deficiencies in the theory of centrifugation limit the use of sedimentation theory as a basis for the design of centrifuges, and evaluation and design needs to be based on pilot studies.

The major type of centrifuge used for the dewatering of water plant sludge is the scroll-discharge, solid bowl decanter centrifuge. The solid bowl centrifuge (also called *scroll* or *decanter centrifuge*) is a horizontal unit that utilizes a scroll conveyor inside the centrifuge bowl (see Fig. 22-17). The unit is fed continuously with the solids settling against the bowl wall. The scroll rotates at a slightly different speed than the bowl and conveys the dewatered sludge to the small end of the centrifuge where it is discharged. The water is directed from the central axis of the centrifuge toward the centrifuge's large end where it is discharged. The water exits through adjustable weirs (level rings), which also control the pool depth.

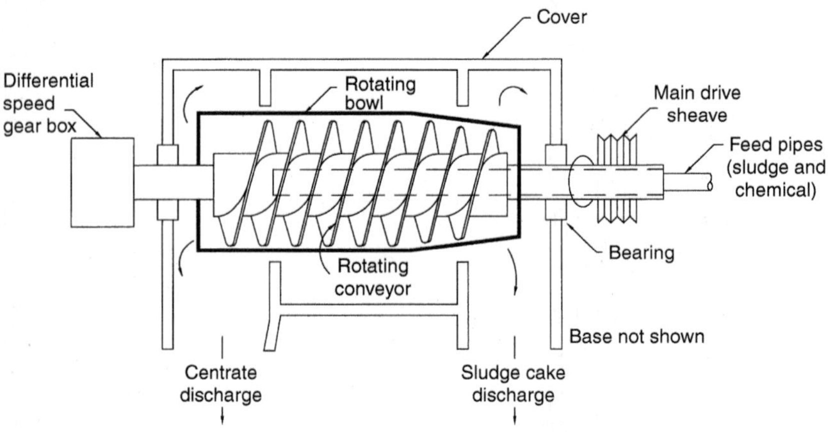

FIGURE 22-17 Schematic of horizontal scroll centrifuge. (*Source: EE&T, 1996.*)

The best procedure for evaluation of centrifuges is pilot tests on prototype equipment. Tests should be conducted on a centrifuge exactly like that to be used in full scale except smaller. Tests should be conducted for operational parameters of concern such as feed flow rate, feed suspended solids concentration, polymer conditioning, bowl speed, pool depth, and scroll speed. Scale-up considering only liquid loading is often referred to as the *Sigma concept*. The Sigma concept is based on Stokes' law describing the settling of discrete particles under the influence of gravity. Gravity is replaced by the centrifugal acceleration and the expression is integrated over the depth of the water pool. One then ends up with a term for the allowable flow (Q) through the centrifuge.

$$Q = \left(\frac{V\omega^2}{g \ln\left(\frac{r_2}{r_1}\right)} \right) \left(\frac{g(\rho_p - \rho)d^2}{18\mu} \right) \tag{22-15}$$

where V = volume of sludge/water in the pool
ω = radial velocity of centrifuge (rad/s)
r_2 = radius from centerline of centrifuge to bowl
r_1 = radius from centerline of centrifuge to pool level
ρ_p = particle density
ρ = fluid density
d = particle diameter
μ = viscosity

Note that the first set of terms on the right hand side of the equation consists of machine variables and the next set of terms relates to sludge variables. Therefore, in scale-up, if it is assumed that the sludge is the same for full scale as in the pilot studies,

$$Q_2 = \frac{Q_1 \Sigma_2}{\Sigma_1} \tag{22-16}$$

where Q_2 is the allowable flow in the full-scale centrifuge based on the optimal flow (Q_1) obtained in the pilot plant, and

$$\Sigma = \frac{V\omega^2}{g \ln\left(\frac{r_2}{r_1}\right)} \tag{22-17}$$

which are variables obtainable for the particular size pilot and full-scale centrifuge. An analysis of the solids-loading limitation is known as the *Beta concept* and is expressed as

$$Q_{s2} = \frac{Q_{S1}\beta_2}{\beta_1} \tag{22-18}$$

where Q_s is the solids throughput in units such as kg/h and

$$\beta = \Delta W S_b N D z \pi \tag{22-19}$$

where ΔW = bowl/conveyor differential speed
S_b = pitch of blades
N = number of leads
D = total bowl diameter
z = pool depth

Again, all beta terms are made up of machine variables and set by the manufacturer for the unit of interest.

In addition to the Sigma and Beta concepts, many centrifuge manufacturers have established scale up factors to predict the performance of their equipment based on pilot test results. In scale-up, multiple scale-up factors should be calculated to determine the limiting condition for design.

Example 22-5 Centrifuge Performance and Scale-Up
A centrifuge pilot study is typically conducted to evaluate performance as well as to allow for scale-up to select the appropriate size for the final installation. Figure 22-18 shows example results from such testing with a pilot centrifuge. Recovery refers to the solids capture and a desired value is over 95 percent. If, for this example, a 24 percent cake was desired, then the pilot centrifuge performed adequately at about 9.084 m^3/h (40 gpm) using a polymer dose of 2000 to 2500 mg/day-kg (4 to 5 lb/day-ton). The pertinent information for the pilot centrifuge, as well as for an example full-scale unit is shown following. Using the Sigma concept, what is the capacity of the full-scale unit?

	Pilot unit	Full-scale unit
Clarification length (cm)	78.74	232.664
Bowl radius (cm)	21.336	36.83
Radius to pool (cm)	12.954	24.13
rpm	3250	3150

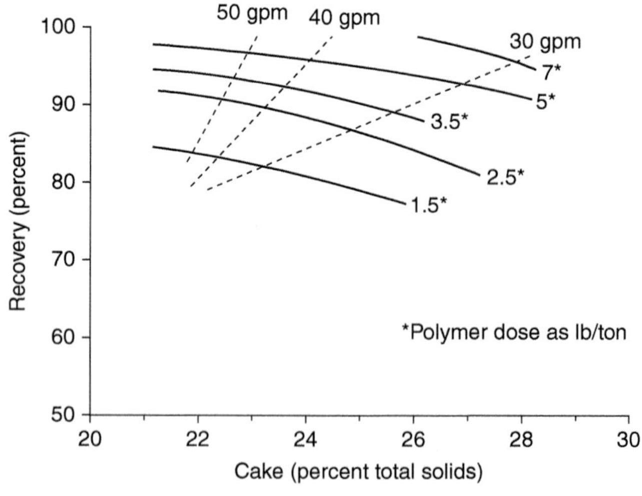

FIGURE 22-18 Centrifuge pilot performance data. Note: multiply gpm by 0.23 to convert to m^3/h; multiply lb/ton by 2 to convert to kg/tonne.

Solution

To calculate the sigma values, Eq. 22-17 needs to be used. The volume V of sludge or water in the pool can be found as

$$V = \pi L\left(r_2^2 - r_1^2\right)$$

where L is the clarification length. Therefore, for the pilot unit,

$$V = \pi(78.74 \text{ in}) \left[(21.336 \text{ cm})^2 - (12.954 \text{ cm})^2\right] = 71{,}000 \text{ cm}^3$$

and ω is found as

$$\omega = (3250 \text{ rpm})\left(\frac{\text{min}}{60 \text{ s}}\right)\left(2\pi\frac{\text{rad}}{\text{revolution}}\right) = 340 \text{ rad/s}$$

Using Eq. 22-17,

$$\Sigma_1 = \frac{(71{,}000 \text{ cm}^3)(340 \text{ rad/s})^2}{(9.81 \text{ m/s}^2)\ (100 \text{ cm/m}) \ln\left({}^{21.336 \text{ cm}}\!/\!_{12.954 \text{ cm}}\right)} = 1.68 \times 10^7 \text{cm}^2 = 16.8 \text{ m}^2$$

Similarly, $\Sigma_2 = 157.7 \text{ m}^2$, such that the Sigma ratio, or scale-up factor, is 157.7/16.8 = 9.39, and, from Eq. 22-16, the design flow for the full-scale centrifuge would be 9.084 m^3/h (9.39) = 85.3 m^3/h. ▲

Filter Presses

The filter press is another process option available for dewatering sludge, and it generally produces the highest final cake concentration of any of the mechanical dewatering devices. Early filter presses were used frequently in Europe for dewatering thin slurries, such as china clays and wastewater sludges. Their practical use for water treatment residuals began around 1965. Experiments commenced in England in 1956, but were disappointing until the advent of the use of polymers as conditioners. The first known uses in the United States were at the Atlanta Waterworks and at the Little Falls Treatment Plant of the Passaic Valley Water Commission.

At the beginning of a filter cycle, sludge is forced into contact with the cloth, which retains the solid matter while passing the liquid filtrate. Very quickly the cloth becomes coated with a cake of sludge solids, and all future filtering occurs through this cake, which increases in depth as succeeding layers build up. The type of cloth does not affect the rate of filtration after the first few minutes, and it can be ignored from a theoretical point of view.

Filter presses are heavy, cumbersome pieces of equipment demanding costly foundations and relatively large buildings. Apart from minor refinements, for several decades filter press design changed little until the advent of the diaphragm filter. The original design of the plate and frame filter (Fig. 22-19) consisted of a series of frames into which sludge is passed under high pressure, up to 1570 kPa (225 psi), in order to dewater the sludge against the outer cloth-covered plate. The depth of the cake was consequently fixed, being governed by the distance between filter plates. Different sizes of plates were manufactured to give cakes of; for example, 19, 25, or 38 mm (¾, 1, or 1½ in) depth. Such filters have been used to dewater sludge in an acceptable manner for many years.

A considerable change in design resulted with the introduction of diaphragm filters (Fig. 22-20). The advantage of this system is that the thickness of the cake is infinitely variable within the limits of the machine dimensions. Sludge is filtered through a cloth for a fixed period of time, perhaps 20 min, at which stage the sludge supply is cut off and water

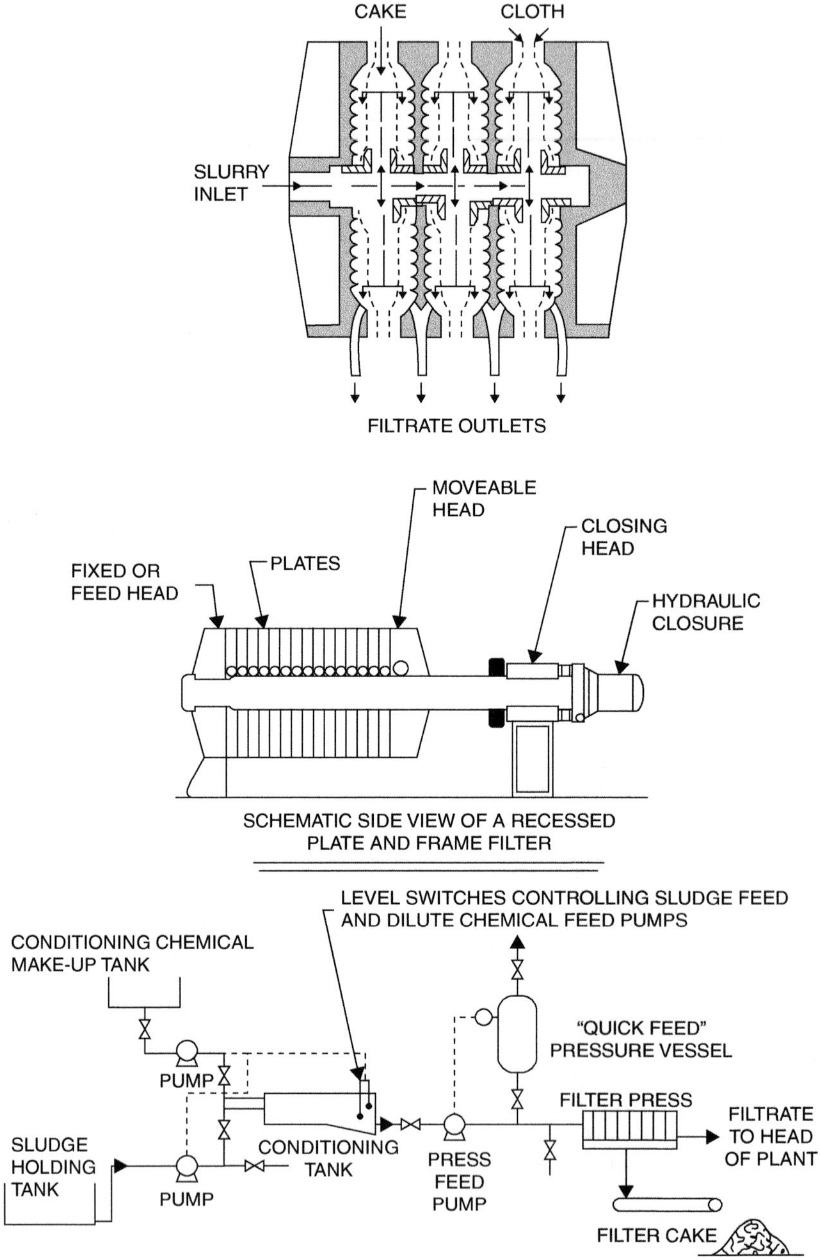

FIGURE 22-19 Plate and frame press schematic. (*Source: Innocenti, 1988.*)

FIGURE 22-20 Diaphragm filter press. (*Courtesy of US Filter – JWI products, Holland, Mich.*)

or compressed air is applied behind an expandable diaphragm that further squeezes water out of the sludge. The cake is dislodged by shaking or by rotating the cloth, depending on the manufacturer's design, and falls into a hopper for disposal. Hanging cakes, where the cake refuses to leave the cloth, an unhappy feature of the older plate and frame press, are consequently eliminated.

SPENT FILTER BACKWASH TREATMENT

All water treatment facilities (e.g., conventional, softening, iron removal, and direct filtration) that have granular media filters as well as low-pressure membranes generate spent filter backwash water during media cleaning operations. Untreated SFBW consists of the washwater and solids carried directly off the top of a filter or cleaned off the membrane during backwash operations. SFBW typically is the largest residual stream in terms of volume, though it is typically low in solids relative to other residual streams.

The volume of SFBW generated at a water treatment plant is dependent on a number of factors including backwash duration, washwater rate, backwashing frequency (filter run length), and number of filters. Filters are typically backwashed every 1 to 4 days, depending on treatment effectiveness, water quality, season, and operator and state regulatory preferences. Many facilities stagger backwashes, so that even when filter run lengths are long, one or more filters are backwashed on a daily basis. Backwash duration and rate also vary with treatment process, water quality, and season.

Naturally, contaminants that are removed by filtration are concentrated in the SFBW. Cornwell et al. (2001) found that TOC levels in SFBW were on average higher by threefold compared to corresponding raw water levels. Observed increases in total trihalomethanes (TTHM) and haloacetic acids (HAA6) were 92- and 24-fold, respectively, in SFBW when raw water is prechlorinated. Average total aluminum and total iron levels in SFBW were 7- and 20-fold higher in SFBW, whereas manganese levels were increased by nearly 13-fold. Table 22-5 compares contaminants in the raw water to SFBW. Corresponding average increases in aluminum and iron concentrations for plants using alum and ferric coagulants were somewhat higher (22-fold and 12-fold, respectively), while the average increase in aluminum levels for plants using polyaluminum chloride coagulants was much higher (55-fold).

TABLE 22-5 Comparison of Contaminant Levels in Raw Water and Spent Filter Backwash Water

Parameter	Raw water		Spent filter backwash water		Multiple increase
	Range	Average	Range	Average	
TOC (mg/L)	0.7–5.4	2.4	0.8–191	8.0	3.3
TTHM (μg/L)	ND–21.8	0.6	ND–198	55.0	91.7
HAA6 (μg/L)	ND–21.5	1.9	ND–211	46.1	24.3
Br (mg/L)	ND–0.68	0.038	ND–0.46	0.033	–0.1
A1 (mg/L)	ND–30	0.72	ND–145.8	14.7	20.4
Fe (mg/L)	ND–56.6	1.2	ND–132	8.7	7.3
Mn (mg/L)	0.01–5.5	0.11	0.01–17.9	1.4	12.7
Zn (mg/L)	ND–0.5	0.03	ND–1.0	0.1	3.3

Source: Cornwell et al., 2001.

Edzwald et al. (2001) showed that *dissolved organic carbon* (*DOC*), dissolved metals, and UV-254 were not elevated in untreated SFBW relative to raw water levels. However, that work did support the Cornwell et al.'s (2001) findings that total metals levels and TOC were elevated compared to the corresponding raw water.

Removal of solids from SFBW can remove many of the contaminants since they tend to be associated with the solids. Solids removal reduces TOC, metals, and DBP precursors.

Research completed by Cornwell et al. (2001) on treatment options for SFBW yielded similar results, and also evaluated the effectiveness of more advanced treatment including *dissolved air flotation* (DAF), clarification plus granular media filtration, and MF on removal of particles. In fact, results demonstrate that when paired with polymer conditioning, both sedimentation and DAF resulted in particle removals of greater than 90 percent. This research also indicated that treatment of SFBW to achieve at least 1 log of *Cryptosporidium* removal would reduce the cyst concentration in the SFBW to at or below raw water background levels, alleviating concerns over recycle or discharge to the environment. Other benefits of a 1-log treatment process could include reduction of contaminants such as TOC, Fe, Mn, and solids. Research also indicated that gravity treatment processes should include the use of a polymer to aid solids settling. The type of polymer applied is important for achieving successful treatment of the recycle stream.

Spent filter backwash is produced over a relatively short time periods (around 15 min per filter) but at a high instantaneous flow rate. Equalization (EQ) is a method whereby the SFBW is captured in a storage device and released over a longer period of time and, therefore, at a lower flow rate. For example, capturing the volume of one backwash and releasing it over the next hour will reduce the flow to about 25 percent of the instantaneous rate. Reduction of the flow rate through equalization can be useful in sizing downstream treatment devices, reducing impacts to the main treatment plant if recycle is used, reducing impacts at wastewater plants or sewer line size requirements, or reducing flow rates into the receiving stream. The allowed rate of release from the equalization basin is established by the downstream control, for example, the sewer line capacity, regulatory or treatment limitations on recycle, stream flow discharge limits, and so on.

The EQ basin storage volume is determined on the basis of a flow balance of SFBW entering the basin and the rate of release of equalized SFBW leaving the basin. The storage curve is typically shown over a 24-hour period of fill-and-draw from the EQ basin. Because the rate of SFBW flow into the EQ basin usually exceeds the rate of discharge from the basin, the volume in the basin increases over time as long as filter backwashing is occurring. The volume in the EQ basin will then decrease until the next cycle of washes begins. The peak of the storage curve is the total EQ basin volume required to meet the backwash assumptions used.

SFBW typically flows by gravity to equalization basins, treatment processes, disposal, or recycle. EQ basins assist by buffering the flow of SFBW from filter washing into the treatment process, so that a controlled flow rate can be maintained. SFBW delivered to a lagoon system or batch clarification process may not require equalization as an intermediate step. SFBW treatment systems could include any of the following:

- Lagoons
- Conventional clarifiers with or without tubes or plates
- Decant basins
- DAF
- High-rate clarification processes
- Membranes

Chemical addition is commonly used to increase the allowed hydraulic loading rate for clarification processes. Polymers are most commonly used to aid separation, but in some cases metal salt coagulants are utilized.

From a quality perspective, the driving force behind the degree of treatment required depends on the fate of the treated SFBW. If the supernatant is to be discharged to a receiving surface water, then the state discharge permit requirements determine the treatment necessary. Generally, the permits contain a suspended solids limit of around 20 to 30 mg/L. The ratio of turbidity to suspended solids is site specific and needs to be evaluated in order to set a treated turbidity goal. Absent data, a turbidity of less than 10 ntu should satisfy a 20-mg/L suspended solids limit and a 10-ntu goal is reasonable and can be attained by a properly designed SFBW treatment system. The discharge permits will also often have a pH limit and a chlorine numerical water quality–based effluent limit (WQBEL) that is most likely to be below the method detection limit (MDL) for chlorine. A typical pH limit of 6 to 9 is generally not a problem for utilities to meet without needing adjustments. It is possible for softening plants to exceed a pH 9 limit, and a lowering of the pH may be required. Removing free chlorine is accomplished by the use of a dechlorinating agent such as sodium bisulfite or ascorbic acid. Therefore, for a SFBW treatment system that is to meet discharge limits, the driving force for the design is normally aimed at meeting a set suspended solids limit.

If the treated SFBW is to be recycled, then generally it is the utilities own goals for the quality that determines the treatment requirements. There are two excellent Awwa Research

Foundation reports available to help determine the goals for SFBW treatment prior to recycle: *Recycle Stream Effects on Water Treatment* (Cornwell and Lee, 1993) and *Treatment Options for* Giardia, Cryptosporidium, *and Other Contaminants in Recycled Backwash Water* (Cornwell et al., 2001). A lot of the contaminants of concern (TOC, DBP precursors, Mn, and metals) are associated with the solids in the SFBW, so again the ultimate goal for the SFBW treatment may well be based on solids removal. In this case, however, the treated turbidity goal may well be lower than for the discharge situation. For low-turbidity raw water, the recycle turbidity goal is often set at below the low water turbidity level which can correspond to <5 ntu or even <1 ntu. For higher-turbidity raw waters, contaminant removal for TOC, Mn, and so on, may set the turbidity goal.

SFBW treatment has traditionally been accomplished with either standard clarifiers or plate or tube settlers. Loading rates for standard clarifiers treating SFBW tend to be quite low in order to achieve low treated turbidities or particle counts. In one laboratory study, Cornwell and Lee (1993) found that an overflow rate of 0.12 m/h (0.05 gpm/ft^2) was required to achieve 90 percent particle removal in the 3- to 10-μm size range without the use of polymer. With polymer, acceptable performance could be achieved with an overflow rate of 0.49 m/h (0.2 gpm/ft^2). These ranges were confirmed in a later study using field scale data although rates up to about 1.2 m/h (0.5 gpm/ft^2) with polymer were successful for turbidity removal and total particle count removal. In a 2009 study on two full-scale plants, a plate loading of 0.61 m/h (0.25 gpm/ft^2) with polymer was successfully used (Cornwell et al., 2010). In that same study, another facility had a plate settler designed to treat a surface hydraulic loading up to 2.9 m/h (1.2 gpm/ft^2) although turbidity performance seemed to drop off above about 1.2 m/h (0.5 gpm/ft^2).

High-rate clarification processes have been investigated for SFBW treatment in order to reduce footprint requirements. High-rate SFBW treatment alternatives include the following: (1) conventional-rate DAF (10 to 15 m/h), (2) high-rate DAF (up to 40 m/h), (3) high-rate solids contact clarification processes (including process trade names DensaDeg®, CONTRAFAST®, Trident® HSC, Actiflo™), and (4) membranes. The actual design of these devices is similar to other applications which are discussed in the various chapters of this book. However, the actual design parameters for each device are unique for SFBW. Table 22-6 shows loading rates for various devices.

RECYCLE

It is common for many utilities to recycle various streams back to the beginning of the treatment process. Recycle is most often done to reduce the volume of residuals for disposal, to conserve water, or because it is not convenient or possible to discharge the liquids to a surface water or sewer. The liquid volume associated with sedimentation underflow and spent filter backwash water can range from 3 to 10 percent of the plant production. In water-scarce areas the water is too valuable a resource to discharge, and returning it to the treatment process is an imperative water conservation measure. Discharge into surface waters is regulated by states under the National Permit Discharge Elimination System (NPDES) program. In some areas such a permit cannot be obtained and recycle is the only option.

Liquid streams can be recycled back to a terminal reservoir, to the beginning of the treatment train, or, in some cases, to an intermediate point in the treatment process. Possible streams that could be recycled include the following:

- Spent filter backwash water–with or without treatment
- Clarifier or sedimentation basin sludge
- Sludge thickener overflow (supernatant)

TABLE 22-6 Summary of SFBW Treatment

City	State	Year	Median untreated turbidity (ntu)	Clarification rate (gpm/ft^2)	Residuals concentration (percent solids)	Polymer needed?	Pilot- or full-scale
				Standard-rate DAF			
Durham	NC	1999	100	up to 5	3.5	Yes	Pilot
Boulder	CO	1999	10	up to 7	3.5	Yes	Pilot
Phoenix	AZ	2000	Not reported	up to 6	4 to 6	*	Pilot
Cleveland	OH	2000	20	up to 6	2 to 3	Yes	Pilot
Boulder	CO	2007	10	up to 3	up to 3	Yes	
				High-rate DAF			
Orem	UT	2007	100	12	4.4	Yes	Pilot
Cleveland	OH	2007	20	16	3	Yes	Pilot
Cleveland	OH	2007	20	15	3.6	Yes	Pilot
				Sand-ballasted coagulation/clarification			
Modesto	CA	2005	20	up to 30	Not reported	H	Pilot
Tempe	AZ	1997	10	up to 20	Not reported	H	Pilot
W. Creek	CA	1997	100	up to 20	0.45	H	Pilot
				Solids contact + tube settlers			
W. Nyack	NJ	2005	~50	~12	2 to 3	H	Pilot
Oradell	NJ	2005	~50	~12	2 to 3	‡	Pilot
Orem	UT	2007	100	12	>5	Yes	Pilot
Cleveland	OH	2007	20	16	3	Yes	Pilot
				Tube settlers			
Orem	UT	2007	100	0.45	~0.5	H	
Orem	UT	2007	100	0.92	~0.5	H	
				Tube settlers + upflow buoyant media			
Orem	UT	2007	100	18	~0.3	Yes	Pilot

*Can operate without polymer, but more stable with polymer.
[H]Impact of treatment without flocculation not reported gpm/ft^2 × 2.44 = m/h.
Source: Cornwell et al., 2010.

- Sludge lagoon decant or overflow
- Dewatering operation liquid wastes
  - Filtrate from belt or filter presses
  - Centrate from centrifuges
  - Leachate from sand drying beds

Recycling of these streams has the potential to upset the treatment process or to affect the quality of the finished water. The impacts could be caused by the solids in some of the recycle streams, constituents in the recycle streams, characteristics of the recycle streams (e.g., pH, alkalinity, etc.), or hydraulic impacts. The principal constituents that could be in recycle streams and be of water quality concern include

- Microbiological contaminants
- *Giardia* and *Cryptosporidium* parasites
- Total organic carbon
- Disinfection by-products
- Turbidity and suspended solids
- Metals, such as manganese, aluminum, iron
- Taste and odor causing compounds

Giardia and *Cryptosporidium* in SFBW can have high cyst concentrations (Cornwell and Lee, 1993). For example, one plant studied had *Giardia* and *Cryptosporidium* concentrations of more than 150 cysts/L in the spent filter backwash water compared with 0.2 to 3 cysts/L in the raw water. Laboratory and full-scale confirmation in that research showed that sedimentation was effective in reducing particles (and cyst levels) in the spent filter backwash prior to recycle. However, low overflow rates (less than 0.12 m/h (0.05 gpm/ft^2)) were required to achieve 70 to 80 percent particle removal in the cyst-size range. A nonionic polymer was effective in increasing particle removals to more than 90 percent at overflow rates of 0.5 to 0.75 m/h (0.2 to 0.3 gpm/ft^2). A mass balance model was developed to estimate the increase of the cyst loading to the plant due to recycle; the model was based on varying recycle ratios and varying degrees of cyst removal from the recycle stream prior to recycle. For example, it was found from the model that if a plant was recycling settled SFBW into the raw water line at a 20 percent ratio and was using a backwash clarifier designed for an overflow rate of 0.6 m/h (0.25 gpm/ft^2) without polymer addition, then the cyst loading to the treatment process would be 20 times that present in the source water. However, if the recycle ratio was reduced to 5 percent and the backwash water clarifier efficiency was increased with the use of polymer or with lower overflow rates, the increased cyst loading due to recycle would only be 1.1 times that of the source water. This analysis, however, discounts the additional disinfection that occurs as the cysts are recycled through the various treatment processes, and no information is available as to whether the cysts are viable. Additional discussion of filter backwash recycle and the impact of prefiltration clarification performance is presented in the backwashing section of Chap. 10.

Nearly all coagulant sludges contain high concentrations of manganese. Quantities of dissolved manganese in the water surrounding sludge can be in the range of 1 to 7 mg/L and on storage reach 20 to 30 mg/L. As sludge accumulates in manually cleaned sedimentation basins, the manganese levels in the clarified water can gradually increase. Therefore, some manganese is released to sludge thickener overflows and recycled to the plant or is released in manually cleaned sedimentation basins to the clarified water. Normally the manganese concentrations are low unless large spikes of waste streams are recycled. However, if sludge accumulation were allowed to occupy a significant portion of the thickener or manually cleaned basin or if a hydraulic upset occurred, a situation could develop where large concentrations of manganese could be recycled or released from the manually cleaned sedimentation basin.

In the research by Cornwell and Lee (1993) and Edzwald et al. (2001), it was generally found that if the solids were removed from the waste streams prior to recycle, *total trihalomethane formation potential* (TTHMFP) in the recycle streams was no higher than in the raw waters. The same was found for *total organic carbon* (TOC). However, without solids removal TTHMFP and TOC levels can be quite high in the waste streams. The recycle streams can contain preformed *trihalomethane* (THM), and therefore, THM concentrations leaving the plant with recycle were sometimes found to be higher than that without recycle. This increase could impact a utility's distribution system THM average.

Treatment of waste streams prior to recycle depends on the types of contaminants to be removed and the degree of removal required. Generally, sedimentation is sufficient to remove the solids leaving the supernatant to be recycled. Polymer or coagulant assistance can increase the removal efficiency of particles. The supernatant stream could also be disinfected prior to recycle using ozone, chlorine dioxide, or perhaps potassium permanganate. If particulate removal above that achievable by sedimentation is required, filters could be added or membrane technology employed. If recycle of streams is necessitated by conservation or discharge issues, it appears that an appropriate technology can be selected to allow for recycle without impacting finished water. However, more research and site-specific investigations are needed to determine appropriate treatment for each situation.

MEMBRANE RESIDUALS

As membrane systems are increasingly used for water utility applications, the management of their residuals has become a growing challenge. As outlined in Chap. 11, membrane processes are classified, among other things, by their pore size. Larger pore-size membranes, such as MF and UF systems, remove primarily particles and suspended solids from feedwater. Tighter membranes, such as NF, RO, ED, and EDR systems, are able to separate dissolved substances from feedwater. The types of residuals produced by both systems vary accordingly. In addition to the residuals produced during water treatment, membrane systems typically require various chemical cleaning methods to reduce fouling, which produce associated chemical wastes. In the next sections, membrane residuals fall into two categories on the basis of the system pore sizes and/or system pressure.

Low-Pressure Membranes

As with granular media filtration processes, low-pressure membrane systems require periodic backwashing. Backwashing counteracts the accumulation of foulants on the membrane surface and within pores, which over time can result in a loss of productivity. These foulants can be inorganic (clay or silt colloids, precipitated metals, etc.), organic (natural organic matter, coagulant aid polymers, etc.), and biological (biofilms, etc.). In addition to backwashing, chemical cleaning (or CIP) methods are used to control fouling and restore productivity.

Backwashing typically occurs every hour or so and is accomplished using air, air and permeate, or permeate alone from the permeate side to the feed side, in the opposite direction of filtration. Chemically enhanced backwash (CEBW) involves the addition of chemicals to increase the backwashing effectiveness by oxidizing or dissolving some of the solids in the fouling layer and by inactivating biological growth.

However, backwashing alone cannot completely prevent the loss of productivity, as over time foulants that cannot be removed via backwashing accumulate on the membrane surface and in the membrane pores. The removal of these foulants requires the system to be removed from service (typically for a half or full day) for chemical cleaning (or CIP). Tank-batched solutions of appropriate oxidant, caustic, acid, and detergent are recirculated individually or in combination to dissolve membrane foulants. Sometimes these solutions are heated to help loosen the foulants. Citric acid, caustic soda, and chlorine are the most common agents used for cleaning low-pressure membranes. CIP is usually performed on a monthly or greater interval.

The nature of the residuals produced by a low-pressure membrane system differ depending on if the membranes are used in combination with other treatment processes (integrated membrane systems) or as a stand-alone membrane system. The backwash waste contains not only the contaminants naturally occurring in the source water but also those added during upstream treatment (e.g., coagulation). Therefore, residuals management alternatives (i.e., disposal, reuse, etc.) may differ accordingly (AWWA, 2003).

Generally, MF/UF systems produce residuals on an intermittent basis. However, in large membrane plants (>40 MLD (10 mgd) the backwash sequence operation is rotational such that intermittent backwash flow rates from individual trains essentially become continuous on a plant basis.

Backwashing Quantities and Characteristics. Specific characteristics of MF and UF backwash and concentrate residuals depend on the quality of the water being treated and the recovery of the membrane. Recovery is the ratio of water produced (i.e., feed flow minus water used for backwash, etc.) to feed flow and can range from 85 to 98 percent; thus, the solids in the feedwater may be concentrated by 7 to 50 times over the feedwater concentration, respectively. Backwash flow rates typically represent greater than 95 percent of the total volume of residuals produced (the remaining portion comes from chemical cleaning procedures). Table 22-7 presents typical spent backwash characteristics. If coagulants or other pretreatment chemicals are applied, the characteristics of the MF/UF residuals become more similar to the residuals from a conventional water treatment plant, while those from spent CEBW will more resemble those from CIP (AWWA, 2003).

TABLE 22-7 Typical Characteristics of Low-Pressure Membrane Spent Backwash

Parameter	Characteristics
Frequency of application	Every 30 to 60 minutes
Volume of waste produced	2 to 15% of plant feed flow rate for recoveries of 85 to 98%
Composition of spent backwash	• Algae, precipitated solids, possible chemical residues if using CEBW. • For recoveries of 85 to 98%, feedwater TSS will be concentrated 7 to 50 times. • pH may be <6 or >9. • Cl_2 residuals may be up to 1000 mg/L. • If coagulant is used, TOC could be 5 times the feed concentration; otherwise TOC can be up to twice that of the feed concentration.

Source: Adapted from AWWA, 2003.

Chemical Cleaning Quantities and Characteristics. CIP wastes reflect the chemicals used in the cleaning process. In addition to the portion of remaining active chemical ingredient, the resulting chemical cleaning waste includes dissolved organic materials, suspended solids, and salts from chemical reactions between the chemicals and foulants (AWWA, 2003). Table 22-8 describes typical CIP waste characteristics.

Chlorine residuals in CEBW and CIP wastes may range from 1 to 1000 mg/L as Cl_2, and pH may be acidic (pH < 6) or basic (pH > 9) depending on the chemicals used. When surfactants are employed, they may cause foaming when the spent cleaning solution is discharged. The need for neutralization chemicals can be minimized by mixing the acid and caustic waste solutions, but this practice may also precipitate additional solids such as calcium carbonate or some iron compounds, which will increase the suspended solids concentration of the waste solution. If citric acid is used in the CIP, the biochemical oxygen demand (BOD) of the spent cleaning solution is likely to be high as well.

Cleaning solution volumes will be 2 to 3 times the volume normally present in the membrane modules and piping (one volume for the actual cleaning solution, and one or two

TABLE 22-8 Typical Characteristics of MF/UF Spent Chemical Cleaning Solutions

Parameter	Characteristics
Frequency of application	Every month
Volume of waste produced	Monthly CIP wastes normally <0.05% of plant feed flow rate
Chemicals commonly used	• NaOCl (500–1000 mg/L as Cl_2) • Citric of hydrochloric acid (pH 1 to 2) • Caustic soda (pH 12 to 13)
Composition of spent CIP	• pH from 2 to 14 • TSS up to 500 mg/L (neutralization may precipitate additional solids) • Low concentrations of surfactants • Cl_2 residuals up to 1000 mg/L • TOC 10 to 30 times the feedwater concentration • BOD_5 up to 5000 to 10,000 mg/L if citric acid is used

Source: Adapted from AWWA, 2003.

additional volumes to rinse the solution), or for immersed systems equal to the tank volume (AWWA, 2003). Rinse waters typically have low levels of contamination and may either be returned and blended with the feedwater, combined with backwash wastes for recycling or disposal, or used to make up the next batch of cleaning solution. While the frequency of chemical cleaning varies as a function of the quality of the feedwater, overall chemical cleaning is performed far less frequently than backwashing and thereby contributes less than 5 percent in the generated residuals volume.

Membrane modules may be shipped to the plant preserved in specialty solutions, such as glycerin, which are typically discharged to the sewer. In the absence of a sewer connection, these wastes need to be collected and hauled to an appropriate disposal site (AWWA, 2003).

Residuals Management. Membrane residual treatment and disposal alternatives depend on site-specific factors such as local climate, land availability, size of the MF/UF installation, capabilities of the operating staff, and feasibility and local requirements for different disposal options. Smaller installations tend to use simple methods capable of handling both types of residuals (backwash and cleaning wastes), such as discharge to the sanitary sewer, settling lagoons or percolation/evaporation ponds, and equalization basins to store, mix, and neutralize chemical solutions (AWWA, 2003). It is believed that large installations will increasingly look toward recycling of backwash and reuse of chemical cleaning solutions to significantly reduce volumes and associated disposal costs (MacPhee et al., 2002). The comparative economics for each disposal alternative is influenced by the cost and availability of land, the local value of water, the availability of a sewer system to accept the residuals, and the demand (amount and seasonal variation) for irrigation water (AWWA, 2003).

Surface Discharge. Discharges to receiving surface waters are regulated under the National Pollutant Discharge Elimination System (NPDES), according to water quality standards that protect the designated usage of the receiving waters. Surface water discharge is the most common form of backwash waste disposal regardless of plant size (Kenna and Zander, 2000). Simple settling basins or lagoons typically provide sufficient treatment to allow subsequent surface water discharge for most installations.

Underground Injection. The underground injection control (UIC) program administers groundwater discharge permits (e.g., deep well injection) and provides safeguards to maintain current and future underground sources of drinking water. An NPDES permit may also be involved if a secondary disposal method is required or if the receiving groundwater is hydraulically linked to surface water.

Sewer Discharge. Indirect discharge to *a publicly owned treatment works* (POTW) may be controlled under the *Industrial Pretreatment Program* (IPP) if the POTW practices surface water discharge and has its own NPDES permit (AWWA, 2003). Under the IPP, a POTW is required to enact local sewer ordinances and to enforce standards laid out in the IPP to protect the POTW's NPDES permit from any violation caused by indirect discharge. Other federal and local regulation may apply depending on the method of waste disposal used by the receiving wastewater treatment plant. Most plants do not provide any treatment of the spent backwash prior to discharge to a sewer, but discharges may need to be equalized to meet available sewer capacity (AWWA, 2003).

Land Disposal. Percolation ponds, leaching fields, and other land disposal methods that carry a risk of liquid discharge reaching any groundwaters are affected by the same regulations governing underground discharge (i.e., NPDES and UIC programs). When high concentrations of solids are present, land application of membrane residuals is regulated by the Solids Waste Disposal Act. If disposal is via a landfill, additional Resource Conservation and Recovery Act (RCRA) requirements apply.

Prior to land application, solid residuals must also undergo a TCLP test. Residuals that fail the TCLP test are deemed hazardous wastes and are subject to additional regulations, but in nearly all cases, low-pressure membrane process residuals do not fall into this category. Spray irrigation is a common, low-cost alternative for land application that provides benefits for both the water utility disposing of the residuals and the organization needing irrigation water. To be successful, spray irrigation requires year round demand, low concern for aesthetics, and proper soil conditions at the application site to avoid potential groundwater contamination. Spray irrigation may not be feasible for large MF or UF plants because of the large residuals volumes.

Evaporation Ponds. Although there is no federal regulation for evaporation ponds, as with land disposal, NPDES or UIC permits may be required if there is a chance of intermittent discharge to surface or groundwater. For this disposal method, optimal conditions include low flows and an arid climate. Adequate space should be provided for accumulating several years of solids within the evaporation ponds to reduce the frequency of required cleaning (AWWA, 2003).

Backwash Recycle. The Filter Backwash Recycling Rule (FBRR), which regulates the recycling of backwash and other recycled flows in conventional treatment plants, does not address membrane residual recycling unless membrane residuals are commingled with other conventional process residuals prior to recycling (AWWA, 2003). At large installations, recycling of spent backwash water is generally accompanied by some type of settling process (Kenna and Zander, 2000). At installations where backwash residuals contain a higher concentration of contaminants such as *Cryptosporidium*, additional treatment may be required to safely recycle backwash (MacPhee et al., 2002). The volume of residuals sent for disposal may be reduced by more than 90 percent by recycling the clarified backwash water and neutralized CEBW wastes to the front of the treatment process. This reduction in volume may make hauling the concentrated backwash solids off-site more economically feasible. The following recommendations were made for treatment of spent MF/UF backwash prior to recycle (Le Gouellec et al., 2004):

- MF backwash streams can vary widely in turbidity (<10 to >125 ntu).
- Addition of a metal salt coagulant is necessary if the raw water is not coagulated ahead of the membrane in the main treatment train.

- Ferric chloride generally outperformed aluminum based coagulants. pH adjustment was not needed.
- Sedimentation and DAF are both effective means of clarification for spent backwash.
- A sedimentation hydraulic overflow rate of 0.61 to 1.2 m/h (0.25 to 0.50 gpm/ft^2) is appropriate for settling if the right dose of a primary coagulant is used. Addition of a polymer with the metal salt coagulant may enhance performance.
- Bench-scale testing is an excellent tool for screening coagulant, polymer, pH adjustment, and overflow rate requirements for treatment of MF/UF backwash streams intended for recycle.
- In general, a sedimentation treatment system for MF/UF backwash would include equalization to provide a relatively constant backwash quality and quantity for sedimentation and allow chemical dosing to be set on the basis of average conditions.
- Under suitable treatment conditions, an average of at least 1-log *Cryptosporidium* removal can be achieved through sedimentation of spent backwash.
- Removals of *Bacillus subtilis* and particle counts are similar to those of *Cryptosporidium*, suggesting that these parameters could be used as surrogates for gauging treatment performance for pathogen removal.
- To limit potential recycle impacts, backwash turbidity levels, and particle counts after treatment should be close to those of the raw (influent) water.

CIP Disposal. CIP residuals disposal methods for low-pressure membrane plants depend largely on the frequency of cleaning and the type and strength of chemicals used (AWWA, 2003). For example, large plants that perform frequent chemical cleanings may not be able to use surface water discharge for the spent CIP, unless the CIP wastes can be blended with spent backwash for chemical neutralization. CIP wastes are typically redirected to the sewer after pH adjustment and/or dechlorination if needed, or blended with the backwash residuals prior to joint disposal (AWWA, 2003).

Desalting Membranes

As discussed in Chap. 11, use of desalting membrane technologies is increasing as utilities recognize that such processes can extend water resources by treating seawater, brackish water, and reclaimed water into drinking water. With the continuing installation of desalting plants, residuals management has become a growing problem especially since the principal residual from these systems (i.e., concentrate) typically contains high levels of total dissolved solids (TDS) including heavy metals and radionuclides, which makes its disposal more challenging than for residuals from low-pressure (MF and UF) membrane systems.

The separation of dissolved substances from water can be made under a pressure differential (RO and NF) or under an electrical potential (ED/EDR). Typical overall recoveries for NF plants are 75 to 85 percent for brackish water plants depending on the types and concentrations of potential scaling species (sparingly soluble salts and silica) in the feedwater. Recoveries for seawater plants range from 40 to 60 percent because of material resistance limit imposed by the high osmotic pressure.

Fouling. As with low-pressure membrane systems, over time desalting membrane systems experience a loss in productivity from foulants accumulating on the membrane surface or within the feed channel between the layers of membrane material. As described in the

2004 report "Current Perspectives on Residuals Management for Desalting Membranes" (AWWA, 2004), there are several categories of foulants, including

- *Inorganic.* Clays and silts, in colloidal or suspended form; metal precipitates (e.g., aluminum, iron, and manganese oxides); precipitates from sparingly soluble salts, (e.g., barium sulfate, calcium carbonate, calcium sulfate, etc.); and silica
- *Organic.* Natural organic matter (humic and fulvic acids, etc.)
- *Biological.* Microorganisms that accumulate or actively colonize the membrane surface, as well as the exopolymeric substances (films or slimes) that such organisms produce for the purpose of attachment or protection (i.e., biofilm)

Although these categories encompass the range of desalting membrane foulants, fouling usually consists of a combination of factors, with the relative contribution of each foulant category a function of feedwater characteristics, membrane type, and system operating conditions (AWWA, 2004).

Residuals Quantities and Characteristics. The residuals of NF, RO, and ED systems result from the separation process itself, which generates concentrate, and the CIP process, which produces spent cleaning solution waste (AWWA, 2004). Concentrate contains both dissolved and particulate contaminants at levels that are a direct function of their feedwater concentration and recovery of the system. Thus, the final characteristics of the concentrate from the desalination membrane systems depend on what constituents are removed or added by the upstream treatment process. Spent cleaning solutions are typically generated semiannually, and generally contain high concentrations of the cleaning chemicals and removed contaminants. Concentrate is continuously produced by the membrane separation process; thus, the volume of concentrate that is produced is much greater than that of intermittently generated cleaning waste.

Since the characteristics of the concentrate from NF, brackish water RO (BWRO), and ED/EDR systems can be considered similar, residuals management is about the same. The exception is that ED/EDR systems produce a specialized waste stream of limited flow called *electrode waste* that contains significant levels of hydrogen and chlorine gases. These gases are typically stripped from the electrode waste stream using a degasifier. Although free chlorine may be present in the poststripper waste stream, it is diluted when the stripped electrode waste is mixed with the concentrate flow (AWWA, 2004).

The typical addition of acid and/or antiscalants as part of the pretreatment in desalting membrane systems helps to control scaling from sparingly soluble salts. Chemical cleaning is achieved by using appropriate acidic, caustic, and selected detergent solutions, either individually or in combination, to loosen and dissolve membrane foulants. These solutions are tank batched and circulated directly through the membrane train. Heating the chemical solution can be used to help loosen accumulated materials. Chemical cleaning is typically performed every 6 to 12 months depending on feedwater quality and operating characteristics. Some systems with low fouling feedwaters can operate indefinitely without the need for chemical cleaning, particularly if designed at conservative flux rates (AWWA, 2004).

In contrast to the concentrate, chemical-cleaning residuals from ED/EDR systems are different from those from NF and RO systems.

- NF and RO systems are typically cleaned with acid (mineral or citric) and alkaline (typically containing detergents/surfactants and sometimes chelants) solutions.
- ED/EDR systems are typically cleaned with concentrated sodium chloride solutions and sometimes with chlorine to remove biological foulants (NF and RO systems generally do not use chlorine since it can damage the membranes in these systems).

Concentrate. The concentration of contaminants in the concentrate is dependent on the rejection characteristics of the membrane and is directly proportional to recovery. In contrast to the spent backwash from low-pressure membrane systems, the concentrate from desalting membrane systems contains low particulate concentrations, typically less than 10 mg/L total suspended solids. This is because feedwater turbidity for these systems is required to be less than 1.0 ntu to avoid plugging the flow channels. Likewise, while scale-inhibiting agents added to the membrane feedwater to control precipitation is also concentrated in the desalting membrane process, the levels of these constituents in concentrate are typically less than 30 mg/L.

The majority of residuals present in desalting membrane concentrate are dissolved solids. For ED/EDR, ions will be concentrated in the waste stream to the same levels than described for NF and RO, but the concentrate will also contain free chlorine from the generation of chlorine gas at the electrode. For salinity, the brine concentration can be estimated on he basis of the feedwater quality and a concentration factor that is related to the water recovery as shown in the following. These equations assume a membrane process with 100 percent salt rejection and, therefore, provide a conservative (high) estimate of residuals concentrations:

$$C_b = \frac{C_f}{1-Y} \tag{22-20}$$

where C_b is the brine concentration, C_f the feed concentration, and Y the water recovery. For example, recoveries of 50 and 90 percent would correspond to concentration factors of 2 and 10, respectively.

The quantity of concentrate is a function of the amount of feedwater that is purified (or converted to permeate), as defined by the product water recovery. NF salt rejection is typically less than for RO; therefore for a given feed and recovery, NF concentrate will be less saline than RO concentrate. Additionally, NF rejects monovalent ions (e.g., sodium and chloride) to a lesser degree than multivalent ions (e.g., calcium and sulfate), and thereby enriches the concentrate in multivalent ions compared to that in the feedwater.

Chemical Cleaning (CIP) Waste. Each membrane supplier has developed its own specific solutions or formulations based on laboratory and field experience with their membrane product, and many companies catering to the membrane industry have developed proprietary cleaners in an attempt to optimize cleaning efficiency for specific fouling situations (AWWA, 2004). Table 22-9 presents a listing of standard cleaning solutions to remove various categories of foulants.

TABLE 22-9 Typical Chemical Cleaning Solutions for Desalting Membranes

Foulant type	Cleaning solution
Inorganic salts/ precipitates	• 0.2% HCl • 0.5% H_3PO_4 • 2% citric acid
Metal oxides	• 2% citric acid • 1% Na_2SO_4
Inorganic colloids (silt)	• 0.1% NaOH and 0.05% sodium dodecyl benzene sulfate at pH 12
Silica (and metal silicates)	• 0.1% NaOH, 0.05% sodium dodecyl benzene sulfate at pH 12 • Ammonium bifluoride
Biofilms and organics	• Hypochlorite, hydrogen peroxide, 0.1% NaOH, 0.05% sodium dodecyl benzene sulfate at pH 12 • 1% sodium tripolyphosphate, 1% trisodium phosphate, 1% sodium EDTA

Source: Adapted from AWWA, 2004.

The volume of cleaning solution used and the amount of generated waste are function of

- Source water characteristics (type and level of foulants present)
- Membrane process and module/system characteristics
- Size/capacity of membrane train
- System operating conditions (flux and recovery)
- Cleaning system design (piping diameter and lengths)
- Frequency of cleaning

Typically the volume of the spent cleaning solution is a small percentage of the flow treated (significantly less than 0.1 percent). The total volume of spent cleaning solution includes also the rinses between each cleaning step.

Chemical cleaning is typically a multistage process. Usually a low-pH solution is applied first, followed by a high-pH solution. The trains are also cleaned in steps. The typical approach is to clean first the modules in 1/2 of the vessels in the first stage, followed by the other half of the first stage, and finally all modules in the second stage. Spent cleaning solutions are composed of the removed foulants and the chemicals used to remove them (e.g., detergents, surfactants, acid, caustic, etc.). The spent cleaning solution may need to be treated before disposal, most often by simple pH adjustment (AWWA, 2004).

Residuals Management. In addition to capital- and operation-cost considerations, decisions regarding residuals management should include the assessment of the long-term viability of any selected approach. Regulatory changes in particular have the potential to adversely impact operation of a desalting plant. Disposal of desalination membrane residuals is generally regulated by the same guidelines governing industrial residuals, which results in a longer and costlier permitting process than if the nature of membrane concentrate were taken into consideration at the beginning of the permitting process. Unlike many industrial wastes, membrane concentrate is not characterized by process added chemicals but by the nature of the raw water fed to the process (AWWA, 2004).

Unlike residuals from MF/UF membranes, management and disposal of desalination membrane concentrate can be particularly challenging because volume and a high concentration of dissolved solids can limit the number of options permitted or cost effective, particularly in geographical regions where surface water discharge is unavailable. The most common method of NF, RO, and EDR concentrate management is surface water discharge, followed by sanitary sewer discharge and deep well injection (Mickley, 2001). The two methods typically used for spent chemical cleaning solutions are sewer disposal and blending with concentrate. Prior to blending, it may be necessary to neutralize the acidity or alkalinity of the cleaning solution to prevent unwanted reactions or to ensure that the blended residual is compatible with regulations governing the concentrate discharge. For certain ED/EDR cleaning wastes, residual chlorine may need to be neutralized with a reducing agent.

Spent cleaning solution treatment sometimes is accomplished in the cleaning tank itself at the end of each cleaning step for small facilities. However, for larger facilities, treatment is usually accomplished in separate facilities because of the relatively low cost of these facilities and the significant benefit of reducing membrane system downtime for the overall cleaning process. Commonly, one or two spent cleaning tanks are used with pumps that recirculate the spent cleaning solution from and back to the tank(s) or transfer the treated solution to the disposal point. For pH adjustment, either acid (e.g., sulfuric acid) or base (e.g., sodium hydroxide) is added to the recirculating fluid until the desired pH neutralization point is reached. This can be monitored and controlled manually or automatically (AWWA, 2004).

Surface Water Discharge. The discharge of NF, RO, and ED residuals to a surface water body (i.e., lakes, rivers, lagoons, ocean, etc.) is the most common management practice, primarily because this method has the lowest cost (unless pipeline conveyance distances are long) and the residuals volume improves stream flow. However, the residuals must be compatible with the environment of the receiving water body. Discharge to surface waters requires an NPDES permit, which includes an assessment of potential impacts on the receiving stream from both salinity and specific concentrate constituents. Typically salinity is the limiting factor for surface water discharges, as it is unusual for discharges to be permitted if the concentrate would increase the receiving stream salinity by more than 10 percent. Some facilities address this by combining the concentrate with liquid process streams from other municipal or industrial facilities (e.g., with WWTP effluent, cooling water, etc.).

In general, toxicity is not a concern for desalting membrane system residuals because the principal effect of membrane processes is an increase in the concentration of naturally occurring constituents, which typically have no toxicity. However, concentrates from facilities treating brackish groundwater may be too saline to discharge to fresh surface waters. Therefore such plants are limited to areas with available brackish receiving streams (e.g., estuary, tidal river, etc.). In some states, whole effluent toxicity tests (bioassays) are part of the NPDES permit. The concentrate produced by some groundwaters treated by RO have been found to fail these tests on sensitive test organisms because of major ion toxicity, which is due to excessively high (or low) concentrations of common ions (Mickley, 2000). Most cases in Florida have been associated with high calcium levels. Some of these situations have been also complicated by toxicity due to high levels of fluoride. Because major ion toxicity is different from that due to contaminants such as heavy metals or pesticides, the state of Florida has made allowances in cases where only major ion toxicity is present in membrane concentrates.

Dissolved oxygen (DO) levels may also be a concern for concentrates from desalination of groundwaters, which can have low DO levels. Concentrates from such system must be reaerated prior to discharge to avoid negative impacts on receiving stream biota. Occasionally the groundwater may also contain hydrogen sulfide, which must also be suitably removed from the concentrate prior to discharge. As mentioned earlier, ED/EDR concentrates can contain free chlorine. The free chlorine must be neutralized prior to surface water discharge, which requires that the reducing agent be compatible with the receiving stream (AWWA, 2004).

Sewer. Disposal to the sewer represents an indirect discharge to surface waters and requires permission from the receiving wastewater utility, which may impose pretreatment standards since the discharge permit of the wastewater treatment plant (WWTP) could be affected. If the concentrate has high salinity and its flow contribution is significant, impacts on the biological efficiency of the wastewater plant due to concentrate salinity should also be considered. Additionally, the capacity of the sewer line and WWTP must be addressed. These capacity and quality criteria may limit the amount of desalting residuals discharged to sewer. Sanitary sewer discharge of small volumes usually represents a low-cost disposal method with limited permitting requirements.

Land Application. Disposal of concentrate by land application is generally achieved via irrigation or through groundwater recharge measures such as percolation ponds, rapid infiltration basins, and so on. The selected land application option must be compatible with state regulations protecting groundwater and vegetation (in the case of irrigation). Factors associated with land application include the availability and cost of land, percolation rates, irrigation needs, water quality tolerance of target vegetation to salinity, and ability to meet groundwater quality standards. An assessment of the compatibility with target vegetation should be conducted, including assessment of the sodium adsorption ratio (SAR), trace metals uptake, and other vegetative and percolation factors (AWWA, 2004).

Regulations governing groundwater quality and protection of drinking water aquifers should be investigated as early as possible to confirm the acceptability of this alternative. Usually dilution of the concentrate is required to meet groundwater standards. Where salinity levels are excessive, special salt tolerant species (halophytes) could be considered for irrigation. In general, land applications are used only for smaller volume concentrates since this disposal option is frequently limited by availability of land, dilution water, and/or climate in locations where land application is not possible year around (AWWA, 2004).

Deep Well Injection. Deep well injection and other subsurface injection alternatives such as bore holes are regulated by USEPA's UIC program, which is typically administered at the state level. Regulatory considerations include the transmissivity and salinity of the receiving aquifer and the presence of a structurally isolating and confining layer between the receiving aquifer and any overlying Underground Source of Drinking Water (USDW), which is defined as any water bearing formation containing less than 10,000 mg/L TDS (AWWA, 2004). Deep wells are not feasible in areas subject to earthquakes or where faults are present that can provide a direct hydraulic connection between the receiving aquifer and an overlying potable aquifer.

The cost for deep well injection is higher than surface water disposal, sewer disposal, and land application in cases where these alternative methods do not require long transmission pipelines. In addition to the disposal well(s), one or more small-bore monitoring wells in the local area are also typically required to confirm that vertical movement of water has not occurred. Disposal to deep wells is usually restricted to larger volume concentrates where economies of scale make the disposal option more affordable. A backup means of disposal must be available for use during periodic maintenance testing of the well. Nearly all of the plants using subsurface injection are located in the state of Florida as geologic characteristics make this option inappropriate in many areas of the United States (Mickley, 2001).

Evaporation Pond. Solar evaporation is a viable alternative in relatively warm, dry climates with high evaporation rates, flat terrain, and low land costs. Construction of evaporation ponds typically requires an impervious lining and monitoring wells, depending on state requirements. With little economy of scale, evaporation ponds are usually used only for small volume concentrates. Evaporation ponds are typically designed to accommodate concentrate for the projected life of the desalination facility, so provision for the precipitation of salts must be incorporated into the depth requirements of the pond (AWWA, 2004). In addition, the ultimate fate of the concentrated salts and the future regulatory implications should be considered for any evaporation pond project.

Other. Other concentrate management and disposal alternatives such as blending with WWTP effluent or with power plant cooling water are subsets of the alternatives presented. Blending provides a dilution effect that can support the implementation of a desalting facility.

Zero liquid discharge systems such as thermal evaporators, brine concentrators, vapor compressors, crystallizers, and spray dryers are available to reduce residuals to a solid product for landfill disposal. However, the cost for these thermal systems is typically much higher than the cost for the desalination membrane facility, both from a capital and operating (energy) perspective, making this disposal option unfeasible except for small concentrate flows (AWWA, 2004). Use of high-recovery RO systems in front of the thermal evaporators can reduce the cost for waters of limited hardness where chemical and solids disposal costs are not excessive.

Zero liquid discharge technologies other than thermal evaporation to remove dissolved solids from concentrate are currently being researched, such as fluidized bed crystallization, adsorption with activated alumina, and forward osmosis. These methods may potentially provide cost and energy savings compared to traditional zero liquid discharge systems; for example, Bond and Veerapaneni (2007) described cost savings of 50 to 70 percent and energy savings of 70 to 75 percent using a fluidized bed crystallization process to treat an RO concentrate stream rather than a benchmark zero liquid discharge approach including a brine concentrator and evaporation pond. However, many of these non-thermal concentrate treatment approaches have not yet been established as mature technologies. Figure 22-21 presents the relative maturity level of various concentrate treatment systems.

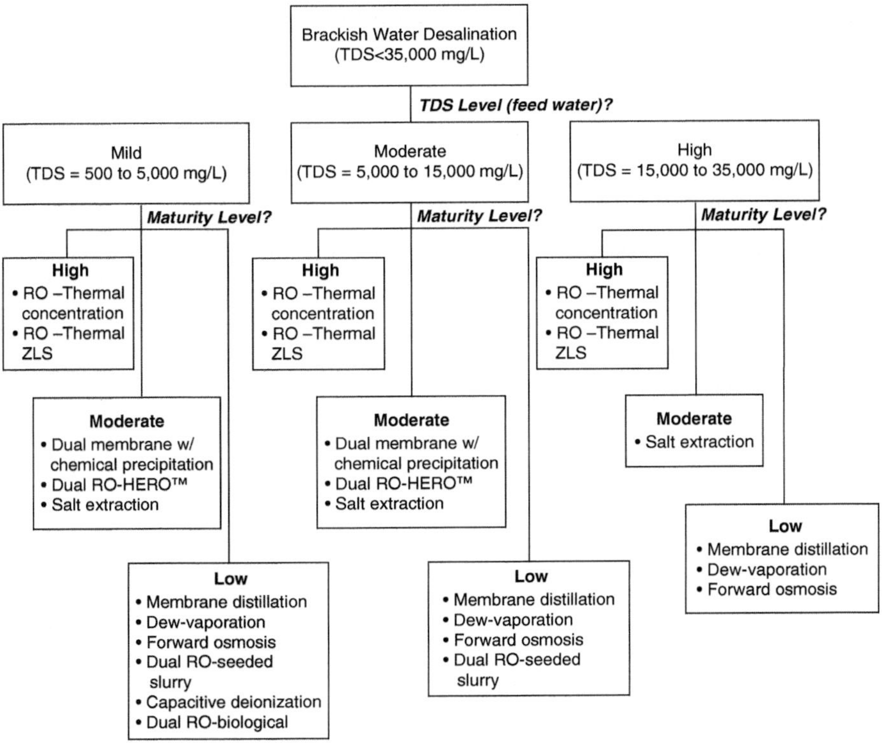

Notes:

1. Maturity is prescribed in terms of applications and experience relative to the municipal water industry. If mature and installed in other applications (e.g. industrial), the maturity relative to municipal application is considered moderate. In general, the developing/alternative desalination technologies in bench or pilot scale testing stage are considered low in maturity.
2. Dual membrane with chemical precipitation implies primary RO, followed by concentrate treatment via chemical precipitation and a secondary membrane process (i.e. RO or ED/EDR). The configuration based on secondary ED/EDR was developed and tested in this study.
3. Thermal ZLD implies thermal concentration and thermal crystallization.
4. Dual RO-HERO™ is considered mature in industrial applications.
5. Salt extraction is considered moderately mature based on existing international/industrial installations.

FIGURE 22-21 Flowchart for general guidance on relative maturity and TDS applicability of recovery enhancement and concentrate minimization technologies. (*Source: Sethi et al., 2009.*)

Two recent Water Research Foundation reports provide an excellent overview of the subject: *Desalination Product Water Recovery and Concentrate Volume Minimization* (Sethi et al., 2009) and *Inland Membrane Concentrate Treatment Strategies for Water Reclamation Systems* (Fox et al., 2009).

Research is ongoing to evaluate the technical and cost feasibility of beneficial use desalination membrane concentrates, including use as a feed source for sodium hypochlorite generation or for energy recovery through heat generation for solar energy ponds (AWWA, 2004). The Water Reuse Foundation report *Beneficial and Nontraditional Uses of Concentrate* (Jordahl, 2006) provides additional information on beneficial use options for desalination membrane concentrate. These and other research may ultimately provide additional alternatives for management of desalination residuals.

ION-EXCHANGE AND INORGANIC ADSORPTION PROCESS RESIDUALS

As described in Chap. 12, ion-exchange and adsorption processes can be used to remove cations, anions, and ionic complexes from water. Both ion-exchange and adsorption processes pass raw water through a vessel containing a solid media. In ion exchange, the removed ions exchange onto the media for other ions of lesser attraction to the media. These released ions end up in the finished water. The process continues until the media is exhausted, at which point the targeted contaminant begins to breakthrough to the finished water. When the concentration of the contaminant breakthrough exceeds a target finished water concentration, the media is regenerated by passing a volume of water containing a high concentration of an exchangeable ion past the media. The media releases the contaminant ions and accepts the exchangeable ions. Therefore the spent regenerant contains the contaminant ions as well as excess exchangeable ions. Ion-exchange media can be a strong acid cation (SAC) exchanger, weak acid cation (WAC) exchanger or anion exchange media (see Chap. 12).

Adsorption processes work similarly to ion-exchange processes, although the adsorption of the contaminant to the solids media does not release ions into the finished water. Adsorption media can also be regenerated to regain adsorption capacity, although occasionally it may be easier to dispose of the target contaminants through disposal of the contaminated solids; in this case, utilities may forgo the regeneration process and simply discard the exhausted media.

The primary cations removed by ion exchange or adsorption include

- Hardness using usually SAC, but occasionally WAC
- Barium using SAC
- Radium using SAC
- Radon using activated alumina, specialty resins, and GAC

The anions that are removed by ion exchange or adsorption include

- Fluoride using activated alumina
- Nitrate using anion exchange
- Arsenic using activated alumina, ion exchange, iron impregnated resins
- Perchlorate using anion exchange
- Selenium using activated alumina or anion exchange
- Chromium using anion exchange

By far the most common ion-exchange process used is softening for the removal of calcium and magnesium.

Softening

Ion exchange has been used for many years as a softening process. Most large plants in the past that utilized ion exchange for softening were located near coastal areas so that brine wastes were discharged to the ocean. Although many of these plants have been abandoned due to corrosion and high maintenance costs, ion exchange is still being practiced by small treatment systems. Its chief advantage over lime softening is the ease of operation. In areas where disposal of lime sludges is a problem, ion exchange may also be favorable if a suitable brine waste discharge point can be found. Wastes are produced from an ion-exchange process from the three phases of bed cleaning: backwash, regeneration, and rinse. Ion-exchange contactors are normally backwashed at a rate of 12.2 to 14.7 m/h (5 to 6 gpm/ft^2) for about 10 min, generating 2037 to 2444 L of backwash waste for each square meter of contactor area. The regenerate brine waste volume depends on the type of exchange media used, the amount of cation exchange capacity that is available, and the efficiency of regeneration. (The units used for ion exchange are in terms of kg of hardness expressed as $CaCO_3$. Hardness refers to any divalent or higher cation.) The capacity of synthetic ion-exchange resins depends on the regenerant driving force used, as shown in Table 22-10.

TABLE 22-10 Ion Exchange Softening Resin Capacity

Salt used (kg/m^3)	Exchange capacity (kg/m^3)	Regenerate use kg NaCl/kg removed
85	49.4–56.5	1.8
142	63.6–70.6	2.4
213	67.1–77.7	3.4

The theoretical salt (NaCl) demand for regeneration is 1.18 kg NaCl/kg hardness removed. Most plants operate between 114- and 142-kg/m^3 (8- and 10-lb/ft^3) regeneration, whereas small homeowner systems operate at 85 kg/m^3 (6 lb/ft^3). Therefore, for every kg of hardness removed approximately 11 to 57 L (3 to 15 gal) of brine are produced. The strength of the initial regenerant solution is normally between 8 and 18 percent NaCl. Increased regenerant strength reduces waste production, as shown in Table 22-11. This brine contains the excess salt in the brine and the cations removed from the resin.

TABLE 22-11 Volume of Spent Regenerate Produced in Ion Exchange Softening

	Volume of regenerant used for indicated solution strength, L/kg removed			
Regenerant use kg NaCl/kg removed	6 percent NaCl	10 percent NaCl	12 percent NaCl	16 percent NaCl
1.8	30.3	18.2	15.1	11.4
2.4	40.1	24.2	20.1	15.1
3.4	56.8	34.1	28.4	21.2

Following regeneration, the media is rinsed to remove any remaining regenerant. The rinse process is conducted in a downward flow mode for about 30 min and uses 2673 to 5347 L/m^3 (20 to 40 gal/ft^3) of resin volume.

The volume of brine solution produced can also be determined in terms of bed volumes (BV). A bed volume as its name implies is the equivalent volume occupied by the media. Therefore, 1 m^3 of resin would have a BV of about 1000 L. The number of bed volumes that a column can operate prior to exhaustion can be found by dividing the exchange capacity by the hardness of the raw water.

Table 22-12 shows data collected by Snoeyink (1984) on quantities of residuals produced by softening ion-exchange plants at six sites. The amount of rinse water used at each plant was in the range of 2673 to 5347 L/m^3 (20 to 40 gal/ft^3) of resin. The TDS concentration of the residual can be estimated as the excess salt plus the cations removed. All of the chloride should be present and the excess sodium not used in exchange. The theoretical NaCl demand is 1.18 kg/kg, and NaCl is 40 percent sodium; therefore, the theoretical sodium demand is 0.472 kg/kg.

TABLE 22-12 Residuals Quantities Produced by Softening Ion Exchange Plants

Plant	Gallons wastewater/1,000 gal. water processed*	Raw water total hardness (mg/L as $CaCO_3$)	Gallons regenerant/ft^3 resin	Gallons rinse/ft^3 resin	Concentration of brine percent	Gallons processed/ft^3 resin-cycle	Dosage (lb salt/ft^3 resin)
Crystal Lake Plant 6	21.9	233	7.3	18.4	11	1220	6.6
Crystal Lake Plant 8	17.2	244	5.1	19.0	15	1400	6.5
Eldon	71.9	75	3.9	61.7	17	750	5.6
Grinnel	49.5	388	14.5	35.0	6	1000	7.2
Holstein	53.5	885	5.7	19.7	14	475	6.6
Estherville	82.8	915	4.4	24.7	15	350	5.5

*Includes backwash water, regenerant waste and rinse water.
Source: Snoeyink, 1984.

Fluoride Removal by Activated Alumina

As indicated earlier, there are several ions other than hardness that can be successfully removed by ion-exchange or adsorption processes. Of these, probably the use of activated alumina for fluoride removal is the most widely practiced. Activated alumina also removes arsenic and selenium. Activated alumina is a mixture of amorphous and gamma aluminum oxide prepared by low-temperature dehydration of aluminum hydroxide.

To regenerate the alumina once it is exhausted with fluoride, a dilute solution of 0.25- to 0.5-M sodium hydroxide is used. Because alumina is both a cation and anion exchanger, Na^+ is exchanged for H^+, which immediately combines with hydroxide to form water (Clifford, 1999).

In order to restore the fluoride removal capacity, the alumina that has been regenerated with sodium hydroxide is acidified by contacting it with excess acid. As an alternative to rinsing the basic alumina with acid, the source water pH can be reduced to 5.5 to 6.0. This avoids the production of an acid rinse residual.

Table 22-13 shows a typical fluoride capacity for activated alumina as a function of the fluoride concentration in order to reach a final fluoride level of 1.4 mg/L.

TABLE 22-13 Fluoride Capacity of Activated Alumina

Raw water fluoride (mg/L)	Raw water TDS (mg/L)	Run length BV	Fluoride capacity (g/m^3)
2.0	810	2300	3700
3.0	1350	1200	3000
5.0	1210	1150	4600

Source: Clifford, 1999.

Similar to ion-exchange processes, prior to regeneration the exhausted contactor is backwashed to remove particles and to restratify the media. Backwashing is accomplished at a flow rate of 19.6 to 22.0 m/h (8 to 9 gpm/ft^2) for 5 to 10 min. This backwash results in about eight bed volumes (BVs) of spent backwash water being produced.

A slow rinse follows both the sodium hydroxide regeneration and the acid rinse. The rinse takes place at about 1.22 m/h (0.5 gpm/ft^3) and is equivalent to about two BV for each rinse. The sodium hydroxide regeneration requires about 1.22-m/h (0.5-gpm/ft^3) flow rate for around 75 min, or about five BV. A 1 percent NaOH concentration is typically used. The acid rinse is also done at a flow rate of 1.22 m/h (0.5 gpm/ft^3) with the use of about 1.5 BV at a 2 percent sulfuric acid concentration. A final rinse of about 30 to 50 BV is used prior to placing the activated alumina back in service.

Snoeyink (1984) presented characteristics of residuals produced at two activated alumina plants as shown in Tables 22-14 and 22-15. In both cases the sodium hydroxide regeneration was done in both an upflow and downflow mode. The acid and final rinse stages were combined by rinsing using raw water that had been lowered to a pH of about 3.2 using sulfuric acid. Also shown are the percentage each residual represents to the total volume of raw water treated.

Removal of Nitrate and Perchlorate Using Anion Exchange

Nitrate and perchlorate can be removed by anion exchange resins. Sodium chloride is used as the regenerate for these systems. Table 22-16 shows the amount of residual produced at several operating nitrate and perchlorate plants as summarized by Min et al. (2004).

TABLE 22-14 Residual Characteristics of Activated Alumina Used to Remove Fluoride at Palo Verde, Ariz*

	F^- (mg/L)	Al (mg/L)	As (mg/L)	$SO^=_4$ (mg/L)	Na (mg/L)	TDS (mg/L)	Mg (mg/L)	Turbidity (ntu)	Ca (mg/L)	Volume percent of flow treated
Backwash	1.8	230	0.210	220	225	880	2.8	17	53	0.6
NaOH upflow regeneration	3.2	13	0.017	184	750	1344	4.3	6.3	30	0.12
Upflow rinse	250	150	0.140	640	740	2629	1.8	20	6.9	0.9
NaOH downflow regeneration	23.5	11	0.030	156	410	1264	0.2	1.0	7.5	0.12
Acid and final rinse	8.4	150	0.475	580	1240	2879	2.9	18	6.8	0.7
Residual composite 3010 gal	91.5	156.4	0.246	469.2	743.5	2130	2.4	17	19.7	2.5

*The terms have been modified from the original source and the residual volume percentages were calculated.
Source: Snoeyink, 1984.

TABLE 22-15 Residual Characteristics of Activated Alumina Used to Remove Fluoride at Gila Bend, Ariz*

	F^- (mg/L)	Al (mg/L)	As (mg/L)	$SO^=_4$ (mg/L)	Na (mg/L)	TDS (mg/L)	Cl^- (mg/L)	Mg (mg/L)	Turbidity (ntu)	Silica (mg/L)	Ca (mg/L)	Cr (mg/L)	Fe (mg/L)	Volume percent of flow treated
Backwash	5.4	2.1	<.002	219	438	1278	590	0.4	300.0	9.5	33.0	0.013	14.5	0.4
NaOH upflow regeneration	5.1	0.42	<.001	206	444	1310	590	0.4	140.0	7.7	32.0	0.016	NA	0.17
Rinse A	6.1	0.22	0.076	739	741	2338	700	0.3	45.0	NA	30.5	0.026	NA	0.08
Rinse B	192	193.3	0.49	721.5	1319	3704	605	<0.1	105.5	5.2	2.0	0.031	NA	0.9
NaOH downflow regeneration	33.4	8.0	0.050	187	528	1453	580	NA	1.0	2.3	4.7	0.021	NA	0.1
Acid and final rinse	5.6	20.1	0.054	162.6	454	1380	570	<0.1	0.38	3.3	1.4	0.010	NA	3.0
Residual composite	41.9	50.7	<0.13	285	624	1832	581	NA	49	NA	5.6	0.015	NA	4.65
Raw water	5.0	0.08	0.014	174	432	1260	570	0.4	0.5	10.6	35.0	0.017	0.11	

*The terms have been modified from the original source and the residual volume percentages were calculated.

Source: Snoeyink, 1984.

TABLE 22-16 Brine Characteristics at Full-Scale Perchlorate and Nitrate Treatment Facilities

Utility	Treatment type	Capacity (gpm)	Contaminants (concentration)	Residual type	Quantity	Residual handling	Permits
A (CA)	Regenerable ion exchange	9000	Nitrate (31-84 mg/L) Perchlorate (~14 μg/L)	Liquid brine	~250,000 gal/day of brine	Brine discharge	NA
B (CA)	Regenerable ion exchange	2500	Nitrate (15 mg/L) Perchlorate (35 μg/L)	Liquid brine	25 gpm of brine	Brine discharge until 2006 Perchlorate brine treatment by 2006	Perchlorate discharge
D (CA)	Regenerable ion exchange	1350	Nitrate (45 mg/L)	Liquid brine	1,700 gal/day of brine	Brine trucking to a WWTP by the equipment provider	NPDES pending Brine permit to WWTP
G (CA)	Regenerable ion exchange	1000	Nitrate (50 mg/L)	Liquid brine	6,000 gal/day	Brine discharge to a WWTP (pending)	NPDES discharge (pending)
H (CA)	Regenerable ion exchange	400–2100	Nitrate (55–70 mg/L)	Liquid brine	140,000 gal/day	Brine discharge to a WWTP	NA
I (CA)	Regenerable ion exchange	3000	Nitrate (34–90 mg/L)	Liquid brine	6,000 gal/day	Regional WWTP	Discharge batch
J (CA)	Regenerable ion exchange	7900	Perchlorate (NA)	Liquid brine	80 gpm of perchlorate brine	Brine discharge until 2006 Brine treatment by 2006	NA

NA = Information not available.
Source: Min et al., 2004.

Approximate values for nitrate and perchlorate in the brine residual are also shown, as well as the disposal method used.

Min et al. (2004) also reported quality analysis on a regenerate brine collected from La Puenta Valley County Water District (LPVCWD), as presented in Table 22-17. The six samples presented in the table were collected consecutively from a pseudo-continuous, countercurrent, ion-exchange process operated by LPVCWD to remove perchlorate. The considerable variability in the spent brine quality could challenge subsequent treatment processes.

TABLE 22-17 Analysis of Individual La Puente Ion Exchange Effluent Brine Samples

	Drum number (in order collected)					
Analysis (mg/L)	No. 1	No. 2	No. 3	No. 4	No. 5	No. 6
Initial analyses						
ClO_4^-	2.913	3.345	3.116	4.315	4.252	3.089
Conduct. (mS/cm)	81.2	95.1	82.9	106.7	107.8	76.1
TDS (%)	5.90	7.09	6.04	8.17	8.28	5.48
Additional analyses						
NO_3^-	1336	489	400	3155	2132	1809
SO_4^{2-}	1560	1252	6001	633	800	517
Cl^-	22638	21623	18671	35522	39166	17074
pH	7.25	0.89	7.18	6.83	4.91	1.24

Source: Min et al., 2004.

The principal options for treating or disposing of nitrate and perchlorate brines are biological degradation followed by discharge, thermal degradation followed by discharge, or direct discharge to a wastewater plant or body of water if allowed. Significant research has been conducted on both degradation processes, but actual applications are limited.

Disposal of Brines

Economical disposal of brine and other ion-exchange and adsorption process waste streams (spent backwash and rinse water) requires a readily available disposal location. Therefore, most large plants that utilize ion-exchange softening are located in coastal communities so that ocean brine disposal is practiced. Ion exchange has been used in small water supply systems in other parts of the country, and residuals have most often been discharged to municipal wastewater systems or to receiving streams.

One of the most common methods to dispose of the residuals is to discharge to a wastewater plant. The effects of changing concentrations of salts in the wastewater due to slug discharge of high TDS brine is of considerable significance. A concentration change of 100 to 200 mg/L of NaCl is rather low and would have no impact on the biological process, whereas a concentration change of 35,000 mg/L NaCl (almost an undiluted discharge of ion-exchange brine) would cause considerable stress to the biological organisms. In terms of impact to biological treatment processes, the effects of brine discharges from water treatment plants should have an insignificant impact on the wastewater treatment process as long as some precautions for equalization and dilution are followed. Toxicity of specific compounds within the waste streams should be carefully monitored and diluted sufficiently to eliminate toxic effects.

In addition to effects on the aerobic biological process, salts can affect anaerobic digestion. Toxicity is normally associated with the cation rather than the anion of the salt. While discussion of cation effects is beyond the scope of this chapter, for additional information a recommended source is the book by Rittmann and McCarty (2001).

Although brine discharge to wastewater systems may be feasible from a biological treatment perspective, there may be other factors that limit the feasibility of this disposal method. If the wastewater is being treated for reuse applications, wastewater utilities may be limited in their ability to accept brine discharges due to effluent TDS limits.

Discharge of the brine to brackish coastal water and to seawater is also a widely practiced method for disposal. Since the brines are primarily a highTDS residual consisting mostly of salt, the permitting process has been fairly easy. Membrane concentrates have been more difficult to permit in some states due to the requirements to pass acute toxicity tests. These tests could also be applied to ion-exchange brines which would make permitting of their disposal more difficult. Mickley (2000) has described extensively the requirements and problems related to toxicity testing on membrane concentrates.

In general, there are no economically feasible technical options for specifically removing excess ions from concentrate within the existing economics of the municipal drinking water industry. In site-specific instances, there may be means of providing enhanced dilution of the concentrate prior to or at final discharge through blending or the use of end-of-the-pipe diffusers (Mickley, 2000).

RESIDUALS CONTAINING ARSENIC

All water systems must supply finished water with less than 0.01 mg/L arsenic according to the Safe Drinking Water Act, (USEPA, 2003); see Chap. 1 on standards and Chap. 2 on health effects. For systems that utilize sources with arsenic concentrations above this regulatory limit, some form of treatment is utilized for arsenic removal. These technologies produce a number of different types of residuals, including both solids and liquids. The feasibility of certain arsenic removal treatment technologies for a given utility may be strongly influenced by the available disposal options for the arsenic residuals. When selecting treatment options, it is critical to evaluate the types of residuals that would be generated, their expected arsenic concentrations, and the residuals pretreatment strategies that would be required prior to final disposal.

Table 22-18 shows the options available to dispose of these materials produced by treatment technologies. In general, arsenic in the residual material can limit the options available for disposal. The material must be analyzed for leaching potential using the TCLP procedure (EPA Test Method 1311) (USEPA, 1991a). The arsenic in the leachate must not exceed 5 mg/L, or it will be categorized as a hazardous waste. In general, removal technologies that produce residual solids will not fail the TCLP; adsorption media will not fail because the arsenic is tightly bound, and coagulant sludges generally do not have levels high enough to fail this test. However, iron or alum coagulation processes treating waters specifically for arsenic can have high enough arsenic concentrations in the residuals to fail the TCLP. In California, the Waste Extraction Test (WET) is used to determine the toxicity of a material, and so, to determine if the waste must be handled as a hazardous material (State of California, 2005). California's WET procedure is similar to that described in the TCLP, but it uses more stringent conditions. Therefore, some sludges could pass the TCLP, but fail the WET.

If the material does fail the TCLP, then it can only be disposed of at a hazardous waste facility. It may be difficult to find such facilities that will accept small quantities of materials, and disposal costs at facilities that do accept these wastes can be high. Transportation to these facilities is regulated under the Hazardous Materials Transportation Act, which covers how the material is packaged, the manifests that must accompany it, and the records that must be kept.

TABLE 22-18 Summary of Residuals and Management Methods for Arsenic Treatment Technologies

Treatment technology	Form of residual	Type of residual	Possible disposal methods
Anion exchange	Liquid	Regeneration streams • Spent backwash • Spent regenerant • Spent rinse stream	Sanitary sewer Direct discharge Evaporation ponds/lagoon
	Solid	Spent resin	Landfill Hazardous waste landfill Return to vendor
Activated alumina	Liquid	Regeneration streams • Spent backwash • Spent regenerant (caustic) • Spent neutralization (acid) • Spent rinse Liquid filtrate (when brine streams are precipitated)	Sanitary sewer Direct discharge Evaporation ponds/lagoon
	Solid	Spent alumina	Landfill Hazardous waste landfill Land application
		Sludge (when brine streams are precipitated)	
Media adsorption	Liquid	Spent backwash Spent regenerant Spent rinse stream	Sanitary sewer Direct discharge Evaporation ponds/lagoon
	Solid	Spent media Solids from backwash	Landfill or Hazardous waste landfill
Iron and manganese removal processes	Liquid	Filter backwash	Direct discharge Sanitary sewer Evaporation ponds/lagoons
	Solid	Sludge (if separated from backwash water)	Sanitary sewer Land application Landfill
		Spent media	Landfill
Membrane processes	Liquid	Brine (reject and backwash streams)	Direct discharge Sanitary sewer Deep well injection Evaporation ponds/lagoon

Source: SAIC, 2000.

Regulations

Arsenic residuals management methods are classified into two groups depending on whether they are liquid residuals or solid/sludge residuals. In the case of liquid residuals, disposal methods may include direct discharge to a water body, indirect discharge via sanitary sewer system, underground injection, management in lagoons, and possible land application. Solid/sludge residuals may be disposed through landfill land application, storage in lagoons, and reactivation, recovery, or disposal of spent resin. Federal and state regulations apply to handling, transportation, and disposal of arsenic-laden residuals. Most states have the administrative authority to implement federal regulations, and are then required to establish and administer regulations meeting the requirements of these acts. Waste disposal regulations are primarily the responsibility of the states, but the state regulations must meet or exceed the federal regulations.

If the water treatment residual has failed the TCLP test because of arsenic concentrations, then on-site lagoons would be regulated by RCRA. Under RCRA, USEPA has established criteria that prohibit practices that contaminate surface or groundwater (40 CFR 257) and also established comprehensive design and operation standards applicable to surface impoundments (USEPA, 1991b).

Residuals Quantities and Characteristics

Table 22-19 provides a summary of example arsenic concentrations in water treatment residuals (Amy et al., 2000). The residuals volumes and arsenic concentrations shown in the table for various types of residuals were calculated assuming a raw water arsenic content and arsenic removal for each treatment technology. Arsenic concentrations in residuals volumes generated in processes listed in Table 22-19 ranged from 0.098 mg/L for membrane technologies to approximately 10 mg/L for activated alumina and ion exchange. On a dry weight basis, theoretical arsenic concentrations ranged from 165 to more than 14,000 mg/kg.

Table 22-20 shows a summary of the key water quality results obtained in USEPA (2000) for residuals samples collected from various arsenic treatment techniques. Results showed that about 99 percent of the arsenic in the ion-exchange regenerant samples was in the dissolved form. Almost none of the arsenic in the activated alumina regenerant stream and the SFBW samples was dissolved. The concentrate samples collected from the RO and NF plants had low arsenic concentration.

The concentration factor, or degree to which arsenic levels were concentrated in the residuals by the various treatment processes, depends on the arsenic removal process utilized. USEPA (2000) reported arsenic concentration factors as low as 12 for SFBW/activated carbon filter samples and as high as 270 for ion-exchange residuals. Arsenic concentration was greater for brine streams than for waste streams from other processes.

TABLE 22-19 Summary of Example Residuals Quantity from Arsenic Processes

Treatment technology	Volume of residuals produced (gal/MG)	As concentration in residuals volume (mg/L)	Quantity of solids produced (lb/MG)	As concentration in solids (mg/kg dry weight)
Conventional coagulation	4300	9.25	180	1850
Softening	9600	4.2	2000	165
Ion exchange	4000	10	5.2	14,250
Activated alumina	4200	9.52	23.4 (calculated)	14,250 (calculated)
Iron-based sorbents	21,000	1.9	23.4 (calculated)	14,250 (calculated)
NF/RO	664,000	0.098	NA	NA
Coagulation/MF	52,600	0.76	112.6	2957

NA = Not applicable.
lb/MG × 0.12 = kg/ML.
Source: Amy et al., 2000.

TABLE 22-20 Arsenic Residuals Sample Characterization

	Untreated residuals characteristics									
Sample type	pH	Alk*	Total hardness	TDS (mg/L)	Total As (mg/L)	Dissolved As (mg/L)	Total Fe (mg/L)	Total Mn (mg/L)	Conductivity (S/cm)	Sulfate (mg/L)
AA regenerant	7.1	268	13	10,240	2.63	0.12	0.83	0.09	22,640	16,338
SFBW (A)	7.6	430	365	460	1.41	<0.002	78.5	7.52	900	4.82
SFBW/ACF (B)	8.1	197	400	341	1.74	0.03	45.9	3.75	680	97.3
RO (A)	7.9	2800	460	14,300	<0.002	<0.002	0.65	0.23	28,500	544
RO (B)	7.3	600	840	11,750	<0.002	<0.002	0.86	1.1 1	23,800	—
NF (A)	7.1	325	1560	1765	0.013	0.007	2.16	0.14	3515	1075
NF (B)	6.6	210	1750	1533	0.005	0.009	0.46	0.08	3080	1190
Ion Ex (B)	9.7	7000	86	6240	24.8	24.7	<0.01	<0.005	8100	910
Ion Ex (A)	9.0	950	90	4100	10.5	10.3	0.49	—	12,440	—

*mg/L as $CaCO_3$.
AA = Activated Alumina.
Source: Adapted from USEPA, 2000.

Treatment of Arsenic-Containing Liquid Residuals

Liquid wastes from arsenic removal processes often have arsenic levels higher than allowed for discharge to streams or land. In this case the arsenic must be removed prior to discharge. Figure 22-22 illustrates some of the possible treatment processes for brines generated from activated alumina and ion-exchange regenerants as well as for concentrates from NF and RO. The treatment processes for the liquid wastes are essentially the same as for treatment of the source water: either adsorption, or ion exchange, or precipitation. Extensive testing has been done to identify methods to treat these arsenic residuals (USEPA, 2000; MacPhee, 2001). A summary of the best treatment technology determined for each residual type is presented in Table 22-21.

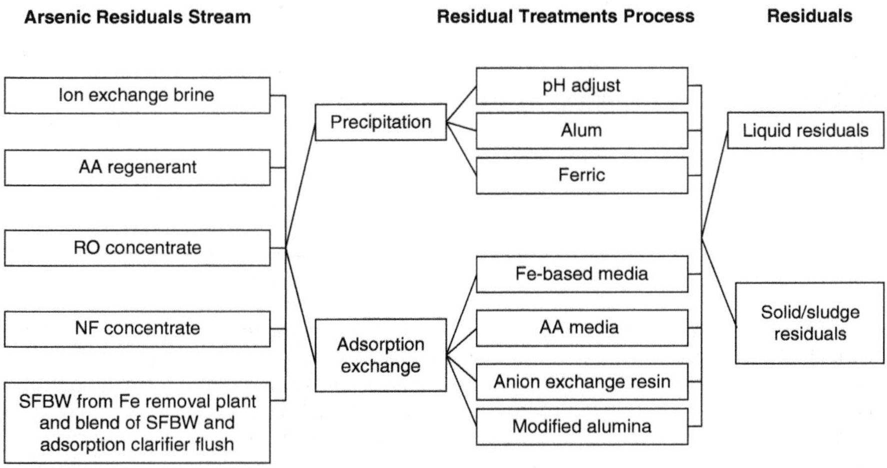

FIGURE 22-22 Arsenic residuals treatment options. (*Source: Adapted from USEPA, 2000.*)

TABLE 22-21 Summary of Treatment Results for Removing Arsenic from Liquid Arsenic Waste

Sample type	Best treatment conditions determined from testing*	Total As remaining (mg/L)
AA regenerant	None	0.154
Ion Ex (A)	None	1.28
Ion Ex (B)	None	18.7
RO (A)	Ferric chloride precipitation	0.041
RO (B)	Iron media adsorption	0.018
NF (A)	Ferric chloride precipitation, iron-based media or AA adsorption	0.009, 0.030
NF (B)	Iron media adsorption, ferric chloride precipitation	<0.002, 0.005
SFBW (A) (settled)	Ferric chloride precipitation	0.013
SFBW/ACF (B)		
(unsettled)	Gravity settling (no chemical addition)	0.043
(settled)	Iron media, ion exchange, or AA adsorption	<0.002

*Goal was to reach 0.05 m/L arsenic which is the discharge standard for many states.
AA = Activated Alumina.
Source: Adapted from MacPhee et al., 2001.

Handling of Solids from Arsenic Removal

As discussed, several solid residuals can be produced by arsenic removal systems. Spent exhausted media can be produced by activated alumina and iron-based sorbents systems. Generally a contract can be established for the media vendor to take back the spent material. They will often supply replacement material at a cost less than normal. If the media cannot be returned to the vendor, then it needs to undergo TCLP or WET (for systems in California) testing and be disposed of properly.

Other solids produced can be coagulant solids from treating liquid residuals or coagulant solids from treatment processes removing arsenic from the raw water such as coagulation followed by filtration or microfiltraiton. These residuals can go to a sewer, be dewatered and disposed of, or be stored in a lagoon.

RESIDUALS CONTAINING RADIOACTIVITY

The revised Radionuclides Rule came into effect on December 8, 2003, mandating that all community water systems (CWSs) come into compliance with this regulation. This rule retained the maximum contaminant levels (MCLs) for combined radium-226/228, gross alpha particle activity and combined beta particle and photon radioactivity and set a new MCL for uranium. The MCLs are presented in Chap. 1.

Systems can use a number of treatment alternatives to come into compliance with this regulation. Best available technologies (BATs) for radionuclide removal, and the residuals associated with these processes, are shown in Table 22-22. Treating water for naturally occurring radionuclides will result in residual streams that are classified as "technologically enhanced naturally occurring radioactive materials" (TENORM). TENORM is defined here as naturally occurring materials, such as rocks, minerals, soils, and water whose radionuclide concentrations or potential for exposure to humans or the environment is enhanced as a result of human activities (e.g., water treatment).

TABLE 22-22 Summary of Treatment Technologies for Removal of Naturally Occurring Radionuclides in Water

Treatment technology	Contaminant removed	Removal efficiency (percent)	Wastes produced	Waste concentrations
Cation exchange	Radium	85–97	Rinse and backwash water Regenerant brine	8 to 9 pCi/L-Ra[H] 50 to 3,500 pCi/L- Ra[H] 22 to 94 pCi/L[I]
Anion exchange	Uranium	95	Rinse and backwash water Brine regenerant solution	2 to 6e+06 pCi/L[H]-U 35 to 4.5e+06 pCi/L[H]-U 1.3 to 11 pCi/L[†]
Lime softening	Radium Uranium	90 85 – 90[‡]	Sludge (at clarifier) Sludge (dry) Filter backwash	76 to 4577 pCi/L-Ra 1 to 21.6 pCi/g-Ra 1 to 10 pCi/g-U 6.3 to 21.9 pCi/L-Ra
Reverse osmosis	Radium Uranium	90+ -ND	Reject water	7 to 43 pCi/L-Ra 200 to 750 pCi/L-U
Electrodialysis	Radium Uranium	90 ND	Reject water	No data
Iron removal Oxidation	Radium	0 to 70[§]	Solids and supernatant from filtration backwash	12 to 1980 pCi/L-Ra
Greensand			Green sand media	28 to 250 pCi/g-Ra
Selective sorbents	Radium Uranium	90+	Selective sorbents (radium selective and activated alumina)	Up to 3.6 pCi/g-Ra
Coagulation/ Filtration	Uranium	50 to 85	Sludge	10,000 to 30,000 pCi/ L-U

[H]Peak values.

[†]Average for given waste forms.

[‡]May be increased to 99 percent by the presence or addition of magnesium carbonate to the water.

[§]May be increased to 90 percent by passing the water through a detention tank after the addition of potassium permanganate prior to filtration.

Source: USEPA, 2005a: data extracted from USEPA 1982, 1986, 1994, 1995; Wade Miller Associates, 1991; and Reid 1985.

Characterization

The Spreadsheet Program to Ascertain Residuals Radionuclide Concentrations (SPARRC) can be used to estimate the waste volumes and concentrations of radionuclides in the residuals from water treatment plants removing naturally occurring radionuclides. The current version of SPAARC (available at www.npdespermits.com/sparrc) incorporates predictive algorithms to estimate radionuclides and cocontaminant removals and focuses on a sound estimate of residual radionuclides concentrations and cooccurring pollutants rather than sizing and designing drinking water treatment technologies. The program allows the operator to select the type of treatment process, as well as input and output parameters such as water flows, doses of coagulant and polymer, and filter capacities. Table 22-23 shows the processes considered.

TABLE 22-23 Software Program to Ascertain Residuals Radionuclide Concentration Elements

Technology	Radionuclides	Co-contaminant
Coagulation/Filtration	Uranium	Arsenic
Lime softening	Radium and uranium	None
IX	Radium, barium, and uranium	None
RO	Radium and uranium	None
Activated alumina	Uranium	Arsenic
Greensand filtration	Radium and barium	None

Source: USEPA, 2005a.

Regulations

No federal agency currently has the legislative authority to regulate the disposal of wastes generated by water treatment facilities on the basis of the residuals' naturally occurring radionuclide content. The U.S. Nuclear Regulatory Commission (USNRC) regulates licensed radionuclide materials such as would be generated from a nuclear power plant. The Atomic Energy Act (AEA) of 1954 and its subsequent amendments provided the legislative authority for this agency to do so. The USNRC regulates the handling and disposal of both high-level and low-level radioactive wastes that are generated by various human-made processes. High-level radioactive waste includes such things as spent fuel rods from nuclear power plants. Low-level radioactive waste includes such things as clothing and building materials used in nuclear processes. The regulations governing both high-level and low-level radioactive wastes deal only with human-made radioactivity and source material.

Uranium and thorium are considered source material (42 USC 2014(z)) and so are subject to NRC or Agreement State licensing and regulation. Water treatment systems that produce residuals containing uranium or thorium must comply with this licensing requirement. However, source material of an unimportant quantity (10 CFR 40.13) is *exempt* from NRC or Agreement State regulation if the uranium or thorium makes up less than 0.05 percent by weight (or approximately 335 pCi/g for natural uranium) of the material. These limits apply to both liquid and solid residuals.

The USEPA regulates the disposal of hazardous solid waste under the Resource Conservation and Recovery Act (RCRA). But radioactivity in solid wastes is not one of the classifications used by RCRA to determine if a particular substance is hazardous. Activated carbons and sludges from water treatment could not be categorized as hazardous under RCRA because of radionuclide activity. No other federal law specifically regulates disposal of waste containing elevated levels of naturally occurring radioactive materials at this time.

State TENORM Regulations

Most states do not specifically regulate water plant wastes on the basis of elevated levels of radioactivity. A few states have enacted restrictions on the disposal of solid wastes containing specific naturally occurring radionuclides.

The states of Illinois and Wisconsin have developed disposal criteria of water treatment plant wastes containing radium. Other states, including New Hampshire, have disposal criteria for wastes containing high levels of NORM and TENORM radiation from uranium and radium. *Naturally occurring radioactive materials* (NORM) include primordial radionuclides that are naturally present in the rocks and minerals of the Earth's crust and cosmogenic radionuclides produced by interactions of cosmic nucleons with target atoms in the atmosphere and in the Earth. TENORM refers to NORM that has been concentrated or changed due to the exposure to the environment and by processes such as water treatment.

Most states do not have specific TENORM regulations and regulate it the same ways as all other sources of radiation. In the absence of specific regulations or guidance, water suppliers would be required to dispose of radioactive residuals in accordance with existing solid waste or, where applicable, hazardous waste requirements.

The U.S. Department of Transportation (USDOT) regulates the shipment of any radioactive waste (USDOT, 1976). DOT is a possible regulatory authority if the waste is shipped off-site for disposal. The waste can be considered radioactive by USDOT if (1) a state authority has designated the waste as radioactive or (2) the radioactivity exceeds USDOT established levels. DOT defines a radioactive waste as a material that has a specific activity greater than 2000 pCi/g. Also, if a state designates a waste as radioactive then USDOT regulations apply. In such cases, shipment must be according to 49 CFR Part 172.392, which requires that the waste be packaged in leak-proof containers with acceptable levels of external radiation and transported in appropriately marked vehicles.

Ultimate Disposal Options

Table 22-24 summarizes the disposal options for radioactive wastes and notes possible constraints for these methods due to radionuclides.

TABLE 22-24 Common Disposal Considerations for Residuals Containing Radioactivity

Direct discharge	• System must have a NPDES permit. • Systems must meet state radionuclides limits.
Discharge to a POTW	• Systems must meet the requirements of the POTW and meet state permitting requirements.
Underground injection	• Class I hazardous injection wells may be a disposal option for radioactive or hazardous wastes. • Systems should check with their state to determine whether the state has more stringent UIC requirements. • USEPA has the authority to take action on any residential waste disposal system if the system introduces contaminants into a USDW whose presence or likely presence causes an imminent and substantial endangerment to public health.
Landfill disposal	• Systems must check with their states to determine whether landfilling is an acceptable means of disposal for hazardous and nonhazardous solid waste containing radionuclides.

Source: USEPA, 2005a.

Mixed Waste

Mixed waste is regulated under RCRA and the Atomic Energy Act (AEA) of 1954. Mixed waste "contains both hazardous waste and source . . . or byproduct material subject to the Atomic Energy Act of 1954" (42 USC 6903.41). Therefore, systems generating waste containing uranium or thorium (source material) as well as hazardous waste could potentially have a mixed waste. If wastes contain licensable amounts of source material (any concentration exceeding the "unimportant quantity" in 10 CFR 40.13 (a)) and hazardous waste, these wastes must be disposed of at a facility authorized to accept mixed waste. Because there are limited disposal pathways, generation of a mixed waste should be avoided if at all possible. (USEPA, 2005).

ULTIMATE DISPOSAL AND UTILIZATION OF SOLIDS

The treatment and disposal of water treatment plant residuals are rapidly becoming integral parts of operating water treatment plant facilities as local, state, and federal regulations are requiring more stringent standards for traditionally practiced residuals management programs. The discharge of untreated residuals to most surface waters is severely restricted under the National Pollutant Discharge Elimination System (NPDES) of the Clean Water Act. Discharge of residuals to the sanitary sewer is equally becoming more restrictive through wastewater pretreatment standards, wastewater treatment plant available capacity, digester ability to handle the mostly inorganic residuals, and the wastewater plants' effluent standards. Sanitary landfill disposal of dewatered residuals has become costly, and the residuals take up valuable space in a disposal system that is already predicting severe shortages in the near future. As a result, beneficial use programs for residuals are increasingly being considered by utilities, not only as a cost-effective alternative but also as a publicly acceptable management practice.

The desired residuals management goal for any water utility is to operate an economically efficient and environmentally attractive residuals management plan by developing one or more long-term agreements that will allow for the proper utilization of residuals in a beneficial application. The beneficial use markets that have exhibited the greatest potential for success are the following:

- *Commercial products.* The use of residuals in the making of commercial products has been successful in such areas as turf farming and topsoil blending. In both cases, the residuals are a substitute for natural soil material and offer economical benefits to the producer. Details on beneficial use options for water plant residuals is found in Cornwell et al. (2000).
- *Co-use with biosolids.* Incorporating residuals in biosolids management programs such as land application and composting can be beneficial. Blended products tend to have lower metal concentrations making the product more marketable. Also, for utilities that operate both a water and wastewater facility, permitting, record keeping, and monitoring requirements are reduced when all the residuals are managed under one program. Co-composting of biosolids and residuals has been successfully practiced at blends up to 25 percent residuals.
- *Land application.* Land application of residuals to agricultural or forested land is a feasible beneficial use alternative. Land application of lime residuals has been practiced for many years due to the value of lime addition to farmland (Russell, 1975). Land application of coagulant residuals has also been successful whether blended with biosolids or applied alone.

Application of residuals to turf farms provides a residuals management option which can benefit a farmer's harvest and supplements the farmed areas with a new soil base. Turf grass has a relatively low nutrient demand but requires significant moisture levels, particularly for initial growth phases. Dewatered residuals applied to a turf farm at the beginning of the seeding process can provide excellent water retention capabilities. The preferred method of application is with a manure-type spreader. The application rate is a function of the type of grass, supplemental fertilization, and residuals quality. A 25-mm (1-in) application depth of residuals at 20 percent solids concentration equals a solids loading rate of 38.1 dry metric tons per hectare (17 dry tons per acre) or a mass loading rate of 1.7 percent.

Greenhouse and leaching experiments undertaken by the University of Buffalo (Van Benschoten et al., 1991) showed the success of using residuals in grass growing. Residuals

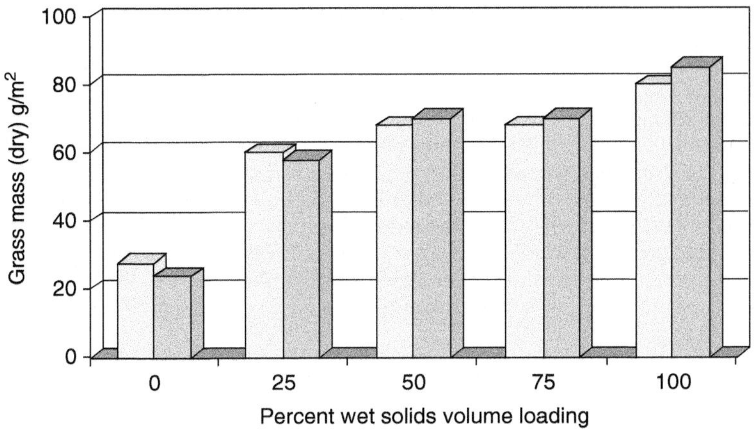

FIGURE 22-23 University of Buffalo greenhouse study: Grass mass versus percent residuals loading. Note: each bar shows the sum of three harvests and represents a unique test. *(Source: Van Benschoten et al., 1991.)*

mixed with native soils at 25, 50, 75, and 100 percent loadings clearly enhanced turf grass growth in the greenhouse studies. As shown in Fig. 22-23, measured turf grass growth in residuals/soil mixtures was up to 3 to 4 times the growth observed in native soils. This improvement in yield translates into a turf farmer being able to bring his fields to harvest sooner and therefore provide a marketable turf product sooner than experienced under current harvesting conditions. Increases in labile phosphorus (as determined in Bray-1 phosphorus tests), indicative of a beneficial increase in soil of phosphorus available to plants, followed the recorded increases in grass mass.

Commercial producers of topsoil utilize a variety of raw soil products to develop a marketable product for nurseries, homeowners, professional landscapers, and so on. In this process, the raw soils are screened and blended with some organic material before being sold as a product. Residuals can be blended during the topsoil production process to increase the nutrient value and water retention capabilities. To a topsoil producer, the residuals are a raw material that can be obtained at little or no cost and increases the profit margin on the final product. The amount of residuals added to the topsoil may be 10 percent or less and is a function of consistency, quality, and availability. The acceptable quality of the residuals is determined by individual topsoil producers. Earthgro Corporation in Pennsylvania is a commercial topsoil producer and has established various classes of quality at which residuals can be accepted as shown in Table 22-25.

TABLE 22-25 Topsoil Blending: Metal Content Limits (ppm)

Parameter	Preapproved	Requires review	Not accepted
Cadmium	<2.0	2.0–25.0	>25.0
Chromium	<1000	—	>1000
Copper	<100	100–1000	>1000
Lead	<200	200–400	>400
Mercury	<0.3	0.3–10.0	>10.0
Nickel	<200	—	>200
Zinc	<300	300–1200	>1200

Blending residuals with topsoil can be a mutually beneficial operation for the utility and the producers. Key elements for long-term success are reasonably consistent solids quality, reliable delivery of dewatered residuals, and a strong contractual arrangement between all parties.

Residuals can be mixed with biosolids and subsequently be a coproduct in the overall end-use management strategy of the biosolids. For a utility that operates both water and wastewater facilities, this type of residuals management program has certain benefits including (1) it avoids separate permitting and monitoring of the residuals, (2) it provides a beneficial and often cost-effective end-use avenue for the residuals, and (3) it reduces most of the metal concentrations in the biosolids product because of the diluting effect of the residuals. Even in situations where the water and wastewater facilities are owned and operated by separate entities, the residuals can enhance the quality of the biosolids product and provide a source of revenue. Residuals can be incorporated with biosolids in a variety of methods, including

- Discharge of liquid residuals to the sanitary sewer
- Discharge of liquid residuals at the wastewater plant influent
- Discharge of liquid residuals at the wastewater plant solids handling stage
- Blending of dewatered residuals with dewatered biosolids

Inorganic water plant residuals can significantly impact primary settler overflow quality, digester space, and digester efficiency. Residuals should also not degrade the end-use biosolids product quality such as lowering nutrient values or increasing/introducing higher metal concentrations. This is particularly true for land application and composting processes.

The land application of residuals can be an environmentally safe and cost-effective method for residuals management. Due to the variable characteristics of different raw water sources, allowable residuals application rates must be considered according to each individual residuals composition. The primary concerns associated with land application of coagulant residuals are due to the presence of heavy metals and the potential for phosphorus binding. Alum residuals, when added to soil, have the capabilities to absorb inorganic P which prohibits plants to extract P from the soil for plant growth. However, with good soil management and crop selection, P depletion can be prevented with proper loading rates and P fertilization. The aluminum concentration of WTP residuals can range from 5 to 15 percent of the total dry solids mass, which is 50 to 100 percent higher than the concentration of the aluminum in most soils (Elliott and Dempsey, 1991). Elliott and Dempsey (1991) determined that a 22.4 metric ton per hectare (10 ton per acre) residuals loading rate with residuals Al concentration of 30 percent, the soil Al level was reported to only increase by 0.3 percent. Aluminum phytotoxicity is dependent on Al solubility and is not a problem when the pH range in the soils is maintained between pH 6 and 6.5 (Elliott and Dempsey, 1991).

The addition of alum residuals to soils could also change the soils physical structure, or bulk density. A soil with a high bulk density is more compact which is unfavorable to plant growth because root penetration is restricted. A low bulk density (less compacted soil) has more pore space for air and water which is beneficial for plant growth (Tisdale and Nelson, 1975). Rengasamy et al. (1980) found that soil mixed with alum residuals increased soil aggregation and moisture retention and, as a result, increased the dry yield of maize. Bugbee and Frink (1985) demonstrated that improvements in aeration and moisture retention promoted by the addition of alum residuals were made to offset the phosphorus deficiency in lettuce.

The impact of land application on various crops has been investigated by Virginia Polytechnic Institute and State University. Novak (1993) studied the impact of alum and PACl residuals on corn when applied at 1.3 to 2.5 percent by dry weight (29.1 to 56.0 metric tons per hectare (13 to 25 tons per acre)). Crop yields from the treated plots were not statistically different from the untreated plots. Mutter et al. (1994) studied the impact of

PACl residuals on wheat when applied at 2, 4, and 8 percent. No negative effects on wheat grain or biomass yield were observed at these loading rates. Although soil aluminum levels were significantly increased at these loading rates, after two crop rotations the soil Al concentrations were found to be similar to background Al levels. The wheat leaf tissue was increased slightly in Al concentration as compared with the control. This increase in Al concentration, however, was not found to be significant.

A detailed description of the processes involved for land application of water residuals are outlined in *Land Application of Water Treatment Plant Sludges* (Elliott et al., 1990). The AwwaRF land application guide provides a good source of technical information concerning the principles and design associated with land-applying residuals. Implementation logistics and residuals quality requirements are also summarized in the report. A number of land application research studies conducted using coagulant residuals have demonstrated neutral or slightly positive impacts on crop growth (Lin and Green, 1990; Geertsema, 1994). Some of the benefits associated with the addition of coagulant residuals to agronomic soils include

- Improvement to soil structure
- Soil pH adjustment
- Addition of trace minerals
- Increased moisture holding capacity
- Soil aeration

Land application of softening sludge has been practiced for years. Farmers were allowed to cart away dewatered softening sludge from a plant in Ohio as early as the 1950s (Reeves, 1980).

In farming regions, the application of nitrogen fertilizers causes a reduction in soil pH. If optimum pH conditions do not exist, crop yields are reduced. Therefore, farmers must apply sufficient quantities of calcium carbonate as a means of counteracting the fertilizer applications. For each 45.4 kg (100 lb) of ammonia fertilizer, 1.4 to 1.8 kg (3 to 4 lb) of limestone must be applied (Russell, 1980).

Several state agencies have evaluated the neutralizing power of softening sludges versus commercial limestone. In 1969, the Ohio Department of Health reported that the total neutralizing power of lime sludge is greater than that of marketed liming materials (Ohio, 1969). To bring the soil pH into the desirable range, 2721 kg/4047 m^2 (3 tons/acre) lime, or about 9070 kg (10 tons) of lime sludge at a 30-percent solids concentration, are required. Subsequent lime applications are required to maintain the desired pH. In Illinois, a calcium carbonate equivalent test (CCE) performed on several softening sludges indicated that the softening sludges were superior to agricultural limestones available locally. Since softening sludges contain a high quantity of calcium carbonate and offer a high degree of neutralization, this resource should be used when it is practical for soil conditioning. The addition of softening sludge also increases the porosity of tight soils, making them more workable for agricultural purposes (Russell, 1980).

Landfilling of water plant solids is the disposal choice when beneficial options cannot be obtained. Landfilling of water plant residuals has three basic categories.

- Sludge monofills
- Codisposal in a municipal or industrial landfill
- Hazardous waste landfills

Since water plant solids are usually not found to be hazardous, the last category is not used. Codisposal into a municipal or industrial waste landfill is a matter of contacting the available landfills in the area and determining these requirements and their tipping fee. Generally landfills will require that the sludge be tested to prove it is nonhazardous, and it must be of a high enough solids concentration that there is no free water. Due to the costs of codisposal landfills, some utilities are building monofills. Monofills are disposal sites specifically constructed to hold water plant residuals. The two major types of sludge monofills are trench filling and area filling.

In trench landfills, sludge is placed entirely below the original ground surface. The trench depth is dependent typically on the depth to groundwater and bedrock to maintain sufficient soil buffers between the sludge and substrata. Trench depth is a function of sidewall stability and equipment limitations as well. The trenching method encompasses both narrow and wide trench type disposal areas, which range in width from about 1 to 15 m (several ft to 50 ft) .

Unlike the trench landfilling technique, where sludge is placed below ground, in area filling techniques sludge is generally placed above the original surface. Area filling may be accomplished in one of three ways.

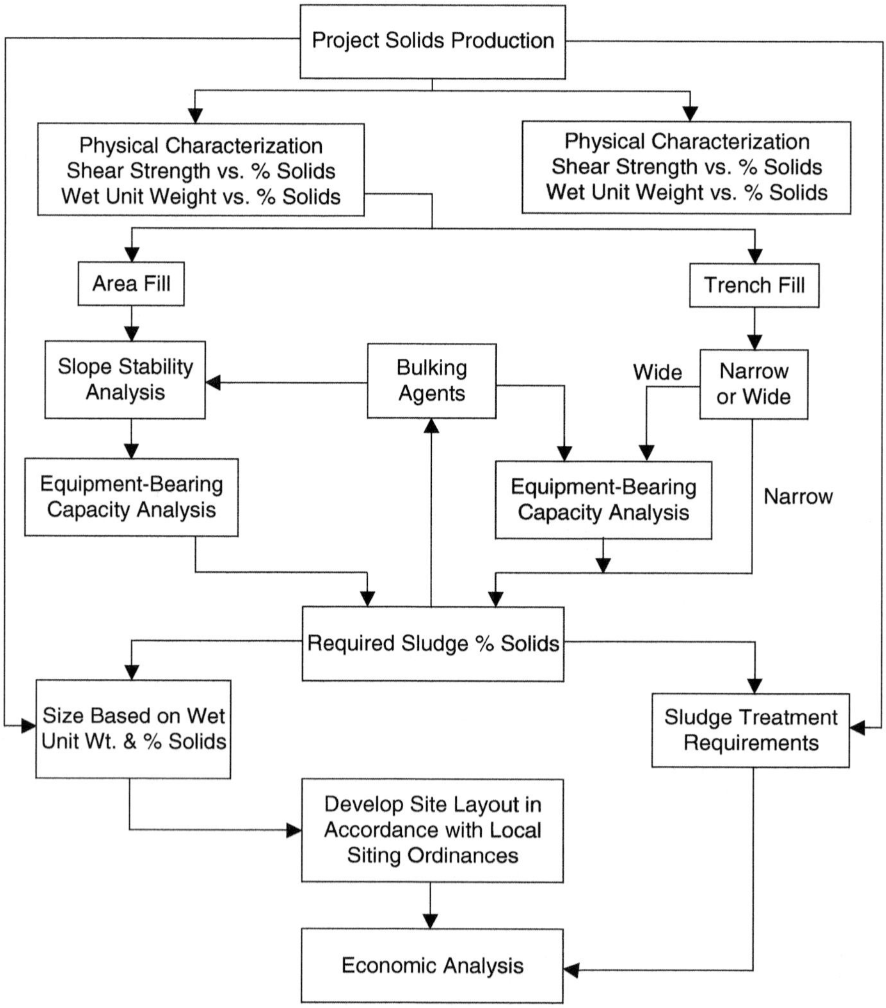

FIGURE 22-24 Considerations for sludge monofill design.

- *Area mound.* Sludge is mixed with soil such that it becomes stable enough to be stacked in mounds.
- *Area layer.* Sludge is spread evenly in layers over a large tract of land.
- *Diked containment.* Earthen dikes are constructed above ground to form a containment structure into which the sludge can be disposed.

Although solids content is not limited in area fill landfills, the requirement that sludge must be capable of supporting equipment due to the lack of sidewall containment necessitates reasonably good sludge stability and bearing capacity. These characteristics are typically achieved through good dewatering, dewatering followed by air drying, or mixing sludge with bulking agents. A combination of these methods can also be employed. Areas with high water tables and those with bedrock close to the surface are particularly amenable to area fill methods of sludge monofilling.

In the area fill layer disposal method, sludge is spread in 15.2- to 30.5-cm (6- to 12-in) layers. This in turn allows additional air drying of the sludge to achieve higher solids concentration and shear strengths. This method would seem favorable for coagulant sludges, which are typically difficult to dewater. The layering method eliminates the need for a separate air-drying area outside the monofill, provided the monofill cell is large enough. The area fill layering technique usually results in relatively stable fill areas when completed and, therefore, requires less extensive equipment and manpower efforts for maintenance than the area fill mound technique.

Figure 22-24 presents a schematic diagram that highlights the essential planning considerations for disposal of water plant sludge in a monofill. The diagram shows the determining factors and the effects of each decision in the planning process as indicated on the figure. Sludge treatment requirements and project economics are affected not only by the quantity of sludge produced but also by the sludge's physical and chemical characteristics.

ABBREVIATIONS

AEA	Atomic Energy Act
BAT	best available technology
BOD	biochemical oxygen demand
BV	bed volume
CEBW	chemically enhanced backwash
CIP	clean-in-place
CST	capillary suction time
CWS	community water systems
DAF	dissolved air flotation
DO	dissolved oxygen
DOC	dissolved organic carbon
ED	electrodialysis
EDR	electrodialysis reversal
EQ	equalization
FBRR	Filter Backwash Recycling Rule
GAC	granular activated carbon
IPP	Industrial Pretreatment Program
MCL	maximum contaminant level
MDL	method detection limit
MF	microfiltration
NF	nanofiltration

NOAA	National Oceanic and Atmospheric Administration
NORM	naturally occurring radioactive materials
NPDES	National Permit Discharge Elimination System
ntu	nephelometric turbidity unit
PAC	powdered activated carbon
PACl	polyaluminum chloride
POTW	publicly owned treatment works
RCRA	Resource Conservation and Recovery Act
RO	reverse osmosis
SAC	strong acid cation
SAR	sodium adsorption ratio
SFBW	spent filter backwash water
SPARRC	Spreadsheet Program to Ascertain Radionuclides Residuals Concentration
SR	specific resistance
TCLP	toxicity characteristic leach procedure
TDS	total dissolved solids
TENORM	technologically enhanced naturally occurring radioactive materials
TOC	total organic carbon
TTF	time to filter
TTHMFP	total trihalomethane formation potential
UF	ultrafiltration
UIC	underground injection control
USDOT	U.S. Department of Transportation
USDW	underground source of drinking water
USEPA	U.S. Environmental Protection Agency
USNRC	U.S. Nuclear Regulatory Commission
WAC	weak acid cation
WET	waste extraction test
WQBEL	water quality based effluent limit

NOTATION FOR EQUATIONS

A	additional chemicals added such as polymer, clay or activated carbon
Al	alum dose as 17.1 percent Al_2O_3
$CaCH$	calcium carbonate hardness removed as $CaCO_3$
$CaNCH$	calcium noncarbonate hardness removed as $CaCO_3$
d	particle diameter
D	total bowl diameter
D_d	drained depth
ΔD_e	required depth change due to evaporation
D_i	initial depth of residuals at loading
$d(z)$	thickness of the sludge layer
$D(z)$	total depth of sludge that can be frozen
DOC_r	dissolved organic carbon removed
E	net pan evaporation
F	latent heat of fusion
Fe	iron dose as Fe
h	convection coefficient
K	conductivity coefficient
M	mass of dry solids

MgCH	magnesium carbonate hardness removed as $CaCO_3$
MgNCH	magnesium noncarbonate hardness removed as $CaCO_3$
N	number of leads
P	percent solids expressed as a decimal
Q	plant flow
r_1	radius from centerline of centrifuge to pool level
r_2	radius from centerline of centrifuge to bowl
s	specific gravity of the sludge
S	sludge production
S_b	pitch of blades
SS	raw water suspended solids
SS_d	drained solids concentration
SS_f	desired final solids concentration
SS_i	initial solids concentration
$t(f)$	freezing time
T	average ambient temperature
T_f	freezing point temperature
Tu	raw water turbidity
V	volume of sludge
V	volume of sludge/water in the pool of a centrifuge
ΔW	bowl/conveyor differential speed
Y	sand bed yield
z	pool depth in a centrifuge
μ	viscosity
ρ	fluid density
ρ_p	particle density
ρ_f	density of frozen sludge
χ	filterability constant
Φ	dimensionless instrument constant
ω	radial velocity of centrifuge (radians)

REFERENCES

Amy, G., M. Edwards, P. Brandhuber, L. McNeill, M. Benjamin, F. Vagliasindi, K. Carlson, and J. Chwirka (2000) "Residuals Generation, Handling and Disposal." In *Arsenic Treatment Options and Residuals Handling Issues*, Awwa Research Foundation report. Denver, CO: Water Research Foundation.

AWWA (1996) *Management of Water Treatment Plant Residuals,* USEPA/625/R-95/008. Denver, CO: American Water Works Association.

AWWA (2003) "Residuals Management for Low-Pressure Membranes," Committee Report, *Journal AWWA*, vol. 95, no. 6, pp. 68–82.

AWWA (2004) "Current Perspectives on Residuals Management for Desalting Membranes," Committee Report, *Journal AWWA*, vol. 96, no. 12, pp. 73–87.

AWWA M37 (1992) *Operational Control of Coagulation and Filtration Processes*, Denver, CO:. American Water Works Association.

Bond, R., and S. Veerapaneni (2007) *Zero Liquid Discharge for Inland Desalination*. Awwa Research Foundation report. Denver, CO: Water Research Foundation.

Bugbee, G.J., and C.R. Frink. (1985) "Alum Sludge as Soil Amendment: Effects on Soil Properties and Plant Growth," *Connecticut Agricultural Experiment Station*. Bulletin 827.

Clifford, D.A. (1999) "Ion Exchange and Inorganic Adsorption." In *Water Quality and Treatment: A Handbook of Community Water Supplies*, 5th ed. New York: McGraw-Hill.

Cornwell, D.A., and J. Susan (1979) "Characteristics of Acid-Treated Alum Sludges," *Journal AWWA*, vol. 71, no.10, pp. 604–608.

Cornwell, D.A. (1981) "Management of Water Treatment Plant Sludges," *Sludge and its Ultimate Disposal*. Ann Arbor.

Cornwell, D.A., and H. Koppers (1990) *Slib, Schlamm, Sludge*, Awwa Research Foundation report. Denver, CO: Water Research Foundation.

Cornwell, D.A., C. Vandermeyden, G. Dillow, and M. Wang (1992) Sludge Physical Characterization. *Landfilling of Water Treatment Plant Coagulant Sludges*, Awwa Research Foundation report. Denver, CO: Water Research Foundation.

Cornwell, D.A., and R.G. Lee (1993) *Recycle Stream Effects on Water Treatment*, Awwa Research Foundation report. Denver, Colo.: Water Research Foundation.

Cornwell, D.A., and C. Vandermeyden (1999) "Sizing Residuals Drying Beds," *Journal AWWA*, vol. 91, no. 11, pp. 94–105.

Cornwell, D.A., R.N. Mutter, and C. Vandermeyden (2000) *Commercial Application and Marketing of Water Plant Residuals* report. Denver, CO: Water Research Foundation.

Cornwell, D.A., M.J. MacPhee, N.E. McTigue, H. Arora, G. DiGiovanni, M. LeChevallier, and J.S. Taylor (2001) *Treatment Options for* Giardia, Cryptosporidium, *and Other Contaminants in Recycled Backwash Water*, Awwa Research Foundation report. Denver, CO: Water Research Foundation.

Cornwell, D.A. (2006) *Water Treatment Residuals Engineering*. Awwa Research Foundation report. Denver, CO: Water Research Foundation.

Cornwell, D.A., J.E. Tobiason, and R. Brown (2010) *Innovative Applications of Treatment Processes for Spent Filter Backwash*. Denver, CO: Water Research Foundation.

Edzwald, J.K., J.E. Tobiason, M.B. Kelley, G.S. Kaminski, H.J. Dunn, and P.B. Gallant (2001) *Impacts of Filter Backwash Recycle on Clarification and Filtration*, Awwa Research Foundation report. Denver, CO: Water Research Foundation.

EE&T, Inc. (1992) "Alternative Evaluation and Preliminary Design Studies," Prepared for Charlotte–Mecklenburg Utility Department, Charlotte, NC.

Elliott, H.A., B.A. Dempsey, D.W. Hamilton, and J.R. DeWolfe (1990) *Land Application of Water Treatment Sludges: Impacts and Management*, Awwa Research Foundation report. Denver, CO: Water Research Foundation.

Elliott, H.A., and B.A. Dempsey (1991) "Agronomic Effects of Land Application of Water Treatment Sludges," *Journal AWWA*, vol. 83 no. 4, pp. 126–131.

Fox, P., M. Abbaszadegan, F. Mohammadesmaeili, and M. Kabiri-Badr (2009) *Inland Membrane Concentrate Treatment Strategies for Water Reclamation Systems*. Denver, CO: Water Research Foundation.

Geertsema, W.S., W.R. Knocke, J.T. Novak, and D. Dove (1994) "Long Term Effects of Sludge Application to Land," *Journal AWWA*, vol. 86, no. 11, pp. 68–74.

Innocenti, P. (1998) "Techniques for Handling Water Treatment Sludge," *Upflow*, vol. 142, no. 2, pp. 1.

Jordahl, J. (2006) *Beneficial and Nontraditional Uses of Concentrate*. Alexandria, VA: Water Reuse Foundation.

Kenna, E., and A. Zander (2000) *Current Management of Membrane Plant Concentrate*, Awwa Research Foundation report. Denver, CO: Water Research Foundation.

Knocke, W.R., and D.L. Wakeland (1983) "Fundamental Characteristics of Water Treatment Plant Sludges," *Journal AWWA*, vol. 75, no. 10, pp. 516–523.

Kynch, G. (1952) "A Theory of Sedimentation," *Trans Faraday Society*, vol. 48, pp. 166.

Le Gouellec, Y., D.A. Cornwell, and M.J. MacPhee (2004) "Treating Microfiltration Backwash," *Journal AWWA*, vol. 96, no. 1, pp. 72–83.

Lin, S.D., and D.C. Green (1990) Two-Year Study of Alum Sludge Application to Corn and Soybean Farmland. Report of Investigation 113. Champaign, IL: Illinois State Water Survey.

MacPhee, M.J., G.E. Chales, and D.A. Cornwell (2001) *Treatment of Arsenic Residuals from Drinking Water Removal Processes*. Under contract to the U.S. Environmental Protection Agency, Washington, DC, EPA/600/R-01/033.

MacPhee, M.J., D.A. Cornwell, and R. Brown (2002) *Trace Contaminants in Water Treatment Chemicals*, Awwa Research Foundation report. Denver, CO: Water Research Foundation.

Martel, C.J. (1989) "Development and Design of Sludge Freezing Beds," *JEED*, vol. 115 no. 4, pp. 799–808.

Martel, C.J., and C.J. Diener (1991) "A Pilot-Scale Study of Alum Sludge Dewatering in a Freezing Bed," *Journal AWWA*, vol. 83, no. 12, pp. 51–55.

Mickley, M. (2001) *Membrane Concentrate Disposal: Practices and Regulation*, Report No. 69, U.S. Department of Interior, Bureau of Reclamation, Technical Service Center, Water Treatment Engineering and Research Group; available online at www.usbr.gov/pmts/water/media/pdfs/report069.pdf (accessed July 7, 2004).

Mickley, M. (2000) *Major Ion Toxicity in Membrane Concentrate*, Awwa Research Foundation report. Denver, CO: Water Research Foundation.

Migneault, W.H. (1988) "Potential for Brickmaking. Freshwater Utility Recycles Its Sludge," *BioCycle*, vol. 4, p. 63.

Min, J., L. Boulos, J. Brown, D. Cornwell, Y. Le Gouellec, E. coppola, S. Baxley, J. Rine, J. Hering, and N. Vural (2004) *Innovative Treatment Alternatives to Minimize Residuals Containing Nitate, Perchlorate, and Arsenic*, Awwa Research Foundation report. Denver, CO: Water Research Foundation.

Mutter, R., et al. (1994) *An Assessment of Cropland Application of Alum Sludge*. Master's Thesis, VPI & SU, Blacksburg, VA.

Novak, J. (1993) *Demonstration of Cropland Application of Alum Sludges*, Awwa Research Foundation report. Denver, CO: Water Research Foundation.

Ohio Department of Health (1996) *Supplement to Report on Waste Sludge and Filter Wastewater Disposal From Water Softening Plants*. Columbus, OH: Ohio Environmental Protection Agency.

Parker, P.J., and A.G. Collins (2000) "Optimization of Freeze–Thaw Conditioning," *Journal AWWA*, vol. 92, no. 5, pp. 77–85.

Parker, P.J., A.G. Collins, and J.P. Depmsey (1998) "Alum Residual Floc Interactions with an Advancing Ice/Water Interface," *Jour. Envir. Eng*., vol. 124, no. 3, pp. 249–253.

Reeves, T. (1980) The Mahoning Valley Sanitary District Plans to Spread Slurry on Farmland. *Farm and Dairy.*

Rengasamy, P., J.M. Oades, and T.W. Hancock (1980) "Improvement of Soil Structure and Plant Growth by Addition of Alum Sludge," *Communications in Soil Science and Plant Analysis*, vol. 11, no. 6, pp. 533–545.

Rittman, M.E., and P.L. McCarty (2001) *Environmental Biotechnology: Principles and Applications.* New York: McGraw-Hill.

Rolan, A.T. (1976) "Evaluation of Water Plant Sludge Disposal Methods," Proceedings of the Annual Meeting. *Jour. North Carolina Section*, vol. 12, no. 1, pp. 56.

Rolan, A.T. (1980) "Determination of Design Loading for Sand Drying Beds," *Jour. North Carolina Section*, AWWA and North Carolina WPCA, vol. L5, no. 1, p. 25.

Russell, G.A. (1975) "From Lagooning to Farmland Application The Next Step in Lime Sludge Disposal," *Journal AWWA*, vol. 67, no. 10, pp. 585–588.

Russell, G.A. (1980) "Agricultural Application of Lime Softening Residue," paper presented at the Missouri AWWA Section Meeting, April 1980, and Illinois AWWA Section Meeting, May 1980.

Sethi, S., S. Walker, P. Xu, and J. Drewes (2009) *Desalination Product Water Recovery and Concentrate Volume Minimization*. Denver, CO: Water Research Foundation.

State of California (2005) *Waste Extraction Test (WET) Procedure,* Title II, Chap. 11, App. II, Sacramento, CA.

Snoeyink, V.L., C.K. Jongeward, A.G. Myers, and S.K. Richter (1984) "Characteristics and Handling of Wastes From Groundwater Treatment Systems," paper presented at the AWWA Annual Conference, Dallas, TX, June 10.

Talmage, W., and B. Fitch (1955) "Determining Thickener Unit Areas," *Ind. & Eng. Chem*., vol. 47, no. 38.

Tisdale, S.L., and W.L. Nelson (1975) *Soil Fertility and Fertilizers,* 3d ed. New York: Macmillian Publishing Co., pp. 105–189.

U.S. Department of Transportation (1976) 49 CFR Part 172. Hazardous Material Transportation Uniform Safety Act. April 15, 1976. 41 Fed. Reg. 15996. Washington, DC:USDOT.

USEPA (1978) *Technology Transfer Process Design Manual–Municipal Sludge Landfills.* EPA-62511-78-010. Cincinnati, OH.

USEPA (1991a) *Filter Backwash and Recycling Rule,* 40 CFR 141.76(a), Washington, DC.

USEPA (1991b) *Solid Waste Disposal Facility Criteria, Final Rule,* 40 DFR Parts 257 and 258, 56 *Federal Register* 50978, Washington, DC. 22-102

USEPA (2000) *Regulations on the Disposal of Arsenic Residuals from Drinking Water Treatment Plants*, EPA/600/R-00/25.

USEPA (2003) *Arsenic Treatment Technology Evaluation Handbook for Small Systems,* EPA 816-R-03-014. www.epa.gov/ogwdw/arsenic/pdfs/handbook_arsenic_treatment-tech.pdf.

USEPA (2005) *A Regulator's Guide to the Management of Radioactive Residuals from Drinking Water Treatment Technologies*. http://epa.gov/rpdweb/docs/tenorm/816-R-05.004.pdf.

U.S. Nuclear Regulatory Commission (1961) 10 CFR Part 40. Domestic Licensing of Source Material. www.nrc.gov/reading-rm/doc-collections/cfr/part040.

Van Benschoten, J.E., J.N. Jensen, and A.R. Griffin (1991) *Land Application of Water Treatment Plant Sludge*. Prepared for the Erie County Water Authority.

Vandermeyden, C., D.A. Cornwell, and K. Schenkelberg (1997) *Non-mechanical Dewatering of Water Plant Residuals*, Awwa Research Foundation report. Denver, CO: Water Research Foundation.

Vesilind, P.A. (1979) *Treatment and Disposal of Wastewater Sludges.* Ann Arbor, MI:Ann Arbor Science.

Vesilind, P.A. (1988) "Capillary Suction Time as Fundamental Measure of Sludge Dewaterability," *Jour. WPCF*, vol. 60, no. 2, pp. 215–219.

Vesilind, P.A., and C.J. Martel (1990) "Freezing of Water and Wastewater Sludges," *Jour. Envir. Eng.*, vol. 116, no. 5, pp. 854–862.

APPENDIX A
CHEMICAL ELEMENTS

TABLE A.1 Atomic Numbers and Masses from IUPAC†

Actinium	Ac	89	(227)	Krypton	Kr	36	83.798
Aluminum	Al	13	26.9815386	Lanthanum	La	57	138.90547
Americium	Am	95	(243)	Lawrencium	Lr	103	(262)
Antimony	Sb	51	121.760	Lead	Pb	82	207.2
Argon	Ar	18	39.948	Lithium	Li	3	6.941
Arsenic	As	33	74.92160	Lutetium	Lu	71	174.967
Astatine	At	85	(210)	Magnesium	Mg	12	24.305
Barium	Ba	56	137.327	Manganese	Mn	25	54.9380
Berkelium	Bk	97	(247)	Mendelevium	Md	101	(258)
Beryllium	Be	4	9.012182	Mercury	Hg	80	200.59
Bismuth	Bi	83	208.98040	Molybdenum	Mo	42	95.94
Boron	B	5	10.811	Neodymium	Nd	60	144.242
Bromine	Br	35	79.904	Neon	Ne	10	20.1797
Cadmium	Cd	48	112.411	Neptunium	Np	93	(237)
Calcium	Ca	20	40.078	Nickel	Ni	28	58.693
Californium	Cf	98	(251)	Niobium	Nb	41	92.9064
Carbon	C	6	12.0107	Nitrogen	N	7	14.0067
Cerium	Ce	58	140.116	Nobelium	No	102	(259)
Cesium	Cs	55	132.90541	Osmium	Os	76	190.23
Chlorine	Cl	17	35.453	Oxygen	O	8	15.9994
Chromium	Cr	24	51.996	Palladium	Pd	46	106.42
Cobalt	Co	27	58.933195	Phosphorus	P	15	30.97376
Copper	Cu	29	63.546	Platinum	Pt	78	195.084
Curium	Cm	96	(247)	Plutonium	Pu	94	(244)
Dysprosium	Dy	66	162.500	Polonium	Po	84	(209)
Einsteinium	Es	99	(252)	Potassium	K	19	39.0983
Erbium	Er	68	167.259	Praseodymium	Pr	59	140.9077
Europium	Eu	63	151.964	Promethium	Pm	61	(145)
Fermium	Fm	100	(257)	Protactinium	Pa	91	231.0359
Fluorine	F	9	18.998403	Radium	Ra	88	(226)
Francium	Fr	87	(223)	Radon	Rn	86	(222)
Gadolinium	Gd	64	157.25	Rhenium	Re	75	186.207
Gallium	Ga	31	69.723	Rhodium	Rh	45	102.9055
Germanium	Ge	32	72.64	Rubidium	Rb	37	85.4678
Gold	Au	79	196.96658	Ruthenium	Ru	44	101.07
Hafnium	Hf	72	178.49	Samarium	Sm	62	150.36
Helium	He	2	4.00260	Scandium	Sc	21	44.9559
Holmium	Ho	67	164.9303	Selenium	Se	34	78.96
Hydrogen	H	1	1.0079	Silicon	Si	14	28.0855
Indium	In	49	114.818	Silver	Ag	47	107.8682
Iodine	I	53	126.9045	Sodium	Na	11	22.989769
Iridium	Ir	77	192.217	Strontium	Sr	38	87.62
Iron	Fe	26	55.847	Sulfur	S	16	32.065

(*Continued*)

TABLE A.1 Atomic Numbers and Masses from IUPAC† (*Continued*)

Tantalum	Ta	73	180.9479	Tungsten	W	74	183.84
Technetium	Tc	43	(98)	Uranium	U	92	238.0289
Tellurium	Te	52	127.60	Vanadium	V	23	50.9415
Terbium	Tb	65	158.9254	Xenon	Xe	54	131.293
Thallium	Tl	81	204.383	Ytterbium	Yb	70	173.04
Thorium	Th	90	232.0381	Yttrium	Y	39	88.9059
Thulium	Tm	69	168.9342	Zinc	Zn	30	65.409
Tin	Sn	50	118.71	Zirconium	Zr	40	91.224
Titanium	Ti	22	47.867				

†IUPAC, *Pure Applied Chemistry,* vol. 75, no. 8, pp. 1107–1122, 2003, and revisions, August 2005. A value in parentheses is the mass number of the longest-lived isotope of the element.

APPENDIX B
USEFUL CONSTANTS

TABLE B.1 Physical and Chemical Constants

Avogadro's number	6.022×10^{23} mol^{-1}
Boltzmann's constant, k_B	1.3805×10^{-23} J K^{-1}
Dielectric constant (water, 25°C, ε)	78.54
Earth gravitation, g	9.806 m s^{-2}
Einstein	1 mole of photons = 6.022×10^{23} photons
Electron charge	1.602×10^{-19} C
Faraday constant, F	96.485 kJ equiv^{-1} V^{-1} = 9.6485×10^{4} C equiv^{-1}
Gas constant, R	1.987 cal mol^{-1} K^{-1} = 8.314 J mol^{-1} K^{-1}, 0.08206 L atm mol^{-1} K^{-1}
Ice point	273.15 K
Molar volume of ideal gas at 273.15 K, 1 atm	22.414 L mol^{-1}
Speed of light in vacuum	2.998×10^{8} m s^{-1}
Planck constant, h	6.626×10^{-34} J s

APPENDIX C
CONVERSION FACTORS

TABLE C-1 Unit Conversion Factors, SI Units to U.S. Customary Units, and U.S. Customary Units to SI Units

SI unit name	Symbol	→	←	Symbol	U.S. customary unit name
		To convert, multiply in direction shown by arrows			
		Acceleration			
Meters per square second	m/s^2	3.2808	0.3048	ft/s^2	Feet per square second
Meters per square second	m/s^2	39.3701	0.0254	in/s^2	Inches per square second
		Area			
Hectare (10,000 m^2)	ha	2.4711	0.4047	ac	Acre
Square centimeter	cm^2	0.1550	6.4516	in^2	Square inch
Square kilometer	km^2	0.3861	2.5900	mi^2	Square mile
Square kilometer	km^2	247.1054	4.047×10^{-2}	ac	Acre
Square meter	m^2	10.7639	9.2903×10^{-2}	ft^2	Square foot
Square meter	m^2	1.1960	0.8361	yd^2	Square yard
		Energy			
Kilojoule	kJ	0.9478	1.0551	Btu	British thermal unit
Joule	J	$2.7778\ 10^{-7}$	3.6×10^6	kW·h	Kilowatt-hour
Joule	J	0.7376	1.356	$ft \cdot lb_f$	Foot-pound (force)
Joule	J	1.0000	1.0000	W·s	Watt-second
Joule	J	0.2388	4.1876	cal	Calorie
Kilojoule	kJ	$2.7778\ 10^{-4}$	3600	kW·h	Kilowatt-hour
Kilojoule	kJ	0.2778	3.600	W·h	Watt-hour
Megajoule	MJ	0.3725	2.6845	hp·h	Horsepower-hour
		Force			
Newton	N	0.2248	4.4482	lb_f	Pound force
		Flow rate			
Cubic meters per day	m^3/d	264.2	3.785×10^{-3}	gal/d	Gallons per day
Cubic meters per day	m^3/d	2.642×10^{-4}	3.785×10^3	mgd	Million gallons per day
Cubic meters per second	m^3/s	35.3147	0.02832	ft^3/s	Cubic feet per second
Cubic meters per second	m^3/s	22.827	0.0438	mgd	Million gallons per day
Cubic meters per second	m^3/s	15850.3	6.3090×10^{-5}	gal/min	Gallons per minute
Liters per second	L/s	22827	4.381×10^{-5}	gal/d	Gallons per day
Liters per second	L/s	$2.2825\ 10^{-2}$	43.8126	mgd	Million gallons per day
Liters per second	L/s	15.852	0.0631	gal/min	Gallons per minute

(*Continued*)

TABLE C-1 Unit Conversion Factors, SI Units to U.S. Customary Units, and U.S. Customary Units to SI Units (*Continued*)

SI unit name	To convert, multiply in direction shown by arrows				U.S. customary unit name
	Symbol	→	←	Symbol	
		Length			
Centimeter	cm	0.3937	2.540	in	Inch
Kilometer	km	0.6214	1.6093	mi	Mile
Meter	m	39.3701	2.54×10^{-2}	in	Inch
Meter	m	3.2808	0.3048	ft	Foot
Meter	m	1.0936	0.9144	yd	Yard
Millimeter	mm	0.03937	25.4	in	Inch
		Mass			
Gram	g	0.0353	28.3495	oz	Ounce
Gram	g	0.0022	$4.5359\ 10^2$	lb	Pound
Kilogram	kg	2.2046	0.45359	lb	Pound
		Power			
Kilowatt	kW	0.9478	1.0551	Btu/s	British thermal units per second
Kilowatt	kW	1.3410	0.7457	hp	Horsepower
Watt	W	0.7376	1.3558	ft-lb_f/sec	Foot-pounds (force) per second
		Pressure (force/area)			
Pascal (newtons per square meter)	Pa(N/m^2)	1.4504×10^{-4}	6.8948×10^3	lb/in^2	Pounds (force) per square inch
Pascal (newtons per square meter)	Pa(N/m^2)	2.0885×10^{-2}	47.8803	lb/ft^2	Pounds (force) per square foot
Pascal (newtons per square meter	Pa(N/m^2)	2.9613×10^{-4}	3.3768×10^3	in Hg	Inches of mercury (60°F)
Pascal (newtons per square meter)	Pa (N/m^2)	4.0187×10^{-3}	2.4884×10^2	in H_2O	Inches of water (60°F)
Kilopascal (kilonewtons per square meter)	kPa (kN/m^2)	0.1450	6.8948	lb/in^2	Pounds (force) per square inch
Kilopascal (kilonewtons per square meter)	kPa (kN/m^2)	9.8688×10^{-3}	1.0133×10^2	atm	Atmosphere (standard)
		Temperature			
Degree Celsius	°C	1.8(°C) + 32	0.555(°F – 32)	°F	Degree Fahrenheit
Kelvin	K	1.8(K) – 459.67	0.555(°F + 459.67)	°F	Degree Fahrenheit
		Velocity			
Meters per second	m/s	2.2369	0.44704	mi/h	Miles per hour
Meters per second	m/s	3.2808	0.3048	ft/s	Feet per second
		Volume			
Cubic centimeter	cm^3	0.0610	16.3781	in^3	Cubic inch
Cubic meter	m^3	35.3147	2.8317×10^{-2}	ft^3	Cubic foot

(*Continued*)

TABLE C-1 Unit Conversion Factors, SI Units to U.S. Customary Units, and U.S. Customary Units to SI Units (*Continued*)

SI unit name	To convert, multiply in direction shown by arrows				U.S. customary unit name
	Symbol	→	←	Symbol	
		Volume			
Cubic meter	m^3	1.3079	0.7646	yd^3	Cubic yard
Cubic meter	m^3	264.1720	3.7854×10^{-3}	gal	Gallon
Cubic meter	m^3	8.1071×10^{-4}	1.2335×10^3	acre-ft	Acre-foot
Liter	L	0.2642	3.7854	gal	Gallon
Liter	L	0.0353	28.3168	ft^3	Cubic foot
Liter	L	33.8150	2.9573×10^{-2}	oz	Ounce (U.S. fluid)

Note: cm is not a SI unit but is included because of its common usage.
UV Dose Conversions
1 mW-sec/cm^2 = 1 mJ/cm^2 = 10 J/m^2

TABLE C-2 Common water treatment conversion factors, SI units to U.S. customary units, and U.S. customary units to SI units

SI unit name	To convert, multiply in direction shown by arrows				U.S. customary unit name
	Symbol	→	←	Symbol	
		Concentration			
Kilogram per cubic meter	kg/m^3	8.34×10^3	1.2×10^{-4}	lb/Mgal	Pounds per million gallons
Milligram per liter	mg/L	8.34	0.12	lb/Mgal	Pounds per million gallons
		Flow rate			
Cubic meters per day	m^3/d	2.642×10^{-4}	3.785×10^3	mgd	Million gallons per day
Megaliters per day	ML/d	0.2642	3.785	mgd	Million gallons per day
Liters per second	L/s	15.852	0.0631	gal/min	Gallons per minute
		Hydraulic loading rate			
Cubic meter per square meter-hour	m^3/m^2-h	0.4098	2.44	gal/min/ft^2	Gallons per minute per square foot
Meter per hour	m/h	0.4098	2.44	gal/min/ft^2	
		Mass loading			
Kilogram per day	Kg/d	2.2046	0.45359	lb/d	Pound per day
		Pressure (force/area)			
Kilopascal	kPa	0.1450	6.8948	lb/in^2	Pounds per sq inch (psi)
Bar	bar	14.504	0.06895	lb/in^2	Pounds per sq inch (psi)

APPENDIX D
PROPERTIES OF WATER AND GASES

TABLE D-1 Physical Properties of Water (SI units)[†]

Temperature °C	Density ρ kg/m^3	Dynamic viscosity $\mu \times 10^3$ N-s/m^2	Kinematic viscosity $\upsilon \times 10^6$ m^2/s	Surface tension σ N/m	Dielectric constant ε	Specific heat c_p J/mole*K	Vapor pressure ρ_v kN/m^2
0	999.84	1.793	1.793	0.07565	87.9	76.023	0.612
4	999.98	1.570	1.570	0.07508	86.7	75.087	0.814
5	999.97	1.521	1.521	0.07494	86.4	75.760	0.873
10	999.70	1.307	1.307	0.07422	84.0	75.584	1.23
15	999.10	1.138	1.139	0.07349	82.2	75.462	1.71
20	998.21	1.002	1.004	0.07274	80.2	75.382	2.34
25	997.05	0.891	0.893	0.07197	78.5	75.333	3.17
30	995.65	0.798	0.801	0.07119	76.6	75.305	4.25
40	992.22	0.653	0.658	0.06960	73.2	75.297	7.38
50	988.03	0.547	0.553	0.06794	69.9	75.332	12.3
60	983.20	0.466	0.474	0.06624	66.7	75.396	19.9
70	977.78	0.404	0.413	0.06448	63.7	75.488	31.2
80	971.82	0.354	0.365	0.06267	60.9	75.608	47.4
90	965.35	0.314	0.326	0.06082	58.1	75.759	70.1
100	958.40	0.282	0.294	0.05891	55.5	75.946	101.3

[†]From National Institute of Standards and Technology (NITS); available online at *Chemistry WebBook*, http://webbook.nist.gov/chemistry/. And from *CRC Handbook of Chemistry and Physics*, 85th ed. Boca Raton, FL, CRC Press, 2004.

TABLE D-2 Solubility of Oxygen in Freshwater (Zero Chlorinity) for Water Saturated Air for 101.3 kPa (1 atm)[†]

Temperature °C	Oxygen solubility mg/L	Temperature °C	Oxygen solubility mg/L
0	14.62	16	9.87
1	14.22	17	9.67
2	13.83	18	9.47
3	13.46	19	9.28
4	13.11	20	9.09
5	12.77	21	8.92
6	12.45	22	8.74
7	12.14	23	8.58
8	11.84	24	8.42
9	11.56	25	8.26
10	11.29	26	8.11
11	11.03	27	7.97
12	10.78	28	7.83
13	10.54	29	7.69
14	10.31	30	7.56
15	10.08	31	7.43

[†]From 21st ed. of *Standard Methods* (APHA, AWWA, and WEF, 2005).

TABLE D-3 Physical Properties of Selected Gases (standard conditions: 0°C, 1 atm)

Gas	Formula	Molecular weight	Specific weight[†] lb/ft^3	Density[‡] g/L
Air (dry)	—	28.97	0.0807	1.293[§]
Ammonia	NH_3	17.03	0.0481	0.7713
Carbon dioxide	CO_2	44.00	0.1233	1.976
Chlorine	Cl_2	70.91	0.2011	3.222[§]
Hydrogen	H_2	2.016	0.0056	0.08986
Hydrogen sulfide	H_2S	34.08	0.0958	1.536
Methane	CH_4	16.03	0.0448	0.7175
Nitrogen	N_2	28.02	0.0780	1.250
Oxygen	O_2	32.00	0.0892	1.429
Ozone	O_3	48.0	0.1338	2.144[§]

[†]Specific weights calculated from densities.

[‡]From *National Institute of Standards and Technology* (NITS); available online at *Chemistry WebBook,* http://webbook.nist.gov/chemistry/.

[§]From Degrémont. *Water Treatment Handbook,* Vol. 1, 6th ed., Paris, Lavoisier Publishing, 1991.

TABLE D-4 Moist Air Bubble Densities† (for Use in Dissolved Air Flotation Applications)

Temperature (°C)	ρ_b (kg/m^3)
0	1.29
4	1.27
10	1.24
15	1.22
20	1.19
25	1.17
30	1.15

†Calculated from dry air densities and maximum humidity ratio (mass of water vapor in moist air to mass of dry air) for 101.3 kPa total pressure.

INDEX

Note: f. indicates figure; t. indicates table.

AA. *See* Activated alumina adsorption
A-C pipes. *See* Asbestos-cement pipe
Acanthamoeba, 2.17–2.18
Acetanilides, 2.53
Acid cleaners, 11.92
Acid-base chemistry, 3.11–3.13
  inorganic carbon chemistry and, 3.18–3.19
Acid-base equilibria, 3.12–3.13
Acidity, 3.23–3.24
Acidobacterium, 21.10
Acinetobacter, 21.10–21.11
Acrylamide, 2.58
Actinomyces, 21.19
Activated alumina adsorption (AA), 12.3
  defluoridation system design for, 12.41–12.42, 12.41f., 12.42t.–12.43t.
  fluoride breakthrough curves with, 12.41, 12.41f.
  fluoride capacity of, 12.42–12.45, 12.42t., 12.44f., 22.57, 22.57t.
  fluoride removal with, 12.7–12.8, 12.41–12.46
  materials in, 12.7
  pH sensitivity of, 12.7
  regeneration of fluoride spent alumina columns in, 12.45–12.46
  residuals from fluoride removal by, 22.57, 22.58t.–22.59t.
  selectivity sequences for, 12.14
Activated carbon. *See* Granular activated carbon; Powdered activated carbon
Activated carbon fibers, 14.8, 14.88–14.89
Activated silica, 8.48
Activation control, 20.13
Acute poisoning, 2.3
Adenoviruses, 2.14
AdOx kinetic model, 18.29–18.30, 18.30f.
Adsorbable organic iodine, 15.17
Adsorbates, 14.2
  in GAC performance estimation, 14.60–14.62
  types of, 14.60
Adsorbent resins, 14.87–14.88
Adsorbents, 14.2
  activated carbon fibers, 14.8, 14.88–14.89
  carbonaceous resins, 14.7–14.8
  characteristics if, 14.4–14.8
  synthetic adsorbent resins, 14.7
  zeolites, 14.8, 14.89
Adsorption. *See also* GAC adsorption systems; Granular activated carbon; Powdered activated carbon
  adsorbates, 14.2
  adsorbent characteristics in, 14.4–14.8
  adsorbents, 14.2
  advantages/disadvantages of, 12.2t.
  air stripping off-gas control using, 6.36–6.41, 6.37f.–6.38f., 6.39t.
  apparent density in, 14.5
  attenuation in, 14.37
  bond, 14.28
  breakthrough capacity in, 14.32, 14.33f.
  breakthrough concentration in, 14.32
  breakthrough curves in, 14.31–14.34, 14.33f.–14.34f.
  bulk solution transport and, 14.28
  competitive adsorption in multisolute systems, 14.14–14.19, 14.15f., 14.18f., 14.18t.–14.19t.
  competitive adsorption in natural waters, 14.21–14.26, 14.22f.–14.23f., 14.25f.
  coprecipitation and, 13.4
  for DBP postformation control, 19.39–19.40
  for DBP precursor control, 19.34
  decolorizing index in, 14.7

Adsorption (*Cont.*):
desorption and, 14.36–14.37
of DOC, 8.29–8.30
equilibrium, 14.8–14.27
equivalent background compound and, 14.23–14.26, 14.23f.
external (film) resistance to transport and, 14.28
Freundlich equation for, 14.9–14.12, 14.9f., 14.10t.–14.11t.
GAC and, 10.15, 14.6–14.7
GAC capacity of, 14.62
granular media, 12.7–12.9
HSDM for, 14.30–14.31, 14.31f.
of humic substances on GAC, 14.27
internal (intraparticle) transport and, 14.28
iodine number in, 14.7
isotherms, 14.8–14.12, 14.9f., 14.10t.–14.11t., 14.12f.
IX kinetics compared to, 12.18
kinetic parameter estimation for, 14.29–14.30
kinetics, 14.28–14.36
mass transfer zone in, 14.31–14.34, 14.32f.
MIB capacity for, 14.22, 14.22f., 14.25f.
micropollutant capacity in natural waters of, 14.24, 14.25f., 14.26
molasses number in, 14.7
of NOM, 14.19–14.21, 14.20f.
of NOM by HMS coagulants, 8.28–8.30
NOM preloading effect on kinetic parameters of, 14.34–14.36, 14.35f.–14.36f.
NOM preloading in natural waters and, 14.26–14.27, 14.26f.–14.27f.
overview of, 14.2–14.4
PAC and, 14.6–14.7
of polymers, 8.13–8.15
rate-limiting step in, 14.28
residuals and, 22.54–22.62
solvent-motivated, 14.13
surface area in, 14.4
surface chemistry of, 14.13
of synthetic resins, 14.87–14.88
of THMs, 14.53
transport mechanisms in, 14.28–14.29
treatment process selection and, 5.19–5.21
water quality and, 14.3

Adsorptive media. *See also* Ion exchange
arsenic and residuals of, 12.54–12.55
arsenic capacity of, 12.49, 12.52–12.54, 12.52f., 12.53t.
arsenic removal by, 12.46–12.55, 12.47f.–12.48f., 12.48t., 12.50t.–12.51t., 12.52.f
arsenic removal process of, 12.47–12.49, 12.48f., 12.48t.
Advanced oxidation processes (AOPs), 5.7, 5.25, 7.23–7.29, 16.5
adOx kinetic model for, 18.29–18.30, 18.30f.
for DBP precursor control, 19.29–19.30, 19.30f.
steady state OH radical model for, 18.28–18.29
UV/Cl_2 AOP mechanism, 18.33
UV/H_2O_2 AOP mechanism, 18.24, 18.25t.–18.26t., 18.27–18.32
UV/O_3 AOP mechanism, 18.33
UV/TiO_2 AOP mechanism, 18.34
Aeration. *See also* Bubble aeration
corrosion control with, 20.75
DBP removal with packed towers and, 19.42, 19.42f.
softening, 13.52
spray aerators, 6.56–6.60, 6.57f.
surface, 6.52–6.56, 6.52f.–6.54f.
Aeromonas hydrophyla, 2.8
in distribution systems, 21.16–21.17
Aesthetic concerns
with color, 2.74–2.75
with hardness, 2.75
with mineralization, 2.75
with staining, 2.75
with taste and odor, 2.73–2.74
with turbidity, 2.74–2.75
water quality and, 2.2, 2.72–2.75
Air binding, 10.49
Air saturation systems, 9.70–9.72, 9.71f.–9.72f.
Air scour
backwash water flow rates compared to, 10.56–10.57, 10.57t.
in backwashing of rapid granular bed filtration, 10.54t., 10.55–10.57
delivery systems for, 10.56
Air stripping. *See also* Packed towers
adsorption for off-gas control in, 6.36–6.41, 6.37f.–6.38f., 6.39t.
low-profile, 6.15
sieve tray columns in, 6.15, 6.17
treatment process selection and, 5.5
types of, 6.1–6.2
of VOCs, 6.42–6.45

Air-dissolving vessel, 9.48
Air-pressure testing, 11.35
Alachlor, 2.53
Aldicarb, 2.55
Aldicarb sulfone, 2.55
Aldicarb sulfoxide, 2.55
Algae
  DAF removing, 5.33, 5.33f.
  photosynthesis and, 3.38–3.39
  taste and odor causing, 3.36, 3.37t.–3.38t., 3.44
Alkalinity
  buffer intensity and, 20.26–20.29, 20.27f.–20.28f.
  causes of, 3.21
  classifications for, 3.23
  concentrations, 3.4
  hardness and, 13.13
  inorganic carbon chemistry and, 3.22–3.23
  measuring, 3.22
  membrane processes and recovery of, 11.89, 11.90f.
  total, 20.26
Alum
  acidulated, 8.21
  sludge, 22.4–22.5
Aluminum, 2.33–2.34, 8.2, 8.3
  corrosion and, 20.35–20.36, 20.36f.
  dosage predictions for, 8.22–8.30
  hydrolysis, 8.16–8.17, 8.16f., 8.23
  required dosage of, 8.42
  residual, 8.36–8.41, 8.37f.–8.40f.
Aluminum hydroxide, 8.17–8.19, 8.17f., 8.18t.
  precipitate solubility of, 8.36–8.41, 8.37f.–8.40f.
Ames test, 2.27
Ammonia
  chlorine reactions with, 7.19–7.20, 17.10–17.15, 17.12f., 17.15t.
  corrosion and, 20.25
  HAA formation and impact of, 19.38, 19.39f.
  THM formation and impact of, 19.38, 19.38f.
Ammonia-oxidizing bacteria, 21.20
Ampholyte polymers, 8.45
Anionic polyelectrolytes, 8.47
Anionic polymers, 8.45
Anodes, 20.3–20.4, 20.6
Anodic current, 20.4
Anthrax, 2.76
Anthropogenic chemicals, 3.24, 3.25t.
  in rivers and streams, 3.33–3.34
Antiscalants, 13.3
AOPs. *See* Advanced oxidation processes
Apparent density, 14.5
Aquifer storage and recovery (ASR), 5.22
  for DBP control, 19.28
Aquifers. *See also* Groundwater
  IPR through direct injection into portable, 16.31–16.32
  MAR, 15.2–15.3
  protection of, 3.71
  SAT for, 15.12–15.15, 15.12f.
ARR. *See* Artificial recharge and recovery
Arrhenius equation, 7.10
Arsenic, 2.34–2.35
  adsorptive media capacity for, 12.49, 12.52–12.54, 12.52f., 12.53t.
  adsorptive media processes for removal of, 12.47–12.49, 12.48f., 12.48t.
  adsorptive media removal of, 12.46–12.55, 12.47f.–12.48f., 12.48t., 12.50t.–12.51t., 12.52.f
  adsorptive media residuals with, 12.54–12.55
  breakthrough curves in IX removal of, 12.56, 12.56f.
  concentration and IX run length for, 12.58–12.59
  downflow v. upflow regeneration in IX removal of, 12.60–12.61, 12.60f.
  in groundwater, 3.40, 3.42t.
  IX modeling for removal of, 12.76–12.77, 12.76f.
  IX removal of nitrate combined with, 12.57–12.58, 12.58t.
  IX removing, 12.55–12.61, 12.56f., 12.58f., 12.58t.–12.59t., 12.60f.
  leaching, 22.15, 22.15f.
  leakage during exhaustion in IX removal of, 12.59
  MF/UF integrated with coagulants removing, 11.61, 11.61f.
  oxidation of arsenite to arsenate, 12.46–12.47
  regulations, 1.19, 22.63–22.64
  residuals with, 22.62–22.66, 22.63t.–22.66t.
  resins for IX removal of, 12.56–12.57
  solids in removal of, 22.66
  spent resin regeneration in IX removal of, 12.59–12.60
  sulfate effecting IX and run length for, 12.59, 12.59t.
  treatment of residuals with, 22.65, 22.65f., 22.66t.
  treatment of spent IX regenerant with, 12.61

Artificial recharge and recovery (ARR), 15.7–15.10. *See also* Dune infiltration
attributes/limitations of, 15.10
in Aurora, CO, 15.27–15.28
engineering factor comparisons for, 15.8–15.9, 15.9t.
in Graz, Austria, 15.27
infiltration rates and operation of, 15.9
LBF/RBF compared to, 15.7, 15.8f.
in Maribor, Slovenia, 15.26
performance of, 15.9–15.10
Asbestos, 2.35
Asbestos-cement pipe (A-C pipe), 20.64–20.65
ASR. *See* Aquifer storage and recovery
Astroviruses, 2.14
Asymmetric membranes, 11.10
Atomic numbers and masses, A.1t.–A.2t.
Atrazine, 2.52–2.53
GAC adsorption systems, breakthrough curves for, 14.54–14.56, 14.55f.–14.56f.
isotherms, 14.23–14.24, 14.23f.
PAC usage rate for removal of, 14.81–14.82
Atterberg limit test, 22.10–22.11
Aurora, CO (ARR/RBF systems), 15.27–15.28
Australia
regulations for IPR, 16.38–16.39
standards of, 1.32–1.33
Available head loss, 10.11

Bacillus subtillis, 11.34
Back diffusion constant, 11.79
Backwashing. *See also* Spent filter backwash water
air scour delivery systems in, 10.56
air scour flow rates compared to, 10.56–10.57, 10.57t.
air-scour assisted, 10.54t., 10.55–10.57
CEB, 11.91
Cryptosporidium recycling and, 10.69
filtration and, 10.11
of GAC adsorption systems, 14.41
GAC and, 10.20–10.21, 10.68
Giardia lamblia recycling and, 10.69
gulf streaming in, 10.63
intermixing of adjacent layers during, 10.63
jet action in, 10.63, 10.65f.
movement of gravel during, 10.64–10.65, 10.65f.
mudballs and, 10.65f., 10.66
rapid granular bed filtration, expansion of filter bed during, 10.60–10.62
rapid granular bed filtration, wash water required for, 10.57–10.58
Backwashing (*Cont.*):
rapid granular bed filtration methods for, 10.53–10.57, 10.54t.
residuals and, 22.44, 22.44t.
sand boils in, 10.63, 10.64f.
skimming during, 10.62–10.63
stratification during, 10.62–10.63
surface wash plus fluidized-bed backwash in, 10.54–10.55
TOC and, 14.41
troughs, 10.57
upflow wash with full fluidization in, 10.53–10.54
wash water v. filtered water chemistry in, 10.69–10.70
wastewater management with, 10.68–10.69
Bacteria, 2.7–2.11. *See also* Coliforms
Acidobacterium, 21.10
Aeromonas hydrophyla, 2.8
ammonia-oxidizing, 21.20
Campylobacter, 2.8–2.9
chloramine inactivation of, 17.28
chlorine dioxide inactivation of, 17.28
chlorine inactivation of, 17.27–17.28
cyanobacteria, 2.18–2.19
distribution systems, identifying, 21.9–21.10
enteric, 2.7–2.8
enteric-like, 2.8
Escherichia coli, 2.6, 2.20–2.21
Helicobacter pylori, 2.9
heterotrophic, 2.22, 21.10–21.13, 21.11t., 21.12f.
Klebsiella, 21.10
Legionella, 2.3, 2.7, 2.9–2.10
Mycobacterium, 2.10–2.11
nitrite-oxidizing, 21.20
Nitrospirae, 21.10
Nocardia, 21.10–21.11
opportunistic, 2.11
ozone inactivation of, 17.28–17.29
Planctomyces, 21.10
Pseudomonas aeruginosa, 21.10–21.12
Rhodococcus, 21.10–21.11
Salmonella, 2.6–2.8, 2.27
Shigella, 2.7
Sphingomonas, 21.10–21.11
Staphylococcus, 21.11
UV disinfection inactivation of, 17.29, 17.29t.
VBNC bacteria, 21.14, 21.15t.
Verrucomicrobium, 21.10
Vibrio chloerae, 2.7–2.8
Yersinia enterocolitica, 2.8

Ballasted flocculation, 5.10, 8.49
Barium, IX removing, 12.33–12.34
Basicity, 8.20–8.21
Batch reactors
  flocculation kinetics in, 8.55
  kinetics in, 4.29–4.31, 4.30f.
  reaction rate expressions in, 4.28–4.33
Batch thickeners, 22.17–22.21, 22.18f.
BDCM. *See* Bromodichloromethane
Beer-Lambert law, 18.7–18.8
Belt filter presses, 22.31–22.32
Benzene, 2.45–2.46
Berlin, Germany (LBF system), 15.24–15.25
Biochemical cycle, 3.27, 3.27f.
Biodegradable organic material formation, 7.28, 19.14
Biodosimetry UV test, 18.9–18.10, 18.9f.
Biofilms, 21.7–21.8, 21.8f.
Biofiltration, 14.42
  DBP removal with, 19.40–19.41, 19.42f.
Biological activated carbon, 14.42
Biological nutrient removal (BNR), 16.10–16.11, 16.10t.
Bioterrorism Act (1992), 1.7
Blended phosphates, 20.44
Blue water, 20.46
BNR. *See* Biological nutrient removal
Body feed
  precoat filtration and, 10.87, 10.92
  precoat filtration and concentration of, 10.93–10.94, 10.94f.
Boron, 2.35–2.36
Brasses, 20.53–20.54
Brine disposal
  in IX removal of nitrate, 12.39–12.40
  locations for, 22.61–22.62
  from softening plants in IX, 12.33
Brines, 22.3
Bromate, 2.66
  control of, 19.40f.
  GAC reactions with, 14.59
  ozonation and, 19.19
Bromide
  chlorine reactions with, 7.20
  ozone reactions with, 7.28–7.29, 7.29f.
  THM reactions with, 19.20–19.22, 19.20f.
Brominated acetic acids, 2.68–2.69
Bromine, 2.65–2.66
  incorporation factor, 19.21
  substitution factor, 19.21
  THMs and, 19.21–19.22, 19.21f.–19.22f.
Bromodichloromethane (BDCM), 2.67
Bromoform, 2.67–2.68
Bronzes, 20.53–20.54
Brownian diffusion, 8.51–8.52, 10.31, 11.42, 11.43f.
Brunauer-Emmett-Teller isotherm equation, 14.4
Bubble aeration, 6.2, 6.41–6.52
  ozone absorption problem for, 6.50–6.52
  sample calculation for, 6.47–6.50
  schematic of, 6.42f.
  schematic of mixed, 6.43f.
  schematic of single, 6.43f.
  tank schematic of, 6.44f.
  tanks in series development for, 6.44–6.45, 6.44f.
Bubble-point testing, 11.35–11.36
Buffer intensity, 20.26–20.29, 20.27f.–20.28f.
Bulk solution transport, 14.28
Bureau of Water Hygiene CWSS, 1.3–1.4

$C \times T$ approach (disinfection regulation), 17.39–17.41, 17.40t.
Cadmium, 2.36
Cake filtration, 10.3
Calcium carbonate equilibria, 13.25–13.29, 13.28t.
Calcium carbonate precipitation potential (CCPP), 20.34
Calcium hypochlorite, 7.17–7.18
  basic chemistry of, 17.5–17.8
Caliciviruses, 2.11–2.12
California state regulations for IPR, 16.37–16.38
Campylobacter, 2.8–2.9
  IPR and, 16.7, 16.7t.
Canada, standards of, 1.32
Cancer, pesticides and risk of, 2.51–2.52
Candy tanks, 9.3, 9.3f.–9.4f.
Capillary suction time test (CST test), 22.12, 22.13f., 22.13t.
Carbamates, 2.55
Carbon dioxide, solubility of, 3.10–3.11
Carbon tetrachloride, 2.46–2.47
Carbonaceous resins, 14.7–14.8, 14.8
Carbonate hardness, 13.13–13.14
  lime softening removing, 13.18–13.19
Carbonate saturometer, 20.35
Carbonic acid, 13.16
  lime softening removing, 13.20–13.21
Carcinogenicity, 2.25–2.26
  of DBPs, 2.61t.–2.63t.

Carcinogens. *See also* Arsenic
DWEL and, 2.29
health-based drinking water concentration for, 2.30
LOAEL/NOAEL, 2.28–2.29
possible, 2.30–2.31
RfD for, 2.28–2.29
risk assessment for, 2.29–2.31
threshold for, 2.29–2.30
Catalysis, 7.13
Cathodes, 20.3–20.4, 20.6
Cathodic current, 20.4
Cationic polyelectrolytes, 8.46–8.47, 8.47f.
Cationic polymers, 8.45
with HMS coagulants, 8.49
CCPP. *See* Calcium carbonate precipitation potential
CCPs. *See* Critical control points
CCR. *See* Consumer Confidence Report
CDC. *See* Centers for Disease Control and Prevention
CEB. *See* Chemically enhanced backwash
Cement. *See* Asbestos-cement pipe
Cement-mortar lined pipe corrosion, 20.64–20.65
Centers for Disease Control and Prevention (CDC), 1.2
Centrifuges, 22.32–22.35, 22.32f., 22.34f.
CFD. *See* Computational fluid dynamics
CFSTR. *See* Continuous flow stirred tank reactor
CFSTRs-in-series model, 4.17–4.18, 4.18f.
reaction rate expressions in, 4.38–4.39, 4.39f.
tracer test data and, 4.18–4.19, 4.20f.
Charge concentrations, 3.4
on particles v. NOM, 3.61–3.62
Chemical constants, B.1t.
Chemical oxidation. *See also* Advanced oxidation processes; Chlorine dioxide; Chlorine/chlorination; Mixed oxidants; Ozone/ozonation; Potassium permanganate
Arrhenius equation for, 7.10
catalysis and, 7.13
as coagulation/flocculation aid, 7.46
color removal by, 7.42–7.43
of cyanotoxins, 7.45
for DBP control, 19.28–19.31, 19.29f.–19.31f.
electrochemical potentials in, 7.2–7.4, 7.3t.
of hydrogen sulfide, 7.42, 7.42t.
ionic reactions in, 7.11–7.12, 7.12t.
of iron, 7.31–7.36, 7.32t., 7.34f.
Chemical oxidation (*Cont.*):
of manganese, 7.31–7.36, 7.32t., 7.34f.
mixed oxidants, 5.7, 7.31
molecularity and, 7.9
of MTBE, 7.46
oxidation state in, 7.4
oxidation-reduction reactions in, 7.4–7.8
pH dependence in, 7.10–7.11, 7.11f.
radical reactions in, 7.12–7.13, 7.13t.
reaction kinetics and, 7.8–7.10
reaction pathways in, 7.14–7.15
reaction rate constant and, 7.9
of SOCs, 7.43–7.46
standard half-cell potentials for, 7.3t.
taste and odor destruction with, 7.37–7.39, 7.41
temperature dependence in, 7.10
thermodynamic principles of, 7.2–7.8
treatment process selection and, 5.5–5.7
uses of, 7.1–7.2
Chemical precipitation. *See* Precipitation
Chemical warfare, 2.76
Chemically enhanced backwash (CEB), 11.91
Chemicals. *See also* Dissolved chemicals; Inorganic constituents; Organic constituents; Volatile organic chemicals
acid-base chemistry, 3.11–3.13
anthropogenic, 3.24, 3.25t., 3.33–3.34
carcinogenicity of, 2.25–2.26
classifying, 2.28
dose-response in, 2.23–2.24
electroneutrality of, 3.5–3.6
equilibrium of, 3.7–3.17
genotoxicity of, 2.25, 2.27
inorganic carbon chemistry, 3.17–3.21
IPR assessing unidentified, 16.19
mutagenicity of, 2.25, 2.27
naturally occurring, 3.24, 3.25t.
oncogenicity of, 2.25
perfluorinated, 2.57–2.58
principles/concepts of, 3.2–3.24
redox chemistry, 3.14–3.17
stoichiometry of, 3.6–3.7
teratogenicity of, 2.25
in treatment additives, linings and coatings, 2.58–2.59
Chick's law, 17.21–17.23, 17.22f.
Chick-Watson law, 4.31–4.33, 17.22–17.23
Chitin, 8.48

Chloramine/chloramination, 2.64
application of, 17.33–17.34
bacteria inactivation by, 17.28
breakpoint reaction in, 17.11–17.13, 17.12f.
chlorine reaction with ammonia in, 7.19–7.20, 17.10–17.15, 17.12f., 17.15t.
chlorine reaction with inorganic constituents in, 17.16, 17.17t.
chlorine reaction with organic constituents in, 7.18–7.19, 17.16
DBP formation and impact of, 19.38, 19.38f.–19.39f.
DBPs and, 19.4t.–19.6t., 19.10–19.13, 19.10f., 19.12f.
formation of, 7.19–7.20
free available chlorine compared to, 21.24, 21.25f.
history of, 17.3
Legionella control with, 21.24–21.25
NDMA and, 19.11, 19.12f.
Chlorate, 2.64–2.65
formation of, 7.21–7.23
GAC reactions with, 14.59
Chloride
calcium hypochlorite, 7.17–7.18
corrosion from, 20.29–20.30
Chlorinated volatile organic chemicals, 2.46–2.49
Chlorine dioxide, 5.6–5.7, 7.34f.
application of, 17.36
bacteria inactivation by, 17.28
basic chemistry of, 17.9–17.10
chlorate formation, 7.21–7.23
chlorite formation, 7.21–7.23
DBPs and, 2.64–2.65, 19.4t.–19.6t., 19.13, 19.29, 19.29f.
demand reactions of, 17.18
GAC reactions with, 14.59
generation of, 7.21, 17.35–17.36
Henry's law and, 17.9
history of, 17.3
in taste and odor control, 7.21
Chlorine/chlorination, 2.64, 5.6
ammonia reactions with, 7.19–7.20, 17.10–17.15, 17.12f., 17.15t.
bacteria inactivation by, 17.27–17.28
basic chemistry of, 17.5–17.8
biological growth in treatment plants controlled by, 7.47
breakpoint reaction in, 17.11–17.13, 17.12f.
bromide reactions with, 7.20
chlorine demand kinetics, 17.17–17.18
Chlorine/chlorination (*Cont.*):
chlorine-chlorite process, 17.35–17.36
contact tank hydraulics in, 17.34
contact time in, 17.33t.
DBPs and, 19.4t.–19.6t., 19.7–19.10, 19.8f.–19.9f.
DBPs and delaying addition of, 19.36, 19.37f.
dechlorination, 17.33
design/application of, 17.31–17.34, 17.32t.
disinfection, UV disinfection compared to, 18.3, 18.3t.
disinfection efficiency and speciation of, 17.10–17.11, 17.11f.
forms of, 7.15–7.16
free available chlorine, 7.16, 17.6, 21.24, 21.25f.
in GAC adsorption systems, 14.44
GAC reactions with, 14.58–14.59
gaseous, 7.17–7.18
Henry's law and, 17.6–17.7
history of, 17.3
hypochlorite ion, 7.16–7.17, 7.16f.
hypochlorous acid, 7.16–7.17, 7.16f.
inorganic constituent reactions with, 17.16, 17.17t.
iodide reactions with, 7.20
liquid, 7.17–7.18
molecular chlorine, 7.16–7.17, 7.16f.
NOM studies with, 19.23
organic constituents reactions with, 7.18–7.19, 17.16
pH and, 17.6–17.8, 17.7f., 20.30, 20.31f.
postchlorination, 17.33
prechlorination, 17.32–17.33
residuals of, 17.30–17.31
sodium hypochlorite, 7.17–7.18
source water and, 17.32–17.33
species of, 7.16–7.17, 7.16f.
superchlorination, 17.33
TOC and decay of, 17.17–17.18
total available chlorine, 17.10
UV/Cl_2 AOP mechanism, 18.33
Chlorite, 2.64–2.65
chlorine-chlorite process, 17.35–17.36
formation of, 7.21–7.23
GAC reactions with, 14.59
3-Chloro-4-(dichloromethyl)-5-hydroxy-2(5h)-furanone (MX), 2.69
Chloroform, 2.67
Chlorophenoxyacetic acids, 2.54
Chlorpyrifos, 2.55
Chromium, 2.36–2.37

Chromophores, 18.20
CIP waste. *See* Clean-in-place waste
Clarifier sludge, 22.3
Clastogenicity, 2.25
Clean Water Act (CWA), 16.5–16.6
Clean-bed head loss, 10.36–10.38
Clean-in-place solutions, 11.91
Clean-in-place waste (CIP waste), 22.4, 22.44–22.45, 22.45t., 22.49–22.50, 22.49t.
Clearwells, 4.46
Clostridium perfringens, 2.22, 17.20
Coagulants. *See also* HMS coagulants; Polyelectrolyte coagulants
  acidity of, 8.33–8.36, 834t.
  alum, 8.2
  aluminum, 8.3
  combinations of, 8.49
  demand, 8.22
  dosage estimations for, 8.41–8.45
  effective acid content of, 8.42
  electrokinetic measurements in control/monitoring of, 8.68–8.70
  jar tests for, 8.64–8.66
  metal salts plus additives, 8.21
  metal salts plus strong acid, 8.21
  MF and, 8.2
  MF/UF integration removing arsenic with, 11.61, 11.61f.
  polyaluminum chloride, 8.20
  polyaluminum hydroxychloride, 8.20
  polyiron chloride, 8.20
  prehydrolyzed metal salts, 8.20–8.21
  residuals and, 22.4–22.16, 22.11–22.12, 22.11f.
  selecting appropriate, 8.41–8.45
  silica, 8.7, 8.8f., 8.48
  simple metal salts, 8.20
Coagulation. *See also* Enhanced coagulation
  chemical oxidation as aid to, 7.46
  DAF and, 9.56–9.58, 9.57f., 9.73–9.75, 9.74f.
  DBP precursor control with, 19.31, 19.32f., 19.33
  definition of, 8.2–8.3
  electrokinetic measurements in control/monitoring of, 8.68–8.70
  electrophoretic mobility and, 8.47, 8.47f.
  electrophoretic mobility in control/monitoring of, 8.68
  lime softening and, 13.53, 13.69–13.70, 13.69f.
  MF/UF integration with, 11.58–11.60, 11.58f.–11.59f.
Coagulation (*Cont.*):
  ozone and, 8.49
  in precipitative softening, 13.59–13.60
  rapid granular bed filtration and pretreatment, 10.43–10.45, 10.44f.
  SCD measurements in monitoring/control of, 8.68–8.69, 8.69f.
  sedimentation and, 9.41–9.42
  temperature's effect on, 8.63–8.64
  treatment process selection and, 5.7–5.9
Coarse-bed filtration, 9.87
COCODAF® dissolved air flotation, 9.69, 9.69f.
Cold polymictic lakes, 3.34
Coliforms. *See also* Fecal coliforms; Total coliform
  in distribution systems, 21.13, 21.13t., 21.27
  fermentation tube procedures and, 17.20
  in GAC/PAC, 21.22
  as indicators, 2.19–2.21, 17.20
  membrane filter procedures and, 17.20
Coliphages, 2.21
Collimated beam, 18.8–18.9, 18.8f.
Collision efficiency factor, 8.51
Colloids
  destabilization of, 8.14–8.15
  DLVO theory of stability for, 8.7
  like-charge electrostatic attraction of, 8.12
  mass transport of, 11.42–11.43, 11.43f.
Color
  aesthetic concerns with, 2.74–2.75
  chemical oxidation removing, 7.42–7.43
  corrosion and, 20.38
  DAF removing, 5.33, 5.33f.
  iron corrosion and, 20.41–20.42
  IX removing, 12.61–12.67, 12.63f.–12.65f.
  sludge production and, 22.6
Community Water System Survey (CWSS), 1.3–1.4
Compaction density, 22.14
Competitive adsorption
  in multisolute systems, 14.14–14.19, 14.15f., 14.18f., 14.18t.–14.19t.
  in natural waters, 14.21–14.26, 14.22f.–14.23f., 14.25f.
Composite membranes, 11.10–11.11, 11.11f.
Compound analysis, 20.71
Computational fluid dynamics (CFD), 4.24–4.26
  disinfection and, 17.41
  sedimentation and, 9.26, 9.88
  tracer tests and, 4.25–4.26

Concentration cell corrosion, 20.10
Concentration polarization
  layer, 11.38–11.41
  precipitative fouling and, 11.41
Concentrations
  adsorption and breakthrough, 14.32
  alkalinity, 3.4
  of bodyfeed and precoat filtration, 10.93–10.94, 10.94f.
  for carcinogens, based on health, 2.30
  charge, 3.4
  flocculation and particle, 4.45–4.46
  gas phase, 3.4
  of inorganic carbon chemistry in groundwater/surface water, 3.18t.
  ionic strength and, 3.5, 3.5t.
  of particles, 3.49–3.52, 3.51t.
  residuals and total metal, 22.15–22.16, 22.15t.
  of TOC, 3.62t.
  of TOC in GAC adsorption systems, 14.44–14.48, 14.47f.
  water quality evaluation with, 3.4
Concrete pipe corrosion, 20.64–20.65
Connections, 20.3
Consumer Confidence Report (CCR), 1.7, 1.19
  requirements of, 1.32
Contact time, 17.33t.
Contaminant Candidate List (CCL), 1.8–1.9, 1.10f., 1.11t.–1.13t., 2.31
Contaminants. *See also* Microbial contaminants
  coprecipitation removing, 13.4–13.5
  current regulations for, 1.19, 1.20t.–1.25t.
  endocrine disruptors, 2.56
  membrane processes size ranges of, 11.4, 11.5f., 11.6t.–11.7t., 11.8
  nanoparticles, 2.58
  pharmaceuticals, 2.56–2.57
  precipitation removing, 13.3
  in raw water compared to SFBW, 22.38, 22.38t.
  risk assessment of, 2.27–2.33
  trace organic, 15.17, 15.18t.
  treatment process selection, removal of, 5.29, 5.30t.–5.31t., 5.32
  UCMRs, 1.14–1.15
  USEPA regulatory determinations for, 1.9–1.10, 1.13, 1.14t.
Continuous flow reactors, 4.2–4.3
  CFSTR, 4.3
  conceptual models for, 4.22–4.24
  dispersion model for, 4.20–4.22, 4.21f.
  exit age distribution estimating removal in, 4.40–4.43, 4.42f.
  flocculation in ideal, 8.55–8.63, 8.57f.–8.58f. 8.60f., 8.57t.
  hydraulic characteristics of, 4.43–4.47, 4.44t.
  mathematical models for non-ideal, 4.16–4.24
  open/closed boundaries in, 4.21
  plug flow reactor, 4.2
Continuous flow stirred tank reactor (CFSTR), 4.3
  CFSTRs-in-series model, 4.17–4.18, 4.18f.
  PFR's reaction rate expressions compared to, 4.36, 4.38, 4.38t.
  reaction rate expressions in, 4.36, 4.37t.
  schematic of, 4.36f.
  tracer tests for, 4.13–4.16, 4.13f.–4.15f.
Continuous flow thickeners, 22.17–22.21, 22.18f.
Convective diffusion, 10.31–10.32, 10.31f.
Conventional treatment
  LBF compared to, 15.5t.
  of surface water, 5.33–5.34, 5.34f.
Copper, 2.37–2.38
  corrosion and, 20.30, 20.77–20.78
  pitting corrosion and, 20.50–20.52
  uniform corrosion of, 20.46–20.50, 20.47f., 20.49f.
Coprecipitation
  adsorption and, 13.4
  contaminants removed by, 13.4–13.5
  fluoride removal by, 13.74
  inclusion/occlusion and, 13.4
  metal ions removed by, 13.71–13.72
  principles of, 13.3–13.5
Cornwall, Ontario (drinking water case study), 18.32
Corrosion, 20.2–20.3
  A-C pipes and, 20.64–20.65
  activation control in, 20.13
  aeration for control of, 20.75
  alkalinity and, 20.25
  aluminum and, 20.35–20.36, 20.36f.
  ammonia and, 20.25
  anodes, 20.3–20.4, 20.6
  anodic current in, 20.4
  assessment of, 20.66–20.73
  brasses and, 20.53–20.54
  bronzes and, 20.53–20.54

Corrosion (*Cont.*):
buffer intensity and, 20.26–20.29, 20.27f.–20.28f.
cathodes, 20.3–20.4, 20.6
cathodic current in, 20.4
cement-mortar lined pipes and, 20.64–20.65
chemical factors affecting, 20.25–20.39, 20.80–20.81
chloride causing, 20.29–20.30
color and, 20.38
color from iron and, 20.41–20.42
compound analysis for, 20.71
concentration cell, 20.10
concrete pipes and, 20.64–20.65
connections in, 20.3
control of, 20.73–20.77
copper and, 20.30, 20.77–20.78
copper and pitting, 20.50–20.52
copper and uniform, 20.46–20.50, 20.47f., 20.49f.
coupon weight-loss testing for, 20.67
crevice, 20.10
dealloying, 20.7–20.8
DIC and, 20.25, 20.27f.
differential oxygenation, 20.10
distribution systems and controlling, 21.48–21.49
divalent lead and, 20.54–20.59, 20.55f., 20.57f.–20.58f.
duplex film inhibiting, 20.46
electrochemical rate measurements for, 20.67
electrolytic solution in, 20.3
electron microscopy methods for analyzing, 20.71
elemental spectroscopy after decomposition for analyzing, 20.69–20.70, 20.70t.
engineering considerations for controlling, 20.74
erosion, 20.10
flow velocity and, 20.21
galvanic, 20.6–20.7
galvanized steel and, 20.63–20.64
graphitization and, 20.10
hardness and, 20.32–20.35
human exposure evaluations for, 20.77–20.78
hydrogen sulfide accelerating, 20.35
immersion testing for, 20.68
iron and, 20.9, 20.40–20.45
kinetics of, 20.12–20.13
Larson ratio and, 20.29
lead and, 20.54–20.63
limestone contactors controlling, 20.75

Corrosion (*Cont.*):
LSI and, 20.33–20.34
magnesium and, 20.35
manganese and, 20.36, 20.37f.
manufacturing processes and, 20.24
mass balance equation and, 20.14–20.15
materials selection for controlling, 20.73–20.74
membrane cell theory and, 20.51
metallic precipitates and, 20.35–20.38, 20.36f.–20.37f.
microbiologically influenced, 20.8–20.10
mixed-potential theory and, 20.13
Nernst equation and, 20.4–20.5
NOM and, 20.8–20.9, 20.38
nonprotective scales and, 20.11
optical microscopic techniques for analyzing, 20.71
ORP and, 20.30, 20.31f., 20.32
orthophosphate and, 20.38, 20.43, 20.48–20.49, 20.49f.
oxygen and, 20.32
particulate lead and, 20.54
pH and, 20.38–20.39, 20.43, 20.46–20.47
photomicroscopy techniques for analyzing, 20.71
physical factors affecting, 20.20–20.24
physical inspect of, 20.66
pinhole leak and, 20.4
pipe linings, coating and paints for controlling, 20.76–20.77
pipe rig systems for, 20.67–20.68
on pipe surfaces, 20.11–20.12
of pipes in distribution systems, 21.26
pitting, 20.7–20.10, 20.7f., 20.50–20.52
planned interval tests for, 20.67
polyphosphates and, 20.39, 20.43–20.44
potential-pH diagrams (Pourbaix diagrams) and, 20.18–20.20, 20.19f.–20.20f.
protective scales and, 20.11
rate measurements of, 20.66–20.67
secondary effects of control measures for, 20.75–20.76
selective leaching in, 20.7–20.8
siderite model of, 20.41
silicates and, 20.39, 20.44–20.45
solubility diagrams for, 20.13–20.18, 20.16f.–20.18f.
statistical testing criteria for, 20.79–20.80
stray current, 20.10
sulfate causing, 20.9, 20.29–20.30, 20.48
TDS and, 20.39

Corrosion (*Cont.*):
temperature influencing, 20.21–20.24, 20.23f.–20.24f.
tetravalent lead and, 20.60–20.63, 20.60f.–20.61f.
thermodynamic driving forces of, 20.12–20.13
transport control in, 20.13
tuberculation and, 20.4, 20.8
types of, 20.5–20.7
uniform, 20.6, 20.46–20.50, 20.47f., 20.49f.
water sampling for controlling, 20.77–20.81, 20.80t.
wet chemical procedures for analyzing, 20.71
XAFS/XANES for identifying, 20.73
x-ray diffraction for, 20.72–20.73, 20.73f.
x-ray fluorescence spectrometry for analyzing, 20.70–20.71
zinc and, 20.35
County Sanitation Districts of Los Angeles County, 16.29–16.30
Coupon weight-loss testing, 20.67
Crevice corrosion, 20.10
Critical control points (CCPs), 16.19–16.21
Cross-connections, 21.29, 21.60–21.63, 21.62t.
Cryptosporidium, 2.15–2.16, 10.41, 17.20
backwashing and recycling of, 10.69
DAF and removal of, 9.81–9.83, 9.82f.–9.83f.
DE filtration removing, 10.8
filtration and removal of, 10.5–10.8, 10.6t.
IPR and, 16.7–16.8, 16.7t.
membrane processes removing, 11.33–11.34, 11.33f.
MF/UF removing, 11.33–11.34, 11.33f.
Milwaukee outbreak of, 2.3
as ovoid particles, 8.3–8.4
precoat filtration removing, 10.88–10.89
pressure filtration removing, 10.8
rapid granular bed filtration and, 10.22
in SFBW, 22.42
SSF removing, 5.12, 10.8, 10.78–10.79
CST test. *See* Capillary suction time test
Cumulative age distribution, in tracer tests, 4.4, 4.5f., 4.6, 4.7f., 4.8, 4.12f.
Cuprosolvency, 20.6
Cuprous oxide, 20.51
CWA. *See* Clean Water Act
CWSS. *See* Community Water System Survey
Cyanobacteria, 2.18–2.19
Cyanotoxins
chemical oxidation of, 7.45
PAC removing, 14.84
Cyclospora, 2.17
2,4-D. *See* 2,4-Dichlorophenoxyacetic acid
Dacthal, 2.54
DAF. *See* Dissolved air flotation
DALYs. *See* Disability-adjusted life years
Dan Region, Israel (SAT system), 15.28
Darcy-Weisbach equation, 10.13, 10.37
Dayton, Ohio's magnesium-carbonate process, 13.51–13.52
DBCM. *See* Dibromochloromethane
DBCP. *See* 1,2-Dibromo-3-chloropropane
DBPs. *See* Disinfection by-products
DCA. *See* Dichloroacetic acid
DE filtration. *See* Diatomaceous earth filtration
Dealloying, 20.7–20.8
Dechlorination, 17.33
Decolorizing index, 14.7
Department of Homeland Security (DHS), 1.7
Depth filtration, 10.3
Desorption, 14.36–14.37
Developmental toxicity, 2.25
Dewatering lagoons, 22.25
DHS. *See* Department of Homeland Security
Diatomaceous earth filtration (DE filtration), 10.91. *See also* Precoat filtration
Cryptosporidium and *Giardia lamblia* removal by, 10.8
Diazinon, 2.55
1,2-Dibromo-3-chloropropane (DBCP), 2.55–2.56
Dibromochloromethane (DBCM), 2.67
DIC. *See* Dissolved inorganic carbon
Dichloroacetic acid (DCA), 2.68
Dichlorobenzenes, 2.47
1,2-Dichloroethane, 2.47
1,1-Dichloroethylene, 2.47–2.48
1,2-Dichloroethylenes, 2.48
Dichloromethane, 2.48
2,4-Dichlorophenoxyacetic acid (2,4-D), 2.54
Differential oxygenation corrosion, 20.10
Differential settling, 8.52
Diffuse layers, 8.8–8.11, 8.9f.–8.10f.
Diffused aeration. *See* Bubble aeration
Dimensionless dispersion number, 17.38
Dimictic lakes, 3.35–3.36
Direct filtration, 10.4, 22.3
advantages of, 10.70
disadvantages of, 10.70–10.71
effluent turbidity in, 10.73
filtration rates for, 10.73–10.74
in Las Vegas, 10.73
Los Angeles plant for, 10.73
pretreatment for, 10.72–10.73
source waters for, 10.71–10.72

Direct potable reuse (DPR), 16.2. *See also* Water reuse
  in Windhoek's Goreangab Reclamation Plant, Namibia, 16.32–16.33
Disability-adjusted life years (DALYs), 16.18
Discrete particle settling
  drag force and, 9.5–9.7, 9.6f.
  flocculation and, 9.7–9.9
  particle shape effecting, 9.7
  predicting settling efficiency, 9.11
  Reynolds number and, 9.6, 9.6f., 9.13
  in sedimentation, 9.5–9.11, 9.6f.
  settlement in tanks, 9.9–9.11, 9.10f.
  settling velocity, 9.9–9.11, 9.10f.
  terminal settling velocity in, 9.5–9.7
Disinfectants. *See also* Microbial/Disinfection By-Product
  alternative primary, 19.38
  alternative secondary, 19.38, 19.38f.–19.39f.
  background of, 2.59–2.60
  basic chemistry of, 17.5–17.10
  decay of, 4.33, 4.33f.
  microbial growth and residuals of, 21.23–21.25
  regulations for, 1.19
  residuals and, 17.30–17.31, 21.51–21.53
Disinfection, 17.1–17.2. *See also* UV disinfection
  $C \times T$ approach for regulation of, 17.39–17.41, 17.40t.
  CFD and, 17.41
  Chick's law and, 17.21–17.23, 17.22f.
  Chick-Watson law of, 4.31–4.33, 17.22–17.23
  chlorine speciation and efficiency of, 17.10–17.11, 17.11f.
  chlorine v. UV disinfection for, 18.3, 18.3t.
  contact time in, 17.33t.
  DBP control, modifying, 19.36–19.39
  distribution systems and secondary, 21.49, 21.50f. 21.50f.
  history of, 17.3–17.4
  Hom's law and, 17.24–17.25
  IDDF for, 17.38
  indicators and, 17.19–17.21
  kinetics, 17.21–17.27, 17.22f., 17.37–17.38
  microbial risk assessment for, 17.19–17.21
  microorganism physical/physiological factors influencing efficiency of, 17.30
  models for decay of, 17.26–17.27
  power law models for, 17.26
  preoxidation in, 17.32–17.33
  regulation issues for, 17.4–17.5, 18.3
Disinfection (*Cont.*):
  residence time and, 17.37–17.38
  secondary, 17.31
  SWTR and, 17.4–17.5
  technique comparisons for, 17.41–17.42, 17.42t.
  temperature's impact on, 17.27
  terminal, 17.33
  treatment process selection and, 5.22–5.24
  trends in, 17.2, 17.2t.
Disinfection by-products (DBPs), 2.2, 11.4, 19.1–19.2
  adsorption for postformation control of, 19.39–19.40
  adsorption for precursor control of, 19.34
  AOPs for precursor control of, 19.29–19.30, 19.30f.
  ASR for controlling, 19.28
  background of, 2.59–2.60
  BDCM, 2.67
  biofiltration removing, 19.40–19.41, 19.42f.
  bromate, 2.66
  bromide, 19.20–19.22, 19.20f.
  brominated acetic acids, 2.68–2.69
  bromine, 2.65–2.66, 19.21–19.22, 19.21f.–19.22f.
  bromoform, 2.67–2.68
  carcinogenicity of, 2.61t.–2.63t.
  chemical oxidation for control of, 19.28–19.31, 19.29f.–19.31f.
  chemistry changes for control of, 19.38–19.39
  chloramine and, 19.4t.–19.6t., 19.10–19.13, 19.10f., 19.12f.
  chloramines, 2.64
  chlorate, 2.64–2.65
  chlorination addition delayed for controlling, 19.36, 19.37f.
  chlorination and, 19.4t.–19.6t., 19.7–19.10, 19.8f.–19.9f.
  chlorine, 2.64
  chlorine dioxide and, 2.64–2.65, 19.4t.–19.6t., 19.13, 19.29, 19.29f.
  chlorite, 2.64–2.65
  chloroform, 2.67
  coagulation for precursor control of, 19.31, 19.32f., 19.33
  DBCM, 2.67
  DCA, 2.68
  disinfectant dose and, 19.17–19.19, 19.17f.–19.18f.
  disinfection modification for, 19.36–19.39

Disinfection by-products (DBPs) (*Cont.*):
DOC and formation of, 19.24–19.25, 19.25f.–19.26f.
evaluation of health effects of, 2.60, 2.61t.–2.63t., 2.63–2.64
factors influencing formation of, 19.16–19.27
formation potential test for, 19.23
GAC adsorption systems removing, 14.43–14.44, 14.46–14.48, 19.39–19.40, 19.41f.
genotoxicity of, 2.61t.–2.63t.
granular media filters for precursor control of, 19.34–19.35
HAAs, 2.68, 19.7–19.8
halogenated, 7.2, 19.16–19.17
inorganic, 2.64–2.66
iodide, 19.20–19.21
iodine, 2.66, 19.21–19.22, 19.21f.–19.22f.
in IPR, 16.12t.–16.13t., 16.15–16.16
IX for precursor control of, 19.33–19.34, 19.33f.
major groups of, 19.3, 19.4t.–19.6t., 19.7
membrane separation and, 19.43
models for formation of, 19.27
MX, 2.69
NF/RO membrane process removing, 11.29–11.30, 11.30t., 19.35–19.36
nitrosamines, 2.69
NOM and, 19.3
occurrence of, 2.61t.–2.63t.
organic, 2.66–2.69
ozonation and, 19.4t.–19.6t., 19.13–19.14, 19.14f.–19.15f., 19.29, 19.29f.
ozone, 2.66
PAC removing precursors of, 14.85
packed tower aeration removing, 19.42, 19.42f.
pH and, 19.19
postformation removal of, 19.39–19.43
precursor material and, 19.22–19.25
precursor material removal for controlling, 19.27–19.36
reaction time and, 19.16–19.17, 19.16f.
regulation of, 2.60, 19.45–19.46
residuals and, 19.17–19.19
seasonal effects on, 19.25–19.27
simulated distribution system test for, 19.24
source control of precursors to, 19.28
spatial temporal variability of, 19.46
stability of, 19.43–19.45
TCA, 2.68
temperature and, 19.20

Disinfection by-products (DBPs) (*Cont.*):
THMs, 2.66–2.67, 19.7–19.8
uniform formation conditions test for, 19.23
UV and, 19.4t.–19.6t., 19.14–19.15
UV disinfection and, 18.15
Disinfection By-Products Rule, 2.60
Dispersed air flotation, 9.46
Dissolved air flotation (DAF)
air quantity required for, 9.76–9.77, 9.77f.
air release devices for, 9.77–9.78
air saturation systems and, 9.70–9.72, 9.71f.–9.72f.
air solubility in pressurized recycle water exiting saturator in, 9.50–9.51, 9.50f.
air solubility in water in, 9.48–9.49, 9.49f., 9.49t.
air/solids ratio on float in, 9.78
algae/color/TOC removal with, 5.33, 5.33f.
bubble rise velocities in, 9.59–9.60, 9.59f.–9.60f.
bubble suspension in contact zone in, 9.51–9.54, 9.53t.
circular tanks for, 9.68
coagulation and, 9.56–9.58, 9.57f., 9.73–9.75, 9.74f.
COCODAF® design for, 9.69, 9.69f.
compactness of, 9.87
contact zone in, 9.47–9.48, 9.47f., 9.54
contact zone model in, 9.55–9.59, 9.56t., 9.57f.
conventional systems for, 9.65–9.68, 9.66t.
costs of, 9.86–9.87
countercurrent flotation tank for, 9.69, 9.69f.
Cryptosporidium removal and, 9.81–9.83, 9.82f.–9.83f.
degree of agitation in, 9.75–9.76
efficiency factors for, 9.72–9.79
emerging technology for, 9.88–9.89
examples of, 9.52–9.53, 9.62, 9.64–9.65
filtration combined with, 9.69–9.70, 9.83–9.84, 9.84f.
float removal and, 9.78–9.79
for floc separation in flocculation, 8.65
floc-bubble aggregate rise velocities in, 9.60–9.62, 9.61f.
flocculation and, 9.47, 9.56–9.58, 9.57f., 9.75–9.76, 9.75t., 9.76f.
flocculation time and, 9.75–9.76, 9.75t.
fundamentals of, 9.48–9.65
Giardia lamblia removal and, 9.81–9.83, 9.82f.–9.83f.
high-rate systems for, 9.65–9.68, 9.66t., 9.67f.

Dissolved air flotation (DAF) (*Cont.*):
history of, 9.45–9.46
hydraulic flocculation and, 9.76, 9.76f.
microflotation in, 9.46
models for, 9.54–9.59
moist bubble densities for, D.3t.
NOM removal by, 9.83–9.84, 9.84f.
nomenclature of, 9.89–9.92
performance case studies for, 9.79–9.84
pressure flotation in, 9.46
rectangular tanks for, 9.68–9.69
schematic of, 9.47f.
sedimentation compared with, 9.84–9.88, 9.85t.
separation zone in, 9.48, 9.54, 9.63–9.65, 9.63f.
sludge removal and, 9.88
solids loading and, 9.86
source water influence on float in, 9.78–9.79
tank types for, 9.68–9.70, 9.69f.
tank variables for, 9.58–9.59
treatment of low turbidity, low color waters (case study), 9.80
treatment of turbid lowland (high-alkalinity) river water (case study), 9.79–9.80
treatment of waters with algae (case study), 9.81, 9.81t., 9.82f.
treatment of waters with natural color and humic substances (case study), 9.80, 9.80t.
treatment process selection and, 5.10–5.11
turbidity and, 9.83–9.84, 9.84f.
vacuum flotation in, 9.46
of wastewater, 9.46
Dissolved chemicals, 3.24, 3.26
groundwater, composition of, 3.41t.
Dissolved inorganic carbon (DIC), 16.16
buffer intensity and, 20.26–20.29, 20.27f.–20.28f.
corrosion and, 20.25, 20.27f.
in surface water, 3.29
types of, 3.18, 3.18t.
Dissolved organic carbon (DOC), 3.34, 3.59, 3.61
adsorption of, 8.29–8.30
DBP formation and, 19.24–19.25, 19.25f.–19.26f.
in IPR, 16.11, 16.12t.–16.13t.
IX removing, 12.61–12.67, 12.63f.–12.65f.
magnetic ion exchange for removal of, 12.66–12.67
measurements of, 3.62–3.63, 3.66–3.67
natural treatment systems removing, 15.17
POC and, 13.64
precipitative softening removing, 13.65–13.67
UV as surrogate for, 3.63–3.65
Dissolved organic matter (DOM)
categories of, 14.2–14.3, 14.2f.
as micropollutants, 14.2
Dissolved oxygen, 20.32
Distillation, 5.16
Distribution systems
actinomyces in, 21.19
Aeromonas hydrophyla in, 21.16–21.17
bacteria identification in, 21.9–21.10
bacterial nutrients in, 21.25–21.26
biofilms compared to suspended microbes in, 21.7–21.8, 21.8f.
coliform bacteria in, 21.13, 21.13t.
coliforms in, 21.13, 21.13t., 21.27
corrosion control in, 21.48–21.49
cross-connections in, 21.29, 21.60–21.63, 21.62t.
design standards for, 21.46–21.47
engineering and design of, 21.43–21.47
engineering transient control strategies for, 21.60
epidemiological studies on risks in, 21.4–21.7
fecal coliforms in, 21.13
free-living protozoa in, 21.18
fungi in, 21.18–21.19
HCGI and, 21.4–21.7
heterotrophic bacteria in, 21.10–21.13, 21.11t., 21.12f.
hydraulic models for, 21.44–21.45, 21.44f.
integrated system management for, 21.43, 21.43f.
intrusion events in, 21.29–21.30
iron in, 21.20
Legionella in, 21.14–21.16
low/negative-pressure transients in, 21.30
maintenance control strategies for, 21.60, 21.61f.
manganese in, 21.20
microbial contaminant control for, 21.47–21.65
microbial contaminants in, 21.7–21.21
microbial growth in, 21.22–21.28
monitoring, 21.34–21.43
Mycobacterium in, 21.17
nitrification in, 21.28
nitrogen in, 21.19–21.20
nutrient loading in, 21.48
on-line monitoring in, 21.37–21.38
operational control strategies for, 21.58–21.60, 21.59f.

Distribution systems (*Cont.*):
- operation/maintenance approaches to microbial contaminants in, 21.50–21.55, 21.51f., 21.51t., 21.53t., 21.55t.
- pathogen occurrence in, 21.9
- physical integrity of, 21.55–21.65, 21.56t.–21.57t.
- pipe corrosion in, 21.26
- recontamination of, 21.28–21.34, 21.30t., 21.33t.
- residence time in, 21.27
- sampling frequency for, 21.41
- sampling optimization for, 21.41
- sampling plan design, theoretical approach, for, 21.38–21.41
- secondary disinfection in, 21.49, 21.50f. 21.50f.
- stressed organisms in, 21.14
- sulfur in, 21.21
- TEVA for, 21.42
- treatment process selection and, 5.25–5.27, 21.21–21.22
- trench work in, 21.64
- UDF in, 21.52–21.53
- USEPA regulation requirements for, 21.35–21.36, 21.35t.–21.36t.
- VBNC bacteria in, 21.14, 21.15t.
- water age control in, 21, 21.51–21.52, 53t.
- water main installation and repair for, 21.63–21.65
- water quality and design practices of, 21.46–21.47
- water quality in, 5.25–5.27
- water quality models for, 21.44–21.45, 21.44f.
- water quality parameter measurements in, 21.36–21.38
- water security in, 21.41–21.42
- waterborne disease outbreaks and, 21.2–21.4

Diuron, 2.54
Divalent lead corrosion, 20.54–20.59, 20.55f., 20.57f.–20.58f.
DLVO theory of colloid stability, 8.7
DNA, 18.11, 18.11f.
DOC. *See* Dissolved organic carbon
DOM. *See* Dissolved organic matter
Dose-response relationship, 2.23–2.24
DPR. *See* Direct potable reuse
Drag force, 9.5–9.7, 9.6f.
Drinking water advisories (DWAs), 1.18–1.19
Drinking water equivalent level (DWEL), 2.29
Drinking water standards. *See* Standards
Drinking Water State Revolving Fund (DWSRF), 1.28–1.29
Driving force, 11.21–11.22, 11.22t.
Dune infiltration, 15.8–15.9
- in Leiduin, the Netherlands, 15.26–15.27

Duplex film, 20.46
Düsseldorf, Germany (RBF system), 15.25–15.26
DWAs. *See* Drinking water advisories
DWEL. *See* Drinking water equivalent level
DWSRF. *See* Drinking Water State Revolving Fund

ED. *See* Electrodialysis
EDB. *See* Ethylene dibromide
EDR. *See* Electrodialysis reversal
Effluent organic matter (EfOM), 14.3
EfOM. *See* Effluent organic matter
Electrical resistance, 3.56–3.57
Electrochemical potentials, 7.2–7.4, 7.3t.
Electrochemical rate measurements (corrosion), 20.67
Electrodialysis (ED), 5.16
- cast membrane sheets in, 11.23, 11.24f.
- cell, 11.23–11.24, 11.23f.
- cell pairs, 11.24, 11.25f.
- mass transport in, 11.50–11.53
- membrane stacks in, 11.24, 11.26, 11.26f.
- MF/UF integration with, 11.59–11.60, 11.59f.
- modules in, 11.24, 11.26, 11.26f.
- pilot plant testing for, 11.95–11.96
- process of, 11.22, 11.22f.
- use of, 11.4, 11.8

Electrodialysis reversal (EDR), 11.2
- mass transport in, 11.51–11.53, 11.53f.–11.54f.
- pilot plant testing for, 11.95–11.96
- uses of, 11.4, 11.8

Electrolytic flotation, 9.46
Electrolytic solution, 20.3
Electromagnetic spectrum, 18.4, 18.4f.
Electromotive force (EMF), 3.15–3.16
Electron microscopy, 20.71
Electroneutrality, 3.5–3.6
Electrophoretic mobility
- coagulation and, 8.47, 8.47f.
- coagulation control/monitoring and measurements of, 8.68
- jar test interpretations for, 6.69–6.70, 6.70f.

Electrostatic stabilization
electrical double layer in, 8.8–8.11, 8.9f.–8.10f.
Gouy-Chapman model in, 8.8–8.9, 8.9f.
like-charge electrostatic attraction of colloids in, 8.12
of particles, 8.7–8.12
secondary minimum aggregation in, 8.11–8.12
surface charge origins in, 8.7–8.8, 8.8f.
Elemental spectroscopy after decomposition for, 20.69–20.70, 20.70t.
EMCT. *See* Equilibrium multicomponent chromatographic theory
EMF. *See* Electromotive force
Empty-bed contact time
GAC adsorption systems and, 14.40–14.41, 14.41f.
IX and, 12.25, 12.27
NOM breakthrough and, 14.40–14.41, 14.41f.
Endocrine disruptors, 2.56
ozonation of, 7.45–7.46
Enhanced coagulation, 8.5–8.6, 8.5t.
Enhanced Surface Water Treatment Rules, 2.15
Enmeshment, 8.24
by hydroxide precipitates, 8.62–8.63
Entamoeba histolytica, 2.16–2.17
Enteric bacteria, 2.7–2.8
Enteric-like bacteria, 2.8
Enterococcus, 2.21
Enteroviruses, 2.13–2.14
Environmental buffer, in IPR, 16.23
Environmental considerations, for treatment process selection, 5.28–5.29
Epichlorohydrin, 2.59
Equilibrium, gas transfer, 6.2–6.10
Equilibrium constant, 3.8–3.9, 6.5
Equilibrium multicomponent chromatographic theory (EMCT), 12.2
IX modeling using, 12.74–12.82, 12.76f.–12.82f.
Equivalent background compound, 14.23–14.26, 14.23f.
Ergun equation, 10.13, 10.37–10.38, 10.58–10.59
Erosion corrosion, 20.10
Escherichia coli, 2.6
as indicator, 2.20–2.21
Ethylbenzene, 2.45–2.46
Ethylene dibromide (EDB), 2.55–2.56
Ethylene dichloride, 2.47
Ethylene thiourea (ETU), 2.56
ETU. *See* Ethylene thiourea
European Union, standards of, 1.33
Excess lime softening, 13.30f., 13.34–13.46
equilibria calculations for, 13.39–13.46
stoichiometric calculations for, 13.35–13.39
Exhaustion rate, 12.27
Exit age distribution
continuous flow reactors estimating removal with, 4.40–4.43, 4.42f.
in tracer tests, 4.4, 4.5f., 4.6, 4.7f., 4.8, 4.12f.
External (film) resistance to transport, 14.28

Fabricated self-supporting underdrain system, 10.24
FACA. *See* Federal Advisory Committee Act
False-floor underdrain with nozzles, 10.25, 10.25f.
Fecal coliforms
in distribution systems, 21.13
as indicators, 2.20
Federal Advisory Committee Act (FACA), 1.25–1.26
Fermentation tube procedures, 17.20
Ferric hydroxide, 8.17–8.19, 8.18f., 8.18t.
precipitate solubility of, 8.41
Ferric iron salts, 8.22–8.30
Film theory model, 11.78–11.79
Filter backwash, 1.19
Filter elements, 10.87, 10.89, 10.90f.–10.91f.
Filter presses, 22.35–22.37, 22.36f.–22.37f.
Filtration. *See also* Diatomaceous earth filtration; Direct filtration; Granular media filters; Gravity filters; In-line filtration; Lakebank filtration; Membrane filtration; Precoat filtration; Pressure filtration; Rapid granular bed filtration; Slow sand filtration
in air, 10.27, 10.27f.
available head loss in, 10.11
backwashing and, 10.11
cake, 10.3
configurations of, 10.4–10.5
conventional, 10.4
Cryptosporidium removal and, 10.5–10.8, 10.6t.
DAF combined with, 9.69–9.70, 9.83–9.84, 9.84f.
depth, 10.3
filter cycle, 10.11
filter run, 10.11
flow control systems for, 10.49–10.53, 10.50f.
Giardia lamblia removal and, 10.5–10.8, 10.6t.
GWR and, 10.6–10.7
in lime softening, 13.61
mechanical control systems for, 10.49

Filtration (*Cont.*):
  microorganism removal by, 10.7–10.8
  nonmechanical control systems for, 10.49, 10.52
  Proportional-level declining rate control systems for, 10.52
  proportional-level equal rate control systems for, 10.51
  proportional-level influent flow splitting for, 10.51
  rates of, 10.21–10.22
  regulatory requirements for, 10.5–10.7, 10.6.t.
  secondary roles of, 10.4–10.5
  as sedimentation alternative, 9.88–9.89
  straining, 10.3
  SWTR and requirements for, 10.5–10.6, 10.6t.
  terminal head loss in, 10.11
  types of, 10.3–10.4
  vacuum filtration, 10.3, 22.30–22.31, 22.30f.
  variable-level declining rate control systems for, 10.51–10.52
  variable-level influent flow splitting for, 10.50–10.51
FISH. *See* Fluorescence in situ hybridization
Flat-bottom clarifiers, 9.3, 9.3f., 9.37
Float. *See* Residuals; Sludge
Floc ballasting, 9.37–9.39, 9.38f.
  compactness of, 9.87
  inclined settling and, 9.30
Floc-blanket process, 5.10
  ballasted floc systems for, 9.37–9.39, 9.38f.
  clarification mechanism for, 9.23
  costs of, 9.86–9.87
  effective depth in, 9.34
  examples of, 9.17, 9.18f., 9.36, 9.41
  fine sand ballasting in, 9.37–9.38, 9.38f.
  flat-bottomed tanks in, 9.37
  floc volume fraction for, 9.19–9.20
  floc-blanket reactor-PAC-UF process, 14.80–14.81, 14.80f.
  history of, 9.2–9.3
  hopper-bottomed tanks, 9.33–9.35, 9.33f.–9.35f.
  inclined settling with, 9.37
  lime softening and, 13.54–13.55
  performance prediction for, 9.23–9.25, 9.24f.
  sludge recycling in, 9.37
  sludge removal in, 9.35
  solids loading in, 9.84, 9.86
  tanks for, 9.33–9.39, 9.33f.–9.35f., 9.38f.
  upflow velocity and, 9.24f., 9.25
  wind effects on, 9.45
Floc-bubble aggregates, 9.60–9.62, 9.61f.
Flocculant aids, 9.43–9.45, 9.44f.
Flocculant polymers, 8.46
Flocculation, 4.27–4.28. *See also* Coagulation; Rapid mixing; Sedimentation
  ballasted, 5.10, 8.49
  in batch/plug flow reactor kinetics, 8.55
  Brownian diffusion in, 8.51–8.52
  chemical oxidation as aid to, 7.46
  collision efficiency factor in, 8.51
  DAF and, 9.47, 9.56–9.58, 9.57f., 9.75–9.76, 9.75t., 9.76f.
  DAF for floc separation in, 8.65
  definition of, 8.2–8.3
  by differential settling, 8.52
  discrete particle settling and, 9.7–9.9
  floc density and enmeshment by hydroxide precipitates in, 8.62–8.63
  floc density in, 8.60–8.62
  floc disaggregation and, 8.59, 8.60f.
  floc size in, 8.59–8.62
  G value concept and, 8.53–8.54
  hydraulic, 9.76, 9.76f.
  in ideal continuous flow reactors, 8.55–8.63, 8.57f.–8.58f. 8.60f., 8.57t.
  index, 8.66
  jar tests and floc particle separation in, 8.65
  kinetics, 4.28
  lime softening and, 13.52–13.53
  mathematical models of, 8.61
  MF/UF integration with, 11.58–11.60, 11.58f.–11.59f.
  orthokinetic, 8.52
  particle concentration in, 4.45–4.46
  purpose of, 8.50
  sedimentation and, 4.46, 9.7–9.9, 9.42–9.43, 9.42f.–9.43f.
  Stokes' law and, 8.52, 9.6
  sweep, 8.24
  temperature's effect on, 8.63–8.64
  transport in laminar shear and, 8.52
  transport mechanisms in, 8.50–8.54
  treatment process selection and, 5.7–5.9
  turbidity, 8.66–8.67, 8.67f.
  in turbulent transport, 8.52–8.53
Florida state regulations for IPR, 16.38
Flotation. *See* Dissolved air flotation
Flow velocity, 20.21

Flow-through curves, 9.26
Fluidization
clarifiers as collectors in, 9.17–9.19
head loss in granular media filters and, 10.58, 10.59f.
point of incipient, 10.58–10.60
sedimentation and, 9.14–9.20
Fluidized bed crystalizers, 13.56–13.58, 13.57f.–13.58f.
Fluorescence in situ hybridization (FISH), 21.10
Fluoride, 2.38
AA and breakthrough curves for, 12.41, 12.41f.
AA capacity for, 12.42–12.45, 12.42t., 12.44f., 22.57, 22.57t.
AA removal of, 12.7–12.8, 12.41–12.46
coprecipitation removing, 13.74
regeneration of fluoride spent alumina columns in AA, 12.45–12.46
residuals from AA removal of, 22.57, 22.58t.–22.59t.
Flushing, 21.53–21.54
Foam flotation, 9.45
Formation potential test, 19.23
Fort Benton, Montana (UV disinfection), 18.17
Fouling indexes, 11.9t., 11.67–11.69
Fractal dimension, 8.62
Fractal geometry, 8.61
Free available chlorine, 7.16, 17.6
chloramine compared to, 21.24, 21.25f.
Freeze-thaw beds, 22.25–22.26
Freundlich equation, 14.9–14.12, 14.9f., 14.10t.–14.11t.
Froth flotation, 9.45
Froude number, 9.27
Fulvic acids, of NOM, 3.60–3.61, 3.61f.
Fungi, 21.18–21.19
Fungicides, 2.56

G value concept, 8.53–8.54
GAC. *See* Granular activated carbon
GAC adsorption systems
atrazine breakthrough curves in, 14.54–14.56, 14.55f.–14.56f.
backwashing of, 14.41
bed life/carbon usage of, 14.63–14.64
bench-scale column tests for, 14.64–14.65
biological activity in, 14.42–14.44, 14.43f., 14.43t.
characteristics of, 14.37–14.39
chlorine in, 14.44
constant diffusivity design equation for, 14.64–14.65
GAC adsorption systems (*Cont.*):
data analysis for, 14.67–14.73, 14.67f.
DBP precursor removal by, 14.43–14.44, 14.46–14.48
DBP removal with, 19.39–19.40, 19.41f.
design conditions for, 14.47, 14.47t.
empirical models for, 14.74–14.77, 14.75f.–14.76f., 14.75t.
empty bed contact time and, 14.40–14.41, 14.41f.
field scale efficiency of, 14.46–14.58, 14.54t.
flow arrangements for, 14.38–14.39
HLR in, 14.40–14.41
initial TOC concentration in, 14.44–14.48, 14.47f.
mathematical models for, 14.73–14.74
parallel-column analysis for, 14.68–14.72, 14.69f.–14.71f.
particle size in, 14.39–14.40
performance of, 14.57–14.58
pilot-plant tests for, 14.66–14.67
preozonation and, 14.42–14.44, 14.43f., 14.44t.
pretreatment for, 14.44–14.46, 14.45f.
removal efficiency factors for, 14.39–14.46
series-column analysis for, 14.72–14.73, 14.72f.
single-stage, 14.37–14.38
SOC removal by, 14.51–14.56
taste and odor removal by, 14.48–14.51, 14.49f.–14.50f.
uniformity coefficient in, 14.39–14.40
VOC removal by, 14.51–14.53, 14.52f., 14.52t.
Galvanic corrosion, 20.6–20.7
Galvanized steel corrosion, 20.63–20.64
Gang stirrer, 8.64
Gas phase concentrations, 3.4
Gas transfer
equilibrium, 6.2–6.10
Henry's law constant plots, 6.4f.
Henry's law constant temperature and, 6.8t.–6.9t.
Henry's law constants for gases in water compounds, 6.8t.
Henry's law constants for organic compounds, 6.6t.–6.7t.
Henry's law for, 3.10t., 5.5, 6.2–6.10
Henry's law unit conversions, 6.5t.
mass transfer in, 6.10, 6.11f., 6.12–6.14
salting-out (Setschenow) coefficients, 6.9–6.10, 6.10t.
theory of, 6.2–6.14

Gaseous residuals, 22.4
Gases, physical properties of, D.2t.
Genotoxicity, 2.25, 2.27
  of DBPs, 2.61t.–2.63t.
Geosmin, 2.74
  ozone removing, 7.38–7.39, 7.40f.–7.41f.
Giardia lamblia, 2.15, 17.20
  backwashing and recycling of, 10.69
  $C \times T$ approach for regulation of disinfection for, 17.39–17.41, 17.40t.
  DAF and removal of, 9.81–9.83, 9.82f.–9.83f.
  DE filtration removing, 10.8
  filtration and removal of, 10.5–10.8, 10.6t.
  IPR and, 16.7–16.8, 16.7t.
  membrane processes removing, 11.33–11.34, 11.33f.
  MF/UF removing, 11.33–11.34, 11.33f.
  as ovoid particles, 8.3–8.4
  precoat filtration removing, 10.88–10.89
  pressure filtration removing, 10.8
  rapid granular bed filtration and, 10.22
  in SFBW, 22.42
  SSF removing, 5.12, 10.8, 10.78–10.79
Giardia muris, 10.41
Glyphosate, 2.54
Gouy-Chapman model, 8.8–8.9, 8.9f.
Graetz-Leveque correlation, 11.41
Granular activated carbon (GAC), 5.11–5.13. *See also* GAC adsorption systems
  adsorption capacity of, 14.62
  adsorption of humic substances on, 14.27
  adsorption qualities of, 10.15, 14.6–14.7
  backwashing and, 10.20–10.21, 10.68
  breakthrough curves for, 14.31–14.34, 14.33f.–14.44f.
  bromate reactions with, 14.59
  chemical characteristics of, 14.6
  chlorate reactions with, 14.59
  chlorine dioxide reactions with, 14.59
  chlorine reactions with, 14.58–14.59
  chlorite reactions with, 14.59
  coliforms in, 21.22
  fouling, 14.26–14.27
  isotherm in performance estimation for, 14.59–14.64, 14.61f.
  off-gas control with, 6.36–6.41, 6.37f.–6.38f., 6.39t.
  PAC compared to, 5.20–5.21, 14.3
  perchlorate reactions with, 14.59
  physical characteristics of, 14.4–14.6, 14.5t.
  preloading effect and, 14.53
  production of, 14.4
Granular activated carbon (GAC) (*Cont.*):
  in rapid granular bed filtration, 10.20–10.21
  reactivation by-products and, 14.87
  reactivation furnaces and, 14.86
  reactivation performance and, 14.86–14.87
  reactivation principles and, 14.85–14.86
  turbidity reduction with, 14.57–14.58
  water vapor isotherms for, 6.38f.
Granular media adsorption, 12.7–12.9
Granular media filters, 4.46, 10.3
  bed depth and media size in, 10.18
  configurations for, 10.15–10.21, 10.16f.
  Darcy-Weisbach equation for, 10.13, 10.37
  for DBP precursor control, 19.34–19.35
  Ergun equation for, 10.13, 10.37–10.38, 10.58–10.59
  fixed-bed porosity in, 10.14, 10.14t.
  grain density (specific gravity) in, 10.13, 10.14t.
  grain hardness in, 10.13–10.14
  grain shape in, 10.13, 10.14t.
  grain size and distribution in, 10.12–10.13, 10.12f.
  grain sphericity in, 10.13, 10.14t.
  head loss for fixed-bed flow in, 10.59
  head loss in fluidized, 10.58, 10.59f.
  head loss loss development in, 10.38–10.39
  Kozeny equation for, 10.13, 10.37–10.38
  media properties in, 10.12–10.15, 10.14t.
  media types in, 10.11
  performance of, 10.9, 10.10f., 10.11
  point of incipient fluidization and, 10.58–10.60
  Reynolds number, 10.58, 10.61
  schematic of, 10.9f.–10.10f.
  sieve analysis in, 10.12, 10.12f., 10.14–10.15
  treatment process selection and, 5.11–5.13
Graphitization, 20.10
Gravity filters, 10.3
  declining-rate control systems for, 10.50
  equal-rate control systems for, 10.49–10.50
  mechanical control systems for, 10.49
  nonmechanical control systems for, 10.49, 10.52
  pressure filtration compared with, 10.75–10.76
  proportional-level declining rate control systems for, 10.52
  proportional-level equal rate control systems for, 10.51
  proportional-level influent flow splitting for, 10.51
  rate control systems for, 10.49–10.52, 10.50f.
  variable-level declining rate control systems for, 10.51–10.52
  variable-level influent flow splitting for, 10.50–10.51

Graz, Austria (ARR system), 15.27
Green rust, 20.41
Green water, 20.46
Grotthuss-Draper law, 18.18
Ground Water Rule (GWR), 10.6–10.7
Groundwater
  arsenic in, 3.40, 3.42t.
  augmentation, 16.29–16.31
  dissolved chemicals composition in, 3.41t.
  hydrogeochemical cycle in, 3.39, 3.40f.
  inorganic carbon chemistry concentrations in, 3.18t.
  LBF/RBF source water mixing with, 15.6
  lime softening for, 5.36, 5.36f.
  nitrate in, 3.40
  recharge, 16.23
  regulations, 1.19
  SAT and recharge of, 15.13
  subsurface treatment of, 15.10–15.11
  treatment process selection and, 5.3–5.4
  treatment process selection for iron/manganese removal in, 5.35, 5.35f.
  UV disinfection applications in, 18.16
Groundwater Replenishment System, Orange County, California, 16.31–16.32
Gulf streaming, 10.63
GWR. *See* Ground Water Rule
Gypsum, 3.14

HAAs. *See* Haloacetic acids
Haber-Weiss mechanism, 18.24, 18.25t.–18.26t., 18.27–18.34
Hagen-Poiseuille equation, 11.55
Halides, 13.74
Haloacetic acids (HAAs), 2.44–2.45, 2.59, 2.68, 19.7–19.8
  ammonia's impact on formation of, 19.38, 19.39f.
  biodegradation of, 19.20
  reaction time and, 19.16–19.17, 19.16f.
  regulations for, 19.45–19.46
  spatial temporal variability of, 19.46
  stability of, 19.43–19.45, 19.44f.–19.45f.
Haloform reaction, 7.14–7.15, 7.15f.
Halogens, 7.11–7.12
Hardness, 2.38–2.39
  aesthetic concerns with, 2.75
  alkalinity and, 13.13
  carbonate, 13.13–13.14
  corrosion and, 20.32–20.35
  lime softening and, 13.12–13.18
Hardness (*Cont.*):
  lime softening removing carbonate, 13.18–13.19
  lime softening removing noncarbonate, 13.19–13.20
  low/moderate, 5.33–5.44
  magnesium hydroxide and, 13.15, 13.15t.
  noncarbonate hardness, 13.13–13.14
  of particles, 14.6
  removal reasons for, 13.14
  treatment objectives for, 13.14–13.16, 13.15t.
  types of, 13.13–13.14
  units of expression for, 13.12–13.13
  U.S. distribution of, 13.16–13.18, 13.17f.
  water supplies classification by, 13.16, 13.16t.
HAV. *See* Hepatitis A virus
Hazen's law, 9.39–9.40
HCGI. *See* Highly credible gastrointestinal illness
Head loss
  available, 10.11
  calculations example for, 10.38
  clean-bed, 10.36–10.38
  for fluidized granular media bed, 10.58, 10.59f.
  granular media filters and development of, 10.38–10.39
  in granular media filters with fixed bed flow, 10.59
  models for, 10.36–10.38
  negative head and, 10.49, 10.49f.
  rapid granular bed filtration and development of, 10.46–10.48, 10.47f.–10.48f.
  terminal, 10.11
Health advisories, 1.16, 1.18–1.19
Health reference level (HRL), 1.13
Health-based drinking water concentration, for carcinogens, 2.30
Heat, 17.4
Heavy metals, 15.19
Helicobacter pylori, 2.9
Henry's law, 6.18, 6.18f.
  chlorine and, 17.6–17.7
  chlorine dioxide and, 17.9
  estimating, 6.3
  gas transfer, constant plots, 6.4f.
  gas transfer, constants for gases in water compounds, 6.8t.
  gas transfer, constants for organix compounds, 6.6t.–6.7t.
  gas transfer, temperature and, 6.8t.–6.9t.
  gas transfer, unit conversions, 6.5t.

Henry's law (*Cont.*):
gas transfer and, 3.10t., 5.5, 6.2–6.10
ionic strength and, 6.9
pH and, 6.9–6.10
salting-out (Setschenow) coefficients, 6.9–6.10, 6.10t.
solubility of gas in water by, 9.49
surfactants and, 6.10
VOCs and, 3.9–3.10
Hepatitis A virus (HAV), 2.12–2.13
Hepatitis E virus (HEV), 2.13, 2.21
Herbicides, 2.52–2.54
acetanilides, 2.53
chlorophenoxyacetic acids, 2.54
triazine, 2.52–2.53
Heterogeneous buffer system, 20.28
Heterotrophic bacteria
in distribution systems, 21.10–21.13, 21.11t., 21.12f.
in drinking water, 21.25
as indicator, 2.22
Heterotrophic plate count organisms, 17.20
HEV. *See* Hepatitis E virus
Highest occupied molecular orbital (HOMO), 18.18–18.19, 18.18f.
Highly credible gastrointestinal illness (HCGI), 21.4–21.7
High-rate DAF systems, 9.65–9.68, 9.67f.
conventional DAF systems compared to, 9.66t.
Hinckley Reservoir example problem, 8.43–8.45
Hindered settling, 9.4
compression point in, 9.14
equations for, 9.12–9.14
particle interaction in, 9.11
Richardson and Zaki equation for, 9.13
sedimentation and, 9.11–9.14
settling rate prediction for, 9.14, 9.15f.
solids flux in, 9.11–9.12, 9.12f.
HLR. *See* Hydraulic loading rate
HMS coagulants, 8.5, 8.14–8.19
acidity of, 8.33–8.36, 834t.
acidulated alum, 8.21
aluminum hydroxide, 8.17–8.19, 8.17f., 8.18t.
cationic polymers with, 8.49
effective dosage of, 8.25
ferric hydroxide, 8.17–8.19, 8.18f., 8.18t.
hydrolysis, 8.16–8.17, 8.16f.
hydrolysis products, 8.16–8.19, 8.17f.–8.18f., 8.18t.
impurities in, 8.21–8.22
iron as impurity of, 8.21–8.22
HMS coagulants (*Cont.*):
jar tests for, 8.25–8.28, 8.26t., 8.28t.
metal salts plus additives, 8.21
metal salts plus strong acid, 8.21
models for predicting dosages of, 8.22–8.30
NOM adsorption and, 8.28–8.30
NOM removal by, 8.24–8.25
polyaluminum chloride, 8.20
polyaluminum hydroxychloride, 8.20
polyiron chloride, 8.20
prehydrolyzed metal salts, 8.20–8.21
residual aluminum/iron and, 8.36–8.41, 8.37f.–8.40f.
simple metal salts, 8.20
titration curves for, 8.34–8.36, 8.35f., 8.39f.
Hollow fine fiber membranes, 11.4f., 11.10, 11.12–11.13
RO and, 11.10, 11.18, 11.19f.
HOMO. *See* Highest occupied molecular orbital
Homogeneous buffer system, 20.26
Homogeneous Surface Diffusion Model (HSDM), 14.30–14.31, 14.31f.
Hom's law, 17.24–17.25
Hoover process, 13.51
Horizontal-flow tanks, 9.26–9.28, 9.28f.–9.29f.
Hot-process softening, 13.52
HRL. *See* Health reference level
HSDM. *See* Homogeneous Surface Diffusion Model
Hueco Bolson Recharge Project, El Paso, Texas, 16.32
Human exposure evaluations for corrosion, 20.77–20.78
Humic acids, 7.42–7.43
of NOM, 3.60–3.61, 3.60f.
Humic substances, 7.42–7.43
adsorption on GAC of, 14.27
Hydraulic loading rate (HLR), 10.3
in GAC adsorption systems, 14.40–14.41
in natural treatment systems, 15.21, 15.22t.
in SAT systems, 15.14, 15.15t.
Hydrocarbons, 19.23
Hydrodynamic retardation, 8.7
Hydrogen IX softening, 12.33
Hydrogen peroxide
UV ozonation and formation of, 7.27–7.28
UV/H_2O_2 AOP mechanism, 18.24, 18.25t.–18.26t., 18.27–18.32
Hydrogen sulfide
chemical oxidation of, 7.42, 7.42t.
corrosion accelerated by, 20.35

Hydrogeochemical cycle, 3.26–3.27, 3.26f.
in groundwater, 3.39, 3.40f.
Hydrolysis
aluminum, 8.16–8.17, 8.16f., 8.23
basicity, 8.20–8.21
iron, 8.16–8.17
products, 8.16–8.19, 8.17f.–8.18f., 8.18t.
Hydrolyzing metal salt coagulants. *See* HMS coagulants
Hydroxyl radical, 7.25, 7.27–7.28
Hyperfiltration, 11.11
Hypochlorite ion, 7.16–7.17, 7.16f.
Hypochlorous acid, 7.16–7.17, 7.16f.

IDDF. *See* Integrated Disinfection Design Framework
Impediment modulus, 10.29
Inclined settling, 9.2
cocurrent settling in, 9.21, 9.21f., 9.32
compactness of, 9.87
countercurrent settling, 9.20, 9.21f., 9.31, 9.31f.
crossflow settling in, 9.21–9.22, 9.21f., 9.32, 9.32f.
example for, 9.22–9.23
floc ballasting and, 9.30
floc blanket process and, 9.37
flow geometries for, 9.20–9.22
rapid start-up of, 9.87
sludge removal and, 9.88
tanks for, 9.29–9.32, 9.30f–9.32f.
Indicators
Clostridium perfringens as, 2.22
coliforms as, 2.19–2.21, 17.20
coliphages as, 2.21
disinfection and, 17.19–17.21
enterococcus as, 2.21
Escherichia coli as, 2.20–2.21
fecal coliforms as, 2.20
heterotrophic bacteria as, 2.22
MPA for, 2.22
particle counts as, 2.23
total coliform as, 2.19–2.20
turbidity as, 2.22–2.23
of water quality, 2.19–2.23
Indicators compounds, 16.34, 16.35t.
Indirect potable reuse (IPR). *See also* Water reuse
Australia regulations for, 16.38–16.39
BNR in, 16.10–16.11, 16.10t.
California state regulations for, 16.37–16.38
Campylobacter and, 16.7, 16.7t.
Indirect potable reuse (IPR) (*Cont.*):
CCPs for, 16.19–16.21
components of planned, 16.2
in County Sanitation Districts of Los Angeles County, 16.29–16.30
Cryptosporidium and, 16.7–16.8, 16.7t.
DALYs and, 16.18
DBPs in, 16.12t.–16.13t., 16.15–16.16
by direct injection into portable aquifer, 16.31–16.32
dissolved inorganic chemicals in, 16.16
DOC in, 16.11, 16.12t.–16.13t.
environmental buffer in, 16.23
evolution of, 16.3t.–16.4t., 16.5
Florida state regulations for, 16.38
Giardia lamblia and, 16.7–16.8, 16.7t.
groundwater augmentation and, 16.29–16.31
groundwater recharge and, 16.23
in Groundwater Replenishment System, Orange County, California, 16.31–16.32
health risk assessment for, 16.19–16.22, 16.20t–16.21t., 16.22f.
in Hueco Bolson Recharge Project, El Paso, Texas, 16.32
indicators compounds and, 16.34, 16.35t.
in Montebello Forebay Spreading Operations, California, 16.30
multiple-barrier approach in, 16.24–16.27, 16.25f.
nitrosamines in, 16.15–16.16
nutrients and, 16.8, 16.10–16.11
ozonation in, 16.15
pathogens and, 16.6–16.8, 16.7t., 16.9f.–16.10f.
PDFs and, 16.21–16.22, 16.22f.
perfluorinated chemicals in, 16.14–16.15
pesticides in, 16.12t.–16.13t., 16.14
PFOA/PFOS in, 16.14–16.15
pharmaceuticals in, 16.12t.–16.13t., 16.14
posttreatment stabilization in, 16.16
potential hazardous events with, 16.40t.
in Prairie Waters Project, City of Aurora, Colorado, 16.30–16.31
probabilistic exposure assessment for, 16.21–16.22, 16.22f.
public perception to, 16.39–16.42, 16.40t.
real-time monitoring and operating strategies for, 16.33–16.34
recycled water contribution and, 16.37
reliability/health risk considerations of, 16.17–16.22
Salmonella and, 16.7, 16.7t.

Indirect potable reuse (IPR) (*Cont.*):
Shigella and, 16.7, 16.7t.
source control in, 16.6
source water characteristics for, 16.5–16.16
surface water augmentation and, 16.23, 16.27–16.29
surrogate parameters for, 16.34, 16.36t.
suspended solids in, 16.16
TDS and, 16.8
THMs in, 16.15
TOC in, 16.11, 16.12t.–16.13t.
tolerable level of risk in, 16.18
treatment goals from health perspective for, 16.17–16.18
treatment process selection for, 16.5
unidentified chemical assessment for, 16.19
U.S. Federal regulations for, 16.36–16.37
UV/H_2O_2 AOP mechanism applications for, 18.30–18.31
VOC in, 16.11, 16.12t.–16.13t., 16.14
in Water Replenishment District of Southern California, 16.29
in Western Corridor Recycled Water Project, Queensland, Australia, 16.41
Induction period, 13.2
In-line filtration, 10.4, 10.70
Inorganic carbon chemistry, 3.17–3.21
acid-base chemistry and, 3.18–3.19
alkalinity and, 3.22–3.23
closed system of, 3.20, 3.21t.
groundwater/surface water concentrations of, 3.18t.
open system of, 3.19–3.20, 3.21t.
Inorganic constituents
aluminum, 2.33–2.34
arsenic, 2.34–2.35
asbestos, 2.35
boron, 2.35–2.36
cadmium, 2.36
chlorine reactions with, 17.16, 17.17t.
chromium, 2.36–2.37
copper, 2.37–2.38
fluoride, 2.38
hardness, 2.38–2.39
iron, 2.39
lead, 2.39–2.40
lime softening removing, 13.70–13.74, 13.71t.
manganese, 2.40
mercury, 2.40–2.41
nitrate, 2.41–2.42
nitrite, 2.41–2.42
Inorganic constituents (*Cont.*):
perchlorate, 2.42–2.43
precipitative softening removing, 13.70–13.74
selenium, 2.43
sodium, 2.43–2.44
sulfate, 2.44
taste and odor of, 2.73–2.74
Inorganic DBPs, 2.64–2.66
Insecticides, 2.54–2.56
Integrated Disinfection Design Framework (IDDF), 17.38
Integrated membrane systems, 16.5
Internal (intraparticle) transport, 14.28
Interstate Quarantine Act, 1.2–1.3
Intrusion events, 21.29–21.30
Iodide, 19.20–19.21
chlorine reactions with, 7.20
Iodine, 2.66
adsorbable organic, 15.17
THMs and, 19.21–19.22, 19.21f.–19.22f.
Iodine number, 14.7
Ion exchange (IX), 12.2–12.3. *See also* Adsorptive media
adsorption kinetics compared to, 12.18
advantages/disadvantages of, 12.2t.
arsenic breakthrough curves in, 12.56, 12.56f.
arsenic concentration on run length in, 12.58–12.59
arsenic leakage during exhaustion with, 12.59
arsenic removal by, 12.55–12.61, 12.56f., 12.58f., 12.58t.–12.59t., 12.60f.
arsenic spent resin regeneration in, 12.59–12.60
arsenic/nitrate combined removal by, 12.57–12.58, 12.58t.
barium removal by, 12.33–12.34
bed regeneration in, 12.3
bed size in, 12.25, 12.27
binary, 12.19–12.20, 12.20f.
breakthrough curves in, 12.20–12.22, 12.20f.–12.21f.
breakthrough detection in, 12.22–12.23
brine disposal from softening plants in, 12.33
brine disposal in nitrate removal by, 12.39–12.40
capacity in, 12.4
cocurrent regeneration in, 12.27
color removal by, 12.61–12.67, 12.63f.–12.65f.
columns in parallel for, 12.23f., 12.25
columns in series for, 12.24–12.25, 12.24f.
countercurrent regeneration in, 12.27

Ion exchange (IX) (*Cont.*):
  cross-linking, 12.4, 12.5f.
  DBP precursor control with, 19.33–19.34, 19.33f.
  DOC removal by, 12.61–12.67, 12.63f.–12.65f.
  downflow v. upflow regeneration for arsenic in removal by, 12.60–12.61, 12.60f.
  empty-bed contact time in, 12.25, 12.27
  exhaustion rate in, 12.27
  fixed-bed columns in, 12.27
  functionality in, 12.4
  future of, 12.4
  history of, 12.4
  hydrogen IX softening, 12.33
  inorganic anions removal process summaries for, 12.84t.–12.85t.
  inorganic cation removal process summaries for, 12.83t.–12.84t.
  isotherm plots for, 12.14–12.17, 12.15f., 12.17t.
  magnetic, 12.66–12.67
  modeling using EMCT for, 12.74–12.82, 12.76f.–12.82f.
  models for removal of arsenic by, 12.76–12.77, 12.76f.
  models for removal of nitrate by, 12.77–12.79, 12.77t.–12.78t., 12.78f.–12.79f.
  models for removal of perchlorate by, 12.79–12.82, 12.80f.–12.82f.
  multicolumn processes for, 12.24–12.25
  multicomponent, 12.20–12.22, 12.20f.–12.21f.
  nitrate removal by, 12.35–12.40, 12.36f., 12.38f., 12.40f.
  nitrate-laden resin regeneration in, 12.38–12.39, 12.38f.
  operating capacity in, 12.25, 12.26t.
  partial regeneration for, 12.23–12.24
  perchlorate removal by (anion exchange), 12.73–12.74
  pH as breakthrough device for, 12.22–12.23
  pH effecting uranium removal by, 12.69–12.70
  porosity in, 12.4
  precipitative softening v., 13.61–13.63
  process selection for, 13.61–13.63
  pure rates of, 12.17–12.18
  quats and, 12.10
  radium removal by, 12.34–12.35
  radium/uranium combined removal by, 12.72–12.73

Ion exchange (IX) (*Cont.*):
  regenerant reuse in, 12.24
  residuals and, 22.54–22.62
  residuals from nitrate/perchlorate removal by, 22.57, 22.60t., 22.61
  resin characteristics in, 12.25, 12.26t.
  resin choices for nitrate removal by, 12.37–12.38
  resins for arsenic removal by, 12.56–12.57
  selectivity coefficients and, 12.11–12.12, 12.11f.
  selectivity sequences in, 12.12–12.14, 12.13t.
  separation factor in, 12.12
  service flow rate in, 12.27
  single-column service cycle in, 12.23, 12.23f.
  sodium IX softening, 12.27–12.33
  softening, 5.17
  softening capacity in, 12.25, 12.26t.
  softening residuals in, 22.54–22.55, 22.55t.–22.56t.
  special purpose resins in, 12.9–12.11, 12.10t.
  spent brine reuse in, 12.27
  strong-acid cation exchange resins in, 12.4–12.5
  strong-base anion exchange resins in, 12.6
  sulfate effecting arsenic run length for, 12.59, 12.59t.
  TOC removal and, 12.10
  treatment of spent arsenic regenerant with, 12.61
  treatment process selection and, 5.16–5.17
  uranium removal by (anion exchange), 12.67–12.73
  uranium spent resin regeneration in, 12.71–12.72, 12.72f.
  uses of, 12.3–12.4
  water quality in nitrate removal by, 12.35, 12.36f., 12.37
  weak-acid cation exchange resins in, 12.5–12.6
  weak-base anion exchange resins in, 12.6–12.7
Ionic reactions, 7.11–7.12, 7.12t.
Ionic strength, 3.5, 3.5t.
  Henry's law and, 6.9
IPR. *See* Indirect potable reuse
Iron, 2.39
  chemical oxidation of, 7.31–7.36, 7.32t., 7.34f.
  color from corrosion and, 20.41–20.42
  corrosion and, 20.9, 20.40–20.45
  in distribution systems, 21.20

Iron (*Cont.*):
green rust and, 20.41
groundwater, treatment process selection for removal of, 5.35, 5.35f.
as HMS coagulant impurity, 8.21–8.22
hydrolysis, 8.16–8.17
in lakes and reservoirs, 3.39
pH and, 20.43
release, 20.40, 20.42
residual, 8.36–8.41, 8.37f.–8.40f.
sequestration of, 20.45–20.46
Isospora, 2.16
Isotherms
adsorption, 14.8–14.12, 14.9f., 14.10t.–14.11t., 14.12f.
atrazine, 14.23–14.24, 14.23f.
Brunauer-Emmett-Teller isotherm equation, 14.4
GAC, 6.38f.
in GAC performance estimation, 14.59–14.64, 14.61f.
IX and, 12.14–12.17, 12.15f., 12.17t.
IX. *See* Ion exchange

Jablonski diagram, 18.19, 18.19f.
Jar tests
for coagulants, 8.64–8.66
electrophoretic mobility measurement interpretations for, 8.69–8.70, 8.70f.
floc particle separation in, 8.65
for HMS coagulants, 8.25–8.28, 8.26t., 8.28t.
NOM removal and, 8.30–8.33, 8.31f.–8.32f.
PAC dose estimation with, 14.81
for rapid mixing, 8.64–8.66
SCD measurement interpretations for, 8.69–8.70, 8.70f.
Jet action, 10.63, 10.65f.

Kedem-Katchalsky equation, 11.37
Klebsiella, 21.10
Kolmogoroff microscale, 8.53
Kozeny equation, 10.13, 10.37–10.38

Lakebank filtration (LBF), 15.3
ARR compared to, 15.7, 15.8f.
attributes/limitations of, 15.6–15.7
in Berlin, Germany, 15.24–15.25
conventional treatment compared to, 15.5t.
design parameter effects on efficiency of, 15.21, 15.21t.
hydrogeological conditions in, 15.4–15.6
pumping rates in, 15.6

Lakebank filtration (LBF) (*Cont.*):
removal mechanisms of, 15.4, 15.4f.
source water mixing with groundwater in, 15.6
in United States and Europe, 15.24t.
Lakes and reservoirs, 3.34–3.39
classes of, 3.34–3.35
cycling of nutrients in, 3.39
iron/manganese in, 3.39
stratification and mixing of, 3.35–3.36, 3.35f.
water quality in, 3.36, 3.37t.–3.38t., 3.38–3.39
Laminar flow, 9.26
Laminar shear
G value concept and, 8.53–8.54
transport in, 8.52
Langelier saturation index (LSI), 13.8, 13.8t., 20.23
corrosion and, 20.33–20.34
equation defining, 20.33
Larson ratio, 20.29
Las Vegas, direct filtration in, 10.73
Latice ions, 13.2
LBF. *See* Lakebank filtration
Lead, 2.39–2.40
corrosion and, 20.54–20.63
divalent, 20.54–20.59, 20.55f., 20.57f.–20.58f.
leaching, 20.59
particulate, 20.54
pipes and regulations for, 20.53–20.54
tetravalent, 20.60–20.63, 20.60f.–20.61f.
Lead and Copper Rule, 2.37–2.38
Legionella, 2.3, 2.7, 2.9–2.10
chloramine controlling, 21.24–21.25
in distribution systems, 21.14–21.16
waterborne disease outbreaks of, 21.3–21.4, 21.3f.
Leiduin, the Netherlands (dune infiltration), 15.26–15.27
Light-blockage particle counter, 3.54–3.56
Light-scattering particle counter, 3.56
Like-charge electrostatic attraction of colloids, 8.12
Lime sludge, 22.3, 22.11–22.12, 22.11f.
Lime softening
aeration softening and, 13.52
carbonate hardness removal by, 13.18–13.19
carbonic acid removal with, 13.20–13.21
chemistry of, 13.18–13.29
coagulation and, 13.53, 13.69–13.70, 13.69f.

Lime softening (*Cont.*):
equilibrium models estimating doses for, 13.24–13.25
excess, 13.30f., 13.34–13.46
filtration in, 13.61
final stabilization of, 13.22–13.23
floc-blanket process and, 13.54–13.55
flocculation and, 13.52–13.53
for groundwater, 5.36, 5.36f.
hardness and, 13.12–13.18
Hoover process in, 13.51
hot-process softening and, 13.52
inorganic constituents removed by, 13.70–13.74, 13.71t.
magnesium-carbonate process in, 13.51–13.52
monitoring and control of, 13.60–13.61
noncarbonate hardness removal by, 13.19–13.20
pH adjustment in, 13.21
process variations facilitating lime recovery in, 13.50–13.52
process variations for, 13.29–13.30, 13.30f.
radionuclides removed by, 13.72, 13.73f.
rapid mixing and, 13.56
recarbonation of, 13.21–13.22
recirculation and, 13.53–13.54
silica removed by, 13.72, 13.73f.
with sodium hydroxide, 13.20
split-treatment softening, 13.47–13.50
stoichiometry basis for dose estimates of, 13.23–13.24
straight, 13.30–13.31
straight lime-soda softening, 13.30f., 13.31–13.34
three-stage, 13.51
TOC removal by, 13.65–13.69, 13.65f., 13.68t., 13.69f.
two-stage, 13.46–13.47, 13.51
Limestone contactors, 20.75
Limiting current density, 11.51–11.52
Limiting salt, 11.71–11.76, 11.74t.
Linear solution diffusion model, 11.76–11.78, 11.77f.–1178f.
Lingin, 19.23
Linton and Sherwood correlation, 11.41
London-van der Waals force, 8.7
Los Angeles County, County Sanitation Districts of, 16.29–16.30
Los Angeles (California) Department of Water and Power, 10.73
Louisville, KY (RBF system), 15.25
Lowest observed adverse effect level (LOAEL), 2.28–2.29
Lowest unoccupied molecular orbital (LUMO), 18.18–18.19, 18.18f.
LSI. *See* Langelier saturation index
LUMO. *See* Lowest unoccupied molecular orbital

Magnesium
corrosion and, 20.35
magnesium-carbonate process, 13.51–13.52
Magnesium hydroxide
hardness and, 13.15, 13.15t.
lime recovery interference from, 13.50
precipitation and equilibria of, 13.9–13.10, 13.27–13.29, 13.28t.
Magnetic ion exchange, 12.66–12.67, 19.33, 19.33f.
Maintenance wash, 5.15
Managed aquifer recharge (MAR), 15.2–15.3
Manganese, 2.40
chemical oxidation of, 7.31–7.36, 7.32t., 7.34f.
corrosion and, 20.36, 20.37f.
in distribution systems, 21.20
groundwater, treatment process selection for removal of, 5.35, 5.35f.
in lakes and reservoirs, 3.39
Manifold-lateral underdrain system, 10.24, 10.24f.
MAR. *See* Managed aquifer recharge
Marble test, 20.34
Maribor, Slovenia (ARR/RBF systems), 15.26
Mass balance equation, 20.14–20.15
Mass flux, 6.12–6.14
Mass fractal dimension, 8.62
Mass transfer
in gas transfer, 6.10, 6.11f., 6.12–6.14
phase controlling, 6.13–6.14, 6.14t.
volumetric, 6.13
zone (adsorption), 14.31–14.34, 14.32f.
Mass transport
of colloids and particles, 11.42–11.43, 11.43f.
concentration polarization and precipitative fouling in, 11.41
concentration polarization layer in, 11.38–11.41
in ED, 11.50–11.53
in EDR, 11.51–11.53, 11.53f.–11.54f.
global rejection and, 11.44
local rejection and, 11.44
mechanical sieving in, 11.44–11.45

Mass transport (*Cont.*):
membrane processes and, 11.37–11.50
osmotic pressure in, 11.38, 11.47–11.48
permeate flux in, 11.38–11.40
polarization factor in, 11.40
RO and, 11.46–11.50, 11.46f.
separation mechanisms and, 11.43–11.44
solute, 11.9
solute transport/rejection in, 11.46–11.50, 11.46f.
temperature correction factors for, 11.55–11.56
transmembrane pressure in, 11.39–11.40
Maximum contaminant level goals (MCLGs), 1.15, 16.17–16.18
enforceability of, 2.27
for NPDWRs, 1.20t.–1.25t.
Maximum contaminant levels (MCLs), 1.4, 16.17–16.18
for NPDWRs, 1.20t.–1.25t.
MCLGs. *See* Maximum contaminant level goals
MCLs. *See* Maximum contaminant levels
M/DBP. *See* Microbial/Disinfection By-Product
Mechanical dewatering, 5.27. *See also* Non-mechanical dewatering
belt filter presses for, 22.31–22.32
centrifuges for, 22.32–22.35, 22.32f., 22.34f.
filter presses for, 22.35–22.37, 22.36f.–22.37f.
residuals and, 22.29–22.37
vacuum filtration for, 22.30–22.31, 22.30f.
Mechanical sieving, 11.44–11.45
Membrane backwash, 22.4
Membrane cell theory, 20.51
Membrane films, 11.11
Membrane filtration
procedures, 17.20
regulatory environment and, 11.4, 11.8
Membrane processes. *See also* Electrodialysis; Microfiltration; Nanofiltration; Reverse osmosis; Ultralfiltration
air-pressure testing for, 11.35
alkalinity recovery in, 11.89, 11.90f.
array design example for, 11.83–11.86
array modeling for, 11.76–11.83, 11.77f.–11.78f., 11.81f.
asymmetric membranes, 11.10
back-diffusion constant in, 11.79
benefits of, 11.2
bubble-point testing for, 11.35–11.36
charge repulsion in, 11.28
classification by geometry for, 11.12–11.15
Membrane processes (*Cont.*):
classification by material for, 11.9–11.12
composite membranes, 11.10–11.11, 11.11f.
concentrate streams in, 11.65, 11.93–11.94
concentration polarization layer and precipitative fouling in, 11.41
concentration polarization layer in, 11.38–11.41
contaminant size ranges for, 11.4, 11.5f., 11.6t.–11.7t., 11.8
coupling model for, 11.79–11.80
Cryptosporidium removed by, 11.33–11.34, 11.33f.
DBP removal by, 11.29–11.30, 11.30t.
desalting, 22.47–22.53, 22.49t.
design criteria for, 11.57–11.91
dissolved solute interaction in, 11.27–11.28
dissolved solutes influence on electrokinetic properties of, 11.28
driving force classifications for, 11.21–11.22, 11.22t.
DuPont equation in, 11.56
fiber breakage in, 11.36
film theory model for, 11.78–11.79
finger structure in, 11.10
fouling control in, 11.70
fouling indexes and, 11.9t., 11.67–11.69
Giardia lamblia removed by, 11.33–11.34, 11.33f.
global rejection and, 11.44
Hagen-Poiseuille equation in, 11.55
limiting salt and, 11.71–11.76, 11.74t.
linear solution diffusion model for, 11.76–11.78, 11.77f.–1178f.
local rejection and, 11.44
low-pressure, 22.43–22.47
mass transport in, 11.37–11.50
mass transport of colloids/particles in, 11.42–11.43, 11.43f.
mechanical sieving in, 11.44–11.45
membrane films, 11.11
membrane fouling in, 11.28
membrane integrity testing for, 11.34–11.35, 11.34t.
modeling linear arrays for, 11.80–11.83, 11.81f.
modified fouling index and, 11.67, 11.68f.
module repair consideration in, 11.36
MWC and, 11.8–11.9
osmotic pressure in, 11.38, 11.47–11.48
pathogen removal through, 11.33–11.34, 11.33f.

Membrane processes (*Cont.*):
permeate flux in, 11.38–11.40
permeate streams in, 11.65, 11.67
pH and, 11.28
phase inversion membranes, 11.10–11.12
pilot plant testing for, 11.94–11.96
polarization factor in, 11.40
polyamide membranes, 11.11
pore size in, 11.9
posttreatment for, 11.86–11.91, 11.87t.–11.88t.
pretreatment in, 11.69–11.71
process integrity monitoring for, 11.36
recycling residuals from, 22.46–22.47
residuals and, 22.43–22.53, 22.44t.–22.45t.
scaling control in, 11.71–11.76
separation mechanisms and, 11.43–11.44
silt density index and, 11.67
SOC/pesticide removal in, 11.30, 11.31t.–11.32t.
solute transport/rejection in, 11.46–11.50, 11.46f.
sonic sensor testing for, 11.36
sponge structure in, 11.10
static microfilters for, 11.76
surface characterization techniques for, 11.26–11.27, 11.27t.
symmetric membranes, 11.10
temperature correction factors in, 11.55–11.56
terminology for, 11.2t.–11.3t.
theoretical normalized flux equation for, 11.55
thin-film composite membranes, 11.11–11.12
transmembrane pressure in, 11.39–11.40
treatment process selection and, 5.13–5.16
two stage system for, 11.81–11.83, 11.81f.
waste disposal in, 11.91–11.94, 11.93t.
Membrane stacks, 11.24, 11.26, 11.26f.
Mercury, 2.40–2.41
Mercury vapor UV lamps, 18.5–18.6, 18.5t., 18.6f.
Metal ions, 13.71–13.72
Metal solubility, 3.14–3.15
Methyl chloroform, 2.48–2.49
Methylene chloride, 2.48
2-Methylisoborneol (MIB), 2.74
Methyl-*tert*-butyl-ether (MTBE), 2.50
chemical oxidation of, 7.46
Metolachlor, 2.53
Metribuzin, 2.52–2.53
MF. *See* Microfiltration
MIB, 2.74
adsorption capacity of, 14.22, 14.22f., 14.25f.
ozone removing, 7.38–7.39, 7.40f.–7.41f.
PAC usage rate for removal of, 14.82, 14.82f.
MIC. *See* Microbiologically influenced corrosion
Microbial contaminants. *See also* Contaminants; Distribution systems
best available treatment for, 21.48–21.49
in distribution systems, 21.7–21.21
distribution systems controlling, 21.47–21.65
operation/maintenance approaches to distribution systems and, 21.50–21.55, 21.51f., 21.51t., 21.53t., 21.55t.
pathways for, 21.30–21.31
in storage facilities, 21.31–21.32, 21.54–21.55, 21.55t., 21.65
in water main repairs and installations, 21.32–21.34, 21.33t., 21.63–21.65
Microbial growth
disinfectant residuals and, 21.23–21.25
in distribution systems, 21.22–21.28
temperature and, 21.23
treatment types for, 21.23
Microbial risk assessment, 2.31–2.33
for disinfection, 17.19–17.21
Microbial/Disinfection By-Product (M/DBP), 1.25–1.26
Microbiologically influenced corrosion (MIC), 20.8–20.10
Microfiltration (MF), 5.13–5.15, 11.2, 16.5
coagulants and, 8.2
coagulants removing arsenic with, 11.61, 11.61f.
coagulation, flocculation, sedimentation integration with, 11.58–11.60, 11.58f.–11.59f.
concentrate streams in, 11.93–11.94
configuration of, 11.12–11.15, 11.12f.–11.15f., 11.17t.
crossflow operation in, 11.13, 11.14f.
Cryptosporidium removed by, 11.33–11.34, 11.33f.
dead end operation in, 11.13, 11.14f.
flow patterns of, 11.13, 11.14f.–11.15f., 11.15
Giardia lamblia removed by, 11.33–11.34, 11.33f.
mechanical sieving in, 11.44–11.45
NF/RO/EDR integration with, 11.59–11.60, 11.59f.
oxidation reactions integrated with, 11.60–11.61, 11.60f.

Microfiltration (MF) (*Cont.*):
PAC with, 14.80
pilot plant testing for, 11.95
process residuals of, 11.91–11.92
residuals removed by, 22.3–22.4
seawater RO integrated with, 11.61–11.62
UF integrated process design/applications with, 11.57–11.65
use of, 11.8
uses of, 11.4
Microflotation, 9.46
Microorganisms. *See also* Pathogens
disinfection efficiency influenced by physical/ physiological factors of, 17.30
filtration and removal of, 10.7–10.8
natural treatment systems removing, 15.16, 15.16t.
SSF removing, 10.78–10.79
taste and odor from, 2.74
UV and, 18.1–18.2
waterborne disease and, 2.3, 2.4t.
Micropollutants
adsorption capacity for, 14.24, 14.25f., 14.26
DOM as, 14.2
Microscopic particulate analysis (MPA), 2.22
Milwaukee, Wisconsin, *Cryptosporidium* outbreak in, 2.3
Mineralization, 2.75
Minimum fluidization velocity, 10.60
Mixed oxidants, 5.7, 7.31
Mixed waste, 22.69
Mixed-media filter, 10.19
Mixed-potential theory, 20.13
Models/modeling
adOx kinetic model, 18.29–18.30, 18.30f.
coupling model in, 11.79–11.80
for DAF, 9.54–9.59
DAF contact zone, 9.55–9.59, 9.56t., 9.57f.
for DBP formation, 19.27
for disinfection decay, 17.26–17.27
disinfection power law models, 17.26
distribution systems and hydraulic, 21.44–21.45, 21.44f.
distribution systems and water quality, 21.44–21.45, 21.44f.
film theory model in, 11.78–11.79
for flocculation, 8.61
GAC adsorption systems and empirical, 14.74–14.77, 14.75f.–14.76f., 14.75t.
GAC adsorption systems and mathematical, 14.73–14.74
Gouy-Chapman model, 8.8–8.9, 8.9f.
Models/modeling (*Cont.*):
for head loss, 10.36–10.38
for HMS coagulant dosage predictions, 8.22–8.30
HSDM for adsorption, 14.30–14.31, 14.31f.
for IX removal of arsenic, 12.76–12.77, 12.76f.
for IX removal of nitrate, 12.77–12.79, 12.77t.–12.78t., 12.78f.–12.79f.
for IX removal of perchlorate, 12.79–12.82, 12.80f.–12.82f.
for IX using EMCT, 12.74–12.82, 12.76f.–12.82f.
lime softening/recarbonation dose estimation with equilibrium, 13.24–13.25
linear arrays, 11.80–11.83, 11.81f.
linear solution diffusion model in, 11.76–11.78, 11.77f.–1178f.
membrane processes and array, 11.76–11.83, 11.77f.–11.78f., 11.81f.
PAC and mathematical, 14.82–14.83
for precoat filtration, 10.94–10.95
for rapid granular bed filtration, 10.26–10.39
siderite model of corrosion, 20.41
for SSF, 10.29–10.31
steady state OH radical model, 18.28–18.29
Modified fouling index, 11.67, 11.68f.
Moist bubble densities, for DAF, D.3t.
Molar absorption coefficient, 18.20, 18.20f., 18.21t.
Molasses number, 14.7
Molecular chlorine, 7.16–7.17, 7.16f.
Molecular weight cutoff (MWC), 11.8–11.9
Molecularity, 7.9
Monochloroethene, 2.49–2.50
Monochromatic UV photolysis, 18.21–18.23, 18.21t.
Monomers, 8.45
Montebello Forebay Spreading Operations, California, 16.30
Moringa oleifera seed extract, 8.48
Mother liquor, 13.3
MPA. *See* Microscopic particulate analysis
MTBE. *See* Methyl-*tert*-butyl-ether
Mudballs, 10.65f., 10.66
Multiple-barrier treatment approach, 16.24–16.27, 16.25f.
UV disinfection for, 18.16
Mutagenicity, 2.25, 2.27
MWC. *See* Molecular weight cutoff
MX, 2.69
Mycobacterium, 2.10–2.11
in distribution systems, 21.17

Naegleria fowleri, 2.17, 17.20
Nanofiltration (NF), 5.13, 5.15–5.16, 11.2
  concentrate streams in, 11.65, 11.93–11.94
  configuration of, 11.18–11.21
  DBP removal by, 11.29–11.30, 11.30t., 19.35–19.36
  dissolved solutes influence on membrane electrokinetic properties and, 11.28
  fouling control in, 11.70
  fouling indexes and, 11.9t., 11.67–11.69
  limiting salt and, 11.71–11.76, 11.74t.
  MF/UF integration with, 11.59–11.60, 11.59f.
  modified fouling index and, 11.67, 11.68f.
  permeate streams in, 11.65, 11.67
  pilot plant testing for, 11.95–11.96
  posttreatment for, 11.86–11.91, 11.87t.–11.88t.
  pretreatment in, 11.69–11.71
  process concepts of, 11.65, 11.65f.–11.66f., 11.67
  process integrity monitoring for, 11.36
  residuals removed by, 22.4
  scaling control in, 11.71–11.76
  silt density index and, 11.67
  sonic sensor testing for, 11.36
  static microfilters for, 11.76
  SW configurations for, 11.18, 11.19f.–11.20f., 11.20–11.21
  uses of, 11.4, 11.8
Nanoparticles, 2.58, 8.12
NAS. *See* National Academy of Sciences
National Academy of Sciences (NAS), 1.30
National Drinking Water Advisory Council (NDWAC), 1.9
  consultation from, 1.30–1.31
National Infrastructure Protection Plan (NIPP), 1.7
National Interim Primary Drinking Water Regulations (NIPDWRs), 1.4
National Permit Discharge Elimination System (NPDES), 22.40
National Primary Drinking Water Regulations (NPDWRs), 1.4, 1.5t., 1.15–1.16
  current regulations for, 1.19, 1.20t.–1.25t.
  MCLGs/MCLs for, 1.19, 1.20t.–1.25t.
  review, 1.16, 1.17f.
National Research Council (NRC), 1.9, 1.13
National Secondary Drinking Water Regulations, 1.16, 1.18t.
Natural organic matter (NOM), 3.2, 14.3
  adsorption, preloading in natural waters of, 14.26–14.27, 14.26f.–14.27f.
  adsorption kinetic parameters, preloading effect of, 14.34–14.36, 14.35f.–14.36f.
  adsorption of, 14.19–14.21, 14.20f.
  breakthrough of, 14.33–14.34, 14.34f.
  charge concentrations on particles v., 3.61–3.62
  chlorination studies with, 19.23
  corrosion and, 20.8–20.9, 20.38
  DAF for, 9.83–9.84, 9.84f.
  DBPs and, 19.3
  empty-bed contact time and, 14.40–14.41, 14.41f.
  enhanced coagulation and, 8.5–8.6, 8.5t.
  fractions, 3.59–3.60
  fulvic acids of, 3.60–3.61, 3.61f.
  HMS coagulants and adsorption of, 8.28–8.30
  HMS coagulants for removal of, 8.24–8.25
  humic acids of, 3.60–3.61, 3.60f.
  jar test results for removal of, 8.30–8.33, 8.31f.–8.32f.
  measurements of, 3.66–3.67
  natural treatment systems removing bulk, 15.16–15.17
  PAC removing precursors of, 14.85
  precipitative softening removing, 13.63–13.70
  reaction products of, 19.2, 19.2f.
  sources of, 3.58–3.59
  SUVA and, 3.65–3.66, 3.65t., 3.66f.
  types of, 3.25t., 3.59–3.62
  water quality, effects of, 3.58, 3.58t.
Natural treatment systems. *See also* Lakebank filtration; Riverbank filtration
  ARR, 15.7–15.10
  background of, 15.2
  bulk NOM removal by, 15.16–15.17
  case studies for, 15.23–15.28
  clogging in, 15.22–15.23
  DOC removed by, 15.17
  heavy metals removed by, 15.19
  HLRs in, 15.21, 15.22t.
  MAR, 15.2–15.3
  microorganism removal in, 15.16, 15.16t.
  nitrogen removal by, 15.18
  particles in, 15.15–15.16
  phosphorus removal by, 15.18–15.19
  production well type/number/spacing/distance in, 15.20
  redox conditions in, 15.21–15.22
  SAT, 15.12–15.15, 15.12f.

Natural treatment systems (*Cont.*):
subsurface groundwater treatment in, 15.10–15.11
trace organic contaminants removed by, 15.17, 15.18t.
treatment process selection and, 5.21–5.22
turbidity in, 15.15–15.16
water quality improvements in, 15.15–15.19
Naturally occurring chemicals, 3.24, 3.25t.
NDMA. *See* Nitrosodimethylamine
NDWAC. *See* National Drinking Water Advisory Council
Negative head, 10.49, 10.49f.
Nematicides, 2.55–2.56
Nephelometer, 8.4
Nernst equation, 3.15–3.16, 20.4–20.5
Neutralization processes, 20.75
New York City, NY (UV disinfection), 18.17
NF. *See* Nanofiltration
NIPDWRs. *See* National Interim Primary Drinking Water Regulations
NIPP. *See* National Infrastructure Protection Plan
Nitrate, 2.41–2.42
brine disposal in IX removal of, 12.39–12.40
in groundwater, 3.40
IX modeling for removal of, 12.77–12.79, 12.77t.–12.78t., 12.78f.–12.79f.
IX removal of arsenic combined with, 12.57–12.58, 12.58t.
IX removing, 12.35–12.40, 12.36f., 12.38f., 12.40f.
nitrate-laden resin regeneration in IX removal of, 12.38–12.39, 12.38f.
residuals from IX removing, 22.57, 22.60t., 22.61
resin choices for IX removal of, 12.37–12.38
strong-base anion exchange resins and removal of, 12.37
water quality effects on IX removal of, 12.35, 12.36f., 12.37
Nitredox method, 15.10
Nitrification
in distribution systems, 21.28
process of, 21.19–21.20
Nitrite, 2.41–2.42
Nitrite-oxidizing bacteria, 21.20
Nitrogen, 15.18
in distribution systems, 21.19–21.20
Nitrogen-enriched air, 9.50
Nitrosamines, 2.69
in IPR, 16.15–16.16
Nitrosodimethylamine (NDMA), 18.21–18.22
chloramine and, 19.11, 19.12f.
Nitrosomonas, 21.20
Nitrospirae, 21.10
No observed adverse effect level (NOAEL), 2.28–2.29
Nocardia, 21.10–21.11
NOM. *See* Natural organic matter
Noncarbonate hardness, 13.13–13.14
lime softening removing, 13.19–13.20
Noncarcinogenic risk assessment, 2.28–2.29
Nonionic polyelectrolytes, 8.47
Nonionic polymers, 8.45
Non-mechanical dewatering
dewatering lagoons for, 22.25
freeze-thaw beds for, 22.25–22.26
of residuals, 22.21–22.29
sand drying beds for, 22.21–22.22, 22.23f., 22.24, 22.26–22.29
solar drying beds for, 22.24
Norovirus, 2.11
NPDES. *See* National Permit Discharge Elimination System
NPDWRs. *See* National Primary Drinking Water Regulations
NRC. *See* National Research Council
Nutrients
BNR, 16.10–16.11, 16.10t.
distribution systems and bacterial, 21.25–21.26
distribution systems and loading, 21.48
IPR and, 16.8, 16.10–16.11
lakes and reservoirs, cycling of, 3.39

Odor. *See* Taste and odor
Office of Management and Budget (OMB), 1.31
OMB. *See* Office of Management and Budget
Oncogenicity, 2.25
Organic constituents, 2.44–2.45. *See also* Dissolved organic carbon; Total organic carbon; Trihalomethanes
acrylamide, 2.58
chlorine reactions with, 7.18–7.19, 17.16
epichlorohydrin, 2.59
fungicides, 2.56
HAAs, 2.44–2.45, 2.59
herbicides, 2.52–2.54
insecticides, 2.54–2.56
mixtures of, 2.57
PAHs, 2.59
pesticides, 2.50–2.52
precipitative softening removing, 13.70
SOCs, 1.4

Organic constituents (*Cont.*):
  taste and odor of, 2.73–2.74
  THMs, 2.44–2.45, 2.59
  VOCs, 2.45–2.50
Organic DBPs, 2.66–2.69
Organochlorines, 2.55
Organophosphates, 2.55
ORP. *See* Oxidation-reduction potential
Orthokinetic flocculation, 8.52
Orthophosphate, 20.38, 20.43, 20.48–20.49, 20.49f.
Osmotic pressure, 11.38, 11.47–11.48
Ostwald ripening, 13.3
Oxidation reactions, 3.15
  MF/UF integrated with, 11.60–11.61, 11.60f.
Oxidation state, 7.4
Oxidation-reduction potential (ORP), 3.16–3.17
  corrosion and, 20.30, 20.31f., 20.32
Oxidation-reduction reactions, 7.4–7.8
Oxidation-reductions chemistry. *See* Redox chemistry
Oxygen
  corrosion and, 20.32
  solubility in freshwater of, D.2t.
Ozone/ozonation, 2.66, 5.6
  auto-decomposition of, 7.25
  bacteria inactivation by, 17.28–17.29
  basic chemistry of, 17.10
  biodegradable organic material formation and, 7.28, 19.14
  bromate formation and, 19.19
  bromide reactions with, 7.28–7.29, 7.29f.
  bubble aeration and absorption of, 6.50–6.52
  coagulation and, 8.49
  DBPs and, 19.4t.–19.6t., 19.13–19.14, 19.14f.–19.15f., 19.29, 19.29f.
  demand reactions of, 17.18–17.19
  design/application of, 17.36–17.37
  of endocrine disruptors, 7.45–7.46
  GAC adsorption systems and pretreatment of, 14.42–14.44, 14.43f., 14.44t.
  generation of, 7.23, 7.24f.
  geosmin removal by, 7.38–7.39, 7.40f.–7.41f.
  history of, 17.3–17.4
  hydrogen peroxide formation and, 7.27–7.28
  hydroxyl radical and, 7.25, 7.27–7.28
  indicators compounds and, 16.34, 16.35t.
  in IPR, 16.15
  MIB removal by, 7.38–7.39, 7.40f.–7.41f.
  pH and, 19.19
  of pharmaceuticals, 7.45–7.46
  reaction pathways in, 7.25–7.27, 7.26f.
  UV/O_3 AOP mechanism, 18.33
PAC. *See* Powdered activated carbon
Packed towers
  applications of, 6.14–6.15, 6.17
  cascade, 6.14–6.15
  cocurrent, 6.14–6.15
  cross-flow, 6.14–6.15
  DBP removal with aeration by, 19.42, 19.42f.
  design procedure for, 6.28–6.36
  diameter determination for, 6.26–6.27
  dissolved solids impacting performance of, 6.36
  flooding and pressure drop in, 6.26f.
  governing equations of, 6.17–6.28
  height expression of, 6.17–6.26
  HTU (height of transfer unit) in, 6.20
  K_La value determination for, 6.21–6.25, 6.23t.
  K_La value estimation for, 6.25–6.26
  minimum air-to-water ratio for, 6.19–6.20
  NTU (number of transfer units) in, 6.20, 6.21f.
  operating line for, 6.18, 6.18f.
  operating power requirements for, 6.27–6.28
  packing materials for, 6.14, 6.16f., 6.17t.
  schematic of, 6.15f., 6.18f.
  TCA calculation and, 6.29–6.36
PAHs. *See* Polynuclear aromatic hydrocarbons
Particle counts
  as indicator, 2.23
  instruments for, 8.4–8.5
  rapid granular bed filtration and instruments for, 10.43, 10.44f.
Particles. *See also* Discrete particle settling
  characteristics of, 3.45–3.49
  charge concentrations on NOM v., 3.61–3.62
  colloidal, 8.4
  concentration of, 3.49–3.52, 3.51t., 8.4–8.5
  counting, 3.54–3.57
  density of, 3.46–3.47
  density wetted in water of, 14.5
  destabilization mechanisms of, 8.14–8.15
  diffuse layers of, 8.8–8.11, 8.9f.–8.10f.
  dissolved, 8.4
  DLVO theory of colloid stability and, 8.7
  double-layer compression of, 8.14
  electrostatic stabilization of, 8.7–8.12
  external fields and, 8.15
  flocculation and concentration of, 4.45–4.46
  GAC adsorption systems and size of, 14.39–14.40
  Gouy-Chapman model for, 8.8–8.9, 8.9f.
  hardness of, 14.6
  hindered settling, interaction of, 9.11

Particles (*Cont.*):
hydrodynamic retardation of, 8.7
interparticle bridging of, 8.15
London-van der Waals force and, 8.7
mass transport of, 11.42–11.43, 11.43f.
in natural treatment systems, 15.15–15.16
ovoid, 8.3–8.4
PAC and size of, 14.77–14.78, 14.78f.
rapid granular bed filtration, removal efficiency as function of size of, 10.32–10.35, 10.33f.–10.34f.
raw water sources of, 3.43–3.44
repulsive and attractive forces in, 8.7, 8.10–8.11
secondary minimum aggregation in, 8.11–8.12
sedimentation and shape of, 9.7
shape of, 3.47
size distribution of, 3.52–3.53, 3.53f.–3.55f., 3.56t.
size of, 3.45–3.46, 3.46f., 8.3–8.4
sources/types of, 3.43–3.45, 3.43t.
stability measurements for, 3.57
stability of suspensions of, 8.6–8.14
steric stabilization of, 8.12–8.14, 8.13f.
surface charge neutralization for, 8.14–8.15, 8.22–8.24
surface charge of, 3.47–3.49
suspended, 8.4
treatment system sources of, 3.44–3.45
turbidity of, 3.49, 3.50f., 8.4–8.5
UV disinfection and interference of, 18.15
water quality and importance of, 3.42–3.43
Particulate lead corrosion, 20.54
Particulate organic carbon (POC), 13.64
Partnership for Safe Water, 1.26
Pathogens, 2.6–2.7. *See also Cryptosporidium*; *Giardia lamblia*
Acanthamoeba, 2.17–2.18
adenoviruses, 2.14
astroviruses, 2.14
bacteria, 2.7–2.11
caliciviruses, 2.11–2.12
Campylobacter, 2.8–2.9
cyanobacteria, 2.18–2.19
Cyclospora, 2.17
in distribution systems, 21.9
emerging, 2.6
Entamoeba histolytica, 2.16–2.17
enteroviruses, 2.13–2.14
Escherichia coli, 2.6, 2.20–2.21
HAV, 2.12–2.13
Pathogens (*Cont.*):
Helicobacter pylori, 2.9
HEV, 2.13, 2.21
IPR and, 16.6–16.8, 16.7t., 16.9f.–16.10f.
Isospora, 2.16
Legionella, 2.3, 2.7, 2.9–2.10
membrane processes removing, 11.33–11.34, 11.33f.
Naegleria fowleri, 2.17
norovirus, 2.11
opportunistic bacterial, 2.11
protozoa, 2.15–2.18
raw water sources of, 3.44
Rotaviruses, 2.12
Salmonella, 2.6–2.8, 2.27
Shigella, 2.7
Toxoplasma, 2.16
UV disinfection doses for inactivation of, 18.12, 18.12f.
Vibrio chloerae, 2.7–2.8
viruses, 2.11–2.14
in wastewater, 17.21
Yersinia enterocolitica, 2.8
PCR. *See* Polymerase chain reaction
PDFs. *See* Probability density functions
Pellet reactors, 13.56–13.58, 13.57f.–13.58f.
Perchlorate, 2.42–2.43
GAC reactions with, 14.59
IX modeling for removal of, 12.79–12.82, 12.80f.–12.82f.
IX removal of (by anion exchange), 12.73–12.74
residuals from IX removing, 22.57, 22.60t., 22.61
Perchloroethylene, 2.48
Perfluorinated chemicals, 2.57–2.58
in IPR, 16.14–16.15
Perfluorooctanoic acid (PFOA), 16.14–16.15
Perfluorooctanoic sulfonate (PFOS), 16.14–16.15
Perlite, 10.91–10.92
Personal care products, 7.45–7.46. *See also* Micropollutants
Pesticides, 2.50–2.52. *See also* Micropollutants
alachlor, 2.53
aldicarb, 2.55
aldicarb sulfone, 2.55
aldicarb sulfoxide, 2.55
atrazine, 2.52–2.53
cancer risk of, 2.51–2.52
carbamates, 2.55
chlorpyrifos, 2.55

Pesticides (*Cont.*):
2,4-D, 2.54
dacthal, 2.54
DBCP, 2.55–2.56
diazinon, 2.55
diuron, 2.54
EDB, 2.55–2.56
ETU, 2.56
glyphosate, 2.54
herbicides, 2.52–2.54
insecticides, 2.54–2.56
in IPR, 16.12t.–16.13t., 16.14
membrane processes removing, 11.30, 11.31t.–11.32t.
metolachlor, 2.53
metribuzin, 2.52–2.53
organochlorines, 2.55
organophosphates, 2.55
PAC removing, 14.83
prometon, 2.52–2.53
simazine, 2.52–2.53
sources of, 2.51
PFOA. *See* Perfluorooctanoic acid
PFOS. *See* Perfluorooctanoic sulfonate
PFRs. *See* Plug flow reactors
pH, 3.3, 3.4t.
AA's sensitivity to, 12.7
chemical oxidation and, 7.10–7.11, 7.11f.
chlorination and, 17.6–17.8, 17.7f., 20.30, 20.31f.
corrosion and, 20.38–20.39, 20.43, 20.46–20.47
DBPs and, 19.19
Henry's law and, 6.9–6.10
iron and, 20.43
as IX breakthrough device, 12.22–12.23
IX removal of uranium effected by, 12.69–12.70
lime softening and adjustment of, 13.21
membrane processes and, 11.28
of minimum solubility, 8.38–8.41, 8.40f.
ozonation and, 19.19
potential-pH diagrams (Pourbaix diagrams), 20.18–20.20, 20.19f.–20.20f.
target, 8.42
temperature's effects on, 20.23–20.24, 20.23f.
Pharmaceuticals, 2.56–2.57
in IPR, 16.12t.–16.13t., 16.14
ozonation of, 7.45–7.46
Phase inversion membranes, 11.10–11.12
Phosphates
blended, 20.44
precipitative softening removing, 13.74
Phosphorus, 3.4
natural treatment systems removing, 15.18–15.19
in water reuse, 16.11
Photochemical reaction, 18.17
Photolyase, 18.11
Photolysis. *See* UV photolysis
Photomicroscopy, 20.71
Photons, 18.17
Photoreactivation, 18.11–18.12
Photosynthesis, algae and, 3.38–3.39
Physical constants, B.1t.
Physical integrity, of distribution systems, 21.55–21.65, 21.56t.–21.57t.
Physical properties
of selected gases, D.2t.
of water, D.1t.
Pinhole leak, 20.4
Pipes/piping
A-C pipe corrosion, 20.64–20.65
cement-mortar lined pipe corrosion, 20.64–20.65
concrete pipe corrosion, 20.64–20.65
corrosion and rig systems of, 20.67–20.68
corrosion control with linings, coating and paints in, 20.76–20.77
corrosion on, 20.11–20.12
distribution systems and corrosion of, 21.26
lead regulations for, 20.53–20.54
as PFRs, 4.44–4.45
water age control in, 21, 21.51–21.52, 53t.
Pitting corrosion, 20.7–20.10, 20.7f.
copper and, 20.50–20.52
Planctomyces, 21.10
Planned interval tests, 20.67
Plug flow conditions, 9.26
Plug flow reactors (PFRs), 4.2
CFSTR's reaction rate expressions compared to, 4.36, 4.38, 4.38t.
dispersion model, reaction rate expressions of, 4.39, 4.40f.
dispersion model for, 4.20–4.22, 4.21f.
flocculation kinetics in, 8.55
pipes modeled as, 4.44–4.45
reaction rate expressions in, 4.34, 4.35t.
schematic of, 4.34f.
tracer tests for, 4.11–4.13, 4.12f.
Plumbosolvency, 20.6
POC. *See* Particulate organic carbon
Point of incipient fluidization, 10.58–10.60
Poisoning, acute, 2.3
Polarization factor, 11.40

Polyacrylamide polymers, 8.47
Polyaluminum chloride, 8.20, 22.9
Polyaluminum hydroxychloride, 8.20
Polyamide membranes, 11.11
Polychromatic UV photolysis, 18.23–18.24
Polyelectrolyte coagulants
ampholyte polymers, 8.45
anionic polyelectrolytes, 8.47
anionic polymers, 8.45
cationic polyelectrolytes, 8.46–8.47, 8.47f.
cationic polymers, 8.45
chitin, 8.48
flocculant polymers, 8.46
impurities in, 8.47
monomers, 8.45
Moringa oleifera seed extract, 8.48
from natural organic compounds, 8.47
nonionic polyelectrolytes, 8.47
nonionic polymers, 8.45
polyacrylamide polymers, 8.47
primary coagulant polymers, 8.46
quaternary amines, 8.46
restabilization of, 8.46–8.47
sedimentation and, 9.43–9.45, 9.44f.
types of, 8.45–8.46
usage of, 8.45–8.46
Polyiron chloride, 8.20
Polymerase chain reaction (PCR), 21.10
Polymers
adsorption of, 8.13–8.15
ampholyte, 8.45
anionic, 8.45
cationic, 8.45, 8.49
configuration of, 8.12–8.13, 8.13f.
flocculant, 8.46
nonionic, 8.45
polyacrylamide, 8.47
primary coagulant, 8.46
as rapid granular bed filtration aid, 10.21
Polynuclear aromatic hydrocarbons (PAHs), 2.59
Polyphosphate, 20.39, 20.43–20.44
Postprecipitation, 13.4
Potable reuse. *See* Direct potable reuse; Indirect potable reuse
Potassium ferrate, 7.30
Potassium permanganate, 5.6, 7.29–7.30, 7.34f., 17.4
taste and odor elimination with, 7.39
Potential-pH diagrams (Pourbaix diagrams), 20.18–20.20, 20.19f.–20.20f.
Pourbaix diagrams, 20.18–20.20, 20.19f.–20.20f.
Powdered activated carbon (PAC), 3.45, 11.58–11.59
adsorption qualities of, 14.6–14.7
advantages/disadvantages of, 14.77, 14.78t.
application of, 14.77–14.81
atrazine removal and usage rate of, 14.81–14.82
chemical characteristics of, 14.6
coliforms in, 21.22
cyanotoxins removed by, 14.84
DBP precursors removed by, 14.85
floc-blanket reactor-PAC-UF process, 14.80–14.81, 14.80f.
GAC compared to, 5.20–5.21, 14.3
in Hueco Bolson Recharge Project, El Paso, Texas, 16.32
jar tests estimating dose of, 14.81
mathematical models for dose of, 14.82–14.83
with MF/UF, 14.80
MIB removal and usage rate of, 14.82, 14.82f.
NOM precursors removed by, 14.85
particle size in, 14.77–14.78, 14.78f.
pesticide removal by, 14.83
physical characteristics of, 14.4–14.6, 14.5t.
production of, 14.4
taste and odor removal with, 14.83–14.84
Power law models, 17.26
Power law relation, 8.61
Prairie Waters Project, City of Aurora, Colorado, 16.30–16.31
Precipitation. *See also* Coprecipitation
calcium carbonate equilibria and, 13.25–13.29, 13.28t.
common-ion effect in, 13.9
contaminant removal by, 13.3
coordination in, 13.2
error sources in calculating, 13.10–13.11
magnesium hydroxide equilibria in, 13.9–13.10, 13.27–13.29, 13.28t.
nucleation in, 13.2–13.3
postprecipitation, 13.4
principles of, 13.2–13.3
reactions, 3.14–3.15
solubility equilibria in, 13.5, 13.6t.–13.8t., 13.8–13.12
solubility product constants in, 13.5, 13.6t–13.8t.
treatment process selection and, 5.17–5.19

Precipitative softening. *See also* Lime softening
coagulation in, 13.59–13.60
conventional systems for, 13.52–13.54
DOC removal by, 13.65–13.67
inorganic constituents removed by, 13.70–13.74
IX v., 13.61–13.63
metal ions removed by, 13.71–13.72
monitoring and control of, 13.60–13.61
NOM removal by, 13.63–13.70
organic constituents removed by, 13.70
pellet reactors for, 13.56–13.58, 13.57f.–13.58f.
phosphate removal with, 13.74
process selection for, 13.61–13.63
residuals of, 13.70
solids contact clarifiers for, 13.54–13.56, 13.54f.–13.55f.
TOC removal by, 13.64–13.69, 13.65f., 13.69f.
Precoat filtration, 10.2, 10.87–10.88
advantages of, 10.88–10.89
applications of, 10.88–10.89
body feed and, 10.87, 10.92
body feed concentration and, 10.93–10.94, 10.94f.
Cryptosporidium removal with, 10.88–10.89
disadvantages of, 10.89
filter elements in, 10.87, 10.89, 10.90f.–10.91f.
filter media in, 10.91–10.92
filtration rate effecting, 10.94
Giardia lamblia removal with, 10.88–10.89
grades of filter media for, 10.88
mathematical model for, 10.94–10.95
performance of, 10.88–10.89
precoating in, 10.92, 10.92f.
pressure filter vessels in, 10.91, 10.91f.
schematic of, 10.88f.
septum in, 10.87, 10.89, 10.90f.
spent cake removal in, 10.92–10.93
SWTR and, 10.87
theoretical aspects of, 10.93–10.95
vacuum filter vessels in, 10.90–10.91
Prehydrolyzed metal salts, 8.20–8.21
Premix clarifiers, 9.32–9.33
Premix-recirculation clarifiers, 9.32–9.33, 9.33f.
Pressure
equilibrium constant and effects of, 3.9, 6.5
low/negative-pressure transients in distribution systems, 21.30
membrane processes and low, 22.43–22.47
packed towers and drop in, 6.26f.
Pressure filtration, 10.3
applications of, 10.76
configuration of, 10.74–10.75, 10.74f.
Cryptosporidium and *Giardia lamblia* removal by, 10.8
gravity filters compared with, 10.75–10.76
operation of, 10.75
rate control of, 10.75–10.76
in small water systems, 10.76
Pressure flotation, 9.46
Primary coagulant polymers, 8.46
Probability density functions (PDFs), 16.21–16.22, 16.22f.
Process integrity monitoring, 11.36
Prometon, 2.52–2.53
Protozoa, 2.15–2.18. *See also Cryptosporidium; Giardia lamblia*
Acanthamoeba, 2.17–2.18
Cyclospora, 2.17
Entamoeba histolytica, 2.16–2.17
free-living, 2.17–2.18, 21.18
Isospora, 2.16
Naegleria fowleri, 2.17
parasitic, 2.15–2.17
Toxoplasma, 2.16
Pseudomonas aeruginosa, 21.10–21.12
Public Notification Rule, 1.31–1.32
pX notation, 3.3, 3.4t.

Quantum yield, 18.20–18.21, 18.21t.
Quaternary amines, 8.46
Quats, 12.10

Radical reactions, 7.12–7.13, 7.13t.
Radioactive residuals, 22.66–22.69, 22.67t.–22.69t.
characterization of, 22.67, 22.68t.
disposal options for, 22.69, 22.69t.
regulations for, 22.68
Radionuclides
health effects of, 2.70–2.71
lime softening removing, 13.72, 13.73f.
particle types of, 2.69
radium, 2.71
radon, 2.72
standards for, 2.71
uranium, 2.72
Radium, 2.71
IX removal of uranium combined with, 12.72–12.73
IX removing, 12.34–12.35

Radon, 2.72
regulations, 1.19
Rainwater, 3.28–3.29, 3.28t.
Rapid granular bed filtration
air binding in, 10.49
air scour delivery systems in, 10.56
air-scour assisted backwashing of, 10.54t., 10.55–10.57
available head loss in, 10.11
backwash in, 10.11
backwash water and air scour flow rates in, 10.56–10.57, 10.57t.
backwashing and expansion of filter bed in, 10.60–10.62
backwashing methods for, 10.53–10.57, 10.54t.
backwashing troughs in, 10.57
backwashing wash water required for, 10.57–10.58
bed depth and media size in, 10.18
Brownian diffusion, 10.31
configurations for, 10.15–10.21, 10.16f.
continuous turbidity monitoring and, 10.43, 10.44f.
Cryptosporidium and, 10.22
diffusion mechanism in, 10.31–10.32, 10.31f.
dirty filter, restarting, and, 10.45–10.46
dirty filter media and, 10.66–10.68
dirty filter rate increases and, 10.45, 10.46f.
dual-media filters in, 10.18–10.20
effluent quality pattern in, 10.39–10.40, 10.40f.
fabricated self-supporting underdrain system for, 10.24
false-floor underdrain with nozzles for, 10.25, 10.25f.
filter cycle, 10.11
filter run in, 10.11
filtering-to-waste in, 10.41–10.42
filtration rates and, 10.21–10.22
GAC in, 10.20–10.21
Giardia lamblia and, 10.22
grain sizes in, 10.16–10.17, 10.17t.
gross production per filter run in, 10.23–10.24, 10.23f.
head loss development in, 10.46–10.48, 10.47f.–10.48f.
initial performance of, 10.40–10.41, 10.41f.
interception mechanism in, 10.31–10.32, 10.31f.
intermixing of adjacent layers during backwashing in, 10.63
Rapid granular bed filtration (*Cont.*):
macroscopic (phenomenological) approach to, 10.29–10.31
manifold-lateral underdrain system, 10.24, 10.24f.
media properties in, 10.12–10.15, 10.14t.
media types in, 10.11
microscopic (fundamental) approach to, 10.31–10.35
microscopic/macroscopic models combined for dirty media in, 10.36
mineral deposits and, 10.66–10.68
models for, 10.26–10.39
movement of gravel during backwashing in, 10.64–10.65, 10.65f.
mudballs and, 10.65f., 10.66
negative head effects in, 10.49, 10.49f.
particle size for removal efficiency in, 10.32–10.35, 10.33f.–10.34f.
particle-counting instruments and, 10.43, 10.44f.
performance of, 10.9, 10.10f., 10.11
polymers aiding, 10.21
pre-treatment coagulation for, 10.43–10.45, 10.44f.
run length in, 10.23–10.24, 10.23f.
schematic of, 10.9f.–10.10f.
sedimentation mechanism in, 10.31–10.32, 10.31f.
single collector efficiency in, 10.32, 10.33t.
single-media filters in, 10.18–10.20
skimming during backwashing in, 10.62–10.63
steady state period in, 10.42
stratification during backwashing in, 10.62–10.63
support gravel in, 10.26
surface wash plus fluidized-bed backwash in, 10.54–10.55
terminal head loss in, 10.11
transport mechanisms for, 10.31–10.32, 10.31f.
triple-media filters in, 10.19–10.20
UFRV in, 10.23–10.24, 10.23f.
underdrain failures in, 10.65–10.66
underdrain systems in, 10.24–10.26
upflow filter in, 10.15–10.16
upflow wash with full fluidization in, 10.53–10.54
Rapid mixing, 5.8. *See also* Flocculation
jar tests for, 8.64–8.66
lime softening and, 13.56
purpose of, 8.50

Rapid-mix tank, 4.45
Rate-limiting step, 7.9, 20.13
  in adsorption, 14.28
Raw waters
  contaminants in SFBW compared to, 22.38, 22.38t.
  turbidity of, 9.41
RBF. *See* Riverbank filtration
Reaction kinetics, 7.8–7.10
Reaction pathways, 7.14–7.15
Reaction rate constant, 7.9
Reaction rate expressions, 4.26–4.28
  in batch reactors, 4.28–4.33
  in CFSTR, 4.36, 4.37t.
  of CFSTRs compared to PFRs, 4.36, 4.38, 4.38t.
  in CFSTRs-in-series model, 4.38–4.39, 4.39f.
  in PFR with dispersion model, 4.39, 4.40f.
  in PFRs, 4.34, 4.35t.
Reactor hydraulics, 4.26
Recarbonation
  equilibrium models estimating doses for, 13.24–13.25
  of lime softening, 13.21–13.22
  single-stage, 13.21
  stoichiometry basis for dose estimates of, 13.23–13.24
  two-stage, 13.21–13.22, 13.46–13.47
Recycled water contribution, 16.37
Recycling residuals, 22.40–22.43
  membrane processes and, 22.46–22.47
Red water, 20.41
Redox chemistry, 3.14–3.17
Reference dose (RfD), 2.28–2.29
Regulations. *See also* National Interim Primary Drinking Water Regulations; National Primary Drinking Water Regulations; National Secondary Drinking Water Regulations; Safe Drinking Water Act; Standards
  arsenic, 1.19, 22.63–22.64
  of Australia for IPR, 16.38–16.39
  $C \times T$ approach for disinfection and, 17.39–17.41, 17.40t.
  of California state for IPR, 16.37–16.38
  contaminants, current, 1.19, 1.20t.–1.25t.
  of DBPs, 2.60, 19.45–19.46
  disinfection issues with, 17.4–17.5, 18.3
  for distribution systems by USEPA, 21.35–21.36, 21.35t.–21.36t.
  early history of, 1.2–1.3
  of Florida state for IPR, 16.38
Regulations (*Cont.*):
  future of, 1.33–1.34
  for HAAs, 19.45–19.46
  Internet as resource for, 1.34–1.35
  Lead and Copper Rule, 2.37–2.38
  of lead for pipes, 20.53–20.54
  M/DBP, 1.25–1.26
  membrane processes and, 11.4, 11.8
  for mixed waste, 22.69
  for radioactive residuals, 22.68
  TENORM, 22.66, 22.68–22.69
  for THMs, 19.45–19.46
  traditional/negotiated, 1.25–1.26
  TTHM Rule forming, 1.4
  of U.S. Federal government for IPR, 16.36–16.37
  of USDOT, 22.69
Relative source contribution factor, 2.29
Reproductive toxicity, 2.25
Reservoirs. *See* Lakes and reservoirs
Residence time
  disinfection and, 17.37–17.38
  in distribution systems, 21.27
  sedimentation and, 9.26
Residuals. *See also* Brine disposal; Sludge
  adsorption and, 22.54–22.62
  of adsorptive media and arsenic, 12.54–12.55
  aluminum, 8.36–8.41, 8.37f.–8.40f.
  with arsenic, 22.62–22.66, 22.63t.–22.66t.
  Atterberg limit test and, 22.10–22.11
  backwashing for, 22.44, 22.44t.
  batch thickeners and, 22.17–22.21, 22.18f.
  belt filter presses for, 22.31–22.32
  brine, 22.3
  centrifuges for, 22.32–22.35, 22.32f., 22.34f.
  of chlorine, 17.30–17.31
  CIP waste and, 22.4, 22.44–22.45, 22.45t., 22.49–22.50, 22.49t.
  clarifier sludge, 22.3
  coagulants and, 22.4–22.16, 22.11–22.12, 22.11f.
  compaction density and, 22.14
  concentration of, 22.49
  continuous flow thickeners and, 22.17–22.21, 22.18f.
  CST test for, 22.12, 22.13f., 22.13t.
  DBPs and, 19.17–19.19
  deep well injection and, 22.52
  desalting membrane processes for, 22.47–22.53, 22.49t.
  dewatering lagoons for, 22.25
  disinfectants and, 17.30–17.31, 21.51–21.53

Residuals (*Cont.*):
of disinfectants and microbial growth, 21.23–21.25
disposal methods for, 22.61–22.62
evaporation ponds and, 22.46
extraction tests for, 22.14–22.15
filter presses for, 22.35–22.37, 22.36f.–22.37f.
of fluoride removal by AA, 22.57, 22.58t.–22.59t.
fouling and, 22.47–22.48
freeze-thaw beds for, 22.25–22.26
gaseous, 22.4
gravity sludge thickeners and, 22.17–22.21, 22.17f.–22.18f.
iron, 8.36–8.41, 8.37f.–8.40f.
IX and, 22.54–22.62
of IX softening, 22.54–22.55, 22.55t.–22.56t.
land application of, 22.72–22.73
land disposal of, 22.46, 22.51–22.52
lime sludge, 22.3
low-pressure membrane processes for, 22.43–22.47
management of, 2.50–2.53, 2.53f., 22.44–22.45, 22.45t.
mechanical dewatering for, 22.29–22.37
membrane backwash, 22.4
membrane processes and, 22.43–22.53, 22.44t.–22.45t.
MF/UF removing, 22.3–22.4
from nitrate/perchlorate removal by IX, 22.57, 22.60t., 22.61
non-mechanical dewatering of, 22.21–22.29
of precipitative softening, 13.70
quantity of, 22.4–22.10, 22.48–22.49
radioactive, 22.66–22.69, 22.67t.–22.69t.
recycling, 22.40–22.43
recycling of membrane processes with, 22.46–22.47
RO/NF removing, 22.4
sand drying beds for, 22.21–22.22, 22.23f., 22.24, 22.26–22.29
sewer discharge and, 22.46, 22.51
SFBW and, 22.37–22.40
shear strength and, 22.14, 22.14f.
solar drying beds for, 22.24
SR test for, 22.12, 22.13f., 22.13t.
surface discharge of, 22.45
suspended solids concentration and, 22.7–22.10
TCLP for, 22.14–22.15
thickening of, 22.16–22.21

Residuals (*Cont.*):
topsoil blending of, 22.71, 22.71t.
total metal concentrations for, 22.15–22.16, 22.15t.
treatment for arsenic in, 22.65, 22.65f., 22.66t.
treatment process selection and, 5.27
in trench landfills, 22.74–22.75
TTF test for, 22.12, 22.13f., 22.13t.
types of, 22.2–22.3, 22.2t.
UF process, 11.91–11.92
ultimate disposal of, 22.70–22.75
underground injection control for, 22.45–22.46
vacuum filtration for, 22.30–22.31, 22.30f.

Restabilization, 8.46–8.47

Reverse osmosis (RO), 5.13, 5.15–5.16, 11.2, 16.5
concentrate streams in, 11.65
configuration of, 11.18–11.21, 11.19f.–11.20f.
DBP removal by, 11.29–11.30, 11.30t., 19.35–19.36
dissolved solutes influence on membrane electrokinetic properties and, 11.28
fouling control in, 11.70
fouling indexes and, 11.9t., 11.67–11.69
hollow fine fiber membranes and, 11.10, 11.18, 11.19f.
limiting salt and, 11.71–11.76, 11.74t.
mass transport and, 11.46–11.50, 11.46f.
MF/UF integration with, 11.59–11.60, 11.59f.
MF/UF integration with seawater, 11.61–11.62
modified fouling index and, 11.67, 11.68f.
organic fouling in, 11.71
permeate streams in, 11.65, 11.67
pilot plant testing for, 11.95–11.96
posttreatment for, 11.86–11.91, 11.87t.–11.88t.
pretreatment in, 11.69–11.71
process concepts of, 11.65, 11.65f.–11.66f., 11.67
process integrity monitoring for, 11.36
residuals removed by, 22.4
scaling control in, 11.71–11.76
silt density index and, 11.67
sonic sensor testing for, 11.36
static microfilters for, 11.76
SW configurations for, 11.18, 11.19f.–11.20f., 11.20–11.21
use of, 11.8
uses of, 11.4

Reynolds number
discrete particle settling, 9.6, 9.6f., 9.13
granular media filters and, 10.58, 10.61
sedimentation and, 9.6, 9.6f.
RfD. *See* Reference dose
Rhodococcus, 21.10–21.11
Richardson and Zaki equation, 9.13
Risk assessment
acceptable, 2.32
carcinogenic, 2.29–2.31
of contaminants, 2.27–2.33
estimation of, 2.30
Internet resources for, 2.76–2.77
microbial, 2.31–2.33, 17.19–17.21
noncarcinogenic, 2.28–2.29
USEPA approach to, 2.27–2.28
Risk management, 1.8–1.19. *See also* National Primary Drinking Water Regulations
CCL, 1.8–1.9, 1.10f., 1.11t.–1.13t., 2.31
DWAs for, 1.18–1.19
health advisories for, 1.16, 1.18–1.19
MCLGs for, 1.15
UCMRs for, 1.14–1.15
Riverbank filtration (RBF), 5.21–5.22, 15.3
ARR compared to, 15.7, 15.8f.
attributes/limitations of, 15.6–15.7
in Aurora, CO, 15.27–15.28
design parameter effects on efficiency of, 15.21, 15.21t.
in Düsseldorf, Germany, 15.25–15.26
hydrogeological conditions in, 15.4–15.6
in Louisville, KY, 15.25
in Maribor, Slovenia, 15.26
pumping rates in, 15.6
removal mechanisms of, 15.4, 15.4f.
slow sand filtration, SAT compared with, 15.19, 15.19t.
source water mixing with groundwater in, 15.6
in United States and Europe, 15.24t.
well distance/spacing in, 15.6
Rivers and streams, 3.29–3.30
anthropogenic chemicals in, 3.33–3.34
downstream, 3.32t., 3.33
upland, 3.30, 3.31t.–3.32t., 3.33
RO. *See* Reverse osmosis
Rotaviruses, 2.12

SAB. *See* Science Advisory Board
Safe Drinking Water Act (SDWA), 1.2. *See also* Consumer Confidence Report; Risk management; State agencies
amendments (1986), 1.6–1.7
Safe Drinking Water Act (SDWA) (*Cont.*):
amendments (1996), 1.7, 1.28–1.29
CWA compared to, 16.5–16.6
DHS and, 1.7
evolution of, 1.3–1.7
future of, 1.33–1.34
in 1974, 1.4, 1.6
NPDWRs, 1.4, 1.5t., 1.15–1.16
origins and passage of, 1.3–1.4
public notification requirements of, 1.31–1.32
regulatory processes, 1.8f.
Salmonella, 2.6–2.8, 2.27
IPR and, 16.7, 16.7t.
Salting-out coefficients, 6.9–6.10, 6.10t.
Sand boils, 10.63, 10.64f.
Sand drying beds, 22.21–22.22, 22.23f., 22.24, 22.26–22.29
SAT. *See* Soil aquifer treatment
Saturator, 9.48
air composition within, 9.50–9.51, 9.50f.
Scales, protective/nonprotective, 20.11
Scaling control, 11.71–11.76
Scanning electron microscope, 20.71
SCD. *See* Streaming current detectors
Schmutzdecke, 5.12, 10.77–10.78, 10.84
Science Advisory Board (SAB), 1.30
SDVB resin. *See* Styrene-divinylbenzene resin
SDWA. *See* Safe Drinking Water Act
Secondary disinfection, 17.31
Secondary minimum aggregation, 8.11–8.12
Sedimentation. *See also* Floc-blanket process; Inclined settling
baffling, 9.40
candelabra flow distribution in, 9.3, 9.3f.
Candy tanks, 9.3, 9.3f.–9.4f.
CFD and, 9.26, 9.88
circular tanks for, 9.28, 9.29f.
coagulation and, 9.41–9.42
compactness of, 9.87
compression point for, 9.14
costs of, 9.86–9.87
DAF compared with, 9.84–9.88, 9.85t.
discrete particle settling in, 9.5–9.11, 9.6f.
drag force and, 9.5–9.7, 9.6f.
emerging technology for, 9.88–9.89
filtration as alternative to, 9.88–9.89
flat-bottom clarifiers for, 9.3, 9.3f., 9.37
flocculant aids and, 9.43–9.45, 9.44f.
flocculation and, 4.46, 9.7–9.9, 9.42–9.43, 9.42f.–9.43f.
flow-through curves for, 9.26
fluidization and, 9.14–9.20

Sedimentation (*Cont.*):
  Froude number and, 9.27
  high-gravity setting for, 5.10
  hindered settling and, 9.11–9.14
  horizontal flow, 5.10
  horizontal-flow tanks for, 9.26–9.28, 9.28f.–9.29f.
  inlets and outlets for, 9.40
  laminar flow and, 9.26
  lateral flow distribution in, 9.3, 9.4f.
  MF/UF integration with, 11.58–11.60, 11.58f.–11.59f.
  modern history of, 9.2–9.3
  multiusory tanks for, 9.28, 9.28f.
  nomenclature of, 9.89–9.92
  particle interaction in hindered settling and, 9.11
  particle shape effecting, 9.7
  plug flow and, 9.26
  polyelectrolyte coagulants and, 9.43–9.45, 9.44f.
  predicting settling efficiency, 9.11
  premix clarifiers, 9.32–9.33
  premix-recirculation clarifiers, 9.32–9.33, 9.33f.
  rapid granular bed filtration and, 10.31–10.32, 10.31f.
  rapid start-up of, 9.87
  rectangular tanks for, 9.26–9.27
  residence time and, 9.26
  Reynolds number and, 9.6, 9.6f.
  seasonal water quality and, 9.40
  settlement in tanks in, 9.9–9.11, 9.10f.
  settling types in, 9.4
  settling velocity, 9.9–9.11, 9.10f.
  sludge removal and, 9.88
  solar radiation and, 9.45
  solids contact clarifiers, 5.10, 9.32–9.33, 9.33f.
  solids flux in, 9.11–9.12, 9.12f.
  solids loading and, 9.84, 9.86
  subsidence in, 9.4
  surface loading tanks for, 9.39
  tank size/shape/depth for, 9.39–.940
  terminal settling velocity in, 9.5–9.7
  tracer tests and, 9.26
  treatment process selection and, 5.9–5.11
  wind effects on, 9.45
Selective leaching, 20.7–20.8
Selectivity reversal, 13.62
Selenium, 2.43
Septum, 10.87, 10.89, 10.90f.
Service flow rate, 12.27
Setschenow coefficients, 6.9–6.10, 6.10t.
Settling. *See also* Discrete particle settling; Hindered settling; Inclined settling
  efficiency, 9.11
  in tanks, 9.9–9.11, 9.10f.
  types of, 9.4
  velocity, 9.9–9.11, 9.10f.
SFBW. *See* Spent filter backwash water
Shear strength, 22.14, 22.14f.
Shigella, 2.7
  IPR and, 16.7, 16.7t.
Siderite model, 20.41
Sieve analysis, 10.12, 10.12f., 10.14–10.15
Sieve tray columns, 6.15, 6.17
Silica
  activated, 8.48
  lime softening removing, 13.72, 13.73f.
  surface charge of, 8.7, 8.8f.
  zeolites and, 14.89
Silicates, 20.39, 20.44–20.45
Silt density index, 11.61–11.62, 11.67
Simazine, 2.52–2.53
Simulated distribution system test, 19.24
Singapore's NEWater Project, 16.28
Skimming, 10.62–10.63
Slow sand filtration (SSF)
  biological activity in, 10.77–10.78
  cleaning, 10.84–10.85
  Cryptosporidium removal with, 5.12, 10.8, 10.78–10.79
  design of, 10.80–10.82
  filtration mechanisms of, 10.77–10.79
  filtration rates/control in, 10.81–10.82, 10.81f.
  Giardia lamblia removal with, 5.12, 10.8, 10.78–10.79
  history of, 10.76–10.77
  maintenance for, 10.84–10.86
  microorganism removal in, 10.78–10.79
  model for, 10.29–10.31
  modifications for enhanced performance of, 10.83–10.84
  performance of, 10.78–10.79, 10.86
  RBF, SAT compared with, 15.19, 15.19t.
  resanding, 10.85–10.86
  schmutzdecke in, 5.12, 10.77–10.78, 10.84
  scraping in, 10.84–10.85
  in small water systems, 10.86–10.87
  source water quality in, 10.82–10.83
  startup of, 10.84

Sludge. *See also* Residuals
air/solids ratio on float in DAF, 9.78
alum, 22.4–22.5
batch thickeners and, 22.17–22.21, 22.18f.
cake, 22.3
clarifier, 22.3
color and production of, 22.6
compaction density and, 22.14
continuous flow thickeners and, 22.17–22.21, 22.18f.
float removal in DAF, 9.78–9.79
floc-blanket process, recycling of, 9.37
floc-blanket process, removal of, 9.35
gravity sludge thickeners and, 22.17–22.21, 22.17f.–22.18f.
land application of, 22.73–22.74
lime, 22.3, 22.11–22.12, 22.11f.
macro/micro properties of, 22.11
monofills, 22.73–22.74, 22.74f.
shear strength and, 22.14, 22.14f.
source water influence on float in DAF, 9.78–9.79
terminology of, 22.2
thickening of, 22.3, 22.16–22.21
treatment process selection and removal of, 9.88
in trench landfills, 22.74–22.75
volume calculations for, 22.16–22.17
Small water systems
pressure filtration in, 10.76
SSF in, 10.86–10.87
UV disinfection in, 18.16
SOCs. *See* Synthetic organic chemicals
Sodium, 2.43–2.44
Sodium hydroxide, lime softening with, 13.20
Sodium hypochlorite, 7.17–7.18
basic chemistry of, 17.5–17.8
risks of, 17.32
Sodium IX softening
design example for, 12.28–12.33
strong-acid cation exchange resins in, 12.27–12.28
Softening. *See also* Lime softening; Precipitative softening
IX, 5.17
IX and capacity of, 12.25, 12.26t.
residuals from IX, 22.54–22.55, 22.55t.–22.56t.
of surface water, 5.34–5.35, 5.34f.
Softening membranes, 13.61–13.63
Soil aquifer treatment (SAT), 15.12–15.15, 15.12f.
comparisons of zones for, 15.12, 15.13t.
in Dan Region, Israel, 15.28
groundwater recharge and, 15.13
HLR in, 15.14, 15.15t.
RBF, SSF compared with, 15.19, 15.19t.
requirements for, 15.14
water depth in, 15.14–15.15
Soil fumigants, 2.55–2.56
Solar drying beds, 22.24
Solar radiation, 9.45
Solids contact clarifiers
for precipitative softening, 13.54–13.56, 13.54f.–13.55f.
for sedimentation, 5.10, 9.32–9.33, 9.33f.
Solids flux, 9.11–9.12, 9.12f.
Solubility diagrams, 20.13–20.18, 20.16f.–20.18f.
Solubility equilibria, 13.5, 13.8–13.12
Solubility product constants, 13.5, 13.6t–13.8t.
Solubility reactions, 3.14–3.15
Solvent-motivated adsorption, 14.13
Solvents, 2.73, 2.73t.
Sonic sensor testing, 11.36
Source control, 16.6
Source water. *See also* Groundwater; Surface water
biochemical cycle in, 3.27, 3.27f.
composition of, 3.24–3.42
DAF, float and influence of, 9.78–9.79
for direct filtration, 10.71–10.72
hydrogeochemical cycle in, 3.26–3.27, 3.26f.
IPR, characteristics of, 16.5–16.16
lakes and reservoirs, 3.34–3.39
LBF/RBF groundwater mixing with, 15.6
nonpoint source pollution of, 3.69
point-source control of, 3.69
prechlorination of, 17.32–17.33
protection of, 3.69
rainwater, 3.28–3.29, 3.28t.
rivers and streams, 3.29–3.30
selection of, 3.67–3.69
SSF and water quality of, 10.82–10.83
treatment process selection and variables of, 5.4
treatment process selection and water quality of, 5.2–5.4
treatment process selection incorporation in various, 5.32–5.36
Specific deposit, 10.29

Specific resistance test (SR test), 22.12, 22.13f., 22.13t.
Specific UV absorbance (SUVA), 3.65–3.66, 3.65t., 3.66f.
  TOC/UV removal estimation with, 8.27–8.28, 8.27f., 8.28t.
Spectrophotometric absorbance, 18.7–18.8
Spent adsorption material, 22.3
Spent filter backwash water (SFBW), 22.3
  characteristics of, 22.10, 22.10t.
  contaminants in raw waters compared to, 22.38, 22.38t.
  Cryptosporidium in, 22.42
  giardia lamblia in, 22.42
  residuals and, 22.37–22.40
  summary of, 22.41t.
  treatment of, 22.39–22.40
Spent rinse water, 22.3
Sphingomonas, 21.10–21.11
Spiral wound membranes (SW), 11.18, 11.19f.–11.20f., 11.20–11.21
Split-treatment softening, 13.47–13.50
Spray aerators, 6.56–6.57
  governing equations for, 6.57–6.60
  sample calculation for, 6.58–6.60
  schematic of, 6.57f.
SR test. *See* Specific resistance test
SSF. *See* Slow sand filtration
Stage 1 Disinfection By-Products Rule, 8.5–8.6, 8.5t.
Staining, aesthetic concerns with, 2.75
Standard half-cell potentials, 7.3t.
Standards. *See also* National Interim Primary Drinking Water Regulations; National Primary Drinking Water Regulations; Regulations; U.S. Environmental Protection Agency
  Australia, 1.32–1.33
  Canada, 1.32
  distribution systems and design, 21.46–21.47
  European Union, 1.33
  future of, 1.33–1.34
  international, 1.32–1.33
  processes for setting, 1.8–1.19
  for radionuclides, 2.71
  of state agencies, 1.29–1.30
  Treasury, 1.3
  USPHS, 1.3
  WHO, 1.33
Staphylococcus, 21.11
Stark-Einstein law, 18.18
State agencies
  new programs for, 1.28–1.29
  primacy of, 1.27
  role of, 1.26–1.30
  standards of, 1.29–1.30
  USEPA co-regulating with, 1.26–1.27
Static microfilters, 11.76
Steady state OH radical model, 18.28–18.29
Steel, galvanized, 20.63–20.64
Steric stabilization, 8.12–8.14, 8.13f.
Stoichiometry, 3.6–3.7
  excess lime softening calculations by, 13.35–13.39
  lime softening dose estimates based on, 13.23–13.24
  recarbonation dose estimates based on, 13.23–13.24
Stokes' law, 8.52, 9.6
Storage facilities, microbial contaminants in, 21.31–21.32, 21.54–21.55, 21.55t., 21.65
Straight lime softening, 13.30–13.31
Straight lime-soda softening, 13.30f., 13.31–13.34
Straining filtration, 10.3
Stratification, 10.62–10.63
Stray current corrosion, 20.10
Streaming current detectors (SCD)
  coagulation control/monitoring and measurements of, 8.68–8.69, 8.69f.
  jar test interpretation for, 8.69–8.70, 8.70f.
Streams. *See* Rivers and streams
Strong-acid cation exchange resins, 12.4–12.5
  adsorption rates of, 12.18
  selectivity sequences in, 12.12–12.13, 12.13t.
  in sodium IX softening, 12.27–12.28
Strong-base anion exchange resins, 12.4–12.5, 12.76–12.77
  adsorption rates of, 12.18
  nitrate removal and, 12.37
  selectivity sequences in, 12.12–12.13, 12.13t.
Styrene-divinylbenzene resin (SDVB resin), 14.87–14.88
Subnatant, 9.48
Subsidence, 9.4
Subsurface groundwater treatment, 15.10–15.11
Sulfate, 2.44
  corrosion from, 20.9, 20.29–20.30, 20.48
  IX removal of arsenic, effect of, 12.59, 12.59t.
  regulations, 1.19
Sulfur, 21.21
Superchlorination, 17.33

Supervisory control and data acquisition, 21.42
Surface aeration
  brush type, 6.52
  governing equations of, 6.53–6.56
  sample calculation for, 6.55–6.66
  schematic of, 6.52f.
  single tank schematic of, 6.53f.
  tanks in series schematic of, 6.54f.
  turbine type, 6.52
  wastewater and, 6.52
Surface charge
  electrostatic stabilization and origins of, 8.7–8.8, 8.8f.
  neutralization, 8.14–8.15, 8.22–8.24
  of particles, 3.47–3.49
  of silica, 8.7, 8.8f.
Surface water
  augmentation, 16.23, 16.27–16.29
  conventional treatment of, 5.33–5.34, 5.34f.
  DIC in, 3.29
  inorganic carbon chemistry concentrations in, 3.18t.
  softening of, 5.34–5.35, 5.34f.
  treatment process selection and, 5.2–5.3
  treatment process selection for high quality, 5.32, 5.32f.
Surface Water Treatment Rule (SWTR)
  $C \times T$ approach for regulation of disinfections under, 17.39–17.41, 17.40t.
  disinfection and, 17.4–17.5
  filtration requirements and, 10.5–10.6, 10.6t.
  precoat filtration and, 10.87
Surfactants, 6.10
Surrogate parameters, 16.34, 16.36t.
Suspended microbes, 21.7–21.8, 21.8f.
Suspended solids, 16.16
  residuals and concentration of, 22.7–22.10
SUVA. *See* Specific UV absorbance
SW. *See* Spiral wound membranes
Sweep flocculation, 8.24
SWTR. *See* Surface Water Treatment Rule
Symmetric membranes, 11.10
Synthetic adsorbent resins, 14.7
Synthetic organic chemicals (SOCs), 1.4, 11.4
  chemical oxidation of, 7.43–7.46
  GAC adsorption systems removing, 14.51–14.56
  membrane processes removing, 11.30, 11.31t.–11.32t.
  types of, 14.2–14.3
System toxicity, 2.25
Taste and odor
  aesthetic concerns with, 2.73–2.74
  algae and, 3.36, 3.37t.–3.38t., 3.44
  chemical oxidation for destruction of, 7.37–7.39, 7.41
  chlorine dioxide for controlling, 7.21
  GAC adsorption systems removing, 14.48–14.51, 14.49f.–14.50f.
  of inorganic/organic constituents, 2.73–2.74
  from microorganisms, 2.74
  PAC removing, 14.83–14.84
  potassium permanganate eliminating, 7.39
  solvents, thresholds for, 2.73, 2.73t.
TCA. *See* Trichloroacetic acid
TCE. *See* Trichloroethylene
TCLP. *See* Toxicity characteristic leach procedure
TCR. *See* Total Coliform Rule
TCRDSAC. *See* Total Coliform Rule Distribution System Advisory Committee
TDS. *See* Total dissolved solids
Technologically enhanced naturally occurring radioactive materials (TENORM), 22.66, 22.68–22.69
Temperature
  chemical oxidation and, 7.10
  coagulation and effect of, 8.63–8.64
  corrosion influenced by, 20.21–20.24, 20.23f.–20.24f.
  DBPs and, 19.20
  disinfection and impact of, 17.27
  equilibrium constant and effects of, 3.9, 6.5
  flocculation and effect of, 8.63–8.64
  gas transfer/Henry's law and, 6.8t.–6.9t.
  mass transport and correction factors in, 11.55–11.56
  membrane processes and correction factors in, 11.55–11.56
  microbial growth and, 21.23
  pH effected by, 20.23–20.24, 20.23f.
TENORM. *See* Technologically enhanced naturally occurring radioactive materials
Teratogenicity, 2.25
Terminal disinfection, 17.33
Terminal head loss, 10.11
Terminal settling velocity, 9.5–9.7
Terpenoids, 19.23
Tetrachloroethylene, 2.48
Tetravalent lead corrosion, 20.60–20.63, 20.60f.–20.61f.
TEVA. *See* Threat Ensemble Vulnerability Assessment

Theoretical normalized flux equation, 11.55
Thermodynamic principles, chemical oxidation, 7.2–7.8
Thin-film composite membranes, 11.11–11.12
THMs. *See* Trihalomethanes
Thornton, Colorado (UV disinfection), 18.17
Threat Ensemble Vulnerability Assessment (TEVA), 21.42
Threshold inhibitors, 13.3
Time to filter test (TTF test), 22.12, 22.13f., 22.13t.
Titanium dioxide, 18.34
Titration curves, 8.34–8.36, 8.35f., 8.39f.
TOC. *See* Total organic carbon
Toluene, 2.45–2.46
Total alkalinity, 20.26
Total available chlorine, 17.10
Total coliform, as indicator, 2.19–2.20
Total Coliform Rule (TCR), 1.16, 2.20
  revising, 21.35
  TCRDSAC, 1.26
Total Coliform Rule Distribution System Advisory Committee (TCRDSAC), 1.26
Total dissolved solids (TDS), 11.4
  corrosion and, 20.39
  IPR and, 16.8
  selectivity reversal with, 13.62
Total organic carbon (TOC), 3.23
  backwashing and, 14.41
  chlorine decay and, 17.17–17.18
  concentrations of, 3.62t.
  DAF removing, 5.33, 5.33f.
  GAC adsorption systems and initial concentration of, 14.44–14.48, 14.47f.
  in IPR, 16.11, 16.12t.–16.13t.
  IX and, 12.10
  lime softening removing, 13.65–13.69, 13.65f., 13.68t., 13.69f.
  measurements of, 3.62–3.63, 3.66–3.67
  POC and, 13.64
  precipitative softening removing, 13.64–13.69, 13.65f., 13.69f.
  SUVA estimating removal of, 8.27–8.28, 8.27f., 8.28t.
  THM formation potential and, 19.24
  UV as surrogate for, 3.63–3.65, 3.64t.
Total oxidants, 17.10
Total Trihalomethanes Rule (TTHM Rule), 1.4
Toxicity characteristic leach procedure (TCLP), 22.14–22.15
Toxicodynamics, 2.24
Toxicokinetics, 2.24
Toxicology, 2.23–2.27. *See also* Chemicals
Toxoplasma, 2.16
Trace organic contaminants, 15.17, 15.18t.
Tracer tests
  CFD and, 4.25–4.26
  for CFSTR, 4.13–4.16, 4.13f.–4.15f.
  CFSTRs-in-series model fit to data from, 4.18–4.19, 4.20f.
  cumulative age distribution in, 4.4, 4.5f., 4.6, 4.7f., 4.8, 4.12f.
  discrete data in, 4.8–4.11, 4.9f.–4.10f.
  exit age distribution in, 4.4, 4.5f., 4.6, 4.7f., 4.8, 4.12f.
  for PFRs, 4.11–4.13, 4.12f.
  sedimentation and, 9.26
  step v. pulse, 4.11
  types of, 4.3–4.4
Transmembrane pressure, 11.39–11.40
Transport control, 20.13
Transport mechanisms
  in adsorption, 14.28–14.29
  Brownian diffusion, 8.51–8.52, 10.31, 11.42, 11.43f.
  differential settling, 8.52
  flocculation and, 8.50–8.54
  G value concept for, 8.53–8.54
  in laminar shear, 8.52
  orthokinetic flocculation, 8.52
  for rapid granular bed filtration, 10.31–10.32, 10.31f.
  Stokes' law, 8.52, 9.6
  turbulent, 8.52–8.53
Treasury Standards, 1.3
Treatment process selection
  adsorption and, 5.19–5.21
  aeration and, 5.5
  air stripping and, 5.5
  chemical oxidation and, 5.5–5.7
  coagulation and, 5.7–5.9
  compactness for, 9.87
  contaminant removal in, 5.29, 5.30t.–5.31t., 5.32
  conventional, 5.33–5.34, 5.34f.
  costs of DAF and sedimentation for, 9.86–9.87
  DAF and, 5.10–5.11
  DAF compared to sedimentation in, 9.83–9.84, 9.84f.
  disinfection and, 5.22–5.24
  distribution systems considerations in, 5.25–5.27, 21.21–21.22
  environmental considerations for, 5.28–5.29

Treatment process selection (*Cont.*):
flexibility in, 5.28
flocculation and, 5.7–5.9
granular media filters and, 5.11–5.13
groundwater considerations for, 5.3–5.4
for high quality surface water, 5.32, 5.32f.
for IPR, 16.5
for iron/manganese removal from groundwater, 5.35, 5.35f.
IX and, 5.16–5.17
membrane processes and, 5.13–5.16
natural treatment systems and, 5.21–5.22
pilot-scale testing for, 5.28
precipitation and, 5.17–5.19
rapid start-up for, 9.87
residuals and, 5.27
sedimentation and, 5.9–5.11
sludge removal and, 9.88
solids loading for, 9.84, 9.86
source water incorporation of, 5.32–5.36
source water quality considerations for, 5.2–5.4
source water variables in, 5.4
surface water and, 5.2–5.3
UV and, 5.24–5.25
UV/H_2O_2 AOP mechanism, 18.31–18.32
Treatment systems. *See* Natural treatment systems
Triazine herbicides, 2.52–2.53
Trichloroacetic acid (TCA), 2.68
packed towers calculating, 6.29–6.36
Trichloroethane, 2.49
1,1,1-Trichloroethane, 2.48–2.49
Trichloroethylene (TCE), 2.49
Trihalomethanes (THMs), 2.44–2.45, 2.59, 2.66–2.67, 19.7–19.8
adsorption of, 14.53
ammonia's impact on formation of, 19.38, 19.38f.
bromide reactions with, 19.20–19.22, 19.20f.
bromine/iodine and, 19.21–19.22, 19.21f.–19.22f.
discovery of, 1.4
haloform reaction and, 7.14–7.15, 7.15f.
in IPR, 16.15
reaction time and, 19.16–19.17, 19.16f.
regulations for, 19.45–19.46
spatial temporal variability of, 19.46
stability of, 19.43–19.45, 19.44f.
TOC and formation potential of, 19.24
TTHM Rule, 1.4
UV controlling precursors to, 19.30–19.31, 19.31f.
TTF test. *See* Time to filter test
TTHM Rule. *See* Total Trihalomethanes Rule
Tuberculation, 20.4, 20.8
Turbidimeter, 8.4
Turbidity
aesthetic concerns with, 2.74–2.75
DAF and, 9.83–9.84, 9.84f.
direct filtration and effluent, 10.73
flocculation, 8.66–8.67, 8.67f.
GAC reducing, 14.57–14.58
as indicator, 2.22–2.23
in natural treatment systems, 15.15–15.16
of particles, 3.49, 3.50f., 8.4–8.5
rapid granular bed filtration and continuous monitoring of, 10.43, 10.44f.
of raw waters, 9.41
Turbulent transport, 8.52–8.53

UDF. *See* Unidirectional flushing
UF. *See* Ultrafiltration
UFRV. *See* Unit filter run volume
Ultrafiltration (UF), 5.13–5.15, 11.2, 16.5
coagulants removing arsenic with, 11.61, 11.61f.
coagulation, flocculation, sedimentation integration with, 11.58–11.60, 11.58f.–11.59f.
configuration of, 11.12–11.15, 11.15f.–11.17f., 11.17t.
crossflow operation in, 11.13, 11.14f.
Cryptosporidium removed by, 11.33–11.34, 11.33f.
dead end operation in, 11.13, 11.14f.
floc-blanket reactor-PAC-UF process, 14.80–14.81, 14.80f.
flow patterns of, 11.13, 11.14f.–11.15f., 11.15
Giardia lamblia removed by, 11.33–11.34, 11.33f.
mechanical sieving in, 11.44–11.45
MF integrated process design/applications with, 11.57–11.65
NF/RO/EDR integration with, 11.59–11.60, 11.59f.
oxidation reactions integrated with, 11.60–11.61, 11.60f.
PAC with, 14.80
pilot plant testing for, 11.95
process residuals of, 11.91–11.92
residuals removed by, 22.3–22.4
seawater RO integrated with, 11.61–11.62
uses of, 11.4, 11.8

Ultraviolet light (UV). *See also* UV disinfection; UV photolysis
absorbance of, 18.14
Beer-Lambert law and, 18.7–18.8
bench-scale testing for, 18.8–18.9
biodosimetry test for, 18.9–18.10, 18.9f.
collimated beam and, 18.8–18.9, 18.8f.
DBPs and, 19.4t.–19.6t., 19.14–19.15
Disinfection Guidance Manual for, 18.5, 18.9
as DOC surrogate, 3.63–3.65
dose calculations for, 18.6–18.10
in electromagnetic spectrum, 18.4, 18.4f.
emerging lamps for, 18.6
energy measurements for, 18.7
fundamentals of, 18.4–18.10
hydrogen peroxide formation and, 7.27–7.28
irradiance calculations for, 18.7–18.8
lamp types for, 18.4–18.6, 18.5t., 18.6f.
mercury vapor lamps for, 18.5–18.6, 18.5t., 18.6f.
microorganisms and, 18.1–18.2
reactors for, 18.10, 18.10f.
spectrophotometric absorbance and, 18.7–18.8
SUVA, 3.65–3.66, 3.65t., 3.66f.
SUVA estimating removal of, 8.27–8.28, 8.27f., 8.28t.
for THM precursor control, 19.30–19.31, 19.31f.
as TOC surrogate, 3.63–3.65, 3.64t.
treatment process selection and, 5.24–5.25
UV/Cl_2 AOP mechanism, 18.33
UV/H_2O_2 AOP mechanism, 18.24, 18.25t.–18.26t., 18.27–18.32
UV/O_3 AOP mechanism, 18.33
UV/TiO_2 AOP mechanism, 18.34
water quality effecting transmission of, 18.14–18.15
Underground injection control, 22.45–22.46
Unidirectional flushing (UDF), 21.52–21.53
Uniform corrosion, 20.6
of copper, 20.46–20.50, 20.47f., 20.49f.
Uniform formation conditions test, 19.23
Unit conversion factors, C.1t.–C.3t.
Unit filter run volume (UFRV), 10.23–10.24, 10.23f.
Unregulated contaminant monitoring regulations (UCMRs), 1.14–1.15
Upflow filter, 10.15–10.16
Upper Occuquan Service Authority, Virginia, 16.27–16.28
Uranium, 2.72
chemistry/speciation of, 12.67
IX removal of (by anion exchange), 12.67–12.73
IX removal of radium combined with, 12.72–12.73
pH effecting IX removal of, 12.69–12.70
regenerability of spent resin in IX removal of, 12.71–12.72, 12.72f.
U.S. Department of Transportation (USDOT), 22.69
U.S. Environmental Protection Agency (USEPA), 1.2. *See also* National Primary Drinking Water Regulations; Risk management
contaminants, regulatory determinations of, 1.9–1.10, 1.13, 1.14t.
future of, 1.33–1.34
grants of, 1.27
health advisories of, 1.16, 1.18–1.19
MCLGs, 1.15, 16.17–16.18
MCLs, 1.4, 16.17–16.18
National Secondary Drinking Water Regulations, 1.16, 1.18t.
peer review/outside consultation for, 1.30–1.31
public notification requirements of, 1.31–1.32
regions of, 1.28f.
regulations for distribution systems by, 21.35–21.36, 21.35t.–21.36t.
risk assessment approach of, 2.27–2.28
SAB of, 1.30
Stage 1 Disinfection By-Products Rule of, 8.5–8.6, 8.5t.
state agencies co-regulating with, 1.26–1.27
traditional/negotiated rulemaking processes of, 1.25–1.26
UCMRs for, 1.14–1.15
U.S. Federal regulations for IPR, 16.36–16.37
U.S. Public Health Services (USPHS)
Bureau of Water Hygiene CWSS, 1.3–1.4
standards, 1.3
U.S. Treasury Department, 1.3
USEPA. *See* U.S. Environmental Protection Agency
USPHS. *See* U.S. Public Health Services
UV. *See* Ultraviolet light

UV disinfection, 18.10
advantages/disadvantages of, 18.3t.
application/design of, 17.37
bacteria inactivation by, 17.29, 17.29t.
chlorine for disinfection compared to, 18.3, 18.3t.
DBPs and, 18.15
demand reactions of, 17.19
dose monitoring for, 18.14
in Fort Benton, Montana, 18.17
fouling interfering with, 18.14–18.15
fundamentals of, 18.11–18.13
groundwater application of, 18.16
history of use, 17.4
integration of, 18.15–18.16
microbial inactivation mechanisms of, 8.11f., 18.11
for multiple-barrier treatment approach, 18.16
in New York City, NY, 18.17
particles interfering with, 18.15
pathogen inactivation and doses in, 18.12, 18.12f.
photoreactivation and, 18.11–18.12
reactor performance for, 18.13
reactor validation for, 18.13–18.14
resurgence of, 18.2–18.3
in Seattle Public Utilities, 18.17
in small water systems, 18.16
in Thornton, Colorado, 18.17
for unfiltered systems, 18.16
virus inactivation and, 18.16
water quality effecting transmission of UV in, 18.14–18.15
UV photolysis, 18.17
chemical pollutants treated with, 18.21–18.24
Grotthuss-Draper law and, 18.18
HOMO and, 18.18–18.19, 18.18f.
Jablonski diagram and, 18.19, 18.19f.
light and matter interactions in, 18.19–18.21, 18.19f.–18.20f.
LUMO and, 18.18–18.19, 18.18f.
molar absorption coefficient and, 18.20, 18.20f., 18.21t.
monochromatic, 18.21–18.23, 18.21t.
NDMA and, 18.21–18.22
polychromatic, 18.23–18.24
quantum yield and, 18.20–18.21, 18.21t.
Stark-Einstein law and, 18.18
UV/Cl_2 AOP mechanism, 18.33
UV/H_2O_2 AOP mechanism, 18.24, 18.25t.–18.26t., 18.27–18.32
adOx kinetic model for, 18.29–18.30, 18.30f.
drinking water applications of, 18.31–18.32
IPR applications of, 18.30–18.31
steady state OH radical model for, 18.28–18.29
UV/O_3 AOP mechanism, 18.33
UV/TiO_2 AOP mechanism, 18.34

Vacuum filtration, 10.3
for mechanical dewatering of residuals, 22.30–22.31, 22.30f.
Vacuum flotation, 9.46
Vadnais Lake example problem, 8.43–8.44
VBNC bacteria. *See* Viable but nonculturable bacteria
Verrucomicrobium, 21.10
Viable but nonculturable bacteria (VBNC bacteria), 21.14, 21.15t.
Vibrio chloerae, 2.7–2.8
Vinyl chloride, 2.49–2.50
Vinylidene chloride, 2.47–2.48
Viruses, 2.11–2.14
adenoviruses, 2.14
astroviruses, 2.14
caliciviruses, 2.11–2.12
enteroviruses, 2.13–2.14
HAV, 2.12–2.13
HEV, 2.13, 2.21
norovirus, 2.11
Rotaviruses, 2.12
UV disinfection and inactivation of, 18.16
VOCs. *See* Volatile organic chemicals
Volatile organic chemicals (VOCs), 2.45–2.50
air stripping of, 6.42–6.45
benzene, 2.45–2.46
carbon tetrachloride, 2.46–2.47
chlorinated, 2.46–2.49
dichlorobenzenes, 2.47
1,2-dichloroethane, 2.47
1,1-dichloroethylene, 2.47–2.48
1,2-dichloroethylenes, 2.48
dichloromethane, 2.48
ethylbenzene, 2.45–2.46
ethylene dichloride, 2.47
GAC adsorption systems removing, 14.51–14.53, 14.52f., 14.52t.
Henry's law and, 3.9–3.10
in IPR, 16.11, 16.12t.–16.13t., 16.14
methyl chloroform, 2.48–2.49
methylene chloride, 2.48

Volatile organic chemicals (VOCs) (*Cont.*):
monochloroethene, 2.49–2.50
MTBE, 2.50
off-gas control with, 6.36–6.41, 6.37f.–6.38f., 6.39t.
perchloroethylene, 2.48
TCE, 2.49
tetrachloroethylene, 2.48
toluene, 2.45–2.46
trichloroethane, 2.49
1,1,1-trichloroethane, 2.48–2.49
vinyl chloride, 2.49–2.50
vinylidene chloride, 2.47–2.48
xylene, 2.45–2.46
Vyredox method, 15.10

Warm monomictic lakes, 3.35
Warm polymictic lakes, 3.35
Wastewater, 3.68–3.69
backwashing management of, 10.68–10.69
DAF of, 9.46
pathogens in, 17.21
surface aeration and, 6.52
upflow filter treatment for, 10.15–10.16
Water age, 21.51–21.52, 21.53t.
Water dissociation, 3.3–3.4
Water properties
physical, 3.3
polar nature, 3.2
Water quality
adsorption and, 14.3
aesthetic concerns with, 2.2
concentrations measuring, 3.4
distribution system design practices and, 21.46–21.47
distribution system models for, 21.44–21.45, 21.44f.
in distribution systems, 5.25–5.27
distribution systems, measuring parameters for, 21.36–21.38
health concerns with, 2.2–2.3
indicators of, 2.19–2.23
IX removal of nitrate and effects of, 12.35, 12.36f., 12.37
in lakes and reservoirs, 3.36, 3.37t.–3.38t., 3.38–3.39
natural treatment systems and improvements in, 15.15–15.19
NOM effects on, 3.58, 3.58t.
on-line monitoring for, 21.37–21.38
particles importance to, 3.42–3.43
sampling techniques for, 21.36–21.37
Water quality (*Cont.*):
sedimentation and seasonal, 9.40
in SSF of source water, 10.82–10.83
treatment process selection, source water considerations with, 5.2–5.4
UV transmission and effects of, 18.14–18.15
Water Replenishment District of Southern California, 16.29
Water Research Foundation, 1.6
Water reuse, 3.68. *See also* Direct potable reuse; Indirect potable reuse
BNR in, 16.10–16.11, 16.10t.
phosphorus in, 16.11
Water security, 21.41–21.42
Water supplies, hardness classification for, 13.16, 13.16t.
Water treatment conversion factors, C.3t.
Water treatment residuals. *See* Residuals
Waterborne disease. *See also Cryptosporidium*; *Escherichia coli*; *Giardia lamblia*; *Legionella*; *Salmonella*; *Shigella*
distribution systems and outbreaks of, 21.2–21.4
Legionella outbreaks and, 21.3–21.4, 21.3f.
microorganisms and, 2.3, 2.4t.
outbreaks, 2.3, 2.4t., 2.5, 2.11–2.12
reporting, 2.5
United States outbreaks of, 21.2f.–21.3f.
in water main repairs and installations, 21.32–21.34, 21.33t., 21.63–21.65
Water's physical properties, D.1t.
Watersheds, protection of, 3.70
Weak-acid cation exchange resins, 12.5–12.6, 12.12–12.13
adsorption rates of, 12.18–12.19
Weak-base anion exchange resins, 12.6–12.7, 12.12–12.13
adsorption rates of, 12.18–12.19
Western Corridor Project, Queensland, Australia, 16.28–16.29
Western Corridor Recycled Water Project, Queensland, Australia, 16.41
Wet chemical procedures, 20.71
White water blanket, 9.51, 9.54
WHO. *See* World Health Organization
Windhoek's Goreangab Reclamation Plant, Namibia, 16.32–16.33
World Health Organization (WHO)
DALYs established by, 16.18
standards of, 1.33

XAFS. *See* X-ray absorption fine structure
XANES. *See* X-ray absorption near-edge spectroscopy
X-ray absorption fine structure (XAFS), 20.73
X-ray absorption near-edge spectroscopy (XANES), 20.73
X-ray diffraction, 20.72–20.73, 20.73f.
X-ray fluorescence spectrometry, 20.70–20.71
X-ray photoelectron spectroscopy, 20.44
Xylene, 2.45–2.46

Yellow water, 20.41
Yersinia enterocolitica, 2.8

Zeolites, 14.8
  high-silica, 14.89
Zinc, 20.35